$107.95 per copy (in United States).
Price subject to change without prior notice.

0046

RSMeans

Repair & Remodeling Cost Data

W9-DGQ-137

27th Annual Edition

2006

Senior Editor
Robert W. Mewis, CCC

Contributing Editors
Christopher Babbitt
Ted Baker
Barbara Balboni
Robert A. Bastoni
John H. Chiang, PE
John Kane
Robert J. Kuchta
Robert C. McNichols
Melville J. Mossman, PE
John J. Moylan
Jeannene D. Murphy
Stephen C. Plotner
Eugene R. Spencer
Marshall J. Stetson
Phillip R. Waier, PE

Senior Engineering Operations Manager
John H. Ferguson, PE

Senior Vice President & General Manager
John Ware

Vice President of Sales
John M. Shea

Production Manager
Michael Kokernak

Technical Support
Wayne D. Anderson
Thomas J. Dion
Jonathan Forgit
Mary Lou Geary
Gary L. Hoitt
Genevieve Medeiros
Paula Reale-Camelio
Kathryn S. Rodriguez
Sheryl A. Rose
Laurie Thom

Book & Cover Design
Norman R. Forgit

RSMeans
Construction Publishers & Consultants
63 Smiths Lane
Kingston, MA 02364-0800
(781) 422-5000

Printed in the United States of America.
ISSN 0271-5945
ISBN 0-87629-795-5

This book is recyclable.

This book is printed on recycled stock.

First Printing

Foreword

RSMeans is a product line of Reed Construction Data, Inc., a leading provider of construction information, products, and services in North America and globally. Reed Construction Data's project information products include more than 100 regional editions, national construction data, sales leads, and local plan rooms in major business centers. Reed Construction Data's PlansDirect provides surveys, plans, and specifications. The First Source suite of products consists of *First Source for Products,* SPEC-DATA™, MANU-SPEC™, CADBlocks, Manufacturer Catalogs, and First Source Exchange (www.firstsourceexchange.com) for the selection of nationally available building products. Reed Construction Data also publishes ProFile, a database of more than 20,000 U.S. architectural firms. RSMeans provides construction cost data, training, and consulting services in print, CD-ROM, and online. Reed Construction Data, headquartered in Atlanta, is owned by Reed Business Information (www.reedconstructiondata.com), a leading provider of critical information and marketing solutions to business professionals in the media, manufacturing, electronics, construction, and retail industries. Its market-leading properties include more than 135 business-to-business publications and over 125 Webzines and Web portals as well as online services, custom publishing, directories, research, and direct-marketing lists. Reed Business Information is a member of the Reed Elsevier plc group (NYSE: RUK and ENL)—a world-leading publisher and information provider operating in the science and medical, legal, education, and business-to-business industry sectors.

Our Mission

Since 1942, RSMeans has been actively engaged in construction cost publishing and consulting throughout North America.

Today, over 60 years after RSMeans began, our primary objective remains the same: to provide you, the construction and facilities professional, with the most current and comprehensive construction cost data possible.

Whether you are a contractor, an owner, an architect, an engineer, a facilities manager, or anyone else who needs a fast and reliable construction cost estimate, you'll find this publication to be a highly useful and necessary tool.

Today, with the constant flow of new construction methods and materials, it's difficult to find the time to look at and evaluate all the different construction cost possibilities. In addition, because labor and material costs keep changing, last year's cost information is not a reliable basis for today's estimate or budget.

That's why so many construction professionals turn to RSMeans. We keep track of the costs for you, along with a wide range of other key information, from city cost indexes . . . to productivity rates . . . to crew composition . . . to contractor's overhead and profit rates.

RSMeans performs these functions by collecting data from all facets of the industry and organizing it in a format that is instantly accessible to you. From the preliminary budget to the detailed unit price estimate, you'll find the data in this book useful for all phases of construction cost determination.

The Staff, the Organization, and Our Services

When you purchase one of RSMeans' publications, you are, in effect, hiring the services of a full-time staff of construction and engineering professionals.

Our thoroughly experienced and highly qualified staff works daily at collecting, analyzing, and disseminating comprehensive cost information for your needs. These staff members have years of practical construction experience and engineering training prior to joining the firm. As a result, you can count on them not only for the cost figures, but also for additional background reference information that will help you create a realistic estimate.

The RSMeans organization is always prepared to help you solve construction problems through its five major divisions: Construction and Cost Data Publishing, Electronic Products and Services, Consulting Services, Insurance Services, and Educational Services.

Besides a full array of construction cost estimating books, RSMeans also publishes a number of other reference works for the construction industry. Subjects include construction estimating and project and business management; special topics such as HVAC, roofing, plumbing, and hazardous waste remediation; and a library of facility management references.

In addition, you can access all of our construction cost data through your computer with *Means CostWorks 2006* CD-ROM, an electronic tool that offers over 50,000 lines of RSMeans detailed construction cost data, along with assembly and whole building cost data. You can also access RSMeans cost information from our Web site at www.rsmeans.com.

What's more, you can increase your knowledge and improve your construction estimating and management performance with an RSMeans Construction Seminar or In-House Training Program. These two-day seminar programs offer unparalleled opportunities for everyone in your organization to get updated on a wide variety of construction-related issues.

RSMeans also is a worldwide provider of construction cost management and analysis services for commercial and government owners and of claims and valuation services for insurers.

In short, RSMeans can provide you with the tools and expertise for constructing accurate and dependable construction estimates and budgets in a variety of ways.

Robert Snow Means Established a Tradition of Quality That Continues Today

Robert Snow Means spent years building RSMeans, making certain he always delivered a quality product.

Today, at RSMeans, we do more than talk about the quality of our data and the usefulness of our books. We stand behind all of our data, from historical cost indexes to construction materials and techniques to current costs.

If you have any questions about our products or services, please call us toll-free at 1-800-334-3509. Our customer service representatives will be happy to assist you. You can also visit our Web site at www.rsmeans.com.

Table of Contents

Related RSMeans Products and Services

Robert W. Mewis, CCC, Senior Editor of this cost data book, suggests the following RSMeans products and services as **companion** information resources to *Repair & Remodeling Cost Data*:

Construction Cost Data Books

Building Construction Cost Data 2006
Facilities Construction Cost Data 2006
Facilities Maintenance & Repair Cost Data 2006

Reference Books

ADA Compliance Pricing Guide, 2nd Edition
Building Security: Strategies & Costs
Designing & Building with the IBC
Estimating Building Costs
Estimating Handbook, 2nd Edition
Job Order Contracting
Plan Reading & Material Takeoff
Project Scheduling and Management for Construction
Repair & Remodeling Estimating, 4th Edition

Seminars and In-House Training

Advanced Project Management
Means CostWorks Training
Means Data for Job Order Contracting (JOC)
Plan Reading & Material Takeoff
Scheduling & Project Management
Repair & Remodeling Estimating

RSMeans on the Internet

Visit RSMeans at **http://www.rsmeans.com.** The site contains useful ***interactive*** cost and reference material. Request or download ***FREE*** estimating software demos. Visit our bookstore for convenient ordering and to learn more about new publications and companion products.

RSMeans Data on CD-ROM

Get the information found in RSMeans traditional cost books on RSMeans' new, easy-to-use CD-ROM, *CostWorks 2006. Means CostWorks* users can now enhance their Costlist with the *Means CostWorks Estimator*. For more information see the special full-color brochure inserted in this book.

RSMeans Business Solutions

RSMeans Business Solutions provides construction cost-related services including: Customized Databases, Benchmark Studies, Estimating Audits, Feasibility Studies, Litigation Support, and Predictive Cost Models.

New! RSMeans for Job Order Contracting (JOC)

New methods and "best practices" for estimating and project management to reduce administrative costs for a $147 billion+ renovation market in the U.S. Finish renovation projects faster in schools and universities, on government building projects, and in medical facilities.

* RSMeans Job Order Contracting (JOC) Cost Data
* JOCWorks Software (Basic, Advanced, PRO)
* Consultation in Contracting Methods
* RSMeans Data™ Licensing Program

Construction Costs for Software Applications

Over 25 unit price and assemblies cost databases are available through a number of leading estimating and facilities management software providers (listed below). For more information see the "yellow" pages at the back of this publication.

RSMeansData™ is also available to federal, state, and local government agencies as multi-year, multi-seat licenses.

- 3D International
- 4Clicks-Solutions, LLC
- Aepco, Inc.
- Applied Flow Technology
- ArenaSoft Estimating
- Best Software–Timberline
- BSD – Building Systems Design, Inc.
- CMS – Construction Management Software
- Corecon Technologies, Inc.
- CorVet Systems
- Estimating Systems, Inc.
- Maximus Asset Solutions
- MC^2 – Management Computer Controls
- Shaw Beneco Enterprises, Inc.
- US Cost, Inc.
- VFA–Vanderweil Facility Advisers
- WinEstimator, Inc.

How the Book Is Built: An Overview

A Powerful Construction Tool

You have in your hands one of the most powerful construction tools available today. A successful project is built on the foundation of an accurate and dependable estimate. This book will enable you to construct just such an estimate.

For the casual user the book is designed to be:

- quickly and easily understood so you can get right to your estimate.
- filled with valuable information so you can understand the necessary factors that go into the cost estimate.

For the regular user, the book is designed to be:

- a handy desk reference that can be quickly referred to for key costs.
- a comprehensive, fully reliable source of current construction costs and productivity rates so you'll be prepared to estimate any project.
- a source book for preliminary project cost, product selections, and alternate materials and methods.

To meet all of these requirements we have organized the book into the following clearly defined sections.

How To Use the Book: The Details

This section contains an in-depth explanation of how the book is arranged . . . and how you can use it to determine a reliable construction cost estimate. It includes information about how we develop our cost figures and how to completely prepare your estimate.

Unit Price Section

All cost data has been divided into the 16 divisions according to the MasterFormat system of classification and numbering as developed by the Construction Specifications Institute (CSI) and Construction Specifications Canada (CSC). For a listing of these divisions and an outline of their subdivisions, see the Unit Price Section Table of Contents.

Estimating tips are included at the beginning of each division.

Assemblies Section

The cost data in this section has been organized in an "Assemblies" format. These assemblies are the functional elements of a building and are arranged according to the 7 divisions of the UNIFORMAT II classification system. For a complete explanation of a typical "Assemblies" page, see "How To Use the Assemblies Cost Tables."

Reference Section

This section includes information on Equipment Rental Costs, Crew Listings, Historical Cost Indexes, City Cost Indexes, Location Factors, Reference Tables, Change Orders, Square Foot Costs, and a listing of Abbreviations.

Equipment Rental Costs: This section contains the average costs to rent and operate hundreds of pieces of construction equipment.

Crew Listings: This section lists all the crews referenced in the book. For the purposes of this book, a crew is composed of more than one trade classification and/or the addition of power equipment to any trade classification. Power equipment is included in the cost of the crew. Costs are shown both with the bare labor rates and with the installing contractor's overhead and profit added. For each, the total crew cost per eight-hour day and the composite cost per labor-hour are listed.

Historical Cost Indexes: These indexes provide you with data to adjust construction costs over time.

City Cost Indexes: All costs in this book are U.S. national averages. Costs vary because of the regional economy. You can adjust costs by CSI Division to over 316 locations throughout the U.S. and Canada by using the data in this section.

Location Factors: You can adjust total project costs to over 900 locations throughout the U.S. and Canada by using the data in this section.

Reference Tables: At the beginning of selected major classifications in the Unit Price and Assemblies sections are "reference numbers" shown in a bold box. These numbers refer you to related information in the Reference Section. In this section, you'll find reference tables, explanations, and estimating information that support how we develop the unit price data, technical data, and estimating procedures.

Change Orders: This section includes information on the factors that influence the pricing of change orders.

Square Foot Costs: Formerly Division 17, this section contains costs for 59 different building types that allow you to make a rough estimate for the overall cost of a project or its major components.

Abbreviations: A listing of abbreviations used throughout this book, along with the terms they represent, is included in this section.

Index

A comprehensive listing of all terms and subjects in this book will help you quickly find what you need when you are not sure where it falls in MasterFormat.

The Scope of This Book

This book is designed to be as comprehensive and as easy to use as possible. To that end we have made certain assumptions and limited its scope in three key ways:

1. We have established material prices based on a national average.
2. We have computed labor costs based on a 30-city national average of union wage rates.
3. We have targeted the data for projects of a certain size range.

For a more detailed explanation of how the cost data is developed, see "How To Use the Book: The Details."

Project Size

This book is aimed primarily at residential, commercial and industrial repair/remodeling projects costing $10,000 to $1,000,000.

With reasonable exercise of judgment the figures can be used for any building work. *However, for civil engineering structures such as bridges, dams, highways, or the like, please refer to **RSMeans Heavy Construction Cost Data.***

How to Use the Book: The Details

What's Behind the Numbers? The Development of Cost Data

The staff at RSMeans continuously monitors developments in the construction industry in order to ensure reliable, thorough, and up-to-date cost information.

While *overall* construction costs may vary relative to general economic conditions, price fluctuations within the industry are dependent upon many factors. Individual price variations may, in fact, be opposite to overall economic trends. Therefore, costs are continually monitored and complete updates are published yearly. Also, new items are frequently added in response to changes in materials and methods.

Costs—$ (U.S.)

All costs represent U.S. national averages and are given in U.S. dollars. The RSMeans City Cost Indexes can be used to adjust costs to a particular location. The City Cost Indexes for Canada can be used to adjust U.S. national averages to local costs in Canadian dollars. No exchange rate conversion is necessary.

Material Costs

The RSMeans staff contacts manufacturers, dealers, distributors, and contractors all across the U.S. and Canada to determine national average material costs. If you have access to current material costs for your specific location, you may wish to make adjustments to reflect differences from the national average. Included within material costs are fasteners for a normal installation. RSMeans engineers use manufacturers' recommendations, written specifications, and/or standard construction practice for size and spacing of fasteners. Adjustments to material costs may be required for your specific application or location. Material costs do not include sales tax.

Labor Costs

Labor costs are based on the average of wage rates from 30 major U.S. cities. Rates are determined from labor union agreements or prevailing wages for construction trades for the current year. Rates, along with overhead and profit markups, are listed on the inside back cover of this book.

- If wage rates in your area vary from those used in this book, or if rate increases are expected within a given year, labor costs should be adjusted accordingly.

Labor costs reflect productivity based on actual working conditions. These figures include time spent during a normal workday on tasks other than actual installation, such as material receiving and handling, mobilization at site, site movement, breaks, and cleanup.

Productivity data is developed over an extended period so as not to be influenced by abnormal variations and reflects a typical average.

Equipment Costs

Equipment costs include not only rental, but also operating costs for equipment under normal use. The operating costs include parts and labor for routine servicing such as repair and replacement of pumps, filters, and worn lines. Normal operating expendables, such as fuel, lubricants, tires, and electricity (where applicable), are also included. Extraordinary operating expendables with highly variable wear patterns, such as diamond bits and blades, are excluded. These costs are included under materials. Equipment rental rates are obtained from industry sources throughout North America—contractors, suppliers, dealers, manufacturers, and distributors.

Crew Equipment Cost/Day—The power equipment required for each crew is included in the crew cost. The daily cost for crew equipment is based on dividing the weekly bare rental rate by 5 (number of working days per week) and then adding the hourly operating cost times 8 (hours per day). This "Crew Equipment Cost/Day" is listed in the Reference Section.

Mobilization/Demobilization—The cost to move construction equipment from an equipment yard or rental company to the job site and back again is not included in equipment costs. Mobilization (to the site) and demobilization (from the site) costs can be found in the Unit Price Section. If a piece of equipment is already at the job site, it is not appropriate to utilize mob/demob costs again in an estimate.

General Conditions

Cost data in this book is presented in two ways: Bare Costs and Total Cost including O&P (Overhead and Profit). General Conditions, when applicable, should also be added to the Total Cost including O&P. The costs for General Conditions are listed in Division 1 of the Unit Price Section and the Reference Section of this book. General Conditions for the *Installing Contractor* may range from 0% to 10% of the Total Cost including O&P. For the *General* or *Prime Contractor*, costs for General Conditions may range from 5% to 15% of the Total Cost, including O&P, with a figure of 10% as the most typical allowance.

Overhead and Profit

Total Cost, including O&P, for the *Installing Contractor* is shown in the last column on both the Unit Price and the Assemblies pages of this book. This figure is the sum of the bare material cost plus 10% for profit, the base labor cost plus total overhead and profit, and the bare equipment cost plus 10% for profit. Details for the calculation of Overhead and Profit on labor are shown on the inside back cover and in the Reference Section of this book. (See the "How to Use the Unit Price Pages" for an example of this calculation.)

Factors Affecting Costs

Costs can vary depending upon a number of variables. Here's how we have handled the main factors affecting costs.

Quality—The prices for materials and the workmanship upon which productivity is based represent sound construction work. They are also in line with U.S. government specifications.

Overtime—We have made no allowance for overtime. If you anticipate premium time or work beyond normal working hours, be sure to make an appropriate adjustment to your labor costs.

Productivity—The productivity, daily output, and labor-hour figures for each line item are based on working an eight-hour day in daylight hours in moderate temperatures. For work that extends beyond normal work hours or is performed under adverse conditions, productivity may decrease. (See the section in "How To Use the Unit Price Pages" for more on productivity.)

Size of Project—The size, scope of work, and type of construction project will have a significant impact on cost. Economies of scale can reduce costs for large projects. Unit costs can often run higher for small projects. Costs in this book are intended for the size and type of project as previously described in "How the Book Is Built: An Overview." Costs for projects of a significantly different size or type should be adjusted accordingly.

Location—Material prices in this book are for metropolitan areas. However, in dense urban areas, traffic and site storage limitations may increase costs. Beyond a 20-mile radius of large cities, extra trucking or transportation charges may also increase the material costs slightly. On the other hand, lower wage rates may be in effect. Be sure to consider both of these factors when preparing an estimate, particularly if the job site is located in a central city or remote rural location.

In addition, highly specialized subcontract items may require travel and per-diem expenses for mechanics.

Other Factors—
- season of year
- contractor management
- weather conditions
- local union restrictions
- building code requirements
- availability of:
  - adequate energy
  - skilled labor
  - building materials
- owner's special requirements/restrictions
- safety requirements
- environmental considerations

Unpredictable Factors—General business conditions influence "in-place" costs of all items. Substitute materials and construction methods may have to be employed. These may affect the installed cost and/or life cycle costs. Such factors may be difficult to evaluate and cannot necessarily be predicted on the basis of the job's location in a particular section of the country. Thus, where these factors apply, you may find significant but unavoidable cost variations for which you will have to apply a measure of judgment to your estimate.

Rounding of Costs

In general, all unit prices in excess of $5.00 have been rounded to make them easier to use and still maintain adequate precision of the results. The rounding rules we have chosen are in the following table.

Prices from . . .	Rounded to the nearest . . .
$.01 to $5.00	$.01
$5.01 to $20.00	$.05
$20.01 to $100.00	$.50
$100.01 to $300.00	$1.00
$300.01 to $1,000.00	$5.00
$1,000.01 to $10,000.00	$25.00
$10,000.01 to $50,000.00	$100.00
$50,000.01 and above	$500.00

Final Checklist

Estimating can be a straightforward process provided you remember the basics. Here's a checklist of some of the steps you should remember to complete before finalizing your estimate.

Did you remember to . . .
- factor in the City Cost Index for your locale?
- take into consideration which items have been marked up and by how much?
- mark up the entire estimate sufficiently for your purposes?
- read the background information on techniques and technical matters that could impact your project time span and cost?
- include all components of your project in the final estimate?
- make use of Minimum Labor/Equipment Charges for Small Quantities (see the following page for more details)?
- double check your figures for accuracy?
- call RSMeans if you have any questions about your estimate or the data you've found in our publications?

Remember, RSMeans stands behind its publications. If you have any questions about your estimate . . . about the costs you've used from our books . . . or even about the technical aspects of the job that may affect your estimate, feel free to call the RSMeans editors at 1-800-334-3509.

Using Minimum Labor/Equipment Charges for Small Quantities

Estimating small construction or repair tasks often creates situations in which the quantity of work to be performed is very small. When this occurs, the labor and/or equipment costs to perform the work may be too low to allow for the crew to get to the job, receive instructions, find materials, get set up, perform the work, clean up, and get to the next job. In these situations, the estimator should compare the developed labor and/or equipment costs for performing the work (e.g., quantity x labor and/or equipment costs) with the *"minimum labor/equipment charge"* within that Unit Price section of the book.

If the labor and/or equipment costs developed by the estimator are LOWER THAN the *"minimum labor/equipment charge"* listed at the bottom of specific sections of Unit Price costs, the estimator should adjust the developed costs upward to the *"minimum labor/equipment charge."* The proper use of a *"minimum labor/equipment charge"* results in having enough money in the estimate to cover the contractor's higher cost of performing a very small amount of work during a partial workday.

A *"minimum labor/equipment charge"* should be used only when the task being estimated is the only task the crew will perform at the job site that day. If, however, the crew will be able to perform other tasks at the job site that day, the use of a *"minimum labor/equipment charge"* is not appropriate.

08500 | Windows

		08550 | Wood Windows	CREW	DAILY OUTPUT	LABOR-HOURS	UNIT	2006 BARE COSTS MAT.	LABOR	EQUIP.	TOTAL	TOTAL INCL O&P	
200	0010	**CASEMENT WINDOW** Including frame, screen, and grills										200
	0100	Avg. quality, bldrs. model, 2'-0" x 3'-0" H, dbl. insulated glass	1 Carp	10	.800	Ea.	186	28.50		214.50	252	
	0150	Low E glass		10	.800		225	28.50		253.50	295	
	0200	2'-0" x 4'-6" high, double insulated glass		9	.889		242	31.50		273.50	320	
	0250	Low E glass		9	.889		320	31.50		351.50	410	
	0300	2'-4" x 6'-0" high, double insulated glass		8	1		355	35.50		390.50	450	
	0350	Low E glass		8	1		385	35.50		420.50	485	
	0522	Vinyl clad, premium, double insulated glass, 2'-0" x 3'-0"		10	.800		264	28.50		292.50	335	
	8100	Metal clad, deluxe, dbl. insul. glass, 2'-0" x 3'-0" high		10	.800		206	28.50		234.50	274	
	8120	2'-0" x 4'-0" high		9	.889		248	31.50		279.50	325	
	8140	2'-0" x 5'-0" high		8	1		282	35.50		317.50	370	
	8300	For installation, add per leaf						15%				
	9000	Minimum labor/equipment charge	1 Carp	3	2.667	Job		95		95	157	

Example:

Establish the bid price to install two casement windows. Assume installation of 2′ x 4′ metal clad windows with insulating glass [Unit Price line number 08550-200-8120], and that this is the only task this crew will perform at the job site that day.

Solution:

Step One — Develop the Bare Labor Cost for this task:

Bare Labor Cost = 2 windows @ $31.50/each = $63.00

Step Two — Evaluate the *"minimum labor/equipment charge"* for this Unit Price section against the developed Bare Labor Cost for this task:

"minimum labor/equipment charge" = $95.00 (compare with $63.00)

Step Three — Choose to adjust the developed labor cost upward to the *"minimum labor/equipment charge."*

Step Four — Develop the bid price for this task (including O&P):

Add together the marked-up Bare Material Cost for this task and the marked-up *"minimum labor/equipment charge"* for this Unit Price section.

2 x ($248.00 + 10%) + ($95.00 + 65.7%)
= 2 x ($248.00 + $24.80) + ($95.00 + $62.42)
= 2 x ($272.80) + $157.42
= $545.60 + $157.42
= $703.02

ANSWER: $703.02 is the correct bid price to use. This sum takes into consideration the Material Cost (with 10% for profit) for these two windows, plus the *"minimum labor/equipment charge"* (with O&P included) for this section of the Unit Price book.

Unit Price Section

Table of Contents

How to Use the Unit Price Pages

The following is a detailed explanation of a sample entry in the Unit Price Section. Next to each bold number below is the described item with the appropriate component of the sample entry following in parentheses. Some prices are listed as bare costs; others as costs that include overhead and profit of the installing contractor. In most cases, if the work is to be subcontracted, the general contractor will need to add an additional markup (RSMeans suggests using 10%) to the figures in the column "Total Incl. O&P."

1 *Division Number/Title (03300/Cast-In-Place Concrete)*

Use the Unit Price Section Table of Contents to locate specific items. The sections are classified according to the CSI MasterFormat (1995 Edition).

2 *Line Numbers (03310 240 3920)*

Each unit price line item has been assigned a unique 12-digit code based on the CSI MasterFormat classification.

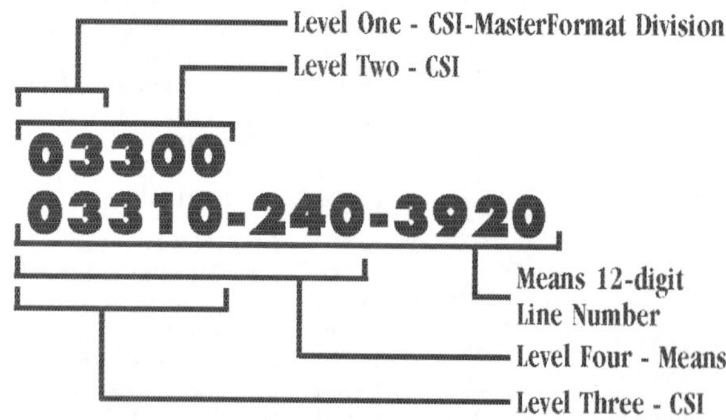

3 *Description (Concrete-In-Place, etc.)*

Each line item is described in detail. Sub-items and additional sizes are indented beneath the appropriate line items. The first line or two after the main item (in boldface) may contain descriptive information that pertains to all line items beneath this boldface listing.

4 *Reference Number Information*

R033053 -10 You'll see reference numbers shown in bold rectangles at the beginning of some sections. These refer to related items in the Reference Section, visually identified by a vertical gray bar on the page edges.

The relation may be: (1) an estimating procedure that should be read before estimating, (2) an alternate pricing method, or (3) technical information.

The "R" designates the Reference Section. The numbers refer to the MasterFormat classification system.

It is strongly recommended that you review all reference numbers that appear within the section in which you are working.

Note: **Not all reference numbers appear in all Means publications.**

03300 | Cast-In-Place Concrete

03310 | Structural Concrete				CREW	DAILY OUTPUT	LABOR-HOURS	UNIT	2006 BARE COSTS MAT.	LABOR	EQUIP.	TOTAL	TOTAL INCL O&P	
240	0010	**CONCRETE IN PLACE**	R033053 -10										240
	0020	Including forms (4 uses), co[illegible]lacemet, rei[illegible]g											
	0050	steel and finishing unless othe[illegible]ndicated	R033053 -50										240
	0300	Beams, 5 kip per L.F., 10' span		C-14A	15.62	12.804	C.Y.	287	455		788	1,125	
	3800	Footings, spread under 1 C.Y.		C-14C	38.07	2.942	[illegible]	175	99.50	.56	275.06	[illegible]	
	3850	Over 5 C.Y.			81.04	1.382		242	47	.26	289.26	[illegible]	
	3900	Footings, strip, 18" x 9", unreinfo[illegible]			40	2.800		109	95	.53	204.53	278	
	3920	18" x 9", reinforced			35	3.200		130	108	.61	238.61	325	
	3925	20" x 10", unreinforced			45	2.489		106	84.50	.47	190.97	257	
	3930	20" x 10", reinforced	R033053 -10	C-14C	40	2.800	C.Y.	122	95	.53	217.53	293	
	3935	24" x 12", unreinforced			55	2.036		103	69	.39	172.39	228	
	3940	24" x 12", reinforced	R033053 -50		48	2.333		121	79	.44	200.44	264	
	3945	36" x 12", unreinforced			70	[illegible]		99.50	54	.30	153.80	199	

5 *Crew (C-14C)*

The "Crew" column designates the typical trade or crew used to install the item. If an installation can be accomplished by one trade and requires no power equipment, that trade and the number of workers are listed (for example, "2 Carpenters"). If an installation requires a composite crew, a crew code designation is listed (for example, "C-14C"). You'll find full details on all composite crews in the Crew Listings.

- For a complete list of all trades utilized in this book and their abbreviations, see the inside back cover.

Crews

Crew No.	Bare Costs		Incl. Subs O & P		Cost Per Labor-Hour	
Crew C-14C	Hr.	Daily	Hr.	Daily	Bare Costs	Incl. O&P
1 Carpenter Foreman (out)	$37.55	$300.40	$62.20	$497.60	$33.85	$56.21
6 Carpenters	35.55	1706.40	58.90	2827.20		
2 Rodmen (reinf.)	39.50	632.00	67.80	1084.80		
4 Laborers	27.40	876.80	45.40	1452.80		
1 Cement Finisher	34.40	275.20	54.20	433.60		
1 Gas Engine Vibrator		21.40		23.55	.19	.21
112 L.H., Daily Totals		$3812.20		$6319.55	$34.04	$56.42

Productivity: Daily Output (35)/Labor-Hours (3.200)

The "Daily Output" represents the typical number of units the designated crew will install in a normal 8-hour day. To find out the number of days the given crew would require to complete the installation, divide your quantity by the daily output. For example:

Quantity	÷	Daily Output	=	Duration
100 C.Y.	÷	35/ Crew Day	=	2.86 Crew Days

The "Labor-Hours" figure represents the number of labor-hours required to install one unit of work. To find out the number of labor-hours required for your particular task, multiply the quantity of the item times the number of labor-hours shown. For example:

Quantity	x	Productivity Rate	=	Duration
100 C.Y.	x	3.200 Labor-Hours/ C.Y.	=	320 Labor-Hours

Unit (C.Y.)

The abbreviated designation indicates the unit of measure upon which the price, production, and crew are based (C.Y. = Cubic Yard). For a complete listing of abbreviations refer to the Abbreviations Listing in the Reference Section of this book.

Bare Costs:

Mat. (Bare Material Cost) (130.00)

The unit material cost is the "bare" material cost with no overhead and profit included. *Costs shown reflect national average material prices for January of the current year and include delivery to the job site. No sales taxes are included.*

Labor (108.00)

The unit labor cost is derived by multiplying bare labor-hour costs for Crew C-14C by labor-hour units. The bare labor-hour cost is found in the Crew Section under C-14C. (If a trade is listed, the hourly labor cost—the wage rate—is found on the inside back cover.)

Labor-Hour Cost Crew C-14C	x	Labor-Hour Units	=	Labor
$33.85	x	3.200	=	$108.45

Equip. (Equipment) (.61)

Equipment costs for each crew are listed in the description of each crew. Tools or equipment whose value justifies purchase or ownership by a contractor are considered overhead as shown on the inside back cover. The unit equipment cost is derived by multiplying the bare equipment hourly cost by the labor-hour units.

Equipment Cost Crew C-14C	x	Labor-Hour Units	=	Equip.
.19	x	3.20	=	.61

Total (238.61)

The total of the bare costs is the arithmetic total of the three previous columns: mat., labor, and equip.

Material	+	Labor	+	Equip.	=	Total
$130	+	$108.00	+	$.61	=	$238.61

Total Costs Including O&P

This figure is the sum of the bare material cost plus 10% for profit; the bare labor cost plus total overhead and profit (per the inside back cover or, if a crew is listed, from the crew listings); and the bare equipment cost plus 10% for profit.

Material is Bare Material Cost + 10% = 130 + 13.00	=	$143.00
Labor for Crew C-14C = Labor-Hour Cost (56.21) x Labor-Hour Units (3.20)	=	$179.87
Equip. is Bare Equip. Cost + 10% = .61 + .06	=	$ .67
Total (Rounded)	=	$325

Division 1 General Requirements

Estimating Tips

The General Requirements of any contract are very important to both the bidder and the owner. These lay the ground rules under which the contract will be executed and have a significant influence on the cost of operations. Therefore, it is extremely important to thoroughly read and understand the General Requirements, both before preparing an estimate and when the estimate is complete, to ascertain that nothing in the contract is overlooked. Caution should be exercised when applying items listed in Division 1 to an estimate. Many of the items are included in the unit prices listed in the other divisions such as markups on labor and company overhead.

01200 Price & Payment Procedures

- When estimating historic preservation projects (depending on the condition of the existing structure and the owner's requirements), a 15%-20% contingency or allowance is recommended, regardless of the stage of the drawings.

01300 Administrative Requirements

- Before determining a final cost estimate, it is a good practice to review all the items listed in Subdivision 01300 to make final adjustments for items that may need customizing to specific job conditions.
- Historic preservation projects may require specialty labor and methods, as well as extra time, to protect existing materials that must be preserved and/or restored. Some additional expenses may be incurred in architectural fees for facility surveys and other special inspections and analyses.
- Requirements for initial and periodic submittals can represent a significant cost to the General Requirements of a job. Thoroughly check the submittal specifications when estimating a project to determine any costs that should be included.

01400 Quality Requirements

- All projects will require some degree of Quality Control. This cost is not included in the unit cost of construction listed in each division. Depending upon the terms of the contract, the various costs of inspection and testing can be the responsibility of either the owner or the contractor. Be sure to include the required costs in your estimate.

01500 Temporary Facilities & Controls

- Barricades, access roads, safety nets, scaffolding, security, and many more requirements for the execution of a safe project are elements of direct cost. These costs can easily be overlooked when preparing an estimate. When looking through the major classifications of this subdivision, determine which items apply to each division in your estimate.

01740 Execution Requirements

- When preparing an estimate, thoroughly read the specifications to determine the requirements for Contract Closeout. Final cleaning, record documentation, operation and maintenance data, warranties and bonds, and spare parts and maintenance materials can all be elements of cost for the completion of a contract. Do not overlook these in your estimate.

01830 Operations & Maintenance

- If maintenance and repair are included in your contract, they require special attention. Estimating the cost to remove and replace any unit usually requires a site visit to determine the accessibility and the specific difficulty at that location. Obstructions, dust control, safety, and often, overtime hours must be considered when preparing your estimate.

Reference Numbers

Reference numbers are shown in bold squares at the beginning of some major classifications. These numbers refer to related items in the Reference Section. The reference information may be an estimating procedure, an alternate pricing method, or technical information.

Note: Not all subdivisions listed here necessarily appear in this publication.

01100 | Summary

		01103 | Models & Renderings	CREW	DAILY OUTPUT	LABOR-HOURS	UNIT	2006 BARE COSTS MAT.	LABOR	EQUIP.	TOTAL	TOTAL INCL O&P	
500	0010	**RENDERINGS** Color, matted, 20" x 30", eye level,										500
	0020	1 building, minimum				Ea.	1,850			1,850	2,050	
	0050	Average					2,650			2,650	2,925	
	0100	Maximum					4,250			4,250	4,675	
	1000	5 buildings, minimum					3,725			3,725	4,075	
	1100	Maximum				↓	7,425			7,425	8,175	
		01107 | Professional Consultant										
100	0011	**ARCHITECTURAL FEES** R011110-10										100
	0020	For new construction										
	0060	Minimum				Project					4.90%	
	0090	Maximum									16%	
	0100	For alteration work, to $500,000, add to fee									50%	
	0150	Over $500,000, add to fee				↓					25%	
200	0010	**CONSTRUCTION MANAGEMENT FEES**										200
	0060	For work to $10,000				Project					10%	
	0070	To $25,000									9%	
	0090	To $100,000									6%	
	0100	To $500,000									5%	
	0110	To $1,000,000				↓					4%	
300	0010	**ENGINEERING FEES** R011110-30										300
	0020	Educational planning consultant, minimum				Project					.50%	
	0100	Maximum				"					2.50%	
	0400	Elevator & conveying systems, minimum				Contrct					2.50%	
	0500	Maximum									5%	
	1000	Mechanical (plumbing & HVAC), minimum									4.10%	
	1100	Maximum				↓					10.10%	
	1200	Structural, minimum				Project					1%	
	1300	Maximum				"					2.50%	
700	0010	**TOPOGRAPHICAL SURVEYS**										700
	0020	Minimum	A-7	3.30	7.273	Acre	16.80	283	17.65	317.45	495	
	0100	Maximum	A-8	.60	53.333		50.50	2,000	97	2,147.50	3,425	
	0300	Lot location and lines, minimum, for large quantities	A-7	2	12		26.50	465	29	520.50	815	
	0320	Average	"	1.25	19.200		47.50	745	46.50	839	1,300	
	0400	Maximum, for small quantities	A-8	1	32	↓	75.50	1,200	58	1,333.50	2,100	
	1100	Crew for layout of building, trenching or pipe laying, 2 person crew	A-6	1	16	Day		565	58	623	985	
	1200	3 person crew	A-7	1	24			930	58.50	988.50	1,575	
	1400	Crew for roadway layout, 4 person crew	A-8	1	32	↓		1,200	58	1,258	2,025	

01200 | Price & Payment Procedures

		01250 | Contract Modification Procedures	CREW	DAILY OUTPUT	LABOR-HOURS	UNIT	2006 BARE COSTS MAT.	LABOR	EQUIP.	TOTAL	TOTAL INCL O&P	
200	0010	**CONTINGENCIES**										200
	0020	For estimate at conceptual stage				Project					25%	
	0050	Schematic stage									20%	
	0100	Preliminary working drawing stage (Design Dev.)									15%	
	0150	Final working drawing stage				↓					8%	

01200 | Price & Payment Procedures

01250 | Contract Modification Procedures

			CREW	DAILY OUTPUT	LABOR-HOURS	UNIT	2006 BARE COSTS MAT.	LABOR	EQUIP.	TOTAL	TOTAL INCL O&P	
400	0010	**FACTORS** Cost adjustments R012153-10										400
	0100	Add to construction costs for particular job requirements										
	0500	Cut & patch to match existing construction, add, minimum				Costs	2%	3%				
	0550	Maximum					5%	9%				
	0800	Dust protection, add, minimum					1%	2%				
	0850	Maximum					4%	11%				
	1100	Equipment usage curtailment, add, minimum					1%	1%				
	1150	Maximum					3%	10%				
	1400	Material handling & storage limitation, add, minimum					1%	1%				
	1450	Maximum					6%	7%				
	1700	Protection of existing work, add, minimum					2%	2%				
	1750	Maximum					5%	7%				
	2000	Shift work requirements, add, minimum						5%				
	2050	Maximum						30%				
	2300	Temporary shoring and bracing, add, minimum					2%	5%				
	2350	Maximum					5%	12%				
	2400	Work inside prisons and high security areas, add, minimum						30%				
	2450	Maximum				↓		50%				
500	0010	**JOB CONDITIONS** Modifications to total										500
	0020	project cost summaries										
	0100	Economic conditions, favorable, deduct				Project					2%	
	0200	Unfavorable, add									5%	
	0300	Hoisting conditions, favorable, deduct									2%	
	0400	Unfavorable, add									5%	
	0500	General Contractor management, experienced, deduct									2%	
	0600	Inexperienced, add									10%	
	0700	Labor availability, surplus, deduct									1%	
	0800	Shortage, add									10%	
	0900	Material storage area, available, deduct									1%	
	1000	Not available, add									2%	
	1100	Subcontractor availability, surplus, deduct									5%	
	1200	Shortage, add									12%	
	1300	Work space, available, deduct									2%	
	1400	Not available, add				↓					5%	

01290 | Payment Procedures

			CREW	DAILY OUTPUT	LABOR-HOURS	UNIT	MAT.	LABOR	EQUIP.	TOTAL	TOTAL INCL O&P	
800	0010	**TAXES** R012909-80										800
	0020	Sales tax, state, average				%	4.83%					
	0050	Maximum R012909-85					7%					
	0200	Social Security, on first $90,000 of wages						7.65%				
	0300	Unemployment, combined Federal and State, minimum						.80%				
	0350	Average						6.20%				
	0400	Maximum				↓		11.76%				

01300 | Administrative Requirements

01310 | Project Management/Coordination

			CREW	DAILY OUTPUT	LABOR-HOURS	UNIT	2006 BARE COSTS MAT.	LABOR	EQUIP.	TOTAL	TOTAL INCL O&P	
150	0010	**PERMITS**										150
	0020	Rule of thumb, most cities, minimum				Job					.50%	

01300 | Administrative Requirements

		01310 | Project Management/Coordination	CREW	DAILY OUTPUT	LABOR-HOURS	UNIT	2006 BARE COSTS MAT.	LABOR	EQUIP.	TOTAL	TOTAL INCL O&P	
150	0100	Maximum				Job					2%	150
350	0010	**INSURANCE**										350
	0020	Builders risk, standard, minimum (R013113-40)				Job					.24%	
	0050	Maximum									.64%	
	0200	All-risk type, minimum (R013113-60)									.25%	
	0250	Maximum									.62%	
	0400	Contractor's equipment floater, minimum				Value					.50%	
	0450	Maximum				″					1.50%	
	0600	Public liability, average				Job					2.02%	
	0810	Workers' compensation & employer's liability										
	2000	Range of 35 trades in 50 states, excl. wrecking, min.				Payroll		2.40%				
	2100	Average						16.30%				
	2200	Maximum						137.60%				
400	0010	**MAIN OFFICE EXPENSE** Average for General Contractors										400
	0020	As a percentage of their annual volume										
	0030	Annual volume to $50,000, minimum				% Vol.				20%		
	0040	Maximum								30%		
	0060	To $100,000, minimum								17%		
	0070	Maximum								22%		
	0080	To $250,000, minimum								16%		
	0090	Maximum								19%		
	0110	To $500,000, minimum								14%		
	0120	Maximum								16%		
	0130	To $1,000,000, minimum								8%		
	0140	Maximum								10%		
500	0010	**MARK-UP** For General Contractors for change										500
	0100	of scope of job as bid										
	0200	Extra work, by subcontractors, add				%					10%	
	0250	By General Contractor, add									15%	
	0400	Omitted work, by subcontractors, deduct all but									5%	
	0450	By General Contractor, deduct all but									7.50%	
	0600	Overtime work, by subcontractors, add									15%	
	0650	By General Contractor, add									10%	
	1000	Installing contractors, on his own labor, minimum						53.70%				
	1100	Maximum						97.30%				
600	0010	**OVERHEAD** (R013113-50)										600
	0020	As percent of direct costs, minimum				%				5%		
	0050	Average								16%		
	0100	Maximum								30%		
620	0010	**OVERHEAD & PROFIT** Allowance to add to items in this (R013113-50)										620
	0020	book that do not include Subs' O&P, average				%				30%		
	0100	Allowance to add to items in this book that										
	0110	do include Subs' O&P, minimum				%					5%	
	0150	Average									10%	
	0200	Maximum									15%	
	0290	Typical, by size of project, under $50,000								40%		
	0310	$50,000 to $100,000								35%		
	0320	$100,000 to $500,000								25%		
	0330	$500,000 to $1,000,000								20%		
700	0010	**FIELD PERSONNEL**										700
	0020	Clerk, average				Week		335		335	515	
	0180	Project manager, minimum						1,500		1,500	2,300	
	0200	Average						1,700		1,700	2,600	
	0220	Maximum						1,950		1,950	3,000	
	0240	Superintendent, minimum						1,450		1,450	2,225	

01300 | Administrative Requirements

	01310 | Project Management/Coordination	CREW	DAILY OUTPUT	LABOR-HOURS	UNIT	2006 BARE COSTS MAT.	LABOR	EQUIP.	TOTAL	TOTAL INCL O&P	
700	0260 Average				Week		1,575		1,575	2,425	700
	0280 Maximum				↓		1,800		1,800	2,750	

01400 | Quality Requirements

	01450 | Quality Control	CREW	DAILY OUTPUT	LABOR-HOURS	UNIT	2006 BARE COSTS MAT.	LABOR	EQUIP.	TOTAL	TOTAL INCL O&P	
500	0010 **TESTING** and Inspectional Services										500
	1800 Compressive test, cylinder, delivered to lab, ASTM C 39				Ea.					13	
	1900 Picked up by lab, minimum									15	
	1950 Average									20	
	2000 Maximum									30	
	2200 Compressive strength, cores (not incl. drilling), ASTM C 42									40	
	2300 Patching core holes				↓					24	
	4730 Soil testing										
	4735 Soil density, nuclear method, ASTM D2922				Ea.					38.67	
	4740 Sand cone method ASTM D1556									30.17	
	4750 Moisture content, ASTM D 2216									10	
	4780 Permeability test, double ring infiltrometer									550	
	4800 Permeability, var. or constant head, undist., ASTM D 2434									250	
	4850 Recompacted									275	
	4900 Proctor compaction, 4" standard mold, ASTM D 698									135	
	4950 6" modified mold									75	
	5100 Shear tests, triaxial, minimum									450	
	5150 Maximum									600	
	5550 Technician for inspection, per day, earthwork									220	
	5650 Bolting									280	
	5750 Roofing									255	
	5790 Welding				↓					270	
	5820 Non-destructive testing, dye penetrant				Day					330	
	5840 Magnetic particle									330	
	5860 Radiography									495	
	5880 Ultrasonic				↓					340	
	6000 Welding certification, minimum				Ea.					100	
	6100 Maximum				"					275	
	7000 Underground storage tank										
	7500 Volumetric tightness test, <=12,000 gal				Ea.					400	
	7510 <=30,000 gal				"					660	
	7600 Vadose zone (soil gas) sampling, 10-40 samples, min.				Day					1,500	
	7610 Maximum				"					2,500	
	7700 Ground water monitoring incl. drilling 3 wells, min.				Total					5,000	
	7710 Maximum				"					7,000	
	8000 X-ray concrete slabs				Ea.					200	

01500 | Temporary Facilities & Controls

		01510 | Temporary Utilities	CREW	DAILY OUTPUT	LABOR-HOURS	UNIT	2006 BARE COSTS MAT.	LABOR	EQUIP.	TOTAL	TOTAL INCL O&P	
800	0010	**TEMPORARY UTILITIES**										800
	0100	Heat, incl. fuel and operation, per week, 12 hrs. per day	1 Skwk	100	.080	CSF Flr	7.20	2.92		10.12	12.75	
	0200	24 hrs. per day	"	60	.133		10.80	4.87		15.67	19.85	
	0350	Lighting, incl. service lamps, wiring & outlets, minimum	1 Elec	34	.235		2.39	9.90		12.29	17.85	
	0360	Maximum	"	17	.471		5.20	19.75		24.95	36.50	
	0400	Power for temp lighting only, per month, min/month 6.6 KWH								.75	1.18	
	0450	Maximum/month 23.6 KWH								2.85	3.14	
	0600	Power for job duration incl. elevator, etc., minimum								47	51.70	
	0650	Maximum								110	121	

		01520 | Construction Facilities	CREW	DAILY OUTPUT	LABOR-HOURS	UNIT	MAT.	LABOR	EQUIP.	TOTAL	TOTAL INCL O&P	
500	0010	**OFFICE**										500
	0020	Trailer, furnished, no hookups, 20′ x 8′, buy	2 Skwk	1	16	Ea.	7,550	585		8,135	9,250	
	0250	Rent per month					167			167	183	
	0300	32′ x 8′, buy	2 Skwk	.70	22.857		10,400	835		11,235	12,800	
	0350	Rent per month					183			183	201	
	0400	50′ x 10′, buy	2 Skwk	.60	26.667		18,100	975		19,075	21,600	
	0450	Rent per month					293			293	320	
	0500	50′ x 12′, buy	2 Skwk	.50	32		23,300	1,175		24,475	27,500	
	0550	Rent per month					340			340	375	
	0700	For air conditioning, rent per month, add					41			41	45	
	0800	For delivery, add per mile				Mile	3.50			3.50	3.85	
	1200	Storage boxes, 20′ x 8′, buy	2 Skwk	1.80	8.889	Ea.	3,250	325		3,575	4,100	
	1250	Rent per month					75			75	82.50	
	1300	40′ x 8′, buy	2 Skwk	1.40	11.429		4,275	415		4,690	5,375	
	1350	Rent per month					109			109	120	

		01530 | Temporary Construction	CREW	DAILY OUTPUT	LABOR-HOURS	UNIT	MAT.	LABOR	EQUIP.	TOTAL	TOTAL INCL O&P	
700	0010	**PROTECTION**										700
	0020	Stair tread, 2″ x 12″ planks, 1 use	1 Carp	75	.107	Tread	4.80	3.79		8.59	11.60	
	0100	Exterior plywood, 1/2″ thick, 1 use		65	.123		1.84	4.38		6.22	9.30	
	0200	3/4″ thick, 1 use		60	.133		2.46	4.74		7.20	10.55	
900	0010	**WINTER PROTECTION** Reinforced plastic on wood										900
	0100	framing to close openings	2 Clab	750	.021	S.F.	.39	.58		.97	1.40	
	0200	Tarpaulins hung over scaffolding, 8 uses, not incl. scaffolding		1500	.011		.19	.29		.48	.69	
	0250	Tarpaulin polyester reinf. w/integral fastening system 11 mils thick		1600	.010		.77	.27		1.04	1.30	
	0300	Prefab. fiberglass panels, steel frame, 8 uses		1200	.013		.75	.37		1.12	1.44	

		01540 | Construction Aids	CREW	DAILY OUTPUT	LABOR-HOURS	UNIT	MAT.	LABOR	EQUIP.	TOTAL	TOTAL INCL O&P	
550	0010	**PUMP STAGING**, Aluminum (R015423-20)										550
	0200	24′ long pole section, buy				Ea.	360			360	400	
	0300	18′ long pole section, buy					278			278	305	
	0400	12′ long pole section, buy					191			191	210	
	0500	6′ long pole section, buy					100			100	110	
	0600	6′ long splice joint section, buy					73.50			73.50	81	
	0700	Pump jack					119			119	131	
	0900	Foldable brace					51.50			51.50	56.50	
	1000	Workbench/back safety rail support					63			63	69	
	1100	Scaffolding planks/workbench, 14″ wide x 24′ long					580			580	640	
	1200	Plank end safety rail					196			196	215	
	1250	Safety net, 22′ long					285			285	315	
	1300	System in place, 50′ working height, per use based on 50 uses	2 Carp	84.80	.189	C.S.F.	5.30	6.70		12	16.95	
	1400	100 uses		84.80	.189		2.66	6.70		9.36	14	

		01540 | Construction Aids	CREW	DAILY OUTPUT	LABOR-HOURS	UNIT	2006 BARE COSTS MAT.	LABOR	EQUIP.	TOTAL	TOTAL INCL O&P	
550	1500	150 uses (R015423-20)	2 Carp	84.80	.189	C.S.F.	1.78	6.70		8.48	13.05	550
750	0010	SCAFFOLDING (R015423-10)										750
	0015	Steel tube, reg., rent/mo., no plank, incl. erect or dismantle										
	0090	Building exterior, wall face, 1 to 5 stories, 6'-4" x 5' frames	3 Carp	24	1	C.S.F.	25.50	35.50		61	87	
	0200	6 to 12 stories	4 Carp	21.20	1.509		25.50	53.50		79	117	
	0310	13 to 20 stories	5 Carp	20	2		25.50	71		96.50	146	
	0460	Building interior, wall face area, up to 16' high	3 Carp	25	.960		25.50	34		59.50	84.50	
	0560	16' to 40' high		23	1.043		25.50	37		62.50	89.50	
	0800	Building interior floor area, up to 30' high		312	.077	C.C.F.	2.70	2.73		5.43	7.50	
	0900	Over 30' high	4 Carp	275	.116	"	2.70	4.14		6.84	9.80	
	0910	Steel tubular, heavy duty shoring, buy										
	0920	Frames 5' high 2' wide				Ea.	86.50			86.50	95.50	
	0925	5' high 4' wide					98			98	108	
	0930	6' high 2' wide					99.50			99.50	109	
	0935	6' high 4' wide					117			117	128	
	0940	Accessories										
	0945	Cross braces				Ea.	18.50			18.50	20.50	
	0950	U-head, 8" x 8"					20			20	22	
	0955	J-head, 4" x 8"					14.80			14.80	16.25	
	0960	Base plate, 8" x 8"					16.40			16.40	18.05	
	0965	Leveling jack					35			35	39	
	1000	Steel tubular, regular, buy										
	1100	Frames 3' high 5' wide				Ea.	67			67	73.50	
	1150	5' high 5' wide					77.50			77.50	85	
	1200	6'-4" high 5' wide					97			97	107	
	1350	7'-6" high 6' wide					167			167	184	
	1500	Accessories cross braces					17.35			17.35	19.05	
	1550	Guardrail post					17.35			17.35	19.05	
	1600	Guardrail 7' section					8.40			8.40	9.20	
	1650	Screw jacks & plates					27.50			27.50	30.50	
	1700	Sidearm brackets					32.50			32.50	35.50	
	1750	8" casters					38			38	42	
	1800	Plank 2" x 10" x 16'-0"					49			49	54	
	1900	Stairway section					283			283	310	
	1910	Stairway starter bar					33.50			33.50	37	
	1920	Stairway inside handrail					61			61	67.50	
	1930	Stairway outside handrail					84.50			84.50	93	
	1940	Walk-thru frame guardrail					42.50			42.50	46.50	
	2000	Steel tubular, regular, rent/mo.										
	2100	Frames 3' high 5' wide				Ea.	3.94			3.94	4.33	
	2150	5' high 5' wide					3.94			3.94	4.33	
	2200	6'-4" high 5' wide					3.94			3.94	4.33	
	2250	7'-6" high 6' wide					7.35			7.35	8.10	
	2500	Accessories, cross braces					.63			.63	.69	
	2550	Guardrail post					1.05			1.05	1.16	
	2600	Guardrail 7' section					.79			.79	.87	
	2650	Screw jacks & plates					1.58			1.58	1.74	
	2700	Sidearm brackets					1.58			1.58	1.74	
	2750	8" casters					6.30			6.30	6.95	
	2800	Outrigger for rolling tower					3.15			3.15	3.47	
	2850	Plank 2" x 10" x 16'-0"					5.25			5.25	5.80	
	2900	Stairway section					10.50			10.50	11.55	
	2910	Stairway starter bar					.11			.11	.12	
	2920	Stairway inside handrail					5.25			5.25	5.80	
	2930	Stairway outside handrail					5.25			5.25	5.80	

	01540 Construction Aids		CREW	DAILY OUTPUT	LABOR-HOURS	UNIT	2006 BARE COSTS MAT.	LABOR	EQUIP.	TOTAL	TOTAL INCL O&P	
750	2940	Walk-thru frame guardrail R015423-10				Ea.	2.10			2.10	2.31	750
	3000	Steel tubular, heavy duty shoring, rent/mo.										
	3250	5′ high 2′ & 4′ wide				Ea.	5.25			5.25	5.80	
	3300	6′ high 2′ & 4′ wide					5.25			5.25	5.80	
	3500	Accessories, cross braces					1.05			1.05	1.16	
	3600	U - head, 8″ x 8″					1.05			1.05	1.16	
	3650	J - head, 4″ x 8″					1.05			1.05	1.16	
	3700	Base plate, 8″ x 8″					1.05			1.05	1.16	
	3750	Leveling jack					2.10			2.10	2.31	
	5700	Planks, 2x10x16′-0″, labor only, erect/remove to 50′ H	3 Carp	144	.167			5.95		5.95	9.80	
	5800	Over 50′ high	4 Carp	160	.200			7.10		7.10	11.80	
755	0010	**SCAFFOLDING SPECIALTIES**										755
	1200	Sidewalk bridge, heavy duty steel posts & beams, including										
	1210	parapet protection & waterproofing										
	1220	8′ to 10′ wide, 2 posts	3 Carp	15	1.600	L.F.	33.50	57		90.50	131	
	1230	3 posts	"	10	2.400	"	51.50	85.50		137	198	
	1500	Sidewalk bridge using tubular steel										
	1512	scaffold frames, including planking	3 Carp	55	.436	L.F.	4.96	15.50		20.46	31	
	1600	For 2 uses per month, deduct from all above					50%					
	1700	For 1 use every 2 months, add to all above					100%					
	1900	Catwalks, 20″ wide, no guardrails, 7′ span, buy				Ea.	144			144	159	
	2000	10′ span, buy				"	185			185	203	
	2800	Hand winch-operated masons scaffolding, no plank										
	2810	plank moving not required										
	2900	98′ long, 10′-6″ high, buy				Ea.	30,300			30,300	33,300	
	3000	Rent per month					1,200			1,200	1,325	
	3100	28′-6″ high, buy					37,200			37,200	40,900	
	3200	Rent per month					1,475			1,475	1,625	
	3400	196′ long, 28′-6″ high, buy					72,000			72,000	79,000	
	3500	Rent per month					2,875			2,875	3,150	
	3600	64′-6″ high, buy					98,500			98,500	108,500	
	3700	Rent per month					3,950			3,950	4,325	
	3720	Putlog, standard, 8′ span, with hangers, buy					70.50			70.50	77.50	
	3730	Rent per month					10.50			10.50	11.55	
	3750	12′ span, buy					106			106	117	
	3755	Rent per month					15.75			15.75	17.35	
	3760	Trussed type, 16′ span, buy					243			243	267	
	3770	Rent per month					21			21	23	
	3790	22′ span, buy					291			291	320	
	3795	Rent per month					31.50			31.50	34.50	
	4000	7 step					660			660	725	
	4100	Rolling towers, buy, 5′ wide, 7′ long, 10′ high					1,325			1,325	1,450	
	4200	For 5′ high added sections, to buy, add					219			219	241	
	4300	Complete incl. wheels, railings, outriggers,										
	4350	21′ high, to buy				Ea.	2,225			2,225	2,450	
	4400	Rent/month				"	167			167	184	
760	0010	**STAGING AIDS** and fall protection equipment										760
	0100	Sidewall staging bracket, tubular, buy				Ea.	34			34	37	
	0110	Cost each per day, based on 250 days' use				Day	.14			.14	.15	
	0200	Guard post, buy				Ea.	16.50			16.50	18.15	
	0210	Cost each per day, based on 250 days' use				Day	.07			.07	.07	
	0300	End guard chains, buy per pair				Pair	27.50			27.50	30.50	
	0310	Cost per set per day, based on 250 days' use				Day	.14			.14	.15	
	1010	Cost each per day, based on 250 days' use				"	.03			.03	.03	

01540 | Construction Aids

		01540 Construction Aids	CREW	DAILY OUTPUT	LABOR-HOURS	UNIT	2006 BARE COSTS MAT.	LABOR	EQUIP.	TOTAL	TOTAL INCL O&P	
760	1100	Wood bracket, buy				Ea.	14.20			14.20	15.60	760
	1110	Cost each per day, based on 250 days' use				Day	.06			.06	.06	
	2010	Cost per pair per day, based on 250 days' use				"	.36			.36	.39	
	2100	Steel siderail jack, buy per pair				Pair	69.50			69.50	76	
	2110	Cost per pair per day, based on 250 days' use				Day	.28			.28	.30	
	3010	Cost each per day, based on 250 days' use				"	.19			.19	.21	
	3100	Aluminum scaffolding plank, 20" wide x 24' long, buy				Ea.	725			725	795	
	3110	Cost each per day, based on 250 days' use				Day	2.89			2.89	3.18	
	4000	Nylon full-body harness, lanyard and rope grab				Ea.	218			218	239	
	4010	Cost each per day, based on 250 days' use				Day	.87			.87	.96	
	4100	Rope for safety line, 5/8" x 100' nylon, buy				Ea.	45			45	49.50	
	4110	Cost each per day, based on 250 days' use				Day	.18			.18	.20	
	4200	Permanent U-Bolt roof anchor, buy				Ea.	32.50			32.50	36	
	4300	Temporary (one use) roof ridge anchor, buy				"	26.50			26.50	29	
	5000	Installation (setup and removal) of staging aids										
	5010	Sidewall staging bracket	2 Carp	64	.250	Ea.		8.90		8.90	14.75	
	5020	Guard post with 2 wood rails	"	64	.250			8.90		8.90	14.75	
	5030	End guard chains, set	1 Carp	64	.125			4.44		4.44	7.35	
	5100	Roof shingling bracket		96	.083			2.96		2.96	4.91	
	5200	Ladder jack		64	.125			4.44		4.44	7.35	
	5300	Wood plank, 2x10x16'	2 Carp	80	.200			7.10		7.10	11.80	
	5310	Aluminum scaffold plank, 20" x 24'	"	40	.400			14.20		14.20	23.50	
	5410	Safety rope	1 Carp	40	.200			7.10		7.10	11.80	
	5420	Permanent U-Bolt roof anchor (install only)	2 Carp	40	.400			14.20		14.20	23.50	
	5430	Temporary roof ridge anchor (install only)	1 Carp	64	.125			4.44		4.44	7.35	
780	0010	**SWING STAGING**, 500 lb cap., 2' wide to 24' long, hand-operated hoist										780
	0020	steel cable type, with 60' cables, buy				Ea.	4,675			4,675	5,150	
	0030	Rent per month				"	470			470	515	
	0600	Lightweight (not for masons) 24' long for 150' height										
	0610	manual type, buy				Ea.	4,950			4,950	5,450	
	0620	Rent per month					495			495	545	
	0700	Powered, electric or air, to 150' high, buy					17,500			17,500	19,300	
	0710	Rent per month					1,225			1,225	1,350	
	0780	To 300' high, buy					20,600			20,600	22,600	
	0800	Rent per month					1,450			1,450	1,575	
	1000	Bosun's chair or work basket 3' x 3.5', to 300' high, electric, buy					7,725			7,725	8,500	
	1010	Rent per month					540			540	595	
	2200	Move swing staging (setup and remove)	E-4	2	16	Move		645	44.50	689.50	1,250	
800	0010	**TARPAULINS**										800
	0020	Cotton duck, 10 oz. to 13.13 oz. per S.Y., minimum				S.F.	.50			.50	.55	
	0050	Maximum					.58			.58	.64	
	0100	Polyvinyl-coated nylon, 14 oz. to 18 oz., minimum					.48			.48	.53	
	0150	Maximum					.68			.68	.75	
	0200	Reinforced polyethylene 3 mils thick, white					.11			.11	.12	
	0300	4 mils thick, white, clear or black					.14			.14	.15	
	0400	5.5 mils thick, clear					.20			.20	.22	
	0500	White, fire retardant					.18			.18	.20	
	0600	7.5 mils, oil resistant, fire retardant					.19			.19	.21	
	0700	8.5 mils, black					.24			.24	.26	
	0720	Steel reinforced polyethylene, 4 mils thick					.53			.53	.58	
	0730	Polyester reinforced w/integral fastening system 11 mils thick					1.07			1.07	1.18	
	0740	Mylar polyester, non-reinforced, 7 mils thick					1.17			1.17	1.29	
820	0010	**SMALL TOOLS** R013113-50										820
	0020	As % of contractor's work, minimum				Total					.50%	

GENERAL REQUIREMENTS 1

01500 | Temporary Facilities & Controls

01540 | Construction Aids

			CREW	DAILY OUTPUT	LABOR-HOURS	UNIT	2006 BARE COSTS MAT.	LABOR	EQUIP.	TOTAL	TOTAL INCL O&P	
820	0100	Maximum R013113-50				Total					2%	820

01550 | Vehicular Access & Parking

			CREW	DAILY OUTPUT	LABOR-HOURS	UNIT	MAT.	LABOR	EQUIP.	TOTAL	TOTAL INCL O&P	
700	0010	**ROADS AND SIDEWALKS** Temporary										700
	2200	Sidewalks, 2″ x 12″ planks, 2 uses	1 Carp	350	.023	S.F.	.80	.81		1.61	2.23	
	2300	Exterior plywood, 2 uses, 1/2″ thick	↓	750	.011	↓	.31	.38		.69	.97	
	2400	5/8″ thick	↓	650	.012	↓	.33	.44		.77	1.10	
	2500	3/4″ thick	↓	600	.013	↓	.41	.47		.88	1.24	

01560 | Barriers & Enclosures

			CREW	DAILY OUTPUT	LABOR-HOURS	UNIT	MAT.	LABOR	EQUIP.	TOTAL	TOTAL INCL O&P	
100	0010	**BARRICADES**										100
	0020	5′ high, 3 rail @ 2″ x 8″, fixed	2 Carp	20	.800	L.F.	5.10	28.50		33.60	52.50	
	0150	Movable	"	30	.533	"	4.38	18.95		23.33	36.50	
	0300	Stock units, 6′ high, 8′ wide, plain, buy				Ea.	435			435	480	
	0350	With reflective tape, buy				"	525			525	580	
	0400	Break-a-way 3″ PVC pipe barricade										
	0410	with 3 ea. 1′ x 4′ reflectorized panels, buy				Ea.	305			305	335	
	0500	Plywood with steel legs, 32″ wide				↓	72			72	79	
	0600	Telescoping Christmas tree, 9′ high, 5 flags, buy				↓	122			122	134	
	0800	Traffic cones, PVC, 18″ high				↓	6.25			6.25	6.85	
	0850	28″ high				↓	19			19	21	
	0900	Barrels, 55 gal., with flasher	1 Clab	96	.083	↓	54	2.28		56.28	63.50	
	1000	Guardrail, wooden, 3′ high, 1″ x 6″, on 2″ x 4″ posts	2 Carp	200	.080	L.F.	1.04	2.84		3.88	5.85	
	1100	2″ x 6″, on 4″ x 4″ posts	"	165	.097	↓	1.91	3.45		5.36	7.80	
	1200	Portable metal with base pads, buy				↓	15.95			15.95	17.55	
	1250	Typical installation, assume 10 reuses	2 Carp	600	.027	↓	1.65	.95		2.60	3.39	
	1300	Barricade tape, polyethelyne, 7 mil, 3″ wide x 500′ long roll				Ea.	25			25	27.50	
250	0010	**TEMPORARY FENCING**										250
	0020	Chain link, 11 ga., 5′ high	2 Clab	400	.040	L.F.	4.79	1.10		5.89	7.05	
	0100	6′ high	↓	300	.053	↓	4.62	1.46		6.08	7.50	
	0200	Rented chain link, 6′ high, to 1,000′ (up to 12 mo.)	↓	400	.040	↓	3.03	1.10		4.13	5.15	
	0250	Over 1,000′ (up to 12 mo.)	↓	300	.053	↓	2.19	1.46		3.65	4.83	
	0350	Plywood, painted, 2″ x 4″ frame, 4′ high	A-4	135	.178	↓	5.15	6.10		11.25	15.70	
	0400	4″ x 4″ frame, 8′ high	"	110	.218	↓	9.85	7.50		17.35	23	
	0500	Wire mesh on 4″ x 4″ posts, 4′ high	2 Carp	100	.160	↓	8.15	5.70		13.85	18.35	
	0550	8′ high	"	80	.200	↓	12.60	7.10		19.70	25.50	
400	0010	**TEMPORARY CONSTRUCTION** See also Division 01530										400
800	0010	**WATCHMAN**										800
	0020	Service, monthly basis, uniformed person, minimum				Hr.					20	
	0100	Maximum				↓					50	
	0200	Person and command dog, minimum				↓					28	
	0300	Maximum				↓					60	
	0500	Sentry dog, leased, with job patrol (yard dog), 1 dog				Week					250	
	0600	2 dogs				"					320	
	0800	Purchase, trained sentry dog, minimum				Ea.					1,500	
	0900	Maximum				"					3,000	

01580 | Project Signs

			CREW	DAILY OUTPUT	LABOR-HOURS	UNIT	MAT.	LABOR	EQUIP.	TOTAL	TOTAL INCL O&P	
700	0010	**SIGNS**										700
	0020	Hi Intensity reflectorized, no posts, buy				S.F.	16.55			16.55	18.20	

01700 | Execution Requirements

	01740 | Cleaning		CREW	DAILY OUTPUT	LABOR-HOURS	UNIT	2006 BARE COSTS MAT.	LABOR	EQUIP.	TOTAL	TOTAL INCL O&P	
500	0010	CLEANING UP										500
	0020	After job completion, allow, minimum				Job					.30%	
	0040	Maximum				″					1%	
	0052	Cleanup of floor area, continuous, per day, during const.	A-5	16	1.125	M.S.F.	1.70	31	2.01	34.71	55	
	0100	Final by GC at end of job	″	11.50	1.565	″	2.71	43	2.80	48.51	77	

01800 | Facility Operation

	01810 | Commissioning		CREW	DAILY OUTPUT	LABOR-HOURS	UNIT	2006 BARE COSTS MAT.	LABOR	EQUIP.	TOTAL	TOTAL INCL O&P	
100	0010	**COMMISSIONING** Including documentation of design intent										100
	0100	performance verification, O&M, training, min				Project					.50%	
	0150	Maximum				″					.75%	

	01840 | Moving Equipment		CREW	DAILY OUTPUT	LABOR-HOURS	UNIT	MAT.	LABOR	EQUIP.	TOTAL	TOTAL INCL O&P	
100	0010	**MOVING EQUIPMENT**, Remove and reset, 100′ distance,										100
	0020	No obstructions, no assembly or leveling unless noted										
	0100	Annealing furnace, 24′ overall	B-67	4	4	Ea.		145	60.50	205.50	295	
	0200	Annealing oven, small		14	1.143			41.50	17.30	58.80	84	
	0240	Very large		1	16			580	242	822	1,175	
	0400	Band saw, small		12	1.333			48	20	68	98	
	0440	Large		8	2			72.50	30.50	103	148	
	0500	Blue print copy machine		7	2.286			82.50	34.50	117	168	
	0600	Bonding mill, 6″		7	2.286			82.50	34.50	117	168	
	0620	12″		6	2.667			96.50	40.50	137	197	
	0640	18″		4	4			145	60.50	205.50	295	
	0660	24″		2	8			289	121	410	590	
	0700	Boring machine (jig)	B-68	7	3.429			125	34.50	159.50	235	
	0800	Bridgeport mill, standard	B-67	14	1.143			41.50	17.30	58.80	84	
	1000	Calibrator, 6 unit	″	14	1.143			41.50	17.30	58.80	84	
	1100	Comparitor, bench top	2 Clab	14	1.143			31.50		31.50	52	
	1140	Floor mounted	B-67	7	2.286			82.50	34.50	117	168	
	1200	Computer, desk top	2 Clab	25	.640			17.55		17.55	29	
	1300	Copy machine	″	25	.640			17.55		17.55	29	
	1500	Deflasher	B-67	14	1.143			41.50	17.30	58.80	84	
	1600	Degreaser, small		14	1.143			41.50	17.30	58.80	84	
	1640	Large 24′ overall		1	16			580	242	822	1,175	
	1700	Desk with chair	2 Clab	25	.640			17.55		17.55	29	
	1800	Dial press	B-67	7	2.286			82.50	34.50	117	168	
	1900	Drafting table	2 Clab	14	1.143			31.50		31.50	52	
	2000	Drill press, bench top	″	14	1.143			31.50		31.50	52	
	2040	Floor mounted	B-67	14	1.143			41.50	17.30	58.80	84	
	2080	Industrial radial	″	7	2.286			82.50	34.50	117	168	
	2100	Dust collector, portable	2 Clab	25	.640			17.55		17.55	29	
	2140	Stationary, small	B-67	7	2.286			82.50	34.50	117	168	
	2180	Stationary, large	″	2	8			289	121	410	590	
	2300	Electric discharge machine	B-68	7	3.429			125	34.50	159.50	235	
	2400	Environmental chamber walls, including assembly	4 Clab	18	1.778	L.F.		48.50		48.50	80.50	
	2600	File cabinet	2 Clab	25	.640	Ea.		17.55		17.55	29	

01840 | Moving Equipment			CREW	DAILY OUTPUT	LABOR-HOURS	UNIT	MAT.	LABOR	EQUIP.	TOTAL	TOTAL INCL O&P	
100	2800	Grinder/sander, pedestal mount	B-67	14	1.143	Ea.		41.50	17.30	58.80	84	100
	3000	Hack saw, power	2 Clab	24	.667			18.25		18.25	30.50	
	3100	Hydraulic press	B-67	14	1.143			41.50	17.30	58.80	84	
	3500	Laminar flow tables	"	14	1.143			41.50	17.30	58.80	84	
	3600	Lathe, bench	2 Clab	14	1.143			31.50		31.50	52	
	3640	6"	B-67	14	1.143			41.50	17.30	58.80	84	
	3680	10"		13	1.231			44.50	18.65	63.15	90.50	
	3720	12"		12	1.333			48	20	68	98	
	4000	Milling machine		8	2			72.50	30.50	103	148	
	4100	Molding press, 25 ton		5	3.200			116	48.50	164.50	237	
	4140	60 ton		4	4			145	60.50	205.50	295	
	4180	100 ton		2	8			289	121	410	590	
	4220	150 ton		1.50	10.667			385	162	547	790	
	4260	200 ton		1	16			580	242	822	1,175	
	4300	300 ton		.75	21.333			770	325	1,095	1,575	
	4700	Oil pot stand		14	1.143			41.50	17.30	58.80	84	
	5000	Press, 10 ton		14	1.143			41.50	17.30	58.80	84	
	5040	15 ton		12	1.333			48	20	68	98	
	5080	20 ton		10	1.600			58	24	82	118	
	5120	30 ton		8	2			72.50	30.50	103	148	
	5160	45 ton		6	2.667			96.50	40.50	137	197	
	5200	60 ton		4	4			145	60.50	205.50	295	
	5240	75 ton		2.50	6.400			231	97	328	470	
	5280	100 ton	↓	2	8	↓		289	121	410	590	
	5500	Raised floor, including assembly	2 Carp	250	.064	S.F.		2.28		2.28	3.77	
	5600	Rolling mill, 6"	B-67	7	2.286	Ea.		82.50	34.50	117	168	
	5640	9"		6	2.667			96.50	40.50	137	197	
	5680	12"		4	4			145	60.50	205.50	295	
	5720	13"		3.50	4.571			165	69.50	234.50	335	
	5760	18"		2	8			289	121	410	590	
	5800	25"		1	16			580	242	822	1,175	
	6000	Sander, floor stand		14	1.143			41.50	17.30	58.80	84	
	6100	Screw machine		7	2.286			82.50	34.50	117	168	
	6200	Shaper, 16"		14	1.143			41.50	17.30	58.80	84	
	6300	Shear, power assist		4	4			145	60.50	205.50	295	
	6400	Slitter, 6"		14	1.143			41.50	17.30	58.80	84	
	6440	8"		13	1.231			44.50	18.65	63.15	90.50	
	6480	10"		12	1.333			48	20	68	98	
	6520	12"		11	1.455			52.50	22	74.50	108	
	6560	16"		10	1.600			58	24	82	118	
	6600	20"		8	2			72.50	30.50	103	148	
	6640	24"		6	2.667			96.50	40.50	137	197	
	6800	Snag and tap machine		7	2.286			82.50	34.50	117	168	
	6900	Solder machine (auto)	↓	7	2.286			82.50	34.50	117	168	
	7000	Storage cabinet metal, small	2 Clab	36	.444			12.20		12.20	20	
	7040	Large		25	.640			17.55		17.55	29	
	7100	Storage rack open, small		14	1.143			31.50		31.50	52	
	7140	Large	↓	7	2.286			62.50		62.50	104	
	7200	Surface bench, small	B-67	14	1.143			41.50	17.30	58.80	84	
	7240	Large	"	5	3.200			116	48.50	164.50	237	
	7300	Surface grinder, large wet	B-68	5	4.800			175	48.50	223.50	330	
	7500	Time check machine	2 Clab	14	1.143			31.50		31.50	52	
	8000	Welder, 30 KVA (bench)		14	1.143			31.50		31.50	52	
	8100	Work bench with chair	↓	25	.640	↓		17.55		17.55	29	

Column group "2006 BARE COSTS" covers MAT., LABOR, EQUIP., TOTAL.

Division 2 Site Construction

Estimating Tips

02200 Site Preparation

- If possible visit the site and take an inventory of the type, quantity, and size of the trees. Certain trees may have a landscape resale value or firewood value. Stump disposal can be very expensive, particularly if they cannot be buried at the site. Consider using a bulldozer in lieu of hand cutting trees.
- Estimators should visit the site to determine the need for haul road access, storage of materials, and security considerations. When estimating for access roads on unstable soil, consider using a geotextile stabilization fabric. It can greatly reduce the quantity of crushed stone or gravel. Sites of limited size and access can cause cost overruns due to lost productivity. Theft and damage is another consideration if the location is isolated. A temporary fence or security guards may be required. Investigate the site thoroughly.

02210 Subsurface Investigation

In preparing estimates on structures involving earthwork or foundations, all information concerning soil characteristics should be obtained. Look particularly for hazardous waste, evidence of prior dumping of debris, and previous stream beds.

02220 Selective Demolition

The costs shown for selective demolition do not include rubbish handling or disposal. These items should be estimated separately using Means data or other sources.

- Historic preservation often requires that the contractor remove materials from the existing structure, rehab them, and replace them. The estimator must be aware of any related measures and precautions that must be taken when doing selective demolition and cutting and patching. Requirements may include special handling and storage as well as security.
- In addition to Section 02220, you can find selective demolition items in each division. Example: Roofing demolition is in Division 7.

02300 Earthwork

- Estimating the actual cost of performing earthwork requires careful consideration of the variables involved. This includes items such as type of soil, whether water will be encountered, dewatering, whether banks need bracing, disposal of excavated earth, and length of haul to fill or spoil sites, etc. If the project has large quantities of cut or fill, consider raising or lowering the site to reduce costs, while paying close attention to the effect on site drainage and utilities.
- If the project has large quantities of fill, creating a borrow pit on the site can significantly lower the costs.
- It is very important to consider what time of year the project is scheduled for completion. Bad weather can create large cost overruns from dewatering, site repair, and lost productivity from cold weather.

02500 Utility Services
02600 Drainage & Containment

- Never assume that the water, sewer, and drainage lines will go in at the early stages of the project. Consider the site access needs before dividing the site in half with open trenches, loose pipe, and machinery obstructions. Always inspect the site to establish that the site drawings are complete. Check off all existing utilities on your drawing as you locate them. If you find any discrepancies, mark up the site plan for further research. Differing site conditions can be very costly if discovered later in the project.
- See also Section 02955 for restoration of pipe where removal/replacement may be undesirable. Use of new types of piping materials can reduce the overall project cost. Owners/design engineers should consider the installing construction as a valuable source of current information on piping products that could lead to significant utility cost savings.

02700 Bases, Ballasts, Pavements/Appurtenances

- When estimating paving, keep in mind the project schedule. If an asphaltic paving project is in a colder climate and runs through to the spring, consider placing the base course in the autumn and then topping it in the spring just prior to completion. This could save considerable costs in spring repair. Keep in mind that prices for asphalt and concrete are generally higher in the cold seasons.
- See also Sections 02960/02965.

02900 Planting

- The timing of planting and guarantee specifications often dictate the costs for establishing tree and shrub growth and a stand of grass or ground cover. Establish the work performance schedule to coincide with the local planting season. Maintenance and growth guarantees can add from 20% to 100% to the total landscaping cost. The cost to replace trees and shrubs can be as high as 5% of the total cost depending on the planting zone, soil conditions, and time of year.

02960 & 02965 Flexible Pavement Surfacing Recovery

- Recycling of asphalt pavement is becoming very popular and is an alternative to removal and replacement of asphalt pavement. It can be a good value engineering proposal if removed pavement can be recycled either at the site or another site that is reasonably close to the project site.

Reference Numbers

Reference numbers are shown in bold squares at the beginning of some major classifications. These numbers refer to related items in the Reference Section. The reference information may be an estimating procedure, an alternate pricing method, or technical information.

Note: Not all subdivisions listed here necessarily appear in this publication.

02050 | Basic Site Materials & Methods

02055 | Soils

		02055 Soils	CREW	DAILY OUTPUT	LABOR-HOURS	UNIT	2006 BARE COSTS MAT.	LABOR	EQUIP.	TOTAL	TOTAL INCL O&P	
150	0010	**BORROW**										150
	0020	Spread, 200 H.P. dozer, no compaction, 2 mi. RT haul										
	0200	Common borrow	B-15	600	.047	C.Y.	6.10	1.43	3.12	10.65	12.50	
	0700	Screened loam		600	.047		21.50	1.43	3.12	26.05	30	
	0800	Topsoil, weed free		600	.047		21	1.43	3.12	25.55	29	
	0900	For 5 mile haul, add	B-34B	200	.040			1.13	2.38	3.51	4.47	

02060 | Aggregate

		02060 Aggregate	CREW	DAILY OUTPUT	LABOR-HOURS	UNIT	MAT.	LABOR	EQUIP.	TOTAL	TOTAL INCL O&P	
150	0010	**BORROW**										150
	0020	Spread, with 200 H.P. dozer, no compaction, 2 mi RT haul										
	0100	Bank run gravel	B-15	600	.047	C.Y.	18.15	1.43	3.12	22.70	25.50	
	0300	Crushed stone (1.40 tons per CY) , 1-1/2″		600	.047		27.50	1.43	3.12	32.05	36.50	
	0320	3/4″		600	.047		27.50	1.43	3.12	32.05	36.50	
	0340	1/2″		600	.047		27	1.43	3.12	31.55	36	
	0360	3/8″		600	.047		27	1.43	3.12	31.55	35.50	
	0400	Sand, washed, concrete		600	.047		27	1.43	3.12	31.55	35.50	
	0500	Dead or bank sand		600	.047		4.21	1.43	3.12	8.76	10.40	
	0600	Select structural fill		600	.047		8.05	1.43	3.12	12.60	14.60	
	0900	For 5 mile haul, add	B-34B	200	.040			1.13	2.38	3.51	4.47	

02080 | Utility Materials

		02080 Utility Materials	CREW	DAILY OUTPUT	LABOR-HOURS	UNIT	MAT.	LABOR	EQUIP.	TOTAL	TOTAL INCL O&P	
400	0010	**UTILITY BOXES** Precast concrete, 6″ thick										400
	0050	5′ x 10′ x 6′ high, I.D.	B-13	2	28	Ea.	1,550	840	325	2,715	3,450	
	0350	Hand hole, precast concrete, 1-1/2″ thick										
	0400	1′-0″ x 2′-0″ x 1′-9″, I.D., light duty	B-1	4	6	Ea.	265	168		433	570	
	0450	4′-6″ x 3′-2″ x 2′-0″, O.D., heavy duty	B-6	3	8	″	815	240	75.50	1,130.50	1,375	
600	0010	**UTILITY ACCESSORIES** R312316-40										600
	0400	Underground tape, detectable, reinforced, alum. foil core, 2″	1 Clab	150	.053	C.L.F.	1.46	1.46		2.92	4.03	
	0500	6″		140	.057	″	3.65	1.57		5.22	6.60	
	9000	Minimum labor/equipment charge		4	2	Job		55		55	91	

02100 | Site Remediation

02110 | Hazmat Removal & Handling

		02110 Hazmat Removal & Handling	CREW	DAILY OUTPUT	LABOR-HOURS	UNIT	2006 BARE COSTS MAT.	LABOR	EQUIP.	TOTAL	TOTAL INCL O&P	
300	0010	**HAZARDOUS WASTE CLEANUP/PICKUP/DISPOSAL**										300
	0100	For contractor equipment, i.e., dozer,										
	0110	Front end loader, dump truck, etc., see Reference Section										
	1000	Solid pickup										
	1100	55 gal. drums				Ea.					220	
	1120	Bulk material, minimum				Ton					165	
	1130	Maximum				″					550	
	1200	Transportation to disposal site										
	1220	Truckload = 80 drums or 25 C.Y. or 18 tons										
	1260	Minimum				Mile					2.50	
	1270	Maximum				″					4.40	
	3000	Liquid pickup, vacuum truck, stainless steel tank										
	3100	Minimum charge, 4 hours										
	3110	1 compartment, 2200 gallon				Hr.					110	

02100 | Site Remediation

		02110 | Hazmat Removal & Handling	CREW	DAILY OUTPUT	LABOR-HOURS	UNIT	2006 BARE COSTS MAT.	LABOR	EQUIP.	TOTAL	TOTAL INCL O&P	
300	3120	2 compartment, 5000 gallon				Hr.					110	300
	3400	Transportation in 6900 gallon bulk truck				Mile					4.75	
	3410	In teflon lined truck				"					5.50	
	5000	Heavy sludge or dry vacuumable material				Hr.					110	
	6000	Dumpsite disposal charge, minimum				Ton					110	
	6020	Maximum				"					440	

		02115 | Underground Storage Tank Removal	CREW	DAILY OUTPUT	LABOR-HOURS	UNIT	MAT.	LABOR	EQUIP.	TOTAL	TOTAL INCL O&P	
200	0010	REMOVAL OF UNDERGROUND STORAGE TANKS R026510-20										200
	0011	Petroleum storage tanks, non-leaking										
	0100	Excavate & load onto trailer										
	0110	3000 gal. to 5000 gal. tank	B-14	4	12	Ea.		350	56.50	406.50	630	
	0120	6000 gal to 8000 gal tank	B-3A	3	13.333	↓		390	241	631	905	
	0130	9000 gal to 12000 gal tank	"	2	20	↓		585	360	945	1,350	
	0190	Known leaking tank add				%				100%	100%	
	0200	Remove sludge, water and remaining product from tank bottom										
	0201	of tank with vacuum truck										
	0300	3000 gal to 5000 gal tank	A-13	5	1.600	Ea.		56.50	100	156.50	199	
	0310	6000 gal to 8000 gal tank	↓	4	2	↓		70.50	125	195.50	249	
	0320	9000 gal to 12000 gal tank	↓	3	2.667	↓		94	167	261	330	
	0390	Dispose of sludge off-site, average				Gal.					4.40	
	0400	Insert inert solid CO2 "dry ice" into tank										
	0401	For cleaning/transporting tanks (1.5 lbs./100 gal. cap)	1 Clab	500	.016	Lb.	1.37	.44		1.81	2.24	
	1020	Haul tank to certified salvage dump, 100 miles round trip										
	1023	3000 gal. to 5000 gal. tank				Ea.				550	690	
	1026	6000 gal. to 8000 gal. tank				↓				650	825	
	1029	9,000 gal. to 12,000 gal. tank				↓				875	1,100	
	1100	Disposal of contaminated soil to landfill										
	1110	Minimum				C.Y.					110	
	1111	Maximum				"					330	
	1120	Disposal of contaminated soil to										
	1121	bituminous concrete batch plant										
	1130	Minimum				C.Y.					55	
	1131	Maximum				"					110	
	2010	Decontamination of soil on site incl poly tarp on top/bottom										
	2011	Soil containment berm, and chemical treatment										
	2020	Minimum	B-11C	100	.160	C.Y.	5.70	5.15	2.26	13.11	17.05	
	2021	Maximum	"	100	.160	↓	7.40	5.15	2.26	14.81	18.90	
	2050	Disposal of decontaminated soil, minimum				↓					66	
	2055	Maximum				↓					135	

02200 | Site Preparation

		02210 | Subsurface Investigation	CREW	DAILY OUTPUT	LABOR-HOURS	UNIT	2006 BARE COSTS MAT.	LABOR	EQUIP.	TOTAL	TOTAL INCL O&P	
120	0010	BORING AND EXPLORATORY DRILLING										120
	0020	Borings, initial field stake out & determination of elevations	A-6	1	16	Day		565	58	623	985	
	0100	Drawings showing boring details				Total		185		185	270	
	0200	Report and recommendations from P.E.				↓		415		415	595	
	0300	Mobilization and demobilization, minimum	B-55	4	6	↓		165	202	367	495	
	0350	For over 100 miles, per added mile	↓	450	.053	Mile		1.47	1.80	3.27	4.38	

2 SITE CONSTRUCTION

	02210 | Subsurface Investigation		Crew	Daily Output	Labor-Hours	Unit	2006 Bare Costs Mat.	Labor	Equip.	Total	Total Incl O&P	
120	0600	Auger holes in earth, no samples, 2-1/2" diameter	B-55	78.60	.305	L.F.		8.40	10.30	18.70	25	120
	0650	4" diameter		67.50	.356			9.75	11.95	21.70	29.50	
	0800	Cased borings in earth, with samples, 2-1/2" diameter		55.50	.432		15.30	11.90	14.55	41.75	52.50	
	0850	4" diameter		32.60	.736		24	20	25	69	87.50	
	1000	Drilling in rock, "BX" core, no sampling	B-56	34.90	.458			14.35	30.50	44.85	56.50	
	1050	With casing & sampling		31.70	.505		15.30	15.80	33.50	64.60	79.50	
	1200	"NX" core, no sampling		25.92	.617			19.30	41	60.30	76.50	
	1250	With casing and sampling		25	.640		18.75	20	42.50	81.25	100	
	1400	Drill rig and crew with truck mounted auger	B-55	1	24	Day		660	810	1,470	1,975	
	1450	With crawler type drill	B-56	1	16	"		500	1,075	1,575	1,975	
	1500	For inner city borings add, minimum									10%	
	1510	Maximum									20%	
200	0010	**CORE DRILLING**										200
	0020	Reinf. conc slab, up to 6" thick, incl. bit, layout & set up										
	0100	1" diameter core	B-89A	28	.571	Ea.	2.59	18.25	4.13	24.97	37.50	
	0150	Each added inch thick, add		300	.053		.46	1.70	.39	2.55	3.73	
	0300	3" diameter core		23	.696		5.75	22	5	32.75	48.50	
	0350	Each added inch thick, add		186	.086		1.04	2.75	.62	4.41	6.35	
	0500	4" diameter core		19	.842		5.75	27	6.10	38.85	57.50	
	0550	Each added inch thick, add		170	.094		1.31	3.01	.68	5	7.15	
	0700	6" diameter core		14	1.143		9.50	36.50	8.25	54.25	79.50	
	0750	Each added inch thick, add		140	.114		1.61	3.65	.83	6.09	8.70	
	0900	8" diameter core		11	1.455		12.95	46.50	10.50	69.95	102	
	0950	Each added inch thick, add		95	.168		2.18	5.40	1.22	8.80	12.60	
	1100	10" diameter core		10	1.600		17.30	51	11.55	79.85	116	
	1150	Each added inch thick, add		80	.200		2.86	6.40	1.44	10.70	15.25	
	1300	12" diameter core		9	1.778		21	57	12.85	90.85	131	
	1350	Each added inch thick, add		68	.235		3.43	7.50	1.70	12.63	18	
	1500	14" diameter core		7	2.286		25	73	16.50	114.50	166	
	1550	Each added inch thick, add		55	.291		4.36	9.30	2.10	15.76	22.50	
	1700	18" diameter core		4	4		32.50	128	29	189.50	278	
	1750	Each added inch thick, add		28	.571		5.70	18.25	4.13	28.08	41	
	1760	For horizontal holes, add to above								30%	30%	
	1770	Prestressed hollow core plank, 6" thick										
	1780	1" diameter core	B-89A	52	.308	Ea.	1.72	9.85	2.22	13.79	20.50	
	1790	Each added inch thick, add		350	.046		.30	1.46	.33	2.09	3.09	
	1800	3" diameter core		50	.320		3.79	10.20	2.31	16.30	23.50	
	1810	Each added inch thick, add		240	.067		.63	2.13	.48	3.24	4.72	
	1820	4" diameter core		48	.333		5.05	10.65	2.41	18.11	25.50	
	1830	Each added inch thick, add		216	.074		.87	2.37	.53	3.77	5.45	
	1840	6" diameter core		44	.364		6.25	11.60	2.63	20.48	29	
	1850	Each added inch thick, add		175	.091		1.04	2.92	.66	4.62	6.65	
	1860	8" diameter core		32	.500		8.35	16	3.61	27.96	39.50	
	1870	Each added inch thick, add		118	.136		1.45	4.33	.98	6.76	9.85	
	1880	10" diameter core		28	.571		11.30	18.25	4.13	33.68	47	
	1890	Each added inch thick, add		99	.162		1.56	5.15	1.17	7.88	11.50	
	1900	12" diameter core		22	.727		13.75	23	5.25	42	59	
	1910	Each added inch thick, add		85	.188		2.29	6	1.36	9.65	13.90	
	1950	Minimum charge for above, 3" diameter core		7	2.286	Total		73	16.50	89.50	138	
	2000	4" diameter core		6.80	2.353			75	17	92	143	
	2050	6" diameter core		6	2.667			85	19.25	104.25	161	
	2100	8" diameter core		5.50	2.909			93	21	114	176	
	2150	10" diameter core		4.75	3.368			108	24.50	132.50	204	
	2200	12" diameter core		3.90	4.103			131	29.50	160.50	249	
	2250	14" diameter core		3.38	4.734			151	34	185	287	
	2300	18" diameter core		3.15	5.079			162	36.50	198.50	310	

02210 | Subsurface Investigation

			Crew	Daily Output	Labor-Hours	Unit	2006 Bare Costs Mat.	Labor	Equip.	Total	Total Incl O&P	
200	3010	Bits for core drill, diamond, premium, 1″ diameter				Ea.	116			116	127	200
	3020	3″ diameter					286			286	315	
	3040	4″ diameter					320			320	350	
	3050	6″ diameter					510			510	560	
	3080	8″ diameter					695			695	765	
	3120	12″ diameter					1,100			1,100	1,200	
	3180	18″ diameter					2,275			2,275	2,500	
	3240	24″ diameter				↓	3,025			3,025	3,325	

02220 | Site Demolition

			Crew	Daily Output	Labor-Hours	Unit	2006 Bare Costs Mat.	Labor	Equip.	Total	Total Incl O&P	
110	0010	**BUILDING DEMOLITION** Large urban projects, incl. 20 mi. haul (R024119-10)										110
	0011	No foundation or dump fees, C.F. is vol. of building standing										
	0012	Steel	B-8	21500	.003	C.F.		.09	.11	.20	.27	
	0050	Concrete		15300	.004			.13	.16	.29	.38	
	0080	Masonry		20100	.003			.10	.12	.22	.29	
	0100	Mixture of types, average	↓	20100	.003			.10	.12	.22	.29	
	0500	Small bldgs, or single bldgs, no salvage included, steel	B-3	14800	.003			.10	.12	.22	.29	
	0600	Concrete		11300	.004			.13	.15	.28	.38	
	0650	Masonry		14800	.003			.10	.12	.22	.29	
	0700	Wood	↓	14800	.003	↓		.10	.12	.22	.29	
	1000	Single family, one story house, wood, minimum				Ea.				2,525	2,975	
	1020	Maximum								4,400	5,275	
	1200	Two family, two story house, wood, minimum								3,300	3,950	
	1220	Maximum								6,375	7,700	
	1300	Three family, three story house, wood, minimum								4,400	5,275	
	1320	Maximum				↓				7,700	9,250	
	1400	Gutting building, see division 02220-340										
	5000	For buildings with no interior walls, deduct				Ea.				50%		
130	0010	**BLDG. FOOTINGS AND FOUNDATIONS DEMOLITION** (R024119-10)										130
	0200	Floors, concrete slab on grade,										
	0240	4″ thick, plain concrete	B-9C	500	.080	S.F.		2.22	.29	2.51	4	
	0280	Reinforced, wire mesh		470	.085			2.37	.31	2.68	4.26	
	0300	Rods		400	.100			2.78	.37	3.15	5	
	0400	6″ thick, plain concrete		375	.107			2.97	.39	3.36	5.35	
	0420	Reinforced, wire mesh		340	.118			3.27	.43	3.70	5.90	
	0440	Rods	↓	300	.133	↓		3.71	.49	4.20	6.70	
	1000	Footings, concrete, 1′ thick, 2′ wide	B-5	300	.187	L.F.		5.65	3.14	8.79	12.70	
	1080	1′-6″ thick, 2′ wide		250	.224			6.80	3.77	10.57	15.25	
	1120	3′ wide		200	.280			8.50	4.71	13.21	19.05	
	1140	2′ thick, 3′ wide	↓	175	.320			9.70	5.40	15.10	22	
	1200	Average reinforcing, add								10%	10%	
	1220	Heavy reinforcing, add				↓				20%	20%	
	2000	Walls, block, 4″ thick	1 Clab	180	.044	S.F.		1.22		1.22	2.02	
	2040	6″ thick		170	.047			1.29		1.29	2.14	
	2080	8″ thick		150	.053			1.46		1.46	2.42	
	2100	12″ thick	↓	150	.053			1.46		1.46	2.42	
	2200	For horizontal reinforcing, add								10%	10%	
	2220	For vertical reinforcing, add								20%	20%	
	2400	Concrete, plain concrete, 6″ thick	B-9	160	.250			6.95	.92	7.87	12.50	
	2420	8″ thick		140	.286			7.95	1.05	9	14.30	
	2440	10″ thick		120	.333			9.25	1.23	10.48	16.70	
	2500	12″ thick	↓	100	.400			11.10	1.47	12.57	20	
	2600	For average reinforcing, add								10%	10%	
	2620	For heavy reinforcing, add				↓				20%	20%	
	9000	Minimum labor/equipment charge	A-1	2	4	Job		110	24.50	134.50	209	

For expanded coverage of these items see ***Means Site Work & Landscape Cost Data 2006***

02200 | Site Preparation

02220 | Site Demolition

			Crew	Daily Output	Labor-Hours	Unit	2006 Bare Costs Mat.	2006 Bare Costs Labor	2006 Bare Costs Equip.	2006 Bare Costs Total	Total Incl O&P
210	0010	**MINOR BUILDING DECONSTRUCTION**, for salvage, avg house R024119-10									
	0225	2 story, pre 1970 house, 1400 SF, lbr for all salv mat, min	6 Clab	25	1.920	SF Flr.		52.50		52.50	87
	0230	Maximum	″	15	3.200	″		87.50		87.50	145
	0235	Salvage of carpet, tackless	2 Clab	2400	.007	S.F.		.18		.18	.30
	0240	Wood floors, incl denailing and packaging	3 Clab	270	.089	″		2.44		2.44	4.04
	0260	Wood doors and trim, standard	1 Clab	16	.500	Ea.		13.70		13.70	22.50
	0280	Base or cove mouldings, incl denailing and packaging		500	.016	L.F.		.44		.44	.73
	0320	Closet shelving and trim, incl denailing and packaging		18	.444	Set		12.20		12.20	20
	0340	Kitchen cabinets, uppers and lowers, prefab type	2 Clab	24	.667	L.F.		18.25		18.25	30.50
	0360	Kitchen cabinets, uppers and lowers, built-in		12	1.333	″		36.50		36.50	60.50
	0370	Bath fixt, incl toilet, tub, vanity, and med cabinet		3	5.333	Set		146		146	242
220	0010	**FENCING DEMOLITION** R024119-10									
	1600	Fencing, barbed wire, 3 strand	2 Clab	430	.037	L.F.		1.02		1.02	1.69
	1650	5 strand	″	280	.057			1.57		1.57	2.59
	1700	Chain link, posts & fabric, remove only, 8′ to 10′ high	B-6	445	.054			1.62	.51	2.13	3.19
230	0010	**HYDRODEMOLITION** R024119-10									
	0015	Hydrodemolition, concrete pavement, 4000 PSI, 2″ depth	B-5	500	.112	S.F.		3.40	1.88	5.28	7.60
	0120	4″ depth		450	.124			3.78	2.09	5.87	8.45
	0130	6″ depth		400	.140			4.25	2.35	6.60	9.50
	0410	6000 PSI, 2″ depth		410	.137			4.14	2.30	6.44	9.30
	0420	4″ depth		350	.160			4.85	2.69	7.54	10.85
	0430	6″ depth		300	.187			5.65	3.14	8.79	12.70
	0510	8000 PSI, 2″ depth		330	.170			5.15	2.85	8	11.55
	0520	4″ depth		280	.200			6.05	3.36	9.41	13.60
	0530	6″ depth		240	.233			7.10	3.92	11.02	15.85
240	0010	**MINOR SITE DEMOLITION** R024119-10									
	0015	No hauling, abandon catch basin or manhole	B-6	7	3.429	Ea.		103	32.50	135.50	203
	0020	Remove existing catch basin or manhole, masonry		4	6			180	56.50	236.50	355
	0030	Catch basin or manhole frames and covers, stored		13	1.846			55.50	17.40	72.90	109
	0040	Remove and reset		7	3.429			103	32.50	135.50	203
	0100	Roadside delineators, remove only	B-80	175	.183			5.45	3.04	8.49	12.25
	0110	Remove and reset	″	100	.320			9.55	5.30	14.85	21.50
	0400	Minimum labor/equipment charge	B-6	4	6	Job		180	56.50	236.50	355
	0800	Guiderail, corrugated steel, remove only	B-80A	100	.240	L.F.		6.60	1.76	8.36	12.85
	0850	Remove and reset	″	40	.600	″		16.45	4.40	20.85	32
	0860	Guide posts, remove only	B-80B	120	.267	Ea.		7.85	1.96	9.81	14.95
	0870	Remove and reset	B-55	50	.480	″		13.20	16.15	29.35	39.50
	0890	Minimum labor/equipment charge	2 Clab	4	4	Job		110		110	182
	0900	Hydrants, fire, remove only	B-21A	5	8	Ea.		275	107	382	560
	0950	Remove and reset	″	2	20	″		685	268	953	1,400
	0990	Minimum labor/equipment charge	2 Plum	2	8	Job		340		340	530
	1000	Masonry walls, block or tile, solid, remove	B-5	1800	.031	C.F.		.94	.52	1.46	2.12
	1100	Cavity wall		2200	.025			.77	.43	1.20	1.73
	1200	Brick, solid		900	.062			1.89	1.05	2.94	4.23
	1300	With block back-up		1130	.050			1.50	.83	2.33	3.37
	1400	Stone, with mortar		900	.062			1.89	1.05	2.94	4.23
	1500	Dry set		1500	.037			1.13	.63	1.76	2.54
	1600	Median barrier, precast concrete, remove and store	B-3	430	.112	L.F.		3.30	4.06	7.36	9.85
	1610	Remove and reset	″	390	.123	″		3.64	4.48	8.12	10.90
	1650	Minimum labor/equipment charge	A-1	4	2	Job		55	12.25	67.25	105
	2900	Pipe removal, sewer/water, no excavation, 12″ diameter	B-6	175	.137	L.F.		4.11	1.29	5.40	8.10
	2930	15″-18″ diameter		150	.160			4.80	1.51	6.31	9.45
	2960	21″-24″ diameter		120	.200			6	1.89	7.89	11.80
	3000	27″-36″ diameter		90	.267			8	2.51	10.51	15.75
	3200	Steel, welded connections, 4″ diameter		160	.150			4.50	1.41	5.91	8.85

02220 | Site Demolition

		02220 Site Demolition	CREW	DAILY OUTPUT	LABOR-HOURS	UNIT	MAT.	LABOR	EQUIP.	TOTAL	TOTAL INCL O&P	
							2006 BARE COSTS					
240	3300	10" diameter (R024119-10)	B-6	80	.300	L.F.		9	2.83	11.83	17.75	240
	3390	Minimum labor/equipment charge		3	8	Job		240	75.50	315.50	475	
	3500	Railroad track removal, ties and track	B-13	330	.170	L.F.		5.10	1.97	7.07	10.45	
	3600	Ballast	B-14	500	.096	C.Y.		2.79	.45	3.24	5.05	
	3700	Remove and re-install, ties & track using new bolts & spikes		50	.960	L.F.		28	4.52	32.52	50.50	
	3800	Turnouts using new bolts and spikes		1	48	Ea.		1,400	226	1,626	2,525	
	3890	Minimum labor/equipment charge		5	9.600	Job		279	45	324	505	
	4000	Sidewalk removal, bituminous, 2-1/2" thick	B-6	325	.074	S.Y.		2.22	.70	2.92	4.37	
	4050	Brick, set in mortar		185	.130			3.89	1.22	5.11	7.70	
	4100	Concrete, plain, 4"		160	.150			4.50	1.41	5.91	8.85	
	4200	Mesh reinforced		150	.160			4.80	1.51	6.31	9.45	
	4290	Minimum labor/equipment charge	B-39	12	4	Job		116	12.30	128.30	205	
250	0010	**DEMOLISH, REMOVE PAVEMENT AND CURB** (R024119-10)										250
	5010	Pavement removal, bituminous roads, 3" thick	B-38	690	.058	S.Y.		1.81	1.21	3.02	4.26	
	5050	4" to 6" thick		420	.095			2.97	1.99	4.96	7	
	5100	Bituminous driveways		640	.063			1.95	1.31	3.26	4.60	
	5200	Concrete to 6" thick, hydraulic hammer, mesh reinforced		255	.157			4.90	3.28	8.18	11.55	
	5300	Rod reinforced		200	.200			6.25	4.18	10.43	14.70	
	5400	Concrete, 7" to 24" thick, plain		33	1.212	C.Y.		38	25.50	63.50	89.50	
	5500	Reinforced		24	1.667	"		52	35	87	123	
	5590	Minimum labor/equipment charge		6	6.667	Job		208	139	347	490	
	5600	With hand held air equipment, bituminous, to 6" thick	B-39	1900	.025	S.F.		.73	.08	.81	1.29	
	5700	Concrete to 6" thick, no reinforcing		1600	.030			.87	.09	.96	1.53	
	5800	Mesh reinforced		1400	.034			1	.11	1.11	1.75	
	5900	Rod reinforced		765	.063			1.82	.19	2.01	3.20	
	5990	Minimum labor/equipment charge	B-38	6	6.667	Job		208	139	347	490	
	6000	Curbs, concrete, plain	B-6	360	.067	L.F.		2	.63	2.63	3.94	
	6100	Reinforced		275	.087			2.62	.82	3.44	5.15	
	6200	Granite		360	.067			2	.63	2.63	3.94	
	6300	Bituminous		528	.045			1.36	.43	1.79	2.69	
	6390	Minimum labor/equipment charge		6	4	Job		120	37.50	157.50	237	
310	0010	**SELECTIVE DEMOLITION, CUTOUT** (R024119-10)										310
	0020	Concrete, elev. slab, light reinforcement, under 6 CF	B-9C	65	.615	C.F.		17.10	2.26	19.36	31	
	0050	Light reinforcing, over 6 C.F.	"	75	.533	"		14.85	1.96	16.81	26.50	
	0200	Slab on grade to 6" thick, not reinforced, under 8 S.F.	B-9	85	.471	S.F.		13.10	1.73	14.83	23.50	
	0250	8 - 16 S.F.	"	175	.229	"		6.35	.84	7.19	11.50	
	0255	For over 16 SF see 02220-130-0400										
	0600	Walls, not reinforced, under 6 C.F.	B-9	60	.667	C.F.		18.55	2.45	21	33	
	0650	6 - 12 C.F.	"	80	.500	"		13.90	1.84	15.74	25	
	0655	For over 12 CF see 02220-130-2500										
	1000	Concrete, elevated slab, bar reinforced, under 6 C.F.	B-9C	45	.889	C.F.		24.50	3.27	27.77	44.50	
	1050	Bar reinforced, over 6 C.F.	"	50	.800	"		22	2.94	24.94	40	
	1200	Slab on grade to 6" thick, bar reinforced, under 8 S.F.	B-9	75	.533	S.F.		14.85	1.96	16.81	26.50	
	1250	8 - 16 S.F.	"	150	.267	"		7.40	.98	8.38	13.40	
	1255	For over 16 SF see 02220-130-0440										
	1400	Walls, bar reinforced, under 6 C.F.	B-9C	50	.800	C.F.		22	2.94	24.94	40	
	1450	6 - 12 CF	"	70	.571	"		15.90	2.10	18	29	
	1455	For over 12 CF see 02220-130-2500 & 2600										
	2000	Brick, to 4 S.F. opening, not including toothing										
	2040	4" thick	B-9C	30	1.333	Ea.		37	4.91	41.91	67	
	2060	8" thick		18	2.222			62	8.20	70.20	111	
	2080	12" thick		10	4			111	14.70	125.70	200	
	2400	Concrete block, to 4 S.F. opening, 2" thick		35	1.143			32	4.21	36.21	57	
	2420	4" thick		30	1.333			37	4.91	41.91	67	
	2440	8" thick		27	1.481			41	5.45	46.45	74	

02200 | Site Preparation

02220 | Site Demolition

		02220 | Site Demolition	CREW	DAILY OUTPUT	LABOR-HOURS	UNIT	MAT.	LABOR	EQUIP.	TOTAL	TOTAL INCL O&P	
							2006 BARE COSTS					
310	2460	12" thick (R024119-10)	B-9C	24	1.667	Ea.		46.50	6.15	52.65	84	310
	2600	Gypsum block, to 4 S.F. opening, 2" thick	B-9	80	.500			13.90	1.84	15.74	25	
	2620	4" thick		70	.571			15.90	2.10	18	29	
	2640	8" thick		55	.727			20	2.68	22.68	36.50	
	2800	Terra cotta, to 4 S.F. opening, 4" thick		70	.571			15.90	2.10	18	29	
	2840	8" thick		65	.615			17.10	2.26	19.36	31	
	2880	12" thick		50	.800			22	2.94	24.94	40	
	4000	For toothing masonry, see Division 04910-800										
	6000	Walls, interior, not including re-framing,										
	6010	openings to 5 S.F.										
	6100	Drywall to 5/8" thick	1 Clab	24	.333	Ea.		9.15		9.15	15.15	
	6200	Paneling to 3/4" thick		20	.400			10.95		10.95	18.15	
	6300	Plaster, on gypsum lath		20	.400			10.95		10.95	18.15	
	6340	On wire lath		14	.571			15.65		15.65	26	
	7000	Wood frame, not including re-framing, openings to 5 S.F.										
	7200	Floors, sheathing and flooring to 2" thick	1 Clab	5	1.600	Ea.		44		44	72.50	
	7310	Roofs, sheathing to 1" thick, not including roofing		6	1.333			36.50		36.50	60.50	
	7410	Walls, sheathing to 1" thick, not including siding		7	1.143			31.50		31.50	52	
	8500	Minimum labor/equipment charge		4	2	Job		55		55	91	
320	0010	**SELECTIVE DEMOLITION, DISPOSAL ONLY** (R024119-10)										320
	0015	Urban bldg w/salvage value allowed										
	0020	Including loading and 5 mile haul to dump										
	0200	Steel frame	B-3	430	.112	C.Y.		3.30	4.06	7.36	9.85	
	0300	Concrete frame		365	.132			3.89	4.79	8.68	11.60	
	0400	Masonry construction		445	.108			3.19	3.93	7.12	9.50	
	0500	Wood frame		247	.194			5.75	7.05	12.80	17.20	
330	0010	**SELECTIVE DEMOLITION, DUMP CHARGES** (R024119-10)										330
	0020	Dump charges, typical urban city, tipping fees only										
	0100	Building construction materials				Ton					70	
	0200	Trees, brush, lumber									50	
	0300	Rubbish only									60	
	0500	Reclamation station, usual charge									85	
340	0010	**SELECTIVE DEMOLITION, GUTTING** (R024119-10)										340
	0020	Building interior, including disposal, dumpter fees not incl.										
	0500	Residential building										
	0560	Minimum	B-16	400	.080	SF Flr.		2.25	1.19	3.44	5	
	0580	Maximum	"	360	.089	"		2.50	1.32	3.82	5.60	
	0900	Commercial building										
	1000	Minimum	B-16	350	.091	SF Flr.		2.57	1.36	3.93	5.75	
	1020	Maximum		250	.128	"		3.60	1.91	5.51	8.05	
	3000	Minimum labor/equipment charge		4	8	Job		225	119	344	500	
350	0010	**SELECTIVE DEMOLITION, RUBBISH HANDLING** (R024119-10)										350
	0020	The following are to be added to the demolition prices										
	0400	Chute, circular, prefabricated steel, 18" diameter	B-1	40	.600	L.F.	28.50	16.85		45.35	59.50	
	0440	30" diameter	"	30	.800	"	38.50	22.50		61	79	
	0725	Dumpster, weekly rental, 1 dump/week, 20 C.Y. capacity (8 Tons)				Week	420			420	462	
	0800	30 C.Y. capacity (10 Tons)					605			605	665	
	0840	40 C.Y. capacity (13 Tons)					775			775	825	
	0900	Alternate pricing for dumpsters										
	0910	Delivery, average for all sizes				Ea.	50			50	55	
	0920	Haul, average for all sizes					150			150	165	
	0930	Rent per day, average for all sizes					5			5	5.50	
	0940	Rent per month, average for all sizes					45			45	49.50	

		02220 | Site Demolition	CREW	DAILY OUTPUT	LABOR-HOURS	UNIT	2006 BARE COSTS MAT.	2006 BARE COSTS LABOR	2006 BARE COSTS EQUIP.	2006 BARE COSTS TOTAL	TOTAL INCL O&P	
350	0950	Disposal fee per ton, average for all sizes (R024119-10)				Ton	40			40	44	350
	1000	Dust partition, 6 mil polyethylene, 1" x 3" frame	2 Carp	2000	.008	S.F.	.18	.28		.46	.67	
	1080	2" x 4" frame	"	2000	.008	"	.32	.28		.60	.82	
	2000	Load, haul, and dump, 50' haul	2 Clab	24	.667	C.Y.		18.25		18.25	30.50	
	2040	100' haul		16.50	.970			26.50		26.50	44	
	2080	Over 100' haul, add per 100 L.F.		35.50	.451			12.35		12.35	20.50	
	2120	In elevators, per 10 floors, add	↓	140	.114			3.13		3.13	5.20	
	3000	Loading & trucking, including 2 mile haul, chute loaded	B-16	45	.711			20	10.60	30.60	44.50	
	3040	Hand loading truck, 50' haul	"	48	.667			18.75	9.95	28.70	42	
	3080	Machine loading truck	B-17	120	.267			7.90	4.61	12.51	17.90	
	3120	Wheeled 50' and ramp dump loaded	2 Clab	24	.667			18.25		18.25	30.50	
	5000	Haul, per mile, up to 8 C.Y. truck	B-34B	1165	.007			.19	.41	.60	.77	
	5100	Over 8 C.Y. truck	"	1550	.005	↓		.15	.31	.46	.58	
360	0010	**SELECTIVE DEMOLITION, SAW CUTTING** (R024119-10)										360
	0015	Asphalt, up to 3" deep	B-89	1050	.015	L.F.	.28	.48	.28	1.04	1.39	
	0020	Each additional inch of depth		1800	.009		.06	.28	.17	.51	.70	
	0400	Concrete slabs, mesh reinforcing, up to 3" deep		980	.016		.38	.51	.30	1.19	1.57	
	0420	Each additional inch of depth	↓	1600	.010		.13	.31	.19	.63	.84	
	0800	Concrete walls, hydraulic saw, plain, per inch of depth	B-89B	250	.064		.34	2.01	1.86	4.21	5.65	
	0820	Rod reinforcing, per inch of depth		150	.107		.47	3.35	3.09	6.91	9.25	
	1200	Masonry walls, hydraulic saw, brick, per inch of depth		300	.053		.34	1.67	1.55	3.56	4.75	
	1220	Block walls, solid, per inch of depth	↓	250	.064		.36	2.01	1.86	4.23	5.65	
	2000	Brick or masonry w/hand held saw, per inch of depth	A-1	125	.064		.29	1.75	.39	2.43	3.66	
	5000	Wood sheathing to 1" thick, on walls	1 Carp	200	.040			1.42		1.42	2.36	
	5020	On roof	"	250	.032	↓		1.14		1.14	1.88	
	9000	Minimum labor/equipment charge	A-1	2	4	Job		110	24.50	134.50	209	
	9950	See also Div. 02210-200 core drilling										
370	0010	**SELECTIVE DEMOLITION, TORCH CUTTING** (R024119-10)										370
	0020	Steel, 1" thick plate	1 Clab	360	.022	L.F.	.18	.61		.79	1.21	
	0040	1" diameter bar	"	210	.038	Ea.		1.04		1.04	1.73	
	1000	Oxygen lance cutting, reinforced concrete walls										
	1040	12" to 16" thick walls	1 Clab	10	.800	L.F.		22		22	36.50	
	1080	24" thick walls	"	6	1.333	"		36.50		36.50	60.50	
	1090	Minimum labor/equipment charge	E-25	2	4	Job		168	40.50	208.50	360	
	1100	See also division 05090-920										
		02250 | Shoring & Underpinning										
100	0010	**GROUTING, PRESSURE**										100
	0020	Grouting, pressure, cement & sand, 1:1 mix, minimum	B-61	124	.323	Bag	9.50	9.45	2.10	21.05	28.50	
	0100	Maximum		51	.784	"	9.50	23	5.10	37.60	53.50	
	0200	Cement and sand, 1:1 mix, minimum		250	.160	C.F.	19.05	4.70	1.04	24.79	30	
	0300	Maximum		100	.400		28.50	11.75	2.60	42.85	53.50	
	0400	Epoxy cement grout, minimum		137	.292		141	8.55	1.90	151.45	172	
	0500	Maximum	↓	57	.702	↓	141	20.50	4.57	166.07	195	
	0600	Structural epoxy grout				Gal.	49.50			49.50	54.50	
	0700	Alternate pricing method: (Add for materials)										
	0710	5 person crew and equipment	B-61	1	40	Day		1,175	260	1,435	2,200	
400	0010	**SHEET PILING**										400
	0020	Sheet piling steel, not incl. wales, 22 psf, 15' excav., left in place	B-40	10.81	5.920	Ton	935	210	255	1,400	1,650	
	0100	Drive, extract & salvage (R314116-40)		6	10.667	"	415	380	460	1,255	1,575	
	1200	15' deep excavation, 22 psf, left in place		983	.065	S.F.	10.85	2.31	2.80	15.96	18.85	
	1300	Drive, extract & salvage	↓	545	.117	"	4.66	4.17	5.05	13.88	17.60	
	2100	Rent steel sheet piling and wales, first month				Ton	222			222	244	
	2200	Per added month				"	22			22	24.50	
	3900	Wood, solid sheeting, incl. wales, braces and spacers,										

02200 | Site Preparation

	02250 | Shoring & Underpinning		CREW	DAILY OUTPUT	LABOR-HOURS	UNIT	2006 BARE COSTS MAT.	LABOR	EQUIP.	TOTAL	TOTAL INCL O&P	
400	3910	drive, extract & salvage, 8' deep excavation (R314116-40)	B-31	330	.121	S.F.	1.78	3.57	.41	5.76	8.30	400
	4520	Left in place, 8' deep, 55 S.F./hr.		440	.091	"	3.20	2.68	.31	6.19	8.30	
	4990	Minimum labor/equipment charge		2	20	Job		590	68	658	1,050	
	5000	For treated lumber add cost of treatment to lumber										
	5010	See Division 06070-400										
500	0010	**SHORING**										500
	0020	Shoring, existing building, with timber, no salvage allowance	B-51	2.20	21.818	M.B.F.	800	605	58.50	1,463.50	1,950	
	1000	On cribbing with 35 ton screw jacks, per box and jack		3.60	13.333	Jack	48	370	35.50	453.50	705	
	1090	Minimum labor/equipment charge		2	24	Ea.		665	64.50	729.50	1,175	
	1100	Masonry openings in walls, see Div. 02220-310										
800	0010	**UNDERPINNING FOUNDATIONS** Including excavation,										800
	0020	forming, reinforcing, concrete and equipment										
	0100	5' to 16' below grade, 100 to 500 C.Y.	B-52	2.30	24.348	C.Y.	240	790	173	1,203	1,750	
	0200	Over 500 C.Y.		2.50	22.400		216	730	159	1,105	1,625	
	0400	16' to 25' below grade, 100 to 500 C.Y.		2	28		264	910	199	1,373	2,000	
	0500	Over 500 C.Y.		2.10	26.667		250	870	189	1,309	1,900	
	0700	26' to 40' below grade, 100 to 500 C.Y.		1.60	35		288	1,150	248	1,686	2,475	
	0800	Over 500 C.Y.		1.80	31.111		264	1,000	221	1,485	2,200	
	0900	For under 50 C.Y., add					10%	40%				

02300 | Earthwork

	02305 | Equipment		CREW	DAILY OUTPUT	LABOR-HOURS	UNIT	2006 BARE COSTS MAT.	LABOR	EQUIP.	TOTAL	TOTAL INCL O&P	
250	0010	**MOBILIZATION OR DEMOB.** (One or the other, unless noted) (R015433-10)										250
	0015	Up to 25 mi haul dist (50 mi RT for mob/demob crew)										
	0020	Dozer, loader, backhoe, excav., grader, paver, roller, 70 to 150 H.P.	B-34N	4	2	Ea.		56.50	112	168.50	216	
	0100	Above 150 HP	B-34K	3	2.667			75.50	175	250.50	315	
	0300	Scraper, towed type (incl. tractor), 6 C.Y. capacity		3	2.667			75.50	175	250.50	315	
	0400	10 C.Y.		2.50	3.200			90.50	210	300.50	380	
	0600	Self-propelled scraper, 15 C.Y.		2.50	3.200			90.50	210	300.50	380	
	0700	24 C.Y.		2	4			113	262	375	475	
	0900	Shovel or dragline, 3/4 C.Y.		3.60	2.222			63	146	209	263	
	1000	1-1/2 C.Y.		3	2.667			75.50	175	250.50	315	
	1100	Small equipment, placed in rear of, or towed by pickup truck	A-3A	8	1			27.50	10.65	38.15	56.50	
	1150	Equip up to 70 HP, on flatbed trailer behind pickup truck	A-3D	4	2			55	43.50	98.50	138	
	2000	Mob & demob truck-mounted crane up to 75 ton, driver only	1 Eqhv	3.60	2.222			84.50		84.50	134	
	2100	Crane, truck-mounted, over 75 ton	A-3E	2.50	6.400			213	34	247	380	
	2200	Crawler-mounted, up to 75 ton	A-3F	2	8			266	278	544	730	
	2300	Over 75 ton	A-3G	1.50	10.667			355	405	760	1,000	
	2500	For each additional 5 miles haul distance, add						10%	10%			
	3000	For large pieces of equipment, allow for assembly/knockdown										
	3100	For mob/demob of micro-tunneling equip, see section 02441-400										
	3200	For mob/demob of pile driving equip, see section 02455-650										

	02315 | Excavation and Fill		CREW	DAILY OUTPUT	LABOR-HOURS	UNIT	MAT.	LABOR	EQUIP.	TOTAL	TOTAL INCL O&P	
110	0010	**BACKFILL, GENERAL** (R312323-30)										110
	0015	By hand, no compaction, light soil	1 Clab	14	.571	L.C.Y.		15.65		15.65	26	

02315 | Excavation and Fill

			CREW	DAILY OUTPUT	LABOR-HOURS	UNIT	2006 BARE COSTS MAT.	LABOR	EQUIP.	TOTAL	TOTAL INCL O&P	
110	0100	Heavy soil (R312323-30)	1 Clab	11	.727	L.C.Y.		19.95		19.95	33	110
	0300	Compaction in 6″ layers, hand tamp, add to above		20.60	.388	E.C.Y.		10.65		10.65	17.65	
	0400	Roller compaction operator walking, add	B-10A	100	.120			4.03	1.26	5.29	7.85	
	0500	Air tamp, add	B-9D	190	.211			5.85	.97	6.82	10.75	
	0600	Vibrating plate, add	A-1D	60	.133			3.65	.44	4.09	6.55	
	0800	Compaction in 12″ layers, hand tamp, add to above	1 Clab	34	.235			6.45		6.45	10.70	
	1000	Air tamp, add	B-9	285	.140			3.90	.52	4.42	7	
	1100	Vibrating plate, add	A-1E	90	.089			2.44	.38	2.82	4.46	
	1200	Trench, dozer, no compaction, 60 HP	B-10L	425	.028	L.C.Y.		.95	.75	1.70	2.35	
	1300	Dozer backfilling, bulk, up to 300′ haul, no compaction	B-10B	1200	.010	″		.34	.77	1.11	1.38	
	1400	Air tamped, add	B-11B	80	.200	E.C.Y.		6.25	2.45	8.70	12.80	
	1900	Dozer backfilling, trench, up to 300′ haul, no compaction	B-10B	900	.013	L.C.Y.		.45	1.02	1.47	1.85	
	2000	Air tamped, add	B-11B	80	.200	E.C.Y.		6.25	2.45	8.70	12.80	
	2350	Spreading in 8″ layers, small dozer	B-10B	1060	.011	L.C.Y.		.38	.87	1.25	1.57	
	2450	Compacting with vibrating plate, 8″ lifts	A-1D	73	.110	E.C.Y.		3	.36	3.36	5.40	
120	0010	**BACKFILL, STRUCTURAL** Dozer or F.E. loader										120
	0020	From existing stockpile, no compaction										
	2000	80 H.P., 50′ haul, sand & gravel	B-10L	1100	.011	L.C.Y.		.37	.29	.66	.91	
	2020	Common earth		975	.012			.41	.33	.74	1.02	
	2040	Clay		850	.014			.47	.38	.85	1.17	
	2400	300′ haul, sand & gravel		370	.032			1.09	.86	1.95	2.69	
	2420	Common earth		330	.036			1.22	.97	2.19	3.02	
	2440	Clay		290	.041			1.39	1.10	2.49	3.44	
	3000	105 H.P., 50′ haul, sand & gravel	B-10W	1350	.009			.30	.34	.64	.86	
	3020	Common earth		1225	.010			.33	.38	.71	.95	
	3040	Clay		1100	.011			.37	.42	.79	1.05	
	3300	300′ haul, sand & gravel		465	.026			.87	1	1.87	2.48	
	3320	Common earth		415	.029			.97	1.12	2.09	2.79	
	3340	Clay		370	.032			1.09	1.25	2.34	3.12	
310	0010	**COMPACTION, GENERAL**										310
	5000	Riding, vibrating roller, 6″ lifts, 2 passes	B-10Y	3000	.004	E.C.Y.		.13	.14	.27	.37	
	5020	3 passes		2300	.005			.18	.18	.36	.48	
	5040	4 passes		1900	.006			.21	.22	.43	.58	
	5060	12″ lifts, 2 passes		5200	.002			.08	.08	.16	.21	
	5080	3 passes		3500	.003			.12	.12	.24	.31	
	5100	4 passes		2600	.005			.16	.16	.32	.43	
	5600	Sheepsfoot or wobbly wheel roller, 6″ lifts, 2 passes	B-10G	2400	.005			.17	.35	.52	.66	
	5620	3 passes		1735	.007			.23	.49	.72	.91	
	5640	4 passes		1300	.009			.31	.65	.96	1.22	
	5680	12″ lifts, 2 passes		5200	.002			.08	.16	.24	.30	
	5700	3 passes		3500	.003			.12	.24	.36	.45	
	5720	4 passes		2600	.005			.16	.33	.49	.61	
	7000	Walk behind, vibrating plate 18″ wide, 6″ lifts, 2 passes	A-1D	200	.040			1.10	.13	1.23	1.97	
	7020	3 passes		185	.043			1.18	.14	1.32	2.12	
	7040	4 passes		140	.057			1.57	.19	1.76	2.80	
	7200	12″ lifts, 2 passes	A-1E	560	.014			.39	.06	.45	.72	
	7220	3 passes		375	.021			.58	.09	.67	1.07	
	7240	4 passes		280	.029			.78	.12	.90	1.44	
	7500	Vibrating roller 24″ wide, 6″ lifts, 2 passes	B-10A	420	.029			.96	.30	1.26	1.87	
	7520	3 passes		280	.043			1.44	.45	1.89	2.79	
	7540	4 passes		210	.057			1.92	.60	2.52	3.73	
	7600	12″ lifts, 2 passes		840	.014			.48	.15	.63	.93	
	7620	3 passes		560	.021			.72	.22	.94	1.40	
	7640	4 passes		420	.029			.96	.30	1.26	1.87	
	8000	Rammer tamper, 6″ to 11″, 4″ lifts, 2 passes	A-1F	130	.062			1.69	.28	1.97	3.10	

02300 | Earthwork

	02315 | Excavation and Fill		Crew	Daily Output	Labor-Hours	Unit	2006 Bare Costs Mat.	Labor	Equip.	Total	Total Incl O&P	
310	8050	3 passes	A-1F	97	.082	E.C.Y.		2.26	.38	2.64	4.16	310
	8100	4 passes		65	.123			3.37	.57	3.94	6.25	
	8200	8″ lifts, 2 passes		260	.031			.84	.14	.98	1.56	
	8250	3 passes		195	.041			1.12	.19	1.31	2.07	
	8300	4 passes		130	.062			1.69	.28	1.97	3.10	
	8400	13″ to 18″, 4″ lifts, 2 passes	A-1G	390	.021			.56	.10	.66	1.03	
	8450	3 passes		290	.028			.76	.13	.89	1.39	
	8500	4 passes		195	.041			1.12	.19	1.31	2.07	
	8600	8″ lifts, 2 passes		780	.010			.28	.05	.33	.52	
	8650	3 passes		585	.014			.37	.06	.43	.69	
	8700	4 passes		390	.021			.56	.10	.66	1.03	
	9000	Water, 3000 gal. truck, 3 mile haul	B-45	1888	.008		.20	.28	.28	.76	.97	
	9010	6 mile haul		1444	.011		.20	.36	.37	.93	1.21	
	9020	12 mile haul		1000	.016		.20	.52	.53	1.25	1.64	
	9030	6000 gal. wagon, 3 mile haul	B-59	2000	.004		.20	.11	.16	.47	.58	
	9040	6 mile haul	"	1600	.005		.20	.14	.20	.54	.67	
320	0010	**COMPACTION, STRUCTURAL** R312323-30										320
	0020	Steel wheel tandem roller, 5 tons	B-10E	8	1.500	Hr.		50.50	14.30	64.80	96	
	0050	Air tamp, 6″ to 8″ lifts, common fill	B-9	250	.160	E.C.Y.		4.45	.59	5.04	8	
	0060	Select fill	"	300	.133			3.71	.49	4.20	6.70	
	0600	Vibratory plate, 8″ lifts, common fill	A-1D	200	.040			1.10	.13	1.23	1.97	
	0700	Select fill	"	216	.037			1.01	.12	1.13	1.81	
	9000	Minimum labor/equipment charge	1 Clab	4	2	Job		55		55	91	
424	0010	**EXCAVATING, BULK BANK MEASURE** Common earth piled R312316-40										424
	0020	For loading onto trucks, add								15%	15%	
	0200	Backhoe, hydraulic, crawler mtd., 1 C.Y. cap. = 75 C.Y./hr. R312316-45	B-12A	600	.027	B.C.Y.		.87	.93	1.80	2.44	
	1200	Front end loader, track mtd., 1-1/2 C.Y. cap. = 70 C.Y./hr.	B-10N	560	.021			.72	.56	1.28	1.77	
	1500	Wheel mounted, 3/4 C.Y. cap. = 45 C.Y./hr.	B-10R	360	.033			1.12	.54	1.66	2.38	
	9000	Minimum labor/equipment charge	B-10L	2	6	Job		202	159	361	500	
432	0010	**EXCAVATING, BULK, DOZER** Open site										432
	2000	80 H.P., 50′ haul, sand & gravel	B-10L	460	.026	B.C.Y.		.88	.69	1.57	2.16	
	2200	150′ haul, sand & gravel		230	.052			1.75	1.39	3.14	4.33	
	2400	300′ haul, sand & gravel		120	.100			3.36	2.66	6.02	8.30	
	3000	105 H.P., 50′ haul, sand & gravel	B-10W	700	.017			.58	.66	1.24	1.65	
	3200	150′ haul, sand & gravel		310	.039			1.30	1.49	2.79	3.72	
	3300	300′ haul, sand & gravel		140	.086			2.88	3.31	6.19	8.25	
	4000	200 H.P., 50′ haul, sand & gravel	B-10B	1400	.009			.29	.66	.95	1.18	
	4200	150′ haul, sand & gravel		595	.020			.68	1.55	2.23	2.78	
	4400	300′ haul, sand & gravel		310	.039			1.30	2.97	4.27	5.35	
	5040	Clay	B-10M	1025	.012			.39	1.17	1.56	1.91	
	5400	300′ haul, sand & gravel	"	470	.026			.86	2.55	3.41	4.17	
462	0010	**EXCAVATION, STRUCTURAL**										462
	0015	Hand, pits to 6′ deep, sandy soil	1 Clab	8	1	B.C.Y.		27.50		27.50	45.50	
	0100	Heavy soil or clay		4	2			55		55	91	
	0300	Pits 6′ to 12′ deep, sandy soil		5	1.600			44		44	72.50	
	0500	Heavy soil or clay		3	2.667			73		73	121	
	0700	Pits 12′ to 18′ deep, sandy soil		4	2			55		55	91	
	0900	Heavy soil or clay		2	4			110		110	182	
	1100	Hand loading trucks from stock pile, sandy soil		12	.667			18.25		18.25	30.50	
	1300	Heavy soil or clay		8	1			27.50		27.50	45.50	
	1500	For wet or muck hand excavation, add to above				%				50%	50%	
	6000	Machine excavation, for spread and mat footings, elevator pits,										
	6001	and small building foundations										
	6030	Common earth, hydraulic backhoe, 1/2 C.Y. bucket	B-12E	55	.291	B.C.Y.		9.55	6.10	15.65	22	
	6035	3/4 C.Y. bucket	B-12F	90	.178			5.80	5.30	11.10	15.25	

	02315 | Excavation and Fill		CREW	DAILY OUTPUT	LABOR-HOURS	UNIT	2006 BARE COSTS MAT.	LABOR	EQUIP.	TOTAL	TOTAL INCL O&P	
462	6040	1 C.Y. bucket	B-12A	108	.148	B.C.Y.		4.85	5.20	10.05	13.50	462
	6050	1-1/2 C.Y. bucket	B-12B	144	.111			3.64	5	8.64	11.35	
	6060	2 C.Y. bucket	B-12C	200	.080			2.62	4.60	7.22	9.25	
	6070	Sand and gravel, 3/4 C.Y. bucket	B-12F	100	.160			5.25	4.78	10.03	13.70	
	6080	1 C.Y. bucket	B-12A	120	.133			4.37	4.67	9.04	12.20	
	6090	1-1/2 C.Y. bucket	B-12B	160	.100			3.28	4.52	7.80	10.25	
	6100	2 C.Y. bucket	B-12C	220	.073			2.38	4.18	6.56	8.45	
	6110	Clay, till, or blasted rock, 3/4 C.Y. bucket	B-12F	80	.200			6.55	5.95	12.50	17.10	
	6120	1 C.Y. bucket	B-12A	95	.168			5.50	5.90	11.40	15.40	
	6130	1-1/2 C.Y. bucket	B-12B	130	.123			4.03	5.55	9.58	12.60	
	6140	2 C.Y. bucket	B-12C	175	.091	↓		2.99	5.25	8.24	10.65	
	9000	Minimum labor/equipment charge	1 Clab	4	2	Job		55		55	91	
490	0010	**HAULING**, excavated or borrow, loose cubic yards										490
	0012	no loading included, highway haulers										
	0020	6 C.Y. dump truck, 1/4 mile round trip, 5.0 loads/hr.	B-34A	195	.041	L.C.Y.		1.16	1.68	2.84	3.73	
	0200	4 mile round trip, 1.8 loads/hr.	"	70	.114			3.24	4.67	7.91	10.40	
	0310	12 C.Y. dump truck, 1/4 mile round trip 3.7 loads/hr.	B-34B	288	.028			.79	1.66	2.45	3.10	
	0500	4 mile round trip, 1.6 loads/hr.	"	125	.064			1.81	3.81	5.62	7.15	
	0600	16.5 C.Y. dump trailer, 1 mile round trip, 2.6 loads/hr.	B-34C	280	.029			.81	1.45	2.26	2.91	
	1100	4 mile round trip, 1.6 loads/hr.	"	172	.047			1.32	2.35	3.67	4.74	
	1150	20 C.Y. dump trailer, 1 mile round trip, 2.5 loads/hr.	B-34D	325	.025			.70	1.28	1.98	2.55	
	1240	4 mile round trip, 1.5 loads/hr.	"	195	.041			1.16	2.14	3.30	4.24	
	1300	Hauling in medium traffic, add								20%	20%	
	1400	Heavy traffic, add								30%	30%	
	1600	Grading at dump, or embankment if required, by dozer	B-10B	1000	.012	↓		.40	.92	1.32	1.66	
	1800	Spotter at fill or cut, if required	1 Clab	8	1	Hr.		27.50		27.50	45.50	
	4700	Highway hauling beyond 20 miles, per loaded mile, minimum				Mile				1.20	1.32	
	4750	Maximum				"				2.40	2.64	
520	0010	**FILL**, spread dumped material, no compaction										520
	0020	By dozer, no compaction	B-10B	1000	.012	L.C.Y.		.40	.92	1.32	1.66	
	0100	By hand	1 Clab	12	.667	"		18.25		18.25	30.50	
	9000	Minimum labor/equipment charge	"	4	2	Job		55		55	91	
610	0010	**EXCAVATING, TRENCH** or continuous footing, common earth										610
	0020	No sheeting or dewatering included										
	1400	By hand with pick and shovel 2′ to 6′ deep, light soil	1 Clab	8	1	B.C.Y.		27.50		27.50	45.50	
	1500	Heavy soil	"	4	2	"		55		55	91	
	1700	For tamping backfilled trenches, air tamp, add	A-1G	100	.080	E.C.Y.		2.19	.37	2.56	4.04	
	1900	Vibrating plate, add	B-18	180	.133	"		3.74	.19	3.93	6.40	
	2100	Trim sides and bottom for concrete pours, common earth	↓	1500	.016	S.F.		.45	.02	.47	.77	
	2300	Hardpan	↓	600	.040	"		1.12	.06	1.18	1.92	
	9000	Minimum labor/equipment charge	1 Clab	4	2	Job		55		55	91	
620	0010	**EXCAVATING, UTILITY TRENCH** Common earth										620
	0050	Trenching with chain trencher, 12 H.P., operator walking										
	0100	4″ wide trench, 12″ deep	B-53	800	.010	L.F.		.35	.06	.41	.63	
	0150	18″ deep		750	.011			.38	.06	.44	.66	
	0200	24″ deep		700	.011			.40	.07	.47	.71	
	0300	6″ wide trench, 12″ deep		650	.012			.43	.07	.50	.76	
	0350	18″ deep		600	.013			.47	.08	.55	.83	
	0400	24″ deep		550	.015			.51	.09	.60	.91	
	0450	36″ deep		450	.018			.63	.11	.74	1.11	
	0600	8″ wide trench, 12″ deep		475	.017			.59	.10	.69	1.05	
	0650	18″ deep		400	.020			.70	.12	.82	1.24	
	0700	24″ deep		350	.023			.80	.14	.94	1.42	
	0750	36″ deep		300	.027	↓		.94	.16	1.10	1.65	
	0900	Minimum labor/equipment charge	↓	2	4	Job		141	24	165	248	

02300 | Earthwork

02315 | Excavation and Fill

		02315 Excavation and Fill	CREW	DAILY OUTPUT	LABOR-HOURS	UNIT	MAT.	LABOR	EQUIP.	TOTAL	TOTAL INCL O&P	
							2006 BARE COSTS					
620	1000	Backfill by hand including compaction, add										620
	1050	4" wide trench, 12" deep	A-1G	800	.010	L.F.		.27	.05	.32	.50	
	1100	18" deep		530	.015			.41	.07	.48	.77	
	1150	24" deep		400	.020			.55	.09	.64	1.01	
	1300	6" wide trench, 12" deep		540	.015			.41	.07	.48	.75	
	1350	18" deep		405	.020			.54	.09	.63	1	
	1400	24" deep		270	.030			.81	.14	.95	1.50	
	1450	36" deep		180	.044			1.22	.21	1.43	2.25	
	1600	8" wide trench, 12" deep		400	.020			.55	.09	.64	1.01	
	1650	18" deep		265	.030			.83	.14	.97	1.52	
	1700	24" deep		200	.040			1.10	.19	1.29	2.02	
	1750	36" deep		135	.059			1.62	.27	1.89	2.99	
	2000	Chain trencher, 40 H.P. operator riding										
	2050	6" wide trench and backfill, 12" deep	B-54	1200	.007	L.F.		.23	.18	.41	.57	
	2100	18" deep		1000	.008			.28	.22	.50	.68	
	2150	24" deep		975	.008			.29	.22	.51	.70	
	2200	36" deep		900	.009			.31	.24	.55	.76	
	2250	48" deep		750	.011			.38	.29	.67	.91	
	2300	60" deep		650	.012			.43	.33	.76	1.05	
	2400	8" wide trench and backfill, 12" deep		1000	.008			.28	.22	.50	.68	
	2450	18" deep		950	.008			.30	.23	.53	.72	
	2500	24" deep		900	.009			.31	.24	.55	.76	
	2550	36" deep		800	.010			.35	.27	.62	.86	
	2600	48" deep		650	.012			.43	.33	.76	1.05	
	2700	12" wide trench and backfill, 12" deep		975	.008			.29	.22	.51	.70	
	2750	18" deep		860	.009			.33	.25	.58	.80	
	2800	24" deep		800	.010			.35	.27	.62	.86	
	2850	36" deep		725	.011			.39	.30	.69	.94	
	3000	16" wide trench and backfill, 12" deep		835	.010			.34	.26	.60	.82	
	3050	18" deep		750	.011			.38	.29	.67	.91	
	3100	24" deep		700	.011			.40	.31	.71	.98	
	3200	Compaction with vibratory plate, add								50%	50%	
	5100	Hand excavate and trim for pipe bells after trench excavation										
	5200	8" pipe	1 Clab	155	.052	L.F.		1.41		1.41	2.34	
	5300	18" pipe	"	130	.062	"		1.69		1.69	2.79	
	9000	Minimum labor/equipment charge	A-1G	4	2	Job		55	9.25	64.25	101	
640	0010	**UTILITY BEDDING** For pipe and conduit, not incl. compaction										640
	0050	Crushed or screened bank run gravel	B-6	150	.160	L.C.Y.	20.50	4.80	1.51	26.81	32	
	0100	Crushed stone 3/4" to 1/2"		150	.160		27.50	4.80	1.51	33.81	40	
	0200	Sand, dead or bank		150	.160		4.21	4.80	1.51	10.52	14.10	
	0500	Compacting bedding in trench	A-1D	90	.089	E.C.Y.		2.44	.29	2.73	4.36	
	0600	If material source exceeds 2 miles, add for extra mileage.										
	0610	See 02315-490 for hauling mileage add.										

02360 | Soil Treatment

		02360 Soil Treatment	CREW	DAILY OUTPUT	LABOR-HOURS	UNIT	MAT.	LABOR	EQUIP.	TOTAL	TOTAL INCL O&P	
200	0010	**TERMITE PRETREATMENT**										200
	0020	Slab and walls, residential	1 Skwk	1200	.007	SF Flr.	.29	.24		.53	.72	
	0100	Commercial, minimum		2496	.003		.31	.12		.43	.53	
	0200	Maximum		1645	.005		.47	.18		.65	.80	
	0390	Minimum labor/equipment charge		4	2	Job		73		73	119	
	0400	Insecticides for termite control, minimum		14.20	.563	Gal.	11.85	20.50		32.35	46.50	
	0500	Maximum		11	.727	"	20.50	26.50		47	66	

02300 | Earthwork

02370 | Erosion & Sedimentation Control

		02370 Erosion & Sedimentation Control	CREW	DAILY OUTPUT	LABOR-HOURS	UNIT	2006 BARE COSTS MAT.	LABOR	EQUIP.	TOTAL	TOTAL INCL O&P	
700	0010	**SYNTHETIC EROSION CONTROL**										700
	0020	Jute mesh, 100 SY per roll, 4′ wide, stapled	B-80A	2400	.010	S.Y.	.75	.27	.07	1.09	1.36	
	0100	Plastic netting, stapled, 2″ x 1″ mesh, 20 mil	B-1	2500	.010		.68	.27		.95	1.20	
	0200	Polypropylene mesh, stapled, 6.5 oz./S.Y.		2500	.010		1.40	.27		1.67	1.99	
	0300	Tobacco netting, or jute mesh #2, stapled	↓	2500	.010	↓	.08	.27		.35	.54	
	1000	Silt fence, polypropylene, 3′ high, ideal conditions	2 Clab	1600	.010	L.F.	.34	.27		.61	.82	
	1100	Adverse conditions	″	950	.017	″	.34	.46		.80	1.13	
	1200	Place and remove hay bales	A-2	3	8	Ton	56	220	43	319	470	
	1250	Hay bales, staked	″	2500	.010	L.F.	2.25	.26	.05	2.56	2.96	

02400 | Tunneling, Boring & Jacking

02441 | Microtunneling

		02441 Microtunneling	CREW	DAILY OUTPUT	LABOR-HOURS	UNIT	2006 BARE COSTS MAT.	LABOR	EQUIP.	TOTAL	TOTAL INCL O&P	
400	0010	**MICROTUNNELING** Not including excavation, backfill, shoring,										400
	0020	or dewatering, average 50′/day, slurry method										
	0100	24″ to 48″ outside diameter, minimum				L.F.					700	
	0110	Adverse conditions, add				%					50%	
	1000	Rent microtunneling machine, average monthly lease				Month					94,000	
	1010	Operating technician				Day					700	
	1100	Mobilization and demobilization, minimum				Job					47,000	
	1110	Maximum				″					450,000	

02445 | Boring or Jacking Conduits

		02445 Boring or Jacking Conduits	CREW	DAILY OUTPUT	LABOR-HOURS	UNIT	2006 BARE COSTS MAT.	LABOR	EQUIP.	TOTAL	TOTAL INCL O&P	
300	0010	**HORIZONTAL BORING** Casing only, 100′ minimum,										300
	0020	not incl. jacking pits or dewatering										
	0100	Roadwork, 1/2″ thick wall, 24″ diameter casing	B-42	20	3.200	L.F.	72.50	100	56.50	229	310	
	0200	36″ diameter		16	4		115	125	71	311	415	
	0300	48″ diameter		15	4.267		169	133	75.50	377.50	490	
	0500	Railroad work, 24″ diameter		15	4.267		72.50	133	75.50	281	385	
	0600	36″ diameter		14	4.571		115	143	81	339	455	
	0700	48″ diameter	↓	12	5.333		169	166	94.50	429.50	570	
	0900	For ledge, add								155	190	
	1000	Small diameter boring, 3″, sandy soil	B-82	900	.018		19.95	.56	.07	20.58	23	
	1040	Rocky soil	″	500	.032	↓	19.95	1	.13	21.08	24	

02450 | Foundation & Load Bearing Elements

02455 | Driven Piles

		02455 Driven Piles	CREW	DAILY OUTPUT	LABOR-HOURS	UNIT	2006 BARE COSTS MAT.	LABOR	EQUIP.	TOTAL	TOTAL INCL O&P	
100	0010	**CAST IN PLACE CONCRETE PILES**, 200 piles, 60′ long										100
	0020	unless specified otherwise, not incl. pile caps or mobilization										
	0800	Cast in place friction pile, 50′ long, fluted,										
	0810	tapered steel, 4000 psi concrete, no reinforcing										

02450 | Foundation & Load Bearing Elements

02455 | Driven Piles

		02455 Driven Piles	CREW	DAILY OUTPUT	LABOR-HOURS	UNIT	2006 BARE COSTS MAT.	LABOR	EQUIP.	TOTAL	TOTAL INCL O&P	
100	0900	12″ diameter, 7 ga.	B-19	600	.107	V.L.F.	18.35	3.79	2.62	24.76	29	100
	1200	18″ diameter, 7 ga.	″	480	.133	″	27.50	4.74	3.27	35.51	42	
	1300	End bearing, fluted, constant diameter,										
	1320	4000 psi concrete, no reinforcing										
	1340	12″ diameter, 7 ga.	B-19	600	.107	V.L.F.	19.15	3.79	2.62	25.56	30	
	1400	18″ diameter, 7 ga.	″	480	.133	″	30.50	4.74	3.27	38.51	45	
450	0010	**PRESTRESSED CONCRETE PILES**, 200 piles										450
	0020	unless specified otherwise, not incl. pile caps or mobilization										
	3100	Precast, prestressed, 40′ long, 10″ thick, square	B-19	700	.091	V.L.F.	9.10	3.25	2.24	14.59	17.80	
	3200	12″ thick, square		680	.094		11.45	3.34	2.31	17.10	20.50	
	3400	14″ thick, square		600	.107		13.55	3.79	2.62	19.96	24	
	4000	18″ thick, square	B-19A	520	.123		25.50	4.37	3.86	33.73	40	
600	0010	**STEEL PILES**, Not including mobilization or demobilization										600
	0100	Step tapered, round, concrete filled										
	0110	8″ tip, 60 ton capacity, 30′ depth	B-19	760	.084	V.L.F.	7.60	2.99	2.07	12.66	15.55	
	0120	60′ depth		740	.086		8.55	3.07	2.12	13.74	16.80	
	0250	"H" Sections, 50′ long, HP8 x 36		640	.100		12.80	3.55	2.46	18.81	22.50	
	1300	HP14 X 102 	B-19A	510	.125		37	4.46	3.94	45.40	52	
	2600	Pipe piles, 50′ lg. 8″ diam., 29 lb. per L.F., no concrete	B-19	500	.128		13.85	4.55	3.14	21.54	26	
	2700	Concrete filled		460	.139		14.70	4.94	3.42	23.06	28	
	3500	14″ diameter, 46 lb. per L.F., no concrete		430	.149		23	5.30	3.65	31.95	37.50	
	3600	Concrete filled		355	.180		24.50	6.40	4.43	35.33	42.50	
	4100	18″ diameter, 59 lb. per L.F., no concrete		355	.180		33.50	6.40	4.43	44.33	52	
	4200	Concrete filled		310	.206		33	7.35	5.05	45.40	54	
650	0010	**TIMBER PILES**, Friction or end bearing, not including										650
	0050	mobilization or demobilization										
	0100	Untreated piles, up to 30′ long, 12″ butts, 8″ points	B-19	625	.102	V.L.F.	6.55	3.64	2.51	12.70	15.95	
	0200	30′ to 39′ long, 12″ butts, 8″ points 	″	700	.091	″	6.55	3.25	2.24	12.04	15	
	0800	Treated piles, 12 lb. per C.F.,										
	0810	friction or end bearing, ASTM class B										
	1000	Up to 30′ long, 12″ butts, 8″ points	B-19	625	.102	V.L.F.	9.70	3.64	2.51	15.85	19.40	
	1100	30′ to 39′ long, 12″ butts, 8″ points		700	.091		9.45	3.25	2.24	14.94	18.20	
	2700	Mobilization for 10,000 L.F. pile job, add		3300	.019			.69	.48	1.17	1.66	
	2800	25,000 L.F. pile job, add		8500	.008			.27	.18	.45	.64	
800	0010	**PILING SPECIAL COSTS**										800
	0011	Pile caps, see Division 03310-240										
	0500	Cutoffs, concrete piles, plain	1 Pile	5.50	1.455	Ea.		50.50		50.50	85.50	
	0600	With steel thin shell, add		38	.211			7.30		7.30	12.40	
	0700	Steel pile or "H" piles		19	.421			14.60		14.60	25	
	0800	Wood piles		38	.211			7.30		7.30	12.40	
	1000	Testing, any type piles, test load is twice the design load										
	1050	50 ton design load, 100 ton test				Ea.				15,000	16,050	
	1100	100 ton design load, 200 ton test								19,250	20,300	
	1200	200 ton design load, 400 ton test								26,200	28,800	
900	0010	**MOBILIZATION**										900
	0020	Set up & remove, air compressor, 600 C.F.M.	A-5	3.30	5.455	Ea.		150	9.75	159.75	258	
	0100	1200 C.F.M.	″	2.20	8.182			224	14.65	238.65	385	
	0200	Crane, with pile leads and pile hammer, 75 ton	B-19	.60	106			3,800	2,625	6,425	9,125	
	0300	150 ton	″	.36	177			6,325	4,375	10,700	15,200	

02465 | Bored Piles

		02465 Bored Piles	CREW	DAILY OUTPUT	LABOR-HOURS	UNIT	MAT.	LABOR	EQUIP.	TOTAL	TOTAL INCL O&P	
800	0010	**DRILLED CAISSONS** Incl. excav., concrete, 50 lbs. reinf.										800
	0020	per C.Y., not incl. mobiliz., boulder removal, disposal										
	0100	Open style, machine drilled, to 50′ deep, in stable ground, R316326-60										
	0110	no casings or ground water, 18″ diam., 0.065 C.Y./L.F.	B-43	200	.240	V.L.F.	7	7.30	16.90	31.20	38	

02465 | Bored Piles

		02465 Bored Piles	CREW	DAILY OUTPUT	LABOR-HOURS	UNIT	2006 BARE COSTS MAT.	LABOR	EQUIP.	TOTAL	TOTAL INCL O&P	
800	0200	24″ diameter, 0.116 C.Y./L.F. R316326-60	B-43	190	.253	V.L.F.	12.55	7.65	17.80	38	46	800
	0210	4′ bell diameter , add		20	2.400	Ea.	124	73	169	366	440	
	0500	48″ diameter, 0.465 C.Y./L.F.		100	.480	V.L.F.	50.50	14.55	34	99.05	116	
	0510	9′ bell diameter , add		2	24	Ea.	1,000	730	1,700	3,430	4,125	
	1200	Open style, machine drilled, to 50′ deep, in wet ground, pulled										
	1220	pulled casing and pumping										
	1400	24″ diameter, 0.116 C.Y./L.F.	B-48	125	.448	V.L.F.	12.55	13.90	29	55.45	68.50	
	1410	4′ bell diameter , add	″	19.80	2.828	Ea.	124	88	184	396	480	
	1700	48″ diameter, 0.465 C.Y./L.F.	B-49	55	1.600	V.L.F.	50.50	52	78	180.50	226	
	1710	9′ bell diameter , add	″	3.30	26.667	Ea.	1,000	865	1,300	3,165	3,925	
	2300	Open style, machine drilled, to 50′ deep, in soft rocks and										
	2320	medium hard shales										
	2500	24″ diameter, 0.116 C.Y./L.F.	B-49	30	2.933	V.L.F.	12.55	95.50	143	251.05	325	
	2510	4′ bell diameter, add		10.90	8.073	Ea.	124	262	395	781	995	
	2800	48″ diameter, 0.465 C.Y./L.F.		10	8.800	V.L.F.	50.50	286	430	766.50	990	
	2810	9′ bell diameter , add		1.10	80	Ea.	750	2,600	3,900	7,250	9,350	
	3600	For rock excavation, sockets, add, minimum		120	.733	C.F.		24	35.50	59.50	78	
	3650	Average		95	.926			30	45	75	98.50	
	3700	Maximum		48	1.833			59.50	89.50	149	196	
	3900	For 50′ to 100′ deep, add				V.L.F.				7%	7%	
	4000	For 100′ to 150′ deep, add								25%	25%	
	4100	For 150′ to 200′ deep, add								30%	30%	
	4200	For casings left in place, add				Lb.	.72			.72	.79	
	4300	For other than 50 lb. reinf. per C.Y., add or deduct				″	.77			.77	.85	
	4400	For steel "I" beam cores, add	B-49	8.30	10.602	Ton	1,525	345	515	2,385	2,800	
	4500	Load and haul excess excavation, 2 miles	B-34B	178	.045	L.C.Y.		1.27	2.68	3.95	5	
	4600	For mobilization, 50 mile radius, rig to 36″	B-43	2	24	Ea.		730	1,700	2,430	3,025	
	4650	Rig to 84″	B-48	1.75	32			995	2,075	3,070	3,875	
	4700	For low headroom, add								50%		
	4750	For difficult access, add								25%		

02500 | Utility Services

02510 | Water Distribution

		02510 Water Distribution	CREW	DAILY OUTPUT	LABOR-HOURS	UNIT	2006 BARE COSTS MAT.	LABOR	EQUIP.	TOTAL	TOTAL INCL O&P	
710	0010	**TAPPING, CROSSES AND SLEEVES**										710
	4000	Drill and tap pressurized main (labor only)										
	4100	6″ main, 1″ to 2″ service	Q-1	3	5.333	Ea.		205		205	320	
	4150	8″ main, 1″ to 2″ service	″	2.75	5.818	″		224		224	350	
	4500	Tap and insert gate valve										
	4600	8″ main, 4″ branch	B-21	3.20	8.750	Ea.		281	47.50	328.50	515	
	4651	Piping, drill, tap & insert gate valve, 8″ main, 6″ branch		2.70	10.370			335	56.50	391.50	605	
	4800	12″ main, 6″ branch		2.35	11.915			380	65	445	695	
730	0010	**WATER SUPPLY, DUCTILE IRON PIPE** cement lined R331113-50										730
	0020	Not including excavation or backfill										
	2000	Pipe, class 50 water piping, 18′ lengths R221113-80										
	2020	Mechanical joint, 4″ diameter	B-21A	200	.200	L.F.	13.50	6.85	2.68	23.03	29	
	3000	Tyton, push-on joint, 4″ diameter		400	.100		7.60	3.44	1.34	12.38	15.35	
	3020	6″ diameter		333.33	.120		8.90	4.12	1.61	14.63	18.10	
	8000	Fittings, mechanical joint										
	8006	90° bend, 4″ diameter	B-20A	16	2	Ea.	174	67		241	298	

02500 | Utility Services

02510 | Water Distribution

	Code	Description	Crew	Daily Output	Labor-Hours	Unit	2006 Bare Costs Mat.	Labor	Equip.	Total	Total Incl O&P
730	8020	6″ diameter (R331113-50)	B-20A	12.80	2.500	Ea.	210	83.50		293.50	365
	8200	Wye or tee, 4″ diameter		10.67	2.999		268	100		368	455
	8220	6″ diameter (R221113-80)		8.53	3.751		360	125		485	595
	8398	45° bends, 4″ diameter		16	2		157	67		224	280
	8450	Decreaser, 6″ x 4″ diameter		14.22	2.250		191	75		266	330
	8460	8″ x 6″ diameter		11.64	2.749		279	92		371	450
	8550	Butterfly valves with boxes, cast iron									
	8560	4″ diameter	B-20	6	4	Ea.	760	124		884	1,050
	9600	Steel sleeve and tap, 4″ diameter		3	8		430	249		679	880
	9620	6″ diameter		2	12		505	375		880	1,175

02520 | Wells

	Code	Description	Crew	Daily Output	Labor-Hours	Unit	2006 Bare Costs Mat.	Labor	Equip.	Total	Total Incl O&P
510	0010	**WELLS & ACCESSORIES,** domestic (R221113-50)									
	0100	Drilled, 4″ to 6″ diameter	B-23	120	.333	L.F.		9.25	29.50	38.75	48
	0200	8″ diameter	″	95.20	.420	″		11.70	37.50	49.20	60.50
	1400	Remove & reset pump, minimum	B-21	4	7	Ea.		225	38	263	410
	1420	Maximum	″	2	14	″		450	76.50	526.50	820
	1500	Pumps, installed in wells to 100′ deep, 4″ submersible									
	1510	1/2 H.P.	Q-1	3.22	4.969	Ea.	345	191		536	675
	1520	3/4 H.P.		2.66	6.015		400	231		631	800
	1600	1 H.P.		2.29	6.987		425	269		694	880
	1700	1-1/2 H.P.	Q-22	1.60	10		1,100	385	400	1,885	2,250
	1800	2 H.P.		1.33	12.030		1,275	460	480	2,215	2,650
	1900	3 H.P.		1.14	14.035		1,475	540	560	2,575	3,075
	2000	5 H.P.		1.14	14.035		2,000	540	560	3,100	3,650
	2050	Remove and install motor only, 4 H.P.		1.14	14.035		755	540	560	1,855	2,275
	5000	Wells to 180 ft. deep, 4″ submersible, 1 HP	B-21	1.10	25.455		1,925	815	139	2,879	3,600
	5500	2 HP		1.10	25.455		2,325	815	139	3,279	4,050
	6000	3 HP		1	28		3,175	900	153	4,228	5,125
	7000	5 HP		.90	31.111		3,825	1,000	170	4,995	6,025
	9000	Minimum labor/equipment charge		1.80	15.556	Job		500	85	585	910

02530 | Sanitary Sewerage

	Code	Description	Crew	Daily Output	Labor-Hours	Unit	2006 Bare Costs Mat.	Labor	Equip.	Total	Total Incl O&P
730	0010	**SEWAGE COLLECTION, CONCRETE PIPE**									
	0020	See 02630-530 for sewage/drainage collection, concrete pipe									
780	0010	**SEWAGE COLLECTION, POLYVINYL CHLORIDE PIPE**									
	0020	Not including excavation or backfill									
	2000	10′ lengths, S.D.R. 35, B&S, 4″ diameter	B-20	375	.064	L.F.	2.53	1.99		4.52	6.05
	2040	6″ diameter		350	.069		5.30	2.13		7.43	9.35
	2080	8″ diameter		335	.072		8.05	2.23		10.28	12.50
	2120	10″ diameter	B-21	330	.085		11.65	2.72	.46	14.83	17.75
	4000	Piping, DWV PVC,no exc/bkfill, 10′ L, Sch 40, 4″ dia	B-20	375	.064		2.05	1.99		4.04	5.55
	4010	6″ dia		350	.069		4.43	2.13		6.56	8.40
	4020	8″ dia		335	.072		13.15	2.23		15.38	18.10

02540 | Septic Tank Systems

	Code	Description	Crew	Daily Output	Labor-Hours	Unit	2006 Bare Costs Mat.	Labor	Equip.	Total	Total Incl O&P
400	0010	**SEPTIC TANKS**									
	0015	Septic tanks, not incl exc or piping, precast, 1,000 gal	B-21	8	3.500	Ea.	585	112	19.10	716.10	850
	0020	1,250 gallon		8	3.500		820	112	19.10	951.10	1,100
	0060	1,500 gallon		7	4		965	128	22	1,115	1,275
	0100	2,000 gallon		5	5.600		1,175	180	30.50	1,385.50	1,600
	0140	2,500 gallon		5	5.600		1,550	180	30.50	1,760.50	2,025
	0180	4,000 gallon		4	7		4,525	225	38	4,788	5,375
	0220	5,000 gal., 4 piece	B-13	3	18.667		7,225	560	217	8,002	9,100

02540 | Septic Tank Systems

		02540 Septic Tank Systems	CREW	DAILY OUTPUT	LABOR-HOURS	UNIT	2006 BARE COSTS MAT.	LABOR	EQUIP.	TOTAL	TOTAL INCL O&P	
400	0300	15,000 gallon, 4 piece	B-13B	1.70	32.941	Ea.	16,400	985	560	17,945	20,200	400
	0400	25,000 gallon, 4 piece		1.10	50.909		34,600	1,525	865	36,990	41,500	
	0500	40,000 gallon, 4 piece		.80	70		40,800	2,100	1,200	44,100	49,500	
	0520	50,000 gallon, 5 piece	B-13C	.60	93.333		46,900	2,800	2,650	52,350	59,000	
	0540	75,000 gallon, cast in place	C-14C	.25	448		57,000	15,200	85	72,285	88,500	
	0560	100,000 gallon	"	.15	746		70,500	25,300	142	95,942	119,500	
	0600	High density polyethylene, 1,000 gallon	B-21	6	4.667		910	150	25.50	1,085.50	1,275	
	0700	1,500 gallon		4	7		1,175	225	38	1,438	1,675	
	0900	Galley, 4′ x 4′ x 4′		16	1.750		216	56	9.55	281.55	340	
	1000	Distribution boxes, concrete, 7 outlets	2 Clab	16	1		109	27.50		136.50	166	
	1100	9 outlets	"	8	2		295	55		350	415	
	1150	Leaching field chambers, 13′ x 3′-7″ x 1′-4″, standard	B-13	16	3.500		590	105	40.50	735.50	865	
	1200	Heavy duty, 8′ x 4′ x 1′-6″		14	4		410	120	46.50	576.50	695	
	1300	13′ x 3′-9″ x 1′-6″		12	4.667		1,450	140	54	1,644	1,875	
	1350	20′ x 4′ x 1′-6″		5	11.200		955	335	130	1,420	1,750	
	1420	Leaching pit, 6′, dia, 3′ deep complete					680			680	745	
	1600	Leaching pit, 6′-6″ diameter, 6′ deep	B-21	5	5.600		655	180	30.50	865.50	1,050	
	1620	8′ deep		4	7		770	225	38	1,033	1,250	
	1700	8′ diameter, H-20 load, 6′ deep		4	7		1,050	225	38	1,313	1,575	
	1720	8′ deep		3	9.333		1,300	300	51	1,651	2,000	
	2000	Velocity reducing pit, precast conc., 6′ diameter, 3′ deep		4.70	5.957		325	191	32.50	548.50	710	
	2200	Excavation for septic tank, 3/4 C.Y. backhoe	B-12F	145	.110	C.Y.		3.61	3.29	6.90	9.40	
	2400	4′ trench for disposal field, 3/4 C.Y. backhoe	"	335	.048	L.F.		1.56	1.43	2.99	4.09	
	2600	Gravel fill, run of bank	B-6	150	.160	C.Y.	16.50	4.80	1.51	22.81	27.50	
	2800	Crushed stone, 3/4″	"	150	.160	"	23	4.80	1.51	29.31	34.50	

02550 | Piped Energy Distribution

		02550 Piped Energy Distribution	CREW	DAILY OUTPUT	LABOR-HOURS	UNIT	MAT.	LABOR	EQUIP.	TOTAL	TOTAL INCL O&P	
464	0010	**PIPING, GAS SERVICE & DISTRIBUTION, POLYETHYLENE**										464
	0020	not including excavation or backfill										
	1000	60 psi coils, comp cplg @ 100′, 1/2″ diameter, SDR 9.3	B-20A	608	.053	L.F.	.51	1.76		2.27	3.37	
	1040	1-1/4″ diameter, SDR 11		544	.059		.95	1.97		2.92	4.18	
	1100	2″ diameter, SDR 11		488	.066		1.17	2.19		3.36	4.79	
	1160	3″ diameter, SDR 11		408	.078		2.43	2.62		5.05	6.85	
	1500	60 PSI 40′ joints with coupling, 3″ diameter, SDR 11	B-21A	408	.098		2.43	3.37	1.31	7.11	9.45	
	1540	4″ diameter, SDR 11		352	.114		5.60	3.90	1.52	11.02	14.05	
	1600	6″ diameter, SDR 11		328	.122		17.35	4.19	1.64	23.18	27.50	
	1640	8″ diameter, SDR 11		272	.147		23.50	5.05	1.97	30.52	36	
	9000	Minimum labor/equipment charge	B-20	2	12	Job		375		375	615	
550	0010	**GAS STATION PRODUCT LINE**										550
	0020	Primary containment pipe, fiberglass-reinforced										
	0030	Plastic pipe 15′ & 30′ lengths										
	0040	2″ diameter	Q-6	425	.056	L.F.	3.49	2.27		5.76	7.35	
	0050	3″ diameter		400	.060		4.57	2.41		6.98	8.80	
	0060	4″ diameter		375	.064		5.90	2.57		8.47	10.50	
	0100	Fittings										
	0110	Elbows, 90° & 45°, bell-ends, 2″	Q-6	24	1	Ea.	35.50	40		75.50	102	
	0120	3″ diameter		22	1.091		37	44		81	109	
	0130	4″ diameter		20	1.200		49	48		97	129	
	0200	Tees, bell ends, 2″		21	1.143		43	46		89	119	
	0210	3″ diameter		18	1.333		43	53.50		96.50	131	
	0220	4″ diameter		15	1.600		59	64.50		123.50	165	
	0230	Flanges bell ends, 2″		24	1		14.15	40		54.15	78	
	0240	3″ diameter		22	1.091		17.85	44		61.85	87.50	
	0250	4″ diameter		20	1.200		24.50	48		72.50	102	

For expanded coverage of these items see *Means Site Work & Landscape Cost Data 2006*

02500 | Utility Services

02550 | Piped Energy Distribution

		02550 Piped Energy Distribution	CREW	DAILY OUTPUT	LABOR-HOURS	UNIT	2006 BARE COSTS MAT.	LABOR	EQUIP.	TOTAL	TOTAL INCL O&P	
550	0260	Sleeve couplings, 2"	Q-6	21	1.143	Ea.	9	46		55	81.50	550
	0270	3" diameter		18	1.333		12.80	53.50		66.30	97.50	
	0280	4" diameter		15	1.600		17.65	64.50		82.15	119	
	0290	Threaded adapters 2"		21	1.143		11.85	46		57.85	84.50	
	0300	3" diameter		18	1.333		21	53.50		74.50	107	
	0310	4" diameter		15	1.600		28	64.50		92.50	131	
	0320	Reducers, 2"		27	.889		15.75	35.50		51.25	73	
	0330	3" diameter		22	1.091		18.25	44		62.25	88	
	0340	4" diameter		20	1.200		23.50	48		71.50	101	
	1010	Gas station product line for secondary containment (double wall)										
	1100	Fiberglass reinforced plastic pipe 25' lengths										
	1120	Pipe, plain end, 3"	Q-6	375	.064	L.F.	6	2.57		8.57	10.60	
	1130	4" diameter		350	.069		9.85	2.76		12.61	15.10	
	1140	5" diameter		325	.074		12.80	2.97		15.77	18.65	
	1150	6" diameter		300	.080		13.15	3.21		16.36	19.50	
	1200	Fittings										
	1230	Elbows, 90° & 45°, 3"	Q-6	18	1.333	Ea.	43	53.50		96.50	131	
	1240	4" diameter		16	1.500		75.50	60.50		136	177	
	1250	5" diameter		14	1.714		175	69		244	299	
	1260	6" diameter		12	2		177	80.50		257.50	320	
	1270	Tees, 3"		15	1.600		64	64.50		128.50	171	
	1280	4" diameter		12	2		94	80.50		174.50	228	
	1290	5" diameter		9	2.667		190	107		297	375	
	1300	6" diameter		6	4		199	161		360	470	
	1310	Couplings, 3"		18	1.333		30.50	53.50		84	117	
	1320	4" diameter		16	1.500		78.50	60.50		139	180	
	1330	5" diameter		14	1.714		163	69		232	286	
	1340	6" diameter		12	2		169	80.50		249.50	310	
	1350	Cross-over nipples, 3"		18	1.333		6.90	53.50		60.40	91	
	1360	4" diameter		16	1.500		8.10	60.50		68.60	102	
	1370	5" diameter		14	1.714		12.05	69		81.05	120	
	1380	6" diameter		12	2		12.65	80.50		93.15	139	
	1400	Telescoping, reducers, concentric 4" x 3"		18	1.333		23	53.50		76.50	109	
	1410	5" x 4"		17	1.412		60	56.50		116.50	154	
	1420	6" x 5"		16	1.500		144	60.50		204.50	253	

02580 | Elec/Communication Structures

		02580 Elec/Communication Structures	CREW	DAILY OUTPUT	LABOR-HOURS	UNIT	MAT.	LABOR	EQUIP.	TOTAL	TOTAL INCL O&P	
420	0010	**ELECTRIC & TELEPHONE UNDERGROUND**, Not including excavation,										420
	0200	backfill and cast in place concrete										
	4200	Underground duct, banks ready for concrete fill, min. of 7.5"										
	4400	between conduits, ctr. to ctr.(for wire & cable see Div. 16120)										
	4600	2 @ 2" diameter	2 Elec	240	.067	L.F.	1.50	2.80		4.30	5.95	
	4800	4 @ 2" diameter		120	.133		3	5.60		8.60	11.95	
	5600	4 @ 4" diameter		80	.200		6.30	8.40		14.70	19.90	
	6200	Rigid galvanized steel, 2 @ 2" diameter		180	.089		14.80	3.73		18.53	22	
	6400	4 @ 2" diameter		90	.178		29.50	7.45		36.95	44	
	7400	4 @ 4" diameter		34	.471		96.50	19.75		116.25	137	
	9990	Minimum labor/equipment charge	1 Elec	3.50	2.286	Job		96		96	148	

 Important: See the Reference Section for supporting data - Crews, Rental Equipment, City Cost Indexes and Reference Data

02600 | Drainage & Containment

		02630 | Storm Drainage	CREW	DAILY OUTPUT	LABOR-HOURS	UNIT	2006 BARE COSTS MAT.	LABOR	EQUIP.	TOTAL	TOTAL INCL O&P	
110	0010	**CATCH BASIN GRATES AND FRAMES** not including footing, excavation										110
	1600	Frames & covers, C.I., 24″ square, 500 lb.	B-6	7.80	3.077	Ea.	268	92.50	29	389.50	475	
	1700	26″ D shape, 600 lb.	″	7	3.429	″	460	103	32.50	595.50	710	
	3320	Frames and covers, existing, raised for paving, 2″, including										
	3340	row of brick, concrete collar, up to 12″ wide frame	B-6	18	1.333	Ea.	39.50	40	12.55	92.05	122	
120	0010	**REMOVE AND REPLACE CATCH BASIN COVER**										120
	0023	Remove catch basin cover	1 Clab	80	.100	Ea.		2.74		2.74	4.54	
	0033	Re-place catch basin cover	″	80	.100	″		2.74		2.74	4.54	
400	0010	**STORM DRAINAGE MANHOLES, FRAMES & COVERS** not including										400
	0020	footing, excavation, backfill (See line items for frame & cover)										
	0050	Brick, 4′ inside diameter, 4′ deep	D-1	1	16	Ea.	360	515		875	1,225	
	0100	6′ deep		.70	22.857		505	735		1,240	1,750	
	0150	8′ deep		.50	32		645	1,025		1,670	2,375	
	0200	For depths over 8′, add		4	4	V.L.F.	182	129		311	410	
	1110	Precast, 4′ I.D., 4′ deep	B-22	4.10	7.317	Ea.	710	238	56	1,004	1,225	
	1120	6′ deep		3	10		915	325	76.50	1,316.50	1,625	
	1130	8′ deep		2	15		1,075	490	114	1,679	2,100	
	1140	For depths over 8′, add		16	1.875	V.L.F.	150	61	14.30	225.30	280	
510	0010	**PIPING, STORM DRAINAGE, CORRUGATED METAL** R221113-50										510
	0020	Not including excavation or backfill										
	2000	Corrugated metal pipe, galvanized and coated										
	2020	Bituminous coated with paved invert, 20′ lengths										
	2040	8″ diameter, 16 ga.	B-14	330	.145	L.F.	9.30	4.22	.69	14.21	17.95	
	2080	12″ diameter, 16 ga.	″	210	.229	″	13.40	6.65	1.08	21.13	27	
530	0010	**SEWAGE/DRAINAGE COLLECTION, CONCRETE PIPE**										530
	0020	Not including excavation or backfill										
	1000	Non-reinforced pipe, extra strength, B&S or T&G joints										
	1010	6″ diameter	B-14	265.04	.181	L.F.	4.46	5.25	.85	10.56	14.50	
	1020	8″ diameter		224	.214		4.91	6.20	1.01	12.12	16.70	
	1040	12″ diameter		200	.240		6.70	6.95	1.13	14.78	20	
	2000	Reinforced culvert, class 3, no gaskets										
	2010	12″ diameter	B-14	150	.320	L.F.	12.10	9.30	1.51	22.91	30.50	
	2020	15″ diameter	″	150	.320	″	15.50	9.30	1.51	26.31	34	

02700 | Bases, Ballasts, Pavements & Appurtenances

		02720 | Unbound Base Courses & Ballasts	CREW	DAILY OUTPUT	LABOR-HOURS	UNIT	2006 BARE COSTS MAT.	LABOR	EQUIP.	TOTAL	TOTAL INCL O&P	
200	0010	**AGGREGRATE BASE COURSE** For roadways and large areas										200
	0050	Crushed 3/4″ stone base, compacted, 3″ deep	B-36C	5200	.008	S.Y.	2.69	.26	.51	3.46	3.93	
	0100	6″ deep		5000	.008		5.40	.27	.53	6.20	6.90	
	0200	9″ deep		4600	.009		8.05	.29	.58	8.92	10	
	0300	12″ deep		4200	.010		10.75	.32	.63	11.70	13.05	
	0301	Crushed 1-1/2″ stone base, compacted to 4″ deep	B-36B	6000	.011		4.25	.35	.50	5.10	5.80	
	0302	6″ deep		5400	.012		6.40	.38	.56	7.34	8.25	
	0303	8″ deep		4500	.014		8.50	.46	.67	9.63	10.85	
	0304	12″ deep		3800	.017		12.75	.55	.79	14.09	15.80	
	0310	Minimum labor/equipment charge	B-36	1	40	Job		1,250	1,150	2,400	3,325	

SITE CONSTRUCTION 2

02700 | Bases, Ballasts, Pavements & Appurtenances

02720 | Unbound Base Courses & Ballasts

		02720 Unbound Base Courses & Ballasts	CREW	DAILY OUTPUT	LABOR-HOURS	UNIT	2006 BARE COSTS MAT.	LABOR	EQUIP.	TOTAL	TOTAL INCL O&P
200	0350	Bank run gravel, spread and compacted									
	0370	6" deep	B-32	6000	.005	S.Y.	3.54	.18	.26	3.98	4.47
	0390	9" deep		4900	.007		5.30	.22	.32	5.84	6.55
	0400	12" deep		4200	.008		7.05	.26	.38	7.69	8.65
	6900	For small and irregular areas, add						50%	50%		
	6990	Minimum labor/equipment charge	1 Clab	3	2.667	Job		73		73	121
	7000	Prepare and roll sub-base, small areas to 2500 S.Y.	B-32A	1500	.016	S.Y.		.54	.65	1.19	1.58
	8000	Large areas over 2500 S.Y.	B-32	3700	.009	"		.30	.43	.73	.94
	9000	Minimum labor/equipment charge	1 Clab	4	2	Job		55		55	91

02740 | Flexible Pavement

		02740 Flexible Pavement	CREW	DAILY OUTPUT	LABOR-HOURS	UNIT	MAT.	LABOR	EQUIP.	TOTAL	TOTAL INCL O&P
310	0010	**ASPHALTIC CONCRETE PAVEMENT, HIGHWAYS**									
	0020	and large paved areas									
	0080	Binder course, 1-1/2" thick	B-25	7725	.011	S.Y.	2.89	.34	.27	3.50	4.04
	0120	2" thick		6345	.014		3.85	.42	.33	4.60	5.30
	0160	3" thick		4905	.018		5.70	.54	.43	6.67	7.65
	0300	Wearing course, 1" thick	B-25B	10575	.009		2.04	.28	.22	2.54	2.94
	0340	1-1/2" thick		7725	.012		3.11	.38	.30	3.79	4.37
	0380	2" thick		6345	.015		4.18	.46	.36	5	5.75
	0420	2-1/2" thick		5480	.018		5.15	.54	.42	6.11	7
	0460	3" thick		4900	.020		6.15	.60	.47	7.22	8.25
	0500	Open graded friction course	B-25C	5000	.010		1.85	.30	.37	2.52	2.93
	0800	Alternate method of figuring paving costs									
	0810	Binder course, 1-1/2" thick	B-25	630	.140	Ton	38	4.21	3.36	45.57	52
	0811	2" thick		690	.128		38	3.84	3.07	44.91	51
	0812	3" thick		800	.110		38	3.31	2.65	43.96	50
	0813	4" thick		850	.104		38	3.12	2.49	43.61	49.50
	0850	Wearing course, 1" thick	B-25B	575	.167		41	5.10	4.03	50.13	57.50
	0851	1-1/2" thick		630	.152		41	4.67	3.67	49.34	56.50
	0852	2" thick		690	.139		41	4.27	3.35	48.62	55.50
	0853	2-1/2" thick		745	.129		41	3.95	3.11	48.06	55
	0854	3" thick		800	.120		41	3.68	2.89	47.57	54
	1000	Pavement replacement over trench, 2" thick	B-37	90	.533	S.Y.	4.25	15.50	1.27	21.02	31.50
	1050	4" thick		70	.686		8.40	19.90	1.63	29.93	43.50
	1080	6" thick		55	.873		13.40	25.50	2.08	40.98	58.50
	1200	For paving projects 300 tons or less add for trucking									
	1300	See 02315-490-0010 for hauling costs									
315	0011	**ASPHALTIC CONCRETE PAVEMENT, LOTS & DRIVEWAYS**									
	0020	6" stone base, 2" binder course, 1" topping	B-25C	9000	.005	S.F.	1.42	.16	.21	1.79	2.06
	0300	Binder course, 1-1/2" thick		35000	.001		.34	.04	.05	.43	.51
	0400	2" thick		25000	.002		.45	.06	.07	.58	.67
	0500	3" thick		15000	.003		.69	.10	.12	.91	1.06
	0600	4" thick		10800	.004		.90	.14	.17	1.21	1.40
	0800	Sand finish course, 3/4" thick		41000	.001		.18	.04	.05	.27	.31
	0900	1" thick		34000	.001		.22	.04	.05	.31	.37
	1000	Fill pot holes, hot mix, 2" thick	B-16	4200	.008		.47	.21	.11	.79	.98
	1100	4" thick		3500	.009		.68	.26	.14	1.08	1.32
	1120	6" thick		3100	.010		.92	.29	.15	1.36	1.66
	1140	Cold patch, 2" thick	B-51	3000	.016		.47	.44	.04	.95	1.30
	1160	4" thick		2700	.018		.90	.49	.05	1.44	1.86
	1180	6" thick		1900	.025		1.40	.70	.07	2.17	2.77

02750 | Rigid Pavement

		02750 Rigid Pavement	CREW	DAILY OUTPUT	LABOR-HOURS	UNIT	MAT.	LABOR	EQUIP.	TOTAL	TOTAL INCL O&P
300	0010	**PLAIN CEMENT CONCRETE PAVEMENT**									
	0015	Including joints, finishing and curing									

		02750 | Rigid Pavement	CREW	DAILY OUTPUT	LABOR-HOURS	UNIT	2006 BARE COSTS MAT.	LABOR	EQUIP.	TOTAL	TOTAL INCL O&P	
300	0020	Fixed form, 12′ pass, unreinforced, 6″ thick	B-26	3000	.029	S.Y.	21.50	.91	.90	23.31	26	300
	0100	8″ thick		2750	.032		29.50	.99	.98	31.47	35	
	0400	12″ thick		1800	.049		40	1.52	1.50	43.02	48	
	0505	Minimum labor/equipment charge	1 Cefi	1	8	Job		275		275	435	
	0700	Finishing, broom finish small areas	2 Cefi	120	.133	S.Y.		4.59		4.59	7.25	
	1000	Curing, with sprayed membrane by hand	2 Clab	1500	.011	″	.50	.29		.79	1.03	
		02760 | Paving Specialties										
300	0010	PAINTED TRAFFIC LINES AND MARKINGS										300
	0020	Acrylic waterborne, white or yellow, 4″ wide	B-78	20000	.002	L.F.	.17	.07	.02	.26	.32	
	0200	6″ wide		11000	.004		.16	.12	.04	.32	.42	
	0500	8″ wide		10000	.005		.22	.13	.04	.39	.50	
	0600	12″ wide		4000	.012		.41	.33	.10	.84	1.11	
	0620	Arrows or gore lines		2300	.021	S.F.	.64	.58	.17	1.39	1.85	
	0640	Temporary paint, white or yellow		15000	.003	L.F.	.19	.09	.03	.31	.39	
	0660	Removal	1 Clab	300	.027			.73		.73	1.21	
	0680	Temporary tape	2 Clab	1500	.011		1.85	.29		2.14	2.52	
	0710	Thermoplastic, white or yellow, 4″ wide	B-79	15000	.003		.74	.07	.06	.87	1	
	0730	6″ wide		14000	.003		1.07	.08	.06	1.21	1.38	
	0740	8″ wide		12000	.003		1.44	.09	.07	1.60	1.82	
	0750	12″ wide		6000	.007		2.15	.19	.15	2.49	2.83	
	0760	Arrows		660	.061	S.F.	1.98	1.69	1.34	5.01	6.45	
	0770	Gore lines		2500	.016		1.31	.45	.36	2.12	2.57	
	0780	Letters		660	.061		1.64	1.69	1.34	4.67	6.05	
500	0010	PAVEMENT MARKINGS										500
	0790	Layout of pavement marking	A-2	25000	.001	L.F.		.03	.01	.04	.05	
	0800	Parking stall, paint, white	B-78	440	.109	Stall	3.43	3.03	.88	7.34	9.75	
	1000	Street letters and numbers	″	1600	.030	S.F.	.66	.83	.24	1.73	2.38	
		02770 | Curbs and Gutters										
100	0010	BITUMINOUS CONCRETE CURBS										100
	0012	Curbs, asphaltic, machine formed, 8″ wide, 6″ high, 40 L.F./ton	B-27	1000	.032	L.F.	.82	.89	.23	1.94	2.64	
	0100	8″ wide, 8″ high, 30 L.F. per ton		900	.036		.94	.99	.26	2.19	2.96	
	0150	Asphaltic berm, 12″ W, 3″-6″ H, 35 L.F./ton, before pavement		700	.046		1.07	1.28	.33	2.68	3.65	
	0200	12″ W, 1-1/2″ to 4″ H, 60 L.F. per ton, laid with pavement	B-2	1050	.038		.65	1.06		1.71	2.47	
300	0010	CEMENT CONCRETE CURBS										300
	0300	Concrete, wood forms, 6″ x 18″, straight	C-2A	500	.096	L.F.	2.89	3.30		6.19	8.55	
	0400	6″ x 18″, radius	″	200	.240		3	8.25		11.25	16.85	
	0415	Machine formed, 6″ x 18″, straight	B-69A	2000	.024		3.80	.73	.30	4.83	5.70	
	0416	6″ x 18″, radius	″	900	.053		3.96	1.62	.66	6.24	7.75	
	0550	Precast, 6″ x 18″, straight	B-29	700	.080		7.95	2.40	1.20	11.55	13.95	
	0600	6″ x 18″, radius	″	325	.172		9.05	5.15	2.59	16.79	21.50	
500	0010	STONE CURBS										500
	1000	Granite, split face, straight, 5″ x 16″	D-13	500	.096	L.F.	10.20	3.27	1.07	14.54	17.70	
	1100	6″ x 18″	″	450	.107		13.40	3.63	1.19	18.22	22	
	1300	Radius curbing, 6″ x 18″, over 10′ radius	B-29	260	.215		16.40	6.45	3.24	26.09	32	
	1400	Corners, 2′ radius		80	.700	Ea.	55	21	10.50	86.50	106	
	1600	Edging, 4-1/2″ x 12″, straight		300	.187	L.F.	5.10	5.60	2.80	13.50	17.80	
	1800	Curb inlets, (guttermouth) straight		41	1.366	Ea.	122	41	20.50	183.50	224	
	2000	Indian granite (belgian block)										
	2100	Jumbo, 10-1/2″ x 7-1/2″ x 4″, grey	D-1	150	.107	L.F.	1.86	3.43		5.29	7.60	
	2150	Pink		150	.107		2.41	3.43		5.84	8.20	
	2200	Regular, 9″ x 4-1/2″ x 4-1/2″, grey		160	.100		1.68	3.22		4.90	7.05	
	2250	Pink		160	.100		2.31	3.22		5.53	7.75	

		02770 | Curbs and Gutters	Crew	Daily Output	Labor-Hours	Unit	2006 Bare Costs Mat.	Labor	Equip.	Total	Total Incl O&P	
500	2300	Cubes, 4″ x 4″ x 4″, grey	D-1	175	.091	L.F.	1.60	2.94		4.54	6.55	500
	2350	Pink		175	.091		1.68	2.94		4.62	6.60	
	2400	6″ x 6″ x 6″, pink		155	.103		4.17	3.32		7.49	10	
	2500	Alternate pricing method for indian granite										
	2550	Jumbo, 10-1/2″ x 7-1/2″ x 4″ (30 lb), grey				Ton	106			106	116	
	2600	Pink					140			140	154	
	2650	Regular, 9″ x 4-1/2″ x 4-1/2″ (20 lb), grey					119			119	131	
	2700	Pink					162			162	178	
	2750	Cubes, 4″ x 4″ x 4″ (15 lb), grey					194			194	214	
	2800	Pink					216			216	238	
	2850	6″ x 6″ x 6″ (25 lb), pink					162			162	178	
	2900	For pallets, add					16.50			16.50	18.15	
		02775 | Sidewalks										
275	0010	**SIDEWALKS, DRIVEWAYS, & PATIOS** No base										275
	0020	Asphaltic concrete, 2″ thick	B-37	720	.067	S.Y.	4.01	1.94	.16	6.11	7.75	
	0100	2-1/2″ thick	″	660	.073	″	5.10	2.11	.17	7.38	9.25	
	0110	Bedding for brick or stone, mortar, 1″ thick	D-1	300	.053	S.F.	.65	1.71		2.36	3.50	
	0120	2″ thick	″	200	.080		1.63	2.57		4.20	5.95	
	0130	Sand, 2″ thick	B-18	8000	.003		.16	.08		.24	.32	
	0140	4″ thick	″	4000	.006		.32	.17	.01	.50	.65	
	0300	Concrete, 3000 psi, CIP, 6 x 6 - W1.4 x W1.4 mesh,										
	0310	broomed finish, no base, 4″ thick	B-24	600	.040	S.F.	1.45	1.30		2.75	3.71	
	0350	5″ thick		545	.044		1.93	1.43		3.36	4.46	
	0400	6″ thick		510	.047		2.26	1.53		3.79	4.97	
	0450	For bank run gravel base, 4″ thick, add	B-18	2500	.010		.42	.27	.01	.70	.94	
	0520	8″ thick, add	″	1600	.015		.85	.42	.02	1.29	1.66	
	0550	Exposed aggregate finish, add to above, minimum	B-24	1875	.013		.11	.42		.53	.80	
	0600	Maximum	″	455	.053		.35	1.71		2.06	3.18	
	0950	Concrete tree grate, 5′ square	B-6	25	.960	Ea.	345	29	9.05	383.05	435	
	0960	Cast iron tree grate with frame, 2 piece, round, 5′ diameter		25	.960		950	29	9.05	988.05	1,100	
	0980	Square, 5′ side		25	.960		980	29	9.05	1,018.05	1,125	
	1700	Redwood, prefabricated, 4′ x 4′ sections	2 Carp	316	.051	S.F.	8.10	1.80		9.90	11.90	
	1750	Redwood planks, 1″ thick, on sleepers	″	240	.067	″	5.65	2.37		8.02	10.20	
	2250	Stone dust, 4″ thick	B-62	900	.027	S.Y.	2.94	.80	.17	3.91	4.72	
	9000	Minimum labor/equipment charge	D-1	2	8	Job		257		257	415	
		02780 | Unit Pavers										
100	0010	**ASPHALT BLOCKS**										100
	0020	Rectangular, 6″ x 12″ x 1-1/4″, w/bed & neopr. adhesive	D-1	135	.119	S.F.	4.64	3.81		8.45	11.30	
	0100	3″ thick		130	.123		6.50	3.96		10.46	13.55	
	0300	Hexagonal tile, 8″ wide, 1-1/4″ thick		135	.119		4.64	3.81		8.45	11.30	
	0400	2″ thick		130	.123		6.50	3.96		10.46	13.55	
	0500	Square, 8″ x 8″, 1-1/4″ thick		135	.119		4.64	3.81		8.45	11.30	
	0600	2″ thick		130	.123		6.50	3.96		10.46	13.55	
	9000	Minimum labor/equipment charge	1 Bric	2	4	Job		146		146	237	
200	0010	**BRICK PAVING**										200
	0012	4″ x 8″ x 1-1/2″, without joints (4.5 brick/S.F.)	D-1	110	.145	S.F.	2.57	4.68		7.25	10.45	
	0100	Grouted, 3/8″ joint (3.9 brick/S.F.)		90	.178		3.01	5.70		8.71	12.55	
	0200	4″ x 8″ x 2-1/4″, without joints (4.5 bricks/S.F.)		110	.145		3.33	4.68		8.01	11.25	
	0300	Grouted, 3/8″ joint (3.9 brick/S.F.)		90	.178		3.07	5.70		8.77	12.65	
	0500	Bedding, asphalt, 3/4″ thick	B-25	5130	.017		.38	.52	.41	1.31	1.71	
	0540	Course washed sand bed, 1″ thick	B-18	5000	.005		.18	.13	.01	.32	.43	
	0580	Mortar, 1″ thick	D-1	300	.053		.54	1.71		2.25	3.38	
	0620	2″ thick		200	.080		1.09	2.57		3.66	5.35	
	1500	Brick on 1″ thick sand bed laid flat, 4.5 per S.F.		100	.160		2.53	5.15		7.68	11.15	

02700 | Bases, Ballasts, Pavements & Appurtenances

		02780 | Unit Pavers	CREW	DAILY OUTPUT	LABOR-HOURS	UNIT	2006 BARE COSTS MAT.	LABOR	EQUIP.	TOTAL	TOTAL INCL O&P	
200	2000	Brick pavers, laid on edge, 7.2 per S.F.	D-1	70	.229	S.F.	2.48	7.35		9.83	14.60	200
	2500	For 4" thick concrete bed and joints, add	↓	595	.027		1.03	.86		1.89	2.53	
	2800	For steam cleaning, add	A-1H	950	.008	↓	.05	.23	.06	.34	.51	
	9000	Minimum labor/equipment charge	1 Bric	2	4	Job		146		146	237	
600	0010	**STONE PAVERS**										600
	1300	Slate, natural cleft, irregular, 3/4" thick	D-1	92	.174	S.F.	5.25	5.60		10.85	14.85	
	1350	Random rectangular, gauged, 1/2" thick		105	.152		11.40	4.90		16.30	20.50	
	1400	Random rectangular, butt joint, gauged, 1/4" thick	↓	150	.107		12.25	3.43		15.68	19.05	
	1450	For sand rubbed finish, add					5.70			5.70	6.30	
	1550	Granite blocks, 3-1/2" x 3-1/2" x 3-1/2"	D-1	92	.174		6.40	5.60		12	16.05	
	1600	4" to 12" long, 3" to 5" wide, 3" to 5" thick	"	98	.163	↓	5.30	5.25		10.55	14.35	
700	0010	**STEPS** Incl. excav., borrow & concrete base, where applicable										700
	0100	Brick steps	B-24	35	.686	LF Riser	10.35	22.50		32.85	47.50	
	0200	Railroad ties	2 Clab	25	.640		3.04	17.55		20.59	32.50	
	0300	Bluestone treads, 12" x 2" or 12" x 1-1/2"	B-24	30	.800	↓	23	26		49	68	
	0490	Minimum labor/equipment charge	D-1	2	8	Job		257		257	415	
	0500	Concrete, cast in place, see Division 03310-240										
	0600	Precast concrete, see Division 03480-800										

		02785 | Flexible Pavement Coating	CREW	DAILY OUTPUT	LABOR-HOURS	UNIT	MAT.	LABOR	EQUIP.	TOTAL	TOTAL INCL O&P	
250	0010	**FOG SEAL**										250
	0012	Sealcoating, 2 coat coal tar pitch emulsion over 10,000 SY	B-45	5000	.003	S.Y.	.53	.10	.11	.74	.87	
	0030	1000 to 10,000 S.Y.	"	3000	.005		.53	.17	.18	.88	1.06	
	0100	Under 1000 S.Y.	B-1	1050	.023		.53	.64		1.17	1.64	
	0300	Petroleum resistant, over 10,000 S.Y.	B-45	5000	.003		.66	.10	.11	.87	1.02	
	0320	1000 to 10,000 S.Y.	"	3000	.005		.66	.17	.18	1.01	1.21	
	0400	Under 1000 S.Y.	B-1	1050	.023		.66	.64		1.30	1.79	
	0600	Non-skid pavement renewal, over 10,000 S.Y.	B-45	5000	.003		.77	.10	.11	.98	1.14	
	0620	1000 to 10,000 S.Y.	"	3000	.005		.77	.17	.18	1.12	1.33	
	0700	Under 1000 S.Y.	B-1	1050	.023		.77	.64		1.41	1.91	
	0800	Prepare and clean surface for above	A-2	8545	.003	↓		.08	.02	.10	.15	
	1000	Hand seal asphalt curbing	B-1	4420	.005	L.F.	.39	.15		.54	.68	
	1900	Asphalt surface treatment, single course, small area										
	1901	0.30 gal/S.Y. asphalt material, 20#/S.Y. aggregate	B-91	5000	.013	S.Y.	.91	.42	.30	1.63	1.99	
	1910	Roadway or large area		10000	.006		.84	.21	.15	1.20	1.41	
	1950	Asphalt surface treatment, dbl. course for small area		3000	.021		1.68	.69	.49	2.86	3.50	
	1960	Roadway or large area		6000	.011		1.51	.35	.25	2.11	2.49	
	1980	Asphalt surface treatment,single course, for shoulders	↓	7500	.009	↓	.97	.28	.20	1.45	1.74	
300	0010	**LATEX MODIFIED EMULSION**										300
	3600	Waterproofing, membrane, tar and fabric, small area	B-63	233	.172	S.Y.	6.20	4.97	.66	11.83	15.75	
	3640	Large area		1435	.028		5.70	.81	.11	6.62	7.75	
	3680	Preformed rubberized asphalt, small area		100	.400		8.40	11.60	1.54	21.54	30	
	3720	Large area	↓	367	.109		7.65	3.16	.42	11.23	14	
	3780	Rubberized asphalt (latex) seal	B-45	5000	.003	↓	1.21	.10	.11	1.42	1.62	
	9000	Minimum labor/equipment charge	1 Clab	2	4	Job		110		110	182	
400	0010	**SAND SEAL**										400
	2080	Sand sealing, sharp sand, asphalt emulsion, small area	B-91	10000	.006	S.Y.	.66	.21	.15	1.02	1.22	
	2120	Roadway or large area	"	18000	.004	"	.57	.12	.08	.77	.90	
	3000	Sealing random cracks, min 1/2" wide, to 1-1/2", 1,000 L.F.	B-77	2800	.014	L.F.	.52	.40	.14	1.06	1.38	
	3040	10,000 L.F.		4000	.010	"	.40	.28	.09	.77	1	
	3080	Alternate method, 1,000 L.F.		200	.200	Gal.	7.85	5.55	1.90	15.30	19.95	
	3120	10,000 L.F.	↓	325	.123	"	6.30	3.43	1.17	10.90	13.90	
	3200	Multi-cracks (flooding), 1 coat, small area	B-92	460	.070	S.Y.	2.36	1.94	.71	5.01	6.60	

02700 | Bases, Ballasts, Pavements & Appurtenances

	02785	Flexible Pavement Coating	CREW	DAILY OUTPUT	LABOR-HOURS	UNIT	2006 BARE COSTS MAT.	LABOR	EQUIP.	TOTAL	TOTAL INCL O&P	
400	3240	Large area	B-92	2850	.011	S.Y.	2.10	.31	.11	2.52	2.96	400
	3280	2 coat, small area		230	.139		7.75	3.88	1.42	13.05	16.55	
	3320	Large area		1425	.022		7.05	.63	.23	7.91	9.05	
	3360	Alternate method, small area		115	.278	Gal.	8.10	7.75	2.84	18.69	25	
	3400	Large area		715	.045	"	7.60	1.25	.46	9.31	10.95	
500	0010	**SLURRY SEAL**										500
	0100	Slurry seal, type I, 8 lbs agg./S.Y., 1 coat, small or irregular area	B-90	2800	.023	S.Y.	.90	.68	.68	2.26	2.85	
	0150	Roadway or large area		10000	.006		.90	.19	.19	1.28	1.51	
	0200	Type II, 12 lbs aggregate/S.Y., 2 coats,small or irregular area		2000	.032		1.80	.95	.95	3.70	4.58	
	0250	Roadway or large area		8000	.008		1.80	.24	.24	2.28	2.63	
	0300	Type III, 20 lbs aggregate/S.Y., 2 coats, small or irregular area		1800	.036		2.12	1.06	1.06	4.24	5.25	
	0350	Roadway or large area		6000	.011		2.12	.32	.32	2.76	3.21	
	0400	Slurry seal, thermoplastic coal-tar, type I, small or irregular area		2400	.027		1.87	.80	.79	3.46	4.22	
	0450	Roadway or large area		8000	.008		1.87	.24	.24	2.35	2.71	
	0500	Type II, small or irregular area		2400	.027		2.39	.80	.79	3.98	4.79	
	0550	Roadway or large area		7800	.008		2.39	.25	.24	2.88	3.30	
	0600	Average mobilization cost				Ea.				3,750	3,750	
600	0010	**SURFACE TREATMENT**										600
	3000	Pavement overlay, polypropylene										
	3040	6 oz. per S.Y., ideal conditions	B-63	10000	.004	S.Y.	1.25	.12	.02	1.39	1.59	
	3080	Adverse conditions		1000	.040		1.68	1.16	.15	2.99	3.91	
	3120	4 oz. per S.Y., ideal conditions		10000	.004		.85	.12	.02	.99	1.15	
	3160	Adverse conditions		1000	.040		1.10	1.16	.15	2.41	3.28	
	3200	Tack coat, emulsion, .05 gal per S.Y., 1000 S.Y	B-45	2500	.006		.24	.21	.21	.66	.82	
	3240	10,000 S.Y.		10000	.002		.19	.05	.05	.29	.35	
	3270	.10 gal per S.Y., 1000 S.Y.		2500	.006		.45	.21	.21	.87	1.06	
	3275	.10 gal per S.Y., 10,000 S.Y.		10000	.002		.45	.05	.05	.55	.64	
	3280	.15 gal per S.Y., 1000 S.Y.		2500	.006		.66	.21	.21	1.08	1.29	
	3320	10,000 S.Y.		10000	.002		.53	.05	.05	.63	.72	

02800 | Site Improvements and Amenities

	02810	Irrigation System	CREW	DAILY OUTPUT	LABOR-HOURS	UNIT	2006 BARE COSTS MAT.	LABOR	EQUIP.	TOTAL	TOTAL INCL O&P	
300	0010	**SPRINKLER IRRIGATION SYSTEM** For lawns										300
	0100	Golf course with fully automatic system	C-17	.05	1600	9 holes	86,000	59,000		145,000	191,000	
	0200	24′ diam. head at 15′ O.C incl. piping, auto oper., minimum	B-20	70	.343	Head	18.95	10.65		29.60	38.50	
	0300	Maximum		40	.600		43.50	18.65		62.15	79	
	0500	60′ diam. head at 40′ O.C. incl. piping, auto oper., minimum		28	.857		57.50	26.50		84	107	
	0600	Maximum		23	1.043		161	32.50		193.50	231	
	0800	Residential system, custom, 1″ supply		2000	.012	S.F.	.28	.37		.65	.93	
	0900	1-1/2″ supply		1800	.013	"	.33	.41		.74	1.04	
	02815	**Outdoor Fountains**										
100	0010	**YARD FOUNTAINS**										100
	0100	Outdoor fountain, 48″ high with bowl and figures	2 Clab	2	8	Ea.	172	219		391	555	
	02820	**Fences & Gates**										
130	0010	**FENCE, CHAIN LINK INDUSTRIAL**, schedule 40										130
	0020	3 strands barb wire, 2″ post @ 10′ O.C., set in concrete, 6′ H										

02820 | Fences & Gates

		02820 Fences & Gates	CREW	DAILY OUTPUT	LABOR-HOURS	UNIT	2006 BARE COSTS MAT.	LABOR	EQUIP.	TOTAL	TOTAL INCL O&P	
130	0200	9 ga. wire, galv. steel	B-80C	240	.100	L.F.	12.75	2.75	.56	16.06	19.15	130
	0300	Aluminized steel		240	.100		16.35	2.75	.56	19.66	23	
	0500	6 ga. wire, galv. steel		240	.100		20	2.75	.56	23.31	27	
	0600	Aluminized steel		240	.100		22.50	2.75	.56	25.81	30	
	0800	6 ga. wire, 6′ high but omit barbed wire, galv. steel		250	.096		19.40	2.64	.54	22.58	26.50	
	0900	Aluminized steel		250	.096		27	2.64	.54	30.18	35	
	0920	8′ H, 6 ga. wire, 2-1/2″ line post, galv. steel		180	.133		31	3.66	.75	35.41	41	
	0940	Aluminized steel		180	.133		38	3.66	.75	42.41	48.50	
	1100	Add for corner posts, 3″ diam., galv. steel		40	.600	Ea.	88.50	16.50	3.35	108.35	128	
	1200	Aluminized steel		40	.600		106	16.50	3.35	125.85	148	
	1300	Add for braces, galv. steel		80	.300		24	8.25	1.68	33.93	42	
	1350	Aluminized steel		80	.300		32	8.25	1.68	41.93	51	
	1400	Gate for 6′ high fence, 1-5/8″ frame, 3′ wide, galv. steel		10	2.400		141	66	13.40	220.40	280	
	1500	Aluminized steel		10	2.400		174	66	13.40	253.40	315	
	2000	5′-0″ high fence, 9 ga., no barbed wire, 2″ line post,										
	2010	10′ O.C., 1-5/8″ top rail										
	2100	Galvanized steel	B-80C	300	.080	L.F.	11.05	2.20	.45	13.70	16.25	
	2200	Aluminized steel		300	.080	"	13.05	2.20	.45	15.70	18.45	
	2400	Gate, 4′ wide, 5′ high, 2″ frame, galv. steel		10	2.400	Ea.	161	66	13.40	240.40	300	
	2500	Aluminized steel		10	2.400	"	181	66	13.40	260.40	325	
	3100	Overhead slide gate, chain link, 6′ high, to 18′ wide		38	.632	L.F.	142	17.35	3.53	162.88	188	
	3110	Cantilever type	B-80	48	.667		61	19.95	11.10	92.05	112	
	3120	8′ high		24	1.333		88.50	40	22	150.50	187	
	3130	10′ high		18	1.778		105	53	29.50	187.50	234	
	3150	Tennis courts, 11 ga. wire, 1-3/4″ mesh, 2-1/2″ line posts										
	3170	1-5/8″ top rail, 3″ corner and gate posts										
	3190	10′ high, galvanized steel	B-80	190	.168	L.F.	16.10	5.05	2.80	23.95	29	
	3210	Vinyl covered 9 ga. wire		190	.168	"	17.75	5.05	2.80	25.60	31	
	3240	Corner posts for above, 3″ diameter, 10′ high		30	1.067	Ea.	169	32	17.75	218.75	258	
	3300	Residential, 11 ga. wire, 1-5/8″ line post @ 10′ O.C.										
	3310	1-3/8″ top rail										
	3350	4′ high, galvanized steel	B-80C	475	.051	L.F.	8.85	1.39	.28	10.52	12.35	
	3400	Aluminized		475	.051	"	8.95	1.39	.28	10.62	12.45	
	3600	Gate, 3′ wide, 1-3/8″ frame, galv. steel		10	2.400	Ea.	72.50	66	13.40	151.90	203	
	3700	Aluminized		10	2.400	"	82	66	13.40	161.40	214	
	3900	3′ high, galvanized steel		620	.039	L.F.	5.35	1.06	.22	6.63	7.90	
	4000	Aluminized		620	.039	"	6.55	1.06	.22	7.83	9.25	
	4200	Gate, 3′ wide, 1-3/8″ frame, galv. steel		12	2	Ea.	59.50	55	11.20	125.70	168	
	4300	Aluminized		12	2	"	67	55	11.20	133.20	177	
	7001	Snow fence on steel posts 10′ O.C., 4′ high	B-1	500	.048	L.F.	2.38	1.35		3.73	4.85	
	8000	Components only, fabric, galvanized, 4′ high	B-80C	585	.041		3.45	1.13	.23	4.81	5.90	
	8040	6′ high		430	.056		4.26	1.53	.31	6.10	7.55	
	8200	Posts, line post, 4′ high		34	.706	Ea.	11.75	19.40	3.95	35.10	49	
	8240	6′ high		26	.923	"	13	25.50	5.15	43.65	62	
	8400	Top rails, 1-3/8″ O.D.		1000	.024	L.F.	1.65	.66	.13	2.44	3.06	
	8440	1-5/8″ O.D.		1000	.024	"	1.86	.66	.13	2.65	3.29	
	9000	Minimum labor/equipment charge	B-80	2	16	Job		480	266	746	1,075	
410	0010	**FENCES, MISC. METAL**										410
	0012	Chicken wire, posts @ 4′, 1″ mesh, 4′ high	B-80C	410	.059	L.F.	1.76	1.61	.33	3.70	4.95	
	0100	2″ mesh, 6′ high		350	.069		1.59	1.88	.38	3.85	5.25	
	0200	Galv. steel, 12 ga., 2″ x 4″ mesh, posts 5′ O.C., 3′ high		300	.080		2.39	2.20	.45	5.04	6.75	
	0300	5′ high		300	.080		3.19	2.20	.45	5.84	7.60	
	0400	14 ga., 1″ x 2″ mesh, 3′ high		300	.080		2.54	2.20	.45	5.19	6.90	
	0500	5′ high		300	.080		3.51	2.20	.45	6.16	7.95	
	1000	Kennel fencing, 1-1/2″ mesh, 6′ long, 3′-6″ wide, 6′-2″ high	2 Clab	4	4	Ea.	400	110		510	620	

02800 | Site Improvements and Amenities

02820 | Fences & Gates

		02820 Fences & Gates	CREW	DAILY OUTPUT	LABOR-HOURS	UNIT	2006 BARE COSTS MAT.	LABOR	EQUIP.	TOTAL	TOTAL INCL O&P
410	1050	12′ long	2 Clab	4	4	Ea.	480	110		590	705
	1200	Top covers, 1-1/2″ mesh, 6′ long		15	1.067		81	29		110	138
	1250	12′ long		12	1.333		130	36.50		166.50	204
	4500	Security fence, prison grade, set in concrete, 12′ high	B-80	25	1.280	L.F.	35	38.50	21.50	95	125
	4600	16′ high	″	20	1.600		42	48	26.50	116.50	154
	5300	Tubular picket, steel, 6′ sections, 1-9/16″ posts, 4′ high	B-80C	300	.080		25	2.20	.45	27.65	31
	5400	2″ posts, 5′ high		240	.100		34.50	2.75	.56	37.81	42.50
	5600	2″ posts, 6′ high		200	.120		39	3.30	.67	42.97	48.50
	5700	Staggered picket 1-9/16″ posts, 4′ high		300	.080		22.50	2.20	.45	25.15	28.50
	5800	2″ posts, 5′ high		240	.100		36.50	2.75	.56	39.81	45.50
	5900	2″ posts, 6′ high		200	.120		38	3.30	.67	41.97	48
	6200	Gates, 4′ high, 3′ wide	B-1	10	2.400	Ea.	215	67.50		282.50	350
	6300	5′ high, 3′ wide		10	2.400		279	67.50		346.50	415
	6400	6′ high, 3′ wide		10	2.400		287	67.50		354.50	425
	6500	4′ wide		10	2.400		335	67.50		402.50	480
520	0010	**FENCE, WOOD RAIL**									
	0012	Picket, No. 2 cedar, Gothic, 2 rail, 3′ high	B-1	160	.150	L.F.	5.40	4.21		9.61	12.95
	0050	Gate, 3′-6″ wide	B-80C	9	2.667	Ea.	46.50	73.50	14.90	134.90	188
	0600	Open rail, rustic, No. 1 cedar, 2 rail, 3′ high		160	.150	L.F.	4.84	4.12	.84	9.80	13
	0650	Gate, 3′ wide		9	2.667	Ea.	54	73.50	14.90	142.40	197
	1200	Stockade, No. 2 cedar, treated wood rails, 6′ high		160	.150	L.F.	6.75	4.12	.84	11.71	15.15
	1250	Gate, 3′ wide		9	2.667	Ea.	55.50	73.50	14.90	143.90	198
	3300	Board, shadow box, 1″ x 6″, treated pine, 6′ high		160	.150	L.F.	10	4.12	.84	14.96	18.70
	3400	No. 1 cedar, 6′ high		150	.160		19.70	4.40	.89	24.99	29.50
	3900	Basket weave, No. 1 cedar, 6′ high		160	.150		19.50	4.12	.84	24.46	29
	3950	Gate, 3′-6″ wide	B-1	8	3	Ea.	133	84		217	286
	4000	Treated pine, 6′ high		150	.160	L.F.	11.55	4.49		16.04	20
	4200	Gate, 3′-6″ wide		9	2.667	Ea.	62	75		137	193
	8000	Posts only, 4′ high		40	.600		14.45	16.85		31.30	44
	8040	6′ high		30	.800		19.95	22.50		42.45	59
	9000	Minimum labor/equipment charge	1 Clab	2	4	Job		110		110	182
	9015	Fence rail, made from recycled plastic, various colors, 2 rail	B-1	150	.160	L.F.	4.47	4.49		8.96	12.35
	9018	3 rail		150	.160		5.95	4.49		10.44	14
	9020	4 rail		150	.160		7.20	4.49		11.69	15.40
	9030	Fence pole, made from recycled plastic, various colors, 7′		96	.250	Ea.	17.20	7		24.20	30.50
	9040	Stockade fence, made from recycled plastic, various colors, 4′ high	B-80C	160	.150	L.F.	30	4.12	.84	34.96	40.50
	9050	6′ high		160	.150	″	36.50	4.12	.84	41.46	48
	9060	6′ pole		96	.250	Ea.	18.35	6.85	1.40	26.60	33
	9070	9′ pole		96	.250	″	27.50	6.85	1.40	35.75	43.50
	9080	Picket fence, made from recycled plastic, various colors, 3′ high		160	.150	L.F.	30	4.12	.84	34.96	40.50
	9090	4′ high		160	.150	″	36.50	4.12	.84	41.46	48
	9100	3′ high gate		8	3	Ea.	57.50	82.50	16.75	156.75	217
	9110	4′ high gate		8	3		66.50	82.50	16.75	165.75	227
	9120	5′ high pole		96	.250		10.30	6.85	1.40	18.55	24
	9130	6′ high pole		96	.250		12.60	6.85	1.40	20.85	26.50
	9140	Pole cap only					3.44			3.44	3.78
	9150	Keeper pins only					.23			.23	.25

02830 | Retaining Walls

		02830 Retaining Walls	CREW	DAILY OUTPUT	LABOR-HOURS	UNIT	MAT.	LABOR	EQUIP.	TOTAL	TOTAL INCL O&P
800	0010	**STONE RETAINING WALLS**									
	0015	Including excavation, concrete footing and									
	0020	stone 3′ below grade. Price is exposed face area.									
	0200	Decorative random stone, to 6′ high, 1′-6″ thick, dry set	D-1	35	.457	S.F.	35.50	14.70		50.20	63.50
	0300	Mortar set		40	.400		35.50	12.85		48.35	60.50
	0500	Cut stone, to 6′ high, 1′-6″ thick, dry set		35	.457		35.50	14.70		50.20	63.50

02800 | Site Improvements and Amenities

02830 | Retaining Walls

	Line	02830 Retaining Walls	Crew	Daily Output	Labor-Hours	Unit	2006 Bare Costs Mat.	Labor	Equip.	Total	Total Incl O&P	
800	0600	Mortar set	D-1	40	.400	S.F.	35.50	12.85		48.35	60.50	800
	0800	Random stone, 6′ to 10′ high, 2′ thick, dry set		45	.356		35.50	11.45		46.95	58	
	0900	Mortar set		50	.320		35.50	10.30		45.80	56	
	1100	Cut stone, 6′ to 10′ high, 2′ thick, dry set		45	.356		35.50	11.45		46.95	58	
	1200	Mortar set		50	.320		35.50	10.30		45.80	56	
	9000	Minimum labor/equipment charge		2	8	Job		257		257	415	

02840 | Walk/Road/Parking Appurtenances

	Line	02840 Walk/Road/Parking Appurtenances	Crew	Daily Output	Labor-Hours	Unit	Mat.	Labor	Equip.	Total	Total Incl O&P	
200	0010	GUIDE/GUARD RAIL										200
	0012	Corrugated stl, galv. stl posts, 6′-3″ O.C.	B-80	850	.038	L.F.	16.45	1.13	.63	18.21	20.50	
	0200	End sections, galvanized, flared		50	.640	Ea.	68	19.15	10.65	97.80	118	
	0300	Wrap around end		50	.640	″	102	19.15	10.65	131.80	156	
	0400	Timber guide rail, 4″ x 8″ with 6″ x 8″ wood posts, treated		960	.033	L.F.	21.50	1	.55	23.05	26	
400	0010	FENDERS										400
	0015	Bumper rails for garages, 12 Ga. rail, 6″ wide, with steel										
	0020	posts 12′-6″ O.C., minimum	E-4	190	.168	L.F.	13.15	6.80	.47	20.42	27.50	
	0030	Average		165	.194		16.45	7.85	.54	24.84	33.50	
	0100	Maximum		140	.229		19.70	9.25	.64	29.59	39.50	
	9000	Minimum labor/equipment charge		2	16	Job		645	44.50	689.50	1,250	
800	0010	PARKING BUMPERS										800
	0020	Parking barriers, timber w/saddles, treated type										
	0100	4″ x 4″ for cars	B-2	520	.077	L.F.	2.54	2.14		4.68	6.35	
	0200	6″ x 6″ for trucks		520	.077	″	5.30	2.14		7.44	9.40	
	1000	Wheel stops, precast concrete incl. dowels, 6″ x 10″ x 6′-0″		120	.333	Ea.	40	9.25		49.25	59.50	
	1100	8″ x 13″ x 6′-0″		120	.333		46.50	9.25		55.75	66.50	
	1300	Pipe bollards, conc filled/paint, 8′ L x 4′ D hole, 6″ diam.	B-6	20	1.200		300	36	11.30	347.30	400	
	1400	8″ diam.		15	1.600		455	48	15.10	518.10	595	
	1500	12″ diam.		12	2		595	60	18.85	673.85	770	
	1540	Bollards, precast concrete, free standing, round, 15″ Dia x 36″ H		14	1.714		127	51.50	16.15	194.65	241	
	1550	16″ Dia x 30″ H		14	1.714		127	51.50	16.15	194.65	241	
	1560	18″ Dia x 34″ H		14	1.714		261	51.50	16.15	328.65	390	
	1570	18″ Dia x 72″ H		14	1.714		320	51.50	16.15	387.65	450	
	1572	26″ Dia x 18″ H		12	2		320	60	18.85	398.85	470	
	1574	26″ Dia x 30″ H		12	2		470	60	18.85	548.85	640	
	1578	26″ Dia x 50″ H		12	2		635	60	18.85	713.85	820	
	1580	Square, 22″ x 32″ H		12	2		340	60	18.85	418.85	490	
	1585	22″ x 41″ H		12	2		600	60	18.85	678.85	780	
	1590	Hexagon, 26″ x 41″ H		10	2.400		625	72	22.50	719.50	825	
	1592	Bollards, steel, 3′ H, retractable, incl hydraulic controls, min		4	6		27,800	180	56.50	28,036.50	31,000	
	1594	Max		2	12		31,000	360	113	31,473	34,800	
	1600	Bollards, recycled plastic, 5″ x 5″ x 5′ L		20	1.200		47	36	11.30	94.30	122	
	1610	Wheel stops, recycled plastic, yellow, 4″ x 6″ x 6′ L		20	1.200		47	36	11.30	94.30	122	
	1620	Speed bumps, recycled plastic, yellow, 3″H x 10″ W x 6′ L		20	1.200		101	36	11.30	148.30	182	
	1630	3″H x 10″ W x 9′ L		20	1.200		139	36	11.30	186.30	224	

02870 | Site Furnishings

	Line	02870 Site Furnishings	Crew	Daily Output	Labor-Hours	Unit	Mat.	Labor	Equip.	Total	Total Incl O&P	
310	0010	BENCHES										310
	0012	Seating, benches, park, precast conc, w/backs, wood rails, 4′ long	2 Clab	5	3.200	Ea.	335	87.50		422.50	515	
	0100	8′ long		4	4		700	110		810	950	
	0300	Fiberglass, without back, one piece, 4′ long		10	1.600		440	44		484	560	
	0400	8′ long		7	2.286		905	62.50		967.50	1,100	
	0500	Steel barstock pedestals w/backs, 2″ x 3″ wood rails, 4′ long		10	1.600		835	44		879	990	
	0510	8′ long		7	2.286		985	62.50		1,047.50	1,175	
	0520	3″ x 8″ wood plank, 4′ long		10	1.600		840	44		884	1,000	

2 SITE CONSTRUCTION

02800 | Site Improvements and Amenities

02870 | Site Furnishings

		02870 Site Furnishings	CREW	DAILY OUTPUT	LABOR-HOURS	UNIT	2006 BARE COSTS MAT.	LABOR	EQUIP.	TOTAL	TOTAL INCL O&P	
310	0530	8′ long	2 Clab	7	2.286	Ea.	875	62.50		937.50	1,075	310
	0540	Backless, 4″ x 4″ wood plank, 4′ square		10	1.600		815	44		859	970	
	0550	8′ long		7	2.286		775	62.50		837.50	955	
	0600	Aluminum pedestals, with backs, aluminum slats, 8′ long		8	2		405	55		460	535	
	0610	15′ long		5	3.200		400	87.50		487.50	585	
	0620	Portable, aluminum slats, 8′ long		8	2		355	55		410	480	
	0630	15′ long		5	3.200		520	87.50		607.50	715	
	0800	Cast iron pedestals, back & arms, wood slats, 4′ long		8	2		310	55		365	430	
	0820	8′ long		5	3.200		895	87.50		982.50	1,125	
	0840	Backless, wood slats, 4′ long		8	2		535	55		590	680	
	0860	8′ long		5	3.200		530	87.50		617.50	730	
	1700	Steel frame, fir seat, 10′ long		10	1.600		190	44		234	282	
	9000	Minimum labor/equipment charge		2	8	Job		219		219	365	

02890 | Traffic Signs & Signals

		02890 Traffic Signs & Signals	CREW	DAILY OUTPUT	LABOR-HOURS	UNIT	MAT.	LABOR	EQUIP.	TOTAL	TOTAL INCL O&P	
100	0010	SIGNS										100
	0012	Stock, 24″ x 24″, no posts, .080″ alum. reflectorized	B-80	70	.457	Ea.	55	13.65	7.60	76.25	91	
	0100	High intensity		70	.457		55	13.65	7.60	76.25	91	
	0300	30″ x 30″, reflectorized		70	.457		112	13.65	7.60	133.25	153	
	0400	High intensity		70	.457		112	13.65	7.60	133.25	153	
	0600	Guide and directional signs, 12″ x 18″, reflectorized		70	.457		35.50	13.65	7.60	56.75	70	
	0700	High intensity		70	.457		34	13.65	7.60	55.25	68	
	0900	18″ x 24″, stock signs, reflectorized		70	.457		40	13.65	7.60	61.25	74.50	
	1000	High intensity		70	.457		40	13.65	7.60	61.25	74.50	
	1200	24″ x 24″, stock signs, reflectorized		70	.457		49.50	13.65	7.60	70.75	85	
	1300	High intensity		70	.457		49.50	13.65	7.60	70.75	85	
	1500	Add to above for steel posts, galvanized, 10′-0″ upright, bolted		200	.160		17.90	4.78	2.66	25.34	30.50	
	1600	12′-0″ upright, bolted		140	.229		24	6.85	3.80	34.65	41.50	
	1800	Highway road signs, aluminum, over 20 S. F., reflectorized		350	.091	S.F.	23	2.73	1.52	27.25	31	
	2000	High intensity		350	.091		23	2.73	1.52	27.25	31	
	2200	Highway, suspended over road, 80 S.F. min., reflectorized		165	.194		23	5.80	3.23	32.03	38	
	2300	High intensity		165	.194		23	5.80	3.23	32.03	38	

02900 | Planting

02905 | Plants, Planting, Transplanting

		02905 Plants, Planting, Transplanting	CREW	DAILY OUTPUT	LABOR-HOURS	UNIT	2006 BARE COSTS MAT.	LABOR	EQUIP.	TOTAL	TOTAL INCL O&P	
725	0010	PLANTING										725
	0012	Moving shrubs on site, 12″ ball	B-62	28	.857	Ea.		25.50	5.50	31	48	
	0100	24″ ball	″	22	1.091			32.50	7	39.50	60.50	
	0300	Moving trees on site, 36″ ball	B-6	3.75	6.400			192	60.50	252.50	375	
	0400	60″ ball	″	1	24			720	226	946	1,425	
925	0010	TREE REMOVAL										925
	0100	Dig & lace, shrubs, broadleaf evergreen, 18″-24″	B-1	55	.436	Ea.		12.25		12.25	20.50	
	0200	2′-3′	″	35	.686			19.25		19.25	32	
	0300	3′-4′	B-6	30	.800			24	7.55	31.55	47.50	
	0400	4′-5′	″	20	1.200			36	11.30	47.30	71	
	1000	Deciduous, 12″-15″	B-1	110	.218			6.10		6.10	10.15	

02900 | Planting

02905 | Plants, Planting, Transplanting

			CREW	DAILY OUTPUT	LABOR-HOURS	UNIT	2006 BARE COSTS MAT.	LABOR	EQUIP.	TOTAL	TOTAL INCL O&P	
925	1100	18"-24"	B-1	65	.369	Ea.		10.35		10.35	17.15	925
	1200	2'-3'	↓	55	.436			12.25		12.25	20.50	
	1300	3'-4'	B-6	50	.480			14.40	4.53	18.93	28.50	
	2000	Evergreeen, 18"-24"	B-1	55	.436			12.25		12.25	20.50	
	2100	2'-0" to 2'-6"		50	.480			13.45		13.45	22.50	
	2200	2'-6" to 3'-0"		35	.686			19.25		19.25	32	
	2300	3'-0" to 3'-6"		20	1.200			33.50		33.50	56	
	3000	Trees, deciduous, small, 2'-3'	↓	55	.436			12.25		12.25	20.50	
	3100	3'-4'	B-6	50	.480			14.40	4.53	18.93	28.50	
	3200	4'-5'		35	.686			20.50	6.45	26.95	40.50	
	3300	5'-6'		30	.800			24	7.55	31.55	47.50	
	4000	Shade, 5'-6'		50	.480			14.40	4.53	18.93	28.50	
	4100	6'-8'		35	.686			20.50	6.45	26.95	40.50	
	4200	8'-10'		25	.960			29	9.05	38.05	57	
	4300	2" caliper		12	2			60	18.85	78.85	118	
	5000	Evergreen, 4'-5'		35	.686			20.50	6.45	26.95	40.50	
	5100	5'-6'		25	.960			29	9.05	38.05	57	
	5200	6'-7'		19	1.263			38	11.90	49.90	74.50	
	5300	7'-8'		15	1.600			48	15.10	63.10	94.50	
	5400	8'-10'	↓	11	2.182	↓		65.50	20.50	86	129	

02910 | Plant Preparation

			CREW	DAILY OUTPUT	LABOR-HOURS	UNIT	MAT.	LABOR	EQUIP.	TOTAL	TOTAL INCL O&P	
710	0010	**LAWN BED PREPARATION**										710
	0100	Rake topsoil, site material, harley rock rake, ideal	B-6	33	.727	M.S.F.		22	6.85	28.85	43	
	0200	Adverse	"	7	3.429			103	32.50	135.50	203	
	0300	Screened loam, york rake and finish, ideal	B-62	24	1			30	6.40	36.40	56	
	0400	Adverse	"	20	1.200			36	7.70	43.70	67	
	1000	Remove topsoil & stock pile on site, 75 HP dozer, 6" deep, 50' haul	B-10L	30	.400			13.45	10.65	24.10	33	
	1050	300' haul		6.10	1.967			66	52.50	118.50	164	
	1100	12" deep, 50' haul		15.50	.774			26	20.50	46.50	64	
	1150	300' haul	↓	3.10	3.871			130	103	233	320	
	1200	200 HP dozer, 6" deep, 50' haul	B-10B	125	.096			3.23	7.35	10.58	13.25	
	1250	300' haul		30.70	.391			13.15	30	43.15	54	
	1300	12" deep, 50' haul		62	.194			6.50	14.85	21.35	27	
	1350	300' haul	↓	15.40	.779	↓		26	60	86	108	
	1400	Alternate method, 75 HP dozer, 50' haul	B-10L	860	.014	C.Y.		.47	.37	.84	1.16	
	1450	300' haul	"	114	.105			3.54	2.80	6.34	8.75	
	1500	200 HP dozer, 50' haul	B-10B	2660	.005			.15	.35	.50	.62	
	1600	300' haul	"	570	.021	↓		.71	1.62	2.33	2.91	
	1800	Rolling topsoil, hand push roller	1 Clab	3200	.003	S.F.		.07		.07	.11	
	1850	Tractor drawn roller	B-66	10666	.001	"		.03	.02	.05	.06	
	2000	Root raking and loading, residential, no boulders	B-6	53.30	.450	M.S.F.		13.50	4.25	17.75	26.50	
	2100	With boulders		32	.750			22.50	7.05	29.55	44.50	
	2200	Municipal, no boulders		200	.120			3.60	1.13	4.73	7.10	
	2300	With boulders	↓	120	.200			6	1.89	7.89	11.80	
	2400	Large commercial, no boulders	B-10B	400	.030			1.01	2.30	3.31	4.14	
	2500	With boulders	"	240	.050			1.68	3.84	5.52	6.90	
	3000	Scarify subsoil, residential, skid steer loader w/scarifiers, 50 HP	B-66	32	.250			8.80	5.60	14.40	20	
	3050	Municipal, skid steer loader w/scarifiers, 50 HP	"	120	.067			2.35	1.49	3.84	5.35	
	3100	Large commercial, 75 HP, dozer w/scarifier	B-10L	240	.050	↓		1.68	1.33	3.01	4.15	
	3500	Screen topsoil from stockpile, vibrating screen, wet material (organic)	B-10P	200	.060	C.Y.		2.02	3.97	5.99	7.60	
	3550	Dry material	"	300	.040			1.34	2.65	3.99	5.05	
	3600	Mixing with conditioners, manure and peat	B-10R	550	.022	↓		.73	.35	1.08	1.56	
	3650	Mobilization add for 2 days or less operation	B-34K	3	2.667	Job		75.50	175	250.50	315	
	3800	Spread conditioned topsoil, 6" deep, by hand	B-1	360	.067	S.Y.	4.65	1.87		6.52	8.20	
	3850	300 HP dozer	B-10M	27	.444	M.S.F.	505	14.95	44.50	564.45	630	

02900 | Planting

2 SITE CONSTRUCTION

02910 | Plant Preparation

		02910 Plant Preparation	CREW	DAILY OUTPUT	LABOR-HOURS	UNIT	2006 BARE COSTS MAT.	LABOR	EQUIP.	TOTAL	TOTAL INCL O&P	
710	4000	Spread soil conditioners, alum. sulfate, 1#/S.Y., hand push spreader	1 Clab	17500	.001	S.Y.	15.30	.01		15.31	16.85	710
	4050	Tractor spreader	B-66	700	.011	M.S.F.	1,700	.40	.26	1,700.66	1,875	
	4100	Fertilizer, 0.2#/S.Y., push spreader	1 Clab	17500	.001	S.Y.	.08	.01		.09	.11	
	4150	Tractor spreader	B-66	700	.011	M.S.F.	8.90	.40	.26	9.56	10.70	
	4200	Ground limestone, 1#/S.Y., push spreader	1 Clab	17500	.001	S.Y.	.09	.01		.10	.12	
	4250	Tractor spreader	B-66	700	.011	M.S.F.	10	.40	.26	10.66	11.90	
	4300	Lusoil, 3#/S.Y., push spreader	1 Clab	17500	.001	S.Y.	.47	.01		.48	.54	
	4350	Tractor spreader	B-66	700	.011	M.S.F.	52	.40	.26	52.66	58.50	
	4400	Manure, 18#/S.Y., push spreader	1 Clab	2500	.003	S.Y.	2.79	.09		2.88	3.22	
	4450	Tractor spreader	B-66	280	.029	M.S.F.	310	1.01	.64	311.65	340	
	4500	Perlite, 1″ deep, push spreader	1 Clab	17500	.001	S.Y.	8.60	.01		8.61	9.50	
	4550	Tractor spreader	B-66	700	.011	M.S.F.	960	.40	.26	960.66	1,050	
	4600	Vermiculite, push spreader	1 Clab	17500	.001	S.Y.	2.63	.01		2.64	2.91	
	4650	Tractor spreader	B-66	700	.011	M.S.F.	292	.40	.26	292.66	320	
	5000	Spread topsoil, skid steer loader and hand dress	B-62	270	.089	C.Y.	21	2.67	.57	24.24	28	
	5100	Articulated loader and hand dress	B-100	320	.038	↓	21	1.26	1.74	24	27	
	5200	Articulated loader and 75HP dozer	B-10M	500	.024	↓	21	.81	2.39	24.20	27	
	5300	Road grader and hand dress	B-11L	1000	.016	↓	21	.51	.46	21.97	24.50	
	6000	Tilling topsoil, 20 HP tractor, disk harrow, 2″ deep	B-66	450	.018	M.S.F.		.63	.40	1.03	1.43	
	6050	4″ deep	↓	360	.022	↓		.78	.50	1.28	1.79	
	6100	6″ deep	↓	270	.030	↓		1.04	.66	1.70	2.38	
	6150	26″ rototiller, 2″ deep	A-1J	1250	.006	S.Y.		.18	.08	.26	.38	
	6200	4″ deep	↓	1000	.008	↓		.22	.10	.32	.47	
	6250	6″ deep	↓	750	.011	↓		.29	.14	.43	.63	
720	0010	**PLANT BED PREPARATION, SHRUB & TREE**										720
	0100	Backfill planting pit, by hand, on site topsoil	2 Clab	18	.889	C.Y.		24.50		24.50	40.50	
	0200	Prepared planting mix	″	24	.667	↓		18.25		18.25	30.50	
	0300	Skid steer loader, on site topsoil	B-62	340	.071	↓		2.12	.45	2.57	3.94	
	0400	Prepared planting mix	″	410	.059	↓		1.76	.38	2.14	3.27	
	1000	Excavate planting pit, by hand, sandy soil	2 Clab	16	1	↓		27.50		27.50	45.50	
	1100	Heavy soil or clay	″	8	2	↓		55		55	91	
	1200	1/2 C.Y. backhoe, sandy soil	B-11C	150	.107	↓		3.42	1.51	4.93	7.15	
	1300	Heavy soil or clay	″	115	.139	↓		4.46	1.97	6.43	9.35	
	2000	Mix planting soil, incl. loam, manure, peat, by hand	2 Clab	60	.267	↓	36.50	7.30		43.80	52.50	
	2100	Skid steer loader	B-62	150	.160	↓	36.50	4.80	1.03	42.33	49.50	
	3000	Pile sod, skid steer loader	″	2800	.009	S.Y.		.26	.06	.32	.48	
	3100	By hand	2 Clab	400	.040	↓		1.10		1.10	1.82	
	4000	Remove sod, F.E. loader	B-10S	2000	.006	↓		.20	.12	.32	.45	
	4100	Sod cutter	B-12K	3200	.005	↓		.16	.30	.46	.59	
	4200	By hand	2 Clab	240	.067	↓		1.83		1.83	3.03	
810	0010	**LOAM & TOPSOIL**										810
	0700	Furnish and place, truck dumped, screened, 4″ deep	B-10S	1300	.009	S.Y.	2.71	.31	.19	3.21	3.68	
	0800	6″ deep	↓	820	.015	″	3.46	.49	.29	4.24	4.92	
	0810	Minimum labor/equipment charge	↓	2	6	Ea.		202	121	323	460	
	0900	Fine grading and seeding, incl. lime, fertilizer & seed,										
	1000	With equipment	B-14	1000	.048	S.Y.	.36	1.39	.23	1.98	2.94	
	2000	Minimum labor/equipment charge	1 Clab	4	2	Job		55		55	91	

02915 | Shrub and Tree Transplanting

		02915 Shrub and Tree Transplanting	CREW	DAILY OUTPUT	LABOR-HOURS	UNIT	MAT.	LABOR	EQUIP.	TOTAL	TOTAL INCL O&P	
200	0010	**GROUND COVER**										200
	0012	Plants, pachysandra, in prepared beds	B-1	15	1.600	C	26	45		71	103	
	0200	Vinca minor, 1 yr, bare root	↓	12	2	″	27	56		83	123	
	0600	Stone chips, in 50 lb. bags, Georgia marble	↓	520	.046	Bag	2.44	1.30		3.74	4.83	
	0700	Onyx gemstone	↓	260	.092	↓	17.70	2.59		20.29	23.50	
	0800	Quartz	↓	260	.092	↓	6.60	2.59		9.19	11.55	

02900 | Planting

		02915 | Shrub and Tree Transplanting	CREW	DAILY OUTPUT	LABOR-HOURS	UNIT	2006 BARE COSTS MAT.	LABOR	EQUIP.	TOTAL	TOTAL INCL O&P	
200	0900	Pea gravel, truckload lots	B-1	28	.857	Ton	25	24		49	67.50	200
		02920 | Lawns & Grasses										
310	0010	**SEEDING, GENERAL** R329219 -50										310
	0020	Mechanical seeding, 215 lb./acre	B-66	1.50	5.333	Acre	520	188	119	827	1,000	
	0100	44 lb./M.S.Y.	″	2500	.003	S.Y.	.15	.11	.07	.33	.43	
	0600	Limestone hand push spreader, 50 lbs. per M.S.F.	1 Clab	180	.044	M.S.F.	3.46	1.22		4.68	5.85	
	9000	Minimum labor/equipment charge	″	4	2	Job		55		55	91	
400	0010	**SODDING**										400
	0020	Sodding, 1″ deep, bluegrass sod, on level ground, over 8 MSF	B-63	22	1.818	M.S.F.	223	52.50	7	282.50	340	
	0200	4 M.S.F.		17	2.353		249	68	9.05	326.05	395	
	0300	1000 S.F.		13.50	2.963		271	86	11.40	368.40	455	
	0500	Sloped ground, over 8 M.S.F.		6	6.667		223	193	25.50	441.50	590	
	0600	4 M.S.F.		5	8		249	232	31	512	690	
	0700	1000 S.F.		4	10		271	290	38.50	599.50	815	
	1000	Bent grass sod, on level ground, over 6 M.S.F.		20	2		500	58	7.70	565.70	655	
	1100	3 M.S.F.		18	2.222		555	64.50	8.55	628.05	725	
	1200	Sodding 1000 S.F. or less		14	2.857		630	82.50	11	723.50	845	
	1500	Sloped ground, over 6 M.S.F.		15	2.667		500	77	10.25	587.25	690	
	1600	3 M.S.F.		13.50	2.963		555	86	11.40	652.40	765	
	1700	1000 S.F.		12	3.333		630	96.50	12.85	739.35	865	
		02930 | Exterior Plants										
310	0010	**SHRUBS AND TREES** Evergreen, in prepared beds, B & B										310
	0100	Arborvitae pyramidal, 4′-5′	B-17	30	1.067	Ea.	44	31.50	18.45	93.95	120	
	0150	Globe, 12″-15″	B-1	96	.250		11.05	7		18.05	24	
	0300	Cedar, blue, 8′-10′	B-17	18	1.778		183	52.50	30.50	266	320	
	0500	Hemlock, canadian, 2-1/2′-3′	B-1	36	.667		23	18.70		41.70	56.50	
	0550	Holly, Savannah, 8′ - 10′ H		9.68	2.479		530	69.50		599.50	695	
	0600	Juniper, andorra, 18″-24″		80	.300		15.90	8.40		24.30	31.50	
	0620	Wiltoni, 15″-18″		80	.300		15.25	8.40		23.65	30.50	
	0640	Skyrocket, 4-1/2′-5′	B-17	55	.582		48.50	17.20	10.05	75.75	92.50	
	0660	Blue pfitzer, 2′-2-1/2′	B-1	44	.545		25	15.30		40.30	53	
	0680	Ketleerie, 2-1/2′-3′		50	.480		32	13.45		45.45	57.50	
	0700	Pine, black, 2-1/2′-3′		50	.480		39	13.45		52.45	65.50	
	0720	Mugo, 18″-24″		60	.400		34	11.25		45.25	56	
	0740	White, 4′-5′	B-17	75	.427		50.50	12.65	7.35	70.50	84	
	0800	Spruce, blue, 18″-24″	B-1	60	.400		38	11.25		49.25	60.50	
	0840	Norway, 4′-5′	B-17	75	.427		85.50	12.65	7.35	105.50	123	
	0900	Yew, denisforma, 12″-15″	B-1	60	.400		23	11.25		34.25	43.50	
	1000	Capitata, 18″-24″		30	.800		19.65	22.50		42.15	58.50	
	1100	Hicksi, 2′-2-1/2′		30	.800		29	22.50		51.50	69	
320	0010	**SHRUBS** Broadleaf evergreen, planted in prepared beds										320
	0100	Andromeda, 15″-18″, container	B-1	96	.250	Ea.	22.50	7		29.50	36.50	
	0200	Azalea, 15″ - 18″, container		96	.250		25	7		32	39	
	0300	Barberry, 9″-12″, container		130	.185		10.05	5.20		15.25	19.65	
	0400	Boxwood, 15″-18″, B & B		96	.250		27	7		34	41	
	0500	Euonymus, emerald gaiety, 12″ to 15″, container		115	.209		16.20	5.85		22.05	27.50	
	0600	Holly, 15″-18″, B & B		96	.250		15.80	7		22.80	29	
	0900	Mount laurel, 18″ - 24″, B & B		80	.300		50	8.40		58.40	69	
	1000	Paxistema, 9 - 12″ high		130	.185		16.20	5.20		21.40	26.50	
	1100	Rhododendron, 18″-24″, container		48	.500		28	14.05		42.05	54	

02900 | Planting

	02930 | Exterior Plants		Crew	Daily Output	Labor-Hours	Unit	2006 Bare Costs Mat.	Labor	Equip.	Total	Total Incl O&P	
320	1200	Rosemary, 1 gal container	B-1	600	.040	Ea.	59	1.12		60.12	67	320
	2000	Deciduous, amelanchier, 2'-3', B & B		57	.421		80.50	11.80		92.30	108	
	2100	Azalea, 15"-18", B & B		96	.250		21.50	7		28.50	35	
	2300	Bayberry, 2'-3', B & B		57	.421		24.50	11.80		36.30	46.50	
	2600	Cotoneaster, 15"-18", B & B		80	.300		14.70	8.40		23.10	30	
	2800	Dogwood, 3'-4', B & B	B-17	40	.800		23	23.50	13.80	60.30	79	
	2900	Euonymus, alatus compacta, 15" to 18", container	B-1	80	.300		19.30	8.40		27.70	35	
	3200	Forsythia, 2'-3', container	"	60	.400		17	11.25		28.25	37.50	
	3300	Hibiscus, 3'-4', B & B	B-17	75	.427		13.25	12.65	7.35	33.25	43	
	3400	Honeysuckle, 3'-4', B & B	B-1	60	.400		18.95	11.25		30.20	39.50	
	3500	Hydrangea, 2'-3', B & B	"	57	.421		22	11.80		33.80	43.50	
	3600	Lilac, 3'-4', B & B	B-17	40	.800		23	23.50	13.80	60.30	78.50	
	3900	Privet, bare root, 18"-24"	B-1	80	.300		11.75	8.40		20.15	27	
	4100	Quince, 2'-3', B & B	"	57	.421		19.50	11.80		31.30	41	
	4200	Russian olive, 3'-4', B & B	B-17	75	.427		21	12.65	7.35	41	52	
	4400	Spirea, 3'-4', B & B	B-1	70	.343		24	9.60		33.60	42.50	
	4500	Viburnum, 3'-4', B & B	B-17	40	.800		25	23.50	13.80	62.30	81	
410	0010	**TREES** Deciduous, in prep. beds, balled & burlapped (B&B)										410
	0100	Ash, 2" caliper	B-17	8	4	Ea.	113	118	69	300	395	
	0200	Beech, 5'-6'		50	.640		221	18.95	11.05	251	287	
	0300	Birch, 6'-8', 3 stems		20	1.600		122	47.50	27.50	197	242	
	0500	Crabapple, 6'-8'		20	1.600		159	47.50	27.50	234	283	
	0600	Dogwood, 4'-5'		40	.800		68	23.50	13.80	105.30	128	
	0700	Eastern redbud 4'-5'		40	.800		133	23.50	13.80	170.30	200	
	0800	Elm, 8'-10'		20	1.600		112	47.50	27.50	187	231	
	0900	Ginkgo, 6'-7'		24	1.333		164	39.50	23	226.50	271	
	1000	Hawthorn, 8'-10', 1" caliper		20	1.600		125	47.50	27.50	200	246	
	1100	Honeylocust, 10'-12', 1-1/2" caliper		10	3.200		149	94.50	55.50	299	380	
	1300	Larch, 8'		32	1		96.50	29.50	17.30	143.30	173	
	1400	Linden, 8'-10', 1" caliper		20	1.600		104	47.50	27.50	179	222	
	1500	Magnolia, 4'-5'		20	1.600		71.50	47.50	27.50	146.50	186	
	1600	Maple, red, 8'-10', 1-1/2" caliper		10	3.200		159	94.50	55.50	309	390	
	1700	Mountain ash, 8'-10', 1" caliper		16	2		164	59	34.50	257.50	315	
	1800	Oak, 2-1/2"-3" caliper		6	5.333		248	158	92	498	630	
	2100	Planetree, 9'-11', 1-1/4" caliper		10	3.200		102	94.50	55.50	252	330	
	2200	Plum, 6'-8', 1" caliper		20	1.600		90.50	47.50	27.50	165.50	207	
	2300	Poplar, 9'-11', 1-1/4" caliper		10	3.200		47	94.50	55.50	197	267	
	2500	Sumac, 2'-3'		75	.427		22.50	12.65	7.35	42.50	53.50	
	2700	Tulip, 5'-6'		40	.800		49.50	23.50	13.80	86.80	108	
	2800	Willow, 6'-8', 1" caliper		20	1.600		60.50	47.50	27.50	135.50	175	
	9000	Minimum labor/equipment charge	1 Clab	4	2	Job		55		55	91	

	02945 | Planting Accessories		Crew	Daily Output	Labor-Hours	Unit	Mat.	Labor	Equip.	Total	Total Incl O&P	
120	0010	**EDGING**										120
	0050	Aluminum alloy, including stakes, 1/8" x 4", mill finish	B-1	390	.062	L.F.	2.43	1.73		4.16	5.55	
	0051	Black paint		390	.062		2.82	1.73		4.55	5.95	
	0052	Black anodized		390	.062		3.26	1.73		4.99	6.45	
	0100	Brick, set horizontally, 1-1/2 bricks per L.F.	D-1	370	.043		1.03	1.39		2.42	3.39	
	0150	Set vertically, 3 bricks per L.F.	"	135	.119		2.76	3.81		6.57	9.25	
	0200	Corrugated aluminum, roll, 4" wide	1 Carp	650	.012		.37	.44		.81	1.14	
	0250	6" wide	"	550	.015		.46	.52		.98	1.37	
	0600	Railroad ties, 6" x 8"	2 Carp	170	.094		2.69	3.35		6.04	8.50	
	0650	7" x 9"		136	.118		2.99	4.18		7.17	10.25	
	0750	2" x 4"		330	.048		2.20	1.72		3.92	5.30	
	0800	Steel edge strips, incl. stakes, 1/4" x 5"	B-1	390	.062		3.62	1.73		5.35	6.85	

02900 | Planting

		02945 | Planting Accessories	CREW	DAILY OUTPUT	LABOR-HOURS	UNIT	2006 BARE COSTS MAT.	LABOR	EQUIP.	TOTAL	TOTAL INCL O&P	
120	0850	3/16″ x 4″	B-1	390	.062	L.F.	2.86	1.73		4.59	6	120
	9000	Minimum labor/equipment charge	1 Carp	4	2	Job		71		71	118	
300	0010	**PLANTERS**										300
	0012	Concrete, sandblasted, precast, 48″ diameter, 24″ high	2 Clab	15	1.067	Ea.	585	29		614	690	
	0100	Fluted, precast, 7′ diameter, 36″ high		10	1.600		985	44		1,029	1,150	
	0300	Fiberglass, circular, 36″ diameter, 24″ high		15	1.067		415	29		444	510	
	0400	60″ diameter, 24″ high	↓	10	1.600	↓	705	44		749	850	
	9000	Minimum labor/equipment charge	1 Clab	2	4	Job		110		110	182	
510	0010	**TREE GUYING**										510
	0015	Tree guying Including stakes, guy wire and wrap										
	0100	Less than 3″ caliper, 2 stakes	2 Clab	35	.457	Ea.	15.50	12.55		28.05	38	
	0200	3″ to 4″ caliper, 3 stakes	″	21	.762	″	18.30	21		39.30	54.50	
	1000	Including arrowhead anchor, cable, turnbuckles and wrap										
	1100	Less than 3″ caliper, 3″ anchors	2 Clab	20	.800	Ea.	48.50	22		70.50	90	
	1200	3″ to 6″ caliper, 4″ anchors		15	1.067		69.50	29		98.50	125	
	1300	6″ caliper, 6″ anchors		12	1.333		86	36.50		122.50	156	
	1400	8″ caliper, 8″ anchors	↓	9	1.778	↓	98.50	48.50		147	190	

02950 | Site Restoration & Rehabilitation

		02955 | Restoration of Underground Piping	CREW	DAILY OUTPUT	LABOR-HOURS	UNIT	2006 BARE COSTS MAT.	LABOR	EQUIP.	TOTAL	TOTAL INCL O&P	
230	0010	**LINING PIPE** with cement, incl. bypass and cleaning										230
	0020	Less than 10,000 L.F., urban, 6″ to 10″	C-17E	130	.615	L.F.	7.20	22.50	.62	30.32	45.50	
	0050	10″ to 12″		125	.640		8.85	23.50	.65	33	49	
	0070	12″ to 16″		115	.696		9.10	25.50	.70	35.30	53	
	0100	16″ to 20″		95	.842		10.70	31	.85	42.55	63.50	
	0200	24″ to 36″		90	.889		11.50	33	.90	45.40	67	
	0300	48″ to 72″		80	1		18.35	37	1.01	56.36	81.50	
	0500	Rural, 6″ to 10″		180	.444		7.20	16.40	.45	24.05	35.50	
	0550	10″ to 12″		175	.457		8.85	16.85	.46	26.16	38	
	0570	12″ to 16″		160	.500		9.25	18.45	.51	28.21	41	
	0600	16″ to 20″		135	.593		9.80	22	.60	32.40	47.50	
	0700	24″ to 36″		125	.640		11.70	23.50	.65	35.85	52	
	0800	48″ to 72″		100	.800		18.35	29.50	.81	48.66	69.50	
	1000	Greater than 10,000 L.F., urban, 6″ to 10″		160	.500		7.20	18.45	.51	26.16	38.50	
	1050	10″ to 12″		155	.516		8.75	19.05	.52	28.32	41	
	1070	12″ to 16″		140	.571		9.10	21	.58	30.68	45	
	1100	16″ to 20″		120	.667		9.80	24.50	.67	34.97	51.50	
	1200	24″ to 36″		115	.696		11.70	25.50	.70	37.90	55.50	
	1300	48″ to 72″		95	.842		18.35	31	.85	50.20	72	
	1500	Rural, 6″ to 10″		215	.372		7.20	13.75	.38	21.33	31	
	1550	10″ to 12″		210	.381		8.85	14.05	.38	23.28	33	
	1570	12″ to 16″		185	.432		9.10	15.95	.44	25.49	36.50	
	1600	16″ to 20″		150	.533		9.80	19.70	.54	30.04	43.50	
	1700	24″ to 36″		140	.571		11.70	21	.58	33.28	48	
	1800	48″ to 72″	↓	120	.667	↓	18.50	24.50	.67	43.67	61	

02950 | Site Restoration & Rehabilitation

		02960 | Flex. Pavement Surfacing Recovery	CREW	DAILY OUTPUT	LABOR-HOURS	UNIT	2006 BARE COSTS MAT.	LABOR	EQUIP.	TOTAL	TOTAL INCL O&P	
100	0010	PAVEMENT MILLING AND PAVEMENT COLD PLANING										100
	5200	Cold planing & cleaning, 1" to 3" asphalt pavmt., over 25,000 S.Y.	B-71	6000	.009	S.Y.		.30	.90	1.20	1.47	
	5280	5,000 S.Y. to 10,000 S.Y.	"	4000	.014	"		.44	1.35	1.79	2.20	
	5300	Asphalt pavement removal from conc. base, no haul										
	5320	Rip, load & sweep 1" to 3"	B-70	8000	.007	S.Y.		.22	.16	.38	.53	
	5330	3" to 6" deep	"	5000	.011			.35	.25	.60	.85	
	5340	Profile grooving, asphalt pavement load & sweep, 1" deep	B-71	12500	.004			.14	.43	.57	.70	
	5350	3" deep		9000	.006			.20	.60	.80	.98	
	5360	6" deep	↓	5000	.011	↓		.35	1.08	1.43	1.76	
		02965 | Flex./Bit. Pavement Recycling										
200	0010	COLD IN-PLACE RECYCLED BITUMINOUS PAVEMENT COURSES										200
	5000	Reclamation, pulverizing and blending with existing base										
	5040	Aggregate base, 4" thick pavement, over 15,000 S.Y.	B-73	2400	.027	S.Y.		.89	1.69	2.58	3.29	
	5080	5,000 S.Y. to 15,000 S.Y.		2200	.029			.97	1.84	2.81	3.59	
	5120	8" thick pavement, over 15,000 S.Y.		2200	.029			.97	1.84	2.81	3.59	
	5160	5,000 S.Y. to 15,000 S.Y.	↓	2000	.032	↓		1.07	2.03	3.10	3.95	
	5180	Add for mobilization and demobilization for crew B-73				Ea.				1,700	2,150	
400	0010	HOT IN-PLACE RECYCLED BITUMINOUS PAVEMENT COURSES										400
	5500	Recycle asphalt pavement at site										
	5520	Remove, rejuvenate and spread 4" deep	B-72	2500	.026	S.Y.	2.55	.83	3.56	6.94	8.05	
	5521	6" deep	"	2000	.032	"	3.75	1.03	4.45	9.23	10.70	
		02990 | Structure Moving										
300	0010	MOVING BUILDINGS One day move, up to 24' wide										300
	0020	Reset on new foundation, patch & hook-up, average move				Total					9,300	
	0040	Wood or steel frame bldg., based on ground floor area	B-4	185	.259	S.F.		7.25	1.77	9.02	13.90	
	0060	Masonry bldg., based on ground floor area	"	137	.350			9.75	2.39	12.14	18.80	
	0200	For 24' to 42' wide, add				↓					15%	
	0220	For each additional day on road, add	B-4	1	48	Day		1,350	330	1,680	2,550	
	0240	Construct new basement, move building, 1 day										
	0300	move, patch & hook-up, based on ground floor area	B-3	155	.310	S.F.	6.45	9.15	11.25	26.85	34.50	

Division 3 Concrete

Estimating Tips

General

- Carefully check all the plans and specifications. Concrete often appears on drawings other than structural drawings, including mechanical and electrical drawings for equipment pads. The cost of cutting and patching is often difficult to estimate. See Subdivisions 02220 and 03055 for demolition costs.
- Always obtain concrete prices from suppliers near the job site. A volume discount can often be negotiated depending upon competition in the area. Remember to add for waste, particularly for slabs and footings on grade.

03100 Concrete Forms & Accessories

- A primary cost for concrete construction is forming. Most jobs today are constructed with prefabricated forms. The selection of the forms best suited for the job and the total square feet of forms required for efficient concrete forming and placing are key elements in estimating concrete construction. Enough forms must be available for erection to make efficient use of the concrete placing equipment and crew.
- Concrete accessories for forming and placing depend upon the systems used. Study the plans and specifications to assure that all special accessory requirements have been included in the cost estimate such as anchor bolts, inserts, and hangers.

03200 Concrete Reinforcement

- Ascertain that the reinforcing steel supplier has included all accessories, cutting, bending, and an allowance for lapping, splicing, and waste. A good rule of thumb is 10% for lapping, splicing, and waste. Also, 10% waste should be allowed for welded wire fabric.

03300 Cast-in-Place Concrete

- When estimating structural concrete, pay particular attention to requirements for concrete additives, curing methods, and surface treatments. Special consideration for climate, hot or cold, must be included in your estimate. Be sure to include requirements for concrete placing equipment and concrete finishing.

03400 Precast Concrete
03500 Cementitious Decks & Toppings

- The cost of hauling precast concrete structural members is often an important factor. For this reason, it is important to get a quote from the nearest supplier. It may become economically feasible to set up precasting beds on the site if the hauling costs are prohibitive.

Reference Numbers

Reference numbers are shown in bold squares at the beginning of some major classifications. These numbers refer to related items in the Reference Section. The reference information may be an estimating procedure, an alternate pricing method, or technical information.

Note: Not all subdivisions listed here necessarily appear in this publication.

03050 | Basic Concrete Materials & Methods

3 CONCRETE

		03055 | Selective Demolition	CREW	DAILY OUTPUT	LABOR-HOURS	UNIT	2006 BARE COSTS MAT.	LABOR	EQUIP.	TOTAL	TOTAL INCL O&P	
110	0010	SELECTIVE CONCRETE DEMOLITION R024119-10										110
	0012	Excludes saw cutting, torch cutting, loading or hauling										
	0050	Break up into small pieces, minimum reinforcing	B-9	24	1.667	C.Y.		46.50	6.15	52.65	84	
	0060	Average reinforcing		16	2.500			69.50	9.20	78.70	125	
	0070	Maximum reinforcing		8	5			139	18.40	157.40	251	
	0150	Remove whole pieces, up to 2 tons per piece	E-18	36	1.111	Ea.		44	26.50	70.50	109	
	0160	2 - 5 tons per piece		30	1.333			53	31.50	84.50	131	
	0170	5 - 10 tons per piece		24	1.667			66	39.50	105.50	164	
	0180	10 - 15 tons per piece		18	2.222			88	52.50	140.50	218	
	0250	Precast unit embedded in masonry, up to 1 CF	D-1	16	1			32		32	52	
	0260	1 - 2 CF		12	1.333			43		43	69.50	
	0270	2 - 5 CF		10	1.600			51.50		51.50	83.50	
	0280	5 - 10 CF		8	2			64.50		64.50	104	
	1910	Minimum labor/equipment charge	B-9	2	20	Job		555	73.50	628.50	1,000	
		03060 | Basic Concrete Materials										
100	0010	CONCRETE ADMIXTURES & SURFACE TREATMENTS										100
	0040	Abrasives, aluminum oxide, over 20 tons				Lb.	1.18			1.18	1.30	
	0070	Under 1 ton					1.28			1.28	1.41	
	0100	Silicon carbide, black, over 20 tons					1.64			1.64	1.80	
	0120	Under 1 ton					1.77			1.77	1.95	
	0200	Air entraining agent, .7 to 1.5 oz. per bag, 55 gallon lots				Gal.	8.95			8.95	9.85	
	0220	5 gallon lots					8.25			8.25	9.05	
	0300	Bonding agent, acrylic latex, 250 S.F. per gallon					17.80			17.80	19.55	
	0320	Epoxy resin, 80 S.F. per gallon					39			39	43	
	0400	Calcium chloride, 50 lb. bags				Ton	415			415	455	
	0420	Less than truckload lots				Bag	29.50			29.50	32.50	
	0500	Carbon black, liquid, 2 to 8 lbs. per bag of cement				Lb.	4.27			4.27	4.70	
	0600	Colors, integral, 2 to 10 lb. per bag of cement, minimum					2.32			2.32	2.55	
	0610	Average					3.12			3.12	3.43	
	0620	Maximum					3.95			3.95	4.35	
	0700	Curing compound, solvent based, 400 S.F./gal, 55 gal. lots				Gal.	12.35			12.35	13.60	
	0720	5 gallon lots					14.20			14.20	15.60	
	0800	Water based, 250 S.F./gal, 55 gallon lots					11.35			11.35	12.50	
	0820	5 gallon lots					13			13	14.30	
	0900	Dustproofing compound, (200-600 S.F./gal.), 55 gallon lots					3.84			3.84	4.22	
	0920	5 gallon lots					5.70			5.70	6.30	
	1000	Epoxy dustproof coating, colors, (300-400 S.F. per coat),										
	1010	or transparent, (400-600 S.F. per coat)				Gal.	29.50			29.50	32	
	1100	Hardeners, metallic, 55 lb. bags, natural (grey)				Lb.	.70			.70	.77	
	1200	Colors					.97			.97	1.07	
	1300	Non-metallic, 55 lb. bags, natural grey					.30			.30	.33	
	1320	Colors					.55			.55	.61	
	1600	Sealer, hardener and dustproofer, epoxy-based, 200 SF/gal, min				Gal.	29.50			29.50	32	
	1620	Maximum					32			32	35.50	
	1630	Solvent-based, 250 SF/gal, minimum					12.35			12.35	13.60	
	1640	Maximum					14.20			14.20	15.60	
	1650	Water based, 350 S.F., minimum					11.35			11.35	12.50	
	1660	Maximum					13			13	14.30	
	1700	Colors (300-400 S.F. per gallon)					49			49	53.50	
	1800	Set accelerator for below freezing, 1 to 1-1/2 gal. per C.Y.					10			10	11	
	1900	Set retarder, 2 to 4 fl. oz. per bag of cement					13.30			13.30	14.65	
	2000	Waterproofing, integral 1 lb. per bag of cement				Lb.	.98			.98	1.08	
	2100	Powdered metallic, 40 lbs. per 100 S.F., minimum					1.22			1.22	1.34	
	2120	Maximum					2.44			2.44	2.68	

03050 | Basic Concrete Materials & Methods

03060 | Basic Concrete Materials

		03060 Basic Concrete Materials	CREW	DAILY OUTPUT	LABOR-HOURS	UNIT	2006 BARE COSTS MAT.	LABOR	EQUIP.	TOTAL	TOTAL INCL O&P	
870	0010	WINTER PROTECTION										870
	0012	For heated ready mix, add, minimum				C.Y.	4.73			4.73	5.20	
	0050	Maximum				"	5.90			5.90	6.50	
	0100	Temporary heat to protect concrete, 24 hours, minimum	2 Clab	50	.320	M.S.F.	108	8.75		116.75	134	
	0150	Maximum	"	25	.640	"	141	17.55		158.55	184	
	0200	Temporary shelter for slab on grade, wood frame/polyethylene sheeting										
	0201	Build or remove, minimum	2 Carp	10	1.600	M.S.F.	284	57		341	405	
	0210	Maximum	"	3	5.333	"	340	190		530	690	
	0300	See also Division 03390-200										

03100 | Concrete Forms & Accessories

03110 | Structural C.I.P. Forms

		03110 Structural C.I.P. Forms	CREW	DAILY OUTPUT	LABOR-HOURS	UNIT	2006 BARE COSTS MAT.	LABOR	EQUIP.	TOTAL	TOTAL INCL O&P	
405	0010	FORMS IN PLACE, BEAMS AND GIRDERS R031113-40										405
	0020	See also Elevated Slabs, Division 03110-420										
	0500	Exterior spandrel, job-built plywood, 12" wide, 1 use R031113-60	C-2	225	.213	SFCA	3.76	7.35		11.11	16.35	
	0650	4 use		310	.155		1.22	5.35		6.57	10.20	
	1000	18" wide, 1 use		250	.192		3.22	6.65		9.87	14.55	
	1150	4 use		315	.152		1.05	5.25		6.30	9.85	
	1500	24" wide, 1 use		265	.181		2.95	6.25		9.20	13.60	
	1650	4 use		325	.148		.96	5.10		6.06	9.50	
	2000	Interior beam, job-built plywood, 12" wide, 1 use		300	.160		4.10	5.50		9.60	13.65	
	2150	4 use		377	.127		1.33	4.40		5.73	8.75	
	2500	24" wide, 1 use		320	.150		3.01	5.20		8.21	11.90	
	2650	4 use		395	.122		.97	4.20		5.17	8	
	3000	Encasing steel beam, hung, job-built plywood, 1 use		325	.148		2.94	5.10		8.04	11.70	
	3150	4 use	↓	430	.112	↓	.96	3.85		4.81	7.45	
	9000	Minimum labor/equipment charge	2 Carp	2	8	Job		284		284	470	
410	0010	FORMS IN PLACE, COLUMNS R031113-40										410
	0500	Round fiberglass, 4 use per mo., rent, 12" diameter	C-1	160	.200	L.F.	6.40	6.70		13.10	18.10	
	0550	16" diameter R031113-60		150	.213		7.65	7.15		14.80	20.50	
	0600	18" diameter		140	.229		8.50	7.65		16.15	22	
	0650	24" diameter		135	.237		10.60	7.95		18.55	25	
	0700	28" diameter		130	.246		11.80	8.25		20.05	26.50	
	0800	30" diameter		125	.256		12.35	8.60		20.95	28	
	0850	36" diameter		120	.267		16.45	8.95		25.40	33	
	1500	Round fiber tube, 1 use, 8" diameter		155	.206		1.51	6.90		8.41	13.10	
	1550	10" diameter		155	.206		1.94	6.90		8.84	13.60	
	1600	12" diameter		150	.213		2.32	7.15		9.47	14.40	
	1650	14" diameter		145	.221		3.03	7.40		10.43	15.60	
	1700	16" diameter		140	.229		3.54	7.65		11.19	16.60	
	1750	20" diameter		135	.237		5.65	7.95		13.60	19.35	
	1800	24" diameter		130	.246		7.30	8.25		15.55	21.50	
	1850	30" diameter		125	.256		10.40	8.60		19	25.50	
	1900	36" diameter		115	.278		13.15	9.30		22.45	30	
	1950	42" diameter		100	.320		31.50	10.70		42.20	53	
	2000	48" diameter	↓	85	.376	↓	40	12.60		52.60	65	
	2200	For seamless type, add					15%					

	03110 | Structural C.I.P. Forms		CREW	DAILY OUTPUT	LABOR-HOURS	UNIT	2006 BARE COSTS MAT.	LABOR	EQUIP.	TOTAL	TOTAL INCL O&P	
410	3000	Round, steel, 4 use per mo., rent, regular duty, 12″ diam. R031113-40	C-1	145	.221	L.F.	10.45	7.40		17.85	24	410
	3050	16″ diameter		125	.256		11.75	8.60		20.35	27	
	3100	Heavy duty, 20″ diameter R031113-60		105	.305		12.90	10.20		23.10	31	
	3150	24″ diameter		85	.376		14.15	12.60		26.75	36.50	
	3200	30″ diameter		70	.457		16.25	15.30		31.55	43.50	
	3250	36″ diameter		60	.533		17.40	17.85		35.25	48.50	
	3300	48″ diameter		50	.640		24	21.50		45.50	62	
	3350	60″ diameter		45	.711		32	24		56	74.50	
	4000	Column capitals, steel, 4 uses/mo., 24″ col, 4′ cap diameter		12	2.667	Ea.	198	89.50		287.50	365	
	4050	5′ cap diameter		11	2.909		226	97.50		323.50	410	
	4100	6′ cap diameter		10	3.200		310	107		417	520	
	4150	7′ cap diameter		9	3.556		380	119		499	610	
	4500	For second and succeeding months, deduct					50%					
	5000	Job-built plywood, 8″ x 8″ columns, 1 use	C-1	165	.194	SFCA	2.18	6.50		8.68	13.15	
	5050	2 use		195	.164		1.25	5.50		6.75	10.50	
	5100	3 use		210	.152		.87	5.10		5.97	9.40	
	5150	4 use		215	.149		.72	4.99		5.71	9.05	
	5500	12″ x 12″ columns, 1 use		180	.178		2.19	5.95		8.14	12.25	
	5550	2 use		210	.152		1.20	5.10		6.30	9.75	
	5600	3 use		220	.145		.87	4.87		5.74	9.05	
	5650	4 use		225	.142		.71	4.77		5.48	8.70	
	6000	16″ x 16″ columns, 1 use		185	.173		2.23	5.80		8.03	12.05	
	6050	2 use		215	.149		1.17	4.99		6.16	9.55	
	6100	3 use		230	.139		.89	4.66		5.55	8.75	
	6150	4 use		235	.136		.73	4.56		5.29	8.35	
	6500	24″ x 24″ columns, 1 use		190	.168		2.54	5.65		8.19	12.15	
	6550	2 use		216	.148		1.40	4.96		6.36	9.80	
	6600	3 use		230	.139		1.02	4.66		5.68	8.85	
	6650	4 use		238	.134		.83	4.51		5.34	8.35	
	7000	36″ x 36″ columns, 1 use		200	.160		2.26	5.35		7.61	11.40	
	7050	2 use		230	.139		1.27	4.66		5.93	9.15	
	7100	3 use		245	.131		.90	4.38		5.28	8.25	
	7150	4 use		250	.128		.74	4.29		5.03	7.90	
	7500	Steel framed plywood, 4 use per mo., rent, 8″ x 8″		340	.094		3.26	3.15		6.41	8.85	
	7550	10″ x 10″		350	.091		2.45	3.06		5.51	7.80	
	7600	12″ x 12″		370	.086		3.11	2.90		6.01	8.20	
	7650	16″ x 16″		400	.080		3.41	2.68		6.09	8.20	
	7700	20″ x 20″		420	.076		1.64	2.55		4.19	6.05	
	7750	24″ x 24″		440	.073		1.54	2.44		3.98	5.75	
	7755	30″ x 30″		440	.073		1.38	2.44		3.82	5.55	
	9000	Minimum labor/equipment charge	2 Carp	2	8	Job		284		284	470	
420	0010	**FORMS IN PLACE, ELEVATED SLABS** R031113-40										420
	1000	Flat plate, job-built plywood, to 15′ high, 1 use	C-2	470	.102	S.F.	4.31	3.53		7.84	10.60	
	1150	4 use R031113-60		560	.086		1.40	2.96		4.36	6.45	
	1500	15′ to 20′ high ceilings, 4 use		495	.097		1.66	3.35		5.01	7.40	
	2000	Flat slab, drop panels, job-built plywood, to 15′ high, 1 use		449	.107		4.79	3.69		8.48	11.35	
	2150	4 use		544	.088		1.56	3.05		4.61	6.75	
	2250	15′ to 20′ high ceilings, 4 use		480	.100		4.20	3.45		7.65	10.30	
	3500	Floor slab, with 20″ metal pans, 1 use		415	.116		6.10	3.99		10.09	13.30	
	3650	4 use		500	.096		2.86	3.31		6.17	8.65	
	4500	With 30″ fiberglass domes, 1 use		405	.119		6.30	4.09		10.39	13.70	
	4550	4 use		470	.102		3.04	3.53		6.57	9.20	
	5000	Box out for slab openings, over 16″ deep, 1 use		190	.253	SFCA	3.22	8.70		11.92	18	
	5050	2 use		240	.200	″	1.77	6.90		8.67	13.40	
	5500	Shallow slab box outs, to 10 S.F.		42	1.143	Ea.	10.70	39.50		50.20	77.50	

	03110 | Structural C.I.P. Forms			CREW	DAILY OUTPUT	LABOR-HOURS	UNIT	2006 BARE COSTS MAT.	LABOR	EQUIP.	TOTAL	TOTAL INCL O&P	
420	5550	Over 10 S.F. (use perimeter)	R031113-40	C-2	600	.080	L.F.	1.43	2.76		4.19	6.15	420
	6000	Bulkhead forms for slab, with keyway, 1 use, 2 piece			500	.096		1.59	3.31		4.90	7.25	
	6100	3 piece (see also edge forms)	R031113-60		460	.104		1.74	3.60		5.34	7.85	
	6200	Bulkhead form, 4-1/2" high, exp metal, keyway & stakes		C-1	1200	.027		.84	.89		1.73	2.40	
	6210	5-1/2" high			1100	.029		.90	.97		1.87	2.61	
	6215	7-1/2" high			960	.033		1.08	1.12		2.20	3.04	
	6220	9-1/2" high			840	.038		1.88	1.28		3.16	4.19	
	6500	Curb forms, wood, 6" to 12" high, on elevated slabs, 1 use			180	.178	SFCA	1.54	5.95		7.49	11.55	
	6550	2 use			205	.156		.85	5.25		6.10	9.60	
	6600	3 use			220	.145		.62	4.87		5.49	8.80	
	6650	4 use			225	.142		.50	4.77		5.27	8.45	
	7000	Edge forms to 6" high, on elevated slab, 4 use			500	.064	L.F.	.19	2.14		2.33	3.76	
	7500	Depressed area forms to 12" high, 4 use			300	.107		.72	3.57		4.29	6.70	
	7550	12" to 24" high, 4 use			175	.183		.97	6.15		7.12	11.20	
	8000	Perimeter deck and rail for elevated slabs, straight			90	.356		14.60	11.90		26.50	36	
	8050	Curved			65	.492		20	16.50		36.50	49.50	
	8500	Void forms, round fiber, 3" diameter			450	.071		.64	2.38		3.02	4.65	
	8650	8" diameter			375	.085		1.82	2.86		4.68	6.75	
	9000	Minimum labor/equipment charge		2 Carp	2	8	Job		284		284	470	
425	0010	**FORMS IN PLACE, EQUIPMENT FOUNDATIONS** job built	R031113-40										425
	0020	1 use		C-2	160	.300	SFCA	2.55	10.35		12.90	19.95	
	0050	2 use	R031113-60		190	.253		1.40	8.70		10.10	16	
	0100	3 use			200	.240		1.02	8.30		9.32	14.85	
	0150	4 use			205	.234		.83	8.10		8.93	14.30	
	9000	Minimum labor/equipment charge		1 Carp	3	2.667	Job		95		95	157	
430	0010	**FORMS IN PLACE, FOOTINGS**	R031113-40										430
	0020	Continuous wall, plywood, 1 use		C-1	375	.085	SFCA	2.62	2.86		5.48	7.60	
	0150	4 use	R031113-60	"	485	.066	"	.85	2.21		3.06	4.60	
	1500	Keyway, 4 use, tapered wood, 2" x 4"		1 Carp	530	.015	L.F.	.20	.54		.74	1.11	
	1550	2" x 6"		"	500	.016	"	.30	.57		.87	1.27	
	3000	Pile cap, square or rectangular, job-built plywood, 1 use		C-1	290	.110	SFCA	2.45	3.70		6.15	8.85	
	3150	4 use			383	.084		.80	2.80		3.60	5.50	
	5000	Spread footings, job-built lumber, 1 use			305	.105		1.83	3.52		5.35	7.85	
	5150	4 use			414	.077		.59	2.59		3.18	4.94	
	9000	Minimum labor/equipment charge		1 Carp	3	2.667	Job		95		95	157	
435	0010	**FORMS IN PLACE, GRADE BEAM**	R031113-40										435
	0020	Job-built plywood, 1 use		C-2	530	.091	SFCA	1.74	3.13		4.87	7.10	
	0150	4 use	R031113-60	"	605	.079	"	.57	2.74		3.31	5.15	
	9000	Minimum labor/equipment charge		2 Carp	2	8	Job		284		284	470	
445	0010	**FORMS IN PLACE, SLAB ON GRADE**	R031113-40										445
	1000	Bulkhead forms w/keyway, wood, 6" high, 1 use		C-1	510	.063	L.F.	.97	2.10		3.07	4.55	
	1400	Bulkhead form, 4-1/2" high, exp metal, incl keyway & stakes	R031113-60		1200	.027		.84	.89		1.73	2.40	
	1410	5-1/2" high			1100	.029		.90	.97		1.87	2.61	
	1420	7-1/2" high			960	.033		1.08	1.12		2.20	3.04	
	1430	9-1/2" high			840	.038		1.88	1.28		3.16	4.19	
	2000	Curb forms, wood, 6" to 12" high, on grade, 1 use			215	.149	SFCA	2.23	4.99		7.22	10.70	
	2150	4 use			275	.116	"	.72	3.90		4.62	7.25	
	3000	Edge forms, wood, 4 use, on grade, to 6" high			600	.053	L.F.	.29	1.79		2.08	3.28	
	3050	7" to 12" high			435	.074	SFCA	.76	2.47		3.23	4.92	
	3500	For depressed slabs, 4 use, to 12" high			300	.107	L.F.	.69	3.57		4.26	6.65	
	3550	To 24" high			175	.183		.91	6.15		7.06	11.15	
	4000	For slab blockouts, to 12" high, 1 use			200	.160		.74	5.35		6.09	9.70	
	4050	To 24" high, 1 use			120	.267		.94	8.95		9.89	15.85	
	4100	Plastic (extruded), to 6" high, multiple use, on grade			800	.040		.37	1.34		1.71	2.63	
	5000	Screed, 24 ga. metal key joint, see Div 03150-250											

	03110 | Structural C.I.P. Forms		Crew	Daily Output	Labor-Hours	Unit	2006 Bare Costs Mat.	Labor	Equip.	Total	Total Incl O&P	
445	5020 Wood, incl. wood stakes, 1″ x 3″	R031113-40	C-1	900	.036	L.F.	.57	1.19		1.76	2.59	445
	5050 2″ x 4″			900	.036	″	.66	1.19		1.85	2.69	
	6000 Trench forms in floor, wood, 1 use	R031113-60		160	.200	SFCA	2.50	6.70		9.20	13.85	
	6150 4 use			185	.173	″	.75	5.80		6.55	10.45	
	8760 Void form, corrugated fiberboard, 6″ x 12″, 10′ long			240	.133	S.F.	.77	4.47		5.24	8.25	
	9000 Minimum labor/equipment charge		1 Carp	2	4	Job		142		142	236	
455	0010 **FORMS IN PLACE, WALLS**	R031113-10										455
	0100 Box out for wall openings, to 16″ thick, to 10 S.F.		C-2	24	2	Ea.	23.50	69		92.50	140	
	0150 Over 10 S.F. (use perimeter)	R031113-40	″	280	.171	L.F.	2.02	5.90		7.92	12	
	0250 Brick shelf, 4″ w, add to wall forms, use wall area abv shelf											
	0260 1 use	R031113-60	C-2	240	.200	SFCA	2.20	6.90		9.10	13.85	
	0350 4 use			300	.160	″	.88	5.50		6.38	10.10	
	0500 Bulkhead, with keyway, 1 use, 2 piece			265	.181	L.F.	3.17	6.25		9.42	13.85	
	0550 3 piece			175	.274		4	9.45		13.45	20	
	0600 Bulkhead forms with keyway, 1 Pc. expanded metal, 8″ wall		C-1	1000	.032		1.08	1.07		2.15	2.97	
	0610 10″ wall			800	.040		1.88	1.34		3.22	4.29	
	0620 12″ wall			525	.061		2.28	2.04		4.32	5.90	
	0700 Buttress, to 8′ high, 1 use		C-2	350	.137	SFCA	4.01	4.74		8.75	12.25	
	0850 4 use			480	.100	″	1.32	3.45		4.77	7.15	
	1000 Corbel or haunch, to 12″ wide, add to wall forms, 1 use			150	.320	L.F.	2.14	11.05		13.19	20.50	
	1150 4 use			180	.267	″	.69	9.20		9.89	16	
	2000 Wall, below grade, job-built plywood, to 8′ high, 1 use			300	.160	SFCA	2.47	5.50		7.97	11.85	
	2150 4 use			435	.110		.93	3.81		4.74	7.30	
	2400 Over 8′ to 16′ high, 1 use			280	.171		4.97	5.90		10.87	15.25	
	2420 2 use			345	.139		2	4.80		6.80	10.15	
	2430 3 use			375	.128		1.68	4.42		6.10	9.15	
	2440 4 use			395	.122		1.51	4.20		5.71	8.60	
	2445 Exterior wall, 8′ to 16′ high, 1 use			280	.171		2.17	5.90		8.07	12.20	
	2550 4 use			395	.122		.70	4.20		4.90	7.70	
	2700 Over 16′ high, 1 use			235	.204		2.42	7.05		9.47	14.35	
	2850 4 use			330	.145		.79	5		5.79	9.15	
	3000 For architectural finish, add			1820	.026		5.45	.91		6.36	7.50	
	4000 Radial, smooth curved, job-built plywood, 1 use			245	.196		2.27	6.75		9.02	13.70	
	4150 4 use			335	.143		.74	4.95		5.69	9	
	4200 Below grade, job-built plywood, 1 use			225	.213		3.13	7.35		10.48	15.65	
	4210 2 use			225	.213		1.73	7.35		9.08	14.10	
	4220 3 use			225	.213		1.40	7.35		8.75	13.75	
	4230 4 use			225	.213		1.02	7.35		8.37	13.30	
	4600 Retaining wall, battered, job-built plyw'd, to 8′ high, 1 use			300	.160		1.81	5.50		7.31	11.15	
	4750 4 use			390	.123		.54	4.25		4.79	7.65	
	4900 Over 8′ to 16′ high, 1 use			240	.200		1.98	6.90		8.88	13.65	
	5050 4 use			320	.150		.64	5.20		5.84	9.30	
	5750 Liners for forms (add to wall forms), A.B.S. plastic											
	5800 Aged wood, 4″ wide, 1 use		1 Carp	250	.032	SFCA	6.45	1.14		7.59	8.95	
	5820 2 use			400	.020		3.54	.71		4.25	5.05	
	5840 4 use			750	.011		2.09	.38		2.47	2.93	
	5900 Fractured rope rib, 1 use			250	.032		3.84	1.14		4.98	6.10	
	6000 4 use			750	.011		1.25	.38		1.63	2	
	6100 Ribbed look, 1/2″ & 3/4″ deep, 1 use			300	.027		5.45	.95		6.40	7.55	
	6200 4 use			800	.010		1.77	.36		2.13	2.54	
	6300 Rustic brick pattern, 1 use			250	.032		3.84	1.14		4.98	6.10	
	6400 4 use			750	.011		1.25	.38		1.63	2	
	6500 Striated, random, 3/8″ x 3/8″ deep, 1 use			300	.027		3.84	.95		4.79	5.80	
	6600 4 use			800	.010		1.25	.36		1.61	1.96	
	7500 Lintel or sill forms, 1 use			30	.267		2.93	9.50		12.43	18.95	
	7560 4 use			37	.216		.95	7.70		8.65	13.80	

	03110 | Structural C.I.P. Forms		CREW	DAILY OUTPUT	LABOR-HOURS	UNIT	2006 BARE COSTS MAT.	LABOR	EQUIP.	TOTAL	TOTAL INCL O&P	
455	7800	Modular prefabricated plywood, to 8′ high, 1 use R031113-10	C-2	1180	.041	SFCA	1.91	1.40		3.31	4.43	455
	7860	4 use		1260	.038		.63	1.32		1.95	2.87	
	8000	To 16′ high, 1 use R031113-40		715	.067		2.50	2.32		4.82	6.60	
	8060	4 use		790	.061		.83	2.10		2.93	4.39	
	8100	Over 16′ high, 1 use R031113-60		715	.067		3	2.32		5.32	7.15	
	8160	4 use		790	.061		1	2.10		3.10	4.58	
	8600	Pilasters, 1 use		270	.178		2.87	6.15		9.02	13.30	
	8660	4 use		385	.125		.93	4.31		5.24	8.20	
	9475	For elevated walls, add						10%				
	9480	For battered walls, 1 side battered, add					10%	10%				
	9485	For battered walls, 2 sides battered, add					15%	15%				
	9900	Minimum labor/equipment charge	2 Carp	2	8	Job		284		284	470	
500	0010	**GAS STATION FORMS** Curb fascia, with template,										500
	0050	12 ga. steel, left in place, 9″ high	1 Carp	50	.160	L.F.	9.95	5.70		15.65	20.50	
	1000	Sign or light bases, 18″ diameter, 9″ high		9	.889	Ea.	62.50	31.50		94	122	
	1050	30″ diameter, 13″ high		8	1	″	99.50	35.50		135	169	
	1990	Minimum labor/equipment charge		2	4	Job		142		142	236	
	2000	Island forms, 10′ long, 9″ high, 3′- 6″ wide	C-1	10	3.200	Ea.	278	107		385	485	
	2050	4′ wide		9	3.556		287	119		406	510	
	2500	20′ long, 9″ high, 4′ wide		6	5.333		460	179		639	805	
	2550	5′ wide		5	6.400		480	214		694	885	
	9000	Minimum labor/equipment charge	1 Carp	3	2.667	Job		95		95	157	
800	0010	**SCAFFOLDING** See Division 01540-750										800

	03150 | Concrete Accessories		CREW	DAILY OUTPUT	LABOR-HOURS	UNIT	MAT.	LABOR	EQUIP.	TOTAL	TOTAL INCL O&P	
085	0012	**ANCHOR BOLTS** See Division 04080-070										085
250	0010	**EXPANSION JOINTS**										250
	0020	Keyed, cold, 24 ga, incl. stakes, 3-1/2″ high	1 Carp	200	.040	L.F.	.69	1.42		2.11	3.12	
	0050	4-1/2″ high		200	.040		.84	1.42		2.26	3.28	
	0100	5-1/2″ high		195	.041		.90	1.46		2.36	3.41	
	0150	7-1/2″ high		190	.042		1.08	1.50		2.58	3.67	
	0300	Poured asphalt, plain, 1/2″ x 1″	1 Clab	450	.018		.39	.49		.88	1.24	
	0350	1″ x 2″		400	.020		1.55	.55		2.10	2.62	
	0500	Neoprene, liquid, cold applied, 1/2″ x 1″		450	.018		1.85	.49		2.34	2.85	
	0550	1″ x 2″		400	.020		6.25	.55		6.80	7.80	
	0700	Polyurethane, poured, 2 part, 1/2″ x 1″		400	.020		1.27	.55		1.82	2.31	
	0750	1″ x 2″		350	.023		5.10	.63		5.73	6.65	
	0900	Rubberized asphalt, hot or cold applied, 1/2″ x 1″		450	.018		.37	.49		.86	1.22	
	0950	1″ x 2″		400	.020		1.36	.55		1.91	2.41	
	1100	Hot applied, fuel resistant, 1/2″ x 1″		450	.018		1.04	.49		1.53	1.95	
	1150	1″ x 2″		400	.020		2	.55		2.55	3.11	
	2000	Premolded, bituminous fiber, 1/2″ x 6″	1 Carp	375	.021		.37	.76		1.13	1.67	
	2050	1″ x 12″		300	.027		1.51	.95		2.46	3.23	
	2250	Cork with resin binder, 1/2″ x 6″		375	.021		1.68	.76		2.44	3.11	
	2300	1″ x 12″		300	.027		6.10	.95		7.05	8.30	
	2500	Neoprene sponge, closed cell, 1/2″ x 6″		375	.021		1.42	.76		2.18	2.82	
	2550	1″ x 12″		300	.027		6.50	.95		7.45	8.70	
	2750	Polyethylene foam, 1/2″ x 6″		375	.021		.47	.76		1.23	1.78	
	2800	1″ x 12″		300	.027		1.69	.95		2.64	3.43	
	3000	Polyethylene backer rod, 3/8″ diameter		460	.017		.03	.62		.65	1.05	
	3050	3/4″ diameter		460	.017		.07	.62		.69	1.09	
	3100	1″ diameter		460	.017		.11	.62		.73	1.14	
	3500	Polyurethane foam, with polybutylene, 1/2″ x 1/2″		475	.017		.92	.60		1.52	2	
	3550	1″ x 1″		450	.018		1.82	.63		2.45	3.05	

	03150 | Concrete Accessories		CREW	DAILY OUTPUT	LABOR-HOURS	UNIT	2006 BARE COSTS MAT.	LABOR	EQUIP.	TOTAL	TOTAL INCL O&P	
250	3750	Polyurethane foam, regular, closed cell, 1/2" x 6"	1 Carp	375	.021	L.F.	.62	.76		1.38	1.94	250
	3800	1" x 12"		300	.027		1.24	.95		2.19	2.93	
	4000	Polyvinyl chloride foam, closed cell, 1/2" x 6"		375	.021		2.42	.76		3.18	3.92	
	4050	1" x 12"		300	.027		6.15	.95		7.10	8.35	
	4250	Rubber, gray sponge, 1/2" x 6"		375	.021		2.65	.76		3.41	4.18	
	4300	1" x 12"	↓	300	.027	↓	10.25	.95		11.20	12.80	
	4500	Lead wool for joints, 1 ton lots				Lb.	1.90			1.90	2.09	
	5000	For installation in walls, add						75%				
	5250	For installation in boxouts, add						25%				
600	0010	**SHORES** Erect and strip, by hand, horizontal members										600
	0500	Aluminum joists and stringers	2 Carp	60	.267	Ea.		9.50		9.50	15.70	
	0600	Steel, adjustable beams		45	.356			12.65		12.65	21	
	0700	Wood joists		50	.320			11.40		11.40	18.85	
	0800	Wood stringers		30	.533			18.95		18.95	31.50	
	1000	Vertical members to 10' high		55	.291			10.35		10.35	17.15	
	1050	To 13' high		50	.320			11.40		11.40	18.85	
	1100	To 16' high		45	.356	↓		12.65		12.65	21	
	1500	Reshoring	↓	1400	.011	S.F.	.42	.41		.83	1.13	
	1600	Flying truss system	C-17D	9600	.009	SFCA		.32	.06	.38	.59	
	1760	Horizontal, aluminum joists, 6-1/4" high x 5' to 21' span, buy				L.F.	33.50			33.50	37	
	1770	Beams, 7-1/4" high x 4' to 30' span				"	47			47	51.50	
	1810	Horizontal, steel beam, adjustable, 4' to 7' span				Ea.	535			535	590	
	1830	6' to 10' span					690			690	760	
	1920	9' to 15' span					875			875	965	
	1940	12' to 20' span				↓	895			895	985	
	1970	Steel stringer, W8x10, 4' to 16' span, buy				L.F.	19.25			19.25	21	
	3000	Rent for job duration, aluminum joist @ 2' O.C., per mo				SF Flr.	.84			.84	.93	
	3050	Steel W8x10					.48			.48	.53	
	3060	Steel adjustable				↓	1.73			1.73	1.90	
	3500	#1 post shore, steel, 5'-7" to 9'-6" high, 10000# cap., buy				Ea.	355			355	390	
	3550	#2 post shore, 7'-3" to 12'-10" high, 7800# capacity					380			380	420	
	3600	#3 post shore, 8'-10" to 16'-1" high, 3800# capacity				↓	415			415	455	
	5010	Frame shoring systems, steel, 12000#/leg, buy										
	5040	Frame, 2' wide x 6' high				Ea.	330			330	365	
	5250	X-brace					52.50			52.50	57.50	
	5550	Base plate					41.50			41.50	45.50	
	5600	Screw jack					132			132	145	
	5650	U-head, 8" x 8"				↓	21			21	23	
660	0010	**STAIR TREAD INSERTS**										660
	0015	Cast iron, abrasive, 3" wide	1 Carp	90	.089	L.F.	8.05	3.16		11.21	14.10	
	0020	4" wide		80	.100		10.70	3.56		14.26	17.70	
	0040	6" wide		75	.107		16.10	3.79		19.89	24	
	0050	9" wide		70	.114		24	4.06		28.06	33.50	
	0100	12" wide	↓	65	.123		32	4.38		36.38	43	
	0300	Cast aluminum, compared to cast iron, deduct					10%					
	0500	Extruded aluminum safety tread, 3" wide	1 Carp	75	.107		9.15	3.79		12.94	16.35	
	0550	4" wide		75	.107		12.20	3.79		15.99	19.75	
	0600	6" wide		75	.107		18.30	3.79		22.09	26.50	
	0650	9" wide to resurface stairs	↓	70	.114	↓	27.50	4.06		31.56	37	
	1700	Cement fill for pan-type metal treads, plain	1 Cefi	115	.070	S.F.	1.61	2.39		4	5.55	
	1750	Non-slip	"	100	.080	"	1.77	2.75		4.52	6.30	
860	0010	**WATERSTOP**										860
	0020	PVC, ribbed 3/16" thick, 4" wide	1 Carp	155	.052	L.F.	.79	1.83		2.62	3.91	

03100 | Concrete Forms & Accessories

	03150 | Concrete Accessories	CREW	DAILY OUTPUT	LABOR-HOURS	UNIT	2006 BARE COSTS MAT.	LABOR	EQUIP.	TOTAL	TOTAL INCL O&P	
860	0050 6" wide	1 Carp	145	.055	L.F.	1.30	1.96		3.26	4.68	860
	0500 Ribbed, PVC, with center bulb, 6" wide, 3/16" thick		135	.059		1.19	2.11		3.30	4.80	
	0550 3/8" thick		130	.062		1.86	2.19		4.05	5.65	

03200 | Concrete Reinforcement

	03210 | Reinforcing Steel	CREW	DAILY OUTPUT	LABOR-HOURS	UNIT	2006 BARE COSTS MAT.	LABOR	EQUIP.	TOTAL	TOTAL INCL O&P	
600	0010 **REINFORCING IN PLACE** A615 Grade 60, incl. access. labor (R032110-10)										600
	0102 Beams & Girders, #3 to #7	4 Rodm	3200	.010	Lb.	.44	.40		.84	1.17	
	0152 #8 to #18 (R032110-20)		5400	.006		.44	.23		.67	.89	
	0202 Columns, #3 to #7		3000	.011		.44	.42		.86	1.21	
	0252 #8 to #18 (R032110-25)		4600	.007		.44	.27		.71	.96	
	0402 Elevated slabs, #4 to #7		5800	.006		.47	.22		.69	.88	
	0502 Footings, #4 to #7		4200	.008		.44	.30		.74	1.01	
	0552 #8 to #18		7200	.004		.44	.18		.62	.79	
	0602 Slab on grade, #3 to #7		4200	.008		.43	.30		.73	.99	
	0702 Walls, #3 to #7		6000	.005		.44	.21		.65	.85	
	0752 #8 to #18		8000	.004		.44	.16		.60	.76	
	2000 Unloading & sorting, add to above	C-5	100	.560	Ton		21.50	6.50	28	43.50	
	2200 Crane cost for handling, add to above, minimum		135	.415			16	4.82	20.82	32.50	
	2210 Average		92	.609			23.50	7.05	30.55	47.50	
	2220 Maximum		35	1.600			61.50	18.60	80.10	125	
	2400 Dowels, 2 feet long, deformed, #3	2 Rodm	520	.031	Ea.	.35	1.22		1.57	2.48	
	2410 #4		480	.033		.63	1.32		1.95	2.95	
	2420 #5		435	.037		.98	1.45		2.43	3.57	
	2430 #6		360	.044		1.41	1.76		3.17	4.56	
	2450 Longer and heavier dowels, add		725	.022	Lb.	.47	.87		1.34	2.02	
	2500 Smooth dowels, 12" long, 1/4" or 3/8" diameter		140	.114	Ea.	.72	4.51		5.23	8.55	
	2520 5/8" diameter		125	.128		1.25	5.05		6.30	10.10	
	2530 3/4" diameter		110	.145		1.55	5.75		7.30	11.55	
	2600 Dowel sleeves for CIP concrete, 2-part system										
	2610 Sleeve base, plastic, for #5 bar, fasten to edge form	1 Rodm	200	.040	Ea.	.36	1.58		1.94	3.11	
	2615 Sleeve, plastic, for #5 bar x 9" long, snap onto base		400	.020		1.03	.79		1.82	2.49	
	2620 Sleeve base, for #6 bar		175	.046		.36	1.81		2.17	3.50	
	2625 Sleeve, for #6 bar		350	.023		1.06	.90		1.96	2.72	
	2630 Sleeve base, for #8 bar		150	.053		.44	2.11		2.55	4.10	
	2635 Sleeve, for #8 bar		300	.027		1.18	1.05		2.23	3.11	
	2700 Dowel caps, visual warning only, plastic, #3 to #8	2 Rodm	800	.020		.27	.79		1.06	1.66	
	2720 #7 to #14		750	.021		.35	.84		1.19	1.83	
	2750 Impalement protective, plastic, #3 to #7		800	.020		1.59	.79		2.38	3.11	
	2760 #7 to #11		775	.021		2.21	.82		3.03	3.83	
	2770 #11 to #16		750	.021		2.25	.84		3.09	3.93	
	3000 For epoxy dowel anchoring, see Div 05090-300										
	9000 Minimum labor/equipment charge	1 Rodm	4	2	Job		79		79	136	

	03220 | Welded Wire Fabric	CREW	DAILY OUTPUT	LABOR-HOURS	UNIT	MAT.	LABOR	EQUIP.	TOTAL	TOTAL INCL O&P	
200	0010 **WELDED WIRE FABRIC** ASTM A185 (R032205-30)										200
	0050 Sheets										

03200 | Concrete Reinforcement

03220 | Welded Wire Fabric

		03220 Welded Wire Fabric	CREW	DAILY OUTPUT	LABOR-HOURS	UNIT	2006 BARE COSTS MAT.	LABOR	EQUIP.	TOTAL	TOTAL INCL O&P	
200	0100	6 x 6 - W1.4 x W1.4 (10 x 10) 21 lb. per C.S.F. R032205-30	2 Rodm	35	.457	C.S.F.	12	18.05		30.05	44	200
	0200	6 x 6 - W2.1 x W2.1 (8 x 8) 30 lb. per C.S.F.		31	.516		14.30	20.50		34.80	51	
	0300	6 x 6 - W2.9 x W2.9 (6 x 6) 42 lb. per C.S.F.		29	.552		17.15	22		39.15	56.50	
	0400	6 x 6 - W4 x W4 (4 x 4) 58 lb. per C.S.F.		27	.593		23.50	23.50		47	65.50	
	0500	4 x 4 - W1.4 x W1.4 (10 x 10) 31 lb. per C.S.F.		31	.516		17.70	20.50		38.20	54.50	
	0600	4 x 4 - W2.1 x W2.1 (8 x 8) 44 lb. per C.S.F.		29	.552		21	22		43	60.50	
	0650	4 x 4 - W2.9 x W2.9 (6 x 6) 61 lb. per C.S.F.		27	.593		27.50	23.50		51	70.50	
	0700	4 x 4 - W4 x W4 (4 x 4) 85 lb. per C.S.F.	↓	25	.640	↓	39	25.50		64.50	86	
	0750	Rolls										
	0800	2 x 2 - #14 galv., 21 lb/C.S.F., beam & column wrap	2 Rodm	6.50	2.462	C.S.F.	30.50	97		127.50	201	
	0900	2 x 2 - #12 galv. for gunite reinforcing	"	6.50	2.462	"	35	97		132	206	
	9000	Minimum labor/equipment charge	1 Rodm	4	2	Job		79		79	136	

03240 | Fibrous Reinforcing

		03240 Fibrous Reinforcing	CREW	DAILY OUTPUT	LABOR-HOURS	UNIT	MAT.	LABOR	EQUIP.	TOTAL	TOTAL INCL O&P	
300	0010	**FIBROUS REINFORCING**										300
	0100	Synthetic fibers, add to concrete				Lb.	3.97			3.97	4.37	
	0110	1-1/2 lb. per C.Y.				C.Y.	6.15			6.15	6.75	
	0150	Steel fibers, add to concrete				Lb.	.46			.46	.51	
	0155	25 lb. per C.Y.				C.Y.	11.50			11.50	12.65	
	0160	50 lb. per C.Y.					23			23	25.50	
	0170	75 lb. per C.Y.					35.50			35.50	39	
	0180	100 lb. per C.Y.				↓	46			46	50.50	

03300 | Cast-In-Place Concrete

03310 | Structural Concrete

		03310 Structural Concrete	CREW	DAILY OUTPUT	LABOR-HOURS	UNIT	2006 BARE COSTS MAT.	LABOR	EQUIP.	TOTAL	TOTAL INCL O&P	
220	0010	**CONCRETE, READY MIX** Normal weight R033105-20										220
	0015	Excludes all additives and treatments										
	0020	2000 psi R033105-30				C.Y.	83.50			83.50	92	
	0100	2500 psi					85			85	93.50	
	0150	3000 psi R033105-40					87			87	95.50	
	0200	3500 psi					89			89	98	
	0300	4000 psi R033105-50					91			91	100	
	0350	4500 psi					93			93	102	
	0400	5000 psi					96			96	106	
	0411	6000 psi					109			109	120	
	0412	8000 psi					179			179	196	
	0413	10,000 psi					253			253	279	
	0414	12,000 psi					305			305	335	
	1000	For high early strength cement, add					10%					
	2000	For all lightweight aggregate, add				↓	45%					
	3000	For integral colors, 2500 psi (5 bag mix)										
	3100	Red, yellow or brown, 1.8 lb. per bag, add				C.Y.	21			21	23	
	3200	9.4 lb. per bag, add					109			109	120	
	3400	Black, 1.8 lb. per bag, add					28			28	31	
	3500	7.5 lb. per bag, add					117			117	129	
	3700	Green, 1.8 lb. per bag, add					35.50			35.50	39	
	3800	7.5 lb. per bag, add				↓	148			148	163	
240	0010	**CONCRETE IN PLACE** R033053-10										240
	0020	Including forms (4 uses), concrete, placement, reinforcing										

03310 | Structural Concrete		CREW	DAILY OUTPUT	LABOR-HOURS	UNIT	2006 BARE COSTS MAT.	LABOR	EQUIP.	TOTAL	TOTAL INCL O&P
240 0050	steel and finishing unless otherwise indicated (R033053-50)									
0300	Beams, 5 kip per L.F., 10′ span	C-14A	15.62	12.804	C.Y.	287	455	46	788	1,125
0350	25′ span (R033053-60)	″	18.55	10.782		298	385	39	722	1,000
0500	Chimney foundations, industrial, minimum	C-14C	32.22	3.476		129	118	.66	247.66	340
0510	Maximum (R033105-80)	″	23.71	4.724		152	160	.90	312.90	435
0700	Columns, square, 12″ x 12″, minimum reinforcing	C-14A	11.96	16.722		305	595	60.50	960.50	1,400
0720	Average reinforcing (R033105-85)		10.13	19.743		485	705	71.50	1,261.50	1,775
0740	Maximum reinforcing		9.03	22.148		725	790	80	1,595	2,200
0800	16″ x 16″, minimum reinforcing		16.22	12.330		243	440	44.50	727.50	1,050
0820	Average reinforcing		12.57	15.911		410	565	57.50	1,032.50	1,450
0840	Maximum reinforcing		10.25	19.512		640	695	70.50	1,405.50	1,925
0900	24″ x 24″, minimum reinforcing		23.66	8.453		207	300	30.50	537.50	760
0920	Average reinforcing		17.71	11.293		370	400	41	811	1,125
0940	Maximum reinforcing		14.15	14.134		585	505	51	1,141	1,525
1000	36″ x 36″, minimum reinforcing		33.69	5.936		182	211	21.50	414.50	575
1020	Average reinforcing		23.32	8.576		325	305	31	661	895
1040	Maximum reinforcing		17.82	11.223		545	400	40.50	985.50	1,300
1200	16″ diameter, minimum reinforcing		31.49	6.351		236	226	23	485	660
1220	Average reinforcing		19.12	10.460		415	370	38	823	1,125
1240	Maximum reinforcing		13.77	14.524		630	515	52.50	1,197.50	1,600
1300	20″ diameter, minimum reinforcing		41.04	4.873		238	174	17.60	429.60	570
1320	Average reinforcing		24.05	8.316		400	296	30	726	965
1340	Maximum reinforcing		17.01	11.758		630	420	42.50	1,092.50	1,425
1400	24″ diameter, minimum reinforcing		51.85	3.857		223	137	13.90	373.90	490
1420	Average reinforcing		27.06	7.391		400	263	26.50	689.50	905
1440	Maximum reinforcing		18.29	10.935		620	390	39.50	1,049.50	1,375
1500	36″ diameter, minimum reinforcing		75.04	2.665		224	95	9.60	328.60	415
1520	Average reinforcing		37.49	5.335		380	190	19.25	589.25	755
1540	Maximum reinforcing		22.84	8.757		600	310	31.50	941.50	1,225
1900	Elevated slabs, flat slab with drops, 125 psf Sup. Load, 20′ span	C-14B	38.45	5.410		242	192	18.75	452.75	610
1950	30′ span		50.99	4.079		250	145	14.15	409.15	530
2100	Flat plate, 125 psf Sup. Load, 15′ span		30.24	6.878		220	245	24	489	675
2150	25′ span		49.60	4.194		226	149	14.55	389.55	510
2300	Waffle const., 30″ domes, 125 psf Sup. Load, 20′ span		37.07	5.611		330	200	19.45	549.45	715
2350	30′ span		44.07	4.720		294	168	16.40	478.40	620
2500	One way joists, 30″ pans, 125 psf Sup. Load, 15′ span		27.38	7.597		410	270	26.50	706.50	930
2550	25′ span		31.15	6.677		375	237	23	635	830
2700	One way beam & slab, 125 psf Sup. Load, 15′ span		20.59	10.102		245	360	35	640	905
2750	25′ span		28.36	7.334		225	261	25.50	511.50	705
2900	Two way beam & slab, 125 psf Sup. Load, 15′ span		24.04	8.652		232	310	30	572	800
2950	25′ span		35.87	5.799		196	206	20	422	575
3100	Elevated slabs including finish, not									
3110	including forms or reinforcing									
3150	Regular concrete, 4″ slab	C-8	2613	.021	S.F.	1.18	.66	.27	2.11	2.67
3200	6″ slab		2585	.022		1.74	.67	.27	2.68	3.31
3250	2-1/2″ thick floor fill		2685	.021		.76	.65	.26	1.67	2.17
3300	Lightweight, 110# per C.F., 2-1/2″ thick floor fill		2585	.022		1.04	.67	.27	1.98	2.54
3400	Cellular concrete, 1-5/8″ fill, under 5000 S.F.		2000	.028		.70	.87	.35	1.92	2.57
3450	Over 10,000 S.F.		2200	.025		.66	.79	.32	1.77	2.36
3500	Add per floor for 3 to 6 stories high		31800	.002			.05	.02	.07	.11
3520	For 7 to 20 stories high		21200	.003			.08	.03	.11	.17
3800	Footings, spread under 1 C.Y.	C-14C	38.07	2.942	C.Y.	175	99.50	.56	275.06	360
3850	Over 5 C.Y.		81.04	1.382		242	47	.26	289.26	345
3900	Footings, strip, 18″ x 9″, unreinforced		40	2.800		109	95	.53	204.53	278
3920	18″ x 9″, reinforced		35	3.200		130	108	.61	238.61	325
3925	20″ x 10″, unreinforced		45	2.489		106	84.50	.47	190.97	257

CONCRETE 3

	03310 Structural Concrete		CREW	DAILY OUTPUT	LABOR-HOURS	UNIT	MAT.	LABOR	EQUIP.	TOTAL	TOTAL INCL O&P	
240	3930	20" x 10", reinforced R033053-10	C-14C	40	2.800	C.Y.	122	95	.53	217.53	293	240
	3935	24" x 12", unreinforced		55	2.036		103	69	.39	172.39	228	
	3940	24" x 12", reinforced R033053-50		48	2.333		121	79	.44	200.44	264	
	3945	36" x 12", unreinforced		70	1.600		99.50	54	.30	153.80	199	
	3950	36" x 12", reinforced R033053-60		60	1.867		115	63	.35	178.35	231	
	4000	Foundation mat, under 10 C.Y.		38.67	2.896		179	98	.55	277.55	360	
	4050	Over 20 C.Y. R033105-80	↓	56.40	1.986		156	67	.38	223.38	283	
	4200	Grade walls, 8" thick, 8' high	C-14D	45.83	4.364		157	154	15.75	326.75	445	
	4250	14' high R033105-85		27.26	7.337		203	259	26.50	488.50	680	
	4260	12" thick, 8' high		64.32	3.109		140	110	11.25	261.25	350	
	4270	14' high		40.01	4.999		159	176	18.05	353.05	485	
	4300	15" thick, 8' high		80.02	2.499		132	88	9	229	300	
	4350	12' high		51.26	3.902		140	138	14.10	292.10	400	
	4500	18' high	↓	48.85	4.094	↓	155	144	14.80	313.80	425	
	4520	Handicap access ramp, railing both sides, 3' wide	C-14H	14.58	3.292	L.F.	204	115	1.48	320.48	415	
	4525	5' wide		12.22	3.928		212	137	1.77	350.77	460	
	4530	With 6" curb and rails both sides, 3' wide		8.55	5.614		211	196	2.53	409.53	560	
	4535	5' wide	↓	7.31	6.566	↓	215	230	2.95	447.95	620	
	4650	Slab on grade, not including finish, 4" thick	C-14E	60.75	1.449	C.Y.	106	50.50	.35	156.85	202	
	4700	6" thick	"	92	.957	"	102	33.50	.23	135.73	168	
	4751	Slab on grade, incl. troweled finish, not incl. forms										
	4760	or reinforcing, over 10,000 S.F., 4" thick	C-14F	3425	.021	S.F.	1.16	.68	.01	1.85	2.37	
	4820	6" thick		3350	.021		1.69	.69	.01	2.39	2.98	
	4840	8" thick		3184	.023		2.31	.73	.01	3.05	3.73	
	4900	12" thick		2734	.026		3.47	.85	.01	4.33	5.20	
	4950	15" thick	↓	2505	.029	↓	4.36	.93	.01	5.30	6.30	
	5000	Slab on grade, incl. textured finish, not incl. forms										
	5001	or reinforcing, 4" thick	C-14G	2873	.019	S.F.	1.13	.62	.01	1.76	2.24	
	5010	6" thick		2590	.022		1.77	.69	.01	2.47	3.06	
	5020	8" thick	↓	2320	.024	↓	2.31	.77	.01	3.09	3.78	
	5200	Lift slab in place above the foundation, incl. forms,										
	5210	reinforcing, concrete and columns, minimum	C-14B	2113	.098	S.F.	5.35	3.50	.34	9.19	12.05	
	5250	Average		1650	.126		5.80	4.48	.44	10.72	14.30	
	5300	Maximum	↓	1500	.139	↓	6.25	4.93	.48	11.66	15.65	
	5500	Lightweight, ready mix, including screed finish only,										
	5510	not including forms or reinforcing										
	5550	1:4 for structural roof decks	C-14B	260	.800	C.Y.	122	28.50	2.78	153.28	184	
	5600	1:6 for ground slab with radiant heat	C-14F	92	.783		123	25.50	.23	148.73	176	
	5650	1:3:2 with sand aggregate, roof deck	C-14B	260	.800		120	28.50	2.78	151.28	182	
	5700	Ground slab	C-14F	107	.673		120	21.50	.20	141.70	167	
	5900	Pile caps, incl. forms and reinf., sq. or rect., under 5 C.Y.	C-14C	54.14	2.069		151	70	.39	221.39	282	
	5950	Over 10 C.Y.		75	1.493		139	50.50	.28	189.78	237	
	6000	Triangular or hexagonal, under 5 C.Y.		53	2.113		107	71.50	.40	178.90	236	
	6050	Over 10 C.Y.	↓	85	1.318		120	44.50	.25	164.75	206	
	6200	Retaining walls, gravity, 4' high see division 02830-800	C-14D	66.20	3.021		125	107	10.90	242.90	325	
	6250	10' high		125	1.600		117	56.50	5.80	179.30	229	
	6300	Cantilever, level backfill loading, 8' high		70	2.857		136	101	10.30	247.30	330	
	6350	16' high	↓	91	2.198	↓	130	77.50	7.95	215.45	280	
	6800	Stairs, not including safety treads, free standing, 3'-6" wide	C-14H	83	.578	LF Nose	5.30	20	.26	25.56	39.50	
	6850	Cast on ground		125	.384	"	4.03	13.45	.17	17.65	26.50	
	7000	Stair landings, free standing		200	.240	S.F.	4.36	8.40	.11	12.87	18.80	
	7050	Cast on ground	↓	475	.101	"	3.05	3.54	.05	6.64	9.25	
	9000	Minimum labor/equipment charge	2 Carp	1	16	Job		570		570	940	
700	0010	PLACING CONCRETE R033105-70										700
	0020	Includes labor and equipment to place and vibrate										

	03310 | Structural Concrete	CREW	DAILY OUTPUT	LABOR-HOURS	UNIT	2006 BARE COSTS MAT.	LABOR	EQUIP.	TOTAL	TOTAL INCL O&P
700 0050	Beams, elevated, small beams, pumped (R033105-70)	C-20	60	1.067	C.Y.		31.50	12.40	43.90	65
0100	With crane and bucket	C-7	45	1.600			48	22.50	70.50	103
0200	Large beams, pumped	C-20	90	.711			21	8.25	29.25	43.50
0250	With crane and bucket	C-7	65	1.108			33	15.55	48.55	71
0400	Columns, square or round, 12" thick, pumped	C-20	60	1.067			31.50	12.40	43.90	65
0450	With crane and bucket	C-7	40	1.800			54	25.50	79.50	116
0600	18" thick, pumped	C-20	90	.711			21	8.25	29.25	43.50
0650	With crane and bucket	C-7	55	1.309			39.50	18.40	57.90	84
0800	24" thick, pumped	C-20	92	.696			20.50	8.10	28.60	42.50
0850	With crane and bucket	C-7	70	1.029			31	14.45	45.45	66
1000	36" thick, pumped	C-20	140	.457			13.55	5.30	18.85	28
1050	With crane and bucket	C-7	100	.720			21.50	10.10	31.60	46
1400	Elevated slabs, less than 6" thick, pumped	C-20	140	.457			13.55	5.30	18.85	28
1450	With crane and bucket	C-7	95	.758			22.50	10.65	33.15	48.50
1500	6" to 10" thick, pumped	C-20	160	.400			11.90	4.65	16.55	24.50
1550	With crane and bucket	C-7	110	.655			19.65	9.20	28.85	42
1600	Slabs over 10" thick, pumped	C-20	180	.356			10.55	4.13	14.68	22
1650	With crane and bucket	C-7	130	.554			16.60	7.80	24.40	35.50
1900	Footings, continuous, shallow, direct chute	C-6	120	.400			11.55	.36	11.91	19.35
1950	Pumped	C-20	150	.427			12.65	4.96	17.61	26
2000	With crane and bucket	C-7	90	.800			24	11.25	35.25	51.50
2100	Footings, continuous, deep, direct chute	C-6	140	.343			9.90	.31	10.21	16.60
2150	Pumped	C-20	160	.400			11.90	4.65	16.55	24.50
2200	With crane and bucket	C-7	110	.655			19.65	9.20	28.85	42
2400	Footings, spread, under 1 C.Y., direct chute	C-6	55	.873			25	.78	25.78	42.50
2450	Pumped	C-20	65	.985			29	11.45	40.45	60
2500	With crane and bucket	C-7	45	1.600			48	22.50	70.50	103
2600	Over 5 C.Y., direct chute	C-6	120	.400			11.55	.36	11.91	19.35
2650	Pumped	C-20	150	.427			12.65	4.96	17.61	26
2700	With crane and bucket	C-7	100	.720			21.50	10.10	31.60	46
2900	Foundation mats, over 20 C.Y., direct chute	C-6	350	.137			3.96	.12	4.08	6.65
2950	Pumped	C-20	400	.160			4.75	1.86	6.61	9.80
3000	With crane and bucket	C-7	300	.240			7.20	3.37	10.57	15.40
3200	Grade beams, direct chute	C-6	150	.320			9.25	.28	9.53	15.45
3250	Pumped	C-20	180	.356			10.55	4.13	14.68	22
3300	With crane and bucket	C-7	120	.600			18	8.40	26.40	39
3500	High rise, for more than 5 stories, pumped, add per story	C-20	2100	.030			.91	.35	1.26	1.87
3510	With crane and bucket, add per story	C-7	2100	.034			1.03	.48	1.51	2.20
3700	Pile caps, under 5 C.Y., direct chute	C-6	90	.533			15.40	.47	15.87	26
3750	Pumped	C-20	110	.582			17.25	6.75	24	35.50
3800	With crane and bucket	C-7	80	.900			27	12.65	39.65	58
3850	Pile cap, 5 C.Y. to 10 C.Y., direct chute	C-6	175	.274			7.95	.24	8.19	13.25
3900	Pumped	C-20	200	.320			9.50	3.72	13.22	19.60
3950	With crane and bucket	C-7	150	.480			14.40	6.75	21.15	31
4000	Over 10 C.Y., direct chute	C-6	215	.223			6.45	.20	6.65	10.80
4050	Pumped	C-20	240	.267			7.90	3.10	11	16.35
4100	With crane and bucket	C-7	185	.389			11.65	5.45	17.10	25
4300	Slab on grade, 4" thick, direct chute	C-6	110	.436			12.60	.39	12.99	21
4350	Pumped	C-20	130	.492			14.60	5.70	20.30	30.50
4400	With crane and bucket	C-7	110	.655			19.65	9.20	28.85	42
4600	Over 6" thick, direct chute	C-6	165	.291			8.40	.26	8.66	14.10
4650	Pumped	C-20	185	.346			10.25	4.02	14.27	21
4700	With crane and bucket	C-7	145	.497			14.90	6.95	21.85	31.50
4900	Walls, 8" thick, direct chute	C-6	90	.533			15.40	.47	15.87	26
4950	Pumped	C-20	100	.640			19	7.45	26.45	39
5000	With crane and bucket	C-7	80	.900			27	12.65	39.65	58

03300 | Cast-In-Place Concrete

		03310 | Structural Concrete	CREW	DAILY OUTPUT	LABOR-HOURS	UNIT	2006 BARE COSTS MAT.	LABOR	EQUIP.	TOTAL	TOTAL INCL O&P	
700	5050	12" thick, direct chute (R033105-70)	C-6	100	.480	C.Y.		13.85	.43	14.28	23.50	700
	5100	Pumped	C-20	110	.582			17.25	6.75	24	35.50	
	5200	With crane and bucket	C-7	90	.800			24	11.25	35.25	51.50	
	5300	15" thick, direct chute	C-6	105	.457			13.20	.41	13.61	22	
	5350	Pumped	C-20	120	.533			15.85	6.20	22.05	33	
	5400	With crane and bucket	C-7	95	.758			22.50	10.65	33.15	48.50	
	5600	Wheeled concrete dumping, add to placing costs above										
	5610	Walking cart, 50' haul, add	C-18	32	.281	C.Y.		7.75	1.63	9.38	14.65	
	5620	150' haul, add		24	.375			10.35	2.18	12.53	19.55	
	5700	250' haul, add		18	.500			13.80	2.90	16.70	26	
	5800	Riding cart, 50' haul, add	C-19	80	.113			3.11	.93	4.04	6.15	
	5810	150' haul, add		60	.150			4.14	1.24	5.38	8.20	
	5900	250' haul, add		45	.200			5.50	1.65	7.15	10.95	
	9000	Minimum labor/equipment charge	C-6	2	24	Job		695	21.50	716.50	1,175	
		03350 | Concrete Finishing										
300	0010	**FINISHING FLOORS**										300
	0020	Monolithic, screed finish	1 Cefi	900	.009	S.F.		.31		.31	.48	
	0100	Screed and bull float (darby) finish		725	.011			.38		.38	.60	
	0150	Screed, float, and broom finish		630	.013			.44		.44	.69	
	0200	Screed, float, and hand trowel		600	.013			.46		.46	.72	
	0250	Machine trowel		550	.015			.50		.50	.79	
	0370	Minimum labor/equipment charge		4	2	Job		69		69	108	
	0400	Integral topping and finish, using 1:1:2 mix, 3/16" thick	C-10B	1000	.040	S.F.	.06	1.21	.18	1.45	2.22	
	0450	1/2" thick		950	.042		.17	1.27	.19	1.63	2.45	
	0500	3/4" thick		850	.047		.26	1.42	.21	1.89	2.82	
	0600	1" thick		750	.053		.35	1.61	.23	2.19	3.25	
	0800	Granolithic topping, laid after, 1:1:1-1/2 mix, 1/2" thick		590	.068		.19	2.05	.30	2.54	3.86	
	0820	3/4" thick		580	.069		.29	2.08	.30	2.67	4.02	
	0850	1" thick		575	.070		.39	2.10	.31	2.80	4.17	
	0950	2" thick		500	.080		.77	2.42	.35	3.54	5.15	
	9100	Minimum labor/equipment charge	C-10	2	12	Job		385		385	615	
325	0010	**CONTROL JOINTS, SAW CUT**										325
	0100	Sawcut in green concrete										
	0120	1" depth	C-27	2000	.008	L.F.	.08	.28	.05	.41	.58	
	0140	1-1/2" depth		1800	.009		.13	.31	.06	.50	.69	
	0160	2" depth		1600	.010		.17	.34	.07	.58	.79	
	0200	Clean out control joint of debris	C-28	6000	.001			.05		.05	.07	
	0300	Joint sealant										
	0320	Backer rod, polyethylene, 1/4" diameter	1 Cefi	460	.017	L.F.	.02	.60		.62	.96	
	0340	Sealant, polyurethane										
	0360	1/4" x 1/4" (308 LF/Gal)	1 Cefi	270	.030	L.F.	.16	1.02		1.18	1.79	
	0380	1/4" x 1/2" (154 LF/Gal)	"	255	.031	"	.32	1.08		1.40	2.05	
350	0010	**FINISHING WALLS**										350
	0020	Break ties and patch voids	1 Cefi	540	.015	S.F.	.03	.51		.54	.83	
	0050	Burlap rub with grout	"	450	.018		.03	.61		.64	.99	
	0300	Bush hammer, green concrete	B-39	1000	.048			1.39	.15	1.54	2.45	
	0350	Cured concrete	"	650	.074			2.14	.23	2.37	3.77	
	0700	Sandblast, light penetration	C-10	1100	.022		.24	.70		.94	1.39	
	0750	Heavy penetration		375	.064		.48	2.05		2.53	3.81	
	9000	Minimum labor/equipment charge		2	12	Job		385		385	615	
600	0010	**SLAB TEXTURE STAMPING**										600
	0020	Approx. 3 S.F.- 5 S.F. each, buy, minimum				Ea.	99			99	109	
	0030	Average					139			139	153	
	0120	Maximum					169			169	186	

03300 | Cast-In-Place Concrete

03350 | Concrete Finishing

		03350 Concrete Finishing	CREW	DAILY OUTPUT	LABOR-HOURS	UNIT	2006 BARE COSTS MAT.	LABOR	EQUIP.	TOTAL	TOTAL INCL O&P	
600	0200	Commonly used chemicals for texture systems										600
	0210	Hardener, colored powder				S.F.	.51			.51	.56	
	0220	Release agent, colored powder					.08			.08	.09	
	0230	Curing & sealing compound, solvent based					.05			.05	.05	
	0300	Broadcasting hardener & release agent, stamping	2 Cefi	1000	.016			.55		.55	.87	

03370 | Specially Placed Concrete

		03370 Specially Placed Concrete	CREW	DAILY OUTPUT	LABOR-HOURS	UNIT	MAT.	LABOR	EQUIP.	TOTAL	TOTAL INCL O&P	
300	0010	**GUNITE (DRY-MIX)**										300
	0020	Applied in 1″ layers, no mesh included	C-8	2000	.028	S.F.	.27	.87	.35	1.49	2.09	
	0300	Typical in place, including mesh, 2″ thick, minimum	C-16	1000	.072		.88	2.37	.70	3.95	5.65	
	0350	Maximum		500	.144		.88	4.74	1.40	7.02	10.30	
	0500	4″ thick, minimum		750	.096		1.41	3.16	.93	5.50	7.80	
	0550	Maximum		350	.206		1.41	6.75	2	10.16	14.90	
	0900	Prepare old walls, no scaffolding, minimum	C-10	1000	.024			.77		.77	1.23	
	0950	Maximum	″	275	.087			2.80		2.80	4.47	
	1100	For high finish requirement or close tolerance, add, minimum						50%				
	1150	Maximum						110%				
	9000	Minimum labor/equipment charge	C-10	1	24	Job		770		770	1,225	

03390 | Concrete Curing

		03390 Concrete Curing	CREW	DAILY OUTPUT	LABOR-HOURS	UNIT	MAT.	LABOR	EQUIP.	TOTAL	TOTAL INCL O&P	
200	0010	**WATER CURING**										200
	0015	With burlap, 4 uses assumed, 7.5 oz.	2 Clab	55	.291	C.S.F.	7.05	7.95		15	21	
	0100	10 oz.		55	.291		12.65	7.95		20.60	27	
	0200	Curing blanket, burlap/poly, 2-ply		70	.229		15.05	6.25		21.30	27	
	0300	Sprayed membrane curing compound		95	.168		5.55	4.61		10.16	13.75	
	0710	Electrically, heated pads, 15 watts/S.F., 20 uses, minimum				S.F.	.19			.19	.21	
	0800	Maximum				″	.32			.32	.35	
	9000	Minimum labor/equipment charge	1 Clab	5	1.600	Job		44		44	72.50	

03400 | Precast Concrete

03410 | Plant-Precast Structural Concrete

		03410 Plant-Precast Structural Concrete	CREW	DAILY OUTPUT	LABOR-HOURS	UNIT	2006 BARE COSTS MAT.	LABOR	EQUIP.	TOTAL	TOTAL INCL O&P	
100	0010	**PRECAST BEAMS** R034105-30										100
	0011	L-shaped, 20′ span, 12″ x 20″	C-11	32	2.250	Ea.	1,050	88	48	1,186	1,375	
	1000	Inverted tee beams, add to above, small beams					15%					
	1050	Large beams					20%					
	1200	Rectangular, 20′ span, 12″ x 20″	C-11	32	2.250		780	88	48	916	1,075	
	1250	18″ x 36″		24	3		885	117	64	1,066	1,250	
	1300	24″ x 44″		22	3.273		1,100	128	70	1,298	1,525	
	1400	30′ span, 12″ x 36″		24	3		1,200	117	64	1,381	1,575	
	1450	18″ x 44″		20	3.600		1,500	141	77	1,718	2,000	
	1500	24″ x 52″		16	4.500		1,925	176	96	2,197	2,550	
	1600	40′ span, 12″ x 52″		20	3.600		2,200	141	77	2,418	2,775	
	1650	18″ x 52″		16	4.500		2,625	176	96	2,897	3,325	
	1700	24″ x 52″		12	6		2,875	235	128	3,238	3,750	
	2000	"T" shaped, 20′ span, 12″ x 20″		32	2.250		1,325	88	48	1,461	1,675	
	2050	18″ x 36″		24	3		1,500	117	64	1,681	1,925	
	2100	24″ x 44″		22	3.273		1,875	128	70	2,073	2,375	

03400 | Precast Concrete

	03410 | Plant-Precast Structural Concrete		CREW	DAILY OUTPUT	LABOR-HOURS	UNIT	2006 BARE COSTS MAT.	LABOR	EQUIP.	TOTAL	TOTAL INCL O&P	
100	2200	30′ span, 12″ x 36″ R034105-30	C-11	24	3	Ea.	2,025	117	64	2,206	2,500	100
	2250	18″ x 44″		20	3.600		2,550	141	77	2,768	3,150	
	2300	24″ x 52″		16	4.500		3,275	176	96	3,547	4,025	
	2500	40′ span, 12″ x 52″		20	3.600		3,750	141	77	3,968	4,450	
	2550	18″ x 52″		16	4.500		4,475	176	96	4,747	5,350	
	2600	24″ x 52″		12	6		4,900	235	128	5,263	5,975	
210	0010	**PRECAST COLUMNS** R034105-30										210
	0020	Rectangular to 12′ high, small columns	C-11	120	.600	L.F.	42.50	23.50	12.85	78.85	104	
	0050	Large columns	″	96	.750	″	74.50	29.50	16.05	120.05	153	
620	0010	**PRECAST SLAB PLANKS** R034105-30										620
	0020	Prestressed roof/floor members, grouted, solid, 4″ thick	C-11	2400	.030	S.F.	4.44	1.17	.64	6.25	7.70	
	0050	6″ thick		2800	.026		5.50	1.01	.55	7.06	8.45	
	0100	Hollow, 8″ thick		3200	.023		5.75	.88	.48	7.11	8.50	
	0150	10″ thick		3600	.020		6.10	.78	.43	7.31	8.60	
	0200	12″ thick		4000	.018		6.90	.70	.38	7.98	9.30	
	03450 | Plant-Precast Architectural Concrete											
850	0011	**WALL PANELS** R034513-10										850
	2200	Fiberglass reinforced cement with urethane core										
	2210	R20, 8′ x 8′, minimum	E-2	750	.075	S.F.	16.55	2.90	1.90	21.35	25.50	
	2220	Maximum	″	600	.093	″	23	3.63	2.38	29.01	34.50	
855	0010	**PRECAST WINDOW SILLS**										855
	0600	Precast concrete, 4″ tapers to 3″, 9″ wide	D-1	70	.229	L.F.	9.35	7.35		16.70	22	
	0650	11″ wide		60	.267		12.50	8.55		21.05	27.50	
	0700	13″ wide, 3 1/2″ tapers to 2 1/2″, 12″ wall		50	.320		13.50	10.30		23.80	31.50	
	03480 | Precast Concrete Specialties											
400	0010	**LINTELS**										400
	0800	Precast concrete, 4″ wide, 8″ high, to 5′ long	D-10	28	1.429	Ea.	26	48	19.15	93.15	128	
	0850	5′-12′ long		24	1.667		66.50	56	22.50	145	188	
	1000	6″ wide, 8″ high, to 5′ long		26	1.538		32.50	52	20.50	105	142	
	1050	5′-12′ long		22	1.818		90	61.50	24.50	176	225	
	1200	8″ wide, 8″ high, to 5′ long		24	1.667		31.50	56	22.50	110	150	
	1250	5′-12′ long		20	2		80.50	67.50	27	175	227	
	1400	10″ wide, 8″ high U-Shape, to 14′ long		18	2.222		168	75	30	273	340	
	1450	12″ wide, 8″ high U-Shape, to 19′ long		16	2.500		219	84.50	33.50	337	415	
800	0010	**PRECAST STAIRS**										800
	0020	Precast concrete treads on steel stringers, 3′ wide	C-12	75	.640	Riser	101	22.50	8.50	132	157	
	0300	Front entrance, 5′ wide with 48″ platform, 2 risers		16	3	Flight	315	105	40	460	565	
	0350	5 risers		12	4		515	140	53	708	855	
	0500	6′ wide, 2 risers		15	3.200		335	112	42.50	489.50	600	
	0550	5 risers		11	4.364		530	153	58	741	900	
	1200	Basement entrance stairs, steel bulkhead doors, minimum	B-51	22	2.182		700	60.50	5.85	766.35	875	
	1250	Maximum	″	11	4.364		1,200	121	11.70	1,332.70	1,550	

03500 | Cementitious Decks & Underlayments

03510 | Cementitious Roof Deck

			CREW	DAILY OUTPUT	LABOR-HOURS	UNIT	2006 BARE COSTS MAT.	LABOR	EQUIP.	TOTAL	TOTAL INCL O&P	
200	0010	**WOOD FIBER** Lightweight cement system R051223-50										200
	0050	Plank, beveled, 1″ thick	2 Carp	1000	.016	S.F.	1.85	.57		2.42	2.98	
	0100	Plank, T & G, 1-1/2″ thick		975	.016		1.97	.58		2.55	3.14	
	0150	2″ thick		950	.017		2.61	.60		3.21	3.86	
	0200	2-1/2″ thick		925	.017		2.86	.62		3.48	4.17	
	0300	3-1/2″ thick		875	.018		4.79	.65		5.44	6.35	
	0350	4″ thick		850	.019		5.35	.67		6.02	7	
	1000	Bulb tee, sub-purlin and grout, 6′ span, add	E-1	5000	.005		1.65	.19	.02	1.86	2.16	
	1100	8′ span	″	4200	.006		1.87	.22	.02	2.11	2.48	

03520 | Lightweight Concrete Roof Insulation

			CREW	DAILY OUTPUT	LABOR-HOURS	UNIT	MAT.	LABOR	EQUIP.	TOTAL	TOTAL INCL O&P	
250	0010	**INSULATING ROOF FILL**, lightweight cellular concrete R035216-10										250
	0020	Portland cement and foaming agent	C-8	50	1.120	C.Y.	84.50	34.50	14	133	164	
	0100	Poured vermiculite or perlite, field mix,										
	0110	1:6 field mix	C-8	50	1.120	C.Y.	92.50	34.50	14	141	173	
	0200	Ready mix, 1:6 mix, roof fill, 2″ thick		10000	.006	S.F.	.51	.17	.07	.75	.92	
	0250	3″ thick		7700	.007	″	.77	.23	.09	1.09	1.31	

CONCRETE 3

03900 | Concrete Restoration & Cleaning

03920 | Concrete Resurfacing

			CREW	DAILY OUTPUT	LABOR-HOURS	UNIT	2006 BARE COSTS MAT.	LABOR	EQUIP.	TOTAL	TOTAL INCL O&P	
600	0010	**PATCHING CONCRETE**										600
	0100	Floors, 1/4″ thick, small areas, regular grout	1 Cefi	170	.047	S.F.	.96	1.62		2.58	3.61	
	0150	Epoxy grout	″	100	.080	″	6.20	2.75		8.95	11.20	
	0300	Slab on grade, cut outs, up to 50 C.F.	2 Cefi	50	.320	C.F.	6.45	11		17.45	24.50	
	2000	Walls, including chipping, cleaning and epoxy grout										
	2100	1/4″ deep	1 Cefi	65	.123	S.F.	7.50	4.23		11.73	14.90	
	2150	1/2″ deep		50	.160		15.05	5.50		20.55	25	
	2200	3/4″ deep		40	.200		22.50	6.90		29.40	36	
	2510	Underlayment, P.C based self-leveling, 4100 psi, pumped, 1/4″	C-8	20000	.003		1.35	.09	.04	1.48	1.66	
	2520	1/2″		19000	.003		2.69	.09	.04	2.82	3.15	
	2530	3/4″		18000	.003		4.04	.10	.04	4.18	4.64	
	2540	1″		17000	.003		5.40	.10	.04	5.54	6.10	
	2550	1-1/2″		15000	.004		8.05	.12	.05	8.22	9.15	
	2560	Hand mix, 1/2″	C-18	4000	.002		2.69	.06	.01	2.76	3.07	
	2610	Topping, P.C. based self-level/dry 6100 psi, pumped, 1/4″	C-8	20000	.003		2.08	.09	.04	2.21	2.47	
	2620	1/2″		19000	.003		4.16	.09	.04	4.29	4.77	
	2630	3/4″		18000	.003		6.25	.10	.04	6.39	7.05	
	2660	1″		17000	.003		8.35	.10	.04	8.49	9.35	
	2670	1-1/2″		15000	.004		12.50	.12	.05	12.67	14	
	2680	Hand mix, 1/2″	C-18	4000	.002		4.16	.06	.01	4.23	4.69	
	9000	Minimum labor/equipment charge	1 Cefi	4.50	1.778	Job		61		61	96.50	

03930 | Concrete Rehabilitation

			CREW	DAILY OUTPUT	LABOR-HOURS	UNIT	MAT.	LABOR	EQUIP.	TOTAL	TOTAL INCL O&P	
300	0010	**CRACK REPAIR**, including chipping, sand blasting and cleaning										300
	0100	Epoxy injection, 1/8″ wide, 12″ deep	B-9	80	.500	L.F.	4.36	13.90	1.84	20.10	30	
	0110	1/4″ wide, 12″ deep		60	.667		8.75	18.55	2.45	29.75	43	
	0200	Latex injection, 1/8″ wide, 12″ deep		100	.400		1.75	11.10	1.47	14.32	22	
	0210	1/4″ wide, 12″ deep		75	.533		3.50	14.85	1.96	20.31	30.50	

Division Notes

		CREW	DAILY OUTPUT	LABOR-HOURS	UNIT	2000 BARE COSTS MAT.	LABOR	EQUIP.	TOTAL	TOTAL INCL O&P

Division 4 Masonry

Estimating Tips

04050 Basic Masonry Materials & Methods

- The terms *mortar* and *grout* are often used interchangeably, and incorrectly. Mortar is used to bed masonry units, seal the entry of air and moisture, provide architectural appearance, and allow for size variations in the units. Grout is used primarily in reinforced masonry construction and is used to bond the masonry to the reinforcing steel. Common mortar types are M(2500 psi), S(1800 psi), N(750 psi), and O(350 psi), and conform to ASTM C270. Grout is either fine or coarse and conforms to ASTM C476, and in-place strengths generally exceed 2500 psi. Mortar and grout are different components of masonry construction and are placed by entirely different methods. An estimator should be aware of their unique uses and costs.
- Waste, specifically the loss/droppings of mortar and the breakage of brick and block, is included in all masonry assemblies in this division. A factor of 25% is added for mortar and 3% for brick and concrete masonry units.
- Scaffolding or staging is not included in any of the Division 4 costs. Refer to section 01540 for scaffolding and staging costs.

04800 Masonry Assemblies

- The most common types of unit masonry are brick and concrete masonry. The major classifications of brick are building brick (ASTM C62), facing brick (ASTM C216), glazed brick, fire brick, and pavers. Many varieties of texture and appearance can exist within these classifications, and the estimator would be wise to check local custom and availability within the project area. For repair and remodeling jobs, matching the existing brick may be the most important criteria.
- Brick and concrete block are priced by the piece and then converted into a price per square foot of wall. Openings less than two square feet are generally ignored by the estimator because any savings in units used is offset by the cutting and trimming required.
- It is often difficult and expensive to find and purchase small lots of historic brick. Costs can vary widely. Many design issues affect costs, selection of mortar mix, and repairs or replacement of masonry materials. Cleaning techniques must be reflected in the estimate.
- All masonry walls, whether interior or exterior, require bracing. The cost of bracing walls during construction should be included by the estimator, and this bracing must remain in place until permanent bracing is complete. Permanent bracing of masonry walls is accomplished by masonry itself, in the form of pilasters or abutting wall corners, or by anchoring the walls to the structural frame. Accessories in the form of anchors, anchor slots, and ties are used, but their supply and installation can be by different trades. For instance, anchor slots on spandrel beams and columns are supplied and welded in place by the steel fabricator, but the ties from the slots into the masonry are installed by the bricklayer. Regardless of the installation method the estimator must be certain that these accessories are accounted for in pricing.

Reference Numbers

Reference numbers are shown in bold squares at the beginning of some major classifications. These numbers refer to related items in the Reference Section. The reference information may be an estimating procedure, an alternate pricing method, or technical information.

Note: Not all subdivisions listed here necessarily appear in this publication.

4 MASONRY

	04055 | Selective Demolition	CREW	DAILY OUTPUT	LABOR-HOURS	UNIT	2006 BARE COSTS MAT.	LABOR	EQUIP.	TOTAL	TOTAL INCL O&P	
110 0010	**SELECTIVE DEMOLITION, MASONRY** R024119-10										110
0300	Concrete block walls, unreinforced, 2" thick	2 Clab	1200	.013	S.F.		.37		.37	.61	
0310	4" thick		1150	.014			.38		.38	.63	
0320	6" thick		1100	.015			.40		.40	.66	
0330	8" thick		1050	.015			.42		.42	.69	
0340	10" thick		1000	.016			.44		.44	.73	
0360	12" thick		950	.017			.46		.46	.76	
0380	Reinforced alternate courses, 2" thick		1130	.014			.39		.39	.64	
0390	4" thick		1080	.015			.41		.41	.67	
0400	6" thick		1035	.015			.42		.42	.70	
0410	8" thick		990	.016			.44		.44	.73	
0420	10" thick		940	.017			.47		.47	.77	
0430	12" thick		890	.018			.49		.49	.82	
0440	Reinforced alternate courses & vertically 48" OC, 4" thick		900	.018			.49		.49	.81	
0450	6" thick		850	.019			.52		.52	.85	
0460	8" thick		800	.020			.55		.55	.91	
0480	10" thick		750	.021			.58		.58	.97	
0490	12" thick		700	.023			.63		.63	1.04	
1000	Chimney, 16" x 16", soft old mortar	1 Clab	55	.145	C.F.		3.99		3.99	6.60	
1020	Hard mortar		40	.200			5.50		5.50	9.10	
1030	16" x 20", soft old mortar		55	.145			3.99		3.99	6.60	
1040	Hard mortar		40	.200			5.50		5.50	9.10	
1050	16" x 24", soft old mortar		55	.145			3.99		3.99	6.60	
1060	Hard mortar		40	.200			5.50		5.50	9.10	
1080	20" x 20", soft old mortar		55	.145			3.99		3.99	6.60	
1100	Hard mortar		40	.200			5.50		5.50	9.10	
1110	20" x 24", soft old mortar		55	.145			3.99		3.99	6.60	
1120	Hard mortar		40	.200			5.50		5.50	9.10	
1140	20" x 32", soft old mortar		55	.145			3.99		3.99	6.60	
1160	Hard mortar		40	.200			5.50		5.50	9.10	
1200	48" x 48", soft old mortar		55	.145			3.99		3.99	6.60	
1220	Hard mortar		40	.200			5.50		5.50	9.10	
1250	Metal, high temp steel jacket, 24" diameter	E-2	130	.431	V.L.F.		16.75	11	27.75	42	
1260	60" diameter	"	60	.933			36.50	24	60.50	91	
1280	Flue lining, up to 12" x 12"	1 Clab	200	.040			1.10		1.10	1.82	
1282	Up to 24" x 24"		150	.053			1.46		1.46	2.42	
2000	Columns, 8" x 8", soft old mortar		48	.167			4.57		4.57	7.55	
2020	Hard mortar		40	.200			5.50		5.50	9.10	
2060	16" x 16", soft old mortar		16	.500			13.70		13.70	22.50	
2100	Hard mortar		14	.571			15.65		15.65	26	
2140	24" x 24", soft old mortar		8	1			27.50		27.50	45.50	
2160	Hard mortar		6	1.333			36.50		36.50	60.50	
2200	36" x 36", soft old mortar		4	2			55		55	91	
2220	Hard mortar		3	2.667			73		73	121	
2230	Alternate pricing method, soft old mortar		30	.267	C.F.		7.30		7.30	12.10	
2240	Hard mortar		23	.348	"		9.55		9.55	15.80	
3000	Copings, precast or masonry, to 8" wide										
3020	Soft old mortar	1 Clab	180	.044	L.F.		1.22		1.22	2.02	
3040	Hard mortar	"	160	.050	"		1.37		1.37	2.27	
3100	To 12" wide										
3120	Soft old mortar	1 Clab	160	.050	L.F.		1.37		1.37	2.27	
3140	Hard mortar	"	140	.057	"		1.57		1.57	2.59	
4000	Fireplace, brick, 30" x 24" opening										
4020	Soft old mortar	1 Clab	2	4	Ea.		110		110	182	
4040	Hard mortar		1.25	6.400			175		175	291	
4100	Stone, soft old mortar		1.50	5.333			146		146	242	

Important: See the Reference Section for supporting data - Crews, Rental Equipment, City Cost Indexes and Reference Data

04055 | Selective Demolition

			Crew	Daily Output	Labor-Hours	Unit	2006 Bare Costs Mat.	Labor	Equip.	Total	Total Incl O&P	
110	4120	Hard mortar (R024119-10)	1 Clab	1	8	Ea.		219		219	365	110
	5000	Veneers, brick, soft old mortar		140	.057	S.F.		1.57		1.57	2.59	
	5020	Hard mortar		125	.064			1.75		1.75	2.91	
	5100	Granite and marble, 2″ thick		180	.044			1.22		1.22	2.02	
	5120	4″ thick		170	.047			1.29		1.29	2.14	
	5140	Stone, 4″ thick		180	.044			1.22		1.22	2.02	
	5160	8″ thick		175	.046			1.25		1.25	2.08	
	5400	Alternate pricing method, stone, 4″ thick		60	.133	C.F.		3.65		3.65	6.05	
	5420	8″ thick		85	.094	″		2.58		2.58	4.27	
	9000	Minimum labor/equipment charge		2	4	Job		110		110	182	

04060 | Masonry Mortar

			Crew	Daily Output	Labor-Hours	Unit	Mat.	Labor	Equip.	Total	Total Incl O&P	
200	0010	**CEMENT**										200
	0020	Gypsum, 80 lb. bag, T.L. lots				Bag	16.70			16.70	18.35	
	0050	L.T.L. lots					22.50			22.50	25	
	0100	Masonry, 70 lb. bag, T.L. lots					6.45			6.45	7.10	
	0150	L.T.L. lots					6.85			6.85	7.50	
	0200	White, 70 lb. bag, T.L. lots					13.40			13.40	14.75	
	0250	L.T.L. lots					14.30			14.30	15.75	
400	0010	**LIME**										400
	0020	Masons, hydrated, 50 lb. bag, T.L. lots				Bag	6.55			6.55	7.20	
	0050	L.T.L. lots					7.40			7.40	8.15	
	0200	Finish, double hydrated, 50 lb. bag, T.L. lots					8.40			8.40	9.25	
	0250	L.T.L. lots					9.30			9.30	10.25	
500	0010	**MORTAR**										500
	0100	Type M, 1:1:6 mix	1 Brhe	143	.056	C.F.	3.44	1.55		4.99	6.30	
	0200	Type N, 1:3 mix		143	.056		3.12	1.55		4.67	5.95	
	0300	Type O, 1:3 mix		143	.056		5.25	1.55		6.80	8.30	
	0400	Type PM, 1:1:6 mix, 2500 psi		143	.056		5.55	1.55		7.10	8.65	
	0500	Type S, 1/2:1:4 mix		143	.056		5.90	1.55		7.45	9	
	2000	With portland cement and lime										
	2100	Type M, 1:1/4:3 mix	1 Brhe	143	.056	C.F.	6.60	1.55		8.15	9.75	
	2200	Type N, 1:1:6 mix, 750 psi		143	.056		5.55	1.55		7.10	8.60	
	2300	Type O, 1:2:9 mix (Pointing Mortar)		143	.056		5.90	1.55		7.45	9	
	2400	Type PL, 1:1/2:4 mix, 2500 psi		143	.056		6.05	1.55		7.60	9.15	
	2500	Type K, 1:3:12 mix, 75 psi		143	.056		5.25	1.55		6.80	8.25	
	2600	Type S, 1:1/2:4 mix, 1800 psi		143	.056		6.95	1.55		8.50	10.15	
	2650	Pre-mixed, type S or N					4.44			4.44	4.89	
	2700	Mortar for glass block	1 Brhe	143	.056		8.15	1.55		9.70	11.45	
	2800	Gypsum cement mortar					8.15			8.15	9	
	2900	Mortar for Fire Brick, 80 lb. bag, T.L. Lots				Bag	21			21	23	

04070 | Masonry Grout

			Crew	Daily Output	Labor-Hours	Unit	Mat.	Labor	Equip.	Total	Total Incl O&P	
420	0010	**GROUTING** Bond bms. & lintels, 8″ dp., pumped, not incl. block										420
	0020	8″ thick, 0.2 C.F. per L.F.	D-4	1400	.023	L.F.	.82	.73	.10	1.65	2.18	
	0050	10″ thick, 0.25 C.F. per L.F.		1200	.027		.91	.85	.12	1.88	2.50	
	0060	12″ thick, 0.3 C.F. per L.F.		1040	.031		1.09	.98	.14	2.21	2.93	
	0200	Concrete block cores, solid, 4″ thk., by hand, 0.067 C.F./S.F. of wall	D-8	1100	.036	S.F.	.24	1.20		1.44	2.22	
	0210	6″ thick, pumped, 0.175 C.F. per S.F.	D-4	720	.044		.64	1.41	.20	2.25	3.20	
	0250	8″ thick, pumped, 0.258 C.F. per S.F.		680	.047		.94	1.50	.21	2.65	3.67	
	0300	10″ thick, pumped, 0.340 C.F. per S.F.		660	.048		1.24	1.54	.22	3	4.08	
	0350	12″ thick, pumped, 0.422 C.F. per S.F.		640	.050		1.54	1.59	.23	3.36	4.50	
	0500	Cavity walls, 2″ space, pumped, 0.167 C.F./S.F. of wall		1700	.019		.61	.60	.09	1.30	1.72	

		04070 | Masonry Grout	CREW	DAILY OUTPUT	LABOR-HOURS	UNIT	2006 BARE COSTS MAT.	LABOR	EQUIP.	TOTAL	TOTAL INCL O&P	
420	0550	3″ space, 0.250 C.F./S.F.	D-4	1200	.027	S.F.	.91	.85	.12	1.88	2.50	420
	0600	4″ space, 0.333 C.F. per S.F.		1150	.028		1.21	.89	.13	2.23	2.91	
	0700	6″ space, 0.500 C.F. per S.F.		800	.040		1.82	1.27	.18	3.27	4.26	
	0800	Door frames, 3′ x 7′ opening, 2.5 C.F. per opening		60	.533	Opng.	9.10	16.95	2.42	28.47	40	
	0850	6′ x 7′ opening, 3.5 C.F. per opening		45	.711	"	12.75	22.50	3.22	38.47	54	
	2000	Grout, C476, for bond beams, lintels and CMU cores		350	.091	C.F.	3.65	2.91	.41	6.97	9.15	
	9000	Minimum labor/equipment charge	1 Bric	2	4	Job		146		146	237	

		04080 | Anchorage & Reinforcement	CREW	DAILY OUTPUT	LABOR-HOURS	UNIT	MAT.	LABOR	EQUIP.	TOTAL	TOTAL INCL O&P	
070	0010	**ANCHOR BOLTS**										070
	0020	Hooked, with nut and washer, 1/2″ diam., 8″ long	1 Bric	200	.040	Ea.	.68	1.46		2.14	3.12	
	0030	12″ long		190	.042		1.29	1.54		2.83	3.92	
	0040	5/8″ diameter, 8″ long		180	.044		1.01	1.62		2.63	3.75	
	0050	12″ long		170	.047		1.11	1.72		2.83	4.01	
	0060	3/4″ diameter, 8″ long		160	.050		1.54	1.83		3.37	4.66	
	0070	12″ long		150	.053		1.92	1.95		3.87	5.25	
200	0010	**REINFORCING**										200
	0015	Steel bars A615, placed horiz., #3 & #4 bars	1 Bric	450	.018	Lb.	.43	.65		1.08	1.52	
	0020	#5 & #6 bars		800	.010		.43	.37		.80	1.06	
	0050	Placed vertical, #3 & #4 bars		350	.023		.43	.84		1.27	1.83	
	0060	#5 & #6 bars		650	.012		.43	.45		.88	1.20	
	0500	Joint reinforcing, ladder type, mill std galvanized										
	0600	9 ga. sides, 9 ga. ties, 4″ wall	1 Bric	30	.267	C.L.F.	9.40	9.75		19.15	26	
	0650	6″ wall		30	.267		10.65	9.75		20.40	27.50	
	0700	8″ wall		25	.320		11	11.70		22.70	31	
	0750	10″ wall		20	.400		11.70	14.60		26.30	36.50	
	0800	12″ wall		20	.400		12.35	14.60		26.95	37	
	1000	Truss type										
	1100	9 ga. sides, 9 ga. ties, 4″ wall	1 Bric	30	.267	C.L.F.	11.50	9.75		21.25	28.50	
	1150	6″ wall		30	.267		11.95	9.75		21.70	29	
	1200	8″ wall		25	.320		12.60	11.70		24.30	33	
	1250	10″ wall		20	.400		13.25	14.60		27.85	38	
	1300	12″ wall		20	.400		14.05	14.60		28.65	39	
	1500	3/16″ sides, 9 ga. ties, 4″ wall		30	.267		16.85	9.75		26.60	34.50	
	1550	6″ wall		30	.267		17	9.75		26.75	34.50	
	1600	8″ wall		25	.320		17.60	11.70		29.30	38.50	
	1650	10″ wall		20	.400		18.30	14.60		32.90	43.50	
	1700	12″ wall		20	.400		19.15	14.60		33.75	44.50	
	2000	3/16″ sides, 3/16″ ties, 4″ wall		30	.267		19.70	9.75		29.45	37.50	
	2050	6″ wall		30	.267		19.80	9.75		29.55	38	
	2100	8″ wall		25	.320		20.50	11.70		32.20	41.50	
	2150	10″ wall		20	.400		21.50	14.60		36.10	47	
	2200	12″ wall		20	.400		22.50	14.60		37.10	48	
	2500	Cavity truss type, galvanized										
	2600	9 ga. sides, 9 ga. ties, 4″ wall	1 Bric	25	.320	C.L.F.	28.50	11.70		40.20	50.50	
	2650	6″ wall		25	.320		29	11.70		40.70	50.50	
	2700	8″ wall		20	.400		29.50	14.60		44.10	56	
	2750	10″ wall		15	.533		30.50	19.50		50	65	
	2800	12″ wall		15	.533		31.50	19.50		51	66	
	3000	3/16″ sides, 9 ga. ties, 4″ wall		25	.320		35	11.70		46.70	57.50	
	3050	6″ wall		25	.320		35.50	11.70		47.20	58	
	3100	8″ wall		20	.400		36	14.60		50.60	63	
	3150	10″ wall		15	.533		36.50	19.50		56	72	
	3200	12″ wall		15	.533		38	19.50		57.50	73	

		04080 | Anchorage & Reinforcement	CREW	DAILY OUTPUT	LABOR-HOURS	UNIT	2006 BARE COSTS MAT.	LABOR	EQUIP.	TOTAL	TOTAL INCL O&P	
200	3500	For hot dip galvanizing, add					80%					200
650	0010	WALL TIES										650
	0020	For brick veneer, galv., corrugated, 7/8" x 7", 22 Ga.	1 Bric	10.50	.762	C	8.35	28		36.35	54	
	0100	24 Ga.		10.50	.762		7.20	28		35.20	53	
	0150	16 Ga.		10.50	.762		18.25	28		46.25	65	
	0200	Buck anchors, galv., corrugated, 16 gauge, 2" bend, 8" x 2"		10.50	.762		121	28		149	178	
	0250	8" x 3"		10.50	.762		107	28		135	163	
	0300	Adjustable, rectangular, 4-1/8" wide										
	0350	Anchor and tie, 3/16" wire, mill galv.										
	0400	2-3/4" eye, 3-1/4" tie	1 Bric	1.05	7.619	M	430	278		708	920	
	0500	4-3/4" tie		1.05	7.619		475	278		753	970	
	0520	5-1/2" tie		1.05	7.619		500	278		778	1,000	
	0550	4-3/4" eye, 3-1/4" tie		1.05	7.619		480	278		758	975	
	0570	4-3/4" tie		1.05	7.619		540	278		818	1,050	
	0580	5-1/2" tie		1.05	7.619		565	278		843	1,075	
	0670	3/16" diameter		10.50	.762	C	25.50	28		53.50	73	
	0680	1/4" diameter		10.50	.762		33	28		61	81.50	
	0850	8" long, 3/16" diameter		10.50	.762		28.50	28		56.50	76.50	
	0855	1/4" diameter		10.50	.762		78.50	28		106.50	132	
	1000	Rectangular type, galvanized, 1/4" diameter, 2" x 6"		10.50	.762		35.50	28		63.50	84	
	1050	4" x 6"		10.50	.762		39.50	28		67.50	88.50	
	1100	3/16" diameter, 2" x 6"		10.50	.762		28.50	28		56.50	76	
	1150	4" x 6"		10.50	.762		30.50	28		58.50	78.50	
	1200	Mesh wall tie, 1/2" mesh, hot dip galvanized										
	1400	16 gauge, 12" long, 3" wide	1 Bric	9	.889	C	73.50	32.50		106	134	
	1420	6" wide		9	.889		112	32.50		144.50	176	
	1440	12" wide		8.50	.941		180	34.50		214.50	254	
	1500	Rigid partition anchors, plain, 8" long, 1" x 1/8"		10.50	.762		69	28		97	121	
	1550	1" x 1/4"		10.50	.762		114	28		142	170	
	1580	1-1/2" x 1/8"		10.50	.762		92	28		120	146	
	1600	1-1/2" x 1/4"		10.50	.762		187	28		215	251	
	1650	2" x 1/8"		10.50	.762		119	28		147	176	
	1700	2" x 1/4"		10.50	.762		219	28		247	286	
	2000	Column flange ties, wire, galvanized										
	2300	3/16" diameter, up to 3" wide	1 Bric	10.50	.762	C	73.50	28		101.50	126	
	2350	To 5" wide		10.50	.762		80	28		108	133	
	2400	To 7" wide		10.50	.762		85.50	28		113.50	139	
	2600	To 9" wide		10.50	.762		92	28		120	146	
	2650	1/4" diameter, up to 3" wide		10.50	.762		99	28		127	154	
	2700	To 5" wide		10.50	.762		109	28		137	165	
	2800	To 7" wide		10.50	.762		118	28		146	174	
	2850	To 9" wide		10.50	.762		128	28		156	186	
	2900	For hot dip galvanized, add					35%					
	4000	Channel slots, 1-3/8" x 1/2" x 8"										
	4100	12 gauge, plain	1 Bric	10.50	.762	C	188	28		216	251	
	4150	16 gauge, galvanized	"	10.50	.762	"	104	28		132	159	
	4200	Channel slot anchors										
	4300	16 gauge, galvanized, 1-1/4" x 3-1/2"				C	41.50			41.50	45.50	
	4350	1-1/4" x 5-1/2"					48			48	53	
	4400	1-1/4" x 7-1/2"					54.50			54.50	60	
	4500	1/8" plain, 1-1/4" x 3-1/2"					67			67	73.50	
	4550	1-1/4" x 5-1/2"					72			72	79	
	4600	1-1/4" x 7-1/2"					78			78	86	
	4700	For corrugation, add					39			39	43	
	4750	For hot dip galvanized, add					35%					

04050 | Basic Masonry Materials and Methods

4 MASONRY

		04080 | Anchorage & Reinforcement	CREW	DAILY OUTPUT	LABOR-HOURS	UNIT	2006 BARE COSTS MAT.	LABOR	EQUIP.	TOTAL	TOTAL INCL O&P	
650	5000	Dowels										650
	5100	Plain, 1/4" diameter, 3" long				C	28			28	31	
	5150	4" long					31			31	34	
	5200	6" long					38			38	42	
	5300	3/8" diameter, 3" long					37			37	40.50	
	5350	4" long					46			46	50.50	
	5400	6" long					53			53	58.50	
	5500	1/2" diameter, 3" long					54			54	59.50	
	5550	4" long					64			64	70.50	
	5600	6" long					85			85	93.50	
	5700	5/8" diameter, 3" long					74			74	81.50	
	5750	4" long					91			91	100	
	5800	6" long					125			125	138	
	6000	3/4" diameter, 3" long					92			92	101	
	6100	4" long					118			118	130	
	6150	6" long					167			167	184	
	6300	For hot dip galvanized, add					35%					

		04090 | Masonry Accessories	CREW	DAILY OUTPUT	LABOR-HOURS	UNIT	MAT.	LABOR	EQUIP.	TOTAL	TOTAL INCL O&P	
170	0010	**CONTROL JOINT**										170
	0020	Rubber, 4" and wider wall	1 Bric	400	.020	L.F.	1.80	.73		2.53	3.17	
	0050	PVC, 4" wall		400	.020		1.10	.73		1.83	2.40	
	0100	Rubber, 6" wall		320	.025		2.31	.91		3.22	4.02	
	0120	PVC, 6" wall		320	.025		1.48	.91		2.39	3.11	
	0140	Rubber, 8" and wider wall		280	.029		2.99	1.04		4.03	4.98	
	0160	PVC, 8" wall		280	.029		1.68	1.04		2.72	3.54	
	0180	12" wall		240	.033		3.46	1.22		4.68	5.80	
860	0010	**VENT BOX**										860
	0020	Extruded aluminum, 4" deep, 2-3/8" x 8-1/8"	1 Bric	30	.267	Ea.	22.50	9.75		32.25	40.50	
	0050	5" x 8-1/8"		25	.320		28.50	11.70		40.20	50	
	0100	2-1/4" x 25"		25	.320		58	11.70		69.70	82.50	
	0150	5" x 16-1/2"		22	.364		52.50	13.30		65.80	79.50	
	0200	6" x 16-1/2"		22	.364		68	13.30		81.30	96.50	
	0250	7-3/4" x 16-1/2"		20	.400		66.50	14.60		81.10	97	
	0400	For baked enamel finish, add					35%					
	0500	For cast aluminum, painted, add					60%					
	1000	Stainless steel ventilators, 6" x 6"	1 Bric	25	.320		98.50	11.70		110.20	127	
	1050	8" x 8"		24	.333		104	12.20		116.20	134	
	1100	12" x 12"		23	.348		120	12.70		132.70	153	
	1150	12" x 6"		24	.333		105	12.20		117.20	135	
	1200	Foundation block vent, galv., 1-1/4" thk, 8" high, 16" long, no damper		30	.267		19.25	9.75		29	37	
	1250	For damper, add					6.40			6.40	7.05	

04200 | Masonry Units

		04210 | Clay Masonry Units	CREW	DAILY OUTPUT	LABOR-HOURS	UNIT	2006 BARE COSTS MAT.	LABOR	EQUIP.	TOTAL	TOTAL INCL O&P	
100	0010	**COMMON BUILDING BRICK** C62, TL lots, material only R042110-20										100
	0020	Standard, minimum				M	315			315	345	

04200 | Masonry Units

	04210 Clay Masonry Units		CREW	DAILY OUTPUT	LABOR-HOURS	UNIT	2006 BARE COSTS MAT.	LABOR	EQUIP.	TOTAL	TOTAL INCL O&P	
100	0050	Average (select) R042110-20				M	385			385	425	100
300	0010	**FACE BRICK** C216, TL lots, material only R042110-20										300
	0300	Standard modular, 4" x 2-2/3" x 8", minimum				M	400			400	440	
	0350	Maximum R042110-50					515			515	565	
	0450	Economy, 4" x 4" x 8", minimum					775			775	855	
	0500	Maximum					1,000			1,000	1,100	
	0510	Economy, 4" x 4" x 12", minimum					1,075			1,075	1,200	
	0520	Maximum					1,400			1,400	1,550	
	0550	Jumbo, 6" x 4" x 12", minimum					1,225			1,225	1,350	
	0600	Maximum					1,550			1,550	1,700	
	0610	Jumbo, 8" x 4" x 12", minimum					1,225			1,225	1,350	
	0620	Maximum					1,550			1,550	1,700	
	0650	Norwegian, 4" x 3-1/5" x 12", minimum					785			785	860	
	0700	Maximum					1,025			1,025	1,125	
	0710	Norwegian, 6" x 3-1/5" x 12", minimum					1,175			1,175	1,300	
	0720	Maximum					1,525			1,525	1,675	
	0850	Standard glazed, plain colors, 4" x 2-2/3" x 8", minimum					1,225			1,225	1,350	
	0900	Maximum					1,600			1,600	1,750	
	1000	Deep trim shades, 4" x 2-2/3" x 8", minimum					1,250			1,250	1,375	
	1050	Maximum					1,400			1,400	1,525	
	1080	Jumbo utility, 4" x 4" x 12"					1,125			1,125	1,250	
	1120	4" x 8" x 8"					1,175			1,175	1,275	
	1140	4" x 8" x 16"					2,450			2,450	2,700	
	1260	Engineer, 4" x 3-1/5" x 8", minimum					500			500	550	
	1270	Maximum					530			530	580	
	1350	King, 4" x 2-3/4" x 10", minimum					465			465	515	
	1360	Maximum					500			500	550	
	1400	Norman, 4" x 2-3/4" x 12"					465			465	515	
	1450	Roman, 4" x 2" x 12"					465			465	515	
	1500	SCR, 6" x 2-2/3" x 12"					465			465	515	
	1550	Double, 4" x 5-1/3" x 8"					465			465	515	
	1600	Triple, 4" x 5-1/3" x 12"					465			465	515	
	1770	Standard modular, double glazed, 4" x 2-2/3" x 8"					1,050			1,050	1,150	
	1850	Jumbo, colored glazed ceramic, 6" x 4" x 12"					1,550			1,550	1,700	
	2050	Jumbo utility, glazed, 4" x 4" x 12"					1,175			1,175	1,275	
	2100	4" x 8" x 8"					1,600			1,600	1,775	
	2150	4" x 16" x 8"					2,775			2,775	3,050	
	2170	For less than truck load lots, add					10			10	11	
	2180	For buff or gray brick, add					15			15	16.50	
	3050	Used brick, minimum					375			375	410	
	3100	Maximum					490			490	540	
	3150	Add for brick to match existing work, minimum					5%					
	3200	Maximum					50%					

04500 | Refractories

	04580 Refractory Brick		CREW	DAILY OUTPUT	LABOR-HOURS	UNIT	2006 BARE COSTS MAT.	LABOR	EQUIP.	TOTAL	TOTAL INCL O&P	
260	0010	**FIRE CLAY**										260
	0020	Gray, high duty, 100 lb. bag				Bag	46			46	50.50	

04500 | Refractories

		04580 | Refractory Brick	CREW	DAILY OUTPUT	LABOR-HOURS	UNIT	2006 BARE COSTS MAT.	LABOR	EQUIP.	TOTAL	TOTAL INCL O&P	
260	0050	100 lb. drum, premixed (400 brick per drum)				Drum	56.50			56.50	62.50	260

4 MASONRY

04700 | Simulated Masonry

		04710 | Simulated Brick	CREW	DAILY OUTPUT	LABOR-HOURS	UNIT	2006 BARE COSTS MAT.	LABOR	EQUIP.	TOTAL	TOTAL INCL O&P	
600	0010	SIMULATED BRICK										600
	0020	Aluminum, baked on colors	1 Carp	200	.040	S.F.	2.73	1.42		4.15	5.35	
	0050	Fiberglass panels		200	.040		3	1.42		4.42	5.65	
	0100	Urethane pieces cemented in mastic		150	.053		5.20	1.90		7.10	8.85	
	0150	Vinyl siding panels	↓	200	.040	↓	2.07	1.42		3.49	4.64	

04800 | Masonry Assemblies

		04810 | Unit Masonry Assemblies	CREW	DAILY OUTPUT	LABOR-HOURS	UNIT	2006 BARE COSTS MAT.	LABOR	EQUIP.	TOTAL	TOTAL INCL O&P	
050	0010	**AUTOCLAVED AERATED CONCRETE BLOCK** Scaffolding not incl										050
	0050	Solid, 4" x 12" x 24", incl mortar	D-8	600	.067	S.F.	1.87	2.20		4.07	5.60	
	0060	6" x 12" x 24"		600	.067		2.51	2.20		4.71	6.35	
	0070	8" x 8" x 24"		575	.070		3.15	2.30		5.45	7.20	
	0080	10" x 12" x 24"		575	.070		4.11	2.30		6.41	8.25	
	0090	12" x 12" x 24"	↓	550	.073	↓	4.78	2.40		7.18	9.15	
100	0010	**BRICK VENEER** Scaffolding not included, truck load lots R042110-20										100
	0015	Material costs incl. 3% brick and 25% mortar waste										
	2000	Standard, sel. common, 4" x 2-2/3" x 8", (6.75/S.F.) R042110-50	D-8	230	.174	S.F.	3.28	5.75		9.03	12.90	
	2020	Standard, red, 4" x 2-2/3" x 8", running bond (6.75/SF)		220	.182		3.28	6		9.28	13.35	
	2050	Full header every 6th course (7.88/S.F.)		185	.216		3.82	7.15		10.97	15.80	
	2100	English, full header every 2nd course (10.13/S.F.)		140	.286		4.90	9.45		14.35	20.50	
	2150	Flemish, alternate header every course (9.00/S.F.)		150	.267		4.36	8.80		13.16	19.10	
	2200	Flemish, alt. header every 6th course (7.13/S.F.)		205	.195		3.46	6.45		9.91	14.25	
	2250	Full headers throughout (13.50/S.F.)		105	.381		6.50	12.60		19.10	27.50	
	2300	Rowlock course (13.50/S.F.)		100	.400		6.50	13.20		19.70	28.50	
	2350	Rowlock stretcher (4.50/S.F.)		310	.129		2.20	4.26		6.46	9.30	
	2400	Soldier course (6.75/S.F.)		200	.200		3.28	6.60		9.88	14.30	
	2450	Sailor course (4.50/S.F.)		290	.138		2.20	4.56		6.76	9.80	
	2600	Buff or gray face, running bond, (6.75/S.F.)		220	.182		3.47	6		9.47	13.55	
	2700	Glazed face brick, running bond		210	.190		9.05	6.30		15.35	20	
	2750	Full header every 6th course (7.88/S.F.)		170	.235		10.55	7.75		18.30	24	
	3000	Jumbo, 6" x 4" x 12" running bond (3.00/S.F.)		435	.092		4.03	3.04		7.07	9.35	
	3050	Norman, 4" x 2-2/3" x 12" running bond, (4.5/S.F.)		320	.125		4.43	4.13		8.56	11.55	
	3100	Norwegian, 4" x 3-1/5" x 12" (3.75/S.F.)		375	.107		3.32	3.52		6.84	9.35	
	3150	Economy, 4" x 4" x 8" (4.50/S.F.)		310	.129		3.94	4.26		8.20	11.25	
	3200	Engineer, 4" x 3-1/5" x 8" (5.63/S.F.)		260	.154		3.33	5.10		8.43	11.90	
	3250	Roman, 4" x 2" x 12" (6.00/S.F.)	↓	250	.160	↓	5.10	5.30		10.40	14.15	

04810 | Unit Masonry Assemblies

		04810 Unit Masonry Assemblies	CREW	DAILY OUTPUT	LABOR-HOURS	UNIT	2006 BARE COSTS MAT.	LABOR	EQUIP.	TOTAL	TOTAL INCL O&P	
100	3300	SCR, 6" x 2-2/3" x 12" (4.50/S.F.) R042110-20	D-8	310	.129	S.F.	4.63	4.26		8.89	12	100
	3350	Utility, 4" x 4" x 12" (3.00/S.F.)		450	.089		3.74	2.94		6.68	8.90	
	3400	For cavity wall construction, add R042110-50						15%				
	3450	For stacked bond, add						10%				
	3500	For interior veneer construction, add						15%				
	3550	For curved walls, add						30%				
	9000	Minimum labor/equipment charge	D-1	2	8	Job		257		257	415	
110	0010	**OVERSIZED BRICK**, scaffolding not included										110
	0100	Veneer, 4" x 2.25" x 16"	D-8	387	.103	S.F.	4.48	3.41		7.89	10.50	
	0105	4" x 2.75" x 16"		412	.097		4.05	3.21		7.26	9.65	
	0110	4" x 4" x 16"		460	.087		3.96	2.87		6.83	9	
	0120	4" x 8" x 16"		533	.075		9.35	2.48		11.83	14.25	
	0125	Loadbearing, 6" x 4" x 16", grouted and reinforced		387	.103		6.85	3.41		10.26	13.10	
	0130	8" x 4" x 16", grouted and reinforced		327	.122		7.20	4.04		11.24	14.45	
	0135	6" x 8" x 16", grouted and reinforced		440	.091		6.55	3		9.55	12.05	
	0140	8" x 8" x 16", grouted and reinforced		400	.100		7.30	3.30		10.60	13.35	
	0145	Curtainwall / reinforced veneer, 6" x 4" x 16"		387	.103		10.45	3.41		13.86	17.05	
	0150	8" x 4" x 16"		327	.122		12.60	4.04		16.64	20.50	
	0155	6" x 8" x 16"		440	.091		10.55	3		13.55	16.45	
	0160	8" x 8" x 16"		400	.100		12.70	3.30		16	19.30	
	0200	For 1 to 3 slots in face, add					25%					
	0210	For 4 to 7 slots in face, add					15%					
	0220	For bond beams, add					20%					
	0230	For bullnose shapes, add					20%					
	0240	For open end knockout, add					10%					
	0250	For white or gray color group, add					10%					
	0260	For 135 degree corner, add					250%					
160	0010	**CHIMNEY** See Div. 03310-240 for foundation, add to prices below										160
	0100	Brick, 16" x 16", 8" flue, scaff. not incl.	D-1	18.20	.879	V.L.F.	17.45	28.50		45.95	65	
	0150	16" x 20" with one 8" x 12" flue		16	1		27	32		59	81.50	
	0200	16" x 24" with two 8" x 8" flues		14	1.143		38.50	36.50		75	102	
	0250	20" x 20" with one 12" x 12" flue		13.70	1.168		31.50	37.50		69	95.50	
	0300	20" x 24" with two 8" x 12" flues		12	1.333		44	43		87	118	
	0350	20" x 32" with two 12" x 12" flues		10	1.600		55	51.50		106.50	144	
170	0010	**COLUMNS** Face brick, includes mortar, scaffolding not included										170
	0050	8" x 8", 9 brick per course	D-1	56	.286	V.L.F.	4.21	9.20		13.41	19.55	
	0100	12" x 8", 13.5 brick		37	.432		6.30	13.90		20.20	29.50	
	0200	12" x 12", 20 brick		25	.640		9.35	20.50		29.85	44	
	0300	16" x 12", 27 brick		19	.842		12.65	27		39.65	58	
	0400	16" x 16", 36 brick		14	1.143		16.85	36.50		53.35	78	
	0500	20" x 16", 45 brick		11	1.455		21	47		68	99	
	0600	20" x 20", 56 brick		9	1.778		26	57		83	122	
	0700	24" x 20", 68 brick		7	2.286		32	73.50		105.50	154	
	0800	24" x 24", 81 brick		6	2.667		38	85.50		123.50	181	
	1000	36" x 36", 182 brick		3	5.333		85	171		256	370	
	9000	Minimum labor/equipment charge		2	8	Job		257		257	415	
172	0010	**CONCRETE BLOCK, BACK-UP,** C90, 2000 psi										172
	0020	Normal weight, 8" x 16" units, tooled joint 1 side										
	0050	Not-reinforced, 2000 psi, 2" thick	D-8	475	.084	S.F.	1.01	2.78		3.79	5.60	
	0200	4" thick		460	.087		1.21	2.87		4.08	6	
	0300	6" thick		440	.091		1.76	3		4.76	6.80	
	0350	8" thick		400	.100		1.93	3.30		5.23	7.45	
	0400	10" thick		330	.121		2.41	4		6.41	9.15	
	0450	12" thick	D-9	310	.155		2.78	4.98		7.76	11.10	

MASONRY 4

04810 | Unit Masonry Assemblies

		04810 Unit Masonry Assemblies	CREW	DAILY OUTPUT	LABOR-HOURS	UNIT	2006 BARE COSTS MAT.	LABOR	EQUIP.	TOTAL	TOTAL INCL O&P	
172	1000	Reinforced, alternate courses, 4″ thick	D-8	450	.089	S.F.	1.29	2.94		4.23	6.20	172
	1100	6″ thick		430	.093		1.85	3.07		4.92	7	
	1150	8″ thick		395	.101		2.02	3.34		5.36	7.65	
	1200	10″ thick		320	.125		2.51	4.13		6.64	9.45	
	1250	12″ thick	D-9	300	.160		2.88	5.15		8.03	11.50	
	9000	Minimum labor/equipment charge	D-1	2	8	Job		257		257	415	
175	0010	**CONCRETE BLOCK BOND BEAM** C90, 2000 psi										175
	0020	Not including grout or reinforcing										
	0130	8″ high, 8″ thick	D-8	565	.071	L.F.	2	2.34		4.34	6	
	0150	12″ thick	D-9	510	.094		2.78	3.03		5.81	7.95	
	0525	Lightweight, 6″ thick	D-8	592	.068		1.84	2.23		4.07	5.65	
	2000	Including grout and 2 #5 bars										
	2100	Regular block, 8″ high, 8″ thick	D-8	300	.133	L.F.	3.77	4.40		8.17	11.30	
	2150	12″ thick	D-9	250	.192	″	5	6.15		11.15	15.50	
	9000	Minimum labor/equipment charge	D-1	2	8	Job		257		257	415	
182	0010	**CONCRETE BLOCK, DECORATIVE** C90, 2000 psi										182
	1000	Fluted high strength										
	1100	8″ x 16″ x 4″ thick, flutes 1 side,	D-8	345	.116	S.F.	3.21	3.83		7.04	9.75	
	1150	Flutes 2 sides		335	.119		3.90	3.94		7.84	10.70	
	1200	8″ thick		300	.133		5.05	4.40		9.45	12.75	
	1250	For special colors, add					.32			.32	.35	
	1400	Deep grooved, smooth face										
	1450	8″ x 16″ x 4″ thick	D-8	345	.116	S.F.	2.10	3.83		5.93	8.50	
	1500	8″ thick	″	300	.133	″	3.63	4.40		8.03	11.15	
	1600											
	4000	Slump block										
	4100	4″ face height x 16″ x 4″ thick	D-1	165	.097	S.F.	3.42	3.12		6.54	8.80	
	4150	6″ thick		160	.100		4.82	3.22		8.04	10.50	
	4200	8″ thick		155	.103		5.20	3.32		8.52	11.10	
	4250	10″ thick		140	.114		9.75	3.67		13.42	16.65	
	4300	12″ thick		130	.123		10.15	3.96		14.11	17.60	
	5000	Split rib profile units, 1″ deep ribs, 8 ribs										
	5100	8″ x 16″ x 4″ thick	D-8	345	.116	S.F.	2.47	3.83		6.30	8.90	
	5150	6″ thick		325	.123		2.87	4.07		6.94	9.75	
	5200	8″ thick		300	.133		3.31	4.40		7.71	10.80	
	5250	12″ thick	D-9	275	.175		3.94	5.60		9.54	13.45	
	9000	Minimum labor/equipment charge	D-1	2	8	Job		257		257	415	
184	0010	**CONCRETE BLOCK, EXTERIOR** C90, 2000 psi										184
	0020	Reinforced alt courses, tooled joints 2 sides										
	0100	Normal weight, 8″ x 16″ x 6″ thick	D-8	395	.101	S.F.	1.97	3.34		5.31	7.60	
	0200	8″ thick		360	.111		2.90	3.67		6.57	9.15	
	0250	10″ thick		290	.138		3.51	4.56		8.07	11.25	
	0300	12″ thick	D-9	250	.192		3.59	6.15		9.74	13.95	
	9000	Minimum labor/equipment charge	D-1	2	8	Job		257		257	415	
186	0010	**CONCRETE BLOCK FOUNDATION WALL** C90/C145										186
	0050	Normal-weight, cut joints, horiz joint reinf, no vert reinf										
	0200	Hollow, 8″ x 16″ x 6″ thick	D-8	455	.088	S.F.	2.13	2.90		5.03	7.05	
	0250	8″ thick		425	.094		2.31	3.11		5.42	7.60	
	0300	10″ thick		350	.114		2.81	3.78		6.59	9.20	
	0350	12″ thick	D-9	300	.160		3.18	5.15		8.33	11.85	
	1000	Reinforced, #4 vert @ 48″										
	1100	Hollow, 8″ x 16″ block, 4″ thick	D-8	455	.088	S.F.	2.02	2.90		4.92	6.95	
	1125	6″ thick		445	.090		2.86	2.97		5.83	7.95	
	1150	8″ thick		415	.096		3.34	3.18		6.52	8.80	

	04810 | Unit Masonry Assemblies		CREW	DAILY OUTPUT	LABOR-HOURS	UNIT	2006 BARE COSTS MAT.	LABOR	EQUIP.	TOTAL	TOTAL INCL O&P	
186	1200	10″ thick	D-8	340	.118	S.F.	4.13	3.89		8.02	10.85	186
	1250	12″ thick	D-9	290	.166	↓	4.81	5.30		10.11	13.95	
	9000	Minimum labor/equipment charge	D-1	2	8	Job		257		257	415	
188	0010	**CONCRETE BLOCK INSULATION INSERTS**										188
	0100	Inserts, styrofoam, plant installed, add to block prices										
	0200	8″ x 16″ units, 6″ thick				S.F.	1.02			1.02	1.12	
	0250	8″ thick					1.02			1.02	1.12	
	0300	10″ thick					1.20			1.20	1.32	
	0350	12″ thick					1.26			1.26	1.39	
	0500	8″ x 8″ units, 8″ thick					.84			.84	.92	
	0550	12″ thick				↓	1.02			1.02	1.12	
189	0010	**CONCRETE BLOCK, INTERLOCKING**										189
	0100	Not including grout or reinforcing										
	0200	8″ x 16″ units, 2,000 psi, 8″ thick	D-1	245	.065	S.F.	2.14	2.10		4.24	5.75	
	0300	12″ thick	″	220	.073		3.17	2.34		5.51	7.25	
	0400	Including grout & reinforcing, 8″ thick	D-4	245	.131		6.15	4.15	.59	10.89	14.10	
	0450	12″ thick	″	220	.145	↓	7.35	4.63	.66	12.64	16.25	
	9000	Minimum labor/equipment charge	D-1	2	8	Job		257		257	415	
190	0010	**CONCRETE BLOCK, LINTELS** C90, normal weight										190
	0100	Including grout and horizontal reinforcing										
	0200	8″ x 8″ x 8″, 1 #4 bar	D-4	300	.107	L.F.	3.99	3.39	.48	7.86	10.35	
	0250	2 #4 bars		295	.108		4.18	3.45	.49	8.12	10.70	
	1000	12″ x 8″ x 8″, 1 #4 bar		275	.116		5.60	3.70	.53	9.83	12.70	
	1150	2 #5 bars	↓	270	.119	↓	6	3.77	.54	10.31	13.25	
	9000	Minimum labor/equipment charge	D-1	2	8	Job		257		257	415	
210	0010	**CONCRETE BLOCK, PARTITIONS**, scaffolding not included										210
	1000	Lightweight block, tooled joints, 2 sides, hollow										
	1100	Not reinforced, 8″ x 16″ x 4″ thick	D-8	440	.091	S.F.	1.34	3		4.34	6.35	
	1150	6″ thick		410	.098		1.84	3.22		5.06	7.25	
	1200	8″ thick		385	.104		2.26	3.43		5.69	8.05	
	1250	10″ thick	↓	370	.108		2.95	3.57		6.52	9.05	
	1300	12″ thick	D-9	350	.137		3.02	4.41		7.43	10.45	
	1500	Reinforced alternate courses, 4″ thick	D-8	435	.092		1.96	3.04		5	7.10	
	1600	6″ thick		405	.099		1.92	3.26		5.18	7.40	
	1650	8″ thick		380	.105		2.34	3.48		5.82	8.25	
	1700	10″ thick	↓	365	.110		3.04	3.62		6.66	9.20	
	1750	12″ thick	D-9	345	.139	↓	3.11	4.47		7.58	10.65	
	4000	Regular block, tooled joints, 2 sides, hollow										
	4100	Not reinforced, 8″ x 16″ x 4″ thick	D-8	430	.093	S.F.	1.14	3.07		4.21	6.25	
	4150	6″ thick		400	.100		1.69	3.30		4.99	7.20	
	4200	8″ thick		375	.107		1.85	3.52		5.37	7.75	
	4250	10″ thick	↓	360	.111		2.34	3.67		6.01	8.55	
	4300	12″ thick	D-9	340	.141		2.70	4.54		7.24	10.30	
	4500	Reinforced alternate courses, 8″ x 16″ x 4″ thick	D-8	425	.094		1.22	3.11		4.33	6.40	
	4550	6″ thick		395	.101		1.78	3.34		5.12	7.40	
	4600	8″ thick		370	.108		1.95	3.57		5.52	7.95	
	4650	10″ thick	↓	355	.113		2.70	3.72		6.42	9	
	4700	12″ thick	D-9	335	.143	↓	2.81	4.61		7.42	10.55	
	9000	Minimum labor/equipment charge	D-1	2	8	Job		257		257	415	
225	0010	**CONCRETE BRICK** C55, grade N, type I										225
	0100	Regular, 4 x 2-1/4 x 8	D-8	220	.182	Ea.	.35	6		6.35	10.15	
	0125	Rusticated, 4 x 2-1/4 x 8		220	.182		.39	6		6.39	10.20	
	0150	Frog, 4 x 2-1/4 x 8	↓	220	.182	↓	.38	6		6.38	10.15	

	04810 | Unit Masonry Assemblies		CREW	DAILY OUTPUT	LABOR-HOURS	UNIT	2006 BARE COSTS MAT.	LABOR	EQUIP.	TOTAL	TOTAL INCL O&P	
225	0200	Double, 4 x 4-7/8 x 8	D-8	180	.222	Ea.	.61	7.35		7.96	12.55	225
250	0010	**COPING** Stock units										250
	0050	Precast concrete, 10″ wide, 4″ tapers to 3-1/2″, 8″ wall	D-1	75	.213	L.F.	15	6.85		21.85	27.50	
	0100	12″ wide, 3-1/2″ tapers to 3″, 10″ wall		70	.229		10.50	7.35		17.85	23.50	
	0150	16″ wide, 4″ tapers to 3-1/2″, 14″ wall		60	.267		14.50	8.55		23.05	30	
	0300	Limestone for 12″ wall, 4″ thick		90	.178		13.50	5.70		19.20	24	
	0350	6″ thick		80	.200		15.75	6.45		22.20	28	
	0500	Marble, to 4″ thick, no wash, 9″ wide		90	.178		19.25	5.70		24.95	30.50	
	0550	12″ wide		80	.200		29	6.45		35.45	42.50	
	0700	Terra cotta, 9″ wide		90	.178		4.75	5.70		10.45	14.50	
	0750	12″ wide		80	.200		7.80	6.45		14.25	19.05	
	0800	Aluminum, for 12″ wall		80	.200		11.25	6.45		17.70	23	
	9000	Minimum labor/equipment charge		2	8	Job		257		257	415	
260	0010	**CORNICES** Brick, on existing building										260
	0110	Face bricks, 12 brick/S.F., minimum	D-1	30	.533	SF Face	5	17.15		22.15	33.50	
	0150	15 brick/S.F., maximum		23	.696	″	6	22.50		28.50	43	
	9000	Minimum labor/equipment charge		1.50	10.667	Job		345		345	555	
325	0010	**GLASS BLOCK**										325
	0100	Plain, 4″ thick, under 1,000 S.F., 6″ x 6″	D-8	115	.348	S.F.	17.95	11.50		29.45	38.50	
	0150	8″ x 8″		160	.250		11.95	8.25		20.20	26.50	
	0160	end block		160	.250		34	8.25		42.25	50.50	
	0170	90 deg corner		160	.250		32.50	8.25		40.75	49.50	
	0180	45 deg corner		160	.250		14.05	8.25		22.30	29	
	0200	12″ x 12″		175	.229		14.25	7.55		21.80	28	
	0210	4″ x 8″		160	.250		8	8.25		16.25	22	
	0220	6″ x 8″		160	.250		9.40	8.25		17.65	24	
	0300	1,000 to 5,000 S.F., 6″ x 6″		135	.296		17.60	9.80		27.40	35.50	
	0350	8″ x 8″		190	.211		11.75	6.95		18.70	24	
	0400	12″ x 12″		215	.186		14	6.15		20.15	25.50	
	0410	4″ x 8″		215	.186		3.73	6.15		9.88	14.05	
	0420	6″ x 8″		215	.186		4.13	6.15		10.28	14.50	
	0700	For solar reflective blocks, add					100%					
	1000	Thinline, plain, 3-1/8″ thick, under 1,000 S.F., 6″ x 6″	D-8	115	.348	S.F.	19.70	11.50		31.20	40	
	1050	8″ x 8″		160	.250		9.90	8.25		18.15	24.50	
	1400	For cleaning block after installation (both sides), add		1000	.040		.10	1.32		1.42	2.25	
	9000	Minimum labor/equipment charge	D-1	2	8	Job		257		257	415	
350	0010	**GLAZED CONCRETE BLOCK** C744										350
	0100	Single face, 8″ x 16″ units, 2″ thick	D-8	360	.111	S.F.	6.45	3.67		10.12	13.05	
	0200	4″ thick		345	.116		6.55	3.83		10.38	13.45	
	0250	6″ thick		330	.121		7.05	4		11.05	14.25	
	0300	8″ thick		310	.129		7.55	4.26		11.81	15.25	
	0350	10″ thick		295	.136		8.50	4.48		12.98	16.60	
	0400	12″ thick	D-9	280	.171		8.90	5.50		14.40	18.70	
	0700	Double face, 8″ x 16″ units, 4″ thick	D-8	340	.118		10.25	3.89		14.14	17.55	
	0750	6″ thick		320	.125		11.25	4.13		15.38	19.10	
	0800	8″ thick		300	.133		11.75	4.40		16.15	20	
	1500	Cove base, 8″ x 16″, 2″ thick		315	.127	L.F.	6.45	4.19		10.64	13.90	
	1550	4″ thick		285	.140		6.20	4.64		10.84	14.35	
	1600	6″ thick		265	.151		6.80	4.99		11.79	15.55	
	1650	8″ thick		245	.163		7.45	5.40		12.85	16.90	
	9000	Minimum labor/equipment charge	D-1	2	8	Job		257		257	415	
500	0010	**TERRA COTTA COPING**										500
	0020	Split type, not glazed, 9″ wide	D-1	90	.178	L.F.	7.50	5.70		13.20	17.50	

04810 | Unit Masonry Assemblies

			CREW	DAILY OUTPUT	LABOR-HOURS	UNIT	2006 BARE COSTS MAT.	LABOR	EQUIP.	TOTAL	TOTAL INCL O&P	
500	0100	13" wide	D-1	80	.200	L.F.	11.30	6.45		17.75	23	500
	0200	Split type, glazed, 9" wide		90	.178		12.70	5.70		18.40	23.50	
	0250	13" wide		80	.200		16.60	6.45		23.05	28.50	
	0500	Partition or back-up blocks, scored, in C.L. lots										
	0700	Non-load bearing 12" x 12", 3" thick, special order	D-8	550	.073	S.F.	12.35	2.40		14.75	17.45	
	0850	8" thick		400	.100		7.45	3.30		10.75	13.55	
	1000	Load bearing, 12" x 12", 4" thick, in walls		500	.080		5.10	2.64		7.74	9.90	
	1400	8" thick, in walls		400	.100		7.45	3.30		10.75	13.55	
	9000	Minimum labor/equipment charge	D-1	2	8	Job		257		257	415	
650	0016	**WALLS** R042110-20										650
	0800	4" wall, face, 4" x 2-2/3" x 8"	D-8	215	.186	S.F.	3.23	6.15		9.38	13.50	
	0850	4" thick, as back up, 6.75 bricks per S.F. R042110-50		240	.167		2.64	5.50		8.14	11.85	
	0900	8" thick wall, 13.50 brick per S.F.		135	.296		5.50	9.80		15.30	22	
	1000	12" thick wall, 20.25 bricks per S.F.		95	.421		8.25	13.90		22.15	31.50	
	1050	16" thick wall, 27.00 bricks per S.F.		75	.533		11.20	17.60		28.80	41	
	1200	Reinforced, 4" x 2-2/3" x 8" , 4" wall		205	.195		2.64	6.45		9.09	13.35	
	1250	8" thick wall, 13.50 brick per S.F.		130	.308		5.50	10.15		15.65	22.50	
	1300	12" thick wall, 20.25 bricks per S.F.		90	.444		8.25	14.70		22.95	33	
	1350	16" thick wall, 27.00 bricks per S.F.		70	.571		11.20	18.85		30.05	43	
	9000	Minimum labor/equipment charge	D-1	2	8	Job		257		257	415	
700	0010	**STRUCTURAL BRICK** C652, Grade SW, incl mortar, scaffolding not incl										700
	0100	Standard unit, 4-5/8" x 2-3/4" x 9-5/8"	D-8	245	.163	S.F.	4.67	5.40		10.07	13.90	
	0120	Bond beam		225	.178		8.85	5.85		14.70	19.30	
	0140	V cut bond beam		225	.178		9.15	5.85		15	19.60	
	0160	Stretcher quoin, 5-5/8" x 2-3/4" x 9-5/8"		245	.163		8.80	5.40		14.20	18.45	
	0180	Corner quoin		245	.163		9.90	5.40		15.30	19.65	
	0200	Corner, 45 deg, 4-5/8" x 2-3/4" x 10-7/16"		235	.170		10.25	5.60		15.85	20.50	
750	0010	**STRUCTURAL FACING TILE** Scaffolding not incl, standard colors										750
	0020	6T series, 5-1/3" x 12", 2.3 pieces per S.F., glazed 1 side, 2" thick	D-8	225	.178	S.F.	6	5.85		11.85	16.15	
	0100	4" thick		220	.182		7.05	6		13.05	17.50	
	0150	Glazed 2 sides		195	.205		11.20	6.80		18	23.50	
	0250	6" thick		210	.190		10.95	6.30		17.25	22	
	0300	Glazed 2 sides		185	.216		12.80	7.15		19.95	25.50	
	0400	8" thick		180	.222		13.65	7.35		21	27	
	9000	Minimum labor/equipment charge	D-1	2	8	Job		257		257	415	

04850 | Stone Assemblies

			CREW	DAILY OUTPUT	LABOR-HOURS	UNIT	MAT.	LABOR	EQUIP.	TOTAL	TOTAL INCL O&P	
100	0010	**BLUESTONE** Cut to size										100
	0500	Sills, natural cleft, 10" wide to 6' long, 1-1/2" thick	D-11	70	.343	L.F.	12.35	11.75		24.10	32.50	
	0550	2" thick		63	.381		14.05	13.05		27.10	36.50	
	0600	Smooth finish, 1-1/2" thick		70	.343		18.50	11.75		30.25	39.50	
	0650	2" thick		63	.381		22	13.05		35.05	45	
	0800	Thermal finish, 1-1/2" thick		70	.343		20	11.75		31.75	41	
	0850	2" thick		63	.381		22	13.05		35.05	45	
	1000	Stair treads, natural cleft, 12" wide, 6' long, 1-1/2" thick	D-10	115	.348		18.20	11.75	4.66	34.61	44	
	1050	2" thick		105	.381		19.10	12.85	5.10	37.05	47	
	1100	Smooth finish, 1-1/2" thick		115	.348		22	11.75	4.66	38.41	48	
	1150	2" thick		105	.381		22	12.85	5.10	39.95	50	
	1300	Thermal finish		115	.348		27.50	11.75	4.66	43.91	54	
	1350	2" thick		105	.381		31	12.85	5.10	48.95	60	
	9000	Minimum labor/equipment charge	D-1	2.50	6.400	Job		206		206	335	
300	0010	**GRANITE** Cut to size										300
	0050	Veneer, polished face, 3/4" to 1-1/2" thick										

04800 | Masonry Assemblies

04850 | Stone Assemblies

			CREW	DAILY OUTPUT	LABOR-HOURS	UNIT	2006 BARE COSTS MAT.	LABOR	EQUIP.	TOTAL	TOTAL INCL O&P	
300	0150	Low price, gray, light gray, etc.	D-10	130	.308	S.F.	22.50	10.40	4.13	37.03	46	300
	0180	Medium price, pink, brown, etc.		130	.308		24.50	10.40	4.13	39.03	48.50	
	0220	High price, red, black, etc.	↓	130	.308	↓	37.50	10.40	4.13	52.03	63	
	2500	Steps, copings, etc., finished on more than one surface										
	2550	Minimum	D-10	50	.800	C.F.	84	27	10.75	121.75	148	
	2600	Maximum	"	50	.800	"	135	27	10.75	172.75	203	
	3500	Curbing, city street type, See Division 02770-300										
	3800	Radius curbs, over 5′ radius, add				L.F.	50%					
	3850	Under 5′ radius, add				"	100%					
	9000	Minimum labor/equipment charge	D-1	2	8	Job		257		257	415	
400	0010	**LIMESTONE,** Cut to size										400
	0020	Veneer facing panels										
	0750	5″ thick, 5′ x 14′ panels	D-10	275	.145	S.F.	50.50	4.91	1.95	57.36	65.50	
	1000	Sugarcube finish, 2″ Thick, 3′ x 5′ panels		275	.145		25	4.91	1.95	31.86	37.50	
	1050	3″ Thick, 4′ x 9′ panels		275	.145		37.50	4.91	1.95	44.36	51.50	
	1200	4″ Thick, 5′ x 11′ panels		275	.145		39.50	4.91	1.95	46.36	53.50	
	1400	Sugarcube, textured finish, 4-1/2″ thick, 5′ x 12′		275	.145		43.50	4.91	1.95	50.36	58	
	1450	5″ thick, 5′ x 14′ panels		275	.145	↓	51.50	4.91	1.95	58.36	66.50	
	2000	Coping, sugarcube finish, top & 2 sides		30	1.333	C.F.	125	45	17.90	187.90	230	
	2100	Sills, lintels, jambs, trim, stops, sugarcube finish, average		20	2		125	67.50	27	219.50	277	
	2150	Detailed		20	2	↓	125	67.50	27	219.50	277	
	2300	Steps, extra hard, 14″ wide, 6″ rise	↓	50	.800	L.F.	22.50	27	10.75	60.25	80.50	
	3000	Quoins, plain finish, 6″x12″x12″	D-12	25	1.280	Ea.	125	42		167	206	
	3050	6″x16″x24″	"	25	1.280	"	167	42		209	251	
	9000	Minimum labor/equipment charge	D-1	2	8	Job		257		257	415	
500	0011	**MARBLE**, ashlar, split face, 4″ + or - thick, random										500
	0040	lengths 1′ to 4′ & heights 2″ to 7-1/2″, average	D-8	175	.229	S.F.	14.55	7.55		22.10	28.50	
	0100	Base, polished, 3/4″ or 7/8″ thick, polished, 6″ high	D-10	65	.615	L.F.	13.25	21	8.25	42.50	57	
	0300	Carvings or bas relief, from templates, average		80	.500	S.F.	118	16.85	6.70	141.55	164	
	0350	Maximum	↓	80	.500	"	276	16.85	6.70	299.55	340	
	1000	Facing, polished finish, cut to size, 3/4″ to 7/8″ thick										
	1050	Average	D-10	130	.308	S.F.	19.45	10.40	4.13	33.98	43	
	1100	Maximum	"	130	.308	"	45	10.40	4.13	59.53	71	
	2500	Flooring, polished tiles, 12″ x 12″ x 3/8″ thick										
	2510	Thin set, average	D-11	90	.267	S.F.	9.75	9.15		18.90	25.50	
	2600	Maximum		90	.267		150	9.15		159.15	180	
	2700	Mortar bed, average		65	.369		9.90	12.65		22.55	31.50	
	2740	Maximum	↓	65	.369	↓	150	12.65		162.65	186	
	3500	Thresholds, 3′ long, 7/8″ thick, 4″ to 5″ wide, plain	D-12	24	1.333	Ea.	14.05	43.50		57.55	86	
	3550	Beveled		24	1.333	"	16.30	43.50		59.80	88.50	
	3700	Window stools, polished, 7/8″ thick, 5″ wide	↓	85	.376	L.F.	13.10	12.30		25.40	34.50	
	9000	Minimum labor/equipment charge	D-1	2	8	Job		257		257	415	
600	0011	**ROUGH STONE WALL**, Dry										600
	0100	Random fieldstone, under 18″ thick	D-12	60	.533	C.F.	8.90	17.40		26.30	38.50	
	0150	Over 18″ thick	"	63	.508	"	10.70	16.60		27.30	39	
	9000	Minimum labor/equipment charge	D-1	2	8	Job		257		257	415	
700	0011	**SANDSTONE OR BROWNSTONE**										700
	0100	Sawed face veneer, 2-1/2″ thick, to 2′ x 4′ panels	D-10	130	.308	S.F.	16.25	10.40	4.13	30.78	39	
	0150	4′ thick, to 3′-6″ x 8′panels		100	.400		16.25	13.50	5.35	35.10	46	
	0300	Split face, random sizes	↓	100	.400	↓	11.70	13.50	5.35	30.55	41	
	9000	Minimum labor/equipment charge	D-1	2.50	6.400	Job		206		206	335	
800	0010	**SLATE** Pennsylvania, blue gray to gray black; Vermont,										800
	0050	Unfading green, mottled green & purple, gray & purple										

04800 | Masonry Assemblies

04850 | Stone Assemblies

		04850 Stone Assemblies	CREW	DAILY OUTPUT	LABOR-HOURS	UNIT	2006 BARE COSTS MAT.	LABOR	EQUIP.	TOTAL	TOTAL INCL O&P	
800	0100	Virginia, blue black										800
	3100	Stair landings, 1″ thick, black, clear	D-1	65	.246	S.F.	15	7.90		22.90	29.50	
	3200	Ribbon	″	65	.246	″	13.50	7.90		21.40	27.50	
	3500	Stair treads, sand finish, 1″ thick x 12″ wide										
	3550	Under 3 L.F.	D-10	85	.471	L.F.	17.75	15.90	6.30	39.95	52	
	3600	3 L.F. to 6 L.F.	″	120	.333	″	17.90	11.25	4.47	33.62	43	
	3700	Ribbon, sand finish, 1″ thick x 12″ wide										
	3750	To 6 L.F.	D-10	120	.333	L.F.	14.75	11.25	4.47	30.47	39.50	
	4000	Stools or sills, sand finish, 1″ thick, 6″ wide	D-12	160	.200		9.25	6.55		15.80	21	
	4200	10″ wide		90	.356		14	11.60		25.60	34.50	
	4400	2″ thick, 6″ wide		140	.229		15.25	7.45		22.70	29	
	4600	10″ wide	↓	90	.356		23.50	11.60		35.10	44.50	
	4800	For lengths over 3′, add				↓	25%					
	9000	Minimum labor/equipment charge	D-1	2.50	6.400	Job		206		206	335	
900	0010	**WINDOW SILL**										900
	0020	Bluestone, thermal top, 10″ wide, 1-1/2″ thick	D-1	85	.188	S.F.	14.85	6.05		20.90	26	
	0050	2″ thick		75	.213	″	17.35	6.85		24.20	30	
	0100	Cut stone, 5″ x 8″ plain		48	.333	L.F.	10.20	10.70		20.90	28.50	
	0200	Face brick on edge, brick, 8″ wide		80	.200		2.15	6.45		8.60	12.80	
	0400	Marble, 9″ wide, 1″ thick		85	.188		7.50	6.05		13.55	18.05	
	0900	Slate, colored, unfading, honed, 12″ wide, 1″ thick		85	.188		15.25	6.05		21.30	26.50	
	0950	2″ thick	↓	70	.229	↓	21.50	7.35		28.85	35.50	
	9000	Minimum labor/equipment charge	1 Bric	2	4	Job		146		146	237	

04880 | Masonry Fireplaces

		04880 Masonry Fireplaces	CREW	DAILY OUTPUT	LABOR-HOURS	UNIT	MAT.	LABOR	EQUIP.	TOTAL	TOTAL INCL O&P	
600	0010	**FIREPLACE** For prefabricated fireplace, see Div. 10305-100										600
	0100	Brick fireplace, not incl. foundations or chimneys										
	0110	30″ x 29″ opening, incl. chamber, plain brickwork	D-1	.40	40	Ea.	415	1,275		1,690	2,525	
	0200	Fireplace box only (110 brick)	″	2	8	″	136	257		393	565	
	0300	For elaborate brickwork and details, add					35%	35%				
	0400	For hearth, brick & stone, add	D-1	2	8	Ea.	153	257		410	585	
	0410	For steel angle, damper, cleanouts, add		4	4		107	129		236	325	
	0600	Plain brickwork, incl. metal circulator		.50	32	↓	790	1,025		1,815	2,550	
	0800	Face brick only, standard size, 8″ x 2-2/3″ x 4″	↓	.30	53.333	M	415	1,725		2,140	3,225	
	0900	Stone fireplace, fieldstone, add				SF Face	10.90			10.90	12	
	1000	Cut stone, add				″	12			12	13.20	
	9000	Minimum labor/equipment charge	D-1	2	8	Job		257		257	415	

04900 | Masonry Restoration and Cleaning

04910 | Unit Masonry Restoration

		04910 Unit Masonry Restoration	CREW	DAILY OUTPUT	LABOR-HOURS	UNIT	2006 BARE COSTS MAT.	LABOR	EQUIP.	TOTAL	TOTAL INCL O&P	
200	0010	**CAULKING MASONRY** 1/2″ x 1/2″ joint										200
	0050	Re-caulk only, oil base	1 Bric	225	.036	L.F.	.24	1.30		1.54	2.37	
	0100	Acrylic latex		205	.039		.26	1.43		1.69	2.60	
	0200	Polyurethane		200	.040		.34	1.46		1.80	2.75	
	0300	Silicone		195	.041		.50	1.50		2	2.98	
	1000	Cut out and re-caulk, oil base		145	.055		.24	2.02		2.26	3.53	
	1050	Acrylic latex		130	.062		.26	2.25		2.51	3.94	
	1100	Polyurethane	↓	125	.064	↓	.34	2.34		2.68	4.18	

04900 | Masonry Restoration and Cleaning

		04910 | Unit Masonry Restoration	CREW	DAILY OUTPUT	LABOR-HOURS	UNIT	2006 BARE COSTS MAT.	LABOR	EQUIP.	TOTAL	TOTAL INCL O&P	
200	1150	Silicone	1 Bric	120	.067	L.F.	.50	2.44		2.94	4.50	200
	9000	Minimum labor/equipment charge		4	2	Job		73		73	119	
600	0010	**NEEDLE BEAM MASONRY** Incl. wood shoring 10′ x 10′ opening										600
	0400	Block, concrete, 8″ thick	B-9	7.10	5.634	Ea.	36.50	157	20.50	214	325	
	0420	12″ thick		6.70	5.970		42.50	166	22	230.50	345	
	0800	Brick, 4″ thick with 8″ backup block		5.70	7.018		42.50	195	26	263.50	400	
	1000	Brick, solid, 8″ thick		6.20	6.452		36.50	179	23.50	239	365	
	1040	12″ thick		4.90	8.163		42.50	227	30	299.50	455	
	1080	16″ thick		4.50	8.889		53.50	247	32.50	333	505	
	2000	Add for additional floors of shoring	B-1	6	4		36.50	112		148.50	227	
	9000	Minimum labor/equipment charge	″	2	12	Job		335		335	560	
720	0010	**POINTING MASONRY**										720
	0300	Cut and repoint brick, hard mortar, running bond	1 Bric	80	.100	S.F.	.43	3.66		4.09	6.40	
	0320	Common bond		77	.104		.43	3.80		4.23	6.60	
	0360	Flemish bond		70	.114		.45	4.18		4.63	7.30	
	0400	English bond		65	.123		.45	4.50		4.95	7.80	
	0600	Soft old mortar, running bond		100	.080		.44	2.92		3.36	5.25	
	0620	Common bond		96	.083		.43	3.05		3.48	5.40	
	0640	Flemish bond		90	.089		.45	3.25		3.70	5.75	
	0680	English bond		82	.098		.45	3.57		4.02	6.30	
	0700	Stonework, hard mortar		140	.057	L.F.	.57	2.09		2.66	4.02	
	0720	Soft old mortar		160	.050	″	.57	1.83		2.40	3.60	
	1000	Repoint, mask and grout method, running bond		95	.084	S.F.	.57	3.08		3.65	5.60	
	1020	Common bond		90	.089		.57	3.25		3.82	5.90	
	1040	Flemish bond		86	.093		.57	3.40		3.97	6.15	
	1060	English bond		77	.104		.57	3.80		4.37	6.80	
	2000	Scrub coat, sand grout on walls, minimum		120	.067		2.77	2.44		5.21	7	
	2020	Maximum		98	.082		1.99	2.98		4.97	7.05	
	9000	Minimum labor/equipment charge		3	2.667	Job		97.50		97.50	158	
800	0010	**TOOTHING MASONRY**										800
	0500	Brickwork, soft old mortar	1 Clab	40	.200	V.L.F.		5.50		5.50	9.10	
	0520	Hard mortar		30	.267			7.30		7.30	12.10	
	0700	Blockwork, soft old mortar		70	.114			3.13		3.13	5.20	
	0720	Hard mortar		50	.160			4.38		4.38	7.25	
	9000	Minimum labor/equipment charge		4	2	Job		55		55	91	
		04930 | Unit Masonry Cleaning										
220	0010	**MASONRY CLEANING**										220
	0200	Chemical cleaning, new construction, brush and wash, minimum	D-1	1000	.016	S.F.	.04	.51		.55	.87	
	0220	Average		800	.020		.06	.64		.70	1.10	
	0240	Maximum		600	.027		.08	.86		.94	1.47	
	0260	Light restoration, minimum		800	.020		.07	.64		.71	1.12	
	0270	Average		400	.040		.11	1.29		1.40	2.21	
	0280	Maximum		330	.048		.15	1.56		1.71	2.69	
	0300	Heavy restoration, minimum		600	.027		.08	.86		.94	1.48	
	0310	Average		400	.040		.12	1.29		1.41	2.22	
	0320	Maximum		250	.064		.16	2.06		2.22	3.51	
	0400	High pressure water only, minimum	B-9	2000	.020			.56	.07	.63	1	
	0420	Average		1500	.027			.74	.10	.84	1.34	
	0440	Maximum		1000	.040			1.11	.15	1.26	2	
	0800	High pressure water and chemical, minimum		1800	.022		.09	.62	.08	.79	1.21	
	0820	Average		1200	.033		.13	.93	.12	1.18	1.82	
	0840	Maximum		800	.050		.18	1.39	.18	1.75	2.69	
	1200	Sandblast, wet system, minimum		1750	.023		.16	.64	.08	.88	1.32	
	1220	Average		1100	.036		.24	1.01	.13	1.38	2.09	

	04930 | Unit Masonry Cleaning		CREW	DAILY OUTPUT	LABOR-HOURS	UNIT	2006 BARE COSTS MAT.	LABOR	EQUIP.	TOTAL	TOTAL INCL O&P	
220	1240	Maximum	B-9	700	.057	S.F.	.32	1.59	.21	2.12	3.22	220
	1400	Dry system, minimum		2500	.016		.16	.44	.06	.66	.98	
	1420	Average		1750	.023		.24	.64	.08	.96	1.41	
	1440	Maximum	↓	1000	.040		.32	1.11	.15	1.58	2.36	
	1800	For walnut shells, add					.41			.41	.45	
	1820	For corn chips, add					.41			.41	.45	
	2000	Steam cleaning, minimum	B-9	3000	.013			.37	.05	.42	.66	
	2020	Average		2500	.016			.44	.06	.50	.80	
	2040	Maximum	↓	1500	.027			.74	.10	.84	1.34	
	4000	Add for masking doors and windows				↓					.80	
	4200	Add for pedestrian protection				Job					10%	
	4400	Add for wire cut face brick				S.F.	.15			.15	.16	
	9000	Minimum labor/equipment charge	D-4	2	16	Job		510	72.50	582.50	900	

For information about Means Estimating Seminars, see yellow pages 12 and 13 in back of book

Division Notes

		CREW	DAILY OUTPUT	LABOR-HOURS	UNIT	2000 BARE COSTS MAT.	LABOR	EQUIP.	TOTAL	TOTAL INCL O&P

Division 5 Metals

Estimating Tips

05050 Basic Metal Materials & Methods

- Nuts, bolts, washers, connection angles, and plates can add a significant amount to both the tonnage of a structural steel job as well as the estimated cost. As a rule of thumb, add 10% to the total weight to account for these accessories.
- Type 2 steel construction, commonly referred to as "simple construction," consists generally of field-bolted connections with lateral bracing supplied by other elements of the building, such as masonry walls or x-bracing. The estimator should be aware, however, that shop connections may be accomplished by welding or bolting. The method may be particular to the fabrication shop and may have an impact on the estimated cost.

05200 Metal Joists

- In any given project the total weight of open web steel joists is determined by the loads to be supported and the design. However, economies can be realized in minimizing the amount of labor used to place the joists. This is done by maximizing the joist spacing and therefore minimizing the number of joists required to be installed on the job. Certain spacings and locations may be required by the design, but in other cases maximizing the spacing and keeping it as uniform as possible will keep the costs down.

05300 Metal Deck

- The takeoff and estimating of metal deck involves more than simply the area of the floor or roof and the type of deck specified or shown on the drawings. Many different sizes and types of openings may exist. Small openings for individual pipes or conduits may be drilled after the floor/roof is installed, but larger openings may require special deck lengths as well as reinforcing or structural support. The estimator should determine who will be supplying this reinforcing. Additionally, some deck terminations are part of the deck package, such as screed angles and pour stops, and others will be part of the steel contract, such as angles attached to structural members and cast-in-place angles and plates. The estimator must ensure that all pieces are accounted for in the complete estimate.

05500 Metal Fabrications

- The most economical steel stairs are those that use common materials, standard details, and most importantly, a uniform and relatively simple method of field assembly. Commonly available A36 channels and plates are very good choices for the main stringers of the stairs, as are angles and tees for the carrier members. Risers and treads are usually made by specialty shops, and it is most economical to use a typical detail in as many places as possible. The stairs should be pre-assembled and shipped directly to the site. The field connections should be simple and straightforward to be accomplished efficiently and with a minimum of equipment and labor.

Reference Numbers

Reference numbers are shown in bold squares at the beginning of some major classifications. These numbers refer to related items in the Reference Section. The reference information may be an estimating procedure, an alternate pricing method, or technical information.

Note: Not all subdivisions listed here necessarily appear in this publication.

05060 | Selective Demolition

		05060 Selective Demolition	CREW	DAILY OUTPUT	LABOR-HOURS	UNIT	2006 BARE COSTS MAT.	LABOR	EQUIP.	TOTAL	TOTAL INCL O&P	
110	0010	SELECTIVE METALS DEMOLITION R024119-10										110
	0015	Excludes shores, bracing, cutting, loading, hauling, dumping										
	0020	Remove nuts only up to 3/4" diameter	1 Sswk	480	.017	Ea.		.67		.67	1.25	
	0030	7/8" to 1-1/4" diameter		240	.033			1.33		1.33	2.49	
	0040	1-3/8" to 2" diameter		160	.050			2		2	3.74	
	0060	Unbolt and remove structural bolts up to 3/4" diameter		240	.033			1.33		1.33	2.49	
	0070	7/8" to 2" diameter		160	.050			2		2	3.74	
	0140	Light weight framing members, remove whole or cut up, up to 20 lb		240	.033			1.33		1.33	2.49	
	0150	21 - 40 lb	2 Sswk	210	.076			3.04		3.04	5.70	
	0160	41 - 80 lb	3 Sswk	180	.133			5.35		5.35	9.95	
	0170	81 - 120 lb	4 Sswk	150	.213			8.50		8.50	15.95	
	0230	Structural members, remove whole or cut up, up to 500 lb	E-19	48	.500			19.50	19.85	39.35	57	
	0240	1/4 - 2 tons	E-18	36	1.111			44	26.50	70.50	109	
	0250	2 - 5 tons	E-24	30	1.067			42	21.50	63.50	99	
	0260	5 - 10 tons	E-20	24	2.667			104	47.50	151.50	240	
	0270	10 - 15 tons	E-2	18	3.111			121	79.50	200.50	305	
	0340	Fabricated item, remove whole or cut up, up to 20 lb	1 Sswk	96	.083			3.33		3.33	6.20	
	0350	21 - 40 lb	2 Sswk	84	.190			7.60		7.60	14.25	
	0360	41 - 80 lb	3 Sswk	72	.333			13.30		13.30	25	
	0370	81 - 120 lb	4 Sswk	60	.533			21.50		21.50	40	
	0380	121 - 500 lb	E-19	48	.500			19.50	19.85	39.35	57	
	0390	501 - 1000 lb	"	36	.667			26	26.50	52.50	75.50	
	2950	Minimum labor/equipment charge	B-13	2	28	Job		840	325	1,165	1,725	

05090 | Metal Fastenings

		05090 Metal Fastenings	CREW	DAILY OUTPUT	LABOR-HOURS	UNIT	MAT.	LABOR	EQUIP.	TOTAL	TOTAL INCL O&P	
300	0010	CHEMICAL ANCHORS, Includes layout & drilling										300
	1430	Chemical anchor, w/rod & epoxy cartridge, 3/4" diam. x 9-1/2" long	B-89A	27	.593	Ea.	12.80	18.95	4.28	36.03	50	
	1435	1" diameter x 11-3/4" long		24	.667		25	21.50	4.81	51.31	68	
	1440	1-1/4" diameter x 14" long		21	.762		47	24.50	5.50	77	98	
	1445	1-3/4" diameter x 15" long		20	.800		89.50	25.50	5.80	120.80	147	
	1450	18" long		17	.941		107	30	6.80	143.80	175	
	1455	2" diameter x 18" long		16	1		137	32	7.20	176.20	210	
	1460	24" long		15	1.067		179	34	7.70	220.70	261	
	1500	Chemical anchoring, epoxy cartridge, excludes layout, drilling, fastener										
	1530	For fastener 3/4" dia x 6" embedment	B-89A	27	.593	Ea.	5.05	18.95	4.28	28.28	41.50	
	1535	1" dia x 8" embedment		24	.667		7.55	21.50	4.81	33.86	48.50	
	1540	1-1/4" dia x 10" embedment		21	.762		15.10	24.50	5.50	45.10	62.50	
	1545	1-3/4" dia x 12" embedment		20	.800		25	25.50	5.80	56.30	76	
	1550	14" embedment		17	.941		30	30	6.80	66.80	90.50	
	1555	2" dia x 12" embedment		16	1		40.50	32	7.20	79.70	105	
	1560	18" embedment		15	1.067		50.50	34	7.70	92.20	120	
340	0010	DRILLING										340
	0050	Up to 4" deep in conc/brick floor/wall, incl. bit & layout, no anchor										
	0100	Holes, 1/4" diameter	1 Carp	75	.107	Ea.	.09	3.79		3.88	6.40	
	0150	For each additional inch of depth, add		430	.019		.02	.66		.68	1.13	
	0200	3/8" diameter		63	.127		.09	4.51		4.60	7.60	
	0250	For each additional inch of depth, add		340	.024		.02	.84		.86	1.41	
	0300	1/2" diameter		50	.160		.09	5.70		5.79	9.50	
	0350	For each additional inch of depth, add		250	.032		.02	1.14		1.16	1.90	
	0400	5/8" diameter		48	.167		.16	5.95		6.11	10	
	0450	For each additional inch of depth, add		240	.033		.04	1.18		1.22	2	
	0500	3/4" diameter		45	.178		.19	6.30		6.49	10.65	
	0550	For each additional inch of depth, add		220	.036		.05	1.29		1.34	2.19	
	0600	7/8" diameter		43	.186		.23	6.60		6.83	11.20	
	0650	For each additional inch of depth, add		210	.038		.06	1.35		1.41	2.30	

05090 | Metal Fastenings

			CREW	DAILY OUTPUT	LABOR-HOURS	UNIT	2006 BARE COSTS MAT.	LABOR	EQUIP.	TOTAL	TOTAL INCL O&P	
340	0700	1" diameter	1 Carp	40	.200	Ea.	.27	7.10		7.37	12.10	340
	0750	For each additional inch of depth, add		190	.042		.07	1.50		1.57	2.55	
	0800	1-1/4" diameter		38	.211		.38	7.50		7.88	12.80	
	0850	For each additional inch of depth, add		180	.044		.10	1.58		1.68	2.72	
	0900	1-1/2" diameter		35	.229		.58	8.15		8.73	14.10	
	0950	For each additional inch of depth, add		165	.048		.14	1.72		1.86	3.02	
	1000	For ceiling installations, add						40%				
	1100	Drilling & layout for drywall/plaster walls, up to 1" deep, no anchor										
	1200	Holes, 1/4" diameter	1 Carp	150	.053	Ea.	.01	1.90		1.91	3.15	
	1300	3/8" diameter		140	.057		.01	2.03		2.04	3.38	
	1400	1/2" diameter		130	.062		.01	2.19		2.20	3.63	
	1500	3/4" diameter		120	.067		.02	2.37		2.39	3.96	
	1600	1" diameter		110	.073		.03	2.59		2.62	4.32	
	1700	1-1/4" diameter		100	.080		.05	2.84		2.89	4.76	
	1800	1-1/2" diameter		90	.089		.07	3.16		3.23	5.35	
	1900	For ceiling installations, add						40%				
	1910	Drilling & layout for steel, up to 1/4" deep, no anchor										
	1920	Holes, 1/4" diameter	1 Sswk	112	.071	Ea.	.12	2.85		2.97	5.50	
	1925	For each additional 1/4" depth, add		336	.024		.12	.95		1.07	1.92	
	1930	3/8" diameter		104	.077		.14	3.07		3.21	5.90	
	1935	For each additional 1/4" depth, add		312	.026		.14	1.02		1.16	2.07	
	1940	1/2" diameter		96	.083		.16	3.33		3.49	6.35	
	1945	For each additional 1/4" depth, add		288	.028		.16	1.11		1.27	2.25	
	1950	5/8" diameter		88	.091		.26	3.63		3.89	7.10	
	1955	For each additional 1/4" depth, add		264	.030		.26	1.21		1.47	2.54	
	1960	3/4" diameter		80	.100		.29	4		4.29	7.75	
	1965	For each additional 1/4" depth, add		240	.033		.29	1.33		1.62	2.81	
	1970	7/8" diameter		72	.111		.34	4.44		4.78	8.70	
	1975	For each additional 1/4" depth, add		216	.037		.34	1.48		1.82	3.15	
	1980	1" diameter		64	.125		.39	4.99		5.38	9.80	
	1985	For each additional 1/4" depth, add		192	.042		.39	1.66		2.05	3.54	
	1990	For drilling up, add						40%				
	2000	Minimum labor/equipment charge	1 Carp	4	2	Job		71		71	118	
380	0010	**EXPANSION ANCHORS**										380
	0100	Anchors for concrete, brick or stone, no layout and drilling										
	0200	Expansion shields, zinc, 1/4" diameter, 1-5/16" long, single	1 Carp	90	.089	Ea.	1.06	3.16		4.22	6.40	
	0300	1-3/8" long, double		85	.094		1.16	3.35		4.51	6.85	
	0400	3/8" diameter, 1-1/2" long, single		85	.094		1.74	3.35		5.09	7.45	
	0500	2" long, double		80	.100		2.15	3.56		5.71	8.25	
	0600	1/2" diameter, 2-1/16" long, single		80	.100		2.88	3.56		6.44	9.05	
	0700	2-1/2" long, double		75	.107		2.78	3.79		6.57	9.35	
	0800	5/8" diameter, 2-5/8" long, single		75	.107		4.12	3.79		7.91	10.85	
	0900	2-3/4" long, double		70	.114		4.12	4.06		8.18	11.30	
	1000	3/4" diameter, 2-3/4" long, single		70	.114		6.10	4.06		10.16	13.50	
	1100	3-15/16" long, double		65	.123		8.15	4.38		12.53	16.25	
	1500	Self drilling anchor, snap-off, for 1/4" diameter bolt		26	.308		.91	10.95		11.86	19.10	
	1600	3/8" diameter bolt		23	.348		1.32	12.35		13.67	22	
	1700	1/2" diameter bolt		20	.400		2.03	14.20		16.23	25.50	
	1800	5/8" diameter bolt		18	.444		3.39	15.80		19.19	29.50	
	1900	3/4" diameter bolt		16	.500		5.70	17.80		23.50	36	
	2100	Hollow wall anchors for gypsum wall board, plaster or tile										
	2300	1/8" diameter, short	1 Carp	160	.050	Ea.	.29	1.78		2.07	3.27	
	2400	Long		150	.053		.34	1.90		2.24	3.51	
	2500	3/16" diameter, short		150	.053		.61	1.90		2.51	3.81	
	2600	Long		140	.057		.65	2.03		2.68	4.09	

	05090 | Metal Fastenings		Crew	Daily Output	Labor-Hours	Unit	2006 Bare Costs Mat.	Labor	Equip.	Total	Total Incl O&P	
380	2700	1/4" diameter, short	1 Carp	140	.057	Ea.	.74	2.03		2.77	4.18	380
	2800	Long		130	.062		.84	2.19		3.03	4.54	
	3000	Toggle bolts, bright steel, 1/8" diameter, 2" long		85	.094		.27	3.35		3.62	5.85	
	3100	4" long		80	.100		.41	3.56		3.97	6.35	
	3200	3/16" diameter, 3" long		80	.100		.46	3.56		4.02	6.40	
	3300	6" long		75	.107		.64	3.79		4.43	7	
	3400	1/4" diameter, 3" long		75	.107		.51	3.79		4.30	6.85	
	3500	6" long		70	.114		.73	4.06		4.79	7.55	
	3600	3/8" diameter, 3" long		70	.114		.99	4.06		5.05	7.85	
	3700	6" long		60	.133		1.73	4.74		6.47	9.75	
	3800	1/2" diameter, 4" long		60	.133		2.57	4.74		7.31	10.70	
	3900	6" long		50	.160		4.23	5.70		9.93	14.05	
	4000	Nailing anchors										
	4100	Nylon nailing anchor, 1/4" diameter, 1" long	1 Carp	3.20	2.500	C	19.80	89		108.80	169	
	4200	1-1/2" long		2.80	2.857		25.50	102		127.50	196	
	4300	2" long		2.40	3.333		42.50	119		161.50	243	
	4400	Metal nailing anchor, 1/4" diameter, 1" long		3.20	2.500		30.50	89		119.50	181	
	4500	1-1/2" long		2.80	2.857		41.50	102		143.50	214	
	4600	2" long		2.40	3.333		52.50	119		171.50	254	
	8000	Wedge anchors, not including layout or drilling										
	8050	Carbon steel, 1/4" diameter, 1-3/4" long	1 Carp	150	.053	Ea.	.46	1.90		2.36	3.65	
	8100	3 1/4" long		140	.057		.61	2.03		2.64	4.04	
	8150	3/8" diameter, 2-1/4" long		145	.055		.69	1.96		2.65	4.01	
	8200	5" long		140	.057		1.21	2.03		3.24	4.70	
	8250	1/2" diameter, 2-3/4" long		140	.057		1.06	2.03		3.09	4.53	
	8300	7" long		125	.064		1.81	2.28		4.09	5.75	
	8350	5/8" diameter, 3-1/2" long		130	.062		2.09	2.19		4.28	5.90	
	8400	8-1/2" long		115	.070		4.45	2.47		6.92	9	
	8450	3/4" diameter, 4-1/4" long		115	.070		2.52	2.47		4.99	6.90	
	8500	10" long		95	.084		5.75	2.99		8.74	11.25	
	8550	1" diameter, 6" long		100	.080		8.45	2.84		11.29	14	
	8575	9" long		85	.094		11	3.35		14.35	17.65	
	8600	12" long		75	.107		11.85	3.79		15.64	19.35	
	8650	1-1/4" diameter, 9" long		70	.114		15.40	4.06		19.46	23.50	
	8700	12" long		60	.133		19.70	4.74		24.44	29.50	
	8750	For type 303 stainless steel, add					350%					
	8800	For type 316 stainless steel, add					450%					
	8950	Self-drilling concrete screw, hex washer head, 3/16" dia x 1-3/4" long	1 Carp	300	.027	Ea.	.35	.95		1.30	1.96	
	8960	2-1/4" long		250	.032		.54	1.14		1.68	2.47	
	8970	Phillips flat head, 3/16" dia x 1-3/4" long		300	.027		.36	.95		1.31	1.97	
	8980	2-1/4" long		250	.032		.53	1.14		1.67	2.46	
	9000	Minimum labor/equipment charge		4	2	Job		71		71	118	
460	0010	**LAG SCREWS**										460
	0020	Steel, 1/4" diameter, 2" long	1 Carp	200	.040	Ea.	.09	1.42		1.51	2.46	
	0200	1/2" diameter, 3" long		130	.062		.42	2.19		2.61	4.08	
	0300	5/8" diameter, 3" long		120	.067		.82	2.37		3.19	4.83	
540	0010	**MACHINERY ANCHORS**, heavy duty, incl. sleeve, floating base nut,										540
	0020	lower stud & coupling nut, fiber plug, connecting stud, washer & nut.										
	0030	For flush mounted embedment in poured concrete heavy equip. pads.										
	0200	1/2" diameter stud & bolt	E-16	40	.400	Ea.	58	16.40	2.23	76.63	97	
	0300	5/8" diameter		35	.457		64	18.70	2.55	85.25	108	
	0500	3/4" diameter		30	.533		74	22	2.98	98.98	126	
	0600	7/8" diameter		25	.640		81	26	3.57	110.57	142	
	0800	1" diameter		20	.800		85	33	4.46	122.46	160	
	0900	1-1/4" diameter		15	1.067		113	43.50	5.95	162.45	212	

05090 | Metal Fastenings

		05090 Metal Fastenings	CREW	DAILY OUTPUT	LABOR-HOURS	UNIT	2006 BARE COSTS MAT.	LABOR	EQUIP.	TOTAL	TOTAL INCL O&P
580	0010	POWDER ACTUATED TOOLS & FASTENERS									
	0020	Stud driver, .22 caliber, buy, minimum				Ea.	340			340	370
	0100	Maximum				"	545			545	600
	0300	Powder charges for above, low velocity				C	17.40			17.40	19.15
	0400	Standard velocity					25			25	27.50
	0600	Drive pins & studs, 1/4" & 3/8" diam., to 3" long, minimum	1 Carp	4.80	1.667		13.05	59.50		72.55	112
	0700	Maximum	"	4	2		51	71		122	174
	0800	Pneumatic stud driver for 1/8" diameter studs				Ea.	2,350			2,350	2,600
	0900	Drive pins for above, 1/2" to 3/4" long	1 Carp	1	8	M	540	284		824	1,075
860	0010	WELD STUDS									
	0020	1/4" diameter, 2-11/16" long	E-10	1120	.014	Ea.	.27	.59	.24	1.10	1.64
	0100	4-1/8" long		1080	.015		.25	.61	.25	1.11	1.68
	0200	3/8" diameter, 4-1/8" long		1080	.015		.29	.61	.25	1.15	1.72
	0300	6-1/8" long		1040	.015		.38	.63	.26	1.27	1.87
	9000	Minimum labor/equipment charge	1 Sswk	2	4	Job		160		160	299
900	0010	WELDING STRUCTURAL (R050521-20)									
	0020	Field welding, 1/8" E6011, cost per welder, no oper. engr	E-14	8	1	Hr.	3.80	42	11.15	56.95	95
	0200	With 1/2 operating engineer	E-13	8	1.500		3.80	59.50	11.15	74.45	122
	0300	With 1 operating engineer	E-12	8	2		3.80	77	11.15	91.95	150
	0500	With no operating engineer, 2# weld rod per ton	E-14	8	1	Ton	3.80	42	11.15	56.95	95
	0600	8# E6011 per ton	"	2	4		15.20	168	44.50	227.70	380
	0800	With one operating engineer per welder, 2# E6011 per ton	E-12	8	2		3.80	77	11.15	91.95	150
	0900	8# E6011 per ton	"	2	8		15.20	310	44.50	369.70	600
	1200	Continuous fillet, stick welding, incl. equipment									
	1300	Single pass, 1/8" thick, 0.1#/L.F.	E-14	150	.053	L.F.	.19	2.24	.59	3.02	5.05
	1400	3/16" thick, 0.2#/L.F.		75	.107		.38	4.47	1.19	6.04	10.10
	1500	1/4" thick, 0.3#/L.F.		50	.160		.57	6.70	1.78	9.05	15.15
	1610	5/16" thick, 0.4#/L.F.		38	.211		.76	8.85	2.35	11.96	19.90
	1800	3 passes, 3/8" thick, 0.5#/L.F.		30	.267		.95	11.20	2.97	15.12	25.50
	2010	4 passes, 1/2" thick, 0.7#/L.F.		22	.364		1.33	15.25	4.05	20.63	34.50
	2200	5 to 6 passes, 3/4" thick, 1.3#/L.F.		12	.667		2.47	28	7.45	37.92	63.50
	2400	8 to 11 passes, 1" thick, 2.4#/L.F.		6	1.333		4.56	56	14.85	75.41	126
	2600	For all position welding, add, minimum						20%			
	2700	Maximum						300%			
	2900	For semi-automatic welding, deduct, minimum						5%			
	3000	Maximum						15%			
	4000	Cleaning and welding plates, bars, or rods									
	4010	to existing beams, columns, or trusses	E-14	12	.667	L.F.	.95	28	7.45	36.40	61.50
	9000	Minimum labor/equipment charge	"	4	2	Job		84	22.50	106.50	182
920	0010	STEEL CUTTING									
	0020	Hand burning, incl. preparation, torch cutting & grinding, no staging									
	0100	Steel to 1/2" thick	E-25	320	.025	L.F.		1.05	.25	1.30	2.24
	0150	3/4" thick		260	.031			1.29	.31	1.60	2.75
	0200	1" thick		200	.040			1.68	.41	2.09	3.59
	9000	Minimum labor/equipment charge		2	4	Job		168	40.50	208.50	360

		05120 | Structural Steel	CREW	DAILY OUTPUT	LABOR-HOURS	UNIT	2006 BARE COSTS MAT.	LABOR	EQUIP.	TOTAL	TOTAL INCL O&P	
180	0010	CANOPY FRAMING										180
	0020	6″ and 8″ members	E-4	3000	.011	Lb.	1.14	.43	.03	1.60	2.09	
	9000	Minimum labor/equipment charge	1 Sswk	1	8	Job		320		320	600	
220	0010	CEILING SUPPORTS										220
	1000	Entrance door/folding partition supports	E-4	60	.533	L.F.	19	21.50	1.49	41.99	63	
	1100	Linear accelerator door supports		14	2.286		86.50	92.50	6.40	185.40	275	
	1200	Lintels or shelf angles, hung, exterior hot dipped galv.		267	.120		12.95	4.85	.33	18.13	23.50	
	1250	Two coats primer paint instead of galv.		267	.120		11.25	4.85	.33	16.43	22	
	1400	Monitor support, ceiling hung, expansion bolted		4	8	Ea.	300	325	22.50	647.50	960	
	1450	Hung from pre-set inserts		6	5.333		325	216	14.90	555.90	775	
	1600	Motor supports for overhead doors		4	8		153	325	22.50	500.50	800	
	1700	Partition support for heavy folding partitions, without pocket		24	1.333	L.F.	43	54	3.72	100.72	153	
	1750	Supports at pocket only		12	2.667		86.50	108	7.45	201.95	305	
	2000	Rolling grilles & fire door supports		34	.941		37	38	2.63	77.63	115	
	2100	Spider-leg light supports, expansion bolted to ceiling slab		8	4	Ea.	124	162	11.15	297.15	455	
	2150	Hung from pre-set inserts		12	2.667	″	133	108	7.45	248.45	355	
	2400	Toilet partition support		36	.889	L.F.	43	36	2.48	81.48	117	
	2500	X-ray travel gantry support		12	2.667	″	148	108	7.45	263.45	375	
260	0010	COLUMNS, STRUCTURAL R051223-10										260
	0020	Shop fab'd for 100-ton, 1-2 story project, bolted conn's.										
	0800	Steel, concrete filled, extra strong pipe, 3-1/2″ diameter	E-2	660	.085	L.F.	31.50	3.30	2.16	36.96	43	
	0830	4″ diameter		780	.072		35	2.79	1.83	39.62	45.50	
	0890	5″ diameter		1020	.055		41.50	2.14	1.40	45.04	51.50	
	0930	6″ diameter		1200	.047		55	1.82	1.19	58.01	65	
	0940	8″ diameter		1100	.051		55	1.98	1.30	58.28	65.50	
	1500	Steel pipe, extra strong, no concrete, 3″ to 5″ diameter		16000	.004	Lb.	.95	.14	.09	1.18	1.39	
	3300	Structural tubing, square, A500GrB, 4″ to 6″ square, light section		11270	.005	″	.95	.19	.13	1.27	1.54	
	8090	For projects 75 to 99 tons, add				All	10%					
	8092	50 to 74 tons, add					20%					
	8094	25 to 49 tons, add					30%	10%				
	8096	10 to 24 tons, add					50%	25%				
	8098	2 to 9 tons, add					75%	50%				
	8099	Less than 2 tons, add					100%	100%				
	9000	Minimum labor/equipment charge	1 Sswk	1	8	Job		320		320	600	
300	0010	CURB EDGING										300
	0020	Steel angle w/anchors, on forms, 1″ x 1″, 0.8#/L.F.	E-4	350	.091	L.F.	1.61	3.70	.26	5.57	8.95	
	0300	4″ x 4″ angles, 8.2#/L.F.		275	.116	″	10.55	4.71	.32	15.58	21	
	9000	Minimum labor/equipment charge		4	8	Job		325	22.50	347.50	630	
440	0010	LIGHTWEIGHT FRAMING R051223-35										440
	0200	For load-bearing steel studs see Division 05410-400										
	0400	Angle framing, field fabricated, 4″ and larger R051223-45	E-3	440	.055	Lb.	.55	2.22	.20	2.97	4.97	
	0450	Less than 4″ angles		265	.091		.57	3.68	.34	4.59	7.90	
	0600	Channel framing, field fabricated, 8″ and larger		500	.048		.57	1.95	.18	2.70	4.48	
	0650	Less than 8″ channels		335	.072		.57	2.91	.27	3.75	6.35	
	1000	Continuous slotted channel framing system, shop fab, min	2 Sswk	2400	.007		2.95	.27		3.22	3.74	
	1200	Maximum	″	1600	.010		3.33	.40		3.73	4.41	
	1250	Plate & bar stock for reinforcing beams and trusses					1.05			1.05	1.15	
	1300	Cross bracing, rods, shop fabricated, 3/4″ diameter	E-3	700	.034		1.14	1.39	.13	2.66	3.99	
	1310	7/8″ diameter		850	.028		1.14	1.15	.11	2.40	3.51	
	1320	1″ diameter		1000	.024		1.14	.97	.09	2.20	3.17	
	1330	Angle, 5″ x 5″ x 3/8″		2800	.009		1.14	.35	.03	1.52	1.94	
	1350	Hanging lintels, shop fabricated, average		850	.028		1.14	1.15	.11	2.40	3.51	
	1380	Roof frames, shop fabricated, 3′-0″ square, 5′ span	E-2	4200	.013		1.14	.52	.34	2	2.55	
	1400	Tie rod, not upset, 1-1/2″ to 4″ diameter, with turnbuckle	2 Sswk	800	.020		1.24	.80		2.04	2.85	

05100 | Structural Metal Framing

	05120 | Structural Steel		CREW	DAILY OUTPUT	LABOR-HOURS	UNIT	2006 BARE COSTS MAT.	LABOR	EQUIP.	TOTAL	TOTAL INCL O&P	
440	1420	No turnbuckle (R051223-35)	2 Sswk	700	.023	Lb.	1.19	.91		2.10	3.02	440
	1500	Upset, 1-3/4" to 4" diameter, with turnbuckle		800	.020		1.24	.80		2.04	2.85	
	1520	No turnbuckle (R051223-45)		700	.023	↓	1.19	.91		2.10	3.02	
	9000	Minimum labor/equipment charge	↓	2	8	Job		320		320	600	
480	0010	**LINTELS**										480
	0020	Plain steel angles, under 500 lb.	1 Bric	550	.015	Lb.	.73	.53		1.26	1.66	
	0100	500 to 1000 lb.		640	.013		.71	.46		1.17	1.52	
	0200	1,000 to 2,000 lb.		640	.013		.69	.46		1.15	1.50	
	0300	2,000 to 4,000 lb.	↓	640	.013		.67	.46		1.13	1.48	
	0500	For built-up angles and plates, add to above					.24			.24	.26	
	0700	For engineering, add to above					.10			.10	.10	
	0900	For galvanizing, add to above, under 500 lb.					.30			.30	.33	
	0950	500 to 2,000 lb.					.27			.27	.30	
	1000	Over 2,000 lb.				↓	.22			.22	.25	
	2000	Steel angles, 3-1/2" x 3", 1/4" thick, 2'-6" long	1 Bric	47	.170	Ea.	10.25	6.20		16.45	21.50	
	2100	4'-6" long		26	.308		18.45	11.25		29.70	39	
	2500	3-1/2" x 3-1/2" x 5/16", 5'-0" long		18	.444		27.50	16.25		43.75	56.50	
	2600	4" x 3-1/2", 1/4" thick, 5'-0" long		21	.381		23.50	13.90		37.40	48.50	
	2700	9'-0" long		12	.667		42.50	24.50		67	86	
	2800	4" x 3-1/2" x 5/16", 7'-0" long		12	.667		41	24.50		65.50	84.50	
	2900	5" x 3-1/2" x 5/16", 10'-0" long	↓	8	1	↓	66	36.50		102.50	132	
	3500	For precast concrete lintels, see Div. 03480-400										
	9000	Minimum labor/equipment charge	1 Bric	4	2	Job		73		73	119	
520	0010	**PIPE SUPPORT FRAMING**										520
	0020	Under 10#/L.F.	E-4	3900	.008	Lb.	1.27	.33	.02	1.62	2.05	
	0200	10.1 to 15#/L.F.		4300	.007		1.25	.30	.02	1.57	1.96	
	0400	15.1 to 20#/L.F.		4800	.007		1.24	.27	.02	1.53	1.88	
	0600	Over 20#/L.F.	↓	5400	.006	↓	1.22	.24	.02	1.48	1.81	
560	0010	**PLATES** (R051223-80)										560
	0020	For connections & stiffener plates, shop fabricated										
	0050	1/8" thick (5.1 Lb./S.F.)				S.F.	4.85			4.85	5.35	
	0100	1/4" thick (10.2 Lb./S.F.)					9.70			9.70	10.65	
	0300	3/8" thick (15.3 Lb./S.F.)					14.55			14.55	16	
	0400	1/2" thick (20.4 Lb./S.F.)					19.40			19.40	21.50	
	0450	3/4" thick (30.6 Lb./S.F.)					29			29	32	
	0500	1" thick (40.8 Lb.S.F.)				↓	39			39	42.50	
680	0010	**STRUCTURAL STEEL PROJECTS** (R050515-30)										680
	0020	Shop fab'd for 100-ton, 1-2 story project, bolted conn's.										
	0700	Offices, hospitals, etc., steel bearing, 1 to 2 stories (R050521-20)	E-5	10.30	7.767	Ton	1,900	305	147	2,352	2,825	
	1300	Industrial bldgs., 1 story, beams & girders, steel bearing		12.90	6.202		1,900	244	118	2,262	2,675	
	1400	Masonry bearing (R051223-10)		10	8		1,900	315	152	2,367	2,850	
	1600	1 story with roof trusses, steel bearing		10.60	7.547		2,250	298	143	2,691	3,175	
	1700	Masonry bearing (R051223-15)	↓	8.30	9.639		2,250	380	183	2,813	3,375	
	5390	For projects 75 to 99 tons, add					10%					
	5392	50 to 74 tons, add (R051223-20)					20%					
	5394	25 to 49 tons, add					30%	10%				
	5396	10 to 24 tons, add (R051223-25)					50%	25%				
	5398	2 to 9 tons, add					75%	50%				
	5399	Less than 2 tons, add (R051223-30)				↓	100%	100%				
720	0010	**STRUCTURAL STEEL** (R050516-30)										720
	0020	Shop fab'd for 100-ton, 1-2 story project, bolted conn's.										
	0050	Beams, W 6 x 9 (R050521-20)	E-2	720	.078	L.F.	10.25	3.03	1.98	15.26	18.95	
	0100	W 8 x 10	↓	720	.078	↓	11.40	3.03	1.98	16.41	20	

05100 | Structural Metal Framing

		05120 | Structural Steel		CREW	DAILY OUTPUT	LABOR-HOURS	UNIT	2006 BARE COSTS MAT.	LABOR	EQUIP.	TOTAL	TOTAL INCL O&P
720	0150	W 10 x 15	R051223-10	E-2	720	.078	L.F.	17.10	3.03	1.98	22.11	26.50
	0200	Columns, W 6 x 15			540	.104		18.55	4.03	2.64	25.22	30.50
	0250	W 8 x 31	R051223-15		540	.104		38.50	4.03	2.64	45.17	52
	0500	Girders, W 12 x 22			900	.062		25	2.42	1.59	29.01	33.50
	0550	W 14 x 26	R051223-20		900	.062		29.50	2.42	1.59	33.51	38.50
	0600	W 16 x 31			900	.062		35.50	2.42	1.59	39.51	45
	0700	Joists (bar joists, H or K series), span to 30′	R051223-25	E-7	30000	.003	Lb.	.73	.11	.05	.89	1.05
	0750	Span to 50′		"	20000	.004	"	.72	.16	.08	.96	1.17
	7990	For projects 75 to 99 tons, add	R051223-30				All	10%				
	7992	50 to 75 tons, add						20%				
	7994	25 to 49 tons, add						30%	10%			
	7996	10 to 24 tons, add						50%	25%			
	7998	2 to 9 tons, add						75%	50%			
	7999	Less than 2 tons, add						100%	100%			
	9000	Minimum labor/equipment charge		E-2	2	28	Job		1,100	715	1,815	2,725

5 METALS

05200 | Metal Joists

		05210 | Steel Joists	CREW	DAILY OUTPUT	LABOR-HOURS	UNIT	2006 BARE COSTS MAT.	LABOR	EQUIP.	TOTAL	TOTAL INCL O&P
600	0010	**OPEN WEB JOISTS**									
	0020	K series, 40-ton job lots, horiz. bridging, spans to 30′, minimum	E-7	15	5.333	Ton	1,175	210	107	1,492	1,800
	0050	Average		12	6.667		1,325	263	134	1,722	2,075
	0080	Maximum		9	8.889		1,575	350	178	2,103	2,575
	0410	Span 30′ to 50′, minimum		17	4.706		1,150	186	94.50	1,430.50	1,725
	0440	Average		17	4.706		1,300	186	94.50	1,580.50	1,875
	0460	Maximum		10	8		1,375	315	161	1,851	2,275
	1010	CS series, horizontal bridging									
	1020	Spans to 30′, minimum	E-7	15	5.333	Ton	1,225	210	107	1,542	1,850
	1040	Average		12	6.667		1,350	263	134	1,747	2,125
	1060	Maximum		9	8.889		1,600	350	178	2,128	2,575
	2000	LH series, bolted cross bridging									
	2020	Spans to 96′, minimum	E-7	16	5	Ton	1,325	197	100	1,622	1,950
	2040	Average		13	6.154		1,450	243	124	1,817	2,175
	2080	Maximum		11	7.273		1,725	287	146	2,158	2,575
	3010	DLH series, bolted cross bridging									
	3020	Spans to 144′ (shipped in 2 pieces), minimum	E-7	16	5	Ton	1,450	197	100	1,747	2,075
	3040	Average		13	6.154		1,550	243	124	1,917	2,275
	3100	Maximum		11	7.273		1,875	287	146	2,308	2,750
	4010	SLH series, bolted cross bridging									
	4020	Spans to 200′ (shipped in 3 pieces), minimum	E-7	16	5	Ton	1,425	197	100	1,722	2,050
	4040	Average		13	6.154		1,600	243	124	1,967	2,325
	4060	Maximum		11	7.273		1,900	287	146	2,333	2,775
	6100	For less than 40-ton job lots									
	6102	For 30 to 39 tons, add					10%				
	6104	20 to 29 tons, add					20%				
	6106	10 to 19 tons, add					30%				
	6107	5 to 9 tons, add					50%	25%			
	6108	1 to 4 tons, add					75%	50%			
	6109	Less than 1 ton, add					100%	100%			

05300 | Metal Deck

05310 | Steel Deck

		05310 Steel Deck	CREW	DAILY OUTPUT	LABOR-HOURS	UNIT	2006 BARE COSTS MAT.	LABOR	EQUIP.	TOTAL	TOTAL INCL O&P	
300	0010	**METAL DECKING** Steel decking R053100-10										300
	0200	Cellular units, galv, 2" deep, 20-20 gauge, over 15 squares	E-4	1460	.022	S.F.	5.85	.89	.06	6.80	8.20	
	0400	3" deep, galvanized, 20-20 gauge		1375	.023		6.45	.94	.06	7.45	8.95	
	1000	4-1/2" deep, galvanized, 20-18 gauge		1100	.029		9	1.18	.08	10.26	12.20	
	1500	For acoustical deck, add					15%					
	1900	For multi-story or congested site, add						50%				
	2100	Open type, galv., 1-1/2" deep wide rib, 22 gauge, under 50 squares	E-4	4500	.007	S.F.	1.47	.29	.02	1.78	2.17	
	2400	Over 500 squares		5100	.006		1.06	.25	.02	1.33	1.65	
	5200	Non-cellular composite deck, galv., 2" deep, 22 gauge		3860	.008		1.40	.34	.02	1.76	2.20	
	5300	20 gauge		3600	.009		1.55	.36	.02	1.93	2.41	
	5400	18 gauge		3380	.009		1.97	.38	.03	2.38	2.92	
	5500	16 gauge		3200	.010		2.46	.40	.03	2.89	3.50	
	5700	3" deep, galv., 22 gauge		3200	.010		1.53	.40	.03	1.96	2.47	
	5800	20 gauge		3000	.011		1.71	.43	.03	2.17	2.72	
	5900	18 gauge		2850	.011		2.10	.45	.03	2.58	3.19	
	6000	16 gauge		2700	.012		2.80	.48	.03	3.31	4.02	
	6100	Slab form, steel, 28 gauge, 9/16" deep, uncoated		4000	.008		.98	.32	.02	1.32	1.71	
	6200	Galvanized		4000	.008		.87	.32	.02	1.21	1.59	
	9000	Minimum labor/equipment charge	1 Sswk	1	8	Job		320		320	600	

05400 | Cold-Formed Metal Framing

05410 | Load-Bearing Metal Studs

		05410 Load-Bearing Metal Studs	CREW	DAILY OUTPUT	LABOR-HOURS	UNIT	2006 BARE COSTS MAT.	LABOR	EQUIP.	TOTAL	TOTAL INCL O&P	
100	0010	**BRACING**, shear wall X-bracing, per 10' x 10' bay, one face										100
	0120	Metal strap, 20 ga x 4" wide	2 Carp	18	.889	Ea.	18	31.50		49.50	72.50	
	0130	6" wide		18	.889		28	31.50		59.50	83	
	0160	18 ga x 4" wide		16	1		26	35.50		61.50	87.50	
	0170	6" wide		16	1		38.50	35.50		74	101	
	0410	Continuous strap bracing, per horizontal row on both faces										
	0420	Metal strap, 20 ga x 2" wide, studs 12" O.C.	1 Carp	7	1.143	C.L.F.	46	40.50		86.50	119	
	0430	16" O.C.		8	1		46	35.50		81.50	110	
	0440	24" O.C.		10	.800		46	28.50		74.50	98	
	0450	18 ga x 2" wide, studs 12" O.C.		6	1.333		64	47.50		111.50	149	
	0460	16" O.C.		7	1.143		64	40.50		104.50	138	
	0470	24" O.C.		8	1		64	35.50		99.50	129	
120	0010	**BRIDGING**, solid between studs w/ 1-1/4" leg track, per stud bay										120
	0200	Studs 12" O.C., 18 ga x 2-1/2" wide	1 Carp	125	.064	Ea.	.77	2.28		3.05	4.62	
	0210	3-5/8" wide		120	.067		.93	2.37		3.30	4.95	
	0220	4" wide		120	.067		.99	2.37		3.36	5	
	0230	6" wide		115	.070		1.29	2.47		3.76	5.50	
	0240	8" wide		110	.073		1.63	2.59		4.22	6.10	
	0300	16 ga x 2-1/2" wide		115	.070		.97	2.47		3.44	5.15	
	0310	3-5/8" wide		110	.073		1.18	2.59		3.77	5.60	
	0320	4" wide		110	.073		1.26	2.59		3.85	5.65	
	0330	6" wide		105	.076		1.61	2.71		4.32	6.25	
	0340	8" wide		100	.080		2.06	2.84		4.90	7	
	1200	Studs 16" O.C., 18 ga x 2-1/2" wide		125	.064		.99	2.28		3.27	4.86	
	1210	3-5/8" wide		120	.067		1.19	2.37		3.56	5.25	
	1220	4" wide		120	.067		1.27	2.37		3.64	5.35	

	05410	Load-Bearing Metal Studs	CREW	DAILY OUTPUT	LABOR-HOURS	UNIT	2006 BARE COSTS MAT.	LABOR	EQUIP.	TOTAL	TOTAL INCL O&P	
120	1230	6" wide	1 Carp	115	.070	Ea.	1.65	2.47		4.12	5.90	120
	1240	8" wide		110	.073		2.10	2.59		4.69	6.60	
	1300	16 ga x 2-1/2" wide		115	.070		1.24	2.47		3.71	5.45	
	1310	3-5/8" wide		110	.073		1.52	2.59		4.11	5.95	
	1320	4" wide		110	.073		1.62	2.59		4.21	6.05	
	1330	6" wide		105	.076		2.06	2.71		4.77	6.75	
	1340	8" wide		100	.080		2.64	2.84		5.48	7.60	
	2200	Studs 24" O.C., 18 ga x 2-1/2" wide		125	.064		1.43	2.28		3.71	5.35	
	2210	3-5/8" wide		120	.067		1.72	2.37		4.09	5.80	
	2220	4" wide		120	.067		1.84	2.37		4.21	5.95	
	2230	6" wide		115	.070		2.39	2.47		4.86	6.75	
	2240	8" wide		110	.073		3.03	2.59		5.62	7.60	
	2300	16 ga x 2-1/2" wide		115	.070		1.79	2.47		4.26	6.05	
	2310	3-5/8" wide		110	.073		2.20	2.59		4.79	6.70	
	2320	4" wide		110	.073		2.34	2.59		4.93	6.85	
	2330	6" wide		105	.076		2.98	2.71		5.69	7.75	
	2340	8" wide		100	.080		3.82	2.84		6.66	8.90	
	3000	Continuous bridging, per row										
	3100	16 ga x 1-1/2" channel thru studs 12" O.C.	1 Carp	6	1.333	C.L.F.	43	47.50		90.50	126	
	3110	16" O.C.		7	1.143		43	40.50		83.50	115	
	3120	24" O.C.		8.80	.909		43	32.50		75.50	101	
	4100	2" x 2" angle x 18 ga, studs 12" O.C.		7	1.143		62.50	40.50		103	137	
	4110	16" O.C.		9	.889		62.50	31.50		94	122	
	4120	24" O.C.		12	.667		62.50	23.50		86	109	
	4200	16 ga, studs 12" O.C.		5	1.600		80.50	57		137.50	183	
	4210	16" O.C.		7	1.143		80.50	40.50		121	156	
	4220	24" O.C.		10	.800		80.50	28.50		109	136	
300	0010	**FRAMING, BOXED HEADERS/BEAMS**										300
	0200	Double, 18 ga x 6" deep	2 Carp	220	.073	L.F.	4.49	2.59		7.08	9.20	
	0210	8" deep		210	.076		4.99	2.71		7.70	10	
	0220	10" deep		200	.080		6.05	2.84		8.89	11.35	
	0230	12 " deep		190	.084		6.65	2.99		9.64	12.25	
	0300	16 ga x 8" deep		180	.089		5.75	3.16		8.91	11.55	
	0310	10" deep		170	.094		6.85	3.35		10.20	13.10	
	0320	12 " deep		160	.100		7.50	3.56		11.06	14.15	
	0400	14 ga x 10" deep		140	.114		8	4.06		12.06	15.55	
	0410	12 " deep		130	.123		8.75	4.38		13.13	16.90	
	1210	Triple, 18 ga x 8" deep		170	.094		7.20	3.35		10.55	13.50	
	1220	10" deep		165	.097		8.65	3.45		12.10	15.20	
	1230	12 " deep		160	.100		9.55	3.56		13.11	16.40	
	1300	16 ga x 8" deep		145	.110		8.30	3.92		12.22	15.65	
	1310	10" deep		140	.114		9.90	4.06		13.96	17.65	
	1320	12 " deep		135	.119		10.85	4.21		15.06	18.90	
	1400	14 ga x 10" deep		115	.139		10.90	4.95		15.85	20	
	1410	12 " deep		110	.145		12.10	5.15		17.25	22	
400	0010	**FRAMING, STUD WALLS** w/ top & bottom track, no openings,										400
	0020	headers, beams, bridging or bracing										
	4100	8' high walls, 18 ga x 2-1/2" wide, studs 12" O.C.	2 Carp	54	.296	L.F.	7.55	10.55		18.10	26	
	4110	16" O.C.		77	.208		6.05	7.40		13.45	18.90	
	4120	24" O.C.		107	.150		4.51	5.30		9.81	13.75	
	4130	3-5/8" wide, studs 12" O.C.		53	.302		8.95	10.75		19.70	27.50	
	4140	16" O.C.		76	.211		7.15	7.50		14.65	20.50	
	4150	24" O.C.		105	.152		5.35	5.40		10.75	14.90	
	4160	4" wide, studs 12" O.C.		52	.308		9.40	10.95		20.35	28.50	
	4170	16" O.C.		74	.216		7.55	7.70		15.25	21	

05410 | Load-Bearing Metal Studs

		Description	CREW	DAILY OUTPUT	LABOR-HOURS	UNIT	2006 BARE COSTS MAT.	LABOR	EQUIP.	TOTAL	TOTAL INCL O&P	
400	4180	24" O.C.	2 Carp	103	.155	L.F.	5.65	5.50		11.15	15.35	400
	4190	6" wide, studs 12" O.C.		51	.314		11.90	11.15		23.05	31.50	
	4200	16" O.C.		73	.219		9.55	7.80		17.35	23.50	
	4210	24" O.C.		101	.158		7.15	5.65		12.80	17.25	
	4220	8" wide, studs 12" O.C.		50	.320		14.55	11.40		25.95	35	
	4230	16" O.C.		72	.222		11.70	7.90		19.60	26	
	4240	24" O.C.		100	.160		8.85	5.70		14.55	19.10	
	4300	16 ga x 2-1/2" wide, studs 12" O.C.		47	.340		8.85	12.10		20.95	29.50	
	4310	16" O.C.		68	.235		7	8.35		15.35	21.50	
	4320	24" O.C.		94	.170		5.15	6.05		11.20	15.70	
	4330	3-5/8" wide, studs 12" O.C.		46	.348		10.55	12.35		22.90	32	
	4340	16" O.C.		66	.242		8.35	8.60		16.95	23.50	
	4350	24" O.C.		92	.174		6.15	6.20		12.35	17.05	
	4360	4" wide, studs 12" O.C.		45	.356		11.15	12.65		23.80	33.50	
	4370	16" O.C.		65	.246		8.85	8.75		17.60	24.50	
	4380	24" O.C.		90	.178		6.55	6.30		12.85	17.65	
	4390	6" wide, studs 12" O.C.		44	.364		14.05	12.95		27	37	
	4400	16" O.C.		64	.250		11.15	8.90		20.05	27	
	4410	24" O.C.		88	.182		8.25	6.45		14.70	19.80	
	4420	8" wide, studs 12" O.C.		43	.372		17.25	13.25		30.50	41	
	4430	16" O.C.		63	.254		13.75	9.05		22.80	30	
	4440	24" O.C.		86	.186		10.20	6.60		16.80	22	
	5100	10' high walls, 18 ga x 2-1/2" wide, studs 12" O.C.		54	.296		9.05	10.55		19.60	27.50	
	5110	16" O.C.		77	.208		7.15	7.40		14.55	20	
	5120	24" O.C.		107	.150		5.25	5.30		10.55	14.60	
	5130	3-5/8" wide, studs 12" O.C.		53	.302		10.75	10.75		21.50	29.50	
	5140	16" O.C.		76	.211		8.50	7.50		16	22	
	5150	24" O.C.		105	.152		6.25	5.40		11.65	15.90	
	5160	4" wide, studs 12" O.C.		52	.308		11.30	10.95		22.25	30.50	
	5170	16" O.C.		74	.216		8.95	7.70		16.65	22.50	
	5180	24" O.C.		103	.155		6.60	5.50		12.10	16.40	
	5190	6" wide, studs 12" O.C.		51	.314		14.25	11.15		25.40	34	
	5200	16" O.C.		73	.219		11.30	7.80		19.10	25.50	
	5210	24" O.C.		101	.158		8.35	5.65		14	18.55	
	5220	8" wide, studs 12" O.C.		50	.320		17.40	11.40		28.80	38	
	5230	16" O.C.		72	.222		13.85	7.90		21.75	28.50	
	5240	24" O.C.		100	.160		10.25	5.70		15.95	20.50	
	5300	16 ga x 2-1/2" wide, studs 12" O.C.		47	.340		10.65	12.10		22.75	32	
	5310	16" O.C.		68	.235		8.35	8.35		16.70	23	
	5320	24" O.C.		94	.170		6.05	6.05		12.10	16.75	
	5330	3-5/8" wide, studs 12" O.C.		46	.348		12.75	12.35		25.10	34.50	
	5340	16" O.C.		66	.242		10	8.60		18.60	25.50	
	5350	24" O.C.		92	.174		7.25	6.20		13.45	18.25	
	5360	4" wide, studs 12" O.C.		45	.356		13.50	12.65		26.15	36	
	5370	16" O.C.		65	.246		10.60	8.75		19.35	26	
	5380	24" O.C.		90	.178		7.70	6.30		14	18.90	
	5390	6" wide, studs 12" O.C.		44	.364		16.95	12.95		29.90	40	
	5400	16" O.C.		64	.250		13.35	8.90		22.25	29.50	
	5410	24" O.C.		88	.182		9.70	6.45		16.15	21.50	
	5420	8" wide, studs 12" O.C.		43	.372		21	13.25		34.25	45	
	5430	16" O.C.		63	.254		16.40	9.05		25.45	33	
	5440	24" O.C.		86	.186		11.95	6.60		18.55	24	
	6190	12' high walls, 18 ga x 6" wide, studs 12" O.C.		41	.390		16.60	13.85		30.45	41.50	
	6200	16" O.C.		58	.276		13.05	9.80		22.85	30.50	
	6210	24" O.C.		81	.198		9.55	7		16.55	22	
	6220	8" wide, studs 12" O.C.		40	.400		20.50	14.20		34.70	46	

05400 | Cold-Formed Metal Framing

05410 | Load-Bearing Metal Studs

			CREW	DAILY OUTPUT	LABOR-HOURS	UNIT	2006 BARE COSTS MAT.	LABOR	EQUIP.	TOTAL	TOTAL INCL O&P	
400	6230	16″ O.C.	2 Carp	57	.281	L.F.	16	10		26	34	400
	6240	24″ O.C.		80	.200		11.70	7.10		18.80	24.50	
	6390	16 ga x 6″ wide, studs 12″ O.C.		35	.457		19.85	16.25		36.10	49	
	6400	16″ O.C.		51	.314		15.50	11.15		26.65	35.50	
	6410	24″ O.C.		70	.229		11.15	8.15		19.30	25.50	
	6420	8″ wide, studs 12″ O.C.		34	.471		24.50	16.75		41.25	54.50	
	6430	16″ O.C.		50	.320		19.05	11.40		30.45	40	
	6440	24″ O.C.		69	.232		13.75	8.25		22	29	
	6530	14 ga x 3-5/8″ wide, studs 12″ O.C.		34	.471		18.80	16.75		35.55	48	
	6540	16″ O.C.		48	.333		14.65	11.85		26.50	36	
	6550	24″ O.C.		65	.246		10.55	8.75		19.30	26	
	6560	4″ wide, studs 12″ O.C.		33	.485		19.90	17.25		37.15	50.50	
	6570	16″ O.C.		47	.340		15.55	12.10		27.65	37	
	6580	24″ O.C.		64	.250		11.15	8.90		20.05	27	
	6730	12 ga x 3-5/8″ wide, studs 12″ O.C.		31	.516		26	18.35		44.35	59	
	6740	16″ O.C.		43	.372		20	13.25		33.25	44	
	6750	24″ O.C.		59	.271		14.15	9.65		23.80	31.50	
	6760	4″ wide, studs 12″ O.C.		30	.533		28	18.95		46.95	62.50	
	6770	16″ O.C.		42	.381		21.50	13.55		35.05	46	
	6780	24″ O.C.		58	.276		15.20	9.80		25	33	
	7390	16′ high walls, 16 ga x 6″ wide, studs 12″ O.C.		33	.485		25.50	17.25		42.75	56.50	
	7400	16″ O.C.		48	.333		19.85	11.85		31.70	41.50	
	7410	24″ O.C.		67	.239		14.05	8.50		22.55	29.50	
	7420	8″ wide, studs 12″ O.C.		32	.500		31.50	17.80		49.30	64	
	7430	16″ O.C.		47	.340		24.50	12.10		36.60	47	
	7440	24″ O.C.		66	.242		17.25	8.60		25.85	33.50	
	7560	14 ga x 4″ wide, studs 12″ O.C.		31	.516		26	18.35		44.35	59	
	7570	16″ O.C.		45	.356		19.90	12.65		32.55	43	
	7580	24″ O.C.		61	.262		14.10	9.30		23.40	31	
	7590	6″ wide, studs 12″ O.C.		30	.533		32.50	18.95		51.45	67	
	7600	16″ O.C.		44	.364		25	12.95		37.95	49	
	7610	24″ O.C.		60	.267		17.70	9.50		27.20	35	
	7760	12 ga x 4″ wide, studs 12″ O.C.		29	.552		36.50	19.60		56.10	72.50	
	7770	16″ O.C.		40	.400		28	14.20		42.20	54.50	
	7780	24″ O.C.		55	.291		19.45	10.35		29.80	38.50	
	7790	6″ wide, studs 12″ O.C.		28	.571		46	20.50		66.50	84	
	7800	16″ O.C.		39	.410		35	14.60		49.60	62.50	
	7810	24″ O.C.		54	.296		24.50	10.55		35.05	44.50	
	8590	20′ high walls, 14 ga x 6″ wide, studs 12″ O.C.		29	.552		39.50	19.60		59.10	76	
	8600	16″ O.C.		42	.381		30.50	13.55		44.05	56	
	8610	24″ O.C.		57	.281		21.50	10		31.50	40	
	8620	8″ wide, studs 12″ O.C.		28	.571		48.50	20.50		69	87	
	8630	16″ O.C.		41	.390		37.50	13.85		51.35	64	
	8640	24″ O.C.		56	.286		26	10.15		36.15	46	
	8790	12 ga x 6″ wide, studs 12″ O.C.		27	.593		56.50	21		77.50	97.50	
	8800	16″ O.C.		37	.432		43.50	15.35		58.85	73	
	8810	24″ O.C.		51	.314		30	11.15		41.15	51.50	
	8820	8″ wide, studs 12″ O.C.		26	.615		69	22		91	113	
	8830	16″ O.C.		36	.444		53	15.80		68.80	84	
	8840	24″ O.C.		50	.320		36.50	11.40		47.90	59	
	9000	Minimum labor/equipment charge		4	4	Job		142		142	236	

05420 | Cold-Formed Metal Joists

			CREW	DAILY OUTPUT	LABOR-HOURS	UNIT	MAT.	LABOR	EQUIP.	TOTAL	TOTAL INCL O&P	
100	0010	**BRACING**, continuous, per row, top & bottom										100
	0120	Flat strap, 20 ga x 2″ wide, joists at 12″ O.C.	1 Carp	4.67	1.713	C.L.F.	48.50	61		109.50	154	

	05420 Cold-Formed Metal Joists		CREW	DAILY OUTPUT	LABOR-HOURS	UNIT	2006 BARE COSTS MAT.	LABOR	EQUIP.	TOTAL	TOTAL INCL O&P	
100	0130	16" O.C.	1 Carp	5.33	1.501	C.L.F.	46.50	53.50		100	140	100
	0140	24" O.C.		6.66	1.201		45	42.50		87.50	121	
	0150	18 ga x 2" wide, joists at 12" O.C.		4	2		63	71		134	188	
	0160	16" O.C.		4.67	1.713		62	61		123	170	
	0170	24" O.C.	↓	5.33	1.501	↓	61	53.50		114.50	156	
120	0010	**BRIDGING**, solid between joists w/ 1-1/4" leg track, per joist bay										120
	0230	Joists 12" O.C., 18 ga track x 6" wide	1 Carp	80	.100	Ea.	1.29	3.56		4.85	7.30	
	0240	8" wide		75	.107		1.63	3.79		5.42	8.10	
	0250	10" wide		70	.114		2.02	4.06		6.08	8.95	
	0260	12" wide		65	.123		2.33	4.38		6.71	9.80	
	0330	16 ga track x 6" wide		70	.114		1.61	4.06		5.67	8.50	
	0340	8" wide		65	.123		2.06	4.38		6.44	9.50	
	0350	10" wide		60	.133		2.54	4.74		7.28	10.65	
	0360	12" wide		55	.145		2.91	5.15		8.06	11.75	
	0440	14 ga track x 8" wide		60	.133		2.59	4.74		7.33	10.70	
	0450	10" wide		55	.145		3.18	5.15		8.33	12.05	
	0460	12" wide		50	.160		3.68	5.70		9.38	13.45	
	0550	12 ga track x 10" wide		45	.178		4.67	6.30		10.97	15.60	
	0560	12" wide		40	.200		5.30	7.10		12.40	17.60	
	1230	16" O.C., 18 ga track x 6" wide		80	.100		1.65	3.56		5.21	7.70	
	1240	8" wide		75	.107		2.10	3.79		5.89	8.60	
	1250	10" wide		70	.114		2.59	4.06		6.65	9.60	
	1260	12" wide		65	.123		2.99	4.38		7.37	10.55	
	1330	16 ga track x 6" wide		70	.114		2.06	4.06		6.12	9	
	1340	8" wide		65	.123		2.64	4.38		7.02	10.15	
	1350	10" wide		60	.133		3.25	4.74		7.99	11.45	
	1360	12" wide		55	.145		3.73	5.15		8.88	12.65	
	1440	14 ga track x 8" wide		60	.133		3.32	4.74		8.06	11.50	
	1450	10" wide		55	.145		4.08	5.15		9.23	13.05	
	1460	12" wide		50	.160		4.72	5.70		10.42	14.60	
	1550	12 ga track x 10" wide		45	.178		6	6.30		12.30	17.05	
	1560	12" wide		40	.200		6.80	7.10		13.90	19.25	
	2230	24" O.C., 18 ga track x 6" wide		80	.100		2.39	3.56		5.95	8.55	
	2240	8" wide		75	.107		3.03	3.79		6.82	9.65	
	2250	10" wide		70	.114		3.75	4.06		7.81	10.85	
	2260	12" wide		65	.123		4.32	4.38		8.70	12	
	2330	16 ga track x 6" wide		70	.114		2.98	4.06		7.04	10.05	
	2340	8" wide		65	.123		3.82	4.38		8.20	11.45	
	2350	10" wide		60	.133		4.70	4.74		9.44	13	
	2360	12" wide		55	.145		5.40	5.15		10.55	14.50	
	2440	14 ga track x 8" wide		60	.133		4.80	4.74		9.54	13.15	
	2450	10" wide		55	.145		5.90	5.15		11.05	15.05	
	2460	12" wide		50	.160		6.85	5.70		12.55	16.90	
	2550	12 ga track x 10" wide		45	.178		8.65	6.30		14.95	20	
	2560	12" wide	↓	40	.200	↓	9.80	7.10		16.90	22.50	
200	0010	**FRAMING, BAND JOIST** (track) fastened to bearing wall										200
	0220	18 ga track x 6" deep	2 Carp	1000	.016	L.F.	1.05	.57		1.62	2.10	
	0230	8" deep		920	.017		1.33	.62		1.95	2.49	
	0240	10" deep		860	.019		1.65	.66		2.31	2.91	
	0320	16 ga track x 6" deep		900	.018		1.31	.63		1.94	2.49	
	0330	8" deep		840	.019		1.68	.68		2.36	2.97	
	0340	10" deep		780	.021		2.07	.73		2.80	3.49	
	0350	12" deep		740	.022		2.37	.77		3.14	3.88	
	0430	14 ga track x 8" deep		750	.021		2.11	.76		2.87	3.58	
	0440	10" deep	↓	720	.022	↓	2.59	.79		3.38	4.16	

	05420 Cold-Formed Metal Joists		Crew	Daily Output	Labor-Hours	Unit	2006 Bare Costs Mat.	Labor	Equip.	Total	Total Incl O&P
200	0450	12" deep	2 Carp	700	.023	L.F.	3	.81		3.81	4.65
	0540	12 ga track x 10" deep		670	.024		3.81	.85		4.66	5.60
	0550	12" deep		650	.025		4.32	.88		5.20	6.20
300	0010	**FRAMING, BOXED HEADERS/BEAMS**									
	0200	Double, 18 ga x 6" deep	2 Carp	220	.073	L.F.	4.49	2.59		7.08	9.20
	0210	8" deep		210	.076		4.99	2.71		7.70	10
	0220	10" deep		200	.080		6.05	2.84		8.89	11.35
	0230	12" deep		190	.084		6.65	2.99		9.64	12.25
	0300	16 ga x 8" deep		180	.089		5.75	3.16		8.91	11.55
	0310	10" deep		170	.094		6.85	3.35		10.20	13.10
	0320	12" deep		160	.100		7.50	3.56		11.06	14.15
	0400	14 ga x 10" deep		140	.114		8	4.06		12.06	15.55
	0410	12" deep		130	.123		8.75	4.38		13.13	16.90
	0500	12 ga x 10" deep		110	.145		10.55	5.15		15.70	20
	0510	12" deep		100	.160		11.70	5.70		17.40	22.50
	1210	Triple, 18 ga x 8" deep		170	.094		7.20	3.35		10.55	13.50
	1220	10" deep		165	.097		8.65	3.45		12.10	15.20
	1230	12" deep		160	.100		9.55	3.56		13.11	16.40
	1300	16 ga x 8" deep		145	.110		8.30	3.92		12.22	15.65
	1310	10" deep		140	.114		9.90	4.06		13.96	17.65
	1320	12" deep		135	.119		10.85	4.21		15.06	18.90
	1400	14 ga x 10" deep		115	.139		11.60	4.95		16.55	21
	1410	12" deep		110	.145		12.75	5.15		17.90	22.50
	1500	12 ga x 10" deep		90	.178		15.45	6.30		21.75	27.50
	1510	12" deep		85	.188		17.15	6.70		23.85	30
410	0010	**FRAMING, JOISTS**, no band joists (track), web stiffeners, headers,									
	0020	beams, bridging or bracing									
	0030	Joists (2" flange) and fasteners, materials only									
	0220	18 ga x 6" deep				L.F.	1.39			1.39	1.52
	0230	8" deep					1.65			1.65	1.81
	0240	10" deep					1.93			1.93	2.13
	0320	16 ga x 6" deep					1.70			1.70	1.87
	0330	8" deep					2.04			2.04	2.24
	0340	10" deep					2.37			2.37	2.61
	0350	12" deep					2.70			2.70	2.97
	0430	14 ga x 8" deep					2.57			2.57	2.83
	0440	10" deep					2.96			2.96	3.26
	0450	12" deep					3.37			3.37	3.71
	0540	12 ga x 10" deep					4.32			4.32	4.75
	0550	12" deep					4.91			4.91	5.40
	1010	Installation of joists to band joists, beams & headers, labor only									
	1220	18 ga x 6" deep	2 Carp	110	.145	Ea.		5.15		5.15	8.55
	1230	8" deep		90	.178			6.30		6.30	10.45
	1240	10" deep		80	.200			7.10		7.10	11.80
	1320	16 ga x 6" deep		95	.168			6		6	9.90
	1330	8" deep		70	.229			8.15		8.15	13.45
	1340	10" deep		60	.267			9.50		9.50	15.70
	1350	12" deep		55	.291			10.35		10.35	17.15
	1430	14 ga x 8" deep		65	.246			8.75		8.75	14.50
	1440	10" deep		45	.356			12.65		12.65	21
	1450	12" deep		35	.457			16.25		16.25	27
	1540	12 ga x 10" deep		40	.400			14.20		14.20	23.50
	1550	12" deep		30	.533			18.95		18.95	31.50
	9000	Minimum labor/equipment charge		4	4	Job		142		142	236

05420 | Cold-Formed Metal Joists

		05420 Cold-Formed Metal Joists	CREW	DAILY OUTPUT	LABOR-HOURS	UNIT	2006 BARE COSTS MAT.	LABOR	EQUIP.	TOTAL	TOTAL INCL O&P	
500	0010	**FRAMING, WEB STIFFENERS** at joist bearing, fabricated from										500
	0020	stud piece (1-5/8" flange) to stiffen joist (2" flange)										
	2120	For 6" deep joist, with 18 ga x 2-1/2" stud	1 Carp	120	.067	Ea.	1.67	2.37		4.04	5.75	
	2130	3-5/8" stud		110	.073		1.85	2.59		4.44	6.30	
	2140	4" stud		105	.076		1.79	2.71		4.50	6.45	
	2150	6" stud		100	.080		1.95	2.84		4.79	6.85	
	2160	8" stud		95	.084		2	2.99		4.99	7.15	
	2220	8" deep joist, with 2-1/2" stud		120	.067		1.83	2.37		4.20	5.95	
	2230	3-5/8" stud		110	.073		1.99	2.59		4.58	6.45	
	2240	4" stud		105	.076		1.95	2.71		4.66	6.65	
	2250	6" stud		100	.080		2.13	2.84		4.97	7.05	
	2260	8" stud		95	.084		2.30	2.99		5.29	7.50	
	2320	10" deep joist, with 2-1/2" stud		110	.073		2.59	2.59		5.18	7.10	
	2330	3-5/8" stud		100	.080		2.84	2.84		5.68	7.85	
	2340	4" stud		95	.084		2.81	2.99		5.80	8.05	
	2350	6" stud		90	.089		3.04	3.16		6.20	8.60	
	2360	8" stud		85	.094		3.09	3.35		6.44	8.95	
	2420	12" deep joist, with 2-1/2" stud		110	.073		2.74	2.59		5.33	7.30	
	2430	3-5/8" stud		100	.080		2.97	2.84		5.81	8	
	2440	4" stud		95	.084		2.91	2.99		5.90	8.15	
	2450	6" stud		90	.089		3.19	3.16		6.35	8.75	
	2460	8" stud		85	.094		3.43	3.35		6.78	9.35	
	3130	For 6" deep joist, with 16 ga x 3-5/8" stud		100	.080		1.93	2.84		4.77	6.85	
	3140	4" stud		95	.084		1.91	2.99		4.90	7.05	
	3150	6" stud		90	.089		2.10	3.16		5.26	7.55	
	3160	8" stud		85	.094		2.21	3.35		5.56	8	
	3230	8" deep joist, with 3-5/8" stud		100	.080		2.14	2.84		4.98	7.05	
	3240	4" stud		95	.084		2.10	2.99		5.09	7.25	
	3250	6" stud		90	.089		2.33	3.16		5.49	7.80	
	3260	8" stud		85	.094		2.49	3.35		5.84	8.30	
	3330	10" deep joist, with 3-5/8" stud		85	.094		2.92	3.35		6.27	8.75	
	3340	4" stud		80	.100		2.98	3.56		6.54	9.20	
	3350	6" stud		75	.107		3.25	3.79		7.04	9.85	
	3360	8" stud		70	.114		3.38	4.06		7.44	10.45	
	3430	12" deep joist, with 3-5/8" stud		85	.094		3.19	3.35		6.54	9.05	
	3440	4" stud		80	.100		3.13	3.56		6.69	9.35	
	3450	6" stud		75	.107		3.48	3.79		7.27	10.15	
	3460	8" stud		70	.114		3.72	4.06		7.78	10.85	
	4230	For 8" deep joist, with 14 ga x 3-5/8" stud		90	.089		2.77	3.16		5.93	8.30	
	4240	4" stud		85	.094		2.84	3.35		6.19	8.65	
	4250	6" stud		80	.100		3.07	3.56		6.63	9.25	
	4260	8" stud		75	.107		3.29	3.79		7.08	9.90	
	4330	10" deep joist, with 3-5/8" stud		75	.107		3.89	3.79		7.68	10.60	
	4340	4" stud		70	.114		3.88	4.06		7.94	11	
	4350	6" stud		65	.123		4.25	4.38		8.63	11.95	
	4360	8" stud		60	.133		4.44	4.74		9.18	12.75	
	4430	12" deep joist, with 3-5/8" stud		75	.107		4.14	3.79		7.93	10.85	
	4440	4" stud		70	.114		4.23	4.06		8.29	11.40	
	4450	6" stud		65	.123		4.58	4.38		8.96	12.30	
	4460	8" stud		60	.133		4.91	4.74		9.65	13.25	
	5330	For 10" deep joist, with 12 ga x 3-5/8" stud		65	.123		4.11	4.38		8.49	11.75	
	5340	4" stud		60	.133		4.24	4.74		8.98	12.50	
	5350	6" stud		55	.145		4.67	5.15		9.82	13.70	
	5360	8" stud		50	.160		5.15	5.70		10.85	15.05	
	5430	12" deep joist, with 3-5/8" stud		65	.123		4.55	4.38		8.93	12.25	
	5440	4" stud	▼	60	.133	▼	4.47	4.74		9.21	12.75	

05400 | Cold-Formed Metal Framing

	05420 | Cold-Formed Metal Joists		CREW	DAILY OUTPUT	LABOR-HOURS	UNIT	2006 BARE COSTS MAT.	LABOR	EQUIP.	TOTAL	TOTAL INCL O&P	
500	5450	6″ stud	1 Carp	55	.145	Ea.	5.10	5.15		10.25	14.15	500
	5460	8″ stud		50	.160		5.85	5.70		11.55	15.85	

	05425 | Cold-Formed Roof Framing		CREW	DAILY OUTPUT	LABOR-HOURS	UNIT	MAT.	LABOR	EQUIP.	TOTAL	TOTAL INCL O&P	
100	0010	**FRAMING, BRACING**										100
	0020	Continuous bracing, per row										
	0100	16 ga x 1-1/2″ channel thru rafters/trusses @ 16″ O.C.	1 Carp	4.50	1.778	C.L.F.	43	63		106	153	
	0120	24″ O.C.		6	1.333		43	47.50		90.50	126	
	0300	2″ x 2″ angle x 18 ga, rafters/trusses @ 16″ O.C.		6	1.333		62.50	47.50		110	148	
	0320	24″ O.C.		8	1		62.50	35.50		98	128	
	0400	16 ga, rafters/trusses @ 16″ O.C.		4.50	1.778		80.50	63		143.50	194	
	0420	24″ O.C.		6.50	1.231		80.50	44		124.50	161	
200	0010	**FRAMING, BRIDGING**										200
	0020	Solid, between rafters w/ 1-1/4″ leg track, per rafter bay										
	1200	Rafters 16″ O.C., 18 ga x 4″ deep	1 Carp	60	.133	Ea.	1.27	4.74		6.01	9.25	
	1210	6″ deep		57	.140		1.65	4.99		6.64	10.05	
	1220	8″ deep		55	.145		2.10	5.15		7.25	10.85	
	1230	10″ deep		52	.154		2.59	5.45		8.04	11.90	
	1240	12″ deep		50	.160		2.99	5.70		8.69	12.70	
	2200	24″ O.C., 18 ga x 4″ deep		60	.133		1.84	4.74		6.58	9.85	
	2210	6″ deep		57	.140		2.39	4.99		7.38	10.90	
	2220	8″ deep		55	.145		3.03	5.15		8.18	11.90	
	2230	10″ deep		52	.154		3.75	5.45		9.20	13.15	
	2240	12″ deep		50	.160		4.32	5.70		10.02	14.15	
500	0010	**FRAMING, PARAPETS**										500
	0100	3′ high installed on 1st story, 18 ga x 4″ wide studs, 12″ O.C.	2 Carp	100	.160	L.F.	4.71	5.70		10.41	14.60	
	0110	16″ O.C.		150	.107		4	3.79		7.79	10.70	
	0120	24″ O.C.		200	.080		3.30	2.84		6.14	8.35	
	0200	6″ wide studs, 12″ O.C.		100	.160		6	5.70		11.70	16	
	0210	16″ O.C.		150	.107		5.10	3.79		8.89	11.90	
	0220	24″ O.C.		200	.080		4.22	2.84		7.06	9.35	
	1100	Installed on 2nd story, 18 ga x 4″ wide studs, 12″ O.C.		95	.168		4.71	6		10.71	15.10	
	1110	16″ O.C.		145	.110		4	3.92		7.92	10.90	
	1120	24″ O.C.		190	.084		3.30	2.99		6.29	8.60	
	1200	6″ wide studs, 12″ O.C.		95	.168		6	6		12	16.50	
	1210	16″ O.C.		145	.110		5.10	3.92		9.02	12.10	
	1220	24″ O.C.		190	.084		4.22	2.99		7.21	9.60	
	2100	Installed on gable, 18 ga x 4″ wide studs, 12″ O.C.		85	.188		4.71	6.70		11.41	16.30	
	2110	16″ O.C.		130	.123		4	4.38		8.38	11.65	
	2120	24″ O.C.		170	.094		3.30	3.35		6.65	9.20	
	2200	6″ wide studs, 12″ O.C.		85	.188		6	6.70		12.70	17.70	
	2210	16″ O.C.		130	.123		5.10	4.38		9.48	12.85	
	2220	24″ O.C.		170	.094		4.22	3.35		7.57	10.20	
550	0010	**FRAMING, ROOF RAFTERS**										550
	0100	Boxed ridge beam, double, 18 ga x 6″ deep	2 Carp	160	.100	L.F.	4.49	3.56		8.05	10.85	
	0110	8″ deep		150	.107		4.99	3.79		8.78	11.80	
	0120	10″ deep		140	.114		6.05	4.06		10.11	13.40	
	0130	12″ deep		130	.123		6.65	4.38		11.03	14.55	
	0200	16 ga x 6″ deep		150	.107		5.10	3.79		8.89	11.90	
	0210	8″ deep		140	.114		5.75	4.06		9.81	13.05	
	0220	10″ deep		130	.123		6.85	4.38		11.23	14.80	
	0230	12″ deep		120	.133		7.50	4.74		12.24	16.10	
	1100	Rafters, 2″ flange, material only, 18 ga x 6″ deep					1.39			1.39	1.52	
	1110	8″ deep					1.65			1.65	1.81	
	1120	10″ deep					1.93			1.93	2.13	

5 METALS

	05425 | Cold-Formed Roof Framing		CREW	DAILY OUTPUT	LABOR-HOURS	UNIT	2006 BARE COSTS MAT.	LABOR	EQUIP.	TOTAL	TOTAL INCL O&P	
550	1130	12" deep				L.F.	2.25			2.25	2.47	550
	1200	16 ga x 6" deep					1.70			1.70	1.87	
	1210	8" deep					2.04			2.04	2.24	
	1220	10" deep					2.37			2.37	2.61	
	1230	12" deep					2.70			2.70	2.97	
	2100	Installation only, ordinary rafter to 4:12 pitch, 18 ga x 6" deep	2 Carp	35	.457	Ea.		16.25		16.25	27	
	2110	8" deep		30	.533			18.95		18.95	31.50	
	2120	10" deep		25	.640			23		23	37.50	
	2130	12" deep		20	.800			28.50		28.50	47	
	2200	16 ga x 6" deep		30	.533			18.95		18.95	31.50	
	2210	8" deep		25	.640			23		23	37.50	
	2220	10" deep		20	.800			28.50		28.50	47	
	2230	12" deep		15	1.067			38		38	63	
	8100	Add to labor, ordinary rafters on steep roofs						25%				
	8110	Dormers & complex roofs						50%				
	8200	Hip & valley rafters to 4:12 pitch						25%				
	8210	Steep roofs						50%				
	8220	Dormers & complex roofs						75%				
	8300	Hip & valley jack rafters to 4:12 pitch						50%				
	8310	Steep roofs						75%				
	8320	Dormers & complex roofs						100%				
	9000	Minimum labor/equipment charge	2 Carp	4	4	Job		142		142	236	
600	0010	**FRAMING, ROOF TRUSSES**										600
	0020	Fabrication of trusses on ground, Fink (W) or King Post, to 4:12 pitch										
	0120	18 ga x 4" chords, 16' span	2 Carp	12	1.333	Ea.	52.50	47.50		100	137	
	0130	20' span		11	1.455		66	51.50		117.50	158	
	0140	24' span		11	1.455		79	51.50		130.50	173	
	0150	28' span		10	1.600		92	57		149	195	
	0160	32' span		10	1.600		105	57		162	210	
	0250	6" chords, 28' span		9	1.778		116	63		179	232	
	0260	32' span		9	1.778		132	63		195	250	
	0270	36' span		8	2		149	71		220	282	
	0280	40' span		8	2		165	71		236	300	
	1120	5:12 to 8:12 pitch, 18 ga x 4" chords, 16' span		10	1.600		60	57		117	160	
	1130	20' span		9	1.778		75	63		138	188	
	1140	24' span		9	1.778		90	63		153	205	
	1150	28' span		8	2		105	71		176	234	
	1160	32' span		8	2		120	71		191	250	
	1250	6" chords, 28' span		7	2.286		132	81.50		213.50	280	
	1260	32' span		7	2.286		151	81.50		232.50	300	
	1270	36' span		6	2.667		170	95		265	345	
	1280	40' span		6	2.667		189	95		284	365	
	2120	9:12 to 12:12 pitch, 18 ga x 4" chords, 16' span		8	2		75	71		146	201	
	2130	20' span		7	2.286		94	81.50		175.50	238	
	2140	24' span		7	2.286		113	81.50		194.50	259	
	2150	28' span		6	2.667		132	95		227	300	
	2160	32' span		6	2.667		150	95		245	320	
	2250	6" chords, 28' span		5	3.200		165	114		279	370	
	2260	32' span		5	3.200		189	114		303	395	
	2270	36' span		4	4		212	142		354	470	
	2280	40' span		4	4		236	142		378	495	
	4900	Minimum labor/equipment charge		4	4	Job		142		142	236	
	5120	Erection only of roof trusses, to 4:12 pitch, 16' span	F-6	48	.833	Ea.		27.50	13.25	40.75	59.50	
	5130	20' span		46	.870			28.50	13.85	42.35	62	
	5140	24' span		44	.909			30	14.45	44.45	65	
	5150	28' span		42	.952			31	15.15	46.15	67.50	

05400 | Cold-Formed Metal Framing

05425 | Cold-Formed Roof Framing

			CREW	DAILY OUTPUT	LABOR-HOURS	UNIT	2006 BARE COSTS MAT.	LABOR	EQUIP.	TOTAL	TOTAL INCL O&P	
600	5160	32′ span	F-6	40	1	Ea.		33	15.90	48.90	71.50	600
	5170	36′ span		38	1.053			34.50	16.75	51.25	75	
	5180	40′ span		36	1.111			36.50	17.70	54.20	79	
	5220	5:12 to 8:12 pitch, 16′ span		42	.952			31	15.15	46.15	67.50	
	5230	20′ span		40	1			33	15.90	48.90	71.50	
	5240	24′ span		38	1.053			34.50	16.75	51.25	75	
	5250	28′ span		36	1.111			36.50	17.70	54.20	79	
	5260	32′ span		34	1.176			38.50	18.75	57.25	83.50	
	5270	36′ span		32	1.250			41	19.90	60.90	89	
	5280	40′ span		30	1.333			43.50	21	64.50	95	
	5320	9:12 to 12:12 pitch, 16′ span		36	1.111			36.50	17.70	54.20	79	
	5330	20′ span		34	1.176			38.50	18.75	57.25	83.50	
	5340	24′ span		32	1.250			41	19.90	60.90	89	
	5350	28′ span		30	1.333			43.50	21	64.50	95	
	5360	32′ span		28	1.429			47	22.50	69.50	102	
	5370	36′ span		26	1.538			50.50	24.50	75	110	
	5380	40′ span		24	1.667			54.50	26.50	81	119	
	9000	Minimum labor/equipment charge		2	20	Job		655	320	975	1,425	
650	0010	**FRAMING, SOFFITS & CANOPIES**										650
	0130	Continuous ledger track @ wall, studs @ 16″ O.C., 18 ga x 4″ wide	2 Carp	535	.030	L.F.	.85	1.06		1.91	2.69	
	0140	6″ wide		500	.032		1.10	1.14		2.24	3.09	
	0150	8″ wide		465	.034		1.40	1.22		2.62	3.57	
	0160	10″ wide		430	.037		1.73	1.32		3.05	4.09	
	0230	Studs @ 24″ O.C., 18 ga x 4″ wide		800	.020		.81	.71		1.52	2.07	
	0240	6″ wide		750	.021		1.05	.76		1.81	2.42	
	0250	8″ wide		700	.023		1.33	.81		2.14	2.82	
	0260	10″ wide		650	.025		1.65	.88		2.53	3.26	
	1000	Horizontal soffit and canopy members, material only										
	1030	1-5/8″ flange studs, 18 ga x 4″ deep				L.F.	1.13			1.13	1.24	
	1040	6″ deep					1.42			1.42	1.56	
	1050	8″ deep					1.72			1.72	1.89	
	1140	2″ flange joists, 18 ga x 6″ deep					1.58			1.58	1.74	
	1150	8″ deep					1.88			1.88	2.07	
	1160	10″ deep					2.21			2.21	2.43	
	4030	Installation only, 18 ga, 1-5/8″ flange x 4″ deep	2 Carp	130	.123	Ea.		4.38		4.38	7.25	
	4040	6″ deep		110	.145			5.15		5.15	8.55	
	4050	8″ deep		90	.178			6.30		6.30	10.45	
	4140	2″ flange, 18 ga x 6″ deep		110	.145			5.15		5.15	8.55	
	4150	8″ deep		90	.178			6.30		6.30	10.45	
	4160	10″ deep		80	.200			7.10		7.10	11.80	
	6010	Clips to attach facia to rafter tails, 2″ x 2″ x 18 ga angle	1 Carp	120	.067		.74	2.37		3.11	4.75	
	6020	16 ga angle	″	100	.080		.95	2.84		3.79	5.75	
	9000	Minimum labor/equipment charge	2 Carp	4	4	Job		142		142	236	

5 METALS

05500 | Metal Fabrications

05514 | Ladders

			CREW	DAILY OUTPUT	LABOR-HOURS	UNIT	2006 BARE COSTS MAT.	LABOR	EQUIP.	TOTAL	TOTAL INCL O&P	
500	0010	**LADDER**, shop fabricated										500
	0020	Steel, 20″ wide, bolted to concrete, with cage	E-4	50	.640	V.L.F.	68	26	1.79	95.79	125	

05500 | Metal Fabrications

	05514 | Ladders		Crew	Daily Output	Labor-Hours	Unit	2006 Bare Costs Mat.	Labor	Equip.	Total	Total Incl O&P	
500	0100	Without cage	E-4	85	.376	V.L.F.	31.50	15.25	1.05	47.80	64.50	500
	0300	Aluminum, bolted to concrete, with cage		50	.640		91.50	26	1.79	119.29	151	
	0400	Without cage		85	.376		53	15.25	1.05	69.30	88	
	1350	Alternating tread stair, 56/68°, steel, standard paint color	2 Sswk	50	.320		151	12.80		163.80	190	
	1360	Non-standard paint color		50	.320		171	12.80		183.80	212	
	1370	Galvanized steel		50	.320		170	12.80		182.80	211	
	1380	Stainless steel		50	.320		253	12.80		265.80	305	
	1390	68°, aluminum		50	.320		185	12.80		197.80	228	
	9000	Minimum labor/equipment charge	E-4	2	16	Job		645	44.50	689.50	1,250	
	05517 | Metal Stairs											
300	0010	**FIRE ESCAPE**, shop fabricated										300
	0200	2′ wide balcony, 1″ x 1/4″ bars 1-1/2″ O.C.	1 Sswk	5	1.600	L.F.	45.50	64		109.50	170	
	0400	1st story cantilevered stair, standard	″	.09	88.889	Ea.	1,900	3,550		5,450	8,750	
	0500	Cable counterweight				″	1,775			1,775	1,950	
	0700	Platform & fixed stair, 36″ x 40″	1 Sswk	.17	47.059	Flight	845	1,875		2,720	4,450	
	0900	For 3′-6″ wide escapes, add to above					100%	150%				
350	0010	**FIRE ESCAPE STAIRS**										350
	0020	One story, disappearing, stainless steel	2 Sswk	20	.800	V.L.F.	184	32		216	262	
	0100	Portable ladder				Ea.	55.50			55.50	61.50	
700	0010	**STAIR**, shop fabricated, steel stringers, safety nosing on treads										700
	0020	Grating tread and pipe railing, 3′-6″ wide	E-4	35	.914	Riser	218	37	2.55	257.55	310	
	0100	4′-0″ wide		30	1.067		283	43	2.98	328.98	395	
	0200	Cement fill metal pan, picket rail, 3′-6″ wide		35	.914		325	37	2.55	364.55	430	
	0300	4′-0″ wide		30	1.067		370	43	2.98	415.98	490	
	0350	Wall rail, both sides, 3′-6″ wide		53	.604		250	24.50	1.68	276.18	320	
	0500	Checkered plate tread, industrial, 3′-6″ wide		28	1.143		218	46	3.19	267.19	330	
	0550	Circular, for tanks, 3′-0″ wide		33	.970		240	39	2.71	281.71	340	
	0600	For isolated stairs, add						100%				
	0800	Custom steel stairs, 3′-6″ wide, minimum	E-4	35	.914		325	37	2.55	364.55	430	
	0810	Average		30	1.067		435	43	2.98	480.98	565	
	0900	Maximum		20	1.600		545	64.50	4.46	613.96	725	
	1100	For 4′ wide stairs, add					5%	5%				
	1300	For 5′ wide stairs, add					10%	10%				
	1500	Landing, steel pan, conventional	E-4	160	.200	S.F.	43.50	8.10	.56	52.16	64	
	1810	Spiral aluminum, 5′-0″ diameter, stock units		45	.711	Riser	395	29	1.98	425.98	490	
	1820	Custom units		45	.711		745	29	1.98	775.98	875	
	1900	Spiral, cast iron, 4′-0″ diameter, ornamental, minimum		45	.711		350	29	1.98	380.98	445	
	1920	Maximum		25	1.280		480	52	3.57	535.57	630	
	2000	Spiral, steel, industrial checkered plate, 4′ diameter		45	.711		350	29	1.98	380.98	445	
	2200	Stock units, 6′-0″ diameter		40	.800		425	32.50	2.23	459.73	535	
	3110	Spiral steel, stock units, primed, flat metal tread, 3′-6″ dia	2 Carp	1.60	10	Flight	1,050	355		1,405	1,775	
	3120	4′-0″ dia		1.45	11.034		1,225	390		1,615	2,000	
	3130	4′-6″ dia		1.35	11.852		1,350	420		1,770	2,175	
	3140	5′-0″ dia		1.25	12.800		1,450	455		1,905	2,350	
	3210	Galvanized, 3′-6″ dia		1.60	10		1,850	355		2,205	2,650	
	3220	4′-0″ dia		1.45	11.034		2,075	390		2,465	2,950	
	3230	4′-6″ dia		1.35	11.852		2,275	420		2,695	3,200	
	3240	5′-0″ dia		1.25	12.800		2,450	455		2,905	3,450	
	3310	Checkered plate tread, 3′-6″ dia		1.45	11.034		1,300	390		1,690	2,075	
	3320	4′-0″ dia		1.35	11.852		1,475	420		1,895	2,325	
	3330	4′-6″ dia		1.25	12.800		1,625	455		2,080	2,525	
	3340	5′-0″ dia		1.15	13.913		1,750	495		2,245	2,750	
	3410	Galvanized, 3′-6″ dia		1.45	11.034		2,125	390		2,515	2,975	

05517 | Metal Stairs

			CREW	DAILY OUTPUT	LABOR-HOURS	UNIT	2006 BARE COSTS MAT.	LABOR	EQUIP.	TOTAL	TOTAL INCL O&P	
700	3420	4'-0" dia	2 Carp	1.35	11.852	Flight	2,375	420		2,795	3,325	700
	3430	4'-6" dia		1.25	12.800		2,575	455		3,030	3,600	
	3440	5'-0" dia		1.15	13.913		2,775	495		3,270	3,900	
	3510	Red oak tread on flat metal, 3'-6" dia		1.35	11.852		1,775	420		2,195	2,650	
	3520	4'-0" dia		1.25	12.800		1,975	455		2,430	2,900	
	3530	4'-6" dia		1.15	13.913		2,125	495		2,620	3,175	
	3540	5'-0" dia	↓	1.05	15.238	↓	2,300	540		2,840	3,425	
	3900	Industrial ships ladder, 3' W, grating treads, 2 line pipe rail	E-4	30	1.067	Riser	142	43	2.98	187.98	240	
	4000	Aluminum		30	1.067	"	218	43	2.98	263.98	325	
	9000	Minimum labor/equipment charge	↓	2	16	Job		645	44.50	689.50	1,250	

05520 | Handrails & Railings

			CREW	DAILY OUTPUT	LABOR-HOURS	UNIT	2006 BARE COSTS MAT.	LABOR	EQUIP.	TOTAL	TOTAL INCL O&P	
700	0010	**RAILING, PIPE**, shop fabricated										700
	0020	Aluminum, 2 rail, satin finish, 1-1/4" diameter	E-4	160	.200	L.F.	19.75	8.10	.56	28.41	37.50	
	0030	Clear anodized		160	.200		24.50	8.10	.56	33.16	43	
	0040	Dark anodized		160	.200		27.50	8.10	.56	36.16	46.50	
	0080	1-1/2" diameter, satin finish		160	.200		23.50	8.10	.56	32.16	42	
	0090	Clear anodized		160	.200		26.50	8.10	.56	35.16	45	
	0100	Dark anodized		160	.200		29	8.10	.56	37.66	48	
	0140	Aluminum, 3 rail, 1-1/4" diam., satin finish		137	.234		30	9.45	.65	40.10	52	
	0150	Clear anodized		137	.234		38	9.45	.65	48.10	60	
	0160	Dark anodized		137	.234		42	9.45	.65	52.10	64.50	
	0200	1-1/2" diameter, satin finish		137	.234		36.50	9.45	.65	46.60	58.50	
	0210	Clear anodized		137	.234		41	9.45	.65	51.10	63.50	
	0220	Dark anodized		137	.234		45	9.45	.65	55.10	68	
	0500	Steel, 2 rail, on stairs, primed, 1-1/4" diameter		160	.200		16.20	8.10	.56	24.86	33.50	
	0520	1-1/2" diameter		160	.200		17.80	8.10	.56	26.46	35.50	
	0540	Galvanized, 1-1/4" diameter		160	.200		22.50	8.10	.56	31.16	40.50	
	0560	1-1/2" diameter		160	.200		25	8.10	.56	33.66	43.50	
	0580	Steel, 3 rail, primed, 1-1/4" diameter		137	.234		24	9.45	.65	34.10	45	
	0600	1-1/2" diameter		137	.234		25.50	9.45	.65	35.60	46.50	
	0620	Galvanized, 1-1/4" diameter		137	.234		34	9.45	.65	44.10	56	
	0640	1-1/2" diameter		137	.234		40	9.45	.65	50.10	62.50	
	0700	Stainless steel, 2 rail, 1-1/4" diam. #4 finish		137	.234		58.50	9.45	.65	68.60	82.50	
	0720	High polish		137	.234		94	9.45	.65	104.10	121	
	0740	Mirror polish		137	.234		118	9.45	.65	128.10	148	
	0760	Stainless steel, 3 rail, 1-1/2" diam., #4 finish		120	.267		88	10.80	.74	99.54	118	
	0770	High polish		120	.267		145	10.80	.74	156.54	181	
	0780	Mirror finish		120	.267		177	10.80	.74	188.54	216	
	0900	Wall rail, alum. pipe, 1-1/4" diam., satin finish		213	.150		11.30	6.10	.42	17.82	24	
	0905	Clear anodized		213	.150		13.75	6.10	.42	20.27	27	
	0910	Dark anodized		213	.150		16.65	6.10	.42	23.17	30	
	0915	1-1/2" diameter, satin finish		213	.150		12.50	6.10	.42	19.02	25.50	
	0920	Clear anodized		213	.150		15.70	6.10	.42	22.22	29	
	0925	Dark anodized		213	.150		19.40	6.10	.42	25.92	33.50	
	0930	Steel pipe, 1-1/4" diameter, primed		213	.150		9.85	6.10	.42	16.37	22.50	
	0935	Galvanized		213	.150		14.25	6.10	.42	20.77	27.50	
	0940	1-1/2" diameter		176	.182		10.10	7.35	.51	17.96	25.50	
	0945	Galvanized		213	.150		14.30	6.10	.42	20.82	27.50	
	0955	Stainless steel pipe, 1-1/2" diam., #4 finish		107	.299		46.50	12.10	.83	59.43	74.50	
	0960	High polish		107	.299		95	12.10	.83	107.93	128	
	0965	Mirror polish	↓	107	.299	↓	112	12.10	.83	124.93	146	
	9000	Minimum labor/equipment charge	1 Sswk	2	4	Job		160		160	299	
780	0010	**RAILINGS, INDUSTRIAL** Welded, shop fabricated										780
	0020	2 rail, 3'-6" high, 1-1/2" pipe	E-4	255	.125	L.F.	21.50	5.10	.35	26.95	33.50	

05500 | Metal Fabrications

	05520 Handrails & Railings		Crew	Daily Output	Labor-Hours	Unit	2006 Bare Costs Mat.	Labor	Equip.	Total	Total Incl O&P	
780	0200	For 4" high kick plate, 10 gauge, add				L.F.	4.48			4.48	4.93	780
	0500	For curved rails, add				↓	30%	30%				
	9000	Minimum labor/equipment charge	1 Sswk	2	4	Job		160		160	299	

	05560 Metal Castings		Crew	Daily Output	Labor-Hours	Unit	Mat.	Labor	Equip.	Total	Total Incl O&P	
200	0010	**CONSTRUCTION CASTINGS**										200
	0020	Manhole covers and frames see Division 02630-110										
	0100	Column bases, cast iron, 16" x 16", approx. 65 lbs.	E-4	46	.696	Ea.	107	28	1.94	136.94	173	
	0200	32" x 32", approx. 256 lbs.		23	1.391	"	400	56.50	3.88	460.38	550	
	0600	Miscellaneous C.I. castings, light sections, less than 150 lbs		3200	.010	Lb.	1.66	.40	.03	2.09	2.62	
	1300	Special low volume items	↓	3200	.010	"	2.89	.40	.03	3.32	3.97	

	05580 Formed Metal Fabrications		Crew	Daily Output	Labor-Hours	Unit	Mat.	Labor	Equip.	Total	Total Incl O&P	
200	0010	**ALUMINUM COLUMNS**										200
	0020	Aluminum, extruded, stock units, no cap or base, 6" diameter	E-4	240	.133	L.F.	8.70	5.40	.37	14.47	20	
	0100	8" diameter	"	170	.188		11.70	7.60	.53	19.83	27.50	
	0500	For square columns, add to column prices above				↓	50%					
600	0010	**LAMP POSTS**										600
	0020	Aluminum, 7' high, stock units, post only	1 Carp	16	.500	Ea.	32	17.80		49.80	65	
	0100	Mild steel, plain		16	.500	"	28	17.80		45.80	60.50	
	9000	Minimum labor/equipment charge	↓	4	2	Job		71		71	118	
900	0010	**WINDOW GUARDS**, shop fabricated										900
	0015	Expanded metal, steel angle frame, permanent	E-4	350	.091	S.F.	19.65	3.70	.26	23.61	28.50	
	0025	Steel bars, 1/2" x 1/2", spaced 5" O.C.	"	290	.110	"	13.60	4.46	.31	18.37	23.50	
	0030	Hinge mounted, add				Opng.	39.50			39.50	43.50	
	0040	Removable type, add				"	25			25	27.50	
	0050	For galvanized guards, add				S.F.	35%					
	0070	For pivoted or projected type, add					105%	40%				
	0100	Mild steel, stock units, economy	E-4	405	.079		5.35	3.20	.22	8.77	12.10	
	0200	Deluxe		405	.079	↓	10.95	3.20	.22	14.37	18.30	
	0400	Woven wire, stock units, 3/8" channel frame, 3' x 5' opening		40	.800	Opng.	144	32.50	2.23	178.73	222	
	0500	4' x 6' opening	↓	38	.842		231	34	2.35	267.35	320	
	0800	Basket guards for above, add					197			197	217	
	1000	Swinging guards for above, add				↓	68			68	74.50	
	9000	Minimum labor/equipment charge	1 Sswk	2	4	Job		160		160	299	

05700 | Ornamental Metal

	05720 Ornamental Handrails & Railings		Crew	Daily Output	Labor-Hours	Unit	2006 Bare Costs Mat.	Labor	Equip.	Total	Total Incl O&P	
700	0010	**RAILINGS, ORNAMENTAL**, shop fabricated										700
	0020	Aluminum, bronze or stainless, minimum	1 Sswk	24	.333	L.F.	27.50	13.30		40.80	55.50	
	0100	Maximum		9	.889		261	35.50		296.50	355	
	0200	Aluminum ornamental rail, minimum		15	.533		27.50	21.50		49	70.50	
	0300	Maximum		8	1		82.50	40		122.50	166	
	0400	Hand-forged wrought iron, minimum		12	.667		86	26.50		112.50	145	
	0500	Maximum		8	1		258	40		298	360	
	0600	Composite metal/wood/glass, minimum		6	1.333		165	53.50		218.50	281	
	0700	Maximum		5	1.600	↓	330	64		394	485	
	9000	Minimum labor/equipment charge	↓	2	4	Job		160		160	299	

	05810 Exp. Joint Cover Assemblies		CREW	DAILY OUTPUT	LABOR-HOURS	UNIT	2006 BARE COSTS MAT.	LABOR	EQUIP.	TOTAL	TOTAL INCL O&P	
350	0010	**EXPANSION JOINT ASSEMBLIES** Custom units										350
	0200	Floor cover assemblies, 1" space, aluminum	1 Sswk	38	.211	L.F.	16.20	8.40		24.60	33.50	
	0300	Bronze		38	.211		33	8.40		41.40	52	
	0500	2" space, aluminum		38	.211		19.60	8.40		28	37.50	
	0600	Bronze		38	.211		35.50	8.40		43.90	55	
	0800	Wall and ceiling assemblies, 1" space, aluminum		38	.211		9.70	8.40		18.10	26.50	
	0900	Bronze		38	.211		31	8.40		39.40	50	
	1100	2" space, aluminum		38	.211		16.40	8.40		24.80	34	
	1200	Bronze		38	.211		38	8.40		46.40	57.50	
	1400	Floor to wall assemblies, 1" space, aluminum		38	.211		14.75	8.40		23.15	32	
	1500	Bronze or stainless		38	.211		37.50	8.40		45.90	57	
	1700	Gym floor angle covers, aluminum, 3" x 3" angle		46	.174		12.80	6.95		19.75	27	
	1800	3" x 4" angle		46	.174		15.10	6.95		22.05	29.50	
	2000	Roof closures, aluminum, flat roof, low profile, 1" space		57	.140		29.50	5.60		35.10	43	
	2100	High profile		57	.140		36	5.60		41.60	50.50	
	2300	Roof to wall, low profile, 1" space		57	.140		16.30	5.60		21.90	28.50	
	2400	High profile		57	.140	▼	21	5.60		26.60	33.50	
	9000	Minimum labor/equipment charge	▼	2	4	Job		160		160	299	

For information about Means Estimating Seminars, see yellow pages 12 and 13 in back of book

Division 6 Wood & Plastics

Estimating Tips

06050 Basic Wood & Plastic Materials & Methods

- Common to any wood-framed structure are the accessory connector items such as screws, nails, adhesives, hangers, connector plates, straps, angles, and holdowns. For typical wood-framed buildings, such as residential projects, the aggregate total for these items can be significant, especially in areas where seismic loading is a concern. For floor and wall framing, the material cost is based on 10 to 25 lbs. per MBF. Holdowns, hangers, and other connectors should be taken off by the piece.

06100 Rough Carpentry

- Lumber is a traded commodity and therefore sensitive to supply and demand in the marketplace. Even in "budgetary" estimating of wood-framed projects, it is advisable to call local suppliers for the latest market pricing.
- Common quantity units for wood framed projects are "thousand board feet" (MBF). A board foot is a volume of wood, 1″ x 1′ x 1′, or 144 cubic inches. Board-foot quantities are generally calculated using nominal material dimensions—dressed sizes are ignored. Board foot per lineal foot of any stick of lumber can be calculated by dividing the nominal cross sectional area by 12. As an example, 2,000 lineal feet of 2 x 12 equates to 4 MBF by dividing the nominal area, 2 x 12, by 12, which equals 2, and multiplying by 2,000 to give 4,000 board feet. This simple rule applies to all nominal dimensioned lumber.
- Waste is an issue of concern at the quantity takeoff for any area of construction. Framing lumber is sold in even foot lengths, i.e., 10′, 12′, 14′, 16′, and depending on spans, wall heights and the grade of lumber, waste is inevitable. A rule of thumb for lumber waste is 5% to 10% depending on material quality and the complexity of the framing.
- Wood in various forms and shapes is used in many projects, even where the main structural framing is steel, concrete, or masonry. Plywood as a back-up partition material and 2x boards used as blocking and cant strips around roof edges are two common examples. The estimator should ensure that the costs of all wood materials are included in the final estimate.

06200 Finish Carpentry

- It is necessary to consider the grade of workmanship when estimating labor costs for erecting millwork and interior finish. In practice, there are three grades: premium, custom, and economy. The Means daily output for base and case moldings is in the range of 200 to 250 L.F. per carpenter per day. This is appropriate for most average custom-grade projects. For premium projects an adjustment to productivity of 25% to 50% should be made depending on the complexity of the job.

Reference Numbers

Reference numbers are shown in bold squares at the beginning of some major classifications. These numbers refer to related items in the Reference Section. The reference information may be an estimating procedure, an alternate pricing method, or technical information.

Note: Not all subdivisions listed here necessarily appear in this publication.

	06052 | Selective Demolition	CREW	DAILY OUTPUT	LABOR-HOURS	UNIT	2006 BARE COSTS MAT.	LABOR	EQUIP.	TOTAL	TOTAL INCL O&P	
110 0010	SELECTIVE DEMOLITION, WOOD FRAMING R024119-10										110
0100	Timber connector, nailed, small	1 Clab	96	.083	Ea.		2.28		2.28	3.78	
0110	Medium		60	.133			3.65		3.65	6.05	
0120	Large		48	.167			4.57		4.57	7.55	
0130	Bolted, small		48	.167			4.57		4.57	7.55	
0140	Medium		32	.250			6.85		6.85	11.35	
0150	Large		24	.333			9.15		9.15	15.15	
2958	Beams, 2″ x 6″	2 Clab	1100	.015	L.F.		.40		.40	.66	
2960	2″ x 8″		825	.019			.53		.53	.88	
2965	2″ x 10″		665	.024			.66		.66	1.09	
2970	2″ x 12″		550	.029			.80		.80	1.32	
2972	2″ x 14″		470	.034			.93		.93	1.55	
2975	4″ x 8″	B-1	413	.058			1.63		1.63	2.70	
2980	4″ x 10″		330	.073			2.04		2.04	3.38	
2985	4″ x 12″		275	.087			2.45		2.45	4.06	
3000	6″ x 8″		275	.087			2.45		2.45	4.06	
3040	6″ x 10″		220	.109			3.06		3.06	5.05	
3080	6″ x 12″		185	.130			3.64		3.64	6.05	
3120	8″ x 12″		140	.171			4.81		4.81	7.95	
3160	10″ x 12″		110	.218			6.10		6.10	10.15	
3162	Alternate pricing method		1.10	21.818	M.B.F.		610		610	1,025	
3170	Blocking, in 16″ OC wall framing, 2″ x 4″	1 Clab	600	.013	L.F.		.37		.37	.61	
3172	2″ x 6″		400	.020			.55		.55	.91	
3174	In 24″ OC wall framing, 2″ x 4″		600	.013			.37		.37	.61	
3176	2″ x 6″		400	.020			.55		.55	.91	
3178	Alt method, wood blocking removal from wood framimg		.40	20	M.B.F.		550		550	910	
3179	Wood blocking removal from steel framimg		.36	22.222	″		610		610	1,000	
3180	Bracing, let in, 1″ x 3″, studs 16″ OC		1050	.008	L.F.		.21		.21	.35	
3181	Studs 24″ OC		1080	.007			.20		.20	.34	
3182	1″ x 4″, studs 16″ OC		1050	.008			.21		.21	.35	
3183	Studs 24″ OC		1080	.007			.20		.20	.34	
3184	1″ x 6″, studs 16″ OC		1050	.008			.21		.21	.35	
3185	Studs 24″ OC		1080	.007			.20		.20	.34	
3186	2″ x 3″, studs 16″ OC		800	.010			.27		.27	.45	
3187	Studs 24″ OC		830	.010			.26		.26	.44	
3188	2″ x 4″, studs 16″ OC		800	.010			.27		.27	.45	
3189	Studs 24″ OC		830	.010			.26		.26	.44	
3190	2″ x 6″, studs 16″ OC		800	.010			.27		.27	.45	
3191	Studs 24″ OC		830	.010			.26		.26	.44	
3192	2″ x 8″, studs 16″ OC		800	.010			.27		.27	.45	
3193	Studs 24″ OC		830	.010			.26		.26	.44	
3194	"T" shaped metal bracing, studs at 16″ OC		1060	.008			.21		.21	.34	
3195	Studs at 24″ OC		1200	.007			.18		.18	.30	
3196	Metal straps, studs at 16″ OC		1200	.007			.18		.18	.30	
3197	Studs at 24″ OC		1240	.006			.18		.18	.29	
3200	Columns, round, 8′ to 14′ tall		40	.200	Ea.		5.50		5.50	9.10	
3202	Dimensional lumber sizes	2 Clab	1.10	14.545	M.B.F.		400		400	660	
3250	Blocking, between joists	1 Clab	320	.025	Ea.		.69		.69	1.14	
3252	Bridging, metal strap, between joists		320	.025	Pr.		.69		.69	1.14	
3254	Wood, between joists		320	.025	″		.69		.69	1.14	
3260	Door buck, studs, header & access, 8′ high 2″ x 4″ wall, 3′ wide		32	.250	Ea.		6.85		6.85	11.35	
3261	4′ wide		32	.250			6.85		6.85	11.35	
3262	5′ wide		32	.250			6.85		6.85	11.35	
3263	6′ wide		32	.250			6.85		6.85	11.35	
3264	8′ wide		30	.267			7.30		7.30	12.10	
3265	10′ wide		30	.267			7.30		7.30	12.10	

	06052 Selective Demolition		CREW	DAILY OUTPUT	LABOR-HOURS	UNIT	2006 BARE COSTS MAT.	LABOR	EQUIP.	TOTAL	TOTAL INCL O&P	
110	3266	12′ wide R024119 -10	1 Clab	30	.267	Ea.		7.30		7.30	12.10	110
	3267	2″ x 6″ wall, 3′ wide		32	.250			6.85		6.85	11.35	
	3268	4′ wide		32	.250			6.85		6.85	11.35	
	3269	5′ wide		32	.250			6.85		6.85	11.35	
	3270	6′ wide		32	.250			6.85		6.85	11.35	
	3271	8′ wide		30	.267			7.30		7.30	12.10	
	3272	10′ wide		30	.267			7.30		7.30	12.10	
	3273	12′ wide		30	.267			7.30		7.30	12.10	
	3274	Window buck, studs, header & access, 8′ high 2″ x 4″ wall, 2′ wide		24	.333			9.15		9.15	15.15	
	3275	3′ wide		24	.333			9.15		9.15	15.15	
	3276	4′ wide		24	.333			9.15		9.15	15.15	
	3277	5′ wide		24	.333			9.15		9.15	15.15	
	3278	6′ wide		24	.333			9.15		9.15	15.15	
	3279	7′ wide		24	.333			9.15		9.15	15.15	
	3280	8′ wide		22	.364			9.95		9.95	16.50	
	3281	10′ wide		22	.364			9.95		9.95	16.50	
	3282	12′ wide		22	.364			9.95		9.95	16.50	
	3283	2″ x 6″ wall, 2′ wide		24	.333			9.15		9.15	15.15	
	3284	3′ wide		24	.333			9.15		9.15	15.15	
	3285	4′ wide		24	.333			9.15		9.15	15.15	
	3286	5′ wide		24	.333			9.15		9.15	15.15	
	3287	6′ wide		24	.333			9.15		9.15	15.15	
	3288	7′ wide		24	.333			9.15		9.15	15.15	
	3289	8′ wide		22	.364			9.95		9.95	16.50	
	3290	10′ wide		22	.364			9.95		9.95	16.50	
	3291	12′ wide		22	.364			9.95		9.95	16.50	
	3400	Fascia boards, 1″ x 6″		500	.016	L.F.		.44		.44	.73	
	3440	1″ x 8″		450	.018			.49		.49	.81	
	3480	1″ x 10″		400	.020			.55		.55	.91	
	3490	2″ x 6″		450	.018			.49		.49	.81	
	3500	2″ x 8″		400	.020			.55		.55	.91	
	3510	2″ x 10″		350	.023			.63		.63	1.04	
	3610	Furring, on wood walls or ceiling		4000	.002	S.F.		.05		.05	.09	
	3620	On masonry or concrete walls or ceiling		1200	.007	″		.18		.18	.30	
	3800	Headers over openings, 2 @ 2″ x 6″		110	.073	L.F.		1.99		1.99	3.30	
	3840	2 @ 2″ x 8″		100	.080			2.19		2.19	3.63	
	3880	2 @ 2″ x 10″		90	.089			2.44		2.44	4.04	
	3885	Alternate pricing method		.26	30.651	M.B.F.		840		840	1,400	
	3920	Joists, 1″ x 4″		1250	.006	L.F.		.18		.18	.29	
	3930	1″ x 6″		1135	.007			.19		.19	.32	
	3940	1″ x 8″		1000	.008			.22		.22	.36	
	3950	1″ x 10″		895	.009			.25		.25	.41	
	3960	1″ x 12″		765	.010			.29		.29	.47	
	4200	2″ x 4″	2 Clab	1000	.016			.44		.44	.73	
	4230	2″ x 6″		970	.016			.45		.45	.75	
	4240	2″ x 8″		940	.017			.47		.47	.77	
	4250	2″ x 10″		910	.018			.48		.48	.80	
	4280	2″ x 12″		880	.018			.50		.50	.83	
	4281	2″ x 14″		850	.019			.52		.52	.85	
	4282	Composite joists, 9-1/2″		960	.017			.46		.46	.76	
	4283	11-7/8″		930	.017			.47		.47	.78	
	4284	14″		897	.018			.49		.49	.81	
	4285	16″		865	.019			.51		.51	.84	
	4290	Wood joists, alternate pricing method		1.50	10.667	M.B.F.		292		292	485	
	4500	Open web joist, 12″ deep		500	.032	L.F.		.88		.88	1.45	
	4505	14″ deep		475	.034			.92		.92	1.53	

06052 | Selective Demolition

R024119-10

		06052 Selective Demolition	CREW	DAILY OUTPUT	LABOR-HOURS	UNIT	2006 BARE COSTS MAT.	LABOR	EQUIP.	TOTAL	TOTAL INCL O&P	
110	4510	16" deep	2 Clab	450	.036	L.F.		.97		.97	1.61	110
	4520	18" deep		425	.038			1.03		1.03	1.71	
	4530	24" deep		400	.040			1.10		1.10	1.82	
	4550	Ledger strips, 1" x 2"	1 Clab	1200	.007			.18		.18	.30	
	4560	1" x 3"		1200	.007			.18		.18	.30	
	4570	1" x 4"		1200	.007			.18		.18	.30	
	4580	2" x 2"		1100	.007			.20		.20	.33	
	4590	2" x 4"		1000	.008			.22		.22	.36	
	4600	2" x 6"		1000	.008			.22		.22	.36	
	4601	2" x 8 or 2" x 10"		800	.010			.27		.27	.45	
	4602	4" x 6"		600	.013			.37		.37	.61	
	4604	4" x 8"		450	.018			.49		.49	.81	
	5400	Posts, 4" x 4"	2 Clab	800	.020			.55		.55	.91	
	5405	4" x 6"		550	.029			.80		.80	1.32	
	5410	4" x 8"		440	.036			1		1	1.65	
	5425	4" x 10"		390	.041			1.12		1.12	1.86	
	5430	4" x 12"		350	.046			1.25		1.25	2.08	
	5440	6" x 6"		400	.040			1.10		1.10	1.82	
	5445	6" x 8"		350	.046			1.25		1.25	2.08	
	5450	6" x 10"		320	.050			1.37		1.37	2.27	
	5455	6" x 12"		290	.055			1.51		1.51	2.50	
	5480	8" x 8"		300	.053			1.46		1.46	2.42	
	5500	10" x 10"		240	.067			1.83		1.83	3.03	
	5660	Tongue and groove floor planks		2	8	M.B.F.		219		219	365	
	5750	Rafters, ordinary, 16" OC, 2" x 4"		880	.018	S.F.		.50		.50	.83	
	5755	2" x 6"		840	.019			.52		.52	.86	
	5760	2" x 8"		820	.020			.53		.53	.89	
	5770	2" x 10"		820	.020			.53		.53	.89	
	5780	2" x 12"		810	.020			.54		.54	.90	
	5785	24" OC, 2" x 4"		1170	.014			.37		.37	.62	
	5786	2" x 6"		1117	.014			.39		.39	.65	
	5787	2" x 8"		1091	.015			.40		.40	.67	
	5788	2" x 10"		1091	.015			.40		.40	.67	
	5789	2" x 12"		1077	.015			.41		.41	.67	
	5795	Rafters, ordinary, 2" x 4" (alternate method)		862	.019	L.F.		.51		.51	.84	
	5800	2" x 6" (alternate method)		850	.019			.52		.52	.85	
	5840	2" x 8" (alternate method)		837	.019			.52		.52	.87	
	5855	2" x 10" (alternate method)		825	.019			.53		.53	.88	
	5865	2" x 12" (alternate method)		812	.020			.54		.54	.89	
	5870	Sill plate, 2" x 4"	1 Clab	1170	.007			.19		.19	.31	
	5871	2" x 6"		780	.010			.28		.28	.47	
	5872	2" x 8"		586	.014			.37		.37	.62	
	5873	Alternate pricing method		.78	10.256	M.B.F.		281		281	465	
	5885	Ridge board, 1" x 4"	2 Clab	900	.018	L.F.		.49		.49	.81	
	5886	1" x 6"		875	.018			.50		.50	.83	
	5887	1" x 8"		850	.019			.52		.52	.85	
	5888	1" x 10"		825	.019			.53		.53	.88	
	5889	1" x 12"		800	.020			.55		.55	.91	
	5890	2" x 4"		900	.018			.49		.49	.81	
	5892	2" x 6"		875	.018			.50		.50	.83	
	5894	2" x 8"		850	.019			.52		.52	.85	
	5896	2" x 10"		825	.019			.53		.53	.88	
	5898	2" x 12"		800	.020			.55		.55	.91	
	5900	Hip & valley rafters, 2" x 6"		500	.032			.88		.88	1.45	
	5940	2" x 8"		420	.038			1.04		1.04	1.73	
	6050	Rafter tie, 1" x 4"		1250	.013			.35		.35	.58	

06050 | Basic Wood / Plastic Materials / Methods

	06052 Selective Demolition		CREW	DAILY OUTPUT	LABOR-HOURS	UNIT	2006 BARE COSTS MAT.	LABOR	EQUIP.	TOTAL	TOTAL INCL O&P	
110	6052	1″ x 6″ R024119-10	2 Clab	1135	.014	L.F.		.39		.39	.64	110
	6054	2″ x 4″		1000	.016			.44		.44	.73	
	6056	2″ x 6″		970	.016			.45		.45	.75	
	6070	Sleepers, on concrete, 1″ x 2″	1 Clab	4700	.002			.05		.05	.08	
	6075	1″ x 3″		4000	.002			.05		.05	.09	
	6080	2″ x 4″		3000	.003			.07		.07	.12	
	6085	2″ x 6″		2600	.003			.08		.08	.14	
	6086	Sheathing from roof, 5/16″	2 Clab	1600	.010	S.F.		.27		.27	.45	
	6088	3/8″		1525	.010			.29		.29	.48	
	6090	1/2″		1400	.011			.31		.31	.52	
	6092	5/8″		1300	.012			.34		.34	.56	
	6094	3/4″		1200	.013			.37		.37	.61	
	6096	Board sheathing from roof		1400	.011			.31		.31	.52	
	6100	Sheathing, from walls, 1/4″		1200	.013			.37		.37	.61	
	6110	5/16″		1175	.014			.37		.37	.62	
	6120	3/8″		1150	.014			.38		.38	.63	
	6130	1/2″		1125	.014			.39		.39	.65	
	6140	5/8″		1100	.015			.40		.40	.66	
	6150	3/4″		1075	.015			.41		.41	.68	
	6152	Board sheathing from walls		1500	.011			.29		.29	.48	
	6158	Subfloor, with boards		1050	.015			.42		.42	.69	
	6160	Plywood, 1/2″ thick		768	.021			.57		.57	.95	
	6162	5/8″ thick		760	.021			.58		.58	.96	
	6164	3/4″thick		750	.021			.58		.58	.97	
	6165	1-1/8″ thick		720	.022			.61		.61	1.01	
	6166	Underlayment, particle board, 3/8″ thick	1 Clab	780	.010			.28		.28	.47	
	6168	1/2″ thick		768	.010			.29		.29	.47	
	6170	5/8″ thick		760	.011			.29		.29	.48	
	6172	3/4″ thick		750	.011			.29		.29	.48	
	6200	Stairs and stringers, minimum	2 Clab	40	.400	Riser		10.95		10.95	18.15	
	6240	Maximum	″	26	.615	″		16.85		16.85	28	
	6300	Components, tread	1 Clab	110	.073	Ea.		1.99		1.99	3.30	
	6320	Riser		80	.100	″		2.74		2.74	4.54	
	6390	Stringer, 2″ x 10″		260	.031	L.F.		.84		.84	1.40	
	6400	2″ x 12″		260	.031			.84		.84	1.40	
	6410	3″ x 10″		250	.032			.88		.88	1.45	
	6420	3″ x 12″		250	.032			.88		.88	1.45	
	6590	Wood studs, 2″ x 3″	2 Clab	3076	.005			.14		.14	.24	
	6600	2″ x 4″		2000	.008			.22		.22	.36	
	6640	2″ x 6″		1600	.010			.27		.27	.45	
	6720	Wall framing, including studs plates and blocking, 2″ x 4″	1 Clab	600	.013	S.F.		.37		.37	.61	
	6740	2″ x 6″		480	.017	″		.46		.46	.76	
	6750	Headers, 2″ x 4″		1125	.007	L.F.		.19		.19	.32	
	6755	2″ x 6″		1125	.007			.19		.19	.32	
	6760	2″ x 8″		1050	.008			.21		.21	.35	
	6765	2″ x 10″		1050	.008			.21		.21	.35	
	6770	2″ x 12″		1000	.008			.22		.22	.36	
	6780	4″ x 10″		525	.015			.42		.42	.69	
	6785	4″ x 12″		500	.016			.44		.44	.73	
	6790	6″ x 8″		560	.014			.39		.39	.65	
	6795	6″ x 10″		525	.015			.42		.42	.69	
	6797	6″ x 12″		500	.016			.44		.44	.73	
	8000	Soffit, T & G wood		520	.015	S.F.		.42		.42	.70	
	8010	Hardboard, vinyl or aluminum		640	.013			.34		.34	.57	
	8030	Plywood	2 Carp	315	.051			1.81		1.81	2.99	
	9000	Minimum labor/equipment charge	1 Clab	4	2	Job		55		55	91	

WOOD & PLASTICS 6

06052 | Selective Demolition

			CREW	DAILY OUTPUT	LABOR-HOURS	UNIT	2006 BARE COSTS MAT.	LABOR	EQUIP.	TOTAL	TOTAL INCL O&P	
110	9500	See Div. 02220-350 for rubbish handling (R024119-10)										110
120	0010	SELECTIVE DEMOLITION, MILLWORK AND TRIM (R024119-10)										120
	1000	Cabinets, wood, base cabinets, per L.F.	2 Clab	80	.200	L.F.		5.50		5.50	9.10	
	1020	Wall cabinets, per L.F.	"	80	.200	"		5.50		5.50	9.10	
	1060	Remove and reset, base cabinets	2 Carp	18	.889	Ea.		31.50		31.50	52.50	
	1070	Wall cabinets	"	20	.800	"		28.50		28.50	47	
	1100	Steel, painted, base cabinets	2 Clab	60	.267	L.F.		7.30		7.30	12.10	
	1120	Wall cabinets		60	.267	"		7.30		7.30	12.10	
	1200	Casework, large area		320	.050	S.F.		1.37		1.37	2.27	
	1220	Selective		200	.080	"		2.19		2.19	3.63	
	1500	Counter top, minimum		200	.080	L.F.		2.19		2.19	3.63	
	1510	Maximum		120	.133			3.65		3.65	6.05	
	1550	Remove and reset, minimum	2 Carp	50	.320			11.40		11.40	18.85	
	1560	Maximum	"	40	.400			14.20		14.20	23.50	
	2000	Paneling, 4′ x 8′ sheets	2 Clab	2000	.008	S.F.		.22		.22	.36	
	2100	Boards, 1″ x 4″		700	.023			.63		.63	1.04	
	2120	1″ x 6″		750	.021			.58		.58	.97	
	2140	1″ x 8″		800	.020			.55		.55	.91	
	3000	Trim, baseboard, to 6″ wide		1200	.013	L.F.		.37		.37	.61	
	3040	Greater than 6″ and up to 12″ wide		1000	.016			.44		.44	.73	
	3080	Remove and reset, minimum	2 Carp	400	.040			1.42		1.42	2.36	
	3090	Maximum	"	300	.053			1.90		1.90	3.14	
	3100	Ceiling trim	2 Clab	1000	.016			.44		.44	.73	
	3120	Chair rail		1200	.013			.37		.37	.61	
	3140	Railings with balusters		240	.067			1.83		1.83	3.03	
	3160	Wainscoting		700	.023	S.F.		.63		.63	1.04	
	9000	Minimum labor/equipment charge	1 Clab	4	2	Job		55		55	91	

06070 | Lumber Treatment

			CREW	DAILY OUTPUT	LABOR-HOURS	UNIT	MAT.	LABOR	EQUIP.	TOTAL	TOTAL INCL O&P	
400	0011	LUMBER TREATMENT										400
	0400	Fire retardant, wet				M.B.F.	310			310	340	
	0500	KDAT					297			297	325	
	0700	Salt treated, water borne, .40 lb. retention					142			142	156	
	0800	Oil borne, 8 lb. retention					166			166	183	
	1000	Kiln dried lumber, 1″ & 2″ thick, softwoods					95			95	104	
	1100	Hardwoods					101			101	111	
	1500	For small size 1″ stock, add					12.80			12.80	14.10	
	1700	For full size rough lumber, add					20%					
600	0010	PLYWOOD TREATMENT										600
	0020	Fire retardant, 1/4″ thick				M.S.F.	237			237	261	
	0030	3/8″ thick					261			261	287	
	0050	1/2″ thick					279			279	305	
	0070	5/8″ thick					297			297	325	
	0100	3/4″ thick					325			325	360	
	0200	For KDAT, add					71			71	78	
	0500	Salt treated water borne, .25 lb., wet, 1/4″ thick					130			130	143	
	0530	3/8″ thick					136			136	150	
	0550	1/2″ thick					142			142	156	
	0570	5/8″ thick					155			155	170	
	0600	3/4″ thick					161			161	177	
	0800	For KDAT add					71			71	78	
	0900	For .40 lb., per C.F. retention, add					59.50			59.50	65.50	
	1000	For certification stamp, add					35			35	38.50	

	06090	Wood & Plastic Fastenings	CREW	DAILY OUTPUT	LABOR-HOURS	UNIT	2006 BARE COSTS MAT.	LABOR	EQUIP.	TOTAL	TOTAL INCL O&P	
600	0010	**NAILS**, material only, based upon 50# box purchase										600
	0020	Copper nails, plain				Lb.	6.60			6.60	7.25	
	0400	Stainless steel, plain					5.95			5.95	6.55	
	0500	Box, 3d to 20d, bright					.94			.94	1.03	
	0520	Galvanized					1.38			1.38	1.52	
	0600	Common, 3d to 60d, plain					.78			.78	.86	
	0700	Galvanized					1.02			1.02	1.12	
	0800	Aluminum					4.45			4.45	4.90	
	1000	Annular or spiral thread, 4d to 60d, plain					1.88			1.88	2.07	
	1200	Galvanized					1.95			1.95	2.15	
	1400	Drywall nails, plain					.79			.79	.87	
	1600	Galvanized					1.54			1.54	1.69	
	1800	Finish nails, 4d to 10d, plain					1.05			1.05	1.16	
	2000	Galvanized					1.32			1.32	1.45	
	2100	Aluminum					4.10			4.10	4.51	
	2300	Flooring nails, hardened steel, 2d to 10d, plain					1.61			1.61	1.77	
	2400	Galvanized					2.25			2.25	2.48	
	2500	Gypsum lath nails, 1-1/8″, 13 ga. flathead, blued					1.54			1.54	1.69	
	2600	Masonry nails, hardened steel, 3/4″ to 3″ long, plain					1.48			1.48	1.63	
	2700	Galvanized					1.55			1.55	1.71	
	2900	Roofing nails, threaded, galvanized					1.32			1.32	1.45	
	3100	Aluminum					4.80			4.80	5.30	
	3300	Compressed lead head, threaded, galvanized					1.50			1.50	1.65	
	3600	Siding nails, plain shank, galvanized					1.45			1.45	1.60	
	3800	Aluminum					4.11			4.11	4.52	
	5000	Add to prices above for cement coating					.10			.10	.11	
	5200	Zinc or tin plating					.13			.13	.14	
	5500	Vinyl coated sinkers, 8d to 16d				↓	.57			.57	.63	
650	0010	**NAILS**, pneumatic tools										650
	0020	Framing, per carton of 5000, 2″				Ea.	37			37	41	
	0100	2-3/8″					42.50			42.50	46.50	
	0200	Per carton of 4000, 3″					37.50			37.50	41.50	
	0300	3-1/4″					40			40	44	
	0400	Per carton of 5000, 2-3/8″, galv.					57.50			57.50	63.50	
	0500	Per carton of 4000, 3″, galv.					65			65	71.50	
	0600	3-1/4″, galv.					80.50			80.50	88.50	
	0700	Roofing, per carton of 7200, 1″					35			35	38.50	
	0800	1-1/4″					32.50			32.50	36	
	0900	1-1/2″					37.50			37.50	41.50	
	1000	1-3/4″				↓	45.50			45.50	50	
750	0010	**WOOD SCREWS**										750
	0020	Steel, #8 x 1″ long				C	3.29			3.29	3.62	
	0100	Brass					11.50			11.50	12.65	
	0600	#10, 2″ long, steel					6.30			6.30	6.90	
	0700	Brass					23			23	25.50	
	1500	#12, 3″ long, steel				↓	11.50			11.50	12.65	
800	0010	**TIMBER CONNECTORS** Add up cost of each part for total										800
	0020	cost of connection										
	0100	Connector plates, steel, with bolts, straight	2 Carp	75	.213	Ea.	22.50	7.60		30.10	37.50	
	0110	Tee		50	.320		33	11.40		44.40	55.50	
	0120	T- Strap, 14 gauge, 12″ x 8″ x 2″		50	.320		33	11.40		44.40	55.50	
	0150	Anchor plate, 7 gauge, 9″ x 7″	↓	75	.213		22.50	7.60		30.10	37.50	
	0200	Bolts, machine, sq. hd. with nut & washer, 1/2″ diameter, 4″ long	1 Carp	140	.057		.38	2.03		2.41	3.79	
	0300	7-1/2″ long		130	.062		.67	2.19		2.86	4.36	
	0500	3/4″ diameter, 7-1/2″ long		130	.062		1.83	2.19		4.02	5.65	
	0610	Machine bolts, w/ nut, washer, 3/4″ dia, 15″ L, HD's & beam hangers	↓	95	.084	↓	3.36	2.99		6.35	8.65	

	06090 | Wood & Plastic Fastenings		CREW	DAILY OUTPUT	LABOR-HOURS	UNIT	2006 BARE COSTS MAT.	LABOR	EQUIP.	TOTAL	TOTAL INCL O&P	
800	0800	Drilling bolt holes in timber, 1/2″ diameter	1 Carp	450	.018	Inch		.63		.63	1.05	800
	0900	1″ diameter		350	.023	″		.81		.81	1.35	
	1100	Framing anchors, 2 or 3 dimensional, 10 gauge, no nails incl.		175	.046	Ea.	.56	1.63		2.19	3.31	
	1150	Framing anchors, 18 gauge, 4 1/2″ x 2 3/4″		175	.046		.56	1.63		2.19	3.31	
	1160	Framing anchors, 18 gauge, 4 1/2″ x 3″		175	.046		.56	1.63		2.19	3.31	
	1170	Clip anchors plates, 18 gauge, 12″ x 1 1/8″		175	.046		.56	1.63		2.19	3.31	
	1250	Holdowns, 3 gauge base, 10 gauge body		8	1		18.75	35.50		54.25	79.50	
	1260	Holdowns, 7 gauge 11 1/16″ x 3 1/4″		8	1		18.75	35.50		54.25	79.50	
	1270	Holdowns, 7 gauge 14 3/8″ x 3 1/8″		8	1		18.75	35.50		54.25	79.50	
	1275	Holdowns, 12 gauge 8″ x 2 1/2″		8	1		18.75	35.50		54.25	79.50	
	1300	Joist and beam hangers, 18 ga. galv., for 2″ x 4″ joist		175	.046		.70	1.63		2.33	3.46	
	1400	2″ x 6″ to 2″ x 10″ joist		165	.048		.80	1.72		2.52	3.74	
	1600	16 ga. galv., 3″ x 6″ to 3″ x 10″ joist		160	.050		3.16	1.78		4.94	6.45	
	1700	3″ x 10″ to 3″ x 14″ joist		160	.050		3.67	1.78		5.45	7	
	1800	4″ x 6″ to 4″ x 10″ joist		155	.052		2.78	1.83		4.61	6.10	
	1900	4″ x 10″ to 4″ x 14″ joist		155	.052		3.77	1.83		5.60	7.20	
	2000	Two-2″ x 6″ to two-2″ x 10″ joists		150	.053		3.04	1.90		4.94	6.50	
	2100	Two-2″ x 10″ to two-2″ x 14″ joists		150	.053		3.04	1.90		4.94	6.50	
	2300	3/16″ thick, 6″ x 8″ joist		145	.055		6.60	1.96		8.56	10.50	
	2400	6″ x 10″ joist		140	.057		7.80	2.03		9.83	11.90	
	2500	6″ x 12″ joist		135	.059		9.40	2.11		11.51	13.80	
	2700	1/4″ thick, 6″ x 14″ joist		130	.062		11.65	2.19		13.84	16.40	
	2900	Plywood clips, extruded aluminum H clip, for 3/4″ panels					.18			.18	.20	
	3000	Galvanized 18 ga. back-up clip					.17			.17	.19	
	3200	Post framing, 16 ga. galv. for 4″ x 4″ base, 2 piece	1 Carp	130	.062		6.55	2.19		8.74	10.80	
	3300	Cap		130	.062		3.21	2.19		5.40	7.15	
	3500	Rafter anchors, 18 ga. galv., 1-1/2″ wide, 5-1/4″ long		145	.055		.55	1.96		2.51	3.86	
	3600	10-3/4″ long		145	.055		1.09	1.96		3.05	4.45	
	3800	Shear plates, 2-5/8″ diameter		120	.067		1.96	2.37		4.33	6.10	
	3900	4″ diameter		115	.070		4.50	2.47		6.97	9.05	
	4000	Sill anchors, embedded in concrete or block, 18-5/8″ long		115	.070		1.32	2.47		3.79	5.55	
	4100	Spike grids, 4″ x 4″, flat or curved		120	.067		.49	2.37		2.86	4.47	
	4400	Split rings, 2-1/2″ diameter		120	.067		1.58	2.37		3.95	5.65	
	4500	4″ diameter		110	.073		2.45	2.59		5.04	7	
	4550	Tie plate, 20 gauge, 7″ x 3 1/8″		110	.073		2.45	2.59		5.04	7	
	4560	Tie plate, 20 gauge, 5″ x 4 1/8″		110	.073		2.45	2.59		5.04	7	
	4575	Twist straps, 18 gauge, 12″ x 1 1/4″		110	.073		2.45	2.59		5.04	7	
	4580	Twist straps, 18 gauge, 16″ x 1 1/4″		110	.073		2.45	2.59		5.04	7	
	4600	Strap ties, 20 ga., 2 -1/16″ wide, 12 13/16″ long		180	.044		1.43	1.58		3.01	4.19	
	4700	Strap ties, 16 ga., 1-3/8″ wide, 12″ long		180	.044		1.43	1.58		3.01	4.19	
	4800	24″ long		160	.050		1.97	1.78		3.75	5.10	
	5000	Toothed rings, 2-5/8″ or 4″ diameter		90	.089		1.35	3.16		4.51	6.75	
	5200	Truss plates, nailed, 20 gauge, up to 32′ span		17	.471	Truss	9.75	16.75		26.50	38.50	
	5400	Washers, 2″ x 2″ x 1/8″				Ea.	.31			.31	.34	
	5500	3″ x 3″ x 3/16″				″	.80			.80	.88	
	9000	Minimum labor/equipment charge	1 Carp	4	2	Job		71		71	118	
825	0010	**ROUGH HARDWARE**, average % of carpentry material										825
	0020	Minimum					.50%					
	0200	Maximum					1.50%					
850	0010	**BRACING**										850
	0300	Let-in, "T" shaped, 22 ga. galv. steel, studs at 16″ O.C.	1 Carp	5.80	1.379	C.L.F.	53	49		102	140	
	0400	Studs at 24″ O.C.		6	1.333		53	47.50		100.50	137	
	0500	16 ga. galv. steel straps, studs at 16″ O.C.		6	1.333		80	47.50		127.50	167	
	0600	Studs at 24″ O.C.		6.20	1.290		80	46		126	164	

6 WOOD & PLASTICS

		06110 | Wood Framing	CREW	DAILY OUTPUT	LABOR-HOURS	UNIT	2006 BARE COSTS MAT.	LABOR	EQUIP.	TOTAL	TOTAL INCL O&P	
100	0010	**BLOCKING**										100
	2600	Miscellaneous, to wood construction										
	2620	2″ x 4″	1 Carp	.17	47.059	M.B.F.	560	1,675		2,235	3,400	
	2625	Pneumatic nailed		.21	38.095		560	1,350		1,910	2,875	
	2660	2″ x 8″		.27	29.630		640	1,050		1,690	2,450	
	2665	Pneumatic nailed		.33	24.242		640	860		1,500	2,125	
	2720	To steel construction										
	2740	2″ x 4″	1 Carp	.14	57.143	M.B.F.	560	2,025		2,585	4,000	
	2780	2″ x 8″		.21	38.095	″	640	1,350		1,990	2,950	
	9000	Minimum labor/equipment charge		4	2	Job		71		71	118	
150	0012	**BRACING** Let-in, with 1″ x 6″ boards, studs @ 16″ O.C.		150	.053	L.F.	.55	1.90		2.45	3.75	150
	0202	Studs @ 24″ O.C.		230	.035	″	.55	1.24		1.79	2.66	
200	0012	**BRIDGING** Wood, for joists 16″ O.C., 1″ x 3″		130	.062	Pr.	.42	2.19		2.61	4.09	200
	0017	Pneumatic nailed		170	.047		.42	1.67		2.09	3.24	
	0102	2″ x 3″ bridging		130	.062		.57	2.19		2.76	4.25	
	0107	Pneumatic nailed		170	.047		.57	1.67		2.24	3.40	
	0302	Steel, galvanized, 18 ga., for 2″ x 10″ joists at 12″ O.C.		130	.062		1.18	2.19		3.37	4.91	
	0402	24″ O.C.		140	.057		1.42	2.03		3.45	4.93	
	0902	Compression type, 16″ O.C., 2″ x 8″ joists		200	.040		1.38	1.42		2.80	3.88	
	1002	2″ x 12″ joists		200	.040		1.38	1.42		2.80	3.88	
505	0010	**FRAMING, BEAMS & GIRDERS** R061110-30										505
	1002	Single, 2″ x 6″	2 Carp	700	.023	L.F.	.59	.81		1.40	2	
	1007	Pneumatic nailed		812	.020		.59	.70		1.29	1.81	
	1022	2″ x 8″		650	.025		.85	.88		1.73	2.39	
	1027	Pneumatic nailed		754	.021		.85	.75		1.60	2.19	
	1042	2″ x 10″		600	.027		1.29	.95		2.24	2.99	
	1047	Pneumatic nailed		696	.023		1.29	.82		2.11	2.77	
	1062	2″ x 12″		550	.029		1.60	1.03		2.63	3.47	
	1067	Pneumatic nailed		638	.025		1.60	.89		2.49	3.24	
	1082	2″ x 14″		500	.032		2.18	1.14		3.32	4.28	
	1087	Pneumatic nailed		580	.028		2.18	.98		3.16	4.03	
	1102	3″ x 8″		550	.029		2.59	1.03		3.62	4.56	
	1122	3″ x 10″		500	.032		3.26	1.14		4.40	5.45	
	1142	3″ x 12″		450	.036		3.92	1.26		5.18	6.40	
	1162	3″ x 14″		400	.040		4.73	1.42		6.15	7.55	
	1170	4″ x 6″	F-3	1100	.036		1.87	1.31	.58	3.76	4.85	
	1182	4″ x 8″		1000	.040		2.59	1.44	.64	4.67	5.90	
	1202	4″ x 10″		950	.042		3.26	1.52	.67	5.45	6.80	
	1222	4″ x 12″		900	.044		4.04	1.60	.71	6.35	7.85	
	1242	4″ x 14″		850	.047		4.95	1.70	.75	7.40	9.05	
	2002	Double, 2″ x 6″	2 Carp	625	.026		1.18	.91		2.09	2.81	
	2007	Pneumatic nailed		725	.022		1.18	.78		1.96	2.60	
	2022	2″ x 8″		575	.028		1.71	.99		2.70	3.52	
	2027	Pneumatic nailed		667	.024		1.71	.85		2.56	3.29	
	2042	2″ x 10″		550	.029		2.58	1.03		3.61	4.55	
	2047	Pneumatic nailed		638	.025		2.58	.89		3.47	4.32	
	2062	2″ x 12″		525	.030		3.20	1.08		4.28	5.30	
	2067	Pneumatic nailed		610	.026		3.20	.93		4.13	5.05	
	2082	2″ x 14″		475	.034		4.36	1.20		5.56	6.75	
	2087	Pneumatic nailed		551	.029		4.36	1.03		5.39	6.50	
	3002	Triple, 2″ x 6″		550	.029		1.78	1.03		2.81	3.66	
	3007	Pneumatic nailed		638	.025		1.78	.89		2.67	3.43	
	3022	2″ x 8″		525	.030		2.56	1.08		3.64	4.62	
	3027	Pneumatic nailed		609	.026		2.56	.93		3.49	4.37	
	3042	2″ x 10″		500	.032		3.87	1.14		5.01	6.15	
	3047	Pneumatic nailed		580	.028		3.87	.98		4.85	5.90	

	06110 | Wood Framing		CREW	DAILY OUTPUT	LABOR-HOURS	UNIT	2006 BARE COSTS MAT.	LABOR	EQUIP.	TOTAL	TOTAL INCL O&P	
505	3062	2″ x 12″ R061110-30	2 Carp	475	.034	L.F.	4.80	1.20		6	7.30	505
	3067	Pneumatic nailed		551	.029		4.80	1.03		5.83	7	
	3082	2″ x 14″		450	.036		6.55	1.26		7.81	9.30	
	3087	Pneumatic nailed	↓	522	.031	↓	6.55	1.09		7.64	9	
	9000	Minimum labor/equipment charge	1 Carp	2	4	Job		142		142	236	
510	0010	**FRAMING, CEILINGS**										510
	6002	Suspended, 2″ x 3″	2 Carp	1000	.016	L.F.	.38	.57		.95	1.36	
	6052	2″ x 4″		900	.018		.37	.63		1	1.46	
	6102	2″ x 6″		800	.020		.59	.71		1.30	1.83	
	6152	2″ x 8″	↓	650	.025	↓	.85	.88		1.73	2.39	
	9000	Minimum labor/equipment charge	1 Carp	4	2	Job		71		71	118	
515	0010	**FRAMING, COLUMNS**										515
	0100	4″ x 4″	2 Carp	390	.041	L.F.	1.25	1.46		2.71	3.80	
	0150	4″ x 6″		275	.058		1.87	2.07		3.94	5.50	
	0200	4″ x 8″		220	.073		2.59	2.59		5.18	7.15	
	0250	6″ x 6″		215	.074		4.54	2.65		7.19	9.35	
	0300	6″ x 8″		175	.091		6.05	3.25		9.30	12.05	
	0350	6″ x 10″	↓	150	.107	↓	10.90	3.79		14.69	18.30	
	9000	Minimum labor/equipment charge	1 Carp	2	4	Job		142		142	236	
520	0010	**HEAVY MILL TIMBER FRAMING**										520
	0020	Beams, single 6″ x 10″	2 Carp	1.10	14.545	M.B.F.	2,150	515		2,665	3,225	
	0100	Single 8″ x 16″		1.20	13.333	″	2,700	475		3,175	3,750	
	0202	Built from 2″ lumber, multiple 2″ x 14″		900	.018	B.F.	.93	.63		1.56	2.08	
	0212	Built from 3″ lumber, multiple 3″ x 6″		700	.023		1.28	.81		2.09	2.76	
	0222	Multiple 3″ x 8″		800	.020		1.30	.71		2.01	2.60	
	0232	Multiple 3″ x 10″		900	.018		1.30	.63		1.93	2.48	
	0242	Multiple 3″ x 12″		1000	.016		1.31	.57		1.88	2.38	
	0252	Built from 4″ lumber, multiple 4″ x 6″		800	.020		.94	.71		1.65	2.21	
	0262	Multiple 4″ x 8″		900	.018		.97	.63		1.60	2.12	
	0272	Multiple 4″ x 10″	↓	1000	.016	↓	.98	.57		1.55	2.02	
	0281											
	0282	Multiple 4″ x 12″	2 Carp	1100	.015	B.F.	1.01	.52		1.53	1.97	
	0292	Columns, structural grade, 1500f, 4″ x 4″		450	.036	L.F.	2.34	1.26		3.60	4.66	
	0302	6″ x 6″		225	.071		6.40	2.53		8.93	11.25	
	0402	8″ x 8″		240	.067		12.45	2.37		14.82	17.65	
	0502	10″ x 10″		90	.178		19.85	6.30		26.15	32.50	
	0602	12″ x 12″		70	.229	↓	28.50	8.15		36.65	45	
	0802	Floor planks, 2″ thick, T & G, 2″ x 6″		1050	.015	B.F.	1.88	.54		2.42	2.97	
	0902	2″ x 10″		1100	.015		1.88	.52		2.40	2.93	
	1102	3″ thick, 3″ x 6″		1050	.015		1.35	.54		1.89	2.39	
	1202	3″ x 10″		1100	.015		1.35	.52		1.87	2.35	
	1402	Girders, structural grade, 12″ x 12″		800	.020		1.94	.71		2.65	3.31	
	1502	10″ x 16″	↓	1000	.016	↓	1.89	.57		2.46	3.02	
	2050	Roof planks, see division 06150-600										
	2302	Roof purlins, 4″ thick, structural grade	2 Carp	1050	.015	B.F.	1.32	.54		1.86	2.35	
	2502	Roof trusses, add timber connectors, division 06090-800	″	450	.036	″	1.29	1.26		2.55	3.51	
	9000	Minimum labor/equipment charge	1 Carp	2	4	Job		142		142	236	
530	0010	**FRAMING, JOISTS** R061110-30										530
	2002	Joists, 2″ x 4″	2 Carp	1250	.013	L.F.	.37	.46		.83	1.16	
	2007	Pneumatic nailed		1438	.011		.37	.40		.77	1.07	
	2100	2″ x 6″		1250	.013		.59	.46		1.05	1.40	
	2105	Pneumatic nailed		1438	.011		.59	.40		.99	1.31	
	2152	2″ x 8″		1100	.015		.85	.52		1.37	1.80	
	2157	Pneumatic nailed		1265	.013		.85	.45		1.30	1.69	
	2202	2″ x 10″	↓	900	.018	↓	1.29	.63		1.92	2.47	

	06110 | Wood Framing		CREW	DAILY OUTPUT	LABOR-HOURS	UNIT	2006 BARE COSTS MAT.	LABOR	EQUIP.	TOTAL	TOTAL INCL O&P	
530	2207	Pneumatic nailed R061110-30	2 Carp	1035	.015	L.F.	1.29	.55		1.84	2.33	530
	2252	2" x 12"		875	.018		1.60	.65		2.25	2.84	
	2257	Pneumatic nailed		1006	.016		1.60	.57		2.17	2.70	
	2302	2" x 14"		770	.021		2.18	.74		2.92	3.62	
	2307	Pneumatic nailed		886	.018		2.18	.64		2.82	3.46	
	2352	3" x 6"		925	.017		1.92	.62		2.54	3.13	
	2402	3" x 10"		780	.021		3.26	.73		3.99	4.80	
	2452	3" x 12"		600	.027		3.92	.95		4.87	5.90	
	2502	4" x 6"		800	.020		1.87	.71		2.58	3.24	
	2552	4" x 10"		600	.027		3.26	.95		4.21	5.15	
	2602	4" x 12"		450	.036		4.04	1.26		5.30	6.55	
	2607	Sister joist, 2" x 6"		800	.020		.59	.71		1.30	1.83	
	2608	Pneumatic nailed		960	.017		.59	.59		1.18	1.63	
	2612	2" x 8"		640	.025		.85	.89		1.74	2.41	
	2613	Pneumatic nailed		768	.021		.85	.74		1.59	2.17	
	2617	2" x 10"		535	.030		1.29	1.06		2.35	3.18	
	2618	Pneumatic nailed		642	.025		1.29	.89		2.18	2.89	
	2622	2" x 12"		455	.035		1.60	1.25		2.85	3.83	
	2627	Pneumatic nailed		546	.029		1.60	1.04		2.64	3.49	
	3000	Composite wood joist 9-1/2" deep		.90	17.778	M.L.F.	1,825	630		2,455	3,050	
	3010	11-1/2" deep		.88	18.182		1,950	645		2,595	3,225	
	3020	14" deep		.82	19.512		2,250	695		2,945	3,625	
	3030	16" deep		.78	20.513		2,550	730		3,280	4,000	
	4000	Open web joist 12" deep		.88	18.182		1,825	645		2,470	3,075	
	4010	14" deep		.82	19.512		2,125	695		2,820	3,475	
	4020	16" deep		.78	20.513		2,200	730		2,930	3,625	
	4030	18" deep		.74	21.622		2,250	770		3,020	3,750	
	6000	Composite rim joist, 1-1/4" x 9-1/2"		90	.178		2,000	6.30		2,006.30	2,200	
	6010	1-1/4" x 11-1/2"		.88	18.182		2,225	645		2,870	3,525	
	6020	1-1/4" x 14-1/2"		.82	19.512		2,800	695		3,495	4,225	
	6030	1-1/4" x 16-1/2"		.78	20.513		3,200	730		3,930	4,700	
	9000	Minimum labor/equipment charge	1 Carp	4	2	Job		71		71	118	
545	0010	**FRAMING, MISCELLANEOUS**										545
	2002	Firestops, 2" x 4"	2 Carp	780	.021	L.F.	.37	.73		1.10	1.62	
	2007	Pneumatic nailed		952	.017		.37	.60		.97	1.40	
	2102	2" x 6"		600	.027		.59	.95		1.54	2.22	
	2107	Pneumatic nailed		732	.022		.59	.78		1.37	1.94	
	5002	Nailers, treated, wood construction, 2" x 4"		800	.020		.49	.71		1.20	1.71	
	5007	Pneumatic nailed		960	.017		.49	.59		1.08	1.51	
	5102	2" x 6"		750	.021		.75	.76		1.51	2.08	
	5107	Pneumatic nailed		900	.018		.75	.63		1.38	1.87	
	5122	2" x 8"		700	.023		.98	.81		1.79	2.42	
	5127	Pneumatic nailed		840	.019		.98	.68		1.66	2.19	
	5202	Steel construction, 2" x 4"		750	.021		.49	.76		1.25	1.79	
	5222	2" x 6"		700	.023		.75	.81		1.56	2.17	
	5242	2" x 8"		650	.025		.98	.88		1.86	2.52	
	7002	Rough bucks, treated, for doors or windows, 2" x 6"		400	.040		.75	1.42		2.17	3.18	
	7007	Pneumatic nailed		480	.033		.75	1.18		1.93	2.78	
	7102	2" x 8"		380	.042		.98	1.50		2.48	3.55	
	7107	Pneumatic nailed		456	.035		.98	1.25		2.23	3.14	
	8000	Stair stringers, 2" x 10"		130	.123		1.29	4.38		5.67	8.65	
	8100	2" x 12"		130	.123		1.60	4.38		5.98	9	
	8150	3" x 10"		125	.128		3.26	4.55		7.81	11.15	
	8200	3" x 12"		125	.128		3.92	4.55		8.47	11.85	
	8870	Composite LSL, 1-1/4" x 11-1/2"		130	.123		2.23	4.38		6.61	9.70	
	8880	1-1/4" x 14-1/2"		130	.123		2.80	4.38		7.18	10.35	

For expanded coverage of these items see *Means Interior Cost Data 2006*

	06110	Wood Framing	CREW	DAILY OUTPUT	LABOR-HOURS	UNIT	2006 BARE COSTS MAT.	LABOR	EQUIP.	TOTAL	TOTAL INCL O&P	
545	9000	Minimum labor/equipment charge	1 Carp	4	2	Job		71		71	118	545
550	0010	**PARTITIONS** Wood stud with single bottom plate and										550
	0020	double top plate, no waste, std. & better lumber										
	0182	2" x 4" studs, 8' high, studs 12" O.C.	2 Carp	80	.200	L.F.	4.50	7.10		11.60	16.75	
	0187	12" O.C., pneumatic nailed		96	.167		4.50	5.95		10.45	14.75	
	0202	16" O.C.		100	.160		3.68	5.70		9.38	13.45	
	0207	16" O.C., pneumatic nailed		120	.133		3.68	4.74		8.42	11.90	
	0302	24" O.C.		125	.128		2.86	4.55		7.41	10.70	
	0307	24" O.C., pneumatic nailed		150	.107		2.86	3.79		6.65	9.45	
	0382	10' high, studs 12" O.C.		80	.200		5.30	7.10		12.40	17.65	
	0387	12" O.C., pneumatic nailed		96	.167		5.30	5.95		11.25	15.65	
	0402	16" O.C.		100	.160		4.30	5.70		10	14.15	
	0407	16" O.C., pneumatic nailed		120	.133		4.30	4.74		9.04	12.60	
	0502	24" O.C.		125	.128		3.27	4.55		7.82	11.15	
	0507	24" O.C., pneumatic nailed		150	.107		3.27	3.79		7.06	9.90	
	0582	12' high, studs 12" O.C.		65	.246		6.15	8.75		14.90	21.50	
	0587	12" O.C., pneumatic nailed		78	.205		6.15	7.30		13.45	18.85	
	0602	16" O.C.		80	.200		4.91	7.10		12.01	17.20	
	0607	16" O.C., pneumatic nailed		96	.167		4.91	5.95		10.86	15.20	
	0701	24" O.C.		100	.160		3.68	5.70		9.38	13.45	
	0702											
	0706	24" O.C., pneumatic nailed	2 Carp	120	.133	L.F.	3.68	4.74		8.42	11.90	
	0782	2" x 6" studs, 8' high, studs 12" O.C.		70	.229		7.15	8.15		15.30	21.50	
	0787	12" O.C., pneumatic nailed		84	.190		7.15	6.75		13.90	19.10	
	0802	16" O.C.		90	.178		5.85	6.30		12.15	16.90	
	0807	16" O.C., pneumatic nailed		108	.148		5.85	5.25		11.10	15.20	
	0902	24" O.C.		115	.139		4.56	4.95		9.51	13.20	
	0907	24" O.C., pneumatic nailed		138	.116		4.56	4.12		8.68	11.85	
	0982	10' high, studs 12" O.C.		70	.229		8.45	8.15		16.60	23	
	0987	12" O.C., pneumatic nailed		84	.190		8.45	6.75		15.20	20.50	
	1002	16" O.C.		90	.178		6.85	6.30		13.15	17.95	
	1007	16" O.C., pneumatic nailed		108	.148		6.85	5.25		12.10	16.25	
	1102	24" O.C.		115	.139		5.20	4.95		10.15	13.95	
	1107	24" O.C., pneumatic nailed		138	.116		5.20	4.12		9.32	12.60	
	1182	12' high, studs 12" O.C.		55	.291		9.75	10.35		20.10	28	
	1187	12" O.C., pneumatic nailed		66	.242		9.75	8.60		18.35	25	
	1202	16" O.C.		70	.229		7.80	8.15		15.95	22	
	1207	16" O.C., pneumatic nailed		84	.190		7.80	6.75		14.55	19.80	
	1302	24" O.C.		90	.178		5.85	6.30		12.15	16.90	
	1307	24" O.C., pneumatic nailed		108	.148		5.85	5.25		11.10	15.20	
	1402	For horizontal blocking, 2" x 4", add		600	.027		.41	.95		1.36	2.02	
	1502	2" x 6", add		600	.027		.65	.95		1.60	2.29	
	1600	For openings, add		250	.064			2.28		2.28	3.77	
	1702	Headers for above openings, material only, add				B.F.	.70			.70	.77	
	9000	Minimum labor/equipment charge	1 Carp	4	2	Job		71		71	118	
552	0010	**FRAMING, PORCH OR DECK**										552
	0100	Treated lumber, posts or columns, 4" x 4"	2 Carp	390	.041	L.F.	1.41	1.46		2.87	3.97	
	0110	4" x 6"		275	.058		2.49	2.07		4.56	6.15	
	0120	4" x 8"		220	.073		3.32	2.59		5.91	7.95	
	0130	Girder, single, 4" x 4"		675	.024		1.41	.84		2.25	2.95	
	0140	4" x 6"		600	.027		2.49	.95		3.44	4.31	
	0150	4" x 8"		525	.030		3.32	1.08		4.40	5.45	
	0160	Double, 2" x 4"		625	.026		.99	.91		1.90	2.60	
	0170	2" x 6"		600	.027		1.53	.95		2.48	3.25	
	0180	2" x 8"		575	.028		1.99	.99		2.98	3.83	

06110 | Wood Framing

		06110 Wood Framing	CREW	DAILY OUTPUT	LABOR-HOURS	UNIT	2006 BARE COSTS MAT.	LABOR	EQUIP.	TOTAL	TOTAL INCL O&P	
552	0190	2″ x 10″	2 Carp	550	.029	L.F.	3.03	1.03		4.06	5.05	552
	0200	2″ x 12″		525	.030		3.96	1.08		5.04	6.15	
	0210	Triple, 2″ x 4″		575	.028		1.49	.99		2.48	3.28	
	0220	2″ x 6″		550	.029		2.29	1.03		3.32	4.23	
	0230	2″ x 8″		525	.030		2.99	1.08		4.07	5.10	
	0240	2″ x 10″		500	.032		4.55	1.14		5.69	6.90	
	0250	2″ x 12″		475	.034		5.95	1.20		7.15	8.55	
	0260	Ledger, bolted 4′ O.C., 2″ x 4″		400	.040		.59	1.42		2.01	3.01	
	0270	2″ x 6″		395	.041		.85	1.44		2.29	3.33	
	0280	2″ x 8″		390	.041		1.08	1.46		2.54	3.61	
	0300	2″ x 12″		380	.042		2.06	1.50		3.56	4.74	
	0310	Joists, 2″ x 4″		1250	.013		.50	.46		.96	1.30	
	0320	2″ x 6″		1250	.013		.77	.46		1.23	1.60	
	0330	2″ x 8″		1100	.015		1	.52		1.52	1.96	
	0340	2″ x 10″		900	.018		1.53	.63		2.16	2.73	
	0350	2″ x 12″		875	.018		1.83	.65		2.48	3.09	
	0440	Balusters, square, 2″ x 2″		220	.073		.31	2.59		2.90	4.62	
	0450	Turned, 2″ x 2″		140	.114		.41	4.06		4.47	7.20	
	0460	Stair stringer, 2″ x 10″		130	.123		1.53	4.38		5.91	8.95	
	0470	2″ x 12″		130	.123		1.83	4.38		6.21	9.25	
	0480	Stair treads, 1″ x 4″		140	.114		.64	4.06		4.70	7.45	
	0490	2″ x 4″		140	.114		.50	4.06		4.56	7.30	
	0500	2″ x 6″		160	.100		.75	3.56		4.31	6.75	
	0510	5/4″ x 6″		160	.100		.60	3.56		4.16	6.55	
	0520	Turned handrail post, 4″ x 4″		64	.250	Ea.	13.05	8.90		21.95	29	
	0530	Lattice panel, 4′ x 8′		1600	.010	S.F.	.70	.36		1.06	1.36	
	0540	Cedar, posts or columns, 4″ x 4″		390	.041	L.F.	3.18	1.46		4.64	5.90	
	0550	4″ x 6″		275	.058		4.83	2.07		6.90	8.75	
	0560	4″ x 8″		220	.073		7.25	2.59		9.84	12.25	
	0800	Decking, 1″ x 4″		550	.029		1.36	1.03		2.39	3.21	
	0810	2″ x 4″		600	.027		2.63	.95		3.58	4.46	
	0820	2″ x 6″		640	.025		4.24	.89		5.13	6.15	
	0830	5/4″ x 6″		640	.025		3.79	.89		4.68	5.65	
	0840	Railings and trim, 1″ x 4″		600	.027		1.36	.95		2.31	3.07	
	0860	2″ x 4″		600	.027		2.63	.95		3.58	4.46	
	0870	2″ x 6″		600	.027		4.24	.95		5.19	6.25	
	0920	Stair treads, 1″ x 4″		140	.114		1.36	4.06		5.42	8.25	
	0930	2″ x 4″		140	.114		2.63	4.06		6.69	9.65	
	0940	2″ x 6″		160	.100		4.24	3.56		7.80	10.55	
	0950	5/4″ x 6″		160	.100		3.79	3.56		7.35	10.05	
	0980	Redwood, posts or columns, 4″ x 4″		390	.041		6.35	1.46		7.81	9.40	
	0990	4″ x 6″		275	.058		10.65	2.07		12.72	15.20	
	1000	4″ x 8″		220	.073		18.55	2.59		21.14	25	
	1280	Railings and trim, 1″ x 4″		600	.027		1.48	.95		2.43	3.20	
	1310	2″ x 6″		600	.027		5.75	.95		6.70	7.90	
	1420	Alternative decking, wood / plastic composite, 5/4″ x 6″		640	.025		1.93	.89		2.82	3.59	
	1430	Vinyl, 1-1/2″ x 5-1/2″		640	.025		3.48	.89		4.37	5.30	
	1440	1″ x 4″ square edge fir		550	.029		1.30	1.03		2.33	3.14	
	1450	1″ x 4″ tongue and groove fir		450	.036		1.30	1.26		2.56	3.52	
	1460	1″ x 4″ mahogany		550	.029		1.30	1.03		2.33	3.14	
	1470	Accessories, joist hangers, 2″ x 4″	1 Carp	160	.050	Ea.	.70	1.78		2.48	3.72	
	1480	2″ x 6″ through 2″ x 12″	″	150	.053		.80	1.90		2.70	4.02	
	1530	Post footing, incl excav, backfill, tube form & concrete, 4′ deep, 8″ dia	F-7	12	2.667		10.40	84		94.40	150	
	1540	10″ diameter		11	2.909		14.70	91.50		106.20	168	
	1550	12″ diameter		10	3.200		19.70	101		120.70	189	

06100 | Rough Carpentry

6 WOOD & PLASTICS

	06110 Wood Framing		CREW	DAILY OUTPUT	LABOR-HOURS	UNIT	2006 BARE COSTS MAT.	LABOR	EQUIP.	TOTAL	TOTAL INCL O&P	
555	0010	FRAMING, ROOFS										555
	2000	Fascia boards, 2" x 8" R061110-30	2 Carp	225	.071	L.F.	.85	2.53		3.38	5.15	
	2100	2" x 10"		180	.089		1.29	3.16		4.45	6.65	
	5000	Rafters, to 4 in 12 pitch, 2" x 6", ordinary		1000	.016		.59	.57		1.16	1.59	
	5060	2" x 8", ordinary		950	.017		.85	.60		1.45	1.93	
	5250	Composite rafter, 9-1/2" deep		575	.028		1.82	.99		2.81	3.64	
	5260	11-1/2" deep		575	.028		1.96	.99		2.95	3.79	
	5300	Hip and valley rafters, 2" x 6", ordinary		760	.021		.59	.75		1.34	1.89	
	5360	2" x 8", ordinary		720	.022		.85	.79		1.64	2.25	
	5540	Hip and valley jacks, 2" x 6", ordinary		600	.027		.59	.95		1.54	2.22	
	5600	2" x 8", ordinary		490	.033		.85	1.16		2.01	2.86	
	5761	For slopes steeper than 4 in 12, add						30%				
	5770	For dormers or complex roofs, add						50%				
	5780	Rafter tie, 1" x 4", #3	2 Carp	800	.020	L.F.	.35	.71		1.06	1.57	
	5800	Ridge board, #2 or better, 1" x 6"		600	.027		.70	.95		1.65	2.34	
	5820	1" x 8"		550	.029		.89	1.03		1.92	2.68	
	5840	1" x 10"		500	.032		1.30	1.14		2.44	3.31	
	5860	2" x 6"		500	.032		.59	1.14		1.73	2.53	
	5880	2" x 8"		450	.036		.85	1.26		2.11	3.03	
	5900	2" x 10"		400	.040		1.29	1.42		2.71	3.78	
	5920	Roof cants, split, 4" x 4"		650	.025		1.25	.88		2.13	2.83	
	5940	6" x 6"		600	.027		4.54	.95		5.49	6.55	
	5960	Roof curbs, untreated, 2" x 6"		520	.031		.59	1.09		1.68	2.46	
	5980	2" x 12"		400	.040		1.60	1.42		3.02	4.12	
	6000	Sister rafters, 2" x 6"		800	.020		.59	.71		1.30	1.83	
	6020	2" x 8"		640	.025		.85	.89		1.74	2.41	
	6040	2" x 10"		535	.030		1.29	1.06		2.35	3.18	
	6060	2" x 12"		455	.035		1.60	1.25		2.85	3.83	
	9000	Minimum labor/equipment charge	1 Carp	4	2	Job		71		71	118	
560	0010	FRAMING, SILLS										560
	2002	Ledgers, nailed, 2" x 4"	2 Carp	755	.021	L.F.	.37	.75		1.12	1.66	
	2052	2" x 6"		600	.027		.59	.95		1.54	2.22	
	2102	Bolted, not including bolts, 3" x 6"		325	.049		1.92	1.75		3.67	5	
	2152	3" x 12"		233	.069		3.92	2.44		6.36	8.35	
	2602	Mud sills, redwood, construction grade, 2" x 4"		895	.018		2.20	.64		2.84	3.47	
	2622	2" x 6"		780	.021		3.30	.73		4.03	4.84	
	4002	Sills, 2" x 4"		600	.027		.37	.95		1.32	1.98	
	4052	2" x 6"		550	.029		.59	1.03		1.62	2.36	
	4082	2" x 8"		500	.032		.85	1.14		1.99	2.82	
	4202	Treated, 2" x 4"		550	.029		.49	1.03		1.52	2.24	
	4222	2" x 6"		500	.032		.75	1.14		1.89	2.70	
	4242	2" x 8"		450	.036		.98	1.26		2.24	3.16	
	4402	4" x 4"		450	.036		1.39	1.26		2.65	3.62	
	4422	4" x 6"		350	.046		2.46	1.63		4.09	5.40	
	4462	4" x 8"		300	.053		3.27	1.90		5.17	6.75	
	4481	4" x 10"		260	.062		4.10	2.19		6.29	8.15	
	9000	Minimum labor/equipment charge	1 Carp	4	2	Job		71		71	118	
565	0010	FRAMING, SLEEPERS										565
	0100	On concrete, treated, 1" x 2"	2 Carp	2350	.007	L.F.	.10	.24		.34	.51	
	0150	1" x 3"		2000	.008		.19	.28		.47	.68	
	0200	2" x 4"		1500	.011		.49	.38		.87	1.16	
	0250	2" x 6"		1300	.012		.75	.44		1.19	1.55	
	9000	Minimum labor/equipment charge	1 Carp	4	2	Job		71		71	118	
570	0010	FRAMING, SOFFITS & CANOPIES										570
	1002	Canopy or soffit framing , 1" x 4"	2 Carp	900	.018	L.F.	.46	.63		1.09	1.56	

	06110 | Wood Framing		CREW	DAILY OUTPUT	LABOR-HOURS	UNIT	2006 BARE COSTS MAT.	LABOR	EQUIP.	TOTAL	TOTAL INCL O&P	
570	1042	1" x 8"	2 Carp	750	.021	L.F.	.89	.76		1.65	2.23	570
	1102	2" x 4"		620	.026		.37	.92		1.29	1.93	
	1142	2" x 8"		500	.032		.85	1.14		1.99	2.82	
	1202	3" x 4"		500	.032		1.09	1.14		2.23	3.08	
	1242	3" x 10"		300	.053		3.26	1.90		5.16	6.75	
	9000	Minimum labor/equipment charge	1 Carp	4	2	Job		71		71	118	
575	0010	FRAMING, TREATED LUMBER										575
	0100	2" x 4"				M.B.F.	730			730	800	
	0110	2" x 6"					750			750	825	
	0120	2" x 8"					730			730	805	
	0130	2" x 10"					895			895	985	
	0140	2" x 12"					975			975	1,075	
	0200	4" x 4"					1,050			1,050	1,150	
	0210	4" x 6"					1,225			1,225	1,350	
	0220	4" x 8"					1,225			1,225	1,350	
590	0010	FRAMING, WALLS R061110-30										590
	0100	Door buck, studs, header, access, 8' H, 2"x4" wall, 3' W	1 Carp	32	.250	Ea.	15.75	8.90		24.65	32	
	0110	4' wide		32	.250		16.95	8.90		25.85	33.50	
	0120	5' wide		32	.250		21	8.90		29.90	38	
	0130	6' wide		32	.250		22.50	8.90		31.40	40	
	0140	8' wide		30	.267		33	9.50		42.50	52	
	0150	10' wide		30	.267		44.50	9.50		54	64.50	
	0160	12' wide		30	.267		65.50	9.50		75	87.50	
	0170	2" x 6" wall, 3' wide		32	.250		23	8.90		31.90	40	
	0180	4' wide		32	.250		24	8.90		32.90	41.50	
	0190	5' wide		32	.250		28	8.90		36.90	45.50	
	0200	6' wide		32	.250		29.50	8.90		38.40	47.50	
	0210	8' wide		30	.267		40	9.50		49.50	60	
	0220	10' wide		30	.267		51.50	9.50		61	72.50	
	0230	12' wide		30	.267		72.50	9.50		82	95	
	0240	Window buck, studs, header & access, 8' high 2" x 4" wall, 2' wide		24	.333		16.55	11.85		28.40	38	
	0250	3' wide		24	.333		19.45	11.85		31.30	41	
	0260	4' wide		24	.333		21.50	11.85		33.35	43	
	0270	5' wide		24	.333		25.50	11.85		37.35	47.50	
	0280	6' wide		24	.333		28	11.85		39.85	50.50	
	0300	8' wide		22	.364		41	12.95		53.95	66.50	
	0310	10' wide		22	.364		53.50	12.95		66.45	80.50	
	0320	12' wide		22	.364		76.50	12.95		89.45	106	
	0330	2" x 6" wall, 2' wide		24	.333		25.50	11.85		37.35	47.50	
	0340	3' wide		24	.333		28.50	11.85		40.35	51	
	0350	4' wide		24	.333		31	11.85		42.85	53.50	
	0360	5' wide		24	.333		35	11.85		46.85	58	
	0370	6' wide		24	.333		38.50	11.85		50.35	62	
	0380	7' wide		24	.333		48.50	11.85		60.35	72.50	
	0390	8' wide		22	.364		52.50	12.95		65.45	79.50	
	0400	10' wide		22	.364		66	12.95		78.95	94	
	0410	12' wide		22	.364		90	12.95		102.95	121	
	2002	Headers over openings, 2" x 6"	2 Carp	360	.044	L.F.	.59	1.58		2.17	3.27	
	2007	2" x 6", pneumatic nailed		432	.037		.59	1.32		1.91	2.83	
	2052	2" x 8"		340	.047		.85	1.67		2.52	3.71	
	2057	2" x 8", pneumatic nailed		408	.039		.85	1.39		2.24	3.25	
	2100	2" x 10"		320	.050		1.29	1.78		3.07	4.37	
	2105	2" x 10", pneumatic nailed		384	.042		1.29	1.48		2.77	3.87	
	2152	2" x 12"		300	.053		1.60	1.90		3.50	4.90	
	2157	2" x 12", pneumatic nailed		360	.044		1.60	1.58		3.18	4.38	

06110 | Wood Framing

		06110 Wood Framing	CREW	DAILY OUTPUT	LABOR-HOURS	UNIT	2006 BARE COSTS MAT.	LABOR	EQUIP.	TOTAL	TOTAL INCL O&P	
590	2202	4" x 12" (R061110-30)	2 Carp	190	.084	L.F.	4.04	2.99		7.03	9.40	590
	2207	4" x 12", pneumatic nailed		228	.070		4.04	2.49		6.53	8.55	
	2252	6" x 12"		140	.114		9.60	4.06		13.66	17.30	
	3000	Radius, 2" x 6"		270	.059		.93	2.11		3.04	4.51	
	3010	2" x 6", pneumatic nailed		324	.049		.93	1.76		2.69	3.93	
	3020	2" x 8"		255	.063		1.33	2.23		3.56	5.15	
	3030	2" x 8", pneumatic nailed		296	.054		1.33	1.92		3.25	4.65	
	3040	2" x 10"		240	.067		2.02	2.37		4.39	6.15	
	3050	2" x 10", pneumatic nailed		285	.056		2.02	2		4.02	5.55	
	3060	2" x 12"		225	.071		2.50	2.53		5.03	6.95	
	3070	2" x 12", pneumatic nailed		270	.059		2.50	2.11		4.61	6.25	
	5002	Plates, untreated, 2" x 3"		850	.019		.38	.67		1.05	1.53	
	5007	2" x 3", pneumatic nailed		1020	.016		.38	.56		.94	1.34	
	5022	2" x 4"		800	.020		.37	.71		1.08	1.59	
	5027	2" x 4", pneumatic nailed		960	.017		.37	.59		.96	1.39	
	5041	2" x 6"		750	.021		.59	.76		1.35	1.91	
	5046	2" x 6", pneumatic nailed		900	.018		.59	.63		1.22	1.70	
	5122	Studs, 8' high wall, 2" x 3"		1200	.013		.38	.47		.85	1.21	
	5127	2" x 3", pneumatic nailed		1440	.011		.38	.40		.78	1.07	
	5142	2" x 4"		1100	.015		.37	.52		.89	1.27	
	5147	2" x 4", pneumatic nailed		1320	.012		.37	.43		.80	1.12	
	5162	2" x 6"		1000	.016		.59	.57		1.16	1.59	
	5167	2" x 6", pneumatic nailed		1200	.013		.59	.47		1.06	1.44	
	5182	3" x 4"		800	.020		1.09	.71		1.80	2.38	
	5187	3" x 4", pneumatic nailed	▼	960	.017	▼	1.09	.59		1.68	2.18	
	8200	For 12' high walls, deduct						5%				
	8220	For stub wall, 6' high, add						20%				
	8240	3' high, add						40%				
	8250	For second story & above, add						5%				
	8300	For dormer & gable, add						15%				
	9000	Minimum labor/equipment charge	1 Carp	4	2	Job		71		71	118	
600	0010	**FURRING**										600
	0012	Wood strips, 1" x 2", on walls, on wood	1 Carp	550	.015	L.F.	.19	.52		.71	1.07	
	0017	Pneumatic nailed		710	.011		.19	.40		.59	.87	
	0302	On masonry		495	.016		.19	.57		.76	1.16	
	0402	On concrete		260	.031		.19	1.09		1.28	2.02	
	0602	1" x 3", on walls, on wood		550	.015		.28	.52		.80	1.17	
	0607	Pneumatic nailed		710	.011		.28	.40		.68	.97	
	0702	On masonry		495	.016		.28	.57		.85	1.26	
	0802	On concrete		260	.031		.28	1.09		1.37	2.12	
	0852	On ceilings, on wood		350	.023		.28	.81		1.09	1.66	
	0857	Pneumatic nailed		450	.018		.28	.63		.91	1.36	
	0902	On masonry		320	.025		.28	.89		1.17	1.78	
	0952	On concrete		210	.038	▼	.28	1.35		1.63	2.55	
	9000	Minimum labor/equipment charge	▼	4	2	Job		71		71	118	
700	0010	**GROUNDS**										700
	0020	For casework, 1" x 2" wood strips, on wood	1 Carp	330	.024	L.F.	.19	.86		1.05	1.64	
	0102	On masonry		285	.028		.19	1		1.19	1.86	
	0202	On concrete		250	.032		.19	1.14		1.33	2.09	
	0402	For plaster, 3/4" deep, on wood		450	.018		.19	.63		.82	1.26	
	0502	On masonry		225	.036		.19	1.26		1.45	2.30	
	0602	On concrete		175	.046		.19	1.63		1.82	2.90	
	0702	On metal lath		200	.040	▼	.19	1.42		1.61	2.57	
	9000	Minimum labor/equipment charge	▼	4	2	Job		71		71	118	

06100 | Rough Carpentry

06120 | Structural Panels

		06120 Structural Panels	CREW	DAILY OUTPUT	LABOR-HOURS	UNIT	2006 BARE COSTS MAT.	LABOR	EQUIP.	TOTAL	TOTAL INCL O&P	
900	0010	**STRUCTURAL INSULATED PANELS**										900
	0100	Structural insul. panels, 7/16" OSB both faces, EPS insul, 3-5/8" T	F-3	2075	.019	S.F.	3	.70	.31	4.01	4.78	
	0110	5-5/8" thick		1725	.023		3.25	.84	.37	4.46	5.35	
	0120	7-3/8" thick		1425	.028		3.50	1.01	.45	4.96	6	
	0130	9-3/8" thick		1125	.036		3.75	1.28	.57	5.60	6.85	
	0140	7/16" OSB one face, EPS insul, 3-5/8" thick		2175	.018		1.56	.66	.29	2.51	3.13	
	0150	5-5/8" thick		1825	.022		1.84	.79	.35	2.98	3.70	
	0160	7-3/8" thick		1525	.026		2.09	.95	.42	3.46	4.31	
	0170	9-3/8" thick		1225	.033		2.40	1.18	.52	4.10	5.15	
	0180	11-3/8" thick		925	.043		2.57	1.56	.69	4.82	6.15	
	0190	7/16" OSB - 1/2" GWB faces , EPS insul, 3-5/8" T		2075	.019		2.50	.70	.31	3.51	4.23	
	0200	5-5/8" thick		1725	.023		2.78	.84	.37	3.99	4.84	
	0210	7-3/8" thick		1425	.028		3.09	1.01	.45	4.55	5.55	
	0220	9-3/8" thick		1125	.036		3.43	1.28	.57	5.28	6.50	
	0230	11-3/8" thick		825	.048		3.65	1.75	.77	6.17	7.75	
	0240	7/16" OSB - 1/2" MRGWB faces , EPS insul, 3-5/8" T		2075	.019		2.50	.70	.31	3.51	4.23	
	0250	5-5/8" thick		1725	.023		2.78	.84	.37	3.99	4.84	
	0300	For 1/2" GWB added to OSB skin, add					.60			.60	.66	
	0310	For 1/2" MRGWB added to OSB skin, add					.75			.75	.83	
	0320	For one T1-11 skin, add to OSB-OSB					.75			.75	.83	
	0330	For one 19/32" CDX skin, add to OSB-OSB					.72			.72	.79	

06150 | Wood Decking

		06150 Wood Decking	CREW	DAILY OUTPUT	LABOR-HOURS	UNIT	MAT.	LABOR	EQUIP.	TOTAL	TOTAL INCL O&P	
600	0010	**ROOF DECKS**										600
	0020	For laminated decks, see division 06170-550										
	0400	Cedar planks, 3" thick	2 Carp	320	.050	S.F.	6.40	1.78		8.18	10	
	0500	4" thick		250	.064		8.65	2.28		10.93	13.25	
	0702	Douglas fir, 3" thick		320	.050		2.40	1.78		4.18	5.60	
	0802	4" thick		250	.064		3.21	2.28		5.49	7.30	
	1002	Hemlock, 3" thick		320	.050		2.40	1.78		4.18	5.60	
	1102	4" thick		250	.064		3.20	2.28		5.48	7.30	
	1302	Western white spruce, 3" thick		320	.050		2.31	1.78		4.09	5.50	
	1402	4" thick		250	.064		3.08	2.28		5.36	7.15	
	9000	Minimum labor/equipment charge	1 Carp	2	4	Job		142		142	236	

06160 | Sheathing

		06160 Sheathing	CREW	DAILY OUTPUT	LABOR-HOURS	UNIT	MAT.	LABOR	EQUIP.	TOTAL	TOTAL INCL O&P	
800	0010	**SHEATHING** Plywood on roof, CDX (R061110-30)										800
	0032	5/16" thick	2 Carp	1600	.010	S.F.	.62	.36		.98	1.27	
	0037	Pneumatic nailed (R061636-20)		1952	.008		.62	.29		.91	1.16	
	0052	3/8" thick		1525	.010		.46	.37		.83	1.12	
	0057	Pneumatic nailed		1860	.009		.46	.31		.77	1.01	
	0102	1/2" thick		1400	.011		.61	.41		1.02	1.35	
	0103	Pneumatic nailed		1708	.009		.61	.33		.94	1.23	
	0202	5/8" thick		1300	.012		.66	.44		1.10	1.46	
	0207	Pneumatic nailed		1586	.010		.66	.36		1.02	1.32	
	0302	3/4" thick		1200	.013		.82	.47		1.29	1.69	
	0307	Pneumatic nailed		1464	.011		.82	.39		1.21	1.54	
	0502	Plywood on walls with exterior CDX, 3/8" thick		1200	.013		.46	.47		.93	1.29	
	0507	Pneumatic nailed		1488	.011		.46	.38		.84	1.13	
	0602	1/2" thick		1125	.014		.61	.51		1.12	1.52	
	0607	Pneumatic nailed		1395	.011		.61	.41		1.02	1.36	
	0702	5/8" thick		1050	.015		.66	.54		1.20	1.63	
	0707	Pneumatic nailed		1302	.012		.66	.44		1.10	1.45	
	0802	3/4" thick		975	.016		.82	.58		1.40	1.87	

06160 | Sheathing

		06160 Sheathing	Crew	Daily Output	Labor-Hours	Unit	2006 Bare Costs Mat.	Labor	Equip.	Total	Total Incl O&P
800	0807	Pneumatic nailed (R061110-30)	2 Carp	1209	.013	S.F.	.82	.47		1.29	1.68
	1000	For shear wall construction, add						20%			
	1200	For structural 1 exterior plywood, add (R061636-20)				S.F.	10%				
	1402	With boards, on roof 1" x 6" boards, laid horizontal	2 Carp	725	.022		1.20	.78		1.98	2.62
	1502	Laid diagonal		650	.025		1.20	.88		2.08	2.77
	1702	1" x 8" boards, laid horizontal		875	.018		1.11	.65		1.76	2.31
	1802	Laid diagonal		725	.022		1.11	.78		1.89	2.53
	2000	For steep roofs, add						40%			
	2200	For dormers, hips and valleys, add					5%	50%			
	2402	Boards on walls, 1" x 6" boards, laid regular	2 Carp	650	.025		1.20	.88		2.08	2.77
	2502	Laid diagonal		585	.027		1.20	.97		2.17	2.93
	2702	1" x 8" boards, laid regular		765	.021		1.11	.74		1.85	2.46
	2802	Laid diagonal		650	.025		1.11	.88		1.99	2.68
	2852	Gypsum, weatherproof, 1/2" thick		1050	.015		.62	.54		1.16	1.58
	2902	Sealed, 4/10" thick		1100	.015		.45	.52		.97	1.36
	3000	Wood fiber, regular, no vapor barrier, 1/2" thick		1200	.013		.55	.47		1.02	1.40
	3100	5/8" thick		1200	.013		.72	.47		1.19	1.58
	3300	No vapor barrier, in colors, 1/2" thick		1200	.013		.78	.47		1.25	1.65
	3400	5/8" thick		1200	.013		.96	.47		1.43	1.85
	3600	With vapor barrier one side, white, 1/2" thick		1200	.013		.55	.47		1.02	1.40
	3700	Vapor barrier 2 sides, 1/2" thick		1200	.013		.77	.47		1.24	1.64
	3800	Asphalt impregnated, 25/32" thick		1200	.013		.28	.47		.75	1.10
	3850	Intermediate, 1/2" thick		1200	.013		.18	.47		.65	.99
	9000	Minimum labor/equipment charge	1 Carp	2	4	Job		142		142	236
850	0010	**SUBFLOOR** (R061636-20)									
	0011	Plywood, CDX, 1/2" thick	2 Carp	1500	.011	SF Flr.	.61	.38		.99	1.31
	0017	Pneumatic nailed		1860	.009		.61	.31		.92	1.19
	0102	5/8" thick		1350	.012		.66	.42		1.08	1.43
	0107	Pneumatic nailed		1674	.010		.66	.34		1	1.29
	0202	3/4" thick		1250	.013		.82	.46		1.28	1.65
	0207	Pneumatic nailed		1550	.010		.82	.37		1.19	1.51
	0302	1-1/8" thick, 2-4-1 including underlayment		1050	.015		1.19	.54		1.73	2.21
	0452	1" x 8" S4S, laid regular		1000	.016		1.11	.57		1.68	2.17
	0462	Laid diagonal		850	.019		1.11	.67		1.78	2.34
	0502	1" x 10" S4S, laid regular		1100	.015		1.36	.52		1.88	2.36
	0602	Laid diagonal		900	.018		1.36	.63		1.99	2.55
	9000	Minimum labor/equipment charge	1 Carp	4	2	Job		71		71	118
900	0010	**UNDERLAYMENT** (R061636-20)									
	0030	Plywood, underlayment grade, 3/8" thick	2 Carp	1500	.011	SF Flr.	.78	.38		1.16	1.49
	0080	Pneumatic nailed		1860	.009		.78	.31		1.09	1.37
	0102	1/2" thick		1450	.011		.87	.39		1.26	1.61
	0107	Pneumatic nailed		1798	.009		.87	.32		1.19	1.48
	0202	5/8" thick		1400	.011		1.22	.41		1.63	2.01
	0207	Pneumatic nailed		1736	.009		1.22	.33		1.55	1.88
	0302	3/4" thick		1300	.012		1.32	.44		1.76	2.18
	0306	Pneumatic nailed		1612	.010		1.32	.35		1.67	2.03
	0502	Particle board, 3/8" thick		1500	.011		.33	.38		.71	.99
	0507	Pneumatic nailed		1860	.009		.33	.31		.64	.87
	0602	1/2" thick		1450	.011		.37	.39		.76	1.06
	0607	Pneumatic nailed		1798	.009		.37	.32		.69	.93
	0802	5/8" thick		1400	.011		.41	.41		.82	1.12
	0807	Pneumatic nailed		1736	.009		.41	.33		.74	.99
	0902	3/4" thick		1300	.012		.56	.44		1	1.35
	0907	Pneumatic nailed		1612	.010		.56	.35		.91	1.20
	1102	Hardboard, underlayment grade, 4' x 4', .215" thick		1500	.011		.39	.38		.77	1.06

06100 | Rough Carpentry

06160 | Sheathing

		06160 Sheathing	CREW	DAILY OUTPUT	LABOR-HOURS	UNIT	MAT.	LABOR	EQUIP.	TOTAL	TOTAL INCL O&P	
							2006 BARE COSTS					
900	9000	Minimum labor/equipment charge R061636-20	1 Carp	4	2	Job		71		71	118	900

06170 | Prefabricated Structural Wood

		06170 Prefabricated Structural Wood	CREW	DAILY OUTPUT	LABOR-HOURS	UNIT	MAT.	LABOR	EQUIP.	TOTAL	TOTAL INCL O&P	
550	0010	**LAMINATED ROOF DECK**										550
	0020	Pine or hemlock, 3" thick	2 Carp	425	.038	S.F.	3.02	1.34		4.36	5.55	
	0100	4" thick		325	.049		4.02	1.75		5.77	7.30	
	0300	Cedar, 3" thick		425	.038		3.69	1.34		5.03	6.30	
	0400	4" thick		325	.049		4.70	1.75		6.45	8.05	
	0600	Fir, 3" thick		425	.038		3.08	1.34		4.42	5.60	
	0700	4" thick		325	.049		3.85	1.75		5.60	7.15	
	9000	Minimum labor/equipment charge	1 Carp	3	2.667	Job		95		95	157	
600	0010	**STRUCTURAL JOISTS** Fabricated "I" joists with wood flanges,										600
	0100	Plywood webs, incl. bridging & blocking, panels 24" O.C.										
	1200	15' to 24' span, 50 psf live load	F-5	2400	.013	SF Flr.	2.07	.48		2.55	3.08	
	1300	55 psf live load		2250	.014		2.23	.51		2.74	3.30	
	1400	24' to 30' span, 45 psf live load		2600	.012		2.57	.44		3.01	3.56	
	1500	55 psf live load		2400	.013		2.91	.48		3.39	4	
	1600	Tubular steel open webs, 45 psf, 24" O.C., 40' span	F-3	6250	.006		2.07	.23	.10	2.40	2.77	
	1700	55' span		7750	.005		2.01	.19	.08	2.28	2.61	
	1800	70' span		9250	.004		2.61	.16	.07	2.84	3.21	
	1900	85 psf live load, 26' span		2300	.017		2.43	.63	.28	3.34	4	
980	0010	**ROOF TRUSSES**										980
	0020	For timber connectors, see div. 06090-800										
	0100	Fink (W) or King post type, 2'-0" O.C.										
	0200	Metal plate connected, 4 in 12 slope										
	0210	24' to 29' span	F-3	3000	.013	SF Flr.	1.84	.48	.21	2.53	3.04	
	0300	30' to 43' span		3000	.013		2.04	.48	.21	2.73	3.26	
	0400	44' to 60' span		3000	.013		2.25	.48	.21	2.94	3.50	
	0700	Glued and nailed, add					50%					

06180 | Glued-Laminated Construction

		06180 Glued-Laminated Construction	CREW	DAILY OUTPUT	LABOR-HOURS	UNIT	MAT.	LABOR	EQUIP.	TOTAL	TOTAL INCL O&P	
400	0010	**LAMINATED FRAMING** Not including decking										400
	0020	30 lb., short term live load, 15 lb. dead load										
	0200	Straight roof beams, 20' clear span, beams 8' O.C.	F-3	2560	.016	SF Flr.	1.78	.56	.25	2.59	3.15	
	0300	Beams 16' O.C.		3200	.013		1.28	.45	.20	1.93	2.37	
	0500	40' clear span, beams 8' O.C.		3200	.013		3.40	.45	.20	4.05	4.70	
	0600	Beams 16' O.C.		3840	.010		2.77	.38	.17	3.32	3.85	
	0800	60' clear span, beams 8' O.C.	F-4	2880	.017		5.85	.59	.33	6.77	7.70	
	0900	Beams 16' O.C.	"	3840	.013		4.35	.44	.25	5.04	5.80	
	1100	Tudor arches, 30' to 40' clear span, frames 8' O.C.	F-3	1680	.024		7.60	.86	.38	8.84	10.20	
	1200	Frames 16' O.C.	"	2240	.018		5.95	.64	.28	6.87	7.90	
	1400	50' to 60' clear span, frames 8' O.C.	F-4	2200	.022		8.20	.77	.43	9.40	10.80	
	1500	Frames 16' O.C.		2640	.018		7	.64	.36	8	9.15	
	1700	Radial arches, 60' clear span, frames 8' O.C.		1920	.025		7.70	.89	.50	9.09	10.45	
	1800	Frames 16' O.C.		2880	.017		5.90	.59	.33	6.82	7.80	
	2000	100' clear span, frames 8' O.C.		1600	.030		7.95	1.06	.59	9.60	11.15	
	2100	Frames 16' O.C.		2400	.020		7	.71	.40	8.11	9.30	
	2300	120' clear span, frames 8' O.C.		1440	.033		10.55	1.18	.66	12.39	14.30	
	2400	Frames 16' O.C.		1920	.025		9.65	.89	.50	11.04	12.60	
	2600	Bowstring trusses, 20' O.C., 40' clear span	F-3	2400	.017		4.76	.60	.27	5.63	6.55	
	2700	60' clear span	F-4	3600	.013		4.27	.47	.26	5	5.75	
	2800	100' clear span		4000	.012		6.05	.43	.24	6.72	7.60	
	2900	120' clear span		3600	.013		6.50	.47	.26	7.23	8.20	
	3100	For premium appearance, add to S.F. prices					5%					
	3300	For industrial type, deduct					15%					

06100 | Rough Carpentry

06180 | Glued-Laminated Construction

		06180 Glued-Laminated Construction	CREW	DAILY OUTPUT	LABOR-HOURS	UNIT	2006 BARE COSTS MAT.	LABOR	EQUIP.	TOTAL	TOTAL INCL O&P	
400	3500	For stain and varnish, add				SF Flr.	5%					400
	3900	For 3/4" laminations, add to straight					25%					
	4100	Add to curved				↓	15%					
	4300	Alternate pricing method: (use nominal footage of										
	4310	components). Straight beams, camber less than 6"	F-3	3.50	11.429	M.B.F.	2,625	410	182	3,217	3,775	
	4400	Columns, including hardware		2	20		2,825	720	320	3,865	4,650	
	4600	Curved members, radius over 32'		2.50	16		2,900	575	255	3,730	4,425	
	4700	Radius 10' to 32'	↓	3	13.333		2,875	480	212	3,567	4,175	
	4900	For complicated shapes, add maximum					100%					
	5100	For pressure treating, add to straight					35%					
	5200	Add to curved				↓	45%					
	6000	Laminated veneer members, southern pine or western species										
	6050	1-3/4" wide x 5-1/2" deep	2 Carp	480	.033	L.F.	3.02	1.18		4.20	5.30	
	6100	9-1/2" deep		480	.033		3.76	1.18		4.94	6.10	
	6150	14" deep		450	.036		5.60	1.26		6.86	8.25	
	6200	18" deep	↓	450	.036	↓	7.60	1.26		8.86	10.45	
	6300	Parallel strand members, southern pine or western species										
	6350	1-3/4" wide x 9-1/4" deep	2 Carp	480	.033	L.F.	3.56	1.18		4.74	5.90	
	6400	11-1/4" deep		450	.036		4.38	1.26		5.64	6.90	
	6450	14" deep		400	.040		5.20	1.42		6.62	8.10	
	6500	3-1/2" wide x 9-1/4" deep		480	.033		8.65	1.18		9.83	11.45	
	6550	11-1/4" deep		450	.036		10.70	1.26		11.96	13.90	
	6600	14" deep		400	.040		12.70	1.42		14.12	16.35	
	6650	7" wide x 9-1/4" deep		450	.036		18.05	1.26		19.31	22	
	6700	11-1/4" deep		420	.038		22.50	1.35		23.85	27	
	6750	14" deep	↓	400	.040	↓	27	1.42		28.42	32	
	9000	Minimum labor/equipment charge	F-3	2.50	16	Job		575	255	830	1,225	

06200 | Finish Carpentry

06220 | Millwork

		06220 Millwork	CREW	DAILY OUTPUT	LABOR-HOURS	UNIT	2006 BARE COSTS MAT.	LABOR	EQUIP.	TOTAL	TOTAL INCL O&P	
200	0010	**MOLDINGS, BASE**										200
	0500	Base, stock pine, 9/16" x 3-1/2"	1 Carp	240	.033	L.F.	2.21	1.18		3.39	4.39	
	0550	9/16" x 4-1/2"		200	.040		2.40	1.42		3.82	5	
	0561	Base shoe, oak, 3/4" x 1"		240	.033	↓	1.13	1.18		2.31	3.20	
	9000	Minimum labor/equipment charge	↓	4	2	Job		71		71	118	
400	0010	**MOLDINGS, CASINGS**										400
	0090	Apron, stock pine, 5/8" x 2"	1 Carp	250	.032	L.F.	1.04	1.14		2.18	3.02	
	0110	5/8" x 3-1/2"		220	.036		1.25	1.29		2.54	3.52	
	0300	Band, stock pine, 11/16" x 1-1/8"		270	.030		.57	1.05		1.62	2.38	
	0350	11/16" x 1-3/4"		250	.032		1.12	1.14		2.26	3.11	
	0700	Casing, stock pine, 11/16" x 2-1/2"		240	.033		1.28	1.18		2.46	3.37	
	0750	11/16" x 3-1/2"		215	.037		1.83	1.32		3.15	4.20	
	0760	Door & window casing, exterior, 1-1/4" x 2"		200	.040		2.13	1.42		3.55	4.70	
	0770	Finger jointed, 1-1/4" x 2"		200	.040	↓	1.04	1.42		2.46	3.50	
	9000	Minimum labor/equipment charge	↓	4	2	Job		71		71	118	
450	0010	**MOLDINGS, CEILINGS**										450
	0600	Bed, stock pine, 9/16" x 1-3/4"	1 Carp	270	.030	L.F.	.88	1.05		1.93	2.72	

	06220 | Millwork		Crew	Daily Output	Labor-Hours	Unit	2006 Bare Costs Mat.	Labor	Equip.	Total	Total Incl O&P	
450	0650	9/16" x 2"	1 Carp	240	.033	L.F.	1.19	1.18		2.37	3.27	450
	1200	Cornice molding, stock pine, 9/16" x 1-3/4"		330	.024		.82	.86		1.68	2.33	
	1300	9/16" x 2-1/4"		300	.027		1.02	.95		1.97	2.69	
	2400	Cove scotia, stock pine, 9/16" x 1-3/4"		270	.030		.65	1.05		1.70	2.47	
	2500	11/16" x 2-3/4"		255	.031		1.25	1.12		2.37	3.23	
	2600	Crown, stock pine, 9/16" x 3-5/8"		250	.032		1.86	1.14		3	3.93	
	2700	11/16" x 4-5/8"		220	.036		3.11	1.29		4.40	5.55	
	9000	Minimum labor/equipment charge		4	2	Job		71		71	118	
500	0010	**MOLDINGS, EXTERIOR**										500
	1500	Cornice, boards, pine, 1" x 2"	1 Carp	330	.024	L.F.	.36	.86		1.22	1.83	
	1700	1" x 6"		250	.032		1.24	1.14		2.38	3.24	
	2000	1" x 12"		180	.044		1.89	1.58		3.47	4.70	
	2200	Three piece, built-up, pine, minimum		80	.100		2.31	3.56		5.87	8.45	
	2300	Maximum		65	.123		5	4.38		9.38	12.75	
	3000	Corner board, sterling pine, 1" x 4"		200	.040		.69	1.42		2.11	3.12	
	3100	1" x 6"		200	.040		1.50	1.42		2.92	4.01	
	3350	Fascia, sterling pine, 1" x 6"		250	.032		1.50	1.14		2.64	3.53	
	3370	1" x 8"		225	.036		1.90	1.26		3.16	4.18	
	3400	Trim, exterior, sterling pine, back band		250	.032		.74	1.14		1.88	2.69	
	3500	Casing		250	.032		1.96	1.14		3.10	4.04	
	3600	Crown		250	.032		1.89	1.14		3.03	3.96	
	3700	Porch rail with balusters		22	.364		15.40	12.95		28.35	38.50	
	3800	Screen		395	.020		1.19	.72		1.91	2.50	
	3850											
	4100	Verge board, sterling pine, 1" x 4"	1 Carp	200	.040	L.F.	.78	1.42		2.20	3.22	
	4200	1" x 6"		200	.040		1.08	1.42		2.50	3.55	
	4300	2" x 6"		165	.048		1.75	1.72		3.47	4.79	
	4400	2" x 8"		165	.048		2.32	1.72		4.04	5.40	
	4700	For redwood trim, add					200%					
	9000	Minimum labor/equipment charge	1 Carp	4	2	Job		71		71	118	
700	0010	**MOLDINGS, TRIM**										700
	0200	Astragal, stock pine, 11/16" x 1-3/4"	1 Carp	255	.031	L.F.	1.16	1.12		2.28	3.13	
	0250	1-5/16" x 2-3/16"		240	.033		2.71	1.18		3.89	4.94	
	0800	Chair rail, stock pine, 5/8" x 2-1/2"		270	.030		1.44	1.05		2.49	3.33	
	0900	5/8" x 3-1/2"		240	.033		2.20	1.18		3.38	4.38	
	1000	Closet pole, stock pine, 1-1/8" diameter		200	.040		.87	1.42		2.29	3.32	
	1100	Fir, 1-5/8" diameter		200	.040		1.60	1.42		3.02	4.12	
	3300	Half round, stock pine, 1/4" x 1/2"		270	.030		.23	1.05		1.28	2	
	3350	1/2" x 1"		255	.031		.60	1.12		1.72	2.51	
	3400	Handrail, fir, single piece, stock, hardware not included										
	3450	1-1/2" x 1-3/4"	1 Carp	80	.100	L.F.	1.77	3.56		5.33	7.85	
	3470	Pine, 1-1/2" x 1-3/4"		80	.100		1.77	3.56		5.33	7.85	
	3500	1-1/2" x 2-1/2"		76	.105		1.63	3.74		5.37	8	
	3600	Lattice, stock pine, 1/4" x 1-1/8"		270	.030		.41	1.05		1.46	2.20	
	3700	1/4" x 1-3/4"		250	.032		.45	1.14		1.59	2.38	
	3800	Miscellaneous, custom, pine, 1" x 1"		270	.030		.34	1.05		1.39	2.12	
	3900	1" x 3"		240	.033		.69	1.18		1.87	2.72	
	4100	Birch or oak, nominal 1" x 1"		240	.033		.53	1.18		1.71	2.54	
	4200	Nominal 1" x 3"		215	.037		1.81	1.32		3.13	4.18	
	4400	Walnut, nominal 1" x 1"		215	.037		.87	1.32		2.19	3.15	
	4500	Nominal 1" x 3"		200	.040		2.61	1.42		4.03	5.25	
	4700	Teak, nominal 1" x 1"		215	.037		1.23	1.32		2.55	3.54	
	4800	Nominal 1" x 3"		200	.040		3.52	1.42		4.94	6.25	
	4900	Quarter round, stock pine, 1/4" x 1/4"		275	.029		.22	1.03		1.25	1.95	
	4950	3/4" x 3/4"		255	.031		.64	1.12		1.76	2.55	
	5600	Wainscot moldings, 1-1/8" x 9/16", 2' high, minimum		76	.105	S.F.	10.50	3.74		14.24	17.75	

		06220 | Millwork	CREW	DAILY OUTPUT	LABOR-HOURS	UNIT	2006 BARE COSTS MAT.	LABOR	EQUIP.	TOTAL	TOTAL INCL O&P	
700	5700	Maximum	1 Carp	65	.123	S.F.	19.65	4.38		24.03	29	700
	9000	Minimum labor/equipment charge		4	2	Job		71		71	118	
800	0010	**MOLDINGS, WINDOW AND DOOR**										800
	2800	Door moldings, stock, decorative, 1-1/8″ wide, plain	1 Carp	17	.471	Set	43.50	16.75		60.25	75.50	
	2900	Detailed		17	.471	″	87.50	16.75		104.25	124	
	2960	Clear pine door jamb, no stops, 11/16″ x 4-9/16″		240	.033	L.F.	2.30	1.18		3.48	4.49	
	3150	Door trim set, 1 head and 2 sides, pine, 2-1/2 wide		5.90	1.356	Opng.	22	48		70	104	
	3170	3-1/2″ wide		5.30	1.509	″	31	53.50		84.50	123	
	3250	Glass beads, stock pine, 3/8″ x 1/2″		275	.029	L.F.	.31	1.03		1.34	2.05	
	3270	3/8″ x 7/8″		270	.030		.41	1.05		1.46	2.20	
	4850	Parting bead, stock pine, 3/8″ x 3/4″		275	.029		.35	1.03		1.38	2.10	
	4870	1/2″ x 3/4″		255	.031		.44	1.12		1.56	2.33	
	5000	Stool caps, stock pine, 11/16″ x 3-1/2″		200	.040		2.20	1.42		3.62	4.78	
	5100	1-1/16″ x 3-1/4″		150	.053		3.48	1.90		5.38	6.95	
	5300	Threshold, oak, 3′ long, inside, 5/8″ x 3-5/8″		32	.250	Ea.	7.75	8.90		16.65	23.50	
	5400	Outside, 1-1/2″ x 7-5/8″		16	.500	″	35	17.80		52.80	68	
	5900	Window trim sets, including casings, header, stops,										
	5910	stool and apron, 2-1/2″ wide, minimum	1 Carp	13	.615	Opng.	30	22		52	69	
	5950	Average		10	.800		35	28.50		63.50	85.50	
	6000	Maximum		6	1.333		62.50	47.50		110	147	
	9000	Minimum labor/equipment charge		4	2	Job		71		71	118	
900	0010	**SOFFITS** R061110-30										900
	0200	Soffits, pine, 1″ x 4″	2 Carp	420	.038	L.F.	.35	1.35		1.70	2.63	
	0210	1″ x 6″		420	.038		.55	1.35		1.90	2.85	
	0220	1″ x 8″		420	.038		.70	1.35		2.05	3.01	
	0230	1″ x 10″		400	.040		1.08	1.42		2.50	3.55	
	0240	1″ x 12″		400	.040		1.10	1.42		2.52	3.56	
	0250	STK cedar, 1″ x 4″		420	.038		.56	1.35		1.91	2.86	
	0260	1″ x 6″		420	.038		.83	1.35		2.18	3.15	
	0270	1″ x 8″		420	.038		1.09	1.35		2.44	3.44	
	0280	1″ x 10″		400	.040		2.13	1.42		3.55	4.70	
	0290	1″ x 12″		400	.040		2.89	1.42		4.31	5.55	
	1000	Exterior AC plywood, 1/4″ thick		420	.038	S.F.	.75	1.35		2.10	3.07	
	1050	3/8″ thick		420	.038		.78	1.35		2.13	3.10	
	1100	1/2″ thick		420	.038		.87	1.35		2.22	3.20	
	1150	Polyvinyl chloride, white, solid	1 Carp	230	.035		.82	1.24		2.06	2.95	
	1160	Perforated	″	230	.035		.82	1.24		2.06	2.95	
	9000	Minimum labor/equipment charge	2 Carp	5	3.200	Job		114		114	188	
		06250 | Prefinished Paneling										
200	0010	**PANELING, HARDBOARD**										200
	0050	Not incl. furring or trim, hardboard, tempered, 1/8″ thick	2 Carp	500	.032	S.F.	.32	1.14		1.46	2.23	
	0100	1/4″ thick		500	.032		.48	1.14		1.62	2.41	
	0300	Tempered pegboard, 1/8″ thick		500	.032		.39	1.14		1.53	2.31	
	0400	1/4″ thick		500	.032		.51	1.14		1.65	2.44	
	0600	Untempered hardboard, natural finish, 1/8″ thick		500	.032		.35	1.14		1.49	2.27	
	0700	1/4″ thick		500	.032		.34	1.14		1.48	2.25	
	0900	Untempered pegboard, 1/8″ thick		500	.032		.35	1.14		1.49	2.27	
	1000	1/4″ thick		500	.032		.39	1.14		1.53	2.31	
	1200	Plastic faced hardboard, 1/8″ thick		500	.032		.58	1.14		1.72	2.52	
	1300	1/4″ thick		500	.032		.77	1.14		1.91	2.73	
	1500	Plastic faced pegboard, 1/8″ thick		500	.032		.55	1.14		1.69	2.49	
	1600	1/4″ thick		500	.032		.68	1.14		1.82	2.63	
	1800	Wood grained, plain or grooved, 1/4″ thick, minimum		500	.032		.51	1.14		1.65	2.44	

06200 | Finish Carpentry

06250 | Prefinished Paneling

			CREW	DAILY OUTPUT	LABOR-HOURS	UNIT	2006 BARE COSTS MAT.	LABOR	EQUIP.	TOTAL	TOTAL INCL O&P	
200	1900	Maximum	2 Carp	425	.038	S.F.	1.08	1.34		2.42	3.41	200
	2100	Moldings for hardboard, wood or aluminum, minimum		500	.032	L.F.	.35	1.14		1.49	2.27	
	2200	Maximum		425	.038	"	.98	1.34		2.32	3.30	
	9000	Minimum labor/equipment charge	1 Carp	2	4	Job		142		142	236	
500	0010	PANELING, PLYWOOD R061636-20										500
	2400	Plywood, prefinished, 1/4" thick, 4' x 8' sheets										
	2410	with vertical grooves. Birch faced, minimum	2 Carp	500	.032	S.F.	.84	1.14		1.98	2.80	
	2420	Average		420	.038		1.28	1.35		2.63	3.65	
	2430	Maximum		350	.046		1.87	1.63		3.50	4.75	
	2600	Mahogany, African		400	.040		2.39	1.42		3.81	4.99	
	2700	Philippine (Lauan)		500	.032		1.03	1.14		2.17	3.01	
	2900	Oak or Cherry, minimum		500	.032		2	1.14		3.14	4.08	
	3000	Maximum		400	.040		3.07	1.42		4.49	5.75	
	3200	Rosewood		320	.050		4.36	1.78		6.14	7.75	
	3400	Teak		400	.040		3.07	1.42		4.49	5.75	
	3600	Chestnut		375	.043		4.54	1.52		6.06	7.50	
	3800	Pecan		400	.040		1.96	1.42		3.38	4.52	
	3900	Walnut, minimum		500	.032		2.62	1.14		3.76	4.76	
	3950	Maximum		400	.040		4.96	1.42		6.38	7.80	
	4000	Plywood, prefinished, 3/4" thick, stock grades, minimum		320	.050		1.19	1.78		2.97	4.26	
	4100	Maximum		224	.071		5.10	2.54		7.64	9.85	
	4300	Architectural grade, minimum		224	.071		3.78	2.54		6.32	8.35	
	4400	Maximum		160	.100		5.80	3.56		9.36	12.25	
	4600	Plywood, "A" face, birch, V.C., 1/2" thick, natural		450	.036		1.79	1.26		3.05	4.06	
	4700	Select		450	.036		1.96	1.26		3.22	4.25	
	4900	Veneer core, 3/4" thick, natural		320	.050		1.89	1.78		3.67	5.05	
	5000	Select		320	.050		2.13	1.78		3.91	5.30	
	5200	Lumber core, 3/4" thick, natural		320	.050		2.84	1.78		4.62	6.05	
	5500	Plywood, knotty pine, 1/4" thick, A2 grade		450	.036		1.55	1.26		2.81	3.80	
	5600	A3 grade		450	.036		1.96	1.26		3.22	4.25	
	5800	3/4" thick, veneer core, A2 grade		320	.050		2.01	1.78		3.79	5.15	
	5900	A3 grade		320	.050		2.26	1.78		4.04	5.45	
	6100	Aromatic cedar, 1/4" thick, plywood		400	.040		1.98	1.42		3.40	4.54	
	6200	1/4" thick, particle board		400	.040		.96	1.42		2.38	3.42	
	9000	Minimum labor/equipment charge	1 Carp	2	4	Job		142		142	236	

06260 | Board Paneling

			CREW	DAILY OUTPUT	LABOR-HOURS	UNIT	MAT.	LABOR	EQUIP.	TOTAL	TOTAL INCL O&P	
400	0010	PANELING, BOARDS										400
	6400	Wood board paneling, 3/4" thick, knotty pine	2 Carp	300	.053	S.F.	1.40	1.90		3.30	4.68	
	6500	Rough sawn cedar		300	.053		1.79	1.90		3.69	5.10	
	6700	Redwood, clear, 1" x 4" boards		300	.053		4.18	1.90		6.08	7.75	
	6900	Aromatic cedar, closet lining, boards		275	.058		3.22	2.07		5.29	6.95	
	9000	Minimum labor/equipment charge	1 Carp	2	4	Job		142		142	236	

06270 | Closet/Utility Wood Shelving

			CREW	DAILY OUTPUT	LABOR-HOURS	UNIT	MAT.	LABOR	EQUIP.	TOTAL	TOTAL INCL O&P	
200	0010	SHELVING										200
	0020	Pine, clear grade, no edge band, 1" x 8"	1 Carp	115	.070	L.F.	1.81	2.47		4.28	6.10	
	0100	1" x 10"		110	.073		2.31	2.59		4.90	6.80	
	0200	1" x 12"		105	.076		3.10	2.71		5.81	7.90	
	0600	Plywood, 3/4" thick with lumber edge, 12" wide		75	.107		1.66	3.79		5.45	8.15	
	0700	24" wide		70	.114		2.98	4.06		7.04	10.05	
	0900	Bookcase, clear grade pine, shelves 12" O.C., 8" deep, /SF shelf		70	.114	S.F.	5.90	4.06		9.96	13.20	
	1000	12" deep shelves		65	.123	"	10.10	4.38		14.48	18.35	
	1200	Adjustable closet rod and shelf, 12" wide, 3' long		20	.400	Ea.	8.70	14.20		22.90	33	
	1300	8' long		15	.533	"	22	18.95		40.95	55.50	

WOOD & PLASTICS 6

06200 | Finish Carpentry

	06270 | Closet/Utility Wood Shelving		Crew	Daily Output	Labor-Hours	Unit	2006 Bare Costs Mat.	Labor	Equip.	Total	Total Incl O&P	
200	1500	Prefinished shelves with supports, stock, 8″ wide	1 Carp	75	.107	L.F.	3.83	3.79		7.62	10.50	200
	1600	10″ wide		70	.114	″	4.26	4.06		8.32	11.45	
	9000	Minimum labor/equipment charge	↓	4	2	Job		71		71	118	

06400 | Architectural Woodwork

	06410 | Custom Cabinets		Crew	Daily Output	Labor-Hours	Unit	2006 Bare Costs Mat.	Labor	Equip.	Total	Total Incl O&P	
100	0010	**CABINETS** Corner china cabinets, stock pine,										100
	0020	80″ high, unfinished, minimum	2 Carp	6.60	2.424	Ea.	455	86		541	645	
	0100	Maximum	″	4.40	3.636	″	1,000	129		1,129	1,325	
	0300	Built-in drawer units, pine, 18″ deep, 32″ high, unfinished										
	0400	Minimum	2 Carp	53	.302	L.F.	120	10.75		130.75	150	
	0500	Maximum	″	40	.400	″	146	14.20		160.20	185	
	0700	Kitchen base cabinets, hardwood, not incl. counter tops,										
	0710	24″ deep, 35″ high, prefinished										
	0800	One top drawer, one door below, 12″ wide	2 Carp	24.80	.645	Ea.	138	23		161	190	
	0820	15″ wide		24	.667		188	23.50		211.50	247	
	0840	18″ wide		23.30	.687		205	24.50		229.50	266	
	0860	21″ wide		22.70	.705		213	25		238	276	
	0880	24″ wide		22.30	.717		245	25.50		270.50	315	
	1000	Four drawers, 12″ wide		24.80	.645		325	23		348	400	
	1020	15″ wide		24	.667		251	23.50		274.50	315	
	1040	18″ wide		23.30	.687		280	24.50		304.50	350	
	1060	24″ wide		22.30	.717		305	25.50		330.50	380	
	1200	Two top drawers, two doors below, 27″ wide		22	.727		272	26		298	340	
	1220	30″ wide		21.40	.748		291	26.50		317.50	365	
	1240	33″ wide		20.90	.766		300	27		327	375	
	1260	36″ wide		20.30	.788		315	28		343	390	
	1280	42″ wide		19.80	.808		335	28.50		363.50	420	
	1300	48″ wide		18.90	.847		360	30		390	445	
	1500	Range or sink base, two doors below, 30″ wide		21.40	.748		240	26.50		266.50	310	
	1520	33″ wide		20.90	.766		258	27		285	330	
	1540	36″ wide		20.30	.788		270	28		298	345	
	1560	42″ wide		19.80	.808		288	28.50		316.50	365	
	1580	48″ wide	↓	18.90	.847		300	30		330	380	
	1800	For sink front units, deduct					52			52	57.50	
	2000	Corner base cabinets, 36″ wide, standard	2 Carp	18	.889		395	31.50		426.50	490	
	2100	Lazy Susan with revolving door	″	16.50	.970	↓	385	34.50		419.50	475	
	4000	Kitchen wall cabinets, hardwood, 12″ deep with two doors										
	4050	12″ high, 30″ wide	2 Carp	24.80	.645	Ea.	151	23		174	204	
	4100	36″ wide		24	.667		176	23.50		199.50	233	
	4400	15″ high, 30″ wide		24	.667		158	23.50		181.50	214	
	4420	33″ wide		23.30	.687		177	24.50		201.50	236	
	4440	36″ wide		22.70	.705		179	25		204	239	
	4450	42″ wide		22.70	.705		208	25		233	270	
	4700	24″ high, 30″ wide		23.30	.687		197	24.50		221.50	257	
	4720	36″ wide		22.70	.705		218	25		243	281	
	4740	42″ wide		22.30	.717		240	25.50		265.50	305	
	5000	30″ high, one door, 12″ wide	↓	22	.727	↓	133	26		159	189	

	06410 | Custom Cabinets		CREW	DAILY OUTPUT	LABOR-HOURS	UNIT	2006 BARE COSTS MAT.	LABOR	EQUIP.	TOTAL	TOTAL INCL O&P	
100	5020	15" wide	2 Carp	21.40	.748	Ea.	150	26.50		176.50	209	100
	5040	18" wide		20.90	.766		164	27		191	226	
	5060	24" wide		20.30	.788		185	28		213	251	
	5300	Two doors, 27" wide		19.80	.808		229	28.50		257.50	300	
	5320	30" wide		19.30	.829		222	29.50		251.50	293	
	5340	36" wide		18.80	.851		253	30.50		283.50	330	
	5360	42" wide		18.50	.865		276	31		307	355	
	5380	48" wide		18.40	.870		310	31		341	390	
	6000	Corner wall, 30" high, 24" wide		18	.889		152	31.50		183.50	220	
	6050	30" wide		17.20	.930		180	33		213	253	
	6100	36" wide		16.50	.970		195	34.50		229.50	272	
	6500	Revolving Lazy Susan		15.20	1.053		298	37.50		335.50	385	
	7000	Broom cabinet, 84" high, 24" deep, 18" wide		10	1.600		405	57		462	540	
	7500	Oven cabinets, 84" high, 24" deep, 27" wide		8	2	↓	590	71		661	770	
	7750	Valance board trim	↓	396	.040	L.F.	8.40	1.44		9.84	11.65	
	9000	For deluxe models of all cabinets, add					40%					
	9500	For custom built in place, add					25%	10%				
	9550	Rule of thumb, kitchen cabinets not including										
	9560	appliances & counter top, minimum	2 Carp	30	.533	L.F.	95.50	18.95		114.45	137	
	9600	Maximum	"	25	.640	"	250	23		273	315	
	9700	Minimum labor/equipment charge	1 Carp	3	2.667	Job		95		95	157	
210	0010	**CASEWORK, FRAMES**										210
	0050	Base cabinets, counter storage, 36" high, one bay										
	0100	18" wide	1 Carp	2.70	2.963	Ea.	107	105		212	293	
	0400	Two bay, 36" wide		2.20	3.636		163	129		292	395	
	1100	Three bay, 54" wide		1.50	5.333		194	190		384	530	
	2800	Book cases, one bay, 7' high, 18" wide		2.40	3.333		126	119		245	335	
	3500	Two bay, 36" wide		1.60	5		183	178		361	495	
	4100	Three bay, 54" wide		1.20	6.667		300	237		537	725	
	5100	Coat racks, one bay, 7' high, 24" wide		4.50	1.778		126	63		189	243	
	5300	Two bay, 48" wide		2.75	2.909		175	103		278	365	
	5800	Three bay, 72" wide		2.10	3.810		258	135		393	505	
	6100	Wall mounted cabinet, one bay, 24" high, 18" wide		3.60	2.222		69.50	79		148.50	207	
	6800	Two bay, 36" wide		2.20	3.636		101	129		230	325	
	7400	Three bay, 54" wide		1.70	4.706		126	167		293	415	
	8400	30" high, one bay, 18" wide		3.60	2.222		75.50	79		154.50	214	
	9000	Two bay, 36" wide		2.15	3.721		100	132		232	330	
	9400	Three bay, 54" wide		1.60	5		125	178		303	430	
	9800	Wardrobe, 7' high, single, 24" wide		2.70	2.963		139	105		244	325	
	9880	Partition & adjustable shelves, 48" wide		1.70	4.706		176	167		343	470	
	9950	Partition, adjustable shelves & drawers, 48" wide		1.40	5.714	↓	265	203		468	625	
	9970	Minimum labor/equipment charge	↓	4	2	Job		71		71	118	
220	0010	**CABINET DOORS**										220
	2000	Glass panel, hardwood frame										
	2200	12" wide, 18" high	1 Carp	34	.235	Ea.	22	8.35		30.35	38.50	
	2600	30" high		32	.250		25.50	8.90		34.40	43	
	4450	18" wide, 18" high		32	.250		24	8.90		32.90	41	
	4550	30" high	↓	29	.276	↓	26.50	9.80		36.30	45.50	
	5000	Hardwood, raised panel										
	5100	12" wide, 18" high	1 Carp	16	.500	Ea.	33.50	17.80		51.30	66.50	
	5200	30" high		15	.533		50	18.95		68.95	86.50	
	5500	18" wide, 18" high		15	.533		37	18.95		55.95	72	
	5600	30" high	↓	14	.571	↓	45	20.50		65.50	83	
	6000	Plastic laminate on particle board										

For expanded coverage of these items see *Means Interior Cost Data 2006*

06400 | Architectural Woodwork

		06410 | Custom Cabinets	CREW	DAILY OUTPUT	LABOR-HOURS	UNIT	2006 BARE COSTS MAT.	LABOR	EQUIP.	TOTAL	TOTAL INCL O&P	
220	6100	12" wide, 18" high	1 Carp	25	.320	Ea.	10.35	11.40		21.75	30.50	220
	6140	30" high		23	.348		17.25	12.35		29.60	39.50	
	6500	18" wide, 18" high		24	.333		15.50	11.85		27.35	36.50	
	6600	30" high		22	.364		26	12.95		38.95	50	
	9000	Minimum labor/equipment charge		4	2	Job		71		71	118	
230	0010	CABINET HARDWARE										230
	1000	Catches, minimum	1 Carp	235	.034	Ea.	.84	1.21		2.05	2.93	
	1040	Maximum	"	80	.100	"	5.20	3.56		8.76	11.65	
	2000	Door/drawer pulls, handles										
	2200	Handles and pulls, projecting, metal, minimum	1 Carp	160	.050	Ea.	3.90	1.78		5.68	7.25	
	2240	Maximum		68	.118		7.15	4.18		11.33	14.85	
	2300	Wood, minimum		160	.050		3.90	1.78		5.68	7.25	
	2340	Maximum		68	.118		7.15	4.18		11.33	14.85	
	2600	Flush, metal, minimum		160	.050		3.90	1.78		5.68	7.25	
	2640	Maximum		68	.118		7.15	4.18		11.33	14.85	
	3000	Drawer tracks/glides, minimum		48	.167	Pr.	6.60	5.95		12.55	17.10	
	3040	Maximum		24	.333		19.25	11.85		31.10	40.50	
	4000	Cabinet hinges, minimum		160	.050		2.25	1.78		4.03	5.45	
	4040	Maximum		68	.118		7.85	4.18		12.03	15.60	
240	0010	DRAWERS										240
	0100	Solid hardwood front										
	1000	4" high, 12" wide	1 Carp	17	.471	Ea.	3.50	16.75		20.25	31.50	
	1200	18" wide	"	16	.500	"	4.95	17.80		22.75	35	
	2800	Plastic laminate on particle board front										
	3000	4" high, 12" wide	1 Carp	17	.471	Ea.	4.39	16.75		21.14	32.50	
	3200	18" wide	"	16	.500	"	6.60	17.80		24.40	37	
	5400	Plywood, flush panel front										
	6000	4" high, 12" wide	1 Carp	17	.471	Ea.	5.95	16.75		22.70	34	
	6200	18" wide		16	.500	"	8.90	17.80		26.70	39.50	
	9000	Minimum labor/equipment charge		4	2	Job		71		71	118	
400	0010	VANITIES										400
	8000	Vanity bases, 2 doors, 30" high, 21" deep, 24" wide	2 Carp	20	.800	Ea.	184	28.50		212.50	249	
	8050	30" wide		16	1		211	35.50		246.50	291	
	8100	36" wide		13.33	1.200		282	42.50		324.50	380	
	8150	48" wide		11.43	1.400		335	50		385	455	
	9000	For deluxe models of all vanities, add to above					40%					
	9500	For custom built in place, add to above					25%	10%				
500	0010	CABINETS, OUTDOOR										500
	0020	Cabinet, base, sink/range, 36"	2 Carp	20.30	.788	Ea.	1,300	28		1,328	1,500	
	0100	Cabinet, base, 36"		20.30	.788		1,875	28		1,903	2,125	
	0200	Cabinet, filler strip, 1" x 30"		158	.101		26	3.60		29.60	34.50	
	0210	Cabinet, filler strip, 2" x 30"		158	.101		33.50	3.60		37.10	42.50	

		06415 | Countertops	CREW	DAILY OUTPUT	LABOR-HOURS	UNIT	MAT.	LABOR	EQUIP.	TOTAL	TOTAL INCL O&P	
100	0010	COUNTER TOP										100
	0020	Stock plastic laminate, 24" wide w/ backsplash, minimum	1 Carp	30	.267	L.F.	8.85	9.50		18.35	25.50	
	0100	Maximum		25	.320		16	11.40		27.40	36.50	
	0300	Custom plastic, 7/8" thick, aluminum molding, no splash		30	.267		17.70	9.50		27.20	35	
	0400	Cove splash		30	.267		23	9.50		32.50	41	
	0600	1-1/4" thick, no splash		28	.286		20	10.15		30.15	39	
	0700	Square splash		28	.286		25.50	10.15		35.65	45	
	0900	Square edge, plastic face, 7/8" thick, no splash		30	.267		22	9.50		31.50	40	

06415 | Countertops

		06415 Countertops	CREW	DAILY OUTPUT	LABOR-HOURS	UNIT	2006 BARE COSTS MAT.	LABOR	EQUIP.	TOTAL	TOTAL INCL O&P	
100	1000	With splash	1 Carp	30	.267	L.F.	28.50	9.50		38	46.50	100
	1200	For stainless channel edge, 7/8" thick, add					2.32			2.32	2.55	
	1300	1-1/4" thick, add					2.73			2.73	3	
	1500	For solid color suede finish, add					2.27			2.27	2.50	
	1700	For end splash, add				Ea.	15.15			15.15	16.65	
	1900	For cut outs, standard, add, minimum	1 Carp	32	.250		3.03	8.90		11.93	18.10	
	2000	Maximum		8	1		5.05	35.50		40.55	64.50	
	2100	Postformed, including backsplash and front edge		30	.267	L.F.	9.10	9.50		18.60	25.50	
	2110	Mitred, add		12	.667	Ea.		23.50		23.50	39.50	
	2200	Built-in place, 25" wide, plastic laminate		25	.320	L.F.	12.10	11.40		23.50	32	
	2300	Ceramic tile mosaic		25	.320		26.50	11.40		37.90	48	
	2500	Marble, stock, with splash, 1/2" thick, minimum	1 Bric	17	.471		32.50	17.20		49.70	63.50	
	2700	3/4" thick, maximum	"	13	.615		81	22.50		103.50	126	
	2900	Maple, solid, laminated, 1-1/2" thick, no splash	1 Carp	28	.286		53	10.15		63.15	75	
	3000	With square splash		28	.286		63	10.15		73.15	86	
	3200	Stainless steel		24	.333	S.F.	121	11.85		132.85	153	
	3400	Recessed cutting block with trim, 16" x 20" x 1"		8	1	Ea.	61.50	35.50		97	127	
	9000	Minimum labor/equipment charge		3.75	2.133	Job		76		76	126	

06430 | Stairs & Railings

		06430 Stairs & Railings	CREW	DAILY OUTPUT	LABOR-HOURS	UNIT	MAT.	LABOR	EQUIP.	TOTAL	TOTAL INCL O&P	
500	0010	RAILINGS										500
	0020	Custom design, architectural grade, hardwood, minimum	1 Carp	38	.211	L.F.	5.60	7.50		13.10	18.55	
	0100	Maximum		30	.267		46.50	9.50		56	67	
	0300	Stock interior railing with spindles 6" O.C., 4' long		40	.200		42	7.10		49.10	58	
	0400	8' long		48	.167		21	5.95		26.95	33	
	9000	Minimum labor/equipment charge		3	2.667	Job		95		95	157	
620	0010	STAIRS, PREFABRICATED										620
	0100	Box stairs, prefabricated, 3'-0" wide										
	0110	Oak treads, up to 14 risers	2 Carp	39	.410	Riser	71.50	14.60		86.10	103	
	0600	With pine treads for carpet, up to 14 risers	"	39	.410	"	46	14.60		60.60	74.50	
	1100	For 4' wide stairs, add				Flight	25%					
	1550	Stairs, prefabricated stair handrail with balusters	1 Carp	30	.267	L.F.	64	9.50		73.50	85.50	
	1700	Basement stairs, prefabricated, pine treads										
	1710	Pine risers, 3' wide, up to 14 risers	2 Carp	52	.308	Riser	46	10.95		56.95	68.50	
	4000	Residential, wood, oak treads, prefabricated		1.50	10.667	Flight	930	380		1,310	1,650	
	4200	Built in place		.44	36.364	"	1,925	1,300		3,225	4,275	
	4400	Spiral, oak, 4'-6" diameter, unfinished, prefabricated,										
	4500	incl. railing, 9' high	2 Carp	1.50	10.667	Flight	4,400	380		4,780	5,475	
	9000	Minimum labor/equipment charge	"	3	5.333	Job		190		190	315	
630	0010	STAIR PARTS										630
	0020	Balusters, turned, 3" high, pine, minimum	1 Carp	28	.286	Ea.	3.70	10.15		13.85	21	
	0100	Maximum		26	.308		19	10.95		29.95	39	
	0300	30" high birch balusters, minimum		28	.286		6.30	10.15		16.45	24	
	0400	Maximum		26	.308		27.50	10.95		38.45	48	
	0600	42" high, pine balusters, minimum		27	.296		4.90	10.55		15.45	23	
	0700	Maximum		25	.320		27.50	11.40		38.90	49	
	0900	42" high birch balusters, minimum		27	.296		10.70	10.55		21.25	29	
	1000	Maximum		25	.320		38.50	11.40		49.90	61.50	
	1050	Baluster, stock pine, 1-1/4" x 1-1/4"		240	.033	L.F.	3.12	1.18		4.30	5.40	
	1100	1-3/4" x 1-3/4"		220	.036	"	8.90	1.29		10.19	11.95	
	1200	Newels, 3-1/4" wide, starting, minimum		7	1.143	Ea.	38.50	40.50		79	110	
	1300	Maximum		6	1.333		320	47.50		367.50	430	
	1500	Landing, minimum		5	1.600		104	57		161	208	
	1600	Maximum		4	2		350	71		421	505	
	1800	Railings, oak, built-up, minimum		60	.133	L.F.	31.50	4.74		36.24	43	

For expanded coverage of these items see *Means Interior Cost Data 2006*

06400 | Architectural Woodwork

	06430 | Stairs & Railings		Crew	Daily Output	Labor-Hours	Unit	2006 Bare Costs Mat.	Labor	Equip.	Total	Total Incl O&P	
630	1900	Maximum	1 Carp	55	.145	L.F.	43.50	5.15		48.65	56.50	630
	2100	Add for sub rail		110	.073		5.30	2.59		7.89	10.15	
	2300	Risers, beech, 3/4″ x 7-1/2″ high		64	.125		5.80	4.44		10.24	13.70	
	2400	Fir, 3/4″ x 7-1/2″ high		64	.125		1.60	4.44		6.04	9.10	
	2600	Oak, 3/4″ x 7-1/2″ high		64	.125		6.10	4.44		10.54	14.05	
	2800	Pine, 3/4″ x 7-1/2″ high		66	.121		3.05	4.31		7.36	10.50	
	2850	Skirt board, pine, 1″ x 10″		55	.145		2.90	5.15		8.05	11.75	
	2900	1″ x 12″		52	.154		3.45	5.45		8.90	12.85	
	3000	Treads, oak, 1-1/4″ x 10″ wide, 3′ long		18	.444	Ea.	29	15.80		44.80	58	
	3100	4′ long, oak		17	.471		116	16.75		132.75	156	
	3300	1-1/4″ x 11-1/2″ wide, 3′ long, oak		18	.444		31	15.80		46.80	60	
	3400	6′ long, oak		14	.571		186	20.50		206.50	239	
	3600	Beech treads, add					40%					
	3800	For mitered return nosings, add				L.F.	3.90			3.90	4.29	
	9000	Minimum labor/equipment charge	1 Carp	3	2.667	Job		95		95	157	

	06440 | Wood Ornaments		Crew	Daily Output	Labor-Hours	Unit	Mat.	Labor	Equip.	Total	Total Incl O&P	
150	0010	**BEAMS, DECORATIVE**										150
	0020	Rough sawn cedar, non-load bearing, 4″ x 4″	2 Carp	180	.089	L.F.	1.35	3.16		4.51	6.75	
	0100	4″ x 6″		170	.094		2.60	3.35		5.95	8.40	
	0200	4″ x 8″		160	.100		3.34	3.56		6.90	9.55	
	0300	4″ x 10″		150	.107		4.64	3.79		8.43	11.40	
	0400	4″ x 12″		140	.114		5.60	4.06		9.66	12.95	
	0500	8″ x 8″		130	.123		7.85	4.38		12.23	15.90	
	1100	Beam connector plates see div. 06090-800										
	9000	Minimum labor/equipment charge	1 Carp	3	2.667	Job		95		95	157	
350	0010	**GRILLES** and panels, hardwood, sanded										350
	0020	2′ x 4′ to 4′ x 8′, custom designs, unfinished, minimum	1 Carp	38	.211	S.F.	12.10	7.50		19.60	26	
	0050	Average		30	.267		26.50	9.50		36	44.50	
	0100	Maximum		19	.421		40.50	14.95		55.45	69.50	
	0300	As above, but prefinished, minimum		38	.211		12.10	7.50		19.60	26	
	0400	Maximum		19	.421		45.50	14.95		60.45	75	
	9000	Minimum labor/equipment charge		2	4	Job		142		142	236	
400	0010	**LOUVERS**										400
	0020	Redwood, 2′-0″ diameter, full circle	1 Carp	16	.500	Ea.	136	17.80		153.80	180	
	0100	Half circle		16	.500		130	17.80		147.80	173	
	0200	Octagonal		16	.500		104	17.80		121.80	144	
	0300	Triangular, 5/12 pitch, 5′-0″ at base		16	.500		220	17.80		237.80	272	
	9000	Minimum labor/equipment charge		3.50	2.286	Job		81.50		81.50	135	
500	0010	**FIREPLACE MANTELS**										500
	0015	6″ molding, 6′ x 3′-6″ opening, minimum	1 Carp	5	1.600	Opng.	139	57		196	247	
	0100	Maximum		5	1.600		173	57		230	284	
	0300	Prefabricated pine, colonial type, stock, deluxe		2	4		2,175	142		2,317	2,625	
	0400	Economy		3	2.667		375	95		470	565	
	9000	Minimum labor/equipment charge		3	2.667	Job		95		95	157	
550	0010	**FIREPLACE MANTEL BEAMS**										550
	0020	Rough texture wood, 4″ x 8″	1 Carp	36	.222	L.F.	5.15	7.90		13.05	18.75	
	0100	4″ x 10″		35	.229	″	6.20	8.15		14.35	20.50	
	0300	Laminated hardwood, 2-1/4″ x 10-1/2″ wide, 6′ long		5	1.600	Ea.	98.50	57		155.50	202	
	0400	8′ long		5	1.600	″	137	57		194	244	
	0600	Brackets for above, rough sawn		12	.667	Pr.	9.05	23.50		32.55	49.50	
	0700	Laminated		12	.667	″	13.70	23.50		37.20	54.50	
	9000	Minimum labor/equipment charge		4	2	Job		71		71	118	

06440 | Wood Ornaments

			Crew	Daily Output	Labor-Hours	Unit	2006 Bare Costs Mat.	Labor	Equip.	Total	Total Incl O&P	
700	0010	**COLUMNS**										700
	0050	Aluminum, round colonial, 6" diameter	2 Carp	80	.200	V.L.F.	18.10	7.10		25.20	31.50	
	0100	8" diameter		62.25	.257		25	9.15		34.15	42.50	
	0200	10" diameter		55	.291		30	10.35		40.35	50	
	0250	Fir, stock units, hollow round, 6" diameter		80	.200		20.50	7.10		27.60	34.50	
	0300	8" diameter		80	.200		24.50	7.10		31.60	38.50	
	0350	10" diameter		70	.229		31	8.15		39.15	47.50	
	0400	Solid turned, to 8' high, 3-1/2" diameter		80	.200		8.90	7.10		16	21.50	
	0500	4-1/2" diameter		75	.213		12.75	7.60		20.35	26.50	
	0600	5-1/2" diameter		70	.229		17.80	8.15		25.95	33	
	0800	Square columns, built-up, 5" x 5"		65	.246		16.55	8.75		25.30	32.50	
	0900	Solid, 3-1/2" x 3-1/2"		130	.123		7.65	4.38		12.03	15.65	
	1600	Hemlock, tapered, T & G, 12" diam, 10' high		100	.160		38	5.70		43.70	51.50	
	1700	16' high		65	.246		67.50	8.75		76.25	88.50	
	1900	10' high, 14" diameter		100	.160		98	5.70		103.70	117	
	2000	18' high		65	.246		93	8.75		101.75	117	
	2200	18" diameter, 12' high		65	.246		131	8.75		139.75	159	
	2300	20' high		50	.320		126	11.40		137.40	158	
	2500	20" diameter, 14' high		40	.400		155	14.20		169.20	195	
	2600	20' high	↓	35	.457		160	16.25		176.25	203	
	2800	For flat pilasters, deduct				↓	33%					
	3000	For splitting into halves, add				Ea.	114			114	125	
	4000	Rough sawn cedar posts, 4" x 4"	2 Carp	250	.064	V.L.F.	3.70	2.28		5.98	7.85	
	4100	4" x 6"		235	.068		6.70	2.42		9.12	11.35	
	4200	6" x 6"		220	.073		12.50	2.59		15.09	18.05	
	4300	8" x 8"	↓	200	.080	↓	28.50	2.84		31.34	36	
	9000	Minimum labor/equipment charge	1 Carp	3	2.667	Job		95		95	157	

06445 | Simulated Wood Ornaments

			Crew	Daily Output	Labor-Hours	Unit	Mat.	Labor	Equip.	Total	Total Incl O&P	
100	0010	**MILLWORK, HIGH DENSITY POLYMER**										100
	0100	Base, 9/16" x 3-3/16"	1 Carp	230	.035	L.F.	1.34	1.24		2.58	3.52	
	0200	Casing, fluted, 5/8" x 3-1/4"		215	.037		3.12	1.32		4.44	5.60	
	0300	Chair rail, 9/16" x 2-1/4"		260	.031		2.04	1.09		3.13	4.05	
	0600	Cove, 13/16" x 3-3/4"		260	.031		3.90	1.09		4.99	6.10	
	0700	Crown, 3/4" x 3-13/16"		260	.031		5.85	1.09		6.94	8.25	
	0800	Half round, 15/16" x 2"	↓	240	.033	↓	17.50	1.18		18.68	21	

06470 | Screen, Blinds & Shutters

			Crew	Daily Output	Labor-Hours	Unit	Mat.	Labor	Equip.	Total	Total Incl O&P	
100	0010	**SHUTTERS, EXTERIOR**										100
	0012	Aluminum, louvered, 1'-4" wide, 3'-0" long	1 Carp	10	.800	Pr.	45.50	28.50		74	97	
	0400	6'-8" long		9	.889		91.50	31.50		123	154	
	1000	Pine, louvered, primed, each 1'-2" wide, 3'-3" long		10	.800		86.50	28.50		115	143	
	1100	4'-7" long		10	.800		117	28.50		145.50	176	
	1250	Each 1'-4" wide, 3'-0" long		10	.800		88.50	28.50		117	145	
	1350	5'-3" long		10	.800		133	28.50		161.50	193	
	1500	Each 1'-6" wide, 3'-3" long		10	.800		92	28.50		120.50	148	
	1600	4'-7" long		10	.800		129	28.50		157.50	189	
	1610	Door blinds, 6'-9" long 1'-3" wide		9	.889		161	31.50		192.50	230	
	1615	1' - 6" wide		9	.889		178	31.50		209.50	249	
	1620	Hemlock, louvered, 1'-2" wide, 5'-7" long		10	.800		142	28.50		170.50	203	
	1630	Each 1'-4" wide, 2'-2" long		10	.800		88.50	28.50		117	145	
	1640	3'-0" long		10	.800		88.50	28.50		117	145	
	1650	3'-3" long		10	.800		95.50	28.50		124	152	
	1660	3'-11" long	↓	10	.800	↓	108	28.50		136.50	166	

06400 | Architectural Woodwork

06470 | Screen, Blinds & Shutters

		06470 Screen, Blinds & Shutters	CREW	DAILY OUTPUT	LABOR-HOURS	UNIT	2006 BARE COSTS MAT.	LABOR	EQUIP.	TOTAL	TOTAL INCL O&P	
100	1670	4′-3″ long	1 Carp	10	.800	Pr.	106	28.50		134.50	163	100
	1680	5′-3″ long		10	.800		132	28.50		160.50	192	
	1690	5′-11″ long		10	.800		149	28.50		177.50	211	
	1700	Door blinds, 6′-9″ long, each 1′-3″ wide		9	.889		150	31.50		181.50	218	
	1710	1′-6″ wide		9	.889		161	31.50		192.50	230	
	1720	Hemlock, solid raised panel, each 1′-4″ wide, 3′-3″ long		10	.800		143	28.50		171.50	204	
	1730	3′-11″ long		10	.800		167	28.50		195.50	231	
	1740	4′-3″ long		10	.800		181	28.50		209.50	246	
	1750	4′-7″ long		10	.800		195	28.50		223.50	262	
	1760	4′-11″ long		10	.800		212	28.50		240.50	281	
	1770	5′-11″ long		10	.800		242	28.50		270.50	315	
	1800	Door blinds, 6′-9″ long, each 1′-3″ wide		9	.889		272	31.50		303.50	355	
	1900	1′-6″ wide		9	.889		297	31.50		328.50	380	
	2500	Polystyrene, solid raised panel, each 1′-4″ wide, 3′-3″ long		10	.800		38.50	28.50		67	89	
	2700	4′-7″ long		10	.800		48	28.50		76.50	100	
	4500	Polystyrene, louvered, each 1′-2″ wide, 3′-3″ long		10	.800		26	28.50		54.50	75.50	
	4600	4′-7″ long		10	.800		32	28.50		60.50	82.50	
	4750	5′-3″ long		10	.800		37	28.50		65.50	87.50	
	4850	6′-8″ long		9	.889		44	31.50		75.50	101	
	6000	Vinyl, louvered, each 1′-2″ x 4′-7″ long		10	.800		32	28.50		60.50	82	
	6200	Each 1′-4″ x 6′-8″ long		9	.889		45	31.50		76.50	102	
	9000	Minimum labor/equipment charge		4	2	Job		71		71	118	
200	0010	**SHUTTERS, INTERIOR** Wood, louvered,										200
	0200	Two panel, 27″ wide, 36″ high	1 Carp	5	1.600	Set	108	57		165	213	
	0300	33″ wide, 36″ high		5	1.600		142	57		199	250	
	0500	47″ wide, 36″ high		5	1.600		189	57		246	300	
	1000	Four panel, 27″ wide, 36″ high		5	1.600		128	57		185	235	
	1100	33″ wide, 36″ high		5	1.600		157	57		214	267	
	1300	47″ wide, 36″ high		5	1.600		223	57		280	340	

06600 | Plastic Fabrications

06620 | Non-Structural Plastics

		06620 Non-Structural Plastics	CREW	DAILY OUTPUT	LABOR-HOURS	UNIT	2006 BARE COSTS MAT.	LABOR	EQUIP.	TOTAL	TOTAL INCL O&P	
810	0010	**SOLID SURFACE COUNTERTOPS**, Acrylic polymer										810
	0020	Pricing for orders of 100 L.F. or greater										
	0100	25″ wide, solid colors	2 Carp	28	.571	L.F.	44	20.50		64.50	82	
	0200	Patterned colors		28	.571		56	20.50		76.50	95	
	0300	Premium patterned colors		28	.571		70	20.50		90.50	111	
	0400	With silicone attached 4″ backsplash, solid colors		27	.593		48.50	21		69.50	88.50	
	0500	Patterned colors		27	.593		61.50	21		82.50	103	
	0600	Premium patterned colors		27	.593		76.50	21		97.50	119	
	0700	With hard seam attached 4″ backsplash, solid colors		23	.696		48.50	24.50		73	94.50	
	0800	Patterned colors		23	.696		61.50	24.50		86	109	
	0900	Premium patterned colors		23	.696		76.50	24.50		101	125	
	1000	Pricing for order of 51 - 99 L.F.										
	1100	25″ wide, solid colors	2 Carp	24	.667	L.F.	51	23.50		74.50	95.50	
	1200	Patterned colors		24	.667		64.50	23.50		88	111	
	1300	Premium patterned colors		24	.667		80.50	23.50		104	128	
	1400	With silicone attached 4″ backsplash, solid colors		23	.696		55.50	24.50		80	103	

06600 | Plastic Fabrications

	06620 Non-Structural Plastics		Crew	Daily Output	Labor-Hours	Unit	2006 Bare Costs Mat.	Labor	Equip.	Total	Total Incl O&P	
810	1500	Patterned colors	2 Carp	23	.696	L.F.	70.50	24.50		95	119	810
	1600	Premium patterned colors		23	.696		88	24.50		112.50	138	
	1700	With hard seam attached 4″ backsplash, solid colors		20	.800		55.50	28.50		84	109	
	1800	Patterned colors		20	.800		70.50	28.50		99	125	
	1900	Premium patterned colors	↓	20	.800	↓	88	28.50		116.50	144	
	2000	Pricing for order of 1 - 50 L.F.										
	2100	25″ wide, solid colors	2 Carp	20	.800	L.F.	59.50	28.50		88	113	
	2200	Patterned colors		20	.800		75.50	28.50		104	130	
	2300	Premium patterned colors		20	.800		94.50	28.50		123	151	
	2400	With silicone attached 4″ backsplash, solid colors		19	.842		65.50	30		95.50	122	
	2500	Patterned colors		19	.842		83	30		113	141	
	2600	Premium patterned colors		19	.842		103	30		133	164	
	2700	With hard seam attached 4″ backsplash, solid colors		15	1.067		65.50	38		103.50	135	
	2800	Patterned colors		15	1.067		83	38		121	154	
	2900	Premium patterned colors	↓	15	1.067	↓	103	38		141	177	
	3000	Sinks, pricing for order of 100 or greater units										
	3100	Single bowl, hard seamed, solid colors, 13″ x 17″	1 Carp	3	2.667	Ea.	298	95		393	485	
	3200	10″ x 15″		7	1.143		138	40.50		178.50	220	
	3300	Cutouts for sinks	↓	8	1	↓		35.50		35.50	59	
	3400	Sinks, pricing for order of 51 - 99 units										
	3500	Single bowl, hard seamed, solid colors, 13″ x 17″	1 Carp	2.55	3.137	Ea.	345	112		457	560	
	3600	10″ x 15″		6	1.333		159	47.50		206.50	253	
	3700	Cutouts for sinks	↓	7	1.143	↓		40.50		40.50	67.50	
	3800	Sinks, pricing for order of 1 - 50 units										
	3900	Single bowl, hard seamed, solid colors, 13″ x 17″	1 Carp	2	4	Ea.	405	142		547	680	
	4000	10″ x 15″		4.55	1.758		186	62.50		248.50	310	
	4100	Cutouts for sinks		5.25	1.524			54		54	90	
	4200	Cooktop cutouts, pricing for 100 or greater units		4	2		22	71		93	142	
	4300	51 - 99 units		3.40	2.353		25.50	83.50		109	167	
	4400	1 - 50 units	↓	3	2.667	↓	29.50	95		124.50	190	
850	0010	**VANITY TOPS**										850
	0015	Solid surface, center bowl, 17″ x 19″	1 Carp	12	.667	Ea.	182	23.50		205.50	240	
	0020	19″ x 25″		12	.667		220	23.50		243.50	282	
	0030	19″ x 31″		12	.667		267	23.50		290.50	335	
	0040	19″ x 37″		12	.667		310	23.50		333.50	380	
	0050	22″ x 25″		10	.800		194	28.50		222.50	260	
	0060	22″ x 31″		10	.800		227	28.50		255.50	296	
	0070	22″ x 37″		10	.800		264	28.50		292.50	335	
	0080	22″ x 43″		10	.800		300	28.50		328.50	375	
	0090	22″ x 49″		10	.800		335	28.50		363.50	410	
	0110	22″ x 55″		8	1		380	35.50		415.50	475	
	0120	22″ x 61″		8	1		430	35.50		465.50	535	
	0220	Double bowl, 22″ x 61″		8	1		490	35.50		525.50	595	
	0230	Double bowl, 22″ x 73″	↓	8	1	↓	675	35.50		710.50	805	
	0240	For aggregate colors, add					35%					
	0250	For faucets and fittings see 15410-300										

For information about Means Estimating Seminars, see yellow pages 12 and 13 in back of book

Division Notes

		CREW	DAILY OUTPUT	LABOR-HOURS	UNIT	2000 BARE COSTS MAT.	LABOR	EQUIP.	TOTAL	TOTAL INCL O&P

Division 7 Thermal & Moisture Protection

Estimating Tips

07100 Dampproofing & Waterproofing

- Be sure of the job specifications before pricing this subdivision. The difference in cost between waterproofing and dampproofing can be great. Waterproofing will hold back standing water. Dampproofing prevents the transmission of water vapor. Also included in this section are vapor retarding membranes.

07200 Thermal Protection

- Insulation and fireproofing products are measured by area, thickness, volume or R value. Specifications may give only what the specific R value should be in a certain situation. The estimator may need to choose the type of insulation to meet that R value.

07300 Shingles, Roof Tiles & Roof Coverings
07400 Roofing & Siding Panels

- Many roofing and siding products are bought and sold by the square. One square is equal to an area that measures 100 square feet.

This simple change in unit of measure could create a large error if the estimator is not observant. Accessories necessary for a complete installation must be figured into any calculations for both material and labor.

07500 Membrane Roofing
07600 Flashing & Sheet Metal
07700 Roof Specialties & Accessories

- The items in these subdivisions compose a roofing system. No one component completes the installation, and all must be estimated. Built-up or single-ply membrane roofing systems are made up of many products and installation trades. Wood blocking at roof perimeters or penetrations, parapet coverings, reglets, roof drains, gutters, downspouts, sheet metal flashing, skylights, smoke vents, and roof hatches all need to be considered along with the roofing material. Several different installation trades will need to work together on the roofing system. Inherent difficulties in the scheduling and coordination of various trades must be accounted for when estimating labor costs.

07900 Joint Sealers

- To complete the weather-tight shell the sealants and caulkings must be estimated. Where different materials meet—at expansion joints, at flashing penetrations, and at hundreds of other locations throughout a construction project—they provide another line of defense against water penetration. Often, an entire system is based on the proper location and placement of caulking or sealants. The detailed drawings that are included as part of a set of architectural plans show typical locations for these materials. When caulking or sealants are shown at typical locations, this means the estimator must include them for all the locations where this detail is applicable. Be careful to keep different types of sealants separate, and remember to consider backer rods and primers if necessary.

Reference Numbers

Reference numbers are shown in bold squares at the beginning of some major classifications. These numbers refer to related items in the Reference Section. The reference information may be an estimating procedure, an alternate pricing method, or technical information.

Note: Not all subdivisions listed here necessarily appear in this publication.

07050 | Basic Thermal & Moisture Protection Materials & Methods

07060 | Selective Demolition

			CREW	DAILY OUTPUT	LABOR-HOURS	UNIT	2006 BARE COSTS MAT.	LABOR	EQUIP.	TOTAL	TOTAL INCL O&P	
110	0010	**SELECTIVE DEMO, THERMAL & MOISTURE PROTECTION** R024119-10										110
	0200	Waterproofing demo, scrape off, to 1/2" thick	2 Clab	2000	.008	S.F.		.22		.22	.36	
	0210	Over 1/2" thick		1750	.009	"		.25		.25	.42	
	0250	Protection / drain board		3900	.004	B.F.		.11		.11	.19	
	1000	Deck, roof, concrete plank	B-13	1680	.033	S.F.		1	.39	1.39	2.06	
	1100	Gypsum plank		3900	.014			.43	.17	.60	.88	
	1150	Metal decking		3500	.016			.48	.19	.67	.98	
	1200	Wood, boards, tongue and groove, 2" x 6"	2 Clab	960	.017			.46		.46	.76	
	1220	2" x 10"		1040	.015			.42		.42	.70	
	1280	Standard planks, 1" x 6"		1080	.015			.41		.41	.67	
	1320	1" x 8"		1160	.014			.38		.38	.63	
	1340	1" x 12"		1200	.013			.37		.37	.61	
	1350	Plywood, to 1" thick		2000	.008			.22		.22	.36	
	2000	Gutters, aluminum or wood, edge hung	1 Clab	240	.033	L.F.		.91		.91	1.51	
	2100	Built-in		100	.080	"		2.19		2.19	3.63	
	2200	Insulation removal, loose fitting		3000	.003	C.F.		.07		.07	.12	
	2250	Air barrier		3500	.002	S.F.		.06		.06	.10	
	2300	Batts or blankets		1400	.006	C.F.		.16		.16	.26	
	2350	Rigid board		3450	.002	B.F.		.06		.06	.11	
	2500	Roof accessories, plumbing vent flashing		14	.571	Ea.		15.65		15.65	26	
	2600	Adjustable metal chimney flashing		9	.889	"		24.50		24.50	40.50	
	2650	Coping, sheet metal, up to 12" wide		240	.033	L.F.		.91		.91	1.51	
	2660	Concrete, up to 12" wide	2 Clab	160	.100	"		2.74		2.74	4.54	
	3000	Roofing, built-up, 5 ply roof, no gravel	B-2	1600	.025	S.F.		.70		.70	1.15	
	3001	Including gravel		890	.045			1.25		1.25	2.07	
	3100	Gravel removal, minimum		5000	.008			.22		.22	.37	
	3120	Maximum		2000	.020			.56		.56	.92	
	3400	Roof insulation board, up to 2" thick		3900	.010			.29		.29	.47	
	3405	Over 2" thick		7800	.005	B.F.		.14		.14	.24	
	3450	Roll roofing, cold adhesive	1 Clab	12	.667	Sq.		18.25		18.25	30.50	
	4000	Shingles, asphalt strip, 1 layer	B-2	3500	.011	S.F.		.32		.32	.53	
	4100	Slate		2500	.016			.44		.44	.74	
	4300	Wood		2200	.018			.51		.51	.84	
	4500	Skylight to 10 S.F.	1 Clab	8	1	Ea.		27.50		27.50	45.50	
	5000	Siding, metal, horizontal		444	.018	S.F.		.49		.49	.82	
	5020	Vertical		400	.020			.55		.55	.91	
	5200	Wood, boards, vertical		400	.020			.55		.55	.91	
	5220	Clapboards, horizontal		380	.021			.58		.58	.96	
	5240	Shingles		350	.023			.63		.63	1.04	
	5260	Textured plywood		725	.011			.30		.30	.50	
	9000	Minimum labor/equipment charge		2	4	Job		110		110	182	

07100 | Dampproofing and Waterproofing

07110 | Dampproofing

			CREW	DAILY OUTPUT	LABOR-HOURS	UNIT	2006 BARE COSTS MAT.	LABOR	EQUIP.	TOTAL	TOTAL INCL O&P	
100	0010	**BITUMINOUS ASPHALT COATING** For foundation										100
	0030	Brushed on, below grade, 1 coat	1 Rofc	665	.012	S.F.	.09	.37		.46	.76	
	0100	2 coat		500	.016		.18	.49		.67	1.07	
	0300	Sprayed on, below grade, 1 coat, 25.6 S.F./gal.		830	.010		.09	.30		.39	.63	

07100 | Dampproofing and Waterproofing

07110 | Dampproofing

		07110 Dampproofing	Crew	Daily Output	Labor-Hours	Unit	2006 Bare Costs Mat.	Labor	Equip.	Total	Total Incl O&P	
100	0400	2 coat, 20.5 S.F./gal.	1 Rofc	500	.016	S.F.	.17	.49		.66	1.07	100
	0600	Troweled on, asphalt with fibers, 1/16" thick		500	.016		.18	.49		.67	1.08	
	0700	1/8" thick		400	.020		.32	.61		.93	1.45	
	1000	1/2" thick		350	.023		1.05	.70		1.75	2.41	
	9000	Minimum labor/equipment charge		3	2.667	Job		81.50		81.50	147	

07130 | Sheet Waterproofing

		07130 Sheet Waterproofing	Crew	Daily Output	Labor-Hours	Unit	Mat.	Labor	Equip.	Total	Total Incl O&P	
200	0010	ELASTOMERIC WATERPROOFING										200
	0090	EPDM, plain, 45 mils thick	2 Rofc	580	.028	S.F.	.99	.84		1.83	2.61	
	0100	60 mils thick		570	.028		1.07	.86		1.93	2.72	
	0300	Nylon reinforced sheets, 45 mils thick		580	.028		1.50	.84		2.34	3.17	
	0400	60 mils thick		570	.028		1.89	.86		2.75	3.62	
	0600	Vulcanizing splicing tape for above, 2" wide				C.L.F.	41			41	45	
	0700	4" wide				"	55			55	60.50	
	0900	Adhesive, bonding, 60 SF per gal				Gal.	14.60			14.60	16.05	
	1000	Splicing, 75 SF per gal				"	31			31	34.50	
	1200	Neoprene sheets, plain, 45 mils thick	2 Rofc	580	.028	S.F.	1.41	.84		2.25	3.07	
	1300	60 mils thick		570	.028		2.21	.86		3.07	3.97	
	1500	Nylon reinforced, 45 mils thick		580	.028		1.80	.84		2.64	3.50	
	1600	60 mils thick		570	.028		2.05	.86		2.91	3.80	
	1800	120 mils thick		500	.032		3.37	.98		4.35	5.45	
	1900	Adhesive, splicing, 150 S.F. per gal. per coat				Gal.	18.15			18.15	19.95	
	2100	Fiberglass reinforced, fluid applied, 1/8" thick	2 Rofc	500	.032	S.F.	1.62	.98		2.60	3.54	
	2200	Polyethylene and rubberized asphalt sheets, 1/8" thick		550	.029		.54	.89		1.43	2.19	
	2210	Asphaltic hardboard protection board, 1/8" thick		500	.032		.31	.98		1.29	2.10	
	2220	1/4" thick		450	.036		.55	1.09		1.64	2.57	
	2400	Polyvinyl chloride sheets, plain, 10 mils thick		580	.028		.16	.84		1	1.70	
	2500	20 mils thick		570	.028		.25	.86		1.11	1.82	
	2700	30 mils thick		560	.029		.35	.87		1.22	1.96	
	3000	Adhesives, trowel grade, 40-100 SF per gal				Gal.	23			23	25.50	
	3100	Brush grade, 100-250 SF per gal.				"	23			23	25.50	
	3300	Bitumen modified polyurethane, fluid applied, 55 mils thick	2 Rofc	665	.024	S.F.	.67	.74		1.41	2.06	
	3600	Vinyl plastic, sprayed on, 25 to 40 mils thick		475	.034	"	1	1.03		2.03	2.95	
	9000	Minimum labor/equipment charge		2	8	Job		245		245	440	
500	0010	MEMBRANE WATERPROOFING										500
	0012	On slabs, 1 ply, felt, mopped	G-1	3000	.019	S.F.	.19	.53	.11	.83	1.29	
	0300	On slabs, 2 ply, felt, mopped		2500	.022		.39	.64	.13	1.16	1.73	
	0400	On slabs, 2 ply, glass fiber fabric, mopped		1650	.034		.46	.97	.20	1.63	2.47	
	0600	On slabs, 3 ply, felt, mopped		2100	.027		.58	.76	.16	1.50	2.19	
	0700	On slabs, 3 ply, glass fiber fabric, mopped		1550	.036		.62	1.03	.22	1.87	2.78	
	0710	Asphaltic hardboard protection board, 1/8" thick	2 Rofc	500	.032		.31	.98		1.29	2.10	
	1000	For adhered 1/4" EPS protection board, add		3500	.005		.16	.14		.30	.43	
	1050	3/8" thick, add		3500	.005		.18	.14		.32	.45	
	1060	1/2" thick, add		3500	.005		.20	.14		.34	.47	
	1070	Fiberglass fabric, black, 20/10 mesh		116	.138	Sq.	9.95	4.22		14.17	18.55	
	1080	White, 20/10 mesh		116	.138	"	10.25	4.22		14.47	18.90	
	1100	1/16" urethane, troweled		200	.080	S.F.	.66	2.45		3.11	5.15	
	1200	Roller applied		120	.133	"	.60	4.08		4.68	8	
	9000	Minimum labor/equipment charge		2	8	Job		245		245	440	

07160 | Cement. & Reactive Waterproofing

		07160 Cement. & Reactive Waterproofing	Crew	Daily Output	Labor-Hours	Unit	Mat.	Labor	Equip.	Total	Total Incl O&P	
150	0010	**CEMENTITIOUS WATERPROOFING** One coat cement base										150
	0020	1/8" application, sprayed on	G-2	1000	.024	S.F.	1.55	.70	.12	2.37	2.98	

07100 | Dampproofing and Waterproofing

07160 | Cement. & Reactive Waterproofing

		07160 Cement. & Reactive Waterproofing	CREW	DAILY OUTPUT	LABOR-HOURS	UNIT	2006 BARE COSTS MAT.	LABOR	EQUIP.	TOTAL	TOTAL INCL O&P	
150	0030	2 coat, cementitious/metallic slurry, troweled, 1/4" thick	1 Cefi	2.48	3.226	C.S.F.	36	111		147	215	150
	0040	3 coat, 3/8" thick		1.84	4.348		54	150		204	296	
	0050	4 coat, 1/2" thick		1.20	6.667		72	229		301	440	

07190 | Water Repellents

		07190 Water Repellents	CREW	DAILY OUTPUT	LABOR-HOURS	UNIT	MAT.	LABOR	EQUIP.	TOTAL	TOTAL INCL O&P	
700	0010	**RUBBER COATING**										700
	0020	Water base liquid, roller applied	2 Rofc	7000	.002	S.F.	.57	.07		.64	.76	
	0200	Silicone or stearate, sprayed on CMU, 1 coat	1 Rofc	4000	.002		.32	.06		.38	.47	
	0300	2 coats		3000	.003		.65	.08		.73	.86	
	9000	Minimum labor/equipment charge		3	2.667	Job		81.50		81.50	147	

07200 | Thermal Protection

07210 | Building Insulation

		07210 Building Insulation	CREW	DAILY OUTPUT	LABOR-HOURS	UNIT	2006 BARE COSTS MAT.	LABOR	EQUIP.	TOTAL	TOTAL INCL O&P	
150	0010	**BLOWN-IN INSULATION** Ceilings, with open access										150
	0020	Cellulose, 3-1/2" thick, R13	G-4	5000	.005	S.F.	.19	.13	.05	.37	.48	
	0030	5-3/16" thick, R19		3800	.006		.28	.18	.06	.52	.66	
	0050	6-1/2" thick, R22		3000	.008		.35	.22	.08	.65	.84	
	0100	8-11/16" thick, R30		2600	.009		.48	.26	.09	.83	1.05	
	1000	Fiberglass, 5" thick, R11		3800	.006		.17	.18	.06	.41	.54	
	1050	6" thick, R13		3000	.008		.21	.22	.08	.51	.68	
	1100	8-1/2" thick, R19		2200	.011		.29	.31	.10	.70	.94	
	1200	10" thick, R22		1800	.013		.34	.37	.13	.84	1.13	
	1300	12" thick, R26		1500	.016		.41	.45	.15	1.01	1.36	
	2000	Mineral wool, 4" thick, R12		3500	.007		.19	.19	.06	.44	.60	
	2050	6" thick, R17		2500	.010		.21	.27	.09	.57	.78	
	2100	9" thick, R23		1750	.014		.30	.38	.13	.81	1.11	
	2500	Wall installation, incl. drilling & patching from outside, two 1"										
	2510	diam. holes @ 16" O.C., top & mid-point of wall, add to above										
	2700	For masonry	G-4	415	.058	S.F.	.06	1.62	.55	2.23	3.36	
	2800	For wood siding		840	.029		.06	.80	.27	1.13	1.70	
	2900	For stucco/plaster		665	.036		.06	1.01	.34	1.41	2.12	
	9000	Minimum labor/equipment charge		4	6	Job		168	56.50	224.50	340	
350	0010	**FLOOR INSULATION, NONRIGID** Including										350
	0020	spring type wire fasteners										
	2000	Fiberglass, blankets or batts, paper or foil backing										
	2100	1 side, 3-1/2" thick, R11	1 Carp	700	.011	S.F.	.36	.41		.77	1.07	
	2150	6" thick, R19		600	.013		.49	.47		.96	1.33	
	2200	8-1/2" thick, R30		550	.015		.78	.52		1.30	1.72	
	9000	Minimum labor/equipment charge		4	2	Job		71		71	118	
500	0010	**POURED INSULATION**, Cellulose fiber										500
	0020	R3.8 per inch	1 Carp	200	.040	C.F.	.58	1.42		2	3	
	0040	Ceramic type (perlite), R3.2 per inch		200	.040		1.70	1.42		3.12	4.23	
	0080	Fiberglass wool, R4 per inch		200	.040		.41	1.42		1.83	2.81	
	0100	Mineral wool, R3 per inch		200	.040		.35	1.42		1.77	2.75	
	0300	Polystyrene, R4 per inch		200	.040		2.89	1.42		4.31	5.55	

07200 | Thermal Protection

	07210 | Building Insulation		CREW	DAILY OUTPUT	LABOR-HOURS	UNIT	2006 BARE COSTS MAT.	LABOR	EQUIP.	TOTAL	TOTAL INCL O&P	
500	0400	Vermiculite or perlite, R2.7 per inch	1 Carp	200	.040	C.F.	1.70	1.42		3.12	4.23	500
	9000	Minimum labor/equipment charge	↓	4	2	Job		71		71	118	
550	0010	**MASONRY INSULATION** Vermiculite or perlite, poured										550
	0100	In cores of concrete block, 4″ thick wall, .115 CF/SF	D-1	4800	.003	S.F.	.20	.11		.31	.39	
	0200	6″ thick wall, .175 CF/SF		3000	.005		.30	.17		.47	.61	
	0300	8″ thick wall, .258 CF/SF		2400	.007		.44	.21		.65	.83	
	0400	10″ thick wall, .340 CF/SF		1850	.009		.58	.28		.86	1.09	
	0500	12″ thick wall, .422 CF/SF	↓	1200	.013	↓	.72	.43		1.15	1.49	
	0550	For sand fill, deduct from above					70%					
	0600	Poured cavity wall, vermiculite or perlite, water repellant	D-1	250	.064	C.F.	1.70	2.06		3.76	5.20	
	0700	Foamed in place, urethane in 2-5/8″ cavity	G-2	1035	.023	S.F.	.41	.68	.11	1.20	1.67	
	0800	For each 1″ added thickness, add	″	2372	.010	″	.12	.30	.05	.47	.66	
600	0010	**PERIMETER INSULATION**										600
	0600	Polystyrene, expanded, 1″ thick, R4	1 Carp	680	.012	S.F.	.25	.42		.67	.97	
	0700	2″ thick, R8		675	.012	″	.53	.42		.95	1.28	
	9000	Minimum labor/equipment charge		4	2	Job		71		71	118	
700	0012	**REFLECTIVE INSULATION**, aluminum foil on reinforced scrim		1900	.004	S.F.	.14	.15		.29	.41	700
	0102	Reinforced with woven polyolefin		1900	.004		.17	.15		.32	.44	
	0502	With single bubble air space, R8.8		1500	.005		.28	.19		.47	.62	
	0602	With double bubble air space, R9.8		1500	.005	↓	.30	.19		.49	.64	
	9000	Minimum labor/equipment charge	↓	4	2	Job		71		71	118	
800	0010	**SPRAYED INSULATION**										800
	0020	Fibrous/cementitious, finished wall, 1″ thick, R3.7	G-2	2050	.012	S.F.	.23	.34	.06	.63	.87	
	0100	Attic, 5.2″ thick, R19	″	1550	.015	″	.37	.45	.08	.90	1.23	
	0300	Foam type, incl. preparation										
	0600	3 #/CF, 1″ thick, R3.8	G-2	770	.031	S.F.	.50	.91	.15	1.56	2.20	
	0700	2″ thick, R7.5	″	475	.051	″	1.02	1.48	.25	2.75	3.79	
	1000	Coating, 50-70 mils, elastomeric aromatic urethane										
	1100	Finish coat, 5 mils, elastomeric aliphatic, 200 S.F./gal.				Gal.	40.50			40.50	44.50	
	9000	Minimum labor/equipment charge	G-2	2	12	Job		350	58.50	408.50	635	
900	0010	**WALL INSULATION, RIGID**										900
	0040	Fiberglass, 1.5#/CF, unfaced, 1″ thick, R4.1	1 Carp	1000	.008	S.F.	.38	.28		.66	.89	
	0060	1-1/2″ thick, R6.2		1000	.008		.49	.28		.77	1.01	
	0080	2″ thick, R8.3		1000	.008		.59	.28		.87	1.12	
	0120	3″ thick, R12.4		800	.010		.68	.36		1.04	1.34	
	0370	3#/CF, unfaced, 1″ thick, R4.3		1000	.008		.43	.28		.71	.94	
	0390	1-1/2″ thick, R6.5		1000	.008		.82	.28		1.10	1.37	
	0400	2″ thick, R8.7		890	.009		.99	.32		1.31	1.62	
	0420	2-1/2″ thick, R10.9		800	.010		1.21	.36		1.57	1.92	
	0440	3″ thick, R13		800	.010		1.44	.36		1.80	2.17	
	0520	Foil faced, 1″ thick, R4.3		1000	.008		.95	.28		1.23	1.52	
	0540	1-1/2″ thick, R6.5		1000	.008		1.29	.28		1.57	1.89	
	0560	2″ thick, R8.7		890	.009		1.61	.32		1.93	2.30	
	0580	2-1/2″ thick, R10.9		800	.010		1.90	.36		2.26	2.68	
	0600	3″ thick, R13		800	.010		2.07	.36		2.43	2.87	
	0670	6#/CF, unfaced, 1″ thick, R4.3		1000	.008		.92	.28		1.20	1.48	
	0690	1-1/2″ thick, R6.5		890	.009		1.42	.32		1.74	2.09	
	0700	2″ thick, R8.7		800	.010		2	.36		2.36	2.79	
	0721	2-1/2″ thick, R10.9		800	.010		2.19	.36		2.55	3	
	0741	3″ thick, R13		730	.011		2.62	.39		3.01	3.53	
	0821	Foil faced, 1″ thick, R4.3		1000	.008		1.30	.28		1.58	1.90	
	0840	1-1/2″ thick, R6.5		890	.009		1.87	.32		2.19	2.59	
	0850	2″ thick, R8.7		800	.010		2.44	.36		2.80	3.27	
	0880	2-1/2″ thick, R10.9	↓	800	.010	↓	2.93	.36		3.29	3.81	

07210 | Building Insulation

		07210 Building Insulation	CREW	DAILY OUTPUT	LABOR-HOURS	UNIT	2006 BARE COSTS MAT.	LABOR	EQUIP.	TOTAL	TOTAL INCL O&P
900	0900	3" thick, R13	1 Carp	730	.011	S.F.	3.50	.39		3.89	4.50
	1500	Foamglass, 1-1/2" thick, R4.5		800	.010		1.25	.36		1.61	1.97
	1550	3" thick, R9		730	.011		3.01	.39		3.40	3.96
	1600	Isocyanurate, 4' x 8' sheet, foil faced, both sides									
	1610	1/2" thick, R3.9	1 Carp	800	.010	S.F.	.29	.36		.65	.91
	1620	5/8" thick, R4.5		800	.010		.48	.36		.84	1.12
	1630	3/4" thick, R5.4		800	.010		.38	.36		.74	1.01
	1640	1" thick, R7.2		800	.010		.52	.36		.88	1.16
	1650	1-1/2" thick, R10.8		730	.011		.56	.39		.95	1.27
	1660	2" thick, R14.4		730	.011		.74	.39		1.13	1.46
	1670	3" thick, R21.6		730	.011		1.78	.39		2.17	2.61
	1680	4" thick, R28.8		730	.011		2.19	.39		2.58	3.06
	1700	Perlite, 1" thick, R2.77		800	.010		.28	.36		.64	.90
	1750	2" thick, R5.55		730	.011		.54	.39		.93	1.24
	1900	Extruded polystyrene, 25 PSI compressive strength, 1" thick, R5		800	.010		.47	.36		.83	1.11
	1940	2" thick R10		730	.011		.93	.39		1.32	1.67
	1960	3" thick, R15		730	.011		1.30	.39		1.69	2.08
	2100	Expanded polystyrene, 1" thick, R3.85		800	.010		.22	.36		.58	.83
	2120	2" thick, R7.69		730	.011		.55	.39		.94	1.26
	2140	3" thick, R11.49		730	.011		.68	.39		1.07	1.40
	9000	Minimum labor/equipment charge		4	2	Job		71		71	118
950	0010	**WALL OR CEILING INSUL., NON-RIGID**									
	0040	Fiberglass, kraft faced, batts or blankets									
	0060	3-1/2" thick, R11, 11" wide	1 Carp	1150	.007	S.F.	.29	.25		.54	.73
	0080	15" wide		1600	.005		.29	.18		.47	.61
	0100	23" wide		1600	.005		.29	.18		.47	.61
	0140	6" thick, R19, 11" wide		1000	.008		.42	.28		.70	.93
	0160	15" wide		1350	.006		.42	.21		.63	.81
	0180	23" wide		1600	.005		.42	.18		.60	.75
	0200	9" thick, R30, 15" wide		1150	.007		.71	.25		.96	1.19
	0220	23" wide		1350	.006		.71	.21		.92	1.13
	0240	12" thick, R38, 15" wide		1000	.008		.85	.28		1.13	1.41
	0260	23" wide		1350	.006		.85	.21		1.06	1.29
	0400	Fiberglass, foil faced, batts or blankets									
	0420	3-1/2" thick, R11, 15" wide	1 Carp	1600	.005	S.F.	.47	.18		.65	.81
	0440	23" wide		1600	.005		.47	.18		.65	.81
	0460	6" thick, R19, 15" wide		1350	.006		.51	.21		.72	.91
	0480	23" wide		1600	.005		.51	.18		.69	.85
	0500	9" thick, R30, 15" wide		1150	.007		.78	.25		1.03	1.27
	0550	23" wide		1350	.006		.78	.21		.99	1.21
	0800	Fiberglass, unfaced, batts or blankets									
	0820	3-1/2" thick, R11, 15" wide	1 Carp	1350	.006	S.F.	.33	.21		.54	.71
	0830	23" wide		1600	.005		.33	.18		.51	.65
	0860	6" thick, R19, 15" wide		1150	.007		.56	.25		.81	1.03
	0880	23" wide		1350	.006		.56	.21		.77	.97
	0900	9" thick, R30, 15" wide		1000	.008		.66	.28		.94	1.20
	0920	23" wide		1150	.007		.66	.25		.91	1.14
	0940	12" thick, R38, 15" wide		1000	.008		.83	.28		1.11	1.38
	0960	23" wide		1150	.007		.83	.25		1.08	1.32
	1300	Mineral fiber batts, kraft faced									
	1320	3-1/2" thick, R12	1 Carp	1600	.005	S.F.	.30	.18		.48	.62
	1340	6" thick, R19		1600	.005		.40	.18		.58	.73
	1380	10" thick, R30		1350	.006		.59	.21		.80	1
	1850	Friction fit wire insulation supports, 16" O.C.		960	.008	Ea.	.07	.30		.37	.57
	1900	For foil backing, add				S.F.	.04			.04	.04

		07210 | Building Insulation	CREW	DAILY OUTPUT	LABOR-HOURS	UNIT	2006 BARE COSTS MAT.	LABOR	EQUIP.	TOTAL	TOTAL INCL O&P	
950	9000	Minimum labor/equipment charge	1 Carp	4	2	Job		71		71	118	950
		07220 | Roof and Deck Insulation										
700	0010	ROOF DECK INSULATION										700
	0020	Fiberboard low density, 1/2" thick R1.39	1 Rofc	1000	.008	S.F.	.19	.24		.43	.65	
	0030	1" thick R2.78		800	.010		.38	.31		.69	.97	
	0080	1 1/2" thick R4.17		800	.010		.57	.31		.88	1.18	
	0100	2" thick R5.56		800	.010		.76	.31		1.07	1.39	
	0110	Fiberboard high density, 1/2" thick R1.3		1000	.008		.20	.24		.44	.66	
	0120	1" thick R2.5		800	.010		.40	.31		.71	.99	
	0130	1-1/2" thick R3.8		800	.010		.60	.31		.91	1.21	
	0200	Fiberglass, 3/4" thick R2.78		1000	.008		.50	.24		.74	.99	
	0400	15/16" thick R3.70		1000	.008		.67	.24		.91	1.18	
	0460	1-1/16" thick R4.17		1000	.008		.84	.24		1.08	1.36	
	0600	1-5/16" thick R5.26		1000	.008		1.15	.24		1.39	1.71	
	0650	2-1/16" thick R8.33		800	.010		1.23	.31		1.54	1.90	
	0700	2-7/16" thick R10		800	.010		1.40	.31		1.71	2.09	
	1500	Foamglass, 1-1/2" thick R4.5		800	.010		1.30	.31		1.61	1.98	
	1530	3" thick R9		700	.011		2.65	.35		3	3.55	
	1600	Tapered for drainage		600	.013	B.F.	1.13	.41		1.54	1.97	
	1650	Perlite, 1/2" thick R1.32		1050	.008	S.F.	.28	.23		.51	.73	
	1655	3/4" thick R2.08		800	.010		.30	.31		.61	.88	
	1660	1" thick R2.78		800	.010		.32	.31		.63	.90	
	1670	1-1/2" thick R4.17		800	.010		.40	.31		.71	.99	
	1680	2" thick R5.56		700	.011		.64	.35		.99	1.33	
	1685	2-1/2" thick R6.67		700	.011		.79	.35		1.14	1.50	
	1690	Tapered for drainage		800	.010	B.F.	.61	.31		.92	1.22	
	1700	Polyisocyanurate, 2#/CF density, 3/4" thick, R5.1		1500	.005	S.F.	.60	.16		.76	.95	
	1705	1" thick R7.14		1400	.006		.65	.17		.82	1.03	
	1715	1-1/2" thick R10.87		1250	.006		.69	.20		.89	1.11	
	1725	2" thick R14.29		1100	.007		.91	.22		1.13	1.40	
	1735	2-1/2" thick R16.67		1050	.008		1	.23		1.23	1.52	
	1745	3" thick R21.74		1000	.008		1.46	.24		1.70	2.05	
	1755	3-1/2" thick R25		1000	.008		1.48	.24		1.72	2.07	
	1765	Tapered for drainage		1400	.006	B.F.	.76	.17		.93	1.15	
	1900	Extruded Polystyrene										
	1910	15 PSI compressive strength, 1" thick, R5	1 Rofc	1500	.005	S.F.	.43	.16		.59	.76	
	1920	2" thick, R10		1250	.006		.66	.20		.86	1.08	
	1930	3" thick R15		1000	.008		.86	.24		1.10	1.39	
	1932	4" thick R20		1000	.008		1.34	.24		1.58	1.91	
	1934	Tapered for drainage		1500	.005	B.F.	.44	.16		.60	.77	
	1940	25 PSI compressive strength, 1" thick R5		1500	.005	S.F.	.63	.16		.79	.98	
	1942	2" thick R10		1250	.006		1.21	.20		1.41	1.68	
	1944	3" thick R15		1000	.008		1.84	.24		2.08	2.46	
	1946	4" thick R20		1000	.008		2.45	.24		2.69	3.14	
	1948	Tapered for drainage		1500	.005	B.F.	.50	.16		.66	.84	
	1950	40 psi compressive strength, 1" thick R5		1500	.005	S.F.	.44	.16		.60	.77	
	1952	2" thick R10		1250	.006		.86	.20		1.06	1.30	
	1954	3" thick R15		1000	.008		1.26	.24		1.50	1.83	
	1956	4" thick R20		1000	.008		1.69	.24		1.93	2.30	
	1958	Tapered for drainage		1400	.006	B.F.	.63	.17		.80	1	
	1960	60 PSI compressive strength, 1" thick R5		1450	.006	S.F.	.53	.17		.70	.88	
	1962	2" thick R10		1200	.007		.94	.20		1.14	1.40	
	1964	3" thick R15		975	.008		1.40	.25		1.65	1.99	
	1966	4" thick R20		950	.008		1.95	.26		2.21	2.61	

	07220 | Roof and Deck Insulation		Crew	Daily Output	Labor-Hours	Unit	2006 Bare Costs Mat.	Labor	Equip.	Total	Total Incl O&P	
700	1968	Tapered for drainage	1 Rofc	1400	.006	B.F.	.76	.17		.93	1.15	700
	2010	Expanded polystyrene, 1#/CF density, 3/4" thick R2.89		1500	.005	S.F.	.25	.16		.41	.57	
	2020	1" thick R3.85		1500	.005		.25	.16		.41	.57	
	2100	2" thick R7.69		1250	.006		.53	.20		.73	.93	
	2110	3" thick R11.49		1250	.006		.61	.20		.81	1.02	
	2120	4" thick R15.38		1200	.007		.74	.20		.94	1.18	
	2130	5" thick R19.23		1150	.007		.92	.21		1.13	1.39	
	2140	6" thick R23.26		1150	.007	↓	1.08	.21		1.29	1.57	
	2150	Tapered for drainage	↓	1500	.005	B.F.	.44	.16		.60	.77	
	2400	Composites with 2" EPS										
	2410	1" fiberboard	1 Rofc	950	.008	S.F.	.98	.26		1.24	1.54	
	2420	7/16" oriented strand board		800	.010		1.16	.31		1.47	1.83	
	2430	1/2" plywood		800	.010		1.25	.31		1.56	1.93	
	2440	1" perlite	↓	800	.010	↓	1.03	.31		1.34	1.68	
	2450	Composites with 1-1/2" polyisocyanurate										
	2460	1" fiberboard	1 Rofc	800	.010	S.F.	1.34	.31		1.65	2.02	
	2470	1" perlite		850	.009		1.41	.29		1.70	2.07	
	2480	7/16" oriented strand board		800	.010	↓	1.62	.31		1.93	2.33	
	9000	Minimum labor/equipment charge	↓	3.25	2.462	Job		75.50		75.50	135	

	07240 | Ext. Insulation Finish Systems (EIFS)		Crew	Daily Output	Labor-Hours	Unit	Mat.	Labor	Equip.	Total	Total Incl O&P	
100	0010	**EXTERIOR INSULATION FINISH SYSTEM**										100
	0095	Field applied, 1" EPS insulation	J-1	295	.136	S.F.	2.07	4.15	.36	6.58	9.35	
	0100	With 1/2" cement board sheathing		220	.182		2.83	5.55	.48	8.86	12.65	
	0105	2" EPS insulation		295	.136		2.40	4.15	.36	6.91	9.75	
	0110	With 1/2" cement board sheathing		220	.182		3.16	5.55	.48	9.19	13	
	0115	3" EPS insulation		295	.136		2.53	4.15	.36	7.04	9.90	
	0120	With 1/2" cement board sheathing		220	.182		3.29	5.55	.48	9.32	13.15	
	0125	4" EPS insulation		295	.136		2.95	4.15	.36	7.46	10.35	
	0130	With 1/2" cement board sheathing		220	.182		4.47	5.55	.48	10.50	14.45	
	0140	Premium finish add		1265	.032		.29	.97	.08	1.34	1.97	
	0150	Heavy duty reinforcement add	↓	914	.044	↓	1.73	1.34	.12	3.19	4.19	
	0160	2.5#/S.Y. metal lath substrate add	1 Lath	75	.107	S.Y.	2.22	3.50		5.72	8	
	0170	3.4#/S.Y. metal lath substrate add	"	75	.107	"	2.32	3.50		5.82	8.10	
	0180	Color or texture change,	J-1	1265	.032	S.F.	.75	.97	.08	1.80	2.48	
	0190	With substrate leveling base coat	1 Plas	530	.015		.75	.49		1.24	1.62	
	0210	With substrate sealing base coat	1 Pord	1224	.007	↓	.07	.21		.28	.41	
	0370	V groove shape in panel face				L.F.	.54			.54	.59	
	0380	U groove shape in panel face				"	.71			.71	.78	
	0433	Crack repair, acrylic rubber, fluid applied, 20 mils thick	1 Plas	350	.023	S.F.	1.36	.74		2.10	2.70	
	0437	50 mils thick, reinforced	"	200	.040	"	2.53	1.30		3.83	4.87	
	0440	For higher than one story, add						25%				

	07260 | Vapor Retarders		Crew	Daily Output	Labor-Hours	Unit	Mat.	Labor	Equip.	Total	Total Incl O&P	
100	0010	**BUILDING PAPER**										100
	0020	Aluminum and kraft laminated, foil 1 side	1 Carp	37	.216	Sq.	4.78	7.70		12.48	18	
	0100	Foil 2 sides		37	.216		7.95	7.70		15.65	21.50	
	0400	Asphalt felt sheathing paper, 15#	↓	37	.216	↓	3.41	7.70		11.11	16.50	
	0450	Housewrap, exterior, spun bonded polypropylene										
	0470	Small roll	1 Carp	3800	.002	S.F.	.18	.08		.26	.32	
	0480	Large roll	"	4000	.002	"	.11	.07		.18	.24	
	0500	Material only, 3' x 111.1' roll				Ea.	61			61	67	
	0520	9' x 111.1' roll				"	110			110	121	
	0600	Polyethylene vapor barrier, standard, .002" thick	1 Carp	37	.216	Sq.	.96	7.70		8.66	13.80	

07200 | Thermal Protection

07260 | Vapor Retarders

		07260 Vapor Retarders	CREW	DAILY OUTPUT	LABOR-HOURS	UNIT	2006 BARE COSTS MAT.	LABOR	EQUIP.	TOTAL	TOTAL INCL O&P	
100	0700	.004" thick	1 Carp	37	.216	Sq.	2.74	7.70		10.44	15.75	100
	0900	.006" thick		37	.216		4.09	7.70		11.79	17.25	
	1200	.010" thick		37	.216		5.10	7.70		12.80	18.35	
	1300	Clear reinforced, fire retardant, .008" thick		37	.216		8.55	7.70		16.25	22	
	1350	Cross laminated type, .003" thick		37	.216		6.65	7.70		14.35	20	
	1400	.004" thick		37	.216		7.40	7.70		15.10	21	
	1500	Red rosin paper, 5 sq rolls, 4 lb per square		37	.216		1.73	7.70		9.43	14.65	
	1600	5 lbs. per square		37	.216		2.19	7.70		9.89	15.15	
	1800	Reinf. waterproof, .002" polyethylene backing, 1 side		37	.216		5	7.70		12.70	18.25	
	1900	2 sides		37	.216		6.65	7.70		14.35	20	
	2100	Roof deck vapor barrier, class 1 metal decks	1 Rofc	37	.216		11.75	6.60		18.35	25	
	2200	For all other decks	″	37	.216		8.35	6.60		14.95	21	
	2400	Waterproofed kraft with sisal or fiberglass fibers, minimum	1 Carp	37	.216		5.40	7.70		13.10	18.70	
	2500	Maximum		37	.216		13.50	7.70		21.20	27.50	
	9950	Minimum labor/equipment charge		4	2	Job		71		71	118	

07300 | Shingles, Roof Tiles and Roof Coverings

07310 | Shingles

		07310 Shingles	CREW	DAILY OUTPUT	LABOR-HOURS	UNIT	2006 BARE COSTS MAT.	LABOR	EQUIP.	TOTAL	TOTAL INCL O&P	
050	0010	**ALUMINUM SHINGLES**										050
	0020	Mill finish, .019 thick	1 Carp	5	1.600	Sq.	158	57		215	268	
	0100	.020" thick	″	5	1.600		177	57		234	288	
	0300	For colors, add					15.15			15.15	16.65	
	0600	Ridge cap, .024" thick	1 Carp	170	.047	L.F.	1.82	1.67		3.49	4.77	
	0700	End wall flashing, .024" thick		170	.047		1.61	1.67		3.28	4.54	
	0900	Valley section, .024" thick		170	.047		2.48	1.67		4.15	5.50	
	1000	Starter strip, .024" thick		400	.020		1.35	.71		2.06	2.67	
	1200	Side wall flashing, .024" thick		170	.047		1.62	1.67		3.29	4.55	
	9000	Minimum labor/equipment charge		3	2.667	Job		95		95	157	
100	0010	**ASPHALT SHINGLES**										100
	0100	Standard strip shingles										
	0150	Inorganic, class A, 210-235 lb/sq	1 Rofc	5.50	1.455	Sq.	38.50	44.50		83	122	
	0155	Pneumatic nailed		7	1.143		38.50	35		73.50	105	
	0200	Organic, class C, 235-240 lb/sq		5	1.600		43	49		92	136	
	0205	Pneumatic nailed		6.25	1.280		43	39		82	118	
	0250	Standard, laminated multi-layered shingles										
	0300	Class A, 240-260 lb/sq	1 Rofc	4.50	1.778	Sq.	47	54.50		101.50	150	
	0305	Pneumatic nailed		5.63	1.422		47	43.50		90.50	130	
	0350	Class C, 260-300 lb/square, 4 bundles/square		4	2		50	61		111	165	
	0355	Pneumatic nailed		5	1.600		50	49		99	143	
	0400	Premium, laminated multi-layered shingles										
	0450	Class A, 260-300 lb, 4 bundles/sq	1 Rofc	3.50	2.286	Sq.	63	70		133	195	
	0455	Pneumatic nailed		4.37	1.831		63	56		119	170	
	0500	Class C, 300-385 lb/square, 5 bundles/square		3	2.667		75.50	81.50		157	230	
	0505	Pneumatic nailed		3.75	2.133		75.50	65.50		141	200	
	0800	#15 felt underlayment		64	.125		3.41	3.83		7.24	10.65	
	0825	#30 felt underlayment		58	.138		7.20	4.22		11.42	15.50	
	0850	Self adhering polyethylene and rubberized asphalt underlayment		22	.364		46	11.15		57.15	70.50	
	0900	Ridge shingles		330	.024	L.F.	1.33	.74		2.07	2.79	

07310	Shingles		Crew	Daily Output	Labor-Hours	Unit	2006 Bare Costs Mat.	Labor	Equip.	Total	Total Incl O&P	
100	0905	Pneumatic nailed	1 Rofc	412.50	.019	L.F.	1.33	.59		1.92	2.53	100
	1000	For steep roofs (7 to 12 pitch or greater), add						50%				
	9000	Minimum labor/equipment charge	1 Rofc	3	2.667	Job		81.50		81.50	147	
500	0010	**FIBER CEMENT SHINGLES**										500
	0012	Field shingles, 16" x 9.35", 500 lb per square	1 Rofc	2.20	3.636	Sq.	292	111		403	520	
	0200	Shakes, 16" x 9.35", 550 lb per square		2.20	3.636	"	265	111		376	490	
	0300	Hip & ridge, 4.75 x 14"		1	8	C.L.F.	720	245		965	1,225	
	0400	Hexagonal, 16" x 16"		3	2.667	Sq.	198	81.50		279.50	365	
	0500	Square, 16" x 16"		3	2.667		178	81.50		259.50	340	
	2000	For steep roofs (7/12 pitch or greater), add						50%				
	9000	Minimum labor/equipment charge	1 Rofc	3	2.667	Job		81.50		81.50	147	
800	0010	**SLATE**, Buckingham, Virginia, black										800
	0100	3/16" - 1/4" thick	1 Rots	1.75	4.571	Sq.	585	140		725	890	
	0200	1/4" thick		1.75	4.571		615	140		755	925	
	0900	Pennsylvania black, Bangor, #1 clear		1.75	4.571		475	140		615	775	
	1200	Vermont, unfading, green, mottled green		1.75	4.571		365	140		505	655	
	1300	Semi-weathering green & gray		1.75	4.571		350	140		490	635	
	1400	Purple		1.75	4.571		390	140		530	680	
	1500	Black or gray		1.75	4.571		360	140		500	645	
	2500	Slate roof repair, extensive replacement		1	8		575	245		820	1,075	
	2600	Repair individual pieces, scattered		19	.421	Ea.	5.70	12.90		18.60	29.50	
	9000	Minimum labor/equipment charge		3	2.667	Job		81.50		81.50	147	
900	0010	**STEEL SHINGLES**										900
	0012	Galvanized, 26 gauge	1 Rots	2.20	3.636	Sq.	183	111		294	400	
	0200	24 gauge	"	2.20	3.636		192	111		303	410	
	0300	For colored galvanized shingles, add					50			50	55	
	0500	For 1" factory applied polystyrene insulation, add					36.50			36.50	40.50	
	9000	Minimum labor/equipment charge	1 Rots	3	2.667	Job		81.50		81.50	147	
980	0010	**WOOD SHINGLES** R061110-30										980
	0012	16" No. 1 red cedar shingles, 5" exposure, on roof	1 Carp	2.50	3.200	Sq.	187	114		301	395	
	0015	Pneumatic nailed		3.25	2.462		187	87.50		274.50	350	
	0200	7-1/2" exposure, on walls		2.05	3.902		125	139		264	365	
	0205	Pneumatic nailed		2.67	2.996		125	107		232	315	
	0300	18" No. 1 red cedar perfections, 5-1/2" exposure, on roof		2.75	2.909		168	103		271	355	
	0305	Pneumatic nailed		3.57	2.241		168	79.50		247.50	315	
	0600	Resquared, and rebutted, 5-1/2" exposure, on roof		3	2.667		210	95		305	390	
	0605	Pneumatic nailed		3.90	2.051		210	73		283	350	
	0900	7-1/2" exposure, on walls		2.45	3.265		154	116		270	360	
	0905	Pneumatic nailed		3.18	2.516		154	89.50		243.50	315	
	1000	Add to above for fire retardant shingles, 16" long					35			35	38.50	
	1050	18" long					35			35	38.50	
	1060	Preformed ridge shingles	1 Carp	400	.020	L.F.	1.65	.71		2.36	3	
	1100	Hand-split red cedar shakes, 1/2" thick x 24" long, 10" exp. on roof		2.50	3.200	Sq.	116	114		230	315	
	1105	Pneumatic nailed		3.25	2.462		116	87.50		203.50	272	
	1110	3/4" thick x 24" long, 10" exp. on roof		2.25	3.556		116	126		242	335	
	1115	Pneumatic nailed		2.92	2.740		116	97.50		213.50	288	
	1200	1/2" thick, 18" long, 8-1/2" exp. on roof		2	4		89.50	142		231.50	335	
	1205	Pneumatic nailed		2.60	3.077		89.50	109		198.50	280	
	1210	3/4" thick x 18" long, 8 1/2" exp. on roof		1.80	4.444		89.50	158		247.50	360	
	1215	Pneumatic nailed		2.34	3.419		89.50	122		211.50	300	
	1255	10" exp. on walls		2	4		101	142		243	345	
	1260	10" exposure on walls, pneumatic nailed		2.60	3.077		101	109		210	292	
	1700	Add to above for fire retardant shakes, 24" long					50			50	55	
	1800	18" long					45			45	49.50	

	07310 | Shingles		Crew	Daily Output	Labor-Hours	Unit	2006 Bare Costs Mat.	Labor	Equip.	Total	Total Incl O&P	
980	1810	Ridge shakes (R061110-30)	1 Carp	350	.023	L.F.	2.35	.81		3.16	3.94	980
	2000	White cedar shingles, 16" long, extras, 5" exposure, on roof		2.40	3.333	Sq.	163	119		282	375	
	2005	Pneumatic nailed		3.12	2.564		163	91		254	330	
	2050	5" exposure on walls		2	4		163	142		305	415	
	2055	Pneumatic nailed		2.60	3.077		163	109		272	360	
	2100	7-1/2" exposure, on walls		2	4		116	142		258	365	
	2105	Pneumatic nailed		2.60	3.077		116	109		225	310	
	2150	"B" grade, 5" exposure on walls		2	4		146	142		288	395	
	2155	Pneumatic nailed		2.60	3.077		146	109		255	340	
	2300	For 15# organic felt underlayment on roof, 1 layer, add		64	.125		3.41	4.44		7.85	11.10	
	2400	2 layers, add		32	.250		6.80	8.90		15.70	22.50	
	2600	For steep roofs (7/12 pitch or greater), add to above						50%				
	2700	Panelized systems, No.1 cedar shingles on 5/16" CDX plywood										
	2800	On walls, 8' strips, 7" or 14" exposure	2 Carp	700	.023	S.F.	3.20	.81		4.01	4.87	
	3500	On roofs, 8' strips, 7" or 14" exposure	1 Carp	3	2.667	Sq.	320	95		415	505	
	3505	Pneumatic nailed		4	2	"	320	71		391	470	
	9000	Minimum labor/equipment charge		3	2.667	Job		95		95	157	

	07320 | Roof Tiles		Crew	Daily Output	Labor-Hours	Unit	Mat.	Labor	Equip.	Total	Total Incl O&P	
100	0010	**ALUMINUM ROOF TILES**										100
	0020	Accessories included, .032" thick, mission tile	1 Carp	2.50	3.200	Sq.	565	114		679	815	
	0200	Spanish tiles		3	2.667	"	390	95		485	585	
	9000	Minimum labor/equipment charge		3	2.667	Job		95		95	157	
200	0010	**CLAY TILE** ASTM C1167, GR 1, severe weathering, acces. incl.										200
	0200	Lanai tile or Classic tile, 158 pc per sq	1 Rots	1.65	4.848	Sq.	405	148		553	715	
	0300	Americana, 158 pc per sq, most colors		1.65	4.848		525	148		673	840	
	0350	Green, gray or brown		1.65	4.848		505	148		653	825	
	0400	Blue		1.65	4.848		505	148		653	825	
	0600	Spanish tile, 171 pc per sq, red		1.80	4.444		268	136		404	540	
	0800	Blend		1.80	4.444		430	136		566	720	
	0900	Glazed white		1.80	4.444		515	136		651	810	
	1100	Mission tile, 192 pc per sq, machine scored finish, red		1.15	6.957		650	213		863	1,100	
	1700	French tile, 133 pc per sq, smooth finish, red		1.35	5.926		590	181		771	975	
	1750	Blue or green		1.35	5.926		705	181		886	1,100	
	1800	Norman black 317 pc per sq		1	8		830	245		1,075	1,350	
	2200	Williamsburg tile, 158 pc per sq, aged cedar		1.35	5.926		505	181		686	880	
	2250	Gray or green		1.35	5.926		505	181		686	880	
	3000	For steep roofs (7/12 pitch or greater), add to above						50%				
	9000	Minimum labor/equipment charge	1 Rots	3	2.667	Job		81.50		81.50	147	
300	0010	**CONCRETE TILE** Including installation of accessories										300
	0020	Corrugated, 13" x 16-1/2", 90 per sq, 950 lb per sq										
	0050	Earthtone colors, nailed to wood deck	1 Rots	1.35	5.926	Sq.	93.50	181		274.50	430	
	0150	Blues		1.35	5.926		94.50	181		275.50	430	
	0200	Greens		1.35	5.926		94.50	181		275.50	430	
	0250	Premium colors		1.35	5.926		158	181		339	500	
	0500	Shakes, 13" x 16-1/2", 90 per sq, 950 lb per sq										
	0600	All colors, nailed to wood deck	1 Rots	1.50	5.333	Sq.	191	163		354	505	
	1500	Accessory pieces, ridge & hip, 10" x 16-1/2", 8 lbs. each				Ea.	2.32			2.32	2.55	
	1700	Rake, 6-1/2" x 16-3/4", 9 lbs. each					2.32			2.32	2.55	
	1800	Mansard hip, 10" x 16-1/2", 9.2 lbs. each					2.32			2.32	2.55	
	1900	Hip starter, 10" x 16-1/2", 10.5 lbs. each					9.80			9.80	10.75	
	2000	3 or 4 way apex, 10" each side, 11.5 lbs. each					10.55			10.55	11.60	
	9000	Minimum labor/equipment charge	1 Rots	3	2.667	Job		81.50		81.50	147	

07400 | Roofing and Siding Panels

		07410 Metal Roof and Wall Panels	Crew	Daily Output	Labor-Hours	Unit	2006 Bare Costs Mat.	Labor	Equip.	Total	Total Incl O&P	
100	0010	**ALUMINUM ROOFING PANELS**										100
	0020	Corrugated or ribbed, .0155″ thick, natural	G-3	1200	.027	S.F.	.69	.93		1.62	2.26	
	0300	Painted		1200	.027		1	.93		1.93	2.60	
	0400	Corrugated, .018″ thick, on steel frame, natural finish		1200	.027		.90	.93		1.83	2.49	
	0600	Painted		1200	.027		1.11	.93		2.04	2.72	
	0700	Corrugated, on steel frame, natural, .024″ thick		1200	.027		1.30	.93		2.23	2.93	
	0800	Painted		1200	.027		1.57	.93		2.50	3.23	
	0900	.032″ thick, natural		1200	.027		1.54	.93		2.47	3.19	
	1200	Painted		1200	.027		2.13	.93		3.06	3.84	
	9000	Minimum labor/equipment charge	1 Rofc	3	2.667	Job		81.50		81.50	147	
500	0010	**MANSARD** Colored aluminum, with battens, .032″ thick										500
	0600	Stock units, straight surfaces	1 Shee	115	.070	S.F.	2.39	2.93		5.32	7.30	
	0700	Concave or convex surfaces		75	.107	″	2.63	4.50		7.13	10.05	
	0800	For framing, to 5′ high, add		115	.070	L.F.	2.63	2.93		5.56	7.55	
	0900	Soffits, to 1′ wide		125	.064	S.F.	1.30	2.70		4	5.70	
	9000	Minimum labor/equipment charge		2.50	3.200	Job		135		135	215	
700	0010	**STEEL ROOFING PANELS**										700
	0012	Corrugated or ribbed, on steel framing, 30 ga galv	G-3	1100	.029	S.F.	1.05	1.01		2.06	2.80	
	0100	28 ga		1050	.030		1.10	1.06		2.16	2.92	
	0300	26 ga		1000	.032		1.53	1.11		2.64	3.48	
	0400	24 ga		950	.034		1.89	1.17		3.06	3.97	
	0600	Colored, 28 ga		1050	.030		1.08	1.06		2.14	2.90	
	0700	26 ga		1000	.032		1.65	1.11		2.76	3.62	
	0710	Flat profile, 1-3/4″ standing seams, 10″ wide, standard finish, 26 ga		1000	.032		3.22	1.11		4.33	5.35	
	0715	24 ga		950	.034		3.74	1.17		4.91	6	
	0720	22 ga		900	.036		4.61	1.24		5.85	7.05	
	0725	Zinc aluminum alloy finish, 26 ga		1000	.032		2.52	1.11		3.63	4.57	
	0730	24 ga		950	.034		3.01	1.17		4.18	5.20	
	0735	22 ga		900	.036		3.45	1.24		4.69	5.80	
	0740	12″ wide, standard finish, 26 ga		1000	.032		3.21	1.11		4.32	5.35	
	0745	24 ga		950	.034		4.22	1.17		5.39	6.55	
	0750	Zinc aluminum alloy finish, 26 ga		1000	.032		3.65	1.11		4.76	5.80	
	0755	24 ga		950	.034		3	1.17		4.17	5.20	
	0840	Flat profile, 1″ x 3/8″ batten, 12″ wide, standard finish, 26 ga		1000	.032		2.83	1.11		3.94	4.91	
	0845	24 ga		950	.034		3.32	1.17		4.49	5.55	
	0850	22 ga		900	.036		3.98	1.24		5.22	6.40	
	0855	Zinc aluminum alloy finish, 26 ga		1000	.032		2.72	1.11		3.83	4.79	
	0860	24 ga		950	.034		3.03	1.17		4.20	5.20	
	0865	22 ga		900	.036		3.50	1.24		4.74	5.85	
	0870	16-1/2″ wide, standard finish, 24 ga		950	.034		3.27	1.17		4.44	5.50	
	0875	22 ga		900	.036		3.66	1.24		4.90	6.05	
	0880	Zinc aluminum alloy finish, 24 ga		950	.034		2.85	1.17		4.02	5.05	
	0885	22 ga		900	.036		3.19	1.24		4.43	5.50	
	0890	Flat profile, 2″ x 2″ batten, 12″ wide, standard finish, 26 ga		1000	.032		3.25	1.11		4.36	5.40	
	0895	24 ga		950	.034		3.88	1.17		5.05	6.15	
	0900	22 ga		900	.036		4.75	1.24		5.99	7.25	
	0905	Zinc aluminum alloy finish, 26 ga		1000	.032		3.03	1.11		4.14	5.15	
	0910	24 ga		950	.034		3.46	1.17		4.63	5.70	
	0915	22 ga		900	.036		4.03	1.24		5.27	6.45	
	0920	16-1/2″ wide, standard finish, 24 ga		950	.034		3.58	1.17		4.75	5.85	
	0925	22 ga		900	.036		4.18	1.24		5.42	6.60	
	0930	Zinc aluminum alloy finish, 24 ga		950	.034		3.24	1.17		4.41	5.45	
	0935	22 ga		900	.036		3.68	1.24		4.92	6.05	
	9000	Minimum labor/equipment charge	1 Rofc	2	4	Job		122		122	220	

07400 | Roofing and Siding Panels

07420 | Plastic Roof and Wall Panels

		07420 Plastic Roof and Wall Panels	CREW	DAILY OUTPUT	LABOR-HOURS	UNIT	2006 BARE COSTS MAT.	LABOR	EQUIP.	TOTAL	TOTAL INCL O&P	
770	0010	**FIBERGLASS CORRUGATED ROOFING PANELS**										770
	0012	Corrugated, 8 oz per SF	G-3	1000	.032	S.F.	2.24	1.11		3.35	4.26	
	0100	12 oz per SF		1000	.032		3.23	1.11		4.34	5.35	
	0300	Corrugated siding, 6 oz per SF		880	.036		1.94	1.26		3.20	4.17	
	0400	8 oz per SF		880	.036		2.24	1.26		3.50	4.50	
	0500	Fire retardant		880	.036		3.09	1.26		4.35	5.45	
	0600	12 oz. siding, textured		880	.036		3.14	1.26		4.40	5.50	
	0700	Fire retardant		880	.036		4.14	1.26		5.40	6.60	
	0900	Flat panels, 6 oz per SF, clear or colors		880	.036		1.73	1.26		2.99	3.94	
	1100	Fire retardant, class A		880	.036		3	1.26		4.26	5.35	
	1300	8 oz per SF, clear or colors		880	.036		2.24	1.26		3.50	4.50	
	9000	Minimum labor/equipment charge	1 Rofc	2	4	Job		122		122	220	

07460 | Siding

		07460 Siding	CREW	DAILY OUTPUT	LABOR-HOURS	UNIT	MAT.	LABOR	EQUIP.	TOTAL	TOTAL INCL O&P	
100	0010	**ALUMINUM SIDING PANELS**										100
	0012	Corrugated, on steel framing, .019 thick, natural finish	G-3	775	.041	S.F.	.74	1.44		2.18	3.13	
	0100	Painted		775	.041		.96	1.44		2.40	3.38	
	0400	Farm type, .021″ thick on steel frame, natural		775	.041		.86	1.44		2.30	3.27	
	0600	Painted		775	.041		1.05	1.44		2.49	3.48	
	0700	Industrial type, corrugated, on steel, .024″ thick, mill		775	.041		1.09	1.44		2.53	3.52	
	0900	Painted		775	.041		1.31	1.44		2.75	3.76	
	1000	.032″ thick, mill		775	.041		1.59	1.44		3.03	4.07	
	1200	Painted		775	.041		2.02	1.44		3.46	4.54	
	1300	V-Beam, on steel frame, .032″ thick, mill		775	.041		1.86	1.44		3.30	4.37	
	1500	Painted		775	.041		2.21	1.44		3.65	4.75	
	1600	.040″ thick, mill		775	.041		2.26	1.44		3.70	4.81	
	1800	Painted		775	.041		2.65	1.44		4.09	5.25	
	3800	Horizontal, colored clapboard, 8″ wide, plain	2 Carp	515	.031		1.21	1.10		2.31	3.16	
	3810	Insulated		515	.031		1.22	1.10		2.32	3.17	
	3830	8″ embossed, painted		515	.031		1.35	1.10		2.45	3.32	
	3840	Insulated		515	.031		1.50	1.10		2.60	3.48	
	3860	12″ painted, smooth		600	.027		1.25	.95		2.20	2.95	
	3870	Insulated		600	.027		1.45	.95		2.40	3.17	
	3890	12″ embossed, painted		600	.027		1.40	.95		2.35	3.11	
	3900	Insulated		515	.031		1.56	1.10		2.66	3.55	
	4000	Vertical board & batten, colored, non-insulated		515	.031		1.21	1.10		2.31	3.16	
	4200	For simulated wood design, add					.09			.09	.10	
	4300	Corners for above, outside	2 Carp	515	.031	V.L.F.	1.86	1.10		2.96	3.88	
	4500	Inside corners	″	515	.031	″	1.10	1.10		2.20	3.04	
	4520	For simulated wood design, add				S.F.	.09			.09	.10	
	9000	Minimum labor/equipment charge	1 Carp	3	2.667	Job		95		95	157	
300	0010	**FASCIA** Aluminum, reverse board and batten,										300
	0100	.032″ thick, colored, no furring included	1 Shee	145	.055	S.F.	2.95	2.33		5.28	6.95	
	0200	Residential type, aluminum	1 Carp	200	.040	L.F.	1.17	1.42		2.59	3.65	
	0220	Vinyl	″	200	.040	″	.88	1.42		2.30	3.33	
	0300	Steel, galv and enameled, stock, no furring, long panels	1 Shee	145	.055	S.F.	3	2.33		5.33	7	
	0600	Short panels		115	.070	″	4	2.93		6.93	9.05	
	9000	Minimum labor/equipment charge		4	2	Job		84.50		84.50	134	
500	0010	**FIBER CEMENT SIDING**										500
	0020	Lap siding, 5/16″ thick, 6″ wide, smooth texture	2 Carp	415	.039	S.F.	1.23	1.37		2.60	3.63	
	0025	Woodgrain texture		415	.039		1.23	1.37		2.60	3.63	
	0030	7-1/2″ wide, smooth texture		425	.038		1.08	1.34		2.42	3.40	
	0035	Woodgrain texture		425	.038		1.08	1.34		2.42	3.40	
	0040	8″ wide, smooth texture		425	.038		1.07	1.34		2.41	3.39	

		07460 | Siding	CREW	DAILY OUTPUT	LABOR-HOURS	UNIT	2006 BARE COSTS MAT.	LABOR	EQUIP.	TOTAL	TOTAL INCL O&P	
500	0045	Roughsawn texture	2 Carp	425	.038	S.F.	1.07	1.34		2.41	3.39	500
	0050	9-1/2″ wide, smooth texture		440	.036		1.03	1.29		2.32	3.28	
	0055	Woodgrain texture		440	.036		1.03	1.29		2.32	3.28	
	0060	12″ wide, smooth texture		455	.035		.99	1.25		2.24	3.16	
	0065	Woodgrain texture		455	.035		.99	1.25		2.24	3.16	
	0070	Panel siding, 5/16″ thick, smooth texture		750	.021		.89	.76		1.65	2.24	
	0075	Stucco texture		750	.021		.89	.76		1.65	2.24	
	0080	Grooved woodgrain texture		750	.021		.89	.76		1.65	2.24	
	0085	V - grooved woodgrain texture		750	.021		.89	.76		1.65	2.24	
	0090	Wood starter strip		400	.040	L.F.	.19	1.42		1.61	2.57	
600	0010	VINYL SIDING										600
	0020	Clapboard profile, woodgrain texture, .048 thick, double 4	1 Carp	255	.031	S.F.	.74	1.12		1.86	2.66	
	0100	with 3/8″ insulation		255	.031		.84	1.12		1.96	2.77	
	0200	Soffit and fascia		205	.039		1.60	1.39		2.99	4.06	
	0300	Window and door trim moldings		185	.043	L.F.	.40	1.54		1.94	2.99	
	0500	Corner posts, outside corner		205	.039		1.44	1.39		2.83	3.88	
	0600	Inside corner		205	.039		.67	1.39		2.06	3.04	
	9000	Minimum labor/equipment charge		3	2.667	Job		95		95	157	
750	0010	SOFFIT										750
	0012	Aluminum, residential, .020″ thick	1 Carp	210	.038	S.F.	1.24	1.35		2.59	3.60	
	0100	Baked enamel on steel, 16 or 18 gauge		105	.076		4.60	2.71		7.31	9.55	
	0300	Polyvinyl chloride, white, solid		230	.035		.82	1.24		2.06	2.95	
	0400	Perforated		230	.035		.82	1.24		2.06	2.95	
	0500	For colors, add					.09			.09	.10	
	9000	Minimum labor/equipment charge	1 Carp	3	2.667	Job		95		95	157	
800	0010	STEEL SIDING										800
	0020	Beveled, vinyl coated, 8″ wide, including fasteners	1 Carp	265	.030	S.F.	1.50	1.07		2.57	3.43	
	0050	10″ wide		275	.029	″	1.60	1.03		2.63	3.47	
	0060	Minimum labor/equipment charge		3	2.667	Job		95		95	157	
	0070											
	0080	Galv, corrugated or ribbed, on steel frame, 30 gauge	G-3	800	.040	S.F.	1	1.39		2.39	3.35	
	0100	28 gauge		795	.040		1.05	1.40		2.45	3.42	
	0300	26 gauge		790	.041		1.47	1.41		2.88	3.90	
	0400	24 gauge		785	.041		1.48	1.42		2.90	3.92	
	0600	22 gauge		770	.042		1.70	1.45		3.15	4.21	
	0700	Colored, corrugated/ribbed, on steel frame, 10 yr fnsh, 28 ga.		800	.040		1.56	1.39		2.95	3.97	
	0900	26 gauge		795	.040		1.63	1.40		3.03	4.05	
	1000	24 gauge		790	.041		1.89	1.41		3.30	4.36	
	1020	20 gauge		785	.041		2.40	1.42		3.82	4.93	
	9000	Minimum labor/equipment charge	1 Carp	3	2.667	Job		95		95	157	
900	0010	WOOD SIDING, BOARDS										900
	3200	Wood, cedar bevel, A grade, 1/2″ x 6″	1 Carp	250	.032	S.F.	3.20	1.14		4.34	5.40	
	3300	1/2″ x 8″		275	.029		3.11	1.03		4.14	5.15	
	3500	3/4″ x 10″, clear grade		300	.027		3.63	.95		4.58	5.55	
	3600	"B" grade		300	.027		2.80	.95		3.75	4.65	
	3800	Cedar, rough sawn, 1″ x 4″, A grade, natural		240	.033		2.91	1.18		4.09	5.15	
	3900	Stained		240	.033		3.28	1.18		4.46	5.55	
	4100	1″ x 12″, board & batten, #3 & Btr., natural		260	.031		2.20	1.09		3.29	4.23	
	4200	Stained		260	.031		2.56	1.09		3.65	4.63	
	4400	1″ x 8″ channel siding, #3 & Btr., natural		250	.032		2.14	1.14		3.28	4.23	
	4500	Stained		250	.032		2.43	1.14		3.57	4.55	
	4700	Redwood, clear, beveled, vertical grain, 1/2″ x 4″		200	.040		3.40	1.42		4.82	6.10	
	4750	1/2″ x 6″		225	.036		2.86	1.26		4.12	5.25	
	4800	1/2″ x 8″		250	.032		2.31	1.14		3.45	4.42	

07400 | Roofing and Siding Panels

07460 | Siding

		07460 Siding	CREW	DAILY OUTPUT	LABOR-HOURS	UNIT	2006 BARE COSTS MAT.	LABOR	EQUIP.	TOTAL	TOTAL INCL O&P	
900	5000	3/4″ x 10″	1 Carp	300	.027	S.F.	3.77	.95		4.72	5.70	900
	5200	Channel siding, 1″ x 10″, B grade		285	.028		2.43	1		3.43	4.32	
	5250	Redwood, T&G boards, B grade, 1″ x 4″	2 Carp	300	.053		2.90	1.90		4.80	6.35	
	5270	1″ x 8″	″	375	.043		2.50	1.52		4.02	5.25	
	5400	White pine, rough sawn, 1″ x 8″, natural	1 Carp	275	.029		.73	1.03		1.76	2.51	
	5500	Stained		275	.029		1.08	1.03		2.11	2.90	
	9000	Minimum labor/equipment charge		2	4	Job		142		142	236	
950	0010	**WOOD PRODUCT SIDING** R061636-20										950
	0030	Lap siding, hardboard, 7/16″ x 8″, primed										
	0051	Wood grain texture finish	L-2	552	.029	S.F.	1.13	.90		2.03	2.73	
	0100	Panels, 7/16″ thick, smooth, textured or grooved, primed	2 Carp	700	.023		.84	.81		1.65	2.27	
	0200	Stained		700	.023		.96	.81		1.77	2.41	
	0700	Particle board, overlaid, 3/8″ thick		750	.021		.69	.76		1.45	2.02	
	0900	Plywood, medium density overlaid, 3/8″ thick		750	.021		.84	.76		1.60	2.18	
	1000	1/2″ thick		700	.023		.95	.81		1.76	2.40	
	1100	3/4″ thick		650	.025		1.28	.88		2.16	2.86	
	1600	Texture 1-11, cedar, 5/8″ thick, natural		675	.024		2.38	.84		3.22	4.02	
	1700	Factory stained		675	.024		2.21	.84		3.05	3.83	
	1900	Texture 1-11, fir, 5/8″ thick, natural		675	.024		1.01	.84		1.85	2.51	
	2000	Factory stained		675	.024		1.08	.84		1.92	2.59	
	2050	Texture 1-11, S.Y.P., 5/8″ thick, natural		675	.024		.81	.84		1.65	2.29	
	2100	Factory stained		675	.024		.88	.84		1.72	2.37	
	2200	Rough sawn cedar, 3/8″ thick, natural		675	.024		1.08	.84		1.92	2.59	
	2300	Factory stained		675	.024		1.20	.84		2.04	2.72	
	2500	Rough sawn fir, 3/8″ thick, natural		675	.024		.58	.84		1.42	2.04	
	2600	Factory stained		675	.024		.65	.84		1.49	2.12	
	2800	Redwood, textured siding, 5/8″ thick		675	.024		1.80	.84		2.64	3.38	
	3000	Polyvinyl chloride coated, 3/8″ thick		750	.021		.88	.76		1.64	2.23	
	9000	Minimum labor/equipment charge	1 Carp	2	4	Job		142		142	236	

07500 | Membrane Roofing

07510 | Built-Up Bituminous Roofing

		07510 Built-Up Bituminous Roofing	CREW	DAILY OUTPUT	LABOR-HOURS	UNIT	2006 BARE COSTS MAT.	LABOR	EQUIP.	TOTAL	TOTAL INCL O&P	
050	0010	**BUILT- UP ROOFING COMPONENTS**										050
	0012	Asphalt saturated felt, #30, 2 square per roll	1 Rofc	58	.138	Sq.	7.20	4.22		11.42	15.50	
	0200	#15, 4 sq per roll, plain or perforated, not mopped		58	.138		3.41	4.22		7.63	11.35	
	0300	Roll roofing, smooth, #65		15	.533		6.85	16.30		23.15	37	
	0500	#90		15	.533		19.05	16.30		35.35	50.50	
	0520	Mineralized		15	.533		16.55	16.30		32.85	47.50	
	0540	D.C. (Double coverage), 19″ selvage edge		10	.800		32	24.50		56.50	79	
	0580	Adhesive (lap cement)				Gal.	3.97			3.97	4.37	
	0600	Steep, flat or dead level asphalt, 10 ton lots, bulk				Ton	280			280	310	
	0800	Packaged				″	410			410	450	
	9000	Minimum labor/equipment charge	1 Rofc	4	2	Job		61		61	110	
300	0010	**BUILT-UP ROOFING**										300
	0120	Asphalt flood coat with gravel/slag surfacing, not including										
	0140	Insulation, flashing or wood nailers										
	0200	Asphalt base sheet, 3 plies #15 asphalt felt, mopped	G-1	22	2.545	Sq.	49.50	73	15.30	137.80	202	

	07510	Built-Up Bituminous Roofing	CREW	DAILY OUTPUT	LABOR-HOURS	UNIT	2006 BARE COSTS MAT.	LABOR	EQUIP.	TOTAL	TOTAL INCL O&P
300	0350	On nailable decks	G-1	21	2.667	Sq.	55.50	76	16.05	147.55	216
	0500	4 plies #15 asphalt felt, mopped		20	2.800		71.50	80	16.85	168.35	242
	0550	On nailable decks		19	2.947		65	84.50	17.75	167.25	242
	0700	Coated glass base sheet, 2 plies glass (type IV), mopped		22	2.545		53	73	15.30	141.30	206
	0850	3 plies glass, mopped		20	2.800		63.50	80	16.85	160.35	232
	0950	On nailable decks		19	2.947		60	84.50	17.75	162.25	237
	1100	4 plies glass fiber felt (type IV), mopped		20	2.800		77	80	16.85	173.85	247
	1150	On nailable decks		19	2.947		70	84.50	17.75	172.25	248
	1200	Coated & saturated base sheet, 3 plies #15 asph. felt, mopped		20	2.800		57	80	16.85	153.85	226
	1250	On nailable decks		19	2.947		53.50	84.50	17.75	155.75	230
	1300	4 plies #15 asphalt felt, mopped		22	2.545		66.50	73	15.30	154.80	221
	2000	Asphalt flood coat, smooth surface									
	2200	Asphalt base sheet & 3 plies #15 asphalt felt, mopped	G-1	24	2.333	Sq.	54.50	66.50	14.05	135.05	195
	2400	On nailable decks		23	2.435		51	69.50	14.65	135.15	198
	2600	4 plies #15 asphalt felt, mopped		24	2.333		63.50	66.50	14.05	144.05	205
	2700	On nailable decks		23	2.435		60.50	69.50	14.65	144.65	208
	2900	Coated glass fiber base sheet, mopped, and 2 plies of									
	2910	glass fiber felt (type IV)	G-1	25	2.240	Sq.	48.50	64	13.50	126	183
	3100	On nailable decks		24	2.333		46	66.50	14.05	126.55	186
	3200	3 plies, mopped		23	2.435		59	69.50	14.65	143.15	206
	3300	On nailable decks		22	2.545		55.50	73	15.30	143.80	209
	3800	4 plies glass fiber felt (type IV), mopped		23	2.435		69	69.50	14.65	153.15	217
	3900	On nailable decks		22	2.545		65.50	73	15.30	153.80	220
	4000	Coated & saturated base sheet, 3 plies #15 asph. felt, mopped		24	2.333		52.50	66.50	14.05	133.05	193
	4200	On nailable decks		23	2.435		49.50	69.50	14.65	133.65	195
	4300	4 plies #15 organic felt, mopped		22	2.545		62	73	15.30	150.30	216
	4500	Coal tar pitch with gravel/slag surfacing									
	4600	4 plies #15 tarred felt, mopped	G-1	21	2.667	Sq.	118	76	16.05	210.05	284
	4800	3 plies glass fiber felt (type IV), mopped	″	19	2.947	″	96	84.50	17.75	198.25	277
	5000	Coated glass fiber base sheet, and 2 plies of									
	5010	glass fiber felt, (type IV), mopped	G-1	19	2.947	Sq.	95.50	84.50	17.75	197.75	276
	5300	On nailable decks		18	3.111		85	89	18.75	192.75	274
	5600	4 plies glass fiber felt (type IV), mopped		21	2.667		133	76	16.05	225.05	300
	5800	On nailable decks		20	2.800		123	80	16.85	219.85	298
400	0010	**CANTS**									
	0012	Lumber, treated, 4″ x 4″ cut diagonally	1 Rofc	325	.025	L.F.	1.38	.75		2.13	2.87
	0100	Foamglass		325	.025		2.15	.75		2.90	3.72
	0300	Mineral or fiber, trapezoidal, 1″x 4″ x 48″		325	.025		.17	.75		.92	1.54
	0400	1-1/2″ x 5-5/8″ x 48″		325	.025		.29	.75		1.04	1.67
	9000	Minimum labor/equipment charge		4	2	Job		61		61	110
700	0010	**ROOFING FELTS**									
	0012	Glass fibered roofing felt, #15, not mopped	1 Rofc	58	.138	Sq.	4.49	4.22		8.71	12.55
	0300	Base sheet, #45, channel vented		58	.138		19.15	4.22		23.37	28.50
	0400	#50, coated		58	.138		10.35	4.22		14.57	19
	0500	Cap, mineral surfaced		58	.138		19.15	4.22		23.37	28.50
	0600	Flashing membrane, #65		16	.500		29.50	15.30		44.80	59.50
	0800	Coal tar fibered, #15, no mopping		58	.138		8.70	4.22		12.92	17.15
	0900	Asphalt felt, #15, 4 sq per roll, no mopping		58	.138		3.41	4.22		7.63	11.35
	1100	#30, 2 sq per roll		58	.138		7.20	4.22		11.42	15.50
	1200	Double coated, #33		58	.138		6.85	4.22		11.07	15.10
	1400	#40, base sheet		58	.138		9.40	4.22		13.62	17.95
	1450	Coated and saturated		58	.138		7.50	4.22		11.72	15.85
	1500	Tarred felt, organic, #15, 4 sq rolls		58	.138		10.85	4.22		15.07	19.50
	1550	#30, 2 sq roll		58	.138		18.55	4.22		22.77	28
	1700	Add for mopping above felts, per ply, asphalt, 24 lb per sq	G-1	192	.292		4.92	8.35	1.76	15.03	22.50
	1800	Coal tar mopping, 30 lb per sq		186	.301		9.25	8.60	1.81	19.66	27.50

07510 | Built-Up Bituminous Roofing

		07510 Built-Up Bituminous Roofing	Crew	Daily Output	Labor-Hours	Unit	2006 Bare Costs Mat.	Labor	Equip.	Total	Total Incl O&P	
700	1900	Flood coat, with asphalt, 60 lb per sq	G-1	60	.933	Sq.	12.30	26.50	5.60	44.40	68	700
	2000	With coal tar, 75 lb per sq		56	1		23	28.50	6	57.50	83.50	
	9000	Minimum labor/equipment charge	1 Rofc	4	2	Job		61		61	110	

07520 | Cold Applied Bituminous Roofing

		07520 Cold Applied Bituminous Roofing	Crew	Daily Output	Labor-Hours	Unit	Mat.	Labor	Equip.	Total	Total Incl O&P	
200	0010	COLD APPLIED BUILT UP ROOF										200
	0020	3 ply system, installation only (components listed below)	G-5	50	.800	Sq.		22	2.93	24.93	43	
	0100	Spunbond poly. fabric, 1.35 oz/SY, 36"W, 10.8 Sq/roll				Ea.	132			132	146	
	0200	49" wide, 14.6 Sq./roll					183			183	201	
	0300	2.10 oz./S.Y., 36" wide, 10.8 Sq./roll					199			199	219	
	0400	49" wide, 14.6 Sq./roll					270			270	297	
	0500	Base & finish coat, 3 gal./Sq., 5 gal./can				Gal.	3.51			3.51	3.86	
	0600	Coating, ceramic granules, 1/2 Sq./bag				Ea.	12.40			12.40	13.65	
	0700	Aluminum, 2 gal./Sq.				Gal.	10			10	11	
	0800	Emulsion, fibered or non-fibered, 4 gal./Sq.				"	4.59			4.59	5.05	

07530 | Elastomeric Membrane Roofing

		07530 Elastomeric Membrane Roofing	Crew	Daily Output	Labor-Hours	Unit	Mat.	Labor	Equip.	Total	Total Incl O&P	
350	0010	ELASTOMERIC ROOFING										350
	0100	For Elastomeric waterproofing, see Division 07130-200										
	0110	Acrylic rubber, fluid applied, 20 mils thick	G-5	2000	.020	S.F.	1.80	.56	.07	2.43	3.06	
	0120	50 mils, reinforced		1200	.033		2.80	.93	.12	3.85	4.87	
	0130	For walking surface, add		900	.044		.85	1.23	.16	2.24	3.34	
	0300	Hypalon neoprene, fluid applied, 20 mil thick, not-reinforced	G-1	1135	.049		2.05	1.41	.30	3.76	5.10	
	0600	Non-woven polyester, reinforced		960	.058		2.07	1.67	.35	4.09	5.65	
	0700	5 coat neoprene deck, 60 mil thick, under 10,000 SF		325	.172		4.56	4.93	1.04	10.53	15	
	0900	Over 10,000 SF		625	.090		4.25	2.56	.54	7.35	9.85	
	1300	Vinyl plastic traffic deck, sprayed, 2 to 4 mils thick		625	.090		1.33	2.56	.54	4.43	6.65	
	1500	Vinyl and neoprene membrane traffic deck		1550	.036		1.41	1.03	.22	2.66	3.65	
	9000	Minimum labor/equipment charge	1 Rofc	2	4	Job		122		122	220	
800	0010	SINGLE-PLY MEMBRANE										800
	0800	Chlorosulfonated polyethylene-hypalon (CSPE), 45 mils,										
	0900	0.29 P.S.F., fully adhered	G-5	26	1.538	Sq.	123	42.50	5.65	171.15	219	
	1200	Mechanically attached	"	35	1.143	"	125	32	4.18	161.18	200	
	3500	Ethylene propylene diene monomer (EPDM), 45 mils, 0.28 P.S.F.										
	3600	Loose-laid & ballasted with stone (10 P.S.F.)	G-5	51	.784	Sq.	61	22	2.87	85.87	109	
	3700	Mechanically attached		35	1.143		53.50	32	4.18	89.68	120	
	3800	Fully adhered with adhesive		26	1.538		77	42.50	5.65	125.15	168	
	4500	60 mils, 0.40 P.S.F.										
	4600	Loose-laid & ballasted with stone (10 P.S.F.)	G-5	51	.784	Sq.	75.50	22	2.87	100.37	125	
	4700	Mechanically attached		35	1.143		66.50	32	4.18	102.68	135	
	4800	Fully adhered with adhesive		26	1.538		90	42.50	5.65	138.15	183	
	4810	45 mil, .28 PSF, membrane only					31.50			31.50	34.50	
	4820	60 mil, .40 PSF, membrane only					43.50			43.50	48	
	4850	Seam tape for membrane, 3" x 100' roll				Ea.	41			41	45	
	4900	Batten strips, 10' sections					2.95			2.95	3.25	
	4910	Cover tape for batten strips, 6" x 100' roll					144			144	158	
	4930	Plate anchors				M	73.50			73.50	80.50	
	4970	Adhesive for fully adhered systems, 60 S.F./gal.				Gal.	14.60			14.60	16.05	
	7500	Polyisobutylene (PIB), 100 mils, 0.57 P.S.F.										
	7600	Loose-laid & ballasted with stone/gravel (10 P.S.F.)	G-5	51	.784	Sq.	128	22	2.87	152.87	183	
	7700	Partially adhered with adhesive		35	1.143		160	32	4.18	196.18	238	
	7800	Hot asphalt attachment		35	1.143		153	32	4.18	189.18	230	
	7900	Fully adhered with contact cement		26	1.538		165	42.50	5.65	213.15	265	

07540 | PVC Single Ply Membrane

		07540 PVC Single Ply Membrane	CREW	DAILY OUTPUT	LABOR-HOURS	UNIT	2006 BARE COSTS MAT.	LABOR	EQUIP.	TOTAL	TOTAL INCL O&P	
800	0010	POLYVINYL CHLORIDE ROOFING										800
	8200	Heat welded seams										
	8700	Reinforced, 48 mils, 0.33 P.S.F.										
	8750	Loose-laid & ballasted with stone/gravel (12 P.S.F.)	G-5	51	.784	Sq.	97.50	22	2.87	122.37	149	
	8800	Mechanically attached		35	1.143		88.50	32	4.18	124.68	159	
	8850	Fully adhered with adhesive		26	1.538		119	42.50	5.65	167.15	213	
	8860	Reinforced, 60 mils, .40 P.S.F.										
	8870	Loose-laid & ballasted with stone/gravel (12 P.S.F.)	G-5	51	.784	Sq.	90	22	2.87	114.87	141	
	8880	Mechanically attached		35	1.143		81	32	4.18	117.18	151	
	8890	Fully adhered with adhesive		26	1.538		111	42.50	5.65	159.15	205	

07550 | Modified Bit. Membrane Roofing

		07550 Modified Bit. Membrane Roofing	CREW	DAILY OUTPUT	LABOR-HOURS	UNIT	MAT.	LABOR	EQUIP.	TOTAL	TOTAL INCL O&P	
500	0010	MODIFIED BITUMEN ROOFING R075213-30										500
	0020	Base sheet, #15 glass fiber felt, nailed to deck	1 Rofc	58	.138	Sq.	5.25	4.22		9.47	13.35	
	0030	Spot mopped to deck	G-1	295	.190		6.95	5.45	1.14	13.54	18.65	
	0040	Fully mopped to deck	"	192	.292		9.40	8.35	1.76	19.51	27.50	
	0050	#15 organic felt, nailed to deck	1 Rofc	58	.138		4.15	4.22		8.37	12.15	
	0060	Spot mopped to deck	G-1	295	.190		5.85	5.45	1.14	12.44	17.45	
	0070	Fully mopped to deck	"	192	.292		8.35	8.35	1.76	18.46	26	
	0080	SBS modified, granule surf cap sheet, poly rein., mopped										
	1500	Glass fiber reinforced, mopped, 160 mils	G-1	2000	.028	S.F.	.45	.80	.17	1.42	2.13	
	1600	Smooth surface cap sheet, mopped, 145 mils		2100	.027		.45	.76	.16	1.37	2.05	
	1700	Smooth surface flashing, 145 mils		1260	.044		.45	1.27	.27	1.99	3.07	
	1800	150 mils		1260	.044		.44	1.27	.27	1.98	3.05	
	1900	Granular surface flashing, 150 mils		1260	.044		.49	1.27	.27	2.03	3.11	
	2000	160 mils		1260	.044		.71	1.27	.27	2.25	3.35	
	2100	APP mod., smooth surf. cap sheet, poly. reinf., torched, 160 mils	G-5	2100	.019		.44	.53	.07	1.04	1.51	
	2150	170 mils		2100	.019		.50	.53	.07	1.10	1.58	
	2200	Granule surface cap sheet, poly. reinf., torched, 180 mils		2000	.020		.55	.56	.07	1.18	1.69	
	2250	Smooth surface flashing, torched, 160 mils		1260	.032		.44	.88	.12	1.44	2.20	
	2300	170 mils		1260	.032		.50	.88	.12	1.50	2.27	
	2350	Granule surface flashing, torched, 180 mils		1260	.032		.55	.88	.12	1.55	2.33	
	2400	Fibrated aluminum coating	1 Rofc	3800	.002		.10	.06		.16	.23	

07580 | Roll Roofing

		07580 Roll Roofing	CREW	DAILY OUTPUT	LABOR-HOURS	UNIT	MAT.	LABOR	EQUIP.	TOTAL	TOTAL INCL O&P	
200	0010	ROLL ROOFING										200
	0100	Asphalt, mineral surface										
	0200	1 ply #15 organic felt, 1 ply mineral surfaced										
	0300	Selvage roofing, lap 19", nailed & mopped	G-1	27	2.074	Sq.	42	59.50	12.50	114	167	
	0400	3 plies glass fiber felt (type IV), 1 ply mineral surfaced										
	0500	Selvage roofing, lapped 19", mopped	G-1	25	2.240	Sq.	65	64	13.50	142.50	201	
	0600	Coated glass fiber base sheet, 2 plies of glass fiber										
	0700	Felt (type IV), 1 ply mineral surfaced selvage										
	0800	Roofing, lapped 19", mopped	G-1	25	2.240	Sq.	71	64	13.50	148.50	208	
	0900	On nailable decks	"	24	2.333	"	66	66.50	14.05	146.55	208	
	1000	3 plies glass fiber felt (type III), 1 ply mineral surfaced										
	1100	Selvage roofing, lapped 19", mopped	G-1	25	2.240	Sq.	65	64	13.50	142.50	201	

07590 | Roof Maintenance and Repairs

		07590 Roof Maintenance and Repairs	CREW	DAILY OUTPUT	LABOR-HOURS	UNIT	MAT.	LABOR	EQUIP.	TOTAL	TOTAL INCL O&P	
300	0010	ROOF COATINGS										300
	0012	Asphalt, brush grade, material only				Gal.	2.95			2.95	3.25	
	0200	Asphalt base, fibered aluminum coating					9.80			9.80	10.80	
	0300	Asphalt primer, 5 gallon					4.21			4.21	4.63	
	0600	Coal tar pitch, 200 lb. barrels				Ton	615			615	680	
	0700	Tar roof cement, 5 gal. lots				Gal.	6.50			6.50	7.15	

 Important: See the Reference Section for supporting data - Crews, Rental Equipment, City Cost Indexes and Reference Data

07500 | Membrane Roofing

07590 | Roof Maintenance and Repairs

			Crew	Daily Output	Labor-Hours	Unit	2006 Bare Costs Mat.	Labor	Equip.	Total	Total Incl O&P	
300	0800	Glass fibered roof & patching cement, 5 gallon				Gal.	4.64			4.64	5.10	300
	0900	Reinforcing glass membrane, 450 S.F./roll				Ea.	45.50			45.50	50	
	1000	Neoprene roof coating, 5 gal, 2 gal/sq				Gal.	22			22	24	
	1100	Roof patch & flashing cement, 5 gallon					19.15			19.15	21	
	1200	Roof resaturant, glass fibered, 3 gal/sq					7.30			7.30	8.05	
	1300	Mineral rubber, 3 gal/sq					4.71			4.71	5.20	

07600 | Flashing and Sheet Metal

07610 | Sheet Metal Roofing

			Crew	Daily Output	Labor-Hours	Unit	2006 Bare Costs Mat.	Labor	Equip.	Total	Total Incl O&P	
300	0010	**COPPER ROOFING**, field fabricated										300
	0012	Batten seam, over 10 sq, 16 oz, 130 lb/sq	1 Shee	1.10	7.273	Sq.	440	305		745	975	
	0200	18 oz, 145 lb per sq		1	8		490	335		825	1,075	
	0300	20 oz, 160 lb per sq		1	8		545	335		880	1,125	
	0400	Standing seam, over 10 squares, 16 oz, 125 lb per sq		1.30	6.154		425	259		684	880	
	0600	18 oz, 140 lb per sq		1.20	6.667		475	281		756	965	
	0700	20 oz, 150 lb per sq		1.10	7.273		510	305		815	1,050	
	0900	Flat seam, over 10 squares, 16 oz, 115 lb per sq		1.20	6.667		390	281		671	875	
	1000	20 oz, 145 lb per sq		1.10	7.273		490	305		795	1,025	
	1200	For abnormal conditions or small areas, add					25%	100%				
	1300	For lead-coated copper, add					25%					
	9000	Minimum labor/equipment charge	1 Shee	2	4	Job		169		169	268	
500	0010	**LEAD ROOFING**, field fabricated										500
	0020	5 lb. per SF, batten seam	1 Shee	1.20	6.667	Sq.	395	281		676	880	
	0100	Flat seam		1.30	6.154	"	395	259		654	850	
	9000	Minimum labor/equipment charge		2	4	Job		169		169	268	
700	0010	**STAINLESS STEEL ROOFING**										700
	0020	Type 304, batten seam, 28 gauge	1 Shee	1.20	6.667	Sq.	400	281		681	885	
	0100	26 gauge	"	1.15	6.957		495	293		788	1,000	
	0200	For standing seam construction, deduct					2%					
	0500	For flat seam construction, deduct					3%					
	0800	For lead or terne coated stainless, 28 gauge, add					91			91	100	
	0900	For 26 gauge, add					121			121	133	
	9000	Minimum labor/equipment charge	1 Shee	2.75	2.909	Job		123		123	195	
900	0010	**ZINC / COPPER ALLOY ROOFING**, field fabricated										900
	0012	Batten seam, .020 thick	1 Shee	1.20	6.667	Sq.	560	281		841	1,075	
	0100	.027" thick		1.15	6.957		680	293		973	1,200	
	0300	.032" thick		1.10	7.273		765	305		1,070	1,325	
	0400	.040" thick		1.05	7.619		900	320		1,220	1,500	
	0600	For standing seam construction, deduct					2%					
	0700	For flat seam construction, deduct					3%					
	9000	Minimum labor/equipment charge	1 Shee	2.75	2.909	Job		123		123	195	

07620 | Sheet Metal Flashing and Trim

			Crew	Daily Output	Labor-Hours	Unit	Mat.	Labor	Equip.	Total	Total Incl O&P	
100	0010	**SHEET METAL CLADDING**										100
	0100	Aluminum, up to 6 bends, .032" thick, window casing	1 Carp	180	.044	S.F.	.71	1.58		2.29	3.40	
	0200	Window sill		72	.111	L.F.	.71	3.95		4.66	7.35	
	0300	Door casing		180	.044	S.F.	.71	1.58		2.29	3.40	
	0400	Fascia		250	.032		.71	1.14		1.85	2.66	
	0500	Rake trim		225	.036		.71	1.26		1.97	2.87	

	07620 | Sheet Metal Flashing and Trim		Crew	Daily Output	Labor-Hours	Unit	2006 Bare Costs Mat.	Labor	Equip.	Total	Total Incl O&P	
100	0700	.024″ thick, window casing	1 Carp	180	.044	S.F.	1.30	1.58		2.88	4.05	100
	0800	Window sill		72	.111	L.F.	1.30	3.95		5.25	8	
	0900	Door casing		180	.044	S.F.	1.30	1.58		2.88	4.05	
	1000	Fascia		250	.032		1.30	1.14		2.44	3.31	
	1100	Rake trim		225	.036		1.30	1.26		2.56	3.52	
	1200	Vinyl coated aluminum, up to 6 bends, window casing		180	.044		.82	1.58		2.40	3.52	
	1300	Window sill		72	.111	L.F.	.82	3.95		4.77	7.45	
	1400	Door casing		180	.044	S.F.	.82	1.58		2.40	3.52	
	1500	Fascia		250	.032		.82	1.14		1.96	2.78	
	1600	Rake trim		225	.036		.82	1.26		2.08	2.99	
	07650 | Flexible Flashing											
600	0010	**FLASHING**, including up to 4 bends										600
	0020	Aluminum, mill finish, .013″ thick	1 Rofc	145	.055	S.F.	.38	1.69		2.07	3.45	
	0030	.016″ thick		145	.055		.56	1.69		2.25	3.65	
	0060	.019″ thick		145	.055		.71	1.69		2.40	3.81	
	0100	.032″ thick		145	.055		1.17	1.69		2.86	4.32	
	0200	.040″ thick		145	.055		1.58	1.69		3.27	4.77	
	0300	.050″ thick		145	.055		2.01	1.69		3.70	5.25	
	0325	Mill finish 5″ x 7″ step flashing, .016″ thick		1920	.004	Ea.	.12	.13		.25	.36	
	0350	Mill finish 12″ x 12″ step flashing, .016″ thick		1600	.005	″	.48	.15		.63	.81	
	0400	Painted finish, add				S.F.	.26			.26	.29	
	0500	Fabric-backed 2 sides, .004″ thick	1 Rofc	330	.024		1.01	.74		1.75	2.44	
	0700	.005″ thick		330	.024		1.19	.74		1.93	2.64	
	0750	Mastic-backed, self adhesive		460	.017		2.46	.53		2.99	3.67	
	0800	Mastic-coated 2 sides, .004″ thick		330	.024		1.01	.74		1.75	2.44	
	1000	.005″ thick		330	.024		1.19	.74		1.93	2.64	
	1100	.016″ thick		330	.024		1.31	.74		2.05	2.77	
	1300	Asphalt flashing cement, 5 gallon				Gal.	4.77			4.77	5.25	
	1600	Copper, 16 oz, sheets, under 1000 lbs.	1 Rofc	115	.070	S.F.	2.76	2.13		4.89	6.85	
	1700	Over 4000 lbs.		155	.052		3.06	1.58		4.64	6.20	
	1900	20 oz sheets, under 1000 lbs.		110	.073		4.10	2.23		6.33	8.50	
	2000	Over 4000 lbs.		145	.055		3.80	1.69		5.49	7.20	
	2200	24 oz sheets, under 1000 lbs.		105	.076		4.92	2.33		7.25	9.60	
	2300	Over 4000 lbs.		135	.059		4.56	1.81		6.37	8.25	
	2500	32 oz sheets, under 1000 lbs.		100	.080		6.55	2.45		9	11.60	
	2600	Over 4000 lbs.		130	.062		6.10	1.88		7.98	10.10	
	2700	W shape for valleys, 16 oz, 24″ wide		100	.080	L.F.	6.20	2.45		8.65	11.20	
	2800	Copper, paperbacked 1 side, 2 oz		330	.024	S.F.	.90	.74		1.64	2.32	
	2900	3 oz		330	.024		1.18	.74		1.92	2.63	
	3100	Paperbacked 2 sides, 2 oz		330	.024		.91	.74		1.65	2.33	
	3150	3 oz		330	.024		1.17	.74		1.91	2.62	
	3200	5 oz		330	.024		1.75	.74		2.49	3.26	
	3250	7 oz		330	.024		2.86	.74		3.60	4.48	
	3400	Mastic-backed 2 sides, copper, 2 oz		330	.024		1.07	.74		1.81	2.51	
	3500	3 oz		330	.024		1.32	.74		2.06	2.78	
	3700	5 oz		330	.024		1.94	.74		2.68	3.46	
	3800	Fabric-backed 2 sides, copper, 2 oz		330	.024		1.14	.74		1.88	2.58	
	4000	3 oz		330	.024		1.47	.74		2.21	2.95	
	4100	5 oz		330	.024		2	.74		2.74	3.53	
	4300	Copper-clad stainless steel, .015″ thick, under 500 lbs.		115	.070		3.32	2.13		5.45	7.50	
	4400	Over 2000 lbs.		155	.052		3.20	1.58		4.78	6.35	
	4600	.018″ thick, under 500 lbs.		100	.080		4.39	2.45		6.84	9.25	
	4700	Over 2000 lbs.		145	.055		3.21	1.69		4.90	6.55	
	4900	Fabric, asphalt-saturated cotton, specification grade		35	.229	S.Y.	1.93	7		8.93	14.65	
	5000	Utility grade		35	.229		1.22	7		8.22	13.90	

07600 | Flashing and Sheet Metal

	07650 | Flexible Flashing	CREW	DAILY OUTPUT	LABOR-HOURS	UNIT	2006 BARE COSTS MAT.	LABOR	EQUIP.	TOTAL	TOTAL INCL O&P	
600	5200 Open-mesh fabric, saturated, 40 oz per S.Y.	1 Rofc	35	.229	S.Y.	1.35	7		8.35	14.05	600
	5300 Close-mesh fabric, saturated, 17 oz per S.Y.		35	.229		1.42	7		8.42	14.10	
	5500 Fiberglass, resin-coated		35	.229		1.16	7		8.16	13.85	
	5600 Asphalt-coated, 40 oz per S.Y.		35	.229		7.90	7		14.90	21.50	
	5800 Lead, 2.5 lb. per SF, up to 12" wide		135	.059	S.F.	3.03	1.81		4.84	6.60	
	5900 Over 12" wide		135	.059		3.69	1.81		5.50	7.30	
	6100 Lead-coated copper, fabric-backed, 2 oz		330	.024		1.54	.74		2.28	3.02	
	6200 5 oz		330	.024		1.77	.74		2.51	3.28	
	6400 Mastic-backed 2 sides, 2 oz		330	.024		1.20	.74		1.94	2.65	
	6500 5 oz		330	.024		1.49	.74		2.23	2.97	
	6700 Paperbacked 1 side, 2 oz		330	.024		1.04	.74		1.78	2.47	
	6800 3 oz		330	.024		1.22	.74		1.96	2.67	
	7000 Paperbacked 2 sides, 2 oz		330	.024		1.07	.74		1.81	2.51	
	7100 5 oz		330	.024		1.75	.74		2.49	3.26	
	7300 Polyvinyl chloride, black, .010" thick		285	.028		.17	.86		1.03	1.73	
	7400 .020" thick		285	.028		.24	.86		1.10	1.80	
	7600 .030" thick		285	.028		.32	.86		1.18	1.89	
	7700 .056" thick		285	.028		.76	.86		1.62	2.38	
	7900 Black or white for exposed roofs, .060" thick		285	.028		1.66	.86		2.52	3.37	
	8060 PVC tape, 5" x 45 mils, for joint covers, 100 L.F./roll				Ea.	85			85	93.50	
	8100 Rubber, butyl, 1/32" thick	1 Rofc	285	.028	S.F.	.75	.86		1.61	2.37	
	8200 1/16" thick		285	.028		1.12	.86		1.98	2.77	
	8300 Neoprene, cured, 1/16" thick		285	.028		1.58	.86		2.44	3.28	
	8400 1/8" thick		285	.028		3.20	.86		4.06	5.05	
	8500 Shower pan, bituminous membrane, 7 oz		155	.052		1.16	1.58		2.74	4.12	
	8550 3 ply copper and fabric, 3 oz		155	.052		1.74	1.58		3.32	4.75	
	8600 7 oz		155	.052		3.61	1.58		5.19	6.80	
	8650 Copper, 16 oz		100	.080		3.33	2.45		5.78	8.05	
	8700 Lead on copper and fabric, 5 oz		155	.052		1.77	1.58		3.35	4.79	
	8800 7 oz		155	.052		3.14	1.58		4.72	6.30	
	8900 Stainless steel sheets, 32 ga, .010" thick		155	.052		2.70	1.58		4.28	5.80	
	9000 28 ga, .015" thick		155	.052		3.35	1.58		4.93	6.55	
	9100 26 ga, .018" thick		155	.052		4.06	1.58		5.64	7.30	
	9200 24 ga, .025" thick		155	.052		5.30	1.58		6.88	8.65	
	9290 For mechanically keyed flashing, add					40%					
	9300 Stainless steel, paperbacked 2 sides, .005" thick	1 Rofc	330	.024	S.F.	2.38	.74		3.12	3.95	
	9320 Steel sheets, galvanized, 20 gauge		130	.062		.91	1.88		2.79	4.38	
	9340 30 gauge		160	.050		.39	1.53		1.92	3.18	
	9400 Terne coated stainless steel, .015" thick, 28 ga		155	.052		5.05	1.58		6.63	8.40	
	9500 .018" thick, 26 ga		155	.052		5.70	1.58		7.28	9.15	
	9600 Zinc and copper alloy (brass), .020" thick		155	.052		3.55	1.58		5.13	6.75	
	9700 .027" thick		155	.052		4.76	1.58		6.34	8.10	
	9800 .032" thick		155	.052		5.55	1.58		7.13	8.95	
	9900 .040" thick		155	.052		6.80	1.58		8.38	10.30	
	9950 Minimum labor/equipment charge		3	2.667	Job		81.50		81.50	147	

THERMAL & MOISTURE PROTECTION 7

7 THERMAL & MOISTURE PROTECTION

	07710 | Manufactured Roof Specialties		CREW	DAILY OUTPUT	LABOR-HOURS	UNIT	2006 BARE COSTS MAT.	LABOR	EQUIP.	TOTAL	TOTAL INCL O&P	
400	0010	DOWNSPOUTS										400
	0020	Aluminum 2" x 3", .020" thick, embossed	1 Shee	190	.042	L.F.	.85	1.77		2.62	3.76	
	0100	Enameled		190	.042		1.26	1.77		3.03	4.21	
	0300	Enameled, .024" thick, 2" x 3"		180	.044		1.77	1.87		3.64	4.93	
	0400	3" x 4"		140	.057		2.50	2.41		4.91	6.60	
	0600	Round, corrugated aluminum, 3" diameter, .020" thick		190	.042		1.36	1.77		3.13	4.32	
	0700	4" diameter, .025" thick		140	.057		2.03	2.41		4.44	6.05	
	0900	Wire strainer, round, 2" diameter		155	.052	Ea.	1.85	2.18		4.03	5.50	
	1000	4" diameter		155	.052		2	2.18		4.18	5.65	
	1200	Rectangular, perforated, 2" x 3"		145	.055		2.25	2.33		4.58	6.20	
	1300	3" x 4"		145	.055		3.25	2.33		5.58	7.30	
	1500	Copper, round, 16 oz., stock, 2" diameter		190	.042	L.F.	4.55	1.77		6.32	7.80	
	1600	3" diameter		190	.042		4.40	1.77		6.17	7.65	
	1800	4" diameter		145	.055		5.20	2.33		7.53	9.40	
	1900	5" diameter		130	.062		7.70	2.59		10.29	12.60	
	2100	Rectangular, corrugated copper, stock, 2" x 3"		190	.042		3.88	1.77		5.65	7.10	
	2200	3" x 4"		145	.055		4.60	2.33		6.93	8.75	
	2400	Rectangular, plain copper, stock, 2" x 3"		190	.042		5.25	1.77		7.02	8.60	
	2500	3" x 4"		145	.055		5.95	2.33		8.28	10.25	
	2700	Wire strainers, rectangular, 2" x 3"		145	.055	Ea.	2.85	2.33		5.18	6.85	
	2800	3" x 4"		145	.055		4.45	2.33		6.78	8.60	
	3000	Round, 2" diameter		145	.055		2.65	2.33		4.98	6.60	
	3100	3" diameter		145	.055		3.70	2.33		6.03	7.75	
	3300	4" diameter		145	.055		5.75	2.33		8.08	10.05	
	3400	5" diameter		115	.070		8.30	2.93		11.23	13.80	
	3600	Lead-coated copper, round, stock, 2" diameter		190	.042	L.F.	5.50	1.77		7.27	8.85	
	3700	3" diameter		190	.042		7.35	1.77		9.12	10.90	
	3900	4" diameter		145	.055		9.70	2.33		12.03	14.40	
	4000	5" diameter, corrugated		130	.062		13.80	2.59		16.39	19.30	
	4200	6" diameter, corrugated		105	.076		14.85	3.21		18.06	21.50	
	4300	Rectangular, corrugated, stock, 2" x 3"		190	.042		7.80	1.77		9.57	11.40	
	4500	Plain, stock, 2" x 3"		190	.042		7.35	1.77		9.12	10.90	
	4600	3" x 4"		145	.055		7.95	2.33		10.28	12.45	
	4800	Steel, galvanized, round, corrugated, 2" or 3" diam, 28 ga		190	.042		1.34	1.77		3.11	4.29	
	4900	4" diameter, 28 gauge		145	.055		1.40	2.33		3.73	5.25	
	5100	5" diameter, 28 gauge		130	.062		2.34	2.59		4.93	6.70	
	5200	26 gauge		130	.062		2.35	2.59		4.94	6.70	
	5400	6" diameter, 28 gauge		105	.076		3.32	3.21		6.53	8.75	
	5500	26 gauge		105	.076		3.25	3.21		6.46	8.70	
	5700	Rectangular, corrugated, 28 gauge, 2" x 3"		190	.042		1.26	1.77		3.03	4.21	
	5800	3" x 4"		145	.055		1.26	2.33		3.59	5.10	
	6000	Rectangular, plain, 28 gauge, galvanized, 2" x 3"		190	.042		1.78	1.77		3.55	4.78	
	6100	3" x 4"		145	.055		1.58	2.33		3.91	5.45	
	6300	Epoxy painted, 24 gauge, corrugated, 2" x 3"		190	.042		1.55	1.77		3.32	4.53	
	6400	3" x 4"		145	.055		1.78	2.33		4.11	5.65	
	6600	Wire strainers, rectangular, 2" x 3"		145	.055	Ea.	1.57	2.33		3.90	5.45	
	6700	3" x 4"		145	.055		3.13	2.33		5.46	7.15	
	6900	Round strainers, 2" or 3" diameter		145	.055		1.57	2.33		3.90	5.45	
	7000	4" diameter		145	.055		3.13	2.33		5.46	7.15	
	7200	5" diameter		145	.055		2.20	2.33		4.53	6.10	
	7300	6" diameter		115	.070		2.63	2.93		5.56	7.55	
	9000	Minimum labor/equipment charge		4	2	Job		84.50		84.50	134	
450	0010	DRIP EDGE										450
	0020	Aluminum, .016" thick, 5" wide, mill finish	1 Carp	400	.020	L.F.	.30	.71		1.01	1.51	
	0100	White finish		400	.020		.35	.71		1.06	1.57	
	0200	8" wide, mill finish		400	.020		.42	.71		1.13	1.64	

07700 | Roof Specialties and Accessories

	07710 | Manufactured Roof Specialties	CREW	DAILY OUTPUT	LABOR-HOURS	UNIT	2006 BARE COSTS MAT.	LABOR	EQUIP.	TOTAL	TOTAL INCL O&P	
450	0300 Ice belt, 28″ wide, mill finish	1 Carp	100	.080	L.F.	3.76	2.84		6.60	8.85	450
	0310 Vented, mill finish		400	.020		1.32	.71		2.03	2.63	
	0320 Painted finish		400	.020		1.43	.71		2.14	2.75	
	0400 Galvanized, 5″ wide		400	.020		.40	.71		1.11	1.62	
	0500 8″ wide, mill finish		400	.020		.49	.71		1.20	1.72	
	0510 Rake edge, aluminum, 1-1/2″ x 1-1/2″		400	.020		.15	.71		.86	1.35	
	0520 3-1/2″ x 1-1/2″		400	.020		.19	.71		.90	1.39	
	9000 Minimum labor/equipment charge		4	2	Job		71		71	118	
500	0010 **DOWNSPOUT ELBOWS**										500
	0020 Aluminum, 2″ x 3″, embossed	1 Shee	100	.080	Ea.	3.30	3.37		6.67	9	
	0100 Enameled		100	.080		3.30	3.37		6.67	9	
	0200 3″ x 4″, .025″ thick, embossed		100	.080		3.30	3.37		6.67	9	
	0300 Enameled		100	.080		3.20	3.37		6.57	8.85	
	0400 Round corrugated, 3″, embossed, .020″ thick		100	.080		1.98	3.37		5.35	7.55	
	0500 4″, .025″ thick		100	.080		4.29	3.37		7.66	10.05	
	0600 Copper, 16 oz. round, 2″ diameter		100	.080		11.15	3.37		14.52	17.65	
	0700 3″ diameter		100	.080		8.15	3.37		11.52	14.35	
	0800 4″ diameter		100	.080		9.50	3.37		12.87	15.80	
	1000 2″ x 3″ corrugated		100	.080		6.15	3.37		9.52	12.10	
	1100 3″ x 4″ corrugated		100	.080		7.20	3.37		10.57	13.30	
	9000 Minimum labor/equipment charge		4	2	Job		84.50		84.50	134	
550	0010 **GRAVEL STOP**										550
	0020 Aluminum, .050″ thick, 4″ face height, mill finish	1 Shee	145	.055	L.F.	4.28	2.33		6.61	8.40	
	0080 Duranodic finish		145	.055		4.13	2.33		6.46	8.25	
	0100 Painted		145	.055		4.77	2.33		7.10	8.95	
	0300 6″ face height		135	.059		4.59	2.50		7.09	9	
	0350 Duranodic finish		135	.059		4.87	2.50		7.37	9.30	
	0400 Painted		135	.059		5.65	2.50		8.15	10.20	
	0600 8″ face height		125	.064		5.75	2.70		8.45	10.60	
	0650 Duranodic finish		125	.064		5.60	2.70		8.30	10.45	
	0700 Painted		125	.064		5.70	2.70		8.40	10.55	
	0900 12″ face height, .080 thick, 2 piece		100	.080		7.65	3.37		11.02	13.75	
	0950 Duranodic finish		100	.080		7.05	3.37		10.42	13.10	
	1000 Painted		100	.080		8.30	3.37		11.67	14.50	
	1500 Polyvinyl chloride, 6″ face height		135	.059		3.67	2.50		6.17	8	
	1600 9″ face height		125	.064		4.33	2.70		7.03	9.05	
	1800 Stainless steel, 24 ga., 6″ face height		135	.059		8	2.50		10.50	12.75	
	1900 12″ face height		100	.080		16.80	3.37		20.17	24	
	2100 20 ga., 6″ face height		135	.059		9.05	2.50		11.55	13.95	
	2200 12″ face height		100	.080		19.15	3.37		22.52	26.50	
	9000 Minimum labor/equipment charge		3.50	2.286	Job		96.50		96.50	153	
650	0010 **GUTTERS**										650
	0012 Aluminum, stock units, 5″ K type, .027″ thick, plain	1 Shee	120	.067	L.F.	1.33	2.81		4.14	5.95	
	0100 Enameled		120	.067		1.25	2.81		4.06	5.85	
	0300 5″ K type type, .032″ thick, plain		120	.067		2.35	2.81		5.16	7.05	
	0400 Enameled		120	.067		2.36	2.81		5.17	7.05	
	0700 Copper, half round, 16 oz, stock units, 4″ wide		120	.067		4.10	2.81		6.91	9	
	0900 5″ wide		120	.067		5.25	2.81		8.06	10.20	
	1000 6″ wide		115	.070		5.70	2.93		8.63	10.90	
	1200 K type, 16 oz, stock, 4″ wide		120	.067		5.05	2.81		7.86	10	
	1300 5″ wide		120	.067		5.50	2.81		8.31	10.50	
	1500 Lead coated copper, half round, stock, 4″ wide		120	.067		7.50	2.81		10.31	12.70	
	1600 6″ wide		115	.070		11.40	2.93		14.33	17.15	
	1800 K type, stock, 4″ wide		120	.067		8.10	2.81		10.91	13.35	
	1900 5″ wide		120	.067		8.90	2.81		11.71	14.20	

	07710 Manufactured Roof Specialties		CREW	DAILY OUTPUT	LABOR-HOURS	UNIT	2006 BARE COSTS MAT.	LABOR	EQUIP.	TOTAL	TOTAL INCL O&P	
650	2100	Stainless steel, half round or box, stock, 4″ wide	1 Shee	120	.067	L.F.	4.65	2.81		7.46	9.55	650
	2200	5″ wide		120	.067		5	2.81		7.81	9.95	
	2400	Steel, galv, half round or box, 28 ga, 5″ wide, plain		120	.067		1.18	2.81		3.99	5.75	
	2500	Enameled		120	.067		1.12	2.81		3.93	5.70	
	2700	26 ga, stock, 5″ wide		120	.067		.95	2.81		3.76	5.50	
	2800	6″ wide	↓	120	.067		1.60	2.81		4.41	6.25	
	3000	Vinyl, O.G., 4″ wide	1 Carp	110	.073		.85	2.59		3.44	5.20	
	3100	5″ wide		110	.073		1	2.59		3.59	5.40	
	3200	4″ half round, stock units	↓	110	.073	↓	.68	2.59		3.27	5.05	
	3250	Joint connectors				Ea.	1.36			1.36	1.50	
	3300	Wood, clear treated cedar, fir or hemlock, 3″ x 4″	1 Carp	100	.080	L.F.	6.30	2.84		9.14	11.60	
	3400	4″ x 5″	″	100	.080	″	7.30	2.84		10.14	12.70	
	9000	Minimum labor/equipment charge	1 Shee	3.75	2.133	Job		90		90	143	
700	0010	**GUTTER GUARD**										700
	0020	6″ wide strip, aluminum mesh	1 Carp	500	.016	L.F.	.37	.57		.94	1.35	
	0100	Vinyl mesh		500	.016	″	.40	.57		.97	1.38	
	9000	Minimum labor/equipment charge	↓	4	2	Job		71		71	118	
750	0010	**REGLET**										750
	0020	Aluminum, .025″ thick, in concrete parapet	1 Carp	225	.036	L.F.	1.02	1.26		2.28	3.21	
	0100	Copper, 10 oz.		225	.036		1.77	1.26		3.03	4.04	
	0300	16 oz.		225	.036		2.35	1.26		3.61	4.68	
	0400	Galvanized steel, 24 gauge		225	.036		.75	1.26		2.01	2.92	
	0600	Stainless steel, .020″ thick		225	.036		1.74	1.26		3	4	
	0700	Zinc and copper alloy, 20 oz.	↓	225	.036		1.95	1.26		3.21	4.24	
	0900	Counter flashing for above, 12″ wide, .032″ aluminum	1 Shee	150	.053		1.32	2.25		3.57	5.05	
	1000	Copper, 10 oz.		150	.053		3.70	2.25		5.95	7.65	
	1200	16 oz.		150	.053		4.11	2.25		6.36	8.10	
	1300	Galvanized steel, .020″ thick		150	.053		.69	2.25		2.94	4.34	
	1500	Stainless steel, .020″ thick		150	.053		3.07	2.25		5.32	6.95	
	1600	Zinc and copper alloy, 20 oz.	↓	150	.053	↓	3.46	2.25		5.71	7.40	
	9000	Minimum labor/equipment charge	1 Carp	3	2.667	Job		95		95	157	
800	0010	**EXPANSION JOINT**										800
	0300	Butyl or neoprene center with foam insulation, metal flanges										
	0400	Aluminum, .032″ thick for openings to 2-1/2″	1 Rofc	165	.048	L.F.	8.15	1.48		9.63	11.60	
	0600	For joint openings to 3-1/2″		165	.048		9.55	1.48		11.03	13.15	
	0610	For joint openings to 5″		165	.048		11.45	1.48		12.93	15.25	
	0620	For joint openings to 8″		165	.048		18.05	1.48		19.53	22.50	
	0700	Copper, 16 oz. for openings to 2-1/2″		165	.048		11.45	1.48		12.93	15.25	
	0900	For joint openings to 3-1/2″		165	.048		13.15	1.48		14.63	17.15	
	0910	For joint openings to 5″		165	.048		15.40	1.48		16.88	19.60	
	0920	For joint openings to 8″		165	.048		23	1.48		24.48	28	
	1000	Galvanized steel, 26 ga. for openings to 2-1/2″		165	.048		6.90	1.48		8.38	10.25	
	1200	For joint openings to 3-1/2″		165	.048		8.15	1.48		9.63	11.60	
	1210	For joint openings to 5″		165	.048		10.20	1.48		11.68	13.85	
	1220	For joint openings to 8″		165	.048		17.35	1.48		18.83	22	
	1300	Lead-coated copper, 16 oz. for openings to 2-1/2″		165	.048		20.50	1.48		21.98	25	
	1500	For joint openings to 3-1/2″		165	.048		24	1.48		25.48	29	
	1600	Stainless steel, .018″, for openings to 2-1/2″		165	.048		10.20	1.48		11.68	13.85	
	1800	For joint openings to 3-1/2″		165	.048		11.60	1.48		13.08	15.45	
	1810	For joint openings to 5″		165	.048		14.35	1.48		15.83	18.45	
	1820	For joint openings to 8″		165	.048		22	1.48		23.48	26.50	
	1900	Neoprene, double-seal type with thick center, 4-1/2″ wide		125	.064		9.80	1.96		11.76	14.30	
	1950	Polyethylene bellows, with galv steel flat flanges		100	.080		3.80	2.45		6.25	8.60	
	1960	With galvanized angle flanges	↓	100	.080		4.19	2.45		6.64	9	
	2000	Roof joint with extruded aluminum cover, 2″	1 Shee	115	.070	↓	26	2.93		28.93	33.50	

7 THERMAL & MOISTURE PROTECTION

		07710 | Manufactured Roof Specialties	CREW	DAILY OUTPUT	LABOR-HOURS	UNIT	2006 BARE COSTS MAT.	LABOR	EQUIP.	TOTAL	TOTAL INCL O&P	
800	2100	Roof joint, plastic curbs, foam center, standard	1 Rofc	100	.080	L.F.	9.60	2.45		12.05	14.95	800
	2200	Large		100	.080		12.80	2.45		15.25	18.45	
	2300	Transitions, regular, minimum		10	.800	Ea.	82.50	24.50		107	135	
	2350	Maximum		4	2		105	61		166	226	
	2400	Large, minimum		9	.889		121	27		148	182	
	2450	Maximum		3	2.667		127	81.50		208.50	286	
	2500	Roof to wall joint with extruded aluminum cover	1 Shee	115	.070	L.F.	22.50	2.93		25.43	29	
	2650											
	2700	Wall joint, closed cell foam on PVC cover, 9″ wide	1 Rofc	125	.064	L.F.	3.41	1.96		5.37	7.25	
	2800	12″ wide	″	115	.070	″	3.86	2.13		5.99	8.10	
	9000	Minimum labor/equipment charge	1 Shee	3	2.667	Job		112		112	179	
		07720 | Roof Accessories										
480	0010	**PITCH POCKETS**										480
	0100	Adjustable, 4″ to 7″, welded corners, 4″ deep	1 Rofc	48	.167	Ea.	10.75	5.10		15.85	21	
	0200	Side extenders, 6″	″	240	.033	″	1.80	1.02		2.82	3.81	
500	0010	**ROOF VENTS**										500
	0020	Mushroom shape, for built-up roofs, aluminum	1 Rofc	30	.267	Ea.	24	8.15		32.15	41	
	0100	PVC, 6″ high		30	.267	″	27.50	8.15		35.65	45	
	9000	Minimum labor/equipment charge		2.75	2.909	Job		89		89	160	
550	0010	**RIDGE VENT**										550
	0100	Aluminum strips, mill finish	1 Rofc	160	.050	L.F.	1.21	1.53		2.74	4.08	
	0150	Painted finish		160	.050	″	2.08	1.53		3.61	5.05	
	0200	Connectors		48	.167	Ea.	1.99	5.10		7.09	11.35	
	0300	End caps		48	.167	″	.96	5.10		6.06	10.20	
	0400	Galvanized strips		160	.050	L.F.	2.07	1.53		3.60	5.05	
	0430	Molded polyethylene, shingles not included		160	.050	″	2.70	1.53		4.23	5.70	
	0440	End plugs		48	.167	Ea.	.96	5.10		6.06	10.20	
	0450	Flexible roll, shingles not included		160	.050	L.F.	1.99	1.53		3.52	4.93	
560	0010	**SNOW GUARDS**										560
	0100	Slate & asphalt shingle roofs	1 Rofc	160	.050	Ea.	7.75	1.53		9.28	11.25	
	0200	Standing seam metal roofs		48	.167		12.25	5.10		17.35	22.50	
	0300	Surface mount for metal roofs		48	.167		6.75	5.10		11.85	16.60	
	0400	Double rail pipe type, including pipe		130	.062	L.F.	17.75	1.88		19.63	23	
700	0010	**ROOF HATCHES** With curb, 1″ fiberglass insulation, 2′-6″ x 3′-0″										700
	0500	Aluminum curb and cover	G-3	10	3.200	Ea.	550	111		661	785	
	0520	Galvanized steel curb and aluminum cover		10	3.200		380	111		491	600	
	0540	Galvanized steel curb and cover		10	3.200		445	111		556	670	
	0600	2′-6″ x 4′-6″, aluminum curb and cover		9	3.556		650	124		774	915	
	0800	Galvanized steel curb and aluminum cover		9	3.556		540	124		664	795	
	0900	Galvanized steel curb and cover		9	3.556		685	124		809	950	
	1100	4′ x 4′ aluminum curb and cover		8	4		670	139		809	960	
	1120	Galvanized steel curb and aluminum cover		8	4		540	139		679	820	
	1140	Galvanized steel curb and cover		8	4		625	139		764	915	
	1200	2′-6″ x 8′-0″, aluminum curb and cover		6.60	4.848		1,375	169		1,544	1,775	
	1400	Galvanized steel curb and aluminum cover		6.60	4.848		1,050	169		1,219	1,425	
	1500	Galvanized steel curb and cover		6.60	4.848		940	169		1,109	1,300	
	1800	For plexiglass panels, 2′-6″ x 3′-0″, add to above					375			375	410	
	9000	Minimum labor/equipment charge	2 Carp	2	8	Job		284		284	470	
800	0010	**ROOF WALKWAY**										800
	0020	Asphalt impregnated, 3′ x 6′ x 1/2″ thick	1 Rofc	400	.020	S.F.	1.20	.61		1.81	2.42	

07700 | Roof Specialties and Accessories

	07720 | Roof Accessories		CREW	DAILY OUTPUT	LABOR-HOURS	UNIT	2006 BARE COSTS MAT.	LABOR	EQUIP.	TOTAL	TOTAL INCL O&P	
800	0100	3′ x 3′ x 3/4″ thick	1 Rofc	400	.020	S.F.	2.30	.61		2.91	3.63	800
	0600	100% recycled rubber, 3′ x 4′ x 3/8″		400	.020	L.F.	4.24	.61		4.85	5.75	
	0610	3′ x 4′ x 1/2″		400	.020		4.59	.61		5.20	6.15	
	0620	3′ x 4′ x 3/4″		400	.020		5.75	.61		6.36	7.45	
	9000	Minimum labor/equipment charge		2.75	2.909	Job		89		89	160	
850	0010	**SMOKE HATCHES** Unlabeled, not including hand winch operator										850
	0200	For 3′-0″ long, add to roof hatches from Division 07720-700				Ea.	25%	5%				
	0250	For 4′-0″ long, add to roof hatches from Division 07720-700					20%	5%				
	0300	For 8′-0″ long, add to roof hatches from Division 07720-700					10%	5%				
860	0010	**SMOKE VENT,** insulated, 4′ x 4′										860
	0100	Aluminum cover and frame	G-3	13	2.462	Ea.	1,150	85.50		1,235.50	1,400	
	0200	Galvanized steel cover and frame		13	2.462		1,050	85.50		1,135.50	1,300	
	0300	4′ x 8′ aluminum cover and frame		8	4		1,550	139		1,689	1,925	
	0400	Galvanized steel cover and frame		8	4		1,350	139		1,489	1,725	
	9000	Minimum labor/equipment charge	2 Carp	2	8	Job		284		284	470	
865	0010	**VENTS**										865
	0100	Soffit or eave, aluminum, mill finish, strips, 2-1/2″ wide	1 Carp	200	.040	L.F.	.33	1.42		1.75	2.72	
	0200	3″ wide		200	.040		.34	1.42		1.76	2.73	
	0300	Enamel finish, 3″ wide		200	.040		.47	1.42		1.89	2.88	
	0400	Mill finish, rectangular, 4″ x 16″		72	.111	Ea.	1.47	3.95		5.42	8.15	
	0500	8″ x 16″		72	.111	″	1.76	3.95		5.71	8.50	
870	0010	**VENTS, ONE-WAY**										870
	0020	Plastic, for insulated decks, 1 per M.S.F., minimum	1 Rofc	40	.200	Ea.	12.90	6.10		19	25	
	0100	Maximum		20	.400		29.50	12.25		41.75	54.50	
	0300	Aluminum		30	.267		12.90	8.15		21.05	29	
	0800	Polystyrene baffles, 12″ wide for 16″ O.C. rafter spacing	1 Carp	90	.089		.31	3.16		3.47	5.60	
	0900	For 24″ O.C. rafter spacing		110	.073		.52	2.59		3.11	4.85	
	9000	Minimum labor/equipment charge		3	2.667	Job		95		95	157	

07800 | Fire and Smoke Protection

	07812 | Cementitious Fireproofing		CREW	DAILY OUTPUT	LABOR-HOURS	UNIT	2006 BARE COSTS MAT.	LABOR	EQUIP.	TOTAL	TOTAL INCL O&P	
600	0010	**SPRAYED** Mineral fiber or cementitious for fireproofing,										600
	0050	not incl tamping or canvas protection										
	0100	1″ thick, on flat plate steel	G-2	3000	.008	S.F.	.43	.23	.04	.70	.89	
	0200	Flat decking		2400	.010		.43	.29	.05	.77	1	
	0400	Beams		1500	.016		.43	.47	.08	.98	1.32	
	0500	Corrugated or fluted decks		1250	.019		.64	.56	.09	1.29	1.71	
	0700	Columns, 1-1/8″ thick		1100	.022		.48	.64	.11	1.23	1.69	
	0800	2-3/16″ thick		700	.034		.91	1	.17	2.08	2.81	
	0850	For tamping, add						10%				
	0900	For canvas protection, add	G-2	5000	.005	S.F.	.06	.14	.02	.22	.33	
	9000	Minimum labor/equipment charge	″	3	8	Job		234	39	273	425	

	07840 | Firestopping		CREW	DAILY OUTPUT	LABOR-HOURS	UNIT	MAT.	LABOR	EQUIP.	TOTAL	TOTAL INCL O&P	
100	0010	**FIRESTOPPING** R078413-30										100
	0100	Metallic piping, non insulated										

07800 | Fire and Smoke Protection

07840 | Firestopping		CREW	DAILY OUTPUT	LABOR-HOURS	UNIT	2006 BARE COSTS MAT.	LABOR	EQUIP.	TOTAL	TOTAL INCL O&P
100 0110	Through walls, 2" diameter (R078413-30)	1 Carp	16	.500	Ea.	9.95	17.80		27.75	40.50
0120	4" diameter		14	.571		15.20	20.50		35.70	50.50
0130	6" diameter		12	.667		20.50	23.50		44	62
0140	12" diameter		10	.800		36.50	28.50		65	87
0150	Through floors, 2" diameter		32	.250		6.05	8.90		14.95	21.50
0160	4" diameter		28	.286		8.70	10.15		18.85	26.50
0170	6" diameter		24	.333		11.40	11.85		23.25	32
0180	12" diameter		20	.400		19.25	14.20		33.45	44.50
0190	Metallic piping, insulated									
0200	Through walls, 2" diameter	1 Carp	16	.500	Ea.	14.15	17.80		31.95	45
0210	4" diameter		14	.571		19.40	20.50		39.90	55
0220	6" diameter		12	.667		24.50	23.50		48	66.50
0230	12" diameter		10	.800		40.50	28.50		69	91.50
0240	Through floors, 2" diameter		32	.250		10.20	8.90		19.10	26
0250	4" diameter		28	.286		12.85	10.15		23	31
0260	6" diameter		24	.333		15.60	11.85		27.45	37
0270	12" diameter		20	.400		19.25	14.20		33.45	44.50
0280	Non metallic piping, non insulated									
0290	Through walls, 2" diameter	1 Carp	12	.667	Ea.	41	23.50		64.50	85
0300	4" diameter		10	.800		51.50	28.50		80	104
0310	6" diameter		8	1		72	35.50		107.50	138
0330	Through floors, 2" diameter		16	.500		32	17.80		49.80	65
0340	4" diameter		6	1.333		40	47.50		87.50	123
0350	6" diameter		6	1.333		48	47.50		95.50	131
0370	Ductwork, insulated & non insulated, round									
0380	Through walls, 6" diameter	1 Carp	12	.667	Ea.	21	23.50		44.50	62.50
0390	12" diameter		10	.800		41.50	28.50		70	93
0400	18" diameter		8	1		67.50	35.50		103	134
0410	Through floors, 6" diameter		16	.500		11.45	17.80		29.25	42
0420	12" diameter		14	.571		21	20.50		41.50	56.50
0430	18" diameter		12	.667		36.50	23.50		60	79.50
0440	Ductwork, insulated & non insulated, rectangular									
0450	With stiffener/closure angle, through walls, 6" x 12"	1 Carp	8	1	Ea.	17.35	35.50		52.85	78
0460	12" x 24"		6	1.333		23	47.50		70.50	104
0470	24" x 48"		4	2		65.50	71		136.50	190
0480	With stiffener/closure angle, through floors, 6" x 12"		10	.800		9.35	28.50		37.85	57.50
0490	12" x 24"		8	1		16.85	35.50		52.35	77.50
0500	24" x 48"		6	1.333		33	47.50		80.50	115
0510	Multi trade openings									
0520	Through walls, 6" x 12"	1 Carp	2	4	Ea.	36.50	142		178.50	276
0530	12" x 24"	"	1	8		147	284		431	630
0540	24" x 48"	2 Carp	1	16		585	570		1,155	1,575
0550	48" x 96"	"	.75	21.333		2,350	760		3,110	3,850
0560	Through floors, 6" x 12"	1 Carp	2	4		36.50	142		178.50	276
0570	12" x 24"	"	1	8		147	284		431	630
0580	24" x 48"	2 Carp	.75	21.333		585	760		1,345	1,900
0590	48" x 96"	"	.50	32		2,350	1,150		3,500	4,475
0600	Structural penetrations, through walls									
0610	Steel beams, W8 x 10	1 Carp	8	1	Ea.	23	35.50		58.50	84
0620	W12 x 14		6	1.333		36.50	47.50		84	119
0630	W21 x 44		5	1.600		73	57		130	174
0640	W36 x 135		3	2.667		177	95		272	350
0650	Bar joists, 18" deep		6	1.333		33.50	47.50		81	115
0660	24" deep		6	1.333		41.50	47.50		89	125
0670	36" deep		5	1.600		62.50	57		119.50	163
0680	48" deep		4	2		73	71		144	198

THERMAL & MOISTURE PROTECTION 7

07800 | Fire and Smoke Protection

07840 | Firestopping

			CREW	DAILY OUTPUT	LABOR-HOURS	UNIT	2006 BARE COSTS MAT.	LABOR	EQUIP.	TOTAL	TOTAL INCL O&P	
100	0690	Construction joints, floor slab at exterior wall R078413-30										100
	0700	Precast, brick, block or drywall exterior										
	0710	2" wide joint	1 Carp	125	.064	L.F.	5.20	2.28		7.48	9.45	
	0720	4" wide joint	"	75	.107	"	10.40	3.79		14.19	17.75	
	0730	Metal panel, glass or curtain wall exterior										
	0740	2" wide joint	1 Carp	40	.200	L.F.	12.30	7.10		19.40	25.50	
	0750	4" wide joint	"	25	.320	"	16.80	11.40		28.20	37.50	
	0760	Floor slab to drywall partition										
	0770	Flat joint	1 Carp	100	.080	L.F.	5.10	2.84		7.94	10.30	
	0780	Fluted joint		50	.160		10.40	5.70		16.10	21	
	0790	Etched fluted joint	↓	75	.107	↓	6.75	3.79		10.54	13.75	
	0800	Floor slab to concrete/masonry partition										
	0810	Flat joint	1 Carp	75	.107	L.F.	11.45	3.79		15.24	18.90	
	0820	Fluted joint	"	50	.160	"	13.55	5.70		19.25	24.50	
	0830	Concrete/CMU wall joints										
	0840	1" wide	1 Carp	100	.080	L.F.	6.25	2.84		9.09	11.55	
	0850	2" wide		75	.107		11.45	3.79		15.24	18.90	
	0860	4" wide	↓	50	.160	↓	22	5.70		27.70	33.50	
	0870	Concrete/CMU floor joints										
	0880	1" wide	1 Carp	200	.040	L.F.	3.12	1.42		4.54	5.80	
	0890	2" wide		150	.053		5.70	1.90		7.60	9.45	
	0900	4" wide	↓	100	.080	↓	10.90	2.84		13.74	16.70	

07900 | Joint Sealers

07920 | Joint Sealants

			CREW	DAILY OUTPUT	LABOR-HOURS	UNIT	2006 BARE COSTS MAT.	LABOR	EQUIP.	TOTAL	TOTAL INCL O&P	
800	0010	**CAULKING AND SEALANTS**										800
	0020	Acoustical sealant, elastomeric, cartridges				Ea.	2.28			2.28	2.51	
	0032	Backer rod, polyethylene, 1/4" diameter	1 Bric	460	.017	L.F.	.02	.64		.66	1.05	
	0052	1/2" diameter		460	.017		.04	.64		.68	1.07	
	0072	3/4" diameter		460	.017		.07	.64		.71	1.10	
	0092	1" diameter	↓	460	.017	↓	.11	.64		.75	1.15	
	0100	Acrylic latex caulk, white										
	0200	11 fl. oz cartridge				Ea.	1.88			1.88	2.07	
	0500	1/4" x 1/2"	1 Bric	248	.032	L.F.	.15	1.18		1.33	2.08	
	0600	1/2" x 1/2"		250	.032		.31	1.17		1.48	2.24	
	0800	3/4" x 3/4"		230	.035		.69	1.27		1.96	2.82	
	0900	3/4" x 1"		200	.040		.92	1.46		2.38	3.38	
	1000	1" x 1"	↓	180	.044	↓	1.15	1.62		2.77	3.91	
	1400	Butyl based, bulk				Gal.	23			23	25	
	1500	Cartridges				"	28			28	30.50	
	1700	Bulk, in place 1/4" x 1/2", 154 L.F./gal.	1 Bric	230	.035	L.F.	.15	1.27		1.42	2.22	
	1800	1/2" x 1/2", 77 L.F./gal.	"	180	.044	"	.30	1.62		1.92	2.97	
	2000	Latex acrylic based, bulk				Gal.	24			24	26.50	
	2100	Cartridges				"	29.50			29.50	32.50	
	2200	Bulk in place, 1/4" x 1/2", 154 L.F./gal.	1 Bric	230	.035	L.F.	.16	1.27		1.43	2.23	
	2250											
	2300	Polysulfide compounds, 1 component, bulk				Gal.	45			45	49.50	
	2400	Cartridges				"	48			48	53	
	2600	1 or 2 component, in place, 1/4" x 1/4", 308 L.F./gal.	1 Bric	145	.055	L.F.	.15	2.02		2.17	3.43	

07900 | Joint Sealers

	07920 | Joint Sealants	CREW	DAILY OUTPUT	LABOR-HOURS	UNIT	2006 BARE COSTS MAT.	LABOR	EQUIP.	TOTAL	TOTAL INCL O&P	
800	2700 1/2" x 1/4", 154 L.F./gal.	1 Bric	135	.059	L.F.	.29	2.17		2.46	3.83	800
	2900 3/4" x 3/8", 68 L.F./gal.		130	.062		.66	2.25		2.91	4.38	
	3000 1" x 1/2", 38 L.F./gal.	↓	130	.062	↓	1.19	2.25		3.44	4.95	
	3200 Polyurethane, 1 or 2 component				Gal.	49			49	54	
	3300 Cartridges				"	47			47	51.50	
	3500 Bulk, in place, 1/4" x 1/4"	1 Bric	150	.053	L.F.	.16	1.95		2.11	3.34	
	3600 1/2" x 1/4"		145	.055		.32	2.02		2.34	3.62	
	3800 3/4" x 3/8", 68 L.F./gal.		130	.062		.72	2.25		2.97	4.44	
	3900 1" x 1/2"	↓	110	.073	↓	1.27	2.66		3.93	5.70	
	4100 Silicone rubber, bulk				Gal.	35.50			35.50	39	
	4200 Cartridges				"	38			38	42	
	4300 Bulk in place, 1/4" x 1/2", 154 L.F./gal.	1 Bric	235	.034	L.F.	.23	1.24		1.47	2.27	
	4350										
	4400 Neoprene gaskets, closed cell, adhesive, 1/8" x 3/8"	1 Bric	240	.033	L.F.	.21	1.22		1.43	2.21	
	4500 1/4" x 3/4"		215	.037		.49	1.36		1.85	2.75	
	4700 1/2" x 1"		200	.040		1.44	1.46		2.90	3.95	
	4800 3/4" x 1-1/2"	↓	165	.048	↓	3	1.77		4.77	6.15	
	5500 Resin epoxy coating, 2 component, heavy duty				Gal.	26.50			26.50	29	
	5802 Tapes, sealant, P.V.C. foam adhesive, 1/16" x 1/4"				L.F.	.05			.05	.05	
	5902 1/16" x 1/2"					.07			.07	.08	
	5952 1/16" x 1"					.12			.12	.13	
	6002 1/8" x 1/2"				↓	.08			.08	.09	
	6200 Urethane foam, 2 component, handy pack, 1 C.F.				Ea.	28.50			28.50	31	
	6300 50.0 C.F. pack				C.F.	14.45			14.45	15.90	
	9000 Minimum labor/equipment charge	1 Bric	4	2	Job		73		73	119	

For information about Means Estimating Seminars, see yellow pages 12 and 13 in back of book

Division Notes

		CREW	DAILY OUTPUT	LABOR-HOURS	UNIT	2000 BARE COSTS MAT.	LABOR	EQUIP.	TOTAL	TOTAL INCL O&P

Division 8
Doors & Windows

Estimating Tips

08100 Metal Doors & Frames

- Most metal doors and frames look alike, but there may be significant differences among them. When estimating these items be sure to choose the line item that most closely compares to the specification or door schedule requirements regarding:
  - type of metal
  - metal gauge
  - door core material
  - fire rating
  - finish

08200 Wood & Plastic Doors

- Wood and plastic doors vary considerably in price. The primary determinant is the veneer material. Lauan, birch, and oak are the most common veneers. Other variables include the following:
  - hollow or solid core
  - fire rating
  - flush or raised panel
  - finish
- If the specifications require compliance with AWI (Architectural Woodwork Institute) standards or acoustical standards, the cost of the door may increase substantially. All wood doors are priced pre-mortised for hinges and pre-drilled for cylindrical locksets.
- Frequently, doors, frames, and windows are unique in old buildings. Specified replacement units could be stock, custom (similar to the original), or exact reproduction. The estimator should work closely with a window consultant to determine any extra costs that may be associated with the unusual installation requirements.

08300 Specialty Doors

- There are many varieties of special doors, and they are usually priced per each. Add frames, hardware, or operators required for a complete installation.

08510 Steel Windows

- Most metal windows are delivered preglazed. However, some metal windows are priced without glass. Refer to 08800 Glazing for glass pricing. The grade C indicates commercial grade windows, usually ASTM C-35.

08550 Wood Windows

- All wood windows are priced preglazed. The two glazing options priced are single pane float glass and insulating glass 1/2″ thick. Add the cost of screens and grills if required.

08700 Hardware

- Hardware costs add considerably to the cost of a door. The most efficient method to determine the hardware requirements for a project is to review the door schedule. This schedule, in conjunction with the specifications, is all you should need to take off the door hardware.
- Door hinges are priced by the pair, with most doors requiring 1-1/2 pairs per door. The hinge prices do not include installation labor because it is included in door installation. Hinges are classified according to the frequency of use.

08800 Glazing

- Different openings require different types of glass. The three most common types are:
  - float
  - tempered
  - insulating
- Most exterior windows are glazed with insulating glass. Entrance doors and window walls, where the glass is less than 18″ from the floor, are generally glazed with tempered glass. Interior windows and some residential windows are glazed with float glass.
- Energy efficient coatings are also available

08900 Glazed Curtain Wall

- Glazed curtain walls consist of the metal tube framing and the glazing material. The cost data in this subdivision is presented for the metal tube framing alone or the composite wall. If your estimate requires a detailed takeoff of the framing, be sure to add the glazing cost.

Reference Numbers

Reference numbers are shown in bold squares at the beginning of some major classifications. These numbers refer to related items in the Reference Section. The reference information may be an estimating procedure, an alternate pricing method, or technical information.

Note: Not all subdivisions listed here necessarily appear in this publication.

08060 | Selective Demolition

		08060 Selective Demolition	CREW	DAILY OUTPUT	LABOR-HOURS	UNIT	2006 BARE COSTS MAT.	LABOR	EQUIP.	TOTAL	TOTAL INCL O&P	
110	0010	SELECTIVE DEMOLITION, DOORS R024119-10										110
	0200	Doors, exterior, 1-3/4" thick, single, 3' x 7' high	1 Clab	16	.500	Ea.		13.70		13.70	22.50	
	0220	Double, 6' x 7' high		12	.667			18.25		18.25	30.50	
	0500	Interior, 1-3/8" thick, single, 3' x 7' high		20	.400			10.95		10.95	18.15	
	0520	Double, 6' x 7' high		16	.500			13.70		13.70	22.50	
	0700	Bi-folding, 3' x 6'-8" high		20	.400			10.95		10.95	18.15	
	0720	6' x 6'-8" high		18	.444			12.20		12.20	20	
	0900	Bi-passing, 3' x 6'-8" high		16	.500			13.70		13.70	22.50	
	0940	6' x 6'-8" high		14	.571			15.65		15.65	26	
	1500	Remove and reset, minimum	1 Carp	8	1			35.50		35.50	59	
	1520	Maximum		6	1.333			47.50		47.50	78.50	
	2000	Frames, including trim, metal		8	1			35.50		35.50	59	
	2200	Wood	2 Carp	32	.500			17.80		17.80	29.50	
	2201	Alternate pricing method	1 Carp	200	.040	L.F.		1.42		1.42	2.36	
	2950	Minimum labor/equipment charge	1 Clab	4	2	Job		55		55	91	
	3000	Special doors, counter doors	2 Carp	6	2.667	Ea.		95		95	157	
	3100	Double acting		10	1.600			57		57	94	
	3200	Floor door (trap type), or access type		8	2			71		71	118	
	3300	Glass, sliding, including frames		12	1.333			47.50		47.50	78.50	
	3400	Overhead, commercial, 12' x 12' high		4	4			142		142	236	
	3440	up to 20' x 16' high		3	5.333			190		190	315	
	3445	up to 35' x 30' high		1	16			570		570	940	
	3500	Residential, 9' x 7' high		8	2			71		71	118	
	3540	16' x 7' high		7	2.286			81.50		81.50	135	
	3600	Remove and reset, minimum		4	4			142		142	236	
	3620	Maximum		2.50	6.400			228		228	375	
	3700	Roll-up grille		5	3.200			114		114	188	
	3800	Revolving door		2	8			284		284	470	
	3900	Storefront swing door		3	5.333			190		190	315	
	5032	Remove skylight, plstc domes,flush/curb mtd	G-3	395	.081	S.F.		2.82		2.82	4.56	
	7100	Remove double swing pneumatic doors, openers and sensors	2 Skwk	.50	32	Opng.		1,175		1,175	1,900	
	7570	Remove shock absorbing door	2 Sswk	1.90	8.421	"		335		335	630	
	9000	Minimum labor/equipment charge	1 Carp	4	2	Job		71		71	118	
120	0010	SELECTIVE DEMOLITION, WINDOWS R024119-10										120
	0200	Aluminum, including trim, to 12 S.F.	1 Clab	16	.500	Ea.		13.70		13.70	22.50	
	0240	To 25 S.F.		11	.727			19.95		19.95	33	
	0280	To 50 S.F.		5	1.600			44		44	72.50	
	0320	Storm windows/scree, to 12 S.F.		27	.296			8.10		8.10	13.45	
	0360	To 25 S.F.		21	.381			10.45		10.45	17.30	
	0400	To 50 S.F.		16	.500			13.70		13.70	22.50	
	0600	Glass, minimum		200	.040	S.F.		1.10		1.10	1.82	
	0620	Maximum		150	.053	"		1.46		1.46	2.42	
	1000	Steel, including trim, to 12 S.F.		13	.615	Ea.		16.85		16.85	28	
	1020	To 25 S.F.		9	.889			24.50		24.50	40.50	
	1040	To 50 S.F.		4	2			55		55	91	
	2000	Wood, including trim, to 12 S.F.		22	.364			9.95		9.95	16.50	
	2020	To 25 S.F.		18	.444			12.20		12.20	20	
	2060	To 50 S.F.		13	.615			16.85		16.85	28	
	2065	To 180 S.F.		8	1			27.50		27.50	45.50	
	5020	Remove and reset window, minimum	1 Carp	6	1.333			47.50		47.50	78.50	
	5040	Average		4	2			71		71	118	
	5080	Maximum		2	4			142		142	236	
	9000	Minimum labor/equipment charge	1 Clab	4	2	Job		55		55	91	

08110 | Steel Doors and Frames

			CREW	DAILY OUTPUT	LABOR-HOURS	UNIT	2006 BARE COSTS MAT.	LABOR	EQUIP.	TOTAL	TOTAL INCL O&P
200	0010	COMMERCIAL STEEL DOORS R081313-20									
	0015	Flush, full panel, hollow core									
	0020	1-3/8" thick, 20 ga., 2'-0"x 6'-8"	2 Carp	20	.800	Ea.	246	28.50		274.50	320
	0040	2'-8" x 6'-8"		18	.889		254	31.50		285.50	330
	0060	3'-0" x 6'-8"		17	.941		259	33.50		292.50	340
	0100	3'-0" x 7'-0"		17	.941		241	33.50		274.50	320
	0120	For vision lite, add					84			84	92.50
	0140	For narrow lite, add					85.50			85.50	94
	0320	Half glass, 20 ga., 2'-0" x 6'-8"	2 Carp	20	.800		365	28.50		393.50	450
	0340	2'-8" x 6'-8"		18	.889		375	31.50		406.50	470
	0360	3'-0" x 6'-8"		17	.941		380	33.50		413.50	475
	0400	3'-0" x 7'-0"		17	.941		380	33.50		413.50	475
	0410	1-3/8" thick, 18 ga., 2'-0"x 6'-8"		20	.800		295	28.50		323.50	370
	0420	3'-0" x 6'-8"		17	.941		299	33.50		332.50	385
	0425	3'-0" x 7'-0"		17	.941		305	33.50		338.50	390
	0450	For vision lite, add					84			84	92.50
	0452	For narrow lite, add					85.50			85.50	94
	0460	Half glass, 18 ga., 2'-0" x 6'-8"	2 Carp	20	.800		415	28.50		443.50	505
	0465	2'-8" x 6'-8"		18	.889		430	31.50		461.50	530
	0470	3'-0" x 6'-8"		17	.941		420	33.50		453.50	520
	0475	3'-0" x 7'-0"		17	.941		425	33.50		458.50	525
	0500	Hollow core, 1-3/4" thick, full panel, 20 ga., 2'-8" x 6'-8"		18	.889		277	31.50		308.50	360
	0520	3'-0" x 6'-8"		17	.941		259	33.50		292.50	340
	0640	3'-0" x 7'-0"		17	.941		305	33.50		338.50	395
	0680	4'-0" x 7'-0"		15	1.067		430	38		468	535
	0700	4'-0" x 8'-0"		13	1.231		490	44		534	615
	1000	18 ga., 2'-8" x 6'-8"		17	.941		204	33.50		237.50	281
	1020	3'-0" x 6'-8"		16	1		199	35.50		234.50	278
	1120	3'-0" x 7'-0"		17	.941		345	33.50		378.50	430
	1180	4'-0" x 7'-0"		14	1.143		415	40.50		455.50	525
	1200	4'-0" x 8'-0"		17	.941		490	33.50		523.50	595
	1212	For vision lite, add					84			84	92.50
	1214	For narrow lite, add					85.50			85.50	94
	1230	Half glass, 20 ga., 2'-8" x 6'-8"	2 Carp	20	.800		380	28.50		408.50	465
	1240	3'-0" x 6'-8"		18	.889		380	31.50		411.50	475
	1260	3'-0" x 7'-0"		18	.889		390	31.50		421.50	485
	1320	18 ga., 2'-8" x 6'-8"		18	.889		425	31.50		456.50	525
	1340	3'-0" x 6'-8"		17	.941		420	33.50		453.50	520
	1360	3'-0" x 7'-0"		17	.941		430	33.50		463.50	530
	1380	4'-0" x 7'-0"		15	1.067		535	38		573	655
	1400	4'-0" x 8'-0"		14	1.143		615	40.50		655.50	745
	1720	Insulated, 1-3/4" thick, full panel, 18 ga., 3'-0" x 6'-8"		15	1.067		370	38		408	475
	1740	2'-8" x 7'-0"		16	1		390	35.50		425.50	485
	1760	3'-0" x 7'-0"		15	1.067		380	38		418	485
	1800	4'-0" x 8'-0"		13	1.231		565	44		609	695
	1805	For vision lite, add					84			84	92.50
	1810	For narrow lite, add					85.50			85.50	94
	1820	Half glass, 18 ga., 3'-0" x 6'-8"	2 Carp	16	1		495	35.50		530.50	605
	1840	2'-8" x 7'-0"		17	.941		510	33.50		543.50	615
	1860	3'-0" x 7'-0"		16	1		565	35.50		600.50	680
	1900	4'-0" x 8'-0"		14	1.143		685	40.50		725.50	825
	2000	For bottom louver, add					138			138	152
	2020	For baked enamel finish, add					30%	15%			
	2040	For galvanizing, add					15%				
	9000	Minimum labor/equipment charge	1 Carp	4	2	Job		71		71	118

08110 | Steel Doors and Frames

		08110 Steel Doors and Frames	CREW	DAILY OUTPUT	LABOR-HOURS	UNIT	2006 BARE COSTS MAT.	LABOR	EQUIP.	TOTAL	TOTAL INCL O&P	
250	0010	**DOOR FRAMES**										250
	0020	Steel channels with anchors and bar stops										
	0100	6" channel @ 8.2#/L.F., 3' x 7' door, weighs 150#	E-4	13	2.462	Ea.	171	99.50	6.85	277.35	380	
	0200	8" channel @ 11.5#/L.F., 6' x 8' door, weighs 275#		9	3.556		315	144	9.90	468.90	625	
	0300	8' x 12' door, weighs 400#		6.50	4.923		455	199	13.75	667.75	885	
	0800	For frames without bar stops, light sections, deduct					15%					
	0900	Heavy sections, deduct					10%					
	9000	Minimum labor/equipment charge	E-4	4	8	Job		325	22.50	347.50	630	
300	0010	**FIRE DOOR** R081313-20										300
	0015	Steel, flush, "B" label, 90 minute										
	0020	Full panel, 20 ga., 2'-0" x 6'-8"	2 Carp	20	.800	Ea.	310	28.50		338.50	385	
	0040	2'-8" x 6'-8"		18	.889		320	31.50		351.50	405	
	0060	3'-0" x 6'-8"		17	.941		325	33.50		358.50	410	
	0080	3'-0" x 7'-0"		17	.941		335	33.50		368.50	425	
	0140	18 ga., 3'-0" x 6'-8"		16	1		365	35.50		400.50	460	
	0160	2'-8" x 7'-0"		17	.941		385	33.50		418.50	475	
	0180	3'-0" x 7'-0"		16	1		365	35.50		400.50	465	
	0200	4'-0" x 7'-0"		15	1.067		480	38		518	590	
	0220	For "A" label, 3 hour, 18 ga., use same price as "B" label										
	0240	For vision lite, add				Ea.	104			104	115	
	0520	Flush, "B" label 90 min., composite, 20 ga., 2'-0" x 6'-8"	2 Carp	18	.889		400	31.50		431.50	495	
	0540	2'-8" x 6'-8"		17	.941		410	33.50		443.50	505	
	0560	3'-0" x 6'-8"		16	1		410	35.50		445.50	515	
	0580	3'-0" x 7'-0"		16	1		425	35.50		460.50	525	
	0640	Flush, "A" label 3 hour, composite, 18 ga., 3'-0" x 6'-8"		15	1.067		345	38		383	445	
	0660	2'-8" x 7'-0"		16	1		365	35.50		400.50	460	
	0680	3'-0" x 7'-0"		15	1.067		360	38		398	460	
	0700	4'-0" x 7'-0"		14	1.143		465	40.50		505.50	580	
	9000	Minimum labor/equipment charge	1 Carp	4	2	Job		71		71	118	
600	0010	**RESIDENTIAL STEEL DOOR**										600
	0020	Prehung, insulated, exterior										
	0030	Embossed, full panel, 2'-8" x 6'-8"	2 Carp	17	.941	Ea.	216	33.50		249.50	293	
	0040	3'-0" x 6'-8"		15	1.067		217	38		255	300	
	0060	3'-0" x 7'-0"		15	1.067		280	38		318	375	
	0070	5'-4" x 6'-8", double		8	2		440	71		511	605	
	0220	Half glass, 2'-8" x 6'-8"		17	.941		261	33.50		294.50	345	
	0240	3'-0" x 6'-8"		16	1		261	35.50		296.50	345	
	0260	3'-0" x 7'-0"		16	1		315	35.50		350.50	410	
	0270	5'-4" x 6'-8", double		8	2		540	71		611	715	
	0720	Raised plastic face, full panel, 2'-8" x 6'-8"		16	1		256	35.50		291.50	340	
	0740	3'-0" x 6'-8"		15	1.067		258	38		296	345	
	0760	3'-0" x 7'-0"		15	1.067		261	38		299	350	
	0780	5'-4" x 6'-8", double		8	2		485	71		556	650	
	0820	Half glass, 2'-8" x 6'-8"		17	.941		287	33.50		320.50	370	
	0840	3'-0" x 6'-8"		16	1		290	35.50		325.50	380	
	0860	3'-0" x 7'-0"		16	1		320	35.50		355.50	410	
	0880	5'-4" x 6'-8", double		8	2		630	71		701	815	
	1320	Flush face, full panel, 2'-6" x 6'-8"		16	1		216	35.50		251.50	297	
	1340	3'-0" x 6'-8"		15	1.067		216	38		254	300	
	1360	3'-0" x 7'-0"		15	1.067		278	38		316	370	
	1380	5'-4" x 6'-8", double		8	2		405	71		476	565	
	1420	Half glass, 2'-8" x 6'-8"		17	.941		270	33.50		303.50	355	
	1440	3'-0" x 6'-8"		16	1		274	35.50		309.50	360	
	1460	3'-0" x 7'-0"		16	1		315	35.50		350.50	405	
	1480	5'-4" x 6'-8", double		8	2		530	71		601	700	

08110 | Steel Doors and Frames

		08110 Steel Doors and Frames	Crew	Daily Output	Labor-Hours	Unit	2006 Bare Costs Mat.	2006 Bare Costs Labor	2006 Bare Costs Equip.	2006 Bare Costs Total	Total Incl O&P
600	1500	Sidelight, full lite, 1′-0″ x 6′-8″with grille				Ea.	211			211	232
	1510	1′-0″ x 6′-8″, low e					242			242	267
	1520	1′-0″ x 6′-8″, half lite					189			189	208
	1530	1′-0″ x 6′-8″, half lite, low e					230			230	253
	2300	Interior, residential, closet, bi-fold, 6′-8″ x 2′-0″ wide	2 Carp	16	1		140	35.50		175.50	213
	2330	3′-0″ wide		16	1		157	35.50		192.50	232
	2360	4′-0″ wide		15	1.067		238	38		276	325
	2400	5′-0″ wide		14	1.143		276	40.50		316.50	375
	2420	6′-0″ wide		13	1.231		310	44		354	415
	9000	Minimum labor/equipment charge	1 Carp	4	2	Job		71		71	118
820	0010	**STEEL FRAMES, KNOCK DOWN**									
	0020	16 ga., up to 5-3/4″ jamb depth									
	0025	6′-8″ high, 3′-0″ wide, single	2 Carp	16	1	Ea.	121	35.50		156.50	192
	0028	3′-6″ wide, single		16	1		107	35.50		142.50	177
	0030	4′-0″ wide, single		16	1		107	35.50		142.50	177
	0040	6′-0″ wide, double		14	1.143		143	40.50		183.50	226
	0045	8′-0″ wide, double		14	1.143		149	40.50		189.50	232
	0100	7′-0″ high, 3′-0″ wide, single		16	1		124	35.50		159.50	195
	0110	3′-6″ wide, single		16	1		110	35.50		145.50	180
	0112	4′-0″ wide, single		16	1		188	35.50		223.50	266
	0140	6′-0″ wide, double		14	1.143		150	40.50		190.50	233
	0145	8′-0″ wide, double		14	1.143		140	40.50		180.50	222
	1000	16 ga., up to 4-7/8″ deep, 7′-0″ H, 3′-0″ W, single		16	1		121	35.50		156.50	192
	1140	6′-0″ wide, double		14	1.143		148	40.50		188.50	231
	2800	14 ga., up to 3-7/8″ deep, 7′-0″ high, 3′-0″ wide, single		16	1		124	35.50		159.50	195
	2840	6′-0″ wide, double		14	1.143		152	40.50		192.50	235
	3000	14 ga., up to 5-3/4″ deep, 6′-8″ high, 3′-0″ wide, single		16	1		134	35.50		169.50	207
	3002	3′-6″ wide, single		16	1		131	35.50		166.50	203
	3005	4′-0″ wide, single		16	1		131	35.50		166.50	203
	3600	up to 5-3/4″ jamb depth, 7′-0″ high, 4′-0″ wide, single		15	1.067		103	38		141	177
	3620	6′-0″ wide, double		12	1.333		162	47.50		209.50	257
	3640	8′-0″ wide, double		12	1.333		136	47.50		183.50	229
	3700	8′-0″ high, 4′-0″ wide, single		15	1.067		136	38		174	213
	3740	8′-0″ wide, double		12	1.333		169	47.50		216.50	265
	4000	6-3/4″ deep, 7′-0″ high, 4′-0″ wide, single		15	1.067		144	38		182	222
	4020	6′-0″ wide, double		12	1.333		165	47.50		212.50	261
	4040	8′-0″, wide double		12	1.333		191	47.50		238.50	289
	4100	8′-0″ high, 4′-0″ wide, single		15	1.067		162	38		200	241
	4140	8′-0″ wide, double		12	1.333		189	47.50		236.50	287
	4400	8-3/4″ deep, 7′-0″ high, 4′-0″ wide, single		15	1.067		152	38		190	230
	4440	8′-0″ wide, double		12	1.333		200	47.50		247.50	299
	4500	8′-0″ high, 4′-0″ wide, single		15	1.067		178	38		216	258
	4540	8′-0″ wide, double		12	1.333		208	47.50		255.50	310
	4900	For welded frames, add					44.50			44.50	49
	5400	14 ga., "B" label, up to 5-3/4″ deep, 7′-0″ high, 4′-0″ wide, single	2 Carp	15	1.067		160	38		198	239
	5440	8′-0″ wide, double		12	1.333		187	47.50		234.50	285
	5800	6-3/4″ deep, 7′-0″ high, 4′-0″ wide, single		15	1.067		140	38		178	217
	5840	8′-0″ wide, double		12	1.333		207	47.50		254.50	305
	6200	8-3/4″ deep, 7′-0″ high, 4′-0″ wide, single		15	1.067		170	38		208	250
	6240	8′-0″ wide, double		12	1.333		220	47.50		267.50	320
	6300	For "A" label use same price as "B" label									
	6400	For baked enamel finish, add					30%	15%			
	6500	For galvanizing, add					15%				
	6600	For hospital stop, add				Ea.	263			263	289
	7900	Transom lite frames, fixed, add	2 Carp	155	.103	S.F.	42.50	3.67		46.17	53
	8000	Movable, add	"	130	.123	"	51.50	4.38		55.88	64

08100 | Metal Doors and Frames

	08110 Steel Doors and Frames		Crew	Daily Output	Labor-Hours	Unit	2006 Bare Costs Mat.	Labor	Equip.	Total	Total Incl O&P	
820	9000	Minimum labor/equipment charge	1 Carp	4	2	Job		71		71	118	820

	08180 Metal Screen and Storm Doors		Crew	Daily Output	Labor-Hours	Unit	2006 Bare Costs Mat.	Labor	Equip.	Total	Total Incl O&P	
100	0010	**STORM DOORS & FRAMES** Aluminum, residential,										100
	0020	combination storm and screen										
	0400	Clear anodic coating, 6′-8″ x 2′-6″ wide	2 Carp	15	1.067	Ea.	162	38		200	242	
	0420	2′-8″ wide		14	1.143		186	40.50		226.50	273	
	0440	3′-0″ wide	↓	14	1.143	↓	186	40.50		226.50	273	
	0500	For 7′ door height, add					5%					
	1000	Mill finish, 6′-8″ x 2′-6″ wide	2 Carp	15	1.067	Ea.	216	38		254	300	
	1020	2′-8″ wide		14	1.143		216	40.50		256.50	305	
	1040	3′-0″ wide	↓	14	1.143		234	40.50		274.50	325	
	1100	For 7′-0″ door, add					5%					
	1500	White painted, 6′-8″ x 2′-6″ wide	2 Carp	15	1.067		216	38		254	300	
	1520	2′-8″ wide		14	1.143		220	40.50		260.50	310	
	1540	3′-0″ wide	↓	14	1.143		229	40.50		269.50	320	
	1600	For 7′-0″ door, add				↓	5%					
	2000	Wood door & screen, see Division 08210-930										
	9000	Minimum labor/equipment charge	1 Carp	4	2	Job		71		71	118	

08200 | Wood and Plastic Doors

	08210 Wood Doors		Crew	Daily Output	Labor-Hours	Unit	2006 Bare Costs Mat.	Labor	Equip.	Total	Total Incl O&P	
450	0010	**KALAMEIN**										450
	0020	Interior, flush type, 3′ x 7′	2 Carp	4.30	3.721	Opng.	176	132		308	410	
	9000	Minimum labor/equipment charge	1 Carp	2	4	Job		142		142	236	
720	0010	**PRE-HUNG DOORS**										720
	0300	Exterior, wood, comb. storm & screen, 6′-9″ x 2′-6″ wide	2 Carp	15	1.067	Ea.	283	38		321	375	
	0320	2′-8″ wide		15	1.067		283	38		321	375	
	0340	3′-0″ wide	↓	15	1.067		291	38		329	385	
	0360	For 7′-0″ high door, add				↓	24.50			24.50	27	
	0370	For aluminum storm doors, see Division 08180-100										
	1600	Entrance door, flush, birch, solid core										
	1620	4-5/8″ solid jamb, 1-3/4″ x 6′-8″ x 2′-8″ wide	2 Carp	16	1	Ea.	288	35.50		323.50	375	
	1640	3′-0″ wide	″	16	1		296	35.50		331.50	385	
	1680	For 7′-0″ high door, add				↓	17.50			17.50	19.25	
	2000	Entrance door, colonial, 6 panel pine										
	2020	4-5/8″ solid jamb, 1-3/4″ x 6′-8″ x 2′-8″ wide	2 Carp	16	1	Ea.	520	35.50		555.50	630	
	2040	3′-0″ wide	″	16	1		520	35.50		555.50	630	
	2060	For 7′-0″ high door, add					48			48	52.50	
	2200	For 5-5/8″ solid jamb, add				↓	38.50			38.50	42	
	2990											
	4000	Interior, passage door, 4-5/8″ solid jamb										
	4400	Lauan, flush, solid core, 1-3/8″ x 6′-8″ x 2′-6″ wide	2 Carp	20	.800	Ea.	178	28.50		206.50	242	
	4420	2′-8″ wide		20	.800		178	28.50		206.50	242	
	4440	3′-0″ wide		19	.842		191	30		221	260	
	4600	Hollow core, 1-3/8″ x 6′-8″ x 2′-6″ wide		20	.800		120	28.50		148.50	179	
	4620	2′-8″ wide	↓	20	.800	↓	118	28.50		146.50	177	

08210 | Wood Doors

		08210 Wood Doors	CREW	DAILY OUTPUT	LABOR-HOURS	UNIT	2006 BARE COSTS MAT.	LABOR	EQUIP.	TOTAL	TOTAL INCL O&P	
720	4640	3'-0" wide	2 Carp	19	.842	Ea.	121	30		151	183	720
	4700	For 7'-0" high door, add					23			23	25.50	
	5000	Birch, flush, solid core, 1-3/8" x 6'-8" x 2'-6" wide	2 Carp	20	.800		166	28.50		194.50	229	
	5020	2'-8" wide		20	.800		187	28.50		215.50	253	
	5040	3'-0" wide		19	.842		200	30		230	270	
	5200	Hollow core, 1-3/8" x 6'-8" x 2'-6" wide		20	.800		137	28.50		165.50	198	
	5220	2'-8" wide		20	.800		144	28.50		172.50	205	
	5240	3'-0" wide		19	.842		144	30		174	208	
	5280	For 7'-0" high door, add					19.95			19.95	22	
	5500	Hardboard paneled, 1-3/8" x 6'-8" x 2'-6" wide	2 Carp	20	.800		138	28.50		166.50	199	
	5520	2'-8" wide		20	.800		145	28.50		173.50	206	
	5540	3'-0" wide		19	.842		143	30		173	207	
	6000	Pine paneled, 1-3/8" x 6'-8" x 2'-6" wide		20	.800		240	28.50		268.50	310	
	6020	2'-8" wide		20	.800		261	28.50		289.50	335	
	6040	3'-0" wide		19	.842		266	30		296	340	
	6500	For 5-5/8" solid jamb, add					11.80			11.80	12.95	
	6520	For split jamb, deduct					13.95			13.95	15.35	
	9000	Minimum labor/equipment charge	1 Carp	4	2	Job		71		71	118	
850	0010	**TIN CLAD**										850
	0020	3 ply, 6' x 7', double sliding, doors only	2 Carp	1	16	Opng.	1,375	570		1,945	2,475	
	1000	For electric operator, add	1 Elec	2	4	"	2,450	168		2,618	2,950	
	9000	Minimum labor/equipment charge	2 Carp	1	16	Job		570		570	940	
900	0010	**WOOD DOOR, ARCHITECTURAL**										900
	0015	Flush, int., 1-3/8", 7 ply, hollow core,										
	0020	Lauan face, 2'-0" x 6'-8"	2 Carp	17	.941	Ea.	29	33.50		62.50	87.50	
	0040	2'-6" x 6'-8"		17	.941		33	33.50		66.50	92	
	0080	3'-0" x 6'-8"		17	.941		39	33.50		72.50	98.50	
	0100	4'-0" x 6'-8"		16	1		69.50	35.50		105	136	
	0120	Birch face, 2'-0" x 6'-8"		17	.941		46	33.50		79.50	106	
	0140	2'-6" x 6'-8"		17	.941		44.50	33.50		78	105	
	0180	3'-0" x 6'-8"		17	.941		50	33.50		83.50	111	
	0200	4'-0" x 6'-8"		16	1		95	35.50		130.50	163	
	0220	Oak face, 2'-0" x 6'-8"		17	.941		72.50	33.50		106	136	
	0240	2'-6" x 6'-8"		17	.941		77.50	33.50		111	141	
	0280	3'-0" x 6'-8"		17	.941		83	33.50		116.50	147	
	0300	4'-0" x 6'-8"		16	1		105	35.50		140.50	175	
	0320	Walnut face, 2'-0" x 6'-8"		17	.941		147	33.50		180.50	217	
	0340	2'-6" x 6'-8"		17	.941		150	33.50		183.50	220	
	0380	3'-0" x 6'-8"		17	.941		155	33.50		188.50	227	
	0400	4'-0" x 6'-8"		16	1		176	35.50		211.50	253	
	0430	For 7'-0" high, add					14.45			14.45	15.90	
	0440	For 8'-0" high, add					20.50			20.50	22.50	
	0480	For prefinishing, clear, add					32			32	35	
	0500	For prefinishing, stain, add					43.50			43.50	47.50	
	1320	M.D. overlay on hardboard, 2'-0" x 6'-8"	2 Carp	17	.941		89.50	33.50		123	154	
	1340	2'-6" x 6'-8"		17	.941		89.50	33.50		123	154	
	1380	3'-0" x 6'-8"		17	.941		106	33.50		139.50	173	
	1400	4'-0" x 6'-8"		16	1		146	35.50		181.50	220	
	1420	For 7'-0" high, add					7.95			7.95	8.75	
	1440	For 8'-0" high, add					21			21	23.50	
	1720	H.P. plastic laminate, 2'-0" x 6'-8"	2 Carp	16	1		213	35.50		248.50	294	
	1740	2'-6" x 6'-8"		16	1		213	35.50		248.50	294	
	1780	3'-0" x 6'-8"		15	1.067		246	38		284	335	
	1800	4'-0" x 6'-8"		14	1.143		340	40.50		380.50	445	
	1820	For 7'-0" high, add					8.35			8.35	9.20	
	1840	For 8'-0" high, add					21.50			21.50	23.50	

	08210	Wood Doors	CREW	DAILY OUTPUT	LABOR-HOURS	UNIT	2006 BARE COSTS MAT.	LABOR	EQUIP.	TOTAL	TOTAL INCL O&P
900	2020	5 ply particle core, lauan face, 2′-6″ x 6′-8″	2 Carp	15	1.067	Ea.	73	38		111	144
	2040	3′-0″ x 6′-8″		14	1.143		76	40.50		116.50	151
	2080	3′-0″ x 7′-0″		13	1.231		85	44		129	166
	2100	4′-0″ x 7′-0″		12	1.333		98	47.50		145.50	187
	2120	Birch face, 2′-6″ x 6′-8″		15	1.067		83	38		121	155
	2140	3′-0″ x 6′-8″		14	1.143		91	40.50		131.50	168
	2180	3′-0″ x 7′-0″		13	1.231		93	44		137	175
	2200	4′-0″ x 7′-0″		12	1.333		113	47.50		160.50	204
	2220	Oak face, 2′-6″ x 6′-8″		15	1.067		92	38		130	164
	2240	3′-0″ x 6′-8″		14	1.143		101	40.50		141.50	179
	2280	3′-0″ x 7′-0″		13	1.231		104	44		148	187
	2300	4′-0″ x 7′-0″		12	1.333		127	47.50		174.50	219
	2320	Walnut face, 2′-0″ x 6′-8″		15	1.067		102	38		140	175
	2340	2′-6″ x 6′-8″		14	1.143		116	40.50		156.50	196
	2380	3′-0″ x 6′-8″		13	1.231		131	44		175	217
	2400	4′-0″ x 6′-8″		12	1.333		171	47.50		218.50	267
	2440	For 8′-0″ high, add					25			25	27.50
	2460	For 8′-0″ high walnut, add					13.65			13.65	15.05
	2480	For solid wood core, add					30.50			30.50	33.50
	2720	For prefinishing, clear, add					19.95			19.95	22
	2740	For prefinishing, stain, add					44.50			44.50	49
	2750										
	3320	M.D. overlay on hardboard, 2′-6″ x 6′-8″	2 Carp	14	1.143	Ea.	89.50	40.50		130	166
	3340	3′-0″ x 6′-8″		13	1.231		93.50	44		137.50	176
	3380	3′-0″ x 7′-0″		12	1.333		95.50	47.50		143	184
	3400	4′-0″ x 7′-0″		10	1.600		117	57		174	222
	3440	For 8′-0″ height, add					26			26	28.50
	3460	For solid wood core, add					34			34	37.50
	3720	H.P. plastic laminate, 2′-6″ x 6′-8″	2 Carp	13	1.231		131	44		175	217
	3740	3′-0″ x 6′-8″		12	1.333		148	47.50		195.50	242
	3780	3′-0″ x 7′-0″		11	1.455		154	51.50		205.50	255
	3800	4′-0″ x 7′-0″		8	2		187	71		258	325
	3840	For 8′-0″ height, add					26			26	28.50
	3860	For solid wood core, add					32			32	35
	4000	Exterior, flush, solid wood stave core, birch, 1-3/4″ x 7′-0″ x 2′-6″	2 Carp	15	1.067		158	38		196	236
	4020	2′-8″ wide		15	1.067		165	38		203	244
	4040	3′-0″ wide		14	1.143		176	40.50		216.50	262
	4100	Oak faced 1-3/4″ x 7′-0″ x 2′-6″ wide		15	1.067		174	38		212	254
	4120	2′-8″ wide		15	1.067		186	38		224	267
	4140	3′-0″ wide		14	1.143		198	40.50		238.50	286
	4200	Walnut faced, 1-3/4″ x 7′-0″ x 2′-6″ wide		15	1.067		255	38		293	345
	4220	2′-8″ wide		15	1.067		267	38		305	355
	4240	3′-0″ wide		14	1.143		278	40.50		318.50	375
	4300	For 6′-8″ high door, deduct from 7′-0″ door					14.20			14.20	15.60
	9000	Minimum labor/equipment charge	1 Carp	4	2	Job		71		71	118
910	0010	**WOOD DOORS, DECORATOR**									
	3000	Solid wood, 1-3/4″ thick stile and rail									
	3020	Mahogany, 3′-0″ x 7′-0″, minimum	2 Carp	14	1.143	Ea.	945	40.50		985.50	1,125
	3030	Maximum		10	1.600		1,300	57		1,357	1,550
	3040	3′-6″ x 8′-0″, minimum		10	1.600		895	57		952	1,075
	3050	Maximum		8	2		1,625	71		1,696	1,925
	3100	Pine, 3′-0″ x 7′-0″, minimum		14	1.143		425	40.50		465.50	540
	3110	Maximum		10	1.600		705	57		762	870
	3120	3′-6″ x 8′-0″, minimum		10	1.600		655	57		712	815
	3130	Maximum		8	2		1,075	71		1,146	1,325

08210 | Wood Doors

		08210 Wood Doors	CREW	DAILY OUTPUT	LABOR-HOURS	UNIT	2006 BARE COSTS MAT.	LABOR	EQUIP.	TOTAL	TOTAL INCL O&P	
910	3200	Red oak, 3'-0" x 7'-0", minimum	2 Carp	14	1.143	Ea.	1,475	40.50		1,515.50	1,700	910
	3210	Maximum		10	1.600		1,750	57		1,807	2,025	
	3220	3'-6" x 8'-0", minimum		10	1.600		1,600	57		1,657	1,875	
	3230	Maximum		8	2		2,850	71		2,921	3,275	
	4000	Hand carved door, mahogany										
	4020	3'-0" x 7'-0", minimum	2 Carp	14	1.143	Ea.	1,450	40.50		1,490.50	1,650	
	4030	Maximum		11	1.455		2,825	51.50		2,876.50	3,200	
	4040	3'-6" x 8'-0", minimum		10	1.600		1,800	57		1,857	2,100	
	4050	Maximum		8	2		2,775	71		2,846	3,200	
	4200	Red oak, 3'-0" x 7'-0", minimum		14	1.143		4,400	40.50		4,440.50	4,925	
	4210	Maximum		11	1.455		12,100	51.50		12,151.50	13,400	
	4220	3'-6" x 8'-0", minimum		10	1.600		4,950	57		5,007	5,550	
	4280	For 6'-8" high door, deduct from 7'-0" door					31.50			31.50	34.50	
	4400	For custom finish, add					335			335	370	
	4600	Side light, mahogany, 7'-0" x 1'-6" wide, minimum	2 Carp	18	.889		810	31.50		841.50	945	
	4610	Maximum		14	1.143		2,375	40.50		2,415.50	2,700	
	4620	8'-0" x 1'-6" wide, minimum		14	1.143		1,525	40.50		1,565.50	1,750	
	4630	Maximum		10	1.600		1,725	57		1,782	2,000	
	4640	Side light, oak, 7'-0" x 1'-6" wide, minimum		18	.889		935	31.50		966.50	1,075	
	4650	Maximum		14	1.143		1,675	40.50		1,715.50	1,900	
	4660	8'-0" x 1-6" wide, minimum		14	1.143		880	40.50		920.50	1,050	
	4670	Maximum		10	1.600		1,675	57		1,732	1,925	
	6520	Interior cafe doors, 2'-6" opening, stock, panel pine		16	1		188	35.50		223.50	265	
	6540	3'-0" opening		16	1		196	35.50		231.50	274	
	6550	Louvered pine										
	6560	2'-6" opening	2 Carp	16	1	Ea.	164	35.50		199.50	239	
	8000	3'-0" opening		16	1		175	35.50		210.50	251	
	8010	2'-6" opening, hardwood		16	1		282	35.50		317.50	370	
	8020	3'-0" opening		16	1		310	35.50		345.50	405	
	8800	Pre-hung doors see Division 08210-720										
	9000	Minimum labor/equipment charge	1 Carp	4	2	Job		71		71	118	
920	0010	**WOOD DOORS, PANELED**										920
	0020	Interior, six panel, hollow core, 1-3/8" thick										
	0040	Molded hardboard, 2'-0" x 6'-8"	2 Carp	17	.941	Ea.	47.50	33.50		81	108	
	0060	2'-6" x 6'-8"		17	.941		51	33.50		84.50	112	
	0070	2'-8" x 6'-8"		17	.941		53.50	33.50		87	115	
	0080	3'-0" x 6'-8"		17	.941		56.50	33.50		90	118	
	0140	Embossed print, molded hardboard, 2'-0" x 6'-8"		17	.941		51	33.50		84.50	112	
	0160	2'-6" x 6'-8"		17	.941		51	33.50		84.50	112	
	0180	3'-0" x 6'-8"		17	.941		56.50	33.50		90	118	
	0540	Six panel, solid, 1-3/8" thick, pine, 2'-0" x 6'-8"		15	1.067		124	38		162	199	
	0560	2'-6" x 6'-8"		14	1.143		139	40.50		179.50	221	
	0580	3'-0" x 6'-8"		13	1.231		160	44		204	249	
	1020	Two panel, bored rail, solid, 1-3/8" thick, pine, 1'-6" x 6'-8"		16	1		226	35.50		261.50	310	
	1040	2'-0" x 6'-8"		15	1.067		297	38		335	390	
	1060	2'-6" x 6'-8"		14	1.143		340	40.50		380.50	445	
	1340	Two panel, solid, 1-3/8" thick, fir, 2'-0" x 6'-8"		15	1.067		124	38		162	199	
	1360	2'-6" x 6'-8"		14	1.143		139	40.50		179.50	221	
	1380	3'-0" x 6'-8"		13	1.231		340	44		384	450	
	1740	Five panel, solid, 1-3/8" thick, fir, 2'-0" x 6'-8"		15	1.067		222	38		260	305	
	1760	2'-6" x 6'-8"		14	1.143		355	40.50		395.50	460	
	1780	3'-0" x 6'-8"		13	1.231		355	44		399	465	
	9000	Minimum labor/equipment charge	1 Carp	4	2	Job		71		71	118	
930	0010	**WOOD DOORS, RESIDENTIAL**										930
	0200	Exterior, combination storm & screen, pine										

08200 | Wood and Plastic Doors

	08210 | Wood Doors		CREW	DAILY OUTPUT	LABOR-HOURS	UNIT	2006 BARE COSTS MAT.	LABOR	EQUIP.	TOTAL	TOTAL INCL O&P	
930	0260	2'-8" wide	2 Carp	10	1.600	Ea.	272	57		329	395	930
	0280	3'-0" wide		9	1.778		278	63		341	410	
	0300	7'-1" x 3'-0" wide		9	1.778		298	63		361	435	
	0400	Full lite, 6'-9" x 2'-6" wide		11	1.455		290	51.50		341.50	405	
	0420	2'-8" wide		10	1.600		290	57		347	415	
	0440	3'-0" wide		9	1.778		299	63		362	435	
	0500	7'-1" x 3'-0" wide		9	1.778		320	63		383	460	
	0700	Dutch door, pine, 1-3/4" x 6'-8" x 2'-8" wide, minimum		12	1.333		675	47.50		722.50	820	
	0720	Maximum		10	1.600		710	57		767	880	
	0800	3'-0" wide, minimum		12	1.333		685	47.50		732.50	835	
	0820	Maximum		10	1.600		750	57		807	920	
	1000	Entrance door, colonial, 1-3/4" x 6'-8" x 2'-8" wide		16	1		355	35.50		390.50	455	
	1020	6 panel pine, 3'-0" wide		15	1.067		400	38		438	505	
	1100	8 panel pine, 2'-8" wide		16	1		595	35.50		630.50	715	
	1120	3'-0" wide		15	1.067		535	38		573	655	
	1200	For tempered safety glass lites, (min of 2)add					62.50			62.50	68.50	
	1300	Flush, birch, solid core, 1-3/4" x 6'-8" x 2'-8" wide	2 Carp	16	1		94.50	35.50		130	163	
	1320	3'-0" wide		15	1.067		98	38		136	171	
	1350	7'-0" x 2'-8" wide		16	1		104	35.50		139.50	173	
	1360	3'-0" wide		15	1.067		111	38		149	185	
	1380	For tempered safety glass lites, add					93			93	102	
	1550	For handcarved door, see division 08210-910										
	2700	Interior, closet, bi-fold, w/hardware, no frame or trim incl.										
	2720	Flush, birch, 6'-6" or 6'-8" x 2'-6" wide	2 Carp	13	1.231	Ea.	48.50	44		92.50	126	
	2740	3'-0" wide		13	1.231		52.50	44		96.50	131	
	2760	4'-0" wide		12	1.333		96	47.50		143.50	184	
	2780	5'-0" wide		11	1.455		97	51.50		148.50	193	
	2800	6'-0" wide		10	1.600		103	57		160	208	
	2820	Flush, hardboard, primed, 6'-8" x 2'-6" wide		13	1.231		40.50	44		84.50	117	
	2840	3'-0" wide		13	1.231		43.50	44		87.50	121	
	2860	4'-0" wide		12	1.333		85	47.50		132.50	172	
	2880	5'-0" wide		11	1.455		92.50	51.50		144	187	
	2900	6'-0" wide		10	1.600		104	57		161	208	
	3000	Raised panel pine, 6'-6" or 6'-8" x 2'-6" wide		13	1.231		154	44		198	242	
	3020	3'-0" wide		13	1.231		196	44		240	288	
	3040	4'-0" wide		12	1.333		300	47.50		347.50	410	
	3060	5'-0" wide		11	1.455		360	51.50		411.50	480	
	3080	6'-0" wide		10	1.600		395	57		452	530	
	3200	Louvered, pine 6'-6" or 6'-8" x 2'-6" wide		13	1.231		104	44		148	187	
	3220	3'-0" wide		13	1.231		157	44		201	246	
	3240	4'-0" wide		12	1.333		194	47.50		241.50	292	
	3260	5'-0" wide		11	1.455		220	51.50		271.50	330	
	3280	6'-0" wide		10	1.600		243	57		300	360	
	4400	Bi-passing closet, incl. hardware and frame, no trim incl.										
	4420	Flush, lauan, 6'-8" x 4'-0" wide	2 Carp	12	1.333	Opng.	164	47.50		211.50	260	
	4440	5'-0" wide		11	1.455		180	51.50		231.50	284	
	4460	6'-0" wide		10	1.600		193	57		250	305	
	4600	Flush, birch, 6'-8" x 4'-0" wide		12	1.333		201	47.50		248.50	300	
	4620	5'-0" wide		11	1.455		203	51.50		254.50	310	
	4640	6'-0" wide		10	1.600		241	57		298	360	
	4800	Louvered, pine, 6'-8" x 4'-0" wide		12	1.333		400	47.50		447.50	520	
	4820	5'-0" wide		11	1.455		385	51.50		436.50	505	
	4840	6'-0" wide		10	1.600		490	57		547	635	
	5000	Paneled, pine, 6'-8" x 4'-0" wide		12	1.333		380	47.50		427.50	500	
	5020	5'-0" wide		11	1.455		400	51.50		451.50	525	
	5040	6'-0" wide		10	1.600		470	57		527	615	

	08210 | Wood Doors	CREW	DAILY OUTPUT	LABOR-HOURS	UNIT	2006 BARE COSTS MAT.	LABOR	EQUIP.	TOTAL	TOTAL INCL O&P	
930	6100 Folding accordion, closet, including track and frame										930
	6120 Vinyl, 2 layer, stock (see also Division 10651-100)	2 Carp	400	.040	S.F.	3.09	1.42		4.51	5.75	
	6200 Rigid PVC	"	400	.040	"	4.63	1.42		6.05	7.45	
	6220 For custom partition, add					25%	10%				
	7310 Passage doors, flush, no frame included										
	7320 Hardboard, hollow core, 1-3/8" x 6'-8" x 1'-6" wide	2 Carp	18	.889	Ea.	40	31.50		71.50	96.50	
	7330 2'-0" wide		18	.889		40.50	31.50		72	97	
	7340 2'-6" wide		18	.889		44.50	31.50		76	102	
	7350 2'-8" wide		18	.889		47	31.50		78.50	105	
	7360 3'-0" wide		17	.941		49.50	33.50		83	110	
	7420 Lauan, hollow core, 1-3/8" x 6'-8" x 1'-6" wide		18	.889		28	31.50		59.50	83.50	
	7440 2'-0" wide		18	.889		27	31.50		58.50	82.50	
	7450 2'-4" wide		18	.889		30.50	31.50		62	86	
	7460 2'-6" wide		18	.889		30.50	31.50		62	86	
	7480 2'-8" wide		18	.889		32	31.50		63.50	88	
	7500 3'-0" wide		17	.941		33.50	33.50		67	92.50	
	7700 Birch, hollow core, 1-3/8" x 6'-8" x 1'-6" wide		18	.889		35.50	31.50		67	91.50	
	7720 2'-0" wide		18	.889		40	31.50		71.50	96.50	
	7740 2'-6" wide		18	.889		44.50	31.50		76	102	
	7760 2'-8" wide		18	.889		46	31.50		77.50	103	
	7780 3'-0" wide		17	.941		50	33.50		83.50	111	
	8000 Pine louvered, 1-3/8" x 6'-8" x 1'-6" wide		19	.842		100	30		130	160	
	8020 2'-0" wide		18	.889		125	31.50		156.50	191	
	8040 2'-6" wide		18	.889		137	31.50		168.50	203	
	8060 2'-8" wide		18	.889		144	31.50		175.50	211	
	8080 3'-0" wide	↓	17	.941	↓	154	33.50		187.50	226	
	8090										
	8300 Pine paneled, 1-3/8" x 6'-8" x 1'-6" wide	2 Carp	19	.842	Ea.	108	30		138	169	
	8320 2'-0" wide		18	.889		125	31.50		156.50	191	
	8330 2'-4" wide		18	.889		137	31.50		168.50	203	
	8340 2'-6" wide		18	.889		140	31.50		171.50	207	
	8360 2'-8" wide		18	.889		151	31.50		182.50	220	
	8380 3'-0" wide		17	.941		158	33.50		191.50	230	
	8804 Pocket door,6 panel pine, 2'-6" x 6'-8"		10.50	1.524		204	54		258	315	
	8814 2'-8" x 6'-8"		10.50	1.524		215	54		269	325	
	8824 3'-0" x 6'-8"	↓	10.50	1.524	↓	224	54		278	335	
	9900 Minimum labor/equipment charge	1 Carp	4	2	Job		71		71	118	
950	0010 **WOOD FIRE DOORS**										950
	0020 Particle core, 7 face plys, "B" label,										
	0040 1 hour, birch face, 1-3/4" x 2'-6" x 6'-8"	2 Carp	14	1.143	Ea.	298	40.50		338.50	400	
	0080 3'-0" x 6'-8"		13	1.231		310	44		354	415	
	0090 3'-0" x 7'-0"		12	1.333		320	47.50		367.50	435	
	0100 4'-0" x 7'-0"		12	1.333		430	47.50		477.50	550	
	0140 Oak face, 2'-6" x 6'-8"		14	1.143		298	40.50		338.50	400	
	0180 3'-0" x 6'-8"		13	1.231		310	44		354	415	
	0190 3'-0" x 7'-0"		12	1.333		325	47.50		372.50	435	
	0200 4'-0" x 7'-0"		12	1.333		420	47.50		467.50	545	
	0240 Walnut face, 2'-6" x 6'-8"		14	1.143		390	40.50		430.50	500	
	0280 3'-0" x 6'-8"		13	1.231		400	44		444	515	
	0290 3'-0" x 7'-0"		12	1.333		420	47.50		467.50	540	
	0300 4'-0" x 7'-0"		12	1.333		565	47.50		612.50	700	
	0440 M.D. overlay on hardboard, 2'-6" x 6'-8"		15	1.067		262	38		300	350	
	0480 3'-0" x 6'-8"		14	1.143		272	40.50		312.50	370	
	0490 3'-0" x 7'-0"		13	1.231		287	44		331	390	
	0500 4'-0" x 7'-0"	↓	12	1.333	↓	350	47.50		397.50	465	

For expanded coverage of these items see *Means Interior Cost Data 2006*

		08210 Wood Doors	CREW	DAILY OUTPUT	LABOR-HOURS	UNIT	2006 BARE COSTS MAT.	LABOR	EQUIP.	TOTAL	TOTAL INCL O&P	
950	0540	H.P. plastic laminate, 2'-6" x 6'-8"	2 Carp	13	1.231	Ea.	350	44		394	460	950
	0590	3'-0" x 7'-0"		11	1.455		370	51.50		421.50	495	
	0600	4'-0" x 7'-0"		10	1.600		475	57		532	615	
	0740	90 minutes, birch face, 1-3/4" x 2'-6" x 6'-8"		14	1.143		256	40.50		296.50	350	
	0780	3'-0" x 6'-8"		13	1.231		262	44		306	360	
	0790	3'-0" x 7'-0"		12	1.333		315	47.50		362.50	430	
	0800	4'-0" x 7'-0"		12	1.333		380	47.50		427.50	495	
	0840	Oak face, 2'-6" x 6'-8"		14	1.143		269	40.50		309.50	365	
	0880	3'-0" x 6'-8"		13	1.231		279	44		323	380	
	0890	3'-0" x 7'-0"		12	1.333		293	47.50		340.50	400	
	0900	4'-0" x 7'-0"		12	1.333		405	47.50		452.50	525	
	0940	Walnut face, 2'-6" x 6'-8"		14	1.143		370	40.50		410.50	475	
	0980	3'-0" x 6'-8"		13	1.231		380	44		424	490	
	0990	3'-0" x 7'-0"		12	1.333		395	47.50		442.50	515	
	1000	4'-0" x 7'-0"		12	1.333		570	47.50		617.50	705	
	1140	M.D. overlay on hardboard, 2'-6" x 6'-8"		15	1.067		292	38		330	385	
	1180	3'-0" x 6'-8"		14	1.143		300	40.50		340.50	400	
	1190	3'-0" x 7'-0"		13	1.231		310	44		354	420	
	1200	4'-0" x 7'-0"		12	1.333		425	47.50		472.50	545	
	1240	For 8'-0" height, add					53.50			53.50	59	
	1260	For 8'-0" height walnut, add					71			71	78	
	1340	H.P. plastic laminate, 2'-6" x 6'-8"	2 Carp	13	1.231		365	44		409	475	
	1380	3'-0" x 6'-8"		12	1.333		380	47.50		427.50	500	
	1390	3'-0" x 7'-0"		11	1.455		385	51.50		436.50	510	
	1400	4'-0" x 7'-0"		10	1.600		500	57		557	645	
	2200	Custom architectural "B" label, flush, 1-3/4" thick, birch,										
	2210	Solid core										
	2220	2'-6" x 7'-0"	2 Carp	15	1.067	Ea.	248	38		286	335	
	2260	3'-0" x 7'-0"		14	1.143		257	40.50		297.50	350	
	2300	4'-0" x 7'-0"		13	1.231		360	44		404	470	
	2420	4'-0" x 8'-0"		11	1.455		450	51.50		501.50	580	
	2480	For oak veneer, add					50%					
	2500	For walnut veneer, add					75%					
	9000	Minimum labor/equipment charge	1 Carp	4	2	Job		71		71	118	
960	0010	**WOOD FRAMES**										960
	0400	Exterior frame, incl. ext. trim, pine, 5/4 x 4-9/16" deep	2 Carp	375	.043	L.F.	5.10	1.52		6.62	8.10	
	0420	5-3/16" deep		375	.043		7.95	1.52		9.47	11.25	
	0440	6-9/16" deep		375	.043		7.90	1.52		9.42	11.15	
	0600	Oak, 5/4 x 4-9/16" deep		350	.046		9.45	1.63		11.08	13.05	
	0620	5-3/16" deep		350	.046		10.60	1.63		12.23	14.35	
	0640	6-9/16" deep		350	.046		11.80	1.63		13.43	15.70	
	0800	Walnut, 5/4 x 4-9/16" deep		350	.046		11.10	1.63		12.73	14.90	
	0820	5-3/16" deep		350	.046		16	1.63		17.63	20.50	
	0840	6-9/16" deep		350	.046		18.95	1.63		20.58	23.50	
	1000	Sills, 8/4 x 8" deep, oak, no horns		100	.160		17.55	5.70		23.25	29	
	1020	2" horns		100	.160		15.45	5.70		21.15	26.50	
	1040	3" horns		100	.160		16	5.70		21.70	27	
	1100	8/4 x 10" deep, oak, no horns		90	.178		22.50	6.30		28.80	35	
	1120	2" horns		90	.178		21	6.30		27.30	33.50	
	1140	3" horns		90	.178		21	6.30		27.30	33.50	
	2000	Exterior, colonial, frame & trim, 3' opng., in-swing, minimum		22	.727	Ea.	310	26		336	385	
	2010	Average		21	.762		460	27		487	555	
	2020	Maximum		20	.800		1,050	28.50		1,078.50	1,200	
	2100	5'-4" opening, in-swing, minimum		17	.941		355	33.50		388.50	445	
	2120	Maximum		15	1.067		1,050	38		1,088	1,225	
	2140	Out-swing, minimum		17	.941		360	33.50		393.50	450	

08210 | Wood Doors

			CREW	DAILY OUTPUT	LABOR-HOURS	UNIT	2006 BARE COSTS MAT.	LABOR	EQUIP.	TOTAL	TOTAL INCL O&P	
960	2160	Maximum	2 Carp	15	1.067	Ea.	1,100	38		1,138	1,275	960
	2400	6'-0" opening, in-swing, minimum		16	1		340	35.50		375.50	430	
	2420	Maximum		10	1.600		1,100	57		1,157	1,300	
	2460	Out-swing, minimum		16	1		365	35.50		400.50	460	
	2480	Maximum		10	1.600		1,275	57		1,332	1,525	
	2600	For two sidelights, add, minimum		30	.533	Opng.	345	18.95		363.95	410	
	2620	Maximum		20	.800	"	1,125	28.50		1,153.50	1,275	
	2700	Custom birch frame, 3'-0" opening		16	1	Ea.	199	35.50		234.50	278	
	2750	6'-0" opening		16	1		300	35.50		335.50	390	
	2900	Exterior, modern, plain trim, 3' opng., in-swing, minimum		26	.615		33.50	22		55.50	73.50	
	2920	Average		24	.667		40	23.50		63.50	83.50	
	2940	Maximum		22	.727		48.50	26		74.50	96.50	
	3000	Interior frame, pine, 11/16" x 3-5/8" deep		375	.043	L.F.	4.31	1.52		5.83	7.25	
	3020	4-9/16" deep		375	.043		5.80	1.52		7.32	8.90	
	3200	Oak, 11/16" x 3-5/8" deep		350	.046		4	1.63		5.63	7.10	
	3220	4-9/16" deep		350	.046		4.31	1.63		5.94	7.45	
	3240	5-3/16" deep		350	.046		4.47	1.63		6.10	7.60	
	3400	Walnut, 11/16" x 3-5/8" deep		350	.046		6.75	1.63		8.38	10.15	
	3420	4-9/16" deep		350	.046		7.10	1.63		8.73	10.50	
	3440	5-3/16" deep		350	.046		7.40	1.63		9.03	10.85	
	3600	Pocket door frame		16	1	Ea.	62.50	35.50		98	128	
	3800	Threshold, oak, 5/8" x 3-5/8" deep		200	.080	L.F.	2.60	2.84		5.44	7.55	
	3820	4-5/8" deep		190	.084		3.20	2.99		6.19	8.50	
	3840	5-5/8" deep		180	.089		5.65	3.16		8.81	11.45	
	4000	For casing see division 06220-400 & 06220-800										
	9000	Minimum labor/equipment charge	1 Carp	4	2	Job		71		71	118	

08220 | Plastic Doors

			CREW	DAILY OUTPUT	LABOR-HOURS	UNIT	MAT.	LABOR	EQUIP.	TOTAL	TOTAL INCL O&P	
100	0010	**FIBERGLASS DOORS**										100
	0020	Exterior, fiberglass, door, 2'-8" wide x 6'-8" high	2 Carp	15	1.067	Ea.	305	38		343	400	
	0040	3'-0" wide x 6'-8" high		15	1.067		305	38		343	400	
	0060	3'-0" wide x 7'-0" high		15	1.067		375	38		413	480	
	0080	3'-0" wide x 6'-8" high, with two lites		15	1.067		370	38		408	470	
	0100	3'-0" wide x 7'-0" high, with two lites		15	1.067		440	38		478	550	
	0110	Half glass, 3'-0" wide x 6'-8" high		15	1.067		415	38		453	520	
	0120	3'-0" wide x 6'-8" high, low e		15	1.067		440	38		478	545	
	0130	3'-0" wide x 7'-0" high		15	1.067		485	38		523	595	
	0140	3'-0" wide x 7'-0" high, low e		15	1.067		510	38		548	625	
	0150	Side lights, 1'-0" wide x 6'-8" high,					281			281	310	
	0160	1'-0" wide x 6'-8" high, low e					293			293	325	
	0180	1'-0" wide x 6'-8" high, full glass					320			320	355	
	0190	1'-0" wide x 6'-8" high, low e					345			345	380	

08260 | Sliding Wood and Plastic Doors

			CREW	DAILY OUTPUT	LABOR-HOURS	UNIT	MAT.	LABOR	EQUIP.	TOTAL	TOTAL INCL O&P	
700	0010	**GLASS, SLIDING, VINYL**										700
	0012	Vinyl clad, 1" insul. glass, 6'-0" x 6'-10" high	2 Carp	4	4	Opng.	1,250	142		1,392	1,625	
	0030	6'-0" x 8'-0" high		4	4	Ea.	1,900	142		2,042	2,325	
	0100	8'-0" x 6'-10" high		4	4	Opng.	1,950	142		2,092	2,375	
	0500	3 leaf, 9'-0" x 6'-10" high		3	5.333		1,775	190		1,965	2,275	
	0600	12'-0" x 6'-10" high		3	5.333		2,200	190		2,390	2,750	
	9000	Minimum labor/equipment charge	1 Carp	4	2	Job		71		71	118	
900	0010	**GLASS, SLIDING, WOOD**										900
	0020	Wood, 5/8" tempered insul. glass, 6' wide, premium	2 Carp	4	4	Ea.	1,150	142		1,292	1,500	
	0100	Economy		4	4		785	142		927	1,100	
	0150	8' wide, wood, premium		3	5.333		1,350	190		1,540	1,800	

08200 | Wood and Plastic Doors

	08260	Sliding Wood and Plastic Doors	CREW	DAILY OUTPUT	LABOR-HOURS	UNIT	2006 BARE COSTS MAT.	LABOR	EQUIP.	TOTAL	TOTAL INCL O&P	
900	0200	Economy	2 Carp	3	5.333	Ea.	900	190		1,090	1,300	900
	0250	12′ wide, wood, vinyl clad		2.50	6.400		2,900	228		3,128	3,550	
	0300	Economy	↓	2.50	6.400	↓	2,050	228		2,278	2,625	
	0350	Aluminum, 5/8″ tempered insulated glass, 6′ wide										
	0400	Premium	2 Carp	4	4	Ea.	1,350	142		1,492	1,700	
	0450	Economy		4	4		705	142		847	1,000	
	0500	8′ wide, premium		3	5.333		1,525	190		1,715	2,000	
	0550	Economy		3	5.333		1,300	190		1,490	1,750	
	0600	12′ wide, premium		2.50	6.400		2,475	228		2,703	3,075	
	0650	Economy	↓	2.50	6.400	↓	1,475	228		1,703	1,975	
	1000	Replacement doors, wood										
	1050	6′ wide, vinyl clad	2 Carp	4	4	Ea.	900	142		1,042	1,225	
	9000	Minimum labor/equipment charge	1 Carp	2	4	Job		142		142	236	

08300 | Specialty Doors

	08310	Access Doors and Panels	CREW	DAILY OUTPUT	LABOR-HOURS	UNIT	2006 BARE COSTS MAT.	LABOR	EQUIP.	TOTAL	TOTAL INCL O&P	
100	0010	ACCESS DOORS										100
	1000	Fire rated door with lock										
	1100	Metal, 12″ x 12″	1 Carp	10	.800	Ea.	143	28.50		171.50	204	
	1150	18″ x 18″		9	.889		186	31.50		217.50	258	
	1200	24″ x 24″		9	.889		225	31.50		256.50	300	
	1250	24″ x 36″		8	1		300	35.50		335.50	390	
	1300	24″ x 48″		8	1		375	35.50		410.50	470	
	1350	36″ x 36″		7.50	1.067		450	38		488	560	
	1400	48″ x 48″		7.50	1.067		575	38		613	700	
	1600	Stainless steel, 12″ x 12″		10	.800		255	28.50		283.50	325	
	1650	18″ x 18″		9	.889		370	31.50		401.50	460	
	1700	24″ x 24″		9	.889		455	31.50		486.50	555	
	1750	24″ x 36″	↓	8	1	↓	580	35.50		615.50	700	
	2000	Flush door for finishing										
	2100	Metal 8″ x 8″	1 Carp	10	.800	Ea.	41	28.50		69.50	92	
	2150	12″ x 12″	″	10	.800	″	46	28.50		74.50	97.50	
	3000	Recessed door for acoustic tile										
	3100	Metal, 12″ x 12″	1 Carp	4.50	1.778	Ea.	63	63		126	174	
	3150	12″ x 24″		4.50	1.778		82	63		145	196	
	3200	24″ x 24″		4	2		110	71		181	239	
	3250	24″ x 36″	↓	4	2	↓	141	71		212	274	
	4000	Recessed door for drywall										
	4100	Metal 12″ x 12″	1 Carp	6	1.333	Ea.	70	47.50		117.50	156	
	4150	12″ x 24″		5.50	1.455		104	51.50		155.50	200	
	4200	24″ x 36″	↓	5	1.600	↓	166	57		223	276	
	6000	Standard door										
	6100	Metal, 8″ x 8″	1 Carp	10	.800	Ea.	36.50	28.50		65	87	
	6150	12″ x 12″		10	.800		41	28.50		69.50	92	
	6200	18″ x 18″		9	.889		57	31.50		88.50	115	
	6250	24″ x 24″		9	.889		73.50	31.50		105	134	
	6300	24″ x 36″		8	1		109	35.50		144.50	179	
	6350	36″ x 36″	↓	8	1	↓	133	35.50		168.50	205	

08300 | Specialty Doors

08310 | Access Doors and Panels

		08310 Access Doors and Panels	Crew	Daily Output	Labor-Hours	Unit	2006 Bare Costs Mat.	Labor	Equip.	Total	Total Incl O&P	
100	6500	Stainless steel, 8" x 8"	1 Carp	10	.800	Ea.	72.50	28.50		101	127	100
	6550	12" x 12"		10	.800		96.50	28.50		125	153	
	6600	18" x 18"		9	.889		179	31.50		210.50	250	
	6650	24" x 24"		9	.889		234	31.50		265.50	310	
	7010	Aluminum cover	G-3	11	2.909		535	101		636	750	
	9000	Minimum labor/equipment charge	1 Carp	4	2	Job		71		71	118	
150	0010	**BULKHEAD CELLAR DOORS**										150
	0020	Steel, not incl. sides, 44" x 62"	1 Carp	5.50	1.455	Ea.	214	51.50		265.50	320	
	0100	52" x 73"		5.10	1.569		238	56		294	355	
	0500	With sides and foundation plates, 57" x 45" x 24"		4.70	1.702		280	60.50		340.50	410	
	0600	42" x 49" x 51"		4.30	1.860		335	66		401	480	
	9000	Minimum labor/equipment charge		2	4	Job		142		142	236	
300	0010	**FLOOR, COMMERCIAL**										300
	0020	Aluminum tile, steel frame, one leaf, 2' x 2' opng.	2 Sswk	3.50	4.571	Opng.	385	183		568	760	
	0050	3'-6" x 3'-6" opening		3.50	4.571		695	183		878	1,100	
	0500	Double leaf, 4' x 4' opening		3	5.333		1,025	213		1,238	1,525	
	0550	5' x 5' opening		3	5.333		1,500	213		1,713	2,050	
	9000	Minimum labor/equipment charge		2	8	Job		320		320	600	
350	0010	**FLOOR, INDUSTRIAL**										350
	0020	Steel 300 psf L.L., single leaf, 2' x 2', 175#	2 Sswk	6	2.667	Opng.	565	107		672	820	
	0050	3' x 3' opening, 300#		5.50	2.909		780	116		896	1,075	
	0300	Double leaf, 4' x 4' opening, 455#		5	3.200		1,175	128		1,303	1,550	
	0350	5' x 5' opening, 645#		4.50	3.556		1,525	142		1,667	1,975	
	1000	Aluminum, 300 psf L.L., single leaf, 2' x 2', 60#		6	2.667		555	107		662	810	
	1050	3' x 3' opening, 100#		5.50	2.909		865	116		981	1,175	
	1500	Double leaf, 4' x 4' opening, 160#		5	3.200		1,375	128		1,503	1,750	
	1550	5' x 5' opening, 235#		4.50	3.556		1,800	142		1,942	2,275	
	9000	Minimum labor/equipment charge		2	8	Job		320		320	600	

08330 | Coiling Doors and Grilles

		08330 Coiling Doors and Grilles	Crew	Daily Output	Labor-Hours	Unit	Mat.	Labor	Equip.	Total	Total Incl O&P	
130	0010	**COUNTER DOORS** (security item)										130
	0020	Manual, incl. frm and hdwe, galv. stl., 4' roll-up, 6' long	2 Carp	2	8	Opng.	950	284		1,234	1,525	
	0300	Galvanized steel, UL label		1.80	8.889		1,050	315		1,365	1,675	
	0600	Stainless steel, 4' high roll-up, 6' long		2	8		1,450	284		1,734	2,075	
	0700	10' long		1.80	8.889		2,100	315		2,415	2,825	
	2000	Aluminum, 4' high, 4' long		2.20	7.273		920	259		1,179	1,425	
	2020	6' long		2	8		1,025	284		1,309	1,600	
	2040	8' long		1.90	8.421		1,175	299		1,474	1,800	
	2060	10' long		1.80	8.889		1,400	315		1,715	2,075	
	2080	14' long		1.40	11.429		2,025	405		2,430	2,925	
	2100	6' high, 4' long		2	8		1,025	284		1,309	1,600	
	2120	6' long		1.60	10		1,200	355		1,555	1,925	
	2140	10' long		1.40	11.429		1,625	405		2,030	2,475	
	9000	Minimum labor/equipment charge	1 Carp	2	4	Job		142		142	236	
640	0010	**COILING GRILLE**										640
	2020	Aluminum, manual operated, mill finish	2 Sswk	82	.195	S.F.	23	7.80		30.80	40	
	2040	Bronze anodized		82	.195	"	36.50	7.80		44.30	54.50	
	2060	Steel, manual operated, 10' x 10' high		1	16	Opng.	2,050	640		2,690	3,475	
	2080	15' x 8' high		.80	20	"	2,425	800		3,225	4,175	
	3000	For safety edge bottom bar, electric, add				L.F.	42.50			42.50	47	
	8000	For motor operation, add	2 Sswk	5	3.200	Opng.	1,050	128		1,178	1,400	
	9000	Minimum labor/equipment charge	"	1	16	Job		640		640	1,200	
720	0010	**ROLLING SERVICE DOORS** Steel, manual, 20 ga., incl. hardware										720
	0120	8' x 8' high, class A fire door	2 Sswk	1.40	11.429	Ea.	1,100	455		1,555	2,050	
	0130	12' x 12' high, standard		1.20	13.333		1,375	535		1,910	2,500	
	0140	12' x 12' high, class A fire door		1	16		1,925	640		2,565	3,325	

For expanded coverage of these items see *Means Interior Cost Data 2006*

08300 | Specialty Doors

08330 | Coiling Doors and Grilles

		08330 Coiling Doors and Grilles	Crew	Daily Output	Labor-Hours	Unit	2006 Bare Costs Mat.	Labor	Equip.	Total	Total Incl O&P	
720	0160	10′ x 20′ high, standard	2 Sswk	.50	32	Ea.	1,625	1,275		2,900	4,200	720
	0180	10′ x 20′ high, class A fire door		.40	40		2,900	1,600		4,500	6,200	
	3000	For 18 ga. doors, add				S.F.	.82			.82	.90	
	3300	For enamel finish, add				″	.98			.98	1.08	
	3600	For safety edge bottom bar, pneumatic, add				L.F.	12.85			12.85	14.15	
	4000	For weatherstripping, extruded rubber, jambs, add					8.35			8.35	9.15	
	4100	Hood, add					5.30			5.30	5.85	
	4200	Sill, add					3.03			3.03	3.33	
	4500	Motor operators, to 14′ x 14′ opening	2 Sswk	5	3.200	Ea.	780	128		908	1,100	
	4700	For fire door, additional fusible link, add				″	15.15			15.15	16.65	
	9000	Minimum labor/equipment charge	2 Sswk	1	16	Job		640		640	1,200	

08340 | Special Function Doors

		08340 Special Function Doors	Crew	Daily Output	Labor-Hours	Unit	Mat.	Labor	Equip.	Total	Total Incl O&P	
100	0010	COLD STORAGE										100
	0020	Single, 20 ga. galvanized steel										
	0300	Horizontal sliding, 5′ x 7′, manual operation, 3.5″ thick	2 Carp	2	8	Ea.	2,600	284		2,884	3,350	
	0400	4″ thick		2	8		3,175	284		3,459	3,950	
	0500	6″ thick		2	8		2,850	284		3,134	3,625	
	0800	5′ x 7′, power operation, 2″ thick		1.90	8.421		4,750	299		5,049	5,725	
	0900	4″ thick		1.90	8.421		4,825	299		5,124	5,800	
	1000	6″ thick		1.90	8.421		5,500	299		5,799	6,550	
	1300	9′ x 10′, manual operation, 2″ insulation		1.70	9.412		3,825	335		4,160	4,775	
	1400	4″ insulation		1.70	9.412		3,950	335		4,285	4,900	
	1500	6″ insulation		1.70	9.412		4,775	335		5,110	5,800	
	1800	Power operation, 2″ insulation		1.60	10		6,600	355		6,955	7,850	
	1900	4″ insulation		1.60	10		6,750	355		7,105	8,025	
	2000	6″ insulation		1.70	9.412		7,650	335		7,985	8,975	
	2300	For stainless steel face, add					20%					
	3000	Hinged, lightweight, 3′ x 7′-0″, galvanized 1 face, 2″ thick	2 Carp	2	8	Ea.	1,200	284		1,484	1,800	
	3050	4″ thick		1.90	8.421		1,400	299		1,699	2,025	
	3300	Aluminum doors, 3′ x 7′-0″, 4″ thick		1.90	8.421		1,125	299		1,424	1,750	
	3350	6″ thick		1.40	11.429		2,025	405		2,430	2,900	
	3600	Stainless steel, 3′ x 7′-0″, 4″ thick		1.90	8.421		1,450	299		1,749	2,100	
	3650	6″ thick		1.40	11.429		2,425	405		2,830	3,350	
	3900	Painted, 3′ x 7′-0″, 4″ thick		1.90	8.421		1,050	299		1,349	1,650	
	3950	6″ thick		1.40	11.429		1,975	405		2,380	2,850	
	5000	Bi-parting, electric operated										
	5010	6′ x 8′ opening, galv. faces, 4″ thick for cooler	2 Carp	.80	20	Opng.	6,200	710		6,910	8,000	
	5050	For freezer, 4″ thick		.80	20		6,825	710		7,535	8,700	
	5300	For door buck framing and door protection, add		2.50	6.400		505	228		733	935	
	6000	Galvanized batten door, galvanized hinges, 4′ x 7′		2	8		1,525	284		1,809	2,150	
	6050	6′ x 8′		1.80	8.889		2,075	315		2,390	2,825	
	6500	Fire door, 3 hr., 6′ x 8′, single slide		.80	20		7,175	710		7,885	9,050	
	6550	Double, bi-parting		.70	22.857		10,700	815		11,515	13,200	
	9000	Minimum labor/equipment charge	1 Carp	2	4	Job		142		142	236	
610	0010	ACOUSTICAL DOORS										610
	0020	Including framed seals, 3′ x 7′, wood, 27 STC rating	2 Carp	1.50	10.667	Ea.	385	380		765	1,050	
	0100	Steel, 40 STC rating		1.50	10.667		1,450	380		1,830	2,225	
	0200	45 STC rating		1.50	10.667		1,925	380		2,305	2,750	
	0300	48 STC rating		1.50	10.667		2,425	380		2,805	3,300	
	0400	52 STC rating		1.50	10.667		2,925	380		3,305	3,825	
	9000	Minimum labor/equipment charge	1 Carp	4	2	Job		71		71	118	
700	0010	DOUBLE ACTING, SWING										700
	0020	Including frame, closer, hardware and vision panel										

08300 | Specialty Doors

	08340 | Special Function Doors		CREW	DAILY OUTPUT	LABOR-HOURS	UNIT	2006 BARE COSTS MAT.	LABOR	EQUIP.	TOTAL	TOTAL INCL O&P	
700	1000	.063″ aluminum, 7′-0″ high, 4′-0″ wide	2 Carp	4.20	3.810	Pr.	1,825	135		1,960	2,225	700
	1050	6′-0″ wide	″	4	4	″	2,400	142		2,542	2,875	
	2000	Solid core wood, 3/4″ thick, metal frame, stainless steel										
	2010	base plate, 7′ high opening, 4′ wide	2 Carp	4	4	Pr.	2,175	142		2,317	2,625	
	2050	7′ wide		3.80	4.211	″	2,400	150		2,550	2,900	
	9000	Minimum labor/equipment charge		2	8	Job		284		284	470	
710	0010	**GLASS DOOR, SWING**										710
	0020	Including hardware, 1/2″ thick, tempered, 3′ x 7′ opening	2 Glaz	2	8	Opng.	1,800	274		2,074	2,425	
	0100	6′ x 7′ opening		1.40	11.429	″	3,500	390		3,890	4,475	
	9000	Minimum labor/equipment charge		2	8	Job		274		274	440	
720	0010	**SHOCK ABSORBING DOORS**										720
	0020	Rigid, no frame, 1-1/2″ thick, 5′ x 7′	2 Sswk	1.90	8.421	Opng.	1,275	335		1,610	2,025	
	0100	8′ x 8′		1.80	8.889		1,800	355		2,155	2,650	
	0500	Flexible, no frame, insulated, .16″ thick, economy, 5′ x 7′		2	8		1,575	320		1,895	2,325	
	0600	Deluxe		1.90	8.421		2,375	335		2,710	3,225	
	1000	8′ x 8′ opening, economy		2	8		2,450	320		2,770	3,275	
	1100	Deluxe		1.90	8.421		3,150	335		3,485	4,075	
	9000	Minimum labor/equipment charge		2	8	Job		320		320	600	
	08360 | Overhead Doors											
550	0010	**OVERHEAD, COMMERCIAL** Frames not included										550
	1000	Stock, sectional, heavy duty, wood, 1-3/4″ thick, 8′ x 8′ high	2 Carp	2	8	Ea.	605	284		889	1,125	
	1200	12′ x 12′ high		1.50	10.667		1,325	380		1,705	2,075	
	1300	Chain hoist, 14′ x 14′ high		1.30	12.308		1,925	440		2,365	2,850	
	1600	20′ x 16′ high		.65	24.615		3,975	875		4,850	5,800	
	2100	For medium duty custom door, deduct					5%	5%				
	2150	For medium duty stock doors, deduct					10%	5%				
	2300	Fiberglass and aluminum, heavy duty, sectional, 12′ x 12′ high	2 Carp	1.50	10.667	Ea.	1,925	380		2,305	2,725	
	2450	Chain hoist, 20′ x 20′ high	″	.50	32		4,775	1,150		5,925	7,125	
	2900	For electric trolley operator, 1/3 H.P., to 12′ x 12′, add	1 Carp	2	4		700	142		842	1,000	
	2950	Over 12′ x 12′, 1/2 H.P., add	″	1	8		785	284		1,069	1,325	
	9000	Minimum labor/equipment charge	2 Carp	1.50	10.667	Job		380		380	630	
600	0010	**RESIDENTIAL GARAGE DOORS** Including hardware, no frame										600
	0050	Hinged, wood, custom, double door, 9′ x 7′	2 Carp	4	4	Ea.	390	142		532	660	
	0070	16′ x 7′		3	5.333		660	190		850	1,050	
	0200	Overhead, sectional, incl. hardware, fiberglass, 9′ x 7′, standard		5.28	3.030		610	108		718	850	
	0220	Deluxe		5.28	3.030		775	108		883	1,025	
	0300	16′ x 7′, standard		6	2.667		1,100	95		1,195	1,350	
	0320	Deluxe		6	2.667		1,350	95		1,445	1,650	
	0500	Hardboard, 9′ x 7′, standard		8	2		405	71		476	565	
	0520	Deluxe		8	2		540	71		611	715	
	0600	16′ x 7′, standard		6	2.667		790	95		885	1,025	
	0620	Deluxe		6	2.667		920	95		1,015	1,175	
	0700	Metal, 9′ x 7′, standard		5.28	3.030		475	108		583	700	
	0720	Deluxe		8	2		635	71		706	820	
	0800	16′ x 7′, standard		3	5.333		605	190		795	980	
	0820	Deluxe		6	2.667		965	95		1,060	1,225	
	0900	Wood, 9′ x 7′, standard		8	2		500	71		571	670	
	0920	Deluxe		8	2		1,425	71		1,496	1,700	
	1000	16′ x 7′, standard		6	2.667		1,000	95		1,095	1,250	
	1020	Deluxe		6	2.667		2,100	95		2,195	2,450	
	1800	Door hardware, sectional	1 Carp	4	2		220	71		291	360	
	1810	Door tracks only		4	2		102	71		173	230	
	1820	One side only		7	1.143		71	40.50		111.50	146	
	3000	Swing-up, including hardware, fiberglass, 9′ x 7′, standard	2 Carp	8	2		690	71		761	875	
	3020	Deluxe		8	2		720	71		791	915	

DOORS & WINDOWS 8

08360 | Overhead Doors

			CREW	DAILY OUTPUT	LABOR-HOURS	UNIT	2006 BARE COSTS MAT.	LABOR	EQUIP.	TOTAL	TOTAL INCL O&P	
600	3100	16′ x 7′, standard	2 Carp	6	2.667	Ea.	870	95		965	1,100	600
	3120	Deluxe		6	2.667		895	95		990	1,150	
	3200	Hardboard, 9′ x 7′, standard		8	2		315	71		386	465	
	3220	Deluxe		8	2		420	71		491	580	
	3300	16′ x 7′, standard		6	2.667		440	95		535	640	
	3320	Deluxe		6	2.667		655	95		750	875	
	3400	Metal, 9′ x 7′, standard		8	2		345	71		416	500	
	3420	Deluxe		8	2		610	71		681	790	
	3500	16′ x 7′, standard		6	2.667		540	95		635	750	
	3520	Deluxe		6	2.667		870	95		965	1,125	
	3600	Wood, 9′ x 7′, standard		8	2		380	71		451	535	
	3620	Deluxe		8	2		665	71		736	850	
	3700	16′ x 7′, standard		6	2.667		655	95		750	875	
	3720	Deluxe	↓	6	2.667		925	95		1,020	1,175	
	3900	Door hardware only, swing up	1 Carp	4	2		109	71		180	238	
	3920	One side only		7	1.143		61.50	40.50		102	135	
	4000	For electric operator, economy, add		8	1		287	35.50		322.50	375	
	4100	Deluxe, including remote control	↓	8	1	↓	420	35.50		455.50	520	
	4500	For transmitter/receiver control , add to operator				Total	86.50			86.50	95.50	
	4600	Transmitters, additional				″	31.50			31.50	35	
	6000	Replace section, on sectional door, fiberglass, 9′ x 7′	1 Carp	4	2	Ea.	171	71		242	305	
	6020	16′ x 7′		3.50	2.286		257	81.50		338.50	420	
	6200	Hardboard, 9′ x 7′		4	2		89.50	71		160.50	217	
	6220	16′ x 7′		3.50	2.286		171	81.50		252.50	325	
	6300	Metal, 9′ x 7′		4	2		145	71		216	277	
	6320	16′ x 7′		3.50	2.286		238	81.50		319.50	395	
	6500	Wood, 9′ x 7′		4	2		88	71		159	215	
	6520	16′ x 7′		3.50	2.286	↓	171	81.50		252.50	325	
	9000	Minimum labor/equipment charge	↓	2.50	3.200	Job		114		114	188	

08370 | Vertical Lift Doors

			CREW	DAILY OUTPUT	LABOR-HOURS	UNIT	MAT.	LABOR	EQUIP.	TOTAL	TOTAL INCL O&P	
950	0010	VERTICAL LIFT DOORS										950
	0020	Motorized, 14 ga. stl, incl., frm and ctrl pnl										
	0050	16′ x 16′ high	L-10	.50	48	Ea.	17,600	1,925	1,275	20,800	24,200	
	0100	10′ x 20′ high		1.30	18.462		23,900	740	490	25,130	28,200	
	0120	15′ x 20′ high		1.30	18.462		29,300	740	490	30,530	34,100	
	0140	20′ x 20′ high		1	24		34,400	960	635	35,995	40,200	
	0160	25′ x 20′ high		1	24		38,100	960	635	39,695	44,300	
	0170	32′ x 24′ high		.75	32		35,100	1,275	850	37,225	41,800	
	0180	20′ x 25′ high		1	24		39,400	960	635	40,995	45,700	
	0200	25′ x 25′ high		.70	34.286		44,400	1,375	910	46,685	52,500	
	0220	25′ x 30′ high		.70	34.286		52,000	1,375	910	54,285	60,500	
	0240	30′ x 30′ high		.70	34.286		59,500	1,375	910	61,785	69,000	
	0260	35′ x 30′ high	↓	.70	34.286	↓	65,500	1,375	910	67,785	75,500	

	08410 | Metal-Framed Storefronts		Crew	Daily Output	Labor-Hours	Unit	2006 Bare Costs Mat.	Labor	Equip.	Total	Total Incl O&P	
110	0010	**ALUMINUM-FRAMED STOREFRONTS,** No glazing										110
	0020	Entrance, 3′ x 7′ opening, clear anodized finish	2 Sswk	7	2.286	Opng.	254	91.50		345.50	450	
	0100	Bronze finish		7	2.286		340	91.50		431.50	540	
	0500	6′ x 7′ opening, clear finish		6	2.667		310	107		417	540	
	0520	Bronze finish		6	2.667		350	107		457	585	
	1000	With 3′ high transoms, 3′ x 10′ opening, clear finish		6.50	2.462		370	98.50		468.50	595	
	1050	Bronze finish		6.50	2.462		410	98.50		508.50	635	
	1100	Black finish		6.50	2.462		485	98.50		583.50	715	
	1500	With 3′ high transoms, 6′ x 10′ opening, clear finish		5.50	2.909		450	116		566	710	
	1550	Bronze finish		5.50	2.909		485	116		601	745	
	1600	Black finish		5.50	2.909		575	116		691	845	
	9000	Minimum labor/equipment charge		4	4	Job		160		160	299	
120	0010	**ALUMINUM DOORS** Commercial entrance, no glazing										120
	0020	Incl. hinges, push/pull, deadlock, cyl., threshold										
	0800	Narrow stile, no glazing, standard hardware, pair of 2′-6″ x 7′-0″	2 Carp	1.70	9.412	Pr.	770	335		1,105	1,400	
	1000	3′-0″ x 7′-0″, single		3	5.333	Ea.	510	190		700	875	
	1200	Pair of 3′-0″ x 7′-0″		1.70	9.412	Pr.	1,025	335		1,360	1,675	
	1500	3′-6″ x 7′-0″, single		3	5.333	Ea.	545	190		735	915	
	2000	Medium stile, pair of 2′-6″ x 7′-0″		1.70	9.412	Pr.	955	335		1,290	1,600	
	2100	3′-0″ x 7′-0″, single		3	5.333	Ea.	670	190		860	1,050	
	2200	Pair of 3′-0″ x 7′-0″		1.70	9.412	Pr.	1,300	335		1,635	2,000	
	2300	3′-6″ x 7′-0″, single		5.33	3.002	Ea.	790	107		897	1,050	
	5000	Flush panel doors, pair of 2′-6″ x 7′-0″	2 Sswk	2	8	Pr.	945	320		1,265	1,650	
	5050	3′-0″ x 7′-0″, single		2.50	6.400	Ea.	470	256		726	1,000	
	5100	Pair of 3′-0″ x 7′-0″		2	8	Pr.	945	320		1,265	1,650	
	5150	3′-6″ x 7′-0″, single		2.50	6.400	Ea.	560	256		816	1,100	
130	0010	**ALUMINUM DOORS & FRAMES** Entrance, narrow stile, including										130
	0015	Standard hardware, clear finish, not incl. glass, 2′-6″ x 7′-0″ opng.	2 Sswk	2	8	Ea.	695	320		1,015	1,350	
	0020	3′-0″ x 7′-0″ opening		2	8		560	320		880	1,225	
	0030	3′-6″ x 7′-0″ opening		2	8		500	320		820	1,150	
	0100	3′-0″ x 10′-0″ opening, 3′ high transom		1.80	8.889		785	355		1,140	1,525	
	0200	3′-6″ x 10′-0″ opening, 3′ high transom		1.80	8.889		770	355		1,125	1,525	
	0280	5′-0″ x 7′-0″ opening		2	8		850	320		1,170	1,550	
	0300	6′-0″ x 7′-0″ opening		1.30	12.308	Pr.	865	490		1,355	1,875	
	0400	6′-0″ x 10′-0″ opening, 3′ high transom		1.10	14.545		730	580		1,310	1,875	
	0420	7′-0″ x 7′-0″ opening		1	16		830	640		1,470	2,125	
	0500	Wide stile, 2′-6″ x 7′-0″ opening		2	8	Ea.	680	320		1,000	1,350	
	0520	3′-0″ x 7′-0″ opening		2	8		695	320		1,015	1,375	
	0540	3′-6″ x 7′-0″ opening		2	8		740	320		1,060	1,425	
	0560	5′-0″ x 7′-0″ opening		2	8		1,200	320		1,520	1,925	
	0580	6′-0″ x 7′-0″ opening		1.30	12.308	Pr.	1,150	490		1,640	2,200	
	0600	7′-0″ x 7′-0″ opening		1	16	″	1,250	640		1,890	2,575	
	1100	For full vision doors, with 1/2″ glass, add				Leaf	55%					
	1200	For non-standard size, add					67%					
	1300	Light bronze finish, add					36%					
	1400	Dark bronze finish, add					18%					
	1500	For black finish, add					36%					
	1600	Concealed panic device, add					1,025			1,025	1,125	
	1700	Electric striker release, add				Opng.	262			262	288	
	1800	Floor check, add				Leaf	775			775	855	
	1900	Concealed closer, add				″	515			515	570	
	2000	Flush 3′ x 7′ Insulated, 12″x 12″ lite, clear finish	2 Sswk	2	8	Ea.	975	320		1,295	1,675	
	9000	Minimum labor/equipment charge	2 Carp	4	4	Job		142		142	236	
140	0010	**STOREFRONT SYSTEMS** Aluminum frame, clear 3/8″ plate glass,										140
	0020	incl. 3′ x 7′ door with hardware (400 sq. ft. max. wall)										

08400 | Entrances and Storefronts

	08410 | Metal-Framed Storefronts		CREW	DAILY OUTPUT	LABOR-HOURS	UNIT	2006 BARE COSTS MAT.	LABOR	EQUIP.	TOTAL	TOTAL INCL O&P	
140	0500	Wall height to 12′ high, commercial grade	2 Glaz	150	.107	S.F.	13.85	3.65		17.50	21	140
	0600	Institutional grade		130	.123		17.40	4.22		21.62	26	
	0700	Monumental grade		115	.139		26.50	4.77		31.27	36.50	
	1000	6′ x 7′ door with hardware, commercial grade		135	.119		14.15	4.06		18.21	22	
	1100	Institutional grade		115	.139		19.35	4.77		24.12	29	
	1200	Monumental grade	↓	100	.160		36	5.50		41.50	48.50	
	1500	For bronze anodized finish, add					15%					
	1600	For black anodized finish, add					30%					
	1700	For stainless steel framing, add to monumental				↓	75%					
	9000	Minimum labor/equipment charge	2 Glaz	1	16	Job		550		550	885	
300	0010	**STAINLESS STEEL AND GLASS** Entrance unit, narrow stiles										300
	0020	3′ x 7′ opening, including hardware, minimum	2 Sswk	1.60	10	Opng.	4,825	400		5,225	6,050	
	0050	Average		1.40	11.429		5,225	455		5,680	6,600	
	0100	Maximum	↓	1.20	13.333		5,600	535		6,135	7,150	
	1000	For solid bronze entrance units, statuary finish, add					60%					
	1100	Without statuary finish, add				↓	45%					
	2000	Balanced doors, 3′ x 7′, economy	2 Sswk	.90	17.778	Ea.	6,550	710		7,260	8,525	
	2100	Premium		.70	22.857	"	11,300	915		12,215	14,100	
	9000	Minimum labor/equipment charge	↓	2	8	Job		320		320	600	

	08460 | Automatic Entrance Doors		CREW	DAILY OUTPUT	LABOR-HOURS	UNIT	MAT.	LABOR	EQUIP.	TOTAL	TOTAL INCL O&P	
600	0010	**SLIDING ENTRANCE** 12′ x 7′-6″ opng., 5′ x 7′ door, 2 way traf.,										600
	0020	mat activated, panic pushout, incl. operator & hardware,										
	0030	not including glass or glazing	2 Glaz	.70	22.857	Opng.	6,200	785		6,985	8,075	
	9000	Minimum labor/equipment charge	"	.70	22.857	Job		785		785	1,250	

	08480 | Balanced Entrance Doors		CREW	DAILY OUTPUT	LABOR-HOURS	UNIT	MAT.	LABOR	EQUIP.	TOTAL	TOTAL INCL O&P	
150	0010	**BALANCED DOORS**										150
	0020	Hardware & frame, alum. & glass, 3′ x 7′, econ.	2 Sswk	.90	17.778	Ea.	5,300	710		6,010	7,150	
	0150	Premium		.70	22.857	"	6,600	915		7,515	8,950	
	9000	Minimum labor/equipment charge	↓	1	16	Job		640		640	1,200	

	08490 | Sliding Storefronts		CREW	DAILY OUTPUT	LABOR-HOURS	UNIT	MAT.	LABOR	EQUIP.	TOTAL	TOTAL INCL O&P	
100	0010	**SLIDING PANELS**										100
	0020	Mall fronts, aluminum & glass, 15′ x 9′ high	2 Glaz	1.30	12.308	Opng.	2,425	420		2,845	3,350	
	0100	24′ x 9′ high		.70	22.857		3,525	785		4,310	5,125	
	0200	48′ x 9′ high, with fixed panels	↓	.90	17.778		6,575	610		7,185	8,200	
	0500	For bronze finish, add				↓	17%					
	9000	Minimum labor/equipment charge	2 Glaz	1	16	Job		550		550	885	

8 DOORS & WINDOWS

08500 | Windows

	08510 | Steel Windows		CREW	DAILY OUTPUT	LABOR-HOURS	UNIT	2006 BARE COSTS MAT.	LABOR	EQUIP.	TOTAL	TOTAL INCL O&P	
700	0010	**SCREENS**										700
	0020	For metal sash, aluminum or bronze mesh, flat screen	2 Sswk	1200	.013	S.F.	3.61	.53		4.14	4.97	
	0500	Wicket screen, inside window		1000	.016		5.50	.64		6.14	7.25	
	0800	Security screen, aluminum frame with stainless steel cloth		1200	.013		19.45	.53		19.98	22.50	
	0900	Steel grate, painted, on steel frame		1600	.010		10.45	.40		10.85	12.25	
	1000	For solar louvers, add	↓	160	.100	↓	20	4		24	29.50	

08510 | Steel Windows

		08510 Steel Windows	CREW	DAILY OUTPUT	LABOR-HOURS	UNIT	2006 BARE COSTS MAT.	LABOR	EQUIP.	TOTAL	TOTAL INCL O&P	
700	4000	See also Division 05580-900										700
750	0010	**STEEL SASH** Custom units, glazing and trim not included										750
	0100	Casement, 100% vented	2 Sswk	200	.080	S.F.	42	3.20		45.20	52	
	0200	50% vented		200	.080		38	3.20		41.20	48	
	0300	Fixed		200	.080		26	3.20		29.20	35	
	1000	Projected, commercial, 40% vented		200	.080		44	3.20		47.20	54	
	1100	Intermediate, 50% vented		200	.080		48	3.20		51.20	58.50	
	1500	Industrial, horizontally pivoted		200	.080		45.50	3.20		48.70	56	
	1600	Fixed		200	.080		26	3.20		29.20	34.50	
	2000	Industrial security sash, 50% vented		200	.080		49	3.20		52.20	59.50	
	2100	Fixed		200	.080		39.50	3.20		42.70	49.50	
	2500	Picture window		200	.080		24.50	3.20		27.70	33	
	3000	Double hung		200	.080		47	3.20		50.20	58	
	5000	Mullions for above, open interior face		240	.067	L.F.	8.45	2.66		11.11	14.25	
	5100	With interior cover		240	.067	"	13.95	2.66		16.61	20.50	
	5200	Single glazing for above, add	2 Glaz	200	.080	S.F.	5.40	2.74		8.14	10.30	
	6000	Double glazing for above, add		200	.080		10.90	2.74		13.64	16.40	
	6100	Triple glazing for above, add		85	.188		10.35	6.45		16.80	22	
	9000	Minimum labor/equipment charge	1 Sswk	2	4	Job		160		160	299	
770	0010	**STEEL WINDOWS** Stock, including frame, trim and insul. glass R085123-10										770
	1000	Custom units, double hung, 2'-8" x 4'-6" opening	2 Sswk	12	1.333	Ea.	595	53.50		648.50	755	
	1100	2'-4" x 3'-9" opening		12	1.333		490	53.50		543.50	640	
	1500	Commercial projected, 3'-9" x 5'-5" opening		10	1.600		1,050	64		1,114	1,275	
	1600	6'-9" x 4'-1" opening		7	2.286		1,375	91.50		1,466.50	1,700	
	2000	Intermediate projected, 2'-9" x 4'-1" opening		12	1.333		585	53.50		638.50	740	
	2100	4'-1" x 5'-5" opening		10	1.600		1,175	64		1,239	1,425	
	9000	Minimum labor/equipment charge	1 Sswk	3	2.667	Job		107		107	199	

08520 | Aluminum Windows

		08520 Aluminum Windows	CREW	DAILY OUTPUT	LABOR-HOURS	UNIT	MAT.	LABOR	EQUIP.	TOTAL	TOTAL INCL O&P	
100	0010	**ALUMINUM SASH**										100
	0020	Stock, grade C, glaze & trim not incl., casement	2 Sswk	200	.080	S.F.	29.50	3.20		32.70	38	
	0050	Double hung		200	.080		29.50	3.20		32.70	38.50	
	0100	Fixed casement		200	.080		11.70	3.20		14.90	18.85	
	0150	Picture window		200	.080		12.55	3.20		15.75	19.85	
	0200	Projected window		200	.080		27	3.20		30.20	35.50	
	0250	Single hung		200	.080		13.80	3.20		17	21	
	0300	Sliding		200	.080		17.90	3.20		21.10	25.50	
	1000	Mullions for above, tubular		240	.067	L.F.	4.60	2.66		7.26	10.05	
	2950	Single glazing for above, add	2 Glaz	200	.080	S.F.	6	2.74		8.74	11	
	3000	Double glazing for above, add		200	.080		9.65	2.74		12.39	15	
	3100	Triple glazing for above, add		85	.188		11.30	6.45		17.75	23	
	9000	Minimum labor/equipment charge	1 Sswk	2	4	Job		160		160	299	
120	0010	**ALUMINUM WINDOWS** Incl. frame and glazing, Commercial grade										120
	1000	Stock units, casement, 3'-1" x 3'-2" opening	2 Sswk	10	1.600	Ea.	310	64		374	460	
	1050	Add for storms					62.50			62.50	68.50	
	1600	Projected, with screen, 3'-1" x 3'-2" opening	2 Sswk	10	1.600		222	64		286	365	
	1700	Add for storms					59			59	64.50	
	2000	4'-5" x 5'-3" opening	2 Sswk	8	2		310	80		390	495	
	2100	Add for storms					81.50			81.50	90	
	2500	Enamel finish windows, 3'-1" x 3'-2"	2 Sswk	10	1.600		199	64		263	340	
	2600	4'-5" x 5'-3"		8	2		298	80		378	480	
	3000	Single hung, 2' x 3' opening, enameled, standard glazed		10	1.600		149	64		213	284	
	3100	Insulating glass		10	1.600		181	64		245	320	
	3300	2'-8" x 6'-8" opening, standard glazed		8	2		315	80		395	500	

	08520	Aluminum Windows	CREW	DAILY OUTPUT	LABOR-HOURS	UNIT	2006 BARE COSTS MAT.	LABOR	EQUIP.	TOTAL	TOTAL INCL O&P
120	3400	Insulating glass	2 Sswk	8	2	Ea.	405	80		485	600
	3700	3'-4" x 5'-0" opening, standard glazed		9	1.778		204	71		275	355
	3800	Insulating glass		9	1.778		287	71		358	450
	4000	Sliding aluminum, 3' x 2' opening, standard glazed		10	1.600		168	64		232	305
	4100	Insulating glass		10	1.600		186	64		250	325
	4300	5' x 3' opening, standard glazed		9	1.778		213	71		284	365
	4400	Insulating glass		9	1.778		298	71		369	465
	4600	8' x 4' opening, standard glazed		6	2.667		305	107		412	535
	4700	Insulating glass		6	2.667		495	107		602	740
	5000	9' x 5' opening, standard glazed		4	4		465	160		625	810
	5100	Insulating glass		4	4		740	160		900	1,125
	5500	Sliding, with thermal barrier and screen, 6' x 4', 2 track		8	2		630	80		710	845
	5700	4 track		8	2		770	80		850	1,000
	6000	For above units with bronze finish, add					12%				
	6200	For installation in concrete openings, add					5%				
	6400										
	9000	Minimum labor/equipment charge	1 Sswk	3	2.667	Job		107		107	199
500	0010	**JALOUSIES**									
	0020	Aluminum incl. glazing & screens, stock, 1'-7" x 3'-2"	2 Sswk	10	1.600	Ea.	136	64		200	270
	0100	2'-3" x 4'-0"		10	1.600		194	64		258	335
	0200	3'-1" x 2'-0"		10	1.600		153	64		217	289
	0300	3'-1" x 5'-3"		10	1.600		277	64		341	425
	1000	Mullions for above, 2'-0" long		80	.200		10.45	8		18.45	26.50
	1100	5'-3" long		80	.200		17.90	8		25.90	34.50
	9000	Minimum labor/equipment charge	1 Sswk	3.50	2.286	Job		91.50		91.50	171
	08550	**Wood Windows**									
100	0010	**AWNING WINDOW** Including frame, screens and grills									
	0100	Average quality, builders model, 34" x 22", double insulated glass	1 Carp	10	.800	Ea.	233	28.50		261.50	305
	0200	Low E glass		10	.800		246	28.50		274.50	315
	0300	40" x 28", double insulated glass		9	.889		294	31.50		325.50	380
	0400	Low E Glass		9	.889		310	31.50		341.50	400
	0500	48" x 36", double insulated glass		8	1		430	35.50		465.50	530
	0600	Low E glass		8	1		450	35.50		485.50	555
	1000	34" x 22"		10	.800		247	28.50		275.50	320
	1100	40" x 22"		10	.800		270	28.50		298.50	345
	1200	36" x 28"		9	.889		287	31.50		318.50	370
	1300	36" x 36"		9	.889		320	31.50		351.50	410
	1400	48" x 28"		8	1		345	35.50		380.50	440
	1500	60" x 36"		8	1		500	35.50		535.50	610
	2000	Metal clad, deluxe, double insulated glass, 34" x 22"		10	.800		231	28.50		259.50	300
	2100	40" x 22"		10	.800		271	28.50		299.50	345
	2200	36" x 25"		9	.889		251	31.50		282.50	330
	2300	40" x 30"		9	.889		315	31.50		346.50	400
	2400	48" x 28"		8	1		320	35.50		355.50	410
	2500	60" x 36"		8	1		340	35.50		375.50	435
	9000	Minimum labor/equipment charge		4	2	Job		71		71	118
150	0010	**BOW-BAY WINDOW** Including frame, screens and grills,									
	0020	end panels operable									
	1000	Bow type, casement, wood, bldrs mdl, 8' x 5' dbl insltd glass, 4 panel	2 Carp	10	1.600	Ea.	1,500	57		1,557	1,750
	1050	Low E glass		10	1.600		1,200	57		1,257	1,425
	1100	10'-0" x 5'-0", double insulated glass, 6 panels		6	2.667		1,250	95		1,345	1,525
	1200	Low E glass, 6 panels		6	2.667		1,325	95		1,420	1,600
	1300	Vinyl clad, bldrs model, double insulated glass, 6'-0" x 4'-0", 3 panel		10	1.600		1,375	57		1,432	1,600
	1340	9'-0" x 4'-0", 4 panel		8	2		1,375	71		1,446	1,625

	08550 | Wood Windows		CREW	DAILY OUTPUT	LABOR-HOURS	UNIT	2006 BARE COSTS MAT.	LABOR	EQUIP.	TOTAL	TOTAL INCL O&P	
150	1380	10′-0″ x 6′-0″, 5 panels	2 Carp	7	2.286	Ea.	2,200	81.50		2,281.50	2,550	150
	1420	12′-0″ x 6′-0″, 6 panels		6	2.667		2,775	95		2,870	3,200	
	1600	Metal clad, casement, bldrs mdl, 6′-0″ x 4′-0″, dbl insltd gls, 3 panels		10	1.600		855	57		912	1,025	
	1640	9′-0″ x 4′-0″ , 4 panels		8	2		1,200	71		1,271	1,450	
	1680	10′-0″ x 5′-0″, 5 panels		7	2.286		1,650	81.50		1,731.50	1,950	
	1720	12′-0″ x 6′-0″, 6 panels		6	2.667		2,300	95		2,395	2,675	
	2000	Bay window, casement, builders model, 8′ x 5′ dbl insul glass, 4 panels		10	1.600		1,675	57		1,732	1,925	
	2050	Low E glass,		10	1.600		2,025	57		2,082	2,325	
	2100	12′-0″ x 6′-0″ , double insulated glass, 6 panels		6	2.667		2,075	95		2,170	2,450	
	2200	Low E glass		6	2.667		2,050	95		2,145	2,400	
	2300	Vinyl clad, premium, double insulated glass, 8′-0″ x 5′-0″		10	1.600		1,275	57		1,332	1,500	
	2340	10′-0″ x 5′-0″		8	2		1,800	71		1,871	2,125	
	2380	10′-0″ x 6′-0″		7	2.286		1,900	81.50		1,981.50	2,200	
	2420	12′-0″ x 6′-0″		6	2.667		2,250	95		2,345	2,625	
	2600	Metal clad, deluxe, dbl insul. glass, 8′-0″ x 5′-0″ high, 4 panels		10	1.600		1,475	57		1,532	1,725	
	2640	10′-0″ x 5′-0″ high, 5 panels		8	2		1,575	71		1,646	1,875	
	2680	10′-0″ x 6′-0″ high, 5 panels		7	2.286		1,875	81.50		1,956.50	2,175	
	2720	12′-0″ x 6′-0″ high, 6 panels		6	2.667		2,600	95		2,695	3,000	
	3000	Double hung, bldrs. model, bay, 8′ x 4′ high,dbl insulated glass		10	1.600		1,175	57		1,232	1,375	
	3050	Low E glass		10	1.600		1,250	57		1,307	1,475	
	3100	9′-0″ x 5′-0″ high, doublel insulated glass		6	2.667		1,250	95		1,345	1,550	
	3200	Low E glass		6	2.667		1,325	95		1,420	1,625	
	3300	Vinyl clad, premium, double insulated glass, 7′-0″ x 4′-6″		10	1.600		1,225	57		1,282	1,425	
	3340	8′-0″ x 4′-6″		8	2		1,250	71		1,321	1,500	
	3380	8′-0″ x 5′-0″		7	2.286		1,300	81.50		1,381.50	1,550	
	3420	9′-0″ x 5′-0″		6	2.667		1,325	95		1,420	1,625	
	3600	Metal clad, deluxe, dbl insul. glass, 7′-0″ x 4′-0″ high		10	1.600		1,125	57		1,182	1,325	
	3640	8′-0″ x 4′-0″ high		8	2		1,175	71		1,246	1,400	
	3680	8′-0″ x 5′-0″ high		7	2.286		1,200	81.50		1,281.50	1,450	
	3720	9′-0″ x 5′-0″ high		6	2.667		1,275	95		1,370	1,550	
	9000	Minimum labor/equipment charge		2.50	6.400	Job		228		228	375	
200	0010	**CASEMENT WINDOW** Including frame, screen, and grills										200
	0100	Avg. quality, bldrs. model, 2′-0″ x 3′-0″ H, dbl. insulated glass	1 Carp	10	.800	Ea.	186	28.50		214.50	252	
	0150	Low E glass		10	.800		225	28.50		253.50	295	
	0200	2′-0″ x 4′-6″ high, double insulated glass		9	.889		242	31.50		273.50	320	
	0250	Low E glass		9	.889		320	31.50		351.50	410	
	0300	2′-4″ x 6′-0″ high, double insulated glass		8	1		355	35.50		390.50	450	
	0350	Low E glass		8	1		385	35.50		420.50	485	
	0522	Vinyl clad, premium, double insulated glass, 2′-0″ x 3′-0″		10	.800		264	28.50		292.50	335	
	0524	2′-0″ x 4′-0″		9	.889		310	31.50		341.50	395	
	0525	2′-0″ x 5′-0″		8	1		355	35.50		390.50	450	
	0528	2′-0″ x 6′-0″		8	1		400	35.50		435.50	500	
	8100	Metal clad, deluxe, dbl. insul. glass, 2′-0″ x 3′-0″ high		10	.800		206	28.50		234.50	274	
	8120	2′-0″ x 4′-0″ high		9	.889		248	31.50		279.50	325	
	8140	2′-0″ x 5′-0″ high		8	1		282	35.50		317.50	370	
	8160	2′-0″ x 6′-0″ high		8	1		325	35.50		360.50	415	
	8200	For multiple leaf units, deduct for stationary sash										
	8220	2′ high				Ea.	21.50			21.50	23.50	
	8240	4′-6″ high					24.50			24.50	27	
	8260	6′ high					33			33	36.50	
	8300	For installation, add per leaf						15%				
	9000	Minimum labor/equipment charge	1 Carp	3	2.667	Job		95		95	157	
250	0010	**DOUBLE HUNG** Including frame, screens, and grills										250
	0100	Avg. quality, bldrs. model, 2′-0″ x 3′-0″ high, dbl insul. glass	1 Carp	10	.800	Ea.	204	28.50		232.50	272	

08550 | Wood Windows

		08550 Wood Windows	CREW	DAILY OUTPUT	LABOR-HOURS	UNIT	2006 BARE COSTS MAT.	LABOR	EQUIP.	TOTAL	TOTAL INCL O&P	
250	0150	Low E glass	1 Carp	10	.800	Ea.	186	28.50		214.50	252	250
	0200	3′-0″ x 4′-0″ high, double insulated glass		9	.889		266	31.50		297.50	345	
	0250	Low E glass		9	.889		249	31.50		280.50	325	
	0300	4′-0″ x 4′-6″ high, double insulated glass		8	1		310	35.50		345.50	400	
	0350	Low E glass		8	1		335	35.50		370.50	425	
	1000	Vinyl clad, premium, double insulated glass, 2′-6″ x 3′-0″		10	.800		218	28.50		246.50	287	
	1100	3′-0″ x 3′-6″		10	.800		256	28.50		284.50	330	
	1200	3′-0″ x 4′-0″		9	.889		350	31.50		381.50	440	
	1300	3′-0″ x 4′-6″		9	.889		296	31.50		327.50	380	
	1400	3′-0″ x 5′-0″		8	1		310	35.50		345.50	400	
	1500	3′-6″ x 6′-0″		8	1		360	35.50		395.50	460	
	2000	Metal clad, deluxe, dbl. insul. glass, 2′-6″ x 3′-0″ high		10	.800		219	28.50		247.50	287	
	2100	3′-0″ x 3′-6″ high		10	.800		259	28.50		287.50	330	
	2200	3′-0″ x 4′-0″ high		9	.889		275	31.50		306.50	360	
	2300	3′-0″ x 4′-6″ high		9	.889		298	31.50		329.50	380	
	2400	3′-0″ x 5′-0″ high		8	1		320	35.50		355.50	410	
	2500	3′-6″ x 6′-0″ high		8	1		385	35.50		420.50	485	
	9000	Minimum labor/equipment charge		3	2.667	Job		95		95	157	
260	0010	**HALF ROUND WINDOW**, Vinyl clad, double insulated glass, including grill										260
	0800	14″ height x 24″ base	2 Carp	9	1.778	Ea.	345	63		408	485	
	1040	15″ height x 25″ base		8	2		345	71		416	500	
	1060	16″ height x 28″ base		7	2.286		380	81.50		461.50	550	
	1080	17″ height x 29″ base		7	2.286		395	81.50		476.50	570	
	2000	19″ height x 33″ base	1 Carp	6	1.333		425	47.50		472.50	545	
	2100	20″ height x 35″ base		6	1.333		470	47.50		517.50	595	
	2200	21″ height x 37″ base		6	1.333		450	47.50		497.50	575	
	2250	23″ height x 41″ base	2 Carp	6	2.667		490	95		585	695	
	2300	26″ height x 48″ base		6	2.667		510	95		605	715	
	2350	30″ height x 56″ base		6	2.667		600	95		695	815	
	3000	36″ height x 67″base	1 Carp	4	2		1,025	71		1,096	1,250	
	3040	38″ height x 71″ base	2 Carp	5	3.200		955	114		1,069	1,250	
	3050	40″ height x 75″ base	″	5	3.200		1,250	114		1,364	1,575	
	5000	Elliptical, 71″ x 16″	1 Carp	11	.727		785	26		811	910	
	5100	95″ x 21″	″	10	.800		1,100	28.50		1,128.50	1,275	
670	0010	**PICTURE WINDOW** Including frame and grills										670
	0100	Average quality, bldrs. model, 3′-6″ x 4′-0″ high, dbl insulated glass	2 Carp	12	1.333	Ea.	284	47.50		331.50	390	
	0150	Low E glass		12	1.333		310	47.50		357.50	420	
	0200	4′-0″ x 4′-6″ high, double insulated glass		11	1.455		315	51.50		366.50	430	
	0250	Low E glass		11	1.455		345	51.50		396.50	465	
	0300	5′-0″ x 4′-0″ high, double insulated glass		11	1.455		390	51.50		441.50	510	
	0350	Low E glass		11	1.455		435	51.50		486.50	565	
	0400	6′-0″ x 4′-6″ high, double insulated glass		10	1.600		495	57		552	640	
	0450	Low E glass		10	1.600		560	57		617	710	
	1000	Vinyl clad, premium, dbl. insul. glass, 4′-0″ x 4′-0″		12	1.333		455	47.50		502.50	580	
	1100	4′-0″ x 6′-0″		11	1.455		675	51.50		726.50	825	
	1200	5′-0″ x 6′-0″		10	1.600		970	57		1,027	1,175	
	1300	6′-0″ x 6′-0″		10	1.600		990	57		1,047	1,175	
	2000	Metal clad, deluxe, dbl. insul. glass, 4′-0″ x 4′-0″ high		12	1.333		325	47.50		372.50	435	
	2100	4′-0″ x 6′-0″ high		11	1.455		480	51.50		531.50	610	
	2200	5′-0″ x 6′-0″ high		10	1.600		525	57		582	675	
	2300	6′-0″ x 6′-0″ high		10	1.600		605	57		662	760	
	9000	Minimum labor/equipment charge		2.75	5.818	Job		207		207	345	
750	0010	**SLIDING WINDOW** Including frame, screen, and grills										750
	0100	Average quality, bldrs. model, 3′-0″ x 3′-0″ high, double insulated	1 Carp	10	.800	Ea.	153	28.50		181.50	215	
	0120	Low E glass		10	.800		193	28.50		221.50	259	
	0200	4′-0″ x 3′-6″ high, double insulated		9	.889		182	31.50		213.50	253	

08550 | Wood Windows

		08550 Wood Windows	CREW	DAILY OUTPUT	LABOR-HOURS	UNIT	2006 BARE COSTS MAT.	LABOR	EQUIP.	TOTAL	TOTAL INCL O&P	
750	0220	Low E glass	1 Carp	9	.889	Ea.	227	31.50		258.50	305	750
	0300	6'-0" x 5'-0" high, double insulated		8	1		335	35.50		370.50	425	
	0320	Low E glass		8	1		400	35.50		435.50	500	
	1000	Vinyl clad, premium, dbl. insulated glass, 3'-0" x 3'-0"		10	.800		510	28.50		538.50	605	
	1050	4'-0" x 3'-6"		9	.889		630	31.50		661.50	745	
	1100	5'-0" x 4'-0"		9	.889		760	31.50		791.50	890	
	1150	6'-0" x 5'-0"		8	1		965	35.50		1,000.50	1,100	
	2000	Metal clad, deluxe, double insulated glass, 3'-0" x 3'-0" high		10	.800		290	28.50		318.50	365	
	2050	4'-0" x 3'-6" high		9	.889		355	31.50		386.50	445	
	2100	5'-0" x 4'-0" high		9	.889		430	31.50		461.50	525	
	2150	6'-0" x 5'-0" high		8	1		655	35.50		690.50	780	
	9000	Minimum labor/equipment charge		3	2.667	Job		95		95	157	
760	0010	**TRANSOM WINDOWS**										760
	0050	Vinyl clad, premium, double insulated glass, 32" x 8"	1 Carp	16	.500	Ea.	173	17.80		190.80	220	
	0100	36" x 8"		16	.500		184	17.80		201.80	232	
	0110	36" x 12"		16	.500		175	17.80		192.80	223	
	2000	Custom sizes, up to 350 sq. in.		12	.667		224	23.50		247.50	286	
	2100	351 to 750 sq. in.		12	.667		282	23.50		305.50	350	
	2200	751 to 1150 sq. in.		11	.727		345	26		371	425	
	2300	1151 to 1450 sq. in.		11	.727		395	26		421	480	
	2400	1451 to 1850 sq. in.	2 Carp	12	1.333		460	47.50		507.50	585	
	2500	1851 to 2250 sq. in.		12	1.333		520	47.50		567.50	650	
	2600	2251 to 2650 sq. in.		11	1.455		585	51.50		636.50	725	
	2700	2651 to 3050 sq. in.		11	1.455		595	51.50		646.50	740	
	2800	3051 to 3450 sq. in.		11	1.455		650	51.50		701.50	800	
	2900	3451 to 3850 sq. in.		10	1.600		700	57		757	865	
	3000	3851 to 4250 sq. in.		10	1.600		760	57		817	930	
	3100	4251 to 4650 sq. in.		10	1.600		775	57		832	945	
	3200	4651 to 5050 sq. in.		10	1.600		825	57		882	1,000	
	3300	5051 to 5450 sq. in.		9	1.778		880	63		943	1,075	
	3400	5451 to 5850 sq. in.		9	1.778		850	63		913	1,050	
	3600	6251 to 6650 sq. in.		8	2		1,025	71		1,096	1,250	
	3700	6651 to 7050 sq. in.		8	2		1,075	71		1,146	1,325	
780	0010	**TRAPEZOID WINDOWS**										780
	0900	20" base x 44" leg x 53" leg	2 Carp	13	1.231	Ea.	355	44		399	465	
	1000	24" base x 90" leg x 102" leg		8	2		590	71		661	770	
	3000	36" base x 0" leg x 22" leg		12	1.333		375	47.50		422.50	495	
	3010	36" base x 4" leg x 25" leg		13	1.231		395	44		439	510	
	3050	36" base x 26" leg x 48" leg		9	1.778		410	63		473	555	
	3100	36" base x 42" legs, 50" peak		9	1.778		480	63		543	635	
	3200	36" base x 60" leg x 81" leg		11	1.455		625	51.50		676.50	770	
	4320	44" base x 23" leg x 56" leg		11	1.455		495	51.50		546.50	630	
	4350	44" base x 59" leg x 92" leg		10	1.600		740	57		797	910	
	4500	46" base x 15" leg x 46" leg		8	2		375	71		446	530	
	4550	46" base x 16" leg x 48" leg		8	2		395	71		466	555	
	4600	46" base x 50" leg x 80" leg		7	2.286		595	81.50		676.50	790	
	6600	66" base x 12" leg x 42" leg		8	2		500	71		571	675	
	6650	66" base x 12" legs, 28" peak		9	1.778		410	63		473	555	
	6700	68" base x 3" legs, 31" peak		8	2		510	71		581	680	
800	0010	**WINDOW GRILLE OR MUNTIN** Snap-in type										800
	0020	Standard pattern interior grills										
	2000	Wood, awning window, glass size 28" x 16" high	1 Carp	30	.267	Ea.	20	9.50		29.50	38	
	2060	44" x 24" high		32	.250		29.50	8.90		38.40	47.50	
	2100	Casement, glass size, 20" x 36" high		30	.267		25	9.50		34.50	43	
	2180	20" x 56" high		32	.250		36	8.90		44.90	54.50	

08500 | Windows

08550 | Wood Windows

		08550 Wood Windows	CREW	DAILY OUTPUT	LABOR-HOURS	UNIT	MAT.	LABOR	EQUIP.	TOTAL	TOTAL INCL O&P	
							2006 BARE COSTS					
800	2200	Double hung, glass size, 16" x 24" high	1 Carp	24	.333	Set	44	11.85		55.85	68	800
	2280	32" x 32" high		34	.235	"	123	8.35		131.35	150	
	2500	Picture, glass size, 48" x 48" high		30	.267	Ea.	141	9.50		150.50	171	
	2580	60" x 68" high		28	.286	"	108	10.15		118.15	136	
	2600	Sliding, glass size, 14" x 36" high		24	.333	Set	25	11.85		36.85	47	
	2680	36" x 36" high		22	.364	"	38.50	12.95		51.45	63.50	
	9000	Minimum labor/equipment charge		5	1.600	Job		57		57	94	
820	0010	**WOOD SASH** Including glazing but not including trim										820
	0050	Custom, 5'-0" x 4'-0", 1" dbl. glazed, 3/16" thick lites	2 Carp	3.20	5	Ea.	164	178		342	475	
	0100	1/4" thick lites		5	3.200		169	114		283	375	
	0200	1" thick, triple glazed		5	3.200		385	114		499	615	
	0300	7'-0" x 4'-6" high, 1" double glazed, 3/16" thick lites		4.30	3.721		395	132		527	650	
	0400	1/4" thick lites		4.30	3.721		445	132		577	705	
	0500	1" thick, triple glazed		4.30	3.721		505	132		637	775	
	0600	8'-6" x 5'-0" high, 1" double glazed, 3/16" thick lites		3.50	4.571		530	163		693	855	
	0700	1/4" thick lites		3.50	4.571		580	163		743	910	
	0800	1" thick, triple glazed		3.50	4.571		585	163		748	915	
	0900	Window frames only, based on perimeter length				L.F.	3.20			3.20	3.52	
	1200	Window sill, stock, per lineal foot					7.20			7.20	7.90	
	1250	Casing, stock					2.58			2.58	2.84	
	3000	Replacement sash, double hung, double glazing, to 12 S.F.	1 Carp	64	.125	S.F.	17.15	4.44		21.59	26	
	3100	12 S.F. to 20 S.F.		94	.085		17.35	3.03		20.38	24	
	3200	20 S.F. and over		106	.075		15.10	2.68		17.78	21	
	3800	Triple glazing for above, add					2.40			2.40	2.64	
	7000	Sash, single lite, 2'-0" x 2'-0" high	1 Carp	20	.400	Ea.	34	14.20		48.20	60.50	
	7050	2'-6" x 2'-0" high		19	.421		39	14.95		53.95	68	
	7100	2'-6" x 2'-6" high		18	.444		45	15.80		60.80	75.50	
	7150	3'-0" x 2'-0" high		17	.471		48	16.75		64.75	80.50	
	9000	Minimum labor/equipment charge		4	2	Job		71		71	118	
840	0010	**WOOD SCREENS**										840
	0020	Over 3 S.F., 3/4" frames	2 Carp	375	.043	S.F.	3.69	1.52		5.21	6.55	
	0100	1-1/8" frames		375	.043		6.45	1.52		7.97	9.55	
	0200	Rescreen wood frame		500	.032		.54	1.14		1.68	2.47	
	9000	Minimum labor/equipment charge	1 Carp	4	2	Job		71		71	118	

08560 | Plastic Windows

		08560 Plastic Windows	CREW	DAILY OUTPUT	LABOR-HOURS	UNIT	MAT.	LABOR	EQUIP.	TOTAL	TOTAL INCL O&P	
100	0010	**VINYL SINGLE HUNG WINDOWS**										100
	0100	Grids, low E, J fin, ext. jambs, 21" x 53"	2 Carp	18	.889	Ea.	146	31.50		177.50	214	
	0110	21" x 57"		17	.941		150	33.50		183.50	221	
	0120	21" x 65"		16	1		156	35.50		191.50	230	
	0130	25" x 41"		20	.800		138	28.50		166.50	199	
	0140	25" x 49"		18	.889		152	31.50		183.50	221	
	0150	25" x 57"		17	.941		156	33.50		189.50	227	
	0160	25" x 65"		16	1		162	35.50		197.50	237	
	0170	29" x 41"		18	.889		147	31.50		178.50	214	
	0180	29" x 53"		18	.889		157	31.50		188.50	226	
	0190	29" x 57"		17	.941		161	33.50		194.50	233	
	0200	29" x 65"		16	1		167	35.50		202.50	243	
	0210	33" x 41"		20	.800		152	28.50		180.50	214	
	0220	33" x 53"		18	.889		163	31.50		194.50	233	
	0230	33" x 57"		17	.941		167	33.50		200.50	240	
	0240	33" x 65"		16	1		174	35.50		209.50	250	
	0250	37" x 41"		20	.800		160	28.50		188.50	223	
	0260	37" x 53"		18	.889		172	31.50		203.50	242	

08500 | Windows

	08560 | Plastic Windows		CREW	DAILY OUTPUT	LABOR-HOURS	UNIT	2006 BARE COSTS MAT.	LABOR	EQUIP.	TOTAL	TOTAL INCL O&P	
100	0270	37″ x 57″	2 Carp	17	.941	Ea.	175	33.50		208.50	249	100
	0280	37″ x 65″		16	1		182	35.50		217.50	260	
200	0010	VINYL DOUBLE HUNG WINDOWS										200
	0100	Grids, low E, J fin, ext. jambs, 21″ x 53″	2 Carp	18	.889	Ea.	167	31.50		198.50	237	
	0102	21″ x 37″		18	.889		149	31.50		180.50	217	
	0104	21″ x 41″		18	.889		153	31.50		184.50	221	
	0106	21″ x 49″		18	.889		160	31.50		191.50	229	
	0110	21″ x 57″		17	.941		171	33.50		204.50	244	
	0120	21″ x 65″		16	1		177	35.50		212.50	254	
	0128	25″ x 37″		20	.800		157	28.50		185.50	220	
	0130	25″ x 41″		20	.800		161	28.50		189.50	224	
	0140	25″ x 49″		18	.889		166	31.50		197.50	235	
	0145	25″ x 53″		18	.889		172	31.50		203.50	243	
	0150	25″ x 57″		17	.941		173	33.50		206.50	246	
	0160	25″ x 65″		16	1		184	35.50		219.50	261	
	0162	25″ x 69″		16	1		191	35.50		226.50	269	
	0164	25″ x 77″		16	1		202	35.50		237.50	281	
	0168	29″ x 37″		18	.889		162	31.50		193.50	231	
	0170	29″ x 41″		18	.889		166	31.50		197.50	235	
	0172	29″ x 49″		18	.889		174	31.50		205.50	244	
	0180	29″ x 53″		18	.889		178	31.50		209.50	248	
	0190	29″ x 57″		17	.941		181	33.50		214.50	255	
	0200	29″ x 65″		16	1		188	35.50		223.50	266	
	0202	29″ x 69″		16	1		195	35.50		230.50	274	
	0205	29″ x 77″		16	1		207	35.50		242.50	286	
	0208	33″ x 37″		20	.800		167	28.50		195.50	230	
	0210	33″ x 41″		20	.800		170	28.50		198.50	234	
	0215	33″ x 49″		20	.800		179	28.50		207.50	244	
	0220	33″ x 53″		18	.889		183	31.50		214.50	254	
	0230	33″ x 57″		17	.941		187	33.50		220.50	262	
	0240	33″ x 65″		16	1		192	35.50		227.50	270	
	0242	33″ x 69″		16	1		204	35.50		239.50	283	
	0246	33″ x 77″		16	1		214	35.50		249.50	294	
	0250	37″ x 41″		20	.800		174	28.50		202.50	239	
	0255	37″ x 49″		20	.800		183	28.50		211.50	249	
	0260	37″ x 53″		18	.889		191	31.50		222.50	263	
	0270	37″ x 57″		17	.941		195	33.50		228.50	270	
	0280	37″ x 65″		16	1		200	35.50		235.50	279	
	0282	37″ x 69″		16	1		262	35.50		297.50	345	
	0286	37″ x 77″		16	1		274	35.50		309.50	360	
	0300	Solid vinyl, average quality, double insulated glass, 2′-0″ x 3′-0″	1 Carp	10	.800		137	28.50		165.50	198	
	0310	3′-0″ x 4′-0″		9	.889		165	31.50		196.50	234	
	0320	4′-0″ x 4′-6″		8	1		197	35.50		232.50	276	
	0330	Premium, double insulated glass, 2′-6″ x 3′-0″		10	.800		153	28.50		181.50	215	
	0340	3′-0″ x 3′-6″		9	.889		177	31.50		208.50	248	
	0350	3′-0″ x 4″-0″		9	.889		188	31.50		219.50	260	
	0360	3′-0″ x 4′-6″		9	.889		193	31.50		224.50	265	
	0370	3′-0″ x 5′-0″		8	1		198	35.50		233.50	277	
	0380	3′-6″ x 6″-0″		8	1		218	35.50		253.50	298	
300	0010	VINYL CASEMENT WINDOWS										300
	0100	Grids, low E, J fin, ext. jambs, 1 lt, 21″ x 41″	2 Carp	20	.800	Ea.	216	28.50		244.50	285	
	0110	21″ x 47″		20	.800		236	28.50		264.50	305	
	0120	21″ x 53″		20	.800		255	28.50		283.50	325	
	0128	24″ x 35″		19	.842		208	30		238	278	
	0130	24″ x 41″		19	.842		226	30		256	298	

08560 | Plastic Windows

			CREW	DAILY OUTPUT	LABOR-HOURS	UNIT	2006 BARE COSTS MAT.	LABOR	EQUIP.	TOTAL	TOTAL INCL O&P	
300	0140	24″ x 47″	2 Carp	19	.842	Ea.	244	30		274	320	300
	0150	24″ x 53″		19	.842		263	30		293	340	
	0158	28″ x 35″		19	.842		221	30		251	294	
	0160	28″ x 41″		19	.842		239	30		269	315	
	0170	28″ x 47″		19	.842		258	30		288	335	
	0180	28″ x 53″		19	.842		284	30		314	365	
	0184	28″ x 59″		19	.842		290	30		320	370	
	0188	Two lites, 33″ x 35″		18	.889		355	31.50		386.50	445	
	0190	33″ x 41″		18	.889		380	31.50		411.50	475	
	0200	33″ x 47″		18	.889		410	31.50		441.50	505	
	0210	33″ x 53″		18	.889		440	31.50		471.50	535	
	0212	33″ x 59″		18	.889		465	31.50		496.50	565	
	0215	33″ x 72″		18	.889		480	31.50		511.50	585	
	0220	41″ x 41″		18	.889		415	31.50		446.50	515	
	0230	41″ x 47″		18	.889		445	31.50		476.50	545	
	0240	41″ x 53″		17	.941		475	33.50		508.50	575	
	0242	41″ x 59″		17	.941		500	33.50		533.50	605	
	0246	41″ x 72″		17	.941		520	33.50		553.50	630	
	0250	47″ x 41″		17	.941		420	33.50		453.50	520	
	0260	47″ x 47″		17	.941		450	33.50		483.50	550	
	0270	47″ x 53″		17	.941		475	33.50		508.50	575	
	0272	47″ x 59″		17	.941		520	33.50		553.50	625	
	0280	56″ x 41″		15	1.067		450	38		488	560	
	0290	56″ x 47″		15	1.067		475	38		513	585	
	0300	56″ x 53″		15	1.067		520	38		558	635	
	0302	56″ x 59″		15	1.067		540	38		578	660	
	0310	56″ x 72″		15	1.067		590	38		628	710	
	0340	Solid vinyl, premium, double insulated glass, 2′-0″ x 3′-0″ high	1 Carp	10	.800		185	28.50		213.50	250	
	0360	2′-0″ x 4′-0″ high		9	.889		227	31.50		258.50	300	
	0380	2′-0″ x 5′-0″ high		8	1		258	35.50		293.50	345	
400	0010	**VINYL PICTURE WINDOWS**										400
	0100	Grids, low E, J fin, ext. jambs, 33″ x 47″	2 Carp	12	1.333	Ea.	217	47.50		264.50	320	
	0110	35″ x 71″		12	1.333		230	47.50		277.50	330	
	0120	41″ x 47″		12	1.333		252	47.50		299.50	355	
	0130	41″ x 71″		12	1.333		273	47.50		320.50	380	
	0140	47″ x 47″		12	1.333		285	47.50		332.50	395	
	0150	47″ x 71″		11	1.455		299	51.50		350.50	415	
	0160	53″ x 47″		11	1.455		280	51.50		331.50	395	
	0170	53″ x 71″		11	1.455		293	51.50		344.50	410	
	0180	59″ x 47″		11	1.455		320	51.50		371.50	435	
	0190	59″ x 71″		11	1.455		340	51.50		391.50	460	
	0200	71″ x 47″		10	1.600		350	57		407	485	
	0210	71″ x 71″		10	1.600		370	57		427	505	
500	0010	**VINYL HALF ROUND WINDOW,** including grill, j fin, low E, ext. jambs										500
	0100	10″ height x 20″ base	2 Carp	8	2	Ea.	242	71		313	385	
	0110	15″ height x 30″ base		8	2		340	71		411	495	
	0120	17″ height x 34″ base		7	2.286		355	81.50		436.50	525	
	0130	19″ height x 38″ base		7	2.286		405	81.50		486.50	580	
	0140	19″ height x 33″ base		7	2.286		350	81.50		431.50	520	
	0150	24″ height x 48″ base	1 Carp	6	1.333		415	47.50		462.50	535	
	0160	25″ height x 50″ base	″	6	1.333		460	47.50		507.50	585	
	0170	30″ height x 60″ base	2 Carp	6	2.667		560	95		655	770	

08580 | Special Function Windows

			CREW	DAILY OUTPUT	LABOR-HOURS	UNIT	2006 BARE COSTS MAT.	LABOR	EQUIP.	TOTAL	TOTAL INCL O&P
900	0010	**STORM WINDOWS** Aluminum, residential									
	0300	Basement, mill finish, incl. fiberglass screen									
	0320	1'-10" x 1'-0" high	2 Carp	30	.533	Ea.	29.50	18.95		48.45	64
	0340	2'-9" x 1'-6" high		30	.533		32	18.95		50.95	67
	0360	3'-4" x 2'-0" high		30	.533		39	18.95		57.95	74
	1600	Double-hung, combination, storm & screen									
	1700	Custom, clear anodic coating, 2'-0" x 3'-5" high	2 Carp	30	.533	Ea.	76.50	18.95		95.45	116
	1720	2'-6" x 5'-0" high		28	.571		102	20.50		122.50	147
	1740	4'-0" x 6'-0" high		25	.640		217	23		240	276
	1800	White painted, 2'-0" x 3'-5" high		30	.533		91	18.95		109.95	132
	1820	2'-6" x 5'-0" high		28	.571		146	20.50		166.50	195
	1840	4'-0" x 6'-0" high		25	.640		262	23		285	325
	2000	Average quality, clear anodic coating, 2'-0" x 3'-5" high		30	.533		77.50	18.95		96.45	117
	2020	2'-6" x 5'-0" high		28	.571		95	20.50		115.50	138
	2040	4'-0" x 6'-0" high		25	.640		115	23		138	165
	2400	White painted, 2'-0" x 3'-5" high		30	.533		76.50	18.95		95.45	116
	2420	2'-6" x 5'-0" high		28	.571		84.50	20.50		105	127
	2440	4'-0" x 6'-0" high		25	.640		92.50	23		115.50	140
	2600	Mill finish, 2'-0" x 3'-5" high		30	.533		69.50	18.95		88.45	108
	2620	2'-6" x 5'-0" high		28	.571		77.50	20.50		98	119
	2640	4'-0" x 6-8" high		25	.640		87	23		110	134
	4000	Picture window, storm, 1 lite, white or bronze finish									
	4020	4'-6" x 4'-6" high	2 Carp	25	.640	Ea.	117	23		140	166
	4040	5'-8" x 4'-6" high		20	.800		132	28.50		160.50	193
	4400	Mill finish, 4'-6" x 4'-6" high		25	.640		117	23		140	166
	4420	5'-8" x 4'-6" high		20	.800		132	28.50		160.50	193
	4600	3 lite, white or bronze finish									
	4620	4'-6" x 4'-6" high	2 Carp	25	.640	Ea.	142	23		165	194
	4640	5'-8" x 4'-6" high		20	.800		158	28.50		186.50	221
	4800	Mill finish, 4'-6" x 4'-6" high		25	.640		125	23		148	175
	4820	5'-8" x 4'-6" high		20	.800		132	28.50		160.50	193
	5001	Sliding glass door, storm window, 6'-0" x 6'-8", fixed	1 Glaz	1.60	5		335	171		506	645
	5101	Operable	"	2.10	3.810		335	130		465	580
	6000	Sliding window, storm, 2 lite, white or bronze finish									
	6020	3'-4" x 2'-7" high	2 Carp	28	.571	Ea.	105	20.50		125.50	149
	6040	4'-4" x 3'-3" high		25	.640		143	23		166	195
	6060	5'-4" x 6'-0" high		20	.800		229	28.50		257.50	299
	6400	3 lite, white or bronze finish									
	6420	4'-4" x 3'-3" high	2 Carp	25	.640	Ea.	166	23		189	220
	6440	5'-4" x 6'-0" high		20	.800		299	28.50		327.50	375
	6460	6'-0" x 6'-0" high		18	.889		300	31.50		331.50	385
	6800	Mill finish, 4'-4" x 3'-3" high		25	.640		143	23		166	195
	6820	5'-4" x 6'-0" high		20	.800		300	28.50		328.50	375
	6840	6'-0" x 6-0" high		18	.889		310	31.50		341.50	395
	8000	PVC framed									
	8100	Double-hung, combination, storm & screen									
	8120	2'-6" x 3'-5"	2 Carp	15	1.067	Ea.	65.50	38		103.50	135
	8140	4' x 6'	"	12.50	1.280	"	112	45.50		157.50	200
	8600	Single lite picture storm									
	8620	4'-6" x 4'-6"	2 Carp	12.50	1.280	Ea.	76	45.50		121.50	160
	8640	5'-8" x 4'-6"	"	10	1.600	"	83	57		140	185
	9000	Magnetic interior storm window									
	9100	3/16" plate glass	1 Glaz	107	.075	S.F.	4.32	2.56		6.88	8.90
	9410	Minimum labor/equipment charge	1 Carp	4	2	Job		71		71	118

DOORS & WINDOWS 8

08500 | Windows

08590 | Window Restoration & Replace

		08590 Window Restoration & Replace	Crew	Daily Output	Labor-Hours	Unit	2006 Bare Costs Mat.	Labor	Equip.	Total	Total Incl O&P	
600	0010	SOLID VINYL REPLACEMENT WINDOWS R085313-20										600
	0020	Double hung, insulated glass, up to 83 united inches	2 Carp	8	2	Ea.	218	71		289	360	
	0040	84 to 93		8	2		243	71		314	385	
	0060	94 to 101		6	2.667		277	95		372	460	
	0080	102 to 111		6	2.667		305	95		400	495	
	0100	112 to 120		6	2.667		345	95		440	535	
	0120	For each united inch over 120 , add		800	.020	Inch	3.65	.71		4.36	5.20	
	0140	Casement windows, one operating sash , 42 to 60 united inches		8	2	Ea.	191	71		262	330	
	0160	61 to 70		8	2		217	71		288	355	
	0180	71 to 80		8	2		236	71		307	375	
	0200	81 to 96		8	2		250	71		321	395	
	0220	Two operating sash, 58 to 78 united inches		8	2		380	71		451	540	
	0240	79 to 88		8	2		405	71		476	570	
	0260	89 to 98		8	2		445	71		516	610	
	0280	99 to 108		6	2.667		465	95		560	665	
	0300	109 to 121		6	2.667		500	95		595	705	
	0320	Three operating sash, 73 to 108 united inches		8	2		600	71		671	780	
	0340	109 to 118		8	2		635	71		706	815	
	0360	119 to 128		6	2.667		650	95		745	870	
	0380	129 to 138		6	2.667		695	95		790	920	
	0400	139 to 156		6	2.667		735	95		830	965	
	0420	Four operating sash, 98 to 118 united inches		8	2		865	71		936	1,075	
	0440	119 to 128		8	2		925	71		996	1,150	
	0460	129 to 138		6	2.667		980	95		1,075	1,225	
	0480	139 to 148		6	2.667		1,025	95		1,120	1,275	
	0500	149 to 168		6	2.667		1,100	95		1,195	1,350	
	0520	169 to 178		6	2.667		1,200	95		1,295	1,450	
	0540	For venting unit to fixed unit, deduct					17.40			17.40	19.15	
	0560	Fixed picture window, up to 63 united inches	2 Carp	8	2		138	71		209	270	
	0580	64 to 83		8	2		163	71		234	298	
	0600	84 to 101		8	2		210	71		281	350	
	0620	For each united inch over 101, add		900	.018	Inch	2.53	.63		3.16	3.83	
	0640	Picture window opt., low E glazing, up to 101 united inches				Ea.	19.65			19.65	21.50	
	0660	102 to 124					25.50			25.50	28	
	0680	124 and over					38			38	42	
	0700	Options, low E glazing, up to 101 united inches					9.80			9.80	10.80	
	0720	102 to 124					12.65			12.65	13.90	
	0740	124 and over					19.05			19.05	21	
	0760	Muntins, between glazing, square, per lite					1.85			1.85	2.04	
	0780	Diamond shape, per full or partial diamond					2.92			2.92	3.21	
	0800	Celluose fiber insulation, poured into sash balance cavity	1 Carp	36	.222	C.F.	.58	7.90		8.48	13.75	
	0820	Silicone caulking at perimeter	"	800	.010	L.F.	.12	.36		.48	.73	

08600 | Skylights

08620 | Unit Skylights

		08620 Unit Skylights	Crew	Daily Output	Labor-Hours	Unit	2006 Bare Costs Mat.	Labor	Equip.	Total	Total Incl O&P	
800	0010	**SKYLIGHT** Plastic domes, flush or curb mounted, ten or										800
	0100	more units, curb not included										
	0300	Nominal size under 10 S.F., double	G-3	130	.246	S.F.	19.10	8.55		27.65	35	
	0400	Single		160	.200		13.95	6.95		20.90	26.50	

08600 | Skylights

	08620 Unit Skylights		CREW	DAILY OUTPUT	LABOR-HOURS	UNIT	2006 BARE COSTS MAT.	LABOR	EQUIP.	TOTAL	TOTAL INCL O&P	
800	0600	10 S.F. to 20 S.F., double	G-3	315	.102	S.F.	16.85	3.53		20.38	24.50	800
	0700	Single		395	.081		8.70	2.82		11.52	14.10	
	0900	20 S.F. to 30 S.F., double		395	.081		15.40	2.82		18.22	21.50	
	1000	Single		465	.069		10.95	2.39		13.34	15.90	
	1200	30 S.F. to 65 S.F., double		465	.069		11.75	2.39		14.14	16.75	
	1300	Single	↓	610	.052		15	1.82		16.82	19.45	
	1500	For insulated 4″ curbs, double, add					25%					
	1600	Single, add				↓	30%					
	1800	For integral insulated 9″ curbs, double, add					30%					
	1900	Single, add					40%					
	2120	Ventilating insulated plexiglass dome with										
	2130	curb mounting, 36″ x 36″	G-3	12	2.667	Ea.	380	93		473	570	
	2150	52″ x 52″		12	2.667		570	93		663	775	
	2160	28″ x 52″		10	3.200		445	111		556	670	
	2170	36″ x 52″	↓	10	3.200		480	111		591	710	
	2180	For electric opening system, add				↓	285			285	315	
	2200	Field fabricated, factory type, aluminum and wire glass	G-3	120	.267	S.F.	14.70	9.25		23.95	31	
	2300	Insulated safety glass with aluminum frame		160	.200		85.50	6.95		92.45	106	
	2400	Sandwich panels, fiberglass, for walls, 1-9/16″ thick, to 250 SF		200	.160		15.55	5.55		21.10	26	
	2500	250 SF and up		265	.121		13.95	4.20		18.15	22	
	2700	As above, but for roofs, 2-3/4″ thick, to 250 SF		295	.108		22.50	3.77		26.27	30.50	
	2800	250 SF and up	↓	330	.097	↓	18.35	3.37		21.72	25.50	

08700 | Hardware

	08710 Door Hardware		CREW	DAILY OUTPUT	LABOR-HOURS	UNIT	2006 BARE COSTS MAT.	LABOR	EQUIP.	TOTAL	TOTAL INCL O&P	
100	0010	**AUTOMATIC OPENERS COMMERCIAL**										100
	0020	Pneumatic, incl opener, motion sens, control box, tubing, compressor										
	0050	For single swing door, per opening	2 Skwk	.80	20	Ea.	3,725	730		4,455	5,300	
	0100	Pair, per opening		.50	32	Opng.	6,200	1,175		7,375	8,725	
	1000	For single sliding door, per opening		.60	26.667		4,125	975		5,100	6,125	
	1300	Bi-parting pair	↓	.50	32	↓	6,225	1,175		7,400	8,750	
	1420	Electronic door opener incl motion sens, 12V control box, motor										
	1450	For single swing door, per opening	2 Skwk	.80	20	Opng.	3,250	730		3,980	4,775	
	1500	Pair, per opening		.50	32		5,325	1,175		6,500	7,750	
	1600	For single sliding door, per opening		.60	26.667		3,525	975		4,500	5,475	
	1700	Bi-parting pair	↓	.50	32	↓	5,400	1,175		6,575	7,850	
	1750	Handicap actuator buttons, 2, including 12V DC wiring, add	1 Carp	1.50	5.333	Pr.	375	190		565	725	
150	0010	**AVERAGE** Percentage for hardware, total job cost										150
	0025	Minimum									.75%	
	0050	Maximum									3.50%	
	0500	Total hardware for building, average distribution					85%	15%				
	1000	Door hardware, apartment, interior				Door	127			127	140	
	1500	Hospital bedroom, minimum					284			284	315	
	2000	Maximum					625			625	685	
	2250	School, single exterior, incl. lever, not incl. panic device					420			420	460	
	2500	Single interior, regular use, no lever included					280			280	310	
	2600	Heavy use, incl. lever and closer					490			490	540	
	2850	Stairway, single interior				↓	700			700	770	
	3100	Double exterior, with panic device				Pr.	985			985	1,075	

		08710 | Door Hardware	CREW	DAILY OUTPUT	LABOR-HOURS	UNIT	2006 BARE COSTS MAT.	LABOR	EQUIP.	TOTAL	TOTAL INCL O&P	
150	3600	Toilet, public, single interior				Door	154			154	170	150
300	0010	**DOOR CLOSERS**										300
	0015	Door closer Rack and pinion	1 Carp	6.50	1.231	Ea.	134	44		178	220	
	0020	Adjustable backcheck, 3 way mount, all sizes, regular arm		6	1.333		137	47.50		184.50	230	
	0040	Hold open arm		6	1.333		157	47.50		204.50	252	
	0100	Fusible link		6.50	1.231		122	44		166	207	
	0200	Non sized, regular arm		6	1.333		136	47.50		183.50	229	
	0240	Hold open arm		6	1.333		170	47.50		217.50	266	
	0400	4 way mount, non sized, regular arm		6	1.333		187	47.50		234.50	285	
	0440	Hold open arm		6	1.333		201	47.50		248.50	300	
	2000	Backcheck and adjustable power, hinge face mount										
	2010	All sizes, regular arm	1 Carp	6.50	1.231	Ea.	172	44		216	262	
	2040	Hold open arm		6.50	1.231		185	44		229	277	
	2400	Top jamb mount, all sizes, regular arm		6	1.333		172	47.50		219.50	268	
	2440	Hold open arm		6	1.333		185	47.50		232.50	283	
	2800	Top face mount, all sizes, regular arm		6.50	1.231		172	44		216	262	
	2840	Hold open arm		6.50	1.231		185	44		229	276	
	4000	Backcheck, overhead concealed, all sizes, regular arm		5.50	1.455		182	51.50		233.50	286	
	4040	Concealed arm		5	1.600		194	57		251	310	
	4400	Compact overhead, concealed, all sizes, regular arm		5.50	1.455		330	51.50		381.50	450	
	4440	Concealed arm		5	1.600		345	57		402	475	
	4800	Concealed in door, all sizes, regular arm		5.50	1.455		123	51.50		174.50	221	
	4840	Concealed arm		5	1.600		132	57		189	239	
	4900	Floor concealed, all sizes, single acting		2.20	3.636		156	129		285	385	
	4940	Double acting		2.20	3.636		201	129		330	435	
	5000	For cast aluminum cylinder, deduct					16.50			16.50	18.15	
	5040	For delayed action, add					29			29	32	
	5080	For fusible link arm, add					11.95			11.95	13.15	
	5120	For shock absorbing arm, add					36			36	39.50	
	5160	For spring power adjustment, add					27.50			27.50	30.50	
	6000	Closer-holder, hinge face mount, all sizes, exposed arm	1 Carp	6.50	1.231		127	44		171	212	
	7000	Electronic closer-holder, hinge facemount, concealed arm		5	1.600		193	57		250	305	
	7400	With built-in detector		5	1.600		580	57		637	735	
	9000	Minimum labor/equipment charge		4	2	Job		71		71	118	
320	0010	**DEADLOCKS**										320
	0011	Mortise heavy duty outside key(security item)	1 Carp	9	.889	Ea.	131	31.50		162.50	197	
	0020	Double cylinder		9	.889		145	31.50		176.50	212	
	0100	Medium duty, outside key		10	.800		102	28.50		130.50	159	
	0110	Double cylinder		10	.800		127	28.50		155.50	187	
	1000	Tubular, standard duty, outside key		10	.800		55	28.50		83.50	107	
	1010	Double cylinder		10	.800		70.50	28.50		99	125	
	1200	Night latch, outside key		10	.800		69	28.50		97.50	123	
340	0010	**DOORSTOPS**										340
	0020	Holder & bumper, floor or wall	1 Carp	32	.250	Ea.	31.50	8.90		40.40	49.50	
	1300	Wall bumper, 4" diameter, with rubber pad, aluminum		32	.250		9.35	8.90		18.25	25	
	1600	Door bumper, floor type, aluminum		32	.250		4.82	8.90		13.72	20	
	1900	Plunger type, door mounted		32	.250		25.50	8.90		34.40	43	
	9000	Minimum labor/equipment charge		6	1.333	Job		47.50		47.50	78.50	
400	0010	**ENTRANCE LOCKS**										400
	0015	Cylinder, grip handle deadlocking latch	1 Carp	9	.889	Ea.	120	31.50		151.50	185	
	0020	Deadbolt		8	1		146	35.50		181.50	220	
	0100	Push and pull plate, dead bolt		8	1		139	35.50		174.50	212	
	0900	For handicapped lever, add					152			152	167	

	08710 | Door Hardware		CREW	DAILY OUTPUT	LABOR-HOURS	UNIT	2006 BARE COSTS MAT.	LABOR	EQUIP.	TOTAL	TOTAL INCL O&P	
500	0010	HASP										500
	0015	3″	1 Carp	26	.308	Ea.	2.68	10.95		13.63	21	
	0020	4-1/2″		13	.615		3.43	22		25.43	40.50	
	0040	6″		12.50	.640		5.20	23		28.20	43	
520	0010	HINGES R087120-10										520
	0012	Full mortise, avg. freq., steel base, USP, 4-1/2″ x 4-1/2″				Pr.	21			21	23.50	
	0100	5″ x 5″, USP					35.50			35.50	39	
	0200	6″ x 6″, USP					75.50			75.50	83	
	0400	Brass base, 4-1/2″ x 4-1/2″, US10					43.50			43.50	48	
	0500	5″ x 5″, US10					63.50			63.50	70	
	0600	6″ x 6″, US10					108			108	119	
	0800	Stainless steel base, 4-1/2″ x 4-1/2″, US32					65.50			65.50	72	
	0900	For non removable pin, add(security item)				Ea.	2.38			2.38	2.62	
	0910	For floating pin, driven tips, add					2.73			2.73	3	
	0930	For hospital type tip on pin, add					11.80			11.80	12.95	
	0940	For steeple type tip on pin, add					10.30			10.30	11.35	
	0950	Full mortise, high frequency, steel base, 3-1/2″ x 3-1/2″, US26D				Pr.	19.30			19.30	21	
	1000	4-1/2″ x 4-1/2″, USP					51.50			51.50	56.50	
	1100	5″ x 5″, USP					48			48	53	
	1200	6″ x 6″, USP					117			117	129	
	1400	Brass base, 3-1/2″ x 3-1/2″, US4					40.50			40.50	44.50	
	1430	4-1/2″ x 4-1/2″, US10					69			69	76	
	1500	5″ x 5″, US10					103			103	113	
	1600	6″ x 6″, US10					149			149	164	
	1800	Stainless steel base, 4-1/2″ x 4-1/2″, US32					111			111	122	
	1810	5″ x 4-1/2″, US32					154			154	170	
	1930	For hospital type tip on pin, add				Ea.	6.95			6.95	7.65	
	1950	Full mortise, low frequency, steel base, 3-1/2″ x 3-1/2″, US26D				Pr.	10.10			10.10	11.10	
	2000	4-1/2″ x 4-1/2″, USP					9.40			9.40	10.35	
	2100	5″ x 5″, USP					26			26	28.50	
	2200	6″ x 6″, USP					52			52	57	
	2300	4-1/2″ x 4-1/2″, US3					15.20			15.20	16.70	
	2310	5″ x 5″, US3					37.50			37.50	41.50	
	2400	Brass bass, 4-1/2″ x 4-1/2″, US10					36.50			36.50	40	
	2500	5″ x 5″, US10					55.50			55.50	61	
	2800	Stainless steel base, 4-1/2″ x 4-1/2″, US32					63			63	69	
550	0010	KICK PLATE										550
	0020	Stainless steel	1 Carp	15	.533	Ea.	27.50	18.95		46.45	62	
	0500	Bronze		15	.533		35	18.95		53.95	69.50	
	2000	Aluminum, .050, with 3 beveled edges, 10″ x 28″		15	.533		19.80	18.95		38.75	53.50	
	2010	10″ x 30″		15	.533		21.50	18.95		40.45	55	
	2020	10″ x 34″		15	.533		22.50	18.95		41.45	56	
	2040	10″ x 38″		15	.533		25	18.95		43.95	59	
	9000	Minimum labor/equipment charge		6	1.333	Job		47.50		47.50	78.50	
650	0010	LOCKSET Standard duty, cylindrical, with sectional trim(security item)										650
	0020	Non-keyed, passage	1 Carp	12	.667	Ea.	43	23.50		66.50	86.50	
	0100	Privacy		12	.667		53.50	23.50		77	98	
	0400	Keyed, single cylinder function		10	.800		74.50	28.50		103	129	
	0420	Hotel		8	1		106	35.50		141.50	176	
	0500	Lever handled, keyed, single cylinder function		10	.800		132	28.50		160.50	192	
	1000	Heavy duty with sectional trim, non-keyed, passages		12	.667		123	23.50		146.50	176	
	1100	Privacy		12	.667		156	23.50		179.50	212	
	1400	Keyed, single cylinder function		10	.800		184	28.50		212.50	250	
	1420	Hotel		8	1		275	35.50		310.50	360	
	1600	Communicating		10	.800		218	28.50		246.50	287	
	1690	For re-core cylinder, add					31			31	34	

		08710 | Door Hardware	CREW	DAILY OUTPUT	LABOR-HOURS	UNIT	2006 BARE COSTS MAT.	LABOR	EQUIP.	TOTAL	TOTAL INCL O&P	
650	1700	Residential, interior door, minimum	1 Carp	16	.500	Ea.	14.90	17.80		32.70	46	650
	1720	Maximum		8	1		39.50	35.50		75	103	
	1800	Exterior, minimum		14	.571		33	20.50		53.50	70	
	1810	Average		8	1		65	35.50		100.50	131	
	1820	Maximum		8	1		140	35.50		175.50	212	
	9000	Minimum labor/equipment charge		6	1.333	Job		47.50		47.50	78.50	
700	0010	**MORTISE LOCKSET** Comm., wrought knobs & full escutcheon trim										700
	0020	Non-keyed, passage, minimum	1 Carp	9	.889	Ea.	160	31.50		191.50	228	
	0030	Maximum		8	1		258	35.50		293.50	340	
	0040	Privacy, minimum		9	.889		170	31.50		201.50	240	
	0050	Maximum		8	1		278	35.50		313.50	365	
	0100	Keyed, office/entrance/apartment, minimum		8	1		195	35.50		230.50	273	
	0110	Maximum		7	1.143		335	40.50		375.50	440	
	0120	Single cylinder, typical, minimum		8	1		167	35.50		202.50	243	
	0130	Maximum		7	1.143		310	40.50		350.50	410	
	0200	Hotel, minimum		7	1.143		201	40.50		241.50	289	
	0210	Maximum		6	1.333		325	47.50		372.50	440	
	0300	Communication, double cylinder, minimum		8	1		201	35.50		236.50	280	
	0310	Maximum		7	1.143		260	40.50		300.50	355	
	1000	Wrought knobs and sectional trim, non-keyed, passage, minimum		10	.800		106	28.50		134.50	163	
	1010	Maximum		9	.889		209	31.50		240.50	283	
	1040	Privacy, minimum		10	.800		124	28.50		152.50	183	
	1050	Maximum		9	.889		223	31.50		254.50	298	
	1100	Keyed, entrance, office/apartment, minimum		9	.889		184	31.50		215.50	255	
	1110	Maximum		8	1		265	35.50		300.50	350	
	1120	Single cylinder, typical, minimum		9	.889		177	31.50		208.50	247	
	1130	Maximum		8	1		257	35.50		292.50	340	
	2000	Cast knobs and full escutcheon trim										
	2010	Non-keyed, passage, minimum	1 Carp	9	.889	Ea.	225	31.50		256.50	300	
	2020	Maximum		8	1		365	35.50		400.50	460	
	2040	Privacy, minimum		9	.889		271	31.50		302.50	350	
	2050	Maximum		8	1		390	35.50		425.50	490	
	2120	Keyed, single cylinder, typical, minimum		8	1		271	35.50		306.50	355	
	2130	Maximum		7	1.143		425	40.50		465.50	540	
	2200	Hotel, minimum		7	1.143		300	40.50		340.50	400	
	2210	Maximum		6	1.333		540	47.50		587.50	670	
	3000	Cast knob and sectional trim, non-keyed, passage, minimum		10	.800		175	28.50		203.50	240	
	3010	Maximum		10	.800		350	28.50		378.50	430	
	3040	Privacy, minimum		10	.800		199	28.50		227.50	266	
	3050	Maximum		10	.800		350	28.50		378.50	430	
	3100	Keyed, office/entrance/apartment, minimum		9	.889		226	31.50		257.50	300	
	3110	Maximum		9	.889		355	31.50		386.50	445	
	3120	Single cylinder, typical, minimum		9	.889		226	31.50		257.50	300	
	3130	Maximum		9	.889		440	31.50		471.50	535	
	3190	For re-core cylinder, add					30.50			30.50	33.50	
	3800	Cipher lockset(security item)	1 Carp	13	.615		715	22		737	820	
	3900	Keyless, pushbutton type										
	4000	Residential/light commercial, deadbolt, standard	1 Carp	9	.889	Ea.	102	31.50		133.50	165	
	4010	Heavy duty		9	.889		121	31.50		152.50	186	
	4020	Industrial, heavy duty, with deadbolt		9	.889		238	31.50		269.50	315	
	4030	Key override		9	.889		264	31.50		295.50	345	
	4040	Lever activated handle		9	.889		290	31.50		321.50	375	
	4050	Key override		9	.889		320	31.50		351.50	410	
	4060	Double sided pushbutton type		8	1		530	35.50		565.50	645	
	4070	Key override		8	1		570	35.50		605.50	690	

	08710 Door Hardware		Crew	Daily Output	Labor-Hours	Unit	2006 Bare Costs Mat.	Labor	Equip.	Total	Total Incl O&P
750	0010	PANIC DEVICE									
	0015	For rim locks, single door exit only	1 Carp	6	1.333	Ea.	375	47.50		422.50	490
	0020	Outside key and pull		5	1.600		425	57		482	560
	0200	Bar and vertical rod, exit only		5	1.600		540	57		597	690
	0210	Outside key and pull		4	2		645	71		716	830
	0400	Bar and concealed rod		4	2		545	71		616	715
	0600	Touch bar, exit only		6	1.333		430	47.50		477.50	550
	0610	Outside key and pull		5	1.600		515	57		572	665
	0700	Touch bar and vertical rod, exit only		5	1.600		585	57		642	740
	0710	Outside key and pull		4	2		685	71		756	870
	1000	Mortise, bar, exit only		4	2		480	71		551	650
	1600	Touch bar, exit only		4	2		550	71		621	725
	2000	Narrow stile, rim mounted, bar, exit only		6	1.333		580	47.50		627.50	715
	2010	Outside key and pull		5	1.600		630	57		687	785
	2200	Bar and vertical rod, exit only		5	1.600		595	57		652	750
	2210	Outside key and pull		4	2		595	71		666	775
	2400	Bar and concealed rod, exit only		3	2.667		695	95		790	920
	3000	Mortise, bar, exit only		4	2		505	71		576	675
	3600	Touch bar, exit only		4	2		735	71		806	930
	4000	Double doors, exit only		2	4	Pr.	705	142		847	1,000
	4500	Exit & entrance		2	4	"	805	142		947	1,125
	9000	Minimum labor/equipment charge		2.50	3.200	Job		114		114	188
770	0010	DOOR AND WINDOW ACCESSORIES									
	0050	Door closing coordinator, 36" (for paired openings up to 56")	1 Carp	8	1	Ea.	87.50	35.50		123	155
	0060	48" (for paired openings up to 84")		8	1		93.50	35.50		129	162
	0070	56" (for paired openings up to 96")		8	1		103	35.50		138.50	172
780	0010	PUSH-PULL PLATE									
	0100	Push plate, .050 thick, 4" x 16", aluminum	1 Carp	12	.667	Ea.	6.85	23.50		30.35	47
	0500	Bronze		12	.667		17.35	23.50		40.85	58.50
	1500	Pull handle and push bar, aluminum		11	.727		121	26		147	176
	2000	Bronze		10	.800		157	28.50		185.50	219
	3000	Push plate both sides, aluminum		14	.571		14.80	20.50		35.30	50
	3500	Bronze		13	.615		37	22		59	77
	4000	Door pull, designer style, cast aluminum, minimum		12	.667		64.50	23.50		88	111
	5000	Maximum		8	1		325	35.50		360.50	420
	6000	Cast bronze, minimum		12	.667		76	23.50		99.50	123
	7000	Maximum		8	1		355	35.50		390.50	450
	8000	Walnut, minimum		12	.667		58	23.50		81.50	104
	9000	Maximum		8	1		325	35.50		360.50	415
	9800	Minimum labor/equipment charge		5	1.600	Job		57		57	94
790	0010	PEEPHOLE									
	2010	Peephole	1 Carp	32	.250	Ea.	14.10	8.90		23	30.50
800	0010	SPECIAL HINGES									
	0015	Paumelle, high frequency									
	0020	Steel base, 6" x 4-1/2", US10				Pr.	134			134	147
	0100	Bronze base, 5" x 4-1/2", US10					170			170	187
	0200	Paumelle, average frequency, steel base, 4-1/2" x 3-1/2", US10					90.50			90.50	99.50
	0400	Olive knuckle, low frequency, brass base, 6" x 4-1/2", US10					155			155	171
	1000	Electric hinge with concealed conductor, average frequency									
	1010	Steel base, 4-1/2" x 4-1/2", US26D				Pr.	286			286	315
	1100	Bronze base, 4-1/2" x 4-1/2", US26D				"	300			300	330
	1200	Electric hinge with concealed conductor, high frequency									
	1210	Steel base, 4-1/2" x 4-1/2", US26D				Pr.	213			213	235
	1600	Double weight, 800 lb., steel base, removable pin, 5" x 6", USP					125			125	137
	1700	Steel base-welded pin, 5" x 6", USP					141			141	155
	1800	Triple weight, 2000 lb., steel base, welded pin, 5" x 6", USP					143			143	158

For expanded coverage of these items see *Means Interior Cost Data 2006*

08700 | Hardware

08710 | Door Hardware

			Crew	Daily Output	Labor-Hours	Unit	2006 Bare Costs Mat.	Labor	Equip.	Total	Total Incl O&P	
800	2000	Pivot reinf., high frequency, steel base, 7-3/4" door plate, USP				Pr.	166			166	183	800
	2200	Bronze base, 7-3/4" door plate, US10					198			198	218	
	3000	Swing clear, full mortise, full or half surface, high frequency,										
	3010	Steel base, 5" high, USP				Pr.	143			143	157	
	3200	Swing clear, full mortise, average frequency										
	3210	Steel base, 4-1/2" high, USP				Pr.	113			113	125	
	4000	Wide throw, average frequency, steel base, 4-1/2" x 6", USP					86.50			86.50	95.50	
	4200	High frequency, steel base, 4-1/2" x 6", USP					133			133	146	
	4600	Spring hinge, single acting, 6" flange, steel				Ea.	49			49	54	
	4700	Brass					86			86	94.50	
	4900	Double acting, 6" flange, steel					86			86	95	
	4950	Brass					141			141	155	
	5000	T-strap, galvanized, 4"				Pr.	17.20			17.20	18.90	
	5010	6"					27			27	30	
	5020	8"					41.50			41.50	45.50	
	9000	Continuous hinge, steel, full mortise, heavy duty	2 Carp	64	.250	L.F.	11.10	8.90		20	27	

08720 | Weatherstripping & Seals

			Crew	Daily Output	Labor-Hours	Unit	2006 Bare Costs Mat.	Labor	Equip.	Total	Total Incl O&P	
100	0010	**ASTRAGALS** One piece overlapping										100
	0400	Cadmium plated steel, flat, 3/16" x 2"	1 Carp	90	.089	L.F.	3.06	3.16		6.22	8.60	
	0600	Prime coated steel, flat, 1/8" x 3"		90	.089		4.25	3.16		7.41	9.95	
	0800	Stainless steel, flat, 3/32" x 1-5/8"		90	.089		15.95	3.16		19.11	23	
	1000	Aluminum, flat, 1/8" x 2"		90	.089		3.07	3.16		6.23	8.65	
	1200	Nail on, "T" extrusion		120	.067		.68	2.37		3.05	4.68	
	1300	Vinyl bulb insert		105	.076		1.10	2.71		3.81	5.70	
	1600	Screw on, "T" extrusion		90	.089		4.47	3.16		7.63	10.15	
	1700	Vinyl insert		75	.107		3.02	3.79		6.81	9.60	
	2000	"L" extrusion, neoprene bulbs		75	.107		1.74	3.79		5.53	8.20	
	2100	Neoprene sponge insert		75	.107		5.30	3.79		9.09	12.15	
	2200	Magnetic		75	.107		8.70	3.79		12.49	15.90	
	2400	Spring hinged security seal, with cam		75	.107		5.60	3.79		9.39	12.45	
	2600	Spring loaded locking bolt, vinyl insert		45	.178		7.65	6.30		13.95	18.85	
	2800	Neoprene sponge strip, "Z" shaped, aluminum		60	.133		3.63	4.74		8.37	11.85	
	2900	Solid neoprene strip, nail on aluminum strip		90	.089		3.05	3.16		6.21	8.60	
	3000	One piece stile protection										
	3020	Neoprene fabric loop, nail on aluminum strips	1 Carp	60	.133	L.F.	.54	4.74		5.28	8.45	
	3110	Flush mounted aluminum extrusion, 1/2" x 1-1/4"		60	.133		2.89	4.74		7.63	11.05	
	3140	3/4" x 1-3/8"		60	.133		3.67	4.74		8.41	11.90	
	3160	1-1/8" x 1-3/4"		60	.133		5.85	4.74		10.59	14.30	
	3300	Mortise, 9/16" x 3/4"		60	.133		3.11	4.74		7.85	11.25	
	3320	13/16" x 1-3/8"		60	.133		3.37	4.74		8.11	11.55	
	3600	Spring bronze strip, nail on type		105	.076		2.65	2.71		5.36	7.40	
	3620	Screw on, with retainer		75	.107		2.17	3.79		5.96	8.70	
	3800	Flexible stainless steel housing, pile insert, 1/2" door		105	.076		6	2.71		8.71	11.10	
	3820	3/4" door		105	.076		6.75	2.71		9.46	11.95	
	4000	Extruded aluminum retainer, flush mount, pile insert		105	.076		2.10	2.71		4.81	6.80	
	4080	Mortise, felt insert		90	.089		3.74	3.16		6.90	9.35	
	4160	Mortise with spring, pile insert		90	.089		2.92	3.16		6.08	8.45	
	4400	Rigid vinyl retainer, mortise, pile insert		105	.076		2.09	2.71		4.80	6.80	
	4600	Wool pile filler strip, aluminum backing		105	.076		2.10	2.71		4.81	6.80	
	5000	Two piece overlapping astragal, extruded aluminum retainer										
	5010	Pile insert	1 Carp	60	.133	L.F.	2.72	4.74		7.46	10.85	
	5020	Vinyl bulb insert		60	.133		1.76	4.74		6.50	9.80	
	5040	Vinyl flap insert		60	.133		5.45	4.74		10.19	13.85	
	5060	Solid neoprene flap insert		60	.133		5.40	4.74		10.14	13.80	
	5080	Hypalon rubber flap insert		60	.133		5.55	4.74		10.29	13.95	

08700 | Hardware

	08720 Weatherstripping & Seals		Crew	Daily Output	Labor-Hours	Unit	2006 Bare Costs Mat.	Labor	Equip.	Total	Total Incl O&P	
100	5090	Snap on cover, pile insert	1 Carp	60	.133	L.F.	6.30	4.74		11.04	14.80	100
	5400	Magnetic aluminum, surface mounted		60	.133		21.50	4.74		26.24	31.50	
	5500	Interlocking aluminum, 5/8" x 1" neoprene bulb insert		45	.178		3.39	6.30		9.69	14.20	
	5600	Adjustable aluminum, 9/16" x 21/32", pile insert	↓	45	.178		16.10	6.30		22.40	28	
	5790	For vinyl bulb, deduct					.41			.41	.45	
	5800	Magnetic, adjustable, 9/16" x 21/32"	1 Carp	45	.178	↓	20.50	6.30		26.80	33	
	6000	Two piece stile protection										
	6010	Cloth backed rubber loop, 1" gap, nail on aluminum strips	1 Carp	45	.178	L.F.	3.47	6.30		9.77	14.25	
	6040	Screw on aluminum strips		45	.178		5.40	6.30		11.70	16.40	
	6100	1-1/2" gap, screw on aluminum extrusion		45	.178		4.86	6.30		11.16	15.80	
	6240	Vinyl fabric loop, slotted aluminum extrusion, 1" gap		45	.178		1.72	6.30		8.02	12.35	
	6300	1-1/4" gap	↓	45	.178	↓	5.10	6.30		11.40	16.10	
300	0010	**WEATHERSTRIPPING**, Window, double hung, 3' X 5'										300
	0020	Zinc	1 Carp	7.20	1.111	Opng.	12.05	39.50		51.55	79	
	0100	Bronze		7.20	1.111		23.50	39.50		63	91.50	
	0200	Vinyl V strip		7	1.143		3.76	40.50		44.26	71.50	
	0500	As above but heavy duty, zinc		4.60	1.739		15.35	62		77.35	119	
	0600	Bronze		4.60	1.739		26.50	62		88.50	132	
	1000	Doors, wood frame, interlocking, for 3' x 7' door, zinc		3	2.667		13.75	95		108.75	172	
	1100	Bronze		3	2.667		21.50	95		116.50	181	
	1300	6' x 7' opening, zinc		2	4		15.05	142		157.05	253	
	1400	Bronze	↓	2	4	↓	28.50	142		170.50	267	
	1700	Wood frame, spring type, bronze										
	1800	3' x 7' door	1 Carp	7.60	1.053	Opng.	17.70	37.50		55.20	81.50	
	1900	6' x 7' door	"	7	1.143	"	21	40.50		61.50	90.50	
	2200	Metal frame, spring type, bronze										
	2300	3' x 7' door	1 Carp	3	2.667	Opng.	30	95		125	190	
	2400	6' x 7' door	"	2.50	3.200	"	41	114		155	233	
	2500	For stainless steel, spring type, add					133%					
	2700	Metal frame, extruded sections, 3' x 7' door, aluminum	1 Carp	2	4	Opng.	40	142		182	280	
	2800	Bronze		2	4		101	142		243	345	
	3100	6' x 7' door, aluminum		1.20	6.667		51	237		288	450	
	3200	Bronze	↓	1.20	6.667	↓	119	237		356	525	
	3500	Threshold weatherstripping										
	3650	Door sweep, flush mounted, aluminum	1 Carp	25	.320	Ea.	11.80	11.40		23.20	32	
	3700	Vinyl		25	.320		14	11.40		25.40	34.50	
	5000	Garage door bottom weatherstrip, 12' aluminum, clear		14	.571		18.90	20.50		39.40	54.50	
	5010	Bronze		14	.571		72	20.50		92.50	113	
	5050	Bottom protection, 12' aluminum, clear		14	.571		21	20.50		41.50	56.50	
	5100	Bronze		14	.571	↓	89.50	20.50		110	132	
	9000	Minimum labor/equipment charge	↓	3	2.667	Job		95		95	157	
800	0010	**THRESHOLDS**										800
	0011	Threshold 3' long saddles aluminum	1 Carp	48	.167	L.F.	3.62	5.95		9.57	13.80	
	0100	Aluminum, 8" wide, 1/2" thick		12	.667	Ea.	30.50	23.50		54	73	
	0500	Bronze		60	.133	L.F.	32	4.74		36.74	43	
	0600	Bronze, panic threshold, 5" wide, 1/2" thick ♿		12	.667	Ea.	60	23.50		83.50	106	
	0700	Rubber, 1/2" thick, 5-1/2" wide		20	.400		33.50	14.20		47.70	60.50	
	0800	2-3/4" wide		20	.400	↓	14.05	14.20		28.25	39	
	9000	Minimum labor/equipment charge	↓	4	2	Job		71		71	118	
	08750 Window Hardware											
400	0010	**WINDOW HARDWARE**										400
	1000	Handles, surface mounted, aluminum	1 Carp	24	.333	Ea.	1.90	11.85		13.75	21.50	
	1020	Brass		24	.333		2.20	11.85		14.05	22	
	1040	Chrome	↓	24	.333	↓	2.02	11.85		13.87	22	

08700 | Hardware

08750 | Window Hardware

	08750 Window Hardware		Crew	Daily Output	Labor-Hours	Unit	2006 Bare Costs Mat.	Labor	Equip.	Total	Total Incl O&P	
400	1500	Recessed, aluminum	1 Carp	12	.667	Ea.	1.10	23.50		24.60	40.50	400
	1520	Brass		12	.667		1.22	23.50		24.72	41	
	1540	Chrome		12	.667		1.15	23.50		24.65	41	
	2000	Latches, aluminum		20	.400		1.58	14.20		15.78	25	
	2020	Brass		20	.400		1.90	14.20		16.10	25.50	
	2040	Chrome		20	.400		1.78	14.20		15.98	25.50	
	9000	Minimum labor/equipment charge		6	1.333	Job		47.50		47.50	78.50	

08770 | Door/Window Accessories

	08770 Door/Window Accessories		Crew	Daily Output	Labor-Hours	Unit	2006 Bare Costs Mat.	Labor	Equip.	Total	Total Incl O&P	
560	0010	DOOR ACCESSORIES										560
	1000	Knockers, brass, standard	1 Carp	16	.500	Ea.	37	17.80		54.80	70	
	1100	Deluxe		10	.800		115	28.50		143.50	173	
	4000	Security chain, standard		18	.444		6.35	15.80		22.15	33	
	4100	Deluxe		18	.444		38	15.80		53.80	68	
	4500	Rubber door silencers		540	.015		.10	.53		.63	.98	
	9000	Minimum labor/equipment charge		6	1.333	Job		47.50		47.50	78.50	

08800 | Glazing

08810 | Glass

	08810 Glass		Crew	Daily Output	Labor-Hours	Unit	2006 Bare Costs Mat.	Labor	Equip.	Total	Total Incl O&P	
260	0010	FLOAT GLASS										260
	0020	3/16" thick, plain	2 Glaz	130	.123	S.F.	4.67	4.22		8.89	11.95	
	0200	Tempered, clear		130	.123		5.45	4.22		9.67	12.80	
	0300	Tinted		130	.123		6.95	4.22		11.17	14.45	
	0600	1/4" thick, clear, plain		120	.133		5.55	4.57		10.12	13.45	
	0700	Tinted		120	.133		5.45	4.57		10.02	13.35	
	0800	Tempered, clear		120	.133		6.80	4.57		11.37	14.80	
	0900	Tinted		120	.133		9.35	4.57		13.92	17.65	
	1600	3/8" thick, clear, plain		75	.213		8.80	7.30		16.10	21.50	
	1700	Tinted		75	.213		10.90	7.30		18.20	24	
	1800	Tempered, clear		75	.213		13.65	7.30		20.95	27	
	1900	Tinted		75	.213		16.95	7.30		24.25	30.50	
	2200	1/2" thick, clear, plain		55	.291		17.60	9.95		27.55	35.50	
	2300	Tinted		55	.291		19.05	9.95		29	37	
	2400	Tempered, clear		55	.291		20.50	9.95		30.45	38.50	
	2500	Tinted		55	.291		25.50	9.95		35.45	44	
	2800	5/8" thick, clear, plain		45	.356		19.05	12.20		31.25	40.50	
	2900	Tempered, clear		45	.356		22	12.20		34.20	43.50	
	3200	3/4" thick, clear, plain		35	.457		24.50	15.65		40.15	52	
	3300	Tempered, clear		35	.457		28.50	15.65		44.15	56.50	
	3600	1" thick, clear, plain		30	.533		41	18.25		59.25	74.50	
	8900	For low emissivity coating for 3/16" & 1/4" only, add to above					15%					
	9000	Minimum labor/equipment charge	1 Glaz	2	4	Job		137		137	221	
270	0010	FULL VISION										270
	0020	Up to 10' high	H-2	130	.185	S.F.	55	5.90		60.90	70	
	0100	10' to 20' high, minimum		110	.218		58.50	7		65.50	75.50	
	0150	Average		100	.240		63	7.65		70.65	81.50	

		08810 | Glass	CREW	DAILY OUTPUT	LABOR-HOURS	UNIT	2006 BARE COSTS MAT.	LABOR	EQUIP.	TOTAL	TOTAL INCL O&P	
270	0200	Maximum	H-2	80	.300	S.F.	70.50	9.60		80.10	93.50	270
	9000	Minimum labor/equipment charge	1 Glaz	2	4	Job		137		137	221	
300	0010	**GLAZING VARIABLES** R088110-10										300
	0500	For high rise glazing, exterior, add per S.F. per story				S.F.					.09	
	0600	For glass replacement, add				"		100%				
	0700	For gasket settings, add				L.F.	4.08			4.08	4.49	
	0900	For sloped glazing, add				S.F.		25%				
	2000	Fabrication, polished edges, 1/4" thick				Inch	.35			.35	.39	
	2100	1/2" thick					.89			.89	.98	
	2500	Mitered edges, 1/4" thick					.89			.89	.98	
	2600	1/2" thick					1.44			1.44	1.58	
460	0010	**INSULATING GLASS** 2 lites 1/8" float, 1/2" thk, under 15 S.F. R088110-10										460
	0020	Clear	2 Glaz	95	.168	S.F.	7.85	5.75		13.60	17.95	
	0100	Tinted		95	.168		11.55	5.75		17.30	22	
	0200	2 lites 3/16" float, for 5/8" thk unit, 15 to 30 S.F., clear		90	.178		9.55	6.10		15.65	20.50	
	0400	1" thk, dbl. glazed, 1/4" float, 30-70 S.F., clear		75	.213		13.45	7.30		20.75	26.50	
	0500	Tinted		75	.213		16.35	7.30		23.65	30	
	2000	Both lites, light & heat reflective		85	.188		22	6.45		28.45	34.50	
	2500	Heat reflective, film inside, 1" thick unit, clear		85	.188		19.05	6.45		25.50	31.50	
	2600	Tinted		85	.188		20.50	6.45		26.95	33	
	3000	Film on weatherside, clear, 1/2" thick unit		95	.168		13.60	5.75		19.35	24.50	
	3100	5/8" thick unit		90	.178		16.60	6.10		22.70	28	
	3200	1" thick unit		85	.188		18.75	6.45		25.20	31	
	3350	Minimum	1 Glaz	50	.160		10.25	5.50		15.75	20	
	3360	Maximum	"	25	.320		11.10	10.95		22.05	30	
	3370	Reflective or tinted, add					2.82			2.82	3.10	
	9000	Minimum labor/equipment charge	1 Glaz	2	4	Job		137		137	221	
500	0010	**LAMINATED GLASS** (security item)										500
	0020	Clear float .03" vinyl 1/4"	2 Glaz	90	.178	S.F.	9.45	6.10		15.55	20	
	0100	3/8" thick		78	.205		14.90	7.05		21.95	27.50	
	0200	.06" vinyl, 1/2" thick		65	.246		17.40	8.45		25.85	33	
	1000	5/8" thick		90	.178		20.50	6.10		26.60	32.50	
	2000	Bullet-resisting, 1-3/16" thick, to 15 S.F.		16	1		56	34.50		90.50	117	
	2100	Over 15 S.F.		16	1		50.50	34.50		85	111	
	2500	2-1/4" thick, to 15 S.F.		12	1.333		67	45.50		112.50	147	
	2600	Over 15 S.F.		12	1.333		59.50	45.50		105	139	
	0015	3/16" thick, material					4.27			4.27	4.70	
675	0010	**REFLECTIVE GLASS**										675
	0100	1/4" float with fused metallic oxide fixed	2 Glaz	115	.139	S.F.	11.35	4.77		16.12	20	
	0500	1/4" float glass with reflective applied coating		115	.139		9.45	4.77		14.22	18.10	
	2000	Solar film on glass, not including glass, minimum		180	.089		4.57	3.04		7.61	9.95	
	2050	Solar film on glass, not including glass, maximum		225	.071		10.55	2.44		12.99	15.60	
850	0010	**WINDOW GLASS**										850
	0015	1/8" thick, clear float	2 Glaz	480	.033	S.F.	3.78	1.14		4.92	6	
	0500	3/16" thick, clear		480	.033		4.66	1.14		5.80	7	
	0600	Tinted		480	.033		5.25	1.14		6.39	7.65	
	0700	Tempered		480	.033		6.35	1.14		7.49	8.85	
	2000	Replace broken window lite, 1/8" glass (9 S.F. maximum)	1 Glaz	48	.167		4.21	5.70		9.91	13.85	
	2100	1/4" plate (16 S.F. maximum)	"	48	.167		4.64	5.70		10.34	14.30	
	9000	Minimum labor/equipment charge	2 Glaz	5	3.200	Job		110		110	177	
900	0010	**WIRE GLASS** (chicken wire) (security item)										900
	0012	1/4" thick rough obscure	2 Glaz	135	.119	S.F.	12.45	4.06		16.51	20.50	
	1000	Polished wire, 1/4" thick, diamond, clear		135	.119		15.45	4.06		19.51	23.50	
	1500	Pinstripe, obscure		135	.119		15.50	4.06		19.56	23.50	

08800 | Glazing

08830 | Mirrors

				DAILY	LABOR-		2006 BARE COSTS				TOTAL	
			CREW	OUTPUT	HOURS	UNIT	MAT.	LABOR	EQUIP.	TOTAL	INCL O&P	
100	0010	**MIRRORS** No frames, wall type, 1/4" plate glass, polished edge										100
	0100	Up to 5 S.F.	2 Glaz	125	.128	S.F.	6.70	4.38		11.08	14.40	
	0200	Over 5 S.F.		160	.100		6.45	3.43		9.88	12.60	
	0500	Door type, 1/4" plate glass, up to 12 S.F.		160	.100		6.95	3.43		10.38	13.15	
	1000	Float glass, up to 10 S.F., 1/8" thick		160	.100		3.91	3.43		7.34	9.80	
	1100	3/16" thick		150	.107		4.54	3.65		8.19	10.90	
	1500	12" x 12" wall tiles, square edge, clear		195	.082		1.59	2.81		4.40	6.30	
	1600	Veined		195	.082		4.33	2.81		7.14	9.30	
	2000	1/4" thick, stock sizes, one way transparent		125	.128		15.10	4.38		19.48	23.50	
	2010	Bathroom, unframed, laminated		160	.100		11.10	3.43		14.53	17.75	
	2500	Tempered		160	.100		12.25	3.43		15.68	19	

08840 | Plastic Glazing

			CREW	DAILY OUTPUT	LABOR-HOURS	UNIT	MAT.	LABOR	EQUIP.	TOTAL	TOTAL INCL O&P	
600	0010	**PLEXIGLASS ACRYLIC**, Masked										600
	0020	1/8" thick, cut sheets	2 Glaz	170	.094	S.F.	3.45	3.22		6.67	9	
	0200	Full sheets		195	.082		1.80	2.81		4.61	6.50	
	0500	1/4" thick, cut sheets		165	.097		6.10	3.32		9.42	12.05	
	0600	Full sheets		185	.086		3.31	2.96		6.27	8.40	
	0900	3/8" thick, cut sheets		155	.103		11.15	3.54		14.69	17.95	
	1000	Full sheets		180	.089		6	3.04		9.04	11.50	
	1300	1/2" thick, cut sheets		135	.119		12.85	4.06		16.91	20.50	
	1400	Full sheets		150	.107		12.45	3.65		16.10	19.60	
	1700	3/4" thick, cut sheets		115	.139		45.50	4.77		50.27	57.50	
	1800	Full sheets		130	.123		26.50	4.22		30.72	36	
	2100	1" thick, cut sheets		105	.152		51.50	5.20		56.70	65	
	2200	Full sheets		125	.128		32	4.38		36.38	42	
	3000	Colored, 1/8" thick, cut sheets		170	.094		10.60	3.22		13.82	16.85	
	3200	Full sheets		195	.082		6.85	2.81		9.66	12.10	
	3500	1/4" thick, cut sheets		165	.097		11.85	3.32		15.17	18.40	
	3600	Full sheets		185	.086		8.10	2.96		11.06	13.65	
	4000	Mirrors, untinted, cut sheets, 1/8" thick		185	.086		4.98	2.96		7.94	10.25	
	4200	1/4" thick		180	.089		8	3.04		11.04	13.70	
650	0010	**POLYCARBONATE** (security item)										650
	0020	1/8" thick	2 Glaz	170	.094	S.F.	6	3.22		9.22	11.80	
	0500	3/16" thick		165	.097		7.25	3.32		10.57	13.30	
	1000	1/4" thick		155	.103		8	3.54		11.54	14.50	
	1500	3/8" thick		150	.107		14.75	3.65		18.40	22	
	9000	Minimum labor/equipment charge	1 Glaz	2	4	Job		137		137	221	
700	0010	**SECURITY FILM** clear, 32000 psi tensile strength, adh to glass R088110-10										700
	0100	.002" thick, daylight installation	H-2	950	.025	S.F.	.65	.81		1.46	2.03	
	0150	.004" thick, daylight installation		800	.030		.77	.96		1.73	2.41	
	0200	.006" thick, daylight installation		700	.034		.82	1.10		1.92	2.68	
	0210	Install for anchorage		600	.040		.91	1.28		2.19	3.08	
	0400	.007" thick, daylight istallation		600	.040		.87	1.28		2.15	3.04	
	0410	Install for anchorage		500	.048		.97	1.53		2.50	3.55	
	0500	.008" thick, daylight installation		500	.048		1.20	1.53		2.73	3.81	
	0510	Install for anchorage		500	.048		1.33	1.53		2.86	3.96	
	0600	.015" thick, daylight installation		400	.060		2.49	1.92		4.41	5.85	
	0610	Install for anchorage		400	.060		1.33	1.92		3.25	4.59	
	0900	Security Film Anchorage, mechanical attachment and cover plate	H-3	370	.043	L.F.	6.55	1.31		7.86	9.35	
	0950	Security film anchorage, wet glaze structural caulking	1 Glaz	225	.036	"	1.20	1.22		2.42	3.28	
	1000	Adhered security film removal	1 Clab	275	.029	S.F.		.80		.80	1.32	

8 DOORS & WINDOWS

08800 | Glazing

08850 | Glazing Accessories

		08850 Glazing Accessories	Crew	Daily Output	Labor-Hours	Unit	2006 Bare Costs Mat.	Labor	Equip.	Total	Total Incl O&P	
100	0010	**GLAZING GASKETS**										100
	0015	1/4" glass	2 Glaz	200	.080	L.F.	.23	2.74		2.97	4.67	
	0020	3/8"		200	.080		.34	2.74		3.08	4.79	
	0040	1/2"		200	.080		.38	2.74		3.12	4.84	
	0060	3/4"		180	.089		.35	3.04		3.39	5.30	
	0080	1"		180	.089		.40	3.04		3.44	5.35	
	1000	Glazing compound, wood	1 Glaz	58	.138		.53	4.72		5.25	8.20	
	1005	Glazing compound, per window, up to 30 L.F.					.53			.53	.58	
	1006	Glazing compound, per window, up to 30 L.F.				Ea.	15.80			15.80	17.40	
	1020	Metal	1 Glaz	58	.138	L.F.	.53	4.72		5.25	8.20	

08900 | Glazed Curtain Wall

08910 | Glazed Aluminum Curtain Wall

		08910 Glazed Aluminum Curtain Wall	Crew	Daily Output	Labor-Hours	Unit	2006 Bare Costs Mat.	Labor	Equip.	Total	Total Incl O&P	
700	0010	**TUBE FRAMING** For window walls and store fronts, aluminum, stock										700
	0050	Plain tube frame, mill finish, 1-3/4" x 1-3/4"	2 Glaz	103	.155	L.F.	6.55	5.30		11.85	15.75	
	0150	1-3/4" x 4"		98	.163		8.75	5.60		14.35	18.60	
	0200	1-3/4" x 4-1/2"		95	.168		10.20	5.75		15.95	20.50	
	0250	2" x 6"		89	.180		14.65	6.15		20.80	26	
	0350	4" x 4"		87	.184		14.35	6.30		20.65	26	
	0400	4-1/2" x 4-1/2"		85	.188		16.65	6.45		23.10	28.50	
	0450	Glass bead		240	.067		1.86	2.28		4.14	5.75	
	1000	Flush tube frame, mill finish, 1/4" glass, 1-3/4" x 4", open header		80	.200		8.60	6.85		15.45	20.50	
	1050	Open sill		82	.195		7.45	6.70		14.15	18.95	
	1100	Closed back header		83	.193		12	6.60		18.60	24	
	1150	Closed back sill		85	.188		11.40	6.45		17.85	23	
	1200	Vertical mullion, one piece		75	.213		12.70	7.30		20	26	
	1250	Two piece		73	.219		13.65	7.50		21.15	27	
	1300	90° or 180° vertical corner post		75	.213		21.50	7.30		28.80	35.50	
	1400	1-3/4" x 4-1/2", open header		80	.200		10.45	6.85		17.30	22.50	
	1450	Open sill		82	.195		8.65	6.70		15.35	20.50	
	1500	Closed back header		83	.193		12.75	6.60		19.35	24.50	
	1550	Closed back sill		85	.188		12.30	6.45		18.75	24	
	1600	Vertical mullion, one piece		75	.213		13.75	7.30		21.05	27	
	1650	Two piece		73	.219		14.55	7.50		22.05	28	
	1700	90° or 180° vertical corner post		75	.213		14.90	7.30		22.20	28	
	2000	Flush tube frame, mill fin. for ins. glass, 2" x 4-1/2", open header		75	.213		11.45	7.30		18.75	24.50	
	2050	Open sill		77	.208		10.05	7.10		17.15	22.50	
	2100	Closed back header		78	.205		12.50	7.05		19.55	25	
	2150	Closed back sill		80	.200		12.35	6.85		19.20	24.50	
	2200	Vertical mullion, one piece		70	.229		13.85	7.85		21.70	28	
	2250	Two piece		68	.235		14.80	8.05		22.85	29.50	
	2300	90° or 180° vertical corner post		70	.229		14	7.85		21.85	28	
	5000	Flush tube frame, mill fin., thermal brk., 2-1/4"x 4-1/2", open header		74	.216		12.65	7.40		20.05	26	
	5050	Open sill		75	.213		11.05	7.30		18.35	24	
	5100	Vertical mullion, one piece		69	.232		15.25	7.95		23.20	29.50	
	5150	Two piece		67	.239		16.30	8.20		24.50	31	
	5200	90° or 180° vertical corner post		69	.232		14.65	7.95		22.60	29	
	6980	Door stop (snap in)		380	.042		2.60	1.44		4.04	5.20	
	7000	For joints, 90°, clip type, add				Ea.	20			20	22	

08900 | Glazed Curtain Wall

		08910 | Glazed Aluminum Curtain Wall	CREW	DAILY OUTPUT	LABOR-HOURS	UNIT	2006 BARE COSTS MAT.	LABOR	EQUIP.	TOTAL	TOTAL INCL O&P	
700	7050	Screw spline joint, add				Ea.	15.35			15.35	16.85	700
	7100	For joint other than 90°, add				↓	32			32	35.50	
	8000	For bronze anodized aluminum, add					15%					
	8020	For black finish, add					27%					
	8050	For stainless steel materials, add					350%					
	8100	For monumental grade, add					50%					
	8150	For steel stiffener, add	2 Glaz	200	.080	L.F.	7.85	2.74		10.59	13	
	8200	For 2 to 5 stories, add per story				Story		5%				
	9000	Minimum labor/equipment charge	2 Glaz	2	8	Job		274		274	440	
900	0010	**WINDOW WALLS**, Aluminum stock including glazing										900
	0020	Minimum	H-2	160	.150	S.F.	28	4.80		32.80	39	
	0050	Average		140	.171		34.50	5.50		40	47	
	0100	Maximum	↓	110	.218	↓	108	7		115	130	
	0500	For translucent sandwich wall systems, see Div. 07420-770										
	0850	Cost of the above walls depends on material,										
	0860	finish, repetition, and size of units.										
	0870	The larger the opening, the lower the S.F. cost										
	1200	Double glazed acoustical window wall for airports,										
	1220	including 1" thick glass with 2" x 4-1/2" tube frame	H-2	40	.600	S.F.	62.50	19.20		81.70	99.50	
		08950 | Translucent Wall & Roof Asemblies										
100	0010	**SKYROOFS**, Translucent panels, 2-3/4" thick										100
	0020	Under 500 S.F.	G-3	395	.081	SF Hor.	20.50	2.82		23.32	27	
	0100	Over 5000 S.F.		465	.069		18.10	2.39		20.49	24	
	0300	Continuous vaulted, semi-circular, to 8' wide, double glazed		145	.221		45.50	7.70		53.20	62.50	
	0400	Single glazed		160	.200		30.50	6.95		37.45	45.50	
	0600	To 20' wide, single glazed		175	.183		34.50	6.35		40.85	48.50	
	0700	Over 20' wide, single glazed		200	.160		39.50	5.55		45.05	52	
	0900	Motorized opening type, single glazed, 1/3 opening		145	.221		42	7.70		49.70	59	
	1000	Full opening	↓	130	.246	↓	48	8.55		56.55	67	
	1200	Pyramid type units, self-supporting, to 30' clear opening,										
	1300	square or circular, single glazed, minimum	G-3	200	.160	SF Hor.	22	5.55		27.55	33.50	
	1310	Average		165	.194		31.50	6.75		38.25	45.50	
	1400	Maximum		130	.246		45	8.55		53.55	63.50	
	1500	Grid type, 4' to 10' modules, single glass glazed, minimum		200	.160		29	5.55		34.55	41	
	1550	Maximum		128	.250		47.50	8.70		56.20	66	
	1600	Preformed acrylic, minimum		300	.107		35	3.71		38.71	44.50	
	1650	Maximum	↓	175	.183	↓	48.50	6.35		54.85	64	
	9000	Minimum labor/equipment charge	2 Carp	8	2	Job		71		71	118	

Division 9 Finishes

Estimating Tips

General

- Room Finish Schedule: A complete set of plans should contain a room finish schedule. If one is not available, it would be well worth the time and effort to put one together. A room finish schedule should contain the room number, room name (for clarity), floor materials, base materials, wainscot materials, wainscot height, wall materials (for each wall), ceiling materials, ceiling height, and special instructions.
- Surplus Finishes: Review the specifications to determine if there is any requirement to provide certain amounts of extra materials for the owner's maintenance department. In some cases the owner may require a substantial amount of materials, especially when it is a special order item or long lead time item.

09200 Plaster & Gypsum Board

- Lath is estimated by the square yard for both gypsum and metal lath, plus usually 5% allowance for waste. Furring, channels, and accessories are measured by the linear foot. An extra foot should be allowed for each accessory miter or stop.
- Plaster is also estimated by the square yard. Deductions for openings vary by preference, from zero deduction to 50% of all openings over 2 feet in width. Some estimators deduct a percentage of the total yardage for openings. The estimator should allow one extra square foot for each linear foot of horizontal interior or exterior angle located below the ceiling level. Also, double the areas of small radius work.
- Each room should be measured, perimeter times maximum wall height. Floors and ceiling areas are equal to length times width.
- Drywall accessories, studs, track, and acoustical caulking are all measured by the linear foot. Drywall taping is figured by the square foot. Gypsum wallboard is estimated by the square foot. No material deductions should be made for door or window openings under 32 S.F. Coreboard can be obtained in a 1″ thickness for solid wall and shaft work. Additions should be made to price out the inside or outside corners.
- Different types of partition construction should be listed separately on the quantity sheets. There may be walls with studs of various widths, double studded, and similar or dissimilar surface materials. Shaft work is usually different construction from surrounding partitions requiring separate quantities and pricing of the work.

09300 Tile
09400 Terrazzo

- Tile and terrazzo areas are taken off on a square foot basis. Trim and base materials are measured by the linear foot. Accent tiles are listed per each. Two basic methods of installation are used. Mud set is approximately 30% more expensive than the thin set. In terrazzo work, be sure to include the linear footage of embedded decorative strips, grounds, machine rubbing, and power cleanup.

09600 Flooring

- Wood flooring is available in strip, parquet, or block configuration. The latter two types are set in adhesives with quantities estimated by the square foot. The laying pattern will influence labor costs and material waste. In addition to the material and labor for laying wood floors, the estimator must make allowances for sanding and finishing these areas unless the flooring is prefinished.
- Most of the various types of flooring are all measured on a square foot basis. Base is measured by the linear foot. If adhesive materials are to be quantified, they are estimated at a specified coverage rate by the gallon depending upon the specified type and the manufacturer's recommendations.
- Sheet flooring is measured by the square yard. Roll widths vary, so consideration should be given to use the most economical width, as waste must be figured into the total quantity. Consider also the installation methods available, direct glue down or stretched.

09700 Wall Finishes

- Wall coverings are estimated by the square foot. The area to be covered is measured, length by height of wall above baseboards, to calculate the square footage of each wall. This figure is divided by the number of square feet in the single roll which is being used. Deduct, in full, the areas of openings such as doors and windows. Where a pattern match is required allow 25%-30% waste. One gallon of paste should be sufficient to hang 12 single rolls of light- to medium-weight paper.

09800 Acoustical Treatment

- Acoustical systems fall into several categories. The takeoff of these materials should be by the square foot of area with a 5% allowance for waste. Do not forget about scaffolding, if applicable, when estimating these systems.

09900 Paints & Coatings

- A major portion of the work in painting involves surface preparation. Be sure to include cleaning, sanding, filling, and masking costs in the estimate.
- Painting is one area where bids vary to a greater extent than almost any other section of a project. This arises from the many methods of measuring surfaces to be painted. The estimator should check the plans and specifications carefully to be sure of the required number of coats.
- Protection of adjacent surfaces is not included in painting costs. When considering the method of paint application, an important factor is the amount of protection and masking required. These must be estimated separately and may be the determining factor in choosing the method of application.

Reference Numbers

Reference numbers are shown in bold squares at the beginning of some major classifications. These numbers refer to related items in the Reference Section. The reference information may be an estimating procedure, an alternate pricing method, or technical information.

Note: Not all subdivisions listed here necessarily appear in this publication.

09060 | Selective Demolition

			CREW	DAILY OUTPUT	LABOR-HOURS	UNIT	2006 BARE COSTS MAT.	LABOR	EQUIP.	TOTAL	TOTAL INCL O&P	
110	0010	**SELECTIVE DEMOLITION, CEILINGS** R024119-10										110
	0200	Ceiling, drywall, furred and nailed or screwed	2 Clab	800	.020	S.F.		.55		.55	.91	
	0220	On metal frame		760	.021			.58		.58	.96	
	0240	On suspension system, including system		720	.022			.61		.61	1.01	
	1000	Plaster, lime and horse hair, on wood lath, incl. lath		700	.023			.63		.63	1.04	
	1020	On metal lath		570	.028			.77		.77	1.27	
	1100	Gypsum, on gypsum lath		720	.022			.61		.61	1.01	
	1120	On metal lath		500	.032			.88		.88	1.45	
	1200	Suspended ceiling, mineral fiber, 2′ x 2′ or 2′ x 4′		1500	.011			.29		.29	.48	
	1250	On suspension system, incl. system		1200	.013			.37		.37	.61	
	1500	Tile, wood fiber, 12″ x 12″, glued		900	.018			.49		.49	.81	
	1540	Stapled		1500	.011			.29		.29	.48	
	1580	On suspension system, incl. system		760	.021			.58		.58	.96	
	2000	Wood, tongue and groove, 1″ x 4″		1000	.016			.44		.44	.73	
	2040	1″ x 8″		1100	.015			.40		.40	.66	
	2400	Plywood or wood fiberboard, 4′ x 8′ sheets		1200	.013			.37		.37	.61	
	9000	Minimum labor/equipment charge	1 Clab	2	4	Job		110		110	182	
120	0010	**SELECTIVE DEMOLITION, FLOORING** R024119-10										120
	0200	Brick with mortar	2 Clab	475	.034	S.F.		.92		.92	1.53	
	0400	Carpet, bonded, including surface scraping		2000	.008			.22		.22	.36	
	0480	Tackless		9000	.002			.05		.05	.08	
	0600	Composition, acrylic or epoxy		400	.040			1.10		1.10	1.82	
	0700	Concrete, scarify skin	A-1A	225	.036			1.30	1.39	2.69	3.64	
	0800	Resilient, sheet goods	2 Clab	1400	.011			.31		.31	.52	
	0820	For gym floors		900	.018			.49		.49	.81	
	0900	Vinyl composition tile, 12″ x 12″		1000	.016			.44		.44	.73	
	2000	Tile, ceramic, thin set		675	.024			.65		.65	1.08	
	2020	Mud set		625	.026			.70		.70	1.16	
	2200	Marble, slate, thin set		675	.024			.65		.65	1.08	
	2220	Mud set		625	.026			.70		.70	1.16	
	2600	Terrazzo, thin set		450	.036			.97		.97	1.61	
	2620	Mud set		425	.038			1.03		1.03	1.71	
	2640	Cast in place		300	.053			1.46		1.46	2.42	
	3000	Wood, block, on end	1 Carp	400	.020			.71		.71	1.18	
	3200	Parquet		450	.018			.63		.63	1.05	
	3400	Strip flooring, interior, 2-1/4″ x 25/32″ thick		325	.025			.88		.88	1.45	
	3500	Exterior, porch flooring, 1″ x 4″		220	.036			1.29		1.29	2.14	
	3800	Subfloor, tongue and groove, 1″ x 6″		325	.025			.88		.88	1.45	
	3820	1″ x 8″		430	.019			.66		.66	1.10	
	3840	1″ x 10″		520	.015			.55		.55	.91	
	4000	Plywood, nailed		600	.013			.47		.47	.79	
	4100	Glued and nailed		400	.020			.71		.71	1.18	
	8000	Remove flooring, bead blast, minimum	A-1A	1000	.008			.29	.31	.60	.82	
	8100	Maximum		400	.020			.73	.78	1.51	2.05	
	8150	Mastic only		1500	.005			.19	.21	.40	.55	
	9000	Minimum labor/equipment charge	1 Clab	4	2	Job		55		55	91	
130	0010	**SELECTIVE DEMOLITION, WALLS AND PARTITIONS** R024119-10										130
	0100	Brick, 4″ to 12″ thick	B-9C	220	.182	C.F.		5.05	.67	5.72	9.10	
	0200	Concrete block, 4″ thick		1000	.040	S.F.		1.11	.15	1.26	2	
	0280	8″ thick		810	.049			1.37	.18	1.55	2.47	
	0300	Exterior stucco 1″ thick over mesh	B-9	3200	.013			.35	.05	.40	.63	
	1000	Drywall, nailed or screwed	1 Clab	1000	.008			.22		.22	.36	
	1020	Glued and nailed		900	.009			.24		.24	.40	
	1500	Fiberboard, nailed		900	.009			.24		.24	.40	

09050 | Basic Material Finishes and Methods

09060 | Selective Demolition

		09060 Selective Demolition	CREW	DAILY OUTPUT	LABOR-HOURS	UNIT	2006 BARE COSTS MAT.	LABOR	EQUIP.	TOTAL	TOTAL INCL O&P	
130	1520	Glued and nailed (R024119-10)	1 Clab	800	.010	S.F.		.27		.27	.45	130
	1568	Plenum barrier, sheet lead		300	.027			.73		.73	1.21	
	2000	Movable walls, metal, 5′ high		300	.027			.73		.73	1.21	
	2020	8′ high		400	.020			.55		.55	.91	
	2200	Metal or wood studs, finish 2 sides, fiberboard	B-1	520	.046			1.30		1.30	2.15	
	2250	Lath and plaster		260	.092			2.59		2.59	4.29	
	2300	Plasterboard (drywall)		520	.046			1.30		1.30	2.15	
	2350	Plywood		450	.053			1.50		1.50	2.48	
	3000	Plaster, lime and horsehair, on wood lath	1 Clab	400	.020			.55		.55	.91	
	3020	On metal lath		335	.024			.65		.65	1.08	
	3400	Gypsum or perlite, on gypsum lath		410	.020			.53		.53	.89	
	3420	On metal lath		300	.027			.73		.73	1.21	
	3800	Toilet partitions, slate or marble		5	1.600	Ea.		44		44	72.50	
	3820	Hollow metal		8	1	″		27.50		27.50	45.50	
	5000	Wallcovering, vinyl	1 Pape	700	.011	S.F.		.36		.36	.58	
	5040	Designer	″	480	.017	″		.53		.53	.85	
	9000	Minimum labor/equipment charge	1 Clab	4	2	Job		55		55	91	

09100 | Metal Support Assemblies

09110 | Non-Load Bearing Wall Framing

		09110 Non-Load Bearing Wall Framing	CREW	DAILY OUTPUT	LABOR-HOURS	UNIT	2006 BARE COSTS MAT.	LABOR	EQUIP.	TOTAL	TOTAL INCL O&P	
100	0010	METAL STUDS AND TRACK										100
	1600	Non-load bearing, galv, 8′ high, 25 ga. 1-5/8″ wide, 16″ O.C.	1 Carp	619	.013	S.F.	.24	.46		.70	1.02	
	1610	24″ O.C.		950	.008		.18	.30		.48	.70	
	1620	2-1/2″ wide, 16″ O.C.		613	.013		.29	.46		.75	1.08	
	1630	24″ O.C.		938	.009		.22	.30		.52	.74	
	1640	3-5/8″ wide, 16″ O.C.		600	.013		.33	.47		.80	1.15	
	1650	24″ O.C.		925	.009		.25	.31		.56	.78	
	1660	4″ wide, 16″ O.C.		594	.013		.35	.48		.83	1.18	
	1670	24″ O.C.		925	.009		.26	.31		.57	.80	
	1680	6″ wide, 16″ O.C.		588	.014		.51	.48		.99	1.36	
	1690	24″ O.C.		906	.009		.38	.31		.69	.94	
	1700	20 ga. studs, 1-5/8″ wide, 16″ O.C.		494	.016		.39	.58		.97	1.38	
	1710	24″ O.C.		763	.010		.29	.37		.66	.94	
	1720	2-1/2″ wide, 16″ O.C.		488	.016		.46	.58		1.04	1.48	
	1730	24″ O.C.		750	.011		.35	.38		.73	1.01	
	1740	3-5/8″ wide, 16″ O.C.		481	.017		.55	.59		1.14	1.58	
	1750	24″ O.C.		738	.011		.41	.39		.80	1.09	
	1760	4″ wide, 16″ O.C.		475	.017		.60	.60		1.20	1.65	
	1770	24″ O.C.		738	.011		.45	.39		.84	1.14	
	1780	6″ wide, 16″ O.C.		469	.017		.76	.61		1.37	1.84	
	1790	24″ O.C.		725	.011		.57	.39		.96	1.28	
	2000	Non-load bearing, galv, 10′ high, 25 ga. 1-5/8″ wide, 16″ O.C.		495	.016		.22	.57		.79	1.20	
	2100	24″ O.C.		760	.011		.17	.37		.54	.80	
	2200	2-1/2″ wide, 16″ O.C.		490	.016		.27	.58		.85	1.26	
	2250	24″ O.C.		750	.011		.20	.38		.58	.85	
	2300	3-5/8″ wide, 16″ O.C.		480	.017		.31	.59		.90	1.32	
	2350	24″ O.C.		740	.011		.23	.38		.61	.89	
	2400	4″ wide, 16″ O.C.		475	.017		.33	.60		.93	1.35	

FINISHES 9

09100 | Metal Support Assemblies

	09110 | Non-Load Bearing Wall Framing		Crew	Daily Output	Labor-Hours	Unit	2006 Bare Costs Mat.	Labor	Equip.	Total	Total Incl O&P	
100	2450	24" O.C.	1 Carp	740	.011	S.F.	.25	.38		.63	.91	100
	2500	6" wide, 16" O.C.		470	.017		.48	.61		1.09	1.53	
	2550	24" O.C.		725	.011		.36	.39		.75	1.04	
	2600	20 ga. studs, 1-5/8" wide, 16" O.C.		395	.020		.37	.72		1.09	1.59	
	2650	24" O.C.		610	.013		.27	.47		.74	1.07	
	2700	2-1/2" wide, 16" O.C.		390	.021		.44	.73		1.17	1.69	
	2750	24" O.C.		600	.013		.32	.47		.79	1.14	
	2800	3-5/8" wide, 16" O.C.		385	.021		.52	.74		1.26	1.79	
	2850	24" O.C.		590	.014		.38	.48		.86	1.22	
	2900	4" wide, 16" O.C.		380	.021		.57	.75		1.32	1.87	
	2950	24" O.C.		590	.014		.42	.48		.90	1.26	
	3000	6" wide, 16" O.C.		375	.021		.72	.76		1.48	2.05	
	3050	24" O.C.		580	.014		.53	.49		1.02	1.40	
	3060	Non-load bearing, galv, 12' high, 25 ga. 1-5/8" wide, 16" O.C.		413	.019		.21	.69		.90	1.37	
	3070	24" O.C.		633	.013		.16	.45		.61	.91	
	3080	2-1/2" wide, 16" O.C.		408	.020		.26	.70		.96	1.44	
	3090	24" O.C.		625	.013		.19	.46		.65	.96	
	3100	3-5/8" wide, 16" O.C.		400	.020		.30	.71		1.01	1.50	
	3110	24" O.C.		617	.013		.22	.46		.68	1	
	3120	4" wide, 16" O.C.		396	.020		.32	.72		1.04	1.54	
	3130	24" O.C.		617	.013		.23	.46		.69	1.01	
	3140	6" wide, 16" O.C.		392	.020		.46	.73		1.19	1.71	
	3150	24" O.C.		604	.013		.34	.47		.81	1.15	
	3160	20 ga. studs, 1-5/8" wide, 16" O.C.		329	.024		.35	.86		1.21	1.81	
	3170	24" O.C.		508	.016		.26	.56		.82	1.21	
	3180	2-1/2" wide, 16" O.C.		325	.025		.42	.88		1.30	1.91	
	3190	24" O.C.		500	.016		.30	.57		.87	1.27	
	3200	3-5/8" wide, 16" O.C.		321	.025		.50	.89		1.39	2.02	
	3210	24" O.C.		492	.016		.36	.58		.94	1.36	
	3220	4" wide, 16" O.C.		317	.025		.54	.90		1.44	2.09	
	3230	24" O.C.		492	.016		.40	.58		.98	1.40	
	3240	6" wide, 16" O.C.		313	.026		.69	.91		1.60	2.27	
	3250	24" O.C.	↓	483	.017	↓	.50	.59		1.09	1.53	
	5000	Load bearing studs, see division 05410-400										
	9000	Minimum labor/equipment charge	1 Carp	4	2	Job		71		71	118	

	09120 | Ceiling Suspension		Crew	Daily Output	Labor-Hours	Unit	Mat.	Labor	Equip.	Total	Total Incl O&P	
100	0010	**CEILING SUSPENSION SYSTEMS** For gypsum board or plaster										100
	8000	Suspended ceilings, including carriers										
	8200	1-1/2" carriers, 24" O.C. with:										
	8300	7/8" channels, 16" O.C.	1 Lath	165	.048	S.F.	.39	1.59		1.98	2.95	
	8320	24" O.C.		200	.040		.32	1.31		1.63	2.43	
	8400	1-5/8" channels, 16" O.C.		155	.052		.47	1.69		2.16	3.21	
	8420	24" O.C.	↓	190	.042	↓	.37	1.38		1.75	2.60	
	8600	2" carriers, 24" O.C. with:										
	8700	7/8" channels, 16" O.C.	1 Lath	155	.052	S.F.	.44	1.69		2.13	3.17	
	8720	24" O.C.		190	.042		.37	1.38		1.75	2.60	
	8800	1-5/8" channels, 16" O.C.		145	.055		.52	1.81		2.33	3.44	
	8820	24" O.C.	↓	180	.044	↓	.42	1.46		1.88	2.77	

	09130 | Acoustical Suspension		Crew	Daily Output	Labor-Hours	Unit	Mat.	Labor	Equip.	Total	Total Incl O&P	
100	0010	**CEILING SUSPENSION SYSTEMS** For boards and tile										100
	0050	Class A suspension system, 15/16" T bar, 2' x 4' grid	1 Carp	800	.010	S.F.	.52	.36		.88	1.17	
	0300	2' x 2' grid	"	650	.012		.66	.44		1.10	1.45	
	0350	For 9/16" grid, add				↓	.15			.15	.17	

9 FINISHES

09100 | Metal Support Assemblies

	09130 | Acoustical Suspension		CREW	DAILY OUTPUT	LABOR-HOURS	UNIT	2006 BARE COSTS MAT.	LABOR	EQUIP.	TOTAL	TOTAL INCL O&P	
100	0360	For fire rated grid, add				S.F.	.09			.09	.10	100
	0370	For colored grid, add					.19			.19	.21	
	0400	Concealed Z bar suspension system, 12" module	1 Carp	520	.015		.47	.55		1.02	1.43	
	0600	1-1/2" carrier channels, 4' O.C., add	"	470	.017	↓	.10	.61		.71	1.11	
	0700	Carrier channels for ceilings with										
	0900	recessed lighting fixtures, add	1 Carp	460	.017	S.F.	.18	.62		.80	1.22	
	9000	Minimum labor/equipment charge	"	5	1.600	Job		57		57	94	

09200 | Plaster & Gypsum Board

	09205 | Furring & Lathing		CREW	DAILY OUTPUT	LABOR-HOURS	UNIT	2006 BARE COSTS MAT.	LABOR	EQUIP.	TOTAL	TOTAL INCL O&P	
530	0010	**FURRING** Beams & columns, 7/8" galvanized channels,										530
	0030	12" O.C.	1 Lath	155	.052	S.F.	.25	1.69		1.94	2.97	
	0050	16" O.C.		170	.047		.21	1.54		1.75	2.68	
	0070	24" O.C.		185	.043		.14	1.42		1.56	2.40	
	0100	Ceilings, on steel, 7/8" channels, galvanized, 12" O.C.		210	.038		.23	1.25		1.48	2.23	
	0300	16" O.C.		290	.028		.21	.91		1.12	1.67	
	0400	24" O.C.		420	.019		.14	.62		.76	1.14	
	0600	1-5/8" channels, galvanized, 12" O.C.		190	.042		.32	1.38		1.70	2.54	
	0700	16" O.C.		260	.031		.29	1.01		1.30	1.92	
	0900	24" O.C.		390	.021		.19	.67		.86	1.28	
	1000	Walls, 7/8" channels, galvanized, 12" O.C.		235	.034		.23	1.12		1.35	2.02	
	1200	16" O.C.		265	.030		.21	.99		1.20	1.80	
	1300	24" O.C.		350	.023		.14	.75		.89	1.34	
	1500	1-5/8" channels, galvanized, 12" O.C.		210	.038		.32	1.25		1.57	2.33	
	1600	16" O.C.		240	.033		.29	1.09		1.38	2.05	
	1800	24" O.C.		305	.026	↓	.19	.86		1.05	1.58	
	9000	Minimum labor/equipment charge	↓	4	2	Job		65.50		65.50	104	
540	0010	**GYPSUM LATH**										540
	0012	Plain or perforated, nailed, 3/8" thick	1 Lath	85	.094	S.Y.	5.40	3.09		8.49	10.85	
	0100	1/2" thick, nailed		80	.100		4.23	3.28		7.51	9.85	
	0300	Clipped to steel studs, 3/8" thick		75	.107		5.40	3.50		8.90	11.50	
	0400	1/2" thick		70	.114		4.23	3.75		7.98	10.60	
	0600	Firestop gypsum base, to steel studs, 3/8" thick		70	.114		3.69	3.75		7.44	10	
	0700	1/2" thick		65	.123		4.50	4.04		8.54	11.35	
	0900	Foil back, to steel studs, 3/8" thick		75	.107		3.96	3.50		7.46	9.90	
	1000	1/2" thick		70	.114		4.14	3.75		7.89	10.50	
	1500	For ceiling installations, add		216	.037			1.21		1.21	1.93	
	1600	For columns and beams, add		170	.047	↓		1.54		1.54	2.45	
	9000	Minimum labor/equipment charge	↓	4.25	1.882	Job		61.50		61.50	98	
560	0010	**METAL LATH**										560
	0020	Diamond, expanded, 2.5 lb. per S.Y., painted				S.Y.	2.85			2.85	3.14	
	0100	Galvanized, 2.5 lb. per S.Y.					2.37			2.37	2.61	
	0300	3.4 lb. per S.Y., painted					3.71			3.71	4.08	
	0400	Galvanized					3.11			3.11	3.42	
	0600	For 15# asphalt sheathing paper, add					.31			.31	.34	
	0900	Flat rib, 1/8" high, 2.75 lb., painted					3.04			3.04	3.34	
	1000	Foil backed				↓	3.65			3.65	4.02	

For expanded coverage of these items see *Means Interior Cost Data 2006*

09200 | Plaster & Gypsum Board

09205 | Furring & Lathing

		09205 Furring & Lathing	CREW	DAILY OUTPUT	LABOR-HOURS	UNIT	2006 BARE COSTS MAT.	LABOR	EQUIP.	TOTAL	TOTAL INCL O&P	
560	1200	3.4 lb. per S.Y., painted				S.Y.	3.28			3.28	3.61	560
	1300	Galvanized					4.13			4.13	4.54	
	1500	For 15# asphalt sheating paper, add					.31			.31	.34	
	1800	High rib, 3/8″ high, 3.4 lb. per S.Y., painted					4.40			4.40	4.84	
	1900	Galvanized					3.67			3.67	4.04	
	2400	High rib, 3/4″ high, painted, .60 lb. per S.F.				S.F.	.42			.42	.46	
	2500	.75 lb. per S.F.				″	.90			.90	.99	
	2800	Stucco mesh, painted, 3.6 lb.				S.Y.	3.32			3.32	3.65	
	3000	K-lath, perforated, absorbent paper, regular					3.21			3.21	3.53	
	3100	Heavy duty					3.79			3.79	4.17	
	3300	Waterproof, heavy duty, grade B backing					3.71			3.71	4.08	
	3400	Fire resistant backing					4.10			4.10	4.51	
	3600	2.5 lb. diamond painted, on wood framing, on walls	1 Lath	85	.094		2.85	3.09		5.94	8.05	
	3700	On ceilings		75	.107		2.85	3.50		6.35	8.70	
	3900	3.4 lb. diamond painted, on wood framing, on walls		80	.100		3.28	3.28		6.56	8.80	
	4000	On ceilings		70	.114		3.28	3.75		7.03	9.55	
	4200	3.4 lb. diamond painted, wired to steel framing		75	.107		3.28	3.50		6.78	9.15	
	4300	On ceilings		60	.133		3.28	4.37		7.65	10.55	
	4600	Cornices, wired to steel		35	.229		3.28	7.50		10.78	15.50	
	4800	Screwed to steel studs, 2.5 lb.		80	.100		2.85	3.28		6.13	8.35	
	4900	3.4 lb.		75	.107		3.71	3.50		7.21	9.65	
	5100	Rib lath, painted, wired to steel, on walls, 2.5 lb.		75	.107		3.04	3.50		6.54	8.90	
	5200	3.4 lb.		70	.114		4.40	3.75		8.15	10.80	
	5400	4.0 lb.		65	.123		4.52	4.04		8.56	11.35	
	5500	For self-furring lath, add					.08			.08	.09	
	5700	Suspended ceiling system, incl. 3.4 lb. diamond lath, painted	1 Lath	15	.533		11.90	17.50		29.40	41	
	5800	Galvanized	″	15	.533		12.25	17.50		29.75	41.50	
	6000	Hollow metal stud partitions, 3.4 lb. painted lath both sides										
	6010	Non-load bearing, 25 ga., w/rib lath 2-1/2″ studs, 12″ O.C.	1 Lath	20.30	.394	S.Y.	11.85	12.95		24.80	33.50	
	6300	16″ O.C.		21.10	.379		11.15	12.45		23.60	32	
	6350	24″ O.C.		22.70	.352		10.60	11.55		22.15	30	
	6400	3-5/8″ studs, 16″ O.C.		19.50	.410		11.60	13.45		25.05	34.50	
	6600	24″ O.C.		20.40	.392		10.85	12.85		23.70	32.50	
	6700	4″ studs, 16″ O.C.		20.40	.392		11.80	12.85		24.65	33.50	
	6900	24″ O.C.		21.60	.370		11	12.15		23.15	31.50	
	7000	6″ studs, 16″ O.C.		19.50	.410		13.15	13.45		26.60	36	
	7100	24″ O.C.		21.10	.379		12	12.45		24.45	33	
	7200	L.B. partitions, 16 ga., w/rib lath, 2-1/2″ studs, 16″ O.C.		20	.400		11.35	13.10		24.45	33.50	
	7300	3-5/8″ studs, 16 ga.		19.70	.406		12.85	13.30		26.15	35	
	7500	4″ studs, 16 ga.		19.50	.410		13.35	13.45		26.80	36	
	7600	6″ studs, 16 ga.		18.70	.428		15.80	14.05		29.85	40	
	9000	Minimum labor/equipment charge		4.25	1.882	Job		61.50		61.50	98	

09210 | Gypsum Plaster

		09210 Gypsum Plaster	CREW	DAILY OUTPUT	LABOR-HOURS	UNIT	MAT.	LABOR	EQUIP.	TOTAL	TOTAL INCL O&P	
100	0010	**GYPSUM PLASTER**										100
	0020	80# bag, less than 1 ton				Bag	14.45			14.45	15.85	
	0300	2 coats, no lath included, on walls	J-1	105	.381	S.Y.	3.29	11.65	1.01	15.95	23.50	
	0400	On ceilings		92	.435		3.29	13.30	1.15	17.74	26.50	
	0900	3 coats, no lath included, on walls		87	.460		4.66	14.10	1.21	19.97	29	
	1000	On ceilings		78	.513		4.66	15.70	1.35	21.71	32	
	1600	For irregular or curved surfaces, add						30%				
	1800	For columns & beams, add						50%				
	9000	Minimum labor/equipment charge	1 Plas	1	8	Job		260		260	420	
500	0010	**PERLITE OR VERMICULITE PLASTER**										500
	0020	In 100 lb. bags, under 200 bags				Bag	14.55			14.55	16	

09200 | Plaster & Gypsum Board

	09210 | Gypsum Plaster		Crew	Daily Output	Labor-Hours	Unit	2006 Bare Costs Mat.	Labor	Equip.	Total	Total Incl O&P
500	0300	2 coats, no lath included, on walls	J-1	92	.435	S.Y.	3.24	13.30	1.15	17.69	26.50
	0400	On ceilings		79	.506		3.24	15.50	1.34	20.08	30
	0900	3 coats, no lath included, on walls		74	.541		5.85	16.55	1.43	23.83	34.50
	1000	On ceilings		63	.635		5.85	19.45	1.68	26.98	40
	1700	For irregular or curved surfaces, add to above						30%			
	1800	For columns and beams, add to above						50%			
	1900	For soffits, add to ceiling prices						40%			
	9000	Minimum labor/equipment charge	1 Plas	1	8	Job		260		260	420
900	0010	THIN COAT									
	0012	1 coat veneer, not incl. lath	J-1	3600	.011	S.F.	.07	.34	.03	.44	.66
	1000	In 50 lb. bags				Bag	9.85			9.85	10.80

	09220 | Portland Cement Plaster		Crew	Daily Output	Labor-Hours	Unit	Mat.	Labor	Equip.	Total	Total Incl O&P
200	0010	STUCCO									
	0015	3 coats 1" thick, float finish, with mesh, on wood frame	J-2	63	.762	S.Y.	5.35	23.50	1.68	30.53	45.50
	0100	On masonry construction, no mesh incl.	J-1	67	.597		2.05	18.30	1.58	21.93	33.50
	0300	For trowel finish, add	1 Plas	170	.047			1.53		1.53	2.46
	0400	For 3/4" thick, on masonry, deduct	J-1	880	.045		.57	1.39	.12	2.08	3.01
	0600	For coloring and special finish, add, minimum		685	.058		.37	1.79	.15	2.31	3.47
	0700	Maximum		200	.200		1.29	6.15	.53	7.97	11.90
	0900	For soffits, add	J-2	155	.310		1.99	9.60	.68	12.27	18.40
	1000	Exterior stucco, with bonding agent, 3 coats, on walls, no mesh incl.	J-1	200	.200		3.33	6.15	.53	10.01	14.15
	1200	Ceilings		180	.222		3.33	6.80	.59	10.72	15.30
	1300	Beams		80	.500		3.33	15.30	1.32	19.95	29.50
	1500	Columns		100	.400		3.33	12.25	1.06	16.64	24.50
	1550	Minimum labor/equipment charge	1 Plas	1	8	Job		260		260	420
	1600	Mesh, painted, nailed to wood, 1.8 lb.	1 Lath	60	.133	S.Y.	4.17	4.37		8.54	11.55
	1800	3.6 lb.		55	.145		3.32	4.77		8.09	11.20
	1900	Wired to steel, painted, 1.8 lb.		53	.151		4.17	4.95		9.12	12.45
	2100	3.6 lb.		50	.160		3.32	5.25		8.57	12
	9000	Minimum labor/equipment charge		4	2	Job		65.50		65.50	104

	09250 | Gypsum Board		Crew	Daily Output	Labor-Hours	Unit	Mat.	Labor	Equip.	Total	Total Incl O&P
200	0010	CEMENTITIOUS BACKERBOARD									
	0070	Cementitious backerboard, on floor, 3' x 4'x 1/2" sheets	2 Carp	525	.030	S.F.	1.04	1.08		2.12	2.95
	0080	3' x 5' x 1/2" sheets		525	.030		1.04	1.08		2.12	2.95
	0090	3' x 6' x 1/2" sheets		525	.030		.76	1.08		1.84	2.63
	0100	3' x 4'x 5/8" sheets		525	.030		1.08	1.08		2.16	2.99
	0110	3' x 5' x 5/8" sheets		525	.030		1.06	1.08		2.14	2.97
	0120	3' x 6' x 5/8" sheets		525	.030		1.07	1.08		2.15	2.97
	0150	On wall, 3' x 4'x 1/2" sheets		350	.046		1.04	1.63		2.67	3.84
	0160	3' x 5' x 1/2" sheets		350	.046		1.04	1.63		2.67	3.84
	0170	3' x 6' x 1/2" sheets		350	.046		.76	1.63		2.39	3.52
	0180	3' x 4'x 5/8" sheets		350	.046		1.08	1.63		2.71	3.88
	0190	3' x 5' x 5/8" sheets		350	.046		1.06	1.63		2.69	3.86
	0200	3' x 6' x 5/8" sheets		350	.046		1.07	1.63		2.70	3.86
	0250	On counter, 3' x 4'x 1/2" sheets		180	.089		1.04	3.16		4.20	6.40
	0260	3' x 5' x 1/2" sheets		180	.089		1.04	3.16		4.20	6.40
	0270	3' x 6' x 1/2" sheets		180	.089		.76	3.16		3.92	6.10
	0300	3' x 4'x 5/8" sheets		180	.089		1.08	3.16		4.24	6.45
	0310	3' x 5' x 5/8" sheets		180	.089		1.06	3.16		4.22	6.40
	0320	3' x 6' x 5/8" sheets		180	.089		1.07	3.16		4.23	6.40
300	0010	BLUEBOARD For use with thin coat									
	0100	plaster application (see division 09210-900)									

09200 | Plaster & Gypsum Board

		09250 | Gypsum Board	CREW	DAILY OUTPUT	LABOR-HOURS	UNIT	2006 BARE COSTS MAT.	LABOR	EQUIP.	TOTAL	TOTAL INCL O&P	
300	1000	3/8" thick, on walls or ceilings, standard, no finish included	2 Carp	1900	.008	S.F.	.23	.30		.53	.75	300
	1100	With thin coat plaster finish		875	.018		.30	.65		.95	1.41	
	1400	On beams, columns, or soffits, standard, no finish included		675	.024		.26	.84		1.10	1.69	
	1450	With thin coat plaster finish		475	.034		.34	1.20		1.54	2.35	
	3000	1/2" thick, on walls or ceilings, standard, no finish included		1900	.008		.27	.30		.57	.80	
	3100	With thin coat plaster finish		875	.018		.34	.65		.99	1.46	
	3300	Fire resistant, no finish included		1900	.008		.27	.30		.57	.80	
	3400	With thin coat plaster finish		875	.018		.34	.65		.99	1.46	
	3450	On beams, columns, or soffits, standard, no finish included		675	.024		.31	.84		1.15	1.74	
	3500	With thin coat plaster finish		475	.034		.38	1.20		1.58	2.40	
	3700	Fire resistant, no finish included		675	.024		.31	.84		1.15	1.74	
	3800	With thin coat plaster finish		475	.034		.38	1.20		1.58	2.40	
	5000	5/8" thick, on walls or ceilings, fire resistant, no finish included		1900	.008		.29	.30		.59	.82	
	5100	With thin coat plaster finish		875	.018		.36	.65		1.01	1.48	
	5500	On beams, columns, or soffits, no finish included		675	.024		.33	.84		1.17	1.77	
	5600	With thin coat plaster finish		475	.034		.41	1.20		1.61	2.43	
	6000	For high ceilings, over 8' high, add		3060	.005			.19		.19	.31	
	6500	For over 3 stories high, add per story	↓	6100	.003	↓		.09		.09	.15	
	9000	Minimum labor/equipment charge	1 Carp	2	4	Job		142		142	236	
500	0010	**CEILINGS** Gypsum drywall, fire rated, finished										500
	0100	Screwed to grid, channel or joists, 1/2" thick	2 Carp	765	.021	S.F.	.30	.74		1.04	1.56	
	0200	5/8" thick		765	.021		.30	.74		1.04	1.56	
	0300	Over 8' high, 1/2" thick		615	.026		.30	.93		1.23	1.86	
	0400	5/8" thick	↓	615	.026	↓	.30	.93		1.23	1.86	
	0600	Grid suspension system, direct hung										
	0700	1-1/2" C.R.C., with 7/8" hi hat furring channel, 16" O.C.	2 Carp	600	.027	S.F.	1	.95		1.95	2.67	
	0800	24" O.C.		900	.018		.91	.63		1.54	2.05	
	0900	3-5/8" C.R.C., with 7/8" hi hat furring channel, 16" O.C.		600	.027		1.06	.95		2.01	2.74	
	1000	24" O.C.	↓	900	.018	↓	.92	.63		1.55	2.06	
700	0010	**DRYWALL** Gypsum plasterboard, nailed or screwed R092910-10										700
	0100	to studs unless otherwise noted										
	0150	3/8" thick, on walls, standard, no finish included	2 Carp	2000	.008	S.F.	.24	.28		.52	.73	
	0200	On ceilings, standard, no finish included		1800	.009		.24	.32		.56	.78	
	0250	On beams, columns, or soffits, no finish included		675	.024		.24	.84		1.08	1.66	
	0300	1/2" thick, on walls, standard, no finish included		2000	.008		.27	.28		.55	.77	
	0350	Taped and finished (level 4 finish)		965	.017		.30	.59		.89	1.32	
	0390	With compound skim coat (level 5 finish)		775	.021		.34	.73		1.07	1.60	
	0400	Fire resistant, no finish included		2000	.008		.30	.28		.58	.80	
	0450	Taped and finished (level 4 finish)		965	.017		.33	.59		.92	1.35	
	0490	With compound skim coat (level 5 finish)		775	.021		.37	.73		1.10	1.63	
	0500	Water resistant, no finish included		2000	.008		.25	.28		.53	.75	
	0550	Taped and finished (level 4 finish)		965	.017		.28	.59		.87	1.29	
	0590	With compound skim coat (level 5 finish)		775	.021		.32	.73		1.05	1.58	
	0600	Prefinished, vinyl, clipped to studs		900	.018		.56	.63		1.19	1.67	
	1000	On ceilings, standard, no finish included		1800	.009		.27	.32		.59	.82	
	1050	Taped and finished (level 4 finish)		765	.021		.30	.74		1.04	1.57	
	1090	With compound skim coat (level 5 finish)		610	.026		.34	.93		1.27	1.92	
	1100	Fire resistant, no finish included		1800	.009		.30	.32		.62	.85	
	1150	Taped and finished (level 4 finish)		765	.021		.33	.74		1.07	1.60	
	1195	With compound skim coat (level 5 finish)		610	.026		.37	.93		1.30	1.95	
	1200	Water resistant, no finish included		1800	.009		.25	.32		.57	.80	
	1250	Taped and finished (level 4 finish)		765	.021		.28	.74		1.02	1.54	
	1290	With compound skim coat (level 5 finish)		610	.026		.32	.93		1.25	1.90	
	1500	On beams, columns, or soffits, standard, no finish included		675	.024		.31	.84		1.15	1.74	
	1550	Taped and finished (level 4 finish)	↓	475	.034	↓	.30	1.20		1.50	2.32	

9 FINISHES

09200 | Plaster & Gypsum Board

	09250 | Gypsum Board		CREW	DAILY OUTPUT	LABOR-HOURS	UNIT	2006 BARE COSTS MAT.	LABOR	EQUIP.	TOTAL	TOTAL INCL O&P	
700	1590	With compound skim coat (level 5 finish) R092910-10	2 Carp	540	.030	S.F.	.34	1.05		1.39	2.13	700
	1600	Fire resistant, no finish included		675	.024		.35	.84		1.19	1.78	
	1650	Taped and finished (level 4 finish)		475	.034		.33	1.20		1.53	2.35	
	1690	With compound skim coat (level 5 finish)		540	.030		.37	1.05		1.42	2.16	
	1700	Water resistant, no finish included		675	.024		.29	.84		1.13	1.72	
	1750	Taped and finished (level 4 finish)		475	.034		.28	1.20		1.48	2.29	
	1790	With compound skim coat (level 5 finish)		540	.030		.32	1.05		1.37	2.11	
	2000	5/8" thick, on walls, standard, no finish included		2000	.008		.30	.28		.58	.80	
	2050	Taped and finished (level 4 finish)		965	.017		.33	.59		.92	1.35	
	2090	With compound skim coat (level 5 finish)		775	.021		.37	.73		1.10	1.63	
	2100	Fire resistant, no finish included		2000	.008		.30	.28		.58	.80	
	2150	Taped and finished (level 4 finish)		965	.017		.33	.59		.92	1.35	
	2195	With compound skim coat (level 5 finish)		775	.021		.37	.73		1.10	1.63	
	2200	Water resistant, no finish included		2000	.008		.27	.28		.55	.77	
	2250	Taped and finished (level 4 finish)		965	.017		.30	.59		.89	1.32	
	2290	With compound skim coat (level 5 finish)		775	.021		.34	.73		1.07	1.60	
	2300	Prefinished, vinyl, clipped to studs		900	.018		.65	.63		1.28	1.77	
	3000	On ceilings, standard, no finish included		1800	.009		.30	.32		.62	.85	
	3050	Taped and finished (level 4 finish)		765	.021		.33	.74		1.07	1.60	
	3090	With compound skim coat (level 5 finish)		615	.026		.37	.93		1.30	1.94	
	3100	Fire resistant, no finish included		1800	.009		.30	.32		.62	.85	
	3150	Taped and finished (level 4 finish)		765	.021		.33	.74		1.07	1.60	
	3190	With compound skim coat (level 5 finish)		615	.026		.37	.93		1.30	1.94	
	3200	Water resistant, no finish included		1800	.009		.27	.32		.59	.82	
	3250	Taped and finished (level 4 finish)		765	.021		.30	.74		1.04	1.57	
	3290	With compound skim coat (level 5 finish)		615	.026		.34	.93		1.27	1.91	
	3500	On beams, columns, or soffits, no finish included		675	.024		.35	.84		1.19	1.78	
	3550	Taped and finished (level 4 finish)		475	.034		.38	1.20		1.58	2.40	
	3590	With compound skim coat (level 5 finish)		380	.042		.43	1.50		1.93	2.95	
	3600	Fire resistant, no finish included		675	.024		.35	.84		1.19	1.78	
	3650	Taped and finished (level 4 finish)		475	.034		.38	1.20		1.58	2.40	
	3690	With compound skim coat (level 5 finish)		380	.042		.37	1.50		1.87	2.89	
	3700	Water resistant, no finish included		675	.024		.31	.84		1.15	1.74	
	3750	Taped and finished (level 4 finish)		475	.034		.35	1.20		1.55	2.37	
	3790	With compound skim coat (level 5 finish)		380	.042		.34	1.50		1.84	2.86	
	4000	Fireproofing, beams or columns, 2 layers, 1/2" thick, incl finish		330	.048		.63	1.72		2.35	3.56	
	4050	5/8" thick		300	.053		.67	1.90		2.57	3.88	
	4100	3 layers, 1/2" thick		225	.071		.93	2.53		3.46	5.20	
	4150	5/8" thick		210	.076		1	2.71		3.71	5.60	
	5050	For 1" thick coreboard on columns		480	.033		.48	1.18		1.66	2.49	
	5100	For foil-backed board, add					.10			.10	.11	
	5200	For work over 8' high, add	2 Carp	3060	.005			.19		.19	.31	
	5270	For textured spray, add	2 Lath	1600	.010		.05	.33		.38	.58	
	5300	For over 3 stories high, add per story	2 Carp	6100	.003			.09		.09	.15	
	5350	For finishing inner corners, add		950	.017	L.F.	.08	.60		.68	1.07	
	5355	For finishing outer corners, add		1250	.013		.20	.46		.66	.97	
	5500	For acoustical sealant, add per bead	1 Carp	500	.016		.03	.57		.60	.97	
	5550	Sealant, 1 quart tube				Ea.	4.89			4.89	5.40	
	5600	Sound deadening board, 1/4" gypsum	2 Carp	1800	.009	S.F.	.27	.32		.59	.82	
	5650	1/2" wood fiber	"	1800	.009	"	.38	.32		.70	.94	
	9000	Minimum labor/equipment charge	1 Carp	2	4	Job		142		142	236	

	09260 | Gypsum Board Systems		CREW	DAILY OUTPUT	LABOR-HOURS	UNIT	MAT.	LABOR	EQUIP.	TOTAL	TOTAL INCL O&P	
100	0010	**PARTITION WALL** Stud wall, 8' to 12' high										100
	0050	1/2", interior, gypsum board, std, tape & finish 2 sides										

FINISHES 9

	09260 Gypsum Board Systems		CREW	DAILY OUTPUT	LABOR-HOURS	UNIT	2006 BARE COSTS MAT.	LABOR	EQUIP.	TOTAL	TOTAL INCL O&P	
100	0500	Installed on and incl., 2″ x 4″ wood studs, 16″ O.C.	2 Carp	310	.052	S.F.	1.03	1.83		2.86	4.17	100
	1000	Metal studs, NLB, 25 ga., 16″ O.C., 3-5/8″ wide		350	.046		.92	1.63		2.55	3.70	
	1200	6″ wide		330	.048		1.09	1.72		2.81	4.06	
	1400	Water resistant, on 2″ x 4″ wood studs, 16″ O.C.		310	.052		.99	1.83		2.82	4.13	
	1600	Metal studs, NLB, 25 ga., 16″ O.C., 3-5/8″ wide		350	.046		.88	1.63		2.51	3.66	
	1800	6″ wide		330	.048		1.05	1.72		2.77	4.02	
	2000	Fire res., 2 layers, 1-1/2 hr., on 2″ x 4″ wood studs, 16″ O.C.		210	.076		1.69	2.71		4.40	6.35	
	2200	Metal studs, NLB, 25 ga., 16″ O.C., 3-5/8″ wide		250	.064		1.58	2.28		3.86	5.50	
	2400	6″ wide		230	.070		1.75	2.47		4.22	6.05	
	2600	Fire & water res., 2 layers, 1-1/2 hr., 2″ x 4″ studs, 16″ O.C.		210	.076		1.69	2.71		4.40	6.35	
	2800	Metal studs, NLB, 25 ga., 16″ O.C., 3-5/8″ wide		250	.064		1.58	2.28		3.86	5.50	
	3000	6″ wide		230	.070		1.75	2.47		4.22	6.05	
	3200	5/8″, interior, gypsum board, std, tape & finish 2 sides										
	3400	Installed on and including 2″ x 4″ wood studs, 16″ O.C.	2 Carp	300	.053	S.F.	1.09	1.90		2.99	4.34	
	3600	24″ O.C.		330	.048		.99	1.72		2.71	3.95	
	3800	Metal studs, NLB, 25 ga., 16″ O.C., 3-5/8″ wide		340	.047		.98	1.67		2.65	3.85	
	4000	6″ wide		320	.050		1.15	1.78		2.93	4.22	
	4200	24″ O.C., 3-5/8″ wide		360	.044		.90	1.58		2.48	3.61	
	4400	6″ wide		340	.047		1.03	1.67		2.70	3.90	
	4800	Water resistant, on 2″ x 4″ wood studs, 16″ O.C.		300	.053		1.03	1.90		2.93	4.27	
	5000	24″ O.C.		330	.048		.93	1.72		2.65	3.89	
	5200	Metal studs, NLB, 25 ga. 16″ O.C., 3-5/8″ wide		340	.047		.92	1.67		2.59	3.78	
	5400	6″ wide		320	.050		1.09	1.78		2.87	4.15	
	5600	24″ O.C., 3-5/8″ wide		360	.044		.84	1.58		2.42	3.54	
	5800	6″ wide		340	.047		.97	1.67		2.64	3.83	
	6000	Fire res., 2 layers, 2 hr., on 2″ x 4″ wood studs, 16″ O.C.		205	.078		1.59	2.77		4.36	6.35	
	6200	24″ O.C.		235	.068		1.59	2.42		4.01	5.75	
	6400	Metal studs, NLB, 25 ga., 16″ O.C., 3-5/8″ wide		245	.065		1.60	2.32		3.92	5.60	
	6600	6″ wide		225	.071		1.75	2.53		4.28	6.10	
	6800	24″ O.C., 3-5/8″ wide		265	.060		1.50	2.15		3.65	5.20	
	7000	6″ wide		245	.065		1.63	2.32		3.95	5.65	
	7200	Fire & water res., 2 layers, 2 hr., 2″ x 4″ studs, 16″ O.C.		205	.078		1.69	2.77		4.46	6.45	
	7400	24″ O.C.		235	.068		1.59	2.42		4.01	5.75	
	7600	Metal studs, NLB, 25 ga., 16″ O.C., 3-5/8″ wide		245	.065		1.58	2.32		3.90	5.60	
	7800	6″ wide		225	.071		1.75	2.53		4.28	6.10	
	8000	24″ O.C., 3-5/8″ wide		265	.060		1.50	2.15		3.65	5.20	
	8200	6″ wide		245	.065		1.63	2.32		3.95	5.65	
	8600	1/2″ blueboard, mesh tape both sides										
	8620	Installed on and including 2″ x 4″ wood studs, 16″ O.C.	2 Carp	300	.053	S.F.	1.09	1.90		2.99	4.34	
	8640	Metal studs, NLB, 25 ga., 16″ O.C., 3-5/8″ wide		340	.047		.98	1.67		2.65	3.85	
	8660	6″ wide		320	.050		1.15	1.78		2.93	4.22	
	9000	Exterior, 1/2″ gypsum sheathing, 1/2″ gypsum finished, interior,										
	9100	including foil faced insulation, metal studs, 20 ga.										
	9200	16″ O.C., 3-5/8″ wide	2 Carp	290	.055	S.F.	1.91	1.96		3.87	5.35	
	9400	6″ wide	″	270	.059	″	2.12	2.11		4.23	5.80	
800	0010	**SHAFT WALL** Cavity type on 25 ga. J track & C-H studs, 24″ O.C.										800
	0030	1″ thick coreboard wall liner on shaft side										
	0040	2-hour assembly with double layer										
	0060	5/8″ fire rated gypsum board on room side	2 Carp	220	.073	S.F.	1.08	2.59		3.67	5.45	
	0100	3-hour assembly with triple layer										
	0300	5/8″ fire rated gypsum board on room side	2 Carp	180	.089	S.F.	1.38	3.16		4.54	6.75	
	0400	4-hour assembly, 1″ coreboard, 5/8″ fire rated gypsum board										
	0600	and 3/4″ galv. metal furring channels, 24″ O.C., with										
	0700	Double layer 5/8″ fire rated gypsum board on room side	2 Carp	110	.145	S.F.	1.22	5.15		6.37	9.90	
	0900	For taping & finishing, add per side	1 Carp	1050	.008	″	.03	.27		.30	.49	

09200 | Plaster & Gypsum Board

	09260 | Gypsum Board Systems		Crew	Daily Output	Labor-Hours	Unit	2006 Bare Costs Mat.	Labor	Equip.	Total	Total Incl O&P	
800	1000	For insulation, see div. 07210										800

	09270 | Drywall Accessories		Crew	Daily Output	Labor-Hours	Unit	Mat.	Labor	Equip.	Total	Total Incl O&P	
100	0010	**ACCESSORIES, DRYWALL**										100
	0020	Casing bead, galvanized steel	1 Carp	2.90	2.759	C.L.F.	18.10	98		116.10	182	
	0100	Vinyl		3	2.667		18.35	95		113.35	177	
	0300	Corner bead, galvanized steel, 1" x 1"		4	2		13.95	71		84.95	133	
	0400	Corner bead, galvanized steel, 1-1/4" x 1-1/4"		3.50	2.286		20.50	81.50		102	158	
	0600	Vinyl corner bead		4	2		18.55	71		89.55	139	
	0900	Furring channel, galv. steel, 7/8" deep, standard		2.60	3.077		24	109		133	207	
	1000	Resilient		2.55	3.137		23.50	112		135.50	211	
	1100	J trim, galvanized steel, 1/2" wide		3	2.667		18.45	95		113.45	178	
	1120	5/8" wide		2.95	2.712		16.65	96.50		113.15	178	
	1500	Z stud, galvanized steel, 1-1/2" wide		2.60	3.077		34.50	109		143.50	219	
	9000	Minimum labor/equipment charge		3	2.667	Job		95		95	157	

	09280 | Gypsum Wallboard Repairs		Crew	Daily Output	Labor-Hours	Unit	Mat.	Labor	Equip.	Total	Total Incl O&P	
100	0010	**GYPSUM WALLBOARD REPAIRS**										100
	0100	Fill and sand, pin / nail holes	1 Carp	960	.008	Ea.		.30		.30	.49	
	0110	Screw head pops		480	.017			.59		.59	.98	
	0120	Dents, up to 2" square		48	.167		.01	5.95		5.96	9.80	
	0130	2" to 4" square		24	.333		.03	11.85		11.88	19.70	
	0140	Cut square, patch, sand and finish, holes, up to 2" square		12	.667		.03	23.50		23.53	39.50	
	0150	2" to 4" square		11	.727		.07	26		26.07	43	
	0160	4" to 8" square		10	.800		.19	28.50		28.69	47	
	0170	8" to 12" square		8	1		.37	35.50		35.87	59.50	
	9000	Minimum labor/equipment charge		2	4	Job		142		142	236	

09300 | Tile

	09310 | Ceramic Tile		Crew	Daily Output	Labor-Hours	Unit	2006 Bare Costs Mat.	Labor	Equip.	Total	Total Incl O&P	
100	0010	**CERAMIC TILE**										100
	0600	Cove base, 4-1/4" x 4-1/4" high, mud set	D-7	91	.176	L.F.	3.19	5.35		8.54	11.85	
	0700	Thin set		128	.125		3.21	3.80		7.01	9.50	
	0900	6" x 4-1/4" high, mud set		100	.160		2.94	4.86		7.80	10.85	
	1000	Thin set		137	.117		2.94	3.55		6.49	8.80	
	1200	Sanitary cove base, 6" x 4-1/4" high, mud set		93	.172		3.26	5.25		8.51	11.80	
	1300	Thin set		124	.129		3.70	3.92		7.62	10.20	
	1500	6" x 6" high, mud set		84	.190		4.06	5.80		9.86	13.50	
	1600	Thin set		117	.137		4.06	4.15		8.21	10.95	
	2400	Bullnose trim, 4-1/4" x 4-1/4", mud set		82	.195		2.82	5.95		8.77	12.40	
	2500	Thin set		128	.125		2.61	3.80		6.41	8.80	
	2700	6" x 4-1/4" bullnose trim, mud set		84	.190		2.32	5.80		8.12	11.60	
	2800	Thin set		124	.129		2.32	3.92		6.24	8.70	
	3000	Floors, natural clay, random or uniform, thin set, color group 1		183	.087	S.F.	3.69	2.66		6.35	8.20	
	3100	Color group 2		183	.087		3.98	2.66		6.64	8.55	
	3300	Porcelain type, 1 color, color group 2, 1" x 1"		183	.087		4.27	2.66		6.93	8.85	
	3310	2" x 2" or 2" x 1", thin set		190	.084		4.48	2.56		7.04	8.95	
	3350	For random blend, 2 colors, add					.79			.79	.87	

For expanded coverage of these items see *Means Interior Cost Data 2006*

09300 | Tile

	09310 | Ceramic Tile		CREW	DAILY OUTPUT	LABOR-HOURS	UNIT	MAT.	LABOR	EQUIP.	TOTAL	TOTAL INCL O&P	
100	3360	4 colors, add				S.F.	1.12			1.12	1.23	100
	4300	Specialty tile, 4-1/4" x 4-1/4" x 1/2", decorator finish	D-7	183	.087		8.95	2.66		11.61	14	
	4500	Add for epoxy grout, 1/16" joint, 1" x 1" tile		800	.020		.55	.61		1.16	1.56	
	4600	2" x 2" tile		820	.020		.50	.59		1.09	1.48	
	4800	Pregrouted sheets, walls, 4-1/4" x 4-1/4", 6" x 4-1/4"										
	4810	and 8-1/2" x 4-1/4", 4 S.F. sheets, silicone grout	D-7	240	.067	S.F.	4.23	2.03		6.26	7.85	
	5100	Floors, unglazed, 2 S.F. sheets,										
	5110	urethane adhesive	D-7	180	.089	S.F.	4.21	2.70		6.91	8.85	
	5400	Walls, interior, thin set, 4-1/4" x 4-1/4" tile		190	.084		1.99	2.56		4.55	6.20	
	5500	6" x 4-1/4" tile		190	.084		2.35	2.56		4.91	6.60	
	5700	8-1/2" x 4-1/4" tile		190	.084		3.32	2.56		5.88	7.65	
	5800	6" x 6" tile		200	.080		2.73	2.43		5.16	6.80	
	5810	8" x 8" tile		225	.071		3.64	2.16		5.80	7.40	
	5820	12" x 12" tile		300	.053		3.04	1.62		4.66	5.90	
	5830	16" x 16" tile		500	.032		3.29	.97		4.26	5.15	
	6000	Decorated wall tile, 4-1/4" x 4-1/4", minimum		270	.059		3.65	1.80		5.45	6.85	
	6100	Maximum		180	.089		39	2.70		41.70	47	
	6600	Crystalline glazed, 4-1/4" x 4-1/4", mud set, plain		100	.160		3.33	4.86		8.19	11.25	
	6700	4-1/4" x 4-1/4", scored tile		100	.160		4.13	4.86		8.99	12.15	
	6900	6"x 6" plain		93	.172		4.40	5.25		9.65	13.05	
	7000	For epoxy grout, 1/16" joints, 4-1/4" tile, add		800	.020		.33	.61		.94	1.31	
	7200	For tile set in dry mortar, add		1735	.009			.28		.28	.44	
	7300	For tile set in portland cement mortar, add		290	.055			1.68		1.68	2.63	
	9500	Minimum labor/equipment charge		3.25	4.923	Job		150		150	234	

	09330 | Quarry Tile		CREW	DAILY OUTPUT	LABOR-HOURS	UNIT	MAT.	LABOR	EQUIP.	TOTAL	TOTAL INCL O&P	
100	0010	**QUARRY TILE** Base, cove or sanitary, 2" or 5" high, mud set										100
	0100	1/2" thick	D-7	110	.145	L.F.	4.59	4.42		9.01	12	
	0300	Bullnose trim, red, mud set, 6" x 6" x 1/2" thick		120	.133		3.92	4.05		7.97	10.65	
	0400	4" x 4" x 1/2" thick		110	.145		4.43	4.42		8.85	11.80	
	0600	4" x 8" x 1/2" thick, using 8" as edge		130	.123		3.89	3.74		7.63	10.15	
	0700	Floors, mud set, 1,000 S.F. lots, red, 4" x 4" x 1/2" thick		120	.133	S.F.	3.89	4.05		7.94	10.65	
	0900	6" x 6" x 1/2" thick		140	.114		3.14	3.47		6.61	8.90	
	1000	4" x 8" x 1/2" thick		130	.123		3.89	3.74		7.63	10.15	
	1300	For waxed coating, add					.62			.62	.68	
	1500	For colors other than green, add					.37			.37	.41	
	1600	For abrasive surface, add					.44			.44	.48	
	1800	Brown tile, imported, 6" x 6" x 3/4"	D-7	120	.133		4.62	4.05		8.67	11.45	
	1900	8" x 8" x 1"		110	.145		5.25	4.42		9.67	12.70	
	2100	For thin set mortar application, deduct		700	.023			.69		.69	1.09	
	2700	Stair tread, 6" x 6" x 3/4", plain		50	.320		4.82	9.70		14.52	20.50	
	2800	Abrasive		47	.340		4.91	10.35		15.26	21.50	
	3000	Wainscot, 6" x 6" x 1/2", thin set, red		105	.152		3.67	4.63		8.30	11.30	
	3100	Colors other than green		105	.152		4.09	4.63		8.72	11.75	
	3300	Window sill, 6" wide, 3/4" thick		90	.178	L.F.	4.60	5.40		10	13.50	
	3400	Corners		80	.200	Ea.	5.15	6.10		11.25	15.25	
	9000	Minimum labor/equipment charge		3.25	4.923	Job		150		150	234	

	09370 | Metal Tile		CREW	DAILY OUTPUT	LABOR-HOURS	UNIT	MAT.	LABOR	EQUIP.	TOTAL	TOTAL INCL O&P	
100	0010	**METAL TILE** 4' x 4' sheet, 24 ga., tile pattern, nailed										100
	0200	Stainless steel	2 Carp	512	.031	S.F.	23	1.11		24.11	27.50	
	0400	Aluminized steel	"	512	.031	"	12.50	1.11		13.61	15.60	
	9000	Minimum labor/equipment charge	1 Carp	4	2	Job		71		71	118	

	09410 Portland Cement Terrazzo		Crew	Daily Output	Labor-Hours	Unit	2006 Bare Costs Mat.	Labor	Equip.	Total	Total Incl O&P
100	0010	**PORTLAND CEMENT TERRAZZO**, cast-in-place									
	0020	Cove base, 6" high, 16ga. zinc	1 Mstz	20	.400	L.F.	3	13.65		16.65	25
	0100	Curb, 6" high and 6" wide	↓	6	1.333	↓	4.78	45.50		50.28	77
	0300	Divider strip for floors, 14 ga., 1-1/4" deep, zinc		375	.021		1.08	.73		1.81	2.33
	0400	Brass		375	.021		1.94	.73		2.67	3.27
	0600	Heavy top strip 1/4" thick, 1-1/4" deep, zinc		300	.027		1.53	.91		2.44	3.11
	1200	For thin set floors, 16 ga., 1/2" x 1/2", zinc	↓	350	.023	↓	.66	.78		1.44	1.95
	1500	Floor, bonded to concrete, 1-3/4" thick, gray cement	J-3	130	.123	S.F.	2.55	3.79	1.81	8.15	10.75
	1600	White cement, mud set	↓	130	.123	↓	2.90	3.79	1.81	8.50	11.15
	1800	Not bonded, 3" total thickness, gray cement		115	.139		3.19	4.29	2.05	9.53	12.45
	1900	White cement, mud set	↓	115	.139	↓	3.48	4.29	2.05	9.82	12.80
	9000	Minimum labor/equipment charge	1 Mstz	1	8	Job		273		273	430

	09420 Precast Terrazzo		Crew	Daily Output	Labor-Hours	Unit	Mat.	Labor	Equip.	Total	Total Incl O&P
900	0010	**TERRAZZO, PRECAST**									
	0020	Base, 6" high, straight	1 Mstz	35	.229	L.F.	9.65	7.80		17.45	23
	0100	Cove	↓	30	.267	↓	10.90	9.10		20	26.50
	0300	8" high base, straight		30	.267		9.75	9.10		18.85	25
	0400	Cove	↓	25	.320		14.35	10.90		25.25	33
	0600	For white cement, add					.39			.39	.43
	0700	For 16 ga. zinc toe strip, add					1.43			1.43	1.57
	0900	Curbs, 4" x 4" high	1 Mstz	19	.421		27.50	14.35		41.85	53
	1000	8" x 8" high	"	15	.533	↓	31.50	18.20		49.70	63
	1200	Floor tiles, non-slip, 1" thick, 12" x 12"	D-1	29	.552	S.F.	16.20	17.75		33.95	47
	1300	1-1/4" thick, 12" x 12"	↓	29	.552	↓	18.10	17.75		35.85	49
	1500	16" x 16"		23	.696		19.70	22.50		42.20	58
	1600	1-1/2" thick, 16" x 16"	↓	21	.762		18	24.50		42.50	59.50
	1800	For Venetian terrazzo, add					5.40			5.40	5.95
	1900	For white cement, add				↓	.50			.50	.55
	2400	Stair treads, 1-1/2" thick, non-slip, three line pattern	2 Mstz	70	.229	L.F.	36.50	7.80		44.30	52
	2500	Nosing and two lines	↓	70	.229	↓	36.50	7.80		44.30	52
	2700	2" thick treads, straight		60	.267		38.50	9.10		47.60	57
	2800	Curved		50	.320		50.50	10.90		61.40	72.50
	3000	Stair risers, 1" thick, to 6" high, straight sections		60	.267		8.80	9.10		17.90	24
	3100	Cove		50	.320		12.85	10.90		23.75	31.50
	3300	Curved, 1" thick, to 6" high, vertical		48	.333		17.55	11.35		28.90	37
	3400	Cove		38	.421		33.50	14.35		47.85	59.50
	3600	Stair tread and riser, single piece, straight, minimum		60	.267		46.50	9.10		55.60	65.50
	3700	Maximum		40	.400		60	13.65		73.65	87.50
	3900	Curved tread and riser, minimum		40	.400		65	13.65		78.65	93
	4000	Maximum		32	.500		81.50	17.05		98.55	117
	4200	Stair stringers, notched, 1" thick		25	.640		26.50	22		48.50	63
	4300	2" thick		22	.727	↓	31.50	25		56.50	73.50
	4500	Stair landings, structural, non-slip, 1-1/2" thick		85	.188	S.F.	29	6.40		35.40	42
	4600	3" thick	↓	75	.213	↓	41	7.25		48.25	56.50
	4800	Wainscot, 12" x 12" x 1" tiles	1 Mstz	12	.667		5.85	22.50		28.35	42
	4900	16" x 16" x 1-1/2" tiles	"	8	1	↓	12.65	34		46.65	67.50
	9500	Minimum labor/equipment charge	1 Tilf	2	4	Job		137		137	215

	09450 Cast-in-Place Terrazzo		Crew	Daily Output	Labor-Hours	Unit	Mat.	Labor	Equip.	Total	Total Incl O&P
200	0010	**TILE OR TERRAZZO BASE**									
	0020	Scratch coat only	1 Mstz	150	.053	S.F.	.38	1.82		2.20	3.27
	0500	Scratch and brown coat only	↓	75	.107	"	.72	3.64		4.36	6.50
	9000	Minimum labor/equipment charge	↓	1	8	Job		273		273	430

		09510 | Acoustical Ceilings	CREW	DAILY OUTPUT	LABOR-HOURS	UNIT	2006 BARE COSTS MAT.	LABOR	EQUIP.	TOTAL	TOTAL INCL O&P	
700	0010	**SUSPENDED ACOUSTIC CEILING TILES**, Not including										700
	0100	suspension system										
	0300	Fiberglass boards, film faced, 2′ x 2′ or 2′ x 4′, 5/8″ thick	1 Carp	625	.013	S.F.	.52	.46		.98	1.32	
	0400	3/4″ thick		600	.013		1.19	.47		1.66	2.10	
	0500	3″ thick, thermal, R11		450	.018		1.31	.63		1.94	2.49	
	0600	Glass cloth faced fiberglass, 3/4″ thick		500	.016		1.72	.57		2.29	2.83	
	0700	1″ thick		485	.016		1.90	.59		2.49	3.06	
	0820	1-1/2″ thick, nubby face		475	.017		2.36	.60		2.96	3.59	
	1110	Mineral fiber tile, lay-in, 2′ x 2′ or 2′ x 4′, 5/8″ thick, fine texture		625	.013		.45	.46		.91	1.25	
	1115	Rough textured		625	.013		1.10	.46		1.56	1.96	
	1125	3/4″ thick, fine textured		600	.013		1.22	.47		1.69	2.13	
	1130	Rough textured		600	.013		1.53	.47		2	2.47	
	1135	Fissured		600	.013		1.81	.47		2.28	2.78	
	1150	Tegular, 5/8″ thick, fine textured		470	.017		1.08	.61		1.69	2.19	
	1155	Rough textured		470	.017		1.40	.61		2.01	2.54	
	1165	3/4″ thick, fine textured		450	.018		1.53	.63		2.16	2.73	
	1170	Rough textured		450	.018		1.73	.63		2.36	2.95	
	1175	Fissured		450	.018		2.69	.63		3.32	4.01	
	1180	For aluminum face, add					4.91			4.91	5.40	
	1185	For plastic film face, add					.81			.81	.89	
	1190	For fire rating, add					.36			.36	.40	
	1300	Mirror faced panels, 15/16″ thick, 2′ x 2′	1 Carp	500	.016		10.20	.57		10.77	12.20	
	1900	Eggcrate, acrylic, 1/2″ x 1/2″ x 1/2″ cubes		500	.016		1.48	.57		2.05	2.57	
	2100	Polystyrene eggcrate, 3/8″ x 3/8″ x 1/2″ cubes		510	.016		1.24	.56		1.80	2.28	
	2200	1/2″ x 1/2″ x 1/2″ cubes		500	.016		1.66	.57		2.23	2.77	
	2400	Luminous panels, prismatic, acrylic		400	.020		1.80	.71		2.51	3.16	
	2500	Polystyrene		400	.020		.92	.71		1.63	2.19	
	2700	Flat white acrylic		400	.020		3.13	.71		3.84	4.62	
	2800	Polystyrene		400	.020		2.14	.71		2.85	3.53	
	3000	Drop pan, white, acrylic		400	.020		4.59	.71		5.30	6.25	
	3100	Polystyrene		400	.020		3.84	.71		4.55	5.40	
	3600	Perforated aluminum sheets, .024″ thick, corrugated, painted		490	.016		1.83	.58		2.41	2.97	
	3700	Plain		500	.016		3.10	.57		3.67	4.35	
	3720	Mineral fiber, 24″ x 24″ or 48″, reveal edge, painted, 5/8″ thick		600	.013		1.02	.47		1.49	1.91	
	3740	3/4″ thick		575	.014		1.66	.49		2.15	2.65	
	9000	Minimum labor/equipment charge		4	2	Job		71		71	118	
760	0010	**SUSPENDED CEILINGS, COMPLETE** Including standard										760
	0100	suspension system but not incl. 1-1/2″ carrier channels										
	0600	Fiberglass ceiling board, 2′ x 4′ x 5/8″, plain faced,	1 Carp	500	.016	S.F.	1.04	.57		1.61	2.09	
	0700	Offices, 2′ x 4′ x 3/4″		380	.021		1.71	.75		2.46	3.12	
	0800	Mineral fiber, on 15/16″ T bar susp. 2′ x 2′ x 3/4″ lay-in board		345	.023		1.88	.82		2.70	3.43	
	0810	2′ x 4′ x 5/8″ tile		380	.021		.97	.75		1.72	2.31	
	0820	Tegular, 2′ x 2′ x 5/8″ tile on 9/16″ grid		250	.032		1.89	1.14		3.03	3.95	
	0830	2′ x 4′ x 3/4″ tile		275	.029		2.20	1.03		3.23	4.13	
	0900	Luminous panels, prismatic, acrylic		255	.031		2.32	1.12		3.44	4.41	
	1200	Metal pan with acoustic pad, steel		75	.107		3.44	3.79		7.23	10.10	
	1300	Painted aluminum		75	.107		2.35	3.79		6.14	8.90	
	1500	Aluminum, degreased finish		75	.107		3.98	3.79		7.77	10.70	
	1600	Stainless steel		75	.107		7.45	3.79		11.24	14.50	
	1800	Tile, Z bar suspension, 5/8″ mineral fiber tile		150	.053		1.61	1.90		3.51	4.91	
	1900	3/4″ mineral fiber tile		150	.053		1.72	1.90		3.62	5.05	
	2402	For strip lighting, see division 16510-440										
	2500	For rooms under 500 S.F., add				S.F.		25%				
	9000	Minimum labor/equipment charge	1 Carp	2	4	Job		142		142	236	
900	0010	**CEILING TILE**, Stapled or cemented										900
	0100	12″ x 12″ or 12″ x 24″, not including furring										

09500 | Ceilings

	09510 | Acoustical Ceilings		CREW	DAILY OUTPUT	LABOR-HOURS	UNIT	2006 BARE COSTS MAT.	LABOR	EQUIP.	TOTAL	TOTAL INCL O&P	
900	0600	Mineral fiber, vinyl coated, 5/8" thick	1 Carp	1000	.008	S.F.	1.47	.28		1.75	2.09	900
	0700	3/4" thick		1000	.008		1.41	.28		1.69	2.02	
	0900	Fire rated, 3/4" thick, plain faced		1000	.008		1.27	.28		1.55	1.87	
	1000	Plastic coated face		1000	.008		1.36	.28		1.64	1.97	
	1200	Aluminum faced, 5/8" thick, plain		1000	.008		1.12	.28		1.40	1.70	
	3000	Wood fiber tile, 1/2" thick		400	.020		.76	.71		1.47	2.02	
	3100	3/4" thick	↓	400	.020		1.07	.71		1.78	2.36	
	3300	For flameproofing, add					.09			.09	.10	
	3400	For sculptured 3 dimensional, add					.26			.26	.29	
	3900	For ceiling primer, add					.12			.12	.13	
	4000	For ceiling cement, add				↓	.34			.34	.37	
	9000	Minimum labor/equipment charge	1 Carp	4	2	Job		71		71	118	

09600 | Flooring

	09620 | Specialty Flooring		CREW	DAILY OUTPUT	LABOR-HOURS	UNIT	2006 BARE COSTS MAT.	LABOR	EQUIP.	TOTAL	TOTAL INCL O&P	
100	0010	**ATHLETIC FLOORING**										100
	3700	Polyethylene, in rolls, no base incl., landscape surfaces	1 Tilf	275	.029	S.F.	2.57	1		3.57	4.39	
	3800	Nylon action surface, 1/8" thick		275	.029		2.76	1		3.76	4.60	
	3900	1/4" thick		275	.029		3.98	1		4.98	5.95	
	4000	3/8" thick		275	.029		5	1		6	7.05	
	5500	Polyvinyl chloride, sheet goods for gyms, 1/4" thick		80	.100		5	3.43		8.43	10.85	
	5600	3/8" thick	↓	60	.133	↓	5.65	4.57		10.22	13.35	

	09631 | Brick Flooring		CREW	DAILY OUTPUT	LABOR-HOURS	UNIT	MAT.	LABOR	EQUIP.	TOTAL	TOTAL INCL O&P	
100	0010	**BRICK FLOORING**										100
	0020	Acid proof shales, red, 8" x 3-3/4" x 1-1/4" thick	D-7	.43	37.209	M	760	1,125		1,885	2,600	
	0050	2-1/4" thick	D-1	.40	40		825	1,275		2,100	2,975	
	0200	Acid proof clay brick, 8" x 3-3/4" x 2-1/4" thick	"	.40	40	↓	785	1,275		2,060	2,950	
	0260	Cast ceramic, pressed, 4" x 8" x 1/2", unglazed	D-7	100	.160	S.F.	5.15	4.86		10.01	13.25	
	0270	Glazed		100	.160		6.85	4.86		11.71	15.15	
	0280	Hand molded flooring, 4" x 8" x 3/4", unglazed		95	.168		6.80	5.10		11.90	15.50	
	0290	Glazed		95	.168		8.50	5.10		13.60	17.35	
	0300	8" hexagonal, 3/4" thick, unglazed		85	.188		7.45	5.70		13.15	17.15	
	0310	Glazed	↓	85	.188		13.45	5.70		19.15	24	
	0400	Heavy duty industrial, cement mortar bed, 2" thick, not incl. brick	D-1	80	.200		.70	6.45		7.15	11.20	
	0450	Acid proof joints, 1/4" wide	"	65	.246		1.18	7.90		9.08	14.15	
	0500	Pavers, 8" x 4", 1" to 1-1/4" thick, red	D-7	95	.168		3	5.10		8.10	11.30	
	0510	Ironspot	"	95	.168		4.23	5.10		9.33	12.65	
	0540	1-3/8" to 1-3/4" thick, red	D-1	95	.168		2.89	5.40		8.29	12	
	0560	Ironspot		95	.168		4.18	5.40		9.58	13.40	
	0580	2-1/4" thick, red		90	.178		2.94	5.70		8.64	12.50	
	0590	Ironspot	↓	90	.178	↓	4.56	5.70		10.26	14.25	
	0800	For sidewalks and patios with pavers, see division 02780-200										
	0870	For epoxy joints, add	D-1	600	.027	S.F.	2.24	.86		3.10	3.85	
	0880	For Furan underlayment, add	"	600	.027		1.85	.86		2.71	3.43	
	0890	For waxed surface, steam cleaned, add	A-1H	1000	.008	↓	.16	.22	.06	.44	.60	
	9000	Minimum labor/equipment charge	1 Bric	2	4	Job		146		146	237	

		09635 | Marble Flooring	CREW	DAILY OUTPUT	LABOR-HOURS	UNIT	2006 BARE COSTS MAT.	LABOR	EQUIP.	TOTAL	TOTAL INCL O&P	
100	0010	**MARBLE**										100
	0020	Thin gauge tile, 12" x 6", 3/8", White Carara	D-7	60	.267	S.F.	9.10	8.10		17.20	22.50	
	0100	Travertine		60	.267		10	8.10		18.10	23.50	
	0200	12" x 12" x 3/8", thin set, floors		60	.267		6.30	8.10		14.40	19.65	
	0300	On walls		52	.308		8.90	9.35		18.25	24.50	
	9000	Minimum labor/equipment charge		3	5.333	Job		162		162	254	

		09637 | Stone Flooring	CREW	DAILY OUTPUT	LABOR-HOURS	UNIT	MAT.	LABOR	EQUIP.	TOTAL	TOTAL INCL O&P	
100	0010	**SLATE TILE**										100
	0020	Vermont, 6" x 6" x 1/4" thick, thin set	D-7	180	.089	S.F.	4.26	2.70		6.96	8.90	
	9000	Minimum labor/equipment charge	"	3	5.333	Job		162		162	254	

		09643 | Wood Block Flooring	CREW	DAILY OUTPUT	LABOR-HOURS	UNIT	MAT.	LABOR	EQUIP.	TOTAL	TOTAL INCL O&P	
100	0010	**WOOD BLOCK FLOORING**										100
	0020	End grain flooring, coated, 2" thick	1 Carp	295	.027	S.F.	2.98	.96		3.94	4.88	
	0400	Natural finish, 1" thick, fir		125	.064		3.08	2.28		5.36	7.15	
	0600	1-1/2" thick, pine		125	.064		3.03	2.28		5.31	7.10	
	0700	2" thick, pine		125	.064		3.55	2.28		5.83	7.70	
	9000	Minimum labor/equipment charge		2	4	Job		142		142	236	

		09647 | Wood Parquet Flooring	CREW	DAILY OUTPUT	LABOR-HOURS	UNIT	MAT.	LABOR	EQUIP.	TOTAL	TOTAL INCL O&P	
100	0010	**WOOD PARQUET** flooring										100
	5200	Parquetry, standard, 5/16" thick, not incl. finish, oak, minimum	1 Carp	160	.050	S.F.	3.43	1.78		5.21	6.70	
	5300	Maximum		100	.080		5.30	2.84		8.14	10.55	
	5500	Teak, minimum		160	.050		4.53	1.78		6.31	7.95	
	5600	Maximum		100	.080		7.90	2.84		10.74	13.40	
	5650	13/16" thick, select grade oak, minimum		160	.050		8.75	1.78		10.53	12.55	
	5700	Maximum		100	.080		13.25	2.84		16.09	19.30	
	5800	Custom parquetry, including finish, minimum		100	.080		14.65	2.84		17.49	21	
	5900	Maximum		50	.160		19.40	5.70		25.10	31	
	6700	Parquetry, prefinished white oak, 5/16" thick, minimum		160	.050		3.50	1.78		5.28	6.80	
	6800	Maximum		100	.080		7.15	2.84		9.99	12.60	
	7000	Walnut or teak, parquetry, minimum		160	.050		4.91	1.78		6.69	8.35	
	7100	Maximum		100	.080		8.55	2.84		11.39	14.15	
	7200	Acrylic wood parquet blocks, 12" x 12" x 5/16",										
	7210	irradiated, set in epoxy	1 Carp	160	.050	S.F.	7.15	1.78		8.93	10.80	

		09648 | Wood Strip Flooring	CREW	DAILY OUTPUT	LABOR-HOURS	UNIT	MAT.	LABOR	EQUIP.	TOTAL	TOTAL INCL O&P	
100	0010	**WOOD** R061110-30										100
	0020	Fir, vertical grain, 1" x 4", not incl. finish, B & better	1 Carp	255	.031	S.F.	2.51	1.12		3.63	4.61	
	0100	C grade & better		255	.031		2.36	1.12		3.48	4.45	
	0300	Flat grain, 1" x 4", not incl. finish, B & better		255	.031		2.87	1.12		3.99	5	
	0400	C & better		255	.031		2.76	1.12		3.88	4.89	
	4000	Maple, strip, 25/32" x 2-1/4", not incl. finish, select		170	.047		4.78	1.67		6.45	8	
	4100	#2 & better		170	.047		2.95	1.67		4.62	6	
	4300	33/32" x 3-1/4", not incl. finish, #1 grade		170	.047		3.73	1.67		5.40	6.85	
	4400	#2 & better		170	.047		3.32	1.67		4.99	6.40	
	4600	Oak, white or red, 25/32" x 2-1/4", not incl. finish										
	4700	#1 common	1 Carp	170	.047	S.F.	3	1.67		4.67	6.05	
	4900	Select quartered, 2-1/4" wide		170	.047		2.78	1.67		4.45	5.85	
	5000	Clear		170	.047		3.72	1.67		5.39	6.85	
	6100	Prefinished, white oak, prime grade, 2-1/4" wide		170	.047		6.20	1.67		7.87	9.55	
	6200	3-1/4" wide		185	.043		7.95	1.54		9.49	11.30	
	6400	Ranch plank		145	.055		7.70	1.96		9.66	11.70	

09648 | Wood Strip Flooring

			CREW	DAILY OUTPUT	LABOR-HOURS	UNIT	2006 BARE COSTS MAT.	LABOR	EQUIP.	TOTAL	TOTAL INCL O&P	
100	6500	Hardwood blocks, 9″ x 9″, 25/32″ thick (R061110-30)	1 Carp	160	.050	S.F.	5.30	1.78		7.08	8.75	100
	7400	Yellow pine, 3/4″ x 3-1/8″, T & G, C & better, not incl. finish		200	.040		2.28	1.42		3.70	4.87	
	7500	Refinish wood floor, sand, 2 cts poly, wax, soft wood, min.	1 Clab	400	.020		.71	.55		1.26	1.69	
	7600	Hard wood, max		130	.062		1.07	1.69		2.76	3.97	
	7800	Sanding and finishing, 2 coats polyurethane		295	.027		.71	.74		1.45	2.01	
	7900	Subfloor and underlayment, see division 06160										
	8015	Transition molding, 2 1/4″ wide, 5′ long	1 Carp	19.20	.417	Ea.	12.90	14.80		27.70	38.50	
	8300	Floating floor, wood composition strip, complete.	1 Clab	133	.060	S.F.	3.98	1.65		5.63	7.10	
	8310	Floating floor components, T & G wood composite strips					3.45			3.45	3.80	
	8320	Film					.14			.14	.15	
	8330	Foam					.23			.23	.25	
	8340	Adhesive					.21			.21	.23	
	8350	Installation kit					.16			.16	.18	
	8360	Trim, 2″ wide x 3′ long				L.F.	2.31			2.31	2.54	
	8370	Reducer moulding				″	3.99			3.99	4.39	
	9000	Minimum labor/equipment charge	1 Carp	2	4	Job		142		142	236	

09651 | Resilient Base & Access.

			CREW	DAILY OUTPUT	LABOR-HOURS	UNIT	MAT.	LABOR	EQUIP.	TOTAL	TOTAL INCL O&P	
100	0010	**STAIR TREADS AND RISERS** See index for materials other										100
	0100	than rubber and vinyl										
	0300	Rubber, molded tread, 12″ wide, 5/16″ thick, black	1 Tilf	115	.070	L.F.	9.45	2.38		11.83	14.15	
	0400	Colors		115	.070		8.95	2.38		11.33	13.60	
	0600	1/4″ thick, black		115	.070		8.75	2.38		11.13	13.40	
	0700	Colors		115	.070		8.75	2.38		11.13	13.40	
	0900	Grip strip safety tread, colors, 5/16″ thick		115	.070		13	2.38		15.38	18.05	
	1000	3/16″ thick		120	.067		9.15	2.28		11.43	13.70	
	1200	Landings, smooth sheet rubber, 1/8″ thick		120	.067	S.F.	4.20	2.28		6.48	8.20	
	1300	3/16″ thick		120	.067	″	5.35	2.28		7.63	9.50	
	1500	Nosings, 3″ wide, 3/16″ thick, black		140	.057	L.F.	2.90	1.96		4.86	6.25	
	1600	Colors		140	.057		2.79	1.96		4.75	6.15	
	1800	Risers, 7″ high, 1/8″ thick, flat		250	.032		3.39	1.10		4.49	5.45	
	1900	Coved		250	.032		3.03	1.10		4.13	5.05	
	2100	Vinyl, molded tread, 12″ wide, colors, 1/8″ thick		115	.070		3.85	2.38		6.23	8	
	2200	1/4″ thick		115	.070		5.65	2.38		8.03	9.95	
	2300	Landing material, 1/8″ thick		200	.040	S.F.	3.95	1.37		5.32	6.50	
	2400	Riser, 7″ high, 1/8″ thick, coved		175	.046	L.F.	2.25	1.57		3.82	4.93	
	2500	Tread and riser combined, 1/8″ thick		80	.100	″	6.55	3.43		9.98	12.55	
	9000	Minimum labor/equipment charge		3	2.667	Job		91.50		91.50	143	
200	0010	**RESILIENT BASE**										200
	0800	Base, cove, rubber or vinyl, .080″ thick										
	1100	Standard colors, 2-1/2″ high	1 Tilf	315	.025	L.F.	.52	.87		1.39	1.93	
	1150	4″ high		315	.025		.53	.87		1.40	1.94	
	1200	6″ high		315	.025		.92	.87		1.79	2.37	
	1450	1/8″ thick, standard colors, 2-1/2″ high		315	.025		.60	.87		1.47	2.02	
	1500	4″ high		315	.025		.60	.87		1.47	2.02	
	1550	6″ high		315	.025		.95	.87		1.82	2.41	
	1600	Corners, 2-1/2″ high		315	.025	Ea.	1.17	.87		2.04	2.65	
	1630	4″ high		315	.025		1.75	.87		2.62	3.29	
	1660	6″ high		315	.025		2	.87		2.87	3.56	

09653 | Resilient Sheet Flooring

			CREW	DAILY OUTPUT	LABOR-HOURS	UNIT	MAT.	LABOR	EQUIP.	TOTAL	TOTAL INCL O&P	
100	0010	**RESILIENT SHEET FLOORING**										100
	5900	Rubber, sheet goods, 36″ wide, 1/8″ thick	1 Tilf	120	.067	S.F.	4.25	2.28		6.53	8.25	
	5950	3/16″ thick		100	.080		6	2.74		8.74	10.90	
	6000	1/4″ thick		90	.089		7	3.04		10.04	12.45	

09653 Resilient Sheet Flooring			CREW	DAILY OUTPUT	LABOR-HOURS	UNIT	2006 BARE COSTS MAT.	LABOR	EQUIP.	TOTAL	TOTAL INCL O&P	
100	8000	Vinyl sheet goods, backed, .065" thick, minimum	1 Tilf	250	.032	S.F.	2.21	1.10		3.31	4.15	100
	8050	Maximum		200	.040		2.71	1.37		4.08	5.15	
	8100	.080" thick, minimum		230	.035		2.39	1.19		3.58	4.50	
	8150	Maximum		200	.040		3.40	1.37		4.77	5.90	
	8200	.125" thick, minimum		230	.035		2.69	1.19		3.88	4.83	
	8250	Maximum	↓	200	.040	↓	4.25	1.37		5.62	6.85	
	8700	Adhesive cement, 1 gallon does 200 to 300 S.F.				Gal.	16.75			16.75	18.40	
	8800	Asphalt primer, 1 gallon per 300 S.F.					10.30			10.30	11.35	
	8900	Emulsion, 1 gallon per 140 S.F.					13.10			13.10	14.40	
	8950	Latex underlayment, liquid, fortified				↓	33			33	36.50	

09658 Resilient Tile Flooring			CREW	DAILY OUTPUT	LABOR-HOURS	UNIT	MAT.	LABOR	EQUIP.	TOTAL	TOTAL INCL O&P	
100	0010	RESILIENT TILE FLOORING										100
	2200	Cork tile, standard finish, 1/8" thick	1 Tilf	315	.025	S.F.	4.24	.87		5.11	6	
	2250	3/16" thick		315	.025		4.91	.87		5.78	6.75	
	2300	5/16" thick		315	.025		5.45	.87		6.32	7.35	
	2350	1/2" thick		315	.025		6.45	.87		7.32	8.45	
	2500	Urethane finish, 1/8" thick		315	.025		5.20	.87		6.07	7.05	
	2550	3/16" thick		315	.025		5.60	.87		6.47	7.50	
	2600	5/16" thick		315	.025		7.05	.87		7.92	9.10	
	2650	1/2" thick		315	.025		9.85	.87		10.72	12.20	
	6050	Tile, marbleized colors, 12" x 12", 1/8" thick		400	.020		4.86	.69		5.55	6.40	
	6100	3/16" thick		400	.020		6.50	.69		7.19	8.20	
	6300	Special tile, plain colors, 1/8" thick		400	.020		5	.69		5.69	6.55	
	6350	3/16" thick		400	.020		6.80	.69		7.49	8.55	
	7000	Vinyl composition tile, 12" x 12", 1/16" thick		500	.016		.86	.55		1.41	1.81	
	7050	Embossed		500	.016		1.35	.55		1.90	2.35	
	7100	Marbleized		500	.016		1.35	.55		1.90	2.35	
	7150	Solid		500	.016		1.50	.55		2.05	2.51	
	7200	3/32" thick, embossed		500	.016		1.03	.55		1.58	1.99	
	7250	Marbleized		500	.016		1.50	.55		2.05	2.51	
	7300	Solid		500	.016		1.96	.55		2.51	3.02	
	7350	1/8" thick, marbleized		500	.016		1.17	.55		1.72	2.15	
	7400	Solid		500	.016		2.05	.55		2.60	3.12	
	7450	Conductive		500	.016		3.89	.55		4.44	5.15	
	7500	Vinyl tile, 12" x 12", .050" thick, minimum		500	.016		2.10	.55		2.65	3.17	
	7550	Maximum		500	.016		3.95	.55		4.50	5.20	
	7600	1/8" thick, minimum		500	.016		2.75	.55		3.30	3.89	
	7650	Solid colors		500	.016		5.05	.55		5.60	6.40	
	7700	Marbleized or Travertine pattern		500	.016		3.95	.55		4.50	5.20	
	7750	Florentine pattern		500	.016		4.42	.55		4.97	5.70	
	7800	Maximum		500	.016	↓	9.10	.55		9.65	10.85	
	9500	Minimum labor/equipment charge	↓	4	2	Job		68.50		68.50	107	

09662 Static Control Flooring			CREW	DAILY OUTPUT	LABOR-HOURS	UNIT	MAT.	LABOR	EQUIP.	TOTAL	TOTAL INCL O&P	
100	0010	CONDUCTIVE RESILIENT FLOORING										100
	1700	Conductive flooring, rubber tile, 1/8" thick	1 Tilf	315	.025	S.F.	3.50	.87		4.37	5.20	
	1800	Homogeneous vinyl tile, 1/8" thick	"	315	.025	"	4.75	.87		5.62	6.60	

09670 Fluid Applied Flooring			CREW	DAILY OUTPUT	LABOR-HOURS	UNIT	MAT.	LABOR	EQUIP.	TOTAL	TOTAL INCL O&P	
100	0010	RESILIENT TILE UNDERLAYMENT										100
	3600	Latex underlayment, 1/8" thk., cementitious for resilient flooring	1 Tilf	160	.050	S.F.	1.55	1.71		3.26	4.39	

09600 | Flooring

		09673 | Composition Flooring	CREW	DAILY OUTPUT	LABOR-HOURS	UNIT	2006 BARE COSTS MAT.	LABOR	EQUIP.	TOTAL	TOTAL INCL O&P	
100	0010	**COMPOSITION FLOORING**										100
	0020	Cementitous acrylic, 1/4" thick	C-6	520	.092	S.F.	1.30	2.67	.08	4.05	5.90	
	0200	Methyl methachrylate, 1/4" thick	C-8A	3000	.016		5	.48		5.48	6.30	
	0600	Epoxy, with colored quartz chips, broadcast, minimum	C-6	675	.071		2.30	2.06	.06	4.42	5.95	
	0700	Maximum		490	.098		2.79	2.83	.09	5.71	7.80	
	0900	Trowelled, minimum		560	.086		2.97	2.48	.08	5.53	7.40	
	1000	Maximum		480	.100		4.33	2.89	.09	7.31	9.60	
	1200	Heavy duty epoxy topping, 1/4" thick,										
	1300	500 to 1,000 S.F.	C-6	420	.114	S.F.	5.15	3.30	.10	8.55	11.15	
	1500	1,000 to 2,000 S.F.		450	.107		4.06	3.08	.09	7.23	9.60	
	1600	Over 10,000 S.F.		480	.100		3.76	2.89	.09	6.74	9	
	1800	Epoxy terrazzo, 1/4" thick, chemical resistant, minimum	J-3	200	.080		5.45	2.46	1.18	9.09	11.15	
	1900	Maximum	"	150	.107		8.35	3.29	1.57	13.21	16.10	
		09680 | Carpet										
600	0010	**CARPET PAD**, commercial grade										600
	9000	Sponge rubber pad, minimum	1 Tilf	150	.053	S.Y.	3.58	1.83		5.41	6.80	
	9100	Maximum		150	.053		8.25	1.83		10.08	11.95	
	9200	Felt pad, minimum		150	.053		3.57	1.83		5.40	6.80	
	9300	Maximum		150	.053		6.70	1.83		8.53	10.25	
	9400	Bonded urethane pad, minimum		150	.053		3.95	1.83		5.78	7.20	
	9500	Maximum		150	.053		6.75	1.83		8.58	10.30	
	9600	Prime urethane pad, minimum		150	.053		2.27	1.83		4.10	5.35	
	9700	Maximum		150	.053		4.19	1.83		6.02	7.45	
800	0010	**CARPET** Commercial grades, direct cement										800
	0700	Nylon, level loop, 26 oz., light to medium traffic	1 Tilf	75	.107	S.Y.	17.60	3.65		21.25	25	
	0720	28 oz., light to medium traffic		75	.107		16.55	3.65		20.20	24	
	0900	32 oz., medium traffic		75	.107		25	3.65		28.65	33.50	
	1100	40 oz., medium to heavy traffic		75	.107		37	3.65		40.65	46.50	
	2920	Nylon plush, 30 oz., medium traffic		57	.140		18.40	4.81		23.21	27.50	
	3000	36 oz., medium traffic		75	.107		24	3.65		27.65	32.50	
	3100	42 oz., medium to heavy traffic		70	.114		22	3.91		25.91	30.50	
	3200	46 oz., medium to heavy traffic		70	.114		33	3.91		36.91	42.50	
	3300	54 oz., heavy traffic		70	.114		37.50	3.91		41.41	47	
	4500	50 oz., medium to heavy traffic		75	.107		84.50	3.65		88.15	99	
	4700	Patterned, 32 oz., medium to heavy traffic		70	.114		83.50	3.91		87.41	98	
	4900	48 oz., heavy traffic		70	.114		85	3.91		88.91	99.50	
	5000	For less than full roll, add					25%					
	5100	For small rooms, less than 12' wide, add						25%				
	5200	For large open areas (no cuts), deduct						25%				
	5600	For bound carpet baseboard, add	1 Tilf	300	.027	L.F.	1.35	.91		2.26	2.92	
	5610	For stairs, not incl. price of carpet, add	"	30	.267	Riser		9.15		9.15	14.30	
	5620	For borders and patterns, add to labor						18%				
	8950	For tackless, stretched installation, add padding to above										
	9850	For "branded" fiber, add				S.Y.	25%					
	9900	Carpet cleaning machine, rent				Day					35	
	9910	Minimum labor/equipment charge	1 Tilf	3	2.667	Job		91.50		91.50	143	
900	0010	**CARPET TILE**										900
	0100	Tufted nylon, 18" x 18", hard back, 20 oz.	1 Tilf	150	.053	S.Y.	20.50	1.83		22.33	25.50	
	0110	26 oz.		150	.053		35.50	1.83		37.33	42	
	0200	Cushion back, 20 oz.		150	.053		26	1.83		27.83	31.50	
	0210	26 oz.		150	.053		40.50	1.83		42.33	47.50	

09700 | Wall Finishes

	09720 | Wall Coverings		CREW	DAILY OUTPUT	LABOR-HOURS	UNIT	2006 BARE COSTS MAT.	LABOR	EQUIP.	TOTAL	TOTAL INCL O&P	
100	0010	**WALL COVERING** Including sizing, add 10-30% waste @ takeoff										100
	0050	Aluminum foil	1 Pape	275	.029	S.F.	.87	.93		1.80	2.45	
	0100	Copper sheets, .025" thick, vinyl backing		240	.033		4.66	1.06		5.72	6.85	
	0300	Phenolic backing		240	.033		6.05	1.06		7.11	8.35	
	0600	Cork tiles, light or dark, 12" x 12" x 3/16"		240	.033		3.74	1.06		4.80	5.80	
	0700	5/16" thick		235	.034		3.19	1.08		4.27	5.25	
	0900	1/4" basketweave		240	.033		4.90	1.06		5.96	7.10	
	1000	1/2" natural, non-directional pattern		240	.033		6.25	1.06		7.31	8.60	
	1100	3/4" natural, non-directional pattern		240	.033		10.25	1.06		11.31	13	
	1200	Granular surface, 12" x 36", 1/2" thick		385	.021		1.06	.66		1.72	2.23	
	1300	1" thick		370	.022		1.37	.69		2.06	2.62	
	1500	Polyurethane coated, 12" x 12" x 3/16" thick		240	.033		3.31	1.06		4.37	5.35	
	1600	5/16" thick		235	.034		4.71	1.08		5.79	6.95	
	1800	Cork wallpaper, paperbacked, natural		480	.017		1.88	.53		2.41	2.92	
	1900	Colors		480	.017		2.33	.53		2.86	3.41	
	2100	Flexible wood veneer, 1/32" thick, plain woods		100	.080		1.98	2.55		4.53	6.25	
	2200	Exotic woods		95	.084		3.01	2.68		5.69	7.60	
	2400	Gypsum-based, fabric-backed, fire										
	2500	resistant for masonry walls, minimum, 21 oz./S.Y.	1 Pape	800	.010	S.F.	.68	.32		1	1.26	
	2600	Average		720	.011		1.03	.35		1.38	1.70	
	2700	Maximum, (small quantities)		640	.013		1.14	.40		1.54	1.89	
	2750	Acrylic, modified, semi-rigid PVC, .028" thick	2 Carp	330	.048		.95	1.72		2.67	3.91	
	2800	.040" thick	"	320	.050		1.25	1.78		3.03	4.33	
	3000	Vinyl wall covering, fabric-backed, lightweight, (12-15 oz./S.Y.)	1 Pape	640	.013		.59	.40		.99	1.29	
	3300	Medium weight, type 2, (20-24 oz./S.Y.)		480	.017		.73	.53		1.26	1.65	
	3400	Heavy weight, type 3, (28 oz./S.Y.)		435	.018		1.17	.59		1.76	2.23	
	3600	Adhesive, 5 gal. lots, (18SY/Gal.)				Gal.	8.95			8.95	9.80	
	3700	Wallpaper, average workmanship, solid pattern, low cost paper	1 Pape	640	.013	S.F.	.30	.40		.70	.97	
	3900	basic patterns (matching required), avg. cost paper		535	.015		.65	.48		1.13	1.48	
	4000	Paper at $85 per double roll, quality workmanship		435	.018		1.53	.59		2.12	2.62	
	4200	Grass cloths with lining paper, minimum		400	.020		.65	.64		1.29	1.74	
	4300	Maximum		350	.023		2.09	.73		2.82	3.47	
	5990	Wallpaper removal, 1 layer, minimum		800	.010		.06	.32		.38	.58	
	6000	Wallpaper removal, 3 layer, maximum		400	.020		.11	.64		.75	1.14	
	9000	Minimum labor/equipment charge		2	4	Job		127		127	205	

	09770 | Special Wall Surfaces		CREW	DAILY OUTPUT	LABOR-HOURS	UNIT	MAT.	LABOR	EQUIP.	TOTAL	TOTAL INCL O&P	
400	0010	**FIBERGLASS REINFORCED PLASTIC** panels, .090" thick, on walls										400
	0020	Adhesive mounted, embossed surface	2 Carp	640	.025	S.F.	1.52	.89		2.41	3.14	
	0030	Smooth surface		640	.025		1.78	.89		2.67	3.43	
	0040	Fire rated, embossed surface		640	.025		1.98	.89		2.87	3.65	
	0050	Nylon rivet mounted, on drywall, embossed surface		480	.033		1.41	1.18		2.59	3.51	
	0060	Smooth surface		480	.033		1.66	1.18		2.84	3.79	
	0070	Fire rated, embossed surface		480	.033		1.85	1.18		3.03	4	
	0080	On masonry, embossed surface		320	.050		1.41	1.78		3.19	4.50	
	0090	Smooth surface		320	.050		1.78	1.78		3.56	4.91	
	0100	Fire rated, embossed surface		320	.050		1.88	1.78		3.66	5	
	0110	Nylon rivet and adhesive mounted, on drywall, embossed surface		240	.067		1.66	2.37		4.03	5.75	
	0120	Smooth surface		240	.067		1.88	2.37		4.25	6	
	0130	Fire rated, embossed surface		240	.067		2.09	2.37		4.46	6.25	
	0140	On masonry, embossed surface		190	.084		1.66	2.99		4.65	6.80	
	0150	Smooth surface		190	.084		1.88	2.99		4.87	7.05	
	0160	Fire rated, embossed surface		190	.084		2.09	2.99		5.08	7.25	
	0170	For moldings add	1 Carp	250	.032	L.F.	.28	1.14		1.42	2.19	
	0180	On ceilings, for lay in grid system, embossed surface		400	.020	S.F.	1.52	.71		2.23	2.85	

		09770 | Special Wall Surfaces	CREW	DAILY OUTPUT	LABOR-HOURS	UNIT	2006 BARE COSTS MAT.	LABOR	EQUIP.	TOTAL	TOTAL INCL O&P	
400	0190	Smooth surface	1 Carp	400	.020	S.F.	1.78	.71		2.49	3.14	400
	0200	Fire rated, embossed surface	↓	400	.020	↓	1.98	.71		2.69	3.36	
700	0010	**PANEL SYSTEM**										700
	0100	Raised panel, eng. wood core w/ wood veneer, std., paint grade	2 Carp	300	.053	S.F.	10.20	1.90		12.10	14.35	
	0110	Oak veneer		300	.053		16.90	1.90		18.80	21.50	
	0120	Maple veneer		300	.053		21.50	1.90		23.40	27	
	0130	Cherry veneer		300	.053		27.50	1.90		29.40	33	
	0300	Class I fire rated, paint grade		300	.053		12.20	1.90		14.10	16.55	
	0310	Oak veneer		300	.053		20.50	1.90		22.40	25.50	
	0320	Maple veneer		300	.053		26	1.90		27.90	31.50	
	0330	Cherry veneer		300	.053		33	1.90		34.90	39	
	0510	Beadboard, 5/8" MDF, standard, primed		300	.053		7.20	1.90		9.10	11.05	
	0520	Oak veneer, unfinished		300	.053		12.50	1.90		14.40	16.90	
	0530	Maple veneer, unfinished		300	.053		14.80	1.90		16.70	19.45	
	0610	Rustic paneling, 5/8" MDF, standard, maple veneer, unfinished	↓	300	.053	↓	19	1.90		20.90	24	
	5000	For prefinished paneling, see division 06250-500 & 06250-200										
750	0010	**SLATWALL PANELS AND ACCESSORIES**										750
	0100	Slatwall panel, 4' x 8' x 3/4" T, MDF, paint grade	1 Carp	500	.016	S.F.	1.41	.57		1.98	2.49	
	0110	Melamine finish		500	.016		2.14	.57		2.71	3.29	
	0120	High pressure plastic laminate finish	↓	500	.016		3.42	.57		3.99	4.70	
	0130	Aluminum channel inserts, add				↓	2.68			2.68	2.95	
	0200	Accessories, corner forms, 8' L				L.F.	3.82			3.82	4.20	
	0210	T-connector, 8' L					5.30			5.30	5.80	
	0220	J-mold, 8' L					1.35			1.35	1.49	
	0230	Edge cap, 8' L					.88			.88	.97	
	0240	Finish end cap, 8' L				↓	3.23			3.23	3.55	
	0300	Display hook, 4" L				Ea.	1			1	1.10	
	0310	6" L					1.11			1.11	1.22	
	0320	8" L					1.18			1.18	1.30	
	0330	10" L					1.30			1.30	1.43	
	0340	12" L					1.41			1.41	1.55	
	0350	Acrylic, 4" L					.82			.82	.90	
	0360	6" L					.95			.95	1.05	
	0370	8" L					1			1	1.10	
	0380	10" L					1.12			1.12	1.23	
	0400	Waterfall hanger, metal, 12" - 16"					6.65			6.65	7.30	
	0410	Acrylic					9.50			9.50	10.45	
	0500	Shelf bracket, metal, 8"					5.15			5.15	5.70	
	0510	10"					5.55			5.55	6.10	
	0520	12"					5.90			5.90	6.50	
	0530	14"					6.65			6.65	7.30	
	0540	16"					7.40			7.40	8.15	
	0550	Acrylic, 8"					3.08			3.08	3.39	
	0560	10"					3.45			3.45	3.80	
	0570	12"					3.82			3.82	4.20	
	0580	14"					4.31			4.31	4.74	
	0600	Shelf, acrylic, 12" x 16" x 1/4"					26			26	28.50	
	0610	12" x 24" x 1/4"				↓	43			43	47.50	

09800 | Acoustical Treatment

09820 | Acoustical Insul/Sealants

			CREW	DAILY OUTPUT	LABOR-HOURS	UNIT	2006 BARE COSTS MAT.	LABOR	EQUIP.	TOTAL	TOTAL INCL O&P	
500	0010	**SOUND ATTENUATION**										500
	0020	Blanket, 1″ thick	1 Carp	925	.009	S.F.	.25	.31		.56	.79	
	0500	1-1/2″ thick		920	.009		.25	.31		.56	.79	
	1000	2″ thick		915	.009		.31	.31		.62	.85	
	1500	3″ thick		910	.009		.42	.31		.73	.98	
	3400	Urethane plastic foam, open cell, on wall, 2″ thick	2 Carp	2050	.008		2.82	.28		3.10	3.56	
	3500	3″ thick		1550	.010		3.74	.37		4.11	4.72	
	3600	4″ thick		1050	.015		5.25	.54		5.79	6.70	
	3700	On ceiling, 2″ thick		1700	.009		2.81	.33		3.14	3.64	
	3800	3″ thick		1300	.012		3.74	.44		4.18	4.84	
	3900	4″ thick		900	.018		5.25	.63		5.88	6.85	
	4000	Nylon matting 0.4″ thick, with carbon black spinerette										
	4010	plus polyester fabric, on floor	J-4	4000	.004	S.F.	2.09	.12		2.21	2.49	
	4200	Fiberglass reinf. backer board underlayment, 7/16″ thick, on floor	″	800	.020	″	1.75	.61		2.36	2.88	
	9000	Minimum labor/equipment charge	1 Carp	5	1.600	Job		57		57	94	

09840 | Acoustical Wall Treatment

			CREW	DAILY OUTPUT	LABOR-HOURS	UNIT	MAT.	LABOR	EQUIP.	TOTAL	TOTAL INCL O&P	
100	0010	**SOUND ABSORBING PANELS** Perforated steel facing, painted with										100
	0100	fiberglass or mineral filler, no backs, 2-1/4″ thick, modular										
	0200	space units, ceiling or wall hung, white or colored	1 Carp	100	.080	S.F.	9.15	2.84		11.99	14.80	
	0300	Fiberboard sound deadening panels, 1/2″ thick	″	600	.013	″	.30	.47		.77	1.12	
	0500	Fiberglass panels, 4′ x 8′ x 1″ thick, with										
	0600	glass cloth face for walls, cemented	1 Carp	155	.052	S.F.	6.20	1.83		8.03	9.85	
	0700	1-1/2″ thick, dacron covered, inner aluminum frame,										
	0710	wall mounted	1 Carp	300	.027	S.F.	7.60	.95		8.55	9.90	
	0900	Mineral fiberboard panels, fabric covered, 30″x 108″,										
	1000	3/4″ thick, concealed spline, wall mounted	1 Carp	150	.053	S.F.	5.45	1.90		7.35	9.15	
	9000	Minimum labor/equipment charge	″	4	2	Job		71		71	118	

09900 | Paints & Coatings

09910 | Paints

			CREW	DAILY OUTPUT	LABOR-HOURS	UNIT	2006 BARE COSTS MAT.	LABOR	EQUIP.	TOTAL	TOTAL INCL O&P	
100	0010	**CABINETS AND CASEWORK**										100
	1000	Primer coat, oil base, brushwork	1 Pord	650	.012	S.F.	.05	.39		.44	.68	
	2000	Paint, oil base, brushwork, 1 coat		650	.012		.06	.39		.45	.70	
	3000	Stain, brushwork, wipe off		650	.012		.05	.39		.44	.69	
	4000	Shellac, 1 coat, brushwork		650	.012		.06	.39		.45	.69	
	4500	Varnish, 3 coats, brushwork, sand after 1st coat		325	.025		.17	.78		.95	1.44	
	5000	For latex paint, deduct					10%					
300	0010	**DOORS AND WINDOWS, EXTERIOR**										300
	0100	Door frames & trim, only										
	0110	Brushwork, primer	1 Pord	512	.016	L.F.	.05	.50		.55	.86	
	0120	Finish coat, exterior latex		512	.016		.06	.50		.56	.86	
	0130	Primer & 1 coat, exterior latex		300	.027		.11	.85		.96	1.48	
	0135	2 coats, exterior latex, both sides		15	.533	Ea.	5.20	16.90		22.10	32.50	
	0140	Primer & 2 coats, exterior latex		265	.030	L.F.	.17	.96		1.13	1.73	
	0150	Doors, flush, both sides, incl. frame & trim										

	09910 | Paints		CREW	DAILY OUTPUT	LABOR-HOURS	UNIT	2006 BARE COSTS MAT.	LABOR	EQUIP.	TOTAL	TOTAL INCL O&P
300	0160	Roll & brush, primer	1 Pord	10	.800	Ea.	4.01	25.50		29.51	45
	0170	Finish coat, exterior latex		10	.800		4.42	25.50		29.92	45.50
	0180	Primer & 1 coat, exterior latex		7	1.143		8.45	36		44.45	67.50
	0190	Primer & 2 coats, exterior latex		5	1.600		12.85	50.50		63.35	95.50
	0200	Brushwork, stain, sealer & 2 coats polyurethane		4	2		17	63.50		80.50	121
	0210	Doors, French, both sides, 10-15 lite, incl. frame & trim									
	0220	Brushwork, primer	1 Pord	6	1.333	Ea.	2	42.50		44.50	70
	0230	Finish coat, exterior latex		6	1.333		2.21	42.50		44.71	70.50
	0240	Primer & 1 coat, exterior latex		3	2.667		4.21	84.50		88.71	141
	0250	Primer & 2 coats, exterior latex		2	4		6.30	127		133.30	211
	0260	Brushwork, stain, sealer & 2 coats polyurethane		2.50	3.200		6.10	101		107.10	170
	0270	Doors, louvered, both sides, incl. frame & trim									
	0280	Brushwork, primer	1 Pord	7	1.143	Ea.	4.01	36		40.01	62.50
	0290	Finish coat, exterior latex		7	1.143		4.42	36		40.42	63
	0300	Primer & 1 coat, exterior latex		4	2		8.45	63.50		71.95	111
	0310	Primer & 2 coats, exterior latex		3	2.667		12.60	84.50		97.10	150
	0320	Brushwork, stain, sealer & 2 coats polyurethane		4.50	1.778		17	56.50		73.50	109
	0330	Doors, panel, both sides, incl. frame & trim									
	0340	Roll & brush, primer	1 Pord	6	1.333	Ea.	4.01	42.50		46.51	72.50
	0350	Finish coat, exterior latex		6	1.333		4.42	42.50		46.92	73
	0360	Primer & 1 coat, exterior latex		3	2.667		8.45	84.50		92.95	145
	0370	Primer & 2 coats, exterior latex		2.50	3.200		12.60	101		113.60	177
	0380	Brushwork, stain, sealer & 2 coats polyurethane		3	2.667		17	84.50		101.50	155
	0400	Windows, per ext. side, based on 15 SF									
	0410	1 to 6 lite									
	0420	Brushwork, primer	1 Pord	13	.615	Ea.	.79	19.50		20.29	32.50
	0430	Finish coat, exterior latex		13	.615		.87	19.50		20.37	32.50
	0440	Primer & 1 coat, exterior latex		8	1		1.66	31.50		33.16	53
	0450	Primer & 2 coats, exterior latex		6	1.333		2.49	42.50		44.99	70.50
	0460	Stain, sealer & 1 coat varnish		7	1.143		2.42	36		38.42	60.50
	0470	7 to 10 lite									
	0480	Brushwork, primer	1 Pord	11	.727	Ea.	.79	23		23.79	38
	0490	Finish coat, exterior latex		11	.727		.87	23		23.87	38
	0500	Primer & 1 coat, exterior latex		7	1.143		1.66	36		37.66	60
	0510	Primer & 2 coats, exterior latex		5	1.600		2.49	50.50		52.99	84
	0520	Stain, sealer & 1 coat varnish		6	1.333		2.42	42.50		44.92	70.50
	0530	12 lite									
	0540	Brushwork, primer	1 Pord	10	.800	Ea.	.79	25.50		26.29	41.50
	0550	Finish coat, exterior latex		10	.800		.87	25.50		26.37	41.50
	0560	Primer & 1 coat, exterior latex		6	1.333		1.66	42.50		44.16	70
	0570	Primer & 2 coats, exterior latex		5	1.600		2.49	50.50		52.99	84
	0580	Stain, sealer & 1 coat varnish		6	1.333		2.40	42.50		44.90	70.50
	0590	For oil base paint, add					10%				
310	0010	**DOORS & WINDOWS, INTERIOR LATEX**									
	0100	Doors flush, both sides, incl. frame & trim									
	0110	Roll & brush, primer	1 Pord	10	.800	Ea.	3.63	25.50		29.13	44.50
	0120	Finish coat, latex		10	.800		3.66	25.50		29.16	44.50
	0130	Primer & 1 coat latex		7	1.143		7.30	36		43.30	66
	0140	Primer & 2 coats latex		5	1.600		10.75	50.50		61.25	93.50
	0160	Spray, both sides, primer		20	.400		3.82	12.70		16.52	24.50
	0170	Finish coat, latex		20	.400		3.84	12.70		16.54	24.50
	0180	Primer & 1 coat latex		11	.727		7.70	23		30.70	45.50
	0190	Primer & 2 coats latex		8	1		11.35	31.50		42.85	63.50
	0200	Doors, French, both sides, 10-15 lite, incl. frame & trim									
	0210	Roll & brush, primer	1 Pord	6	1.333	Ea.	1.81	42.50		44.31	70

		09910 | Paints	Crew	Daily Output	Labor-Hours	Unit	2006 Bare Costs Mat.	Labor	Equip.	Total	Total Incl O&P	
310	0220	Finish coat, latex	1 Pord	6	1.333	Ea.	1.83	42.50		44.33	70	310
	0230	Primer & 1 coat latex		3	2.667		3.64	84.50		88.14	140	
	0240	Primer & 2 coats latex		2	4		5.35	127		132.35	210	
	0260	Doors, louvered, both sides, incl. frame & trim										
	0270	Roll & brush, primer	1 Pord	7	1.143	Ea.	3.63	36		39.63	62	
	0280	Finish coat, latex		7	1.143		3.66	36		39.66	62	
	0290	Primer & 1 coat, latex		4	2		7.05	63.50		70.55	110	
	0300	Primer & 2 coats, latex		3	2.667		10.95	84.50		95.45	148	
	0320	Spray, both sides, primer		20	.400		3.82	12.70		16.52	24.50	
	0330	Finish coat, latex		20	.400		3.84	12.70		16.54	24.50	
	0340	Primer & 1 coat, latex		11	.727		7.70	23		30.70	45.50	
	0350	Primer & 2 coats, latex		8	1		11.60	31.50		43.10	64	
	0360	Doors, panel, both sides, incl. frame & trim										
	0370	Roll & brush, primer	1 Pord	6	1.333	Ea.	3.82	42.50		46.32	72	
	0380	Finish coat, latex		6	1.333		3.66	42.50		46.16	72	
	0390	Primer & 1 coat, latex		3	2.667		7.30	84.50		91.80	144	
	0400	Primer & 2 coats, latex		2.50	3.200		10.95	101		111.95	175	
	0420	Spray, both sides, primer		10	.800		3.82	25.50		29.32	44.50	
	0430	Finish coat, latex		10	.800		3.84	25.50		29.34	44.50	
	0440	Primer & 1 coat, latex		5	1.600		7.70	50.50		58.20	90	
	0450	Primer & 2 coats, latex		4	2		11.60	63.50		75.10	115	
	0460	Windows, per interior side, based on 15 SF										
	0470	1 to 6 lite										
	0480	Brushwork, primer	1 Pord	13	.615	Ea.	.72	19.50		20.22	32.50	
	0490	Finish coat, enamel		13	.615		.72	19.50		20.22	32.50	
	0500	Primer & 1 coat enamel		8	1		1.44	31.50		32.94	52.50	
	0510	Primer & 2 coats enamel		6	1.333		2.16	42.50		44.66	70.50	
	0530	7 to 10 lite										
	0540	Brushwork, primer	1 Pord	11	.727	Ea.	.72	23		23.72	38	
	0550	Finish coat, enamel		11	.727		.72	23		23.72	38	
	0560	Primer & 1 coat enamel		7	1.143		1.44	36		37.44	59.50	
	0570	Primer & 2 coats enamel		5	1.600		2.16	50.50		52.66	84	
	0590	12 lite										
	0600	Brushwork, primer	1 Pord	10	.800	Ea.	.72	25.50		26.22	41.50	
	0610	Finish coat, enamel		10	.800		.72	25.50		26.22	41.50	
	0620	Primer & 1 coat enamel		6	1.333		1.44	42.50		43.94	69.50	
	0630	Primer & 2 coats enamel		5	1.600		2.16	50.50		52.66	84	
	0650	For oil base paint, add					10%					
320	0010	**DOORS AND WINDOWS, INTERIOR ALKYD (OIL BASE)**										320
	0500	Flush door & frame, 3′ x 7′, oil, primer, brushwork	1 Pord	10	.800	Ea.	2.20	25.50		27.70	43	
	1000	Paint, 1 coat		10	.800		2.04	25.50		27.54	42.50	
	1400	Stain, brushwork, wipe off		18	.444		1.06	14.10		15.16	23.50	
	1600	Shellac, 1 coat, brushwork		25	.320		1.18	10.15		11.33	17.60	
	1800	Varnish, 3 coats, brushwork, sand after 1st coat		9	.889		3.61	28		31.61	49	
	2000	Panel door & frame, 3′ x 7′, oil, primer, brushwork		6	1.333		1.82	42.50		44.32	70	
	2200	Paint, 1 coat		6	1.333		2.04	42.50		44.54	70	
	2600	Stain, brushwork, panel door, 3′ x 7′, not incl. frame		16	.500		1.06	15.85		16.91	26.50	
	2800	Shellac, 1 coat, brushwork		22	.364		1.18	11.55		12.73	19.80	
	3000	Varnish, 3 coats, brushwork, sand after 1st coat		7.50	1.067		3.61	34		37.61	58.50	
	4400	Windows, including frame and trim, per side										
	4600	Colonial type, 6/6 lites, 2′ x 3′, oil, primer, brushwork	1 Pord	14	.571	Ea.	.29	18.10		18.39	29.50	
	5800	Paint, 1 coat		14	.571		.32	18.10		18.42	29.50	
	6200	3′ x 5′ opening, 6/6 lites, primer coat, brushwork		12	.667		.72	21		21.72	35	
	6400	Paint, 1 coat		12	.667		.80	21		21.80	35	
	6800	4′ x 8′ opening, 6/6 lites, primer coat, brushwork		8	1		1.54	31.50		33.04	52.50	
	7000	Paint, 1 coat		8	1		1.72	31.50		33.22	53	

09900 | Paints & Coatings

		09910 | Paints	CREW	DAILY OUTPUT	LABOR-HOURS	UNIT	2006 BARE COSTS MAT.	LABOR	EQUIP.	TOTAL	TOTAL INCL O&P	
320	8000	Single lite type, 2′ x 3′, oil base, primer coat, brushwork	1 Pord	33	.242	Ea.	.29	7.70		7.99	12.65	320
	8200	Paint, 1 coat		33	.242		.32	7.70		8.02	12.70	
	8600	3′ x 5′ opening, primer coat, brushwork		20	.400		.72	12.70		13.42	21.50	
	8800	Paint, 1 coat		20	.400		.80	12.70		13.50	21.50	
	9200	4′ x 8′ opening, primer coat, brushwork		14	.571		1.54	18.10		19.64	30.50	
	9400	Paint, 1 coat		14	.571		1.72	18.10		19.82	31	
400	0010	**FENCES**										400
	0100	Chain link or wire metal, one side, water base										
	0110	Roll & brush, first coat	1 Pord	960	.008	S.F.	.06	.26		.32	.48	
	0120	Second coat		1280	.006		.05	.20		.25	.38	
	0130	Spray, first coat		2275	.004		.06	.11		.17	.24	
	0140	Second coat		2600	.003		.06	.10		.16	.22	
	0150	Picket, water base										
	0160	Roll & brush, first coat	1 Pord	865	.009	S.F.	.06	.29		.35	.54	
	0170	Second coat		1050	.008		.06	.24		.30	.46	
	0180	Spray, first coat		2275	.004		.06	.11		.17	.25	
	0190	Second coat		2600	.003		.06	.10		.16	.23	
	0200	Stockade, water base										
	0210	Roll & brush, first coat	1 Pord	1040	.008	S.F.	.06	.24		.30	.46	
	0220	Second coat		1200	.007		.06	.21		.27	.41	
	0230	Spray, first coat		2275	.004		.06	.11		.17	.25	
	0240	Second coat		2600	.003		.06	.10		.16	.23	
500	0010	**FLOORS, INTERIOR**										500
	0100	Concrete										
	0110	Brushwork, latex, block filler										
	0120	1st coat	1 Pord	975	.008	S.F.	.12	.26		.38	.56	
	0130	2nd coat		1150	.007		.08	.22		.30	.44	
	0140	3rd coat		1300	.006		.07	.20		.27	.38	
	0150	Roll, latex, block filler										
	0160	1st coat	1 Pord	2600	.003	S.F.	.17	.10		.27	.34	
	0170	2nd coat		3250	.002		.10	.08		.18	.24	
	0180	3rd coat		3900	.002		.07	.07		.14	.18	
	0190	Spray, latex, block filler										
	0200	1st coat	1 Pord	2600	.003	S.F.	.14	.10		.24	.32	
	0210	2nd coat		3250	.002		.08	.08		.16	.22	
	0220	3rd coat		3900	.002		.06	.07		.13	.17	
620	0010	**MISCELLANEOUS, EXTERIOR**										620
	0100	Railing, ext., decorative wood, incl. cap & baluster										
	0110	newels & spindles @ 12″ O.C.										
	0120	Brushwork, stain, sand, seal & varnish										
	0130	First coat	1 Pord	90	.089	L.F.	.48	2.82		3.30	5.05	
	0140	Second coat	″	120	.067	″	.48	2.11		2.59	3.92	
	0150	Rough sawn wood, 42″ high, 2″ x 2″ verticals, 6″ O.C.										
	0160	Brushwork, stain, each coat	1 Pord	90	.089	L.F.	.15	2.82		2.97	4.69	
	0170	Wrought iron, 1″ rail, 1/2″ sq. verticals										
	0180	Brushwork, zinc chromate, 60″ high, bars 6″ O.C.										
	0190	Primer	1 Pord	130	.062	L.F.	.51	1.95		2.46	3.69	
	0200	Finish coat		130	.062		.16	1.95		2.11	3.30	
	0210	Additional coat		190	.042		.19	1.33		1.52	2.34	
	0220	Shutters or blinds, single panel, 2′ x 4′, paint all sides										
	0230	Brushwork, primer	1 Pord	20	.400	Ea.	.58	12.70		13.28	21	
	0240	Finish coat, exterior latex		20	.400		.47	12.70		13.17	21	
	0250	Primer & 1 coat, exterior latex		13	.615		.92	19.50		20.42	32.50	
	0260	Spray, primer		35	.229		.84	7.25		8.09	12.60	
	0270	Finish coat, exterior latex		35	.229		.99	7.25		8.24	12.75	
	0280	Primer & 1 coat, exterior latex		20	.400		.91	12.70		13.61	21.50	

For expanded coverage of these items see *Means Interior Cost Data 2006*

	09910 | Paints		CREW	DAILY OUTPUT	LABOR-HOURS	UNIT	2006 BARE COSTS MAT.	LABOR	EQUIP.	TOTAL	TOTAL INCL O&P	
620	0290	For louvered shutters, add				S.F.	10%					620
	0300	Stair stringers, exterior, metal										
	0310	Roll & brush, zinc chromate, to 14", each coat	1 Pord	320	.025	L.F.	.05	.79		.84	1.33	
	0320	Rough sawn wood, 4" x 12"										
	0330	Roll & brush, exterior latex, each coat	1 Pord	215	.037	L.F.	.07	1.18		1.25	1.97	
	0340	Trellis/lattice, 2" x 2" @ 3" O.C. with 2" x 8" supports										
	0350	Spray, latex, per side, each coat	1 Pord	475	.017	S.F.	.07	.53		.60	.94	
	0450	Decking, Ext., sealer, alkyd, brushwork, sealer coat		1140	.007		.06	.22		.28	.42	
	0460	1st coat		1140	.007		.06	.22		.28	.42	
	0470	2nd coat		1300	.006		.04	.20		.24	.36	
	0500	Paint, alkyd, brushwork, primer coat		1140	.007		.07	.22		.29	.44	
	0510	1st coat		1140	.007		.07	.22		.29	.44	
	0520	2nd coat		1300	.006		.05	.20		.25	.37	
	0600	Sand paint, alkyd, brushwork, 1 coat	↓	150	.053	↓	.09	1.69		1.78	2.81	
630	0010	**MISCELLANEOUS, INTERIOR**										630
	2400	Floors, conc./wood, oil base, primer/sealer coat, brushwork	2 Pord	1950	.008	S.F.	.06	.26		.32	.49	
	2450	Roller		5200	.003		.06	.10		.16	.23	
	2600	Spray		6000	.003		.06	.08		.14	.21	
	2650	Paint 1 coat, brushwork		1950	.008		.06	.26		.32	.48	
	2800	Roller		5200	.003		.06	.10		.16	.22	
	2850	Spray		6000	.003		.06	.08		.14	.21	
	3000	Stain, wood floor, brushwork, 1 coat		4550	.004		.05	.11		.16	.24	
	3200	Roller		5200	.003		.05	.10		.15	.22	
	3250	Spray		6000	.003		.05	.08		.13	.20	
	3400	Varnish, wood floor, brushwork		4550	.004		.06	.11		.17	.24	
	3450	Roller		5200	.003		.06	.10		.16	.23	
	3600	Spray	↓	6000	.003		.06	.08		.14	.21	
	3800	Grilles, per side, oil base, primer coat, brushwork	1 Pord	520	.015		.10	.49		.59	.89	
	3850	Spray		1140	.007		.10	.22		.32	.47	
	3920	Paint 2 coats, brushwork		325	.025		.21	.78		.99	1.48	
	3940	Spray	↓	650	.012	↓	.24	.39		.63	.89	
	5000	Pipe, thru 4" diameter, primer or sealer coat, oil base, brushwork	2 Pord	1250	.013	L.F.	.06	.41		.47	.72	
	5100	Spray		2165	.007		.06	.23		.29	.45	
	5350	Paint 2 coats, brushwork		775	.021		.11	.65		.76	1.17	
	5400	Spray		1240	.013		.13	.41		.54	.80	
	6300	Thru 16" diameter, primer or sealer coat, brushwork		310	.052		.25	1.64		1.89	2.91	
	6450	Spray		540	.030		.26	.94		1.20	1.80	
	6500	Paint 2 coats, brushwork		195	.082		.45	2.60		3.05	4.68	
	6550	Spray	↓	310	.052	↓	.50	1.64		2.14	3.18	
	7000	Trim, wood, incl. puttying, under 6" wide										
	7200	Primer coat, oil base, brushwork	1 Pord	650	.012	L.F.	.02	.39		.41	.66	
	7250	Paint, 1 coat, brushwork		650	.012		.03	.39		.42	.66	
	7450	3 coats		325	.025		.08	.78		.86	1.34	
	7500	Over 6" wide, primer coat, brushwork		650	.012		.05	.39		.44	.68	
	7550	Paint, 1 coat, brushwork		650	.012		.05	.39		.44	.69	
	7650	3 coats		325	.025	↓	.15	.78		.93	1.42	
	8000	Cornice, simple design, primer coat, oil base, brushwork		650	.012	S.F.	.05	.39		.44	.68	
	8250	Paint, 1 coat		650	.012		.05	.39		.44	.69	
	8350	Ornate design, primer coat		350	.023		.05	.72		.77	1.21	
	8400	Paint, 1 coat		350	.023		.05	.72		.77	1.22	
	8600	Balustrades, primer coat, oil base, brushwork	↓	520	.015	↓	.05	.49		.54	.83	
	8800											
	8900	Trusses and wood frames, primer coat, oil base, brushwork	1 Pord	800	.010	S.F.	.05	.32		.37	.56	
	8950	Spray		1200	.007		.05	.21		.26	.40	
	9220	Paint 2 coats, brushwork		500	.016		.10	.51		.61	.92	
	9240	Spray	↓	600	.013	↓	.12	.42		.54	.81	

	09910 | Paints		CREW	DAILY OUTPUT	LABOR-HOURS	UNIT	2006 BARE COSTS MAT.	LABOR	EQUIP.	TOTAL	TOTAL INCL O&P	
630	9260	Stain, brushwork, wipe off	1 Pord	600	.013	S.F.	.05	.42		.47	.74	630
	9280	Varnish, 3 coats, brushwork		275	.029		.17	.92		1.09	1.67	
	9350	For latex paint, deduct					10%					
700	0010	**SIDING EXTERIOR**, Alkyd (oil base)										700
	0450	Steel siding, oil base, paint 1 coat, brushwork	2 Pord	2015	.008	S.F.	.06	.25		.31	.46	
	0500	Spray		4550	.004		.08	.11		.19	.27	
	0800	Paint 2 coats, brushwork		1300	.012		.11	.39		.50	.75	
	1000	Spray		4550	.004		.14	.11		.25	.33	
	1200	Stucco, rough, oil base, paint 2 coats, brushwork		1300	.012		.11	.39		.50	.75	
	1400	Roller		1625	.010		.12	.31		.43	.63	
	1600	Spray		2925	.005		.12	.17		.29	.42	
	1800	Texture 1-11 or clapboard, oil base, primer coat, brushwork		1300	.012		.10	.39		.49	.74	
	2000	Spray		4550	.004		.10	.11		.21	.29	
	2400	Paint 2 coats, brushwork		810	.020		.16	.63		.79	1.19	
	2600	Spray		2600	.006		.18	.20		.38	.51	
	3400	Stain 2 coats, brushwork		950	.017		.10	.53		.63	.97	
	4000	Spray		3050	.005		.11	.17		.28	.39	
	4200	Wood shingles, oil base primer coat, brushwork		1300	.012		.09	.39		.48	.73	
	4400	Spray		3900	.004		.08	.13		.21	.30	
	5000	Paint 2 coats, brushwork		810	.020		.13	.63		.76	1.16	
	5200	Spray		2275	.007		.13	.22		.35	.50	
	6500	Stain 2 coats, brushwork		950	.017		.10	.53		.63	.97	
	7000	Spray		2660	.006		.14	.19		.33	.46	
	8000	For latex paint, deduct					10%					
	8100	For work over 12′ H, from pipe scaffolding, add						15%				
	8200	For work over 12′ H, from extension ladder, add						25%				
	8300	For work over 12′ H, from swing staging, add						35%				
710	0010	**SIDING, MISC.**, latex paint										710
	0100	Aluminum siding										
	0110	Brushwork, primer	2 Pord	2275	.007	S.F.	.05	.22		.27	.42	
	0120	Finish coat, exterior latex		2275	.007		.04	.22		.26	.41	
	0130	Primer & 1 coat exterior latex		1300	.012		.10	.39		.49	.74	
	0140	Primer & 2 coats exterior latex		975	.016		.14	.52		.66	1	
	0150	Mineral Fiber shingles										
	0160	Brushwork, primer	2 Pord	1495	.011	S.F.	.10	.34		.44	.65	
	0170	Finish coat, industrial enamel		1495	.011		.10	.34		.44	.65	
	0180	Primer & 1 coat enamel		810	.020		.20	.63		.83	1.23	
	0190	Primer & 2 coats enamel		540	.030		.30	.94		1.24	1.84	
	0200	Roll, primer		1625	.010		.11	.31		.42	.62	
	0210	Finish coat, industrial enamel		1625	.010		.11	.31		.42	.62	
	0220	Primer & 1 coat enamel		975	.016		.22	.52		.74	1.08	
	0230	Primer & 2 coats enamel		650	.025		.33	.78		1.11	1.61	
	0240	Spray, primer		3900	.004		.08	.13		.21	.30	
	0250	Finish coat, industrial enamel		3900	.004		.09	.13		.22	.31	
	0260	Primer & 1 coat enamel		2275	.007		.17	.22		.39	.55	
	0270	Primer & 2 coats enamel		1625	.010		.27	.31		.58	.79	
	0280	Waterproof sealer, first coat		4485	.004		.07	.11		.18	.25	
	0290	Second coat		5235	.003		.06	.10		.16	.23	
	0300	Rough wood incl. shingles, shakes or rough sawn siding										
	0310	Brushwork, primer	2 Pord	1280	.013	S.F.	.12	.40		.52	.77	
	0320	Finish coat, exterior latex		1280	.013		.07	.40		.47	.72	
	0330	Primer & 1 coat exterior latex		960	.017		.19	.53		.72	1.06	
	0340	Primer & 2 coats exterior latex		700	.023		.26	.72		.98	1.45	
	0350	Roll, primer		2925	.005		.16	.17		.33	.45	
	0360	Finish coat, exterior latex		2925	.005		.09	.17		.26	.38	

For expanded coverage of these items see *Means Interior Cost Data 2006*

	09910 | Paints		CREW	DAILY OUTPUT	LABOR-HOURS	UNIT	2006 BARE COSTS MAT.	LABOR	EQUIP.	TOTAL	TOTAL INCL O&P	
710	0370	Primer & 1 coat exterior latex	2 Pord	1790	.009	S.F.	.24	.28		.52	.73	710
	0380	Primer & 2 coats exterior latex		1300	.012		.33	.39		.72	1	
	0390	Spray, primer		3900	.004		.13	.13		.26	.36	
	0400	Finish coat, exterior latex		3900	.004		.07	.13		.20	.29	
	0410	Primer & 1 coat exterior latex		2600	.006		.20	.20		.40	.53	
	0420	Primer & 2 coats exterior latex		2080	.008		.27	.24		.51	.69	
	0430	Waterproof sealer, first coat		4485	.004		.12	.11		.23	.31	
	0440	Second coat	↓	4485	.004	↓	.07	.11		.18	.25	
	0450	Smooth wood incl. butt, T&G, beveled, drop or B&B siding										
	0460	Brushwork, primer	2 Pord	2325	.007	S.F.	.08	.22		.30	.44	
	0470	Finish coat, exterior latex		1280	.013		.07	.40		.47	.72	
	0480	Primer & 1 coat exterior latex		800	.020		.16	.63		.79	1.19	
	0490	Primer & 2 coats exterior latex		630	.025		.23	.81		1.04	1.54	
	0500	Roll, primer		2275	.007		.09	.22		.31	.46	
	0510	Finish coat, exterior latex		2275	.007		.08	.22		.30	.45	
	0520	Primer & 1 coat exterior latex		1300	.012		.17	.39		.56	.82	
	0530	Primer & 2 coats exterior latex		975	.016		.25	.52		.77	1.12	
	0540	Spray, primer		4550	.004		.07	.11		.18	.26	
	0550	Finish coat, exterior latex		4550	.004		.07	.11		.18	.26	
	0560	Primer & 1 coat exterior latex		2600	.006		.14	.20		.34	.47	
	0570	Primer & 2 coats exterior latex		1950	.008		.21	.26		.47	.65	
	0580	Waterproof sealer, first coat		5230	.003		.07	.10		.17	.23	
	0590	Second coat	↓	5980	.003		.07	.09		.16	.21	
	0600	For oil base paint, add				↓	10%					
800	0010	**TRIM, EXTERIOR**										800
	0100	Door frames & trim (see Doors, interior or exterior)										
	0110	Fascia, latex paint, one coat coverage										
	0120	1″ x 4″, brushwork	1 Pord	640	.013	L.F.	.02	.40		.42	.66	
	0130	Roll		1280	.006		.02	.20		.22	.34	
	0140	Spray		2080	.004		.01	.12		.13	.22	
	0150	1″ x 6″ to 1″ x 10″, brushwork		640	.013		.06	.40		.46	.71	
	0160	Roll		1230	.007		.07	.21		.28	.40	
	0170	Spray		2100	.004		.05	.12		.17	.24	
	0180	1″ x 12″, brushwork		640	.013		.06	.40		.46	.71	
	0190	Roll		1050	.008		.07	.24		.31	.46	
	0200	Spray	↓	2200	.004	↓	.05	.12		.17	.24	
	0210	Gutters & downspouts, metal, zinc chromate paint										
	0220	Brushwork, gutters, 5″, first coat	1 Pord	640	.013	L.F.	.06	.40		.46	.70	
	0230	Second coat		960	.008		.05	.26		.31	.48	
	0240	Third coat		1280	.006		.04	.20		.24	.37	
	0250	Downspouts, 4″, first coat		640	.013		.06	.40		.46	.70	
	0260	Second coat		960	.008		.05	.26		.31	.48	
	0270	Third coat	↓	1280	.006	↓	.04	.20		.24	.37	
	0280	Gutters & downspouts, wood										
	0290	Brushwork, gutters, 5″, primer	1 Pord	640	.013	L.F.	.05	.40		.45	.70	
	0300	Finish coat, exterior latex		640	.013		.05	.40		.45	.70	
	0310	Primer & 1 coat exterior latex		400	.020		.11	.63		.74	1.14	
	0320	Primer & 2 coats exterior latex		325	.025		.17	.78		.95	1.44	
	0330	Downspouts, 4″, primer		640	.013		.05	.40		.45	.70	
	0340	Finish coat, exterior latex		640	.013		.05	.40		.45	.70	
	0350	Primer & 1 coat exterior latex		400	.020		.11	.63		.74	1.14	
	0360	Primer & 2 coats exterior latex	↓	325	.025	↓	.08	.78		.86	1.34	
	0370	Molding, exterior, up to 14″ wide										
	0380	Brushwork, primer	1 Pord	640	.013	L.F.	.06	.40		.46	.71	
	0390	Finish coat, exterior latex		640	.013		.06	.40		.46	.71	
	0400	Primer & 1 coat exterior latex	↓	400	.020	↓	.13	.63		.76	1.17	

09900 | Paints & Coatings

09910 | Paints

		09910 Paints	Crew	Daily Output	Labor-Hours	Unit	2006 Bare Costs Mat.	Labor	Equip.	Total	Total Incl O&P	
800	0410	Primer & 2 coats exterior latex	1 Pord	315	.025	L.F.	.13	.81		.94	1.44	800
	0420	Stain & fill		1050	.008		.06	.24		.30	.46	
	0430	Shellac		1850	.004		.07	.14		.21	.29	
	0440	Varnish		1275	.006		.07	.20		.27	.40	
910	0350	**WALLS, MASONRY (CMU), EXTERIOR**										910
	0360	Concrete masonry units (CMU), smooth surface										
	0370	Brushwork, latex, first coat	1 Pord	640	.013	S.F.	.04	.40		.44	.68	
	0380	Second coat		960	.008		.03	.26		.29	.45	
	0390	Waterproof sealer, first coat		736	.011		.24	.34		.58	.81	
	0400	Second coat		1104	.007		.24	.23		.47	.63	
	0410	Roll, latex, paint, first coat		1465	.005		.04	.17		.21	.33	
	0420	Second coat		1790	.004		.03	.14		.17	.26	
	0430	Waterproof sealer, first coat		1680	.005		.24	.15		.39	.50	
	0440	Second coat		2060	.004		.24	.12		.36	.46	
	0450	Spray, latex, paint, first coat		1950	.004		.03	.13		.16	.25	
	0460	Second coat		2600	.003		.03	.10		.13	.19	
	0470	Waterproof sealer, first coat		2245	.004		.24	.11		.35	.44	
	0480	Second coat		2990	.003		.24	.09		.33	.40	
	0490	Concrete masonry unit (CMU), porous										
	0500	Brushwork, latex, first coat	1 Pord	640	.013	S.F.	.07	.40		.47	.72	
	0510	Second coat		960	.008		.04	.26		.30	.46	
	0520	Waterproof sealer, first coat		736	.011		.24	.34		.58	.81	
	0530	Second coat		1104	.007		.24	.23		.47	.63	
	0540	Roll latex, first coat		1465	.005		.05	.17		.22	.34	
	0550	Second coat		1790	.004		.03	.14		.17	.27	
	0560	Waterproof sealer, first coat		1680	.005		.24	.15		.39	.50	
	0570	Second coat		2060	.004		.24	.12		.36	.46	
	0580	Spray latex, first coat		1950	.004		.04	.13		.17	.25	
	0590	Second coat		2600	.003		.03	.10		.13	.19	
	0600	Waterproof sealer, first coat		2245	.004		.24	.11		.35	.44	
	0610	Second coat		2990	.003		.24	.09		.33	.40	
920	0010	**WALLS AND CEILINGS, INTERIOR**										920
	0100	Concrete, dry wall or plaster, oil base, primer or sealer coat										
	0200	Smooth finish, brushwork	1 Pord	1150	.007	S.F.	.05	.22		.27	.40	
	0240	Roller		1350	.006		.05	.19		.24	.35	
	0280	Spray		2750	.003		.03	.09		.12	.19	
	0300	Sand finish, brushwork		975	.008		.05	.26		.31	.47	
	0340	Roller		1150	.007		.05	.22		.27	.41	
	0380	Spray		2275	.004		.04	.11		.15	.22	
	0800	Paint 2 coats, smooth finish, brushwork		680	.012		.10	.37		.47	.71	
	0840	Roller		800	.010		.11	.32		.43	.63	
	0880	Spray		1625	.005		.09	.16		.25	.35	
	0900	Sand finish, brushwork		605	.013		.10	.42		.52	.78	
	0940	Roller		1020	.008		.11	.25		.36	.52	
	0980	Spray		1700	.005		.09	.15		.24	.34	
	1600	Glaze coating, 5 coats, spray, clear		900	.009		.60	.28		.88	1.11	
	1640	Multicolor		900	.009		.85	.28		1.13	1.39	
	1700	For latex paint, deduct					10%					
	1800	For ceiling installations, add						25%				
	2000	Masonry or concrete block, oil base, primer or sealer coat										
	2100	Smooth finish, brushwork	1 Pord	1224	.007	S.F.	.05	.21		.26	.38	
	2180	Spray		2400	.003		.07	.11		.18	.24	
	2200	Sand finish, brushwork		1089	.007		.07	.23		.30	.45	
	2280	Spray		2400	.003		.07	.11		.18	.24	
	2800	Paint 2 coats, smooth finish, brushwork		756	.011		.15	.34		.49	.70	

09910 | Paints

		09910 Paints	Crew	Daily Output	Labor-Hours	Unit	2006 Bare Costs Mat.	Labor	Equip.	Total	Total Incl O&P	
920	2880	Spray	1 Pord	1360	.006	S.F.	.14	.19		.33	.45	920
	2900	Sand finish, brushwork		672	.012		.15	.38		.53	.77	
	2980	Spray		1360	.006		.14	.19		.33	.45	
	3600	Glaze coating, 5 coats, spray, clear		900	.009		.60	.28		.88	1.11	
	3620	Multicolor		900	.009		.85	.28		1.13	1.39	
	4000	Block filler, 1 coat, brushwork		425	.019		.12	.60		.72	1.10	
	4100	Silicone, water repellent, 2 coats, spray		2000	.004		.25	.13		.38	.48	
	4120	For latex paint, deduct					10%					
	8200	For work 8 - 15′ H, add						10%				
	8300	For work over 15′ H, add						20%				
940	0010	**DRY FALL PAINTING**										940
	0100	Walls										
	0200	Wallboard and smooth plaster, one coat, brush	1 Pord	910	.009	S.F.	.04	.28		.32	.50	
	0210	Roll		1560	.005		.04	.16		.20	.31	
	0220	Spray		2600	.003		.04	.10		.14	.21	
	0230	Two coats, brush		520	.015		.09	.49		.58	.88	
	0240	Roll		877	.009		.09	.29		.38	.56	
	0250	Spray		1560	.005		.09	.16		.25	.36	
	0260	Concrete or textured plaster, one coat, brush		747	.011		.04	.34		.38	.60	
	0270	Roll		1300	.006		.04	.20		.24	.36	
	0280	Spray		1560	.005		.04	.16		.20	.31	
	0290	Two coats, brush		422	.019		.09	.60		.69	1.07	
	0300	Roll		747	.011		.09	.34		.43	.65	
	0310	Spray		1300	.006		.09	.20		.29	.41	
	0320	Concrete block, one coat, brush		747	.011		.04	.34		.38	.60	
	0330	Roll		1300	.006		.04	.20		.24	.36	
	0340	Spray		1560	.005		.04	.16		.20	.31	
	0350	Two coats, brush		422	.019		.09	.60		.69	1.07	
	0360	Roll		747	.011		.09	.34		.43	.65	
	0370	Spray		1300	.006		.09	.20		.29	.41	
	0380	Wood, one coat, brush		747	.011		.04	.34		.38	.60	
	0390	Roll		1300	.006		.04	.20		.24	.36	
	0400	Spray		877	.009		.04	.29		.33	.51	
	0410	Two coats, brush		487	.016		.09	.52		.61	.94	
	0420	Roll		747	.011		.09	.34		.43	.65	
	0430	Spray		650	.012		.09	.39		.48	.73	
	0440	Ceilings										
	0450	Wallboard and smooth plaster, one coat, brush	1 Pord	600	.013	S.F.	.04	.42		.46	.73	
	0460	Roll		1040	.008		.04	.24		.28	.44	
	0470	Spray		1560	.005		.04	.16		.20	.31	
	0480	Two coats, brush		341	.023		.09	.74		.83	1.29	
	0490	Roll		650	.012		.09	.39		.48	.73	
	0500	Spray		1300	.006		.09	.20		.29	.41	
	0510	Concrete or textured plaster, one coat, brush		487	.016		.04	.52		.56	.89	
	0520	Roll		877	.009		.04	.29		.33	.51	
	0530	Spray		1560	.005		.04	.16		.20	.31	
	0540	Two coats, brush		276	.029		.09	.92		1.01	1.58	
	0550	Roll		520	.015		.09	.49		.58	.88	
	0560	Spray		1300	.006		.09	.20		.29	.41	
	0570	Structural steel, bar joists or metal deck, one coat, spray		1560	.005		.04	.16		.20	.31	
	0580	Two coats, spray		1040	.008		.09	.24		.33	.49	

09930 | Stains/Transp. Finishes

		09930 Stains/Transp. Finishes	Crew	Daily Output	Labor-Hours	Unit	Mat.	Labor	Equip.	Total	Total Incl O&P	
100	0010	**VARNISH**										100
	0012	1 coat + sealer, on wood trim, no sanding included	1 Pord	400	.020	S.F.	.07	.63		.70	1.10	

9 FINISHES

09930 | Stains/Transp. Finishes

			Crew	Daily Output	Labor-Hours	Unit	2006 Bare Costs Mat.	Labor	Equip.	Total	Total Incl O&P	
100	0100	Hardwood floors, 2 coats, no sanding included, roller	1 Pord	1890	.004	S.F.	.13	.13		.26	.36	100
	9000	Minimum labor/equipment charge	↓	4	2	Job		63.50		63.50	102	

09963 | Glazed Coatings

			Crew	Daily Output	Labor-Hours	Unit	2006 Bare Costs Mat.	Labor	Equip.	Total	Total Incl O&P	
200	0010	**WALL COATINGS**										200
	0100	Acrylic glazed coatings, minimum	1 Pord	525	.015	S.F.	.26	.48		.74	1.07	
	0200	Maximum		305	.026		.54	.83		1.37	1.93	
	0300	Epoxy coatings, minimum		525	.015		.33	.48		.81	1.14	
	0400	Maximum		170	.047		1.02	1.49		2.51	3.52	
	2400	Sprayed perlite or vermiculite, 1/16" thick, minimum		2935	.003		.22	.09		.31	.38	
	2500	Maximum		640	.013		.62	.40		1.02	1.32	
	2700	Vinyl plastic wall coating, minimum		735	.011		.28	.34		.62	.86	
	2800	Maximum		240	.033		.68	1.06		1.74	2.45	
	3000	Urethane on smooth surface, 2 coats, minimum		1135	.007		.21	.22		.43	.59	
	3100	Maximum	↓	665	.012	↓	.46	.38		.84	1.12	

09990 | Paint Restoration

			Crew	Daily Output	Labor-Hours	Unit	2006 Bare Costs Mat.	Labor	Equip.	Total	Total Incl O&P	
500	0010	**SCRAPE AFTER FIRE DAMAGE**										500
	0050	Boards, 1" x 4"	1 Pord	336	.024	L.F.		.75		.75	1.21	
	0060	1" x 6"		260	.031			.98		.98	1.57	
	0070	1" x 8"		207	.039			1.23		1.23	1.97	
	0080	1" x 10"		174	.046			1.46		1.46	2.34	
	0500	Framing, 2" x 4"		265	.030			.96		.96	1.54	
	0510	2" x 6"		221	.036			1.15		1.15	1.84	
	0520	2" x 8"		190	.042			1.33		1.33	2.14	
	0530	2" x 10"		165	.048			1.54		1.54	2.47	
	0540	2" x 12"		144	.056			1.76		1.76	2.83	
	1000	Heavy framing, 3" x 4"		226	.035			1.12		1.12	1.80	
	1010	4" x 4"		210	.038			1.21		1.21	1.94	
	1020	4" x 6"		191	.042			1.33		1.33	2.13	
	1030	4" x 8"		165	.048			1.54		1.54	2.47	
	1040	4" x 10"		144	.056			1.76		1.76	2.83	
	1060	4" x 12"		131	.061	↓		1.94		1.94	3.11	
	2900	For sealing, minimum		825	.010	S.F.	.13	.31		.44	.63	
	2920	Maximum	↓	460	.017	"	.26	.55		.81	1.18	
	3000	For sandblasting, see division 04930-220										
	3020											
	9000	Minimum labor/equipment charge	1 Pord	3	2.667	Job		84.50		84.50	136	
800	0010	**SANDING** and puttying interior trim, compared to										800
	0100	Painting 1 coat, on quality work				L.F.		100%				
	0300	Medium work						50%				
	0400	Industrial grade				↓		25%				
	0500	Surface protection, placement and removal										
	0510	Basic drop cloths	1 Pord	6400	.001	S.F.		.04		.04	.06	
	0520	Masking with paper		800	.010		.03	.32		.35	.54	
	0530	Volume cover up (using plastic sheathing, or building paper)	↓	16000	.001	↓		.02		.02	.03	
900	0010	**SURFACE PREPARATION, EXTERIOR**										900
	0015	Doors, per side, not incl. frames or trim										
	0020	Scrape & sand										
	0030	Wood, flush	1 Pord	616	.013	S.F.		.41		.41	.66	
	0040	Wood, detail		496	.016			.51		.51	.82	
	0050	Wood, louvered		280	.029			.91		.91	1.45	
	0060	Wood, overhead	↓	616	.013	↓		.41		.41	.66	
	0070	Wire brush										

09990 | Paint Restoration

		09990 Paint Restoration	CREW	DAILY OUTPUT	LABOR-HOURS	UNIT	2006 BARE COSTS MAT.	LABOR	EQUIP.	TOTAL	TOTAL INCL O&P	
900	0080	Metal, flush	1 Pord	640	.013	S.F.		.40		.40	.64	900
	0090	Metal, detail		520	.015			.49		.49	.78	
	0100	Metal, louvered		360	.022			.70		.70	1.13	
	0110	Metal or fibr., overhead		640	.013			.40		.40	.64	
	0120	Metal, roll up		560	.014			.45		.45	.73	
	0130	Metal, bulkhead		640	.013			.40		.40	.64	
	0140	Power wash, based on 2500 lb. operating pressure										
	0150	Metal, flush	B-9	2240	.018	S.F.		.50	.07	.57	.89	
	0160	Metal, detail		2120	.019			.52	.07	.59	.95	
	0170	Metal, louvered		2000	.020			.56	.07	.63	1	
	0180	Metal or fibr., overhead		2400	.017			.46	.06	.52	.84	
	0190	Metal, roll up		2400	.017			.46	.06	.52	.84	
	0200	Metal, bulkhead		2200	.018			.51	.07	.58	.91	
	0400	Windows, per side, not incl. trim										
	0410	Scrape & sand										
	0420	Wood, 1-2 lite	1 Pord	320	.025	S.F.		.79		.79	1.27	
	0430	Wood, 3-6 lite		280	.029			.91		.91	1.45	
	0440	Wood, 7-10 lite		240	.033			1.06		1.06	1.70	
	0450	Wood, 12 lite		200	.040			1.27		1.27	2.04	
	0460	Wood, Bay / Bow		320	.025			.79		.79	1.27	
	0470	Wire brush										
	0480	Metal, 1-2 lite	1 Pord	480	.017	S.F.		.53		.53	.85	
	0490	Metal, 3-6 lite		400	.020			.63		.63	1.02	
	0500	Metal, Bay / Bow		480	.017			.53		.53	.85	
	0510	Power wash, based on 2500 lb. operating pressure										
	0520	1-2 lite	B-9	4400	.009	S.F.		.25	.03	.28	.46	
	0530	3-6 lite		4320	.009			.26	.03	.29	.47	
	0540	7-10 lite		4240	.009			.26	.03	.29	.47	
	0550	12 lite		4160	.010			.27	.04	.31	.48	
	0560	Bay / Bow		4400	.009			.25	.03	.28	.46	
	0600	Siding, scrape and sand, light=10-30%, med.=30-70%										
	0610	Heavy=70-100%, % of surface to sand										
	0650	Texture 1-11, light	1 Pord	480	.017	S.F.		.53		.53	.85	
	0660	Med.		440	.018			.58		.58	.93	
	0670	Heavy		360	.022			.70		.70	1.13	
	0680	Wood shingles, shakes, light		440	.018			.58		.58	.93	
	0690	Med.		360	.022			.70		.70	1.13	
	0700	Heavy		280	.029			.91		.91	1.45	
	0710	Clapboard, light		520	.015			.49		.49	.78	
	0720	Med.		480	.017			.53		.53	.85	
	0730	Heavy		400	.020			.63		.63	1.02	
	0740	Wire brush										
	0750	Aluminum, light	1 Pord	600	.013	S.F.		.42		.42	.68	
	0760	Med.		520	.015			.49		.49	.78	
	0770	Heavy		440	.018			.58		.58	.93	
	0780	Pressure wash, based on 2500 lb.. operating pressure										
	0790	Stucco	B-9	3080	.013	S.F.		.36	.05	.41	.65	
	0800	Aluminum or vinyl		3200	.013			.35	.05	.40	.63	
	0810	Siding, masonry, brick & block		2400	.017			.46	.06	.52	.84	
	1300	Miscellaneous, wire brush										
	1310	Metal, pedestrian gate	1 Pord	100	.080	S.F.		2.54		2.54	4.07	
	8000	For Chemical Washing, see Division 04930										
	8010	For Steam Cleaning, see Division 04930										
910	0010	**SURFACE PREPARATION, INTERIOR**										910
	0020	Doors										

09990 | Paint Restoration

			CREW	DAILY OUTPUT	LABOR-HOURS	UNIT	MAT.	LABOR	EQUIP.	TOTAL	TOTAL INCL O&P	
910	0030	Scrape & sand										910
	0040	Wood, flush	1 Pord	616	.013	S.F.		.41		.41	.66	
	0050	Wood, detail		496	.016			.51		.51	.82	
	0060	Wood, louvered	↓	280	.029	↓		.91		.91	1.45	
	0070	Wire brush										
	0080	Metal, flush	1 Pord	640	.013	S.F.		.40		.40	.64	
	0090	Metal, detail		520	.015			.49		.49	.78	
	0100	Metal, louvered	↓	360	.022	↓		.70		.70	1.13	
	0110	Hand wash										
	0120	Wood, flush	1 Pord	2160	.004	S.F.		.12		.12	.19	
	0130	Wood, detailed		2000	.004			.13		.13	.20	
	0140	Wood, louvered		1360	.006			.19		.19	.30	
	0150	Metal, flush		2160	.004			.12		.12	.19	
	0160	Metal, detail		2000	.004			.13		.13	.20	
	0170	Metal, louvered	↓	1360	.006	↓		.19		.19	.30	
	0400	Windows, per side, not incl. trim										
	0410	Scrape & sand										
	0420	Wood, 1-2 lite	1 Pord	360	.022	S.F.		.70		.70	1.13	
	0430	Wood, 3-6 lite		320	.025			.79		.79	1.27	
	0440	Wood, 7-10 lite		280	.029			.91		.91	1.45	
	0450	Wood, 12 lite		240	.033			1.06		1.06	1.70	
	0460	Wood, Bay / Bow	↓	360	.022	↓		.70		.70	1.13	
	0470	Wire brush										
	0480	Metal, 1-2 lite	1 Pord	520	.015	S.F.		.49		.49	.78	
	0490	Metal, 3-6 lite		440	.018			.58		.58	.93	
	0500	Metal, Bay / Bow	↓	520	.015	↓		.49		.49	.78	
	0600	Walls, sanding, light=10-30%										
	0610	Med.=30-70%, heavy=70-100%, % of surface to sand										
	0650	Walls, sand										
	0660	Drywall, gypsum, plaster, light	1 Pord	3077	.003	S.F.		.08		.08	.13	
	0670	Drywall, gypsum, plaster, med.		2160	.004			.12		.12	.19	
	0680	Drywall, gypsum, plaster, heavy		923	.009			.27		.27	.44	
	0690	Wood, T&G, light		2400	.003			.11		.11	.17	
	0700	Wood, T&G, med.		1600	.005			.16		.16	.25	
	0710	Wood, T&G, heavy	↓	800	.010	↓		.32		.32	.51	
	0720	Walls, wash										
	0730	Drywall, gypsum, plaster	1 Pord	3200	.003	S.F.		.08		.08	.13	
	0740	Wood, T&G		3200	.003			.08		.08	.13	
	0750	Masonry, brick & block, smooth		2800	.003			.09		.09	.15	
	0760	Masonry, brick & block, coarse	↓	2000	.004	↓		.13		.13	.20	
	8000	For Chemical Washing, see Division 04930										
	8010	For Steam Cleaning, see Division 04930										

For information about Means Estimating Seminars, see yellow pages 12 and 13 in back of book

Division Notes

		CREW	DAILY OUTPUT	LABOR-HOURS	UNIT	2000 BARE COSTS MAT.	LABOR	EQUIP.	TOTAL	TOTAL INCL O&P

Division 10 Specialties

Estimating Tips

General

- The items in this division are usually priced per square foot or each.
- Many items in Division 10 require some type of support system or special anchors that are not usually furnished with the item. The required anchors must be added to the estimate in the appropriate division.
- Some items in Division 10, such as lockers, may require assembly before installation. Verify the amount of assembly required. Assembly can often exceed installation time.

10150 Compartments & Cubicles

- Support angles and blocking are not included in the installation of toilet compartments, shower/dressing compartments, or cubicles. Appropriate line items from Divisions 5 or 6 may need to be added to support the installations.
- Toilet partitions are priced by the stall. A stall consists of a side wall, pilaster, and door with hardware. Toilet tissue holders and grab bars are extra.

10600 Partitions

- The required acoustical rating of a folding partition can have a significant impact on costs. Verify the sound transmission coefficient rating of the panel priced to the specification requirements.

10800 Toilet/Bath/Laundry Accessories

- Grab bar installation does not include supplemental blocking or backing to support the required load. When grab bars are installed at an existing facility, provisions must be made to attach the grab bars to solid structure.

Reference Numbers

Reference numbers are shown in bold squares at the beginning of some major classifications. These numbers refer to related items in the Reference Section. The reference information may be an estimating procedure, an alternate pricing method, or technical information.

Note: Not all subdivisions listed here necessarily appear in this publication.

10100 | Visual Display Boards

10110 | Chalkboards

		10110 Chalkboards	CREW	DAILY OUTPUT	LABOR-HOURS	UNIT	2006 BARE COSTS MAT.	LABOR	EQUIP.	TOTAL	TOTAL INCL O&P
240	0010	CHALKBOARDS Porcelain enamel steel									
	0100	Freestanding, reversible									
	0120	Economy, wood frame, 4′ x 6′									
	0140	Chalkboard both sides				Ea.	545			545	600
	0160	Chalkboard one side, cork other side				"	455			455	500
	0200	Standard, lightweight satin finished aluminum, 4′ x 6′									
	0220	Chalkboard both sides				Ea.	565			565	620
	0240	Chalkboard one side, cork other side				"	490			490	540
	0300	Deluxe, heavy duty extruded aluminum, 4′ x 6′									
	0320	Chalkboard both sides				Ea.	1,175			1,175	1,300
	0340	Chalkboard one side, cork other side				"	1,075			1,075	1,175
	3900	Wall hung									
	4000	Aluminum frame and chalktrough									
	4300	3′ x 5′	2 Carp	15	1.067	Ea.	218	38		256	305
	4600	4′ x 12′	"	13	1.231	"	480	44		524	605
	4700	Wood frame and chalktrough									
	4800	3′ x 4′	2 Carp	16	1	Ea.	155	35.50		190.50	230
	5300	4′ x 8′	"	13	1.231	"	310	44		354	415
	5400	Liquid chalk, white porcelain enamel, wall hung									
	5420	Deluxe units, aluminum trim and chalktrough									
	5450	4′ x 4′	2 Carp	16	1	Ea.	191	35.50		226.50	269
	5550	4′ x 12′	"	12	1.333	"	455	47.50		502.50	580
	5700	Wood trim and chalktrough									
	5900	4′ x 4′	2 Carp	16	1	Ea.	440	35.50		475.50	545
	6200	4′ x 8′	↓	14	1.143	"	615	40.50		655.50	745
	9000	Minimum labor/equipment charge	↓	3	5.333	Job		190		190	315

10120 | Tack & Visual Aid Boards

		10120 Tack & Visual Aid Boards	CREW	DAILY OUTPUT	LABOR-HOURS	UNIT	MAT.	LABOR	EQUIP.	TOTAL	TOTAL INCL O&P
940	0011	BULLETIN BOARD									
	2120	Prefabricated, 1/4″ cork, 3′ x 5′ with aluminum frame	2 Carp	16	1	Ea.	114	35.50		149.50	184
	2140	Wood frame	"	16	1	"	147	35.50		182.50	221
	2300	Glass enclosed cabinets, alum., cork panel, hinged doors									
	2600	4′ x 7′, 3 door	2 Carp	10	1.600	Ea.	1,400	57		1,457	1,650
	9000	Minimum labor/equipment charge	"	4	4	Job		142		142	236

10150 | Compartments & Cubicles

10160 | Metal Toilet Compartments

		10160 Metal Toilet Compartments	CREW	DAILY OUTPUT	LABOR-HOURS	UNIT	2006 BARE COSTS MAT.	LABOR	EQUIP.	TOTAL	TOTAL INCL O&P
100	0010	TOILET PARTITIONS, METAL									
	0110	Cubicles, ceiling hung									
	0200	Painted metal	2 Carp	4	4	Ea.	475	142		617	755
	0500	Stainless steel	"	4	4	↓	1,250	142		1,392	1,600
	0600	For handicap units, incl. 52″ grab bars, add				↓	415			415	455
	0900	Floor and ceiling anchored									
	1000	Painted metal	2 Carp	5	3.200	Ea.	470	114		584	705
	1300	Stainless steel	"	5	3.200	↓	1,450	114		1,564	1,800
	1400	For handicap units, incl. 52″ grab bars, add				↓	287			287	315
	1610	Floor mounted									
	1700	Painted metal	2 Carp	7	2.286	Ea.	530	81.50		611.50	715
	2000	Stainless steel	"	7	2.286	↓	1,550	81.50		1,631.50	1,825

10150 | Compartments & Cubicles

10160 | Metal Toilet Compartments

	Line	Description	Crew	Daily Output	Labor-Hours	Unit	2006 Bare Costs Mat.	Labor	Equip.	Total	Total Incl O&P	
100	2100	For handicap units, incl. 52" grab bars, add				Ea.	287			287	315	100
	2200	For juvenile units, deduct				↓	39.50			39.50	43.50	
	2450	Floor mounted, headrail braced										
	2500	Painted metal	2 Carp	6	2.667	Ea.	470	95		565	670	
	2800	Stainless steel	"	6	2.667		1,450	95		1,545	1,725	
	2900	For handicap units, incl. 52" grab bars, add					305			305	335	
	3000	Wall hung partitions, painted metal	2 Carp	7	2.286		595	81.50		676.50	790	
	3300	Stainless steel	"	7	2.286		1,425	81.50		1,506.50	1,675	
	3400	For handicap units, incl. 52" grab bars, add				↓	305			305	335	
	4000	Screens, entrance, floor mounted, 58" high, 48" wide										
	4200	Painted metal	2 Carp	15	1.067	Ea.	212	38		250	297	
	4500	Stainless steel	"	15	1.067	"	765	38		803	910	
	4650	Urinal screen, 18" wide										
	4700	Painted metal	2 Carp	8	2	Ea.	204	71		275	345	
	5000	Stainless steel	"	8	2	"	550	71		621	725	
	5100	Floor mounted, head rail braced										
	5300	Painted metal	2 Carp	8	2	Ea.	198	71		269	335	
	5600	Stainless steel	"	8	2	"	640	71		711	825	
	5750	Pilaster, flush										
	5800	Painted metal	2 Carp	10	1.600	Ea.	269	57		326	390	
	6100	Stainless steel		10	1.600		485	57		542	625	
	6300	Urinal screen, post braced, painted metal		10	1.600		294	57		351	420	
	6600	Stainless steel	↓	10	1.600	↓	480	57		537	625	
	6700	Wall hung, bracket supported										
	6800	Painted metal	2 Carp	10	1.600	Ea.	281	57		338	405	
	7100	Stainless steel		10	1.600		440	57		497	580	
	7400	Flange supported, painted metal		10	1.600		218	57		275	335	
	7700	Stainless steel		10	1.600		510	57		567	655	
	7800	Wedge type, painted metal		10	1.600		257	57		314	375	
	8100	Stainless steel	↓	10	1.600	↓	525	57		582	670	
	9000	Minimum labor/equipment charge	1 Carp	2.50	3.200	Job		114		114	188	

10165 | Plastic Laminate Toilet Compartment

	Line	Description	Crew	Daily Output	Labor-Hours	Unit	Mat.	Labor	Equip.	Total	Total Incl O&P	
100	0010	TOILET PARTITIONS, PLASTIC LAMINATE										100
	0110	Cubicles, ceiling hung										
	0300	Plastic laminate on particle board	2 Carp	4	4	Ea.	575	142		717	865	
	0600	For handicap units, incl. 52" grab bars, add				"	415			415	455	
	0900	Floor and ceiling anchored										
	1100	Plastic laminate on particle board	2 Carp	5	3.200	Ea.	665	114		779	920	
	1400	For handicap units, incl. 52" grab bars, add				"	287			287	315	
	1610	Floor mounted										
	1800	Plastic laminate on particle board	2 Carp	7	2.286	Ea.	570	81.50		651.50	760	
	2450	Floor mounted, headrail braced										
	2600	Plastic laminate on particle board	2 Carp	6	2.667	Ea.	770	95		865	1,000	
	3400	For handicap units, incl. 52" grab bars, add					305			305	335	
	4300	Entrance screen, floor mtd., plas. lam., 58" high, 48" wide	2 Carp	15	1.067		435	38		473	540	
	4800	Urinal screen, 18" wide, ceiling braced, plastic laminate		8	2		293	71		364	440	
	5400	Floor mounted, headrail braced		8	2		355	71		426	510	
	5900	Pilaster, flush, plastic laminate		10	1.600		360	57		417	490	
	6400	Post braced, plastic laminate	↓	10	1.600	↓	360	57		417	490	
	6700	Wall hung, bracket supported										
	6900	Plastic laminate on particle board	2 Carp	10	1.600	Ea.	167	57		224	277	
	7450	Flange supported										
	7500	Plastic laminate on particle board	2 Carp	10	1.600	Ea.	395	57		452	530	

10150 | Compartments & Cubicles

10 SPECIALTIES

10170 | Plastic Toilet Compartments

		10170 Plastic Toilet Compartments	CREW	DAILY OUTPUT	LABOR-HOURS	UNIT	2006 BARE COSTS MAT.	LABOR	EQUIP.	TOTAL	TOTAL INCL O&P	
100	0010	**TOILET PARTITIONS, PLASTIC**										100
	0110	Cubicles, ceiling hung										
	0250	Phenolic	2 Carp	4	4	Ea.	955	142		1,097	1,275	
	0600	For handicap units, incl. 52″ grab bars, add				″	415			415	455	
	0900	Floor and ceiling anchored										
	1050	Phenolic	2 Carp	5	3.200	Ea.	985	114		1,099	1,275	
	1400	For handicap units, incl. 52″ grab bars, add				″	287			287	315	
	1610	Floor mounted										
	1750	Phenolic	2 Carp	7	2.286	Ea.	1,075	81.50		1,156.50	1,325	
	2100	For handicap units, incl. 52″ grab bars, add					287			287	315	
	2200	For juvenile units, deduct					39.50			39.50	43.50	
	2450	Floor mounted, headrail braced										
	2550	Phenolic	2 Carp	6	2.667	Ea.	950	95		1,045	1,200	

10180 | Stone Toilet Compartments

		10180 Stone Toilet Compartments	CREW	DAILY OUTPUT	LABOR-HOURS	UNIT	MAT.	LABOR	EQUIP.	TOTAL	TOTAL INCL O&P	
100	0010	**TOILET PARTITIONS, STONE**										100
	0100	Cubicles, ceiling hung, marble	2 Marb	2	8	Ea.	1,550	279		1,829	2,150	
	0600	For handicap units, incl. 52″ grab bars, add					415			415	455	
	0800	Floor & ceiling anchored, marble	2 Marb	2.50	6.400		1,675	223		1,898	2,200	
	1400	For handicap units, incl. 52″ grab bars, add					287			287	315	
	1600	Floor mounted, marble	2 Marb	3	5.333		995	186		1,181	1,400	
	2400	Floor mounted, headrail braced, marble	″	3	5.333		950	186		1,136	1,350	
	2900	For handicap units, incl. 52″ grab bars, add					305			305	335	
	4100	Entrance screen, floor mounted marble, 58″ high, 48″ wide	2 Marb	9	1.778		625	62		687	785	
	4600	Urinal screen, 18″ wide, ceiling braced, marble	D-1	6	2.667		625	85.50		710.50	825	
	5100	Floor mounted, head rail braced										
	5200	Marble	D-1	6	2.667	Ea.	545	85.50		630.50	740	
	5700	Pilaster, flush, marble		9	1.778		695	57		752	860	
	6200	Post braced, marble		9	1.778		690	57		747	850	
	9000	Minimum labor/equipment charge	1 Carp	2.50	3.200	Job		114		114	188	

10185 | Shower/Dressing Compartments

		10185 Shower/Dressing Compartments	CREW	DAILY OUTPUT	LABOR-HOURS	UNIT	MAT.	LABOR	EQUIP.	TOTAL	TOTAL INCL O&P	
100	0010	**PARTITIONS, SHOWER** Floor mounted, no plumbing										100
	0100	Cabinet, incl. base, no door, painted steel, 1″ thick walls	2 Shee	5	3.200	Ea.	765	135		900	1,050	
	0300	With door, fiberglass		4.50	3.556		620	150		770	920	
	0600	Galvanized and painted steel, 1″ thick walls		5	3.200		805	135		940	1,100	
	0800	Stall, 1″ thick wall, no base, enameled steel		5	3.200		875	135		1,010	1,175	
	1500	Circular fiberglass, cabinet 36″ diameter,		4	4		635	169		804	970	
	1700	One piece, 36″ diameter, less door		4	4		535	169		704	860	
	1800	With door		3.50	4.571		880	193		1,073	1,275	
	4100	Shower doors, economy plastic, 24″ wide	1 Shee	9	.889		109	37.50		146.50	180	
	4200	Tempered glass door, economy		8	1		184	42		226	269	
	4700	Deluxe, tempered glass, chrome on brass frame, minimum		8	1		269	42		311	365	
	4800	Maximum		1	8		735	335		1,070	1,350	
	9990	Minimum labor/equipment charge		2.50	3.200	Job		135		135	215	

10200 | Louvers & Vents

10210 | Wall Louvers

		10210 Wall Louvers	CREW	DAILY OUTPUT	LABOR-HOURS	UNIT	2006 BARE COSTS MAT.	LABOR	EQUIP.	TOTAL	TOTAL INCL O&P	
800	0010	**LOUVERS**										800
	0020	Aluminum with screen, residential, 8″ x 8″	1 Carp	38	.211	Ea.	10.05	7.50		17.55	23.50	
	0100	12″ x 12″	↓	38	.211	↓	11.80	7.50		19.30	25.50	
	0200	12″ x 18″		35	.229		15.35	8.15		23.50	30.50	
	0250	14″ x 24″		30	.267		20	9.50		29.50	37.50	
	0300	18″ x 24″		27	.296		24.50	10.55		35.05	44.50	
	0500	24″ x 30″		24	.333		30.50	11.85		42.35	53	
	0700	Triangle, adjustable, small		20	.400		28	14.20		42.20	54.50	
	0800	Large		15	.533		45.50	18.95		64.45	81.50	
	2100	Midget, aluminum, 3/4″ deep, 1″ diameter		85	.094		.91	3.35		4.26	6.55	
	2150	3″ diameter		60	.133		1.98	4.74		6.72	10.05	
	2200	4″ diameter		50	.160		2.98	5.70		8.68	12.70	
	2250	6″ diameter	↓	30	.267	↓	3.52	9.50		13.02	19.55	
	2300	Ridge vent strip, mill finish	1 Shee	155	.052	L.F.	2.66	2.18		4.84	6.40	
	2400	Under eaves vent, aluminum, mill finish, 16″ x 4″	1 Carp	48	.167	Ea.	1.96	5.95		7.91	11.95	
	2500	16″ x 8″	↓	48	.167	↓	2.17	5.95		8.12	12.20	
	7000	Vinyl gable vent, 8″ x 8″		38	.211		10.90	7.50		18.40	24.50	
	7020	12″ x 12″		38	.211		25	7.50		32.50	40	
	7080	12″ x 18″		35	.229		30	8.15		38.15	46.50	
	7200	18″ x 24″		30	.267	↓	34	9.50		43.50	53	
	9000	Minimum labor/equipment charge	↓	2.50	3.200	Job		114		114	188	

10260 | Wall & Corner Guards

10265 | Wall & Corner Guards

		10265 Wall & Corner Guards	CREW	DAILY OUTPUT	LABOR-HOURS	UNIT	2006 BARE COSTS MAT.	LABOR	EQUIP.	TOTAL	TOTAL INCL O&P	
200	0010	**CORNER GUARDS**										200
	0020	Steel angle w/anchors, 1″ x 1″ x 1/4″, 1.5#/L.F.	2 Carp	160	.100	L.F.	5.50	3.56		9.06	11.95	
	0100	2″ x 2″ x 1/4″ angles, 3.2#/L.F.	″	150	.107	″	9.95	3.79		13.74	17.25	
	9000	Minimum labor/equipment charge	1 Carp	2	4	Job		142		142	236	

10270 | Access Flooring

10275 | Access Flooring

		10275 Access Flooring	CREW	DAILY OUTPUT	LABOR-HOURS	UNIT	2006 BARE COSTS MAT.	LABOR	EQUIP.	TOTAL	TOTAL INCL O&P	
150	0010	**PEDESTAL ACCESS FLOORS** Computer room application, metal										150
	0020	Particle board or steel panels, no covering, under 6,000 S.F.	2 Carp	400	.040	S.F.	14.50	1.42		15.92	18.30	
	0300	Metal covered, over 6,000 S.F.	↓	450	.036	↓	8.40	1.26		9.66	11.35	
	0400	Aluminum, 24″ panels	↓	500	.032		31	1.14		32.14	36	
	0600	For carpet covering, add					6			6	6.60	
	0700	For vinyl floor covering, add					6.05			6.05	6.70	
	0900	For high pressure laminate covering, add					4.05			4.05	4.46	
	0910	For snap on stringer system, add	2 Carp	1000	.016	↓	1.38	.57		1.95	2.46	
	0950	Office applications, to 8″ high, steel panels,										
	0960	no covering, over 6,000 S.F.	2 Carp	500	.032	S.F.	9.70	1.14		10.84	12.55	

10270 | Access Flooring

	10275	Access Flooring	CREW	DAILY OUTPUT	LABOR-HOURS	UNIT	2006 BARE COSTS MAT.	LABOR	EQUIP.	TOTAL	TOTAL INCL O&P	
150	1050	Pedestals, 6″ to 12″	2 Carp	85	.188	Ea.	8.50	6.70		15.20	20.50	150

10300 | Fireplaces & Stoves

	10305	Manufactured Fireplaces	CREW	DAILY OUTPUT	LABOR-HOURS	UNIT	2006 BARE COSTS MAT.	LABOR	EQUIP.	TOTAL	TOTAL INCL O&P	
100	0010	**FIREPLACE, PREFABRICATED** Free standing or wall hung										100
	0100	with hood & screen, minimum	1 Carp	1.30	6.154	Ea.	980	219		1,199	1,425	
	0150	Average		1	8		1,375	284		1,659	1,975	
	0200	Maximum		.90	8.889		3,350	315		3,665	4,225	
	0500	Chimney dbl. wall, all stainless, over 8′-6″, 7″ diam., add		33	.242	V.L.F.	54	8.60		62.60	74	
	0600	10″ diameter, add		32	.250		55	8.90		63.90	75.50	
	0700	12″ diameter, add		31	.258		75	9.15		84.15	97.50	
	0800	14″ diameter, add		30	.267		99	9.50		108.50	125	
	1000	Simulated brick chimney top, 4′ high, 16″ x 16″		10	.800	Ea.	202	28.50		230.50	269	
	1100	24″ x 24″		7	1.143	″	375	40.50		415.50	485	
	1500	Simulated logs, gas fired, 40,000 BTU, 2′ long, minimum		7	1.143	Set	510	40.50		550.50	630	
	1600	Maximum		6	1.333		680	47.50		727.50	830	
	1700	Electric, 1,500 BTU, 1′-6″ long, minimum		7	1.143		140	40.50		180.50	222	
	1800	11,500 BTU, maximum		6	1.333		300	47.50		347.50	410	
	2000	Fireplace, built-in, 36″ hearth, radiant		1.30	6.154	Ea.	540	219		759	955	
	2100	Recirculating, small fan		1	8		860	284		1,144	1,425	
	2150	Large fan		.90	8.889		1,500	315		1,815	2,175	
	2200	42″ hearth, radiant		1.20	6.667		730	237		967	1,200	
	2300	Recirculating, small fan		.90	8.889		925	315		1,240	1,550	
	2350	Large fan		.80	10		1,500	355		1,855	2,250	
	2400	48″ hearth, radiant		1.10	7.273		1,400	259		1,659	1,975	
	2500	Recirculating, small fan		.80	10		1,750	355		2,105	2,525	
	2550	Large fan		.70	11.429		2,700	405		3,105	3,650	
	3000	See through, including doors		.80	10		2,300	355		2,655	3,125	
	3200	Corner (2 wall)		1	8		2,300	284		2,584	3,000	

	10310	Fireplace Specialties & Accessories	CREW	DAILY OUTPUT	LABOR-HOURS	UNIT	2006 BARE COSTS MAT.	LABOR	EQUIP.	TOTAL	TOTAL INCL O&P	
100	0010	**FIREPLACE ACCESSORIES**										100
	0020	Chimney screens, galv., 13″ x 13″ flue	1 Bric	8	1	Ea.	43	36.50		79.50	107	
	0050	Galv., 24″ x 24″ flue		5	1.600		100	58.50		158.50	205	
	0200	Stainless steel, 13″ x 13″ flue		8	1		264	36.50		300.50	350	
	0250	20″ x 20″ flue		5	1.600		360	58.50		418.50	490	
	0400	Cleanout doors and frames, cast iron, 8″ x 8″		12	.667		30	24.50		54.50	72.50	
	0450	12″ x 12″		10	.800		36.50	29		65.50	87.50	
	0500	18″ x 24″		8	1		105	36.50		141.50	175	
	0550	Cast iron frame, steel door, 24″ x 30″		5	1.600		227	58.50		285.50	345	
	0800	Damper, rotary control, steel, 30″ opening		6	1.333		64	48.50		112.50	149	
	0850	Cast iron, 30″ opening		6	1.333		71	48.50		119.50	157	
	0880	36″ opening		6	1.333		77	48.50		125.50	164	
	0900	48″ opening		6	1.333		110	48.50		158.50	200	
	0920	60″ opening		6	1.333		247	48.50		295.50	350	
	1000	84″ opening, special order		5	1.600		635	58.50		693.50	795	
	1050	96″ opening, special order		4	2		645	73		718	830	

10300 | Fireplaces & Stoves

10310 | Fireplace Specialties & Accessories

		10310 Fireplace Specialties & Accessories	CREW	DAILY OUTPUT	LABOR-HOURS	UNIT	2006 BARE COSTS MAT.	LABOR	EQUIP.	TOTAL	TOTAL INCL O&P	
100	1200	Steel plate, poker control, 60" opening	1 Bric	8	1	Ea.	225	36.50		261.50	310	100
	1250	84" opening, special opening		5	1.600		410	58.50		468.50	545	
	1400	"Universal" type, chain operated, 32" x 20" opening		8	1		175	36.50		211.50	253	
	1450	48" x 24" opening		5	1.600		261	58.50		319.50	380	
	1600	Dutch Oven door and frame, cast iron, 12" x 15" opening		13	.615		92	22.50		114.50	138	
	1650	Copper plated, 12" x 15" opening		13	.615		177	22.50		199.50	231	
	1800	Fireplace forms, no accessories, 32" opening		3	2.667		510	97.50		607.50	720	
	1900	36" opening		2.50	3.200		650	117		767	905	
	2000	40" opening		2	4		860	146		1,006	1,175	
	2100	78" opening		1.50	5.333		1,250	195		1,445	1,700	
	2400	Squirrel and bird screens, galvanized, 8" x 8" flue		16	.500		37.50	18.30		55.80	70.50	
	2450	13" x 13" flue		12	.667	↓	40.50	24.50		65	84	
	9000	Minimum labor/equipment charge	↓	3.50	2.286	Job		83.50		83.50	136	

10320 | Stoves

		10320 Stoves	CREW	DAILY OUTPUT	LABOR-HOURS	UNIT	MAT.	LABOR	EQUIP.	TOTAL	TOTAL INCL O&P	
100	0010	**WOODBURNING STOVES**										100
	0015	Cast iron, minimum	2 Carp	1.30	12.308	Ea.	910	440		1,350	1,725	
	0020	Average		1	16		1,400	570		1,970	2,500	
	0030	Maximum	↓	.80	20		2,350	710		3,060	3,750	
	0050	For gas log lighter, add				↓	40			40	44	

10340 | Manufactured Exterior Specialties

10342 | Cupolas

		10342 Cupolas	CREW	DAILY OUTPUT	LABOR-HOURS	UNIT	2006 BARE COSTS MAT.	LABOR	EQUIP.	TOTAL	TOTAL INCL O&P	
100	0010	**CUPOLA**										100
	0020	Stock units, pine, painted, 18" sq., 28" high, alum. roof	1 Carp	4.10	1.951	Ea.	150	69.50		219.50	280	
	0100	Copper roof		3.80	2.105		175	75		250	315	
	0300	23" square, 33" high, aluminum roof		3.70	2.162		325	77		402	485	
	0400	Copper roof		3.30	2.424		440	86		526	630	
	0600	30" square, 37" high, aluminum roof		3.70	2.162		460	77		537	630	
	0700	Copper roof		3.30	2.424		495	86		581	690	
	0900	Hexagonal, 31" wide, 46" high, copper roof		4	2		620	71		691	800	
	1000	36" wide, 50" high, copper roof	↓	3.50	2.286		680	81.50		761.50	885	
	1200	For deluxe stock units, add to above					25%					
	1400	For custom built units, add to above				↓	50%	50%				
	9000	Minimum labor/equipment charge	1 Carp	2.75	2.909	Job		103		103	171	

10344 | Weathervanes

		10344 Weathervanes	CREW	DAILY OUTPUT	LABOR-HOURS	UNIT	MAT.	LABOR	EQUIP.	TOTAL	TOTAL INCL O&P	
800	0010	**WEATHERVANES**										800
	0020	Residential types, minimum	1 Carp	8	1	Ea.	130	35.50		165.50	202	
	0100	Maximum		2	4	"	2,300	142		2,442	2,775	
	9000	Minimum labor/equipment charge	↓	4	2	Job		71		71	118	

10400 | Identification Devices

		10410 | Directories	CREW	DAILY OUTPUT	LABOR-HOURS	UNIT	2006 BARE COSTS MAT.	LABOR	EQUIP.	TOTAL	TOTAL INCL O&P	
100	0010	**DIRECTORY BOARDS**										100
	0050	Plastic, glass covered, 30" x 20"	2 Carp	3	5.333	Ea.	242	190		432	580	
	0100	36" x 48"	↓	2	8	↓	770	284		1,054	1,325	
	0900	Outdoor, weatherproof, black plastic, 36" x 24"		2	8		685	284		969	1,225	
	1000	36" x 36"		1.50	10.667		790	380		1,170	1,500	
	9000	Minimum labor/equipment charge	1 Carp	1	8	Job		284		284	470	

		10430 | Exterior Signage	CREW	DAILY OUTPUT	LABOR-HOURS	UNIT	MAT.	LABOR	EQUIP.	TOTAL	TOTAL INCL O&P	
200	0012	**PLAQUES**										200
	3910	20" x 30", up to 450 letters, cast alum.	2 Carp	4	4	Ea.	820	142		962	1,150	
	4000	Cast bronze	↓	4	4	↓	1,125	142		1,267	1,475	
	4200	30" x 36", up to 900 letters cast aluminum		3	5.333		1,725	190		1,915	2,225	
	4300	Cast bronze		3	5.333		2,225	190		2,415	2,775	
	5100	Exit signs, 24 ga. alum., 14" x 12" surface mounted	1 Carp	30	.267		33.50	9.50		43	52.50	
	5200	10" x 7"	"	20	.400		19.55	14.20		33.75	45	
	6400	Replacement sign faces, 6" or 8"	1 Clab	50	.160		25	4.38		29.38	35	
	9000	Minimum labor/equipment charge	1 Carp	4	2	Job		71		71	118	

		10440 | Interior Signage	CREW	DAILY OUTPUT	LABOR-HOURS	UNIT	MAT.	LABOR	EQUIP.	TOTAL	TOTAL INCL O&P	
200	0010	**INTERIOR SIGNAGE**										200
	1010	Flexible door sign, adhesive back, w/Braille, 5/8" letters, 4" x 4"	1 Clab	32	.250	Ea.	24	6.85		30.85	37.50	
	1050	6" x 6"	↓	32	.250	↓	34.50	6.85		41.35	49	
	1100	8" x 2"		32	.250		25	6.85		31.85	38.50	
	1150	8" x 4"		32	.250		31	6.85		37.85	45.50	
	1200	8" x 8"		32	.250		46	6.85		52.85	62	
	1250	12" x 2"		32	.250		32	6.85		38.85	46.50	
	1300	12" x 6"		32	.250		45	6.85		51.85	61	
	1350	12" x 12"		32	.250		92	6.85		98.85	112	
	1500	Graphic symbols, 2" x 2"		32	.250		11	6.85		17.85	23.50	
	1550	6" x 6"		32	.250		28	6.85		34.85	42.50	
	1600	8" x 8"		32	.250		28	6.85		34.85	42.50	

10 SPECIALTIES

10500 | Lockers

		10505 | Metal Lockers	CREW	DAILY OUTPUT	LABOR-HOURS	UNIT	2006 BARE COSTS MAT.	LABOR	EQUIP.	TOTAL	TOTAL INCL O&P	
500	0011	**LOCKERS** Steel, baked enamel										500
	0110	Single tier box locker, 12" x 15" x 72"	1 Shee	8	1	Ea.	165	42		207	249	
	0120	18" x 15" x 72"	↓	8	1	↓	178	42		220	263	
	0130	12" x 18" x 72"		8	1		170	42		212	254	
	0140	18" x 18" x 72"		8	1		193	42		235	279	
	0410	Double tier, 12" x 15" x 36"		21	.381		175	16.05		191.05	219	
	0420	18" x 15" x 36"		21	.381		237	16.05		253.05	286	
	0430	12" x 18" x 36"		21	.381		184	16.05		200.05	228	
	0440	18" x 18" x 36"		21	.381		197	16.05		213.05	243	
	0500	Two person, 18" x 15" x 72"		8	1		263	42		305	355	
	0510	18" x 18" x 72"		8	1		270	42		312	365	
	0520	Duplex, 15" x 15" x 72"		8	1		258	42		300	350	
	0530	15" x 21" x 72"		8	1		300	42		342	400	
	1100	Wire meshed wardrobe, floor. mtd., open front varsity type		7.50	1.067		166	45		211	254	

10500 | Lockers

	10505 | Metal Lockers		CREW	DAILY OUTPUT	LABOR-HOURS	UNIT	2006 BARE COSTS MAT.	LABOR	EQUIP.	TOTAL	TOTAL INCL O&P	
500	2400	16-person locker unit with clothing rack										500
	2500	72 wide x 15" deep x 72" high	1 Shee	8	1	Ea.	470	42		512	585	
	2550	18" deep		8	1	"	400	42		442	505	
	3250	Rack w/ 24 wire mesh baskets		1.50	5.333	Set	325	225		550	715	
	3260	30 baskets		1.25	6.400		256	270		526	710	
	3270	36 baskets		.95	8.421		325	355		680	920	
	3280	42 baskets	↓	.80	10	↓	560	420		980	1,275	
	3600	For hanger rods, add				Ea.	1.79			1.79	1.97	
	9000	Minimum labor/equipment charge	1 Shee	2.50	3.200	Job		135		135	215	

10520 | Fire Protection Specialties

	10525 | Fire Prot. Specialties		CREW	DAILY OUTPUT	LABOR-HOURS	UNIT	2006 BARE COSTS MAT.	LABOR	EQUIP.	TOTAL	TOTAL INCL O&P	
200	0010	**FIRE EQUIPMENT CABINETS** Not equipped, 20 ga. steel box,										200
	0040	recessed, D.S. glass in door, box size given										
	1000	Portable extinguisher, single, 8" x 12" x 27", alum. door & frame	Q-12	8	2	Ea.	96	75.50		171.50	222	
	1100	Steel door and frame	"	8	2	"	70	75.50		145.50	194	
	3000	Hose rack assy., 1-1/2" valve & 100' hose, 24" x 40" x 5-1/2"										
	3200	Steel door and frame	Q-12	6	2.667	Ea.	154	100		254	325	
	4000	Hose rack assy., 2-1/2" x 1-1/2" valve, 100' hose, 24" x 40" x 8"										
	4200	Steel door and frame	Q-12	6	2.667	Ea.	160	100		260	330	
	5000	Hose rack assy., 2-1/2" x 1-1/2" valve, 100' hose										
	5010	and extinguisher, 30" x 40" x 8"										
	5200	Steel door and frame	Q-12	5	3.200	Ea.	176	120		296	380	
300	0010	**FIRE EXTINGUISHERS**										300
	0120	CO2, portable with swivel horn, 5 lb.				Ea.	130			130	143	
	0140	With hose and "H" horn, 10 lb.				"	160			160	176	
	1000	Dry chemical, pressurized										
	1040	Standard type, portable, painted, 2-1/2 lb.				Ea.	27.50			27.50	30.50	
	1080	10 lb.					80			80	88	
	1100	20 lb.					110			110	121	
	1120	30 lb.					157			157	173	
	2000	ABC all purpose type, portable, 2-1/2 lb.					30			30	33	
	2080	9-1/2 lb.				↓	68			68	75	

10530 | Protective Covers

	10535 | Awnings & Canopies		CREW	DAILY OUTPUT	LABOR-HOURS	UNIT	2006 BARE COSTS MAT.	LABOR	EQUIP.	TOTAL	TOTAL INCL O&P	
050	0010	**AWNINGS, FABRIC**										050
	0020	Including acrylic canvas and frame, standard design										
	0100	Door and window, slope, 3' high, 4' wide	1 Carp	4.50	1.778	Ea.	585	63		648	750	
	0110	6' wide	↓	3.50	2.286	↓	755	81.50		836.50	965	

	10535 | Awnings & Canopies		CREW	DAILY OUTPUT	LABOR-HOURS	UNIT	2006 BARE COSTS MAT.	LABOR	EQUIP.	TOTAL	TOTAL INCL O&P	
050	0120	8′ wide	1 Carp	3	2.667	Ea.	925	95		1,020	1,175	050
	0200	Quarter round convex, 4′ wide		3	2.667		910	95		1,005	1,150	
	0210	6′ wide		2.25	3.556		1,175	126		1,301	1,500	
	0220	8′ wide		1.80	4.444		1,450	158		1,608	1,850	
	0300	Dome, 4′ wide		7.50	1.067		350	38		388	455	
	0310	6′ wide		3.50	2.286		790	81.50		871.50	1,000	
	0320	8′ wide		2	4		1,400	142		1,542	1,775	
	0350	Elongated dome, 4′ wide		1.33	6.015		1,325	214		1,539	1,800	
	0360	6′ wide		1.11	7.207		1,575	256		1,831	2,175	
	0370	8′ wide		1	8		1,850	284		2,134	2,500	
	1000	Entry or walkway, peak, 12′ long, 4′ wide	2 Carp	.90	17.778		4,175	630		4,805	5,625	
	1010	6′ wide		.60	26.667		6,425	950		7,375	8,650	
	1020	8′ wide		.40	40		8,875	1,425		10,300	12,100	
	1100	Radius with dome end, 4′ wide		1.10	14.545		3,175	515		3,690	4,325	
	1110	6′ wide		.70	22.857		5,075	815		5,890	6,950	
	1120	8′ wide		.50	32		7,225	1,150		8,375	9,825	
	2000	Retractable lateral arm awning, manual										
	2010	To 12′ wide, 8′ - 6″ projection	2 Carp	1.70	9.412	Ea.	940	335		1,275	1,575	
	2020	To 14′ wide, 8′ - 6″ projection		1.10	14.545		1,100	515		1,615	2,050	
	2030	To 19′ wide, 8′ - 6″ projection		.85	18.824		1,500	670		2,170	2,750	
	2040	To 24′ wide, 8′ - 6″ projection		.67	23.881		1,875	850		2,725	3,475	
	2050	Motor for above, add	1 Carp	2.67	3		810	107		917	1,075	
	3000	Patio/deck canopy with frame										
	3010	12′ wide, 12′ projection	2 Carp	2	8	Ea.	1,325	284		1,609	1,950	
	3020	16′ wide, 14′ projection	″	1.20	13.333		2,075	475		2,550	3,050	
	9000	For fire retardant canvas, add					7%					
	9010	For lettering or graphics, add					35%					
	9020	For painted or coated acrylic canvas, deduct					8%					
	9030	For translucent or opaque vinyl canvas, add					10%					
	9040	For 6 or more units, deduct					20%	15%				
100	0010	**CANOPIES, RESIDENTIAL** Prefabricated										100
	0500	Carport, free standing, baked enamel, alum., .032″, 40 psf										
	0520	16′ x 8′, 4 posts	2 Carp	3	5.333	Ea.	3,425	190		3,615	4,075	
	0600	20′ x 10′, 6 posts		2	8		3,575	284		3,859	4,400	
	0605	30′ x 10′, 8 posts		2	8		5,350	284		5,634	6,375	
	1000	Door canopies, extruded alum., .032″, 42″ projection, 4′ wide	1 Carp	8	1		400	35.50		435.50	500	
	1020	6′ wide	″	6	1.333		485	47.50		532.50	615	
	1040	8′ wide	2 Carp	9	1.778		625	63		688	795	
	1060	10′ wide		7	2.286		730	81.50		811.50	940	
	1080	12′ wide		5	3.200		865	114		979	1,150	
	1200	54″ projection, 4′ wide	1 Carp	8	1		515	35.50		550.50	625	
	1220	6′ wide	″	6	1.333		650	47.50		697.50	795	
	1240	8′ wide	2 Carp	9	1.778		865	63		928	1,050	
	1260	10′ wide		7	2.286		975	81.50		1,056.50	1,200	
	1280	12′ wide		5	3.200		1,100	114		1,214	1,425	
	1300	Painted, add					20%					
	1310	Bronze anodized, add					50%					
	3000	Window awnings, aluminum, window 3′ high, 4′ wide	1 Carp	10	.800		235	28.50		263.50	305	
	3020	6′ wide	″	8	1		274	35.50		309.50	360	
	3040	9′ wide	2 Carp	9	1.778		445	63		508	595	
	3060	12′ wide	″	5	3.200		610	114		724	860	
	3100	Window, 4′ high, 4′ wide	1 Carp	10	.800		288	28.50		316.50	360	
	3120	6′ wide	″	8	1		385	35.50		420.50	485	
	3140	9′ wide	2 Carp	9	1.778		525	63		588	680	
	3160	12′ wide	″	5	3.200		675	114		789	935	
	3200	Window, 6′ high, 4′ wide	1 Carp	10	.800		435	28.50		463.50	520	

10530 | Protective Covers

10535 | Awnings & Canopies

			CREW	DAILY OUTPUT	LABOR-HOURS	UNIT	2006 BARE COSTS MAT.	LABOR	EQUIP.	TOTAL	TOTAL INCL O&P	
100	3220	6′ wide	1 Carp	8	1	Ea.	600	35.50		635.50	720	100
	3240	9′ wide	2 Carp	9	1.778		820	63		883	1,000	
	3260	12′ wide	"	5	3.200		1,125	114		1,239	1,450	
	3400	Roll-up aluminum, 2′-6″ wide	1 Carp	14	.571		101	20.50		121.50	145	
	3420	3′ wide		12	.667		121	23.50		144.50	173	
	3440	4′ wide		10	.800		156	28.50		184.50	218	
	3460	6′ wide		8	1		192	35.50		227.50	270	
	3480	9′ wide	2 Carp	9	1.778		279	63		342	410	
	3500	12′ wide	"	5	3.200		345	114		459	565	
	3600	Window awnings, canvas, 24″ drop, 3′ wide	1 Carp	30	.267	L.F.	36.50	9.50		46	56	
	3620	4′ wide		40	.200		33	7.10		40.10	48.50	
	3700	30″ drop, 3′ wide		30	.267		51.50	9.50		61	72	
	3720	4′ wide		40	.200		44.50	7.10		51.60	60.50	
	3740	5′ wide		45	.178		40	6.30		46.30	54.50	
	3760	6′ wide		48	.167		37	5.95		42.95	50.50	
	3780	8′ wide		48	.167		30.50	5.95		36.45	43.50	
	3800	10′ wide		50	.160		28	5.70		33.70	40.50	
	3900	Repair canvas, minimum		8	1	Ea.		35.50		35.50	59	
	3920	Maximum		4	2	"		71		71	118	
	9000	Minimum labor/equipment charge		3	2.667	Job		95		95	157	
200	0010	CANOPIES										200
	0020	Wall hung, .032″, aluminum, prefinished, 8′ x 10′	K-2	1.30	18.462	Ea.	1,900	675	136	2,711	3,450	
	0300	8′ x 20′		1.10	21.818		3,775	795	160	4,730	5,775	
	1000	12′ x 20′		1	24		4,825	875	176	5,876	7,100	
	1050											
	2300	Aluminum entrance canopies, flat soffit, .032″										
	2500	3′-6″ x 4′-0″, clear anodized	2 Carp	4	4	Ea.	795	142		937	1,100	
	9000	Minimum labor/equipment charge	"	2	8	Job		284		284	470	

10550 | Postal Specialties

10555 | Mail Delivery Systems

			CREW	DAILY OUTPUT	LABOR-HOURS	UNIT	2006 BARE COSTS MAT.	LABOR	EQUIP.	TOTAL	TOTAL INCL O&P	
600	0010	MAIL BOXES										600
	0020	Horiz., key lock, 5″H x 6″W x 15″D, alum., rear load	1 Carp	34	.235	Ea.	33	8.35		41.35	50	
	0100	Front loading		34	.235		37	8.35		45.35	54.50	
	0200	Double, 5″H x 12″W x 15″D, rear loading		26	.308		60	10.95		70.95	84	
	0300	Front loading		26	.308		63	10.95		73.95	87.50	
	0500	Quadruple, 10″H x 12″W x 15″D, rear loading		20	.400		112	14.20		126.20	147	
	0600	Front loading		20	.400		114	14.20		128.20	149	
	1600	Vault type, horizontal, for apartments, 4″ x 5″		34	.235		36.50	8.35		44.85	54	
	1700	Alphabetical directories, 120 names		10	.800		114	28.50		142.50	173	
	1900	Letter slot, residential		20	.400		60	14.20		74.20	89.50	
	2000	Post office type		8	1		220	35.50		255.50	300	
	9000	Minimum labor/equipment charge		5	1.600	Job		57		57	94	
700	0010	MAIL CHUTES										700
	0020	Aluminum & glass, 14-1/4″ wide, 4-5/8″ deep	2 Shee	4	4	Floor	650	169		819	985	
	0100	8-5/8″ deep		3.80	4.211	"	790	177		967	1,150	
	0600	Lobby collection boxes, aluminum		5	3.200	Ea.	1,800	135		1,935	2,200	
	0700	Bronze or stainless		4.50	3.556	"	2,300	150		2,450	2,775	

10600 | Partitions

10610 | Folding Gates

			CREW	DAILY OUTPUT	LABOR-HOURS	UNIT	2006 BARE COSTS MAT.	LABOR	EQUIP.	TOTAL	TOTAL INCL O&P	
100	0010	**SECURITY GATES** For roll up type, see division 08330-130										100
	0300	Scissors type folding gate, ptd. steel, single, 6-1/2' high, 5-1/2'wide	2 Sswk	4	4	Opng.	142	160		302	455	
	0350	6-1/2' wide		4	4		138	160		298	450	
	0400	7-1/2' wide		4	4		166	160		326	480	
	0600	Double gate, 8' high, 8' wide		2.50	6.400		217	256		473	720	
	0650	10' wide		2.50	6.400		246	256		502	750	
	0700	12' wide		2	8		340	320		660	975	
	0750	14' wide		2	8		390	320		710	1,025	
	0900	Door gate, folding steel, 4' wide, 61" high		4	4		72.50	160		232.50	380	
	1000	71" high		4	4		75	160		235	380	
	1200	81" high		4	4		81.50	160		241.50	390	
	1300	Window gates, 2' to 4' wide, 31" high		4	4		43	160		203	345	
	1500	55" high		3.75	4.267		86	170		256	415	
	1600	79" high	↓	3.50	4.571	↓	99.50	183		282.50	450	

10615 | Demountable Partitions

			CREW	DAILY OUTPUT	LABOR-HOURS	UNIT	MAT.	LABOR	EQUIP.	TOTAL	TOTAL INCL O&P	
100	0010	**PARTITIONS, MOVABLE OFFICE** Demountable, add for doors										100
	0100	Do not deduct door openings from total L.F.										
	0900	Demountable gypsum system on 2" to 2-1/2"										
	1000	steel studs, 9' high, 3" to 3-3/4" thick										
	1200	Vinyl clad gypsum	2 Carp	48	.333	L.F.	48	11.85		59.85	72	
	1300	Fabric clad gypsum		44	.364		119	12.95		131.95	153	
	1500	Steel clad gypsum	↓	40	.400	↓	129	14.20		143.20	166	
	1600	1.75 system, aluminum framing, vinyl clad hardboard,										
	1800	paper honeycomb core panel, 1-3/4" to 2-1/2" thick										
	1900	9' high	2 Carp	48	.333	L.F.	80.50	11.85		92.35	108	
	2100	7' high		60	.267		72	9.50		81.50	94.50	
	2200	5' high	↓	80	.200	↓	61	7.10		68.10	79	
	2250	Unitized gypsum system										
	2300	Unitized panel, 9' high, 2" to 2-1/2" thick										
	2350	Vinyl clad gypsum	2 Carp	48	.333	L.F.	103	11.85		114.85	134	
	2400	Fabric clad gypsum	"	44	.364	"	170	12.95		182.95	209	
	2500	Unitized mineral fiber system										
	2510	Unitized panel, 9' high, 2-1/4" thick, aluminum frame										
	2550	Vinyl clad mineral fiber	2 Carp	48	.333	L.F.	103	11.85		114.85	133	
	2600	Fabric clad mineral fiber	"	44	.364	"	153	12.95		165.95	191	
	2800	Movable steel walls, modular system										
	2900	Unitized panels, 9' high, 48" wide										
	3100	Baked enamel, pre-finished	2 Carp	60	.267	L.F.	117	9.50		126.50	144	
	3200	Fabric clad steel	"	56	.286	"	168	10.15		178.15	202	
	5500	For acoustical partitions, add, minimum				S.F.	1.88			1.88	2.07	
	5550	Maximum				"	8.75			8.75	9.65	
	5700	For doors, see Div. 08100 & 08200										
	5800	For door hardware, see Div. 08700										
	6100	In-plant modular office system, w/prehung hollow core door										
	6200	3" thick polystyrene core panels										
	6250	12' x 12', 2 wall	2 Clab	3.80	4.211	Ea.	2,675	115		2,790	3,150	
	6300	4 wall		1.90	8.421		3,725	231		3,956	4,475	
	6350	16' x 16', 2 wall		3.60	4.444		4,175	122		4,297	4,800	
	6400	4 wall	↓	1.80	8.889	↓	5,375	244		5,619	6,300	
	9000	Minimum labor/equipment charge	2 Carp	3	5.333	Job		190		190	315	

10630 | Port. Partitions/Screens/Panels

			CREW	DAILY OUTPUT	LABOR-HOURS	UNIT	MAT.	LABOR	EQUIP.	TOTAL	TOTAL INCL O&P	
100	0010	**PARTITIONS, PORTABLE** Divider panels, free standing, fiber core										100
	0020	Fabric face straight										

10 SPECIALTIES

10600 | Partitions

	10630 | Port. Partitions/Screens/Panels		CREW	DAILY OUTPUT	LABOR-HOURS	UNIT	2006 BARE COSTS MAT.	LABOR	EQUIP.	TOTAL	TOTAL INCL O&P	
100	0100	3'-0" long, 4'-0" high	2 Carp	100	.160	L.F.	106	5.70		111.70	126	100
	0200	5'-0" high		90	.178		107	6.30		113.30	128	
	0500	6'-0" high		75	.213		122	7.60		129.60	147	
	0900	5'-0" long, 4'-0" high		175	.091		80.50	3.25		83.75	94	
	1000	5'-0" high		150	.107		98.50	3.79		102.29	115	
	1500	6"-0" high		125	.128		94.50	4.55		99.05	112	
	1600	6'-0" long, 5'-0" high		162	.099		79	3.51		82.51	93	
	3100	Curved, 3'-0" long, 5'-0" high		90	.178		83	6.30		89.30	102	
	3150	6'-0" high		75	.213		95.50	7.60		103.10	118	
	3200	Economical panels, fabric face, 4'-0" long, 5'-0" high		132	.121		39.50	4.31		43.81	50.50	
	3250	6'-0" high		112	.143		42.50	5.10		47.60	55	
	3300	5'-0" long, 5'-0" high		150	.107		34.50	3.79		38.29	44.50	
	3350	6'-0" high		125	.128		36	4.55		40.55	47	
	3380	3'-0" curved, 5'-0" high		90	.178		83	6.30		89.30	102	
	3390	6'-0" high		75	.213		95.50	7.60		103.10	118	
	3450	Acoustical panels, 60 to 90 NRC, 3'-0" long, 5'-0" high		90	.178		95.50	6.30		101.80	115	
	3550	6'-0" high		75	.213		91.50	7.60		99.10	114	
	3600	5'-0" long, 5'-0" high		150	.107		66.50	3.79		70.29	79.50	
	3650	6'-0" high		125	.128		73.50	4.55		78.05	88.50	
	3700	6'-0" long, 5'-0" high		162	.099		64.50	3.51		68.01	76.50	
	3750	6'-0" high		138	.116		64.50	4.12		68.62	78	
	3800	Economy acoustical panels, 40 NRC, 4'-0" long, 5'-0" high		132	.121		39.50	4.31		43.81	50.50	
	3850	6'-0" high		112	.143		42	5.10		47.10	54.50	
	3900	5'-0" long, 6'-0" high		125	.128		36	4.55		40.55	47	
	3950	6'-0" long, 5'-0" high		162	.099		29.50	3.51		33.01	38.50	
	9000	Minimum labor/equipment charge		3	5.333	Job		190		190	315	
	10651 | Accordion Folding Partitions											
100	0010	**PARTITIONS, FOLDING ACCORDION**										100
	0100	Vinyl covered, over 150 S.F., frame not included										
	0300	Residential, 1.25 lb. per S.F., 8' maximum height	2 Carp	300	.053	S.F.	17.35	1.90		19.25	22	
	0400	Commercial, 1.75 lb. per S.F., 8' maximum height		225	.071		19.85	2.53		22.38	26	
	0900	Acoustical, 3 lb. per S.F., 17' maximum height		100	.160		23	5.70		28.70	35	
	1200	5 lb. per S.F., 20' maximum height		95	.168		32	6		38	45	
	1500	Vinyl clad wood or steel, electric operation, 5.0 psf		160	.100		43.50	3.56		47.06	53.50	
	1900	Wood, non-acoustic, birch or mahogany, to 10' high		300	.053		23.50	1.90		25.40	28.50	
	9000	Minimum labor/equipment charge		4	4	Job		142		142	236	
	10653 | Folding Panel Partitions											
200	0010	**PARTITIONS, FOLDING LEAF** Acoustic, wood										200
	0100	Vinyl faced, to 18' high, 6 psf, minimum	2 Carp	60	.267	S.F.	44	9.50		53.50	64	
	0150	Average		45	.356		52.50	12.65		65.15	79	
	0200	Maximum		30	.533		68	18.95		86.95	107	
	0400	Formica or hardwood finish, minimum		60	.267		45.50	9.50		55	65.50	
	0500	Maximum		30	.533		48.50	18.95		67.45	85	
	0600	Wood, low acoustical type, 4.5 psf, to 14' high		50	.320		33	11.40		44.40	55.50	
	9000	Minimum labor/equipment charge		4	4	Job		142		142	236	

10670 | Storage Shelving

	10674 | Storage Shelving	CREW	DAILY OUTPUT	LABOR-HOURS	UNIT	2006 BARE COSTS MAT.	LABOR	EQUIP.	TOTAL	TOTAL INCL O&P	
500	0010 SHELVING										500
	0020 Metal, industrial, cross-braced, 3′ wide, 12″ deep	1 Sswk	175	.046	SF Shlf	9.65	1.83		11.48	14	
	0100 24″ deep		330	.024		6.90	.97		7.87	9.40	
	2200 Wide span, 1600 lb. capacity per shelf, 6′ wide, 24″ deep		380	.021		7.25	.84		8.09	9.55	
	2400 36″ deep	↓	440	.018	↓	6.55	.73		7.28	8.55	
	9000 Minimum labor/equipment charge	1 Carp	4	2	Job		71		71	118	

10750 | Telephone Specialties

	10755 | Telephone Enclosures	CREW	DAILY OUTPUT	LABOR-HOURS	UNIT	2006 BARE COSTS MAT.	LABOR	EQUIP.	TOTAL	TOTAL INCL O&P	
400	0010 TELEPHONE ENCLOSURE										400
	0300 Shelf type, wall hung, minimum	2 Carp	5	3.200	Ea.	1,025	114		1,139	1,325	
	0400 Maximum		5	3.200		2,625	114		2,739	3,100	
	0600 Booth type, painted steel, indoor or outdoor, minimum		1.50	10.667		3,200	380		3,580	4,150	
	0700 Maximum (stainless steel)		1.50	10.667		10,700	380		11,080	12,400	
	1900 Outdoor, drive-up type, wall mounted		4	4		860	142		1,002	1,175	
	2000 Post mounted, stainless steel posts		3	5.333	↓	1,325	190		1,515	1,775	
	9000 Minimum labor/equipment charge	↓	4	4	Job		142		142	236	

10 SPECIALTIES

10800 | Toilet/Bath/Laundry Accessories

	10810 | Toilet Accessories	CREW	DAILY OUTPUT	LABOR-HOURS	UNIT	2006 BARE COSTS MAT.	LABOR	EQUIP.	TOTAL	TOTAL INCL O&P	
100	0010 COMMERCIAL TOILET ACCESSORIES										100
	0200 Curtain rod, stainless steel, 5′ long, 1″ diameter	1 Carp	13	.615	Ea.	30.50	22		52.50	70	
	0300 1-1/4″ diameter		13	.615		30	22		52	70	
	0400 Diaper changing station, horizontal, wall mounted, plastic	↓	10	.800	↓	172	28.50		200.50	236	
	0500 Dispenser units, combined soap & towel dispensers,										
	0510 mirror and shelf, flush mounted	1 Carp	10	.800	Ea.	350	28.50		378.50	430	
	0600 Towel dispenser and waste receptacle,										
	0610 18 gallon capacity	1 Carp	10	.800	Ea.	269	28.50		297.50	345	
	0800 Grab bar, straight, 1-1/4″ diameter, stainless steel, 18″ long ♿		24	.333		20	11.85		31.85	41.50	
	0900 24″ long		23	.348		19.80	12.35		32.15	42.50	
	1000 30″ long		22	.364		22	12.95		34.95	46	
	1100 36″ long		20	.400		21	14.20		35.20	46.50	
	1105 42″ long		20	.400		21.50	14.20		35.70	47	
	1200 1-1/2″ diameter, 24″ long		23	.348		52	12.35		64.35	77.50	
	1300 36″ long		20	.400		57	14.20		71.20	86	
	1310 42″ long		18	.444		48	15.80		63.80	79	
	1500 Tub bar, 1-1/4″ diameter, 24″ x 36″		14	.571		89.50	20.50		110	132	
	1600 Plus vertical arm		12	.667		90.50	23.50		114	139	
	1900 End tub bar, 1″ diameter, 90° angle, 16″ x 32″		12	.667		124	23.50		147.50	177	
	2010 Tub/shower/toilet, 2-wall, 36″ x 24″		12	.667		80	23.50		103.50	128	
	2300 Hand dryer, surface mounted, electric, 115 volt, 20 amp		4	2		510	71		581	680	
	2400 230 volt, 10 amp	↓	4	2	↓	510	71		581	680	

10800 | Toilet/Bath/Laundry Accessories

	10810 Toilet Accessories		Crew	Daily Output	Labor-Hours	Unit	2006 Bare Costs Mat.	Labor	Equip.	Total	Total Incl O&P	
100	2600	Hat and coat strip, stainless steel, 4 hook, 36" long	1 Carp	24	.333	Ea.	50.50	11.85		62.35	75	100
	2700	6 hook, 60" long		20	.400		88	14.20		102.20	121	
	3000	Mirror, with stainless steel 3/4" square frame, 18" x 24"		20	.400		66	14.20		80.20	96	
	3100	36" x 24"		15	.533		147	18.95		165.95	194	
	3200	48" x 24"		10	.800		193	28.50		221.50	259	
	3300	72" x 24"		6	1.333		196	47.50		243.50	295	
	3500	With 5" stainless steel shelf, 18" x 24"		20	.400		157	14.20		171.20	196	
	3600	36" x 24"		15	.533		199	18.95		217.95	251	
	3700	48" x 24"		10	.800		229	28.50		257.50	299	
	3800	72" x 24"		6	1.333		340	47.50		387.50	455	
	4100	Mop holder strip, stainless steel, 5 holders, 48" long		20	.400		86.50	14.20		100.70	119	
	4200	Napkin/tampon dispenser, recessed		15	.533		330	18.95		348.95	390	
	4300	Robe hook, single, regular		36	.222		5.30	7.90		13.20	18.95	
	4400	Heavy duty, concealed mounting		36	.222		12.15	7.90		20.05	26.50	
	4600	Soap dispenser, chrome, surface mounted, liquid		20	.400		47	14.20		61.20	75	
	4700	Powder		20	.400		44	14.20		58.20	72	
	5000	Recessed stainless steel, liquid		10	.800		140	28.50		168.50	201	
	5100	Powder		10	.800		183	28.50		211.50	248	
	5300	Soap tank, stainless steel, 1 gallon		10	.800		167	28.50		195.50	231	
	5400	5 gallon		5	1.600		230	57		287	345	
	5600	Shelf, stainless steel, 5" wide, 18 ga., 24" long		24	.333		45.50	11.85		57.35	69.50	
	5700	48" long		16	.500		89.50	17.80		107.30	128	
	5800	8" wide shelf, 18 ga., 24" long		22	.364		53.50	12.95		66.45	80	
	5900	48" long		14	.571		114	20.50		134.50	159	
	6000	Toilet seat cover dispenser, stainless steel, recessed		20	.400		111	14.20		125.20	146	
	6050	Surface mounted		15	.533		27.50	18.95		46.45	61.50	
	6100	Toilet tissue dispenser, surface mounted, SS, single roll		30	.267		11.70	9.50		21.20	28.50	
	6200	Double roll		24	.333		16.30	11.85		28.15	37.50	
	6290	Toilet seat	1 Plum	40	.200		19.85	8.55		28.40	35.50	
	6400	Towel bar, stainless steel, 18" long	1 Carp	23	.348		35	12.35		47.35	59	
	6500	30" long		21	.381		71.50	13.55		85.05	101	
	6700	Towel dispenser, stainless steel, surface mounted		16	.500		37.50	17.80		55.30	71	
	6800	Flush mounted, recessed		10	.800		209	28.50		237.50	277	
	7000	Towel holder, hotel type, 2 guest size		20	.400		16.15	14.20		30.35	41.50	
	7200	Towel shelf, stainless steel, 24" long, 8" wide		20	.400		62	14.20		76.20	91.50	
	7400	Tumbler holder, tumbler only		30	.267		26	9.50		35.50	44	
	7500	Soap, tumbler & toothbrush		30	.267		23.50	9.50		33	41.50	
	7700	Wall urn ash receiver, surface mount, 11" long		12	.667		117	23.50		140.50	169	
	8000	Waste receptacles, stainless steel, with top, 13 gallon		10	.800		231	28.50		259.50	300	
	8100	36 gallon		8	1		425	35.50		460.50	525	
	9000	Minimum labor/equipment charge		5	1.600	Job		57		57	94	

	10820 Bath Accessories		Crew	Daily Output	Labor-Hours	Unit	Mat.	Labor	Equip.	Total	Total Incl O&P	
400	0010	**MEDICINE CABINETS**										400
	0020	With mirror, st. st. frame, 16" x 22", unlighted	1 Carp	14	.571	Ea.	73	20.50		93.50	114	
	0100	Wood frame		14	.571		101	20.50		121.50	145	
	0300	Sliding mirror doors, 20" x 16" x 4-3/4", unlighted		7	1.143		90.50	40.50		131	167	
	0400	24" x 19" x 8-1/2", lighted		5	1.600		143	57		200	251	
	0600	Triple door, 30" x 32", unlighted, plywood body		7	1.143		214	40.50		254.50	305	
	0700	Steel body		7	1.143		282	40.50		322.50	380	
	0900	Oak door, wood body, beveled mirror, single door		7	1.143		138	40.50		178.50	220	
	1000	Double door		6	1.333		335	47.50		382.50	450	
	1200	Hotel cabinets, stainless, with lower shelf, unlighted		10	.800		181	28.50		209.50	246	
	1300	Lighted		5	1.600		269	57		326	390	
	9000	Minimum labor/equipment charge		4	2	Job		71		71	118	

Division Notes

		CREW	DAILY OUTPUT	LABOR-HOURS	UNIT	2000 BARE COSTS MAT.	LABOR	EQUIP.	TOTAL	TOTAL INCL O&P

Division 11 Equipment

Estimating Tips

General

- The items in this division are usually priced per square foot or each. Many of these items are purchased by the owner for installation by the contractor. Check the specifications for responsibilities and include time for receiving, storage, installation, and mechanical and electrical hook-ups in the appropriate divisions.
- Many items in Division 11 require some type of support system that is not usually furnished with the item. Examples of these systems include blocking for the attachment of casework and support angles for ceiling-hung projection screens. The required blocking or supports must be added to the estimate in the appropriate division.
- Some items in Division 11 may require assembly or electrical hookups. Verify the amount of assembly required or the need for a hard electrical connection and add the appropriate costs.

Reference Numbers

Reference numbers are shown in bold squares at the beginning of some major classifications. These numbers refer to related items in the Reference Section. The reference information may be an estimating procedure, an alternate pricing method, or technical information.

Note: Not all subdivisions listed here necessarily appear in this publication.

11010 | Maintenance Equipment

	11013 Floor/Wall Cleaning Equipment		CREW	DAILY OUTPUT	LABOR-HOURS	UNIT	2006 BARE COSTS MAT.	LABOR	EQUIP.	TOTAL	TOTAL INCL O&P	
800	0010	VACUUM CLEANING										800
	0020	Central, 3 inlet, residential	1 Skwk	.90	8.889	Total	640	325		965	1,225	
	0200	Commercial		.70	11.429		1,200	415		1,615	2,000	
	0400	5 inlet system, residential		.50	16		970	585		1,555	2,025	
	0600	7 inlet system, commercial		.40	20		1,075	730		1,805	2,400	
	4010	Rule of thumb: First 1200 S.F., installed									1,180	
	4020	For each additional S.F., add				S.F.					.19	

11020 | Security & Vault Equipment

	11021 Safes and Vault Doors		CREW	DAILY OUTPUT	LABOR-HOURS	UNIT	2006 BARE COSTS MAT.	LABOR	EQUIP.	TOTAL	TOTAL INCL O&P	
600	0010	SAFE										600
	0015	Office, 4 hr. rating, 30″ x 18″ x 18″ inside				Ea.	3,275			3,275	3,600	
	0800	Money, "B" label, 9″ x 14″ x 14″					415			415	455	
	0900	Tool resistive, 24″ x 24″ x 20″					2,800			2,800	3,075	
	1050	Tool and torch resistive, 24″ x 24″ x 20″					6,875			6,875	7,550	

11030 | Teller & Service Equipment

	11038 Bank Equipment		CREW	DAILY OUTPUT	LABOR-HOURS	UNIT	2006 BARE COSTS MAT.	LABOR	EQUIP.	TOTAL	TOTAL INCL O&P	
150	0010	BANK EQUIPMENT										150
	0020	Alarm system, police	2 Elec	1.60	10	Ea.	4,250	420		4,670	5,325	
	0400	Bullet resistant teller window, 44″ x 60″	1 Glaz	.60	13.333		2,825	455		3,280	3,825	
	0500	48″ x 60″	"	.60	13.333		3,500	455		3,955	4,600	
	3000	Counters for banks, frontal only	2 Carp	1	16	Station	1,575	570		2,145	2,700	
	3100	Complete with steel undercounter		.50	32	"	3,100	1,150		4,250	5,275	
	5400	Partitions, bullet-resistant, 1-3/16″ glass, 8′ high		10	1.600	L.F.	171	57		228	282	
	5600	Pass thru, bullet-res. window, painted steel, 24″ x 36″	2 Sswk	1.60	10	Ea.	1,925	400		2,325	2,875	

11040 | Ecclesiastical Equipment

	11041 Ecclesiastical Equipment		CREW	DAILY OUTPUT	LABOR-HOURS	UNIT	2006 BARE COSTS MAT.	LABOR	EQUIP.	TOTAL	TOTAL INCL O&P	
250	0010	CHURCH EQUIPMENT										250
	0020	Altar, wood, custom design, plain	1 Carp	1.40	5.714	Ea.	1,925	203		2,128	2,425	
	0050	Deluxe	"	.20	40	"	9,225	1,425		10,650	12,600	
	0150	Baptistry, fiberglass, 3′-6″ deep, x 13′-7″ long,										

11040 | Ecclesiastical Equipment

	11041 | Ecclesiastical Equipment		CREW	DAILY OUTPUT	LABOR-HOURS	UNIT	2006 BARE COSTS MAT.	LABOR	EQUIP.	TOTAL	TOTAL INCL O&P	
250	0160	steps at both ends, incl. plumbing, minimum	L-8	1	20	Ea.	2,900	740		3,640	4,400	250
	0200	Maximum	″	.70	28.571	↓	5,675	1,050		6,725	7,975	
	0250	Add for filter, heater and lights				↓	1,250			1,250	1,375	
	1500	Pews, bench type, hardwood, minimum	1 Carp	20	.400	L.F.	67	14.20		81.20	97.50	
	1550	Maximum	″	15	.533	↓	133	18.95		151.95	179	
	1570	For kneeler, add				↓	16			16	17.60	
	4000	Steeples, translucent fiberglass, 30″ square, 15′ high	F-3	2	20	Ea.	3,775	720	320	4,815	5,675	
	4150	25′ high	″	1.80	22.222	↓	4,425	800	355	5,580	6,575	
	4600	Aluminum, baked finish, 14′ high, 16″ square					1,875			1,875	2,050	
	4640	35′ high, 8′ base					17,800			17,800	19,600	
	4680	152′ high, custom				↓	406,500			406,500	447,000	
	5000	Wall cross, aluminum, extruded, 2″ x 2″ section	1 Carp	34	.235	L.F.	36.50	8.35		44.85	54.50	
	5150	4″ x 4″ section	↓	29	.276	↓	52.50	9.80		62.30	74.50	
	5300	Bronze, extruded, 1″ x 2″ section	↓	31	.258	↓	72	9.15		81.15	94.50	
	5350	2-1/2″ x 2-1/2″ section	↓	34	.235	↓	109	8.35		117.35	134	

11050 | Library Equipment

	11051 | Library Equipment		CREW	DAILY OUTPUT	LABOR-HOURS	UNIT	2006 BARE COSTS MAT.	LABOR	EQUIP.	TOTAL	TOTAL INCL O&P	
400	0010	LIBRARY EQUIPMENT										400
	0020	Bookshelf, mtl, 90″ high, 10″ shelf, dbl face	1 Carp	11.50	.696	L.F.	144	24.50		168.50	199	
	0300	Single face	″	12	.667	″	160	23.50		183.50	216	
	0600	For 8″ shelving, subtract from above					10%					
	0800	For 42″ high with countertop, subtract from above					20%					
	2500	Carrels, hardwood, 36″ x 24″, minimum	1 Carp	5	1.600	Ea.	665	57		722	825	
	2650	Maximum	↓	4	2	↓	855	71		926	1,050	
	2660	Metal, minimum	↓	5	1.600	↓	235	57		292	350	
	2670	Maximum	↓	4	2	↓	665	71		736	850	

11060 | Theater & Stage Equipment

	11063 | Stage Equipment		CREW	DAILY OUTPUT	LABOR-HOURS	UNIT	2006 BARE COSTS MAT.	LABOR	EQUIP.	TOTAL	TOTAL INCL O&P	
600	0010	STAGE EQUIPMENT										600
	0050	Control boards with dimmers and breakers, minimum	1 Elec	1	8	Ea.	11,500	335		11,835	13,200	
	0150	Maximum	″	.20	40	″	106,000	1,675		107,675	119,000	
	0160											
	0500	Curtain track, straight, light duty	2 Carp	20	.800	L.F.	21.50	28.50		50	70.50	
	0700	Curved sections	↓	12	1.333	″	141	47.50		188.50	234	
	1000	Curtains, velour, medium weight	↓	600	.027	S.F.	6.85	.95		7.80	9.05	
	5000	Stages, portable with steps, folding legs, stock, 8″ high				SF Stg.	22.50			22.50	25	
	5100	16″ high				↓	22			22	24	
	5200	32″ high				↓	31			31	34	

11060 | Theater & Stage Equipment

	11063 | Stage Equipment		CREW	DAILY OUTPUT	LABOR-HOURS	UNIT	2006 BARE COSTS MAT.	LABOR	EQUIP.	TOTAL	TOTAL INCL O&P	
600	5300	40″ high				SF Stg.	51			51	56.50	600

11100 | Mercantile Equipment

	11102 | Barber Shop Equipment		CREW	DAILY OUTPUT	LABOR-HOURS	UNIT	2006 BARE COSTS MAT.	LABOR	EQUIP.	TOTAL	TOTAL INCL O&P	
150	0010	BARBER EQUIPMENT										150
	0020	Chair, hydraulic, movable, minimum	1 Carp	24	.333	Ea.	430	11.85		441.85	495	
	0050	Maximum	″	16	.500		2,750	17.80		2,767.80	3,050	
	0200	Wall hung styling station with mirrors, minimum	L-2	8	2		415	62		477	560	
	0300	Maximum	″	4	4		1,825	124		1,949	2,225	
	0500	Sink, hair washing basin, rough plumbing not incl.	1 Plum	8	1		283	42.50		325.50	375	
	1000	Sterilizer, liquid solution for tools					128			128	140	
	1100	Total equipment, rule of thumb, per chair, minimum	L-8	1	20		1,475	740		2,215	2,825	
	1150	Maximum	″	1	20		4,075	740		4,815	5,700	

	11103 | Cash Register/Checking		CREW	DAILY OUTPUT	LABOR-HOURS	UNIT	MAT.	LABOR	EQUIP.	TOTAL	TOTAL INCL O&P	
200	0010	CHECKOUT COUNTER										200
	0020	Supermarket conveyor, single belt	2 Clab	10	1.600	Ea.	2,300	44		2,344	2,600	
	0100	Double belt, power take-away		9	1.778		3,850	48.50		3,898.50	4,325	
	0400	Double belt, power take-away, incl. side scanning		7	2.286		4,800	62.50		4,862.50	5,375	
	0800	Warehouse or bulk type		6	2.667		5,625	73		5,698	6,300	
	1000	Scanning system, 2 lanes, w/registers, scan gun & memory				System	14,000			14,000	15,400	
	1100	10 lanes, single processor, full scan, with scales				″	133,500			133,500	147,000	
	2000	Register, restaurant, minimum				Ea.	570			570	625	
	2100	Maximum					2,500			2,500	2,750	
	2150	Store, minimum					570			570	625	
	2200	Maximum					2,500			2,500	2,750	

	11104 | Display Cases & Systems		CREW	DAILY OUTPUT	LABOR-HOURS	UNIT	MAT.	LABOR	EQUIP.	TOTAL	TOTAL INCL O&P	
700	0010	REFRIGERATED FOOD CASES										700
	0030	Dairy, multi-deck, 12′ long	Q-5	3	5.333	Ea.	8,600	207		8,807	9,775	
	0100	For rear sliding doors, add					1,225			1,225	1,350	
	0200	Delicatessen case, service deli, 12′ long, single deck	Q-5	3.90	4.103		5,775	159		5,934	6,625	
	0300	Multi-deck, 18 S.F. shelf display		3	5.333		7,075	207		7,282	8,125	
	0400	Freezer, self-contained, chest-type, 30 C.F.		3.90	4.103		4,225	159		4,384	4,900	
	0500	Glass door, upright, 78 C.F.		3.30	4.848		8,025	188		8,213	9,150	
	0600	Frozen food, chest type, 12′ long		3.30	4.848		5,800	188		5,988	6,700	
	0700	Glass door, reach-in, 5 door		3	5.333		11,100	207		11,307	12,500	
	0800	Island case, 12′ long, single deck		3.30	4.848		6,575	188		6,763	7,525	
	0900	Multi-deck		3	5.333		13,900	207		14,107	15,600	
	1000	Meat case, 12′ long, single deck		3.30	4.848		4,775	188		4,963	5,550	
	1050	Multi-deck		3.10	5.161		8,225	200		8,425	9,350	
	1100	Produce, 12′ long, single deck		3.30	4.848		6,325	188		6,513	7,250	
	1200	Multi-deck		3.10	5.161		7,050	200		7,250	8,050	

11110 | Commercial Laundry & Dry Cleaning Equipment

	11119 Laundry Cleaning		CREW	DAILY OUTPUT	LABOR-HOURS	UNIT	2006 BARE COSTS MAT.	LABOR	EQUIP.	TOTAL	TOTAL INCL O&P	
450	0010	**LAUNDRY EQUIPMENT** Not incl. rough-in										450
	0500	Dryers, gas fired residential, 16 lb. capacity, average	1 Plum	3	2.667	Ea.	650	114		764	890	
	1000	Commercial, 30 lb. capacity, coin operated, single		3	2.667		2,750	114		2,864	3,200	
	1100	Double stacked		2	4		5,825	171		5,996	6,675	
	1500	Industrial, 30 lb. capacity		2	4		2,575	171		2,746	3,100	
	1600	50 lb. capacity	↓	1.70	4.706		3,125	201		3,326	3,725	
	2000	Dry cleaners, electric, 20 lb. capacity	L-1	.20	80		32,800	3,400		36,200	41,400	
	2050	25 lb. capacity		.17	94.118		42,700	3,975		46,675	53,000	
	2100	30 lb. capacity		.15	106		45,000	4,525		49,525	56,500	
	2150	60 lb. capacity	↓	.09	177		70,500	7,525		78,025	89,000	
	3500	Folders, blankets & sheets, minimum	1 Elec	.17	47.059		29,900	1,975		31,875	36,000	
	3700	King size with automatic stacker		.10	80		51,000	3,350		54,350	61,000	
	3800	For conveyor delivery, add		.45	17.778		6,100	745		6,845	7,850	
	4500	Ironers, institutional, 110", single roll	↓	.20	40		27,500	1,675		29,175	32,900	
	4700	Lint collector, ductwork not included, 8,000 to 10,000 C.F.M.	Q-10	.30	80		8,025	3,150		11,175	13,800	
	5000	Washers, residential, 4 cycle, average	1 Plum	3	2.667		695	114		809	940	
	5300	Commercial, coin operated, average	"	3	2.667		1,125	114		1,239	1,400	
	6000	Combination washer/extractor, 20 lb. capacity	L-6	1.50	8		4,000	340		4,340	4,925	
	6100	30 lb. capacity		.80	15		7,775	635		8,410	9,525	
	6200	50 lb. capacity		.68	17.647		9,650	750		10,400	11,800	
	6300	75 lb. capacity		.30	40		19,800	1,700		21,500	24,400	
	6350	125 lb. capacity	↓	.16	75	↓	24,100	3,175		27,275	31,400	

11130 | Audio-Visual Equipment

	11136 Projection Screens		CREW	DAILY OUTPUT	LABOR-HOURS	UNIT	2006 BARE COSTS MAT.	LABOR	EQUIP.	TOTAL	TOTAL INCL O&P	
500	0010	**PROJECTION SCREENS** Wall or ceiling hung, matte white										500
	0100	Manually operated, economy	2 Carp	500	.032	S.F.	5.15	1.14		6.29	7.55	
	0300	Intermediate		450	.036		6	1.26		7.26	8.70	
	0400	Deluxe		400	.040	↓	8.30	1.42		9.72	11.50	
	9000	Minimum labor/equipment charge	↓	3	5.333	Job		190		190	315	
600	0011	**MOVIE EQUIPMENT**										600
	0020	Changeover, minimum				Ea.	410			410	450	
	3000	Projection screens, rigid, in wall, acrylic, 1/4" thick	2 Glaz	195	.082	S.F.	37	2.81		39.81	45	
	3100	1/2" thick	"	130	.123	"	43	4.22		47.22	54	
	3700	Sound systems, incl. amplifier, mono, minimum	1 Elec	.90	8.889	Ea.	2,900	375		3,275	3,775	
	3800	Dolby/Super Sound, maximum		.40	20		15,900	840		16,740	18,800	
	4100	Dual system, 2 channel, front surround, minimum		.70	11.429		4,075	480		4,555	5,225	
	4200	Dolby/Super Sound, 4 channel, maximum	↓	.40	20		14,600	840		15,440	17,300	
	5700	Seating, painted steel, upholstered, minimum	2 Carp	35	.457		116	16.25		132.25	155	
	5800	Maximum	"	28	.571	↓	370	20.50		390.50	445	

11140 | Vehicle Service Equipment

11141 | Service Station Equipment

			CREW	DAILY OUTPUT	LABOR-HOURS	UNIT	2006 BARE COSTS MAT.	LABOR	EQUIP.	TOTAL	TOTAL INCL O&P
100	0010	**COMPRESSED AIR EQUIPMENT**									
	0030	Compressors, electric, 1-1/2 H.P., standard controls	L-4	1.50	16	Ea.	281	530		811	1,175
	0550	Dual controls		1.50	16		475	530		1,005	1,400
	0600	5 H.P., 115/230 volt, standard controls		1	24		1,650	795		2,445	3,125
	0650	Dual controls	↓	1	24	↓	1,775	795		2,570	3,250
200	0010	**FUEL DISPENSING EQUIPMENT**									
	1100	Product dispenser with vapor recovery for 6 nozzles, installed, not									
	1110	including piping to storage tanks				Ea.	18,900			18,900	20,800
300	0010	**LUBRICATION EQUIPMENT**									
	3000	Lube equipment, 3 reel type, with pumps, not including piping	L-4	.50	48	Set	7,175	1,600		8,775	10,500
400	0010	**SPRAY PAINTING EQUIPMENT**									
	4000	Spray painting booth, 26′ long, complete	L-4	.40	60	Ea.	13,300	2,000		15,300	17,900

11150 | Parking Control Equipment

11156 | Parking Equipment

			CREW	DAILY OUTPUT	LABOR-HOURS	UNIT	2006 BARE COSTS MAT.	LABOR	EQUIP.	TOTAL	TOTAL INCL O&P
600	0010	**PARKING EQUIPMENT**									
	5000	Barrier gate with programmable controller	2 Elec	3	5.333	Ea.	3,200	224		3,424	3,875
	5020	Industrial	"	3	5.333		4,350	224		4,574	5,150
	5100	Card reader	1 Elec	2	4		1,750	168		1,918	2,175
	5120	Proximity with customer display	2 Elec	1	16		5,300	670		5,970	6,850
	5200	Cashier booth, average	B-22	1	30		9,225	975	229	10,429	12,100
	5300	Collector station, pay on foot	2 Elec	.20	80		105,500	3,350		108,850	121,000
	5320	Credit card only		.50	32		19,500	1,350		20,850	23,500
	5500	Exit verifier	↓	1	16		16,800	670		17,470	19,400
	5600	Fee computer	1 Elec	1.50	5.333		12,800	224		13,024	14,400
	5700	Full sign, 4″ letters	"	2	4		1,175	168		1,343	1,525
	5800	Inductive loop	2 Elec	4	4		162	168		330	435
	5900	Ticket spitter with time/date stamp, standard		2	8		6,075	335		6,410	7,225
	5920	Mag stripe encoding	↓	2	8		17,800	335		18,135	20,100
	5950	Vehicle detector, microprocessor based	1 Elec	3	2.667		375	112		487	585
	6000	Parking control software, minimum		.50	16		21,100	670		21,770	24,200
	6020	Maximum	↓	.20	40	↓	87,500	1,675		89,175	99,000

11160 | Loading Dock Equipment

11161 | Loading Dock Equipment

			CREW	DAILY OUTPUT	LABOR-HOURS	UNIT	2006 BARE COSTS MAT.	LABOR	EQUIP.	TOTAL	TOTAL INCL O&P
200	0010	**DOCK BUMPERS** Bolts not included									
	0012	2″ x 6″ to 4″ x 8″, average	1 Carp	300	.027	B.F.	.60	.95		1.55	2.23
400	0010	**LOADING DOCK**									
	0020	Bumpers, rubber blocks 4-1/2″ thk, 10″ H, 14″ long	1 Carp	26	.308	Ea.	44.50	10.95		55.45	67

11160 | Loading Dock Equipment

	11161 | Loading Dock Equipment		CREW	DAILY OUTPUT	LABOR-HOURS	UNIT	2006 BARE COSTS				TOTAL INCL O&P	
							MAT.	LABOR	EQUIP.	TOTAL		
400	0200	24″ long	1 Carp	22	.364	Ea.	84	12.95		96.95	114	400
	0300	36″ long		17	.471		85.50	16.75		102.25	122	
	0500	12″ high, 14″ long	↓	25	.320		88.50	11.40		99.90	116	
	2200	Dock boards, heavy duty, 60″ x 60″, aluminum, 5,000 lb. cap.					1,250			1,250	1,375	
	4200	Platform lifter, 6′ x 6′, portable, 3,000 lb. capacity					8,075			8,075	8,875	
	6200	Shelters, fabric, for truck or train, scissor arms, minimum	1 Carp	1	8		1,300	284		1,584	1,900	
	6300	Maximum	″	.50	16	↓	1,900	570		2,470	3,025	

11170 | Solid Waste Handling Equipment

	11179 | Waste Handling Equipment		CREW	DAILY OUTPUT	LABOR-HOURS	UNIT	2006 BARE COSTS				TOTAL INCL O&P	
							MAT.	LABOR	EQUIP.	TOTAL		
150	0010	WASTE HANDLING										150
	0020	Compactors, 115 volt, 250#/hr., chute fed	L-4	1	24	Ea.	9,950	795		10,745	12,200	
	0100	Hand fed		2.40	10		7,150	330		7,480	8,425	
	1000	Heavy duty industrial compactor, 0.5 C.Y. capacity		1	24		6,375	795		7,170	8,300	
	1050	1.0 C.Y. capacity	↓	1	24		9,800	795		10,595	12,100	
	1400	For handling hazardous waste materials, 55 gallon drum packer, std.					15,100			15,100	16,600	
	1410	55 gallon drum packer w/HEPA filter					18,800			18,800	20,700	
	1420	55 gallon drum packer w/charcoal & HEPA filter					25,100			25,100	27,600	
	1430	All of the above made explosion proof, add				↓	11,500			11,500	12,600	

11190 | Detention Equipment

	11191 | Detention Equipment		CREW	DAILY OUTPUT	LABOR-HOURS	UNIT	2006 BARE COSTS				TOTAL INCL O&P	
							MAT.	LABOR	EQUIP.	TOTAL		
150	0010	DETENTION EQUIPMENT										150
	0020											
	2000	Cells, prefab., 5′ to 6′ wide, 7′ to 8′ high, 7′ to 8′ deep,										
	2010	bar front, cot, not incl. plumbing	E-4	1.50	21.333	Ea.	8,000	865	59.50	8,924.50	10,500	

11400 | Food Service Equipment

	11405 | Food Storage Equipment		CREW	DAILY OUTPUT	LABOR-HOURS	UNIT	2006 BARE COSTS				TOTAL INCL O&P	
							MAT.	LABOR	EQUIP.	TOTAL		
110	0010	FOOD STORAGE EQUIPMENT										110
	2350	Cooler, reach-in, beverage, 6′ long	Q-1	6	2.667	Ea.	3,775	102		3,877	4,300	
	4300	Freezers, reach-in, 44 C.F.		4	4		8,625	154		8,779	9,725	
	4500	68 C.F.	↓	3	5.333	↓	9,825	205		10,030	11,100	

11405 | Food Storage Equipment

			CREW	DAILY OUTPUT	LABOR-HOURS	UNIT	2006 BARE COSTS MAT.	LABOR	EQUIP.	TOTAL	TOTAL INCL O&P	
110	4600	Freezer, pre-fab, 8′ x 8′ w/refrigeration	2 Carp	.45	35.556	Ea.	7,950	1,275		9,225	10,900	110
	4620	8′ x 12′		.35	45.714		9,775	1,625		11,400	13,500	
	4640	8′ x 16′		.25	64		12,300	2,275		14,575	17,300	
	4660	8′ x 20′		.17	94.118		15,000	3,350		18,350	22,200	
	4680	Reach-in, 1 compartment	Q-1	4	4		2,125	154		2,279	2,600	
	4700	2 compartment		3	5.333		4,325	205		4,530	5,075	
	8300	Refrigerators, reach-in type, 44 C.F.		5	3.200		5,200	123		5,323	5,925	
	8310	With glass doors, 68 C.F.		4	4		7,575	154		7,729	8,600	
	8320	Refrigerator, reach-in, 1 compartment	R-18	7.80	3.333		1,675	108		1,783	2,025	
	8330	2 compartment		6.20	4.194		2,200	136		2,336	2,650	
	8340	3 compartment		5.60	4.643		3,375	151		3,526	3,975	
	8350	Pre-fab, with refrigeration, 8′ x 8′	2 Carp	.45	35.556		5,275	1,275		6,550	7,900	
	8360	8′ x 12′		.35	45.714		6,800	1,625		8,425	10,200	
	8370	8′ x 16′		.25	64		8,650	2,275		10,925	13,300	
	8380	8′ x 20′		.17	94.118		10,600	3,350		13,950	17,300	
	8390	Pass-thru/roll-in, 1 compartment	R-18	7.80	3.333		3,400	108		3,508	3,925	
	8400	2 compartment		6.24	4.167		4,875	135		5,010	5,575	
	8410	3 compartment		5.60	4.643		6,175	151		6,326	7,050	
	8420	Walk-in, alum, door & floor only, no refrig, 6′ x 6′ x 7′-6″	2 Carp	1.40	11.429		6,700	405		7,105	8,050	
	8430	10′ x 6′ x 7′-6″		.55	29.091		9,600	1,025		10,625	12,300	
	8440	12′ x 14′ x 7′-6″		.25	64		13,300	2,275		15,575	18,500	
	8450	12′ x 20′ x 7′-6″		.17	94.118		16,400	3,350		19,750	23,600	
	8460	Refrigerated cabinets, mobile					2,825			2,825	3,100	
	8470	Refrigerator/freezer, reach-in, 1 compartment	R-18	5.60	4.643		4,350	151		4,501	5,025	
	8480	2 compartment	″	4.80	5.417		6,275	176		6,451	7,175	
	8600	Stainless steel shelving, louvered 4-tier, 20″ x 3′	1 Clab	6	1.333		975	36.50		1,011.50	1,125	
	8605	20″ x 4′		6	1.333		1,375	36.50		1,411.50	1,575	
	8610	20″ x 6′		6	1.333		2,000	36.50		2,036.50	2,250	
	8615	24″ x 3′		6	1.333		1,275	36.50		1,311.50	1,450	
	8620	24″ x 4′		6	1.333		1,475	36.50		1,511.50	1,675	
	8625	24″ x 6′		6	1.333		2,150	36.50		2,186.50	2,400	
	8630	Flat 4-tier, 20″ x 3′		6	1.333		1,125	36.50		1,161.50	1,300	
	8635	20″ x 4′		6	1.333		1,350	36.50		1,386.50	1,525	
	8640	20″ x 5′		6	1.333		1,550	36.50		1,586.50	1,750	
	8645	24″ x 3′		6	1.333		1,175	36.50		1,211.50	1,350	
	8650	24″ x 4′		6	1.333		1,400	36.50		1,436.50	1,575	
	8655	24″ x 6′		6	1.333		2,900	36.50		2,936.50	3,250	
	8700	Galvanized shelving, louvered 4-tier, 20″ x 3′		6	1.333		495	36.50		531.50	600	
	8705	20″ x 4′		6	1.333		550	36.50		586.50	665	
	8710	20″ x 6′		6	1.333		570	36.50		606.50	690	
	8715	24″ x 3′		6	1.333		450	36.50		486.50	555	
	8720	24″ x 4′		6	1.333		590	36.50		626.50	710	
	8725	24″ x 6′		6	1.333		845	36.50		881.50	990	
	8730	Flat 4-tier, 20″ x 3′		6	1.333		350	36.50		386.50	445	
	8735	20″ x 4′		6	1.333		395	36.50		431.50	495	
	8740	20″ x 6′		6	1.333		710	36.50		746.50	840	
	8745	24″ x 3′		6	1.333		340	36.50		376.50	435	
	8750	24″ x 4′		6	1.333		535	36.50		571.50	645	
	8755	24″ x 6′		6	1.333		755	36.50		791.50	890	
	8760	Stainless steel dunnage rack, 24″ x 3′		8	1		680	27.50		707.50	795	
	8765	24″ x 4′		8	1		705	27.50		732.50	820	
	8770	Galvanized dunnage rack, 24″ x 3′		8	1		111	27.50		138.50	169	
	8775	24″ x 4′		8	1		144	27.50		171.50	204	
800	0010	**WINE CELLAR**, refrigerated, Redwood interior, carpeted, walk-in type										800
	0020	6′-8″ high, including racks										

11400 | Food Service Equipment

11405 | Food Storage Equipment

			Crew	Daily Output	Labor-Hours	Unit	2006 Bare Costs Mat.	Labor	Equip.	Total	Total Incl O&P	
800	0200	80 "W x 48"D for 900 bottles	2 Carp	1.50	10.667	Ea.	3,125	380		3,505	4,050	800
	0250	80" W x 72" D for 1300 bottles		1.33	12.030		4,125	430		4,555	5,250	
	0300	80" W x 94" D for 1900 bottles		1.17	13.675		5,350	485		5,835	6,700	
	0400	80" W x 124" D for 2500 bottles		1	16		6,475	570		7,045	8,075	

11410 | Food Preparation Equipment

			Crew	Daily Output	Labor-Hours	Unit	Mat.	Labor	Equip.	Total	Total Incl O&P	
110	0010	FOOD PREPARATION EQUIPMENT										110
	1700	Choppers, 5 pounds	R-18	7	3.714	Ea.	2,400	120		2,520	2,850	
	1720	16 pounds		5	5.200		2,700	169		2,869	3,250	
	1740	35 to 40 pounds		4	6.500		3,000	211		3,211	3,625	
	1840	Coffee brewer, 5 burners	1 Plum	3	2.667		1,150	114		1,264	1,425	
	1850	Coffee urn, twin 6 gallon urns		2	4		2,650	171		2,821	3,175	
	1860	Single, 3 gallon		3	2.667		1,675	114		1,789	2,025	
	3000	Fast food equipment, total package, minimum	6 Skwk	.08	600		139,500	21,900		161,400	189,500	
	3100	Maximum	"	.07	685		190,000	25,000		215,000	250,000	
	3800	Food mixers, 20 quarts	L-7	7	4		3,150	137		3,287	3,700	
	3850	40 quarts		5.40	5.185		6,775	177		6,952	7,750	
	3900	60 quarts		5	5.600		9,875	191		10,066	11,200	
	4040	80 quarts		3.90	7.179		11,200	245		11,445	12,800	
	4080	130 quarts		2.20	12.727		17,200	435		17,635	19,600	
	4100	Floor type, 20 quarts		15	1.867		2,550	63.50		2,613.50	2,900	
	4120	60 quarts		14	2		7,225	68.50		7,293.50	8,025	
	4140	80 quarts		12	2.333		9,875	79.50		9,954.50	10,900	
	4160	140 quarts		8.60	3.256		22,200	111		22,311	24,600	
	6700	Peelers, small	R-18	8	3.250		1,150	105		1,255	1,450	
	6720	Large	"	6	4.333		3,650	141		3,791	4,250	
	6800	Pulper/extractor, close coupled, 5 HP	1 Plum	1.90	4.211		3,250	180		3,430	3,850	
	8580	Slicer with table	R-18	9	2.889		4,175	93.50		4,268.50	4,750	

11415 | Food Delivery Carts and Conveyors

			Crew	Daily Output	Labor-Hours	Unit	Mat.	Labor	Equip.	Total	Total Incl O&P	
110	0010	FOOD DELIVERY CARTS										110
	1650	Cabinet, heated, 1 compartment, reach-in	R-18	5.60	4.643	Ea.	2,325	151		2,476	2,825	
	1655	Pass-thru roll-in		5.60	4.643		3,975	151		4,126	4,600	
	1660	2 compartment, reach-in		4.80	5.417		5,925	176		6,101	6,775	
	1670	Mobile					2,625			2,625	2,875	
	6850	Mobile rack w/pan slide					985			985	1,075	
	9180	Tray and silver dispenser, mobile	1 Clab	16	.500		770	13.70		783.70	870	

11420 | Food Cooking Equipment

			Crew	Daily Output	Labor-Hours	Unit	Mat.	Labor	Equip.	Total	Total Incl O&P	
110	0010	COOKING EQUIPMENT										110
	0020	Bake oven, gas, one section	Q-1	8	2	Ea.	4,700	77		4,777	5,275	
	0300	Two sections		7	2.286		9,450	88		9,538	10,500	
	0600	Three sections		6	2.667		14,000	102		14,102	15,600	
	0900	Electric convection, single deck	L-7	4	7		4,875	239		5,114	5,750	
	1300	Broiler, without oven, standard	Q-1	8	2		2,950	77		3,027	3,375	
	1550	Infra-red	L-7	4	7		6,475	239		6,714	7,525	
	4750	Fryer, with twin baskets, modular model	Q-1	7	2.286		1,675	88		1,763	1,975	
	5000	Floor model, on 6" legs	"	5	3.200		1,575	123		1,698	1,950	
	5100	Extra single basket, large					94			94	103	
	5300	Griddle, SS, 24" plate, w/4" legs, elec, 208V, 3 phase, 3' long	Q-1	7	2.286		915	88		1,003	1,125	
	5550	4' long	"	6	2.667		1,225	102		1,327	1,500	
	6200	Iced tea brewer	1 Plum	3.44	2.326		625	99.50		724.50	845	
	6350	Kettle, w/steam jacket, tilting, w/positive lock, SS, 20 gallons	L-7	7	4		5,100	137		5,237	5,850	
	6600	60 gallons	"	6	4.667		7,250	159		7,409	8,225	
	6900	Range, restaurant type, 6 burners and 1 standard oven, 36" wide	Q-1	7	2.286		1,875	88		1,963	2,200	

11420 | Food Cooking Equipment

			CREW	DAILY OUTPUT	LABOR-HOURS	UNIT	2006 BARE COSTS MAT.	LABOR	EQUIP.	TOTAL	TOTAL INCL O&P	
110	6950	Convection	Q-1	7	2.286	Ea.	2,650	88		2,738	3,050	110
	7150	2 standard ovens, 24″ griddle, 60″ wide		6	2.667		4,650	102		4,752	5,250	
	7200	1 standard, 1 convection oven		6	2.667		5,625	102		5,727	6,350	
	7450	Heavy duty, single 34″ standard oven, open top		5	3.200		4,125	123		4,248	4,750	
	7500	Convection oven		5	3.200		5,675	123		5,798	6,450	
	7700	Griddle top		6	2.667		4,925	102		5,027	5,575	
	7750	Convection oven		6	2.667		6,500	102		6,602	7,300	
	8850	Steamer, electric 27 KW	L-7	7	4		8,975	137		9,112	10,100	
	9100	Electric, 10 KW or gas 100,000 BTU	″	5	5.600		4,325	191		4,516	5,075	
	9150	Toaster, conveyor type, 16-22 slices per minute					1,050			1,050	1,175	
	9160	Pop-up, 2 slot					575			575	630	
	9200	For deluxe models of above equipment, add					75%					
	9400	Rule of thumb: Equipment cost based										
	9410	on kitchen work area										
	9420	Office buildings, minimum	L-7	77	.364	S.F.	62	12.40		74.40	88.50	
	9450	Maximum		58	.483		105	16.50		121.50	142	
	9550	Public eating facilities, minimum		77	.364		81.50	12.40		93.90	110	
	9600	Maximum		46	.609		132	21		153	179	
	9750	Hospitals, minimum		58	.483		83.50	16.50		100	119	
	9800	Maximum		39	.718		140	24.50		164.50	194	

11425 | Hood and Ventilation Equipment

			CREW	DAILY OUTPUT	LABOR-HOURS	UNIT	MAT.	LABOR	EQUIP.	TOTAL	TOTAL INCL O&P	
110	0010	**HOOD & VENTILATION EQUIPMENT**										110
	2970	Exhaust hood, sst, gutter on all sides, 4′ x 4′ x 2′	1 Carp	1.80	4.444	Ea.	3,250	158		3,408	3,825	
	2980	4′x 4′ x 7′	″	1.60	5		5,050	178		5,228	5,850	
	7950	Hood fire protection system, minimum	Q-1	3	5.333		3,450	205		3,655	4,100	
	8050	Maximum	″	1	16		25,300	615		25,915	28,900	

11430 | Food Dispensing Equipment

			CREW	DAILY OUTPUT	LABOR-HOURS	UNIT	MAT.	LABOR	EQUIP.	TOTAL	TOTAL INCL O&P	
110	0010	**FOOD DISPENSING EQUIPMENT**										110
	1050	Butter pat dispenser	1 Clab	13	.615	Ea.	780	16.85		796.85	885	
	1100	Bread dispenser, counter top		13	.615		740	16.85		756.85	845	
	1900	Cup and glass dispenser, drop in		4	2		1,050	55		1,105	1,275	
	1920	Disposable cup, drop in		16	.500		320	13.70		333.70	380	
	2650	Dish dispenser, drop in, 12″		11	.727		1,200	19.95		1,219.95	1,350	
	2660	Mobile		10	.800		2,025	22		2,047	2,250	
	3300	Food warmer, counter, 1.2 KW					525			525	580	
	3550	1.6 KW					1,550			1,550	1,700	
	3600	Well, hot food, built-in, rectangular, 12″ x 20″	R-30	10	2.600		360	86		446	535	
	3610	Circular, 7 qt		10	2.600		325	86		411	500	
	3620	Refrigerated, 2 compartments		10	2.600		1,850	86		1,936	2,175	
	3630	3 compartments		9	2.889		1,850	96		1,946	2,200	
	3640	4 compartments		8	3.250		2,575	108		2,683	3,000	
	4720	Frost cold plate		9	2.889		12,700	96		12,796	14,200	
	5700	Hot chocolate dispenser	1 Plum	4	2		855	85.50		940.50	1,075	
	6250	Jet spray dispenser	R-18	4.50	5.778		2,650	187		2,837	3,200	
	6300	Juice dispenser, concentrate	″	4.50	5.778		925	187		1,112	1,325	
	6690	Milk dispenser, bulk, 2 flavor	R-30	8	3.250		1,175	108		1,283	1,475	
	6695	3 flavor	″	8	3.250		1,625	108		1,733	1,950	
	8800	Serving counter, straight	1 Carp	40	.200	L.F.	620	7.10		627.10	695	
	8820	Curved section	″	30	.267	″	775	9.50		784.50	870	
	8825	Solid surface, see section 06620-810										
	8830	Soft serve ice cream machine, medium	R-18	11	2.364	Ea.	8,300	76.50		8,376.50	9,275	
	8840	Large	″	9	2.889	″	12,300	93.50		12,393.50	13,700	

11400 | Food Service Equipment

11435 | Ice Machines

		11435 Ice Machines	CREW	DAILY OUTPUT	LABOR-HOURS	UNIT	2006 BARE COSTS MAT.	LABOR	EQUIP.	TOTAL	TOTAL INCL O&P	
110	0010	ICE MACHINES										110
	5800	Ice cube maker, 50 pounds per day	Q-1	6	2.667	Ea.	1,400	102		1,502	1,675	
	5900	250 pounds per day		1.20	13.333		1,850	510		2,360	2,825	
	6050	500 pounds per day		4	4		2,700	154		2,854	3,225	
	6060	With bin		1.20	13.333		2,200	510		2,710	3,225	
	6090	1000 pounds per day, with bin		1	16		6,625	615		7,240	8,225	
	6100	Ice flakers, 300 pounds per day		1.60	10		3,025	385		3,410	3,925	
	6120	600 pounds per day		.95	16.842		3,325	645		3,970	4,650	
	6130	1000 pounds per day		.75	21.333		4,950	820		5,770	6,700	
	6140	2000 pounds per day		.65	24.615		14,800	945		15,745	17,700	
	6160	Ice storage bin, 500 pound capacity	Q-5	1	16		1,800	620		2,420	2,950	
	6180	1000 pound	"	.56	28.571		3,150	1,100		4,250	5,175	

11440 | Cleaning and Disposal Equipment

		11440 Cleaning and Disposal Equipment	CREW	DAILY OUTPUT	LABOR-HOURS	UNIT	2006 BARE COSTS MAT.	LABOR	EQUIP.	TOTAL	TOTAL INCL O&P	
110	0010	CLEANING & DISPOSAL EQUIPMENT										110
	2700	Dishwasher, commercial, rack type										
	2720	10 to 12 racks per hour	Q-1	3.20	5	Ea.	3,875	192		4,067	4,575	
	2750	Semi-automatic 38 to 50 racks per hour	"	1.30	12.308		6,950	475		7,425	8,350	
	2800	Automatic, 190 to 230 racks per hour	L-6	.35	34.286		8,625	1,450		10,075	11,700	
	2820	235 to 275 racks per hour		.25	48		19,900	2,050		21,950	25,100	
	2840	8,750 to 12,500 dishes per hour		.10	120		43,200	5,100		48,300	55,500	
	2950	Dishwasher hood, canopy type	L-3A	10	1.200	L.F.	465	47		512	585	
	2960	Pant leg type	"	2.50	4.800	Ea.	6,525	188		6,713	7,475	
	5200	Garbage disposal 1.5 HP, 100 GPH	L-1	4.80	3.333		1,400	141		1,541	1,775	
	5210	3 HP, 120 GPH		4.60	3.478		2,250	147		2,397	2,700	
	5220	5 HP, 250 GPH		4.50	3.556		3,225	151		3,376	3,775	
	6750	Pot sink, 3 compartment	1 Plum	7.25	1.103	L.F.	610	47		657	750	
	6760	Pot washer, small		1.60	5	Ea.	16,400	214		16,614	18,400	
	6770	Large		1.20	6.667		36,900	285		37,185	41,000	
	9170	Trash compactor, small, up to 125 lb. compacted weight	L-4	4	6		17,500	199		17,699	19,500	
	9175	Large, up to 175 lb. compacted weight	"	3	8		21,300	265		21,565	23,900	

11450 | Residential Equipment

11454 | Residential Appliances

		11454 Residential Appliances	CREW	DAILY OUTPUT	LABOR-HOURS	UNIT	2006 BARE COSTS MAT.	LABOR	EQUIP.	TOTAL	TOTAL INCL O&P	
500	0010	RESIDENTIAL APPLIANCES										500
	0020	Cooking range, 30" free standing, 1 oven, minimum	2 Clab	10	1.600	Ea.	249	44		293	345	
	0050	Maximum	"	4	4		1,550	110		1,660	1,875	
	0350	Built-in, 30" wide, 1 oven, minimum	1 Elec	6	1.333		460	56		516	590	
	0400	Maximum	2 Carp	2	8		1,350	284		1,634	1,975	
	0900	Counter top cook tops, 4 burner, standard, minimum	1 Elec	6	1.333		192	56		248	298	
	0950	Maximum		3	2.667		535	112		647	765	
	1250	Microwave oven, minimum		4	2		85.50	84		169.50	223	
	1300	Maximum		2	4		405	168		573	705	
	1750	Compactor, residential size, 4 to 1 compaction, minimum	1 Carp	5	1.600		430	57		487	565	
	1800	Maximum	"	3	2.667		485	95		580	685	
	2750	Dishwasher, built-in, 2 cycles, minimum	L-1	4	4		256	169		425	545	
	2800	Maximum		2	8		299	340		639	855	
	3300	Garbage disposal, sink type, minimum		10	1.600		44	68		112	153	

11450 | Residential Equipment

	11454 | Residential Appliances		CREW	DAILY OUTPUT	LABOR-HOURS	UNIT	2006 BARE COSTS MAT.	LABOR	EQUIP.	TOTAL	TOTAL INCL O&P	
500	3350	Maximum	L-1	10	1.600	Ea.	150	68		218	270	500
	4150	Hood for range, 2 speed, vented, 30″ wide, minimum	L-3	5	3.200		38.50	124		162.50	243	
	4200	Maximum		3	5.333		615	207		822	1,025	
	4300	42″ wide, minimum		5	3.200		232	124		356	455	
	4330	Custom		5	3.200		645	124		769	910	
	4350	Maximum	↓	3	5.333		785	207		992	1,200	
	4400	Ventless hood, 2 speed, 30″ wide, minimum	1 Elec	8	1		39	42		81	108	
	4450	Maximum		7	1.143		134	48		182	221	
	4510	36″ wide, minimum		7	1.143		40.50	48		88.50	119	
	4550	Maximum		6	1.333		290	56		346	405	
	4580	42″ wide, minimum		6	1.333		208	56		264	315	
	4600	Maximum	↓	5	1.600		278	67		345	410	
	4650	For vented 1 speed, deduct from maximum				↓	40.50			40.50	44.50	
	4700											
	5380	Oven, built in, standard	1 Elec	4	2	Ea.	395	84		479	565	
	5390	Deluxe	″	2	4		1,975	168		2,143	2,425	
	5450	Refrigerator, no frost, 6 C.F.	2 Clab	15	1.067		295	29		324	375	
	5500	Refrigerator, no frost, 10 C.F. to 12 C.F. minimum		10	1.600		455	44		499	575	
	5600	Maximum	↓	6	2.667		705	73		778	895	
	6400	Sump pump cellar drainer, pedestal, 1/3 H.P., molded PVC base	1 Plum	3	2.667		88.50	114		202.50	275	
	6450	Solid brass	″	2	4	↓	185	171		356	470	
	6460	Sump pump, see also division 15440-940										
	7350	Water softener, automatic, to 30 grains per gallon	2 Plum	5	3.200	Ea.	445	137		582	700	
	7400	To 100 grains per gallon	″	4	4	″	620	171		791	945	
550	0010	**DISAPPEARING STAIRWAY** No trim included										550
	0020	One piece, yellow pine, 8′-0″ ceiling	2 Carp	4	4	Ea.	1,100	142		1,242	1,425	
	0030	9′-0″ ceiling		4	4		1,125	142		1,267	1,475	
	0040	10′-0″ ceiling		3	5.333		1,175	190		1,365	1,625	
	0050	11′-0″ ceiling		3	5.333		1,325	190		1,515	1,775	
	0060	12′-0″ ceiling	↓	3	5.333		1,425	190		1,615	1,875	
	0100	Custom grade, pine, 8′-6″ ceiling, minimum	1 Carp	4	2		117	71		188	247	
	0150	Average		3.50	2.286		118	81.50		199.50	265	
	0200	Maximum	↓	3	2.667	↓	195	95		290	370	
	0250											
	0500	Heavy duty, pivoted, from 7′-7″ to 12′-10″ floor to floor	1 Carp	3	2.667	Ea.	385	95		480	575	
	0600	16′-0″ ceiling		2	4		1,275	142		1,417	1,625	
	0800	Economy folding, pine, 8′-6″ ceiling		4	2		107	71		178	235	
	0900	9′-6″ ceiling	↓	4	2		116	71		187	246	
	1000	Fire escape, galvanized steel, 8′-0″ to 10′-4″ ceiling	2 Carp	1	16		1,350	570		1,920	2,450	
	1010	10′-6″ to 13′-6″ ceiling		1	16		1,700	570		2,270	2,825	
	1100	Automatic electric, aluminum, floor to floor height, 8′ to 9′		1	16		7,000	570		7,570	8,650	
	1500	11′ to 12′		.90	17.778		7,075	630		7,705	8,825	
	1700	14′ to 15′	↓	.70	22.857	↓	8,025	815		8,840	10,200	
	9000	Minimum labor/equipment charge	1 Carp	2	4	Job		142		142	236	

	11460 | Unit Kitchens		CREW	DAILY OUTPUT	LABOR-HOURS	UNIT	MAT.	LABOR	EQUIP.	TOTAL	TOTAL INCL O&P	
100	0010	**UNIT KITCHENS**										100
	1500	Combination range, refrigerator and sink, 30″ wide, minimum	L-1	2	8	Ea.	755	340		1,095	1,350	
	1550	Maximum	″	1	16	″	1,500	680		2,180	2,700	
	1640	Combination range, refrigerator, sink, microwave										
	1660	oven and ice maker	L-1	.80	20	Ea.	4,150	845		4,995	5,875	

11470 | Darkroom Equipment

11471 | Darkroom Processing

			CREW	DAILY OUTPUT	LABOR-HOURS	UNIT	2006 BARE COSTS MAT.	LABOR	EQUIP.	TOTAL	TOTAL INCL O&P	
700	0010	DARKROOM EQUIPMENT										700
	0020	Developing sink, 5″ deep, 24″ x 48″	Q-1	2	8	Ea.	4,450	305		4,755	5,350	
	0050	48″ x 52″		1.70	9.412		4,500	360		4,860	5,500	
	0200	10″ deep, 24″ x 48″		1.70	9.412		5,550	360		5,910	6,650	
	0250	24″ x 108″		1.50	10.667		2,000	410		2,410	2,825	
	0500	Dryers, dehumidified filtered air, 36″ x 25″ x 68″ high	L-7	6	4.667		4,325	159		4,484	5,025	
	0550	48″ x 25″ x 68″ high		5	5.600		7,975	191		8,166	9,100	
	2000	Processors, automatic, color print, minimum		4	7		11,200	239		11,439	12,700	
	2050	Maximum		.60	46.667		11,800	1,600		13,400	15,600	
	2300	Black and white print, minimum		2	14		8,500	480		8,980	10,100	
	2350	Maximum		.80	35		55,500	1,200		56,700	63,000	
	2600	Manual processor, 16″ x 20″ maximum print size		2	14		8,575	480		9,055	10,200	
	2650	20″ x 24″ maximum print size		1	28		8,075	955		9,030	10,500	
	3000	Viewing lites, 20″ x 24″		6	4.667		350	159		509	645	
	3100	20″ x 24″ with color correction		6	4.667		495	159		654	805	
	3500	Washers, round, minimum sheet 11″ x 14″	Q-1	2	8		2,725	305		3,030	3,475	
	3550	Maximum sheet 20″ x 24″		1	16		3,225	615		3,840	4,475	
	3800	Square, minimum sheet 20″ x 24″		1	16		2,875	615		3,490	4,125	
	3900	Maximum sheet 50″ x 56″		.80	20		4,475	770		5,245	6,125	
	4500	Combination tank sink, tray sink, washers, with										
	4510	dry side tables, average	Q-1	.45	35.556	Ea.	8,525	1,375		9,900	11,500	

11472 | Revolving Darkroom Doors

			CREW	DAILY OUTPUT	LABOR-HOURS	UNIT	MAT.	LABOR	EQUIP.	TOTAL	TOTAL INCL O&P	
370	0010	DARKROOM DOORS										370
	0015	Revolving, standard, 2 way, 36″ diameter	2 Carp	3.10	5.161	Opng.	1,850	183		2,033	2,350	
	0020	41″ diameter		3.10	5.161		2,000	183		2,183	2,500	
	0050	3 way, 51″ diameter		1.40	11.429		2,475	405		2,880	3,400	
	1000	4 way, 49″ diameter		1.40	11.429		2,950	405		3,355	3,900	
	2000	Hinged safety, 2 way, 41″ diameter		2.30	6.957		2,375	247		2,622	3,000	
	2500	3 way, 51″ diameter		1.40	11.429		2,975	405		3,380	3,950	
	3000	Pop out safety, 2 way, 41″ diameter		3.10	5.161		2,900	183		3,083	3,500	
	4000	3 way, 51″ diameter		1.40	11.429		2,900	405		3,305	3,875	
	5000	Wheelchair-type, pop out, 51″ diameter		1.40	11.429		3,000	405		3,405	3,975	
	5020	72″ diameter		.90	17.778		6,125	630		6,755	7,800	

11480 | Athletic Recreational & Therapeutic Equipment

11484 | Exercise Equipment

			CREW	DAILY OUTPUT	LABOR-HOURS	UNIT	2006 BARE COSTS MAT.	LABOR	EQUIP.	TOTAL	TOTAL INCL O&P	
400	0010	HEALTH CLUB EQUIPMENT										400
	0020	Abdominal rack, 2 board capacity				Ea.	435			435	480	
	0050	Abdominal board, upholstered					490			490	540	
	0200	Bicycle trainer, minimum					760			760	835	
	0300	Deluxe, electric					3,875			3,875	4,250	
	0400	Bar bell set, chrome plated steel, 25 lbs.					228			228	251	
	0420	100 lbs.					345			345	380	
	0450	200 lbs.					665			665	735	
	0500	Weight plates, cast iron, per lb.				Lb.	4.67			4.67	5.15	
	0520	Storage rack, 10 station				Ea.	735			735	810	

11480 | Athletic Recreational & Therapeutic Equipment

	11484 | Exercise Equipment		CREW	DAILY OUTPUT	LABOR-HOURS	UNIT	2006 BARE COSTS MAT.	LABOR	EQUIP.	TOTAL	TOTAL INCL O&P	
400	0600	Circuit training apparatus, 12 machines minimum	2 Clab	1.25	12.800	Set	25,700	350		26,050	28,900	400
	0700	Average		1	16		31,600	440		32,040	35,400	
	0800	Maximum	↓	.75	21.333		37,400	585		37,985	42,200	
	0820	Dumbbell set, cast iron, with rack and 5 pair				↓	720			720	790	
	0900	Squat racks	2 Clab	5	3.200	Ea.	715	87.50		802.50	930	
	1600	For saunas, see Div. 13035-800										
	1640	For steam baths, see Div. 13035-940										

	11486 | Gymnasium Equipment		CREW	DAILY OUTPUT	LABOR-HOURS	UNIT	MAT.	LABOR	EQUIP.	TOTAL	TOTAL INCL O&P	
700	0010	**SCHOOL EQUIPMENT**										700
	0200	For exterior equipment, see Div. 02880										
	0300	For chalkboards & bulletin boards, see Div. 10120-940 & 10110-240										
	0400	For lockers, see Div. 10505-500										
	1000	Basketball backstops, wall mtd., 6′ extended, fixed, minimum	L-2	1	16	Ea.	1,075	495		1,570	2,000	
	1100	Maximum		1	16		1,575	495		2,070	2,550	
	1200	Swing up, minimum		1	16		1,275	495		1,770	2,225	
	1250	Maximum		1	16		5,000	495		5,495	6,325	
	1300	Portable, manual, heavy duty, spring operated		1.90	8.421		10,600	261		10,861	12,100	
	1400	Ceiling suspended, stationary, minimum		.78	20.513		1,900	635		2,535	3,125	
	1450	Fold up, with accessories, maximum	↓	.40	40		5,700	1,250		6,950	8,325	
	1600	For electrically operated, add	1 Elec	1	8	↓	1,875	335		2,210	2,575	
	2000	Benches, folding, in wall, 14′ table, 2 benches	L-4	2	12	Set	600	400		1,000	1,300	
	3000	Bleachers, telescoping, manual to 15 tier, minimum	F-5	65	.492	Seat	71	17.75		88.75	108	
	3100	Maximum		60	.533		107	19.25		126.25	149	
	3300	16 to 20 tier, minimum		60	.533		170	19.25		189.25	219	
	3400	Maximum		55	.582		213	21		234	269	
	3600	21 to 30 tier, minimum		50	.640		177	23		200	233	
	3700	Maximum	↓	40	.800		213	29		242	282	
	3900	For integral power operation, add, minimum	2 Elec	300	.053		35.50	2.24		37.74	42.50	
	4000	Maximum	"	250	.064	↓	57	2.69		59.69	66.50	
	4100	Boxing ring, elevated, 22′ x 22′	L-4	.10	240	Ea.	8,075	7,950		16,025	21,900	
	4110	For cellular plastic foam padding, add		.10	240		2,550	7,950		10,500	15,800	
	4120	Floor level, including posts and ropes only, 20′ x 20′		.80	30		1,500	995		2,495	3,275	
	4130	Canvas, 30′ x 30′	↓	5	4.800		965	159		1,124	1,325	
	4150	Exercise equipment, bicycle trainer					405			405	445	
	4180	Chinning bar, adjustable, wall mounted	1 Carp	5	1.600		287	57		344	410	
	4200	Exercise ladder, 16′ x 1′-7″, suspended	L-2	3	5.333		1,250	165		1,415	1,650	
	4210	High bar, floor plate attached	1 Carp	4	2		1,075	71		1,146	1,325	
	4240	Parallel bars, adjustable		4	2		2,375	71		2,446	2,725	
	4270	Uneven parallel bars, adjustable	↓	4	2	↓	2,825	71		2,896	3,225	
	4280	Wall mounted, adjustable	L-2	1.50	10.667	Set	815	330		1,145	1,450	
	4300	Rope, ceiling mounted, 18′ long	1 Carp	3.66	2.186	Ea.	212	77.50		289.50	360	
	4330	Side horse, vaulting		5	1.600		1,700	57		1,757	1,975	
	4360	Treadmill, motorized, deluxe, training type	↓	5	1.600		5,450	57		5,507	6,075	
	4390	Weight lifting multi-station, minimum	2 Clab	1	16	↓	2,575	440		3,015	3,550	
	4500	Gym divider curtain, mesh top, vinyl bottom, manual	L-4	500	.048	S.F.	7.75	1.59		9.34	11.10	
	4700	Electric roll up	L-7	400	.070		11.30	2.39		13.69	16.35	
	5500	Gym mats, 2″ thick, naugahyde covered					3.53			3.53	3.88	
	5600	Vinyl/nylon covered					5.30			5.30	5.85	
	5800	Wall pads, 1-1/2″ thick	2 Carp	640	.025		5.80	.89		6.69	7.85	
	6000	Wrestling mats, 1″ thick, heavy duty				↓	4.73			4.73	5.20	
	7000	Scoreboards, baseball, minimum	R-3	1.30	15.385	Ea.	2,975	635	117	3,727	4,375	
	7200	Maximum		.05	400		14,600	16,600	3,050	34,250	45,100	
	7300	Football, minimum		.86	23.256		3,950	965	177	5,092	6,050	
	7400	Maximum	↓	.20	100	↓	12,600	4,150	765	17,515	21,000	

11480 | Athletic Recreational & Therapeutic Equipment

11486 | Gymnasium Equipment

			CREW	DAILY OUTPUT	LABOR-HOURS	UNIT	2006 BARE COSTS MAT.	LABOR	EQUIP.	TOTAL	TOTAL INCL O&P	
700	7500	Basketball (one side), minimum	R-3	2.07	9.662	Ea.	2,050	400	73.50	2,523.50	2,950	700
	7600	Maximum		.30	66.667		14,300	2,750	510	17,560	20,500	
	7700	Hockey-basketball (four sides), minimum		.25	80		5,675	3,325	610	9,610	12,000	
	7800	Maximum		.15	133		23,600	5,525	1,025	30,150	35,700	

11500 | Industrial & Process Equipment

11520 | Industrial Equipment

			CREW	DAILY OUTPUT	LABOR-HOURS	UNIT	2006 BARE COSTS MAT.	LABOR	EQUIP.	TOTAL	TOTAL INCL O&P	
850	0010	VOCATIONAL SHOP EQUIPMENT										850
	0020	Benches, work, wood, average	2 Carp	5	3.200	Ea.	450	114		564	685	
	0100	Metal, average		5	3.200		385	114		499	615	
	0400	Combination belt & disc sander, 6″		4	4		710	142		852	1,025	
	0700	Drill press, floor mounted, 12″, 1/2 H.P.		4	4		350	142		492	620	
	0800	Dust collector, not incl. ductwork, 6″ diameter	1 Shee	1.10	7.273		2,725	305		3,030	3,500	
	1000	Grinders, double wheel, 1/2 H.P.	2 Carp	5	3.200		192	114		306	400	
	1300	Jointer, 4″, 3/4 H.P.		4	4		1,000	142		1,142	1,325	
	1600	Kilns, 16 C.F., to 2000°		4	4		2,150	142		2,292	2,600	
	1900	Lathe, woodworking, 10″, 1/2 H.P.		4	4		915	142		1,057	1,225	
	2200	Planer, 13″ x 6″		4	4		1,725	142		1,867	2,125	
	2500	Potter's wheel, motorized		4	4		980	142		1,122	1,300	
	2800	Saws, band, 14″, 3/4 H.P.		4	4		750	142		892	1,050	
	3100	Metal cutting band saw, 14″		4	4		1,975	142		2,117	2,400	
	3400	Radial arm saw, 10″, 2 H.P.		4	4		695	142		837	1,000	
	3700	Scroll saw, 24″		4	4		1,450	142		1,592	1,825	
	4000	Table saw, 10″, 3 H.P.		4	4		1,700	142		1,842	2,100	
	4300	Welder AC arc, 30 amp capacity		4	4		2,125	142		2,267	2,575	

11600 | Laboratory Equipment

11620 | Laboratory Equipment

			CREW	DAILY OUTPUT	LABOR-HOURS	UNIT	2006 BARE COSTS MAT.	LABOR	EQUIP.	TOTAL	TOTAL INCL O&P	
110	0010	LABORATORY EQUIPMENT										110
	1400	Safety equipment, eye wash, hand held				Ea.	310			310	340	
	1450	Deluge shower				″	595			595	655	
	8000	Alternate pricing method: as percent of lab furniture										
	8050	Installation, not incl. plumbing & duct work				% Furn.					22%	
	8100	Plumbing, final connections, simple system									10%	
	8110	Moderately complex system									15%	
	8120	Complex system									20%	
	8150	Electrical, simple system									10%	
	8160	Moderately complex system									20%	
	8170	Complex system									35%	

For information about Means Estimating Seminars, see yellow pages 12 and 13 in back of book

Division Notes

		CREW	DAILY OUTPUT	LABOR-HOURS	UNIT	2000 BARE COSTS MAT.	LABOR	EQUIP.	TOTAL	TOTAL INCL O&P

Division 12 Furnishings

Estimating Tips

General

- The items in this division are usually priced per square foot or each. Most of these items are purchased by the owner and placed by the supplier. Do not assume the items in Division 12 will be purchased and installed by the supplier. Check the specifications for responsibilities and include receiving, storage, installation, and mechanical and electrical hookups in the appropriate divisions.
- Some items in this division require some type of support system that is not usually furnished with the item. Examples of these systems include blocking for the attachment of casework and heavy drapery rods. The required blocking must be added to the estimate in the appropriate division.

Reference Numbers

Reference numbers are shown in bold squares at the beginning of some major classifications. These numbers refer to related items in the Reference Section. The reference information may be an estimating procedure, an alternate pricing method, or technical information.

Note: Not all subdivisions listed here necessarily appear in this publication.

12300 | Manufactured Casework

12350 | Specialty Casework

				DAILY	LABOR-		2006 BARE COSTS				TOTAL	
			CREW	OUTPUT	HOURS	UNIT	MAT.	LABOR	EQUIP.	TOTAL	INCL O&P	
500	0010	**LABORATORY CASEWORK**										500
	0020	Cabinets, base, door units, metal	2 Carp	18	.889	L.F.	167	31.50		198.50	236	
	0300	Drawer units		18	.889		360	31.50		391.50	450	
	0700	Tall storage cabinets, open, 7′ high		20	.800		350	28.50		378.50	430	
	0900	With glazed doors		20	.800		435	28.50		463.50	520	
	1300	Wall cabinets, metal, 12-1/2″ deep, open		20	.800		109	28.50		137.50	167	
	1500	With doors		20	.800		227	28.50		255.50	297	
	1550	Counter tops, not incl. base cabinets, acidproof, minimum		82	.195	S.F.	26.50	6.95		33.45	40.50	
	1600	Maximum		70	.229		36	8.15		44.15	53	
	1650	Stainless steel		82	.195		78	6.95		84.95	97	

12400 | Furnishings & Accessories

12492 | Blinds and Shades

				DAILY	LABOR-		2006 BARE COSTS				TOTAL	
			CREW	OUTPUT	HOURS	UNIT	MAT.	LABOR	EQUIP.	TOTAL	INCL O&P	
100	0010	**BLINDS, INTERIOR**										100
	0020	Horizontal, 1″ aluminum slats, solid color, stock	1 Carp	590	.014	S.F.	2.89	.48		3.37	3.98	
	3000	Wood folding panels with movable louvers, 7″ x 20″ each		17	.471	Pr.	42	16.75		58.75	74	
	4000	Fixed louver type, stock units, 8″ x 20″ each		17	.471		63	16.75		79.75	97	
	4450	18″ x 40″ each		17	.471		131	16.75		147.75	172	
600	0011	**SHADES** Basswood roll-up, stain finish, 3/8″ slats		300	.027	S.F.	9.85	.95		10.80	12.35	600
	0030	Double layered, heat reflective		685	.012		6.90	.42		7.32	8.30	
	0250	7/8″ slats		300	.027		9.30	.95		10.25	11.75	
	0950	Mylar, single layer, non-heat reflective		685	.012		4.63	.42		5.05	5.80	
	1050	Double layered, heat reflective		685	.012		8	.42		8.42	9.50	
	1150	Triple layered, heat reflective		685	.012		9.25	.42		9.67	10.85	
	5000	Thermal, roll up, R-4		44	.182		8.90	6.45		15.35	20.50	
	5030	R-10.7		44	.182		10.25	6.45		16.70	22	
	5050	Magnetic clips, set of 20				Set	19.15			19.15	21	

12500 | Furniture

12510 | Office Furniture

				DAILY	LABOR-		2006 BARE COSTS				TOTAL	
			CREW	OUTPUT	HOURS	UNIT	MAT.	LABOR	EQUIP.	TOTAL	INCL O&P	
600	0010	**OFFICE CASE GOODS**										600
	0020	Desks, 29″ high, double pedestal, 30″ x 60″, metal, minimum				Ea.	330			330	360	
	0030	Maximum				″	1,050			1,050	1,150	

12540 | Hospitality Furniture

				DAILY	LABOR-		2006 BARE COSTS				TOTAL	
			CREW	OUTPUT	HOURS	UNIT	MAT.	LABOR	EQUIP.	TOTAL	INCL O&P	
500	0010	**FURNITURE, HOTEL**										500
	0020	Standard quality set, minimum				Room	1,500			1,500	1,650	
	0200	Maximum				″	7,900			7,900	8,675	

12500 | Furniture

12540 | Hospitality Furniture

		12540 Hospitality Furniture	CREW	DAILY OUTPUT	LABOR-HOURS	UNIT	2006 BARE COSTS MAT.	LABOR	EQUIP.	TOTAL	TOTAL INCL O&P	
700	0010	**FURNITURE, RESTAURANT**										700
	0020	Bars, built-in, front bar	1 Carp	5	1.600	L.F.	220	57		277	335	
	0200	Back bar		5	1.600	"	160	57		217	270	
	0500	Booth unit, molded plastic, stub wall and 2 seats, minimum		2	4	Set	298	142		440	565	
	0600	Maximum		1.50	5.333	"	1,025	190		1,215	1,450	
	0800	Booth seat, upholstered, foursome, single (end) minimum		5	1.600	Ea.	855	57		912	1,025	
	0900	Maximum		4	2		1,550	71		1,621	1,825	
	1000	Foursome, double, minimum		4	2		855	71		926	1,050	
	1100	Maximum		3	2.667		1,550	95		1,645	1,850	
	1300	Circle booth, upholstered, 1/4 circle, minimum		3	2.667		560	95		655	770	
	1400	Maximum		2	4		1,275	142		1,417	1,625	
	1500	3/4 circle, minimum		1.50	5.333		1,675	190		1,865	2,150	
	1600	Maximum	▼	1	8	▼	3,550	284		3,834	4,400	

12560 | Institutional Furniture

		12560 Institutional Furniture	CREW	DAILY OUTPUT	LABOR-HOURS	UNIT	MAT.	LABOR	EQUIP.	TOTAL	TOTAL INCL O&P	
310	0010	**DORMITORY FURNITURE**										310
	0020	Beds, free standing, minimum				Ea.	300			300	330	
	0100	Maximum				"	620			620	680	
	2050	Rule of thumb: Total cost for furniture, minimum				Student					2,075	
	2150	Maximum				"					4,050	
400	0010	**FURNITURE, HOSPITAL**										400
	0020	Beds, manual, minimum				Ea.	775			775	855	
	0100	Maximum				"	2,375			2,375	2,600	
	1100	Patient wall systems, not incl. plumbing, minimum				Room	960			960	1,050	
	1200	Maximum				"	1,775			1,775	1,950	

For information about Means Estimating Seminars, see yellow pages 12 and 13 in back of book

Division Notes

		CREW	DAILY OUTPUT	LABOR-HOURS	UNIT	2000 BARE COSTS MAT.	LABOR	EQUIP.	TOTAL	TOTAL INCL O&P

Division 13
Special Construction

Estimating Tips

General

- The items and systems in this division are usually estimated, purchased, supplied, and installed as a unit by one or more subcontractors. The estimator must ensure that all parties are operating from the same set of specifications and assumptions and that all necessary items are estimated and will be provided. Many times the complex items and systems are covered but the more common ones such as excavation or a crane are overlooked for the very reason that everyone assumes nobody could miss them. The estimator should be the central focus and be able to ensure that all systems are complete.
- Another area where problems can develop in this division is at the interface between systems. The estimator must ensure, for instance, that anchor bolts, nuts, and washers are estimated and included for the air-supported structures and pre-engineered buildings to be bolted to their foundations. Utility supply is a common area where essential items or pieces of equipment can be missed or overlooked due to the fact that each subcontractor may feel it is another's responsibility. The estimator should also be aware of certain items which may be supplied as part of a package but installed by others and ensure that the installing contractor's estimate includes the cost of installation. Conversely, the estimator must also ensure that items are not costed by two different subcontractors, resulting in an inflated overall estimate.

13120 Pre-Engineered Structures

- The foundations and floor slab, as well as rough mechanical and electrical, should be estimated, as this work is required for the assembly and erection of the structure. Generally, as noted in the book, the pre-engineered building comes as a shell and additional features, such as windows and doors, must be included by the estimator. Here again, the estimator must have a clear understanding of the scope of each portion of the work and all the necessary interfaces.

13200 Storage Tanks

- The prices in this subdivision for above- and below-ground storage tanks do not include foundations or hold-down slabs. The estimator should refer to Divisions 2 and 3 for foundation system pricing. In addition to the foundations, required tank accessories, such as tank gauges, leak detection devices, and additional manholes and piping, must be added to the tank prices.

Reference Numbers

Reference numbers are shown in bold squares at the beginning of some major classifications. These numbers refer to related items in the Reference Section. The reference information may be an estimating procedure, an alternate pricing method, or technical information.

Note: Not all subdivisions listed here necessarily appear in this publication.

		13005 | Selective Demolition	CREW	DAILY OUTPUT	LABOR-HOURS	UNIT	2006 BARE COSTS MAT.	LABOR	EQUIP.	TOTAL	TOTAL INCL O&P	
011	0010	SELECTIVE DEMOLITION, AIR SUPPORTED STRUCTURES										011
	0020	Tank covers, scrim, dbl layer, vinyl poly w/ hdw, blower & controls										
	0050	Round and rectangular R024119-10	B-2	9000	.004	S.F.		.12		.12	.20	
	0100	Warehouse structures										
	0120	Poly/vinyl fabric, 28 oz, incl tension cables & inflation system	4 Clab	9000	.004	SF Flr.		.10		.10	.16	
	0150	Reinforced vinyl, 12 oz., 3000 S.F.	″	5000	.006			.18		.18	.29	
	0200	12,000 to 24,000 S.F.	8 Clab	20000	.003			.09		.09	.15	
	0250	Tedlar vinyl fabric, 28 oz. w/liner, to 3000 S.F.	4 Clab	5000	.006			.18		.18	.29	
	0300	12,000 to 24,000 S.F.	8 Clab	20000	.003			.09		.09	.15	
	0350	Greenhouse/shelter, woven polyethylene with liner										
	0400	3000 S.F.	4 Clab	5000	.006	SF Flr.		.18		.18	.29	
	0450	12,000 to 24,000 S.F.	8 Clab	20000	.003			.09		.09	.15	
	0500	Tennis/gymnasium, poly/vinyl fabric, 28 oz., incl thermal liner	4 Clab	9000	.004			.10		.10	.16	
	0600	Stadium/Convention Ctr, teflon coated fiberglass, incl thermal liner	9 Clab	40000	.002			.05		.05	.08	
	0700	Doors, air lock, 15′ long, 10′ x 10′	2 Carp	1.50	10.667	Ea.		380		380	630	
	0720	15′ x 15′		.80	20			710		710	1,175	
	0750	Revolving personnel door, 6′ dia. x 6′-6″ high		1.50	10.667			380		380	630	
012	0010	SELECTIVE DEMOLITION, GARDEN HOUSES										012
	0020	Garden house, prefab, wood, excl foundation, average	2 Clab	400	.040	SF Flr.		1.10		1.10	1.82	
013	0010	**SELECTIVE DEMOLITION, GREENHOUSES**, excl foundation										013
	0020	Greenhouse, resi-type, free standing, 9′ long x 8′ wide	2 Clab	160	.100	SF Flr.		2.74		2.74	4.54	
	0030	11′ wide		170	.094			2.58		2.58	4.27	
	0040	14′ wide		220	.073			1.99		1.99	3.30	
	0050	17′ wide		320	.050			1.37		1.37	2.27	
	0060	Lean-to type, 4′ wide		64	.250			6.85		6.85	11.35	
	0070	7′ wide		120	.133			3.65		3.65	6.05	
	0080	Geodesic hemishere, 1/8″ plexiglass glazing, 8′ dia		4	4	Ea.		110		110	182	
	0090	24′ dia		.80	20			550		550	910	
	0100	48′ dia		.40	40			1,100		1,100	1,825	
035	0010	SELECTIVE DEMOLITION, SPECIAL PURPOSE ROOMS										035
	0100	Audiometric rooms, under 500 S.F. surface	4 Carp	200	.160	SF Surf		5.70		5.70	9.40	
	0110	Over 500 S.F. surface	″	240	.133	″		4.74		4.74	7.85	
	0200	Clean rooms, 12′ x 12′ soft wall, class 100	1 Carp	.30	26.667	Ea.		950		950	1,575	
	0210	Class 1000		.30	26.667			950		950	1,575	
	0220	Class 10,000		.35	22.857			815		815	1,350	
	0230	Class 100,000		.35	22.857			815		815	1,350	
	0300	Darkrooms, shell complete, 8′ high	2 Carp	220	.073	SF Flr.		2.59		2.59	4.28	
	0310	12′ high		110	.145	″		5.15		5.15	8.55	
	0350	Darkrooms doors, mini-cylindrical, revolving		4	4	Ea.		142		142	236	
	0400	Music room, practice modular		140	.114	SF Surf		4.06		4.06	6.75	
	0500	Refrigeration structures and finishes										
	0510	Wall finish, 2 coat portland cement plaster, 1/2″ thick	1 Clab	200	.040	S.F.		1.10		1.10	1.82	
	0520	Fiberglass panels, 1/8″ thick		400	.020			.55		.55	.91	
	0530	Ceiling finish, polystyrene plastic, 1″ to 2″ thick		500	.016			.44		.44	.73	
	0540	4″ thick		450	.018			.49		.49	.81	
	0550	Refrigerator, prefab aluminum walk-in, 7′-6″ high, 6′ x 6′ OD	2 Carp	100	.160	SF Flr.		5.70		5.70	9.40	
	0560	10′ x 10′ OD		160	.100			3.56		3.56	5.90	
	0570	Over 150 S.F.		200	.080			2.84		2.84	4.71	
	0600	Sauna, prefabricated, including heater & controls, 7′ high, to 30 S.F.		120	.133			4.74		4.74	7.85	
	0610	To 40 S.F.		140	.114			4.06		4.06	6.75	
	0620	To 60 S.F.		175	.091			3.25		3.25	5.40	
	0630	To 100 S.F.		220	.073			2.59		2.59	4.28	
	0640	To 130 S.F.		250	.064			2.28		2.28	3.77	
036	0010	SELECTIVE DEMOLITION, SOUND CONTROL										036
	0120	Acoustical enclosure, 4″ T walls & ceiling panels, 8 lbs/SF	3 Carp	144	.167	SF Surf		5.95		5.95	9.80	

13005 | Selective Demolition

	13005 | Selective Demolition		CREW	DAILY OUTPUT	LABOR-HOURS	UNIT	2006 BARE COSTS MAT.	LABOR	EQUIP.	TOTAL	TOTAL INCL O&P	
036	0130	10.5 lbs/SF	3 Carp	128	.188	SF Surf		6.65		6.65	11.05	036
	0140	Reverb chamber, parallel walls, 4″ thick		120	.200			7.10		7.10	11.80	
	0150	Skewed walls, parallel roof, 4″ thick		110	.218			7.75		7.75	12.85	
	0160	Skewed walls/roof, 4″ layer/air space		96	.250			8.90		8.90	14.75	
	0170	Sound-absorbing panels, painted metal, 2′-6″ x 8′, under 1,000 SF		430	.056			1.98		1.98	3.29	
	0180	Over 1,000 SF		480	.050			1.78		1.78	2.95	
	0190	Flexible transparent curtain, clear	3 Shee	430	.056			2.35		2.35	3.74	
	0192	50% clear, 50% foam		430	.056			2.35		2.35	3.74	
	0194	25% clear, 75% foam		430	.056			2.35		2.35	3.74	
	0196	100% foam		430	.056			2.35		2.35	3.74	
	0200	Audio-masking system, incl speakers, amplfr, signal gnrtr										
	0205	Ceiling mounted, 5,000 SF	2 Elec	4800	.003	S.F.		.14		.14	.22	
	0210	10,000 SF		5600	.003			.12		.12	.19	
	0230	10,000 SF		8800	.002			.08		.08	.12	
	0020	Shielding lead, lined door frame, excl hdwe, 1/16″ thick	1 Clab	4.80	1.667	Ea.		45.50		45.50	75.50	
	0030	Lead sheets, 1/16″ thick	2 Clab	270	.059	S.F.		1.62		1.62	2.69	
	0050	Lead shielding, 1/4″ thick		270	.059			1.62		1.62	2.69	
	0060	1/2″ thick		240	.067			1.83		1.83	3.03	
	0070	Lead glass, 1/4″ thick, 2.0 mm LE, 12″ x 16″	2 Glaz	26	.615	Ea.		21		21	34	
	0080	24″ x 36″		16	1			34.50		34.50	55	
	0090	36″ x 60″		4	4			137		137	221	
	0100	Lead glass window frame, w/ 1/16″ lead & voice passage, 36″ x 60″		4	4			137		137	221	
	0110	Lead glass window frame, 24″ x 36″		16	1			34.50		34.50	55	
	0120	Lead gypsum board, 5/8″ thick with 1/16″ lead	2 Clab	320	.050	S.F.		1.37		1.37	2.27	
	0130	1/8″ lead		280	.057			1.57		1.57	2.59	
	0140	1/32″ lead		400	.040			1.10		1.10	1.82	
	0150	Butt joints, 1/8″ lead or thicker, 2″ x 7′ long batten strip		480	.033	Ea.		.91		.91	1.51	
	0160	X-ray protection, average radiography room, to 300 SF, 1/16″ lead, min		.50	32	Total		875		875	1,450	
	0170	Maximum		.30	53.333			1,450		1,450	2,425	
	0180	Deep therapy X-ray room, 250 KV cap, to 300 SF, 1/4″ lead, min		.20	80			2,200		2,200	3,625	
	0190	Maximum		.12	133			3,650		3,650	6,050	
	0880	Radio frequency shielding, prefab or screen-type copper or steel, min		360	.044	SF Surf		1.22		1.22	2.02	
	0890	Average		310	.052			1.41		1.41	2.34	
	0895	Maximum		290	.055			1.51		1.51	2.50	
039	0010	**SELECTIVE DEMOLITION, GEODESIC DOMES**										039
	0050	Shell only, interlocking plywood panels, 30′ diameter	F-5	3.20	10	Ea.		360		360	595	
	0060	34′ diameter		2.30	13.913			500		500	830	
	0070	39′ diameter		2	16			575		575	955	
	0080	45′ diameter		2.20	14.545			525		525	870	
	0090	55′ diameter		2	16			575		575	955	
	0100	60′ diameter		2	16			575		575	955	
	0110	65′ diameter		1.60	20			720		720	1,200	
040	0010	**SELECTIVE DEMOLITION, TENSION STRUCTURES**										040
	0020	Steel/alum frame, fabric shell, 60′ clear span, 6,000 SF	B-41	2000	.022	SF Flr.		.63	.12	.75	1.16	
	0030	12,000 SF		2200	.020			.57	.11	.68	1.06	
	0040	80′ clear span, 20,800 SF		2440	.018			.51	.10	.61	.96	
	0050	100′ clear span, 10,000 SF	L-5	4350	.013			.51	.15	.66	1.10	
	0060	26,000 SF	″	4600	.012			.49	.14	.63	1.05	
041	0010	**SELECTIVE DEMOLITION, HANGARS**										041
	0020	Prefab, steel T type, galv roof & walls, incl doors, excl fndtn	E-2	2550	.022	SF Flr.		.85	.56	1.41	2.15	
	0030	Circular type, prefab, steel frame, plastic skin, incl foundation, 80′ dia	″	.50	112	Total		4,350	2,850	7,200	11,000	
	0020	Concrete stave, indstrl, conical/sloping bott, excl fndtn, 12′ dia, 35′ h	D-8	.22	181	Ea.		6,000		6,000	9,750	
	0030	16′ dia, 45′ h		.16	250			8,250		8,250	13,400	
	0040	25′ dia, 75′ h		.10	400			13,200		13,200	21,400	
	0050	Steel, factory fabricated, 30,000 gal cap, painted or epoxy lined	L-5	2	28			1,125	325	1,450	2,400	
	0020	Diving stand, stainless steel, 3 meter	2 Clab	3	5.333			146		146	242	

	13005	Selective Demolition	CREW	DAILY OUTPUT	LABOR-HOURS	UNIT	2006 BARE COSTS MAT.	LABOR	EQUIP.	TOTAL	TOTAL INCL O&P	
041	0030	1 meter	2 Clab	5	3.200	Ea.		87.50		87.50	145	041
	0040	Diving board, 16′ long, aluminum		5.40	2.963			81		81	135	
	0050	Fiberglass		5.40	2.963			81		81	135	
	0070	Ladders, heavy duty, stainless steel, 2 tread		14	1.143			31.50		31.50	52	
	0080	4 tread		12	1.333			36.50		36.50	60.50	
	0090	Lifeguard chair, stainless steel, fixed		5	3.200			87.50		87.50	145	
	0100	Slide, tubular, fiberglass, aluminum handrails & ladder, 5′, straight		4	4			110		110	182	
	0110	8′, curved		6	2.667			73		73	121	
	0120	10′, curved		3	5.333			146		146	242	
	0130	12′ straight, with platform		2.50	6.400			175		175	291	
	0140	Removable access ramp, stainless steel		4	4			110		110	182	
	0150	Removable stairs, stainless steel, collapsible		4	4			110		110	182	
	0020	Air terminal & base, copper, 3/8″ dia x 10″, to 75′ h	1 Clab	16	.500			13.70		13.70	22.50	
	0030	1/2″ dia x 12″, over 75′ h		16	.500			13.70		13.70	22.50	
	0050	Aluminum, 1/2″ dia x 12″, to 75′ h		16	.500			13.70		13.70	22.50	
	0060	5/8″ dia x 12″, over 75′ h		16	.500			13.70		13.70	22.50	
	0070	Cable, copper, 220 lb per thousand feet, to 75′ high		640	.013	L.F.		.34		.34	.57	
	0080	375 lb per thousand feet, over 75′ high		460	.017			.48		.48	.79	
	0090	Aluminum, 101 lb per thousand feet, to 75′ high		560	.014			.39		.39	.65	
	0100	199 lb per thousand feet, over 75′ high		480	.017			.46		.46	.76	
	0110	Arrester, 175 V AC, to ground		16	.500	Ea.		13.70		13.70	22.50	
	0120	650 V AC, to ground		13	.615	″		16.85		16.85	28	
128	0010	**SELECTIVE DEMOLITION, PRE-ENGINEERED STEEL BUILDINGS**										128
	0500	Pre-engd steel bldgs, rigid frame, clear span & multi post, excl salvage										
	0550	3,500 to 7,500 S.F	L-11	1350	.024	SF Flr.		.76	1.05	1.81	2.45	
	0600	7,501 to 12,500 S.F		2000	.016			.51	.71	1.22	1.65	
	0650	12,500 S.F or greater		2200	.015			.47	.65	1.12	1.50	
	0700	Pre-engd steel building components										
	0710	Entrance canopy, including frame 4′ x 4′	E-24	8	4	Ea.		157	81.50	238.50	370	
	0720	4′ x 8′	″	7	4.571			179	93	272	420	
	0730	H.M doors, self framing, single leaf	2 Skwk	8	2			73		73	119	
	0740	Double leaf		5	3.200			117		117	191	
	0760	Gutter, eave type		600	.027	L.F.		.97		.97	1.59	
	0770	Sash, single slide, double slide or fixed		24	.667	Ea.		24.50		24.50	40	
	0780	Skylight, fiberglass, to 30 S.F.		16	1			36.50		36.50	59.50	
	0785	Roof vents, circular, 12″ to 24″ diameter		12	1.333			48.50		48.50	79.50	
	0790	Continuous, 10′ long		8	2			73		73	119	
	0900	Shelters, aluminum frame										
	0910	Aluminum frame, acrylic glazing, 3′ x 9′ x 8′ high	2 Skwk	2	8	Ea.		292		292	480	
	0920	9′ x 12′ x 8′ high	″	1.50	10.667	″		390		390	635	
	0930	Silos, concrete stave industrial, not incl foundations										
201	0010	**SELECTIVE DEMOLITION, STORAGE TANKS**										201
	0500	Steel tank, single wall, above ground, not incl fdn, pumps or piping										
	0510	Single wall, 275 gallon	Q-1	3	5.333	Ea.		205		205	320	
	0520	550 thru 2,000 gallon	B-34P	2	12			430	201	631	900	
	0530	5,000 thru 10,000 gallon	B-34Q	2	12			435	535	970	1,275	
	0540	15,000 thru 30,000 gallon	B-34N	2	4			113	223	336	430	
	0600	Steel tank, double wall, above ground not incl fdn, pumps & piping										
	0620	500 thru 2,000 gallon	B-34P	2	12	Ea.		430	201	631	900	

13030 | Special Purpose Rooms

		13035 | Special Purpose Rooms	CREW	DAILY OUTPUT	LABOR-HOURS	UNIT	2006 BARE COSTS MAT.	LABOR	EQUIP.	TOTAL	TOTAL INCL O&P	
700	0010	**REFRIGERATION**										700
	0020	Curbs, 12" high, 4" thick, concrete	2 Carp	58	.276	L.F.	3.52	9.80		13.32	20	
	6300	Rule of thumb for complete units, w/o doors & refrigeration, cooler	↓	146	.110	SF Flr.	103	3.90		106.90	119	
	6400	Freezer	↓	109.60	.146	"	121	5.20		126.20	142	
800	0010	**SAUNA**										800
	0020	Prefabricated, incl. heater & controls, 7' high, 6' x 4', C/C	L-7	2.20	12.727	Ea.	3,675	435		4,110	4,750	
	1700	Door only, cedar, 2'x6', with tempered insulated glass window	2 Carp	3.40	4.706	↓	480	167		647	805	
	1800	Prehung, incl. jambs, pulls & hardware	"	12	1.333	↓	490	47.50		537.50	620	
	2500	Heaters only (incl. above), wall mounted, to 200 C.F.				↓	475			475	520	
	4480	For additional equipment, see div. 11484-400										
900	0010	**SPORT COURT**										900
	0020	Floors, No. 2 & better maple, 25/32" thick				SF Flr.					6.65	
	0300	Squash, regulation court in existing building, minimum				Court	15,300			15,300	16,800	
	0400	Maximum				"	28,500			28,500	31,400	
	0450	Rule of thumb for components:										
	0470	Walls	3 Carp	.15	160	Court	11,700	5,700		17,400	22,300	
	0500	Floor	"	.25	96	↓	5,225	3,425		8,650	11,400	
	0550	Lighting	2 Elec	.60	26.667	↓	1,650	1,125		2,775	3,550	
940	0010	**STEAM BATH**										940
	0020	Heater, timer & head, single, to 140 C.F.	1 Plum	1.20	6.667	Ea.	1,075	285		1,360	1,625	
	0500	To 300 C.F.	"	1.10	7.273	↓	1,200	310		1,510	1,800	
	2000	Multiple, motels, apts., 2 baths, w/ blow-down assm., 500 C.F.	Q-1	1.30	12.308	↓	4,125	475		4,600	5,250	
	2500	4 baths	"	.70	22.857	↓	4,500	880		5,380	6,325	

13080 | Sound, Vibration & Seismic Control

		13081 | Sound Control	CREW	DAILY OUTPUT	LABOR-HOURS	UNIT	2006 BARE COSTS MAT.	LABOR	EQUIP.	TOTAL	TOTAL INCL O&P	
100	0010	**AUDIO MASKING** Acoustical enclosure, 4" thick wall and ceiling panels										100
	0020	8# per S.F., up to 12' span	3 Carp	72	.333	SF Surf	29	11.85		40.85	51.50	
	0300	Better quality panels, 10.5# per S.F.	↓	64	.375	↓	33	13.35		46.35	58	
	0400	Reverb-chamber, 4" thick, parallel walls	↓	60	.400	↓	41	14.20		55.20	68.50	

13090 | Radiation Protection

		13091 | X-Ray/Radio Freq Protection	CREW	DAILY OUTPUT	LABOR-HOURS	UNIT	2006 BARE COSTS MAT.	LABOR	EQUIP.	TOTAL	TOTAL INCL O&P	
600	0010	**SHIELDING LEAD**										600
	0300	Lead sheets, 1/16" thick	2 Lath	135	.119	S.F.	5.40	3.89		9.29	12.10	
	0400	1/8" thick	↓	120	.133	↓	11.05	4.37		15.42	19.10	
	0500	Lead shielding, 1/4" thick	↓	135	.119	↓	19.65	3.89		23.54	27.50	
	0550	1/2" thick	↓	120	.133	↓	40	4.37		44.37	51	
	0600	Lead glass, 1/4" thick, 2.0 mm LE, 12" x 16"	2 Glaz	13	1.231	Ea.	269	42		311	365	
	0700	24" x 36"	"	8	2	"	1,050	68.50		1,118.50	1,275	
	1200	X-ray protection, average radiography or fluoroscopy										

13090 | Radiation Protection

	13091 | X-Ray/Radio Freq Protection		Crew	Daily Output	Labor-Hours	Unit	2006 Bare Costs Mat.	Labor	Equip.	Total	Total Incl O&P	
600	1210	room, up to 300 S.F. floor, 1/16" lead, minimum	2 Lath	.25	64	Total	6,000	2,100		8,100	9,925	600
	1500	Maximum, 7'-0" walls	"	.15	106	"	7,500	3,500		11,000	13,800	
	1600	Deep therapy X-ray room, 250 KV capacity,										
	1800	up to 300 S.F. floor, 1/4" lead, minimum	2 Lath	.08	200	Total	18,500	6,550		25,050	30,700	
	1900	Maximum, 7'-0" walls		.06	266	"	23,900	8,750		32,650	40,200	
	1999	Minimum labor/equipment charge		4.50	3.556	Job		117		117	185	
	2000	X-ray viewing panels, clear lead plastic										
	2010	7 mm thick, 0.3 mm LE, 2.3 lbs/S.F.	H-3	139	.115	S.F.	147	3.49		150.49	168	
	2020	12 mm thick, 0.5 mm LE, 3.9 lbs/S.F.		82	.195		199	5.90		204.90	229	
	2030	18 mm thick, 0.8mm LE, 5.9 lbs/S.F.		54	.296		216	9		225	253	
	2040	22 mm thick, 1.0 mm LE, 7.2 lbs/S.F.		44	.364		221	11.05		232.05	261	
	2050	35 mm thick, 1.5 mm LE, 11.5 lbs/S.F.		28	.571		248	17.35		265.35	300	
	2060	46 mm thick, 2.0 mm LE, 15.0 lbs/S.F.		21	.762		325	23		348	400	
	2090	For panels 12 S.F. to 48 S.F., add crating charge				Ea.					50	
	4000	X-ray barriers, modular, panels mounted within framework for										
	4002	attaching to floor, wall or ceiling, upper portion is clear lead										
	4005	plastic window panels 48"H, lower portion is opaque leaded										
	4008	steel panels 36"H, structural supports not incl.										
	4010	1-section barrier, 36"W x 84"H overall										
	4020	0.5 mm LE panels	H-3	6.40	2.500	Ea.	3,325	76		3,401	3,775	
	4030	0.8 mm LE panels		6.40	2.500		3,525	76		3,601	4,000	
	4040	1.0 mm LE panels		5.33	3.002		3,650	91		3,741	4,175	
	4050	1.5 mm LE panels		5.33	3.002		3,900	91		3,991	4,425	
	4060	2-section barrier, 72"W x 84"H overall										
	4070	0.5 mm LE panels	H-3	4	4	Ea.	6,875	121		6,996	7,775	
	4080	0.8 mm LE panels		4	4		7,275	121		7,396	8,200	
	4090	1.0 mm LE panels		3.56	4.494		7,425	136		7,561	8,400	
	5000	1.5 mm LE panels		3.20	5		8,025	152		8,177	9,075	
	5010	3-section barrier, 108"W x 84"H overall										
	5020	0.5 mm LE panels	H-3	3.20	5	Ea.	9,850	152		10,002	11,000	
	5030	0.8 mm LE panels		3.20	5		10,300	152		10,452	11,600	
	5040	1.0 mm LE panels		2.67	5.993		10,600	182		10,782	11,900	
	5050	1.5 mm LE panels		2.46	6.504		11,500	197		11,697	12,900	
	7000	X-ray barriers, mobile, mounted within framework w/casters on										
	7005	bottom, clear lead plastic window panels on upper portion,										
	7010	opaque on lower, 30"W x 75"H overall, incl. framework										
	7020	24"H upper w/0.5 mm LE, 48"H lower w/0.8 mm LE	1 Carp	16	.500	Ea.	2,250	17.80		2,267.80	2,500	
	7030	48"W x 75"H overall, incl. framework										
	7040	36"H upper w/0.5 mm LE, 36"H lower w/0.8 mm LE	1 Carp	16	.500	Ea.	4,000	17.80		4,017.80	4,425	
	7050	36"H upper w/1.0 mm LE, 36"H lower w/1.5 mm LE	"	16	.500	"	4,825	17.80		4,842.80	5,325	
	7060	72"W x 75"H overall, incl. framework										
	7070	36"H upper w/0.5 mm LE, 36"H lower w/0.8 mm LE	1 Carp	16	.500	Ea.	4,725	17.80		4,742.80	5,225	
	7080	36"H upper w/1.0 mm LE, 36"H lower w/1.5 mm LE	"	16	.500	"	5,975	17.80		5,992.80	6,600	

13120 | Pre-Engineered Structures

	13128 | Pre-Engineered Structures		Crew	Daily Output	Labor-Hours	Unit	2006 Bare Costs Mat.	Labor	Equip.	Total	Total Incl O&P	
045	0010	GUARD HOUSE										045
	0100	Prefab conc w/bullet resistant doors & windows, roof and wiring										

13120 | Pre-Engineered Structures

	13128 Pre-Engineered Structures	CREW	DAILY OUTPUT	LABOR-HOURS	UNIT	2006 BARE COSTS MAT.	LABOR	EQUIP.	TOTAL	TOTAL INCL O&P	
045	0110 8′ x 8′, Level III	L-10	1	24	Ea.	20,000	960	635	21,595	24,400	045
	0120 8′ x 8′, Level IV	"	1	24	"	25,000	960	635	26,595	29,900	
540	0010 **GREENHOUSE** Shell only, stock units, not incl. 2′ stub walls,										540
	0020 foundation, floors, heat or compartments										
	0300 Residential type, free standing, 8′-6″ long x 7′-6″ wide	2 Carp	59	.271	SF Flr.	43.50	9.65		53.15	64	
	0400 10′-6″ wide		85	.188		33.50	6.70		40.20	48	
	0600 13′-6″ wide		108	.148		30	5.25		35.25	42	
	0700 17′-0″ wide		160	.100		33.50	3.56		37.06	43	
	0900 Lean-to type, 3′-10″ wide		34	.471		38.50	16.75		55.25	70	
	1000 6′-10″ wide		58	.276		30	9.80		39.80	49.50	
	1050 8′-0″ wide	↓	60	.267	↓	27.50	9.50		37	45.50	
	1060										
	1100 Wall mounted, to existing window, 3′ x 3′	1 Carp	4	2	Ea.	420	71		491	580	
	1120 4′ x 5′	"	3	2.667	"	625	95		720	840	
	3900 For cooling, add, minimum				SF Flr.	2.63			2.63	2.89	
	4000 Maximum					6.50			6.50	7.15	
	4200 For heaters, 13.6 MBH, add					4.99			4.99	5.50	
	4300 60 MBH, add				↓	1.87			1.87	2.06	
	4500 For benches, 2′ x 3′-6″, add				SF Hor.	22.50			22.50	24.50	
	4600 3′ x 10′, add				S.F.	12.10			12.10	13.30	
	4800 For controls, add, minimum				Total	2,250			2,250	2,475	
	4900 Maximum				"	13,300			13,300	14,700	
	5100 For humidification equipment, add				M.C.F.	5.75			5.75	6.30	
	5200 For vinyl shading, add				S.F.	1.21			1.21	1.33	
600	0010 **KIOSKS**										600
	0020 Round, 5′ diameter, 8′ high, 1/4″ fiberglass wall				Ea.	5,600			5,600	6,175	
	0100 1″ insulated double wall, fiberglass					6,375			6,375	7,025	
	0500 Rectangular, 5′ x 9′, 7′-6″ high, 1/4″ fiberglass wall					8,150			8,150	8,975	
	0600 1″ insulated double wall, fiberglass				↓	9,700			9,700	10,700	
880	0010 **SWIMMING POOL ENCLOSURE** Translucent, free standing,										880
	0020 not including foundations, heat or light										
	0200 Economy, minimum	2 Carp	200	.080	SF Hor.	11.90	2.84		14.74	17.80	
	0300 Maximum		100	.160		35	5.70		40.70	48	
	0400 Deluxe, minimum		100	.160		41	5.70		46.70	55	
	0600 Maximum	↓	70	.229	↓	184	8.15		192.15	215	

13150 | Swimming Pools

	13151 Swimming Pools	CREW	DAILY OUTPUT	LABOR-HOURS	UNIT	2006 BARE COSTS MAT.	LABOR	EQUIP.	TOTAL	TOTAL INCL O&P	
200	0011 **SWIMMING POOLS**, Outdoor, incl. equip. & houses, minimum				SF Surf					36	200
	0300 Maximum									68	
	0400 Residential, incl. equipment, permanent type, minimum									12	
	0700 Maximum									25	
	0900 Municipal, including equipment only, over 5000 S.F., minimum									25	
	1000 Maximum									51	
	1300 Motel or apt., incl. equipment only, under 5000 S.F., minimum									22	
	1400 Maximum				↓					34	

		13281 | Hazardous Material Remediation	CREW	DAILY OUTPUT	LABOR-HOURS	UNIT	2006 BARE COSTS MAT.	LABOR	EQUIP.	TOTAL	TOTAL INCL O&P	
120	0010	**BULK ASBESTOS REMOVAL**										120
	0020	Includes disposable tools and 2 suits and 1respirator filter/day/worker										
	0100	Beams, W 10 x 19	A-9	235	.272	L.F.	.76	10.65		11.41	18.25	
	0110	W 12 x 22		210	.305		.85	11.90		12.75	20.50	
	0120	W 14 x 26		180	.356		.99	13.90		14.89	23.50	
	0130	W 16 x 31		160	.400		1.12	15.65		16.77	26.50	
	0140	W 18 x 40		140	.457		1.27	17.90		19.17	30.50	
	0150	W 24 x 55		110	.582		1.62	23		24.62	39	
	0160	W 30 x 108		85	.753		2.10	29.50		31.60	50.50	
	0170	W 36 x 150		72	.889		2.48	35		37.48	59	
	0200	Boiler insulation		480	.133	S.F.	.43	5.20		5.63	8.95	
	0210	With metal lath add				%				50%		
	0300	Boiler breeching or flue insulation	A-9	520	.123	S.F.	.34	4.81		5.15	8.25	
	0310	For active boiler, add				%				100%		
	0400	Duct or AHU insulation	A-10B	440	.073	S.F.	.20	2.85		3.05	4.87	
	0500	Duct vibration isolation joints, up to 24 Sq. In. duct	A-9	56	1.143	Ea.	3.19	44.50		47.69	76.50	
	0520	25 Sq. In. to 48 Sq. In. duct		48	1.333		3.72	52		55.72	89	
	0530	49 Sq. In. to 76 Sq. In. duct		40	1.600		4.46	62.50		66.96	107	
	0600	Pipe insulation, air cell type, up to 4″ diameter pipe		900	.071	L.F.	.20	2.78		2.98	4.76	
	0610	4″ to 8″ diameter pipe		800	.080		.22	3.13		3.35	5.35	
	0620	10″ to 12″ diameter pipe		700	.091		.25	3.58		3.83	6.15	
	0630	14″ to 16″ diameter pipe		550	.116		.32	4.55		4.87	7.75	
	0650	Over 16″ diameter pipe		650	.098	S.F.	.27	3.85		4.12	6.60	
	0700	With glove bag up to 3″ diameter pipe		300	.213	L.F.	3.15	8.35		11.50	17.05	
	1000	Pipe fitting insulation up to 4″ diameter pipe		320	.200	Ea.	.56	7.80		8.36	13.35	
	1100	6″ to 8″ diameter pipe		304	.211		.59	8.25		8.84	14.10	
	1110	10″ to 12″ diameter pipe		192	.333		.93	13.05		13.98	22.50	
	1120	14″ to 16″ diameter pipe		128	.500		1.39	19.55		20.94	33.50	
	1130	Over 16″ diameter pipe		176	.364	S.F.	1.01	14.20		15.21	24	
	1200	With glove bag, up to 8″ diameter pipe		100	.640	L.F.	6.55	25		31.55	48	
	2000	Scrape foam fireproofing from flat surface		2400	.027	S.F.	.07	1.04		1.11	1.78	
	2100	Irregular surfaces		1200	.053		.15	2.09		2.24	3.56	
	3000	Remove cementitious material from flat surface		1800	.036		.10	1.39		1.49	2.38	
	3100	Irregular surface		1400	.046		.13	1.79		1.92	3.06	
	4000	Scrape acoustical coating/fireproofing, from ceiling		3200	.020		.06	.78		.84	1.34	
	5000	Remove VAT from floor by hand		2400	.027		.07	1.04		1.11	1.78	
	5100	By machine	A-11	4800	.013		.04	.52	.01	.57	.90	
	5150	For 2 layers, add				%				50%		
	6000	Remove contaminated soil from crawl space by hand	A-9	400	.160	C.F.	.45	6.25		6.70	10.70	
	6100	With large production vacuum loader	A-12	700	.091	″	.25	3.58	.71	4.54	6.90	
	7000	Radiator backing, not including radiator removal	A-9	1200	.053	S.F.	.15	2.09		2.24	3.56	
	8000	Cement-asbestos transite board	2 Asbe	1000	.016		.12	.62		.74	1.16	
	8100	Transite shingle siding	A-10B	750	.043		.20	1.67		1.87	2.95	
	8200	Shingle roofing	″	2000	.016		.07	.63		.70	1.10	
	8250	Built-up, no gravel, non-friable	B-2	1400	.029		.07	.79		.86	1.40	
	8300	Asbestos millboard	2 Asbe	1000	.016		.08	.62		.70	1.11	
	9000	For type C (supplied air) respirator equipment, add				%					10%	
125	0010	**ASBESTOS ABATEMENT WORK AREA** Containment and preparation.										125
	0100	Pre-cleaning, HEPA vacuum and wet wipe, flat surfaces	A-10	12000	.005	S.F.	.01	.21		.22	.36	
	0200	Protect carpeted area, 2 layers 6 mil poly on 3/4″ plywood	″	1000	.064		1.50	2.50		4	5.75	
	0300	Separation barrier, 2″ x 4″ @ 16″, 1/2″ plywood ea. side, 8′ high	2 Carp	400	.040		1.25	1.42		2.67	3.74	
	0310	12′ high		320	.050		1.40	1.78		3.18	4.49	
	0320	16′ high		200	.080		1.50	2.84		4.34	6.35	
	0400	Personnel decontam. chamber, 2″ x 4″ @ 16″, 3/4″ ply ea. side		280	.057		2.50	2.03		4.53	6.10	
	0450	Waste decontam. chamber, 2″ x 4″ studs @ 16″, 3/4″ ply ea. side		360	.044		3	1.58		4.58	5.90	

	13281 Hazardous Material Remediation		CREW	DAILY OUTPUT	LABOR-HOURS	UNIT	2006 BARE COSTS MAT.	LABOR	EQUIP.	TOTAL	TOTAL INCL O&P
125	0500	Cover surfaces with polyethelene sheeting									
	0501	Including glue and tape									
	0550	Floors, each layer, 6 mil	A-10	8000	.008	S.F.	.11	.31		.42	.63
	0551	4 mil		9000	.007		.06	.28		.34	.52
	0560	Walls, each layer, 6 mil		6000	.011		.11	.42		.53	.80
	0561	4 mil		7000	.009		.06	.36		.42	.65
	0570	For heights above 12′, add						20%			
	0575	For heights above 20′, add						30%			
	0580	For fire retardant poly, add					100%				
	0590	For large open areas, deduct					10%	20%			
	0600	Seal floor penetrations with foam firestop to 36 Sq. In.	2 Carp	200	.080	Ea.	6.25	2.84		9.09	11.60
	0610	36 Sq. In. to 72 Sq. In.		125	.128		12.50	4.55		17.05	21.50
	0615	72 Sq. In. to 144 Sq. In.		80	.200		25	7.10		32.10	39.50
	0620	Wall penetrations, to 36 square inches		180	.089		6.25	3.16		9.41	12.15
	0630	36 Sq. In. to 72 Sq. In.		100	.160		12.50	5.70		18.20	23
	0640	72 Sq. In. to 144 Sq. In.		60	.267		25	9.50		34.50	43
	0800	Caulk seams with latex	1 Carp	230	.035	L.F.	.15	1.24		1.39	2.22
	0900	Set up neg. air machine, 1-2k C.F.M. /25 M.C.F. volume	1 Asbe	4.30	1.860	Ea.		72.50		72.50	119
130	0010	**DEMOLITION IN ASBESTOS CONTAMINATED AREA**									
	0200	Ceiling, including suspension system, plaster and lath	A-9	2100	.030	S.F.	.08	1.19		1.27	2.03
	0210	Finished plaster, leaving wire lath		585	.109		.30	4.28		4.58	7.35
	0220	Suspended acoustical tile		3500	.018		.05	.72		.77	1.23
	0230	Concealed tile grid system		3000	.021		.06	.83		.89	1.43
	0240	Metal pan grid system		1500	.043		.12	1.67		1.79	2.85
	0250	Gypsum board		2500	.026		.07	1		1.07	1.71
	0260	Lighting fixtures up to 2′ x 4′		72	.889	Ea.	2.48	35		37.48	59
	0400	Partitions, non load bearing									
	0410	Plaster, lath, and studs	A-9	690	.093	S.F.	.81	3.63		4.44	6.80
	0450	Gypsum board and studs	"	1390	.046	"	.13	1.80		1.93	3.08
	9000	For type C (supplied air) respirator equipment, add				%					10%
135	0010	**ASBESTOS ABATEMENT EQUIPMENT** and supplies, buy R028213-20									
	0200	Air filtration device, 2000 C.F.M.				Ea.	2,500			2,500	2,750
	0250	Large volume air sampling pump, minimum					365			365	400
	0260	Maximum					680			680	745
	0300	Airless sprayer unit, 2 gun					2,000			2,000	2,200
	0350	Light stand, 500 watt					250			250	275
	0400	Personal respirators									
	0410	Negative pressure, 1/2 face, dual operation, min.				Ea.	22.50			22.50	25
	0420	Maximum					24			24	26.50
	0450	P.A.P.R., full face, minimum					400			400	440
	0460	Maximum					700			700	770
	0470	Supplied air, full face, incl. air line, minimum					450			450	495
	0480	Maximum					600			600	660
	0500	Personnel sampling pump, minimum					450			450	495
	0510	Maximum					750			750	825
	1500	Power panel, 20 unit, incl. G.F.I.					1,800			1,800	1,975
	1600	Shower unit, including pump and filters					1,125			1,125	1,250
	1700	Supplied air system (type C)					10,000			10,000	11,000
	1750	Vacuum cleaner, HEPA, 16 gal., stainless steel, wet/dry					1,000			1,000	1,100
	1760	55 gallon					2,200			2,200	2,425
	1800	Vacuum loader, 9-18 ton/hr					90,000			90,000	99,000
	1900	Water atomizer unit, including 55 gal. drum					230			230	253
	2000	Worker protection, whole body, foot, head cover & gloves, plastic					5.60			5.60	6.15
	2500	Respirator, single use					10.30			10.30	11.35
	2550	Cartridge for respirator					11.10			11.10	12.20
	2570	Glove bag, 7 mil, 50″ x 64″					8.50			8.50	9.35

13280 | Hazardous Material Remediation

	13281	Hazardous Material Remediation	CREW	DAILY OUTPUT	LABOR-HOURS	UNIT	2006 BARE COSTS MAT.	LABOR	EQUIP.	TOTAL	TOTAL INCL O&P	
135	2580	10 mil, 44" x 60" R028213-20				Ea.	8.40			8.40	9.25	135
	3000	HEPA vacuum for work area, minimun					1,050			1,050	1,150	
	3050	Maximum					2,875			2,875	3,175	
	6000	Disposable polyethelene bags, 6 mil, 3 C.F.					1.15			1.15	1.27	
	6300	Disposable fiber drums, 3 C.F.					6.50			6.50	7.15	
	6400	Pressure sensitive caution lables, 3" x 5"					1.10			1.10	1.21	
	6450	11" x 17"					6.50			6.50	7.15	
	6500	Negative air machine, 1800 C.F.M.					775			775	855	
140	0010	**DECONTAMINATION CONTAINMENT AREA DEMOLITION** and clean-up										140
	0100	Spray exposed substrate with surfactant (bridging)										
	0200	Flat surfaces	A-9	6000	.011	S.F.	.36	.42		.78	1.08	
	0250	Irregular surfaces		4000	.016	"	.31	.63		.94	1.36	
	0300	Pipes, beams, and columns		2000	.032	L.F.	.56	1.25		1.81	2.66	
	1000	Spray encapsulate polyethelene sheeting		8000	.008	S.F.	.31	.31		.62	.85	
	1100	Roll down polyethelene sheeting		8000	.008	"		.31		.31	.51	
	1500	Bag polyethelene sheeting		400	.160	Ea.	.77	6.25		7.02	11.05	
	2000	Fine clean exposed substrate, with nylon brush		2400	.027	S.F.		1.04		1.04	1.70	
	2500	Wet wipe substrate		4800	.013			.52		.52	.85	
	2600	Vacuum surfaces, fine brush		6400	.010			.39		.39	.64	
	3000	Structural demolition										
	3100	Wood stud walls	A-9	2800	.023	S.F.		.89		.89	1.46	
	3500	Window manifolds, not incl. window replacement		4200	.015			.60		.60	.97	
	3600	Plywood carpet protection		2000	.032			1.25		1.25	2.04	
	4000	Remove custom decontamination facility	A-10A	8	3	Ea.	15	118		133	209	
	4100	Remove portable decontamination facility	3 Asbe	12	2	"	12.50	78		90.50	141	
	5000	HEPA vacuum, shampoo carpeting	A-9	4800	.013	S.F.	.05	.52		.57	.91	
	9000	Final cleaning of protected surfaces	A-10A	8000	.003	"		.12		.12	.19	
145	0010	**OSHA TESTING**										145
	0100	Certified technician, minimum				Day					300	
	0110	Maximum				"					500	
	0200	Personal sampling, PCM analysis, NIOSH 7400, minimum	1 Asbe	8	1	Ea.	2.75	39		41.75	66.50	
	0210	Maximum	"	4	2	"	3	78		81	130	
	0300	Industrial hygenist, minimum				Day					400	
	0310	Maximum				"					550	
	1000	Cleaned area samples	1 Asbe	8	1	Ea.	2.63	39		41.63	66.50	
	1100	PCM air sample analysis, NIOSH 7400, minimum		8	1		30	39		69	96.50	
	1110	Maximum		4	2		3.09	78		81.09	130	
	1200	TEM air sample analysis, NIOSH 7402, minimum								100	125	
	1210	Maximum								400	500	
150	0010	**ENCAPSULATION WITH SEALANTS**										150
	0100	Ceilings and walls, minimum	A-9	21000	.003	S.F.	.27	.12		.39	.49	
	0110	Maximum		10600	.006		.42	.24		.66	.85	
	0200	Columns and beams, minimum		13300	.005		.27	.19		.46	.61	
	0210	Maximum		5325	.012		.47	.47		.94	1.29	
	0300	Pipes to 12" diameter including minor repairs, minimum		800	.080	L.F.	.37	3.13		3.50	5.50	
	0310	Maximum		400	.160	"	1.04	6.25		7.29	11.35	
155	0010	**WASTE PACKAGING, HANDLING, & DISPOSAL**										155
	0100	Collect and bag bulk material, 3 C.F. bags, by hand	A-9	400	.160	Ea.	1.15	6.25		7.40	11.45	
	0200	Large production vacuum loader	A-12	880	.073		.80	2.84	.57	4.21	6.15	
	1000	Double bag and decontaminate	A-9	960	.067		2.30	2.61		4.91	6.80	
	2000	Containerize bagged material in drums, per 3 C.F. drum	"	800	.080		6.50	3.13		9.63	12.25	
	3000	Cart bags 50' to dumpster	2 Asbe	400	.040			1.56		1.56	2.55	
	5000	Disposal charges, not including haul, minimum				C.Y.					50	
	5020	Maximum				"					175	

13 SPECIAL CONSTRUCTION

13280 | Hazardous Material Remediation

	13281 Hazardous Material Remediation		CREW	DAILY OUTPUT	LABOR-HOURS	UNIT	2006 BARE COSTS MAT.	LABOR	EQUIP.	TOTAL	TOTAL INCL O&P	
155	5100	Remove refrigerant from system	1 Plum	40	.200	Lb.		8.55		8.55	13.30	155
	9000	For type C (supplied air) respirator equipment, add				%					10%	
440	0010	**REMOVAL** Existing lead paint, by chemicals, per application R028319-60										440
	0050	Baseboard, to 6" wide	1 Pord	64	.125	L.F.	1.57	3.96		5.53	8.10	
	0070	To 12" wide		32	.250	"	3.09	7.95		11.04	16.15	
	0200	Balustrades, one side		28	.286	S.F.	3.50	9.05		12.55	18.40	
	1400	Cabinets, simple design		32	.250		3.06	7.95		11.01	16.10	
	1420	Ornate design		25	.320		3.94	10.15		14.09	20.50	
	1600	Cornice, simple design		60	.133		1.65	4.23		5.88	8.60	
	1620	Ornate design		20	.400		4.86	12.70		17.56	26	
	2800	Doors, one side, flush		84	.095		1.19	3.02		4.21	6.15	
	2820	Two panel		80	.100		1.23	3.17		4.40	6.45	
	2840	Four panel		45	.178		2.17	5.65		7.82	11.45	
	2880	For trim, one side, add		64	.125	L.F.	1.57	3.96		5.53	8.10	
	3000	Fence, picket, one side		30	.267	S.F.	3.29	8.45		11.74	17.15	
	3200	Grilles, one side, simple design		30	.267		3.29	8.45		11.74	17.15	
	3220	Ornate design		25	.320		3.94	10.15		14.09	20.50	
	4400	Pipes, to 4" diameter		90	.089	L.F.	1.13	2.82		3.95	5.75	
	4420	To 8" diameter		50	.160		1.96	5.05		7.01	10.30	
	4440	To 12" diameter		36	.222		2.75	7.05		9.80	14.35	
	4460	To 16" diameter		20	.400		4.89	12.70		17.59	26	
	4500	For hangers, add		40	.200	Ea.	2.45	6.35		8.80	12.90	
	4800	Siding		90	.089	S.F.	1.13	2.82		3.95	5.75	
	5000	Trusses, open		55	.145	SF Face	1.79	4.61		6.40	9.35	
	6200	Windows, one side only, double hung, 1/1 light, 24" x 48" high		4	2	Ea.	24.50	63.50		88	129	
	6220	30" x 60" high		3	2.667		33	84.50		117.50	172	
	6240	36" x 72" high		2.50	3.200		39.50	101		140.50	207	
	6280	40" x 80" high		2	4		49.50	127		176.50	258	
	6400	Colonial window, 6/6 light, 24" x 48" high		2	4		49.50	127		176.50	259	
	6420	30" x 60" high		1.50	5.333		65.50	169		234.50	345	
	6440	36" x 72" high		1	8		98.50	254		352.50	515	
	6480	40" x 80" high		1	8		98.50	254		352.50	515	
	6600	8/8 light, 24" x 48" high		2	4		49.50	127		176.50	259	
	6620	40" x 80" high		1	8		98.50	254		352.50	515	
	6800	12/12 light, 24" x 48" high		1	8		98.50	254		352.50	515	
	6820	40" x 80" high		.75	10.667		131	340		471	690	
	6840	Window frame & trim items, included in pricing above										
	9000	Minimum labor/equipment charge	1 Pord	3	2.667	Job		84.50		84.50	136	
460	0010	**LEAD PAINT ENCAPSULATION**, water based polymer coating,14 mil DFT										460
	0020	Interior, brushwork, trim, under 6"	1 Pord	240	.033	L.F.	2.37	1.06		3.43	4.31	
	0030	6" to 12" wide		180	.044		3.15	1.41		4.56	5.75	
	0040	Balustrades		300	.027		1.90	.85		2.75	3.45	
	0050	Pipe to 4" diameter		500	.016		1.14	.51		1.65	2.06	
	0060	To 8" diameter		375	.021		1.51	.68		2.19	2.75	
	0070	To 12" diameter		250	.032		2.27	1.01		3.28	4.13	
	0080	To 16" diameter		170	.047		3.34	1.49		4.83	6.05	
	0090	Cabinets, ornate design		200	.040	S.F.	2.86	1.27		4.13	5.20	
	0100	Simple design		250	.032	"	2.27	1.01		3.28	4.13	
	0110	Doors, 3'x 7', both sides, incl. frame & trim										
	0120	Flush	1 Pord	6	1.333	Ea.	29	42.50		71.50	100	
	0130	French, 10-15 lite		3	2.667		5.80	84.50		90.30	142	
	0140	Panel		4	2		35	63.50		98.50	141	
	0150	Louvered		2.75	2.909		32	92		124	183	
	0160	Windows, per interior side, per 15 S.F.										
	0170	1 to 6 lite	1 Pord	14	.571	Ea.	20	18.10		38.10	51	
	0180	7 to 10 lite		7.50	1.067		22	34		56	79	

13280 | Hazardous Material Remediation

	13281 Hazardous Material Remediation		Crew	Daily Output	Labor-Hours	Unit	2006 Bare Costs Mat.	Labor	Equip.	Total	Total Incl O&P
460	0190	12 lite	1 Pord	5.75	1.391	Ea.	30	44		74	104
	0200	Radiators		8	1		71	31.50		102.50	129
	0210	Grilles, vents		275	.029	S.F.	2.07	.92		2.99	3.76
	0220	Walls, roller, drywall or plaster		1000	.008		.57	.25		.82	1.04
	0230	With spunbonded reinforcing fabric		720	.011		.65	.35		1	1.29
	0240	Wood		800	.010		.71	.32		1.03	1.29
	0250	Ceilings, roller, drywall or plaster		900	.009		.65	.28		.93	1.17
	0260	Wood		700	.011		.81	.36		1.17	1.47
	0270	Exterior, brushwork, gutters and downspouts		300	.027	L.F.	1.90	.85		2.75	3.45
	0280	Columns		400	.020	S.F.	1.41	.63		2.04	2.57
	0290	Spray, siding		600	.013	"	.95	.42		1.37	1.73
	0300	Miscellaneous									
	0310	Electrical conduit, brushwork, to 2" diameter	1 Pord	500	.016	L.F.	1.14	.51		1.65	2.06
	0320	Brick, block or concrete, spray		500	.016	S.F.	1.14	.51		1.65	2.06
	0330	Steel, flat surfaces and tanks to 12"		500	.016		1.14	.51		1.65	2.06
	0340	Beams, brushwork		400	.020		1.41	.63		2.04	2.57
	0350	Trusses		400	.020		1.41	.63		2.04	2.57
600	0010	**MOLD ABATEMENT WORK AREA** Containment and prep.									
	0100	Pre-cleaning, HEPA vacuum and wet wipe, flat surfaces	A-10	12000	.005	S.F.	.01	.21		.22	.36
	0300	Separation barrier, 2" x 4" @ 16", 1/2" plywood ea. side, 8' high	2 Carp	400	.040		1.25	1.42		2.67	3.74
	0310	12' high		320	.050		1.40	1.78		3.18	4.49
	0320	16' high		200	.080		1.50	2.84		4.34	6.35
	0400	Personnel decontam. chamber, 2" x 4" @ 16", 3/4" ply ea. side		280	.057		2.50	2.03		4.53	6.10
	0450	Waste decontam. chamber, 2" x 4" studs @ 16", 3/4" ply each side		360	.044		3	1.58		4.58	5.90
	0500	Cover surfaces with polyethelene sheeting									
	0501	Including glue and tape									
	0550	Floors, each layer, 6 mil	A-10	8000	.008	S.F.	.11	.31		.42	.63
	0551	4 mil		9000	.007		.06	.28		.34	.52
	0560	Walls, each layer, 6 mil		6000	.011		.09	.42		.51	.78
	0561	4 mil		7000	.009		.07	.36		.43	.66
	0570	For heights above 12', add						20%			
	0575	For heights above 20', add						30%			
	0580	For fire retardant poly, add					100%				
	0590	For large open areas, deduct					10%	20%			
	0600	Seal floor penetrations with foam firestop to 36 sq in	2 Carp	200	.080	Ea.	6.25	2.84		9.09	11.60
	0610	36 sq in to 72 sq in		125	.128		12.50	4.55		17.05	21.50
	0615	72 sq in to 144 sq in		80	.200		25	7.10		32.10	39.50
	0620	Wall penetrations, to 36 square inches		180	.089		6.25	3.16		9.41	12.15
	0630	36 Sq. in. to 72 sq. in.		100	.160		12.50	5.70		18.20	23
	0640	72 Sq. in. to 144 sq. in.		60	.267		25	9.50		34.50	43
	0800	Caulk seams with latex	1 Carp	230	.035	L.F.	.15	1.24		1.39	2.22
	0900	Set up neg. air machine, 1-2k C.F.M. /25 M.C.F. volume	1 Asbe	4.30	1.860	Ea.		72.50		72.50	119
610	0010	**DEMOLITION IN MOLD CONTAMINATED AREA**									
	0200	Ceiling, including suspension system, plaster and lath	A-9	2100	.030	S.F.	.08	1.19		1.27	2.03
	0210	Finished plaster, leaving wire lath		585	.109		.30	4.28		4.58	7.35
	0220	Suspended acoustical tile		3500	.018		.05	.72		.77	1.23
	0230	Concealed tile grid system		3000	.021		.06	.83		.89	1.43
	0240	Metal pan grid system		1500	.043		.12	1.67		1.79	2.85
	0250	Gypsum board		2500	.026		.07	1		1.07	1.71
	0255	Plywood		2500	.026		.07	1		1.07	1.71
	0260	Lighting fixtures up to 2' x 4'		72	.889	Ea.	2.48	35		37.48	59
	0400	Partitions, non load bearing									
	0410	Plaster, lath, and studs	A-9	690	.093	S.F.	.81	3.63		4.44	6.80
	0450	Gypsum board and studs		1390	.046		.13	1.80		1.93	3.08

13280 | Hazardous Material Remediation

	13281	Hazardous Material Remediation	CREW	DAILY OUTPUT	LABOR-HOURS	UNIT	2006 BARE COSTS MAT.	LABOR	EQUIP.	TOTAL	TOTAL INCL O&P	
610	0465	Carpet & pad	A-9	1390	.046	S.F.	.13	1.80		1.93	3.08	610
	0600	Pipe insulation, air cell type, up to 4" diameter pipe		900	.071	L.F.	.20	2.78		2.98	4.76	
	0610	4" to 8" diameter pipe		800	.080		.22	3.13		3.35	5.35	
	0620	10" to 12" diameter pipe		700	.091		.25	3.58		3.83	6.15	
	0630	14" to 16" diameter pipe		550	.116		.32	4.55		4.87	7.75	
	0650	Over 16" diameter pipe		650	.098	S.F.	.27	3.85		4.12	6.60	
	9000	For type C (supplied air) respirator equipment, add				%					10%	
750	0010	**MOLD ABATEMENT METHODS**										750
	0020	Initial inspection, average 3 bedroom home				Total					260	
	0030	Initial inspection, average 5 bedroom home									360	
	0040	Air sample test, each									260	
	0050	Swab sample test, each									155	
	0060	Tape sample test, each									155	
	0070	After remediation air test, each									260	
	0080	Mold abatement plan, average 3 bedroom home									1,275	
	0090	Mold abatement plan, average 5 bedroom home									1,550	
	0100	Packup & removal of contents, average 3 bedroom home, excl storage									10,300	
	0110	Packup & removal of contents, average 5 bedroom home, excl storage									25,800	
	0120	For demolition in mold contaminated areas, see Div. 13281-130										
	0130	For personal protection equipment, see Div. 13281-135										

13600 | Solar and Wind Energy Equipment

	13630	Solar Collector Components	CREW	DAILY OUTPUT	LABOR-HOURS	UNIT	2006 BARE COSTS MAT.	LABOR	EQUIP.	TOTAL	TOTAL INCL O&P	
200	0010	**SOLAR ENERGY**										200
	0020	System/Package prices, not including connecting										
	0030	pipe, insulation, or special heating/plumbing fixtures										
	0500	Hot water, standard package, low temperature										
	0580	2 collectors, circulator, fittings, 120 gal. tank	Q-1	.40	40	Ea.	2,275	1,525		3,800	4,900	
	2250	Controller, liquid temperature	1 Plum	5	1.600	"	87.50	68.50		156	202	
	2300	Circulators, air										
	2310	Blowers										
	2520	Space & DHW system, less duct work	Q-9	.50	32	Ea.	1,475	1,225		2,700	3,550	
	2580	8" diameter, 150 CFM		16	1		39.50	38		77.50	104	
	2660	Shutter/damper		12	1.333		39	50.50		89.50	124	
	2670	Shutter motor		16	1		92.50	38		130.50	163	
	2800	Circulators, liquid, 1/25 HP, 5.3 GPM	Q-1	14	1.143		147	44		191	231	
	2820	1/20 HP, 17 GPM	"	12	1.333		154	51		205	250	
	3000	Collector panels, air with aluminum absorber plate										
	3010	Wall or roof mount										
	3040	Flat black, plastic glazing										
	3080	4' x 8'	Q-9	6	2.667	Ea.	635	101		736	855	
	3300	Collector panels, liquid with copper absorber plate										
	3320	Black chrome, tempered glass glazing										
	3330	Alum. frame, 4' x 8', 5/32" single glazing	Q-1	9.50	1.684	Ea.	630	64.50		694.50	790	
	3450	Flat black, alum. frame, 3.5' x 7.5'		9	1.778		535	68.50		603.50	695	
	3500	4' x 8'		5.50	2.909		605	112		717	840	
	3600	Liquid, full wetted, plastic, alum. frame, 3' x 10'		5	3.200		195	123		318	405	

	13630 | Solar Collector Components		CREW	DAILY OUTPUT	LABOR-HOURS	UNIT	2006 BARE COSTS MAT.	LABOR	EQUIP.	TOTAL	TOTAL INCL O&P	
200	3650	Collector panel mounting, flat roof or ground rack	Q-1	7	2.286	Ea.	48	88		136	190	200
	3670	Roof clamps	↓	70	.229	Set	2.35	8.80		11.15	16.25	
	3700	Roof strap, teflon	1 Plum	205	.039	L.F.	9.50	1.67		11.17	13.05	
	3900	Differential controller with two sensors										
	3930	Thermostat, hard wired	1 Plum	8	1	Ea.	95	42.50		137.50	171	
	4050	Pool valve system	"	2.50	3.200	"	215	137		352	450	
	4150	Sensors										
	4220	Freeze prevention	1 Plum	32	.250	Ea.	22	10.70		32.70	40.50	
	4260											
	4300	Heat exchanger										
	4315	includes coil, blower, circulator										
	4316	and controller for DHW and space hot air										
	4380	70 MBH	Q-1	3.50	4.571	Ea.	335	176		511	640	
	4400	80 MBH	"	3	5.333	"	445	205		650	810	
	4580	Fluid to fluid package includes two circulating pumps										
	4590	expansion tank, check valve, relief valve										
	4600	controller, high temperature cutoff and sensors	Q-1	2.50	6.400	Ea.	695	246		941	1,150	
	4650	Heat transfer fluid										
	4700	Propylene glycol, inhibited anti-freeze	1 Plum	28	.286	Gal.	8.80	12.20		21	28.50	
	4800	Solar storage tanks, knocked down										
	5020	7′ x 10′-6″ = 459 C.F./3000 gallons	Q-10	.80	30	Ea.	13,100	1,175		14,275	16,300	
	5210	7′ x 14′ = 613 C.F./4000 gallons		.60	40		15,800	1,575		17,375	19,800	
	5230	10′-6″ x 14′ = 919 C.F./6000 gallons	↓	.40	60		19,100	2,350		21,450	24,800	
	5250	14′ x 17′-6″ = 1531 C.F./10,000 gallons	Q-11	.30	106	↓	25,000	4,275		29,275	34,300	
	7000	Solar control valves and vents										
	7070	Air eliminator, automatic 3/4″ size	1 Plum	32	.250	Ea.	28.50	10.70		39.20	48	
	7090	Air vent, automatic, 1/8″ fitting		32	.250		11.25	10.70		21.95	29	
	7100	Manual, 1/8″ NPT		32	.250		2.48	10.70		13.18	19.35	
	7120	Backflow preventer, 1/2″ pipe size		16	.500		58.50	21.50		80	97	
	7130	3/4″ pipe size		16	.500		62.50	21.50		84	102	
	7150	Balancing valve, 3/4″ pipe size		20	.400		26	17.10		43.10	55	
	7180	Draindown valve, 1/2″ copper tube		9	.889		202	38		240	281	
	7200	Flow control valve, 1/2″ pipe size		22	.364		58.50	15.55		74.05	88.50	
	7220	Expansion tank, up to 5 gal.		32	.250		61.50	10.70		72.20	84	
	7400	Pressure gauge, 2″ dial		32	.250		22	10.70		32.70	41	
	7450	Relief valve, temp. and pressure 3/4″ pipe size	↓	30	.267	↓	9.10	11.40		20.50	27.50	
	7500	Solenoid valve, normally closed										
	7750	Vacuum relief valve, 3/4″ pipe size	1 Plum	32	.250	Ea.	26.50	10.70		37.20	45.50	
	7800	Thermometers										
	8250	Water storage tank with heat exchanger and electric element										
	8270	66 gal. with 2″ x 2 lb. density insulation	1 Plum	1.60	5	Ea.	730	214		944	1,125	
	8300	80 gal. with 2″ x 2 lb. density insulation	"	1.60	5	"	835	214		1,049	1,250	
	8500	Water storage module, plastic										
	8600	Tubular, 12″ diameter, 4′ high	1 Carp	48	.167	Ea.	91	5.95		96.95	110	
	8610	12″ diameter, 8′ high	"	40	.200		140	7.10		147.10	166	
	8650	Cap, 12″ diameter				↓	16			16	17.60	
	9000	Minimum labor/equipment charge	1 Plum	2	4	Job		171		171	266	

13700 | Security Access and Surveillance

13710 | Security Access

		13710 Security Access	CREW	DAILY OUTPUT	LABOR-HOURS	UNIT	2006 BARE COSTS MAT.	LABOR	EQUIP.	TOTAL	TOTAL INCL O&P	
300	0010	**ACCESS CONTROL**										300
	0020	Card type, 1 time zone, minimum				Ea.	320			320	350	
	0040	Maximum					1,075			1,075	1,175	
	0060	3 time zones, minimum					785			785	865	
	0080	Maximum					1,800			1,800	2,000	
	0100	System with printer, and control console, 3 zones				Total	8,800			8,800	9,700	
	0120	6 zones				″	11,600			11,600	12,800	
	0140	For each door, minimum, add				Ea.	1,300			1,300	1,425	
	0160	Maximum, add					1,925			1,925	2,125	
	0220	Hand geometry scanner, mem of 512 users, excl striker/powr	1 Elec	3	2.667		1,600	112		1,712	1,925	
	0230	Memory upgrade for, adds 9,700 user profiles		8	1		225	42		267	315	
	0240	Adds 32,500 user profiles		8	1		450	42		492	560	
	0250	Prison type, memory of 256 users, excl striker, power		3	2.667		2,150	112		2,262	2,550	
	0260	Memory upgrade for, adds 3,300 user profiles		8	1		160	42		202	241	
	0270	Adds 9,700 user profiles		8	1		350	42		392	450	
	0280	Adds 27,900 user profiles		8	1		450	42		492	560	
	0290	All weather, mem of 512 users, excl striker/pwr		3	2.667		3,225	112		3,337	3,725	
	0300	Facial & fingerprint scanner, combination unit, excl striker/power		3	2.667		4,200	112		4,312	4,800	
	0310	Access for, for initial setup, excl striker/power		3	2.667		100	112		212	283	
400	0010	**ACCESS CONTROL**										400
	0200	Video cameras, wireless, hidden in exit signs, clocks, etc, incl receiver	1 Elec	3	2.667	Ea.	300	112		412	505	
	0210	Accessories for, VCR, single camera		3	2.667		500	112		612	725	
	0220	For multiple cameras		3	2.667		1,500	112		1,612	1,825	
	0230	Video cameras, wireless, for under vehicle searching, complete		2	4		3,500	168		3,668	4,100	
	0240	Metal detector, hand-held, wand type, unit only									90	
	0250	Metal detector, walk through portal type, single zone	1 Elec	2	4		4,000	168		4,168	4,650	
	0260	Multi zone		2	4		5,000	168		5,168	5,750	
	0270	Explosives detector, walk through portal type		2	4		3,500	168		3,668	4,100	
	0280	Explosives detector, hand-held, battery operated									28,000	
	0290	X-ray machine, desk top, for mail/small packages/letters	1 Elec	4	2		3,000	84		3,084	3,425	
	0300	Conveyor type, incl monitor, min		2	4		14,000	168		14,168	15,700	
	0310	Maximum		2	4		25,000	168		25,168	27,800	
	0320	X-ray machine, large unit, for airports, incl monitor, min	2 Elec	1	16		35,000	670		35,670	39,500	
	0330	Maximum	″	.50	32		60,000	1,350		61,350	68,000	

13720 | Detection & Alarm

		13720 Detection & Alarm	CREW	DAILY OUTPUT	LABOR-HOURS	UNIT	MAT.	LABOR	EQUIP.	TOTAL	TOTAL INCL O&P	
065	0010	**DETECTION SYSTEMS**, not including wires & conduits										065
	0100	Burglar alarm, battery operated, mechanical trigger	1 Elec	4	2	Ea.	254	84		338	410	
	0200	Electrical trigger		4	2		305	84		389	465	
	0400	For outside key control, add		8	1		72	42		114	144	
	0600	For remote signaling circuitry, add		8	1		114	42		156	191	
	0800	Card reader, flush type, standard		2.70	2.963		850	124		974	1,125	
	1000	Multi-code		2.70	2.963		1,100	124		1,224	1,400	
	1200	Door switches, hinge switch		5.30	1.509		53.50	63.50		117	157	
	1400	Magnetic switch		5.30	1.509		63	63.50		126.50	167	
	1600	Exit control locks, horn alarm		4	2		315	84		399	480	
	1800	Flashing light alarm		4	2		355	84		439	525	
	2000	Indicating panels, 1 channel		2.70	2.963		335	124		459	560	
	2200	10 channel	2 Elec	3.20	5		1,150	210		1,360	1,575	
	2400	20 channel		2	8		2,250	335		2,585	3,000	
	2600	40 channel		1.14	14.035		4,075	590		4,665	5,400	
	2800	Ultrasonic motion detector, 12 volt	1 Elec	2.30	3.478		210	146		356	455	
	3000	Infrared photoelectric detector		2.30	3.478		173	146		319	415	
	3200	Passive infrared detector		2.30	3.478		259	146		405	510	

13700 | Security Access and Surveillance

	13720 Detection & Alarm		CREW	DAILY OUTPUT	LABOR-HOURS	UNIT	2006 BARE COSTS MAT.	LABOR	EQUIP.	TOTAL	TOTAL INCL O&P	
065	3400	Glass break alarm switch	1 Elec	8	1	Ea.	43.50	42		85.50	112	065
	3420	Switchmats, 30″ x 5′		5.30	1.509		77.50	63.50		141	183	
	3440	30″ x 25′		4	2		186	84		270	335	
	3460	Police connect panel		4	2		223	84		307	375	
	3480	Telephone dialer		5.30	1.509		350	63.50		413.50	485	
	3500	Alarm bell		4	2		71	84		155	207	
	3520	Siren		4	2		134	84		218	276	
	3540	Microwave detector, 10′ to 200′		2	4		610	168		778	935	
	3560	10′ to 350′		2	4		1,775	168		1,943	2,225	
	3594	Fire, alarm control panel										
	3600	4 zone	2 Elec	2	8	Ea.	940	335		1,275	1,550	
	3800	8 zone		1	16		1,400	670		2,070	2,575	
	4000	12 zone		.67	23.988		1,825	1,000		2,825	3,550	
	4200	Battery and rack	1 Elec	4	2		690	84		774	890	
	4400	Automatic charger		8	1		445	42		487	555	
	4600	Signal bell		8	1		49.50	42		91.50	119	
	4800	Trouble buzzer or manual station		8	1		37	42		79	105	
	5000	Detector, rate of rise		8	1		33.50	42		75.50	102	
	5100	Fixed temperature		8	1		28	42		70	95.50	
	5200	Smoke detector, ceiling type		6.20	1.290		75	54		129	166	
	5400	Duct type		3.20	2.500		250	105		355	435	
	5600	Strobe and horn		5.30	1.509		95	63.50		158.50	203	
	5800	Fire alarm horn		6.70	1.194		36.50	50		86.50	118	
	6000	Door holder, electro-magnetic		4	2		77.50	84		161.50	215	
	6200	Combination holder and closer		3.20	2.500		430	105		535	635	
	6400	Code transmitter		4	2		690	84		774	890	
	6600	Drill switch		8	1		86.50	42		128.50	160	
	6800	Master box		2.70	2.963		3,100	124		3,224	3,600	
	7000	Break glass station		8	1		50	42		92	120	
	7800	Remote annunciator, 8 zone lamp		1.80	4.444		175	187		362	480	
	8000	12 zone lamp	2 Elec	2.60	6.154		300	258		558	730	
	8200	16 zone lamp	″	2.20	7.273		300	305		605	800	
	8400	Standpipe or sprinkler alarm, alarm device	1 Elec	8	1		125	42		167	203	
	8600	Actuating device		8	1		290	42		332	385	
	9410	Minimum labor/equipment charge		4	2	Job		84		84	129	

13800 | Building Automation & Control

	13834 Electric/Electronic Control		CREW	DAILY OUTPUT	LABOR-HOURS	UNIT	2006 BARE COSTS MAT.	LABOR	EQUIP.	TOTAL	TOTAL INCL O&P	
200	0010	CONTROL SYSTEMS, ELECTRONIC										200
	9000	Minimum labor/equipment charge	1 Plum	8	1	Job		42.50		42.50	66.50	

	13838 Pneumatic/Electric Controls		CREW	DAILY OUTPUT	LABOR-HOURS	UNIT	MAT.	LABOR	EQUIP.	TOTAL	TOTAL INCL O&P	
200	0010	CONTROL COMPONENTS										200
	5000	Thermostats										
	5030	Manual	1 Shee	8	1	Ea.	27	42		69	96.50	
	5040	1 set back, electric, timed		8	1		81	42		123	156	
	5050	2 set back, electric, timed		8	1		178	42		220	263	
	5200	24 hour, automatic, clock		8	1		106	42		148	184	

13800 | Building Automation & Control

	13838 | Pneumatic/Electric Controls	CREW	DAILY OUTPUT	LABOR-HOURS	UNIT	2006 BARE COSTS MAT.	LABOR	EQUIP.	TOTAL	TOTAL INCL O&P
200	6000 Valves, motorized zone									
	6100 Sweat connections, 1/2" C x C	1 Stpi	20	.400	Ea.	136	17.20		153.20	176
	6110 3/4" C x C		20	.400		145	17.20		162.20	187
	6120 1" C x C		19	.421		170	18.15		188.15	215
	9000 Minimum labor/equipment charge	1 Plum	4	2	Job		85.50		85.50	133

13900 | Fire Suppression

	13910 | Basic Fire Protection Matl/Methd	CREW	DAILY OUTPUT	LABOR-HOURS	UNIT	2006 BARE COSTS MAT.	LABOR	EQUIP.	TOTAL	TOTAL INCL O&P
400	0010 FIRE HOSE AND EQUIPMENT									
	2200 Hose, less couplings									
	2260 Synthetic jacket, lined, 300 lb. test, 1-1/2" diameter	Q-12	2600	.006	L.F.	1.85	.23		2.08	2.40
	2280 2-1/2" diameter	"	2200	.007	"	3.07	.27		3.34	3.81
	2600 Hose rack, swinging, for 1-1/2" diameter hose,									
	2620 Enameled steel, 50′ & 75′ lengths of hose	Q-12	20	.800	Ea.	33	30		63	83.50
	3750 Hydrants, wall, w/caps, single, flush, polished brass									
	3800 2-1/2" x 2-1/2"	Q-12	5	3.200	Ea.	135	120		255	335
	4350 Double, projecting, polished brass									
	4400 2-1/2" x 2-1/2" x 4"	Q-12	5	3.200	Ea.	148	120		268	350
	5600 Nozzles, brass									
	5620 Adjustable fog, 3/4" booster line				Ea.	70.50			70.50	77.50
	5640 1-1/2" leader line					91			91	100
	5660 2-1/2" direct connection					178			178	196
	7140 Standpipe connections, wall, w/plugs & chains									
	7280 Double, flush, polished brass									
	7300 2-1/2" x 2-1/2" x 4"	Q-12	5	3.200	Ea.	330	120		450	550
	9900 Minimum labor/equipment charge	1 Plum	2	4	Job		171		171	266
800	0010 FIRE VALVES									
	3000 Gate, hose, wheel handle, N.R.S., rough brass, 1-1/2"	1 Spri	12	.667	Ea.	112	28		140	167
	3040 2-1/2", 300 lb.		7	1.143	"	149	48		197	238
	9990 Minimum labor/equipment charge		4	2	Job		83.50		83.50	130

	13930 | Wet-Pipe Fire Supp. Sprinklers	CREW	DAILY OUTPUT	LABOR-HOURS	UNIT	MAT.	LABOR	EQUIP.	TOTAL	TOTAL INCL O&P
400	0010 SPRINKLER SYSTEM COMPONENTS									
	0800 Air compressor for dry pipe system, automatic, complete									
	0820 280 gal. system capacity, 3/4 HP	1 Spri	1.30	6.154	Ea.	745	257		1,002	1,225
	1220 Water motor, complete with gong	"	4	2	"	188	83.50		271.50	335
	1800 Firecycle system, controls, includes panel,									
	1820 batteries, solenoid valves and pressure switches	Q-13	1	32	Ea.	8,800	1,275		10,075	11,700
	1980 Detector	1 Spri	16	.500	"	325	21		346	390
	2600 Sprinkler heads, not including supply piping									
	2640 Dry, pendent, 1/2" orifice, 3/4" or 1" NPT									
	2660 1/2" to 6" length	1 Spri	14	.571	Ea.	44.50	24		68.50	85.50
	2700 15-1/4" to 18" length		14	.571		58.50	24		82.50	102
	3600 Foam-water, pendent or upright, 1/2" NPT		12	.667		61	28		89	111
	3700 Standard spray, pendent or upright, brass, 135° to 286° F									
	3740 1/2" NPT, 1/2" orifice	1 Spri	16	.500	Ea.	5.10	21		26.10	38
	3790 For riser & feeder piping, see div. 15107									
	6500 Check, swing, C.I. body, brass fittings, auto. ball drip									

13900 | Fire Suppression

	13930 | Wet-Pipe Fire Supp. Sprinklers		CREW	DAILY OUTPUT	LABOR-HOURS	UNIT	2006 BARE COSTS MAT.	LABOR	EQUIP.	TOTAL	TOTAL INCL O&P	
400	6520	4" size	Q-12	3	5.333	Ea.	163	201		364	495	400
	8200	Dry pipe valve, incl. trim and gauges, 3" size	↓	2	8	↓	1,250	300		1,550	1,850	
	8220	4" size	↓	1	16	↓	1,300	600		1,900	2,375	
	9990	Minimum labor/equipment charge	1 Spri	3	2.667	Job		111		111	174	

For information about Means Estimating Seminars, see yellow pages 12 and 13 in back of book

Division 14
Conveying Systems

Estimating Tips

General

- Many products in Division 14 will require some type of support or blocking for installation not included with the item itself. Examples are supports for conveyors or tube systems, attachment points for lifts, and footings for hoists or cranes. Add these supports in the appropriate division.

14100 Dumbwaiters
14200 Elevators

- Dumbwaiters and elevators are estimated and purchased in a method similar to buying a car. The manufacturer has a base unit with standard features. Added to this base unit price will be whatever options the owner or specifications require. Increased load capacity, additional vertical travel, additional stops, higher speed, and cab finish options are items to be considered. When developing an estimate for dumbwaiters and elevators, remember that some items needed by the installers may have to be included as part of the general contract.

  Examples are:
  - shaftway
  - rail support brackets
  - machine room
  - electrical supply
  - sill angles
  - electrical connections
  - pits
  - roof penthouses
  - pit ladders

  Check the job specifications and drawings before pricing.
- Installation of elevators and handicapped lifts in historic structures can require significant additional costs. The associated structural requirements may involve cutting into and repairing finishes, mouldings, flooring, etc. The estimator must account for these special conditions.

14300 Escalators & Moving Walks

- Escalators and moving walks are specialty items installed by specialty contractors. There are numerous options associated with these items. For specific options contact a manufacturer or contractor. In a method similar to estimating dumbwaiters and elevators, you should verify the extent of general contract work and add items as necessary.

14400 Lifts
14500 Material Handling
14600 Hoists & Cranes

- Products such as correspondence lifts, conveyors, chutes, pneumatic tube systems, material handling cranes, and hoists, as well as other items specified in this subdivision, may require trained installers. The general contractor might not have any choice as to who will perform the installation or when it will be performed. Long lead times are often required for these products, making early decisions in scheduling necessary.

Reference Numbers

Reference numbers are shown in bold squares at the beginning of some major classifications. These numbers refer to related items in the Reference Section. The reference information may be an estimating procedure, an alternate pricing method, or technical information.

Note: Not all subdivisions listed here necessarily appear in this publication.

14100 | Dumbwaiters

	14110 | Manual Dumbwaiters		CREW	DAILY OUTPUT	LABOR-HOURS	UNIT	2006 BARE COSTS MAT.	LABOR	EQUIP.	TOTAL	TOTAL INCL O&P	
400	0010	**DUMBWAITERS**, manual										400
	0020	2 stop, minimum	2 Elev	.75	21.333	Ea.	2,450	1,075		3,525	4,350	
	0100	Maximum		.50	32	"	5,600	1,600		7,200	8,625	
	0300	For each additional stop, add		.75	21.333	Stop	890	1,075		1,965	2,625	

	14120 | Electric Dumbwaiters		CREW	DAILY OUTPUT	LABOR-HOURS	UNIT	MAT.	LABOR	EQUIP.	TOTAL	TOTAL INCL O&P	
400	0010	**DUMBWAITERS**, electric										400
	0020	2 stop, minimum	2 Elev	.13	123	Ea.	6,200	6,150		12,350	16,300	
	0100	Maximum		.11	145	"	18,600	7,275		25,875	31,700	
	0600	For each additional stop, add		.54	29.630	Stop	2,725	1,475		4,200	5,275	
	0750	Correspondence lift, 1 floor, 2 stop, 45 lb capacity	2 Elec	.20	80	Ea.	7,025	3,350		10,375	12,900	

14200 | Elevators

	14210 | Electric Traction Elevators		CREW	DAILY OUTPUT	LABOR-HOURS	UNIT	2006 BARE COSTS MAT.	LABOR	EQUIP.	TOTAL	TOTAL INCL O&P	
100	0010	**ELEVATOR SYSTEMS**										100
	7000	Residential, cab type, 1 floor, 2 stop, minimum	2 Elev	.20	80	Ea.	8,500	4,000		12,500	15,500	
	7100	Maximum		.10	160		14,300	8,000		22,300	28,100	
	7200	2 floor, 3 stop, minimum		.12	133		12,600	6,650		19,250	24,200	
	7300	Maximum		.06	266		20,600	13,300		33,900	43,100	
200	0010	**ELEVATORS** R142000-10										200
	0020	For multi-story buildings, housing project, minimum				% total					2.50%	
	0100	Maximum R142000-20									4.50%	
	0300	Office building, minimum									2.50%	
	0400	Maximum R142000-30									10%	
	0425	Electric freight, base unit, 4000 lb, 200 fpm, 4 stp, std. fin.	2 Elev	.05	320	Ea.	73,500	16,000		89,500	105,000	
	0450	For 5000 lb capacity, add R142000-40					5,250			5,250	5,775	
	0500	For 6000 lb capacity, add					9,275			9,275	10,200	
	0525	For 7000 lb capacity, add					12,400			12,400	13,700	
	0550	For 8000 lb capacity, add					17,000			17,000	18,700	
	0575	For 10000 lb capacity, add					20,300			20,300	22,300	
	0600	For 12000 lb capacity, add					24,800			24,800	27,300	
	0625	For 16000 lb capacity, add					29,800			29,800	32,800	
	0650	For 20000 lb capacity, add					32,900			32,900	36,200	
	0675	For increased speed, 250 fpm, add					9,875			9,875	10,900	
	0700	300 fpm, geared electric, add					12,300			12,300	13,500	
	0725	350 fpm, geared electric, add					14,500			14,500	16,000	
	0750	400 fpm, geared electric, add					16,400			16,400	18,100	
	0775	500 fpm, gearless electric, add					21,000			21,000	23,100	
	0800	600 fpm, gearless electric, add					23,200			23,200	25,500	
	0825	700 fpm, gearless electric, add					27,000			27,000	29,700	
	0850	800 fpm, gearless electric, add					30,000			30,000	33,000	
	0875	For class "B" loading, add					1,775			1,775	1,950	
	0900	For class "C-1" loading, add					4,400			4,400	4,850	
	0925	For class "C-2" loading, add					5,250			5,250	5,775	
	0950	For class "C-3" loading, add					7,225			7,225	7,950	
	0975	For travel over 40 V.L.F., add	2 Elev	7.25	2.207	V.L.F.	105	110		215	285	
	1000	For number of stops over 4, add		.27	59.259	Stop	1,875	2,950		4,825	6,625	

14200 | Elevators

14210 | Electric Traction Elevators

		14210 Electric Traction Elevators	CREW	DAILY OUTPUT	LABOR-HOURS	UNIT	2006 BARE COSTS MAT.	LABOR	EQUIP.	TOTAL	TOTAL INCL O&P	
200	1625	Electric pass., base unit, 2000 lb, 200 fpm, 4 stop, std. fin. R142000-10	2 Elev	.05	320	Ea.	68,500	16,000		84,500	100,000	200
	1650	For 2500 lb capacity, add					2,800			2,800	3,075	
	1675	For 3000 lb capacity, add R142000-20					4,275			4,275	4,700	
	1700	For 3500 lb capacity, add					5,950			5,950	6,525	
	1725	For 4000 lb capacity, add R142000-30					6,350			6,350	7,000	
	1750	For 4500 lb capacity, add					8,350			8,350	9,200	
	1775	For 5000 lb capacity, add R142000-40					10,600			10,600	11,700	
	1800	For increased speed, 250 fpm, geared electric, add					2,350			2,350	2,575	
	1825	300 fpm, geared electric, add					4,850			4,850	5,325	
	1850	350 fpm, geared electric, add					5,750			5,750	6,325	
	1875	400 fpm, geared electric, add					8,150			8,150	8,975	
	1900	500 fpm, gearless electric, add					38,300			38,300	42,100	
	1925	600 fpm, gearless electric, add					40,500			40,500	44,500	
	1950	700 fpm, gearless electric, add					44,300			44,300	48,700	
	1975	800 fpm, gearless electric, add					48,800			48,800	53,500	
	2000	For travel over 40 V.L.F., add	2 Elev	7.25	2.207	V.L.F.	107	110		217	288	
	2025	For number of stops over 4, add		.27	59.259	Stop	2,400	2,950		5,350	7,200	
	2400	Electric hospital, base unit, 4000 lb, 200 fpm, 4 stop, std fin.		.05	320	Ea.	75,000	16,000		91,000	107,000	
	2425	For 4500 lb capacity, add					5,175			5,175	5,700	
	2450	For 5000 lb capacity, add					6,775			6,775	7,450	
	2475	For increased speed, 250 fpm, geared electric, add					2,525			2,525	2,775	
	2500	300 fpm, geared electric, add					4,800			4,800	5,275	
	2525	350 fpm, geared electric, add					5,850			5,850	6,450	
	2550	400 fpm, geared electric, add					8,175			8,175	8,975	
	2575	500 fpm, gearless electric, add					36,600			36,600	40,300	
	2600	600 fpm, gearless electric, add					40,500			40,500	44,500	
	2625	700 fpm, gearless electric, add					44,200			44,200	48,600	
	2650	800 fpm, gearless electric, add					48,800			48,800	53,500	
	2675	For travel over 40 V.L.F., add	2 Elev	7.25	2.207	V.L.F.	107	110		217	288	
	2700	For number of stops over 4, add	″	.27	59.259	Stop	2,975	2,950		5,925	7,850	

14240 | Hydraulic Elevators

		14240 Hydraulic Elevators	CREW	DAILY OUTPUT	LABOR-HOURS	UNIT	MAT.	LABOR	EQUIP.	TOTAL	TOTAL INCL O&P	
200	0010	**HYDRAULIC ELEVATORS**										200
	1025	Hydraulic freight, base unit, 2000 lb, 50 fpm, 2 stop, std. fin.	2 Elev	.10	160	Ea.	36,100	8,000		44,100	52,000	
	1050	For 2500 lb capacity, add					2,625			2,625	2,875	
	1075	For 3000 lb capacity, add					4,150			4,150	4,550	
	1100	For 3500 lb capacity, add					6,275			6,275	6,900	
	1125	For 4000 lb capacity, add					6,700			6,700	7,375	
	1150	For 4500 lb capacity, add					7,725			7,725	8,500	
	1175	For 5000 lb capacity, add					10,500			10,500	11,600	
	1200	For 6000 lb capacity, add					11,100			11,100	12,200	
	1225	For 7000 lb capacity, add					17,000			17,000	18,700	
	1250	For 8000 lb capacity, add					19,000			19,000	20,900	
	1275	For 10000 lb capacity, add					20,200			20,200	22,200	
	1300	For 12000 lb capacity, add					24,400			24,400	26,800	
	1325	For 16000 lb capacity, add					31,800			31,800	35,000	
	1350	For 20000 lb capacity, add					35,300			35,300	38,800	
	1375	For increased speed, 100 fpm, add					755			755	830	
	1400	125 fpm, add					1,450			1,450	1,600	
	1425	150 fpm, add					2,725			2,725	3,000	
	1450	175 fpm, add					4,150			4,150	4,575	
	1475	For class "B" loading, add					1,725			1,725	1,900	
	1500	For class "C-1" loading, add					4,350			4,350	4,800	
	1525	For class "C-2" loading, add					5,225			5,225	5,725	
	1550	For class "C-3" loading, add					7,175			7,175	7,875	
	1575	For travel over 20 V.L.F., add	2 Elev	7.25	2.207	V.L.F.	350	110		460	555	

14200 | Elevators

14240 | Hydraulic Elevators

			CREW	DAILY OUTPUT	LABOR-HOURS	UNIT	2006 BARE COSTS MAT.	LABOR	EQUIP.	TOTAL	TOTAL INCL O&P	
200	1600	For number of stops over 2, add	2 Elev	.27	59.259	Stop	510	2,950		3,460	5,125	200
	2050	Hyd. pass., base unit, 1500 lb, 100 fpm, 2 stop, std. fin.	↓	.10	160	Ea.	30,600	8,000		38,600	46,000	
	2075	For 2000 lb capacity, add					975			975	1,075	
	2100	For 2500 lb capacity, add					2,000			2,000	2,200	
	2125	For 3000 lb capacity, add					3,625			3,625	4,000	
	2150	For 3500 lb capacity, add					5,375			5,375	5,925	
	2175	For 4000 lb capacity, add					6,475			6,475	7,125	
	2200	For 4500 lb capacity, add					7,800			7,800	8,600	
	2225	For 5000 lb capacity, add					11,800			11,800	13,000	
	2250	For increased speed, 125 fpm, add					1,150			1,150	1,275	
	2275	150 fpm, add					2,300			2,300	2,525	
	2300	175 fpm, add					3,725			3,725	4,100	
	2325	200 fpm, add				↓	5,425			5,425	5,975	
	2350	For travel over 12 V.L.F., add	2 Elev	7.25	2.207	V.L.F.	350	110		460	555	
	2375	For number of stops over 2, add		.27	59.259	Stop	3,050	2,950		6,000	7,925	
	2725	Hydraulic hospital, base unit, 4000 lb, 100 fpm, 2 stop, std. fin.	↓	.10	160	Ea.	42,500	8,000		50,500	59,000	
	2750	For 4000 lb capacity, add					5,125			5,125	5,650	
	2775	For 4500 lb capacity, add					6,025			6,025	6,625	
	2800	For 5000 lb capacity, add					8,775			8,775	9,650	
	2825	For increased speed, 125 fpm, add					1,425			1,425	1,575	
	2850	150 fpm, add					2,400			2,400	2,650	
	2875	175 fpm, add					3,875			3,875	4,250	
	2900	200 fpm, add				↓	5,650			5,650	6,200	
	2925	For travel over 12 V.L.F., add	2 Elev	7.25	2.207	V.L.F.	234	110		344	430	
	2950	For number of stops over 2, add	"	.27	59.259	Stop	2,950	2,950		5,900	7,825	

14270 | Custom Elevator Cabs

			CREW	DAILY OUTPUT	LABOR-HOURS	UNIT	MAT.	LABOR	EQUIP.	TOTAL	TOTAL INCL O&P	
200	0010	CAB FINISHES										200
	3325	Passenger elevator cab finishes (based on 3500 lb cab size)										
	3350	Acrylic panel ceiling				Ea.	440			440	485	
	3375	Aluminum eggcrate ceiling					510			510	560	
	3400	Stainless steel doors					1,975			1,975	2,175	
	3425	Carpet flooring					370			370	405	
	3450	Epoxy flooring					305			305	335	
	3475	Quarry tile flooring					470			470	515	
	3500	Slate flooring					655			655	720	
	3525	Textured rubber flooring					119			119	131	
	3550	Stainless steel walls					2,650			2,650	2,925	
	3575	Stainless steel returns at door				↓	560			560	615	
	4450	Hospital elevator cab finishes (based on 3500 lb cab size)										
	4475	Aluminum eggcrate ceiling				Ea.	510			510	560	
	4500	Stainless steel doors					1,625			1,625	1,800	
	4525	Epoxy flooring					305			305	335	
	4550	Quarry tile flooring					465			465	515	
	4575	Textured rubber flooring					119			119	131	
	4600	Stainless steel walls					2,650			2,650	2,925	
	4625	Stainless steel returns at door				↓	560			560	615	

14280 | Elevator Equipment and Controls

			CREW	DAILY OUTPUT	LABOR-HOURS	UNIT	MAT.	LABOR	EQUIP.	TOTAL	TOTAL INCL O&P	
200	0010	ELEVATOR CONTROLS AND DOORS										200
	2975	Passenger elevator options										
	3000	2 car group automatic controls	2 Elev	.66	24.242	Ea.	2,450	1,200		3,650	4,550	
	3025	3 car group automatic controls		.44	36.364		3,675	1,825		5,500	6,825	
	3050	4 car group automatic controls		.33	48.485		6,075	2,425		8,500	10,400	
	3075	5 car group automatic controls	↓	.26	61.538	↓	8,275	3,075		11,350	13,900	

	14280 | Elevator Equipment and Controls	CREW	DAILY OUTPUT	LABOR-HOURS	UNIT	2006 BARE COSTS MAT.	LABOR	EQUIP.	TOTAL	TOTAL INCL O&P	
200	3100 6 car group automatic controls	2 Elev	.22	72.727	Ea.	12,500	3,625		16,125	19,400	200
	3125 Intercom service		3	5.333		325	266		591	765	
	3150 Duplex car selective collective		.66	24.242		2,700	1,200		3,900	4,825	
	3175 Center opening 1 speed doors		2	8		1,400	400		1,800	2,175	
	3200 Center opening 2 speed doors		2	8		1,675	400		2,075	2,475	
	3225 Rear opening doors (opposite front)		2	8		3,900	400		4,300	4,900	
	3250 Side opening 2 speed doors		2	8		3,700	400		4,100	4,700	
	3275 Automatic emergency power switching		.66	24.242		875	1,200		2,075	2,825	
	3300 Manual emergency power switching		8	2		335	100		435	525	
	3625 Hall finishes, stainless steel doors					925			925	1,025	
	3650 Stainless steel frames					925			925	1,025	
	3675 12 month maintenance contract									2,640	
	3700 Signal devices, hall lanterns	2 Elev	8	2		345	100		445	535	
	3725 Position indicators, up to 3		9.40	1.702		239	85		324	395	
	3750 Position indicators, per each over 3		32	.500		66	25		91	111	
	3775 High speed heavy duty door opener					1,675			1,675	1,850	
	3800 Variable voltage, O.H. gearless machine, min.	2 Elev	.16	100		24,600	5,000		29,600	34,800	
	3815 Maximum		.07	228		54,500	11,400		65,900	77,500	
	3825 Basement installed geared machine		.33	48.485		12,200	2,425		14,625	17,100	
	3850 Freight elevator options										
	3875 Doors, bi-parting	2 Elev	.66	24.242	Ea.	4,175	1,200		5,375	6,475	
	3900 Power operated door and gate	"	.66	24.242		16,900	1,200		18,100	20,500	
	3925 Finishes, steel plate floor					735			735	810	
	3950 14 ga. 1/4" x 4' steel plate walls					1,800			1,800	1,975	
	3975 12 month maintenance contract									1,992	
	4000 Signal devices, hall lanterns	2 Elev	8	2		345	100		445	535	
	4025 Position indicators, up to 3		9.40	1.702		239	85		324	395	
	4050 Position indicators, per each over 3		32	.500		66	25		91	111	
	4075 Variable voltage basement installed geared machine		.66	24.242		13,600	1,200		14,800	16,800	
	4100 Hospital elevator options										
	4125 2 car group automatic controls	2 Elev	.66	24.242	Ea.	2,450	1,200		3,650	4,550	
	4150 3 car group automatic controls		.44	36.364		3,675	1,825		5,500	6,825	
	4175 4 car group automatic controls		.33	48.485		6,075	2,425		8,500	10,400	
	4200 5 car group automatic controls		.26	61.538		8,300	3,075		11,375	13,900	
	4225 6 car group automatic controls		.22	72.727		12,500	3,625		16,125	19,400	
	4250 Intercom service		3	5.333		325	266		591	765	
	4275 Duplex car selective collective		.66	24.242		2,700	1,200		3,900	4,825	
	4300 Center opening 1 speed doors		2	8		1,400	400		1,800	2,175	
	4325 Center opening 2 speed doors		2	8		1,850	400		2,250	2,650	
	4350 Rear opening doors (opposite front)		2	8		3,900	400		4,300	4,900	
	4375 Side opening 2 speed doors		2	8		5,925	400		6,325	7,125	
	4400 Automatic emergency power switching		.66	24.242		875	1,200		2,075	2,825	
	4425 Manual emergency power switching		8	2		335	100		435	525	
	4675 Hall finishes, stainless steel doors					925			925	1,025	
	4700 Stainless steel frames					925			925	1,025	
	4725 12 month maintenance contract									3,985	
	4750 Signal devices, hall lanterns	2 Elev	8	2		345	100		445	535	
	4775 Position indicators, up to 3		9.40	1.702		239	85		324	395	
	4800 Position indicators, per each over 3		32	.500		66	25		91	111	
	4825 High speed heavy duty door opener					1,675			1,675	1,850	
	4850 Variable voltage, O.H. gearless machine, min.	2 Elev	.16	100		24,600	5,000		29,600	34,800	
	4865 Maximum		.07	228		54,500	11,400		65,900	77,500	
	4875 Basement installed geared machine		.33	48.485		12,200	2,425		14,625	17,100	
	5000 Drilling for piston, casing included, 18" diameter	B-48	80	.700	V.L.F.	34	21.50	45.50	101	123	

14400 | Lifts

		14420 | Wheelchair Lifts	CREW	DAILY OUTPUT	LABOR-HOURS	UNIT	2006 BARE COSTS MAT.	LABOR	EQUIP.	TOTAL	TOTAL INCL O&P	
100	0010	**CHAIR / WHEELCHAIR LIFT**										100
	7700	Stair climber (chair lift), single seat, minimum	2 Elev	1	16	Ea.	4,100	800		4,900	5,750	
	7800	Maximum		.20	80		5,650	4,000		9,650	12,400	
	8000	Wheelchair lift, minimum		1	16		5,625	800		6,425	7,425	
	8500	Maximum		.50	32		13,300	1,600		14,900	17,200	
	8700	Stair lift, minimum		1	16		11,100	800		11,900	13,500	
	8900	Maximum	↓	.20	80	↓	17,600	4,000		21,600	25,600	
		14450 | Vehicle Lifts										
700	0010	**HYDRAULIC LIFT AUTO/TRUCK**										700
	0200	Double post, 9000 lb frame	L-4	1.15	20.870	Ea.	5,150	690		5,840	6,800	
	0500	Four post, 50000 lb frame		.18	133		56,500	4,425		60,925	69,500	
	2200	Hoists, single post, 8,000# capacity, swivel arms		.40	60		4,525	2,000		6,525	8,225	
	2400	Two posts, adjustable frames, 11,000# capacity		.25	96		5,800	3,175		8,975	11,600	
	2500	24,000# capacity		.15	160		7,750	5,300		13,050	17,200	
	2700	7,500# capacity, frame supports		.50	48		6,450	1,600		8,050	9,700	
	2800	Four post, roll on ramp		.50	48		5,800	1,600		7,400	9,000	
	2810	Hydraulic lifts, above ground, 2 post, clear floor, 6000 lb cap		2.67	8.989		6,075	298		6,373	7,175	
	2815	9000 lb capacity		2.29	10.480		14,400	345		14,745	16,400	
	2820	15,000 lb capacity		2	12		16,300	400		16,700	18,600	
	2825	30,000 lb capacity		1.60	15		36,000	495		36,495	40,400	
	2830	4 post, ramp style, 25,000 lb capacity		2	12		14,100	400		14,500	16,200	
	2835	35,000 lb capacity		1	24		65,500	795		66,295	73,500	
	2840	50,000 lb capacity		1	24		73,500	795		74,295	82,000	
	2845	75,000 lb capacity	↓	1	24		85,000	795		85,795	95,500	
	2850	For drive thru tracks, add, minimum					905			905	995	
	2855	Maximum					1,550			1,550	1,700	
	2860	Ramp extensions, 3′(set of 2)					745			745	815	
	2865	Rolling jack platform					2,575			2,575	2,850	
	2870	Elec/hyd jacking beam					6,900			6,900	7,600	
	2880	Scissor lift, portable, 6000 lb capacity				↓	6,775			6,775	7,450	
		14460 | Correspondence & Parcel Lifts										
200	0010	**CORRESPONDENCE LIFT**										200
	0020	1 floor, 2 stop, 25 lb capacity, electric	2 Elev	.20	80	Ea.	5,025	4,000		9,025	11,700	
	0100	Hand, 5 lb capacity	"	.20	80	"	1,900	4,000		5,900	8,250	

14500 | Material Transport

		14560 | Chutes	CREW	DAILY OUTPUT	LABOR-HOURS	UNIT	2006 BARE COSTS MAT.	LABOR	EQUIP.	TOTAL	TOTAL INCL O&P	
250	0010	**CHUTES** Linen or refuse, incl. sprinklers, 12′ floor height										250
	0020											
	0050	Aluminized steel, 16 ga., 18″ diameter	2 Shee	3.50	4.571	Floor	870	193		1,063	1,250	
	0100	24″ diameter		3.20	5		945	211		1,156	1,375	
	0400	Galvanized steel, 16 ga., 18″ diameter		3.50	4.571		770	193		963	1,150	
	0500	24″ diameter		3.20	5		870	211		1,081	1,300	
	0800	Stainless steel, 18″ diameter		3.50	4.571		1,800	193		1,993	2,275	
	0900	24″ diameter	↓	3.20	5	↓	1,900	211		2,111	2,425	

14500 | Material Transport

14560 | Chutes

			CREW	DAILY OUTPUT	LABOR-HOURS	UNIT	2006 BARE COSTS MAT.	LABOR	EQUIP.	TOTAL	TOTAL INCL O&P	
250	9000	Minimum labor/equipment charge	1 Shee	1	8	Job		335		335	535	250

14580 | Pneumatic Tube Systems

			CREW	DAILY OUTPUT	LABOR-HOURS	UNIT	2006 BARE COSTS MAT.	LABOR	EQUIP.	TOTAL	TOTAL INCL O&P	
800	0010	**PNEUMATIC TUBE SYSTEM** Single tube, 2 stations, blower										800
	0020	100′ long, stock										
	0100	3″ diameter	2 Stpi	.12	133	Total	5,075	5,750		10,825	14,500	
	0300	4″ diameter	″	.09	177	″	5,725	7,650		13,375	18,200	
	0400	Twin tube, two stations or more, conventional system										
	0700	3″ round	2 Stpi	46	.348	L.F.	11.75	14.95		26.70	36.50	
	1050	Add for blower		2	8	System	4,025	345		4,370	4,950	
	1110	Plus for each round station, add		7.50	2.133	Ea.	450	92		542	640	
	1200	Alternate pricing method: base cost, minimum		.75	21.333	Total	4,925	920		5,845	6,825	
	1300	Maximum		.25	64	″	9,775	2,750		12,525	15,000	
	1500	Plus total system length, add, minimum		93.40	.171	L.F.	6.30	7.35		13.65	18.40	
	1600	Maximum	↓	37.60	.426	″	18.70	18.30		37	49	

14600 | Hoists & Cranes

14630 | Bridge Cranes

			CREW	DAILY OUTPUT	LABOR-HOURS	UNIT	2006 BARE COSTS MAT.	LABOR	EQUIP.	TOTAL	TOTAL INCL O&P	
300	0010	**CRANE RAIL**										300
	0020	Box beam bridge, no equipment included	E-4	3400	.009	Lb.	.91	.38	.03	1.32	1.74	
	0210	Running track only, 104 lb per yard, 20′ piece	″	160	.200	L.F.	16.70	8.10	.56	25.36	34	
700	0010	**OVERHEAD BRIDGE CRANES**										700
	0100	1 girder, 20′ span, 3 ton	M-3	1	34	Ea.	19,100	1,375	151	20,626	23,300	
	0125	5 ton		1	34		20,900	1,375	151	22,426	25,300	
	0150	7.5 ton		1	34		24,900	1,375	151	26,426	29,700	
	0175	10 ton		.80	42.500		32,900	1,700	189	34,789	39,100	
	0200	15 ton		.80	42.500		42,300	1,700	189	44,189	49,400	
	0225	30′ span, 3 ton		1	34		19,900	1,375	151	21,426	24,200	
	0250	5 ton		1	34		21,900	1,375	151	23,426	26,300	
	0275	7.5 ton		1	34		26,200	1,375	151	27,726	31,100	
	0300	10 ton		.80	42.500		34,100	1,700	189	35,989	40,400	
	0325	15 ton	↓	.80	42.500		44,200	1,700	189	46,089	51,500	
	0350	2 girder, 40′ span, 3 ton	M-4	.50	72		32,900	2,875	460	36,235	41,100	
	0375	5 ton		.50	72		34,400	2,875	460	37,735	42,800	
	0400	7.5 ton		.50	72		37,700	2,875	460	41,035	46,500	
	0425	10 ton		.40	90		44,100	3,575	570	48,245	54,500	
	0450	15 ton		.40	90		60,000	3,575	570	64,145	72,000	
	0475	25 ton		.30	120		71,000	4,775	765	76,540	86,500	
	0500	50′ span, 3 ton		.50	72		37,400	2,875	460	40,735	46,200	
	0525	5 ton		.50	72		38,900	2,875	460	42,235	47,800	
	0550	7.5 ton		.50	72		41,600	2,875	460	44,935	51,000	
	0575	10 ton		.40	90		48,000	3,575	570	52,145	59,000	
	0600	15 ton		.40	90		63,000	3,575	570	67,145	75,000	
	0625	25 ton	↓	.30	120	↓	74,500	4,775	765	80,040	90,000	

Division Notes

		CREW	DAILY OUTPUT	LABOR-HOURS	UNIT	2000 BARE COSTS MAT.	LABOR	EQUIP.	TOTAL	TOTAL INCL O&P

Division 15 Mechanical

Estimating Tips

15100 Building Services Piping

This subdivision is primarily basic pipe and related materials. The pipe may be used by any of the mechanical disciplines, i.e., plumbing, fire protection, heating, and air conditioning.

- The piping section lists the add to labor for elevated pipe installation. These adds apply to all elevated pipe, fittings, valves, insulation, etc., that are placed above 10′ high. CAUTION: the correct percentage may vary for the same pipe. For example, the percentage add for the basic pipe installation should be based on the maximum height that the craftsman must install for that particular section. If the pipe is to be located 14′ above the floor but it is suspended on threaded rod from beams, the bottom flange of which is 18′ high (4′ rods), then the height is actually 18′ and the add is 20%. The pipe coverer, however, does not have to go above the 14′ and so his or her add should be 10%.
- Most pipe is priced first as straight pipe with a joint (coupling, weld, etc.) every 10′ and a hanger usually every 10′. There are exceptions with hanger spacing such as for cast iron pipe (5′) and plastic pipe (3 per 10′). Following each type of pipe there are several lines listing sizes and the amount to be subtracted to delete couplings and hangers. This is for pipe that is to be buried or supported together on trapeze hangers. The reason that the couplings are deleted is that these runs are usually long and frequently longer lengths of pipe are used. By deleting the couplings the estimator is expected to look up and add back the correct reduced number of couplings.
- When preparing an estimate it may be necessary to approximate the fittings. Fittings usually run between 25% and 50% of the cost of the pipe. The lower percentage is for simpler runs, and the higher number is for complex areas such as mechanical rooms.
- For historic restoration projects, the systems must be as invisible as possible and pathways must be sought for pipes, conduits, and ductwork. While installations in accessible spaces (such as basements and attics) are relatively straightforward to estimate, labor costs may be more difficult to determine when delivery systems must be concealed.

15400 Plumbing Fixtures & Equipment

- Plumbing fixture costs usually require two lines, the fixture itself and its "rough-in, supply and waste."
- In the Assemblies Section (Plumbing D2010) for the desired fixture, the System Components Group at the center of the page shows the fixture on the first line. The rest of the list (fittings, pipe, tubing, etc.) will total up to what we refer to in the Unit Price section as "Rough-in, supply, waste and vent." Note that for most fixtures we allow a nominal 5′ of tubing to reach from the fixture to a main or riser.
- Remember that gas- and oil-fired units need venting.

15500 Heat Generation Equipment

- When estimating the cost of an HVAC system, check to see who is responsible for providing and installing the temperature control system. It is possible to overlook controls, assuming that they would be included in the electrical estimate.
- When looking up a boiler be careful on specified capacity. Some manufacturers rate their products on output while others use input.
- Include HVAC insulation for pipe, boiler, and duct (wrap and liner).
- Be careful when looking up mechanical items to get the correct pressure rating and connection type (thread, weld, flange).

15700 Heating/Ventilation/ Air Conditioning Equipment

- Combination heating and cooling units are sized by the air conditioning requirements. (See Reference No. R236000-20 for preliminary sizing guide.)
- A ton of air conditioning is nominally 400 CFM.
- Rectangular duct is taken off by the linear foot for each size, but its cost is usually estimated by the pound. Remember that SMACNA standards now base duct on internal pressure.
- Prefabricated duct is estimated and purchased like pipe: straight sections and fittings.
- Note that cranes or other lifting equipment are not included on any lines in Division 15. For example, if a crane is required to lift a heavy piece of pipe into place high above a gym floor, or to put a rooftop unit on the roof of a four-story building, etc., it must be added. Due to the potential for extreme variation—from nothing additional required to a major crane or helicopter—we feel that including a nominal amount for "lifting contingency" would be useless and detract from the accuracy of the estimate. When using equipment rental from Means do not forget to include the cost of the operator(s).

Reference Numbers

Reference numbers are shown in bold squares at the beginning of some major classifications. These numbers refer to related items in the Reference Section. The reference information may be an estimating procedure, an alternate pricing method, or technical information.

Note: Not all subdivisions listed here necessarily appear in this publication.

Note: **i2 Trade Service,** *in part, has been used as a reference source for some of the material prices used in Division 15.*

15051 | Mechanical General

			Crew	Daily Output	Labor-Hours	Unit	2006 Bare Costs Mat.	Labor	Equip.	Total	Total Incl O&P	
700	0010	**PIPING** See also Divisions 02500 & 02600 for site work										700
	1000	Add to labor for elevated installation										
	1080	10′ to 15′ high						10%				
	1100	15′ to 20′ high						20%				
	1120	20′ to 25′ high						25%				
	1140	25′ to 30′ high						35%				
	1160	30′ to 35′ high						40%				
	1180	35′ to 40′ high						50%				
	1200	Over 40′ high						55%				

15055 | Selective Mech Demolition

			Crew	Daily Output	Labor-Hours	Unit	Mat.	Labor	Equip.	Total	Total Incl O&P	
300	0010	**HVAC DEMOLITION** R220105-10										300
	0100	Air conditioner, split unit, 3 ton	Q-5	2	8	Ea.		310		310	480	
	0150	Package unit, 3 ton R024119-10	Q-6	3	8	"		320		320	500	
	0298	Boilers										
	0300	Electric, up thru 148 kW	Q-19	2	12	Ea.		480		480	740	
	0310	150 thru 518 kW	"	1	24			955		955	1,475	
	0320	550 thru 2000 kW	Q-21	.40	80			3,250		3,250	5,050	
	0330	2070 kW and up	"	.30	106			4,325		4,325	6,725	
	0340	Gas and/or oil, up thru 150 MBH	Q-7	2.20	14.545			595		595	930	
	0350	160 thru 2000 MBH		.80	40			1,650		1,650	2,550	
	0360	2100 thru 4500 MBH		.50	64			2,625		2,625	4,075	
	0370	4600 thru 7000 MBH		.30	106			4,375		4,375	6,800	
	0380	7100 thru 12,000 MBH		.16	200			8,200		8,200	12,800	
	0390	12,200 thru 25,000 MBH		.12	266			10,900		10,900	17,000	
	1000	Ductwork, 4″ high, 8″ wide	1 Clab	200	.040	L.F.		1.10		1.10	1.82	
	1020	10″ wide		190	.042			1.15		1.15	1.91	
	1040	14″ wide		180	.044			1.22		1.22	2.02	
	1100	6″ high, 8″ wide		165	.048			1.33		1.33	2.20	
	1120	12″ wide		150	.053			1.46		1.46	2.42	
	1140	18″ wide		135	.059			1.62		1.62	2.69	
	1200	10″ high, 12″ wide		125	.064			1.75		1.75	2.91	
	1220	18″ wide		115	.070			1.91		1.91	3.16	
	1240	24″ wide		110	.073			1.99		1.99	3.30	
	1300	12″-14″ high, 16″-18″ wide		85	.094			2.58		2.58	4.27	
	1320	24″ wide		75	.107			2.92		2.92	4.84	
	1340	48″ wide		71	.113			3.09		3.09	5.10	
	1400	18″ high, 24″ wide		67	.119			3.27		3.27	5.40	
	1420	36″ wide		63	.127			3.48		3.48	5.75	
	1440	48″ wide		59	.136			3.72		3.72	6.15	
	1500	30″ high, 36″ wide		56	.143			3.91		3.91	6.50	
	1520	48″ wide		53	.151			4.14		4.14	6.85	
	1540	72″ wide		50	.160			4.38		4.38	7.25	
	1550	Duct heater, electric strip	1 Elec	8	1	Ea.		42		42	64.50	
	1850	Minimum labor/equipment charge	1 Clab	3	2.667	Job		73		73	121	
	2200	Furnace, electric	Q-20	2	10	Ea.		385		385	610	
	2300	Gas or oil, under 120 MBH	Q-9	4	4			152		152	241	
	2340	Over 120 MBH	"	3	5.333			202		202	320	
	2800	Heat pump, package unit, 3 ton	Q-5	2.40	6.667			258		258	400	
	2840	Split unit, 3 ton		2	8			310		310	480	
	3000	Mechanical equipment, light items. Unit is weight, not cooling.		.90	17.778	Ton		690		690	1,075	
	3600	Heavy items		1.10	14.545	"		565		565	875	
	9000	Minimum labor/equipment charge	Q-6	3	8	Job		320		320	500	
600	0010	**PLUMBING DEMOLITION** R220105-10										600
	1020	Fixtures, including 10′ piping										

15 MECHANICAL

15055 | Selective Mech Demolition

		15055 Selective Mech Demolition	CREW	DAILY OUTPUT	LABOR-HOURS	UNIT	2006 BARE COSTS MAT.	LABOR	EQUIP.	TOTAL	TOTAL INCL O&P	
600	1100	Bath tubs, cast iron (R220105-10)	1 Plum	4	2	Ea.		85.50		85.50	133	600
	1120	Fiberglass		6	1.333			57		57	88.50	
	1140	Steel (R024119-10)		5	1.600			68.50		68.50	106	
	1200	Lavatory, wall hung		10	.800			34		34	53	
	1220	Counter top		8	1			42.50		42.50	66.50	
	1300	Sink, single compartment		8	1			42.50		42.50	66.50	
	1320	Double compartment		7	1.143			49		49	76	
	1400	Water closet, floor mounted		8	1			42.50		42.50	66.50	
	1420	Wall mounted		7	1.143			49		49	76	
	1500	Urinal, floor mounted		4	2			85.50		85.50	133	
	1520	Wall mounted		7	1.143			49		49	76	
	1600	Water fountains, free standing		8	1			42.50		42.50	66.50	
	1620	Wall or deck mounted		6	1.333			57		57	88.50	
	2000	Piping, metal, up thru 1-1/2" diameter		200	.040	L.F.		1.71		1.71	2.66	
	2050	2" thru 3-1/2" diameter		150	.053			2.28		2.28	3.54	
	2100	4" thru 6" diameter	2 Plum	100	.160			6.85		6.85	10.60	
	2150	8" thru 14" diameter	"	60	.267			11.40		11.40	17.70	
	2153	16" thru 20 " diameter	Q-18	70	.343			13.80	.99	14.79	22.50	
	2155	24" thru 26 " diameter		55	.436			17.55	1.26	18.81	29	
	2156	30" thru 36 " diameter		40	.600			24	1.73	25.73	39.50	
	2250	Water heater, 40 gal.	1 Plum	6	1.333	Ea.		57		57	88.50	
	6000	Remove and reset fixtures, minimum		6	1.333			57		57	88.50	
	6100	Maximum		4	2			85.50		85.50	133	
	9000	Minimum labor/equipment charge		2	4	Job		171		171	266	

15080 | Mechanical Insulation

		15080 Mechanical Insulation	CREW	DAILY OUTPUT	LABOR-HOURS	UNIT	MAT.	LABOR	EQUIP.	TOTAL	TOTAL INCL O&P	
200	0010	**DUCT INSULATION**										200
	0100	Rule of thumb, as a percentage of total mechanical costs				Job				10%		
	0110	Insulation req'd is based on the surface size/area to be covered										
	3000	Ductwork										
	3020	Blanket type, fiberglass, flexible										
	3030	Fire resistant liner, black coating one side										
	3050	1/2" thick, 2 lb. density	Q-14	380	.042	S.F.	.43	1.48		1.91	2.88	
	3060	1" thick, 1-1/2 lb. density	"	350	.046	"	.56	1.61		2.17	3.24	
	3140	FRK vapor barrier wrap, .75 lb. density										
	3160	1" thick	Q-14	350	.046	S.F.	.28	1.61		1.89	2.93	
	3170	1-1/2" thick		320	.050		.30	1.76		2.06	3.20	
	3180	2" thick		300	.053		.43	1.87		2.30	3.53	
	3190	3" thick		260	.062		.53	2.16		2.69	4.11	
	3200	4" thick		242	.066		.83	2.32		3.15	4.70	
	3490	Board type, fiberglass liner, 3 lb. density										
	3500	Fire resistant, black pigmented, 1 side										
	3520	1" thick	Q-14	150	.107	S.F.	1.61	3.75		5.36	7.85	
	3540	1-1/2" thick	"	130	.123	"	1.97	4.33		6.30	9.20	
	9600	Minimum labor/equipment charge	1 Stpi	4	2	Job		86		86	134	
400	0010	**EQUIPMENT INSULATION**										400
	0100	Rule of thumb, as a percentage of total mechanical costs				Job				10%		
	0110	Insulation req'd is based on the surface size/area to be covered										
	2900	Domestic water heater wrap kit										
	2920	1-1/2" with vinyl jacket, 20-60 gal.	1 Plum	8	1	Ea.	16.05	42.50		58.55	84	
	2925	50 to 80 gallons	"	8	1	"	19.15	42.50		61.65	87.50	
600	0010	**PIPING INSULATION**										600
	0100	Rule of thumb, as a percentage of total mechanical costs				Job				10%		
	0110	Insulation req'd is based on the surface size/area to be covered										
	2930	Insulated protectors, (ADA)										

	15080 | Mechanical Insulation		CREW	DAILY OUTPUT	LABOR-HOURS	UNIT	2006 BARE COSTS MAT.	LABOR	EQUIP.	TOTAL	TOTAL INCL O&P	
600	2935	For exposed piping under sinks or lavatories.										600
	2940	Vinyl coated foam, velcro tabs										
	2945	P Trap, 1-1/4" or 1-1/2"	1 Plum	32	.250	Ea.	16.80	10.70		27.50	35	
	2960	Valve and supply cover										
	2965	1/2", 3/8", and 7/16" pipe size	1 Plum	32	.250	Ea.	16.80	10.70		27.50	35	
	2970	Extension drain cover										
	2975	1-1/4", or 1-1/2" pipe size	1 Plum	32	.250	Ea.	17.60	10.70		28.30	36	
	2980	Tailpiece offset (wheelchair)										
	2985	1-1/4" pipe size	1 Plum	32	.250	Ea.	19.60	10.70		30.30	38	
	4000	Pipe covering (price copper tube one size less than IPS)										
	6120	2" wall, 1/2" iron pipe size	Q-14	145	.110	L.F.	2.92	3.88		6.80	9.55	
	6210	4" iron pipe size	"	125	.128	"	5.30	4.50		9.80	13.20	
	6600	Fiberglass, with all service jacket										
	6840	1" wall, 1/2" iron pipe size	Q-14	240	.067	L.F.	.89	2.34		3.23	4.80	
	6860	3/4" iron pipe size		230	.070		.97	2.45		3.42	5.05	
	6870	1" iron pipe size		220	.073		1.04	2.56		3.60	5.30	
	6880	1-1/4" iron pipe size		210	.076		1.12	2.68		3.80	5.60	
	6890	1-1/2" iron pipe size		210	.076		1.21	2.68		3.89	5.70	
	6900	2" iron pipe size		200	.080		1.31	2.81		4.12	6.05	
	7080	1-1/2" wall, 1/2" iron pipe size		230	.070		1.67	2.45		4.12	5.85	
	7140	2" iron pipe size		190	.084		2.26	2.96		5.22	7.30	
	7160	3" iron pipe size		170	.094		2.54	3.31		5.85	8.20	
	7879	Rubber tubing, flexible closed cell foam										
	8100	1/2" wall, 1/4" iron pipe size	1 Asbe	90	.089	L.F.	.36	3.47		3.83	6.05	
	8120	3/8" iron pipe size		90	.089		.39	3.47		3.86	6.10	
	8130	1/2" iron pipe size		89	.090		.44	3.51		3.95	6.25	
	8140	3/4" iron pipe size		89	.090		.49	3.51		4	6.30	
	8150	1" iron pipe size		88	.091		.54	3.55		4.09	6.40	
	8170	1-1/2" iron pipe size		87	.092		.76	3.59		4.35	6.70	
	8180	2" iron pipe size		86	.093		.97	3.63		4.60	7	
	8200	3" iron pipe size		85	.094		1.37	3.68		5.05	7.50	
	8220	4" iron pipe size		80	.100		2.03	3.91		5.94	8.60	
	8300	3/4" wall, 1/4" iron pipe size		90	.089		.56	3.47		4.03	6.25	
	8330	1/2" iron pipe size		89	.090		.73	3.51		4.24	6.55	
	8340	3/4" iron pipe size		89	.090		.89	3.51		4.40	6.75	
	8350	1" iron pipe size		88	.091		1.01	3.55		4.56	6.90	
	8360	1-1/4" iron pipe size		87	.092		1.36	3.59		4.95	7.35	
	8370	1-1/2" iron pipe size		87	.092		1.54	3.59		5.13	7.55	
	8380	2" iron pipe size		86	.093		1.81	3.63		5.44	7.95	
	8444	1" wall, 1/2" iron pipe size		86	.093		1.41	3.63		5.04	7.50	
	8445	3/4" iron pipe size		84	.095		1.71	3.72		5.43	7.95	
	8446	1" iron pipe size		84	.095		1.99	3.72		5.71	8.25	
	8447	1-1/4" iron pipe size		82	.098		2.25	3.81		6.06	8.70	
	8448	1-1/2" iron pipe size		82	.098		2.61	3.81		6.42	9.05	
	8449	2" iron pipe size		80	.100		3.49	3.91		7.40	10.20	
	8450	2-1/2" iron pipe size		80	.100		4.55	3.91		8.46	11.35	
	8456	Rubber insulation tape, 1/8" x 2" x 30'				Ea.	11.40			11.40	12.50	
	9600	Minimum labor/equipment charge	1 Plum	4	2	Job		85.50		85.50	133	

15 MECHANICAL

15100 | Building Services Piping

15106 | Glass Pipe & Fittings

			CREW	DAILY OUTPUT	LABOR-HOURS	UNIT	2006 BARE COSTS MAT.	LABOR	EQUIP.	TOTAL	TOTAL INCL O&P	
120	0010	**PIPE, GLASS** Borosilicate, couplings & hangers 10′ O.C.										120
	0020	Drainage										
	1100	1-1/2″ diameter	Q-1	52	.308	L.F.	9.30	11.80		21.10	28.50	
	1120	2″ diameter		44	.364		12.40	13.95		26.35	35	
	1140	3″ diameter		39	.410		16.60	15.75		32.35	43	
	1160	4″ diameter		30	.533		30	20.50		50.50	65	
	1180	6″ diameter		26	.615		54	23.50		77.50	96.50	
	9000	Minimum labor/equipment charge	1 Plum	4	2	Job		85.50		85.50	133	
160	0010	**PIPE, GLASS, FITTINGS**										160
	0020	Drainage, beaded ends										
	0040	Coupling & labor required at joints not incl. in fitting										
	0050	price. Add 1 per joint for installed price										
	0070	90° Bend or sweep, 1-1/2″				Ea.	20.50			20.50	22.50	
	0090	2″					26			26	28.50	
	0100	3″					42.50			42.50	47	
	0110	4″					68.50			68.50	75	
	0120	6″ (sweep only)					208			208	229	
	0350	Tee, single sanitary, 1-1/2″					33			33	36.50	
	0370	2″					33			33	36.50	
	0380	3″					49.50			49.50	54.50	
	0390	4″					88.50			88.50	97.50	
	0400	6″					238			238	262	
	0500	Coupling, stainless steel, TFE seal ring										
	0520	1-1/2″	Q-1	32	.500	Ea.	14.30	19.20		33.50	46	
	0530	2″		30	.533		18.10	20.50		38.60	52	
	0540	3″		25	.640		24.50	24.50		49	65	
	0550	4″		23	.696		42	26.50		68.50	87.50	
	0560	6″		20	.800		94.50	30.50		125	152	
	9000	Minimum labor/equipment charge	1 Plum	4	2	Job		85.50		85.50	133	

15107 | Metal Pipe & Fittings

			CREW	DAILY OUTPUT	LABOR-HOURS	UNIT	MAT.	LABOR	EQUIP.	TOTAL	TOTAL INCL O&P	
220	0010	**PIPE, BRASS** Plain end,										220
	0900	Field threaded, coupling & clevis hanger 10′ O.C.										
	0920	Regular weight										
	1120	1/2″ diameter	1 Plum	48	.167	L.F.	4.39	7.10		11.49	15.90	
	1140	3/4″ diameter		46	.174		5.95	7.45		13.40	18.05	
	1160	1″ diameter		43	.186		8.60	7.95		16.55	22	
	1180	1-1/4″ diameter	Q-1	72	.222		13.05	8.55		21.60	27.50	
	1200	1-1/2″ diameter		65	.246		15.50	9.45		24.95	32	
	1220	2″ diameter		53	.302		21.50	11.60		33.10	41.50	
	9000	Minimum labor/equipment charge	1 Plum	4	2	Job		85.50		85.50	133	
260	0010	**PIPE, BRASS, FITTINGS** Rough bronze, threaded										260
	0020											
	1000	Standard wt., 90° Elbow										
	1100	1/2″	1 Plum	12	.667	Ea.	9.05	28.50		37.55	54.50	
	1120	3/4″		11	.727		12.10	31		43.10	62	
	1140	1″		10	.800		19.65	34		53.65	74.50	
	1160	1-1/4″	Q-1	17	.941		32	36		68	91	
	1180	1-1/2″	″	16	1		39.50	38.50		78	104	
	1500	45° Elbow, 1/8″	1 Plum	13	.615		11.10	26.50		37.60	53.50	
	1580	1/2″		12	.667		11.10	28.50		39.60	57	
	1600	3/4″		11	.727		15.80	31		46.80	66	
	1620	1″		10	.800		27	34		61	82.50	
	1640	1-1/4″	Q-1	17	.941		42.50	36		78.50	103	
	1660	1-1/2″	″	16	1		54	38.50		92.50	119	

15107 | Metal Pipe & Fittings

		15107 Metal Pipe & Fittings	CREW	DAILY OUTPUT	LABOR-HOURS	UNIT	2006 BARE COSTS MAT.	LABOR	EQUIP.	TOTAL	TOTAL INCL O&P	
260	2000	Tee, 1/8″	1 Plum	9	.889	Ea.	10.60	38		48.60	70.50	260
	2080	1/2″		8	1		10.60	42.50		53.10	78	
	2100	3/4″		7	1.143		15.10	49		64.10	92.50	
	2120	1″		6	1.333		27	57		84	119	
	2140	1-1/4″	Q-1	10	1.600		50.50	61.50		112	151	
	2160	1-1/2″	″	9	1.778		53	68.50		121.50	164	
	2500	Coupling, 1/8″	1 Plum	26	.308		8	13.15		21.15	29.50	
	2580	1/2″		15	.533		8	23		31	44.50	
	2600	3/4″		14	.571		10.60	24.50		35.10	49.50	
	2620	1″		13	.615		18.10	26.50		44.60	61	
	2640	1-1/4″	Q-1	22	.727		30.50	28		58.50	77	
	2660	1-1/2″		20	.800		39.50	30.50		70	91.50	
	2680	2″		18	.889		65	34		99	125	
	9000	Minimum labor/equipment charge	1 Plum	4	2	Job		85.50		85.50	133	
320	0010	**PIPE, CAST IRON** Soil, on hangers 5′ O.C. (R221113-50)										320
	0020	Single hub, service wt., lead & oakum joints 10′ O.C.										
	2120	2″ diameter	Q-1	63	.254	L.F.	5	9.75		14.75	20.50	
	2140	3″ diameter		60	.267		6.95	10.25		17.20	23.50	
	2160	4″ diameter		55	.291		8.90	11.20		20.10	27	
	2180	5″ diameter	Q-2	76	.316		12.25	12.60		24.85	33	
	2200	6″ diameter	″	73	.329		15.05	13.10		28.15	37	
	2220	8″ diameter	Q-3	59	.542		23	22		45	60	
	4000	No hub, couplings 10′ O.C.										
	4100	1-1/2″ diameter	Q-1	71	.225	L.F.	5.25	8.65		13.90	19.20	
	4120	2″ diameter		67	.239		5.40	9.20		14.60	20	
	4140	3″ diameter		64	.250		7.20	9.60		16.80	23	
	4160	4″ diameter		58	.276		9.15	10.60		19.75	26.50	
	4180	5″ diameter	Q-2	83	.289		13	11.50		24.50	32	
	9000	Minimum labor/equipment charge	1 Plum	4	2	Job		85.50		85.50	133	
360	0010	**PIPE, CAST IRON, FITTINGS** Soil (R221113-50)										360
	0040	Hub and spigot, service weight, lead & oakum joints										
	0080	1/4 bend, 2″	Q-1	16	1	Ea.	10.35	38.50		48.85	71.50	
	0120	3″		14	1.143		13.80	44		57.80	83.50	
	0140	4″		13	1.231		21.50	47.50		69	97	
	0160	5″	Q-2	18	1.333		30	53		83	116	
	0180	6″	″	17	1.412		37.50	56.50		94	129	
	0200	8″	Q-3	11	2.909		113	118		231	310	
	0340	1/8 bend, 2″	Q-1	16	1		7.35	38.50		45.85	68	
	0350	3″		14	1.143		11.55	44		55.55	81	
	0360	4″		13	1.231		16.90	47.50		64.40	92	
	0380	5″	Q-2	18	1.333		23.50	53		76.50	108	
	0400	6″	″	17	1.412		28.50	56.50		85	119	
	0420	8″	Q-3	11	2.909		85	118		203	278	
	0500	Sanitary tee, 2″	Q-1	10	1.600		14.50	61.50		76	111	
	0540	3″		9	1.778		23.50	68.50		92	132	
	0620	4″		8	2		28.50	77		105.50	152	
	0700	5″	Q-2	12	2		57	79.50		136.50	187	
	0800	6″	″	11	2.182		64.50	87		151.50	206	
	0880	8″	Q-3	7	4.571		171	186		357	475	
	0954	10″ x 6″	″	8	4		247	163		410	525	
	5990	No hub										
	6000	Cplg. & labor required at joints not incl. in fitting										
	6010	price. Add 1 coupling per joint for installed price										
	6020	1/4 Bend, 1-1/2″				Ea.	5.80			5.80	6.40	
	6060	2″					6.35			6.35	7	

15 MECHANICAL

15107 | Metal Pipe & Fittings

		15107 Metal Pipe & Fittings	CREW	DAILY OUTPUT	LABOR-HOURS	UNIT	2006 BARE COSTS MAT.	LABOR	EQUIP.	TOTAL	TOTAL INCL O&P	
360	6080	3″ R221113-50				Ea.	8.75			8.75	9.65	360
	6120	4″					12.65			12.65	13.95	
	6184	1/4 Bend, long sweep, 1-1/2″					13.65			13.65	15.05	
	6186	2″					13.65			13.65	15.05	
	6188	3″					16.20			16.20	17.80	
	6189	4″					26			26	28.50	
	6190	5″					47.50			47.50	52.50	
	6191	6″					58			58	64	
	6192	8″					141			141	155	
	6193	10″					253			253	279	
	6200	1/8 Bend, 1-1/2″					4.85			4.85	5.35	
	6210	2″					5.40			5.40	5.95	
	6212	3″					7.30			7.30	8.05	
	6214	4″					9.25			9.25	10.20	
	6380	Sanitary Tee, tapped, 1-1/2″					10.70			10.70	11.75	
	6382	2″ x 1-1/2″					9.65			9.65	10.60	
	6384	2″					10.75			10.75	11.85	
	6386	3″ x 2″					15			15	16.50	
	6388	3″					28			28	30.50	
	6390	4″ x 1-1/2″					13.35			13.35	14.70	
	6392	4″ x 2″					15.05			15.05	16.55	
	6394	6″ x 1-1/2″					31			31	34	
	6396	6″ x 2″					32			32	35	
	6459	Sanitary Tee, 1-1/2″					8.05			8.05	8.85	
	6460	2″					8.05			8.05	8.85	
	6470	3″					10.65			10.65	11.70	
	6472	4″					16.50			16.50	18.15	
	8000	Coupling, standard (by CISPI Mfrs.)										
	8020	1-1/2″	Q-1	48	.333	Ea.	6	12.80		18.80	26.50	
	8040	2″		44	.364		6	13.95		19.95	28	
	8080	3″		38	.421		7.15	16.20		23.35	33	
	8120	4″		33	.485		8.45	18.65		27.10	38.50	
	9000	Minimum labor/equipment charge	1 Plum	4	2	Job		85.50		85.50	133	
420	0010	**PIPE, COPPER** Solder joints										420
	1000	Type K tubing, couplings & clevis hangers 10′ O.C.										
	1100	1/4″ diameter	1 Plum	84	.095	L.F.	1.66	4.07		5.73	8.15	
	1200	1″ diameter		66	.121		5.05	5.20		10.25	13.60	
	1260	2″ diameter		40	.200		11.90	8.55		20.45	26.50	
	2000	Type L tubing, couplings & hangers 10′ O.C.										
	2100	1/4″ diameter	1 Plum	88	.091	L.F.	1.10	3.88		4.98	7.25	
	2120	3/8″ diameter		84	.095		1.53	4.07		5.60	8	
	2140	1/2″ diameter		81	.099		1.77	4.22		5.99	8.50	
	2160	5/8″ diameter		79	.101		2.48	4.32		6.80	9.40	
	2180	3/4″ diameter		76	.105		2.60	4.49		7.09	9.85	
	2200	1″ diameter		68	.118		3.65	5		8.65	11.80	
	2220	1-1/4″ diameter		58	.138		5.05	5.90		10.95	14.75	
	2240	1-1/2″ diameter		52	.154		6.50	6.55		13.05	17.30	
	2260	2″ diameter		42	.190		10.05	8.15		18.20	23.50	
	2280	2-1/2″ diameter	Q-1	62	.258		15.55	9.90		25.45	32.50	
	2300	3″ diameter		56	.286		21	11		32	40	
	2320	3-1/2″ diameter		43	.372		28.50	14.30		42.80	53.50	
	2340	4″ diameter		39	.410		35.50	15.75		51.25	63.50	
	2360	5″ diameter		34	.471		100	18.10		118.10	138	
	2380	6″ diameter	Q-2	40	.600		124	24		148	173	
	2410	For other than full hard temper, add					21%					

MECHANICAL 15

	15107	Metal Pipe & Fittings	Crew	Daily Output	Labor-Hours	Unit	2006 Bare Costs Mat.	Labor	Equip.	Total	Total Incl O&P	
420	2590	For silver solder, add						15%				420
	3000	Type M tubing, couplings & hangers 10′ O.C.										
	3140	1/2″ diameter	1 Plum	84	.095	L.F.	1.39	4.07		5.46	7.85	
	3180	3/4″ diameter		78	.103		2.14	4.38		6.52	9.15	
	3200	1″ diameter		70	.114		2.80	4.88		7.68	10.70	
	4000	Type DWV tubing, couplings & hangers 10′ O.C.										
	4100	1-1/4″ diameter	1 Plum	60	.133	L.F.	4.45	5.70		10.15	13.75	
	4120	1-1/2″ diameter		54	.148		5.55	6.35		11.90	15.95	
	4140	2″ diameter		44	.182		7.40	7.75		15.15	20	
	4160	3″ diameter	Q-1	58	.276		13.15	10.60		23.75	31	
	4180	4″ diameter		40	.400		23	15.35		38.35	49.50	
	4200	5″ diameter		36	.444		68.50	17.10		85.60	102	
	4220	6″ diameter	Q-2	42	.571		98	23		121	144	
	9000	Minimum labor/equipment charge	1 Plum	4	2	Job		85.50		85.50	133	
460	0010	**PIPE, COPPER, FITTINGS** Wrought unless otherwise noted										460
	0040	Solder joints, copper x copper										
	0070	90° elbow, 1/4″	1 Plum	22	.364	Ea.	1.70	15.55		17.25	26	
	0100	1/2″		20	.400		.51	17.10		17.61	27	
	0120	3/4″		19	.421		1.14	18		19.14	29.50	
	0130	1″		16	.500		2.82	21.50		24.32	36	
	0250	45° elbow, 1/4″		22	.364		3.18	15.55		18.73	27.50	
	0270	3/8″		22	.364		2.59	15.55		18.14	27	
	0280	1/2″		20	.400		.94	17.10		18.04	27.50	
	0290	5/8″		19	.421		4.93	18		22.93	33.50	
	0300	3/4″		19	.421		1.64	18		19.64	30	
	0310	1″		16	.500		4.13	21.50		25.63	37.50	
	0320	1-1/4″		15	.533		5.90	23		28.90	42	
	0330	1-1/2″		13	.615		7.10	26.50		33.60	49	
	0340	2″		11	.727		11.80	31		42.80	61.50	
	0350	2-1/2″	Q-1	13	1.231		25	47.50		72.50	101	
	0360	3″		13	1.231		37.50	47.50		85	115	
	0370	3-1/2″		10	1.600		66	61.50		127.50	168	
	0380	4″		9	1.778		79.50	68.50		148	194	
	0390	5″		6	2.667		310	102		412	500	
	0400	6″	Q-2	9	2.667		485	106		591	700	
	0450	Tee, 1/4″	1 Plum	14	.571		3.57	24.50		28.07	42	
	0470	3/8″		14	.571		2.73	24.50		27.23	41	
	0480	1/2″		13	.615		.87	26.50		27.37	42	
	0490	5/8″		12	.667		5.95	28.50		34.45	51	
	0500	3/4″		12	.667		2.11	28.50		30.61	47	
	0510	1″		10	.800		6.50	34		40.50	60	
	0520	1-1/4″		9	.889		9.35	38		47.35	69.50	
	0530	1-1/2″		8	1		14.45	42.50		56.95	82.50	
	0540	2″		7	1.143		22.50	49		71.50	101	
	0550	2-1/2″	Q-1	8	2		45.50	77		122.50	170	
	0560	3″		7	2.286		69.50	88		157.50	214	
	0570	3-1/2″		6	2.667		201	102		303	380	
	0580	4″		5	3.200		168	123		291	375	
	0590	5″		4	4		550	154		704	845	
	0600	6″	Q-2	6	4		750	159		909	1,075	
	0612	Tee, reducing on the outlet, 1/4″	1 Plum	15	.533		6.15	23		29.15	42.50	
	0613	3/8″		15	.533		5.40	23		28.40	41.50	
	0614	1/2″		14	.571		4.74	24.50		29.24	43	
	0615	5/8″		13	.615		9.55	26.50		36.05	51.50	
	0616	3/4″		12	.667		2.97	28.50		31.47	48	
	0617	1″		11	.727		6.90	31		37.90	56	

15107 | Metal Pipe & Fittings

			CREW	DAILY OUTPUT	LABOR-HOURS	UNIT	2006 BARE COSTS MAT.	LABOR	EQUIP.	TOTAL	TOTAL INCL O&P	
460	0618	1-1/4"	1 Plum	10	.800	Ea.	10.10	34		44.10	64	460
	0619	1-1/2"		9	.889		10.75	38		48.75	71	
	0620	2"		8	1		17	42.50		59.50	85	
	0621	2-1/2"	Q-1	9	1.778		53	68.50		121.50	165	
	0622	3"		8	2		58.50	77		135.50	184	
	0623	4"		6	2.667		113	102		215	284	
	0624	5"		5	3.200		520	123		643	765	
	0625	6"	Q-2	7	3.429		715	137		852	995	
	0626	8"	"	6	4		2,900	159		3,059	3,450	
	0630	Tee, reducing on the run, 1/4"	1 Plum	15	.533		7.20	23		30.20	43.50	
	0631	3/8"		15	.533		9.65	23		32.65	46	
	0632	1/2"		14	.571		6.50	24.50		31	45	
	0633	5/8"		13	.615		10.15	26.50		36.65	52	
	0634	3/4"		12	.667		5.25	28.50		33.75	50.50	
	0635	1"		11	.727		8.20	31		39.20	57.50	
	0636	1-1/4"		10	.800		13.30	34		47.30	67.50	
	0637	1-1/2"		9	.889		23	38		61	84	
	0638	2"		8	1		29.50	42.50		72	98.50	
	0639	2-1/2"	Q-1	9	1.778		68.50	68.50		137	182	
	0640	3"		8	2		101	77		178	231	
	0641	4"		6	2.667		224	102		326	405	
	0642	5"		5	3.200		520	123		643	765	
	0643	6"	Q-2	7	3.429		790	137		927	1,075	
	0644	8"	"	6	4		3,050	159		3,209	3,600	
	0650	Coupling, 1/4"	1 Plum	24	.333		.40	14.25		14.65	22.50	
	0670	3/8"		24	.333		.52	14.25		14.77	22.50	
	0680	1/2"		22	.364		.40	15.55		15.95	24.50	
	0690	5/8"		21	.381		1.22	16.25		17.47	27	
	0700	3/4"		21	.381		.76	16.25		17.01	26.50	
	0710	1"		18	.444		1.56	19		20.56	31	
	0715	1-1/4"		17	.471		2.90	20		22.90	34.50	
	0716	1-1/2"		15	.533		3.84	23		26.84	39.50	
	0718	2"		13	.615		6.40	26.50		32.90	48	
	0721	2-1/2"	Q-1	15	1.067		13.60	41		54.60	78.50	
	0722	3"		13	1.231		20.50	47.50		68	96	
	0724	3-1/2"		8	2		39	77		116	163	
	0726	4"		7	2.286		43	88		131	185	
	0728	5"		6	2.667		106	102		208	275	
	0731	6"	Q-2	8	3		175	120		295	380	
	2000	DWV, solder joints, copper x copper										
	2030	90° Elbow, 1-1/4"	1 Plum	13	.615	Ea.	5.20	26.50		31.70	47	
	2050	1-1/2"		12	.667		7.80	28.50		36.30	53	
	2070	2"		10	.800		10.20	34		44.20	64	
	2090	3"	Q-1	10	1.600		25	61.50		86.50	123	
	2100	4"	"	9	1.778		120	68.50		188.50	238	
	2250	Tee, Sanitary, 1-1/4"	1 Plum	9	.889		9.50	38		47.50	69.50	
	2270	1-1/2"		8	1		12.80	42.50		55.30	80.50	
	2290	2"		7	1.143		14.95	49		63.95	92.50	
	2310	3"	Q-1	7	2.286		54	88		142	197	
	2330	4"	"	6	2.667		138	102		240	310	
	2400	Coupling, 1-1/4"	1 Plum	14	.571		2.43	24.50		26.93	40.50	
	2420	1-1/2"		13	.615		3.04	26.50		29.54	44.50	
	2440	2"		11	.727		4.20	31		35.20	53	
	2460	3"	Q-1	11	1.455		8.15	56		64.15	96	
	2480	4"	"	10	1.600		26	61.50		87.50	124	
	9000	Minimum labor/equipment charge	1 Plum	4	2	Job		85.50		85.50	133	

		15107 | Metal Pipe & Fittings	Crew	Daily Output	Labor-Hours	Unit	2006 Bare Costs Mat.	Labor	Equip.	Total	Total Incl O&P	
500	0010	**PIPE, CORROSION RESISTANT** No couplings or hangers										500
	0020	Iron alloy, drain, mechanical joint										
	1000	1-1/2" diameter	Q-1	70	.229	L.F.	22.50	8.80		31.30	38.50	
	1100	2" diameter		66	.242		26	9.30		35.30	43	
	1120	3" diameter		60	.267		36	10.25		46.25	56	
	1140	4" diameter		52	.308		46.50	11.80		58.30	69.50	
	2980	Plastic, epoxy, fiberglass filament wound, B&S joint										
	3000	2" diameter	Q-1	62	.258	L.F.	7.20	9.90		17.10	23.50	
	3100	3" diameter		51	.314		8.30	12.05		20.35	28	
	3120	4" diameter		45	.356		12.15	13.65		25.80	34.50	
	3160	8" diameter	Q-2	38	.632		30.50	25		55.50	72.50	
	3200	12" diameter	"	28	.857		50	34		84	108	
	9800	Minimum labor/equipment charge	1 Plum	4	2	Job		85.50		85.50	133	
560	0010	**PIPE, CORROSION RESISTANT, FITTINGS**										560
	0030	Iron alloy										
	0050	Mechanical joint										
	0060	1/4 Bend, 1-1/2"	Q-1	12	1.333	Ea.	47	51		98	131	
	0080	2"		10	1.600		76.50	61.50		138	180	
	0090	3"		9	1.778		92	68.50		160.50	207	
	0100	4"		8	2		105	77		182	236	
	0160	Tee and Y, sanitary, straight										
	0170	1-1/2"	Q-1	8	2	Ea.	51	77		128	176	
	0180	2"		7	2.286		68	88		156	212	
	0190	3"		6	2.667		105	102		207	275	
	0200	4"		5	3.200		194	123		317	405	
	0360	Coupling, 1-1/2"		14	1.143		35.50	44		79.50	108	
	0380	2"		12	1.333		41	51		92	125	
	0390	3"		11	1.455		42.50	56		98.50	134	
	0400	4"		10	1.600		48.50	61.50		110	149	
	3000	Epoxy, filament wound										
	3030	Quick-lock joint										
	3040	90° Elbow, 2"	Q-1	28	.571	Ea.	67.50	22		89.50	109	
	3060	3"		16	1		78	38.50		116.50	146	
	3070	4"		13	1.231		106	47.50		153.50	191	
	3190	Tee, 2"		19	.842		161	32.50		193.50	228	
	3200	3"		11	1.455		195	56		251	300	
	3210	4"		9	1.778		234	68.50		302.50	365	
	9000	Minimum labor/equipment charge	1 Plum	4	2	Job		85.50		85.50	133	
620	0010	**PIPE, STEEL** R221113-50										620
	0020	All pipe sizes are to Spec. A-53 unless noted otherwise										
	0050	Schedule 40, threaded, with couplings, and clevis type										
	0060	hangers sized for covering, 10′ O.C.										
	0540	Black, 1/4" diameter	1 Plum	66	.121	L.F.	2.04	5.20		7.24	10.30	
	0560	1/2" diameter		63	.127		2.05	5.40		7.45	10.70	
	0570	3/4" diameter		61	.131		2.40	5.60		8	11.35	
	0580	1" diameter		53	.151		3.47	6.45		9.92	13.80	
	0590	1-1/4" diameter	Q-1	89	.180		4.56	6.90		11.46	15.75	
	0600	1-1/2" diameter		80	.200		5.35	7.70		13.05	17.85	
	0610	2" diameter		64	.250		7.10	9.60		16.70	23	
	0620	2-1/2" diameter		50	.320		11	12.30		23.30	31	
	0630	3" diameter		43	.372		14.30	14.30		28.60	37.50	
	0640	3-1/2" diameter		40	.400		19.35	15.35		34.70	45.50	
	0650	4" diameter		36	.444		21	17.10		38.10	49.50	
	0670	6" diameter	Q-2	31	.774		46	31		77	99	

15107 | Metal Pipe & Fittings

				DAILY	LABOR-		2006 BARE COSTS				TOTAL	
			CREW	OUTPUT	HOURS	UNIT	MAT.	LABOR	EQUIP.	TOTAL	INCL O&P	
620	0680	8" diameter (R221113-50)	Q-2	27	.889	L.F.	71	35.50		106.50	133	620
	1290	Galvanized, 1/4" diameter	1 Plum	66	.121		2.42	5.20		7.62	10.70	
	1310	1/2" diameter		63	.127		2.78	5.40		8.18	11.50	
	1320	3/4" diameter		61	.131		3.32	5.60		8.92	12.35	
	1330	1" diameter	▼	53	.151		4.51	6.45		10.96	14.95	
	1350	1-1/2" diameter	Q-1	80	.200		7.10	7.70		14.80	19.75	
	1360	2" diameter		64	.250		9.45	9.60		19.05	25.50	
	1370	2-1/2" diameter		50	.320		15.60	12.30		27.90	36.50	
	1380	3" diameter		43	.372		19.90	14.30		34.20	44	
	1400	4" diameter	▼	36	.444		29	17.10		46.10	58	
	1420	6" diameter	Q-2	31	.774		53	31		84	106	
	1430	8" diameter	"	27	.889	▼	81.50	35.50		117	145	
	2000	Welded, sch. 40, on yoke & roll hangers, sized for covering,										
	2040	Black, 1" diameter	Q-15	93	.172	L.F.	5.05	6.60	.74	12.39	16.65	
	2070	2" diameter		61	.262		8.90	10.10	1.13	20.13	26.50	
	2090	3" diameter		43	.372		14	14.30	1.61	29.91	39	
	2110	4" diameter		37	.432		19.25	16.60	1.87	37.72	49	
	2120	5" diameter	▼	32	.500		25.50	19.20	2.16	46.86	60.50	
	2130	6" diameter	Q-16	36	.667	▼	34.50	26.50	1.92	62.92	81.50	
	9990	Minimum labor/equipment charge	1 Plum	3	2.667	Job		114		114	177	
640	0010	**PIPE, STEEL, FITTINGS** Threaded (R221113-50)										640
	0020	Cast Iron										
	0040	Standard weight, black										
	0060	90° Elbow, straight										
	0070	1/4"	1 Plum	16	.500	Ea.	4.47	21.50		25.97	38	
	0080	3/8"		16	.500		6.40	21.50		27.90	40	
	0090	1/2"		15	.533		2.84	23		25.84	38.50	
	0100	3/4"		14	.571		2.96	24.50		27.46	41.50	
	0110	1"	▼	13	.615		3.50	26.50		30	45	
	0130	1-1/2"	Q-1	20	.800		6.85	30.50		37.35	55.50	
	0140	2"		18	.889		10.70	34		44.70	65	
	0150	2-1/2"		14	1.143		25.50	44		69.50	97	
	0160	3"		10	1.600		42	61.50		103.50	142	
	0180	4"	▼	6	2.667		78	102		180	245	
	0200	6"	Q-2	7	3.429		215	137		352	450	
	0210	8"	"	6	4	▼	525	159		684	830	
	0500	Tee, straight										
	0510	1/4"	1 Plum	10	.800	Ea.	7	34		41	60.50	
	0520	3/8"		10	.800		6.80	34		40.80	60.50	
	0540	3/4"		9	.889		5.15	38		43.15	64.50	
	0550	1"	▼	8	1		4.58	42.50		47.08	71.50	
	0570	1-1/2"	Q-1	13	1.231		10.85	47.50		58.35	85.50	
	0580	2"		11	1.455		15.15	56		71.15	104	
	0590	2-1/2"		9	1.778		39.50	68.50		108	150	
	0600	3"		6	2.667		60.50	102		162.50	226	
	0620	4"	▼	4	4		118	154		272	370	
	0640	6"	Q-2	4	6		355	239		594	760	
	0650	8"	"	3	8	▼	805	320		1,125	1,375	
	0700	Standard weight, galvanized cast iron										
	0720	90° Elbow, straight										
	0730	1/4"	1 Plum	16	.500	Ea.	10.50	21.50		32	44.50	
	0740	3/8"		16	.500		10.50	21.50		32	44.50	
	0750	1/2"		15	.533		8.40	23		31.40	45	
	0760	3/4"		14	.571		11.75	24.50		36.25	51	
	0770	1"	▼	13	.615		13.55	26.50		40.05	56	
	0790	1-1/2"	Q-1	20	.800	▼	29	30.50		59.50	80	

For expanded coverage of these items see *Means Mechanical or Plumbing Cost Data 2006*

15107 | Metal Pipe & Fittings

	15107 Metal Pipe & Fittings	CREW	DAILY OUTPUT	LABOR-HOURS	UNIT	2006 BARE COSTS MAT.	LABOR	EQUIP.	TOTAL	TOTAL INCL O&P
640 0800	2" (R221113-50)	Q-1	18	.889	Ea.	43	34		77	100
0810	2-1/2"		14	1.143		88	44		132	165
0820	3"		10	1.600		134	61.50		195.50	243
0840	4"		6	2.667		245	102		347	430
0860	6"	Q-2	7	3.429		640	137		777	910
0870	8"	"	6	4		1,550	159		1,709	1,975
1100	Tee, straight									
1110	1/4"	1 Plum	10	.800	Ea.	20.50	34		54.50	76
1120	3/8"		10	.800		13.40	34		47.40	67.50
1140	3/4"		9	.889		16.85	38		54.85	77.50
1150	1"		8	1		18.30	42.50		60.80	86.50
1170	1-1/2"	Q-1	13	1.231		42.50	47.50		90	120
1180	2"		11	1.455		53	56		109	145
1190	2-1/2"		9	1.778		114	68.50		182.50	231
1200	3"		6	2.667		295	102		397	485
1220	4"		4	4		340	154		494	615
1240	6"	Q-2	4	6		1,050	239		1,289	1,525
1250	8"	"	3	8		2,375	320		2,695	3,125
5000	Malleable iron, 150 lb.									
5020	Black									
5040	90° elbow, straight									
5060	1/4"	1 Plum	16	.500	Ea.	2.17	21.50		23.67	35.50
5070	3/8"		16	.500		2.17	21.50		23.67	35.50
5090	3/4"		14	.571		1.82	24.50		26.32	40
5100	1"		13	.615		3.16	26.50		29.66	44.50
5120	1-1/2"	Q-1	20	.800		6.85	30.50		37.35	55.50
5130	2"		18	.889		11.80	34		45.80	66
5140	2-1/2"		14	1.143		26.50	44		70.50	97.50
5150	3"		10	1.600		38.50	61.50		100	138
5170	4"		6	2.667		82.50	102		184.50	250
5190	6"	Q-2	7	3.429		279	137		416	515
5450	Tee, straight									
5470	1/4"	1 Plum	10	.800	Ea.	3.16	34		37.16	56.50
5480	3/8"		10	.800		3.16	34		37.16	56.50
5500	3/4"		9	.889		2.90	38		40.90	62
5510	1"		8	1		4.94	42.50		47.44	72
5520	1-1/4"	Q-1	14	1.143		8	44		52	77.50
5530	1-1/2"		13	1.231		9.95	47.50		57.45	84.50
5540	2"		11	1.455		17	56		73	106
5550	2-1/2"		9	1.778		36.50	68.50		105	147
5560	3"		6	2.667		54	102		156	219
5570	3-1/2"		5	3.200		130	123		253	335
5580	4"		4	4		130	154		284	380
5600	6"	Q-2	4	6		410	239		649	825
5650	Coupling									
5670	1/4"	1 Plum	19	.421	Ea.	2.69	18		20.69	31
5680	3/8"		19	.421		2.69	18		20.69	31
5690	1/2"		19	.421		2.07	18		20.07	30.50
5700	3/4"		18	.444		2.44	19		21.44	32
5710	1"		15	.533		3.69	23		26.69	39.50
5730	1-1/2"	Q-1	24	.667		6.40	25.50		31.90	47
5740	2"		21	.762		9.45	29.50		38.95	56
5750	2-1/2"		18	.889		26	34		60	82
5760	3"		14	1.143		35.50	44		79.50	108
5780	4"		10	1.600		71	61.50		132.50	174
5800	6"	Q-2	8	3		204	120		324	410

	15107 | Metal Pipe & Fittings		Crew	Daily Output	Labor-Hours	Unit	2006 Bare Costs Mat.	2006 Bare Costs Labor	2006 Bare Costs Equip.	2006 Bare Costs Total	Total Incl O&P
640	6000	For galvanized elbows, tees, and couplings add (R221113-50)				Ea.	20%				
	9990	Minimum labor/equipment charge	1 Plum	4	2	Job		85.50		85.50	133
660	0010	**PIPE, STEEL, FITTINGS** Flanged, welded and special type									
	3000	Weld joint, butt, carbon steel, standard weight									
	3040	90° elbow, long radius									
	3050	1/2" pipe size	Q-15	16	1	Ea.	25	38.50	4.32	67.82	92.50
	3060	3/4" pipe size		16	1		25	38.50	4.32	67.82	92.50
	3070	1" pipe size		16	1		12.15	38.50	4.32	54.97	78
	3100	2" pipe size		10	1.600		13.10	61.50	6.90	81.50	118
	3120	3" pipe size		7	2.286		19.20	88	9.85	117.05	169
	3130	4" pipe size		5	3.200		31.50	123	13.80	168.30	241
	3136	5" pipe size		4	4		66	154	17.30	237.30	330
	3140	6" pipe size	Q-16	5	4.800		69	191	13.80	273.80	390
	3350	Tee, straight									
	3360	1/2" pipe size	Q-15	10	1.600	Ea.	63	61.50	6.90	131.40	173
	3370	3/4" pipe size		10	1.600		63	61.50	6.90	131.40	173
	3380	1" pipe size		10	1.600		32	61.50	6.90	100.40	138
	3410	2" pipe size		6	2.667		31.50	102	11.50	145	207
	3430	3" pipe size		4	4		47.50	154	17.30	218.80	310
	3440	4" pipe size		3	5.333		67	205	23	295	420
	3446	5" pipe size		2.50	6.400		111	246	27.50	384.50	535
	3450	6" pipe size	Q-16	3	8		115	320	23	458	650
	9990	Minimum labor/equipment charge	Q-15	3	5.333	Job		205	23	228	345
690	0010	**PIPE, GROOVED-JOINT STEEL FITTINGS & VALVES**									
	0012	Fittings are ductile iron. Steel fittings noted.									
	0020	Pipe includes coupling & clevis type hanger 10' O.C.									
	1000	Schedule 40, black									
	1040	3/4" diameter	1 Plum	71	.113	L.F.	2.98	4.81		7.79	10.80
	1050	1" diameter		63	.127		3.62	5.40		9.02	12.45
	1060	1-1/4" diameter		58	.138		4.78	5.90		10.68	14.40
	1070	1-1/2" diameter		51	.157		5.55	6.70		12.25	16.50
	1080	2" diameter		40	.200		7.15	8.55		15.70	21
	1090	2-1/2" diameter	Q-1	57	.281		10.15	10.80		20.95	28
	1100	3" diameter		50	.320		12.75	12.30		25.05	33
	1110	4" diameter		45	.356		17.30	13.65		30.95	40
	1120	5" diameter		37	.432		22.50	16.60		39.10	50.50
	4000	Elbow, 90° or 45°, painted									
	4030	3/4" diameter	1 Plum	50	.160	Ea.	24.50	6.85		31.35	37.50
	4040	1" diameter		50	.160		13.15	6.85		20	25
	4050	1-1/4" diameter		40	.200		13.15	8.55		21.70	28
	4060	1-1/2" diameter		33	.242		13.15	10.35		23.50	30.50
	4070	2" diameter		25	.320		13.15	13.65		26.80	36
	4080	2-1/2" diameter	Q-1	40	.400		13.15	15.35		28.50	38.50
	4100	4" diameter		25	.640		25.50	24.50		50	66
	4110	5" diameter		20	.800		61	30.50		91.50	116
	4250	For galvanized elbows, add					26%				
	4690	Tee, painted									
	4700	3/4" diameter	1 Plum	38	.211	Ea.	26.50	9		35.50	43
	4740	1" diameter		33	.242		20	10.35		30.35	38
	4750	1-1/4" diameter		27	.296		20	12.65		32.65	41.50
	4760	1-1/2" diameter		22	.364		20	15.55		35.55	46
	4770	2" diameter		17	.471		20	20		40	53.50
	4780	2-1/2" diameter	Q-1	27	.593		20	23		43	57.50
	4800	4" diameter		17	.941		43	36		79	104
	4810	5" diameter		13	1.231		101	47.50		148.50	185

		15107 | Metal Pipe & Fittings	CREW	DAILY OUTPUT	LABOR-HOURS	UNIT	2006 BARE COSTS MAT.	LABOR	EQUIP.	TOTAL	TOTAL INCL O&P	
690	4900	For galvanized tees, add				Ea.	24%					690
	9990	Minimum labor/equipment charge	1 Plum	4	2	Job		85.50		85.50	133	
		15108 | Plastic Pipe & Fittings										
520	0010	PIPE, PLASTIC										520
	0020	Fiberglass reinforced, couplings 10′ O.C., hangers 3 per 10′										
	0240	2″ diameter	Q-1	58	.276	L.F.	9.10	10.60		19.70	26.50	
	0280	4″ diameter		47	.340		18.55	13.10		31.65	41	
	0300	6″ diameter	↓	38	.421	↓	30.50	16.20		46.70	59	
	1800	PVC, couplings 10′ O.C., hangers 3 per 10′										
	1820	Schedule 40										
	1860	1/2″ diameter	1 Plum	54	.148	L.F.	1.14	6.35		7.49	11.10	
	1870	3/4″ diameter		51	.157		1.23	6.70		7.93	11.75	
	1880	1″ diameter		46	.174		1.38	7.45		8.83	13.05	
	1890	1-1/4″ diameter		42	.190		1.57	8.15		9.72	14.40	
	1900	1-1/2″ diameter	↓	36	.222		1.69	9.50		11.19	16.60	
	1910	2″ diameter	Q-1	59	.271		1.98	10.40		12.38	18.40	
	1920	2-1/2″ diameter		56	.286		2.70	11		13.70	20	
	1930	3″ diameter		53	.302		3.47	11.60		15.07	22	
	1940	4″ diameter		48	.333		4.39	12.80		17.19	24.50	
	1950	5″ diameter		43	.372		5.75	14.30		20.05	28.50	
	1960	6″ diameter	↓	39	.410	↓	7.65	15.75		23.40	33	
	4100	DWV type, schedule 40, couplings 10′ O.C., hangers 3 per 10′										
	4120	ABS										
	4140	1-1/4″ diameter	1 Plum	42	.190	L.F.	1.26	8.15		9.41	14.05	
	4150	1-1/2″ diameter	″	36	.222		1.29	9.50		10.79	16.15	
	4160	2″ diameter	Q-1	59	.271	↓	1.42	10.40		11.82	17.75	
	4400	PVC										
	4410	1-1/4″ diameter	1 Plum	42	.190	L.F.	1.37	8.15		9.52	14.15	
	4460	2″ diameter	Q-1	59	.271		1.58	10.40		11.98	17.95	
	4470	3″ diameter		53	.302		2.96	11.60		14.56	21.50	
	4480	4″ diameter		48	.333		3.84	12.80		16.64	24	
	4490	6″ diameter	↓	39	.410	↓	7.10	15.75		22.85	32.50	
	5360	CPVC, couplings 10′ O.C., hangers 3 per 10′										
	5380	Schedule 40										
	5460	1/2″ diameter	1 Plum	54	.148	L.F.	2.53	6.35		8.88	12.65	
	5470	3/4″ diameter		51	.157		3.32	6.70		10.02	14.05	
	5480	1″ diameter		46	.174		4.07	7.45		11.52	16.05	
	5490	1-1/4″ diameter		42	.190		4.71	8.15		12.86	17.85	
	5500	1-1/2″ diameter	↓	36	.222		5.25	9.50		14.75	20.50	
	5510	2″ diameter	Q-1	59	.271		6.50	10.40		16.90	23.50	
	5520	2-1/2″ diameter		56	.286		10.45	11		21.45	28.50	
	5530	3″ diameter	↓	53	.302	↓	12.75	11.60		24.35	32	
	5560											
	9900	Minimum labor/equipment charge	1 Plum	4	2	Job		85.50		85.50	133	
560	0010	PIPE, PLASTIC, FITTINGS										560
	2700	PVC (white), schedule 40, socket joints										
	2760	90° elbow, 1/2″	1 Plum	33.30	.240	Ea.	.33	10.25		10.58	16.30	
	2810	2″	Q-1	36.40	.440	″	1.97	16.90		18.87	28.50	
	4500	DWV, ABS, non pressure, socket joints										
	4540	1/4 Bend, 1-1/4″	1 Plum	20.20	.396	Ea.	3.15	16.90		20.05	30	
	4560	1-1/2″	″	18.20	.440		2.41	18.75		21.16	31.50	
	4570	2″	Q-1	33.10	.483	↓	3.73	18.60		22.33	33	
	4800	Tee, sanitary										
	4820	1-1/4″	1 Plum	13.50	.593	Ea.	3.57	25.50		29.07	43.50	

	15108 | Plastic Pipe & Fittings	CREW	DAILY OUTPUT	LABOR-HOURS	UNIT	2006 BARE COSTS MAT.	LABOR	EQUIP.	TOTAL	TOTAL INCL O&P	
560	4830	1-1/2"	1 Plum	12.10	.661	Ea.	3.29	28		31.29	47.50	560
	4840	2"	Q-1	20	.800		4.81	30.50		35.31	53.50	
	5000	DWV, PVC, schedule 40, socket joints										
	5040	1/4 bend, 1-1/4"	1 Plum	20.20	.396	Ea.	4.50	16.90		21.40	31.50	
	5060	1-1/2"	"	18.20	.440		1.71	18.75		20.46	31	
	5070	2"	Q-1	33.10	.483		2.52	18.60		21.12	32	
	5080	3"		20.80	.769		7.20	29.50		36.70	54	
	5090	4"		16.50	.970		12.50	37.50		50	72	
	5100	6"		10.10	1.584		47.50	61		108.50	147	
	5105	8"	Q-2	9.30	2.581		96.50	103		199.50	266	
	5110	1/4 bend, long sweep, 1-1/2"	1 Plum	18.20	.440		3.74	18.75		22.49	33	
	5112	2"	Q-1	33.10	.483		4.03	18.60		22.63	33.50	
	5114	3"		20.80	.769		9.50	29.50		39	56.50	
	5116	4"		16.50	.970		17.70	37.50		55.20	77.50	
	5215	8"	Q-2	9.30	2.581		96	103		199	266	
	5250	Tee, sanitary 1-1/4"	1 Plum	13.50	.593		4.12	25.50		29.62	44	
	5254	1-1/2"	"	12.10	.661		2.95	28		30.95	47.50	
	5255	2"	Q-1	20	.800		4.14	30.50		34.64	52.50	
	5256	3"		13.90	1.151		9.75	44		53.75	80	
	5257	4"		11	1.455		16.95	56		72.95	106	
	5259	6"		6.70	2.388		91	92		183	243	
	5261	8"	Q-2	6.20	3.871		190	154		344	450	
	5264	2" x 1-1/2"	Q-1	22	.727		4.67	28		32.67	48.50	
	5266	3" x 1-1/2"		15.50	1.032		5.90	39.50		45.40	68	
	5268	4" x 3"		12.10	1.322		22.50	51		73.50	104	
	5271	6" x 4"		6.90	2.319		76	89		165	223	
	5314	Combination Y & 1/8 bend, 1-1/2"	1 Plum	12.10	.661		5.35	28		33.35	50	
	5315	2"	Q-1	20	.800		7.15	30.50		37.65	56	
	5317	3"		13.90	1.151		13	44		57	83.50	
	5318	4"		11	1.455		22	56		78	112	
	5324	Combination Y & 1/8 bend, reducing										
	5325	2" x 2" x 1-1/2"	Q-1	22	.727	Ea.	8.05	28		36.05	52.50	
	5327	3" x 3" x 1-1/2"		15.50	1.032		14	39.50		53.50	77	
	5328	3" x 3" x 2"		15.30	1.046		11.65	40		51.65	75.50	
	5329	4" x 4" x 2"		12.20	1.311		19.50	50.50		70	100	
	5331	Wye, 1-1/4"	1 Plum	13.50	.593		4	25.50		29.50	44	
	5332	1-1/2"	"	12.10	.661		3.95	28		31.95	48.50	
	5333	2"	Q-1	20	.800		5.05	30.50		35.55	53.50	
	5334	3"		13.90	1.151		14.45	44		58.45	85	
	5335	4"		11	1.455		21	56		77	110	
	5336	6"		6.70	2.388		44	92		136	192	
	5337	8"	Q-2	6.20	3.871		73.50	154		227.50	320	
	5341	2" x 1-1/2"	Q-1	22	.727		5.90	28		33.90	50	
	5342	3" x 1-1/2"		15.50	1.032		8.30	39.50		47.80	70.50	
	5343	4" x 3"		12.10	1.322		16.75	51		67.75	97.50	
	5344	6" x 4"		6.90	2.319		52	89		141	197	
	5345	8" x 6"	Q-2	6.40	3.750		56	149		205	294	
	5347	Double wye, 1-1/2"	1 Plum	9.10	.879		8	37.50		45.50	67.50	
	5348	2"	Q-1	16.60	.964		9.15	37		46.15	67.50	
	5349	3"		10.40	1.538		23.50	59		82.50	118	
	5350	4"		8.25	1.939		48	74.50		122.50	169	
	5354	2" x 1-1/2"		16.80	.952		8.40	36.50		44.90	66	
	5355	3" x 2"		10.60	1.509		17.60	58		75.60	109	
	5356	4" x 3"		8.45	1.893		38	73		111	155	
	5357	6" x 4"		7.25	2.207		79	85		164	219	
	5410	Reducer bushing, 2" x 1-1/4"		36.50	.438		2.56	16.85		19.41	29	

		15108 Plastic Pipe & Fittings	CREW	DAILY OUTPUT	LABOR-HOURS	UNIT	MAT.	LABOR	EQUIP.	TOTAL	TOTAL INCL O&P	
560	5412	3" x 1-1/2"	Q-1	27.30	.586	Ea.	5.85	22.50		28.35	41.50	560
	5414	4" x 2"		18.20	.879		10.80	34		44.80	64.50	
	5416	6" x 4"	↓	11.10	1.441		31	55.50		86.50	120	
	5418	8" x 6"	Q-2	10.20	2.353	↓	62	94		156	215	
	5500	CPVC, Schedule 80, threaded joints										
	5540	90° Elbow, 1/4"	1 Plum	32	.250	Ea.	8.25	10.70		18.95	25.50	
	5560	1/2"		30.30	.264		4.79	11.25		16.04	23	
	5570	3/4"		26	.308		7.15	13.15		20.30	28.50	
	5580	1"		22.70	.352		10.05	15.05		25.10	34.50	
	5590	1-1/4"		20.20	.396		19.35	16.90		36.25	48	
	5600	1-1/2"	↓	18.20	.440		21	18.75		39.75	52	
	5610	2"	Q-1	33.10	.483		28	18.60		46.60	60	
	5620	2-1/2"		24.20	.661		86.50	25.50		112	135	
	5630	3"	↓	20.80	.769		93	29.50		122.50	148	
	6000	Coupling, 1/4"	1 Plum	32	.250		10.50	10.70		21.20	28	
	6020	1/2"		30.30	.264		8.65	11.25		19.90	27	
	6030	3/4"		26	.308		14	13.15		27.15	36	
	6040	1"		22.70	.352		15.85	15.05		30.90	41	
	6050	1-1/4"		20.20	.396		16.85	16.90		33.75	45	
	6060	1-1/2"	↓	18.20	.440		18.10	18.75		36.85	49	
	6070	2"	Q-1	33.10	.483		21.50	18.60		40.10	52.50	
	6080	2-1/2"		24.20	.661		38	25.50		63.50	81.50	
	6090	3"	↓	20.80	.769	↓	44.50	29.50		74	95	
	9900	Minimum labor/equipment charge	1 Plum	4	2	Job		85.50		85.50	133	

		15110 Valves	CREW	DAILY OUTPUT	LABOR-HOURS	UNIT	MAT.	LABOR	EQUIP.	TOTAL	TOTAL INCL O&P	
160	0010	VALVES, BRONZE										160
	1020	Angle, 150 lb., rising stem, threaded										
	1070	3/4"	1 Plum	20	.400	Ea.	77.50	17.10		94.60	112	
	1080	1"		19	.421		110	18		128	149	
	1100	1-1/2"		13	.615		218	26.50		244.50	281	
	1110	2"	↓	11	.727	↓	345	31		376	430	
	1380	Ball, 150 psi, threaded										
	1460	3/4"	1 Plum	20	.400	Ea.	13.35	17.10		30.45	41	
	1750	Check, swing, class 150, regrinding disc, threaded										
	1850	1/2"	1 Plum	24	.333	Ea.	30	14.25		44.25	55	
	1860	3/4"		20	.400		40.50	17.10		57.60	71	
	1870	1"		19	.421		60	18		78	94	
	1890	1-1/2"		13	.615		99	26.50		125.50	150	
	1900	2"	↓	11	.727	↓	146	31		177	209	
	2850	Gate, N.R.S., soldered, 125 psi										
	2920	1/2"	1 Plum	24	.333	Ea.	19.50	14.25		33.75	43.50	
	2940	3/4"		20	.400		22.50	17.10		39.60	51	
	2950	1"		19	.421		32	18		50	63	
	2970	1-1/2"		13	.615		56.50	26.50		83	103	
	2980	2"	↓	11	.727	↓	76.50	31		107.50	133	
	4850	Globe, class 150, rising stem, threaded										
	4950	1/2"	1 Plum	24	.333	Ea.	42	14.25		56.25	68	
	4960	3/4"	"	20	.400	"	57	17.10		74.10	89	
	5600	Relief, pressure & temperature, self-closing, ASME, threaded										
	5640	3/4"	1 Plum	28	.286	Ea.	85.50	12.20		97.70	113	
	5650	1"		24	.333		124	14.25		138.25	159	
	5660	1-1/4"		20	.400		249	17.10		266.10	300	
	5670	1-1/2"		18	.444		475	19		494	555	
	5680	2"	↓	16	.500	↓	520	21.50		541.50	605	
	6400	Pressure, water, ASME, threaded										

15110 | Valves

			Crew	Daily Output	Labor-Hours	Unit	2006 Bare Costs Mat.	Labor	Equip.	Total	Total Incl O&P	
160	6440	3/4″	1 Plum	28	.286	Ea.	49	12.20		61.20	73	160
	6450	1″		24	.333		53.50	14.25		67.75	81	
	6470	1-1/2″		18	.444		232	19		251	286	
	6900	Reducing, water pressure										
	6940	1/2″	1 Plum	24	.333	Ea.	145	14.25		159.25	182	
	6960	1″	″	19	.421	″	225	18		243	275	
	8350	Tempering, water, sweat connections										
	8400	1/2″	1 Plum	24	.333	Ea.	52	14.25		66.25	79	
	8440	3/4″	″	20	.400	″	63.50	17.10		80.60	96	
	8650	Threaded connections										
	8700	1/2″	1 Plum	24	.333	Ea.	63.50	14.25		77.75	91.50	
	8740	3/4″		20	.400	″	242	17.10		259.10	293	
	9000	Minimum labor/equipment charge		4	2	Job		85.50		85.50	133	
200	0010	**VALVES, IRON BODY**										200
	1020	Butterfly, wafer type, gear actuator, 200 lb.										
	1030	2″	1 Plum	14	.571	Ea.	120	24.50		144.50	170	
	1060	4″	Q-1	5	3.200	″	160	123		283	365	
	1650	Gate, 125 lb., N.R.S.										
	2150	Flanged										
	2240	2-1/2″	Q-1	5	3.200	Ea.	355	123		478	580	
	2260	3″		4.50	3.556		400	137		537	650	
	2280	4″		3	5.333		570	205		775	950	
	3550	OS&Y, 125 lb., flanged										
	3680	4″	Q-1	3	5.333	Ea.	415	205		620	775	
	3690	5″	Q-2	3.40	7.059		685	281		966	1,175	
	3700	6″	″	3	8		685	320		1,005	1,250	
	9000	Minimum labor/equipment charge	1 Plum	3	2.667	Job		114		114	177	
500	0010	**VALVES, PLASTIC**										500
	0020											
	1150	Ball, PVC, socket or threaded, single union										
	1280	2″	1 Plum	17	.471	Ea.	42	20		62	78	
	1650	CPVC, socket or threaded, single union										
	1770	2″	1 Plum	17	.471	Ea.	122	20		142	166	
	3150	Ball check, PVC, socket or threaded										
	3290	2″	1 Plum	17	.471	Ea.	98	20		118	140	
	9000	Minimum labor/equipment charge	″	3.50	2.286	Job		97.50		97.50	152	

15120 | Piping Specialties

			Crew	Daily Output	Labor-Hours	Unit	Mat.	Labor	Equip.	Total	Total Incl O&P	
250	0010	**DIELECTRIC UNIONS** Standard gaskets for water and air										250
	0020	250 psi maximum pressure										
	0280	Female IPT to sweat, straight										
	0340	3/4″ pipe size	1 Plum	20	.400	Ea.	3.63	17.10		20.73	30.50	
	0780	Female IPT to female IPT, straight										
	0800	1/2″ pipe size	1 Plum	24	.333	Ea.	5.35	14.25		19.60	28	
	0840	3/4″ pipe size		20	.400	″	5.45	17.10		22.55	32.50	
	9000	Minimum labor/equipment charge		4	2	Job		85.50		85.50	133	
320	0010	**EXPANSION TANKS**										320
	2000	Steel, liquid expansion, ASME, painted, 15 gallon capacity	Q-5	17	.941	Ea.	365	36.50		401.50	455	
	2040	30 gallon capacity		12	1.333		420	51.50		471.50	540	
	2080	60 gallon capacity		8	2		550	77.50		627.50	725	
	2120	100 gallon capacity		6	2.667		805	103		908	1,050	
	3000	Steel ASME expansion, rubber diaphragm, 19 gal. cap. accept.		12	1.333		1,700	51.50		1,751.50	1,950	
	3020	31 gallon capacity		8	2		1,875	77.50		1,952.50	2,200	
	3040	61 gallon capacity		6	2.667		1,975	103		2,078	2,325	

15120 | Piping Specialties

		15120 Piping Specialties	CREW	DAILY OUTPUT	LABOR-HOURS	UNIT	2006 BARE COSTS MAT.	LABOR	EQUIP.	TOTAL	TOTAL INCL O&P
320	3080	119 gallon capacity	Q-5	4	4	Ea.	2,850	155		3,005	3,375
	9000	Minimum labor/equipment charge	↓	4	4	Job		155		155	241
350	0010	**FLEXIBLE CONNECTORS**, Corrugated, 7/8" O.D., 1/2" I.D.									
	0050	Gas, seamless brass, steel fittings									
	0200	12" long	1 Plum	36	.222	Ea.	12.35	9.50		21.85	28.50
	0220	18" long		36	.222		15.30	9.50		24.80	31.50
	0240	24" long		34	.235		18.05	10.05		28.10	35.50
	0260	30" long		34	.235	↓	19.55	10.05		29.60	37
	9000	Minimum labor/equipment charge	↓	4	2	Job		85.50		85.50	133
580	0010	**MIXING VALVE** Automatic water tempering									
	0040	1/2" size	1 Stpi	19	.421	Ea.	415	18.15		433.15	485
	0050	3/4" size		18	.444	"	415	19.15		434.15	485
	9000	Minimum labor/equipment charge	↓	5	1.600	Job		69		69	107
760	0010	**STEAM TRAP**									
	0020										
	0030	Cast iron body, threaded									
	0040	Inverted bucket									
	0050	1/2" pipe size	1 Stpi	12	.667	Ea.	112	28.50		140.50	168
	0100	1" pipe size		9	.889		300	38.50		338.50	390
	0120	1-1/4" pipe size	↓	8	1	↓	450	43		493	560
	1000	Float & thermostatic, 15 psi									
	1010	3/4" pipe size	1 Stpi	16	.500	Ea.	89.50	21.50		111	132
	1020	1" pipe size		15	.533		107	23		130	154
	1040	1-1/2" pipe size		9	.889		191	38.50		229.50	270
	1060	2" pipe size		6	1.333	↓	350	57.50		407.50	475
	9000	Minimum labor/equipment charge	↓	4	2	Job		86		86	134
820	0010	**STRAINERS, Y TYPE** Bronze body									
	0020										
	0050	Screwed, 150 lb., 1/4" pipe size	1 Stpi	24	.333	Ea.	12.80	14.35		27.15	36.50
	0100	1/2" pipe size		20	.400		16.85	17.20		34.05	45.50
	0120	3/4" pipe size		19	.421		18.45	18.15		36.60	48.50
	0140	1" pipe size		17	.471		19.85	20.50		40.35	53.50
	0160	1-1/2" pipe size		14	.571		42.50	24.50		67	85.50
	0180	2" pipe size		13	.615		57	26.50		83.50	104
	0182	3" pipe size	↓	12	.667		360	28.50		388.50	440
	0220	3" pipe size	Q-5	16	1		600	39		639	720
	0240	4" pipe size	"	15	1.067		1,375	41.50		1,416.50	1,575
	1000	Flanged, 150 lb., 1-1/2" pipe size	1 Stpi	11	.727		315	31.50		346.50	395
	1020	2" pipe size	"	8	1		365	43		408	470
	1030	2-1/2" pipe size	Q-5	5	3.200		560	124		684	815
	1040	3" pipe size		4.50	3.556		695	138		833	975
	1060	4" pipe size	↓	3	5.333		1,050	207		1,257	1,475
	1100	6" pipe size	Q-6	3	8		2,800	320		3,120	3,575
	1106	8" pipe size	"	2.60	9.231	↓	3,425	370		3,795	4,350
	1500	For 300 lb rating, add					40%				
	9000	Minimum labor/equipment charge	1 Stpi	3.75	2.133	Job		92		92	143
940	0010	**WATER SUPPLY METERS**									
	2000	Domestic/commercial, bronze									
	2020	Threaded									
	2060	5/8" diameter, to 20 GPM	1 Plum	16	.500	Ea.	40	21.50		61.50	77
	2080	3/4" diameter, to 30 GPM		14	.571		67.50	24.50		92	113
	2100	1" diameter, to 50 GPM	↓	12	.667	↓	94	28.50		122.50	148
	2300	Threaded/flanged									
	2340	1-1/2" diameter, to 100 GPM	1 Plum	8	1	Ea.	330	42.50		372.50	430

15100 | Building Services Piping

		15120 | Piping Specialties	CREW	DAILY OUTPUT	LABOR-HOURS	UNIT	2006 BARE COSTS MAT.	LABOR	EQUIP.	TOTAL	TOTAL INCL O&P	
940	9000	Minimum labor/equipment charge	1 Plum	3.25	2.462	Job		105		105	163	940

		15140 | Domestic Water Piping	CREW	DAILY OUTPUT	LABOR-HOURS	UNIT	MAT.	LABOR	EQUIP.	TOTAL	TOTAL INCL O&P	
100	0010	**BACKFLOW PREVENTER** Includes valves										100
	0020	and four test cocks, corrosion resistant, automatic operation										
	1000	Double check principle										
	1010	Threaded, with ball valves										
	1020	3/4" pipe size	1 Plum	16	.500	Ea.	148	21.50		169.50	196	
	1030	1" pipe size		14	.571		164	24.50		188.50	218	
	1040	1-1/2" pipe size		10	.800		295	34		329	380	
	1050	2" pipe size		7	1.143		365	49		414	475	
	1080	Threaded, with gate valves										
	1100	3/4" pipe size	1 Plum	16	.500	Ea.	580	21.50		601.50	675	
	1120	1" pipe size		14	.571		620	24.50		644.50	720	
	1140	1-1/2" pipe size		10	.800		755	34		789	885	
	1160	2" pipe size		7	1.143		925	49		974	1,100	
	4000	Reduced pressure principle										
	4100	Threaded, bronze, valves are ball										
	4120	3/4" pipe size	1 Plum	16	.500	Ea.	220	21.50		241.50	275	
	4140	1" pipe size		14	.571		236	24.50		260.50	298	
	4160	1-1/2" pipe size		10	.800		445	34		479	545	
	4180	2" pipe size		7	1.143		500	49		549	625	
	4500	Minimum labor/equipment charge		4	2	Job		85.50		85.50	133	
	5000	Flanged, valves are OS&Y										
	5060	2-1/2" pipe size	Q-1	5	3.200	Ea.	2,550	123		2,673	3,000	
	5080	3" pipe size		4.50	3.556		2,675	137		2,812	3,150	
	5100	4" pipe size		3	5.333		3,375	205		3,580	4,025	
	5120	6" pipe size	Q-2	3	8		4,875	320		5,195	5,875	
	5600	Flanged, iron, valves are OS&Y										
	5660	2-1/2" pipe size	Q-1	5	3.200	Ea.	2,050	123		2,173	2,450	
	5680	3" pipe size		4.50	3.556		2,150	137		2,287	2,575	
	5700	4" pipe size		3	5.333		2,700	205		2,905	3,300	
	5720	6" pipe size	Q-2	3	8		3,900	320		4,220	4,800	
	5800	Rebuild 4" diameter reduced pressure BFP	1 Plum	4	2		220	85.50		305.50	375	
	5810	6" diameter		2.66	3.008		234	128		362	455	
	5820	8" diameter		2	4		289	171		460	585	
	5830	10" diameter		1.60	5		335	214		549	700	
	9010	Minimum labor/equipment charge		2	4	Job		171		171	266	
600	0010	**VACUUM BREAKERS** Hot or cold water										600
	1030	Anti-siphon, brass										
	1060	1/2" size	1 Plum	24	.333	Ea.	27	14.25		41.25	51.50	
	1080	3/4" size		20	.400		32	17.10		49.10	61.50	
	1100	1" size		19	.421		50	18		68	82.50	
	1120	1-1/4" size		15	.533		87	23		110	132	
	1140	1-1/2" size		13	.615		103	26.50		129.50	154	
	1160	2" size		11	.727		160	31		191	225	
	9000	Minimum labor/equipment charge		4	2	Job		85.50		85.50	133	
800	0010	**WATER HAMMER ARRESTORS / SHOCK ABSORBERS**										800
	0490	Copper										
	0500	3/4" male I.P.S. For 1 to 11 fixtures	1 Plum	12	.667	Ea.	15.55	28.50		44.05	61.50	
	0600	1" male I.P.S., For 12 to 32 fixtures		8	1		39.50	42.50		82	110	
	0700	1-1/4" male I.P.S. For 33 to 60 fixtures		8	1		46	42.50		88.50	117	
	0800	1-1/2" male I.P.S. For 61 to 113 fixtures		8	1		62.50	42.50		105	136	

15100 | Building Services Piping

15140 | Domestic Water Piping

			Crew	Daily Output	Labor-Hours	Unit	2006 Bare Costs Mat.	Labor	Equip.	Total	Total Incl O&P	
800	0900	2" male I.P.S.For 114 to 154 fixtures	1 Plum	8	1	Ea.	100	42.50		142.50	177	800
	1000	2-1/2" male I.P.S. For 155 to 330 fixtures		4	2	↓	285	85.50		370.50	450	
	9000	Minimum labor/equipment charge	↓	3.50	2.286	Job		97.50		97.50	152	

15150 | Sanitary Waste and Vent Piping

			Crew	Daily Output	Labor-Hours	Unit	2006 Bare Costs Mat.	Labor	Equip.	Total	Total Incl O&P	
200	0010	CLEANOUTS										200
	0060	Floor type										
	0080	Round or square, scoriated nickel bronze top										
	0100	2" pipe size	1 Plum	10	.800	Ea.	102	34		136	165	
	0140	4" pipe size	"	6	1.333	"	153	57		210	257	
	0980	Round top, recessed for terrazzo										
	1000	2" pipe size	1 Plum	9	.889	Ea.	102	38		140	171	
	1100	4" pipe size	"	4	2		153	85.50		238.50	300	
	1120	5" pipe size	Q-1	6	2.667	↓	194	102		296	370	
	9000	Minimum labor/equipment charge	1 Plum	3	2.667	Job		114		114	177	
250	0010	CLEANOUT TEE										250
	0100	Cast iron, B&S, with countersunk plug										
	0220	3" pipe size	1 Plum	3.60	2.222	Ea.	151	95		246	315	
	0240	4" pipe size	"	3.30	2.424		188	104		292	370	
	0500	For round smooth access cover, same price				↓						
	4000	Plastic, tees and adapters. Add plugs										
	4010	ABS, DWV										
	4020	Cleanout tee, 1-1/2" pipe size	1 Plum	15	.533	Ea.	6.05	23		29.05	42	
	9000	Minimum labor/equipment charge	"	2.75	2.909	Job		124		124	193	
300	0010	FLOOR AND AREA DRAINS										300
	0400	Deck, auto park, C.I., 13" top										
	0440	3", 4", 5", and 6" pipe size	Q-1	8	2	Ea.	745	77		822	940	
	0480	For galvanized body, add				"	355			355	390	
	0500											
	2000	Floor, medium duty, C.I., deep flange, 7" dia top										
	2040	2" and 3" pipe size	Q-1	12	1.333	Ea.	104	51		155	195	
	2080	For galvanized body, add					44			44	48.50	
	2120	For polished bronze top, add				↓	58			58	64	
	2500	Heavy duty, cleanout & trap w/bucket, C.I., 15" top										
	2540	2", 3", and 4" pipe size	Q-1	6	2.667	Ea.	3,325	102		3,427	3,800	
	2560	For galvanized body, add					770			770	845	
	2580	For polished bronze top, add				↓	270			270	297	
350	0010	FLOOR RECEPTORS For connection to 2", 3" & 4" diameter pipe										350
	0200	12-1/2" square top, 25 sq in open area	Q-1	10	1.600	Ea.	505	61.50		566.50	650	
	0300	For grate with 4" diam. x 3-3/4" high funnel, add					89.50			89.50	98.50	
	0400	For grate with 6" diameter x 6" high funnel, add					116			116	127	
	0700	For acid-resisting bucket, add					142			142	156	
	0900	For stainless steel mesh bucket liner, add					112			112	123	
	2000	12-5/8" diameter top, 40 sq. in. open area	Q-1	10	1.600	↓	400	61.50		461.50	535	
	2100	For options, add same prices as square top										
	3000	8" x 4" rectangular top, 7.5 sq. in. open area	Q-1	14	1.143	Ea.	385	44		429	495	
	3100	For trap primer connections, add				"	45.50			45.50	50	
	9000	Minimum labor/equipment charge	Q-1	3	5.333	Job		205		205	320	
400	0010	INTERCEPTORS										400
	0150	Grease, cast iron, 4 GPM, 8 lb. fat capacity	1 Plum	4	2	Ea.	615	85.50		700.50	815	
	0200	7 GPM, 14 lb. fat capacity		4	2		855	85.50		940.50	1,075	
	1000	10 GPM, 20 lb. fat capacity	↓	4	2	↓	1,000	85.50		1,085.50	1,225	

15 MECHANICAL

		15150 | Sanitary Waste and Vent Piping	CREW	DAILY OUTPUT	LABOR-HOURS	UNIT	2006 BARE COSTS MAT.	LABOR	EQUIP.	TOTAL	TOTAL INCL O&P	
400	1160	100 GPM, 200 lb. fat capacity	Q-1	2	8	Ea.	7,600	305		7,905	8,825	400
	1240	300 GPM, 600 lb. fat capacity	"	1	16	↓	16,800	615		17,415	19,500	
	9000	Minimum labor/equipment charge	1 Plum	3	2.667	Job		114		114	177	
800	0010	**TRAPS**										800
	0030	Cast iron, service weight										
	0050	Running P trap, without vent										
	1100	2"	Q-1	16	1	Ea.	28.50	38.50		67	91.50	
	1150	4"	"	13	1.231		82.50	47.50		130	165	
	1160	6"	Q-2	17	1.412		360	56.50		416.50	485	
	3000	P trap, B&S, 2" pipe size	Q-1	16	1		19.60	38.50		58.10	81.50	
	3040	3" pipe size	"	14	1.143	↓	29.50	44		73.50	101	
	3900											
	4700	Copper, drainage, drum trap										
	4840	3" x 6" swivel, 1-1/2" pipe size	1 Plum	16	.500	Ea.	69	21.50		90.50	109	
	5100	P trap, standard pattern										
	5200	1-1/4" pipe size	1 Plum	18	.444	Ea.	32	19		51	64.50	
	5240	1-1/2" pipe size		17	.471		31	20		51	65.50	
	5260	2" pipe size		15	.533		48	23		71	88	
	5280	3" pipe size		11	.727	↓	115	31		146	175	
	9000	Minimum labor/equipment charge	↓	3	2.667	Job		114		114	177	
900	0010	**VENT FLASHING, CAPS**										900
	0120	Vent caps										
	0140	Cast iron										
	0180	2-1/2" - 3-5/8" pipe	1 Plum	21	.381	Ea.	35	16.25		51.25	63.50	
	0190	4" - 4-1/8" pipe	"	19	.421	"	41.50	18		59.50	74	
	0900	Vent flashing										
	1000	Aluminum with lead ring										
	1050	3" pipe	1 Plum	17	.471	Ea.	7.90	20		27.90	40	
	1060	4" pipe	"	16	.500	"	9.55	21.50		31.05	43.50	
	1350	Copper with neoprene ring										
	1440	2" pipe	1 Plum	18	.444	Ea.	14.85	19		33.85	46	
	1450	3" pipe		17	.471		17.40	20		37.40	50.50	
	1460	4" pipe		16	.500	↓	19.20	21.50		40.70	54	
	9000	Minimum labor/equipment charge	↓	4	2	Job		85.50		85.50	133	

		15160 | Storm Drainage Piping	CREW	DAILY OUTPUT	LABOR-HOURS	UNIT	MAT.	LABOR	EQUIP.	TOTAL	TOTAL INCL O&P	
500	0010	**STORM AREA DRAINS**										500
	0140	Cornice, C.I., 45° or 90° outlet										
	0200	3" and 4" pipe size	Q-1	12	1.333	Ea.	171	51		222	268	
	0260	For galvanized body, add					34.50			34.50	37.50	
	0280	For polished bronze dome, add				↓	29.50			29.50	32.50	
	3860	Roof, flat metal deck, C.I. body, 12" C.I. dome										
	3890	3" pipe size	Q-1	14	1.143	Ea.	204	44		248	294	
	3900	4" pipe size	"	13	1.231	"	204	47.50		251.50	299	
	4620	Main, all aluminum, 12" low profile dome										
	4640	2", 3" and 4" pipe size	Q-1	14	1.143	Ea.	226	44		270	320	
	9000	Minimum labor/equipment charge	1 Plum	4	2	Job		85.50		85.50	133	

		15180 | Heating and Cooling Piping	CREW	DAILY OUTPUT	LABOR-HOURS	UNIT	MAT.	LABOR	EQUIP.	TOTAL	TOTAL INCL O&P	
200	0010	**PUMPS, CIRCULATING** Heated or chilled water application										200
	0100											
	0600	Bronze, sweat connections, 1/40 HP, in line										
	0640	3/4" size	Q-1	16	1	Ea.	134	38.50		172.50	208	

15100 | Building Services Piping

15180 | Heating and Cooling Piping

	15180 Heating and Cooling Piping		Crew	Daily Output	Labor-Hours	Unit	2006 Bare Costs Mat.	Labor	Equip.	Total	Total Incl O&P	
200	1000	Flange connection, 3/4" to 1-1/2" size										200
	1040	1/12 HP	Q-1	6	2.667	Ea.	365	102		467	565	
	1060	1/8 HP		6	2.667		630	102		732	855	
	1100	1/3 HP		6	2.667		705	102		807	935	
	1140	2" size, 1/6 HP		5	3.200		905	123		1,028	1,175	
	1180	2-1/2" size, 1/4 HP		5	3.200		1,150	123		1,273	1,475	
	1220	3" size, 1/4 HP		4	4	↓	1,200	154		1,354	1,575	
	9000	Minimum labor/equipment charge	↓	3.25	4.923	Job		189		189	294	

15200 | Process Piping

15210 | Process Air/Gas Piping

	15210 Process Air/Gas Piping		Crew	Daily Output	Labor-Hours	Unit	2006 Bare Costs Mat.	Labor	Equip.	Total	Total Incl O&P	
100	0010	**AIR COMPRESSORS**										100
	5250	Air, reciprocating air cooled, splash lubricated, tank mounted										
	5300	Single stage, 1 phase, 140 psi										
	5303	1/2 HP, 30 gal tank	1 Stpi	3	2.667	Ea.	890	115		1,005	1,150	
	5305	3/4 HP, 30 gal tank		2.60	3.077		920	132		1,052	1,200	
	5307	1 HP, 30 gal tank	↓	2.20	3.636	↓	1,150	157		1,307	1,500	

15400 | Plumbing Fixtures & Equipment

15410 | Plumbing Fixtures

	15410 Plumbing Fixtures		Crew	Daily Output	Labor-Hours	Unit	2006 Bare Costs Mat.	Labor	Equip.	Total	Total Incl O&P	
200	0010	**CARRIERS/SUPPORTS** For plumbing fixtures										200
	0020											
	3000	Lavatory, concealed arm										
	3050	Floor mounted, single										
	3100	High back fixture	1 Plum	6	1.333	Ea.	177	57		234	284	
	3200	Flat slab fixture		6	1.333		223	57		280	335	
	3220	Paraplegic	↓	6	1.333	↓	290	57		347	410	
	6980	Water closet, siphon jet										
	7000	Horizontal, adjustable, caulk										
	7040	Single, 4" pipe size	1 Plum	5.33	1.501	Ea.	360	64		424	495	
	7060	5" pipe size		5.33	1.501		475	64		539	620	
	7100	Double, 4" pipe size		5	1.600		675	68.50		743.50	850	
	7120	5" pipe size	↓	5	1.600	↓	830	68.50		898.50	1,025	
	8200	Water closet, residential										
	8220	Vertical centerline, floor mount										
	8240	Single, 3" caulk, 2" or 3" vent	1 Plum	6	1.333	Ea.	297	57		354	415	
	8260	4" caulk, 2" or 4" vent		6	1.333	"	385	57		442	510	
	9990	Minimum labor/equipment charge	↓	3.50	2.286	Job		97.50		97.50	152	
300	0010	**FAUCETS/FITTINGS**										300
	0020											
	0150	Bath, faucets, diverter spout combination, sweat	1 Plum	8	1	Ea.	69.50	42.50		112	143	
	0200	For integral stops, IPS unions, add				↓	73			73	80.50	

	15410 | Plumbing Fixtures	CREW	DAILY OUTPUT	LABOR-HOURS	UNIT	2006 BARE COSTS MAT.	LABOR	EQUIP.	TOTAL	TOTAL INCL O&P	
300	0420 Bath, press-bal mix valve w/diverter, spout, shower hd, arm/flange	1 Plum	8	1	Ea.	116	42.50		158.50	195	300
	0500 Drain, central lift, 1-1/2" IPS male		20	.400		38	17.10		55.10	68.50	
	0600 Trip lever, 1-1/2" IPS male		20	.400		38.50	17.10		55.60	69	
	0810 Bidet										
	0812 Fitting, over the rim, swivel spray/pop-up drain	1 Plum	8	1	Ea.	138	42.50		180.50	219	
	0840 Flush valves, with vacuum breaker										
	0850 Water closet										
	0860 Exposed, rear spud	1 Plum	8	1	Ea.	118	42.50		160.50	196	
	0870 Top spud		8	1		109	42.50		151.50	187	
	0880 Concealed, rear spud		8	1		154	42.50		196.50	237	
	0890 Top spud		8	1		127	42.50		169.50	207	
	0900 Wall hung		8	1		136	42.50		178.50	217	
	0920 Urinal										
	0930 Exposed, stall	1 Plum	8	1	Ea.	108	42.50		150.50	186	
	0940 Wall, (washout)		8	1		108	42.50		150.50	186	
	0950 Pedestal, top spud		8	1		110	42.50		152.50	188	
	0960 Concealed, stall		8	1		124	42.50		166.50	203	
	0970 Wall (washout)		8	1		123	42.50		165.50	203	
	0971 Automatic flush sensor and operator for										
	0972 urinals or water closets	1 Plum	5.33	1.501	Ea.	355	64		419	490	
	1000 Kitchen sink faucets, top mount, cast spout		10	.800		51.50	34		85.50	110	
	1100 For spray, add		24	.333		11.05	14.25		25.30	34	
	1200 Wall type, swing tube spout		10	.800		64	34		98	124	
	1300 Single control lever handle										
	1310 With pull out spray										
	1320 Polished chrome	1 Plum	10	.800	Ea.	167	34		201	236	
	1330 Polished brass		10	.800		200	34		234	273	
	1340 White		10	.800		184	34		218	255	
	1348 With spray thru escutcheon										
	1350 Polished chrome	1 Plum	10	.800	Ea.	117	34		151	181	
	1360 Polished brass		10	.800		140	34		174	207	
	1370 White		10	.800		128	34		162	194	
	2000 Laundry faucets, shelf type, IPS or copper unions		12	.667		42.50	28.50		71	91	
	2100 Lavatory faucet, centerset, without drain		10	.800		37.50	34		71.50	94	
	2120 With pop-up drain		6.66	1.201		52	51.50		103.50	138	
	2210 Porcelain cross handles and pop-up drain										
	2220 Polished chrome	1 Plum	6.66	1.201	Ea.	110	51.50		161.50	201	
	2230 Polished brass	"	6.66	1.201	"	165	51.50		216.50	262	
	2260 Single lever handle and pop-up drain										
	2270 Black nickel	1 Plum	6.66	1.201	Ea.	212	51.50		263.50	315	
	2280 Polished brass		6.66	1.201		212	51.50		263.50	315	
	2290 Polished chrome		6.66	1.201		141	51.50		192.50	235	
	2800 Self-closing, center set		10	.800		179	34		213	250	
	2810 Automatic sensor and operator, with faucet head		6.15	1.301		300	55.50		355.50	415	
	2850 Medical, bedpan cleanser, with pedal valve,		12	.667		470	28.50		498.50	560	
	2860 With screwdriver stop valve		12	.667		244	28.50		272.50	315	
	2870 With self-closing spray valve		12	.667		182	28.50		210.50	245	
	2900 Faucet, gooseneck spout, wrist handles, grid drain		10	.800		109	34		143	173	
	2940 Mixing valve, knee action, screwdriver stops		4	2		365	85.50		450.50	535	
	3000 Service sink faucet, cast spout, pail hook, hose end		14	.571		72	24.50		96.50	117	
	4000 Shower by-pass valve with union		18	.444		50.50	19		69.50	85	
	4100 Shower arm with flange and head		22	.364		79	15.55		94.55	111	
	4140 Shower, hand held, pin mount, massage action, chrome		22	.364		51.50	15.55		67.05	81	
	4142 Polished brass		22	.364		98	15.55		113.55	132	
	4144 Shower, hand held, wall mtd, adj. spray, 2 wall mounts, chrome		20	.400		110	17.10		127.10	148	
	4146 Polished brass		20	.400		220	17.10		237.10	269	

MECHANICAL 15

		15410 | Plumbing Fixtures	CREW	DAILY OUTPUT	LABOR-HOURS	UNIT	2006 BARE COSTS MAT.	LABOR	EQUIP.	TOTAL	TOTAL INCL O&P	
300	4148	Shower, hand held head, bar mounted 24", adj. spray, chrome	1 Plum	20	.400	Ea.	144	17.10		161.10	185	300
	4150	Polished brass		20	.400		330	17.10		347.10	390	
	4200	Shower thermostatic mixing valve, concealed		8	1		250	42.50		292.50	340	
	4204	For inlet strainer, check, and stops, add					33			33	36	
	4220	Shower pressure balancing mixing valve,										
	4230	With shower head, arm, flange and diverter tub spout										
	4240	Chrome	1 Plum	6.14	1.303	Ea.	138	55.50		193.50	238	
	4250	Polished brass		6.14	1.303		193	55.50		248.50	299	
	4260	Satin		6.14	1.303		193	55.50		248.50	299	
	4270	Polished chrome/brass		6.14	1.303		158	55.50		213.50	261	
	5000	Sillcock, compact, brass, IPS or copper to hose		24	.333		5.50	14.25		19.75	28	
	6000	Stop and waste valves, bronze										
	6100	Angle, solder end 1/2"	1 Plum	24	.333	Ea.	4.87	14.25		19.12	27.50	
	6110	3/4"		20	.400		5.35	17.10		22.45	32.50	
	6300	Straightway, solder end 3/8"		24	.333		4.23	14.25		18.48	26.50	
	6310	1/2"		24	.333		2.71	14.25		16.96	25	
	6330	1"		19	.421		4.74	18		22.74	33	
	6400	Straightway, threaded 3/8"		24	.333		3.87	14.25		18.12	26.50	
	6410	1/2"		24	.333		3.67	14.25		17.92	26	
	6420	3/4"		20	.400		4.38	17.10		21.48	31.50	
	6430	1"		19	.421		8.90	18		26.90	38	
	7800	Water closet, wax gasket		96	.083		.99	3.56		4.55	6.65	
	7820	Gasket toilet tank to bowl		32	.250		1.24	10.70		11.94	17.95	
	8000	Water supply stops, polished chrome plate										
	8200	Angle, 3/8"	1 Plum	24	.333	Ea.	4.73	14.25		18.98	27	
	8300	1/2"		22	.364		4.73	15.55		20.28	29	
	8400	Straight, 3/8"		26	.308		7.05	13.15		20.20	28.50	
	8500	1/2"		24	.333		7.05	14.25		21.30	30	
	8600	Water closet, angle, w/flex riser, 3/8"		24	.333		11.05	14.25		25.30	34	
	9000	Minimum labor/equipment charge		4	2	Job		85.50		85.50	133	
		15411 | Commercial/Indust Fixtures										
500	0010	**HYDRANTS**										500
	0050	Wall type, moderate climate, bronze, encased										
	0200	3/4" IPS connection	1 Plum	16	.500	Ea.	395	21.50		416.50	470	
	1000	Non-freeze, bronze, exposed										
	1100	3/4" IPS connection, 4" to 9" thick wall	1 Plum	14	.571	Ea.	268	24.50		292.50	335	
	1120	10" to 14" thick wall	"	12	.667		290	28.50		318.50	365	
	1280	For anti-siphon type, add					70			70	77	
	9000	Minimum labor/equipment charge	1 Plum	3	2.667	Job		114		114	177	
700	0010	**URINALS**										700
	3000	Wall hung, vitreous china, with hanger & self-closing valve										
	3100	Siphon jet type	Q-1	3	5.333	Ea.	315	205		520	665	
	3120	Blowout type		3	5.333		350	205		555	705	
	3300	Rough-in, supply, waste & vent		2.83	5.654		136	217		353	490	
	5000	Stall type, vitreous china, includes valve		2.50	6.400		520	246		766	955	
	6980	Rough-in, supply, waste and vent		1.99	8.040		184	310		494	685	
	9000	Minimum labor/equipment charge		4	4	Job		154		154	239	
810	0010	**HANDWASHER-DRYER MODULE**										810
	0100	Wall mounted										
	0120	With electric dryer										
	0140	Sensor operated	Q-1	8	2	Ea.	2,300	77		2,377	2,650	
	0160	Sensor operated (ADA)		8	2		2,175	77		2,252	2,525	
	0180	Sensor operated (ADA), surface mounted		8	2		2,825	77		2,902	3,225	
	0200	With paper towels										
	0220	Sensor operated	Q-1	8	2	Ea.	2,250	77		2,327	2,600	

15400 | Plumbing Fixtures & Equipment

	15411 | Commercial/Indust Fixtures	CREW	DAILY OUTPUT	LABOR-HOURS	UNIT	2006 BARE COSTS MAT.	LABOR	EQUIP.	TOTAL	TOTAL INCL O&P	
840	0010 **WASH FOUNTAINS**										840
	1900 Group, foot control										
	2000 Precast terrazzo, circular, 36″ diam., 5 or 6 persons	Q-2	3	8	Ea.	2,975	320		3,295	3,750	
	2100 54″ diameter for 8 or 10 persons		2.50	9.600		3,600	385		3,985	4,550	
	2400 Semi-circular, 36″ diam. for 3 persons		3	8		2,650	320		2,970	3,425	
	2500 54″ diam. for 4 or 5 persons		2.50	9.600		3,275	385		3,660	4,200	
	5610 Group, infrared control, barrier free										
	5614 Precast terrazzo										
	5620 Semi-circular 36″ diam. for 3 persons	Q-2	3	8	Ea.	3,900	320		4,220	4,800	
	5630 46″ diam. for 4 persons		2.80	8.571		4,375	340		4,715	5,325	
	5640 Circular, 54″ diam. for 8 persons, button control		2.50	9.600		6,400	385		6,785	7,625	
	5700 Rough-in, supply, waste and vent for above wash fountains	Q-1	1.82	8.791		176	340		516	720	
	9000 Minimum labor/equipment charge	Q-2	3	8	Job		320		320	495	

	15412 | Drinking Fountains	CREW	DAILY OUTPUT	LABOR-HOURS	UNIT	MAT.	LABOR	EQUIP.	TOTAL	TOTAL INCL O&P	
200	0010 **DRINKING FOUNTAIN** For connection to cold water supply										200
	1000 Wall mounted, non-recessed										
	2700 Stainless steel, single bubbler, no back	1 Plum	4	2	Ea.	855	85.50		940.50	1,075	
	2740 With back		4	2		975	85.50		1,060.50	1,200	
	2780 Dual handle & wheelchair projection type		4	2		540	85.50		625.50	730	
	2820 Dual level for handicapped type		3.20	2.500		1,025	107		1,132	1,300	
	3980 For rough-in, supply and waste, add		2.21	3.620		75.50	155		230.50	325	
	4000 Wall mounted, semi-recessed										
	4200 Poly-marble, single bubbler	1 Plum	4	2	Ea.	680	85.50		765.50	880	
	4600 Stainless steel, satin finish, single bubbler	″	4	2	″	870	85.50		955.50	1,100	
	6000 Wall mounted, fully recessed										
	6400 Poly-marble, single bubbler	1 Plum	4	2	Ea.	765	85.50		850.50	975	
	6800 Stainless steel, single bubbler		4	2		700	85.50		785.50	900	
	7580 For rough-in, supply and waste, add		1.83	4.372		75.50	187		262.50	375	
	7590										
	7600 Floor mounted, pedestal type										
	8600 Enameled iron, heavy duty service, 2 bubblers	1 Plum	2	4	Ea.	1,100	171		1,271	1,475	
	8880 For freeze-proof valve system, add		2	4		335	171		506	635	
	8900 For rough-in, supply and waste, add		1.83	4.372		75.50	187		262.50	375	
	9000 Minimum labor/equipment charge		2	4	Job		171		171	266	

	15413 | Electric Water Coolers	CREW	DAILY OUTPUT	LABOR-HOURS	UNIT	MAT.	LABOR	EQUIP.	TOTAL	TOTAL INCL O&P	
900	0010 **WATER COOLER**										900
	0100 Wall mounted, non-recessed										
	0140 4 GPH	Q-1	4	4	Ea.	500	154		654	790	
	0160 8 GPH, barrier free, sensor operated		4	4		630	154		784	935	
	1000 Dual height, 8.2 GPH		3.80	4.211		805	162		967	1,125	
	1040 14.3 GPH		3.80	4.211		855	162		1,017	1,200	
	3300 Semi-recessed, 8.1 GPH		4	4		840	154		994	1,150	
	4600 Floor mounted, flush-to-wall										
	4640 4 GPH	1 Plum	3	2.667	Ea.	520	114		634	750	
	4980 For stainless steel cabinet, add					112			112	123	
	5000 Dual height, 8.2 GPH	1 Plum	2	4		870	171		1,041	1,225	
	9000 Minimum labor/equipment charge	″	2	4	Job		171		171	266	

	15414 | Emergency Fixtures	CREW	DAILY OUTPUT	LABOR-HOURS	UNIT	MAT.	LABOR	EQUIP.	TOTAL	TOTAL INCL O&P	
200	0010 **INDUSTRIAL SAFETY FIXTURES** Rough-in not included										200
	0020										
	1000 Eye wash fountain										
	1400 Plastic bowl, pedestal mounted	Q-1	4	4	Ea.	196	154		350	455	

15414 | Emergency Fixtures

		15414 Emergency Fixtures	Crew	Daily Output	Labor-Hours	Unit	2006 Bare Costs Mat.	Labor	Equip.	Total	Total Incl O&P	
200	5000	Shower, single head, drench, ball valve, pull, freestanding	Q-1	4	4	Ea.	273	154		427	540	200
	5200	Horizontal or vertical supply		4	4		249	154		403	515	
	6000	Multi-nozzle, eye/face wash combination		4	4		505	154		659	800	
	6400	Multi-nozzle, 12 spray, shower only		4	4		1,300	154		1,454	1,700	
	6600	For freeze-proof, add		6	2.667		287	102		389	475	
	9000	Minimum labor/equipment charge		3	5.333	Job		205		205	320	

15418 | Resi/Comm/Industrial Fixtures

		15418 Resi/Comm/Industrial Fixtures	Crew	Daily Output	Labor-Hours	Unit	2006 Bare Costs Mat.	Labor	Equip.	Total	Total Incl O&P	
100	0010	**BATHS** R224000-40										100
	0100	Tubs, recessed porcelain enamel on cast iron, with trim										
	0180	48″ x 42″	Q-1	4	4	Ea.	1,625	154		1,779	2,025	
	0220	72″ x 36″	″	3	5.333	″	1,675	205		1,880	2,150	
	0300	Mat bottom										
	0380	5′ long	Q-1	4.40	3.636	Ea.	730	140		870	1,025	
	0560	Corner 48″ x 44″		4.40	3.636		1,625	140		1,765	2,000	
	2000	Enameled formed steel, 4′-6″ long		5.80	2.759		335	106		441	530	
	4600	Module tub & showerwall surround, molded fiberglass										
	4610	5′ long x 34″ wide x 76″ high	Q-1	4	4	Ea.	495	154		649	785	
	4750	Handicap with 1-1/2″ OD grab bar, antiskid bottom										
	4760	60″ x 32-3/4″ x 72″ high	Q-1	4	4	Ea.	670	154		824	980	
	4770	60″ x 30″ x 71″ high with molded seat	″	3.50	4.571	″	835	176		1,011	1,200	
	6000	Whirlpool, bath with vented overflow, molded fiberglass										
	6100	66″ x 48″ x 24″	Q-1	1	16	Ea.	2,700	615		3,315	3,925	
	6400	72″ x 36″ x 24″		1	16		2,675	615		3,290	3,875	
	6500	60″ x 30″ x 21″		1	16		2,300	615		2,915	3,500	
	6600	72″ x 42″ x 22″		1	16		3,475	615		4,090	4,775	
	6700	83″ x 65″		.30	53.333		4,700	2,050		6,750	8,350	
	6710	For color add					10%					
	6711	For designer colors and trim add					25%					
	7000	Redwood hot tub system										
	7050	4′ diameter x 4′ deep	Q-1	1	16	Ea.	1,750	615		2,365	2,875	
	7150	6′ diameter x 4′ deep		.80	20		2,575	770		3,345	4,025	
	7200	8′ diameter x 4′ deep		.80	20		3,950	770		4,720	5,550	
	9600	Rough-in, supply, waste and vent, for all above tubs, add		2.07	7.729		166	297		463	640	
	9900	Minimum labor/equipment charge		3	5.333	Job		205		205	320	
400	0010	**LAUNDRY SINKS** With trim										400
	0020	Porcelain enamel on cast iron, black iron frame										
	0050	24″ x 21″, single compartment	Q-1	6	2.667	Ea.	345	102		447	540	
	0100	26″ x 21″, single compartment	″	6	2.667	″	340	102		442	535	
	2000	Molded stone, on wall hanger or legs										
	2020	22″ x 23″, single compartment	Q-1	6	2.667	Ea.	124	102		226	295	
	2100	45″ x 21″, double compartment	″	5	3.200	″	257	123		380	475	
	3000	Plastic, on wall hanger or legs										
	3020	18″ x 23″, single compartment	Q-1	6.50	2.462	Ea.	94	94.50		188.50	251	
	3100	20″ x 24″, single compartment		6.50	2.462		126	94.50		220.50	285	
	3200	36″ x 23″, double compartment		5.50	2.909		151	112		263	340	
	3300	40″ x 24″, double compartment		5.50	2.909		224	112		336	420	
	5000	Stainless steel, counter top, 22″ x 17″ single compartment		6	2.667		54.50	102		156.50	219	
	5200	33″ x 22″, double compartment		5	3.200		66.50	123		189.50	265	
	9600	Rough-in, supply, waste and vent, for all laundry sinks		2.14	7.477		130	287		417	590	
	9810	Minimum labor/equipment charge	1 Plum	3	2.667	Job		114		114	177	
450	0010	**LAVATORIES** With trim, white unless noted otherwise R224000-40										450
	0020											
	0500	Vanity top, porcelain enamel on cast iron										
	0600	20″ x 18″	Q-1	6.40	2.500	Ea.	212	96		308	385	

	15418 | Resi/Comm/Industrial Fixtures		CREW	DAILY OUTPUT	LABOR-HOURS	UNIT	2006 BARE COSTS MAT.	LABOR	EQUIP.	TOTAL	TOTAL INCL O&P	
450	0640	33″ x 19″ oval (R224000-40)	Q-1	6.40	2.500	Ea.	445	96		541	635	450
	0860	For color, add					25%					
	1000	Cultured marble, 19″ x 17″, single bowl	Q-1	6.40	2.500	Ea.	163	96		259	330	
	1120	25″ x 22″, single bowl		6.40	2.500		175	96		271	340	
	1900	Stainless steel, self-rimming, 25″ x 22″, single bowl, ledge		6.40	2.500		247	96		343	420	
	1960	17″ x 22″, single bowl		6.40	2.500		240	96		336	415	
	2600	Steel, enameled, 20″ x 17″, single bowl		5.80	2.759		138	106		244	315	
	2900	Vitreous china, 20″ x 16″, single bowl		5.40	2.963		222	114		336	420	
	2960	20″ x 17″, single bowl		5.40	2.963		147	114		261	340	
	3580	Rough-in, supply, waste and vent for all above lavatories		2.30	6.957		113	267		380	540	
	4000	Wall hung										
	4040	Porcelain enamel on cast iron, 16″ x 14″, single bowl	Q-1	8	2	Ea.	345	77		422	500	
	4180	20″ x 18″, single bowl	″	8	2	″	241	77		318	385	
	4580	For color, add					30%					
	6000	Vitreous china, 18″ x 15″, single bowl with backsplash	Q-1	7	2.286	Ea.	204	88		292	360	
	6960	Rough-in, supply, waste and vent for above lavatories	″	1.66	9.639	″	310	370		680	915	
	9000	Minimum labor/equipment charge	1 Plum	3	2.667	Job		114		114	177	
500	0011	**SHOWERS**, Stall, with drain only										500
	0020											
	1520	32″ square	Q-1	5	3.200	Ea.	345	123		468	570	
	1530	36″ square		4.80	3.333		425	128		553	665	
	1540	Terrazzo receptor, 32″ square		5	3.200		730	123		853	995	
	1580	36″ corner angle		4.80	3.333		800	128		928	1,075	
	3000	Fiberglass, one piece, with 3 walls, 32″ x 32″ square		5.50	2.909		420	112		532	635	
	3100	36″ x 36″ square		5.50	2.909		465	112		577	685	
	4200	Rough-in, supply, waste and vent for above showers		2.05	7.805		244	300		544	735	
	5500	Head, water economizer	1 Plum	24	.333		65	14.25		79.25	93.50	
	9000	Minimum labor/equipment charge	Q-1	4	4	Job		154		154	239	
600	0010	**SINKS** With faucets and drain										600
	2000	Kitchen, countertop, P.E. on C.I., 24″ x 21″ single bowl	Q-1	5.60	2.857	Ea.	232	110		342	425	
	2100	31″ x 22″ single bowl		5.60	2.857		217	110		327	410	
	2200	32″ x 21″ double bowl		4.80	3.333		315	128		443	545	
	3000	Stainless steel, self rimming, 19″ x 18″ single bowl		5.60	2.857		380	110		490	585	
	3100	25″ x 22″ single bowl		5.60	2.857		420	110		530	635	
	3200	33″ x 22″ double bowl		4.80	3.333		610	128		738	870	
	3300	43″ x 22″ double bowl		4.80	3.333		705	128		833	975	
	4000	Steel, enameled, with ledge, 24″ x 21″ single bowl		5.60	2.857		114	110		224	296	
	4100	32″ x 21″ double bowl		4.80	3.333		150	128		278	365	
	4960	For color sinks except stainless steel, add					10%					
	4980	For rough-in, supply, waste and vent, counter top sinks	Q-1	2.14	7.477		130	287		417	590	
	5790	For rough-in, supply, waste & vent, sinks		1.85	8.649		130	330		460	660	
	6650	Service, floor, corner, P.E. on C.I., 28″ x 28″		4.40	3.636		585	140		725	860	
	6790	For rough-in, supply, waste & vent, floor service sinks		1.64	9.756		305	375		680	920	
	9000	Minimum labor/equipment charge		4	4	Job		154		154	239	
900	0010	**WATER CLOSETS** (R224000-40)										900
	0030	For automatic flush, see 15410-300-0972										
	0100											
	0150	Tank type, vitreous china, incl. seat, supply pipe w/stop										
	0200	Wall hung, one piece	Q-1	5.30	3.019	Ea.	390	116		506	610	
	0400	Two piece, close coupled		5.30	3.019		480	116		596	705	
	0960	For rough-in, supply, waste, vent and carrier		2.73	5.861		445	225		670	835	
	1000	Floor mounted, one piece		5.30	3.019		495	116		611	725	
	1100	Two piece, close coupled		5.30	3.019		169	116		285	365	
	1960	For color, add					30%					

MECHANICAL 15

	15418 | Resi/Comm/Industrial Fixtures		Crew	Daily Output	Labor-Hours	Unit	2006 Bare Costs Mat.	Labor	Equip.	Total	Total Incl O&P	
900	1980	For rough-in, supply, waste and vent (R224000-40)	Q-1	3.05	5.246	Ea.	191	202		393	525	900
	3000	Bowl only, with flush valve, seat										
	3100	Wall hung	Q-1	5.80	2.759	Ea.	350	106		456	550	
	3200	For rough-in, supply, waste and vent, single WC		2.56	6.250	"	460	240		700	880	
	9000	Minimum labor/equipment charge		4	4	Job		154		154	239	
	15440 | Plumbing Pumps											
800	0010	**PUMPS, SEWAGE EJECTOR** With operating and level controls										800
	0100	Simplex system incl. tank, cover, pump 15' head										
	0500	37 gal PE tank, 12 GPM, 1/2 HP, 2" discharge	Q-1	3.20	5	Ea.	385	192		577	725	
	0510	3" discharge		3.10	5.161		415	198		613	765	
	0530	87 GPM, .7 HP, 2" discharge		3.20	5		585	192		777	940	
	0540	3" discharge		3.10	5.161		635	198		833	1,000	
	0600	45 gal. coated stl tank, 12 GPM, 1/2 HP, 2" discharge		3	5.333		680	205		885	1,075	
	0610	3" discharge		2.90	5.517		710	212		922	1,100	
	0630	87 GPM, .7 HP, 2" discharge		3	5.333		870	205		1,075	1,275	
	0640	3" discharge		2.90	5.517		920	212		1,132	1,350	
	0660	134 GPM, 1 HP, 2" discharge		2.80	5.714		945	220		1,165	1,400	
	0680	3" discharge		2.70	5.926		995	228		1,223	1,450	
	0700	70 gal. PE tank, 12 GPM, 1/2 HP, 2" discharge		2.60	6.154		750	236		986	1,200	
	0710	3" discharge		2.40	6.667		800	256		1,056	1,275	
	0730	87 GPM, 0.7 HP, 2" discharge		2.50	6.400		960	246		1,206	1,425	
	0740	3" discharge		2.30	6.957		1,025	267		1,292	1,550	
	0760	134 GPM, 1 HP, 2" discharge		2.20	7.273		1,050	279		1,329	1,575	
	0770	3" discharge		2	8		1,125	305		1,430	1,700	
	9000	Minimum labor/equipment charge		2.50	6.400	Job		246		246	380	
940	0010	**PUMPS, SUBMERSIBLE** Sump										940
	7000	Sump pump, automatic										
	7100	Plastic, 1-1/4" discharge, 1/4 HP	1 Plum	6	1.333	Ea.	114	57		171	214	
	7140	1/3 HP		5	1.600		138	68.50		206.50	257	
	7180	1-1/2" discharge, 1/2 HP		4	2		188	85.50		273.50	340	
	7500	Cast iron, 1-1/4" discharge, 1/4 HP		6	1.333		131	57		188	233	
	7540	1/3 HP		6	1.333		155	57		212	259	
	7560	1/2 HP		5	1.600		187	68.50		255.50	310	
	9000	Minimum labor/equipment charge		4	2	Job		85.50		85.50	133	
	15480 | Domestic Water Heaters											
200	0010	**WATER HEATERS**										200
	0050											
	1000	Residential, electric, glass lined tank, 5 yr, 10 gal., single element	1 Plum	2.30	3.478	Ea.	227	149		376	480	
	1060	30 gallon, double element		2.20	3.636		290	155		445	560	
	1080	40 gallon, double element		2	4		310	171		481	610	
	1100	52 gallon, double element		2	4		370	171		541	670	
	1120	66 gallon, double element		1.80	4.444		430	190		620	770	
	1140	80 gallon, double element		1.60	5		555	214		769	940	
	2000	Gas fired, foam lined tank, 10 yr, vent not incl.,										
	2040	30 gallon	1 Plum	2	4	Ea.	630	171		801	955	
	2060	40 gallon		1.90	4.211		635	180		815	975	
	2100	75 gallon		1.50	5.333		820	228		1,048	1,250	
	3000	Oil fired, glass lined tank, 5 yr, vent not included, 30 gallon		2	4		795	171		966	1,150	
	3040	50 gallon		1.80	4.444		1,375	190		1,565	1,800	
	4000	Commercial, 100° rise. NOTE: for each size tank, a range of										
	4010	heaters between the ones shown are available										

15400 | Plumbing Fixtures & Equipment

15480 | Domestic Water Heaters

		15480 Domestic Water Heaters	CREW	DAILY OUTPUT	LABOR-HOURS	UNIT	2006 BARE COSTS MAT.	LABOR	EQUIP.	TOTAL	TOTAL INCL O&P
200	4020	Electric									
	4100	5 gal., 3 kW, 12 GPH, 208V	1 Plum	2	4	Ea.	1,800	171		1,971	2,250
	4160	50 gal., 36 kW, 148 GPH, 208V	"	1.80	4.444		4,250	190		4,440	4,975
	4480	400 gal., 210 kW, 860 GPH, 480V	Q-1	1	16	↓	25,100	615		25,715	28,600
	6000	Gas fired, flush jacket, std. controls, vent not incl.									
	6040	75 MBH input, 73 GPH	1 Plum	1.40	5.714	Ea.	1,575	244		1,819	2,100
	6060	98 MBH input, 95 GPH		1.40	5.714		2,775	244		3,019	3,425
	6180	200 MBH input, 192 GPH	↓	.60	13.333	↓	4,300	570		4,870	5,625
	8000	Oil fired, glass lined, UL listed, std. cntrls, vent not incl.									
	8060	140 gal., 140 MBH input, 134 GPH	Q-1	2.13	7.512	Ea.	1,150	289		1,439	1,700
	8080	140 gal., 199 MBH input, 191 GPH		2	8		11,800	305		12,105	13,500
	8160	140 gal., 540 MBH input, 519 GPH	↓	.96	16.667		16,100	640		16,740	18,700
	8280	201 gal., 1250 MBH input, 1200 GPH	Q-2	1.22	19.672	↓	25,100	785		25,885	28,900
	9000	Minimum labor/equipment charge	1 Plum	1.75	4.571	Job		195		195	305

15500 | Heat Generation Equipment

15510 | Heating Boilers and Accessories

		15510 Heating Boilers and Accessories	CREW	DAILY OUTPUT	LABOR-HOURS	UNIT	2006 BARE COSTS MAT.	LABOR	EQUIP.	TOTAL	TOTAL INCL O&P
120	0010	**BURNERS**									
	0990	Residential, conversion, gas fired, LP or natural									
	1000	Gun type, atmospheric input 72 to 200 MBH	Q-1	2.50	6.400	Ea.	655	246		901	1,100
	1020	120 to 360 MBH		2	8		725	305		1,030	1,275
	1040	280 to 800 MBH	↓	1.70	9.412	↓	1,400	360		1,760	2,100
	3000	Flame retention oil fired assembly, input									
	3040	2.0 to 5.0 GPH	Q-1	2	8	Ea.	525	305		830	1,050
300	0010	**BOILERS, ELECTRIC, ASME** Standard controls and trim									
	1000	Steam, 6 KW, 20.5 MBH	Q-19	1.20	20	Ea.	2,900	795		3,695	4,425
	1160	60 KW, 205 MBH		1	24		4,925	955		5,880	6,875
	1300	296 KW, 1010 MBH	↓	.45	53.333		11,200	2,125		13,325	15,700
	1400	592 KW, 2020 MBH	Q-21	.34	94.118		21,300	3,825		25,125	29,300
	1540	1110 KW, 3788 MBH	"	.19	168		29,800	6,850		36,650	43,300
	2000	Hot water, 7.5 KW, 25.6 MBH	Q-19	1.30	18.462		3,050	735		3,785	4,525
	2040	30 KW, 102 MBH		1.20	20		3,250	795		4,045	4,800
	2060	45 KW, 164 MBH		1.20	20		3,650	795		4,445	5,250
	2070	60 KW, 205 MBH		1.20	20		3,850	795		4,645	5,475
	2080	75 KW, 256 MBH		1.10	21.818		4,150	870		5,020	5,925
	2100	90 KW, 307 MBH	↓	1.10	21.818	↓	4,475	870		5,345	6,275
	9000	Minimum labor/equipment charge	Q-20	1	20	Job		775		775	1,225
400	0010	**BOILERS, GAS FIRED** Natural or propane, standard controls									
	1000	Cast iron, with insulated jacket									
	2000	Steam, gross output, 81 MBH	Q-7	1.40	22.857	Ea.	1,575	940		2,515	3,175
	2060	163 MBH		.90	35.556		2,325	1,450		3,775	4,825
	2200	440 MBH		.51	62.500		4,200	2,575		6,775	8,600
	2320	1,875 MBH		.30	106		12,900	4,375		17,275	21,000
	2400	3570 MBH		.18	181		20,400	7,450		27,850	34,000
	2540	6,970 MBH	↓	.10	320	↓	68,500	13,100		81,600	96,000
	2800										
	3000	Hot water, gross output, 80 MBH	Q-7	1.46	21.918	Ea.	1,500	900		2,400	3,050

15510 | Heating Boilers and Accessories

			CREW	DAILY OUTPUT	LABOR-HOURS	UNIT	2006 BARE COSTS MAT.	LABOR	EQUIP.	TOTAL	TOTAL INCL O&P	
400	3020	100 MBH	Q-7	1.35	23.704	Ea.	1,700	975		2,675	3,375	400
	3040	122 MBH		1.10	29.091		1,850	1,200		3,050	3,875	
	3060	163 MBH		1	32		2,225	1,325		3,550	4,500	
	3200	440 MBH		.58	54.983		3,975	2,250		6,225	7,875	
	3320	2,000 MBH		.26	125		13,400	5,125		18,525	22,800	
	3540	6,970 MBH		.09	359		69,000	14,800		83,800	99,000	
	7000	For tankless water heater, add					10%					
	7050	For additional zone valves up to 312 MBH add					124			124	136	
	7060											
	9900	Minimum labor/equipment charge	Q-6	1	24	Job		965		965	1,500	
460	0010	**BOILERS, GAS/OIL** Combination with burners and controls										460
	1000	Cast iron with insulated jacket										
	2000	Steam, gross output, 720 MBH	Q-7	.43	74.074	Ea.	7,600	3,050		10,650	13,100	
	2140	2,700 MBH		.19	165		17,800	6,800		24,600	30,100	
	2380	6,970 MBH		.09	372		76,500	15,300		91,800	107,500	
	2900	Hot water, gross output										
	2910	200 MBH	Q-6	.62	39.024	Ea.	6,150	1,575		7,725	9,225	
	2920	300 MBH		.49	49.080		6,150	1,975		8,125	9,850	
	2930	400 MBH		.41	57.971		7,200	2,325		9,525	11,600	
	2940	500 MBH		.36	67.039		7,775	2,700		10,475	12,800	
	3000	584 MBH	Q-7	.44	72.072		10,400	2,950		13,350	16,000	
	3060	1,460 MBH		.28	113		15,300	4,650		19,950	24,000	
	3300	13,500 MBH, 403.3 BHP		.04	727		111,500	29,800		141,300	169,000	
500	0010	**BOILERS, OIL FIRED** Standard controls, flame retention burner										500
	1000	Cast iron, with insulated flush jacket										
	2000	Steam, gross output, 109 MBH	Q-7	1.20	26.667	Ea.	2,250	1,100		3,350	4,175	
	2060	207 MBH		.90	35.556		3,050	1,450		4,500	5,650	
	2180	1,084 MBH		.38	85.106		7,725	3,500		11,225	13,900	
	2200	1,360 MBH		.33	98.160		9,050	4,025		13,075	16,200	
	2240	2,175 MBH		.24	133		12,300	5,500		17,800	22,100	
	2340	4,360 MBH		.15	214		23,000	8,800		31,800	39,000	
	2460	6,970 MBH		.09	363		73,000	14,900		87,900	103,500	
	3000	Hot water, same price as steam										
	5000	Steel, insulated jacket, burner										
	7000	Hot water, gross output, 103 MBH	Q-6	1.60	15	Ea.	1,400	605		2,005	2,450	
	7020	122 MBH		1.45	16.506		2,625	665		3,290	3,900	
	7040	137 MBH		1.36	17.595		2,850	705		3,555	4,250	
	7060	168 MBH		1.30	18.405		3,300	740		4,040	4,775	
	7080	225 MBH		1.22	19.704		3,400	790		4,190	4,950	
	7100	315 MBH		.96	25.105		4,075	1,000		5,075	6,050	
	7120	420 MBH		.70	34.483		5,225	1,375		6,600	7,875	
	9000	Minimum labor/equipment charge		1.75	13.714	Job		550		550	855	

15530 | Furnaces

			CREW	DAILY OUTPUT	LABOR-HOURS	UNIT	MAT.	LABOR	EQUIP.	TOTAL	TOTAL INCL O&P	
200	0010	**FURNACE COMPONENTS AND COMBINATIONS**										200
	0080	Coils, A/C evaporator, for gas or oil furnaces										
	0090	Add-on, with holding charge										
	0100	Upflow										
	0120	1-1/2 ton cooling	Q-5	4	4	Ea.	114	155		269	365	
	0130	2 ton cooling		3.70	4.324		136	168		304	410	
	0140	3 ton cooling		3.30	4.848		172	188		360	480	
	0150	4 ton cooling		3	5.333		237	207		444	580	
	0160	5 ton cooling		2.70	5.926		280	230		510	665	
	0300	Downflow										

	15530 | Furnaces		CREW	DAILY OUTPUT	LABOR-HOURS	UNIT	2006 BARE COSTS MAT.	LABOR	EQUIP.	TOTAL	TOTAL INCL O&P	
200	0330	2-1/2 ton cooling	Q-5	3	5.333	Ea.	167	207		374	505	200
	0340	3-1/2 ton cooling		2.60	6.154		199	238		437	590	
	0350	5 ton cooling		2.20	7.273		280	282		562	750	
	0600	Horizontal										
	0630	2 ton cooling	Q-5	3.90	4.103	Ea.	145	159		304	405	
	0640	3 ton cooling		3.50	4.571		164	177		341	455	
	0650	4 ton cooling		3.20	5		207	194		401	530	
	0660	5 ton cooling		2.90	5.517		275	214		489	630	
	2000	Cased evaporator coils for air handlers										
	2100	1-1/2 ton cooling	Q-5	4.40	3.636	Ea.	202	141		343	440	
	2110	2 ton cooling		4.10	3.902		205	151		356	460	
	2120	2-1/2 ton cooling		3.90	4.103		214	159		373	485	
	2130	3 ton cooling		3.70	4.324		237	168		405	520	
	2140	3-1/2 ton cooling		3.50	4.571		248	177		425	545	
	2150	4 ton cooling		3.20	5		296	194		490	625	
	2160	5 ton cooling		2.90	5.517		340	214		554	705	
	3010	Air handler, modular										
	3100	With cased evaporator cooling coil										
	3120	1-1/2 ton cooling	Q-5	3.80	4.211	Ea.	510	163		673	815	
	3130	2 ton cooling		3.50	4.571		535	177		712	865	
	3140	2-1/2 ton cooling		3.30	4.848		580	188		768	925	
	3150	3 ton cooling		3.10	5.161		625	200		825	1,000	
	3160	3-1/2 ton cooling		2.90	5.517		775	214		989	1,175	
	3170	4 ton cooling		2.50	6.400		830	248		1,078	1,300	
	3180	5 ton cooling		2.10	7.619		885	295		1,180	1,425	
	3500	With no cooling coil										
	3520	1-1/2 ton coil size	Q-5	12	1.333	Ea.	360	51.50		411.50	475	
	3530	2 ton coil size		10	1.600		385	62		447	520	
	3540	2-1/2 ton coil size		10	1.600		425	62		487	565	
	3554	3 ton coil size		9	1.778		475	69		544	625	
	3560	3-1/2 ton coil size		9	1.778		530	69		599	685	
	3570	4 ton coil size		8.50	1.882		680	73		753	860	
	3580	5 ton coil size		8	2		730	77.50		807.50	925	
	4000	With heater										
	4120	5 kW, 17.1 MBH	Q-5	16	1	Ea.	400	39		439	500	
	4130	7.5 kW, 25.6 MBH		15.60	1.026		440	39.50		479.50	540	
	4140	10 kW, 34.2 MBH		15.20	1.053		530	41		571	650	
	4150	12.5 KW, 42.7 MBH		14.80	1.081		650	42		692	780	
400	0010	**FURNACES** Hot air heating, blowers, standard controls										400
	0020	not including gas, oil or flue piping										
	1000	Electric, UL listed										
	1020	10.2 MBH	Q-20	5	4	Ea.	325	155		480	605	
	1100	34.1 MBH	″	4.40	4.545	″	425	176		601	745	
	3000	Gas, AGA certified, upflow, direct drive models										
	3020	45 MBH input	Q-9	4	4	Ea.	455	152		607	740	
	3040	60 MBH input		3.80	4.211		605	160		765	920	
	3060	75 MBH input		3.60	4.444		640	169		809	975	
	3100	100 MBH input		3.20	5		680	190		870	1,050	
	3120	125 MBH input		3	5.333		785	202		987	1,175	
	3130	150 MBH input		2.80	5.714		910	217		1,127	1,350	
	3140	200 MBH input		2.60	6.154		2,100	233		2,333	2,675	
	6000	Oil, UL listed, atomizing gun type burner										
	6020	56 MBH output	Q-9	3.60	4.444	Ea.	745	169		914	1,100	
	6040	95 MBH output		3.40	4.706		790	178		968	1,150	
	6060	134 MBH output		3.20	5		1,100	190		1,290	1,500	
	6080	151 MBH output		3	5.333		1,225	202		1,427	1,650	

For expanded coverage of these items see *Means Mechanical or Plumbing Cost Data 2006*

		15530 | Furnaces	Crew	Daily Output	Labor-Hours	Unit	2006 Bare Costs Mat.	Labor	Equip.	Total	Total Incl O&P	
400	6100	200 MBH input	Q-9	2.60	6.154	Ea.	2,225	233		2,458	2,825	400
	9000	Minimum labor/equipment charge		2.75	5.818	Job		221		221	350	
		15540 | Fuel-Fired Heaters										
300	0010	**DUCT FURNACES** Includes burner, controls, stainless steel										300
	0020	heat exchanger. Gas fired, electric ignition										
	1000	Outdoor installation, with vent cap										
	1120	225 MBH output	Q-5	2.50	6.400	Ea.	3,800	248		4,048	4,550	
	1160	375 MBH output		1.60	10		5,050	390		5,440	6,150	
	1180	450 MBH output		1.40	11.429		5,275	445		5,720	6,500	
900	0010	**SPACE HEATERS** Cabinet, grilles, fan, controls, burner,										900
	0020	thermostat, no piping. For flue see 15550										
	1000	Gas fired, floor mounted										
	1100	60 MBH output	Q-5	10	1.600	Ea.	620	62		682	780	
	1180	180 MBH output		6	2.667		925	103		1,028	1,175	
	2000	Suspension mounted, propeller fan, 20 MBH output		8.50	1.882		480	73		553	640	
	2040	60 MBH output		7	2.286		590	88.50		678.50	790	
	2100	130 MBH output		5	3.200		875	124		999	1,150	
	2240	320 MBH output		2	8		1,800	310		2,110	2,450	
	2500	For powered venter and adapter, add					305			305	335	
	5000	Wall furnace, 17.5 MBH output	Q-5	6	2.667		545	103		648	760	
	5020	24 MBH output		5	3.200		575	124		699	825	
	5040	35 MBH output		4	4		770	155		925	1,100	
	9000	Minimum labor/equipment charge		3.50	4.571	Job		177		177	275	
		15550 | Breechings, Chimneys & Stacks										
200	0010	**DRAFT CONTROL DEVICES**										200
	1000	Barometric,gas fired system only, 6″ size for 5″ and 6″ pipes	1 Shee	20	.400	Ea.	45	16.85		61.85	76.50	
	1040	8″ size, for 7″ and 8″ pipes	"	18	.444	"	62	18.75		80.75	98.50	
	2000	All fuel, oil, oil/gas, coal										
	2020	10″ for 9″ and 10″ pipes	1 Shee	15	.533	Ea.	107	22.50		129.50	154	
	3260	For thermal switch for above, add	"	24	.333		44.50	14.05		58.55	71.50	
	5000	Vent damper, bi-metal, gas, 3″ diameter	Q-9	24	.667		25	25.50		50.50	67.50	
	5010	4″ diameter		24	.667		25	25.50		50.50	67.50	
	5020	5″ diameter		23	.696		25	26.50		51.50	69.50	
	5030	6″ diameter		22	.727		25	27.50		52.50	71.50	
	5040	7″ diameter		21	.762		26.50	29		55.50	75.50	
	5050	8″ diameter		20	.800		29	30.50		59.50	80	
	9000	Minimum labor/equipment charge	1 Shee	4	2	Job		84.50		84.50	134	
440	0010	**VENT CHIMNEY** Prefab metal, U.L. listed										440
	0020	Gas, double wall, galvanized steel										
	0080	3″ diameter	Q-9	72	.222	V.L.F.	3.96	8.45		12.41	17.75	
	0100	4″ diameter		68	.235		4.95	8.90		13.85	19.65	
	0120	5″ diameter		64	.250		5.75	9.50		15.25	21.50	
	0140	6″ diameter		60	.267		6.70	10.10		16.80	23.50	
	0160	7″ diameter		56	.286		9.85	10.85		20.70	28	
	0180	8″ diameter		52	.308		10.95	11.65		22.60	30.50	
	0200	10″ diameter		48	.333		23	12.65		35.65	45.50	
	0220	12″ diameter		44	.364		31	13.80		44.80	56	
	0260	16″ diameter		40	.400		70.50	15.15		85.65	102	
	0300	20″ diameter	Q-10	36	.667		107	26		133	160	
	0340	24″ diameter	"	32	.750		167	29.50		196.50	231	
	7800	All fuel, double wall, stainless steel, 6″ diameter	Q-9	60	.267		30.50	10.10		40.60	49.50	
	7802	7″ diameter		56	.286		39.50	10.85		50.35	61	
	7804	8″ diameter		52	.308		46	11.65		57.65	69.50	

15500 | Heat Generation Equipment

15550 | Breechings, Chimneys & Stacks

			CREW	DAILY OUTPUT	LABOR-HOURS	UNIT	2006 BARE COSTS MAT.	LABOR	EQUIP.	TOTAL	TOTAL INCL O&P	
440	7806	10″ diameter	Q-9	48	.333	V.L.F.	67.50	12.65		80.15	94.50	440
	7808	12″ diameter		44	.364		90.50	13.80		104.30	122	
	7810	14″ diameter	↓	42	.381	↓	119	14.45		133.45	154	
	8000	All fuel, double wall, stainless steel fittings										
	8010	Roof support 6″ diameter	Q-9	30	.533	Ea.	89	20		109	130	
	8020	7″ diameter		28	.571		100	21.50		121.50	145	
	8030	8″ diameter		26	.615		112	23.50		135.50	160	
	8040	10″ diameter		24	.667		129	25.50		154.50	182	
	8050	12″ diameter		22	.727		155	27.50		182.50	215	
	8060	14″ diameter		21	.762		197	29		226	263	
	8100	Elbow 15°, 6″ diameter		30	.533		142	20		162	188	
	8120	7″ diameter		28	.571		160	21.50		181.50	211	
	8140	8″ diameter		26	.615		182	23.50		205.50	237	
	8160	10″ diameter		24	.667		238	25.50		263.50	300	
	8180	12″ diameter		22	.727		288	27.50		315.50	360	
	8200	14″ diameter		21	.762		345	29		374	425	
	8300	Insulated tee with insulated tee cap, 6″ diameter		30	.533		134	20		154	180	
	8340	7″ diameter		28	.571		176	21.50		197.50	229	
	8360	8″ diameter		26	.615		199	23.50		222.50	255	
	8380	10″ diameter		24	.667		279	25.50		304.50	345	
	8400	12″ diameter		22	.727		390	27.50		417.50	475	
	8420	14″ diameter		21	.762		515	29		544	610	
	8500	Joist shield, 6″ diameter		30	.533		41	20		61	77	
	8510	7″ diameter		28	.571		44.50	21.50		66	83.50	
	8520	8″ diameter		26	.615		54.50	23.50		78	97	
	8530	10″ diameter		24	.667		74	25.50		99.50	121	
	8540	12″ diameter		22	.727		92	27.50		119.50	145	
	8550	14″ diameter		21	.762		114	29		143	172	
	8600	Round top, 6″ diameter		30	.533		45.50	20		65.50	82	
	8620	7″ diameter		28	.571		62	21.50		83.50	103	
	8640	8″ diameter		26	.615		83.50	23.50		107	129	
	8660	10″ diameter		24	.667		150	25.50		175.50	205	
	8680	12″ diameter		22	.727		214	27.50		241.50	279	
	8700	14″ diameter		21	.762		285	29		314	360	
	8800	Adjustable roof flashing, 6″ diameter		30	.533		54	20		74	91.50	
	8820	7″ diameter		28	.571		62	21.50		83.50	103	
	8840	8″ diameter		26	.615		67	23.50		90.50	111	
	8860	10″ diameter		24	.667		86.50	25.50		112	135	
	8880	12″ diameter		22	.727		111	27.50		138.50	166	
	8900	14″ diameter		21	.762	↓	139	29		168	199	
	9990	Minimum labor/equipment charge	↓	3	5.333	Job		202		202	320	

MECHANICAL 15

15600 | Refrigeration Equipment

15610 | Refrigeration Compressors

			CREW	DAILY OUTPUT	LABOR-HOURS	UNIT	2006 BARE COSTS MAT.	LABOR	EQUIP.	TOTAL	TOTAL INCL O&P	
400	0010	ROTARY / RECIPROCATING REFRIGERANT COMPRESSOR										400
	1000	10 ton	Q-5	1	16	Ea.	7,325	620		7,945	9,025	
	1100	20 ton	Q-6	.72	33.333		11,000	1,350		12,350	14,200	
	1400	50 ton	″	.20	120	↓	15,100	4,825		19,925	24,100	

15600 | Refrigeration Equipment

	15610 | Refrigeration Compressors		CREW	DAILY OUTPUT	LABOR-HOURS	UNIT	2006 BARE COSTS MAT.	LABOR	EQUIP.	TOTAL	TOTAL INCL O&P	
400	1600	130 ton	Q-7	.21	152	Ea.	19,000	6,250		25,250	30,600	400

	15620 | Packaged Water Chillers		CREW	DAILY OUTPUT	LABOR-HOURS	UNIT	MAT.	LABOR	EQUIP.	TOTAL	TOTAL INCL O&P	
100	0010	**ABSORPTION WATER CHILLERS**										100
	3000	Gas fired, air cooled										
	3180	3 ton	Q-5	1.30	12.308	Ea.	6,625	475		7,100	8,050	
	9000	Minimum labor/equipment charge	"	1	16	Job		620		620	965	
600	0010	**CENTRIFUGAL/SCREW/RECIP. WATER CHILLERS,** W/ standard controls										600
	0110	Screw, liquid chiller, air cooled, insulated evaporator										
	0120	130 ton	Q-7	.14	228	Ea.	63,500	9,375		72,875	84,000	
	0124	160 ton		.13	246		78,000	10,100		88,100	101,000	
	0490	Packaged w/integral air cooled condenser, 15 ton cool		.37	86.486		15,300	3,550		18,850	22,300	
	0500	20 ton cooling		.34	94.955		20,300	3,900		24,200	28,400	
	0520	40 ton cooling		.30	108		27,500	4,425		31,925	37,100	
	0680	Scroll water cooled, single compressor, hermetic, tower not incl.										
	0760	10 ton cooling	Q-6	.36	67.039	Ea.	6,300	2,700		9,000	11,100	
	0840	35 ton cooling	Q-7	.33	97.859	"	12,200	4,025		16,225	19,700	
	0980	Water cooled, multiple compr., semi-hermetic, tower not incl.										
	1100	50 ton cooling	Q-7	.28	113	Ea.	29,200	4,675		33,875	39,500	
	1160	100 ton cooling		.18	179		52,000	7,375		59,375	69,000	
	1200	140 ton cooling		.16	202		69,500	8,300		77,800	89,500	
	9000	Minimum labor/equipment charge	Q-6	1	24	Job		965		965	1,500	

	15640 | Packaged Cooling Towers		CREW	DAILY OUTPUT	LABOR-HOURS	UNIT	MAT.	LABOR	EQUIP.	TOTAL	TOTAL INCL O&P	
400	0010	**COOLING TOWERS** Packaged units										400
	0070	Galvanized steel										
	0080	Induced draft, crossflow										
	0100	Vertical, belt drive, 61 tons	Q-6	90	.267	TonAC	87	10.70		97.70	112	
	0150	100 ton		100	.240		69	9.65		78.65	91	
	0200	115 ton		109	.220		67	8.85		75.85	88	
	0250	131 ton		120	.200		59.50	8.05		67.55	78	
	0260	162 ton		132	.182		50	7.30		57.30	66.50	
	1000	For higher capacities, use multiples										
	1500	Induced air, double flow										
	1900	Vertical, gear drive, 167 ton	Q-6	126	.190	TonAC	92.50	7.65		100.15	114	
	2000	297 ton		129	.186		66	7.50		73.50	84	
	2100	582 ton		132	.182		54.50	7.30		61.80	71.50	
	2200	1016 ton		150	.160		51.50	6.45		57.95	66.50	
	3500	For pumps and piping, add		38	.632		43.50	25.50		69	87.50	
	4000	For absorption systems, add					75%	75%				
	5000	Fiberglass										
	5010	Draw thru										
	5100	60 ton	Q-6	1.50	16	Ea.	3,275	645		3,920	4,600	
	5120	125 ton		.99	24.242		6,700	975		7,675	8,900	
	5140	300 ton		.43	55.814		15,700	2,250		17,950	20,800	
	5160	600 ton		.22	109		28,200	4,375		32,575	37,800	
	5180	1000 ton		.15	160		48,400	6,425		54,825	63,000	
	6000	Stainless steel										
	6010	Induced draft, crossflow, horizontal, belt drive										
	6100	57 ton	Q-6	1.50	16	Ea.	8,775	645		9,420	10,700	
	6120	91 ton		.99	24.242		12,600	975		13,575	15,400	
	6140	111 ton		.43	55.814		14,900	2,250		17,150	19,900	
	6160	126 ton		.22	109		16,100	4,375		20,475	24,500	
	6170	Induced draft, crossflow, vertical, gear drive										

 Important: See the Reference Section for supporting data - Crews, Rental Equipment, City Cost Indexes and Reference Data

15600 | Refrigeration Equipment

15640 | Packaged Cooling Towers

				DAILY	LABOR-		2006 BARE COSTS				TOTAL	
			CREW	OUTPUT	HOURS	UNIT	MAT.	LABOR	EQUIP.	TOTAL	INCL O&P	
400	6172	167 ton	Q-6	.75	32	Ea.	25,900	1,275		27,175	30,500	400
	6174	297 ton		.43	55.814		33,300	2,250		35,550	40,100	
	6176	582 ton		.23	104		52,500	4,200		56,700	64,000	
	6180	1016 ton		.15	160	↓	91,500	6,425		97,925	111,000	
	9000	Minimum labor/equipment charge	↓	1	24	Job		965		965	1,500	

15670 | Refrigerant Condensing Units

			CREW	DAILY OUTPUT	LABOR-HOURS	UNIT	MAT.	LABOR	EQUIP.	TOTAL	TOTAL INCL O&P	
300	0010	CONDENSING UNITS										300
	0030	Air cooled, compressor, standard controls										
	0050	1.5 ton	Q-5	2.50	6.400	Ea.	635	248		883	1,075	
	0200	2.5 ton		1.70	9.412		770	365		1,135	1,400	
	0300	3 ton		1.30	12.308		910	475		1,385	1,750	
	0400	4 ton		.90	17.778		1,175	690		1,865	2,375	
	0500	5 ton		.60	26.667	↓	1,425	1,025		2,450	3,175	
	9000	Minimum labor/equipment charge	↓	1.50	10.667	Job		415		415	645	

15700 | Heating/Ventilating/Air Conditioning Equipment

15705 | Curbs/Pads/Stands Prefab

				DAILY	LABOR-		2006 BARE COSTS				TOTAL	
			CREW	OUTPUT	HOURS	UNIT	MAT.	LABOR	EQUIP.	TOTAL	INCL O&P	
600	0010	CURBS/PADS PREFABRICATED										600
	6000	Pad, fiberglass reinforced concrete with polystyrene foam core										
	6050	Condenser, 2″ thick, 20″ x 38″	1 Shee	8	1	Ea.	8.90	42		50.90	77	
	6090	24″ x 36″	″	12	.667		12.65	28		40.65	58.50	
	6280	36″ x 36″	Q-9	8	2		18.20	76		94.20	141	
	6300	36″ x 40″		7	2.286		20.50	86.50		107	161	
	6320	36″ x 48″	↓	7	2.286	↓	24.50	86.50		111	165	

15720 | Air Handling Units

			CREW	DAILY OUTPUT	LABOR-HOURS	UNIT	MAT.	LABOR	EQUIP.	TOTAL	TOTAL INCL O&P	
500	0010	MAKE-UP AIR UNIT										500
	0020	Indoor suspension, natural/LP gas, direct fired,										
	0030	standard control. For flue see division 15550-440										
	0040	70°F temperature rise, MBH is input										
	0100	2000 CFM, 168 MBH	Q-6	3	8	Ea.	6,950	320		7,270	8,125	
	0220	12,000 CFM, 1005 MBH	″	1	24		10,900	965		11,865	13,500	
	0300	24,000 CFM, 2007 MBH	Q-7	1	32		14,200	1,325		15,525	17,700	
	0400	50,000 CFM, 4180 MBH	″	.80	40	↓	19,500	1,650		21,150	24,000	
	9000	Minimum labor/equipment charge	Q-6	2.75	8.727	Job		350		350	545	

15730 | Unitary Air Conditioning Equip

			CREW	DAILY OUTPUT	LABOR-HOURS	UNIT	MAT.	LABOR	EQUIP.	TOTAL	TOTAL INCL O&P	
200	0010	COMPUTER ROOM UNITS										200
	1000	Air cooled, includes remote condenser but not										
	1020	interconnecting tubing or refrigerant										
	1160	6 ton	Q-5	.30	53.333	Ea.	22,200	2,075		24,275	27,600	
	1240	10 ton	″	.25	64		26,100	2,475		28,575	32,600	
	1320	20 ton	Q-6	.29	82.759	↓	34,800	3,325		38,125	43,500	
600	0010	ROOF TOP AIR CONDITIONERS Standard controls, curb, economizer										600
	0020											

	15730 | Unitary Air Conditioning Equip		CREW	DAILY OUTPUT	LABOR-HOURS	UNIT	2006 BARE COSTS MAT.	LABOR	EQUIP.	TOTAL	TOTAL INCL O&P
600	1000	Single zone, electric cool, gas heat									
	1090	2 ton cooling, 55 MBH heating	Q-5	.93	17.204	Ea.	2,900	665		3,565	4,225
	1100	3 ton cooling, 60 MBH heating		.70	22.857		2,925	885		3,810	4,600
	1120	4 ton cooling, 95 MBH heating		.61	26.403		4,250	1,025		5,275	6,275
	1160	10 ton cooling, 200 MBH heating	Q-6	.67	35.982		8,725	1,450		10,175	11,900
	1220	30 ton cooling, 540 MBH heating	Q-7	.47	68.376		28,300	2,800		31,100	35,500
	1240	40 ton cooling, 675 MBH heating	"	.35	91.168		36,900	3,750		40,650	46,400
	2000	Multizone, electric cool, gas heat, economizer									
	2120	20 ton cooling, 360 MBH heating	Q-7	.53	60.038	Ea.	87,500	2,475		89,975	100,000
	2180	30 ton cooling, 540 MBH heating		.37	85.562		113,000	3,500		116,500	129,500
	2210	50 ton cooling, 540 MBH heating		.23	142		129,500	5,825		135,325	151,000
	2220	70 ton cooling, 1500 MBH heating		.16	198		140,000	8,150		148,150	166,500
	2240	80 ton cooling, 1500 MBH heating		.14	228		160,000	9,375		169,375	190,500
	2280	105 ton cooling, 1500 MBH heating		.11	290		185,500	11,900		197,400	222,500
	2400	For hot water heat coil, deduct					5%				
	2500	For steam heat coil, deduct					2%				
	2600	For electric heat, deduct					3%	5%			
	9000	Minimum labor/equipment charge	Q-5	1.50	10.667	Job		415		415	645
800	0010	**WINDOW UNIT AIR CONDITIONERS**									
	4500	10,000 BTUH	1 Carp	6	1.333	Ea.	350	47.50		397.50	465
	4520	12,000 BTUH	L-2	8	2	"	630	62		692	795
	9000	Minimum labor/equipment charge	1 Carp	2	4	Job		142		142	236
840	0010	**SELF-CONTAINED SINGLE PACKAGE**									
	0100	Air cooled, for free blow or duct, not incl. remote condenser									
	0200	3 ton cooling	Q-5	1	16	Ea.	2,975	620		3,595	4,250
	0210	4 ton cooling	"	.80	20		3,225	775		4,000	4,750
	0240	10 ton cooling	Q-7	1	32		6,025	1,325		7,350	8,675
	0280	30 ton cooling	"	.80	40		14,700	1,650		16,350	18,800
	1000	Water cooled for free blow or duct, not including tower									
	1010	Constant volume									
	1100	3 ton cooling	Q-6	1	24	Ea.	2,825	965		3,790	4,600
	1120	5 ton cooling	"	1	24		3,650	965		4,615	5,525
	1140	10 ton cooling	Q-7	.90	35.556		7,225	1,450		8,675	10,200
	1180	30 ton cooling	"	.70	45.714		32,100	1,875		33,975	38,300
	1240	60 ton cooling	Q-8	.30	106		59,000	4,375	230	63,605	72,000
	9000	Minimum labor/equipment charge	Q-5	1	16	Job		620		620	965
900	0010	**SPLIT DUCTLESS SYSTEM**									
	0100	Cooling only, single zone									
	0110	Wall mount									
	0120	3/4 ton cooling	Q-5	2	8	Ea.	1,000	310		1,310	1,575
	0130	1 ton cooling		1.80	8.889		1,150	345		1,495	1,775
	0140	1-1/2 ton cooling		1.60	10		1,400	390		1,790	2,150
	0150	2 ton cooling		1.40	11.429		1,800	445		2,245	2,675
	1000	Ceiling mount									
	1020	2 ton cooling	Q-5	1.40	11.429	Ea.	1,225	445		1,670	2,050
	1030	3 ton cooling	"	1.20	13.333	"	3,300	515		3,815	4,425
	2000	T-Bar mount									
	2010	2 ton cooling	Q-5	1.40	11.429	Ea.	2,450	445		2,895	3,400
	2020	3 ton cooling		1.20	13.333		2,950	515		3,465	4,050
	2030	3-1/2 ton cooling		1.10	14.545		3,550	565		4,115	4,775
	3000	Multizone									
	3010	Wall mount									
	3020	2 @ 3/4 ton cooling	Q-5	1.80	8.889	Ea.	1,200	345		1,545	1,850
	5000	Cooling / Heating									
	5010	Wall mount									
	5110	1 ton cooling	Q-5	1.70	9.412	Ea.	890	365		1,255	1,550

15700 | Heating/Ventilating/Air Conditioning Equipment

		15730 | Unitary Air Conditioning Equip	Crew	Daily Output	Labor-Hours	Unit	2006 Bare Costs Mat.	Labor	Equip.	Total	Total Incl O&P	
900	5120	1-1/2 ton cooling	Q-5	1.50	10.667	Ea.	1,425	415		1,840	2,225	900
	5300	Ceiling mount										
	5310	3 ton cooling	Q-5	1	16	Ea.	3,825	620		4,445	5,175	
	7000	Accessories for all split ductless systems										
	7010	Add for ambient frost control	Q-5	8	2	Ea.	120	77.50		197.50	253	
	7020	Add for tube / wiring kit										
	7030	15′ kit	Q-5	32	.500	Ea.	31.50	19.40		50.90	65	
	7040	35′ kit	"	24	.667	"	102	26		128	152	

		15740 | Heat Pumps	Crew	Daily Output	Labor-Hours	Unit	Mat.	Labor	Equip.	Total	Total Incl O&P	
100	0010	**AIR-SOURCE HEAT PUMPS** (Not including interconnecting tubing)										100
	1000	Air to air, split system, not including curbs, pads, or ductwork										
	1020	2 ton cooling, 8.5 MBH heat @ 0° F	Q-5	1.20	13.333	Ea.	1,625	515		2,140	2,600	
	1040	3 ton cooling, 13 MBH heat @ 0° F		.80	20		1,975	775		2,750	3,375	
	1080	7.5 ton cooling, 33 MBH heat @ 0° F	↓	.30	53.333		6,250	2,075		8,325	10,100	
	1120	15 ton cooling, 64 MBH heat @ 0° F	Q-6	.26	92.308		11,900	3,700		15,600	18,900	
	1130	20 ton cooling, 85 MBH heat @ 0° F		.20	120		15,600	4,825		20,425	24,700	
	1140	25 ton cooling, 119 MBH heat @ 0° F	↓	.20	120	↓	18,800	4,825		23,625	28,200	
	1500	Single package, not including curbs, pads, or plenums										
	1520	2 ton cooling, 6.5 MBH heat @ 0° F	Q-5	1.50	10.667	Ea.	2,525	415		2,940	3,425	
	1560	3 ton cooling, 10 MBH heat @ 0° F	"	1.20	13.333	"	3,025	515		3,540	4,125	
800	0010	**WATER-SOURCE HEAT PUMPS** (Not including interconnecting tubing)										800
	2000	Water source to air, single package										
	2100	1 ton cooling, 13 MBH heat @ 75°F	Q-5	2	8	Ea.	1,125	310		1,435	1,725	
	2140	2 ton cooling, 19 MBH heat @ 75°F		1.70	9.412		1,300	365		1,665	2,000	
	2220	5 ton cooling, 29 MBH heat @ 75°F		.90	17.778	↓	2,000	690		2,690	3,250	
	9000	Minimum labor/equipment charge	↓	1.75	9.143	Job		355		355	550	

		15750 | Humidity Control Equipment	Crew	Daily Output	Labor-Hours	Unit	Mat.	Labor	Equip.	Total	Total Incl O&P	
500	0010	**HUMIDIFIERS**										500
	0520	Steam, room or duct, filter, regulators, auto. controls, 220 V										
	0560	22 lb. per hour	Q-5	5	3.200	Ea.	2,400	124		2,524	2,825	
	0580	33 lb. per hour		4	4		2,450	155		2,605	2,950	
	0620	100 lb. per hour		3	5.333	↓	3,600	207		3,807	4,300	
	9000	Minimum labor/equipment charge	↓	3.50	4.571	Job		177		177	275	

		15760 | Terminal Heating & Cooling Units	Crew	Daily Output	Labor-Hours	Unit	Mat.	Labor	Equip.	Total	Total Incl O&P	
100	0010	**COILS, FLANGED**										100
	1500	Hot water heating, 1 row, 24″ x 48″	Q-5	4	4	Ea.	1,150	155		1,305	1,500	
	9000	Minimum labor/equipment charge	1 Plum	1	8	Job		340		340	530	
200	0010	**DUCT HEATERS** Electric, 480 V, 3 Ph										200
	0020	Finned tubular insert, 500°F										
	0100	8″ wide x 6″ high, 4.0 kW	Q-20	16	1.250	Ea.	570	48.50		618.50	700	
	0120	12″ high, 8.0 kW		15	1.333		940	51.50		991.50	1,100	
	0140	18″ high, 12.0 kW		14	1.429		1,325	55.50		1,380.50	1,550	
	0160	24″ high, 16.0 kW		13	1.538		1,700	59.50		1,759.50	1,975	
	0180	30″ high, 20.0 kW		12	1.667		2,075	64.50		2,139.50	2,375	
	0300	12″ wide x 6″ high, 6.7 kW		15	1.333		600	51.50		651.50	745	
	0320	12″ high, 13.3 kW		14	1.429		975	55.50		1,030.50	1,175	
	0340	18″ high, 20.0 kW	↓	13	1.538	↓	1,375	59.50		1,434.50	1,600	
	8000	To obtain BTU multiply kW by 3413										
	9900	Minimum labor/equipment charge	1 Shee	4	2	Job		84.50		84.50	134	
250	0010	**ELECTRIC HEATING**, not incl. conduit or feed wiring										250
	1300	Baseboard heaters, 2′ long, 375 watt	1 Elec	8	1	Ea.	33	42		75	101	

15760 | Terminal Heating & Cooling Units

		15760 Terminal Heating & Cooling Units	CREW	DAILY OUTPUT	LABOR-HOURS	UNIT	2006 BARE COSTS MAT.	LABOR	EQUIP.	TOTAL	TOTAL INCL O&P
250	1400	3′ long, 500 watt	1 Elec	8	1	Ea.	39	42		81	108
	1600	4′ long, 750 watt		6.70	1.194		46	50		96	129
	1800	5′ long, 935 watt		5.70	1.404		55	59		114	152
	2000	6′ long, 1125 watt		5	1.600		61	67		128	171
	2400	8′ long, 1500 watt		4	2		76.50	84		160.50	214
	2600	9′ long, 1680 watt		3.60	2.222		85.50	93.50		179	238
	2800	10′ long, 1875 watt		3.30	2.424		95	102		197	262
	2950	Wall heaters with fan, 120 to 277 volt									
	3190	1500 watt	1 Elec	4	2	Ea.	101	84		185	240
	3230	2500 watt		3.50	2.286		229	96		325	400
	3250	4000 watt		2.70	2.963		350	124		474	575
	3600	Thermostats, integral		16	.500		18	21		39	52.50
	3800	Line voltage, 1 pole		8	1		22.50	42		64.50	89
	3810	2 pole		8	1		29	42		71	96.50
	9990	Minimum labor/equipment charge		4	2	Job		84		84	129
300	0010	**FAN COIL AIR CONDITIONING** Cabinet mounted, filters, controls									
	0020										
	0100	Chilled water, 1/2 ton cooling	Q-5	8	2	Ea.	600	77.50		677.50	780
	0120	1 ton cooling		6	2.667		720	103		823	955
	0160	2.5 ton cooling		5	3.200		1,550	124		1,674	1,900
	0180	3 ton cooling		4	4		1,700	155		1,855	2,125
	0262	For hot water coil, add					40%	10%			
	0940	Direct expansion, for use w/air cooled condensing, 1.5 ton cooling	Q-5	5	3.200	Ea.	470	124		594	715
	1000	5 ton cooling	"	3	5.333		875	207		1,082	1,275
	1040	10 ton cooling	Q-6	2.60	9.231		2,450	370		2,820	3,275
	1060	20 ton cooling	"	.70	34.286		4,300	1,375		5,675	6,900
	1510	For condensing unit add see division 15670.									
	3000	Chilled water, horizontal unit, housing, 2 pipe, controls									
	3100	1/2 ton cooling	Q-5	8	2	Ea.	930	77.50		1,007.50	1,150
	3110	1 ton cooling		6	2.667		1,125	103		1,228	1,400
	3120	1.5 ton cooling		5.50	2.909		1,350	113		1,463	1,650
	3130	2 ton cooling		5.25	3.048		1,600	118		1,718	1,950
	3140	3 ton cooling		4	4		1,875	155		2,030	2,300
	3150	3.5 ton cooling		3.80	4.211		2,050	163		2,213	2,525
	3160	4 ton cooling		3.80	4.211		2,050	163		2,213	2,525
	4000	With electric heat, 2 pipe									
	4100	1/2 ton cooling	Q-5	8	2	Ea.	1,125	77.50		1,202.50	1,375
	4105	3/4 ton cooling		7	2.286		1,250	88.50		1,338.50	1,525
	4110	1 ton cooling		6	2.667		1,375	103		1,478	1,675
	4120	1.5 ton cooling		5.50	2.909		1,525	113		1,638	1,850
	4130	2 ton cooling		5.25	3.048		1,925	118		2,043	2,300
	4135	2.5 ton cooling		4.80	3.333		2,675	129		2,804	3,125
	4140	3 ton cooling		4	4		3,550	155		3,705	4,150
	4150	3.5 ton cooling		3.80	4.211		4,250	163		4,413	4,925
	4160	4 ton cooling		3.60	4.444		4,900	172		5,072	5,650
	9000	Minimum labor/equipment charge		3.75	4.267	Job		165		165	257
600	0010	**HYDRONIC HEATING** Terminal units, not incl. main supply pipe									
	1000	Radiation									
	1150	Fin tube, wall hung, 14″ slope top cover, with damper									
	1200	1-1/4″ copper tube, 4-1/4″ alum. fin	Q-5	38	.421	L.F.	27.50	16.30		43.80	56
	1250	1-1/4″ steel tube, 4-1/4″ steel fin		36	.444		25	17.20		42.20	54.50
	1255	2″ steel tube, 4-1/4″ steel fin		32	.500		27.50	19.40		46.90	60.50
	1310	Baseboard, pkgd, 1/2″ copper tube, alum. fin, 7″ high		60	.267		6.25	10.35		16.60	23
	1320	3/4″ copper tube, alum. fin, 7″ high		58	.276		6.70	10.70		17.40	24

15700 | Heating/Ventilating/Air Conditioning Equipment

		15760 | Terminal Heating & Cooling Units	CREW	DAILY OUTPUT	LABOR-HOURS	UNIT	2006 BARE COSTS MAT.	LABOR	EQUIP.	TOTAL	TOTAL INCL O&P
600	1340	1" copper tube, alum. fin, 8-7/8" high	Q-5	56	.286	L.F.	13.80	11.05		24.85	32.50
	1360	1-1/4" copper tube, alum. fin, 8-7/8" high	↓	54	.296	↓	20.50	11.50		32	40.50
	1500	Note: fin tube may also require corners, caps, etc.									
	3000	Radiators, cast iron									
	3100	Free standing or wall hung, 6 tube, 25" high	Q-5	96	.167	Section	24	6.45		30.45	36.50
	9000	Minimum labor/equipment charge	"	3.20	5	Job		194		194	300
800	0010	**UNIT HEATERS**									
	3950	Unit heaters, propeller, 115 V 2 psi steam, 60° F entering air									
	4000	Horizontal, 12 MBH	Q-5	12	1.333	Ea.	278	51.50		329.50	385
	4060	43.9 MBH		8	2		425	77.50		502.50	590
	4240	286.9 MBH		2	8		1,325	310		1,635	1,925
	4250	326.0 MBH		1.90	8.421		1,625	325		1,950	2,275
	4260	364 MBH		1.80	8.889		1,700	345		2,045	2,400
	4270	404 MBH	↓	1.60	10		2,200	390		2,590	3,025
	4300	For vertical diffuser, add					154			154	169
	4310	Vertical flow, 40 MBH	Q-5	11	1.455		415	56.50		471.50	550
	4326	131.0 MBH	"	4	4		655	155		810	960
	4354	420 MBH,(460 V)	Q-6	1.80	13.333		1,575	535		2,110	2,550
	4358	500 MBH, (460 V)		1.71	14.035		2,225	565		2,790	3,325
	4362	570 MBH, (460 V)		1.40	17.143		3,275	690		3,965	4,675
	4366	620 MBH, (460 V)		1.30	18.462		3,750	740		4,490	5,275
	4370	960 MBH, (460 V)	↓	1.10	21.818	↓	6,500	875		7,375	8,525

		15780 | Energy Recovery Equipment	CREW	DAILY OUTPUT	LABOR-HOURS	UNIT	MAT.	LABOR	EQUIP.	TOTAL	TOTAL INCL O&P
100	0010	**HEAT RECOVERY PACKAGES**									
	0100	Air to air									
	9000	Minimum labor/equipment charge	1 Shee	4	2	Job		84.50		84.50	134

MECHANICAL 15

15800 | Air Distribution

		15810 | Ducts	CREW	DAILY OUTPUT	LABOR-HOURS	UNIT	2006 BARE COSTS MAT.	LABOR	EQUIP.	TOTAL	TOTAL INCL O&P
100	0010	**METAL DUCTWORK** R233100-20									
	0020	Fabricated rectangular, includes fittings, joints, supports,									
	0100	Aluminum, alloy 3003-H14, under 100 lb.	Q-10	75	.320	Lb.	2.50	12.60		15.10	23
	0110	100 to 500 lb.		80	.300		2	11.80		13.80	21
	0120	500 to 1,000 lb.		95	.253		1.65	9.95		11.60	17.60
	0500	Galvanized steel, under 200 lb.		235	.102		.95	4.02		4.97	7.45
	0520	200 to 500 lb.		245	.098		.81	3.85		4.66	7.05
	0540	500 to 1,000 lb.		255	.094		.71	3.70		4.41	6.70
	0560	1,000 to 2,000 lb.		265	.091		.54	3.56		4.10	6.25
	0580	Over 5,000 lb.	↓	285	.084	↓	.48	3.31		3.79	5.80
	0590										
	1000	Stainless steel, type 304, under 100 lb.	Q-10	165	.145	Lb.	2.50	5.70		8.20	11.85
	1020	100 to 500 lb.		175	.137		2.30	5.40		7.70	11.15
	1030	500 to 1,000 lb.	↓	190	.126	↓	2.20	4.97		7.17	10.30
	9990	Minimum labor/equipment charge	1 Shee	3	2.667	Job		112		112	179
300	0010	**FIBEROUS GLASS DUCTWORK** R233100-20									
	9990	Minimum labor/equipment charge	1 Shee	3	2.667	Job		112		112	179

15800 | Air Distribution

15810 | Ducts

			Crew	Daily Output	Labor-Hours	Unit	2006 Bare Costs Mat.	Labor	Equip.	Total	Total Incl O&P	
500	0010	**FLEXIBLE DUCTS** R233100-20										500
	1300	Flexible, coated fiberglass fabric on corr. resist. metal helix										
	1400	pressure to 12" (WG) UL-181										
	1500	Non-insulated, 3" diameter	Q-9	400	.040	L.F.	1.08	1.52		2.60	3.60	
	1540	5" diameter		320	.050		1.35	1.90		3.25	4.51	
	1560	6" diameter		280	.057		1.60	2.17		3.77	5.20	
	1580	7" diameter		240	.067		1.90	2.53		4.43	6.10	
	1600	8" diameter		200	.080		2.20	3.03		5.23	7.25	
	1640	10" diameter		160	.100		2.76	3.79		6.55	9.10	
	1660	12" diameter		120	.133		3.34	5.05		8.39	11.70	
	1900	Insulated, 1" thick, PE jacket, 3" diameter		380	.042		2.22	1.60		3.82	4.98	
	1910	4" diameter		340	.047		2.22	1.79		4.01	5.30	
	1920	5" diameter		300	.053		2.39	2.02		4.41	5.85	
	1940	6" diameter		260	.062		2.54	2.33		4.87	6.50	
	1960	7" diameter		220	.073		2.95	2.76		5.71	7.65	
	1980	8" diameter		180	.089		3.15	3.37		6.52	8.80	
	2020	10" diameter		140	.114		3.71	4.34		8.05	11	
	2040	12" diameter		100	.160		4.62	6.05		10.67	14.75	
	2060	14" diameter		80	.200		5.60	7.60		13.20	18.20	
	9990	Minimum labor/equipment charge	1 Shee	3	2.667	Job		112		112	179	
700	0010	**GLASS FIBER-REINFORCED PLASTIC DUCTS** R233100-20										700
	9990	Minimum labor/equipment charge	1 Shee	3	2.667	Job		112		112	179	

15820 | Duct Accessories

			Crew	Daily Output	Labor-Hours	Unit	Mat.	Labor	Equip.	Total	Total Incl O&P	
300	0010	**DUCT ACCESSORIES**										300
	0050	Air extractors, 12" x 4"	1 Shee	24	.333	Ea.	16.50	14.05		30.55	40.50	
	0100	8" x 6"		22	.364		16.50	15.35		31.85	42.50	
	0200	20" x 8"		16	.500		37.50	21		58.50	74.50	
	0240	18" x 10"		14	.571		36.50	24		60.50	78.50	
	0280	24" x 12"		10	.800		50.50	33.50		84	109	
	3000	Fire damper, curtain type, 1-1/2 hr rated, vertical, 6" x 6"		24	.333		14.75	14.05		28.80	39	
	3020	8" x 6"		22	.364		15.10	15.35		30.45	41	
	3240	16" x 14"		18	.444		32.50	18.75		51.25	66	
	3400	24" x 20"		8	1		45.50	42		87.50	117	
	5990	Multi-blade dampers, opposed blade, 8" x 6"		24	.333		20	14.05		34.05	44.50	
	5994	8" x 8"		22	.364		21	15.35		36.35	47.50	
	5996	10" x 10"		21	.381		24	16.05		40.05	52	
	6000	12" x 12"		21	.381		27	16.05		43.05	55.50	
	6020	12" x 18"		18	.444		37	18.75		55.75	70.50	
	6030	14" x 10"		20	.400		26.50	16.85		43.35	56	
	6031	14" x 14"		17	.471		33	19.85		52.85	67.50	
	6033	16" x 12"		17	.471		33	19.85		52.85	67.50	
	6035	16" x 16"		16	.500		41	21		62	78.50	
	6037	18" x 16"		15	.533		45	22.50		67.50	85.50	
	6038	18" x 18"		15	.533		49	22.50		71.50	89.50	
	6070	20" x 16"		14	.571		49	24		73	92	
	6072	20" x 20"		13	.615		58.50	26		84.50	106	
	6074	22" x 18"		14	.571		58.50	24		82.50	103	
	6076	24" x 16"		11	.727		57.50	30.50		88	113	
	6078	24" x 20"		8	1		68	42		110	142	
	6080	24" x 24"		8	1		79	42		121	154	
	6110	26" x 26"		6	1.333		85.50	56		141.50	184	
	6133	30" x 30"	Q-9	6.60	2.424		124	92		216	282	
	6135	32" x 32"		6.40	2.500		144	95		239	310	
	6180	48" x 36"		5.60	2.857		236	108		344	430	
	7500	Variable volume modulating motorized damper, incl. elect. mtr.										

15 MECHANICAL

15820 | Duct Accessories

			CREW	DAILY OUTPUT	LABOR-HOURS	UNIT	2006 BARE COSTS MAT.	LABOR	EQUIP.	TOTAL	TOTAL INCL O&P	
300	7504	8″ x 6″	1 Shee	15	.533	Ea.	111	22.50		133.50	158	300
	7506	10″ x 6″		14	.571		111	24		135	161	
	7510	10″ x 10″		13	.615		114	26		140	168	
	7520	12″ x 12″		12	.667		118	28		146	174	
	7522	12″ x 16″		11	.727		121	30.50		151.50	182	
	7524	16″ x 10″		12	.667		118	28		146	175	
	7526	16″ x 14″		10	.800		123	33.50		156.50	190	
	7528	16″ x 18″		9	.889		131	37.50		168.50	204	
	7542	18″ x 18″		8	1		136	42		178	217	
	7544	20″ x 14″		8	1		138	42		180	219	
	7546	20″ x 18″		7	1.143		147	48		195	239	
	7560	24″ x 12″		8	1		140	42		182	221	
	7562	24″ x 18″		7	1.143		156	48		204	249	
	7564	24″ x 24″		6	1.333		162	56		218	268	
	7568	28″ x 10″		7	1.143		234	48		282	335	
	7590	30″ x 14″		5	1.600		246	67.50		313.50	380	
	7600	30″ x 18″		4	2		262	84.50		346.50	425	
	7610	30″ x 24″		3.80	2.105		299	88.50		387.50	470	
	7700	For thermostat, add		8	1		40	42		82	111	
	8000	Multi-blade dampers, parallel blade										
	8100	8″ x 8″	1 Shee	24	.333	Ea.	59	14.05		73.05	87.50	
	8140	16″ x 10″		20	.400		78	16.85		94.85	113	
	8160	18″ x 12″		18	.444		85	18.75		103.75	124	
	8220	28″ x 16″		10	.800		110	33.50		143.50	175	
	9000	Silencers, noise control for air flow, duct				MCFM	49			49	54	
	9900	Minimum labor/equipment charge	1 Shee	4	2	Job		84.50		84.50	134	

15830 | Fans

			CREW	DAILY OUTPUT	LABOR-HOURS	UNIT	MAT.	LABOR	EQUIP.	TOTAL	TOTAL INCL O&P	
100	0010	**FANS**										100
	0020	Air conditioning and process air handling										
	0030	Axial flow, compact, low sound, 2.5″ S.P.										
	0050	3,800 CFM, 5 HP	Q-20	3.40	5.882	Ea.	3,925	228		4,153	4,650	
	0120	15,600 CFM, 10 HP	″	1.60	12.500	″	6,875	485		7,360	8,325	
	0200	In-line centrifugal, supply/exhaust booster										
	0220	aluminum wheel/hub, disconnect switch, 1/4″ S.P.										
	0240	500 CFM, 10″ diameter connection	Q-20	3	6.667	Ea.	940	258		1,198	1,425	
	0280	1,520 CFM, 16″ diameter connection		2	10		1,075	385		1,460	1,800	
	0320	3,480 CFM, 20″ diameter connection		.80	25		1,400	970		2,370	3,075	
	0326	5,080 CFM, 20″ diameter connection		.75	26.667		1,525	1,025		2,550	3,300	
	2500	Ceiling fan, right angle, extra quiet, 0.10″ S.P.										
	2540	210 CFM	Q-20	19	1.053	Ea.	189	41		230	273	
	2580	885 CFM		16	1.250		470	48.50		518.50	595	
	2620	2,960 CFM		11	1.818		870	70.50		940.50	1,075	
	2640	For wall or roof cap, add	1 Shee	16	.500		160	21		181	210	
	4500	Corrosive fume resistant, plastic										
	4600	roof ventilators, centrifugal, V belt drive, motor										
	4620	1/4″ S.P., 250 CFM, 1/4 HP	Q-20	6	3.333	Ea.	2,525	129		2,654	3,000	
	4640	895 CFM, 1/3 HP		5	4		2,750	155		2,905	3,275	
	4660	1630 CFM, 1/2 HP		4	5		3,250	194		3,444	3,875	
	4680	2240 CFM, 1 HP		3	6.667		3,400	258		3,658	4,125	
	5000	Utility set, centrifugal, V belt drive, motor										
	5020	1/4″ S.P., 1200 CFM, 1/4 HP	Q-20	6	3.333	Ea.	2,850	129		2,979	3,350	
	5040	1520 CFM, 1/3 HP		5	4		2,850	155		3,005	3,400	
	5060	1850 CFM, 1/2 HP		4	5		2,875	194		3,069	3,450	
	5080	2180 CFM, 3/4 HP		3	6.667		2,900	258		3,158	3,600	
	6650	Residential, bath exhaust, grille, back draft damper										

	15830 | Fans		CREW	DAILY OUTPUT	LABOR-HOURS	UNIT	2006 BARE COSTS MAT.	LABOR	EQUIP.	TOTAL	TOTAL INCL O&P	
100	6660	50 CFM	Q-20	24	.833	Ea.	30	32.50		62.50	84	100
	6670	110 CFM		22	.909		50	35		85	111	
	6680	Light combination, squirrel cage, 100 watt, 70 CFM		24	.833		63	32.50		95.50	120	
	6700	Light/heater combination, ceiling mounted										
	6710	70 CFM, 1450 watt	Q-20	24	.833	Ea.	74.50	32.50		107	133	
	6800	Heater combination, recessed, 70 CFM		24	.833		36	32.50		68.50	90.50	
	6820	With 2 infrared bulbs		23	.870		53.50	33.50		87	112	
	6900	Kitchen exhaust, grille, complete, 160 CFM		22	.909		63.50	35		98.50	126	
	6910	180 CFM		20	1		54	38.50		92.50	121	
	6920	270 CFM		18	1.111		97.50	43		140.50	175	
	6940	Residential roof jacks and wall caps										
	6944	Wall cap with back draft damper										
	6946	3" & 4" dia. round duct	1 Shee	11	.727	Ea.	13.05	30.50		43.55	63.50	
	6948	6" dia. round duct	"	11	.727	"	31	30.50		61.50	83	
	6958	Roof jack with bird screen and back draft damper										
	6960	3" & 4" dia. round duct	1 Shee	11	.727	Ea.	12.55	30.50		43.05	63	
	6962	3-1/4" x 10" rectangular duct	"	10	.800	"	22.50	33.50		56	78.50	
	6980	Transition										
	6982	3-1/4" x 10" to 6" dia. round	1 Shee	20	.400	Ea.	13.85	16.85		30.70	42.50	
	7100	Direct drive, 320 CFM, 11" sq. damper	Q-20	7	2.857		360	111		471	570	
	7120	600 CFM, 11" sq. damper	"	6	3.333		365	129		494	605	
	8000	Ventilation, residential										
	8020	Attic, roof type										
	8030	Aluminum dome, damper & curb										
	8040	6" diameter, 300 CFM	1 Elec	16	.500	Ea.	266	21		287	325	
	8050	7" diameter, 450 CFM		15	.533		291	22.50		313.50	355	
	8060	9" diameter, 900 CFM		14	.571		465	24		489	550	
	8080	12" diameter, 1000 CFM (gravity)		10	.800		330	33.50		363.50	415	
	8090	16" diameter, 1500 CFM (gravity)		9	.889		400	37.50		437.50	500	
	8100	20" diameter, 2500 CFM (gravity)		8	1		490	42		532	600	
	8110	26" diameter, 4000 CFM (gravity)		7	1.143		590	48		638	725	
	8120	32" diameter, 6500 CFM (gravity)		6	1.333		810	56		866	980	
	8130	38" diameter, 8000 CFM (gravity)		5	1.600		1,200	67		1,267	1,425	
	8140	50" diameter, 13,000 CFM (gravity)		4	2		1,750	84		1,834	2,050	
	8160	Plastic, ABS dome										
	8180	1050 CFM	1 Elec	14	.571	Ea.	97	24		121	144	
	8200	1600 CFM	"	12	.667	"	145	28		173	203	
	8240	Attic, wall type, with shutter, one speed										
	8250	12" diameter, 1000 CFM	1 Elec	14	.571	Ea.	210	24		234	268	
	8260	14" diameter, 1500 CFM		12	.667		227	28		255	293	
	8270	16" diameter, 2000 CFM		9	.889		257	37.50		294.50	340	
	8290	Whole house, wall type, with shutter, one speed										
	8300	30" diameter, 4800 CFM	1 Elec	7	1.143	Ea.	550	48		598	680	
	8310	36" diameter, 7000 CFM		6	1.333		600	56		656	745	
	8320	42" diameter, 10,000 CFM		5	1.600		670	67		737	845	
	8330	48" diameter, 16,000 CFM		4	2		835	84		919	1,050	
	8340	For two speed, add					50			50	55	
	8350	Whole house, lay-down type, with shutter, one speed										
	8360	30" diameter, 4500 CFM	1 Elec	8	1	Ea.	585	42		627	710	
	8370	36" diameter, 6500 CFM		7	1.143		630	48		678	770	
	8380	42" diameter, 9000 CFM		6	1.333		690	56		746	845	
	8390	48" diameter, 12,000 CFM		5	1.600		785	67		852	970	
	8440	For two speed, add					37.50			37.50	41.50	
	8450	For 12 hour timer switch, add	1 Elec	32	.250		37.50	10.50		48	57.50	
	9900	Minimum labor/equipment charge	"	4	2	Job		84		84	129	

15 MECHANICAL

	15840 | Air Terminal Units		Crew	Daily Output	Labor-Hours	Unit	2006 Bare Costs Mat.	Labor	Equip.	Total	Total Incl O&P	
500	0010	**CONSTANT VOLUME MIXING BOXES**										500
	5180	Mixing box, includes electric or pneumatic motor										
	5190	Recommend use with attenuator, see 15820-300-9000										
	5200	Constant volume, 150 to 270 CFM	Q-9	12	1.333	Ea.	545	50.50		595.50	680	
	5210	270 to 600 CFM	"	11	1.455	"	560	55		615	705	
700	0010	**VARIABLE VOLUME MIXING BOXES**										700
	5500	VAV Cool only, pneum. press indep. 300 to 600 CFM	Q-9	11	1.455	Ea.	335	55		390	455	
	5510	500 to 1000 CFM		9	1.778		345	67.50		412.50	485	
	5520	800 to 1600 CFM		9	1.778		360	67.50		427.50	500	
	5530	1100 to 2000 CFM		8	2		370	76		446	530	
	5540	1500 to 3000 CFM		7	2.286		400	86.50		486.50	580	
	5550	2000 to 4000 CFM		6	2.667		410	101		511	610	
	5560	For electric, w/thermostat, press. dependent, add					20			20	22	
	15850 | Air Outlets & Inlets											
300	0010	**DIFFUSERS** Aluminum, opposed blade damper unless noted										300
	0100	Ceiling, linear, also for sidewall										
	0120	2" wide	1 Shee	32	.250	L.F.	30.50	10.55		41.05	50.50	
	0160	4" wide		26	.308		39.50	12.95		52.45	64	
	0180	6" wide		24	.333		47.50	14.05		61.55	75	
	0200	8" wide		22	.364		54.50	15.35		69.85	84.50	
	0500	Perforated, 24" x 24" lay-in panel size, 6" x 6"		16	.500	Ea.	72	21		93	113	
	0520	8" x 8"		15	.533		74	22.50		96.50	117	
	0530	9" x 9"		14	.571		77	24		101	124	
	0540	10" x 10"		14	.571		79.50	24		103.50	126	
	0560	12" x 12"		12	.667		82	28		110	135	
	0590	16" x 16"		11	.727		112	30.50		142.50	173	
	0600	18" x 18"		10	.800		120	33.50		153.50	186	
	0610	20" x 20"		10	.800		137	33.50		170.50	205	
	0620	24" x 24"		9	.889		159	37.50		196.50	235	
	1000	Rectangular, 1 to 4 way blow, 6" x 6"		16	.500		44.50	21		65.50	82.50	
	1010	8" x 8"		15	.533		53	22.50		75.50	94.50	
	1014	9" x 9"		15	.533		57	22.50		79.50	99	
	1016	10" x 10"		15	.533		67.50	22.50		90	110	
	1020	12" x 6"		15	.533		77.50	22.50		100	121	
	1040	12" x 9"		14	.571		84.50	24		108.50	132	
	1060	12" x 12"		12	.667		78.50	28		106.50	131	
	1070	14" x 6"		13	.615		84	26		110	134	
	1074	14" x 14"		12	.667		103	28		131	158	
	1150	18" x 18"		9	.889		145	37.50		182.50	219	
	1170	24" x 12"		10	.800		136	33.50		169.50	204	
	2000	T bar mounting, 24" x 24" lay-in frame, 6" x 6"		16	.500		94.50	21		115.50	138	
	2020	9" x 9"		14	.571		105	24		129	154	
	2040	12" x 12"		12	.667		135	28		163	193	
	2060	15" x 15"		11	.727		173	30.50		203.50	239	
	2080	18" x 18"		10	.800		190	33.50		223.50	264	
	6000	For steel diffusers instead of aluminum, deduct					10%					
	9000	Minimum labor/equipment charge	1 Shee	4	2	Job		84.50		84.50	134	
500	0010	**GRILLES**										500
	0020	Aluminum										
	1000	Air return, 6" x 6"	1 Shee	26	.308	Ea.	14.30	12.95		27.25	36.50	
	1020	10" x 6"		24	.333		16.90	14.05		30.95	41	
	1080	16" x 8"		22	.364		25.50	15.35		40.85	52.50	
	1100	12" x 12"		22	.364		25.50	15.35		40.85	52.50	

15800 | Air Distribution

	15850 | Air Outlets & Inlets		Crew	Daily Output	Labor-Hours	Unit	2006 Bare Costs Mat.	Labor	Equip.	Total	Total Incl O&P	
500	1120	24" x 12"	1 Shee	18	.444	Ea.	45	18.75		63.75	79.50	500
	1300	48" x 24"		12	.667		174	28		202	236	
	9000	Minimum labor/equipment charge		4	2	Job		84.50		84.50	134	
700	0010	**REGISTERS**										700
	0020											
	0980	Air supply										
	1000	Ceiling/wall, O.B. damper, anodized aluminum										
	1010	One or two way deflection, adj. curved face bars										
	1140	14" x 8"	1 Shee	17	.471	Ea.	34.50	19.85		54.35	69.50	
	3000	Baseboard, hand adj. damper, enameled steel										
	3012	8" x 6"	1 Shee	26	.308	Ea.	13.05	12.95		26	35	
	3020	10" x 6"		24	.333		14.70	14.05		28.75	38.50	
	3040	12" x 5"		23	.348		16.80	14.65		31.45	42	
	3060	12" x 6"		23	.348		15.50	14.65		30.15	40.50	
	3080	12" x 8"		22	.364		22.50	15.35		37.85	49	
	3100	14" x 6"		20	.400		16.80	16.85		33.65	45.50	
	9000	Minimum labor/equipment charge		4	2	Job		84.50		84.50	134	
800	0010	**VENTILATORS** Base & damper										800
	9000	Minimum labor/equipment charge	1 Plum	2	4	Job		171		171	266	
	15860 | Air Cleaning Devices											
100	0010	**AIR FILTERS**										100
	0020											
	0050	Activated charcoal type, full flow				MCFM	600			600	660	
	0060	Activated charcoal type, full flow, impregnated media 12" deep					175			175	193	
	0070	Activated charcoal type, HEPA filter & frame for field erection					175			175	193	
	0080	Activated charcoal type, HEPA filter-diffuser, ceiling install.					250			250	275	
	2000	Electronic air cleaner, duct mounted										
	2150	400 - 1000 CFM	1 Shee	2.30	3.478	Ea.	695	147		842	1,000	
	2200	1000 - 1400 CFM		2.20	3.636		725	153		878	1,050	
	2250	1400 - 2000 CFM		2.10	3.810		800	161		961	1,125	
	2950	Mechanical media filtration units										
	3000	High efficiency type, with frame, non-supported				MCFM	45			45	49.50	
	3100	Supported type					55			55	60.50	
	4000	Medium efficiency, extended surface					5.50			5.50	6.05	
	4500	Permanent washable					20			20	22	
	5000	Renewable disposable roll					120			120	132	
	5500	Throwaway glass or paper media type				Ea.	4.60			4.60	5.05	
	9000	Minimum labor/equipment charge	1 Shee	2.25	3.556	Job		150		150	238	

For information about Means Estimating Seminars, see yellow pages 12 and 13 in back of book

Division 16 Electrical

Estimating Tips

16060 Grounding & Bonding

- When taking off grounding system, identify separately the type and size of wire and list each unique type of ground connection.

16100 Wiring Methods

- Conduit should be taken off in three main categories: power distribution, branch power, and branch lighting, so the estimator can concentrate on systems and components, therefore making it easier to ensure all items have been accounted for.
- For cost modifications for elevated conduit installation, add the percentages to labor according to the height of installation and only to the quantities exceeding the different height levels, not to the total conduit quantities.
- Remember that aluminum wiring of equal ampacity is larger in diameter than copper and may require larger conduit.
- If more than three wires at a time are being pulled, deduct percentages from the labor hours of that grouping of wires.
- The estimator should take the weights of materials into consideration when completing a takeoff. Topics to consider include: How will the materials be supported? What methods of support are available? How high will the support structure have to reach? Will the final support structure be able to withstand the total burden? Is the support material included or separate from the fixture, equipment, and material specified?

16200 Electrical Power

- Do not overlook the costs for equipment used in the installation. If scaffolding or highlifts are available in the field, contractors may use them in lieu of the proposed ladders and rolling staging.

16400 Low-Voltage Distribution

- Supports and concrete pads may be shown on drawings for the larger equipment, or the support system may be only a piece of plywood for the back of a panelboard. In either case, it must be included in the costs.

16500 Lighting

- Fixtures should be taken off room by room, using the fixture schedule, specifications, and the ceiling plan. For large concentrations of lighting fixtures in the same area, deduct the percentages from labor hours.

16700 Communications
16800 Sound & Video

- When estimating material costs for special systems, it is always prudent to obtain manufacturers' quotations for equipment prices and special installation requirements which will affect the total costs.

Reference Numbers

Reference numbers are shown in bold squares at the beginning of some major classifications. These numbers refer to related items in the Reference Section. The reference information may be an estimating procedure, an alternate pricing method, or technical information.

Note: Not all subdivisions listed here necessarily appear in this publication.

Note: **i2 Trade Service,** *in part, has been used as a reference source for some of the material prices used in Division 16.*

16055 | Selective Demolition

			CREW	DAILY OUTPUT	LABOR-HOURS	UNIT	2006 BARE COSTS MAT.	LABOR	EQUIP.	TOTAL	TOTAL INCL O&P	
300	0010	ELECTRICAL DEMOLITION R260105-30										300
	0020	Conduit to 15′ high, including fittings & hangers										
	0100	Rigid galvanized steel, 1/2″ to 1″ diameter R024119-10	1 Elec	242	.033	L.F.		1.39		1.39	2.14	
	0120	1-1/4″ to 2″	″	200	.040			1.68		1.68	2.59	
	0140	2-1/2″ to 3-1/2″	2 Elec	302	.053			2.23		2.23	3.43	
	0200	Electric metallic tubing (EMT), 1/2″ to 1″	1 Elec	394	.020			.85		.85	1.31	
	0220	1-1/4″ to 1-1/2″		326	.025			1.03		1.03	1.59	
	0240	2″ to 3″		236	.034			1.42		1.42	2.19	
	0400	Wiremold raceway, including fittings & hangers										
	0420	No. 3000	1 Elec	250	.032	L.F.		1.34		1.34	2.07	
	0440	No. 4000	″	217	.037	″		1.55		1.55	2.39	
	0500	Channels, steel, including fittings & hangers										
	0520	3/4″ x 1-1/2″	1 Elec	308	.026	L.F.		1.09		1.09	1.68	
	0540	1-1/2″ x 1-1/2″	″	269	.030	″		1.25		1.25	1.92	
	0600	Copper bus duct, indoor, 3 phase										
	0610	Including hangers & supports										
	0620	225 amp	2 Elec	135	.119	L.F.		4.98		4.98	7.65	
	0640	400 amp		106	.151			6.35		6.35	9.75	
	0660	600 amp		86	.186			7.80		7.80	12.05	
	0680	1000 amp		60	.267			11.20		11.20	17.25	
	0700	1600 amp		40	.400			16.80		16.80	26	
	0720	3000 amp		10	1.600			67		67	104	
	0800	Plug-in switches, 600V 3 ph, incl. disconnecting										
	0820	wire, conduit terminations, 30 amp	1 Elec	15.50	.516	Ea.		21.50		21.50	33.50	
	0840	60 amp		13.90	.576			24		24	37	
	0850	100 amp		10.40	.769			32.50		32.50	50	
	0860	200 amp		6.20	1.290			54		54	83.50	
	0940	1200 amp	2 Elec	2	8			335		335	520	
	0960	1600 amp	″	1.70	9.412			395		395	610	
	1010	Safety switches, 250 or 600V, incl. disconnection										
	1050	of wire & conduit terminations										
	1100	30 amp	1 Elec	12.30	.650	Ea.		27.50		27.50	42	
	1120	60 amp		8.80	.909			38		38	59	
	1140	100 amp		7.30	1.096			46		46	71	
	1160	200 amp		5	1.600			67		67	104	
	1210	Panel boards, incl. removal of all breakers,										
	1220	conduit terminations & wire connections										
	1230	3 wire, 120/240V, 100A, to 20 circuits	1 Elec	2.60	3.077	Ea.		129		129	199	
	1240	200 amps, to 42 circuits	2 Elec	2.60	6.154			258		258	400	
	1260	4 wire, 120/208V, 125A, to 20 circuits	1 Elec	2.40	3.333			140		140	216	
	1270	200 amps, to 42 circuits	2 Elec	2.40	6.667			280		280	430	
	1300	Transformer, dry type, 1 ph, incl. removal of										
	1320	supports, wire & conduit terminations										
	1340	1 kVA	1 Elec	7.70	1.039	Ea.		43.50		43.50	67	
	1360	5 kVA	″	4.70	1.702			71.50		71.50	110	
	1420	75 kVA	2 Elec	2.50	6.400			269		269	415	
	1440	3 Phase to 600V, primary										
	1460	3 kVA	1 Elec	3.85	2.078	Ea.		87.50		87.50	134	
	1480	15 kVA	2 Elec	4.20	3.810			160		160	246	
	1500	30 kVA	″	3.50	4.571			192		192	296	
	1530	112.5 kVA	R-3	2.90	6.897			286	52.50	338.50	500	
	1560	500 kVA		1.40	14.286			590	109	699	1,025	
	1570	750 kVA		1.10	18.182			755	139	894	1,325	
	1600	Pull boxes & cabinets, sheet metal, incl. removal										
	1620	of supports and conduit terminations										
	1640	6″ x 6″ x 4″	1 Elec	31.10	.257	Ea.		10.80		10.80	16.65	

16055 | Selective Demolition

		Description		Crew	Daily Output	Labor-Hours	Unit	2006 Bare Costs Mat.	2006 Bare Costs Labor	2006 Bare Costs Equip.	2006 Bare Costs Total	Total Incl O&P	
300	1660	12" x 12" x 4"	R024119-10	1 Elec	23.30	.343	Ea.		14.40		14.40	22	300
	1720	Junction boxes, 4" sq. & oct.			80	.100			4.20		4.20	6.45	
	1740	Handy box	R260105-30		107	.075			3.14		3.14	4.84	
	1760	Switch box			107	.075			3.14		3.14	4.84	
	1780	Receptacle & switch plates			257	.031			1.31		1.31	2.01	
	1800	Wire, THW-THWN-THHN, removed from											
	1810	in place conduit, to 15' high											
	1830	#14		1 Elec	65	.123	C.L.F.		5.15		5.15	7.95	
	1840	#12			55	.145			6.10		6.10	9.40	
	1850	#10			45.50	.176			7.40		7.40	11.40	
	1880	#4		2 Elec	53	.302			12.70		12.70	19.55	
	1890	#3			50	.320			13.45		13.45	20.50	
	1910	1/0			33.20	.482			20		20	31	
	1920	2/0			29.20	.548			23		23	35.50	
	1930	3/0			25	.640			27		27	41.50	
	1980	400 kcmil			17	.941			39.50		39.50	61	
	1990	500 kcmil			16.20	.988			41.50		41.50	64	
	2000	Interior fluorescent fixtures, incl. supports											
	2010	& whips, to 15' high											
	2100	Recessed drop-in 2' x 2', 2 lamp		2 Elec	35	.457	Ea.		19.20		19.20	29.50	
	2120	2' x 4', 2 lamp			33	.485			20.50		20.50	31.50	
	2140	2' x 4', 4 lamp			30	.533			22.50		22.50	34.50	
	2160	4' x 4', 4 lamp			20	.800			33.50		33.50	52	
	2180	Surface mount, acrylic lens & hinged frame											
	2200	1' x 4', 2 lamp		2 Elec	44	.364	Ea.		15.25		15.25	23.50	
	2220	2' x 2', 2 lamp			44	.364			15.25		15.25	23.50	
	2260	2' x 4', 4 lamp			33	.485			20.50		20.50	31.50	
	2280	4' x 4', 4 lamp			23	.696			29		29	45	
	2300	Strip fixtures, surface mount											
	2320	4' long, 1 lamp		2 Elec	53	.302	Ea.		12.70		12.70	19.55	
	2340	4' long, 2 lamp			50	.320			13.45		13.45	20.50	
	2360	8' long, 1 lamp			42	.381			16		16	24.50	
	2380	8' long, 2 lamp			40	.400			16.80		16.80	26	
	2400	Pendant mount, industrial, incl. removal											
	2410	of chain or rod hangers, to 15' high											
	2420	4' long, 2 lamp		2 Elec	35	.457	Ea.		19.20		19.20	29.50	
	2440	8' long, 2 lamp		"	27	.593	"		25		25	38.50	
	2460	Interior incandescent, surface, ceiling											
	2470	or wall mount, to 12' high											
	2480	Metal cylinder type, 75 Watt		2 Elec	62	.258	Ea.		10.85		10.85	16.70	
	2500	150 Watt		"	62	.258	"		10.85		10.85	16.70	
	2520	Metal halide, high bay											
	2540	400 Watt		2 Elec	15	1.067	Ea.		45		45	69	
	2560	1000 Watt			12	1.333			56		56	86.50	
	2580	150 Watt, low bay			20	.800			33.50		33.50	52	
	2600	Exterior fixtures, incandescent, wall mount											
	2620	100 Watt		2 Elec	50	.320	Ea.		13.45		13.45	20.50	
	2640	Quartz, 500 Watt			33	.485			20.50		20.50	31.50	
	2660	1500 Watt			27	.593			25		25	38.50	
	2680	Wall pack, mercury vapor											
	2700	175 Watt		2 Elec	25	.640	Ea.		27		27	41.50	
	2720	250 Watt		"	25	.640	"		27		27	41.50	
	9000	Minimum labor/equipment charge		1 Elec	4	2	Job		84		84	129	

16050 | Basic Electrical Materials & Methods

16060 | Grounding & Bonding

	16060 Grounding & Bonding		CREW	DAILY OUTPUT	LABOR-HOURS	UNIT	2006 BARE COSTS MAT.	LABOR	EQUIP.	TOTAL	TOTAL INCL O&P	
800	0010	GROUNDING										800
	0030	Rod, copper clad, 8' long, 1/2" diameter	1 Elec	5.50	1.455	Ea.	13.70	61		74.70	109	
	0040	5/8" diameter		5.50	1.455		15	61		76	111	
	0050	3/4" diameter		5.30	1.509		28.50	63.50		92	129	
	0080	10' long, 1/2" diameter		4.80	1.667		17.70	70		87.70	127	
	0090	5/8" diameter		4.60	1.739		22	73		95	137	
	0100	3/4" diameter		4.40	1.818		31.50	76.50		108	153	
	0130	15' long, 3/4" diameter		4	2		87	84		171	225	
	0260	Wire ground bare armored, #8-1 conductor		2	4	C.L.F.	80.50	168		248.50	350	
	0280	#4-1 conductor		1.60	5		149	210		359	490	
	0390	Bare copper wire, #8 stranded		11	.727		15.90	30.50		46.40	64.50	
	0400	#6		10	.800		28.50	33.50		62	83.50	
	0600	#2	2 Elec	10	1.600		66	67		133	177	
	0800	3/0		6.60	2.424		147	102		249	320	
	1000	4/0		5.70	2.807		184	118		302	385	
	1200	250 kcmil	3 Elec	7.20	3.333		218	140		358	455	
	1800	Water pipe ground clamps, heavy duty										
	2000	Bronze, 1/2" to 1" diameter	1 Elec	8	1	Ea.	14.40	42		56.40	80.50	
	2100	1-1/4" to 2" diameter		8	1		18.95	42		60.95	85.50	
	2200	2-1/2" to 3" diameter		6	1.333		41	56		97	132	
	2800	Brazed connections, #6 wire		12	.667		12.35	28		40.35	56.50	
	3000	#2 wire		10	.800		16.55	33.50		50.05	70	
	3100	3/0 wire		8	1		25	42		67	92	
	3200	4/0 wire		7	1.143		28.50	48		76.50	106	
	3400	250 kcmil wire		5	1.600		33	67		100	141	
	3600	500 kcmil wire		4	2		41	84		125	174	
	9000	Minimum labor/equipment charge		4	2	Job		84		84	129	

16100 | Wiring Methods

16120 | Conductors & Cables

	16120 Conductors & Cables		CREW	DAILY OUTPUT	LABOR-HOURS	UNIT	2006 BARE COSTS MAT.	LABOR	EQUIP.	TOTAL	TOTAL INCL O&P	
120	0010	ARMORED CABLE										120
	0020											
	0050	600 volt, copper (BX), #14, 2 conductor, solid	1 Elec	2.40	3.333	C.L.F.	58.50	140		198.50	281	
	0100	3 conductor, solid		2.20	3.636		93	153		246	335	
	0152	#12, 2 conductor, solid		2.10	3.810		59.50	160		219.50	310	
	0202	3 conductor, solid		1.80	4.444		95	187		282	390	
	0252	#10, 2 conductor, solid		1.80	4.444		108	187		295	405	
	0302	3 conductor, solid		1.50	5.333		149	224		373	510	
	0352	#8, 3 conductor, solid		1.20	6.667		268	280		548	725	
	9010	600 volt, copper (MC) steel clad, #14, 2 wire		2.40	3.333		61.50	140		201.50	284	
	9020	3 wire		2.20	3.636		95.50	153		248.50	340	
	9040	#12, 2 wire		2.30	3.478		62.50	146		208.50	294	
	9050	3 wire		2	4		96.50	168		264.50	365	
	9070	#10, 2 wire		2	4		111	168		279	380	
	9080	3 wire		1.60	5		171	210		381	515	
	9100	#8, 2 wire, stranded		1.80	4.444		210	187		397	520	
	9110	3 wire, stranded		1.30	6.154		325	258		583	755	
	9900	Minimum labor/equipment charge		4	2	Job		84		84	129	

16120 | Conductors & Cables

			CREW	DAILY OUTPUT	LABOR-HOURS	UNIT	2006 BARE COSTS MAT.	LABOR	EQUIP.	TOTAL	TOTAL INCL O&P	
500	0010	**MINERAL INSULATED CABLE** 600 volt										500
	0100	1 conductor, #12	1 Elec	1.60	5	C.L.F.	236	210		446	585	
	1500	2 conductor, #12		1.40	5.714		485	240		725	905	
	1600	#10		1.20	6.667		590	280		870	1,075	
	1800	#8		1.10	7.273		735	305		1,040	1,275	
	2000	#6		1.05	7.619		940	320		1,260	1,525	
	2100	#4	2 Elec	2	8		1,225	335		1,560	1,875	
	2200	3 conductor, #12	1 Elec	1.20	6.667		615	280		895	1,100	
	2400	#10		1.10	7.273		710	305		1,015	1,250	
	2600	#8		1.05	7.619		855	320		1,175	1,450	
	2800	#6		1	8		1,100	335		1,435	1,750	
	3000	#4	2 Elec	1.80	8.889		1,475	375		1,850	2,175	
	3100	4 conductor, #12	1 Elec	1.20	6.667		650	280		930	1,150	
	3200	#10		1.10	7.273		775	305		1,080	1,325	
	3400	#8		1	8		1,000	335		1,335	1,625	
	3600	#6		.90	8.889		1,275	375		1,650	1,975	
	3620	7 conductor, #12		1.10	7.273		825	305		1,130	1,375	
	3640	#10		1	8		1,025	335		1,360	1,650	
	3800	M.I. terminations, 600 volt, 1 conductor, #12		8	1	Ea.	9.60	42		51.60	75	
	5500	2 conductor, #12		6.70	1.194		9.60	50		59.60	88	
	5600	#10		6.40	1.250		14.40	52.50		66.90	97	
	5800	#8		6.20	1.290		14.40	54		68.40	99.50	
	6000	#6		5.70	1.404		14.40	59		73.40	107	
	6200	#4		5.30	1.509		32	63.50		95.50	133	
	6400	3 conductor, #12		5.70	1.404		14.40	59		73.40	107	
	6500	#10		5.50	1.455		14.40	61		75.40	110	
	6600	#8		5.20	1.538		14.40	64.50		78.90	115	
	6800	#6		4.80	1.667		14.40	70		84.40	124	
	7200	#4		4.60	1.739		32	73		105	148	
	7400	4 conductor, #12		4.60	1.739		15.95	73		88.95	131	
	7500	#10		4.40	1.818		15.95	76.50		92.45	136	
	7600	#8		4.20	1.905		15.95	80		95.95	141	
	8400	#6		4	2		33.50	84		117.50	166	
	8500	7 conductor, #12		3.50	2.286		15.95	96		111.95	166	
	8600	#10		3	2.667		33.50	112		145.50	210	
	8800	Crimping tool, plier type					45.50			45.50	50	
	9000	Stripping tool					105			105	116	
	9200	Hand vise					51			51	56	
	9500	Minimum labor/equipment charge	1 Elec	4	2	Job		84		84	129	
550	0010	**NON-METALLIC SHEATHED CABLE** 600 volt										550
	0100	Copper with ground wire, (Romex)										
	0152	#14, 2 wire	1 Elec	2.50	3.200	C.L.F.	18.25	134		152.25	227	
	0202	3 wire		2.30	3.478		29	146		175	257	
	0252	#12, 2 wire		2.20	3.636		27	153		180	265	
	0302	3 wire		2	4		48.50	168		216.50	315	
	0352	#10, 2 wire		2	4		48.50	168		216.50	310	
	0402	3 wire		1.40	5.714		60.50	240		300.50	435	
	0452	#8, 3 conductor		1.30	6.154		138	258		396	550	
	0502	#6, 3 wire		1.20	6.667		214	280		494	665	
	0550	SE type SER aluminum cable, 3 RHW and										
	0602	1 bare neutral, 3 #8 & 1 #8	1 Elec	1.50	5.333	C.L.F.	133	224		357	490	
	0652	3 #6 & 1 #6	"	1.30	6.154		150	258		408	565	
	0702	3 #4 & 1 #6	2 Elec	2.20	7.273		168	305		473	655	
	0752	3 #2 & 1 #4		2	8		248	335		583	790	
	0802	3 #1/0 & 1 #2		1.80	8.889		375	375		750	985	

ELECTRICAL 16

		16120 | Conductors & Cables	CREW	DAILY OUTPUT	LABOR-HOURS	UNIT	2006 BARE COSTS MAT.	LABOR	EQUIP.	TOTAL	TOTAL INCL O&P	
550	0852	3 #2/0 & 1 #1	2 Elec	1.60	10	C.L.F.	440	420		860	1,125	550
	0902	3 #4/0 & 1 #2/0	↓	1.40	11.429	↓	630	480		1,110	1,425	
	6500	Service entrance cap for copper SEU										
	6700	150 amp	1 Elec	10	.800	Ea.	13.95	33.50		47.45	67.50	
	6800	200 amp		8	1	"	29	42		71	96.50	
	9000	Minimum labor/equipment charge	↓	4	2	Job		84		84	129	
900	0010	**WIRE** R260533-22										900
	0020	600 volt type THW, copper, solid, #14	1 Elec	13	.615	C.L.F.	4.25	26		30.25	44.50	
	0030	#12		11	.727		6.40	30.50		36.90	54	
	0040	#10	↓	10	.800	↓	9.70	33.50		43.20	62.50	
	0051	Wire, 600 volt, stranded										
	0140	#8	1 Elec	8	1	C.L.F.	19.70	42		61.70	86	
	0160	#6	"	6.50	1.231		32.50	51.50		84	116	
	0180	#4	2 Elec	10.60	1.509		50	63.50		113.50	153	
	0200	#3		10	1.600		62.50	67		129.50	173	
	0220	#2		9	1.778		79	74.50		153.50	202	
	0240	#1		8	2		100	84		184	239	
	0260	1/0		6.60	2.424		111	102		213	279	
	0280	2/0		5.80	2.759		130	116		246	320	
	0300	3/0		5	3.200		166	134		300	390	
	0350	4/0	↓	4.40	3.636		206	153		359	460	
	0400	250 kcmil	3 Elec	6	4		245	168		413	530	
	0420	300 kcmil		5.70	4.211		291	177		468	590	
	0450	350 kcmil		5.40	4.444		340	187		527	665	
	0480	400 kcmil		5.10	4.706		390	198		588	735	
	0490	500 kcmil		4.80	5		480	210		690	855	
	0510	750 kcmil	↓	3.30	7.273		770	305		1,075	1,325	
	0530	Aluminum, stranded, #8	1 Elec	9	.889		17.90	37.50		55.40	77	
	0540	#6	"	8	1		24.50	42		66.50	91.50	
	0560	#4	2 Elec	13	1.231		30.50	51.50		82	113	
	0580	#2		10.60	1.509		41	63.50		104.50	143	
	0600	#1		9	1.778		60.50	74.50		135	182	
	0620	1/0		8	2		72.50	84		156.50	209	
	0640	2/0		7.20	2.222		85.50	93.50		179	238	
	0680	3/0		6.60	2.424		106	102		208	274	
	0700	4/0	↓	6.20	2.581		118	108		226	297	
	0720	250 kcmil	3 Elec	8.70	2.759		144	116		260	335	
	0740	300 kcmil		8.10	2.963		199	124		323	410	
	0760	350 kcmil		7.50	3.200		202	134		336	430	
	0780	400 kcmil		6.90	3.478		237	146		383	485	
	0800	500 kcmil		6	4		277	168		445	565	
	0850	600 kcmil		5.70	4.211		330	177		507	635	
	0880	700 kcmil	↓	5.10	4.706	↓	380	198		578	725	
	9000	Minimum labor/equipment charge	1 Elec	4	2	Job		84		84	129	

		16132 | Conduit & Tubing	CREW	DAILY OUTPUT	LABOR-HOURS	UNIT	MAT.	LABOR	EQUIP.	TOTAL	TOTAL INCL O&P	
205	0010	**CONDUIT** To 15′ high, includes 2 terminations, 2 elbows and R260533-22										205
	0020	11 beam clamps & couplings per 100 L.F.										
	1161	Field bends, 45° to 90°, 1/2″ diameter	1 Elec	53	.151	Ea.		6.35		6.35	9.75	
	1162	3/4″ diameter		47	.170			7.15		7.15	11	
	1163	1″ diameter		44	.182			7.65		7.65	11.75	
	1164	1-1/4″ diameter		23	.348			14.60		14.60	22.50	
	1165	1-1/2″ diameter		21	.381			16		16	24.50	
	1166	2″ diameter		16	.500			21		21	32.50	
	1991	Field bends, 45° to 90°, 1/2″ diameter		44	.182			7.65		7.65	11.75	
	1992	3/4″ diameter	↓	40	.200	↓		8.40		8.40	12.95	

	16132 | Conduit & Tubing		CREW	DAILY OUTPUT	LABOR-HOURS	UNIT	2006 BARE COSTS MAT.	LABOR	EQUIP.	TOTAL	TOTAL INCL O&P
205	1993	1" diameter (R260533-22)	1 Elec	36	.222	Ea.		9.35		9.35	14.40
	1994	1-1/4" diameter		19	.421			17.70		17.70	27
	1995	1-1/2" diameter		18	.444			18.65		18.65	29
	1996	2" diameter		13	.615			26		26	40
	2500	Steel, intermediate conduit (IMC), 1/2" diameter		100	.080	L.F.	2.14	3.36		5.50	7.55
	2530	3/4" diameter		90	.089		2.53	3.73		6.26	8.55
	2550	1" diameter		70	.114		3.65	4.80		8.45	11.40
	2570	1-1/4" diameter		65	.123		4.80	5.15		9.95	13.25
	2600	1-1/2" diameter		60	.133		5.60	5.60		11.20	14.80
	2630	2" diameter		50	.160		7.15	6.70		13.85	18.20
	2650	2-1/2" diameter		40	.200		14.65	8.40		23.05	29
	2670	3" diameter	2 Elec	60	.267		19.10	11.20		30.30	38.50
	2700	3-1/2" diameter		54	.296		23.50	12.45		35.95	45
	2730	4" diameter		50	.320		27	13.45		40.45	50
	2731	Field bends, 45° to 90°, 1/2" diameter	1 Elec	44	.182	Ea.		7.65		7.65	11.75
	2732	3/4" diameter		40	.200			8.40		8.40	12.95
	2733	1" diameter		36	.222			9.35		9.35	14.40
	2734	1-1/4" diameter		19	.421			17.70		17.70	27
	2735	1-1/2" diameter		18	.444			18.65		18.65	29
	2736	2" diameter		13	.615			26		26	40
	5000	Electric metallic tubing (EMT), 1/2" diameter		170	.047	L.F.	.60	1.98		2.58	3.70
	5020	3/4" diameter		130	.062		.98	2.58		3.56	5.05
	5040	1" diameter		115	.070		1.74	2.92		4.66	6.40
	5060	1-1/4" diameter		100	.080		2.67	3.36		6.03	8.15
	5080	1-1/2" diameter		90	.089		3.37	3.73		7.10	9.45
	5100	2" diameter		80	.100		4.27	4.20		8.47	11.15
	5120	2-1/2" diameter		60	.133		8.15	5.60		13.75	17.60
	5140	3" diameter	2 Elec	100	.160		10.95	6.70		17.65	22.50
	5160	3-1/2" diameter		90	.178		14.05	7.45		21.50	27
	5180	4" diameter		80	.200		15.95	8.40		24.35	30.50
	5200	Field bends, 45° to 90°, 1/2" diameter	1 Elec	89	.090	Ea.		3.78		3.78	5.80
	5220	3/4" diameter		80	.100			4.20		4.20	6.45
	5240	1" diameter		73	.110			4.60		4.60	7.10
	5260	1-1/4" diameter		38	.211			8.85		8.85	13.60
	5280	1-1/2" diameter		36	.222			9.35		9.35	14.40
	5300	2" diameter		26	.308			12.90		12.90	19.90
	5320	Offsets, 1/2" diameter		65	.123			5.15		5.15	7.95
	5340	3/4" diameter		62	.129			5.40		5.40	8.35
	5360	1" diameter		53	.151			6.35		6.35	9.75
	5380	1-1/4" diameter		30	.267			11.20		11.20	17.25
	5400	1-1/2" diameter		28	.286			12		12	18.50
	7600	EMT, "T" fittings with covers, 1/2" diameter, set screw		16	.500		13.40	21		34.40	47.50
	9990	Minimum labor/equipment charge		4	2	Job		84		84	129
230	0010	**CONDUIT IN CONCRETE SLAB** Including terminations,									
	0020	fittings and supports									
	3230	PVC, schedule 40, 1/2" diameter	1 Elec	270	.030	L.F.	.57	1.24		1.81	2.55
	3250	3/4" diameter		230	.035		.67	1.46		2.13	2.99
	3270	1" diameter		200	.040		.88	1.68		2.56	3.56
	3300	1-1/4" diameter		170	.047		1.26	1.98		3.24	4.43
	3330	1-1/2" diameter		140	.057		1.52	2.40		3.92	5.35
	3350	2" diameter		120	.067		1.95	2.80		4.75	6.45
	4350	Rigid galvanized steel, 1/2" diameter		200	.040		2.23	1.68		3.91	5.05
	4400	3/4" diameter		170	.047		2.56	1.98		4.54	5.85
	4450	1" diameter		130	.062		3.88	2.58		6.46	8.25
	4500	1-1/4" diameter		110	.073		5.15	3.05		8.20	10.35

ELECTRICAL 16

	16132 | Conduit & Tubing		Crew	Daily Output	Labor-Hours	Unit	2006 Bare Costs Mat.	Labor	Equip.	Total	Total Incl O&P
230	4600	1-1/2" diameter	1 Elec	100	.080	L.F.	5.95	3.36		9.31	11.75
	4800	2" diameter		90	.089		7.65	3.73		11.38	14.15
	9000	Minimum labor/equipment charge		4	2	Job		84		84	129
240	0010	**CONDUIT IN TRENCH** Includes terminations and fittings									
	0020	Does not include excavation or backfill, see div. 02315									
	0200	Rigid galvanized steel, 2" diameter	1 Elec	150	.053	L.F.	7.35	2.24		9.59	11.55
	0400	2-1/2" diameter	"	100	.080		14.30	3.36		17.66	21
	0600	3" diameter	2 Elec	160	.100		17.60	4.20		21.80	26
	0800	3-1/2" diameter		140	.114		22.50	4.80		27.30	32
	1000	4" diameter		100	.160		24.50	6.70		31.20	37.50
	1200	5" diameter		80	.200		53.50	8.40		61.90	72
	1400	6" diameter		60	.267		78.50	11.20		89.70	103
	9000	Minimum labor/equipment charge	1 Elec	4	2	Job		84		84	129
260	0010	**CUTTING AND DRILLING**									
	0100	Hole drilling to 10' high, concrete wall									
	0110	8" thick, 1/2" pipe size	R-31	12	.667	Ea.	3.51	28	5.40	36.91	53
	0120	3/4" pipe size		12	.667		3.51	28	5.40	36.91	53
	0130	1" pipe size		9.50	.842		7.25	35.50	6.80	49.55	70
	0140	1-1/4" pipe size		9.50	.842		7.25	35.50	6.80	49.55	70
	0150	1-1/2" pipe size		9.50	.842		7.25	35.50	6.80	49.55	70
	0160	2" pipe size		4.40	1.818		7.85	76.50	14.70	99.05	143
	0170	2-1/2" pipe size		4.40	1.818		7.85	76.50	14.70	99.05	143
	0180	3" pipe size		4.40	1.818		7.85	76.50	14.70	99.05	143
	0190	3-1/2" pipe size		3.30	2.424		8.40	102	19.60	130	188
	0200	4" pipe size		3.30	2.424		8.40	102	19.60	130	188
	0500	12" thick, 1/2" pipe size		9.40	.851		5.35	35.50	6.90	47.75	68.50
	0520	3/4" pipe size		9.40	.851		5.35	35.50	6.90	47.75	68.50
	0540	1" pipe size		7.30	1.096		10.25	46	8.85	65.10	92
	0560	1-1/4" pipe size		7.30	1.096		10.25	46	8.85	65.10	92
	0570	1-1/2" pipe size		7.30	1.096		10.25	46	8.85	65.10	92
	0580	2" pipe size		3.60	2.222		12	93.50	17.95	123.45	177
	0590	2-1/2" pipe size		3.60	2.222		12	93.50	17.95	123.45	177
	0600	3" pipe size		3.60	2.222		12	93.50	17.95	123.45	177
	0610	3-1/2" pipe size		2.80	2.857		13.60	120	23	156.60	226
	0630	4" pipe size		2.50	3.200		13.60	134	26	173.60	251
	0650	16" thick, 1/2" pipe size		7.60	1.053		7.20	44	8.50	59.70	85.50
	0670	3/4" pipe size		7	1.143		7.20	48	9.25	64.45	92
	0690	1" pipe size		6	1.333		13.25	56	10.75	80	113
	0710	1-1/4" pipe size		5.50	1.455		13.25	61	11.75	86	122
	0730	1-1/2" pipe size		5.50	1.455		13.25	61	11.75	86	122
	0750	2" pipe size		3	2.667		16.15	112	21.50	149.65	214
	0770	2-1/2" pipe size		2.70	2.963		16.15	124	24	164.15	236
	0790	3" pipe size		2.50	3.200		16.15	134	26	176.15	253
	0810	3-1/2" pipe size		2.30	3.478		18.85	146	28	192.85	277
	0830	4" pipe size		2	4		18.85	168	32.50	219.35	315
	0850	20" thick, 1/2" pipe size		6.40	1.250		9.05	52.50	10.10	71.65	102
	0870	3/4" pipe size		6	1.333		9.05	56	10.75	75.80	108
	0890	1" pipe size		5	1.600		16.25	67	12.95	96.20	136
	0910	1-1/4" pipe size		4.80	1.667		16.25	70	13.45	99.70	141
	0930	1-1/2" pipe size		4.60	1.739		16.25	73	14.05	103.30	146
	0950	2" pipe size		2.70	2.963		20.50	124	24	168.50	241
	0970	2-1/2" pipe size		2.40	3.333		20.50	140	27	187.50	268
	0990	3" pipe size		2.20	3.636		20.50	153	29.50	203	290
	1010	3-1/2" pipe size		2	4		24	168	32.50	224.50	320
	1030	4" pipe size		1.70	4.706		24	198	38	260	375

16132 | Conduit & Tubing

			CREW	DAILY OUTPUT	LABOR-HOURS	UNIT	2006 BARE COSTS MAT.	LABOR	EQUIP.	TOTAL	TOTAL INCL O&P	
260	1050	24" thick, 1/2" pipe size	R-31	5.50	1.455	Ea.	10.85	61	11.75	83.60	119	260
	1070	3/4" pipe size		5.10	1.569		10.85	66	12.65	89.50	127	
	1090	1" pipe size		4.30	1.860		19.25	78	15.05	112.30	158	
	1110	1-1/4" pipe size		4	2		19.25	84	16.15	119.40	168	
	1130	1-1/2" pipe size		4	2		19.25	84	16.15	119.40	168	
	1150	2" pipe size		2.40	3.333		24.50	140	27	191.50	273	
	1170	2-1/2" pipe size		2.20	3.636		24.50	153	29.50	207	295	
	1190	3" pipe size		2	4		24.50	168	32.50	225	320	
	1210	3-1/2" pipe size		1.80	4.444		29.50	187	36	252.50	360	
	1230	4" pipe size		1.50	5.333		29.50	224	43	296.50	425	
	1500	Brick wall, 8" thick, 1/2" pipe size		18	.444		3.51	18.65	3.59	25.75	37	
	1520	3/4" pipe size		18	.444		3.51	18.65	3.59	25.75	37	
	1540	1" pipe size		13.30	.602		7.25	25.50	4.86	37.61	52.50	
	1560	1-1/4" pipe size		13.30	.602		7.25	25.50	4.86	37.61	52.50	
	1580	1-1/2" pipe size		13.30	.602		7.25	25.50	4.86	37.61	52.50	
	1600	2" pipe size		5.70	1.404		7.85	59	11.35	78.20	112	
	1620	2-1/2" pipe size		5.70	1.404		7.85	59	11.35	78.20	112	
	1640	3" pipe size		5.70	1.404		7.85	59	11.35	78.20	112	
	1660	3-1/2" pipe size		4.40	1.818		8.40	76.50	14.70	99.60	143	
	1680	4" pipe size		4	2		8.40	84	16.15	108.55	156	
	1700	12" thick, 1/2" pipe size		14.50	.552		5.35	23	4.46	32.81	46.50	
	1720	3/4" pipe size		14.50	.552		5.35	23	4.46	32.81	46.50	
	1740	1" pipe size		11	.727		10.25	30.50	5.90	46.65	65	
	1760	1-1/4" pipe size		11	.727		10.25	30.50	5.90	46.65	65	
	1780	1-1/2" pipe size		11	.727		10.25	30.50	5.90	46.65	65	
	1800	2" pipe size		5	1.600		12	67	12.95	91.95	131	
	1820	2-1/2" pipe size		5	1.600		12	67	12.95	91.95	131	
	1840	3" pipe size		5	1.600		12	67	12.95	91.95	131	
	1860	3-1/2" pipe size		3.80	2.105		13.60	88.50	17	119.10	170	
	1880	4" pipe size		3.30	2.424		13.60	102	19.60	135.20	194	
	1900	16" thick, 1/2" pipe size		12.30	.650		7.20	27.50	5.25	39.95	55.50	
	1920	3/4" pipe size		12.30	.650		7.20	27.50	5.25	39.95	55.50	
	1940	1" pipe size		9.30	.860		13.25	36	6.95	56.20	78	
	1960	1-1/4" pipe size		9.30	.860		13.25	36	6.95	56.20	78	
	1980	1-1/2" pipe size		9.30	.860		13.25	36	6.95	56.20	78	
	2000	2" pipe size		4.40	1.818		16.15	76.50	14.70	107.35	152	
	2010	2-1/2" pipe size		4.40	1.818		16.15	76.50	14.70	107.35	152	
	2030	3" pipe size		4.40	1.818		16.15	76.50	14.70	107.35	152	
	2050	3-1/2" pipe size		3.30	2.424		18.85	102	19.60	140.45	200	
	2070	4" pipe size		3	2.667		18.85	112	21.50	152.35	218	
	2090	20" thick, 1/2" pipe size		10.70	.748		9.05	31.50	6.05	46.60	65	
	2110	3/4" pipe size		10.70	.748		9.05	31.50	6.05	46.60	65	
	2130	1" pipe size		8	1		16.25	42	8.10	66.35	91.50	
	2150	1-1/4" pipe size		8	1		16.25	42	8.10	66.35	91.50	
	2170	1-1/2" pipe size		8	1		16.25	42	8.10	66.35	91.50	
	2190	2" pipe size		4	2		20.50	84	16.15	120.65	169	
	2210	2-1/2" pipe size		4	2		20.50	84	16.15	120.65	169	
	2230	3" pipe size		4	2		20.50	84	16.15	120.65	169	
	2250	3-1/2" pipe size		3	2.667		24	112	21.50	157.50	223	
	2270	4" pipe size		2.70	2.963		24	124	24	172	245	
	2290	24" thick, 1/2" pipe size		9.40	.851		10.85	35.50	6.90	53.25	74.50	
	2310	3/4" pipe size		9.40	.851		10.85	35.50	6.90	53.25	74.50	
	2330	1" pipe size		7.10	1.127		19.25	47.50	9.10	75.85	104	
	2350	1-1/4" pipe size		7.10	1.127		19.25	47.50	9.10	75.85	104	
	2370	1-1/2" pipe size		7.10	1.127		19.25	47.50	9.10	75.85	104	
	2390	2" pipe size	↓	3.60	2.222	↓	24.50	93.50	17.95	135.95	191	

16132 | Conduit & Tubing

		16132 Conduit & Tubing	CREW	DAILY OUTPUT	LABOR-HOURS	UNIT	2006 BARE COSTS MAT.	LABOR	EQUIP.	TOTAL	TOTAL INCL O&P
260	2410	2-1/2" pipe size	R-31	3.60	2.222	Ea.	24.50	93.50	17.95	135.95	191
	2430	3" pipe size		3.60	2.222		24.50	93.50	17.95	135.95	191
	2450	3-1/2" pipe size		2.80	2.857		29.50	120	23	172.50	243
	2470	4" pipe size		2.50	3.200		29.50	134	26	189.50	268
	2480										
	3000	Knockouts to 8' high, metal boxes & enclosures									
	3020	With hole saw, 1/2" pipe size	1 Elec	53	.151	Ea.		6.35		6.35	9.75
	3040	3/4" pipe size		47	.170			7.15		7.15	11
	3050	1" pipe size		40	.200			8.40		8.40	12.95
	3060	1-1/4" pipe size		36	.222			9.35		9.35	14.40
	3070	1-1/2" pipe size		32	.250			10.50		10.50	16.20
	3080	2" pipe size		27	.296			12.45		12.45	19.15
	3090	2-1/2" pipe size		20	.400			16.80		16.80	26
	4010	3" pipe size		16	.500			21		21	32.50
	4030	3-1/2" pipe size		13	.615			26		26	40
	4050	4" pipe size		11	.727			30.50		30.50	47
	4070	With hand punch set, 1/2" pipe size		40	.200			8.40		8.40	12.95
	4090	3/4" pipe size		32	.250			10.50		10.50	16.20
	4110	1" pipe size		30	.267			11.20		11.20	17.25
	4130	1-1/4" pipe size		28	.286			12		12	18.50
	4150	1-1/2" pipe size		26	.308			12.90		12.90	19.90
	4170	2" pipe size		20	.400			16.80		16.80	26
	4190	2-1/2" pipe size		17	.471			19.75		19.75	30.50
	4200	3" pipe size		15	.533			22.50		22.50	34.50
	4220	3-1/2" pipe size		12	.667			28		28	43
	4240	4" pipe size		10	.800			33.50		33.50	52
	4260	With hydraulic punch, 1/2" pipe size		44	.182			7.65		7.65	11.75
	4280	3/4" pipe size		38	.211			8.85		8.85	13.60
	4300	1" pipe size		38	.211			8.85		8.85	13.60
	4320	1-1/4" pipe size		38	.211			8.85		8.85	13.60
	4340	1-1/2" pipe size		38	.211			8.85		8.85	13.60
	4360	2" pipe size		32	.250			10.50		10.50	16.20
	4380	2-1/2" pipe size		27	.296			12.45		12.45	19.15
	4400	3" pipe size		23	.348			14.60		14.60	22.50
	4420	3-1/2" pipe size		20	.400			16.80		16.80	26

16133 | Multi-outlet Assemblies

		16133 Multi-outlet Assemblies	CREW	DAILY OUTPUT	LABOR-HOURS	UNIT	MAT.	LABOR	EQUIP.	TOTAL	TOTAL INCL O&P
800	0010	SURFACE RACEWAY									
	0090	Metal, straight section									
	0100	No. 500	1 Elec	100	.080	L.F.	.81	3.36		4.17	6.10
	0110	No. 700		100	.080		.91	3.36		4.27	6.20
	0400	No. 1500, small pancake		90	.089		1.65	3.73		5.38	7.55
	0600	No. 2000, base & cover, blank		90	.089		1.61	3.73		5.34	7.50
	0800	No. 3000, base & cover, blank		75	.107		3.25	4.48		7.73	10.50
	1000	No. 4000, base & cover, blank		65	.123		5.25	5.15		10.40	13.75
	1200	No. 6000, base & cover, blank		50	.160		8.80	6.70		15.50	20
	2400	Fittings, elbows, No. 500		40	.200	Ea.	1.47	8.40		9.87	14.55
	2800	Elbow cover, No. 2000		40	.200		2.80	8.40		11.20	16.05
	2880	Tee, No. 500		42	.190		2.82	8		10.82	15.40
	2900	No. 2000		27	.296		8.80	12.45		21.25	29
	3000	Switch box, No. 500		16	.500		9.55	21		30.55	43
	3400	Telephone outlet, No. 1500		16	.500		10.60	21		31.60	44
	3600	Junction box, No. 1500		16	.500		7.40	21		28.40	40.50
	3800	Plugmold wired sections, No. 2000									
	4000	1 circuit, 6 outlets, 3 ft. long	1 Elec	8	1	Ea.	27	42		69	94

16133 | Multi-outlet Assemblies

			CREW	DAILY OUTPUT	LABOR-HOURS	UNIT	2006 BARE COSTS MAT.	LABOR	EQUIP.	TOTAL	TOTAL INCL O&P	
800	4100	2 circuits, 8 outlets, 6 ft. long	1 Elec	5.30	1.509	Ea.	45	63.50		108.50	147	800
	9300	Non-metallic, straight section										
	9310	7/16″ x 7/8″, base & cover, blank	1 Elec	160	.050	L.F.	1.25	2.10		3.35	4.62	
	9320	Base & cover w/ adhesive		160	.050		1.38	2.10		3.48	4.76	
	9340	7/16″ x 1-5/16″, base & cover, blank		145	.055		1.38	2.32		3.70	5.10	
	9350	Base & cover w/ adhesive		145	.055		1.60	2.32		3.92	5.35	
	9370	11/16″ x 2-1/4″, base & cover, blank		130	.062		1.94	2.58		4.52	6.10	
	9380	Base & cover w/ adhesive		130	.062		2.24	2.58		4.82	6.45	
	9400	Fittings, elbows, 7/16″ x 7/8″		50	.160	Ea.	1.38	6.70		8.08	11.85	
	9410	7/16″ x 1-5/16″		45	.178		1.44	7.45		8.89	13.10	
	9420	11/16″ x 2-1/4″		40	.200		1.56	8.40		9.96	14.65	
	9430	Tees, 7/16″ x 7/8″		35	.229		1.82	9.60		11.42	16.80	
	9440	7/16″ x 1-5/16″		32	.250		1.87	10.50		12.37	18.25	
	9450	11/16″ x 2-1/4″		30	.267		1.89	11.20		13.09	19.35	
	9460	Cover clip, 7/16″ x 7/8″		80	.100		.36	4.20		4.56	6.85	
	9470	7/16″ x 1-5/16″		72	.111		.33	4.67		5	7.55	
	9480	11/16″ x 2-1/4″		64	.125		.55	5.25		5.80	8.70	
	9490	Blank end, 7/16″ x 7/8″		50	.160		.52	6.70		7.22	10.90	
	9510	11/16″ x 2-1/4″		40	.200		.88	8.40		9.28	13.90	
	9520	Round fixture box 5.5″ dia x 1″		25	.320		8.70	13.45		22.15	30	
	9530	Device box, 1 gang		30	.267		3.85	11.20		15.05	21.50	
	9540	2 gang		25	.320		5.75	13.45		19.20	27	
	9990	Minimum labor/equipment charge		5	1.600	Job		67		67	104	

16134 | Wireway & Aux Gutters

			CREW	DAILY OUTPUT	LABOR-HOURS	UNIT	MAT.	LABOR	EQUIP.	TOTAL	TOTAL INCL O&P	
150	0010	**WIREWAY** to 15′ high										150
	0100	Screw cover, NEMA 1 w/ fittings and supports, 2-1/2″ x 2-1/2″	1 Elec	45	.178	L.F.	10.95	7.45		18.40	23.50	
	0200	4″ x 4″	″	40	.200		11.60	8.40		20	26	
	0400	6″ x 6″	2 Elec	60	.267		19.70	11.20		30.90	39	
	0600	8″ x 8″	″	40	.400		33	16.80		49.80	62.50	
	4475	Screw cover, NEMA 3R w/ fittings and supports, 4″ x 4″	1 Elec	36	.222		24.50	9.35		33.85	41.50	
	4480	6″ x 6″	2 Elec	55	.291		33	12.20		45.20	55	
	4485	8″ x 8″		36	.444		53	18.65		71.65	87	
	4490	12″ x 12″		18	.889		73.50	37.50		111	139	

16136 | Boxes

			CREW	DAILY OUTPUT	LABOR-HOURS	UNIT	MAT.	LABOR	EQUIP.	TOTAL	TOTAL INCL O&P	
600	0010	**OUTLET BOXES**										600
	0020	Pressed steel, octagon, 4″	1 Elec	20	.400	Ea.	2.21	16.80		19.01	28.50	
	0060	Covers, blank		64	.125		.92	5.25		6.17	9.10	
	0100	Extension rings		40	.200		3.64	8.40		12.04	16.95	
	0152	Square 4″		18	.444		2.24	18.65		20.89	31.50	
	0200	Extension rings		40	.200		3.71	8.40		12.11	17.05	
	0250	Covers, blank		64	.125		1.04	5.25		6.29	9.25	
	0300	Plaster rings		64	.125		2.03	5.25		7.28	10.35	
	0652	Switchbox		24	.333		3.54	14		17.54	25.50	
	1102	Concrete, floor, 1 gang		4.80	1.667		67.50	70		137.50	182	
	9000	Minimum labor/equipment charge		4	2	Job		84		84	129	
620	0010	**OUTLET BOXES, PLASTIC**										620
	0050	4″ diameter, round with 2 mounting nails	1 Elec	25	.320	Ea.	2.48	13.45		15.93	23	
	0102	Bar hanger mounted		23	.348		4.52	14.60		19.12	27.50	
	0202	Square with 2 mounting nails		23	.348		3.72	14.60		18.32	26.50	
	0300	Plaster ring		64	.125		1.55	5.25		6.80	9.80	
	0402	Switch box with 2 mounting nails, 1 gang		27	.296		1.69	12.45		14.14	21	

16100 | Wiring Methods

	16136 Boxes		Crew	Daily Output	Labor-Hours	Unit	2006 Bare Costs Mat.	Labor	Equip.	Total	Total Incl O&P	
620	0502	2 gang	1 Elec	23	.348	Ea.	3.45	14.60		18.05	26.50	620
	0602	3 gang		18	.444		5.45	18.65		24.10	35	
	0702	Old work box		27	.296		3.32	12.45		15.77	23	
	9000	Minimum labor/equipment charge		4	2	Job		84		84	129	
700	0010	PULL BOXES & CABINETS										700
	0100	Sheet metal, pull box, NEMA 1, type SC, 6" W x 6" H x 4" D	1 Elec	8	1	Ea.	10.80	42		52.80	76.50	
	0200	8" W x 8" H x 4" D		8	1		14.80	42		56.80	81	
	0300	10" W x 12" H x 6" D		5.30	1.509		26	63.50		89.50	127	
	0400	16" W x 20" H x 8" D		4	2		98.50	84		182.50	237	
	0500	20" W x 24" H x 8" D		3.20	2.500		115	105		220	289	
	0600	24" W x 36" H x 8" D		2.70	2.963		163	124		287	370	
	0650	Hinged cabinets, NEMA 1, 6" W x 6" H x 4" D		8	1		11	42		53	76.50	
	0802	12" W x 16" H x 6" D		4	2		36	84		120	169	
	1000	20" W x 20" H x 6" D		3.60	2.222		72.50	93.50		166	224	
	1200	20" W x 20" H x 8" D		3.20	2.500		145	105		250	320	
	1400	24" W x 36" H x 8" D		2.70	2.963		233	124		357	450	
	1600	24" W x 42" H x 8" D		2	4		355	168		523	650	
	7000	Cabinets, current transformer										
	7050	Single door, 24" H x 24" W x 10" D	1 Elec	1.60	5	Ea.	139	210		349	480	
	7100	30" H x 24" W x 10" D		1.30	6.154		153	258		411	570	
	7150	36" H x 24" W x 10" D		1.10	7.273		161	305		466	645	
	7200	30" H x 30" W x 10" D		1	8		167	335		502	705	
	7250	36" H x 30" W x 10" D		.90	8.889		230	375		605	830	
	7300	36" H x 36" W x 10" D		.80	10		244	420		664	915	
	7500	Double door, 48" H x 36" W x 10" D		.60	13.333		440	560		1,000	1,350	
	7550	24" H x 24" W x 12" D		1	8		230	335		565	775	
	9990	Minimum labor/equipment charge		2	4	Job		168		168	259	

	16139 Residential Wiring		Crew	Daily Output	Labor-Hours	Unit	Mat.	Labor	Equip.	Total	Total Incl O&P	
700	0010	RESIDENTIAL WIRING										700
	0020	20' avg. runs and #14/2 wiring incl. unless otherwise noted										
	1000	Service & panel, includes 24' SE-AL cable, service eye, meter,										
	1010	Socket, panel board, main bkr., ground rod, 15 or 20 amp										
	1020	1-pole circuit breakers, and misc. hardware										
	1100	100 amp, with 10 branch breakers	1 Elec	1.19	6.723	Ea.	490	282		772	975	
	1110	With PVC conduit and wire		.92	8.696		535	365		900	1,150	
	1120	With RGS conduit and wire		.73	10.959		710	460		1,170	1,500	
	1150	150 amp, with 14 branch breakers		1.03	7.767		760	325		1,085	1,350	
	1170	With PVC conduit and wire		.82	9.756		850	410		1,260	1,575	
	1180	With RGS conduit and wire		.67	11.940		1,200	500		1,700	2,075	
	1200	200 amp, with 18 branch breakers	2 Elec	1.80	8.889		995	375		1,370	1,675	
	1220	With PVC conduit and wire		1.46	10.959		1,100	460		1,560	1,900	
	1230	With RGS conduit and wire		1.24	12.903		1,525	540		2,065	2,525	
	1800	Lightning surge suppressor for above services, add	1 Elec	32	.250		43.50	10.50		54	64	
	2000	Switch devices										
	2100	Single pole, 15 amp, Ivory, with a 1-gang box, cover plate,										
	2110	Type NM (Romex) cable	1 Elec	17.10	.468	Ea.	8.45	19.65		28.10	40	
	2120	Type MC (BX) cable		14.30	.559		22	23.50		45.50	60	
	2130	EMT & wire		5.71	1.401		28	59		87	122	
	2150	3-way, #14/3, type NM cable		14.55	.550		12.10	23		35.10	49	
	2170	Type MC cable		12.31	.650		30	27.50		57.50	75.50	
	2180	EMT & wire		5	1.600		30.50	67		97.50	138	
	2200	4-way, #14/3, type NM cable		14.55	.550		25.50	23		48.50	64	
	2220	Type MC cable		12.31	.650		44	27.50		71.50	90	
	2230	EMT & wire		5	1.600		44	67		111	153	

16139 | Residential Wiring

				DAILY	LABOR-		2006 BARE COSTS				TOTAL	
			CREW	OUTPUT	HOURS	UNIT	MAT.	LABOR	EQUIP.	TOTAL	INCL O&P	
700	2250	S.P., 20 amp, #12/2, type NM cable	1 Elec	13.33	.600	Ea.	14.50	25		39.50	55	700
	2270	Type MC cable		11.43	.700		26.50	29.50		56	74.50	
	2280	EMT & wire		4.85	1.649		34.50	69.50		104	145	
	2290	S.P. rotary dimmer, 600W, no wiring		17	.471		16.90	19.75		36.65	49	
	2300	S.P. rotary dimmer, 600W, type NM cable		14.55	.550		20.50	23		43.50	58	
	2320	Type MC cable		12.31	.650		34	27.50		61.50	79.50	
	2330	EMT & wire		5	1.600		41	67		108	149	
	2350	3-way rotary dimmer, type NM cable		13.33	.600		18.30	25		43.30	59	
	2370	Type MC cable		11.43	.700		31.50	29.50		61	80.50	
	2380	EMT & wire		4.85	1.649		39	69.50		108.50	150	
	2400	Interval timer wall switch, 20 amp, 1-30 min., #12/2										
	2410	Type NM cable	1 Elec	14.55	.550	Ea.	40	23		63	79	
	2420	Type MC cable		12.31	.650		47.50	27.50		75	94	
	2430	EMT & wire		5	1.600		60	67		127	170	
	2500	Decorator style										
	2510	S.P., 15 amp, type NM cable	1 Elec	17.10	.468	Ea.	12	19.65		31.65	43.50	
	2520	Type MC cable		14.30	.559		25.50	23.50		49	64	
	2530	EMT & wire		5.71	1.401		31.50	59		90.50	126	
	2550	3-way, #14/3, type NM cable		14.55	.550		15.65	23		38.65	53	
	2570	Type MC cable		12.31	.650		34	27.50		61.50	79	
	2580	EMT & wire		5	1.600		34	67		101	142	
	2600	4-way, #14/3, type NM cable		14.55	.550		29.50	23		52.50	67.50	
	2620	Type MC cable		12.31	.650		47.50	27.50		75	94	
	2630	EMT & wire		5	1.600		47.50	67		114.50	157	
	2650	S.P., 20 amp, #12/2, type NM cable		13.33	.600		18.05	25		43.05	59	
	2670	Type MC cable		11.43	.700		30	29.50		59.50	78.50	
	2680	EMT & wire		4.85	1.649		38	69.50		107.50	149	
	2700	S.P., slide dimmer, type NM cable		17.10	.468		27	19.65		46.65	60.50	
	2720	Type MC cable		14.30	.559		40.50	23.50		64	80.50	
	2730	EMT & wire		5.71	1.401		47.50	59		106.50	143	
	2770	Type MC cable		14.30	.559		37.50	23.50		61	77	
	2780	EMT & wire		5.71	1.401		44.50	59		103.50	140	
	2800	3-way touch dimmer, type NM cable		13.33	.600		44	25		69	87.50	
	2820	Type MC cable		11.43	.700		57.50	29.50		87	109	
	2830	EMT & wire		4.85	1.649		64.50	69.50		134	178	
	3100	S.P. switch/15 amp recpt., Ivory, 1-gang box, plate										
	3110	Type NM cable	1 Elec	11.43	.700	Ea.	17.50	29.50		47	65	
	3120	Type MC cable		10	.800		31	33.50		64.50	86	
	3130	EMT & wire		4.40	1.818		38	76.50		114.50	160	
	3150	S.P. switch/pilot light, type NM cable		11.43	.700		18.20	29.50		47.70	65.50	
	3170	Type MC cable		10	.800		31.50	33.50		65	87	
	3180	EMT & wire		4.43	1.806		38.50	76		114.50	160	
	3190	2-S.P. switches, 2-#14/2, no wiring		14	.571		7	24		31	44.50	
	3200	2-S.P. switches, 2-#14/2, type NM cables		10	.800		20.50	33.50		54	74.50	
	3220	Type MC cable		8.89	.900		42	38		80	104	
	3230	EMT & wire		4.10	1.951		40	82		122	170	
	3250	3-way switch/15 amp recpt., #14/3, type NM cable		10	.800		24.50	33.50		58	79	
	3270	Type MC cable		8.89	.900		42.50	38		80.50	105	
	3280	EMT & wire		4.10	1.951		43	82		125	173	
	3300	2-3 way switches, 2-#14/3, type NM cables		8.89	.900		32.50	38		70.50	94	
	3320	Type MC cable		8	1		63.50	42		105.50	135	
	3330	EMT & wire		4	2		47.50	84		131.50	182	
	3350	S.P. switch/20 amp recpt., #12/2, type NM cable		10	.800		28	33.50		61.50	82.50	
	3370	Type MC cable		8.89	.900		35.50	38		73.50	97	
	3380	EMT & wire		4.10	1.951		48	82		130	179	
	3400	Decorator style										

	16139 | Residential Wiring	CREW	DAILY OUTPUT	LABOR-HOURS	UNIT	2006 BARE COSTS MAT.	LABOR	EQUIP.	TOTAL	TOTAL INCL O&P
700 3410	S.P. switch/15 amp recpt., type NM cable	1 Elec	11.43	.700	Ea.	21	29.50		50.50	68.50 700
3420	Type MC cable		10	.800		34.50	33.50		68	90
3430	EMT & wire		4.40	1.818		41.50	76.50		118	164
3450	S.P. switch/pilot light, type NM cable		11.43	.700		21.50	29.50		51	69.50
3470	Type MC cable		10	.800		35	33.50		68.50	90.50
3480	EMT & wire		4.40	1.818		42	76.50		118.50	165
3500	2-S.P. switches, 2-#14/2, type NM cables		10	.800		24	33.50		57.50	78.50
3520	Type MC cable		8.89	.900		45.50	38		83.50	108
3530	EMT & wire		4.10	1.951		43.50	82		125.50	174
3550	3-way/15 amp recpt., #14/3, type NM cable		10	.800		28	33.50		61.50	83
3580	EMT & wire		4.10	1.951		46.50	82		128.50	177
3650	2-3 way switches, 2-#14/3, type NM cables		8.89	.900		36	38		74	98
3670	Type MC cable		8	1		67	42		109	139
3680	EMT & wire		4	2		51.50	84		135.50	186
3700	S.P. switch/20 amp recpt., #12/2, type NM cable		10	.800		31.50	33.50		65	86.50
3720	Type MC cable		8.89	.900		39	38		77	101
3730	EMT & wire		4.10	1.951		51.50	82		133.50	183
4000	Receptacle devices									
4010	Duplex outlet, 15 amp recpt., Ivory, 1-gang box, plate									
4015	Type NM cable	1 Elec	14.55	.550	Ea.	6.80	23		29.80	43
4020	Type MC cable		12.31	.650		20	27.50		47.50	64.50
4030	EMT & wire		5.33	1.501		26.50	63		89.50	126
4050	With #12/2, type NM cable		12.31	.650		8.55	27.50		36.05	51.50
4070	Type MC cable		10.67	.750		20.50	31.50		52	71
4080	EMT & wire		4.71	1.699		28.50	71.50		100	142
4100	20 amp recpt., #12/2, type NM cable		12.31	.650		16.20	27.50		43.70	60
4120	Type MC cable		10.67	.750		28	31.50		59.50	79.50
4130	EMT & wire		4.71	1.699		36.50	71.50		108	150
4140	For GFI see line 4300 below									
4150	Decorator style, 15 amp recpt., type NM cable	1 Elec	14.55	.550	Ea.	10.35	23		33.35	47
4170	Type MC cable		12.31	.650		24	27.50		51.50	68
4180	EMT & wire		5.33	1.501		30	63		93	130
4200	With #12/2, type NM cable		12.31	.650		12.10	27.50		39.60	55.50
4220	Type MC cable		10.67	.750		24	31.50		55.50	75
4230	EMT & wire		4.71	1.699		32	71.50		103.50	146
4250	20 amp recpt. #12/2, type NM cable		12.31	.650		19.75	27.50		47.25	63.50
4270	Type MC cable		10.67	.750		31.50	31.50		63	83.50
4280	EMT & wire		4.71	1.699		40	71.50		111.50	154
4300	GFI, 15 amp recpt., type NM cable		12.31	.650		35.50	27.50		63	81
4320	Type MC cable		10.67	.750		48.50	31.50		80	102
4330	EMT & wire		4.71	1.699		55	71.50		126.50	171
4350	GFI with #12/2, type NM cable		10.67	.750		37	31.50		68.50	89
4370	Type MC cable		9.20	.870		49	36.50		85.50	111
4380	EMT & wire		4.21	1.900		57	80		137	186
4400	20 amp recpt., #12/2 type NM cable		10.67	.750		38.50	31.50		70	91
4420	Type MC cable		9.20	.870		50.50	36.50		87	112
4430	EMT & wire		4.21	1.900		59	80		139	188
4500	Weather-proof cover for above receptacles, add		32	.250		4.40	10.50		14.90	21
4550	Air conditioner outlet, 20 amp-240 volt recpt.									
4560	30′ of #12/2, 2 pole circuit breaker									
4570	Type NM cable	1 Elec	10	.800	Ea.	48.50	33.50		82	106
4580	Type MC cable		9	.889		63.50	37.50		101	128
4590	EMT & wire		4	2		67.50	84		151.50	204
4600	Decorator style, type NM cable		10	.800		52.50	33.50		86	110
4620	Type MC cable		9	.889		67.50	37.50		105	132
4630	EMT & wire		4	2		72	84		156	208

	16139 | Residential Wiring	Crew	Daily Output	Labor-Hours	Unit	2006 Bare Costs Mat.	2006 Bare Costs Labor	2006 Bare Costs Equip.	2006 Bare Costs Total	Total Incl O&P	
700 4650	Dryer outlet, 30 amp-240 volt recpt., 20′ of #10/3										700
4660	2 pole circuit breaker										
4670	Type NM cable	1 Elec	6.41	1.248	Ea.	57.50	52.50		110	145	
4680	Type MC cable		5.71	1.401		70.50	59		129.50	168	
4690	EMT & wire		3.48	2.299		71.50	96.50		168	228	
4700	Range outlet, 50 amp-240 volt recpt., 30′ of #8/3										
4710	Type NM cable	1 Elec	4.21	1.900	Ea.	89	80		169	221	
4720	Type MC cable		4	2		133	84		217	275	
4730	EMT & wire		2.96	2.703		91	114		205	275	
4750	Central vacuum outlet, Type NM cable		6.40	1.250		53.50	52.50		106	140	
4770	Type MC cable		5.71	1.401		73.50	59		132.50	171	
4780	EMT & wire		3.48	2.299		75	96.50		171.50	232	
4800	30 amp-110 volt locking recpt., #10/2 circ. bkr.										
4810	Type NM cable	1 Elec	6.20	1.290	Ea.	60	54		114	150	
4820	Type MC cable		5.40	1.481		85.50	62		147.50	190	
4830	EMT & wire		3.20	2.500		82.50	105		187.50	253	
4900	Low voltage outlets										
4910	Telephone recpt., 20′ of 4/C phone wire	1 Elec	26	.308	Ea.	7.95	12.90		20.85	28.50	
4920	TV recpt., 20′ of RG59U coax wire, F type connector	″	16	.500	″	13.20	21		34.20	47	
4950	Door bell chime, transformer, 2 buttons, 60′ of bellwire										
4970	Economy model	1 Elec	11.50	.696	Ea.	55	29		84	106	
4980	Custom model		11.50	.696		86	29		115	140	
4990	Luxury model, 3 buttons		9.50	.842		232	35.50		267.50	310	
6000	Lighting outlets										
6050	Wire only (for fixture), type NM cable	1 Elec	32	.250	Ea.	4.71	10.50		15.21	21.50	
6070	Type MC cable		24	.333		14.30	14		28.30	37	
6080	EMT & wire		10	.800		19.35	33.50		52.85	73.50	
6100	Box (4″), and wire (for fixture), type NM cable		25	.320		10.45	13.45		23.90	32	
6120	Type MC cable		20	.400		20	16.80		36.80	48	
6130	EMT & wire		11	.727		25	30.50		55.50	74.50	
6200	Fixtures (use with lines 6050 or 6100 above)										
6210	Canopy style, economy grade	1 Elec	40	.200	Ea.	26	8.40		34.40	41.50	
6220	Custom grade		40	.200		48	8.40		56.40	66	
6250	Dining room chandelier, economy grade		19	.421		78	17.70		95.70	113	
6270	Luxury grade		15	.533		510	22.50		532.50	595	
6310	Kitchen fixture (fluorescent), economy grade		30	.267		57	11.20		68.20	80	
6320	Custom grade		25	.320		165	13.45		178.45	203	
6350	Outdoor, wall mounted, economy grade		30	.267		27.50	11.20		38.70	48	
6360	Custom grade		30	.267		104	11.20		115.20	131	
6370	Luxury grade		25	.320		235	13.45		248.45	280	
6410	Outdoor PAR floodlights, 1 lamp, 150 watt		20	.400		26.50	16.80		43.30	55	
6420	2 lamp, 150 watt each		20	.400		43	16.80		59.80	73.50	
6430	For infrared security sensor, add		32	.250		90	10.50		100.50	115	
6450	Outdoor, quartz-halogen, 300 watt flood		20	.400		39	16.80		55.80	69	
6600	Recessed downlight, round, pre-wired, 50 or 75 watt trim		30	.267		36	11.20		47.20	57	
6610	With shower light trim		30	.267		45	11.20		56.20	67	
6620	With wall washer trim		28	.286		54	12		66	78	
6630	With eye-ball trim		28	.286		54	12		66	78	
6640	For direct contact with insulation, add					1.80			1.80	1.98	
6700	Porcelain lamp holder	1 Elec	40	.200		3.80	8.40		12.20	17.15	
6710	With pull switch		40	.200		4.13	8.40		12.53	17.50	
6750	Fluorescent strip, 1-20 watt tube, wrap around diffuser, 24″		24	.333		53	14		67	80	
6760	1-40 watt tube, 48″		24	.333		65	14		79	93	
6770	2-40 watt tubes, 48″		20	.400		79	16.80		95.80	113	
6780	With residential ballast		20	.400		89.50	16.80		106.30	125	
6800	Bathroom heat lamp, 1-250 watt		28	.286		38.50	12		50.50	61	

ELECTRICAL 16

16139 | Residential Wiring		CREW	DAILY OUTPUT	LABOR-HOURS	UNIT	2006 BARE COSTS MAT.	LABOR	EQUIP.	TOTAL	TOTAL INCL O&P	
700 6810	2-250 watt lamps	1 Elec	28	.286	Ea.	63	12		75	88	700
6820	For timer switch, see line 2400										
6900	Outdoor post lamp, incl. post, fixture, 35′ of #14/2										
6910	Type NMC cable	1 Elec	3.50	2.286	Ea.	186	96		282	350	
6920	Photo-eye, add		27	.296		30	12.45		42.45	52	
6950	Clock dial time switch, 24 hr., w/enclosure, type NM cable		11.43	.700		56.50	29.50		86	108	
6970	Type MC cable		11	.727		70	30.50		100.50	124	
6980	EMT & wire		4.85	1.649		76	69.50		145.50	191	
7000	Alarm systems										
7050	Smoke detectors, box, #14/3, type NM cable	1 Elec	14.55	.550	Ea.	30	23		53	68.50	
7070	Type MC cable		12.31	.650		44	27.50		71.50	90.50	
7080	EMT & wire		5	1.600		44	67		111	153	
7090	For relay output to security system, add					12.15			12.15	13.35	
8000	Residential equipment										
8050	Disposal hook-up, incl. switch, outlet box, 3′ of flex										
8060	20 amp-1 pole circ. bkr., and 25′ of #12/2										
8070	Type NM cable	1 Elec	10	.800	Ea.	24.50	33.50		58	79	
8080	Type MC cable		8	1		38	42		80	107	
8090	EMT & wire		5	1.600		47.50	67		114.50	156	
8100	Trash compactor or dishwasher hook-up, incl. outlet box,										
8110	3′ of flex, 15 amp-1 pole circ. bkr., and 25′ of #14/2										
8120	Type NM cable	1 Elec	10	.800	Ea.	17.85	33.50		51.35	71.50	
8130	Type MC cable		8	1		33.50	42		75.50	102	
8140	EMT & wire		5	1.600		41	67		108	150	
8150	Hot water sink dispensor hook-up, use line 8100										
8200	Vent/exhaust fan hook-up, type NM cable	1 Elec	32	.250	Ea.	4.71	10.50		15.21	21.50	
8220	Type MC cable		24	.333		14.30	14		28.30	37	
8230	EMT & wire		10	.800		19.35	33.50		52.85	73.50	
8250	Bathroom vent fan, 50 CFM (use with above hook-up)										
8260	Economy model	1 Elec	15	.533	Ea.	22.50	22.50		45	59.50	
8270	Low noise model		15	.533		32	22.50		54.50	69.50	
8280	Custom model		12	.667		117	28		145	172	
8300	Bathroom or kitchen vent fan, 110 CFM										
8310	Economy model	1 Elec	15	.533	Ea.	60.50	22.50		83	101	
8320	Low noise model	"	15	.533	"	79	22.50		101.50	122	
8350	Paddle fan, variable speed (w/o lights)										
8360	Economy model (AC motor)	1 Elec	10	.800	Ea.	105	33.50		138.50	168	
8370	Custom model (AC motor)		10	.800		182	33.50		215.50	252	
8380	Luxury model (DC motor)		8	1		360	42		402	460	
8390	Remote speed switch for above, add		12	.667		26	28		54	71.50	
8500	Whole house exhaust fan, ceiling mount, 36″, variable speed										
8510	Remote switch, incl. shutters, 20 amp-1 pole circ. bkr.										
8520	30′ of #12/2, type NM cable	1 Elec	4	2	Ea.	685	84		769	885	
8530	Type MC cable		3.50	2.286		705	96		801	925	
8540	EMT & wire		3	2.667		715	112		827	960	
8600	Whirlpool tub hook-up, incl. timer switch, outlet box										
8610	3′ of flex, 20 amp-1 pole GFI circ. bkr.										
8620	30′ of #12/2, type NM cable	1 Elec	5	1.600	Ea.	100	67		167	214	
8630	Type MC cable		4.20	1.905		109	80		189	243	
8640	EMT & wire		3.40	2.353		118	99		217	281	
8650	Hot water heater hook-up, incl. 1-2 pole circ. bkr., box;										
8660	3′ of flex, 20′ of #10/2, type NM cable	1 Elec	5	1.600	Ea.	26.50	67		93.50	133	
8670	Type MC cable		4.20	1.905		44	80		124	171	
8680	EMT & wire		3.40	2.353		42.50	99		141.50	199	
9000	Heating/air conditioning										
9050	Furnace/boiler hook-up, incl. firestat, local on-off switch										

		16139 | Residential Wiring	CREW	DAILY OUTPUT	LABOR-HOURS	UNIT	2006 BARE COSTS MAT.	LABOR	EQUIP.	TOTAL	TOTAL INCL O&P	
700	9060	Emergency switch, and 40′ of type NM cable	1 Elec	4	2	Ea.	46	84		130	180	700
	9070	Type MC cable	↓	3.50	2.286	↓	67.50	96		163.50	223	
	9080	EMT & wire	↓	1.50	5.333	↓	79.50	224		303.50	435	
	9100	Air conditioner hook-up, incl. local 60 amp disc. switch										
	9110	3′ sealtite, 40 amp, 2 pole circuit breaker										
	9130	40′ of #8/2, type NM cable	1 Elec	3.50	2.286	Ea.	169	96		265	335	
	9140	Type MC cable	↓	3	2.667	↓	240	112		352	435	
	9150	EMT & wire	↓	1.30	6.154	↓	190	258		448	610	
	9200	Heat pump hook-up, 1-40 & 1-100 amp 2 pole circ. bkr.										
	9210	Local disconnect switch, 3′ sealtite										
	9220	40′ of #8/2 & 30′ of #3/2										
	9230	Type NM cable	1 Elec	1.30	6.154	Ea.	425	258		683	870	
	9240	Type MC cable	↓	1.08	7.407	↓	585	310		895	1,125	
	9250	EMT & wire	↓	.94	8.511	↓	460	355		815	1,050	
	9500	Thermostat hook-up, using low voltage wire										
	9520	Heating only	1 Elec	24	.333	Ea.	6	14		20	28	
	9530	Heating/cooling	"	20	.400	"	7.15	16.80		23.95	34	

		16140 | Wiring Devices	CREW	DAILY OUTPUT	LABOR-HOURS	UNIT	MAT.	LABOR	EQUIP.	TOTAL	TOTAL INCL O&P	
910	0010	WIRING DEVICES										910
	0200	Toggle switch, quiet type, single pole, 15 amp	1 Elec	40	.200	Ea.	4.88	8.40		13.28	18.30	
	0500	20 amp	↓	27	.296	↓	7.15	12.45		19.60	27	
	0550	Rocker, 15 amp		40	.200		5.40	8.40		13.80	18.90	
	0560	20 amp		27	.296		12.70	12.45		25.15	33	
	0600	3 way, 15 amp		23	.348		7	14.60		21.60	30	
	0850	Rocker, 15 amp		23	.348		7.60	14.60		22.20	31	
	0860	20 amp		18	.444		18.40	18.65		37.05	49	
	0900	4 way, 15 amp		15	.533		21	22.50		43.50	57.50	
	1030	Rocker, 15 amp		15	.533		29.50	22.50		52	67	
	1040	20 amp		11	.727		45	30.50		75.50	96.50	
	1650	Dimmer switch, 120 volt, incandescent, 600 watt, 1 pole		16	.500		11.35	21		32.35	45	
	2460	Receptacle, duplex, 120 volt, grounded, 15 amp		40	.200		1.17	8.40		9.57	14.25	
	2470	20 amp		27	.296		8.85	12.45		21.30	29	
	2490	Dryer, 30 amp		15	.533		10.80	22.50		33.30	46.50	
	2500	Range, 50 amp		11	.727		11.40	30.50		41.90	59.50	
	2600	Wall plates, stainless steel, 1 gang		80	.100		1.80	4.20		6	8.45	
	2800	2 gang		53	.151		4.10	6.35		10.45	14.25	
	3200	Lampholder, keyless		26	.308		9.70	12.90		22.60	30.50	
	3400	Pullchain with receptacle		22	.364	↓	9.20	15.25		24.45	33.50	
	9000	Minimum labor/equipment charge	↓	4	2	Job		84		84	129	

		16150 | Wiring Connections	CREW	DAILY OUTPUT	LABOR-HOURS	UNIT	MAT.	LABOR	EQUIP.	TOTAL	TOTAL INCL O&P	
275	0010	MOTOR CONNECTIONS										275
	0020	Flexible conduit and fittings, 115 volt, 1 phase, up to 1 HP motor	1 Elec	8	1	Ea.	4.97	42		46.97	70	
	9000	Minimum labor/equipment charge	"	4	2	Job		84		84	129	

16200 | Electrical Power

	16210 | Electrical Utility Services		CREW	DAILY OUTPUT	LABOR-HOURS	UNIT	2006 BARE COSTS MAT.	LABOR	EQUIP.	TOTAL	TOTAL INCL O&P	
600	0010	METER CENTERS AND SOCKETS										600
	0100	Sockets, single position, 4 terminal, 100 amp	1 Elec	3.20	2.500	Ea.	36	105		141	202	
	0200	150 amp		2.30	3.478		40	146		186	270	
	0300	200 amp		1.90	4.211		53.50	177		230.50	330	
	0400	20 amp		3.20	2.500		83	105		188	253	
	0500	Double position, 4 terminal, 100 amp		2.80	2.857		141	120		261	340	
	0600	150 amp		2.10	3.810		160	160		320	420	
	0700	200 amp		1.70	4.706		345	198		543	685	
	9000	Minimum labor/equipment charge		3	2.667	Job		112		112	173	

	16220 | Motors & Generators		CREW	DAILY OUTPUT	LABOR-HOURS	UNIT	MAT.	LABOR	EQUIP.	TOTAL	TOTAL INCL O&P	
900	0010	VARIABLE FREQUENCY DRIVES/ADJUSTABLE FREQUENCY DRIVES										900
	0100	Enclosed (NEMA 1), 460 volt, for 3 HP motor size	1 Elec	.80	10	Ea.	1,500	420		1,920	2,300	
	0110	5 HP motor size		.80	10		1,625	420		2,045	2,450	
	0120	7.5 HP motor size		.67	11.940		1,925	500		2,425	2,900	
	0130	10 HP motor size		.67	11.940		1,925	500		2,425	2,900	
	0140	15 HP motor size	2 Elec	.89	17.978		2,225	755		2,980	3,625	
	0150	20 HP motor size		.89	17.978		3,300	755		4,055	4,800	
	0160	25 HP motor size		.67	23.881		3,825	1,000		4,825	5,775	
	0170	30 HP motor size		.67	23.881		4,675	1,000		5,675	6,700	
	0180	40 HP motor size		.67	23.881		6,875	1,000		7,875	9,100	
	0190	50 HP motor size		.53	30.189		7,475	1,275		8,750	10,200	
	0200	60 HP motor size	R-3	.56	35.714		8,175	1,475	273	9,923	11,600	
	0210	75 HP motor size		.56	35.714		11,200	1,475	273	12,948	14,900	
	0220	100 HP motor size		.50	40		11,400	1,650	305	13,355	15,500	
	0230	125 HP motor size		.50	40		12,700	1,650	305	14,655	16,800	
	0240	150 HP motor size		.50	40		16,200	1,650	305	18,155	20,700	
	0250	200 HP motor size		.42	47.619		20,000	1,975	365	22,340	25,500	
	1100	Custom-engineered, 460 volt, for 3 HP motor size	1 Elec	.56	14.286		2,475	600		3,075	3,650	
	1110	5 HP motor size		.56	14.286		2,600	600		3,200	3,775	
	1120	7.5 HP motor size		.47	17.021		3,125	715		3,840	4,550	
	1130	10 HP motor size		.47	17.021		3,125	715		3,840	4,550	
	1140	15 HP motor size	2 Elec	.62	25.806		3,425	1,075		4,500	5,450	
	1150	20 HP motor size		.62	25.806		4,675	1,075		5,750	6,825	
	1160	25 HP motor size		.47	34.043		5,075	1,425		6,500	7,775	
	1170	30 HP motor size		.47	34.043		6,925	1,425		8,350	9,800	
	1180	40 HP motor size		.47	34.043		7,725	1,425		9,150	10,700	
	1190	50 HP motor size		.37	43.243		7,725	1,825		9,550	11,300	
	1200	60 HP motor size	R-3	.39	51.282		11,500	2,125	390	14,015	16,400	
	1210	75 HP motor size		.39	51.282		13,300	2,125	390	15,815	18,400	
	1220	100 HP motor size		.35	57.143		13,900	2,375	435	16,710	19,500	
	1230	125 HP motor size		.35	57.143		14,700	2,375	435	17,510	20,300	
	1240	150 HP motor size		.35	57.143		16,400	2,375	435	19,210	22,300	
	1250	200 HP motor size		.29	68.966		22,700	2,850	525	26,075	30,000	
	2000	For complex & special design systems to meet specific										
	2010	requirements, obtain quote from vendor.										

	16230 | Generator Assemblies		CREW	DAILY OUTPUT	LABOR-HOURS	UNIT	MAT.	LABOR	EQUIP.	TOTAL	TOTAL INCL O&P	
450	0010	GENERATOR SET										450
	0020	Gas or gasoline operated, includes battery,										
	0050	charger, muffler & transfer switch										
	0200	3 phase 4 wire, 277/480 volt, 7.5 kW	R-3	.83	24.096	Ea.	6,000	1,000	184	7,184	8,350	
	0300	11.5 kW		.71	28.169		8,500	1,175	215	9,890	11,400	
	0400	20 kW		.63	31.746		10,000	1,325	242	11,567	13,300	

16 ELECTRICAL

16200 | Electrical Power

16230 | Generator Assemblies			CREW	DAILY OUTPUT	LABOR-HOURS	UNIT	2006 BARE COSTS MAT.	LABOR	EQUIP.	TOTAL	TOTAL INCL O&P	
450	0500	35 kW	R-3	.55	36.364	Ea.	12,000	1,500	277	13,777	15,800	450

16270 | Transformers			CREW	DAILY OUTPUT	LABOR-HOURS	UNIT	MAT.	LABOR	EQUIP.	TOTAL	TOTAL INCL O&P	
200	0010	DRY TYPE TRANSFORMER										200
	0050	Single phase, 240/480 volt primary, 120/240 volt secondary										
	0100	1 kVA	1 Elec	2	4	Ea.	219	168		387	500	
	0300	2 kVA		1.60	5		330	210		540	685	
	0500	3 kVA		1.40	5.714		405	240		645	820	
	0700	5 kVA	↓	1.20	6.667		560	280		840	1,050	
	0900	7.5 kVA	2 Elec	2.20	7.273		780	305		1,085	1,325	
	1100	10 kVA		1.60	10		955	420		1,375	1,700	
	1300	15 kVA	↓	1.20	13.333	↓	1,300	560		1,860	2,300	
	2300	3 phase, 480 volt primary 120/208 volt secondary										
	2310	Ventilated, 3 kVA	1 Elec	1	8	Ea.	725	335		1,060	1,325	
	2700	6 kVA		.80	10		1,000	420		1,420	1,750	
	2900	9 kVA	↓	.70	11.429		1,125	480		1,605	2,000	
	3100	15 kVA	2 Elec	1.10	14.545		1,500	610		2,110	2,600	
	3300	30 kVA		.90	17.778		1,775	745		2,520	3,100	
	3500	45 kVA	↓	.80	20	↓	2,125	840		2,965	3,625	
	9000	Minimum labor/equipment charge	1 Elec	1	8	Job		335		335	520	

16400 | Low-Voltage Distribution

16410 | Encl. Switches & Circuit Breakers			CREW	DAILY OUTPUT	LABOR-HOURS	UNIT	2006 BARE COSTS MAT.	LABOR	EQUIP.	TOTAL	TOTAL INCL O&P	
200	0010	CIRCUIT BREAKERS (in enclosure)										200
	0100	Enclosed (NEMA 1), 600 volt, 3 pole, 30 amp	1 Elec	3.20	2.500	Ea.	500	105		605	710	
	0200	60 amp		2.80	2.857		500	120		620	735	
	0400	100 amp		2.30	3.478		570	146		716	850	
	0600	225 amp	↓	1.50	5.333		1,325	224		1,549	1,800	
	0700	400 amp	2 Elec	1.60	10	↓	2,250	420		2,670	3,125	
	9000	Minimum labor/equipment charge	1 Elec	4	2	Job		84		84	129	
800	0010	SAFETY SWITCHES										800
	0100	General duty 240 volt, 3 pole NEMA 1, fusible, 30 amp	1 Elec	3.20	2.500	Ea.	82	105		187	252	
	0200	60 amp		2.30	3.478		138	146		284	375	
	0300	100 amp		1.90	4.211		238	177		415	535	
	0400	200 amp	↓	1.30	6.154		510	258		768	965	
	0500	400 amp	2 Elec	1.80	8.889	↓	1,300	375		1,675	2,000	
	9990	Minimum labor/equipment charge	1 Elec	3	2.667	Job		112		112	173	
840	0010	TIME SWITCHES										840
	0100	Single pole, single throw, 24 hour dial	1 Elec	4	2	Ea.	87	84		171	225	
	0200	24 hour dial with reserve power		3.60	2.222		370	93.50		463.50	550	
	0300	Astronomic dial		3.60	2.222		149	93.50		242.50	310	
	0400	Astronomic dial with reserve power		3.30	2.424		480	102		582	685	
	0500	7 day calendar dial		3.30	2.424		134	102		236	305	
	0600	7 day calendar dial with reserve power		3.20	2.500		520	105		625	730	
	0700	Photo cell 2000 watt	↓	8	1	↓	16.05	42		58.05	82	

16400 | Low-Voltage Distribution

16410 | Encl. Switches & Circuit Breakers

		16410 Encl. Switches & Circuit Breakers	CREW	DAILY OUTPUT	LABOR-HOURS	UNIT	2006 BARE COSTS MAT.	LABOR	EQUIP.	TOTAL	TOTAL INCL O&P	
840	1080	Load management device, 4 loads	1 Elec	2	4	Ea.	845	168		1,013	1,200	840
	1100	Load management device, 8 loads	↓	1	8	↓	1,375	335		1,710	2,050	
	9000	Minimum labor/equipment charge	↓	3.50	2.286	Job		96		96	148	

16420 | Enclosed Controllers

		16420 Enclosed Controllers	CREW	DAILY OUTPUT	LABOR-HOURS	UNIT	MAT.	LABOR	EQUIP.	TOTAL	TOTAL INCL O&P	
800	0010	**RELAYS** Enclosed (NEMA 1)										800
	0100	2 pole, 12 amp	1 Elec	5	1.600	Ea.	83	67		150	195	
	0200	4 pole, 10 amp	"	4.50	1.778	"	110	74.50		184.50	236	

16440 | Swbds, Panels & Control Centers

		16440 Swbds, Panels & Control Centers	CREW	DAILY OUTPUT	LABOR-HOURS	UNIT	MAT.	LABOR	EQUIP.	TOTAL	TOTAL INCL O&P	
700	0010	**PANELBOARD & LOAD CENTER CIRCUIT BREAKERS**										700
	0050	Bolt-on, 10,000 amp IC, 120 volt, 1 pole										
	0100	15 to 50 amp	1 Elec	10	.800	Ea.	13	33.50		46.50	66.50	
	0200	60 amp	↓	8	1	↓	13	42		55	79	
	0300	70 amp	↓	8	1	↓	24.50	42		66.50	91.50	
	0350	240 volt, 2 pole										
	0400	15 to 50 amp	1 Elec	8	1	Ea.	28.50	42		70.50	96	
	0500	60 amp	↓	7.50	1.067	↓	28.50	45		73.50	101	
	0600	80 to 100 amp	↓	5	1.600	↓	73.50	67		140.50	185	
	0700	3 pole, 15 to 60 amp	↓	6.20	1.290	↓	90.50	54		144.50	183	
	0800	70 amp	↓	5	1.600	↓	114	67		181	229	
	0900	80 to 100 amp	↓	3.60	2.222	↓	129	93.50		222.50	286	
	1000	22,000 amp I.C., 240 volt, 2 pole, 70 - 225 amp	↓	2.70	2.963	↓	555	124		679	800	
	1100	3 pole, 70 - 225 amp	↓	2.30	3.478	↓	620	146		766	905	
	1200	14,000 amp I.C., 277 volts, 1 pole, 15 - 30 amp	↓	8	1	↓	34.50	42		76.50	103	
	1300	22,000 amp I.C., 480 volts, 2 pole, 70 - 225 amp	↓	2.70	2.963	↓	555	124		679	800	
	1400	3 pole, 70 - 225 amp	↓	2.30	3.478	↓	685	146		831	980	
	2060	Plug-in tandem, 120/240 V, 2-15 A, 1 pole	↓	11	.727	↓	24	30.50		54.50	73.50	
	2070	1-15 A & 1-20 A	↓	11	.727	↓	24	30.50		54.50	73.50	
	2080	2-20 A	↓	11	.727	↓	24	30.50		54.50	73.50	
	9000	Minimum labor/equipment charge	↓	3	2.667	Job		112		112	173	
720	0010	**PANELBOARDS** (Commercial use)										720
	0050	NQOD, w/20 amp 1 pole bolt-on circuit breakers										
	0100	3 wire, 120/240 volts, 100 amp main lugs										
	0150	10 circuits	1 Elec	1	8	Ea.	420	335		755	985	
	0200	14 circuits	↓	.88	9.091	↓	495	380		875	1,125	
	0250	18 circuits	↓	.75	10.667	↓	540	450		990	1,275	
	0300	20 circuits	↓	.65	12.308	↓	610	515		1,125	1,475	
	0600	4 wire, 120/208 volts, 100 amp main lugs, 12 circuits	↓	1	8	↓	475	335		810	1,050	
	0650	16 circuits	↓	.75	10.667	↓	550	450		1,000	1,300	
	0700	20 circuits	↓	.65	12.308	↓	640	515		1,155	1,500	
	0750	24 circuits	↓	.60	13.333	↓	695	560		1,255	1,625	
	0800	30 circuits	↓	.53	15.094	↓	800	635		1,435	1,850	
	0850	225 amp main lugs, 32 circuits	2 Elec	.90	17.778	↓	900	745		1,645	2,150	
	0900	34 circuits	↓	.84	19.048	↓	925	800		1,725	2,250	
	0950	36 circuits	↓	.80	20	↓	945	840		1,785	2,350	
	1000	42 circuits	↓	.68	23.529	↓	1,050	990		2,040	2,700	
	1200	NEHB,w/20 amp, 1 pole bolt-on circuit breakers										
	1250	4 wire, 277/480 volts, 100 amp main lugs, 12 circuits	1 Elec	.88	9.091	Ea.	915	380		1,295	1,600	
	1300	20 circuits	"	.60	13.333	↓	1,350	560		1,910	2,350	
	1350	225 amp main lugs, 24 circuits	2 Elec	.90	17.778	↓	1,550	745		2,295	2,875	
	1400	30 circuits	↓	.80	20	↓	1,875	840		2,715	3,350	
	1450	36 circuits	↓	.72	22.222	↓	2,175	935		3,110	3,850	

16400 | Low-Voltage Distribution

16440 | Swbds, Panels & Control Centers

			CREW	DAILY OUTPUT	LABOR-HOURS	UNIT	2006 BARE COSTS MAT.	LABOR	EQUIP.	TOTAL	TOTAL INCL O&P	
720	1600	NQOD panel, w/20 amp, 1 pole, circuit breakers										720
	2000	4 wire, 120/208 volts with main circuit breaker										
	2050	100 amp main, 24 circuits	1 Elec	.47	17.021	Ea.	875	715		1,590	2,050	
	2100	30 circuits	"	.40	20		990	840		1,830	2,400	
	2200	225 amp main, 32 circuits	2 Elec	.72	22.222		1,675	935		2,610	3,275	
	2250	42 circuits		.56	28.571		1,825	1,200		3,025	3,850	
	2300	400 amp main, 42 circuits		.48	33.333		2,475	1,400		3,875	4,875	
	2350	600 amp main, 42 circuits	↓	.40	40	↓	3,675	1,675		5,350	6,625	
	2400	NEHB, with 20 amp, 1 pole circuit breaker										
	2450	4 wire, 277/480 volts with main circuit breaker										
	2500	100 amp main, 24 circuits	1 Elec	.42	19.048	Ea.	1,800	800		2,600	3,200	
	2550	30 circuits	"	.38	21.053		2,100	885		2,985	3,675	
	2600	225 amp main, 30 circuits	2 Elec	.72	22.222		2,650	935		3,585	4,375	
	2650	42 circuits	"	.56	28.571	↓	3,275	1,200		4,475	5,450	
	9000	Minimum labor/equipment charge	1 Elec	1	8	Job		335		335	520	

16500 | Lighting

16510 | Interior Luminaires

			CREW	DAILY OUTPUT	LABOR-HOURS	UNIT	2006 BARE COSTS MAT.	LABOR	EQUIP.	TOTAL	TOTAL INCL O&P	
440	0010	**INTERIOR LIGHTING FIXTURES** Including lamps, mounting										440
	0030	hardware and connections										
	0100	Fluorescent, C.W. lamps, troffer, recess mounted in grid, RS										
	0200	Acrylic lens, 1'W x 4'L, two 40 watt	1 Elec	5.70	1.404	Ea.	48.50	59		107.50	145	
	0300	2'W x 2'L, two U40 watt		5.70	1.404		52	59		111	148	
	0600	2'W x 4'L, four 40 watt		4.70	1.702		59	71.50		130.50	175	
	0910	Acrylic lens, 1'W x 4'L, two 32 watt		5.70	1.404		56	59		115	153	
	0930	2'W x 2'L, two U32 watt		5.70	1.404		80	59		139	179	
	0940	2'W x 4'L, two 32 watt		5.30	1.509		72	63.50		135.50	177	
	0950	2'W x 4'L, three 32 watt		5	1.600		76.50	67		143.50	188	
	0960	2'W x 4'L, four 32 watt	↓	4.70	1.702	↓	78.50	71.50		150	197	
	1000	Surface mounted, RS										
	1030	Acrylic lens with hinged & latched door frame										
	1100	1'W x 4'L, two 40 watt	1 Elec	7	1.143	Ea.	73.50	48		121.50	155	
	1200	2'W x 2'L, two U40 watt		7	1.143		79	48		127	161	
	1500	2'W x 4'L, four 40 watt	↓	5.30	1.509	↓	93.50	63.50		157	201	
	2100	Strip fixture										
	2130	Surface mounted										
	2200	4' long, one 40 watt RS	1 Elec	8.50	.941	Ea.	28.50	39.50		68	92	
	2300	4' long, two 40 watt RS	"	8	1		30.50	42		72.50	98	
	2600	8' long, one 75 watt, SL	2 Elec	13.40	1.194		42.50	50		92.50	125	
	2700	8' long, two 75 watt, SL	"	12.40	1.290	↓	51.50	54		105.50	140	
	3000	Pendent mounted, industrial, white porcelain enamel										
	3100	4' long, two 40 watt, RS	1 Elec	5.70	1.404	Ea.	47	59		106	143	
	3200	4' long, two 60 watt, HO	"	5	1.600		74.50	67		141.50	186	
	3300	8' long, two 75 watt, SL	2 Elec	8.80	1.818	↓	88.50	76.50		165	216	
	4220	Metal halide, integral ballast, ceiling, recess mounted										
	4230	prismatic glass lens, floating door										
	4240	2'W x 2'L, 250 watt	1 Elec	3.20	2.500	Ea.	291	105		396	480	
	4250	2'W x 2'L, 400 watt	2 Elec	5.80	2.759	↓	330	116		446	540	

	16510 | Interior Luminaires		CREW	DAILY OUTPUT	LABOR-HOURS	UNIT	2006 BARE COSTS MAT.	LABOR	EQUIP.	TOTAL	TOTAL INCL O&P	
440	4260	Surface mounted, 2'W x 2'L, 250 watt	1 Elec	2.70	2.963	Ea.	288	124		412	505	440
	4270	400 watt	2 Elec	4.80	3.333		340	140		480	590	
	4280	High bay, aluminum reflector,										
	4290	Single unit, 400 watt	2 Elec	4.60	3.478	Ea.	345	146		491	605	
	4300	Single unit, 1000 watt		4	4		495	168		663	805	
	4310	Twin unit, 400 watt		3.20	5		690	210		900	1,075	
	4320	Low bay, aluminum reflector, 250W DX lamp	1 Elec	3.20	2.500		340	105		445	530	
	4340	High pressure sodium integral ballast ceiling, recess mounted										
	4350	prismatic glass lens, floating door										
	4360	2'W x 2'L, 150 watt lamp	1 Elec	3.20	2.500	Ea.	350	105		455	545	
	4370	2'W x 2'L, 400 watt lamp	2 Elec	5.80	2.759		420	116		536	640	
	4380	Surface mounted, 2'W x 2'L, 150 watt lamp	1 Elec	2.70	2.963		395	124		519	625	
	4390	400 watt lamp	2 Elec	4.80	3.333		445	140		585	700	
	4400	High bay, aluminum reflector,										
	4410	Single unit, 400 watt lamp	2 Elec	4.60	3.478	Ea.	320	146		466	575	
	4430	Single unit, 1000 watt lamp	"	4	4		460	168		628	765	
	4440	Low bay, aluminum reflector, 150 watt lamp	1 Elec	3.20	2.500		275	105		380	465	
	4450	Incandescent, high hat can, round alzak reflector, prewired										
	4470	100 watt	1 Elec	8	1	Ea.	60.50	42		102.50	131	
	4480	150 watt		8	1		87.50	42		129.50	161	
	4500	300 watt		6.70	1.194		204	50		254	305	
	4600	Square glass lens with metal trim, prewired										
	4630	100 watt	1 Elec	6.70	1.194	Ea.	47	50		97	129	
	4700	200 watt		6.70	1.194		82	50		132	168	
	6010	Vapor tight, incandescent, ceiling mounted, 200 watt		6.20	1.290		53.50	54		107.50	142	
	6100	Fluorescent, surface mounted, 2 lamps, 4'L, RS, 40 watt		3.20	2.500		97	105		202	269	
	6850	Vandalproof, surface mounted, fluorescent, two 40 watt		3.20	2.500		211	105		316	395	
	6860	Incandescent, one 150 watt		8	1		54.50	42		96.50	125	
	6900	Mirror light, fluorescent, RS, acrylic enclosure, two 40 watt		8	1		87.50	42		129.50	161	
	6910	One 40 watt		8	1		69.50	42		111.50	141	
	6920	One 20 watt		12	.667		55	28		83	104	
	7500	Ballast replacement, by weight of ballast, to 15' high										
	7520	Indoor fluorescent, less than 2 lbs.	1 Elec	10	.800	Ea.		33.50		33.50	52	
	7540	Two 40W, watt reducer, 2 to 5 lbs.		9.40	.851		23	35.50		58.50	80.50	
	7560	Two F96 slimline, over 5 lbs.		8	1		39	42		81	107	
	7580	Vaportite ballast, less than 2 lbs.		9.40	.851			35.50		35.50	55	
	7600	2 lbs. to 5 lbs.		8.90	.899			38		38	58	
	7620	Over 5 lbs.		7.60	1.053			44		44	68	
	7630	Electronic ballast for two tubes		8	1		34.50	42		76.50	103	
	7640	Dimmable ballast one lamp		8	1		49.50	42		91.50	119	
	7650	Dimmable ballast two-lamp		7.60	1.053		80	44		124	156	
	9000	Minimum labor/equipment charge		3	2.667	Job		112		112	173	
800	0010	**RESIDENTIAL FIXTURES**										800
	0400	Fluorescent, interior, surface, circline, 32 watt & 40 watt	1 Elec	20	.400	Ea.	77.50	16.80		94.30	112	
	0500	2' x 2', two U 40 watt		8	1		103	42		145	179	
	0700	Shallow under cabinet, two 20 watt		16	.500		44	21		65	81	
	0900	Wall mounted, 4'L, one 40 watt, with baffle		10	.800		119	33.50		152.50	183	
	2000	Incandescent, exterior lantern, wall mounted, 60 watt		16	.500		31.50	21		52.50	67.50	
	2100	Post light, 150W, with 7' post		4	2		110	84		194	250	
	2500	Lamp holder, weatherproof with 150W PAR		16	.500		19.50	21		40.50	54	
	2550	With reflector and guard		12	.667		54.50	28		82.50	103	
	2600	Interior pendent, globe with shade, 150 watt		20	.400		128	16.80		144.80	167	
	9000	Minimum labor/equipment charge		4	2	Job		84		84	129	

Get your Means CostWorks 2006 today!

Fax : 1-800-632-6732, phone: 1-800-334-3509, or at www.rsmeans.com/cwcw.asp
For information on network pricing call 1-800-334-3509.

Catalog No.	Means CostWorks 2006 CD Subscription (12 Months)	Price	Cost
65066	Assemblies Cost Data	$219.95	
65016	Building Construction Cost Data	$154.95	
65316	Metric Construction Cost Data	$182.95	
65156	Open Shop Building Construction Cost Data	$154.95	
65116	Concrete & Masonry Cost Data	$154.95	
65036	Electrical Cost Data	$154.95	
65166	Heavy Construction Cost Data	$154.95	
65096	Interior Cost Data	$154.95	
65186	Light Commercial Cost Data	$192.95	
65026	Mechanical Cost Data	$154.95	
65216	Plumbing Cost Data	$154.95	
65286	Site Work & Landscape Cost Data	$154.95	
65056	Square Foot Costs	$219.95	
65606*	CostWorks Estimator	$299.95	
65206	Facilities Construction Cost Data	$356.95	
65046	Repair & Remodeling Cost Data	$154.95	
65176	Residential Cost Data	$177.95	
65306	Facilities Maintenance & Repair Cost Data	$356.95	

CD Package Subscription (12 Months)*

Catalog No.	Title	Price	Cost
65516	Builder's Package	$410.95	
65526	Building Professional's Package	$481.95	
65536	Facility Manager's Package	$619.95	
65546	Design Professional's Package	$481.95	

(Packages include FREE Means Estimator)

Subtotal	
MA residents please add 5% sales tax	
Shipping & Handling	**$6.00**
Total	

Name: ______________________

Company: ______________________

Street: ______________________

City: ______________ State: ______ Zip: ________

Phone: (   ) ______________ Fax: (   ) ______________

E-mail: ______________________

☐ Bill me ☐ Please charge my credit card

☐ Visa ☐ MasterCard ☐ American Express ☐ Discover

Card # ______________ Expiration Date: __________

Cardholder Signature: ______________________

All items shipped on a single CD with software code to unlock specific selections purchased.
Buy additional keys any time you need to unlock additional cost data books!

CWCW-2006

From the planning stages through the punch list . . .

maximize your estimating & budgeting efforts quickly & efficiently with a *Means CostWorks 2006* subscription.

As building requirements grow more complex, the estimating and budgeting process has never been more demanding. Tighter deadlines and more stringent cost controls mean you must find ways to work more efficiently. Let ***Means CostWorks 2006***, available in CD format, help you quickly and easily perform in-depth cost research . . . and take your estimating and budgeting to the next level.

Purchase your titles separately or in value-priced packages. With your purchase you receive access to regular cost updates, access to the Means HOTLINE, and the Wage Rate Utility (with selected titles).

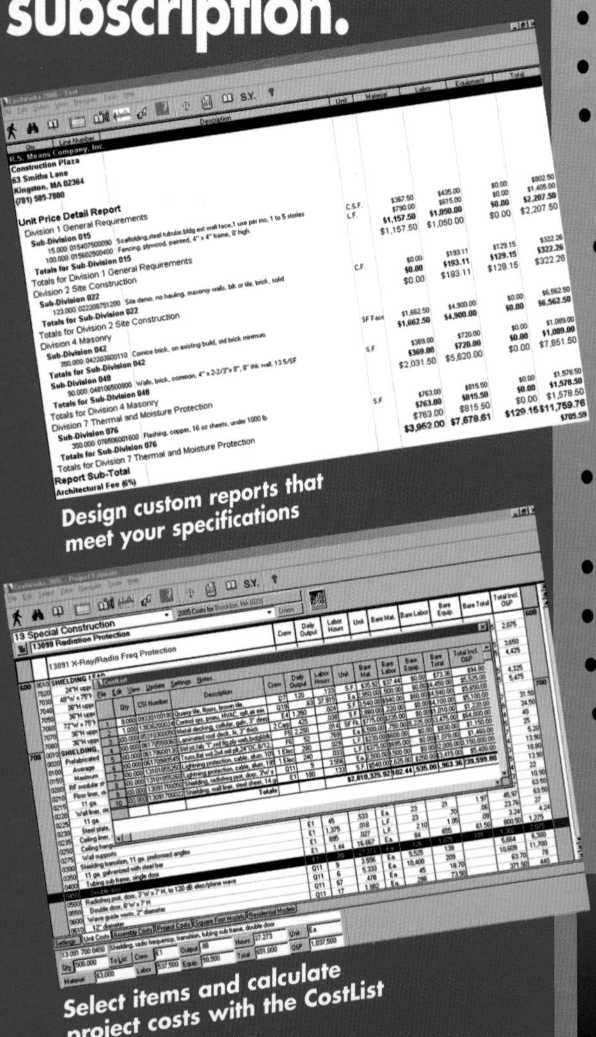

Design custom reports that meet your specifications

Select items and calculate project costs with the CostList

Your *Means CostWorks 2006* subscription now includes, **Free:**

- Access to Means Hotline
- The *RSMeans Quarterly Update Service*
- Wage Rate Utility – allows you to reference Union and Open Shop rates (included with selected titles)
- Means Estimator with your purchase of a *Means CostWorks 2006* package

The same Cost Data you know and trust . . . delivered electronically in a CD format!

- Localize costs to your geographic area to ensure accuracy
- Easily search and organize detailed cost data
- Determine best-cost solutions for each project
- Change and calculate results automatically
- Receive free online cost-factor updates with the *RSMeans Quarterly Update Service*
- Available as a 12-month subscription

Quick and easy . . . Get this powerful tool before you estimate your next project!

Subscribe to *Means CostWorks 2006 . . .* Join the thousands of industry professionals who rely on RSMeans Cost Data each year!

Means CostWorks 2006 subscriptions offer a choice of 17 detailed cost titles that cover every job costing requirement you can imagine. They turn your desktop or laptop PC into a comprehensive cost library. Each title addresses a specific area of the commercial or residential building and maintenance industry.

You get the productivity-enhancing tools to actually put this valuable information to work.

Save hundreds on specially priced CD Package subscriptions! Choose the one that's right for you . . .

Builder's Package
Includes:
- Residential Cost Data
- Light Commercial Cost Data
- Site Work & Landscape Cost Data
- Concrete & Masonry Cost Data
- Open Shop Building Construction Cost Data

Just $410.95
Over $300 off the price of buying each book separately!

Building Professional's Package
Includes:
- Building Construction Cost Data
- Mechanical Cost Data
- Plumbing Cost Data
- Electrical Cost Data
- Square Foot Costs

Just $481.95
Over $250 off the price of buying each book separately!

Facility Manager's Package
Includes:
- Facilities Construction Cost Data
- Facilities Maintenance & Repair Cost Data
- Repair & Remodeling Cost Data
- Building Construction Cost Data

Just $619.95
Over $300 off the price of buying each book separately!

Design Professional's Package
Includes:
- Building Construction Cost Data
- Square Foot Costs
- Interior Cost Data
- Assemblies Cost Data

Just $481.95
Over $200 off the price of buying each book separately!

All 12-month CD Package subscriptions include the Means Estimator FREE!

16520 | Exterior Luminaires

			CREW	DAILY OUTPUT	LABOR-HOURS	UNIT	2006 BARE COSTS MAT.	LABOR	EQUIP.	TOTAL	TOTAL INCL O&P	
300	0010	**EXTERIOR FIXTURES** With lamps										300
	0400	Quartz, 500 watt	1 Elec	5.30	1.509	Ea.	54	63.50		117.50	157	
	1100	Wall pack, low pressure sodium, 35 watt		4	2		214	84		298	365	
	1150	55 watt		4	2		255	84		339	410	
	1160	High pressure sodium, 70 watt		4	2		200	84		284	350	
	1170	150 watt		4	2		230	84		314	380	
	1180	Metal Halide, 175 watt		4	2		255	84		339	410	
	1190	250 watt		4	2		265	84		349	420	
	1200	Floodlights with ballast and lamp,										
	1400	pole mounted, pole not included										
	2250	Low pressure sodium, 55 watt	1 Elec	2.70	2.963	Ea.	485	124		609	725	
	2270	90 watt		2	4	"	535	168		703	850	
	9000	Minimum labor/equipment charge		3.75	2.133	Job		89.50		89.50	138	

16530 | Emergency Lighting

			CREW	DAILY OUTPUT	LABOR-HOURS	UNIT	MAT.	LABOR	EQUIP.	TOTAL	TOTAL INCL O&P	
320	0010	**EXIT AND EMERGENCY LIGHTING**										320
	0080	Exit light ceiling or wall mount, incandescent, single face	1 Elec	8	1	Ea.	42	42		84	111	
	0100	Double face	"	6.70	1.194	"	46	50		96	128	
	0300	Emergency light units, battery operated										
	0350	Twin sealed beam light, 25 watt, 6 volt each										
	0500	Lead battery operated	1 Elec	4	2	Ea.	122	84		206	263	
	0700	Nickel cadmium battery operated		4	2	"	580	84		664	770	
	9000	Minimum labor/equipment charge		4	2	Job		84		84	129	

16550 | Special Purpose Lighting

			CREW	DAILY OUTPUT	LABOR-HOURS	UNIT	MAT.	LABOR	EQUIP.	TOTAL	TOTAL INCL O&P	
820	0010	**TRACK LIGHTING**										820
	0080	Track, 1 circuit, 4′ section	1 Elec	6.70	1.194	Ea.	40.50	50		90.50	122	
	0100	8′ section		5.30	1.509		68	63.50		131.50	173	
	0300	3 circuits, 4′ section		6.70	1.194		54.50	50		104.50	138	
	0400	8′ section		5.30	1.509		84.50	63.50		148	191	
	9000	Minimum labor/equipment charge		3	2.667	Job		112		112	173	

16585 | Lamps

			CREW	DAILY OUTPUT	LABOR-HOURS	UNIT	MAT.	LABOR	EQUIP.	TOTAL	TOTAL INCL O&P	
600	0010	**LAMPS**										600
	0080	Fluorescent, rapid start, cool white, 2′ long, 20 watt	1 Elec	1	8	C	310	335		645	860	
	0100	4′ long, 40 watt		.90	8.889		305	375		680	910	
	0170	4′ long, 34 watt energy saver		.90	8.889		296	375		671	900	
	0176	2′ long, T8, 17 W engergy saver		1	8		440	335		775	1,000	
	0178	3′ long, T8, 25 W energy saver		.90	8.889		450	375		825	1,075	
	0180	4′ long, T8, 32 watt energy saver		.90	8.889		273	375		648	875	
	0560	Twin tube compact lamp		.90	8.889		535	375		910	1,175	
	0570	Double twin tube compact lamp		.80	10		1,300	420		1,720	2,075	
	0600	Mercury vapor, mogul base, deluxe white, 100 watt		.30	26.667		2,725	1,125		3,850	4,700	
	0800	400 watt		.30	26.667		2,875	1,125		4,000	4,900	
	1000	Metal halide, mogul base, 175 watt		.30	26.667		3,400	1,125		4,525	5,450	
	1100	250 watt		.30	26.667		3,850	1,125		4,975	5,950	
	1350	High pressure sodium, 70 watt		.30	26.667		3,950	1,125		5,075	6,075	
	1370	150 watt		.30	26.667		4,225	1,125		5,350	6,375	
	1500	Low pressure sodium, 35 watt		.30	26.667		5,975	1,125		7,100	8,300	
	1600	90 watt		.30	26.667		7,600	1,125		8,725	10,100	
	1800	Incandescent, interior, A21, 100 watt		1.60	5		140	210		350	480	
	1900	A21, 150 watt		1.60	5		143	210		353	485	
	2300	R30, 75 watt		1.30	6.154		535	258		793	990	
	2500	Exterior, PAR 38, 75 watt		1.30	6.154		1,100	258		1,358	1,600	
	2600	PAR 38, 150 watt		1.30	6.154		1,200	258		1,458	1,725	

16500 | Lighting

16585 | Lamps

			CREW	DAILY OUTPUT	LABOR-HOURS	UNIT	2006 BARE COSTS MAT.	LABOR	EQUIP.	TOTAL	TOTAL INCL O&P	
600	9000	Minimum labor/equipment charge	1 Elec	4	2	Job		84		84	129	600

16800 | Sound & Video

16810 | Sound and Video Circuits

			CREW	DAILY OUTPUT	LABOR-HOURS	UNIT	2006 BARE COSTS MAT.	LABOR	EQUIP.	TOTAL	TOTAL INCL O&P	
750	0010	**SOUND AND VIDEO CABLES & FITTINGS**										750
	1250	Nonshielded, #22-2 conductor	1 Elec	10	.800	C.L.F.	10.30	33.50		43.80	63.50	

16820 | Sound Reinforcement

			CREW	DAILY OUTPUT	LABOR-HOURS	UNIT	2006 BARE COSTS MAT.	LABOR	EQUIP.	TOTAL	TOTAL INCL O&P	
300	0010	**DOORBELL SYSTEM** Incl. transformer, button & signal										300
	0020											
	1000	Door chimes, 2 notes, minimum	1 Elec	16	.500	Ea.	22	21		43	57	
	1020	Maximum		12	.667		118	28		146	172	
	1100	Tube type, 3 tube system		12	.667		167	28		195	227	
	1180	4 tube system		10	.800		268	33.50		301.50	345	
	1900	For transformer & button, minimum add		5	1.600		12.70	67		79.70	118	
	1960	Maximum, add		4.50	1.778		38	74.50		112.50	157	
	3000	For push button only, minimum		24	.333		2.48	14		16.48	24	
	3100	Maximum		20	.400		20	16.80		36.80	48	
	9000	Minimum labor/equipment charge		4	2	Job		84		84	129	
800	0010	**PUBLIC ADDRESS SYSTEM**										800
	0100	Conventional, office	1 Elec	5.33	1.501	Speaker	107	63		170	215	
	0200	Industrial		2.70	2.963	"	207	124		331	420	
	9000	Minimum labor/equipment charge		3.50	2.286	Job		96		96	148	
840	0010	**SOUND SYSTEM** not including rough-in wires, cables & conduits										840
	2020	11 station capacity	2 Elec	4	4	Ea.	760	168		928	1,100	
	3600	House telephone, talking station	1 Elec	1.60	5		370	210		580	735	
	3800	Press to talk, release to listen	"	5.30	1.509		86.50	63.50		150	193	
	4000	System-on button					51.50			51.50	57	
	4200	Door release	1 Elec	4	2		92.50	84		176.50	231	
	4400	Combination speaker and microphone		8	1		157	42		199	238	
	4600	Termination box		3.20	2.500		49.50	105		154.50	217	
	4800	Amplifier or power supply		5.30	1.509		570	63.50		633.50	725	
	5000	Vestibule door unit		16	.500	Name	105	21		126	148	
	5200	Strip cabinet		27	.296	Ea.	197	12.45		209.45	236	
	5400	Directory		16	.500	"	93	21		114	135	
	9000	Minimum labor/equipment charge		3.50	2.286	Job		96		96	148	

16850 | Television Equipment

			CREW	DAILY OUTPUT	LABOR-HOURS	UNIT	2006 BARE COSTS MAT.	LABOR	EQUIP.	TOTAL	TOTAL INCL O&P	
600	0010	**T.V. SYSTEMS** not including rough-in wires, cables & conduits										600
	5000	T.V. Antenna only, minimum	1 Elec	6	1.333	Ea.	37.50	56		93.50	128	
	5100	Maximum	1 Elec	4	2	Ea.	157	84		241	300	

Assemblies Section

Table of Contents

How to Use the Assemblies Cost Tables

The following is a detailed explanation of a sample Assemblies Cost Table. Most Assembly Tables are separated into three parts: 1) an illustration of the system to be estimated; 2) the components and related costs of a typical system; and 3) the costs for similar systems with dimensional and/or size variations. For costs of the components that comprise these systems or "assemblies," refer to the Unit Price Section. Next to each bold number below is the item being described with the appropriate component of the sample entry following in parenthesis. In most cases, if the work is to be subcontracted, the general contractor will need to add an additional markup (RSMeans suggests using 10%) to the "Total" figures.

1 System/Line Numbers (B1010 263 1700)

Each Assemblies Cost Line has been assigned a unique identification number based on the UNIFORMAT II classification system.

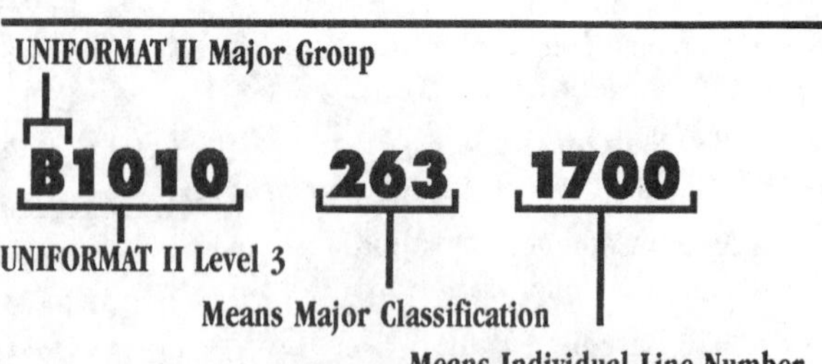

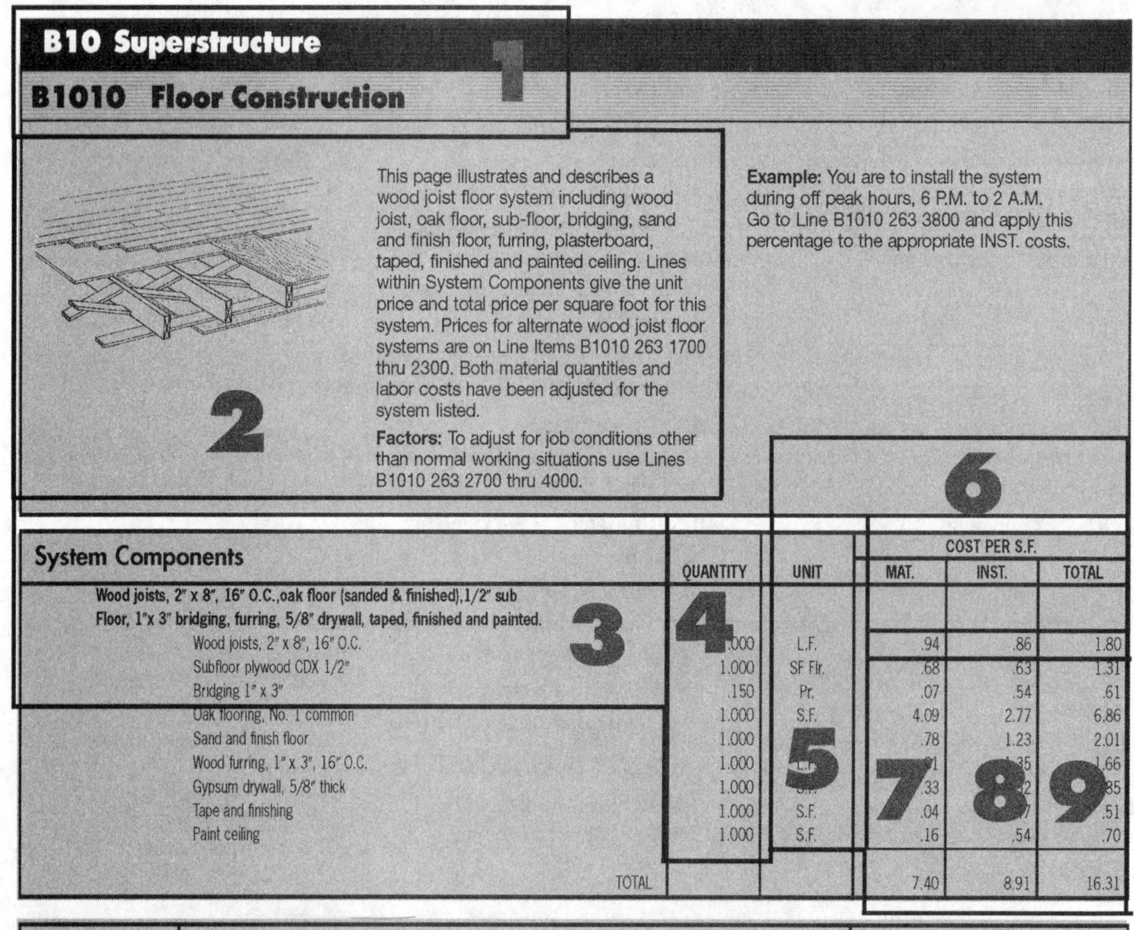

B10 Superstructure

B1010 Floor Construction

1

This page illustrates and describes a wood joist floor system including wood joist, oak floor, sub-floor, bridging, sand and finish floor, furring, plasterboard, taped, finished and painted ceiling. Lines within System Components give the unit price and total price per square foot for this system. Prices for alternate wood joist floor systems are on Line Items B1010 263 1700 thru 2300. Both material quantities and labor costs have been adjusted for the system listed.

Factors: To adjust for job conditions other than normal working situations use Lines B1010 263 2700 thru 4000.

Example: You are to install the system during off peak hours, 6 P.M. to 2 A.M. Go to Line B1010 263 3800 and apply this percentage to the appropriate INST. costs.

2

6

System Components	QUANTITY	UNIT	COST PER S.F. MAT.	COST PER S.F. INST.	COST PER S.F. TOTAL
Wood joists, 2" x 8", 16" O.C.,oak floor (sanded & finished),1/2" sub Floor, 1"x 3" bridging, furring, 5/8" drywall, taped, finished and painted.					
Wood joists, 2" x 8", 16" O.C.	[illegible]	L.F.	.94	.86	1.80
Subfloor plywood CDX 1/2"	1.000	SF Flr.	.68	.63	1.31
Bridging 1" x 3"	.150	Pr.	.07	.54	.61
Oak flooring, No. 1 common	1.000	S.F.	4.09	2.77	6.86
Sand and finish floor	1.000	S.F.	.78	1.23	2.01
Wood furring, 1" x 3", 16" O.C.	1.000	[illegible]	[illegible]	1.35	1.66
Gypsum drywall, 5/8" thick	1.000	[illegible]	.33	[illegible]	[illegible]
Tape and finishing	1.000	S.F.	.04	[illegible]	.51
Paint ceiling	1.000	S.F.	.16	.54	.70
TOTAL			7.40	8.91	16.31

3 4 5 7 8 9

B1010 263	Floor-Ceiling, Wood Joists	COST PER S.F. MAT.	COST PER S.F. INST.	COST PER S.F. TOTAL
1600	For alternate wood joist systems:			
1700	16" on center, 2" x 6" joists	7.10	8.80	15.90
1800	2" x 10"	7.90	9.10	17
1900	2" x 12"	8.20	9.15	17.35
2000	2" x 14"	8.85	9.25	18.10
2100	12" on center, 2" x 10" joists	8.25	9.35	17.60
2200	2" x 12"	8.65	9.40	18.05
2300	2" x 14"	9.45	9.60	19.05

2 *Illustration*

At the top of most assembly tables are an illustration, a brief description, and the design criteria used to develop the cost.

3 *System Components*

The components of a typical system are listed separately to show what has been included in the development of the total system price. The table below contains prices for other similar systems with dimensional and/or size variations.

4 *Quantity*

This is the number of line item units required for one system unit. For example, we assume that it will take .15 pair of 1″ × 3″ bridging on a square foot basis.

5 *Unit of Measure for Each Item*

The abbreviated designation indicates the unit of measure, as defined by industry standards, upon which the price of the component is based. For example, wood joists are priced by L.F. (linear foot) while plywood is priced by S.F. (square foot).

6 *Unit of Measure for Each System (Cost per S.F.)*

Costs shown in the three right-hand columns have been adjusted by the component quantity and unit of measure for the entire system. In this example, "Cost per S.F." is the unit of measure for this system, or assembly.

7 *Materials (7.10)*

This column contains the Materials Cost of each component. These cost figures are bare costs plus 10% for profit.

8 *Installation (8.80)*

Installation includes labor and equipment plus the installing contractor's overhead and profit. Equipment costs are the bare rental costs plus 10% for profit. The labor overhead and profit are defined on the inside back cover of this book.

9 *Total (15.90)*

The figure in this column is the sum of the material and installation costs.

Material Cost	+	Installation Cost	=	Total
$7.10	+	$8.80	=	$15.90

A SUBSTRUCTURE

A1010 Standard Foundations

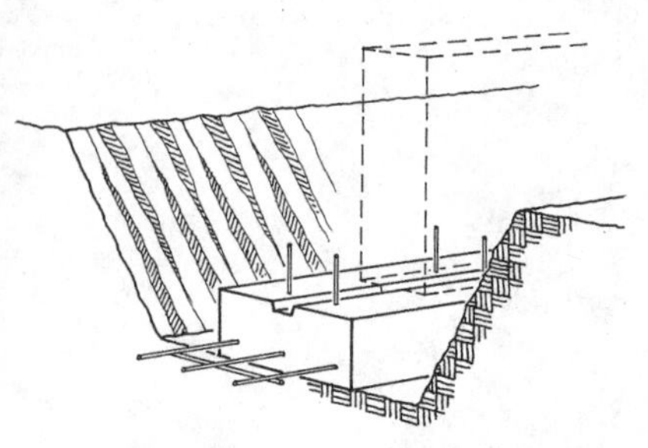

This page illustrates and describes a strip footing system including concrete, forms, reinforcing, keyway and dowels. Lines within System Components give the unit price and total price per linear foot for this system. Prices for alternate strip footing systems are on Line Items A1010 120 1500 thru 2500. Both material quantities and labor costs have been adjusted for the system listed.

Factors: To adjust for job conditions other than normal working situations use Lines A1010 120 2900 thru 4000.

Example: You are to install this footing, and due to a lack of accessibility, only hand tools can be used. Material handling is also a problem. Go to Lines A1010 120 3400 and 3600 and apply these percentages to the appropriate MAT. and INST. costs.

System Components	QUANTITY	UNIT	COST PER L.F. MAT.	COST PER L.F. INST.	COST PER L.F. TOTAL
Strip footing, 2′-0″ wide x 1′-0″ thick, 2000 psi concrete including forms Reinforcing, keyway, and dowels.					
Concrete, 2000 psi	.074	C.Y.	6.81		6.81
Placing concrete	.074	C.Y.		1.43	1.43
Forms, footing, 4 uses	2.000	S.F.	1.88	7.32	9.20
Reinforcing	3.170	Lb.	1.55	1.65	3.20
Keyway, 2″ x 4″, 4 uses	1.000	L.F.	.22	.89	1.11
Dowels, #4 bars, 2′ long, 24″ O.C.	.500	Ea.	.35	1.13	1.48
TOTAL			10.81	12.42	23.23

A1010 120	Strip Footing	COST PER L.F. MAT.	COST PER L.F. INST.	COST PER L.F. TOTAL
1400	Above system with the following:			
1500	2′-0″ wide x 1′ thick, 3000 psi concrete	11.05	12.45	23.50
1600	4000 psi concrete	11.40	12.45	23.85
1800	For alternate footing systems:			
1900	2′-6″ wide x 1′ thick, 2000 psi concrete	12.95	13.25	26.20
2000	3000 psi concrete	13.30	13.25	26.55
2100	4000 psi concrete	13.70	13.25	26.95
2300	3′-0″ wide x 1′ thick, 2000 psi concrete	14.90	13.95	28.85
2400	3000 psi concrete	15.30	13.95	29.25
2500	4000 psi concrete	15.80	13.95	29.75
2700				
2800				
2900	Cut & patch to match existing construction, add, minimum	2%	3%	
3000	Maximum	5%	9%	
3100	Dust protection, add, minimum	1%	2%	
3200	Maximum	4%	11%	
3300	Equipment usage curtailment, add, minimum	1%	1%	
3400	Maximum	3%	10%	
3500	Material handling & storage limitation, add, minimum	1%	1%	
3600	Maximum	6%	7%	
3700	Protection of existing work, add, minimum	2%	2%	
3800	Maximum	5%	7%	
3900	Shift work requirements, add, minimum		5%	
4000	Maximum		30%	

Important: See the Reference Section for critical supporting data - Reference Numbers and City Cost Indexes

A1010 Standard Foundations

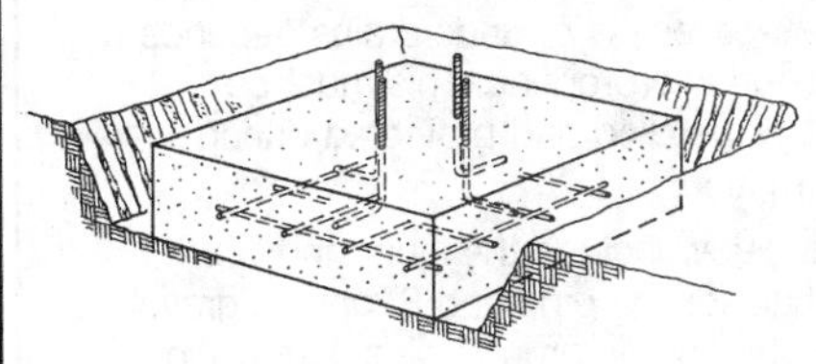

This page illustrates and describes a spread footing system including concrete, forms, reinforcing and anchor bolts. Lines within System Components give the unit price and total price on a cost each basis for this system. Prices for alternate spread footing systems are on Line Items A1010 230 1300 thru 2300. Both material quantities and labor costs have been adjusted for the system listed.

Factors: To adjust for job conditions other than normal working situations use Lines A1010 230 2900 thru 4000.

Example: You are to install the system in an existing occupied building. Access to the site and protection to the building are mandatory. Go to Lines A1010 230 3600 and 3800 and apply these percentages to the appropriate MAT. and INST. costs.

System Components	QUANTITY	UNIT	COST EACH MAT.	COST EACH INST.	COST EACH TOTAL
Interior column footing, 3′ square 1′ thick, 2000 psi concrete					
Including forms, reinforcing, and anchor bolts.					
Concrete, 2000 psi	.330	C.Y.	30.36		30.36
Placing concrete	.330	C.Y.		13.98	13.98
Forms, footing, 4 uses	12.000	SFCA	7.80	51.48	59.28
Reinforcing	11.000	Lb.	5.39	5.72	11.11
Anchor bolts, 3/4″ diameter	2.000	Ea.	1.50	4.74	6.24
TOTAL			45.05	75.92	120.97

A1010 230	Spread Footing	COST EACH MAT.	COST EACH INST.	COST EACH TOTAL
1200	Above system with the following:			
1300	3′ square x 1′ thick, 3000 psi concrete	46	76	122
1400	4000 psi concrete	47.50	76	123.50
1600	For alternate footing systems:			
1700	4′ square x 1′ thick, 2000 psi concrete	76	109	185
1800	3000 psi concrete	78	109	187
1900	4000 psi concrete	80.50	109	189.50
2100	5′ square x 1′-3″ thick, 2000 psi concrete	143	181	324
2200	3000 psi concrete	147	181	328
2300	4000 psi concrete	152	181	333
2600				
2700				
2900	Cut & patch to match existing construction, add, minimum	2%	3%	
3000	Maximum	5%	9%	
3100	Dust protection, add, minimum	1%	2%	
3200	Maximum	4%	11%	
3300	Equipment usage curtailment, add, minimum	1%	1%	
3400	Maximum	3%	10%	
3500	Material handling & storage limitation, add, minimum	1%	1%	
3600	Maximum	6%	7%	
3700	Protection of existing work, add, minimum	2%	2%	
3800	Maximum	5%	7%	
3900	Shift work requirements, add, minimum		5%	
4000	Maximum		30%	

A1010 Standard Foundations

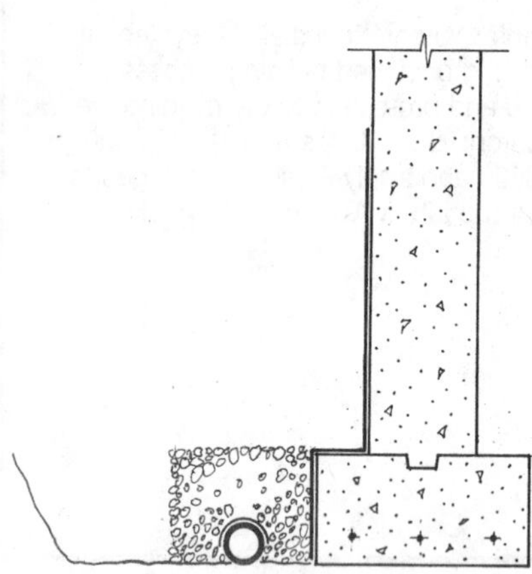

General: Footing drains can be placed either inside or outside of foundation walls depending upon the source of water to be intercepted. If the source of subsurface water is principally from grade or a subsurface stream above the bottom of the footing, outside drains should be used. For high water tables, use inside drains or both inside and outside.

The effectiveness of underdrains depends on good waterproofing. This must be carefully installed and protected during construction.

Costs below include the labor and materials for the pipe and 6″ only of gravel or crushed stone around pipe. Excavation and backfill are not included.

System Components	QUANTITY	UNIT	COST PER L.F. MAT.	INST.	TOTAL
Foundation underdrain, outside, PVC 4″ diameter.					
PVC pipe 4″ diam. S.D.R. 35	1.000	L.F.	2.78	3.28	6.06
Pipe bedding, graded gravel 3/4″ to 1/2″	.070	C.Y.	2.14	.67	2.81
TOTAL			4.92	3.95	8.87

A1010 310	Foundation Underdrain	COST PER L.F. MAT.	INST.	TOTAL
0900	For alternate drain systems:			
1000	Foundation underdrain, outside only, PVC, 4″ diameter	4.92	3.95	8.87
1100	6″ diameter	8.60	4.37	12.97
1200	Bituminous fiber, 4″ diameter	4.92	3.95	8.87
1300	6″ diameter	8.60	4.37	12.97
1400	Porous concrete, 6″ diameter	5.70	4.76	10.46
1450	8″ diameter	7	6.30	13.30
1500	12″ diameter	12.55	7.25	19.80
1600	Corrugated metal, 16 ga. asphalt coated, 6″ diameter	6.10	7.50	13.60
1650	8″ diameter	8.15	7.90	16.05
1700	10″ diameter	9.90	8.25	18.15
2000	Vitrified clay, C-211, 4″ diameter	4.52	7	11.52
2050	6″ diameter	6.70	8.90	15.60
2100	12″ diameter	16.70	10.70	27.40
3000	Outside and inside, PVC, 4″ diameter	9.85	7.90	17.75
3100	6″ diameter	17.20	8.75	25.95
3200	Bituminous fiber, 4″ diameter	9.85	7.90	17.75
3300	6″ diameter	17.20	8.75	25.95
3400	Porous concrete, 6″ diameter	11.35	9.50	20.85
3450	8″ diameter	13.95	12.65	26.60
3500	12″ diameter	25	14.50	39.50
3600	Corrugated metal, 16 ga., asphalt coated, 6″ diameter	12.15	15	27.15
3650	8″ diameter	16.25	15.80	32.05
3700	10″ diameter	19.85	16.55	36.40
4000	Vitrified clay, C-211, 4″ diameter	9.05	13.95	23
4050	6″ diameter	13.35	17.80	31.15
4100	12″ diameter	33.50	21.50	55
4700	Cut & patch to match existing construction, add, minimum	2%	3%	
4800	Maximum	5%	9%	
4900	Dust protection, add, minimum	1%	2%	
5000	Maximum	4%	11%	
5100	Equipment usage curtailment, add, minimum	1%	1%	

A1010 Standard Foundations

A1010 310	Foundation Underdrain	COST PER L.F.		
		MAT.	INST.	TOTAL
5200	Maximum	3%	10%	
5300	Material handling & storage limitation, add, minimum	1%	1%	
5400	Maximum	6%	7%	
5500	Protection of existing work, add, minimum	2%	2%	
5600	Maximum	5%	7%	
5700	Shift work requirements, add, minimum		5%	
5800	Maximum		30%	
5900	Temporary shoring and bracing, add, minimum	2%	5%	
6000	Maximum	5%	12%	

A1030 Slab on Grade

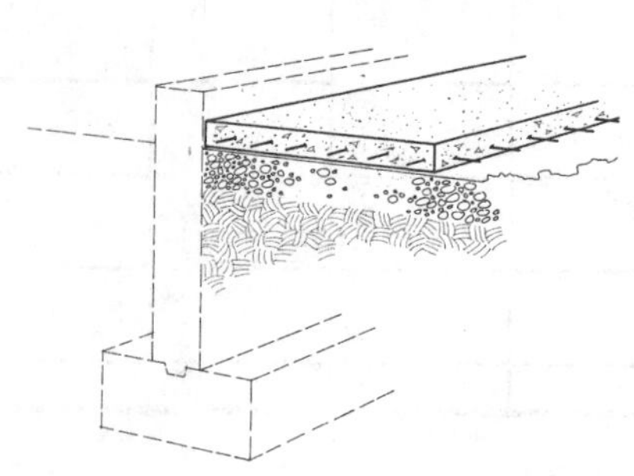

This page illustrates and describes a slab on grade system including slab, bank run gravel, bulkhead forms, placing concrete, welded wire fabric, vapor barrier, steel trowel finish and curing paper. Lines within System Components give the unit price and total price per square foot for this system. Prices for alternate slab on grade systems are on Line Items A1030 110 1500 thru 2600. Both material quantities and labor costs have been adjusted for the system listed.

Factors: To adjust for job conditions other than normal working situations use Lines A1030 110 2900 thru 4000.

Example: You are to install the system at a site where protection of the existing building is required. Go to Line A1030 110 3800 and apply these percentages to the appropriate MAT. and INST. costs.

System Components	QUANTITY	UNIT	COST PER S.F.		
			MAT.	INST.	TOTAL
Ground slab, 4″ thick, 3000 psi concrete, 4″ granular base, vapor barrier					
Welded wire fabric, screed and steel trowel finish.					
Concrete, 4″ thick, 3000 psi concrete	.012	C.Y.	1.15		1.15
Bank run gravel, 4″ deep	.074	C.Y.	.24	.07	.31
Polyethylene vapor barrier, 10 mil.	.011	C.S.F.	.06	.14	.20
Bulkhead forms, expansion material	.100	L.F.	.03	.30	.33
Welded wire fabric, 6 x 6 - #10/10	.011	C.S.F.	.15	.34	.49
Place concrete	.012	C.Y.		.26	.26
Screed & steel trowel finish	1.000	S.F.		.79	.79
TOTAL			1.63	1.90	3.53

A1030 110	Interior Slab on Grade	COST PER S.F.		
		MAT.	INST.	TOTAL
1400	Above system with the following:			
1500	4″ thick slab, 3000 psi concrete, 6″ deep bank run gravel	1.82	1.89	3.71
1600	12″ deep bank run gravel	2.25	1.93	4.18
1700				
1800				
1900				
2000	For alternate slab systems:			
2100	5″ thick slab, 3000 psi concrete, 6″ deep bank run gravel	2.10	1.95	4.05
2200	12″ deep bank run gravel	2.53	1.99	4.52
2300				
2400				
2500	6″ thick slab, 3000 psi concrete, 6″ deep bank run gravel	2.48	2.03	4.51
2600	12″ deep bank run gravel	2.91	2.07	4.98
2700				
2900	Cut & patch to match existing construction, add, minimum	2%	3%	
3000	Maximum	5%	9%	
3100	Dust protection, add, minimum	1%	2%	
3200	Maximum	4%	11%	
3300	Equipment usage curtailment, add, minimum	1%	1%	
3400	Maximum	3%	10%	
3500	Material handling & storage limitation, add, minimum	1%	1%	
3600	Maximum	6%	7%	
3700	Protection of existing work, add, minimum	2%	2%	
3800	Maximum	5%	7%	
3900	Shift work requirements, add, minimum		5%	
4000	Maximum		30%	

A2010 Basement Excavation

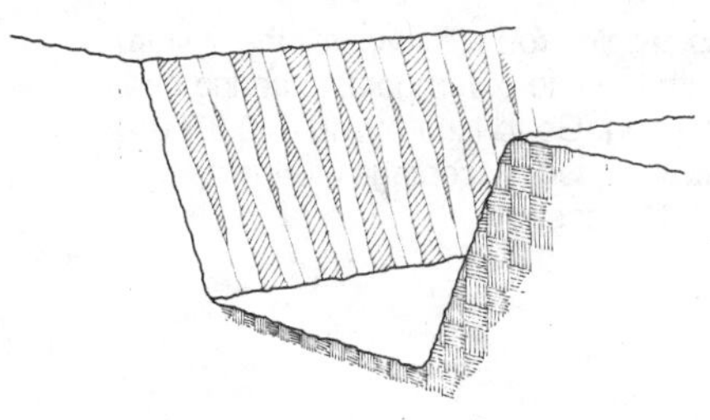

This page illustrates and describes continuous footing and trench excavation systems including a wheel mounted backhoe, operator, equipment rental, fuel, oil, mobilization, no hauling or backfill. Lines within System Components give the unit price and total price per linear foot for this system. Prices for alternate continuous footing and trench excavation systems are on Line Items A2010 130 1300 thru 2700. Both material quantities and labor costs have been adjusted for the system listed.

Factors: To adjust for job conditions other than normal working situations use Lines A2010 130 2900 thru 4000.

Example: You are to install the system and supply dust protection. Go to Line A2010 130 3000 and apply this percentage to the appropriate TOTAL costs.

System Components	QUANTITY	UNIT	COST PER L.F. EQUIP.	LABOR	TOTAL
Continuous footing or trench excav. w/ 3/4 C.Y. wheel mntd backhoe Including operator, equipment rental, fuel, oil and mobilization. Trench Is 4′ wide at bottom, 3′ deep w/sides sloped 1 to 2 and 200′ long. Costs Are based on a production rate of 240 C.Y./day. No hauling/backfill incl.					
Equipment operator	4.000	Hr.		232.50	232.50
3/4 C.Y. backhoe, wheel mounted	.500	Day	76.45		76.45
Operating expense (fuel, oil)	4.000	Hr.	47.96		47.96
Mobilization	1.000	Ea.	160	103	263
TOTAL			284.41	335.50	619.91
COST PER L.F.		L.F.	1.42	1.68	3.10

A2010 130	Excavation, Footings or Trench	COST PER L.F. EQUIP.	LABOR	TOTAL
1200	For alternate trench sizes, 4′ bottom with sloped sides:			
1300	2′ deep, 50′ long	4.91	6.70	11.61
1400	100′ long	2.54	3.36	5.90
1500	300′ long	.93	1.12	2.05
1600	4′ deep, 50′ long	5.10	6.70	11.80
1700	100′ long	2.72	3.36	6.08
1800	300′ long	1.39	1.99	3.38
1900	6′ deep, 50′ long	5.35	6.70	12.05
2000	100′ long	2.91	4.23	7.14
2100	300′ long	2.19	3.44	5.63
2200	8′ deep, 50′ long	5.70	6.70	12.40
2300	100′ long	4.09	5.70	9.79
2400	300′ long	3.02	3.44	6.46
2500	10′ deep, 50′ long	6.95	9.05	16
2600	100′ long	5.20	7.70	12.90
2700	300′ long	4.01	6.85	10.86
2800				
2900	Dust protection, add, minimum		2%	2%
3000	Maximum		11%	11%
3100	Equipment usage curtailment, add, minimum		1%	1%
3200	Maximum		10%	10%
3300	Material handling & storage limitation, add, minimum		1%	1%
3400	Maximum		7%	7%
3500	Protection of existing work, add, minimum		2%	2%
3600	Maximum		7%	7%
3700	Shift work requirements, add, minimum		5%	5%
3800	Maximum		30%	30%
3900	Temporary shoring and bracing, add, minimum		5%	5%
4000	Maximum		12%	12%

A2010 Basement Excavation

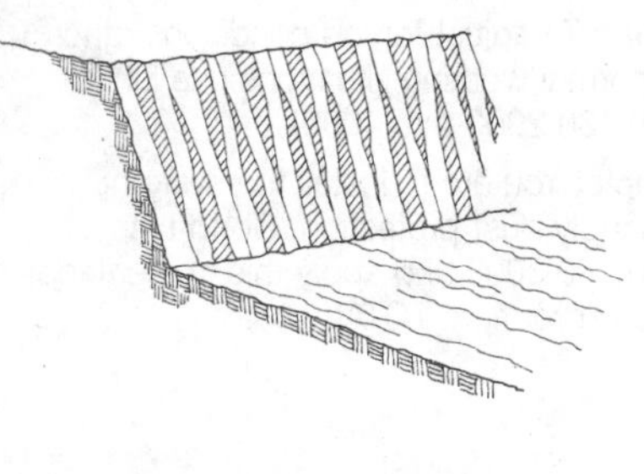

This page illustrates and describes foundation excavation systems including a backhoe-loader, operator, equipment rental, fuel, oil, mobilization, hauling material, no backfilling. Lines within System Components give the unit price and total price per cubic yard for this system. Prices for alternate foundation excavation systems are on Line Items A2010 140 1600 thru 2200. Both material quantities and labor costs have been adjusted for the system listed.

Factors: To adjust for job conditions other than normal working situations use Lines A2010 140 2900 thru 4000.

Example: You are to install the system with the use of temporary shoring and bracing. Go to Line A2010 140 3900 and apply these percentages to the appropriate TOTAL costs.

System Components	QUANTITY	UNIT	COST PER C.Y. EQUIP.	LABOR	TOTAL
Foundation excav. w/ 3/4 C.Y. backhoe-loader, incl. operator, equip Rental, fuel, oil, and mobilization. Hauling of excavated material is Included. Prices based on one day production of 360 C.Y. in medium soil Without backfilling.					
Equipment operator	8.000	Hr.		465	465
Backhoe-loader, 3/4 C.Y.	1.000	Day	152.90		152.90
Operating expense (fuel, oil)	8.000	Hr.	95.92		95.92
Hauling, 12 C.Y. trucks, 1 mile round trip	360.000	C.Y.	727.20	511.20	1,238.40
Mobilization	1.000	Ea.	160	103	263
TOTAL			1,136.02	1,079.20	2,215.22
COST PER C.Y.		C.Y.	3.15	2.99	6.15

A2010 140	Excavation, Foundation	COST PER C.Y. EQUIP.	LABOR	TOTAL
1400	For alternate size excavations:			
1500				
1600	100 C.Y.	3.90	5.05	8.95
1700	200 C.Y.	3.52	3.25	6.77
1800	300 C.Y.	3.27	3.06	6.33
1900	400 C.Y.	3.10	2.96	6.06
2000	500 C.Y.	3.04	2.93	5.97
2100	600 C.Y.	2.97	2.88	5.85
2200	700 C.Y.	2.94	2.83	5.77
2300				
2400				
2500				
2900	Dust protection, add, minimum		2%	2%
3000	Maximum		11%	11%
3100	Equipment usage curtailment, add, minimum		1%	1%
3200	Maximum		10%	10%
3300	Material handling & storage limitation, add, minimum		1%	1%
3400	Maximum		7%	7%
3500	Protection of existing work, add, minimum		2%	2%
3600	Maximum		7%	7%
3700	Shift work requirements, add, minimum		5%	5%
3800	Maximum		30%	30%
3900	Temporary shoring and bracing, add, minimum		5%	5%
4000	Maximum		12%	12%

Important: See the Reference Section for critical supporting data - Reference Numbers and City Cost Indexes

A2020 Basement Walls

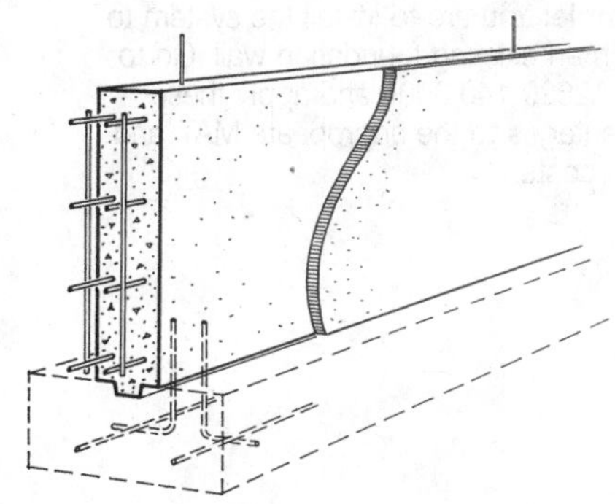

This page illustrates and describes a concrete wall system including concrete, placing concrete, forms, reinforcing, insulation, waterproofing and anchor bolts. Lines within System Components give the unit price and total price per linear foot for this system. Prices for alternate concrete wall systems are on Line Items A2020 120 1500 thru 2600. Both material quantities and labor costs have been adjusted for the system listed.

Factors: To adjust for job conditions other than normal working situations use Lines A2020 120 2900 thru 4000.

Example: You are to install this wall system where delivery of material is difficult. Go to Line A2020 120 3600 and apply these percentages to the appropriate MAT. and INST. costs.

System Components	QUANTITY	UNIT	COST PER L.F. MAT.	COST PER L.F. INST.	COST PER L.F. TOTAL
Cast in place concrete foundation wall, 8″ thick, 3′ high, 2500 psi					
Concrete including forms, reinforcing, waterproofing, and anchor bolts.					
Concrete, 2500 psi, 8″ thick, 3′ high	.070	C.Y.	6.55		6.55
Forms, wall, 4 uses	6.000	S.F.	6.12	37.80	43.92
Reinforcing	6.000	Lb.	2.94	2.16	5.10
Placing concrete	.070	C.Y.		2.74	2.74
Waterproofing	3.000	S.F.	.57	2.64	3.21
Rigid insulaton, 1″ polystyrene	3.000	S.F.	.84	2.07	2.91
Anchor bolts, 1/2″ diameter, 4′ O.C.	.250	Ea.	.36	.63	.99
TOTAL			17.38	48.04	65.42

A2020 120	Concrete Wall	COST PER L.F. MAT.	COST PER L.F. INST.	COST PER L.F. TOTAL
1400	For alternate wall systems:			
1500	8″ thick, 2500 psi concrete, 4′ high	23.50	64.50	88
1600	6′ high	35.50	95.50	131
1700	8′ high	47	128	175
1800	3500 psi concrete, 4′ high	24	64.50	88.50
1900	6′ high	36	95.50	131.50
2000	8′ high	48	128	176
2100	12″ thick, 2500 psi concrete, 4′ high	30.50	67.50	98
2200	6′ high	44	100	144
2300	8′ high	58.50	132	190.50
2400	3500 psi concrete, 4′ high	31	67.50	98.50
2500	8′ high	59.50	132	191.50
2600	10′ high	74	166	240
2700				
2900	Cut & patch to match existing construction, add, minimum	2%	3%	
3000	Maximum	5%	9%	
3100	Dust protection, add, minimum	1%	2%	
3200	Maximum	4%	11%	
3300	Equipment usage curtailment, add, minimum	1%	1%	
3400	Maximum	3%	10%	
3500	Material handling & storage limitation, add, minimum	1%	1%	
3600	Maximum	6%	7%	
3700	Protection of existing work, add, minimum	2%	2%	
3800	Maximum	5%	7%	
3900	Shift work requirements, add, minimum		5%	
4000	Maximum		30%	

A2020 Basement Walls

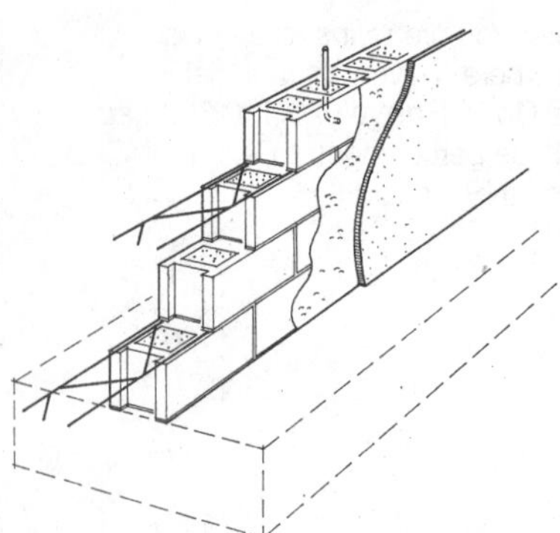

This page illustrates and describes a concrete block wall system including concrete block, masonry reinforcing, parging, waterproofing insulation and anchor bolts. Lines within System Components give the unit price and total price per linear foot for this system. Prices for alternate concrete block wall systems are on Line Items A2020 140 1300 thru 2600. Both material quantities and labor costs have been adjusted for the system listed.

Factors: To adjust for job conditions other than normal working situations use Lines A2020 140 2900 thru 4000.

Example: You are to install the system to match an existing foundation wall. Go to Line A2020 140 3000 and apply these percentages to the appropriate MAT. and INST. costs.

System Components	QUANTITY	UNIT	COST PER L.F. MAT.	COST PER L.F. INST.	COST PER L.F. TOTAL
Concrete block, 8″ thick, masonry reinforcing, parged and Waterproofed, insulation and anchor bolts, wall 2′-8″ high.					
Concrete block, 8″ x 8″ x 16″	2.670	S.F.	6.78	13.48	20.26
Masonry reinforcing	2.000	L.F.	.38	.32	.70
Parging	.193	S.Y.	.71	2.02	2.73
Waterproofing	2.670	S.F.	.51	2.35	2.86
Insulation, 1″ rigid polystyrene	2.670	S.F.	.75	1.84	2.59
Anchor bolts, 1/2″ diameter, 4′ O.C.	.250	Ea.	.36	.63	.99
TOTAL			9.49	20.64	30.13

A2020 140	Concrete Block Wall	COST PER L.F. MAT.	COST PER L.F. INST.	COST PER L.F. TOTAL
1200	For alternate wall systems:			
1300	8″ thick block, 4′ high	14.05	30.50	44.55
1400	6′ high	21	46	67
1500	8′ high	27.50	61	88.50
1600	Grouted solid, 4′ high	18.15	41	59.15
1700	6′ high	27	61.50	88.50
1800	8′ high	36	81.50	117.50
2100	12″ thick block, 4′ high	17.95	44	61.95
2200	6′ high	26.50	66	92.50
2300	8′ high	35.50	87.50	123
2400	Grouted solid, 4′ high	24.50	55	79.50
2500	6′ high	37	83	120
2600	8′ high	49	110	159
2700				
2900	Cut & patch to match existing construction, add, minimum	2%	3%	
3000	Maximum	5%	9%	
3100	Dust protection, add, minimum	1%	2%	
3200	Maximum	4%	11%	
3300	Equipment usage curtailment, add, minimum	1%	1%	
3400	Maximum	3%	10%	
3500	Material handling & storage limitation, add, minimum	1%	1%	
3600	Maximum	6%	7%	
3700	Protection of existing work, add, minimum	2%	2%	
3800	Maximum	5%	7%	
3900	Shift work requirements, add, minimum		5%	
4000	Maximum		30%	

For information about Means Estimating Seminars, see yellow pages 12 and 13 in back of book

B SHELL

B1010 Floor Construction

This page illustrates and describes a reinforced concrete slab system including concrete, placing concrete, formwork, reinforcing, steel trowel finish, V.A. floor tile and acoustical spray ceiling finish. Lines within System Components give the unit price per square foot for this system. Prices for alternate reinforced concrete slab systems are on Line Items B1010 225 1500 thru 1800. Both material quantities and labor costs have been adjusted for the system listed.

Factors: To adjust for job conditions other than normal working situations use Lines B1010 225 2700 thru 4000.

Example: You are to install the system to match an existing floor system. Go to Line B1010 225 2800 and apply these percentages to the appropriate MAT. and INST. costs.

System Components	QUANTITY	UNIT	COST PER S.F.		
			MAT.	INST.	TOTAL
Flat slab system, reinforced concrete with vinyl tile floor, sprayed					
Acoustical ceiling finish, not including columns.					
Concrete, 4000 psi, 6" thick	.020	C.Y.	2		2
Placing concrete	.020	C.Y.		.56	.56
Formwork, 4 use	1.000	S.F.	1.54	4.90	6.44
Edge form	.120	S.F.	.07	.95	1.02
Reinforcing steel	3.000	Lb.	1.47	1.56	3.03
Steel trowel finish	1.000	S.F.		.72	.72
Vinyl asbestos floor tile	1.000	S.F.	3.03	.86	3.89
Acoustical spray ceiling finish	1.000	S.F.	.24	.14	.38
TOTAL			8.35	9.69	18.04

B1010 225	Floor - Ceiling, Concrete Slab	COST PER S.F.		
		MAT.	INST.	TOTAL
1400	For alternate slab systems:			
1500	Concrete, 4000 psi, 7" thick	8.75	10.15	18.90
1600	8" thick	9.35	10.65	20
1700	9" thick	9.85	11.10	20.95
1800	10" thick	10.45	11.70	22.15
1900				
2000				
2100				
2200				
2300				
2400				
2500				
2700	Cut & patch to match existing construction, add, minimum	2%	3%	
2800	Maximum	5%	9%	
2900	Dust protection, add, minimum	1%	2%	
3000	Maximum	4%	11%	
3100	Equipment usage curtailment, add, minimum	1%	1%	
3200	Maximum	3%	10%	
3300	Material handling & storage limitation, add, minimum	1%	1%	
3400	Maximum	6%	7%	
3500	Protection of existing work, add, minimum	2%	2%	
3600	Maximum	5%	7%	
3700	Shift work requirements, add, minimum		5%	
3800	Maximum		30%	
3900	Temporary shoring and bracing, add, minimum	2%	5%	
4000	Maximum	5%	12%	

B1010 Floor Construction

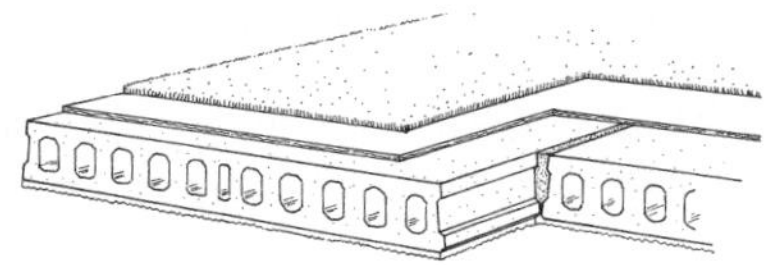

This page illustrates and describes a hollow core prestressed concrete panel system including hollow core slab, grout, carpet, carpet padding, and sprayed ceiling. Lines within System Components give the unit price and total price per square foot for this system. Prices for alternate hollow core prestressed concrete panel systems are on Line Items B1010 232 1300 thru 1600. Both material and labor costs have been adjusted for the system listed.

Factors: To adjust for job conditions other than normal working situations use Lines B1010 232 2700 thru 3800.

Example: You are to install the system where dust control is a major concern. Go to Line B1010 232 2800 and apply these percentages to the appropriate MAT. and INST. costs.

System Components	QUANTITY	UNIT	COST PER S.F. MAT.	INST.	TOTAL
Precast hollow core plank with carpeted floors, padding					
And sprayed textured ceiling.					
Hollow core plank, 4″ thick with grout topping	1.000	S.F.	4.88	2.84	7.72
Nylon carpet, 26 oz. medium traffic	.110	S.Y.	3.03	.63	3.66
Carpet padding, minimum quality	.110	S.Y.	.43	.31	.74
Sprayed texture ceiling	1.000	S.F.	.06	.52	.58
TOTAL			8.40	4.30	12.70

B1010 232	Floor - Ceiling, Conc. Panel	COST PER S.F. MAT.	INST.	TOTAL
1200	For alternate floor systems:			
1300	Hollow core concrete plank, with grout, 6″ thick	9.55	3.88	13.43
1400	8″ thick	9.85	3.59	13.44
1500	10″ thick	10.20	3.35	13.55
1600	12″ thick	11.10	3.16	14.26
1700				
1800				
1900				
2000				
2100				
2200				
2300				
2400				
2500				
2700	Dust protection, add, minimum	1%	2%	
2800	Maximum	4%	11%	
2900	Equipment usage curtailment, add, minimum	1%	1%	
3000	Maximum	3%	10%	
3100	Material handling & storage limitation, add, minimum	1%	1%	
3200	Maximum	6%	7%	
3300	Protection of existing work, add, minimum	2%	2%	
3400	Maximum	5%	7%	
3500	Shift work requirements, add, minimum		5%	
3600	Maximum		30%	
3700	Temporary shoring and bracing, add, minimum	2%	5%	
3800	Maximum	5%	12%	

B1010 Floor Construction

This page illustrates and describes a hollow core plank system including precast hollow core plank, concrete topping, V.A. tile, concealed suspension system and ceiling tile. Lines within System Components give the unit price and total price per square foot for this system. Prices for alternate hollow core plank systems are on Line Items B1010 233 1500 thru 1800. Both material quantities and labor costs have been adjusted for the system listed.

Factors: To adjust for job conditions other than normal working situations use Lines B1010 233 2900 thru 4000.

Example: You are to install the system where equipment usage is a problem. Go to Line B1010 233 3200 and apply these percentages to the appropriate MAT. and INST. costs.

System Components	QUANTITY	UNIT	COST PER S.F. MAT.	COST PER S.F. INST.	COST PER S.F. TOTAL
Precast hollow core plank with 2″ topping, vinyl composition floor tile					
And suspended acoustical tile ceiling.					
Hollow core concrete plank, 4″ thick	1.000	S.F.	4.88	2.84	7.72
Concrete topping, 2″ thick	1.000	S.F.	.85	4.30	5.15
Vinyl composition tile, .08″ thick	1.000	S.F.	1.13	.86	1.99
Concealed Z bar suspension system, 12″ module	1.000	S.F.	.52	.91	1.43
Ceiling tile, mineral fiber, 3/4″ thick	1.000	S.F.	1.55	.47	2.02
TOTAL			8.93	9.38	18.31

B1010 233	Floor - Ceiling, Conc. Plank	COST PER S.F. MAT.	COST PER S.F. INST.	COST PER S.F. TOTAL
1400	For alternate floor systems:			
1500	Hollow core concrete plank, 6″ thick	10.10	8.95	19.05
1600	8″ thick	10.40	8.65	19.05
1700	10″ thick	10.75	8.40	19.15
1800	12″ thick	11.65	8.25	19.90
1900				
2000				
2100				
2200				
2300				
2400				
2500				
2600				
2700				
2900	Dust protection, add, minimum	1%	2%	
3000	Maximum	4%	11%	
3100	Equipment usage curtailment, add, minimum	1%	1%	
3200	Maximum	3%	10%	
3300	Material handling & storage limitation, add, minimum	1%	1%	
3400	Maximum	6%	7%	
3500	Protection of existing work, add, minimum	2%	2%	
3600	Maximum	5%	7%	
3700	Shift work requirements, add, minimum		5%	
3800	Maximum		30%	
3900	Temporary shoring and bracing, add, minimum	2%	5%	
4000	Maximum	5%	12%	

B1010 Floor Construction

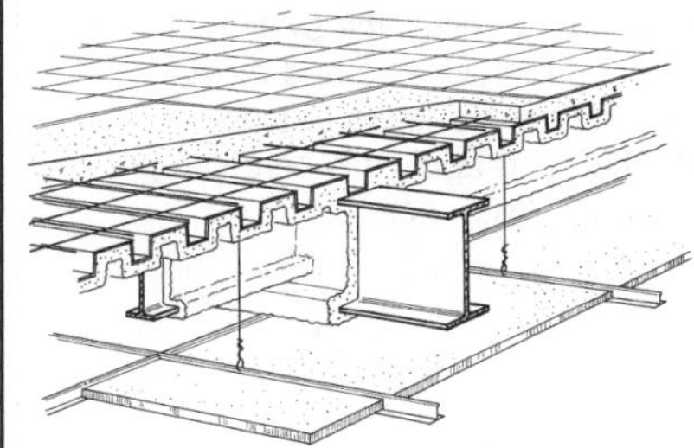

This page illustrates and describes a structural steel w/metal decking and concrete system including steel beams, steel decking, shear studs, concrete, placing concrete, edge form, steel trowel finish, curing, wire fabric, fireproofing, beams and decking, tile floor, and suspended ceiling. Lines within System Components give the unit price and total price per square foot for this system. Prices for alternate structural steel w/metal decking and concrete systems are on Line Items B1010 243 1900 thru 2400. Both material quantities and labor costs have been adjusted for the system listed.

Factors: To adjust for job conditions other than normal working situations use Lines B1010 243 2900 thru 4000.

Example: You are to install the system where material handling and storage are a problem. Go to Line B1010 243 3400 and apply these percentages to the appropriate MAT. and INST. costs.

System Components	QUANTITY	UNIT	COST PER S.F. MAT.	COST PER S.F. INST.	COST PER S.F. TOTAL
Composite structural beams, 20 ga. 3″ deep steel decking, shear studs, 3000 psi, concrete, placing concrete, edge forms, steel trowel finish, Welded wire fabric fireproofing, tile floor and suspended ceiling.					
Steel deck, 20 gage, 3″ deep, galvanized	1.000	S.F.	1.88	.84	2.72
Structural steel framing	.004	Ton	8.40	2.30	10.70
Shear studs, 3/4″	.250	Ea.	.10	.37	.47
Concrete, 3000 psi, 5″ thick	.015	C.Y.	1.43		1.43
Place concrete	.015	C.Y.		.42	.42
Edge form	.140	L.F.	.03	.50	.53
Steel trowel finish	1.000	S.F.		.72	.72
Welded wire fabric 6 x 6 - #10/10	.011	C.S.F.	.15	.34	.49
Fireproofing, sprayed	1.880	S.F.	.88	1.60	2.48
Vinyl composition floor tile	1.000	S.F.	.95	.86	1.81
Suspended acoustical ceiling	1.000	S.F.	1.88	1.24	3.12
TOTAL			15.70	9.19	24.89

B1010 243	Floor - Ceiling, Struc. Steel	COST PER S.F. MAT.	COST PER S.F. INST.	COST PER S.F. TOTAL
1800	For alternate floor systems:			
1900	Composite deck, galvanized, 3″ deep, 22 ga.	15.50	9.15	24.65
2000	18 ga.	16.15	9.25	25.40
2100	16 ga.	16.90	9.30	26.20
2200	Composite deck, galvanized, 2″ deep, 22 ga.	15.35	9.05	24.40
2300	20 ga.	15.55	9.05	24.60
2400	16 ga.	16.55	9.15	25.70
2500				
2900	Dust protection, add, minimum	1%	2%	
3000	Maximum	4%	11%	
3100	Equipment usage curtailment, add, minimum	1%	1%	
3200	Maximum	3%	10%	
3300	Material handling & storage limitation, add, minimum	1%	1%	
3400	Maximum	6%	7%	
3500	Protection of existing work, add, minimum	2%	2%	
3600	Maximum	5%	7%	
3700	Shift work requirements, add, minimum		5%	
3800	Maximum		30%	
3900	Temporary shoring and bracing, add, minimum	2%	5%	
4000	Maximum	5%	12%	

B1010 Floor Construction

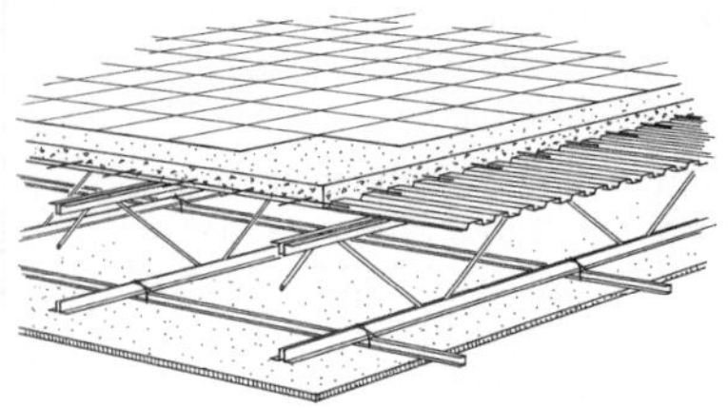

This page illustrates and describes an open web joist and steel slab-form system including open web steel joist, slab form, concrete, placing concrete, wire fabric, steel trowel finish, tile floor and plasterboard. Lines within System Components give the unit price and total price per square foot for this system. Prices for alternate open web joists and steel slab-form systems are on Line Items B1010 245 1900 thru 2200. Both material quantities and labor costs have been adjusted for the systems listed.

Factors: To adjust for job conditions other than normal working situations use Lines B1010 245 2900 thru 4000.

Example: You are to install the system in a congested commercial area and most work will be done at night. Go to Line B1010 245 3800 and apply this percentage to the appropriate INST. cost.

System Components	QUANTITY	UNIT	COST PER S.F. MAT.	INST.	TOTAL
Open web steel joists, 24″ O.C., slab form, 3000 psi concrete, welded Wire fabric, steel trowel finish, floor tile, 5/8″ drywall ceiling.					
Open web steel joists, 12″ deep, 5.2#/L.F., 24″ O.C.	2.630	Lb.	2.08	1	3.08
Slab form, 28 gage, 9/16″ deep, galvanized	1.000	S.F.	.96	.63	1.59
Concrete, 3000 psi, 2-1/2″ thick	.008	C.Y.	.76		.76
Placing concrete	.008	C.Y.		.23	.23
Welded wire fabric 6 x 6 - #10/10	.011	C.S.F.	.15	.34	.49
Steel trowel finish	1.000	S.F.		.72	.72
Vinyl composition floor tile	1.000	S.F.	.95	.86	1.81
Ceiling furring, 3/4″ channels, 24″ O.C.	1.000	S.F.	.15	.99	1.14
Gypsum drywall, 5/8″ thick, finished	1.000	S.F.	.37	1.23	1.60
Paint ceiling	1.000	S.F.	.16	.54	.70
TOTAL			5.58	6.54	12.12

B1010 245	Floor - Ceiling, Steel Joists	COST PER S.F. MAT.	INST.	TOTAL
1800	For alternate floor systems:			
1900	Open web joists, 16″ deep, 6.6#/L.F.	6.10	6.80	12.90
2000	20″ deep, 8.4# L.F.	6.80	7.15	13.95
2100	24″ deep, 11.5# L.F.	8.05	7.75	15.80
2200	26″ deep, 12.8# L.F.	8.55	8	16.55
2300				
2400				
2500				
2600				
2700				
2900	Dust protection, add, minimum	1%	2%	
3000	Maximum	4%	11%	
3100	Equipment usage curtailment, add, minimum	1%	1%	
3200	Maximum	3%	10%	
3300	Material handling & storage limitation, add, minimum	1%	1%	
3400	Maximum	6%	7%	
3500	Protection of existing work, add, minimum	2%	2%	
3600	Maximum	5%	7%	
3700	Shift work requirements, add, minimum		5%	
3800	Maximum		30%	
3900	Temporary shoring and bracing, add, minimum	2%	5%	
4000	Maximum	5%	12%	

 Important: See the Reference Section for critical supporting data - Reference Numbers and City Cost Indexes

B1010 Floor Construction

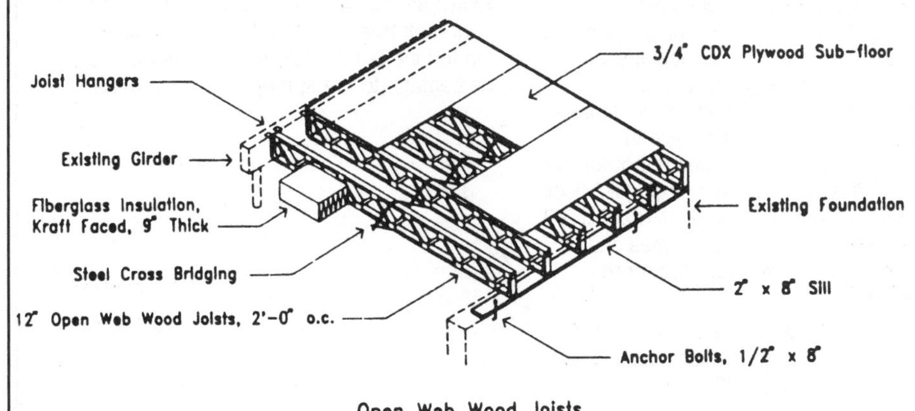

Open Web Wood Joists

This page illustrates and describes floor framing systems including sills, joist hangers, joists, bridging, subfloor and insulation. Lines within systems components give the unit price and total price per square foot for this system. Prices for alternate joist types are on line items B1010 262 0100 through 0700. Both material quantities and labor costs have been adjusted for the system listed.

Factors: To adjust for job conditions other than normal working situations use lines B1010 262 2700 through 4000.

Example: You are to install the system where delivery of material is difficult. Go to line B1010 262 3400 and apply these percentages to the appropriate MAT. and INST. costs.

System Components	QUANTITY	UNIT	COST PER S.F. MAT.	INST.	TOTAL
Open web wood joists, sill, butt to girder, 5/8" subfloor, insul.					
Anchor bolt, hook type, with nut and washer, 1/2" x 8" long	.250	Ea.	.02	.06	.08
Joist & beam hanger, 18 ga. galvanized	.750	Ea.	.07	.21	.28
Bridging, steel, compression, for 2" x 12" joists	.001	C.Pr.	.15	.24	.39
Framing, open web wood joists, 12" deep	.001	M.L.F.	2	1.08	3.08
Framing, sills, 2" x 8"	.001	M.B.F.	.71	1.40	2.11
Sub-floor, plywood, CDX, 3/4" thick	1.000	S.F.	.73	.70	1.43
Insulation, fiberglass, kraft face, 9" thick x 23" wide	1.000	S.F.	.78	.35	1.13
TOTAL			4.46	4.04	8.50

B1010 262	Open Web Wood Joists	COST PER S.F. MAT.	INST.	TOTAL
0080	For alternate floor systems:			
0100	12" open web joists, sill, butt to girder, 5/8"subflr.	4.46	4.04	8.50
0200	14" joists	4.79	4.11	8.90
0300	16" joists	4.89	4.16	9.05
0400	18" joists	4.94	4.24	9.18
0450	Wood struc."I"joists,24"O.C.,to 24'span,50 PSF LL, butt to girder	4.74	3.76	8.50
0500	55 PSF LL	4.91	3.81	8.72
0600	to 30'span, 45 PSF LL	5.30	3.70	9
0700	55 PSF LL	5.65	3.76	9.41
2700	Cut & patch to match existing construction, add, minimum	2%	3%	
2800	Maximum	5%	9%	
2900	Dust protection, add, minimum	1%	2%	
3000	Maximum	4%	11%	
3100	Equipment usage curtailment, add, minimum	1%	1%	
3200	Maximum	3%	10%	
3300	Material handling & storage limitation, add, minimum	1%	1%	
3400	Maximum	6%	7%	
3500	Protection of existing work, add, minimum	2%	2%	
3600	Maximum	5%	7%	
3700	Shift work requirements, add, minimum		5%	
3800	Maximum		30%	
3900	Temporary shoring and bracing, add, minimum	2%	5%	
4000	Maximum	5%	12%	

B1010 Floor Construction

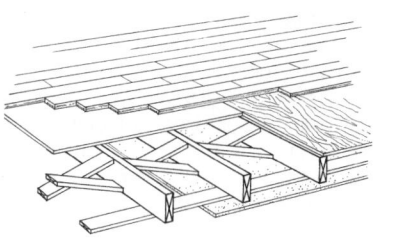

This page illustrates and describes a wood joist floor system including wood joist, oak floor, sub-floor, bridging, sand and finish floor, furring, plasterboard, taped, finished and painted ceiling. Lines within System Components give the unit price and total price per square foot for this system. Prices for alternate wood joist floor systems are on Line Items B1010 263 1700 thru 2300. Both material quantities and labor costs have been adjusted for the system listed.

Factors: To adjust for job conditions other than normal working situations use Lines B1010 263 2700 thru 4000.

Example: You are to install the system during off peak hours, 6 P.M. to 2 A.M. Go to Line B1010 263 3800 and apply this percentage to the appropriate INST. costs.

System Components	QUANTITY	UNIT	COST PER S.F. MAT.	INST.	TOTAL
Wood joists, 2″ x 8″, 16″ O.C.,oak floor (sanded & finished),1/2″ sub Floor, 1″x 3″ bridging, furring, 5/8″ drywall, taped, finished and painted.					
Wood joists, 2″ x 8″, 16″ O.C.	1.000	L.F.	.94	.86	1.80
Subfloor plywood CDX 1/2″	1.000	SF Flr.	.68	.63	1.31
Bridging 1″ x 3″	.150	Pr.	.07	.54	.61
Oak flooring, No. 1 common	1.000	S.F.	4.09	2.77	6.86
Sand and finish floor	1.000	S.F.	.78	1.23	2.01
Wood furring, 1″ x 3″, 16″ O.C.	1.000	L.F.	.31	1.35	1.66
Gypsum drywall, 5/8″ thick	1.000	S.F.	.33	.52	.85
Tape and finishing	1.000	S.F.	.04	.47	.51
Paint ceiling	1.000	S.F.	.16	.54	.70
TOTAL			7.40	8.91	16.31

B1010 263	Floor-Ceiling, Wood Joists	COST PER S.F. MAT.	INST.	TOTAL
1600	For alternate wood joist systems:			
1700	16″ on center, 2″ x 6″ joists	7.10	8.80	15.90
1800	2″ x 10″	7.90	9.10	17
1900	2″ x 12″	8.20	9.15	17.35
2000	2″ x 14″	8.85	9.25	18.10
2100	12″ on center, 2″ x 10″ joists	8.25	9.35	17.60
2200	2″ x 12″	8.65	9.40	18.05
2300	2″ x 14″	9.45	9.60	19.05
2400				
2500				
2700	Cut & patch to match existing construction, add, minimum	2%	3%	
2800	Maximum	5%	9%	
2900	Dust protection, add, minimum	1%	2%	
3000	Maximum	4%	11%	
3100	Equipment usage curtailment, add, minimum	1%	1%	
3200	Maximum	3%	10%	
3300	Material handling & storage limitation, add, minimum	1%	1%	
3400	Maximum	6%	7%	
3500	Protection of existing work, add, minimum	2%	2%	
3600	Maximum	5%	7%	
3700	Shift work requirements, add, minimum		5%	
3800	Maximum		30%	
3900	Temporary shoring and bracing, add, minimum	2%	5%	
4000	Maximum	5%	12%	

B1020 Roof Coverings

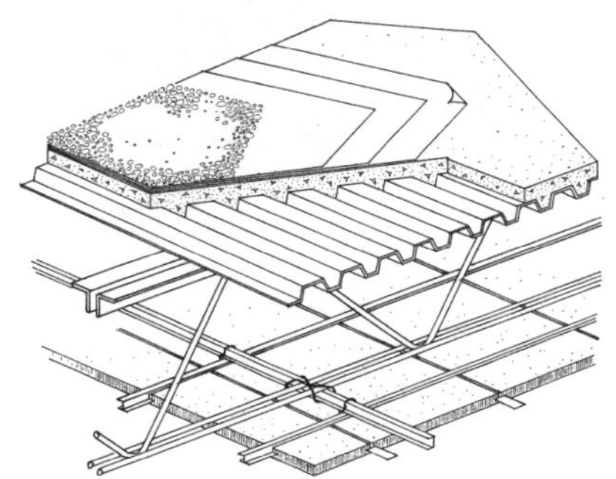

This page illustrates and describes a flat roof bar joist system including asphalt and gravel roof, lightweight concrete, metal decking, open web joists, and suspended acoustic ceiling. Lines within System Components give the unit price and total price per square foot for this system. Prices for alternate flat roof bar joist systems are on Line Items B1020 106 1300 thru 1600. Both material quantities and labor costs have been adjusted for the system listed.

Factors: To adjust for job conditions other than normal working situations use Lines B1020 106 2700 thru 4000.

Example: You are to install the system where material handling and storage present a problem. Go to Line B1020 106 3300 and apply these percentages to the appropriate MAT. and INST. costs.

System Components	QUANTITY	UNIT	COST PER S.F.		
			MAT.	INST.	TOTAL
Three ply asphalt and gravel roof on lightweight concrete, metal decking on Web joist 4′ O.C., with suspended acoustic ceiling.					
Open web joist, 10″ deep, 4′ O.C.	1.250	Lb.	.99	.47	1.46
Metal decking, 1-1/2″ deep, 22 Ga., galvanized	1.000	S.F.	1.61	.56	2.17
Gravel roofing, 3 ply asphalt	.010	C.S.F.	.73	1.48	2.21
Perlite roof fill, 3″ thick	1.000	S.F.	.85	.46	1.31
Suspended ceiling, 3/4″ mineral fiber on "Z" bar suspension	1.000	S.F.	1.89	3.14	5.03
TOTAL			6.07	6.11	12.18

B1020 106	Steel Joist Roof & Ceiling	COST PER S.F.		
		MAT.	INST.	TOTAL
1200	For alternate roof systems:			
1300	Open web bar joist, 16″ deep, 5#/L.F.	6.40	6.30	12.70
1400	18″ deep, 6.6#/L.F.	6.65	6.40	13.05
1500	20″ deep, 9.6#/L.F.	7	6.55	13.55
1600	24″ deep, 11.52#/L.F.	7.35	6.75	14.10
1700				
1800				
1900				
2000				
2100				
2200				
2300				
2400				
2500				
2700	Cut & patch to match existing construction, add, minimum	2%	3%	
2800	Maximum	5%	9%	
2900	Dust protection, add, minimum	1%	2%	
3000	Maximum	4%	11%	
3100	Equipment usage curtailment, add, minimum	1%	1%	
3200	Maximum	3%	10%	
3300	Material handling & storage limitation, add, minimum	1%	1%	
3400	Maximum	6%	7%	
3500	Protection of existing work, add, minimum	2%	2%	
3600	Maximum	5%	7%	
3700	Shift work requirements, add, minimum		5%	
3800	Maximum		30%	
3900	Temporary shoring and bracing, add, minimum	2%	5%	
4000	Maximum	5%	12%	

B1020 Roof Coverings

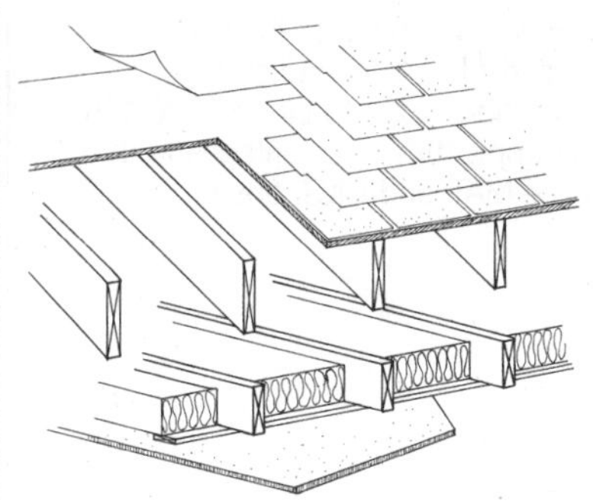

This page illustrates and describes a wood frame roof system including rafters, ceiling joists, sheathing, building paper, asphalt shingles, roof trim, furring, insulation, plaster and paint. Lines within System Components give the unit price and total price per square foot for this system. Prices for alternate wood frame roof systems are on Line Items B1020 202 1900 thru 2700. Both material quantities and labor costs have been adjusted for the system listed.

Factors: To adjust for job conditions other than normal working situations use Lines B1020 202 3300 thru 4000.

Example: You are to install the system while protecting existing work. Go to Line B1020 202 3800 and apply these percentages to the appropriate MAT. and INST. costs.

System Components	QUANTITY	UNIT	COST PER S.F. MAT.	INST.	TOTAL
Wood frame roof system, 4 in 12 pitch, including rafters, sheathing, Shingles, insulation, drywall, thin coat plaster, and painting.					
Rafters, 2″ x 6″, 16″ O.C., 4 in 12 pitch	1.080	L.F.	.70	1.02	1.72
Ceiling joists, 2″ x 6″, 16 O.C.	1.000	L.F.	.65	.75	1.40
Sheathing, 1/2″ CDX	1.080	S.F.	.73	.72	1.45
Building paper, 15# felt	.011	C.S.F.	.04	.14	.18
Asphalt shingles, 240#	.011	C.S.F.	.52	.97	1.49
Roof trim	.100	L.F.	.21	.21	.42
Furring, 1″ x 3″, 16″ O.C.	1.000	L.F.	.31	1.35	1.66
Fiberglass insulation, 6″ batts	1.000	S.F.	.56	.35	.91
Gypsum board, 1/2″ thick	1.000	S.F.	.30	.52	.82
Thin coat plaster	1.000	S.F.	.08	.58	.66
Paint, roller, 2 coats	1.000	S.F.	.16	.54	.70
TOTAL			4.26	7.15	11.41

B1020 202	Wood Frame Roof & Ceiling	COST PER S.F. MAT.	INST.	TOTAL
1800	For alternate roof systems:			
1900	Rafters 16″ O.C., 2″ x 8″	4.58	7.20	11.78
2000	2″ x 10″	5.10	7.75	12.85
2100	2″ x 12″	5.45	7.90	13.35
2200	Rafters 24″ O.C., 2″ x 6″	3.92	6.70	10.62
2300	2″ x 8″	4.15	6.75	10.90
2400	2″ x 10″	4.54	7.15	11.69
2500	2″ x 12″	4.81	7.25	12.06
2600	Roof pitch, 6 in 12, add	3%	10%	
2700	8 in 12, add	5%	12%	
2800				
2900				
3000				
3100				
3300	Cut & patch to match existing construction, add, minimum	2%	3%	
3400	Maximum	5%	9%	
3500	Material handling & storage limitation, add, minimum	1%	1%	
3600	Maximum	6%	7%	
3700	Protection of existing work, add, minimum	2%	2%	
3800	Maximum	5%	7%	
3900	Shift work requirements, add, minimum		5%	
4000	Maximum		30%	

B1020 Roof Construction

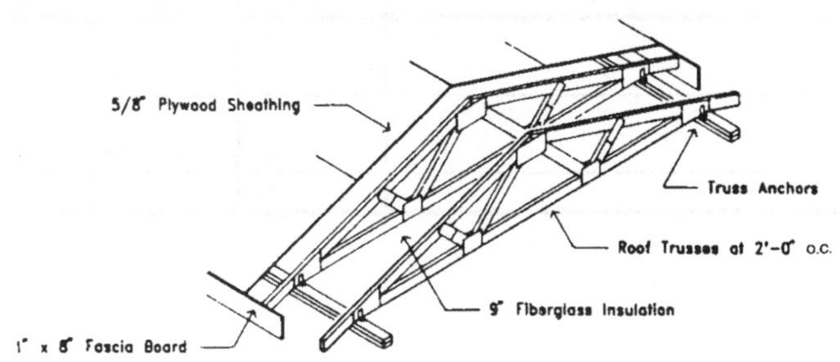

Wood Fabricated Roof Truss System

This page illustrates and describes a sloped roof truss framing system including trusses, truss anchors, sheathing, fascia and insulation. Lines B1020 210 1500 through 1800 are for a flat roof framing system including flat roof trusses, truss anchors, sheathing, fascia and insulation. Lines within system components give the unit price and total price per square foot for this system. Prices for alternate sizes and systems are given in line items B1020 210 0100 through 1800. Both material quantities and labor costs have been adjusted for the system listed.

Factors: To adjust for job conditions other than normal working situations use lines B1020 210 2700 through 4000.

Example: You are to install the system where crane placement and movement will be difficult. Go to line B1020 210 3200 and apply these percentages to the appropriate MAT. and INST. costs.

System Components	QUANTITY	UNIT	COST PER S.F. MAT.	INST.	TOTAL
Roof truss, 2′ O.C., 4/12 slope, 12′ span, 5/8 sheath., fascia,insul.					
Timber connectors, rafter anchors, galv., 1 1/2″ x 5 1/4″	.084	Ea.	.05	.27	.32
Plywood sheathing, CDX, 5/8″ thick	1.054	S.F.	.77	.77	1.54
Truss, 2′ O.C., metal plate connectors, 12′ span	.042	Ea.	1.45	1.47	2.92
Molding, fascia trim, 1″ x 8″	.167	L.F.	.35	.35	.70
Insulation, fiberglass, kraft face, 9″ thick x 23″ wide	1.000	S.F.	.78	.35	1.13
TOTAL			3.40	3.21	6.61

B1020 210	Roof Truss, Wood	COST PER S.F. MAT.	INST.	TOTAL
0090	For alternate roof systems:			
0100	Roof truss, 2′O.C., 4/12p, 1′ovhg, 12′span, 5/8″sheath, fascia, insul.	3.40	3.21	6.61
0200	20′ span	3.21	2.74	5.95
0300	24′ span	3.04	2.57	5.61
0400	26′ span	5.20	3.45	8.65
0500	28′ span	3.01	2.51	5.52
0600	30′ span	3.16	2.47	5.63
0700	32′ span	3.49	2.43	5.92
0800	34′ span	3.73	2.41	6.14
0900	8/12 pitch, 20′ span	4	2.96	6.96
1000	24′ span	3.95	2.77	6.72
1100	26′ span	5.90	3.75	9.65
1200	28′ span	3.93	2.70	6.63
1300	32′ span	4	2.65	6.65
1400	36′ span	4.07	2.60	6.67
1500	Flat roof frame, fab.strl.joists, 2′ O.C., 15′-24′ span, 50 PSF LL	4	2.17	6.17
1600	55 PSF LL	4.17	2.22	6.39
1700	24′-30′span, 45 PSF LL	4.51	2.08	6.59
1800	55 PSF LL	4.89	2.14	7.03
2700	Cut & patch to match existing construction, add, minimum	2%	3%	
2800	Maximum	5%	9%	
2900	Dust protection, add, minimum	1%	2%	
3000	Maximum	4%	11%	
3100	Equipment usage curtailment, add, minimum	1%	1%	
3200	Maximum	3%	10%	
3300	Material handling & storage limitation, add, minimum	1%	1%	
3400	Maximum	6%	7%	
3500	Protection of existing work, add, minimum	2%	2%	

B1020 Roof Construction

B1020 210	Roof Truss, Wood	COST PER S.F.		
		MAT.	INST.	TOTAL
3600	Maximum	5%	7%	
3700	Shift work requirements, add, minimum		5%	
3800	Maximum		30%	
3900	Temporary shoring and bracing, add, minimum	2%	5%	
4000	Maximum	5%	12%	

Important: See the Reference Section for critical supporting data - Reference Numbers and City Cost Indexes

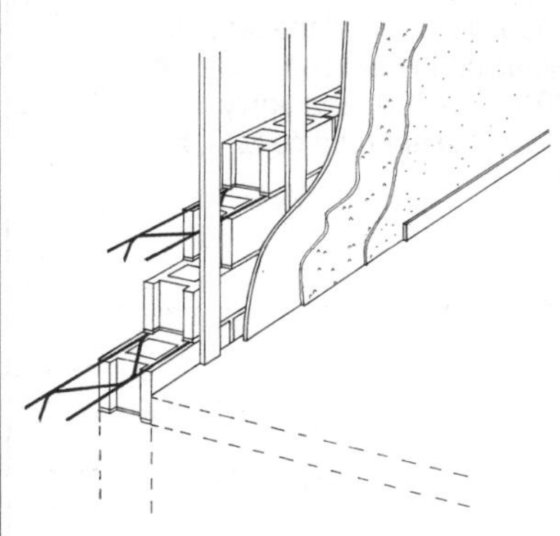

This page illustrates and describes a masonry concrete block system including concrete block wall, pointed, reinforcing, waterproofing, gypsum plaster on gypsum lath, and furring. Lines within System Components give the unit price and total price per square foot for this system. Prices for alternate masonry concrete block wall systems are on Line Items B2010 108 1500 thru 2200. Both material quantities and labor costs have been adjusted for the system listed.

Factors: To adjust for job conditions other than normal working situations use Lines B2010 108 2700 thru 4000.

Example: You are to install the system and match existing construction at several locations. Go to Line B2010 108 2800 and apply these percentages to the appropriate MAT. and INST. costs.

System Components	QUANTITY	UNIT	COST PER S.F. MAT.	COST PER S.F. INST.	COST PER S.F. TOTAL
Concrete block, 8″ thick, reinforced every 2 courses, waterproofing gypsum Plaster over gypsum lath on 1″ x 3″ furring, interior painting & baseboard.					
Concrete block, 8″ x 8″ x 16″, reinforced	1.000	S.F.	3.67	5.15	8.82
Silicone waterproofing, 2 coats	1.000	S.F.	.71	.15	.86
Bituminous coating, 1/16″ thick	1.000	S.F.	.20	.88	1.08
Furring, 1″ x 3″, 16″ O.C.	1.000	L.F.	.31	1.81	2.12
Gypsum lath, 3/8″ thick	.110	S.Y.	.65	.54	1.19
Gypsum plaster, 2 coats	.110	S.Y.	.40	2.19	2.59
Painting, 2 coats	1.000	S.F.	.11	.42	.53
Baseboard wood, 9/16″ x 2-5/8″	.100	L.F.	.24	.20	.44
Paint baseboard, primer + 1 coat enamel	.100	L.F.	.01	.05	.06
TOTAL			6.30	11.39	17.69

B2010 108	Masonry Wall, Concrete Block	COST PER S.F. MAT.	COST PER S.F. INST.	COST PER S.F. TOTAL
1400	For alternate exterior wall systems:			
1500	8″ thick block, fluted 2 sides	8.25	13.35	21.60
1600	Deep grooved	6.60	13.35	19.95
1700	Slump block	8.35	11.60	19.95
1800	Split rib	6.25	13.35	19.60
1900				
2000	12″ thick block, regular	5.60	13.55	19.15
2100	Slump block	13.85	12.60	26.45
2200	Split rib	6.95	15.30	22.25
2300				
2700	Cut & patch to match existing construction, add, minimum	2%	3%	
2800	Maximum	5%	9%	
2900	Dust protection, add, minimum	1%	2%	
3000	Maximum	4%	11%	
3100	Equipment usage curtailment, add, minimum	1%	1%	
3200	Maximum	3%	10%	
3300	Material handling & storage limitation, add, minimum	1%	1%	
3400	Maximum	6%	7%	
3500	Protection of existing work, add, minimum	2%	2%	
3600	Maximum	5%	7%	
3700	Shift work requirements, add, minimum		5%	
3800	Maximum		30%	
3900	Temporary shoring and bracing, add, minimum	2%	5%	
4000	Maximum	5%	12%	

B2010 Exterior Walls

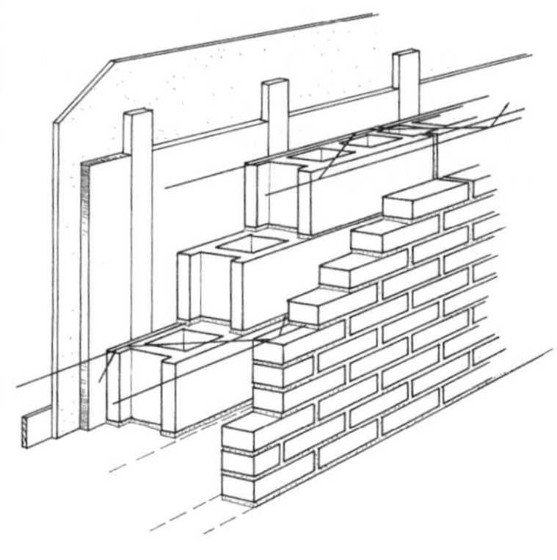

This page illustrates and describes a masonry wall, brick-stone system including brick, concrete block, durawall, insulation, plasterboard, taped and finished, furring, baseboard and painting interior. Lines within System Components give the unit price and total price per square foot for this system. Prices for alternate masonry wall, brick-stone systems are on Line Item B2010 129 1500 thru 2500. Both material quantities and labor costs have been adjusted for the system listed.

Factors: To adjust for job conditions other than normal working situations use Lines B2010 129 3100 thru 4200.

Example: You are to install the system without damaging the existing work. Go to Line B2010 129 3900 and apply these percentages to the appropriate MAT. and INST. costs.

System Components	QUANTITY	UNIT	COST PER S.F.		
			MAT.	INST.	TOTAL
Face brick, 4"thick, concrete block back-up, reinforce every second course, 3/4"insulation, furring, 1/2"drywall, taped, finish, and painted, baseboard					
Face brick, 4" brick	1.000	S.F.	3.61	9.75	13.36
Concrete back-up block, reinforced 8" thick	1.000	S.F.	2.22	5.45	7.67
3/4" rigid polystyrene insulation	1.000	S.F.	.52	.59	1.11
Furring, 1" x 3", wood, 16" O.C.	1.000	L.F.	.31	.95	1.26
Drywall, 1/2" thick	1.000	S.F.	.30	.47	.77
Taping & finishing	1.000	S.F.	.04	.47	.51
Painting, 2 coats	1.000	S.F.	.16	.54	.70
Baseboard, wood, 9/16" x 2-5/8"	.100	L.F.	.24	.20	.44
Paint baseboard, primer + 1 coat enamel	.100	L.F.	.01	.05	.06
TOTAL			7.41	18.47	25.88

B2010 131	Masonry Wall, Brick - Stone	COST PER S.F.		
		MAT.	INST.	TOTAL
1400	For alternate exterior wall systems:			
1500	Face brick, Norman, 4" x 2-2/3" x 12" (4.5 per S.F.)	8.65	15.40	24.05
1600	Roman, 4" x 2" x 12" (6.0 per S.F.)	9.40	17.25	26.65
1700	Engineer, 4" x 3-1/5" x 8" (5.63 per S.F.)	7.45	16.95	24.40
1800	S.C.R., 6" x 2-2/3" x 12" (4.5 per S.F.)	8.90	15.60	24.50
1900	Jumbo, 6" x 4" x 12" (3.0 per S.F.)	8.25	13.65	21.90
2000	Norwegian, 6" x 3-1/5" x 12" (3.75 per S.F.)	7.45	14.40	21.85
2100				
2200				
2300	Stone, veneer, fieldstone, 6" thick	8.70	15.65	24.35
2400	Marble, 2" thick	41.50	27.50	69
2500	Limestone, 2" thick	31.50	18.75	50.25
3100	Cut & patch to match existing construction, add, minimum	2%	3%	
3200	Maximum	5%	9%	
3300	Dust protection, add, minimum	1%	2%	
3400	Maximum	4%	11%	
3500	Equipment usage curtailment, add, minimum	1%	1%	
3600	Maximum	3%	10%	
3700	Material handling & storage limitation, add, minimum	1%	1%	
3800	Maximum	6%	7%	
3900	Protection of existing work, add, minimum	2%	2%	
4000	Maximum	5%	7%	
4100	Shift work requirements, add, minimum		5%	
4200	Maximum		30%	

B2010 Exterior Walls

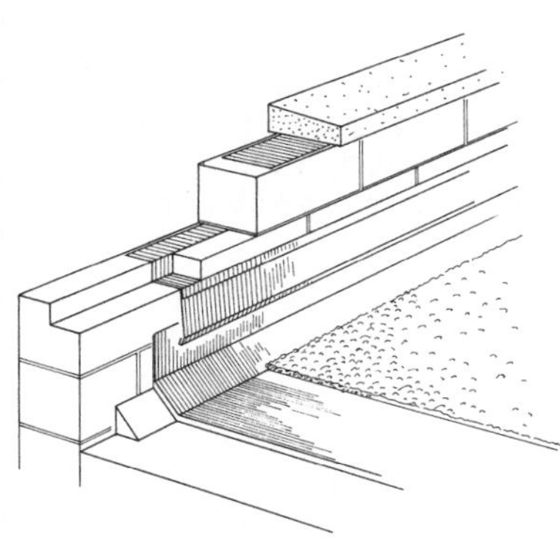

This page illustrates and describes a parapet wall system including wall, coping, flashing and cant strip. Lines within System Components give the unit price and cost per lineal foot for this system. Prices for alternate parapet wall systems are on Lines B2010 142 1300 thru 2400. Both material quantities and labor costs have been adjusted for the system listed.

Factors: To adjust for job conditions other than normal working situations, use Lines B2010 142 2900 thru 4000.

Example: You are to install the system without damaging the adjacent property. Go to Line B2010 142 3600 and apply these percentages to the appropriate MAT. and INST. costs.

System Components	QUANTITY	UNIT	COST PER L.F.		
			MAT.	INST.	TOTAL
Concrete block parapet, incl. reinf., coping, flashing & cant strip, 2′high					
8″ concrete block	2.000	S.F.	7.34	10.30	17.64
Masonry reinforcing	1.000	L.F.	.19	.16	.35
Coping, precast	1.000	L.F.	11.55	11.90	23.45
Roof cant	1.000	L.F.	2.37	1.35	3.72
Through wall flashing	1.000	L.F.	1.95	2.09	4.04
Cap flashing	1.000	L.F.	3.04	3.83	6.87
TOTAL			26.44	29.63	56.07

B2010 142	Parapet Wall	COST PER L.F.		
		MAT.	INST.	TOTAL
1200	For alternate systems:			
1300	Concrete block			
1400	12″ thick	34	38.50	72.50
1500	Split rib block, 8″ thick	26.50	33.50	60
1600	12″ thick	32	39.50	71.50
1700	Brick, common brick, 8″ wall	31	51	82
1800	12″ wall	37	64	101
1900	4″ brick, 4″ backup block	29	49	78
2000	8″ backup block	30.50	50	80.50
2100	Stucco on masonry	26.50	33	59.50
2200	On wood frame	11.35	27	38.35
2300	Wood, T 1-11 siding	23.50	32.50	56
2400	Boards, 1″ x 6″ cedar	25	37	62
2500				
2600				
2700				
2900	Dust protection, add, minimum	1%	2%	
3000	Maximum	4%	11%	
3100	Equipment usage curtailment, add, minimum	1%	1%	
3200	Maximum	3%	10%	
3300	Material handling & storage limitation, add, minimum	1%	1%	
3400	Maximum	6%	7%	
3500	Protection of existing work, add, minimum	2%	2%	
3600	Maximum	5%	7%	
3700	Shift work requirements, add, minimum		5%	
3800	Maximum		30%	
3900	Temporary shoring and bracing, add, minimum	2%	5%	
4000	Maximum	5%	12%	

B2010 Exterior Walls

This page illustrates and describes a masonry cleaning and restoration system, including staging, cleaning, and repointing. Lines within System Components give the unit price and cost per square foot for this system. Prices for alternate systems are on Lines B2010 143 1300 thru 2500. Both material quantities and labor costs have been adjusted for the system listed.

Factors: To adjust for conditions other than normal working situations, use Lines B2010 143 2900 thru 4200.

Example: You are to clean a wall and be concerned about dust control. Go to line B2010 143 3100 and apply these percentages to the appropriate MAT. and INST. costs.

System Components	QUANTITY	UNIT	COST PER S.F. MAT.	COST PER S.F. INST.	COST PER S.F. TOTAL
Repoint existing building, brick, running bond, high pressure cleaning, Water only, soft old mortar.					
Scaffold, building exterior	.010	C.S.F.	.28	.59	.87
Cleaning, high pressure water only	1.000	S.F.		1.47	1.47
Repoint, brick running bond	1.000	S.F.	.49	4.74	5.23
TOTAL			.77	6.80	7.57

B2010 143	Masonry Restoration - Cleaning	COST PER S.F. MAT.	COST PER S.F. INST.	COST PER S.F. TOTAL
1200	For alternate masonry surfaces:			
1300	Brick, common bond	.75	7	7.75
1400	Flemish bond	.78	7.30	8.08
1500	English bond	.78	7.85	8.63
1600	Add for wire cut face brick	.16		.16
1700	Stone, 2′ x 2′ blocks	.91	5.05	5.96
1800	2′ x 4′ blocks	.75	4.29	5.04
2000	Add to above prices for alternate cleaning systems:			
2100	Chemical brush and wash	.03	.52	.55
2200	High pressure chemical and water	.04	.50	.54
2300	Sandblasting, wet system	.20	1.36	1.56
2400	Dry system	.19	.80	.99
2500	Steam cleaning		.40	.40
2600				
2900	Cut & patch to match existing construction, add, minimum	2%	3%	
3000	Maximum	5%	9%	
3100	Dust protection, add, minimum	1%	2%	
3200	Maximum	4%	11%	
3300	Equipment usage curtailment, add, minimum	1%	1%	
3400	Maximum	3%	10%	
3500	Material handling & storage limitation, add, minimum	1%	1%	
3600	Maximum	6%	7%	
3800				
3900	Protection of existing work, add, minimum	2%	2%	
4000	Maximum	5%	7%	
4100	Shift work requirements, add, minimum		5%	
4200	Maximum		30%	

B2010 Exterior Walls

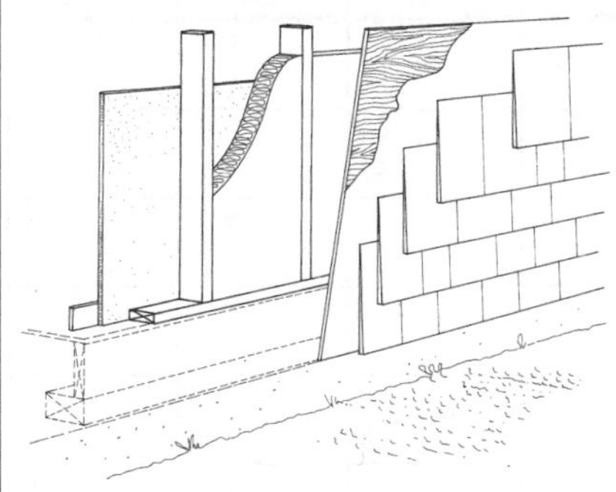

This page illustrates and describes a wood frame exterior wall system including wood studs, sheathing, felt, insulation, plasterboard, taped and finished, baseboard and painted interior. Lines within System Components give the unit price and total price per square foot for this system. Prices for alternate wood frame exterior wall systems are on Line Items B2010 150 1700 thru 2700. Both material quantities and labor costs have been adjusted for the system listed.

Factors: To adjust for job conditions other than normal working situations use Lines B2010 150 3100 thru 4200.

Example: You are to install the system with need for complete temporary bracing. Go to Line B2010 150 4200 and apply these percentages to the appropriate MAT. and INST. costs.

System Components	QUANTITY	UNIT	COST PER S.F. MAT.	COST PER S.F. INST.	COST PER S.F. TOTAL
Wood stud wall, cedar shingle siding, building paper, plywood sheathing, Insulation, 5/8" drywall, taped, finished and painted, baseboard.					
2" x 4" wood studs, 16" O.C.	.100	L.F.	.47	.94	1.41
1/2" CDX sheathing	1.000	S.F.	.68	.67	1.35
18" No. 1 red cedar shingles, 7-1/2" exposure	.008	C.S.F.	1.09	1.67	2.76
15# felt paper	.010	C.S.F.	.04	.13	.17
3-1/2" fiberglass insulation	1.000	S.F.	.52	.29	.81
5/8" drywall	1.000	S.F.	.33	.47	.80
Drywall finishing adder	1.000	S.F.	.04	.47	.51
Baseboard trim, stock pine, 9/16" x 3-1/2", painted	.100	L.F.	.24	.20	.44
Paint baseboard, primer + 1 coat enamel	.100	L.F.	.01	.05	.06
Paint, 2 coats, interior	1.000	S.F.	.16	.54	.70
TOTAL			3.58	5.43	9.01

B2010 150	Wood Frame Exterior Wall	COST PER S.F. MAT.	COST PER S.F. INST.	COST PER S.F. TOTAL
1600	For alternate exterior wall systems:			
1700	Aluminum siding, horizontal clapboard	3.82	5.60	9.42
1800	Cedar bevel siding, 1/2" x 6", vertical , painted	6	5.65	11.65
1900	Redwood siding 1" x 4" to 1" x 6" vertical, T & G	6.25	6.10	12.35
2000	Board and batten	4.91	5.55	10.46
2100	Ship lap siding	4.84	5.65	10.49
2200	Plywood, grooved (T1-11) fir	3.60	5.15	8.75
2300	Redwood	5.10	5.15	10.25
2400	Southern yellow pine	3.38	5.15	8.53
2500	Masonry on stud wall, stucco, wire and plaster	3.14	8.15	11.29
2600	Stone veneer	7.40	10.70	18.10
2700	Brick veneer, brick $275 per M	6.10	13.50	19.60
3100	Cut & patch to match existing construction, add, minimum	2%	3%	
3200	Maximum	5%	9%	
3300	Dust protection, add, minimum	1%	2%	
3400	Maximum	4%	11%	
3500	Material handling & storage limitation, add, minimum	1%	1%	
3600	Maximum	6%	7%	
3700	Protection of existing work, add, minimum	2%	2%	
3800	Maximum	5%	7%	
3900	Shift work requirements, add, minimum		5%	
4000	Maximum		30%	
4100	Temporary shoring and bracing, add, minimum	2%	5%	
4200	Maximum	5%	12%	

B2010 Exterior Walls

B2010 190	Selective Price Sheet	COST PER S.F.		
		MAT.	INST.	TOTAL
0100	Exterior surface, masonry, concrete block, standard 4" thick	1.25	4.98	6.23
0200	6" thick	1.86	5.35	7.21
0300	8" thick	2.04	5.70	7.74
0400	12" thick	2.97	7.35	10.32
0500	Split rib, 4" thick	2.72	6.20	8.92
0600	8" thick	3.64	7.15	10.79
0700	Brick running bond, standard size, 6.75/S.F.	3.61	9.75	13.36
0800	Buff, 6.75/S.F.	3.82	9.75	13.57
0900	Stucco, on frame	.65	4.42	5.07
1000	On masonry	.25	3.46	3.71
1100	Metal, aluminum, horizontal, plain	1.33	1.83	3.16
1200	Insulated	1.34	1.83	3.17
1300	Vertical, plain	1.33	1.83	3.16
1400	Insulated	1.39	1.91	3.30
1500	Wood, beveled siding, "A" grade cedar, 1/2" x 6"	3.52	1.88	5.40
1600	1/2" x 8"	3.42	1.71	5.13
1700	Shingles, 16" #1 red, 7-1/2" exposure	1.37	2.30	3.67
1800	18" perfections, 7-1/2" exposure	1.69	1.92	3.61
1900	Handsplit, 10" exposure	1.27	1.88	3.15
2000	White cedar, 7-1/2" exposure	1.28	2.36	3.64
2100	Vertical, board & batten, redwood	3.74	2.36	6.10
2200	White pine	.80	1.71	2.51
2300	T. & G. boards, redwood, 1" x 4"	3.19	3.14	6.33
2400	1' x 8"	2.75	2.51	5.26
2500				
2600	Interior surface, drywall, taped & finished, standard, 1/2"	.34	.98	1.32
2700	5/8" thick	.37	.98	1.35
2800	Fire resistant, 1/2" thick	.37	.98	1.35
2900	5/8" thick	.37	.98	1.35
3000	Moisture resistant, 1/2" thick	.31	.98	1.29
3100	5/8" thick	.34	.98	1.32
3200	Core board, 1" thick	.53	1.96	2.49
3300	Plaster, gypsum, 2 coats	.40	2.21	2.61
3400	3 coats	.57	2.65	3.22
3500	Perlite or vermiculite, 2 coats	.40	2.53	2.93
3600	3 coats	.72	3.11	3.83
3700	Gypsum lath, standard, 3/8" thick	.66	.54	1.20
3800	1/2" thick	.52	.58	1.10
3900	Fire resistant, 3/8" thick	.45	.66	1.11
4000	1/2" thick	.55	.71	1.26
4100	Metal lath, diamond, 2.5 lb.	.35	.54	.89
4200	Rib, 3.4 lb.	.54	.66	1.20
4300	Framing metal studs including top and bottom			
4400	Runners, walls 10' high			
4500	24" O.C., non load bearing 20 gauge, 2-1/2" wide	.35	.79	1.14
4600	3-5/8" wide	.42	.80	1.22
4700	4" wide	.46	.80	1.26
4800	6" wide	.59	.81	1.40
4900	Load bearing 18 gauge, 2-1/2" wide	.58	.88	1.46
5000	3-5/8" wide	.69	.90	1.59
5100	4" wide	6.20	9.15	15.35
5200	6" wide	.92	.94	1.86
5300	16" O.C., non load bearing 20 gauge, 2-1/2" wide	.48	1.21	1.69
5400	3-5/8" wide	.57	1.22	1.79
5500	4" wide	.63	1.24	1.87
5600	6" wide	.79	1.26	2.05
5700	Load bearing 18 gauge, 2-1/2" wide	.79	1.23	2.02
5800	3-5/8" wide	.94	1.24	2.18

B2010 Exterior Walls

B2010 190	Selective Price Sheet	COST PER S.F. MAT.	COST PER S.F. INST.	COST PER S.F. TOTAL
5900	4" wide	.99	1.28	2.27
6000	6" wide	1.25	1.29	2.54
6100	Framing wood studs incl. double top plate and			
6200	Single bottom plate, walls 10' high			
6300	24" O.C., 2" x 4"	.36	.76	1.12
6400	2" x 6"	.58	.82	1.40
6500	16" O.C., 2" x 4"	.47	.94	1.41
6600	2" x 6"	.75	1.05	1.80
6700	Sheathing, boards, 1" x 6"	1.32	1.45	2.77
6800	1"x 8"	1.23	1.23	2.46
6900	Plywood, 3/8" thick	.50	.79	1.29
7000	1/2" thick	.68	.84	1.52
7100	5/8" thick	.73	.90	1.63
7200	3/4" thick	.90	.97	1.87
7300	Wood fiber, 5/8" thick	.79	.79	1.58
7400	Gypsum weatherproof, 1/2" thick	.68	.90	1.58
7500	Insulation, fiberglass batts, 3-1/2" thick, R11	.40	.67	1.07
7600	6" thick, R19	.54	.79	1.33
7700	Poured 4" thick, fiberglass wool, R4/inch	.45	2.36	2.81
7800	Mineral wool, R3/inch	.39	2.36	2.75
7900	Polystyrene, R4/inch	3.18	2.36	5.54
8000	Perlite or vermiculite, R2.7/inch	1.87	2.36	4.23
8100	Rigid, fiberglass, R4.3/inch, 1" thick	.47	.47	.94
8200	R8.7/inch, 2" thick	1.09	.53	1.62

B2020 Exterior Windows

This page illustrates and describes an aluminum window system including double hung aluminum window, exterior and interior trim, hardware and insulating glass. Lines within System Components give the unit price and total price on a cost each basis for this system. Prices for alternate aluminum window systems are on Line Items B2020 110 1100 thru 2400. Both material quantities and labor costs have been adjusted for the system listed.

Factors: To adjust for job conditions other than normal working situations use Lines B2020 110 3100 thru 4000.

Example: You are to install the system and cut and patch to match existing construction. Go to Line B2020 110 3200 and apply these percentages to the appropriate MAT. and INST. costs.

System Components	QUANTITY	UNIT	COST EACH MAT.	COST EACH INST.	COST EACH TOTAL
Double hung aluminum window, 2′-4″ x 2′-6″ exterior and interior trim, Hardware glazed with insulating glass.					
Double hung aluminum window, 2′-4″ x 2′ 6″	1.000	Ea.	189.48	34.98	224.46
Exterior and interior trim	1.000	Set	32.50	36.50	69
Hardware	1.000	Set	2.09	19.65	21.74
Insulating glass	5.830	S.F.	61.80	25.77	87.57
TOTAL			285.87	116.90	402.77

B2020 110	Windows - Aluminum	COST EACH MAT.	COST EACH INST.	COST EACH TOTAL
1000	For alternate window systems:			
1100	Aluminum, double hung, 2′-8″ x 4′-6″	560	192	752
1200	3′- 4″ x 5′-6″	860	289	1,149
1300	Casement, 3′-6″ x 2′-4″	380	141	521
1400	4′-6″ x 2′-4″	485	176	661
1500	5′-6″ x 2′-4″	615	232	847
1600	Projected window, 2′-1″ x 3′-0″	285	121	406
1700	3′-5″ x 3′-0″	450	173	623
1800	4′-0″ x 4′-0″	710	265	975
1900	Horizontal sliding 3′-0″ x 3′-0″	305	150	455
2000	3′-6″ x 4′-0″	465	213	678
2100	5′-0″ x 6′-0″	980	410	1,390
2200	Picture window, 3′-8″ x 3′-1″	310	174	484
2300	4′-4″ x 4′-5″	510	266	776
2400	5′-8″ x 4′-9″	730	380	1,110
2500				
2600				
2700				
2800				
2900				
3100	Cut & patch to match existing construction, add, minimum	2%	3%	
3200	Maximum	5%	9%	
3300	Dust protection, add, minimum	1%	2%	
3400	Maximum	4%	11%	
3500	Material handling & storage limitation, add, minimum	1%	1%	
3600	Maximum	6%	7%	
3700	Protection of existing work, add, minimum	2%	2%	
3800	Maximum	5%	7%	
3900	Shift work requirements, add, minimum		5%	
4000	Maximum		30%	

B2020 Exterior Windows

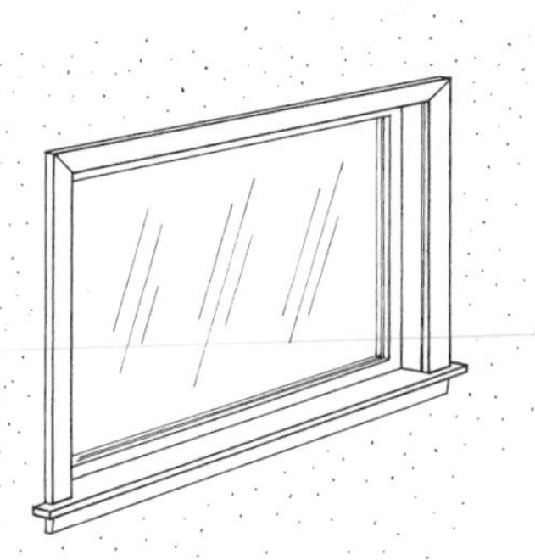

This page illustrates and describes a wood window system including wood picture window, exterior and interior trim, hardware and insulating glass. Lines within System Components give the unit price and total price on a cost-each basis for this system. Prices for alternate wood window systems are on Line Items B2020 112 1300 thru 1800. Both material quantities and labor costs have been adjusted for the system listed.

Factors: To adjust for job conditions other than normal working situations use Lines B2020 112 3100 thru 4000.

Example: You are to install the above system where dust control is vital. Go to Line B2020 112 3400 and apply these percentages to the appropriate MAT. and INST. costs.

System Components	QUANTITY	UNIT	COST EACH		
			MAT.	INST.	TOTAL
Wood picture window 4'-0" x 4'-6", exterior and interior trim, hardware, Glazed with insulating glass.					
4'-0" x 4'-6" wood picture window with insulating glass	1.000	Ea.	380	85.50	465.50
Exterior and interior trim	1.000	Set	68.50	78.50	147
Hardware	1.000	Set	2.09	19.65	21.74
TOTAL			450.59	183.65	634.24

B2020 112	Windows - Wood	COST EACH		
		MAT.	INST.	TOTAL
1200	For alternate window systems:			
1300	Picture window 5'-0" x 4'-0"	550	184	734
1400	6'-0" x 4'-6"	685	192	877
1500	Bow, bay window, 8'-0" x 5'-0", standard	2,300	192	2,492
1600	Deluxe	1,700	192	1,892
1700	Bow, bay window, 12'-0" x 6'-0" standard	2,925	255	3,180
1800	Deluxe	2,375	255	2,630
1900	Double hung, 3'-0" x 4'-0"	315	119	434
2000	4'-0" x 4'-6"	435	157	592
2100	Casement 2'-0" x 3'-0"	283	103	386
2200	2 leaf, 4'-0" x 4'-0"	750	172	922
2300	3 leaf, 6'-0" x 6'-0"	1,350	275	1,625
2400	Awning, 2'-10" x 1'-10"	305	103	408
2500	3'-6" x 2'-4"	385	119	504
2600	4'-0" x 3'-0"	565	157	722
2700	Horizontal sliding 3'-0" x 2'-0"	247	103	350
2800	4'-0" x 3'-6"	291	119	410
2900	6'-0" x 5'-0"	510	157	667
3100	Cut & patch to match existing construction, add, minimum	2%	3%	
3200	Maximum	5%	9%	
3300	Dust protection, add, minimum	1%	2%	
3400	Maximum	4%	11%	
3500	Material handling & storage limitation, add, minimum	1%	1%	
3600	Maximum	6%	7%	
3700	Protection of existing work, add, minimum	2%	2%	
3800	Maximum	5%	7%	
3900	Shift work requirements, add, minimum		5%	
4000	Maximum		30%	

B2020 Exterior Windows

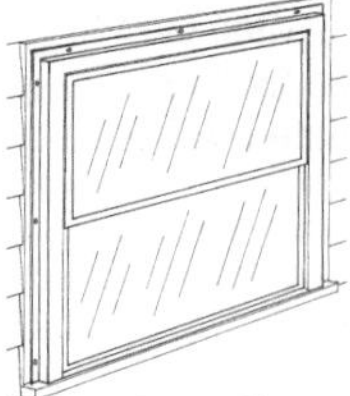

This page illustrates and describes storm window and door systems based on a cost-each price. Prices for alternate storm window and door systems are on Line Items B2020 116 0400 thru 2400. Both material quantities and labor costs have been adjusted for the system listed.

Factors: To adjust for job conditions other than normal working situations use Lines B2020 116 3100 thru 4000 and apply these percentages to the appropriate MAT and INST. costs.

Example: You are to install the system and protect all existing construction. Go to Line B2020 116 3800 and apply these percentages to the appropriate MAT. and INST. costs.

B2020 116	Storm Windows & Doors	COST EACH		
		MAT.	INST.	TOTAL
0350	Storm Window and door systems, single glazed			
0400	Window, custom aluminum anodized, 2′-0″ x 3′-5″	84	31.50	115.50
0500	2′-6″ x 5′-0″	113	33.50	146.50
0600	4′-0″ x 6′-0″	238	37.50	275.50
0700	White painted aluminum, 2′-0″ x 3′-5″	100	31.50	131.50
0800	2′-6″ x 5′-0″	161	33.50	194.50
0900	4′-0″ x 6′-0″	288	37.50	325.50
1000	Average quality aluminum, anodized, 2′-0″ x 3′-5″	85.50	31.50	117
1100	2′-6″ x 5′-0″	104	33.50	137.50
1200	4′-0″ x 6′-0″	127	37.50	164.50
1300	White painted aluminum, 2′-0″ x 3′-5″	84	31.50	115.50
1400	2′-6″ x 5′-0″	93	33.50	126.50
1500	4′-0″ x 6′-0″	102	37.50	139.50
1600	Mill finish, 2′-0″ x 3′-5″	76.50	31.50	108
1700	2′-6″ x 5′-0″	85.50	33.50	119
1800	4′-0″ x 6′-0″	96	37.50	133.50
1900				
2000	Door, aluminum anodized 3′-0″ x 6′-8″	205	67.50	272.50
2100	White painted aluminum, 3′-0″ x 6′-8″	252	67.50	319.50
2200	Mill finish, 3′-0″ x 6′-8″	257	67.50	324.50
2300				
2400	Wood, storm and screen, painted	305	105	410
2500				
2600				
2700				
2800				
3100	Cut & patch to match existing construction, add, minimum	2%	3%	
3200	Maximum	5%	9%	
3300	Dust protection, add, minimum	1%	2%	
3400	Maximum	4%	11%	
3500	Material handling & storage limitation, add, minimum	1%	1%	
3600	Maximum	6%	7%	
3700	Protection of existing work, add, minimum	2%	2%	
3800	Maximum	5%	7%	
3900	Shift work requirements, add, minimum		5%	
4000	Maximum		30%	

SHELL B

B2020 Exterior Windows

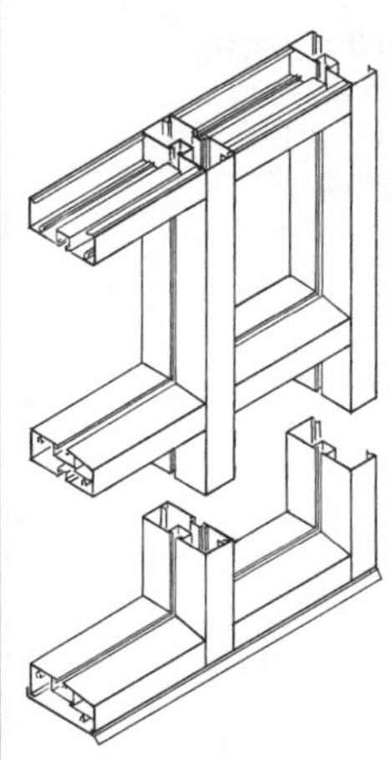

This page illustrates and describes a window wall system including aluminum tube framing, caulking, and glass. Lines within System Components give the unit price and total price per square foot for this system. Prices for alternate window wall systems are on Line Items B2020 124 1400 thru 2600. Both material quantities and labor costs have been adjusted for the system listed.

Factors: To adjust for job conditions other than normal working situations use Lines B2020 124 3100 thru 4000.

Example: You are to install the system with need for complete dust protection. Go to Line B2020 124 3400 and apply these percentages to the appropriate MAT. and INST. costs.

System Components	QUANTITY	UNIT	COST PER S.F.		
			MAT.	INST.	TOTAL
Window wall, including aluminum header, sill, mullions, caulking					
And glass					
Header, mill finish, 1-3/4" x 4-1/2" deep	.167	L.F.	2.34	1.78	4.12
Sill, mill finish, 1-3/4" x 4-1/2" deep	.167	L.F.	2.26	1.74	4
Vertical mullion, 1-3/4" x 4-1/2" deep, 6' O.C.	.191	L.F.	5.73	4.33	10.06
Caulking	.381	L.F.	.06	.73	.79
Glass, 1/4" plate	1.000	S.F.	6.10	7.35	13.45
TOTAL			16.49	15.93	32.42

B2020 124	Aluminum Frame, Window Wall	COST PER S.F.		
		MAT.	INST.	TOTAL
1200	For alternate systems:			
1300				
1400	Mill finish, 2" x 4-1/2" deep, insulating glass	19.10	18.40	37.50
1500	Thermo-break, 2-1/4" x 4-1/2" deep, insulating glass	19.45	18.75	38.20
1600	Bronze finish, 1-3/4" x 4-1/2" deep, 1/4" plate glass	18.35	17.35	35.70
1700	2" x 4-1/2" deep, insulating glass	21	19.95	40.95
1800	Thermo-break, 2-1/4" x 4-1/2" deep, insulating glass	21.50	20.50	42
1900				
2000	Black finish, 1-3/4" x 4-1/2" deep, 1/4" plate glass	19.25	18.05	37.30
2100	2" x 4-1/2" deep, insulating glass	22	20.50	42.50
2200	Thermo-break, 2-1/4" x 4-1/2" deep, insulating glass	22.50	21	43.50
2300				
2400	Stainless steel, 1-3/4" x 4-1/2" deep, 1/4" plate glass	24	22	46
2500	2" x 4-1/2" deep, insulating glass	27	24.50	51.50
2600	Thermo-break, 2-1/4" x 4-1/2" deep, insulating glass	27.50	25	52.50
2700				
3100	Cut & patch to match existing construction, add, minimum	2%	3%	
3200	Maximum	5%	9%	
3300	Dust protection, add, minimum	1%	2%	
3400	Maximum	4%	11%	
3500	Material handling, & storage limitation, add, minimum	1%	1%	
3600	Maximum	6%	7%	
3700	Protection of existing work, add, minimum	2%	2%	
3800	Maximum	5%	7%	
3900	Shift work requirements, add, minimum		5%	
4000	Maximum		30%	

B2030 Exterior Doors

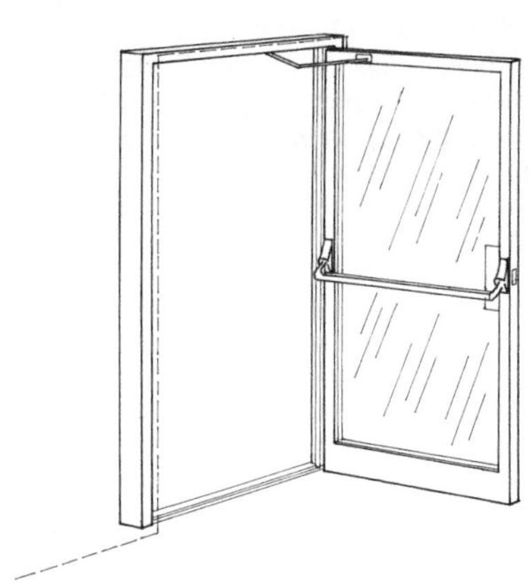

This page illustrates and describes a commercial metal door system, including a single aluminum and glass door, narrow stiles, jamb, hardware, weatherstripping, panic hardware and closer. Lines within System Components give the unit price and total price on a cost each basis for this system. Prices for alternate commercial metal door systems are on Line Items B2030 125 1300 thru 2500. Both material quantities and labor costs have been adjusted for the system listed.

Factors: To adjust for job conditions other than normal working situations, use Lines B2030 125 3100 thru 4000.

Example: You are to install the system and cut and patch to match existing construction. Go to Line B2030 125 3200 and apply these percentages to the appropriate MAT. and INST. costs.

System Components	QUANTITY	UNIT	COST EACH MAT.	COST EACH INST.	COST EACH TOTAL
Single aluminum and glass door, 3′-0″x7′-0″, with narrow stiles, ext. jamb. Weatherstripping, 1/2″ tempered insul. glass, panic hardware, and closer.					
Aluminum door, 3′-0″ x 7′-0″ x 1-3/4″, narrow stiles	1.000	Ea.	620	600	1,220
Tempered insulating glass, 1/2″ thick	20.000	S.F.	450	321	771
Panic hardware	1.000	Set	410	78.50	488.50
Automatic closer	1.000	Ea.	95.50	72.50	168
TOTAL			1,575.50	1,072	2,647.50

B2030 125	Doors, Metal - Commercial	COST EACH MAT.	COST EACH INST.	COST EACH TOTAL
1200	For alternate door systems:			
1300	Single aluminum and glass with transom, 3′-0″ x 10′-0″	2,000	1,275	3,275
1400	Anodized aluminum and glass, 3′-0″ x 7′-0″	1,800	1,275	3,075
1500	With transom, 3′-0″ x 10′-0″	2,300	1,500	3,800
1600	Steel, deluxe, hollow metal 3′-0″ x 7′-0″	1,175	410	1,585
1700	With transom 3′-0″ x 10′-0″	1,600	465	2,065
1800	Fire door, "A" label, 3′-0″ x 7′-0″	1,250	410	1,660
1900	Double, aluminum and glass, 6′-0″ x 7′-0″	2,575	1,825	4,400
2000	With transom, 6′-0″ x 10′-0″	2,775	2,225	5,000
2100	Anodized aluminum and glass 6′-0″ x 7′-0″	2,900	2,150	5,050
2200	With transom, 6′-0″ x 10′-0″	3,050	2,625	5,675
2300	Steel, deluxe, hollow metal, 6′-0″ x 7′-0″	1,950	710	2,660
2400	With transom, 6′-0″ x 10′-0″	2,800	820	3,620
2500	Fire door, "A" label, 6′-0″ x 7′-0″	2,075	705	2,780
2800				
2900				
3100	Cut & patch to match existing construction, add, minimum	2%	3%	
3200	Maximum	5%	9%	
3300	Dust protection, add, minimum	1%	2%	
3400	Maximum	4%	11%	
3500	Material handling & storage limitation, add, minimum	1%	1%	
3600	Maximum	6%	7%	
3700	Protection of existing work, add, minimum	2%	2%	
3800	Maximum	5%	7%	
3900	Shift work requirements, add, minimum		5%	
4000	Maximum		30%	

B2030 Exterior Doors

This page illustrates and describes sliding door systems including a sliding door, frame, and interior and exterior trim with exterior staining. Lines within System Components give the unit price and total price on a cost each basis for this system. Prices for alternate sliding door systems are on Line Items B2030 150 1100 thru 2400. Both material quantities and labor costs have been adjusted for the system listed.

Factors: To adjust for job conditions other than normal working situations use Lines B2030 150 2700 thru 4000.

Example: You are to install the system with temporary shoring and bracing. Go to Line B2030 150 3900 and apply these percentages to the appropriate MAT. and INST. costs.

System Components	QUANTITY	UNIT	COST EACH MAT.	COST EACH INST.	COST EACH TOTAL
Sliding wood door, 6′-0″ x 6′-8″, with wood frame, interior and exterior Trim and exterior staining.					
Sliding wood door, standard, 6′-0″ x 6′-8″, insulated glass	1.000	Ea.	865	236	1,101
Interior & exterior trim	1.000	Set	28.20	39.20	67.40
Stain door	1.000	Ea.	2.15	24	26.15
Stain trim	1.000	Ea.	2	10.20	12.20
TOTAL			897.35	309.40	1,206.75

B2030 150	Doors, Sliding - Patio	COST EACH MAT.	COST EACH INST.	COST EACH TOTAL
1000	For alternate sliding door systems:			
1100	Wood, standard, 8′-0″ x 6′-8″, insulated glass	1,025	395	1,420
1200	12′-0″ x 6′-8″	2,300	475	2,775
1300	Vinyl coated, 6′-0″ x 6′-8″	1,300	310	1,610
1400	8′-0″ x 6′-8″	1,500	395	1,895
1500	12′-0″ x 6′-8″	3,225	475	3,700
1700	Aluminum, standard, 6′-0″ x 6′-8″, insulated glass	840	335	1,175
1800	8′-0″ x 6′-8″	1,500	420	1,920
1900	12′-0″ x 6′-8″	1,675	500	2,175
2000	Anodized, 6′-0″ x 6′-8″	1,550	335	1,885
2100	8′-0″ x 6′-8″	1,750	420	2,170
2200	12′-0″ x 6′-8″	2,775	500	3,275
2300				
2400	Deduct for single glazing	78.50		78.50
2500				
2700	Cut & patch to match existing construction, add, minimum	2%	3%	
2800	Maximum	5%	9%	
2900	Dust protection, add, minimum	1%	2%	
3000	Maximum	4%	11%	
3100	Equipment usage curtailment, add, minimum	1%	1%	
3200	Maximum	3%	10%	
3300	Material handling & storage limitation, add, minimum	1%	1%	
3400	Maximum	6%	7%	
3500	Protection of existing, work, add, minimum	2%	2%	
3600	Maximum	5%	7%	
3700	Shift work requirements, add, minimum		5%	
3800	Maximum		30%	
3900	Temporary shoring and bracing, add, minimum	2%	5%	
4000	Maximum	5%	12%	

B2030 Exterior Doors

This page illustrates and describes residential door systems including a door, frame, trim, hardware, weatherstripping, stained and finished. Lines within System Components give the unit price and total price on a cost each basis for this system. Prices for alternate residential door systems are on Line Items B2030 240 1500 thru 2200. Both material quantities and labor costs have been adjusted for the system listed.

Factors: To adjust for job conditions other than normal working situations use Lines B2030 240 3100 thru 4000.

Example: You are to install the system with a material handling and storage limitation. Go to Line B2030 240 3600 and apply these percentages to the appropriate MAT. and INST. costs.

System Components	QUANTITY	UNIT	COST EACH		
			MAT.	INST.	TOTAL
Single solid wood colonial door, 3′ x 6′-8″ with wood frame and trim, Stained and finished, including hardware and weatherstripping.					
Solid wood colonial door, fir 3′ x 6′-8″ x 1-3/4″, hinges	1.000	Ea.	440	63	503
Hinges, ball bearing	1.000	Ea.	32		32
Exterior frame, with trim	1.000	Set	147.05	42.67	189.72
Interior trim	1.000	Set	24	80	104
Lockset	1.000	Set	36.50	33.50	70
Sill, oak 8″ deep	3.500	Ea.	59.50	32.90	92.40
Weatherstripping	1.000	Set	19.50	62	81.50
Stained and finished	1.000	Ea.	4.60	51	55.60
TOTAL			763.15	365.07	1,128.22

B2030 240	Doors, Residential - Exterior	COST EACH		
		MAT.	INST.	TOTAL
1400	For alternate exterior door system:			
1500	Single doors			
1600	Hollow metal exterior door, plain	610	345	955
1700	Solid core wood door, plain	430	350	780
1800				
1900	Double doors, 6′ x 6′-8″			
2000	Solid colonial double doors, fir	1,275	480	1,755
2100	Hollow metal exterior doors, plain	975	450	1,425
2200	Hollow core wood doors, plain	620	465	1,085
2300				
2800				
2900				
3100	Cut & patch to match existing construction, add, minimum	2%	3%	
3200	Maximum	5%	9%	
3300	Dust protection, add, minimum	1%	2%	
3400	Maximum	4%	11%	
3500	Material handling & storage limitation, add, minimum	1%	1%	
3600	Maximum	6%	7%	
3700	Protection of existing work, add, minimum	2%	2%	
3800	Maximum	5%	7%	
3900	Shift work requirements, add, minimum		5%	
4000	Maximum		30%	

SHELL B

B2030 Exterior Doors

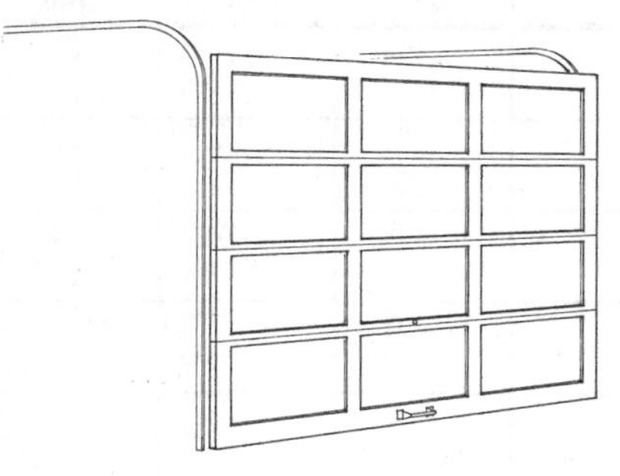

This page illustrates and describes overhead door systems including an overhead door, track, hardware, trim, and electric door opener. Lines within System Components give the unit price and total price on a cost each basis for this system. Prices for alternate overhead door systems are on Line Items B2030 410 1300 thru 2300. Both material quantities and labor costs have been adjusted for the system listed.

Factors: To adjust for job conditions other than normal working situations use Lines B2030 410 3100 thru 4000.

Example: You are to install the system and match existing construction. Go to Line B2030 410 3200 and apply these percentages to the appropriate MAT. and INST. costs.

System Components	QUANTITY	UNIT	COST EACH MAT.	COST EACH INST.	COST EACH TOTAL
Wood overhead door, commercial sectional, including track, door, hardware And trim, electrically operated.					
Commercial, heavy duty wood door, 8′ x 8′ x 1-3/4″ thick	1.000	Ea.	665	470	1,135
Frame, 2 x 8, pressure treated	1.000	Set	25.68	59.52	85.20
Wood trim	1.000	Set	58.32	47.04	105.36
Painting, two coats	1.000	Ea.	18.72	162.72	181.44
Electric trolley operator	1.000	Ea.	770	236	1,006
TOTAL			1,537.72	975.28	2,513

B2030 410	Doors, Overhead	COST EACH MAT.	COST EACH INST.	COST EACH TOTAL
1200	For alternate sectional overhead door systems:			
1300	Commercial, wood, 1-3/4″ thick, 12′ x 12′	2,375	1,375	3,750
1400	14′ x 14′	3,100	1,625	4,725
1500	Fiberglass & aluminum 12′ x 12′	3,050	1,400	4,450
1600	20′ x 20′	6,425	3,550	9,975
1700	Residential, wood, 9′ x 7′	645	375	1,020
1800	16′ x 7′	1,225	555	1,780
1900	Hardboard faced, 9′ x 7′	540	375	915
2000	16′ x 7′	1,000	555	1,555
2100	Fiberglass & aluminum, 9′ x 7′	765	435	1,200
2200	16′ x 7′	1,325	555	1,880
2300	For residential electric opener, add	460	59	519
2400				
2500				
2600				
2700				
2800				
2900				
3100	Cut & patch to match existing construction, add, minumum	2%	3%	
3200	Maximum	5%	9%	
3300	Dust protection, add, minimum	1%	2%	
3400	Maximum	4%	11%	
3500	Material handling & storage limitation, add, minimum	1%	1%	
3600	Maximum	6%	7%	
3700	Protection of existing work, add, minimum	2%	2%	
3800	Maximum	5%	7%	
3900	Shift work requirements, add, minimum		5%	
4000	Maximum		30%	

B2030 810	Selective Price Sheet	COST EACH		
		MAT.	INST.	TOTAL
0100	Door closer, rack and pinion	147	72.50	219.50
0200	Backcheck and adjustable power	151	78.50	229.50
0300	Regular, hinge face mount, all sizes, regular arm	189	72.50	261.50
0400	Hold open arm	204	72.50	276.50
0500	Top jamb mount, all sizes, regular arm	189	78.50	267.50
0600	Hold open arm	204	78.50	282.50
0700	Stop face mount, all sizes, regular arm	189	72.50	261.50
0800	Hold open arm	203	72.50	275.50
0900	Fusible link, hinge face mount, all sizes, regular arm	202	72.50	274.50
1000	Hold open arm	217	72.50	289.50
1100	Top jamb mount, all sizes, regular arm	202	78.50	280.50
1200	Hold open arm	217	78.50	295.50
1300	Stop face mount, all sizes, regular arm	202	72.50	274.50
1400	Hold open arm	216	72.50	288.50
1500				
1600				
1700	Door stops			
1800				
1900	Holder & bumper, floor or wall	34.50	14.75	49.25
2000	Wall bumper	10.30	14.75	25.05
2100	Floor bumper	5.30	14.75	20.05
2200	Plunger type, door mounted	28	14.75	42.75
2300	Hinges, full mortise, material only, per pair			
2400	Low frequency, 4-1/2" x 4-1/2", steel base, USP	10.35		10.35
2500	Brass base, US10	40		40
2600	Stainless steel base, US32	69		69
2700	Average frequency, 4-1/2" x 4-1/2", steel base, USP	23.50		23.50
2800	Brass base, US10	48		48
2900	Stainless steel base, US32	72		72
3000	High frequency, 4-1/2" x 4-1/2", steel base, USP	56.50		56.50
3100	Brass base, US10	76		76
3200	Stainless steel base, US32	122		122
3300	Kick plate			
3400				
3500	6" high, for 3'-0" door, aluminum	30.50	31.50	62
3600	Bronze	38	31.50	69.50
3700	Panic device			
3800				
3900	For rim locks, single door, exit	410	78.50	488.50
4000	Outside key and pull	465	94	559
4100	Bar and vertical rod, exit only	595	94	689
4200	Outside key and pull	710	118	828
4300	Lockset			
4400				
4500	Heavy duty, cylindrical, passage doors	136	39.50	175.50
4600	Classroom	300	59	359
4700	Bedroom, bathroom, and inner office doors	172	39.50	211.50
4800	Apartment, office, and corridor doors	240	47	287
4900	Standard duty, cylindrical, exit doors	82	47	129
5100	Passage doors	58.50	39.50	98
5200	Public restroom, classroom, & office doors	117	59	176
5300	Deadlock, mortise, heavy duty	144	52.50	196.50
5400	Double cylinder	159	52.50	211.50
5500	Entrance lock, cylinder, deadlocking latch	132	52.50	184.50
5600	Deadbolt	161	59	220
5700	Commercial, mortise, wrought knob, keyed, minimum	184	59	243
5800	Maximum	340	67.50	407.50
5900	Cast knob, keyed, minimum	248	52.50	300.50

B2030 Exterior Doors

B2030 810	Selective Price Sheet	COST EACH		
		MAT.	INST.	TOTAL
6000	Maximum	480	52.50	532.50
6100	Push-pull			
6200				
6300	Aluminum	7.55	39.50	47.05
6400	Bronze	19.10	39.50	58.60
6500	Door pull, designer style, minimum	71	39.50	110.50
6600	Maximum	360	59	419
6700	Threshold			
6800				
6900	3′-0″ long door saddles, aluminum, minimum	3.98	9.80	13.78
7000	Maximum	33.50	39.50	73
7100	Bronze, minimum	35	7.85	42.85
7200	Maximum	66	39.50	105.50
7300	Rubber, 1/2″ thick, 5-1/2″ wide	37	23.50	60.50
7400	2-3/4″ wide	15.50	23.50	39
7500	Weatherstripping, per set			
7600				
7700	Doors, wood frame, interlocking for 3′ x 7′ door, zinc	15.15	157	172.15
7800	Bronze	24	157	181
7900	Wood frame, spring type for 3′ x 7′ door, bronze	19.50	62	81.50
8000	Metal frame, spring type for 3′ x 7′ door, bronze	33	157	190
8100	For stainless steel, spring type, add	133%		
8200				
8300	Metal frame, extruded sections, 3′ x 7′ door, aluminum	44	236	280
8400	Bronze	111	236	347

B3010 Roof Coverings

B3010 160	Selective Price Sheet	COST PER S.F.		
		MAT.	INST.	TOTAL
0100	Roofing, built-up, asphalt roll roof, 3 ply organic/mineral surface	.46	1.21	1.67
0200	3 plies glass fiber felt type iv, 1 ply mineral surface	.72	1.30	2.02
0300	Cold applied, 3 ply		.43	.43
0400	Coal tar pitch, 4 ply tarred felt	1.29	1.55	2.84
0500	Mopped, 3 ply glass fiber	1.06	1.71	2.77
0600	4 ply organic felt	1.29	1.55	2.84
0700	Elastomeric, hypalon, neoprene unreinforced	2.26	2.86	5.12
0800	Polyester reinforced	2.28	3.39	5.67
0900	Neoprene, 5 coats 60 mils	5	10	15
1000	Over 10,000 S.F.	4.68	5.20	9.88
1100	PVC, traffic deck sprayed	1.46	5.20	6.66
1200	With neoprene	1.55	2.10	3.65
1300	Shingles, fiber cement, strip, 14" x 30", 325#/sq.	3.20	2	5.20
1500	Shake, 9.35" x 16" 500#/sq.	2.91	2	4.91
1600				
1700	Asphalt, strip, 210-235#/sq.	.42	.80	1.22
1800	235-240#/sq.	.48	.88	1.36
1900	Class A laminated	.52	.98	1.50
2000	Class C laminated	.55	1.10	1.65
2100	Slate, buckingham, 3/16" thick	6.40	2.51	8.91
2200	Black, 1/4" thick	6.75	2.51	9.26
2300	Wood, shingles, 16" no. 1, 5" exp.	2.06	1.88	3.94
2400	Red cedar, 18" perfections	1.85	1.71	3.56
2500	Shakes, 24", 10" exposure	1.27	1.88	3.15
2600	18", 8-1/2" exposure	.99	2.36	3.35
2700	Insulation, ceiling batts, fiberglass, 3-1/2" thick, R11	.32	.29	.61
2800	6" thick, R19	.46	.35	.81
2900	9" thick, R30	.78	.41	1.19
3000	12" thick, R38	.94	.47	1.41
3100	Mineral, 3-1/2" thick, R13	.33	.29	.62
3200	Fiber, 6" thick, R19	.44	.29	.73
3300	Roof deck, fiberboard, 1" thick, R2.78	.42	.55	.97
3400	Mineral, 2" thick, R5.26	.84	.55	1.39
3500	Perlite boards, 3/4" thick, R2.08	.33	.55	.88
3600	2" thick, R5.26	.70	.63	1.33
3700	Polystyrene extruded, R5.26, 1" thick,	.47	.29	.76
3800	2" thick R10	.73	.35	1.08
3900	40 PSI compressive strength, 1" thick R5	.48	.29	.77
4000	Tapered for drainage	.69	.31	1
4100	Foamglass, 1 1/2" thick R4.55	1.43	.55	1.98
4200	3" thick R9.00	2.92	.63	3.55
4300	Ceiling, plaster, gypsum, 2 coats	.40	2.53	2.93
4400	3 coats	.57	3	3.57
4500	Perlite or vermiculite, 2 coats	.40	2.94	3.34
4600	3 coats	.72	3.70	4.42
4700	Gypsum lath, plain 3/8" thick	.66	.54	1.20
4800	1/2" thick	.52	.58	1.10
4900	Firestop, 3/8" thick	.45	.66	1.11
5000	1/2" thick	.55	.71	1.26
5100	Metal lath, rib, 2.75 lb.	.37	.62	.99
5200	3.40 lb.	.54	.66	1.20
5300	Diamond, 2.50 lb.	.35	.62	.97
5400	3.40 lb.	.40	.77	1.17
5500	Drywall, taped and finished standard, 1/2" thick	.34	1.23	1.57
5600	5/8" thick	.37	1.23	1.60
5700	Fire resistant, 1/2" thick	.37	1.23	1.60
5800	5/8" thick	.37	1.23	1.60
5900	Water resist., 1/2" thick	.31	1.23	1.54

B3010 Roof Coverings

B3010 160	Selective Price Sheet	COST PER S.F. MAT.	COST PER S.F. INST.	COST PER S.F. TOTAL
6000	5/8" thick	.34	1.23	1.57
6100	Finish, instead of taping			
6200	For thin coat plaster, add	.08	.58	.66
6300	Finish textured spray, add	.06	.52	.58
6400	Drywall, no finish included, see system A6.9-700			
6500	Tile, stapled glued, mineral fiber plastic coated, 5/8" thick	1.62	.47	2.09
6600	3/4" thick	1.55	.47	2.02
6700	Wood fiber, 1/2" thick	.84	1.18	2.02
6800	3/4" thick	1.18	1.18	2.36
6900	Suspended, fiberglass film faced, 5/8" thick	.57	.75	1.32
7000	3" thick	1.44	1.05	2.49
7100	Mineral fiber 5/8" thick, standard face	.70	.70	1.40
7200	Aluminum faced	1.77	.79	2.56
7300	Wood fiber reveal edge, 1" thick			
7400	3" thick			
7500	Ceiling suspension systems, for tile, "T" bar, class "A", 2' x 4' grid	.58	.59	1.17
7600	2' x 2' grid	.72	.73	1.45
7700	Concealed "Z" bar, 12" module	.52	.91	1.43
7800				
7900	Plaster/drywall, 3/4" channels, steel furring, 16" O.C.	.23	1.44	1.67
8000	24" O.C.	.15	.99	1.14
8100	1-1/2" channels, 16" O.C.	.32	1.60	1.92
8200	24" O.C.	.21	1.07	1.28
8300	Ceiling framing, 2" x 4" studs, 16" O.C.	.31	.79	1.10
8400	24" O.C.	.21	.53	.74

B3020 Roof Openings

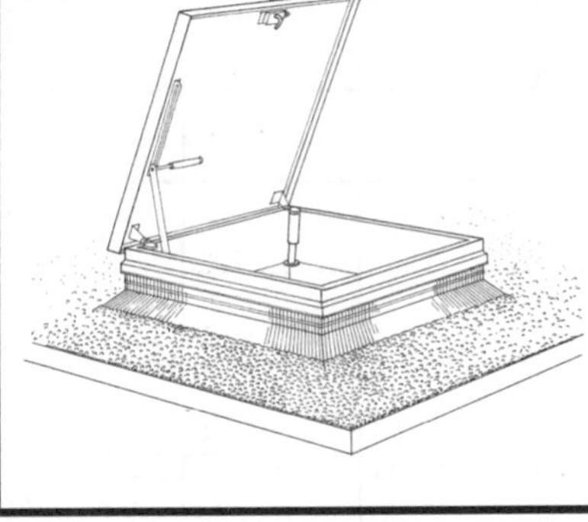

This page illustrates and describes a roof hatch system. Lines within System Components give the unit price and total cost each for this system. Prices for alternate systems are on Lines B3020 230 1500 thru 2400. Both material quantities and labor costs have been adjusted for the system listed.

Factors: To adjust for job conditions other than normal working situtations use Lines B3020 230 2900 thru 4000.

Example: You are to install the system and cut and patch to match existing construction. Use line B3020 230 3000 and apply these percentages to the appropriate MAT. and INST. costs.

System Components	QUANTITY	UNIT	COST EACH MAT.	COST EACH INST.	COST EACH TOTAL
Roof hatch, 2'-6" x 3'-0", aluminum, curb and cover included.					
Through steel construction.					
Roof hatch, 2'-6" x 3'-0", aluminum	1.000	Ea.	605	180	785
Cutout decking	11.000	L.F.		24.64	24.64
Frame opening	44.000	L.F.	27.72	319.88	347.60
Flashing, 16 oz. copper	16.000	S.F.	48.64	61.28	109.92
Cant strip, 4 x 4, foamglass	12.000	L.F.	28.44	16.20	44.64
TOTAL			709.80	602	1,311.80

B3020 230	Roof Hatches, Skylights	COST EACH MAT.	COST EACH INST.	COST EACH TOTAL
1400	For alternate systems:			
1500	Roof hatch, aluminum, curb and cover, 2'-6" x 4'-6"	845	735	1,580
1600	2'-6" x 8'-0"	1,675	1,050	2,725
1700	Skylight, plexiglass dome, with curb mounting, 30" x 32"	525	575	1,100
1800	30" x 45"	620	715	1,335
1900	40" x 45"	715	970	1,685
2000	Smoke hatch, unlabeled, 2'-6" x 3'-0"	860	650	1,510
2100	2'-6" x 4'-6"	1,025	785	1,810
2200	2'-6" x 8'-0"	1,825	1,075	2,900
2300				
2400	For steel ladder, add per vertical linear foot	35	29.50	64.50
2500				
2600				
2700				
2900	Cut & patch to match existing construction, add, minimum	2%	3%	
3000	Maximum	5%	9%	
3100	Dust protection, add, minimum	1%	2%	
3200	Maximum	4%	11%	
3300	Equipment usage curtailment, add, minimum	1%	1%	
3400	Maximum	3%	10%	
3500	Material handling & storage limitation, add, minimum	1%	1%	
3600	Maximum	6%	7%	
3700	Protection of existing work, add, minimum	2%	2%	
3800	Maximum	5%	7%	
3900	Shift work requirements, add, minimum		5%	
4000	Maximum		30%	

Important: See the Reference Section for critical supporting data - Reference Numbers and City Cost Indexes

B3020 Roof Openings

B3020 240	Selective Price Sheet	COST PER L.F. MAT.	INST.	TOTAL
0100	Downspouts per L.F., aluminum, enameled .024" thick, 2" x 3"	1.95	2.98	4.93
0200	3" x 4"	2.75	3.83	6.58
0300	Round .025" thick, 3" diam.	1.50	2.82	4.32
0400	4" diam.	2.23	3.83	6.06
0500	Copper, round 16 oz. stock, 2" diam.	5	2.82	7.82
0600	3" diam.	4.84	2.82	7.66
0700	4" diam.	5.70	3.70	9.40
0800	5" diam.	8.45	4.13	12.58
0900	Rectangular, 2" x 3"	5.80	2.82	8.62
1000	3" x 4"	6.55	3.70	10.25
1100	Lead coated copper, round, 2" diam.	6.05	2.82	8.87
1200	3" diam.	8.10	2.82	10.92
1300	4" diam.	10.70	3.70	14.40
1400	5" diam.	15.15	4.13	19.28
1500	Rectangular, 2" x 3"	8.10	2.82	10.92
1600	3" x 4"	8.75	3.70	12.45
1700	Steel galvanized, round 28 gauge, 3" diam.	1.47	2.82	4.29
1800	4" diam.	1.54	3.70	5.24
1900	5" diam.	2.57	4.13	6.70
2000	6" diam.	3.65	5.10	8.75
2100	Rectangular, 2" x 3"	1.96	2.82	4.78
2200	3" x 4"	1.74	3.70	5.44
2300	Elbows, aluminum, round, 3" diam.	2.18	5.35	7.53
2400	4" diam.	4.72	5.35	10.07
2500	Rectangular, 2" x 3"	3.63	5.35	8.98
2600	3" x 4"	3.63	5.35	8.98
2700	Copper, round 16 oz., 2" diam.	12.30	5.35	17.65
2800	3" diam.	9	5.35	14.35
2900	4" diam.	10.45	5.35	15.80
3100	Rectangular, 2" x 3"	6.75	5.35	12.10
3200	3" x 4"	7.95	5.35	13.30
3300	Drip edge per L.F., aluminum, 5" wide	.33	1.18	1.51
3400	8" wide	.46	1.18	1.64
3500	28" wide	4.14	4.71	8.85
3600				
3700	Steel galvanized, 5" wide	.44	1.18	1.62
3800	8" wide	.54	1.18	1.72
3900				
4000				
4100				
4200				
4300	Flashing 12" wide per S.F., aluminum, mill finish, .013" thick	.42	3.03	3.45
4400	.019" thick	.78	3.03	3.81
4500	.040" thick	1.74	3.03	4.77
4600	.050" thick	2.21	3.03	5.24
4700	Copper, mill finish, 16 oz.	3.04	3.83	6.87
4800	20 oz.	4.51	4	8.51
4900	24 oz.	5.40	4.19	9.59
5000	32 oz.	7.20	4.40	11.60
5100	Lead, 2.5 lb./S.F., 12" wide	3.33	3.26	6.59
5200	Over 12" wide	4.06	3.26	7.32
5300	Lead-coated copper, fabric backed, 2 oz.	1.69	1.33	3.02
5400	5 oz.	1.95	1.33	3.28
5500	Mastic backed, 2 oz.	1.32	1.33	2.65
5600	5 oz.	1.64	1.33	2.97
5700	Paper backed, 2 oz.	1.14	1.33	2.47
5800	3 oz.	1.34	1.33	2.67
5900	Polyvinyl chloride, black, .010" thick	.19	1.54	1.73

B3020 Roof Openings

B3020 240	Selective Price Sheet	COST PER L.F. MAT.	INST.	TOTAL
6000	.020" thick	.26	1.54	1.80
6100	.030" thick	.35	1.54	1.89
6200	.056" thick	.84	1.54	2.38
6300	Steel, galvanized, 20 gauge	1	3.38	4.38
6400	30 gauge	.43	2.75	3.18
6500	Stainless, 32 gauge, .010" thick	2.97	2.84	5.81
6600	28 gauge, .015" thick	3.69	2.84	6.53
6700	26 gauge, .018" thick	4.47	2.84	7.31
6800	24 gauge, .025" thick	5.80	2.84	8.64
6900	Gutters per L.F., aluminum, 5" box, .027" thick	1.46	4.47	5.93
7000	.032" thick	2.59	4.47	7.06
7100	Copper, half round, 4" wide	4.51	4.47	8.98
7200	6" wide	6.25	4.66	10.91
7300	Steel, 26 gauge galvanized, 5" wide	1.05	4.47	5.52
7400	6" wide	1.76	4.47	6.23
7500	Wood, treated hem-fir, 3" x 4"	6.90	4.71	11.61
7600	4" x 5"	8	4.71	12.71
7700	Reglet per L.F., aluminum, .025" thick	1.12	2.09	3.21
7800	Copper, 10 oz.	1.95	2.09	4.04
7900	Steel, galvanized, 24 gauge	.83	2.09	2.92
8000	Stainless, .020" thick	1.91	2.09	4
8100	Counter flashing 12" wide per L.F., aluminum, .032" thick	1.45	3.58	5.03
8200	Copper, 10 oz.	4.07	3.58	7.65
8300	Steel, galvanized, 24 gauge	.76	3.58	4.34
8400	Stainless, .020" thick	3.38	3.58	6.96

For information about Means Estimating Seminars, see yellow pages 12 and 13 in back of book

C INTERIORS

C1010 Partitions

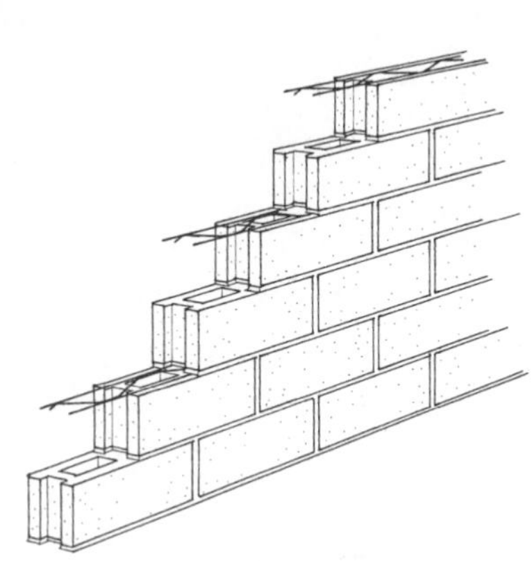

This page illustrates and describes a concrete block wall system including concrete block, horizontal reinforcing alternate courses, mortar, and tooled joints both sides. Lines within System Components give the unit price and total price per square foot for this system. Prices for alternate concrete block wall systems are on Line Items C1010 106 0900 thru 2100. Both material quantities and labor costs have been adjusted for the system listed.

Factors: To adjust for job conditions other than normal working situations use Lines C1010 106 2900 thru 4000.

Example: You are to install the system and protect all existing work. Go to Line C1010 106 3600 and apply these percentages to the appropriate MAT. and INST. costs.

System Components	QUANTITY	UNIT	COST PER S.F.		
			MAT.	INST.	TOTAL
Concrete block partition, including horizontal reinforcing every second Course, mortar, tooled joints both sides.					
8″ x 16″ concrete block, normal weight, 4″ thick	1.000	S.F.	1.25	4.98	6.23
Horizontal reinforcing every second course	.750	L.F.	.14	.12	.26
TOTAL			1.39	5.10	6.49

C1010 106	Partitions, Concrete Block	COST PER S.F.		
		MAT.	INST.	TOTAL
0800	For alternate block partition systems:			
0900	8″ x 16″ concrete block, normal weight, 6″ thick	2	5.45	7.45
1000	8″ thick	2.18	5.80	7.98
1100	10″ thick	2.72	6.05	8.77
1200	12″ thick	3.11	7.45	10.56
1300	8″ x 16″ concrete block, lightweight, 4″ thick	1.61	4.99	6.60
1400	6″ thick	2.16	5.35	7.51
1500	8″ thick	2.63	5.65	8.28
1600	10″ thick	3.39	5.90	9.29
1700	12″ thick	3.46	7.25	10.71
1800	8″ x 16″ glazed concrete block, 4″ thick	7.40	6.30	13.70
1900	8″ thick	8.50	7	15.50
2000	Structural facing tile, 6T series, glazed 2 sides, 4″ thick	12.45	11.10	23.55
2100	6″ thick	14.25	11.70	25.95
2200				
2300				
2400				
2500				
2600				
2900	Cut & patch to match existing construction, add, minimum	2%	3%	
3000	Maximum	5%	9%	
3100	Dust protection, add, minimum	1%	2%	
3200	Maximum	4%	11%	
3300	Material handling & storage limitation, add, minimum	1%	1%	
3400	Maximum	6%	7%	
3500	Protection of existing work, add, minimum	2%	2%	
3600	Maximum	5%	7%	
3700	Shift work requirements, add, minimum		5%	
3800	Maximum		30%	
3900	Temporary shoring and bracing, add, minimum	2%	5%	
4000	Maximum	5%	12%	

C1010 Partitions

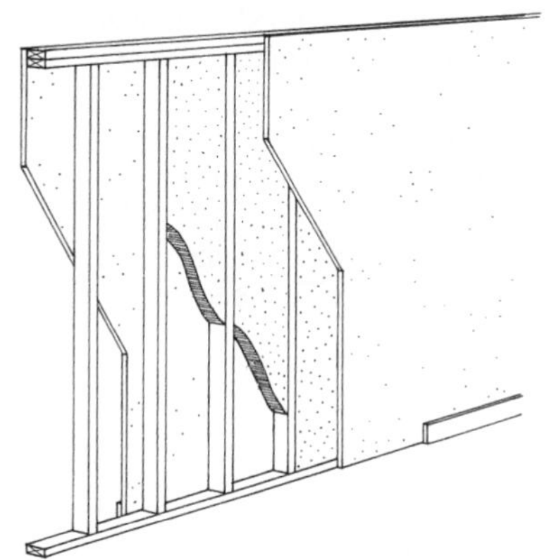

This page illustrates and describes a wood stud partition system including wood studs with plates, gypsum plasterboard – taped and finished, insulation, baseboard and painting. Lines within System Components give the unit price and total price per square foot for this system. Prices for alternate wood stud partition systems are on Line Items C1010 132 1300 thru 2700. Both material quantities and labor costs have been adjusted for the system listed.

Factors: To adjust for job conditions other than normal working situations use Lines C1010 132 2900 thru 4000.

Example: You are to install the system where material handling and storage present a serious problem. Go to Line C1010 132 3400 and apply these percentages to the appropriate MAT. and INST. costs.

System Components	QUANTITY	UNIT	COST PER S.F.		
			MAT.	INST.	TOTAL
Wood stud wall, 2″x4″, 16″ O.C., dbl. top plate, sngl bot. plate, 5/8″ dwl. Taped, finished and painted on 2 faces, insulation, baseboard, wall 8′ high					
Wood studs, 2″ x 4″, 16″ O.C., 8′ high	1.000	S.F.	.51	1.18	1.69
Gypsum drywall, 5/8″ thick	2.000	S.F.	.66	.94	1.60
Taping and finishing	2.000	S.F.	.08	.94	1.02
Insulation, 3-1/2″ fiberglass batts	1.000	S.F.	.52	.29	.81
Baseboard	.200	L.F.	.61	.49	1.10
Paint baseboard, primer + 2 coats	.200	L.F.	.04	.30	.34
Painting, roller, 2 coats	2.000	S.F.	.32	1.08	1.40
TOTAL			2.74	5.22	7.96

C1010 132	Partitions, Wood Stud	COST PER S.F.		
		MAT.	INST.	TOTAL
1200	For alternate wood stud systems:			
1300	2″ x 3″ studs, 8′ high, 16″ O.C.	2.71	5.15	7.86
1400	24″ O.C.	2.60	4.93	7.53
1500	10′ high, 16″ O.C.	2.68	4.93	7.61
1600	24″ O.C.	2.57	4.76	7.33
1700	2″ x 4″ studs, 8′ high, 24″ O.C.	2.62	4.98	7.60
1800	10′ high, 16″ O.C.	2.70	4.98	7.68
1900	24″ O.C.	2.59	4.80	7.39
2000	12′ high, 16″ O.C.	2.66	4.98	7.64
2100	24″ O.C.	2.52	4.79	7.31
2200	2″ x 6″ studs, 8′ high, 16″ O.C.	3.04	5.35	8.39
2300	24″ O.C.	2.86	5.05	7.91
2400	10′ high, 16″ O.C.	2.98	5.10	8.08
2500	24″ O.C.	2.81	4.86	7.67
2600	12′ high, 16″ O.C.	2.92	5.10	8.02
2700	24″ O.C.	2.75	4.88	7.63
2900	Cut & patch to match existing construction, add, minimum	2%	3%	
3000	Maximum	5%	9%	
3100	Dust protection, add, minimum	1%	2%	
3200	Maximum	4%	11%	
3300	Material handling & storage limitation, add, minimum	1%	1%	
3400	Maximum	6%	7%	
3500	Protection of existing work, add, minimum	2%	2%	
3600	Maximum	5%	7%	
3700	Shift work requirements, add, minimum		5%	
3800	Maximum		30%	
3900	Temporary shoring and bracing, add, minimum	2%	5%	
4000	Maximum	5%	12%	

C1010 Partitions

This page illustrates and describes a non-load bearing metal stud partition system including metal studs with runners, gypsum plasterboard, taped and finished, insulation, baseboard and painting. Lines within System Components give the unit price and total price per square foot for this system. Prices for alternate non-load bearing metal stud partition systems are on Line Items C1010 134 1300 thru 2300. Both material quantities and labor costs have been adjusted for the system listed.

Factors: To adjust for job conditions other than normal working situations use Lines C1010 134 2900 thru 4000.

Example: You are to install the system and cut and patch to match existing construction. Go to Line C1010 134 3000 and apply these percentages to the appropriate MAT. and INST. costs.

System Components	QUANTITY	UNIT	COST PER S.F. MAT.	COST PER S.F. INST.	COST PER S.F. TOTAL
Non-load bearing metal studs, including top & bottom runners, 5/8″ drywall, Taped, finished and painted 2 faces, insulation, painted baseboard.					
Metal studs, 25 ga., 3-5/8″ wide, 24″ O.C.	1.000	S.F.	.25	.64	.89
Gypsum drywall, 5/8″ thick	2.000	S.F.	.66	.94	1.60
Taping & finishing	2.000	S.F.	.08	.94	1.02
Insulation, 3-1/2″ fiberglass batts	1.000	S.F.	.52	.29	.81
Baseboard	.200	L.F.	.49	.39	.88
Paint baseboard, primer + 2 coats	.200	L.F.	.03	.24	.27
Painting, roller work, 2 coats	2.000	S.F.	.32	1.08	1.40
TOTAL			2.35	4.52	6.87

C1010 134	Partitions, Metal Stud, NLB	COST PER S.F. MAT.	COST PER S.F. INST.	COST PER S.F. TOTAL
1200	For alternate metal stud systems:			
1300	Non-load bearing, 25 ga., 24″ O.C., 2-1/2″ wide	2.32	4.51	6.83
1400	6″ wide	2.49	4.53	7.02
1500	16″ O.C., 2-1/2″ wide	2.38	4.67	7.05
1600	3-5/8″ wide	2.41	4.68	7.09
1700	6″ wide	2.59	4.69	7.28
1800	20 ga., 24″ O.C., 2-1/2″ wide	2.45	4.67	7.12
1900	3-5/8″ wide	2.56	4.68	7.24
2000	6″ wide	2.69	4.69	7.38
2100	16″ O.C., 2-1/2″ wide	2.54	4.87	7.41
2200	3-5/8″ wide	2.68	4.88	7.56
2300	6″ wide	2.84	4.89	7.73
2400				
2500				
2600				
2700				
2900	Cut & patch to match existing construction, add, minimum	2%	3%	
3000	Maximum	5%	9%	
3100	Dust protection, add, minimum	1%	2%	
3200	Maximum	4%	11%	
3300	Material handling & storage limitation, add, minimum	1%	1%	
3400	Maximum	6%	7%	
3500	Protection of existing work, add, minimum	2%	2%	
3600	Maximum	5%	7%	
3700	Shift work requirements, add, minimum		5%	
3800	Maximum		30%	
3900	Temporary shoring and bracing, add, minimum	2%	5%	
4000	Maximum	5%	12%	

C1010 Partitions

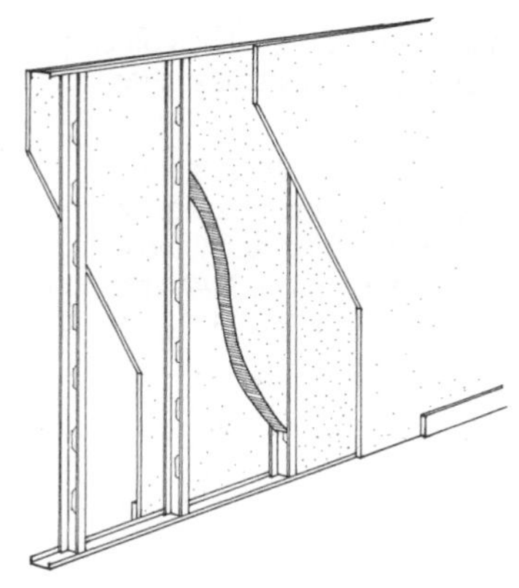

This page illustrates and describes a drywall system including gypsum plasterboard, taped and finished, metal studs with runners, insulation, baseboard and painting. Lines within System Components give the unit price and total price per square foot for this system. Prices for alternate drywall systems are on Line Items C1010 136 1300 thru 1900. Both material quantities and labor costs have been adjusted for the system listed.

Factors: To adjust for job conditions other than normal working situations use Lines C1010 136 2900 thru 4000.

Example: You are to install the system and control dust in the work area. Go to Line C1010 136 3100 and apply these percentages to the appropriate MAT. and INST. costs.

System Components	QUANTITY	UNIT	COST PER S.F. MAT.	COST PER S.F. INST.	COST PER S.F. TOTAL
Gypsum drywall, taped, finished and painted 2 faces, galvanized metal studs					
Including top & bottom runners, insulation, painted baseboard, wall 10′high					
Gypsum drywall, 5/8″ thick, standard	2.000	S.F.	.66	.94	1.60
Taping and finishing	2.000	S.F.	.08	.94	1.02
Metal studs, 20 ga., 3-5/8″ wide, 24″ O.C.	1.000	S.F.	.25	.64	.89
Insulation, 3-1/2″ fiberglass batts	1.000	S.F.	.52	.29	.81
Baseboard	.200	L.F.	.49	.39	.88
Paint baseboard, primer + 2 coats	.200	L.F.	.02	.10	.12
Painting, roller 2 coats	2.000	S.F.	.24	1.02	1.26
TOTAL			2.26	4.32	6.58

C1010 136	Partitions, Drywall	COST PER S.F. MAT.	COST PER S.F. INST.	COST PER S.F. TOTAL
1200	For alternate drywall systems:			
1300	Gypsum drywall, 5/8″ thick, fire resistant	2.26	4.32	6.58
1400	Water resistant	2.20	4.32	6.52
1500	1/2″ thick, standard	2.20	4.32	6.52
1600	Fire resistant	2.26	4.32	6.58
1700	Water resistant	2.16	4.32	6.48
1800	3/8″ thick, vinyl faced, standard	2.84	5.50	8.34
1900	5/8″ thick, vinyl faced, fire resistant	3.04	5.50	8.54
2000				
2100				
2200				
2300				
2400				
2500				
2600				
2700				
2900	Cut & patch to match existing construction, add, minimum	2%	3%	
3000	Maximum	5%	9%	
3100	Dust protection, add, minimum	1%	2%	
3200	Maximum	4%	11%	
3300	Material handling & storage limitation, add, minimum	1%	1%	
3400	Maximum	6%	7%	
3500	Protection of existing work, add, minimum	2%	2%	
3600	Maximum	5%	7%	
3700	Shift work requirements, add, minimum		5%	
3800	Maximum		30%	
3900	Temporary shoring and bracing, add, minimum	2%	5%	
4000	Maximum	5%	12%	

C1010 Partitions

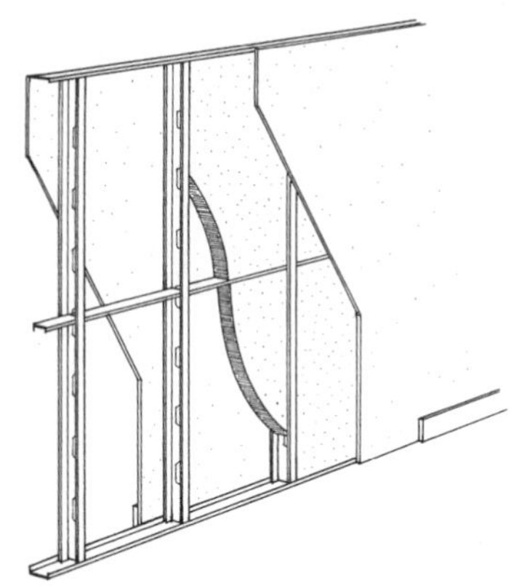

This page illustrates and describes a load bearing metal stud wall system including metal studs, sheetrock–taped and finished, insulation, baseboard and painting. Lines within System Components give the unit price and total price per square foot for this system. Prices for alternate load bearing metal stud wall systems are on Line Items C1010 138 1500 thru 2500. Both material quantities and labor costs have been adjusted for the system listed.

Factors: To adjust for job conditions other than normal working situations use Lines C1010 138 2900 thru 4000.

Example: You are to install the system using temporary shoring and bracing. Go to Line C1010 138 3900 and apply these percentages to the appropriate MAT. and INST. costs.

System Components	QUANTITY	UNIT	COST PER S.F.		
			MAT.	INST.	TOTAL
Load bearing, 18 ga., 3-5/8″, galvanized metal studs, 24″ O.C.,including Top and bottom runners, 1/2″ drywall, taped, finished and painted 2 Faces, 3″ insulation, and painted baseboard, wall 10′ high.					
Metal studs, 24″ O.C., 18 ga., 3-5/8″ wide, galvanized	1.000	S.F.	.69	.90	1.59
Gypsum drywall 1/2″ thick	2.000	S.F.	.60	.94	1.54
Taping and finishing	2.000	S.F.	.08	.94	1.02
Insulation, 3-1/2″ fiberglass batts	1.000	S.F.	.52	.29	.81
Baseboard	.200	L.F.	.49	.39	.88
Paint baseboard, primer + 2 coats	.200	L.F.	.03	.24	.27
Painting, roller 2 coats	2.000	S.F.	.32	1.08	1.40
TOTAL			2.73	4.78	7.51

C1010 138	Partitions, Metal Stud, LB	COST PER S.F.		
		MAT.	INST.	TOTAL
1400	For alternate metal stud systems:			
1500	Load bearing, 18 ga., 24″ O.C., 2-1/2″ wide	2.62	4.76	7.38
1600	6″ wide	2.96	4.82	7.78
1700	16″ O.C. 2-1/2″ wide	2.77	4.98	7.75
1800	3-5/8″ wide	2.90	5	7.90
1900	6″ wide	3.19	5.05	8.24
2000	16 ga., 24″ O.C., 2-1/2″ wide	2.71	4.89	7.60
2100	3-5/8″ wide	2.84	4.91	7.75
2200	6″ wide	3.11	4.95	8.06
2300	16″ O.C., 2-1/2″ wide	2.88	5.15	8.03
2400	3-5/8″ wide	3.04	5.15	8.19
2500	6″ wide	3.37	5.20	8.57
2600				
2700				
2900	Cut & patch to match existing construction, add, minimum	2%	3%	
3000	Maximum	5%	9%	
3100	Dust protection, add, minumum	1%	2%	
3200	Maximum	4%	11%	
3300	Material handling & storage limitation, add, minimum	1%	1%	
3400	Maximum	6%	7%	
3500	Protection of existing work, add, minimum	2%	2%	
3600	Maximum	5%	7%	
3700	Shift work requirements, add, minimum		5%	
3800	Maximum		30%	
3900	Temporary shoring and bracing, add, minimum	2%	5%	
4000	Maximum	5%	12%	

Important: See the Reference Section for critical supporting data - Reference Numbers and City Cost Indexes

C1010 Partitions

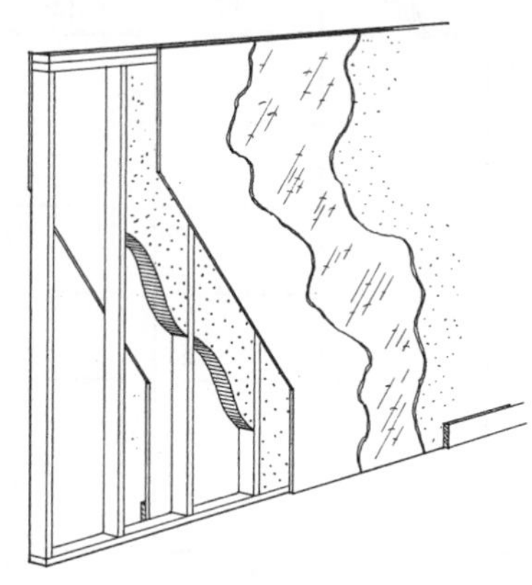

This page illustrates and describes a plaster and lath system including gypsum plaster, gypsum lath, wood studs with plates, insulation, baseboard and painting. Lines within System Components give the unit price and price per square foot for this system. Prices for alternate plaster and lath systems are on Line Items C1010 148 1500 thru 2500. Both material quantities and labor costs have been adjusted for the system listed.

Factors: To adjust for job conditions other than normal working situations use Lines C1010 148 2900 thru 4000.

Example: You are to install the system during evening hours only. Go to Line C1010 148 3800 and apply this percentage to the appropriate INST. costs.

System Components	QUANTITY	UNIT	COST PER S.F. MAT.	INST.	TOTAL
Gypsum plaster, 2 coats, over 3/8″ lath, 2 faces, 2″ x 4″ wood Stud partition, 24″O.C. incl. double top plate,single bottom plate,3-1/2″ Insulation, baseboard and painting.					
Gypsum plaster, 2 coats	.220	S.Y.	.80	4.38	5.18
Lath, gypsum, 3/8″ thick	.220	S.Y.	1.31	1.08	2.39
Wood studs, 2″ x 4″, 24″ O.C.	1.000	S.F.	.36	.76	1.12
Insulation, 3-1/2″ fiberglass batts	1.000	S.F.	.52	.29	.81
Baseboard, 9/16″ x 3-1/2″	.200	L.F.	.49	.39	.88
Paint baseboard, primer + 2 coats	.200	L.F.	.02	.10	.12
Paint, 2 coats	2.000	S.F.	.32	1.08	1.40
TOTAL			3.82	8.08	11.90

C1010 148	Partitions, Plaster & Lath	COST PER S.F. MAT.	INST.	TOTAL
1400	For alternate plaster systems:			
1500	Gypsum plaster, 3 coats	4.14	8.95	13.09
1600	Perlite plaster, 2 coats	3.80	8.75	12.55
1700	3 coats	4.44	9.90	14.34
1800				
1900				
2000	For alternate lath systems:			
2100	Gypsum lath, 1/2″ thick	3.51	8.80	12.31
2200	Foil back, 3/8″ thick	3.45	8.85	12.30
2300	1/2″ thick	3.49	8.95	12.44
2400	Metal lath, 2.5 Lb. diamond	3.18	8.75	11.93
2500	3.4 Lb. diamond	3.28	8.85	12.13
2600				
2700				
2900	Cut & patch to match existing construction, add, minimum	2%	3%	
3000	Maximum	5%	9%	
3100	Dust protection, add, minimum	1%	2%	
3200	Maximum	4%	11%	
3300	Material handling & storage limitation, add, minimum	1%	1%	
3400	Maximum	6%	7%	
3500	Protection of existing work, add, minimum	2%	2%	
3600	Maximum	5%	7%	
3700	Shift work requirements, add, minimum		5%	
3800	Maximum		30%	
3900	Temporary shoring and bracing, add, minimum	2%	5%	
4000	Maximum	5%	12%	

C1010 Partitions

C1010 170	Selective Price Sheet	COST PER S.F. MAT.	INST.	TOTAL
0100	Studs			
0200				
0300	24" O.C. metal, 10' high wall, including			
0400	Top and bottom runners			
0500	Non load bearing, galvanized 25 Ga., 1-5/8" wide	.18	.62	.80
0600	2-1/2" wide	.22	.63	.85
0700	3-5/8" wide	.25	.64	.89
0800	4" wide	.27	.64	.91
0900	6" wide	.39	.65	1.04
1000				
1100	Galvanized 20 Ga., 2-1/2" wide	.35	.79	1.14
1200	3-5/8" wide	.42	.80	1.22
1300	4" wide	.46	.80	1.26
1400	6" wide	.59	.81	1.40
1500	Load bearing, painted 18 Ga., 2-1/2" wide	.58	.88	1.46
1600	3-5/8" wide	.69	.90	1.59
1700	4" wide	.62	.92	1.54
1800	6" wide	.92	.94	1.86
1900	Galvanized 18 Ga., 2-1/2" wide	.58	.88	1.46
2000	3-5/8" wide	.69	.90	1.59
2100	4" wide	.62	.92	1.54
2200	6" wide	.92	.94	1.86
2300	Galvanized 16 Ga., 2-1/2" wide	.67	1.01	1.68
2400	3-5/8" wide	.80	1.03	1.83
2500	4" wide	.85	1.05	1.90
2600	6" wide	1.07	1.07	2.14
2700				
2800				
2900	24" O.C. wood, 10' high wall, including			
3000	Double top plate and shoe			
3100	2" x 4"	.42	.89	1.31
3200	2" x 6"	.67	.95	1.62
3300				
3400				
3500	Furring, 24" O.C. 10' high wall, metal, 3/4" channels	.15	.99	1.14
3600	1-1/2" channels	.21	1.07	1.28
3700	Wood, on wood, 1" x 2" strips	.11	.43	.54
3800	1" x 3" strips	.16	.43	.59
3900	On masonry, 1" x 2" strips	.11	.48	.59
4000	1" x 3" strips	.16	.48	.64
4100	On concrete, 1" x 2" strips	.11	.91	1.02
4200	1" x 3" strips	.16	.91	1.07
4300	Studs			
4400				
4500	16" O.C., metal, 10' high wall, including			
4600	Top and bottom runners			
4700	Non load bearing, galvanized 25 Ga., 1-5/8" wide	.25	.95	1.20
4800	2-1/2" wide	.30	.96	1.26
4900	3-5/8" wide	.34	.98	1.32
5000	4" wide	.36	.99	1.35
5100	6" wide	.53	1	1.53
5200				
5300	Galvanized 20 Ga., 2-1/2" wide	.48	1.21	1.69
5400	3-5/8" wide	.57	1.22	1.79
5500	4" wide	.63	1.24	1.87
5600	6" wide	.79	1.26	2.05
5700	Load bearing, painted 18 Ga., 2-1/2" wide	.79	1.23	2.02
5800	3-5/8" wide	.94	1.24	2.18

Important: See the Reference Section for critical supporting data - Reference Numbers and City Cost Indexes

C1010 Partitions

C1010 170	Selective Price Sheet	COST PER S.F.		
		MAT.	INST.	TOTAL
5900	4" wide	.99	1.28	2.27
6000	6" wide	1.25	1.29	2.54
6100	Galvanized 18 Ga., 2-1/2" wide	.79	1.23	2.02
6200	3-5/8" wide	.94	1.24	2.18
6300	4" wide	.99	1.28	2.27
6400	6" wide	1.25	1.29	2.54
6500	Galvanized 16 Ga., 2-1/2" wide	.92	1.39	2.31
6600	3-5/8" wide	1.10	1.43	2.53
6700	4" wide	1.17	1.45	2.62
6800	6" wide	1.47	1.48	2.95
6900				
7000				
7100	16" O.C., wood, 10' high wall, including			
7200	Double top plate and shoe			
7300	2" x 4"	.52	1.02	1.54
7400	2" x 6"	.66	.92	1.58
7500				
7600				
7700	Furring, 16" O.C. 10' high wall, metal, 3/4" channels	.23	1.44	1.67
7800	1-1/2" channels	.32	1.60	1.92
7900	Wood, on wood, 1" x 2" strips	.16	.65	.81
8000	1" x 3" strips	.23	.65	.88
8100	On masonry, 1" x 2" strips	.16	.71	.87
8200	1" x 3" strips	.23	.71	.94
8300	On concrete, 1" x 2" strips	.16	1.36	1.52
8400	1" x 3" strips	.23	1.36	1.59

C1010 Partitions

C1010 180	Selective Price Sheet, Drywall & Plaster	COST PER S.F.		
		MAT.	INST.	TOTAL
0100	Lath, gypsum perforated			
0200				
0300	Lath, gypsum, perforated, regular, 3/8" thick	.66	.54	1.20
0400	1/2" thick	.52	.66	1.18
0500	Fire resistant, 3/8" thick	.45	.66	1.11
0600	1/2" thick	.55	.71	1.26
0700	Foil back, 3/8" thick	.48	.62	1.10
0800	1/2" thick	.51	.66	1.17
0900				
1000	Metal lath			
1100	Metal lath, diamond, painted, 2.5 lb.	.35	.54	.89
1200	3.4 lb.	.40	.62	1.02
1300	Rib painted, 2.75 lb	.37	.62	.99
1400	3.40 lb	.54	.66	1.20
1500				
1600				
1700	Plaster, gypsum, 2 coats	.40	2.21	2.61
1800	3 coats	.57	2.65	3.22
1900	Perlite/vermiculite, 2 coats	.40	2.53	2.93
2000	3 coats	.72	3.11	3.83
2100	Bondcrete, 1 coat	.41	1.16	1.57
2200				
2500				
2600				
2700	Drywall, standard, 3/8" thick, no finish included	.26	.47	.73
2800	1/2" thick, no finish included	.30	.47	.77
2900	Taped and finished	.34	.98	1.32
3000	5/8" thick, no finish included	.33	.47	.80
3100	Taped and finished	.37	.98	1.35
3200	Fire resistant, 1/2" thick, no finish included	.33	.47	.80
3300	Taped and finished	.37	.98	1.35
3400	5/8" thick, no finish included	.33	.47	.80
3500	Taped and finished	.37	.98	1.35
3600	Water resistant, 1/2" thick, no finish included	.28	.47	.75
3700	Taped and finished	.31	.98	1.29
3800	5/8" thick, no finish included	.30	.47	.77
3900	Taped and finished	.34	.98	1.32
4000	Finish, instead of taping			
4100	For thin coat plaster, add	.08	.58	.66
4200	Finish, textured spray, add	.06	.52	.58

C1010 Partitions

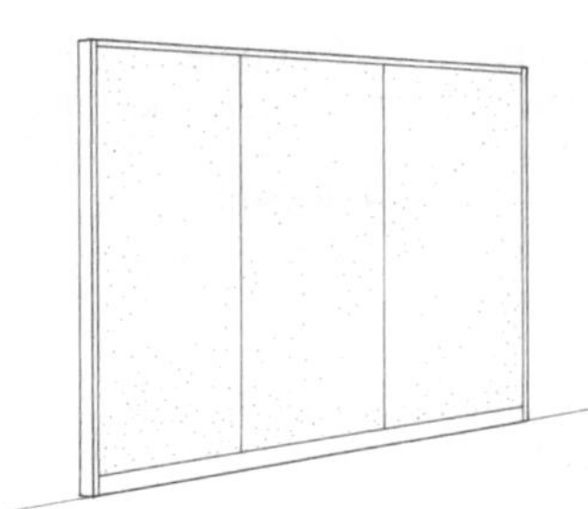

This page illustrates and describes a movable office partition system including demountable office partitions of various styles and sizes based on a cost per square foot basis. Prices for alternate movable office partition systems are on Line Items C1010 210 0300 thru 3600.

C1010 210	Partitions, Movable Office	COST PER S.F. MAT.	COST PER S.F. INST.	COST PER S.F. TOTAL
0250	Office partition, demountable, no deduction for door opening, add for doors			
0300	Air wall, cork finish, semi acoustic, 1-5/8" thick, minimum	35.50	2.90	38.40
0400	Maximum	33.50	4.96	38.46
0500	Acoustic, 2" thick, minimum	29	3.09	32.09
0600	Maximum	50	4.19	54.19
0700				
0800				
0900	In-plant modular office system, w/prehung steel door			
1000	3" thick honeycomb core panels			
1100	12' x 12', 2 wall	5.90	.38	6.28
1200	4 wall	12.30	1.14	13.44
1300				
1400				
1500	Gypsum, demountable, 3" to 3-3/4" thick x 9' high, vinyl clad	5.85	2.18	8.03
1600	Fabric clad	14.55	2.39	16.94
1700	1.75 system, vinyl clad hardboard, paper honeycomb core panel			
1800	1-3/4" to 2-1/2" thick x 9' high	9.80	2.18	11.98
1900	Unitized gypsum panel, 2" to 2-1/2" thick x 9' high, vinyl clad	12.65	2.18	14.83
2000	Fabric clad	21	2.39	23.39
2100				
2200				
2300	Unitized mineral fiber panel system, 2-1/4" thick x 9' high			
2400	Vinyl clad mineral fiber	12.55	2.18	14.73
2500	Fabric clad mineral fiber	18.75	2.39	21.14
2600				
2700	Movable steel walls, modular system			
2800	Unitized panels, 48" wide x 9' high			
2900	Baked enamel, pre-finished	14.20	1.74	15.94
3000	Fabric clad	20.50	1.87	22.37
3100	For acoustical partitions, add, minimum	2.07		2.07
3200	Maximum	9.65		9.65
3300				
3400				
3500	Note: For door prices, see divisions 08100 & 08200			
3600	For door hardware prices, see division 08700			
3700				
3800				
3900				
4000				

C1020 Interior Doors

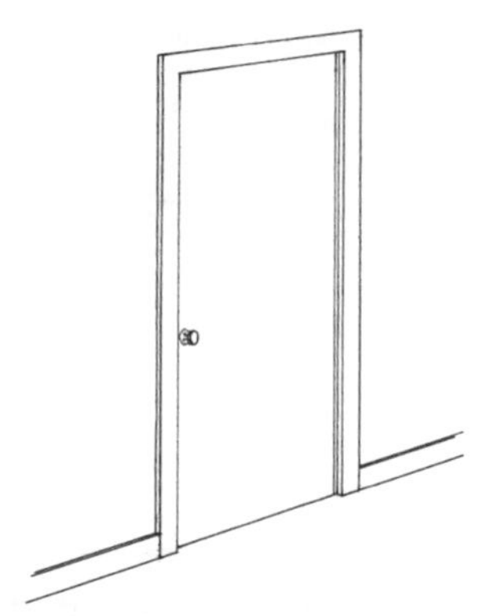

This page illustrates and describes flush interior door systems including hollow core door, jamb, header and trim with hardware. Lines within System Components give the unit price and total price on a cost each basis for this system. Prices for alternate flush interior door systems are on Line items C1020 106 1100 thru 2400. Both material quantities and labor costs have been adjusted for the system listed.

Factors: To adjust for job conditions other than normal working situations use Lines C1020 106 2700 thru 3800.

Example: You are to install the system in an area where dust protection is vital. Go to Line C1020 106 3000 and apply these percentages to the appropriate MAT. and INST. costs.

System Components	QUANTITY	UNIT	COST EACH		
			MAT.	INST.	TOTAL
Single hollow core door, include jamb, header, trim and hardware, painted.					
Hollow core Lauan, 1-3/8" thick, 2'-0" x 6'-8", painted	1.000	Ea.	30	52.50	82.50
Paint door and frame, 1 coat	1.000	Ea.	2.15	24	26.15
Wood jamb, 4-9/16" deep	1.000	Set	102.40	40.16	142.56
Trim, casing	1.000	Set	45.12	62.72	107.84
Hardware, hinges	1.000	Set	10.35		10.35
Hardware, lockset	1.000	Set	16.35	29.50	45.85
TOTAL			206.37	208.88	415.25

C1020 106	Doors, Interior Flush, Wood	COST EACH		
		MAT.	INST.	TOTAL
1000	For alternate door systems:			
1100	Lauan (Mahogany) hollow core, 1-3/8" x 2'-6" x 6'-8"	214	212	426
1200	2'-8" x 6'-8"	219	215	434
1300	3'-0" x 6'-8"	225	222	447
1500	Birch, hollow core, 1-3/8" x 2'-0" x 6'-8"	220	209	429
1600	2'-6" x 6'-8"	230	212	442
1700	2'-8" x 6'-8"	234	215	449
1800	3'-0" x 6'-8"	243	222	465
1900	Solid core, pre-hung, 1-3/8" x 2'-6" x 6'-8"	365	207	572
2000	2'-8" x 6'-8"	390	209	599
2100	3'-0" x 6'-8"	410	216	626
2200				
2400	For metal frame instead of wood, add	50%	20%	
2600				
2700	Cut & patch to match existing construction, add, minimum	2%	3%	
2800	Maximum	5%	9%	
2900	Dust protection, add, minimum	1%	2%	
3000	Maximum	4%	11%	
3100	Equipment usage curtailment, add, minimum	1%	1%	
3200	Maximum	3%	10%	
3300	Material handling & storage limitation, add, minimum	1%	1%	
3400	Maximum	6%	7%	
3500	Protection of existing work, add, minimum	2%	2%	
3600	Maximum	5%	7%	
3700	Shift work requirements, add, minimum		5%	
3800	Maximum		30%	
3900				

C1020 Interior Doors

This page illustrates and describes interior, solid and louvered door systems including a pine panel door, wood jambs, header, and trim with hardware. Lines within System Components give the unit price and total price on a cost each basis for this system. Prices for alternate interior, solid and louvered systems are on Line Items C1020 108 1200 thru 2400. Both material quantities and labor costs have been adjusted for the system listed.

Factors: To adjust for job conditions other than normal working situations use Lines C1020 108 2900 thru 4000.

Example: You are to install the system during night hours only. Go to Line C1020 108 4000 and apply these percentages to the appropriate INST. costs.

System Components	QUANTITY	UNIT	COST EACH		
			MAT.	INST.	TOTAL
Single interior door, including jamb, header, trim and hardware, painted.					
Solid pine panel door, 1-3/8″ thick, 2′-0″ x 6′-8″	1.000	Ea.	138	52.50	190.50
Paint door and frame, 1 coat	1.000	Ea.	2.15	24	26.15
Wooden jamb, 4-5/8″ deep	1.000	Set	102.40	40.16	142.56
Trim, casing	1.000	Set	45.12	62.72	107.84
Hardware, hinges	1.000	Set	10.35		10.35
Hardware, lockset	1.000	Set	16.35	29.50	45.85
TOTAL			314.37	208.88	523.25

C1020 108	Doors, Interior Solid & Louvered	COST EACH		
		MAT.	INST.	TOTAL
1000				
1100	For alternate door systems:			
1200	Solid pine, painted raised panel, 1-3/8″ x 2′-6″ x 6′-8″	335	212	547
1300	2′-8″ x 6′-8″	350	215	565
1400	3′-0″ x 6′-8″	360	222	582
1500				
1600	Louvered pine, painted 1′-6″ x 6′-8″	286	206	492
1700	2′-0″ x 6′-8″	320	212	532
1800	2′-6″ x 6′-8″	335	215	550
1900	3′-0″ x 6′-8″	360	222	582
2200	For prehung door, deduct	5%	30%	
2300				
2400	For metal frame instead of wood, add	50%	20%	
2500				
2900	Cut & patch to match existing construction, add, minimum	2%	3%	
3000	Maximum	5%	9%	
3100	Dust protection, add, minimum	1%	2%	
3200	Maximum	4%	11%	
3300	Equipment usage curtailment, add, minimum	1%	1%	
3400	Maximum	3%	10%	
3500	Material handling & storage limitation, add, minimum	1%	1%	
3600	Maximum	6%	7%	
3700	Protection of existing work, add, minimum	2%	2%	
3800	Maximum	5%	7%	
3900	Shift work requirements, add, minimum		5%	
4000	Maximum		30%	

C1020 Interior Doors

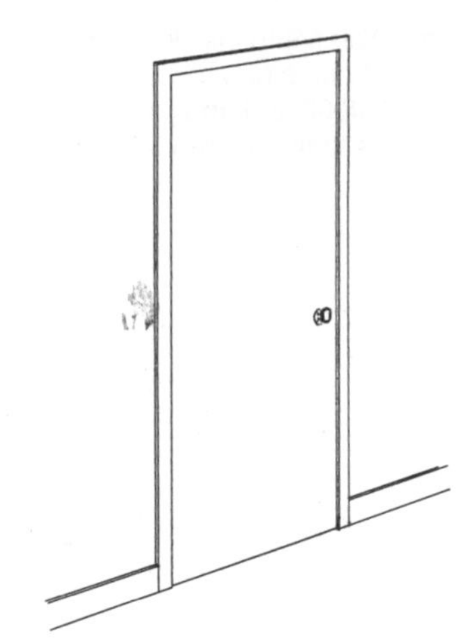

This page illustrates and describes interior metal door systems including a metal door, metal frame and hardware. Lines within System Components give the unit price and total price on a cost each basis for this system. Prices for alternate interior metal door systems are on Line Items C1020 110 1100 thru 2100. Both material quantities and labor costs have been adjusted for the system listed.

Factors: To adjust for job conditions other than normal working situations use Lines C1020 110 2900 thru 4000.

Example: You are to install the system while protecting existing construction. Go to Line C1020 110 3700 and apply these percentages to the appropriate MAT. and INST. costs.

System Components	QUANTITY	UNIT	COST EACH		
			MAT.	INST.	TOTAL
Single metal door, including frame and hardware.					
Hollow metal door, 1-3/8" thick, 2'-6" x 6'-8", painted	1.000	Ea.	340	63	403
Metal frame, 5-3/4" deep	1.000	Set	138	59	197
Paint door and frame, 1 coat	1.000	Ea.	2.15	24	26.15
Hardware, hinges	1.000	Set	70		70
Hardware, passage lockset	1.000	Set	16.35	29.50	45.85
TOTAL			566.50	175.50	742

C1020 110	Doors, Interior Flush, Metal	COST EACH		
		MAT.	INST.	TOTAL
1000	For alternate systems:			
1100	Hollow metal doors, 1-3/8" thick, 2'-8" x 6'-8"	570	176	746
1200	3'-0" x 7'-0"	560	184	744
1300				
1400	Interior fire door, 1-3/8" thick, 2'-6" x 6'-8"	695	176	871
1500	2'-8" x 6'-8"	700	176	876
1600	3'-0" x 7'-0"	700	184	884
1700				
1800	Add to fire doors:			
1900	Baked enamel finish	30%	15%	
2000	Galvanizing	15%		
2200				
2300				
2400				
2900	Cut & patch to match existing construction, add, minimum	2%	3%	
3000	Maximum	5%	9%	
3100	Dust protection, add, minimum	1%	2%	
3200	Maximum	4%	11%	
3300	Equipment usage curtailment, add, minimum	1%	1%	
3400	Maximum	3%	10%	
3500	Material handling & storage limitation, add, minimum	1%	1%	
3600	Maximum	6%	7%	
3700	Protection of existing work, add, minimum	2%	2%	
3800	Maximum	5%	7%	
3900	Shift work requirements, add, minimum		5%	
4000	Maximum		30%	

C1020 Interior Doors

This page illustrates and describes an interior closet door system including an interior closet door, painted, with trim and hardware. Prices for alternate interior closet door systems are on Line Items C1020 112 0500 thru 2200. Both material quantities and labor costs have been adjusted for the system listed.

Factors: To adjust for job conditions other than normal working situations use Lines C1020 112 2900 thru 4000.

Example: You are to install the system and match the existing construction. Go to Line C1020 112 2900 and apply these percentages to the appropriate MAT. and INST. costs.

C1020 112	Doors, Closet	COST PER SET		
		MAT.	INST.	TOTAL
0350	Interior closet door painted, including frame, trim and hardware, prehung.			
0400	Bi-fold doors			
0500	Pine paneled, 3′-0″ x 6′-8″	320	172	492
0600	6′-0″ x 6′-8″	560	231	791
0700	Birch, hollow core, 3′-0″ x 6′-8″	176	187	363
0800	6′-0″ x 6′-8″	264	250	514
0900	Lauan, hollow core, 3′-0″ x 6′-8″	165	172	337
1000	6′-0″ x 6′-8″	241	231	472
1100	Louvered pine, 3′-0″ x 6′-8″	280	172	452
1200	6′-0″ x 6′-8″	395	231	626
1300				
1400	Sliding, bi-passing closet doors			
1500	Pine paneled, 4′-0″ x 6′-8″	535	183	718
1600	6′-0″ x 6′-8″	645	231	876
1700	Birch, hollow core, 4′-0″ x 6′-8″	335	183	518
1800	6′-0″ x 6′-8″	390	231	621
1900	Lauan, hollow core, 4′-0″ x 6′-8″	294	183	477
2000	6′-0″ x 6′-8″	340	231	571
2100	Louvered pine, 4′-0″ x 6′-8″	555	183	738
2200	6′-0″ x 6′-8″	665	231	896
2300				
2400				
2500				
2600				
2700				
2800				
2900	Cut & patch to match existing construction, add, minimum	2%	3%	
3000	Maximum	5%	9%	
3100	Dust protection, add, minimum	1%	2%	
3200	Maximum	4%	11%	
3300	Equipment usage curtailment, add, minimum	1%	1%	
3400	Maximum	3%	10%	
3500	Material handling & storage limitation, add, minimum	1%	1%	
3600	Maximum	6%	7%	
3700	Protection of existing work, add, minimum	2%	2%	
3800	Maximum	5%	7%	
3900	Shift work requirements, add, minimum		5%	
4000	Maximum		30%	

C2010 Stair Construction

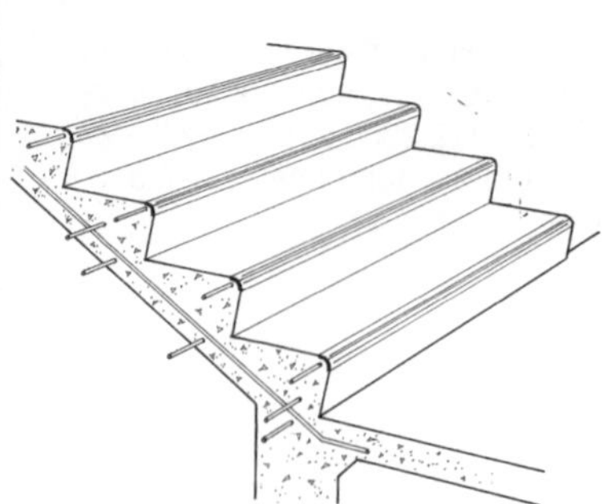

This page illustrates and describes a stair system based on a cost per flight price. Prices for various stair systems are on Line Items C2010 130 0700 thru 3200. Both material quantities and labor costs have been adjusted for the system listed.

Factors: To adjust for job conditions other than normal working situations use Lines C2010 130 3500 thru 4200.

Example: You are to install the system during evenings only. Go to Line C2010 130 4200 and apply this percentage to the appropriate MAT. and INST. costs.

System Components	QUANTITY	UNIT	COST PER FLIGHT		
			MAT.	INST.	TOTAL
Below are various stair systems based on cost per flight of stairs, no side Walls. Stairs are 4′-0″ wide, railings are included unless otherwise noted					

C2010 130	Stairs	COST PER FLIGHT		
		MAT.	INST.	TOTAL
0700	Concrete, cast in place, no nosings, no railings, 12 risers	281	1,625	1,906
0800	24 risers	560	3,250	3,810
0900	Add for 1 intermediate landing	153	450	603
1000	Concrete, cast in place, with nosings, no railings, 12 risers	845	1,925	2,770
1100	24 risers	1,700	3,800	5,500
1200	Add for 1 intermediate landing	200	475	675
1300	Steel, grating tread, safety nosing, 12 risers	3,175	945	4,120
1400	24 risers	6,325	1,900	8,225
1500	Add for intermediate landing	770	252	1,022
1600	Steel, cement fill pan tread, 12 risers	4,750	945	5,695
1700	24 risers	9,500	1,900	11,400
1800	Add for intermediate landing	770	252	1,022
1900	Spiral, industrial, 4′ - 6″ diameter, 12 risers	4,675	675	5,350
2000	24 risers	9,350	1,350	10,700
2100	Wood, box stairs, oak treads, 12 risers	1,900	535	2,435
2200	24 risers	3,775	1,075	4,850
2300	Add for 1 intermediate landing	142	120	262
2400	Wood, basement stairs, no risers, 12 steps	605	217	822
2500	24 steps	1,200	435	1,635
2600	Add for 1 intermediate landing	34	23	57
2700	Wood, open, rough sawn cedar, 12 steps	2,125	435	2,560
2800	24 steps	4,275	865	5,140
2900	Add for 1 intermediate landing	20	24	44
3000	Wood, residential, oak treads, 12 risers	2,350	2,375	4,725
3100	24 risers	4,675	4,725	9,400
3200	Add for 1 intermediate landing	125	109	234
3500	Dust protection, add, minimum	1%	2%	
3600	Maximum	4%	11%	
3700	Material handling & storage limitation, add, minimum	1%	1%	
3800	Maximum	6%	7%	
3900	Protection of existing work, add, minimum	2%	2%	
4000	Maximum	5%	7%	
4100	Shift work requirements, add, minimum		5%	
4200	Maximum		30%	

C3010 Wall Finishes

C3010 210	Selective Price Sheet	COST PER S.F.		
		MAT.	INST.	TOTAL
0100	Painting, on plaster or drywall, brushwork, primer and 1 ct.	.11	.60	.71
0200	Primer and 2 ct.	.16	.95	1.11
0300	Rollerwork, primer and 1 ct.	.12	.51	.63
0400	Primer and 2 ct.	.17	.81	.98
0500	Woodwork incl. puttying, brushwork, primer and 1 ct.	.11	.91	1.02
0600	Primer and 2 ct.	.16	1.20	1.36
0700	Wood trim to 6″ wide, enamel, primer and 1 ct.	.10	.51	.61
0800	Primer and 2 ct.	.15	.65	.80
0900	Cabinets and casework, enamel, primer and 1 ct.	.10	1.02	1.12
1000	Primer and 2 ct.	.15	1.25	1.40
1100	On masonry or concrete, latex, brushwork, primer and 1 ct.	.19	.85	1.04
1200	Primer and 2 ct.	.25	1.22	1.47
1300	For block filler, add	.12	1.04	1.16
1400				
1500	Varnish, wood trim, sealer 1 ct., sanding, puttying, quality work	.14	1.84	1.98
1600	Medium work	.11	1.43	1.54
1700	Without sanding	.14	.22	.36
1800				
1900	Wall coverings, wall paper, at $9.70 per double roll, average workmanship	.33	.64	.97
2000	At $20.00 per double roll, average workmanship	.72	.76	1.48
2100	At $44.00 per double roll, quality workmanship	1.68	.94	2.62
2200				
2300	Grass cloths with lining paper, minimum	.72	1.02	1.74
2400	Maximum	2.30	1.17	3.47
2500	Vinyl, fabric backed, light weight	.65	.64	1.29
2600	Medium weight	.80	.85	1.65
2700	Heavy weight	1.29	.94	2.23
2800				
2900	Cork tiles, 12″ x 12″, 3/16″ thick	4.11	1.70	5.81
3000	5/16″ thick	3.51	1.74	5.25
3100	Granular surface, 12″ x 36″, 1/2″ thick	1.17	1.06	2.23
3200	1″ thick	1.51	1.11	2.62
3300	Aluminum foil	.96	1.49	2.45
3400				
3500	Tile, ceramic, adhesive set, 4-1/4″ x 4-1/4″	2.19	4.01	6.20
3600	6″ x 6″	3	3.81	6.81
3700	Decorated, 4-1/4″ x 4-1/4″, minimum	4.02	2.82	6.84
3800	Maximum	43	4.23	47.23
3900	For epoxy grout, add	.36	.95	1.31
4000	Pregrouted sheets	4.65	3.18	7.83
4100	Glass mosaics, 3/4″ tile on 12″ sheets, minimum	17.30	10.45	27.75
4200	Color group 8	56.50	11.90	68.40
4300	Metal, tile pattern, 4′ x 4′ sheet, 24 ga., nailed			
4400	Stainless steel	25.50	1.84	27.34
4500	Aluminized steel	13.75	1.84	15.59
4600				
4700	Brick, interior veneer, 4″ face brick, running bond, minimum	3.56	9.95	13.51
4800	Maximum	3.56	9.95	13.51
4900	Simulated, urethane pieces, set in mastic	5.70	3.14	8.84
5000	Fiberglass panels	3.30	2.36	5.66
5100	Wall coating, on drywall, thin coat, plain	.08	.58	.66
5200	Stipple	.08	.58	.66
5300	Textured spray	.06	.52	.58
5400				
5500	Paneling not incl. furring or trim, hardboard, tempered, 1/8″ thick	.35	1.88	2.23
5600	1/4″ thick	.53	1.88	2.41
5700	Plastic faced, 1/8″ thick	.64	1.88	2.52
5800	1/4″ thick	.85	1.88	2.73

C3010 Wall Finishes

C3010 210	Selective Price Sheet	COST PER S.F.		
		MAT.	INST.	TOTAL
5900	Woodgrained, 1/4" thick, minimum	.56	1.88	2.44
6000	Maximum	1.19	2.22	3.41
6100	Plywood, 4' x 8' shts. 1/4" thick, prefin., birch faced, min.	.92	1.88	2.80
6200	Maximum	2.06	2.69	4.75
6300	Walnut, minimum	2.88	1.88	4.76
6400	Maximum	5.45	2.36	7.81
6500	Mahogany, African	2.63	2.36	4.99
6600	Philippine	1.13	1.88	3.01
6700	Chestnut	4.99	2.51	7.50
6800	Pecan	2.16	2.36	4.52
6900	Rosewood	4.80	2.95	7.75
7000	Teak	3.38	2.36	5.74
7100	Aromatic cedar, plywood	2.18	2.36	4.54
7200	Particle board	1.06	2.36	3.42
7300	Wood board, 3/4" thick, knotty pine	1.54	3.14	4.68
7400	Rough sawn cedar	1.97	3.14	5.11
7500	Redwood, clear	4.60	3.14	7.74
7600	Aromatic cedar	3.54	3.43	6.97

Important: See the Reference Section for critical supporting data - Reference Numbers and City Cost Indexes

C3020 Floor Finishes

C3020 430	Selective Price Sheet	COST PER S.F.		
		MAT.	INST.	TOTAL
0100	Flooring, carpet, acrylic, 26 oz. light traffic	2.15	.64	2.79
0200	35 oz. heavy traffic	4.50	.64	5.14
0300	Nylon anti-static, 15 oz. light traffic	1.43	.84	2.27
0400	22 oz. medium traffic	2.94	.64	3.58
0500	26 oz. heavy traffic	4.05	.68	4.73
0600	28 oz. heavy traffic	4.55	.68	5.23
0700	Tile, foamed back, needle punch	3.47	.75	4.22
0800	Tufted loop	1.46	.75	2.21
0900	Wool, 36 oz. medium traffic	10.20	.68	10.88
1000	42 oz. heavy traffic	10.40	.68	11.08
1100	Composition, epoxy, with colored chips, minimum	2.53	3.44	5.97
1200	Maximum	3.07	4.75	7.82
1300	Trowelled, minimum	3.27	4.14	7.41
1400	Maximum	4.76	4.84	9.60
1500	Terrazzo, 1/4" thick, chemical resistant, minimum	6	5.15	11.15
1600	Maximum	9.20	6.90	16.10
1700	Resilient, asphalt tile, 1/8" thick	1.17	1.07	2.24
1800	Conductive flooring, rubber, 1/8" thick	3.85	1.36	5.21
1900	Cork tile 1/8" thick, standard finish	4.66	1.36	6.02
2000	Urethane finish	5.70	1.36	7.06
2100	PVC sheet goods for gyms, 1/4" thick	5.50	5.35	10.85
2200	3/8" thick	6.20	7.15	13.35
2300	Vinyl composition 12" x 12" tile, plain, 1/16" thick	.95	.86	1.81
2400	1/8" thick	2.26	.86	3.12
2500	Vinyl tile, 12" x 12" x 1/8" thick, minimum	3.03	.86	3.89
2600	Maximum	10	.86	10.86
2700	Vinyl sheet goods, backed, .093" thick	2.63	1.87	4.50
2900	Slate, random rectangular, 1/4" thick	13.50	5.55	19.05
3000	1/2" thick	12.55	7.95	20.50
3100	Natural cleft, irregular, 3/4" thick	5.80	9.05	14.85
3200	For sand rubbed finish, add	6.30		6.30
3300	Terrazzo, cast in place, bonded 1-3/4" thick, gray cement	2.81	7.95	10.76
3400	White cement	3.19	7.95	11.14
3500	Not bonded 3" thick, gray cement	3.51	8.95	12.46
3600	White cement	3.83	8.95	12.78
3700	Precast, 12" x 12" x 1" thick	17.80	29	46.80
3800	1-1/4" thick	19.90	29	48.90
3900	16" x 16" x 1-1/4" thick	21.50	36.50	58
4000	1-1/2" thick	19.80	39.50	59.30
4100	Marble travertine, standard, 12" x 12" x 3/4" thick	11	12.70	23.70
4200				
4300	Tile, ceramic, natural clay, thin set	4.06	4.16	8.22
4400	Porcelain, thin set	4.93	4.01	8.94
4500	Specialty, decorator finish	9.85	4.16	14.01
4600				
4700	Quarry, red, mud set, 4" x 4" x 1/2" thick	4.28	6.35	10.63
4800	6" x 6" x 1/2" thick	3.45	5.45	8.90
4900	Brown, imported, 6" x 6" x 3/4" thick	5.10	6.35	11.45
5000	8" x 8" x 1" thick	5.75	6.95	12.70
5100	Slate, Vermont, thin set, 6" x 6" x 1/4" thick	4.69	4.23	8.92
5200				
5300	Wood, maple strip, 25/32" x 2-1/4", finished, select	5.40	2.99	8.39
5400	2nd and better	3.39	2.99	6.38
5500	Oak, 25/32" x 2-1/4" finished, clear	3.44	2.99	6.43
5600	No. 1 common	4.23	2.99	7.22
5700	Parquet, standard, 5/16", finished, minimum	3.91	3.17	7.08
5800	Maximum	6	4.93	10.93
5900	Custom, finished, minimum	16.10	4.71	20.81

C3020 Floor Finishes

C3020 430	Selective Price Sheet	COST PER S.F.		
		MAT.	INST.	TOTAL
6000	Maximum	21.50	9.40	30.90
6100	Prefinished, oak, 2-1/4" wide	6.80	2.77	9.57
6200	Ranch plank	8.45	3.25	11.70
6300	Sleepers on concrete, treated, 24" O.C., 1" x 2"	.09	.31	.40
6400	1" x 3"	.18	.40	.58
6500	2" x 4"	.47	.56	1.03
6600	2" x 6"	.72	.64	1.36
6700	Refinish old floors, minimum	.78	.91	1.69
6800	Maximum	1.18	2.79	3.97
6900	Subflooring, plywood, CDX, 1/2" thick	.68	.63	1.31
7000	5/8" thick	.73	.70	1.43
7100	3/4" thick	.90	.75	1.65
7200				
7300	1" x 10" boards, S4S, laid regular	1.50	.86	2.36
7400	Laid diagonal	1.50	1.05	2.55
7500	1" x 8" boards, S4S, laid regular	1.23	.94	2.17
7600	Laid diagonal	1.23	1.11	2.34
7700	Underlayment, plywood, underlayment grade, 3/8" thick	.86	.63	1.49
7800	1/2" thick	.96	.65	1.61
7900	5/8" thick	1.34	.67	2.01
8000	3/4" thick	1.45	.73	2.18
8100	Particle board, 3/8" thick	.36	.63	.99
8200	1/2" thick	.41	.65	1.06
8300	5/8" thick	.45	.67	1.12
8400	3/4" thick	.62	.73	1.35

C3030 Ceiling Finishes

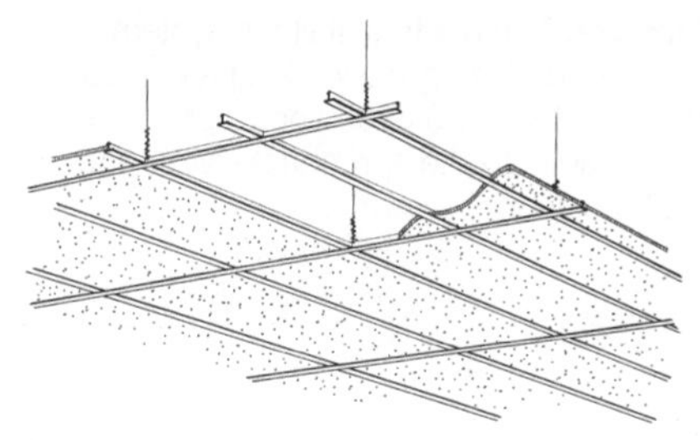

This page illustrates suspended acoustical board systems including acoustic ceiling board, hangers, and T bar suspension. Lines within System Components give the unit price and total price per square foot for this system. Prices for alternate suspended acoustical board systems are on Line Items C3030 210 1100 thru 2400. Both material quantities and labor costs have been adjusted for the system listed.

Factors: To adjust for job conditions other than normal working situations use Lines C3030 210 2900 thru 4000.

Example: You are to install the system and protect existing construction. Go to Line C3030 210 3800 and apply these percentages to the appropriate MAT. and INST. costs.

System Components	QUANTITY	UNIT	COST PER S.F.		
			MAT.	INST.	TOTAL
Suspended acoustical ceiling board installed on exposed grid system.					
Fiberglass boards, film faced, 2′ x 4′, 5/8″ thick	1.000	S.F.	.57	.75	1.32
Hangers, #12 wire	1.000	S.F.	.06	.07	.13
T bar suspension system, 2′ x 4′ grid	1.000	S.F.	.58	.59	1.17
TOTAL			1.21	1.41	2.62

C3030 210	Ceiling, Suspended Acoustical	COST PER S.F.		
		MAT.	INST.	TOTAL
1000	For alternate suspended ceiling systems:			
1100	2′ x 4′ grid, mineral fiber board, aluminum faced, 5/8″ thick	1.85	1.39	3.24
1200	Standard faced	1.34	1.36	2.70
1300	Plastic faced	1.74	1.84	3.58
1400	Fiberglass, film faced, 3″ thick, R11	2.08	1.71	3.79
1500	Grass cloth faced, 3/4″ thick	2.53	1.60	4.13
1600	1″ thick	2.73	1.63	4.36
1700	1-1/2″ thick, nubby face	3.24	1.65	4.89
2200				
2300				
2400	Add for 2′ x 2′ grid system	.14	.15	.29
2500				
2600				
2700				
2900	Cut & patch to match existing construction, add, minimum	2%	3%	
3000	Maximum	5%	9%	
3100	Dust protection, add, minimum	1%	2%	
3200	Maximum	4%	11%	
3300	Equipment usage curtailment, add, minimum	1%	1%	
3400	Maximum	3%	10%	
3500	Material handling & storage limitation, add, minimum	1%	1%	
3600	Maximum	6%	7%	
3700	Protection of existing work, add, minimum	2%	2%	
3800	Maximum	5%	7%	
3900	Shift work requirements, add, minimum		5%	
4000	Maximum		30%	

C3030 Ceiling Finishes

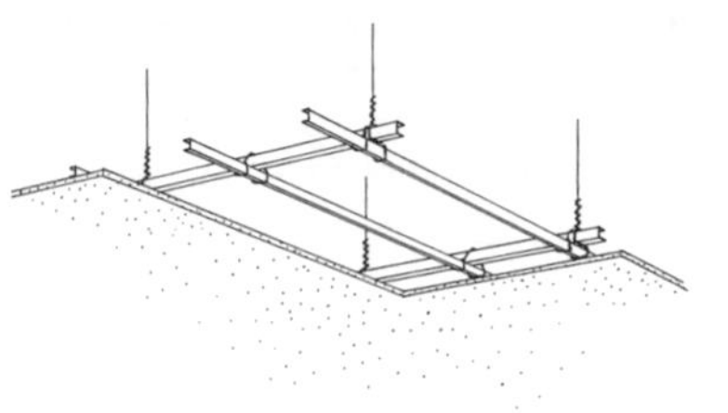

This page illustrates and describes suspended gypsum board systems including gypsum board, metal furring, taping, finished and painted. Lines within System Components give the unit price and total price per square foot for this system. Prices for alternate suspended gypsum board systems are on Line Items C3030 220 1500 thru 1700. Both material quantities and labor costs have been adjusted for the system listed.

Factors: To adjust for job conditions other than normal working situations use Lines C3030 220 2900 thru 4000.

Example: You are to install the system and control dust in the working area. Go to Line C3030 220 3200 and apply these percentages to the appropriate MAT. and INST. costs.

System Components	QUANTITY	UNIT	COST PER S.F. MAT.	COST PER S.F. INST.	COST PER S.F. TOTAL
Suspended ceiling gypsum board,4′ x 8′ x 5/8″ thick,					
On metal furring, taped, finished and painted.					
Gypsum drywall, 4′ x 8′, 5/8″ thick, screwed	1.000	S.F.	.33	.52	.85
Main runners, 1-1/2″ C.R.C., 4′ O.C.	.500	S.F.	.11	.54	.65
25 ga., channels, 2′ O.C.	1.000	S.F.	.15	.99	1.14
Taped and finished	1.000	S.F.	.04	.47	.51
Paint, 2 coats, roller work	1.000	S.F.	.16	.54	.70
TOTAL			.79	3.06	3.85

C3030 220	Ceilings, Suspended Gypsum Board	COST PER S.F. MAT.	COST PER S.F. INST.	COST PER S.F. TOTAL
1400	For alternate drywall ceiling systems:			
1500	Thin coat plaster, 2 coats paint	.67	2.63	3.30
1600	Spray-on sand finish, no paint	.65	2.57	3.22
1700	12″ x 12″ x 3/4″ acoustical wood fiber tile	1.77	3.23	5
1800				
1900				
2000				
2100				
2200				
2300				
2400				
2500				
2600				
2700				
2900	Cut & patch to match existing construction, add, minimum	2%	3%	
3000	Maximum	5%	9%	
3100	Dust protection, add, minimum	1%	2%	
3200	Maximum	4%	11%	
3300	Equipment usage curtailment, add, minimum	1%	1%	
3400	Maximum	3%	10%	
3500	Material handling & storage limitation, add, minimum	1%	1%	
3600	Maximum	6%	7%	
3700	Protection of existing work, add, minimum	2%	2%	
3800	Maximum	5%	7%	
3900	Shift work requirements, add, minimum		5%	
4000	Maximum		30%	

C3030 Ceiling Finishes

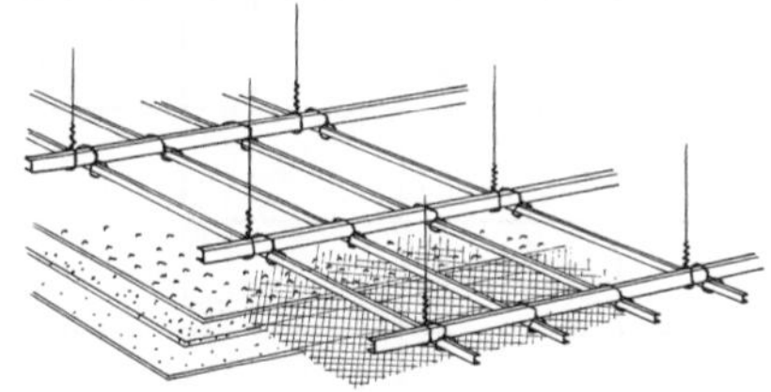

This page illustrates and describes suspended plaster and lath systems including gypsum plaster, lath, furring and runners with ceiling painted. Lines within System Components give the unit price and total price per square foot for this system. Prices for alternate suspended plaster and lath systems are on Line Items C3030 230 1300 thru 2300. Both material quantities and labor costs have been adjusted for the system listed.

Factors: To adjust for job conditions other than normal working situations use Lines C3030 230 2900 thru 4000.

Example: You are to install the system to match existing construction. Go to Line C3030 230 2900 and apply these percentages to the appropriate MAT. and INST. costs.

System Components	QUANTITY	UNIT	COST PER S.F. MAT.	COST PER S.F. INST.	COST PER S.F. TOTAL
Gypsum plaster, 3 coats, on 3.4# rib lath, on 3/4″ C.R.C. furring on 1-1/2″ main runners, ceiling painted.					
Gypsum plaster, 3 coats	.110	S.Y.	.56	2.97	3.53
3.4# rib lath	.110	S.Y.	.53	.65	1.18
Main runners, 1-1/2″ C.R.C. 24″ O.C.	.333	S.F.	.21	1.07	1.28
Furring 3/4″ C.R.C. 16″ O.C.	1.000	S.F.	.23	1.44	1.67
Painting, 2 coats, roller work	1.000	S.F.	.16	.54	.70
TOTAL			1.69	6.67	8.36

C3030 230	Ceiling, Suspended Plaster	COST PER S.F. MAT.	COST PER S.F. INST.	COST PER S.F. TOTAL
1200	For alternate plaster ceiling systems:			
1300	Gypsum plaster, 3 coats, on 2.5# diamond lath	1.51	6.60	8.11
1400	On 3/8″ gypsum lath	1.81	6.60	8.41
1500	2 coats, on 3.4# rib lath	1.67	6.65	8.32
1600	On 2.5# diamond lath	1.35	6.20	7.55
1700	On 3/8″ gypsum lath	1.65	6.20	7.85
1800	Perlite plaster, 3 coats, on 3.4# rib lath	1.84	7.35	9.19
1900	On 2.5# diamond lath	1.66	7.35	9.01
2000	On 3/8″ gypsum lath	1.96	7.35	9.31
2100	2 coats, on 3.4# rib lath	1.52	6.60	8.12
2200	On 2.5# diamond lath	1.34	6.55	7.89
2300	On 3/8″ gypsum lath	1.64	6.55	8.19
2400				
2500				
2600				
2700				
2900	Cut & patch to match existing construction, add, minimum	2%	3%	
3000	Maximum	5%	9%	
3100	Dust protection, add, minimum	1%	2%	
3200	Maximum	4%	11%	
3300	Equipment usage curtailment, add, minimum	1%	1%	
3400	Maximum	3%	10%	
3500	Material handling & storage limitation, add, minimum	1%	1%	
3600	Maximum	6%	7%	
3700	Protection of existing work, add, minimum	2%	2%	
3800	Maximum	5%	7%	
3900	Shift work requirements, add, minimum		5%	
4000	Maximum		30%	

C3030 Ceiling Finishes

C3030 250	Selective Price Sheet	COST PER S.F.		
		MAT.	INST.	TOTAL
0100	Ceiling, plaster, gypsum, 2 coats	.40	2.53	2.93
0200	3 coats	.57	3	3.57
0300	Perlite or vermiculite, 2 coats	.40	2.94	3.34
0400	3 coats	.72	3.70	4.42
0500	Gypsum lath, plain, 3/8" thick	.66	.54	1.20
0600	1/2" thick	.52	.58	1.10
0700	Firestop, 3/8" thick	.45	.66	1.11
0800	1/2" thick	.55	.71	1.26
0900	Metal lath, rib, 2.75 lb.	.37	.62	.99
1000	3.40 lb.	.54	.66	1.20
1100	Diamond, 2.50 lb.	.35	.62	.97
1200	3.40 lb.	.40	.77	1.17
1300				
1400				
1500	Drywall, standard, 1/2" thick, no finish included	.30	.52	.82
1600	Taped and finished	.34	1.23	1.57
1700	5/8" thick, no finish included	.33	.52	.85
1800	Taped and finished	.37	1.23	1.60
1900	Fire resistant, 1/2" thick, no finish included	.33	.52	.85
2000	Taped and finished	.37	1.23	1.60
2100	5/8" thick, no finish included	.33	.52	.85
2200	Taped and finished	.37	1.23	1.60
2300	Water resistant, 1/2" thick, no finish included	.28	.52	.80
2400	Taped and finished	.31	1.23	1.54
2500	5/8" thick, no finish included	.30	.52	.82
2600	Taped and finished	.34	1.23	1.57
2700	Finish, instead of taping			
2800	For thin coat plaster, add	.08	.58	.66
2900	Finish, textured spray, add	.06	.52	.58
3000				
3100				
3200				
3300	Tile, stapled or glued, mineral fiber plastic coated, 5/8" thick	1.62	.47	2.09
3400	3/4" thick	1.55	.47	2.02
3500	Wood fiber, 1/2" thick	.84	1.18	2.02
3600	3/4" thick	1.18	1.18	2.36
3700	Suspended, fiberglass, film faced, 5/8" thick	.57	.75	1.32
3800	3" thick	1.44	1.05	2.49
3900	Mineral fiber, 5/8" thick, standard	.70	.70	1.40
4000	Aluminum	1.77	.79	2.56
4100	Wood fiber, reveal edge, 1" thick			
4200	3" thick			
4300	Framing, metal furring, 3/4" channels, 12" O.C.	.25	1.98	2.23
4400	16" O.C.	.23	1.44	1.67
4500	24" O.C.	.15	.99	1.14
4600				
4700	1-1/2" channels, 12" O.C.	.35	2.19	2.54
4800	16" O.C.	.32	1.60	1.92
4900	24" O.C.	.21	1.07	1.28
5000				
5100				
5200				
5300	Ceiling suspension systems, for tile,			
5400	Concealed "Z" bar, 12" module	.52	.91	1.43
5500	Class A, "T" bar 2'-0" x 4'-0" grid	.58	.59	1.17
5600	2'-0" x 2'-0" grid	.72	.73	1.45
5700	Carrier channels for lighting fixtures, add	.20	1.02	1.22
5800				

C3030 Ceiling Finishes

C3030 250	Selective Price Sheet	COST PER S.F.		
		MAT.	INST.	TOTAL
5900				
6000				
6100	Wood, furring 1" x 3", on wood, 12" O.C.	.31	1.35	1.66
6200	16" O.C..	.23	1.01	1.24
6300	24" O.C.	.16	.68	.84
6400				
6500	On concrete, 12" O.C.	.31	2.24	2.55
6600	16" O.C.	.23	1.68	1.91
6700	24" O.C.	.16	1.12	1.28
6800				
6900	Joists, 2" x 4", 12" O.C.	.53	.97	1.50
7000	16" O.C.	.44	.81	1.25
7100	24" O.C.	.36	.66	1.02
7200	32" O.C.	.27	.50	.77
7300	2" x 6", 12" O.C.	.85	.98	1.83
7400	16" O.C.	.70	.81	1.51
7500	24" O.C.	.56	.65	1.21
7600	32" O.C.	.42	.48	.90

For information about Means Estimating Seminars, see yellow pages 12 and 13 in back of book

D SERVICES

D1010 Elevators and Lifts

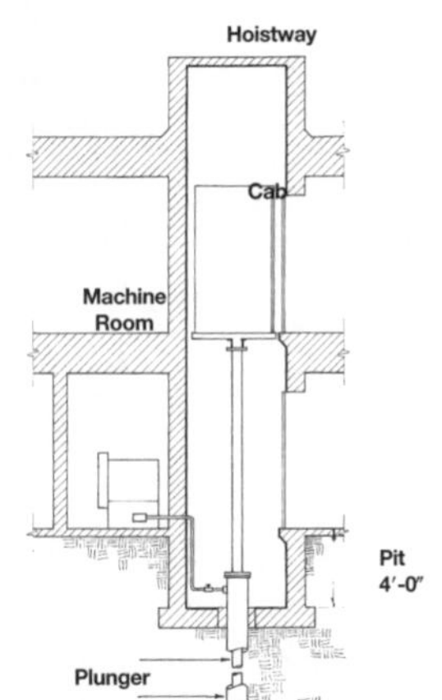

This page illustrates and describes oil hydraulic elevator systems. Prices for various oil hydraulic elevator systems are on Line Items D1010 130 0400 thru 2000. Both material quantities and labor costs have been adjusted for the system listed.

Factors: To adjust for job conditions other than normal working situations use Lines D1010 130 3100 thru 4000.

Example: You are to install the system with limited equipment usage. Go to Line D1010 130 3400 and apply this percentage to the appropriate TOTAL costs.

D1010 130	Oil Hydraulic Elevators	COST EACH		
		MAT.	INST.	TOTAL
0310	Oil-hydraulic elevator systems			
0320	Including piston and piston shaft			
0330				
0400	1500 Lb. passenger, 2 floors	33,700	12,300	46,000
0500	3 floors	42,000	19,000	61,000
0600	4 floors	60,000	26,800	86,800
0700	5 floors	68,000	33,400	101,400
0800	6 floors	75,500	39,800	115,300
0900				
1000	2500 Lb. passenger, 2 floors	35,900	12,300	48,200
1100	3 floors	44,200	19,000	63,200
1200	4 floors	58,500	26,800	85,300
1300	5 floors	66,500	33,400	99,900
1400	6 floors	77,500	39,800	117,300
1500				
1600	4000 Lb. passenger, 2 floors	40,800	12,300	53,100
1700	3 floors	50,500	19,000	69,500
1800	4 floors	67,000	26,800	93,800
1900	5 floors	75,000	33,400	108,400
2000	6 floors	82,500	39,800	122,300
2100				
2200				
2300				
2400				
2500				
2600				
2700				
2800				
2900				
3000				
3100	Dust protection, add, minimum	1%	2%	
3200	Maximum	4%	11%	
3300	Equipment usage curtailment, add, minimum	1%	1%	
3400	Maximum	3%	10%	
3500	Material handling & storage limitation, add, minimum	1%	1%	
3600	Maximum	6%	7%	
3700	Protection of existing work, add, minimum	2%	2%	
3800	Maximum	5%	7%	
3900	Shift work requirements, add, minimum		5%	
4000	Maximum		30%	

D2010 Plumbing Fixtures

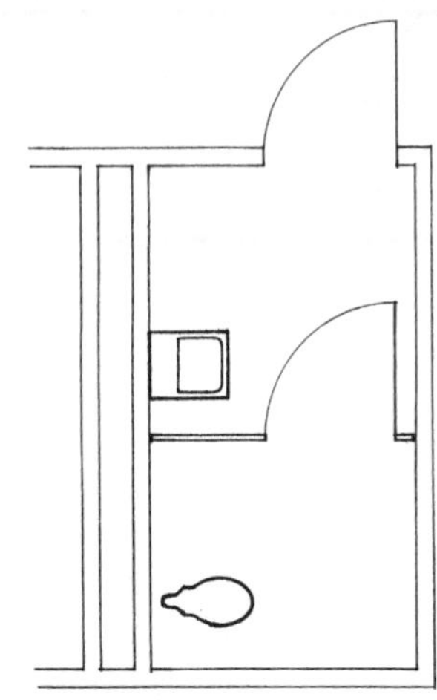

This page illustrates and describes a women's public restroom system including a water closet, lavatory, accessories, and service piping. Lines within System Components give the unit price and total price on a cost each basis for this system. Prices for alternate women's public restroom systems are on Line Items D2010 956 1400 thru 1700. Both material quantities and labor costs have been adjusted for the system listed.

Factors: To adjust for job conditions other than normal working situations use Lines D2010 956 2900 thru 4000.

Example: You are to install the system and protect surrounding area from dust. Go to Line D2010 956 3100 and apply these percentages to the MAT. and INST. costs.

System Components	QUANTITY	UNIT	COST EACH		
			MAT.	INST.	TOTAL
Public women's restroom incl. water closet, lavatory, accessories and Necessary service piping to install this system in one wall.					
Water closet, wall mounted, one piece	1.000	Ea.	385	165	550
Rough-in waste and vent for water closet	1.000	Set	505	375	880
Lavatory, 20" x 18" P.E. cast iron with accessories	1.000	Ea.	265	120	385
Rough-in waste and vent for lavatory	1.000	Set	340	575	915
Toilet partition, painted metal between walls, floor mounted	1.000	Ea.	515	157	672
For handicap unit, add	1.000	Ea.	335		335
Grab bar, 36" long	1.000	Ea.	46	47	93
Mirror, 18" x 24", with stainless steel shelf	1.000	Ea.	172	23.50	195.50
Napkin/tampon dispenser, recessed	1.000	Ea.	360	31.50	391.50
Soap Dispenser, chrome, surface mounted, liquid	1.000	Ea.	51.50	23.50	75
Toilet tissue dispenser, surface mounted, stainless steel	1.000	Ea.	12.85	15.70	28.55
Towel dispenser, surface mounted, stainless steel	1.000	Ea.	41.50	29.50	71
TOTAL			3,028.85	1,562.70	4,591.55

D2010 956	Plumbing - Public Restroom	COST EACH		
		MAT.	INST.	TOTAL
1200				
1300	For alternate size restrooms:			
1400	Two water closets, two lavatories	5,400	2,975	8,375
1500				
1600	For each additional water closet over 2, add	1,475	690	2,165
1700	For each additional lavatory over 2, add	830	740	1,570
1800				
1900				
2400	NOTE: PLUMBING APPROXIMATIONS			
2500	WATER CONTROL: water meter, backflow preventer,			
2600	Shock absorbers, vacuum breakers, mixer....10 to 15% of fixtures			
2700	PIPE AND FITTINGS: .... 30 to 60% of fixtures			
2800				
2900	Cut & patch to match existing construction, add, minimum	2%	3%	
3000	Maximum	5%	9%	
3100	Dust protection, add, minimum	1%	2%	
3200	Maximum	4%	11%	
3300	Equipment usage curtailment, add, minimum	1%	1%	
3400	Maximum	3%	10%	
3500	Material handling & storage limitation, add, minimum	1%	1%	

D2010 Plumbing Fixtures

D2010 956	Plumbing - Public Restroom	COST EACH		
		MAT.	INST.	TOTAL
3600	Maximum	6%	7%	
3700	Protection of existing work, add, minimum	2%	2%	
3800	Maximum	5%	7%	
3900	Shift work requirements, add, minimum		5%	
4000	Maximum		30%	

D2010 Plumbing Fixtures

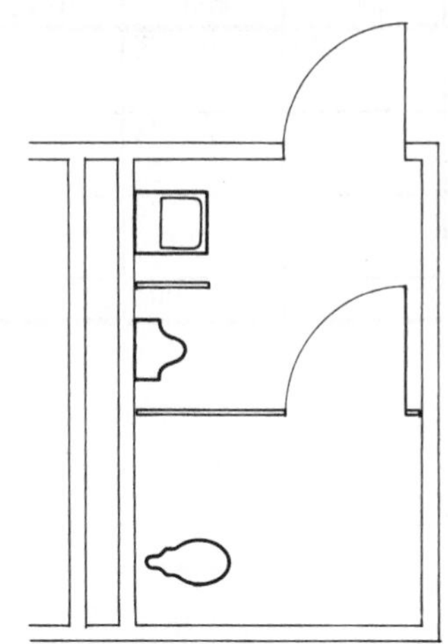

This page illustrates and describes a men's public restroom system including a water closet, urinal, lavatory, accessories and service piping. Lines within System Components give the unit price and total price on a cost each basis for this system. Prices for alternate men's public restroom systems are on Line Items D2010 957 1800 thru 2200. Both material quantities and labor costs have been adjusted for the system listed.

Factors: To adjust for job conditions other than normal working situations use Lines D2010 957 2900 thru 4000.

Example: You are to install the system and match existing construction. Go to Line D2010 957 3000 and apply these percentages to the appropriate MAT. and INST. costs.

System Components	QUANTITY	UNIT	COST EACH		
			MAT.	INST.	TOTAL
Public men's restroom incl. water closet, urinal, lavatory, accessories, And necessary service piping to install this system in one wall.					
Water closet, wall mounted, one piece	1.000	Ea.	385	165	550
Rough-in waste & vent for water closet	1.000	Set	505	375	880
Urinal, wall hung	1.000	Ea.	345	320	665
Rough-in waste & vent for urinal	1.000	Set	150	340	490
Lavatory, 20″ x 18″, P.E. cast iron with accessories	1.000	Ea.	265	120	385
Rough-in waste & vent for lavatory	1.000	Set	340	575	915
Partition, painted mtl., between walls, floor mntd.	1.000	Ea.	515	157	672
For handicap unit, add	1.000	Ea.	335		335
Grab bars	1.000	Ea.	46	47	93
Urinal screen, painted metal, wall mounted	1.000	Ea.	310	94	404
Mirror, 18″ x 24″ with stainless steel shelf	1.000	Ea.	172	23.50	195.50
Soap dispenser, surface mounted, liquid	1.000	Ea.	51.50	23.50	75
Toilet tissue dispenser, surface mounted, stainless steel	1.000	Ea.	12.85	15.70	28.55
Towel dispenser, surface mounted, stainless steel	1.000	Ea.	41.50	29.50	71
TOTAL			3,473.85	2,285.20	5,759.05

D2010 957	Plumbing - Public Restroom	COST EACH		
		MAT.	INST.	TOTAL
1600				
1700	For alternate size restrooms:			
1800	Two water closets, two urinals, two lavatories	6,650	4,450	11,100
1900				
2000	For each additional water closet over 2, add	1,475	690	2,165
2100	For each additional urinal over 2, add	805	755	1,560
2200	For each additional lavatory over 2, add	830	740	1,570
2300				
2400	NOTE: PLUMBING APPROXIMATIONS			
2500	WATER CONTROL: water meter, backflow preventer,			
2600	Shock absorbers, vacuum breakers, mixer....10 to 15% of fixtures			
2700	PIPE AND FITTINGS: .... 30 to 60% of fixtures			
2800				
2900	Cut & patch to match existing construction, add, minimum	2%	3%	
3000	Maximum	5%	9%	
3100	Dust protection, add, minimum	1%	2%	
3200	Maximum	4%	11%	
3300	Equipment usage curtailment, add, minimum	1%	1%	

D2010 Plumbing Fixtures

D2010 957	Plumbing - Public Restroom	COST EACH		
		MAT.	INST.	TOTAL
3400	Maximum	3%	10%	
3500	Material handling, & storage limitation, add, minimum	1%	1%	
3600	Maximum	6%	7%	
3700	Protection of existing work, add, minimum	2%	2%	
3800	Maximum	5%	7%	
3900	Shift work requirements, add, minimum		5%	
4000	Maximum		30%	

D2010 Plumbing Fixtures

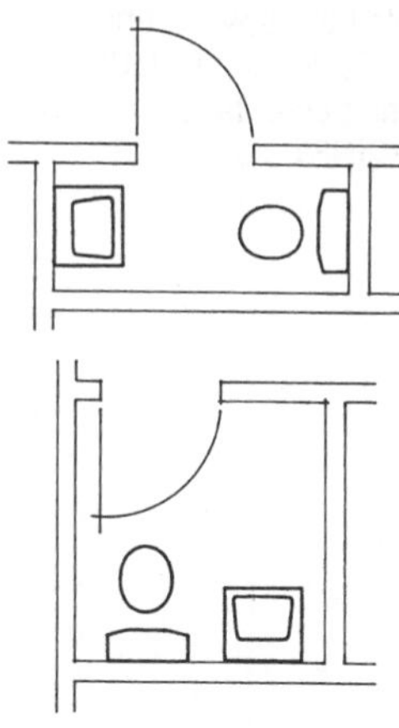

This page illustrates and describes a two fixture lavatory system including a water closet, lavatory, accessories and all service piping. Lines within System Components give the unit price and total price on a cost each basis for this system. Prices for an alternate two fixture lavatory system are on Line Item D2010 958 1900. Both material quantities and labor costs have been adjusted for the system listed.

Factors: To adjust for job conditions other than normal working situations use Lines D2010 958 2900 thru 4000.

Example: You are to install the system while controlling dust in the work area. Go to Line D2010 958 3200 and apply these percentages to the appropriate MAT. and INST. costs.

System Components	QUANTITY	UNIT	COST EACH		
			MAT.	INST.	TOTAL
Two fixture bathroom incl. water closet, lavatory, accessories and Necessary service piping to install this system in 2 walls.					
Water closet, floor mounted, 2 piece, close coupled	1.000	Ea.	186	180	366
Rough in waste & vent for water closet	1.000	Set	210	315	525
Lavatory, 20" x 18", P.E. cast iron with accessories	1.000	Ea.	265	120	385
Rough in waste & vent for lavatory	1.000	Set	340	575	915
Additional service piping					
1/2" copper pipe with sweat solder joints	10.000	L.F.	19.50	65.50	85
2" black schedule 40 steel pipe with threaded couplings	12.000	L.F.	93.60	179.40	273
4" cast iron soil pipe with lead and oakum joints	7.000	L.F.	68.60	121.80	190.40
Accessories					
Toilet tissue dispenser, chrome, single roll	1.000	Ea.	12.85	15.70	28.55
18" long stainless steel towel bar	1.000	Ea.	38.50	20.50	59
Medicine cabinet with mirror, 20" x 16", unlighted	1.000	Ea.	80.50	33.50	114
TOTAL			1,314.55	1,626.40	2,940.95

D2010 958	Plumbing - Two Fixture Bathroom	COST EACH		
		MAT.	INST.	TOTAL
1800				
1900	Above system installed in 1 wall with all necessary service piping	1,225	1,400	2,625
2000				
2400	NOTE: PLUMBING APPROXIMATIONS			
2500	WATER CONTROL: water meter, backflow preventer,			
2600	Shock absorbers, vacuum breakers, mixer....10 to 15% of fixtures			
2700	PIPE AND FITTINGS: .... 30 to 60% of fixtures			
2800				
2900	Cut & patch to match existing construction, add, minimum	2%	3%	
3000	Maximum	5%	9%	
3100	Dust protection, add, minimum	1%	2%	
3200	Maximum	4%	11%	
3300	Equipment usage curtailment, add, minimum	1%	1%	
3400	Maximum	3%	10%	
3500	Material handling & storage limitation, add, minimum	1%	1%	
3600	Maximum	6%	7%	
3700	Protection of existing work, add, minimum	2%	2%	
3800	Maximum	5%	7%	
3900	Shift work requirements, add, minimum		5%	
4000	Maximum		30%	

D2010 Plumbing Fixtures

This page illustrates and describes a three fixture bathroom system including a water closet, tub, lavatory, accessories and service piping. Lines within System Components give the unit price and total price on a cost each basis for this system. Prices for an alternate three fixture bathroom system are on Line Item D2010 959 1700. Both material quantities and labor costs have been adjusted for the system listed.

Factors: To adjust for job conditions other than normal working situations use Lines D2010 959 2900 thru 4000.

Example: You are to install the system and protect all existing work. Go to Line D2010 959 3800 and apply these percentages to the appropriate MAT. and INST. costs.

System Components	QUANTITY	UNIT	COST EACH MAT.	COST EACH INST.	COST EACH TOTAL
Three fixture bathroom incl. water closet, bathtub, lavatory, accessories, And necessary service piping to install this system in 1 wall.					
Water closet, floor mounted, 2 piece, close coupled	1.000	Ea.	186	180	366
Rough-in waste & vent for water closet	1.000	Set	195.30	292.95	488.25
Bathtub, P.E. cast iron 5' long with accessories	1.000	Ea.	805	217	1,022
Rough-in waste & vent for bathtub	1.000	Set	172.90	437	609.90
Lavatory, 20" x 18" P.E. cast iron with accessories	1.000	Ea.	265	120	385
Rough-in waste & vent for lavatory	1.000	Set	340	575	915
Accessories					
Toilet tissue dispenser, chrome, single roll	1.000	Ea.	12.85	15.70	28.55
18" long stainless steel towel bar	2.000	Ea.	77	41	118
Medicine cabinet with mirror, 20" x 16", unlighted	1.000	Ea.	80.50	33.50	114
TOTAL			2,134.55	1,912.15	4,046.70

D2010 959	Plumbing - Three Fixture Bathroom	COST EACH MAT.	COST EACH INST.	COST EACH TOTAL
1600				
1700	Above system installed in one wall with all necessary service piping	2,125	1,875	4,000
2400	NOTE: PLUMBING APPROXIMATIONS			
2500	WATER CONTROL: water meter, backflow preventer,			
2600	Shock absorbers, vacuum breakers, mixer....10 to 15% of fixtures			
2700	PIPE AND FITTINGS: .... 30 to 60% of fixtures			
2800				
2900	Cut & patch to match existing construction, add, minimum	2%	3%	
3000	Maximum	5%	9%	
3100	Dust protection, add, minimum	1%	2%	
3200	Maximum	4%	11%	
3300	Equipment usage curtailment, add, minimum	1%	1%	
3400	Maximum	3%	10%	
3500	Material handling & storage limitation, add, minimum	1%	1%	
3600	Maximum	6%	7%	
3700	Protection of existing work, add, minimum	2%	2%	
3800	Maximum	5%	7%	
3900	Shift work requirements, add, minimum		5%	
4000	Maximum		30%	

D2010 Plumbing Fixtures

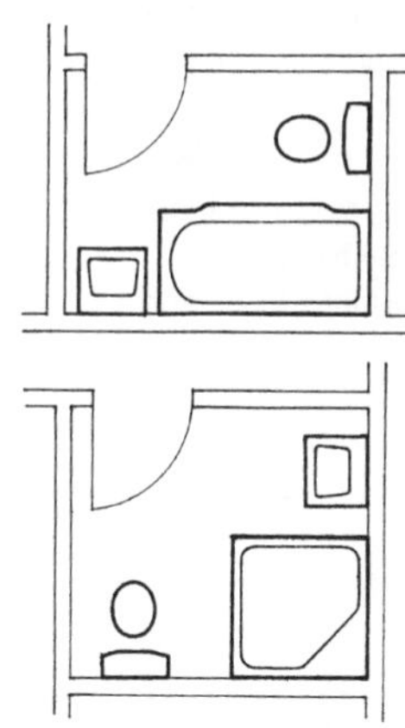

This page illustrates and describes a three fixture bathroom system including a water closet, tub, lavatory, accessories, and service piping. Lines within System Components give the unit price and total price on a cost each basis for this system. Prices for an alternate three fixture bathroom system are on Line Item D2010 960 2000. Both material quantities and labor costs have been adjusted for the system listed.

Factors: To adjust for job conditions other than normal working situations use Lines D2010 960 2900 thru 4000.

Example: You are to install the system and protect the surrounding area from dust. Go to Line D2010 960 3100 and apply these percentages to the appropriate MAT. and INST. costs.

System Components	QUANTITY	UNIT	COST EACH		
			MAT.	INST.	TOTAL
Three fixture bathroom incl. water closet, bathtub, lavatory, accessories,					
And necessary service piping to install this system in 2 walls.					
Water closet, floor mounted, 2 piece, close coupled	1.000	Ea.	186	180	366
Rough-in waste & vent for water closet	1.000	Set	210	315	525
Bathtub, P.E. cast iron, 5′ long with accessories	1.000	Ea.	805	217	1,022
Rough-in waste & vent for bathtub	1.000	Set	182	460	642
Lavatory, 20″ x 18″ P.E. cast iron with accessories	1.000	Ea.	265	120	385
Rough-in waste & vent for lavatory	1.000	Set	340	575	915
Additional service piping					
1-1/4″ copper DWV type tubing, sweat solder joints	6.000	L.F.	29.40	53.10	82.50
2″ black schedule 40 steel pipe with threaded couplings	12.000	L.F.	93.60	179.40	273
Accessories					
Toilet tissue dispenser, chrome, single roll	1.000	Ea.	12.85	15.70	28.55
18″ long stainless steel towel bar	2.000	Ea.	77	41	118
Medicine cabinet with mirror, 20″ x 16″, unlighted	1.000	Ea.	80.50	33.50	114
TOTAL			2,281.35	2,189.70	4,471.05

D2010 960	Plumbing - Three Fixture Bathroom	COST EACH		
		MAT.	INST.	TOTAL
2000	Above system with corner contour tub, P.E. cast iron	3,250	2,200	5,450
2300				
2400	NOTE: PLUMBING APPROXIMATIONS			
2500	WATER CONTROL: water meter, backflow preventer,			
2600	Shock absorbers, vacuum breakers, mixer....10 to 15% of fixtures			
2700	PIPE AND FITTINGS: .... 30 to 60% of fixtures			
2800				
2900	Cut & patch to match existing construction, add, minimum	2%	3%	
3000	Maximum	5%	9%	
3100	Dust protection, add, minimum	1%	2%	
3200	Maximum	4%	11%	
3300	Equipment usage curtailment, add, minimum	1%	1%	
3400	Maximum	3%	10%	
3500	Material handling & storage limitation, add, minimum	1%	1%	
3600	Maximum	6%	7%	
3700	Protection of existing work, add, minimum	2%	2%	
3800	Maximum	5%	7%	
3900	Shift work requirements, add, minimum		5%	
4000	Maximum		30%	

D2010 Plumbing Fixtures

This page illustrates and describes a three fixture bathroom system including a water closet, shower, lavatory, accessories, and service piping. Lines within System Components give the unit price and total price on a cost each basis for this system. Prices for an alternate three fixture bathroom system are on Line Item D2010 961 2200. Both material quantities and labor costs have been adjusted for the system listed.

Factors: To adjust for job conditions other than normal working situations use Lines D2010 961 2900 thru 4000.

Example: You are to install the system and protect existing construction. Go to Line D2010 961 3700 and apply these percentages to the appropriate MAT. and INST. costs.

System Components	QUANTITY	UNIT	COST EACH MAT.	COST EACH INST.	COST EACH TOTAL
Three fixture bathroom incl. water closet, shower, lavatory, accessories,					
And necessary service piping to install this system in 2 walls.					
Water closet, floor mounted, 2 piece, close coupled	1.000	Ea.	186	180	366
Rough-in waste & vent for water closet	1.000	Set	210	315	525
32″ shower, enameled stall, molded stone receptor	1.000	Ea.	380	191	571
Rough-in waste & vent for shower	1.000	Set	268	465	733
Lavatory, 20″ x 18″ P.E. cast iron with accessories	1.000	Ea.	265	120	385
Rough-in waste & vent for lavatory	1.000	Set	340	575	915
Additional service piping					
1-1/4″ Copper DWV type tubing, sweat solder joints	4.000	L.F.	19.60	35.40	55
2″ black schedule 40 steel pipe with threaded couplings	6.000	L.F.	46.80	89.70	136.50
4″ cast iron soil with lead and oakum joints	7.000	L.F.	68.60	121.80	190.40
Accessories					
Toilet tissue dispenser, chrome, single roll	1.000	Ea.	12.85	15.70	28.55
18″ long stainless steel towel bar	2.000	Ea.	77	41	118
Medicine cabinet with mirror, 20″ x 16″, unlighted	1.000	Ea.	80.50	33.50	114
TOTAL			1,954.35	2,183.10	4,137.45

D2010 961	Plumbing - Three Fixture Bathroom	COST EACH MAT.	COST EACH INST.	COST EACH TOTAL
2100	Above system installed with 36″ corner angle shower, enameled steel,			
2200	Molded stone receptor	2,450	2,200	4,650
2300				
2400	NOTE: PLUMBING APPROXIMATIONS			
2500	WATER CONTROL: water meter, backflow preventer,			
2600	Shock absorbers, vacuum breakers, mixer....10 to 15% of fixtures			
2700	PIPE AND FITTINGS: .... 30 to 60% of fixtures			
2800				
2900	Cut & patch to match existing construction, add, minimum	2%	3%	
3000	Maximum	5%	9%	
3100	Dust protection, add, minimum	1%	2%	
3200	Maximum	4%	11%	
3300	Equipment usage curtailment, add, minimum	1%	1%	
3400	Maximum	3%	10%	
3500	Material handling & storage limitation, add, minimum	1%	1%	
3600	Maximum	6%	7%	
3700	Protection of existing work, add, minimum	2%	2%	
3800	Maximum	5%	7%	
3900	Shift work requirements, add, minimum		5%	
4000	Maximum		30%	

D2010 Plumbing Fixtures

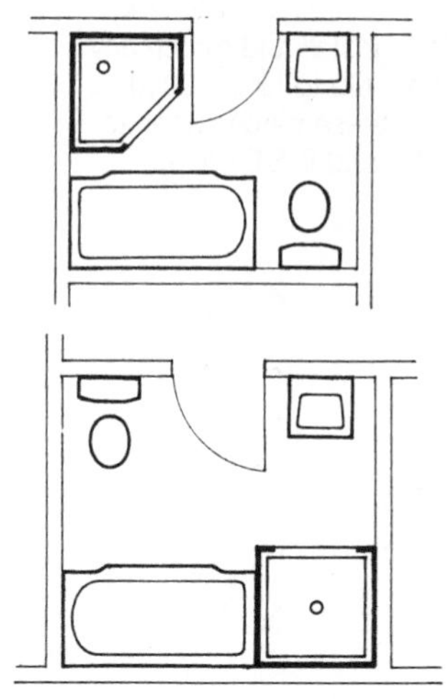

This page illustrates and describes a four fixture bathroom system including a water closet, shower, bathtub, lavatory, accessories, and service piping. Lines within System Components give the unit price and total price on a cost each basis for this system. Prices for an alternate four fixture bathroom system are on Line Item D2010 962 1900. Both material quantities and labor costs have been adjusted for the system listed.

Factors: To adjust for job conditions other than normal working situations use Lines D2010 962 2900 thru 4000.

Example: You are to install the system during weekends and evenings. Go to Line D2010 962 4000 and apply these percentages to the appropriate MAT. and INST. costs.

System Components	QUANTITY	UNIT	COST EACH MAT.	COST EACH INST.	COST EACH TOTAL
Four fixture bathroom incl. water closet, shower, bathtub, lavatory					
Accessories and necessary service piping to install this system in 2 walls.					
Water closet, floor mounted, 2 piece, close coupled	1.000	Ea.	186	180	366
Rough-in waste & vent for water closet	1.000	Set	210	315	525
32″ shower, enameled steel stall, molded stone receptor	1.000	Ea.	380	191	571
Rough-in waste & vent for shower	1.000	Set	268	465	733
Bathtub, P.E. cast iron, 5′ long with accessories	1.000	Ea.	805	217	1,022
Rough-in waste & vent for bathtub	1.000	Set	182	460	642
Lavatory, 20″ x 18″ P.E. cast iron with accessories	1.000	Ea.	265	120	385
Rough-in waste & vent for lavatory	1.000	Set	340	575	915
Accessories					
Toilet tissue dispenser, chrome, single roll	1.000	Ea.	12.85	15.70	28.55
18″ long stainless steel towel bar	2.000	Ea.	77	41	118
Medicine cabinet with mirror, 20″ x 16″ unlighted	1.000	Ea.	80.50	33.50	114
TOTAL			2,806.35	2,613.20	5,419.55

D2010 962	Plumbing - Four Fixture Bathroom	COST EACH MAT.	COST EACH INST.	COST EACH TOTAL
1900	Above system with 36″ corner angle shower, plumbing in 3 walls	3,425	2,875	6,300
2000				
2300				
2400	NOTE: PLUMBING APPROXIMATIONS			
2500	WATER CONTROL: water meter, backflow preventer,			
2600	Shock absorbers, vacuum breakers, mixer....10 to 15% of fixtures			
2700	PIPE AND FITTINGS: .... 30 to 60% of fixtures			
2800				
2900	Cut & patch to match existing construction, add, minimum	2%	3%	
3000	Maximum	5%	9%	
3100	Dust protection, add, minimum	1%	2%	
3200	Maximum	4%	11%	
3300	Equipment usage curtailment, add, minimum	1%	1%	
3400	Maximum	3%	10%	
3500	Material handling & storage limitation, add, minimum	1%	1%	
3600	Maximum	6%	7%	
3700	Protection of existing work, add, minimum	2%	2%	
3800	Maximum	5%	7%	
3900	Shift work requirements, add, minimum		5%	
4000	Maximum		30%	

D2010 Plumbing Fixtures

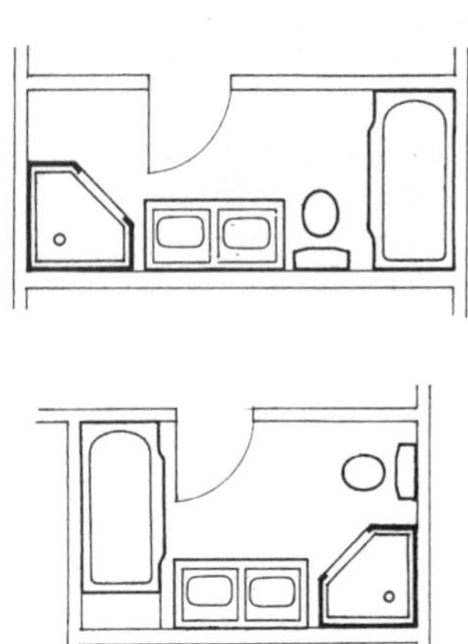

This page illustrates and describes a five fixture bathroom system including a water closet, shower, bathtub, lavatories, accessories, and service piping. Lines within System Components give the unit price and total price on a cost each basis for this system. Prices for an alternate five fixture bathroom system are on Line Item D2010 963 1900. Both material quantities and labor costs have been adjusted for the system listed.

Factors: To adjust for job conditions other than normal working situations use Lines D2010 963 2900 thru 4000.

Example: You are to install and match any existing construction. Go to Line D2010 963 2900 and apply these percentages to the appropriate MAT. and INST. costs.

System Components	QUANTITY	UNIT	COST EACH MAT.	INST.	TOTAL
Five fixture bathroom incl. water closet, shower, bathtub, 2 lavatories, Accessories and necessary service piping to install this system in 1 wall.					
Water closet, floor mounted, 2 piece, close coupled	1.000	Ea.	186	180	366
Rough-in waste & vent for water closet	1.000	Set	210	315	525
36″ corner angle shower, enameled steel stall, molded stone receptor	1.000	Ea.	880	199	1,079
Rough-in waste & vent for shower	1.000	Set	268	465	733
Bathtub, P.E. cast iron, 5′ long with accessories	1.000	Ea.	805	217	1,022
Rough-in waste & vent for bathtub	1.000	Set	182	460	642
Countertop, plastic laminate, with backsplash	2.000	Ea.	124	62.80	186.80
Vanity base, 2 doors, 24″ wide	2.000	Ea.	195	85.50	280.50
Lavatories, 20″ x 18″ cabinet mntd, PECI	2.000	Ea.	468	298	766
Rough-in waste & vent for lavatories	1.600	Set	198.40	664	862.40
Accessories					
Toilet tissue dispenser, chrome, single roll	1.000	Ea.	12.85	15.70	28.55
18″ long stainless steel towel bars	2.000	Ea.	77	41	118
Medicine cabinet with mirror, 20″ x 16″, unlighted	2.000	Ea.	161	67	228
TOTAL			3,767.25	3,070	6,837.25

D2010 963	Plumbing - Five Fixture Bathroom	COST EACH MAT.	INST.	TOTAL
1900	Above system installed in 2 walls with all necessary service piping	3,825	3,175	7,000
2000				
2100				
2200				
2300				
2400	NOTE: PLUMBING APPROXIMATIONS			
2500	WATER CONTROL: water meter, backflow preventer,			
2600	Shock absorbers, vacuum breakers, mixer....10 to 15% of fixtures			
2700	PIPE AND FITTINGS: .... 30 to 60% of fixtures			
2800				
2900	Cut & patch to match existing construction, add, minimum	2%	3%	
3000	Maximum	5%	9%	
3100	Dust protection, add, minimum	1%	2%	
3200	Maximum	4%	11%	
3300	Equipment usage curtailment, add, minimum	1%	1%	
3400	Maximum	3%	10%	
3500	Material handling & storage limitation, add, minimum	1%	1%	
3600	Maximum	6%	7%	

D2010 Plumbing Fixtures

D2010 963	Plumbing - Five Fixture Bathroom	COST EACH		
		MAT.	INST.	TOTAL
3700	Protection of existing work, add, minimum	2%	2%	
3800	Maximum	5%	7%	
3900	Shift work requirements, add, minimum		5%	
4000	Maximum		30%	

D3010 Energy Supply

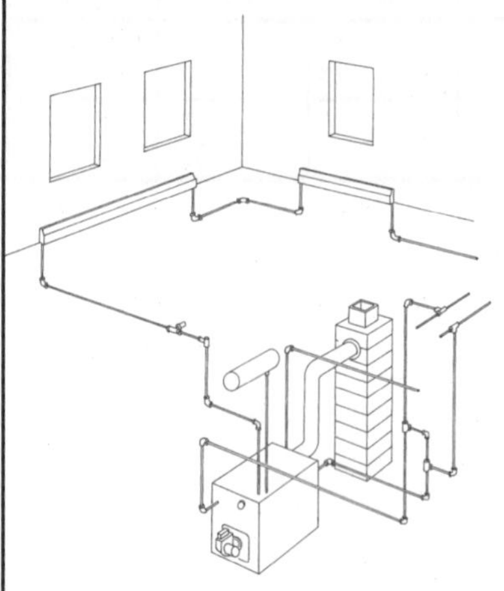

This page illustrates and describes an oil fired hot water baseboard system including an oil fired boiler, fin tube radiation and all fittings and piping. Lines within System Components give the unit price and total price per square foot for this system. Prices for alternate oil fired hot water baseboard systems are on Line Items D3010 540 1700 thru 2100. Both material quantities and labor costs have been adjusted for the system listed.

Factors: To adjust for job conditions other than normal working situations use Lines D3010 540 2900 thru 4000.

Example: You are to install the system while protecting all existing work. Go to Line D3010 540 3700 and apply these percentages to the appropriate MAT. and INST. costs.

System Components	QUANTITY	UNIT	COST PER S.F. MAT.	COST PER S.F. INST.	COST PER S.F. TOTAL
Oil fired hot water baseboard system including boiler, fin tube radiation And all necessary fittings and piping.					
Area to 800 S.F.					
Boiler, cast iron, w/oil piping, 97 MBH	1.000	Ea.	1,906.25	1,168.75	3,075
Copper piping	130.000	L.F.	728	1,189.50	1,917.50
Fin tube radiation	68.000	L.F.	499.80	1,128.80	1,628.60
Circulator	1.000	Ea.	405	159	564
Oil tank	1.000	Ea.	355	193	548
Expansion tank, ASME	1.000	Ea.	415	69	484
TOTAL			4,309.05	3,908.05	8,217.10
COST PER S.F.		S.F.	5.39	4.89	10.28

D3010 540	Heating - Oil Fired Hot Water	COST PER S.F. MAT.	COST PER S.F. INST.	COST PER S.F. TOTAL
1600	For alternate hot water systems:			
1700	Cast iron boiler, area to 1000 S.F.	4.58	4.21	8.79
1800	To 1200 S.F.	3.96	3.81	7.77
1900	To 1600 S.F.	4.73	3.81	8.54
2000	To 2000 S.F.	3.96	3.38	7.34
2100	To 3000 S.F.	2.93	2.85	5.78
2200				
2300				
2700				
2800				
3000	Maximum	5%	9%	
3100	Dust protection, add, minimum	1%	2%	
3200	Maximum	4%	11%	
3300	Equipment usage curtailment, add, minimum	1%	1%	
3400	Maximum	3%	10%	
3500	Material handling & storage limitation, add, minimum	1%	1%	
3600	Maximum	6%	7%	
3700	Protection of existing work, add, minimum	2%	2%	
3800	Maximum	5%	7%	
3900	Shift work requirements, add, minimum		5%	
4000	Maximum		30%	

SERVICES D

Important: See the Reference Section for critical supporting data - Reference Numbers and City Cost Indexes

D3010 Energy Supply

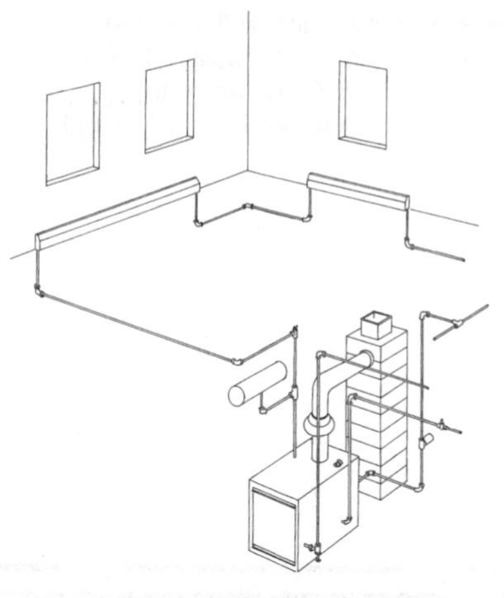

This page illustrates and describes a gas fired hot water baseboard system including a gas fired boiler, fin tube radiation, fittings and piping. Lines within System Components give the unit price and total price per square foot for this system. Prices for alternate gas fired hot water baseboard systems are on Line Items D3010 550 1500 thru 1900. Both material quantities and labor costs have been adjusted for the system listed.

Factors: To adjust for job conditions other than normal working situations use Lines D3010 550 2900 thru 4000.

Example: You are to install the system with minimal equipment usage. Go to Line D3010 550 3400 and apply these percentages to the appropriate MAT. and INST. costs.

System Components

System Components	QUANTITY	UNIT	COST PER S.F. MAT.	INST.	TOTAL
Gas fired hot water baseboard system including boiler, fin tube radiation,					
All necessary fittings and piping.					
Area to 800 S.F.					
Cast iron boiler, insulating jacket, gas piping, 80 MBH	1.000	Ea.	2,062.50	1,750	3,812.50
Copper piping	130.000	L.F.	728	1,189.50	1,917.50
Fin tube radiation	68.000	L.F.	499.80	1,128.80	1,628.60
Circulator, flange connection	1.000	Ea.	405	159	564
Expansion tank, ASME	1.000	Ea.	415	69	484
TOTAL			4,110.30	4,296.30	8,406.60
COST PER S.F.		S.F.	5.14	5.37	10.51

D3010 550 Heating - Gas Fired Hot Water

		COST PER S.F. MAT.	INST.	TOTAL
1400	For alternate hot water systems:			
1500	Cast iron boiler, area to 1000 S.F.	4.32	4.72	9.04
1600	To 1200 S.F.	3.75	4.24	7.99
1700	To 1600 S.F.	4.57	4.13	8.70
1800	To 2000 S.F.	5.15	3.70	8.85
1900	To 3000 S.F.	3.11	3.24	6.35
2000				
2100				
2900	Cut & patch to match existing construction, add, minimum	2%	3%	
3000	Maximum	5%	9%	
3100	Dust protection, add, minimum	1%	2%	
3200	Maximum	4%	11%	
3300	Equipment usage curtailment, add, minimum	1%	1%	
3400	Maximum	3%	10%	
3500	Material handling & storage limitation, add, minimum	1%	1%	
3600	Maximum	6%	7%	
3700	Protection of existing work, add, minimum	2%	2%	
3800	Maximum	5%	7%	
3900	Shift work requirements, add, minimum		5%	
4000	Maximum		30%	

D3010 Energy Supply

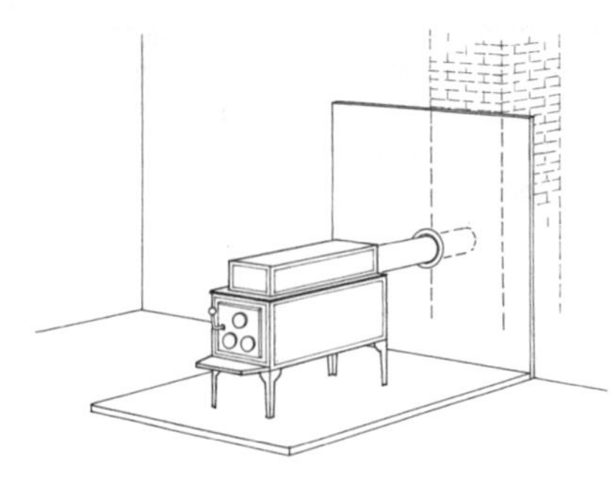

This page illustrates and describes a wood burning stove system including a free standing stove, preformed hearth, masonry chimney, and necessary piping and fittings. Lines within System Components give the unit price and total price on a cost each basis for this system. Prices for alternate wood burning stove systems are on Line Items D3010 991 1600 thru 2000. Both material quantities and labor costs have been adjusted for the system listed.

Factors: To adjust for job conditions other than normal working situations use Lines D3010 991 2900 thru 4000.

Example: You are to install the system and protect existing construction. Go to Line D3010 991 3500 and apply these percentages to the appropriate MAT., and INST. costs.

System Components	QUANTITY	UNIT	COST EACH		
			MAT.	INST.	TOTAL
Cast iron, free standing,wood burning stove with preformed hearth, masonry Chimney, and all necessary piping and fittings to install in chimney.					
Cast iron wood burning stove, stove pipe	1.000	Ea.	1,000	725	1,725
Wall panel or hearth, non-combustible	1.000	Ea.	444.60	132.05	576.65
16" x 16" brick chimney, 8" x 8" flue	20.000	V.L.F.	480	1,150	1,630
Foundation	.500	C.Y.	71	97.87	168.87
TOTAL			1,995.60	2,104.92	4,100.52

D3010 991	Wood Burning Stoves	COST EACH		
		MAT.	INST.	TOTAL
1400	For alternate wood burning systems:			
1500				
1600	Installed in existing fireplace	1,225	795	2,020
1700				
1800	Installed with insulated metal chimney			
1900	System including ceiling package,			
2000	Metal chimney to 10 L.F.	2,000	1,125	3,125
2100				
2200				
2300				
2400				
2500				
2600				
2700				
2900	Dust protection, add, minimum	1%	2%	
3000	Maximum	4%	11%	
3100	Equipment usage curtailment, add, minimum	1%	1%	
3200	Maximum	3%	10%	
3300	Material handling & storage limitation, add, minimum	1%	1%	
3400	Maximum	6%	7%	
3500	Protection of existing work, add, minimum	2%	2%	
3600	Maximum	5%	7%	
3700	Shift work requirements, add, minimum		5%	
3800	Maximum		30%	
3900	Temporary shoring and bracing, add, minimum	2%	5%	
4000	Maximum	5%	12%	

D3010 Energy Supply

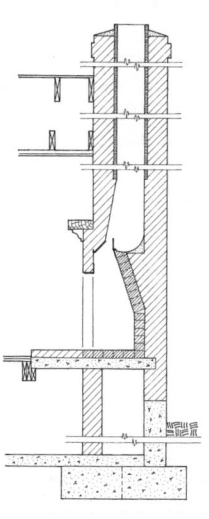

This page illustrates and describes masonry fireplace systems including a brick fireplace, footing, foundation, hearth, firebox, chimney and flue.
Lines within System Components give the unit price and total price on a cost each basis for this system. Prices for alternate masonry fireplace systems are on Line Items D3010 992 1300 thru 1500. Both material quantities and labor costs have been adjusted for the system listed.

Factors: To adjust for job conditions other than normal working situations use Lines D3010 992 2700 thru 4000.

Example: You are to install the system with some temporary shoring and bracing. Go to Line D3010 992 3900 and apply these percentages to the appropriate MAT. and INST. costs.

System Components	QUANTITY	UNIT	COST EACH MAT.	COST EACH INST.	COST EACH TOTAL
Brick masonry fireplace, including footing, foundation, hearth, firebox, Chimney and flue, chimney 12′ above firebox.					
Footing, 4′ x 7′ x 12″ thick, 3000 psi concrete	1.040	C.Y.	142.48	170.46	312.94
Foundation, 12″ concrete block	180.000	S.F.	630	1,503	2,133
Fireplace, brick faced, 6′-0″ wide x 5′-0″ high	1.000	Ea.	455	2,075	2,530
Hearth	1.000	Ea.	168	415	583
Chimney, 20″ x 20″, one 12″ x 12″ flue, 12 V.L.F.	12.000	V.L.F.	414	732	1,146
Mantle, wood	7.000	L.F.	39.55	91.70	131.25
TOTAL			1,849.03	4,987.16	6,836.19

D3010 992	Masonry Fireplace	COST EACH MAT.	COST EACH INST.	COST EACH TOTAL
1200	Above system with the following:			
1300	Chimney, 20″ x 20″, one 12″ x 12″ flue, 18 V.L.F.	2,050	5,350	7,400
1400	24 V.L.F.	2,275	5,725	8,000
1500	Fieldstone face instead of brick	2,200	4,975	7,175
1600				
1700				
1800				
1900				
2000				
2100				
2200				
2300				
2700	Cut & patch to match existing construction, add, minimum	2%	3%	
2800	Maximum	5%	9%	
2900	Dust protection, add, minimum	1%	2%	
3000	Maximum	4%	11%	
3100	Equipment usage curtailment, add, minimum	1%	1%	
3200	Maximum	3%	10%	
3300	Material handling & storage limitation, add, minimum	1%	1%	
3400	Maximum	6%	7%	
3500	Protection of existing work, add, minimum	2%	2%	
3600	Maximum	5%	7%	
3700	Shift work requirements, add, minimum		5%	
3800	Maximum		30%	
3900	Temporary shoring and bracing, add, minimum	2%	5%	
4000	Maximum	5%	12%	

D3020 Heat Generating Systems

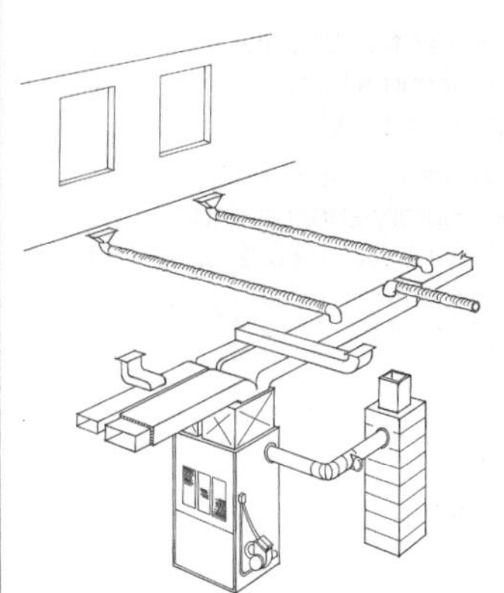

This page illustrates and describes an oil fired forced air system including an oil fired furnace, ductwork, registers and hookups. Lines within System Components give the unit price and total price per square foot for this system. Prices for alternate oil fired forced air systems are on Line Items D3020 122 1700 thru 2800. Both material quantities and labor costs have been adjusted for the system listed.

Factors: To adjust for job conditions other than normal working situations use Lines D3020 122 3100 thru 4200.

Example: You are to install the system during evenings and weekends. Go to Line D3020 122 4200 and apply this percentage to the appropriate INST. cost.

System Components	QUANTITY	UNIT	COST PER S.F. MAT.	INST.	TOTAL
Oil fired hot air heating system including furnace, ductwork, registers And all necessary hookups.					
Area to 800 S.F., heat only					
Furnace, oil, atomizing gun type burner, w/oil piping	1.000	Ea.	1,025	335	1,360
Oil tank, steel, 275 gallon	1.000	Ea.	355	193	548
Duct, galvanized, steel	312.000	Lb.	327.60	1,996.80	2,324.40
Insulation, blanket type, duct work	270.000	S.F.	167.40	707.40	874.80
Flexible duct, 6″ diameter, insulated	100.000	L.F.	279	371	650
Registers, baseboard, gravity, 12″ x 6″	8.000	Ea.	136.40	188	324.40
Return, damper, 36″ x 18″	1.000	Ea.	160	53.50	213.50
TOTAL			2,450.40	3,844.70	6,295.10
COST PER S.F.		S.F.	3.06	4.81	7.87

D3020 122	Heating-Cooling, Oil, Forced Air	COST PER S.F. MAT.	INST.	TOTAL
1600	For alternate heating systems:			
1700	Oil fired, area to 1000 S.F.	2.50	3.90	6.40
1800	To 1200 S.F.	2.21	3.43	5.64
1900	To 1600 S.F.	1.87	2.87	4.74
2000	To 2000 S.F.	2	3.70	5.70
2100	To 3000 S.F.	1.67	2.93	4.60
2200	For combined heating and cooling systems:			
2300	Oil fired, heating and cooling, area to 800 S.F.	4.94	5.70	10.64
2400	To 1000 S.F.	4.50	4.66	9.16
2500	To 1200 S.F.	3.87	4.07	7.94
2600	To 1600 S.F.	3.19	3.37	6.56
2700	To 2000 S.F.	3.12	4.15	7.27
2800	To 3000 S.F.	2.54	3.25	5.79
3100	Cut & patch to match existing construction, add, minimum	2%	3%	
3200	Maximum	5%	9%	
3300	Dust protection, add, minimum	1%	2%	
3400	Maximum	4%	11%	
3500	Equipment usage curtailment, add, minimum	1%	1%	
3600	Maximum	3%	10%	
3700	Material handling & storage limitation, add, minimum	1%	1%	
3800	Maximum	6%	7%	
3900	Protection of existing work, add, minimum	2%	2%	
4000	Maximum	5%	7%	
4100	Shift work requirements, add, minimum		5%	
4200	Maximum		30%	

D30 HVAC

D3020 Heat Generating Systems

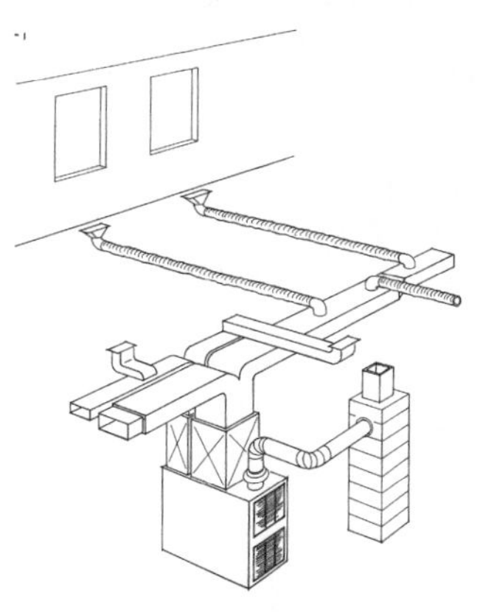

This page illustrates and describes a gas fired forced air system including a gas fired furnace, ductwork, registers and hookups. Lines within System Components give the unit price and total price per square foot for this system. Prices for alternate gas fired forced air systems are on Line Items D3020 124 1500 thru 2600. Both material quantities and labor costs have been adjusted for the system listed.

Factors: To adjust for job conditions other than normal working situations use Lines D3020 124 2900 thru 4000.

Example: You are to install the system with material handling and storage limitations. Go to Line D3020 124 3500 and apply these percentages to the appropriate MAT. and INST. costs.

System Components	QUANTITY	UNIT	COST PER S.F. MAT.	COST PER S.F. INST.	COST PER S.F. TOTAL
Gas fired hot air heating system including furnace, ductwork, registers					
And all necessary hookups.					
Area to 800 S.F., heat only					
Furnace, gas, AGA certified, direct drive, w/gas piping, 44 MBH	1.000	Ea.	625	301.25	926.25
Duct, galvanized steel	312.000	Lb.	327.60	1,996.80	2,324.40
Insulation, blanket type, ductwork	270.000	S.F.	167.40	707.40	874.80
Flexible duct, 6" diameter, insulated	100.000	L.F.	279	371	650
Registers, baseboard, gravity, 12" x 6"	8.000	Ea.	136.40	188	324.40
Return, damper, 36" x 18"	1.000	Ea.	160	53.50	213.50
TOTAL			1,695.40	3,617.95	5,313.35
COST PER S.F.		S.F.	2.12	4.52	6.64

D3020 124	Heating-Cooling, Gas, Forced Air	COST PER S.F. MAT.	COST PER S.F. INST.	COST PER S.F. TOTAL
1400	For alternate heating systems:			
1500	Gas fired, area to 1000 S.F.	1.91	3.68	5.59
1600	To 1200 S.F.	1.71	3.25	4.96
1700	To 1600 S.F.	1.69	3.06	4.75
1800	To 2000 S.F.	1.71	3.61	5.32
1900	To 3000 S.F.	1.37	2.87	4.24
2000	For combined heating and cooling systems:			
2100	Gas fired, heating and cooling, area to 800 S.F.	4	5.45	9.45
2200	To 1000 S.F.	3.90	4.45	8.35
2300	To 1200 S.F.	3.38	3.89	7.27
2400	To 1600 S.F.	3.01	3.56	6.57
2500	To 2000 S.F.	2.83	4.06	6.89
2600	To 3000 S.F.	2.23	3.19	5.42
2900	Cut & patch to match existing construction, add, minimum	2%	3%	
3000	Maximum	5%	9%	
3100	Dust protection, add, minimum	1%	2%	
3200	Maximum	4%	11%	
3300	Equipment usage curtailment, add, minimum	1%	1%	
3400	Maximum	3%	10%	
3500	Material handling & storage limitation, add, minimum	1%	1%	
3600	Maximum	6%	7%	
3700	Protection of existing work, add, minimum	2%	2%	
3800	Maximum	5%	7%	
3900	Shift work requirements, add, minimum		5%	
4000	Maximum		30%	

D SERVICES

D3020 Heat Generating Systems

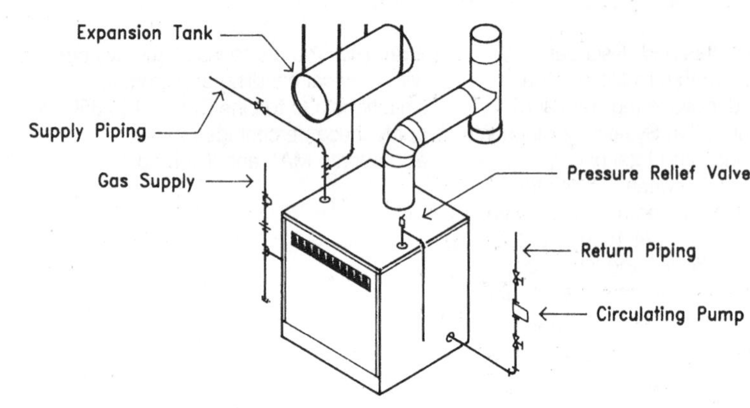

Cast Iron Boiler, Hot Water, Gas Fired

This page illustrates and describes boilers including expansion tank, circulating pump and all service piping. Lines within Systems Components give the material and installation price on a cost each basis for the components. Prices for alternate boiler systems are on Line Items D3020 130 1010 thru D3020 138 1040. Material quantities and labor costs have been adjusted for the system listed.

Factors: To adjust for job conditions other than normal working situations use Line D3020 138 2700 thru 4000.

System Components	QUANTITY	UNIT	COST EACH MAT.	COST EACH INST.	COST EACH TOTAL
Cast iron boiler, including piping, chimney and accessories					
Boilers, gas fired, std controls, CI, insulated, HW, gross output 100 MBH	1.000	Ea.	4,300	1,500	5,800
Pipe, black steel, Sch 40, threaded, W/coupling & hangers, 10′ OC, 3/4″ dia	20.000	L.F.	59.40	195.75	255.15
Pipe, black steel, Sch 40, threaded, W/coupling & hangers, 10′ OC, 1″ dia.	20.000	L.F.	82.13	215	297.13
Elbow, 90°, black, straight, 3/4″ dia.	9.000	Ea.	18	342	360
Elbow, 90°, black, straight, 1″ dia.	6.000	Ea.	20.88	246	266.88
Tee, black, straight, 3/4″ dia.	2.000	Ea.	6.38	118	124.38
Tee, black, straight, 1″ dia.	2.000	Ea.	10.90	133	143.90
Tee, black, reducing, 1″ dia.	2.000	Ea.	12.40	133	145.40
Pipe cap, black, 3/4″ dia.	1.000	Ea.	2.27	16.60	18.87
Union, black with brass seat, 3/4″ dia.	2.000	Ea.	16.80	82	98.80
Union, black with brass seat, 1″ dia.	2.000	Ea.	24.10	89	113.10
Pipe nipples, black, 3/4″ dia.	5.000	Ea.	6.60	21.75	28.35
Pipe nipples, black, 1″ dia.	3.000	Ea.	5.73	15	20.73
Valves, bronze, gate, N.R.S., threaded, class 150, 3/4″ size	2.000	Ea.	53	53	106
Valves, bronze, gate, N.R.S., threaded, class 150, 1″ size	2.000	Ea.	67	56	123
Gas cock, brass, 3/4″ size	1.000	Ea.	15.65	24	39.65
Thermometer, stem type, 9″ case, 8″ stem, 3/4″ NPT	2.000	Ea.	260	38.30	298.30
Tank, steel, liquid expansion, ASME, painted, 15 gallon capacity	1.000	Ea.	400	56.50	456.50
Pump, circulating, bronze, flange connection, 3/4″ to 1-1/2″ size, 1/8 HP	1.000	Ea.	695	159	854
Vent chimney, all fuel, pressure tight, double wall, SS, 6″ dia.	20.000	L.F.	880	322	1,202
Vent chimney, elbow, 90° fixed, 6″ dia.	2.000	Ea.	496	64	560
Vent chimney, Tee, 6″ dia.	2.000	Ea.	328	80	408
Vent chimney, ventilated roof thimble, 6″ dia.	1.000	Ea.	220	37	257
Vent chimney, adjustable roof flashing, 6″ dia.	1.000	Ea.	59.50	32	91.50
Vent chimney, stack cap, 6″ diameter	1.000	Ea.	174	21	195
Insulation, fiberglass pipe covering, 1″ wall, 1″ IPS	20.000	L.F.	22.80	83.40	106.20
TOTAL			8,236.54	4,133.30	12,369.84

D3020 130	Boiler, Cast Iron, Hot Water, Gas	COST EACH MAT.	COST EACH INST.	COST EACH TOTAL
1000	For alternate boilers:			
1010	Boiler, cast iron, gas, hot water, 100 MBH	8,225	4,125	12,350
1020	200 MBH	6,750	4,750	11,500
1030	320 MBH	7,000	5,150	12,150
1040	440 MBH	9,300	6,625	15,925
1050	544 MBH	16,400	10,500	26,900
1060	765 MBH	20,400	11,200	31,600
1070	1088 MBH	22,700	11,900	34,600

 Important: See the Reference Section for critical supporting data - Reference Numbers and City Cost Indexes

D3020 Heat Generating Systems

D3020 130	Boiler, Cast Iron, Hot Water, Gas	COST EACH		
		MAT.	INST.	TOTAL
1080	1530 MBH	26,300	15,900	42,200
1090	2312 MBH	29,800	18,700	48,500

D3020 134	Boiler, Cast Iron, Steam, Gas	COST EACH		
		MAT.	INST.	TOTAL
1010	Boiler, cast iron, gas, steam, 100 MBH	4,825	4,025	8,850
1020	200 MBH	11,700	7,525	19,225
1030	320 MBH	13,000	8,475	21,475
1040	544 MBH	19,500	13,300	32,800
1050	765 MBH	22,000	13,800	35,800
1060	1275 MBH	26,000	15,200	41,200
1070	2675 MBH	33,400	22,200	55,600

D3020 136	Boiler, Cast Iron, Hot Water, Gas/Oil	COST EACH		
		MAT.	INST.	TOTAL
1010	Boiler, cast iron, gas & oil, hot water, 584 MBH	20,300	11,600	31,900
1020	876 MBH	23,900	12,400	36,300
1030	1168 MBH	25,900	14,000	39,900
1040	1460 MBH	29,500	17,000	46,500
1050	2044 MBH	33,800	17,500	51,300
1060	2628 MBH	42,800	21,200	64,000

D3020 138	Boiler, Cast Iron, Steam, Gas/Oil	COST EACH		
		MAT.	INST.	TOTAL
1010	Boiler, cast iron, gas & oil, steam, 810 MBH	21,700	15,000	36,700
1020	1360 MBH	25,100	16,000	41,100
1030	2040 MBH	31,900	20,600	52,500
1040	2700 MBH	34,500	22,900	57,400
2700	Cut & patch to match existing construction, add, minimum	2%	3%	
2800	Maximum	5%	9%	
2900	Dust protection, add, minimum	1%	2%	
3000	Maximum	4%	11%	
3100	Equipment usage curtailment, add, minimum	1%	1%	
3200	Maximum	3%	10%	
3300	Material handling & storage limitation, add, minimum	1%	1%	
3400	Maximum	6%	7%	
3500	Protection of existing work, add, minimum	2%	2%	
3600	Maximum	5%	7%	
3700	Shift work requirements, add, minimum		5%	
3800	Maximum		30%	
3900	Temporary shoring and bracing, add, minimum	2%	5%	
4000	Maximum	5%	12%	

D3020 Heat Generating Systems

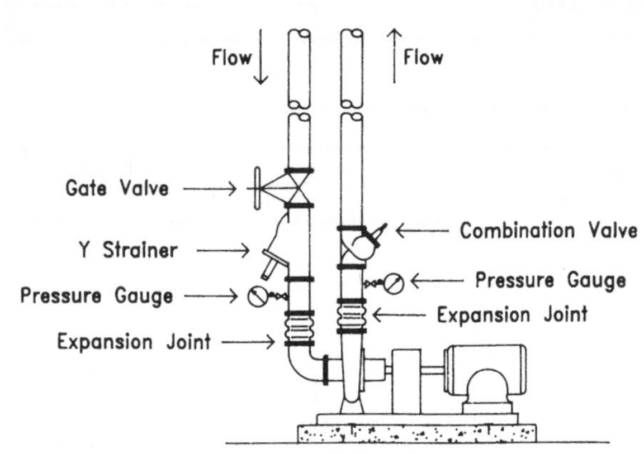

Base Mounted End–Suction Pump

This page illustrates and describes end suction base mounted circulating pumps, including strainer and piping. Lines within system components give the material and installation price on a cost each basis for the components. Prices for alternate circulating pumps are on Line items D3020 330 1010 thru 1050. Material quantities and labor costs have been adjusted for the system listed.

Factors: To adjust for job conditions other than normal working situations uses Lines D3020 330 2700 thru 4000.

System Components	QUANTITY	UNIT	COST EACH MAT.	COST EACH INST.	COST EACH TOTAL
Circulator pump, end suction, including pipe,					
Valves, fittings and guages.					
Pump, circulating, CI, base mounted, 2-1/2" size, 3 HP, to 150 GPM	1.000	Ea.	3,475	530	4,005
Pipe, black steel, Sch. 40, on yoke & roll hangers, 10' O.C., 2-1/2" dia	12.000	L.F.	153.60	265.44	419.04
Elbow, 90°, weld joint, steel, 2-1/2" pipe size	1.000	Ea.	17.25	129.50	146.75
Flange, weld neck, 150 LB, 2-1/2" pipe size	8.000	Ea.	176	518	694
Valve, iron body, gate, 125 lb., N.R.S., flanged, 2-1/2" size	1.000	Ea.	390	191	581
Strainer, Y type, iron body, flanged, 125 lb., 2-1/2" pipe size	1.000	Ea.	84.50	193	277.50
Multipurpose valve, CI body, 2" size	1.000	Ea.	425	67	492
Expansion joint, flanged spool, 6" F to F, 2-1/2" dia	2.000	Ea.	488	155	643
T-O-L, weld joint, socket, 1/4" pipe size, nozzle	2.000	Ea.	11	89.60	100.60
T-O-L, weld joint, socket, 1/2" pipe size, nozzle	2.000	Ea.	11	93.90	104.90
Control gauges, pressure or vacuum, 3-1/2" diameter dial	2.000	Ea.	44	33.50	77.50
Pressure/temperature relief plug, 316 SS, 3/4" OD, 7-1/2" insertion	2.000	Ea.	126	33.50	159.50
Insulation, fiberglass pipe covering, 1-1/2" wall, 2-1/2" IPS	12.000	L.F.	31.92	61.20	93.12
Pump control system	1.000	Ea.	995	495	1,490
Pump balancing	1.000	Ea.		236	236
TOTAL			6,428.27	3,091.64	9,519.91

D3020 330	Circulating Pump Systems, End Suction	COST EACH MAT.	COST EACH INST.	COST EACH TOTAL
1000	For alternate pump sizes:			
1010	Pump, base mtd with motor, end-suction, 2-1/2" size, 3 HP, to 150 GPM	6,425	3,075	9,500
1020	3" size, 5 HP, to 225 GPM	6,450	3,550	10,000
1030	4" size, 7-1/2 HP, to 350 GPM	7,575	4,275	11,850
1040	5" size, 15 HP, to 1000 GPM	10,700	6,450	17,150
1050	6" size, 25 HP, to 1550 GPM	14,600	7,800	22,400
2700	Cut & patch to match existing construction, add, minimum	2%	3%	
2800	Maximum	5%	9%	
2900	Dust protection, add, minimum	1%	2%	
3000	Maximum	4%	11%	
3100	Equipment usage curtailment, add, minimum	1%	1%	
3200	Maximum	3%	10%	
3300	Material handling & storage limitation, add, minimum	1%	1%	
3400	Maximum	6%	7%	
3500	Protection of existing work, add, minimum	2%	2%	
3600	Maximum	5%	7%	

SERVICES D

Important: See the Reference Section for critical supporting data - Reference Numbers and City Cost Indexes

D3020 Heat Generating Systems

D3020 330	Circulating Pump Systems, End Suction	COST EACH		
		MAT.	INST.	TOTAL
3700	Shift work requirements, add, minimum		5%	
3800	Maximum		30%	
3900	Temporary shoring and bracing, add, minimum	2%	5%	
4000	Maximum	5%	12%	

D3040 Distribution Systems

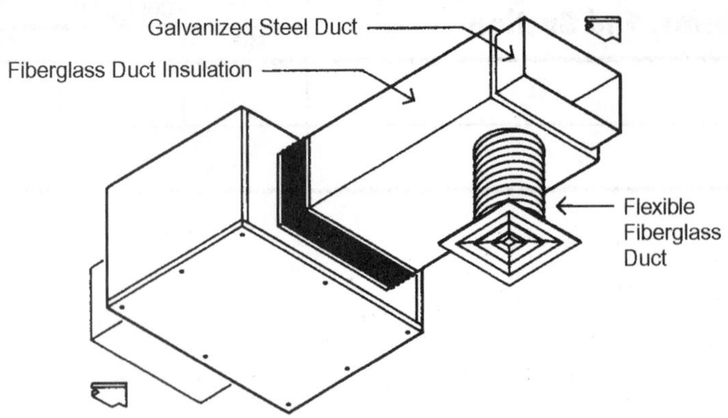

Horizontal Fan Coil Air Conditioning System

This page illustrates and describes fan coil air conditioners including duct work, duct installation, piping and diffusers. Lines within Systems Components give the material and installation price on a cost each basis for the components. Prices for alternate fan coil A/C unit systems are on Line Items D3040 118 1010 thru D3040 126 1120. Material quantities and labor costs have been adjusted for the system listed.

Factors: To adjust for job conditions other than normal working situations use Lines D3040 126 2700 thru 4000.

System Components	QUANTITY	UNIT	COST EACH MAT.	COST EACH INST.	COST EACH TOTAL
Fan coil A/C system, horizontal, with housing, controls, 2 pipe, 1/2 ton					
Fan coil A/C, horizontal housing, filters, chilled water, 1/2 ton cooling	1.000	Ea.	1,025	121	1,146
Pipe, black steel, Sch 40, threaded, W/cplgs & hangers, 10′ OC, 3/4″ dia.	20.000	L.F.	56.76	187.05	243.81
Elbow, 90°, black, straight, 3/4″ dia.	6.000	Ea.	12	228	240
Tee, black, straight, 3/4″ dia.	2.000	Ea.	6.38	118	124.38
Union, black with brass seat, 3/4″ dia.	2.000	Ea.	16.80	82	98.80
Pipe nipples, 3/4″ diam	3.000	Ea.	3.96	13.05	17.01
Valves, bronze, gate, N.R.S., threaded, class 150, 3/4″ size	2.000	Ea.	53	53	106
Circuit setter, balance valve, bronze body, threaded, 3/4″ pipe size	1.000	Ea.	57.50	27	84.50
Insulation, fiberglass pipe covering, 1″ wall, 3/4″ IPS	20.000	L.F.	21.40	79.80	101.20
Ductwork, 12″ x 8″ fabricated, galvanized steel, 12 LF	55.000	Lb.	57.75	352	409.75
Insulation, ductwork, blanket type, fiberglass, 1″ thk, 1-1/2 LB density	40.000	S.F.	24.80	104.80	129.60
Diffusers, aluminum, OB damper, ceiling, perf, 24″x24″ panel size, 6″x6″	2.000	Ea.	158	67	225
Ductwork, flexible, fiberglass fabric, insulated, 1″thk, PE jacket, 6″ dia	16.000	L.F.	44.64	59.36	104
Round volume control damper 6″ dia.	2.000	Ea.	53	49	102
Fan coil unit balancing	1.000	Ea.		69.50	69.50
Re-heat coil balancing	1.000	Ea.		99	99
Diffuser/register, high, balancing	2.000	Ea.		193	193
TOTAL			1,590.99	1,902.56	3,493.55

D3040 118	Fan Coil A/C Unit, Two Pipe	COST EACH MAT.	COST EACH INST.	COST EACH TOTAL
1000	For alternate A/C units:			
1010	Fan coil A/C system, cabinet mounted, controls, 2 pipe, 1/2 Ton	890	1,000	1,890
1020	1 Ton	1,100	1,125	2,225
1030	1-1/2 Ton	1,225	1,150	2,375
1040	2 Ton	1,875	1,375	3,250
1050	3 Ton	2,600	1,450	4,050

D3040 120	Fan Coil A/C Unit, Two Pipe, Electric Heat	COST EACH MAT.	COST EACH INST.	COST EACH TOTAL
1010	Fan coil A/C system, cabinet mntd, elect. ht, controls, 2 pipe, 1/2 Ton	1,325	1,000	2,325
1020	1 Ton	1,600	1,125	2,725
1030	1-1/2 Ton	1,900	1,150	3,050
1040	2 Ton	2,850	1,400	4,250
1050	3 Ton	4,775	1,450	6,225

D3040 Distribution Systems

D3040 122	Fan Coil A/C Unit, Four Pipe	COST EACH		
		MAT.	INST.	TOTAL
1010	Fan coil A/C system, cabinet mounted, controls, 4 pipe, 1/2 Ton	1,400	1,950	3,350
1020	1 Ton	1,650	2,075	3,725
1030	1-1/2 Ton	1,800	2,100	3,900
1040	2 Ton	2,275	2,175	4,450
1050	3 Ton	3,375	2,325	5,700

D3040 124	Fan Coil A/C, Horizontal, Duct Mount, 2 Pipe	COST EACH		
		MAT.	INST.	TOTAL
1010	Fan coil A/C system, horizontal w/housing, controls, 2 pipe, 1/2 Ton	1,600	1,900	3,500
1020	1 Ton	2,100	2,775	4,875
1030	1-1/2 Ton	2,625	3,625	6,250
1040	2 Ton	3,400	4,350	7,750
1050	3 Ton	4,125	5,775	9,900
1060	3-1/2 Ton	4,350	5,975	10,325
1070	4 Ton	4,500	6,775	11,275
1080	5 Ton	5,300	8,625	13,925
1090	6 Ton	5,375	9,250	14,625
1100	7 Ton	5,400	9,600	15,000
1110	8 Ton	5,400	9,800	15,200
1120	10 Ton	6,275	10,200	16,475

D3040 126	Fan Coil A/C, Horiz., Duct Mount, 2 Pipe, Elec. Ht.	COST EACH		
		MAT.	INST.	TOTAL
1010	Fan coil A/C system, horiz. hsng, elect. ht, ctrls, 2 pipe, 1/2 Ton	1,825	1,900	3,725
1020	1 Ton	2,375	2,775	5,150
1030	1-1/2 Ton	2,825	3,625	6,450
1040	2 Ton	3,750	4,350	8,100
1050	3 Ton	5,975	5,950	11,925
1060	3-1/2 Ton	6,750	5,975	12,725
1070	4 Ton	7,600	6,800	14,400
1080	5 Ton	8,675	8,625	17,300
1090	6 Ton	8,750	9,275	18,025
1100	7 Ton	8,900	9,225	18,125
1110	8 Ton	9,725	9,325	19,050
1120	10 Ton	10,300	9,575	19,875
2700	Cut & patch to match existing construction, add, minimum	2%	3%	
2800	Maximum	5%	9%	
2900	Dust protection, add, minimum	1%	2%	
3000	Maximum	4%	11%	
3100	Equipment usage curtailment, add, minimum	1%	1%	
3200	Maximum	3%	10%	
3300	Material handling & storage limitation, add, minimum	1%	1%	
3400	Maximum	6%	7%	
3500	Protection of existing work, add, minimum	2%	2%	
3600	Maximum	5%	7%	
3700	Shift work requirements, add, minimum		5%	
3800	Maximum		30%	
3900	Temporary shoring and bracing, add, minimum	2%	5%	
4000	Maximum	5%	12%	

D3040 Distribution Systems

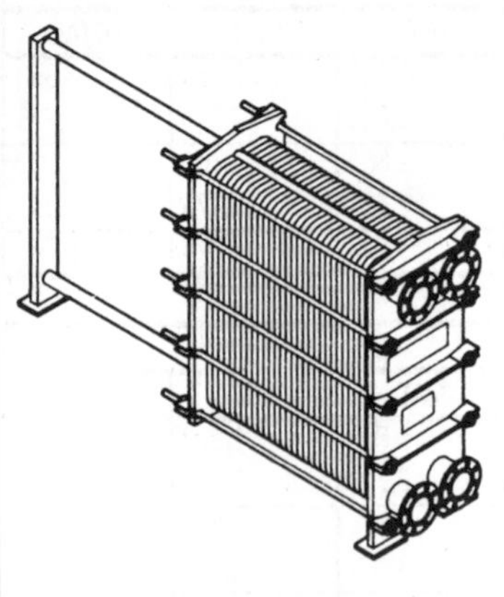

Plate Heat Exchanger

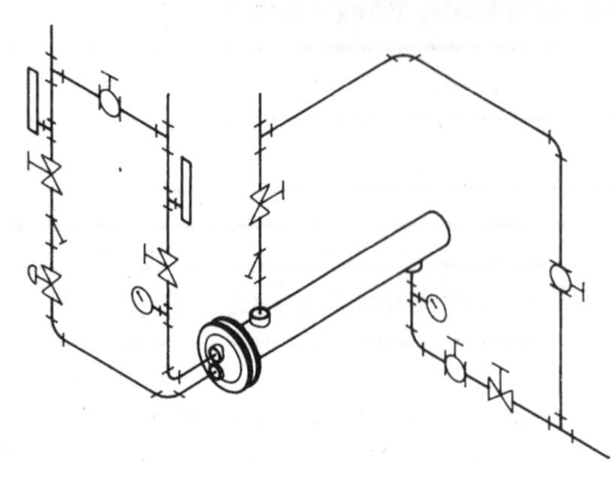

Shell and Tube Heat Exchanger

This page illustrates and describes plate and shell and tube type heat exchangers including pressure gauge and exchanger control system. Lines within Systems Components give the material and installation price on a cost each basis for the components. Prices for alternate heat exchanger systems are on Line Items D3040 610 1010 thru D3040 620 1040. Material quantities and labor costs have been adjusted for the system listed.

Factors To adjust for job conditions other then normal working situations use Lines D3040 620 2700 thru 4000.

System Components	QUANTITY	UNIT	COST EACH MAT.	COST EACH INST.	COST EACH TOTAL
Shell and tube type heat exchanger with related valves and piping.					
Heat exchanger, 4 pass, 3/4" O.D. copper tubes, by steam at 10 psi, 40 GPM	1.000	Ea.	3,000	241	3,241
Pipe, black steel, Sch 40, threaded, W/couplings & hangers, 10' OC, 2" dia.	40.000	L.F.	393.90	754.98	1,148.88
Elbow, 90°, straight, 2" dia.	16.000	Ea.	208	848	1,056
Reducer, black steel, concentric, 2" dia.	2.000	Ea.	29.30	91	120.30
Valves, bronze, gate, rising stem, threaded, class 150, 2" size	6.000	Ea.	654	291	945
Valves, bronze, globe, class 150, rising stem, threaded, 2" size	2.000	Ea.	566	97	663
Strainers, Y type, bronze body, screwed, 150 lb., 2" pipe size	2.000	Ea.	125	82	207
Valve, electric motor actuated, brass, 2 way, screwed, 1-1/2" pipe size	1.000	Ea.	425	41.50	466.50
Union, black with brass seat, 2" dia.	8.000	Ea.	200	448	648
Tee, black, straight, 2" dia.	9.000	Ea.	149.85	783	932.85
Tee, black, reducing run and outlet, 2" dia.	4.000	Ea.	104	348	452
Pipe nipple, black, 2" dia.	21.000	Ea.	3.90	7.48	11.38
Thermometers, stem type, 9" case, 8" stem, 3/4" NPT	2.000	Ea.	260	38.30	298.30
Gauges, pressure or vacuum, 3-1/2" diameter dial	2.000	Ea.	44	33.50	77.50
Insulation, fiberglass pipe covering, 1" wall, 2" IPS	40.000	L.F.	57.60	183.60	241.20
Heat exchanger control system	1.000	Ea.	2,100	1,725	3,825
Coil balancing	1.000	Ea.		99	99
TOTAL			8,320.55	6,112.36	14,432.91

D3040 610	Heat Exchanger, Plate Type	COST EACH MAT.	COST EACH INST.	COST EACH TOTAL
1000	For alternate heat exchangers:			
1010	Plate heat exchanger, 400 GPM	37,700	12,500	50,200
1020	800 GPM	60,500	15,800	76,300
1030	1200 GPM	91,000	21,400	112,400
1040	1800 GPM	123,000	26,600	149,600

D3040 620	Heat Exchanger, Shell & Tube	COST EACH MAT.	COST EACH INST.	COST EACH TOTAL
1010	Shell & tube heat exchanger, 40 GPM	8,325	6,100	14,425
1020	96 GPM	14,800	11,000	25,800
1030	240 GPM	26,800	14,800	41,600
1040	600 GPM	51,000	20,900	71,900
2700	Cut & patch to match existing construction, add, minimum	2%	3%	
2800	Maximum	5%	9%	
2900	Dust protection, add, minimum	1%	2%	
3000	Maximum	4%	11%	

Important: See the Reference Section for critical supporting data - Reference Numbers and City Cost Indexes

D3040 Distribution Systems

D3040 620	Heat Exchanger, Shell & Tube	COST EACH		
		MAT.	INST.	TOTAL
3100	Equipment usage curtailment, add, minimum	1%	1%	
3200	Maximum	3%	10%	
3300	Material handling & storage limitation, add, minimum	1%	1%	
3400	Maximum	6%	7%	
3500	Protection of existing work, add, minimum	2%	2%	
3600	Maximum	5%	7%	
3700	Shift work requirements, add, minimum		5%	
3800	Maximum		30%	
3900	Temporary shoring and bracing, add, minimum	2%	5%	
4000	Maximum	5%	12%	

D3050 Terminal & Package Units

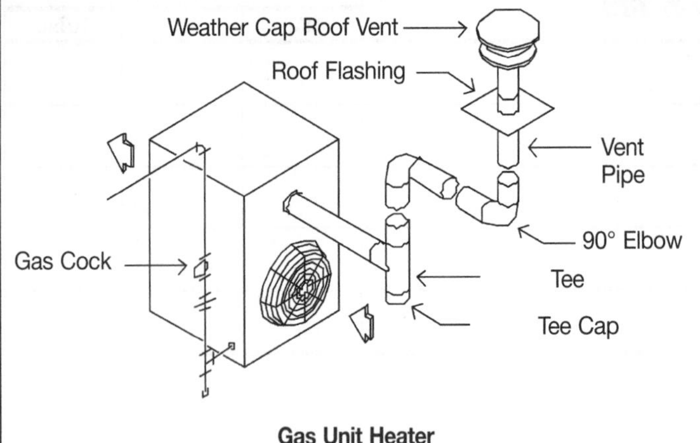

Gas Unit Heater

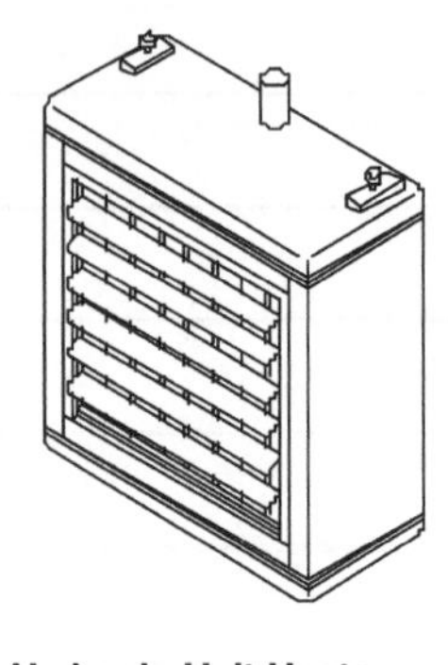
Hydronic Unit Heater

This page illustrates and describes unit heaters including piping and vents. Lines within Systems Components give the material and installation price on a cost each basis for this system. Prices for alternate unit heater systems are on Line items D3050 120 1010 thru D3050 140 1040. Material quantities and labor costs have been adjusted for the system listed.

Factors: To adjust for job conditions other than normal working situations use Lines D3050 140 2700 thru 4000.

System Components	QUANTITY	UNIT	COST EACH MAT.	COST EACH INST.	COST EACH TOTAL
Unit heaters complete with piping, thermostats and chimney as required					
Space heater, propeller fan, 20 MBH output	1.000	Ea.	525	113	638
Pipe, black steel, Sch 40, threaded, W/coupling & hangers, 10′ OC, 3/4″ dia	20.000	L.F.	55.44	182.70	238.14
Elbow, 90°, black steel, straight, 1/2″ dia.	3.000	Ea.	5.01	106.50	111.51
Elbow, 90°, black steel, straight, 3/4″ dia.	3.000	Ea.	6	114	120
Tee, black steel, reducing, 3/4″ dia.	1.000	Ea.	4.92	59	63.92
Union, black with brass seat, 3/4″ dia.	1.000	Ea.	8.40	41	49.40
Pipe nipples, black, 1/2″ dia	2.000	Ea.	4.50	16.90	21.40
Pipe nipples, black, 3/4″ dia	2.000	Ea.	2.64	8.70	11.34
Pipe cap, black, 3/4″ dia.	1.000	Ea.	2.27	16.60	18.87
Gas cock, brass, 3/4″ size	1.000	Ea.	15.65	24	39.65
Thermostat, 1 set back, electric, timed	1.000	Ea.	89	67	156
Wiring, thermostat hook-up	1.000	Ea.	6.60	21.50	28.10
Vent chimney, prefab metal, U.L. listed, gas, double wall, galv. st, 4″ dia	12.000	L.F.	65.40	170.40	235.80
Vent chimney, gas, double wall, galv. steel, elbow 90°, 4″ dia.	2.000	Ea.	44	57	101
Vent chimney, gas, double wall, galv. steel, Tee, 4″ dia.	1.000	Ea.	27.50	37	64.50
Vent chimney, gas, double wall, galv. steel, T cap, 4″ dia.	1.000	Ea.	2.13	23	25.13
Vent chimney, gas, double wall, galv. steel, roof flashing, 4″ dia.	1.000	Ea.	7.35	28.50	35.85
Vent chimney, gas, double wall, galv. steel, top, 4″ dia.	1.000	Ea.	10.45	22	32.45
TOTAL			882.26	1,108.80	1,991.06

D3050 120	Unit Heaters, Gas	COST EACH MAT.	COST EACH INST.	COST EACH TOTAL
1000	For alternate unit heaters:			
1010	Space heater, suspended, gas fired, propeller fan, 20 MBH	880	1,100	1,980
1020	60 MBH	1,025	1,150	2,175
1030	100 MBH	1,275	1,225	2,500
1040	160 MBH	1,550	1,350	2,900
1050	200 MBH	1,850	1,475	3,325
1060	280 MBH	2,425	1,575	4,000
1070	320 MBH	3,150	1,800	4,950

D3050 Terminal & Package Units

D3050 130	Unit Heaters, Hydronic	COST EACH		
		MAT.	INST.	TOTAL
1010	Space heater, suspended, horiz. mount, HW, prop. fan, 20 MBH	1,450	1,175	2,625
1020	60 MBH	1,900	1,325	3,225
1030	100 MBH	2,250	1,450	3,700
1040	150 MBH	2,750	1,600	4,350
1050	200 MBH	2,925	1,800	4,725
1060	300 MBH	3,325	2,100	5,425

D3050 140	Cabinet Unit Heaters, Hydronic	COST EACH		
		MAT.	INST.	TOTAL
1010	Unit heater, cabinet type, horizontal blower, hot water, 20 MBH	1,750	1,150	2,900
1020	60 MBH	2,500	1,275	3,775
1030	100 MBH	2,750	1,400	4,150
1040	120 MBH	2,800	1,450	4,250
2700	Cut & patch to match existing construction, add, minimum	2%	3%	
2800	Maximum	5%	9%	
2900	Dust protection, add, minimum	1%	2%	
3000	Maximum	4%	11%	
3100	Equipment usage curtailment, add, minimum	1%	1%	
3200	Maximum	3%	10%	
3300	Material handling & storage limitation, add, minimum	1%	1%	
3400	Maximum	6%	7%	
3500	Protection of existing work, add, minimum	2%	2%	
3600	Maximum	5%	7%	
3700	Shift work requirements, add, minimum		5%	
3800	Maximum		30%	
3900	Temporary shoring and bracing, add, minimum	2%	5%	
4000	Maximum	5%	12%	

D3050 Terminal & Package Units

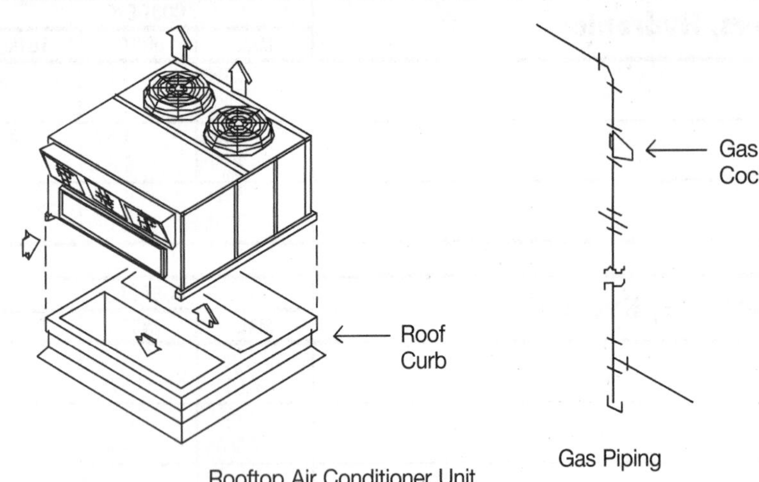

This page illustrates and describes packaged air conditioners including rooftop DX unit, pipes and fittings. Lines within Systems Components give the material and installation price on a cost each basis for the components. Prices for alternate A/C systems are on Line Items D3050 175 1010 thru 1060. Material quantities and labor costs have been adjusted for the system listed.

Factors: To adjust for job conditions other than normal working situations use Lines D3050 175 2700 thru 4000.

System Components	QUANTITY	UNIT	COST EACH		
			MAT.	INST.	TOTAL
Packaged rooftop DX unit, gas, with related piping and valves.					
Roof top A/C, curb, economizer, sgl zone, elec cool, gas ht, 5 ton, 112 MBH	1.000	Ea.	5,125	1,725	6,850
Pipe, black steel, Sch 40, threaded, W/cplgs & hangers, 10′ OC, 1″ dia	20.000	L.F.	80.22	210	290.22
Elbow, 90°, black, straight, 3/4″ dia.	3.000	Ea.	6	114	120
Elbow, 90°, black, straight, 1″ dia.	3.000	Ea.	10.44	123	133.44
Tee, black, reducing, 1″ dia.	1.000	Ea.	6.20	66.50	72.70
Union, black with brass seat, 1″ dia.	1.000	Ea.	12.05	44.50	56.55
Pipe nipple, black, 3/4″ dia	2.000	Ea.	5.28	17.40	22.68
Pipe nipple, black, 1″ dia	2.000	Ea.	3.82	10	13.82
Cap, black, 1″ dia.	1.000	Ea.	2.75	17.70	20.45
Gas cock, brass, 1″ size	1.000	Ea.	20	28	48
Control system, pneumatic, Rooftop A/C unit	1.000	Ea.	2,150	1,900	4,050
Rooftop unit heat/cool balancing	1.000	Ea.		375	375
TOTAL			7,421.76	4,631.10	12,052.86

D3050 175	Rooftop Air Conditioner, Const. Volume	COST EACH		
		MAT.	INST.	TOTAL
1000	For alternate A/C systems:			
1010	A/C, Rooftop, DX cool, gas heat, curb, economizer, fltrs, 5 Ton	7,425	4,625	12,050
1020	7-1/2 Ton	9,775	4,825	14,600
1030	12-1/2 Ton	13,200	5,275	18,475
1040	18 Ton	20,200	5,775	25,975
1050	25 Ton	27,000	6,575	33,575
1060	40 Ton	43,000	8,800	51,800
1010	A/C, Rooftop, DX cool, gas heat, curb, ecmizr, fltrs, VAV, 12-1/2 Ton	15,400	5,625	21,025
1020	18 Ton	22,600	6,200	28,800
1030	25 Ton	30,600	7,125	37,725
1040	40 Ton	50,500	9,675	60,175
1050	60 Ton	71,000	13,200	84,200
1060	80 Ton	86,000	16,400	102,400
2700	Cut & patch to match existing construction, add, minimum	2%	3%	
2800	Maximum	5%	9%	
2900	Dust protection, add, minimum	1%	2%	
3000	Maximum	4%	11%	
3100	Equipment usage curtailment, add, minimum	1%	1%	
3200	Maximum	3%	10%	
3300	Material handling & storage limitation, add, minimum	1%	1%	
3400	Maximum	6%	7%	
3500	Protection of existing work, add, minimum	2%	2%	

Important: See the Reference Section for critical supporting data - Reference Numbers and City Cost Indexes

D3050 Terminal & Package Units

D3050 175	Rooftop Air Conditioner, Const. Volume	COST EACH		
		MAT.	INST.	TOTAL
3600	Maximum	5%	7%	
3700	Shift work requirements, add, minimum		5%	
3800	Maximum		30%	
3900	Temporary shoring and bracing, add, minimum	2%	5%	
4000	Maximum	5%	12%	

D3050 Terminal & Package Units

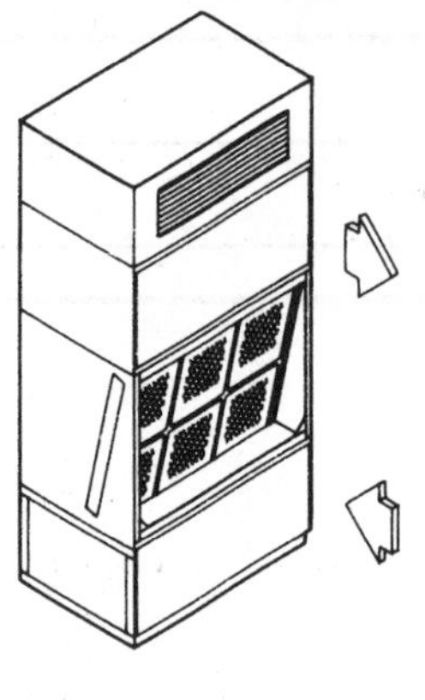
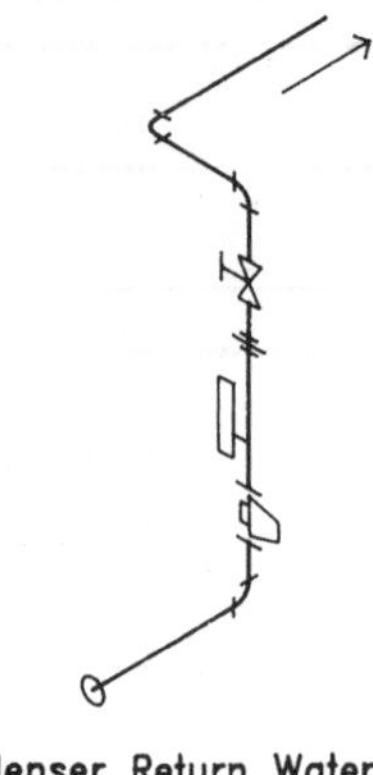

Condenser Supply Water Piping

Condenser Return Water Piping

Self–Contained Air Conditioner Unit, Water Cooled

This page illustrates and describes packaged electrical air conditioning units including multizone control and related piping. Lines within Systems Components give the material and installation price on a cost each basis for the components. Prices for alternate packaged electrical A/C unit systems are on Line Items D3050 210 1010 thru D3050 225 1060. Material quantities and labor costs have been adjusted for the system listed.

Factors: To adjust for job conditions other than normal working situations use Lines D3050 225 2700 thru 4000.

System Components	QUANTITY	UNIT	COST EACH MAT.	COST EACH INST.	COST EACH TOTAL
Packaged water cooled electric air conditioning unit with related piping and valves.					
Self-contained, water cooled, elect. heat, not inc tower, 5 ton, const vol	1.000	Ea.	6,875	1,250	8,125
Pipe, black steel, Sch 40, threaded, W/cplg & hangers, 10′ OC, 1-1/4″ dia.	20.000	L.F.	107.50	231.13	338.63
Elbow, 90°, black, straight, 1-1/4″ dia.	6.000	Ea.	34.20	261	295.20
Tee, black, straight, 1-1/4″ dia.	2.000	Ea.	17.60	137	154.60
Tee, black, reducing, 1-1/4″ dia.	2.000	Ea.	22.60	137	159.60
Thermometers, stem type, 9″ case, 8″ stem, 3/4″ NPT	2.000	Ea.	260	38.30	298.30
Union, black with brass seat, 1-1/4″ dia.	2.000	Ea.	33	91	124
Pipe nipple, black, 1-1/4″ dia	3.000	Ea.	7.50	16.13	23.63
Valves, bronze, gate, N.R.S., threaded, class 150, 1-1/4″ size	2.000	Ea.	103	71	174
Circuit setter, bal valve, bronze body, threaded, 1-1/4″ pipe size	1.000	Ea.	110	35.50	145.50
Insulation, fiberglass pipe covering, 1″ wall, 1-1/4″ IPS	20.000	L.F.	24.60	87.40	112
Control system, pneumatic, A/C Unit with heat	1.000	Ea.	2,150	1,900	4,050
Re-heat coil balancing	1.000	Ea.		99	99
Rooftop unit heat/cool balancing	1.000	Ea.		375	375
TOTAL			9,745	4,729.46	14,474.46

D3050 210	AC Unit, Package, Elec. Ht., Water Cooled	COST EACH MAT.	COST EACH INST.	COST EACH TOTAL
1000	For alternate A/C systems:			
1010	A/C, Self contained, single pkg., water cooled, elect. heat, 5 Ton	9,750	4,725	14,475
1020	10 Ton	15,500	5,825	21,325
1030	20 Ton	21,300	7,475	28,775
1040	30 Ton	30,500	8,075	38,575
1050	40 Ton	37,600	8,925	46,525
1060	50 Ton	47,500	10,700	58,200

D3050 215	AC Unit, Package, Elec. Ht., Water Cooled, VAV	COST EACH MAT.	COST EACH INST.	COST EACH TOTAL
1010	A/C, Self contained, single pkg., water cooled, elect. ht, VAV, 10 Ton	17,600	5,825	23,425
1020	20 Ton	25,000	7,475	32,475
1030	30 Ton	35,600	8,075	43,675
1040	40 Ton	45,100	8,925	54,025
1050	50 Ton	55,500	10,700	66,200
1060	60 Ton	64,000	13,100	77,100

Important: See the Reference Section for critical supporting data - Reference Numbers and City Cost Indexes

D3050 Terminal & Package Units

D3050 220	AC Unit, Package, Hot Water Coil, Water Cooled	COST EACH		
		MAT.	INST.	TOTAL
1010	A/C, Self contn'd, single pkg., water cool, H/W ht, const. vol, 5 Ton	8,375	6,550	14,925
1020	10 Ton	12,800	7,825	20,625
1030	20 Ton	34,800	9,825	44,625
1040	30 Ton	46,000	12,500	58,500
1050	40 Ton	56,500	15,100	71,600
1060	50 Ton	69,500	18,600	88,100

D3050 225	AC Unit, Package, HW Coil, Water Cooled, VAV	COST EACH		
		MAT.	INST.	TOTAL
1010	A/C, Self contn'd, single pkg., water cool, H/W ht, VAV, 10 Ton	14,900	7,350	22,250
1020	20 Ton	21,700	9,575	31,275
1030	30 Ton	29,900	12,000	41,900
1040	40 Ton	39,600	13,000	52,600
1050	50 Ton	47,200	15,400	62,600
1060	60 Ton	56,500	16,700	73,200
2700	Cut & patch to match existing construction, add, minimum	2%	3%	
2800	Maximum	5%	9%	
2900	Dust protection, add, minimum	1%	2%	
3000	Maximum	4%	11%	
3100	Equipment usage curtailment, add, minimum	1%	1%	
3200	Maximum	3%	10%	
3300	Material handling & storage limitation, add, minimum	1%	1%	
3400	Maximum	6%	7%	
3500	Protection of existing work, add, minimum	2%	2%	
3600	Maximum	5%	7%	
3700	Shift work requirements, add, minimum		5%	
3800	Maximum		30%	
3900	Temporary shoring and bracing, add, minimum	2%	5%	
4000	Maximum	5%	12%	

D3050 Terminal & Package Units

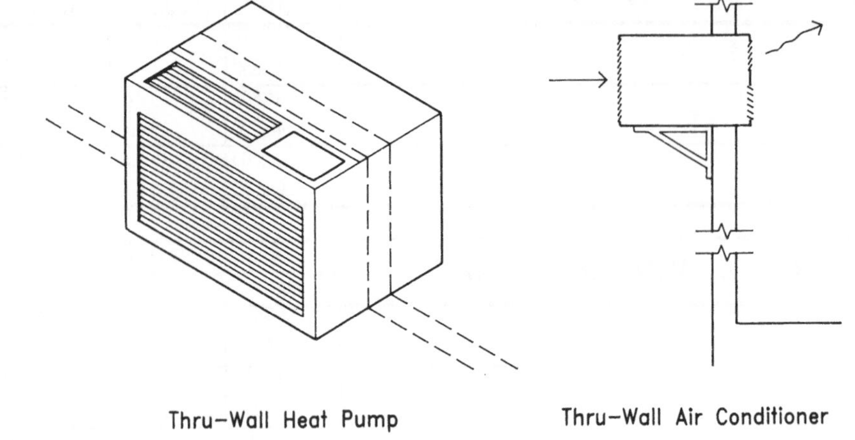

This page illustrates and describes thru-wall air conditioning and heat pump units. Lines within Systems Components give the material and installation price on a cost each basis for the components. Prices for alternate thru-wall unit systems are on Line Items D3050 255 1010 thru 1050. Material quantities and labor costs have been adjusted for the system listed.

Factors: To adjust for job conditions other than normal working conditions use Lines D3050 255 2700 thru 4000.

System Components	QUANTITY	UNIT	COST EACH MAT.	COST EACH INST.	COST EACH TOTAL
Electric thru-wall air conditioning unit.					
A/C unit, thru-wall, electric heat., cabinet, louver, 1/2 ton	1.000	Ea.	1,050	161	1,211
TOTAL			1,050	161	1,211

D3050 255	Thru-Wall A/C Unit	COST EACH MAT.	COST EACH INST.	COST EACH TOTAL
1000	For alternate A/C units:			
1010	A/C Unit, thru-the-wall, sup. elect. heat, cabinet, louver, 1/2 Ton	1,050	161	1,211
1020	3/4 Ton	1,100	193	1,293
1030	1 Ton	1,225	241	1,466
1040	1-1/2 Ton	1,550	400	1,950
1050	2 Ton	1,575	505	2,080
1010	Heat pump, thru-the-wall, cabinet, louver, 1/2 Ton	1,125	121	1,246
1020	3/4 Ton	1,200	161	1,361
1030	1 Ton	1,400	241	1,641
1040	Sup. elect. heat, 1-1/2 Ton	2,450	620	3,070
1050	2 Ton	2,775	645	3,420
2700	Cut & patch to match existing construction, add, minimum	2%	3%	
2800	Maximum	5%	9%	
2900	Dust protection, add, minimum	1%	2%	
3000	Maximum	4%	11%	
3100	Equipment usage curtailment, add, minimum	1%	1%	
3200	Maximum	3%	10%	
3300	Material handling & storage limitation, add, minimum	1%	1%	
3400	Maximum	6%	7%	
3500	Protection of existing work, add, minimum	2%	2%	
3600	Maximum	5%	7%	
3700	Shift work requirements, add, minimum		5%	
3800	Maximum		30%	
3900	Temporary shoring and bracing, add, minimum	2%	5%	
4000	Maximum	5%	12%	

D4010 Sprinklers

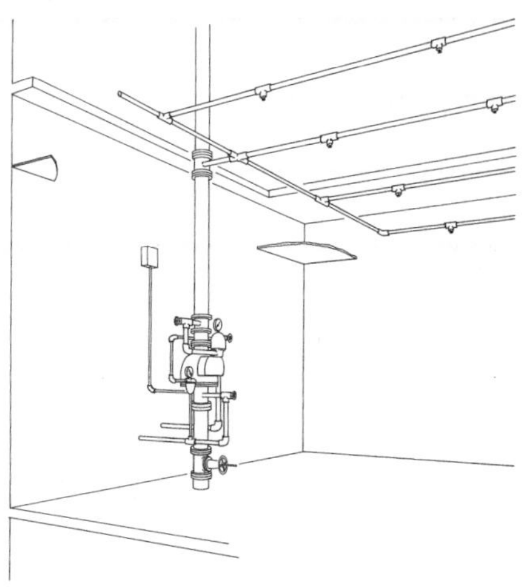

This page illustrates and describes a dry type fire sprinkler system. Lines within System Components give the unit cost of a system for a 2,000 square foot building. Lines D4010 320 1700 thru 2400 give the square foot costs for alternate systems. Both material quantities and labor costs have been adjusted for the system listed.

Factors: To adjust for conditions other than normal working conditions, use Lines D4010 320 2900 thru 4000.

System Components	QUANTITY	UNIT	COST EACH MAT.	COST EACH INST.	COST EACH TOTAL
Pre-action fire sprinkler system, ordinary hazard, open area to 2000 S.F.					
On one floor.					
Air compressor	1.000	Ea.	820	400	1,220
4″ OS & Y valve	1.000	Ea.	455	320	775
Pipe riser 4″ diameter	10.000	L.F.	210	280.50	490.50
Sprinkler head supply piping, 1″	163.000	L.F.	286.50	750	1,036.50
Sprinkler head supply piping, 1-1/4″	163.000	L.F.	275	591.25	866.25
Sprinkler head supply piping, 2-1/2″	163.000	L.F.	459.80	725.80	1,185.60
Pipe fittings, 1″	8.000	Ea.	43.60	532	575.60
Pipe fittings, 1-1/4″	8.000	Ea.	70.40	548	618.40
Pipe fittings, 2-1/2″	4.000	Ea.	162	424	586
Pipe fittings, 4″	2.000	Ea.	286	478	764
Detectors	2.000	Ea.	710	65	775
Sprinkler heads	16.000	Ea.	89.60	520	609.60
Fire department connection	1.000	Ea.	360	188	548
TOTAL			4,227.90	5,822.55	10,050.45

D4010 320	Fire Sprinkler Systems, Dry	COST PER S.F. MAT.	COST PER S.F. INST.	COST PER S.F. TOTAL
1700	Ordinary hazard, one floor, area to 2000 S.F./floor	2.12	2.91	5.03
1800	For each additional floor, add per floor	.94	2.43	3.37
1900	Area to 3200 S.F./ floor	1.72	2.89	4.61
2000	For each additional floor, add per floor	.99	2.58	3.57
2100	Area to 5000 S.F./ floor	1.87	2.93	4.80
2200	For each additional floor, add per floor	1.27	2.72	3.99
2300	Area to 8000 S.F./ floor	1.45	2.46	3.91
2400	For each additional floor, add per floor	1.02	2.33	3.35
2500				
2600				
2700				
2800				
2900	Cut & patch to match existing construction, add, minimum	2%	3%	
3000	Maximum	5%	9%	
3100	Dust protection, add, minimum	1%	2%	
3200	Maximum	4%	11%	
3300	Equipment usage curtailment, add, minimum	1%	1%	
3400	Maximum	3%	10%	

D SERVICES

D4010 Sprinklers

D4010 320		Fire Sprinkler Systems, Dry	COST PER S.F.		
			MAT.	INST.	TOTAL
3500	Material handling & storage limitation, add, minimum		1%	1%	
3600	Maximum		6%	7%	
3700	Protection of existing work, add, minimum		2%	2%	
3800	Maximum		5%	7%	
3900	Shift work requirements, add, minimum			5%	
4000	Maximum			30%	

D4010 Sprinklers

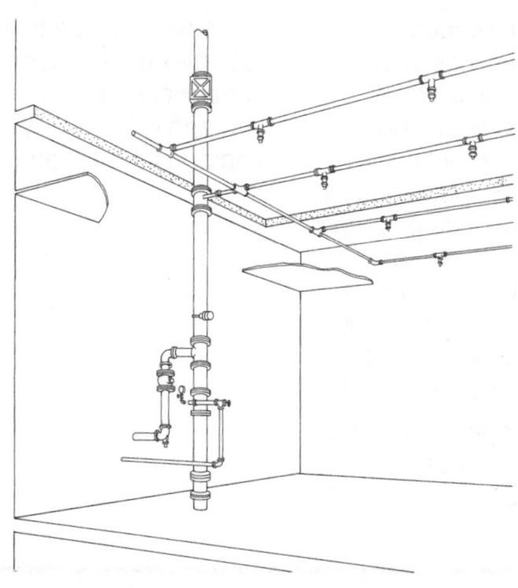

This page illustrates and describes a wet type fire sprinkler system. Lines within System Components give the unit cost of a system for a 2,000 square foot building. Lines D4010 420 1900 thru 2600 give the square foot costs for alternate systems. Both material quantities and labor costs have been adjusted for the system listed.

Factors: To adjust for conditions other than normal working conditions, use Lines D4010 420 2900 thru 4000.

System Components	QUANTITY	UNIT	COST EACH MAT.	COST EACH INST.	COST EACH TOTAL
Wet pipe fire sprinkler system, ordinary hazard, open area to 2000 S.F. On one floor.					
4″ OS & Y valve	1.000	Ea.	455	320	775
Wet pipe alarm valve	1.000	Ea.	945	470	1,415
Water motor alarm	1.000	Ea.	207	130	337
3″ check valve	1.000	Ea.	180	315	495
Pipe riser, 4″ diameter	10.000	L.F.	210	280.50	490.50
Water gauges and trim	1.000	Set	1,425	940	2,365
Electric fire horn	1.000	Ea.	40	77.50	117.50
Sprinkler head supply piping, 1″ diameter	168.000	L.F.	286.50	750	1,036.50
Sprinkler head supply piping, 1-1/2″ diameter	168.000	L.F.	324.50	657.25	981.75
Sprinkler head supply piping, 2-1/2″ diameter	168.000	L.F.	459.80	725.80	1,185.60
Pipe fittings, 1″	8.000	Ea.	43.60	532	575.60
Pipe fittings, 1-1/2″	8.000	Ea.	87.60	588	675.60
Pipe fittings, 2-1/2″	4.000	Ea.	162	424	586
Pipe fittings, 4″	2.000	Ea.	286	478	764
Sprinkler heads	16.000	Ea.	89.60	520	609.60
Fire department connection	1.000	Ea.	360	188	548
TOTAL			5,561.60	7,396.05	12,957.65

D4010 420	Fire Sprinkler Systems, Wet	COST PER S.F. MAT.	COST PER S.F. INST.	COST PER S.F. TOTAL
1900	Ordinary hazard, one floor, area to 2000 S.F./floor	2.76	3.71	6.47
2000	For each additional floor, add per floor	.96	2.48	3.44
2100	Area to 3200 S.F./floor	2.12	3.34	5.46
2200	For each additional floor, add per floor	.99	2.58	3.57
2300	Area to 5000 S.F./ floor	2	3.21	5.21
2400	For each additional floor, add per floor	1.27	2.72	3.99
2500	Area to 8000 S.F./ floor	1.49	2.64	4.13
2600	For each additional floor, add per floor	1.10	2.47	3.57
2700				
2900	Cut & patch to match existing construction, add minimum	2%	3%	
3000	Maximum	5%	9%	
3100	Dust protection, add, minimum	1%	2%	
3200	Maximum	4%	11%	
3300	Equipment usage curtailment, add, minimum	1%	1%	
3400	Maximum	3%	10%	
3500	Material handling & storage limitation, add, minimum	1%	1%	

D5010 Electrical Service/Distribution

This page illustrates and describes commercial service systems including a meter socket, service head and cable, entrance switch, steel conduit, copper wire, panel board, ground rod, wire, and conduit. Lines within System Components give the unit price and total price on a cost each basis for this system. Prices for alternate commercial service systems are on Line Items D5010 210 1500 thru 1700. Both material quantities and labor costs have been adjusted for the system listed.

Factors: To adjust for job conditions other than normal working situations use Lines D5010 210 2900 thru 4000.

Example: You are to install the system with maximum equipment usage curtailment. Go to Line D5010 210 3400 and apply these percentages to the appropriate MAT. and INST. costs.

System Components	QUANTITY	UNIT	COST EACH		
			MAT.	INST.	TOTAL
Commercial electric service including service breakers, metering 120/208 Volt, 3 phase, 4 wire, feeder, and panel board.					
100 Amp Service					
Meter socket	1.000	Ea.	39.50	162	201.50
Service head	.200	C.L.F.	54.40	104	158.40
Service entrance switch	1.000	Ea.	262	272	534
Rigid steel conduit	20.000	L.F.	106	159	265
600 volt copper wire #3	1.000	C.L.F.	69	104	173
Panel board, 24 circuits, 20 Amp breakers	1.000	Ea.	765	865	1,630
Ground rod	1.000	Ea.	15.05	94	109.05
TOTAL			1,310.95	1,760	3,070.95

D5010 210	Commercial Service - 3 Phase	COST EACH		
		MAT.	INST.	TOTAL
1400	For alternate size services:			
1500	120/208 Volt, 3 phase, 4 wire service, 60 Amp	885	1,375	2,260
1600	200 Amp	2,600	3,100	5,700
1700	400 Amp	5,225	3,875	9,100
1800				
1900				
2000				
2100				
2200				
2300				
2400				
2500				
2900	Cut & patch to match existing construction, add, minimum	2%	3%	
3000	Maximum	5%	9%	
3100	Dust protection, add, minimum	1%	2%	
3200	Maximum	4%	11%	
3300	Equipment usage curtailment, add, minimum	1%	1%	
3400	Maximum	3%	10%	
3500	Material handling & storage limitation, add, minimum	1%	1%	
3600	Maximum	6%	7%	
3700	Protection of existing work, add, minimum	2%	2%	
3800	Maximum	5%	7%	
3900	Shift work requirements, add, minimum		5%	
4000	Maximum		30%	

D5010 Electrical Service/Distribution

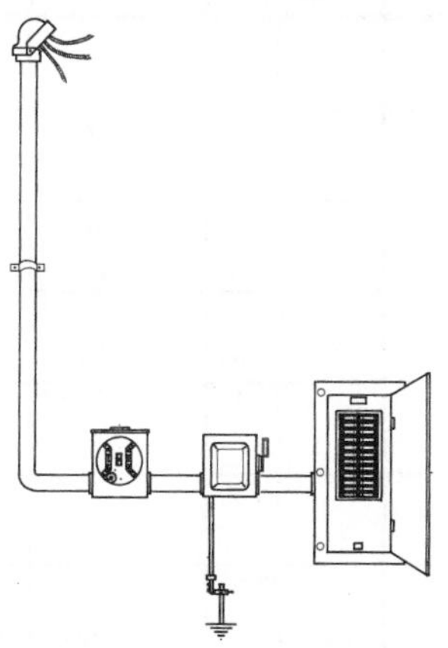

This page illustrates and describes a residential, single phase system including a weather cap, service entrance cable, meter socket, entrance switch, ground rod, ground cable, EMT, and panelboard. Lines with System Components give the unit price and total price on a cost each basis for this system. Prices for an alternate residential, single phase system are also given. Both material quantities and labor costs have been adjusted for the system listed.

Factors: To adjust for job conditions other than normal working situations use Lines D5010 220 2900 thru 4000.

Example: You are to install the system with a minimum equipment usage curtailment. Go to Line D5010 220 3300 and apply these percentages to the appropriate MAT. and INST. costs.

System Components	QUANTITY	UNIT	COST EACH MAT.	COST EACH INST.	COST EACH TOTAL
100 Amp Service, single phase					
Weathercap	1.000	Ea.	10.45	43	53.45
Service entrance cable	.200	C.L.F.	54.40	104	158.40
Meter socket	1.000	Ea.	39.50	162	201.50
Entrance disconnect switch	1.000	Ea.	262	272	534
Ground rod, with clamp	1.000	Ea.	19.45	108	127.45
Ground cable	.100	C.L.F.	16.40	32.50	48.90
Panelboard, 12 circuit	1.000	Ea.	238	430	668
TOTAL			640.20	1,151.50	1,791.70
200 Amp Service , single phase					
Weathercap	1.000	Ea.	32	64.50	96.50
Service entrance cable	.200	C.L.F.	138	148	286
Meter socket	1.000	Ea.	59	272	331
Entrance disconnect switch	1.000	Ea.	565	400	965
Ground rod, with clamp	1.000	Ea.	35	118	153
Ground cable	.100	C.L.F.	14.30	17.80	32.10
3/4" EMT	10.000	L.F.	10.80	39.80	50.60
Panelboard, 24 circuit	1.000	Ea.	545	665	1,210
TOTAL			1,399.10	1,725.10	3,124.20

D5010 220	Residential Service - Single Phase	COST EACH MAT.	COST EACH INST.	COST EACH TOTAL
2800				
2810	100 Amp Service, single phase	640	1,150	1,790
2820	200 Amp Service, single phase	1,400	1,725	3,125
2900	Cut & patch to match existing construction, add, minimum	2%	3%	
3000	Maximum	5%	9%	
3100	Dust protection, add, minimum	1%	2%	
3200	Maximum	4%	11%	
3300	Equipment usage curtailment, add, minimum	1%	1%	
3400	Maximum	3%	10%	
3500	Material handling & storage limitation, add, minimum	1%	1%	
3600	Maximum	6%	7%	
3700	Protection of existing work, add, minimum	2%	2%	
3800	Maximum	5%	7%	
3900	Shift work requirements, add, minimum		5%	
4000	Maximum		30%	

D5020 Lighting and Branch Wiring

D5020 180	Selective Price Sheet	COST EACH		
		MAT.	INST.	TOTAL
0100	Using non-metallic sheathed, cable, air conditioning receptacle	17.85	52	69.85
0200	Disposal wiring	13.45	57.50	70.95
0300	Dryer circuit	38.50	94	132.50
0400	Duplex receptacle	17.85	40	57.85
0500	Fire alarm or smoke detector	68.50	52	120.50
0600	Furnace circuit & switch	22	86.50	108.50
0700	Ground fault receptacle	47.50	64.50	112
0800	Heater circuit	19.45	64.50	83.95
0900	Lighting wiring	19.40	32.50	51.90
1000	Range circuit	83	129	212
1100	Switches single pole	15.95	32.50	48.45
1200	3-way	21.50	43	64.50
1300	Water heater circuit	25.50	104	129.50
1400	Weatherproof receptacle	126	86.50	212.50
1500	Using BX cable, air conditioning receptacle	31	62.50	93.50
1600	Disposal wiring	26	69	95
1700	Dryer circuit	52.50	113	165.50
1800	Duplex receptacle	31	48	79
1900	Fire alarm or smoke detector	68.50	52	120.50
2000	Furnace circuit & switch	36.50	104	140.50
2100	Ground fault receptacle	59	78.50	137.50
2200	Heater circuit	28	78.50	106.50
2300	Lighting wiring	28	39	67
2400	Range circuit	117	157	274
2500	Switches, single pole	29	39	68
2600	3-way	32	52	84
2700	Water heater circuit	44.50	123	167.50
2800	Weatherproof receptacle	134	104	238
2900	Using EMT conduit, air conditioning receptacle	40	77.50	117.50
3000	Disposal wiring	36	86.50	122.50
3100	Dryer circuit	54	140	194
3200	Duplex receptacle	40	59.50	99.50
3300	Fire alarm or smoke detector	85.50	77.50	163
3400	Furnace circuit & switch	44	129	173
3500	Ground fault receptacle	77	96	173
3600	Heater circuit	36.50	96	132.50
3700	Lighting wiring	35	48.50	83.50
3800	Range circuit	85	192	277
3900	Switches, single pole	38	48.50	86.50
4000	3-way	34.50	64.50	99
4100	Water heater circuit	42.50	152	194.50
4200	Weatherproof receptacle	143	129	272
4300	Using aluminum conduit, air conditioning receptacle	44.50	104	148.50
4400	Disposal wiring	40.50	115	155.50
4500	Dryer circuit	69	185	254
4600	Duplex receptacle	43.50	79.50	123
4700	Fire alarm or smoke detector	97.50	104	201.50
4800	Furnace circuit & switch	54.50	173	227.50
4900	Ground fault receptacle	82	129	211
5000	Heater circuit	44.50	129	173.50
5100	Lighting wiring	49.50	64.50	114
5200	Range circuit	107	259	366
5300	Switches, single pole	56	64.50	120.50
5400	3-way	52.50	86.50	139
5500	Water heater circuit	58.50	207	265.50
5600	Weatherproof receptacle	166	173	339
5700	Using galvanized steel conduit	45.50	110	155.50
5800	Disposal wiring	38	123	161

D5020 Lighting and Branch Wiring

D5020 180	Selective Price Sheet	COST EACH		
		MAT.	INST.	TOTAL
5900	Dryer circuit	75	199	274
6000	Duplex receptacle	45.50	85	130.50
6100	Fire alarm or smoke detector	94	110	204
6200	Furnace circuit & switch	51.50	185	236.50
6300	Ground fault receptacle	81	136	217
6400	Heater circuit	44	136	180
6500	Lighting wiring	49	69	118
6600	Range circuit	106	272	378
6700	Switches, single pole	55	69	124
6800	3-way	55.50	89	144.50
6900	Water heater circuit	57.50	216	273.50
7000	Weatherproof receptacle	161	185	346
7100				
7200				
7300				
7400				
7500				
7600				
7700				
7800				
7900				
8000				
8100				
8200				
8300				
8400				

D5020 Lighting and Branch Wiring

This page illustrates and describes fluorescent lighting systems including a fixture, lamp, outlet box and wiring. Lines within System Components give the unit price and total price on a cost each basis for this system. Prices for alternate fluorescent lighting systems are on Line Items D5020 248 1300 thru 1500. Both material quantities and labor costs have been adjusted for the system listed.

Factors: To adjust for job conditions other than normal working situations use Lines D5020 248 2900 thru 4000.

Example: You are to install the system during evening hours. Go to Line D5020 248 3900 and apply this percentage to the appropriate INST. cost.

System Components	QUANTITY	UNIT	COST EACH		
			MAT.	INST.	TOTAL
Fluorescent lighting, including fixture, lamp, outlet box and wiring.					
Recessed lighting fixture, on suspended system	1.000	Ea.	65	110	175
Outlet box	1.000	Ea.	2.46	29	31.46
#12 wire	.660	C.L.F.	4.65	31.02	35.67
Conduit, EMT, 1/2" conduit	20.000	L.F.	13.20	60.80	74
TOTAL			85.31	230.82	316.13

D5020 248	Lighting, Fluorescent	COST EACH		
		MAT.	INST.	TOTAL
1200	For alternate lighting fixtures:			
1300	Surface mounted, 2' x 4', acrylic prismatic diffuser	126	242	368
1400	Strip fixture, 8' long, two 8' lamps	79	228	307
1500	Pendant mounted, industrial, 8' long, with reflectors	120	263	383
1600				
1700				
1800				
1900				
2000				
2100				
2200				
2300				
2900	Cut & patch to match existing construction, add, minimum	2%	3%	
3000	Maximum	5%	9%	
3100	Dust protection, add, minimum	1%	2%	
3200	Maximum	4%	11%	
3300	Equipment usage curtailment, add, minimum	1%	1%	
3400	Maximum	3%	10%	
3500	Material handling & storage limitation, add, minimum	1%	1%	
3600	Maximum	6%	7%	
3700	Protection of existing work, add, minimum	2%	2%	
3800	Maximum	5%	7%	
3900	Shift work requirements, add, minimum		5%	
4000	Maximum		30%	

D5020 Lighting and Branch Wiring

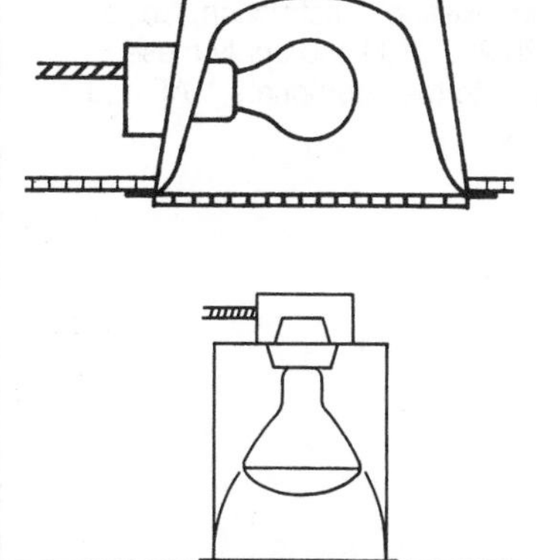

This page illustrates and describes incandescent lighting systems including a fixture, lamp, outlet box, conduit and wiring. Lines within System Components give the unit price and total price on a cost each basis for this system. Prices for an alternate incandescent lighting system are also given. Both material quantities and labor costs have been adjusted for the system listed.

Factors: To adjust for conditions other than normal working situations use Lines D5020 250 2900 thru 4000.

Example: You are to install the system and cut and match existing construction. Go to Line D5020 250 3000 and apply this percentage to the appropriate INST. costs.

System Components	QUANTITY	UNIT	COST EACH MAT.	COST EACH INST.	COST EACH TOTAL
Incandescent light fixture, including lamp, outlet box, conduit and wiring.					
Recessed wide reflector with flat glass lens	1.000	Ea.	90.50	77.50	168
Outlet box	1.000	Ea.	2.46	29	31.46
Armored cable, 3 wire	.200	C.L.F.	20.40	47	67.40
TOTAL			113.36	153.50	266.86
Recessed flood light fixture, including lamp, outlet box, conduit and wiring					
Recessed, R-40 flood lamp with refractor	1.000	Ea.	96	64.50	160.50
150 watt R-40 flood lamp	.010	Ea.	6.40	4	10.40
Outlet box	1.000	Ea.	2.46	29	31.46
Outlet box cover	1.000	Ea.	1.14	8.10	9.24
Romex, 12-2 with ground	.200	C.L.F.	5.90	47	52.90
Conduit, 1/2" EMT	20.000	L.F.	13.20	60.80	74
TOTAL			125.10	213.40	338.50

D5020 250	Lighting, Incandescent	COST EACH MAT.	COST EACH INST.	COST EACH TOTAL
2810	Incandescent light fixture, including lamp, outlet box, conduit and wiring	113	154	267
2820	Recessed flood light fixture, including lamp, outlet box, conduit and wiring	125	213	338
2900	Cut & patch to match existing construction, add, minimum	2%	3%	
3000	Maximum	5%	9%	
3100	Dust protection, add, minimum	1%	2%	
3200	Maximum	4%	11%	
3300	Equipment usage curtailment, add, minimum	1%	1%	
3400	Maximum	3%	10%	
3500	Material handling & storage limitation, add, minimum	1%	1%	
3600	Maximum	6%	7%	
3700	Protection of existing work, add, minimum	2%	2%	
3800	Maximum	5%	7%	
3900	Shift work requirements, add, minimum		5%	
4000	Maximum		30%	

D5020 Lighting and Branch Wiring

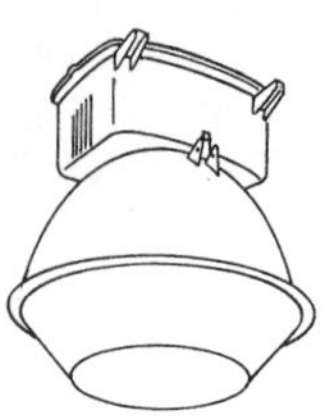

This page illustrates and describes high intensity lighting systems including a lamp, EMT conduit, EMT "T" fitting with cover, and wire. Lines within System Components give the unit price and total price on a cost each basis for this system. Prices for alternate high intensity lighting systems are on Line Items D5020 252 1100 thru 1800. Both material quantities and labor costs have been adjusted for the system listed.

Factors: To adjust for job conditions other than normal working situations use Lines D5020 252 2900 thru 4000.

Example: You are to install the system and protect existing construction. Go to Line D5020 252 3700 and apply these percentages to the appropriate MAT. and INST. costs.

System Components	QUANTITY	UNIT	COST EACH		
			MAT.	INST.	TOTAL
High intensity lighting system consisting of 400 watt metal halide fixture					
And lamp with 1/2" EMT conduit and fittings using #12 wire, high bay.					
Single unit, 400 watt	1.000	Ea.	380	225	605
Electric metallic tubing (EMT), 1/2" diameter	30.000	Ea.	19.80	91.20	111
1/2" EMT "T" fitting with cover	1.000	Ea.	14.75	32.50	47.25
Wire, 600 volt, type THWN-THHN, copper, solid, #12	.600	Ea.	4.23	28.20	32.43
TOTAL			418.78	376.90	795.68

D5020 252	Lighting, High Intensity	COST EACH		
		MAT.	INST.	TOTAL
1000	For alternate high intensity systems:			
1200	High bay: 400 watt, high pressure sodium and lamp	390	375	765
1400	1000 watt, metal halide	585	410	995
1500	High pressure sodium	545	410	955
1700	Low bay: 250 watt, metal halide	410	315	725
1800	150 watt, high pressure sodium	345	315	660
1900				
2000				
2100				
2200				
2300				
2400				
2500				
2600				
2700				
2900	Cut & patch to match existing construction, add, minimum	2%	3%	
3000	Maximum	5%	9%	
3100	Dust protection, add, minimum	1%	2%	
3200	Maximum	4%	11%	
3300	Equipment usage curtailment, add, minimum	1%	1%	
3400	Maximum	3%	10%	
3500	Material handling & storage limitation, add, minimum	1%	1%	
3600	Maximum	6%	7%	
3700	Protection of existing work, add, minimum	2%	2%	
3800	Maximum	5%	7%	
3900	Shift work requirements, add, minimum		5%	
4000	Maximum		30%	

SERVICES D

D5090 Other Electrical Systems

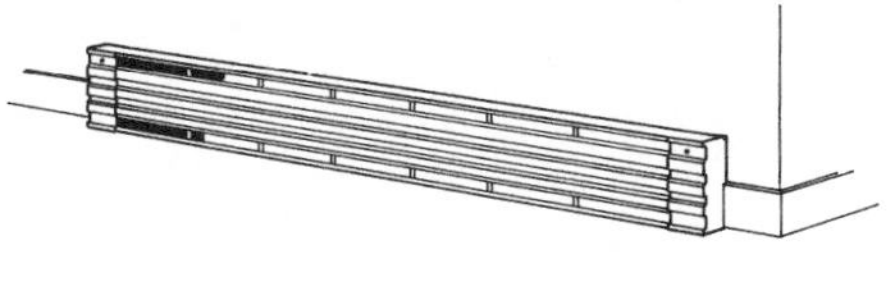

This page illustrates and describes baseboard heat systems including a thermostat, outlet box, breaker, and feed. Lines within System Components give the unit price and total price on a cost each basis for this system. Prices for alternate baseboard heat systems are on Line Items D5090 430 1500 thru 1800. Both material quantities and labor costs have been adjusted for the system listed.

Factors: To adjust for job conditions other than normal working situations use Lines D5090 430 2900 thru 4000.

Example: You are to install the system during evenings and weekends only. Go to Line D5090 430 4000 and apply this percentage to the appropriate INST. cost.

System Components	QUANTITY	UNIT	COST EACH		
			MAT.	INST.	TOTAL
Baseboard heat including thermostat, outlet box, breaker and feed.					
Electric baseboard heater, 4′ long	1.000	Ea.	51	77.50	128.50
Thermostat, integral	1.000	Ea.	19.80	32.50	52.30
Romex, 12-3 with ground	.400	C.L.F.	21.40	103.60	125
Panel board breaker, 20 Amp	1.000	Ea.	31.50	64.50	96
Outlet box	1.000	Ea.	2.46	29	31.46
TOTAL			126.16	307.10	433.26

D5090 530	Heat, Baseboard	COST EACH		
		MAT.	INST.	TOTAL
1400	For alternate baseboard heating systems:			
1500	Electric baseboard, 2′ long	111	294	405
1600	6′ long	142	335	477
1700	8′ long	160	360	520
1800	10′ long	180	385	565
1900				
2000				
2100				
2200				
2300				
2400				
2500				
2600				
2700				
2900	Cut & patch to match existing construction, add, minimum	2%	3%	
3000	Maximum	5%	9%	
3100	Dust protection, add, minimum	1%	2%	
3200	Maximum	4%	11%	
3300	Equipment usage curtailment, add, minimum	1%	1%	
3400	Maximum	3%	10%	
3500	Material handling & storage limitation, add, minimum	1%	1%	
3600	Maximum	6%	7%	
3700	Protection of existing work, add, minimum	2%	2%	
3800	Maximum	5%	7%	
3900	Shift work requirements, add, minimum		5%	
4000	Maximum		30%	

For information about Means Estimating Seminars, see yellow pages 12 and 13 in back of book

E EQUIPMENT & FURNISHINGS

E1090 Other Equipment

This page illustrates and describes kitchen systems including top and bottom cabinets, custom laminated plastic top, single bowl sink, and appliances. Lines within System Components give the unit price and total price on a cost each basis for this system. Prices for alternate kitchen systems are on Line Items E1090 310 1500 and 1600. Both material quantities and labor costs have been adjusted for the system listed.

Factors: To adjust for job conditions other than normal working situations use Lines E1090 310 2900 thru 4000.

Example: You are to install the system and protect the work area from dust. Go to Line E1090 310 3200 and apply these percentages to the appropriate MAT. and INST. costs.

System Components	QUANTITY	UNIT	COST EACH MAT.	COST EACH INST.	COST EACH TOTAL
Kitchen cabinets including wall and base cabinets, custom laminated Plastic top,sink & appliances,no plumbing or electrical rough-in included.					
Prefinished wood cabinets, average quality, wall and base	20.000	L.F.	2,100	630	2,730
Custom laminated plastic counter top	20.000	L.F.	195	314	509
Stainless steel sink, 22″ x 25″	1.000	Ea.	465	171	636
Faucet, top mount	1.000	Ea.	56.50	53	109.50
Dishwasher, built-in	1.000	Ea.	282	262	544
Compactor, built-in	1.000	Ea.	470	94	564
Range hood, vented, 30″ wide	1.000	Ea.	42.50	200	242.50
TOTAL			3,611	1,724	5,335

E1090 310	Kitchens	COST EACH MAT.	COST EACH INST.	COST EACH TOTAL
1400	For alternate kitchen systems:			
1500	Prefinished wood cabinets, high quality	7,200	1,975	9,175
1600	Custom cabinets, built in place, high quality	9,325	2,275	11,600
1700				
1800				
1900				
2000				
2100				
2200	NOTE: No plumbing or electric rough-ins are included in the above			
2300	Prices, for plumbing see Division 15, for electric see Division 16.			
2400				
2500				
2600				
2700				
2900	Cut & patch to match existing construction, add, minimum	2%	3%	
3000	Maximum	5%	9%	
3100	Dust protection, add, minimum	1%	2%	
3200	Maximum	4%	11%	
3300	Equipment usage curtailment, add, minimum	1%	1%	
3400	Maximum	3%	10%	
3500	Material handling & storage limitation, add, minimum	1%	1%	
3600	Maximum	6%	7%	
3700	Protection of existing work, add, minimum	2%	2%	
3800	Maximum	5%	7%	
3900	Shift work requirements, add, minimum		5%	
4000	Maximum		30%	

E1090 Other Equipment

E1090 315	Selective Price Sheet	COST EACH		
		MAT.	INST.	TOTAL
0100	Cabinets standard wood, base, one drawer one door, 12″ wide	152	38	190
0200	15″ wide	207	39.50	246.50
0300	18″ wide	225	40.50	265.50
0400	21″ wide	234	41.50	275.50
0500	24″ wide	270	42.50	312.50
0600				
0700	Two drawers two doors, 27″ wide	299	43	342
0800	30″ wide	320	44	364
0900	33″ wide	330	45	375
1000	36″ wide	345	46.50	391.50
1100	42″ wide	370	47.50	417.50
1200	48″ wide	395	50	445
1300	Drawer base (4 drawers), 12″ wide	360	38	398
1400	15″ wide	276	39.50	315.50
1500	18″ wide	310	40.50	350.50
1600	24″ wide	335	42.50	377.50
1700	Sink or range base, 30″ wide	264	44	308
1800	33″ wide	283	45	328
1900	36″ wide	296	46.50	342.50
2000	42″ wide	315	47.50	362.50
2100	Corner base, 36″ wide	435	52.50	487.50
2200	Lazy susan with revolving door	420	57	477
2300	Cabinets standard wood, wall two doors, 12″ high, 30″ wide	166	38	204
2400	36″ wide	193	39.50	232.50
2500	15″ high, 30″ high	174	39.50	213.50
2600	36″ wide	197	41.50	238.50
2700	24″ high, 30″ wide	216	40.50	256.50
2800	36″ wide	239	41.50	280.50
2900	30″ high, 30″ wide	244	49	293
3000	36″ wide	279	50	329
3100	42″ wide	305	51	356
3200	48″ wide	340	51	391
3300	One door, 30″ high, 12″ wide	146	43	189
3400	15″ wide	165	44	209
3500	18″ wide	181	45	226
3600	24″ wide	204	46.50	250.50
3700	Corner, 30″ high, 24″ wide	167	52.50	219.50
3800	36″ wide	215	57	272
3900	Broom, 84″ high, 24″ deep, 18″ wide	445	94	539
4000	Oven, 84″ high, 24″ deep, 27″ wide	650	118	768
4100	Valance board, 4′ long	37	9.50	46.50
4200	6″ long	55.50	14.30	69.80
4300	Counter tops, laminated plastic, stock 25″ wide w/backsplash, min.	9.75	15.70	25.45
4400	Maximum	17.60	18.85	36.45
4500	Custom, 7/8″ thick, no splasn	19.45	15.70	35.15
4600	Cove splash	25.50	15.70	41.20
4700	1-1/4″ thick, no splash	22	16.85	38.85
4800	Square splash	28	16.85	44.85
4900	Post formed	10	15.70	25.70
5000				
5100	Maple laminated 1-1/2″ thick, no splash	58	16.85	74.85
5200	Square splash	69	16.85	85.85
5300				
5400				
5500	Appliances, range, free standing, minimum	274	72.50	346.50
5600	Maximum	1,700	182	1,882
5700	Built-in, minimum	505	86.50	591.50
5800	Maximum	1,500	470	1,970

E1090 Other Equipment

E1090 315	Selective Price Sheet	COST EACH		
		MAT.	INST.	TOTAL
5900	Counter top range 4 burner, maximum	211	86.50	297.50
6000	Maximum	590	173	763
6100	Compactor, built-in, minimum	470	94	564
6200	Maximum	530	157	687
6300	Dishwasher, built-in, minimum	282	262	544
6400	Maximum	330	525	855
6500	Garbage disposer, sink-pipe, minimum	48	105	153
6600	Maximum	165	105	270
6700	Range hood, 30″ wide, 2 speed, minimum	42.50	200	242.50
6800	Maximum	680	335	1,015
6900	Refrigerator, no frost, 12 cu. ft.	500	72.50	572.50
7000	20 cu. ft.	775	121	896
7100	Plumb. not incl. rough-ins, sinks porc. C.I., single bowl, 21″ x 24″	255	171	426
7200	21″ x 30″	239	171	410
7300	Double bowl, 20″ x 32″	345	199	544
7400				
7500	Stainless steel, single bowl, 19″ x 18″	415	171	586
7600	22″ x 25″	465	171	636
7700				
7800				
7900				
8000				
8100				
8200				
8300				
8400				

For information about Means Estimating Seminars, see yellow pages 12 and 13 in back of book

G BUILDING SITEWORK

G1030 Site Clearing

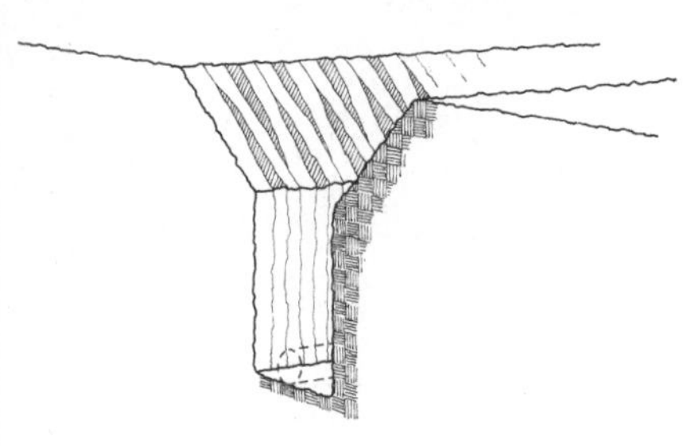

This page illustrates and describes utility trench excavation with backfill systems including a trench, backfill, excavated material, utility pipe or ductwork not included. Lines within System Components give the unit price and total price on a linear foot basis for this system. Prices for alternate utility trench excavation with backfill systems are on Line Items G1030 810 1100 thru 2500. Both material quantities and labor costs have been adjusted for the system listed.

Factors: To adjust for job conditions other than normal working situations use Lines G1030 810 3100 thru 4000.

Example: You are to install the system and protect existing construction. Go to Line G1030 810 3600 and apply these percentages to the appropriate TOTAL cost.

System Components	QUANTITY	UNIT	COST PER L.F. EQUIP.	LABOR	TOTAL
Cont. utility trench excav. with 3/8 C.Y. wheel mtd. backhoe, Backfilling with excav. mat. and comp. in 12″ lifts. Trench is 2′ wide by 4′ dp. Cost of utility piping or ductwork is not incl. Cost based on an Excavation production rate of 150 C.Y. daily no hauling included.					
Machine excavate trench, 2′ wide by 4′ deep	.296	C.Y.	.49	1.63	2.12
Dozer backfill with excavated material	.296	C.Y.	.25	.45	.70
Compact in 12″ lifts, vibrating plate	.296	C.Y.	.06	1.84	1.90
TOTAL			.80	3.92	4.72

G1030 810	Excavation, Utility Trench	COST PER L.F. EQUIP.	LABOR	TOTAL
1000	Alternate size trenches:			
1100	2′ wide x 2′ deep	.40	1.95	2.35
1200	3′ deep	.60	2.94	3.54
1300	With sloping sides, 2′ wide x 5′ deep	2.24	11	13.25
1400	6′ deep	2.99	14.70	17.70
1500	7′ deep	3.85	18.85	22.50
1600	8′ deep	4.80	23.50	28.50
1700	9′ deep	5.85	28.50	34.50
1800	10′ deep	7	34.50	41.50
2000	For hauling excavated material up to 2 miles & backfilling w/ gravel			
2100	Gravel, 2′ wide by 2′ deep, add	3.33	2.94	6.27
2200	4′ deep, add	6.65	5.85	12.50
2300	6′ deep, add	25	22	47
2400	8′ deep, add	40	35	75
2500	10′ deep, add	58.50	51.50	110
2600				
2700				
2800	For shallow, hand excavated trenches, no backfill, to 6′ dp, light soil		45.50	45.50
2900	Heavy soil		91	91
3000				
3100	Equipment usage curtailment, add, minimum	1%	1%	
3200	Maximum	3%	10%	
3300	Material handling & storage limitation, add, minimum	1%	1%	
3400	Maximum	6%	7%	
3500	Protection of existing work, add, minimum	2%	2%	
3600	Maximum	5%	7%	
3700	Shift work requirements, add, minimum		5%	
3800	Maximum		30%	
3900	Temporary shoring and bracing, add, minimum	2%	5%	
4000	Maximum	5%	12%	

G2010 Roadways

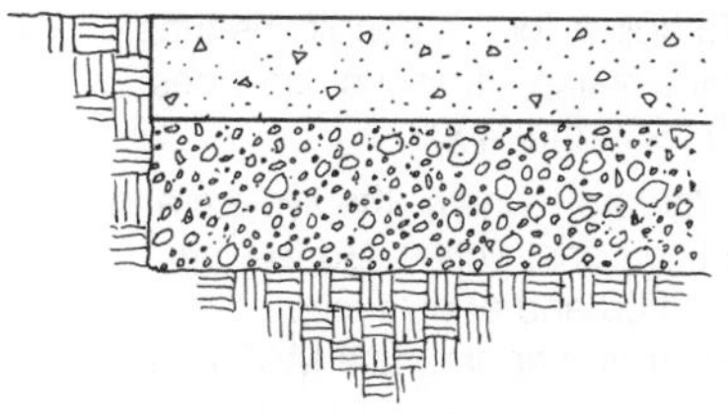

This page illustrates and describes driveway systems including concrete slab, gravel base, broom finish, compaction, and joists. Lines within System Components give the unit price and total price on a cost each basis for this system. Prices for alternate driveway systems are on Line Items G2010 240 1300 thru 2700. Both material quantities and labor costs have been adjusted for the system listed.

Factors: To adjust for job conditions other than normal working situations use Lines G2010 240 2900 thru 4000.

Example: You are to install the system and match existing construction. Go to Line G2010 240 3000 and apply these percentages to the appropriate MAT. and INST. costs.

System Components	QUANTITY	UNIT	COST EACH		
			MAT.	INST.	TOTAL
Complete driveway, 10′ x 30′, 4″ concrete slab, 3000 psi, on 6″ compacted Gravel base, concrete broom finished and cured with 10′ x 10′ joints.					
Grade and compact subgrade	33.000	S.Y.		52.14	52.14
Place and compact 6″ crushed stone base	33.000	S.Y.	194.70	33.33	228.03
Place and remove edgeforms (4 use)	80.000	L.F.	25.60	236.80	262.40
Concrete for slab, 3000 psi	4.000	C.Y.	382		382
Place concrete, 4″ thick slab, direct chute	4.000	C.Y.		83.72	83.72
Screed, float and broom finish slab	4.000	C.Y.		207	207
Spray on membrane curing compound	3.000	S.F.	18.30	22.95	41.25
Saw cut 3″ deep joints	60.000	L.F.	25.20	69	94.20
TOTAL			645.80	704.94	1,350.74

G2010 240	Driveways	COST EACH		
		MAT.	INST.	TOTAL
1200	For alternate driveway systems:			
1300	Concrete: 10′ x 30′ with 6″ concrete on 6″ crushed stone	835	745	1,580
1400	20′ x 30′ with 4″ concrete	1,325	1,325	2,650
1500	With 6″ concrete	1,700	1,400	3,100
1600	10′ wide, for each additional 10′ length over 30′, 4″ thick, add	210	225	435
1700	6″ thick, add	282	241	523
1800	20′ wide, for each additional 10′ length over 30′, 4″ thick, add	425	425	850
1900	6″ thick, add	545	450	995
2000	Asphalt: 10′ x 30′ with 2″ binder, 1″ topping on 6″ cr. st., sealed	425	450	875
2100	3″ binder, 1″ topping	490	460	950
2200	20′ x 30′ with 2″ binder, 1″ topping	845	635	1,480
2300	3″ binding, 1″ topping	980	655	1,635
2400	10′ wide for each add'l. 10′ length over 30′, 2″ b + 1″ t, add	141	62	203
2500	3″ binder, 1″ topping	164	65	229
2600	20′ wide for each add'l. 10′ length over 30′, 2″ b + 1″ t, add	282	124	406
2700	3″ binder, 1″ topping	325	130	455
2800				
2900	Cut & patch to match existing construction, add, minimum	2%	3%	
3000	Maximum	5%	9%	
3100	Dust protection, add, minimum	1%	2%	
3200	Maximum	4%	11%	
3300	Equipment usage curtailment, add, minimum	1%	1%	
3400	Maximum	3%	10%	
3500	Material handling & storage limitation, add, minimum	1%	1%	
3600	Maximum	6%	7%	
3700	Protection of existing work, add, minimum	2%	2%	
3800	Maximum	5%	7%	
3900	Shift work requirements, add, minimum		5%	
4000	Maximum		30%	

G2020 Parking Lots

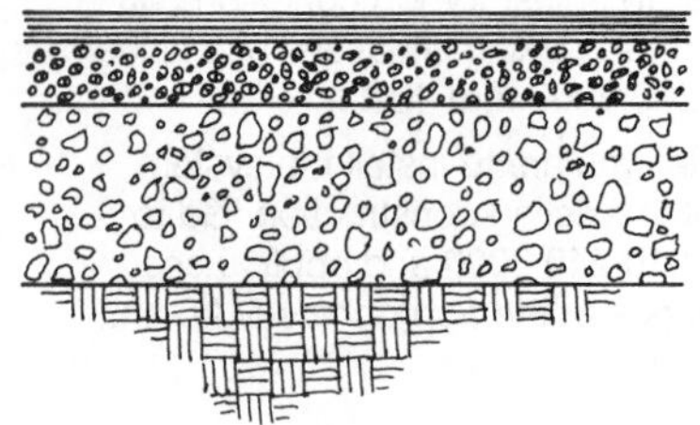

This page illustrates and describes asphalt parking lot systems including asphalt binder, topping, crushed stone base, painted parking stripes and concrete parking blocks. Lines within System Components give the unit price and total price per square yard for this system. Prices for alternate asphalt parking lot systems are on Line Items G2020 230 1500 thru 2500. Both material quantities and labor costs have been adjusted for the system listed.

Factors: To adjust for job conditions other than normal working situations use Lines G2020 230 2900 thru 4000.

Example: You are to install the system and match existing construction. Go to Line G2020 230 2900 and apply these percentages to the appropriate MAT. and INST. costs.

System Components	QUANTITY	UNIT	COST PER S.Y. MAT.	INST.	TOTAL
Parking lot consisting of 2″ asphalt binder and 1″ topping on 6″ Crushed stone base with painted parking stripes and concrete parking blocks					
Fine grade and compact subgrade	1.000	S.Y.		1.58	1.58
6″ crushed stone base, stone	.320	Ton	5.90	1.01	6.91
2″ asphalt binder	1.000	S.Y.	4.24	1.05	5.29
1″ asphalt topping	1.000	S.Y.	2.25	.69	2.94
Paint parking stripes	.500	L.F.	.10	.07	.17
6″ x 10″ x 6′ precast concrete parking blocks	.020	Ea.	.88	.31	1.19
Mobilization of equipment	.005	Ea.		1.32	1.32
TOTAL			13.37	6.03	19.40

G2020 230	Parking Lots, Asphalt	COST PER S.Y. MAT.	INST.	TOTAL
1400	For alternate parking lot systems:			
1500	Above system on 9″ crushed stone	16.35	6.10	22.45
1600	12″ crushed stone	19.30	6.20	25.50
1700	On bank run gravel, 6″ deep	11.35	5.60	16.95
1800	9″ deep	13.30	5.75	19.05
1900	12″ deep	15.25	5.85	21.10
2000	3″ binder plus 1″ topping on 6″ crushed stone	15.45	6.35	21.80
2100	9″ deep crushed stone	18.45	6.40	24.85
2200	12″ deep crushed stone	21.50	6.50	28
2300	On bank run gravel, 6″ deep	13.40	5.90	19.30
2400	9″ deep	15.40	6.05	21.45
2500	12″ deep	17.35	6.15	23.50
2600				
2700				
2900	Cut & patch to match existing construction, add, minimum	2%	3%	
3000	Maximum	5%	9%	
3100	Dust protection, add, minimum	1%	2%	
3200	Maximum	4%	11%	
3300	Equipment usage curtailment, add, minimum	1%	1%	
3400	Maximum	3%	10%	
3500	Material handling & storage limitation, add, minimum	1%	1%	
3600	Maximum	6%	7%	
3700	Protection of existing work, add, minimum	2%	2%	
3800	Maximum	5%	7%	
3900	Shift work requirements, add, minimum		5%	
4000	Maximum		30%	

Important: See the Reference Section for critical supporting data - Reference Numbers and City Cost Indexes

G2020 Parking Lots

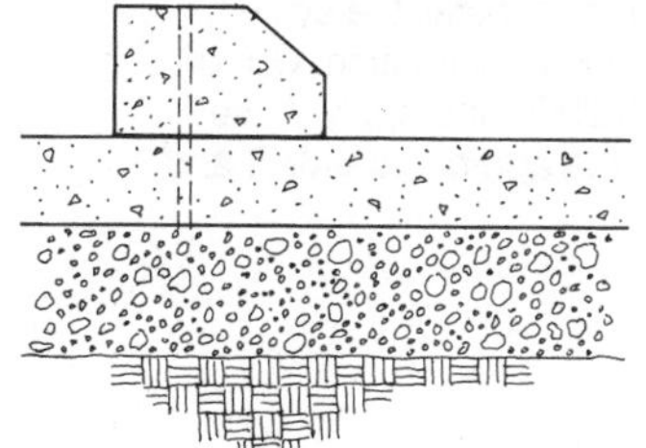

This page illustrates and describes concrete parking lot systems including a concrete slab, unreinforced, compacted gravel base, broom finished, cured, joints, painted parking stripes, and parking blocks. Lines within System Components give the unit price and total price per square yard for this system. Prices for alternate concrete parking lot systems are on Line Items G2020 240 1500 thru 2500. All materials have been adjusted according to the system listed.

Factors: To adjust for job conditions other than normal working situations use Lines G2020 240 2900 thru 4000.

Example: You are to install the system and protect existing construction. Go to Line G2020 240 3800 and apply these percentages to the appropriate MAT. and INST. costs.

System Components	QUANTITY	UNIT	COST PER S.Y.		
			MAT.	INST.	TOTAL
Parking lot, 4″ slab, 3000 psi concrete on 6″ compacted gravel base					
With 12′ x 12′ joints, painted parking strips, conc. parking blocks.					
Grade and compact subgrade	1.000	S.Y.		1.58	1.58
6″ crushed stone base, stone	.320	Ton	5.90	1.01	6.91
Place and remove edge forms, 4 use	.250	L.F.	.08	.74	.82
Concrete for slab, 3000 psi	1.000	S.F.	10.51		10.51
Place concrete, 4″ thick slab, direct chute	1.000	S.F.		2.31	2.31
Spray on membrane curing compound	1.000	S.F.	.55	.48	1.03
Saw cut 3″ deep joints	2.250	L.F.	.95	2.59	3.54
Paint parking stripes	.500	L.F.	.10	.07	.17
6″ x 10″ x 6′ precast concrete parking blocks	.020	Ea.	.88	.31	1.19
TOTAL			18.97	9.09	28.06

G2020 240	Parking Lots, Concrete	COST PER S.Y.		
		MAT.	INST.	TOTAL
1400	For alternate parking lot systems:			
1500	Above system on 3″ crushed stone	18.95	9.10	28.05
1600	9″ crushed stone	22	9.20	31.20
1700	On bank run gravel, 6″ deep	16.95	8.65	25.60
1800	9″ deep	18.90	8.75	27.65
1900	On compacted subgrade only	13.05	8.05	21.10
2000	6″ concrete slab, 3000 psi on 3″ crushed stone	24	10.15	34.15
2100	6″ crushed stone	24	10.15	34.15
2200	9″ crushed stone	27	10.25	37.25
2300	On bank run gravel, 6″ deep	22	9.80	31.80
2400	9″ deep	24	9.90	33.90
2500	On compacted subgrade only	24	10.15	34.15
2600				
2700				
2900	Cut & patch to match existing construction, add, minimum	2%	3%	
3000	Maximum	5%	9%	
3100	Dust protection, add, minimum	1%	2%	
3200	Maximum	4%	11%	
3300	Equipment usage curtailment, add, minimum	1%	1%	
3400	Maximum	3%	10%	
3500	Material handling & storage limitation, add, minimum	1%	1%	
3600	Maximum	6%	7%	
3700	Protection of existing work, add, minimum	2%	2%	
3800	Maximum	5%	7%	
3900	Shift work requirements, add, minimum		5%	
4000	Maximum		30%	

G2030 Pedestrian Paving

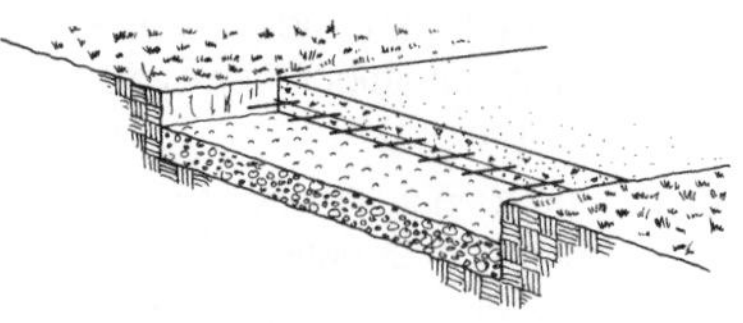

This page illustrates and describes sidewalk systems including concrete, welded wire and broom finish. Lines within System Components give unit price and total price per square foot for this system. Prices for alternate sidewalk systems are on Line Items G2030 105 1700 thru 1900. Both material quantities and labor costs have been adjusted for the system listed.

Factors: To adjust for job conditions other than normal working situations use Lines G2030 105 2900 thru 4000.

Example: You are to install the system and match existing construction. Go to Line G2030 105 2900 and apply these percentages to the appropriate MAT. and INST. costs.

System Components	QUANTITY	UNIT	COST PER S.F.		
			MAT.	INST.	TOTAL
4″ thick concrete sidewalk with welded wire fabric					
3000 psi air entrained concrete, broom finish.		S.F.			
Gravel fill, 4″ deep	.012	C.Y.	.24	.07	.31
Compact fill	.012	C.Y.		.02	.02
Hand grade	1.000	S.F.		1.64	1.64
Edge form	.250	L.F.	.08	.74	.82
Welded wire fabric	.011	S.F.	.15	.34	.49
Concrete, 3000 psi air entrained	.012	C.Y.	1.15		1.15
Place concrete	.012	C.Y.		.26	.26
Broom finish	1.000	S.F.		.69	.69
TOTAL			1.62	3.76	5.38

G2030 105	Sidewalks	COST PER S.F.		
		MAT.	INST.	TOTAL
1600	For alternate sidewalk systems:			
1700	Asphalt (bituminous), 2″ thick	.73	2.10	2.83
1800	Brick, on sand, bed, 4.5 brick per S.F.	2.78	8.35	11.13
1900	Flagstone, slate, 1″ thick, rectangular	7.85	9.05	16.90
2000				
2100				
2200				
2300				
2400				
2500				
2600				
2700				
2900	Cut & patch to match existing construction, add, minimum	2%	3%	
3000	Maximum	5%	9%	
3100	Dust protection, add, minimum	1%	2%	
3200	Maximum	4%	11%	
3300	Equipment usage curtailment, add, minimum	1%	1%	
3400	Maximum	3%	10%	
3500	Material handling & storage limitation, add, minimum	1%	1%	
3600	Maximum	6%	7%	
3700	Protection of existing work, add, minimum	2%	2%	
3800	Maximum	5%	7%	
3900	Shift work requirements, add, minimum		5%	
4000	Maximum		30%	

G2050 Landscaping

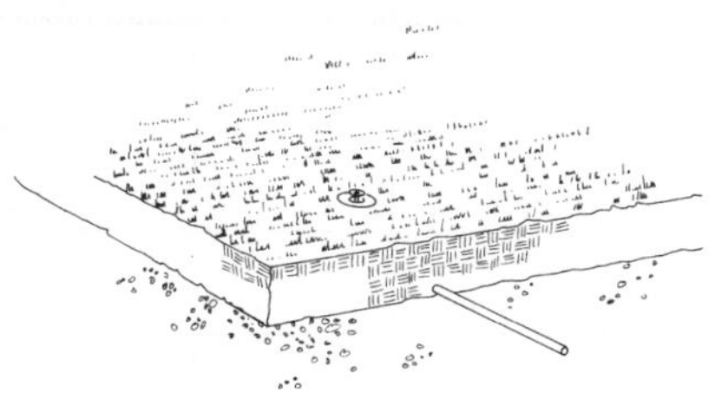

This page describes landscaping—lawn establishment systems including loam, lime, fertilizer, and side and top mulching. Lines within system components give the unit price and total price per square yard for this system. Prices for alternate landscaping—lawn establishment systems are on Line Items G2050 420 1100 and 1200. Both material quantities and labor costs have been adjusted for the system listed.

Factors: To adjust for job conditions other than normal working situations use Lines G2050 420 2900 thru 4000.

Example: You are to install the system and provide dust protection. Go to Line G2050 420 3200 and apply these percentages to the appropriate MAT. and INST. costs.

System Components	QUANTITY	UNIT	COST PER S.Y. MAT.	INST.	TOTAL
Establishing lawns with loam, lime, fertilizer, seed and top mulching					
On rough graded areas.					
Furnish and place loam 4″ deep	.110	C.Y.	2.98	.70	3.68
Fine grade, lime, fertilize and seed	1.000	S.Y.	.40	2.54	2.94
Hay mulch, 1 bale/M.S.F.	1.000	S.Y.	.09	.45	.54
Rolling with hand roller	1.000	S.Y.		.79	.79
TOTAL			3.47	4.48	7.95

G2050 420	Landscaping - Lawn Establishment	COST PER S.Y. MAT.	INST.	TOTAL
1000	For alternate lawn systems:			
1100	Above system with jute mesh in place of hay mulch	4.21	4.56	8.77
1200	Above system with sod in place of seed	5.30	1.59	6.89
1300				
1400				
1500				
1600				
1700				
1800				
1900				
2000				
2100				
2200				
2300				
2400				
2500				
2600				
2700				
2900	Cut & patch to match existing construction, add, minimum	2%	3%	
3000	Maximum	5%	9%	
3100	Dust protection, add, minimum	1%	2%	
3200	Maximum	4%	11%	
3300	Equipment usage curtailment, add, minimum	1%	1%	
3400	Maximum	3%	10%	
3500	Material handling & storage limitation, add, minimum	1%	1%	
3600	Maximum	6%	7%	
3700	Protection of existing work, add, minimum	2%	2%	
3800	Maximum	5%	7%	
3900	Shift work requirements, add, minimum		5%	
4000	Maximum		30%	

G2050 Landscaping

G2050 510	Selective Price Sheet	COST PER UNIT		
		MAT.	INST.	TOTAL
0100	Shrubs and trees, Evergreen, in prepared beds			
0200				
0300	Arborvitae pyramidal, 4'-5'	48	72	120
0400	Globe, 12"-15"	12.15	11.65	23.80
0500	Cedar, blue, 8'-10'	201	120	321
0600	Hemlock, Canadian, 2-1/2'-3'	25.50	31	56.50
0700	Juniper, andora, 18"-24"	17.50	13.95	31.45
0800	Wiltoni, 15"-18"	16.75	13.95	30.70
0900	Skyrocket, 4-1/2'-5'	53.50	39	92.50
1000	Blue pfitzer, 2'-2-1/2'	27.50	25.50	53
1100	Ketleerie, 2-1/2'-3'	35	22.50	57.50
1200	Pine, black, 2-1/2'-3'	43	22.50	65.50
1300	Mugo, 18"-24"	37.50	18.60	56.10
1400	White, 4'-5'	55.50	28.50	84
1500	Spruce, blue, 18"-24"	42	18.60	60.60
1600	Norway, 4'-5'	94.50	28.50	123
1700	Yew, denisforma, 12"-15"	25	18.60	43.60
1800	Capitata, 18"-24"	21.50	37	58.50
1900	Hicksi, 2'-2-1/2'	32	37	69
2000				
2100	Trees, Deciduous, in prepared beds			
2200				
2300	Beech, 5'-6'	244	43	287
2400	Dogwood, 4'-5'	74.50	53.50	128
2500	Elm, 8'-10'	123	108	231
2600	Magnolia, 4'-5'	78.50	108	186.50
2700	Maple, red, 8'-10', 1-1/2" caliper	175	215	390
2800	Oak, 2-1/2"-3" caliper	272	360	632
2900	Willow, 6'-8', 1" caliper	67	108	175
3000				
3100	Shrubs, Broadleaf Evergreen, in prepared beds			
3200				
3300	Andromeda, 15"-18", container	25	11.65	36.65
3400	Azalea, 15"-18", container	27.50	11.65	39.15
3500	Barberry, 9"-12", container	11.05	8.60	19.65
3600	Boxwood, 15"-18", B & B	29.50	11.65	41.15
3700	Euonymus, emerald gaiety, 12"-15", container	17.85	9.70	27.55
3800	Holly, 15"-18", B & B	17.40	11.65	29.05
3900	Mount laurel, 18"-24", B & B	55	13.95	68.95
4000	Privet, 18"-24", B & B	17.85	8.60	26.45
4100	Rhodendron, 18"-24", container	30.50	23.50	54
4200	Rosemary, 1 gal. container	65	1.86	66.86
4300	Deciduous, amalanchier, 2'-3', B & B	88.50	19.60	108.10
4400	Azalea, 15"-18", B & B	23.50	11.65	35.15
4500	Bayberry, 2'-3', B & B	27	19.60	46.60
4600	Cotoneaster, 15"-18", B & B	16.15	13.95	30.10
4700	Dogwood, 3'-4', B & B	25.50	53.50	79
4800	Euonymus, alatus compacta, 15"-18", container	21	13.95	34.95
4900	Forsythia, 2'-3', container	18.70	18.60	37.30
5000	Honeysuckle, 3'-4', B & B	21	18.60	39.60
5100	Hydrangea, 2'-3', B & B	24	19.60	43.60
5200	Lilac, 3'-4', B & B	25	53.50	78.50
5300	Quince, 2'-3', B & B	21.50	19.60	41.10
5400				
5500	Ground Cover			
5600				
5700	Plants, Pachysandra, prepared beds, per hundred	28.50	74.50	103
5800	Vinca minor or English ivy, per hundred	29.50	93	122.50

G2050 Landscaping

G2050 510	Selective Price Sheet	COST PER UNIT		
		MAT.	INST.	TOTAL
5900	Plant bed prep. 18″ dp., mach., per square foot	2.25	.33	2.58
6000	By hand, per square foot	2.25	2.34	4.59
6100	Stone chips, Georgia marble, 50# bags, per bag	2.68	2.15	4.83
6200	Onyx gemstone, per bag	19.45	4.29	23.74
6300	Quartz, per bag	7.25	4.29	11.54
6400	Pea gravel, truck load lots, per cubic yard	27.50	40	67.50
6500	Mulch, polyethylene mulch, per square yard	.19	.36	.55
6600	Wood chips, 2″ deep, per square yard	1.95	1.65	3.60
6700	Peat moss, 1″ deep, per square yard	1.89	.40	2.29
6800	Erosion control per square yard, Jute mesh stapled	.83	.53	1.36
6900	Plastic netting, stapled	.75	.45	1.20
7000	Polypropylene mesh, stapled	1.54	.45	1.99
7100	Tobacco netting, stapled	.09	.45	.54
7200	Lawns per square yard, seeding incl. fine grade, limestone, fert. and seed	.40	2.54	2.94
7300	Sodding, incl. fine grade, level ground	245	94	339
7400	On slope			
7500	Edging, per linear foot, Redwood, untreated, 1″ x 4″	1.21	1.88	3.09
7600	2″ x 4″	2.42	2.86	5.28
7700	Stl. edge strips, 1/4″ x 5″ inc. stakes	4.28	2.86	7.14
7800	3/16″ x 4″	3.42	2.86	6.28
7900	Brick edging, set on edge	2.27	6.20	8.47
8000	Set flat	1.13	2.26	3.39
8100				
8200				
8300				
8400				

G3020 Sanitary Sewer

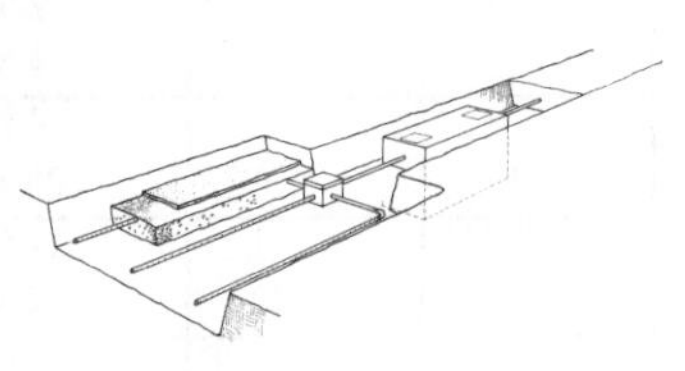

This page illustrates and describes drainage and utilities – septic system including a tank, distribution box, excavation, piping, crushed stone and backfill. Lines within System Components give the unit price and total price on a cost each basis for this system. Prices for alternate drainage and utilities – septic systems are on Line Items G3020 610 1700 thru 2000. Both material quantities and labor costs have been adjusted for the system listed.

Factors: To adjust for job conditions other than normal working situations use Lines G3020 610 3100 thru 4000.

Example: You are to install the system with a material handling limitation. Go to Line G3020 610 3500 and apply these percentages to the appropriate MAT. and INST. costs.

System Components	QUANTITY	UNIT	COST EACH MAT.	COST EACH INST.	COST EACH TOTAL
Septic system including tank, distribution box, excavation, piping, crushed Stone and backfill for a 1000 S.F. leaching field.					
Precast concrete septic tank, 1000 gal. capacity	1.000	Ea.	645	205	850
Concrete distribution box	1.000	Ea.	120	45.50	165.50
Sch. 40 sewer pipe, PVC, 4″ diameter	25.000	L.F.	472.60	557.60	1,030.20
Sch. 40 sewer pipe fittings, PVC, 4″ diameter	8.000	Ea.	48.80	824	872.80
Excavation for septic tank	120.000	C.Y.		169.56	169.56
Excavation for disposal field	120.000	C.Y.		413.09	413.09
Crushed stone backfill	76.000	C.Y.	1,379.40	718.96	2,098.36
Backfill with excavated material	26.000	C.Y.		35.88	35.88
Building paper	6.000	C.S.F.	14.46	76.50	90.96
TOTAL			2,680.26	3,046.09	5,726.35

G3020 610	Septic Systems	COST EACH MAT.	COST EACH INST.	COST EACH TOTAL
1600	For alternate septic systems:			
1700	1000 gal. tank with 2000 S.F. field	4,000	4,325	8,325
1800	With leaching pits	2,475	920	3,395
1900	2000 gal. tank with 2000 S.F. field	4,625	4,450	9,075
2000	With leaching pits	3,225	1,350	4,575
2100				
2200				
2300				
2400				
2500				
2600				
2700				
2800				
2900				
3100	Dust protection, add, minimum	1%	2%	
3200	Maximum	4%	11%	
3300	Equipment usage curtailment, add, minimum	1%	1%	
3400	Maximum	3%	10%	
3500	Material handling & storage limitation, add, minimum	1%	1%	
3600	Maximum	6%	7%	
3700	Protection of existing work, add, minimum	2%	2%	
3800	Maximum	5%	7%	
3900	Shift work requirements, add, minimum		5%	
4000	Maximum		30%	

Important: See the Reference Section for critical supporting data - Reference Numbers and City Cost Indexes

G3060 Fuel Distribution

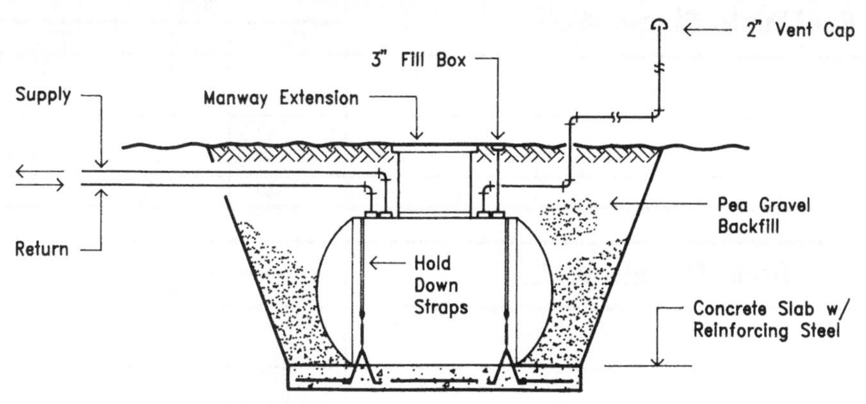

Fiberglass Underground Fuel Storage Tank, Single Wall

This page illustrates and describes fiberglass tanks including hold down slab, reinforcing steel and peastone gravel. Lines within Systems Components give the material and installation price on a cost each basis for the components. Prices for alternate fiberglass tank systems are on Line Items G3060 305 1010 thru G3060 310 1100. Material quantities and labor costs have been adjusted for the system listed.

Factors: To adjust for job conditions other than normal working situations use Lines G3060 310 2700 thru 4000.

System Components	QUANTITY	UNIT	COST EACH MAT.	COST EACH INST.	COST EACH TOTAL
Single wall underground fiberglass storage tanks including, excavation, hold down slab, backfill, manway extension and related piping.					
Tank, fiberglass, underground, single wall, U.L. listed, 550 gal	1.000	Ea.	2,575	360	2,935
Tank, fiberglass, manways, add	1.000	Ea.	1,375		1,375
Tank, for hold-downs 500-4000 gal, add	1.000	Ea.	310	128	438
Foot valve, single poppet, 3/4" dia	1.000	Ea.	41.50	30	71.50
Tubing, copper, Type L, 3/4" dia	120.000	L.F.	343.20	840	1,183.20
Elbow 90°, copper, wrought, cu x cu, 3/4" dia.	8.000	Ea.	10	224	234
Fuel oil specialties, valve, ball chk, globe type, fusible, 3/4" dia	2.000	Ea.	92	54	146
Elbow 90°, black steel, straight, 2" dia.	3.000	Ea.	39	159	198
Union, black steel, with brass seat, 2" dia.	1.000	Ea.	25	56	81
Pipe and nipples, steel, Sch. 40, threaded, black, 2" dia	2.000	Ea.	296.40	568.10	864.50
Vent protector / breather, 2" dia.	1.000	Ea.	17.60	16.75	34.35
Pipe, black steel, welded, Sch. 40, 3" dia.	30.000	L.F.	462	713.10	1,175.10
Elbow 90°, steel, weld joint, butt, 3" dia.	3.000	Ea.	63	443.55	506.55
Flange, weld neck, 150 lb., 3" pipe size	1.000	Ea.	23.50	73.95	97.45
Fuel fill box, locking inner cover, 3" dia.	1.000	Ea.	130	107	237
Remote tank gauging system, 30', 5" pointer travel	1.000	Ea.	4,050	268	4,318
Tank leak detector system, 8 channel, external monitoring	1.000	Ea.	1,825		1,825
Tank leak detection system, probes, well monitoring, liquid phase detection	2.000	Ea.	1,640		1,640
Excavating, trench, NO sheeting or dewatering, 6'-10' D,3/4 CY hyd backhoe	18.000	C.Y.		109.62	109.62
Concrete hold down pad, 6' x 5' x 8" thick	30.000	C.Y.	76.20	37.20	113.40
Reinforcing in hold down pad, #3 to #7	120.000	Lb.	56.40	62.40	118.80
Stone back-fill, hand spread, pea gravel	9.000	C.Y.	576	360	936
Corrosion resistance, wrap & coat, small diam pipe, 1" diam, add	120.000	L.F.	157.20		157.20
Corrosion resistance, wrap & coat, small diameter pipe, 2" diam, add	36.000	L.F.	51.84		51.84
Corrosion resistance wrap & coat, 4" diam, add	30.000	L.F.	52.20		52.20
TOTAL			14,288.04	4,610.67	18,898.71

G3060 305	Fiberglass Fuel Tank, Single Wall	COST EACH MAT.	COST EACH INST.	COST EACH TOTAL
1000	For alternate storage tanks:			
1010	Storage tank, fuel, underground, single wall fiberglass, 550 Gal.	14,300	4,600	18,900
1020	2000 Gal.	18,900	6,300	25,200
1030	4000 Gal.	25,100	8,600	33,700
1040	6000 Gal.	26,900	9,575	36,475
1050	8000 Gal.	31,100	11,700	42,800
1060	10,000 Gal.	36,400	12,700	49,100
1070	15,000 Gal.	44,500	15,800	60,300

G3060 Fuel Distribution

G3060 305	Fiberglass Fuel Tank, Single Wall	COST EACH		
		MAT.	INST.	TOTAL
1080	20,000 Gal.	55,500	19,200	74,700
1090	25,000 Gal.	65,500	22,400	87,900
1100	30,000 Gal.	74,500	24,200	98,700
1110	40,000 Gal.	96,000	30,400	126,400
1120	48,000 Gal.	110,000	35,000	145,000

G3060 310	Fiberglass Fuel Tank, Double Wall	COST EACH		
		MAT.	INST.	TOTAL
1010	Storage tank, fuel, underground, double wall fiberglass, 550 Gal.	17,300	5,275	22,575
1020	2500 Gal.	25,100	7,350	32,450
1030	4000 Gal.	32,000	9,125	41,125
1040	6000 Gal.	34,400	10,300	44,700
1050	8000 Gal.	40,900	12,400	53,300
1060	10,000 Gal.	45,900	14,300	60,200
1070	15,000 Gal.	56,500	16,500	73,000
1080	20,000 Gal.	69,000	20,000	89,000
1090	25,000 Gal.	84,500	23,200	107,700
1100	30,000 Gal.	96,500	24,500	121,000
2700	Cut & patch to match existing construction, add, minimum	2%	3%	
2800	Maximum	5%	9%	
2900	Dust protection, add, minimum	1%	2%	
3000	Maximum	4%	11%	
3100	Equipment usage curtailment, add, minimum	1%	1%	
3200	Maximum	3%	10%	
3300	Material handling & storage limitation, add, minimum	1%	1%	
3400	Maximum	6%	7%	
3500	Protection of existing work, add, minimum	2%	2%	
3600	Maximum	5%	7%	
3700	Shift work requirements, add, minimum		5%	
3800	Maximum		30%	
3900	Temporary shoring and bracing, add, minimum	2%	5%	
4000	Maximum	5%	12%	

G3060 Fuel Distribution

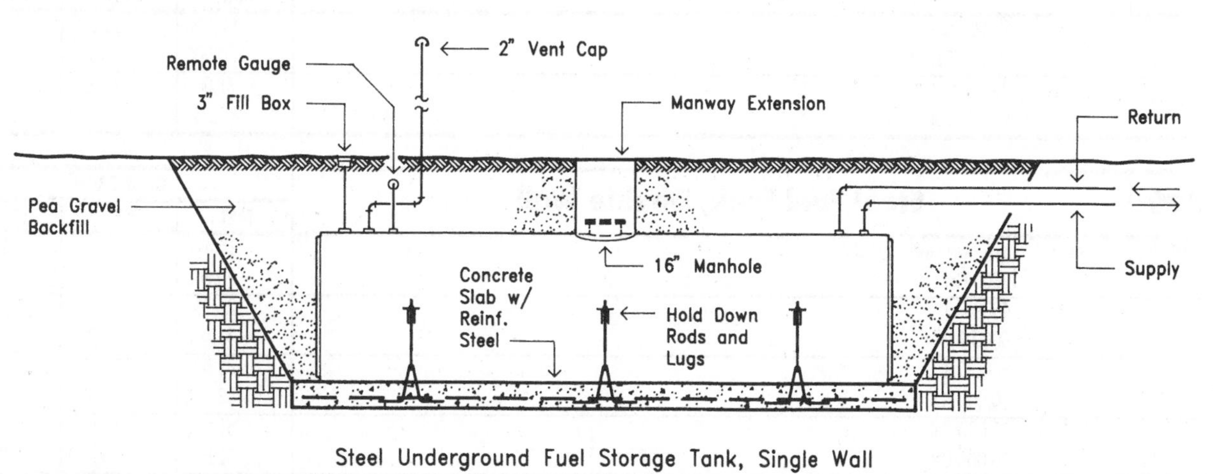

Steel Underground Fuel Storage Tank, Single Wall

System Components	QUANTITY	UNIT	COST EACH		
			MAT.	INST.	TOTAL
Single wall underground steel storage tank including, excavation, hold down slab, backfill, manway extension and related piping.					
Tank, steel, underground, sti-P3, set in place, 500 gal, 7 Ga. shell	1.000	Ea.	1,100	355	1,455
Tank, for manways, add	1.000	Ea.	910		910
Tanks, for hold-downs, 500-2000 gal, add	1.000	Ea.	131	128	259
Foot valve, single poppet, 3/4″ dia	1.000	Ea.	41.50	30	71.50
Tubing, copper, Type L, 3/4″ dia.	120.000	L.F.	343.20	840	1,183.20
Elbow 90°, copper, wrought, cu x cu, 3/4″ dia.	8.000	Ea.	10	224	234
Fuel oil valve, ball chk, globe type, fusible, 3/4″ dia.	2.000	Ea.	92	54	146
Elbow 90°, black steel, straight, 2″ dia.	3.000	Ea.	39	159	198
Union, black steel, with brass seat, 3/4″ dia.	1.000	Ea.	8.40	41	49.40
Pipe and nipples, steel, Sch. 40, threaded, black, 2″ dia	2.000	Ea.	296.40	568.10	864.50
Vent protector / breather, 2″ dia.	1.000	Ea.	17.60	16.75	34.35
Pipe, black steel, welded, Sch. 40, 3″ dia.	30.000	L.F.	462	713.10	1,175.10
Elbow 90°, steel, weld joint, butt, 90° elbow, 3″ dia.	3.000	Ea.	63	443.55	506.55
Flange, steel, weld neck, 150 lb., 3″ pipe size	1.000	Ea.	23.50	73.95	97.45
Fuel fill box, locking inner cover, 3″ dia.	1.000	Ea.	130	107	237
Remote tank gauging system, 30′, 5″ pointer travel	1.000	Ea.	4,050	268	4,318
Tank leak detector system, 8 channel, external monitoring	1.000	Ea.	1,825		1,825
Tank leak detection system, probes, well monitoring, liquid phase detection	2.000	Ea.	1,640		1,640
Excavation, NO sheeting or dewatering,6′-10′ D,3/4 CY hyd backhoe	18.000	C.Y.		109.62	109.62
Concrete hold down pad, 6′ x 5′ x 8″ thick	30.000	C.Y.	76.20	37.20	113.40
Reinforcing in hold down pad, #3 to #7	120.000	Lb.	56.40	62.40	118.80
Stone back-fill, hand spread, pea gravel	9.000	C.Y.	576	360	936
Corrosion resistance, wrap & coat, small diam pipe, 1″ dia., add	120.000	L.F.	157.20		157.20
Corrosion resistance, wrap & coat, small diameter pipe, 2″ dia., add	36.000	L.F.	51.84		51.84
Corrosion resistance wrap & coat, 4″ dia., add	30.000	L.F.	52.20		52.20
TOTAL			12,152.44	4,590.67	16,743.11

G3060 315	Steel Fuel Tank, Single Wall	COST EACH		
		MAT.	INST.	TOTAL
1000	For alternate storage tanks:			
1010	Storage tank, fuel, underground, single wall steel, 550 Gal.	12,200	4,575	16,775
1020	2000 Gal.	16,500	6,300	22,800
1030	5000 Gal.	25,200	8,900	34,100
1040	10,000 Gal.	37,300	13,600	50,900
1050	15,000 Gal.	41,800	15,700	57,500

G3060 Fuel Distribution

G3060 315	Steel Fuel Tank, Single Wall	COST EACH		
		MAT.	INST.	TOTAL
1060	20,000 Gal.	52,500	19,000	71,500
1070	25,000 Gal.	60,500	22,200	82,700
1080	30,000 Gal.	71,000	23,800	94,800
1090	40,000 Gal.	87,500	30,000	117,500

G3060 320	Steel Fuel Tank, Double Wall	COST EACH		
		MAT.	INST.	TOTAL
1010	Storage tank, fuel, underground, double wall steel, 500 Gal.	15,400	5,375	20,775
1020	2000 Gal.	20,600	7,075	27,675
1030	4000 Gal.	27,700	9,225	36,925
1040	6000 Gal.	33,400	10,200	43,600
1050	8000 Gal.	38,500	12,600	51,100
1060	10,000 Gal.	43,200	14,500	57,700
1070	15,000 Gal.	51,000	16,900	67,900
1080	20,000 Gal.	61,500	20,300	81,800
1090	25,000 Gal.	82,000	23,200	105,200
1100	30,000 Gal.	93,500	24,800	118,300
1110	40,000 Gal.	116,000	31,100	147,100
1120	50,000 Gal.	151,000	37,300	188,300

G3060 325	Steel Tank, Above Ground, Single Wall	COST EACH		
		MAT.	INST.	TOTAL
1010	Storage tank, fuel, above ground, single wall steel, 550 Gal.	9,750	3,800	13,550
1020	2000 Gal.	12,800	3,900	16,700
1030	5000 Gal.	15,800	5,950	21,750
1040	10,000 Gal.	28,300	6,350	34,650
1050	15,000 Gal.	29,800	6,525	36,325
1060	20,000 Gal.	35,000	6,725	41,725
1070	25,000 Gal.	38,100	6,900	45,000
1080	30,000 Gal.	44,700	7,175	51,875

G3060 330	Steel Tank, Above Ground, Double Wall	COST EACH		
		MAT.	INST.	TOTAL
1010	Storage tank, fuel, above ground, double wall steel, 500 Gal.	13,000	3,850	16,850
1020	2000 Gal.	17,000	3,925	20,925
1030	4000 Gal.	22,700	4,000	26,700
1040	6000 Gal.	26,400	6,175	32,575
1050	8000 Gal.	30,700	6,350	37,050
1060	10,000 Gal.	32,600	6,450	39,050
1070	15,000 Gal.	43,300	6,675	49,975
1080	20,000 Gal.	47,600	6,900	54,500
1090	25,000 Gal.	55,000	7,100	62,100
1100	30,000 Gal.	59,500	7,375	66,875
2700	Cut & patch to match existing construction, add, minimum	2%	3%	
2800	Maximum	5%	9%	
2900	Dust protection, add, minimum	1%	2%	
3000	Maximum	4%	11%	
3100	Equipment usage curtailment, add, minimum	1%	1%	
3200	Maximum	3%	10%	
3300	Material handling & storage limitation, add, minimum	1%	1%	
3400	Maximum	6%	7%	
3500	Protection of existing work, add, minimum	2%	2%	
3600	Maximum	5%	7%	
3700	Shift work requirements, add, minimum		5%	
3800	Maximum		30%	
3900	Temporary shoring and bracing, add, minimum	2%	5%	
4000	Maximum	5%	12%	

Important: See the Reference Section for critical supporting data - Reference Numbers and City Cost Indexes

G4020 Site Lighting

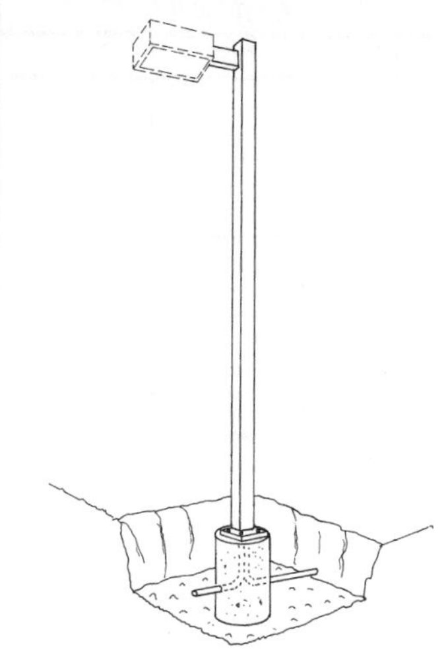

This page illustrates and describes light poles for parking or walkway area lighting. Included are aluminum or steel light poles, single or multiple fixture bracket arms, excavation, concrete footing, backfill and compaction. Lines within system components give the unit price and total price on a cost per each basis. Prices for alternate systems are shown on lines D4020 0200 through 1440. Both material quantities and labor costs have been adjusted for the system listed.

Factors: To adjust for job conditions other than normal working situations use lines D4020 2700 through 3600.
Example: You are to install the system where you must be careful not to damage existing walks or parking areas. Go to line D4020 3300 and apply these percentages to the appropriate MAT. and INST. costs.f

System Components	QUANTITY	UNIT	COST EACH MAT.	COST EACH INST.	COST EACH TOTAL
Light poles, aluminum, 20′ high, 1 arm bracket					
Aluminum light pole, 20′, no concrete base	1.000	Ea.	840	498	1,338
Bracket arm for Aluminum light pole	1.000	Ea.	110	64.50	174.50
Excavation by hand, pits to 6′ deep, heavy soil or clay	2.368	C.Y.		215.49	215.49
Footing, concrete incl forms, reinforcing, spread, under 1 C.Y.	.465	C.Y.	63.71	76.22	139.93
Backfill by hand	1.903	C.Y.		62.80	62.80
Compaction vibrating plate	1.903	C.Y.		8.49	8.49
TOTAL			1,013.71	925.50	1,939.21

G4020 210	Light Pole (Installed)	COST EACH MAT.	COST EACH INST.	COST EACH TOTAL
0195	For alternate light pole systems:			
0200	Light pole, aluminum, 20′ high, 1 arm bracket	1,025	925	1,950
0240	2 arm brackets	1,125	925	2,050
0280	3 arm brackets	1,225	960	2,185
0320	4 arm brackets	1,350	960	2,310
0360	30′ high, 1 arm bracket	1,825	1,175	3,000
0400	2 arm brackets	1,925	1,175	3,100
0440	3 arm brackets	2,025	1,225	3,250
0480	4 arm brackets	2,150	1,225	3,375
0680	40′ high, 1 arm bracket	2,225	1,550	3,775
0720	2 arm brackets	2,350	1,550	3,900
0760	3 arm brackets	2,450	1,575	4,025
0800	4 arm brackets	2,550	1,575	4,125
0840	Steel, 20′ high, 1 arm bracket	1,225	985	2,210
0880	2 arm brackets	1,325	985	2,310
0920	3 arm brackets	1,350	1,025	2,375
0960	4 arm brackets	1,450	1,025	2,475
1000	30′ high, 1 arm bracket	1,425	1,250	2,675
1040	2 arm brackets	1,500	1,250	2,750
1080	3 arm brackets	1,525	1,275	2,800
1120	4 arm brackets	1,625	1,275	2,900
1320	40′ high, 1 arm bracket	1,850	1,700	3,550
1360	2 arm brackets	1,950	1,700	3,650
1400	3 arm brackets	1,975	1,725	3,700
1440	4 arm brackets	2,075	1,725	3,800
2700	Cut & patch to match existing construction, add, minimum	2%	3%	

G4020 Site Lighting

G4020 210	Light Pole (Installed)	COST EACH MAT.	COST EACH INST.	COST EACH TOTAL
2800	Maximum	5%	9%	
2900	Equipment usage curtailment, add, minimum	1%	1%	
3000	Maximum	3%	10%	
3100	Material handling & storage limitation, add, minimum	1%	1%	
3200	Maximum	6%	7%	
3300	Protection of existing work, add, minimum	2%	2%	
3400	Maximum	5%	7%	
3500	Shift work requirements, add, minimum		5%	
3600	Maximum		30%	

For information about Means Estimating Seminars, see yellow pages 12 and 13 in back of book

Reference Section

All the reference information is in one section, making it easy to find what you need to know . . . and easy to use the book on a daily basis. This section is visually identified by a vertical gray bar on the page edges.

In this Reference Section, we've included Equipment Rental Costs, a listing of rental and operating costs; Crew Listings, a full listing of all crews and equipment, and their costs; Historical Cost Indexes for cost comparisons over time; City Cost Indexes and Location Factors for adjusting costs to the region you are in; Reference Tables, where you will find explanations, estimating information and procedures, or technical data; Change Orders, information on pricing changes to contract documents; Square Foot Costs that allow you to make a rough estimate for the overall cost of a project; and an explanation of all the Abbreviations in the book.

Table of Contents

Construction Equipment Rental Costs

Estimating Tips

- This section contains the average costs to rent and operate hundreds of pieces of construction equipment. This is useful information when estimating the time and material requirements of any particular operation in order to establish a unit or total cost. Equipment costs include not only rental, but also operating costs for equipment under normal use.

Rental Costs

- Equipment rental rates are obtained from industry sources throughout North America - contractors, suppliers, dealers, manufacturers, and distributors.
- Rental rates vary throughout the country with larger cities generally having lower rates. Lease plans for new equipment are available for periods in excess of six months with a percentage of payments applying toward purchase.
- Monthly rental rates vary from 2% to 5% of the purchase price of the equipment depending on the anticipated life of the equipment and its wearing parts.
- Weekly rental rates are about 1/3 the monthly rates, and daily rental rates are about 1/3 the weekly rate.

Operating Costs

- The operating costs include parts and labor for routine servicing such as repair and replacement of pumps, filters and worn lines. Normal operating expendables such as fuel, lubricants, tires and electricity (where applicable) are also included.
- Extraordinary operating expendables with highly variable wear patterns such as diamond bits and blades are excluded. These costs can be found as material costs in the unit price section.
- The hourly operating costs listed do not include the operator's wages.

Crew Equipment Cost/Day

- Any power equipment required by a crew is shown in the Crew Listings with a daily cost.
- The daily cost of crew equipment is based on dividing the weekly rental rate by 5 (number of working days in the week), and then adding the hourly operating cost times 8 (the number of hours in a day). This "Crew Equipment Cost/Day" is shown in the far right column of the Equipment Rental pages.
- If equipment is needed for only one or two days, it is best to develop your own cost by including components for daily rent and hourly operating cost. This is important when the listed Crew for a task does not contain the equipment needed, such as a crane for lifting mechanical heating/cooling equipment up onto a roof.

Mobilization/ Demobilization

- The cost to move construction equipment from an equipment yard or rental company to the jobsite and back again is not included in equipment rental costs listed in the Reference section, nor in the bare equipment cost of any Unit Cost line item, nor in any equipment costs shown in the Crew listings.
- Mobilization (to the site) and demobilization (from the site) costs can be found in the Unit Cost section.
- If a piece of equipment is already at the jobsite, it is not appropriate to utilize mob/ demob costs again in an estimate.

01 54 | Construction Aids

01 54 33 | Rental Equipment

		01 54 33 | Rental Equipment		UNIT	HOURLY OPER. COST	RENT PER DAY	RENT PER WEEK	RENT PER MONTH	CREW EQUIPMENT COST/DAY	
10	0010	**CONCRETE EQUIPMENT RENTAL**, without operators	R015433 -10							10
	0150	For batch plant, see div. 01590-500								
	0200	Bucket, concrete lightweight, 1/2 C.Y.	R033105 -70	Ea.	.55	16	48	144	14	
	0300	1 C.Y.			.60	19	57	171	16.20	
	0400	1-1/2 C.Y.			.75	26	78	234	21.60	
	0500	2 C.Y.			.85	30.50	92	276	25.20	
	0580	8 C.Y.			4.55	205	615	1,850	159.40	
	0600	Cart, concrete, self propelled, operator walking, 10 C.F.			2.15	58.50	175	525	52.20	
	0700	Operator riding, 18 C.F.			3.30	80	240	720	74.40	
	0800	Conveyer for concrete, portable, gas, 16″ wide, 26′ long			7.30	117	350	1,050	128.40	
	0900	46′ long			7.65	142	425	1,275	146.20	
	1000	56′ long			7.80	150	450	1,350	152.40	
	1100	Core drill, electric, 2-1/2 H.P., 1″ to 8″ bit diameter			2.00	81	243	730	64.60	
	1150	11 HP, 8″ to 18″ cores			6.61	104	312.80	940	115.45	
	1200	Finisher, concrete floor, gas, riding trowel, 48″ diameter			4.80	86.50	260	780	90.40	
	1300	Gas, manual, 3 blade, 36″ trowel			1.00	16.65	50	150	18	
	1400	4 blade, 48″ trowel			1.45	21	63	189	24.20	
	1500	Float, hand-operated (Bull float) 48″ wide			.08	13.35	40	120	8.65	
	1570	Curb builder, 14 H.P., gas, single screw			9.75	210	630	1,900	204	
	1590	Double screw			10.40	242	725	2,175	228.20	
	1600	Grinder, concrete and terrazzo, electric, floor			2.68	116	349	1,050	91.25	
	1700	Wall grinder			1.35	58.50	175	525	45.80	
	1800	Mixer, powered, mortar and concrete, gas, 6 C.F., 18 H.P.			5.20	107	320	960	105.60	
	1900	10 C.F., 25 H.P.			6.30	128	385	1,150	127.40	
	2000	16 C.F.			6.65	152	455	1,375	144.20	
	2100	Concrete, stationary, tilt drum, 2 C.Y.			5.35	200	600	1,800	162.80	
	2120	Pump, concrete, truck mounted 4″ line 80′ boom			21.70	880	2,635	7,900	700.60	
	2140	5″ line, 110′ boom			28.40	1,175	3,495	10,500	926.20	
	2160	Mud jack, 50 C.F. per hr.			5.79	118	355	1,075	117.30	
	2180	225 C.F. per hr.			7.56	158	473.20	1,425	155.10	
	2190	Shotcrete pump rig, 12 CY/hr			11.80	233	700	2,100	234.40	
	2600	Saw, concrete, manual, gas, 18 H.P.			3.50	35	105	315	49	
	2650	Self-propelled, gas, 30 H.P.			6.75	88.50	265	795	107	
	2700	Vibrators, concrete, electric, 60 cycle, 2 H.P.			.37	8.35	25	75	7.95	
	2800	3 H.P.			.56	12.35	37	111	11.90	
	2900	Gas engine, 5 H.P.			.90	15	45	135	16.20	
	3000	8 H.P.			1.30	18.35	55	165	21.40	
	3050	Vibrating screed, gas engine, 8 H.P.			2.27	66.50	199	595	57.95	
	3100	Concrete transit mixer, hydraulic drive								
	3120	6 x 4, 250 H.P., 8 C.Y., rear discharge			34.85	540	1,625	4,875	603.80	
	3200	Front discharge			40.95	675	2,020	6,050	731.60	
	3300	6 x 6, 285 H.P., 12 C.Y., rear discharge			40.05	635	1,900	5,700	700.40	
	3400	Front discharge			41.85	680	2,045	6,125	743.80	
20	0010	**EARTHWORK EQUIPMENT RENTAL**, without operators	R015433 -10							20
	0040	Aggregate spreader, push type 8′ to 12′ wide		Ea.	1.80	29.50	88	264	32	
	0045	Tailgate type, 8′ wide	R312323 -30	″	1.75	31.50	95	285	33	
	0050	Augers for vertical drilling								
	0055	Earth auger, truck-mounted, for fence & sign posts	R312316 -40	Ea.	8.40	480	1,445	4,325	356.20	
	0060	For borings and monitoring wells			30.95	640	1,920	5,750	631.60	
	0070	Earth auger, portable, trailer mounted	R312316 -45		1.75	23	69	207	27.80	
	0075	Earth auger, truck-mounted, for caissons, water wells, utility poles			160.90	3,475	10,460	31,400	3,379	
	0080	Auger, horizontal boring machine, 12″ to 36″ diameter, 45 H.P.			16.70	190	570	1,700	247.60	
	0090	12″ to 48″ diameter, 65 H.P.			23.50	340	1,020	3,050	392	
	0095	Auger, for fence posts, gas engine, hand held			.35	4.33	13	39	5.40	
	0100	Excavator, diesel hydraulic, crawler mounted, 1/2 C.Y. cap.			16.05	345	1,035	3,100	335.40	
	0120	5/8 C.Y. capacity			19.55	465	1,395	4,175	435.40	
	0140	3/4 C.Y. capacity			22.85	490	1,475	4,425	477.80	

01 54 | Construction Aids

01 54 33 | Rental Equipment

			UNIT	HOURLY OPER. COST	RENT PER DAY	RENT PER WEEK	RENT PER MONTH	CREW EQUIPMENT COST/DAY	
20	0150	1 C.Y. capacity R015433-10	Ea.	27.40	570	1,705	5,125	560.20	20
	0200	1-1/2 C.Y. capacity		33.85	755	2,260	6,775	722.80	
	0300	2 C.Y. capacity R312323-30		43.70	950	2,850	8,550	919.60	
	0320	2-1/2 C.Y. capacity		56.20	1,275	3,840	11,500	1,218	
	0340	3-1/2 C.Y. capacity R312316-40		95.20	2,100	6,305	18,900	2,023	
	0341	Attachments							
	0342	Bucket thumbs R312316-45		2.55	208	625	1,875	145.40	
	0345	Grapples		1.00	219	656	1,975	139.20	
	0350	Gradall type, truck mounted, 3 ton @ 15' radius, 5/8 C.Y.		38.75	885	2,655	7,975	841	
	0370	1 C.Y. capacity		44.65	1,025	3,085	9,250	974.20	
	0400	Backhoe-loader, 40 to 45 H.P., 5/8 C.Y. capacity		8.60	183	550	1,650	178.80	
	0450	45 H.P. to 60 H.P., 3/4 C.Y. capacity		10.90	232	695	2,075	226.20	
	0460	80 H.P., 1-1/4 C.Y. capacity		13.95	277	830	2,500	277.60	
	0470	112 H.P., 1-1/2 C.Y. capacity		18.55	400	1,200	3,600	388.40	
	0480	Attachments							
	0482	Compactor, 20,000 lb		4.35	118	355	1,075	105.80	
	0485	Hydraulic hammer, 750 ft-lbs		2.00	70	210	630	58	
	0486	Hydraulic hammer, 1200 ft-lbs		4.20	133	400	1,200	113.60	
	0500	Brush chipper, gas engine, 6" cutter head, 35 H.P.		6.60	102	305	915	113.80	
	0550	12" cutter head, 130 H.P.		10.25	152	455	1,375	173	
	0600	15" cutter head, 165 H.P.		14.30	162	485	1,450	211.40	
	0750	Bucket, clamshell, general purpose, 3/8 C.Y.		1.00	35	105	315	29	
	0800	1/2 C.Y.		1.10	41.50	125	375	33.80	
	0850	3/4 C.Y.		1.25	51.50	155	465	41	
	0900	1 C.Y.		1.30	55	165	495	43.40	
	0950	1-1/2 C.Y.		2.05	75	225	675	61.40	
	1000	2 C.Y.		2.20	85	255	765	68.60	
	1010	Bucket, dragline, medium duty, 1/2 C.Y.		.60	22.50	67	201	18.20	
	1020	3/4 C.Y.		.60	23.50	71	213	19	
	1030	1 C.Y.		.65	25.50	76	228	20.40	
	1040	1-1/2 C.Y.		1.00	38.50	115	345	31	
	1050	2 C.Y.		1.05	43.50	130	390	34.40	
	1070	3 C.Y.		1.60	58.50	175	525	47.80	
	1200	Compactor, manually guided 2-drum vibratory smooth roller, 7.5 H.P.		4.85	145	435	1,300	125.80	
	1250	Rammer compactor, gas, 1000 lb. blow		1.75	38.50	115	345	37	
	1300	Vibratory plate, gas, 18" plate, 3000 lb. blow		1.65	22	66	198	26.40	
	1350	21" plate, 5000 lb. blow		2.00	31	93	279	34.60	
	1370	Curb builder/extruder, 14 H.P., gas, single screw		9.75	212	635	1,900	205	
	1390	Double screw		10.40	248	745	2,225	232.20	
	1500	Disc harrow attachment, for tractor		.37	61.50	185	555	39.95	
	1750	Extractor, piling, see lines 2500 to 2750							
	1810	Feller buncher, shearing & accumulating trees, 100 H.P.	Ea.	22.60	520	1,555	4,675	491.80	
	1860	Grader, self-propelled, 25,000 lb.		18.25	400	1,200	3,600	386	
	1910	30,000 lb.		21.05	480	1,445	4,325	457.40	
	1920	40,000 lb.		30.55	745	2,230	6,700	690.40	
	1930	55,000 lb.		40.80	1,050	3,130	9,400	952.40	
	1950	Hammer, pavement demo., hyd., gas, self-prop., 1000 to 1250 lb.		18.30	305	915	2,750	329.40	
	2000	Diesel 1300 to 1500 lb.		28.20	610	1,830	5,500	591.60	
	2050	Pile driving hammer, steam or air, 4150 ft.-lb. @ 225 BPM		6.60	272	815	2,450	215.80	
	2100	8750 ft.-lb. @ 145 BPM		8.50	440	1,325	3,975	333	
	2150	15,000 ft.-lb. @ 60 BPM		8.85	475	1,425	4,275	355.80	
	2200	24,450 ft.-lb. @ 111 BPM		11.75	525	1,570	4,700	408	
	2250	Leads, 15,000 ft.-lb. hammers	L.F.	.03	1.45	4.35	13.05	1.10	
	2300	24,450 ft.-lb. hammers and heavier	"	.05	2.47	7.40	22	1.90	
	2350	Diesel type hammer, 22,400 ft.-lb.	Ea.	25.50	605	1,820	5,450	568	
	2400	41,300 ft.-lb.		34.75	655	1,970	5,900	672	
	2450	141,000 ft.-lb.		56.45	1,125	3,380	10,100	1,128	
	2500	Vib. elec. hammer/extractor, 200 KW diesel generator, 34 H.P.		28.70	635	1,910	5,725	611.60	

01 54 33 | Rental Equipment

				UNIT	HOURLY OPER. COST	RENT PER DAY	RENT PER WEEK	RENT PER MONTH	CREW EQUIPMENT COST/DAY
20	2550	80 H.P.	R015433 -10	Ea.	49.50	930	2,785	8,350	953
	2600	150 H.P.			91.90	1,825	5,440	16,300	1,823
	2700	Extractor, steam or air, 700 ft.-lb.	R312323 -30		14.85	435	1,310	3,925	380.80
	2750	1000 ft.-lb.			16.95	540	1,620	4,850	459.60
	2800	Log chipper, up to 22" diam, 600 H.P.	R312316 -40		41.71	1,100	3,331.60	10,000	1,000
	2850	Logger, for skidding & stacking logs, 150 H.P.			36.70	825	2,470	7,400	787.60
	2900	Rake, spring tooth, with tractor	R312316 -45		8.04	214	643	1,925	192.90
	3000	Roller, vibratory, tandem, smooth drum, 20 H.P.			5.40	118	355	1,075	114.20
	3050	35 H.P.			7.75	225	675	2,025	197
	3100	Towed type vibratory compactor, smooth drum, 50 H.P.			24.70	435	1,300	3,900	457.60
	3150	Sheepsfoot, 50 H.P.			25.50	455	1,370	4,100	478
	3170	Landfill compactor, 220 HP			50.30	1,200	3,595	10,800	1,121
	3200	Pneumatic tire roller, 80 H.P.			9.15	310	930	2,800	259.20
	3250	120 H.P.			14.75	540	1,615	4,850	441
	3300	Sheepsfoot vibratory roller, 200 H.P.			38.75	895	2,690	8,075	848
	3320	340 H.P.			53.55	1,300	3,870	11,600	1,202
	3350	Smooth drum vibratory roller, 75 H.P.			15.90	485	1,460	4,375	419.20
	3400	125 H.P.			20.25	600	1,800	5,400	522
	3410	Rotary mower, brush, 60", with tractor			11.50	232	695	2,075	231
	3450	Scrapers, towed type, 9 to 12 C.Y. capacity			4.42	211	632	1,900	161.75
	3500	12 to 17 C.Y. capacity			1.50	281	842	2,525	180.40
	3550	Scrapers, self-propelled, 4 x 4 drive, 2 engine, 14 C.Y. capacity			84.60	1,500	4,500	13,500	1,577
	3600	2 engine, 24 C.Y. capacity			121.45	2,300	6,890	20,700	2,350
	3640	32 - 44 C.Y. capacity			143.20	2,675	8,060	24,200	2,758
	3650	Self-loading, 11 C.Y. capacity			40.95	825	2,475	7,425	822.60
	3700	22 C.Y. capacity			76.90	1,725	5,190	15,600	1,653
	3710	Screening plant 110 H.P. w/ 5' x 10' screen			23.00	380	1,135	3,400	411
	3720	5' x 16' screen			25.05	475	1,430	4,300	486.40
	3850	Shovels, see Cranes division 01590-600							
	3860	Shovel/backhoe bucket, 1/2 C.Y.		Ea.	1.85	56.50	170	510	48.80
	3870	3/4 C.Y.			1.90	63.50	190	570	53.20
	3880	1 C.Y.			2.00	71.50	215	645	59
	3890	1-1/2 C.Y.			2.15	86.50	260	780	69.20
	3910	3 C.Y.			2.45	120	360	1,075	91.60
	3950	Stump chipper, 18" deep, 30 H.P.			4.56	45	135	405	63.50
	4110	Tractor, crawler, with bulldozer, torque converter, diesel 80 H.P.			16.35	315	940	2,825	318.80
	4150	105 H.P.			22.35	475	1,420	4,250	462.80
	4200	140 H.P.			27.50	590	1,765	5,300	573
	4260	200 H.P.			41.10	985	2,960	8,875	920.80
	4310	300 H.P.			53.40	1,275	3,850	11,600	1,197
	4360	410 H.P.			72.30	1,600	4,790	14,400	1,536
	4370	500 H.P.			95.55	2,125	6,395	19,200	2,043
	4380	700 H.P.			142.40	3,350	10,045	30,100	3,148
	4400	Loader, crawler, torque conv., diesel, 1-1/2 C.Y., 80 H.P.			15.15	325	970	2,900	315.20
	4450	1-1/2 to 1-3/4 C.Y., 95 H.P.			17.55	385	1,160	3,475	372.40
	4510	1-3/4 to 2-1/4 C.Y., 130 H.P.			23.75	610	1,835	5,500	557
	4530	2-1/2 to 3-1/4 C.Y., 190 H.P.			35.75	845	2,540	7,625	794
	4560	3-1/2 to 5 C.Y., 275 H.P.			48.55	1,200	3,610	10,800	1,110
	4610	Tractor loader, wheel, torque conv., 4 x 4, 1 to 1-1/4 C.Y., 65 H.P.			9.95	192	575	1,725	194.60
	4620	1-1/2 to 1-3/4 C.Y., 80 H.P.			12.55	235	705	2,125	241.40
	4650	1-3/4 to 2 C.Y., 100 H.P.			14.25	280	840	2,525	282
	4710	2-1/2 to 3-1/2 C.Y., 130 H.P.			15.30	305	915	2,750	305.40
	4730	3 to 4-1/2 C.Y., 170 H.P.			21.00	460	1,380	4,150	444
	4760	5-1/4 to 5-3/4 C.Y., 270 H.P.			34.75	705	2,115	6,350	701
	4810	7 to 8 C.Y., 375 H.P.			59.05	1,250	3,740	11,200	1,220
	4870	12-1/2 C.Y., 690 H.P.			82.40	1,925	5,810	17,400	1,821
	4880	Wheeled, skid steer, 10 C.F., 30 H.P. gas			8.75	140	420	1,250	154
	4890	1 C.Y., 78 H.P., diesel			10.50	197	590	1,775	202

01 54 33 | Rental Equipment

		01 54 33 | Rental Equipment		UNIT	HOURLY OPER. COST	RENT PER DAY	RENT PER WEEK	RENT PER MONTH	CREW EQUIPMENT COST/DAY	
20	4891	Attachments for all skid steer loaders	R015433-10							20
	4892	Auger		Ea.	.40	66.50	200	600	43.20	
	4893	Backhoe	R312323-30		.67	112	336	1,000	72.55	
	4894	Broom			.61	101	303	910	65.50	
	4895	Forks	R312316-40		.22	37	111	335	23.95	
	4896	Grapple			.50	83.50	250	750	54	
	4897	Concrete hammer	R312316-45		.96	159	478	1,425	103.30	
	4898	Tree spade			.95	158	475	1,425	102.60	
	4899	Trencher			.64	106	318	955	68.70	
	4900	Trencher, chain, boom type, gas, operator walking, 12 H.P.			2.70	43.50	130	390	47.60	
	4910	Operator riding, 40 H.P.			8.85	243	730	2,200	216.80	
	5000	Wheel type, diesel, 4' deep, 12" wide			47.60	745	2,230	6,700	826.80	
	5100	Diesel, 6' deep, 20" wide			64.90	1,775	5,295	15,900	1,578	
	5150	Ladder type, diesel, 5' deep, 8" wide			23.25	705	2,115	6,350	609	
	5200	Diesel, 8' deep, 16" wide			57.25	1,675	5,010	15,000	1,460	
	5210	Tree spade, self-propelled			11.10	267	800	2,400	248.80	
	5250	Truck, dump, tandem, 12 ton payload			20.45	272	815	2,450	326.60	
	5300	Three axle dump, 16 ton payload			27.95	420	1,265	3,800	476.60	
	5350	Dump trailer only, rear dump, 16-1/2 C.Y.			4.25	115	345	1,025	103	
	5400	20 C.Y.			4.60	132	395	1,175	115.80	
	5450	Flatbed, single axle, 1-1/2 ton rating			11.85	56.50	170	510	128.80	
	5500	3 ton rating			15.15	91.50	275	825	176.20	
	5550	Off highway rear dump, 25 ton capacity			40.40	980	2,945	8,825	912.20	
	5600	35 ton capacity			41.30	1,000	3,015	9,050	933.40	
	5610	50 ton capacity			53.15	1,250	3,780	11,300	1,181	
	5620	65 ton capacity			56.85	1,375	4,105	12,300	1,276	
	5630	100 ton capacity			73.00	1,775	5,325	16,000	1,649	
	6000	Vibratory plow, 25 H.P., walking			4.65	58.50	175	525	72.20	
40	0010	**GENERAL EQUIPMENT RENTAL**, without operators	R314116-45							40
	0150	Aerial lift, scissor type, to 15' high, 1000 lb. cap., electric		Ea.	2.35	48.50	145	435	47.80	
	0160	To 25' high, 2000 lb. capacity	R314116-40		2.85	70	210	630	64.80	
	0170	Telescoping boom to 40' high, 500 lb. capacity, gas			12.25	278	835	2,500	265	
	0180	To 45' high, 500 lb. capacity	R312323-30		13.05	320	960	2,875	296.40	
	0190	To 60' high, 600 lb. capacity			15.00	425	1,280	3,850	376	
	0195	Air compressor, portable, 6.5 CFM, electric			.40	12.35	37	111	10.60	
	0196	Gasoline			.57	18.65	56	168	15.75	
	0200	Air compressor, portable, gas engine, 60 C.F.M.			4.90	33	99	297	59	
	0300	160 C.F.M.			8.70	46.50	140	420	97.60	
	0400	Diesel engine, rotary screw, 250 C.F.M.			8.60	85	255	765	119.80	
	0500	365 C.F.M.			11.50	112	335	1,000	159	
	0550	450 C.F.M.			13.90	130	390	1,175	189.20	
	0600	600 C.F.M.			23.65	218	655	1,975	320.20	
	0700	750 C.F.M.			25.55	228	685	2,050	341.40	
	0800	For silenced models, small sizes, add								
	0900	Large sizes, add								
	0920	Air tools and accessories								
	0930	Breaker, pavement, 60 lb.		Ea.	.40	9.65	29	87	9	
	0940	80 lb.			.40	10.65	32	96	9.60	
	0950	Drills, hand (jackhammer) 65 lb.			.45	15.35	46	138	12.80	
	0960	Track or wagon, swing boom, 4" drifter			38.85	695	2,090	6,275	728.80	
	0970	5" drifter			45.75	760	2,280	6,850	822	
	0975	Track mounted quarry drill, 6" diameter drill			55.65	965	2,900	8,700	1,025	
	0980	Dust control per drill			.86	14.65	44	132	15.70	
	0990	Hammer, chipping, 12 lb.			.40	20	60	180	15.20	
	1000	Hose, air with couplings, 50' long, 3/4" diameter			.04	6	18	54	3.90	
	1100	1" diameter			.04	6	18	54	3.90	
	1200	1-1/2" diameter			.04	7.35	22	66	4.70	
	1300	2" diameter			.10	17.35	52	156	11.20	

01 54 33 | Rental Equipment

			UNIT	HOURLY OPER. COST	RENT PER DAY	RENT PER WEEK	RENT PER MONTH	CREW EQUIPMENT COST/DAY
40	1400	2-1/2" diameter (R314116-45)	Ea.	.12	19.35	58	174	12.55
	1410	3" diameter		.17	28.50	86	258	18.55
	1450	Drill, steel, 7/8" x 2' (R314116-40)		.05	8.35	25	75	5.40
	1460	7/8" x 6'		.06	9.35	28	84	6.10
	1520	Moil points (R312323-30)		.03	4.33	13	39	2.85
	1525	Pneumatic nailer w/accessories		.41	27.50	82	246	19.70
	1530	Sheeting driver for 60 lb. breaker		.10	7.35	22	66	5.20
	1540	For 90 lb. breaker		.15	10	30	90	7.20
	1550	Spade, 25 lb.		.35	6.65	20	60	6.80
	1560	Tamper, single, 35 lb.		.58	38.50	116	350	27.85
	1570	Triple, 140 lb.		.87	58	174	520	41.75
	1580	Wrenches, impact, air powered, up to 3/4" bolt		.25	7.65	23	69	6.60
	1590	Up to 1-1/4" bolt		.35	16.35	49	147	12.60
	1600	Barricades, barrels, reflectorized, 1 to 50 barrels		.02	3.10	9.30	28	2
	1610	100 to 200 barrels		.01	2.33	7	21	1.50
	1620	Barrels with flashers, 1 to 50 barrels		.02	3.77	11.30	34	2.40
	1630	100 to 200 barrels		.02	3	9	27	1.95
	1640	Barrels with steady burn type C lights		.03	5	15	45	3.25
	1650	Illuminated board, trailer mounted, with generator		.65	115	345	1,025	74.20
	1670	Portable barricade, stock, with flashers, 1 to 6 units		.02	3.77	11.30	34	2.40
	1680	25 to 50 units		.02	3.50	10.50	31.50	2.25
	1690	Butt fusion machine, electric		18.65	330	990	2,975	347.20
	1695	Electro fusion machine		8.55	133	400	1,200	148.40
	1700	Carts, brick, hand powered, 1000 lb. capacity		.30	49.50	148	445	32
	1800	Gas engine, 1500 lb., 7-1/2' lift		3.09	94.50	284	850	81.50
	1822	Dehumidifier, medium, 6 lb/hr, 150 CFM		.75	45.50	137	410	33.40
	1824	Large, 18 lb/hr, 600 CFM		1.48	90.50	272	815	66.25
	1830	Distributor, asphalt, trailer mtd, 2000 gal., 38 H.P. diesel		7.25	252	755	2,275	209
	1840	3000 gal., 38 H.P. diesel		8.55	290	870	2,600	242.40
	1850	Drill, rotary hammer, electric, 1-1/2" diameter		.40	24.50	74	222	18
	1860	Carbide bit for above		.04	6	18	54	3.90
	1865	Rotary, crawler, 250 H.P.		95.45	1,800	5,410	16,200	1,846
	1870	Emulsion sprayer, 65 gal., 5 H.P. gas engine		2.00	75	225	675	61
	1880	200 gal., 5 H.P. engine		5.15	125	375	1,125	116.20
	1920	Floodlight, mercury vapor, or quartz, on tripod						
	1930	1000 watt	Ea.	.30	12	36	108	9.60
	1940	2000 watt		.52	22	66	198	17.35
	1950	Floodlights, trailer mounted with generator, 1 - 300 watt light		2.55	65	195	585	59.40
	1960	2 - 1000 watt lights		3.60	110	330	990	94.80
	2000	4 - 300 watt lights		2.95	76.50	230	690	69.60
	2020	Forklift, wheeled, for brick, 18', 3000 lb., 2 wheel drive, gas		15.25	182	545	1,625	231
	2040	28', 4000 lb., 4 wheel drive, diesel		12.55	237	710	2,125	242.40
	2050	For rough terrain, 8000 lb., 16' lift, 68 H.P.		16.90	375	1,125	3,375	360.20
	2060	For plant, 4 T. capacity, 80 H.P., 2 wheel drive, gas		8.85	88.50	265	795	123.80
	2080	10 T. capacity, 120 H.P., 2 wheel drive, diesel		13.30	162	485	1,450	203.40
	2100	Generator, electric, gas engine, 1.5 KW to 3 KW		1.90	12	36	108	22.40
	2200	5 KW		2.60	17.35	52	156	31.20
	2300	10 KW		4.30	33.50	100	300	54.40
	2400	25 KW		8.35	61.50	185	555	103.80
	2500	Diesel engine, 20 KW		6.40	66.50	200	600	91.20
	2600	50 KW		12.25	71.50	215	645	141
	2700	100 KW		18.15	80	240	720	193.20
	2800	250 KW		51.65	148	445	1,325	502.20
	2850	Hammer, hydraulic, for mounting on boom, to 500 ft.-lb.		1.85	63.50	190	570	52.80
	2860	1000 ft.-lb.		3.25	103	310	930	88
	2900	Heaters, space, oil or electric, 50 MBH		1.25	7.35	22	66	14.40
	3000	100 MBH		2.26	10.65	32	96	24.50
	3100	300 MBH		7.24	33.50	100	300	77.90

01 54 | Construction Aids

01 54 33 | Rental Equipment

			Unit	Hourly Oper. Cost	Rent per Day	Rent per Week	Rent per Month	Crew Equipment Cost/Day	
40	3150	500 MBH (R314116-45)	Ea.	14.60	50	150	450	146.80	40
	3200	Hose, water, suction with coupling, 20' long, 2" diameter		.02	3.33	10	30	2.15	
	3210	3" diameter (R314116-40)		.03	5.65	17	51	3.65	
	3220	4" diameter		.03	6.35	19	57	4.05	
	3230	6" diameter (R312323-30)		.11	18.65	56	168	12.10	
	3240	8" diameter		.27	45.50	137	410	29.55	
	3250	Discharge hose with coupling, 50' long, 2" diameter		.01	2	6	18	1.30	
	3260	3" diameter		.02	3.33	10	30	2.15	
	3270	4" diameter		.02	4.67	14	42	2.95	
	3280	6" diameter		.06	11.65	35	105	7.50	
	3290	8" diameter		.33	54.50	164	490	35.45	
	3295	Insulation blower		.11	7.35	22	66	5.30	
	3300	Ladders, extension type, 16' to 36' long		.20	33	99	297	21.40	
	3400	40' to 60' long		.30	49.50	149	445	32.20	
	3405	Lance for cutting concrete		2.36	77	231	695	65.10	
	3407	Lawn mower, rotary, 22", 5HP		1.05	23.50	71	213	22.60	
	3408	48" self propelled		3.52	126	377	1,125	103.55	
	3410	Level, laser type, for pipe and sewer leveling		1.22	81	243	730	58.35	
	3430	Electronic		.81	54	162	485	38.90	
	3440	Laser type, rotating beam for grade control		1.20	79.50	239	715	57.40	
	3460	Builders level with tripod and rod		.08	12.65	38	114	8.25	
	3500	Light towers, towable, with diesel generator, 2000 watt		2.95	76.50	230	690	69.60	
	3600	4000 watt		3.60	110	330	990	94.80	
	3700	Mixer, powered, plaster and mortar, 6 C.F., 7 H.P.		1.30	21.50	65	195	23.40	
	3800	10 C.F., 9 H.P.		1.55	35	105	315	33.40	
	3850	Nailer, pneumatic		.41	27.50	82	246	19.70	
	3900	Paint sprayers complete, 8 CFM		.69	45.50	137	410	32.90	
	4000	17 CFM		1.12	74.50	223	670	53.55	
	4020	Pavers, bituminous, rubber tires, 8' wide, 50 H.P., diesel		28.05	800	2,405	7,225	705.40	
	4030	10' wide, 150 H.P.		62.65	1,525	4,595	13,800	1,420	
	4050	Crawler, 8' wide, 100 H.P., diesel		58.30	1,650	4,915	14,700	1,449	
	4060	10' wide, 150 H.P.		70.45	1,925	5,810	17,400	1,726	
	4070	Concrete paver, 12' to 24' wide, 250 H.P.		58.75	1,350	4,035	12,100	1,277	
	4080	Placer-spreader-trimmer, 24' wide, 300 H.P.		92.80	2,500	7,475	22,400	2,237	
	4100	Pump, centrifugal gas pump, 1-1/2", 4 MGPH		2.75	40	120	360	46	
	4200	2", 8 MGPH		3.55	43.50	130	390	54.40	
	4300	3", 15 MGPH		3.70	45	135	405	56.60	
	4400	6", 90 MGPH		17.75	163	490	1,475	240	
	4500	Submersible electric pump, 1-1/4", 55 GPM		.36	15	45	135	11.90	
	4600	1-1/2", 83 GPM		.39	17	51	153	13.30	
	4700	2", 120 GPM		.45	21.50	64	192	16.40	
	4800	3", 300 GPM		.95	36.50	110	330	29.60	
	4900	4", 560 GPM		6.34	145	435	1,300	137.70	
	5000	6", 1590 GPM		9.47	213	640	1,925	203.75	
	5100	Diaphragm pump, gas, single, 1-1/2" diameter		.88	40	120	360	31.05	
	5200	2" diameter		2.90	50	150	450	53.20	
	5300	3" diameter		2.90	50	150	450	53.20	
	5400	Double, 4" diameter		3.90	70	210	630	73.20	
	5500	Trash pump, self-priming, gas, 2" diameter		2.80	19.65	59	177	34.20	
	5600	Diesel, 4" diameter		6.85	55	165	495	87.80	
	5650	Diesel, 6" diameter		21.40	122	365	1,100	244.20	
	5655	Grout Pump		8.05	70	210	630	106.40	
	5660	Rollers, see division 01590-200							
	5700	Salamanders, L.P. gas fired, 100,000 BTU	Ea.	2.27	11.35	34	102	24.95	
	5705	50,000 BTU		1.70	8	24	72	18.40	
	5720	Sandblaster, portable, open top, 3 C.F. capacity		.40	20.50	62	186	15.60	
	5730	6 C.F. capacity		.65	29.50	88	264	22.80	
	5740	Accessories for above		.11	18.65	56	168	12.10	

01 54 33 | Rental Equipment

				UNIT	HOURLY OPER. COST	RENT PER DAY	RENT PER WEEK	RENT PER MONTH	CREW EQUIPMENT COST/DAY
40	5750	Sander, floor	R314116-45	Ea.	.75	18.35	55	165	17
	5760	Edger			.55	17.65	53	159	15
	5800	Saw, chain, gas engine, 18" long	R314116-40		1.35	16	48	144	20.40
	5900	36" long			.55	48.50	145	435	33.40
	5950	60" long	R312323-30		.55	50	150	450	34.40
	6000	Masonry, table mounted, 14" diameter, 5 H.P.			1.30	56	168	505	44
	6050	Portable cut-off, 8 H.P.			1.45	29.50	88	264	29.20
	6100	Circular, hand held, electric, 7-1/4" diameter			.20	5	15	45	4.60
	6200	12" diameter			.27	8.35	25	75	7.15
	6250	Wall saw, w/hydraulic power, 10 H.P			2.04	108	323.20	970	80.95
	6275	Shot blaster, walk behind, 20" wide			1.70	495	1,490	4,475	311.60
	6300	Steam cleaner, 100 gallons per hour			2.40	63.50	190	570	57.20
	6310	200 gallons per hour			3.25	76.50	230	690	72
	6340	Tar Kettle/Pot, 400 gallon			3.29	51.50	155	465	57.30
	6350	Torch, cutting, acetylene-oxygen, 150' hose			1.50	7	21	63	16.20
	6360	Hourly operating cost includes tips and gas			8.10				64.80
	6410	Toilet, portable chemical			.11	18	54	162	11.70
	6420	Recycle flush type			.13	22	66	198	14.25
	6430	Toilet, fresh water flush, garden hose,			.15	24.50	74	222	16
	6440	Hoisted, non-flush, for high rise			.13	21.50	65	195	14.05
	6450	Toilet, trailers, minimum			.22	37.50	112	335	24.15
	6460	Maximum			.67	112	337	1,000	72.75
	6465	Tractor, farm with attachment			10.30	223	670	2,000	216.40
	6470	Trailer, office, see division 01520-500							
	6500	Trailers, platform, flush deck, 2 axle, 25 ton capacity		Ea.	4.25	90	270	810	88
	6600	40 ton capacity			5.55	125	375	1,125	119.40
	6700	3 axle, 50 ton capacity			6.00	138	415	1,250	131
	6800	75 ton capacity			7.45	182	545	1,625	168.60
	6810	Trailer mounted cable reel for H.V. line work			4.56	217	652	1,950	166.90
	6820	Trailer mounted cable tensioning rig			8.96	425	1,280	3,850	327.70
	6830	Cable pulling rig			58.52	2,425	7,260	21,800	1,920
	6850	Trailer, storage, see division 01520-500							
	6900	Water tank, engine driven discharge, 5000 gallons		Ea.	5.60	120	360	1,075	116.80
	6925	10,000 gallons			7.70	168	505	1,525	162.60
	6950	Water truck, off highway, 6000 gallons			49.25	700	2,095	6,275	813
	7010	Tram car for H.V. line work, powered, 2 conductor			5.78	118	354	1,050	117.05
	7020	Transit (builder's level) with tripod			.08	12.65	38	114	8.25
	7030	Trench box, 3000 lbs. 6'x8'			.62	104	311	935	67.15
	7040	7200 lbs. 6'x20'			.71	118	354	1,050	76.50
	7050	8000 lbs., 8' x 16'			.93	156	467	1,400	100.85
	7060	9500 lbs., 8'x20'			1.20	200	599	1,800	129.40
	7065	11,000 lbs., 8'x24'			1.35	226	677	2,025	146.20
	7070	12,000 lbs., 10' x 20'			1.77	294	883	2,650	190.75
	7100	Truck, pickup, 3/4 ton, 2 wheel drive			5.75	53.50	160	480	78
	7200	4 wheel drive			5.90	63.50	190	570	85.20
	7250	Crew carrier, 9 passenger			9.40	97.50	292.80	880	133.75
	7290	Tool van, 24,000 G.V.W.			8.80	103	310	930	132.40
	7300	Tractor, 4 x 2, 30 ton capacity, 195 H.P.			13.65	165	495	1,475	208.20
	7410	250 H.P.			18.20	243	730	2,200	291.60
	7500	6 x 2, 40 ton capacity, 240 H.P.			17.45	270	810	2,425	301.60
	7600	6 x 4, 45 ton capacity, 240 H.P.			22.20	293	880	2,650	353.60
	7620	Vacuum truck, hazardous material, 2500 gallon			7.38	310	926	2,775	244.25
	7625	5,000 gallon			11.37	410	1,234.80	3,700	337.90
	7640	Tractor, with A frame, boom and winch, 225 H.P.			15.75	223	670	2,000	260
	7650	Vacuum, H.E.P.A., 16 gal., wet/dry		Ea.	.27	24	72	216	16.55
	7655	55 gal, wet/dry			.60	36	108	325	26.40
	7660	Water tank, portable			1.00	9.65	29	87	13.80
	7690	Large production vacuum loader, 3150 CFM			16.25	615	1,850	5,550	500

01 54 33 | Rental Equipment

				UNIT	HOURLY OPER. COST	RENT PER DAY	RENT PER WEEK	RENT PER MONTH	CREW EQUIPMENT COST/DAY
40	7700	Welder, electric, 200 amp	R314116 -45	Ea.	3.52	42.50	127	380	53.55
	7800	300 amp			5.04	48	144	430	69.10
	7900	Gas engine, 200 amp	R314116 -40		6.65	25.50	76	228	68.40
	8000	300 amp			9.00	28.50	86	258	89.20
	8100	Wheelbarrow, any size	R312323 -30		.06	10.65	32	96	6.90
	8200	Wrecking ball, 4000 lb.			1.95	70	210	630	57.60
50	0010	**HIGHWAY EQUIPMENT RENTAL**, without operators							
	0050	Asphalt batch plant, portable drum mixer, 100 ton/hr.		Ea.	57.45	1,350	4,025	12,100	1,265
	0060	200 ton/hr.			63.80	1,400	4,230	12,700	1,356
	0070	300 ton/hr.			74.25	1,675	5,005	15,000	1,595
	0100	Backhoe attachment, long stick, up to 185 HP, 10.5' long			.31	20.50	61	183	14.70
	0140	Up to 250 HP, 12' long			.33	22	66	198	15.85
	0180	Over 250 HP, 15' long			.44	29	87	261	20.90
	0200	Special dipper arm, up to 100 HP, 32' long			.90	59.50	179	535	43
	0240	Over 100 HP, 33' long			1.12	74.50	224	670	53.75
	0300	Concrete batch plant, portable, electric, 200 CY/Hr			10.42	495	1,490	4,475	381.35
	0500	Grader attachment, ripper/scarifier, rear mounted							
	0520	Up to 135 HP		Ea.	2.85	60	180	540	58.80
	0540	Up to 180 HP			3.40	76.50	230	690	73.20
	0580	Up to 250 HP			3.75	86.50	260	780	82
	0700	Pvmt. removal bucket, for hyd. excavator, up to 90 HP			1.45	45	135	405	38.60
	0740	Up to 200 HP			1.65	65	195	585	52.20
	0780	Over 200 HP			1.75	78.50	235	705	61
	0900	Aggregate spreader, self-propelled, 187 HP			37.80	790	2,375	7,125	777.40
	1000	Chemical spreader, 3 C.Y.			2.25	51.50	155	465	49
	1900	Hammermill, traveling, 250 HP			47.35	1,750	5,220	15,700	1,423
	2000	Horizontal borer, 3" diam, 13 HP gas driven			4.25	53.50	160	480	66
	2200	Hydromulchers, gas power, 3000 gal., for truck mounting			10.65	187	560	1,675	197.20
	2400	Joint & crack cleaner, walk behind, 25 HP			2.10	48.50	145	435	45.80
	2500	Filler, trailer mounted, 400 gal., 20 HP			6.05	182	545	1,625	157.40
	3000	Paint striper, self propelled, double line, 30 HP			5.20	155	465	1,400	134.60
	3200	Post drivers, 6" I-Beam frame, for truck mounting			8.25	415	1,240	3,725	314
	3400	Road sweeper, self propelled, 8' wide, 90 HP			25.40	485	1,450	4,350	493.20
	4000	Road mixer, self-propelled, 130 HP			30.65	595	1,780	5,350	601.20
	4100	310 HP			58.55	2,050	6,150	18,500	1,698
	4200	Cold mix paver, incl pug mill and bitumen tank,							
	4220	165 HP		Ea.	70.45	1,950	5,825	17,500	1,729
	4250	Paver, asphalt, wheel or crawler, 130 H.P., diesel			69.15	1,850	5,540	16,600	1,661
	4300	Paver, road widener, gas 1' to 6', 67 HP			31.70	680	2,035	6,100	660.60
	4400	Diesel, 2' to 14', 88 HP			42.35	995	2,980	8,950	934.80
	4600	Slipform pavers, curb and gutter, 2 track, 75 HP			25.20	655	1,960	5,875	593.60
	4700	4 track, 165 HP			35.25	760	2,280	6,850	738
	4800	Median barrier, 215 HP			35.85	790	2,370	7,100	760.80
	4901	Trailer, low bed, 75 ton capacity			7.90	178	535	1,600	170.20
	5000	Road planer, walk behind, 10" cutting width, 10 HP			2.05	27	81	243	32.60
	5100	Self propelled, 12" cutting width, 64 HP			5.45	150	450	1,350	133.60
	5200	Pavement profiler, 4' to 6' wide, 450 HP			161.05	2,925	8,755	26,300	3,039
	5300	8' to 10' wide, 750 HP			256.35	4,350	13,040	39,100	4,659
	5400	Roadway plate, steel, 1"x8'x20'			.06	10.35	31	93	6.70
	5600	Stabilizer,self-propelled, 150 HP			29.65	565	1,690	5,075	575.20
	5700	310 HP			48.65	1,200	3,570	10,700	1,103
	5800	Striper, thermal, truck mounted 120 gal. paint, 150 H.P.			33.15	485	1,450	4,350	555.20
	6000	Tar kettle, 330 gal., trailer mounted			2.97	36.50	110	330	45.75
	7000	Tunnel locomotive, diesel, 8 to 12 ton			22.10	555	1,670	5,000	510.80
	7005	Electric, 10 ton			21.20	630	1,895	5,675	548.60
	7010	Muck cars, 1/2 C.Y. capacity			1.55	21	63	189	25
	7020	1 C.Y. capacity			1.75	29	87	261	31.40
	7030	2 C.Y. capacity			1.90	33.50	100	300	35.20

01 54 33 | Rental Equipment

		01 54 33 | Rental Equipment	UNIT	HOURLY OPER. COST	RENT PER DAY	RENT PER WEEK	RENT PER MONTH	CREW EQUIPMENT COST/DAY	
50	7040	Side dump, 2 C.Y. capacity	Ea.	2.10	41.50	125	375	41.80	50
	7050	3 C.Y. capacity		2.80	48.50	145	435	51.40	
	7060	5 C.Y. capacity		3.95	60	180	540	67.60	
	7100	Ventilating blower for tunnel, 7-1/2 H.P.		1.25	47	141	425	38.20	
	7110	10 H.P.		1.44	48.50	146	440	40.70	
	7120	20 H.P.		2.33	62	186	560	55.85	
	7140	40 H.P.		4.04	88.50	265	795	85.30	
	7160	60 H.P.		6.16	137	410	1,225	131.30	
	7175	75 H.P.		7.89	182	545	1,625	172.10	
	7180	200 H.P.	↓	17.60	270	810	2,425	302.80	
	7800	Windrow loader, elevating		35.00	905	2,715	8,150	823	
60	0010	**LIFTING** & HOISTING EQUIPMENT RENTAL, without operators (R015433-10)							60
	0120	Aerial lift truck, 2 person, to 80'	Ea.	19.35	600	1,805	5,425	515.80	
	0140	Boom work platform, 40' snorkel (R015433-15)		10.60	222	665	2,000	217.80	
	0150	Crane, flatbed mntd, 3 ton cap.		14.65	197	590	1,775	235.20	
	0200	Crane, climbing, 106' jib, 6000 lb. capacity, 410 FPM (R312316-45)		54.75	1,375	4,150	12,500	1,268	
	0300	101' jib, 10,250 lb. capacity, 270 FPM	↓	60.25	1,750	5,250	15,800	1,532	
	0400	Tower, static, 130' high, 106' jib,							
	0500	6200 lb. capacity at 400 FPM	Ea.	57.95	1,600	4,790	14,400	1,422	
	0600	Crawler mounted, lattice boom, 1/2 C.Y., 15 tons at 12' radius		22.08	595	1,790	5,375	534.65	
	0700	3/4 C.Y., 20 tons at 12' radius		29.44	715	2,140	6,425	663.50	
	0800	1 C.Y., 25 tons at 12' radius		39.25	945	2,830	8,500	880	
	0900	1-1/2 C.Y., 40 tons at 12' radius		43.65	975	2,925	8,775	934.20	
	1000	2 C.Y., 50 tons at 12' radius		54.25	1,375	4,155	12,500	1,265	
	1100	3 C.Y., 75 tons at 12' radius		51.75	1,350	4,060	12,200	1,226	
	1200	100 ton capacity, 60' boom		66.30	1,775	5,300	15,900	1,590	
	1300	165 ton capacity, 60' boom		96.50	2,300	6,890	20,700	2,150	
	1400	200 ton capacity, 70' boom		102.55	2,450	7,375	22,100	2,295	
	1500	350 ton capacity, 80' boom		149.45	3,475	10,450	31,400	3,286	
	1600	Truck mounted, lattice boom, 6 x 4, 20 tons at 10' radius		30.88	1,175	3,500	10,500	947.05	
	1700	25 tons at 10' radius		32.98	1,250	3,740	11,200	1,012	
	1800	8 x 4, 30 tons at 10' radius		42.67	1,325	3,970	11,900	1,135	
	1900	40 tons at 12' radius		36.20	1,400	4,200	12,600	1,130	
	2000	8 x 4, 60 tons at 15' radius		45.96	1,475	4,440	13,300	1,256	
	2050	82 tons at 15' radius		48.65	1,550	4,670	14,000	1,323	
	2100	90 tons at 15' radius		50.65	1,700	5,110	15,300	1,427	
	2200	115 tons at 15' radius		51.45	1,900	5,675	17,000	1,547	
	2300	150 tons at 18' radius		47.40	1,925	5,800	17,400	1,539	
	2350	165 tons at 18' radius		80.90	2,275	6,825	20,500	2,012	
	2400	Truck mounted, hydraulic, 12 ton capacity		35.20	590	1,775	5,325	636.60	
	2500	25 ton capacity		35.35	615	1,840	5,525	650.80	
	2550	33 ton capacity		36.20	645	1,935	5,800	676.60	
	2560	40 ton capacity		34.65	620	1,855	5,575	648.20	
	2600	55 ton capacity		51.45	900	2,700	8,100	951.60	
	2700	80 ton capacity		61.10	950	2,850	8,550	1,059	
	2720	100 ton capacity		83.70	2,425	7,250	21,800	2,120	
	2740	120 ton capacity		90.35	2,625	7,900	23,700	2,303	
	2760	150 ton capacity		110.25	3,325	9,990	30,000	2,880	
	2800	Self-propelled, 4 x 4, with telescoping boom, 5 ton		16.15	293	880	2,650	305.20	
	2900	12-1/2 ton capacity		27.05	535	1,600	4,800	536.40	
	3000	15 ton capacity		28.20	595	1,785	5,350	582.60	
	3050	20 ton capacity		29.05	620	1,855	5,575	603.40	
	3100	25 ton capacity		33.35	715	2,150	6,450	696.80	
	3150	40 ton capacity		48.00	885	2,660	7,975	916	
	3200	Derricks, guy, 20 ton capacity, 60' boom, 75' mast		15.61	340	1,016	3,050	328.10	
	3300	100' boom, 115' mast		24.93	580	1,740	5,225	547.45	
	3400	Stiffleg, 20 ton capacity, 70' boom, 37' mast	↓	17.67	435	1,310	3,925	403.35	

01 54 33 | Rental Equipment

			UNIT	HOURLY OPER. COST	RENT PER DAY	RENT PER WEEK	RENT PER MONTH	CREW EQUIPMENT COST/DAY	
60	3500	100' boom, 47' mast R015433-10	Ea.	27.59	705	2,120	6,350	644.70	60
	3550	Helicopter, small, lift to 1250 lbs. maximum, w/pilot		74.47	2,725	8,210	24,600	2,238	
	3600	Hoists, chain type, overhead, manual, 3/4 ton R015433-15		.10	1.33	4	12	1.60	
	3900	10 ton		.60	10	30	90	10.80	
	4000	Hoist and tower, 5000 lb. cap., portable electric, 40' high R312316-45		4.25	195	585	1,750	151	
	4100	For each added 10' section, add		.09	15.35	46	138	9.90	
	4200	Hoist and single tubular tower, 5000 lb. electric, 100' high		5.74	272	817	2,450	209.30	
	4300	For each added 6'-6" section, add		.15	25.50	77	231	16.60	
	4400	Hoist and double tubular tower, 5000 lb., 100' high		6.15	300	900	2,700	229.20	
	4500	For each added 6'-6" section, add		.17	29	87	261	18.75	
	4550	Hoist and tower, mast type, 6000 lb., 100' high		6.65	310	933	2,800	239.80	
	4570	For each added 10' section, add		.11	18.65	56	168	12.10	
	4600	Hoist and tower, personnel, electric, 2000 lb., 100' @ 125 FPM		13.77	830	2,490	7,475	608.15	
	4700	3000 lb., 100' @ 200 FPM		15.70	935	2,810	8,425	687.60	
	4800	3000 lb., 150' @ 300 FPM		17.40	1,050	3,150	9,450	769.20	
	4900	4000 lb., 100' @ 300 FPM		18.03	1,075	3,210	9,625	786.25	
	5000	6000 lb., 100' @ 275 FPM	↓	19.49	1,125	3,370	10,100	829.90	
	5100	For added heights up to 500', add	L.F.	.01	1.67	5	15	1.10	
	5200	Jacks, hydraulic, 20 ton	Ea.	.05	4.67	14	42	3.20	
	5500	100 ton	"	.30	13.35	40	120	10.40	
	6000	Jacks, hydraulic, climbing with 50' jackrods							
	6010	and control consoles, minimum 3 mo. rental							
	6100	30 ton capacity	Ea.	1.68	112	336	1,000	80.65	
	6150	For each added 10' jackrod section, add		.05	3.33	10	30	2.40	
	6300	50 ton capacity		2.71	180	541	1,625	129.90	
	6350	For each added 10' jackrod section, add		.06	4	12	36	2.90	
	6500	125 ton capacity		7.10	475	1,420	4,250	340.80	
	6550	For each added 10' jackrod section, add		.49	32.50	97	291	23.30	
	6600	Cable jack, 10 ton capacity with 200' cable		1.41	93.50	281	845	67.50	
	6650	For each added 50' of cable, add	↓	.16	10.35	31	93	7.50	
70	0010	**WELLPOINT EQUIPMENT RENTAL**, without operators R312319-90							70
	0020	Based on 2 months rental							
	0100	Combination jetting & wellpoint pump, 60 H.P. diesel	Ea.	10.93	278	833	2,500	254.05	
	0200	High pressure gas jet pump, 200 H.P., 300 psi	"	21.98	237	712	2,125	318.25	
	0300	Discharge pipe, 8" diameter	L.F.	.01	.45	1.35	4.05	.35	
	0350	12" diameter		.01	.67	2	6	.50	
	0400	Header pipe, flows up to 150 G.P.M., 4" diameter		.01	.41	1.22	3.66	.30	
	0500	400 G.P.M., 6" diameter		.01	.48	1.45	4.35	.35	
	0600	800 G.P.M., 8" diameter		.01	.67	2	6	.50	
	0700	1500 G.P.M., 10" diameter		.01	.70	2.10	6.30	.50	
	0800	2500 G.P.M., 12" diameter		.02	1.32	3.97	11.90	.95	
	0900	4500 G.P.M., 16" diameter		.03	1.69	5.08	15.25	1.25	
	0950	For quick coupling aluminum and plastic pipe, add	↓	.03	1.75	5.25	15.75	1.30	
	1100	Wellpoint, 25' long, with fittings & riser pipe, 1-1/2" or 2" diameter	Ea.	.05	3.50	10.49	31.50	2.50	
	1200	Wellpoint pump, diesel powered, 4" diameter, 20 H.P.		5.06	160	480	1,450	136.50	
	1300	6" diameter, 30 H.P.		6.72	199	596	1,800	172.95	
	1400	8" suction, 40 H.P.		9.12	272	817	2,450	236.35	
	1500	10" suction, 75 H.P.		13.06	320	955	2,875	295.50	
	1600	12" suction, 100 H.P.		19.21	510	1,530	4,600	459.70	
	1700	12" suction, 175 H.P.	↓	26.64	560	1,680	5,050	549.10	
80	0010	**MARINE EQUIPMENT RENTAL**, without operators							80
	0200	Barge, 400 Ton, 30' wide x 90' long	Ea.	17.00	240	720	2,150	280	
	0240	800 Ton, 45' wide x 90' long		28.10	345	1,030	3,100	430.80	
	2000	Tugboat, diesel, 100 HP		18.80	170	510	1,525	252.40	
	2040	250 HP		35.50	315	945	2,825	473	
	2080	380 HP	Ea.	80.55	935	2,805	8,425	1,205	

Crew No.	Bare Costs		Incl. Subs O & P		Cost Per Labor-Hour	

Crew A-1	Hr.	Daily	Hr.	Daily	Bare Costs	Incl. O&P
1 Building Laborer	$27.40	$219.20	$45.40	$363.20	$27.40	$45.40
1 Concrete saw, gas manual		49.00		53.90	6.13	6.74
8 L.H., Daily Totals		$268.20		$417.10	$33.53	$52.14

Crew A-1A	Hr.	Daily	Hr.	Daily	Bare Costs	Incl. O&P
1 Skilled Worker	$36.50	$292.00	$59.70	$477.60	$36.50	$59.70
1 Shot Blaster, 20"		311.60		342.75	38.95	42.85
8 L.H., Daily Totals		$603.60		$820.35	$75.45	$102.55

Crew A-1B	Hr.	Daily	Hr.	Daily	Bare Costs	Incl. O&P
1 Building Laborer	$27.40	$219.20	$45.40	$363.20	$27.40	$45.40
1 Concr. saw, gas, self-prop.		107.00		117.70	13.38	14.71
8 L.H., Daily Totals		$326.20		$480.90	$40.78	$60.11

Crew A-1C	Hr.	Daily	Hr.	Daily	Bare Costs	Incl. O&P
1 Building Laborer	$27.40	$219.20	$45.40	$363.20	$27.40	$45.40
1 Brush saw		20.40		22.45	2.55	2.81
8 L.H., Daily Totals		$239.60		$385.65	$29.95	$48.21

Crew A-1D	Hr.	Daily	Hr.	Daily	Bare Costs	Incl. O&P
1 Building Laborer	$27.40	$219.20	$45.40	$363.20	$27.40	$45.40
1 Vibrating plate, gas, 18"		26.40		29.05	3.30	3.63
8 L.H., Daily Totals		$245.60		$392.25	$30.70	$49.03

Crew A-1E	Hr.	Daily	Hr.	Daily	Bare Costs	Incl. O&P
1 Building Laborer	$27.40	$219.20	$45.40	$363.20	$27.40	$45.40
1 Vibrating plate, gas, 21"		34.60		38.05	4.33	4.76
8 L.H., Daily Totals		$253.80		$401.25	$31.73	$50.16

Crew A-1F	Hr.	Daily	Hr.	Daily	Bare Costs	Incl. O&P
1 Building Laborer	$27.40	$219.20	$45.40	$363.20	$27.40	$45.40
1 Rammer/tamper, gas, 8"		37.00		40.70	4.63	5.09
8 L.H., Daily Totals		$256.20		$403.90	$32.03	$50.49

Crew A-1G	Hr.	Daily	Hr.	Daily	Bare Costs	Incl. O&P
1 Building Laborer	$27.40	$219.20	$45.40	$363.20	$27.40	$45.40
1 Rammer/tamper, gas, 8"		37.00		40.70	4.63	5.09
8 L.H., Daily Totals		$256.20		$403.90	$32.03	$50.49

Crew A-1H	Hr.	Daily	Hr.	Daily	Bare Costs	Incl. O&P
1 Building Laborer	$27.40	$219.20	$45.40	$363.20	$27.40	$45.40
1 Pressure washer		57.20		62.90	7.15	7.87
8 L.H., Daily Totals		$276.40		$426.10	$34.55	$53.27

Crew A-1J	Hr.	Daily	Hr.	Daily	Bare Costs	Incl. O&P
1 Building Laborer	$27.40	$219.20	$45.40	$363.20	$27.40	$45.40
1 Rototiller		103.55		113.90	12.94	14.24
8 L.H., Daily Totals		$322.75		$477.10	$40.34	$59.64

Crew A-1K	Hr.	Daily	Hr.	Daily	Bare Costs	Incl. O&P
1 Building Laborer	$27.40	$219.20	$45.40	$363.20	$27.40	$45.40
1 Lawn aerator		22.60		24.85	2.83	3.11
8 L.H., Daily Totals		$241.80		$388.05	$30.23	$48.51

Crew A-1L	Hr.	Daily	Hr.	Daily	Bare Costs	Incl. O&P
1 Building Laborer	$27.40	$219.20	$45.40	$363.20	$27.40	$45.40
1 Power blower/vacuum		22.60		24.85	2.83	3.11
8 L.H., Daily Totals		$241.80		$388.05	$30.23	$48.51

Crew No.	Bare Costs		Incl. Subs O & P		Cost Per Labor-Hour	

Crew A-1M	Hr.	Daily	Hr.	Daily	Bare Costs	Incl. O&P
1 Building Laborer	$27.40	$219.20	$45.40	$363.20	$27.40	$45.40
1 Snow blower		103.55		113.90	12.94	14.24
8 L.H., Daily Totals		$322.75		$477.10	$40.34	$59.64

Crew A-2	Hr.	Daily	Hr.	Daily	Bare Costs	Incl. O&P
2 Laborers	$27.40	$438.40	$45.40	$726.40	$27.47	$45.23
1 Truck Driver (light)	27.60	220.80	44.90	359.20		
1 Light Truck, 1.5 Ton		128.80		141.70	5.37	5.90
24 L.H., Daily Totals		$788.00		$1227.30	$32.84	$51.13

Crew A-2A	Hr.	Daily	Hr.	Daily	Bare Costs	Incl. O&P
2 Laborers	$27.40	$438.40	$45.40	$726.40	$27.47	$45.23
1 Truck Driver (light)	27.60	220.80	44.90	359.20		
1 Light Truck, 1.5 Ton		128.80		141.70		
1 Concrete Saw		107.00		117.70	9.83	10.81
24 L.H., Daily Totals		$895.00		$1345.00	$37.30	$56.04

Crew A-3	Hr.	Daily	Hr.	Daily	Bare Costs	Incl. O&P
1 Truck Driver (heavy)	$28.35	$226.80	$46.15	$369.20	$28.35	$46.15
1 Dump Truck, 12 Ton		326.60		359.25	40.83	44.91
8 L.H., Daily Totals		$553.40		$728.45	$69.18	$91.06

Crew A-3A	Hr.	Daily	Hr.	Daily	Bare Costs	Incl. O&P
1 Truck Driver (light)	$27.60	$220.80	$44.90	$359.20	$27.60	$44.90
1 Pickup Truck (4x4)		85.20		93.70	10.65	11.72
8 L.H., Daily Totals		$306.00		$452.90	$38.25	$56.62

Crew A-3B	Hr.	Daily	Hr.	Daily	Bare Costs	Incl. O&P
1 Equip. Oper. (medium)	$36.70	$293.60	$57.95	$463.60	$32.53	$52.05
1 Truck Driver (heavy)	28.35	226.80	46.15	369.20		
1 Dump Truck, 16 Ton		476.60		524.25		
1 F.E. Loader, 3 C.Y.		305.40		335.95	48.88	53.76
16 L.H., Daily Totals		$1302.40		$1693.00	$81.41	$105.81

Crew A-3C	Hr.	Daily	Hr.	Daily	Bare Costs	Incl. O&P
1 Equip. Oper. (light)	$35.20	$281.60	$55.60	$444.80	$35.20	$55.60
1 Wheeled Skid Steer Loader		202.00		222.20	25.25	27.78
8 L.H., Daily Totals		$483.60		$667.00	$60.45	$83.38

Crew A-3D	Hr.	Daily	Hr.	Daily	Bare Costs	Incl. O&P
1 Truck Driver, Light	$27.60	$220.80	$44.90	$359.20	$27.60	$44.90
1 Pickup Truck (4x4)		85.20		93.70		
1 Flatbed Trailer, 25 Ton		88.00		96.80	21.65	23.82
8 L.H., Daily Totals		$394.00		$549.70	$49.25	$68.72

Crew A-3E	Hr.	Daily	Hr.	Daily	Bare Costs	Incl. O&P
1 Equip. Oper. (crane)	$38.10	$304.80	$60.15	$481.20	$33.23	$53.15
1 Truck Driver (heavy)	28.35	226.80	46.15	369.20		
1 Pickup Truck (4x4)		85.20		93.70	5.33	5.86
16 L.H., Daily Totals		$616.80		$944.10	$38.56	$59.01

Crew A-3F	Hr.	Daily	Hr.	Daily	Bare Costs	Incl. O&P
1 Equip. Oper. (crane)	$38.10	$304.80	$60.15	$481.20	$33.23	$53.15
1 Truck Driver (heavy)	28.35	226.80	46.15	369.20		
1 Pickup Truck (4x4)		85.20		93.70		
1 Tractor, 6x2, 40 Ton Cap.		301.60		331.75		
1 Lowbed Trailer, 75 Ton		170.20		187.20	34.81	38.29
16 L.H., Daily Totals		$1088.60		$1463.05	$68.04	$91.44

Crew No.	Bare Costs		Incl. Subs O & P		Cost Per Labor-Hour	
Crew A-3G	**Hr.**	**Daily**	**Hr.**	**Daily**	**Bare Costs**	**Incl. O&P**
1 Equip. Oper. (crane)	$38.10	$304.80	$60.15	$481.20	$33.23	$53.15
1 Truck Driver (heavy)	28.35	226.80	46.15	369.20		
1 Pickup Truck (4x4)		85.20		93.70		
1 Tractor, 6x4, 45 Ton Cap.		353.60		388.95		
1 Lowbed Trailer, 75 Ton		170.20		187.20	38.06	41.87
16 L.H., Daily Totals		$1140.60		$1520.25	$71.29	$95.02
Crew A-4	**Hr.**	**Daily**	**Hr.**	**Daily**	**Bare Costs**	**Incl. O&P**
2 Carpenters	$35.55	$568.80	$58.90	$942.40	$34.27	$56.23
1 Painter, Ordinary	31.70	253.60	50.90	407.20		
24 L.H., Daily Totals		$822.40		$1349.60	$34.27	$56.23
Crew A-5	**Hr.**	**Daily**	**Hr.**	**Daily**	**Bare Costs**	**Incl. O&P**
2 Laborers	$27.40	$438.40	$45.40	$726.40	$27.42	$45.34
.25 Truck Driver (light)	27.60	55.20	44.90	89.80		
.25 Light Truck, 1.5 Ton		32.20		35.40	1.79	1.97
18 L.H., Daily Totals		$525.80		$851.60	$29.21	$47.31
Crew A-6	**Hr.**	**Daily**	**Hr.**	**Daily**	**Bare Costs**	**Incl. O&P**
1 Instrument Man	$36.50	$292.00	$59.70	$477.60	$35.38	$57.45
1 Rodman/Chainman	34.25	274.00	55.20	441.60		
1 Laser Transit/Level		58.35		64.20	3.64	4.01
16 L.H., Daily Totals		$624.35		$983.40	$39.02	$61.46
Crew A-7	**Hr.**	**Daily**	**Hr.**	**Daily**	**Bare Costs**	**Incl. O&P**
1 Chief Of Party	$45.80	$366.40	$73.60	$588.80	$38.85	$62.83
1 Instrument Man	36.50	292.00	59.70	477.60		
1 Rodman/Chainman	34.25	274.00	55.20	441.60		
1 Laser Transit/Level		58.35		64.20	2.43	2.67
24 L.H., Daily Totals		$990.75		$1572.20	$41.28	$65.50
Crew A-8	**Hr.**	**Daily**	**Hr.**	**Daily**	**Bare Costs**	**Incl. O&P**
1 Chief Of Party	$45.80	$366.40	$73.60	$588.80	$37.70	$60.93
1 Instrument Man	36.50	292.00	59.70	477.60		
2 Rodmen/Chainmen	34.25	548.00	55.20	883.20		
1 Laser Transit/Level		58.35		64.20	1.82	2.00
32 L.H., Daily Totals		$1264.75		$2013.80	$39.52	$62.93
Crew A-9	**Hr.**	**Daily**	**Hr.**	**Daily**	**Bare Costs**	**Incl. O&P**
1 Asbestos Foreman	$39.55	$316.40	$64.50	$516.00	$39.11	$63.80
7 Asbestos Workers	39.05	2186.80	63.70	3567.20		
64 L.H., Daily Totals		$2503.20		$4083.20	$39.11	$63.80
Crew A-10	**Hr.**	**Daily**	**Hr.**	**Daily**	**Bare Costs**	**Incl. O&P**
1 Asbestos Foreman	$39.55	$316.40	$64.50	$516.00	$39.11	$63.80
7 Asbestos Workers	39.05	2186.80	63.70	3567.20		
64 L.H., Daily Totals		$2503.20		$4083.20	$39.11	$63.80
Crew A-10A	**Hr.**	**Daily**	**Hr.**	**Daily**	**Bare Costs**	**Incl. O&P**
1 Asbestos Foreman	$39.55	$316.40	$64.50	$516.00	$39.22	$63.97
2 Asbestos Workers	39.05	624.80	63.70	1019.20		
24 L.H., Daily Totals		$941.20		$1535.20	$39.22	$63.97
Crew A-10B	**Hr.**	**Daily**	**Hr.**	**Daily**	**Bare Costs**	**Incl. O&P**
1 Asbestos Foreman	$39.55	$316.40	$64.50	$516.00	$39.18	$63.90
3 Asbestos Workers	39.05	937.20	63.70	1528.80		
32 L.H., Daily Totals		$1253.60		$2044.80	$39.18	$63.90

Crew No.	Bare Costs		Incl. Subs O & P		Cost Per Labor-Hour	
Crew A-10C	**Hr.**	**Daily**	**Hr.**	**Daily**	**Bare Costs**	**Incl. O&P**
3 Asbestos Workers	$39.05	$937.20	$63.70	$1528.80	$39.05	$63.70
1 Flatbed Truck		128.80		141.70	5.37	5.90
24 L.H., Daily Totals		$1066.00		$1670.50	$44.42	$69.60
Crew A-10D	**Hr.**	**Daily**	**Hr.**	**Daily**	**Bare Costs**	**Incl. O&P**
2 Asbestos Workers	$39.05	$624.80	$63.70	$1019.20	$37.16	$59.70
1 Equip. Oper. (crane)	38.10	304.80	60.15	481.20		
1 Equip. Oper. Oiler	32.45	259.60	51.25	410.00		
1 Hydraulic Crane, 33 Ton		676.60		744.25	21.14	23.26
32 L.H., Daily Totals		$1865.80		$2654.65	$58.30	$82.96
Crew A-11	**Hr.**	**Daily**	**Hr.**	**Daily**	**Bare Costs**	**Incl. O&P**
1 Asbestos Foreman	$39.55	$316.40	$64.50	$516.00	$39.11	$63.80
7 Asbestos Workers	39.05	2186.80	63.70	3567.20		
2 Chipping Hammers		30.40		33.45	.48	.52
64 L.H., Daily Totals		$2533.60		$4116.65	$39.59	$64.32
Crew A-12	**Hr.**	**Daily**	**Hr.**	**Daily**	**Bare Costs**	**Incl. O&P**
1 Asbestos Foreman	$39.55	$316.40	$64.50	$516.00	$39.11	$63.80
7 Asbestos Workers	39.05	2186.80	63.70	3567.20		
1 Large Prod. Vac. Loader		500.00		550.00	7.81	8.59
64 L.H., Daily Totals		$3003.20		$4633.20	$46.92	$72.39
Crew A-13	**Hr.**	**Daily**	**Hr.**	**Daily**	**Bare Costs**	**Incl. O&P**
1 Equip. Oper. (light)	$35.20	$281.60	$55.60	$444.80	$35.20	$55.60
1 Large Prod. Vac. Loader		500.00		550.00	62.50	68.75
8 L.H., Daily Totals		$781.60		$994.80	$97.70	$124.35
Crew B-1	**Hr.**	**Daily**	**Hr.**	**Daily**	**Bare Costs**	**Incl. O&P**
1 Laborer Foreman (outside)	$29.40	$235.20	$48.70	$389.60	$28.07	$46.50
2 Laborers	27.40	438.40	45.40	726.40		
24 L.H., Daily Totals		$673.60		$1116.00	$28.07	$46.50
Crew B-1A	**Hr.**	**Daily**	**Hr.**	**Daily**	**Bare Costs**	**Incl. O&P**
1 Laborer Foreman	$29.40	$235.20	$48.70	$389.60	$28.07	$46.50
2 Laborers	27.40	438.40	45.40	726.40		
2 Cutting Torches		32.40		35.65		
2 Gases		129.60		142.55	6.75	7.43
24 L.H., Daily Totals		$835.60		$1294.20	$34.82	$53.93
Crew B-1B	**Hr.**	**Daily**	**Hr.**	**Daily**	**Bare Costs**	**Incl. O&P**
1 Laborer Foreman	$29.40	$235.20	$48.70	$389.60	$30.58	$49.91
2 Laborers	27.40	438.40	45.40	726.40		
1 Equip. Oper. (crane)	38.10	304.80	60.15	481.20		
2 Cutting Torches		32.40		35.65		
2 Gases		129.60		142.55		
1 Hyd. Crane, 12 Ton		636.60		700.25	24.96	27.45
32 L.H., Daily Totals		$1777.00		$2475.65	$55.54	$77.36
Crew B-2	**Hr.**	**Daily**	**Hr.**	**Daily**	**Bare Costs**	**Incl. O&P**
1 Laborer Foreman (outside)	$29.40	$235.20	$48.70	$389.60	$27.80	$46.06
4 Laborers	27.40	876.80	45.40	1452.80		
40 L.H., Daily Totals		$1112.00		$1842.40	$27.80	$46.06

Crews

Crew No.	Bare Costs		Incl. Subs O & P		Cost Per Labor-Hour	
Crew B-3	**Hr.**	**Daily**	**Hr.**	**Daily**	**Bare Costs**	**Incl. O&P**
1 Laborer Foreman (outside)	$29.40	$235.20	$48.70	$389.60	$29.60	$48.29
2 Laborers	27.40	438.40	45.40	726.40		
1 Equip. Oper. (med.)	36.70	293.60	57.95	463.60		
2 Truck Drivers (heavy)	28.35	453.60	46.15	738.40		
1 F.E. Loader, T.M., 2.5 C.Y.		794.00		873.40		
2 Dump Trucks, 16 Ton		953.20		1048.50	36.40	40.04
48 L.H., Daily Totals		$3168.00		$4239.90	$66.00	$88.33
Crew B-3A	**Hr.**	**Daily**	**Hr.**	**Daily**	**Bare Costs**	**Incl. O&P**
4 Laborers	$27.40	$876.80	$45.40	$1452.80	$29.26	$47.91
1 Equip. Oper. (med.)	36.70	293.60	57.95	463.60		
1 Hyd. Excavator, 1.5 C.Y.		722.80		795.10	18.07	19.88
40 L.H., Daily Totals		$1893.20		$2711.50	$47.33	$67.79
Crew B-3B	**Hr.**	**Daily**	**Hr.**	**Daily**	**Bare Costs**	**Incl. O&P**
2 Laborers	$27.40	$438.40	$45.40	$726.40	$29.96	$48.73
1 Equip. Oper. (med.)	36.70	293.60	57.95	463.60		
1 Truck Driver (heavy)	28.35	226.80	46.15	369.20		
1 Backhoe Loader, 80 H.P.		277.60		305.35		
1 Dump Truck, 16 Ton		476.60		524.25	23.57	25.93
32 L.H., Daily Totals		$1713.00		$2388.80	$53.53	$74.66
Crew B-3C	**Hr.**	**Daily**	**Hr.**	**Daily**	**Bare Costs**	**Incl. O&P**
3 Laborers	$27.40	$657.60	$45.40	$1089.60	$29.73	$48.54
1 Equip. Oper. (med.)	36.70	293.60	57.95	463.60		
1 F.E. Crawler Ldr, 4 C.Y.		1110.00		1221.00	34.69	38.16
32 L.H., Daily Totals		$2061.20		$2774.20	$64.42	$86.70
Crew B-4	**Hr.**	**Daily**	**Hr.**	**Daily**	**Bare Costs**	**Incl. O&P**
1 Laborer Foreman (outside)	$29.40	$235.20	$48.70	$389.60	$27.89	$46.08
4 Laborers	27.40	876.80	45.40	1452.80		
1 Truck Driver (heavy)	28.35	226.80	46.15	369.20		
1 Tractor, 4 x 2, 195 H.P.		208.20		229.00		
1 Platform Trailer		119.40		131.35	6.83	7.51
48 L.H., Daily Totals		$1666.40		$2571.95	$34.72	$53.59
Crew B-5	**Hr.**	**Daily**	**Hr.**	**Daily**	**Bare Costs**	**Incl. O&P**
1 Laborer Foreman (outside)	$29.40	$235.20	$48.70	$389.60	$30.34	$49.46
4 Laborers	27.40	876.80	45.40	1452.80		
2 Equip. Oper. (med.)	36.70	587.20	57.95	927.20		
1 Air Compr., 250 C.F.M.		119.80		131.80		
2 Air Tools & Accessories		18.00		19.80		
2-50 Ft. Air Hoses, 1.5″ Dia.		9.40		10.35		
1 F.E. Loader, T.M., 2.5 C.Y.		794.00		873.40	16.81	18.49
56 L.H., Daily Totals		$2640.40		$3804.95	$47.15	$67.95
Crew B-6	**Hr.**	**Daily**	**Hr.**	**Daily**	**Bare Costs**	**Incl. O&P**
2 Laborers	$27.40	$438.40	$45.40	$726.40	$30.00	$48.80
1 Equip. Oper. (light)	35.20	281.60	55.60	444.80		
1 Backhoe Loader, 48 H.P.		226.20		248.80	9.43	10.37
24 L.H., Daily Totals		$946.20		$1420.00	$39.43	$59.17
Crew B-6B	**Hr.**	**Daily**	**Hr.**	**Daily**	**Bare Costs**	**Incl. O&P**
2 Laborer Foremen (out)	$29.40	$470.40	$48.70	$779.20	$28.07	$46.50
4 Laborers	27.40	876.80	45.40	1452.80		
1 Winch Truck		305.20		335.70		
1 Flatbed Truck		128.80		141.70		
1 Butt Fusion Machine		347.20		381.90	16.28	17.90
48 L.H., Daily Totals		$2128.40		$3091.30	$44.35	$64.40

Crew No.	Bare Costs		Incl. Subs O & P		Cost Per Labor-Hour	
Crew B-7	**Hr.**	**Daily**	**Hr.**	**Daily**	**Bare Costs**	**Incl. O&P**
1 Laborer Foreman (outside)	$29.40	$235.20	$48.70	$389.60	$29.28	$48.04
4 Laborers	27.40	876.80	45.40	1452.80		
1 Equip. Oper. (med.)	36.70	293.60	57.95	463.60		
1 Chipping Machine		173.00		190.30		
1 F.E. Loader, T.M., 2.5 C.Y.		794.00		873.40		
2 Chain Saws, 36″		66.80		73.50	21.54	23.69
48 L.H., Daily Totals		$2439.40		$3443.20	$50.82	$71.73
Crew B-7A	**Hr.**	**Daily**	**Hr.**	**Daily**	**Bare Costs**	**Incl. O&P**
2 Laborers	$27.40	$438.40	$45.40	$726.40	$30.00	$48.80
1 Equip. Oper. (light)	35.20	281.60	55.60	444.80		
1 Rake w/Tractor		192.90		212.20		
2 Chain Saws, 18″		40.80		44.90	9.74	10.71
24 L.H., Daily Totals		$953.70		$1428.30	$39.74	$59.51
Crew B-8	**Hr.**	**Daily**	**Hr.**	**Daily**	**Bare Costs**	**Incl. O&P**
1 Laborer Foreman (outside)	$29.40	$235.20	$48.70	$389.60	$30.84	$49.87
2 Laborers	27.40	438.40	45.40	726.40		
2 Equip. Oper. (med.)	36.70	587.20	57.95	927.20		
1 Equip. Oper. Oiler	32.45	259.60	51.25	410.00		
2 Truck Drivers (heavy)	28.35	453.60	46.15	738.40		
1 Hyd. Crane, 25 Ton		650.80		715.90		
1 F.E. Loader, T.M., 2.5 C.Y.		794.00		873.40		
2 Dump Trucks, 16 Ton		953.20		1048.50	37.47	41.22
64 L.H., Daily Totals		$4372.00		$5829.40	$68.31	$91.09
Crew B-9	**Hr.**	**Daily**	**Hr.**	**Daily**	**Bare Costs**	**Incl. O&P**
1 Laborer Foreman (outside)	$29.40	$235.20	$48.70	$389.60	$27.80	$46.06
4 Laborers	27.40	876.80	45.40	1452.80		
1 Air Compr., 250 C.F.M.		119.80		131.80		
2 Air Tools & Accessories		18.00		19.80		
2-50 Ft. Air Hoses, 1.5″ Dia.		9.40		10.35	3.68	4.05
40 L.H., Daily Totals		$1259.20		$2004.35	$31.48	$50.11
Crew B-9A	**Hr.**	**Daily**	**Hr.**	**Daily**	**Bare Costs**	**Incl. O&P**
2 Laborers	$27.40	$438.40	$45.40	$726.40	$27.72	$45.65
1 Truck Driver (heavy)	28.35	226.80	46.15	369.20		
1 Water Tanker		116.80		128.50		
1 Tractor		208.20		229.00		
2-50 Ft. Disch. Hoses		4.30		4.75	13.72	15.09
24 L.H., Daily Totals		$994.50		$1457.85	$41.44	$60.74
Crew B-9B	**Hr.**	**Daily**	**Hr.**	**Daily**	**Bare Costs**	**Incl. O&P**
2 Laborers	$27.40	$438.40	$45.40	$726.40	$27.72	$45.65
1 Truck Driver (heavy)	28.35	226.80	46.15	369.20		
2-50 Ft. Disch. Hoses		4.30		4.75		
1 Water Tanker		116.80		128.50		
1 Tractor		208.20		229.00		
1 Pressure Washer		50.60		55.65	15.83	17.41
24 L.H., Daily Totals		$1045.10		$1513.50	$43.55	$63.06
Crew B-9C	**Hr.**	**Daily**	**Hr.**	**Daily**	**Bare Costs**	**Incl. O&P**
1 Laborer Foreman (outside)	$29.40	$235.20	$48.70	$389.60	$27.80	$46.06
4 Laborers	27.40	876.80	45.40	1452.80		
1 Air Compr., 250 C.F.M.		119.80		131.80		
2-50 Ft. Air Hoses, 1.5″ Dia.		9.40		10.35		
2 Breaker, Pavement, 60 lb.		18.00		19.80	3.68	4.05
40 L.H., Daily Totals		$1259.20		$2004.35	$31.48	$50.11

Crew No.	Bare Costs		Incl. Subs O & P		Cost Per Labor-Hour	

Crew B-9D	Hr.	Daily	Hr.	Daily	Bare Costs	Incl. O&P
1 Laborer Foreman (Outside)	$29.40	$235.20	$48.70	$389.60	$27.80	$46.06
4 Common Laborers	27.40	876.80	45.40	1452.80		
1 Air Compressor, 250 CFM		119.80		131.80		
2 Air hoses, 1.5" x 50'		9.40		10.35		
2 Air tampers		55.70		61.25	4.63	5.09
40 L.H., Daily Totals		$1296.90		$2045.80	$32.43	$51.15

Crew B-10	Hr.	Daily	Hr.	Daily	Bare Costs	Incl. O&P
1 Equip. Oper. (med.)	$36.70	$293.60	$57.95	$463.60	$33.60	$53.77
.5 Laborer	27.40	109.60	45.40	181.60		
12 L.H., Daily Totals		$403.20		$645.20	$33.60	$53.77

Crew B-10A	Hr.	Daily	Hr.	Daily	Bare Costs	Incl. O&P
1 Equip. Oper. (med.)	$36.70	$293.60	$57.95	$463.60	$33.60	$53.77
.5 Laborer	27.40	109.60	45.40	181.60		
1 Walk behind compactor, 7.5 HP		125.80		138.40	10.48	11.53
12 L.H., Daily Totals		$529.00		$783.60	$44.08	$65.30

Crew B-10B	Hr.	Daily	Hr.	Daily	Bare Costs	Incl. O&P
1 Equip. Oper. (med.)	$36.70	$293.60	$57.95	$463.60	$33.60	$53.77
.5 Laborer	27.40	109.60	45.40	181.60		
1 Dozer, 200 H.P.		920.80		1012.90	76.73	84.41
12 L.H., Daily Totals		$1324.00		$1658.10	$110.33	$138.18

Crew B-10C	Hr.	Daily	Hr.	Daily	Bare Costs	Incl. O&P
1 Equip. Oper. (med.)	$36.70	$293.60	$57.95	$463.60	$33.60	$53.77
.5 Laborer	27.40	109.60	45.40	181.60		
1 Dozer, 200 H.P.		920.80		1012.90		
1 Vibratory Roller, Towed		457.60		503.35	114.87	126.35
12 L.H., Daily Totals		$1781.60		$2161.45	$148.47	$180.12

Crew B-10D	Hr.	Daily	Hr.	Daily	Bare Costs	Incl. O&P
1 Equip. Oper. (med.)	$36.70	$293.60	$57.95	$463.60	$33.60	$53.77
.5 Laborer	27.40	109.60	45.40	181.60		
1 Dozer, 200 H.P.		920.80		1012.90		
1 Sheepsft. Roller, Towed		478.00		525.80	116.57	128.22
12 L.H., Daily Totals		$1802.00		$2183.90	$150.17	$181.99

Crew B-10E	Hr.	Daily	Hr.	Daily	Bare Costs	Incl. O&P
1 Equip. Oper. (med.)	$36.70	$293.60	$57.95	$463.60	$33.60	$53.77
.5 Laborer	27.40	109.60	45.40	181.60		
1 Tandem Roller, 5 Ton		114.20		125.60	9.52	10.47
12 L.H., Daily Totals		$517.40		$770.80	$43.12	$64.24

Crew B-10F	Hr.	Daily	Hr.	Daily	Bare Costs	Incl. O&P
1 Equip. Oper. (med.)	$36.70	$293.60	$57.95	$463.60	$33.60	$53.77
.5 Laborer	27.40	109.60	45.40	181.60		
1 Tandem Roller, 10 Ton		197.00		216.70	16.42	18.06
12 L.H., Daily Totals		$600.20		$861.90	$50.02	$71.83

Crew B-10G	Hr.	Daily	Hr.	Daily	Bare Costs	Incl. O&P
1 Equip. Oper. (med.)	$36.70	$293.60	$57.95	$463.60	$33.60	$53.77
.5 Laborer	27.40	109.60	45.40	181.60		
1 Sheepsft. Roll., 130 H.P.		848.00		932.80	70.67	77.73
12 L.H., Daily Totals		$1251.20		$1578.00	$104.27	$131.50

Crew B-10H	Hr.	Daily	Hr.	Daily	Bare Costs	Incl. O&P
1 Equip. Oper. (med.)	$36.70	$293.60	$57.95	$463.60	$33.60	$53.77
.5 Laborer	27.40	109.60	45.40	181.60		
1 Diaphr. Water Pump, 2"		53.20		58.50		
1-20 Ft. Suction Hose, 2"		2.15		2.35		
2-50 Ft. Disch. Hoses, 2"		2.60		2.85	4.83	5.31
12 L.H., Daily Totals		$461.15		$708.90	$38.43	$59.08

Crew B-10I	Hr.	Daily	Hr.	Daily	Bare Costs	Incl. O&P
1 Equip. Oper. (med.)	$36.70	$293.60	$57.95	$463.60	$33.60	$53.77
.5 Laborer	27.40	109.60	45.40	181.60		
1 Diaphr. Water Pump, 4"		73.20		80.50		
1-20 Ft. Suction Hose, 4"		4.05		4.45		
2-50 Ft. Disch. Hoses, 4"		5.90		6.50	6.93	7.62
12 L.H., Daily Totals		$486.35		$736.65	$40.53	$61.39

Crew B-10J	Hr.	Daily	Hr.	Daily	Bare Costs	Incl. O&P
1 Equip. Oper. (med.)	$36.70	$293.60	$57.95	$463.60	$33.60	$53.77
.5 Laborer	27.40	109.60	45.40	181.60		
1 Centr. Water Pump, 3"		56.60		62.25		
1-20 Ft. Suction Hose, 3"		3.65		4.00		
2-50 Ft. Disch. Hoses, 3"		4.30		4.75	5.38	5.92
12 L.H., Daily Totals		$467.75		$716.20	$38.98	$59.69

Crew B-10K	Hr.	Daily	Hr.	Daily	Bare Costs	Incl. O&P
1 Equip. Oper. (med.)	$36.70	$293.60	$57.95	$463.60	$33.60	$53.77
.5 Laborer	27.40	109.60	45.40	181.60		
1 Centr. Water Pump, 6"		240.00		264.00		
1-20 Ft. Suction Hose, 6"		12.10		13.30		
2-50 Ft. Disch. Hoses, 6"		15.00		16.50	22.26	24.48
12 L.H., Daily Totals		$670.30		$939.00	$55.86	$78.25

Crew B-10L	Hr.	Daily	Hr.	Daily	Bare Costs	Incl. O&P
1 Equip. Oper. (med.)	$36.70	$293.60	$57.95	$463.60	$33.60	$53.77
.5 Laborer	27.40	109.60	45.40	181.60		
1 Dozer, 80 H.P.		318.80		350.70	26.57	29.22
12 L.H., Daily Totals		$722.00		$995.90	$60.17	$82.99

Crew B-10M	Hr.	Daily	Hr.	Daily	Bare Costs	Incl. O&P
1 Equip. Oper. (med.)	$36.70	$293.60	$57.95	$463.60	$33.60	$53.77
.5 Laborer	27.40	109.60	45.40	181.60		
1 Dozer, 300 H.P.		1197.00		1316.70	99.75	109.73
12 L.H., Daily Totals		$1600.20		$1961.90	$133.35	$163.50

Crew B-10N	Hr.	Daily	Hr.	Daily	Bare Costs	Incl. O&P
1 Equip. Oper. (med.)	$36.70	$293.60	$57.95	$463.60	$33.60	$53.77
.5 Laborer	27.40	109.60	45.40	181.60		
1 F.E. Loader, T.M., 1.5 C.Y		315.20		346.70	26.27	28.89
12 L.H., Daily Totals		$718.40		$991.90	$59.87	$82.66

Crew B-10O	Hr.	Daily	Hr.	Daily	Bare Costs	Incl. O&P
1 Equip. Oper. (med.)	$36.70	$293.60	$57.95	$463.60	$33.60	$53.77
.5 Laborer	27.40	109.60	45.40	181.60		
1 F.E. Loader, T.M., 2.25 C.Y.		557.00		612.70	46.42	51.06
12 L.H., Daily Totals		$960.20		$1257.90	$80.02	$104.83

Crew B-10P	Hr.	Daily	Hr.	Daily	Bare Costs	Incl. O&P
1 Equip. Oper. (med.)	$36.70	$293.60	$57.95	$463.60	$33.60	$53.77
.5 Laborer	27.40	109.60	45.40	181.60		
1 F.E. Loader, T.M., 2.5 C.Y.		794.00		873.40	66.17	72.78
12 L.H., Daily Totals		$1197.20		$1518.60	$99.77	$126.55

Crew No.	Bare Costs		Incl. Subs O & P		Cost Per Labor-Hour	
Crew B-10Q	**Hr.**	**Daily**	**Hr.**	**Daily**	**Bare Costs**	**Incl. O&P**
1 Equip. Oper. (med.)	$36.70	$293.60	$57.95	$463.60	$33.60	$53.77
.5 Laborer	27.40	109.60	45.40	181.60		
1 F.E. Loader, T.M., 5 C.Y.		1110.00		1221.00	92.50	101.75
12 L.H., Daily Totals		$1513.20		$1866.20	$126.10	$155.52
Crew B-10R	**Hr.**	**Daily**	**Hr.**	**Daily**	**Bare Costs**	**Incl. O&P**
1 Equip. Oper. (med.)	$36.70	$293.60	$57.95	$463.60	$33.60	$53.77
.5 Laborer	27.40	109.60	45.40	181.60		
1 F.E. Loader, W.M., 1 C.Y.		194.60		214.05	16.22	17.84
12 L.H., Daily Totals		$597.80		$859.25	$49.82	$71.61
Crew B-10S	**Hr.**	**Daily**	**Hr.**	**Daily**	**Bare Costs**	**Incl. O&P**
1 Equip. Oper. (med.)	$36.70	$293.60	$57.95	$463.60	$33.60	$53.77
.5 Laborer	27.40	109.60	45.40	181.60		
1 F.E. Loader, W.M., 1.5 C.Y.		241.40		265.55	20.12	22.13
12 L.H., Daily Totals		$644.60		$910.75	$53.72	$75.90
Crew B-10T	**Hr.**	**Daily**	**Hr.**	**Daily**	**Bare Costs**	**Incl. O&P**
1 Equip. Oper. (med.)	$36.70	$293.60	$57.95	$463.60	$33.60	$53.77
.5 Laborer	27.40	109.60	45.40	181.60		
1 F.E. Loader, W.M.,2.5 C.Y.		305.40		335.95	25.45	28.00
12 L.H., Daily Totals		$708.60		$981.15	$59.05	$81.77
Crew B-10U	**Hr.**	**Daily**	**Hr.**	**Daily**	**Bare Costs**	**Incl. O&P**
1 Equip. Oper. (med.)	$36.70	$293.60	$57.95	$463.60	$33.60	$53.77
.5 Laborer	27.40	109.60	45.40	181.60		
1 F.E. Loader, W.M., 5.5 C.Y.		701.00		771.10	58.42	64.26
12 L.H., Daily Totals		$1104.20		$1416.30	$92.02	$118.03
Crew B-10V	**Hr.**	**Daily**	**Hr.**	**Daily**	**Bare Costs**	**Incl. O&P**
1 Equip. Oper. (med.)	$36.70	$293.60	$57.95	$463.60	$33.60	$53.77
.5 Laborer	27.40	109.60	45.40	181.60		
1 Dozer, 700 H.P.		3148.00		3462.80	262.33	288.57
12 L.H., Daily Totals		$3551.20		$4108.00	$295.93	$342.34
Crew B-10W	**Hr.**	**Daily**	**Hr.**	**Daily**	**Bare Costs**	**Incl. O&P**
1 Equip. Oper. (med.)	$36.70	$293.60	$57.95	$463.60	$33.60	$53.77
.5 Laborer	27.40	109.60	45.40	181.60		
1 Dozer, 105 H.P.		462.80		509.10	38.57	42.42
12 L.H., Daily Totals		$866.00		$1154.30	$72.17	$96.19
Crew B-10X	**Hr.**	**Daily**	**Hr.**	**Daily**	**Bare Costs**	**Incl. O&P**
1 Equip. Oper. (med.)	$36.70	$293.60	$57.95	$463.60	$33.60	$53.77
.5 Laborer	27.40	109.60	45.40	181.60		
1 Dozer, 410 H.P.		1536.00		1689.60	128.00	140.80
12 L.H., Daily Totals		$1939.20		$2334.80	$161.60	$194.57
Crew B-10Y	**Hr.**	**Daily**	**Hr.**	**Daily**	**Bare Costs**	**Incl. O&P**
1 Equip. Oper. (med.)	$36.70	$293.60	$57.95	$463.60	$33.60	$53.77
.5 Laborer	27.40	109.60	45.40	181.60		
1 Vibratory Drum Roller		419.20		461.10	34.93	38.43
12 L.H., Daily Totals		$822.40		$1106.30	$68.53	$92.20
Crew B-11A	**Hr.**	**Daily**	**Hr.**	**Daily**	**Bare Costs**	**Incl. O&P**
1 Equipment Oper. (med.)	$36.70	$293.60	$57.95	$463.60	$32.05	$51.68
1 Laborer	27.40	219.20	45.40	363.20		
1 Dozer, 200 H.P.		920.80		1012.90	57.55	63.31
16 L.H., Daily Totals		$1433.60		$1839.70	$89.60	$114.99

Crew No.	Bare Costs		Incl. Subs O & P		Cost Per Labor-Hour	
Crew B-11B	**Hr.**	**Daily**	**Hr.**	**Daily**	**Bare Costs**	**Incl. O&P**
1 Equipment Oper. (light)	$35.20	$281.60	$55.60	$444.80	$31.30	$50.50
1 Laborer	27.40	219.20	45.40	363.20		
1 Air Powered Tamper		27.85		30.65		
1 Air Compr. 365 C.F.M.		159.00		174.90		
2-50 Ft. Air Hoses, 1.5" Dia.		9.40		10.35	12.27	13.50
16 L.H., Daily Totals		$697.05		$1023.90	$43.57	$64.00
Crew B-11C	**Hr.**	**Daily**	**Hr.**	**Daily**	**Bare Costs**	**Incl. O&P**
1 Equipment Oper. (med.)	$36.70	$293.60	$57.95	$463.60	$32.05	$51.68
1 Laborer	27.40	219.20	45.40	363.20		
1 Backhoe Loader, 48 H.P.		226.20		248.80	14.14	15.55
16 L.H., Daily Totals		$739.00		$1075.60	$46.19	$67.23
Crew B-11K	**Hr.**	**Daily**	**Hr.**	**Daily**	**Bare Costs**	**Incl. O&P**
1 Equipment Oper. (med.)	$36.70	$293.60	$57.95	$463.60	$32.05	$51.68
1 Laborer	27.40	219.20	45.40	363.20		
1 Trencher, 8' D., 16" W.		1460.00		1606.00	91.25	100.38
16 L.H., Daily Totals		$1972.80		$2432.80	$123.30	$152.06
Crew B-11L	**Hr.**	**Daily**	**Hr.**	**Daily**	**Bare Costs**	**Incl. O&P**
1 Equipment Oper. (med.)	$36.70	$293.60	$57.95	$463.60	$32.05	$51.68
1 Laborer	27.40	219.20	45.40	363.20		
1 Grader, 30,000 Lbs.		457.40		503.15	28.59	31.45
16 L.H., Daily Totals		$970.20		$1329.95	$60.64	$83.13
Crew B-11M	**Hr.**	**Daily**	**Hr.**	**Daily**	**Bare Costs**	**Incl. O&P**
1 Equipment Oper. (med.)	$36.70	$293.60	$57.95	$463.60	$32.05	$51.68
1 Laborer	27.40	219.20	45.40	363.20		
1 Backhoe Loader, 80 H.P.		277.60		305.35	17.35	19.09
16 L.H., Daily Totals		$790.40		$1132.15	$49.40	$70.77
Crew B-11S	**Hr.**	**Daily**	**Hr.**	**Daily**	**Bare Costs**	**Incl. O&P**
1 Equipment Operator (med.)	$36.70	$293.60	$57.95	$463.60	$33.60	$53.77
.5 Laborer	27.40	109.60	45.40	181.60		
1 Dozer, 300 H.P.		1197.00		1316.70		
1 Ripper, Beam & 1 Shank		73.20		80.50	105.85	116.44
12 L.H., Daily Totals		$1673.40		$2042.40	$139.45	$170.21
Crew B-11T	**Hr.**	**Daily**	**Hr.**	**Daily**	**Bare Costs**	**Incl. O&P**
1 Equipment Operator (med.)	$36.70	$293.60	$57.95	$463.60	$33.60	$53.77
.5 Laborer	27.40	109.60	45.40	181.60		
1 Dozer, 410 H.P.		1536.00		1689.60		
1 Ripper, Beam & 2 Shanks		82.00		90.20	134.83	148.32
12 L.H., Daily Totals		$2021.20		$2425.00	$168.43	$202.09
Crew B-11W	**Hr.**	**Daily**	**Hr.**	**Daily**	**Bare Costs**	**Incl. O&P**
1 Equipment Operator (med.)	$36.70	$293.60	$57.95	$463.60	$28.97	$47.07
1 Common Laborer	27.40	219.20	45.40	363.20		
10 Truck Drivers, Heavy	28.35	2268.00	46.15	3692.00		
1 Dozer, 200 H.P.		920.80		1012.90		
1 Vib. roller, smth, towed, 23 Ton		457.60		503.35		
10 Dump Truck, 10 Ton		3266.00		3592.60	48.38	53.22
96 L.H., Daily Totals		$7425.20		$9627.65	$77.35	$100.29

Crew No.	Bare Costs		Incl. Subs O & P		Cost Per Labor-Hour	
Crew B-11Y	**Hr.**	**Daily**	**Hr.**	**Daily**	**Bare Costs**	**Incl. O&P**
1 Laborer Foreman (Outside)	$29.40	$235.20	$48.70	$389.60	$30.72	$49.95
5 Common Laborers	27.40	1096.00	45.40	1816.00		
3 Equipment Operators (med.)	36.70	880.80	57.95	1390.80		
1 Dozer, 80 H.P.		318.80		350.70		
2 Walk behind compactors, 7.5 HP		251.60		276.75		
4 Vibratory plates, gas, 21"		138.40		152.25	9.84	10.83
72 L.H., Daily Totals		$2920.80		$4376.10	$40.56	$60.78
Crew B-12A	**Hr.**	**Daily**	**Hr.**	**Daily**	**Bare Costs**	**Incl. O&P**
1 Equip. Oper. (crane)	$38.10	$304.80	$60.15	$481.20	$32.75	$52.78
1 Laborer	27.40	219.20	45.40	363.20		
1 Hyd. Excavator, 1 C.Y.		560.20		616.20	35.01	38.51
16 L.H., Daily Totals		$1084.20		$1460.60	$67.76	$91.29
Crew B-12B	**Hr.**	**Daily**	**Hr.**	**Daily**	**Bare Costs**	**Incl. O&P**
1 Equip. Oper. (crane)	$38.10	$304.80	$60.15	$481.20	$32.75	$52.78
1 Laborer	27.40	219.20	45.40	363.20		
1 Hyd. Excavator, 1.5 C.Y.		722.80		795.10	45.18	49.69
16 L.H., Daily Totals		$1246.80		$1639.50	$77.93	$102.47
Crew B-12C	**Hr.**	**Daily**	**Hr.**	**Daily**	**Bare Costs**	**Incl. O&P**
1 Equip. Oper. (crane)	$38.10	$304.80	$60.15	$481.20	$32.75	$52.78
1 Laborer	27.40	219.20	45.40	363.20		
1 Hyd. Excavator, 2 C.Y.		919.60		1011.55	57.48	63.22
16 L.H., Daily Totals		$1443.60		$1855.95	$90.23	$116.00
Crew B-12D	**Hr.**	**Daily**	**Hr.**	**Daily**	**Bare Costs**	**Incl. O&P**
1 Equip. Oper. (crane)	$38.10	$304.80	$60.15	$481.20	$32.75	$52.78
1 Laborer	27.40	219.20	45.40	363.20		
1 Hyd. Excavator, 3.5 C.Y.		2023.00		2225.30	126.44	139.08
16 L.H., Daily Totals		$2547.00		$3069.70	$159.19	$191.86
Crew B-12E	**Hr.**	**Daily**	**Hr.**	**Daily**	**Bare Costs**	**Incl. O&P**
1 Equip. Oper. (crane)	$38.10	$304.80	$60.15	$481.20	$32.75	$52.78
1 Laborer	27.40	219.20	45.40	363.20		
1 Hyd. Excavator, .5 C.Y.		335.40		368.95	20.96	23.06
16 L.H., Daily Totals		$859.40		$1213.35	$53.71	$75.84
Crew B-12F	**Hr.**	**Daily**	**Hr.**	**Daily**	**Bare Costs**	**Incl. O&P**
1 Equip. Oper. (crane)	$38.10	$304.80	$60.15	$481.20	$32.75	$52.78
1 Laborer	27.40	219.20	45.40	363.20		
1 Hyd. Excavator, .75 C.Y.		477.80		525.60	29.86	32.85
16 L.H., Daily Totals		$1001.80		$1370.00	$62.61	$85.63
Crew B-12G	**Hr.**	**Daily**	**Hr.**	**Daily**	**Bare Costs**	**Incl. O&P**
1 Equip. Oper. (crane)	$38.10	$304.80	$60.15	$481.20	$32.75	$52.78
1 Laborer	27.40	219.20	45.40	363.20		
1 Power Shovel, .5 C.Y.		534.65		588.10		
1 Clamshell Bucket, .5 C.Y.		33.80		37.20	35.53	39.08
16 L.H., Daily Totals		$1092.45		$1469.70	$68.28	$91.86
Crew B-12H	**Hr.**	**Daily**	**Hr.**	**Daily**	**Bare Costs**	**Incl. O&P**
1 Equip. Oper. (crane)	$38.10	$304.80	$60.15	$481.20	$32.75	$52.78
1 Laborer	27.40	219.20	45.40	363.20		
1 Power Shovel, 1 C.Y.		880.00		968.00		
1 Clamshell Bucket, 1 C.Y.		43.40		47.75	57.71	63.48
16 L.H., Daily Totals		$1447.40		$1860.15	$90.46	$116.26

Crew No.	Bare Costs		Incl. Subs O & P		Cost Per Labor-Hour	
Crew B-12I	**Hr.**	**Daily**	**Hr.**	**Daily**	**Bare Costs**	**Incl. O&P**
1 Equip. Oper. (crane)	$38.10	$304.80	$60.15	$481.20	$32.75	$52.78
1 Laborer	27.40	219.20	45.40	363.20		
1 Power Shovel, .75 C.Y.		663.50		729.85		
1 Dragline Bucket, .75 C.Y.		19.00		20.90	42.66	46.92
16 L.H., Daily Totals		$1206.50		$1595.15	$75.41	$99.70
Crew B-12J	**Hr.**	**Daily**	**Hr.**	**Daily**	**Bare Costs**	**Incl. O&P**
1 Equip. Oper. (crane)	$38.10	$304.80	$60.15	$481.20	$32.75	$52.78
1 Laborer	27.40	219.20	45.40	363.20		
1 Gradall, 3 Ton, .5 C.Y.		841.00		925.10	52.56	57.82
16 L.H., Daily Totals		$1365.00		$1769.50	$85.31	$110.60
Crew B-12K	**Hr.**	**Daily**	**Hr.**	**Daily**	**Bare Costs**	**Incl. O&P**
1 Equip. Oper. (crane)	$38.10	$304.80	$60.15	$481.20	$32.75	$52.78
1 Laborer	27.40	219.20	45.40	363.20		
1 Gradall, 3 Ton, 1 C.Y.		974.20		1071.60	60.89	66.98
16 L.H., Daily Totals		$1498.20		$1916.00	$93.64	$119.76
Crew B-12L	**Hr.**	**Daily**	**Hr.**	**Daily**	**Bare Costs**	**Incl. O&P**
1 Equip. Oper. (crane)	$38.10	$304.80	$60.15	$481.20	$32.75	$52.78
1 Laborer	27.40	219.20	45.40	363.20		
1 Power Shovel, .5 C.Y.		534.65		588.10		
1 F.E. Attachment, .5 C.Y.		48.80		53.70	36.46	40.11
16 L.H., Daily Totals		$1107.45		$1486.20	$69.21	$92.89
Crew B-12M	**Hr.**	**Daily**	**Hr.**	**Daily**	**Bare Costs**	**Incl. O&P**
1 Equip. Oper. (crane)	$38.10	$304.80	$60.15	$481.20	$32.75	$52.78
1 Laborer	27.40	219.20	45.40	363.20		
1 Power Shovel, .75 C.Y.		663.50		729.85		
1 F.E. Attachment, .75 C.Y.		53.20		58.50	44.79	49.27
16 L.H., Daily Totals		$1240.70		$1632.75	$77.54	$102.05
Crew B-12N	**Hr.**	**Daily**	**Hr.**	**Daily**	**Bare Costs**	**Incl. O&P**
1 Equip. Oper. (crane)	$38.10	$304.80	$60.15	$481.20	$32.75	$52.78
1 Laborer	27.40	219.20	45.40	363.20		
1 Power Shovel, 1 C.Y.		880.00		968.00		
1 F.E. Attachment, 1 C.Y.		59.00		64.90	58.69	64.56
16 L.H., Daily Totals		$1463.00		$1877.30	$91.44	$117.34
Crew B-12O	**Hr.**	**Daily**	**Hr.**	**Daily**	**Bare Costs**	**Incl. O&P**
1 Equip. Oper. (crane)	$38.10	$304.80	$60.15	$481.20	$32.75	$52.78
1 Laborer	27.40	219.20	45.40	363.20		
1 Power Shovel, 1.5 C.Y.		934.20		1027.60		
1 F.E. Attachment, 1.5 C.Y.		69.20		76.10	62.71	68.98
16 L.H., Daily Totals		$1527.40		$1948.10	$95.46	$121.76
Crew B-12P	**Hr.**	**Daily**	**Hr.**	**Daily**	**Bare Costs**	**Incl. O&P**
1 Equip. Oper. (crane)	$38.10	$304.80	$60.15	$481.20	$32.75	$52.78
1 Laborer	27.40	219.20	45.40	363.20		
1 Crawler Crane, 40 Ton		934.20		1027.60		
1 Dragline Bucket, 1.5 C.Y.		31.00		34.10	60.33	66.36
16 L.H., Daily Totals		$1489.20		$1906.10	$93.08	$119.14
Crew B-12Q	**Hr.**	**Daily**	**Hr.**	**Daily**	**Bare Costs**	**Incl. O&P**
1 Equip. Oper. (crane)	$38.10	$304.80	$60.15	$481.20	$32.75	$52.78
1 Laborer	27.40	219.20	45.40	363.20		
1 Hyd. Excavator, 5/8 C.Y.		435.40		478.95	27.21	29.93
16 L.H., Daily Totals		$959.40		$1323.35	$59.96	$82.71

Crew No.	Bare Costs		Incl. Subs O & P		Cost Per Labor-Hour	
Crew B-12R	**Hr.**	**Daily**	**Hr.**	**Daily**	**Bare Costs**	**Incl. O&P**
1 Equip. Oper. (crane)	$38.10	$304.80	$60.15	$481.20	$32.75	$52.78
1 Laborer	27.40	219.20	45.40	363.20		
1 Hyd. Excavator, 1.5 C.Y.		722.80		795.10	45.18	49.69
16 L.H., Daily Totals		$1246.80		$1639.50	$77.93	$102.47
Crew B-12S	**Hr.**	**Daily**	**Hr.**	**Daily**	**Bare Costs**	**Incl. O&P**
1 Equip. Oper. (crane)	$38.10	$304.80	$60.15	$481.20	$32.75	$52.78
1 Laborer	27.40	219.20	45.40	363.20		
1 Hyd. Excavator, 2.5 C.Y.		1218.00		1339.80	76.13	83.74
16 L.H., Daily Totals		$1742.00		$2184.20	$108.88	$136.52
Crew B-12T	**Hr.**	**Daily**	**Hr.**	**Daily**	**Bare Costs**	**Incl. O&P**
1 Equip. Oper. (crane)	$38.10	$304.80	$60.15	$481.20	$32.75	$52.78
1 Laborer	27.40	219.20	45.40	363.20		
1 Crawler Crane, 75 Ton		1226.00		1348.60		
1 F.E. Attachment, 3 C.Y.		91.60		100.75	82.35	90.59
16 L.H., Daily Totals		$1841.60		$2293.75	$115.10	$143.37
Crew B-12V	**Hr.**	**Daily**	**Hr.**	**Daily**	**Bare Costs**	**Incl. O&P**
1 Equip. Oper. (crane)	$38.10	$304.80	$60.15	$481.20	$32.75	$52.78
1 Laborer	27.40	219.20	45.40	363.20		
1 Crawler Crane, 75 Ton		1226.00		1348.60		
1 Dragline Bucket, 3 C.Y.		47.80		52.60	79.61	87.57
16 L.H., Daily Totals		$1797.80		$2245.60	$112.36	$140.35
Crew B-13	**Hr.**	**Daily**	**Hr.**	**Daily**	**Bare Costs**	**Incl. O&P**
1 Laborer Foreman (outside)	$29.40	$235.20	$48.70	$389.60	$29.94	$48.81
4 Laborers	27.40	876.80	45.40	1452.80		
1 Equip. Oper. (crane)	38.10	304.80	60.15	481.20		
1 Equip. Oper. Oiler	32.45	259.60	51.25	410.00		
1 Hyd. Crane, 25 Ton		650.80		715.90	11.62	12.78
56 L.H., Daily Totals		$2327.20		$3449.50	$41.56	$61.59
Crew B-13A	**Hr.**	**Daily**	**Hr.**	**Daily**	**Bare Costs**	**Incl. O&P**
1 Laborer Foreman	$29.40	$235.20	$48.70	$389.60	$30.61	$49.67
2 Laborers	27.40	438.40	45.40	726.40		
2 Equipment Operators	36.70	587.20	57.95	927.20		
2 Truck Drivers (heavy)	28.35	453.60	46.15	738.40		
1 Crane, 75 Ton		1226.00		1348.60		
1 F.E. Lder, 3.75 C.Y.		1110.00		1221.00		
2 Dump Trucks, 12 Ton		653.20		718.50	53.38	58.72
56 L.H., Daily Totals		$4703.60		$6069.70	$83.99	$108.39
Crew B-13B	**Hr.**	**Daily**	**Hr.**	**Daily**	**Bare Costs**	**Incl. O&P**
1 Laborer Foreman (outside)	$29.40	$235.20	$48.70	$389.60	$29.94	$48.81
4 Laborers	27.40	876.80	45.40	1452.80		
1 Equip. Oper. (crane)	38.10	304.80	60.15	481.20		
1 Equip. Oper. Oiler	32.45	259.60	51.25	410.00		
1 Hyd. Crane, 55 Ton		951.60		1046.75	16.99	18.69
56 L.H., Daily Totals		$2628.00		$3780.35	$46.93	$67.50
Crew B-13C	**Hr.**	**Daily**	**Hr.**	**Daily**	**Bare Costs**	**Incl. O&P**
1 Laborer Foreman (outside)	$29.40	$235.20	$48.70	$389.60	$29.94	$48.81
4 Laborers	27.40	876.80	45.40	1452.80		
1 Equip. Oper. (crane)	38.10	304.80	60.15	481.20		
1 Equip. Oper. Oiler	32.45	259.60	51.25	410.00		
1 Crawler Crane, 100 Ton		1590.00		1749.00	28.39	31.23
56 L.H., Daily Totals		$3266.40		$4482.60	$58.33	$80.04

Crew No.	Bare Costs		Incl. Subs O & P		Cost Per Labor-Hour	
Crew B-14	**Hr.**	**Daily**	**Hr.**	**Daily**	**Bare Costs**	**Incl. O&P**
1 Laborer Foreman (outside)	$29.40	$235.20	$48.70	$389.60	$29.03	$47.65
4 Laborers	27.40	876.80	45.40	1452.80		
1 Equip. Oper. (light)	35.20	281.60	55.60	444.80		
1 Backhoe Loader, 48 H.P.		226.20		248.80	4.71	5.18
48 L.H., Daily Totals		$1619.80		$2536.00	$33.74	$52.83
Crew B-15	**Hr.**	**Daily**	**Hr.**	**Daily**	**Bare Costs**	**Incl. O&P**
1 Equipment Oper. (med)	$36.70	$293.60	$57.95	$463.60	$30.60	$49.41
.5 Laborer	27.40	109.60	45.40	181.60		
2 Truck Drivers (heavy)	28.35	453.60	46.15	738.40		
2 Dump Trucks, 16 Ton		953.20		1048.50		
1 Dozer, 200 H.P.		920.80		1012.90	66.93	73.62
28 L.H., Daily Totals		$2730.80		$3445.00	$97.53	$123.03
Crew B-16	**Hr.**	**Daily**	**Hr.**	**Daily**	**Bare Costs**	**Incl. O&P**
1 Laborer Foreman (outside)	$29.40	$235.20	$48.70	$389.60	$28.14	$46.41
2 Laborers	27.40	438.40	45.40	726.40		
1 Truck Driver (heavy)	28.35	226.80	46.15	369.20		
1 Dump Truck, 16 Ton		476.60		524.25	14.89	16.38
32 L.H., Daily Totals		$1377.00		$2009.45	$43.03	$62.79
Crew B-17	**Hr.**	**Daily**	**Hr.**	**Daily**	**Bare Costs**	**Incl. O&P**
2 Laborers	$27.40	$438.40	$45.40	$726.40	$29.59	$48.14
1 Equip. Oper. (light)	35.20	281.60	55.60	444.80		
1 Truck Driver (heavy)	28.35	226.80	46.15	369.20		
1 Backhoe Loader, 48 H.P.		226.20		248.80		
1 Dump Truck, 12 Ton		326.60		359.25	17.28	19.00
32 L.H., Daily Totals		$1499.60		$2148.45	$46.87	$67.14
Crew B-17A	**Hr.**	**Daily**	**Hr.**	**Daily**	**Bare Costs**	**Incl. O&P**
2 Laborer Foremen	$29.40	$470.40	$48.70	$779.20	$29.82	$49.25
6 Laborers	27.40	1315.20	45.40	2179.20		
1 Skilled Worker Foreman	38.50	308.00	63.00	504.00		
1 Skilled Worker	36.50	292.00	59.70	477.60		
80 L.H., Daily Totals		$2385.60		$3940.00	$29.82	$49.25
Crew B-18	**Hr.**	**Daily**	**Hr.**	**Daily**	**Bare Costs**	**Incl. O&P**
1 Laborer Foreman (outside)	$29.40	$235.20	$48.70	$389.60	$28.07	$46.50
2 Laborers	27.40	438.40	45.40	726.40		
1 Vibrating Compactor		34.60		38.05	1.44	1.59
24 L.H., Daily Totals		$708.20		$1154.05	$29.51	$48.09
Crew B-19	**Hr.**	**Daily**	**Hr.**	**Daily**	**Bare Costs**	**Incl. O&P**
1 Pile Driver Foreman	$36.70	$293.60	$62.15	$497.20	$35.52	$58.61
4 Pile Drivers	34.70	1110.40	58.80	1881.60		
2 Equip. Oper. (crane)	38.10	609.60	60.15	962.40		
1 Equip. Oper. Oiler	32.45	259.60	51.25	410.00		
1 Crane, 40 Ton & Access.		934.20		1027.60		
60 L.F. Pile Leads		66.00		72.60		
1 Hammer, Diesel, 22k Ft-Lb		568.00		624.80	24.55	27.01
64 L.H., Daily Totals		$3841.40		$5476.20	$60.07	$85.62
Crew B-19A	**Hr.**	**Daily**	**Hr.**	**Daily**	**Bare Costs**	**Incl. O&P**
1 Pile Driver Foreman	$36.70	$293.60	$62.15	$497.20	$35.52	$58.61
4 Pile Drivers	34.70	1110.40	58.80	1881.60		
2 Equip. Oper. (crane)	38.10	609.60	60.15	962.40		
1 Equip. Oper. Oiler	32.45	259.60	51.25	410.00		
1 Crawler Crane, 75 Ton		1226.00		1348.60		
60 L.F. Leads, 25K Ft. Lbs.		114.00		125.40		
1 Hammer, Diesel, 41k Ft-Lb		672.00		739.20	31.39	34.53
64 L.H., Daily Totals		$4285.20		$5964.40	$66.91	$93.14

Crew No.	Bare Costs		Incl. Subs O & P		Cost Per Labor-Hour	
Crew B-20	**Hr.**	**Daily**	**Hr.**	**Daily**	**Bare Costs**	**Incl. O&P**
1 Laborer Foreman (out)	$29.40	$235.20	$48.70	$389.60	$31.10	$51.27
1 Skilled Worker	36.50	292.00	59.70	477.60		
1 Laborer	27.40	219.20	45.40	363.20		
24 L.H., Daily Totals		$746.40		$1230.40	$31.10	$51.27

Crew B-20A	Hr.	Daily	Hr.	Daily	Bare Costs	Incl. O&P
1 Laborer Foreman	$29.40	$235.20	$48.70	$389.60	$33.41	$53.40
1 Laborer	27.40	219.20	45.40	363.20		
1 Plumber	42.70	341.60	66.40	531.20		
1 Plumber Apprentice	34.15	273.20	53.10	424.80		
32 L.H., Daily Totals		$1069.20		$1708.80	$33.41	$53.40

Crew B-21	Hr.	Daily	Hr.	Daily	Bare Costs	Incl. O&P
1 Laborer Foreman (out)	$29.40	$235.20	$48.70	$389.60	$32.10	$52.54
1 Skilled Worker	36.50	292.00	59.70	477.60		
1 Laborer	27.40	219.20	45.40	363.20		
.5 Equip. Oper. (crane)	38.10	152.40	60.15	240.60		
.5 S.P. Crane, 5 Ton		152.60		167.85	5.45	6.00
28 L.H., Daily Totals		$1051.40		$1638.85	$37.55	$58.54

Crew B-21A	Hr.	Daily	Hr.	Daily	Bare Costs	Incl. O&P
1 Laborer Foreman	$29.40	$235.20	$48.70	$389.60	$34.35	$54.75
1 Laborer	27.40	219.20	45.40	363.20		
1 Plumber	42.70	341.60	66.40	531.20		
1 Plumber Apprentice	34.15	273.20	53.10	424.80		
1 Equip. Oper. (crane)	38.10	304.80	60.15	481.20		
1 S.P. Crane, 12 Ton		536.40		590.05	13.41	14.75
40 L.H., Daily Totals		$1910.40		$2780.05	$47.76	$69.50

Crew B-21B	Hr.	Daily	Hr.	Daily	Bare Costs	Incl. O&P
1 Laborer Foreman	$29.40	$235.20	$48.70	$389.60	$29.94	$49.01
3 Laborers	27.40	657.60	45.40	1089.60		
1 Equip. Oper. (crane)	38.10	304.80	60.15	481.20		
1 Hyd. Crane, 12 Ton		636.60		700.25	15.92	17.51
40 L.H., Daily Totals		$1834.20		$2660.65	$45.86	$66.52

Crew B-21C	Hr.	Daily	Hr.	Daily	Bare Costs	Incl. O&P
1 Laborer Foreman	$29.40	$235.20	$48.70	$389.60	$29.94	$48.81
4 Laborers	27.40	876.80	45.40	1452.80		
1 Equip. Oper. (crane)	38.10	304.80	60.15	481.20		
1 Equip. Oper. Oiler	32.45	259.60	51.25	410.00		
2 Cutting Torches		32.40		35.65		
2 Gases		129.60		142.55		
1 Crane, 90 Ton		1427.00		1569.70	28.38	31.21
56 L.H., Daily Totals		$3265.40		$4481.50	$58.32	$80.02

Crew B-22	Hr.	Daily	Hr.	Daily	Bare Costs	Incl. O&P
1 Laborer Foreman (out)	$29.40	$235.20	$48.70	$389.60	$32.50	$53.04
1 Skilled Worker	36.50	292.00	59.70	477.60		
1 Laborer	27.40	219.20	45.40	363.20		
.75 Equip. Oper. (crane)	38.10	228.60	60.15	360.90		
.75 S.P. Crane, 5 Ton		228.90		251.80	7.63	8.39
30 L.H., Daily Totals		$1203.90		$1843.10	$40.13	$61.43

Crew B-22A	Hr.	Daily	Hr.	Daily	Bare Costs	Incl. O&P
1 Laborer Foreman (out)	$29.40	$235.20	$48.70	$389.60	$31.43	$51.43
1 Skilled Worker	36.50	292.00	59.70	477.60		
2 Laborers	27.40	438.40	45.40	726.40		
.75 Equipment Oper. (crane)	38.10	228.60	60.15	360.90		
.75 Crane, 5 Ton		228.90		251.80		
1 Generator, 5 KW		31.20		34.30		
1 Butt Fusion Machine		347.20		381.90	15.98	17.58
38 L.H., Daily Totals		$1801.50		$2622.50	$47.41	$69.01

Crew B-22B	Hr.	Daily	Hr.	Daily	Bare Costs	Incl. O&P
1 Skilled Worker	$36.50	$292.00	$59.70	$477.60	$31.95	$52.55
1 Laborer	27.40	219.20	45.40	363.20		
1 Electro Fusion Machine		148.40		163.25	9.28	10.20
16 L.H., Daily Totals		$659.60		$1004.05	$41.23	$62.75

Crew B-23	Hr.	Daily	Hr.	Daily	Bare Costs	Incl. O&P
1 Laborer Foreman (outside)	$29.40	$235.20	$48.70	$389.60	$27.80	$46.06
4 Laborers	27.40	876.80	45.40	1452.80		
1 Drill Rig, Wells		3379.00		3716.90		
1 Light Truck, 3 Ton		176.20		193.80	88.88	97.77
40 L.H., Daily Totals		$4667.20		$5753.10	$116.68	$143.83

Crew B-23A	Hr.	Daily	Hr.	Daily	Bare Costs	Incl. O&P
1 Laborer Foreman (outside)	$29.40	$235.20	$48.70	$389.60	$31.17	$50.68
1 Laborer	27.40	219.20	45.40	363.20		
1 Equip. Operator (medium)	36.70	293.60	57.95	463.60		
1 Drill Rig, Wells		3379.00		3716.90		
1 Pickup Truck, 3/4 Ton		78.00		85.80	144.04	158.45
24 L.H., Daily Totals		$4205.00		$5019.10	$175.21	$209.13

Crew B-23B	Hr.	Daily	Hr.	Daily	Bare Costs	Incl. O&P
1 Laborer Foreman (outside)	$29.40	$235.20	$48.70	$389.60	$31.17	$50.68
1 Laborer	27.40	219.20	45.40	363.20		
1 Equip. Operator (medium)	36.70	293.60	57.95	463.60		
1 Drill Rig, Wells		3379.00		3716.90		
1 Pickup Truck, 3/4 Ton		78.00		85.80		
1 Pump, Cntfgl, 6"		240.00		264.00	154.04	169.45
24 L.H., Daily Totals		$4445.00		$5283.10	$185.21	$220.13

Crew B-24	Hr.	Daily	Hr.	Daily	Bare Costs	Incl. O&P
1 Cement Finisher	$34.40	$275.20	$54.20	$433.60	$32.45	$52.83
1 Laborer	27.40	219.20	45.40	363.20		
1 Carpenter	35.55	284.40	58.90	471.20		
24 L.H., Daily Totals		$778.80		$1268.00	$32.45	$52.83

Crew B-25	Hr.	Daily	Hr.	Daily	Bare Costs	Incl. O&P
1 Laborer Foreman	$29.40	$235.20	$48.70	$389.60	$30.12	$49.12
7 Laborers	27.40	1534.40	45.40	2542.40		
3 Equip. Oper. (med.)	36.70	880.80	57.95	1390.80		
1 Asphalt Paver, 130 H.P		1661.00		1827.10		
1 Tandem Roller, 10 Ton		197.00		216.70		
1 Roller, Pneumatic Wheel		259.20		285.10	24.06	26.47
88 L.H., Daily Totals		$4767.60		$6651.70	$54.18	$75.59

Crew No.	Bare Costs		Incl. Subs O & P		Cost Per Labor-Hour	
Crew B-25B	**Hr.**	**Daily**	**Hr.**	**Daily**	**Bare Costs**	**Incl. O&P**
1 Laborer Foreman	$29.40	$235.20	$48.70	$389.60	$30.67	$49.86
7 Laborers	27.40	1534.40	45.40	2542.40		
4 Equip. Oper. (medium)	36.70	1174.40	57.95	1854.40		
1 Asphalt Paver, 130 H.P.		1661.00		1827.10		
2 Rollers, Steel Wheel		394.00		433.40		
1 Roller, Pneumatic Wheel		259.20		285.10	24.11	26.52
96 L.H., Daily Totals		$5258.20		$7332.00	$54.78	$76.38

Crew B-25C	**Hr.**	**Daily**	**Hr.**	**Daily**	**Bare Costs**	**Incl. O&P**
1 Laborer Foreman	$29.40	$235.20	$48.70	$389.60	$30.83	$50.13
3 Laborers	27.40	657.60	45.40	1089.60		
2 Equip. Oper. (medium)	36.70	587.20	57.95	927.20		
1 Asphalt Paver, 130 H.P.		1661.00		1827.10		
1 Roller, Steel Wheel		197.00		216.70	38.71	42.58
48 L.H., Daily Totals		$3338.00		$4450.20	$69.54	$92.71

Crew B-26	**Hr.**	**Daily**	**Hr.**	**Daily**	**Bare Costs**	**Incl. O&P**
1 Laborer Foreman (outside)	$29.40	$235.20	$48.70	$389.60	$31.01	$50.82
6 Laborers	27.40	1315.20	45.40	2179.20		
2 Equip. Oper. (med.)	36.70	587.20	57.95	927.20		
1 Rodman (reinf.)	39.50	316.00	67.80	542.40		
1 Cement Finisher	34.40	275.20	54.20	433.60		
1 Grader, 30,000 Lbs.		457.40		503.15		
1 Paving Mach. & Equip.		2237.00		2460.70	30.62	33.68
88 L.H., Daily Totals		$5423.20		$7435.85	$61.63	$84.50

Crew B-27	**Hr.**	**Daily**	**Hr.**	**Daily**	**Bare Costs**	**Incl. O&P**
1 Laborer Foreman (outside)	$29.40	$235.20	$48.70	$389.60	$27.90	$46.23
3 Laborers	27.40	657.60	45.40	1089.60		
1 Berm Machine		232.20		255.40	7.26	7.98
32 L.H., Daily Totals		$1125.00		$1734.60	$35.16	$54.21

Crew B-28	**Hr.**	**Daily**	**Hr.**	**Daily**	**Bare Costs**	**Incl. O&P**
2 Carpenters	$35.55	$568.80	$58.90	$942.40	$32.83	$54.40
1 Laborer	27.40	219.20	45.40	363.20		
24 L.H., Daily Totals		$788.00		$1305.60	$32.83	$54.40

Crew B-29	**Hr.**	**Daily**	**Hr.**	**Daily**	**Bare Costs**	**Incl. O&P**
1 Laborer Foreman (outside)	$29.40	$235.20	$48.70	$389.60	$29.94	$48.81
4 Laborers	27.40	876.80	45.40	1452.80		
1 Equip. Oper. (crane)	38.10	304.80	60.15	481.20		
1 Equip. Oper. Oiler	32.45	259.60	51.25	410.00		
1 Gradall, 3 Ton, 1/2 C.Y.		841.00		925.10	15.02	16.52
56 L.H., Daily Totals		$2517.40		$3658.70	$44.96	$65.33

Crew B-30	**Hr.**	**Daily**	**Hr.**	**Daily**	**Bare Costs**	**Incl. O&P**
1 Equip. Oper. (med.)	$36.70	$293.60	$57.95	$463.60	$31.13	$50.08
2 Truck Drivers (heavy)	28.35	453.60	46.15	738.40		
1 Hyd. Excavator, 1.5 C.Y.		722.80		795.10		
2 Dump Trucks, 16 Ton		953.20		1048.50	69.83	76.82
24 L.H., Daily Totals		$2423.20		$3045.60	$100.96	$126.90

Crew B-31	**Hr.**	**Daily**	**Hr.**	**Daily**	**Bare Costs**	**Incl. O&P**
1 Laborer Foreman (outside)	$29.40	$235.20	$48.70	$389.60	$29.43	$48.76
3 Laborers	27.40	657.60	45.40	1089.60		
1 Carpenter	35.55	284.40	58.90	471.20		
1 Air Compr., 250 C.F.M.		119.80		131.80		
1 Sheeting Driver		7.20		7.90		
2-50 Ft. Air Hoses, 1.5" Dia.		9.40		10.35	3.41	3.75
40 L.H., Daily Totals		$1313.60		$2100.45	$32.84	$52.51

Crew No.	Bare Costs		Incl. Subs O & P		Cost Per Labor-Hour	
Crew B-32	**Hr.**	**Daily**	**Hr.**	**Daily**	**Bare Costs**	**Incl. O&P**
1 Laborer	$27.40	$219.20	$45.40	$363.20	$34.38	$54.81
3 Equip. Oper. (med.)	36.70	880.80	57.95	1390.80		
1 Grader, 30,000 Lbs.		457.40		503.15		
1 Tandem Roller, 10 Ton		197.00		216.70		
1 Dozer, 200 H.P.		920.80		1012.90	49.23	54.15
32 L.H., Daily Totals		$2675.20		$3486.75	$83.61	$108.96

Crew B-32A	**Hr.**	**Daily**	**Hr.**	**Daily**	**Bare Costs**	**Incl. O&P**
1 Laborer	$27.40	$219.20	$45.40	$363.20	$33.60	$53.77
2 Equip. Oper. (medium)	36.70	587.20	57.95	927.20		
1 Grader, 30,000 Lbs.		457.40		503.15		
1 Roller, Vibratory, 29,000 Lbs.		522.00		574.20	40.81	44.89
24 L.H., Daily Totals		$1785.80		$2367.75	$74.41	$98.66

Crew B-32B	**Hr.**	**Daily**	**Hr.**	**Daily**	**Bare Costs**	**Incl. O&P**
1 Laborer	$27.40	$219.20	$45.40	$363.20	$33.60	$53.77
2 Equip. Oper. (medium)	36.70	587.20	57.95	927.20		
1 Dozer, 200 H.P.		920.80		1012.90		
1 Roller, Vibratory, 29,000 Lbs.		522.00		574.20	60.12	66.13
24 L.H., Daily Totals		$2249.20		$2877.50	$93.72	$119.90

Crew B-32C	**Hr.**	**Daily**	**Hr.**	**Daily**	**Bare Costs**	**Incl. O&P**
1 Laborer Foreman	$29.40	$235.20	$48.70	$389.60	$32.38	$52.23
2 Laborers	27.40	438.40	45.40	726.40		
3 Equip. Oper. (medium)	36.70	880.80	57.95	1390.80		
1 Grader, 30,000 Lbs.		457.40		503.15		
1 Roller, Steel Wheel		197.00		216.70		
1 Dozer, 200 H.P.		920.80		1012.90	32.82	36.10
48 L.H., Daily Totals		$3129.60		$4239.55	$65.20	$88.33

Crew B-33A	**Hr.**	**Daily**	**Hr.**	**Daily**	**Bare Costs**	**Incl. O&P**
1 Equip. Oper. (med.)	$36.70	$293.60	$57.95	$463.60	$34.04	$54.36
.5 Laborer	27.40	109.60	45.40	181.60		
.25 Equip. Oper. (med.)	36.70	73.40	57.95	115.90		
1 Scraper, Towed, 7 C.Y.		161.75		177.95		
1.25 Dozer, 300 H.P.		1496.25		1645.90	118.43	130.28
14 L.H., Daily Totals		$2134.60		$2584.95	$152.47	$184.64

Crew B-33B	**Hr.**	**Daily**	**Hr.**	**Daily**	**Bare Costs**	**Incl. O&P**
1 Equip. Oper. (med.)	$36.70	$293.60	$57.95	$463.60	$34.04	$54.36
.5 Laborer	27.40	109.60	45.40	181.60		
.25 Equip. Oper. (med.)	36.70	73.40	57.95	115.90		
1 Scraper, Towed, 10 C.Y.		180.40		198.45		
1.25 Dozer, 300 H.P.		1496.25		1645.90	119.76	131.74
14 L.H., Daily Totals		$2153.25		$2605.45	$153.80	$186.10

Crew B-33C	**Hr.**	**Daily**	**Hr.**	**Daily**	**Bare Costs**	**Incl. O&P**
1 Equip. Oper. (med.)	$36.70	$293.60	$57.95	$463.60	$34.04	$54.36
.5 Laborer	27.40	109.60	45.40	181.60		
.25 Equip. Oper. (med.)	36.70	73.40	57.95	115.90		
1 Scraper, Towed, 12 C.Y.		180.40		198.45		
1.25 Dozer, 300 H.P.		1496.25		1645.90	119.76	131.74
14 L.H., Daily Totals		$2153.25		$2605.45	$153.80	$186.10

Crew B-33D	**Hr.**	**Daily**	**Hr.**	**Daily**	**Bare Costs**	**Incl. O&P**
1 Equip. Oper. (med.)	$36.70	$293.60	$57.95	$463.60	$34.04	$54.36
.5 Laborer	27.40	109.60	45.40	181.60		
.25 Equip. Oper. (med.)	36.70	73.40	57.95	115.90		
1 S.P. Scraper, 14 C.Y.		1577.00		1734.70		
.25 Dozer, 300 H.P.		299.25		329.20	134.02	147.42
14 L.H., Daily Totals		$2352.85		$2825.00	$168.06	$201.78

Crews

Crew B-33E	Bare Costs Hr.	Bare Costs Daily	Incl. Subs O & P Hr.	Incl. Subs O & P Daily	Cost Per Labor-Hour Bare Costs	Cost Per Labor-Hour Incl. O&P
1 Equip. Oper. (med.)	$36.70	$293.60	$57.95	$463.60	$34.04	$54.36
.5 Laborer	27.40	109.60	45.40	181.60		
.25 Equip. Oper. (med.)	36.70	73.40	57.95	115.90		
1 S.P. Scraper, 24 C.Y.		2350.00		2585.00		
.25 Dozer, 300 H.P.		299.25		329.20	189.23	208.16
14 L.H., Daily Totals		$3125.85		$3675.30	$223.27	$262.52

Crew B-33F	Hr.	Daily	Hr.	Daily	Bare Costs	Incl. O&P
1 Equip. Oper. (med.)	$36.70	$293.60	$57.95	$463.60	$34.04	$54.36
.5 Laborer	27.40	109.60	45.40	181.60		
.25 Equip. Oper. (med.)	36.70	73.40	57.95	115.90		
1 Elev. Scraper, 11 C.Y.		822.60		904.85		
.25 Dozer, 300 H.P.		299.25		329.20	80.13	88.15
14 L.H., Daily Totals		$1598.45		$1995.15	$114.17	$142.51

Crew B-33G	Hr.	Daily	Hr.	Daily	Bare Costs	Incl. O&P
1 Equip. Oper. (med.)	$36.70	$293.60	$57.95	$463.60	$34.04	$54.36
.5 Laborer	27.40	109.60	45.40	181.60		
.25 Equip. Oper. (med.)	36.70	73.40	57.95	115.90		
1 Elev. Scraper, 20 C.Y.		1653.00		1818.30		
.25 Dozer, 300 H.P.		299.25		329.20	139.45	153.39
14 L.H., Daily Totals		$2428.85		$2908.60	$173.49	$207.75

Crew B-34A	Hr.	Daily	Hr.	Daily	Bare Costs	Incl. O&P
1 Truck Driver (heavy)	$28.35	$226.80	$46.15	$369.20	$28.35	$46.15
1 Dump Truck, 12 Ton		326.60		359.25	40.83	44.91
8 L.H., Daily Totals		$553.40		$728.45	$69.18	$91.06

Crew B-34B	Hr.	Daily	Hr.	Daily	Bare Costs	Incl. O&P
1 Truck Driver (heavy)	$28.35	$226.80	$46.15	$369.20	$28.35	$46.15
1 Dump Truck, 16 Ton		476.60		524.25	59.58	65.53
8 L.H., Daily Totals		$703.40		$893.45	$87.93	$111.68

Crew B-34C	Hr.	Daily	Hr.	Daily	Bare Costs	Incl. O&P
1 Truck Driver (heavy)	$28.35	$226.80	$46.15	$369.20	$28.35	$46.15
1 Truck Tractor, 40 Ton		301.60		331.75		
1 Dump Trailer, 16.5 C.Y.		103.00		113.30	50.58	55.63
8 L.H., Daily Totals		$631.40		$814.25	$78.93	$101.78

Crew B-34D	Hr.	Daily	Hr.	Daily	Bare Costs	Incl. O&P
1 Truck Driver (heavy)	$28.35	$226.80	$46.15	$369.20	$28.35	$46.15
1 Truck Tractor, 40 Ton		301.60		331.75		
1 Dump Trailer, 20 C.Y.		115.80		127.40	52.18	57.39
8 L.H., Daily Totals		$644.20		$828.35	$80.53	$103.54

Crew B-34E	Hr.	Daily	Hr.	Daily	Bare Costs	Incl. O&P
1 Truck Driver (heavy)	$28.35	$226.80	$46.15	$369.20	$28.35	$46.15
1 Truck, Off Hwy., 25 Ton		912.20		1003.40	114.03	125.43
8 L.H., Daily Totals		$1139.00		$1372.60	$142.38	$171.58

Crew B-34F	Hr.	Daily	Hr.	Daily	Bare Costs	Incl. O&P
1 Truck Driver (heavy)	$28.35	$226.80	$46.15	$369.20	$28.35	$46.15
1 Truck, Off Hwy., 22 C.Y.		933.40		1026.75	116.68	128.34
8 L.H., Daily Totals		$1160.20		$1395.95	$145.03	$174.49

Crew B-34G	Hr.	Daily	Hr.	Daily	Bare Costs	Incl. O&P
1 Truck Driver (heavy)	$28.35	$226.80	$46.15	$369.20	$28.35	$46.15
1 Truck, Off Hwy., 34 C.Y.		1181.00		1299.10	147.63	162.39
8 L.H., Daily Totals		$1407.80		$1668.30	$175.98	$208.54

Crew B-34H	Hr.	Daily	Hr.	Daily	Bare Costs	Incl. O&P
1 Truck Driver (heavy)	$28.35	$226.80	$46.15	$369.20	$28.35	$46.15
1 Truck, Off Hwy., 42 C.Y.		1276.00		1403.60	159.50	175.45
8 L.H., Daily Totals		$1502.80		$1772.80	$187.85	$221.60

Crew B-34J	Hr.	Daily	Hr.	Daily	Bare Costs	Incl. O&P
1 Truck Driver (heavy)	$28.35	$226.80	$46.15	$369.20	$28.35	$46.15
1 Truck, Off Hwy., 60 C.Y.		1649.00		1813.90	206.13	226.74
8 L.H., Daily Totals		$1875.80		$2183.10	$234.48	$272.89

Crew B-34K	Hr.	Daily	Hr.	Daily	Bare Costs	Incl. O&P
1 Truck Driver (heavy)	$28.35	$226.80	$46.15	$369.20	$28.35	$46.15
1 Truck Tractor, 240 H.P.		353.60		388.95		
1 Low Bed Trailer		170.20		187.20	65.48	72.02
8 L.H., Daily Totals		$750.60		$945.35	$93.83	$118.17

Crew B-34N	Hr.	Daily	Hr.	Daily	Bare Costs	Incl. O&P
1 Truck Driver (heavy)	$28.35	$226.80	$46.15	$369.20	$28.35	$46.15
1 Dump Truck, 12 Ton		326.60		359.25		
1 Flatbed Trailer, 40 Ton		119.40		131.35	55.75	61.33
8 L.H., Daily Totals		$672.80		$859.80	$84.10	$107.48

Crew B-34P	Hr.	Daily	Hr.	Daily	Bare Costs	Incl. O&P
1 Pipe Fitter	$43.05	$344.40	$66.95	$535.60	$35.78	$56.60
1 Truck Driver (light)	27.60	220.80	44.90	359.20		
1 Equip. Oper. (medium)	36.70	293.60	57.95	463.60		
1 Flatbed Truck, 3 Ton		176.20		193.80		
1 Backhoe Loader, 48 H.P.		226.20		248.80	16.77	18.44
24 L.H., Daily Totals		$1261.20		$1801.00	$52.55	$75.04

Crew B-34Q	Hr.	Daily	Hr.	Daily	Bare Costs	Incl. O&P
1 Pipe Fitter	$43.05	$344.40	$66.95	$535.60	$36.25	$57.33
1 Truck Driver (light)	27.60	220.80	44.90	359.20		
1 Eqip. Oper. (crane)	38.10	304.80	60.15	481.20		
1 Flatbed Trailer, 25 Ton		88.00		96.80		
1 Dump Truck, 12 Ton		326.60		359.25		
1 Hyd. Crane, 25 Ton		650.80		715.90	44.39	48.83
24 L.H., Daily Totals		$1935.40		$2547.95	$80.64	$106.16

Crew B-34R	Hr.	Daily	Hr.	Daily	Bare Costs	Incl. O&P
1 Pipe Fitter	$43.05	$344.40	$66.95	$535.60	$36.25	$57.33
1 Truck Driver (light)	27.60	220.80	44.90	359.20		
1 Equip. Oper. (crane)	38.10	304.80	60.15	481.20		
1 Flatbed Trailer, 25 Ton		88.00		96.80		
1 Dump Truck, 12 Ton		326.60		359.25		
1 Hyd. Crane, 25 Ton		650.80		715.90		
1 Hyd. Excavator, 1 C.Y.		560.20		616.20	67.73	74.51
24 L.H., Daily Totals		$2495.60		$3164.15	$103.98	$131.84

Crew B-34S	Hr.	Daily	Hr.	Daily	Bare Costs	Incl. O&P
2 Pipe Fitters	$43.05	$688.80	$66.95	$1071.20	$38.14	$60.05
1 Truck Driver (heavy)	28.35	226.80	46.15	369.20		
1 Equip. Oper. (crane)	38.10	304.80	60.15	481.20		
1 Flatbed Trailer, 40 Ton		119.40		131.35		
1 Truck Tractor, 40 Ton		301.60		331.75		
1 Truck Crane, 80 Ton		1059.00		1164.90		
1 Hyd. Excavator, 2 C.Y.		919.60		1011.55	74.99	82.49
32 L.H., Daily Totals		$3620.00		$4561.15	$113.13	$142.54

Crew No.	Bare Costs		Incl. Subs O & P		Cost Per Labor-Hour	
Crew B-34T	**Hr.**	**Daily**	**Hr.**	**Daily**	**Bare Costs**	**Incl. O&P**
2 Pipe Fitters	$43.05	$688.80	$66.95	$1071.20	$38.14	$60.05
1 Truck Driver (heavy)	28.35	226.80	46.15	369.20		
1 Equip. Oper. (crane)	38.10	304.80	60.15	481.20		
1 Flatbed Trailer, 40 Ton		119.40		131.35		
1 Truck Tractor, 40 Ton		301.60		331.75		
1 Truck Crane, 80 Ton		1059.00		1164.90	46.25	50.88
32 L.H., Daily Totals		$2700.40		$3549.60	$84.39	$110.93

Crew B-35	Hr.	Daily	Hr.	Daily	Bare Costs	Incl. O&P
1 Laborer Foreman (out)	$29.40	$235.20	$48.70	$389.60	$34.43	$55.27
1 Skilled Worker	36.50	292.00	59.70	477.60		
1 Welder (plumber)	42.70	341.60	66.40	531.20		
1 Laborer	27.40	219.20	45.40	363.20		
1 Equip. Oper. (crane)	38.10	304.80	60.15	481.20		
1 Equip. Oper. Oiler	32.45	259.60	51.25	410.00		
1 Electric Welding Mach.		69.10		76.00		
1 Hyd. Excavator, .75 C.Y.		477.80		525.60	11.39	12.53
48 L.H., Daily Totals		$2199.30		$3254.40	$45.82	$67.80

Crew B-35A	Hr.	Daily	Hr.	Daily	Bare Costs	Incl. O&P
1 Laborer Foreman (out)	$29.40	$235.20	$48.70	$389.60	$33.42	$53.86
2 Laborers	27.40	438.40	45.40	726.40		
1 Skilled Worker	36.50	292.00	59.70	477.60		
1 Welder (plumber)	42.70	341.60	66.40	531.20		
1 Equip. Oper. (crane)	38.10	304.80	60.15	481.20		
1 Equip. Oper. Oiler	32.45	259.60	51.25	410.00		
1 Welder, 300 amp		89.20		98.10		
1 Crane, 75 Ton		1226.00		1348.60	23.49	25.83
56 L.H., Daily Totals		$3186.80		$4462.70	$56.91	$79.69

Crew B-36	Hr.	Daily	Hr.	Daily	Bare Costs	Incl. O&P
1 Laborer Foreman (outside)	$29.40	$235.20	$48.70	$389.60	$31.52	$51.08
2 Laborers	27.40	438.40	45.40	726.40		
2 Equip. Oper. (med.)	36.70	587.20	57.95	927.20		
1 Dozer, 200 H.P.		920.80		1012.90		
1 Aggregate Spreader		32.00		35.20		
1 Tandem Roller, 10 Ton		197.00		216.70	28.75	31.62
40 L.H., Daily Totals		$2410.60		$3308.00	$60.27	$82.70

Crew B-36A	Hr.	Daily	Hr.	Daily	Bare Costs	Incl. O&P
1 Laborer Foreman (outside)	$29.40	$235.20	$48.70	$389.60	$33.00	$53.04
2 Laborers	27.40	438.40	45.40	726.40		
4 Equip. Oper. (med.)	36.70	1174.40	57.95	1854.40		
1 Dozer, 200 H.P.		920.80		1012.90		
1 Aggregate Spreader		32.00		35.20		
1 Roller, Steel Wheel		197.00		216.70		
1 Roller, Pneumatic Wheel		259.20		285.10	25.16	27.68
56 L.H., Daily Totals		$3257.00		$4520.30	$58.16	$80.72

Crew B-36B	Hr.	Daily	Hr.	Daily	Bare Costs	Incl. O&P
1 Laborer Foreman (outside)	$29.40	$235.20	$48.70	$389.60	$32.42	$52.18
2 Laborers	27.40	438.40	45.40	726.40		
4 Equip. Oper. (medium)	36.70	1174.40	57.95	1854.40		
1 Truck Driver, Heavy	28.35	226.80	46.15	369.20		
1 Grader, 30,000 Lbs.		457.40		503.15		
1 F.E. Loader, crl, 1.5 C.Y.		372.40		409.65		
1 Dozer, 300 H.P.		1197.00		1316.70		
1 Roller, Vibratory		522.00		574.20		
1 Truck, Tractor, 240 H.P.		353.60		388.95		
1 Water Tanker, 5000 Gal.		116.80		128.50	47.18	51.89
64 L.H., Daily Totals		$5094.00		$6660.75	$79.60	$104.07

Crew No.	Bare Costs		Incl. Subs O & P		Cost Per Labor-Hour	
Crew B-36C	**Hr.**	**Daily**	**Hr.**	**Daily**	**Bare Costs**	**Incl. O&P**
1 Laborer Foreman (outside)	$29.40	$235.20	$48.70	$389.60	$33.57	$53.74
3 Equip. Oper. (medium)	36.70	880.80	57.95	1390.80		
1 Truck Driver, Heavy	28.35	226.80	46.15	369.20		
1 Grader, 30,000 Lbs.		457.40		503.15		
1 Dozer, 300 H.P.		1197.00		1316.70		
1 Roller, Vibratory		522.00		574.20		
1 Truck, Tractor, 240 H.P.		353.60		388.95		
1 Water Tanker, 5000 Gal.		116.80		128.50	66.17	72.79
40 L.H., Daily Totals		$3989.60		$5061.10	$99.74	$126.53

Crew B-37	Hr.	Daily	Hr.	Daily	Bare Costs	Incl. O&P
1 Laborer Foreman (outside)	$29.40	$235.20	$48.70	$389.60	$29.03	$47.65
4 Laborers	27.40	876.80	45.40	1452.80		
1 Equip. Oper. (light)	35.20	281.60	55.60	444.80		
1 Tandem Roller, 5 Ton		114.20		125.60	2.38	2.62
48 L.H., Daily Totals		$1507.80		$2412.80	$31.41	$50.27

Crew B-38	Hr.	Daily	Hr.	Daily	Bare Costs	Incl. O&P
1 Laborer Foreman (outside)	$29.40	$235.20	$48.70	$389.60	$31.22	$50.61
2 Laborers	27.40	438.40	45.40	726.40		
1 Equip. Oper. (light)	35.20	281.60	55.60	444.80		
1 Equip. Oper. (medium)	36.70	293.60	57.95	463.60		
1 Backhoe Loader, 48 H.P.		226.20		248.80		
1 Hyd.Hammer, (1200 lb)		113.60		124.95		
1 F.E. Loader (170 H.P.)		444.00		488.40		
1 Pavt. Rem. Bucket		52.20		57.40	20.90	22.99
40 L.H., Daily Totals		$2084.80		$2943.95	$52.12	$73.60

Crew B-39	Hr.	Daily	Hr.	Daily	Bare Costs	Incl. O&P
1 Laborer Foreman (outside)	$29.40	$235.20	$48.70	$389.60	$29.03	$47.65
4 Laborers	27.40	876.80	45.40	1452.80		
1 Equip. Oper. (light)	35.20	281.60	55.60	444.80		
1 Air Compr., 250 C.F.M.		119.80		131.80		
2 Air Tools & Accessories		18.00		19.80		
2-50 Ft. Air Hoses, 1.5" Dia.		9.40		10.35	3.07	3.38
48 L.H., Daily Totals		$1540.80		$2449.15	$32.10	$51.03

Crew B-40	Hr.	Daily	Hr.	Daily	Bare Costs	Incl. O&P
1 Pile Driver Foreman (out)	$36.70	$293.60	$62.15	$497.20	$35.52	$58.61
4 Pile Drivers	34.70	1110.40	58.80	1881.60		
2 Equip. Oper. (crane)	38.10	609.60	60.15	962.40		
1 Equip. Oper. Oiler	32.45	259.60	51.25	410.00		
1 Crane, 40 Ton		934.20		1027.60		
1 Vibratory Hammer & Gen.		1823.00		2005.30	43.08	47.39
64 L.H., Daily Totals		$5030.40		$6784.10	$78.60	$106.00

Crew B-40B	Hr.	Daily	Hr.	Daily	Bare Costs	Incl. O&P
1 Laborer Foreman	$29.40	$235.20	$48.70	$389.60	$30.36	$49.38
3 Laborers	27.40	657.60	45.40	1089.60		
1 Equip. Oper. (crane)	38.10	304.80	60.15	481.20		
1 Equip. Oper. Oiler	32.45	259.60	51.25	410.00		
1 Crane, 40 Ton		1130.00		1243.00	23.54	25.90
48 L.H., Daily Totals		$2587.20		$3613.40	$53.90	$75.28

Crew B-41	Hr.	Daily	Hr.	Daily	Bare Costs	Incl. O&P
1 Laborer Foreman (outside)	$29.40	$235.20	$48.70	$389.60	$28.48	$46.94
4 Laborers	27.40	876.80	45.40	1452.80		
.25 Equip. Oper. (crane)	38.10	76.20	60.15	120.30		
.25 Equip. Oper. Oiler	32.45	64.90	51.25	102.50		
.25 Crawler Crane, 40 Ton		233.55		256.90	5.31	5.84
44 L.H., Daily Totals		$1486.65		$2322.10	$33.79	$52.78

Crew B-42	Hr.	Daily	Hr.	Daily	Bare Costs	Incl. O&P
1 Laborer Foreman (outside)	$29.40	$235.20	$48.70	$389.60	$31.19	$52.05
4 Laborers	27.40	876.80	45.40	1452.80		
1 Equip. Oper. (crane)	38.10	304.80	60.15	481.20		
1 Equip. Oper. Oiler	32.45	259.60	51.25	410.00		
1 Welder	39.95	319.60	74.70	597.60		
1 Hyd. Crane, 25 Ton		650.80		715.90		
1 Gas Welding Machine		89.20		98.10		
1 Horiz. Boring Csg. Mch.		392.00		431.20	17.69	19.46
64 L.H., Daily Totals		$3128.00		$4576.40	$48.88	$71.51

Crew B-43	Hr.	Daily	Hr.	Daily	Bare Costs	Incl. O&P
1 Laborer Foreman (outside)	$29.40	$235.20	$48.70	$389.60	$30.36	$49.38
3 Laborers	27.40	657.60	45.40	1089.60		
1 Equip. Oper. (crane)	38.10	304.80	60.15	481.20		
1 Equip. Oper. Oiler	32.45	259.60	51.25	410.00		
1 Drill Rig & Augers		3379.00		3716.90	70.40	77.44
48 L.H., Daily Totals		$4836.20		$6087.30	$100.76	$126.82

Crew B-44	Hr.	Daily	Hr.	Daily	Bare Costs	Incl. O&P
1 Pile Driver Foreman	$36.70	$293.60	$62.15	$497.20	$34.89	$57.88
4 Pile Drivers	34.70	1110.40	58.80	1881.60		
2 Equip. Oper. (crane)	38.10	609.60	60.15	962.40		
1 Laborer	27.40	219.20	45.40	363.20		
1 Crane, 40 Ton, & Access.		934.20		1027.60		
45 L.F. Leads, 15K Ft. Lbs.		49.50		54.45	15.41	16.95
64 L.H., Daily Totals		$3216.50		$4786.45	$50.30	$74.83

Crew B-45	Hr.	Daily	Hr.	Daily	Bare Costs	Incl. O&P
1 Equip. Oper. (med.)	$36.70	$293.60	$57.95	$463.60	$32.53	$52.05
1 Truck Driver (heavy)	28.35	226.80	46.15	369.20		
1 Dist. Tank Truck, 3K Gal.		242.40		266.65		
1 Tractor, 4 x 2, 250 H.P.		291.60		320.75	33.38	36.71
16 L.H., Daily Totals		$1054.40		$1420.20	$65.91	$88.76

Crew B-46	Hr.	Daily	Hr.	Daily	Bare Costs	Incl. O&P
1 Pile Driver Foreman	$36.70	$293.60	$62.15	$497.20	$31.38	$52.66
2 Pile Drivers	34.70	555.20	58.80	940.80		
3 Laborers	27.40	657.60	45.40	1089.60		
1 Chain Saw, 36″ Long		33.40		36.75	.70	.77
48 L.H., Daily Totals		$1539.80		$2564.35	$32.08	$53.43

Crew B-47	Hr.	Daily	Hr.	Daily	Bare Costs	Incl. O&P
1 Blast Foreman	$29.40	$235.20	$48.70	$389.60	$30.67	$49.90
1 Driller	27.40	219.20	45.40	363.20		
1 Equip. Oper. (light)	35.20	281.60	55.60	444.80		
1 Crawler Type Drill, 4″		728.80		801.70		
1 Air Compr., 600 C.F.M.		320.20		352.20		
2-50 Ft. Air Hoses, 3″ Dia.		37.10		40.80	45.26	49.78
24 L.H., Daily Totals		$1822.10		$2392.30	$75.93	$99.68

Crew B-47A	Hr.	Daily	Hr.	Daily	Bare Costs	Incl. O&P
1 Drilling Foreman	$29.40	$235.20	$48.70	$389.60	$33.32	$53.37
1 Equip. Oper. (heavy)	38.10	304.80	60.15	481.20		
1 Oiler	32.45	259.60	51.25	410.00		
1 Quarry Drill		822.00		904.20	34.25	37.68
24 L.H., Daily Totals		$1621.60		$2185.00	$67.57	$91.05

Crew B-47C	Hr.	Daily	Hr.	Daily	Bare Costs	Incl. O&P
1 Laborer	$27.40	$219.20	$45.40	$363.20	$31.30	$50.50
1 Equip. Oper. (light)	35.20	281.60	55.60	444.80		
1 Air Compressor, 750 CFM		341.40		375.55		
2-50′ Air Hoses, 3″		37.10		40.80		
1 Air Track Drill, 4″		728.80		801.70	69.21	76.13
16 L.H., Daily Totals		$1608.10		$2026.05	$100.51	$126.63

Crew B-47E	Hr.	Daily	Hr.	Daily	Bare Costs	Incl. O&P
1 Laborer Foreman	$29.40	$235.20	$48.70	$389.60	$27.90	$46.23
3 Laborers	27.40	657.60	45.40	1089.60		
1 Truck, Flatbed, 3 Ton		176.20		193.80	5.51	6.06
32 L.H., Daily Totals		$1069.00		$1673.00	$33.41	$52.29

Crew B-47G	Hr.	Daily	Hr.	Daily	Bare Costs	Incl. O&P
1 Laborer Foreman	$29.40	$235.20	$48.70	$389.60	$29.85	$48.78
2 Laborers	27.40	438.40	45.40	726.40		
1 Equip. Oper. (light)	35.20	281.60	55.60	444.80		
1 Air Track Drill, 4″		728.80		801.70		
1 Air Compr., 600 C.F.M.		320.20		352.20		
2-50 Ft. Air Hoses, 3″ Dia.		37.10		40.80		
1 Grout Pump		106.40		117.05	37.27	41.00
32 L.H., Daily Totals		$2147.70		$2872.55	$67.12	$89.78

Crew B-48	Hr.	Daily	Hr.	Daily	Bare Costs	Incl. O&P
1 Laborer Foreman (outside)	$29.40	$235.20	$48.70	$389.60	$31.05	$50.27
3 Laborers	27.40	657.60	45.40	1089.60		
1 Equip. Oper. (crane)	38.10	304.80	60.15	481.20		
1 Equip. Oper. Oiler	32.45	259.60	51.25	410.00		
1 Equip. Oper. (light)	35.20	281.60	55.60	444.80		
1 Centr. Water Pump, 6″		240.00		264.00		
1-20 Ft. Suction Hose, 6″		12.10		13.30		
1-50 Ft. Disch. Hose, 6″		7.50		8.25		
1 Drill Rig & Augers		3379.00		3716.90	64.98	71.47
56 L.H., Daily Totals		$5377.40		$6817.65	$96.03	$121.74

Crew B-49	Hr.	Daily	Hr.	Daily	Bare Costs	Incl. O&P
1 Laborer Foreman (outside)	$29.40	$235.20	$48.70	$389.60	$32.48	$52.81
3 Laborers	27.40	657.60	45.40	1089.60		
2 Equip. Oper. (crane)	38.10	609.60	60.15	962.40		
2 Equip. Oper. Oilers	32.45	519.20	51.25	820.00		
1 Equip. Oper. (light)	35.20	281.60	55.60	444.80		
2 Pile Drivers	34.70	555.20	58.80	940.80		
1 Hyd. Crane, 25 Ton		650.80		715.90		
1 Centr. Water Pump, 6″		240.00		264.00		
1-20 Ft. Suction Hose, 6″		12.10		13.30		
1-50 Ft. Disch. Hose, 6″		7.50		8.25		
1 Drill Rig & Augers		3379.00		3716.90	48.74	53.62
88 L.H., Daily Totals		$7147.80		$9365.55	$81.22	$106.43

Crew B-50	Hr.	Daily	Hr.	Daily	Bare Costs	Incl. O&P
2 Pile Driver Foremen	$36.70	$587.20	$62.15	$994.40	$33.75	$56.06
6 Pile Drivers	34.70	1665.60	58.80	2822.40		
2 Equip. Oper. (crane)	38.10	609.60	60.15	962.40		
1 Equip. Oper. Oiler	32.45	259.60	51.25	410.00		
3 Laborers	27.40	657.60	45.40	1089.60		
1 Crane, 40 Ton		934.20		1027.60		
60 L.F. Leads, 15K Ft. Lbs.		66.00		72.60		
1 Hammer, 15K Ft. Lbs.		355.80		391.40		
1 Air Compr., 600 C.F.M.		320.20		352.20		
2-50 Ft. Air Hoses, 3″ Dia.		37.10		40.80		
1 Chain Saw, 36″ Long		33.40		36.75	15.62	17.19
112 L.H., Daily Totals		$5526.30		$8200.15	$49.37	$73.25

Crew No.	Bare Costs		Incl. Subs O & P		Cost Per Labor-Hour	
Crew B-51	**Hr.**	**Daily**	**Hr.**	**Daily**	**Bare Costs**	**Incl. O&P**
1 Laborer Foreman (outside)	$29.40	$235.20	$48.70	$389.60	$27.77	$45.87
4 Laborers	27.40	876.80	45.40	1452.80		
1 Truck Driver (light)	27.60	220.80	44.90	359.20		
1 Light Truck, 1.5 Ton		128.80		141.70	2.68	2.95
48 L.H., Daily Totals		$1461.60		$2343.30	$30.45	$48.82

Crew B-52	Hr.	Daily	Hr.	Daily	Bare Costs	Incl. O&P
1 Carpenter Foreman	$37.55	$300.40	$62.20	$497.60	$32.54	$53.48
1 Carpenter	35.55	284.40	58.90	471.20		
3 Laborers	27.40	657.60	45.40	1089.60		
1 Cement Finisher	34.40	275.20	54.20	433.60		
.5 Rodman (reinf.)	39.50	158.00	67.80	271.20		
.5 Equip. Oper. (med.)	36.70	146.80	57.95	231.80		
.5 F.E. Ldr., T.M., 2.5 C.Y.		397.00		436.70	7.09	7.80
56 L.H., Daily Totals		$2219.40		$3431.70	$39.63	$61.28

Crew B-53	Hr.	Daily	Hr.	Daily	Bare Costs	Incl. O&P
1 Equip. Oper. (light)	$35.20	$281.60	$55.60	$444.80	$35.20	$55.60
1 Trencher, Chain, 12 H.P.		47.60		52.35	5.95	6.55
8 L.H., Daily Totals		$329.20		$497.15	$41.15	$62.15

Crew B-54	Hr.	Daily	Hr.	Daily	Bare Costs	Incl. O&P
1 Equip. Oper. (light)	$35.20	$281.60	$55.60	$444.80	$35.20	$55.60
1 Trencher, Chain, 40 H.P.		216.80		238.50	27.10	29.81
8 L.H., Daily Totals		$498.40		$683.30	$62.30	$85.41

Crew B-54A	Hr.	Daily	Hr.	Daily	Bare Costs	Incl. O&P
.17 Labor Foreman (outside)	$29.40	$39.98	$48.70	$66.23	$35.64	$56.61
1 Equipment Operator (med.)	36.70	293.60	57.95	463.60		
1 Wheel Trencher, 67 H.P.		826.80		909.50	88.33	97.17
9.36 L.H., Daily Totals		$1160.38		$1439.33	$123.97	$153.78

Crew B-54B	Hr.	Daily	Hr.	Daily	Bare Costs	Incl. O&P
.25 Labor Foreman (outside)	$29.40	$58.80	$48.70	$97.40	$35.24	$56.10
1 Equipment Operator (med.)	36.70	293.60	57.95	463.60		
1 Wheel Trencher, 150 H.P.		1578.00		1735.80	157.80	173.58
10 L.H., Daily Totals		$1930.40		$2296.80	$193.04	$229.68

Crew B-55	Hr.	Daily	Hr.	Daily	Bare Costs	Incl. O&P
2 Laborers	$27.40	$438.40	$45.40	$726.40	$27.47	$45.23
1 Truck Driver (light)	27.60	220.80	44.90	359.20		
1 Auger, 4" to 36" Dia		631.60		694.75		
1 Flatbed 3 Ton Truck		176.20		193.80	33.66	37.02
24 L.H., Daily Totals		$1467.00		$1974.15	$61.13	$82.25

Crew B-56	Hr.	Daily	Hr.	Daily	Bare Costs	Incl. O&P
1 Laborer	$27.40	$219.20	$45.40	$363.20	$31.30	$50.50
1 Equip. Oper. (light)	35.20	281.60	55.60	444.80		
1 Crawler Type Drill, 4"		728.80		801.70		
1 Air Compr., 600 C.F.M.		320.20		352.20		
1-50 Ft. Air Hose, 3" Dia.		18.55		20.40	66.73	73.40
16 L.H., Daily Totals		$1568.35		$1982.30	$98.03	$123.90

Crew B-57	Hr.	Daily	Hr.	Daily	Bare Costs	Incl. O&P
1 Laborer Foreman (outside)	$29.40	$235.20	$48.70	$389.60	$31.66	$51.08
2 Laborers	27.40	438.40	45.40	726.40		
1 Equip. Oper. (crane)	38.10	304.80	60.15	481.20		
1 Equip. Oper. (light)	35.20	281.60	55.60	444.80		
1 Equip. Oper. Oiler	32.45	259.60	51.25	410.00		
1 Power Shovel, 1 C.Y.		880.00		968.00		
1 Clamshell Bucket, 1 C.Y.		43.40		47.75		
1 Centr. Water Pump, 6"		240.00		264.00		
1-20 Ft. Suction Hose, 6"		12.10		13.30		
20-50 Ft. Disch. Hoses, 6"		150.00		165.00	27.61	30.38
48 L.H., Daily Totals		$2845.10		$3910.05	$59.27	$81.46

Crew B-58	Hr.	Daily	Hr.	Daily	Bare Costs	Incl. O&P
2 Laborers	$27.40	$438.40	$45.40	$726.40	$30.00	$48.80
1 Equip. Oper. (light)	35.20	281.60	55.60	444.80		
1 Backhoe Loader, 48 H.P.		226.20		248.80		
1 Small Helicopter, w/pilot		2238.00		2461.80	102.68	112.94
24 L.H., Daily Totals		$3184.20		$3881.80	$132.68	$161.74

Crew B-59	Hr.	Daily	Hr.	Daily	Bare Costs	Incl. O&P
1 Truck Driver (heavy)	$28.35	$226.80	$46.15	$369.20	$28.35	$46.15
1 Truck, 30 Ton		208.20		229.00		
1 Water tank, 6000 Gal.		116.80		128.50	40.63	44.69
8 L.H., Daily Totals		$551.80		$726.70	$68.98	$90.84

Crew B-59A	Hr.	Daily	Hr.	Daily	Bare Costs	Incl. O&P
2 Laborers	$27.40	$438.40	$45.40	$726.40	$27.72	$45.65
1 Truck Driver (heavy)	28.35	226.80	46.15	369.20		
1 Water tank, 5K w/pum		116.80		128.50		
1 Truck, 30 Ton		208.20		229.00	13.54	14.90
24 L.H., Daily Totals		$990.20		$1453.10	$41.26	$60.55

Crew B-60	Hr.	Daily	Hr.	Daily	Bare Costs	Incl. O&P
1 Laborer Foreman (outside)	$29.40	$235.20	$48.70	$389.60	$32.16	$51.73
2 Laborers	27.40	438.40	45.40	726.40		
1 Equip. Oper. (crane)	38.10	304.80	60.15	481.20		
2 Equip. Oper. (light)	35.20	563.20	55.60	889.60		
1 Equip. Oper. Oiler	32.45	259.60	51.25	410.00		
1 Crawler Crane, 40 Ton		934.20		1027.60		
45 L.F. Leads, 15K Ft. Lbs.		49.50		54.45		
1 Backhoe Loader, 48 H.P.		226.20		248.80	21.65	23.81
56 L.H., Daily Totals		$3011.10		$4227.65	$53.81	$75.54

Crew B-61	Hr.	Daily	Hr.	Daily	Bare Costs	Incl. O&P
1 Laborer Foreman (outside)	$29.40	$235.20	$48.70	$389.60	$29.36	$48.10
3 Laborers	27.40	657.60	45.40	1089.60		
1 Equip. Oper. (light)	35.20	281.60	55.60	444.80		
1 Cement Mixer, 2 C.Y.		162.80		179.10		
1 Air Compr., 160 C.F.M.		97.60		107.35	6.51	7.16
40 L.H., Daily Totals		$1434.80		$2210.45	$35.87	$55.26

Crew B-62	Hr.	Daily	Hr.	Daily	Bare Costs	Incl. O&P
2 Laborers	$27.40	$438.40	$45.40	$726.40	$30.00	$48.80
1 Equip. Oper. (light)	35.20	281.60	55.60	444.80		
1 Loader, Skid Steer		154.00		169.40	6.42	7.06
24 L.H., Daily Totals		$874.00		$1340.60	$36.42	$55.86

Crew No.	Bare Costs		Incl. Subs O & P		Cost Per Labor-Hour	
Crew B-63	**Hr.**	**Daily**	**Hr.**	**Daily**	**Bare Costs**	**Incl. O&P**
4 Laborers	$27.40	$876.80	$45.40	$1452.80	$28.96	$47.44
1 Equip. Oper. (light)	35.20	281.60	55.60	444.80		
1 Loader, Skid Steer		154.00		169.40	3.85	4.24
40 L.H., Daily Totals		$1312.40		$2067.00	$32.81	$51.68
Crew B-64	**Hr.**	**Daily**	**Hr.**	**Daily**	**Bare Costs**	**Incl. O&P**
1 Laborer	$27.40	$219.20	$45.40	$363.20	$27.50	$45.15
1 Truck Driver (light)	27.60	220.80	44.90	359.20		
1 Power Mulcher (small)		113.80		125.20		
1 Light Truck, 1.5 Ton		128.80		141.70	15.16	16.68
16 L.H., Daily Totals		$682.60		$989.30	$42.66	$61.83
Crew B-65	**Hr.**	**Daily**	**Hr.**	**Daily**	**Bare Costs**	**Incl. O&P**
1 Laborer	$27.40	$219.20	$45.40	$363.20	$27.50	$45.15
1 Truck Driver (light)	27.60	220.80	44.90	359.20		
1 Power Mulcher (large)		185.20		203.70		
1 Light Truck, 1.5 Ton		128.80		141.70	19.63	21.59
16 L.H., Daily Totals		$754.00		$1067.80	$47.13	$66.74
Crew B-66	**Hr.**	**Daily**	**Hr.**	**Daily**	**Bare Costs**	**Incl. O&P**
1 Equip. Oper. (light)	$35.20	$281.60	$55.60	$444.80	$35.20	$55.60
1 Backhoe Ldr. w/Attchmt.		178.80		196.70	22.35	24.59
8 L.H., Daily Totals		$460.40		$641.50	$57.55	$80.19
Crew B-67	**Hr.**	**Daily**	**Hr.**	**Daily**	**Bare Costs**	**Incl. O&P**
1 Millwright	$37.10	$296.80	$58.50	$468.00	$36.15	$57.05
1 Equip. Oper. (light)	35.20	281.60	55.60	444.80		
1 Forklift		242.40		266.65	15.15	16.67
16 L.H., Daily Totals		$820.80		$1179.45	$51.30	$73.72
Crew B-68	**Hr.**	**Daily**	**Hr.**	**Daily**	**Bare Costs**	**Incl. O&P**
2 Millwrights	$37.10	$593.60	$58.50	$936.00	$36.47	$57.53
1 Equip. Oper. (light)	35.20	281.60	55.60	444.80		
1 Forklift		242.40		266.65	10.10	11.11
24 L.H., Daily Totals		$1117.60		$1647.45	$46.57	$68.64
Crew B-69	**Hr.**	**Daily**	**Hr.**	**Daily**	**Bare Costs**	**Incl. O&P**
1 Laborer Foreman (outside)	$29.40	$235.20	$48.70	$389.60	$30.36	$49.38
3 Laborers	27.40	657.60	45.40	1089.60		
1 Equip Oper. (crane)	38.10	304.80	60.15	481.20		
1 Equip Oper. Oiler	32.45	259.60	51.25	410.00		
1 Truck Crane, 80 Ton		1059.00		1164.90	22.06	24.27
48 L.H., Daily Totals		$2516.20		$3535.30	$52.42	$73.65
Crew B-69A	**Hr.**	**Daily**	**Hr.**	**Daily**	**Bare Costs**	**Incl. O&P**
1 Laborer Foreman	$29.40	$235.20	$48.70	$389.60	$30.45	$49.51
3 Laborers	27.40	657.60	45.40	1089.60		
1 Equip. Oper. (medium)	36.70	293.60	57.95	463.60		
1 Concrete Finisher	34.40	275.20	54.20	433.60		
1 Curb Paver		593.60		652.95	12.37	13.60
48 L.H., Daily Totals		$2055.20		$3029.35	$42.82	$63.11
Crew B-69B	**Hr.**	**Daily**	**Hr.**	**Daily**	**Bare Costs**	**Incl. O&P**
1 Laborer Foreman	$29.40	$235.20	$48.70	$389.60	$30.45	$49.51
3 Laborers	27.40	657.60	45.40	1089.60		
1 Equip. Oper. (medium)	36.70	293.60	57.95	463.60		
1 Cement Finisher	34.40	275.20	54.20	433.60		
1 Curb/Gutter Paver		738.00		811.80	15.38	16.91
48 L.H., Daily Totals		$2199.60		$3188.20	$45.83	$66.42

Crew No.	Bare Costs		Incl. Subs O & P		Cost Per Labor-Hour	
Crew B-70	**Hr.**	**Daily**	**Hr.**	**Daily**	**Bare Costs**	**Incl. O&P**
1 Laborer Foreman (outside)	$29.40	$235.20	$48.70	$389.60	$31.67	$51.25
3 Laborers	27.40	657.60	45.40	1089.60		
3 Equip. Oper. (med.)	36.70	880.80	57.95	1390.80		
1 Motor Grader, 30,000 Lb.		457.40		503.15		
1 Grader Attach., Ripper		73.20		80.50		
1 Road Sweeper, S.P.		493.20		542.50		
1 F.E. Loader, 1-3/4 C.Y.		241.40		265.55	22.59	24.85
56 L.H., Daily Totals		$3038.80		$4261.70	$54.26	$76.10
Crew B-71	**Hr.**	**Daily**	**Hr.**	**Daily**	**Bare Costs**	**Incl. O&P**
1 Laborer Foreman (outside)	$29.40	$235.20	$48.70	$389.60	$31.67	$51.25
3 Laborers	27.40	657.60	45.40	1089.60		
3 Equip. Oper. (med.)	36.70	880.80	57.95	1390.80		
1 Pvmt. Profiler, 750 H.P.		4659.00		5124.90		
1 Road Sweeper, S.P.		493.20		542.50		
1 F.E. Loader, 1-3/4 C.Y.		241.40		265.55	96.31	105.95
56 L.H., Daily Totals		$7167.20		$8802.95	$127.98	$157.20
Crew B-72	**Hr.**	**Daily**	**Hr.**	**Daily**	**Bare Costs**	**Incl. O&P**
1 Laborer Foreman (outside)	$29.40	$235.20	$48.70	$389.60	$32.30	$52.09
3 Laborers	27.40	657.60	45.40	1089.60		
4 Equip. Oper. (med.)	36.70	1174.40	57.95	1854.40		
1 Pvmt. Profiler, 750 H.P.		4659.00		5124.90		
1 Hammermill, 250 H.P.		1423.00		1565.30		
1 Windrow Loader		823.00		905.30		
1 Mix Paver 165 H.P.		1729.00		1901.90		
1 Roller, Pneu. Tire, 12 T.		259.20		285.10	138.96	152.85
64 L.H., Daily Totals		$10960.40		$13116.10	$171.26	$204.94
Crew B-73	**Hr.**	**Daily**	**Hr.**	**Daily**	**Bare Costs**	**Incl. O&P**
1 Laborer Foreman (outside)	$29.40	$235.20	$48.70	$389.60	$33.46	$53.66
2 Laborers	27.40	438.40	45.40	726.40		
5 Equip. Oper. (med.)	36.70	1468.00	57.95	2318.00		
1 Road Mixer, 310 H.P.		1698.00		1867.80		
1 Roller, Tandem, 12 Ton		197.00		216.70		
1 Hammermill, 250 H.P.		1423.00		1565.30		
1 Motor Grader, 30,000 Lb.		457.40		503.15		
.5 F.E. Loader, 1-3/4 C.Y.		120.70		132.75		
.5 Truck, 30 Ton		104.10		114.50		
.5 Water Tank 5000 Gal.		58.40		64.25	63.42	69.76
64 L.H., Daily Totals		$6200.20		$7898.45	$96.88	$123.42
Crew B-74	**Hr.**	**Daily**	**Hr.**	**Daily**	**Bare Costs**	**Incl. O&P**
1 Laborer Foreman (outside)	$29.40	$235.20	$48.70	$389.60	$32.54	$52.28
1 Laborer	27.40	219.20	45.40	363.20		
4 Equip. Oper. (med.)	36.70	1174.40	57.95	1854.40		
2 Truck Drivers (heavy)	28.35	453.60	46.15	738.40		
1 Motor Grader, 30,000 Lb.		457.40		503.15		
1 Grader Attach., Ripper		73.20		80.50		
2 Stabilizers, 310 H.P.		2206.00		2426.60		
1 Flatbed Truck, 3 Ton		176.20		193.80		
1 Chem. Spreader, Towed		49.00		53.90		
1 Vibr. Roller, 29,000 Lb.		522.00		574.20		
1 Water Tank 5000 Gal.		116.80		128.50		
1 Truck, 30 Ton		208.20		229.00	59.51	65.46
64 L.H., Daily Totals		$5891.20		$7535.25	$92.05	$117.74

Crew B-75	Bare Costs Hr.	Bare Costs Daily	Incl. Subs O & P Hr.	Incl. Subs O & P Daily	Cost Per Labor-Hour Bare Costs	Cost Per Labor-Hour Incl. O&P
1 Laborer Foreman (outside)	$29.40	$235.20	$48.70	$389.60	$33.14	$53.15
1 Laborer	27.40	219.20	45.40	363.20		
4 Equip. Oper. (med.)	36.70	1174.40	57.95	1854.40		
1 Truck Driver (heavy)	28.35	226.80	46.15	369.20		
1 Motor Grader, 30,000 Lb.		457.40		503.15		
1 Grader Attach., Ripper		73.20		80.50		
2 Stabilizers, 310 H.P.		2206.00		2426.60		
1 Dist. Truck, 3000 Gal.		242.40		266.65		
1 Vibr. Roller, 29,000 Lb.		522.00		574.20	62.52	68.77
56 L.H., Daily Totals		$5356.60		$6827.50	$95.66	$121.92

Crew B-76	Hr.	Daily	Hr.	Daily	Bare Costs	Incl. O&P
1 Dock Builder Foreman	$36.70	$293.60	$62.15	$497.20	$35.43	$58.63
5 Dock Builders	34.70	1388.00	58.80	2352.00		
2 Equip. Oper. (crane)	38.10	609.60	60.15	962.40		
1 Equip. Oper. Oiler	32.45	259.60	51.25	410.00		
1 Crawler Crane, 50 Ton		1265.00		1391.50		
1 Barge, 400 Ton		280.00		308.00		
1 Hammer, 15K Ft. Lbs.		355.80		391.40		
60 L.F. Leads, 15K Ft. Lbs.		66.00		72.60		
1 Air Compr., 600 C.F.M.		320.20		352.20		
2-50 Ft. Air Hoses, 3" Dia.		37.10		40.80	32.32	35.55
72 L.H., Daily Totals		$4874.90		$6778.10	$67.75	$94.18

Crew B-76A	Hr.	Daily	Hr.	Daily	Bare Costs	Incl. O&P
1 Laborer Foreman	$29.40	$235.20	$48.70	$389.60	$29.62	$48.39
5 Laborers	27.40	1096.00	45.40	1816.00		
1 Equip. Oper. (crane)	38.10	304.80	60.15	481.20		
1 Equip. Oper. Oiler	32.45	259.60	51.25	410.00		
1 Crawler Crane, 50 Ton		1265.00		1391.50		
1 Barge, 400 Ton		280.00		308.00	24.14	26.55
64 L.H., Daily Totals		$3440.60		$4796.30	$53.76	$74.94

Crew B-77	Hr.	Daily	Hr.	Daily	Bare Costs	Incl. O&P
1 Laborer Foreman	$29.40	$235.20	$48.70	$389.60	$27.84	$45.96
3 Laborers	27.40	657.60	45.40	1089.60		
1 Truck Driver (light)	27.60	220.80	44.90	359.20		
1 Crack Cleaner, 25 H.P.		45.80		50.40		
1 Crack Filler, Trailer Mtd.		157.40		173.15		
1 Flatbed Truck, 3 Ton		176.20		193.80	9.49	10.43
40 L.H., Daily Totals		$1493.00		$2255.75	$37.33	$56.39

Crew B-78	Hr.	Daily	Hr.	Daily	Bare Costs	Incl. O&P
1 Laborer Foreman	$29.40	$235.20	$48.70	$389.60	$27.77	$45.87
4 Laborers	27.40	876.80	45.40	1452.80		
1 Truck Driver (light)	27.60	220.80	44.90	359.20		
1 Paint Striper, S.P.		134.60		148.05		
1 Flatbed Truck, 3 Ton		176.20		193.80		
1 Pickup Truck, 3/4 Ton		78.00		85.80	8.10	8.91
48 L.H., Daily Totals		$1721.60		$2629.25	$35.87	$54.78

Crew B-78A	Hr.	Daily	Hr.	Daily	Bare Costs	Incl. O&P
1 Equip. Oper. (light)	$35.20	$281.60	$55.60	$444.80	$35.20	$55.60
1 Line Remov. (metal balls) 115 HP		795.80		875.40	99.48	109.42
8 L.H., Daily Totals		$1077.40		$1320.20	$134.68	$165.02

Crew B-78B	Hr.	Daily	Hr.	Daily	Bare Costs	Incl. O&P
2 Laborers	$27.40	$438.40	$45.40	$726.40	$28.27	$46.53
.25 Equip. Oper. (light)	35.20	70.40	55.60	111.20		
1 Pickup Truck, 3/4 Ton		78.00		85.80		
1 Line Remov. (walk behind) 11 HP		44.80		49.30		
.25 Road Sweeper, SP		123.30		135.65	13.67	15.04
18 L.H., Daily Totals		$754.90		$1108.35	$41.94	$61.57

Crew B-79	Hr.	Daily	Hr.	Daily	Bare Costs	Incl. O&P
1 Laborer Foreman	$29.40	$235.20	$48.70	$389.60	$27.84	$45.96
3 Laborers	27.40	657.60	45.40	1089.60		
1 Truck Driver (light)	27.60	220.80	44.90	359.20		
1 Thermo. Striper, T.M.		555.20		610.70		
1 Flatbed Truck, 3 Ton		176.20		193.80		
2 Pickup Truck, 3/4 Ton		156.00		171.60	22.19	24.40
40 L.H., Daily Totals		$2001.00		$2814.50	$50.03	$70.36

Crew B-79A	Hr.	Daily	Hr.	Daily	Bare Costs	Incl. O&P
1.5 Equip. Oper. (light)	$35.20	$422.40	$55.60	$667.20	$35.20	$55.60
.5 Line Remov. (grinder) 115 HP		332.70		365.95		
1 Line Remov. (metal balls) 115 HP		795.80		875.40	94.04	103.45
12 L.H., Daily Totals		$1550.90		$1908.55	$129.24	$159.05

Crew B-80	Hr.	Daily	Hr.	Daily	Bare Costs	Incl. O&P
1 Laborer Foreman	$29.40	$235.20	$48.70	$389.60	$29.90	$48.65
1 Laborer	27.40	219.20	45.40	363.20		
1 Truck Driver (light)	27.60	220.80	44.90	359.20		
1 Equip. Oper. (light)	35.20	281.60	55.60	444.80		
1 Flatbed Truck, 3 Ton		176.20		193.80		
1 Fence Post Auger, T.M.		356.20		391.80	16.64	18.30
32 L.H., Daily Totals		$1489.20		$2142.40	$46.54	$66.95

Crew B-80A	Hr.	Daily	Hr.	Daily	Bare Costs	Incl. O&P
3 Laborers	$27.40	$657.60	$45.40	$1089.60	$27.40	$45.40
1 Flatbed Truck, 3 Ton		176.20		193.80	7.34	8.08
24 L.H., Daily Totals		$833.80		$1283.40	$34.74	$53.48

Crew B-80B	Hr.	Daily	Hr.	Daily	Bare Costs	Incl. O&P
3 Laborers	$27.40	$657.60	$45.40	$1089.60	$29.35	$47.95
1 Equip. Oper. (light)	35.20	281.60	55.60	444.80		
1 Crane, Flatbed Mnt.		235.20		258.70	7.35	8.09
32 L.H., Daily Totals		$1174.40		$1793.10	$36.70	$56.04

Crew B-80C	Hr.	Daily	Hr.	Daily	Bare Costs	Incl. O&P
2 Laborers	$27.40	$438.40	$45.40	$726.40	$27.47	$45.23
1 Truck Driver (light)	27.60	220.80	44.90	359.20		
1 Light Truck, 1.5 Ton		128.80		141.70		
1 Manual fence post auger, gas		5.40		5.95	5.59	6.15
24 L.H., Daily Totals		$793.40		$1233.25	$33.06	$51.38

Crew B-81	Hr.	Daily	Hr.	Daily	Bare Costs	Incl. O&P
1 Laborer	$27.40	$219.20	$45.40	$363.20	$30.82	$49.83
1 Equip. Oper. (med.)	36.70	293.60	57.95	463.60		
1 Truck Driver (heavy)	28.35	226.80	46.15	369.20		
1 Hydromulcher, T.M.		197.20		216.90		
1 Tractor Truck, 4x2		208.20		229.00	16.89	18.58
24 L.H., Daily Totals		$1145.00		$1641.90	$47.71	$68.41

Crew No.	Bare Costs		Incl. Subs O & P		Cost Per Labor-Hour	
Crew B-82	**Hr.**	**Daily**	**Hr.**	**Daily**	**Bare Costs**	**Incl. O&P**
1 Laborer	$27.40	$219.20	$45.40	$363.20	$31.30	$50.50
1 Equip. Oper. (light)	35.20	281.60	55.60	444.80		
1 Horiz. Borer, 6 H.P.		66.00		72.60	4.13	4.54
16 L.H., Daily Totals		$566.80		$880.60	$35.43	$55.04
Crew B-83	**Hr.**	**Daily**	**Hr.**	**Daily**	**Bare Costs**	**Incl. O&P**
1 Tugboat Captain	$36.70	$293.60	$57.95	$463.60	$32.05	$51.68
1 Tugboat Hand	27.40	219.20	45.40	363.20		
1 Tugboat, 250 H.P.		473.00		520.30	29.56	32.52
16 L.H., Daily Totals		$985.80		$1347.10	$61.61	$84.20
Crew B-84	**Hr.**	**Daily**	**Hr.**	**Daily**	**Bare Costs**	**Incl. O&P**
1 Equip. Oper. (med.)	$36.70	$293.60	$57.95	$463.60	$36.70	$57.95
1 Rotary Mower/Tractor		231.00		254.10	28.88	31.76
8 L.H., Daily Totals		$524.60		$717.70	$65.58	$89.71
Crew B-85	**Hr.**	**Daily**	**Hr.**	**Daily**	**Bare Costs**	**Incl. O&P**
3 Laborers	$27.40	$657.60	$45.40	$1089.60	$29.45	$48.06
1 Equip. Oper. (med.)	36.70	293.60	57.95	463.60		
1 Truck Driver (heavy)	28.35	226.80	46.15	369.20		
1 Aerial Lift Truck, 80'		515.80		567.40		
1 Brush Chipper, 130 H.P.		173.00		190.30		
1 Pruning Saw, Rotary		7.15		7.85	17.40	19.14
40 L.H., Daily Totals		$1873.95		$2687.95	$46.85	$67.20
Crew B-86	**Hr.**	**Daily**	**Hr.**	**Daily**	**Bare Costs**	**Incl. O&P**
1 Equip. Oper. (med.)	$36.70	$293.60	$57.95	$463.60	$36.70	$57.95
1 Stump Chipper, S.P.		63.50		69.85	7.94	8.73
8 L.H., Daily Totals		$357.10		$533.45	$44.64	$66.68
Crew B-86A	**Hr.**	**Daily**	**Hr.**	**Daily**	**Bare Costs**	**Incl. O&P**
1 Equip. Oper. (medium)	$36.70	$293.60	$57.95	$463.60	$36.70	$57.95
1 Grader, 30,000 Lbs.		457.40		503.15	57.18	62.89
8 L.H., Daily Totals		$751.00		$966.75	$93.88	$120.84
Crew B-86B	**Hr.**	**Daily**	**Hr.**	**Daily**	**Bare Costs**	**Incl. O&P**
1 Equip. Oper. (medium)	$36.70	$293.60	$57.95	$463.60	$36.70	$57.95
1 Dozer, 200 H.P.		920.80		1012.90	115.10	126.61
8 L.H., Daily Totals		$1214.40		$1476.50	$151.80	$184.56
Crew B-87	**Hr.**	**Daily**	**Hr.**	**Daily**	**Bare Costs**	**Incl. O&P**
1 Laborer	$27.40	$219.20	$45.40	$363.20	$34.84	$55.44
4 Equip. Oper. (med.)	36.70	1174.40	57.95	1854.40		
2 Feller Bunchers, 50 H.P.		983.60		1081.95		
1 Log Chipper, 22" Tree		1000.00		1100.00		
1 Dozer, 105 H.P.		462.80		509.10		
1 Chainsaw, Gas, 36" Long		33.40		36.75	62.00	68.19
40 L.H., Daily Totals		$3873.40		$4945.40	$96.84	$123.63
Crew B-88	**Hr.**	**Daily**	**Hr.**	**Daily**	**Bare Costs**	**Incl. O&P**
1 Laborer	$27.40	$219.20	$45.40	$363.20	$35.37	$56.16
6 Equip. Oper. (med.)	36.70	1761.60	57.95	2781.60		
2 Feller Bunchers, 50 H.P.		983.60		1081.95		
1 Log Chipper, 22" Tree		1000.00		1100.00		
2 Log Skidders, 50 H.P.		1575.20		1732.70		
1 Dozer, 105 H.P.		462.80		509.10		
1 Chainsaw, Gas, 36" Long		33.40		36.75	72.41	79.65
56 L.H., Daily Totals		$6035.80		$7605.30	$107.78	$135.81

Crew No.	Bare Costs		Incl. Subs O & P		Cost Per Labor-Hour	
Crew B-89	**Hr.**	**Daily**	**Hr.**	**Daily**	**Bare Costs**	**Incl. O&P**
1 Equip. Oper. (light)	$35.20	$281.60	$55.60	$444.80	$31.40	$50.25
1 Truck Driver (light)	27.60	220.80	44.90	359.20		
1 Truck, Stake Body, 3 Ton		176.20		193.80		
1 Concrete Saw		107.00		117.70		
1 Water Tank, 65 Gal.		13.80		15.20	18.56	20.42
16 L.H., Daily Totals		$799.40		$1130.70	$49.96	$70.67
Crew B-89A	**Hr.**	**Daily**	**Hr.**	**Daily**	**Bare Costs**	**Incl. O&P**
1 Skilled Worker	$36.50	$292.00	$59.70	$477.60	$31.95	$52.55
1 Laborer	27.40	219.20	45.40	363.20		
1 Core Drill (large)		115.45		127.00	7.22	7.94
16 L.H., Daily Totals		$626.65		$967.80	$39.17	$60.49
Crew B-89B	**Hr.**	**Daily**	**Hr.**	**Daily**	**Bare Costs**	**Incl. O&P**
1 Equip. Oper. (light)	$35.20	$281.60	$55.60	$444.80	$31.40	$50.25
1 Truck Driver, Light	27.60	220.80	44.90	359.20		
1 Wall Saw, Hydraulic, 10 H.P.		80.95		89.05		
1 Generator, Diesel, 100 KW		193.20		212.50		
1 Water Tank, 65 Gal.		13.80		15.20		
1 Flatbed Truck, 3 Ton		176.20		193.80	29.01	31.91
16 L.H., Daily Totals		$966.55		$1314.55	$60.41	$82.16
Crew B-90	**Hr.**	**Daily**	**Hr.**	**Daily**	**Bare Costs**	**Incl. O&P**
1 Laborer Foreman (outside)	$29.40	$235.20	$48.70	$389.60	$29.84	$48.55
3 Laborers	27.40	657.60	45.40	1089.60		
2 Equip. Oper. (light)	35.20	563.20	55.60	889.60		
2 Truck Drivers (heavy)	28.35	453.60	46.15	738.40		
1 Road Mixer, 310 H.P.		1698.00		1867.80		
1 Dist. Truck, 2000 Gal.		209.00		229.90	29.80	32.78
64 L.H., Daily Totals		$3816.60		$5204.90	$59.64	$81.33
Crew B-90A	**Hr.**	**Daily**	**Hr.**	**Daily**	**Bare Costs**	**Incl. O&P**
1 Laborer Foreman	$29.40	$235.20	$48.70	$389.60	$33.00	$53.04
2 Laborers	27.40	438.40	45.40	726.40		
4 Equip. Oper. (medium)	36.70	1174.40	57.95	1854.40		
2 Graders, 30,000 Lbs.		914.80		1006.30		
1 Roller, Steel Wheel		197.00		216.70		
1 Roller, Pneumatic Wheel		259.20		285.10	24.48	26.93
56 L.H., Daily Totals		$3219.00		$4478.50	$57.48	$79.97
Crew B-90B	**Hr.**	**Daily**	**Hr.**	**Daily**	**Bare Costs**	**Incl. O&P**
1 Laborer Foreman	$29.40	$235.20	$48.70	$389.60	$32.38	$52.23
2 Laborers	27.40	438.40	45.40	726.40		
3 Equip. Oper. (medium)	36.70	880.80	57.95	1390.80		
1 Roller, Steel Wheel		197.00		216.70		
1 Roller, Pneumatic Wheel		259.20		285.10		
1 Road Mixer, 310 H.P.		1698.00		1867.80	44.88	49.37
48 L.H., Daily Totals		$3708.60		$4876.40	$77.26	$101.60
Crew B-91	**Hr.**	**Daily**	**Hr.**	**Daily**	**Bare Costs**	**Incl. O&P**
1 Laborer Foreman (outside)	$29.40	$235.20	$48.70	$389.60	$32.42	$52.18
2 Laborers	27.40	438.40	45.40	726.40		
4 Equip. Oper. (med.)	36.70	1174.40	57.95	1854.40		
1 Truck Driver (heavy)	28.35	226.80	46.15	369.20		
1 Dist. Truck, 3000 Gal.		242.40		266.65		
1 Aggreg. Spreader, S.P.		777.40		855.15		
1 Roller, Pneu. Tire, 12 Ton		259.20		285.10		
1 Roller, Steel, 10 Ton		197.00		216.70	23.06	25.37
64 L.H., Daily Totals		$3550.80		$4963.20	$55.48	$77.55

Crews

Crew B-92	Hr.	Daily	Hr.	Daily	Bare Costs	Incl. O&P
1 Laborer Foreman (outside)	$29.40	$235.20	$48.70	$389.60	$27.90	$46.23
3 Laborers	27.40	657.60	45.40	1089.60		
1 Crack Cleaner, 25 H.P.		45.80		50.40		
1 Air Compressor		59.00		64.90		
1 Tar Kettle, T.M.		45.75		50.35		
1 Flatbed Truck, 3 Ton		176.20		193.80	10.21	11.23
32 L.H., Daily Totals		$1219.55		$1838.65	$38.11	$57.46

Crew B-93	Hr.	Daily	Hr.	Daily	Bare Costs	Incl. O&P
1 Equip. Oper. (med.)	$36.70	$293.60	$57.95	$463.60	$36.70	$57.95
1 Feller Buncher, 50 H.P.		491.80		541.00	61.48	67.62
8 L.H., Daily Totals		$785.40		$1004.60	$98.18	$125.57

Crew B-94A	Hr.	Daily	Hr.	Daily	Bare Costs	Incl. O&P
1 Laborer	$27.40	$219.20	$45.40	$363.20	$27.40	$45.40
1 Diaph. Water Pump, 2"		53.20		58.50		
1-20 Ft. Suction Hose, 2"		2.15		2.35		
2-50 Ft. Disch. Hoses, 2"		2.60		2.85	7.24	7.97
8 L.H., Daily Totals		$277.15		$426.90	$34.64	$53.37

Crew B-94B	Hr.	Daily	Hr.	Daily	Bare Costs	Incl. O&P
1 Laborer	$27.40	$219.20	$45.40	$363.20	$27.40	$45.40
1 Diaph. Water Pump, 4"		73.20		80.50		
1-20 Ft. Suction Hose, 4"		4.05		4.45		
2-50 Ft. Disch. Hoses, 4"		5.90		6.50	10.39	11.43
8 L.H., Daily Totals		$302.35		$454.65	$37.79	$56.83

Crew B-94C	Hr.	Daily	Hr.	Daily	Bare Costs	Incl. O&P
1 Laborer	$27.40	$219.20	$45.40	$363.20	$27.40	$45.40
1 Centr. Water Pump, 3"		56.60		62.25		
1-20 Ft. Suction Hose, 3"		3.65		4.00		
2-50 Ft. Disch. Hoses, 3"		4.30		4.75	8.07	8.88
8 L.H., Daily Totals		$283.75		$434.20	$35.47	$54.28

Crew B-94D	Hr.	Daily	Hr.	Daily	Bare Costs	Incl. O&P
1 Laborer	$27.40	$219.20	$45.40	$363.20	$27.40	$45.40
1 Centr. Water Pump, 6"		240.00		264.00		
1-20 Ft. Suction Hose, 6"		12.10		13.30		
2-50 Ft. Disch. Hoses, 6"		15.00		16.50	33.39	36.73
8 L.H., Daily Totals		$486.30		$657.00	$60.79	$82.13

Crew B-95A	Hr.	Daily	Hr.	Daily	Bare Costs	Incl. O&P
1 Equip. Oper. (crane)	$38.10	$304.80	$60.15	$481.20	$32.75	$52.78
1 Laborer	27.40	219.20	45.40	363.20		
1 Hyd. Excavator, 5/8 C.Y.		435.40		478.95	27.21	29.93
16 L.H., Daily Totals		$959.40		$1323.35	$59.96	$82.71

Crew B-95B	Hr.	Daily	Hr.	Daily	Bare Costs	Incl. O&P
1 Equip. Oper. (crane)	$38.10	$304.80	$60.15	$481.20	$32.75	$52.78
1 Laborer	27.40	219.20	45.40	363.20		
1 Hyd. Excavator, 1.5 C.Y.		722.80		795.10	45.18	49.69
16 L.H., Daily Totals		$1246.80		$1639.50	$77.93	$102.47

Crew B-95C	Hr.	Daily	Hr.	Daily	Bare Costs	Incl. O&P
1 Equip. Oper. (crane)	$38.10	$304.80	$60.15	$481.20	$32.75	$52.78
1 Laborer	27.40	219.20	45.40	363.20		
1 Hyd. Excavator, 2.5 C.Y.		1218.00		1339.80	76.13	83.74
16 L.H., Daily Totals		$1742.00		$2184.20	$108.88	$136.52

Crew C-1	Hr.	Daily	Hr.	Daily	Bare Costs	Incl. O&P
3 Carpenters	$35.55	$853.20	$58.90	$1413.60	$33.51	$55.53
1 Laborer	27.40	219.20	45.40	363.20		
32 L.H., Daily Totals		$1072.40		$1776.80	$33.51	$55.53

Crew C-2	Hr.	Daily	Hr.	Daily	Bare Costs	Incl. O&P
1 Carpenter Foreman (out)	$37.55	$300.40	$62.20	$497.60	$34.53	$57.20
4 Carpenters	35.55	1137.60	58.90	1884.80		
1 Laborer	27.40	219.20	45.40	363.20		
48 L.H., Daily Totals		$1657.20		$2745.60	$34.53	$57.20

Crew C-2A	Hr.	Daily	Hr.	Daily	Bare Costs	Incl. O&P
1 Carpenter Foreman (out)	$37.55	$300.40	$62.20	$497.60	$34.33	$56.42
3 Carpenters	35.55	853.20	58.90	1413.60		
1 Cement Finisher	34.40	275.20	54.20	433.60		
1 Laborer	27.40	219.20	45.40	363.20		
48 L.H., Daily Totals		$1648.00		$2708.00	$34.33	$56.42

Crew C-3	Hr.	Daily	Hr.	Daily	Bare Costs	Incl. O&P
1 Rodman Foreman	$41.50	$332.00	$71.25	$570.00	$36.19	$61.11
4 Rodmen (reinf.)	39.50	1264.00	67.80	2169.60		
1 Equip. Oper. (light)	35.20	281.60	55.60	444.80		
2 Laborers	27.40	438.40	45.40	726.40		
3 Stressing Equipment		31.20		34.30		
.5 Grouting Equipment		77.55		85.30	1.70	1.87
64 L.H., Daily Totals		$2424.75		$4030.40	$37.89	$62.98

Crew C-4	Hr.	Daily	Hr.	Daily	Bare Costs	Incl. O&P
1 Rodman Foreman	$41.50	$332.00	$71.25	$570.00	$40.00	$68.66
3 Rodmen (reinf.)	39.50	948.00	67.80	1627.20		
3 Stressing Equipment		31.20		34.30	.98	1.07
32 L.H., Daily Totals		$1311.20		$2231.50	$40.98	$69.73

Crew C-5	Hr.	Daily	Hr.	Daily	Bare Costs	Incl. O&P
1 Rodman Foreman	$41.50	$332.00	$71.25	$570.00	$38.58	$64.84
4 Rodmen (reinf.)	39.50	1264.00	67.80	2169.60		
1 Equip. Oper. (crane)	38.10	304.80	60.15	481.20		
1 Equip. Oper. Oiler	32.45	259.60	51.25	410.00		
1 Hyd. Crane, 25 Ton		650.80		715.90	11.62	12.78
56 L.H., Daily Totals		$2811.20		$4346.70	$50.20	$77.62

Crew C-6	Hr.	Daily	Hr.	Daily	Bare Costs	Incl. O&P
1 Laborer Foreman (outside)	$29.40	$235.20	$48.70	$389.60	$28.90	$47.42
4 Laborers	27.40	876.80	45.40	1452.80		
1 Cement Finisher	34.40	275.20	54.20	433.60		
2 Gas Engine Vibrators		42.80		47.10	.89	.98
48 L.H., Daily Totals		$1430.00		$2323.10	$29.79	$48.40

Crew C-7	Hr.	Daily	Hr.	Daily	Bare Costs	Incl. O&P
1 Laborer Foreman (outside)	$29.40	$235.20	$48.70	$389.60	$29.99	$48.79
5 Laborers	27.40	1096.00	45.40	1816.00		
1 Cement Finisher	34.40	275.20	54.20	433.60		
1 Equip. Oper. (med.)	36.70	293.60	57.95	463.60		
1 Equip. Oper. Oiler	32.45	259.60	51.25	410.00		
2 Gas Engine Vibrators		42.80		47.10		
1 Concrete Bucket, 1 C.Y.		16.20		17.80		
1 Hyd. Crane, 55 Ton		951.60		1046.75	14.04	15.44
72 L.H., Daily Totals		$3170.20		$4624.45	$44.03	$64.23

Crew No.	Bare Costs		Incl. Subs O & P		Cost Per Labor-Hour	

Crew C-8	Hr.	Daily	Hr.	Daily	Bare Costs	Incl. O&P
1 Laborer Foreman (outside)	$29.40	$235.20	$48.70	$389.60	$31.01	$50.18
3 Laborers	27.40	657.60	45.40	1089.60		
2 Cement Finishers	34.40	550.40	54.20	867.20		
1 Equip. Oper. (med.)	36.70	293.60	57.95	463.60		
1 Concrete Pump (small)		700.60		770.65	12.51	13.76
56 L.H., Daily Totals		$2437.40		$3580.65	$43.52	$63.94

Crew C-8A	Hr.	Daily	Hr.	Daily	Bare Costs	Incl. O&P
1 Laborer Foreman (outside)	$29.40	$235.20	$48.70	$389.60	$30.07	$48.88
3 Laborers	27.40	657.60	45.40	1089.60		
2 Cement Finishers	34.40	550.40	54.20	867.20		
48 L.H., Daily Totals		$1443.20		$2346.40	$30.07	$48.88

Crew C-8B	Hr.	Daily	Hr.	Daily	Bare Costs	Incl. O&P
1 Laborer Foreman (outside)	$29.40	$235.20	$48.70	$389.60	$29.66	$48.57
3 Laborers	27.40	657.60	45.40	1089.60		
1 Equipment Operator	36.70	293.60	57.95	463.60		
1 Vibrating Screed		57.95		63.75		
1 Vibratory Roller		522.00		574.20		
1 Dozer, 200 H.P.		920.80		1012.90	37.52	41.27
40 L.H., Daily Totals		$2687.15		$3593.65	$67.18	$89.84

Crew C-8C	Hr.	Daily	Hr.	Daily	Bare Costs	Incl. O&P
1 Laborer Foreman (outside)	$29.40	$235.20	$48.70	$389.60	$30.45	$49.51
3 Laborers	27.40	657.60	45.40	1089.60		
1 Cement Finisher	34.40	275.20	54.20	433.60		
1 Equipment Operator (med.)	36.70	293.60	57.95	463.60		
1 Shotcrete Rig, 12 CY/hr		234.40		257.85	4.88	5.37
48 L.H., Daily Totals		$1696.00		$2634.25	$35.33	$54.88

Crew C-8D	Hr.	Daily	Hr.	Daily	Bare Costs	Incl. O&P
1 Laborer Foreman (outside)	$29.40	$235.20	$48.70	$389.60	$31.60	$50.98
1 Laborer	27.40	219.20	45.40	363.20		
1 Cement Finisher	34.40	275.20	54.20	433.60		
1 Equipment Operator (light)	35.20	281.60	55.60	444.80		
1 Compressor, 250 CFM		119.80		131.80		
2 Hoses, 1", 50'		7.80		8.60	3.99	4.39
32 L.H., Daily Totals		$1138.80		$1771.60	$35.59	$55.37

Crew C-8E	Hr.	Daily	Hr.	Daily	Bare Costs	Incl. O&P
1 Laborer Foreman (outside)	$29.40	$235.20	$48.70	$389.60	$31.60	$50.98
1 Laborer	27.40	219.20	45.40	363.20		
1 Cement Finisher	34.40	275.20	54.20	433.60		
1 Equipment Operator (light)	35.20	281.60	55.60	444.80		
1 Compressor, 250 CFM		119.80		131.80		
2 Hoses, 1", 50'		7.80		8.60		
1 Concrete Pump (small)		700.60		770.65	25.88	28.47
32 L.H., Daily Totals		$1839.40		$2542.25	$57.48	$79.45

Crew C-10	Hr.	Daily	Hr.	Daily	Bare Costs	Incl. O&P
1 Laborer	$27.40	$219.20	$45.40	$363.20	$32.07	$51.27
2 Cement Finishers	34.40	550.40	54.20	867.20		
24 L.H., Daily Totals		$769.60		$1230.40	$32.07	$51.27

Crew C-10B	Hr.	Daily	Hr.	Daily	Bare Costs	Incl. O&P
3 Laborers	$27.40	$657.60	$45.40	$1089.60	$30.20	$48.92
2 Cement Finishers	34.40	550.40	54.20	867.20		
1 Concrete mixer, 10 CF		127.40		140.15		
2 Concrete finishers, 48" dia		48.40		53.25	4.40	4.83
40 L.H., Daily Totals		$1383.80		$2150.20	$34.60	$53.75

Crew No.	Bare Costs		Incl. Subs O & P		Cost Per Labor-Hour	

Crew C-11	Hr.	Daily	Hr.	Daily	Bare Costs	Incl. O&P
1 Struc. Steel Foreman	$41.95	$335.60	$78.45	$627.60	$39.13	$70.89
6 Struc. Steel Workers	39.95	1917.60	74.70	3585.60		
1 Equip. Oper. (crane)	38.10	304.80	60.15	481.20		
1 Equip. Oper. Oiler	32.45	259.60	51.25	410.00		
1 Truck Crane, 150 Ton		1539.00		1692.90	21.38	23.51
72 L.H., Daily Totals		$4356.60		$6797.30	$60.51	$94.40

Crew C-12	Hr.	Daily	Hr.	Daily	Bare Costs	Incl. O&P
1 Carpenter Foreman (out)	$37.55	$300.40	$62.20	$497.60	$34.95	$57.41
3 Carpenters	35.55	853.20	58.90	1413.60		
1 Laborer	27.40	219.20	45.40	363.20		
1 Equip. Oper. (crane)	38.10	304.80	60.15	481.20		
1 Hyd. Crane, 12 Ton		636.60		700.25	13.26	14.59
48 L.H., Daily Totals		$2314.20		$3455.85	$48.21	$72.00

Crew C-13	Hr.	Daily	Hr.	Daily	Bare Costs	Incl. O&P
1 Struc. Steel Worker	$39.95	$319.60	$74.70	$597.60	$38.48	$69.43
1 Welder	39.95	319.60	74.70	597.60		
1 Carpenter	35.55	284.40	58.90	471.20		
1 Gas Welding Machine		89.20		98.10	3.72	4.09
24 L.H., Daily Totals		$1012.80		$1764.50	$42.20	$73.52

Crew C-14	Hr.	Daily	Hr.	Daily	Bare Costs	Incl. O&P
1 Carpenter Foreman (out)	$37.55	$300.40	$62.20	$497.60	$34.57	$57.18
5 Carpenters	35.55	1422.00	58.90	2356.00		
4 Laborers	27.40	876.80	45.40	1452.80		
4 Rodmen (reinf.)	39.50	1264.00	67.80	2169.60		
2 Cement Finishers	34.40	550.40	54.20	867.20		
1 Equip. Oper. (crane)	38.10	304.80	60.15	481.20		
1 Equip. Oper. Oiler	32.45	259.60	51.25	410.00		
1 Crane, 80 Ton, & Tools		1059.00		1164.90	7.35	8.09
144 L.H., Daily Totals		$6037.00		$9399.30	$41.92	$65.27

Crew C-14A	Hr.	Daily	Hr.	Daily	Bare Costs	Incl. O&P
1 Carpenter Foreman (out)	$37.55	$300.40	$62.20	$497.60	$35.61	$59.15
16 Carpenters	35.55	4550.40	58.90	7539.20		
4 Rodmen (reinf.)	39.50	1264.00	67.80	2169.60		
2 Laborers	27.40	438.40	45.40	726.40		
1 Cement Finisher	34.40	275.20	54.20	433.60		
1 Equip. Oper. (med.)	36.70	293.60	57.95	463.60		
1 Gas Engine Vibrator		21.40		23.55		
1 Concrete Pump (small)		700.60		770.65	3.61	3.97
200 L.H., Daily Totals		$7844.00		$12624.20	$39.22	$63.12

Crew C-14B	Hr.	Daily	Hr.	Daily	Bare Costs	Incl. O&P
1 Carpenter Foreman (out)	$37.55	$300.40	$62.20	$497.60	$35.56	$58.96
16 Carpenters	35.55	4550.40	58.90	7539.20		
4 Rodmen (reinf.)	39.50	1264.00	67.80	2169.60		
2 Laborers	27.40	438.40	45.40	726.40		
2 Cement Finishers	34.40	550.40	54.20	867.20		
1 Equip. Oper. (med.)	36.70	293.60	57.95	463.60		
1 Gas Engine Vibrator		21.40		23.55		
1 Concrete Pump (small)		700.60		770.65	3.47	3.82
208 L.H., Daily Totals		$8119.20		$13057.80	$39.03	$62.78

Crews

Crew No.	Bare Costs		Incl. Subs O & P		Cost Per Labor-Hour	
Crew C-14C	**Hr.**	**Daily**	**Hr.**	**Daily**	**Bare Costs**	**Incl. O&P**
1 Carpenter Foreman (out)	$37.55	$300.40	$62.20	$497.60	$33.85	$56.21
6 Carpenters	35.55	1706.40	58.90	2827.20		
2 Rodmen (reinf.)	39.50	632.00	67.80	1084.80		
4 Laborers	27.40	876.80	45.40	1452.80		
1 Cement Finisher	34.40	275.20	54.20	433.60		
1 Gas Engine Vibrator		21.40		23.55	.19	.21
112 L.H., Daily Totals		$3812.20		$6319.55	$34.04	$56.42

Crew C-14D	Hr.	Daily	Hr.	Daily	Bare Costs	Incl. O&P
1 Carpenter Foreman (out)	$37.55	$300.40	$62.20	$497.60	$35.29	$58.44
18 Carpenters	35.55	5119.20	58.90	8481.60		
2 Rodmen (reinf.)	39.50	632.00	67.80	1084.80		
2 Laborers	27.40	438.40	45.40	726.40		
1 Cement Finisher	34.40	275.20	54.20	433.60		
1 Equip. Oper. (med.)	36.70	293.60	57.95	463.60		
1 Gas Engine Vibrator		21.40		23.55		
1 Concrete Pump (small)		700.60		770.65	3.61	3.97
200 L.H., Daily Totals		$7780.80		$12481.80	$38.90	$62.41

Crew C-14E	Hr.	Daily	Hr.	Daily	Bare Costs	Incl. O&P
1 Carpenter Foreman (out)	$37.55	$300.40	$62.20	$497.60	$34.84	$58.33
2 Carpenters	35.55	568.80	58.90	942.40		
4 Rodmen (reinf.)	39.50	1264.00	67.80	2169.60		
3 Laborers	27.40	657.60	45.40	1089.60		
1 Cement Finisher	34.40	275.20	54.20	433.60		
1 Gas Engine Vibrator		21.40		23.55	.24	.27
88 L.H., Daily Totals		$3087.40		$5156.35	$35.08	$58.60

Crew C-14F	Hr.	Daily	Hr.	Daily	Bare Costs	Incl. O&P
1 Laborer Foreman (out)	$29.40	$235.20	$48.70	$389.60	$32.29	$51.63
2 Laborers	27.40	438.40	45.40	726.40		
6 Cement Finishers	34.40	1651.20	54.20	2601.60		
1 Gas Engine Vibrator		21.40		23.55	.30	.33
72 L.H., Daily Totals		$2346.20		$3741.15	$32.59	$51.96

Crew C-14G	Hr.	Daily	Hr.	Daily	Bare Costs	Incl. O&P
1 Laborer Foreman (out)	$29.40	$235.20	$48.70	$389.60	$31.69	$50.90
2 Laborers	27.40	438.40	45.40	726.40		
4 Cement Finishers	34.40	1100.80	54.20	1734.40		
1 Gas Engine Vibrator		21.40		23.55	.38	.42
56 L.H., Daily Totals		$1795.80		$2873.95	$32.07	$51.32

Crew C-14H	Hr.	Daily	Hr.	Daily	Bare Costs	Incl. O&P
1 Carpenter Foreman (out)	$37.55	$300.40	$62.20	$497.60	$34.99	$57.90
2 Carpenters	35.55	568.80	58.90	942.40		
1 Rodman (reinf.)	39.50	316.00	67.80	542.40		
1 Laborer	27.40	219.20	45.40	363.20		
1 Cement Finisher	34.40	275.20	54.20	433.60		
1 Gas Engine Vibrator		21.40		23.55	.45	.49
48 L.H., Daily Totals		$1701.00		$2802.75	$35.44	$58.39

Crew C-15	Hr.	Daily	Hr.	Daily	Bare Costs	Incl. O&P
1 Carpenter Foreman (out)	$37.55	$300.40	$62.20	$497.60	$33.24	$54.71
2 Carpenters	35.55	568.80	58.90	942.40		
3 Laborers	27.40	657.60	45.40	1089.60		
2 Cement Finishers	34.40	550.40	54.20	867.20		
1 Rodman (reinf.)	39.50	316.00	67.80	542.40		
72 L.H., Daily Totals		$2393.20		$3939.20	$33.24	$54.71

Crew C-16	Hr.	Daily	Hr.	Daily	Bare Costs	Incl. O&P
1 Laborer Foreman (outside)	$29.40	$235.20	$48.70	$389.60	$32.90	$54.09
3 Laborers	27.40	657.60	45.40	1089.60		
2 Cement Finishers	34.40	550.40	54.20	867.20		
1 Equip. Oper. (med.)	36.70	293.60	57.95	463.60		
2 Rodmen (reinf.)	39.50	632.00	67.80	1084.80		
1 Concrete Pump (small)		700.60		770.65	9.73	10.70
72 L.H., Daily Totals		$3069.40		$4665.45	$42.63	$64.79

Crew C-17	Hr.	Daily	Hr.	Daily	Bare Costs	Incl. O&P
2 Skilled Worker Foremen	$38.50	$616.00	$63.00	$1008.00	$36.90	$60.36
8 Skilled Workers	36.50	2336.00	59.70	3820.80		
80 L.H., Daily Totals		$2952.00		$4828.80	$36.90	$60.36

Crew C-17A	Hr.	Daily	Hr.	Daily	Bare Costs	Incl. O&P
2 Skilled Worker Foremen	$38.50	$616.00	$63.00	$1008.00	$36.91	$60.36
8 Skilled Workers	36.50	2336.00	59.70	3820.80		
.125 Equip. Oper. (crane)	38.10	38.10	60.15	60.15		
.125 Crane, 80 Ton, & Tools		132.38		145.60	1.63	1.80
81 L.H., Daily Totals		$3122.48		$5034.55	$38.54	$62.16

Crew C-17B	Hr.	Daily	Hr.	Daily	Bare Costs	Incl. O&P
2 Skilled Worker Foremen	$38.50	$616.00	$63.00	$1008.00	$36.93	$60.35
8 Skilled Workers	36.50	2336.00	59.70	3820.80		
.25 Equip. Oper. (crane)	38.10	76.20	60.15	120.30		
.25 Crane, 80 Ton, & Tools		264.75		291.25		
.25 Walk Behind Power Tool		6.05		6.65	3.30	3.63
82 L.H., Daily Totals		$3299.00		$5247.00	$40.23	$63.98

Crew C-17C	Hr.	Daily	Hr.	Daily	Bare Costs	Incl. O&P
2 Skilled Worker Foremen	$38.50	$616.00	$63.00	$1008.00	$36.94	$60.35
8 Skilled Workers	36.50	2336.00	59.70	3820.80		
.375 Equip. Oper. (crane)	38.10	114.30	60.15	180.45		
.375 Crane, 80 Ton & Tools		397.13		436.85	4.78	5.26
83 L.H., Daily Totals		$3463.43		$5446.10	$41.72	$65.61

Crew C-17D	Hr.	Daily	Hr.	Daily	Bare Costs	Incl. O&P
2 Skilled Worker Foremen	$38.50	$616.00	$63.00	$1008.00	$36.96	$60.35
8 Skilled Workers	36.50	2336.00	59.70	3820.80		
.5 Equip. Oper. (crane)	38.10	152.40	60.15	240.60		
.5 Crane, 80 Ton & Tools		529.50		582.45	6.30	6.93
84 L.H., Daily Totals		$3633.90		$5651.85	$43.26	$67.28

Crew C-17E	Hr.	Daily	Hr.	Daily	Bare Costs	Incl. O&P
2 Skilled Worker Foremen	$38.50	$616.00	$63.00	$1008.00	$36.90	$60.36
8 Skilled Workers	36.50	2336.00	59.70	3820.80		
1 Hyd. Jack with Rods		80.65		88.70	1.01	1.11
80 L.H., Daily Totals		$3032.65		$4917.50	$37.91	$61.47

Crew C-18	Hr.	Daily	Hr.	Daily	Bare Costs	Incl. O&P
.125 Laborer Foreman (out)	$29.40	$29.40	$48.70	$48.70	$27.62	$45.77
1 Laborer	27.40	219.20	45.40	363.20		
1 Concrete Cart, 10 C.F.		52.20		57.40	5.80	6.38
9 L.H., Daily Totals		$300.80		$469.30	$33.42	$52.15

Crew C-19	Hr.	Daily	Hr.	Daily	Bare Costs	Incl. O&P
.125 Laborer Foreman (out)	$29.40	$29.40	$48.70	$48.70	$27.62	$45.77
1 Laborer	27.40	219.20	45.40	363.20		
1 Concrete Cart, 18 C.F.		74.40		81.85	8.27	9.09
9 L.H., Daily Totals		$323.00		$493.75	$35.89	$54.86

Crew No.	Bare Costs		Incl. Subs O & P		Cost Per Labor-Hour	
Crew C-20	**Hr.**	**Daily**	**Hr.**	**Daily**	**Bare Costs**	**Incl. O&P**
1 Laborer Foreman (outside)	$29.40	$235.20	$48.70	$389.60	$29.69	$48.48
5 Laborers	27.40	1096.00	45.40	1816.00		
1 Cement Finisher	34.40	275.20	54.20	433.60		
1 Equip. Oper. (med.)	36.70	293.60	57.95	463.60		
2 Gas Engine Vibrators		42.80		47.10		
1 Concrete Pump (small)		700.60		770.65	11.62	12.78
64 L.H., Daily Totals		$2643.40		$3920.55	$41.31	$61.26
Crew C-21	**Hr.**	**Daily**	**Hr.**	**Daily**	**Bare Costs**	**Incl. O&P**
1 Laborer Foreman (outside)	$29.40	$235.20	$48.70	$389.60	$29.69	$48.48
5 Laborers	27.40	1096.00	45.40	1816.00		
1 Cement Finisher	34.40	275.20	54.20	433.60		
1 Equip. Oper. (med.)	36.70	293.60	57.95	463.60		
2 Gas Engine Vibrators		42.80		47.10		
1 Concrete Conveyer		152.40		167.65	3.05	3.36
64 L.H., Daily Totals		$2095.20		$3317.55	$32.74	$51.84
Crew C-22	**Hr.**	**Daily**	**Hr.**	**Daily**	**Bare Costs**	**Incl. O&P**
1 Rodman Foreman	$41.50	$332.00	$71.25	$570.00	$39.68	$67.88
4 Rodmen (reinf.)	39.50	1264.00	67.80	2169.60		
.125 Equip. Oper. (crane)	38.10	38.10	60.15	60.15		
.125 Equip. Oper. Oiler	32.45	32.45	51.25	51.25		
.125 Hyd. Crane, 25 Ton		81.35		89.50	1.94	2.13
42 L.H., Daily Totals		$1747.90		$2940.50	$41.62	$70.01
Crew C-23	**Hr.**	**Daily**	**Hr.**	**Daily**	**Bare Costs**	**Incl. O&P**
2 Skilled Worker Foremen	$38.50	$616.00	$63.00	$1008.00	$36.66	$59.56
6 Skilled Workers	36.50	1752.00	59.70	2865.60		
1 Equip. Oper. (crane)	38.10	304.80	60.15	481.20		
1 Equip. Oper. Oiler	32.45	259.60	51.25	410.00		
1 Crane, 90 Ton		1427.00		1569.70	17.84	19.62
80 L.H., Daily Totals		$4359.40		$6334.50	$54.50	$79.18
Crew C-24	**Hr.**	**Daily**	**Hr.**	**Daily**	**Bare Costs**	**Incl. O&P**
2 Skilled Worker Foremen	$38.50	$616.00	$63.00	$1008.00	$36.66	$59.56
6 Skilled Workers	36.50	1752.00	59.70	2865.60		
1 Equip. Oper. (crane)	38.10	304.80	60.15	481.20		
1 Equip. Oper. Oiler	32.45	259.60	51.25	410.00		
1 Truck Crane, 150 Ton		1539.00		1692.90	19.24	21.16
80 L.H., Daily Totals		$4471.40		$6457.70	$55.90	$80.72
Crew C-25	**Hr.**	**Daily**	**Hr.**	**Daily**	**Bare Costs**	**Incl. O&P**
2 Rodmen (reinf.)	$39.50	$632.00	$67.80	$1084.80	$31.03	$54.15
2 Rodman Helpers	22.55	360.80	40.50	648.00		
32 L.H., Daily Totals		$992.80		$1732.80	$31.03	$54.15
Crew C-27	**Hr.**	**Daily**	**Hr.**	**Daily**	**Bare Costs**	**Incl. O&P**
2 Cement Finishers	$34.40	$550.40	$54.20	$867.20	$34.40	$54.20
1 Concrete Saw		107.00		117.70	6.69	7.36
16 L.H., Daily Totals		$657.40		$984.90	$41.09	$61.56
Crew C-28	**Hr.**	**Daily**	**Hr.**	**Daily**	**Bare Costs**	**Incl. O&P**
1 Cement Finisher	$34.40	$275.20	$54.20	$433.60	$34.40	$54.20
1 Portable Air Compressor		15.75		17.35	1.96	2.16
8 L.H., Daily Totals		$290.95		$450.95	$36.36	$56.36
Crew D-1	**Hr.**	**Daily**	**Hr.**	**Daily**	**Bare Costs**	**Incl. O&P**
1 Bricklayer	$36.55	$292.40	$59.30	$474.40	$32.15	$52.15
1 Bricklayer Helper	27.75	222.00	45.00	360.00		
16 L.H., Daily Totals		$514.40		$834.40	$32.15	$52.15

Crew No.	Bare Costs		Incl. Subs O & P		Cost Per Labor-Hour	
Crew D-2	**Hr.**	**Daily**	**Hr.**	**Daily**	**Bare Costs**	**Incl. O&P**
3 Bricklayers	$36.55	$877.20	$59.30	$1423.20	$33.26	$54.06
2 Bricklayer Helpers	27.75	444.00	45.00	720.00		
.5 Carpenter	35.55	142.20	58.90	235.60		
44 L.H., Daily Totals		$1463.40		$2378.80	$33.26	$54.06
Crew D-3	**Hr.**	**Daily**	**Hr.**	**Daily**	**Bare Costs**	**Incl. O&P**
3 Bricklayers	$36.55	$877.20	$59.30	$1423.20	$33.15	$53.83
2 Bricklayer Helpers	27.75	444.00	45.00	720.00		
.25 Carpenter	35.55	71.10	58.90	117.80		
42 L.H., Daily Totals		$1392.30		$2261.00	$33.15	$53.83
Crew D-4	**Hr.**	**Daily**	**Hr.**	**Daily**	**Bare Costs**	**Incl. O&P**
1 Bricklayer	$36.55	$292.40	$59.30	$474.40	$31.81	$51.23
2 Bricklayer Helpers	27.75	444.00	45.00	720.00		
1 Equip. Oper. (light)	35.20	281.60	55.60	444.80		
1 Grout Pump, 50 C.F./hr		117.30		129.05		
1 Hose & Hopper		15.60		17.15		
1 Accessory		12.10		13.30	4.53	4.98
32 L.H., Daily Totals		$1163.00		$1798.70	$36.34	$56.21
Crew D-5	**Hr.**	**Daily**	**Hr.**	**Daily**	**Bare Costs**	**Incl. O&P**
1 Bricklayer	$36.55	$292.40	$59.30	$474.40	$36.55	$59.30
8 L.H., Daily Totals		$292.40		$474.40	$36.55	$59.30
Crew D-6	**Hr.**	**Daily**	**Hr.**	**Daily**	**Bare Costs**	**Incl. O&P**
3 Bricklayers	$36.55	$877.20	$59.30	$1423.20	$32.29	$52.42
3 Bricklayer Helpers	27.75	666.00	45.00	1080.00		
.25 Carpenter	35.55	71.10	58.90	117.80		
50 L.H., Daily Totals		$1614.30		$2621.00	$32.29	$52.42
Crew D-7	**Hr.**	**Daily**	**Hr.**	**Daily**	**Bare Costs**	**Incl. O&P**
1 Tile Layer	$34.25	$274.00	$53.70	$429.60	$30.38	$47.63
1 Tile Layer Helper	26.50	212.00	41.55	332.40		
16 L.H., Daily Totals		$486.00		$762.00	$30.38	$47.63
Crew D-8	**Hr.**	**Daily**	**Hr.**	**Daily**	**Bare Costs**	**Incl. O&P**
3 Bricklayers	$36.55	$877.20	$59.30	$1423.20	$33.03	$53.58
2 Bricklayer Helpers	27.75	444.00	45.00	720.00		
40 L.H., Daily Totals		$1321.20		$2143.20	$33.03	$53.58
Crew D-9	**Hr.**	**Daily**	**Hr.**	**Daily**	**Bare Costs**	**Incl. O&P**
3 Bricklayers	$36.55	$877.20	$59.30	$1423.20	$32.15	$52.15
3 Bricklayer Helpers	27.75	666.00	45.00	1080.00		
48 L.H., Daily Totals		$1543.20		$2503.20	$32.15	$52.15
Crew D-10	**Hr.**	**Daily**	**Hr.**	**Daily**	**Bare Costs**	**Incl. O&P**
1 Bricklayer Foreman	$38.55	$308.40	$62.55	$500.40	$33.74	$54.40
1 Bricklayer	36.55	292.40	59.30	474.40		
2 Bricklayer Helpers	27.75	444.00	45.00	720.00		
1 Equip. Oper. (crane)	38.10	304.80	60.15	481.20		
1 Truck Crane, 12.5 Ton		536.40		590.05	13.41	14.75
40 L.H., Daily Totals		$1886.00		$2766.05	$47.15	$69.15
Crew D-11	**Hr.**	**Daily**	**Hr.**	**Daily**	**Bare Costs**	**Incl. O&P**
1 Bricklayer Foreman	$38.55	$308.40	$62.55	$500.40	$34.28	$55.62
1 Bricklayer	36.55	292.40	59.30	474.40		
1 Bricklayer Helper	27.75	222.00	45.00	360.00		
24 L.H., Daily Totals		$822.80		$1334.80	$34.28	$55.62

Crew No.	Bare Costs		Incl. Subs O & P		Cost Per Labor-Hour	

Crew D-12	Hr.	Daily	Hr.	Daily	Bare Costs	Incl. O&P
1 Bricklayer Foreman	$38.55	$308.40	$62.55	$500.40	$32.65	$52.96
1 Bricklayer	36.55	292.40	59.30	474.40		
2 Bricklayer Helpers	27.75	444.00	45.00	720.00		
32 L.H., Daily Totals		$1044.80		$1694.80	$32.65	$52.96

Crew D-13	Hr.	Daily	Hr.	Daily	Bare Costs	Incl. O&P
1 Bricklayer Foreman	$38.55	$308.40	$62.55	$500.40	$34.04	$55.15
1 Bricklayer	36.55	292.40	59.30	474.40		
2 Bricklayer Helpers	27.75	444.00	45.00	720.00		
1 Carpenter	35.55	284.40	58.90	471.20		
1 Equip. Oper. (crane)	38.10	304.80	60.15	481.20		
1 Truck Crane, 12.5 Ton		536.40		590.05	11.18	12.29
48 L.H., Daily Totals		$2170.40		$3237.25	$45.22	$67.44

Crew E-1	Hr.	Daily	Hr.	Daily	Bare Costs	Incl. O&P
1 Welder Foreman	$41.95	$335.60	$78.45	$627.60	$39.03	$69.58
1 Welder	39.95	319.60	74.70	597.60		
1 Equip. Oper. (light)	35.20	281.60	55.60	444.80		
1 Gas Welding Machine		89.20		98.10	3.72	4.09
24 L.H., Daily Totals		$1026.00		$1768.10	$42.75	$73.67

Crew E-2	Hr.	Daily	Hr.	Daily	Bare Costs	Incl. O&P
1 Struc. Steel Foreman	$41.95	$335.60	$78.45	$627.60	$38.90	$69.81
4 Struc. Steel Workers	39.95	1278.40	74.70	2390.40		
1 Equip. Oper. (crane)	38.10	304.80	60.15	481.20		
1 Equip. Oper. Oiler	32.45	259.60	51.25	410.00		
1 Crane, 90 Ton		1427.00		1569.70	25.48	28.03
56 L.H., Daily Totals		$3605.40		$5478.90	$64.38	$97.84

Crew E-3	Hr.	Daily	Hr.	Daily	Bare Costs	Incl. O&P
1 Struc. Steel Foreman	$41.95	$335.60	$78.45	$627.60	$40.62	$75.95
1 Struc. Steel Worker	39.95	319.60	74.70	597.60		
1 Welder	39.95	319.60	74.70	597.60		
1 Gas Welding Machine		89.20		98.10	3.72	4.09
24 L.H., Daily Totals		$1064.00		$1920.90	$44.34	$80.04

Crew E-4	Hr.	Daily	Hr.	Daily	Bare Costs	Incl. O&P
1 Struc. Steel Foreman	$41.95	$335.60	$78.45	$627.60	$40.45	$75.64
3 Struc. Steel Workers	39.95	958.80	74.70	1792.80		
1 Gas Welding Machine		89.20		98.10	2.79	3.07
32 L.H., Daily Totals		$1383.60		$2518.50	$43.24	$78.71

Crew E-5	Hr.	Daily	Hr.	Daily	Bare Costs	Incl. O&P
2 Struc. Steel Foremen	$41.95	$671.20	$78.45	$1255.20	$39.42	$71.65
5 Struc. Steel Workers	39.95	1598.00	74.70	2988.00		
1 Equip. Oper. (crane)	38.10	304.80	60.15	481.20		
1 Welder	39.95	319.60	74.70	597.60		
1 Equip. Oper. Oiler	32.45	259.60	51.25	410.00		
1 Crane, 90 Ton		1427.00		1569.70		
1 Gas Welding Machine		89.20		98.10	18.95	20.85
80 L.H., Daily Totals		$4669.40		$7399.80	$58.37	$92.50

Crew E-6	Hr.	Daily	Hr.	Daily	Bare Costs	Incl. O&P
3 Struc. Steel Foremen	$41.95	$1006.80	$78.45	$1882.80	$39.44	$71.83
9 Struc. Steel Workers	39.95	2876.40	74.70	5378.40		
1 Equip. Oper. (crane)	38.10	304.80	60.15	481.20		
1 Welder	39.95	319.60	74.70	597.60		
1 Equip. Oper. Oiler	32.45	259.60	51.25	410.00		
1 Equip. Oper. (light)	35.20	281.60	55.60	444.80		
1 Crane, 90 Ton		1427.00		1569.70		
1 Gas Welding Machine		89.20		98.10		
1 Air Compr., 160 C.F.M.		97.60		107.35		
2 Impact Wrenches		25.20		27.70	12.80	14.09
128 L.H., Daily Totals		$6687.80		$10997.65	$52.24	$85.92

Crew E-7	Hr.	Daily	Hr.	Daily	Bare Costs	Incl. O&P
1 Struc. Steel Foreman	$41.95	$335.60	$78.45	$627.60	$39.42	$71.65
4 Struc. Steel Workers	39.95	1278.40	74.70	2390.40		
1 Equip. Oper. (crane)	38.10	304.80	60.15	481.20		
1 Equip. Oper. Oiler	32.45	259.60	51.25	410.00		
1 Welder Foreman	41.95	335.60	78.45	627.60		
2 Welders	39.95	639.20	74.70	1195.20		
1 Crane, 90 Ton		1427.00		1569.70		
2 Gas Welding Machines		178.40		196.25	20.07	22.07
80 L.H., Daily Totals		$4758.60		$7497.95	$59.49	$93.72

Crew E-8	Hr.	Daily	Hr.	Daily	Bare Costs	Incl. O&P
1 Struc. Steel Foreman	$41.95	$335.60	$78.45	$627.60	$39.17	$70.88
4 Struc. Steel Workers	39.95	1278.40	74.70	2390.40		
1 Welder Foreman	41.95	335.60	78.45	627.60		
4 Welders	39.95	1278.40	74.70	2390.40		
1 Equip. Oper. (crane)	38.10	304.80	60.15	481.20		
1 Equip. Oper. Oiler	32.45	259.60	51.25	410.00		
1 Equip. Oper. (light)	35.20	281.60	55.60	444.80		
1 Crane, 90 Ton		1427.00		1569.70		
4 Gas Welding Machines		356.80		392.50	17.15	18.87
104 L.H., Daily Totals		$5857.80		$9334.20	$56.32	$89.75

Crew E-9	Hr.	Daily	Hr.	Daily	Bare Costs	Incl. O&P
2 Struc. Steel Foremen	$41.95	$671.20	$78.45	$1255.20	$39.44	$71.83
5 Struc. Steel Workers	39.95	1598.00	74.70	2988.00		
1 Welder Foreman	41.95	335.60	78.45	627.60		
5 Welders	39.95	1598.00	74.70	2988.00		
1 Equip. Oper. (crane)	38.10	304.80	60.15	481.20		
1 Equip. Oper. Oiler	32.45	259.60	51.25	410.00		
1 Equip. Oper. (light)	35.20	281.60	55.60	444.80		
1 Crane, 90 Ton		1427.00		1569.70		
5 Gas Welding Machines		446.00		490.60	14.63	16.10
128 L.H., Daily Totals		$6921.80		$11255.10	$54.07	$87.93

Crew E-10	Hr.	Daily	Hr.	Daily	Bare Costs	Incl. O&P
1 Welder Foreman	$41.95	$335.60	$78.45	$627.60	$40.95	$76.58
1 Welder	39.95	319.60	74.70	597.60		
1 Gas Welding Machines		89.20		98.10		
1 Truck, 3 Ton		176.20		193.80	16.59	18.25
16 L.H., Daily Totals		$920.60		$1517.10	$57.54	$94.83

Crew E-11	Hr.	Daily	Hr.	Daily	Bare Costs	Incl. O&P
2 Painters, Struc. Steel	$32.45	$519.20	$63.70	$1019.20	$31.88	$57.10
1 Building Laborer	27.40	219.20	45.40	363.20		
1 Equip. Oper. (light)	35.20	281.60	55.60	444.80		
1 Air Compressor 250 C.F.M.		119.80		131.80		
1 Sand Blaster		15.60		17.15		
1 Sand Blasting Accessories		12.10		13.30	4.61	5.07
32 L.H., Daily Totals		$1167.50		$1989.45	$36.49	$62.17

Crew No.	Bare Costs		Incl. Subs O & P		Cost Per Labor-Hour	
Crew E-12	**Hr.**	**Daily**	**Hr.**	**Daily**	**Bare Costs**	**Incl. O&P**
1 Welder Foreman	$41.95	$335.60	$78.45	$627.60	$38.58	$67.03
1 Equip. Oper. (light)	35.20	281.60	55.60	444.80		
1 Gas Welding Machine		89.20		98.10	5.58	6.13
16 L.H., Daily Totals		$706.40		$1170.50	$44.16	$73.16

Crew E-13	Hr.	Daily	Hr.	Daily	Bare Costs	Incl. O&P
1 Welder Foreman	$41.95	$335.60	$78.45	$627.60	$39.70	$70.83
.5 Equip. Oper. (light)	35.20	140.80	55.60	222.40		
1 Gas Welding Machine		89.20		98.10	7.43	8.18
12 L.H., Daily Totals		$565.60		$948.10	$47.13	$79.01

Crew E-14	Hr.	Daily	Hr.	Daily	Bare Costs	Incl. O&P
1 Welder Foreman	$41.95	$335.60	$78.45	$627.60	$41.95	$78.45
1 Gas Welding Machine		89.20		98.10	11.15	12.27
8 L.H., Daily Totals		$424.80		$725.70	$53.10	$90.72

Crew E-16	Hr.	Daily	Hr.	Daily	Bare Costs	Incl. O&P
1 Welder Foreman	$41.95	$335.60	$78.45	$627.60	$40.95	$76.58
1 Welder	39.95	319.60	74.70	597.60		
1 Gas Welding Machine		89.20		98.10	5.58	6.13
16 L.H., Daily Totals		$744.40		$1323.30	$46.53	$82.71

Crew E-17	Hr.	Daily	Hr.	Daily	Bare Costs	Incl. O&P
1 Structural Steel Foreman	$41.95	$335.60	$78.45	$627.60	$40.95	$76.58
1 Structural Steel Worker	39.95	319.60	74.70	597.60		
1 Power Tool		5.40		5.95	.34	.37
16 L.H., Daily Totals		$660.60		$1231.15	$41.29	$76.95

Crew E-18	Hr.	Daily	Hr.	Daily	Bare Costs	Incl. O&P
1 Structural Steel Foreman	$41.95	$335.60	$78.45	$627.60	$39.70	$72.10
3 Structural Steel Workers	39.95	958.80	74.70	1792.80		
1 Equipment Operator (med.)	36.70	293.60	57.95	463.60		
1 Crane, 20 Ton		947.05		1041.75	23.68	26.04
40 L.H., Daily Totals		$2535.05		$3925.75	$63.38	$98.14

Crew E-19	Hr.	Daily	Hr.	Daily	Bare Costs	Incl. O&P
1 Structural Steel Worker	$39.95	$319.60	$74.70	$597.60	$39.03	$69.58
1 Structural Steel Foreman	41.95	335.60	78.45	627.60		
1 Equip. Oper. (light)	35.20	281.60	55.60	444.80		
1 Power Tool		5.40		5.95		
1 Crane, 20 Ton		947.05		1041.75	39.68	43.65
24 L.H., Daily Totals		$1889.25		$2717.70	$78.71	$113.23

Crew E-20	Hr.	Daily	Hr.	Daily	Bare Costs	Incl. O&P
1 Structural Steel Foreman	$41.95	$335.60	$78.45	$627.60	$39.03	$70.42
5 Structural Steel Workers	39.95	1598.00	74.70	2988.00		
1 Equip. Oper. (crane)	38.10	304.80	60.15	481.20		
1 Oiler	32.45	259.60	51.25	410.00		
1 Power Tool		5.40		5.95		
1 Crane, 40 Ton		1130.00		1243.00	17.74	19.51
64 L.H., Daily Totals		$3633.40		$5755.75	$56.77	$89.93

Crew E-22	Hr.	Daily	Hr.	Daily	Bare Costs	Incl. O&P
1 Skilled Worker Foreman	$38.50	$308.00	$63.00	$504.00	$37.17	$60.80
2 Skilled Workers	36.50	584.00	59.70	955.20		
24 L.H., Daily Totals		$892.00		$1459.20	$37.17	$60.80

Crew No.	Bare Costs		Incl. Subs O & P		Cost Per Labor-Hour	
Crew E-24	**Hr.**	**Daily**	**Hr.**	**Daily**	**Bare Costs**	**Incl. O&P**
3 Structural Steel Workers	$39.95	$958.80	$74.70	$1792.80	$39.14	$70.51
1 Equipment Operator (medium)	36.70	293.60	57.95	463.60		
1-25 Ton Crane		650.80		715.90	20.34	22.37
32 L.H., Daily Totals		$1903.20		$2972.30	$59.48	$92.88

Crew E-25	Hr.	Daily	Hr.	Daily	Bare Costs	Incl. O&P
1 Welder Foreman	$41.95	$335.60	$78.45	$627.60	$41.95	$78.45
1 Cutting Torch		16.20		17.80		
1 Gas		64.80		71.30	10.13	11.14
8 L.H., Daily Totals		$416.60		$716.70	$52.08	$89.59

Crew F-3	Hr.	Daily	Hr.	Daily	Bare Costs	Incl. O&P
4 Carpenters	$35.55	$1137.60	$58.90	$1884.80	$36.06	$59.15
1 Equip. Oper. (crane)	38.10	304.80	60.15	481.20		
1 Hyd. Crane, 12 Ton		636.60		700.25	15.92	17.51
40 L.H., Daily Totals		$2079.00		$3066.25	$51.98	$76.66

Crew F-4	Hr.	Daily	Hr.	Daily	Bare Costs	Incl. O&P
4 Carpenters	$35.55	$1137.60	$58.90	$1884.80	$35.46	$57.83
1 Equip. Oper. (crane)	38.10	304.80	60.15	481.20		
1 Equip. Oper. Oiler	32.45	259.60	51.25	410.00		
1 Hyd. Crane, 55 Ton		951.60		1046.75	19.83	21.81
48 L.H., Daily Totals		$2653.60		$3822.75	$55.29	$79.64

Crew F-5	Hr.	Daily	Hr.	Daily	Bare Costs	Incl. O&P
1 Carpenter Foreman	$37.55	$300.40	$62.20	$497.60	$36.05	$59.73
3 Carpenters	35.55	853.20	58.90	1413.60		
32 L.H., Daily Totals		$1153.60		$1911.20	$36.05	$59.73

Crew F-6	Hr.	Daily	Hr.	Daily	Bare Costs	Incl. O&P
2 Carpenters	$35.55	$568.80	$58.90	$942.40	$32.80	$53.75
2 Building Laborers	27.40	438.40	45.40	726.40		
1 Equip. Oper. (crane)	38.10	304.80	60.15	481.20		
1 Hyd. Crane, 12 Ton		636.60		700.25	15.92	17.51
40 L.H., Daily Totals		$1948.60		$2850.25	$48.72	$71.26

Crew F-7	Hr.	Daily	Hr.	Daily	Bare Costs	Incl. O&P
2 Carpenters	$35.55	$568.80	$58.90	$942.40	$31.48	$52.15
2 Building Laborers	27.40	438.40	45.40	726.40		
32 L.H., Daily Totals		$1007.20		$1668.80	$31.48	$52.15

Crew G-1	Hr.	Daily	Hr.	Daily	Bare Costs	Incl. O&P
1 Roofer Foreman	$32.60	$260.80	$58.60	$468.80	$28.59	$51.37
4 Roofers, Composition	30.60	979.20	55.00	1760.00		
2 Roofer Helpers	22.55	360.80	40.50	648.00		
1 Application Equipment		146.20		160.80		
1 Tar Kettle/Pot		57.30		63.05		
1 Crew Truck		133.75		147.15	6.02	6.62
56 L.H., Daily Totals		$1938.05		$3247.80	$34.61	$57.99

Crew G-2	Hr.	Daily	Hr.	Daily	Bare Costs	Incl. O&P
1 Plasterer	$32.45	$259.60	$52.35	$418.80	$29.25	$47.58
1 Plasterer Helper	27.90	223.20	45.00	360.00		
1 Building Laborer	27.40	219.20	45.40	363.20		
1 Grouting Equipment		117.30		129.05	4.89	5.38
24 L.H., Daily Totals		$819.30		$1271.05	$34.14	$52.96

Crew No.	Bare Costs		Incl. Subs O & P		Cost Per Labor-Hour	
Crew G-2A	**Hr.**	**Daily**	**Hr.**	**Daily**	**Bare Costs**	**Incl. O&P**
1 Roofer, composition	$30.60	$244.80	$55.00	$440.00	$26.85	$46.97
1 Roofer Helper	22.55	180.40	40.50	324.00		
1 Building Laborer	27.40	219.20	45.40	363.20		
1 Spray Equipment		117.30		129.05	4.89	5.38
24 L.H., Daily Totals		$761.70		$1256.25	$31.74	$52.35
Crew G-3	**Hr.**	**Daily**	**Hr.**	**Daily**	**Bare Costs**	**Incl. O&P**
2 Sheet Metal Workers	$42.15	$674.40	$67.05	$1072.80	$34.78	$56.23
2 Building Laborers	27.40	438.40	45.40	726.40		
32 L.H., Daily Totals		$1112.80		$1799.20	$34.78	$56.23
Crew G-4	**Hr.**	**Daily**	**Hr.**	**Daily**	**Bare Costs**	**Incl. O&P**
1 Laborer Foreman (outside)	$29.40	$235.20	$48.70	$389.60	$28.07	$46.50
2 Building Laborers	27.40	438.40	45.40	726.40		
1 Light Truck, 1.5 Ton		128.80		141.70		
1 Air Compr., 160 C.F.M.		97.60		107.35	9.43	10.38
24 L.H., Daily Totals		$900.00		$1365.05	$37.50	$56.88
Crew G-5	**Hr.**	**Daily**	**Hr.**	**Daily**	**Bare Costs**	**Incl. O&P**
1 Roofer Foreman	$32.60	$260.80	$58.60	$468.80	$27.78	$49.92
2 Roofers, Composition	30.60	489.60	55.00	880.00		
2 Roofer Helpers	22.55	360.80	40.50	648.00		
1 Application Equipment		146.20		160.80	3.66	4.02
40 L.H., Daily Totals		$1257.40		$2157.60	$31.44	$53.94
Crew G-6A	**Hr.**	**Daily**	**Hr.**	**Daily**	**Bare Costs**	**Incl. O&P**
2 Roofers, Composition	$30.60	$489.60	$55.00	$880.00	$30.60	$55.00
1 Small Compressor		10.60		11.65		
2 Pneumatic Nailers		39.40		43.35	3.12	3.43
16 L.H., Daily Totals		$539.60		$935.00	$33.72	$58.43
Crew G-7	**Hr.**	**Daily**	**Hr.**	**Daily**	**Bare Costs**	**Incl. O&P**
1 Carpenter	$35.55	$284.40	$58.90	$471.20	$35.55	$58.90
1 Small Compressor		10.60		11.65		
1 Pneumatic Nailer		19.70		21.65	3.78	4.16
8 L.H., Daily Totals		$314.70		$504.50	$39.33	$63.06
Crew H-1	**Hr.**	**Daily**	**Hr.**	**Daily**	**Bare Costs**	**Incl. O&P**
2 Glaziers	$34.25	$548.00	$55.20	$883.20	$37.10	$64.95
2 Struc. Steel Workers	39.95	639.20	74.70	1195.20		
32 L.H., Daily Totals		$1187.20		$2078.40	$37.10	$64.95
Crew H-2	**Hr.**	**Daily**	**Hr.**	**Daily**	**Bare Costs**	**Incl. O&P**
2 Glaziers	$34.25	$548.00	$55.20	$883.20	$31.97	$51.93
1 Building Laborer	27.40	219.20	45.40	363.20		
24 L.H., Daily Totals		$767.20		$1246.40	$31.97	$51.93
Crew H-3	**Hr.**	**Daily**	**Hr.**	**Daily**	**Bare Costs**	**Incl. O&P**
1 Glazier	$34.25	$274.00	$55.20	$441.60	$30.33	$49.40
1 Helper	26.40	211.20	43.60	348.80		
16 L.H., Daily Totals		$485.20		$790.40	$30.33	$49.40
Crew J-1	**Hr.**	**Daily**	**Hr.**	**Daily**	**Bare Costs**	**Incl. O&P**
3 Plasterers	$32.45	$778.80	$52.35	$1256.40	$30.63	$49.41
2 Plasterer Helpers	27.90	446.40	45.00	720.00		
1 Mixing Machine, 6 C.F.		105.60		116.15	2.64	2.90
40 L.H., Daily Totals		$1330.80		$2092.55	$33.27	$52.31

Crew No.	Bare Costs		Incl. Subs O & P		Cost Per Labor-Hour	
Crew J-2	**Hr.**	**Daily**	**Hr.**	**Daily**	**Bare Costs**	**Incl. O&P**
3 Plasterers	$32.45	$778.80	$52.35	$1256.40	$30.99	$49.85
2 Plasterer Helpers	27.90	446.40	45.00	720.00		
1 Lather	32.80	262.40	52.05	416.40		
1 Mixing Machine, 6 C.F.		105.60		116.15	2.20	2.42
48 L.H., Daily Totals		$1593.20		$2508.95	$33.19	$52.27
Crew J-3	**Hr.**	**Daily**	**Hr.**	**Daily**	**Bare Costs**	**Incl. O&P**
1 Terrazzo Worker	$34.10	$272.80	$53.45	$427.60	$30.80	$48.28
1 Terrazzo Helper	27.50	220.00	43.10	344.80		
1 Terrazzo Grinder, Electric		91.25		100.40		
1 Terrazzo Mixer		144.20		158.60	14.72	16.19
16 L.H., Daily Totals		$728.25		$1031.40	$45.52	$64.47
Crew J-4	**Hr.**	**Daily**	**Hr.**	**Daily**	**Bare Costs**	**Incl. O&P**
1 Tile Layer	$34.25	$274.00	$53.70	$429.60	$30.38	$47.63
1 Tile Layer Helper	26.50	212.00	41.55	332.40		
16 L.H., Daily Totals		$486.00		$762.00	$30.38	$47.63
Crew K-1	**Hr.**	**Daily**	**Hr.**	**Daily**	**Bare Costs**	**Incl. O&P**
1 Carpenter	$35.55	$284.40	$58.90	$471.20	$31.58	$51.90
1 Truck Driver (light)	27.60	220.80	44.90	359.20		
1 Truck w/Power Equip.		176.20		193.80	11.01	12.11
16 L.H., Daily Totals		$681.40		$1024.20	$42.59	$64.01
Crew K-2	**Hr.**	**Daily**	**Hr.**	**Daily**	**Bare Costs**	**Incl. O&P**
1 Struc. Steel Foreman	$41.95	$335.60	$78.45	$627.60	$36.50	$66.02
1 Struc. Steel Worker	39.95	319.60	74.70	597.60		
1 Truck Driver (light)	27.60	220.80	44.90	359.20		
1 Truck w/Power Equip.		176.20		193.80	7.34	8.08
24 L.H., Daily Totals		$1052.20		$1778.20	$43.84	$74.10
Crew L-1	**Hr.**	**Daily**	**Hr.**	**Daily**	**Bare Costs**	**Incl. O&P**
1 Electrician	$42.00	$336.00	$64.70	$517.60	$42.35	$65.55
1 Plumber	42.70	341.60	66.40	531.20		
16 L.H., Daily Totals		$677.60		$1048.80	$42.35	$65.55
Crew L-2	**Hr.**	**Daily**	**Hr.**	**Daily**	**Bare Costs**	**Incl. O&P**
1 Carpenter	$35.55	$284.40	$58.90	$471.20	$30.98	$51.25
1 Carpenter Helper	26.40	211.20	43.60	348.80		
16 L.H., Daily Totals		$495.60		$820.00	$30.98	$51.25
Crew L-3	**Hr.**	**Daily**	**Hr.**	**Daily**	**Bare Costs**	**Incl. O&P**
1 Carpenter	$35.55	$284.40	$58.90	$471.20	$38.81	$62.39
.5 Electrician	42.00	168.00	64.70	258.80		
.5 Sheet Metal Worker	42.15	168.60	67.05	268.20		
16 L.H., Daily Totals		$621.00		$998.20	$38.81	$62.39
Crew L-3A	**Hr.**	**Daily**	**Hr.**	**Daily**	**Bare Costs**	**Incl. O&P**
1 Carpenter Foreman (outside)	$37.55	$300.40	$62.20	$497.60	$39.08	$63.82
.5 Sheet Metal Worker	42.15	168.60	67.05	268.20		
12 L.H., Daily Totals		$469.00		$765.80	$39.08	$63.82
Crew L-4	**Hr.**	**Daily**	**Hr.**	**Daily**	**Bare Costs**	**Incl. O&P**
2 Skilled Workers	$36.50	$584.00	$59.70	$955.20	$33.13	$54.33
1 Helper	26.40	211.20	43.60	348.80		
24 L.H., Daily Totals		$795.20		$1304.00	$33.13	$54.33

Crew No.	Bare Costs		Incl. Subs O & P		Cost Per Labor-Hour	
Crew L-5	**Hr.**	**Daily**	**Hr.**	**Daily**	**Bare Costs**	**Incl. O&P**
1 Struc. Steel Foreman	$41.95	$335.60	$78.45	$627.60	$39.97	$73.16
5 Struc. Steel Workers	39.95	1598.00	74.70	2988.00		
1 Equip. Oper. (crane)	38.10	304.80	60.15	481.20		
1 Hyd. Crane, 25 Ton		650.80		715.90	11.62	12.78
56 L.H., Daily Totals		$2889.20		$4812.70	$51.59	$85.94
Crew L-5A	**Hr.**	**Daily**	**Hr.**	**Daily**	**Bare Costs**	**Incl. O&P**
1 Structural Steel Foreman	$41.95	$335.60	$78.45	$627.60	$39.99	$72.00
2 Structural Steel Workers	39.95	639.20	74.70	1195.20		
1 Equip. Oper. (crane)	38.10	304.80	60.15	481.20		
1 Crane,SP, 25 Ton		696.80		766.50	21.78	23.95
32 L.H., Daily Totals		$1976.40		$3070.50	$61.77	$95.95
Crew L-6	**Hr.**	**Daily**	**Hr.**	**Daily**	**Bare Costs**	**Incl. O&P**
1 Plumber	$42.70	$341.60	$66.40	$531.20	$42.47	$65.83
.5 Electrician	42.00	168.00	64.70	258.80		
12 L.H., Daily Totals		$509.60		$790.00	$42.47	$65.83
Crew L-7	**Hr.**	**Daily**	**Hr.**	**Daily**	**Bare Costs**	**Incl. O&P**
2 Carpenters	$35.55	$568.80	$58.90	$942.40	$34.14	$55.87
1 Building Laborer	27.40	219.20	45.40	363.20		
.5 Electrician	42.00	168.00	64.70	258.80		
28 L.H., Daily Totals		$956.00		$1564.40	$34.14	$55.87
Crew L-8	**Hr.**	**Daily**	**Hr.**	**Daily**	**Bare Costs**	**Incl. O&P**
2 Carpenters	$35.55	$568.80	$58.90	$942.40	$36.98	$60.40
.5 Plumber	42.70	170.80	66.40	265.60		
20 L.H., Daily Totals		$739.60		$1208.00	$36.98	$60.40
Crew L-9	**Hr.**	**Daily**	**Hr.**	**Daily**	**Bare Costs**	**Incl. O&P**
1 Laborer Foreman (inside)	$27.90	$223.20	$46.25	$370.00	$31.92	$54.24
2 Building Laborers	27.40	438.40	45.40	726.40		
1 Struc. Steel Worker	39.95	319.60	74.70	597.60		
.5 Electrician	42.00	168.00	64.70	258.80		
36 L.H., Daily Totals		$1149.20		$1952.80	$31.92	$54.24
Crew L-10	**Hr.**	**Daily**	**Hr.**	**Daily**	**Bare Costs**	**Incl. O&P**
1 Structural Steel Foreman	$41.95	$335.60	$78.45	$627.60	$40.00	$71.10
1 Structural Steel Worker	39.95	319.60	74.70	597.60		
1 Equip. Oper. (crane)	38.10	304.80	60.15	481.20		
1 Hyd. Crane, 12 Ton		636.60		700.25	26.53	29.18
24 L.H., Daily Totals		$1596.60		$2406.65	$66.53	$100.28
Crew L-11	**Hr.**	**Daily**	**Hr.**	**Daily**	**Bare Costs**	**Incl. O&P**
2 Wreckers	$27.40	$438.40	$51.35	$821.60	$32.03	$54.61
1 Equip. Oper. (crane)	38.10	304.80	60.15	481.20		
1 Equip. Oper. (light)	35.20	281.60	55.60	444.80		
1 Hyd. Excavator, 2.5 C.Y.		1218.00		1339.80		
1 Skid steer loader		202.00		222.20	44.38	48.81
32 L.H., Daily Totals		$2444.80		$3309.60	$76.41	$103.42
Crew M-1	**Hr.**	**Daily**	**Hr.**	**Daily**	**Bare Costs**	**Incl. O&P**
3 Elevator Constructors	$49.95	$1198.80	$77.00	$1848.00	$47.45	$73.15
1 Elevator Apprentice	39.95	319.60	61.60	492.80		
5 Hand Tools		54.00		59.40	1.69	1.86
32 L.H., Daily Totals		$1572.40		$2400.20	$49.14	$75.01

Crew No.	Bare Costs		Incl. Subs O & P		Cost Per Labor-Hour	
Crew M-3	**Hr.**	**Daily**	**Hr.**	**Daily**	**Bare Costs**	**Incl. O&P**
1 Electrician Foreman (out)	$44.00	$352.00	$67.80	$542.40	$40.11	$62.66
1 Common Laborer	27.40	219.20	45.40	363.20		
.25 Equipment Operator, Medium	36.70	73.40	57.95	115.90		
1 Elevator Constructor	49.95	399.60	77.00	616.00		
1 Elevator Apprentice	39.95	319.60	61.60	492.80		
.25 Crane, SP, 4 x 4, 20 ton		150.85		165.95	4.44	4.88
34 L.H., Daily Totals		$1514.65		$2296.25	$44.55	$67.54
Crew M-4	**Hr.**	**Daily**	**Hr.**	**Daily**	**Bare Costs**	**Incl. O&P**
1 Electrician Foreman (out)	$44.00	$352.00	$67.80	$542.40	$39.76	$62.14
1 Common Laborer	27.40	219.20	45.40	363.20		
.25 Equipment Operator, Crane	38.10	76.20	60.15	120.30		
.25 Equipment Operator, Oiler	32.45	64.90	51.25	102.50		
1 Elevator Constructor	49.95	399.60	77.00	616.00		
1 Elevator Apprentice	39.95	319.60	61.60	492.80		
.25 Crane, Hyd, SP, 4WD, 40 Ton		229.00		251.90	6.36	7.00
36 L.H., Daily Totals		$1660.50		$2489.10	$46.12	$69.14
Crew Q-1	**Hr.**	**Daily**	**Hr.**	**Daily**	**Bare Costs**	**Incl. O&P**
1 Plumber	$42.70	$341.60	$66.40	$531.20	$38.43	$59.75
1 Plumber Apprentice	34.15	273.20	53.10	424.80		
16 L.H., Daily Totals		$614.80		$956.00	$38.43	$59.75
Crew Q-1C	**Hr.**	**Daily**	**Hr.**	**Daily**	**Bare Costs**	**Incl. O&P**
1 Plumber	$42.70	$341.60	$66.40	$531.20	$37.85	$59.15
1 Plumber Apprentice	34.15	273.20	53.10	424.80		
1 Equip. Oper. (medium)	36.70	293.60	57.95	463.60		
1 Trencher, Chain		1460.00		1606.00	60.83	66.92
24 L.H., Daily Totals		$2368.40		$3025.60	$98.68	$126.07
Crew Q-2	**Hr.**	**Daily**	**Hr.**	**Daily**	**Bare Costs**	**Incl. O&P**
2 Plumbers	$42.70	$683.20	$66.40	$1062.40	$39.85	$61.97
1 Plumber Apprentice	34.15	273.20	53.10	424.80		
24 L.H., Daily Totals		$956.40		$1487.20	$39.85	$61.97
Crew Q-3	**Hr.**	**Daily**	**Hr.**	**Daily**	**Bare Costs**	**Incl. O&P**
1 Plumber Foreman (inside)	$43.20	$345.60	$67.20	$537.60	$40.69	$63.28
2 Plumbers	42.70	683.20	66.40	1062.40		
1 Plumber Apprentice	34.15	273.20	53.10	424.80		
32 L.H., Daily Totals		$1302.00		$2024.80	$40.69	$63.28
Crew Q-4	**Hr.**	**Daily**	**Hr.**	**Daily**	**Bare Costs**	**Incl. O&P**
1 Plumber Foreman (inside)	$43.20	$345.60	$67.20	$537.60	$40.69	$63.28
1 Plumber	42.70	341.60	66.40	531.20		
1 Welder (plumber)	42.70	341.60	66.40	531.20		
1 Plumber Apprentice	34.15	273.20	53.10	424.80		
1 Electric Welding Mach.		69.10		76.00	2.16	2.38
32 L.H., Daily Totals		$1371.10		$2100.80	$42.85	$65.66
Crew Q-5	**Hr.**	**Daily**	**Hr.**	**Daily**	**Bare Costs**	**Incl. O&P**
1 Steamfitter	$43.05	$344.40	$66.95	$535.60	$38.75	$60.25
1 Steamfitter Apprentice	34.45	275.60	53.55	428.40		
16 L.H., Daily Totals		$620.00		$964.00	$38.75	$60.25
Crew Q-6	**Hr.**	**Daily**	**Hr.**	**Daily**	**Bare Costs**	**Incl. O&P**
2 Steamfitters	$43.05	$688.80	$66.95	$1071.20	$40.18	$62.48
1 Steamfitter Apprentice	34.45	275.60	53.55	428.40		
24 L.H., Daily Totals		$964.40		$1499.60	$40.18	$62.48

Crew No.	Bare Costs		Incl. Subs O & P		Cost Per Labor-Hour	
Crew Q-7	**Hr.**	**Daily**	**Hr.**	**Daily**	**Bare Costs**	**Incl. O&P**
1 Steamfitter Foreman (inside)	$43.55	$348.40	$67.70	$541.60	$41.03	$63.79
2 Steamfitters	43.05	688.80	66.95	1071.20		
1 Steamfitter Apprentice	34.45	275.60	53.55	428.40		
32 L.H., Daily Totals		$1312.80		$2041.20	$41.03	$63.79
Crew Q-8	**Hr.**	**Daily**	**Hr.**	**Daily**	**Bare Costs**	**Incl. O&P**
1 Steamfitter Foreman (inside)	$43.55	$348.40	$67.70	$541.60	$41.03	$63.79
1 Steamfitter	43.05	344.40	66.95	535.60		
1 Welder (steamfitter)	43.05	344.40	66.95	535.60		
1 Steamfitter Apprentice	34.45	275.60	53.55	428.40		
1 Electric Welding Mach.		69.10		76.00	2.16	2.38
32 L.H., Daily Totals		$1381.90		$2117.20	$43.19	$66.17
Crew Q-9	**Hr.**	**Daily**	**Hr.**	**Daily**	**Bare Costs**	**Incl. O&P**
1 Sheet Metal Worker	$42.15	$337.20	$67.05	$536.40	$37.93	$60.33
1 Sheet Metal Apprentice	33.70	269.60	53.60	428.80		
16 L.H., Daily Totals		$606.80		$965.20	$37.93	$60.33
Crew Q-10	**Hr.**	**Daily**	**Hr.**	**Daily**	**Bare Costs**	**Incl. O&P**
2 Sheet Metal Workers	$42.15	$674.40	$67.05	$1072.80	$39.33	$62.57
1 Sheet Metal Apprentice	33.70	269.60	53.60	428.80		
24 L.H., Daily Totals		$944.00		$1501.60	$39.33	$62.57
Crew Q-11	**Hr.**	**Daily**	**Hr.**	**Daily**	**Bare Costs**	**Incl. O&P**
1 Sheet Metal Foreman (inside)	$42.65	$341.20	$67.85	$542.80	$40.16	$63.89
2 Sheet Metal Workers	42.15	674.40	67.05	1072.80		
1 Sheet Metal Apprentice	33.70	269.60	53.60	428.80		
32 L.H., Daily Totals		$1285.20		$2044.40	$40.16	$63.89
Crew Q-12	**Hr.**	**Daily**	**Hr.**	**Daily**	**Bare Costs**	**Incl. O&P**
1 Sprinkler Installer	$41.80	$334.40	$65.15	$521.20	$37.63	$58.65
1 Sprinkler Apprentice	33.45	267.60	52.15	417.20		
16 L.H., Daily Totals		$602.00		$938.40	$37.63	$58.65
Crew Q-13	**Hr.**	**Daily**	**Hr.**	**Daily**	**Bare Costs**	**Incl. O&P**
1 Sprinkler Foreman (inside)	$42.30	$338.40	$65.95	$527.60	$39.84	$62.10
2 Sprinkler Installers	41.80	668.80	65.15	1042.40		
1 Sprinkler Apprentice	33.45	267.60	52.15	417.20		
32 L.H., Daily Totals		$1274.80		$1987.20	$39.84	$62.10
Crew Q-14	**Hr.**	**Daily**	**Hr.**	**Daily**	**Bare Costs**	**Incl. O&P**
1 Asbestos Worker	$39.05	$312.40	$63.70	$509.60	$35.15	$57.33
1 Asbestos Apprentice	31.25	250.00	50.95	407.60		
16 L.H., Daily Totals		$562.40		$917.20	$35.15	$57.33
Crew Q-15	**Hr.**	**Daily**	**Hr.**	**Daily**	**Bare Costs**	**Incl. O&P**
1 Plumber	$42.70	$341.60	$66.40	$531.20	$38.43	$59.75
1 Plumber Apprentice	34.15	273.20	53.10	424.80		
1 Electric Welding Mach.		69.10		76.00	4.32	4.75
16 L.H., Daily Totals		$683.90		$1032.00	$42.75	$64.50
Crew Q-16	**Hr.**	**Daily**	**Hr.**	**Daily**	**Bare Costs**	**Incl. O&P**
2 Plumbers	$42.70	$683.20	$66.40	$1062.40	$39.85	$61.97
1 Plumber Apprentice	34.15	273.20	53.10	424.80		
1 Electric Welding Mach.		69.10		76.00	2.88	3.17
24 L.H., Daily Totals		$1025.50		$1563.20	$42.73	$65.14

Crew No.	Bare Costs		Incl. Subs O & P		Cost Per Labor-Hour	
Crew Q-17	**Hr.**	**Daily**	**Hr.**	**Daily**	**Bare Costs**	**Incl. O&P**
1 Steamfitter	$43.05	$344.40	$66.95	$535.60	$38.75	$60.25
1 Steamfitter Apprentice	34.45	275.60	53.55	428.40		
1 Electric Welding Mach.		69.10		76.00	4.32	4.75
16 L.H., Daily Totals		$689.10		$1040.00	$43.07	$65.00
Crew Q-17A	**Hr.**	**Daily**	**Hr.**	**Daily**	**Bare Costs**	**Incl. O&P**
1 Steamfitter	$43.05	$344.40	$66.95	$535.60	$38.53	$60.22
1 Steamfitter Apprentice	34.45	275.60	53.55	428.40		
1 Equip. Oper. (crane)	38.10	304.80	60.15	481.20		
1 Truck Crane, 12 Ton		636.60		700.25		
1 Electric Welding Mach.		69.10		76.00	29.40	32.34
24 L.H., Daily Totals		$1630.50		$2221.45	$67.93	$92.56
Crew Q-18	**Hr.**	**Daily**	**Hr.**	**Daily**	**Bare Costs**	**Incl. O&P**
2 Steamfitters	$43.05	$688.80	$66.95	$1071.20	$40.18	$62.48
1 Steamfitter Apprentice	34.45	275.60	53.55	428.40		
1 Electric Welding Mach.		69.10		76.00	2.88	3.17
24 L.H., Daily Totals		$1033.50		$1575.60	$43.06	$65.65
Crew Q-19	**Hr.**	**Daily**	**Hr.**	**Daily**	**Bare Costs**	**Incl. O&P**
1 Steamfitter	$43.05	$344.40	$66.95	$535.60	$39.83	$61.73
1 Steamfitter Apprentice	34.45	275.60	53.55	428.40		
1 Electrician	42.00	336.00	64.70	517.60		
24 L.H., Daily Totals		$956.00		$1481.60	$39.83	$61.73
Crew Q-20	**Hr.**	**Daily**	**Hr.**	**Daily**	**Bare Costs**	**Incl. O&P**
1 Sheet Metal Worker	$42.15	$337.20	$67.05	$536.40	$38.74	$61.20
1 Sheet Metal Apprentice	33.70	269.60	53.60	428.80		
.5 Electrician	42.00	168.00	64.70	258.80		
20 L.H., Daily Totals		$774.80		$1224.00	$38.74	$61.20
Crew Q-21	**Hr.**	**Daily**	**Hr.**	**Daily**	**Bare Costs**	**Incl. O&P**
2 Steamfitters	$43.05	$688.80	$66.95	$1071.20	$40.64	$63.04
1 Steamfitter Apprentice	34.45	275.60	53.55	428.40		
1 Electrician	42.00	336.00	64.70	517.60		
32 L.H., Daily Totals		$1300.40		$2017.20	$40.64	$63.04
Crew Q-22	**Hr.**	**Daily**	**Hr.**	**Daily**	**Bare Costs**	**Incl. O&P**
1 Plumber	$42.70	$341.60	$66.40	$531.20	$38.43	$59.75
1 Plumber Apprentice	34.15	273.20	53.10	424.80		
1 Truck Crane, 12 Ton		636.60		700.25	39.79	43.77
16 L.H., Daily Totals		$1251.40		$1656.25	$78.22	$103.52
Crew Q-22A	**Hr.**	**Daily**	**Hr.**	**Daily**	**Bare Costs**	**Incl. O&P**
1 Plumber	$42.70	$341.60	$66.40	$531.20	$35.59	$56.26
1 Plumber Apprentice	34.15	273.20	53.10	424.80		
1 Laborer	27.40	219.20	45.40	363.20		
1 Equip. Oper. (crane)	38.10	304.80	60.15	481.20		
1 Truck Crane, 12 Ton		636.60		700.25	19.89	21.88
32 L.H., Daily Totals		$1775.40		$2500.65	$55.48	$78.14
Crew Q-23	**Hr.**	**Daily**	**Hr.**	**Daily**	**Bare Costs**	**Incl. O&P**
1 Plumber Foreman	$44.70	$357.60	$69.50	$556.00	$41.37	$64.62
1 Plumber	42.70	341.60	66.40	531.20		
1 Equip. Oper. (medium)	36.70	293.60	57.95	463.60		
1 Power Tool		5.40		5.95		
1 Crane, 20 Ton		947.05		1041.75	39.68	43.65
24 L.H., Daily Totals		$1945.25		$2598.50	$81.05	$108.27

Crew No.	Bare Costs		Incl. Subs O & P		Cost Per Labor-Hour	
Crew R-1	**Hr.**	**Daily**	**Hr.**	**Daily**	**Bare Costs**	**Incl. O&P**
1 Electrician Foreman	$42.50	$340.00	$65.50	$524.00	$36.88	$57.80
3 Electricians	42.00	1008.00	64.70	1552.80		
2 Helpers	26.40	422.40	43.60	697.60		
48 L.H., Daily Totals		$1770.40		$2774.40	$36.88	$57.80
Crew R-1A	**Hr.**	**Daily**	**Hr.**	**Daily**	**Bare Costs**	**Incl. O&P**
1 Electrician	$42.00	$336.00	$64.70	$517.60	$34.20	$54.15
1 Helper	26.40	211.20	43.60	348.80		
16 L.H., Daily Totals		$547.20		$866.40	$34.20	$54.15
Crew R-2	**Hr.**	**Daily**	**Hr.**	**Daily**	**Bare Costs**	**Incl. O&P**
1 Electrician Foreman	$42.50	$340.00	$65.50	$524.00	$37.06	$58.14
3 Electricians	42.00	1008.00	64.70	1552.80		
2 Helpers	26.40	422.40	43.60	697.60		
1 Equip. Oper. (crane)	38.10	304.80	60.15	481.20		
1 S.P. Crane, 5 Ton		305.20		335.70	5.45	6.00
56 L.H., Daily Totals		$2380.40		$3591.30	$42.51	$64.14
Crew R-3	**Hr.**	**Daily**	**Hr.**	**Daily**	**Bare Costs**	**Incl. O&P**
1 Electrician Foreman	$42.50	$340.00	$65.50	$524.00	$41.42	$64.11
1 Electrician	42.00	336.00	64.70	517.60		
.5 Equip. Oper. (crane)	38.10	152.40	60.15	240.60		
.5 S.P. Crane, 5 Ton		152.60		167.85	7.63	8.39
20 L.H., Daily Totals		$981.00		$1450.05	$49.05	$72.50
Crew R-4	**Hr.**	**Daily**	**Hr.**	**Daily**	**Bare Costs**	**Incl. O&P**
1 Struc. Steel Foreman	$41.95	$335.60	$78.45	$627.60	$40.76	$73.45
3 Struc. Steel Workers	39.95	958.80	74.70	1792.80		
1 Electrician	42.00	336.00	64.70	517.60		
1 Gas Welding Machine		89.20		98.10	2.23	2.45
40 L.H., Daily Totals		$1719.60		$3036.10	$42.99	$75.90
Crew R-5	**Hr.**	**Daily**	**Hr.**	**Daily**	**Bare Costs**	**Incl. O&P**
1 Electrician Foreman	$42.50	$340.00	$65.50	$524.00	$36.37	$57.10
4 Electrician Linemen	42.00	1344.00	64.70	2070.40		
2 Electrician Operators	42.00	672.00	64.70	1035.20		
4 Electrician Groundmen	26.40	844.80	43.60	1395.20		
1 Crew Truck		133.75		147.15		
1 Tool Van		132.40		145.65		
1 Pickup Truck, 3/4 Ton		78.00		85.80		
.2 Crane, 55 Ton		190.32		209.35		
.2 Crane, 12 Ton		127.32		140.05		
.2 Auger, Truck Mtd.		675.80		743.40		
1 Tractor w/Winch		260.00		286.00	18.15	19.97
88 L.H., Daily Totals		$4798.39		$6782.20	$54.52	$77.07

Crew No.	Bare Costs		Incl. Subs O & P		Cost Per Labor-Hour	
Crew R-6	**Hr.**	**Daily**	**Hr.**	**Daily**	**Bare Costs**	**Incl. O&P**
1 Electrician Foreman	$42.50	$340.00	$65.50	$524.00	$36.37	$57.10
4 Electrician Linemen	42.00	1344.00	64.70	2070.40		
2 Electrician Operators	42.00	672.00	64.70	1035.20		
4 Electrician Groundmen	26.40	844.80	43.60	1395.20		
1 Crew Truck		133.75		147.15		
1 Tool Van		132.40		145.65		
1 Pickup Truck, 3/4 Ton		78.00		85.80		
.2 Crane, 55 Ton		190.32		209.35		
.2 Crane, 12 Ton		127.32		140.05		
.2 Auger, Truck Mtd.		675.80		743.40		
1 Tractor w/Winch		260.00		286.00		
3 Cable Trailers		500.70		550.75		
.5 Tensioning Rig		163.85		180.25		
.5 Cable Pulling Rig		960.00		1056.00	36.62	40.28
88 L.H., Daily Totals		$6422.94		$8569.20	$72.99	$97.38
Crew R-7	**Hr.**	**Daily**	**Hr.**	**Daily**	**Bare Costs**	**Incl. O&P**
1 Electrician Foreman	$42.50	$340.00	$65.50	$524.00	$29.08	$47.25
5 Electrician Groundmen	26.40	1056.00	43.60	1744.00		
1 Crew Truck		133.75		147.15	2.79	3.07
48 L.H., Daily Totals		$1529.75		$2415.15	$31.87	$50.32
Crew R-8	**Hr.**	**Daily**	**Hr.**	**Daily**	**Bare Costs**	**Incl. O&P**
1 Electrician Foreman	$42.50	$340.00	$65.50	$524.00	$36.88	$57.80
3 Electrician Linemen	42.00	1008.00	64.70	1552.80		
2 Electrician Groundmen	26.40	422.40	43.60	697.60		
1 Pickup Truck, 3/4 Ton		78.00		85.80		
1 Crew Truck		133.75		147.15	4.41	4.85
48 L.H., Daily Totals		$1982.15		$3007.35	$41.29	$62.65
Crew R-9	**Hr.**	**Daily**	**Hr.**	**Daily**	**Bare Costs**	**Incl. O&P**
1 Electrician Foreman	$42.50	$340.00	$65.50	$524.00	$34.26	$54.25
1 Electrician Lineman	42.00	336.00	64.70	517.60		
2 Electrician Operators	42.00	672.00	64.70	1035.20		
4 Electrician Groundmen	26.40	844.80	43.60	1395.20		
1 Pickup Truck, 3/4 Ton		78.00		85.80		
1 Crew Truck		133.75		147.15	3.31	3.64
64 L.H., Daily Totals		$2404.55		$3704.95	$37.57	$57.89
Crew R-10	**Hr.**	**Daily**	**Hr.**	**Daily**	**Bare Costs**	**Incl. O&P**
1 Electrician Foreman	$42.50	$340.00	$65.50	$524.00	$39.48	$61.32
4 Electrician Linemen	42.00	1344.00	64.70	2070.40		
1 Electrician Groundman	26.40	211.20	43.60	348.80		
1 Crew Truck		133.75		147.15		
3 Tram Cars		351.15		386.25	10.10	11.11
48 L.H., Daily Totals		$2380.10		$3476.60	$49.58	$72.43
Crew R-11	**Hr.**	**Daily**	**Hr.**	**Daily**	**Bare Costs**	**Incl. O&P**
1 Electrician Foreman	$42.50	$340.00	$65.50	$524.00	$39.43	$61.41
4 Electricians	42.00	1344.00	64.70	2070.40		
1 Equip. Oper. (crane)	38.10	304.80	60.15	481.20		
1 Common Laborer	27.40	219.20	45.40	363.20		
1 Crew Truck		133.75		147.15		
1 Crane, 12 Ton		636.60		700.25	13.76	15.13
56 L.H., Daily Totals		$2978.35		$4286.20	$53.19	$76.54

Crew No.	Bare Costs		Incl. Sub O & P		Cost Per Labor-Hour	
Crew R-12	**Hr.**	**Daily**	**Hr.**	**Daily**	**Bare Costs**	**Incl. O&P**
1 Carpenter Foreman	$36.05	$288.40	$59.75	$478.00	$33.14	$55.42
4 Carpenters	35.55	1137.60	58.90	1884.80		
4 Common Laborers	27.40	876.80	45.40	1452.80		
1 Equip. Oper. (med.)	36.70	293.60	57.95	463.60		
1 Steel Worker	39.95	319.60	74.70	597.60		
1 Dozer, 200 H.P.		920.80		1012.90		
1 Pickup Truck, 3/4 Ton		78.00		85.80	11.35	12.49
88 L.H., Daily Totals		$3914.80		$5975.50	$44.49	$67.91
Crew R-13	**Hr.**	**Daily**	**Hr.**	**Daily**	**Bare Costs**	**Incl. O&P**
1 Electrician Foreman	$42.50	$340.00	$65.50	$524.00	$40.09	$62.07
3 Electricians	42.00	1008.00	64.70	1552.80		
.25 Equip. Oper. (crane)	38.10	76.20	60.15	120.30		
1 Equipment Oiler	32.45	259.60	51.25	410.00		
.25-1 Hyd. Crane, 33 Ton		169.15		186.05	4.03	4.43
42 L.H., Daily Totals		$1852.95		$2793.15	$44.12	$66.50
Crew R-15	**Hr.**	**Daily**	**Hr.**	**Daily**	**Bare Costs**	**Incl. O&P**
1 Electrician Foreman	$42.50	$340.00	$65.50	$524.00	$40.95	$63.32
4 Electricians	42.00	1344.00	64.70	2070.40		
1 Equipment Operator	35.20	281.60	55.60	444.80		
1 Aerial Lift Truck		265.00		291.50	5.52	6.07
48 L.H., Daily Totals		$2230.60		$3330.70	$46.47	$69.39
Crew R-18	**Hr.**	**Daily**	**Hr.**	**Daily**	**Bare Costs**	**Incl. O&P**
.25 Electrician Foreman	$42.50	$85.00	$65.50	$131.00	$32.44	$51.78
1 Electrician	42.00	336.00	64.70	517.60		
2 Helpers	26.40	422.40	43.60	697.60		
26 L.H., Daily Totals		$843.40		$1346.20	$32.44	$51.78
Crew R-19	**Hr.**	**Daily**	**Hr.**	**Daily**	**Bare Costs**	**Incl. O&P**
.5 Electrician Foreman	$42.50	$170.00	$65.50	$262.00	$42.10	$64.86
2 Electricians	42.00	672.00	64.70	1035.20		
20 L.H., Daily Totals		$842.00		$1297.20	$42.10	$64.86
Crew R-21	**Hr.**	**Daily**	**Hr.**	**Daily**	**Bare Costs**	**Incl. O&P**
1 Electrician Foreman	$42.50	$340.00	$65.50	$524.00	$41.99	$64.73
3 Electricians	42.00	1008.00	64.70	1552.80		
.1 Equip. Oper. (med.)	36.70	29.36	57.95	46.36		
.1 Hyd. Crane 25 Ton		69.68		76.65	2.12	2.34
32. L.H., Daily Totals		$1447.04		$2199.81	$44.11	$67.07
Crew R-22	**Hr.**	**Daily**	**Hr.**	**Daily**	**Bare Costs**	**Incl. O&P**
.66 Electrician Foreman	$42.50	$224.40	$65.50	$345.84	$35.38	$55.76
2 Helpers	26.40	422.40	43.60	697.60		
2 Electricians	42.00	672.00	64.70	1035.20		
37.28 L.H., Daily Totals		$1318.80		$2078.64	$35.38	$55.76
Crew R-30	**Hr.**	**Daily**	**Hr.**	**Daily**	**Bare Costs**	**Incl. O&P**
.25 Electrician	$44.00	$88.00	$67.80	$135.60	$33.17	$53.06
1 Electrician	42.00	336.00	64.70	517.60		
2 Laborers (Semi-Skilled)	27.40	438.40	45.40	726.40		
26 L.H., Daily Totals		$862.40		$1379.60	$33.17	$53.06
Crew R-31	**Hr.**	**Daily**	**Hr.**	**Daily**	**Bare Costs**	**Incl. O&P**
1 Electrician	$42.00	$336.00	$64.70	$517.60	$42.00	$64.70
1 Core Drill, Elec, 2.5 HP		64.60		71.05	8.08	8.88
8 L.H., Daily Totals		$400.60		$588.65	$50.08	$73.58

Historical Cost Indexes

The table below lists both the Means Historical Cost Index based on Jan. 1, 1993 = 100 as well as the computed value of an index based on Jan. 1, 2006 costs. Since the Jan. 1, 2006 figure is estimated, space is left to write in the actual index figures as they become available through either the quarterly "Means Construction Cost Indexes" or as printed in the "Engineering News-Record." To compute the actual index based on Jan. 1, 2006 = 100, divide the Historical Cost Index for a particular year by the actual Jan. 1, 2006 Construction Cost Index. Space has been left to advance the index figures as the year progresses.

Year	Historical Cost Index Jan. 1, 1993 = 100		Current Index Based on Jan. 1, 2006 = 100	
	Est.	Actual	Est.	Actual
Oct 2006				
July 2006				
April 2006				
Jan 2006	155.9		100.0	100.0
July 2005		151.6	97.2	
2004		143.7	92.2	
2003		132.0	84.7	
2002		128.7	82.6	
2001		125.1	80.2	
2000		120.9	77.5	
1999		117.6	75.4	
1998		115.1	73.8	
1997		112.8	72.4	
1996		110.2	70.7	
1995		107.6	69.0	
1994		104.4	67.0	
1993		101.7	65.2	
1992		99.4	63.8	

Year	Historical Cost Index Jan. 1, 1993 = 100	Current Index Based on Jan. 1, 2006 = 100	
	Actual	Est.	Actual
July 1991	96.8	62.1	
1990	94.3	60.5	
1989	92.1	59.1	
1988	89.9	57.6	
1987	87.7	56.2	
1986	84.2	54.0	
1985	82.6	53.0	
1984	82.0	52.6	
1983	80.2	51.4	
1982	76.1	48.8	
1981	70.0	44.9	
1980	62.9	40.3	
1979	57.8	37.1	
1978	53.5	34.3	
1977	49.5	31.8	
1976	46.9	30.1	
1975	44.8	28.7	
1974	41.4	26.6	

Year	Historical Cost Index Jan. 1, 1993 = 100	Current Index Based on Jan. 1, 2006 = 100	
	Actual	Est.	Actual
July 1973	37.7	24.2	
1972	34.8	22.3	
1971	32.1	20.6	
1970	28.7	18.4	
1969	26.9	17.3	
1968	24.9	16.0	
1967	23.5	15.1	
1966	22.7	14.6	
1965	21.7	13.9	
1964	21.2	13.6	
1963	20.7	13.3	
1962	20.2	13.0	
1961	19.8	12.7	
1960	19.7	12.6	
1959	19.3	12.4	
1958	18.8	12.1	
1957	18.4	11.8	
1956	17.6	11.3	

Adjustments to Costs

The Historical Cost Index can be used to convert National Average building costs at a particular time to the approximate building costs for some other time.

Time Adjustment using the Historical Cost Indexes:

$$\frac{\text{Index for Year A}}{\text{Index for Year B}} \text{ x Cost in Year B = Cost in Year A}$$

Example:

Estimate and compare construction costs for different years in the same city.

To estimate the National Average construction cost of a building in 1970, knowing that it cost $900,000 in 2006:

INDEX in 1970 = 28.7

INDEX in 2006 = 155.9

$$\frac{\text{INDEX 1970}}{\text{INDEX 2006}} \text{ x Cost 2006 = Cost 1970}$$

$$\frac{28.7}{155.9} \text{ x \$900,000 = .184 x \$900,000 = \$165,600}$$

The construction cost of the building in 1970 is $165,600.

Note: The City Cost Indexes for Canada can be used to convert U.S. National averages to local costs in Canadian dollars.

How to Use the City Cost Indexes

What you should know before you begin

Means City Cost Indexes (CCI) are an extremely useful tool to use when you want to compare costs from city to city and region to region.

This publication contains average construction cost indexes for 316 major U.S. and Canadian cities and Location Factors covering over 930 three-digit zip code locations.

Keep in mind that a City Cost Index number is a *percentage ratio* of a specific city's cost to the national average cost of the same item at a stated time period.

In other words, these index figures represent relative construction *factors* (or, if you prefer, multipliers) for Material and Installation costs, as well as the weighted average for Total In Place costs for each CSI MasterFormat division. Installation costs include both labor and equipment rental costs. When estimating equipment rental rates only, for a specific location, use 01590 EQUIPMENT RENTAL index.

The 30 City Average Index is the average of 30 major U.S. cities and serves as a National Average.

Index figures for both material and installation are based on the 30 major city average of 100 and represent the cost relationship as of July 1, 2003. The index for each division is computed from representative material and labor quantities for that division. The weighted average for each city is a weighted total of the components listed above it, but does not include relative productivity between trades or cities.

As changes occur in local material prices, labor rates and equipment rental rates, the impact of these changes should be accurately measured by the change in the City Cost Index for each particular city (as compared to the 30 City Average).

Therefore, if you know (or have estimated) building costs in one city today, you can easily convert those costs to expected building costs in another city.

In addition, by using the Historical Cost Index, you can easily convert National Average building costs at a particular time to the approximate building costs for some other time. The City Cost Indexes can then be applied to calculate the costs for a particular city.

Quick Calculations

Location Adjustment Using the City Cost Indexes:

$$\frac{\text{Index for City A}}{\text{Index for City B}} \times \text{Cost in City B} = \text{Cost in City A}$$

Time Adjustment for the National Average Using the Historical Cost Index:

$$\frac{\text{Index for Year A}}{\text{Index for Year B}} \times \text{Cost in Year B} = \text{Cost in Year A}$$

Adjustment from the National Average:

$$\frac{\text{Index for City A}}{100} \times \text{National Average Cost} = \text{Cost in City A}$$

Since each of the other RSMeans publications contains many different items, any *one* item multiplied by the particular city index may give incorrect results. However, the larger the number of items compiled, the closer the results should be to actual costs for that particular city.

The City Cost Indexes for Canadian cities are calculated using Canadian material and equipment prices and labor rates, in Canadian dollars. Therefore, indexes for Canadian cities can be used to convert U.S. National Average prices to local costs in Canadian dollars.

How to use this section

1. Compare costs from city to city.

In using the Means Indexes, remember that an index number is not a fixed number but a *ratio:* It's a percentage ratio of a building component's cost at any stated time to the National Average cost of that same component at the same time period. Put in the form of an equation:

$$\frac{\text{Specific City Cost}}{\text{National Average Cost}} \times 100 = \text{City Index Number}$$

Therefore, when making cost comparisons between cities, do not subtract one city's index number from the index number of another city and read the result as a percentage difference. Instead, divide one city's index number by that of the other city. The resulting number may then be used as a multiplier to calculate cost differences from city to city.

The formula used to find cost differences between cities for the purpose of comparison is as follows:

$$\frac{\text{City A Index}}{\text{City B Index}} \times \text{City B Cost (Known)} = \text{City A Cost (Unknown)}$$

In addition, you can use *Means CCI* to calculate and compare costs division by division between cities using the same basic formula. (Just be sure that you're comparing similar divisions.)

2. Compare a specific city's construction costs with the National Average.

When you're studying construction location feasibility, it's advisable to compare a prospective project's cost index with an index of the National Average cost.

For example, divide the weighted average index of construction costs of a specific city by that of the 30 City Average, which = 100.

$$\frac{\text{City Index}}{100} = \text{\% of National Average}$$

As a result, you get a ratio that indicates the relative cost of construction in that city in comparison with the National Average.

3. Convert U.S. National Average to actual costs in Canadian City.

$$\frac{\text{Index for Canadian City}}{100} \times \text{National Average Cost} = \text{Cost in Canadian City in \$ CAN}$$

4. Adjust construction cost data based on a National Average.

When you use a source of construction cost data which is based on a National Average (such as *Means cost data publications*), it is necessary to adjust those costs to a specific location.

$$\frac{\text{City Index}}{100} \times \frac{\text{"Book" Cost Based on}}{\text{National Average Costs}} = \frac{\text{City Cost}}{\text{(Unknown)}}$$

5. When applying the City Cost Indexes to demolition projects, use the appropriate division index. For example, for removal of existing doors and windows, use the Division 8 index.

What you might like to know about how we developed the Indexes

To create a reliable index, RSMeans researched the building type most often constructed in the United States and Canada. Because it was concluded that no one type of building completely represented the building construction industry, nine different types of buildings were combined to create a composite model.

The exact material, labor and equipment quantities are based on detailed analysis of these nine building types, then each quantity is weighted in proportion to expected usage. These various material items, labor hours, and equipment rental rates are thus combined to form a composite building representing as closely as possible the actual usage of materials, labor and equipment used in the North American Building Construction Industry.

The following structures were chosen to make up that composite model:

1. Factory, 1 story
2. Office, 2–4 story
3. Store, Retail
4. Town Hall, 2–3 story
5. High School, 2–3 story
6. Hospital, 4–8 story
7. Garage, Parking
8. Apartment, 1–3 story
9. Hotel/Motel, 2–3 story

For the purposes of ensuring the timeliness of the data, the components of the index for the composite model have been streamlined. They currently consist of:

- specific quantities of 66 commonly used construction materials;
- specific labor-hours for 21 building construction trades; and
- specific days of equipment rental for 6 types of construction equipment (normally used to install the 66 material items by the 21 trades.)

A sophisticated computer program handles the updating of all costs for each city on a quarterly basis. Material and equipment price quotations are gathered quarterly from over 316 cities in the United States and Canada. These prices and the latest negotiated labor wage rates for 21 different building trades are used to compile the quarterly update of the City Cost Index.

The 30 major U.S. cities used to calculate the National Average are:

Atlanta, GA	Memphis, TN
Baltimore, MD	Milwaukee, WI
Boston, MA	Minneapolis, MN
Buffalo, NY	Nashville, TN
Chicago, IL	New Orleans, LA
Cincinnati, OH	New York, NY
Cleveland, OH	Philadelphia, PA
Columbus, OH	Phoenix, AZ
Dallas, TX	Pittsburgh, PA
Denver, CO	St. Louis, MO
Detroit, MI	San Antonio, TX
Houston, TX	San Diego, CA
Indianapolis, IN	San Francisco, CA
Kansas City, MO	Seattle, WA
Los Angeles, CA	Washington, DC

F.Y.I.: The CSI MasterFormat Divisions

1. General Requirements
2. Site Construction
3. Concrete
4. Masonry
5. Metals
6. Wood & Plastics
7. Thermal & Moisture Protection
8. Doors & Windows
9. Finishes
10. Specialties
11. Equipment
12. Furnishings
13. Special Construction
14. Conveying Systems
15. Mechanical
16. Electrical

The information presented in the CCI is organized according to the Construction Specifications Institute (CSI) MasterFormat.

What the CCI does not indicate

The weighted average for each city is a total of the components listed above weighted to reflect typical usage, but it does *not* include the productivity variations between trades or cities.

In addition, the CCI does not take into consideration factors such as the following:

- managerial efficiency
- competitive conditions
- automation
- restrictive union practices
- unique local requirements
- regional variations due to specific building codes

	DIVISION	UNITED STATES			ALABAMA																	
		30 CITY AVERAGE			BIRMINGHAM			HUNTSVILLE			MOBILE			MONTGOMERY			TUSCALOOSA					
		MAT.	INST.	TOTAL	MAT.	INST.	TOTAL	MAT.	INST.	TOTAL	MAT.	INST.	TOTAL	MAT.	INST.	TOTAL	MAT.	INST.	TOTAL			
01590	EQUIPMENT RENTAL	.0	100.0	100.0	.0	101.3	101.3	.0	101.2	101.2	.0	97.7	97.7	.0	97.7	97.7	.0	101.2	101.2			
02	SITE CONSTRUCTION	100.0	100.0	100.0	83.2	92.9	90.3	81.3	92.9	89.8	94.2	85.8	88.0	92.5	86.4	88.0	81.8	91.6	89.0			
03100	CONCRETE FORMS & ACCESSORIES	100.0	100.0	100.0	95.3	73.4	76.4	97.5	66.7	70.9	97.6	51.4	57.7	96.3	46.7	53.5	97.4	38.5	46.6			
03200	CONCRETE REINFORCEMENT	100.0	100.0	100.0	92.1	87.9	90.0	92.1	74.8	83.3	95.1	51.8	73.1	95.1	86.3	90.6	92.1	86.6	89.3			
03300	CAST-IN-PLACE CONCRETE	100.0	100.0	100.0	93.3	65.9	82.1	88.3	64.5	78.5	93.2	53.0	76.7	94.9	47.8	75.5	91.9	45.7	72.9			
03	CONCRETE	100.0	100.0	100.0	91.7	74.6	83.2	89.4	68.7	79.1	92.3	53.6	73.0	93.0	56.2	74.7	91.2	51.9	71.7			
04	MASONRY	100.0	100.0	100.0	88.5	76.2	80.9	88.1	68.4	75.9	88.8	50.8	65.3	87.9	35.2	55.3	87.0	37.5	56.3			
05	METALS	100.0	100.0	100.0	92.8	94.2	93.3	94.2	89.8	92.8	93.0	77.4	88.0	92.8	90.8	92.1	93.4	91.3	92.7			
06	WOOD & PLASTICS	100.0	100.0	100.0	96.3	73.7	84.5	96.3	66.3	80.7	96.5	51.1	72.9	94.9	46.8	69.8	96.3	37.1	65.5			
07	THERMAL & MOISTURE PROTECTION	100.0	100.0	100.0	96.8	79.3	89.4	96.6	74.3	87.2	96.6	67.5	84.4	96.2	61.2	81.5	96.6	60.0	81.2			
08	DOORS & WINDOWS	100.0	100.0	100.0	98.3	76.5	92.6	98.3	63.2	89.1	98.3	51.4	86.0	98.3	56.8	87.4	98.3	56.7	87.4			
09200	PLASTER & GYPSUM BOARD	100.0	100.0	100.0	103.0	73.5	84.7	100.6	65.9	79.0	100.6	50.3	69.3	100.6	45.9	66.6	100.6	36.0	60.5			
095,098	CEILINGS & ACOUSTICAL TREATMENT	100.0	100.0	100.0	104.3	73.5	85.9	104.3	65.9	81.4	104.3	50.3	72.1	104.3	45.9	69.5	104.3	36.0	63.6			
09600	FLOORING	100.0	100.0	100.0	104.3	48.3	89.2	104.3	51.9	90.1	114.8	52.4	97.9	113.1	27.1	89.8	104.3	39.7	86.8			
097,099	WALL FINISHES, PAINTS & COATINGS	100.0	100.0	100.0	94.8	70.9	80.2	94.8	61.9	74.8	99.6	54.9	72.3	94.8	54.0	69.9	94.8	46.5	65.3			
09	FINISHES	100.0	100.0	100.0	101.1	68.1	83.9	100.6	62.8	80.9	105.3	51.3	77.2	104.3	42.4	72.1	100.6	37.9	67.9			
10 - 14	TOTAL DIV. 10000 - 14000	100.0	100.0	100.0	100.0	85.3	96.9	100.0	83.4	96.5	100.0	64.3	92.4	100.0	76.5	95.0	100.0	74.1	94.5			
15	MECHANICAL	100.0	100.0	100.0	99.9	67.0	86.3	99.9	67.2	86.3	99.9	68.2	86.7	99.9	40.2	75.1	99.9	35.4	73.1			
16	ELECTRICAL	100.0	100.0	100.0	97.5	67.0	82.7	97.5	68.8	83.6	97.5	50.1	74.4	96.7	67.4	82.5	97.0	67.0	82.4			
01 - 16	WEIGHTED AVERAGE	100.0	100.0	100.0	96.5	75.2	87.0	96.3	72.0	85.5	97.2	61.1	81.2	97.0	57.1	79.3	96.3	55.4	78.1			

	DIVISION	ALASKA									ARIZONA								
		ANCHORAGE			FAIRBANKS			JUNEAU			FLAGSTAFF			MESA/TEMPE			PHOENIX		
		MAT.	INST.	TOTAL	MAT.	INST.	TOTAL	MAT.	INST.	TOTAL	MAT.	INST.	TOTAL	MAT.	INST.	TOTAL	MAT.	INST.	TOTAL
01590	EQUIPMENT RENTAL	.0	118.4	118.4	.0	118.4	118.4	.0	118.4	118.4	.0	95.0	95.0	.0	98.3	98.3	.0	98.9	98.9
02	SITE CONSTRUCTION	143.7	133.7	136.4	127.6	133.7	132.1	139.3	133.7	135.2	81.4	100.8	95.6	86.0	103.9	99.1	86.5	104.6	99.8
03100	CONCRETE FORMS & ACCESSORIES	133.2	115.6	118.0	136.3	119.9	122.1	134.7	115.6	118.2	102.7	67.6	72.4	97.3	63.0	67.7	98.4	68.9	73.0
03200	CONCRETE REINFORCEMENT	141.4	108.0	124.4	144.1	108.0	125.7	108.4	108.0	108.2	105.6	74.5	89.8	105.0	74.5	89.5	103.3	74.6	88.7
03300	CAST-IN-PLACE CONCRETE	197.4	115.2	163.6	164.6	115.7	144.5	198.3	115.2	164.1	90.2	76.2	84.4	97.7	69.5	86.1	97.8	76.7	89.1
03	CONCRETE	153.6	113.5	133.7	134.1	115.5	124.9	148.8	113.4	131.3	116.9	71.7	94.5	101.1	67.4	84.4	100.7	72.5	86.7
04	MASONRY	227.3	121.5	161.7	214.0	121.5	156.7	221.8	121.5	159.6	100.1	65.1	78.4	105.1	52.5	72.5	93.3	66.2	76.5
05	METALS	129.2	102.5	120.7	129.3	102.8	120.8	129.3	102.4	120.7	93.8	69.5	86.0	94.3	70.3	86.7	95.8	71.2	87.9
06	WOOD & PLASTICS	119.3	113.4	116.2	119.6	118.8	119.2	119.3	113.4	116.2	108.2	68.2	87.4	99.1	68.4	83.1	100.1	69.7	84.3
07	THERMAL & MOISTURE PROTECTION	172.8	115.6	148.7	170.2	118.7	148.6	170.8	115.6	147.5	99.7	69.1	86.8	99.1	63.4	84.1	99.0	69.2	86.5
08	DOORS & WINDOWS	125.9	111.0	122.0	123.1	113.4	120.6	123.1	111.0	119.9	102.1	68.3	93.3	98.8	68.1	90.8	100.0	69.1	91.9
09200	PLASTER & GYPSUM BOARD	132.8	113.6	120.9	142.5	119.1	128.0	132.8	113.6	120.9	92.3	67.4	76.8	96.8	67.4	78.5	98.9	68.7	80.1
095,098	CEILINGS & ACOUSTICAL TREATMENT	132.4	113.6	121.2	132.4	119.1	124.4	132.4	113.6	121.2	104.8	67.4	82.5	97.4	67.4	79.5	104.2	68.7	83.0
09600	FLOORING	165.0	125.3	154.2	165.0	125.3	154.2	165.0	125.3	154.2	94.5	49.6	82.4	96.5	61.8	87.1	96.8	64.6	88.1
097,099	WALL FINISHES, PAINTS & COATINGS	166.6	114.6	134.9	166.6	128.9	143.6	166.6	114.6	134.9	93.2	56.3	70.7	102.8	56.3	74.5	102.8	58.3	75.7
09	FINISHES	154.3	117.4	135.1	153.7	122.1	137.3	153.0	117.4	134.5	97.4	62.6	79.3	97.8	61.5	78.9	100.0	66.5	82.6
10 - 14	TOTAL DIV. 10000 - 14000	100.0	115.2	103.2	100.0	115.9	103.4	100.0	115.2	103.2	100.0	81.2	96.0	100.0	76.8	95.0	100.0	81.7	96.1
15	MECHANICAL	100.5	108.7	103.9	100.5	116.9	107.3	100.5	105.5	102.6	100.2	76.8	90.5	100.2	70.0	87.7	100.2	76.9	90.6
16	ELECTRICAL	148.4	114.2	131.8	150.2	114.2	132.7	150.2	114.2	132.7	99.5	64.1	82.3	91.4	64.1	78.2	100.5	64.1	82.8
01 - 16	WEIGHTED AVERAGE	133.7	114.7	125.3	130.3	117.6	124.6	132.5	114.0	124.3	100.4	72.3	87.9	97.8	68.9	85.0	98.8	73.6	87.6

	DIVISION	ARIZONA						ARKANSAS											
		PRESCOTT			TUCSON			FORT SMITH			JONESBORO			LITTLE ROCK			PINE BLUFF		
		MAT.	INST.	TOTAL	MAT.	INST.	TOTAL	MAT.	INST.	TOTAL	MAT.	INST.	TOTAL	MAT.	INST.	TOTAL	MAT.	INST.	TOTAL
01590	EQUIPMENT RENTAL	.0	95.0	95.0	.0	98.3	98.3	.0	86.0	86.0	.0	107.8	107.8	.0	86.0	86.0	.0	86.0	86.0
02	SITE CONSTRUCTION	69.6	100.0	91.9	82.9	104.4	98.6	77.2	83.6	81.9	99.8	98.9	99.1	77.0	83.6	81.8	79.2	83.6	82.4
03100	CONCRETE FORMS & ACCESSORIES	98.1	54.6	60.5	97.9	68.6	72.6	97.5	41.0	48.7	84.2	46.3	51.5	91.7	57.6	62.3	76.8	57.4	60.0
03200	CONCRETE REINFORCEMENT	105.6	70.8	87.9	86.8	74.5	80.5	97.3	74.4	85.7	93.0	47.7	70.0	97.5	69.6	83.3	97.4	69.6	83.3
03300	CAST-IN-PLACE CONCRETE	90.1	62.3	78.7	100.5	76.5	90.7	88.9	66.5	79.7	84.7	55.3	72.6	88.9	66.7	79.8	81.5	66.6	75.4
03	CONCRETE	102.7	60.4	81.7	99.3	72.3	85.9	86.9	56.8	72.0	83.5	51.3	67.5	86.5	63.4	75.0	84.3	63.2	73.9
04	MASONRY	100.5	57.1	73.6	95.0	65.1	76.5	96.0	55.2	70.7	91.4	45.5	62.9	94.1	55.2	70.0	115.9	55.2	78.3
05	METALS	93.8	65.1	84.6	95.0	70.1	87.1	96.3	70.7	88.2	90.6	73.8	85.3	92.2	69.4	84.9	95.0	69.1	86.8
06	WOOD & PLASTICS	103.3	52.9	77.1	99.3	69.7	83.9	100.0	39.3	68.4	85.7	47.3	65.7	96.9	61.3	78.4	78.1	61.3	69.4
07	THERMAL & MOISTURE PROTECTION	98.3	60.0	82.2	100.7	65.6	85.9	99.6	47.8	77.8	101.5	50.1	79.9	98.0	50.1	77.8	98.0	50.1	77.9
08	DOORS & WINDOWS	102.1	56.2	90.1	95.9	69.1	88.9	96.9	44.8	83.3	98.5	46.2	84.9	96.9	56.3	86.3	92.4	56.3	83.0
09200	PLASTER & GYPSUM BOARD	89.6	51.6	66.0	97.1	68.7	79.5	83.8	38.4	55.6	92.6	46.2	63.7	83.8	60.9	69.6	76.8	60.9	66.9
095,098	CEILINGS & ACOUSTICAL TREATMENT	103.1	51.6	72.4	97.4	68.7	80.3	89.7	38.4	59.1	87.6	46.2	62.9	89.7	60.9	72.5	85.4	60.9	70.8
09600	FLOORING	92.8	49.4	81.1	96.0	49.6	83.4	115.3	73.1	103.9	77.1	43.6	68.0	116.6	73.1	104.8	104.2	73.1	95.8
097,099	WALL FINISHES, PAINTS & COATINGS	93.2	47.1	65.1	100.5	56.3	73.5	97.0	56.3	72.1	85.5	50.6	64.2	97.0	57.7	73.0	97.0	57.7	73.0
09	FINISHES	95.0	52.0	72.6	97.5	63.5	79.8	95.5	48.0	70.8	87.4	46.4	66.0	95.8	61.1	77.7	90.3	61.1	75.1
10 - 14	TOTAL DIV. 10000 - 14000	100.0	78.6	95.4	100.0	81.7	96.1	100.0	71.9	94.0	100.0	51.2	89.6	100.0	74.6	94.6	100.0	74.6	94.6
15	MECHANICAL	100.2	75.4	89.9	100.2	70.5	87.9	100.1	41.6	75.8	100.2	41.5	75.8	100.1	57.9	82.6	100.1	43.9	76.8
16	ELECTRICAL	99.2	64.1	82.1	93.4	62.0	78.1	95.2	66.6	81.3	102.1	46.6	75.1	94.9	68.4	82.0	94.6	68.4	81.8
01 - 16	WEIGHTED AVERAGE	98.3	66.8	84.3	97.1	71.2	85.6	95.9	56.9	78.6	95.2	53.4	76.6	95.1	63.8	81.2	95.2	60.9	80.0

DIVISION	ARKANSAS			CALIFORNIA														
	TEXARKANA			ANAHEIM			BAKERSFIELD			FRESNO			LOS ANGELES			OAKLAND		
	MAT.	INST.	TOTAL	MAT.	INST.	TOTAL	MAT.	INST.	TOTAL	MAT.	INST.	TOTAL	MAT.	INST.	TOTAL	MAT.	INST.	TOTAL
01590 EQUIPMENT RENTAL	.0	86.7	86.7	.0	102.2	102.2	.0	99.6	99.6	.0	99.6	99.6	.0	98.6	98.6	.0	103.5	103.5
02 SITE CONSTRUCTION	94.6	84.4	87.1	103.2	109.0	107.5	108.0	106.1	106.6	109.5	106.2	107.1	100.5	107.8	105.8	154.4	104.7	117.9
03100 CONCRETE FORMS & ACCESSORIES	84.7	38.4	44.8	105.7	121.7	119.5	97.9	121.2	118.0	101.9	124.6	121.5	105.8	121.6	119.4	106.3	134.8	130.9
03200 CONCRETE REINFORCEMENT	97.0	46.2	71.2	95.4	112.6	104.1	108.6	112.5	110.6	91.9	112.8	102.6	112.8	112.9	112.9	102.9	113.6	108.3
03300 CAST-IN-PLACE CONCRETE	88.9	43.1	70.1	108.1	118.5	112.4	103.2	117.4	109.1	113.0	113.3	113.1	91.3	116.8	101.8	141.7	117.9	131.9
03 CONCRETE	83.8	42.5	63.3	107.3	118.0	112.6	106.5	117.3	111.9	108.9	117.5	113.2	104.0	117.3	110.6	126.7	123.9	125.3
04 MASONRY	96.0	31.9	56.2	93.2	111.0	104.2	109.1	110.3	109.9	112.7	113.5	113.2	95.6	117.3	109.1	154.1	125.1	136.1
05 METALS	87.6	57.7	78.1	110.5	99.8	107.1	105.1	98.9	103.1	110.7	100.4	107.4	111.7	98.7	107.6	104.6	105.1	104.8
06 WOOD & PLASTICS	87.6	41.0	63.3	96.3	121.9	109.6	86.8	122.0	105.1	100.2	125.8	113.5	92.3	121.6	107.6	102.8	136.2	120.2
07 THERMAL & MOISTURE PROTECTION	99.2	41.0	74.7	102.7	115.2	107.9	99.2	109.1	103.4	95.9	111.4	102.4	100.7	117.4	107.7	110.1	128.7	118.0
08 DOORS & WINDOWS	97.4	39.7	82.3	102.7	116.6	106.3	101.5	114.4	104.9	101.9	116.5	105.7	97.0	116.4	102.1	106.1	128.1	111.8
09200 PLASTER & GYPSUM BOARD	81.9	40.1	55.9	98.8	122.4	113.5	98.8	122.4	113.5	96.6	126.3	115.0	104.1	122.4	115.5	101.1	136.5	123.1
095,098 CEILINGS & ACOUSTICAL TREATMENT	92.3	40.1	61.1	120.4	122.4	121.6	120.4	122.4	121.6	120.4	126.3	123.9	123.5	122.4	122.9	116.7	136.5	128.5
09600 FLOORING	107.0	41.6	89.3	122.8	104.8	117.9	117.0	93.7	110.7	133.3	135.6	133.9	113.5	104.8	111.2	116.1	126.1	118.8
097,099 WALL FINISHES, PAINTS & COATINGS	97.0	35.5	59.4	110.9	109.0	109.8	110.5	95.5	101.4	133.7	96.7	111.1	100.3	109.0	105.6	115.4	137.8	129.0
09 FINISHES	94.1	38.8	65.3	114.3	117.5	116.0	114.5	114.1	114.3	120.7	124.5	122.7	112.0	117.3	114.7	117.0	134.3	126.0
10 - 14 TOTAL DIV. 10000 - 14000	100.0	36.6	86.5	100.0	113.0	102.8	100.0	111.9	102.5	100.0	125.3	105.4	100.0	112.3	102.6	100.0	128.8	106.1
15 MECHANICAL	100.1	36.7	73.8	100.2	107.3	103.1	100.2	96.2	98.6	100.2	109.1	103.9	100.1	107.2	103.1	100.3	136.0	115.1
16 ELECTRICAL	96.3	38.8	68.4	89.5	105.3	97.2	89.6	97.2	93.3	88.4	96.6	92.4	97.0	114.2	105.3	103.4	138.9	120.6
01 - 16 WEIGHTED AVERAGE	94.6	44.0	72.1	102.6	110.6	106.1	102.3	105.9	103.9	104.1	110.9	107.1	102.5	112.1	106.8	110.7	127.1	118.0

DIVISION	CALIFORNIA																	
	OXNARD			REDDING			RIVERSIDE			SACRAMENTO			SAN DIEGO			SAN FRANCISCO		
	MAT.	INST.	TOTAL	MAT.	INST.	TOTAL	MAT.	INST.	TOTAL	MAT.	INST.	TOTAL	MAT.	INST.	TOTAL	MAT.	INST.	TOTAL
01590 EQUIPMENT RENTAL	.0	98.2	98.2	.0	99.2	99.2	.0	100.7	100.7	.0	103.0	103.0	.0	97.1	97.1	.0	108.8	108.8
02 SITE CONSTRUCTION	109.5	103.9	105.4	114.3	105.0	107.5	100.8	106.7	105.2	118.7	111.3	113.3	101.2	100.7	100.8	156.7	111.4	123.5
03100 CONCRETE FORMS & ACCESSORIES	104.0	121.8	119.3	103.7	123.5	120.8	106.2	121.6	119.5	104.8	125.2	122.4	108.1	108.7	108.6	106.7	135.6	131.7
03200 CONCRETE REINFORCEMENT	108.6	112.5	110.6	105.3	112.8	109.1	107.6	112.5	110.1	95.2	112.9	104.2	104.9	112.4	108.7	117.2	113.9	115.5
03300 CAST-IN-PLACE CONCRETE	109.6	117.7	112.9	124.1	111.5	118.9	107.1	118.5	111.8	122.1	114.1	118.8	112.0	106.1	109.6	141.6	119.5	132.5
03 CONCRETE	110.0	117.7	113.8	119.7	116.4	118.1	108.8	117.9	113.3	115.9	118.0	116.9	112.9	107.9	110.4	129.0	124.9	127.0
04 MASONRY	113.7	109.6	111.1	116.3	110.8	112.9	86.9	110.6	101.6	123.3	113.6	117.3	99.4	108.6	105.1	154.4	128.5	138.3
05 METALS	104.9	99.5	103.1	109.9	99.6	106.6	110.8	99.7	107.2	100.0	100.0	100.0	108.4	98.8	105.3	110.7	107.1	109.5
06 WOOD & PLASTICS	95.1	122.0	109.1	95.3	125.8	111.1	96.3	121.9	109.6	97.5	126.0	112.4	102.2	105.8	104.1	102.8	136.4	120.3
07 THERMAL & MOISTURE PROTECTION	105.4	113.5	108.8	105.9	109.5	107.4	102.9	114.7	107.9	110.6	113.2	111.7	107.8	103.4	105.9	113.3	130.6	120.6
08 DOORS & WINDOWS	100.4	116.6	104.6	103.2	118.1	107.1	102.7	116.6	106.3	117.6	118.2	117.7	103.8	106.6	104.5	110.5	128.4	115.2
09200 PLASTER & GYPSUM BOARD	98.8	122.4	113.5	97.2	126.3	115.3	97.7	122.4	113.0	98.7	126.3	115.8	97.4	105.7	102.6	103.9	136.5	124.2
095,098 CEILINGS & ACOUSTICAL TREATMENT	120.4	122.4	121.6	127.2	126.3	126.7	116.2	122.4	119.9	116.7	126.3	122.4	104.2	105.7	105.1	125.8	136.5	132.2
09600 FLOORING	117.0	104.8	113.7	116.2	113.9	115.6	121.5	104.8	117.0	118.9	113.9	117.6	110.7	104.8	109.1	116.1	126.1	118.8
097,099 WALL FINISHES, PAINTS & COATINGS	109.9	102.7	105.5	109.9	115.8	113.5	107.4	109.0	108.4	112.8	115.8	114.7	108.2	109.0	108.7	115.4	149.6	136.3
09 FINISHES	114.3	116.9	115.7	116.0	122.1	119.2	112.2	117.5	115.0	115.6	123.1	119.5	107.7	107.8	107.8	119.9	136.1	128.3
10 - 14 TOTAL DIV. 10000 - 14000	100.0	113.3	102.8	100.0	124.4	105.2	100.0	113.0	102.8	100.0	126.1	105.6	100.0	110.9	102.3	100.0	129.4	106.3
15 MECHANICAL	100.2	107.3	103.2	100.2	105.6	102.5	100.1	107.3	103.1	100.2	110.6	104.5	100.3	104.0	101.8	100.3	148.6	120.3
16 ELECTRICAL	95.4	106.8	100.9	98.1	104.2	101.1	89.6	99.7	94.5	97.7	106.1	101.8	98.9	96.4	97.7	103.4	150.5	126.3
01 - 16 WEIGHTED AVERAGE	103.8	110.0	106.5	106.6	110.3	108.3	102.2	109.5	105.5	106.6	113.0	109.4	104.0	103.9	104.0	112.8	132.9	121.7

DIVISION	CALIFORNIA												COLORADO					
	SAN JOSE			SANTA BARBARA			STOCKTON			VALLEJO			COLORADO SPRINGS			DENVER		
	MAT.	INST.	TOTAL	MAT.	INST.	TOTAL	MAT.	INST.	TOTAL	MAT.	INST.	TOTAL	MAT.	INST.	TOTAL	MAT.	INST.	TOTAL
01590 EQUIPMENT RENTAL	.0	100.0	100.0	.0	99.6	99.6	.0	99.2	99.2	.0	103.6	103.6	.0	95.2	95.2	.0	100.3	100.3
02 SITE CONSTRUCTION	147.9	99.8	112.6	109.4	106.1	107.0	107.5	105.6	106.1	115.5	111.4	112.5	93.3	96.8	95.8	91.9	106.0	102.3
03100 CONCRETE FORMS & ACCESSORIES	106.3	134.7	130.8	104.7	121.7	119.4	104.4	124.8	122.0	106.1	133.4	129.6	92.5	85.1	86.1	101.8	85.9	88.1
03200 CONCRETE REINFORCEMENT	95.9	113.6	104.9	108.6	112.6	110.6	109.2	112.9	111.1	102.5	113.7	108.2	112.8	88.5	100.5	112.8	88.6	100.5
03300 CAST-IN-PLACE CONCRETE	129.7	117.5	124.7	109.2	117.6	112.6	109.1	113.3	110.8	127.1	115.3	122.2	95.1	86.5	91.5	88.9	86.9	88.1
03 CONCRETE	117.9	123.8	120.8	109.9	117.6	113.7	109.9	117.6	113.7	119.6	122.2	120.9	106.7	86.3	96.6	100.9	86.8	93.9
04 MASONRY	149.4	125.2	134.4	109.6	110.7	110.3	113.3	113.5	113.4	87.4	125.9	111.3	106.0	81.1	90.6	105.7	83.3	91.8
05 METALS	104.2	106.9	105.1	105.3	99.5	103.5	106.8	100.7	104.8	103.3	102.1	102.9	98.0	89.3	95.2	100.3	89.5	96.9
06 WOOD & PLASTICS	102.0	135.9	119.7	95.1	122.0	109.1	97.1	125.8	112.0	95.9	135.9	116.7	93.3	87.8	90.4	102.9	87.7	95.0
07 THERMAL & MOISTURE PROTECTION	101.9	129.5	113.5	100.3	112.5	105.4	105.6	110.8	107.8	113.1	126.0	118.5	105.9	85.4	97.3	105.4	82.1	95.6
08 DOORS & WINDOWS	92.7	127.9	101.9	101.5	116.6	105.5	100.8	118.1	105.3	118.9	128.1	121.3	98.5	91.3	96.6	99.8	91.2	97.6
09200 PLASTER & GYPSUM BOARD	100.4	136.5	122.8	98.8	122.4	113.5	99.2	126.3	116.0	101.4	136.5	123.2	87.8	87.6	87.7	97.4	87.6	91.3
095,098 CEILINGS & ACOUSTICAL TREATMENT	108.7	136.5	125.3	120.4	122.4	121.6	120.4	126.3	123.9	126.1	136.5	132.3	101.7	87.6	93.3	101.8	87.6	93.3
09600 FLOORING	113.9	126.1	117.2	117.0	96.1	111.3	117.0	114.1	116.2	123.7	126.1	124.4	106.9	85.2	101.0	107.8	96.6	104.8
097,099 WALL FINISHES, PAINTS & COATINGS	111.5	137.8	127.5	109.9	102.7	105.5	109.9	116.6	114.0	112.0	137.8	127.7	105.6	51.7	72.7	105.6	78.4	89.0
09 FINISHES	112.5	134.1	123.7	114.5	115.4	115.0	114.4	123.0	118.9	117.4	133.7	125.9	98.4	81.6	89.7	100.0	87.2	93.3
10 - 14 TOTAL DIV. 10000 - 14000	100.0	128.1	106.0	100.0	113.3	102.8	100.0	125.4	105.4	100.0	127.0	105.8	100.0	89.8	97.8	100.0	89.8	97.8
15 MECHANICAL	100.2	138.0	115.9	100.2	107.3	103.2	100.2	104.4	101.9	100.2	122.9	109.6	100.1	79.9	91.7	100.1	87.4	94.8
16 ELECTRICAL	103.3	140.8	121.5	86.7	110.0	98.0	97.9	123.1	110.1	93.0	123.5	107.9	99.6	88.6	94.3	101.5	94.9	98.3
01 - 16 WEIGHTED AVERAGE	107.3	127.5	116.3	102.6	110.5	106.1	104.4	113.5	108.4	105.5	122.2	112.9	100.4	85.7	93.9	100.6	89.9	95.8

	DIVISION	COLORADO												CONNECTICUT					
		FORT COLLINS			GRAND JUNCTION			GREELEY			PUEBLO			BRIDGEPORT			BRISTOL		
		MAT.	INST.	TOTAL	MAT.	INST.	TOTAL	MAT.	INST.	TOTAL	MAT.	INST.	TOTAL	MAT.	INST.	TOTAL	MAT.	INST.	TOTAL
01590	EQUIPMENT RENTAL	.0	96.7	96.7	.0	100.0	100.0	.0	96.7	96.7	.0	96.9	96.9	.0	101.5	101.5	.0	101.5	101.5
02	SITE CONSTRUCTION	102.2	99.2	100.0	121.8	101.8	107.1	90.1	97.9	95.8	114.1	96.2	101.0	103.3	104.9	104.5	102.4	104.9	104.2
03100	CONCRETE FORMS & ACCESSORIES	101.2	81.4	84.1	106.6	81.8	85.2	98.6	48.1	55.1	103.4	85.4	87.8	102.2	123.2	120.3	102.2	122.7	119.9
03200	CONCRETE REINFORCEMENT	113.7	79.9	96.5	113.7	79.8	96.5	113.6	79.3	96.2	109.5	88.5	98.8	107.5	128.1	117.9	107.5	128.0	117.9
03300	CAST-IN-PLACE CONCRETE	103.2	82.2	94.6	111.2	83.2	99.7	86.3	58.9	75.0	99.7	87.7	94.8	108.1	117.0	111.7	101.2	116.9	107.7
03	CONCRETE	113.1	81.4	97.4	115.4	81.9	98.8	98.4	58.6	78.6	105.3	86.9	96.2	109.3	121.7	115.5	106.0	121.4	113.7
04	MASONRY	118.2	59.1	81.6	144.2	61.5	92.9	112.2	39.2	66.9	105.2	80.2	89.7	109.2	127.8	120.7	101.0	127.8	117.6
05	METALS	96.2	81.3	91.5	96.9	80.5	91.7	96.2	79.6	90.9	98.4	90.3	95.8	100.9	124.6	108.5	100.9	124.1	108.3
06	WOOD & PLASTICS	102.7	85.4	93.7	105.9	85.6	95.3	99.8	46.8	72.2	102.7	88.2	95.1	100.4	122.3	111.8	100.4	122.3	111.8
07	THERMAL & MOISTURE PROTECTION	105.8	70.9	91.1	99.7	67.1	86.0	105.2	59.0	85.7	99.2	84.9	93.2	101.8	126.2	112.1	102.0	122.4	110.6
08	DOORS & WINDOWS	95.8	87.9	93.8	100.5	88.0	97.2	95.8	66.9	88.3	95.2	91.5	94.2	106.9	131.8	113.4	106.9	121.8	110.8
09200	PLASTER & GYPSUM BOARD	95.9	85.3	89.3	109.1	85.3	94.3	94.8	45.6	64.2	87.4	87.6	87.5	105.5	122.0	115.8	105.5	122.0	115.8
095,098	CEILINGS & ACOUSTICAL TREATMENT	95.9	85.3	89.6	98.8	85.3	90.7	95.9	45.6	65.9	111.6	87.6	97.3	98.5	122.0	112.5	98.5	122.0	112.5
09600	FLOORING	107.7	68.4	97.1	115.2	68.4	102.5	106.3	68.4	96.1	111.5	96.6	107.5	99.6	120.6	105.3	99.6	120.6	105.3
097,099	WALL FINISHES, PAINTS & COATINGS	105.6	51.3	72.5	115.2	75.0	90.7	105.6	31.9	60.7	115.2	47.9	74.1	91.5	118.5	108.0	91.5	114.1	105.3
09	FINISHES	98.8	76.0	86.9	106.8	79.4	92.5	97.6	48.5	72.0	105.2	83.8	94.1	102.1	122.0	112.5	102.1	121.5	112.2
10 - 14	TOTAL DIV. 10000 - 14000	100.0	88.0	97.4	100.0	89.3	97.7	100.0	79.3	95.6	100.0	90.7	98.0	100.0	115.4	103.3	100.0	115.3	103.3
15	MECHANICAL	100.1	83.1	93.0	100.1	72.7	88.7	100.1	77.5	90.7	100.1	71.2	88.1	100.1	111.8	104.9	100.1	111.8	104.9
16	ELECTRICAL	96.4	94.2	95.3	90.9	63.9	77.8	96.4	94.1	95.3	91.8	80.6	86.3	97.7	109.6	103.5	97.7	105.8	101.6
01 - 16	WEIGHTED AVERAGE	101.1	82.4	92.8	103.6	76.8	91.7	98.7	70.4	86.1	100.0	83.1	92.5	102.4	117.8	109.2	101.6	116.6	108.2

	DIVISION	CONNECTICUT																	
		HARTFORD			NEW BRITAIN			NEW HAVEN			NORWALK			STAMFORD			WATERBURY		
		MAT.	INST.	TOTAL	MAT.	INST.	TOTAL	MAT.	INST.	TOTAL	MAT.	INST.	TOTAL	MAT.	INST.	TOTAL	MAT.	INST.	TOTAL
01590	EQUIPMENT RENTAL	.0	101.5	101.5	.0	101.5	101.5	.0	102.1	102.1	.0	101.5	101.5	.0	101.5	101.5	.0	101.5	101.5
02	SITE CONSTRUCTION	102.8	104.9	104.4	102.6	104.9	104.3	102.4	105.8	104.9	103.0	104.9	104.4	103.7	105.0	104.6	102.8	104.9	104.4
03100	CONCRETE FORMS & ACCESSORIES	101.1	122.8	119.8	102.5	122.8	120.0	102.0	123.1	120.2	102.2	123.4	120.5	102.2	123.6	120.7	102.2	123.1	120.2
03200	CONCRETE REINFORCEMENT	107.5	128.0	117.9	107.5	128.0	117.9	107.5	128.0	117.9	107.5	128.1	117.9	107.5	128.1	118.0	107.5	128.0	117.9
03300	CAST-IN-PLACE CONCRETE	102.4	116.9	108.3	102.9	116.9	108.6	104.7	117.0	109.7	106.3	125.5	114.2	108.1	125.6	115.3	108.1	117.0	111.7
03	CONCRETE	106.5	121.4	113.9	106.8	121.4	114.1	121.5	121.6	121.5	108.5	124.7	116.5	109.3	124.9	117.0	109.3	121.6	115.4
04	MASONRY	101.1	127.8	117.6	101.1	127.8	117.6	101.2	127.8	117.7	101.4	127.6	117.6	101.5	127.6	117.7	101.5	127.8	117.8
05	METALS	106.0	124.1	111.8	97.3	124.1	105.9	97.5	124.4	106.1	100.9	124.7	108.5	100.9	125.0	108.6	100.9	124.5	108.4
06	WOOD & PLASTICS	100.4	122.3	111.8	100.4	122.3	111.8	100.4	122.3	111.8	100.4	122.3	111.8	100.4	122.3	111.8	100.4	122.3	111.8
07	THERMAL & MOISTURE PROTECTION	100.8	122.4	109.9	102.0	123.4	111.0	102.1	123.4	111.0	102.0	127.4	112.7	102.0	127.4	112.7	102.0	123.4	111.0
08	DOORS & WINDOWS	106.9	121.8	110.8	106.9	121.8	110.8	106.9	131.8	113.4	106.9	131.8	113.4	106.9	131.8	113.4	106.9	131.8	113.4
09200	PLASTER & GYPSUM BOARD	105.5	122.0	115.8	105.5	122.0	115.8	105.5	122.0	115.8	105.5	122.0	115.8	105.5	122.0	115.8	105.5	122.0	115.8
095,098	CEILINGS & ACOUSTICAL TREATMENT	98.5	122.0	112.5	98.5	122.0	112.5	98.5	122.0	112.5	98.5	122.0	112.5	98.5	122.0	112.5	98.5	122.0	112.5
09600	FLOORING	99.6	120.6	105.3	99.6	120.6	105.3	99.6	120.6	105.3	99.6	120.6	105.3	99.6	120.6	105.3	99.6	120.6	105.3
097,099	WALL FINISHES, PAINTS & COATINGS	91.5	114.1	105.3	91.5	114.1	105.3	91.5	118.5	108.0	91.5	118.5	108.0	91.5	118.5	108.0	91.5	121.8	110.0
09	FINISHES	102.1	121.5	112.2	102.1	121.5	112.2	102.2	122.0	112.5	102.1	122.0	112.5	102.2	122.0	112.5	102.0	122.4	112.6
10 - 14	TOTAL DIV. 10000 - 14000	100.0	115.4	103.3	100.0	115.3	103.3	100.0	115.4	103.3	100.0	115.4	103.3	100.0	115.5	103.3	100.0	115.4	103.3
15	MECHANICAL	100.1	111.8	104.9	100.1	111.8	104.9	100.1	111.8	104.9	100.1	111.9	105.0	100.1	111.9	105.0	100.1	111.8	104.9
16	ELECTRICAL	97.3	110.7	103.8	97.7	105.8	101.7	97.6	109.7	103.5	97.7	109.6	103.5	97.7	150.4	123.3	97.2	109.6	103.2
01 - 16	WEIGHTED AVERAGE	102.3	117.3	109.0	101.1	116.6	108.0	102.8	117.8	109.4	101.9	118.2	109.1	102.0	124.0	111.8	101.9	117.7	108.9

	DIVISION	D.C.			DELAWARE			FLORIDA											
		WASHINGTON			WILMINGTON			DAYTONA BEACH			FORT LAUDERDALE			JACKSONVILLE			MELBOURNE		
		MAT.	INST.	TOTAL	MAT.	INST.	TOTAL	MAT.	INST.	TOTAL	MAT.	INST.	TOTAL	MAT.	INST.	TOTAL	MAT.	INST.	TOTAL
01590	EQUIPMENT RENTAL	.0	102.3	102.3	.0	118.4	118.4	.0	97.7	97.7	.0	89.3	89.3	.0	97.7	97.7	.0	97.7	97.7
02	SITE CONSTRUCTION	102.8	89.7	93.2	87.4	111.1	104.8	113.4	86.5	93.7	99.5	73.6	80.5	113.5	87.0	94.1	121.5	87.1	96.2
03100	CONCRETE FORMS & ACCESSORIES	98.5	80.8	83.2	96.2	98.5	98.2	98.5	67.8	72.0	96.0	66.3	70.4	98.2	51.8	58.2	93.9	72.0	75.0
03200	CONCRETE REINFORCEMENT	105.5	92.8	99.0	102.3	95.6	98.9	95.1	84.3	89.6	95.1	70.1	82.4	95.1	49.2	71.8	96.2	84.4	90.2
03300	CAST-IN-PLACE CONCRETE	114.5	86.2	102.9	80.5	91.4	85.0	90.4	69.2	81.7	94.9	67.2	83.5	91.3	57.0	77.2	109.0	75.0	95.0
03	CONCRETE	111.9	86.3	99.2	98.6	96.5	97.5	91.1	72.6	81.9	93.1	68.6	80.9	91.5	54.8	73.3	102.4	76.5	89.5
04	MASONRY	95.0	81.4	86.6	111.1	91.8	99.1	87.6	63.7	72.8	87.8	65.6	74.1	87.2	49.9	64.1	86.6	73.5	78.5
05	METALS	104.9	110.0	106.5	102.4	112.6	105.6	94.6	94.3	94.5	94.5	87.4	92.2	94.1	78.2	89.1	103.2	95.0	100.6
06	WOOD & PLASTICS	96.4	81.1	88.4	94.5	100.0	97.4	97.5	70.3	83.3	92.9	63.8	77.7	97.5	50.5	73.0	92.7	70.3	81.0
07	THERMAL & MOISTURE PROTECTION	103.6	84.8	95.7	101.8	105.0	103.2	99.6	68.0	86.3	99.6	72.6	88.2	99.8	57.1	81.8	100.0	75.6	89.7
08	DOORS & WINDOWS	102.7	90.9	99.6	94.3	102.5	96.5	100.7	68.0	92.1	98.3	61.1	88.6	100.7	47.9	86.9	99.9	73.9	93.1
09200	PLASTER & GYPSUM BOARD	108.8	80.6	91.3	110.7	99.9	104.0	100.6	70.0	81.6	100.2	63.4	77.3	100.6	49.7	69.0	96.7	70.0	80.1
095,098	CEILINGS & ACOUSTICAL TREATMENT	103.7	80.6	90.0	99.9	99.9	99.9	104.3	70.0	83.9	104.3	63.4	79.9	104.3	49.7	71.7	99.2	70.0	81.8
09600	FLOORING	115.9	103.1	112.5	85.0	105.3	90.5	118.7	71.8	106.0	118.7	58.3	102.4	118.7	47.9	99.5	115.2	71.8	103.5
097,099	WALL FINISHES, PAINTS & COATINGS	117.2	84.9	97.4	92.8	93.6	93.3	110.6	69.5	85.6	106.7	48.1	71.0	110.6	46.4	71.4	110.6	94.7	100.9
09	FINISHES	104.6	85.0	94.4	101.6	99.2	100.3	108.7	68.8	87.9	106.7	62.3	83.6	108.7	49.8	78.0	106.3	74.0	89.5
10 - 14	TOTAL DIV. 10000 - 14000	100.0	96.1	99.2	100.0	93.2	98.6	100.0	81.6	96.1	100.0	87.9	97.4	100.0	76.7	95.0	100.0	85.3	96.9
15	MECHANICAL	99.9	91.4	96.4	100.3	113.6	105.8	99.9	70.3	87.6	99.9	67.2	86.3	99.9	49.5	79.0	99.9	75.9	89.9
16	ELECTRICAL	99.2	98.3	98.7	96.2	102.2	99.1	96.6	65.4	81.4	96.6	73.4	85.3	96.4	66.9	82.0	97.5	72.1	85.2
01 - 16	WEIGHTED AVERAGE	102.5	91.2	97.5	99.7	103.9	101.5	98.3	72.8	87.0	97.7	70.3	85.5	98.3	59.6	81.1	100.8	77.8	90.6

	DIVISION	FLORIDA																	
		MIAMI			ORLANDO			PANAMA CITY			PENSACOLA			ST. PETERSBURG			TALLAHASSEE		
		MAT.	INST.	TOTAL	MAT.	INST.	TOTAL	MAT.	INST.	TOTAL	MAT.	INST.	TOTAL	MAT.	INST.	TOTAL	MAT.	INST.	TOTAL
01590	**EQUIPMENT RENTAL**	.0	89.3	89.3	.0	97.7	97.7	.0	97.7	97.7	.0	97.7	97.7	.0	97.7	97.7	.0	97.7	97.7
02	**SITE CONSTRUCTION**	98.8	73.5	80.2	114.1	86.2	93.6	127.7	83.6	95.4	125.4	86.1	96.6	114.4	85.5	93.2	114.8	85.6	93.4
03100	CONCRETE FORMS & ACCESSORIES	96.3	66.4	70.5	98.2	69.4	73.3	97.3	26.9	36.5	87.5	49.5	54.8	95.8	45.8	52.6	98.2	38.0	46.3
03200	CONCRETE REINFORCEMENT	95.1	70.1	82.4	95.1	80.8	87.8	99.3	48.1	73.3	101.8	48.5	74.7	98.5	55.4	76.6	95.1	48.7	71.5
03300	CAST-IN-PLACE CONCRETE	92.4	67.4	82.1	98.3	69.1	86.3	96.0	33.5	70.3	96.0	53.2	78.4	102.3	53.5	82.2	94.7	47.0	75.1
03	**CONCRETE**	91.9	68.7	80.4	92.7	72.6	82.8	99.7	34.9	67.5	98.7	52.3	75.7	97.2	52.0	74.8	93.1	45.1	69.3
04	**MASONRY**	86.2	66.0	73.7	91.3	63.7	74.2	92.0	26.8	51.6	89.1	48.4	63.9	136.3	47.8	81.4	90.2	37.7	57.6
05	**METALS**	99.6	87.7	95.8	100.5	92.7	98.0	94.4	63.8	84.6	94.3	77.5	89.0	97.7	78.3	91.5	87.8	76.6	84.2
06	**WOOD & PLASTICS**	92.9	63.8	77.7	97.5	72.8	84.7	96.3	26.9	60.2	85.8	49.8	67.1	94.6	45.7	69.2	95.9	36.2	64.8
07	**THERMAL & MOISTURE PROTECTION**	103.3	69.4	89.0	99.8	68.3	86.5	100.0	30.5	70.8	99.8	51.3	79.4	99.5	48.2	77.9	99.8	45.2	76.8
08	**DOORS & WINDOWS**	98.3	61.7	88.7	100.7	67.7	92.1	98.3	25.7	79.3	98.3	48.4	85.2	99.3	44.8	85.1	99.3	39.6	83.7
09200	PLASTER & GYPSUM BOARD	100.2	63.4	77.3	103.0	72.7	84.1	98.7	25.5	53.2	94.6	49.0	66.3	97.9	44.8	64.9	100.6	35.0	59.9
095,098	CEILINGS & ACOUSTICAL TREATMENT	104.3	63.4	79.9	104.3	72.7	85.4	98.3	25.5	54.9	98.3	49.0	68.9	98.3	44.8	66.4	104.3	35.0	63.0
09600	FLOORING	126.0	59.0	107.9	118.7	71.8	106.0	118.1	18.4	91.1	112.0	50.9	95.5	117.0	50.7	99.1	118.7	37.2	96.7
097,099	WALL FINISHES, PAINTS & COATINGS	106.7	48.1	71.0	110.6	53.3	75.6	110.6	23.7	57.6	110.6	53.9	76.0	110.6	44.9	70.5	110.6	38.2	66.4
09	**FINISHES**	108.8	62.4	84.7	109.0	68.4	87.9	107.8	24.5	64.4	105.2	50.1	76.5	106.3	46.4	75.1	108.8	36.9	71.4
10 - 14	**TOTAL DIV. 10000 - 14000**	100.0	87.9	97.4	100.0	81.8	96.1	100.0	42.9	87.8	100.0	48.0	88.9	100.0	50.2	89.4	100.0	72.9	94.2
15	**MECHANICAL**	99.9	73.8	89.1	99.9	65.8	85.8	99.9	24.3	68.5	99.9	48.2	78.5	99.9	48.0	78.4	99.9	38.5	74.4
16	**ELECTRICAL**	97.5	71.6	84.9	97.3	43.5	71.2	95.6	32.6	64.9	98.7	62.5	81.1	96.9	47.0	72.6	97.0	40.3	69.4
01 - 16	**WEIGHTED AVERAGE**	98.7	71.4	86.6	99.7	68.6	85.9	99.4	36.8	71.6	99.1	57.2	80.5	101.6	54.2	80.5	97.6	48.2	75.6

	DIVISION	FLORIDA			GEORGIA														
		TAMPA			ALBANY			ATLANTA			AUGUSTA			COLUMBUS			MACON		
		MAT.	INST.	TOTAL	MAT.	INST.	TOTAL	MAT.	INST.	TOTAL	MAT.	INST.	TOTAL	MAT.	INST.	TOTAL	MAT.	INST.	TOTAL
01590	**EQUIPMENT RENTAL**	.0	97.7	97.7	.0	90.1	90.1	.0	92.8	92.8	.0	92.2	92.2	.0	90.1	90.1	.0	102.6	102.6
02	**SITE CONSTRUCTION**	114.6	86.1	93.7	99.7	76.0	82.3	99.4	95.7	96.7	96.0	92.4	93.4	99.6	76.8	82.9	100.4	93.9	95.6
03100	CONCRETE FORMS & ACCESSORIES	99.7	72.7	76.4	97.5	44.6	51.9	98.1	76.6	79.6	95.1	50.2	56.4	97.4	57.1	62.6	96.8	54.1	59.9
03200	CONCRETE REINFORCEMENT	95.1	87.1	91.1	94.7	90.8	92.7	98.9	92.6	95.7	100.0	74.6	87.1	95.1	91.9	93.5	96.3	91.1	93.7
03300	CAST-IN-PLACE CONCRETE	100.0	59.7	83.4	96.6	44.8	75.3	99.0	72.8	88.2	93.5	47.3	74.5	96.2	53.7	78.7	94.9	44.8	74.3
03	**CONCRETE**	95.8	72.0	84.0	93.8	54.9	74.5	100.6	78.3	89.5	96.1	54.3	75.4	93.7	63.8	78.9	93.2	59.3	76.4
04	**MASONRY**	88.7	73.5	79.2	90.6	38.7	58.4	90.7	68.6	77.0	90.9	39.1	58.8	89.8	55.6	68.6	103.8	37.4	62.6
05	**METALS**	97.8	93.0	96.3	93.8	87.3	91.7	91.8	80.4	88.1	90.6	67.8	83.3	93.4	92.0	92.9	89.2	90.6	89.7
06	**WOOD & PLASTICS**	99.0	75.0	86.5	96.3	41.2	67.6	97.0	78.4	87.3	93.9	51.7	71.9	96.3	56.8	75.7	103.8	55.8	78.8
07	**THERMAL & MOISTURE PROTECTION**	99.8	60.6	83.3	96.6	53.4	78.4	101.8	74.9	90.5	101.3	48.4	79.0	96.3	61.0	81.5	95.1	56.1	78.7
08	**DOORS & WINDOWS**	100.7	68.7	92.3	98.3	48.9	85.4	99.6	75.9	93.4	93.8	52.8	83.1	98.3	59.3	88.1	96.7	57.8	86.5
09200	PLASTER & GYPSUM BOARD	100.6	74.9	84.6	100.3	40.2	63.0	114.0	78.3	91.8	112.9	50.8	74.3	100.3	56.2	72.9	105.8	55.1	74.3
095,098	CEILINGS & ACOUSTICAL TREATMENT	104.3	74.9	86.8	103.4	40.2	65.7	115.5	78.3	93.3	116.4	50.8	77.3	103.4	56.2	75.3	99.5	55.1	73.0
09600	FLOORING	118.7	50.7	100.3	118.7	31.1	95.0	83.9	74.6	81.4	82.7	39.6	71.1	118.7	52.4	100.8	92.3	36.5	77.2
097,099	WALL FINISHES, PAINTS & COATINGS	110.6	44.9	70.5	106.7	39.9	65.9	90.7	81.5	85.1	90.7	37.9	58.5	106.7	47.4	70.5	108.5	46.7	70.8
09	**FINISHES**	108.7	66.1	86.5	106.5	40.2	71.9	98.5	76.7	87.2	98.1	46.6	71.3	106.4	54.4	79.3	94.6	49.4	71.0
10 - 14	**TOTAL DIV. 10000 - 14000**	100.0	84.0	96.6	100.0	76.8	95.0	100.0	83.3	96.4	100.0	63.0	92.1	100.0	78.6	95.4	100.0	76.8	95.1
15	**MECHANICAL**	99.9	71.3	88.0	99.9	47.9	78.3	100.1	78.1	91.0	100.1	63.8	85.0	99.9	48.3	78.5	99.9	40.4	75.2
16	**ELECTRICAL**	96.5	47.0	72.4	92.2	65.0	78.9	95.5	80.1	88.0	95.7	48.1	72.6	92.6	49.7	71.7	90.3	66.4	78.7
01 - 16	**WEIGHTED AVERAGE**	99.4	70.8	86.7	97.2	56.3	79.1	97.7	79.0	89.4	96.3	57.5	79.0	97.2	60.4	80.8	95.7	59.1	79.4

	DIVISION	GEORGIA						HAWAII			IDAHO								
		SAVANNAH			VALDOSTA			HONOLULU			BOISE			LEWISTON			POCATELLO		
		MAT.	INST.	TOTAL	MAT.	INST.	TOTAL	MAT.	INST.	TOTAL	MAT.	INST.	TOTAL	MAT.	INST.	TOTAL	MAT.	INST.	TOTAL
01590	**EQUIPMENT RENTAL**	.0	91.2	91.2	.0	90.1	90.1	.0	99.4	99.4	.0	101.7	101.7	.0	94.6	94.6	.0	101.7	101.7
02	**SITE CONSTRUCTION**	100.2	76.9	83.1	109.7	76.2	85.1	139.5	106.4	115.2	78.2	102.4	95.9	82.1	96.7	92.8	79.3	102.4	96.2
03100	CONCRETE FORMS & ACCESSORIES	97.7	51.2	57.6	84.3	44.3	49.8	107.5	142.9	138.1	97.6	81.3	83.5	111.5	68.0	74.0	97.7	80.8	83.1
03200	CONCRETE REINFORCEMENT	96.0	74.8	85.3	97.0	42.3	69.2	108.6	126.2	117.5	108.2	76.7	92.2	115.7	97.3	106.4	108.5	76.2	92.1
03300	CAST-IN-PLACE CONCRETE	93.2	49.9	75.4	94.5	50.5	76.4	201.5	127.9	171.3	94.2	88.1	91.7	102.2	92.7	98.3	93.5	87.9	91.2
03	**CONCRETE**	92.4	56.6	74.6	97.6	47.7	72.9	154.7	133.3	144.1	104.3	82.7	93.6	112.1	82.3	97.3	100.9	82.3	91.7
04	**MASONRY**	92.6	52.3	67.6	95.9	47.6	65.9	133.8	128.9	130.8	133.5	72.3	95.6	134.7	85.0	103.9	129.6	64.5	89.2
05	**METALS**	89.5	82.1	87.2	93.0	70.5	85.8	144.4	110.9	133.7	99.6	76.1	92.1	92.6	87.3	90.9	107.7	75.0	97.3
06	**WOOD & PLASTICS**	109.5	48.7	77.8	81.8	40.2	60.1	98.9	146.9	123.9	95.4	81.3	88.0	103.3	61.8	81.7	95.4	81.3	88.0
07	**THERMAL & MOISTURE PROTECTION**	96.7	51.5	77.7	96.5	56.7	79.7	107.5	128.9	116.5	92.7	78.0	86.5	135.9	79.7	112.2	92.9	68.8	82.7
08	**DOORS & WINDOWS**	99.4	49.2	86.3	93.7	37.3	79.0	105.8	137.9	114.2	94.7	76.6	89.9	114.5	67.6	102.3	94.7	70.1	88.2
09200	PLASTER & GYPSUM BOARD	100.6	47.8	67.8	92.0	39.1	59.2	120.6	148.0	137.6	85.5	80.6	82.5	141.8	60.7	91.4	85.5	80.6	82.5
095,098	CEILINGS & ACOUSTICAL TREATMENT	104.3	47.8	70.6	98.3	39.1	63.0	120.4	148.0	136.9	110.7	80.6	92.7	131.5	60.7	89.3	111.6	80.6	93.1
09600	FLOORING	118.7	46.7	99.2	110.3	37.3	90.5	168.2	132.6	158.6	99.0	53.8	86.7	136.4	91.3	124.2	99.3	53.8	87.0
097,099	WALL FINISHES, PAINTS & COATINGS	107.5	47.4	70.8	106.7	34.6	62.7	104.0	148.6	131.2	102.2	50.3	70.5	127.5	74.7	95.3	102.2	52.4	71.8
09	**FINISHES**	107.0	49.6	77.1	102.1	41.3	70.4	134.5	143.0	138.9	99.0	73.2	85.6	157.5	71.9	112.9	99.4	73.4	85.9
10 - 14	**TOTAL DIV. 10000 - 14000**	100.0	65.9	92.7	100.0	64.7	92.5	100.0	123.5	105.0	100.0	80.8	95.9	100.0	82.8	96.3	100.0	80.8	95.9
15	**MECHANICAL**	99.9	46.1	77.6	99.9	40.6	75.3	100.2	116.5	107.0	100.1	75.5	89.9	101.2	89.1	96.2	100.1	75.4	89.8
16	**ELECTRICAL**	94.1	61.9	78.4	90.8	29.5	61.0	108.9	123.3	115.9	96.0	77.0	86.8	81.3	83.6	82.4	91.1	76.1	83.8
01 - 16	**WEIGHTED AVERAGE**	97.0	57.7	79.6	96.9	47.6	75.0	120.6	124.8	122.5	100.2	78.9	90.7	107.4	83.6	96.8	100.3	77.3	90.1

City Cost Indexes

	DIVISION	IDAHO			ILLINOIS														
		TWIN FALLS			CHICAGO			DECATUR			EAST ST. LOUIS			JOLIET			PEORIA		
		MAT.	INST.	TOTAL	MAT.	INST.	TOTAL	MAT.	INST.	TOTAL	MAT.	INST.	TOTAL	MAT.	INST.	TOTAL	MAT.	INST.	TOTAL
01590	EQUIPMENT RENTAL	.0	101.7	101.7	.0	91.5	91.5	.0	102.4	102.4	.0	109.1	109.1	.0	89.3	89.3	.0	101.6	101.6
02	SITE CONSTRUCTION	85.8	100.3	96.5	99.1	90.8	93.0	89.3	96.0	94.2	105.4	96.4	98.8	99.0	89.6	92.1	97.9	95.1	95.8
03100	CONCRETE FORMS & ACCESSORIES	98.5	35.2	43.9	99.7	137.6	132.4	97.3	102.2	101.5	91.0	110.2	107.6	100.6	138.1	133.0	95.6	105.0	103.7
03200	CONCRETE REINFORCEMENT	110.4	75.9	92.9	100.5	146.9	124.1	97.4	100.7	99.1	99.2	99.0	99.1	100.5	123.7	112.3	97.4	96.4	96.9
03300	CAST-IN-PLACE CONCRETE	95.9	44.2	74.7	113.0	137.9	123.2	101.0	102.6	101.7	94.5	105.3	99.0	112.9	128.4	119.3	98.7	103.8	100.8
03	CONCRETE	108.5	46.9	78.0	110.4	138.5	124.4	98.7	102.4	100.5	89.7	107.3	98.4	110.4	131.0	120.6	97.5	103.3	100.4
04	MASONRY	132.8	37.8	73.9	93.7	136.8	120.4	73.9	106.8	94.3	74.6	111.5	97.5	94.7	127.6	115.1	115.2	111.9	113.1
05	METALS	107.7	72.6	96.5	98.5	124.9	106.9	98.7	106.9	101.3	96.4	117.1	103.0	96.2	112.5	101.4	98.7	106.0	101.0
06	WOOD & PLASTICS	96.0	34.1	63.7	101.6	135.2	119.1	99.7	99.0	99.3	95.3	111.2	103.6	103.2	141.5	123.1	99.7	101.1	100.4
07	THERMAL & MOISTURE PROTECTION	93.7	45.4	73.3	101.2	129.7	113.2	94.7	99.9	96.9	89.5	101.9	94.7	100.6	123.4	110.2	94.9	103.9	98.7
08	DOORS & WINDOWS	97.9	41.9	83.3	103.2	139.4	112.6	96.3	100.1	97.3	86.1	111.7	92.8	101.1	136.6	110.4	96.3	99.1	97.0
09200	PLASTER & GYPSUM BOARD	85.0	32.1	52.1	101.3	136.0	122.9	105.6	98.8	101.4	102.0	111.3	107.8	98.7	142.5	125.9	105.6	100.9	102.7
095,098	CEILINGS & ACOUSTICAL TREATMENT	103.9	32.1	61.1	104.2	136.0	123.2	95.5	98.8	97.4	88.6	111.3	102.2	104.2	142.5	127.1	95.5	100.9	98.7
09600	FLOORING	100.6	53.8	87.9	83.5	127.9	95.6	96.7	105.6	99.1	105.1	118.0	108.6	83.1	127.9	95.2	96.7	107.5	99.6
097,099	WALL FINISHES, PAINTS & COATINGS	102.2	30.9	58.7	78.9	134.3	112.7	90.8	101.7	97.4	98.8	101.9	100.6	76.7	132.0	110.5	90.8	93.0	92.1
09	FINISHES	98.3	37.2	66.5	92.6	135.7	115.0	95.8	102.9	99.5	94.4	110.0	102.6	91.9	136.9	115.3	95.9	104.3	100.3
10 - 14	TOTAL DIV. 10000 - 14000	100.0	45.5	88.4	100.0	123.8	105.1	100.0	88.6	97.6	100.0	90.2	97.9	100.0	120.9	104.5	100.0	88.8	97.6
15	MECHANICAL	100.1	62.1	84.3	100.0	126.1	110.8	100.0	106.2	102.6	99.9	92.3	96.8	100.1	113.4	105.6	100.0	98.4	99.3
16	ELECTRICAL	83.3	69.9	76.8	98.8	117.1	107.7	98.7	89.1	94.0	95.9	89.5	92.8	97.1	103.8	100.4	98.0	86.6	92.4
01 - 16	WEIGHTED AVERAGE	100.8	57.7	81.7	100.1	126.3	111.7	97.0	101.1	98.8	94.5	102.0	97.8	99.4	118.5	107.9	99.1	99.9	99.4

	DIVISION	ILLINOIS						INDIANA											
		ROCKFORD			SPRINGFIELD			ANDERSON			BLOOMINGTON			EVANSVILLE			FORT WAYNE		
		MAT.	INST.	TOTAL	MAT.	INST.	TOTAL	MAT.	INST.	TOTAL	MAT.	INST.	TOTAL	MAT.	INST.	TOTAL	MAT.	INST.	TOTAL
01590	EQUIPMENT RENTAL	.0	101.6	101.6	.0	102.4	102.4	.0	96.7	96.7	.0	85.9	85.9	.0	121.3	121.3	.0	96.7	96.7
02	SITE CONSTRUCTION	97.1	95.7	96.1	95.3	96.0	95.8	88.2	96.4	94.2	77.0	95.8	90.8	82.0	130.1	117.2	89.2	96.5	94.6
03100	CONCRETE FORMS & ACCESSORIES	99.9	117.1	114.8	99.2	102.5	102.0	96.4	79.9	82.2	100.5	80.7	83.4	93.1	82.2	83.7	95.0	77.8	80.2
03200	CONCRETE REINFORCEMENT	97.4	125.2	111.5	97.4	96.8	97.1	94.9	80.7	87.7	84.1	80.4	82.2	92.2	80.4	86.2	94.9	77.5	86.0
03300	CAST-IN-PLACE CONCRETE	101.0	96.1	99.0	93.8	103.0	97.6	101.8	82.0	93.6	100.8	79.5	92.0	96.3	92.0	94.5	108.1	80.4	96.8
03	CONCRETE	98.9	111.6	105.2	95.4	102.0	98.6	96.2	81.3	88.8	102.7	79.9	91.4	102.9	85.4	94.2	99.1	79.2	89.3
04	MASONRY	89.2	107.7	100.7	75.5	107.7	95.4	92.3	84.9	87.7	90.3	80.3	84.1	86.6	86.6	86.6	92.4	84.7	87.6
05	METALS	98.7	121.1	105.8	101.3	105.2	102.5	91.6	88.7	90.7	94.4	75.7	88.4	87.1	86.3	86.8	91.6	87.2	90.2
06	WOOD & PLASTICS	99.7	117.4	108.9	102.3	99.0	100.6	113.1	79.2	95.4	118.2	79.7	98.2	97.3	79.8	88.2	112.9	76.1	93.7
07	THERMAL & MOISTURE PROTECTION	97.9	112.6	104.1	94.4	100.9	97.1	100.3	82.3	92.7	91.4	84.3	88.4	95.5	88.1	92.4	100.1	86.1	94.2
08	DOORS & WINDOWS	96.3	115.7	101.4	96.3	99.1	97.0	98.9	80.9	94.2	103.1	81.2	97.4	95.6	80.4	91.7	98.9	76.0	92.9
09200	PLASTER & GYPSUM BOARD	105.6	117.7	113.1	105.6	98.8	101.4	96.6	79.2	85.8	97.0	79.9	86.4	91.6	78.5	83.4	95.8	76.1	83.6
095,098	CEILINGS & ACOUSTICAL TREATMENT	95.5	117.7	108.7	95.5	98.8	97.4	91.5	79.2	84.2	81.6	79.9	80.6	86.2	78.5	81.6	91.5	76.1	82.3
09600	FLOORING	96.7	89.3	94.7	96.9	105.6	99.3	88.2	92.7	89.4	104.5	92.7	101.3	97.5	88.7	95.1	88.2	80.8	86.2
097,099	WALL FINISHES, PAINTS & COATINGS	90.8	104.9	99.4	90.8	101.7	97.4	91.5	77.0	82.6	90.2	89.0	89.5	96.8	88.0	91.4	91.5	80.1	84.5
09	FINISHES	95.9	110.9	103.7	95.9	103.2	99.7	89.7	81.9	85.7	94.7	83.7	89.0	92.8	83.7	88.1	89.5	78.2	83.6
10 - 14	TOTAL DIV. 10000 - 14000	100.0	98.0	99.6	100.0	88.9	97.6	100.0	90.7	98.0	100.0	90.2	97.9	100.0	95.3	99.0	100.0	89.3	97.7
15	MECHANICAL	100.1	101.9	100.8	100.0	101.9	100.8	99.9	78.8	91.2	99.7	84.9	93.6	99.9	86.8	94.5	99.9	81.2	92.2
16	ELECTRICAL	98.1	107.4	102.6	98.6	86.9	92.9	85.1	90.6	87.8	101.6	88.0	95.0	95.0	87.6	91.4	86.2	82.9	84.6
01 - 16	WEIGHTED AVERAGE	98.1	107.9	102.4	97.3	99.8	98.4	94.9	84.8	90.4	98.2	84.1	91.9	95.3	90.0	93.0	95.4	83.1	89.9

	DIVISION	INDIANA															IOWA		
		GARY			INDIANAPOLIS			MUNCIE			SOUTH BEND			TERRE HAUTE			CEDAR RAPIDS		
		MAT.	INST.	TOTAL	MAT.	INST.	TOTAL	MAT.	INST.	TOTAL	MAT.	INST.	TOTAL	MAT.	INST.	TOTAL	MAT.	INST.	TOTAL
01590	EQUIPMENT RENTAL	.0	96.7	96.7	.0	91.5	91.5	.0	96.2	96.2	.0	106.8	106.8	.0	121.3	121.3	.0	95.6	95.6
02	SITE CONSTRUCTION	88.7	99.2	96.4	87.9	100.1	96.9	77.2	95.8	90.8	88.2	96.2	94.1	83.3	130.0	117.6	89.7	94.6	93.3
03100	CONCRETE FORMS & ACCESSORIES	96.5	106.3	105.0	97.0	84.4	86.1	90.9	79.7	81.2	100.9	81.5	84.2	94.6	82.2	83.9	102.1	80.6	83.6
03200	CONCRETE REINFORCEMENT	94.9	100.3	97.6	94.2	82.2	88.1	93.0	80.7	86.8	94.9	76.7	85.6	92.2	80.9	86.4	93.3	84.5	88.8
03300	CAST-IN-PLACE CONCRETE	106.3	111.1	108.3	97.9	85.0	92.6	106.0	81.4	95.9	99.0	85.1	93.3	93.2	91.2	92.4	108.0	78.2	95.8
03	CONCRETE	98.4	106.9	102.6	98.1	83.8	91.0	101.4	81.0	91.3	91.1	83.2	87.2	105.7	85.2	95.5	100.3	81.1	90.8
04	MASONRY	93.7	107.8	102.5	99.6	84.4	90.1	92.6	84.9	87.8	90.6	84.6	86.9	93.4	85.7	88.6	106.7	79.6	89.9
05	METALS	91.6	103.7	95.5	93.5	78.2	88.6	96.1	89.0	93.8	91.6	101.2	94.7	87.8	87.0	87.6	87.8	92.8	89.4
06	WOOD & PLASTICS	111.1	106.1	108.5	110.1	83.8	96.4	110.3	79.0	94.0	113.0	80.3	96.0	99.6	81.2	90.0	109.0	80.0	93.9
07	THERMAL & MOISTURE PROTECTION	99.6	105.8	102.2	98.5	86.4	93.4	93.7	82.3	88.9	99.4	87.4	94.3	95.6	85.1	91.1	98.9	81.8	91.7
08	DOORS & WINDOWS	98.9	104.4	100.4	106.7	83.8	100.8	98.0	80.8	93.5	92.3	79.8	89.0	96.2	82.0	92.5	99.3	84.7	95.5
09200	PLASTER & GYPSUM BOARD	91.7	106.9	101.1	94.8	83.7	87.9	91.3	79.2	83.8	96.2	80.3	86.3	91.6	79.9	84.3	103.7	79.8	88.9
095,098	CEILINGS & ACOUSTICAL TREATMENT	91.5	106.9	100.7	95.8	83.7	88.6	82.5	79.2	80.5	91.5	80.3	84.9	86.2	79.9	82.5	114.7	79.8	93.9
09600	FLOORING	88.2	119.9	96.8	88.0	92.7	89.3	96.6	92.7	95.5	88.2	85.3	87.4	97.5	97.2	97.4	122.3	55.6	104.2
097,099	WALL FINISHES, PAINTS & COATINGS	91.5	107.9	101.5	91.5	89.0	90.0	90.2	77.0	82.2	91.5	83.8	86.8	96.8	93.3	94.7	105.5	76.5	87.8
09	FINISHES	88.9	109.2	99.5	90.8	86.4	88.5	91.4	81.8	86.4	89.7	82.4	85.9	92.8	86.1	89.4	111.9	74.6	92.5
10 - 14	TOTAL DIV. 10000 - 14000	100.0	94.3	98.8	100.0	92.4	98.4	100.0	90.3	97.9	100.0	90.2	97.9	100.0	94.4	98.8	100.0	90.4	98.0
15	MECHANICAL	99.9	101.3	100.5	99.9	87.7	94.9	99.7	78.7	91.0	99.9	82.8	92.8	99.9	76.9	90.4	100.3	85.9	94.3
16	ELECTRICAL	94.4	104.1	99.1	104.4	90.6	97.7	89.7	74.6	82.4	99.2	88.6	94.1	93.5	88.0	90.9	96.1	85.5	90.9
01 - 16	WEIGHTED AVERAGE	96.2	104.3	99.8	98.8	87.2	93.7	96.2	82.4	90.1	95.3	86.8	91.5	96.0	88.3	92.6	98.9	84.4	92.5

DIVISION		IOWA																	
		COUNCIL BLUFFS			DAVENPORT			DES MOINES			DUBUQUE			SIOUX CITY			WATERLOO		
		MAT.	INST.	TOTAL	MAT.	INST.	TOTAL	MAT.	INST.	TOTAL	MAT.	INST.	TOTAL	MAT.	INST.	TOTAL	MAT.	INST.	TOTAL
01590	**EQUIPMENT RENTAL**	.0	95.3	95.3	.0	99.6	99.6	.0	101.6	101.6	.0	94.2	94.2	.0	99.6	99.6	.0	99.6	99.6
02	**SITE CONSTRUCTION**	95.4	91.4	92.4	87.8	99.1	96.1	80.9	99.9	94.9	87.7	91.6	90.5	97.3	95.5	95.9	88.3	94.5	92.9
03100	CONCRETE FORMS & ACCESSORIES	80.2	56.9	60.1	101.7	93.1	94.3	103.7	80.4	83.6	82.0	69.2	70.9	102.1	69.0	73.5	102.6	48.1	55.6
03200	CONCRETE REINFORCEMENT	95.2	78.9	86.9	93.3	95.5	94.4	93.3	79.6	86.3	92.0	84.2	88.1	93.3	66.8	79.9	93.3	84.0	88.6
03300	CAST-IN-PLACE CONCRETE	112.4	72.1	95.8	104.0	99.0	102.0	108.3	76.5	95.2	105.8	91.0	99.7	107.1	56.5	86.3	108.0	47.1	83.0
03	**CONCRETE**	102.1	67.5	84.9	98.3	96.0	97.2	99.4	79.6	89.6	96.6	80.4	88.6	99.8	65.2	82.6	100.3	55.9	78.3
04	**MASONRY**	106.4	76.0	87.6	102.7	91.8	96.0	100.5	77.9	86.5	107.6	73.9	86.7	99.7	59.4	74.7	101.3	60.8	76.2
05	**METALS**	93.0	89.3	91.8	87.8	101.9	92.3	87.9	91.9	89.2	86.4	91.9	88.2	87.8	82.9	86.2	87.8	90.5	88.7
06	**WOOD & PLASTICS**	84.8	52.8	68.1	109.0	91.5	99.9	110.5	80.3	94.8	86.8	66.9	76.5	109.0	69.3	88.3	109.7	47.8	77.5
07	**THERMAL & MOISTURE PROTECTION**	98.2	66.9	85.0	98.4	92.1	95.7	99.1	78.4	90.4	98.5	70.3	86.6	98.4	57.6	81.2	98.2	52.9	79.1
08	**DOORS & WINDOWS**	98.3	60.4	88.4	99.3	92.4	97.5	99.3	83.5	95.1	98.3	77.6	92.9	99.3	66.4	90.7	95.0	60.8	86.0
09200	PLASTER & GYPSUM BOARD	93.9	51.9	67.8	103.7	91.3	96.0	100.9	79.9	87.8	94.3	66.4	76.9	103.7	68.5	81.9	103.7	46.4	68.1
095,098	CEILINGS & ACOUSTICAL TREATMENT	109.6	51.9	75.1	114.7	91.3	100.8	113.0	79.9	93.2	109.6	66.4	83.8	114.7	68.5	87.1	114.7	46.4	74.0
09600	FLOORING	97.3	45.5	83.3	108.8	84.1	102.1	108.6	41.5	90.4	111.1	38.7	91.5	109.3	58.6	95.6	110.5	58.9	96.5
097,099	WALL FINISHES, PAINTS & COATINGS	97.2	69.5	80.3	101.4	93.3	96.5	101.4	78.7	87.6	104.4	76.5	87.4	102.6	66.0	80.3	102.6	35.5	61.7
09	**FINISHES**	101.2	54.5	76.9	107.7	91.1	99.1	105.9	72.4	88.4	105.7	63.0	83.5	109.0	66.8	87.0	108.3	47.9	76.8
10 - 14	**TOTAL DIV. 10000 - 14000**	100.0	69.3	93.4	100.0	95.1	98.9	100.0	90.9	98.1	100.0	87.6	97.4	100.0	87.6	97.4	100.0	80.5	95.9
15	**MECHANICAL**	100.3	79.0	91.5	100.3	95.3	98.2	100.3	80.2	92.0	100.3	78.9	91.4	100.3	82.4	92.9	100.3	45.2	77.4
16	**ELECTRICAL**	101.0	85.5	93.4	94.7	93.0	93.9	96.6	81.2	89.1	99.6	68.6	84.6	96.1	78.0	87.3	96.1	67.0	82.0
01 - 16	**WEIGHTED AVERAGE**	99.3	75.2	88.6	97.9	94.9	96.5	97.8	82.2	90.9	97.8	77.2	88.7	98.4	74.9	88.0	97.8	62.0	81.9

DIVISION		KANSAS															KENTUCKY		
		DODGE CITY			KANSAS CITY			SALINA			TOPEKA			WICHITA			BOWLING GREEN		
		MAT.	INST.	TOTAL	MAT.	INST.	TOTAL	MAT.	INST.	TOTAL	MAT.	INST.	TOTAL	MAT.	INST.	TOTAL	MAT.	INST.	TOTAL
01590	**EQUIPMENT RENTAL**	.0	103.6	103.6	.0	100.1	100.1	.0	103.6	103.6	.0	101.6	101.6	.0	103.6	103.6	.0	96.0	96.0
02	**SITE CONSTRUCTION**	112.0	92.9	98.0	93.1	91.3	91.8	101.9	93.3	95.6	94.9	90.6	91.8	96.0	93.2	94.0	69.6	100.2	92.0
03100	CONCRETE FORMS & ACCESSORIES	93.4	67.9	71.4	99.3	94.4	95.1	89.2	47.9	53.6	98.6	44.2	51.7	95.4	56.4	61.7	84.1	83.2	83.3
03200	CONCRETE REINFORCEMENT	105.4	57.3	81.0	100.1	87.4	93.7	104.8	79.5	92.0	97.4	95.0	96.2	97.4	80.0	88.5	83.5	71.7	77.5
03300	CAST-IN-PLACE CONCRETE	117.3	56.4	92.2	91.9	96.1	93.6	101.7	48.2	79.7	93.0	52.4	76.3	88.6	57.9	76.0	87.9	81.4	85.2
03	**CONCRETE**	115.8	63.0	89.6	97.0	94.0	95.5	102.8	55.5	79.4	94.9	58.2	76.7	92.6	62.6	77.7	94.4	80.5	87.5
04	**MASONRY**	104.1	57.6	75.3	104.3	97.0	99.8	118.6	50.5	76.4	98.6	62.3	76.1	92.5	65.4	75.7	93.9	78.7	84.5
05	**METALS**	96.1	79.8	90.9	100.9	96.4	99.5	95.9	89.4	93.8	101.3	97.4	100.0	101.3	88.8	97.3	92.1	78.1	87.6
06	**WOOD & PLASTICS**	92.3	75.4	83.5	99.4	93.4	96.3	88.4	48.6	67.7	96.4	40.0	67.0	94.3	55.1	73.9	92.9	82.8	87.6
07	**THERMAL & MOISTURE PROTECTION**	95.4	62.2	81.4	93.2	98.6	95.5	95.0	57.2	79.1	94.9	69.3	84.1	94.6	65.2	82.2	85.8	79.1	83.0
08	**DOORS & WINDOWS**	96.1	63.5	87.6	95.1	90.3	93.9	96.1	52.9	84.8	96.3	59.1	86.6	96.3	61.7	87.3	95.3	75.5	90.1
09200	PLASTER & GYPSUM BOARD	102.3	74.6	85.1	98.8	93.1	95.2	101.6	47.1	67.7	103.6	38.3	63.0	103.6	53.7	72.6	87.5	82.6	84.5
095,098	CEILINGS & ACOUSTICAL TREATMENT	86.9	74.6	79.6	87.8	93.1	91.0	86.9	47.1	63.2	87.8	38.3	58.2	87.8	53.7	67.5	86.2	82.6	84.1
09600	FLOORING	96.0	48.9	83.3	87.2	103.1	91.5	93.9	35.7	78.2	97.5	45.5	83.4	96.7	78.7	91.8	93.2	60.3	84.3
097,099	WALL FINISHES, PAINTS & COATINGS	90.8	48.2	64.8	98.8	86.4	91.2	90.8	38.0	58.5	90.8	67.0	76.3	90.8	61.2	72.7	96.8	74.4	83.1
09	**FINISHES**	94.0	63.1	77.9	92.3	94.0	93.2	92.7	43.4	67.0	93.8	44.5	68.1	93.7	60.7	76.5	91.1	78.1	84.3
10 - 14	**TOTAL DIV. 10000 - 14000**	100.0	51.1	89.6	100.0	80.7	95.9	100.0	62.0	91.9	100.0	65.3	92.6	100.0	65.9	92.7	100.0	60.8	91.6
15	**MECHANICAL**	100.0	64.3	85.2	99.9	95.9	98.2	100.0	64.2	85.2	100.0	71.6	88.2	100.0	65.4	85.7	99.9	86.6	94.4
16	**ELECTRICAL**	99.4	74.8	87.4	104.3	106.8	105.5	99.1	70.8	85.4	103.0	78.7	91.2	101.3	70.8	86.5	93.9	85.5	89.8
01 - 16	**WEIGHTED AVERAGE**	100.4	68.4	86.2	98.9	96.1	97.7	99.2	63.9	83.5	98.7	69.3	85.6	97.9	69.5	85.3	94.6	82.7	89.3

DIVISION		KENTUCKY									LOUISIANA								
		LEXINGTON			LOUISVILLE			OWENSBORO			ALEXANDRIA			BATON ROUGE			LAKE CHARLES		
		MAT.	INST.	TOTAL	MAT.	INST.	TOTAL	MAT.	INST.	TOTAL	MAT.	INST.	TOTAL	MAT.	INST.	TOTAL	MAT.	INST.	TOTAL
01590	**EQUIPMENT RENTAL**	.0	103.5	103.5	.0	96.0	96.0	.0	121.3	121.3	.0	86.7	86.7	.0	86.2	86.2	.0	86.2	86.2
02	**SITE CONSTRUCTION**	77.3	102.2	95.6	67.1	100.0	91.2	81.9	129.3	116.6	102.0	85.5	89.9	107.4	85.0	91.0	108.9	84.8	91.2
03100	CONCRETE FORMS & ACCESSORIES	98.1	70.4	74.2	92.8	82.4	83.8	89.1	63.1	66.6	79.6	43.5	48.5	98.1	59.7	65.0	98.6	59.2	64.6
03200	CONCRETE REINFORCEMENT	92.2	95.4	93.8	92.2	96.7	94.5	83.6	92.6	88.2	98.8	66.2	82.2	94.4	64.3	79.1	94.4	64.3	79.1
03300	CAST-IN-PLACE CONCRETE	95.0	78.9	88.4	91.9	79.2	86.7	90.7	74.4	84.0	92.9	43.0	72.4	83.4	51.0	70.1	88.3	52.3	73.5
03	**CONCRETE**	96.9	78.4	87.7	95.0	84.1	89.6	103.6	73.0	88.4	88.4	48.8	68.7	91.8	58.1	75.0	94.2	58.4	76.4
04	**MASONRY**	90.6	46.7	63.4	92.2	81.4	85.5	90.7	50.5	65.8	113.5	52.4	75.6	98.3	53.9	70.8	96.0	54.9	70.6
05	**METALS**	93.6	88.1	91.8	101.0	89.3	97.3	84.0	86.2	84.7	86.4	75.0	82.8	94.0	71.2	86.7	88.6	71.2	83.0
06	**WOOD & PLASTICS**	104.0	73.9	88.3	102.5	82.8	92.2	93.1	61.7	76.7	81.8	42.5	61.3	104.7	63.0	83.0	102.7	62.3	81.7
07	**THERMAL & MOISTURE PROTECTION**	95.9	77.0	87.9	86.0	81.4	84.0	95.5	66.6	83.3	99.6	53.6	80.2	98.9	58.9	82.1	101.1	58.6	83.2
08	**DOORS & WINDOWS**	96.2	79.5	91.8	96.2	86.4	93.6	93.6	74.8	88.7	99.0	49.6	86.1	103.8	60.0	92.4	103.8	61.2	92.7
09200	PLASTER & GYPSUM BOARD	94.3	72.5	80.8	95.3	82.6	87.4	86.5	59.9	70.0	78.1	41.6	55.4	97.9	62.6	75.9	97.9	61.8	75.5
095,098	CEILINGS & ACOUSTICAL TREATMENT	91.3	72.5	80.1	91.3	82.6	86.1	75.1	59.9	66.0	88.8	41.6	60.7	98.4	62.6	77.0	98.4	61.8	76.6
09600	FLOORING	98.3	41.9	83.0	96.8	73.3	90.5	95.7	60.3	86.1	104.4	65.9	94.0	108.0	64.6	96.2	108.2	50.5	92.6
097,099	WALL FINISHES, PAINTS & COATINGS	96.8	57.6	72.9	96.8	81.5	87.5	96.8	99.6	98.5	97.0	40.6	62.6	98.0	46.6	66.6	98.0	42.9	64.4
09	**FINISHES**	94.9	63.3	78.4	94.3	80.5	87.1	88.8	65.6	76.7	92.3	46.5	68.4	101.9	59.5	79.8	101.9	56.1	78.0
10 - 14	**TOTAL DIV. 10000 - 14000**	100.0	76.7	95.0	100.0	79.9	95.7	100.0	94.6	98.9	100.0	66.7	92.9	100.0	76.7	95.0	100.0	76.7	95.0
15	**MECHANICAL**	99.9	47.0	78.0	99.9	84.6	93.6	99.9	51.7	79.9	100.1	62.5	84.5	100.0	58.0	82.6	100.0	63.5	84.8
16	**ELECTRICAL**	94.4	49.5	72.6	94.4	85.5	90.1	93.5	84.1	88.9	92.0	60.4	76.6	94.0	63.7	79.3	94.0	66.2	80.4
01 - 16	**WEIGHTED AVERAGE**	96.0	65.7	82.5	96.4	85.4	91.5	94.4	73.5	85.1	95.4	59.5	79.5	98.1	62.9	82.5	97.5	64.1	82.7

City Cost Indexes

	DIVISION	LOUISIANA									MAINE								
		MONROE			NEW ORLEANS			SHREVEPORT			AUGUSTA			BANGOR			LEWISTON		
		MAT.	INST.	TOTAL	MAT.	INST.	TOTAL	MAT.	INST.	TOTAL	MAT.	INST.	TOTAL	MAT.	INST.	TOTAL	MAT.	INST.	TOTAL
01590	**EQUIPMENT RENTAL**	.0	86.7	86.7	.0	87.3	87.3	.0	86.7	86.7	.0	101.5	101.5	.0	101.5	101.5	.0	101.5	101.5
02	**SITE CONSTRUCTION**	102.0	85.0	89.5	111.5	87.5	93.9	100.7	84.8	89.1	84.3	102.4	97.6	84.2	100.4	96.1	82.9	100.4	95.7
03100	CONCRETE FORMS & ACCESSORIES	78.9	43.7	48.5	98.2	67.7	71.9	98.5	46.5	53.7	100.9	63.4	68.5	95.0	89.9	90.6	101.0	89.9	91.4
03200	CONCRETE REINFORCEMENT	97.7	66.3	81.8	94.4	67.3	80.6	97.3	66.0	81.4	87.5	111.6	99.7	87.5	111.8	99.8	107.5	111.8	109.7
03300	CAST-IN-PLACE CONCRETE	92.9	46.6	73.8	87.5	75.5	82.6	91.6	50.2	74.6	83.7	61.8	74.7	83.7	63.1	75.3	95.3	63.1	82.1
03	**CONCRETE**	88.2	50.0	69.2	93.7	70.7	82.3	88.3	52.4	70.5	98.6	72.0	85.4	97.7	84.1	90.9	103.1	84.1	93.6
04	**MASONRY**	108.3	56.3	76.1	97.8	62.4	75.8	101.0	49.6	69.1	97.1	50.9	68.5	114.9	58.4	79.9	98.9	58.4	73.8
05	**METALS**	86.4	73.3	82.2	101.2	75.5	93.0	82.7	72.7	79.5	87.7	85.4	87.0	87.3	82.1	85.7	90.4	82.1	87.8
06	**WOOD & PLASTICS**	81.0	42.9	61.2	99.0	70.4	84.1	101.9	46.7	73.1	99.5	61.4	79.7	93.0	96.0	94.5	99.5	96.0	97.7
07	**THERMAL & MOISTURE PROTECTION**	99.6	56.3	81.4	102.2	63.6	85.9	98.7	55.0	80.3	102.2	52.9	81.5	102.1	59.4	84.1	102.0	59.4	84.1
08	**DOORS & WINDOWS**	99.0	54.5	87.4	104.5	70.3	95.6	96.9	52.0	85.2	103.8	60.6	92.5	103.7	81.0	97.8	106.9	81.0	100.2
09200	PLASTER & GYPSUM BOARD	77.7	42.1	55.6	96.8	70.2	80.3	83.8	45.9	60.3	104.0	59.5	76.3	100.4	95.0	97.0	106.2	95.0	99.2
095,098	CEILINGS & ACOUSTICAL TREATMENT	88.8	42.1	60.9	97.5	70.2	81.2	89.7	45.9	63.6	90.0	59.5	71.8	87.5	95.0	91.9	98.5	95.0	96.4
09600	FLOORING	104.0	41.7	87.1	108.6	49.5	92.6	115.1	61.8	100.7	99.6	42.0	84.0	97.1	52.4	85.0	99.6	52.4	86.8
097,099	WALL FINISHES, PAINTS & COATINGS	97.0	48.5	67.4	99.7	67.2	79.9	97.0	40.6	62.6	91.5	35.1	57.1	91.5	32.4	55.4	91.5	32.4	55.4
09	**FINISHES**	92.2	42.8	66.5	102.1	64.2	82.4	96.5	48.0	71.3	98.6	55.2	76.0	96.7	77.9	86.9	100.8	77.9	88.9
10 - 14	**TOTAL DIV. 10000 - 14000**	100.0	61.0	91.7	100.0	79.3	95.6	100.0	70.0	93.6	100.0	59.6	91.4	100.0	79.4	95.6	100.0	79.4	95.6
15	**MECHANICAL**	100.1	53.8	80.9	100.0	65.4	85.6	100.1	62.4	84.4	100.1	76.0	90.1	100.1	77.1	90.5	100.1	77.1	90.5
16	**ELECTRICAL**	93.7	61.3	78.0	95.1	69.0	82.4	92.6	70.2	81.7	97.6	87.8	92.9	95.9	85.0	90.6	97.7	85.0	91.5
01 - 16	**WEIGHTED AVERAGE**	95.3	57.9	78.7	99.8	69.7	86.4	94.6	61.3	79.8	97.4	73.2	86.7	97.7	79.8	89.8	98.9	79.8	90.4

	DIVISION	MAINE			MARYLAND						MASSACHUSETTS								
		PORTLAND			BALTIMORE			HAGERSTOWN			BOSTON			BROCKTON			FALL RIVER		
		MAT.	INST.	TOTAL	MAT.	INST.	TOTAL	MAT.	INST.	TOTAL	MAT.	INST.	TOTAL	MAT.	INST.	TOTAL	MAT.	INST.	TOTAL
01590	**EQUIPMENT RENTAL**	.0	101.5	101.5	.0	103.3	103.3	.0	99.1	99.1	.0	108.4	108.4	.0	103.7	103.7	.0	104.9	104.9
02	**SITE CONSTRUCTION**	81.6	100.4	95.4	97.1	93.1	94.1	86.9	88.7	88.2	89.4	109.1	103.8	87.2	105.0	100.2	86.2	105.1	100.1
03100	CONCRETE FORMS & ACCESSORIES	100.1	89.9	91.3	102.9	74.8	78.6	91.2	76.4	78.4	106.5	137.9	133.6	106.1	124.2	121.7	106.1	123.1	120.8
03200	CONCRETE REINFORCEMENT	107.5	111.8	109.7	93.3	88.1	90.6	82.4	75.8	79.1	106.3	145.1	126.0	107.5	144.9	126.5	107.5	124.2	115.9
03300	CAST-IN-PLACE CONCRETE	86.8	63.1	77.0	99.0	79.0	90.8	84.2	60.4	74.4	108.1	145.5	123.5	102.9	141.7	118.8	99.5	142.2	117.1
03	**CONCRETE**	98.9	84.1	91.5	100.5	79.8	90.3	86.1	71.8	79.0	112.2	141.0	126.5	109.4	133.4	121.3	107.7	129.2	118.4
04	**MASONRY**	96.5	58.4	72.8	92.7	72.5	80.2	96.6	77.0	84.4	118.0	148.1	136.7	114.1	141.8	131.3	113.9	141.7	131.1
05	**METALS**	91.8	82.1	88.7	101.3	98.7	100.5	97.5	90.2	95.2	103.0	127.4	110.8	99.8	124.5	107.6	99.8	116.0	105.0
06	**WOOD & PLASTICS**	99.5	96.0	97.7	100.0	76.0	87.5	87.5	75.8	81.4	103.5	138.9	121.9	102.5	123.5	113.4	102.5	122.5	112.9
07	**THERMAL & MOISTURE PROTECTION**	101.8	59.4	84.0	97.8	80.1	90.4	97.2	71.2	86.3	102.4	142.5	119.3	102.3	135.2	116.1	102.2	130.7	114.2
08	**DOORS & WINDOWS**	106.9	81.0	100.2	94.5	83.2	91.5	91.7	76.8	87.8	102.9	136.8	111.7	100.8	127.7	107.8	100.8	120.1	105.8
09200	PLASTER & GYPSUM BOARD	106.2	95.0	99.2	103.3	75.5	86.1	98.5	75.5	84.2	108.6	139.1	127.6	103.8	123.2	115.8	103.8	121.8	115.0
095,098	CEILINGS & ACOUSTICAL TREATMENT	98.5	95.0	96.4	103.1	75.5	86.7	104.0	75.5	87.0	97.6	139.1	122.3	99.5	123.2	113.6	99.5	121.8	112.8
09600	FLOORING	99.6	52.4	86.8	92.0	78.6	88.4	87.2	79.2	85.0	100.3	163.7	117.4	100.7	163.7	117.8	100.5	163.7	117.6
097,099	WALL FINISHES, PAINTS & COATINGS	91.5	32.4	55.4	95.9	87.4	90.7	95.9	39.7	61.6	94.6	153.1	130.3	94.0	130.1	116.0	94.0	130.1	116.0
09	**FINISHES**	100.8	77.9	88.9	96.1	76.3	85.8	93.6	73.0	82.9	101.1	145.0	124.0	100.9	132.3	117.3	100.9	131.7	117.0
10 - 14	**TOTAL DIV. 10000 - 14000**	100.0	79.4	95.6	100.0	85.8	97.0	100.0	88.6	97.6	100.0	120.4	104.4	100.0	116.9	103.6	100.0	117.6	103.8
15	**MECHANICAL**	100.1	77.1	90.5	100.0	86.1	94.2	100.0	90.0	95.9	100.1	128.8	112.0	100.1	110.1	104.2	100.1	109.9	104.2
16	**ELECTRICAL**	97.8	85.0	91.6	105.1	91.2	98.4	101.7	84.3	93.2	96.8	120.4	108.3	95.8	103.4	99.5	95.4	103.4	99.3
01 - 16	**WEIGHTED AVERAGE**	98.5	79.8	90.2	99.5	84.7	92.9	96.2	81.9	89.8	102.5	132.0	115.6	101.1	121.0	110.0	100.9	119.1	109.0

	DIVISION	MASSACHUSETTS																	
		HYANNIS			LAWRENCE			LOWELL			NEW BEDFORD			PITTSFIELD			SPRINGFIELD		
		MAT.	INST.	TOTAL	MAT.	INST.	TOTAL	MAT.	INST.	TOTAL	MAT.	INST.	TOTAL	MAT.	INST.	TOTAL	MAT.	INST.	TOTAL
01590	**EQUIPMENT RENTAL**	.0	103.7	103.7	.0	103.7	103.7	.0	101.5	101.5	.0	104.9	104.9	.0	101.5	101.5	.0	101.5	101.5
02	**SITE CONSTRUCTION**	83.7	105.0	99.3	87.7	105.0	100.4	86.8	104.9	100.1	85.8	105.1	100.0	87.9	103.6	99.4	87.3	103.8	99.4
03100	CONCRETE FORMS & ACCESSORIES	96.5	122.8	119.2	105.9	123.2	120.9	102.6	123.4	120.6	106.1	123.1	120.8	102.6	99.2	99.7	102.8	107.7	107.0
03200	CONCRETE REINFORCEMENT	86.2	124.1	105.4	106.6	132.5	119.8	107.5	132.6	120.2	107.5	124.2	115.9	89.4	116.9	103.4	107.5	122.2	114.9
03300	CAST-IN-PLACE CONCRETE	93.9	141.5	113.5	103.7	141.7	119.3	94.3	141.8	113.8	96.6	142.2	115.3	102.9	117.8	109.0	98.1	119.4	106.9
03	**CONCRETE**	99.0	128.8	113.8	109.6	130.6	120.1	100.4	130.6	115.4	106.3	129.2	117.7	101.6	108.5	105.1	102.3	113.8	108.0
04	**MASONRY**	113.0	141.8	130.8	113.4	141.8	131.0	98.8	141.5	125.3	113.7	141.7	131.0	99.5	114.2	108.6	99.1	116.8	110.1
05	**METALS**	96.1	115.5	102.3	97.1	119.4	104.2	97.1	116.7	103.3	99.8	116.0	105.0	96.9	104.4	99.3	99.7	106.7	101.9
06	**WOOD & PLASTICS**	92.1	122.1	107.8	102.5	122.1	112.7	101.5	122.1	112.3	102.5	122.5	112.9	101.5	97.3	99.3	101.5	107.2	104.5
07	**THERMAL & MOISTURE PROTECTION**	101.8	131.7	114.4	102.2	135.0	116.0	102.0	134.2	115.6	102.2	130.7	114.2	102.0	110.7	105.7	102.0	112.9	106.6
08	**DOORS & WINDOWS**	97.0	120.0	103.0	100.8	124.5	107.0	106.9	124.5	111.5	100.8	120.1	105.8	106.9	104.9	106.4	106.9	111.7	108.2
09200	PLASTER & GYPSUM BOARD	96.1	121.8	112.1	106.2	121.8	115.9	106.2	121.8	115.9	103.8	121.8	115.0	106.2	96.3	100.0	106.2	106.5	106.4
095,098	CEILINGS & ACOUSTICAL TREATMENT	88.4	121.8	108.4	98.5	121.8	112.4	98.5	121.8	112.4	99.5	121.8	112.8	98.5	96.3	97.2	98.5	106.5	103.3
09600	FLOORING	96.6	163.7	114.7	99.6	163.7	117.0	99.6	163.7	117.0	100.5	163.7	117.6	99.8	128.6	107.6	99.6	128.6	107.4
097,099	WALL FINISHES, PAINTS & COATINGS	94.0	130.1	116.0	91.6	130.1	115.1	91.5	133.1	116.9	94.0	130.1	116.0	91.5	104.5	99.4	93.3	104.5	100.1
09	**FINISHES**	95.6	131.5	114.3	100.6	131.5	116.7	100.6	131.8	116.9	100.8	131.7	116.9	100.6	104.8	102.8	100.7	111.3	106.2
10 - 14	**TOTAL DIV. 10000 - 14000**	100.0	116.8	103.6	100.0	116.8	103.6	100.0	116.8	103.6	100.0	117.6	103.8	100.0	104.7	101.0	100.0	106.9	101.5
15	**MECHANICAL**	100.1	109.8	104.1	100.1	111.4	104.8	100.1	124.6	110.2	100.1	109.9	104.2	100.1	97.0	98.8	100.1	98.4	99.4
16	**ELECTRICAL**	94.0	103.4	98.6	97.3	120.4	108.5	97.7	120.4	108.7	96.7	103.4	100.0	97.7	93.7	95.8	97.7	93.7	95.8
01 - 16	**WEIGHTED AVERAGE**	98.1	119.0	107.4	100.9	122.6	110.5	99.7	125.0	111.0	100.8	119.1	109.0	99.9	103.0	101.3	100.4	105.9	102.8

	DIVISION	MASSACHUSETTS			MICHIGAN														
		WORCESTER			ANN ARBOR			DEARBORN			DETROIT			FLINT			GRAND RAPIDS		
		MAT.	INST.	TOTAL	MAT.	INST.	TOTAL	MAT.	INST.	TOTAL	MAT.	INST.	TOTAL	MAT.	INST.	TOTAL	MAT.	INST.	TOTAL
01590	**EQUIPMENT RENTAL**	.0	101.5	101.5	.0	112.2	112.2	.0	112.2	112.2	.0	97.7	97.7	.0	112.2	112.2	.0	105.1	105.1
02	**SITE CONSTRUCTION**	86.8	104.9	100.1	77.2	95.3	90.5	76.9	95.7	90.7	90.4	97.5	95.6	68.2	94.3	87.3	82.1	87.6	86.1
03100	CONCRETE FORMS & ACCESSORIES	103.2	122.8	120.1	97.4	119.4	116.4	97.3	123.9	120.2	98.8	124.0	120.5	101.0	94.9	95.7	97.7	76.6	79.5
03200	CONCRETE REINFORCEMENT	107.5	143.9	126.0	95.2	122.2	108.9	95.2	123.1	109.4	94.6	123.1	109.1	95.2	121.9	108.8	92.2	83.3	87.7
03300	CAST-IN-PLACE CONCRETE	94.3	141.5	113.7	92.3	107.9	98.7	90.2	121.6	103.1	99.1	121.7	108.4	93.0	96.4	94.4	95.1	96.7	95.8
03	**CONCRETE**	100.4	132.3	116.3	94.0	116.4	105.1	92.9	123.2	108.0	97.2	121.9	109.5	94.5	101.4	97.9	96.9	84.3	90.7
04	**MASONRY**	98.8	141.7	125.4	98.8	103.1	101.5	98.7	122.7	113.6	97.8	122.7	113.2	98.9	89.8	93.3	94.8	54.4	69.7
05	**METALS**	99.7	120.7	106.4	102.4	125.5	109.8	102.5	127.8	110.6	107.1	106.2	106.8	102.5	123.0	109.0	93.9	78.4	89.0
06	**WOOD & PLASTICS**	102.0	122.1	112.5	100.1	122.6	111.8	100.1	124.0	112.5	101.1	124.0	113.0	103.6	96.3	99.8	98.0	77.2	87.2
07	**THERMAL & MOISTURE PROTECTION**	102.0	129.3	113.5	102.8	108.6	105.2	101.9	124.8	111.5	99.5	124.8	110.2	100.8	91.0	96.6	91.4	60.6	78.4
08	**DOORS & WINDOWS**	106.9	127.5	112.3	96.6	117.8	102.1	96.6	118.5	102.3	98.2	121.1	104.2	96.6	101.1	97.8	94.3	68.6	87.6
09200	PLASTER & GYPSUM BOARD	106.2	121.8	115.9	98.1	122.0	112.9	98.1	123.4	113.8	98.1	123.4	113.8	99.6	94.9	96.7	92.9	71.9	79.8
095,098	CEILINGS & ACOUSTICAL TREATMENT	98.5	121.8	112.4	91.3	122.0	109.6	91.3	123.4	110.4	92.3	123.4	110.8	91.3	94.9	93.4	91.3	71.9	79.7
09600	FLOORING	99.6	160.4	116.1	88.0	105.0	92.6	87.6	122.8	97.2	87.7	122.8	97.2	87.8	71.6	83.5	97.0	42.3	82.2
097,099	WALL FINISHES, PAINTS & COATINGS	91.5	130.1	115.1	86.6	99.7	94.6	86.6	113.8	103.2	88.4	113.8	103.9	86.6	85.0	85.6	96.8	41.2	62.8
09	**FINISHES**	100.6	130.8	116.3	92.4	115.4	104.4	92.3	123.0	108.3	93.3	123.0	108.8	92.1	89.0	90.5	94.2	66.3	79.7
10 - 14	**TOTAL DIV. 10000 - 14000**	100.0	110.8	102.3	100.0	110.1	102.2	100.0	112.7	102.7	100.0	112.7	102.7	100.0	90.3	97.9	100.0	94.5	98.8
15	**MECHANICAL**	100.1	108.1	103.4	100.0	99.6	99.8	100.0	117.5	107.3	100.0	120.0	108.3	100.0	99.9	99.9	99.9	55.0	81.3
16	**ELECTRICAL**	97.7	106.8	102.1	94.1	85.8	90.0	94.1	119.2	106.3	95.8	119.1	107.1	94.1	108.0	100.9	94.8	59.8	77.8
01 - 16	**WEIGHTED AVERAGE**	100.1	120.0	109.0	97.4	105.9	101.2	97.3	118.9	106.9	99.1	117.6	107.3	97.2	99.9	98.4	96.0	68.1	83.6

	DIVISION	MICHIGAN												MINNESOTA					
		KALAMAZOO			LANSING			MUSKEGON			SAGINAW			DULUTH			MINNEAPOLIS		
		MAT.	INST.	TOTAL	MAT.	INST.	TOTAL	MAT.	INST.	TOTAL	MAT.	INST.	TOTAL	MAT.	INST.	TOTAL	MAT.	INST.	TOTAL
01590	**EQUIPMENT RENTAL**	.0	105.1	105.1	.0	112.2	112.2	.0	105.1	105.1	.0	112.2	112.2	.0	101.8	101.8	.0	106.2	106.2
02	**SITE CONSTRUCTION**	82.6	88.0	86.5	83.2	93.8	91.0	80.4	87.8	85.8	70.4	93.8	87.5	83.0	102.7	97.4	82.9	108.9	101.9
03100	CONCRETE FORMS & ACCESSORIES	97.2	86.9	88.3	100.9	95.9	96.6	97.9	83.3	85.3	97.4	95.4	95.7	99.7	122.4	119.3	100.4	138.3	133.1
03200	CONCRETE REINFORCEMENT	92.2	85.0	88.5	95.2	121.7	108.7	92.7	84.8	88.7	95.2	121.3	108.5	95.6	110.1	102.9	95.8	128.9	112.6
03300	CAST-IN-PLACE CONCRETE	96.8	98.7	97.6	92.3	91.8	92.1	94.6	95.1	94.8	91.2	92.1	91.6	109.7	108.8	109.3	108.0	126.4	115.6
03	**CONCRETE**	100.0	89.9	95.0	94.2	100.3	97.2	95.0	87.0	91.1	93.4	100.1	96.7	101.0	116.0	108.4	101.8	132.7	117.2
04	**MASONRY**	97.6	84.4	89.4	92.6	91.8	92.1	94.9	77.8	84.3	100.3	86.1	91.5	109.1	118.0	114.6	109.1	137.5	126.7
05	**METALS**	92.4	82.3	89.2	100.8	122.1	107.6	90.1	81.9	87.5	102.5	121.3	108.5	93.0	126.6	103.8	95.8	139.1	109.6
06	**WOOD & PLASTICS**	99.6	85.5	92.3	103.0	96.4	99.6	96.5	82.5	89.2	96.2	96.3	96.2	112.7	123.1	118.1	113.1	137.3	125.7
07	**THERMAL & MOISTURE PROTECTION**	91.0	83.1	87.7	100.8	90.9	96.6	90.2	76.4	84.4	101.2	88.2	95.7	100.4	121.0	109.0	100.3	136.5	115.5
08	**DOORS & WINDOWS**	91.1	80.2	88.2	96.6	101.2	97.8	90.3	79.2	87.4	96.1	101.1	97.4	96.7	124.4	104.0	99.9	142.8	111.1
09200	PLASTER & GYPSUM BOARD	92.9	80.4	85.1	100.5	95.0	97.1	82.1	77.3	79.1	98.1	94.9	96.1	105.9	124.2	117.3	106.1	138.6	126.3
095,098	CEILINGS & ACOUSTICAL TREATMENT	91.3	80.4	84.8	91.3	95.0	93.5	93.0	77.3	83.7	91.3	94.9	93.4	100.3	124.2	114.6	101.2	138.6	123.5
09600	FLOORING	97.0	67.1	88.9	96.3	79.4	91.7	96.2	77.9	91.3	88.0	57.0	79.6	105.0	122.6	109.8	102.4	131.2	110.2
097,099	WALL FINISHES, PAINTS & COATINGS	96.8	83.6	88.7	100.8	91.1	94.9	96.0	59.7	73.9	86.6	78.2	81.5	91.4	112.0	104.0	97.6	127.9	116.1
09	**FINISHES**	94.2	82.6	88.1	96.7	92.3	94.4	92.3	79.4	85.6	92.2	86.1	89.1	102.5	121.8	112.5	102.3	136.2	120.0
10 - 14	**TOTAL DIV. 10000 - 14000**	100.0	97.8	99.5	100.0	91.7	98.2	100.0	96.1	99.2	100.0	91.7	98.2	100.0	102.8	100.6	100.0	109.3	102.0
15	**MECHANICAL**	99.9	84.6	93.6	100.0	90.0	95.8	99.8	86.5	94.3	100.0	92.1	96.7	100.1	114.6	106.1	100.1	119.3	108.1
16	**ELECTRICAL**	93.7	88.6	91.2	92.3	86.1	89.3	94.2	82.5	88.5	93.4	81.2	87.5	103.5	100.6	102.1	105.3	114.3	109.7
01 - 16	**WEIGHTED AVERAGE**	95.8	85.8	91.4	97.2	95.2	96.3	94.4	83.6	89.6	97.0	93.4	95.4	99.5	114.5	106.2	100.5	126.7	112.2

	DIVISION	MINNESOTA									MISSISSIPPI								
		ROCHESTER			SAINT PAUL			ST. CLOUD			BILOXI			GREENVILLE			JACKSON		
		MAT.	INST.	TOTAL	MAT.	INST.	TOTAL	MAT.	INST.	TOTAL	MAT.	INST.	TOTAL	MAT.	INST.	TOTAL	MAT.	INST.	TOTAL
01590	**EQUIPMENT RENTAL**	.0	101.8	101.8	.0	101.8	101.8	.0	101.6	101.6	.0	98.3	98.3	.0	98.3	98.3	.0	98.3	98.3
02	**SITE CONSTRUCTION**	82.1	101.9	96.7	85.1	103.4	98.5	79.9	105.4	98.6	101.5	85.9	90.1	105.5	85.7	91.0	98.1	85.7	89.0
03100	CONCRETE FORMS & ACCESSORIES	100.2	111.1	109.6	91.1	134.3	128.4	83.1	133.0	126.2	95.9	41.2	48.8	80.4	34.0	40.4	95.6	38.8	46.6
03200	CONCRETE REINFORCEMENT	95.6	128.3	112.2	92.3	128.8	110.8	94.3	128.4	111.6	95.1	64.7	79.6	103.1	43.2	72.6	95.1	46.8	70.6
03300	CAST-IN-PLACE CONCRETE	105.9	103.4	104.9	109.0	125.6	115.8	98.0	123.8	108.6	105.1	44.8	80.3	105.5	39.2	78.3	103.1	41.7	77.8
03	**CONCRETE**	99.2	112.7	105.9	103.3	130.7	116.9	90.6	129.3	109.8	97.9	48.8	73.5	101.0	39.6	70.5	96.9	43.2	70.3
04	**MASONRY**	108.4	116.5	113.4	118.8	137.4	130.3	109.5	128.6	121.3	90.1	37.0	57.2	133.3	38.3	74.4	92.0	38.3	58.8
05	**METALS**	92.9	136.6	106.9	92.4	138.7	107.2	90.6	136.1	105.1	90.4	83.0	88.0	88.8	72.8	83.7	90.4	74.6	85.3
06	**WOOD & PLASTICS**	113.1	110.1	111.6	102.7	132.1	118.0	91.2	132.0	112.5	96.2	41.1	67.5	78.4	33.3	54.9	96.3	39.6	66.7
07	**THERMAL & MOISTURE PROTECTION**	100.2	107.6	103.4	100.2	135.8	115.2	98.2	121.3	108.0	96.4	45.9	75.2	96.5	41.2	73.2	96.3	42.4	73.6
08	**DOORS & WINDOWS**	96.7	128.1	104.9	94.1	140.0	106.1	91.6	139.9	104.2	98.3	48.0	85.1	97.7	37.8	82.0	98.7	42.3	83.9
09200	PLASTER & GYPSUM BOARD	105.7	110.9	108.9	100.4	133.5	121.0	87.3	133.5	116.0	103.5	40.0	64.1	92.0	32.1	54.8	103.5	38.5	63.1
095,098	CEILINGS & ACOUSTICAL TREATMENT	99.5	110.9	106.3	97.8	133.5	119.1	72.6	133.5	108.9	104.3	40.0	66.0	98.3	32.1	58.8	104.3	38.5	65.0
09600	FLOORING	104.8	78.5	97.7	97.8	131.2	106.8	96.7	131.2	106.0	118.7	39.6	97.3	108.9	33.8	88.6	118.7	39.1	97.2
097,099	WALL FINISHES, PAINTS & COATINGS	93.7	103.8	99.9	97.6	127.9	116.1	101.3	127.9	117.5	106.7	35.2	63.1	106.7	34.6	62.7	106.7	34.6	62.7
09	**FINISHES**	102.3	104.8	103.6	99.5	133.2	117.0	90.6	132.5	112.4	107.2	39.8	72.1	101.4	33.8	66.2	107.2	38.5	71.4
10 - 14	**TOTAL DIV. 10000 - 14000**	100.0	99.9	100.0	100.0	108.3	101.8	100.0	107.1	101.5	100.0	51.4	89.6	100.0	49.8	89.3	100.0	50.6	89.5
15	**MECHANICAL**	100.1	106.4	102.7	100.1	115.7	106.6	99.7	117.9	107.2	99.9	53.9	80.8	99.9	34.3	72.7	99.9	34.0	72.5
16	**ELECTRICAL**	103.5	87.2	95.6	104.1	114.2	109.0	101.4	114.2	107.7	97.0	61.0	79.5	96.3	37.1	67.6	97.8	37.1	68.3
01 - 16	**WEIGHTED AVERAGE**	99.2	108.5	103.4	99.7	124.5	110.7	95.7	123.3	108.0	97.8	55.5	79.0	99.3	44.6	75.0	97.8	46.1	74.8

DIVISION		MISSISSIPPI			MISSOURI														
		MERIDIAN			CAPE GIRARDEAU			COLUMBIA			JOPLIN			KANSAS CITY			SPRINGFIELD		
		MAT.	INST.	TOTAL	MAT.	INST.	TOTAL	MAT.	INST.	TOTAL	MAT.	INST.	TOTAL	MAT.	INST.	TOTAL	MAT.	INST.	TOTAL
01590	EQUIPMENT RENTAL	.0	98.3	98.3	.0	108.3	108.3	.0	109.1	109.1	.0	107.1	107.1	.0	104.9	104.9	.0	102.8	102.8
02	SITE CONSTRUCTION	97.0	85.7	88.7	94.3	93.3	93.5	103.4	94.6	96.9	103.6	98.8	100.1	97.4	98.1	97.9	97.3	93.8	94.7
03100	CONCRETE FORMS & ACCESSORIES	78.9	33.0	39.3	83.4	77.2	78.1	85.0	72.0	73.8	105.0	73.5	77.8	103.9	106.6	106.3	100.4	73.7	77.4
03200	CONCRETE REINFORCEMENT	101.8	46.3	73.6	94.5	84.5	89.4	101.1	113.3	107.3	111.9	75.7	93.5	106.3	114.2	110.3	97.4	112.9	105.3
03300	CAST-IN-PLACE CONCRETE	99.6	41.7	75.7	87.4	83.3	85.7	90.4	84.3	87.9	105.1	72.6	91.7	97.6	107.7	101.8	103.6	66.9	88.5
03	CONCRETE	93.7	40.6	67.3	87.8	82.4	85.1	86.8	85.8	86.3	101.6	74.6	88.2	97.4	108.7	103.0	100.2	79.7	90.0
04	MASONRY	89.8	28.6	51.8	113.8	81.7	93.9	131.0	86.2	103.2	96.5	61.7	74.9	101.3	106.8	104.7	88.0	82.9	84.8
05	METALS	88.7	73.7	83.9	92.7	109.1	98.0	93.5	121.2	102.4	99.4	89.8	96.3	106.9	114.9	109.5	97.4	107.2	100.5
06	WOOD & PLASTICS	77.1	32.0	53.6	78.5	74.5	76.5	89.1	66.2	77.2	106.9	73.3	89.4	106.2	106.1	106.2	100.0	73.2	86.0
07	THERMAL & MOISTURE PROTECTION	96.1	39.1	72.1	92.7	80.8	87.7	89.6	84.8	87.6	93.5	69.6	83.4	92.3	108.0	98.9	94.1	78.7	87.6
08	DOORS & WINDOWS	97.6	36.6	81.7	93.9	75.5	89.1	90.8	91.5	91.0	92.2	76.2	88.0	98.3	109.1	101.1	96.3	82.2	92.6
09200	PLASTER & GYPSUM BOARD	92.0	30.7	53.9	100.8	73.7	84.0	100.1	65.2	78.4	103.4	72.3	84.1	100.1	106.0	103.7	103.8	72.3	84.3
095,098	CEILINGS & ACOUSTICAL TREATMENT	98.3	30.7	58.0	86.1	73.7	78.7	88.6	65.2	74.6	85.4	72.3	77.6	91.4	106.0	100.1	88.6	72.3	78.9
09600	FLOORING	107.6	27.5	85.9	92.7	74.8	87.9	101.9	91.1	98.9	111.5	50.3	95.0	91.6	98.3	93.4	105.4	50.3	90.5
097,099	WALL FINISHES, PAINTS & COATINGS	106.7	34.2	62.5	101.0	75.7	85.6	98.8	72.2	82.6	92.5	44.7	63.4	97.5	114.6	107.9	93.1	64.0	75.4
09	FINISHES	100.5	31.5	64.5	91.0	75.5	82.9	93.1	72.3	82.3	98.3	66.2	81.6	94.7	105.7	100.4	96.6	67.7	81.6
10 - 14	TOTAL DIV. 10000 - 14000	100.0	49.8	89.3	100.0	66.8	92.9	100.0	90.6	98.0	100.0	77.0	95.1	100.0	97.4	99.4	100.0	88.2	97.5
15	MECHANICAL	99.9	34.0	72.6	100.1	100.9	100.5	99.9	101.8	100.7	100.1	57.5	82.4	99.9	105.4	102.2	100.0	69.7	87.4
16	ELECTRICAL	96.0	61.3	79.1	101.4	114.6	107.8	99.0	92.5	95.8	96.8	62.1	80.0	108.3	109.4	108.8	102.4	67.1	85.2
01 - 16	WEIGHTED AVERAGE	96.0	46.7	74.1	96.4	92.3	94.6	97.1	92.4	95.0	98.6	70.5	86.1	100.9	106.8	103.5	98.4	78.7	89.6

DIVISION		MISSOURI						MONTANA											
		ST. JOSEPH			ST. LOUIS			BILLINGS			BUTTE			GREAT FALLS			HELENA		
		MAT.	INST.	TOTAL	MAT.	INST.	TOTAL	MAT.	INST.	TOTAL	MAT.	INST.	TOTAL	MAT.	INST.	TOTAL	MAT.	INST.	TOTAL
01590	EQUIPMENT RENTAL	.0	103.4	103.4	.0	109.6	109.6	.0	98.6	98.6	.0	98.3	98.3	.0	98.3	98.3	.0	98.3	98.3
02	SITE CONSTRUCTION	98.9	93.7	95.1	94.0	96.4	95.8	87.1	96.0	93.6	94.5	94.6	94.6	98.1	95.4	96.1	100.1	95.4	96.6
03100	CONCRETE FORMS & ACCESSORIES	103.8	85.8	88.2	98.3	105.4	104.4	96.8	63.6	68.2	84.1	61.2	64.3	98.0	64.5	69.1	98.1	65.0	69.5
03200	CONCRETE REINFORCEMENT	105.0	103.3	104.2	86.5	111.3	99.1	93.3	75.7	84.3	101.2	75.5	88.1	93.3	75.6	84.3	96.6	63.3	79.7
03300	CAST-IN-PLACE CONCRETE	97.6	102.8	99.7	87.4	107.4	95.6	119.3	69.6	98.9	122.8	71.1	101.5	129.8	56.8	99.8	132.2	66.6	105.3
03	CONCRETE	97.2	95.6	96.4	87.5	108.2	97.8	105.3	68.9	87.2	105.3	68.3	86.9	110.5	64.9	87.8	112.2	66.1	89.3
04	MASONRY	100.8	93.1	96.0	94.9	111.6	105.3	121.6	70.9	90.1	121.2	65.4	86.6	125.1	74.1	93.5	120.3	66.7	87.1
05	METALS	102.8	106.4	104.0	97.4	123.4	105.7	106.3	86.1	99.8	99.2	85.4	94.8	102.8	85.7	97.3	102.0	78.9	94.6
06	WOOD & PLASTICS	106.9	83.2	94.6	94.4	103.1	98.9	101.2	63.3	81.4	88.5	61.7	74.5	103.9	63.3	82.7	104.0	63.3	82.8
07	THERMAL & MOISTURE PROTECTION	92.7	94.5	93.5	92.6	106.7	98.5	98.8	68.2	85.9	98.3	67.3	85.2	98.9	67.8	85.8	98.9	67.6	85.7
08	DOORS & WINDOWS	97.0	94.0	96.2	93.0	110.7	97.6	99.0	62.7	89.5	95.9	61.7	86.9	99.3	62.7	89.7	98.7	59.5	88.5
09200	PLASTER & GYPSUM BOARD	104.7	82.5	90.9	109.1	103.0	105.3	103.7	62.6	78.2	95.0	61.0	73.8	103.7	62.6	78.2	103.1	62.6	77.9
095,098	CEILINGS & ACOUSTICAL TREATMENT	90.6	82.5	85.7	91.2	103.0	98.2	112.8	62.6	82.8	112.1	61.0	81.6	114.7	62.6	83.6	112.1	62.6	82.6
09600	FLOORING	93.9	80.6	90.3	100.5	97.1	99.6	107.3	49.0	91.5	98.4	39.0	82.3	107.3	52.9	92.6	107.3	51.3	92.1
097,099	WALL FINISHES, PAINTS & COATINGS	93.1	77.3	83.5	101.0	108.4	105.6	97.2	59.7	74.3	97.2	40.6	62.7	97.2	45.8	65.8	97.2	50.8	68.9
09	FINISHES	95.7	83.4	89.3	95.9	103.5	99.9	106.4	60.0	82.2	102.2	54.2	77.2	107.0	59.9	82.4	106.3	60.7	82.5
10 - 14	TOTAL DIV. 10000 - 14000	100.0	92.0	98.3	100.0	99.8	100.0	100.0	67.1	93.0	100.0	66.1	92.8	100.0	68.4	93.3	100.0	56.4	90.7
15	MECHANICAL	100.1	92.4	96.9	100.1	108.3	103.5	100.3	76.5	90.4	100.3	69.5	87.5	100.3	78.2	91.1	100.3	73.9	89.4
16	ELECTRICAL	105.3	93.2	99.4	104.3	114.6	109.3	94.2	78.7	86.6	100.9	75.6	88.6	94.2	73.3	84.0	94.2	74.6	84.7
01 - 16	WEIGHTED AVERAGE	99.9	93.3	97.0	97.0	109.0	102.3	102.1	74.5	89.8	101.2	70.9	87.7	102.7	73.8	89.9	102.5	71.5	88.7

DIVISION		MONTANA			NEBRASKA												NEVADA		
		MISSOULA			GRAND ISLAND			LINCOLN			NORTH PLATTE			OMAHA			CARSON CITY		
		MAT.	INST.	TOTAL	MAT.	INST.	TOTAL	MAT.	INST.	TOTAL	MAT.	INST.	TOTAL	MAT.	INST.	TOTAL	MAT.	INST.	TOTAL
01590	EQUIPMENT RENTAL	.0	98.3	98.3	.0	101.6	101.6	.0	101.6	101.6	.0	101.6	101.6	.0	90.9	90.9	.0	101.7	101.7
02	SITE CONSTRUCTION	77.3	94.5	89.9	100.3	91.5	93.9	91.6	91.5	91.5	101.0	89.7	92.7	80.3	90.3	87.6	61.9	104.0	92.8
03100	CONCRETE FORMS & ACCESSORIES	87.7	57.4	61.5	96.8	49.6	56.1	101.5	45.8	53.4	96.6	45.9	52.8	96.5	73.6	76.7	94.2	100.7	99.8
03200	CONCRETE REINFORCEMENT	103.0	81.1	91.9	106.0	74.7	90.1	97.4	75.6	86.3	107.4	74.4	90.7	102.0	76.4	89.0	113.7	114.1	113.9
03300	CAST-IN-PLACE CONCRETE	90.4	63.1	79.2	118.0	57.4	93.1	106.5	59.9	87.4	118.0	50.8	90.4	110.6	74.1	95.6	109.7	91.4	102.2
03	CONCRETE	84.1	64.9	74.5	109.7	58.4	84.2	101.7	57.7	79.8	110.0	54.3	82.3	103.1	74.5	88.9	109.3	99.9	104.6
04	MASONRY	146.1	58.7	91.9	105.9	48.6	70.4	96.1	64.9	76.7	91.3	39.3	59.1	101.4	78.0	86.9	131.3	80.6	99.9
05	METALS	95.9	86.9	93.0	93.6	83.9	90.5	98.7	85.4	94.5	93.8	81.0	89.7	98.0	77.4	91.5	93.3	102.3	96.2
06	WOOD & PLASTICS	92.9	58.0	74.7	96.3	44.1	69.1	101.0	38.5	68.5	96.2	44.1	69.0	95.9	74.9	84.9	90.7	103.2	97.2
07	THERMAL & MOISTURE PROTECTION	97.6	68.8	85.5	94.5	57.4	78.9	94.8	61.1	80.6	94.5	46.1	74.1	90.2	71.5	82.3	99.8	88.9	95.2
08	DOORS & WINDOWS	95.9	62.0	87.1	90.1	49.0	79.4	95.5	48.6	83.3	89.4	51.3	79.5	98.7	67.5	90.6	94.3	108.8	98.1
09200	PLASTER & GYPSUM BOARD	96.8	57.2	72.2	101.6	42.4	64.8	105.6	36.7	62.8	102.0	42.4	65.0	107.9	74.7	87.3	83.6	103.0	95.7
095,098	CEILINGS & ACOUSTICAL TREATMENT	112.1	57.2	79.3	86.9	42.4	60.4	95.5	36.7	60.4	88.6	42.4	61.1	105.7	74.7	87.2	103.9	103.0	103.4
09600	FLOORING	100.6	61.0	89.9	94.9	34.7	78.6	96.7	41.8	81.8	94.8	36.1	78.9	122.7	46.3	102.1	100.3	63.8	90.5
097,099	WALL FINISHES, PAINTS & COATINGS	97.2	45.8	65.8	90.8	41.3	60.6	90.8	41.7	60.8	90.8	50.0	65.9	148.6	71.0	101.2	102.2	78.7	87.9
09	FINISHES	102.3	56.2	78.3	93.1	44.2	67.6	96.2	42.5	68.2	93.6	43.8	67.6	114.1	67.9	90.0	96.8	91.3	93.9
10 - 14	TOTAL DIV. 10000 - 14000	100.0	51.5	89.7	100.0	68.9	93.4	100.0	68.2	93.2	100.0	51.4	89.6	100.0	71.2	93.9	100.0	107.8	101.7
15	MECHANICAL	100.3	69.3	87.4	100.0	74.8	89.6	100.0	74.9	89.6	100.0	73.8	89.1	99.8	81.1	92.0	100.1	93.0	97.2
16	ELECTRICAL	99.0	78.4	89.0	91.0	70.3	80.9	96.6	70.3	83.8	92.9	43.0	68.7	90.6	86.3	88.5	91.0	110.9	100.7
01 - 16	WEIGHTED AVERAGE	98.9	70.1	86.1	97.5	65.6	83.3	98.2	67.1	84.4	97.1	58.8	80.1	99.3	78.2	89.9	98.6	97.9	98.3

	DIVISION	NEVADA						NEW HAMPSHIRE									NEW JERSEY		
		LAS VEGAS			RENO			MANCHESTER			NASHUA			PORTSMOUTH			CAMDEN		
		MAT.	INST.	TOTAL	MAT.	INST.	TOTAL	MAT.	INST.	TOTAL	MAT.	INST.	TOTAL	MAT.	INST.	TOTAL	MAT.	INST.	TOTAL
01590	**EQUIPMENT RENTAL**	.0	101.7	101.7	.0	101.7	101.7	.0	101.5	101.5	.0	101.5	101.5	.0	101.5	101.5	.0	99.4	99.4
02	**SITE CONSTRUCTION**	61.5	106.2	94.3	61.6	104.0	92.7	87.7	99.9	96.6	89.0	99.9	97.0	83.2	98.9	94.7	92.2	105.0	101.6
03100	CONCRETE FORMS & ACCESSORIES	93.1	108.7	106.6	94.3	100.8	99.9	102.3	62.2	67.7	102.8	62.2	67.8	89.6	57.8	62.2	102.8	125.6	122.4
03200	CONCRETE REINFORCEMENT	108.1	119.3	113.8	108.1	118.5	113.4	107.5	88.6	97.9	107.5	88.6	97.9	85.9	88.5	87.2	107.5	114.7	111.1
03300	CAST-IN-PLACE CONCRETE	103.5	108.8	105.6	111.7	91.4	103.4	103.8	96.3	100.7	95.3	96.3	95.7	90.4	90.0	90.2	82.8	125.3	100.3
03	**CONCRETE**	105.4	110.4	107.8	109.4	100.7	105.1	107.3	79.2	93.3	103.2	79.2	91.3	94.1	75.0	84.6	97.2	122.3	109.6
04	**MASONRY**	122.2	100.4	108.7	130.7	80.6	99.6	99.1	92.2	94.8	99.4	92.2	94.9	94.7	81.9	86.8	91.3	121.9	110.3
05	**METALS**	93.9	107.4	98.2	93.7	104.2	97.1	99.7	85.2	95.1	99.7	85.2	95.0	96.1	83.1	92.0	99.5	101.7	100.2
06	**WOOD & PLASTICS**	90.4	107.7	99.4	90.7	103.2	97.2	101.5	54.2	76.9	101.5	54.2	76.9	86.7	54.2	69.8	101.5	126.0	114.3
07	**THERMAL & MOISTURE PROTECTION**	103.2	101.9	102.7	99.8	88.9	95.2	101.8	87.3	95.7	102.2	87.3	95.9	101.8	90.3	97.0	101.7	122.5	110.5
08	**DOORS & WINDOWS**	94.7	112.4	99.3	94.7	109.8	98.6	106.9	63.1	95.5	106.9	63.1	95.5	107.8	57.3	94.6	106.9	120.2	110.4
09200	PLASTER & GYPSUM BOARD	84.4	107.7	98.8	85.5	103.0	96.4	106.2	52.1	72.6	106.2	52.1	72.6	96.0	52.1	68.7	106.2	125.8	118.4
095,098	CEILINGS & ACOUSTICAL TREATMENT	109.9	107.7	108.6	111.6	103.0	106.5	98.5	52.1	70.8	98.5	52.1	70.8	87.5	52.1	66.3	98.5	125.8	114.8
09600	FLOORING	100.3	99.2	100.0	100.3	63.8	90.5	99.8	102.9	100.7	99.6	102.9	100.5	93.9	102.9	96.3	99.6	126.8	107.0
097,099	WALL FINISHES, PAINTS & COATINGS	102.2	117.1	111.3	102.2	78.7	87.9	91.5	94.9	93.6	91.5	94.9	93.6	91.5	38.9	59.4	91.5	115.5	106.1
09	**FINISHES**	98.3	107.7	103.2	99.0	91.3	95.0	100.8	71.7	85.6	101.0	71.7	85.7	94.6	62.8	78.1	101.3	125.1	113.7
10 - 14	**TOTAL DIV. 10000 - 14000**	100.0	103.3	100.7	100.0	107.8	101.7	100.0	74.6	94.6	100.0	74.6	94.6	100.0	70.7	93.7	100.0	116.6	103.5
15	**MECHANICAL**	100.1	107.1	103.0	100.1	93.1	97.2	100.1	90.3	96.0	100.1	90.3	96.0	100.1	84.8	93.7	100.1	116.6	106.9
16	**ELECTRICAL**	93.0	108.8	100.6	91.0	110.9	100.7	97.8	72.7	85.6	97.6	72.7	85.5	95.9	72.7	84.6	98.0	114.8	106.2
01 - 16	**WEIGHTED AVERAGE**	98.3	107.1	102.2	98.9	98.2	98.6	100.9	82.4	92.7	100.6	82.4	92.5	97.8	78.0	89.0	99.6	116.8	107.2

	DIVISION	NEW JERSEY															NEW MEXICO		
		ELIZABETH			JERSEY CITY			NEWARK			PATERSON			TRENTON			ALBUQUERQUE		
		MAT.	INST.	TOTAL	MAT.	INST.	TOTAL	MAT.	INST.	TOTAL	MAT.	INST.	TOTAL	MAT.	INST.	TOTAL	MAT.	INST.	TOTAL
01590	**EQUIPMENT RENTAL**	.0	101.5	101.5	.0	99.4	99.4	.0	101.5	101.5	.0	101.5	101.5	.0	98.9	98.9	.0	116.3	116.3
02	**SITE CONSTRUCTION**	106.2	105.4	105.6	92.2	104.8	101.4	110.6	105.4	106.8	104.7	104.8	104.8	94.0	105.3	102.2	78.5	111.2	102.5
03100	CONCRETE FORMS & ACCESSORIES	113.1	130.1	127.8	102.8	130.2	126.4	101.0	130.2	126.2	101.6	130.0	126.1	101.7	129.7	125.8	96.0	68.6	72.4
03200	CONCRETE REINFORCEMENT	82.8	114.4	98.8	107.5	114.4	111.0	107.5	114.4	111.0	107.5	114.4	111.0	107.5	115.0	111.3	108.7	68.1	88.1
03300	CAST-IN-PLACE CONCRETE	91.3	131.1	107.7	82.8	131.1	102.7	96.7	131.2	110.9	105.7	131.1	116.1	98.1	125.1	109.2	102.6	74.2	90.9
03	**CONCRETE**	99.6	126.5	113.0	97.2	126.3	111.6	103.8	126.5	115.1	108.1	126.4	117.2	104.5	124.1	114.2	105.3	71.5	88.5
04	**MASONRY**	113.5	127.6	122.3	91.3	127.6	113.8	101.1	127.6	117.6	97.0	127.6	116.0	90.9	121.8	110.1	119.3	64.4	85.3
05	**METALS**	95.8	107.2	99.5	99.6	104.3	101.1	99.6	107.3	102.0	94.4	107.0	98.4	94.4	102.2	96.9	100.1	87.7	96.2
06	**WOOD & PLASTICS**	117.0	131.1	124.4	101.5	131.1	116.9	103.3	131.1	117.8	103.3	131.1	117.8	101.5	131.0	116.9	95.4	70.1	82.2
07	**THERMAL & MOISTURE PROTECTION**	102.5	133.5	115.6	101.7	133.5	115.1	101.9	133.5	115.2	102.3	125.9	112.2	100.6	122.1	109.6	99.0	73.4	88.2
08	**DOORS & WINDOWS**	108.1	124.1	112.3	106.9	124.1	111.4	113.1	124.1	116.0	113.1	124.1	116.0	106.9	123.3	111.2	94.6	72.2	88.8
09200	PLASTER & GYPSUM BOARD	109.6	131.0	122.9	106.2	131.0	121.6	106.2	131.0	121.6	106.2	131.0	121.6	106.2	131.0	121.6	87.5	68.8	75.9
095,098	CEILINGS & ACOUSTICAL TREATMENT	87.5	131.0	113.4	98.5	131.0	117.9	98.5	131.0	117.9	98.5	131.0	117.9	98.5	131.0	117.9	109.0	68.8	85.0
09600	FLOORING	104.5	126.8	110.5	99.6	126.8	107.0	99.8	126.8	107.1	99.6	126.8	107.0	99.8	126.8	107.1	100.3	62.2	90.0
097,099	WALL FINISHES, PAINTS & COATINGS	91.5	133.8	117.3	91.5	133.8	117.3	91.5	133.8	117.3	91.5	133.8	117.3	91.5	115.5	106.2	102.2	58.6	75.6
09	**FINISHES**	101.1	130.1	116.2	101.3	130.1	116.3	101.8	130.1	116.5	101.4	130.1	116.4	101.3	128.0	115.2	99.1	66.2	82.0
10 - 14	**TOTAL DIV. 10000 - 14000**	100.0	118.7	104.0	100.0	118.7	104.0	100.0	118.7	104.0	100.0	118.7	104.0	100.0	117.1	103.6	100.0	71.5	93.9
15	**MECHANICAL**	100.1	121.9	109.1	100.1	125.9	110.8	100.1	124.7	110.3	100.1	125.9	110.8	100.1	127.2	111.3	100.1	72.2	88.5
16	**ELECTRICAL**	95.9	135.0	114.9	99.5	137.6	118.0	99.4	137.6	118.0	99.5	135.0	116.8	98.1	134.7	115.9	84.0	78.1	81.1
01 - 16	**WEIGHTED AVERAGE**	100.8	123.6	110.9	99.8	124.5	110.8	102.1	124.6	112.1	101.4	124.2	111.5	99.6	122.6	109.8	98.5	76.2	88.6

	DIVISION	NEW MEXICO												NEW YORK					
		FARMINGTON			LAS CRUCES			ROSWELL			SANTA FE			ALBANY			BINGHAMTON		
		MAT.	INST.	TOTAL	MAT.	INST.	TOTAL	MAT.	INST.	TOTAL	MAT.	INST.	TOTAL	MAT.	INST.	TOTAL	MAT.	INST.	TOTAL
01590	**EQUIPMENT RENTAL**	.0	116.3	116.3	.0	86.8	86.8	.0	116.3	116.3	.0	116.3	116.3	.0	115.6	115.6	.0	115.8	115.8
02	**SITE CONSTRUCTION**	83.8	111.2	103.9	89.2	86.6	87.3	88.2	111.2	105.1	78.2	111.2	102.4	72.1	106.6	97.4	94.4	90.2	91.3
03100	CONCRETE FORMS & ACCESSORIES	96.0	68.6	72.4	93.4	66.9	70.5	96.0	68.5	72.3	96.0	68.6	72.4	97.4	93.1	93.7	102.5	80.5	83.5
03200	CONCRETE REINFORCEMENT	118.8	68.1	93.0	112.7	56.6	84.2	117.9	57.0	87.0	116.6	68.1	91.9	103.3	99.1	101.2	102.3	86.9	94.5
03300	CAST-IN-PLACE CONCRETE	103.2	74.2	91.3	89.8	64.7	79.5	95.1	74.1	86.5	97.0	74.2	87.6	87.7	100.8	93.1	104.8	95.5	101.0
03	**CONCRETE**	109.1	71.5	90.4	87.2	64.9	76.1	109.4	69.3	89.5	103.8	71.5	87.7	100.6	97.7	99.2	100.5	88.8	94.7
04	**MASONRY**	123.6	64.4	86.9	108.7	58.9	77.9	124.4	64.4	87.2	116.8	64.4	84.3	89.1	97.2	94.2	105.8	88.3	95.0
05	**METALS**	97.8	87.7	94.6	97.7	73.6	90.0	97.8	82.3	92.8	97.8	87.7	94.6	96.4	108.6	100.3	92.9	115.2	100.0
06	**WOOD & PLASTICS**	95.4	70.1	82.2	87.2	68.7	77.6	95.4	70.1	82.2	95.4	70.1	82.2	97.0	92.4	94.6	105.9	79.6	92.2
07	**THERMAL & MOISTURE PROTECTION**	99.6	73.4	88.6	84.3	66.6	76.8	99.9	73.4	88.7	99.2	73.4	88.3	93.8	92.8	93.4	102.5	87.3	96.1
08	**DOORS & WINDOWS**	97.5	72.2	90.9	87.1	68.0	82.1	93.5	68.8	87.1	93.7	72.2	88.1	95.6	88.7	93.8	90.8	77.6	87.3
09200	PLASTER & GYPSUM BOARD	82.7	68.8	74.0	85.1	68.8	75.0	82.7	68.8	74.0	82.7	68.8	74.0	109.8	91.9	98.7	114.6	78.5	92.2
095,098	CEILINGS & ACOUSTICAL TREATMENT	100.5	68.8	81.6	98.5	68.8	80.7	100.5	68.8	81.6	100.5	68.8	81.6	97.9	91.9	94.3	97.9	78.5	86.4
09600	FLOORING	100.3	62.2	90.0	132.4	62.4	113.5	100.3	62.2	90.0	100.3	62.2	90.0	87.1	96.5	89.7	97.7	91.2	95.9
097,099	WALL FINISHES, PAINTS & COATINGS	102.2	58.6	75.6	95.6	58.6	73.0	102.2	58.6	75.6	102.2	58.6	75.6	86.1	78.4	81.4	91.1	82.4	85.8
09	**FINISHES**	96.5	66.2	80.7	110.8	65.2	87.1	96.9	66.2	80.9	96.3	66.2	80.6	97.2	92.0	94.5	98.3	82.2	89.9
10 - 14	**TOTAL DIV. 10000 - 14000**	100.0	71.5	93.9	100.0	67.7	93.1	100.0	71.5	93.9	100.0	71.5	93.9	100.0	95.7	99.1	100.0	92.8	98.5
15	**MECHANICAL**	100.1	72.2	88.5	100.3	71.6	88.4	100.1	72.0	88.4	100.1	72.2	88.5	100.3	91.8	96.8	100.5	84.5	93.9
16	**ELECTRICAL**	82.4	78.1	80.3	84.7	58.9	72.2	83.1	78.1	80.7	84.0	78.1	81.1	103.6	94.1	99.0	101.6	87.4	94.7
01 - 16	**WEIGHTED AVERAGE**	98.7	76.2	88.7	95.8	67.9	83.4	98.7	75.3	88.3	97.5	76.2	88.0	97.8	96.4	97.2	98.5	88.8	94.2

	DIVISION	NEW YORK																	
		BUFFALO			HICKSVILLE			NEW YORK			RIVERHEAD			ROCHESTER			SCHENECTADY		
		MAT.	INST.	TOTAL	MAT.	INST.	TOTAL	MAT.	INST.	TOTAL	MAT.	INST.	TOTAL	MAT.	INST.	TOTAL	MAT.	INST.	TOTAL
01590	**EQUIPMENT RENTAL**	.0	93.1	93.1	.0	117.2	117.2	.0	117.5	117.5	.0	117.2	117.2	.0	116.5	116.5	.0	115.6	115.6
02	**SITE CONSTRUCTION**	96.3	93.7	94.4	112.7	131.1	126.2	136.2	128.7	130.7	113.4	131.1	126.3	74.3	107.4	98.6	71.8	106.6	97.3
03100	CONCRETE FORMS & ACCESSORIES	98.9	115.2	113.0	89.3	154.4	145.5	111.1	181.7	172.0	94.9	154.4	146.2	99.0	102.7	102.2	102.1	92.6	93.9
03200	CONCRETE REINFORCEMENT	104.1	104.6	104.4	103.6	182.3	143.6	111.1	180.5	146.4	105.6	182.3	144.6	105.1	85.9	95.3	102.0	99.1	100.5
03300	CAST-IN-PLACE CONCRETE	115.7	119.1	117.1	98.0	159.1	123.1	118.4	160.9	135.8	99.7	159.1	124.1	110.1	101.6	106.6	96.2	100.1	97.8
03	**CONCRETE**	106.8	113.7	110.2	103.9	159.8	131.7	117.7	172.4	144.9	104.8	159.8	132.1	112.1	100.0	106.1	104.8	97.2	101.0
04	**MASONRY**	104.7	121.1	114.9	109.0	160.5	140.9	102.8	158.4	137.3	115.1	160.5	143.2	102.7	100.2	101.2	90.0	95.9	93.7
05	**METALS**	100.3	94.9	98.6	103.1	138.8	114.5	104.4	145.4	117.5	103.5	138.8	114.8	96.7	107.1	100.0	96.4	108.6	100.3
06	**WOOD & PLASTICS**	102.4	114.6	108.8	87.2	154.5	122.3	108.6	189.9	150.9	93.1	154.5	125.1	99.3	103.7	101.6	102.5	92.4	97.2
07	**THERMAL & MOISTURE PROTECTION**	97.7	110.5	103.1	109.7	149.3	126.4	114.3	161.4	134.1	110.3	149.3	126.7	94.6	98.4	96.2	93.9	92.2	93.2
08	**DOORS & WINDOWS**	92.9	103.7	95.7	88.3	156.0	106.0	97.7	178.7	118.8	88.3	156.0	106.0	96.0	92.6	95.1	95.6	88.7	93.8
09200	PLASTER & GYPSUM BOARD	99.6	114.7	109.0	101.4	156.0	135.3	113.9	191.9	162.4	104.2	156.0	136.4	101.5	103.8	102.9	109.8	91.9	98.7
095,098	CEILINGS & ACOUSTICAL TREATMENT	94.6	114.7	106.6	77.5	156.0	124.3	107.8	191.9	158.0	82.6	156.0	126.4	99.0	103.8	101.9	97.9	91.9	94.3
09600	FLOORING	97.9	119.6	103.8	97.3	86.2	94.3	96.6	173.9	117.5	98.6	86.2	95.2	87.4	104.8	92.1	87.1	96.5	89.7
097,099	WALL FINISHES, PAINTS & COATINGS	95.1	114.3	106.8	114.1	151.4	136.9	96.9	151.6	130.3	114.1	151.4	136.9	93.0	103.4	99.4	86.1	78.4	81.4
09	**FINISHES**	95.5	116.7	106.6	103.2	140.1	122.4	108.0	178.6	144.8	105.2	140.1	123.4	96.7	103.8	100.4	96.9	91.7	94.2
10 - 14	**TOTAL DIV. 10000 - 14000**	100.0	107.3	101.6	100.0	137.8	108.1	100.0	142.4	109.1	100.0	137.8	108.1	100.0	101.1	100.2	100.0	95.2	99.0
15	**MECHANICAL**	99.8	99.6	99.7	99.8	152.0	121.5	100.2	164.6	126.9	99.9	152.0	121.5	99.8	94.7	97.7	100.3	91.1	96.5
16	**ELECTRICAL**	100.2	99.2	99.7	102.0	155.1	127.8	109.9	181.3	144.6	103.5	155.1	128.6	103.9	94.9	99.5	101.1	94.1	97.7
01 - 16	**WEIGHTED AVERAGE**	99.8	105.7	102.4	101.2	149.5	122.7	106.0	164.4	131.9	102.2	149.5	123.2	99.8	99.7	99.8	98.1	95.9	97.1

	DIVISION	NEW YORK															NORTH CAROLINA		
		SYRACUSE			UTICA			WATERTOWN			WHITE PLAINS			YONKERS			ASHEVILLE		
		MAT.	INST.	TOTAL	MAT.	INST.	TOTAL	MAT.	INST.	TOTAL	MAT.	INST.	TOTAL	MAT.	INST.	TOTAL	MAT.	INST.	TOTAL
01590	**EQUIPMENT RENTAL**	.0	115.6	115.6	.0	115.6	115.6	.0	115.6	115.6	.0	117.2	117.2	.0	117.2	117.2	.0	93.3	93.3
02	**SITE CONSTRUCTION**	93.2	107.0	103.3	69.9	105.4	96.0	77.9	107.4	99.6	125.6	125.2	125.4	135.5	125.1	127.8	102.7	72.2	80.3
03100	CONCRETE FORMS & ACCESSORIES	101.3	89.9	91.4	102.6	84.0	86.6	84.7	90.5	89.7	110.7	139.5	135.6	110.9	139.6	135.7	95.1	42.7	49.9
03200	CONCRETE REINFORCEMENT	103.3	96.1	99.6	103.3	84.5	93.8	104.0	85.1	94.4	103.9	181.3	143.2	108.0	181.3	145.2	93.8	46.9	70.0
03300	CAST-IN-PLACE CONCRETE	97.2	97.2	97.2	89.0	93.3	90.7	103.6	83.1	95.2	104.3	132.3	115.8	116.8	132.3	123.1	96.5	49.3	77.1
03	**CONCRETE**	103.6	94.4	99.1	101.5	88.4	95.0	114.3	87.9	101.2	105.9	143.7	124.7	116.5	143.8	130.1	100.3	47.6	74.1
04	**MASONRY**	96.4	99.3	98.2	88.7	91.2	90.2	89.8	91.2	90.7	97.5	132.0	118.9	102.4	132.0	120.7	80.5	40.1	55.5
05	**METALS**	96.3	106.6	99.6	94.4	102.1	96.9	94.5	101.6	96.7	91.6	134.7	105.4	100.5	134.8	111.4	92.4	76.6	87.4
06	**WOOD & PLASTICS**	102.5	87.5	94.7	102.5	83.6	92.7	83.0	92.4	87.9	109.5	140.3	125.5	109.3	140.3	125.4	94.7	43.2	67.9
07	**THERMAL & MOISTURE PROTECTION**	102.2	96.1	99.6	93.8	92.4	93.2	94.0	92.1	93.2	114.8	140.0	125.4	115.1	140.0	125.6	101.3	45.7	77.9
08	**DOORS & WINDOWS**	93.5	84.2	91.1	95.6	78.9	91.2	95.6	81.0	91.8	92.2	148.2	106.8	95.8	148.3	109.5	92.6	43.4	79.7
09200	PLASTER & GYPSUM BOARD	109.8	87.0	95.6	109.8	82.9	93.1	102.0	92.0	95.8	106.6	140.9	127.9	113.5	140.9	130.5	103.4	41.5	64.9
095,098	CEILINGS & ACOUSTICAL TREATMENT	97.9	87.0	91.4	97.9	82.9	89.0	97.9	92.0	94.4	78.0	140.9	115.5	106.1	140.9	126.9	94.1	41.5	62.7
09600	FLOORING	88.6	92.4	89.6	87.1	87.1	87.1	79.3	87.1	81.4	92.8	168.7	113.3	92.4	168.7	113.1	101.8	48.1	87.3
097,099	WALL FINISHES, PAINTS & COATINGS	91.6	86.2	88.3	86.1	84.0	84.8	86.1	70.8	76.8	94.5	151.4	129.2	94.5	175.0	143.6	110.4	37.7	66.1
09	**FINISHES**	98.2	89.6	93.7	96.9	84.3	90.4	94.0	87.6	90.7	97.6	146.4	123.0	106.1	149.1	128.5	97.7	43.1	69.3
10 - 14	**TOTAL DIV. 10000 - 14000**	100.0	96.3	99.2	100.0	92.7	98.4	100.0	78.5	95.4	100.0	133.2	107.1	100.0	133.2	107.1	100.0	59.0	91.3
15	**MECHANICAL**	100.3	89.0	95.6	100.3	87.2	94.9	100.3	74.1	89.5	100.5	131.7	113.4	100.5	131.7	113.4	100.1	42.6	76.2
16	**ELECTRICAL**	101.1	92.2	96.8	97.7	91.8	94.8	101.1	91.4	96.4	99.0	147.1	122.4	106.4	147.1	126.2	97.8	39.4	69.4
01 - 16	**WEIGHTED AVERAGE**	99.0	94.7	97.1	96.9	90.9	94.2	98.5	88.5	94.1	99.5	138.2	116.7	104.5	138.5	119.6	96.8	48.9	75.5

	DIVISION	NORTH CAROLINA																	
		CHARLOTTE			DURHAM			FAYETTEVILLE			GREENSBORO			RALEIGH			WILMINGTON		
		MAT.	INST.	TOTAL	MAT.	INST.	TOTAL	MAT.	INST.	TOTAL	MAT.	INST.	TOTAL	MAT.	INST.	TOTAL	MAT.	INST.	TOTAL
01590	**EQUIPMENT RENTAL**	.0	93.3	93.3	.0	99.6	99.6	.0	99.6	99.6	.0	99.6	99.6	.0	99.6	99.6	.0	93.3	93.3
02	**SITE CONSTRUCTION**	102.9	72.2	80.4	102.7	82.0	87.5	100.7	82.0	87.0	102.6	82.0	87.5	103.7	82.0	87.8	103.6	72.2	80.6
03100	CONCRETE FORMS & ACCESSORIES	102.7	42.8	51.0	97.0	43.0	50.4	92.7	43.1	49.9	97.1	42.9	50.3	99.9	43.0	50.8	96.6	43.1	50.4
03200	CONCRETE REINFORCEMENT	94.2	47.0	70.2	94.2	55.0	74.3	93.3	55.0	73.8	94.2	55.0	74.3	94.2	55.0	74.3	94.5	55.0	74.4
03300	CAST-IN-PLACE CONCRETE	99.0	49.3	78.5	96.8	49.4	77.3	93.4	49.4	75.3	95.9	49.3	76.8	102.4	49.4	80.6	96.1	49.4	76.9
03	**CONCRETE**	101.3	47.7	74.7	99.8	49.3	74.8	97.4	49.4	73.6	99.4	49.2	74.5	102.8	49.3	76.2	100.2	49.4	75.0
04	**MASONRY**	85.9	40.1	57.5	82.8	40.1	56.3	86.6	40.1	57.8	82.6	40.1	56.3	87.4	40.1	58.1	70.4	40.1	51.6
05	**METALS**	94.9	76.7	89.1	97.5	80.3	92.0	97.4	80.3	91.9	98.0	80.1	92.3	95.3	80.3	90.5	91.9	80.3	88.2
06	**WOOD & PLASTICS**	103.9	43.2	72.3	97.0	43.2	69.0	91.8	43.2	66.5	97.0	43.2	69.0	100.3	43.2	70.6	96.5	43.2	68.8
07	**THERMAL & MOISTURE PROTECTION**	101.3	47.3	78.6	101.8	46.2	78.4	101.5	46.9	78.5	101.8	46.2	78.4	101.6	46.3	78.3	101.3	46.9	78.4
08	**DOORS & WINDOWS**	96.6	43.4	82.7	96.6	46.1	83.4	92.7	46.1	80.5	96.6	46.1	83.4	93.4	46.1	81.1	92.7	46.1	80.5
09200	PLASTER & GYPSUM BOARD	109.6	41.5	67.3	109.6	41.5	67.3	103.6	41.5	65.0	109.6	41.5	67.3	109.6	41.5	67.3	104.3	41.5	65.3
095,098	CEILINGS & ACOUSTICAL TREATMENT	100.9	41.5	65.5	100.9	41.5	65.5	94.9	41.5	63.1	100.9	41.5	65.5	100.9	41.5	65.5	94.9	41.5	63.1
09600	FLOORING	105.8	48.1	90.2	106.0	48.1	90.3	101.9	48.1	87.4	106.0	48.1	90.3	106.0	48.1	90.3	102.8	48.1	88.0
097,099	WALL FINISHES, PAINTS & COATINGS	110.4	37.7	66.1	110.4	37.7	66.1	110.4	37.7	66.1	110.4	37.7	66.1	110.4	37.7	66.1	110.4	37.7	66.1
09	**FINISHES**	101.5	43.1	71.1	101.6	43.1	71.2	98.0	43.1	69.4	101.6	43.1	71.2	101.7	43.1	71.2	98.3	43.1	69.6
10 - 14	**TOTAL DIV. 10000 - 14000**	100.0	59.0	91.3	100.0	74.5	94.6	100.0	74.5	94.6	100.0	59.0	91.3	100.0	74.5	94.6	100.0	74.5	94.6
15	**MECHANICAL**	100.1	42.6	76.2	100.1	42.8	76.3	100.1	42.8	76.3	100.1	42.7	76.3	100.1	42.8	76.3	100.1	42.8	76.3
16	**ELECTRICAL**	99.6	38.5	69.9	97.3	37.7	68.3	94.7	37.7	67.0	97.9	37.7	68.6	97.9	37.7	68.6	98.3	37.7	68.8
01 - 16	**WEIGHTED AVERAGE**	98.6	48.9	76.5	98.4	50.7	77.2	97.1	50.8	76.5	98.5	50.3	77.1	98.4	50.7	77.2	96.4	49.9	75.7

City Cost Indexes

DIVISION		NORTH CAROLINA			NORTH DAKOTA													OHIO		
		WINSTON-SALEM			BISMARCK			FARGO			GRAND FORKS			MINOT				AKRON		
		MAT.	INST.	TOTAL	MAT.	INST.	TOTAL	MAT.	INST.	TOTAL	MAT.	INST.	TOTAL	MAT.	INST.	TOTAL		MAT.	INST.	TOTAL
01590	EQUIPMENT RENTAL	.0	99.6	99.6	.0	98.3	98.3	.0	98.3	98.3	.0	98.3	98.3	.0	98.3	98.3		.0	97.4	97.4
02	SITE CONSTRUCTION	102.9	82.0	87.6	90.5	96.5	94.9	89.9	96.5	94.8	98.7	94.0	95.3	96.2	96.5	96.4		101.7	106.7	105.4
03100	CONCRETE FORMS & ACCESSORIES	98.5	42.8	50.4	95.1	49.8	56.0	96.1	50.1	56.4	91.8	43.4	50.1	86.9	56.4	60.6		98.7	96.7	96.9
03200	CONCRETE REINFORCEMENT	94.2	47.0	70.2	101.2	78.4	89.7	93.3	77.6	85.3	99.7	78.6	89.0	103.1	78.5	90.6		94.8	96.8	95.9
03300	CAST-IN-PLACE CONCRETE	99.0	49.3	78.5	105.1	55.2	84.6	110.2	57.3	88.4	107.6	52.6	85.0	107.6	53.8	85.5		98.5	105.1	101.2
03	CONCRETE	101.0	47.7	74.5	99.6	58.2	79.1	110.0	58.9	84.7	105.5	54.3	80.1	104.2	60.7	82.6		98.6	98.6	98.6
04	MASONRY	82.8	40.1	56.3	104.5	65.1	80.1	106.9	48.7	70.8	106.1	66.6	81.6	107.1	66.4	81.9		89.7	102.5	97.7
05	METALS	95.3	76.7	89.3	92.6	83.0	89.5	95.0	81.6	90.7	92.6	79.0	88.2	92.9	83.1	89.8		89.8	81.8	87.2
06	WOOD & PLASTICS	97.0	43.2	69.0	88.3	44.7	65.6	88.3	45.4	66.0	84.7	41.1	62.0	79.8	53.7	66.2		92.3	94.8	93.6
07	THERMAL & MOISTURE PROTECTION	101.8	46.2	78.4	99.2	55.6	80.8	99.8	52.6	79.9	99.8	56.0	81.4	99.5	56.7	81.5		99.6	100.7	100.0
08	DOORS & WINDOWS	96.6	43.9	82.8	99.4	50.4	86.6	99.4	50.7	86.7	99.4	45.5	85.3	99.5	55.3	88.0		104.3	96.7	102.3
09200	PLASTER & GYPSUM BOARD	109.6	41.5	67.3	113.5	43.6	70.0	113.5	44.3	70.5	111.6	39.9	67.0	110.1	52.8	74.5		97.7	94.2	95.6
095,098	CEILINGS & ACOUSTICAL TREATMENT	100.9	41.5	65.5	138.5	43.6	81.9	138.5	44.3	82.3	138.5	39.9	79.7	138.5	52.8	87.4		90.4	94.2	92.7
09600	FLOORING	106.0	48.1	90.3	107.8	75.0	98.9	107.6	44.8	90.6	106.0	44.8	89.4	103.2	77.5	96.3		103.4	94.3	100.9
097,099	WALL FINISHES, PAINTS & COATINGS	110.4	37.7	66.1	97.2	37.5	60.8	97.2	71.0	81.2	97.2	28.7	55.4	97.2	31.4	57.1		107.9	118.5	114.4
09	FINISHES	101.6	43.1	71.2	114.7	52.0	82.0	114.5	49.9	80.9	114.4	41.2	76.3	113.2	57.0	84.0		99.4	98.5	98.9
10 - 14	TOTAL DIV. 10000 - 14000	100.0	59.0	91.3	100.0	69.1	93.4	100.0	69.2	93.4	100.0	45.0	88.3	100.0	70.3	93.7		100.0	86.8	97.2
15	MECHANICAL	100.1	42.7	76.2	100.5	65.3	85.9	100.5	69.3	87.6	100.5	45.2	77.5	100.5	63.4	85.1		99.9	102.0	100.8
16	ELECTRICAL	97.9	37.7	68.6	93.6	69.7	82.0	93.5	68.7	81.4	96.2	69.7	83.3	99.2	70.9	85.4		98.2	94.9	96.6
01 - 16	WEIGHTED AVERAGE	98.2	49.6	76.6	99.3	66.6	84.8	100.9	65.2	85.1	100.5	59.2	82.2	100.6	67.8	86.1		97.8	98.0	97.9

DIVISION		OHIO																	
		CANTON			CINCINNATI			CLEVELAND			COLUMBUS			DAYTON			LORAIN		
		MAT.	INST.	TOTAL	MAT.	INST.	TOTAL	MAT.	INST.	TOTAL	MAT.	INST.	TOTAL	MAT.	INST.	TOTAL	MAT.	INST.	TOTAL
01590	EQUIPMENT RENTAL	.0	97.4	97.4	.0	102.5	102.5	.0	97.7	97.7	.0	96.4	96.4	.0	96.7	96.7	.0	97.4	97.4
02	SITE CONSTRUCTION	101.8	106.8	105.5	74.4	108.5	99.4	101.7	106.9	105.5	85.6	101.8	97.5	73.2	108.1	98.8	101.1	106.0	104.7
03100	CONCRETE FORMS & ACCESSORIES	98.7	85.6	87.4	97.0	88.3	89.5	98.8	102.7	102.1	99.2	86.3	88.1	96.9	79.2	81.6	98.7	103.7	103.0
03200	CONCRETE REINFORCEMENT	94.8	83.6	89.1	98.9	86.0	92.4	95.4	97.0	96.2	76.7	83.2	80.0	98.9	82.1	90.4	94.8	96.9	95.9
03300	CAST-IN-PLACE CONCRETE	99.5	101.8	100.4	85.1	88.8	86.6	96.7	112.5	103.2	88.9	90.9	89.7	77.5	89.1	82.3	93.8	107.5	99.4
03	CONCRETE	99.1	90.1	94.6	92.4	88.1	90.3	97.8	104.0	100.9	91.4	87.0	89.2	88.7	82.9	85.8	96.3	102.6	99.4
04	MASONRY	90.5	92.7	91.9	82.4	94.2	89.7	94.0	107.3	102.3	95.8	91.1	92.9	81.8	88.8	86.1	86.7	99.9	94.9
05	METALS	89.8	75.5	85.2	88.0	88.0	88.0	91.2	85.0	89.2	96.1	80.6	91.1	87.3	77.8	84.2	90.3	82.8	87.9
06	WOOD & PLASTICS	92.6	84.0	88.1	96.8	86.7	91.5	91.5	99.2	95.5	103.7	84.9	93.9	97.9	75.8	86.4	92.3	105.0	98.9
07	THERMAL & MOISTURE PROTECTION	100.1	96.2	98.4	94.2	96.2	95.1	98.3	112.4	104.2	98.1	93.0	96.0	99.4	89.7	95.3	100.0	107.2	103.1
08	DOORS & WINDOWS	98.7	77.2	93.1	96.1	85.5	93.4	95.2	99.1	96.2	99.5	83.5	95.3	96.3	78.6	91.7	98.7	102.2	99.6
09200	PLASTER & GYPSUM BOARD	98.5	83.2	89.0	99.8	86.6	91.6	96.9	98.8	98.1	95.2	84.3	88.5	99.8	75.4	84.7	97.7	104.7	102.0
095,098	CEILINGS & ACOUSTICAL TREATMENT	90.4	83.2	86.1	98.8	86.6	91.6	88.6	98.8	94.7	93.9	84.3	88.2	99.8	75.4	85.3	90.4	104.7	98.9
09600	FLOORING	103.6	85.6	98.7	108.4	99.3	106.0	103.2	108.0	104.5	97.5	91.3	95.8	111.2	86.3	104.4	103.6	106.7	104.4
097,099	WALL FINISHES, PAINTS & COATINGS	107.9	90.4	97.2	107.8	92.8	98.6	107.9	121.9	116.4	98.4	98.5	98.4	107.8	90.3	97.1	107.9	121.9	116.4
09	FINISHES	99.6	85.1	92.1	99.3	90.6	94.8	98.9	105.3	102.2	97.5	88.2	92.7	100.3	80.8	90.2	99.5	106.7	103.3
10 - 14	TOTAL DIV. 10000 - 14000	100.0	75.0	94.7	100.0	92.7	98.4	100.0	105.2	101.1	100.0	92.8	98.5	100.0	90.6	98.0	100.0	103.2	100.7
15	MECHANICAL	99.9	86.3	94.2	99.8	88.5	95.1	99.9	108.1	103.3	100.0	91.7	96.6	100.8	88.0	95.5	99.9	90.8	96.1
16	ELECTRICAL	97.4	91.0	94.3	98.1	81.5	90.1	98.2	107.1	102.6	98.1	88.4	93.4	96.1	88.3	92.3	97.5	88.6	93.2
01 - 16	WEIGHTED AVERAGE	97.3	88.4	93.4	94.9	90.2	92.8	97.2	104.3	100.4	97.3	89.6	93.9	94.6	87.0	91.2	96.9	97.1	97.0

DIVISION		OHIO									OKLAHOMA								
		SPRINGFIELD			TOLEDO			YOUNGSTOWN			ENID			LAWTON			MUSKOGEE		
		MAT.	INST.	TOTAL	MAT.	INST.	TOTAL	MAT.	INST.	TOTAL	MAT.	INST.	TOTAL	MAT.	INST.	TOTAL	MAT.	INST.	TOTAL
01590	EQUIPMENT RENTAL	.0	96.7	96.7	.0	99.5	99.5	.0	97.4	97.4	.0	77.1	77.1	.0	78.2	78.2	.0	86.7	86.7
02	SITE CONSTRUCTION	73.7	107.2	98.3	84.9	102.9	98.1	101.6	107.1	105.6	108.0	88.9	94.0	103.4	90.5	94.0	94.1	84.9	87.4
03100	CONCRETE FORMS & ACCESSORIES	96.9	85.4	87.0	99.2	96.4	96.8	98.7	89.8	91.0	93.5	39.4	46.8	97.4	53.9	59.9	98.8	35.7	44.3
03200	CONCRETE REINFORCEMENT	98.9	82.1	90.4	76.7	95.3	86.2	94.8	96.7	95.8	97.0	82.4	89.6	97.3	82.4	89.7	97.0	38.5	67.3
03300	CAST-IN-PLACE CONCRETE	81.3	89.0	84.5	88.9	105.7	95.8	97.6	103.4	100.0	91.6	50.0	74.5	88.6	50.1	72.8	82.4	38.5	64.4
03	CONCRETE	90.6	85.6	88.1	91.4	98.9	95.1	98.2	95.0	96.6	90.0	51.4	70.8	86.8	57.9	72.4	82.6	38.3	60.6
04	MASONRY	82.1	88.6	86.2	105.7	102.8	103.9	89.9	95.2	93.2	101.6	61.4	76.6	96.0	61.4	74.5	111.2	54.9	76.3
05	METALS	87.3	77.6	84.2	95.9	89.2	93.8	89.8	82.1	87.3	89.9	67.1	82.6	93.3	67.2	84.9	89.8	58.6	79.9
06	WOOD & PLASTICS	99.1	84.5	91.5	103.7	94.2	98.7	92.3	87.4	89.7	96.9	36.4	65.4	100.0	55.8	77.0	102.1	36.0	67.7
07	THERMAL & MOISTURE PROTECTION	99.3	90.5	95.6	100.3	107.8	103.4	100.2	99.0	99.7	99.9	61.6	83.7	99.6	63.6	84.5	99.5	47.6	77.6
08	DOORS & WINDOWS	94.4	80.7	90.8	97.4	93.7	96.4	98.7	93.6	97.4	95.4	50.5	83.7	96.9	61.0	87.5	95.4	35.4	79.7
09200	PLASTER & GYPSUM BOARD	99.8	84.3	90.2	95.2	93.9	94.4	97.7	86.6	90.8	81.4	35.5	52.9	83.8	55.5	66.2	85.3	34.9	54.0
095,098	CEILINGS & ACOUSTICAL TREATMENT	99.8	84.3	90.6	93.9	93.9	93.9	90.4	86.6	88.1	80.3	35.5	53.6	89.7	55.5	69.3	89.7	34.9	57.0
09600	FLOORING	111.2	86.3	104.4	96.6	96.4	96.6	103.6	93.7	100.9	112.4	51.4	95.9	115.3	51.4	98.0	116.5	41.4	96.2
097,099	WALL FINISHES, PAINTS & COATINGS	107.8	90.3	97.1	98.4	106.5	103.4	107.9	100.0	103.1	97.0	65.4	77.7	97.0	65.4	77.7	97.0	35.0	59.2
09	FINISHES	100.3	85.9	92.8	97.2	97.0	97.1	99.6	90.9	95.0	93.8	42.5	67.1	96.9	53.9	74.5	96.9	37.0	65.7
10 - 14	TOTAL DIV. 10000 - 14000	100.0	91.6	98.2	100.0	87.6	97.3	100.0	85.7	97.0	100.0	59.6	91.4	100.0	62.0	91.9	100.0	59.0	91.3
15	MECHANICAL	100.8	88.0	95.4	100.0	102.9	101.2	99.9	92.5	96.8	100.1	65.3	85.6	100.1	65.3	85.6	100.1	29.2	70.6
16	ELECTRICAL	96.1	87.0	91.7	97.9	105.7	101.7	97.5	93.1	95.4	94.9	71.5	83.5	96.4	71.5	84.3	94.5	34.3	65.2
01 - 16	WEIGHTED AVERAGE	94.6	87.9	91.6	97.6	100.0	98.7	97.2	93.3	95.5	96.0	62.0	80.9	96.4	65.3	82.6	95.6	44.0	72.7

	DIVISION	OKLAHOMA						OREGON											
		OKLAHOMA CITY			TULSA			EUGENE			MEDFORD			PORTLAND			SALEM		
		MAT.	INST.	TOTAL	MAT.	INST.	TOTAL	MAT.	INST.	TOTAL	MAT.	INST.	TOTAL	MAT.	INST.	TOTAL	MAT.	INST.	TOTAL
01590	EQUIPMENT RENTAL	.0	78.5	78.5	.0	86.7	86.7	.0	99.6	99.6	.0	99.6	99.6	.0	99.6	99.6	.0	99.6	99.6
02	SITE CONSTRUCTION	104.7	91.1	94.7	100.4	85.9	89.8	108.7	104.7	105.7	118.2	104.7	108.3	109.8	104.7	106.1	108.3	104.7	105.6
03100	CONCRETE FORMS & ACCESSORIES	98.5	46.5	53.6	98.5	46.8	53.9	104.0	105.7	105.5	98.8	105.5	104.6	105.6	106.0	105.9	105.4	105.8	105.8
03200	CONCRETE REINFORCEMENT	97.3	82.4	89.7	97.3	82.3	89.7	105.8	99.3	102.5	103.1	99.3	101.2	106.6	99.5	103.0	106.7	99.5	103.0
03300	CAST-IN-PLACE CONCRETE	95.9	53.3	78.4	89.8	48.8	72.9	106.9	105.4	106.3	110.4	105.3	108.3	114.8	105.5	111.0	110.9	105.4	108.7
03	CONCRETE	90.4	55.7	73.1	87.4	55.1	71.4	108.3	104.0	106.2	114.9	103.9	109.4	112.4	104.2	108.3	110.5	104.1	107.3
04	MASONRY	97.8	63.5	76.5	95.3	63.5	75.6	113.4	103.3	107.1	110.6	103.3	106.1	114.8	105.9	109.3	118.4	105.9	110.6
05	METALS	95.2	67.1	86.2	92.6	81.0	88.9	94.2	96.8	95.0	93.8	96.5	94.7	95.4	97.2	96.0	94.6	97.1	95.4
06	WOOD & PLASTICS	101.9	44.8	72.1	100.7	46.1	72.3	95.2	105.5	100.6	88.6	105.5	97.4	96.3	105.5	101.1	96.3	105.5	101.1
07	THERMAL & MOISTURE PROTECTION	98.9	63.3	83.9	99.5	60.2	82.9	105.2	95.3	101.0	105.8	92.6	100.3	105.1	101.5	103.6	105.1	98.2	102.2
08	DOORS & WINDOWS	96.9	55.0	86.0	96.9	55.3	86.0	98.9	106.2	100.8	101.5	106.2	102.7	96.4	106.2	99.0	98.3	106.2	100.3
09200	PLASTER & GYPSUM BOARD	83.8	44.1	59.1	83.8	45.3	59.9	95.1	105.5	101.5	93.3	105.5	100.8	92.9	105.5	100.7	92.9	105.5	100.7
095,098	CEILINGS & ACOUSTICAL TREATMENT	89.7	44.1	62.5	89.7	45.3	63.2	106.0	105.5	105.7	116.2	105.5	109.8	106.0	105.5	105.7	106.0	105.5	105.7
09600	FLOORING	115.3	51.4	98.0	115.1	53.5	98.4	115.4	98.6	110.9	112.9	98.6	109.1	115.4	98.6	110.9	115.4	98.6	110.9
097,099	WALL FINISHES, PAINTS & COATINGS	97.0	65.4	77.7	97.0	51.7	69.3	115.2	72.0	88.8	115.2	62.6	83.1	115.2	72.0	88.8	115.2	72.0	88.8
09	FINISHES	97.0	48.0	71.4	96.4	47.6	71.0	110.1	100.8	105.2	112.3	99.7	105.8	109.8	100.8	105.1	109.7	100.8	105.0
10 - 14	TOTAL DIV. 10000 - 14000	100.0	61.5	91.8	100.0	62.2	91.9	100.0	106.4	101.4	100.0	106.4	101.4	100.0	106.4	101.4	100.0	106.4	101.4
15	MECHANICAL	100.1	66.4	86.1	100.1	60.3	83.6	100.2	106.1	102.7	100.2	106.1	102.7	100.2	108.9	103.8	100.2	106.1	102.7
16	ELECTRICAL	95.8	71.5	84.0	96.4	44.7	71.3	98.2	100.7	99.4	101.5	90.2	96.0	98.6	105.4	101.9	98.3	100.7	99.5
01 - 16	WEIGHTED AVERAGE	97.1	64.3	82.6	96.2	59.9	80.1	101.7	102.8	102.2	103.3	101.1	102.3	102.2	104.6	103.3	102.2	103.2	102.6

	DIVISION	PENNSYLVANIA																	
		ALLENTOWN			ALTOONA			ERIE			HARRISBURG			PHILADELPHIA			PITTSBURGH		
		MAT.	INST.	TOTAL	MAT.	INST.	TOTAL	MAT.	INST.	TOTAL	MAT.	INST.	TOTAL	MAT.	INST.	TOTAL	MAT.	INST.	TOTAL
01590	EQUIPMENT RENTAL	.0	115.6	115.6	.0	115.6	115.6	.0	115.6	115.6	.0	114.8	114.8	.0	94.7	94.7	.0	115.4	115.4
02	SITE CONSTRUCTION	91.9	106.8	102.8	96.1	106.7	103.9	92.7	107.4	103.5	82.9	105.5	99.5	101.8	95.2	97.0	98.3	109.1	106.2
03100	CONCRETE FORMS & ACCESSORIES	100.5	111.9	110.3	82.7	85.8	85.4	99.8	95.6	96.1	93.2	88.2	88.9	99.3	136.4	131.3	100.6	99.0	99.2
03200	CONCRETE REINFORCEMENT	103.3	101.9	102.6	100.2	93.8	96.9	102.3	90.9	96.5	103.3	98.4	100.8	104.1	137.0	120.8	100.3	109.9	105.2
03300	CAST-IN-PLACE CONCRETE	88.1	101.9	93.7	98.2	85.6	93.0	96.5	84.2	91.5	97.2	93.4	95.6	99.2	129.3	111.6	99.1	95.5	97.6
03	CONCRETE	98.1	107.5	102.8	93.5	88.7	91.1	92.6	92.0	92.3	100.3	93.4	96.9	108.1	133.5	120.7	96.1	101.0	98.6
04	MASONRY	93.7	99.2	97.1	95.8	83.7	88.3	86.0	95.9	92.2	94.4	89.6	91.4	98.8	135.5	121.5	88.6	100.3	95.8
05	METALS	96.9	121.4	104.7	90.5	115.3	98.5	90.7	114.4	98.3	98.8	119.1	105.3	105.0	127.8	112.3	96.3	124.3	105.3
06	WOOD & PLASTICS	101.8	115.2	108.8	79.2	87.6	83.6	98.3	95.1	96.6	95.5	87.1	91.1	97.2	136.3	117.6	100.0	98.5	99.2
07	THERMAL & MOISTURE PROTECTION	102.2	116.2	108.1	101.1	93.0	97.7	100.8	98.2	99.7	105.2	108.6	106.6	102.9	136.8	117.1	101.9	100.3	101.2
08	DOORS & WINDOWS	93.5	110.8	98.0	88.0	93.1	89.4	88.2	93.2	89.5	93.5	94.2	93.7	96.7	140.2	108.1	93.5	108.3	97.4
09200	PLASTER & GYPSUM BOARD	107.4	115.3	112.3	100.3	87.0	92.1	107.4	94.7	99.5	107.4	86.5	94.4	104.1	137.2	124.6	99.1	98.2	98.6
095,098	CEILINGS & ACOUSTICAL TREATMENT	88.5	115.3	104.5	92.8	87.0	89.4	88.5	94.7	92.2	88.5	86.5	87.3	98.3	137.2	121.5	88.6	98.2	94.4
09600	FLOORING	88.6	96.4	90.7	82.6	57.0	75.6	90.8	84.7	89.1	88.8	87.5	88.4	87.0	138.6	101.0	96.6	104.3	98.7
097,099	WALL FINISHES, PAINTS & COATINGS	91.6	90.0	90.6	86.8	107.8	99.6	98.1	91.3	94.0	91.6	87.3	89.0	94.6	138.4	121.3	99.5	113.5	108.0
09	FINISHES	95.5	106.3	101.2	93.9	81.9	87.7	97.0	92.8	94.8	94.5	87.4	90.8	100.7	137.2	119.7	95.9	101.0	98.6
10 - 14	TOTAL DIV. 10000 - 14000	100.0	105.9	101.3	100.0	97.8	99.5	100.0	102.9	100.6	100.0	97.4	99.4	100.0	125.9	105.5	100.0	103.0	100.6
15	MECHANICAL	100.3	109.5	104.1	99.8	90.4	95.9	99.8	94.0	97.4	100.3	94.0	97.7	100.1	133.4	113.9	100.1	98.8	99.6
16	ELECTRICAL	100.7	99.5	100.1	89.8	105.0	97.2	91.3	90.6	91.0	99.7	86.8	93.4	96.1	138.2	116.6	95.9	105.1	100.3
01 - 16	WEIGHTED AVERAGE	98.0	107.4	102.2	94.4	94.5	94.5	94.4	96.7	95.4	98.2	95.3	96.9	101.0	131.1	114.4	97.0	104.2	100.2

	DIVISION	PENNSYLVANIA									PUERTO RICO			RHODE ISLAND			SOUTH CAROLINA		
		READING			SCRANTON			YORK			SAN JUAN			PROVIDENCE			CHARLESTON		
		MAT.	INST.	TOTAL	MAT.	INST.	TOTAL	MAT.	INST.	TOTAL	MAT.	INST.	TOTAL	MAT.	INST.	TOTAL	MAT.	INST.	TOTAL
01590	EQUIPMENT RENTAL	.0	118.2	118.2	.0	115.6	115.6	.0	114.8	114.8	.0	89.1	89.1	.0	103.4	103.4	.0	99.3	99.3
02	SITE CONSTRUCTION	99.2	111.1	107.9	92.4	106.8	103.0	82.2	105.5	99.3	116.4	91.8	98.4	81.9	104.0	98.1	96.5	80.9	85.1
03100	CONCRETE FORMS & ACCESSORIES	98.8	88.6	90.0	100.7	88.7	90.3	81.3	88.1	87.2	94.9	19.5	29.9	104.9	117.7	115.9	97.1	37.3	45.5
03200	CONCRETE REINFORCEMENT	102.3	99.6	100.9	103.3	112.7	108.1	102.3	98.4	100.3	193.0	12.0	101.1	107.5	115.8	111.7	94.2	67.2	80.5
03300	CAST-IN-PLACE CONCRETE	72.5	98.9	83.3	92.0	92.1	92.1	86.7	93.3	89.4	108.0	30.4	76.1	88.5	111.0	97.7	85.3	47.8	69.9
03	CONCRETE	94.9	95.7	95.3	100.0	95.9	98.0	98.1	93.3	95.8	114.5	22.5	68.8	102.3	114.5	108.4	94.3	48.3	71.5
04	MASONRY	100.4	92.1	95.3	94.1	94.5	94.4	93.4	89.6	91.0	220.6	16.8	94.3	108.9	128.4	121.0	90.3	35.1	56.1
05	METALS	101.5	120.2	107.5	98.8	125.3	107.3	96.4	119.0	103.6	114.8	30.4	87.8	99.7	109.9	103.0	92.7	80.6	88.8
06	WOOD & PLASTICS	97.3	86.2	91.6	101.8	85.6	93.3	86.2	87.1	86.6	91.1	19.8	53.9	102.4	116.5	109.8	97.0	36.5	65.5
07	THERMAL & MOISTURE PROTECTION	102.3	110.8	105.8	102.1	102.8	102.4	100.8	108.6	104.0	148.2	23.2	95.6	101.0	116.5	107.5	101.4	42.6	76.7
08	DOORS & WINDOWS	94.3	94.4	94.4	93.5	97.9	94.7	90.4	94.2	91.4	153.5	16.5	117.7	100.8	115.8	104.7	96.6	42.1	82.4
09200	PLASTER & GYPSUM BOARD	107.4	85.7	93.9	109.8	84.9	94.3	103.8	86.5	93.0	258.0	17.4	108.5	103.3	116.1	111.2	109.6	34.6	63.0
095,098	CEILINGS & ACOUSTICAL TREATMENT	87.1	85.7	86.3	97.9	84.9	90.2	90.2	86.5	88.0	334.4	17.4	145.3	97.8	116.1	108.7	100.9	34.6	61.3
09600	FLOORING	85.2	101.6	89.6	88.6	91.1	89.3	84.3	91.2	86.2	199.8	16.4	150.1	100.5	129.7	108.4	106.0	39.2	87.9
097,099	WALL FINISHES, PAINTS & COATINGS	92.8	99.8	97.1	91.6	98.6	95.9	91.6	87.3	89.0	199.0	15.2	86.8	94.0	120.2	110.0	110.4	38.2	66.3
09	FINISHES	97.8	91.1	94.3	98.2	89.4	93.6	93.2	87.4	90.2	250.3	18.7	129.7	100.3	120.7	110.9	101.9	37.2	68.2
10 - 14	TOTAL DIV. 10000 - 14000	100.0	100.2	100.0	100.0	101.8	100.4	100.0	97.4	99.4	100.0	19.6	82.9	100.0	97.4	99.4	100.0	64.6	92.4
15	MECHANICAL	100.3	107.9	103.5	100.3	86.2	94.5	100.3	94.0	97.7	103.7	15.3	67.0	100.1	109.6	104.0	100.1	43.0	76.4
16	ELECTRICAL	97.9	93.5	95.7	100.7	91.7	96.3	94.5	86.8	90.7	129.8	15.3	74.2	96.5	105.4	100.8	97.3	32.4	65.8
01 - 16	WEIGHTED AVERAGE	98.8	100.9	99.7	98.8	96.3	97.7	96.2	95.3	95.8	135.2	25.5	86.5	99.9	112.7	105.6	97.3	47.9	75.3

DIVISION		SOUTH CAROLINA												SOUTH DAKOTA					
		COLUMBIA			FLORENCE			GREENVILLE			SPARTANBURG			ABERDEEN			PIERRE		
		MAT.	INST.	TOTAL	MAT.	INST.	TOTAL	MAT.	INST.	TOTAL	MAT.	INST.	TOTAL	MAT.	INST.	TOTAL	MAT.	INST.	TOTAL
01590	**EQUIPMENT RENTAL**	.0	99.3	99.3	.0	99.3	99.3	.0	99.3	99.3	.0	99.3	99.3	.0	98.3	98.3	.0	98.3	98.3
02	**SITE CONSTRUCTION**	95.8	80.9	84.9	107.4	80.9	88.0	102.3	80.6	86.3	102.1	80.6	86.3	86.1	93.5	91.5	84.4	93.5	91.1
03100	CONCRETE FORMS & ACCESSORIES	101.9	39.5	48.1	82.4	39.4	45.3	96.9	39.2	47.2	100.8	39.2	47.7	95.6	38.8	46.6	94.0	40.5	47.8
03200	CONCRETE REINFORCEMENT	94.2	67.2	80.5	93.9	67.2	80.3	93.8	47.4	70.2	93.8	47.4	70.2	99.9	47.4	73.2	99.4	61.1	79.9
03300	CAST-IN-PLACE CONCRETE	80.5	47.4	66.9	72.7	47.7	62.4	72.7	47.6	62.4	72.7	47.6	62.4	102.6	49.5	80.8	99.6	45.4	77.4
03	**CONCRETE**	92.3	49.2	70.9	94.8	49.2	72.2	93.5	45.5	69.7	93.8	45.5	69.8	98.9	45.6	72.4	96.6	47.4	72.2
04	**MASONRY**	89.1	36.0	56.1	74.6	35.1	50.1	72.5	35.1	49.3	74.5	35.1	50.1	108.2	57.1	76.5	104.8	60.5	77.3
05	**METALS**	92.7	80.5	88.8	91.7	80.4	88.1	91.7	73.0	85.7	91.7	73.0	85.7	98.9	68.9	89.3	98.9	75.1	91.3
06	**WOOD & PLASTICS**	102.9	39.7	70.0	80.6	39.7	59.3	96.9	39.7	67.1	101.3	39.7	69.2	101.2	37.0	67.7	99.2	39.1	67.9
07	**THERMAL & MOISTURE PROTECTION**	101.4	42.4	76.6	101.6	42.9	76.9	101.6	42.9	76.9	101.7	42.9	77.0	98.1	50.9	78.3	98.2	50.1	78.0
08	**DOORS & WINDOWS**	96.6	43.8	82.8	92.7	43.8	79.9	92.6	39.2	78.6	92.6	39.2	78.7	95.1	39.7	80.6	98.3	44.6	84.3
09200	PLASTER & GYPSUM BOARD	109.6	37.9	65.0	98.8	37.9	60.9	104.9	37.9	63.2	106.7	37.9	63.9	105.5	35.6	62.1	103.7	37.8	62.8
095,098	CEILINGS & ACOUSTICAL TREATMENT	100.9	37.9	63.3	94.9	37.9	60.9	94.1	37.9	60.5	94.1	37.9	60.5	113.4	35.6	67.0	110.9	37.8	67.3
09600	FLOORING	105.8	39.2	87.8	95.7	39.2	80.4	103.3	40.0	86.1	105.0	40.0	87.4	108.1	56.1	94.1	107.3	40.0	89.1
097,099	WALL FINISHES, PAINTS & COATINGS	110.4	38.2	66.3	110.4	38.2	66.3	110.4	38.2	66.3	110.4	38.2	66.3	97.2	34.4	58.9	97.2	39.2	61.8
09	**FINISHES**	101.8	39.0	69.1	96.7	39.0	66.7	99.3	39.1	68.0	100.1	39.1	68.3	106.9	41.1	72.6	105.6	39.6	71.2
10 - 14	**TOTAL DIV. 10000 - 14000**	100.0	65.0	92.5	100.0	64.9	92.5	100.0	64.9	92.5	100.0	64.9	92.5	100.0	53.6	90.1	100.0	63.8	92.3
15	**MECHANICAL**	100.1	36.2	73.6	100.1	36.2	73.6	100.1	36.2	73.6	100.1	36.2	73.6	100.2	38.6	74.6	100.2	37.0	73.9
16	**ELECTRICAL**	97.9	34.2	66.9	95.5	19.5	58.6	97.4	31.1	65.2	97.4	31.1	65.2	96.7	59.3	78.5	93.0	54.7	74.4
01 - 16	**WEIGHTED AVERAGE**	97.1	47.3	75.0	95.4	45.2	73.1	95.7	45.4	73.4	95.9	45.4	73.5	99.5	53.1	78.9	98.8	53.5	78.7

DIVISION		SOUTH DAKOTA						TENNESSEE											
		RAPID CITY			SIOUX FALLS			CHATTANOOGA			JACKSON			JOHNSON CITY			KNOXVILLE		
		MAT.	INST.	TOTAL	MAT.	INST.	TOTAL	MAT.	INST.	TOTAL	MAT.	INST.	TOTAL	MAT.	INST.	TOTAL	MAT.	INST.	TOTAL
01590	**EQUIPMENT RENTAL**	.0	98.3	98.3	.0	99.5	99.5	.0	104.5	104.5	.0	104.9	104.9	.0	97.4	97.4	.0	97.4	97.4
02	**SITE CONSTRUCTION**	84.6	93.3	90.9	85.6	95.4	92.8	101.4	96.2	97.6	98.9	95.4	96.3	110.8	84.9	91.8	88.4	85.0	85.9
03100	CONCRETE FORMS & ACCESSORIES	103.1	36.0	45.2	94.1	41.5	48.7	98.3	45.6	52.8	89.0	37.2	44.3	82.9	46.4	51.4	97.2	46.4	53.4
03200	CONCRETE REINFORCEMENT	93.3	61.0	76.9	93.3	61.1	77.0	87.9	42.9	65.0	89.7	42.7	65.8	88.5	43.1	65.4	87.9	43.1	65.1
03300	CAST-IN-PLACE CONCRETE	98.9	43.0	75.9	103.0	48.3	80.5	102.0	47.8	79.7	101.1	41.3	76.5	82.0	52.4	69.8	95.8	48.0	76.1
03	**CONCRETE**	95.9	44.6	70.4	96.1	48.9	72.7	92.9	47.7	70.4	94.7	41.7	68.4	96.9	49.6	73.4	90.0	48.2	69.2
04	**MASONRY**	104.7	53.7	73.1	102.2	60.5	76.3	99.1	43.1	64.4	113.0	33.5	63.7	111.8	42.9	69.1	76.5	42.9	55.7
05	**METALS**	100.7	75.3	92.6	100.8	75.7	92.8	93.5	75.5	87.8	91.0	73.8	85.5	90.9	75.1	85.9	94.1	75.4	88.1
06	**WOOD & PLASTICS**	105.4	34.9	68.7	99.2	39.7	68.2	101.3	46.6	72.8	86.7	37.1	60.8	75.3	47.7	60.9	90.9	47.7	68.4
07	**THERMAL & MOISTURE PROTECTION**	98.6	47.5	77.1	98.2	51.3	78.4	97.9	50.2	77.8	95.1	39.5	71.7	92.2	47.9	73.5	90.3	48.1	72.6
08	**DOORS & WINDOWS**	99.3	42.4	84.4	99.3	45.0	85.1	101.2	46.5	86.9	101.8	39.2	85.5	97.0	47.6	84.1	93.7	47.6	81.6
09200	PLASTER & GYPSUM BOARD	105.5	33.5	60.8	105.5	38.4	63.8	84.6	45.7	60.4	90.1	35.9	56.4	90.0	46.9	63.2	97.3	46.9	66.0
095,098	CEILINGS & ACOUSTICAL TREATMENT	117.7	33.5	67.5	117.7	38.4	70.4	96.4	45.7	66.2	96.2	35.9	60.2	91.9	46.9	65.0	94.5	46.9	66.1
09600	FLOORING	107.3	75.8	98.8	107.3	79.7	99.8	104.4	49.4	89.5	92.8	23.9	74.2	97.7	50.8	85.0	103.5	50.8	89.2
097,099	WALL FINISHES, PAINTS & COATINGS	97.2	39.2	61.8	97.2	39.2	61.8	108.9	45.8	70.4	96.0	31.0	56.3	106.7	53.7	74.4	106.7	53.7	74.4
09	**FINISHES**	107.7	43.4	74.2	107.6	48.1	76.6	98.2	46.3	71.2	95.9	34.1	63.7	98.5	47.9	72.1	94.0	47.9	70.0
10 - 14	**TOTAL DIV. 10000 - 14000**	100.0	61.5	91.8	100.0	64.0	92.3	100.0	51.6	89.7	100.0	48.7	89.1	100.0	59.3	91.3	100.0	59.4	91.3
15	**MECHANICAL**	100.2	34.5	72.9	100.2	36.0	73.6	99.9	44.8	77.1	100.0	55.5	81.5	99.8	59.0	82.9	99.8	59.0	82.9
16	**ELECTRICAL**	93.5	54.8	74.7	92.9	71.9	82.7	104.2	52.8	79.2	100.9	51.1	76.7	92.5	51.1	72.4	97.9	59.7	79.3
01 - 16	**WEIGHTED AVERAGE**	99.4	52.2	78.4	99.2	57.3	80.6	98.6	54.1	78.8	98.3	51.7	77.6	97.3	56.4	79.1	94.7	57.5	78.2

DIVISION		TENNESSEE						TEXAS											
		MEMPHIS			NASHVILLE			ABILENE			AMARILLO			AUSTIN			BEAUMONT		
		MAT.	INST.	TOTAL	MAT.	INST.	TOTAL	MAT.	INST.	TOTAL	MAT.	INST.	TOTAL	MAT.	INST.	TOTAL	MAT.	INST.	TOTAL
01590	**EQUIPMENT RENTAL**	.0	102.3	102.3	.0	105.4	105.4	.0	86.7	86.7	.0	86.7	86.7	.0	85.8	85.8	.0	88.6	88.6
02	**SITE CONSTRUCTION**	93.8	91.7	92.3	93.5	98.7	97.3	100.7	85.2	89.3	101.1	86.0	90.1	89.4	84.9	86.1	96.9	87.1	89.7
03100	CONCRETE FORMS & ACCESSORIES	98.5	67.1	71.4	97.7	65.0	69.5	95.2	42.7	49.9	97.4	54.0	59.9	92.8	57.1	62.0	102.3	53.0	59.8
03200	CONCRETE REINFORCEMENT	102.3	69.3	85.5	86.9	68.2	77.4	94.5	52.6	73.2	94.5	54.0	73.9	98.8	52.0	75.0	92.9	44.1	68.1
03300	CAST-IN-PLACE CONCRETE	88.9	71.3	81.6	96.6	69.2	85.3	93.1	42.2	72.2	97.9	48.2	77.5	87.7	49.1	71.9	87.4	54.5	73.9
03	**CONCRETE**	89.2	70.5	79.9	90.6	68.5	79.6	88.4	45.4	67.0	90.8	52.8	71.9	80.5	54.0	67.3	86.4	52.7	69.7
04	**MASONRY**	89.1	76.4	81.2	84.4	65.7	72.8	98.8	53.8	70.9	100.7	51.3	70.1	92.3	54.7	69.0	100.4	58.8	74.6
05	**METALS**	94.8	94.4	94.7	95.1	91.5	94.0	96.3	66.6	86.8	96.3	67.1	87.0	95.4	65.0	85.7	99.8	64.2	88.4
06	**WOOD & PLASTICS**	96.7	67.9	81.7	97.6	66.2	81.2	96.6	42.0	68.2	98.0	57.1	76.7	93.4	60.9	76.5	110.9	53.4	81.0
07	**THERMAL & MOISTURE PROTECTION**	91.9	74.4	84.5	95.9	66.7	83.6	99.6	49.8	78.6	101.6	48.9	79.4	88.8	54.1	74.2	105.1	58.5	85.5
08	**DOORS & WINDOWS**	100.8	69.3	92.6	98.5	67.4	90.4	92.8	45.1	80.3	92.8	51.4	81.9	94.5	58.8	85.2	97.4	49.2	84.8
09200	PLASTER & GYPSUM BOARD	94.4	67.4	77.7	92.4	65.8	75.9	83.8	41.1	57.3	83.8	56.6	66.9	88.1	60.5	71.0	93.6	52.8	68.3
095,098	CEILINGS & ACOUSTICAL TREATMENT	104.2	67.4	82.3	94.8	65.8	77.5	89.7	41.1	60.7	89.7	56.6	70.0	87.7	60.5	71.5	91.3	52.8	68.3
09600	FLOORING	97.7	45.1	83.4	107.6	77.3	99.4	115.3	68.0	102.5	115.1	63.8	101.2	100.5	48.9	86.6	112.1	72.0	101.2
097,099	WALL FINISHES, PAINTS & COATINGS	99.8	59.4	75.2	110.5	68.1	84.6	95.8	54.6	70.7	95.8	37.8	60.4	92.5	42.5	62.0	94.0	51.0	67.8
09	**FINISHES**	96.5	61.6	78.3	103.4	67.7	84.8	96.4	48.3	71.4	96.4	54.4	74.5	91.4	54.1	72.0	92.8	56.1	73.6
10 - 14	**TOTAL DIV. 10000 - 14000**	100.0	80.4	95.8	100.0	79.4	95.6	100.0	73.3	94.3	100.0	61.9	91.9	100.0	62.1	91.9	100.0	77.3	95.2
15	**MECHANICAL**	99.9	75.5	89.8	99.8	71.0	87.9	100.1	45.1	77.3	100.1	51.5	79.9	99.8	59.5	83.1	100.0	66.1	85.9
16	**ELECTRICAL**	98.3	76.7	87.8	102.3	69.4	86.3	96.5	48.8	73.3	97.0	60.0	79.0	99.5	69.8	85.0	94.4	72.5	83.7
01 - 16	**WEIGHTED AVERAGE**	96.6	76.2	87.5	97.5	73.7	86.9	96.6	53.3	77.4	97.2	58.0	79.8	94.6	61.7	80.0	97.1	64.1	82.4

DIVISION		TEXAS																	
		CORPUS CHRISTI			DALLAS			EL PASO			FORT WORTH			HOUSTON			LAREDO		
		MAT.	INST.	TOTAL	MAT.	INST.	TOTAL	MAT.	INST.	TOTAL	MAT.	INST.	TOTAL	MAT.	INST.	TOTAL	MAT.	INST.	TOTAL
01590	EQUIPMENT RENTAL	.0	94.6	94.6	.0	98.1	98.1	.0	86.7	86.7	.0	86.7	86.7	.0	98.8	98.8	.0	85.8	85.8
02	SITE CONSTRUCTION	124.2	80.2	92.0	126.2	86.3	97.0	101.2	84.8	89.2	100.6	85.6	89.6	124.2	85.6	95.9	89.9	84.7	86.1
03100	CONCRETE FORMS & ACCESSORIES	96.3	38.7	46.6	93.7	59.8	64.5	94.7	47.2	53.7	96.2	59.0	64.1	90.8	68.2	71.3	90.5	39.0	46.1
03200	CONCRETE REINFORCEMENT	97.9	50.1	73.6	96.8	56.2	76.2	94.5	48.9	71.4	94.5	56.1	75.0	94.6	64.2	79.2	98.8	50.3	74.2
03300	CAST-IN-PLACE CONCRETE	104.0	45.5	79.9	96.6	56.6	80.2	95.4	37.0	71.4	91.6	51.6	75.2	90.2	69.2	81.5	80.0	61.3	72.3
03	CONCRETE	90.1	45.2	67.8	88.4	59.5	74.0	89.4	44.8	67.3	87.7	56.6	72.2	85.8	69.3	77.6	80.2	49.8	65.1
04	MASONRY	84.0	50.0	62.9	102.3	61.1	76.8	97.5	51.4	68.9	94.3	61.1	73.7	96.4	65.5	77.2	92.6	53.8	68.5
05	METALS	94.9	77.9	89.5	93.4	82.3	89.9	96.1	62.3	85.3	93.1	68.3	85.2	104.4	89.5	99.6	96.3	63.9	85.9
06	WOOD & PLASTICS	105.9	38.1	70.6	95.8	60.4	77.4	96.6	50.5	72.6	103.3	60.2	80.8	96.1	68.6	81.8	89.4	37.9	62.6
07	THERMAL & MOISTURE PROTECTION	91.3	46.9	72.6	99.5	63.6	84.4	99.2	53.9	80.1	99.9	54.8	80.9	99.4	66.9	85.7	87.9	51.1	72.4
08	DOORS & WINDOWS	101.5	39.9	85.4	105.5	55.9	92.6	92.8	45.8	80.5	86.8	55.8	78.7	107.1	67.4	96.7	95.1	40.6	80.9
09200	PLASTER & GYPSUM BOARD	88.1	36.9	56.3	87.6	59.8	70.3	83.8	49.8	62.7	83.8	59.8	68.9	91.9	68.3	77.2	85.1	36.9	55.2
095,098	CEILINGS & ACOUSTICAL TREATMENT	87.7	36.9	57.4	92.9	59.8	73.2	89.7	49.8	65.9	89.7	59.8	71.9	92.8	68.3	78.2	87.7	36.9	57.4
09600	FLOORING	113.6	48.4	95.9	104.3	59.3	92.1	115.3	69.5	102.9	149.4	47.4	121.8	99.6	65.2	90.3	97.1	48.4	83.9
097,099	WALL FINISHES, PAINTS & COATINGS	105.9	41.9	66.8	104.7	54.1	73.8	95.8	36.3	59.5	97.0	54.0	70.7	102.3	61.8	77.6	92.5	55.5	69.9
09	FINISHES	97.1	40.1	67.4	98.4	59.0	77.8	96.4	50.7	72.6	106.6	56.4	80.5	97.3	66.9	81.5	90.1	41.5	64.8
10 - 14	TOTAL DIV. 10000 - 14000	100.0	74.0	94.5	100.0	78.8	95.5	100.0	59.9	91.5	100.0	78.3	95.4	100.0	82.8	96.3	100.0	71.4	93.9
15	MECHANICAL	99.8	46.0	77.5	99.8	67.8	86.5	100.1	34.4	72.8	100.1	59.7	83.3	99.9	72.4	88.5	99.8	41.5	75.6
16	ELECTRICAL	97.6	54.3	76.6	96.1	65.5	81.2	95.7	56.2	76.5	95.7	63.8	80.2	97.0	71.0	84.3	99.3	62.8	81.5
01 - 16	WEIGHTED AVERAGE	97.3	53.1	77.7	98.3	67.1	84.4	96.6	51.6	76.6	96.2	62.8	81.4	99.6	73.0	87.7	94.6	53.7	76.5

DIVISION		TEXAS															UTAH		
		LUBBOCK			ODESSA			SAN ANTONIO			WACO			WICHITA FALLS			LOGAN		
		MAT.	INST.	TOTAL	MAT.	INST.	TOTAL	MAT.	INST.	TOTAL	MAT.	INST.	TOTAL	MAT.	INST.	TOTAL	MAT.	INST.	TOTAL
01590	EQUIPMENT RENTAL	.0	96.9	96.9	.0	86.7	86.7	.0	88.7	88.7	.0	86.7	86.7	.0	86.7	86.7	.0	100.7	100.7
02	SITE CONSTRUCTION	130.2	83.3	95.8	100.9	85.5	89.6	89.6	89.5	89.5	99.6	85.5	89.3	100.3	85.2	89.2	89.7	100.6	97.7
03100	CONCRETE FORMS & ACCESSORIES	94.8	41.9	49.2	95.1	39.9	47.5	90.5	58.5	62.9	95.8	41.4	48.8	95.8	43.0	50.3	100.9	59.6	65.3
03200	CONCRETE REINFORCEMENT	95.8	52.5	73.8	94.5	52.3	73.1	105.5	52.5	78.6	94.5	51.8	72.8	94.5	53.0	73.4	108.5	77.6	92.8
03300	CAST-IN-PLACE CONCRETE	93.3	48.3	74.8	93.1	44.3	73.1	78.5	68.9	74.6	84.1	53.2	71.4	89.8	47.9	72.6	86.7	70.5	80.0
03	CONCRETE	87.5	48.1	68.0	88.4	44.8	66.7	80.6	61.6	71.1	84.0	48.5	66.4	86.8	47.6	67.3	108.6	67.2	88.0
04	MASONRY	98.2	49.7	68.1	98.8	50.0	68.6	92.4	64.3	75.0	95.7	59.7	73.4	96.1	59.6	73.4	118.6	60.4	82.6
05	METALS	99.9	80.8	93.8	95.7	65.6	86.1	97.0	68.0	87.7	96.2	65.9	86.5	96.2	67.4	87.0	99.4	75.4	91.7
06	WOOD & PLASTICS	97.1	42.1	68.5	96.6	39.9	67.1	89.4	57.3	72.7	102.3	37.1	68.3	102.3	42.0	70.9	87.9	57.9	72.3
07	THERMAL & MOISTURE PROTECTION	91.9	50.8	74.6	99.6	46.2	77.1	87.9	64.8	78.1	100.1	50.6	79.2	100.1	54.2	80.8	99.4	67.7	86.1
08	DOORS & WINDOWS	104.1	43.3	88.2	92.8	41.7	79.4	97.1	56.1	86.4	86.8	38.8	74.3	86.8	45.0	75.9	89.3	59.7	81.6
09200	PLASTER & GYPSUM BOARD	84.5	41.1	57.5	83.8	39.0	56.0	85.1	56.7	67.5	83.8	36.1	54.2	83.8	41.1	57.3	82.9	56.5	66.5
095,098	CEILINGS & ACOUSTICAL TREATMENT	92.3	41.1	61.7	89.7	39.0	59.4	87.7	56.7	69.2	89.7	36.1	57.7	89.7	41.1	60.7	101.4	56.5	74.6
09600	FLOORING	106.6	40.2	88.6	115.3	39.7	94.9	97.1	55.3	85.8	149.6	36.9	119.1	150.0	77.7	130.5	100.3	59.6	89.3
097,099	WALL FINISHES, PAINTS & COATINGS	107.3	34.0	62.5	95.8	34.0	58.0	92.5	55.5	69.9	97.0	35.0	59.2	98.7	55.5	72.4	102.2	47.9	69.0
09	FINISHES	98.4	39.9	67.9	96.4	38.6	66.3	90.1	56.9	72.8	106.6	38.5	71.2	106.9	50.4	77.5	97.6	57.6	76.7
10 - 14	TOTAL DIV. 10000 - 14000	100.0	73.0	94.2	100.0	59.2	91.3	100.0	77.9	95.3	100.0	75.4	94.8	100.0	60.1	91.5	100.0	70.7	93.7
15	MECHANICAL	99.7	48.3	78.3	100.1	37.7	74.2	99.8	72.4	88.4	100.1	55.5	81.6	100.1	51.3	79.8	100.1	71.2	88.1
16	ELECTRICAL	95.1	45.9	71.2	96.5	42.7	70.4	99.5	62.8	81.7	96.5	67.1	82.2	99.6	65.6	83.1	91.0	74.5	83.0
01 - 16	WEIGHTED AVERAGE	98.6	53.5	78.6	96.6	48.5	75.2	95.0	67.0	82.5	96.4	57.5	79.1	97.1	57.9	79.7	99.1	70.5	86.4

DIVISION		UTAH									VERMONT						VIRGINIA		
		OGDEN			PROVO			SALT LAKE CITY			BURLINGTON			RUTLAND			ALEXANDRIA		
		MAT.	INST.	TOTAL	MAT.	INST.	TOTAL	MAT.	INST.	TOTAL	MAT.	INST.	TOTAL	MAT.	INST.	TOTAL	MAT.	INST.	TOTAL
01590	EQUIPMENT RENTAL	.0	100.7	100.7	.0	99.5	99.5	.0	100.7	100.7	.0	101.5	101.5	.0	101.5	101.5	.0	101.2	101.2
02	SITE CONSTRUCTION	78.4	100.6	94.6	86.3	98.6	95.3	78.4	100.5	94.6	80.6	98.8	93.9	80.6	98.8	93.9	113.3	86.2	93.4
03100	CONCRETE FORMS & ACCESSORIES	100.9	59.6	65.3	102.5	59.7	65.6	99.4	59.7	65.2	92.4	50.5	56.2	103.2	50.5	57.8	92.4	75.2	77.5
03200	CONCRETE REINFORCEMENT	113.4	77.6	95.2	117.4	77.7	97.2	110.7	77.7	93.9	107.5	53.9	80.2	107.5	53.9	80.2	83.7	85.9	84.8
03300	CAST-IN-PLACE CONCRETE	88.0	70.5	80.8	86.8	70.5	80.1	95.5	70.5	85.2	99.8	62.9	84.6	95.2	62.9	81.9	99.9	80.8	92.0
03	CONCRETE	99.3	67.2	83.3	108.6	67.2	88.0	117.1	67.2	92.3	104.7	56.0	80.5	103.2	56.0	79.7	102.7	80.2	91.5
04	MASONRY	111.9	60.4	80.0	124.3	60.4	84.7	129.2	60.4	86.6	99.8	60.2	75.2	87.5	60.2	70.6	87.0	73.7	78.7
05	METALS	99.9	75.4	92.0	97.7	75.4	90.6	103.0	75.4	94.2	101.4	65.9	90.0	99.7	66.0	88.9	101.6	98.2	100.5
06	WOOD & PLASTICS	87.9	57.9	72.3	89.5	57.9	73.0	88.1	57.9	72.3	89.0	48.5	67.9	102.0	48.5	74.2	96.0	75.6	85.3
07	THERMAL & MOISTURE PROTECTION	98.2	67.7	85.4	101.2	67.7	87.1	100.9	67.7	86.9	101.7	55.2	82.1	101.5	56.4	82.5	100.6	81.5	92.6
08	DOORS & WINDOWS	89.3	59.7	81.6	93.5	59.7	84.7	89.3	59.7	81.6	106.9	45.2	90.8	106.9	45.2	90.8	96.6	78.0	91.7
09200	PLASTER & GYPSUM BOARD	82.9	56.5	66.5	83.3	56.5	66.7	82.9	56.5	66.5	104.7	46.3	68.4	104.7	46.3	68.4	109.6	74.7	87.9
095,098	CEILINGS & ACOUSTICAL TREATMENT	101.4	56.5	74.6	101.4	56.5	74.6	100.4	56.5	74.2	92.6	46.3	64.9	92.6	46.3	64.9	100.9	74.7	85.3
09600	FLOORING	100.3	59.6	89.3	101.1	59.6	89.9	99.9	59.6	89.0	99.6	68.8	91.3	99.6	68.8	91.3	106.0	91.4	102.0
097,099	WALL FINISHES, PAINTS & COATINGS	102.2	47.9	69.0	102.2	61.1	77.1	102.2	61.1	77.1	91.5	39.6	59.9	91.5	39.6	59.9	121.8	84.2	98.8
09	FINISHES	96.5	57.6	76.2	98.2	59.1	77.8	96.6	59.1	77.1	98.3	52.1	74.2	98.2	52.1	74.2	102.3	78.8	90.1
10 - 14	TOTAL DIV. 10000 - 14000	100.0	70.7	93.7	100.0	70.7	93.7	100.0	70.7	93.7	100.0	92.0	98.3	100.0	92.0	98.3	100.0	86.6	97.1
15	MECHANICAL	100.1	71.2	88.1	100.1	71.2	88.1	100.2	71.2	88.2	100.1	66.1	86.0	100.1	66.1	86.0	99.9	86.3	94.3
16	ELECTRICAL	91.5	74.5	83.3	91.2	74.5	83.1	95.0	74.5	85.0	98.3	68.7	83.9	97.6	68.7	83.6	98.3	95.1	96.8
01 - 16	WEIGHTED AVERAGE	97.4	70.5	85.5	99.6	70.5	86.7	101.3	70.7	87.7	100.5	64.9	84.7	99.5	64.9	84.1	99.9	84.9	93.3

	DIVISION	VIRGINIA																	
		ARLINGTON			NEWPORT NEWS			NORFOLK			PORTSMOUTH			RICHMOND			ROANOKE		
		MAT.	INST.	TOTAL	MAT.	INST.	TOTAL	MAT.	INST.	TOTAL	MAT.	INST.	TOTAL	MAT.	INST.	TOTAL	MAT.	INST.	TOTAL
01590	**EQUIPMENT RENTAL**	.0	99.6	99.6	.0	104.8	104.8	.0	105.5	105.5	.0	104.7	104.7	.0	104.8	104.8	.0	99.6	99.6
02	**SITE CONSTRUCTION**	124.2	82.5	93.6	106.7	84.5	90.4	106.0	85.6	91.1	105.3	83.9	89.6	107.3	84.9	90.9	104.3	80.7	87.0
03100	CONCRETE FORMS & ACCESSORIES	91.1	71.7	74.4	97.2	59.4	64.6	101.7	59.6	65.4	84.9	59.4	62.9	98.2	56.0	61.8	97.0	36.0	44.4
03200	CONCRETE REINFORCEMENT	94.4	62.8	78.3	94.2	74.6	84.2	94.2	74.6	84.2	93.9	74.6	84.1	94.2	72.1	83.0	94.2	72.7	83.3
03300	CAST-IN-PLACE CONCRETE	97.2	76.4	88.6	97.8	54.8	80.1	100.7	54.9	81.9	96.8	54.3	79.4	104.1	54.7	83.8	109.9	46.3	83.7
03	**CONCRETE**	107.5	72.7	90.2	100.4	62.1	81.4	102.1	62.2	82.3	99.0	61.9	80.6	103.4	60.1	81.9	106.2	48.3	77.5
04	**MASONRY**	99.1	67.8	79.7	91.7	53.8	68.2	98.0	53.8	70.6	96.5	53.6	69.9	90.3	49.4	65.0	91.8	38.8	58.9
05	**METALS**	100.4	85.6	95.7	101.7	90.5	98.1	100.7	90.7	97.5	100.7	89.0	97.0	103.6	89.8	99.2	101.4	84.0	95.9
06	**WOOD & PLASTICS**	92.4	75.6	83.6	97.0	63.3	79.4	102.6	63.3	82.1	83.4	63.3	72.9	98.2	57.7	77.1	97.0	35.3	64.9
07	**THERMAL & MOISTURE PROTECTION**	102.2	72.3	89.6	101.4	51.2	80.3	101.2	51.2	80.2	101.4	51.2	80.3	101.1	54.2	81.3	101.4	44.6	77.5
08	**DOORS & WINDOWS**	94.6	71.7	88.6	96.6	60.0	87.0	96.6	62.6	87.7	96.7	62.6	87.8	96.6	55.5	85.9	96.6	44.7	83.1
09200	PLASTER & GYPSUM BOARD	105.8	74.7	86.5	109.6	61.3	79.6	109.6	61.3	79.6	101.8	61.3	76.6	109.6	55.6	76.0	109.6	33.4	62.2
095,098	CEILINGS & ACOUSTICAL TREATMENT	94.9	74.7	82.9	100.9	61.3	77.3	100.9	61.3	77.3	100.9	61.3	77.3	100.9	55.6	73.8	100.9	33.4	60.6
09600	FLOORING	104.2	67.5	94.2	106.0	39.4	88.0	105.8	39.4	87.8	96.9	39.4	81.3	105.8	62.5	94.1	106.0	35.3	86.9
097,099	WALL FINISHES, PAINTS & COATINGS	121.8	84.2	98.8	110.4	42.0	68.7	110.4	63.3	81.7	110.4	63.3	81.7	110.4	54.9	76.5	110.4	32.8	63.0
09	**FINISHES**	100.6	72.6	86.0	101.7	53.8	76.8	101.6	56.3	78.0	97.8	56.3	76.2	101.5	56.9	78.3	101.5	34.8	66.8
10 - 14	**TOTAL DIV. 10000 - 14000**	100.0	68.2	93.2	100.0	78.1	95.3	100.0	78.1	95.3	100.0	78.4	95.4	100.0	77.5	95.2	100.0	71.3	93.9
15	**MECHANICAL**	100.1	82.7	92.8	100.1	63.4	84.8	100.1	63.6	84.9	100.1	63.6	84.9	100.1	67.4	86.5	100.1	39.3	74.9
16	**ELECTRICAL**	95.8	95.2	95.5	97.7	59.2	79.0	97.8	63.1	80.9	96.2	63.1	80.1	98.4	71.5	85.3	97.7	34.2	66.8
01 - 16	**WEIGHTED AVERAGE**	100.5	79.3	91.1	99.6	64.7	84.1	100.0	65.8	84.8	98.9	65.5	84.0	100.3	66.7	85.4	100.2	48.1	77.1

	DIVISION	WASHINGTON																	
		EVERETT			RICHLAND			SEATTLE			SPOKANE			TACOMA			VANCOUVER		
		MAT.	INST.	TOTAL	MAT.	INST.	TOTAL	MAT.	INST.	TOTAL	MAT.	INST.	TOTAL	MAT.	INST.	TOTAL	MAT.	INST.	TOTAL
01590	**EQUIPMENT RENTAL**	.0	104.0	104.0	.0	90.0	90.0	.0	103.7	103.7	.0	90.0	90.0	.0	104.0	104.0	.0	97.1	97.1
02	**SITE CONSTRUCTION**	93.8	115.2	109.5	104.4	89.4	93.3	97.8	113.0	109.0	105.2	89.4	93.6	96.6	115.2	110.3	108.7	100.7	102.8
03100	CONCRETE FORMS & ACCESSORIES	108.0	101.8	102.7	113.4	80.3	84.8	98.4	102.5	101.9	120.4	80.3	85.8	98.4	102.2	101.7	99.2	96.9	97.2
03200	CONCRETE REINFORCEMENT	108.0	94.7	101.2	102.9	87.7	95.1	106.8	94.8	100.7	103.5	87.7	95.5	106.8	94.7	100.6	107.6	94.5	100.9
03300	CAST-IN-PLACE CONCRETE	94.5	106.3	99.3	118.5	83.6	104.1	99.2	108.5	103.0	122.7	83.6	106.6	97.2	108.3	101.8	108.6	100.1	105.1
03	**CONCRETE**	95.7	101.4	98.5	108.0	82.7	95.4	98.4	102.5	100.5	110.7	82.7	96.8	97.4	102.3	99.8	107.2	97.2	102.2
04	**MASONRY**	132.8	103.0	114.4	115.0	83.3	95.3	126.8	103.7	112.5	116.5	83.3	95.9	126.6	103.7	112.4	127.2	100.1	110.4
05	**METALS**	106.0	88.1	100.3	94.1	82.0	90.2	107.7	90.1	102.1	96.7	82.0	92.0	107.7	88.2	101.4	104.8	89.6	100.0
06	**WOOD & PLASTICS**	98.7	101.1	99.9	96.0	79.6	87.5	88.8	101.1	95.2	105.8	79.6	92.2	87.8	101.1	94.7	82.3	96.6	89.7
07	**THERMAL & MOISTURE PROTECTION**	104.6	98.4	102.0	139.9	81.2	115.2	104.6	100.6	102.9	137.7	81.1	113.8	104.2	99.2	102.1	106.8	90.9	100.1
08	**DOORS & WINDOWS**	98.6	98.8	98.7	111.7	76.2	102.4	100.8	98.8	100.3	111.5	76.2	102.2	99.3	98.8	99.1	96.2	95.3	96.0
09200	PLASTER & GYPSUM BOARD	104.7	101.1	102.4	124.7	78.9	96.3	98.2	101.1	100.0	129.5	78.9	98.1	100.2	101.1	100.7	98.7	96.8	97.5
095,098	CEILINGS & ACOUSTICAL TREATMENT	111.3	101.1	105.2	107.7	78.9	90.5	114.5	101.1	106.5	107.7	78.9	90.5	114.7	101.1	106.6	109.7	96.8	102.0
09600	FLOORING	119.1	105.1	115.3	110.2	43.2	92.0	111.0	105.1	109.4	113.9	74.2	103.1	111.6	105.1	109.9	117.5	79.5	107.2
097,099	WALL FINISHES, PAINTS & COATINGS	111.1	90.5	98.5	112.6	74.5	89.3	111.1	92.2	99.6	112.6	74.5	89.3	111.1	92.2	99.6	119.0	68.1	87.9
09	**FINISHES**	113.4	101.3	107.1	123.2	71.9	96.5	111.0	101.6	106.1	125.0	78.3	100.7	111.5	101.6	106.3	109.5	90.3	99.5
10 - 14	**TOTAL DIV. 10000 - 14000**	100.0	102.7	100.6	100.0	83.1	96.4	100.0	103.2	100.7	100.0	83.0	96.4	100.0	103.2	100.7	100.0	80.6	95.9
15	**MECHANICAL**	99.9	101.9	100.7	100.5	98.5	99.7	99.9	112.2	105.0	100.5	87.7	95.2	100.0	100.6	100.3	100.1	104.0	101.7
16	**ELECTRICAL**	103.6	97.7	100.7	95.8	97.6	96.7	103.4	106.5	104.9	94.1	81.8	88.1	103.4	100.6	102.0	114.0	103.1	108.7
01 - 16	**WEIGHTED AVERAGE**	103.6	101.0	102.4	105.0	86.9	96.9	103.8	104.7	104.2	105.7	83.3	95.8	103.5	101.4	102.6	105.4	97.7	102.0

	DIVISION	WASHINGTON			WEST VIRGINIA												WISCONSIN		
		YAKIMA			CHARLESTON			HUNTINGTON			PARKERSBURG			WHEELING			EAU CLAIRE		
		MAT.	INST.	TOTAL	MAT.	INST.	TOTAL	MAT.	INST.	TOTAL	MAT.	INST.	TOTAL	MAT.	INST.	TOTAL	MAT.	INST.	TOTAL
01590	**EQUIPMENT RENTAL**	.0	104.0	104.0	.0	99.6	99.6	.0	99.6	99.6	.0	99.6	99.6	.0	99.6	99.6	.0	100.6	100.6
02	**SITE CONSTRUCTION**	99.8	113.8	110.1	102.1	86.0	90.3	103.5	86.8	91.2	108.3	85.9	91.9	108.9	85.6	91.8	86.5	102.9	98.5
03100	CONCRETE FORMS & ACCESSORIES	98.9	95.8	96.2	104.9	91.7	93.5	98.1	93.7	94.3	87.5	91.7	91.1	89.5	91.6	91.3	98.1	99.3	99.2
03200	CONCRETE REINFORCEMENT	107.2	93.4	100.2	94.2	87.3	90.7	94.2	85.5	89.8	92.8	89.0	90.9	92.2	91.6	91.9	91.8	105.6	98.8
03300	CAST-IN-PLACE CONCRETE	103.9	81.6	94.7	95.7	108.0	100.8	103.2	112.5	107.0	96.5	102.1	98.8	96.5	101.7	98.6	100.3	98.4	99.5
03	**CONCRETE**	102.1	89.9	96.0	99.9	97.1	98.5	103.0	99.2	101.1	103.1	95.4	99.2	103.1	95.7	99.4	96.9	100.4	98.7
04	**MASONRY**	118.8	71.1	89.2	89.0	93.4	91.8	91.2	94.4	93.2	77.0	91.0	85.7	100.1	90.3	94.1	92.8	101.5	98.2
05	**METALS**	105.9	82.8	98.5	101.7	99.4	100.9	101.7	98.8	100.8	100.3	99.8	100.2	100.4	101.1	100.6	91.9	105.2	96.2
06	**WOOD & PLASTICS**	88.2	101.1	95.0	105.6	90.5	97.7	97.0	91.8	94.3	85.8	90.6	88.3	87.7	90.6	89.2	112.2	98.4	105.0
07	**THERMAL & MOISTURE PROTECTION**	104.4	79.5	93.9	101.3	91.1	97.0	101.6	91.6	97.4	101.7	88.4	96.1	101.8	88.6	96.2	97.4	90.9	94.6
08	**DOORS & WINDOWS**	98.8	85.0	95.2	97.8	84.0	94.2	96.6	84.3	93.4	97.2	82.0	93.2	98.1	90.2	96.0	100.9	96.7	99.8
09200	PLASTER & GYPSUM BOARD	100.0	101.1	100.7	108.9	90.1	97.2	107.8	91.4	97.6	102.1	90.2	94.7	102.4	90.2	94.8	105.0	98.7	101.1
095,098	CEILINGS & ACOUSTICAL TREATMENT	109.6	101.1	104.5	98.3	90.1	93.4	94.1	91.4	92.5	93.2	90.2	91.4	93.2	90.2	91.4	98.5	98.7	98.6
09600	FLOORING	113.0	65.0	100.0	105.8	107.1	106.2	105.8	104.3	105.4	100.2	102.9	100.9	101.4	103.4	101.9	90.5	109.4	95.6
097,099	WALL FINISHES, PAINTS & COATINGS	111.1	74.5	88.8	110.4	95.9	101.6	110.4	93.1	99.9	110.4	96.6	102.0	110.4	101.3	104.9	87.7	94.2	91.7
09	**FINISHES**	110.7	88.3	99.0	100.9	95.2	97.9	99.5	95.7	97.5	97.2	94.4	95.8	97.6	94.8	96.2	97.3	100.9	99.2
10 - 14	**TOTAL DIV. 10000 - 14000**	100.0	98.5	99.7	100.0	95.0	98.9	100.0	96.2	99.2	100.0	95.0	98.9	100.0	102.5	100.5	100.0	94.8	98.9
15	**MECHANICAL**	100.0	97.6	99.0	100.1	84.9	93.8	100.1	88.4	95.2	100.1	89.7	95.8	100.1	94.6	97.8	100.2	90.9	96.3
16	**ELECTRICAL**	106.1	97.6	102.0	97.7	87.3	92.6	97.7	95.8	96.8	98.2	96.9	97.5	95.6	96.4	95.9	101.5	83.1	92.6
01 - 16	**WEIGHTED AVERAGE**	103.7	91.8	98.4	99.5	91.0	95.7	99.6	93.4	96.9	98.7	92.6	96.0	99.7	94.2	97.2	97.8	96.2	97.1

DIVISION		WISCONSIN																	
		GREEN BAY			KENOSHA			LA CROSSE			MADISON			MILWAUKEE			RACINE		
		MAT.	INST.	TOTAL	MAT.	INST.	TOTAL	MAT.	INST.	TOTAL	MAT.	INST.	TOTAL	MAT.	INST.	TOTAL	MAT.	INST.	TOTAL
01590	**EQUIPMENT RENTAL**	.0	98.3	98.3	.0	97.9	97.9	.0	100.6	100.6	.0	100.1	100.1	.0	86.3	86.3	.0	100.1	100.1
02	**SITE CONSTRUCTION**	89.2	98.8	96.2	96.5	102.1	100.6	80.6	102.8	96.9	90.8	105.5	101.6	92.2	93.8	93.4	91.2	106.0	102.0
03100	CONCRETE FORMS & ACCESSORIES	109.2	98.5	100.0	109.2	108.7	108.7	82.6	98.8	96.6	99.8	99.4	99.5	102.1	114.2	112.6	99.3	108.6	107.3
03200	CONCRETE REINFORCEMENT	90.1	93.3	91.7	89.1	101.5	95.4	91.5	92.2	91.9	89.2	92.4	90.8	89.2	101.7	95.6	89.2	101.4	95.4
03300	CAST-IN-PLACE CONCRETE	103.8	100.4	102.4	115.6	102.6	110.2	90.2	97.8	93.3	104.2	100.9	102.8	104.2	108.5	105.9	104.2	102.2	103.3
03	**CONCRETE**	98.2	98.4	98.3	103.7	105.1	104.4	88.3	97.5	92.9	97.6	98.7	98.1	97.9	109.0	103.4	97.5	104.9	101.2
04	**MASONRY**	124.3	100.5	109.5	99.2	111.4	106.8	92.0	101.5	97.9	103.2	104.3	103.9	102.7	118.9	112.7	102.6	111.3	108.0
05	**METALS**	94.3	99.5	96.0	95.3	102.7	97.7	91.9	99.0	94.1	96.4	97.4	96.7	98.1	94.1	96.8	96.4	102.7	98.4
06	**WOOD & PLASTICS**	118.1	98.4	107.8	115.9	108.3	111.9	94.7	98.4	96.6	110.2	98.3	104.0	112.9	113.1	113.0	110.4	108.3	109.3
07	**THERMAL & MOISTURE PROTECTION**	99.4	89.0	95.0	96.4	103.7	99.5	96.8	90.0	93.9	95.7	97.3	96.4	94.9	111.9	102.1	96.3	103.2	99.2
08	**DOORS & WINDOWS**	99.1	89.9	96.7	96.9	108.0	99.8	100.9	86.7	97.2	101.9	96.3	100.4	104.2	110.6	105.8	101.9	108.0	103.5
09200	PLASTER & GYPSUM BOARD	97.3	98.7	98.1	90.0	108.9	101.8	98.3	98.7	98.5	98.1	98.7	98.4	99.0	113.7	108.1	98.1	108.9	104.8
095,098	CEILINGS & ACOUSTICAL TREATMENT	91.8	98.7	95.9	80.5	108.9	97.4	96.8	98.7	97.9	86.1	98.7	93.6	89.5	113.7	103.9	86.1	108.9	99.7
09600	FLOORING	108.3	109.4	108.6	109.7	110.7	109.9	82.9	109.5	90.1	93.5	107.2	97.2	96.2	121.1	102.9	93.5	110.7	98.1
097,099	WALL FINISHES, PAINTS & COATINGS	97.2	77.1	84.9	101.5	101.1	101.3	87.7	71.6	77.9	91.5	96.3	94.4	93.8	111.3	104.5	91.4	92.2	91.9
09	**FINISHES**	100.7	98.7	99.7	96.6	109.0	103.1	93.3	98.4	96.0	93.8	101.0	97.5	95.8	115.8	106.2	93.8	107.9	101.2
10 - 14	**TOTAL DIV. 10000 - 14000**	100.0	93.1	98.5	100.0	101.0	100.2	100.0	94.8	98.9	100.0	93.3	98.6	100.0	103.7	100.8	100.0	101.0	100.2
15	**MECHANICAL**	100.5	90.1	96.2	100.0	98.2	99.3	100.2	90.6	96.2	99.8	93.2	97.0	99.8	103.2	101.2	99.8	95.4	97.9
16	**ELECTRICAL**	96.6	90.8	93.8	96.0	96.0	96.0	101.8	91.7	96.9	97.5	95.3	96.4	97.1	103.1	100.0	95.8	98.0	96.9
01 - 16	**WEIGHTED AVERAGE**	99.7	95.2	97.7	98.5	103.1	100.5	96.1	95.7	95.9	98.3	98.1	98.2	99.0	106.2	102.2	98.1	102.9	100.2

DIVISION		WYOMING									CANADA								
		CASPER			CHEYENNE			ROCK SPRINGS			CALGARY, ALBERTA			EDMONTON, ALBERTA			HALIFAX, NOVA SCOTIA		
		MAT.	INST.	TOTAL	MAT.	INST.	TOTAL	MAT.	INST.	TOTAL	MAT.	INST.	TOTAL	MAT.	INST.	TOTAL	MAT.	INST.	TOTAL
01590	**EQUIPMENT RENTAL**	.0	101.7	101.7	.0	101.7	101.7	.0	101.7	101.7	.0	107.0	107.0	.0	107.0	107.0	.0	98.6	98.6
02	**SITE CONSTRUCTION**	79.2	100.0	94.4	79.0	100.0	94.4	82.0	99.6	94.9	116.6	104.8	107.9	120.3	104.8	108.9	99.1	95.5	96.5
03100	CONCRETE FORMS & ACCESSORIES	97.7	42.6	50.1	98.5	42.8	50.5	99.2	30.4	39.8	122.0	74.0	80.6	119.7	74.0	80.3	90.5	64.2	67.8
03200	CONCRETE REINFORCEMENT	115.6	51.7	83.2	108.5	51.9	79.7	117.9	51.5	84.2	161.5	64.2	112.0	161.5	64.2	112.0	154.9	47.8	100.4
03300	CAST-IN-PLACE CONCRETE	97.6	75.8	88.7	97.6	75.9	88.7	98.1	42.5	75.3	191.6	84.3	147.5	207.0	84.3	156.5	178.5	66.1	132.3
03	**CONCRETE**	104.1	56.5	80.5	103.0	56.7	80.0	104.8	39.6	72.4	161.3	76.4	119.1	168.6	76.4	122.8	153.1	62.5	108.1
04	**MASONRY**	107.0	38.1	64.3	104.4	40.2	64.6	170.7	31.1	84.2	181.1	71.9	113.4	182.3	71.9	113.9	172.7	67.5	107.5
05	**METALS**	97.2	62.2	86.1	98.0	62.7	86.7	95.0	61.2	84.2	135.6	83.2	118.9	136.4	83.2	119.4	120.8	70.5	104.7
06	**WOOD & PLASTICS**	95.1	40.5	66.7	95.1	40.5	66.7	97.9	28.7	61.9	116.2	73.6	94.0	112.7	73.6	92.3	84.5	63.5	73.6
07	**THERMAL & MOISTURE PROTECTION**	99.6	53.5	80.2	99.6	54.0	80.4	100.0	44.1	76.5	117.5	77.6	100.7	117.7	77.6	100.8	104.8	65.8	88.3
08	**DOORS & WINDOWS**	93.5	43.2	80.3	94.2	43.2	80.9	100.4	36.8	83.8	91.9	70.6	86.3	91.9	70.6	86.3	82.8	59.0	76.6
09200	PLASTER & GYPSUM BOARD	82.7	38.7	55.4	82.7	38.7	55.4	88.2	26.6	50.0	152.4	72.4	102.7	154.5	72.4	103.5	167.9	62.6	102.5
095,098	CEILINGS & ACOUSTICAL TREATMENT	100.5	38.7	63.7	99.6	38.7	63.3	100.5	26.6	56.4	110.2	72.4	87.7	125.6	72.4	93.9	108.5	62.6	81.1
09600	FLOORING	100.3	41.2	84.3	100.3	66.2	91.1	102.6	40.4	85.8	132.9	82.4	119.2	132.9	82.4	119.2	109.6	59.9	96.2
097,099	WALL FINISHES, PAINTS & COATINGS	102.2	54.9	73.3	102.0	54.9	73.3	102.2	33.3	60.2	109.1	73.8	87.5	109.2	73.8	87.6	109.1	58.2	78.1
09	**FINISHES**	96.5	42.4	68.3	96.2	47.6	70.9	98.0	31.3	63.2	122.3	75.4	97.9	126.5	75.4	99.9	116.5	62.9	88.6
10 - 14	**TOTAL DIV. 10000 - 14000**	100.0	79.2	95.6	100.0	79.3	95.6	100.0	47.3	88.8	140.0	79.8	127.2	140.0	79.8	127.2	140.0	58.1	122.5
15	**MECHANICAL**	100.1	65.9	85.9	100.1	60.0	83.4	100.1	54.9	81.3	101.4	74.7	90.3	101.5	74.7	90.4	101.0	66.4	86.7
16	**ELECTRICAL**	90.6	63.3	77.3	90.6	74.4	82.7	89.9	62.7	76.7	117.6	84.2	101.4	118.3	84.2	101.8	127.4	68.3	98.7
01 - 16	**WEIGHTED AVERAGE**	97.7	59.9	80.9	97.6	61.2	81.4	101.5	51.4	79.2	124.3	79.6	104.4	125.9	79.6	105.3	119.3	68.2	96.6

DIVISION		CANADA																	
		HAMILTON, ONTARIO			KITCHENER, ONTARIO			LAVAL, QUEBEC			LONDON, ONTARIO			MONTREAL, QUEBEC			OSHAWA, ONTARIO		
		MAT.	INST.	TOTAL	MAT.	INST.	TOTAL	MAT.	INST.	TOTAL	MAT.	INST.	TOTAL	MAT.	INST.	TOTAL	MAT.	INST.	TOTAL
01590	**EQUIPMENT RENTAL**	.0	103.1	103.1	.0	103.0	103.0	.0	99.2	99.2	.0	103.2	103.2	.0	101.0	101.0	.0	103.0	103.0
02	**SITE CONSTRUCTION**	112.7	103.2	105.7	101.9	102.6	102.4	94.0	96.5	95.9	112.6	103.0	105.5	93.7	96.8	96.0	114.2	102.8	105.8
03100	CONCRETE FORMS & ACCESSORIES	121.4	84.1	89.2	115.6	77.1	82.4	130.0	78.7	85.8	127.1	79.8	86.3	130.3	79.0	86.1	122.6	80.6	86.4
03200	CONCRETE REINFORCEMENT	175.4	85.7	129.8	110.3	85.6	97.8	157.0	80.5	118.1	135.2	84.2	109.3	159.6	80.6	119.4	174.7	86.2	129.8
03300	CAST-IN-PLACE CONCRETE	157.2	91.8	130.3	146.7	74.0	116.8	140.1	86.6	118.1	157.2	89.5	129.4	137.5	88.2	117.2	169.7	79.2	132.5
03	**CONCRETE**	146.9	87.2	117.2	124.8	77.9	101.5	136.3	82.0	109.3	140.9	84.2	112.7	135.5	82.7	109.3	152.9	81.4	117.4
04	**MASONRY**	168.1	88.5	118.7	163.6	84.5	114.6	161.3	79.1	110.3	168.0	85.8	117.0	164.5	79.1	111.5	166.7	87.8	117.7
05	**METALS**	139.3	86.8	122.5	116.3	86.2	106.7	105.6	83.0	98.4	124.5	86.3	112.3	125.0	83.5	111.7	107.2	86.5	100.6
06	**WOOD & PLASTICS**	117.4	82.9	99.5	110.9	75.5	92.4	131.5	78.7	104.0	117.4	78.1	97.0	131.5	79.0	104.2	118.4	78.9	97.9
07	**THERMAL & MOISTURE PROTECTION**	110.0	85.8	99.8	108.6	81.5	97.2	104.7	83.3	95.7	112.6	84.0	100.6	105.3	83.9	96.3	109.6	82.0	98.0
08	**DOORS & WINDOWS**	91.9	82.7	89.5	83.4	77.1	81.8	91.9	69.8	86.1	92.9	79.1	89.3	91.9	69.9	86.2	90.9	80.2	88.1
09200	PLASTER & GYPSUM BOARD	191.0	82.5	123.6	157.6	74.9	106.2	151.6	78.1	105.9	191.5	77.6	120.7	154.6	78.1	107.1	159.9	78.4	109.3
095,098	CEILINGS & ACOUSTICAL TREATMENT	113.7	82.5	95.1	107.7	74.9	88.1	95.8	78.1	85.2	115.4	77.6	92.8	107.7	78.1	90.0	100.9	78.4	87.5
09600	FLOORING	132.9	87.9	120.7	128.4	87.9	117.5	132.9	88.6	120.9	134.9	87.9	122.2	132.9	88.6	120.9	132.9	90.2	121.3
097,099	WALL FINISHES, PAINTS & COATINGS	109.1	95.8	101.0	109.1	87.7	96.1	109.1	83.3	93.3	109.1	95.8	101.0	109.1	83.3	93.3	109.1	100.6	103.9
09	**FINISHES**	128.6	86.2	106.5	120.1	80.2	99.4	117.5	81.5	98.8	129.8	83.0	105.4	121.0	81.7	100.5	120.7	84.7	102.0
10 - 14	**TOTAL DIV. 10000 - 14000**	140.0	85.3	128.3	140.0	83.3	127.9	140.0	73.8	125.9	140.0	84.1	128.1	140.0	74.6	126.0	140.0	84.5	128.2
15	**MECHANICAL**	101.5	82.4	93.6	101.0	78.8	91.8	101.0	71.8	88.9	101.6	79.0	92.2	101.6	71.9	89.3	101.0	80.4	92.5
16	**ELECTRICAL**	130.4	91.7	111.6	125.2	88.5	107.4	122.1	71.6	97.5	130.7	88.7	110.3	121.1	71.6	97.0	126.3	89.2	108.3
01 - 16	**WEIGHTED AVERAGE**	124.5	87.9	108.2	115.6	83.7	101.4	115.2	78.7	99.0	121.9	85.3	105.6	118.5	79.0	101.0	118.7	85.7	104.1

City Cost Indexes

	DIVISION	CANADA																	
		OTTAWA, ONTARIO			QUEBEC, QUEBEC			REGINA, SASKATCHEWAN			SASKATOON, SASKATCHEWAN			ST CATHARINES, ONTARIO			ST JOHNS, NEWFOUNDLAND		
		MAT.	INST.	TOTAL	MAT.	INST.	TOTAL	MAT.	INST.	TOTAL	MAT.	INST.	TOTAL	MAT.	INST.	TOTAL	MAT.	INST.	TOTAL
01590	**EQUIPMENT RENTAL**	.0	103.0	103.0	.0	101.4	101.4	.0	98.6	98.6	.0	98.6	98.6	.0	100.4	100.4	.0	98.6	98.6
02	**SITE CONSTRUCTION**	112.9	102.8	105.5	95.5	96.9	96.5	112.8	94.5	99.4	107.9	94.6	98.1	102.2	98.6	99.5	114.8	94.2	99.7
03100	CONCRETE FORMS & ACCESSORIES	120.1	80.6	86.1	130.2	79.3	86.3	102.0	54.2	60.8	102.0	54.2	60.7	113.2	78.7	83.4	99.5	55.1	61.2
03200	CONCRETE REINFORCEMENT	173.8	84.1	128.2	148.3	80.6	113.9	126.1	63.3	94.2	119.0	63.3	90.7	111.3	85.6	98.3	164.7	49.5	106.2
03300	CAST-IN-PLACE CONCRETE	159.6	89.4	130.7	151.5	88.6	125.7	162.8	63.5	122.0	148.0	63.5	113.2	140.1	78.3	114.7	179.6	62.6	131.5
03	**CONCRETE**	147.7	84.6	116.3	140.4	82.9	111.9	134.0	59.9	97.2	125.7	59.8	93.0	121.6	80.1	101.0	167.3	57.5	112.8
04	**MASONRY**	168.1	85.4	116.8	165.4	79.1	111.9	168.3	58.9	100.5	168.0	58.9	100.3	163.1	82.0	112.8	163.0	58.3	98.1
05	**METALS**	116.4	86.3	106.8	120.7	83.8	108.9	105.6	70.8	94.5	105.7	70.7	94.5	106.6	86.3	100.1	108.2	68.3	95.5
06	**WOOD & PLASTICS**	116.9	79.7	97.5	132.2	79.1	104.5	95.2	52.8	73.2	93.8	52.8	72.5	108.3	79.5	93.3	94.6	54.2	73.6
07	**THERMAL & MOISTURE PROTECTION**	110.2	82.6	98.6	104.9	84.1	96.1	105.5	60.0	86.4	104.3	59.1	85.3	108.6	82.6	97.6	108.1	57.7	86.9
08	**DOORS & WINDOWS**	91.9	79.9	88.8	91.9	76.8	88.0	86.9	51.9	77.8	86.0	51.9	77.0	83.0	79.6	82.1	98.0	51.9	86.0
09200	PLASTER & GYPSUM BOARD	232.4	79.2	137.2	197.2	78.1	123.2	168.0	51.6	95.7	148.5	51.6	88.3	142.2	79.0	103.0	173.8	53.1	98.8
095,098	CEILINGS & ACOUSTICAL TREATMENT	106.8	79.2	90.3	97.5	78.1	85.9	122.2	51.6	80.1	122.2	51.6	80.1	100.9	79.0	87.8	106.8	53.1	74.8
09600	FLOORING	132.9	86.7	120.4	132.9	88.6	120.9	120.0	56.3	102.7	120.0	56.3	102.7	126.6	87.9	116.1	114.6	50.9	97.4
097,099	WALL FINISHES, PAINTS & COATINGS	109.1	89.6	97.2	109.1	83.3	93.3	109.1	60.7	79.6	109.1	51.8	74.1	109.1	89.6	97.2	109.1	56.1	76.8
09	**FINISHES**	133.3	82.9	107.0	125.0	81.7	102.4	124.0	55.1	88.1	120.8	54.1	86.1	115.6	81.9	98.0	119.4	54.5	85.6
10 - 14	**TOTAL DIV. 10000 - 14000**	140.0	82.3	127.7	140.0	74.7	126.1	140.0	56.0	122.1	140.0	56.0	122.1	140.0	63.4	123.7	140.0	56.7	122.2
15	**MECHANICAL**	101.6	79.3	92.4	101.5	71.9	89.2	101.3	59.7	84.0	101.2	59.7	84.0	101.0	77.3	91.2	101.4	56.5	82.8
16	**ELECTRICAL**	122.1	89.7	106.3	125.9	71.6	99.5	126.4	61.9	95.0	126.5	61.9	95.1	128.2	89.7	109.5	121.0	57.4	90.1
01 - 16	**WEIGHTED AVERAGE**	120.5	85.5	105.0	119.4	79.4	101.6	116.1	62.9	92.5	114.6	62.8	91.6	113.6	83.1	100.0	120.1	60.9	93.8

	DIVISION	CANADA																	
		THUNDER BAY, ONTARIO			TORONTO, ONTARIO			VANCOUVER, B C			WINDSOR, ONTARIO			WINNIPEG, MANITOBA					
		MAT.	INST.	TOTAL	MAT.	INST.	TOTAL	MAT.	INST.	TOTAL	MAT.	INST.	TOTAL	MAT.	INST.	TOTAL	MAT.	INST.	TOTAL
01590	**EQUIPMENT RENTAL**	.0	100.4	100.4	.0	103.2	103.2	.0	111.3	111.3	.0	100.4	100.4	.0	105.3	105.3	.0	.0	.0
02	**SITE CONSTRUCTION**	107.9	98.7	101.1	113.8	103.6	106.3	118.6	107.2	110.3	98.3	98.8	98.7	112.2	100.0	103.3	.0	.0	.0
03100	CONCRETE FORMS & ACCESSORIES	122.6	79.7	85.6	123.0	90.4	94.8	114.5	82.2	86.7	122.6	80.3	86.1	122.2	65.2	73.0	.0	.0	.0
03200	CONCRETE REINFORCEMENT	99.5	85.0	92.1	170.9	86.6	128.0	173.2	78.6	125.1	109.0	84.3	96.5	161.5	54.3	107.0	.0	.0	.0
03300	CAST-IN-PLACE CONCRETE	154.2	89.6	127.6	163.3	98.8	136.7	153.2	91.5	127.9	143.5	90.7	121.8	157.5	70.2	121.6	.0	.0	.0
03	**CONCRETE**	130.4	84.3	107.5	149.2	92.5	121.1	147.9	85.2	116.8	123.5	84.9	104.3	144.6	65.6	105.4	.0	.0	.0
04	**MASONRY**	163.9	82.8	113.6	186.9	96.2	130.7	177.3	86.1	120.8	163.3	89.0	117.2	178.5	61.5	106.0	.0	.0	.0
05	**METALS**	106.4	85.4	99.7	130.4	88.5	117.0	141.5	88.4	124.6	106.5	86.6	100.1	139.2	73.7	118.3	.0	.0	.0
06	**WOOD & PLASTICS**	118.4	79.9	98.3	118.4	88.9	103.0	113.9	80.2	96.3	118.4	78.5	97.6	116.1	65.9	90.0	.0	.0	.0
07	**THERMAL & MOISTURE PROTECTION**	108.8	81.9	97.5	110.2	92.3	102.7	116.6	86.9	104.1	108.6	85.5	98.9	105.3	68.2	89.6	.0	.0	.0
08	**DOORS & WINDOWS**	82.0	79.2	81.2	90.9	87.7	90.1	94.4	78.4	90.2	81.7	79.0	81.0	91.9	59.2	83.4	.0	.0	.0
09200	PLASTER & GYPSUM BOARD	170.8	79.4	114.0	174.9	88.6	121.3	149.2	79.1	105.7	163.5	78.0	110.4	153.9	64.5	98.4	.0	.0	.0
095,098	CEILINGS & ACOUSTICAL TREATMENT	95.8	79.4	86.0	121.3	88.6	101.8	111.1	79.1	92.0	95.8	78.0	85.1	107.7	64.5	81.9	.0	.0	.0
09600	FLOORING	132.9	50.6	110.6	132.9	93.2	122.1	132.9	85.3	120.0	132.9	88.6	120.9	132.9	62.5	113.8	.0	.0	.0
097,099	WALL FINISHES, PAINTS & COATINGS	109.1	91.6	98.4	112.1	100.6	105.1	109.1	95.2	100.6	109.1	91.2	98.2	109.2	53.3	75.1	.0	.0	.0
09	**FINISHES**	120.8	76.4	97.7	128.4	92.2	109.6	122.6	83.8	102.4	119.4	83.1	100.5	122.0	63.7	91.6	.0	.0	.0
10 - 14	**TOTAL DIV. 10000 - 14000**	140.0	64.3	123.9	140.0	87.4	128.8	140.0	83.2	127.9	140.0	65.5	124.1	140.0	59.6	122.9	.0	.0	.0
15	**MECHANICAL**	101.0	78.0	91.5	101.5	88.0	95.9	101.5	81.4	93.2	101.0	80.5	92.5	101.5	65.9	86.7	.0	.0	.0
16	**ELECTRICAL**	125.2	87.9	107.1	129.8	92.3	111.6	132.9	80.8	107.5	133.4	89.7	112.1	131.9	68.6	101.1	.0	.0	.0
01 - 16	**WEIGHTED AVERAGE**	114.8	82.8	100.6	124.2	92.1	109.9	125.7	85.6	107.9	114.6	85.4	101.6	124.1	68.9	99.6	.0	.0	.0

Location Factors

Costs shown in *Means cost data publications* are based on National Averages for materials and installation. To adjust these costs to a specific location, simply multiply the base cost by the factor and divide by 100 for that city. The data is arranged alphabetically by state and postal zip code numbers. For a city not listed, use the factor for a nearby city with similar economic characteristics.

STATE/ZIP	CITY	MAT.	INST.	TOTAL
ALABAMA				
350-352	Birmingham	96.5	75.2	87.0
354	Tuscaloosa	96.3	55.4	78.1
355	Jasper	96.5	52.1	76.8
356	Decatur	96.2	55.9	78.3
357-358	Huntsville	96.3	72.0	85.5
359	Gadsden	96.1	58.8	79.6
360-361	Montgomery	97.0	57.1	79.3
362	Anniston	95.7	46.5	73.9
363	Dothan	96.3	49.0	75.3
364	Evergreen	95.7	50.2	75.5
365-366	Mobile	97.2	61.1	81.2
367	Selma	95.9	51.6	76.2
368	Phenix City	96.6	54.9	78.1
369	Butler	96.1	49.0	75.2
ALASKA				
995-996	Anchorage	133.7	114.7	125.3
997	Fairbanks	130.3	117.6	124.6
998	Juneau	132.5	114.0	124.3
999	Ketchikan	141.9	113.6	129.3
ARIZONA				
850,853	Phoenix	98.8	73.6	87.6
852	Mesa/Tempe	97.8	68.9	85.0
855	Globe	98.1	65.5	83.6
856-857	Tucson	97.1	71.2	85.6
859	Show Low	98.2	68.0	84.8
860	Flagstaff	100.4	72.3	87.9
863	Prescott	98.3	66.8	84.3
864	Kingman	96.7	67.2	83.6
865	Chambers	96.8	68.3	84.1
ARKANSAS				
716	Pine Bluff	95.2	60.9	80.0
717	Camden	93.2	38.9	69.1
718	Texarkana	94.6	44.0	72.1
719	Hot Springs	92.4	38.4	68.4
720-722	Little Rock	95.1	63.8	81.2
723	West Memphis	94.6	53.4	76.3
724	Jonesboro	95.2	53.4	76.6
725	Batesville	93.3	48.4	73.3
726	Harrison	94.5	48.4	74.0
727	Fayetteville	91.9	46.7	71.8
728	Russellville	93.2	46.6	72.5
729	Fort Smith	95.9	56.9	78.6
CALIFORNIA				
900-902	Los Angeles	102.5	112.1	106.8
903-905	Inglewood	98.2	109.0	103.0
906-908	Long Beach	99.6	109.0	103.8
910-912	Pasadena	100.4	109.3	104.4
913-916	Van Nuys	103.5	109.1	106.0
917-918	Alhambra	102.7	109.1	105.5
919-921	San Diego	104.0	103.9	104.0
922	Palm Springs	100.1	107.2	103.2
923-924	San Bernardino	97.9	108.1	102.4
925	Riverside	102.2	109.5	105.5
926-927	Santa Ana	99.9	108.2	103.5
928	Anaheim	102.6	110.6	106.1
930	Oxnard	103.8	110.0	106.5
931	Santa Barbara	102.6	110.5	106.1
932-933	Bakersfield	102.3	105.9	103.9
934	San Luis Obispo	103.4	107.9	105.4
935	Mojave	100.5	103.1	101.7
936-938	Fresno	104.1	110.9	107.1
939	Salinas	104.1	116.4	109.6
940-941	San Francisco	112.8	132.9	121.7
942,956-958	Sacramento	106.6	113.0	109.4
943	Palo Alto	105.5	127.2	115.1
944	San Mateo	108.4	127.8	117.0
945	Vallejo	105.5	122.2	112.9
946	Oakland	110.7	127.1	118.0
947	Berkeley	110.1	126.7	117.5
948	Richmond	109.5	126.1	116.9
949	San Rafael	110.2	122.8	115.8
950	Santa Cruz	109.4	116.4	112.5

STATE/ZIP	CITY	MAT.	INST.	TOTAL
CALIFORNIA (CONT'D)				
951	San Jose	107.3	127.5	116.3
952	Stockton	104.4	113.5	108.4
953	Modesto	104.3	111.0	107.3
954	Santa Rosa	104.1	122.0	112.1
955	Eureka	105.9	101.1	103.8
959	Marysville	105.0	112.5	108.3
960	Redding	106.6	110.3	108.3
961	Susanville	105.7	110.7	107.9
COLORADO				
800-802	Denver	100.6	89.9	95.8
803	Boulder	97.7	86.5	92.7
804	Golden	99.9	85.8	93.6
805	Fort Collins	101.1	82.4	92.8
806	Greeley	98.7	70.4	86.1
807	Fort Morgan	98.4	85.8	92.8
808-809	Colorado Springs	100.4	85.7	93.9
810	Pueblo	100.0	83.1	92.5
811	Alamosa	101.3	79.5	91.7
812	Salida	101.3	80.2	91.9
813	Durango	101.7	81.4	92.7
814	Montrose	100.3	79.6	91.1
815	Grand Junction	103.6	76.8	91.7
816	Glenwood Springs	101.1	83.3	93.2
CONNECTICUT				
060	New Britain	101.1	116.6	108.0
061	Hartford	102.3	117.3	109.0
062	Willimantic	101.8	117.2	108.7
063	New London	98.1	117.3	106.7
064	Meriden	100.1	117.8	108.0
065	New Haven	102.8	117.8	109.4
066	Bridgeport	102.4	117.8	109.2
067	Waterbury	101.9	117.7	108.9
068	Norwalk	101.9	118.2	109.1
069	Stamford	102.0	124.0	111.8
D.C.				
200-205	Washington	102.5	91.2	97.5
DELAWARE				
197	Newark	100.7	103.9	102.1
198	Wilmington	99.7	103.9	101.5
199	Dover	100.7	103.9	102.1
FLORIDA				
320,322	Jacksonville	98.3	59.6	81.1
321	Daytona Beach	98.3	72.8	87.0
323	Tallahassee	97.6	48.2	75.6
324	Panama City	99.4	36.8	71.6
325	Pensacola	99.1	57.2	80.5
326,344	Gainesville	99.6	57.7	81.0
327-328,347	Orlando	99.7	68.6	85.9
329	Melbourne	100.8	77.8	90.6
330-332,340	Miami	98.7	71.4	86.6
333	Fort Lauderdale	97.7	70.3	85.5
334,349	West Palm Beach	96.6	66.6	83.3
335-336,346	Tampa	99.4	70.8	86.7
337	St. Petersburg	101.6	54.2	80.5
338	Lakeland	98.4	70.5	86.0
339,341	Fort Myers	97.7	62.8	82.2
342	Sarasota	99.5	65.1	84.2
GEORGIA				
300-303,399	Atlanta	97.7	79.0	89.4
304	Statesboro	97.1	43.6	73.4
305	Gainesville	96.0	55.7	78.1
306	Athens	95.3	62.2	80.6
307	Dalton	97.4	49.4	76.1
308-309	Augusta	96.3	57.5	79.0
310-312	Macon	95.7	59.1	79.4
313-314	Savannah	97.0	57.7	79.6
315	Waycross	97.0	50.0	76.1
316	Valdosta	96.9	47.6	75.0
317,398	Albany	97.2	56.3	79.1
318-319	Columbus	97.2	60.4	80.8

STATE/ZIP	CITY	MAT.	INST.	TOTAL
HAWAII				
967	Hilo	115.0	124.8	119.3
968	Honolulu	120.6	124.8	122.5
STATES & POSS.				
969	Guam	196.4	52.7	132.5
IDAHO				
832	Pocatello	100.3	77.3	90.1
833	Twin Falls	100.8	57.7	81.7
834	Idaho Falls	98.3	58.8	80.8
835	Lewiston	107.4	83.6	96.8
836-837	Boise	100.2	78.9	90.7
838	Coeur d'Alene	106.6	75.9	93.0
ILLINOIS				
600-603	North Suburban	99.5	125.3	111.0
604	Joliet	99.4	118.5	107.9
605	South Suburban	99.5	125.3	111.0
606-608	Chicago	100.1	126.3	111.7
609	Kankakee	95.7	104.1	99.4
610-611	Rockford	98.1	107.9	102.4
612	Rock Island	95.4	97.0	96.1
613	La Salle	96.7	102.6	99.3
614	Galesburg	96.5	98.9	97.6
615-616	Peoria	99.1	99.9	99.4
617	Bloomington	95.8	99.2	97.3
618-619	Champaign	99.2	100.8	99.9
620-622	East St. Louis	94.5	102.0	97.8
623	Quincy	95.7	94.6	95.2
624	Effingham	95.1	97.1	96.0
625	Decatur	97.0	101.1	98.8
626-627	Springfield	97.3	99.8	98.4
628	Centralia	93.3	101.1	96.8
629	Carbondale	92.9	97.3	94.9
INDIANA				
460	Anderson	94.9	84.8	90.4
461-462	Indianapolis	98.8	87.2	93.7
463-464	Gary	96.2	104.3	99.8
465-466	South Bend	95.3	86.8	91.5
467-468	Fort Wayne	95.4	83.1	89.9
469	Kokomo	93.0	85.7	89.8
470	Lawrenceburg	92.0	81.8	87.5
471	New Albany	93.6	77.5	86.5
472	Columbus	96.1	84.0	90.7
473	Muncie	96.2	82.4	90.1
474	Bloomington	98.2	84.1	91.9
475	Washington	94.0	88.4	91.5
476-477	Evansville	95.3	90.0	93.0
478	Terre Haute	96.0	88.3	92.6
479	Lafayette	95.7	82.4	89.8
IOWA				
500-503,509	Des Moines	97.8	82.2	90.9
504	Mason City	96.0	62.0	80.9
505	Fort Dodge	96.2	58.1	79.3
506-507	Waterloo	97.8	62.0	81.9
508	Creston	96.4	66.0	82.9
510-511	Sioux City	98.4	74.9	88.0
512	Sibley	97.0	51.5	76.8
513	Spencer	98.8	49.9	77.1
514	Carroll	95.8	55.7	78.0
515	Council Bluffs	99.3	75.2	88.6
516	Shenandoah	96.2	52.9	77.0
520	Dubuque	97.8	77.2	88.7
521	Decorah	96.8	52.8	77.3
522-524	Cedar Rapids	98.9	84.4	92.5
525	Ottumwa	96.8	71.1	85.4
526	Burlington	96.0	75.5	86.9
527-528	Davenport	97.9	94.9	96.5
KANSAS				
660-662	Kansas City	98.9	96.1	97.7
664-666	Topeka	98.7	69.3	85.6
667	Fort Scott	97.6	72.3	86.4
668	Emporia	97.4	57.7	79.8
669	Belleville	99.2	63.7	83.5
670-672	Wichita	97.9	69.5	85.3
673	Independence	99.0	67.7	85.1
674	Salina	99.2	63.9	83.5
675	Hutchinson	94.5	49.5	74.5
676	Hays	98.5	67.0	84.5
677	Colby	99.2	53.1	78.8

STATE/ZIP	CITY	MAT.	INST.	TOTAL
KANSAS (CONT'D)				
678	Dodge City	100.4	68.4	86.2
679	Liberal	98.3	44.2	74.3
KENTUCKY				
400-402	Louisville	96.4	85.4	91.5
403-405	Lexington	96.0	65.7	82.5
406	Frankfort	96.3	66.7	83.1
407-409	Corbin	93.3	43.3	71.1
410	Covington	93.8	91.9	93.0
411-412	Ashland	92.5	96.9	94.5
413-414	Campton	94.1	43.7	71.7
415-416	Pikeville	95.0	60.3	79.6
417-418	Hazard	93.5	43.4	71.3
420	Paducah	92.2	88.4	90.5
421-422	Bowling Green	94.6	82.7	89.3
423	Owensboro	94.4	73.5	85.1
424	Henderson	92.0	90.4	91.3
425-426	Somerset	91.6	45.1	70.9
427	Elizabethtown	91.1	84.5	88.2
LOUISIANA				
700-701	New Orleans	99.8	69.7	86.4
703	Thibodaux	97.7	67.2	84.2
704	Hammond	95.0	62.7	80.7
705	Lafayette	97.4	62.3	81.8
706	Lake Charles	97.5	64.1	82.7
707-708	Baton Rouge	98.1	62.9	82.5
710-711	Shreveport	94.6	61.3	79.8
712	Monroe	95.3	57.9	78.7
713-714	Alexandria	95.4	59.5	79.5
MAINE				
039	Kittery	95.2	73.4	85.5
040-041	Portland	98.5	79.8	90.2
042	Lewiston	98.9	79.8	90.4
043	Augusta	97.4	73.2	86.7
044	Bangor	97.7	79.8	89.8
045	Bath	96.3	73.4	86.1
046	Machias	95.9	72.8	85.6
047	Houlton	96.0	76.4	87.3
048	Rockland	95.1	73.4	85.4
049	Waterville	96.3	73.3	86.1
MARYLAND				
206	Waldorf	98.3	72.6	86.9
207-208	College Park	98.4	81.4	90.8
209	Silver Spring	97.6	77.9	88.9
210-212	Baltimore	99.5	84.7	92.9
214	Annapolis	98.9	79.2	90.1
215	Cumberland	95.4	81.2	89.1
216	Easton	96.8	44.8	73.7
217	Hagerstown	96.2	81.9	89.8
218	Salisbury	97.3	53.5	77.8
219	Elkton	94.6	66.7	82.2
MASSACHUSETTS				
010-011	Springfield	100.4	105.9	102.8
012	Pittsfield	99.9	103.0	101.3
013	Greenfield	97.9	105.3	101.2
014	Fitchburg	96.6	119.8	106.9
015-016	Worcester	100.1	120.0	109.0
017	Framingham	96.1	123.8	108.4
018	Lowell	99.7	125.0	111.0
019	Lawrence	100.9	122.6	110.5
020-022, 024	Boston	102.5	132.0	115.6
023	Brockton	101.1	121.0	110.0
025	Buzzards Bay	95.5	119.0	105.9
026	Hyannis	98.1	119.0	107.4
027	New Bedford	100.8	119.1	109.0
MICHIGAN				
480,483	Royal Oak	95.3	107.9	100.9
481	Ann Arbor	97.4	105.9	101.2
482	Detroit	99.1	117.6	107.3
484-485	Flint	97.2	99.9	98.4
486	Saginaw	97.0	93.4	95.4
487	Bay City	96.9	93.5	95.4
488-489	Lansing	97.2	95.2	96.3
490	Battle Creek	95.5	87.9	92.1
491	Kalamazoo	95.8	85.8	91.4
492	Jackson	93.8	92.3	93.2
493,495	Grand Rapids	96.0	68.1	83.6
494	Muskegon	94.4	83.6	89.6

Location Factors

STATE/ZIP	CITY	MAT.	INST.	TOTAL
MICHIGAN (CONT'D)				
496	Traverse City	93.3	73.1	84.3
497	Gaylord	94.5	76.9	86.7
498-499	Iron Mountain	96.3	87.7	92.5
MINNESOTA				
550-551	Saint Paul	99.7	124.5	110.7
553-555	Minneapolis	100.5	126.7	112.2
556-558	Duluth	99.5	114.5	106.2
559	Rochester	99.2	108.5	103.4
560	Mankato	96.3	105.9	100.6
561	Windom	95.1	80.3	88.5
562	Willmar	94.4	87.4	91.3
563	St. Cloud	95.7	123.3	108.0
564	Brainerd	96.0	105.1	100.0
565	Detroit Lakes	97.9	100.6	99.1
566	Bemidji	97.3	101.2	99.0
567	Thief River Falls	96.8	96.1	96.5
MISSISSIPPI				
386	Clarksdale	95.9	30.8	67.0
387	Greenville	99.3	44.6	75.0
388	Tupelo	97.3	37.3	70.6
389	Greenwood	97.2	33.3	68.8
390-392	Jackson	97.8	46.1	74.8
393	Meridian	96.0	46.7	74.1
394	Laurel	97.1	33.4	68.8
395	Biloxi	97.8	55.5	79.0
396	McComb	95.7	52.1	76.4
397	Columbus	97.1	36.7	70.3
MISSOURI				
630-631	St. Louis	97.0	109.0	102.3
633	Bowling Green	95.7	95.8	95.7
634	Hannibal	94.5	86.3	90.9
635	Kirksville	96.8	79.2	89.0
636	Flat River	96.7	97.0	96.8
637	Cape Girardeau	96.4	92.3	94.6
638	Sikeston	94.5	83.2	89.5
639	Poplar Bluff	94.0	83.8	89.5
640-641	Kansas City	100.9	106.8	103.5
644-645	St. Joseph	99.9	93.3	97.0
646	Chillicothe	96.7	71.2	85.4
647	Harrisonville	96.4	101.3	98.6
648	Joplin	98.6	70.5	86.1
650-651	Jefferson City	96.1	89.7	93.3
652	Columbia	97.1	92.4	95.0
653	Sedalia	96.1	84.9	91.2
654-655	Rolla	95.0	78.6	87.7
656-658	Springfield	98.4	78.7	89.6
MONTANA				
590-591	Billings	102.1	74.5	89.8
592	Wolf Point	100.8	68.7	86.6
593	Miles City	98.8	70.0	86.0
594	Great Falls	102.7	73.8	89.9
595	Havre	99.6	68.9	86.0
596	Helena	102.5	71.5	88.7
597	Butte	101.2	70.9	87.7
598	Missoula	98.9	70.1	86.1
599	Kalispell	97.8	69.4	85.2
NEBRASKA				
680-681	Omaha	99.3	78.2	89.9
683-685	Lincoln	98.2	67.1	84.4
686	Columbus	96.1	49.6	75.5
687	Norfolk	97.7	60.6	81.3
688	Grand Island	97.5	65.6	83.3
689	Hastings	97.1	56.7	79.2
690	Mccook	96.9	48.2	75.3
691	North Platte	97.1	58.8	80.1
692	Valentine	99.2	38.5	72.3
693	Alliance	99.0	35.9	71.0
NEVADA				
889-891	Las Vegas	98.3	107.1	102.2
893	Ely	98.7	79.0	90.0
894-895	Reno	98.9	98.2	98.6
897	Carson City	98.6	97.9	98.3
898	Elko	97.6	85.4	92.2
NEW HAMPSHIRE				
030	Nashua	100.6	82.4	92.5
031	Manchester	100.9	82.4	92.7

STATE/ZIP	CITY	MAT.	INST.	TOTAL
NEW HAMPSHIRE (CONT'D)				
032-033	Concord	98.4	82.4	91.3
034	Keene	97.2	50.9	76.6
035	Littleton	97.4	61.6	81.5
036	Charleston	96.6	47.9	75.0
037	Claremont	95.9	47.9	74.6
038	Portsmouth	97.8	78.0	89.0
NEW JERSEY				
070-071	Newark	102.1	124.6	112.1
072	Elizabeth	100.8	123.6	110.9
073	Jersey City	99.8	124.5	110.8
074-075	Paterson	101.4	124.2	111.5
076	Hackensack	99.5	124.5	110.6
077	Long Branch	99.1	125.3	110.8
078	Dover	99.7	124.5	110.7
079	Summit	99.8	123.6	110.4
080,083	Vineland	97.6	117.3	106.3
081	Camden	99.6	116.8	107.2
082,084	Atlantic City	98.4	116.5	106.4
085-086	Trenton	99.6	122.6	109.8
087	Point Pleasant	99.5	122.3	109.6
088-089	New Brunswick	100.0	123.4	110.4
NEW MEXICO				
870-872	Albuquerque	98.5	76.2	88.6
873	Gallup	98.3	76.2	88.5
874	Farmington	98.7	76.2	88.7
875	Santa Fe	97.5	76.2	88.0
877	Las Vegas	96.9	76.2	87.7
878	Socorro	96.5	76.2	87.5
879	Truth/Consequences	96.9	70.8	85.3
880	Las Cruces	95.8	67.9	83.4
881	Clovis	97.0	75.2	87.3
882	Roswell	98.7	75.3	88.3
883	Carrizozo	99.0	76.2	88.9
884	Tucumcari	97.9	75.2	87.8
NEW YORK				
100-102	New York	106.0	164.4	131.9
103	Staten Island	101.6	159.9	127.5
104	Bronx	99.8	159.9	126.5
105	Mount Vernon	99.5	138.5	116.8
106	White Plains	99.5	138.2	116.7
107	Yonkers	104.5	138.5	119.6
108	New Rochelle	100.0	138.0	116.9
109	Suffern	99.8	126.2	111.5
110	Queens	101.4	158.8	126.9
111	Long Island City	103.1	158.8	127.8
112	Brooklyn	103.3	159.1	128.1
113	Flushing	103.3	158.8	128.0
114	Jamaica	101.5	158.8	126.9
115,117,118	Hicksville	101.2	149.5	122.7
116	Far Rockaway	103.4	158.8	128.0
119	Riverhead	102.2	149.5	123.2
120-122	Albany	97.8	96.4	97.2
123	Schenectady	98.1	95.9	97.1
124	Kingston	100.9	117.3	108.2
125-126	Poughkeepsie	100.1	119.4	108.7
127	Monticello	99.5	119.9	108.5
128	Glens Falls	92.7	91.3	92.1
129	Plattsburgh	96.8	88.4	93.1
130-132	Syracuse	99.0	94.7	97.1
133-135	Utica	96.9	90.9	94.2
136	Watertown	98.5	88.5	94.1
137-139	Binghamton	98.5	88.8	94.2
140-142	Buffalo	99.8	105.7	102.4
143	Niagara Falls	97.3	104.7	100.6
144-146	Rochester	99.8	99.7	99.8
147	Jamestown	96.3	87.9	92.6
148-149	Elmira	96.3	85.4	91.4
NORTH CAROLINA				
270,272-274	Greensboro	98.5	50.3	77.1
271	Winston-Salem	98.2	49.6	76.6
275-276	Raleigh	98.4	50.7	77.2
277	Durham	98.4	50.7	77.2
278	Rocky Mount	95.7	35.5	69.0
279	Elizabeth City	96.4	37.8	70.4
280	Gastonia	97.7	48.9	76.0
281-282	Charlotte	98.6	48.9	76.5
283	Fayetteville	97.1	50.8	76.5
284	Wilmington	96.4	49.9	75.7
285	Kinston	94.4	35.1	68.1

STATE/ZIP	CITY	MAT.	INST.	TOTAL
NORTH CAROLINA (CONT'D)				
286	Hickory	94.8	33.9	67.8
287-288	Asheville	96.8	48.9	75.5
289	Murphy	95.7	33.0	67.9
NORTH DAKOTA				
580-581	Fargo	100.9	65.2	85.1
582	Grand Forks	100.5	59.2	82.2
583	Devils Lake	99.9	60.8	82.6
584	Jamestown	100.0	52.6	78.9
585	Bismarck	99.3	66.6	84.8
586	Dickinson	100.7	60.5	82.9
587	Minot	100.6	67.8	86.1
588	Williston	99.3	60.2	81.9
OHIO				
430-432	Columbus	97.3	89.6	93.9
433	Marion	93.9	88.4	91.5
434-436	Toledo	97.6	100.0	98.7
437-438	Zanesville	94.3	86.1	90.6
439	Steubenville	95.2	95.1	95.1
440	Lorain	96.9	97.1	97.0
441	Cleveland	97.2	104.3	100.4
442-443	Akron	97.8	98.0	97.9
444-445	Youngstown	97.2	93.3	95.5
446-447	Canton	97.3	88.4	93.4
448-449	Mansfield	94.6	93.8	94.2
450	Hamilton	94.5	89.6	92.3
451-452	Cincinnati	94.9	90.2	92.8
453-454	Dayton	94.6	87.0	91.2
455	Springfield	94.6	87.9	91.6
456	Chillicothe	93.5	94.2	93.8
457	Athens	96.3	79.9	89.0
458	Lima	96.8	87.9	92.8
OKLAHOMA				
730-731	Oklahoma City	97.1	64.3	82.6
734	Ardmore	94.0	63.7	80.5
735	Lawton	96.4	65.3	82.6
736	Clinton	95.3	62.0	80.6
737	Enid	96.0	62.0	80.9
738	Woodward	94.2	62.1	79.9
739	Guymon	95.2	32.9	67.5
740-741	Tulsa	96.2	59.9	80.1
743	Miami	93.0	67.5	81.7
744	Muskogee	95.6	44.0	72.7
745	Mcalester	92.6	55.1	75.9
746	Ponca City	93.1	62.2	79.4
747	Durant	93.1	61.7	79.2
748	Shawnee	94.6	60.2	79.3
749	Poteau	92.3	64.7	80.1
OREGON				
970-972	Portland	102.2	104.6	103.3
973	Salem	102.2	103.2	102.6
974	Eugene	101.7	102.8	102.2
975	Medford	103.3	101.1	102.3
976	Klamath Falls	103.4	101.1	102.4
977	Bend	102.2	102.9	102.5
978	Pendleton	96.5	102.6	99.2
979	Vale	94.4	93.2	93.9
PENNSYLVANIA				
150-152	Pittsburgh	97.0	104.2	100.2
153	Washington	94.0	102.8	98.0
154	Uniontown	94.2	101.2	97.3
155	Bedford	95.2	92.6	94.0
156	Greensburg	95.3	102.2	98.3
157	Indiana	94.1	100.7	97.0
158	Dubois	95.4	97.3	96.2
159	Johnstown	95.1	98.1	96.5
160	Butler	92.3	102.3	96.7
161	New Castle	92.3	101.0	96.2
162	Kittanning	92.8	103.6	97.6
163	Oil City	92.3	98.1	94.9
164-165	Erie	94.4	96.7	95.4
166	Altoona	94.4	94.5	94.5
167	Bradford	95.7	96.6	96.1
168	State College	95.4	93.9	94.7
169	Wellsboro	96.4	94.2	95.4
170-171	Harrisburg	98.2	95.3	96.9
172	Chambersburg	95.9	89.9	93.3
173-174	York	96.2	95.3	95.8
175-176	Lancaster	94.8	90.6	92.9

STATE/ZIP	CITY	MAT.	INST.	TOTAL
PENNSYLVANIA (CONT'D)				
177	Williamsport	93.4	82.4	88.5
178	Sunbury	95.4	94.4	95.0
179	Pottsville	94.5	94.0	94.3
180	Lehigh Valley	95.9	114.4	104.1
181	Allentown	98.0	107.4	102.2
182	Hazleton	95.3	96.6	95.9
183	Stroudsburg	95.2	101.1	97.8
184-185	Scranton	98.8	96.3	97.7
186-187	Wilkes-Barre	95.2	94.9	95.1
188	Montrose	94.8	95.5	95.1
189	Doylestown	94.7	119.7	105.8
190-191	Philadelphia	101.0	131.1	114.4
193	Westchester	97.6	122.0	108.5
194	Norristown	96.6	126.0	109.7
195-196	Reading	98.8	100.9	99.7
PUERTO RICO				
009	San Juan	135.2	25.5	86.5
RHODE ISLAND				
028	Newport	99.6	112.7	105.4
029	Providence	99.9	112.7	105.6
SOUTH CAROLINA				
290-292	Columbia	97.1	47.3	75.0
293	Spartanburg	95.9	45.4	73.5
294	Charleston	97.3	47.9	75.3
295	Florence	95.4	45.2	73.1
296	Greenville	95.7	45.4	73.4
297	Rock Hill	95.3	33.0	67.6
298	Aiken	96.1	72.0	85.4
299	Beaufort	97.0	36.2	70.0
SOUTH DAKOTA				
570-571	Sioux Falls	99.2	57.3	80.6
572	Watertown	97.9	51.0	77.1
573	Mitchell	96.9	50.8	76.4
574	Aberdeen	99.5	53.1	78.9
575	Pierre	98.8	53.5	78.7
576	Mobridge	97.5	50.4	76.6
577	Rapid City	99.4	52.2	78.4
TENNESSEE				
370-372	Nashville	97.5	73.7	86.9
373-374	Chattanooga	98.6	54.1	78.8
375,380-381	Memphis	96.6	76.2	87.5
376	Johnson City	97.3	56.4	79.1
377-379	Knoxville	94.7	57.5	78.2
382	Mckenzie	96.4	59.4	80.0
383	Jackson	98.3	51.7	77.6
384	Columbia	94.9	55.2	77.3
385	Cookeville	96.2	49.9	75.6
TEXAS				
750	Mckinney	97.6	55.5	78.9
751	Waxahackie	97.5	56.3	79.2
752-753	Dallas	98.3	67.1	84.4
754	Greenville	97.7	40.7	72.4
755	Texarkana	96.6	54.0	77.7
756	Longview	96.9	42.6	72.8
757	Tyler	97.4	56.2	79.1
758	Palestine	93.6	43.0	71.1
759	Lufkin	94.5	46.7	73.3
760-761	Fort Worth	96.2	62.8	81.4
762	Denton	96.5	51.6	76.6
763	Wichita Falls	97.1	57.9	79.7
764	Eastland	95.6	42.0	71.8
765	Temple	94.5	50.3	74.9
766-767	Waco	96.4	57.5	79.1
768	Brownwood	96.5	39.5	71.2
769	San Angelo	96.2	47.7	74.7
770-772	Houston	99.6	73.0	87.7
773	Huntsville	97.7	40.3	72.2
774	Wharton	99.1	45.1	75.1
775	Galveston	97.3	70.6	85.4
776-777	Beaumont	97.1	64.1	82.4
778	Bryan	94.7	64.5	81.3
779	Victoria	99.2	47.6	76.3
780	Laredo	94.6	53.7	76.5
781-782	San Antonio	95.0	67.0	82.5
783-784	Corpus Christi	97.3	53.1	77.7
785	Mc Allen	97.2	47.4	75.1
786-787	Austin	94.6	61.7	80.0

STATE/ZIP	CITY	MAT.	INST.	TOTAL
TEXAS (CONT'D)				
788	Del Rio	96.5	33.2	68.4
789	Giddings	94.0	42.1	71.0
790-791	Amarillo	97.2	58.0	79.8
792	Childress	96.1	52.5	76.7
793-794	Lubbock	98.6	53.5	78.6
795-796	Abilene	96.6	53.3	77.4
797	Midland	98.7	49.9	77.0
798-799,885	El Paso	96.6	51.6	76.6
UTAH				
840-841	Salt Lake City	101.3	70.7	87.7
842,844	Ogden	97.4	70.5	85.5
843	Logan	99.1	70.5	86.4
845	Price	99.6	50.1	77.6
846-847	Provo	99.6	70.5	86.7
VERMONT				
050	White River Jct.	98.6	46.5	75.4
051	Bellows Falls	97.1	48.8	75.6
052	Bennington	97.4	49.1	76.0
053	Brattleboro	97.9	49.3	76.3
054	Burlington	100.5	64.9	84.7
056	Montpelier	97.3	64.9	82.9
057	Rutland	99.5	64.9	84.1
058	St. Johnsbury	98.8	49.1	76.7
059	Guildhall	97.4	48.6	75.7
VIRGINIA				
220-221	Fairfax	99.3	80.6	91.0
222	Arlington	100.5	79.3	91.1
223	Alexandria	99.9	84.9	93.3
224-225	Fredericksburg	98.0	68.4	84.9
226	Winchester	98.6	53.8	78.7
227	Culpeper	98.5	55.3	79.3
228	Harrisonburg	98.8	47.1	75.8
229	Charlottesville	99.1	61.5	82.4
230-232	Richmond	100.3	66.7	85.4
233-235	Norfolk	100.0	65.8	84.8
236	Newport News	99.6	64.7	84.1
237	Portsmouth	98.9	65.5	84.0
238	Petersburg	98.9	66.7	84.6
239	Farmville	98.3	41.1	72.9
240-241	Roanoke	100.2	48.1	77.1
242	Bristol	97.9	49.0	76.2
243	Pulaski	97.7	42.4	73.1
244	Staunton	98.4	45.6	75.0
245	Lynchburg	98.6	52.2	78.0
246	Grundy	98.0	44.4	74.2
WASHINGTON				
980-981,987	Seattle	103.8	104.7	104.2
982	Everett	103.6	101.0	102.4
983-984	Tacoma	103.5	101.4	102.6
985	Olympia	102.3	101.3	101.9
986	Vancouver	105.4	97.7	102.0
988	Wenatchee	103.5	86.6	96.0
989	Yakima	103.7	91.8	98.4
990-992	Spokane	105.7	83.3	95.8
993	Richland	105.0	86.9	96.9
994	Clarkston	103.9	84.3	95.2
WEST VIRGINIA				
247-248	Bluefield	96.8	79.4	89.1
249	Lewisburg	98.3	83.1	91.6
250-253	Charleston	99.5	91.0	95.7
254	Martinsburg	98.1	80.0	90.1
255-257	Huntington	99.6	93.4	96.9
258-259	Beckley	96.6	88.8	93.1
260	Wheeling	99.7	94.2	97.2
261	Parkersburg	98.7	92.6	96.0
262	Buckhannon	98.1	93.4	96.0
263-264	Clarksburg	98.5	93.4	96.2
265	Morgantown	98.6	93.5	96.4
266	Gassaway	97.9	92.2	95.4
267	Romney	97.9	86.5	92.9
268	Petersburg	97.9	89.3	94.1
WISCONSIN				
530,532	Milwaukee	99.0	106.2	102.2
531	Kenosha	98.5	103.1	100.5
534	Racine	98.1	102.9	100.2
535	Beloit	98.0	98.4	98.2
537	Madison	98.3	98.1	98.2

STATE/ZIP	CITY	MAT.	INST.	TOTAL
WISCONSIN (CONT'D)				
538	Lancaster	95.8	94.0	95.0
539	Portage	94.4	97.7	95.9
540	New Richmond	95.5	98.5	96.8
541-543	Green Bay	99.7	95.2	97.7
544	Wausau	94.9	94.3	94.6
545	Rhinelander	98.0	93.7	96.1
546	La Crosse	96.1	95.7	95.9
547	Eau Claire	97.8	96.2	97.1
548	Superior	95.4	102.8	98.7
549	Oshkosh	95.5	95.2	95.4
WYOMING				
820	Cheyenne	97.6	61.2	81.4
821	Yellowstone Nat'l Park	96.8	54.6	78.0
822	Wheatland	98.1	56.0	79.4
823	Rawlins	99.4	53.1	78.8
824	Worland	97.4	51.4	77.0
825	Riverton	98.4	53.1	78.3
826	Casper	97.7	59.9	80.9
827	Newcastle	97.2	53.1	77.6
828	Sheridan	98.1	58.7	80.6
829-831	Rock Springs	101.5	51.4	79.2
CANADIAN FACTORS (reflect Canadian currency)				
ALBERTA				
	Calgary	124.3	79.6	104.4
	Edmonton	125.9	79.6	105.3
	Fort McMurray	116.6	79.1	100.0
	Lethbridge	117.6	78.6	100.3
	Lloydminster	116.7	79.1	100.0
	Medicine Hat	116.7	78.6	99.8
	Red Deer	117.1	78.6	100.0
BRITISH COLUMBIA				
	Kamloops	117.8	81.8	101.8
	Prince George	119.1	81.8	102.5
	Vancouver	125.7	85.6	107.9
	Victoria	119.1	82.3	102.8
MANITOBA				
	Brandon	117.0	68.5	95.5
	Portage la Prairie	117.0	68.5	95.5
	Winnipeg	124.1	68.9	99.6
NEW BRUNSWICK				
	Bathurst	115.1	60.9	91.1
	Dalhousie	115.1	60.9	91.1
	Fredericton	117.0	64.9	93.8
	Moncton	115.4	60.9	91.2
	Newcastle	115.1	60.9	91.1
	Saint John	118.1	64.9	94.5
NEWFOUNDLAND				
	Corner Brook	119.8	60.9	93.7
	St. John's	120.1	60.9	93.8
NORTHWEST TERRITORIES				
	Yellowknife	120.4	77.4	101.3
NOVA SCOTIA				
	Dartmouth	117.8	67.1	95.3
	Halifax	119.3	68.2	96.6
	New Glasgow	115.8	67.1	94.2
	Sydney	113.3	67.1	92.8
	Yarmouth	115.6	67.1	94.1
ONTARIO				
	Barrie	119.9	84.1	104.0
	Brantford	119.1	87.6	105.1
	Cornwall	119.1	84.3	103.6
	Hamilton	124.5	87.9	108.2
	Kingston	120.1	84.7	104.4
	Kitchener	115.6	83.7	101.4
	London	121.9	85.3	105.6
	North Bay	119.1	82.4	102.8
	Oshawa	118.7	85.7	104.1
	Ottawa	120.5	85.5	105.0
	Owen Sound	120.1	82.5	103.4
	Peterborough	119.1	84.1	103.5
	Sarnia	119.3	87.9	105.4
	St. Catharines	113.6	83.1	100.0
	Sudbury	113.7	82.7	99.9

Location Factors

STATE/ZIP	CITY	MAT.	INST.	TOTAL
ONTARIO (CONT'D)	Thunder Bay	114.8	82.8	100.6
	Toronto	124.2	92.1	109.9
	Windsor	114.6	85.4	101.6
PRINCE EDWARD ISLAND	Charlottetown	117.7	56.9	90.7
	Summerside	117.3	56.9	90.5
QUEBEC	Cap-de-la-Madeleine	115.9	78.9	99.5
	Charlesbourg	115.9	78.9	99.5
	Chicoutimi	115.0	78.6	98.8
	Gatineau	115.3	78.7	99.1
	Laval	115.2	78.7	99.0
	Montreal	118.5	79.0	101.0
	Quebec	119.4	79.4	101.6
	Sherbrooke	115.7	78.7	99.3
	Trois Rivieres	116.2	78.9	99.6
SASKATCHEWAN	Moose Jaw	114.2	62.9	91.4
	Prince Albert	113.7	62.7	91.1
	Regina	116.1	62.9	92.5
	Saskatoon	114.6	62.8	91.6
YUKON	Whitehorse	114.0	61.9	90.8

R011105-05 Tips for Accurate Estimating

1. Use pre-printed or columnar forms for orderly sequence of dimensions and locations and for recording telephone quotations.
2. Use only the front side of each paper or form except for certain pre-printed summary forms.
3. Be consistent in listing dimensions: For example, length x width x height. This helps in rechecking to ensure that, the total length of partitions is appropriate for the building area.
4. Use printed (rather than measured) dimensions where given.
5. Add up multiple printed dimensions for a single entry where possible.
6. Measure all other dimensions carefully.
7. Use each set of dimensions to calculate multiple related quantities.
8. Convert foot and inch measurements to decimal feet when listing. Memorize decimal equivalents to .01 parts of a foot (1/8″ equals approximately .01′).
9. Do not "round off" quantities until the final summary.
10. Mark drawings with different colors as items are taken off.
11. Keep similar items together, different items separate.
12. Identify location and drawing numbers to aid in future checking for completeness.
13. Measure or list everything on the drawings or mentioned in the specifications.
14. It may be necessary to list items not called for to make the job complete.
15. Be alert for: Notes on plans such as N.T.S. (not to scale); changes in scale throughout the drawings; reduced size drawings; discrepancies between the specifications and the drawings.
16. Develop a consistent pattern of performing an estimate. For example:
    a. Start the quantity takeoff at the lower floor and move to the next higher floor.
    b. Proceed from the main section of the building to the wings.
    c. Proceed from south to north or vice versa, clockwise or counterclockwise.
    d. Take off floor plan quantities first, elevations next, then detail drawings.
17. List all gross dimensions that can be either used again for different quantities, or used as a rough check of other quantities for verification (exterior perimeter, gross floor area, individual floor areas, etc.).
18. Utilize design symmetry or repetition (repetitive floors, repetitive wings, symmetrical design around a center line, similar room layouts, etc.). Note: Extreme caution is needed here so as not to omit or duplicate an area.
19. Do not convert units until the final total is obtained. For instance, when estimating concrete work, keep all units to the nearest cubic foot, then summarize and convert to cubic yards.
20. When figuring alternatives, it is best to total all items involved in the basic system, then total all items involved in the alternates. Therefore you work with positive numbers in all cases. When adds and deducts are used, it is often confusing whether to add or subtract a portion of an item; especially on a complicated or involved alternate.

R011105-10 Unit Gross Area Requirements

The figures in the table below indicate typical ranges in square feet as a function of the "occupant" unit. This table is best used in the preliminary design stages to help determine the probable size requirement for the total project.

Building Type	Unit	Gross Area in S.F.		
		1/4	Median	3/4
Apartments	Unit	660	860	1,100
Auditorium & Play Theaters	Seat	18	25	38
Bowling Alleys	Lane		940	
Churches & Synagogues	Seat	20	28	39
Dormitories	Bed	200	230	275
Fraternity & Sorority Houses	Bed	220	315	370
Garages, Parking	Car	325	355	385
Hospitals	Bed	685	850	1,075
Hotels	Rental Unit	475	600	710
Housing for the elderly	Unit	515	635	755
Housing, Public	Unit	700	875	1,030
Ice Skating Rinks	Total	27,000	30,000	36,000
Motels	Rental Unit	360	465	620
Nursing Homes	Bed	290	350	450
Restaurants	Seat	23	29	39
Schools, Elementary	Pupil	65	77	90
Junior High & Middle	↓	85	110	129
Senior High	↓	102	130	145
Vocational	↓	110	135	195
Shooting Ranges	Point		450	
Theaters & Movies	Seat		15	

R011105-20 Floor Area Ratios

Table below lists commonly used gross to net area and net to gross area ratios expressed in % for various building types.

Building Type	Gross to Net Ratio	Net to Gross Ratio	Building Type	Gross to Net Ratio	Net to Gross Ratio
Apartment	156	64	School Buildings (campus type)		
Bank	140	72	Administrative	150	67
Church	142	70	Auditorium	142	70
Courthouse	162	61	Biology	161	62
Department Store	123	81	Chemistry	170	59
Garage	118	85	Classroom	152	66
Hospital	183	55	Dining Hall	138	72
Hotel	158	63	Dormitory	154	65
Laboratory	171	58	Engineering	164	61
Library	132	76	Fraternity	160	63
Office	135	75	Gymnasium	142	70
Restaurant	141	70	Science	167	60
Warehouse	108	93	Service	120	83
			Student Union	172	59

The gross area of a building is the total floor area based on outside dimensions.

The net area of a building is the usable floor area for the function intended and excludes such items as stairways, corridors and mechanical rooms. In the case of a commercial building, it might be considered as the "leasable area."

R011105-30 Occupancy Determinations

Description		S.F. Required per Person		
		BOCA	SBC	UBC
Assembly Areas	Fixed Seats	**	6	7
	Movable Seats		15	15
	Concentrated	7		
	Unconcentrated	15		
	Standing Space	3		
Educational	Unclassified			
	Classrooms	20	40	20
	Shop Areas	50	100	50
Institutional	Unclassified		125	
	In-Patient Areas	240		
	Sleeping Areas	120		
Mercantile	Basement	30	30	20
	Ground Floor	30	30	30
	Upper Floors	60	60	50
Office		100	100	100

BOCA=Building Officials & Code Administrators
SBC=Southern Building Code
UBC=Uniform Building Code

** The occupancy load for assembly area with fixed seats shall be determined by the number of fixed seats installed.

R011105-40 Weather Data and Design Conditions

City	Latitude (1)		Winter Temperatures (1)			Winter Degree Days (2)	Summer (Design Dry Bulb) Temperatures and Relative Humidity		
	0	1′	Med. of Annual Extremes	99%	97½%		1%	2½%	5%
UNITED STATES									
Albuquerque, NM	35	0	5.1	12	16	4,400	96/61	94/61	92/61
Atlanta, GA	33	4	11.9	17	22	3,000	94/74	92/74	90/73
Baltimore, MD	39	2	7	14	17	4,600	94/75	91/75	89/74
Birmingham, AL	33	3	13	17	21	2,600	96/74	94/75	92/74
Bismarck, ND	46	5	-32	-23	-19	8,800	95/68	91/68	88/67
Boise, ID	43	3	1	3	10	5,800	96/65	94/64	91/64
Boston, MA	42	2	-1	6	9	5,600	91/73	88/71	85/70
Burlington, VT	44	3	-17	-12	-7	8,200	88/72	85/70	82/69
Charleston, WV	38	2	3	7	11	4,400	92/74	90/73	87/72
Charlotte, NC	35	1	13	18	22	3,200	95/74	93/74	91/74
Casper, WY	42	5	-21	-11	-5	7,400	92/58	90/57	87/57
Chicago, IL	41	5	-8	-3	2	6,600	94/75	91/74	88/73
Cincinnati, OH	39	1	0	1	6	4,400	92/73	90/72	88/72
Cleveland, OH	41	2	-3	1	5	6,400	91/73	88/72	86/71
Columbia, SC	34	0	16	20	24	2,400	97/76	95/75	93/75
Dallas, TX	32	5	14	18	22	2,400	102/75	100/75	97/75
Denver, CO	39	5	-10	-5	1	6,200	93/59	91/59	89/59
Des Moines, IA	41	3	-14	-10	-5	6,600	94/75	91/74	88/73
Detroit, MI	42	2	-3	3	6	6,200	91/73	88/72	86/71
Great Falls, MT	47	3	-25	-21	-15	7,800	91/60	88/60	85/59
Hartford, CT	41	5	-4	3	7	6,200	91/74	88/73	85/72
Houston, TX	29	5	24	28	33	1,400	97/77	95/77	93/77
Indianapolis, IN	39	4	-7	-2	2	5,600	92/74	90/74	87/73
Jackson, MS	32	2	16	21	25	2,200	97/76	95/76	93/76
Kansas City, MO	39	1	-4	2	6	4,800	99/75	96/74	93/74
Las Vegas, NV	36	1	18	25	28	2,800	108/66	106/65	104/65
Lexington, KY	38	0	-1	3	8	4,600	93/73	91/73	88/72
Little Rock, AR	34	4	11	15	20	3,200	99/76	96/77	94/77
Los Angeles, CA	34	0	36	41	43	2,000	93/70	89/70	86/69
Memphis, TN	35	0	10	13	18	3,200	98/77	95/76	93/76
Miami, FL	25	5	39	44	47	200	91/77	90/77	89/77
Milwaukee, WI	43	0	-11	-8	-4	7,600	90/74	87/73	84/71
Minneapolis, MN	44	5	-22	-16	-12	8,400	92/75	89/73	86/71
New Orleans, LA	30	0	28	29	33	1,400	93/78	92/77	90/77
New York, NY	40	5	6	11	15	5,000	92/74	89/73	87/72
Norfolk, VA	36	5	15	20	22	3,400	93/77	91/76	89/76
Oklahoma City, OK	35	2	4	9	13	3,200	100/74	97/74	95/73
Omaha, NE	41	2	-13	-8	-3	6,600	94/76	91/75	88/74
Philadelphia, PA	39	5	6	10	14	4,400	93/75	90/74	87/72
Phoenix, AZ	33	3	27	31	34	1,800	109/71	107/71	105/71
Pittsburgh, PA	40	3	-1	3	7	6,000	91/72	88/71	86/70
Portland, ME	43	4	-10	-6	-1	7,600	87/72	84/71	81/69
Portland, OR	45	4	18	17	23	4,600	89/68	85/67	81/65
Portsmouth, NH	43	1	-8	-2	2	7,200	89/73	85/71	83/70
Providence, RI	41	4	-1	5	9	6,000	89/73	86/72	83/70
Rochester, NY	43	1	-5	1	5	6,800	91/73	88/71	85/70
Salt Lake City, UT	40	5	0	3	8	6,000	97/62	95/62	92/61
San Francisco, CA	37	5	36	38	40	3,000	74/63	71/62	69/61
Seattle, WA	47	4	22	22	27	5,200	85/68	82/66	78/65
Sioux Falls, SD	43	4	-21	-15	-11	7,800	94/73	91/72	88/71
St. Louis, MO	38	4	-3	3	8	5,000	98/75	94/75	91/75
Tampa, FL	28	0	32	36	40	680	92/77	91/77	90/76
Trenton, NJ	40	1	4	11	14	5,000	91/75	88/74	85/73
Washington, DC	38	5	7	14	17	4,200	93/75	91/74	89/74
Wichita, KS	37	4	-3	3	7	4,600	101/72	98/73	96/73
Wilmington, DE	39	4	5	10	14	5,000	92/74	89/74	87/73
ALASKA									
Anchorage	61	1	-29	-23	-18	10,800	71/59	68/58	66/56
Fairbanks	64	5	-59	-51	-47	14,280	82/62	78/60	75/59
CANADA									
Edmonton, Alta.	53	3	-30	-29	-25	11,000	85/66	82/65	79/63
Halifax, N.S.	44	4	-4	1	5	8,000	79/66	76/65	74/64
Montreal, Que.	45	3	-20	-16	-10	9,000	88/73	85/72	83/71
Saskatoon, Sask.	52	1	-35	-35	-31	11,000	89/68	86/66	83/65
St. John's, N.F.	47	4	1	3	7	8,600	77/66	75/65	73/64
Saint John, N.B.	45	2	-15	-12	-8	8,200	80/67	77/65	75/64
Toronto, Ont.	43	4	-10	-5	-1	7,000	90/73	87/72	85/71
Vancouver, B.C.	49	1	13	15	19	6,000	79/67	77/66	74/65
Winnipeg, Man.	49	5	-31	-30	-27	10,800	89/73	86/71	84/70

(1) Handbook of Fundamentals, ASHRAE, Inc., NY 1989
(2) Local Climatological Annual Survey, USDC Env. Science Services Administration, Asheville, NC

R011105-50 Metric Conversion Factors

Description: This table is primarily for converting customary U.S. units in the left hand column to SI metric units in the right hand column. In addition, conversion factors for some commonly encountered Canadian and non-SI metric units are included.

	If You Know		Multiply By		To Find
Length	Inches	x	25.4[a]	=	Millimeters
	Feet	x	0.3048[a]	=	Meters
	Yards	x	0.9144[a]	=	Meters
	Miles (statute)	x	1.609	=	Kilometers
Area	Square inches	x	645.2	=	Square millimeters
	Square feet	x	0.0929	=	Square meters
	Square yards	x	0.8361	=	Square meters
Volume (Capacity)	Cubic inches	x	16,387	=	Cubic millimeters
	Cubic feet	x	0.02832	=	Cubic meters
	Cubic yards	x	0.7646	=	Cubic meters
	Gallons (U.S. liquids)[b]	x	0.003785	=	Cubic meters[c]
	Gallons (Canadian liquid)[b]	x	0.004546	=	Cubic meters[c]
	Ounces (U.S. liquid)[b]	x	29.57	=	Milliliters[c, d]
	Quarts (U.S. liquid)[b]	x	0.9464	=	Liters[c, d]
	Gallons (U.S. liquid)[b]	x	3.785	=	Liters[c, d]
Force	Kilograms force[d]	x	9.807	=	Newtons
	Pounds force	x	4.448	=	Newtons
	Pounds force	x	0.4536	=	Kilograms force[d]
	Kips	x	4448	=	Newtons
	Kips	x	453.6	=	Kilograms force[d]
Pressure, Stress, Strength (Force per unit area)	Kilograms force per square centimeter[d]	x	0.09807	=	Megapascals
	Pounds force per square inch (psi)	x	0.006895	=	Megapascals
	Kips per square inch	x	6.895	=	Megapascals
	Pounds force per square inch (psi)	x	0.07031	=	Kilograms force per square centimeter[d]
	Pounds force per square foot	x	47.88	=	Pascals
	Pounds force per square foot	x	4.882	=	Kilograms force per square meter[d]
Bending Moment Or Torque	Inch-pounds force	x	0.01152	=	Meter-kilograms force[d]
	Inch-pounds force	x	0.1130	=	Newton-meters
	Foot-pounds force	x	0.1383	=	Meter-kilograms force[d]
	Foot-pounds force	x	1.356	=	Newton-meters
	Meter-kilograms force[d]	x	9.807	=	Newton-meters
Mass	Ounces (avoirdupois)	x	28.35	=	Grams
	Pounds (avoirdupois)	x	0.4536	=	Kilograms
	Tons (metric)	x	1000	=	Kilograms
	Tons, short (2000 pounds)	x	907.2	=	Kilograms
	Tons, short (2000 pounds)	x	0.9072	=	Megagrams[e]
Mass per Unit Volume	Pounds mass per cubic foot	x	16.02	=	Kilograms per cubic meter
	Pounds mass per cubic yard	x	0.5933	=	Kilograms per cubic meter
	Pounds mass per gallon (U.S. liquid)[b]	x	119.8	=	Kilograms per cubic meter
	Pounds mass per gallon (Canadian liquid)[b]	x	99.78	=	Kilograms per cubic meter
Temperature	Degrees Fahrenheit	(F-32)/1.8		=	Degrees Celsius
	Degrees Fahrenheit	(F+459.67)/1.8		=	Degrees Kelvin
	Degrees Celsius	C+273.15		=	Degrees Kelvin

[a]The factor given is exact
[b]One U.S. gallon = 0.8327 Canadian gallon
[c]1 liter = 1000 milliliters = 1000 cubic centimeters
1 cubic decimeter = 0.001 cubic meter
[d]Metric but not SI unit
[e]Called "tonne" in England and "metric ton" in other metric countries

R011105-60 Weights and Measures

Measures of Length
1 Mile = 1760 Yards = 5280 Feet
1 Yard = 3 Feet = 36 inches
1 Foot = 12 Inches
1 Mil = 0.001 Inch
1 Fathom = 2 Yards = 6 Feet
1 Rod = 5.5 Yards = 16.5 Feet
1 Hand = 4 Inches
1 Span = 9 Inches
1 Micro-inch = One Millionth Inch or 0.000001 Inch
1 Micron = One Millionth Meter + 0.00003937 Inch

Surveyor's Measure
1 Mile = 8 Furlongs = 80 Chains
1 Furlong = 10 Chains = 220 Yards
1 Chain = 4 Rods = 22 Yards = 66 Feet = 100 Links
1 Link = 7.92 Inches

Square Measure
1 Square Mile = 640 Acres = 6400 Square Chains
1 Acre = 10 Square Chains = 4840 Square Yards = 43,560 Sq. Ft.
1 Square Chain = 16 Square Rods = 484 Square Yards = 4356 Sq. Ft.
1 Square Rod = 30.25 Square Yards = 272.25 Square Feet = 625 Square Lines
1 Square Yard = 9 Square Feet
1 Square Foot = 144 Square Inches
An Acre equals a Square 208.7 Feet per Side

Cubic Measure
1 Cubic Yard = 27 Cubic Feet
1 Cubic Foot = 1728 Cubic Inches
1 Cord of Wood = 4 x 4 x 8 Feet = 128 Cubic Feet
1 Perch of Masonry = 16½ x 1½ x 1 Foot = 24.75 Cubic Feet

Avoirdupois or Commercial Weight
1 Gross or Long Ton = 2240 Pounds
1 Net or Short ton = 2000 Pounds
1 Pound = 16 Ounces = 7000 Grains
1 Ounce = 16 Drachms = 437.5 Grains
1 Stone = 14 Pounds

Shipping Measure
For Measuring Internal Capacity of a Vessel:
1 Register Ton = 100 Cubic Feet

For Measurement of Cargo:
Approximately 40 Cubic Feet of Merchandise is considered a Shipping Ton, unless that bulk would weigh more than 2000 Pounds, in which case Freight Charge may be based upon weight.

40 Cubic Feet = 32.143 U.S. Bushels = 31.16 Imp. Bushels

Liquid Measure
1 Imperial Gallon = 1.2009 U.S. Gallon = 277.42 Cu. In.
1 Cubic Foot = 7.48 U.S. Gallons

R011110-10 Architectural Fees

Tabulated below are typical percentage fees by project size, for good professional architectural service. Fees may vary from those listed depending upon degree of design difficulty and economic conditions in any particular area.

Rates can be interpolated horizontally and vertically. Various portions of the same project requiring different rates should be adjusted proportionately. For alterations, add 50% to the fee for the first $500,000 of project cost and add 25% to the fee for project cost over $500,000.

Architectural fees tabulated below include Structural, Mechanical and Electrical Engineering Fees. They do not include the fees for special consultants such as kitchen planning, security, acoustical, interior design, etc.

Civil Engineering fees are included in the Architectural fee for project sites requiring minimal design such as city sites. However, separate Civil Engineering fees must be added when utility connections require design, drainage calculations are needed, stepped foundations are required, or provisions are required to protect adjacent wetlands.

Building Types	Total Project Size in Thousands of Dollars						
	100	250	500	1,000	5,000	10,000	50,000
Factories, garages, warehouses, repetitive housing	9.0%	8.0%	7.0%	6.2%	5.3%	4.9%	4.5%
Apartments, banks, schools, libraries, offices, municipal buildings	12.2	12.3	9.2	8.0	7.0	6.6	6.2
Churches, hospitals, homes, laboratories, museums, research	15.0	13.6	12.7	11.9	9.5	8.8	8.0
Memorials, monumental work, decorative furnishings	—	16.0	14.5	13.1	10.0	9.0	8.3

R011110-30 Engineering Fees

Typical **Structural Engineering Fees** based on type of construction and total project size. These fees are included in Architectural Fees.

Type of Construction	Total Project Size (in thousands of dollars)			
	$500	$500-$1,000	$1,000-$5,000	Over $5000
Industrial buildings, factories & warehouses	Technical payroll times 2.0 to 2.5	1.60%	1.25%	1.00%
Hotels, apartments, offices, dormitories, hospitals, public buildings, food stores		2.00%	1.70%	1.20%
Museums, banks, churches and cathedrals		2.00%	1.75%	1.25%
Thin shells, prestressed concrete, earthquake resistive		2.00%	1.75%	1.50%
Parking ramps, auditoriums, stadiums, convention halls, hangars & boiler houses		2.50%	2.00%	1.75%
Special buildings, major alterations, underpinning & future expansion	▼	Add to above 0.5%	Add to above 0.5%	Add to above 0.5%

For complex reinforced concrete or unusually complicated structures, add 20% to 50%.

Typical **Mechanical and Electrical Engineering Fees** are based on the size of the subcontract. The fee structure for both are shown below. These fees are included in Architectural Fees.

Type of Construction	Subcontract Size							
	$25,000	$50,000	$100,000	$225,000	$350,000	$500,000	$750,000	$1,000,000
Simple structures	6.4%	5.7%	4.8%	4.5%	4.4%	4.3%	4.2%	4.1%
Intermediate structures	8.0	7.3	6.5	5.6	5.1	5.0	4.9	4.8
Complex structures	10.1	9.0	9.0	8.0	7.5	7.5	7.0	7.0

For renovations, add 15% to 25% to applicable fee.

R012153-10 Repair and Remodeling

Cost figures are based on new construction utilizing the most cost-effective combination of labor, equipment and material with the work scheduled in proper sequence to allow the various trades to accomplish their work in an efficient manner.

The costs for repair and remodeling work must be modified due to the following factors that may be present in any given repair and remodeling project.

1. Equipment usage curtailment due to the physical limitations of the project, with only hand-operated equipment being used.
2. Increased requirement for shoring and bracing to hold up the building while structural changes are being made and to allow for temporary storage of construction materials on above-grade floors.
3. Material handling becomes more costly due to having to move within the confines of an enclosed building. For multi-story construction, low capacity elevators and stairwells may be the only access to the upper floors.
4. Large amount of cutting and patching and attempting to match the existing construction is required. It is often more economical to remove entire walls rather than create many new door and window openings. This sort of trade-off has to be carefully analyzed.
5. Cost of protection of completed work is increased since the usual sequence of construction usually cannot be accomplished.
6. Economies of scale usually associated with new construction may not be present. If small quantities of components must be custom fabricated due to job requirements, unit costs will naturally increase. Also, if only small work areas are available at a given time, job scheduling between trades becomes difficult and subcontractor quotations may reflect the excessive start-up and shut-down phases of the job.
7. Work may have to be done on other than normal shifts and may have to be done around an existing production facility which has to stay in production during the course of the repair and remodeling.
8. Dust and noise protection of adjoining non-construction areas can involve substantial special protection and alter usual construction methods.
9. Job may be delayed due to unexpected conditions discovered during demolition or removal. These delays ultimately increase construction costs.
10. Piping and ductwork runs may not be as simple as for new construction. Wiring may have to be snaked through walls and floors.
11. Matching "existing construction" may be impossible because materials may no longer be manufactured. Substitutions may be expensive.
12. Weather protection of existing structure requires additional temporary structures to protect building at openings.
13. On small projects, because of local conditions, it may be necessary to pay a tradesman for a minimum of four hours for a task that is completed in one hour.

All of the above areas can contribute to increased costs for a repair and remodeling project. Each of the above factors should be considered in the planning, bidding and construction stage in order to minimize the increased costs associated with repair and remodeling jobs.

General Requirements | R0121 | Allowances

R012157-20 Construction Time Requirements

Table at left is average construction time in months for different types of building projects. Table at right is the construction time in months for different size projects. Design time runs 25% to 40% of construction time.

Type Building	Construction Time	Project Value	Construction Time
Industrial Buildings	12 Months	Under $1,400,000	10 Months
Commercial Buildings	15 Months	Up to $3,800,000	15 Months
Research & Development	18 Months	Up to $19,000,000	21 Months
Institutional Buildings	20 Months	over $19,000,000	28 Months

General Requirements | R0129 | Payment Procedures

R012909-80 Sales Tax by State

State sales tax on materials is tabulated below (5 states have no sales tax). Many states allow local jurisdictions, such as a county or city, to levy additional sales tax.

Some projects may be sales tax exempt, particularly those constructed with public funds.

State	Tax (%)	State	Tax (%)	State	Tax (%)	State	Tax (%)
Alabama	4	Illinois	6.25	Montana	0	Rhode Island	7
Alaska	0	Indiana	6	Nebraska	5.5	South Carolina	5
Arizona	5.6	Iowa	5	Nevada	6.5	South Dakota	4
Arkansas	6	Kansas	5.3	New Hampshire	0	Tennessee	7
California	6.25	Kentucky	6	New Jersey	6	Texas	6.25
Colorado	2.9	Louisiana	4	New Mexico	5	Utah	4.75
Connecticut	6	Maine	5	New York	4.25	Vermont	6
Delaware	0	Maryland	5	North Carolina	4.5	Virginia	4
District of Columbia	5.75	Massachusetts	5	North Dakota	5	Washington	6.5
Florida	6	Michigan	6	Ohio	6	West Virginia	6
Georgia	4	Minnesota	6.5	Oklahoma	4.5	Wisconsin	5
Hawaii	4	Mississippi	7	Oregon	0	Wyoming	4
Idaho	6	Missouri	4.225	Pennsylvania	6	Average	4.83 %

Sales Tax by Province (Canada)

GST - a value-added tax, which the government imposes on most goods and services provided in or imported into Canada. PST - a retail sales tax, which five of the provinces impose on the price of most goods and some services. QST - a value-added tax, similar to the federal GST, which Quebec imposes. HST - Three provinces have combined their retail sales tax with the federal GST into one harmonized tax.

Province	PST (%)	QST (%)	GST(%)	HST(%)
Alberta	0	0	7	0
British Columbia	7.5	0	7	0
Manitoba	7	0	7	0
New Brunswick	0	0	0	15
Newfoundland	0	0	0	15
Northwest Territories	0	0	7	0
Nova Scotia	0	0	0	15
Ontario	8	0	7	0
Prince Edward Island	10	0	7	0
Quebec	0	7.5	7	0
Saskatchewan	6	0	7	0
Yukon	0	0	7	0

R012909-85 Unemployment Taxes and Social Security Taxes

State Unemployment Tax rates vary not only from state to state, but also with the experience rating of the contractor. The Federal Unemployment Tax rate is 6.2% of the first $7,000 of wages. This is reduced by a credit of up to 5.4% for timely payment to the state. The minimum Federal Unemployment Tax is 0.8% after all credits.

Social Security (FICA) for 2006 is estimated at time of publication to be 7.65% of wages up to $90,000.

R012909-90 Overtime

One way to improve the completion date of a project or eliminate negative float from a schedule is to compress activity duration times. This can be achieved by increasing the crew size or working overtime with the proposed crew.

To determine the costs of working overtime to compress activity duration times, consider the following examples. Below is an overtime efficiency and cost chart based on a five, six, or seven day week with an eight through twelve hour day. Payroll percentage increases for time and one half and double time are shown for the various working days.

Days per Week	Hours per Day	Production Efficiency					Payroll Cost Factors	
		1 Week	2 Weeks	3 Weeks	4 Weeks	Average 4 Weeks	@ 1-1/2 Times	@ 2 Times
5	8	100%	100%	100%	100%	100 %	100 %	100 %
	9	100	100	95	90	96.25	105.6	111.1
	10	100	95	90	85	91.25	110.0	120.0
	11	95	90	75	65	81.25	113.6	127.3
	12	90	85	70	60	76.25	116.7	133.3
6	8	100	100	95	90	96.25	108.3	116.7
	9	100	95	90	85	92.50	113.0	125.9
	10	95	90	85	80	87.50	116.7	133.3
	11	95	85	70	65	78.75	119.7	139.4
	12	90	80	65	60	73.75	122.2	144.4
7	8	100	95	85	75	88.75	114.3	128.6
	9	95	90	80	70	83.75	118.3	136.5
	10	90	85	75	65	78.75	121.4	142.9
	11	85	80	65	60	72.50	124.0	148.1
	12	85	75	60	55	68.75	126.2	152.4

General Requirements | R0131 | Project Management & Coordination

R013113-40 Builder's Risk Insurance

Builder's Risk Insurance is insurance on a building during construction. Premiums are paid by the owner or the contractor. Blasting, collapse and underground insurance would raise total insurance costs above those listed. Floater policy for materials delivered to the job runs $.75 to $1.25 per $100 value. Contractor equipment insurance runs $.50 to $1.50 per $100 value. Insurance for miscellaneous tools to $1,500 value runs from $3.00 to $7.50 per $100 value.

Tabulated below are New England Builder's Risk insurance rates in dollars per $100 value for $1,000 deductible. For $25,000 deductible, rates can be reduced 13% to 34%. On contracts over $1,000,000, rates may be lower than those tabulated. Policies are written annually for the total completed value in place. For "all risk" insurance (excluding flood, earthquake and certain other perils) add $.025 to total rates below.

Coverage	Frame Construction (Class 1)				Brick Construction (Class 4)				Fire Resistive (Class 6)			
	Range			Average	Range			Average	Range			Average
Fire Insurance	$.350	to	$ .850	$.600	$.158	to	$.189	$.174	$.052	to	$.080	$.070
Extended Coverage	.115	to	.200	.158	.080	to	.105	.101	.081	to	.105	.100
Vandalism	.012	to	.016	.014	.008	to	.011	.011	.008	to	.011	.010
Total Annual Rate	$.477	to	$1.066	$.772	$.246	to	$.305	$.286	$.141	to	$.196	$.180

R013113-50 General Contractor's Overhead

There are two distinct types of overhead on a construction project: Project Overhead and Main Office Overhead. Project Overhead includes those costs at a construction site not directly associated with the installation of construction materials. Examples of Project Overhead costs include the following:

1. Superintendent
2. Construction office and storage trailers
3. Temporary sanitary facilities
4. Temporary utilities
5. Security fencing
6. Photographs
7. Clean up
8. Performance and payment bonds

The above Project Overhead items are also referred to as General Requirements and therefore are estimated in Division 1. Division 1 is the first division listed in the CSI MasterFormat but it is usually the last division estimated. The sum of the costs in Divisions 1 through 16 is referred to as the sum of the direct costs.

All construction projects also include indirect costs. The primary components of indirect costs are the contractor's Main Office Overhead and profit. The amount of the Main Office Overhead expense varies depending on the the following:

1. Owner's compensation
2. Project managers and estimator's wages
3. Clerical support wages
4. Office rent and utilities
5. Corporate legal and accounting costs
6. Advertising
7. Automobile expenses
8. Association dues
9. Travel and entertainment expenses

These costs are usually calculated as a percentage of annual sales volume. This percentage can range from 35% for a small contractor doing less than $500,000 to 5% for a large contractor with sales in excess of $100 million.

R013113-60 Workers' Compensation Insurance Rates by Trade

The table below tabulates the national averages for Workers' Compensation insurance rates by trade and type of building. The average "Insurance Rate" is multiplied by the "% of Building Cost" for each trade. This produces the "Workers' Compensation Cost" by % of total labor cost, to be added for each trade by building type to determine the weighted average Workers' Compensation rate for the building types analyzed.

Trade	Insurance Rate (% Labor Cost) Range	Insurance Rate (% Labor Cost) Average	% of Building Cost Office Bldgs.	% of Building Cost Schools & Apts.	% of Building Cost Mfg.	Workers' Compensation Office Bldgs.	Workers' Compensation Schools & Apts.	Workers' Compensation Mfg.
Excavation, Grading, etc.	4.2 % to 19.5%	10.6%	4.8%	4.9%	4.5%	.51%	.52%	.48%
Piles & Foundations	7.7 to 76.7	22.1	7.1	5.2	8.7	1.57	1.15	1.92
Concrete	5.2 to 35.8	15.8	5.0	14.8	3.7	.79	2.34	.58
Masonry	5.1 to 28.3	14.9	6.9	7.5	1.9	1.03	1.12	.28
Structural Steel	7.8 to 103.8	39.7	10.7	3.9	17.6	4.25	1.55	6.99
Miscellaneous & Ornamental Metals	4.9 to 24.8	12.2	2.8	4.0	3.6	.34	.49	.44
Carpentry & Millwork	7.0 to 53.2	18.4	3.7	4.0	0.5	.68	.74	.09
Metal or Composition Siding	5.3 to 32.1	17.3	2.3	0.3	4.3	.40	.05	.74
Roofing	7.8 to 78.9	32.4	2.3	2.6	3.1	.75	.84	1.00
Doors & Hardware	4.5 to 24.9	11.8	0.9	1.4	0.4	.11	.17	.05
Sash & Glazing	4.6 to 33.7	13.9	3.5	4.0	1.0	.49	.56	.14
Lath & Plaster	4.3 to 40.2	14.0	3.3	6.9	0.8	.46	.97	.11
Tile, Marble & Floors	2.4 to 31.8	9.5	2.6	3.0	0.5	.25	.29	.05
Acoustical Ceilings	2.4 to 29.7	11.4	2.4	0.2	0.3	.27	.02	.03
Painting	4.5 to 33.7	13.3	1.5	1.6	1.6	.20	.21	.21
Interior Partitions	7.0 to 53.2	18.4	3.9	4.3	4.4	.72	.79	.81
Miscellaneous Items	2.6 to 137.6	17.3	5.2	3.7	9.7	.90	.64	1.68
Elevators	2.8 to 42.8	6.9	2.1	1.1	2.2	.14	.08	.15
Sprinklers	3.2 to 22.1	8.6	0.5	—	2.0	.04	—	.17
Plumbing	2.7 to 12.4	8.2	4.9	7.2	5.2	.40	.59	.43
Heat., Vent., Air Conditioning	4.0 to 27.2	11.8	13.5	11.0	12.9	1.59	1.30	1.52
Electrical	2.6 to 12.7	6.8	10.1	8.4	11.1	.69	.57	.75
Total	2.4 % to 137.6%	—	100.0%	100.0%	100.0%	16.58%	14.99%	18.62%
	Overall Weighted Average 16.73%							

Workers' Compensation Insurance Rates by States

The table below lists the weighted average Workers' Compensation base rate for each state with a factor comparing this with the national average of 16.3%.

State	Weighted Average	Factor	State	Weighted Average	Factor	State	Weighted Average	Factor
Alabama	25.7%	158	Kentucky	20.5%	126	North Dakota	13.1%	80
Alaska	24.0	147	Louisiana	28.2	173	Ohio	14.5	89
Arizona	7.5	46	Maine	20.9	128	Oklahoma	14.7	90
Arkansas	13.7	84	Maryland	19.1	117	Oregon	13.0	80
California	18.8	115	Massachusetts	13.4	82	Pennsylvania	14.3	88
Colorado	10.9	67	Michigan	17.4	107	Rhode Island	21.1	129
Connecticut	22.4	137	Minnesota	25.5	156	South Carolina	16.8	103
Delaware	17.3	106	Mississippi	17.9	110	South Dakota	17.1	105
District of Columbia	17.1	105	Missouri	17.7	109	Tennessee	18.1	111
Florida	23.3	143	Montana	20.4	125	Texas	12.7	78
Georgia	22.9	140	Nebraska	22.9	140	Utah	12.3	75
Hawaii	16.9	104	Nevada	13.6	83	Vermont	25.0	153
Idaho	11.7	72	New Hampshire	22.6	139	Virginia	13.5	83
Illinois	18.8	115	New Jersey	12.2	75	Washington	10.9	67
Indiana	7.1	44	New Mexico	17.7	109	West Virginia	13.2	81
Iowa	12.3	75	New York	14.4	88	Wisconsin	14.7	90
Kansas	8.9	55	North Carolina	15.4	94	Wyoming	9.1	56
			Weighted Average for U.S. is	16.7% of payroll = 100%				

Rates in the following table are the base or manual costs per $100 of payroll for Workers' Compensation in each state. Rates are usually applied to straight time wages only and not to premium time wages and bonuses.

The weighted average skilled worker rate for 35 trades is 16.3%. For bidding purposes, apply the full value of Workers' Compensation directly to total labor costs, or if labor is 38%, materials 42% and overhead and profit 20% of total cost, carry 38/80 x 16.3% =7.7% of cost (before overhead and profit) into overhead. Rates vary not only from state to state but also with the experience rating of the contractor.

Rates are the most current available at the time of publication.

R013113-60 Workers' Compensation Insurance Rates by Trade and State (cont.)

State	Carpentry — 3 stories or less	Carpentry — interior cab. work	Carpentry — general	Concrete Work — NOC	Concrete Work — flat (flr., sdwk.)	Electrical Wiring — inside	Excavation — earth NOC	Excavation — rock	Glaziers	Insulation Work	Lathing	Masonry	Painting & Decorating	Pile Driving	Plastering	Plumbing	Roofing	Sheet Metal Work (HVAC)	Steel Erection — door & sash	Steel Erection — inter., ornam.	Steel Erection — structure	Steel Erection — NOC	Tile Work — (interior ceramic)	Waterproofing	Wrecking
	5651	5437	5403	5213	5221	5190	6217	6217	5462	5479	5443	5022	5474	6003	5480	5183	5551	5538	5102	5102	5040	5057	5348	9014	5701
AL	29.57	16.77	31.62	11.24	10.07	8.55	12.90	12.90	26.88	17.64	12.34	23.46	32.15	38.02	32.13	11.79	65.04	21.69	19.37	19.37	56.15	31.21	12.21	6.54	56.15
AK	21.71	15.70	15.84	13.63	12.88	13.24	20.07	20.07	26.84	31.05	11.29	31.31	21.16	51.97	26.78	12.29	41.90	14.82	12.85	12.85	50.64	25.20	9.35	9.63	50.64
AZ	6.88	5.05	11.62	5.56	3.49	3.85	4.74	4.74	7.40	7.69	4.27	5.37	4.87	9.83	5.20	3.90	12.19	5.65	9.50	9.50	18.98	8.11	2.04	2.34	50.31
AR	11.87	9.14	15.06	11.65	6.81	5.31	9.16	9.16	11.83	30.45	7.61	10.44	10.67	16.93	12.24	5.61	22.08	12.99	7.43	7.43	36.81	16.98	6.36	4.67	38.24
CA	28.28	9.07	28.28	13.18	13.18	9.93	7.88	7.88	16.37	23.34	10.90	14.55	19.45	21.18	18.12	11.59	39.79	15.33	13.55	13.55	23.93	21.91	7.74	19.45	21.91
CO	14.58	6.77	10.47	11.38	6.17	4.52	8.41	8.41	7.33	10.76	4.73	10.57	8.54	13.61	8.00	7.15	19.17	10.23	6.98	6.98	27.39	12.94	7.07	4.70	27.39
CT	18.69	17.64	27.54	26.02	12.53	8.70	10.59	10.59	16.86	19.70	19.27	24.01	18.88	25.92	24.21	10.86	43.51	13.48	16.02	16.02	61.01	29.58	9.39	6.44	50.86
DE	17.74	17.74	14.29	14.46	11.19	7.77	11.13	11.13	13.41	14.29	15.42	14.16	18.57	23.83	15.42	9.28	31.57	11.76	14.60	14.60	33.94	14.60	11.81	14.16	33.94
DC	12.26	11.99	12.28	14.08	16.31	7.27	9.99	9.99	23.64	10.47	9.64	19.37	10.29	17.95	12.01	14.00	18.86	8.55	17.12	17.12	46.07	20.00	28.91	4.70	46.07
FL	24.98	19.24	28.32	24.68	12.25	10.16	12.77	12.77	18.92	17.94	13.27	20.39	19.51	47.98	24.53	10.82	37.58	16.40	13.31	13.31	56.10	32.45	10.28	8.90	56.10
GA	32.05	17.69	24.16	15.16	10.44	8.97	16.86	16.86	15.69	22.02	17.16	20.30	17.13	31.16	19.94	10.77	46.74	17.52	13.97	13.97	44.46	48.01	9.64	8.55	44.46
HI	17.38	11.62	28.20	13.74	12.27	7.10	7.73	7.73	20.55	21.19	10.94	17.87	11.33	20.15	15.61	6.06	33.24	8.02	11.11	11.11	31.71	21.70	9.78	11.54	31.71
ID	11.92	6.31	13.47	10.64	5.99	5.25	6.90	6.90	9.58	8.14	6.31	9.91	8.88	15.41	9.45	5.99	29.54	9.11	7.94	7.94	33.81	14.13	7.27	4.09	33.81
IL	17.63	14.46	17.30	27.55	11.47	19.13	9.77	9.77	17.62	12.10	9.88	16.21	10.52	28.78	11.82	9.62	27.47	13.74	15.18	15.18	52.69	23.84	13.78	4.43	52.69
IN	5.29	14.84	6.93	5.29	13.20	2.58	4.34	4.34	4.67	4.81	2.49	5.20	5.25	8.82	4.02	2.73	11.74	4.35	5.20	5.20	18.17	8.49	3.31	2.82	18.17
IA	11.94	5.77	9.18	13.10	7.10	4.40	5.19	5.19	12.22	8.84	5.48	9.22	8.30	10.69	8.48	5.61	17.71	6.88	8.97	8.97	42.71	42.83	4.81	3.70	31.77
KS	9.72	8.65	10.24	7.86	5.92	3.32	4.15	4.15	7.75	7.72	4.21	8.13	6.79	10.74	7.35	5.65	18.03	6.03	7.26	7.26	22.40	11.92	5.40	3.31	23.40
KY	19.74	12.89	18.65	24.60	7.70	8.10	15.00	15.00	21.87	24.34	11.81	10.91	14.25	22.06	15.53	8.42	54.39	17.05	14.51	14.51	50.29	28.09	12.52	5.04	50.26
LA	23.14	24.93	53.17	26.43	15.61	9.88	17.49	17.49	20.14	21.43	24.51	28.33	29.58	31.25	22.37	8.64	77.12	21.20	18.98	18.98	51.76	24.21	13.77	13.78	66.41
ME	13.43	10.42	43.59	24.29	11.12	3.72	12.56	12.56	13.35	15.39	14.03	17.26	16.07	31.44	17.95	9.07	32.07	10.07	15.08	15.08	37.84	61.81	11.66	5.92	37.84
MD	13.71	5.95	10.55	18.82	8.70	5.15	12.91	12.91	25.52	20.07	11.37	13.31	9.37	26.22	14.21	8.58	48.72	12.74	13.66	13.66	56.25	37.81	8.03	7.14	26.80
MA	9.93	6.07	16.09	17.95	8.10	3.69	6.49	6.49	7.55	13.48	5.76	13.04	7.54	14.31	5.34	4.55	38.25	7.47	12.35	12.35	35.13	27.39	9.24	3.32	28.82
MI	19.68	11.72	21.53	19.00	8.83	5.94	9.96	9.96	11.76	13.31	13.59	17.06	13.34	36.96	15.69	7.35	33.24	9.51	9.92	9.92	36.96	27.43	10.97	5.38	39.06
MN	19.53	19.80	41.48	14.65	15.73	7.53	14.95	14.95	15.68	11.90	22.05	19.18	16.95	24.63	22.05	10.88	70.18	14.23	10.58	10.58	104.13	33.30	14.55	6.68	132.92
MS	16.68	14.87	24.71	12.67	9.14	5.85	12.50	12.50	13.10	15.77	8.50	11.88	13.12	19.91	28.30	9.82	38.94	20.61	10.87	10.87	32.50	30.23	10.01	6.49	32.50
MO	24.01	11.40	15.92	15.49	10.23	6.90	9.30	9.30	11.03	17.88	11.24	14.55	13.46	16.69	15.51	9.66	31.18	13.82	11.26	11.26	53.28	36.77	8.63	6.32	53.28
MT	21.18	10.79	19.52	12.88	10.39	6.82	19.69	19.69	13.67	20.72	22.29	15.50	10.31	60.73	12.40	10.22	45.28	12.09	12.45	12.45	40.26	22.93	6.26	5.14	40.26
NE	23.85	14.32	22.60	27.52	12.77	9.52	18.07	18.07	21.82	33.02	12.77	23.10	17.70	23.25	17.17	11.13	38.42	17.27	13.90	13.90	60.90	30.50	10.52	7.63	50.40
NV	20.81	8.10	10.79	9.56	7.96	6.16	10.76	10.76	12.49	10.61	9.45	9.87	9.00	16.26	9.36	7.27	18.09	21.51	10.59	10.59	24.87	29.06	6.55	5.43	4.09
NH	27.93	14.04	25.82	31.58	12.51	7.80	18.33	18.33	13.55	38.86	9.24	22.82	12.51	20.88	15.76	11.83	49.65	16.44	13.62	13.62	52.78	22.94	14.95	6.55	52.78
NJ	12.63	9.48	12.63	10.77	8.25	4.50	8.33	8.33	7.06	12.93	11.17	12.87	10.95	13.78	11.17	6.14	32.47	6.17	10.72	10.72	22.56	12.00	4.75	4.81	26.39
NM	29.00	8.32	20.65	18.60	11.34	7.16	10.46	10.46	14.57	12.61	8.59	12.43	13.86	21.32	11.32	8.92	30.32	10.81	19.90	19.90	40.88	27.34	8.55	7.65	40.88
NY	17.66	6.98	15.91	17.16	13.57	6.32	10.02	10.02	12.02	9.09	5.11	19.20	13.75	16.81	8.43	8.53	33.44	17.47	9.63	9.63	18.80	19.14	9.43	6.09	11.29
NC	16.52	12.25	14.99	13.22	7.32	8.85	9.45	9.45	11.16	11.72	10.68	9.98	10.17	15.13	14.39	8.39	24.40	12.94	7.40	7.40	68.39	20.27	6.05	4.83	68.39
ND	10.49	10.49	10.49	5.94	5.94	4.15	5.93	5.93	10.49	10.49	9.25	8.19	6.57	21.54	9.25	5.46	20.98	5.46	21.45	21.45	21.54	21.54	10.49	20.98	14.48
OH	7.48	9.85	9.98	11.81	11.78	6.16	9.02	9.02	8.65	14.76	38.58	12.85	13.30	23.94	3.76	7.06	24.80	9.48	10.09	10.09	31.53	15.16	9.91	5.99	31.53
OK	13.44	9.11	12.70	12.65	6.63	5.82	10.25	10.25	16.10	19.50	8.46	10.98	9.09	20.29	12.21	7.24	22.32	9.49	13.78	13.78	39.78	24.01	6.95	5.88	39.78
OR	16.72	9.16	15.83	14.23	8.06	4.50	9.48	9.48	14.13	8.79	7.41	14.91	11.59	15.81	11.05	5.30	22.92	9.68	9.53	9.53	29.02	11.69	10.63	4.52	29.02
PA	13.48	13.48	12.50	14.67	10.14	6.58	8.34	8.34	9.68	12.50	11.87	12.38	14.03	17.07	11.87	7.29	27.70	8.07	15.29	15.29	24.67	15.29	8.67	12.38	24.67
RI	19.53	11.65	18.07	18.23	16.24	4.43	10.38	10.38	12.85	22.78	11.97	25.11	24.13	37.66	17.25	8.52	33.92	10.29	14.07	14.07	59.49	37.50	14.36	7.70	78.79
SC	22.26	17.21	19.93	13.51	7.81	9.42	11.95	11.95	14.32	11.08	8.53	13.27	16.47	18.32	18.15	8.83	43.19	10.97	11.47	11.47	24.55	27.29	8.90	5.98	24.55
SD	21.00	17.87	17.14	19.81	7.46	5.25	15.58	15.58	11.54	12.77	8.94	7.73	12.16	23.97	11.77	10.03	22.29	11.34	11.11	11.11	62.70	23.72	7.24	4.06	62.70
TN	30.13	13.71	22.70	18.06	9.32	8.15	13.14	13.14	11.29	13.81	12.27	15.91	13.61	16.19	16.22	10.57	36.23	13.55	12.52	12.52	42.02	23.33	9.07	6.20	42.02
TX	13.47	9.95	13.47	11.11	8.00	7.76	8.79	8.79	9.83	12.47	6.78	12.04	8.63	16.18	8.63	7.35	19.58	16.38	9.69	9.69	31.53	13.20	6.81	7.37	14.13
UT	9.45	6.23	10.52	7.68	7.99	7.05	6.38	6.38	9.55	10.95	12.32	13.75	15.29	17.24	10.60	6.41	26.27	6.30	8.99	8.99	23.51	23.51	6.06	5.83	26.53
VT	24.65	11.60	22.01	36.23	15.66	8.18	13.75	13.75	25.11	29.18	12.53	23.26	13.86	23.35	18.10	11.77	42.54	15.20	15.81	15.81	72.09	57.39	9.47	12.14	72.09
VA	13.36	8.95	11.28	12.51	5.79	5.62	7.58	7.58	9.32	10.10	17.52	9.64	10.83	15.93	9.09	6.69	21.06	8.52	11.01	11.01	49.54	21.40	6.14	3.14	49.54
WA	10.29	10.29	10.29	7.89	7.89	3.28	8.11	8.11	12.02	10.01	10.29	9.82	9.08	18.17	11.05	4.63	18.54	4.64	12.96	12.96	8.65	8.65	10.03	10.23	8.65
WV	12.30	12.30	12.30	27.37	27.37	5.86	8.90	8.90	6.99	13.15	13.15	11.94	12.35	13.34	12.35	5.48	14.63	6.99	10.89	10.89	10.89	10.89	13.15	4.46	48.62
WI	12.40	11.36	16.86	12.29	9.90	4.24	7.44	7.44	13.31	11.62	8.09	18.99	14.32	13.64	12.11	5.42	34.72	7.16	9.18	9.18	37.88	24.06	13.99	5.52	37.88
WY	8.27	8.27	8.27	8.27	8.27	8.27	8.27	8.27	8.27	8.27	8.27	8.27	8.27	8.27	8.27	8.27	8.27	8.27	8.27	8.27	8.27	8.27	8.27	8.27	8.27
AVG.	17.27	11.81	18.39	15.82	10.29	6.84	10.63	10.63	13.87	15.76	11.44	14.92	13.29	22.07	14.00	8.22	32.40	11.76	12.19	12.19	39.65	24.35	9.52	7.04	40.06

R013113-60 Workers' Compensation (cont.) (Canada in Canadian dollars)

Province		Alberta	British Columbia	Manitoba	Ontario	New Brunswick	Newfndld. & Labrador	Northwest Territories	Nova Scotia	Prince Edward Island	Quebec	Saskat-chewan	Yukon
Carpentry—3 stories or less	Rate	9.71	7.19	4.57	4.62	4.83	9.78	4.61	7.67	6.04	15.41	7.33	5.58
	Code	42143	721028	40102	723	422	4226	4-41	4226	401	80110	B1317	202
Carpentry—interior cab. work	Rate	1.60	6.16	4.57	4.62	4.31	6.37	4.61	5.74	4.05	15.41	3.68	5.58
	Code	42133	721021	40102	723	427	4270	4-41	4274	402	80110	B11-27	202
CARPENTRY—general	Rate	9.71	7.19	4.57	4.62	4.83	6.37	4.61	7.67	6.04	15.41	7.33	5.58
	Code	42143	721028	40102	723	422	4299	4-41	4226	401	80110	B1317	202
CONCRETE WORK—NOC	Rate	5.96	7.53	7.60	15.25	4.83	9.78	4.61	4.83	6.04	16.51	7.33	5.58
	Code	42104	721010	40110	748	422	4224	4-41	4224	401	80100	B13-14	203
CONCRETE WORK—flat (flr. sidewalk)	Rate	5.96	7.53	7.60	15.25	4.83	9.78	4.61	4.83	6.04	16.51	7.33	5.58
	Code	42104	721010	40110	748	422	4224	4-41	4224	401	80100	B13-14	203
ELECTRICAL Wiring—inside	Rate	2.50	2.17	2.52	2.94	2.35	3.25	4.13	2.23	4.05	7.64	3.68	5.58
	Code	42124	721019	40203	704	426	4261	4-46	4261	402	80170	B11-05	206
EXCAVATION—earth NOC	Rate	3.31	3.85	3.84	4.20	3.16	5.53	3.71	4.11	4.25	8.36	4.37	5.58
	Code	40604	721031	40706	711	421	4214	4-43	4214	404	80030	R11-06	207
EXCAVATION—rock	Rate	3.31	3.85	3.84	4.20	3.16	5.53	3.71	4.11	4.25	8.36	4.37	5.58
	Code	40604	721031	40706	711	421	4214	4-43	4214	404	80030	R11-06	207
GLAZIERS	Rate	4.17	4.07	4.57	8.12	5.58	7.07	4.61	7.67	4.05	14.36	7.33	5.58
	Code	42121	715020	40109	751	423	4233	4-41	4233	402	80150	B13-04	212
INSULATION WORK	Rate	3.33	5.88	4.57	8.12	5.58	7.07	4.61	7.67	6.04	15.41	6.15	5.58
	Code	42184	721029	40102	751	423	4234	4-41	4234	401	80110	B12-07	202
LATHING	Rate	6.71	8.99	4.57	4.62	4.31	6.37	4.61	5.74	4.05	15.41	7.33	5.58
	Code	42135	721033	40102	723	427	4279	4-41	4271	402	80110	B13-16	202
MASONRY	Rate	5.96	8.99	4.57	11.44	5.58	7.07	4.61	7.67	6.04	16.51	7.33	5.58
	Code	42102	721037	40102	741	423	4231	4-41	4231	401	80100	B13-18	202
PAINTING & DECORATING	Rate	5.20	5.88	3.45	6.41	4.31	6.37	4.61	5.74	4.05	15.41	6.15	5.58
	Code	42111	721041	40105	719	427	4275	4-41	4275	402	80110	B12-01	202
PILE DRIVING	Rate	5.96	5.63	3.84	5.84	4.83	9.78	3.71	4.83	6.04	8.36	7.33	5.58
	Code	42159	722004	40706	732	422	4221	4-43	4221	401	80030	B13-10	202
PLASTERING	Rate	6.71	8.99	5.13	6.41	4.31	6.37	4.61	5.74	4.05	15.41	6.15	5.58
	Code	42135	721042	40108	719	427	4271	4-41	4271	402	80110	B12-21	202
PLUMBING	Rate	2.50	4.67	2.88	3.67	3.09	3.87	4.13	2.23	4.05	7.61	3.68	5.58
	Code	42122	721043	40204	707	424	4241	4-46	4241	402	80160	B11-01	214
ROOFING	Rate	9.88	10.10	6.52	11.60	10.25	9.78	4.61	9.49	6.04	22.49	7.33	5.58
	Code	42118	721036	40403	728	430	4236	4-41	4236	401	80130	B13-20	202
SHEET METAL WORK (HVAC)	Rate	2.50	4.67	6.52	3.67	3.09	3.87	4.13	3.23	4.05	7.61	3.68	7.39
	Code	42117	721043	40402	707	424	4244	4-46	4244	402	80160	B11-07	208
STEEL ERECTION—door & sash	Rate	3.33	13.44	9.89	15.25	4.83	9.78	4.61	7.67	6.04	29.92	7.33	5.58
	Code	42106	722005	40502	748	422	4227	4-41	4227	401	80080	B13-22	202
STEEL ERECTION—inter., ornam.	Rate	3.33	13.44	9.89	15.25	4.83	9.78	4.61	7.67	6.04	29.92	7.33	5.58
	Code	42106	722005	40502	748	422	4227	4-41	4227	401	80080	B13-22	202
STEEL ERECTION—structure	Rate	3.33	13.44	9.89	15.25	4.83	9.78	4.61	7.67	6.04	29.92	7.33	5.58
	Code	42106	722005	40502	748	422	4227	4-41	4227	401	80080	B13-22	202
STEEL ERECTION—NOC	Rate	3.33	13.44	9.89	15.25	4.83	9.78	4.61	7.67	6.04	29.92	7.33	5.58
	Code	42106	722005	40502	748	422	4227	4-41	4227	401	80080	B13-22	202
TILE WORK—inter. (ceramic)	Rate	4.62	5.67	2.17	6.41	4.31	6.37	4.61	5.74	4.05	15.41	7.33	5.58
	Code	42113	721054	40103	719	427	4276	4-41	4276	402	80110	B13-01	202
WATERPROOFING	Rate	5.20	5.88	4.57	4.62	5.58	6.37	4.61	7.67	4.05	22.49	6.15	5.58
	Code	42139	721016	40102	723	423	4299	4-41	4239	402	80130	B12-17	202
WRECKING	Rate	3.31	5.91	6.46	15.25	3.16	5.53	3.71	4.11	6.04	15.41	7.33	5.58
	Code	40604	721005	40106	748	421	4211	4-43	4211	401	80110	B13-09	202

General Requirements | R0131 | Project Management & Coordination

R013113-80 Performance Bond

This table shows the cost of a Performance Bond for a construction job scheduled to be completed in 12 months. Add 1% of the premium cost per month for jobs requiring more than 12 months to complete. The rates are "standard" rates offered to contractors that the bonding company considers financially sound and capable of doing the work. Preferred rates are offered by some bonding companies based upon financial strength of the contractor. Actual rates vary from contractor to contractor and from bonding company to bonding company. Contractors should prequalify through a bonding agency before submitting a bid on a contract that requires a bond.

Contract Amount	Building Construction Class B Projects	Highways & Bridges	
		Class A New Construction	Class A-1 Highway Resurfacing
First $ 100,000 bid	$25.00 per M	$15.00 per M	$9.40 per M
Next 400,000 bid	$ 2,500 plus $15.00 per M	$ 1,500 plus $10.00 per M	$ 940 plus $7.20 per M
Next 2,000,000 bid	8,500 plus 10.00 per M	5,500 plus 7.00 per M	3,820 plus 5.00 per M
Next 2,500,000 bid	28,500 plus 7.50 per M	19,500 plus 5.50 per M	15,820 plus 4.50 per M
Next 2,500,000 bid	47,250 plus 7.00 per M	33,250 plus 5.00 per M	28,320 plus 4.50 per M
Over 7,500,000 bid	64,750 plus 6.00 per M	45,750 plus 4.50 per M	39,570 plus 4.00 per M

General Requirements | R0151 | Temporary Utilities

R015113-65 Temporary Power Equipment

Cost data for the temporary equipment was developed utilizing the following information.

1) Re-usable material-services, transformers, equipment and cords are based on new purchase and prorated to three projects.
2) PVC feeder includes trench and backfill.
3) Connections include disconnects and fuses.
4) Labor units include an allowance for removal.
5) No utility company charges or fees are included.
6) Concrete pads or vaults are not included.
7) Utility company conduits not included.

R015423-10 Steel Tubular Scaffolding

On new construction, tubular scaffolding is efficient up to 60′ high or five stories. Above this it is usually better to use a hung scaffolding if construction permits. Swing scaffolding operations may interfere with tenants. In this case, the tubular is more practical at all heights.

In repairing or cleaning the front of an existing building the cost of tubular scaffolding per S.F. of building front increases as the height increases above the first tier. The first tier cost is relatively high due to leveling and alignment.

The minimum efficient crew for erection is three workers. For heights over 50′, a crew of four is more efficient. Use two or more on top and two at the bottom for handing up or hoisting. Four workers can erect and dismantle about nine frames per hour up to five stories. From five to eight stories they will average six frames per hour. With 7′ horizontal spacing this will run about 400 S.F. and 265 S.F. of wall surface, respectively. Time for placing planks must be added to the above. On heights above 50′, five planks can be placed per labor-hour.

The table below shows the number of pieces required to erect tubular steel scaffolding for 1000 S.F. of building frontage. This area is made up of a scaffolding system that is 12 frames (11 bays) long by 2 frames high.

For jobs under twenty-five frames, add 50% to rental cost. Rental rates will be lower for jobs over three months duration. Large quantities for long periods can reduce rental rates by 20%.

Description of Component	CSI Line Item	Number of Pieces for 1000 S.F. of Building Front	Unit
5′ Wide Standard Frame, 6′-4″ High	01540-750-2200	24	Ea.
Leveling Jack & Plate	01540-750-2650	24	
Cross Brace	01540-750-2500	44	
Side Arm Bracket, 21″	01540-750-2700	12	
Guardrail Post	01540-750-2550	12	
Guardrail, 7′ section	01540-750-2600	22	
Stairway Section	01540-750-2900	2	
Stairway Starter Bar	01540-750-2910	1	
Stairway Inside Handrail	01540-750-2920	2	
Stairway Outside Handrail	01540-750-2930	2	
Walk-Thru Frame Guardrail	01540-750-2940	2	▼

Scaffolding is often used as falsework over 15′ high during construction of cast-in-place concrete beams and slabs. Two foot wide scaffolding is generally used for heavy beam construction. The span between frames depends upon the load to be carried with a maximum span of 5′.

Heavy duty shoring frames with a capacity of 10,000#/leg can be spaced up to 10′ O.C. depending upon form support design and loading.

Scaffolding used as horizontal shoring requires less than half the material required with conventional shoring.

On new construction, erection is done by carpenters.

Rolling towers supporting horizontal shores can reduce labor and speed the job. For maintenance work, catwalks with spans up to 70′ can be supported by the rolling towers.

R015423-20 Pump Staging

Pump staging is generally not available for rent. The table below shows the number of pieces required to erect pump staging for 2400 S.F. of building frontage. This area is made up of a pump jack system that is 3 poles (2 bays) wide by 2 poles high.

Item	CSI Line Item	Number of Pieces for 2400 S.F. of Building Front	Unit
Aluminum pole section, 24′ long	01540-550-0200	6	Ea.
Aluminum splice joint, 6′ long	01540-550-0600	3	
Aluminum foldable brace	01540-550-0900	3	
Aluminum pump jack	01540-550-0700	3	
Aluminum support for workbench/back safety rail	01540-550-1000	3	
Aluminum scaffold plank/workbench, 14″ wide x 24′ long	01540-550-1100	4	
Safety net, 22′ long	01540-550-1250	2	
Aluminum plank end safety rail	01540-550-1200	2	▼

The cost in place for this 2400 S.F. will depend on how many uses are realized during the life of the equipment.

R015433-10 Contractor Equipment

Rental Rates shown elsewhere in the book pertain to late model high quality machines in excellent working condition, rented from equipment dealers. Rental rates from contractors may be substantially lower than the rental rates from equipment dealers depending upon economic conditions; for older, less productive machines, reduce rates by a maximum of 15%. Any overtime must be added to the base rates. For shift work, rates are lower. Usual rule of thumb is 150% of one shift rate for two shifts; 200% for three shifts.

For periods of less than one week, operated equipment is usually more economical to rent than renting bare equipment and hiring an operator.

Costs to move equipment to a job site (mobilization) or from a job site (demobilization) are not included in rental rates, nor in any Equipment costs on any Unit Price line items or crew listings. These costs can be found elsewhere. If a piece of equipment is already at a job site, it is not appropriate to utilize mob/demob costs in an estimate again.

Rental rates vary throughout the country with larger cities generally having lower rates. Lease plans for new equipment are available for periods in excess of six months with a percentage of payments applying toward purchase.

Monthly rental rates vary from 2% to 5% of the cost of the equipment depending on the anticipated life of the equipment and its wearing parts. Weekly rates are about 1/3 the monthly rates and daily rental rates about 1/3 the weekly rate.

The hourly operating costs for each piece of equipment include costs to the user such as fuel, oil, lubrication, normal expendables for the equipment, and a percentage of mechanic's wages chargeable to maintenance. The hourly operating costs listed do not include the operator's wages.

The daily cost for equipment used in the standard crews is figured by dividing the weekly rate by five, then adding eight times the hourly operating cost to give the total daily equipment cost, not including the operator. This figure is in the right hand column of the Equipment listings under Crew Equipment Cost/Day.

Pile Driving rates shown for pile hammer and extractor do not include leads, crane, boiler or compressor. Vibratory pile driving requires an added field specialist during set-up and pile driving operation for the electric model. The hydraulic model requires a field specialist for set-up only. Up to 125 reuses of sheet piling are possible using vibratory drivers. For normal conditions, crane capacity for hammer type and size are as follows.

Crane Capacity	Hammer Type and Size		
	Air or Steam	Diesel	Vibratory
25 ton	to 8,750 ft.-lb.		70 H.P.
40 ton	15,000 ft.-lb.	to 32,000 ft.-lb.	170 H.P.
60 ton	25,000 ft.-lb.		300 H.P.
100 ton		112,000 ft.-lb.	

Cranes should be specified for the job by size, building and site characteristics, availability, performance characteristics, and duration of time required.

Backhoes & Shovels rent for about the same as equivalent size cranes but maintenance and operating expense is higher. Crane operators rate must be adjusted for high boom heights. Average adjustments: for 150′ boom add 2% per hour; over 185′, add 4% per hour; over 210′, add 6% per hour; over 250′, add 8% per hour and over 295′, add 12% per hour.

Tower Cranes of the climbing or static type have jibs from 50′ to 200′ and capacities at maximum reach range from 4,000 to 14,000 pounds. Lifting capacities increase up to maximum load as the hook radius decreases.

Typical rental rates, based on purchase price are about 2% to 3% per month.

Erection and dismantling runs between 500 and 2000 labor hours. Climbing operation takes 10 labor hours per 20′ climb. Crane dead time is about 5 hours per 40′ climb. If crane is bolted to side of the building add cost of ties and extra mast sections. Climbing cranes have from 80′ to 180′ of mast while static cranes have 80′ to 800′ of mast.

Truck Cranes can be converted to tower cranes by using tower attachments. Mast heights over 400′ have been used.

A single 100′ high material **Hoist and Tower** can be erected and dismantled in about 400 labor hours; a double 100′ high hoist and tower in about 600 labor hours. Erection times for additional heights are 3 and 4 labor hours per vertical foot respectively up to 150′, and 4 to 5 labor hours per vertical foot over 150′ high. A 40′ high portable Buck hoist takes about 160 labor hours to erect and dismantle. Additional heights take 2 labor hours per vertical foot to 80′ and 3 labor hours per vertical foot for the next 100′. Most material hoists do not meet local code requirements for carrying personnel.

A 150′ high **Personnel Hoist** requires about 500 to 800 labor hours to erect and dismantle. Budget erection time at 5 labor hours per vertical foot for all trades. Local code requirements or labor scarcity requiring overtime can add up to 50% to any of the above erection costs.

Earthmoving Equipment: The selection of earthmoving equipment depends upon the type and quantity of material, moisture content, haul distance, haul road, time available, and equipment available. Short haul cut and fill operations may require dozers only, while another operation may require excavators, a fleet of trucks, and spreading and compaction equipment. Stockpiled material and granular material are easily excavated with front end loaders. Scrapers are most economically used with hauls between 300′ and 1-1/2 miles if adequate haul roads can be maintained. Shovels are often used for blasted rock and any material where a vertical face of 8′ or more can be excavated. Special conditions may dictate the use of draglines, clamshells, or backhoes. Spreading and compaction equipment must be matched to the soil characteristics, the compaction required and the rate the fill is being supplied.

R015433-15 Heavy Lifting

Hydraulic Climbing Jacks

The use of hydraulic heavy lift systems is an alternative to conventional type crane equipment. The lifting, lowering, pushing, or pulling mechanism is a hydraulic climbing jack moving on a square steel jackrod from 1-5/8″ to 4″ square, or a steel cable. The jackrod or cable can be vertical or horizontal, stationary or movable, depending on the individual application. When the jackrod is stationary, the climbing jack will climb the rod and push or pull the load along with itself. When the climbing jack is stationary, the jackrod is movable with the load attached to the end and the climbing jack will lift or lower the jackrod with the attached load. The heavy lift system is normally operated by a single control lever located at the hydraulic pump.

The system is flexible in that one or more climbing jacks can be applied wherever a load support point is required, and the rate of lift synchronized.

Economic benefits have been demonstrated on projects such as: erection of ground assembled roofs and floors, complete bridge spans, girders and trusses, towers, chimney liners and steel vessels, storage tanks, and heavy machinery. Other uses are raising and lowering offshore work platforms, caissons, tunnel sections and pipelines.

R019313-10 Facility Maintenance - Frequency Table

The following table lists "average" frequency data for selected facility maintenance activities. The frequencies given are for a normal standard of maintenance under average conditions.

Activity	Average Frequency	Notes
Acoustical tile, cleaning		
Heavy smoking	2-3 years	
Non-smoking	10 years	
Carpet, cleaning		Frequency depends on the type of carpet and the occupancy in the building.
Heavy traffic	Weekly	
Light traffic	Every 6 weeks	
Carpet, vacuuming		
Heavy traffic	Daily	6-8 passes by machine in key areas.
Light traffic	Twice weekly	
Offices	Weekly	
Corridor and lobby, policing		Includes picking up loose trash, removing cigarette butts from in and around jardinieres and/or sand urns. Frequency depends on occupancy in the building.
Main	4 times daily	
Secondary	Daily	
Elevator, cleaning		The frequency of elevator cleaning is dependent upon the amount of traffic, size of the elevator car and the occupancy in the building.
Passenger	Daily	
Freight	Weekly	
Elevator lobby, cleaning	Daily	Includes sweeping, mopping and rinsing.
Escalator, cleaning	Daily	Frequency given is for a normal standard of cleaning under average conditions. Approximately 40% of escalator treads are exposed when escalator is stopped.
Floors, mopping		Frequency given is for a normal standard of cleaning under average conditions.
Main corridors	Daily	
Secondary	Weekly	
Floors, sweeping	Daily	Frequency given is for a normal standard of cleaning under average conditions.
Floors, waxing and polishing		Frequency given is for a normal standard of cleaning under average conditions.
Office area	Every 9 weeks	
Open area	Every 9 weeks	
Flower beds		Frequency depends on geographic location.
Fall clean-up	Every year	
Fertilize	2 times per year	
Mulch	Every year	
Police-up	30 times per year	
Weed		
With mulch	15 times per year	
No mulch	25 times per year	
Lawn, mowing	30 times per year	Frequency depends on geographic location.
Fertilize	2 times per year	
Sweep	3 times per year	
Weed control	2 times per year	
Edge-trim		
Walks	30 times per year	
Shrub	10 times per year	
Light fixtures		Frequency depends on occupancy in the building.
Dusting	Every month	
Washing	Every 3 months	
Shrub Areas		Frequency depends on geographic location.
Fertilize	Every year	
Mulch	2 times per year	
Police-up	30 times per year	
Prune	5 times per year	
Weed	10 times per year	
Stairway		Frequency is dependent upon weather conditions, type of building, type of occupancy, and amount of traffic.
Sweeping and dusting	Daily	
Mopping or scrubbing	Weekly	
Toilet, cleaning	Daily	Includes toilet cleaning, collect waste, cleaning wash basins, urinals, water closets, partitions, walls and floors.
Trash collection	Daily	
Trees		Frequency depends on geographic location.
Fertilize	Every year	
Prune	2 times per year	
Pest control		
Spray	3 times per year	
Systemic	Every year	
Urn and Jardiniere cleaning	Daily	Includes the removal of refuse and debris, cleaning and polishing.
Walks, sweeping	30 times per year	
Walls, periodic cleaning	Every 6 months	Damp wipe, spot removal.
Windows, washing	Every month	Frequency given is for a normal standard of cleaning under average conditions.

R019313-20 Facilities Maintenance Labor-Hours

This section lists minimum and maximum cleaning times per unit.
For more information, see *Means Facilities Maintenance Standards* book.

	Unit	
	Minimum	Maximum
Blast Cleaning		
White-metal	100 S.F./Hr	
Near-white	175 S.F./Hr.	
Commercial	370 S.F./Hr.	
Brush-off	870 S.F./Hr.	
Paint Application		
Brushing	125 S.F./Hr.	
Rolling	125 S.F./Hr	
Spraying	500 S.F./Hr.	
Plaster Cleaning		
Wall dusting	2 sec./S.F.	3 sec./S.F.
Vacuuming	4 sec./S.F.	5 sec./S.F.
Spot washing	125 S.F./Hr.	175 S.F./Hr.
Thorough cleaning	275 S.F./Hr.	
Plaster Repair		
Gypsum and lime repair	5 S.Y./Hr.	10 S.Y./Hr.
Ceramic Tile Repair		
General	7 S.F./Hr.	10 S.F./Hr.
Adhesive tile setting	9S.F./Hr	12 S.F./Hr.
Pointing tile joints	10 S.F./Hr.	15 S.F./Hr.
Floor Cleaning		
Manual sweeping	10 min./1000 S.F.	25 min./1000 S.F.
Dust mopping	5 min./1000 S.F.	20 min./1000 S.F.
Buffing	15 min./1000 S.F.	40 min./1000 S.F.
Spray buffin	20 min./1000 S.F.	50 min./1000 S.F.
Damp mopping	15 min./1000 S.F.	30 min./1000 S.F.
Wet mopping	30 min./1000 S.F.	50 min./1000 S.F.
Scrubbing	50 min./1000 S.F.	140 min./1000 S.F.
Scrubbing using electric floor machine		
General	15 min./1000 S.F.	30 min./1000 S.F.
Stripping	100 min./1000 S.F.	200 min./1000 S.F.
Waxing and Buffing using power maching		
Rewaxing	15 min./1000 S.F.	30 min./1000 S.F.
Stripping and rewaxing (two coats)	100 min./1000 S.F.	300 min./1000 S.F.
Waxing and buffing (one coat)	30 min./1000 S.F.	70 min./1000 S.F.
Carpets		
Dry vacuuming	15 min./1000 S.F.	40 min./1000 S.F.
Wet vacuuming	30 min./1000 S.F.	50 min./1000 S.F.
Carpet mopping	20 min./1000 S.F.	40 min./1000 S.F.
Shampooing	175 min./1000 S.F.	250 min./1000 S.F.
Resilient Floor Repair		
Grinding	50 S.F./Hr.	80 S.F./Hr.
Floor Replacement		
Removal (by hand)		
Tiles	100 S.F./Hr.	130 S.F./Hr.
Sheet	120 S.F./Hr	160 S.F./Hr.
Hardwood	40 S.F./Hr.	60 S.F./Hr.
Replacement		
Ceramic	10 S.F./Hr.	20 S.F./Hr.
Resilient	40 S.F./Hr.	70 S.F./Hr.
Hardwood	25 S.F./Hr.	35 S.F./Hr.
Add for related items:		
Replace wood subfloor	80 S.F./Hr.	100 S.F./Hr.
Replace underlayment	75 S.F./Hr.	90 S.F./Hr.
Replace floor moulding	10 S.F./Hr.	30 S.F./Hr.

Unit		
	Minimum	Maximum
Seamless Floor Repair	5 S.F./Hr.	15 S.F./Hr.
Wood Floors		
Sanding	40 S.F./Hr.	60 S.F./Hr.
Sealing	200 S.F./Hr.	300 S.F./Hr.
Waxing*		
Wood Floor Repair		
Loose boards or tiles	50 S.F./Hr.	250S.F./Hr.
Wood strip floor replacement	30 S.F./Hr.	60 S.F./Hr.
Window Washing	300 S.F./Hr.	450 S.F./Hr.
Venetian Blinds, cleaning	15 min./set of blinds	

This section lists average cleaning times per each, or square foot	
	Average
Floor Operations	
Sweeping	
Halls and Corridors	15min./1000 S.F.
General Rooms	30 min./100 S.F.
Dust Mop (unobstructed)	10 min./1000 S.F.
Dust Mop (obstructed)	15 min./1000 S.F.
Damp Mop (unobstructed)	20 min./1000 S.F.
Damp Mop (obstructed)	40 min./1000 S.F.
Wet Mop and Rinse	100 min./1000 S.F.
Hand Scrubbing 12″ brush	300 min./1000 S.F.
Deck Scrubbing	100 min./1000 S.F.
Machine Scrubbing	
12″ diameter	50 min./1000 S.F.
14″diameter	40 min./1000 S.F.
16″ diameter	35 min./1000 S.F.
18″ diameter	31 min./1000 S.F.
19″ diameter	28 min./1000 S.F.
21″ diameter	25 min./1000 S.F.
23″ diameter	23 min./1000 S.F.
24″ diameter	20 min./1000 S.F.
32″ diameter	18 min./1000 S.F.
36″ diameter	15 min./1000 S.F.
Automatic Scrub Machine (24″)	5 min./1000 S.F.
Vacuum (unobstructed)	20 min./1000 S.F.
Vacuum (obstructed)	30 min./1000 S.F.
Waxing	30 min./1000 S.F.
Machine Polish (19″ machine)	15 min./1000 S.F.
Rectangular Machine (48″ plate)	5 min./1000 S.F.
Buff with steel wool	20 min./1000 S.F.
Strip and Rewax	150 min./1000 S.F.
Dry Strip and Rewax	120 min./1000 S.F.
Spray Buffing (unobstructed)	30 min./1000 S.F.
Spray Buffing (obstructed)	45 min./1000 S.F.
Carpeting	
Vacuuming (unobstructed)	20 min./1000 S.F.
Vacuuming (obstructed)	30 min./1000 S.F.
Spot Vacuuming	15 min./1000 S.F.
Shampoo (dry foam)	60 min. 1000 S.F.
Pile Lift	30 min 1000 S.F.
Lockers	.20 min
Radiators	.30 min.
Tables (medium)	.50 min.
Telephones	.15 min.
Towel dispensers	.12 min.

*See waxing and buffing using electric power machine.

R019313-20 Facilities Maintenance Labor-Hours (cont.)

	Average
Towel Disposal Cans	.40 min.
Typewriter and Stand	.50 min
Wash Basin (office)	.60 min
Waste Basin (office)	.50 min.
Window Sill	.20 min.
Venetian Blinds, std. size	3.50 min.
Washrooms	
Cleaning commode	4 min.
Door (spot wash both sides)	1 min.
Mirrors	1 min.
Sanitary napkin dispenser	.50 min.
Urinals	3 min.
Wash basin-soap dispenser	3 min.
General cleaning	120 min./1000 S.F.
Wall Washing	
Painted walls (manual)	240 min./1000 S.F.
Painted walls (machine)	150 min./1000 S.F.
Marble walls (manual)	90 min./1000 S.F.
Ceiling Washing	
Ceiling washing (manual)	300 min./1000 S.F.
Ceiling washing (machine)	180 min./1000 S.F.
Window Washing	
Single pane	125 min./1000 S.F.
Multi-pane	170 min./1000 S.F.
Frosted single pane	190 min./1000 S.F.
Opaque glass	50 min./1000 S.F.
Plate glass	35 min./1000 S.F.
Office partitions (glass)	110 min./1000 S.F.
Dusting Lamps and Lighting Fixtures	
Wall fluorescent fixtures	.15 min.
Desk fluorescent lamp	.30 min.
Table lamp and shade	.60 min.
Floor lamp and shade	.60 min.
Washing Fluorescent Light Fixtures	
Ceiling fixtures (egg crate) 4′ ea.	9 min.
Ceiling fixtures (egg crate) 8′ ea.	12 min.
Dusting	
Air conditioners	.30 min.
Ash trays (desk)	.25 min.
Book cases (3-tier set)	.30 min.
Chairs	.30 min.
Cigarette stands	.40 min.

	Average
Couch	.25 min.
Desks	.80 min.
Desk Trays	.15 min.
File cabinets (4 drawer)	.40 min.
Fabric Upholstery Cleaning	
Whisk or vacuum armless chair	.50 min.
Armchair	1 min.
Couch	2 min.
Shampooing armless chair	4 min.
Armchair	7 min.
Couch	20 min.
Stairway Cleaning	
Sweep and dust, 1 flight, 15 steps	6 min.
Damp mop, 1 flight, 15 steps	5 min.
Scrubbing (hand)	20 min.
Carpentry	
Repair door surface closer	1-1/2 Hrs.
Repair concealed door closer	4 Hrs.
Repair door damage at shop	4 Hrs.
Repair and replace screens	1 Hr.
Repair broken glass	1-1/2 Hrs.
Repair and replace ceiling tile	2 Hrs.
Repair small drywall damage	4 Hrs.
Repair and replace sash balance	4 Hrs.
Change lock on door	1-1/2 Hrs.
Painting	
Repaint 20′ x 15′ room, 1 coat	34 Hrs.
Repaint small bathroom, 1 coat	11-1/2 Hrs.
Repaint exterior window	2-1/2 Hrs.
Plumbing and Steamfitting	
Clear stopped water closet	3 Hrs.
Clear stopped basin	1-1/2 Hrs.
Replace leaking radiator valve	6 Hrs.
Clear external sewer stoppage	9 Hrs.
Install replacement valve, faucet, or trap	10 Hrs.
Electrical	
Replace fluorescent lamp ballast	2 Hrs.
Replace fractional HP motor	10 Hrs.
Replace blown fuse or reset Circuit breaker	1-1/2 Hrs.
Repair exterior light damage	4 Hrs.
Control circuit problems	5 Hrs.

R019313-30 A Review of Major Building Materials–Advantages and Disadvantages

Material	Advantages	Disadvantages
Aluminum	Lightweight High resistance to corrosion Low electrical resistance Good conductor	Softness Limited strength for structural uses Low stiffness High rate of thermal expansion Low rate of fire resistance Relatively high cost
Concrete	High compressive resistance Durable Resistance to moisture, rot, insects, fire and wear Watertight (depending on water-cement ration)	Workability Lack of tensile strength Lack of resistance to many types of chemical exposure, such as salt Hard to remove stains
Copper	High resistance to corrosion Good electrical conductivity Workable Forms its own surface protection Thermal contraction and expansion not high	Costly Strength varies with treatment and mechanical working
Epoxy	Can have varied properties depending on composition Liquid use very applicable Controllable Excellent strength properties Small creep during curing Hard Tough Resistant to abrasion Resistant to corrosion, salts, acids, petroleum products, solvents and other chemicals Adhesion to surfaces good	Color Form
Glass	Considerably strong Non-corrosive	Brittle, subject to shattering under shock Transmits energy at a rapid rate Large sizes expensive
Lead	Good resistance to corrosion	Heavy High coefficient of thermal expansion Difficult to hold in place
Masonry	Available in small units Appearance (available in many textures, sizes and colors) Good insulator	Stains hard to remove Porous (absorption rate high) Shrinkage of mortar Thermal and expansion cracking Others — see concrete
Paper	Readily available Low in cost Used in conjunction with other materials	Susceptible to water damage Susceptible to rot Relatively weak Highly combustible
Plastics — general	Applicable to many uses Relatively low in cost Workable into many shapes High strength Lightweight Non-corrosive Others — see specific plastic entry	Lack of resistance to fire Low stiffness High rate of thermal expansion Low thermal conductivity Some cases of chemical or physical instability with time Non-salvageable Others — see specific plastic entry

R019313-30 A Review of Major Building Materials—Advantages and Disadvantages (cont.)

Material	Advantages	Disadvantages
Plastics — specific		
Acrylics	Transparent Hard Weather resistant Shatter resistant	Easily scratched
Alkyds	Water resistant Touch Good adhesive properties	
Melamines	Hard Durable Abrasive resistant Chemical and heat resistant	
Polyamides (nylon)	Hard Tough Wear resistant	Costly
Polyesters	Weather and chemical resistant Stiff Hard	
Polyethylene	Flexible Tough Translucent Low cost	Easily scratched
Polystyrene	Hard Clear Water and chemical resistant Low cost	Brittle
Vinyls	Tough Wear resistant Stain resistant	
Plywood	Stronger than standard lumber Durable High resistance to impact and load Not as affected by moisture changes as standard wood Workability See Wood for others	Dimensional stability not as good as that of standard structural lumber Poor quality glues are possible Thermal expansion and contraction more of a problem
Steel	Strong Most resistant to aging Most reliable in quality, non-combustible, non-rotting, dimensionally stable with time and moisture change Resistant to staining	Costly Resultant loss of strength when exposed to intense heat, heat, rapid heat gain and loss Corrosive when exposed to moisture and air or other corrosive conditions
Tin	Extremely workable Very resistant to corrosion	Heavy Loss of strength when exposed to heat
Wood	Readily available Relatively low in cost Simple to work with Available in many shapes and forms Good insulating properties	Paintability changes with moisture content Combustible Susceptible to rot and insect infestation Soft and easily damaged (porous) Dimensional changes due to changes in temperature and moisture Strength changes with changing moisture content
Zinc	Fairly high resistance to corrosion Forms its own protective surface Workable	Brittle

Existing Conditions | R0241 | Demolition

R024119-10 Demolition Defined

Whole Building Demolition - Demolition of the whole building with no concern for any particular building element, component, or material type being demolished. This type of demolition is accomplished with large pieces of construction equipment that break up the structure, load it into trucks and haul it to a disposal site, but disposal or dump fees are not included. Demolition of below-grade foundation elements, such as footings, foundation walls, grade beams, slabs on grade, etc., is not included. Certain mechanical equipment containing flammable liquids or ozone-depleting refrigerants, electric lighting elements, communication equipment components, and other building elements may contain hazardous waste, and must be removed, either selectively or carefully, as hazardous waste before the building can be demolished.

Foundation Demolition - Demolition of below-grade foundation footings, foundation walls, grade beams, and slabs on grade. This type of demolition is accomplished by hand or pneumatic hand tools, and does not include saw cutting, or handling, loading, hauling, or disposal of the debris.

Gutting - Removal of building interior finishes and electrical/mechanical systems down to the load-bearing and sub-floor elements of the rough building frame, with no concern for any particular building element, component, or material type being demolished. This type of demolition is accomplished by hand or pneumatic hand tools, and includes loading into trucks, but not hauling, disposal or dump fees, scaffolding, or shoring. Certain mechanical equipment containing flammable liquids or ozone-depleting refrigerants, electric lighting elements, communication equipment components, and other building elements may contain hazardous waste, and must be removed, either selectively or carefully, as hazardous waste, before the building is gutted.

Selective Demolition - Demolition of a selected building element, component, or finish, with some concern for surrounding or adjacent elements, components, or finishes (see the first Subdivision (s) at the beginning of appropriate Divisions). This type of demolition is accomplished by hand or pneumatic hand tools, and does not include handling, loading, storing, hauling, or disposal of the debris, scaffolding, or shoring. "Gutting" methods may be used in order to save time, but damage that is caused to surrounding or adjacent elements, components, or finishes may have to be repaired at a later time.

Careful Removal - Removal of a piece of service equipment, building element or component, or material type, with great concern for both the removed item and surrounding or adjacent elements, components or finishes. The purpose of careful removal may be to protect the removed item for later re-use, preserve a higher salvage value of the removed item, or replace an item while taking care to protect surrounding or adjacent elements, components, connections, or finishes from cosmetic and/or structural damage. An approximation of the time required to perform this type of removal is 1/3 to 1/2 the time it would take to install a new item of like kind (see Reference Numbers R220105-10 and R260105-30). This type of removal is accomplished by hand or pneumatic hand tools, and does not include loading, hauling, or storing the removed item, scaffolding, shoring, or lifting equipment.

Cutout Demolition - Demolition of a small quantity of floor, wall, roof, or other assembly, with concern for the appearance and structural integrity of the surrounding materials. This type of demolition is accomplished by hand or pneumatic hand tools, and does not include saw cutting, handling, loading, hauling, or disposal of debris, scaffolding, or shoring.

Rubbish Handling - Work activities that involve handling, loading or hauling of debris. Generally, the cost of rubbish handling must be added to the cost of all types of demolition, with the exception of whole building demolition.

Minor Site Demolition - Demolition of site elements outside the footprint of a building. This type of demolition is accomplished by hand or pneumatic hand tools, or with larger pieces of construction equipment, and may include loading a removed item onto a truck (check the Crew for equipment used). It does not include saw cutting, hauling or disposal of debris, and, sometimes, handling or loading.

Existing Conditions | R0265 | Underground Storage Tank Removal

R026510-20 Underground Storage Tank Removal

Underground Storage Tank Removal can be divided into two categories: Non-Leaking and Leaking. Prior to removing an underground storage tank, tests should be made, with the proper authorities present, to determine whether a tank has been leaking or the surrounding soil has been contaminated.

To safely remove Liquid Underground Storage Tanks:

1. Excavate to the top of the tank.
2. Disconnect all piping.
3. Open all tank vents and access ports.
4. Remove all liquids and/or sludge.
5. Purge the tank with an inert gas.
6. Provide access to the inside of the tank and clean out the interior using proper personal protective equipment (PPE).
7. Excavate soil surrounding the tank using proper PPE for on-site personnel.
8. Pull and properly dispose of the tank.
9. Clean up the site of all contaminated material.
10. Install new tanks or close the excavation.

R028213-20 Asbestos Removal Process

Asbestos removal is accomplished by a specialty contractor who understands the federal and state regulations regarding the handling and disposal of the material. The process of asbestos removal is divided into many individual steps. An accurate estimate can be calculated only after all the steps have been priced.

The steps are generally as follows:

1. Obtain an asbestos abatement plan from an industrial hygienist.
2. Monitor the air quality in and around the removal area and along the path of travel between the removal area and transport area. This establishes the background contamination.
3. Construct a two part decontamination chamber at entrance to removal area.
4. Install a HEPA filter to create a negative pressure in the removal area.
5. Install wall, floor and ceiling protection as required by the plan, usually 2 layers of fireproof 6 mil polyethylene.
6. Industrial hygienist visually inspects work area to verify compliance with plan.
7. Provide temporary supports for conduit and piping affected by the removal process.
8. Proceed with asbestos removal and bagging process. Monitor air quality as described in Step #2. Discontinue operations when contaminate levels exceed applicable standards.
9. Document the legal disposal of materials in accordance with EPA standards.
10. Thoroughly clean removal area including all ledges, crevices and surfaces.
11. Post abatement inspection by industrial hygienist to verify plan compliance.
12. Provide a certificate from a licensed industrial hygienist attesting that contaminate levels are within acceptable standards before returning area to regular use.

R028319-60 Lead Paint Remediation Methods

Lead paint remediation can be accomplished by the following methods.

1. Abrasive blast
2. Chemical stripping
3. Power tool cleaning with vacuum collection system
4. Encapsulation
5. Remove and replace
6. Enclosure

Each of these methods has strengths and weakness depending on the specific circumstances of the project. The following is an overview of each method.

1. **Abrasive blasting** is usually accomplished with sand or recyclable metallic blast. Before work can begin, the area must be contained to ensure the blast material with lead does not escape to the atmosphere. The use of vacuum blast greatly reduces the containment requirements. Lead abatement equipment that may be associated with this work includes a negative air machine. In addition, it is necessary to have an industrial hygienist monitor the project on a continual basis. When the work is complete, the spent blast sand with lead must be disposed of as a hazardous material. If metallic shot was used, the lead is separated from the shot and disposed of as hazardous material. Worker protection includes disposable clothing and respiratory protection.
2. **Chemical stripping** requires strong chemicals be applied to the surface to remove the lead paint. Before the work can begin, the area under/adjacent to the work area must be covered to catch the chemical and removed lead. After the chemical is applied to the painted surface it is usually covered with paper. The chemical is left in place for the specified period, then the paper with lead paint is pulled or scraped off. The process may require several chemical applications. The paper with chemicals and lead paint adhered to it, plus the containment and loose scrapings collected by a HEPA (High Efficiency Particulate Air Filter) vac, must be disposed of as a hazardous material. The chemical stripping process usually requires a neutralizing agent and several wash downs after the paint is removed. Worker protection includes a neoprene or other compatible protective clothing and respiratory protection with face shield. An industrial hygienist is required intermittently during the process.
3. **Power tool cleaning** is accomplished using shrouded needle blasting guns. The shrouding with different end configurations is held up against the surface to be cleaned. The area is blasted with hardened needles and the shroud captures the lead with a HEPA vac and deposits it in a holding tank. An industrial hygienist monitors the project, protective clothing and a respirator is required until air samples prove otherwise. When the work is complete the lead must be disposed of as a hazardous material.
4. **Encapsulation** is a method that leaves the well bonded lead paint in place after the peeling paint has been removed. Before the work can begin, the area under/adjacent to the work must be covered to catch the scrapings. The scraped surface is then washed with a detergent and rinsed. The prepared surface is covered with approximately 10 mils of paint. A reinforcing fabric can also be embedded in the paint covering. The scraped paint and containment must be disposed of as a hazardous material. Workers must wear protective clothing and respirators.
5. **Remove and replace** is an effective way to remove lead paint from windows, gypsum walls and concrete masonry surfaces. The painted materials are removed and new materials are installed. Workers should wear a respirator and tyvek suit. The demolished materials must be disposed of as hazardous waste if it fails the TCLP (Toxicity Characteristic Leachate Process) test.
6. **Enclosure** is the process that permanently seals lead painted materials in place. This process has many applications such as covering lead painted drywall with new drywall, covering exterior construction with tyvek paper then residing, or covering lead painted structural members with aluminum or plastic. The seams on all enclosing materials must be securely sealed. An industrial hygienist monitors the project, and protective clothing and a respirator is required until air samples prove otherwise.

All the processes require clearance monitoring and wipe testing as required by the hygienist.

R031113-10 Wall Form Materials

Aluminum Forms

Approximate weight is 3 lbs. per S.F.C.A. Standard widths are available from 4″ to 36″ with 36″ most common. Standard lengths of 2′, 4′, 6′ to 8′ are available. Forms are lightweight and fewer ties are needed with the wider widths. The form face is either smooth or textured.

Metal Framed Plywood Forms

Manufacturers claim over 75 reuses of plywood and over 300 reuses of steel frames. Many specials such as corners, fillers, pilasters, etc. are available. Monthly rental is generally about 15% of purchase price for first month and 9% per month thereafter with 90% of rental applied to purchase for the first month and decreasing percentages thereafter. Aluminum framed forms cost 25% to 30% more than steel framed.

After the first month, extra days may be prorated from the monthly charge. Rental rates do not include ties, accessories, cleaning, loss of hardware or freight in and out. Approximate weight is 5 lbs. per S.F. for steel; 3 lbs. per S.F. for aluminum.

Forms can be rented with option to buy.

Plywood Forms, Job Fabricated

There are two types of plywood used for concrete forms.

1. Exterior plyform which is completely waterproof. This is face oiled to facilitate stripping. Ten reuses can be expected with this type with 25 reuses possible.
2. An overlaid type consists of a resin fiber fused to exterior plyform. No oiling is required except to facilitate cleaning. This is available in both high density (HDO) and medium density overlaid (MDO). Using HDO, 50 reuses can be expected with 200 possible.

Plyform is available in 5/8″ and 3/4″ thickness. High density overlaid is available in 3/8″, 1/2″, 5/8″ and 3/4″ thickness.

5/8″ thick is sufficient for most building forms, while 3/4″ is best on heavy construction.

Plywood Forms, Modular, Prefabricated

There are many plywood forming systems without frames. Most of these are manufactured from 1-1/8″ (HDO) plywood and have some hardware attached. These are used principally for foundation walls 8′ or less high. With care and maintenance, 100 reuses can be attained with decreasing quality of surface finish.

Steel Forms

Approximate weight is 6-1/2 lbs. per S.F.C.A. including accessories. Standard widths are available from 2″ to 24″, with 24″ most common. Standard lengths are from 2′ to 8′, with 4′ the most common. Forms are easily ganged into modular units.

Forms are usually leased for 15% of the purchase price per month prorated daily over 30 days.

Rental may be applied to sale price, and usually rental forms are bought. With careful handling and cleaning 200 to 400 reuses are possible.

Straight wall gang forms up to 12′ x 20′ or 8′ x 30′ can be fabricated. These crane handled forms usually lease for approx. 9% per month.

Individual job analysis is available from the manufacturer at no charge.

R031113-30 Slipforms

The slipform method of forming may be used for forming circular silo and multi-celled storage bin type structures over 30′ high, and building core shear walls over eight stories high. The shear walls, usually enclose elevator shafts, stairwells, mechanical spaces, and toilet rooms. Reuse of the form on duplicate structures will reduce the height necessary and spread the cost of building the form. Slipform systems can be used to cast chimneys, towers, piers, dams, underground shafts or other structures capable of being extruded.

Slipforms are usually 4′ high and are raised semi-continuously by jacks climbing on rods which are embedded in the concrete. The jacks are powered by a hydraulic, pneumatic, or electric source and are available in 3, 6, and 22 ton capacities. Interior work decks and exterior scaffolds must be provided for placing inserts, embedded items, reinforcing steel, and concrete. Scaffolds below the form for finishers may be required. The interior work decks are often used as roof slab forms on silos and bin work. Form raising rates will range from 6″ to 20″ per hour for silos; 6″ to 30″ per hour for buildings; and 6″ to 48″ per hour for shaft work.

Reinforcing bars and stressing strands are usually hoisted by crane or gin pole, and the concrete material can be hoisted by crane, winch-powered skip, or pumps. The slipform system is operated on a continuous 24-hour day when a monolithic structure is desired. For least cost, the system is operated only during normal working hours.

Placing concrete will range from 0.5 to 1.5 labor-hours per C.Y. Bucks, blockouts, keyways, weldplates, etc. are extra.

R031113-40 Forms for Reinforced Concrete

Design Economy

Avoid many sizes in proportioning beams and columns.

From story to story avoid changing column dimensions. Gain strength by adding steel or using a richer mix. If a change in size of column is necessary, vary one dimension only to minimize form alterations. Keep beams and columns the same width.

From floor to floor in a multi-story building vary beam depth, not width, as that will leave slab panel form unchanged. It is cheaper to vary the strength of a beam from floor to floor by means of steel area than by 2″ changes in either width or depth.

Cost Factors

Material includes the cost of lumber, cost of rent for metal pans or forms if used, nails, form ties, form oil, bolts and accessories.

Labor includes the cost of carpenters to make up, erect, remove and repair, plus common labor to clean and move. Having carpenters remove forms minimizes repairs.

Improper alignment and condition of forms will increase finishing cost. When forms are heavily oiled, concrete surfaces must be neutralized before finishing. Special curing compounds will cause spillages to spall off in first frost. Gang forming methods will reduce costs on large projects.

Materials Used

Boards are seldom used unless their architectural finish is required. Generally, steel, fiberglass and plywood are used for contact surfaces. Labor on plywood is 10% less than with boards. The plywood is backed up with 2 x 4′s at 12″ to 32″ O.C. Walers are generally 2 - 2 x 4′s. Column forms are held together with steel yokes or bands. Shoring is with adjustable shoring or scaffolding for high ceilings.

Reuse

Floor and column forms can be reused four or possibly five times without excessive repair. Remember to allow for 10% waste on each reuse.

When modular sized wall forms are made, up to twenty uses can be expected with exterior plyform.

When forms are reused, the cost to erect, strip, clean and move will not be affected. 10% replacement of lumber should be included and about one hour of carpenter time for repairs on each reuse per 100 S.F.

The reuse cost for certain accessory items normally rented on a monthly basis will be lower than the cost for the first use.

After fifth use, new material required plus time needed for repair prevent form cost from dropping further and it may go up. Much depends on care in stripping, the number of special bays, changes in beam or column sizes and other factors.

Costs for multiple use of formwork may be developed as follows:

2 Uses

$$\frac{\text{(1st Use + Reuse)}}{2} = \text{avg. cost/2 uses}$$

3 Uses

$$\frac{\text{(1st Use + 2 Reuse)}}{3} = \text{avg. cost/3 uses}$$

4 Uses

$$\frac{\text{(1st use + 3 Reuse)}}{4} = \text{avg. cost/4 uses}$$

R031113-60 Formwork Labor-Hours

Item	Hours Required				Total Hours	Multiple Use		
	Unit	Fabricate	Erect & Strip	Clean & Move	1 Use	2 Use	3 Use	4 Use
Beam and Girder, interior beams, 12″ wide	100 S.F.	6.4	8.3	1.3	16.0	13.3	12.4	12.0
Hung from steel beams	↓	5.8	7.7	1.3	14.8	12.4	11.6	11.2
Beam sides only, 36″ high	↓	5.8	7.2	1.3	14.3	11.9	11.1	10.7
Beam bottoms only, 24″ wide	↓	6.6	13.0	1.3	20.9	18.1	17.2	16.7
Box out for openings	↓	9.9	10.0	1.1	21.0	16.6	15.1	14.3
Buttress forms, to 8′ high	↓	6.0	6.5	1.2	13.7	11.2	10.4	10.0
Centering, steel, 3/4″ rib lath	↓		1.0		1.0			
3/8″ rib lath or slab form	↓		0.9		0.9			
Chamfer strip or keyway	100 L.F.		1.5		1.5	1.5	1.5	1.5
Columns, fiber tube 8″ diameter	↓		20.6		20.6			
12″	↓		21.3		21.3			
16″	↓		22.9		22.9			
20″	↓		23.7		23.7			
24″	↓		24.6		24.6			
30″	↓		25.6		25.6			
Round Steel, 12″ diameter	↓		22.0		22.0	22.0	22.0	22.0
16″	↓		25.6		25.6	25.6	25.6	25.6
20″	↓		30.5		30.5	30.5	30.5	30.5
24″	↓		37.7		37.7	37.7	37.7	37.7
Plywood 8″ x 8″	100 S.F.	7.0	11.0	1.2	19.2	16.2	15.2	14.7
12″ x 12″	↓	6.0	10.5	1.2	17.7	15.2	14.4	14.0
16″ x 16″	↓	5.9	10.0	1.2	17.1	14.7	13.8	13.4
24″ x 24″	↓	5.8	9.8	1.2	16.8	14.4	13.6	13.2
Steel framed plywood 8″ x 8″	↓		10.0	1.0	11.0	11.0	11.0	11.0
12″ x 12″	↓		9.3	1.0	10.3	10.3	10.3	10.3
16″ x 16″	↓		8.5	1.0	9.5	9.5	9.5	9.5
24″ x 24″	↓		7.8	1.0	8.8	8.8	8.8	8.8
Drop head forms, plywood	↓	9.0	12.5	1.5	23.0	19.0	17.7	17.0
Coping forms	↓	8.5	15.0	1.5	25.0	21.3	20.0	19.4
Culvert, box	↓		14.5	4.3	18.8	18.8	18.8	18.8
Curb forms, 6″ to 12″ high, on grade	↓	5.0	8.5	1.2	14.7	12.7	12.1	11.7
On elevated slabs	↓	6.0	10.8	1.2	18.0	15.5	14.7	14.3
Edge forms to 6″ high, on grade	100 L.F.	2.0	3.5	0.6	6.1	5.6	5.4	5.3
7″ to 12″ high	100 S.F.	2.5	5.0	1.0	8.5	7.8	7.5	7.4
Equipment foundations	↓	10.0	18.0	2.0	30.0	25.5	24.0	23.3
Flat slabs, including drops	↓	3.5	6.0	1.2	10.7	9.5	9.0	8.8
Hung from steel	↓	3.0	5.5	1.2	9.7	8.7	8.4	8.2
Closed deck for domes	↓	3.0	5.8	1.2	10.0	9.0	8.7	8.5
Open deck for pans	↓	2.2	5.3	1.0	8.5	7.9	7.7	7.6
Footings, continuous, 12″ high	↓	3.5	3.5	1.5	8.5	7.3	6.8	6.6
Spread, 12″ high	↓	4.7	4.2	1.6	10.5	8.7	8.0	7.7
Pile caps, square or rectangular	↓	4.5	5.0	1.5	11.0	9.3	8.7	8.4
Grade beams, 24″ deep	↓	2.5	5.3	1.2	9.0	8.3	8.0	7.9
Lintel or Sill forms	↓	8.0	17.0	2.0	27.0	23.5	22.3	21.8
Spandrel beams, 12″ wide	↓	9.0	11.2	1.3	21.5	17.5	16.2	15.5
Stairs	↓		25.0	4.0	29.0	29.0	29.0	29.0
Trench forms in floor	↓	4.5	14.0	1.5	20.0	18.3	17.7	17.4
Walls, Plywood, at grade, to 8′ high	↓	5.0	6.5	1.5	13.0	11.0	9.7	9.5
8′ to 16′	↓	7.5	8.0	1.5	17.0	13.8	12.7	12.1
16′ to 20′	↓	9.0	10.0	1.5	20.5	16.5	15.2	14.5
Foundation walls, to 8′ high	↓	4.5	6.5	1.0	12.0	10.3	9.7	9.4
8′ to 16′ high	↓	5.5	7.5	1.0	14.0	11.8	11.0	10.6
Retaining wall to 12′ high, battered	↓	6.0	8.5	1.5	16.0	13.5	12.7	12.3
Radial walls to 12′ high, smooth	↓	8.0	9.5	2.0	19.5	16.0	14.8	14.3
2′ chords	↓	7.0	8.0	1.5	16.5	13.5	12.5	12.0
Prefabricated modular, to 8′ high	↓	—	4.3	1.0	5.3	5.3	5.3	5.3
Steel, to 8′ high	↓	—	6.8	1.2	8.0	8.0	8.0	8.0
8′ to 16′ high	↓	—	9.1	1.5	10.6	10.3	10.2	10.2
Steel framed plywood to 8′ high	↓	—	6.8	1.2	8.0	7.5	7.3	7.2
8′ to 16′ high	↓	—	9.3	1.2	10.5	9.5	9.2	9.0

R032110-10 Reinforcing Steel Weights and Measures

Bar Designation No.**	Nominal Weight Lb./Ft.	U.S. Customary Units Nominal Dimensions*			SI Units Nominal Dimensions*			
		Diameter in.	Cross Sectional Area, in.^{2}	Perimeter in.	Nominal Weight kg/m	Diameter mm	Cross Sectional Area, cm^2	Perimeter mm
3	.376	.375	.11	1.178	.560	9.52	.71	29.9
4	.668	.500	.20	1.571	.994	12.70	1.29	39.9
5	1.043	.625	.31	1.963	1.552	15.88	2.00	49.9
6	1.502	.750	.44	2.356	2.235	19.05	2.84	59.8
7	2.044	.875	.60	2.749	3.042	22.22	3.87	69.8
8	2.670	1.000	.79	3.142	3.973	25.40	5.10	79.8
9	3.400	1.128	1.00	3.544	5.059	28.65	6.45	90.0
10	4.303	1.270	1.27	3.990	6.403	32.26	8.19	101.4
11	5.313	1.410	1.56	4.430	7.906	35.81	10.06	112.5
14	7.650	1.693	2.25	5.320	11.384	43.00	14.52	135.1
18	13.600	2.257	4.00	7.090	20.238	57.33	25.81	180.1

* The nominal dimensions of a deformed bar are equivalent to those of a plain round bar having the same weight per foot as the deformed bar.
** Bar numbers are based on the number of eighths of an inch included in the nominal diameter of the bars.

R032110-20 Metric Rebar Specification - ASTM A615-81

Grade 300 (300 MPa* = 43,560 psi; +8.7% vs. Grade 40)				
Grade 400 (400 MPa* = 58,000 psi; −3.4% vs. Grade 60)				
Bar No.	Diameter mm	Area mm^2	Equivalent in.^{2}	Comparison with U.S. Customary Bars
10M	11.3	100	.16	Between #3 & #4
15M	16.0	200	.31	#5 (.31 in.^{2})
20M	19.5	300	.47	#6 (.44 in.^{2})
25M	25.2	500	.78	#8 (.79 in.^{2})
30M	29.9	700	1.09	#9 (1.00 in.^{2})
35M	35.7	1000	1.55	#11 (1.56 in.^{2})
45M	43.7	1500	2.33	#14 (2.25 in.^{2})
55M	56.4	2500	3.88	#18 (4.00 in.^{2})

* MPa = megapascals

R032110-25 Comparison of U.S. Customary Units and SI Units for Reinforcing Bars

U.S. Customary Units

Bar Designation No.[b]	Nominal Weight, lb/ft	Nominal Dimensions[a]			Deformation Requirements, in.		
		Diameter in.	Cross Sectional Area, in.^{2}	Perimeter in.	Maximum Average Spacing	Minimum Average Height	Maximum Gap (Chord of 12-1/2% of Nominal Perimeter)
3	0.376	0.375	0.11	1.178	0.262	0.015	0.143
4	0.668	0.500	0.20	1.571	0.350	0.020	0.191
5	1.043	0.625	0.31	1.963	0.437	0.028	0.239
6	1.502	0.750	0.44	2.356	0.525	0.038	0.286
7	2.044	0.875	0.60	2.749	0.612	0.044	0.334
8	2.670	1.000	0.79	3.142	0.700	0.050	0.383
9	3.400	1.128	1.00	3.544	0.790	0.056	0.431
10	4.303	1.270	1.27	3.990	0.889	0.064	0.487
11	5.313	1.410	1.56	4.430	0.987	0.071	0.540
14	7.65	1.693	2.25	5.32	1.185	0.085	0.648
18	13.60	2.257	4.00	7.09	1.58	0.102	0.864

SI UNITS

Bar Designation No.[b]	Nominal Weight kg/m	Nominal Dimensions[a]			Deformation Requirements, mm		
		Diameter, mm	Cross Sectional Area, cm^2	Perimeter, mm	Maximum Average Spacing	Minimum Average Height	Maximum Gap (Chord of 12-1/2% of Nominal Perimeter)
3	0.560	9.52	0.71	29.9	6.7	0.38	3.5
4	0.994	12.70	1.29	39.9	8.9	0.51	4.9
5	1.552	15.88	2.00	49.9	11.1	0.71	6.1
6	2.235	19.05	2.84	59.8	13.3	0.96	7.3
7	3.042	22.22	3.87	69.8	15.5	1.11	8.5
8	3.973	25.40	5.10	79.8	17.8	1.27	9.7
9	5.059	28.65	6.45	90.0	20.1	1.42	10.9
10	6.403	32.26	8.19	101.4	22.6	1.62	11.4
11	7.906	35.81	10.06	112.5	25.1	1.80	13.6
14	11.384	43.00	14.52	135.1	30.1	2.16	16.5
18	20.238	57.33	25.81	180.1	40.1	2.59	21.9

[a]Nominal dimensions of a deformed bar are equivalent to those of a plain round bar having the same weight per foot as the deformed bar.

[b]Bar numbers are based on the number of eighths of an inch included in the nominal diameter of the bars.

R032110-40 Weight of Steel Reinforcing Per Square Foot of Wall (PSF)

Reinforced Weights: The table below suggests the weights per square foot for reinforcing steel in walls. Weights are approximate and will be the same for all grades of steel bars. For bars in two directions, add weights for each size and spacing.

C/C Spacing in Inches	Bar Size								
	#3 Wt. (PSF)	#4 Wt. (PSF)	#5 Wt. (PSF)	#6 Wt. (PSF)	#7 Wt. (PSF)	#8 Wt. (PSF)	#9 Wt. (PSF)	#10 Wt. (PSF)	#11 Wt. (PSF)
2″	2.26	4.01	6.26	9.01	12.27				
3″	1.50	2.67	4.17	6.01	8.18	10.68	13.60	17.21	21.25
4″	1.13	2.01	3.13	4.51	6.13	8.10	10.20	12.91	15.94
5″	.90	1.60	2.50	3.60	4.91	6.41	8.16	10.33	12.75
6″	.752	1.34	2.09	3.00	4.09	5.34	6.80	8.61	10.63
8″	.564	1.00	1.57	2.25	3.07	4.01	5.10	6.46	7.97
10″	.451	.802	1.25	1.80	2.45	3.20	4.08	5.16	6.38
12″	.376	.668	1.04	1.50	2.04	2.67	3.40	4.30	5.31
18″	.251	.445	.695	1.00	1.32	1.78	2.27	2.86	3.54
24″	.188	.334	.522	.751	1.02	1.34	1.70	2.15	2.66
30″	.150	.267	.417	.600	.817	1.07	1.36	1.72	2.13
36″	.125	.223	.348	.501	.681	.890	1.13	1.43	1.77
42″	.107	.191	.298	.429	.584	.753	.97	1.17	1.52
48″	.094	.167	.261	.376	.511	.668	.85	1.08	1.33

R032110-50 Minimum Wall Reinforcement Weight (PSF)

This table lists the approximate minimum wall reinforcement weights per S.F. according to the specification of .12% of gross area for vertical bars and .20% of gross area for horizontal bars.

Location	Wall Thickness	Bar Size	Horizontal Steel Spacing C/C	Sq. In. Req'd per S.F.	Total Wt. per S.F.	Bar Size	Vertical Steel Spacing C/C	Sq. In. Req'd per S.F.	Total Wt. per S.F.	Horizontal & Vertical Steel Total Weight per S.F.
Both Faces	10″	#4	18″	.24	.89#	#3	18″	.14	.50#	1.39#
	12″	#4	16″	.29	1.00	#3	16″	.17	.60	1.60
	14″	#4	14″	.34	1.14	#3	13″	.20	.69	1.84
	16″	#4	12″	.38	1.34	#3	11″	.23	.82	2.16
	18″	#5	17″	.43	1.47	#4	18″	.26	.89	2.36
One Face	6″	#3	9″	.15	.50	#3	18″	.09	.25	.75
	8″	#4	12″	.19	.67	#3	11″	.12	.41	1.08
	10″	#5	15″	.24	.83	#4	16″	.14	.50	1.34

R032110-70 Bend, Place and Tie Reinforcing

Placing and tying by rodmen for footings and slabs runs from nine hrs. per ton for heavy bars to fifteen hrs. per ton for light bars. For beams, columns, and walls, production runs from eight hrs. per ton for heavy bars to twenty hrs. per ton for light bars. Overall average for typical reinforced concrete buildings is about fourteen hrs. per ton. These production figures include the time for placing of accessories and usual inserts, but not their material cost (allow 15% of the cost of delivered bent rods). Equipment handling is necessary for the larger-sized bars so that installation costs for the very heavy bars will not decrease proportionately.

Installation costs for splicing reinforcing bars include allowance for equipment to hold the bars in place while splicing as well as necessary scaffolding for iron workers.

R032110-80 Shop-Fabricated Reinforcing Steel

The material prices for reinforcing, shown in the unit cost sections of the book, are for 50 tons or more of shop-fabricated reinforcing steel and include:

1. Mill base price of reinforcing steel
2. Mill grade/size/length extras
3. Mill delivery to the fabrication shop
4. Shop storage and handling
5. Shop drafting/detailing
6. Shop shearing and bending
7. Shop listing
8. Shop delivery to the job site

Both material and installation costs can be considerably higher for small jobs consisting primarily of smaller bars, while material costs may be slightly lower for larger jobs.

R032205-30 Common Stock Styles of Welded Wire Fabric

This table provides some of the basic specifications, sizes, and weights of welded wire fabric used for reinforcing concrete.

New Designation		Old Designation		Steel Area per Foot				Approximate Weight per 100 S.F.	
Spacing — Cross Sectional Area (in.) — (Sq. in. 100)		Spacing — Wire Gauge (in.) — (AS & W)		Longitudinal		Transverse			
				in.	cm	in.	cm	lbs	kg
Rolls	6 x 6 — W1.4 x W1.4	6 x 6 — 10 x 10		.028	.071	.028	.071	21	9.53
	6 x 6 — W2.0 x W2.0	6 x 6 — 8 x 8	1	.040	.102	.040	.102	29	13.15
	6 x 6 — W2.9 x W2.9	6 x 6 — 6 x 6		.058	.147	.058	.147	42	19.05
	6 x 6 — W4.0 x W4.0	6 x 6 — 4 x 4		.080	.203	.080	.203	58	26.91
	4 x 4 — W1.4 x W1.4	4 x 4 — 10 x 10		.042	.107	.042	.107	31	14.06
	4 x 4 — W2.0 x W2.0	4 x 4 — 8 x 8	1	.060	.152	.060	.152	43	19.50
	4 x 4 — W2.9 x W2.9	4 x 4 — 6 x 6		.087	.227	.087	.227	62	28.12
	4 x 4 — W4.0 x W4.0	4 x 4 — 4 x 4		.120	.305	.120	.305	85	38.56
Sheets	6 x 6 — W2.9 x W2.9	6 x 6 — 6 x 6		.058	.147	.058	.147	42	19.05
	6 x 6 — W4.0 x W4.0	6 x 6 — 4 x 4		.080	.203	.080	.203	58	26.31
	6 x 6 — W5.5 x W5.5	6 x 6 — 2 x 2	2	.110	.279	.110	.279	80	36.29
	4 x 4 — W1.4 x W1.4	4 x 4 — 4 x 4		.120	.305	.120	.305	85	38.56

NOTES: 1. Exact W—number size for 8 gauge is W2.1
2. Exact W—number size for 2 gauge is W5.4

R033053-10 Spread Footings

General: A spread footing is used to convert a concentrated load (from one superstructure column, or substructure grade beams) into an allowable area load on supporting soil.

Because of punching action from the column load, a spread footing is usually thicker than strip footings which support wall loads. One or two story commercial or residential buildings should have no less than 1′ thick spread footings. Heavier loads require no less than 2′ thick. Spread footings may be square, rectangular or octagonal in plan.

Spread footings tend to minimize excavation and foundation materials, as well as labor and equipment. Another advantage is that footings and soil conditions can be readily examined. They are the most widely used type of footing, especially in mild climates and for buildings of four stories or under. This is because they are usually more economical than other types, if suitable soil and site conditions exist.

They are used when suitable supporting soil is located within several feet of the surface or line of subsurface excavation. Suitable soil types include sands and gravels, gravels with a small amount of clay or silt, hardpan, chalk, and rock. Pedestals may be used to bring the column base load down to the top of footing. Alternately, undesirable soil between underside of footing and top of bearing level can be removed and replaced with lean concrete mix or compacted granular material.

Depth of footing should be below topsoil, uncompacted fill, muck, etc. It must be lower than frost penetration but should be above the water table. It must not be at the ground surface because of potential surface erosion. If the ground slopes, approximately three horizontal feet of edge protection must remain. Differential footing elevations may overlap soil stresses or cause excavation problems if clear spacing between footings is less than the difference in depth.

Other footing types are usually used for the following reasons:

A. Bearing capacity of soil is low.
B. Very large footings are required, at a cost disadvantage.
C. Soil under footing (shallow or deep) is very compressible, with probability of causing excessive or differential settlement.
D. Good bearing soil is deep.
E. Potential for scour action exists.
F. Varying subsoil conditions within building perimeter.

Cost of spread footings for a building is determined by:

1. The soil bearing capacity.
2. Typical bay size.
3. Total load (live plus dead) per S.F. for roof and elevated floor levels.
4. The size and shape of the building.
5. Footing configuration. Does the building utilize outer spread footings or are there continuous perimeter footings only or a combination of spread footings plus continuous footings?

Soil Bearing Capacity in Kips per S.F.

Bearing Material	Typical Allowable Bearing Capacity
Hard sound rock	120 KSF
Medium hard rock	80
Hardpan overlaying rock	24
Compact gravel and boulder-gravel; very compact sandy gravel	20
Soft rock	16
Loose gravel; sandy gravel; compact sand; very compact sand-inorganic silt	12
Hard dry consolidated clay	10
Loose coarse to medium sand; medium compact fine sand	8
Compact sand-clay	6
Loose fine sand; medium compact sand-inorganic silts	4
Firm or stiff clay	3
Loose saturated sand-clay; medium soft clay	2

R033053-50 Industrial Chimneys

Foundation requirements in C.Y. of concrete for various sized chimneys.

Size Chimney	2 Ton Soil	3 Ton Soil	Size Chimney	2 Ton Soil	3 Ton Soil	Size Chimney	2 Ton Soil	3 Ton Soil
75′ x 3′-0″	13 C.Y.	11 C.Y.	160′ x 6′-6″	86 C.Y.	76 C.Y.	300′ x 10′-0″	325 C.Y.	245 C.Y.
85′ x 5′-6″	19	16	175′ x 7′-0″	108	95	350′ x 12′-0″	422	320
100′ x 5′-0″	24	20	200′ x 6′-0″	125	105	400′ x 14′-0″	520	400
125′ x 5′-6″	43	36	250′ x 8′-0″	230	175	500′ x 18′-0″	725	575

R033053-60 Maximum Depth of Frost Penetration in Inches

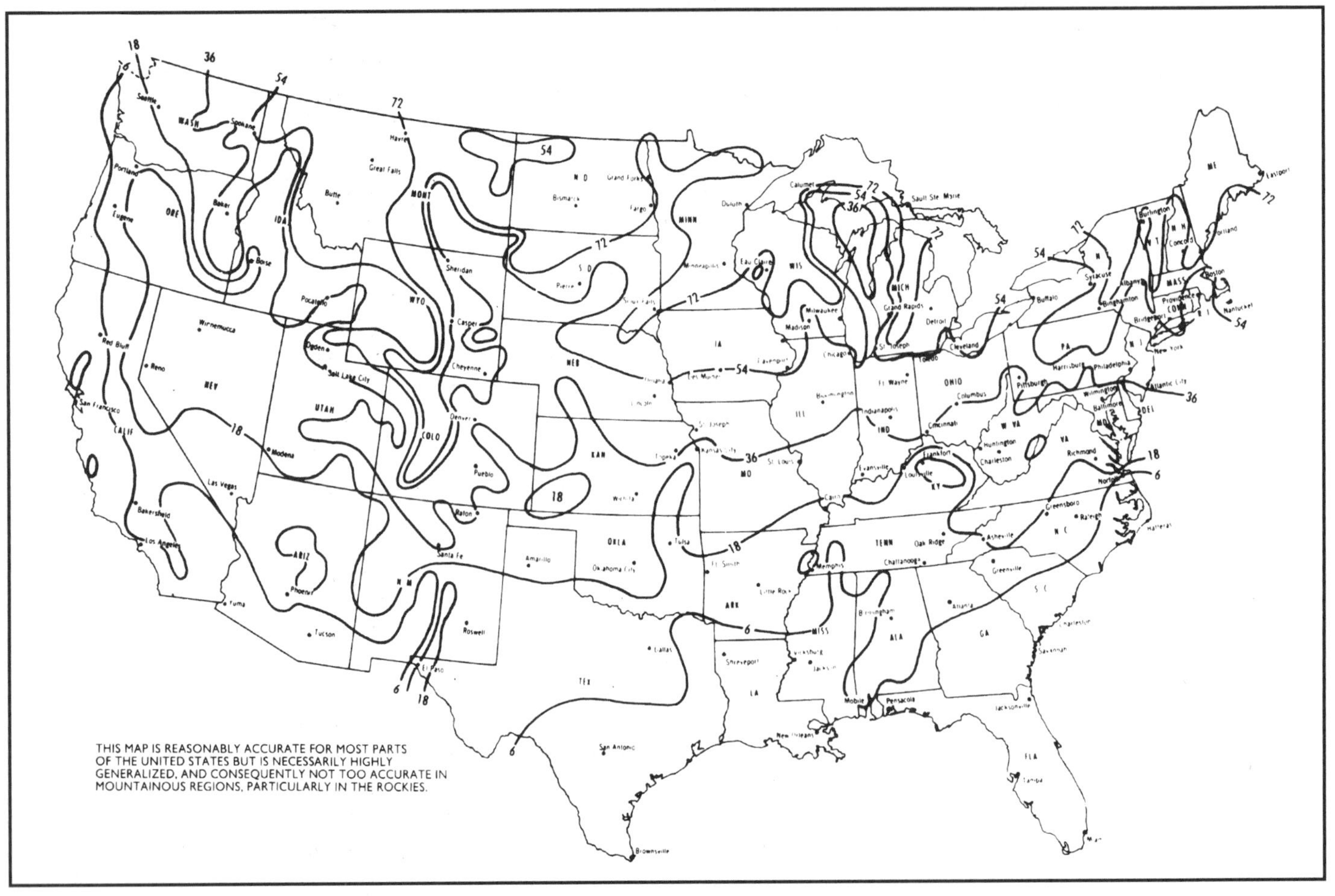

THIS MAP IS REASONABLY ACCURATE FOR MOST PARTS OF THE UNITED STATES BUT IS NECESSARILY HIGHLY GENERALIZED, AND CONSEQUENTLY NOT TOO ACCURATE IN MOUNTAINOUS REGIONS, PARTICULARLY IN THE ROCKIES.

R033105-10 Proportionate Quantities

The tables below show both quantities per S.F. of floor areas as well as form and reinforcing quantities per C.Y. Unusual structural requirements would increase the ratios below. High strength reinforcing would reduce the steel weights. Figures are for 3000 psi concrete and 60,000 psi reinforcing unless specified otherwise.

Type of Construction	Live Load	Span	Per S.F. of Floor Area: Concrete	Per S.F. of Floor Area: Forms	Per S.F. of Floor Area: Reinf.	Per S.F. of Floor Area: Pans	Per C.Y. of Concrete: Forms	Per C.Y. of Concrete: Reinf.	Per C.Y. of Concrete: Pans
Flat Plate	50 psf	15 Ft.	.46 C.F.	1.06 S.F.	1.71lb.		62 S.F.	101 lb.	
		20	.63	1.02	2.40		44	104	
		25	.79	1.02	3.03		35	104	
	100	15	.46	1.04	2.14		61	126	
		20	.71	1.02	2.72		39	104	
		25	.83	1.01	3.47		33	113	
Flat Plate (waffle construction) 20″ domes	50	20	.43	1.00	2.10	.84 S.F.	63	135	53 S.F.
		25	.52	1.00	2.90	.89	52	150	46
		30	.64	1.00	3.70	.87	42	155	37
	100	20	.51	1.00	2.30	.84	53	125	45
		25	.64	1.00	3.20	.83	42	135	35
		30	.76	1.00	4.40	.81	36	160	29
Waffle Construction 30″ domes	50	25	.69	1.06	1.83	.68	42	72	40
		30	.74	1.06	2.39	.69	39	87	39
		35	.86	1.05	2.71	.69	33	85	39
		40	.78	1.00	4.80	.68	35	165	40
Flat Slab (two way with drop panels)	50	20	.62	1.03	2.34		45	102	
		25	.77	1.03	2.99		36	105	
		30	.95	1.03	4.09		29	116	
	100	20	.64	1.03	2.83		43	119	
		25	.79	1.03	3.88		35	133	
		30	.96	1.03	4.66		29	131	
	200	20	.73	1.03	3.03		38	112	
		25	.86	1.03	4.23		32	133	
		30	1.06	1.03	5.30		26	135	
One Way Joists 20″ Pans	50	15	.36	1.04	1.40	.93	78	105	70
		20	.42	1.05	1.80	.94	67	120	60
		25	.47	1.05	2.60	.94	60	150	54
	100	15	.38	1.07	1.90	.93	77	140	66
		20	.44	1.08	2.40	.94	67	150	58
		25	.52	1.07	3.50	.94	55	185	49
One Way Joists 8″ x 16″ filler blocks	50	15	.34	1.06	1.80	.81 Ea.	84	145	64 Ea.
		20	.40	1.08	2.20	.82	73	145	55
		25	.46	1.07	3.20	.83	63	190	49
	100	15	.39	1.07	1.90	.81	74	130	56
		20	.46	1.09	2.80	.82	64	160	48
		25	.53	1.10	3.60	.83	56	190	42
One Way Beam & Slab	50	15	.42	1.30	1.73		84	111	
		20	.51	1.28	2.61		68	138	
		25	.64	1.25	2.78		53	117	
	100	15	.42	1.30	1.90		84	122	
		20	.54	1.35	2.69		68	154	
		25	.69	1.37	3.93		54	145	
	200	15	.44	1.31	2.24		80	137	
		20	.58	1.40	3.30		65	163	
		25	.69	1.42	4.89		53	183	
Two Way Beam & Slab	100	15	.47	1.20	2.26		69	130	
		20	.63	1.29	3.06		55	131	
		25	.83	1.33	3.79		43	123	
	200	15	.49	1.25	2.70		41	149	
		20	.66	1.32	4.04		54	165	
		25	.88	1.32	6.08		41	187	

R033105-10 Proportionate Quantities (cont.)

4000 psi Concrete and 60,000 psi Reinforcing—Form and Reinforcing Quantities per C.Y.					
Item	**Size**	**Forms**	**Reinforcing**	**Minimum**	**Maximum**
Columns (square tied)	10″ x 10″	130 S.F.C.A.	#5 to #11	220 lbs.	875 lbs.
	12″ x 12″	108	#6 to #14	200	955
	14″ x 14″	92	#7 to #14	190	900
	16″ x 16″	81	#6 to #14	187	1082
	18″ x 18″	72	#6 to #14	170	906
	20″ x 20″	65	#7 to #18	150	1080
	22″ x 22″	59	#8 to #18	153	902
	24″ x 24″	54	#8 to #18	164	884
	26″ x 26″	50	#9 to #18	169	994
	28″ x 28″	46	#9 to #18	147	864
	30″ x 30″	43	#10 to #18	146	983
	32″ x 32″	40	#10 to #18	175	866
	34″ x 34″	38	#10 to #18	157	772
	36″ x 36″	36	#10 to #18	175	852
	38″ x 38″	34	#10 to #18	158	765
	40″ x 40″	32 ▼	#10 to #18	143	692

Item	Size	Form	Spiral	Reinforcing	Minimum	Maximum
Columns (spirally reinforced)	12″ diameter	34.5 L.F.	190 lbs.	#4 to #11	165 lbs.	1505 lb.
		34.5	190	#14 & #18	—	1100
	14″	25	170	#4 to #11	150	970
		25	170	#14 & #18	800	1000
	16″	19	160	#4 to #11	160	950
		19	160	#14 & #18	605	1080
	18″	15	150	#4 to #11	160	915
		15	150	#14 & #18	480	1075
	20″	12	130	#4 to #11	155	865
		12	130	#14 & #18	385	1020
	22″	10	125	#4 to #11	165	775
		10	125	#14 & #18	320	995
	24″	9	120	#4 to #11	195	800
		9	120	#14 & #18	290	1150
	26″	7.3	100	#4 to #11	200	729
		7.3	100	#14 & #18	235	1035
	28″	6.3	95	#4 to #11	175	700
		6.3	95	#14 & #18	200	1075
	30″	5.5	90	#4 to #11	180	670
		5.5	90	#14 & #18	175	1015
	32″	4.8	85	#4 to #11	185	615
		4.8	85	#14 & #18	155	955
	34″	4.3	80	#4 to #11	180	600
		4.3	80	#14 & #18	170	855
	36″	3.8	75	#4 to #11	165	570
		3.8	75	#14 & #18	155	865
	40″	3.0	70	#4 to #11	165	500
		3.0	70	#14 & #18	145	765

R033105-10 Proportionate Quantities (cont.)

3000 psi Concrete and 60,000 psi Reinforcing—Form and Reinforcing Quantities per C.Y.						
Item	Type	Loading	Height	C.Y./L.F.	Forms/C.Y.	Reinf./C.Y.
Retaining Walls	Cantilever	Level Backfill	4 Ft.	0.2 C.Y.	49 S.F.	35 lbs.
			8	0.5	42	45
			12	0.8	35	70
			16	1.1	32	85
			20	1.6	28	105
		Highway Surcharge	4	0.3	41	35
			8	0.5	36	55
			12	0.8	33	90
			16	1.2	30	120
			20	1.7	27	155
		Railroad Surcharge	4	0.4	28	45
			8	0.8	25	65
			12	1.3	22	90
			16	1.9	20	100
			20	2.6	18	120
	Gravity, with Vertical Face	Level Backfill	4	0.4	37	None
			7	0.6	27	↓
			10	1.2	20	↓
		Sloping Backfill	4	0.3	31	↓
			7	0.8	21	↓
			10	1.6	15	↓

Item	Span	Live Load in Kips per Linear Foot							
		Under 1 Kip		2 to 3 Kips		4 to 5 Kips		6 to 7 Kips	
		Forms	Reinf.	Forms	Reinf.	Forms	Reinf.	Forms	Reinf.
Beams	10 Ft.	—	—	90 S.F.	170 #	85 S.F.	175 #	75 S.F.	185 #
	16	130 S.F.	165 #	85	180	75	180	65	225
	20	110	170	75	185	62	200	51	200
	26	90	170	65	215	62	215	—	—
	30	85	175	60	200	—	—	—	—

Item	Size	Type	Forms per C.Y.	Reinforcing per C.Y.
Spread Footings	Under 1 C.Y.	1,000 psf soil	24 S.F.	44 lbs.
		5,000	24	42
		10,000	24	52
	1 C.Y. to 5 C.Y.	1,000	14	49
		5,000	14	50
		10,000	14	50
	Over 5 C.Y.	1,000	9	54
		5,000	9	52
		10,000	9	56
Pile Caps (30 Ton Concrete Piles)	Under 5 C.Y.	shallow caps	20	65
		medium	20	50
		deep	20	40
	5 C.Y. to 10 C.Y.	shallow	14	55
		medium	15	45
		deep	15	40
	10 C.Y. to 20 C.Y.	shallow	11	60
		medium	11	45
		deep	12	35
	Over 20 C.Y.	shallow	9	60
		medium	9	45
		deep	10	40

R033105-10 Proportionate Quantities (cont.)

3000 psi Concrete and 60,000 psi Reinforcing — Form and Reinforcing Quantities per C.Y.						
Item	Size	Pile Spacing	50 T Pile	100 T Pile	50 T Pile	100 T Pile
Pile Caps (Steel H Piles)	Under 5 C.Y.	24″ O.C.	24 S.F.	24 S.F.	75 lbs.	90 lbs.
		30″	25	25	80	100
		36″	24	24	80	110
	5 C.Y. to 10 C.Y.	24″	15	15	80	110
		30″	15	15	85	110
		36″	15	15	75	90
	Over 10 C.Y.	24″	13	13	85	90
		30″	11	11	85	95
		36″	10	10	85	90

Item	Height	8″ Thick Forms	8″ Thick Reinf.	10″ Thick Forms	10″ Thick Reinf.	12″ Thick Forms	12″ Thick Reinf.	15″ Thick Forms	15″ Thick Reinf.
Basement Walls	7 Ft.	81 S.F.	44 lbs.	65 S.F.	45 lbs.	54 S.F.	44 lbs.	41 S.F.	43 lbs.
	8	↓	44	↓	45	↓	44	↓	43
	9	↓	46	↓	45	↓	44	↓	43
	10	↓	57	↓	45	↓	44	↓	43
	12	↓	83	↓	50	↓	52	↓	43
	14	↓	116	↓	65	↓	64	↓	51
	16	↓		↓	86	↓	90	↓	65
	18	↓		↓		↓	106	↓	70

R033105-20 Materials for One C.Y. of Concrete

This is an approximate method of figuring quantities of cement, sand and coarse aggregate for a field mix with waste allowance included.

With crushed gravel as coarse aggregate, to determine barrels of cement required, divide 10 by total mix; that is, for 1:2:4 mix, 10 divided by 7 = 1-3/7 barrels. If the coarse aggregate is crushed stone, use 10-1/2 instead of 10 as given for gravel.

To determine tons of sand required, multiply barrels of cement by parts of sand and then by 0.2; that is, for the 1:2:4 mix, as above, 1-3/7 x 2 x .2 = .57 tons.

Tons of crushed gravel are in the same ratio to tons of sand as parts in the mix, or 4/2 x .57 = 1.14 tons.

1 bag cement = 94#
4 bags = 1 barrel

1 C.Y. sand or crushed gravel = 2700#
1 ton sand or crushed gravel = 20 C.F.

1 C.Y. crushed stone = 2575#
1 ton crushed stone = 21 C.F.

Average carload of cement is 692 bags; of sand or gravel is 56 tons.

Do not stack stored cement over 10 bags high.

R033105-30 Metric Equivalents of Cement Content for Concrete Mixes

94 Pound Bags per Cubic Yard	Kilograms per Cubic Meter	94 Pound Bags per Cubic Yard	Kilograms per Cubic Meter
1.0	55.77	7.0	390.4
1.5	83.65	7.5	418.3
2.0	111.5	8.0	446.2
2.5	139.4	8.5	474.0
3.0	167.3	9.0	501.9
3.5	195.2	9.5	529.8
4.0	223.1	10.0	557.7
4.5	251.0	10.5	585.6
5.0	278.8	11.0	613.5
5.5	306.7	11.5	641.3
6.0	334.6	12.0	669.2
6.5	362.5	12.5	697.1

a. If you know the cement content in pounds per cubic yard, multiply by .5933 to obtain kilograms per cubic meter.

b. If you know the cement content in 94 pound bags per cubic yard, multiply by 55.77 to obtain kilograms per cubic meter.

R033105-40 Metric Equivalents of Common Concrete Strengths (to convert other psi values to megapascals, multiply by 0.006895)

U.S. Values psi	SI Value Megapascals	Non-SI Metric Value kgf/cm²*
2000	14	140
2500	17	175
3000	21	210
3500	24	245
4000	28	280
4500	31	315
5000	34	350
6000	41	420
7000	48	490
8000	55	560
9000	62	630
10,000	69	705

* kilograms force per square centimeter

R033105-50 Quantities of Cement, Sand and Stone for One C.Y. of Concrete per Various Mixes

This table can be used to determine the quantities of the ingredients for smaller quantities of site mixed concrete.

Concrete (C.Y.)	Mix = 1:1:1-3/4			Mix = 1:2:2.25			Mix = 1:2.25:3			Mix = 1:3:4		
	Cement (sacks)	Sand (C.Y.)	Stone (C.Y.)	Cement (sacks)	Sand (C.Y.)	Stone (C.Y.)	Cement (sacks)	Sand (C.Y.)	Stone (C.Y.)	Cement (sacks)	Sand (C.Y.)	Stone (C.Y.)
1	10	.37	.63	7.75	.56	.65	6.25	.52	.70	5	.56	.74
2	20	.74	1.26	15.50	1.12	1.30	12.50	1.04	1.40	10	1.12	1.48
3	30	1.11	1.89	23.25	1.68	1.95	18.75	1.56	2.10	15	1.68	2.22
4	40	1.48	2.52	31.00	2.24	2.60	25.00	2.08	2.80	20	2.24	2.96
5	50	1.85	3.15	38.75	2.80	3.25	31.25	2.60	3.50	25	2.80	3.70
6	60	2.22	3.78	46.50	3.36	3.90	37.50	3.12	4.20	30	3.36	4.44
7	70	2.59	4.41	54.25	3.92	4.55	43.75	3.64	4.90	35	3.92	5.18
8	80	2.96	5.04	62.00	4.48	5.20	50.00	4.16	5.60	40	4.48	5.92
9	90	3.33	5.67	69.75	5.04	5.85	56.25	4.68	6.30	45	5.04	6.66
10	100	3.70	6.30	77.50	5.60	6.50	62.50	5.20	7.00	50	5.60	7.40
11	110	4.07	6.93	85.25	6.16	7.15	68.75	5.72	7.70	55	6.16	8.14
12	120	4.44	7.56	93.00	6.72	7.80	75.00	6.24	8.40	60	6.72	8.88
13	130	4.82	8.20	100.76	7.28	8.46	81.26	6.76	9.10	65	7.28	9.62
14	140	5.18	8.82	108.50	7.84	9.10	87.50	7.28	9.80	70	7.84	10.36
15	150	5.56	9.46	116.26	8.40	9.76	93.76	7.80	10.50	75	8.40	11.10
16	160	5.92	10.08	124.00	8.96	10.40	100.00	8.32	11.20	80	8.96	11.84
17	170	6.30	10.72	131.76	9.52	11.06	106.26	8.84	11.90	85	9.52	12.58
18	180	6.66	11.34	139.50	10.08	11.70	112.50	9.36	12.60	90	10.08	13.32
19	190	7.04	11.98	147.26	10.64	12.36	118.76	9.84	13.30	95	10.64	14.06
20	200	7.40	12.60	155.00	11.20	13.00	125.00	10.40	14.00	100	11.20	14.80
21	210	7.77	13.23	162.75	11.76	13.65	131.25	10.92	14.70	105	11.76	15.54
22	220	8.14	13.86	170.05	12.32	14.30	137.50	11.44	15.40	110	12.32	16.28
23	230	8.51	14.49	178.25	12.88	14.95	143.75	11.96	16.10	115	12.88	17.02
24	240	8.88	15.12	186.00	13.44	15.60	150.00	12.48	16.80	120	13.44	17.76
25	250	9.25	15.75	193.75	14.00	16.25	156.25	13.00	17.50	125	14.00	18.50
26	260	9.64	16.40	201.52	14.56	16.92	162.52	13.52	18.20	130	14.56	19.24
27	270	10.00	17.00	209.26	15.12	17.56	168.76	14.04	18.90	135	15.02	20.00
28	280	10.36	17.64	217.00	15.68	18.20	175.00	14.56	19.60	140	15.68	20.72
29	290	10.74	18.28	224.76	16.24	18.86	181.26	15.08	20.30	145	16.24	21.46

R033105-65 Field-Mix Concrete

Presently most building jobs are built with ready-mixed concrete except at isolated locations and some larger jobs requiring over 10,000 C.Y. where land is readily available for setting up a temporary batch plant.

The most economical mix is a controlled mix using local aggregate proportioned by trial to give the required strength with the least cost of material.

R033105-70 Placing Ready-Mixed Concrete

For ground pours allow for 5% waste when figuring quantities.

Prices in the front of the book assume normal deliveries. If deliveries are made before 8 A.M. or after 5 P.M. or on Saturday afternoons add 30%. Negotiated discounts for large volumes are not included in prices in front of book.

For the lower floors without truck access, concrete may be wheeled in rubber-tired buggies, conveyer handled, crane handled or pumped. Pumping is economical if there is top steel. Conveyers are more efficient for thick slabs.

At higher floors the rubber-tired buggies may be hoisted by a hoisting tower and wheeled to location. Placement by a conveyer is limited to three floors and is best for high-volume pours. Pumped concrete is best when building has no crane access. Concrete may be pumped directly as high as thirty-six stories using special pumping techniques. Normal maximum height is about fifteen stories.

Best pumping aggregate is screened and graded bank gravel rather than crushed stone.

Pumping downward is more difficult than pumping upward. Horizontal distance from pump to pour may increase preparation time prior to pour. Placing by cranes, either mobile, climbing or tower types, continues as the most efficient method for high-rise concrete buildings.

R033105-80 Slab on Grade

General: Ground slabs are classified on the basis of use. Thickness is generally controlled by the heaviest concentrated load supported. If load area is greater than 80 sq. in., soil bearing may be important. The base granular fill must be a uniformly compacted material of limited capillarity, such as gravel or crushed rock. Concrete is placed on this surface of the vapor barrier on top of base.

Ground slabs are either single or two course floors. Single course are widely used. Two course floors have a subsequent wear resistant topping.

Reinforcement is provided to maintain tightly closed cracks.

Control joints limit crack locations and provide for differential horizontal movement only. Isolation joints allow both horizontal and vertical differential movement.

Use of Table: Determine appropriate type of slab (A, B, C, or D) by considering type of use or amount of abrasive wear of traffic type.

Determine thickness by maximum allowable wheel load or uniform load, opposite 1st column, thickness. Increase the controlling thickness if details require, and select either plain or reinforced slab thickness and type.

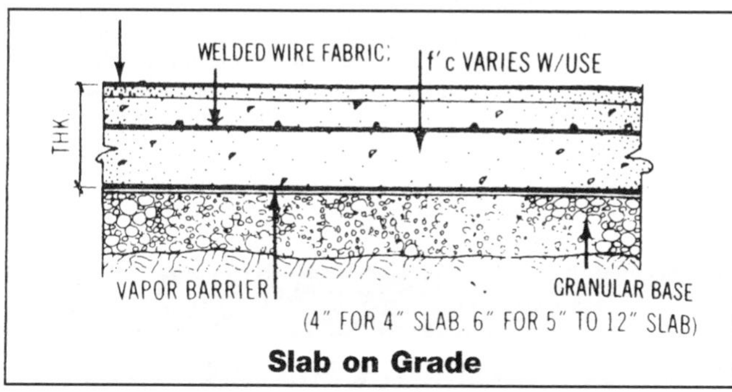

Slab on Grade

Thickness and Loading Assumptions by Type of Use

SLAB THICKNESS (IN.)	TYPE	A	B	C	D	
		A	B	C	D	◄ Slab I.D.
		Non	Light	Normal	Heavy	◄ Industrial
		Little	Light	Moderate	Severe	◄ Abrasion
		Foot Only	Pneumatic Wheels	Solid Rubber Wheels	Steel Tires	◄ Type of Traffic
		Load* (K)	Load* (K)	Load* (K)	Load* (K)	Max. Uniform Load to Slab ▼ (PSF)
4″	Reinf.	4K				100
	Plain					
5″	Reinf.	6K	4K			200
	Plain					
6″	Reinf.		8K	6K	6K	500 to 800
	Plain					
7″	Reinf.			9K	8K	1,500
	Plain					
8″	Reinf.				11K	* Max. Wheel Load in Kips (incl. impact)
	Plain					
10″	Reinf.				14K	
	Plain					
12″	Reinf.					
	Plain					
DESIGN ASSUMPTIONS	Concrete, Chuted	f′c = 3.5 KSI	4 KSI	4.5 KSI	Slab @ 3.5 KSI	ASSUMPTIONS BY SLAB TYPE
	Toppings			1″ Integral	1″ Bonded	
	Finish	Steel Trowel	Steel Trowel	Steel Trowel	Screed & Steel Trowel	
	Compacted Granular Base	4″ deep for 4″ slab thickness 6″ deep for 5″ slab thickness & greater				ASSUMPTIONS FOR ALL SLAB TYPES
	Vapor Barrier	6 mil polyethylene				
	Forms & Joints	Allowances included				
	Rein-forcement	WWF as required ≥ 60,000 psi				

Concrete | R0331 | Structural Concrete

R033105-85 Lift Slabs

The cost advantage of the lift slab method is due to placing all concrete, reinforcing steel, inserts and electrical conduit at ground level and in reduction of formwork. Minimum economical project size is about 30,000 S.F. Slabs may be tilted for parking garage ramps.

It is now used in all types of buildings and has gone up to 22 stories high in apartment buildings. Current trend is to use post-tensioned flat plate slabs with spans from 22′ to 35′. Cylindrical void forms are used when deep slabs are required. One pound of prestressing steel is about equal to seven pounds of conventional reinforcing.

To be considered cured for stressing and lifting, a slab must have attained 75% of design strength. Seven days are usually sufficient with four to five days possible if high early strength cement is used. Slabs can be stacked using two coats of a non-bonding agent to insure that slabs do not stick to each other. Lifting is done by companies specializing in this work. Lift rate is 5′ to 15′ per hour with an average of 10′ per hour. Total areas up to 33,000 S.F. have been lifted at one time. 24 to 36 jacking columns are common. Most economical bay sizes are 24′ to 28′ with four to fourteen stories most efficient. Continuous design reduces reinforcing steel cost. Use of post-tensioned slabs allows larger bay sizes.

Concrete | R0341 | Precast Structural Concrete

R034105-30 Prestressed Precast Concrete Structural Units

Type	Location	Depth	Span in Ft.	Live Load Lb. per S.F.
Double Tee (8′ to 10′)	Floor	28″ to 34″	60 to 80	50 to 80
	Roof	12″ to 24″	30 to 50	40
	Wall	Width 8′	Up to 55′ high	Wind
Multiple Tee (8′)	Roof	8″ to 12″	15 to 40	40
	Floor	8″ to 12″	15 to 30	100
Plank or	Roof or Floor	4″ 6″ 8″ 10″ 12″	Roof / Floor 13 / 12 22 / 18 26 / 25 33 / 29 42 / 32	40 for Roof 100 for Floor
Single Tee (8′ to 10′)	Roof	28″ 32″ 36″ 48″	40 80 100 120	40
AASHO Girder	Bridges	Type 4 5 6	100 110 125	Highway
Box Beam (4′)	Bridges	15″ 27″ 33″	40 to 100	Highway

The majority of precast projects today utilize double tees rather than single tees because of speed and ease of installation. As a result casting beds at manufacturing plants are normally formed for double tees. Single tee projects will therefore require an initial set up charge to be spread over the individual single tee costs.

For floors, a 2″ to 3″ topping is field cast over the shapes. For roofs, insulating concrete or rigid insulation is placed over the shapes.

Member lengths up to 40′ are standard haul, 40′ to 60′ require special permits and lengths over 60′ must be escorted. Over width and/or over length can add up to 100% on hauling costs.

Large heavy members may require two cranes for lifting which would increase erection costs by about 45%. An eight man crew can install 12 to 20 double tees, or 45 to 70 quad tees or planks per day.

Grouting of connections must also be included.

Several system buildings utilizing precast members are available. Heights can go up to 22 stories for apartment buildings. Optimum design ratio is 3 S.F. of surface to 1 S.F. of floor area.

R034136-90 Prestressed Concrete, Post-tensioned

In post-tensioned concrete the steel tendons are tensioned after the concrete has reached about 3/4 of its ultimate strength. The cableways are grouted after tensioning to provide bond between the steel and concrete. If bond is to be prevented, the tendons are coated with a corrosion-preventative grease and wrapped with waterproofed paper or plastic. Bonded tendons are usually used when ultimate strength (beams & girders) are controlling factors.

High strength concrete is used to fully utilize the steel, thereby reducing the size and weight of the member. A plasticizing agent may be added to reduce water content. Maximum size aggregate ranges from 1/2″ to 1-1/2″ depending on the spacing of the tendons.

The types of steel commonly used are bars and strands. Job conditions determine which is best suited. Bars are best for vertical prestresses since they are easy to support. The trend is for steel manufacturers to supply a finished package, cut to length, which reduces field preparation to a minimum.

Bars vary from 3/4″ to 1-3/8″ diameter. Table below gives time in labor-hours per tendon for placing, tensioning and grouting (if required) a 75′ beam. Tendons used in buildings are not usually grouted; tendons for bridges usually are grouted. For strands the table indicates the labor-hours per pound for typical prestressed units 100′ long. Simple span beams usually require one-end stressing regardless of lengths. Continuous beams are usually stressed from two ends. Long slabs are poured from the center outward and stressed in 75′ increments after the initial 150′ center pour.

Labor Hours per Tendon and per Pound of Prestressed Steel

Length	100′ Beam		75′ Beam		100′ Slab	
Type Steel	Strand		Bars		Strand	
Diameter	0.5″		3/4″	1-3/8″	0.5″	0.6″
Number	4	12	1	1	1	1
Force in Kips	100	300	42	143	25	35
Preparation & Placing Cables	3.6	7.4	0.9	2.9	0.9	1.1
Stressing Cables	2.0	2.4	0.8	1.6	0.5	0.5
Grouting, if required	2.5	3.0	0.6	1.3		
Total Labor Hours	8.1	12.8	2.3	5.8	1.4	1.6
Prestressing Steel Weights (Lbs.)	215	640	115	380	53	74
Labor-hours per Lb. Bonded	0.038	0.020	0.020	0.015		
Non-bonded					0.026	0.022

Flat Slab construction — 4000 psi concrete with span-to-depth ratio between 36 and 44. Two way post-tensioned steel averages 1.0 lb. per S.F. for 24′ to 28′ bays (usually strand) and additional reinforcing steel averages .5 lb. per S.F.

Pan and Joist construction — 4000 psi concrete with span-to-depth ratio 28 to 30. Post-tensioned steel averages .8 lb. per S.F. and reinforcing steel about 1.0 lb. per S.F. Placing and stressing averages 40 hours per ton of total material.

Beam construction — 4000 to 5000 psi concrete. Steel weights vary greatly.

Labor cost per pound goes down as the size and length of the tendon increase. The primary economic consideration is the cost per kip for the member.

Post-tensioning becomes feasible for beams and girders over 30′ long; for continuous two-way slabs over 20′ clear; also in transferring upper building loads over longer spans at lower levels. Post-tension suppliers will provide engineering services at no cost to the user. Substantial economies are possible by using post-tensioned Lift Slabs.

Concrete	R0345	Precast Architectural Concrete

R034513-10 Precast Concrete Wall Panels

Panels are either solid or insulated with plain, colored or textured finishes. Transportation is an important cost factor. Prices shown in the unit cost section of the book are based on delivery within 50 miles of a plant including fabricators' overhead and profit. Engineering data is available from fabricators to assist with construction details. Usual minimum job size for economical use of panels is about 5000 S.F. Small jobs can double the prices shown. For large, highly repetitive jobs, deduct up to 15% from the prices shown.

2″ thick panels cost about the same as 3″ thick panels, and maximum panel size is less. For building panels faced with granite, marble or stone, add the material prices from those unit cost sections to the plain panel price shown. There is a growing trend toward aggregate facings and broken rib finish rather than plain gray concrete panels.

No allowance has been made in the unit cost section for supporting steel framework. On one story buildings, panels may rest on grade beams and require only wind bracing and fasteners. On multi-story buildings panels can span from column to column and floor to floor. Plastic-designed steel-framed structures may have large deflections which slow down erection and raise costs.

Large panels are more economical than small panels on a S.F. basis. When figuring areas include all protrusions, returns, etc. Overhangs can triple erection costs. Panels over 45′ have been produced. Larger flat units should be prestressed. Vacuum lifting of smooth finish panels eliminates inserts and can speed erection.

Concrete	R0347	Site-Cast Concrete

R034713-20 Tilt Up Concrete Panels

The advantage of tilt up construction is in the low cost of forms and the placing of concrete and reinforcing. Panels up to 75′ high and 5-1/2″ thick have been tilted using strongbacks. Tilt up has been used for one to five story buildings and is well-suited for warehouses, stores, offices, schools and residences.

The panels are cast in forms on the floor slab. Most jobs use 5-1/2″ thick solid reinforced concrete panels. Sandwich panels with a layer of insulating materials are also used. Where dampness is a factor, lightweight aggregate is used. Optimum panel size is 300 to 500 S.F.

Slabs are usually poured with 3000 psi concrete which permits tilting seven days after pouring. Slabs may be stacked on top of each other and are separated from each other by either two coats of bond breaker or a film of polyethylene. Use of high early-strength cement allows tilting two days after a pour. Tilting up is done with a roller outrigger crane with a capacity of at least 1-1/2 times the weight of the panel at the required reach. Exterior precast columns can be set at the same time as the panels; interior precast columns can be set first and the panels clipped directly to them. The use of cast-in-place concrete columns is diminishing due to shrinkage problems. Structural steel columns are sometimes used if crane rails are planned. Panels can be clipped to the columns or lowered between the flanges. Steel channels with anchors may be used as edge forms for the slab. When the panels are lifted the channels form an integral steel column to take structural loads. Roof loads can be carried directly by the panels for wall heights to 14′.

Requirements of local building codes may be a limiting factor and should be checked. Building floor slabs should be poured first and should be a minimum of 5″ thick with 100% compaction of soil or 6″ thick with less than 100% compaction.

Setting times as fast as nine minutes per panel have been observed, but a safer expectation would be four panels per hour with a crane and a four-man setting crew. If crane erects from inside building, some provision must be made to get crane out after walls are erected. Good yarding procedure is important to minimize delays. Equalizing three-point lifting beams and self-releasing pick-up hooks speed erection. If panels must be carried to their final location, setting time per panel will be increased and erection costs may approach the erection cost range of architectural precast wall panels. Placing panels into slots formed in continuous footers will speed erection.

Reinforcing should be with #5 bars with vertical bars on the bottom. If surface is to be sandblasted, stainless steel chairs should be used to prevent rust staining.

Use of a broom finish is popular since the unavoidable surface blemishes are concealed.

Precast columns run from three to five times the C.Y. price of the panels only.

Concrete	R0352	Lightweight Concrete Roof Insulation

R035216-10 Lightweight Concrete

Lightweight aggregate concrete is usually purchased ready mixed, but it can also be field mixed.

Vermiculite or Perlite comes in bags of 4 C.F. under various trade names. Weight is about 8 lbs. per C.F. For insulating roof fill use 1:6 mix. For structural deck use 1:4 mix over gypsum boards, steeltex, steel centering, etc., supported by closely spaced joists or bulb trees. For structural slabs use 1:3:2 vermiculite sand concrete over steeltex, metal lath, steel centering, etc., on joists spaced 2′-0″ O.C. for maximum L.L. of 80 P.S.F. Use same mix for slab base fill over steel flooring or regular reinforced concrete slab when tile, terrazzo or other finish is to be laid over.

For slabs on grade use 1:3:2 mix when tile, etc., finish is to be laid over. If radiant heating units are installed use a 1:6 mix for a base. After coils are in place, cover with a regular granolithic finish (mix 1:3:2) to a minimum depth of 1-1/2″ over top of units.

Reinforce all slabs with 6 x 6 or 10 x 10 welded wire mesh.

Masonry | R0401 | Maintenance of Masonry

R040130-10 Cleaning Face Brick

On smooth brick a person can clean 70 S.F. an hour; on rough brick 50 S.F. per hour. Use one gallon muriatic acid to 20 gallons of water for 1000 S.F. Do not use acid solution until wall is at least seven days old, but a mild soap solution may be used after two days.

Time has been allowed for clean-up in brick prices.

Masonry | R0405 | Common Work Results for Masonry

R040513-10 Cement Mortar (material only)

Type N - 1:1:6 mix by volume. Use everywhere above grade except as noted below. - 1:3 mix using conventional masonry cement which saves handling two separate bagged materials.

Type M - 1:1/4:3 mix by volume, or 1 part cement, 1/4 (10% by wt.) lime, 3 parts sand. Use for heavy loads and where earthquakes or hurricanes may occur. Also for reinforced brick, sewers, manholes and everywhere below grade.

Mix Proportions by Volume and Compressive Strength of Mortar

Where Used	Allowable Proportions by Volume					Compressive Strength @ 28 days
	Mortar Type	Portland Cement	Masonry Cement	Hydrated Lime	Masonry Sand	
Plain Masonry	M	1	1	—	6	
		1	—	1/4	3	2500 psi
	S	1/2	1	—	4	
		1	—	1/4 to 1/2	4	1800 psi
	N	—	1	—	3	
		1	—	1/2 to 1-1/4	6	750 psi
	O	—	1	—	3	
		1	—	1-1/4 to 2-1/2	9	350 psi
	K	1	—	2-1/2 to 4	12	75 psi
Reinforced Masonry	PM	1	1	—	6	2500 psi
	PL	1	—	1/4 to 1/2	4	2500 psi

Note: The total aggregate should be between 2.25 to 3 times the sum of the cement and lime used.

The labor cost to mix the mortar is included in the productivity and labor cost of unit price lines in unit cost sections for brickwork, blockwork and stonework.

The material cost of mixed mortar is included in the material cost of those same unit price lines and includes the cost of renting and operating a 10 C.F. mixer at the rate of 200 C.F. per day.

There are two types of mortar color used. One type is the inert additive type with about 100 lbs. per M brick as the typical quantity required. These colors are also available in smaller-batch-sized bags (1 lb. to 15 lb.) which can be placed directly into the mixer without measuring. The other type is premixed and replaces the masonry cement. Dark green color has the highest cost.

R040519-50 Masonry Reinforcing

Horizontal joint reinforcing helps prevent wall cracks where wall movement may occur and in many locations is required by code. Horizontal joint reinforcing is generally not considered to be structural reinforcing and an unreinforced wall may still contain joint reinforcing.

Reinforcing strips come in 10′ and 12′ lengths and in truss and ladder shapes, with and without drips. Field labor runs between 2.7 to 5.3 hours per 1000 L.F. for wall thicknesses up to 12″.

The wire meets ASTM A82 for cold drawn steel wire and the typical size is 9 ga. sides and ties with 3/16″ diameter also available. Typical finish is mill galvanized with zinc coating at .10 oz. per S.F. Class I (.40 oz. per S.F.) and Class III (.80 oz per S.F.) are also available, as is hot dipped galvanizing at 1.50 oz. per S.F.

R042110-10 Economy in Bricklaying

Have adequate supervision. Be sure bricklayers are always supplied with materials so there is no waiting. Place best bricklayers at corners and openings.

Use only screened sand for mortar. Otherwise, labor time will be wasted picking out pebbles. Use seamless metal tubs for mortar as they do not leak or catch the trowel. Locate stack and mortar for easy wheeling.

Have brick delivered for stacking. This makes for faster handling, reduces chipping and breakage, and requires less storage space. Many dealers will deliver select common in 2′ x 3′ x 4′ pallets or face brick packaged. This affords quick handling with a crane or forklift and easy tonging in units of ten, which reduces waste.

Use wider bricks for one wythe wall construction. Keep scaffolding away from wall to allow mortar to fall clear and not stain wall.

On large jobs develop specialized crews for each type of masonry unit.

Consider designing for prefabricated panel construction on high rise projects.

Avoid excessive corners or openings. Each opening adds about 50% to labor cost for area of opening.

Bolting stone panels and using window frames as stops reduces labor costs and speeds up erection.

R042110-20 Common and Face Brick

Common building brick manufactured according to ASTM C62 and facing brick manufactured according to ASTM C216 are the two standard bricks available for general building use.

Building brick is made in three grades; SW, where high resistance to damage caused by cyclic freezing is required; MW, where moderate resistance to cyclic freezing is needed; and NW, where little resistance to cyclic freezing is needed. Facing brick is made in only the two grades SW and MW. Additionally, facing brick is available in three types; FBS, for general use; FBX, for general use where a higher degree of precision and lower permissible variation in size than FBS is needed; and FBA, for general use to produce characteristic architectural effects resulting from non-uniformity in size and texture of the units.

In figuring the material cost of brickwork, an allowance of 25% mortar waste and 3% brick breakage was included. If bricks are delivered palletized with 280 to 300 per pallet, or packaged, allow only 1-1/2% for breakage. Packaged or palletized delivery is practical when a job is big enough to have a crane or other equipment available to handle a package of brick. This is so on all industrial work but not always true on small commercial buildings.

The use of buff and gray face is increasing, and there is a continuing trend to the Norman, Roman, Jumbo and SCR brick.

Common red clay brick for backup is not used that often. Concrete block is the most usual backup material with occasional use of sand lime or cement brick. Building brick is commonly used in solid walls for strength and as a fire stop.

Brick panels built on the ground and then crane erected to the upper floors have proven to be economical. This allows the work to be done under cover and without scaffolding.

R042110-50 Brick, Block & Mortar Quantities

Running Bond					
Number of Brick per S.F. of Wall - Single Wythe with 3/8″ Joints				C.F. of Mortar per M Bricks, Waste Included	
Type Brick	Nominal Size (incl. mortar) L H W	Modular Coursing	Number of Brick per S.F.	3/8″ Joint	1/2″ Joint
Standard	8 x 2-2/3 x 4	3C=8″	6.75	10.3	12.9
Economy	8 x 4 x 4	1C=4″	4.50	11.4	14.6
Engineer	8 x 3-1/5 x 4	5C=16″	5.63	10.6	13.6
Fire	9 x 2-1/2 x 4-1/2	2C=5″	6.40	550 # Fireclay	—
Jumbo	12 x 4 x 6 or 8	1C=4″	3.00	23.8	30.8
Norman	12 x 2-2/3 x 4	3C=8″	4.50	14.0	17.9
Norwegian	12 x 3-1/5 x 4	5C=16″	3.75	14.6	18.6
Roman	12 x 2 x 4	2C=4″	6.00	13.4	17.0
SCR	12 x 2-2/3 x 6	3C=8″	4.50	21.8	28.0
Utility	12 x 4 x 4	1C=4″	3.00	15.4	19.6

For Other Bonds Standard Size Add to S.F. Quantities in Table to Left		
Bond Type	Description	Factor
Common	full header every fifth course	+20%
	full header every sixth course	+16.7%
English	full header every second course	+50%
Flemish	alternate headers every course	+33.3%
	every sixth course	+5.6%
Header = W x H exposed		+100%
Rowlock = H x W exposed		+100%
Rowlock stretcher = L x W exposed		+33.3%
Soldier = H x L exposed		—
Sailor = W x L exposed		-33.3%

Concrete Blocks Nominal Size	Approximate Weight per S.F.		Blocks per 100 S.F.	Mortar per M block, waste included	
	Standard	Lightweight		Partitions	Back up
2″ x 8″ x 16″	20 PSF	15 PSF	113	27 C.F.	36 C.F.
4″	30	20		41	51
6″	42	30		56	66
8″	55	38		72	82
10″	70	47		87	97
12″	85	55		102	112

R042210-20 Concrete Block

The material cost of special block such as corner, jamb and head block can be figured at the same price as ordinary block of same size. Labor on specials is about the same as equal-sized regular block.

Bond beam and 16″ high lintel blocks are more expensive than regular units of equal size. Lintel blocks are 8″ long and either 8″ or 16″ high.

Use of motorized mortar spreader box will speed construction of continuous walls.

Hollow non-load-bearing units are made according to ASTM C129 and hollow load-bearing units according to ASTM C90.

R042210-30 Fully Grouted Reinforced Masonry Wall Capacities Per L.F. (Kips & In-Kips)

Earthquake Zones 1, 2 & 3					Allowable Vertical Wall Loads									Allowable Wall Moments Without Vertical Wall Loads	
Thk.	Length Or Height		Type Wall	Rebar	Eccentric Loads				Without Wind or Eccentric Loads		With Wind			Not Wind or Earthquake	
					7.0 in-K/Ft.		3.5 in-K/Ft.				Inspection				
T (Nom.) (in)	h′ (Ft.)	h′/T (in/in)	Brick Grout Conc. M.U.	Size and Spacing (in. O.C.)	Inspection No (K/Ft.)	Inspection Yes (K/Ft.)	Inspection No (K/Ft.)	Inspection Yes (K/Ft.)	Inspection No (K/Ft.)	Inspection Yes (K/Ft.)	No 15 psf (K/Ft.)	No 30 psf (K/Ft.)	Yes 15 & 30 (K/Ft.)	Inspection No (in.-K/Ft.)	Inspection Yes (in.-K/Ft.)
8″	8′	12	4″0″4″	#5@32	10.10	23.35	11.70	24.50	13.30	26.55	13.30	13.30	26.55	6.90	9.65
			Solid CMU		12.75	28.70	14.35	30.30	15.95	31.90	15.95	15.95	31.90	7.80	9.75
	12′	18	4″0″4″		9.30	21.60	10.80	23.05	12.25	24.55	12.25	12.25	24.55	6.90	9.65
			Solid CMU		16.65	26.50	18.15	28.00	14.70	29.45	14.70	14.70	29.45	7.80	9.75
	16′	24	4″0″4″		7.80	18.10	9.05	19.35	10.30	20.60	10.30	9.65	20.60	6.90	9.65
			Solid CMU	↓	9.85	22.20	11.10	23.45	12.35	24.70	12.35	12.35	12.70	7.80	9.75
10″	8′	9.6	4″0″6″	#5@24	14.45	31.50	15.75	32.80	17.05	34.10	17.05	17.05	34.10	11.20	16.20
			Solid CMU		17.85	38.35	19.15	39.65	20.45	40.90	20.45	20.45	40.90	12.65	16.35
			4″2″4″		14.45	31.50	15.75	32.80	17.05	34.10	17.05	17.05	34.10	11.20	16.20
	12′	14.4	4″0″6″		13.90	30.35	15.15	31.60	16.40	32.80	16.40	16.40	32.80	11.20	16.20
			Solid CMU		17.20	36.90	18.45	38.15	19.70	39.40	19.70	19.70	39.40	12.65	16.35
			4″2″4″		13.90	30.35	15.15	31.60	16.40	32.80	16.40	16.40	32.80	11.20	16.20
	16′	19.2	4″0″6″		12.85	28.05	14.00	29.20	15.15	30.35	15.15	15.15	30.35	11.20	16.20
			Solid CMU		15.90	34.10	17.05	35.25	18.20	36.40	13.20	13.20	36.40	12.65	16.35
			4″2″4″		12.85	28.05	14.00	29.20	15.15	30.35	15.15	15.15	30.35	11.20	16.20
	20′	24	4″0″6″		11.15	24.25	12.10	25.25	13.10	26.20	13.10	12.40	26.20	11.20	16.20
			Solid CMU		13.75	29.50	14.75	30.50	15.75	31.50	15.75	15.75	31.50	12.65	16.35
			4″2″4″	↓	11.15	24.25	12.10	25.25	13.10	26.20	13.10	12.40	26.20	11.20	16.20
12″	8′	8	4″0″8″	#8@48	18.60	39.35	19.65	40.40	20.75	41.50	20.75	20.75	41.50	16.65	24.90
			Solid CMU		22.75	47.65	23.80	48.70	24.90	49.80	24.90	24.90	49.80	18.80	25.10
			4″2″6″	#5@20	18.60	39.35	19.65	40.40	20.75	41.50	20.75	20.75	41.50	11.90	18.75
	12′	12	4″0″8″	#8@48	18.20	38.50	19.25	39.55	20.30	40.60	20.30	20.30	40.60	16.65	24.90
			Solid CMU		22.25	46.65	23.30	47.70	24.35	48.75	24.35	24.35	48.75	18.80	25.10
			4″2″6″	#5@20	18.20	38.50	19.25	39.55	20.30	40.60	20.30	20.30	40.60	11.90	18.75
	16′	16	4″0″8″	#8@48	17.45	36.90	18.45	37.90	19.45	38.90	19.45	19.45	38.90	16.65	24.90
			Solid CMU		21.30	44.70	22.35	45.70	23.35	46.70	23.25	23.35	46.70	18.80	25.10
			4″2″6″	#5@20	17.45	36.90	18.45	37.90	19.45	38.90	19.45	19.45	38.90	11.90	18.75
	24′	24	4″0″8″	#8@48	14.30	30.25	15.10	31.10	15.95	31.90	15.95	15.15	31.90	16.65	24.90
			Solid CMU		17.15	36.65	18.30	37.45	19.15	38.30	19.15	19.15	38.30	18.80	25.10
			4″2″6″	#5@20	14.30	30.25	15.10	31.10	15.95	31.90	15.95	15.15	31.90	11.90	18.75
16″	8′	6	4″0″12″	#8@32	26.40	54.40	27.20	55.20	28.00	56.05	28.00	28.00	56.05	31.20	49.10
			Solid CMU		32.00	65.65	32.80	66.45	33.60	67.25	33.60	33.60	67.25	35.25	50.40
			4″2″10″	#5@15	26.40	54.40	27.20	55.20	28.00	56.05	28.00	28.00	56.05	14.75	24.10
	12′	9	4″0″12″	#8@32	26.15	53.95	26.95	54.75	27.75	55.55	27.75	27.75	55.55	31.20	49.10
			Solid CMU		31.70	66.05	32.50	65.85	33.30	66.65	33.00	33.30	66.65	35.25	50.40
			4″2″10″	#5@15	26.15	53.95	26.95	54.75	27.75	55.55	27.75	27.75	55.55	14.75	24.10
	16′	12	4″0″12″	#8@32	25.70	53.05	26.50	53.80	28.30	54.60	27.30	27.30	54.60	31.20	49.10
			Solid CMU		31.20	63.95	31.95	64.75	32.75	65.50	32.75	32.75	65.50	35.25	50.40
			4″2″10″	↓	25.70	53.05	26.50	53.80	27.30	54.60	27.30	27.30	59.60	14.75	24.10
	32′	24	4″0″12″	#8@32	20.35	41.95	20.95	42.55	21.60	43.20	18.75*	18.75*	43.20	31.20	49.10
			Solid CMU		24.65	50.60	25.30	51.20	25.90	51.85	24.55*	24.55*	51.85	35.25	50.40
			4″2″10″	#5@15	20.35	41.95	20.95	42.55	21.60	43.20	18.75*	18.75*	43.20	14.75	24.10

*Zone 3 Only

R042210-30 Unreinforced Masonry Wall Capacities Per L.F. (Kips & In-Kips)

Thk. T (Nom.) (in)	Earthquake Zones 0 & 1 Only			Allowable Vertical Wall Loads					Allowable Wall Moments (Without Vertical Wall Loads)			
	Length Or Height			Eccentric Loads		Without Wind or Eccentric Loads	With Wind		Not Wind or Earthquake		Wind or Earthquake	
									Inspection		Inspection	
	h′ (Ft.)	h′/t (in/in)	Type of Wall	7.0 in-K/Ft. (K/Ft.)	3.5 in-K/Ft. (K/Ft.)	(K/Ft.)	15 psf (K/Ft.)	30 psf (K/Ft.)	No (in-K/Ft.)	Yes (in-K/Ft.)	No (in-K/Ft.)	Yes (in-K/Ft.)
8″	8′	12	Solid Brick	4.85	7.50	10.15	10.15	10.15	1.15	2.30	1.55	3.10
			Solid CM Units	8.85	11.50	14.15	14.15	14.15	.70	1.40	.95	1.85
			Hollow CM Units	7.95	10.60	13.30	13.30	13.30	.70	1.40	.60	1.25
	10′	15	Solid Brick	4.70	7.30	9.85	9.85	9.85	1.15	2.30	1.55	3.10
			Solid CM Units	8.55	11.15	13.75	13.75	13.75	.70	1.40	.90	1.85
			Hollow CM Units	—	4.45	6.30	6.30	6.05	.45	.95	.60	1.25
	12′	18	Solid Brick	—	6.95	9.40	9.40	7.95	1.15	2.30	1.55	3.10
			Solid CM Units	8.15	10.60	13.10	13.10	12.90	.70	1.40	.90	1.85
			Hollow CM Units	—	4.25	6.00	6.00	4.75	.45	.95	.60	1.25
12″	8′	8	Solid Brick	12.30	14.10	15.90	15.90	15.90	2.70	5.40	3.60	7.20
			Solid CM Units	18.50	20.30	22.10	22.10	22.10	1.60	3.20	2.15	4.30
			Hollow CM Units	7.85	9.10	10.40	10.40	10.40	1.10	2.25	1.50	3.00
			Brick & Hollow CMU	5.50	7.80	10.05	10.05	10.05	1.35	2.70	1.80	3.60
	12′	12	Solid Brick	12.05	13.80	15.55	15.55	15.55	2.70	5.40	3.60	7.20
			Solid CM Units	13.45	14.75	16.05	16.05	16.05	1.60	3.20	2.15	4.30
			Hollow CM Units	7.65	8.90	10.15	10.15	10.15	1.10	2.25	1.50	3.00
			Brick & Hollow CMU	5.40	7.65	9.85	9.85	9.05	1.35	2.70	1.80	3.60
	16′	16	Solid Brick	11.55	13.20	14.90	14.90	14.35	2.70	5.40	3.60	7.20
			Solid CM Units	17.35	19.05	20.70	20.70	20.70	1.60	3.20	2.15	4.30
			Hollow CM Units	7.35	8.55	9.75	9.75	9.05	1.10	2.25	1.50	3.00
			Brick & Hollow CMU	5.20	7.30	9.45	9.10	—	1.35	2.70	1.80	3.60
16″	12′	9	Solid Brick	18.60	19.95	21.30	21.30	21.30	4.85	9.75	6.50	13.00
			Solid CM Units	26.95	28.30	29.60	29.60	29.60	2.90	5.85	3.90	7.80
			Hollow CM Units	10.85	12.10	13.30	13.30	13.30	1.50	3.05	2.00	4.05
			Brick & Hollow CMU	9.30	11.10	12.85	12.85	12.85	2.20	4.40	2.90	5.85
	16′	12	Solid Brick	18.30	19.60	20.90	20.90	20.90	4.85	9.75	6.50	13.00
			Solid CM Units	26.50	27.80	29.10	29.10	29.10	2.90	5.85	3.90	7.80
			Hollow CM Units	10.70	11.90	13.10	13.10	13.10	1.50	3.05	2.00	4.05
			Brick & Hollow CMU	9.15	10.90	12.65	12.65	11.10	2.20	4.40	2.90	5.85
	20′	15	Solid Brick	17.80	19.05	20.30	20.30	20.30	4.85	9.75	6.50	13.00
			Solid CM Units	25.75	27.00	28.25	28.25	28.25	2.90	5.85	3.90	7.80
			Hollow CM Units	10.40	11.55	12.70	12.70	10.95	1.50	3.05	2.00	4.05
			Brick & Hollow CMU	8.90	10.60	12.30	12.00	—	2.20	4.40	2.90	5.85
20″	12′	7.2	Solid Brick	24.80	25.85	26.90	26.90	26.90	7.70	15.40	10.25	20.50
			Solid CM Units	35.30	36.40	37.45	37.45	37.45	4.60	9.20	6.15	12.30
			Hollow CM Units	16.15	17.15	18.15	18.15	18.15	2.50	5.05	3.35	6.75
			Brick & Hollow CMU	16.10	17.25	18.45	18.45	18.45	4.05	8.10	5.40	10.80
	16′	9.6	Solid Brick	24.55	25.60	26.70	26.70	26.70	7.70	15.40	10.25	20.50
			Solid CM Units	35.00	36.10	37.15	37.15	37.15	4.60	9.20	6.15	12.30
			Hollow CM Units	16.00	17.00	18.00	18.00	18.00	2.50	5.05	3.35	6.75
			Brick & Hollow CMU	15.95	17.10	18.30	18.30	18.30	4.05	8.10	5.40	10.80
	24′	14.4	Solid Brick	23.70	24.70	25.75	25.75	25.75	7.70	15.40	10.25	20.50
			Solid CM Units	33.80	34.80	35.80	35.80	35.80	4.60	9.20	6.15	12.30
			Hollow CM Units	15.45	16.40	17.40	17.40	16.05	2.50	5.05	3.35	6.75
			Brick & Hollow CMU	15.40	16.50	17.65	17.65	15.25	4.05	8.10	5.40	10.80

R050516-30 Coating Structural Steel

On field-welded jobs, the shop-applied primer coat is necessarily omitted. All painting must be done in the field and usually consists of red oxide rust inhibitive paint or an aluminum paint. The table below shows paint coverage and daily production for field painting.

See Division 05950-650 for hot-dipped galvanizing and for field-applied cold galvanizing and other paints and protective coatings.

See Division 05910-500 for steel surface preparation treatments such as wire brushing, pressure washing and sand blasting.

Type Construction	Surface Area per Ton	Coat	One Gallon Covers		In 8 Hrs. Person Covers		Average per Ton Spray	
			Brush	Spray	Brush	Spray	Gallons	Labor-hours
Light Structural	300 S.F. to 500 S.F.	1st	500 S.F.	455 S.F.	640 S.F.	2000 S.F.	0.9 gals.	1.6 L.H.
		2nd	450	410	800	2400	1.0	1.3
		3rd	450	410	960	3200	1.0	1.0
Medium	150 S.F. to 300 S.F.	All	400	365	1600	3200	0.6	0.6
Heavy Structural	50 S.F. to 150 S.F.	1st	400	365	1920	4000	0.2	0.2
		2nd	400	365	2000	4000	0.2	0.2
		3rd	400	365	2000	4000	0.2	0.2
Weighted Average	225 S.F.	All	400	365	1350	3000	0.6	0.6

R050521-20 Welded Structural Steel

Usual weight reductions with welded design run 10% to 20% compared with bolted or riveted connections. This amounts to about the same total cost compared with bolted structures since field welding is more expensive than bolts. For normal spans of 18′ to 24′ figure 6 to 7 connections per ton.

Trusses — For welded trusses add 4% to weight of main members for connections. Up to 15% less steel can be expected in a welded truss compared to one that is shop bolted. Cost of erection is the same whether shop bolted or welded.

General — Typical electrodes for structural steel welding are E6010, E6011, E60T and E70T. Typical buildings vary between 2# to 8# of weld rod per ton of steel. Buildings utilizing continuous design require about three times as much welding as conventional welded structures. In estimating field erection by welding, it is best to use the average linear feet of weld per ton to arrive at the welding cost per ton. The type, size and position of the weld will have a direct bearing on the cost per linear foot. A typical field welder will deposit 1.8# to 2# of weld rod per hour manually. Using semiautomatic methods can increase production by as much as 50% to 75%.

R050523-10 High Strength Bolts

Common bolts (A307) are usually used in secondary connections (see Division 05090-150).

High strength bolts (A325 and A490) are usually specified for primary connections such as column splices, beam and girder connections to columns, column bracing, connections for supports of operating equipment or of other live loads which produce impact or reversal of stress, and in structures carrying cranes of over 5-ton capacity.

Allow 20 field bolts per ton of steel for a 6 story office building, apartment house or light industrial building. For 6 to 12 stories allow 25 bolts per ton, and above 12 stories, 30 bolts per ton. On power stations, 20 to 25 bolts per ton are needed.

R051223-10 Structural Steel

The bare material prices for structural steel, shown in the unit cost sections of the book, are for 100 tons of shop-fabricated structural steel and include:

1. Mill base price of structural steel
2. Mill scrap/grade/size/length extras
3. Mill delivery to a metals service center (warehouse)
4. Service center storage and handling
5. Service center delivery to a fabrication shop
6. Shop storage and handling
7. Shop drafting/detailing
8. Shop fabrication
9. Shop coat of primer paint
10. Shop listing
11. Shop delivery to the job site

In unit cost sections of the book that contain items for field fabrication of steel components, the bare material cost of steel includes:

1. Mill base price of structural steel
2. Mill scrap/grade/size/length extras
3. Mill delivery to a metals service center (warehouse)
4. Service center storage and handling
5. Service center delivery to the job site

R051223-15 Structural Steel Estimating for Repair and Remodeling Projects

The correct approach to estimating structural steel is dependent upon the amount of steel required for the particular project. If the project is a sizable addition to a building, the data can be used directly from the cost data book. This is not the case however if the project requires a small amount of steel, for instance to reinforce existing roof or floor structural systems. To better understand this, please refer to the unit price line for a W16x31 beam with bolted connections. Assume your project requires the reinforcement of the structural members of the roof system of a 3 story building required for the installation a new piece of HVAC equipment. The project will need 4 pieces of W16x31 that are 30′ long. After pricing this 120 L.F. job using the unit prices for a W16x31, an analysis will reveal that the price is wholly inadequate.

The first problem is apparent if you examine the amount of steel that can be installed per day. The unit price line indicates 900 linear feet per day can be installed by a 5 man crew with an 90 ton crane. This productivity is correct for new construction but certainly is not for repair and remodeling work. Installation of new structural steel considers that each member is installed from the foundation to the roof of the structure in a planned and systematic manner with a crane having unrestricted access to all parts of the project. Additionally each connection is planned and detailed with full field access for fit-up and final bolting. The erection is planned and progresses such that interferences and conflicts with other structural members are minimized, if not completely eliminated. All of these assumptions are clearly not the case with a repair and remodeling job, and a significant decrease in the stated productivity will be observed.

A crane will certainly be needed to lift the members into the general area of the project but in most cases will not be able to place the beams into their final position. An opening in the existing roof may not be large enough to permit the beams to pass through and it may be necessary to bring them into the building through existing windows or doors. Moving the beams to the actual area where they will be installed may involve hand labor and the use of dollies. Finally, hoists and/or jacks may be needed for final positioning.

The connection of new members to existing can often be accomplished by field bolting with accurate field measurements and good planning but in many cases access to both sides of existing members is not possible and field welding becomes the only alternative. In addition to the cost of the actual welding, protection of existing finishes, systems and structure and fire protection must be considered.

New beams can never be installed tight to the existing decks or floors which they must support and the use of shims and tack welding becomes necessary. Additionally, further planning and cost is involved in assuring that existing loads are minimized during the installation and shimming process.

It is apparent that installation of structural steel as part of a repair and remodeling project involves more than simply installing the members and estimating the cost in the same manner as new construction. The best procedure for estimating the total cost is adequate planning and coordination of each process and activity that will be needed. Unit costs for the materials, labor and equipment can then be attached to each needed activity and a final, complete price can be determined.

R051223-20 Steel Estimating Quantities

One estimate on erection is that a crane can handle 35 to 60 pieces per day. Say the average is 45. With usual sizes of beams, girders, and columns, this would amount to about 20 tons per day. The type of connection greatly affects the speed of erection. Moment connections for continuous design slow down production and increase erection costs.

Short open web bar joists can be set at the rate of 75 to 80 per day, with 50 per day being the average for setting long span joists.

After main members are calculated, add the following for usual allowances: base plates 2% to 3%; column splices 4% to 5%; and miscellaneous details 4% to 5%, for a total of 10% to 13% in addition to main members.

The ratio of column to beam tonnage varies depending on type of steels used, typical spans, story heights and live loads.

It is more economical to keep the column size constant and to vary the strength of the column by using high strength steels. This also saves floor space. Buildings have recently gone as high as ten stories with 8″ high strength columns. For light columns under W8X31 lb. sections, concrete filled steel columns are economical.

High strength steels may be used in columns and beams to save floor space and to meet head room requirements. High strength steels in some sizes sometimes require long lead times.

Round, square and rectangular columns, both plain and concrete filled, are readily available and save floor area, but are higher in cost per pound than rolled columns. For high unbraced columns, tube columns may be less expensive.

Below are average minimum figures for the weights of the structural steel frame for different types of buildings using A36 steel, rolled shapes and simple joints. For economy in domes, rise to span ratio = .13. Open web joist framing systems will reduce weights by 10% to 40%. Composite design can reduce steel weight by up to 25% but additional concrete floor slab thickness may be required. Continuous design can reduce the weights up to 20%. There are many building codes with different live load requirements and different structural requirements, such as hurricane and earthquake loadings which can alter the figures.

Structural Steel Weights per S.F. of Floor Area

Type of Building	No. of Stories	Avg. Spans	L.L. #/S.F.	Lbs. Per S.F.	Type of Building	No. of Stories	Avg. Spans	L.L. #/S.F.	Lbs. Per S.F.
		20′x20′		8	Apartments	2-8	20′x20′	40	8
Steel Frame Mfg.	1	30′x30′	40	13		9-25			14
		40′x40′		18		to 10			10
Parking garage	4	Various	80	8.5	Office	20	Various	80	18
Domes	1	200′	30	10		30			26
(Schwedler)*		300′		15		over 50			35

R051223-25 Common Structural Steel Specifications

ASTM A992 (formerly A36, then A572 Grade 50) is the all-purpose carbon grade steel widely used in building and bridge construction.

The other high-strength steels listed below may each have certain advantages over ASTM A992 stuctural carbon steel, depending on the application. They have proven to be economical choices where, due to lighter members, the reduction of dead load and the associated savings in shipping cost can be significant.

ASTM A588 atmospheric weathering, high-strength low-alloy steels can be used in the bare (uncoated) condition, where exposure to normal atmosphere causes a tightly adherant oxide to form on the surface protecting the steel from further oxidation. ASTM A242 corrosion-resistant, high-strength low-alloy steels have enhanced atmospheric corrosion resistance of at least two times that of carbon structural steels with copper, or four times that of carbon structural steels without copper. The reduction or elimination of maintenance resulting from the use of these steels often offsets their higher initial cost.

Steel Type	ASTM Designation	Minimum Yield Stress in KSI	Shapes Available
Carbon	A36	36	All structural shape groups, and plates & bars up thru 8″ thick
	A529	42	Structural shape group 1, and plates & bars up thru 1/2″ thick
High-Strength Low-Alloy Manganese-Vanadium	A441	40	Plates & bars over 4″ up thru 8″ thick
		42	Structural shape groups 4 & 5, and plates & bars over 1-1/2″ up thru 4″ thick
		46	Structural shape group 3, and plates & bars over 3/4″ up thru 1-1/2″ thick
		50	Structural shape groups 1 & 2, and plates & bars up thru 3/4″ thick
High-Strength Low-Alloy Columbium-Vanadium	A572	42	All structural shape groups, and plates & bars up thru 6″ thick
		50	All structural shape groups, and plates & bars up thru 4″ thick
		60	Structural shape groups 1 & 2, and plates & bars up thru 1-1/4″ thick
		65	Structural shape group 1, and plates & bars up thru 1-1/4″ thick
High-Strength Low-Alloy Columbium-Vanadium	A992	50	All structural shape groups
Corrosion-Resistant High-Strength Low-Alloy	A242	42	Structural shape groups 4 & 5, and plates & bars over 1-1/2″ up thru 4″ thick
		46	Structural shape group 3, and plates & bars over 3/4″ up thru 1-1/2″ thick
		50	Structural shape groups 1 & 2, and plates & bars up thru 3/4″ thick
Weathering High-Strength Low-Alloy	A588	42	Plates & bars over 5″ up thru 8″ thick
		46	Plates & bars over 4″ up thru 5″ thick
		50	All structural shape groups, and plates & bars up thru 4″ thick
Quenched and Tempered Low-Alloy	A852	70	Plates & bars up thru 4″ thick
Quenched and Tempered Alloy	A514	90	Plates & bars over 2-1/2″ up thru 6″ thick
		100	Plates & bars up thru 2-1/2″ thick

R051223-30 High Strength Steels

The mill price of high strength steels may be higher than A992 carbon steel but their proper use can achieve overall savings thru total reduced weights. For columns with L/r over 100, A992 steel is best; under 100, high strength steels are economical. For heavy columns, high strength steels are economical when cover plates are eliminated. There is no economy using high strength steels for clip angles or supports or for beams where deflection governs. Thinner members are more economical than thick.

The per ton erection and fabricating costs of the high strength steels will be higher than for A992 since the same number of pieces, but less weight, will be installed.

R051223-35 Common Steel Sections

The upper portion of this table shows the name, shape, common designation and basic characteristics of commonly used steel sections. The lower portion explains how to read the designations used for the above illustrated common sections.

Shape & Designation	Name & Characteristics	Shape & Designation	Name & Characteristics
W	W Shape Parallel flange surfaces	MC	Miscellaneous Channel Infrequently rolled by some producers
S	American Standard Beam (I Beam) Sloped inner flange	L	Angle Equal or unequal legs, constant thickness
M	Miscellaneous Beams Cannot be classified as W, HP or S; infrequently rolled by some producers	T	Structural Tee Cut from W, M or S on center of web
C	American Standard Channel Sloped inner flange	HP	Bearing Pile Parallel flanges and equal flange and web thickness

Common drawing designations follow:

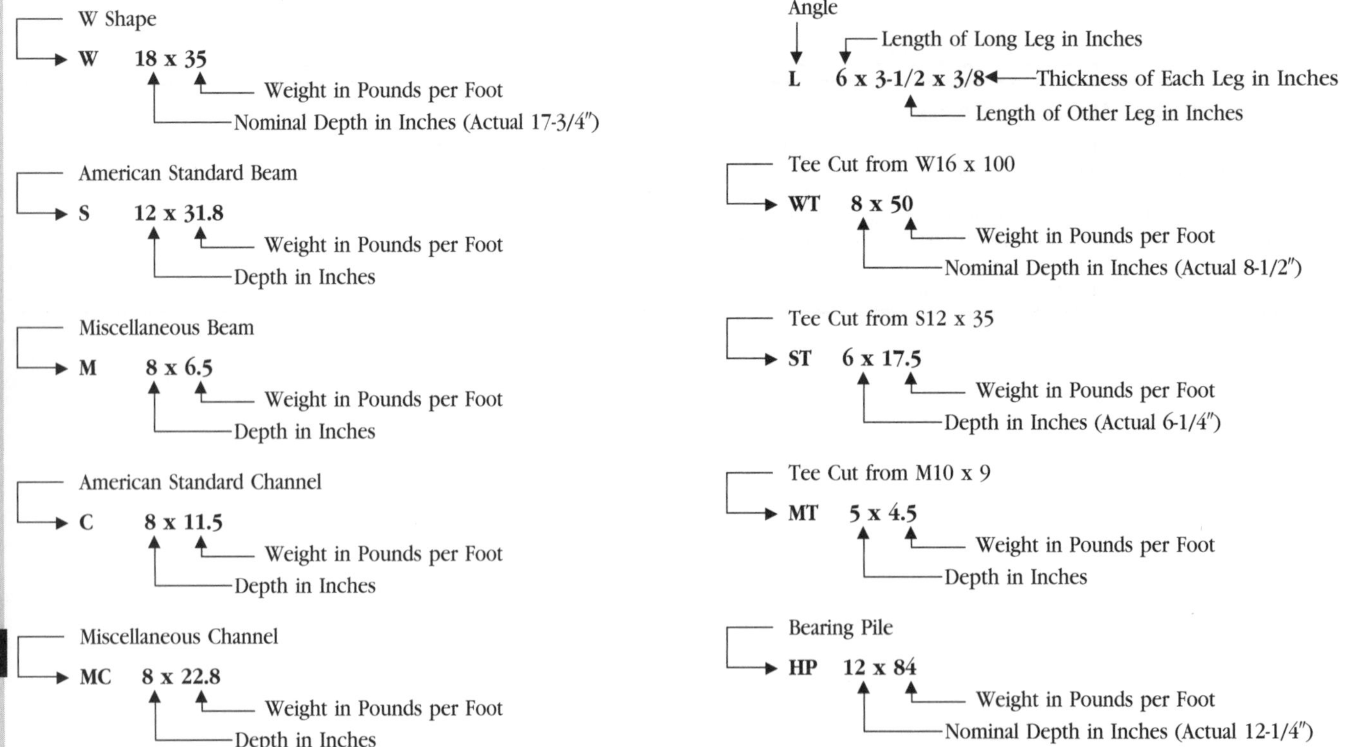

R051223-45 Installation Time for Structural Steel Building Components

The following tables show the expected average installation times for various structural steel shapes. Table A presents installation times for columns, Table B for beams, Table C for light framing and bolts, and Table D for structural steel for various project types.

Table A

Description	Labor-Hours	Unit
Columns		
Steel, Concrete Filled		
3-1/2″ Diameter	.933	Ea.
6-5/8″ Diameter	1.120	Ea.
Steel Pipe		
3″ Diameter	.933	Ea.
8″ Diameter	1.120	Ea.
12″ Diameter	1.244	Ea.
Structural Tubing		
4″ x 4″	.966	Ea.
8″ x 8″	1.120	Ea.
12″ x 8″	1.167	Ea.
W Shape 2 Tier		
W8 x 31	.052	L.F.
W8 x 67	.057	L.F.
W10 x 45	.054	L.F.
W10 x 112	.058	L.F.
W12 x 50	.054	L.F.
W12 x 190	.061	L.F.
W14 x 74	.057	L.F.
W14 x 176	.061	L.F.

Table B

Description	Labor-Hours	Unit	Labor-Hours	Unit
Beams, W Shape				
W6 x 9	.949	Ea.	.093	L.F.
W10 x 22	1.037	Ea.	.085	L.F.
W12 x 26	1.037	Ea.	.064	L.F.
W14 x 34	1.333	Ea.	.069	L.F.
W16 x 31	1.333	Ea.	.062	L.F.
W18 x 50	2.162	Ea.	.088	L.F.
W21 x 62	2.222	Ea.	.077	L.F.
W24 x 76	2.353	Ea.	.072	L.F.
W27 x 94	2.581	Ea.	.067	L.F.
W30 x 108	2.857	Ea.	.067	L.F.
W33 x 130	3.200	Ea.	.071	L.F.
W36 x 300	3.810	Ea.	.077	L.F.

Table C

Description	Labor-Hours	Unit
Light Framing		
Angles 4″ and Larger	.055	lbs.
Less than 4″	.091	lbs.
Channels 8″ and Larger	.048	lbs.
Less than 8″	.072	lbs.
Cross Bracing Angles	.055	lbs.
Rods	.034	lbs.
Hanging Lintels	.069	lbs.
High Strength Bolts in Place		
3/4″ Bolts	.070	Ea.
7/8″ Bolts	.076	Ea.

Table D

Description	Labor-Hours	Unit	Labor-Hours	Unit
Apartments, Nursing Homes, etc.				
1-2 Stories	4.211	Piece	7.767	Ton
3-6 Stories	4.444	Piece	7.921	Ton
7-15 Stories	4.923	Piece	9.014	Ton
Over 15 Stories	5.333	Piece	9.209	Ton
Offices, Hospitals, etc.				
1-2 Stories	4.211	Piece	7.767	Ton
3-6 Stories	4.741	Piece	8.889	Ton
7-15 Stories	4.923	Piece	9.014	Ton
Over 15 Stories	5.120	Piece	9.209	Ton
Industrial Buildings				
1 Story	3.478	Piece	6.202	Ton

R051223-50 Subpurlins

Bulb tee subpurlins are structural members designed to support and reinforce a variety of roof deck systems such as precast cement fiber roof deck tiles, monolithic roof deck systems, and gypsum or lightweight concrete over formboard. Other uses include interstitial service ceiling systems, wall panel systems, and joist anchoring in bond beams. See Unit Price section for pricing on a square foot basis at 32-5/8″ O.C. Maximum span is based on a 3-span condition with a total allowable vertical load of 40 psf.

R051223-80 Dimensions and Weights of Sheet Steel

Gauge No.	Approximate Thickness: Inches (in fractions) Wrought Iron	Approximate Thickness: Inches (in decimal parts) Wrought Iron	Approximate Thickness: Inches (in decimal parts) Steel	Approximate Thickness: Millimeters Steel	Weight: per S.F. in Ounces	Weight: per S.F. in Lbs.	Weight: per Square Meter in Kg.
0000000	1/2″	.5	.4782	12.146	320	20.000	97.650
000000	15/32″	.46875	.4484	11.389	300	18.750	91.550
00000	7/16″	.4375	.4185	10.630	280	17.500	85.440
0000	13/32″	.40625	.3886	9.870	260	16.250	79.330
000	3/8″	.375	.3587	9.111	240	15.000	73.240
00	11/32″	.34375	.3288	8.352	220	13.750	67.130
0	5/16″	.3125	.2989	7.592	200	12.500	61.030
1	9/32″	.28125	.2690	6.833	180	11.250	54.930
2	17/64″	.265625	.2541	6.454	170	10.625	51.880
3	1/4″	.25	.2391	6.073	160	10.000	48.820
4	15/64″	.234375	.2242	5.695	150	9.375	45.770
5	7/32″	.21875	.2092	5.314	140	8.750	42.720
6	13/64″	.203125	.1943	4.935	130	8.125	39.670
7	3/16″	.1875	.1793	4.554	120	7.500	36.320
8	11/64″	.171875	.1644	4.176	110	6.875	33.570
9	5/32″	.15625	.1495	3.797	100	6.250	30.520
10	9/64″	.140625	.1345	3.416	90	5.625	27.460
11	1/8″	.125	.1196	3.038	80	5.000	24.410
12	7/64″	.109375	.1046	2.657	70	4.375	21.360
13	3/32″	.09375	.0897	2.278	60	3.750	18.310
14	5/64″	.078125	.0747	1.897	50	3.125	15.260
15	9/128″	.0713125	.0673	1.709	45	2.813	13.730
16	1/16″	.0625	.0598	1.519	40	2.500	12.210
17	9/160″	.05625	.0538	1.367	36	2.250	10.990
18	1/20″	.05	.0478	1.214	32	2.000	9.765
19	7/160″	.04375	.0418	1.062	28	1.750	8.544
20	3/80″	.0375	.0359	.912	24	1.500	7.324
21	11/320″	.034375	.0329	.836	22	1.375	6.713
22	1/32″	.03125	.0299	.759	20	1.250	6.103
23	9/320″	.028125	.0269	.683	18	1.125	5.490
24	1/40″	.025	.0239	.607	16	1.000	4.882
25	7/320″	.021875	.0209	.531	14	.875	4.272
26	3/160″	.01875	.0179	.455	12	.750	3.662
27	11/640″	.0171875	.0164	.417	11	.688	3.357
28	1/64″	.015625	.0149	.378	10	.625	3.052

R053100-10 Decking Descriptions

General - All Deck Products

Steel deck is made by cold forming structural grade sheet steel into a repeating pattern of parallel ribs. The strength and stiffness of the panels are the result of the ribs and the material properties of the steel. Deck lengths can be varied to suit job conditions, but because of shipping considerations, are usually less than 40 feet. Standard deck width varies with the product used but full sheets are usually 12", 18", 24", 30", or 36". Deck is typically furnished in a standard width with the ends cut square. Any cutting for width, such as at openings or for angular fit, is done at the job site.

Deck is typically attached to the building frame with arc puddle welds, self-drilling screws, or powder or pneumatically driven pins. Sheet to sheet fastening is done with screws, button punching (crimping), or welds.

Composite Floor Deck

After installation and adequate fastening, floor deck serves several purposes. It (a) acts as a working platform, (b) stabilizes the frame, (c) serves as a concrete form for the slab, and (d) reinforces the slab to carry the design loads applied during the life of the building. Composite decks are distinguished by the presence of shear connector devices as part of the deck. These devices are designed to mechanically lock the concrete and deck together so that the concrete and the deck work together to carry subsequent floor loads. These shear connector devices can be rolled-in embossments, lugs, holes, or wires welded to the panels. The deck profile can also be used to interlock concrete and steel.

Composite deck finishes are either galvanized (zinc coated) or phosphatized/painted. Galvanized deck has a zinc coating on both the top and bottom surfaces. The phosphatized/painted deck has a bare (phosphatized) top surface that will come into contact with the concrete. This bare top surface can be expected to develop rust before the concrete is placed. The bottom side of the deck has a primer coat of paint.

Composite floor deck is normally installed so the panel ends do not overlap on the supporting beams. Shear lugs or panel profile shape often prevent a tight metal to metal fit if the panel ends overlap; the air gap caused by overlapping will prevent proper fusion with the structural steel supports when the panel end laps are shear stud welded.

Adequate end bearing of the deck must be obtained as shown on the drawings. If bearing is actually less in the field than shown on the drawings, further investigation is required.

Roof Deck

Roof deck is not designed to act compositely with other materials. Roof deck acts alone in transferring horizontal and vertical loads into the building frame. Roof deck rib openings are usually narrower than floor deck rib openings. This provides adequate support of rigid thermal insulation board.

Roof deck is typically installed to endlap approximately 2" over supports. However, it can be butted (or lapped more than 2") to solve field fit problems. Since designers frequently use the installed deck system as part of the horizontal bracing system (the deck as a diaphragm), any fastening substitution or change should be approved by the designer. Continuous perimeter support of the deck is necessary to limit edge deflection in the finished roof and may be required for diaphragm shear transfer.

Standard roof deck finishes are galvanized or primer painted. The standard factory applied paint for roof deck is a primer paint and is not intended to weather for extended periods of time. Field painting or touching up of abrasions and deterioration of the primer coat or other protective finishes is the responsibility of the contractor.

Cellular Deck

Cellular deck is made by attaching a bottom steel sheet to a roof deck or composite floor deck panel. Cellular deck can be used in the same manner as floor deck. Electrical, telephone, and data wires are easily run through the chase created between the deck panel and the bottom sheet.

When used as part of the electrical distribution system, the cellular deck must be installed so that the ribs line up and create a smooth cell transition at abutting ends. The joint that occurs at butting cell ends must be taped or otherwise sealed to prevent wet concrete from seeping into the cell. Cell interiors must be free of welding burrs, or other sharp intrusions, to prevent damage to wires.

When used as a roof deck, the bottom flat plate is usually left exposed to view. Care must be maintained during erection to keep good alignment and prevent damage.

Cellular deck is sometimes used with the flat plate on the top side to provide a flat working surface. Installation of the deck for this purpose requires special methods for attachment to the frame because the flat plate, now on the top, can prevent direct access to the deck material that is bearing on the structural steel. It may be advisable to treat the flat top surface to prevent slipping.

Cellular deck is always furnished galvanized or painted over galvanized.

Form Deck

Form deck can be any floor or roof deck product used as a concrete form. Connections to the frame are by the same methods used to anchor floor and roof deck. Welding washers are recommended when welding deck that is less than 20 gauge thickness.

Form deck is furnished galvanized, prime painted, or uncoated. Galvanized deck must be used for those roof deck systems where form deck is used to carry a lightweight insulating concrete fill.

Wood, Plastics & Composites | R0611 | Wood Framing

R061110-30 Lumber Product Material Prices

The price of forest products fluctuates widely from location to location and from season to season depending upon economic conditions. The bare material prices in the unit cost sections of the book show the National Average material prices in effect Jan. 1 of this book year. It must be noted that lumber prices in general may change significantly during the year.

Availability of certain items depends upon geographic location and must be checked prior to firm-price bidding.

Wood, Plastics & Composites | R0616 | Sheathing

R061636-20 Plywood

There are two types of plywood used in construction: interior, which is moisture-resistant but not waterproofed, and exterior, which is waterproofed.

The grade of the exterior surface of the plywood sheets is designated by the first letter: A, for smooth surface with patches allowed; B, for solid surface with patches and plugs allowed; C, which may be surface plugged or may have knot holes up to 1″ wide; and D, which is used only for interior type plywood and may have knot holes up to 2-1/2″ wide. "Structural Grade" is specifically designed for engineered applications such as box beams. All CC & DD grades have roof and floor spans marked on them.

Underlayment-grade plywood runs from 1/4″ to 1-1/4″ thick. Thicknesses 5/8″ and over have optional tongue and groove joints which eliminate the need for blocking the edges. Underlayment 19/32″ and over may be referred to as Sturd-i-Floor.

The price of plywood can fluctuate widely due to geographic and economic conditions.

Typical uses for various plywood grades are as follows:

AA-AD Interior — cupboards, shelving, paneling, furniture

BB Plyform — concrete form plywood

CDX — wall and roof sheathing

Structural — box beams, girders, stressed skin panels

AA-AC Exterior — fences, signs, siding, soffits, etc.

Underlayment — base for resilient floor coverings

Overlaid HDO — high density for concrete forms & highway signs

Overlaid MDO — medium density for painting, siding, soffits & signs

303 Siding — exterior siding, textured, striated, embossed, etc.

Thermal & Moisture Protection | R0731 | Shingles & Shakes

R073126-20 Roof Slate

16″, 18″ and 20″ are standard lengths, and slate usually comes in random widths. For standard 3/16″ thickness use 1-1/2″ copper nails. Allow for 3% breakage.

Thermal & Moisture Protection | R0751 | Built-Up Bituminous Roofing

R075113-20 Built-Up Roofing

Asphalt is available in kegs of 100 lbs. each; coal tar pitch in 560 lb. kegs. Prepared roofing felts are available in a wide range of sizes, weights and characteristics. However, the most commonly used are #15 (432 S.F. per roll, 13 lbs. per square) and #30 (216 S.F. per roll, 27 lbs. per square).

Inter-ply bitumen varies from 24 lbs. per sq. (asphalt) to 30 lbs. per sq. (coal tar) per ply, MF4@ 25%. Flood coat bitumen also varies from 60 lbs. per sq. (asphalt) to 75 lbs. per sq. (coal tar), MF4@ 25%. Expendable equipment (mops, brooms, screeds, etc.) runs about 16% of the bitumen cost. For new, inexperienced crews this factor may be much higher.

Rigid insulation board is typically applied in two layers. The first is mechanically attached to nailable decks or spot or solid mopped to non-nailable decks; the second layer is then spot or solid mopped to the first layer. Membrane application follows the insulation, except in protected membrane roofs, where the membrane goes down first and the insulation on top, followed with ballast (stone or concrete pavers). Insulation and related labor costs are NOT included in prices for built-up roofing.

Thermal & Moisture Protection | R0752 | Modified Bituminous Membrane Roofing

R075213-30 Modified Bitumen Roofing

The cost of modified bitumen roofing is highly dependent on the type of installation that is planned. Installation is based on the type of modifier used in the bitumen. The two most popular modifiers are atactic polypropylene (APP) and styrene butadiene styrene (SBS). The modifiers are added to heated bitumen during the manufacturing process to change its characteristics. A polyethylene, polyester or fiberglass reinforcing sheet is then sandwiched between layers of this bitumen. When completed, the result is a pre-assembled, built-up roof that has increased elasticity and weatherablility. Some manufacturers include a surfacing material such as ceramic or mineral granules, metal particles or sand.

The preferred method of adhering SBS-modified bitumen roofing to the substrate is with hot-mopped asphalt (much the same as built-up roofing). This installation method requires a tar kettle/pot to heat the asphalt, as well as the labor, tools and equipment necessary to distribute and spread the hot asphalt.

The alternative method for applying APP and SBS modified bitumen is as follows. A skilled installer uses a torch to melt a small pool of bitumen off the membrane. This pool must form across the entire roll for proper adhesion. The installer must unroll the roofing at a pace slow enough to melt the bitumen, but fast enough to prevent damage to the rest of the membrane.

Modified bitumen roofing provides the advantages of both built-up and single-ply roofing. Labor costs are reduced over those of built-up roofing because only a single ply is necessary. The elasticity of single-ply roofing is attained with the reinforcing sheet and polymer modifiers. Modifieds have some self-healing characteristics and because of their multi-layer construction, they offer the reliability and safety of built-up roofing.

Thermal & Moisture Protection | R0784 | Firestopping

R078413-30 Firestopping

Firestopping is the sealing of structural, mechanical, electrical and other penetrations through fire-rated assemblies. The basic components of firestop systems are safing insulation and firestop sealant on both sides of wall penetrations and the top side of floor penetrations.

Pipe penetrations are assumed to be through concrete, grout, or joint compound and can be sleeved or unsleeved. Costs for the penetrations and sleeves are not included. An annular space of 1″ is assumed. Escutcheons are not included.

Metallic pipe is assumed to be copper, aluminum, cast iron or similar metallic material. Insulated metallic pipe is assumed to be covered with a thermal insulating jacket of varying thickness and materials.

Non-metallic pipe is assumed to be PVC, CPVC, FR Polypropylene or similar plastic piping material. Intumescent firestop sealant or wrap strips are included. Collars on both sides of wall penetrations and a sheet metal plate on the underside of floor penetrations are included.

Ductwork is assumed to be sheet metal, stainless steel or similar metallic material. Duct penetrations are assumed to be through concrete, grout or joint compound. Costs for penetrations and sleeves are not included. An annular space of 1/2″ is assumed.

Multi-trade openings include costs for sheet metal forms, firestop mortar, wrap strips, collars and sealants as necessary.

Structural penetrations joints are assumed to be 1/2″ or less. CMU walls are assumed to be within 1-1/2″ of metal deck. Drywall walls are assumed to be tight to the underside of metal decking.

Metal panel, glass or curtain wall systems include a spandrel area of 5′ filled with mineral wool foil-faced insulation. Fasteners and stiffeners are included.

Openings | R0813 | Metal Doors

R081313-20 Steel Door Selection Guide

Standard steel doors are classified into four levels, as recommended by the Steel Door Institute in the chart below. Each of the four levels offers a range of construction models and designs, to meet architectural requirements for preference and appearance, including full flush, seamless, and stile & rail. Recommended minimum gauge requirements are also included.

For complete standard steel door construction specifications and available sizes, refer to the Steel Door Institute Technical Data Series, ANSI A250.8-98 (SDI-100), and ANSI A250.4-94 Test Procedure and Acceptance Criteria for Physical Endurance of Steel Door and Hardware Reinforcements.

Level		Model	Construction	For Full Flush or Seamless		
				Min. Gauge	Thickness (in)	Thickness (mm)
I	Standard Duty	1 2	Full Flush Seamless	20	0.032	0.8
II	Heavy Duty	1 2	Full Flush Seamless	18	0.042	1.0
III	Extra Heavy Duty	1 2 3	Full Flush Seamless *Stile & Rail	16	0.053	1.3
IV	Maximum Duty	1 2	Full Flush Seamless	14	0.067	1.6

*Stiles & rails are 16 gauge; flush panels, when specified, are 18 gauge

Openings | R0851 | Metal Windows

R085123-10 Steel Sash

Ironworker crew will erect 25 S.F. or 1.3 sash unit per hour, whichever is less.

Mechanic will point 30 L.F. per hour.

Painter will paint 90 S.F. per coat per hour.

Glazier production depends on light size.

Allow 1 lb. special steel sash putty per 16″ x 20″ light.

Openings | R0852 | Wood Windows

R085216-10 Window Estimates

To ensure a complete window estimate, be sure to include the material and labor costs for each window, as well as the material and labor costs for an interior wood trim set.

Openings | R0853 | Plastic Windows

R085313-20 Replacement Windows

Replacement windows are typically measured per United Inch.

United Inches are calculated by rounding the width and height of the window opening up to the nearest inch, then adding the two figures.

The labor cost for replacement windows includes removal of sash, existing sash balance or weights, parting bead where necessary and installation of new window.

Debris hauling and dump fees are not included.

Openings | R0871 | Door Hardware

R087120-10 Hinges

All closer equipped doors should have ball bearing hinges. Lead lined or extremely heavy doors require special strength hinges. Usually 1-1/2 pair of hinges are used per door up to 7′-6″ high openings. Table below shows typical hinge requirements.

Use Frequency	Type Hinge Required	Type of Opening	Type of Structure
High	Heavy weight ball bearing	Entrances Toilet Rooms	Banks, Office buildings, Schools, Stores & Theaters Office buildings and Schools
Average	Standard weight ball bearing	Entrances Corridors Toilet Rooms	Dwellings Office buildings and Schools Stores
Low	Plain bearing	Interior	Dwellings

Door Thickness	Weight of Doors in Pounds per Square Foot				
	White Pine	Oak	Hollow Core	Solid Core	Hollow Metal
1-3/8″	3psf	6psf	1-1/2psf	3-1/2— 4psf	6-1/2psf
1-3/4″	3-1/2	7	2-1/2	4-1/2—5-1/4	6-1/2
2-1/4″	4-1/2	9	—	5-1/2—6-3/4	6-1/2

Openings | R0881 | Glass Glazing

R088110-10 Glazing Productivity

Some glass sizes are estimated by the "united inch" (height + width). The table below shows the number of lights glazed in an eight-hour period by the crew size indicated, for glass up to 1/4″ thick. Square or nearly square lights are more economical on a S.F. basis. Long slender lights will have a high S.F. installation cost. For insulated glass reduce production by 33%. For 1/2″ float glass reduce production by 50%. Production time for glazing with two glaziers per day averages: 1/4″ float glass 120 S.F.; 1/2″ float glass 55 S.F.; 1/2″ insulated glass 95 S.F.; 3/4″ insulated glass 75 S.F.

Glazing Method	United Inches per Light							
	40″	60″	80″	100″	135″	165″	200″	240″
Number of Men in Crew	1	1	1	1	2	3	3	4
Industrial sash, putty	60	45	24	15	18	—	—	—
With stops, putty bed	50	36	21	12	16	8	4	3
Wood stops, rubber	40	27	15	9	11	6	3	2
Metal stops, rubber	30	24	14	9	9	6	3	2
Structural glass	10	7	4	3	—	—	—	—
Corrugated glass	12	9	7	4	4	4	3	—
Storefronts	16	15	13	11	7	6	4	4
Skylights, putty glass	60	36	21	12	16	—	—	—
Thiokol set	15	15	11	9	9	6	3	2
Vinyl set, snap on	18	18	13	12	12	7	5	4
Maximum area per light	2.8 S.F.	6.3 S.F.	11.1 S.F.	17.4 S.F.	31.6 S.F.	47 S.F.	69 S.F.	100 S.F.

R092000-50 Lath, Plaster and Gypsum Board

Gypsum board lath is available in 3/8″ thick x 16″ wide x 4′ long sheets as a base material for multi-layer plaster applications. It is also available as a base for either multi-layer or veneer plaster applications in 1/2″ and 5/8″ thick–4′ wide x 8′, 10′ or 12′ long sheets. Fasteners are screws or blued ring shank nails for wood framing and screws for metal framing.

Metal lath is available in diamond mesh pattern with flat or self-furring profiles. Paper backing is available for applications where excessive plaster waste needs to be avoided. A slotted mesh ribbed lath should be used in areas where the span between structural supports is greater than normal. Most metal lath comes in 27″ x 96″ sheets. Diamond mesh weighs 1.75, 2.5 or 3.4 pounds per square yard, slotted mesh lath weighs 2.75 or 3.4 pounds per square yard. Metal lath can be nailed, screwed or tied in place.

Many **accessories** are available. Corner beads, flat reinforcing strips, casing beads, control and expansion joints, furring brackets and channels are some examples. Note that accessories are not included in plaster or stucco line items.

Plaster is defined as a material or combination of materials that when mixed with a suitable amount of water, forms a plastic mass or paste. When applied to a surface, the paste adheres to it and subsequently hardens, preserving in a rigid state the form or texture imposed during the period of elasticity.

Gypsum plaster is made from ground calcined gypsum. It is mixed with aggregates and water for use as a base coat plaster.

Vermiculite plaster is a fire-retardant plaster covering used on steel beams, concrete slabs and other heavy construction materials. Vermiculite is a group name for certain clay minerals, hydrous silicates or aluminum, magnesium and iron that have been expanded by heat.

Perlite plaster is a plaster using perlite as an aggregate instead of sand. Perlite is a volcanic glass that has been expanded by heat.

Gauging plaster is a mix of gypsum plaster and lime putty that when applied produces a quick drying finish coat.

Veneer plaster is a one or two component gypsum plaster used as a thin finish coat over special gypsum board.

Keenes cement is a white cementitious material manufactured from gypsum that has been burned at a high temperature and ground to a fine powder. Alum is added to accelerate the set. The resulting plaster is hard and strong and accepts and maintains a high polish, hence it is used as a finishing plaster.

Stucco is a Portland cement based plaster used primarily as an exterior finish.

Plaster is used on both interior and exterior surfaces. Generally it is applied in multiple-coat systems. A three-coat system uses the terms scratch, brown and finish to identify each coat. A two-coat system uses base and finish to describe each coat. Each type of plaster and application system has attributes that are chosen by the designer to best fit the intended use.

Gypsum Plaster Quantities for 100 S.Y.	2 Coat, 5/8″ Thick		3 Coat, 3/4″ Thick		
	Base	Finish	Scratch	Brown	Finish
	1:3 Mix	2:1 Mix	1:2 Mix	1:3 Mix	2:1 Mix
Gypsum plaster	1,300 lb.		1,350 lb.	650 lb.	
Sand	1.75 C.Y.		1.85 C.Y.	1.35 C.Y.	
Finish hydrated lime		340 lb.			340 lb.
Gauging plaster		170 lb.			170 lb.

Vermiculite or Perlite Plaster Quantities for 100 S.Y.	2 Coat, 5/8″ Thick		3 Coat, 3/4″ Thick		
	Base	Finish	Scratch	Brown	Finish
Gypsum plaster	1,250 lb.		1,450 lb.	800 lb.	
Vermiculite or perlite	7.8 bags		8.0 bags	3.3 bags	
Finish hydrated lime		340 lb.			340 lb.
Gauging plaster		170 lb.			170 lb.

Stucco-Three-Coat System Quantities for 100 S.Y.	On Wood Frame	On Masonry
Portland cement	29 bags	21 bags
Sand	2.6 C.Y.	2.0 C.Y.
Hydrated lime	180 lb.	120 lb.

R092910-10 Levels of Gypsum Drywall Finish

In the past, contract documents often used phrases such as "industry standard" and "workmanlike finish" to specify the expected quality of gypsum board wall and ceiling installations. The vagueness of these descriptions led to unacceptable work and disputes.

In order to resolve this problem, four major trade associations concerned with the manufacture, erection, finish and decoration of gypsum board wall and ceiling systems have developed an industry-wide *Recommended Levels of Gypsum Board Finish.*

The finish of gypsum board walls and ceilings for specific final decoration is dependent on a number of factors. A primary consideration is the location of the surface and the degree of decorative treatment desired. Painted and unpainted surfaces in warehouses and other areas where appearance is normally not critical may simply require the taping of wallboard joints and 'spotting' of fastener heads. Blemish-free, smooth, monolithic surfaces often intended for painted and decorated walls and ceilings in habitated structures, ranging from single-family dwellings through monumental buildings, require additional finishing prior to the application of the final decoration.

Other factors to be considered in determining the level of finish of the gypsum board surface are (1) the type of angle of surface illumination (both natural and artificial lighting), and (2) the paint and method of application or the type and finish of wallcovering specified as the final decoration. Critical lighting conditions, gloss paints, and thin wallcoverings require a higher level of gypsum board finish than do heavily textured surfaces which are subsequently painted or surfaces which are to be decorated with heavy grade wallcoverings.

The following descriptions were developed jointly by the Association of the Wall and Ceiling Industries-International (AWCI), Ceiling & Interior Systems Construction Association (CISCA), Gypsum Association (GA), and Painting and Decorating Contractors of America (PDCA) as a guide.

Level 0: No taping, finishing, or accessories required. This level of finish may be useful in temporary construction or whenever the final decoration has not been determined.

Level 1: All joints and interior angles shall have tape set in joint compound. Surface shall be free of excess joint compound. Tool marks and ridges are acceptable. Frequently specified in plenum areas above ceilings, in attics, in areas where the assembly would generally be concealed or in building service corridors, and other areas not normally open to public view.

Level 2: All joints and interior angles shall have tape embedded in joint compound and wiped with a joint knife leaving a thin coating of joint compound over all joints and interior angles. Fastener heads and accessories shall be covered with a coat of joint compound. Surface shall be free of excess joint compound. Tool marks and ridges are acceptable. Joint compound applied over the body of the tape at the time of tape embedment shall be considered a separate coat of joint compound and shall satisfy the conditions of this level. Specified where water-resistant gypsum backing board is used as a substrate for tile; may be specified in garages, warehouse storage, or other similar areas where surface appearance is not of primary concern.

Level 3: All joints and interior angles shall have tape embedded in joint compound and one additional coat of joint compound applied over all joints and interior angles. Fastener heads and accessories shall be covered with two separate coats of joint compound. All joint compound shall be smooth and free of tool marks and ridges. Typically specified in appearance areas which are to receive heavy- or medium-texture (spray or hand applied) finishes before final painting, or where heavy-grade wallcoverings are to be applied as the final decoration. This level of finish is not recommended where smooth painted surfaces or light to medium wallcoverings are specified.

Level 4: All joints and interior angles shall have tape embedded in joint compound and two separate coats of joint compound applied over all flat joints and one separate coat of joint compound applied over interior angles. Fastener heads and accessories shall be covered with three separate coats of joint compound. All joint compound shall be smooth and free of tool marks and ridges. This level should be specified where flat paints, light textures, or wallcoverings are to be applied. In critical lighting areas, flat paints applied over light textures tend to reduce joint photographing. Gloss, semi-gloss, and enamel paints are not recommended over this level of finish. The weight, texture, and sheen level of wallcoverings applied over this level of finish should be carefully evaluated. Joints and fasteners must be adequately concealed if the wallcovering material is lightweight, contains limited pattern, has a gloss finish, or any combination of these finishes is present. Unbacked vinyl wallcoverings are not recommended over this level of finish.

Level 5: All joints and interior angles shall have tape embedded in joint compound and two separate coats of joint compound applied over all flat joints and one separate coat of joint compound applied over interior angles. Fastener heads and accessories shall be covered with three separate coats of joint compound. A thin skim coat of joint compound or a material manufactured especially for this purpose, shall be applied to the entire surface. The surface shall be smooth and free of tool marks and ridges. This level of finish is highly recommended where gloss, semi-gloss, enamel, or nontextured flat paints are specified or where severe lighting conditions occur. This highest quality finish is the most effective method to provide a uniform surface and minimize the possibility of joint photographing and of fasteners showing through the final decoration.

R096613-10 Terrazzo Floor

The table below lists quantities required for 100 S.F. of 5/8″ terrazzo topping, either bonded or not bonded.

Description	Bonded to Concrete 1-1/8″ Bed, 1:4 Mix	Not Bonded 2-1/8″ Bed and 1/4″ Sand
Portland cement, 94 lb. Bag	6 bags	8 bags
Sand	10 C.F.	20 C.F.
Divider strips, 4′ squares	50 L.F.	50 L.F.
Terrazzo fill, 50 lb. Bag	12 bags	12 bags
15 Lb. tarred felt		1 C.S.F.
Mesh 2 x 2 #14 galvanized		1 C.S.F.
Crew J-3	0.77 days	0.87 days

2′ x 2′ panels require 1.00 L.F. divider strip per S.F.

3′ x 3′ panels require 0.67 L.F. divider strip per S.F.

4′ x 4′ panels require 0.50 L.F. divider strip per S.F.

5′ x 5′ panels require 0.40 L.F. divider strip per S.F.

6′ x 6′ panels require 0.33 L.F. divider strip per S.F.

R097223-10 Wall Covering

The table below lists the quantities required for 100 S.F. of wall covering.

Description	Medium-Priced Paper	Expensive Paper
Paper	1.6 dbl. rolls	1.6 dbl. rolls
Wall sizing	0.25 gallon	0.25 gallon
Vinyl wall paste	0.6 gallon	0.6 gallon
Apply sizing	0.3 hour	0.3 hour
Apply paper	1.2 hours	1.5 hours

Most wallpapers now come in double rolls only.
To remove old paper, allow 1.3 hours per 100 S.F.

R099100-20 Painting

Item	Coat	One Gallon Covers			In 8 Hours a Laborer Covers			Labor-Hours per 100 S.F.		
		Brush	Roller	Spray	Brush	Roller	Spray	Brush	Roller	Spray
Paint wood siding	prime	250 S.F.	225 S.F.	290 S.F.	1150 S.F.	1300 S.F.	2275 S.F.	.695	.615	.351
	others	270	250	290	1300	1625	2600	.615	.492	.307
Paint exterior trim	prime	400	—	—	650	—	—	1.230	—	—
	1st	475	—	—	800	—	—	1.000	—	—
	2nd	520	—	—	975	—	—	.820	—	—
Paint shingle siding	prime	270	255	300	650	975	1950	1.230	.820	.410
	others	360	340	380	800	1150	2275	1.000	.695	.351
Stain shingle siding	1st	180	170	200	750	1125	2250	1.068	.711	.355
	2nd	270	250	290	900	1325	2600	.888	.603	.307
Paint brick masonry	prime	180	135	160	750	800	1800	1.066	1.000	.444
	1st	270	225	290	815	975	2275	.981	.820	.351
	2nd	340	305	360	815	1150	2925	.981	.695	.273
Paint interior plaster or drywall	prime	400	380	495	1150	2000	3250	.695	.400	.246
	others	450	425	495	1300	2300	4000	.615	.347	.200
Paint interior doors and windows	prime	400	—	—	650	—	—	1.230	—	—
	1st	425	—	—	800	—	—	1.000	—	—
	2nd	450	—	—	975	—	—	.820	—	—

Special Construction | R1311 | Swimming Pools

R131113-20 Swimming Pools

Pool prices given per square foot of surface area include pool structure, filter and chlorination equipment, pumps, related piping, ladders/steps, maintenance kit, skimmer and vacuum system. Decks and electrical service to equipment are not included.

Residential in-ground pool construction can be divided into two categories: vinyl lined and gunite. Vinyl lined pool walls are constructed of different materials including wood, concrete, plastic or metal. The bottom is often graded with sand over which the vinyl liner is installed. Vermiculite or soil cement bottoms may be substituted for an added cost.

Gunite pool construction is used both in residential and municipal installations. These structures are steel reinforced for strength and finished with a white cement limestone plaster.

Municipal pools will have a higher cost because plumbing codes require more expensive materials, chlorination equipment and higher filtration rates.

Municipal pools greater than 1,800 S.F. require gutter systems to control waves. This gutter may be formed into the concrete wall. Often a vinyl/stainless steel gutter or gutter/wall system is specified, which will raise the pool cost.

Competition pools usually require tile bottoms and sides with contrasting lane striping, which will also raise the pool cost.

Special Construction | R1331 | Fabric Structures

R133113-10 Air Supported Structures

Air supported structures are made from fabrics that can be classified into two groups: temporary and permanent. Temporary fabrics include nylon, woven polyethylene, vinyl film, and vinyl coated dacron. These have lifespans that range from five to fifteen plus years. The cost per square foot includes a fabric shell, tension cables, primary and back-up inflation systems and doors. The lower cost structures are used for construction shelters, bulk storage and pond covers. The more expensive are used for recreational structures and warehouses.

Permanent fabrics are teflon coated fiberglass. The life of this structure is twenty plus years. The high cost limits its application to architectural designed structures which call for a clear span covered area, such as stadiums and convention centers. Both temporary and permanent structures are available in translucent fabrics which eliminates the need for daytime lighting.

Areas to be covered vary from 10,000 S.F. to any area up to 1000 foot wide by any length. Height restrictions range from a maximum of 1/2 of width to a minimum of 1/6 of the width. Erection of even the largest of the temporary structures requires no more than a week.

Centrifugal fans provide the inflation necessary to support the structure during application of live loads. Airlocks are usually used at large entrances to prevent loss of static pressure. Some manufacturers employ propeller fans which generate sufficient airflow (30,000 CFM) to eliminate the need for airlocks. These fans may also be automatically controlled to resist high wind conditions, regulate humidity (air changes), and provide cooling and heat.

Insulation can be provided with the addition of a second or even third interior liner, creating a dead air space with an "R" value of four to nine. Some structures allow for the liner to be collapsed into the outer shell to enable the internal heat to melt accumulated snow. For cooling or air conditioning, the exterior face of the liner can be aluminized to reflect the sun's heat.

R133113-90 Seismic Bracing

Sometimes referred to as anti-sway bracing, this support system is required in earthquake areas. The individual components must be assembled to make a required system.

Example

15070-400-0040	C-Clamp 3/8″ rod	2 ea.
15070-400-0330	Rod, continuous thread 3/8″	10 L.F.
15070-400-5030	Field, weld 1″	2 ea.

Additionally, height factors must be taken ainto account. Add the following percentages to labor for elevated installations:

15′ to 20′ high	10%
21′ to 25′ high	20%
26′ to 30′ high	30%
31′ to 35′ high	40%
36′ to 40′ high	50%
41′ to 50′ high	60%

Special Construction | R1334 | Fabricated Engineered Structures

R133419-10 Pre-engineered Steel Buildings

These buildings are manufactured by many companies and normally erected by franchised dealers throughout the U.S. The four basic types are: Rigid Frames, Truss type, Post and Beam and the Sloped Beam type. Most popular roof slope is low pitch of 1″ in 12″. The minimum economical area of these buildings is about 3000 S.F. of floor area. Bay sizes are usually 20′ to 24′ but can go as high as 30′ with heavier girts and purlins. Eave heights are usually 12′ to 24′ with 18′ to 20′ most typical.

Material prices shown in the Unit Price section are bare costs for the building shell only and do not include floors, foundations, anchor bolts, interior finishes or utilities. Costs assume at least three bays of 24′ each, a 1″ in 12″ roof slope, and they are based on 30 psf roof load and 20 psf wind load (wind load is a function of wind speed, building height, and terrain characteristics; this should be determined by a Registered Structural Engineer) and no unusual requirements. Costs include the structural frame, 26 ga. non-insulated colored corrugated or ribbed roofing and siding panels, fasteners, closures, trim and flashing but no allowance for insulation, doors, windows, skylights, gutters or downspouts. Very large projects would generally cost less for materials than the prices shown. For roof panel substitutions and wall panel substitutions, see appropriate Unit Price sections.

Conditions at the site, weather, shape and size of the building, and labor availability will affect the erection cost of the building.

R133423-30 Dome Structures

Steel — The four types are Lamella, Schwedler, Arch and Geodesic. For maximum economy, rise should be about 15 to 20% of diameter. Most common diameters are in the 200′ to 300′ range. Lamella domes weigh about 5 P.S.F. of floor area less than Schwedler domes. Schwedler dome weight in lbs. per S.F. approaches .046 times the diameter. Domes below 125′ diameter weigh .07 times diameter and the cost per ton of steel is higher. See R051223-20 for estimating weight.

Wood — Small domes are of sawn lumber, larger ones are laminated. In larger sizes, triaxial and triangular cost about the same; radial domes cost more. Radial domes are economical in the 60′ to 70′ diameter range. Most economical range of all types is 80′ to 200′ diameters. Diameters can run over 400′. All costs are quoted above the foundation. Prices include 2″ decking and a tension tie ring in place.

Plywood — Stock prefab geodesic domes are available with diameters from 24′ to 60′.

Fiberglass — Aluminum framed translucent sandwich panels with spans from 5′ to 45′ are commercially available.

Aluminum — Stressed skin aluminum panels form geodesic domes with spans ranging from 82′ to 232′. An aluminum space truss, triangulated or nontriangulated, with aluminum or clear acrylic closure panels can be used for clear spans of 40′ to 415′.

R142000-10 Freight Elevators

Capacities run from 2,000 lbs. to over 100,000 lbs. with 3,000 lbs. to 10,000 lbs. most common. Travel speeds are generally lower and control less intricate than on passenger elevators. Unit prices in division 14210-200 are for hydraulic and geared elevators.

R142000-20 Elevator Selective Costs

See R142000-40 for cost development.

	Passenger		Freight		Hospital	
A. Base Unit	Hydraulic	Electric	Hydraulic	Electric	Hydraulic	Electric
Capacity	1,500 lb.	2,000 lb.	2,000 lb.	4,000 lb.	4,000 lb.	4,000 lb.
Speed	100 F.P.M.	200 F.P.M.	100 F.P.M.	200 F.P.M.	100 F.P.M.	200 F.P.M.
#Stops/Travel Ft.	2/12	4/40	2/20	4/40	2/20	4/40
Push Button Oper.	Yes	Yes	Yes	Yes	Yes	Yes
Telephone Box & Wire	"	"	"	"	"	"
Emergency Lighting	"	"	No	No	"	"
Cab	Plastic Lam. Walls	Plastic Lam. Walls	Painted Steel	Painted Steel	Plastic Lam. Walls	Plastic Lam. Walls
Cove Lighting	Yes	Yes	No	No	Yes	Yes
Floor	V.C.T.	V.C.T.	Wood w/Safety Treads	Wood w/Safety Treads	V.C.T.	V.C.T.
Doors, & Speedside Slide	Yes	Yes	Yes	Yes	Yes	Yes
Gates, Manual	No	No	No	No	No	No
Signals, Lighted Buttons	Car and Hall	Car and Hall	Car and Hall	Car and Hall	Car and Hall	Car and Hall
O.H. Geared Machine	N.A.	Yes	N.A.	Yes	N.A.	Yes
Variable Voltage Contr.	"	"	N.A.	"	"	"
Emergency Alarm	Yes	"	Yes	"	Yes	"
Class "A" Loading	N.A.	N.A.	"	"	N.A.	N.A.

R142000-30 Passenger Elevators

Electric elevators are used generally but hydraulic elevators can be used for lifts up to 70′ and where large capacities are required. Hydraulic speeds are limited to 200 F.P.M. but cars are self leveling at the stops. On low rises, hydraulic installation runs about 15% less than standard electric types but on higher rises this installation cost advantage is reduced. Maintenance of hydraulic elevators is about the same as electric type but underground portion is not included in the maintenance contract.

In electric elevators there are several control systems available, the choice of which will be based upon elevator use, size, speed and cost criteria. The two types of drives are geared for low speeds and gearless for 450 F.P.M. and over.

The tables on the preceding pages illustrate typical installed costs of the various types of elevators available.

R142000-40 Elevator Cost Development

To price a new car or truck from the factory, you must start with the manufacturer's basic model, then add or exchange optional equipment and features. The same is true for pricing elevators.

Example:

Requirement: One-passenger elevator, five-story hydraulic, 2,500 lb. capacity, 12′ floor to floor, speed 150 F.P.M., emergency power switching and maintenance contract.

Description	Adjustment
A. Base Elevator: Hydraulic Passenger, 1500 lb. Capacity, 100 fpm, 2 Stops, Standard Finish	1 Ea.
B. Capacity Adjustment (2,500 lb.)	1 Ea.
C. Excess Travel Adjustment: 48′ Total Travel (4 x 12′) minus 12′ Base Unit Travel =	36 V.L.F.
D. Stops Adjustment: 5 Total Stops minus 2 Stops (Base Unit) =	3 Stops
E. Speed Adjustment (150 F.P.M.)	1 Ea.
F. Options:	
1. Intercom Service	1 Ea.
2. Emergency Power Switching, Automatic	1 Ea.
3. Stainless Steel Entrance Doors	5 Ea.
4. Maintenance Contract (12 Months)	1 Ea.
5. Position Indicator for main floor level (none indicated in Base Unit)	1 Ea.

Conveying Equipment | R1431 | Escalators

R143110-10 Escalators

Moving stairs can be used for buildings where 600 or more people are to be carried to the second floor or beyond. Freight cannot be carried on escalators and at least one elevator must be available for this function.

Carrying capacity is 5,000 to 8,000 people per hour. Power requirement is 2 to 3 KW per hour and incline angle is 30°.

Conveying Equipment | R1432 | Moving Walks

R143210-20 Moving Ramps and Walks

These are a specialized form of conveyor 3′ to 6′ wide with capacities of 3,600 to 18,000 persons per hour. Maximum speed is 140 F.P.M. and normal incline is 0° to 15°.

Local codes will determine the maximum angle. Outdoor units would require additional weather protection.

Fire Suppression | R2112 | Fire-Suppression Standpipes

R211226-10 Standpipe Systems

The basis for standpipe system design is National Fire Protection Association NFPA 14, however, the authority having jurisdiction should be consulted for special conditions, local requirements and approval.

Standpipe systems, properly designed and maintained, are an effective and valuable time saving aid for extinquishing fires, especially in the upper stories of tall buildings, the interior of large commercial or industrial malls, or other areas where construction features or access make the laying of temporary hose lines time consuming and/or hazardous. Standpipes are frequently installed with automatic sprinkler systems for maximum protection.

There are three general classes of service for standpipe systems:
Class I for use by fire departments and personnel with special training for heavy streams (2-1/2″ hose connections).
Class II for use by building occupants until the arrival of the fire department (1-1/2″ hose connector with hose).
Class III for use by either fire departments and trained personnel or by the building occupants (both 2-1/2″ and 1-1/2″ hose connections or one 2-1/2″ hose valve with an easily removable 2-1/2″ by 1-1/2″ adapter).

Standpipe systems are also classified by the way water is supplied to the system. The four basic types are:
Type 1: Wet standpipe system having supply valve open and water pressure maintained at all times.
Type 2: Standpipe system so arranged through the use of approved devices as to admit water to the system automatically by opening a hose valve.
Type 3: Standpipe system arranged to admit water to the system through manual operation of approved remote control devices located at each hose station.
Type 4: Dry standpipe having no permanent water supply.

R211226-20 NFPA 14 Basic Standpipe Design

Class	Design-Use	Pipe Size Minimums	Water Supply Minimums
Class I	2 1/2″ hose connection on each floor All areas within 150′ of an exit in every exit stairway Fire Department Trained Personnel	Height to 100′, 4″ dia. Heights above 100′, 6″ dia. (275′ max. except with pressure regulators 400′ max.)	For each standpipe riser 500 GPM flow For common supply pipe allow 500 GPM for first standpipe plus 250 GPM for each additional standpipe (2500 GPM max. total) 30 min. duration 65 PSI at 500 GPM
Class II	1 1/2″ hose connection with hose on each floor All areas within 130′ of hose connection measured along path of hose travel Occupant personnel	Height to 50′, 2″ dia. Height above 50′, 2 1/2″ dia.	For each standpipe riser 100 GPM flow For multiple riser common supply pipe 100 GPM 300 min. duration, 65 PSI at 100 GPM
Class III	Both of above. Class I valved connections will meet Class III with additional 2 1/2″ by 1 1/2″ adapter and 1 1/2″ hose.	Same as Class I	Same as Class I

*Note: Where 2 or more standpipes are installed in the same building or section of building they shall be interconnected at the bottom.

Combined Systems

Combined systems are systems where the risers supply both automatic sprinklers and 2-1/2″ hose connection outlets for fire department use. In such a system the sprinkler spacing pattern shall be in accordance with NFPA 13 while the risers and supply piping will be sized in accordance with NFPA 14. When the building is completely sprinklered the risers may be sized by hydraulic calculation.The minimum size riser for buildings not completely sprinklered is 6″.

The minimum water supply of a completely sprinklered, light hazard, high-rise occupancy building will be 500 GPM while the supply required for other types of completely sprinklered high-rise buildings is 1000 GPM.

General System Requirements

1. Approved valves will be provided at the riser for controlling branch lines to hose outlets.
2. A hose valve will be provided at each outlet for attachment of hose.
3. Where pressure at any standpipe outlet exceeds 100 PSI a pressure reducer must be installed to limit the pressure to 100 PSI. Note that the pressure head due to gravity in 100′ of riser is 43.4 PSI. This must be overcome by city pressure, fire pumps, or gravity tanks to provide adequate pressure at the top of the riser.
4. Each hose valve on a wet system having linen hose shall have an automatic drip connection to prevent valve leakage from entering the hose.
5. Each riser will have a valve to isolate it from the rest of the system.
6. One or more fire department connections as an auxiliary supply shall be provided for each Class I or Class III standpipe system. In buildings having two or more zones, a connection will be provided for each zone.
7. There will be no shutoff valve in the fire department connection, but a check valve will be located in the line before it joins the system.
8. All hose connections street side will be identified on a cast plate or fitting as to purpose.

R211313-10 Sprinkler Systems (Automatic)

Sprinkler systems may be classified by type as follows:

1. **Wet Pipe System.** A system employing automatic sprinklers attached to a piping system containing water and connected to a water supply so that water discharges immediately from sprinklers opened by a fire.
2. **Dry Pipe System.** A system employing automatic sprinklers attached to a piping system containing air under pressure, the release of which as from the opening of sprinklers permits the water pressure to open a valve known as a "dry pipe valve". The water then flows into the piping system and out the opened sprinklers.
3. **Pre-Action System.** A system employing automatic sprinklers attached to a piping system containing air that may or may not be under pressure, with a supplemental heat responsive system of generally more sensitive characteristics than the automatic sprinklers themselves, installed in the same areas as the sprinklers; actuation of the heat responsive system, as from a fire, opens a valve which permits water to flow into the sprinkler piping system and to be discharged from any sprinklers which may be open.
4. **Deluge System.** A system employing open sprinklers attached to a piping system connected to a water supply through a valve which is opened by the operation of a heat responsive system installed in the same areas as the sprinklers. When this valve opens, water flows into the piping system and discharges from all sprinklers attached thereto.
5. **Combined Dry Pipe and Pre-Action Sprinkler System.** A system employing automatic sprinklers attached to a piping system containing air under pressure with a supplemental heat responsive system of generally more sensitive characteristics than the automatic sprinklers themselves, installed in the same areas as the sprinklers; operation of the heat responsive system, as from a fire, actuates tripping devices which open dry pipe valves simultaneously and without loss of air pressure in the system. Operation of the heat responsive system also opens approved air exhaust valves at the end of the feed main which facilitates the filling of the system with water which usually precedes the opening of sprinklers. The heat responsive system also serves as an automatic fire alarm system.
6. **Limited Water Supply System.** A system employing automatic sprinklers and conforming to these standards but supplied by a pressure tank of limited capacity.
7. **Chemical Systems.** Systems using halon, carbon dioxide, dry chemical or high expansion foam as selected for special requirements. Agent may extinguish flames by chemically inhibiting flame propagation, suffocate flames by excluding oxygen, interrupting chemical action of oxygen uniting with fuel or sealing and cooling the combustion center.
8. **Firecycle System.** Firecycle is a fixed fire protection sprinkler system utilizing water as its extinguishing agent. It is a time delayed, recycling, preaction type which automatically shuts the water off when heat is reduced below the detector operating temperature and turns the water back on when that temperature is exceeded. The system senses a fire condition through a closed circuit electrical detector system which controls water flow to the fire automatically. Batteries supply up to 90 hour emergency power supply for system operation. The piping system is dry (until water is required) and is monitored with pressurized air. Should any leak in the system piping occur, an alarm will sound, but water will not enter the system until heat is sensed by a firecycle detector.

Area coverage sprinkler systems may be laid out and fed from the supply in any one of several patterns as shown below. It is desirable, if possible, to utilize a central feed and achieve a shorter flow path from the riser to the furthest sprinkler. This permits use of the smallest sizes of pipe possible with resulting savings.

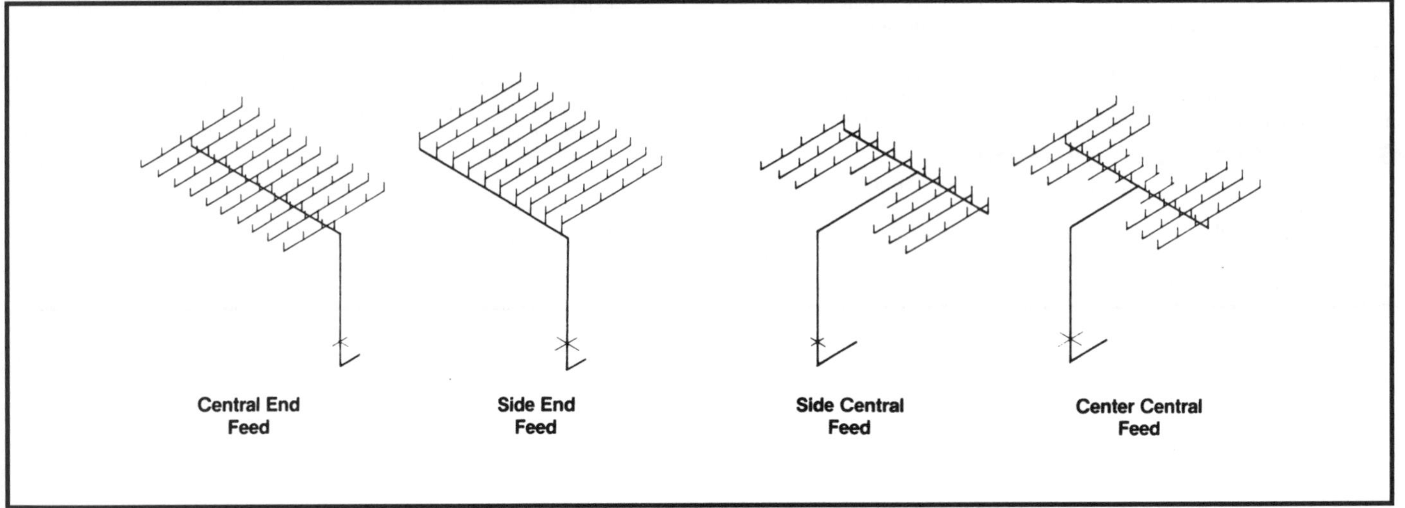

R211313-20 System Classification

System Classification

Rules for installation of sprinkler systems vary depending on the classification of occupancy falling into one of three categories as follows:

Light Hazard Occupancy

The protection area allotted per sprinkler should not exceed 225 S.F., with the maximum distance between lines and sprinklers on lines being 15′. The sprinklers do not need to be staggered. Branch lines should not exceed eight sprinklers on either side of a cross main. Each large area requiring more than 100 sprinklers and without a sub-dividing partition should be supplied by feed mains or risers sized for ordinary hazard occupancy.
Maximum system area = 52,000 S.F.

Included in this group are:

Churches
Clubs
Educational
Hospitals
Institutional
Libraries
(except large stack rooms)
Museums
Nursing Homes
Offices
Residential
Restaurants
Theaters and Auditoriums
(except stages and prosceniums)
Unused Attics

Ordinary Hazard Occupancy

The protection area allotted per sprinkler shall not exceed 130 S.F. of noncombustible ceiling and 130 S.F. of combustible ceiling. The maximum allowable distance between sprinkler lines and sprinklers on line is 15′. Sprinklers shall be staggered if the distance between heads exceeds 12′. Branch lines should not exceed eight sprinklers on either side of a cross main.
Maximum system area = 52,000 S.F.

Included in this group are:

Group 1
Automotive Parking and Showrooms
Bakeries
Beverage manufacturing
Canneries
Dairy Products Manufacturing/Processing
Electronic Plans
Glass and Glass Products Manufacturing
Laundries
Restaurant Service Areas

Group 2
Cereal Mills
Chemical Plants—Ordinary
Confectionery Products
Distilleries
Dry Cleaners
Feed Mills
Horse Stables
Leather Goods Manufacturing
Libraries—Large Stack Room Areas
Machine Shops
Metal Working
Mercantile
Paper and Pulp Mills
Paper Process Plants
Piers and Wharves
Post Offices
Printing and Publishing
Repair Garages
Stages
Textile Manufacturing
Tire Manufacturing
Tobacco Products Manufacturing
Wood Machining
Wood Product Assembly

Extra Hazard Occupancy

The protection area allotted per sprinkler shall not exceed 100 S.F. of noncombustible ceiling and 100 S.F. of combustible ceiling. The maximum allowable distance between lines and between sprinklers on lines is 12′. Sprinklers on alternate lines shall be staggered if the distance between sprinklers on lines exceeds 8′. Branch lines should not exceed six sprinklers on either side of a cross main.
Maximum system area:
Design by pipe schedule = 25,000 S.F.
Design by hydraulic calculation = 40,000 S.F.

Included in this group are:

Group 1
Aircraft hangars
Combustible Hydraulic Fluid Use Area
Die Casting
Metal Extruding
Plywood/Particle Board Manufacturing
Printing (inkls with flash points < 100 degrees F
Rubber Reclaiming, Compounding, Drying, Milling, Vulcanizing
Saw Mills
Textile Picking, Opening, Blending, Garnetting, Carding, Combing of Cotton, Synthetics, Wood Shoddy, or Burlap
Upholstering with Plastic Foams

Group 2
Asphalt Saturating
Flammable Liquids Spraying
Flow Coating
Manufactured/Modular Home Building Assemblies (where finished enclosure is present and has combustible interiors)
Open Oil Quenching
Plastics Processing
Solvent Cleaning
Varnish and Paint Dipping

Fire Suppression | R2113 | Fire-Suppression Sprinkler Systems

R211313-30 Sprinkler Quantities for Various Sizes and Types of Pipe

Sprinkler Quantities: The table below lists the usual maximum number of sprinkler heads for each size of copper and steel pipe for both wet and dry systems. These quantities may be adjusted to meet individual structural needs or local code requirements. Maximum area on any one floor for one system is: light hazard and ordinary hazard 52,000 S.F., extra hazardous 25,000 S.F.

Pipe Size	Light Hazard Occupancy		Ordinary Hazard Occupancy		Extra Hazard Occupancy	
Diameter	Steel Pipe	Copper Pipe	Steel Pipe	Copper Pipe	Steel Pipe	Copper Pipe
1″	2 sprinklers	2 sprinklers	2 sprinklers	2 sprinklers	1 sprinklers	1 sprinklers
1-1/4″	3	3	3	3	2	2
1-1/2″	5	5	5	5	5	5
2″	10	12	10	12	8	8
2-1/2″	30	40	20	25	15	20
3″	60	65	40	45	27	30
3-1/2″	100	115	65	75	40	45
4″			100	115	55	65
5″			160	180	90	100
6″			275	300	150	170

Dry Pipe Systems: A dry pipe system should be installed where a wet pipe system is impractical, as in rooms or buildings which cannot be properly heated.

The use of an approved dry pipe system is more desirable than shutting off the water supply during cold weather.

Not more than 750 gallons of system capacity should be controlled by one dry pipe valve. Where two or more dry pipe valves are used, systems should preferably be divided horizontally.

R211313-40 Adjustment for Sprinkler/Standpipe Installations

Quality/Complexity Multiplier (For all installations)

Economy installation, add . . . . 0 to 5%
Good quality, medium complexity, add . . . . 5 to 15%
Above average quality and complexity, add . . . . 15 to 25%

Plumbing | R2201 | Operation & Maintenance of Plumbing

R220105-10 Demolition (Selective vs. Removal for Replacement)

Demolition can be divided into two basic categories.

One type of demolition involves the removal of material with no concern for its replacement. The labor-hours to estimate this work are found under "Selective Demolition" in the Fire Protection, Plumbing and HVAC Divisions. It is selective in that individual items or all the material installed as a system or trade grouping such as plumbing or heating systems are removed. This may be accomplished by the easiest way possible, such as sawing, torch cutting, or sledge hammer as well as simple unbolting.

The second type of demolition is the removal of some item for repair or replacement. This removal may involve careful draining, opening of unions, disconnecting and tagging of electrical connections, capping of pipes/ducts to prevent entry of debris or leakage of the material contained as well as transport of the item away from its in-place location to a truck/dumpster. An approximation of the time required to accomplish this type of demolition is to use half of the time indicated as necessary to install a new unit. For example; installation of a new pump might be listed as requiring 6 labor-hours so if we had to estimate the removal of the old pump we would allow an additional 3 hours for a total of 9 hours. That is, the complete replacement of a defective pump with a new pump would be estimated to take 9 labor-hours.

R220523-80 Valve Materials

VALVE MATERIALS

Bronze:
Bronze is one of the oldest materials used to make valves. It is most commonly used in hot and cold water systems and other non-corrosive services. It is often used as a seating surface in larger iron body valves to ensure tight closure.

Carbon Steel:
Carbon steel is a high strength material. Therefore, valves made from this metal are used in higher pressure services, such as steam lines up to 600 psi at 850°F. Many steel valves are available with butt-weld ends for economy and are generally used in high pressure steam service as well as other higher pressure non-corrosive services.

Forged Steel:
Valves from tough carbon steel are used in service up to 2000 psi and temperatures up to 1000°F in Gate, Globe and Check valves.

Iron:
Valves are normally used in medium to large pipe lines to control non-corrosive fluid and gases, where pressures do not exceed 250 psi at 450° or 500 psi cold water, oil or gas.

Stainless Steel:
Developed steel alloys can be used in over 90% corrosive services.

Plastic PVC:
This is used in a great variety of valves generally in high corrosive service with lower temperatures and pressures.

VALVE SERVICE PRESSURES

Pressure ratings on valves provide an indication of the safe operating pressure for a valve at some elevated temperature. This temperature is dependent upon the materials used and the fabrication of the valve. When specific data is not available, a good "rule-of-thumb" to follow is the temperature of saturated steam on the primary rating indicated on the valve body. Example: The valve has the number 150S printed on the side indicating 150 psi and hence, a maximum operating temperature of 367°F (temperature of saturated steam and 150 psi).

DEFINITIONS

1. "WOG" – Water, oil, gas (cold working pressures).
2. "SWP" – Steam working pressure.
3. 100% area (full port) – means the area through the valve is equal to or greater than the area of standard pipe.
4. "Standard Opening" – means that the area through the valve is less than the area of standard pipe and therefore these valves should be used only where restriction of flow is unimportant.
5. "Round Port" – means the valve has a full round opening through the plug and body, of the same size and area as standard pipe.
6. "Rectangular Port" – valves have rectangular shaped ports through the plug body. The area of the port is either equal to 100% of the area of standard pipe, or restricted (standard opening). In either case it is clearly marked.
7. "ANSI" – American National Standards Institute.

R220523-90 Valve Selection Considerations

INTRODUCTION: In any piping application, valve performance is critical. Valves should be selected to give the best performance at the lowest cost.

The following is a list of performance characteristics generally expected of valves.

1. Stopping flow or starting it.
2. Throttling flow (Modulation).
3. Flow direction changing.
4. Checking backflow (Permitting flow in only one direction).
5. Relieving or regulating pressure.

In order to properly select the right valve, some facts must be determined.

A. What liquid or gas will flow through the valve?

B. Does the fluid contain suspended particles?

C. Does the fluid remain in liquid form at all times?

D. Which metals does fluid corrode?

E. What are the pressure and temperature limits? (As temperature and pressure rise, so will the price of the valve.)

F. Is there constant line pressure?

G. Is the valve merely an on-off valve?

H. Will checking of backflow be required?

I. Will the valve operate frequently or infrequently?

Valves are classified by design type into such classifications as Gate, Globe, Angle, Check, Ball, Butterfly and Plug. They are also classified by end connection, stem, pressure restrictions and material such as bronze, cast iron, etc. Each valve has a specific use. A quality valve used correctly will provide a lifetime of trouble- free service, but a high quality valve installed in the wrong service may require frequent attention.

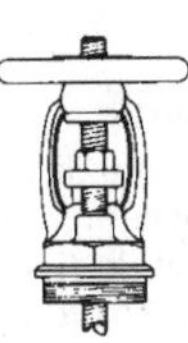

STEM TYPES
(OS & Y)—Rising Stem-Outside Screw and Yoke

Offers a visual indication of whether the valve is open or closed. Recommended where high temperatures, corrosives, and solids in the line might cause damage to inside-valve stem threads. The stem threads are engaged by the yoke bushing so the stem rises through the hand wheel as it is turned.

(R.S.)—Rising Stem-Inside Screw

Adequate clearance for operation must be provided because both the hand wheel and the stem rise.
The valve wedge position is indicated by the position of the stem and hand wheel.

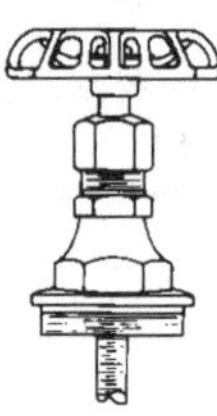

(N.R.S.)—Non-Rising Stem-Inside Screw

A minimum clearance is required for operating this type of valve. Excessive wear or damage to stem threads inside the valve may be caused by heat, corrosion, and solids. Because the hand wheel and stem do not rise, wedge position cannot be visually determined.

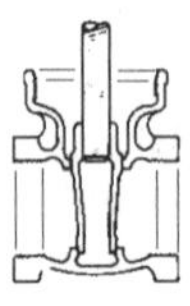

VALVE TYPES
Gate Valves

Provide full flow, minute pressure drop, minimum turbulence and minimum fluid trapped in the line.
They are normally used where operation is infrequent.

Globe Valves

Globe valves are designed for throttling and/or frequent operation with positive shut-off. Particular attention must be paid to the several types of seating materials available to avoid unnecessary wear. The seats must be compatible with the fluid in service and may be composition or metal. The configuration of the Globe valve opening causes turbulence which results in increased resistance. Most bronze Globe valves are rising stem-inside screw, but they are also available on O.S. & Y.

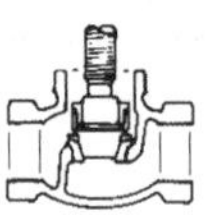

Angle Valves

The fundamental difference between the Angle valve and the Globe valve is the fluid flow through the Angle valve. It makes a 90° turn and offers less resistance to flow than the Globe valve while replacing an elbow. An Angle valve thus reduces the number of joints and installation time.

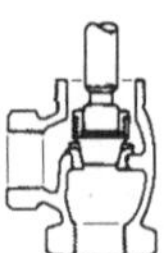

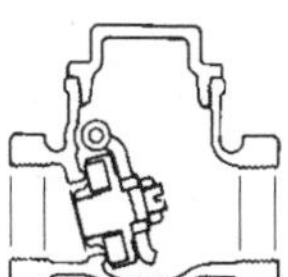
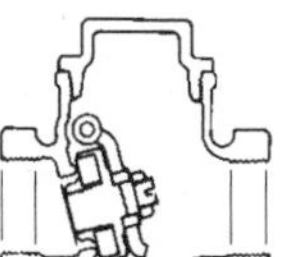
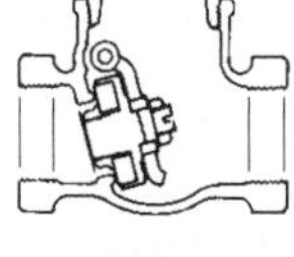

Check Valves

Check valves are designed to prevent backflow by automatically seating when the direction of fluid is reversed.
Swing Check valves are generally installed with Gate-valves, as they provide comparable full flow. Usually recommended for lines where flow velocities are low and should not be used on lines with pulsating flow. Recommended for horizontal installation, or in vertical lines only where flow is upward.

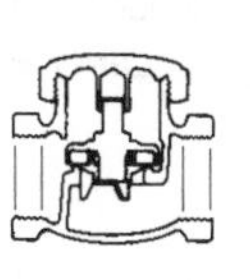
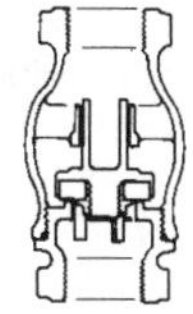

Lift Check Valves

These are commonly used with Globe and Angle valves since they have similar diaphragm seating arrangements and are recommended for preventing backflow of steam, air, gas and water, and on vapor lines with high flow velocities. For horizontal lines, horizontal lift checks should be used and vertical lift checks for vertical lines.

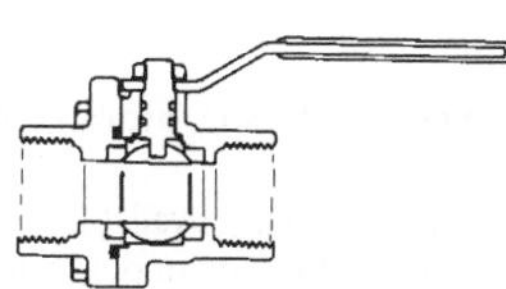

Ball Valves

Ball valves are light and easily installed, yet because of modern elastomeric seats, provide tight closure. Flow is controlled by rotating up to 90° a drilled ball which fits tightly against resilient seals. This ball seats with flow in either direction, and valve handle indicates the degree of opening. Recommended for frequent operation readily adaptable to automation, ideal for installation where space is limited.

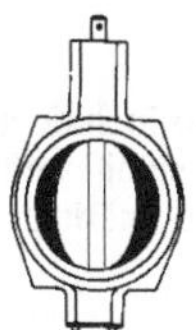

Butterfly Valves

Butterfly valves provide bubble-tight closure with excellent throttling characteristics. They can be used for full-open, closed and for throttling applications.

The Butterfly valve consists of a disc within the valve body which is controlled by a shaft. In its closed position, the valve disc seals against a resilient seat. The disc position throughout the full 90° rotation is visually indicated by the position of the operator.

A Butterfly valve is only a fraction of the weight of a Gate valve and requires no gaskets between flanges in most cases. Recommended for frequent operation and adaptable to automation where space is limited.

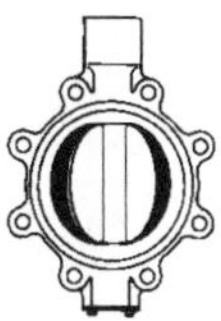

Wafer and Lug type bodies when installed between two pipe flanges, can be easily removed from the line. The pressure of the bolted flanges holds the valve in place.
Locating lugs makes installation easier.

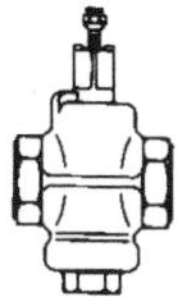

Plug Valves

Lubricated plug valves, because of the wide range of service to which they are adapted, may be classified as all purpose valves. They can be safely used at all pressure and vacuums, and at all temperatures up to the limits of available lubricants. They are the most satisfactory valves for the handling of gritty suspensions and many other destructive, erosive, corrosive and chemical solutions.

R221113-40 Plumbing Approximations for Quick Estimating

Water Control

Water Meter; Backflow Preventer, Shock Absorbers; Vacuum Breakers; Mixer. .... 10 to 15% of Fixtures

Pipe And Fittings .... 30 to 60% of Fixtures

Note: Lower percentage for compact buildings or larger buildings with plumbing in one area.
Larger percentage for large buildings with plumbing spread out.
In extreme cases pipe may be more than 100% of fixtures.
Percentages **do not** include special purpose or process piping.

Plumbing Labor

1 & 2 Story Residential .... Rough-in Labor = 80% of Materials

Apartment Buildings .... Rough-in Labor = 90 to 100% of Materials

Labor for handling and placing fixtures is approximately 25 to 30% of fixtures

Quality/Complexity Multiplier (for all installations)

Economy installation, add.......... 0 to 5%

Good quality, medium complexity, add .... 5 to 15%

Above average quality and complexity, add .... 15 to 25%

R221113-50 Pipe Material Considerations

1. Malleable fittings should be used for gas service.
2. Malleable fittings are used where there are stresses/strains due to expansion and vibration.
3. Cast fittings may be broken as an aid to disassembling of heating lines frozen by long use, temperature and minerals.
4. Cast iron pipe is extensively used for underground and submerged service.
5. Type M (light wall) copper tubing is available in hard temper only and is used for nonpressure and less severe applications than K and L.
6. Type L (medium wall) copper tubing, available hard or soft for interior service.
7. Type K (heavy wall) copper tubing, available in hard or soft temper for use where conditions are severe. For underground and interior service.
8. Hard drawn tubing requires fewer hangers or supports but should not be bent. Silver brazed fittings are recommended, however soft solder is normally used.
9. Type DMV (very light wall) copper tubing designed for drainage, waste and vent plus other non-critical pressure services.

Domestic/Imported Pipe and Fittings Cost

The prices shown in this publication for steel/cast iron pipe and steel, cast iron, malleable iron fittings are based on domestic production sold at the normal trade discounts. The above listed items of foreign manufacture may be available at prices of 1/3 to 1/2 those shown. Some imported items after minor machining or finishing operations are being sold as domestic to further complicate the system.

Caution: Most pipe prices in this book also include a coupling and pipe hangers which for the larger sizes can add significantly to the per foot cost and should be taken into account when comparing "book cost" with quoted supplier's cost.

R221113-70 Piping to 10′ High

When taking off pipe, it is important to identify the different material types and joining procedures, as well as distances between supports and components required for proper support.

During the takeoff, measure through all fittings. Do not subtract the lengths of the fittings, valves, or strainers, etc. This added length plus the final rounding of the totals will compensate for nipples and waste.

When rounding off totals always increase the actual amount to correspond with manufacturer's shipping lengths.

A. Both red brass and yellow brass pipe are normally furnished in 12′ lengths, plain end. The Unit Price section includes in the linear foot costs two field threads and one coupling per 10′ length. A carbon steel clevis type hanger assembly every 10′ is also prorated into the linear foot costs, including both material and labor.

B. Cast iron soil pipe is furnished in either 5′ or 10′ lengths. For pricing purposes, the Unit Price section features 10′ lengths with a joint and a carbon steel clevis hanger assembly every 5′ prorated into the per foot costs of both material and labor.

Three methods of joining are considered: lead and oakum poured joints or push-on gasket type joints for the bell and spigot pipe and a joint clamp for the no-hub soil pipe. The labor and material costs for each of these individual joining procedures are also prorated into the linear costs per foot.

C. Copper tubing covers types K, L, M, and DWV which are furnished in 20′ lengths. Means pricing data is based on a tubing cut each length and a coupling and two soft soldered joints every 10′. A carbon steel, clevis type hanger assembly every 10′ is also prorated into the per foot costs. The prices for refrigeration tubing are for materials only. Labor for full lengths may be based on the type L labor but short cut measures in tight areas can increase the installation labor-hours from 20 to 40%.

D. Corrosion-resistant piping does not lend itself to one particular standard of hanging or support assembly due to its diversity of application and placement. The several varieties of corrosion-resistant piping do not include any material or labor costs for hanger assemblies (See the Unit Price section for appropriate selection).

E. Glass pipe is furnished in standard lengths either 5′ or 10′ long, beaded on one end. Special orders for diverse lengths beaded on both ends are also available. For pricing purposes, R.S. Means features 10′ lengths with a coupling and a carbon steel band hanger assembly every 10′ prorated into the per foot linear costs.

Glass pipe is also available with conical ends and standard lengths ranging from 6″ through 3′ in 6″ increments, then up to 10′ in 12″ increments. Special lengths can be customized for particular installation requirements.

For pricing purposes, Means has based the labor and material pricing on 10′ lengths. Included in these costs per linear foot are the prorated costs for a flanged assembly every 10′ consisting of two flanges, a gasket, two insertable seals, and the required number of bolts and nuts. A carbon steel band hanger assembly based on 10′ center lines has also been prorated into the costs per foot for labor and materials.

F. Plastic pipe of several compositions and joining methods are considered. Fiberglass reinforced pipe (FRP) is priced, based on 10′ lengths (20′ lengths are also available), with coupling and epoxy joints every 10′. FRP is furnished in both "General Service" and "High Strength." A carbon steel clevis hanger assembly, 3 for every 10′, is built into the prorated labor and material costs on a per foot basis.

The PVC and CPVC pipe schedules 40, 80 and 120, plus SDR ratings are all based on 20′ lengths with a coupling installed every 10′, as well as a carbon steel clevis hanger assembly every 3′. The PVC and ABS type DWV piping is based on 10′ lengths with solvent weld couplings every 10′, and with carbon steel clevis hanger assemblies, 3 for every 10′. The rest of the plastic piping in this section is based on flexible 100′ coils and does not include any coupling or supports.

This section ends with PVC drain and sewer piping based on 10′ lengths with bell and spigot ends and 0-ring type, push-on joints.

G. Stainless steel piping includes both weld end and threaded piping, both in the type 304 and 316 specification and in the following schedules, 5, 10, 40, 80, and 160. Although this piping is usually furnished in 20′ lengths, this cost grouping has a joint (either heli-arc butt-welded or threads and coupling) every 10′. A carbon steel clevis type hanger assembly is also included at 10′ intervals and prorated into the linear foot costs.

H. Carbon steel pipe includes both black and galvanized. This section encompasses schedules 40 (standard) and 80 (extra heavy).

Several common methods of joining steel pipe — such as thread and coupled, butt welded, and flanged (150 lb. weld neck flanges) are also included.

For estimating purposes, it is assumed that the piping is purchased in 20′ lengths and that a compatible joint is made up every 10′. These joints are prorated into the labor and material costs per linear foot. The following hanger and support assemblies every 10′ are also included: carbon steel clevis for the T & C pipe, and single rod roll type for both the welded and flanged piping. All of these hangers are oversized to accommodate pipe insulation 3/4″ thick through 5″ pipe size and 1-1/2″ thick from 6″ through 12″ pipe size.

I. Grooved joint steel pipe is priced both black and galvanized, in schedules 10, 40, and 80, furnished in 20′ lengths. This section describes two joining methods: cut groove and roll groove. The schedule 10 piping is roll-grooved, while the heavier schedules are cut-grooved. The labor and material costs are prorated into per linear foot prices, including a coupled joint every 10′, as well as a carbon steel clevis hanger assembly.

Notes:

The pipe hanger assemblies mentioned in the preceding paragraphs include the described hanger; appropriately sized steel, box-type insert and nut; plus a threaded hanger rod. On average, the distance from the pipe center line to the insert face is 2′.

C clamps are used when the pipe is to be supported from steel shapes rather than anchored in the slab. C clamps are slightly less costly than inserts. However, to save time in estimating, it is advisable to use the given line number cost, rather than substituting a C clamp for the insert.

Add to piping labor for elevated installation:

10′ to 15′ high	10%	30′ to 35′ high	40%
15′ to 20′ high	20%	35′ to 40′ high	50%
20′ to 25′ high	25%	Over 40′ high	55%
25′ to 30′ high	35%		

When using the percentage adds for elevated piping installations as shown above, bear in mind that the given heights are for the pipe supports, even though the insert, anchor, or clamp may be several feet higher than the pipe itself.

An allowance has been included in the piping installation time for testing and minor tightening of leaking joints, fittings, stuffing boxes, packing glands, etc. For extraordinary test requirements such as x-rays, prolonged pressure or demonstration tests, a percentage of the piping labor, based on the estimator's experience, must be added to the labor total. A testing service specializing in weld x-rays should be consulted for pricing if it is an estimate requirement. Equipment installation time includes start-up with associated adjustments.

R221316-10 Drainage Requirements

Drainage lines must have a slope to maintain flow for proper operation. This slope should not be less than 1/4″ per foot for 3″ diameter or smaller pipe and not less than 1/8″ per foot for 4″ diameter or larger pipe. The capacity of building drainage systems is calculated on a basis of "drainage fixture units" (d.f.u.) as per the following chart.

Type of Fixture	d.f.u. Value	Type of Fixture	d.f.u. Value
Automatic clothes washer (2″ standpipe)	3	Service sink (trap standard)	3
Bathroom group (water closet, lavatory and		Service sink (P trap)	2
bathtub or shower) tank type closet	6	Urinal, pedestal, syphon jet blowout	6
Bathtub (with or without overhead shower)	2	Urinal, wall hung	4
Clinic sink	6	Urinal, stall washout	4
Combination sink & tray with food disposal	4	Wash sink (cir. or mult.) per faucet set	2
Dental unit or cuspidor	1	Water closet, tank operated	4
Dental lavatory	1	Water closet, valve operated	6
Drinking fountain	1/2	Fixtures not listed above	
Dishwasher, domestic	2	Trap size 1-1/4″ or smaller	1
Floor drains with 2″ waste	3	Trap size 1-1/2″	2
Kitchen sink, domestic with one 1-1/2″ trap	2	Trap size 2″	3
Kitchen sink, domestic with food disposal	2	Trap size 2-1/2″	4
Lavatory with 1-1/4″ waste	1	Trap size 3″	5
Laundry tray (1 or 2 compartment)	2	Trap size 4″	6
Shower stall, domestic	2		

For continuous or nearly continuous flow into the sytem from a pump, air conditioning equipment or other item, allow 2 fixture units for each gallon per minute of flow.

When the "drainage fixture units" (d.f.u.) for each horizontal branch or vertical stack is computed from the table above, the appropriate pipe size for each branch or stack is determined from the table below.

R221316-20 Allowable Fixture Units (d.f.u.) for Branches and Stacks

Pipe Diam.	Horiz. Branch (not incl. drains)	Stack Size for 3 Stories or 3 Levels	Stack size for Over 3 levels	Maximum for 1 Story building Stack
1-1/2″	3	4	8	2
2″	6	10	24	6
2-1/2″	12	20	42	9
3″	20*	48*	72*	20*
4″	160	240	500	90
5″	360	540	1100	200
6″	620	960	1900	350
8″	1400	2200	3600	600
10″	2500	3800	5600	1000
12″	3900	6000	8400	1500
15″	7000			

*Not more than two water closets or bathroom groups within each branch interval nor more than six water closets or bathroom groups on the stack.

Stacks sized for the total may be reduced as load decreases at each story to a minimum diameter of 1/2 the maximum diameter.

R224000-10 Hot Water Consumption Rates

Type of Building	Size Factor	Maximum Hourly Demand	Average Day Demand
Apartment Dwellings	No. of Apartments:		
	Up to 20	12.0 Gal. per apt.	42.0 Gal. per apt.
	21 to 50	10.0 Gal. per apt.	40.0 Gal. per apt.
	51 to 75	8.5 Gal. per apt.	38.0 Gal. per apt.
	76 to 100	7.0 Gal. per apt.	37.0 Gal. per apt.
	101 to 200	6.0 Gal. per apt.	36.0 Gal. per apt.
	201 up	5.0 Gal. per apt.	35.0 Gal. per apt.
Dormitories	Men	3.8 Gal. per man	13.1 Gal. per man
	Women	5.0 Gal. per woman	12.3 Gal. per woman
Hospitals	Per bed	23.0 Gal. per patient	90.0 Gal. per patient
Hotels	Single room with bath	17.0 Gal. per unit	50.0 Gal. per unit
	Double room with bath	27.0 Gal. per unit	80.0 Gal. per unit
Motels	No. of units:		
	Up to 20	6.0 Gal. per unit	20.0 Gal. per unit
	21 to 100	5.0 Gal. per unit	14.0 Gal. per unit
	101 Up	4.0 Gal. per unit	10.0 Gal. per unit
Nursing Homes		4.5 Gal. per bed	18.4 Gal. per bed
Office buildings		0.4 Gal. per person	1.0 Gal. per person
Restaurants	Full meal type	1.5 Gal./max. meals/hr.	2.4 Gal. per meal
	Drive-in snack type	0.7 Gal./max. meals/hr.	0.7 Gal. per meal
Schools	Elementary	0.6 Gal. per student	0.6 Gal. per student
	Secondary & High	1.0 Gal. per student	1.8 Gal. per student

For evaluation purposes, recovery rate and storage capacity are inversely proportional. Water heaters should be sized so that the maximum hourly demand anticipated can be met in addition to allowance for the heat loss from the pipes and storage tank.

R224000-20 Fixture Demands in Gallons Per Fixture Per Hour

Table below is based on 140°F final temperature except for dishwashers in public places (*) where 180°F water is mandatory.

Fixture	Apartment House	Club	Gym	Hospital	Hotel	Indust. Plant	Office	Private Home	School
Bathtubs	20	20	30	20	20			20	
Dishwashers, automatic	15	50-150*		50-150*	50-200*	20-100*		15	20-100*
Kitchen sink	10	20		20	30	20	20	10	20
Laundry, stationary tubs	20	28		28	28			20	
Laundry, automatic wash	75	75		100	150			75	
Private lavatory	2	2	2	2	2	2	2	2	2
Public lavatory	4	6	8	6	8	12	6		15
Showers	30	150	225	75	75	225	30	30	225
Service sink	20	20		20	30	20	20	15	20
Demand factor	0.30	0.30	0.40	0.25	0.25	0.40	0.30	0.30	0.40
Storage capacity factor	1.25	0.90	1.00	0.60	0.80	1.00	2.00	0.70	1.00

To obtain the probable maximum demand multiply the total demands for the fixtures (gal./fixture/hour) by the demand factor. The heater should have a heating capacity in gallons per hour equal to this maximum. The storage tank should have a capacity in gallons equal to the probable maximum demand multiplied by the storage capacity factor.

Plumbing | R2240 Plumbing Fixtures

R224000-30 Minimum Plumbing Fixture Requirements

Minimum Plumbing Fixture Requirements

Type of Building or Occupancy (2)	Water Closets (14) (Fixtures per Person)		Urinals (5,10) (Fixtures per Person)		Lavatories (Fixtures per Person)		Bathtubs or Showers (Fixtures per Person)	Drinking Fountains (Fixtures per Person) (3, 13)
	Male	Female	Male	Female	Male	Female		
Assembly Places-Theatres, Auditoriums, Convention Halls, etc.-for permanent employee use	1: 1 - 15 2: 16 - 35 3: 36 - 55	1: 1 - 15 2: 16 - 35 3: 36 - 55	0: 1 - 9 1: 10 - 50		1 per 40	1 per 40		
	Over 55, add 1 fixture for each additional 40 persons		Add one fixture for each additional 50 males					
Assembly Places-Theatres, Auditoriums, Convention Halls, etc. - for public use	1: 1 - 100 2: 101 - 200 3: 201 - 400	3: 1 - 50 4: 51 - 100 8: 101 - 200 11: 201 - 400	1: 1 - 100 2: 101 - 200 3: 201 - 400 4: 401 - 600		1: 1 - 200 2: 201 - 400 3: 401 - 750	1: 1 - 200 2: 201 - 400 3: 401 - 750		1: 1 - 150 2: 151 - 400 3: 401 - 750
	Over 400, add 1 fixture for each additional 500 males and 1 for each additional 125 females		Over 600, add 1 fixture for each additional 300 males		Over 750, add 1 fixture for each additional 500 persons			Over 750, add one fixture for each additional 500 persons
Dormitories (9) School or Labor	1 per 10	1 per 8	1 per 25		1 per 12	1 per 12	1 per 8	1 per 150 (12)
	Add 1 fixture for each additional 25 males (over 10) and 1 for each additional 20 females (over 8)		Over 150, add 1 fixture for each additional 50 males		Over 12 add 1 fixture for each additional 20 males and 1 for each 15 additional females		For females add 1 bathtub per 30. Over 150, add 1 per 20	
Dormitories-for Staff Use	1: 1 - 15 2: 16 - 35 3: 36 - 55	1: 1 - 15 3: 16 - 35 4: 36 - 55	1 per 50		1 per 40	1 per 40	1 per 8	
	Over 55, add 1 fixture for each additional 40 persons							
Dwellings: Single Dwelling Multiple Dwelling or Apartment House	1 per dwelling 1 per dwelling or apartment unit				1 per dwelling 1 per dwelling or apartment unit		1 per dwelling 1 per dwelling or apartment unit	
Hospital Waiting rooms	1 per room				1 per room			1 per 150 (12)
Hospitals-for employee use	1: 1 - 15 2: 16 - 35 3: 36 - 55	1: 1 - 15 3: 16 - 35 4: 36 - 55	0: 1 - 9 1: 10 - 50		1 per 40	1 per 40		
	Over 55, add 1 fixture for each additional 40 persons		Add 1 fixture for each additional 50 males					
Hospitals: Individual Room Ward Room	1 per room 1 per 8 patients				1 per room 1 per 10 patients		1 per room 1 per 20 patients	1 per 150 (12)
Industrial (6) Warehouses Workshops, Foundries and similar establishments-for employee use	1: 1 -10 2: 11 - 25 3: 26 - 50 4: 51 - 75 5: 76 - 100	1: 1 -10 2: 11 - 25 3: 26 - 50 4: 51 - 75 5: 76 - 100			Up to 100, per 10 persons Over 100, 1 per 15 persons (7, 8)		1 shower for each 15 persons exposed to excessive heat or to skin contamination with poisonous, infectious or irritating material	1 per 150 (12)
	Over 100, add 1 fixture for each additional 30 persons							
Institutional - Other than Hospitals or Penal Institutions (on each occupied floor)	1 per 25	1 per 20	0: 1 - 9 1: 10 - 50		1 per 10	1 per 10	1 per 8	1 per 150 (12)
			Add 1 fixture for each additional 50 males					
Institutional - Other than Hospitals or Penal Institutions (on each occupied floor)-for employee use	1: 1 - 15 2: 16 - 35 3: 36 - 55	1: 1 - 15 3: 16 - 35 4: 36 - 55	0: 1 - 9 1: 10 - 50		1 per 40	1 per 40	1 per 8	1 per 150 (12)
	Over 55, add 1 fixture for each additional 40 persons		Add 1 fixture for each additional 50 males					
Office or Public Buildings	1: 1 - 100 2: 101 - 200 3: 201 - 400	3: 1 - 50 4: 51 - 100 8: 101 - 200 11: 201 - 400	1: 1 - 100 2: 101 - 200 3: 201 - 400 4: 401 - 600		1: 1 - 200 2: 201 - 400 3: 401 - 750	1: 1 - 200 2: 201 - 400 3: 401 - 750		1 per 150 (12)
	Over 400, add 1 fixture for each additional 500 males and 1 for each additional 150 females		Over 600, add 1 fixture for each additional 300 males		Over 750, add 1 fixture for each additional 500 persons			
Office or Public Buildings - for employee use	1: 1 - 15 2: 16 - 35 3: 36 - 55	1: 1 - 15 3: 16 - 35 4: 36 - 55	0: 1 - 9 1: 10 - 50		1 per 40	1 per 40		
	Over 55, add 1 fixture for each additional 40 persons		Add 1 fixture for each additional 50 males					

R224000-30 Minimum Plumbing Fixture Requirements (cont.)

Minimum Plumbing Fixture Requirements

Type of Building or Occupancy	Water Closets (14) (Fixtures per Person)		Urinals (5, 10) (Fixtures per Person)		Lavatories (Fixtures per Person)		Bathtubs or Showers (Fixtures per Person)	Drinking Fountains (Fixtures per Person) (3, 13)
	Male	Female	Male	Female	Male	Female		
Penal Institutions - for employee use	1: 1 - 15 2: 16 - 35 3: 36 - 55	1: 1 - 15 3: 16 - 35 4: 36 - 55	0: 1 - 9 1: 10 - 50		1 per 40	1 per 40		1 per 150 (12)
	Over 55, add 1 fixture for each additional 40 persons		Add 1 fixture for each additional 50 males					
Penal Institutions - for prison use Cell	1 per cell				1 per cell			1 per cellblock floor
Exercise room	1 per exercise room		1 per exercise room		1 per exercise room			1 per exercise room
Restaurants, Pubs and Lounges (11)	1: 1 - 50 2: 51 - 150 3: 151 - 300	1: 1 - 50 2: 51 - 150 4: 151 - 300	1: 1 - 150		1: 1 - 150 2: 151 - 200 3: 201 - 400	1: 1 - 150 2: 151 - 200 3: 201 - 400		
	Over 300, add 1 fixture for each additional 200 persons		Over 150, add 1 fixture for each additional 150 males		Over 400, add 1 fixture for each additional 400 persons			
Schools - for staff use All Schools	1: 1 - 15 2: 16 - 35 3: 36 - 55	1: 1 - 15 3: 16 - 35 4: 36 - 55	1 per 50		1 per 40	1 per 40		
	Over 55, add 1 fixture for each additional 40 persons							
Schools - for student use: Nursery	1: 1 - 20 2: 21 - 50	1: 1 - 20 2: 21 - 50			1: 1 - 25 2: 26 - 50	1: 1 - 25 2: 26 - 50		1 per 150 (12)
	Over 50, add 1 fixture for each additional 50 persons				Over 50, add 1 fixture for each additional 50 persons			
Elementary	1 per 30	1 per 25	1 per 75		1 per 35	1 per 35		1 per 150 (12)
Secondary	1 per 40	1 per 30	1 per 35		1 per 40	1 per 40		1 per 150 (12)
Others (Colleges, Universities, Adult Centers, etc.	1 per 40	1 per 30	1 per 35		1 per 40	1 per 40		1 per 150 (12)
Worship Places Educational and Activities Unit	1 per 150	1 per 75	1 per 150		1 per 2 water closets			1 per 150 (12)
Worship Places Principal Assembly Place	1 per 150	1 per 75	1 per 150		1 per 2 water closets			1 per 150 (12)

Notes:

1. The figures shown are based upon one (1) fixture being the minimum required for the number of persons indicated or any fraction thereof.
2. Building categories not shown on this table shall be considered separately by the Administrative Authority.
3. Drinking fountains shall not be installed in toilet rooms.
4. Laundry trays. One (1) laundry tray or one (1) automatic washer standpipe for each dwelling unit or one (1) laundry trays or one (1) automatic washer standpipes, or combination thereof, for each twelve (12) apartments. Kitchen sinks, one (1) for each dwelling or apartment unit.
5. For each urinal added in excess of the minimum required, one water closet may be deducted. The number of water closets shall not be reduced to less than two-thirds (2/3) of the minimum requirement.
6. As required by ANSI Z4.1-1968, Sanitation in Places of Employment.
7. Where there is exposure to skin contamination with poisonous, infectious, or irritating materials, provide one (1) lavatory for each five (5) persons.
8. Twenty-four (24) lineal inches of wash sink or eighteen (18) inches of a circular basin, when provided with water outlets for such space shall be considered equivalent to one (1) lavatory.
9. Laundry trays, one (1) for each fifty (50) persons. Service sinks, one (1) for each hundred (100) persons.
10. General. In applying this schedule of facilities, consideration shall be given to the accessibility of the fixtures. Conformity purely on a numerical basis may not result in an installation suited to the need of the individual establishment. For example, schools should be provided with toilet facilities on each floor having classrooms.
    a. Surrounding materials, wall and floor space to a point two (2) feet in front of urinal lip and four (4) feet above the floor, and at least two (2) feet to each side of the urinal shall be lined with non-absorbent materials.
    b. Trough urinals shall be prohibited.
11. A restaurant is defined as a business which sells food to be consumed on the premises.
    a. The number of occupants for a drive-in restaurant shall be considered as equal to the number of parking stalls.
    b. Employee toilet facilities shall not to be included in the above restaurant requirements. Hand washing facilities shall be available in the kitchen for employees.
12. Where food is consumed indoors, water stations may be substituted for drinking fountains. Offices, or public buildings for use by more than six (6) persons shall have one (1) drinking fountain for the first one hundred fifty (150) persons and one additional fountain for each three hundred (300) persons thereafter.
13. There shall be a minimum of one (1) drinking fountain per occupied floor in schools, theaters, auditoriums, dormitories, offices of public building.
14. The total number of water closets for females shall be at least equal to the total number of water closets and urinals required for males.

R224000-40 Plumbing Fixture Installation Time

Item	Rough-In	Set	Total Hours	Item	Rough-In	Set	Total Hours
Bathtub	5	5	10	Shower head only	2	1	3
Bathtub and shower, cast iron	6	6	12	Shower drain	3	1	4
Fire hose reel and cabinet	4	2	6	Shower stall, slate		15	15
Floor drain to 4 inch diameter	3	1	4	Slop sink	5	3	8
Grease trap, single, cast iron	5	3	8	Test 6 fixtures			14
Kitchen gas range		4	4	Urinal, wall	6	2	8
Kitchen sink, single	4	4	8	Urinal, pedestal or floor	6	4	10
Kitchen sink, double	6	6	12	Water closet and tank	4	3	7
Laundry tubs	4	2	6	Water closet and tank, wall hung	5	3	8
Lavatory wall hung	5	3	8	Water heater, 45 gals. gas, automatic	5	2	7
Lavatory pedestal	5	3	8	Water heaters, 65 gals. gas, automatic	5	2	7
Shower and stall	6	4	10	Water heaters, electric, plumbing only	4	2	6

Fixture prices in front of book are based on the cost per fixture set in place. The rough-in cost, which must be added for each fixture, includes carrier, if required, some supply, waste and vent pipe connecting fittings and stops. The lengths of rough-in pipe are nominal runs which would connect to the larger runs and stacks. The supply runs and DWV runs and stacks must be accounted for in separate entries. In the eastern half of the United States it is common for the plumber to carry these to a point 5′ outside the building.

R224000-50 Water Cooler Application

Type of Service	Requirement
Office, School or Hospital	12 persons per gallon per hour
Office, Lobby or Department Store	4 or 5 gallons per hour per fountain
Light manufacturing	7 persons per gallon per hour
Heavy manufacturing	5 persons per gallon per hour
Hot heavy manufacturing	4 persons per gallon per hour
Hotel	.08 gallons per hour per room
Theatre	1 gallon per hour per 100 seats

R230500-10 Subcontractors

On the unit cost pages of the R.S. Means Cost Data books, the last column is entitled "Total Incl. O&P". This is normally the cost of the installing contractor. In the HVAC Division, this is the cost of the mechanical contractor. If the particular work being estimated is to be performed by a sub to the mechanical contractor, the mechanical's profit and handling charge (usually 10%) is added to the total of the last column.

R233100-10 Loudness Levels for Moving Air thru Fans, Diffusers, Register, Etc. (Measured in Sones)

Recommended Loudness Levels (Sones)

Area	Very Quiet	Quiet	Noisy
Auditoriums			
Auditorium lobbies	3	4	6
Concert and opera halls	0.8	1	1.5
Courtrooms	2	3	4
Lecture halls	1.5	2	3
Movie theaters	1.5	2	3
Churches and Schools			
Kitchens	4	6	8
Laboratories	2	4	6
Libraries	1.5	2	3
Recreation halls	2	4	8
Sanctuaries	1	1.7	3
Schools and classrooms	1.5	2.5	4
Hospital and Clinics			
Laboratories	2	4	6
Lobbies, waiting rooms	2	4	6
Halls and corridors	2	4	6
Operating rooms	1.5	2.5	4
Private rooms	1	1.7	3
Wards	1.5	2.5	4
Hotels			
Banquet Rooms	1.5	3	6
Individual rooms, suites	1	2	4
Kitchens and laundries	4	7	10
Lobbies	2	4	8
Indoor Sports			
Bowling alleys	3	4	6
Gymnasiums	2	4	6
Swimming pools	4	7	10
Manufacturing Areas			
Assembly lines	5	12	30
Foreman's office	3	5	8
Foundries	10	20	40
General storage	5	10	20
Heavy machinery	10	25	60
Light machinery	5	12	30
Tool maintenance	4	7	10
Stores			
Department stores	3	6	8
Supermarkets	4	7	10
Offices			
Conference rooms	1	1.7	3
Drafting	2	4	8
General open offices	2	4	8
Halls and corridors	2.5	5	10
Professional offices	1.5	3	6
Reception room	1.5	3	6
Tabulation & computation	3	6	12
Public Buildings			
Banks	2	4	6
Court houses	2.5	4	6
Museums	2	3	4
Planetariums	1.5	2	3
Post offices	2.5	4	6
Public libraries	1.5	2	4
Waiting rooms	3	5	8
Restaurants			
Cafeterias	3	6	10
Night clubs	2.5	4	6
Restaurants	2	4	8

R233100-20 Ductwork

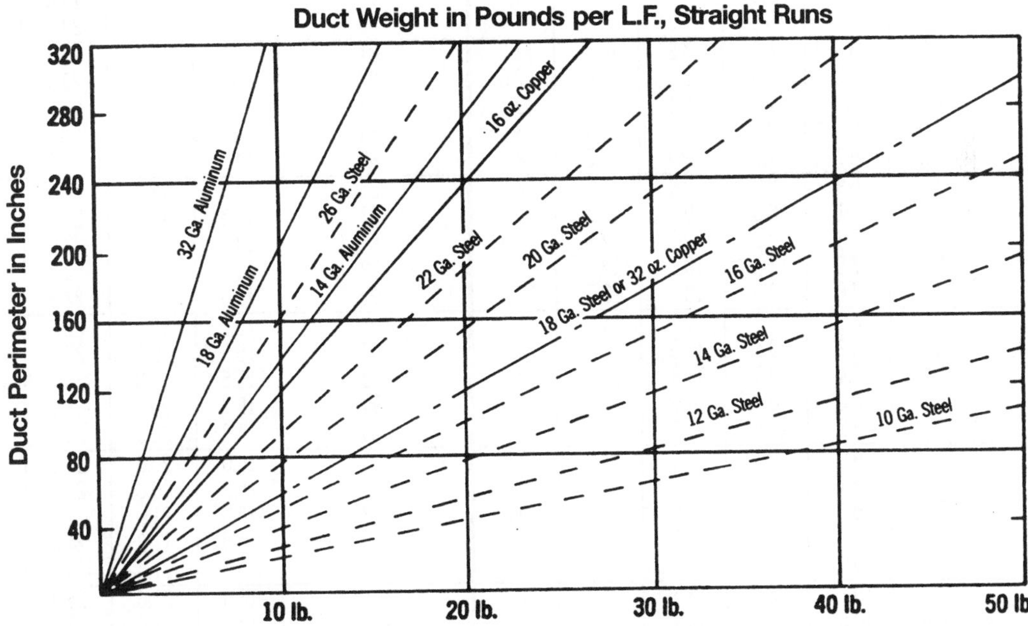

Add to the above for fittings; 90° elbow is 3 L.F.; 45° elbow is 2.5 L.F.; offset is 4 L.F.; transition offset is 6 L.F.; square-to-round transition is 4 L.F.; 90° reducing elbow is 5 L.F. For bracing and waste, add 20% to aluminum and copper, 15% to steel.

R233100-30 Duct Fabrication/Installation

The labor cost for fabricated sheet metal duct includes both the cost of fabrication and installation of the duct. The split is approximately 40% for fabrication, 60% for installation. It is for this reason that the percentage add for elevated installation of fabricated duct is less than the percentage add for prefabricated/preformed duct.

Example: assume a piece of duct cost $100 installed (labor only)

Sheet Metal Fabrication = 40% = $40
Installation = 60% = $60

The add for elevated installation is:

Based on total labor (fabrication & installation) = $100 x 6% = $6.00

Based on installation cost only (Material purchased prefabricated) = $60 x 10% = $6.00

The $6.00 markup (10′ to 15′ high) is the same.

R233100-40 Sheet Metal Calculator (Weight in Lb./Ft. of Length)

Gauge	26	24	22	20	18	16	Gauge	26	24	22	20	18	16
Wt.-Lb./S.F.	.906	1.156	1.406	1.656	2.156	2.656	Wt.-Lb./S.F.	.906	1.156	1.406	1.656	2.156	2.656
SMACNA Max. Dimension – Long Side		30″	54″	84″	85″ Up		SMACNA Max. Dimension – Long Side		30″	54″	84″	85″ Up	
Sum-2 sides							Sum-2 Sides						
2	.3	.40	.50	.60	.80	.90	56	9.3	12.0	14.0	16.2	21.3	25.2
3	.5	.65	.80	.90	1.1	1.4	57	9.5	12.3	14.3	16.5	21.7	25.7
4	.7	.85	1.0	1.2	1.5	1.8	58	9.7	12.5	14.5	16.8	22.0	26.1
5	.8	1.1	1.3	1.5	1.9	2.3	59	9.8	12.7	14.8	17.1	22.4	26.6
6	1.0	1.3	1.5	1.7	2.3	2.7	60	10.0	12.9	15.0	17.4	22.8	27.0
7	1.2	1.5	1.8	2.0	2.7	3.2	61	10.2	13.1	15.3	17.7	23.2	27.5
8	1.3	1.7	2.0	2.3	3.0	3.6	62	10.3	13.3	15.5	18.0	23.6	27.9
9	1.5	1.9	2.3	2.6	3.4	4.1	63	10.5	13.5	15.8	18.3	24.0	28.4
10	1.7	2.2	2.5	2.9	3.8	4.5	64	10.7	13.7	16.0	18.6	24.3	28.8
11	1.8	2.4	2.8	3.2	4.2	5.0	65	10.8	13.9	16.3	18.9	24.7	29.3
12	2.0	2.6	3.0	3.5	4.6	5.4	66	11.0	14.1	16.5	19.1	25.1	29.7
13	2.2	2.8	3.3	3.8	4.9	5.9	67	11.2	14.3	16.8	19.4	25.5	30.2
14	2.3	3.0	3.5	4.1	5.3	6.3	68	11.3	14.6	17.0	19.7	25.8	30.6
15	2.5	3.2	3.8	4.4	5.7	6.8	69	11.5	14.8	17.3	20.0	26.2	31.1
16	2.7	3.4	4.0	4.6	6.1	7.2	70	11.7	15.0	17.5	20.3	26.6	31.5
17	2.8	3.7	4.3	4.9	6.5	7.7	71	11.8	15.2	17.8	20.6	27.0	32.0
18	3.0	3.9	4.5	5.2	6.8	8.1	72	12.0	15.4	18.0	20.9	27.4	32.4
19	3.2	4.1	4.8	5.5	7.2	8.6	73	12.2	15.6	18.3	21.2	27.7	32.9
20	3.3	4.3	5.0	5.8	7.6	9.0	74	12.3	15.8	18.5	21.5	28.1	33.3
21	3.5	4.5	5.3	6.1	8.0	9.5	75	12.5	16.1	18.8	21.8	28.5	33.8
22	3.7	4.7	5.5	6.4	8.4	9.9	76	12.7	16.3	19.0	22.0	28.9	34.2
23	3.8	5.0	5.8	6.7	8.7	10.4	77	12.8	16.5	19.3	22.3	29.3	34.7
24	4.0	5.2	6.0	7.0	9.1	10.8	78	13.0	16.7	19.5	22.6	29.6	35.1
25	4.2	5.4	6.3	7.3	9.5	11.3	79	13.2	16.9	19.8	22.9	30.0	35.6
26	4.3	5.6	6.5	7.5	9.9	11.7	80	13.3	17.1	20.0	23.2	30.4	36.0
27	4.5	5.8	6.8	7.8	10.3	12.2	81	13.5	17.3	20.3	23.5	30.8	36.5
28	4.7	6.0	7.0	8.1	10.6	12.6	82	13.7	17.5	20.5	23.8	31.2	36.9
29	4.8	6.2	7.3	8.4	11.0	13.1	83	13.8	17.8	20.8	24.1	31.5	37.4
30	5.0	6.5	7.5	8.7	11.4	13.5	84	14.0	18.0	21.0	24.4	31.9	37.8
31	5.2	6.7	7.8	9.0	11.8	14.0	85	14.2	18.2	21.3	24.7	32.3	38.3
32	5.3	6.9	8.0	9.3	12.2	14.4	86	14.3	18.4	21.5	24.9	32.7	38.7
33	5.5	7.1	8.3	9.6	12.5	14.9	87	14.5	18.6	21.8	25.2	33.1	39.2
34	5.7	7.3	8.5	9.9	12.9	15.3	88	14.7	18.8	22.0	25.5	33.4	39.6
35	5.8	7.5	8.8	10.2	13.3	15.8	89	14.8	19.0	22.3	25.8	33.8	40.1
36	6.0	7.8	9.0	10.4	13.7	16.2	90	15.0	19.3	22.5	26.1	34.2	40.5
37	6.2	8.0	9.3	10.7	14.1	16.7	91	15.2	19.5	22.8	26.4	34.6	41.0
38	6.3	8.2	9.5	11.0	14.4	17.1	92	15.3	19.7	23.0	26.7	35.0	41.4
39	6.5	8.4	9.8	11.3	14.8	17.6	93	15.5	19.9	23.3	27.0	35.3	41.9
40	6.7	8.6	10.0	11.6	15.2	18.0	94	15.7	20.1	23.5	27.3	35.7	42.3
41	6.8	8.8	10.3	11.9	15.6	18.5	95	15.8	20.3	23.8	27.6	36.1	42.8
42	7.0	9.0	10.5	12.2	16.0	18.9	96	16.0	20.5	24.0	27.8	36.5	43.2
43	7.2	9.2	10.8	12.5	16.3	19.4	97	16.2	20.8	24.3	28.1	36.9	43.7
44	7.3	9.5	11.0	12.8	16.7	19.8	98	16.3	21.0	24.5	28.4	37.2	44.1
45	7.5	9.7	11.3	13.1	17.1	20.3	99	16.5	21.2	24.8	28.7	37.6	44.6
46	7.7	9.9	11.5	13.3	17.5	20.7	100	16.7	21.4	25.0	29.0	38.0	45.0
47	7.8	10.1	11.8	13.6	17.9	21.2	101	16.8	21.6	25.3	29.3	38.4	45.5
48	8.0	10.3	12.0	13.9	18.2	21.6	102	17.0	21.8	25.5	29.6	38.8	45.9
49	8.2	10.5	12.3	14.2	18.6	22.1	103	17.2	22.0	25.8	29.9	39.1	46.4
50	8.3	10.7	12.5	14.5	19.0	22.5	104	17.3	22.3	26.0	30.2	39.5	46.8
51	8.5	11.0	12.8	14.8	19.4	23.0	105	17.5	22.5	26.3	30.5	39.9	47.3
52	8.7	11.2	13.0	15.1	19.8	23.4	106	17.7	22.7	26.5	30.7	40.3	47.7
53	8.8	11.4	13.3	15.4	20.1	23.9	107	17.8	22.9	26.8	31.0	40.7	48.2
54	9.0	11.6	13.5	15.7	20.5	24.3	108	18.0	23.1	27.0	31.3	41.0	48.6
55	9.2	11.8	13.8	16.0	20.9	24.8	109	18.2	23.3	27.3	31.6	41.4	49.1
							110	18.3	23.5	27.5	31.9	41.8	49.5

Example: If duct is 34″ x 20″ x 15′ long, 34″ is greater than 30″ maximum, for 24 ga. so must be 22 ga. 34″ + 20″ = 54″ going across from 54″ find 13.5 lb. per foot. 13.5 x 15′ = 202.5 lbs. For S.F. of surface area 202.5 ÷ 1.406 = 144 S.F.

Note: Figures include an allowance for scrap.

R233100-50 Ductwork Packages (per Ton of Cooling)

System	Sheet Metal	Insulation	Diffusers	Return Register
Roof Top Unit Single Zone	120 Lbs.	52 S.F.	1	1
Roof Top Unit Multizone	240 Lbs.	104 S.F.	2	1
Self-contained Air or Water Cooled	108 Lbs.	—	2	—
Split System Air Cooled	102 Lbs.	—	2	—

Systems reflect most common usage.
Refer to system graphics for duct layout.

Heating, Ventilating & Air Cond. | R2334 | HVAC Fans

R233400-10 Recommended Ventilation Air Changes

Table below lists range of time in minutes per change for various types of facilities.

Facility	Minutes	Facility	Minutes	Facility	Minutes
Assembly Halls	2-10	Dance Halls	2-10	Laundries	1-3
Auditoriums	2-10	Dining Rooms	3-10	Markets	2-10
Bakeries	2-3	Dry Cleaners	1-5	Offices	2-10
Banks	3-10	Factories	2-5	Pool Rooms	2-5
Bars	2-5	Garages	2-10	Recreation Rooms	2-10
Beauty Parlors	2-5	Generator Rooms	2-5	Sales Rooms	2-10
Boiler Rooms	1-5	Gymnasiums	2-10	Theaters	2-8
Bowling Alleys	2-10	Kitchens-Hospitals	2-5	Toilets	2-5
Churches	5-10	Kitchens-Restaurant	1-3	Transformer Rooms	1-5

CFM air required for changes = Volume of room in cubic feet ÷ Minutes per change.

Heating, Ventilating & Air Cond. | R2337 | Air Outlets & Inlets

R233700-60 Diffuser Evaluation

CFM = V × An × K where V = Outlet velocity in feet per minute. An = Neck area in square feet and K = Diffuser delivery factor. An undersized diffuser for a desired CFM will produce a high velocity and noise level. When air moves past people at a velocity in excess of 25 FPM, an annoying draft is felt. An oversized diffuser will result in low velocity with poor mixing. Consideration must be given to avoid vertical stratification or horizontal areas of stagnation.

R235000-10 Heating Systems

Heating Systems

The basic function of a heating system is to bring an enclosed volume up to a desired temperature and then maintain that temperature within a reasonable range. To accomplish this, the selected system must have sufficient capacity to offset transmission losses resulting from the temperature difference on the interior and exterior of the enclosing walls in addition to losses due to cold air infiltration through cracks, crevices and around doors and windows. The amount of heat to be furnished is dependent upon the building size, construction, temperature difference, air leakage, use, shape, orientation and exposure. Air circulation is also an important consideration. Circulation will prevent stratification which could result in heat losses through uneven temperatures at various levels. For example, the most efficient use of unit heaters can usually be achieved by circulating the space volume through the total number of units once every 20 minutes or 3 times an hour. This general rule must, of course, be adapted for special cases such as large buildings with low ratios of heat transmitting surface to cubical volume. The type of occupancy of a building will have considerable bearing on the number of heat transmitting units and the location selected. It is axiomatic, however, that the basis of any successful heating system is to provide the maximum amount of heat at the points of maximum heat loss such as exposed walls, windows, and doors. Large roof areas, wind direction, and wide doorways create problems of excessive heat loss and require special consideration and treatment.

Heat Transmission

Heat transfer is an important parameter to be considered during selection of the exterior wall style, material and window area. A high rate of transfer will permit greater heat loss during the wintertime with the resultant increase in heating energy costs and a greater rate of heat gain in the summer with proportionally greater cooling cost. Several terms are used to describe various aspects of heat transfer. However, for general estimating purposes this book lists U values for systems of construction materials. U is the "overall heat transfer coefficient." It is defined as the heat flow per hour through one square foot when the temperature difference in the air on either side of the structure wall, roof, ceiling or floor is one degree Fahrenheit. The structural segment may be a single homogeneous material or a composite.

Total heat transfer is found using the following equation:

$Q = AU(T_2 - T_1)$ where

- **Q** = Heat flow, BTU per hour
- **A** = Area, square feet
- **U** = Overall heat transfer coefficient
- **$(T_2 - T_1)$** = Difference in temperature of air on each side of the construction component. (Also abbreviated TD)

Note that heat can flow through all surfaces of any building and this flow is in addition to heat gain or loss due to ventilation, infiltration and generation (appliances, machinery, people).

R235000-20 Heating Approximations for Quick Estimating

Oil Piping & Boiler Room Piping

- Small System . . . . . . 20 to 30% of Boiler
- Complex System with Pumps, Headers, Etc. . . . . . . 80 to 110% of Boiler

Breeching With Insulation:

- Small . . . . . . 10 to 15% of Boiler
- Large . . . . . . 15 to 25% of Boiler

Coils: . . . . . . 15 to 30% of Containing Unit

Balancing (Independent) . . . . . . 1/2% of H.V.A.C. Estimate

Quality/Complexity Adjustment: For all heating installations add these adjustments to the estimate to more closely allow for the equipment and conditions of the particular job under consideration.

- Economy installation, add . . . . . . 0 to 5% of System
- Good quality, medium complexity, add . . . . . . 5 to 15% of System
- Above average quality and complexity, add . . . . . . 15 to 25% of System

R235000-30 The Basics of a Heating System

The function of a heating system is to achieve and maintain a desired temperature in a room or building by replacing the amount of heat being dissipated. There are four kinds of heating systems: hot-water, steam, warm-air and electric resistance. Each has certain essential and similar elements with the exception of electric resistance heating.

The basic elements of a heating system are:

A. A **combustion chamber** in which fuel is burned and heat transferred to a conveying medium.

B. The **"fluid"** used for conveying the heat (water, steam or air).

C. **Conductors** or pipes for transporting the fluid to specific desired locations.

D. A means of disseminating the heat, sometimes called **terminal units.**

A. The **combustion chamber** in a furnace heats air which is then distributed. This is called a warm-air system.

The combustion chamber in a boiler heats water which is either distributed as hot water or steam and this is termed a hydronic system.

The maximum allowable working pressures are limited by ASME "Code for Heating Boilers" to 15 PSI for steam and 160 PSI for hot water heating boilers, with a maximum temperature limitation of 250°F. Hot water boilers are generally rated for a working pressure of 30 PSI. High pressure boilers are governed by the ASME "Code for Power Boilers" which is used almost universally for boilers operating over 15 PSIG. High pressure boilers used for a combination of heating/process loads are usually designed for 150 PSIG.

Boiler ratings are usually indicated as either Gross or Net Output. The Gross Load is equal to the Net Load plus a piping and pickup allowance. When this allowance cannot be determined, divide the gross output rating by 1.25 for a value equal to or greater than the net heat loss requirement of the building.

B. Of the three **fluids** used, steam carries the greatest amount of heat per unit volume. This is due to the fact that it gives up its latent heat of vaporization at a temperature considerably above room temperature. Another advantage is that the pressure to produce a positive circulation is readily available. Piping conducts the steam to terminal units and returns condensate to the boiler.

The **steam system** is well adapted to large buildings because of its positive circulation, its comparatively economical installation and its ability to deliver large quantities of heat. Nearly all large office buildings, stores, hotels, and industrial buildings are so heated, in addition to many residences.

Hot water, when used as the heat carrying fluid, gives up a portion of its sensible heat and then returns to the boiler or heating apparatus for reheating. As the heat conveyed by each pound of water is about one-fiftieth of the heat conveyed by a pound of steam, it is necessary to circulate about fifty times as much water as steam by weight (although only one-thirtieth as much by volume). The hot water system is usually, although not necessarily, designed to operate at temperatures below that of the ordinary steam system and so the amount of heat transfer surface must be correspondingly greater. A temperature of 190°F to 200°F is normally the maximum. Circulation in small buildings may depend on the difference in density between hot water and the cool water returning to the boiler; circulating pumps are normally used to maintain a desired rate of flow. Pumps permit a greater degree of flexibility and better control.

In **warm-air** furnace systems, cool air is taken from one or more points in the building, passed over the combustion chamber and flue gas passages and then distributed through a duct system. A disadvantage of this system is that the ducts take up much more building volume than steam or hot water pipes. Advantages of this system are the relative ease with which humidification can be accomplished by the evaporation of water as the air circulates through the heater, and the lack of need for expensive disseminating units as the warm air simply becomes part of the interior atmosphere of the building.

C. Conductors (pipes and ducts) have been lightly treated in the discussion of conveying fluids. For more detailed information such as sizing and distribution methods, the reader is referred to technical publications such as the American Society of Heating, Refrigerating and Air-Conditioning Engineers "Handbook of Fundamentals."

D. Terminal units come in an almost infinite variety of sizes and styles, but the basic principles of operation are very limited. As previously mentioned, warm-air systems require only a simple register or diffuser to mix heated air with that present in the room. Special application items such as radiant coils and infrared heaters are available to meet particular conditions but are not usually considered for general heating needs. Most heating is accomplished by having air flow over coils or pipes containing the heat transporting medium (steam, hot-water, electricity). These units, while varied, may be separated into two general types, (1) radiator/convectors and (2) unit heaters.

Radiator/convectors may be cast, fin-tube or pipe assemblies. They may be direct, indirect, exposed, concealed or mounted within a cabinet enclosure, upright or baseboard style. These units are often collectively referred to as "radiatiors" or "radiation" although none gives off heat either entirely by radiation or by convection but rather a combination of both. The air flows over the units as a gravity "current." It is necessary to have one or more heat-emitting units in each room. The most efficient placement is low along an outside wall or under a window to counteract the cold coming into the room and achieve an even distribution.

In contrast to radiator/convectors which operate most effectively against the walls of smaller rooms, **unit heaters** utilize a fan to move air over heating coils and are very effective in locations of relatively large volume. Unit heaters, while usually suspended overhead, may be floor mounted. They also may take in fresh outside air for ventilation. The heat distributed by unit heaters may be from a remote source and conveyed by a fluid or it may be from the combustion of fuel in each individual heater. In the latter case the only piping required would be for fuel, however, a vent for the products of combustion would be necessary.

The following list gives may of the advantages of unit heaters for applications other than office or residential:

a. Large capacity so smaller number of units are required, **b.** Piping system simplified, **c.** Space saved where they are located overhead out of the way, **d.** Rapid heating directed where needed with effective wide distribution, **e.** Difference between floor and ceiling temperature reduced, **f.** Circulation of air obtained, and ventilation with introduction of fresh air possible, **g.** Heat output flexible and easily controlled.

R235000-35 Heating (42° degrees latitude)

$$\text{Approximate S.F. radiation} = \frac{\text{S.F. sash}}{2} + \frac{\text{S.F. wall + roof — sash}}{20} + \frac{\text{C.F. building}}{200}$$

R235000-50 Factor for Determining Heat Loss for Various Types of Buildings

General: While the most accurate estimates of heating requirements would naturally be based on detailed information about the building being considered, it is possible to arrive at a reasonable approximation using the following procedure:

1. Calculate the cubic volume of the room or building.
2. Select the appropriate factor from Table 1 below. Note that the factors apply only to inside temperatures listed in the first column and to 0°F outside temperature.
3. If the building has bad north and west exposures, multiply the heat loss factor by 1.1.
4. If the outside design temperature is other than 0°F, multiply the factor from Table 1 by the factor from Table 2.
5. Multiply the cubic volume by the factor selected from Table 1. This will give the estimated BTUH heat loss which must be made up to maintain inside temperature.

Table 1 — Building Type	Conditions	Qualifications	Loss Factor*
Factories & Industrial Plants General Office Areas at 70°F	One Story	Skylight in Roof	6.2
		No Skylight in Roof	5.7
	Multiple Story	Two Story	4.6
		Three Story	4.3
		Four Story	4.1
		Five Story	3.9
		Six Story	3.6
	All Walls Exposed	Flat Roof	6.9
		Heated Space Above	5.2
	One Long Warm Common Wall	Flat Roof	6.3
		Heated Space Above	4.7
	Warm Common Walls on Both Long Sides	Flat Roof	5.8
		Heated Space Above	4.1
Warehouses at 60°F	All Walls Exposed	Skylights in Roof	5.5
		No Skylight in Roof	5.1
		Heated Space Above	4.0
	One Long Warm Common Wall	Skylight in Roof	5.0
		No Skylight in Roof	4.9
		Heated Space Above	3.4
	Warm Common Walls on Both Long Sides	Skylight in Roof	4.7
		No Skylight in Roof	4.4
		Heated Space Above	3.0

*Note: This table tends to be conservative particularly for new buildings designed for minimum energy consumption.

Table 2 — Outside Design Temperature Correction Factor (for Degrees Fahrenheit)									
Outside Design Temperature	50	40	30	20	10	0	-10	-20	-30
Correction Factor	.29	.43	.57	.72	.86	1.00	1.14	1.28	1.43

R235000-70 Transmission of Heat

R235000-70 and R235000-80 provide a way to calculate heat transmission of various construction materials from their U values and the TD (Temperature Difference).

1. From the Exterior Enclosure Division or elsewhere, determine U values for the construction desired.
2. Determine the coldest design temperature. The difference between this temperature and the desired interior temperature is the TD (temperature difference).
3. Enter R235000-70 or R235000-80 at correct U Value. Cross horizontally to the intersection with appropriate TD. Read transmission per square foot from bottom of figure.
4. Multiply this value of BTU per hour transmission per square foot of area by the total surface area of that type of construction.

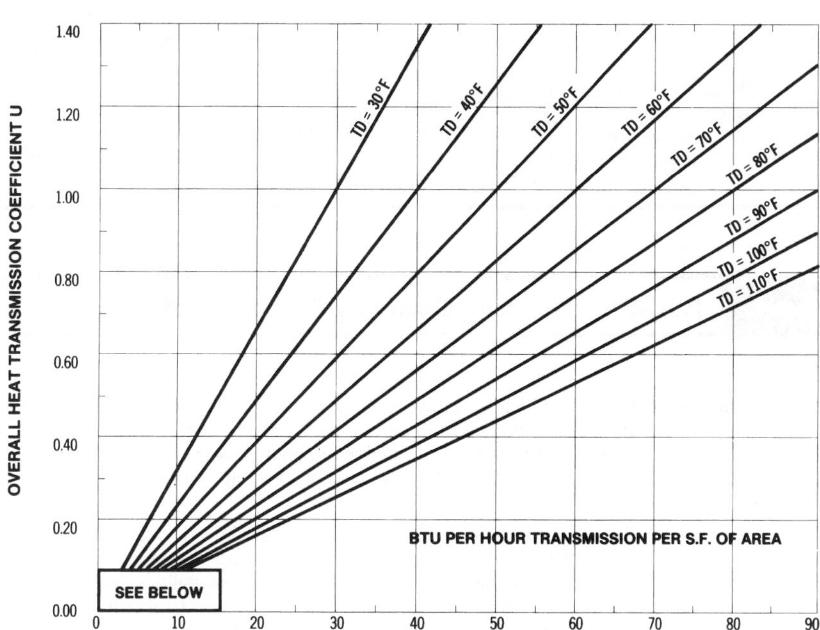

R235000-80 Transmission of Heat (Low Rate)

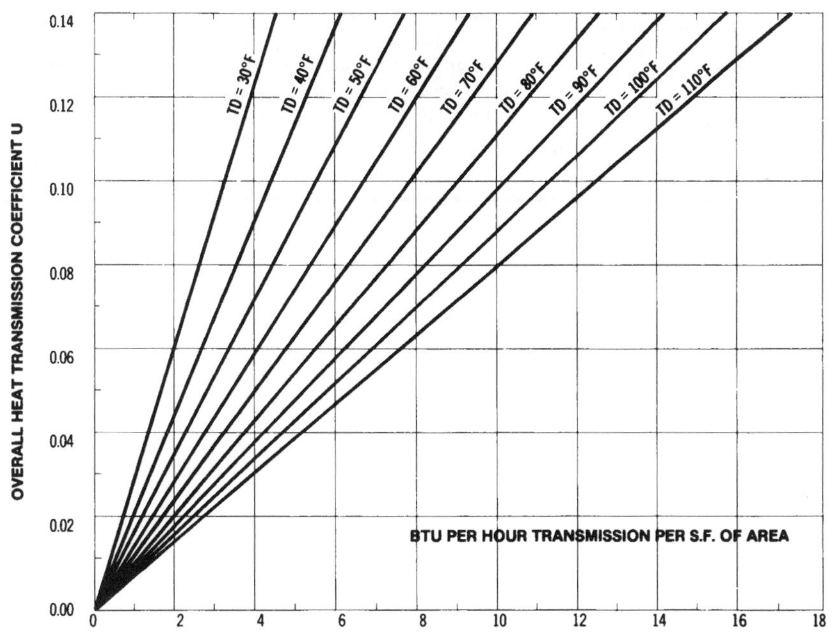

Heating, Ventilating & Air Cond. | R2356 | Solar Energy Heating Equipment

R235616-60 Solar Heating (Space and Hot Water)

Collectors should face as close to due South as possible, however, variations of up to 20 degrees on either side of true South are acceptable. Local climate and collector type may influence the choice between east or west deviations. Obviously they should be located so they are not shaded from the sun's rays. Incline collectors at a slope of latitude minus 5 degrees for domestic hot water and latitude plus 15 degrees for space heating.

Flat plate collectors consist of a number of components as follows: Insulation to reduce heat loss through the bottom and sides of the collector. The enclosure which contains all the components in this assembly is usually weatherproof and prevents dust, wind and water from coming in contact with the absorber plate. The cover plate usually consists of one or more layers of a variety of glass or plastic and reduces the reradiation by creating an air space which traps the heat between the cover and the absorber plates.

The absorber plate must have a good thermal bond with the fluid passages. The absorber plate is usually metallic and treated with a surface coating which improves absorptivity. Black or dark paints or selective coatings are used for this purpose, and the design of this passage and plate combination helps determine a solar system's effectiveness.

Heat transfer fluid passage tubes are attached above and below or integral with an absorber plate for the purpose of transferring thermal energy from the absorber plate to a heat transfer medium. The heat exchanger is a device for transferring thermal energy from one fluid to another.

Piping and storage tanks should be well insulated to minimize heat losses.

Size domestic water heating storage tanks to hold 20 gallons of water per user, minimum, plus 10 gallons per dishwasher or washing machine. For domestic water heating an optimum collector size is approximately 3/4 square foot of area per gallon of water storage. For space heating of residences and small commercial applications the collector is commonly sized between 30% and 50% of the internal floor area. For space heating of large commercial applications, collector areas less than 30% of the internal floor area can still provide significant heat reductions.

A supplementary heat source is recommended for Northern states for December through February.

The solar energy transmission per square foot of collector surface varies greatly with the material used. Initial cost, heat transmittance and useful life are obviously interrelated.

Heating, Ventilating & Air Cond. | R2360 | Central Cooling Equipment

R236000-10 Air Conditioning

General: The purpose of air conditioning is to control the environment of a space so that comfort is provided for the occupants and/or conditions are suitable for the processes or equipment contained therein. The several items which should be evaluated to define system objectives are:

- Temperature Control
- Humidity Control
- Cleanliness
- Odor, smoke and fumes
- Ventilation

Efforts to control the above parameters must also include consideration of the degree or tolerance of variation, the noise level introduced, the velocity of air motion and the energy requirements to accomplish the desired results.

The variation in **temperature** and **humidity** is a function of the sensor and the controller. The controller reacts to a signal from the sensor and produces the appropriate suitable response in either the terminal unit, the conductor of the transporting medium (air, steam, chilled water, etc.), or the source (boiler, evaporating coils, etc.).

The **noise level** is a by-product of the energy supplied to moving components of the system. Those items which usually contribute the most noise are pumps, blowers, fans, compressors and diffusers. The level of noise can be partially controlled through use of vibration pads, isolators, proper sizing, shields, baffles and sound absorbing liners.

Some **air motion** is necessary to prevent stagnation and stratification. The maximum acceptable velocity varies with the degree of heating or cooling which is taking place. Most people feel air moving past them at velocities in excess of 25 FPM as an annoying draft. However, velocities up to 45 FPM may be acceptable in certain cases. Ventilation, expressed as air changes per hour and percentage of fresh air, is usually an item regulated by local codes.

Selection of the system to be used for a particular application is usually a trade-off. In some cases the building size, style, or room available for mechanical use limits the range of possibilities. Prime factors influencing the decision are first cost and total life (operating, maintenance and replacement costs). The accuracy with which each parameter is determined will be an important measure of the reliability of the decision and subsequent satisfactory operation of the installed system.

Heat delivery may be desired from an air conditioning system. Heating capability usually is added as follows: A gas fired burner or hot water/steam/electric coils may be added to the air handling unit directly and heat all air equally. For limited or localized heat requirements the water/steam/electric coils may be inserted into the duct branch supplying the cold areas. Gas fired duct furnaces are also available.

Note: When water or steam coils are used the cost of the piping and boiler must also be added. For a rough estimate use the cost per square foot of the appropriate sized hydronic system with unit heaters. This will provide a cost for the boiler and piping, and the unit heaters of the system would equate to the approximate cost of the heating coils.

R236000-20 Air Conditioning Requirements

BTU's per hour per S.F. of floor area and S.F. per ton of air conditioning.

Type of Building	BTU per S.F.	S.F. per Ton	Type of Building	BTU per S.F.	S.F. per Ton	Type of Building	BTU per S.F.	S.F. per Ton
Apartments, Individual	26	450	Dormitory, Rooms	40	300	Libraries	50	240
Corridors	22	550	Corridors	30	400	Low Rise Office, Exterior	38	320
Auditoriums & Theaters	40	300/18*	Dress Shops	43	280	Interior	33	360
Banks	50	240	Drug Stores	80	150	Medical Centers	28	425
Barber Shops	48	250	Factories	40	300	Motels	28	425
Bars & Taverns	133	90	High Rise Office—Ext. Rms.	46	263	Office (small suite)	43	280
Beauty Parlors	66	180	Interior Rooms	37	325	Post Office, Individual Office	42	285
Bowling Alleys	68	175	Hospitals, Core	43	280	Central Area	46	260
Churches	36	330/20*	Perimeter	46	260	Residences	20	600
Cocktail Lounges	68	175	Hotel, Guest Rooms	44	275	Restaurants	60	200
Computer Rooms	141	85	Corridors	30	400	Schools & Colleges	46	260
Dental Offices	52	230	Public Spaces	55	220	Shoe Stores	55	220
Dept. Stores, Basement	34	350	Industrial Plants, Offices	38	320	Shop'g. Ctrs., Supermarkets	34	350
Main Floor	40	300	General Offices	34	350	Retail Stores	48	250
Upper Floor	30	400	Plant Areas	40	300	Specialty	60	200

*Persons per ton
12,000 BTU = 1 ton of air conditioning

R236000-30 Psychrometric Table

Dewpoint or Saturation Temperature (F)

Relative humidity (%)	32	35	40	45	50	55	60	65	70	75	80	85	90	95	100
100	32	35	40	45	50	55	60	65	70	75	80	85	90	95	100
90	30	33	37	42	47	52	57	62	67	72	77	82	87	92	97
80	27	30	34	39	44	49	54	58	64	68	73	78	83	88	93
70	24	27	31	36	40	45	50	55	60	64	69	74	79	84	88
60	20	24	28	32	36	41	46	51	55	60	65	69	74	79	83
50	16	20	24	28	33	36	41	46	50	55	60	64	69	73	78
40	12	15	18	23	27	31	35	40	45	49	53	58	62	67	71
30	8	10	14	18	21	25	29	33	37	42	46	50	54	59	62
20	6	7	8	9	13	16	20	24	28	31	35	40	43	48	52
10	4	4	5	5	6	8	9	10	13	17	20	24	27	30	34

Dry bulb temperature (F): 32, 35, 40, 45, 50, 55, 60, 65, 70, 75, 80, 85, 90, 95, 100

This table shows the relationship between RELATIVE HUMIDITY, DRY BULB TEMPERATURE AND DEWPOINT.

As an example, assume that the thermometer in a room reads 75°F, and we know that the relative humidity is 50%. The chart shows the dewpoint temperature to be 55°. That is, any surface colder than 55°F will "sweat" or collect condensing moisture. This surface could be the outside of an uninsulated chilled water pipe in the summertime, or the inside surface of a wall or deck in the wintertime. After determining the extreme ambient parameters, the table at the left is useful in determining which surfaces need insulation or vapor barrier protection.

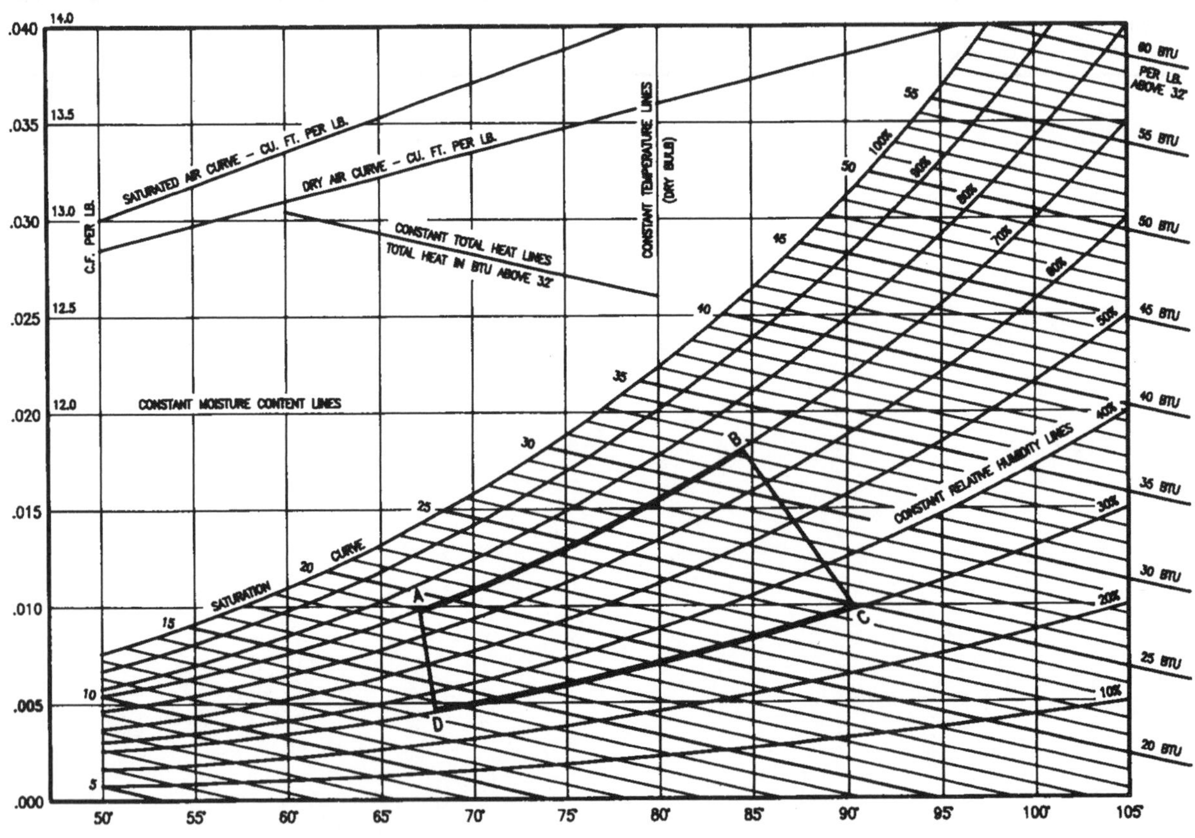

TEMPERATURE DEGREES FAHRENHEIT
TOTAL PRESSURE = 14.696 LB. PER SQ. IN. ABS.

Psychrometric chart showing different variables based on one pound of dry air. Space marked A B C D is temperature-humidity range which is most comfortable for majority of people.

R236000-90 Quality/Complexity Adjustment for Air Conditioning Systems

Economy installation, add	0 to 5%
Good quality, medium complexity, add	5 to 15%
Above average quality and complexity, add	15 to 25%

Add the above adjustments to the estimate to more closely allow for the equipment and conditions of the particular job under consideration.

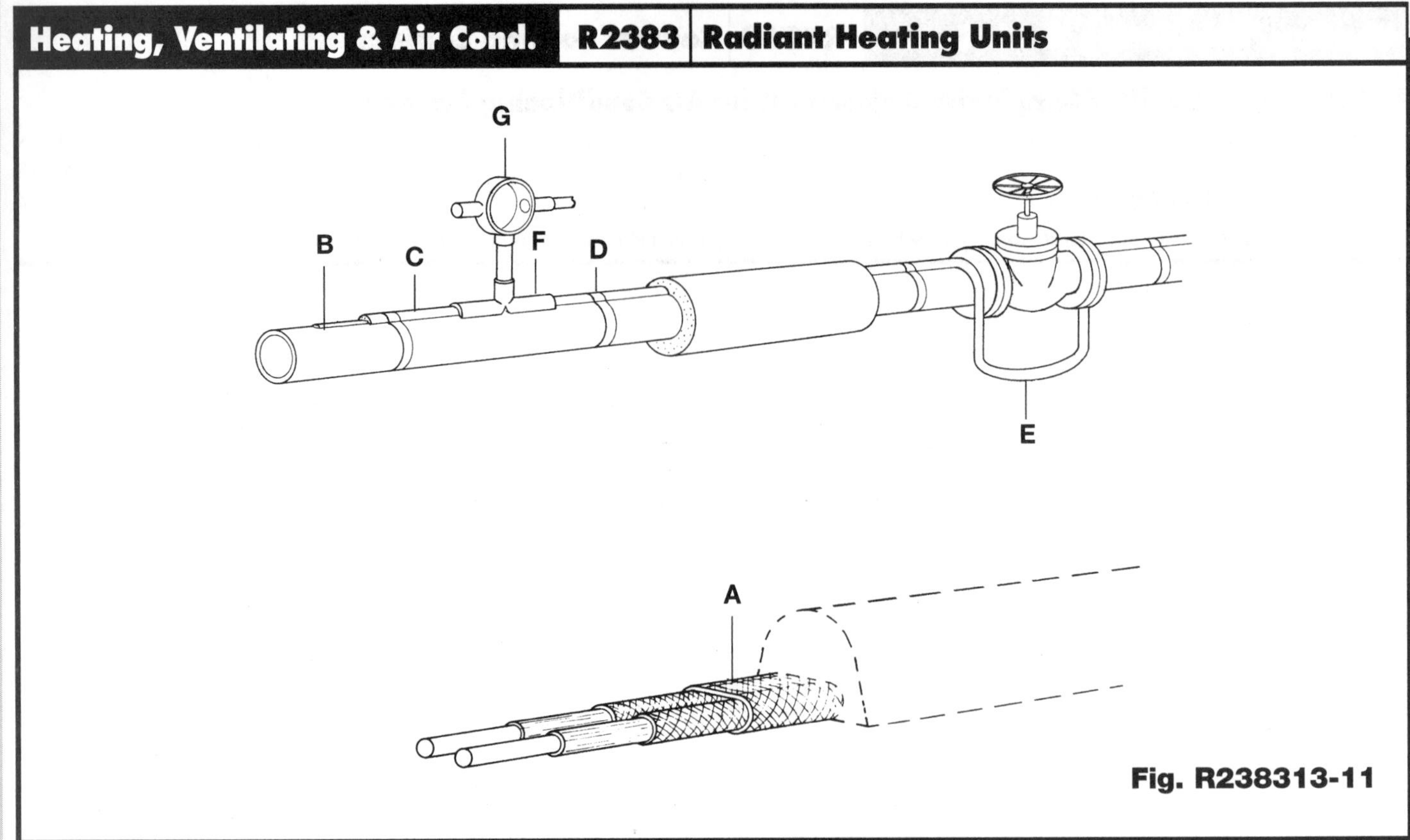

Fig. R238313-11

R238313-10 Heat Trace Systems

Before you can determine the cost of a HEAT TRACE installation the method of attachment must be established. There are (4) common methods:

1. Cable is simply attached to the pipe with polyester tape every 12′.
2. Cable is attached with a continuous cover of 2″ wide aluminum tape.
3. Cable is attached with factory extruded heat transfer cement and covered with metallic raceway with clips every 10′.
4. Cable is attached between layers of pipe insulation using either clips or polyester tape.

In all of the above methods each component of the system must be priced individually.

Example: Components for method 3 must include:

A. Heat trace cable by voltage and watts per linear foot.
B. Heat transfer cement, 1 gallon per 60 linear feet of cover.
C. Metallic raceway by size and type.
D. Raceway clips by size of pipe.

When taking off linear foot lengths of cable add the following for each valve in the system. (E)

SCREWED OR WELDED VALVE:			FLANGED VALVE:			BUTTERFLY VALVES:		
1/2″	=	6″	1/2″	=	1′ -0″	1/2″	=	0′
3/4″	=	9″	3/4″	=	1′ -6″	3/4″	=	0′
1″	=	1′ -0″	1″	=	2′ -0″	1″	=	1′ -0″
1-1/2″	=	1′ -6″	1-1/2″	=	2′ -6″	1-1/2″	=	1′ -6″
2″	=	2′	2″	=	2′ -6″	2″	=	2′ -0″
2-1/2″	=	2′ -6″	2-1/2″	=	3′ -0″	2-1/2″	=	2′ -6″
3″	=	2′ -6″	3″	=	3′ -6″	3″	=	2′ -6″
4″	=	4′ -0″	4″	=	4′ -0″	4″	=	3′ -0″
6″	=	7′ -0″	6″	=	8′ -0″	6″	=	3′ -6″
8″	=	9′ -6″	8″	=	11′ -0″	8″	=	4′ -0″
10″	=	12′ -6″	10″	=	14′ -0″	10″	=	4′ -0″
12″	=	15′ -0″	12″	=	16′ -6″	12″	=	5′ -0″
14″	=	18′ -0″	14″	=	19′ -6″	14″	=	5′ -6″
16″	=	21′ -6″	16″	=	23′ -0″	16″	=	6′ -0″
18″	=	25′ -6″	18″	=	27′ -0″	18″	=	6′ -6″
20″	=	28′ -6″	20″	=	30′ -0″	20″	=	7′ -0″
24″	=	34′ -0″	24″	=	36′ -0″	24″	=	8′ -0″
30″	=	40′ -0″	30″	=	42′ -0″	30″	=	10′ -0″

R238313-10 Heat Trace Systems (cont.)

Add the following quantities of heat transfer cement to linear foot totals for each valve:

Nominal Valve Size	Gallons of Cement per Valve
1/2″	0.14
3/4″	0.21
1″	0.29
1-1/2″	0.36
2″	0.43
2-1/2″	0.70
3″	0.71
4″	1.00
6″	1.43
8″	1.48
10″	1.50
12″	1.60
14″	1.75
16″	2.00
18″	2.25
20″	2.50
24″	3.00
30″	3.75

The following must be added to the list of components to accurately price HEAT TRACE systems:

1. Expediter fitting and clamp fasteners (F)
2. Junction box and nipple connected to expediter fitting (G)
3. Field installed terminal blocks within junction box
4. Ground lugs
5. Piping from power source to expediter fitting
6. Controls
7. Thermostats
8. Branch wiring
9. Cable splices
10. End of cable terminations
11. Branch piping fittings and boxes

Deduct the following percentages from labor if cable lengths in the same area exceed:

150′ to 250′	10%	351′ to 500′	20%
251′ to 350′	15%	Over 500′	25%

Add the following percentages to labor for elevated installations:

15′ to 20′ high	10%	31′ to 35′ high	40%
21′ to 25′ high	20%	36′ to 40′ high	50%
26′ to 30′ high	30%	Over 40′ high	60%

R238313-20 Spiral-Wrapped Heat Trace Cable (Pitch Table)

In order to increase the amount of heat, occasionally heat trace cable is wrapped in a spiral fashion around a pipe; increasing the number of feet of heater cable per linear foot of pipe.

Engineers first determine the heat loss per foot of pipe (based on the insulating material, its thickness, and the temperature differential across it). A ratio is then calculated by the formula:

$$\text{Feet of Heat Trace per Foot of Pipe} = \frac{\text{Watts/Foot of Heat Loss}}{\text{Watts/Foot of the Cable}}$$

The linear distance between wraps (pitch) is then taken from a chart or table. Generally, the pitch is listed on a drawing leaving the estimator to calculate the total length of heat tape required. An approximation may be taken from this table.

Feet of Heat Trace Per Foot of Pipe																
	Nominal Pipe Size in Inches															
Pitch In Inches	1	1¼	1½	2	2½	3	4	6	8	10	12	14	16	18	20	24
3.5	1.80															
4	1.65															
5	1.46	1.60	1.80													
6	1.34	1.45	1.55	1.75												
7	1.25	1.35	1.43	1.57	1.75											
8	1.20	1.28	1.34	1.45	1.60	1.80										
9	1.16	1.23	1.28	1.37	1.51	1.68										
10	1.13	1.19	1.24	1.32	1.44	1.57	1.82									
15	1.06	1.08	1.10	1.15	1.21	1.29	1.42	1.78								
20	1.04	1.05	1.06	1.08	1.13	1.17	1.25	1.49	1.73							
25		1.04	1.04	1.06	1.08	1.11	1.17	1.33	1.51	1.72						
30				1.04	1.05	1.07	1.12	1.24	1.37	1.54	1.70	1.80				
35					1.06	1.06	1.09	1.17	1.28	1.42	1.54	1.64	1.78			
40						1.05	1.07	1.14	1.22	1.33	1.44	1.52	1.64	1.75		
50							1.05	1.09	1.15	1.22	1.29	1.35	1.44	1.53	1.64	1.83
60								1.06	1.11	1.16	1.21	1.25	1.31	1.39	1.46	1.62
70								1.05	1.08	1.12	1.17	1.19	1.24	1.30	1.35	1.47
80									1.06	1.09	1.13	1.15	1.19	1.24	1.28	1.38
90									1.04	1.06	1.10	1.13	1.16	1.19	1.23	1.32
100										1.05	1.08	1.10	1.13	1.15	1.19	1.23

Note: Common practice would normally limit the lower end of the table to 5% of additional heat and above 80% an engineer would likely opt for two (2) parallel cables.

Electrical | R2601 | Operation & Maint. of Electrical Systems

R260105-30 Electrical Demolition (Removal for Replacement)

The purpose of this reference number is to provide a guide to users for electrical "removal for replacement" by applying the rule of thumb: 1/3 of new installation time (typical range from 20% to 50%) for removal. Remember to use reasonable judgment when applying the suggested percentage factor. For example:

Contractors have been requested to remove an existing fluorescent lighting fixture and replace with a new fixture utilizing energy saver lamps and electronic ballast:

In order to fully understand the extent of the project, contractors should visit the job site and estimate the time to perform the renovation work in accordance with applicable national, state and local regulation and codes.

The contractor may need to add extra labor hours to his estimate if he discovers unknown concealed conditions such as: contaminated asbestos ceiling, broken acoustical ceiling tile and need to repair, patch and touch-up paint all the damaged or disturbed areas, tasks normally assigned to general contractors. In addition, the owner could request that the contractors salvage the materials removed and turn over the materials to the owner or dispose of the materials to a reclamation station. The normal removal item is 0.5 labor hour for a lighting fixture and 1.5 labor-hours for new installation time. Revise the estimate times from 2 labor-hours work up to a minimum 4 labor-hours work for just fluorescent lighting fixture.

For removal of large concentrations of lighting fixtures in the same area, apply an "economy of scale" to reduce estimating labor hours.

Electrical | R2605 | Common Work Results for Electrical

R260519-20 Armored Cable

Armored Cable – Quantities are taken off in the same manner as wire.

Bx Type Cable – Productivities are based on an average run of 50′ before terminating at a box fixture, etc. Each 50′ section includes field preparation of (2) ends with hacksaw, identification and tagging of wire. Set up is open coil type without reels attaching cable to snake and pulling across a suspended ceiling or open face wood or steel studding, price does not include drilling of studs.

Cable in Tray – Productivities are based on an average run of 100 L.F. with set up of pulling equipment for (2) 90° bends, attaching cable to pull-in means, identification and tagging of wires, set up of reels. Wire termination to breakers equipment, etc., are not included.

Job Conditions – Productivities are based on new construction to a height of 15′ using rolling staging in an unobstructed area. Material staging is assumed to be within 100′ of work being performed.

R260519-80 Undercarpet Systems

Takeoff Procedure for Power Systems: List components for each fitting type, tap, splice, and bend on your quantity takeoff sheet. Each component must be priced separately. Start at the power supply transition fittings and survey each circuit for the components needed. List the quantities of each component under a specific circuit number. Use the floor plan layout scale to get cable footage.

Reading across the list, combine the totals of each component in each circuit and list the total quantity in the last column. Calculate approximately 5% for scrap for items such as cable, top shield, tape, and spray adhesive. Also provide for final variations that may occur on-site.

Suggested guidelines are:

1. Equal amounts of cable and top shield should be priced.
2. For each roll of cable, price a set of cable splices.
3. For every 1 ft. of cable, price 2-1/2 ft. of hold-down tape.
4. For every 3 rolls of hold-down tape, price 1 can of spray adhesive.

Adjust final figures wherever possible to accommodate standard packaging of the product. This information is available from the distributor.

Each transition fitting requires:

1. 1 base
2. 1 cover
3. 1 transition block

Each floor fitting requires:

1. 1 frame/base kit
2. 1 transition block
3. 2 covers (duplex/blank)

Each tap requires:

1. 1 tap connector for each conductor
2. 1 pair insulating patches
3. 2 top shield connectors

Each splice requires:

1. 1 splice connector for each conductor
2. 1 pair insulating patches
3. 3 top shield connectors

Each cable bend requires:

1. 2 top shield connectors

Each cable dead end (outside of transition block) requires:

1. 1 pair insulating patches

Labor does not include:

1. Patching or leveling uneven floors.
2. Filling in holes or removing projections from concrete slabs.
3. Sealing porous floors.
4. Sweeping and vacuuming floors.
5. Removal of existing carpeting.
6. Carpet square cut-outs.
7. Installation of carpet squares.

Takeoff Procedures for Telephone Systems: After reviewing floor plans identify each transition. Number or letter each cable run from that fitting. Start at the transition fitting and survey each circuit for the components needed. List the cable type, terminations, cable length, and floor fitting type under the specific circuit number. Use the floor plan layout scale to get the cable footage. Add some extra length (next higher increment of 5 feet) to preconnectorized cable.

Transition fittings require:

1. 1 base plate
2. 1 cover
3. 1 transition block

Floor fittings require:

1. 1 frame/base kit
2. 2 covers
3. Modular jacks

Reading across the list, combine the list of components in each circuit and list the total quantity in the last column. Calculate the necessary scrap factors for such items as tape, bottom shield and spray adhesive. Also provide for final variations that may occur on-site.

Adjust final figures whenever possible to accommodate standard packaging. Check that items such as transition fittings, floor boxes, and floor fittings that are to utilize both power and telephone have been priced as combination fittings, so as to avoid duplication.

Make sure to include marking of floors and drilling of fasteners if fittings specified are not the adhesive type.

Labor does not include:

1. Conduit or raceways before transition of floor boxes
2. Telephone cable before transition boxes
3. Terminations before transition boxes
4. Floor preparation as described in power section

Be sure to include all cable folds when pricing labor.

Takeoff Procedure for Data Systems: Start at the transition fittings and take off quantities in the same manner as the telephone system, keeping in mind that data cable does not require top or bottom shields.

The data cable is simply cross-taped on the cable run to the floor fitting.

Data cable can be purchased in either bulk form in which case coaxial connector material and labor must be priced, or in preconnectorized cut lengths.

Data cable cannot be folded and must be notched at 1 inch intervals. A count of all turns must be added to the labor portion of the estimate. (Note: Some manufacturers have prenotched cable.)

Notching required:

1. 90 degree turn requires 8 notches per side.
2. 180 degree turn requires 16 notches per side.

Floor boxes, transition boxes, and fittings are the same as described in the power and telephone procedures.

Since undercarpet systems require special hand tools, be sure to include this cost in proportion to number of crews involved in the installation.

Job Conditions: Productivity is based on new construction in an unobstructed area. Staging area is assumed to be within 200′ of work being performed.

R260519-90 Wire

Wire quantities are taken off by either measuring each cable run or by extending the conduit and raceway quantities times the number of conductors in the raceway. Ten percent should be added for waste and tie-ins. Keep in mind that the unit of measure of wire is C.L.F. not L.F. as in raceways so the formula would read:

$$\frac{\text{(L.F. Raceway x No. of Conductors) x 1.10}}{100} = \text{C.L.F.}$$

Price per C.L.F. of wire includes:
1. Set up wire coils or spools on racks
2. Attaching wire to pull in means
3. Measuring and cutting wire
4. Pulling wire into a raceway
5. Identifying and tagging

Price does not include:
1. Connections to breakers, panelboards or equipment
2. Splices

Job Conditions: Productivity is based on new construction to a height of 15′ using rolling staging in an unobstructed area. Material staging is assumed to be within 100′ of work being performed.

Economy of Scale: If more than three wires at a time are being pulled, deduct the following percentages from the labor of that grouping:

4-5 wires	25%
6-10 wires	30%
11-15 wires	35%
over 15	40%

If a wire pull is less than 100′ in length and is interrupted several times by boxes, lighting outlets, etc., it may be necessary to add the following lengths to each wire being pulled:

Junction box to junction box	2 L.F.
Lighting panel to junction box	6 L.F.
Distribution panel to sub panel	8 L.F.
Switchboard to distribution panel	12 L.F.
Switchboard to motor control center	20 L.F.
Switchboard to cable tray	40 L.F.

Measure of Drops and Riser: It is important when taking off wire quantities to include the wire for drops to electrical equipment. If heights of electrical equipment are not clearly stated, use the following guide:

	Bottom A.F.F.	Top A.F.F.	Inside Cabinet
Safety switch to 100A	5′	6′	2′
Safety switch 400 to 600A	4′	6′	3′
100A panel 12 to 30 circuit	4′	6′	3′
42 circuit panel	3′	6′	4′
Switch box	3′	3′6″	1′
Switchgear	0′	8′	8′
Motor control centers	0′	8′	8′
Transformers - wall mount	4′	8′	2′
Transformers - floor mount	0′	12′	4′

R260519-91 Maximum Circuit Length (approximate) for Various Power Requirements Assuming THW, Copper Wire @ 75° C, Based Upon a 4% Voltage Drop

Maximum Circuit Length: Table R260519-91 indicates typical maximum installed length a circuit can have and still maintain an adequate voltage level at the point of use. The circuit length is similar to the conduit length.

If the circuit length for an ampere load and a copper wire size exceeds the length obtained from Table R260519-91, use the next largest wire size to compensate voltage drop.

Amperes	Wire Size	Maximum Circuit Length in Feet				
		2 Wire, 1 Phase		3 Wire, 3 Phase		
		120V	240V	240V	480V	600V
15	14*	50	105	120	240	300
	14	50	100	120	235	295
20	12*	60	125	145	290	360
	12	60	120	140	280	350
30	10*	65	130	155	305	380
	10	65	130	150	300	375
50	8	60	125	145	285	355
65	6	75	150	175	345	435
85	4	90	185	210	425	530
115	2	110	215	250	500	620
130	1	120	240	275	555	690
150	1/0	130	260	305	605	760
175	2/0	140	285	330	655	820
200	3/0	155	315	360	725	904
230	4/0	170	345	395	795	990
255	250	185	365	420	845	1055
285	300	195	395	455	910	1140
310	350	210	420	485	975	1220
380	500	245	490	565	1130	1415

*Solid Conductor

Note: The circuit length is the one-way distance between the origin and the load.

Example: A 130 ampere load at 480 volts, 3 phase, 3 wire with No. 1 wire can be run a maximum of 555 L.F. and provide satisfactory operation. If the same load is to be wired at the end of a 625 L.F. circuit, then a larger wire must be used.

R260519-92 Minimum Copper and Aluminum Wire Size Allowed for Various Types of Insulation

Minimum Wire Sizes									
	Copper		Aluminum			Copper		Aluminum	
Amperes	THW THWN or XHHW	THHN XHHW *	THW XHHW	THHN XHHW *	Amperes	THW THWN or XHHW	THHN XHHW *	THW XHHW	THHN XHHW *
15A	#14	#14	#12	#12	195	3/0	2/0	250kcmil	4/0
20	#12	#12	#10	#10	200	3/0	3/0	250kcmil	4/0
25	#10	#10	#10	#10	205	4/0	3/0	250kcmil	4/0
30	#10	#10	# 8	# 8	225	4/0	3/0	300kcmil	250kcmil
40	# 8	# 8	# 8	# 8	230	4/0	4/0	300kcmil	250kcmil
45	# 8	# 8	# 6	# 8	250	250kcmil	4/0	350kcmil	300kcmil
50	# 8	# 8	# 6	# 6	255	250kcmil	4/0	400kcmil	300kcmil
55	# 6	# 8	# 4	# 6	260	300kcmil	4/0	400kcmil	350kcmil
60	# 6	# 6	# 4	# 6	270	300kcmil	250kcmil	400kcmil	350kcmil
65	# 6	# 6	# 4	# 4	280	300kcmil	250kcmil	500kcmil	350kcmil
75	# 4	# 6	# 3	# 4	285	300kcmil	250kcmil	500kcmil	400kcmil
85	# 4	# 4	# 2	# 3	290	350kcmil	250kcmil	500kcmil	400kcmil
90	# 3	# 4	# 2	# 2	305	350kcmil	300kcmil	500kcmil	400kcmil
95	# 3	# 4	# 1	# 2	310	350kcmil	300kcmil	500kcmil	500kcmil
100	# 3	# 3	# 1	# 2	320	400kcmil	300kcmil	600kcmil	500kcmil
110	# 2	# 3	1/0	# 1	335	400kcmil	350kcmil	600kcmil	500kcmil
115	# 2	# 2	1/0	# 1	340	500kcmil	350kcmil	600kcmil	500kcmil
120	# 1	# 2	1/0	1/0	350	500kcmil	350kcmil	700kcmil	500kcmil
130	# 1	# 2	2/0	1/0	375	500kcmil	400kcmil	700kcmil	600kcmil
135	1/0	# 1	2/0	1/0	380	500kcmil	400kcmil	750kcmil	600kcmil
150	1/0	# 1	3/0	2/0	385	600kcmil	500kcmil	750kcmil	600kcmil
155	2/0	1/0	3/0	3/0	420	600kcmil	500kcmil		700kcmil
170	2/0	1/0	4/0	3/0	430		500kcmil		750kcmil
175	2/0	2/0	4/0	3/0	435		600kcmil		750kcmil
180	3/0	2/0	4/0	4/0	475		600kcmil		

*Dry Locations Only

Notes:

1. Size #14 to 4/0 is in AWG units (American Wire Gauge).
2. Size 250 to 750 is in kcmil units (Thousand Circular Mils).
3. Use next higher ampere value if exact value is not listed in table.
4. For loads that operate continuously increase ampere value by 25% to obtain proper wire size.
5. Refer to Table R260519-91 for the maximum circuit length for the various size wires.
6. Table R260519-92 has been written for estimating purpose only, based on ambient temperature of 30 °C (86°F); for ambient temperature other than 30 °C (86°F), ampacity correction factors will be applied.

R260519-93 Metric Equivalent, Wire

U.S. vs. European Wire – Approximate Equivalents			
United States		European	
Size AWG or kcmil	Area Cir. Mils.(cmil) mm^2	Size mm^2	Area Cir. Mils.
18	1620/.82	.75	1480
16	2580/1.30	1.0	1974
14	4110/2.08	1.5	2961
12	6530/3.30	2.5	4935
10	10,380/5.25	4	7896
8	16,510/8.36	6	11,844
6	26,240/13.29	10	19,740
4	41,740/21.14	16	31,584
3	52,620/26.65	25	49,350
2	66,360/33.61	–	–
1	83,690/42.39	35	69,090
1/0	105,600/53.49	50	98,700
2/0	133,100/67.42	–	–
3/0	167,800/85.00	70	138,180
4/0	211,600/107.19	95	187,530
250	250,000/126.64	120	236,880
300	300,000/151.97	150	296,100
350	350,000/177.30	–	–
400	400,000/202.63	185	365,190
500	500,000/253.29	240	473,760
600	600,000/303.95	300	592,200
700	700,000/354.60	–	–
750	750,000/379.93	–	–

R260519-94 Size Required and Weight (Lbs./1000 L.F.) of Aluminum and Copper THW Wire by Ampere Load

Amperes	Copper Size	Aluminum Size	Copper Weight	Aluminum Weight
15	14	12	24	11
20	12	10	33	17
30	10	8	48	39
45	8	6	77	52
65	6	4	112	72
85	4	2	167	101
100	3	1	205	136
115	2	1/0	252	162
130	1	2/0	324	194
150	1/0	3/0	397	233
175	2/0	4/0	491	282
200	3/0	250	608	347
230	4/0	300	753	403
255	250	400	899	512
285	300	500	1068	620
310	350	500	1233	620
335	400	600	1396	772
380	500	750	1732	951

R260526-80 Grounding

Grounding When taking off grounding systems, identify separately the type and size of wire.

Example:

Bare copper & size

Bare aluminum & size

Insulated copper & size

Insulated aluminum & size

Count the number of ground rods and their size.

Example:

1. 8′ grounding rod – 5/8″ dia. 20 Ea.
2. 10′ grounding rod – 5/8″ dia. 12 Ea.
3. 15′ grounding rod – 3/4″ dia. 4 Ea.

Count the number of connections; the size of the largest wire will determine the productivity.

Example:

Braze a #2 wire to a #4/0 cable

The 4/0 cable will determine the L.H. and cost to be used.

Include individual connections to:

1. Ground rods
2. Building steel
3. Equipment
4. Raceways

Price does not include:

1. Excavation
2. Backfill
3. Sleeves or raceways used to protect grounding wires
4. Wall penetrations
5. Floor cutting
6. Core drilling

Job Conditions: Productivity is based on a ground floor area, using cable reels in an unobstructed area. Material staging area assumed to be within 100′ of work being performed.

R260533-20 Conduit To 15′ High

List conduit by quantity, size, and type. Do not deduct for lengths occupied by fittings, since this will be allowance for scrap. Example:

A. Aluminum — size
B. Rigid galvanized — size
C. Steel intermediate (IMC) — size
D. Rigid steel, plastic-coated 20 Mil. — size
E. Rigid steel, plastic-coated 40 Mil. — size
F. Electric metallic tubing (EMT) — size
G. PVC Schedule 40 — size

Types (A) thru (E) listed above contain the following per 100 L.F.:

1. (11) Threaded couplings
2. (11) Beam-type hangers
3. (2) Factory sweeps
4. (2) Fiber bushings
5. (4) Locknuts
6. (2) Field threaded pipe terminations
7. (2) Removal of concentric knockouts

Type (F) contains per 100 L.F.:

1. (11) Set screw couplings
2. (11) Beam clamps
3. (2) Field bends on 1/2″ and 3/4″ diameter
4. (2) Factory sweeps for 1″ and above
5. (2) Set screw steel connectors
6. (2) Removal of concentric knockouts

Type (G) contains per 100 L.F.:

1. (11) Field cemented couplings
2. (34) Beam clamps
3. (2) Factory sweeps
4. (2) Adapters
5. (2) Locknuts
6. (2) Removal of concentric knockouts

Labor-hours for all conduit to 15′ include:

1. Unloading by hand
2. Hauling by hand to an area up to 200′ from loading dock
3. Setup of rolling staging
4. Installation of conduit and fittings as described in Conduit models (A) thru (G)

Not included in the material and labor are:

1. Staging rental or purchase
2. Structural modifications
3. Wire
4. Junction boxes
5. Fittings in excess of those described in conduit models (A) thru (G)
6. Painting of conduit

Fittings

Only those fittings listed above are included in the linear foot totals, although they should be listed separately from conduit lengths, without prices, to ensure proper quantities for material procurement.

If the fittings required exceed the quantities included in the model conduit runs, then material and labor costs must be added to the difference. If actual needs per 100 L.F. of conduit are: (2) sweeps, (4) LBs and (1) field bend, then, (4) LBs and (1) field bend must be priced additionally.

R260533-21 Hangers

It is sometimes desirable to substitute an alternate style of hanger if the support being used is not the type described in the conduit models.

One approach is the substitution method:

1. Find the cost of the type hanger described in the conduit model.
2. Calculate the cost of the desired type hanger (it may be necessary to calculate individual components such as drilling, expansion shields, etc.).
3. Calculate the cost difference (delta) between the two types of hangers.
4. Multiply the cost delta by the number of hangers in the model.
5. Divide the total delta cost for hangers in the model by the length of the model to find the delta cost per L.F. for that model.
6. Modify the given unit costs per L.F. for the model by the delta cost per L.F.

Another approach to hanger configurations would be to start with the conduit only and add all the supports and any other items as separate lines. This procedure is most useful if the project involves racking many runs of conduit on a single hanger, for instance, a trapeze type hanger.

Example: Five (5) 2″ RGS conduits, 50 L.F. each, are to be run on trapeze hangers from one pull box to another. The run includes one 90° bend.

1. List the hangers' components to create an assembly cost for each 2′-wide trapeze.
2. List the components for the 50′ conduit run, noting that 6 trapeze supports will be required.

Job Conditions: Productivities are based on new construction to 15′ high, using scaffolding in an unobstructed area. Material storage is assumed to be within 100′ of work being performed.

Add to labor for elevated installations:

15′ to 20′ High–10%	30′ to 35 High–30%
20′ to 25′ High–20%	35′ to 40′ High–35%
25′ to 30′ High–25%	Over 40′ High–40%

Add these percentages to the L.F. labor cost, but not to fittings. Add these percentages only to quantities exceeding the different height levels, rather than the total conduit quantities.

Linear foot price for labor does not include penetrations in walls or floors and must be added to the estimate

R260533-22 Conductors in Conduit

Table below lists maximum number of conductors for various sized conduit using THW, TW or THWN insulations.

Copper Wire Size	Conduit Size																										
	½″			¾″			1″			1-¼″			1-½″			2″			2-½″			3″		3-½″		4″	
	TW	THW	THWN	TW	THW	THWN	TW	THW	THWN	TW	THW	THWN	TW	THW	THWN	TW	THW	THWN	TW	THW	THWN	THW	THWN	THW	THWN	THW	THWN
#14	9	6	13	15	10	24	25	16	39	44	29	69	60	40	94	99	65	154	142	93		143		192			
#12	7	4	10	12	8	18	19	13	29	35	24	51	47	32	70	78	53	114	111	76	164	117		157			
#10	5	4	6	9	6	11	15	11	18	26	19	32	36	26	44	60	43	73	85	61	104	95	160	127		163	
#8	2	1	3	4	3	5	7	5	9	12	10	16	17	13	22	28	22	36	40	32	51	49	79	66	106	85	136
#6		1	1		2	4		4	6		7	11		10	15		16	26		23	37	36	57	48	76	62	98
#4		1	1		1	2		3	4		5	7		7	9		12	16		17	22	27	35	36	47	47	60
#3		1	1		1	1		2	3		4	6		6	8		10	13		15	19	23	29	31	39	40	51
#2		1	1		1	1		2	3		4	5		5	7		9	11		13	16	20	25	27	33	34	43
#1					1	1		1	1		3	3		4	5		6	8		9	12	14	18	19	25	25	32
1/0					1	1		1	1		2	3		3	4		5	7		8	10	12	15	16	21	21	27
2/0					1	1		1	1		1	2		3	3		5	6		7	8	10	13	14	17	18	22
3/0					1	1		1	1		1	1		2	3		4	5		6	7	9	11	12	14	15	18
4/0						1		1	1		1	1		1	2		3	4		5	6	7	9	10	12	13	15
250 kcmil								1	1		1	1		1	1		2	3		4	4	6	7	8	10	10	12
300								1	1		1	1		1	1		2	3		3	4	5	6	7	8	9	11
350									1		1	1		1	1		1	2		3	3	4	5	6	7	8	9
400											1	1		1	1		1	1		2	3	4	5	5	6	7	8
500											1	1		1	1		1	1		1	2	3	4	4	5	6	7
600												1		1	1		1	1		1	1	3	3	4	4	5	5
700														1	1		1	1		1	1	2	3	3	4	4	5
750														1	1		1	1		1	1	2	2	3	3	4	4

R260533-23 Metric Equivalent, Conduit

U.S. vs. European Conduit – Approximate Equivalents			
United States		European	
Trade Size	Inside Diameter Inch/mm	Trade Size	Inside Diameter mm
½	.622/15.8	11	16.4
¾	.824/20.9	16	19.9
1	1.049/26.6	21	25.5
1¼	1.380/35.0	29	34.2
1½	1.610/40.9	36	44.0
2	2.067/52.5	42	51.0
2½	2.469/62.7		
3	3.068/77.9		
3½	3.548/90.12		
4	4.026/102.3		
5	5.047/128.2		
6	6.065/154.1		

R260533-24 Conduit Weight Comparisons (Lbs. per 100 ft.) Empty

Type	1/2″	3/4″	1″	1-1/4″	1-1/2″	2″	2-1/2″	3″	3-1/2″	4″	5″	6″
Rigid Aluminum	28	37	55	72	89	119	188	246	296	350	479	630
Rigid Steel	79	105	153	201	249	332	527	683	831	972	1314	1745
Intermediate Steel (IMC)	60	82	116	150	182	242	401	493	573	638		
Electrical Metallic Tubing (EMT)	29	45	65	96	111	141	215	260	365	390		
Polyvinyl Chloride, Schedule 40	16	22	32	43	52	69	109	142	170	202	271	350
Polyvinyl Chloride Encased Burial						38		67	88	105	149	202
Fibre Duct Encased Burial						127		164	180	206	400	511
Fibre Duct Direct Burial						150		251	300	354		
Transite Encased Burial						160		240	290	330	450	550
Transite Direct Burial						220		310		400	540	640

R260533-25 Conduit Weight Comparisons (Lbs. per 100 ft.) with Maximum Cable Fill*

Type	1/2″	3/4″	1″	1-1/4″	1-1/2″	2″	2-1/2″	3″	3-1/2″	4″	5″	6″
Rigid Galvanized Steel (RGS)	104	140	235	358	455	721	1022	1451	1749	2148	3083	4343
Intermediate Steel (IMC)	84	113	186	293	379	611	883	1263	1501	1830		
Electrical Metallic Tubing (EMT)	54	116	183	296	368	445	641	930	1215	1540		

*Conduit & Heaviest Conductor Combination

R260533-60 Wireway

When "taking off" Wireway, list by size and type.

Example:
1. Screw cover, unflanged + size
2. Screw cover, flanged + size
3. Hinged cover, flanged + size
4. Hinged cover, unflanged + size

Each 10′ length on Wireway contains:
1. 10′ of cover either screw or hinged type
2. (1) Coupling or flange gasket
3. (1) Wall type mount

All fittings must be priced separately.

Substitution of hanger types is done the same as described in R260533-21, "HANGERS," keeping in mind that the wireway model is based on 10′ sections instead of a 100′ conduit run.

Labor-hours for wireway include:
1. Unloading by hand
2. Hauling by hand up to 100′ from loading dock
3. Measuring and marking
4. Mounting wall bracket using (2) anchor type lead fasteners
5. Installing wireway on brackets, to 15′ high (For higher elevations use factors in R260533-20)

Job Conditions: Productivity is based on new construction, to a height of 15′ using rolling staging in an unobstructed area.

Material staging area is assumed to be within 100′ of work being performed.

R260533-65 Outlet Boxes

Outlet boxes should be included on the same takeoff sheet as branch piping or devices to better explain what is included in each circuit.

Each unit price in this section is a stand alone item and contains no other component unless specified. For example, to estimate a duplex outlet, components that must be added are:
1. 4″ square box
2. 4″ plaster ring
3. Duplex receptacle
4. Device cover

The method of mounting outlet boxes is (2) plastic shield fasteners.

Outlet boxes plastic, labor-hours include:
1. Marking box location on wood studding
2. Mounting box

Economy of Scale – For large concentrations of plastic boxes in the same area deduct the following percentages from labor-hour totals:

1	to	10	0%
11	to	25	20%
26	to	50	25%
51	to	100	30%
	over	100	35%

Note: It is important to understand that these percentages are not used on the total job quantities, but only areas where concentrations exceed the levels specified.

R260533-70 Pull Boxes and Cabinets

List cabinets and pull boxes by NEMA type and size.

Example:	**TYPE**	**SIZE**
	NEMA 1	6″W x 6″H x 4″D
	NEMA 3R	6″W x 6″H x 4″D

Labor-hours for wall mount (indoor or outdoor) installations include:
1. Unloading and uncrating
2. Handling of enclosures up to 200′ from loading dock using a dolly or pipe rollers
3. Measuring and marking
4. Drilling (4) anchor type lead fasteners using a hammer drill
5. Mounting and leveling boxes

Note: A plywood backboard is not included.

Labor-hours for ceiling mounting include:
1. Unloading and uncrating
2. Handling boxes up to 100′ from loading dock
3. Measuring and marking
4. Drilling (4) anchor type lead fasteners using a hammer drill
5. Installing and leveling boxes to a height of 15′ using rolling staging

Labor-hours for free standing cabinets include:
1. Unloading and uncrating
2. Handling of cabinets up to 200′ from loading dock using a dolly or pipe rollers
3. Marking of floor
4. Drilling (4) anchor type lead fasteners using a hammer drill
5. Leveling and shimming

Labor-hours for telephone cabinets include:
1. Unloading and uncrating
2. Handling cabinets up to 200′ using a dolly or pipe rollers
3. Measuring and marking
4. Mounting and leveling, using (4) lead anchor type fasteners

R260533-75 Weight Comparisons of Common Size Cast Boxes in Lbs.

Size NEMA 4 or 9	Cast Iron	Cast Aluminum	Size NEMA 7	Cast Iron	Cast Aluminum
6″ x 6″ x 6″	17	7	6″ x 6″ x 6″	40	15
8″ x 6″ x 6″	21	8	8″ x 6″ x 6″	50	19
10″ x 6″ x 6″	23	9	10″ x 6″ x 6″	55	21
12″ x 12″ x 6″	52	20	12″ x 6″ x 6″	100	37
16″ x 16″ x 6″	97	36	16″ x 16″ x 6″	140	52
20″ x 20″ x 6″	133	50	20″ x 20″ x 6″	180	67
24″ x 18″ x 8″	149	56	24″ x 18″ x 8″	250	93
24″ x 24″ x 10″	238	88	24″ x 24″ x 10″	358	133
30″ x 24″ x 12″	324	120	30″ x 24″ x 10″	475	176
36″ x 36″ x 12″	500	185	30″ x 24″ x 12″	510	189

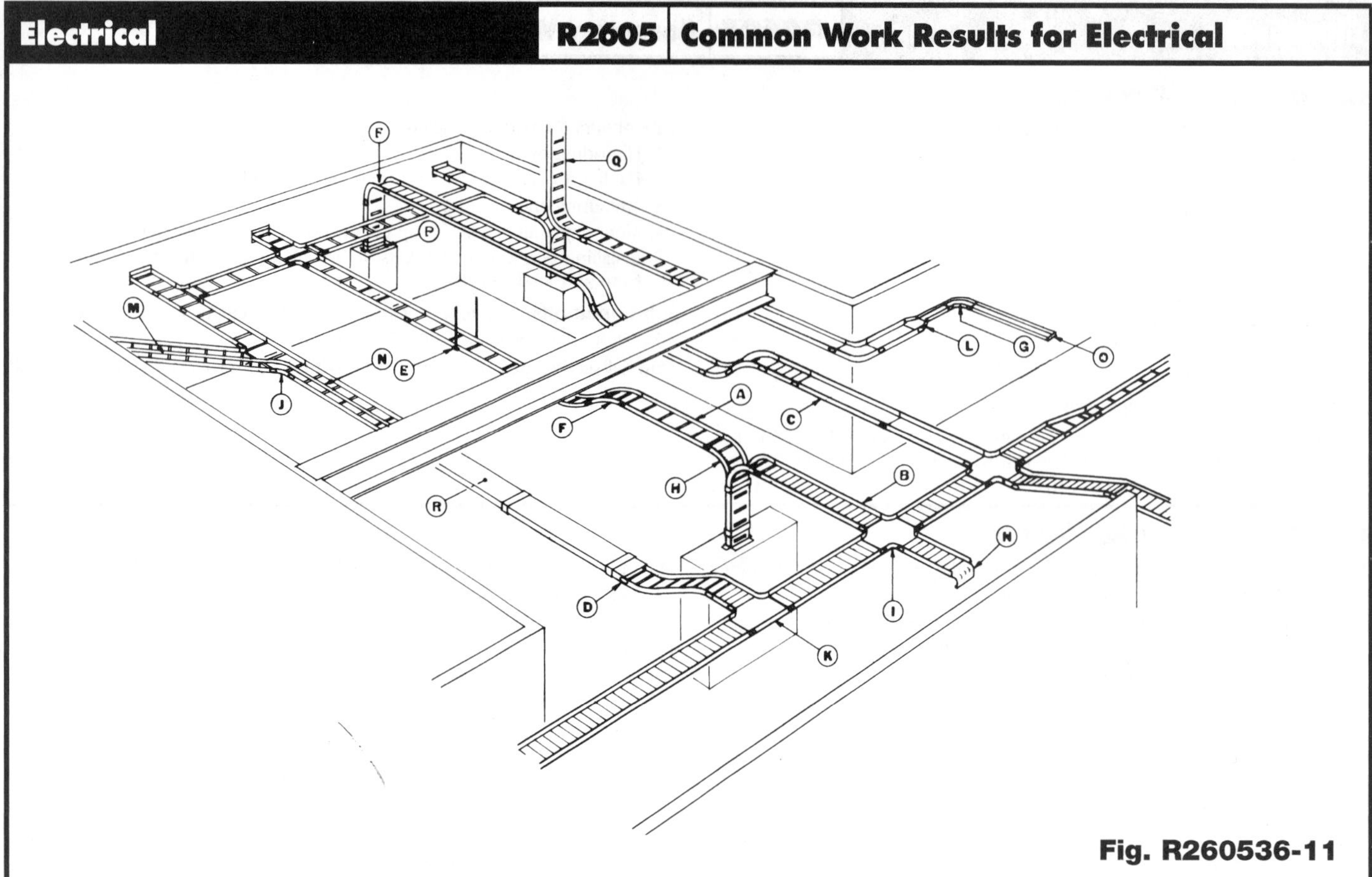

Fig. R260536-11

R260536-10 Cable Tray

Cable Tray - When taking off cable tray it is important to identify separately the different types and sizes involved in the system being estimated. (Fig. R260536-11)

A. – Ladder Type, galvanized or aluminum
B. – Trough Type, galvanized or aluminum
C. – Solid Bottom, galvanized or aluminum

The unit of measure is calculated in linear feet; do not deduct from this footage any length occupied by fittings, this will be the only allowance for scrap. Be sure to include all vertical drops to panels, switch gear, etc.

Hangers – Included in the linear footage of cable tray is

D. – 1 – Pair of connector plates per 12 L.F.

E. – 1 – Pair clamp type hangers and 4′ of 3/8″ threaded rod per 12 L.F.

Not included are structural supports, which must be priced in addition to the hangers.

Fittings – Identify separately the different types of fittings

1.) Ladder Type, galvanized or aluminum
2.) Trough Type, galvanized or aluminum
3.) Solid Bottom Type, galvanized or aluminum

The configuration, radius and rung spacing must also be listed. The unit of measure is "Ea."

F. – Elbow, vertical Ea.
G. – Elbow, horizontal Ea.
H. – Tee, vertical Ea.
I. – Cross, horizontal Ea.
J. – Wye, horizontal Ea.
K. – Tee, horizontal Ea.
L. – Reducing fitting Ea.

Depending on the use of the system other examples of units which must be included are:

M. – Divider strip L.F.
N. – Drop-outs Ea.
O. – End caps Ea.
P. – Panel connectors Ea.

Wire and cable are not included and should be taken off separately, see Division 161.

Job Conditions – Unit prices are based on a new installation to a work plane of 15′ using rolling staging.

Add to labor for elevated installations:

15′ to 20′ High	10%
20′ to 25′ High	20%
25′ to 30′ High	25%
30′ to 35′ High	30%
35′ to 40′ High	35%
Over 40′ High	40%

Add these percentages for L.F. totals but not to fittings. Add percentages to only those quantities that fall in the different elevations, in other words, if the total quantity of cable tray is 200′ but only 75′ is above 15′ then the 10% is added to the 75′ only.

Linear foot costs do not include penetrations through walls and floors which must be added to the estimate.

Cable Tray Covers

Covers – Cable tray covers are taken off in the same manner as the tray itself, making distinctions as to the type of cover. (Fig. R260536-11)

Q. – Vented, galvanized or aluminum
R. – Solid, galvanized or aluminum

Cover configurations are taken off separately noting type, specific radius and widths.

Note: Care should be taken to identify from plans and specifications exactly what is being covered. In many systems only vertical fittings are covered to retain wire and cable.

R260539-30 Conduit In Concrete Slab

List conduit by quantity, size and type.

Example:

A. Rigid galvanized steel + size
B. P.V.C. + size

Rigid galvanized steel (A) contains per 100 L.F.:

1. (20) Ties to slab reinforcing
2. (11) Threaded steel couplings
3. (2) Factory sweeps
4. (2) Field threaded conduit terminations
5. (2) Fiber bushings + locknuts
6. (2) Removal of concentric knockouts

P.V.C. (B) contains per 100 L.F.:

1. (20) Ties to slab reinforcing
2. (11) Field cemented couplings
3. (2) Factory sweeps
4. (2) Adapters
5. (2) Removal of concentric knockouts

R260539-40 Conduit In Trench

Conduit in trench is galvanized steel and contains per 100 L.F.:

1. (11) Threaded couplings
2. (2) Factory sweeps
3. (2) Fiber bushings + (4) locknuts
4. (2) Field threaded conduit terminations
5. (2) Removal of concentric knockouts

Note:

Conduit in Unit Price sections do not include:

1. Floor cutting
2. Excavation or backfill
3. Grouting or patching

Conduit fittings in excess of those listed in the above Conduit model must be added. (Refer to R260533-20 for Procedure example.)

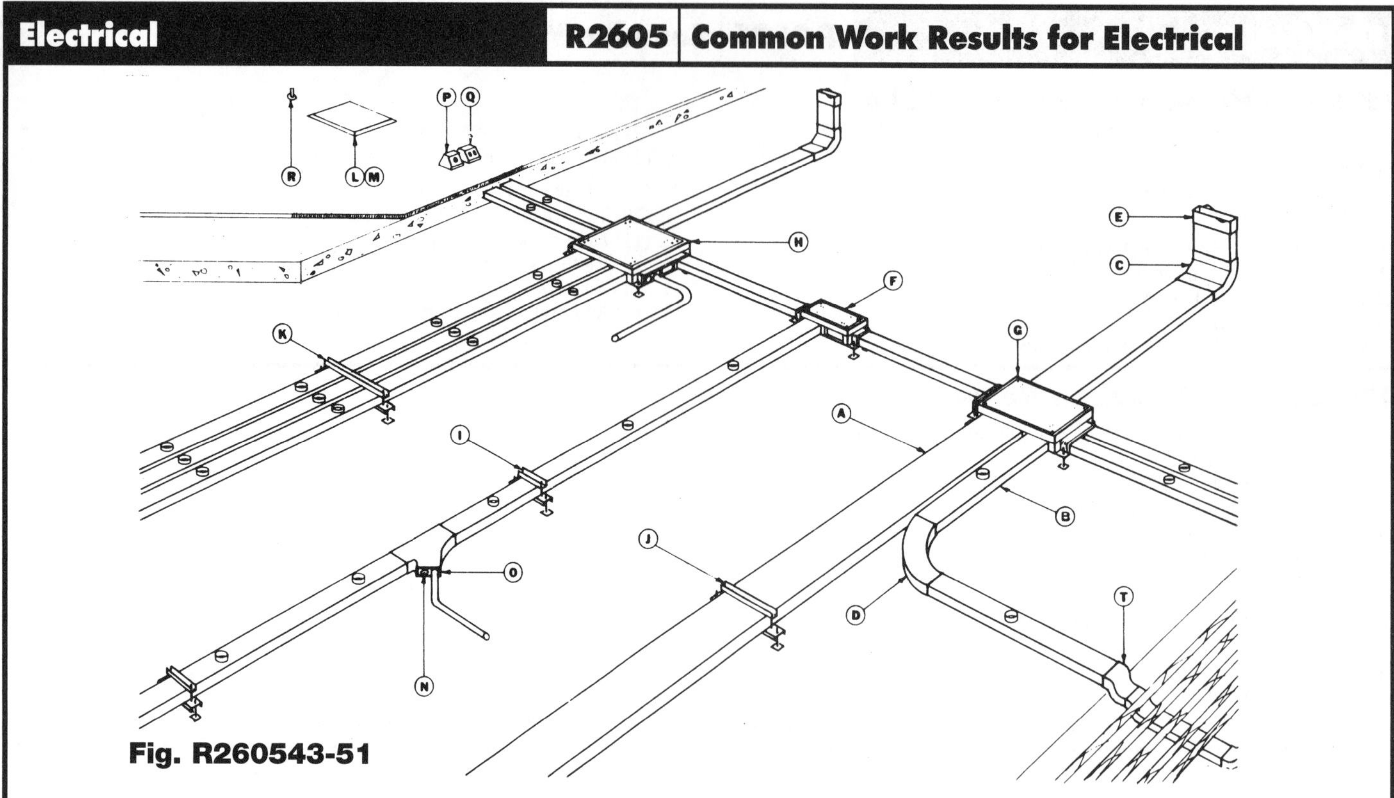

Fig. R260543-51

R260543-50 Underfloor Duct

When pricing Underfloor Duct it is important to identify and list each component, since costs vary significantly from one type of fitting to another. Do not deduct boxes or fittings from linear foot totals; this will be your allowance for scrap.

The first step is to identify the system as either:

FIG. R260543-51 Single Level

FIG. R260543-52 Dual Level

Single Level System

Include on your "takeoff sheet" the following unit price items, making sure to distinquish between Standard and Super duct:

A. Feeder duct (blank) in L.F.
B. Distribution duct (Inserts 2′ on center) in L.F.
C. Elbows (Vertical) Ea.
D. Elbows (Horizontal) Ea.
E. Cabinet connector Ea.
F. Single duct junction box Ea.
G. Double duct junction box Ea.
H. Triple duct junction box Ea.
I. Support, single cell Ea.
J. Support, double cell Ea.
K. Support, triple cell Ea.
L. Carpet pan Ea.
M. Terrazzo pan Ea.
N. Insert to conduit adapter Ea.
O. Conduit adapter Ea.
P. Low tension outlet Ea.
Q. High tension outlet Ea.
R. Galvanized nipple Ea.
S. Wire per C.L.F.
T. Offset (Duct type) Ea.

Dual Level System + Labor
see next page

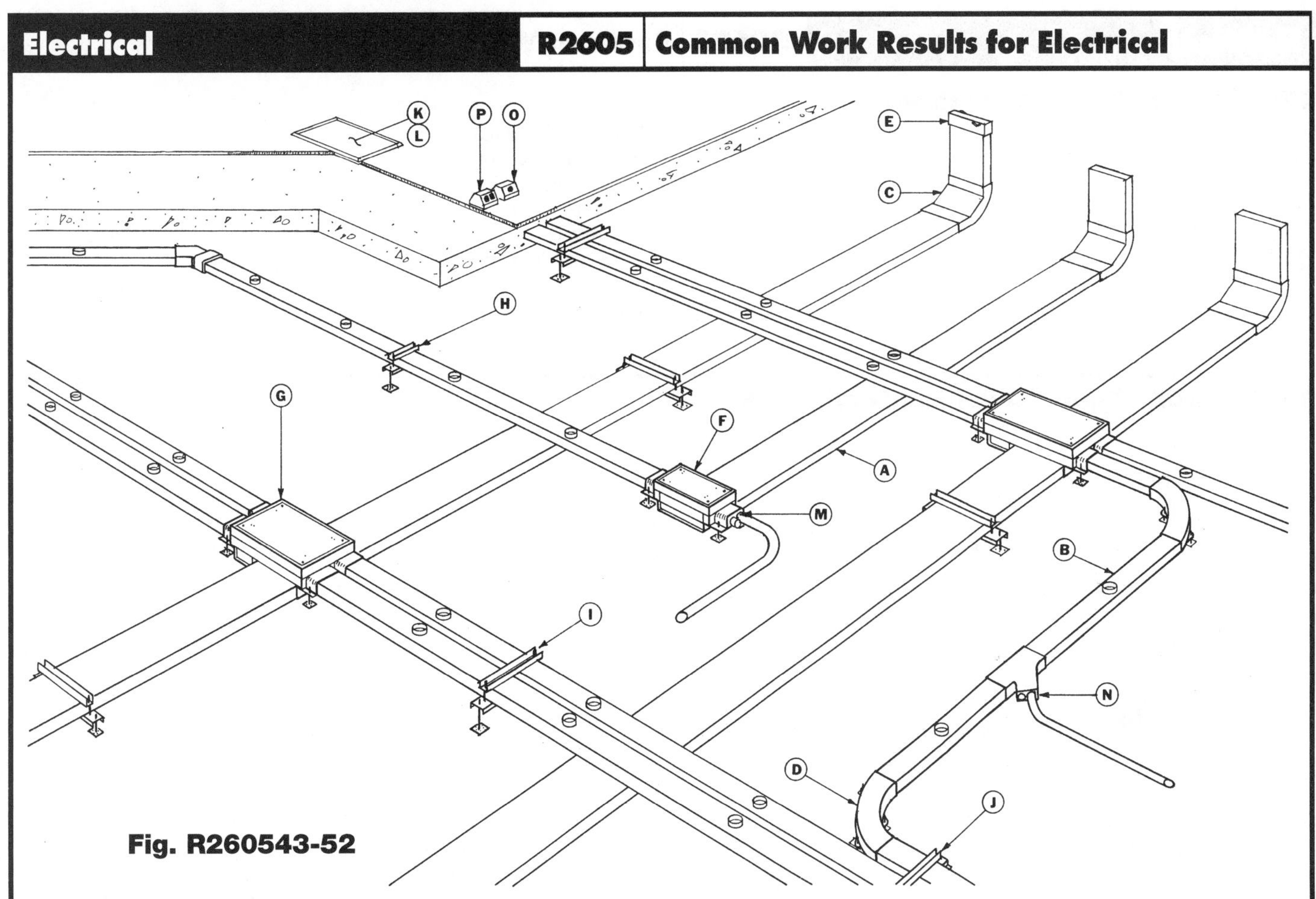

Fig. R260543-52

R260543-50 Underfloor Duct (cont.)

Dual Level

Include the following when "taking off" Dual Level systems:

Distinguish between Standard and Super duct.

A. Feeder duct (blank) in L.F.
B. Distribution duct (Inserts 2′ on center) in L.F.
C. Elbows (Vertical) Ea.
D. Elbows (Horizontal) Ea.
E. Cabinet connector Ea.
F. Single duct, 2 level, junction box Ea.
G. Double duct, 2 level, junction box Ea.
H. Support, single cell Ea.
I. Support, double cell Ea.
J. Support, triple cell Ea.
K. Carpet pan Ea.
L. Terrazzo pan Ea.
M. Insert to conduit adapter Ea.
N. Conduit adapter Ea.
O. Low tension outlet Ea.
P. High tension outlet Ea.
Q. Wire per C.L.F.

Note: Make sure to include risers in linear foot totals. High tension outlets include box, receptacle, covers and related mounting hardware.

Labor-hours for both Single and Dual Level systems include:

1. Unloading and uncrating
2. Hauling up to 200′ from loading dock
3. Measuring and marking
4. Setting raceway and fittings in slab or on grade
5. Leveling raceway and fittings

Labor-hours do not include:

1. Floor cutting
2. Excavation or backfill
3. Concrete pour
4. Grouting or patching
5. Wire or wire pulls
6. Additional outlets after concrete is poured
7. Piping to or from Underfloor Duct

Note: Installation is based on installing up to 150′ of duct. If quantities exceed this, deduct the following percentages:

1. 150′ to 250′ - 10%
2. 250′ to 350′ - 15%
3. 350′ to 500′ - 20%
4. over 500′ - 25%

Deduct these percentages from labor only.

Deduct these percentages from straight sections only.

Do not deduct from fittings or junction boxes.

Job Conditions: Productivity is based on new construction.

Underfloor duct to be installed on first three floors.

Material staging area within 100′ of work being performed.

Area unobstructed and duct not subject to physical damage.

R260580-75 Motor Connections

Motor connections should be listed by size and type of motor. Included in the material and labor cost is:

1. (2) Flex connectors
2. 18″ of flexible metallic wireway
3. Wire identification and termination
4. Test for rotation
5. (2) or (3) Conductors

Price does not include:

1. Mounting of motor
2. Disconnect Switch
3. Motor Starter
4. Controls
5. Conduit or wire ahead of flex

Note: When "Taking off" Motor connections, it is advisable to list connections on the same quantity sheet as Motors, Motor Starters and controls.

R260590-05 Typical Overhead Service Entrance

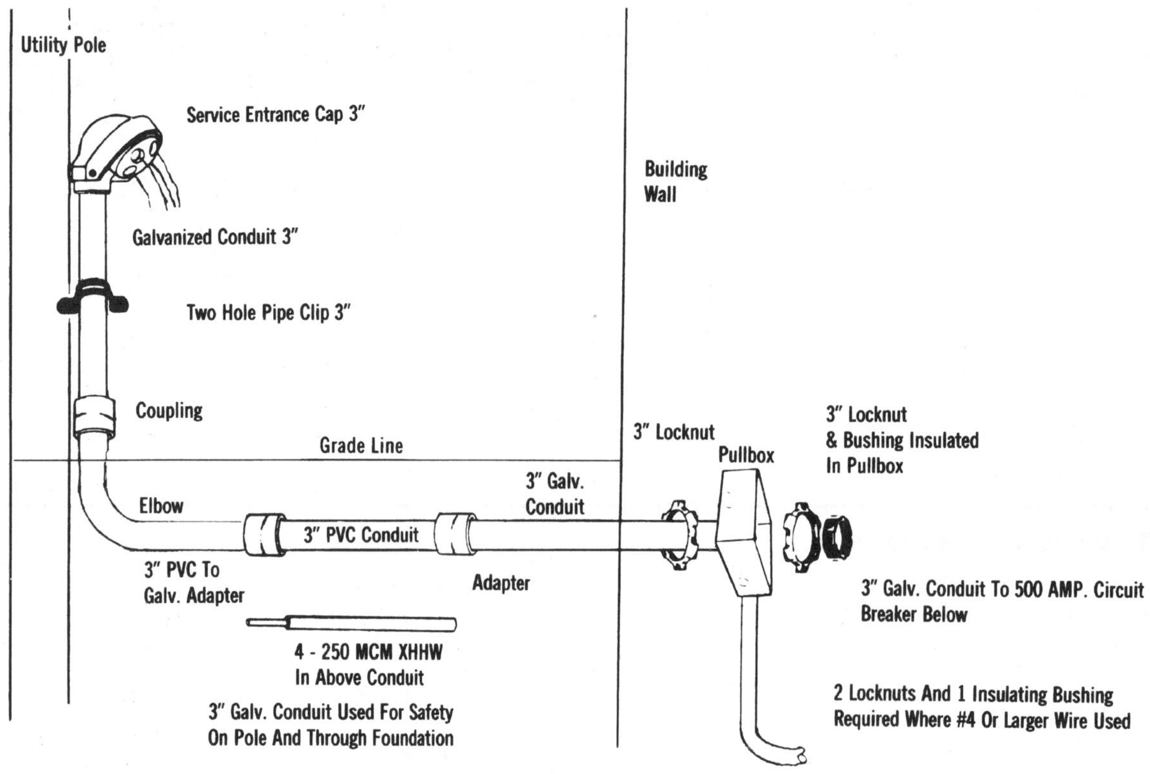

R260913-80 Switchboard Instruments

Switchboard instruments are added to the price of switchboards according to job specifications. This equipment is usually included when ordering "Gear" from the manufacturer and will arrive factory installed.

Included in the labor cost is:
1. Internal wiring connections
2. Wire identification
3. Wire tagging

Transition sections include:
1. Uncrating
2. Hauling sections up to 100′ from loading dock
3. Positioning sections
4. Leveling sections
5. Bolting enclosures
6. Bolting vertical bus bars

Price does not include:
1. Equipment pads
2. Steel channels embedded or grouted in concrete
3. Special knockouts
4. Rigging

Job Conditions: Productivity is based on new construction, equipment to be installed on the first floor within 200′ of the loading dock.

R262213-10 Electric Circuit Voltages

General: The following method provides the user with a simple non-technical means of obtaining comparative costs of wiring circuits. The circuits considered serve the electrical loads of motors, electric heating, lighting and transformers, for example, that require low voltage 60 Hertz alternating current.

The method used here is suitable only for obtaining estimated costs. It is **not** intended to be used as a substitute for electrical engineering design applications.

Conduit and wire circuits can represent from twenty to thirty percent of the total building electrical cost. By following the described steps and using the tables the user can translate the various types of electric circuits into estimated costs.

Wire Size: Wire size is a function of the electric load which is usually listed in one of the following units:

1. Amperes (A)
2. Watts (W)
3. Kilowatts (kW)
4. Volt amperes (VA)
5. Kilovolt amperes (kVA)
6. Horsepower (HP)

These units of electric load must be converted to amperes in order to obtain the size of wire necessary to carry the load. To convert electric load units to amperes one must have an understanding of the voltage classification of the power source and the voltage characteristics of the electrical equipment or load to be energized. The seven A.C. circuits commonly used are illustrated in Figures R262213-11 thru R262213-17 showing the tranformer load voltage and the point of use voltage at the point on the circuit where the load is connected. The difference between the source and point of use voltages is attributed to the circuit voltage drop and is considered to be approximately 4%.

Motor Voltages: Motor voltages are listed by their point of use voltage and not the power source voltage.

For example: 460 volts instead of 480 volts
200 instead of 208 volts
115 volts instead of 120 volts

Lighting and Heating Voltages: Lighting and heating equipment voltages are listed by the power source voltage and not the point of wire voltage.

For example: 480, 277, 120 volt lighting
480 volt heating or air conditioning unit
208 volt heating unit

Transformer Voltages: Transformer primary (input) and secondary (output) voltages are listed by the power source voltage.

For example: Single phase 10 kVA
Primary 240/480 volts
Secondary 120/240 volts

In this case, the primary voltage may be 240 volts with a 120 volts secondary or may be 480 volts with either a 120V or a 240V secondary.

For example: Three phase 10 kVA
Primary 480 volts
Secondary 208Y/120 volts

In this case the transformer is suitable for connection to a circuit with a 3 phase 3 wire or 3 phase 4 wire circuit with a 480 voltage. This application will provide a secondary circuit of 3 phase 4 wire with 208 volts between phase wires and 120 volts between any phase wire and the neutral (white) wire.

R262213-11

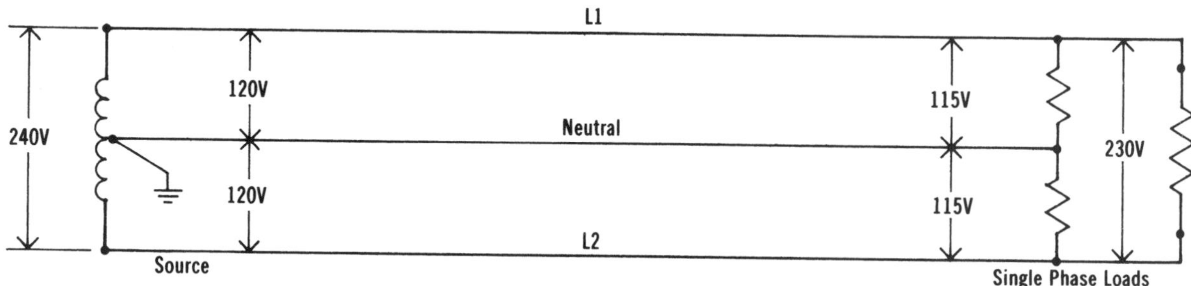

R262213-12

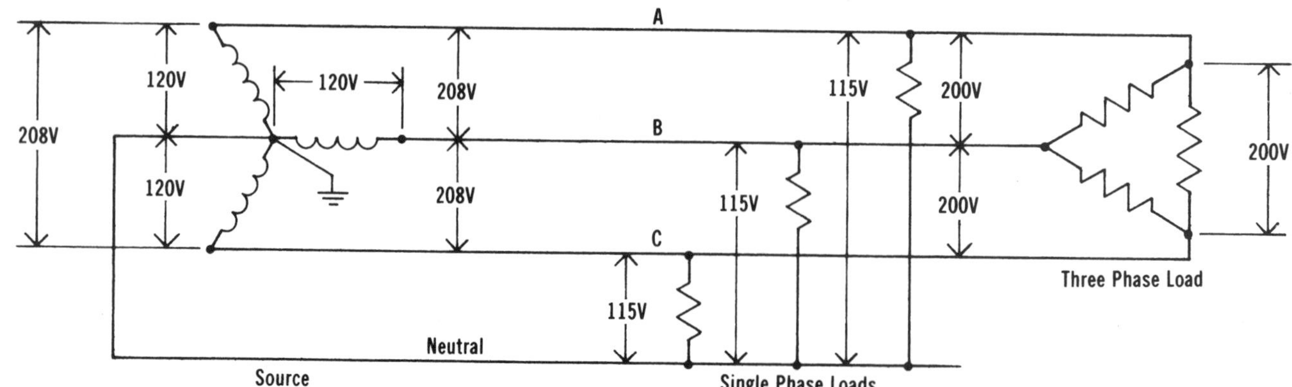

R262213-13 Electric Circuit Voltages (cont.)

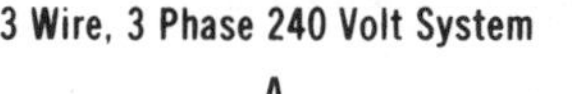

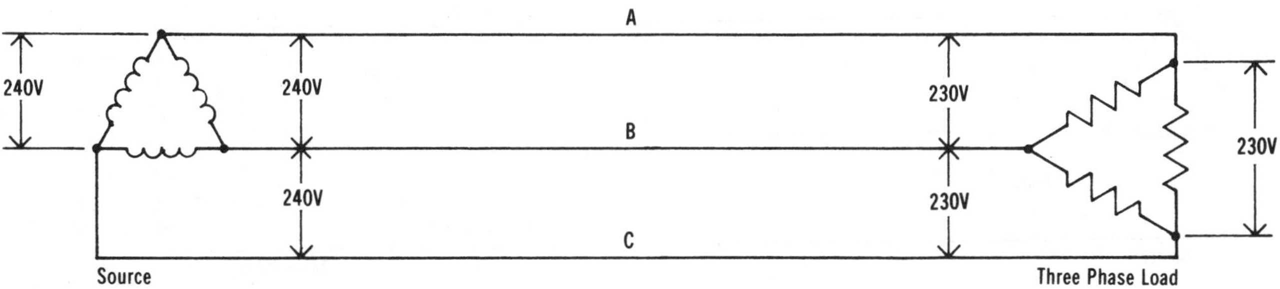

R262213-14

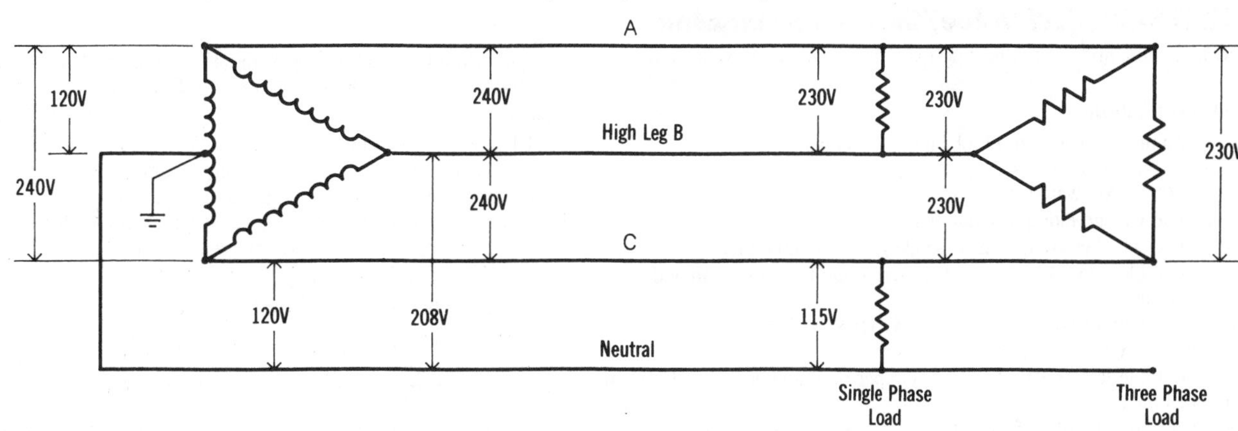

R262213-15

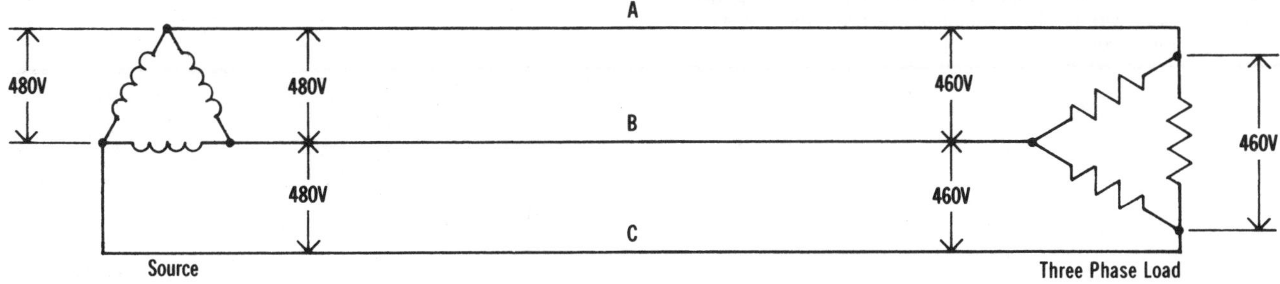

R262213-16

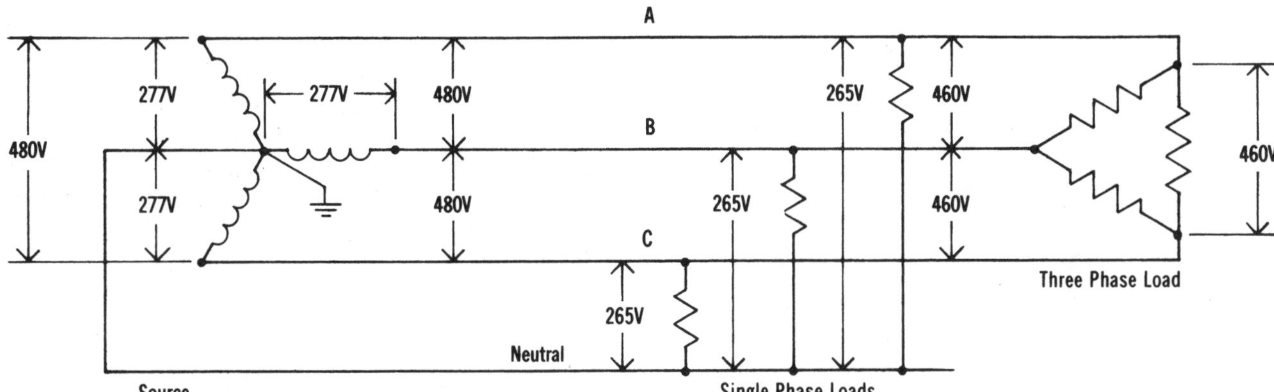

R262213-17 Electric Circuit Voltages (cont.)

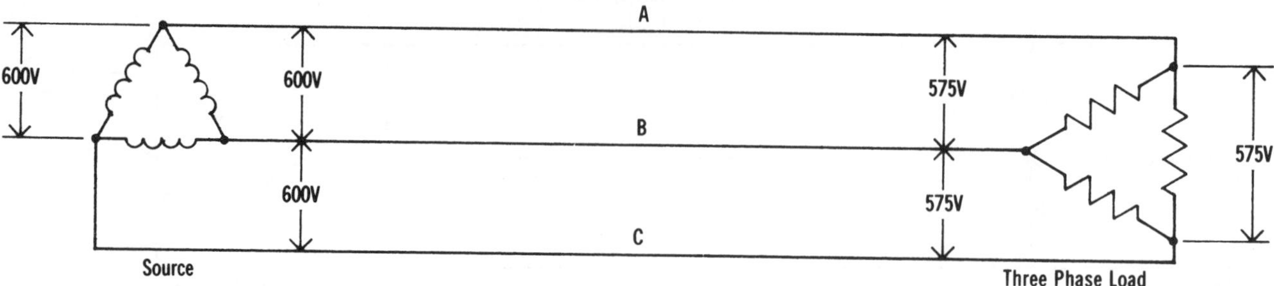

R262213-20 kW Value/Cost Determination

General: Lighting and electric heating loads are expressed in watts and kilowatts.

Cost Determination:

The proper ampere values can be obtained as follows:

1. Convert watts to kilowatts (watts 1000 ÷ kilowatts)
2. Determine voltage rating of equipment.
3. Determine whether equipment is single phase or three phase.
4. Refer to Table R262213-21 to find ampere value from kW, Ton and Btu/hr. values.
5. Determine type of wire insulation – TW, THW, THWN.
6. Determine if wire is copper or aluminum.
7. Refer to Table R260519-92 to obtain copper or aluminum wire size from ampere values.
8. Next refer to Table R260533-22 for the proper conduit size to accommodate the number and size of wires in each particular case.
9. Next refer to unit cost data for the per linear foot cost of the conduit.
10. Next refer to unit cost data for the per linear foot cost of the wire. Multiply cost of wire per L.F. x number of wires in the circuits to obtain total wire cost per L.F.
11. Add values obtained in Step 9 and 10 for total cost per linear foot for conduit and wire x length of circuit = Total Cost.

Notes:

1. 1 Phase refers to single phase, 2 wire circuits.
2. 3 Phase refers to three phase, 3 wire circuits.
3. For circuits which operate continuously for 3 hours or more, multiply the ampere values by 1.25 for a given kw requirement.
4. For kW ratings not listed, add ampere values.

   For example: Find the ampere value of 9 kW at 208 volt, single phase.

   4 kW = 19.2A
   5 kW = 24.0A
   ---
   9 kW = 43.2A

5. "Length of Circuit" refers to the one way distance of the run, not to the total sum of wire lengths.

R262213-21 Ampere Values as Determined by kW Requirements, BTU/HR or Ton, Voltage and Phase Values

			Ampere Values						
			120V	208V		240V		277V	480V
kW	Ton	BTU/HR	1 Phase	1 Phase	3 Phase	1 Phase	3 Phase	1 Phase	3 Phase
0.5	.1422	1,707	4.2A	2.4A	1.4A	2.1A	1.2A	1.8A	0.6A
0.75	.2133	2,560	6.2	3.6	2.1	3.1	1.9	2.7	.9
1.0	.2844	3,413	8.3	4.9	2.8	4.2	2.4	3.6	1.2
1.25	.3555	4,266	10.4	6.0	3.5	5.2	3.0	4.5	1.5
1.5	.4266	5,120	12.5	7.2	4.2	6.3	3.1	5.4	1.8
2.0	.5688	6,826	16.6	9.7	5.6	8.3	4.8	7.2	2.4
2.5	.7110	8,533	20.8	12.0	7.0	10.4	6.1	9.1	3.1
3.0	.8532	10,239	25.0	14.4	8.4	12.5	7.2	10.8	3.6
4.0	1.1376	13,652	33.4	19.2	11.1	16.7	9.6	14.4	4.8
5.0	1.4220	17,065	41.6	24.0	13.9	20.8	12.1	18.1	6.1
7.5	2.1331	25,598	62.4	36.0	20.8	31.2	18.8	27.0	9.0
10.0	2.8441	34,130	83.2	48.0	27.7	41.6	24.0	36.5	12.0
12.5	3.5552	42,663	104.2	60.1	35.0	52.1	30.0	45.1	15.0
15.0	4.2662	51,195	124.8	72.0	41.6	62.4	37.6	54.0	18.0
20.0	5.6883	68,260	166.4	96.0	55.4	83.2	48.0	73.0	24.0
25.0	7.1104	85,325	208.4	120.2	70.0	104.2	60.0	90.2	30.0
30.0	8.5325	102,390		144.0	83.2	124.8	75.2	108.0	36.0
35.0	9.9545	119,455		168.0	97.1	145.6	87.3	126.0	42.1
40.0	11.3766	136,520		192.0	110.8	166.4	96.0	146.0	48.0
45.0	12.7987	153,585			124.8	187.5	112.8	162.0	54.0
50.0	14.2208	170,650			140.0	208.4	120.0	180.4	60.0
60.0	17.0650	204,780			166.4		150.4	216.0	72.0
70.0	19.9091	238,910			194.2		174.6		84.2
80.0	22.7533	273,040			221.6		192.0		96.0
90.0	25.5975	307,170					225.6		108.0
100.0	28.4416	341,300							120.0

R262213-25 kVA Value/Cost Determination

General: Control transformers are listed in VA. Step-down and power transformers are listed in kVA.

Cost Determination:

1. Convert VA to kVA. Volt amperes (VA) ÷ 1000 = Kilovolt amperes (kVA).
2. Determine voltage rating of equipment.
3. Determine whether equipment is single phase or three phase.
4. Refer to Table R262213-26 to find ampere value from kVA value.
5. Determine type of wire insulation – TW, THW, THWN.
6. Determine if wire is copper or aluminum.
7. Refer to Table R260519-92 to obtain copper or aluminum wire size from ampere values.
8. Next refer to Table R260533-22 for the proper conduit size to accommodate the number and size of wires in each particular case.
9. Next refer to unit price data for the per linear foot cost of the conduit.
10. Next refer to unit price data for the per linear foot cost of the wire. Multiply cost of wire per L.F. x number of wires in the circuits to obtain total wire cost.
11. Add values obtained in Step 9 and 10 for total cost per linear foot for conduit and wire x length of circuit = Total Cost.

Example: A transformer rated 10 kVA 480 volts primary, 240 volts secondary, 3 phase has the capacity to furnish the following:

1. Primary amperes = 10 kVA x 1.20 = 12 amperes (from Table R262213-26)
2. Secondary amperes = 10 kVA x 2.40 = 24 amperes (from Table R262213-26)

Note: Transformers can deliver generally 125% of their rated kVA. For instance, a 10 kVA rated transformer can safely deliver 12.5 kVA.

R262213-26 Multiplier Values for kVA to Amperes Determined by Voltage and Phase Values

Volts	Multiplier for Circuits	
	2 Wire, 1 Phase	3 Wire, 3 Phase
115	8.70	
120	8.30	
230	4.30	2.51
240	4.16	2.40
200	5.00	2.89
208	4.80	2.77
265	3.77	2.18
277	3.60	2.08
460	2.17	1.26
480	2.08	1.20
575	1.74	1.00
600	1.66	0.96

R262213-27 Central Air Conditioning Watts per S.F., BTUs per Hour per S.F. of Floor Area and S.F. per Ton of Air Conditioning

Type Building	Watts per S.F.	BTUH per S.F.	S.F. per Ton	Type Building	Watts per S.F.	BTUH per S.F.	S.F. per Ton	Type Building	Watts per S.F.	BTUH per S.F.	S.F. per Ton
Apartments, Individual	3	26	450	Dormitory, Rooms	4.5	40	300	Libraries	5.7	50	240
Corridors	2.5	22	550	Corridors	3.4	30	400	Low Rise Office, Ext.	4.3	38	320
Auditoriums & Theaters	3.3	40	300/18*	Dress Shops	4.9	43	280	Interior	3.8	33	360
Banks	5.7	50	240	Drug Stores	9	80	150	Medical Centers	3.2	28	425
Barber Shops	5.5	48	250	Factories	4.5	40	300	Motels	3.2	28	425
Bars & Taverns	15	133	90	High Rise Off.-Ext. Rms.	5.2	46	263	Office (small suite)	4.9	43	280
Beauty Parlors	7.6	66	180	Interior Rooms	4.2	37	325	Post Office, Int. Office	4.9	42	285
Bowling Alleys	7.8	68	175	Hospitals, Core	4.9	43	280	Central Area	5.3	46	260
Churches	3.3	36	330/20*	Perimeter	5.3	46	260	Residences	2.3	20	600
Cocktail Lounges	7.8	68	175	Hotels, Guest Rooms	5	44	275	Restaurants	6.8	60	200
Computer Rooms	16	141	85	Public Spaces	6.2	55	220	Schools & Colleges	5.3	46	260
Dental Offices	6	52	230	Corridors	3.4	30	400	Shoe Stores	6.2	55	220
Dept. Stores, Basement	4	34	350	Industrial Plants, Offices	4.3	38	320	Shop'g. Ctrs., Sup. Mkts.	4	34	350
Main Floor	4.5	40	300	General Offices	4	34	350	Retail Stores	5.5	48	250
Upper Floor	3.4	30	400	Plant Areas	4.5	40	300	Specialty Shops	6.8	60	200

*Persons per ton

12,000 BTUH = 1 ton of air conditioning

R262213-60 Oil Filled Transformers

Transformers in this section include:

1. Rigging (as required)
2. Rental of crane and operator
3. Setting of oil filled transformer
4. (4) Anchor bolts, nuts and washers in concrete pad

Price does not include:

1. Primary and secondary terminations
2. Transformer pad
3. Equipment grounding
4. Cable
5. Conduit locknuts or bushings

Transformers in Unit Price sections for dry type, back-boost and isolating transformers include:

1. Unloading and uncrating
2. Hauling transformer to within 200′ of loading dock
3. Setting in place
4. Wall mounting hardware
5. Testing

Price does not include:

1. Structural supports
2. Suspension systems
3. Welding or fabrication
4. Primary & secondary terminations

Add the following percentages to the labor for ceiling mounted transformers:

10′ to 15′	= + 15%
15′ to 25′	= + 30%
Over 25′	= + 35%

Job Conditions: Productivities are based on new construction. Installation is assumed to be on the first floor, in an obstructed area to a height of 10′. Material staging area is within 100′ of final transformer location.

Electrical | R2622 | Low Voltage Transformers

R262213-65 Transformer Weight (Lbs.) by kVA

Oil Filled 3 Phase 5/15 KV To 480/277			
kVA	Lbs.	kVA	Lbs.
150	1800	1000	6200
300	2900	1500	8400
500	4700	2000	9700
750	5300	3000	15000
Dry 240/480 To 120/240 Volt			
1 Phase		3 Phase	
kVA	Lbs.	kVA	Lbs.
1	23	3	90
2	36	6	135
3	59	9	170
5	73	15	220
7.5	131	30	310
10	149	45	400
15	205	75	600
25	255	112.5	950
37.5	295	150	1140
50	340	225	1575
75	550	300	1870
100	670	500	2850
167	900	750	4300

Electrical | R2624 | Switchboards & Panelboards

R262416-50 Load Centers and Panelboards

When pricing Load Centers list panels by size and type. List Breakers in a separate column of the "Quantity Sheet," and define by phase and ampere rating.

Material and Labor prices include breakers; for example: a 100A, 3-Wire, 102/240V, 18 circuit panel w/ main breaker, as described in the unit cost section, contains 18 single pole 20A breakers.

If you do not choose to include a full panel of single pole breakers, use the following method to adjust material and labor costs.

Example: In an 18 circuit panel only 16 single pole breakers are desired, requiring that the cost of 2 breakers be subtracted from the panel cost.

1. Go to the appropriate unit cost section of the book to find the unit prices of the given circuit breaker type.
2. Modify those costs as follows: Bare material price x 0.50 Bare labor cost x 0.60.
3. Multiply those modified bare costs by 2 (breakers in this example).
4. Subtract the modified costs for the 2 breakers from the given cost of the panel.

Labor-hours for Load Center installation includes:

1. Unloading, uncrating, and handling enclosures 200′ from unloading area
2. Measuring and marking
3. Drilling (4) lead anchor type fasteners using a hammer drill
4. Mounting and leveling panel to a height of 6′
5. Preparation and termination of feeder cable to lugs or main breaker
6. Branch circuit identification
7. Lacing using tie wraps
8. Testing and load balancing
9. Marking panel directory

Not included in the material and labor are:

1. Modifications to enclosure
2. Structural supports
3. Additional lugs
4. Plywood backboards
5. Painting or lettering

Note: Knockouts are included in the price of terminating pipe runs and need not be added to the Load Center costs.

Job Conditions: Productivity is based on new construction to a height of 6′, in an unobstructed area. Material staging area is assumed to be within 100′ of work being performed.

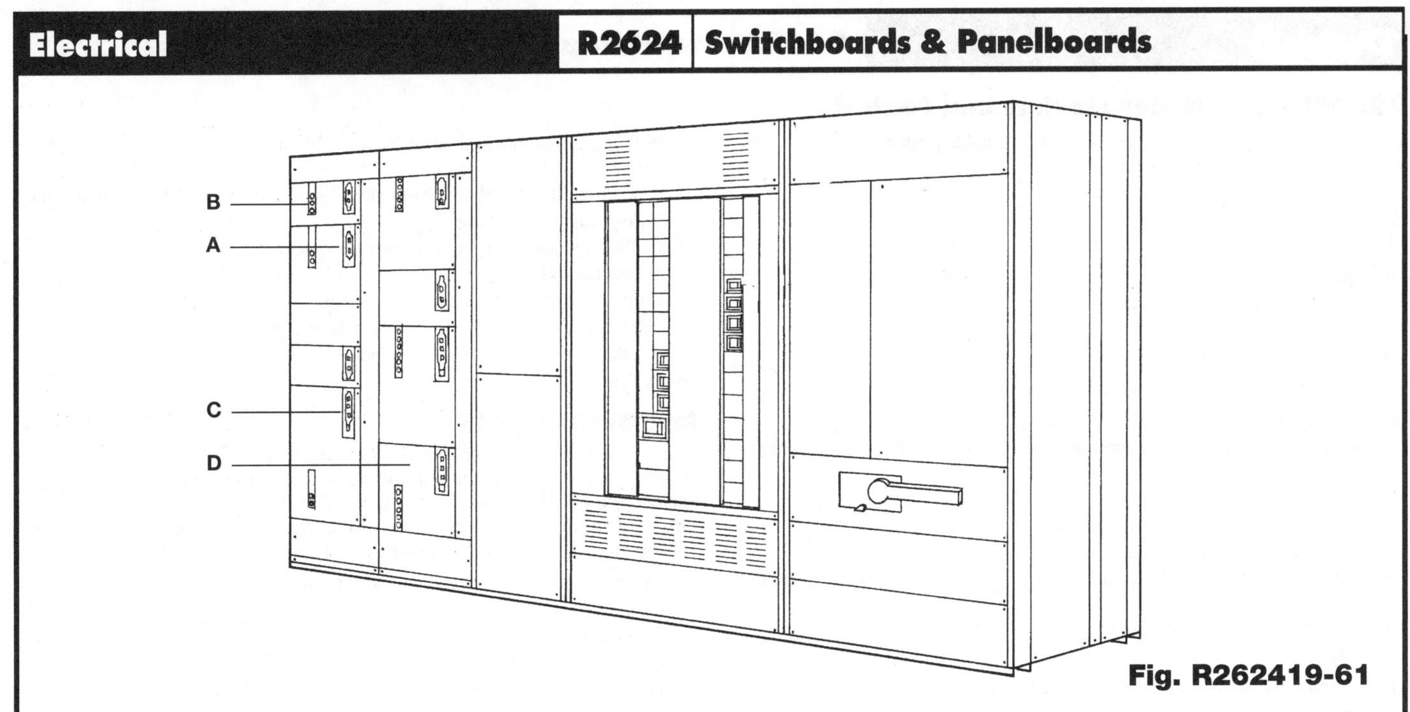

Fig. R262419-61

R262419-60 Motor Control Centers

When taking off Motor Control Centers, list the size, type and height of structures.

Example:

1. 600A, 22,000 RMS, 72″ high
2. 600A, back to back, 72″ high

Next take off individual starters; the number of structures can also be determined by adding the height in inches of starters divided by the height of the structure, and list on the same quantity sheet as the structures. Identify starters by Type, Horsepower rating, Size and Height in inches.

Example:

A. Class I, Type B, FVNR starter, 25 H.P., 18″ high.
Add to the list with Starters, factory installed controls.

Example:

B. Pilot lights
C. Push buttons
D. Auxiliary contacts

Identify starters and structures as either copper or aluminum and by the NEMA type of enclosure.

When pricing starters and structures, be sure to add or deduct adjustments, using lines in the unit price section.

Included in the cost of Motor Control Structures are:

1. Uncrating
2. Hauling to location within 100′ of loading dock
3. Setting structures
4. Leveling
5. Aligning
6. Bolting together structure frames
7. Bolting horizontal bus bars

Labor-hours do not include:

1. Equipment pad
2. Steel channels embedded or grouted in concrete
3. Pull boxes
4. Special knockouts
5. Main switchboard section
6. Transition section
7. Instrumentation
8. External control wiring
9. Conduit or wire

Material for Starters includes:

1. Circuit breaker or fused disconnect
2. Magnetic motor starter
3. Control transformer
4. Control fuse and fuse block

Labor-hours for Starters include:

1. Handling
2. Installing starter within structure
3. Internal wiring connections
4. Lacing within enclosure
5. Testing
6. Phasing

Job Conditions: Productivity is based on new construction. Motor control location assumed to be on first floor in an unobstructed area. Material staging area within 100′ of final location.

Note: Additional labor-hours must be added if M.C.C. is to be installed on other than the first floor, or if rigging is required.

R262419-65 Motor Starters and Controls

Motor starters should be listed on the same "Quantity Sheet" as Motors and Motor Connections. Identify each starter by:

1. Size
2. Voltage
3. Type

Example:

A. FVNR, 480V, 2HP, Size 00
B. FVNR, 480V, 5HP, Size 0, Combination type
C. FVR, 480V, Size 2

The NEMA type of enclosure should also be identified.

Included in the labor-hours are:

1. Unloading, uncrating and handling of starters up to 200′ from loading area
2. Measuring and marking
3. Drilling (4) anchor type lead fasteners, using a hammer drill
4. Mounting and leveling starter
5. Connecting wire or cable to line and load sides of starter (when already lugged)
6. Installation of (3) thermal type heaters
7. Testing

The following is not included unless specified in the unit price description.

1. Control transformer
2. Controls, either factory or field installed
3. Conduit and wire to or from starter
4. Plywood backboard
5. Cable terminations

The following material and labor has been included for Combination type starters:

1. Unloading, uncrating and handling of starter up to 200′ of loading dock
2. Measuring and marking
3. Drilling (4) anchor type lead fasteners using a hammer drill
4. Mounting and leveling
5. Connecting prepared cable conductors
6. Installation of (3) dual element cartridge type fuses
7. Installation of (3) thermal type heaters
8. Test for rotation

MOTOR CONTROLS

When pricing motor controls make sure you consider the type of control system being utilized. If the controls are factory installed and located in the enclosure itself, then you would add to your starter price the items 5200 thru 5800 in the Unit Price section. If control voltage is different from line voltage, add the material and labor cost of a control transformer.

For external control of starters include the following items:

1. Raceways
2. Wire & terminations
3. Control enclosures
4. Fittings
5. Push button stations
6. Indicators

Job Conditions: Productivity is based on new construction to a height of 10′.

Material staging area is assumed to be within 100′ of work being performed.

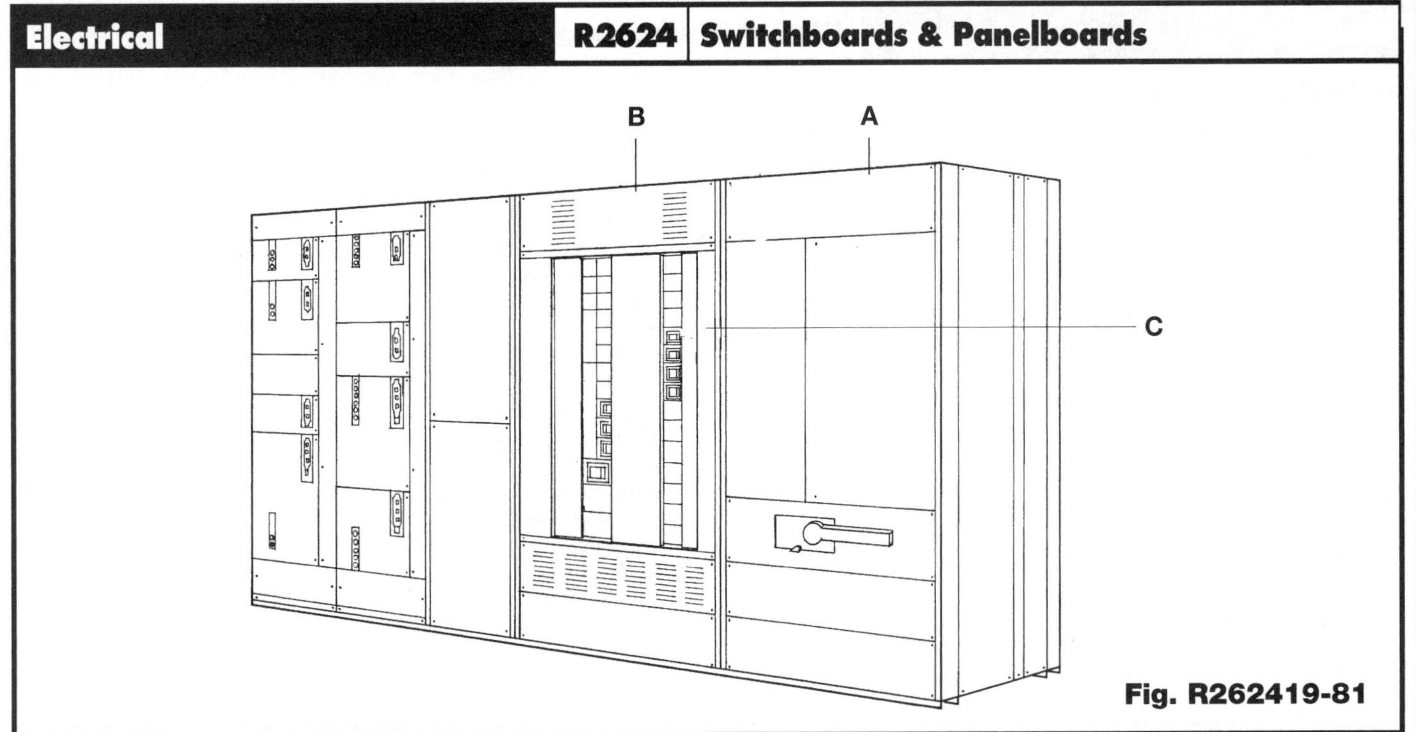

Fig. R262419-81

R262419-80 Distribution Section

After "Taking off" the Switchboard section, include on the same "Quantity sheet" the Distribution section; identify by:

1. Voltage
2. Ampere rating
3. Type

Example:

(Fig. R262419-81) (B) Distribution Section

Included in the labor costs of the Distribution section is:

1. Uncrating
2. Hauling to 200′ of loading dock
3. Setting of distribution panel
4. Leveling & shimming
5. Anchoring of equipment to pad or floor
6. Bolting of horizontal bus bars between
7. Testing of equipment

Not included in the Distribution section is:

1. Breakers (C)
2. Equipment pads
3. Steel channels embedded or grouted in concrete
4. Pull boxes
5. Special knockouts
6. Transition section
7. Conduit or wire

R262419-82 Feeder Section

List quantities on the same sheet as the Distribution section. Identify breakers by: (Fig. R262419-81)

1. Frame type
2. Number of poles
3. Ampere rating

Installation includes:

1. Handling
2. Placing breakers in distribution panel
3. Preparing wire or cable
4. Lacing wire
5. Marking each phase with colored tape
6. Marking panel legend
7. Testing
8. Balancing

R262419-84 Switchgear

It is recommended that "Switchgear" or those items contained in the following sections be quoted from equipment manufacturers as a package price.

Switchboard instruments

Switchboard distribution sections

Switchboard feeder sections

Switchboard Service Disc.

Switchboard In-plant Dist.

Included in these sections are the most common types and sizes of factory assembled equipment.

The recommended procedure for low voltage switchgear would be to price (Fig. R262419-81) (A) Main Switchboard

Identify by:

1. Voltage
2. Amperage
3. Type

Example:

1. 120/208V, 4-wire, 600A, nonfused
2. 277/480V, 4-wire, 600A, nonfused
3. 120/208V, 4-wire, 400A w/fused switch & CT compartment
4. 277/480V, 4-wire, 400A, w/fused switch & CT compartment
5. 120/208V, 4-wire, 800A, w/pressure switch & CT compartment
6. 277/480V, 4-wire, 800A, w/molded CB & CT compartment

Included in the labor costs for Switchboards are:

1. Uncrating
2. Hauling to 200′ of loading dock
3. Setting equipment
4. Leveling and shimming
5. Anchoring
6. Cable identification
7. Testing of equipment

Not included in the Switchboard price is:

1. Rigging
2. Equipment pads
3. Steel channels embedded or grouted in concrete
4. Special knockouts
5. Transition or Auxiliary sections
6. Instrumentation
7. External control
8. Conduit and wire
9. Conductor terminations

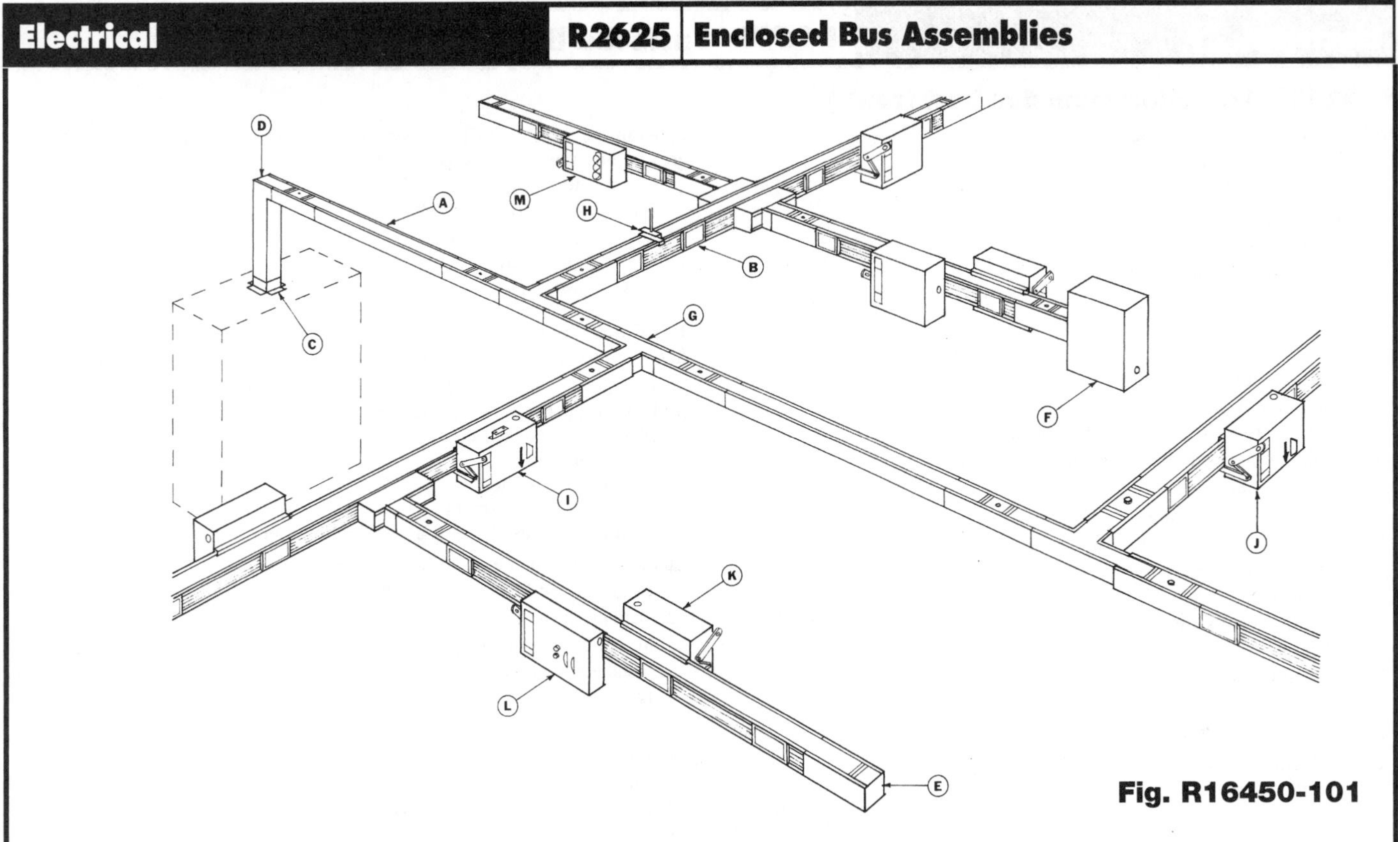

Fig. R16450-101

R262513-10 Aluminum Bus Duct

When taking off bus duct identify the system as either:

1. Aluminum
2. Copper

List straight lengths by type and size

- A. Plug-in — 800 A
- B. Feeder — 800 A

Do not measure thru fittings as you would on conduit, since there is no allowance for scrap in bus duct systems.

If upon taking off linear foot quantities of bus duct you find your quantities are not divisible evenly by 10 ft., then the remainder must be priced as a special item and quoted from the manufactuer. Do not use the bare material cost per L.F. for these special items. You can, however, safely use the bare labor cost per L.F. for the entire length.

Identify fittings by type and ampere rating.

Example:

C. Switchboard stub 800 A

D. Elbows 800 A

E. End box 800 A

F. Cable tap box 800 A

G. Tee Fittings 800 A

H. Hangers

Plug-in Units – List separately plug-in units and identify by type and ampere rating

I. Plug-in switches 600 Volt 3 phase 60 A

J. Plug-in molded case C.B. 60 A

K. Combination starter FVNR NEMA 1

L. Combination contactor & fused switch NEMA 1

M. Combination fusible switch & lighting control 60 A

Labor-hours for feeder and plug-in sections include:

1. Unloading and uncrating
2. Hauling up to 200 ft. from loading dock
3. Measuring and marking
4. Setup of rolling staging
5. Installing hangers
6. Hanging and bolting sections
7. Aligning and leveling
8. Testing

Labor-hours do not include:

1. Modifications to existing structure for hanger supports
2. Threaded rod in excess of 2 ft.
3. Welding
4. Penetrations thru walls
5. Staging rental

Deduct the following percentages from labor only:

150 ft. to 250 ft. — 10%
251 ft. to 350 ft. — 15%
351 ft. to 500 ft. — 20%
Over 500 ft. — 25%

Deduct percentage only if runs are contained in the same area.

Example: If the job entails running 100 ft. in 5 different locations do not deduct 20%, but if the duct is being run in 1 area and the quantity is 500 ft. then you would deduct 20%.

Deduct only from straight lengths, not fittings or plug-in units.

R262513-10 Aluminum Bus Duct (cont.)

Add to labor for elevated installations:

15 ft. to 20 ft. high	10%
21 ft. to 25 ft. high	20%
26 ft. to 30 ft. high	30%
31 ft. to 35 ft. high	40%
36 ft. to 40 ft. high	50%
Over 40 ft. high	60%

Bus Duct Fittings:

Labor-hours for fittings include:
1. Unloading and uncrating
2. Hauling up to 200 ft. from loading dock
3. Installing, fitting, and bolting all ends to in-place sections

Plug-in units include:
1. Unloading and uncrating
2. Hauling up to 200 ft. from loading dock
3. Installing plug-in into in-place duct
4. Setup of rolling staging
5. Connecting load wire to lugs
6. Marking wire
7. Checking phase rotation

Labor-hours for plug-ins do not include:
1. Conduit runs from plug-in
2. Wire from plug-in
3. Conduit termination

Economy of Scale – For large concentrations of plug-in units in the same area deduct the following percentages:

11	to	25	15%
26	to	50	20%
51	to	75	25%
76	to	100	30%
100	and	over	35%

Job Conditions: Productivities are based on new construction in an unobstructed first floor area to a height of 15 ft. using rolling staging.

Material staging area is within 100 ft. of work being performed.
Add to the duct fittings and hangers:
- Plug-in switches (fused)
- Plug-in breakers
- Combination starters
- Combination contactors
- Combination fusible switch and lighting control

Labor-hours for duct and fittings include:
1. Unloading and uncrating
2. Hauling up to 200 ft. from loading dock
3. Measuring and marking
4. Installing duct runs
5. Leveling
6. Sound testing

Labor-hours for plug-ins include:
1. Unloading and uncrating
2. Hauling up to 200 ft. from loading dock
3. Installing plug-ins
4. Preparing wire
5. Wire connections and marking

R262513-15 Weight (Lbs./L.F.) of 4 Pole Aluminum and Copper Bus Duct by Ampere Load

Amperes	Aluminum Feeder	Copper Feeder	Aluminum Plug-In	Copper Plug-In
225			7	7
400			8	13
600	10	10	11	14
800	10	19	13	18
1000	11	19	16	22
1350	14	24	20	30
1600	17	26	25	39
2000	19	30	29	46
2500	27	43	36	56
3000	30	48	42	73
4000	39	67		
5000		78		

R262716-40 Standard Electrical Enclosure Types

NEMA Enclosures

Electrical enclosures serve two basic purposes; they protect people from accidental contact with enclosed electrical devices and connections, and they protect the enclosed devices and connections from specified external conditions. The National Electrical Manufacturers Association (NEMA) has established the following standards. Because these descriptions are not intended to be complete representations of NEMA listings, consultation of NEMA literature is advised for detailed information.

The following definitions and descriptions pertain to NONHAZARDOUS locations.

NEMA Type 1: General purpose enclosures intended for use indoors, primarily to prevent accidental contact of personnel with the enclosed equipment in areas that do not involve unusual conditions.

NEMA Type 2: Dripproof indoor enclosures intended to protect the enclosed equipment against dripping noncorrosive liquids and falling dirt.

NEMA Type 3: Dustproof, raintight and sleet-resistant (ice-resistant) enclosures intended for use outdoors to protect the enclosed equipment against wind-blown dust, rain, sleet, and external ice formation.

NEMA Type 3R: Rainproof and sleet-resistant (ice-resistant) enclosures which are intended for use outdoors to protect the enclosed equipment against rain. These enclosures are constructed so that the accumulation and melting of sleet (ice) will not damage the enclosure and its internal mechanisms.

NEMA Type 3S: Enclosures intended for outdoor use to provide limited protection against wind-blown dust, rain, and sleet (ice) and to allow operation of external mechanisms when ice-laden.

NEMA Type 4: Watertight and dust-tight enclosures intended for use indoors and out – to protect the enclosed equipment against splashing water, see page of water, falling or hose-directed water, and severe external condensation.

NEMA Type 4X: Watertight, dust-tight, and corrosion-resistant indoor and outdoor enclosures featuring the same provisions as Type 4 enclosures, plus corrosion resistance.

NEMA Type 5: Indoor enclosures intended primarily to provide limited protection against dust and falling dirt.

NEMA Type 6: Enclosures intended for indoor and outdoor use – primarily to provide limited protection against the entry of water during occasional temporary submersion at a limited depth.

NEMA Type 6R: Enclosures intended for indoor and outdoor use – primarily to provide limited protection against the entry of water during prolonged submersion at a limited depth.

NEMA Type 11: Enclosures intended for indoor use – primarily to provide, by means of oil immersion, limited protection to enclosed equipment against the corrosive effects of liquids and gases.

NEMA Type 12: Dust-tight and driptight indoor enclosures intended for use indoors in industrial locations to protect the enclosed equipment against fibers, flyings, lint, dust, and dirt, as well as light splashing, see page, dripping, and external condensation of noncorrosive liquids.

NEMA Type 13: Oil-tight and dust-tight indoor enclosures intended primarily to house pilot devices, such as limit switches, foot switches, push buttons, selector switches, and pilot lights, and to protect these devices against lint and dust, see page, external condensation, and sprayed water, oil, and noncorrosive coolant.

The following definitions and descriptions pertain to HAZARDOUS, or CLASSIFIED, locations:

NEMA Type 7: Enclosures intended to use in indoor locations classified as Class 1, Groups A, B, C, or D, as defined in the National Electrical Code.

NEMA Type 9: Enclosures intended for use in indoor locations classified as Class 2, Groups E, F, or G, as defined in the National Electrical Code.

R262726-90 Wiring Devices

Wiring devices should be priced on a separate takeoff form which includes boxes, covers, conduit and wire.

Labor-hours for devices include:

1. Stripping of wire
2. Attaching wire to device using terminators on the device itself, lugs, set screws etc.
3. Mounting of device in box

Labor-hours do not include:

1. Conduit
2. Wire
3. Boxes
4. Plates

Economy of Scale – for large concentrations of devices in the same area deduct the following percentages from labor-hours:

1	to	10	0%
11	to	25	20%
26	to	50	25%
51	to	100	30%
	over	100	35%

R262726-90 Wiring Devices (cont.)

NEMA No.	15 R	20 R	30 R	50 R	60 R
1 125V 2 Pole, 2 Wire	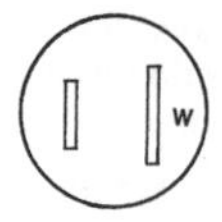 W				
2 250V 2 Pole, 2 Wire		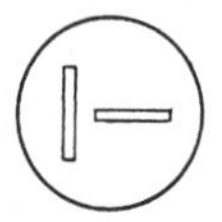	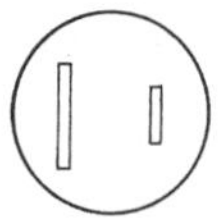		
5 125V 2 Pole, 3 Wire	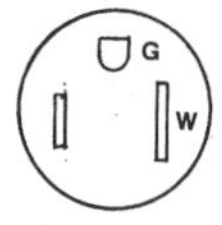 G W	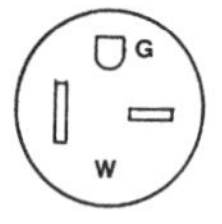 G W	 G W	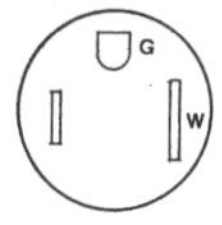 G W	
6 250V 2 Pole, 3 Wire	 G	 G	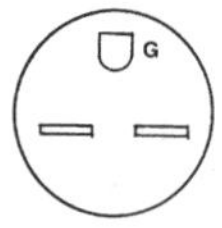 G	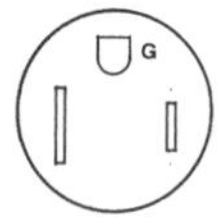 G	
7 277V, AC 2 Pole, 3 Wire	 G W	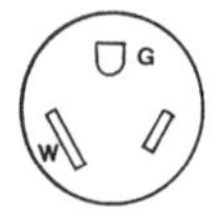 G W	 G W	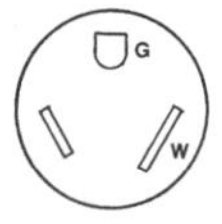 G W	
10 125/250V 3 Pole, 3 Wire		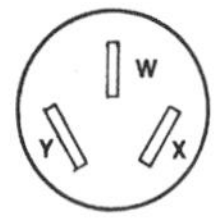 W Y X	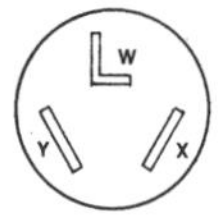 W Y X	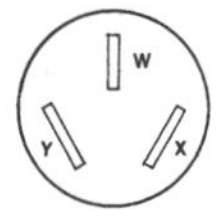 W Y X	
11 3 Phase 250V 3 Pole, 3 Wire	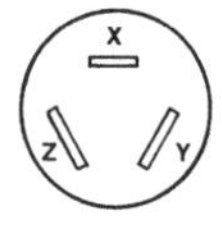 X Z Y	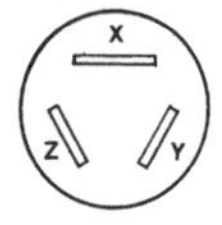 X Z Y	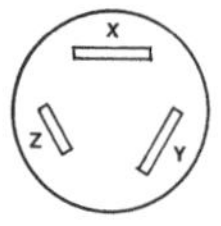 X Z Y	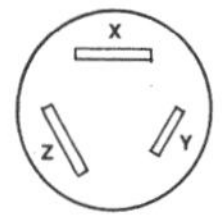 X Z Y	
14 125/250V 3 Pole, 4 Wire	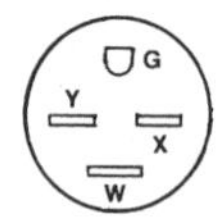 G Y X W	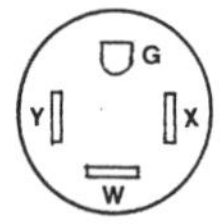 G Y X W	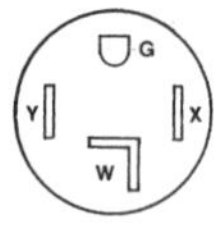 G Y X W	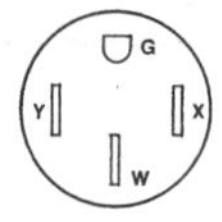 G Y X W	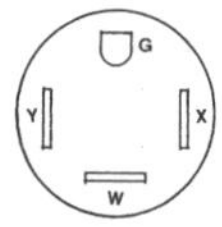 G Y X W
15 3 Phase 250V 3 Pole, 4 Wire	 G Z X Y	 G Z X Y	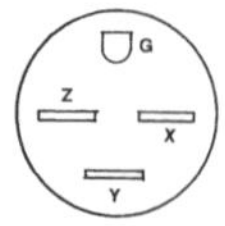 G Z X Y	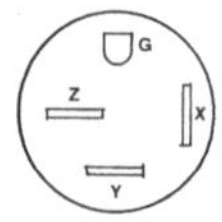 G Z X Y	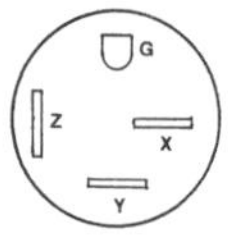 G Z X Y
18 3 Phase 208Y/120V 4 Pole, 4 Wire	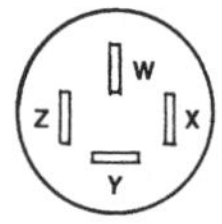 W Z X Y	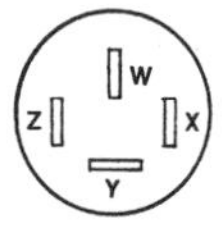 W Z X Y	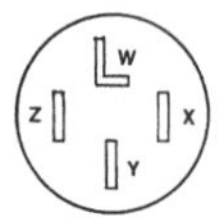 W Z X Y	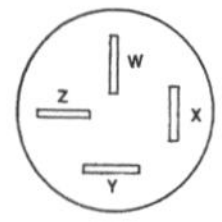 W Z X Y	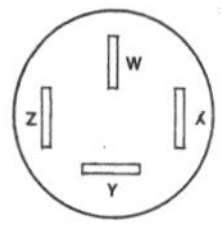 W Z X Y

R262726-90 Wiring Devices (cont.)

NEMA No.	15 R	20 R	30 R	NEMA No.	15 R	20 R	30 R
L 1 125V 2 Pole, 2 Wire	W			**L 13** 3 Phase 600V 3 Pole, 3 Wire			X Y Z
L 2 250V 2 Pole, 2 Wire		15A		**L 14** 125/250V 3 Pole, 4 Wire		X G Y Z	X G W Z
L 5 125 V 2 Pole, 3 Wire	G W	G W	G W	**L 15** 3 Phase 250 V 3 Pole, 4 Wire		X G W Y	X G Y Z
L 6 250 V 2 Pole, 3 Wire	G X Y	G X Y	X G Y	**L 16** 3 Phase 480V 3 Pole, 4 Wire		X G Y Z	X G Y Z
L 7 227 V, AC 2 Pole, 3 Wire	G W	G W	G W	**L 17** 3 Phase 600 V 3 Pole, 4 Wire			X G Y Z
L 8 480 V 2 Pole, 3 Wire		G X Y	G X Y	**L 18** 3 Phase 208Y/120V 4 Pole, 4 Wire		X W Y Z	X W Y Z
L 9 600 V 2 Pole, 3 Wire		G X Y	G X Y	**L 19** 3 Phase 480Y/277V 4 Pole, 4 Wire		X W Y Z	X W Y Z
L 10 125 /250V 3 Pole, 3 Wire		W X Y	X W Y	**L 20** 3 Phase 600Y/347V 4 Pole, 4 Wire		X W Y Z	X W Y Z
L 11 3 Phase 250 V 3 Pole, 3 Wire		Y Z	Y X Z	**L 21** 3 Phase 208Y/120V 4 Pole, 5 Wire		X W G Y Z	X W G Y Z
L 12 3 Phase 480 V 3 Pole, 3 Wire		Y X Z	Y X Z	**L 22** 3 Phase 480Y/277V 4 Pole, 5 Wire		X W G Y Z	X W G Y Z
				L 23 3 Phase 600Y/347V 4 Pole, 5 Wire		X W G Y Z	X W G Y Z

Electrical | R2628 | Low-Voltage Circuit Protective Devices

R262816-80 Safety Switches

List each Safety Switch by type, ampere rating, voltage, single or three phase, fused or nonfused.

Example:

A. General duty, 240V, 3-pole, fused
B. Heavy Duty, 600V, 3-pole, nonfused
C. Heavy Duty, 240V, 2-pole, fused
D. Heavy Duty, 600V, 3-pole, fused

Also include NEMA enclosure type and identify as:

1. Indoor
2. Weatherproof
3. Explosionproof

Installation of Safety Switches includes:

1. Unloading and uncrating
2. Handling disconnects up to 200′ from loading dock
3. Measuring and marking location
4. Drilling (4) anchor type lead fasteners, using a hammer drill
5. Mounting and leveling Safety Switch
6. Installing (3) fuses
7. Phasing and tagging line and load wires

Price does not include:

1. Modifications to enclosure
2. Plywood backboard
3. Conduit or wire
4. Fuses
5. Termination of wires

Job Conditions: Productivities are based on new construction to an installed height of 6′ above finished floor.

Material staging area is assumed to be within 100′ of work in progress.

Electrical | R2632 | Packaged Generator Assemblies

R263213-45 Generator Weight (Lbs.) by kW

3 Phase 4 Wire /480 Volt			
Gas		Diesel	
kW	Lbs.	kW	Lbs.
7.5	600	30	1800
10	630	50	2230
15	960	75	2250
30	1500	100	3840
65	2350	125	4030
85	2570	150	5500
115	4310	175	5650
170	6530	200	5930
		250	6320
		300	7840
		350	8220
		400	10750
		500	11900

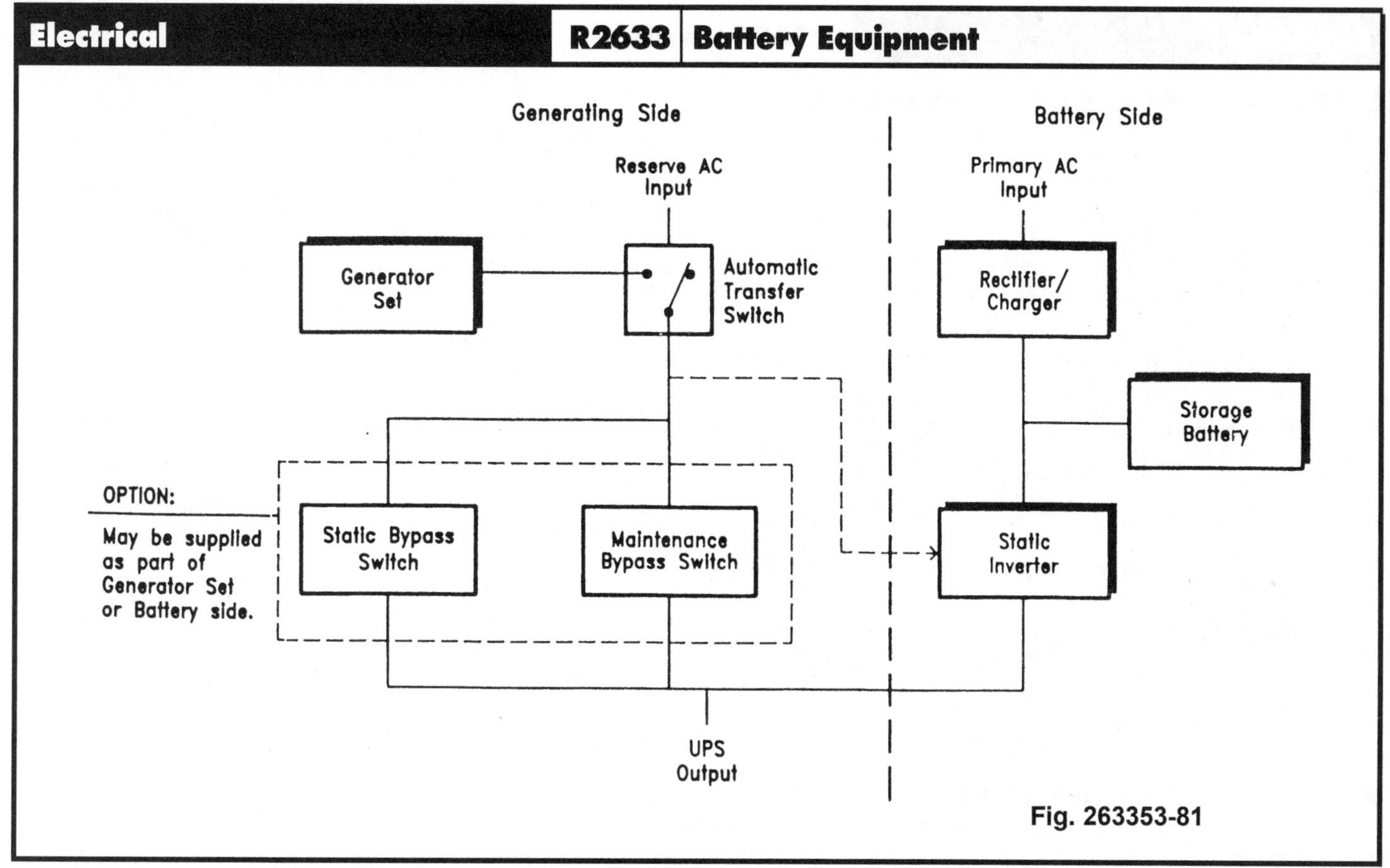

Fig. 263353-81

R263353-80 Uninterruptible Power Supply Systems

General: Uninterruptible Power Supply (UPS) Systems are used to provide power for legally required standby systems. They are also designed to protect computers and provide optional additional coverage for any or all loads in a facility. Figure R263353-81 shows a typical configuration for a UPS System with generator set.

Cost Determination:

It is recommended that UPS System material costs be obtained from equipment manufacturers as a package. Installation costs should be obtained from a vendor or contractor.

The recommended procedure for pricing UPS Systems would be to identify:

1. Frequency – by Hertz (Hz.)
   For example: 50, 60, and 400/415 Hz.
2. Apparent power and real power
   For example: 200 kVA/170 kW
3. Input/Output voltage For example: 120V, 208V or 240V
4. Phase – either single or three-phase
5. Options and accessories – For extended run time more batteries would be required. Other accessories include battery cabinets, battery racks, remote control panel, power distribution unit (PDU), and warranty enhancement plans.

Note:

1. Larger systems can be configured by paralleling two or more standard small size single modules.
   For example: two 15 kVA modules, combined together and configured to 30 kVA.
2. Maximum input current during battery recharge is typically 15% higher than normal current.
3. UPS Systems weights vary depending on options purchased and power rating in kVA.

R263413-30 HP Value/Cost Determination

General: Motors can be powered by any of the seven systems shown in Figure R262213-11 thru Figure R262213-17 provided the motor voltage characteristics are compatible with the power system characteristics.

Cost Determination:

Motor Amperes for the various size H.P. and voltage are listed in Table R263413-31. To find the amperes, locate the required H.P. rating and locate the amperes under the appropriate circuit characteristics.

For example:

A. 100 H.P., 3 phase, 460 volt motor = 124 amperes (Table R263413-31)

B. 10 H.P., 3 phase, 200 volt motor = 32.2 amperes (Table R263413-31)

Motor Wire Size: After the amperes are found in Table R263413-31 the amperes must be increased 25% to compensate for power losses. Next refer to Table R260519-92. Find the appropriate insulation column for copper or aluminum wire to determine the proper wire size.

For example:

A. 100 H.P., 3 phase, 460 volt motor has an ampere value of 124 amperes from Table R263413-31

B. 124A x 1.25 = 155 amperes

C. Refer to Table R260519-92 for THW or THWN wire insulations to find the proper wire size. For a 155 ampere load using copper wire a size 2/0 wire is needed.

D. For the 3 phase motor three wires of 2/0 size are required.

Conduit Size: To obtain the proper conduit size for the wires and type of insulation used, refer to Table R260533-22.

For example: For the 100 H.P., 460V, 3 phase motor, it was determined that three 2/0 wires are required. Assuming THWN insulated copper wire, use Table R260533-22 to determine that three 2/0 wires require 1-1/2″ conduit.

Material Cost of the conduit and wire system depends on:

1. Wire size required
2. Copper or aluminum wire
3. Wire insulation type selected
4. Steel or plastic conduit
5. Type of conduit raceway selected.

Labor Cost of the conduit and wire system depends on:

1. Type and size of conduit
2. Type and size of wires installed
3. Location and height of installation in building or depth of trench
4. Support system for conduit.

R263413-31 Ampere Values Determined by Horsepower, Voltage and Phase Values

H.P.	Amperes							
	Single Phase			Three Phase				
	115V	208V	230V	200V	208V	230V	460V	575V
1/6	4.4A	2.4A	2.2A					
1/4	5.8	3.2	2.9					
1/3	7.2	4.0	3.6					
1/2	9.8	5.4	4.9	2.5A	2.4A	2.2A	1.1A	0.9A
3/4	13.8	7.6	6.9	3.7	3.5	3.2	1.6	1.3
1	16	8.8	8	4.8	4.6	4.2	2.1	1.7
1-1/2	20	11	10	6.9	6.6	6.0	3.0	2.4
2	24	13.2	12	7.8	7.5	6.8	3.4	2.7
3	34	18.7	17	11.0	10.6	9.6	4.8	3.9
5	56	30.8	28	17.5	16.7	15.2	7.6	6.1
7-1/2	80	44	40	25.3	24.2	22	11	9
10	100	55	50	32.2	30.8	28	14	11
15				48.3	46.2	42	21	17
20				62.1	59.4	54	27	22
25				78.2	74.8	68	34	27
30				92	88	80	40	32
40				120	114	104	52	41
50				150	143	130	65	52
60				177	169	154	77	62
75				221	211	192	96	77
100				285	273	248	124	99
125				359	343	312	156	125
150				414	396	360	180	144
200				552	528	480	240	192
250							302	242
300							361	289
350							414	336
400							477	382

R263413-30 Cost Determination (cont.)

Magnetic starters, switches, and motor connection:

To complete the cost picture from H.P. to Costs additional items must be added to the cost of the conduit and wire system to arrive at a total cost.

1. Assembly Table D5020-160 Magnetic Starters Installed Cost lists the various size starters for single phase and three phase motors.
2. Assembly Table D5020-165 Heavy Duty Safety Switches Installed Cost lists safety switches required at the beginning of a motor circuit and also one required in the vicinity of the motor location.
3. Assembly Table D5020-170 Motor Connection lists the various costs for single and three phase motors.

Worksheet to obtain total motor wiring costs:

It is assumed that the motors or motor driven equipment are furnished and installed under other sections for this estimate and the following work is done under this section:

1. Conduit
2. Wire (add 10% for additional wire beyond conduit ends for connections to switches, boxes, starters, etc.)
3. Starters
4. Safety switches
5. Motor connections

Figure R263413-32

Worksheet for Motor Circuits

Item	Type	Size	Quantity	Cost Unit	Cost Total
Wire					
Conduit					
Switch					
Starter					
Switch					
Motor Connection					
Other					
Total Cost					

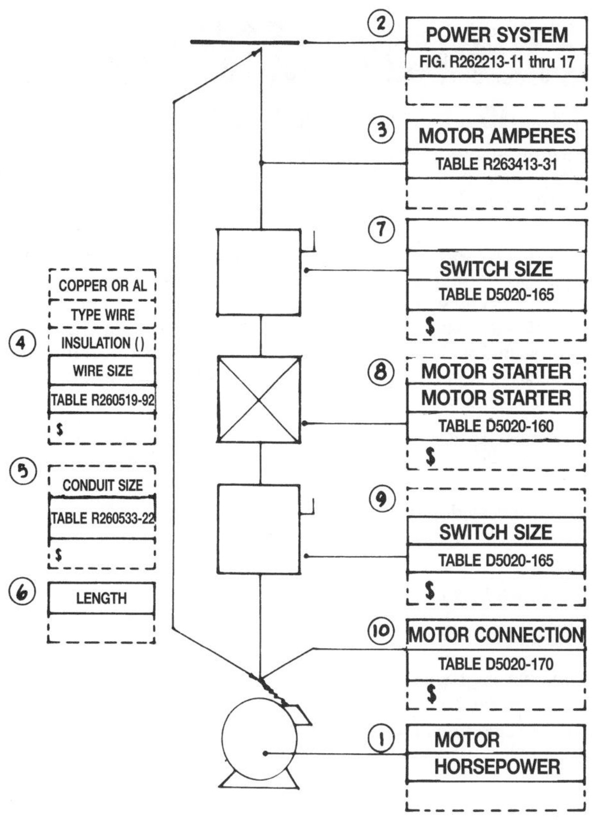

R263413-33 Maximum Horsepower for Starter Size by Voltage

Starter Size	Maximum HP (3φ) 208V	240V	480V	600V
00	1½	1½	2	2
0	3	3	5	5
1	7½	7½	10	10
2	10	15	25	25
3	25	30	50	50
4	40	50	100	100
5		100	200	200
6		200	300	300
7		300	600	600
8		450	900	900
8L		700	1500	1500

Electrical | R2636 | Transfer Switches

R263623-60 Automatic Transfer Switches

When taking off Automatic transfer switches identify by voltage, amperage and number of poles.

Example: Automatic transfer switch, 480V, 3 phase, 30A, NEMA 1 enclosure

Labor-hours for transfer switches include:

1. Unloading, uncrating and handling switches up to 200′ from loading dock
2. Measuring and marking
3. Drilling (4) lead type anchors, using a hammer drill
4. Mounting and leveling
5. Circuit identification
6. Testing and load balancing

Labor-hours do not include:

1. Modifications in enclosure
2. Structural supports
3. Additional lugs
4. Plywood backboard
5. Painting or lettering
6. Conduit runs to or from transfer switch
7. Wire
8. Termination of wires

Electrical | R2651 | Interior Lighting

R265113-40 Interior Lighting Fixtures

When taking off interior lighting fixtures, it is advisable to set up your quantity work sheet to conform to the lighting schedule as it appears on the print. Include the alpha-numeric code plus the symbol on your work sheet.

Take off a particular section or floor of the building and count each type of fixture before going on to another type. It would also be advantageous to include on the same work sheet the pipe, wire, fittings and circuit number associated with each type of lighting fixture. This will help you identify the costs associated with any particular lighting system and in turn make material purchases more specific as to when and how much to order under the classification of lighting.

By taking off lighting first you can get a complete "WALK THRU" of the job. This will become helpful when doing other phases of the project.

Materials for a recessed fixture include:

1. Fixture
2. Lamps
3. 6′ of jack chain
4. (2) S hooks
5. (2) Wire nuts

Labor for interior recessed fixtures include:

1. Unloading by hand
2. Hauling by hand to an area up to 200′ from loading dock
3. Uncrating
4. Layout
5. Installing fixture
6. Attaching jack chain & S hooks
7. Connecting circuit power
8. Reassembling fixture
9. Installing lamps
10. Testing

Material for surface mounted fixtures includes:

1. Fixture
2. Lamps
3. Either (4) lead type anchors, (4) toggle bolts, or (4) ceiling grid clips
4. (2) Wire nuts

Material for pendent mounted fixtures includes:

1. Fixture
2. Lamps
3. (2) Wire nuts
4. Rigid pendents as required by type of fixtures
5. Canopies as required by type of fixture

Labor hours include the following for both surface and pendent fixtures:

1. Unloading by hand
2. Hauling by hand to an area up to 200′ from loading dock
3. Uncrating
4. Layout and marking
5. Drilling (4) holes for either lead anchors or toggle bolts using a hammer drill
6. Installing fixture
7. Leveling fixture
8. Connecting circuit power
9. Installing lamps
10. Testing

Labor for surface or pendent fixtures does not include:

1. Conduit
2. Boxes or covers
3. Connectors
4. Fixture whips
5. Special support
6. Switching
7. Wire

Economy of Scale: For large concentrations of lighting fixtures in the same area deduct the following percentages from labor:

25 to 50	fixtures	15%
51 to 75	fixtures	20%
76 to 100	fixtures	25%
101 and over		30%

Job Conditions: Productivity is based on new construction in a unobstructed first floor location, using rolling staging to 15′ high.

Material staging is assumed to be within 100′ of work being performed.

Add the following percentages to labor for elevated installations:

15′ to 20′	high	10%
21′ to 25′	high	20%
26′ to 30′	high	30%
31′ to 35′	high	40%
36′ to 40′	high	50%
41′ and over		60%

R265723-05 Comparison - Operation of High Intensity Discharge Lamps

Lamp Type	Wattage	Life (Hours)	1 Circuit Wattage	Average Initial Lumens	2 L.L.D.	3 Mean Lumens
M.V.	100 DX	24,000	125	4,000	61%	2,440
L.P.S.	SOX-35	18,000	65	4,800	100%	4,800
H.P.S.	LU-70	12,000	84	5,800	90%	5,220
M.H.	No Equivalent					
M.V.	175 DX	24,000	210	8,500	66%	5,676
M.V.	250 DX	24,000	295	13,000	66%	7,986
L.P.S.	SOX-55	18,000	82	8,000	100%	8,000
H.P.S.	LU-100	12,000	120	9,500	90%	8,550
M.H.	No Equivalent					
M.V.	400 DX	24,000	465	24,000	64%	14,400
L.P.S.	SOX-90	18,000	141	13,500	100%	13,500
H.P.S.	LU-150	16,000	188	16,000	90%	14,400
M.H.	MH-175	7,500	210	14,000	73%	10,200
M.V.	No Equivalent					
L.P.S.	SOX-135	18,000	147	22,500	100%	22,500
H.P.S.	LU-250	20,000	310	25,500	92%	23,205
M.H.	MH-250	7,500	295	20,500	78%	16,000
M.V.	1000 DX	24,000	1,085	63,000	61%	37,820
L.P.S.	SOX-180	18,000	248	33,000	100%	33,000
H.P.S.	LU-400	20,000	480	50,000	90%	45,000
M.H.	MH-400	15,000	465	34,000	72%	24,600
H.P.S.	LU-1000	15,000	1,100	140,000	91%	127,400

1. Includes ballast losses and average lamp watts
2. Lamp lumen depreciation (% of initial light output at 70% rated life)
3. Lamp lumen output at 70% rated life (L.L.D. x initial)
4. Cost of yearly operation = Cicuit Wattage/1000 x annual operating hours x average cost per KWH

M.V. = Mercury Vapor
L.P.S. = Low pressure sodium
H.P.S. = High pressure sodium
L.H. = Metal halide

R265723-10 For Other than Regular Cool White (CW) Lamps

Multiply Material Costs as Follows:					
Regular Lamps	Cool white deluxe (CWX)	x 1.35	Energy Saving Lamps	Cool white (CW/ES)	x 1.35
	Warm white deluxe (WWX)	x 1.35		Cool white deluxe (CWX/ES)	x 1.65
	Warm white (WW)	x 1.30		Warm white (WW/ES)	x 1.55
	Natural (N)	x 2.05		Warm white deluxe (WWX/ES)	x 1.65

R265723-20 Lamp Comparison Chart with Enclosed Floodlight, Ballast, & Lamp for Pole Mounting

Type	Watts	Initial Lumens	Lumens per Watt	Lumens @ 40% Life	Life (Hours)
Incandescent	150	2,880	19	85	750
	300	6,360	21	84	750
	500	10,850	22	80	1,000
	1,000	23,740	24	80	1,000
	1,500	34,400	23	80	1,000
Tungsten	500	10,950	22	97	2,000
Halogen	1,500	35,800	24	97	2,000
Fluorescent	40	3,150	79	88	20,000
Cool	110	9,200	84	87	12,000
White	215	16,000	74	81	12,000
Deluxe	250	12,100	48	86	24,000
Mercury	400	22,500	56	85	24,000
	1,000	63,000	63	75	24,000
Metal	175	14,000	80	77	7,500
Halide	400	34,000	85	75	15,000
	1,000	100,000	100	83	10,000
	1,500	155,000	103	92	1,500
High	70	5,800	83	90	20,000
Pressure	100	9,500	95	90	20,000
Sodium	150	16,000	107	90	24,000
	400	50,000	125	90	24,000
	1,000	140,000	140	90	24,000
Low	55	4,600	131	98	18,000
Pressure	90	12,750	142	98	18,000
Sodium	180	33,000	183	98	18,000

Color: High Pressure Sodium — Slightly Yellow
Low Pressure Sodium — Yellow
Mercury Vapor — Green-Blue
Metal Halide — Blue White

Note: Pole not included.

R265723-25 Energy Efficiency Rating for Luminaires

The energy efficiency program for luminaires recommends the use of a metric called the Luminaire Efficacy Rating (LER). The LER value expresses the total luminaires generated by a lamp, to the watts consumed by the lamp.

Lamp Type	Lumens per Watt
Incandescent	17
Tungsten Halogen	14 - 20
Fluorescent	50 - 104
Metal Halide	64 - 96
High Pressure Sodium	76 - 116

Communications | R2713 | Communications Backbone Cabling

R271323-40 Fiber Optics

Fiber optic systems use optical fiber such as plastic, glass, or fused silica, a transparent material, to transmit radiant power (i.e., light) for control, communication, and signaling applications. The types of fiber optic cables can be nonconductive, conductive, or composite. The composite cables contain fiber optics and current-carrying electrical conductors. The configuration for one of the fiber optic systems is as follows:

The transceiver module acts as transmitting and receiving equipment in a common house, which converts electrical energy to light energy or vice versa.

Pricing the fiber optic system is not an easy task. The performance of the whole system will affect the cost significantly. New specialized tools and techniques decrease the installing cost tremendously. In the fiber optic section of Means Electrical Cost Data, a benchmark for labor-hours and material costs is set up so that users can adjust their costs according to unique project conditions.

Units for Measure: Fiber optic cable is measured in hundred linear feet (C.L.F.) or industry units of measure - meter (m) or kilometer (km). The connectors are counted as units (EA.)

Material Units: Generally, the material costs include only the cable. All the accessories shall be priced separately.

Labor Units: The following procedures are generally included for the installation of fiber optic cables:

- Receiving
- Material handling
- Setting up pulling equipment
- Measuring and cutting cable
- Pulling cable

These additional items are listed and extended: Terminations

Takeoff Procedure: Cable should be taken off by type, size, number of fibers, and number of terminations required. List the lengths of each type of cable on the takeoff sheets. Total and add 10% for waste. Transfer the figures to a cost analysis sheet and extend.

Communications | R2715 | Communications Horizontal Cabling

R271513-75 High Performance Cable

There are several categories used to describe high performance cable. The following information includes a description of categories CAT 3, 5, 5e, 6, and 7, and details classifications of frequency and specific standards. The category standards have evolved under the sponsorship of organizations such as the Telecommunication Industry Association (TIA), the Electronic Industries Alliance (EIA), the American National Standards Institute (ANSI), the International Organization for Standardization (ISO), and the International Electrotechnical Commission (IEC), all of which have catered to the increasing complexities of modern network technology. For network cabling, users must comply with national or international standards. A breakdown of these categories is as follows:

Category 3: Designed to handle frequencies up to 16 MHz.

Category 5: (TIA/EIA 568A) Designed to handle frequencies up to 100 MHz.

Category 5e: Additional transmission performance to exceed Category 5.

Category 6 (draft): Development by TIA and other international groups to handle frequencies of 250 MHz.

Category 7 (draft): Under development to handle a frequency range from 1 to 600 MHz.

R312316-40 Excavating

The selection of equipment used for structural excavation and bulk excavation or for grading is determined by the following factors.

1. Quantity of material
2. Type of material
3. Depth or height of cut
4. Length of haul
5. Condition of haul road
6. Accessibility of site
7. Moisture content and dewatering requirements
8. Availability of excavating and hauling equipment

Some additional costs must be allowed for hand trimming the sides and bottom of concrete pours and other excavation below the general excavation.

When planning excavation and fill, the following should also be considered.

1. Swell factor
2. Compaction factor
3. Moisture content
4. Density requirements

A typical example for scheduling and estimating the cost of excavation of a 15′ deep basement on a dry site when the material must be hauled off the site is outlined below.

Assumptions:

1. Swell factor, 18%
2. No mobilization or demobilization
3. Allowance included for idle time and moving on job
4. No dewatering, sheeting, or bracing
5. No truck spotter or hand trimming

Number of B.C.Y. per truck = 1.5 C.Y. bucket x 8 passes = 12 loose C.Y.

$$= 12 \times \frac{100}{118} = 10.2 \text{ B.C.Y. per truck}$$

Truck Haul Cycle:

Load truck, 8 passes	=	4 minutes
Haul distance, 1 mile	=	9 minutes
Dump time	=	2 minutes
Return, 1 mile	=	7 minutes
Spot under machine	=	1 minute
		23 minute cycle

Fleet Haul Production per day in B.C.Y.

$$4 \text{ trucks} \times \frac{50 \text{ min. hour}}{23 \text{ min. haul cycle}} \times 8 \text{ hrs.} \times 10.2 \text{ B.C.Y.}$$

$$= 4 \times 2.2 \times 8 \times 10.2 = 718 \text{ B.C.Y./day}$$

Add the mobilization and demobilization costs to the total excavation costs. When equipment is rented for more than three days, there is often no mobilization charge by the equipment dealer. On larger jobs outside of urban areas, scrapers can move earth economically provided a dump site or fill area and adequate haul roads are available. Excavation within sheeting bracing or cofferdam bracing is usually done with a clamshell and production is low, since the clamshell may have to be guided by hand between the bracing. When excavating or filling an area enclosed with a wellpoint system, add 10% to 15% to the cost to allow for restricted access. When estimating earth excavation quantities for structures, allow work space outside the building footprint for construction of the foundation and a slope of 1:1 unless sheeting is used.

R312316-45 Excavating Equipment

The table below lists THEORETICAL hourly production in C.Y./hr. bank measure for some typical excavation equipment. Figures assume 50 minute hours, 83% job efficiency, 100% operator efficiency, 90° swing and properly sized hauling units, which must be modified for adverse digging and loading conditions. Actual production costs in the front of the book average about 50% of the theoretical values listed here.

Equipment	Soil Type	B.C.Y. Weight	% Swell	1 C.Y.	1-1/2 C.Y.	2 C.Y.	2-1/2 C.Y.	3 C.Y.	3-1/2 C.Y.	4 C.Y.
Hydraulic Excavator "Backhoe" 15′ Deep Cut	Moist loam, sandy clay	3400 lb.	40%	85	125	175	220	275	330	380
	Sand and gravel	3100	18	80	120	160	205	260	310	365
	Common earth	2800	30	70	105	150	190	240	280	330
	Clay, hard, dense	3000	33	65	100	130	170	210	255	300
Power Shovel Optimum Cut (Ft.)	Moist loam, sandy clay	3400	40	170	245	295	335	385	435	475
				(6.0)	(7.0)	(7.8)	(8.4)	(8.8)	(9.1)	(9.4)
	Sand and gravel	3100	18	165	225	275	325	375	420	460
				(6.0)	(7.0)	(7.8)	(8.4)	(8.8)	(9.1)	(9.4)
	Common earth	2800	30	145	200	250	295	335	375	425
				(7.8)	(9.2)	(10.2)	(11.2)	(12.1)	(13.0)	(13.8)
	Clay, hard, dense	3000	33	120	175	220	255	300	335	375
				(9.0)	(10.7)	(12.2)	(13.3)	(14.2)	(15.1)	(16.0)
Drag Line Optimum Cut (Ft.)	Moist loam, sandy clay	3400	40	130	180	220	250	290	325	385
				(6.6)	(7.4)	(8.0)	(8.5)	(9.0)	(9.5)	(10.0)
	Sand and gravel	3100	18	130	175	210	245	280	315	375
				(6.6)	(7.4)	(8.0)	(8.5)	(9.0)	(9.5)	(10.0)
	Common earth	2800	30	110	160	190	220	250	280	310
				(8.0)	(9.0)	(9.9)	(10.5)	(11.0)	(11.5)	(12.0)
	Clay, hard, dense	3000	33	90	130	160	190	225	250	280
				(9.3)	(10.7)	(11.8)	(12.3)	(12.8)	(13.3)	(12.0)
				Wheel Loaders				Track Loaders		
				3 C.Y.	4 C.Y.	6 C.Y.	8 C.Y.	2-1/4 C.Y.	3 C.Y.	4 C.Y.
Loading Tractors	Moist loam, sandy clay	3400	40	260	340	510	690	135	180	250
	Sand and gravel	3100	18	245	320	480	650	130	170	235
	Common earth	2800	30	230	300	460	620	120	155	220
	Clay, hard, dense	3000	33	200	270	415	560	110	145	200
	Rock, well-blasted	4000	50	180	245	380	520	100	130	180

R312319-90 Wellpoints

A single stage wellpoint system is usually limited to dewatering an average 15′ depth below normal ground water level. Multi-stage systems are employed for greater depth with the pumping equipment installed only at the lowest header level. Ejectors with unlimited lift capacity can be economical when two or more stages of wellpoints can be replaced or when horizontal clearance is restricted, such as in deep trenches or tunneling projects, and where low water flows are expected. Wellpoints are usually spaced on 2-1/2′ to 10′ centers along a header pipe. Wellpoint spacing, header size, and pump size are all determined by the expected flow as dictated by soil conditions.

In almost all soils encountered in wellpoint dewatering, the wellpoints may be jetted into place. Cemented soils and stiff clays may require sand wicks about 12″ in diameter around each wellpoint to increase efficiency and eliminate weeping into the excavation. These sand wicks require 1/2 to 3 C.Y. of washed filter sand and are installed by using a 12″ diameter steel casing and hole puncher jetted into the ground 2′ deeper than the wellpoint. Rock may require predrilled holes.

Labor required for the complete installation and removal of a single stage wellpoint system is in the range of 3/4 to 2 labor-hours per linear foot of header, depending upon jetting conditions, wellpoint spacing, etc.

Continuous pumping is necessary except in some free draining soil where temporary flooding is permissible (as in trenches which are backfilled after each day's work). Good practice requires provision of a stand-by pump during the continuous pumping operation.

Systems for continuous trenching below the water table should be installed three to four times the length of expected daily progress to ensure uninterrupted digging, and header pipe size should not be changed during the job.

For pervious free draining soils, deep wells in place of wellpoints may be economical because of lower installation and maintenance costs. Daily production ranges between two to three wells per day, for 25′ to 40′ depths, to one well per day for depths over 50′.

Detailed analysis and estimating for any dewatering problem is available at no cost from wellpoint manufacturers. Major firms will quote "sufficient equipment" quotes or their affiliates offer lump sum proposals to cover complete dewatering responsibility.

Description for 200′ System with 8″ Header		Quantities
Equipment & Material	Wellpoints 25′ long, 2″ diameter @ 5′ O.C.	40 Each
	Header pipe, 8″ diameter	200 L.F.
	Discharge pipe, 8″ diameter	100 L.F.
	8″ valves	3 Each
	Combination jetting & wellpoint pump (standby)	1 Each
	Wellpoint pump, 8″ diameter	1 Each
	Transportation to and from site	1 Day
	Fuel for 30 days x 60 gal./day	1800 Gallons
	Lubricants for 30 days x 16 lbs./day	480 Lbs.
	Sand for points	40 C.Y.
Labor	Technician to supervise installation	1 Week
	Labor for installation and removal of system	300 Labor-hours
	4 Operators straight time 40 hrs./wk. for 4.33 wks.	693 Hrs.
	4 Operators overtime 2 hrs./wk. for 4.33 wks.	35 Hrs.

R312323-30 Compacting Backfill

Compaction of fill in embankments, around structures, in trenches, and under slabs is important to control settlement. Factors affecting compaction are:

1. Soil gradation
2. Moisture content
3. Equipment used
4. Depth of fill per lift
5. Density required

Production Rate:

$$\frac{1.75' \text{ plate width x 50 F.P.M. x 50 min./hr. x .67' lift}}{27 \text{ C.F. per C.Y.}} = 108.5 \text{ C.Y./hr.}$$

Production Rate for 4 Passes:

$$\frac{108.5 \text{ C.Y.}}{4 \text{ passes}} = 27.125 \text{ C.Y./hr. x 8 hrs.} = 217 \text{ C.Y./day}$$

Example:

Compact granular fill around a building foundation using a 21″ wide x 24″ vibratory plate in 8″ lifts. Operator moves at 50 F.P.M. working a 50 minute hour to develop 95% Modified Proctor Density with 4 passes.

R314116-40 Wood Sheet Piling

Wood sheet piling may be used for depths to 20′ where there is no ground water. If moderate ground water is encountered Tongue & Groove sheeting will help to keep it out. When considerable ground water is present, steel sheeting must be used.

For estimating purposes on trench excavation, sizes are as follows:

Depth	Sheeting	Wales	Braces	B.F. per S.F.
To 8′	3 x 12′s	6 x 8′s, 2 line	6 x 8′s, @ 10′	4.0 @ 8′
8′ x 12′	3 x 12′s	10 x 10′s, 2 line	10 x 10′s, @ 9′	5.0 average
12′ to 20′	3 x 12′s	12 x 12′s, 3 line	12 x 12′s, @ 8′	7.0 average

Sheeting to be toed in at least 2′ depending upon soil conditions. A five person crew with an air compressor and sheeting driver can drive and brace 440 SF/day at 8′ deep, 360 SF/day at 12′ deep, and 320 SF/day at 16′ deep. For normal soils, piling can be pulled in 1/3 the time to install. Pulling difficulty increases with the time in the ground. Production can be increased by high pressure jetting.

R314116-45 Steel Sheet Piling

Limiting weights are 22 to 38#/S.F. of wall surface with 27#/S.F. average for usual types and sizes. (Weights of piles themselves are from 30.7#/L.F. to 57#/L.F. but they are 15″ to 21″ wide.) Lightweight sections 12″ to 28″ wide from 3 ga. to 12 ga. thick are also available for shallow excavations. Piles may be driven two at a time with an impact or vibratory hammer (use vibratory to pull) hung from a crane without leads. A reasonable estimate of the life of steel sheet piling is 10 uses with up to 125 uses possible if a vibratory hammer is used. Used piling costs from 50% to 80% of new piling depending on location and market conditions. Sheet piling and H piles can be rented for about 30% of the delivered mill price for the first month and 5% per month thereafter. Allow 1 labor-hour per pile for cleaning and trimming after driving. These costs increase with depth and hydrostatic head. Vibratory drivers are faster in wet granular soils and are excellent for pile extraction. Pulling difficulty increases with the time in the ground and may cost more than driving. It is often economical to abandon the sheet piling, especially if it can be used as the outer wall form. Allow about 1/3 additional length or more for toeing into ground. Add bracing, waler and strut costs. Waler costs can equal the cost per ton of sheeting.

Earthwork | R3145 | Vibroflotation & Desification

R314513-90 Vibroflotation and Vibro Replacement Soil Compaction

Vibroflotation is a proprietary system of compacting sandy soils in place to increase relative density to about 70%. Typical bearing capacities attained will be 6000 psf for saturated sand and 12,000 psf for dry sand. Usual range is 4000 to 8000 psf capacity. Costs in the front of the book are for a vertical foot of compacted cylinder 6′ to 10′ in diameter.

Vibro replacement is a proprietary system of improving cohesive soils in place to increase bearing capacity. Most silts and clays above or below the water table can be strengthened by installation of stone columns.

The process consists of radial displacement of the soil by vibration. The created hole is then backfilled in stages with coarse granular fill which is thoroughly compacted and displaced into the surrounding soil in the form of a column.

The total project cost would depend on the number and depth of the compacted cylinders. The installing company guarantees relative soil density of the sand cylinders after compaction and the bearing capacity of the soil after the replacement process. Detailed estimating information is available from the installer at no cost.

R316000-20 Pile Caps, Piles and Caissons

General: The function of a reinforced concrete pile cap is to transfer superstructure load from isolated column or pier to each pile in its supporting cluster. To do this, the cap must be thick and rigid, with all piles securely embedded into and bonded to it.

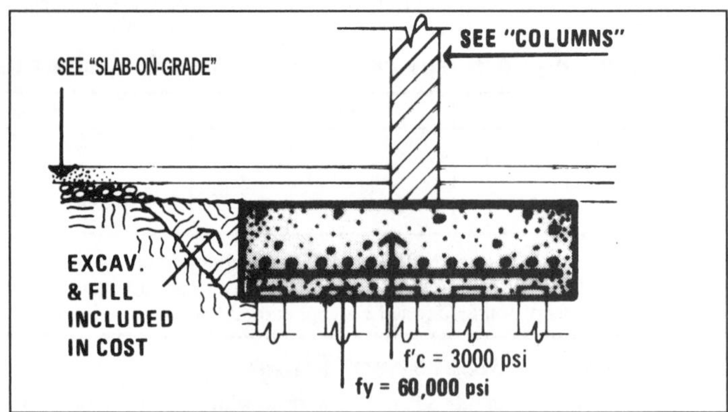

Figure 1.1-331 Section Through Pile Cap

Table 1.1-332 Concrete Quantities for Pile Caps

Load	Number of Piles @ 3'-0" O.C. Per Footing Cluster									
Working (K)	2 (CY)	4 (CY)	6 (CY)	8 (CY)	10 (CY)	12 (CY)	14 (CY)	16 (CY)	18 (CY)	20 (CY)
50	(.9)	(1.9)	(3.3)	(4.9)	(5.7)	(7.8)	(9.9)	(11.1)	(14.4)	(16.5)
100	(1.0)	(2.2)	(3.3)	(4.9)	(5.7)	(7.8)	(9.9)	(11.1)	(14.4)	(16.5)
200	(1.0)	(2.2)	(4.0)	(4.9)	(5.7)	(7.8)	(9.9)	(11.1)	(14.4)	(16.5)
400	(1.1)	(2.6)	(5.2)	(6.3)	(7.4)	(8.2)	(13.7)	(11.1)	(14.4)	(16.5)
800		(2.9)	(5.8)	(7.5)	(9.2)	(13.6)	(17.6)	(15.9)	(19.7)	(22.1)
1200			(5.8)	(8.3)	(9.7)	(14.2)	(18.3)	(20.4)	(21.2)	(22.7)
1600				(9.8)	(11.4)	(14.5)	(19.5)	(20.4)	(24.6)	(27.2)
2000				(9.8)	(11.4)	(16.6)	(24.1)	(21.7)	(26.0)	(28.8)
3000						(17.5)		(26.5)	(30.3)	(32.9)
4000								(30.2)	(30.7)	(36.5)

Table 1.1-333 Concrete Quantities for Pile Caps

Load	Number of Piles @ 4'-6" O.C. Per Footing					
Working (K)	2 (CY)	3 (CY)	4 (CY)	5 (CY)	6 (CY)	7 (CY)
50	(2.3)	(3.6)	(5.6)	(11.0)	(13.7)	(12.9)
100	(2.3)	(3.6)	(5.6)	(11.0)	(13.7)	(12.9)
200	(2.3)	(3.6)	(5.6)	(11.0)	(13.7)	(12.9)
400	(3.0)	(3.6)	(5.6)	(11.0)	(13.7)	(12.9)
800			(6.2)	(11.5)	(14.0)	(12.9)
1200				(13.0)	(13.7)	(13.4)
1600						(14.0)

R316000-20 Pile Caps, Piles and Caissons (cont.)

General: Piles are column-like shafts which receive superstructure loads, overturning forces, or uplift forces. They receive these loads from isolated column or pier foundations (pile caps), foundation walls, grade beams, or foundation mats. The piles then transfer these loads through shallower poor soil strata to deeper soil of adequate support strength and acceptable settlement with load.

Be sure that other foundation types aren't better suited to the job. Consider ground and settlement, as well as loading, when reviewing. Piles usually are associated with difficult foundation problems and substructure condition. Ground conditions determine type of pile (different pile types have been developed to suit ground conditions.) Decide each case by technical study, experience and sound engineering judgment—not rules of thumb. A full investigation of ground conditions, early, is essential to provide maximum information for professional foundation engineering and an acceptable structure.

Piles support loads by end bearing and friction. Both are generally present; however, piles are designated by their principal method of load transfer to soil.

Boring should be taken at expected pile locations. Ground strata (to bedrock or depth of 1-1/2 building width) must be located and identified with appropriate strengths and compressibilities. The sequence of strata determines if end bearing or friction piles are best suited.

End bearing piles have shafts which pass through soft strata or thin hard strata and tip bear on bedrock or penetrate some distance into a dense, adequate soil (sand or gravel.)

Friction piles have shafts which may be entirely embedded in cohesive soil (moist clay), and develop required support mainly by adhesion or "skin-friction" between soil and shaft area.

Piles pass through soil by either one of two ways:

1. Displacement piles force soil out of the way. This may cause compaction, ground heaving, remolding of sensitive soils, damage to adjacent structures, or hard driving.
2. Non-displacement piles have either a hole bored and the pile cast or placed in hole, or open ended pipe (casing) driven and the soil core removed. They tend to eliminate heaving or lateral pressure damage to adjacent structures of piles. Steel "HP" piles are considered of small displacement.

Placement of piles (attitude) is most often vertical; however, they are sometimes battered (placed at a small angle from vertical) to advantageously resist lateral loads. Seldom are piles installed singly but rather in clusters (or groups). Codes require a minimum of three piles per major column load or two per foundation wall or grade beam. Single pile capacity is limited by pile structural strength or support strength of soil. Support capacity of a pile cluster is almost always less than the sum of its individual pile capacities due to overlapping of bearing the friction stresses.

Large rigs for heavy, long piles create large soil surface loads and additional expense on weak ground.

Fewer piles create higher costs per pile.

Pile load tests are frequently required by code, ground situation, or pile type.

R316000-20 Pile Caps, Piles and Caissons (cont.)

General: Caissons, as covered in this section, are drilled cylindrical foundation shafts which function primarily as short column-like compression members. They transfer superstructure loads through inadequate soils to bedrock or hard stratum. They may be either reinforced or unreinforced and either straight or belled out at the bearing level.

Shaft diameters range in size from 20″ to 84″ with the most usual sizes beginning at 34″. If inspection of bottom is required, the minimum diameter practical is 30″. If handwork is required (in addition to mechanical belling, etc.) the minimum diameter is 32″. The most frequently used shaft diameter is probably 36″ with a 5′ or 6′ bell diameter. The maximum bell diameter practical is three times the shaft diameter.

Plain concrete is commonly used, poured directly against the excavated face of soil. Permanent casings add to cost and economically should be avoided. Wet or loose strata are undesirable. The associated installation sometimes involves a mudding operation with bentonite clay slurry to keep walls of excavation stable (costs not included here).

Reinforcement is sometimes used, especially for heavy loads. It is required if uplift, bending moment, or lateral loads exist. A small amount of reinforcement is desirable at the top portion of each caisson, even if the above conditions theoretically are not present. This will provide for construction eccentricities and other possibilities. Reinforcement, if present, should extend below the soft strata. Horizontal reinforcement is not required for belled bottoms.

There are three basic types of caisson bearing details:

1. Belled, which are generally recommended to provide reduced bearing pressure on soil. These are not for shallow depths or poor soils. Good soils for belling include most clays, hardpan, soft shale, and decomposed rock.

   Soils requiring handwork include hard shale, limestone, and sandstone.

   Soils not recommended include sand, gravel, silt, and igneous rock. Compact sand and gravel above water table may stand. Water in the bearing strata is undesirable.
2. Straight shafted, which have no bell but the entire length is enlarged to permit safe bearing pressures. They are most economical for light loads on high bearing capacity soil.
3. Socketed (or keyed), which are used for extremely heavy loads. They involve sinking the shaft into rock for combined friction and bearing support action. Reinforcement of shaft is usually necessary. Wide flange cores are frequently used here.

Advantages include:

A. Shafts can pass through soils that piles cannot
B. No soil heaving or displacement during installation
C. No vibration during installation
D. Less noise than pile driving
E. Bearing strata can be visually inspected & tested

Uses include:

A. Situations where unsuitable soil exists to moderate depth
B. Tall structures
C. Heavy structures
D. Underpinning (extensive use)

See R033053-10 for Soil Bearing Capacities.

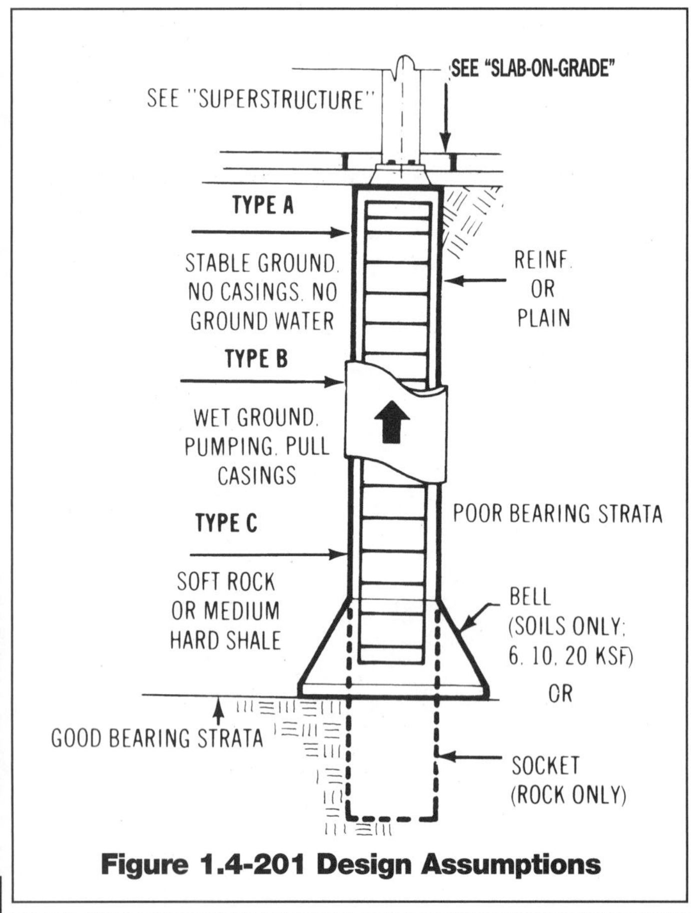

Figure 1.4-201 Design Assumptions

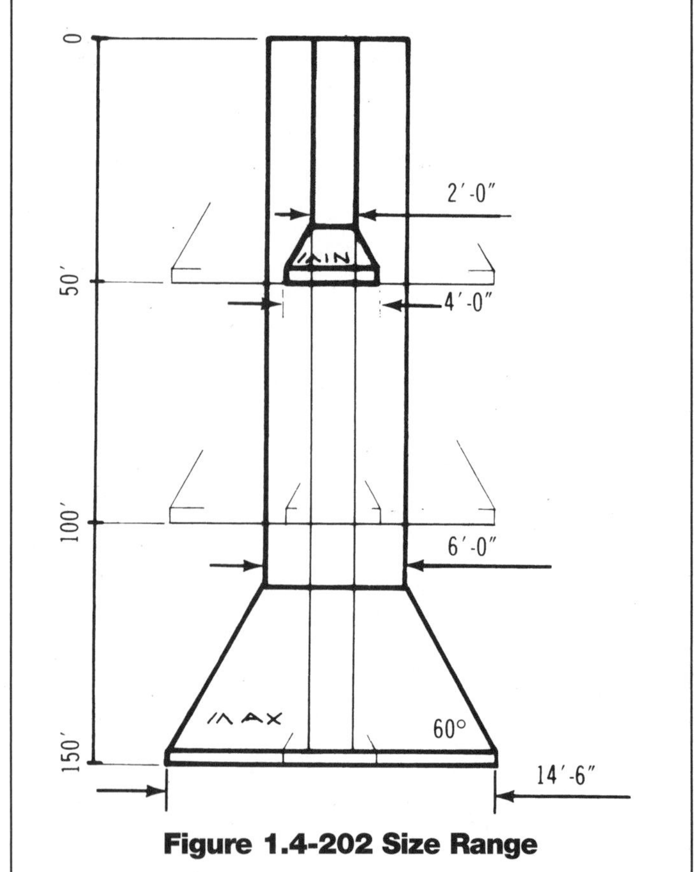

Figure 1.4-202 Size Range

R316326-60 Caissons

The three principal types of cassions are:

(1) Belled Caissons, which except for shallow depths and poor soil conditions, are generally recommended. They provide more bearing than shaft area. Because of its conical shape, no horizontal reinforcement of the bell is required.

(2) Straight Shaft Caissons are used where relatively light loads are to be supported by caissons that rest on high value bearing strata. While the shaft is larger in diameter than for belled types this is more than offset by the saving in time and labor.

(3) Keyed Caissons are used when extremely heavy loads are to be carried. A keyed or socketed caisson transfers its load into rock by a combination of end-bearing and shear reinforcing of the shaft. The most economical shaft often consists of a steel casing, a steel wide flange core and concrete. Allowable compressive stresses of .225 f'c for concrete, 16,000 psi for the wide flange core, and 9,000 psi for the steel casing are commonly used. The usual range of shaft diameter is 18″ to 84″. The number of sizes specified for any one project should be limited due to the problems of casing and auger storage. When hand work is to be performed, shaft diameters should not be less than 32″. When inspection of borings is required a minimum shaft diameter of 30″ is recommended. Concrete caissons are intended to be poured against earth excavation so permanent forms which add to cost should not be used if the excavation is clean and the earth sufficiently impervious to prevent excessive loss of concrete.

Soil Conditions for Belling		
Good	**Requires Handwork**	**Not Recommended**
Clay	Hard Shale	Silt
Sandy Clay	Limestone	Sand
Silty Clay	Sandstone	Gravel
Clayey Silt	Weathered Mica	Igneous Rock
Hard-pan		
Soft Shale		
Decomposed Rock		

R317100-10 Tunnel Excavation

Bored tunnel excavation is common in rock for diameters from 4 feet for sewer and utilities, to 60 feet for vehicles. Production varies from a few linear feet per day to over 200 linear feet per day. In the smaller diameters, the productivity is limited by the restricted area for mucking or the removal of excavated material.

Most of the tunnels in rock today are excavated by boring machines called moles. Preparation for starting the excavation or setting up the mole is very costly. Shafts must be excavated to the invert of the proposed tunnel and the mole must be lowered into the shaft. If excavating a portal tunnel, that is starting at an open face, the cost is reduced considerably both for mobilization and mucking.

In soft ground and mixed material, special bucket excavators and rotary excavators are used inside a shield. Tunnel liners must follow directly behind the shield to support the earth and prevent cave-ins.

Traditional muck haulage operations are performed by rail with locomotives and muck cars. Sometimes conveyors are more economical and require less ventilation of the tunnel.

Ventilation and air compression are other important cost factors to consider in tunnel excavation. Continuous ventilation ducts are sometimes fabricated at the tunnel site.

Tunnel linings are steel, cast in place reinforced concrete, shotcrete, or a combination of these. When required, contact grouting is performed by pumping grout between the lining and the excavation. Intermittent holes are drilled into the lining and separate costs are determined for drilling per hole, grout pump connecting per hole, and grout per cubic foot.

Consolidation grouting and roof bolts may also be required where the excavation is unstable or faulting occurs.

Tunnel boring is usually done 24 hours per day. A typical crew for rock boring is:

Tunneling Crew based on three 8 hour shifts

1 Shifter
1 Walker
1 Machine Operator for mole
1 Oiler
1 Mechanic
3 Locomotives with operators
5 Miners for rails, vent ducts, and roof bolts
1 Electrician
2 Pumps
2 Laborers for hoisting
1 Hoist operator for muck removal
1 Oiler

Surface Crew Based on normal 8 hour shift

2 Shop Mechanics
1 Electrician
1 Shifter
2 Laborers
1 Operator with 18 ton cherry picker
1 Operator with front end loader

EXTERIOR IMPROVEMENTS

Exterior Improvements	R3201	Flexible Paving Surface Treatment

R320113-70 Pavement Maintenance

Routine pavement maintenance should be performed to keep a paved surface from deteriorating under the normal forces of nature and traffic.

The msot important maintenance function is the early detection and repair of minor pavement defects. Cracks and other surface breaks can develop into serious defects if not repaired in their earliest stages. For these reasons a pavement preventive maintenance program should include frequent close inspections of pavement surfaces. When suspicious areas are detected, a detailed investigation should be undertaken to determine the appropriate repair. Where subsurface or pavement deterioration is detected, the Benkelman Beam can be used to make deflection measurements under normal traffic stresses. This is done to determine the extent of the affected area.

Patching or resurfacing work should be done during warm (10°C and above) and dry weather. Adequate compaction is dificult to achieve when hot or warm mixtures are placed on cold pavements. Asphalt and asphalt mixtures usually do not bond well to damp surfaces.

Exterior Improvements	R3292	Turf & Grasses

R329219-50 Seeding

The type of grass is determined by light, shade and moisture content of soil plus intended use. Fertilizer should be disked 4″ before seeding. For steep slopes disk five tons of mulch and lay two tons of hay or straw on surface per acre after seeding. Surface mulch can be staked, lightly disked or tar emulsion sprayed. Material for mulch can be wood chips, peat moss, partially rotted hay or straw, wood fibers and sprayed emulsions. Hemp seed blankets with fertilizer are also available. For spring seeding, watering is necessary. Late fall seeding may have to be reseeded in the spring. Hydraulic seeding, power mulching, and aerial seeding can be used on large areas.

REFERENCE TABLES

R329343-10 Plant Spacing Chart

This chart may be used when plants are to be placed equidistant from each other, staggering their position in each row.

Plant Spacing (Inches)	Row Spacing (Inches)	Plants Per (CSF)	Plant Spacing (Feet)	Row Spacing (Feet)	Plants Per (MSF)
			4	3.46	72
6	5.20	462	5	4.33	46
8	6.93	260	6	5.20	32
10	8.66	166	8	6.93	18
12	10.39	115	10	8.66	12
15	12.99	74	12	10.39	8
18	15.59	51.32	15	12.99	5.13
21	18.19	37.70	20	17.32	2.89
24	20.78	28.87	25	21.65	1.85
30	25.98	18.48	30	25.98	1.28
36	31.18	12.83	40	34.64	0.72

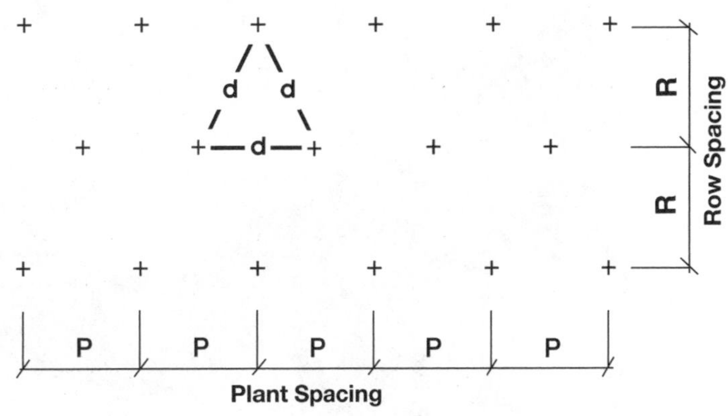

R329343-20 Trees and Plants by Environment and Purposes

Dry, Windy, Exposed Areas
Barberry
Junipers, all varieties
Locust
Maple
Oak
Pines, all varieties
Poplar, Hybrid
Privet
Spruce, all varieties
Sumac, Staghorn

Lightly Wooded Areas
Dogwood
Hemlock
Larch
Pine, White
Rhododendron
Spruce, Norway
Redbud

Total Shade Areas
Hemlock
Ivy, English
Myrtle
Pachysandra
Privet
Spice Bush
Yews, Japanese

Cold Temperatures of Northern U.S. and Canada
Arborvitae, American
Birch, White
Dogwood, Silky
Fir, Balsam
Fir, Douglas
Hemlock
Juniper, Andorra
Juniper, Blue Rug
Linden, Little Leaf
Maple, Sugar
Mountain Ash
Myrtle
Olive, Russian
Pine, Mugho
Pine, Ponderosa
Pine, Red
Pine, Scotch
Poplar, Hybrid
Privet
Rosa Rugosa
Spruce, Dwarf Alberta
Spruce, Black Hills
Spruce, Blue
Spruce, Norway
Spruce, White, Engelman
Yellow Wood

Wet, Swampy Areas
American Arborvitae
Birch, White
Black Gum
Hemlock
Maple, Red
Pine, White
Willow

Poor, Dry, Rocky Soil
Barberry
Crownvetch
Eastern Red Cedar
Juniper, Virginiana
Locust, Black
Locust, Bristly
Locust, Honey
Olive, Russian
Pines, all varieties
Privet
Rosa Rugosa
Sumac, Staghorn

Seashore Planting
Arborvitae, American
Juniper, Tamarix
Locust, Black
Oak, White
Olive, Russian
Pine, Austrian
Pine, Japanese Black
Pine, Mugho
Pine, Scotch
Privet, Amur River
Rosa Rugosa
Yew, Japanese

City Planting
Barberry
Fir, Concolor
Forsythia
Hemlock
Holly, Japanese
Ivy, English
Juniper, Andorra
Linden, Little Leaf
Locust, Honey
Maple, Norway, Silver
Oak, Pin, Red
Olive, Russian
Pachysandra
Pine, Austrian
Pine, White
Privet
Rosa Rugosa
Sumac, Staghorn
Yew, Japanese

Bonsai Planting
Azaleas
Birch, White
Ginkgo
Junipers
Pine, Bristlecone
Pine, Mugho
Spruce,k Engleman
Spruce, Dwarf Alberta

Street Planting
Linden, Little Leaf
Oak, Pin
Ginkgo

Fast Growth
Birch, White
Crownvetch
Dogwood, Silky
Fir, Douglas
Juniper, Blue Pfitzer
Juniper, Blue Rug
Maple, Silver
Olive, Autumn
Pines, Austrian, Ponderosa, Red Scotch and White
Poplar, Hybrid
Privet
Spruce, Norway
Spruce, Serbian
Texus, Cuspidata, Hicksi
Willow

Dense, Impenetrable Hedges
Field Plantings:
- Locust, Bristly,
- Olive, Autumn
- Sumac

Residential Area:
- Barberry, Red or Green
- Juniper, Blue Pfitzer
- Rosa Rugosa

Food for Birds
Ash, Mountain
Barberry
Bittersweet
Cherry, Manchu
Dogwood, Silky
Honesuckle, Rem Red
Hawthorn
Oaks
Olive, Autumn, Russian
Privet
Rosa Rugosa
Sumac

Erosion Control
Crownvetch
Locust, Bristly
Willow

R329343-30 Zones of Plant Hardiness

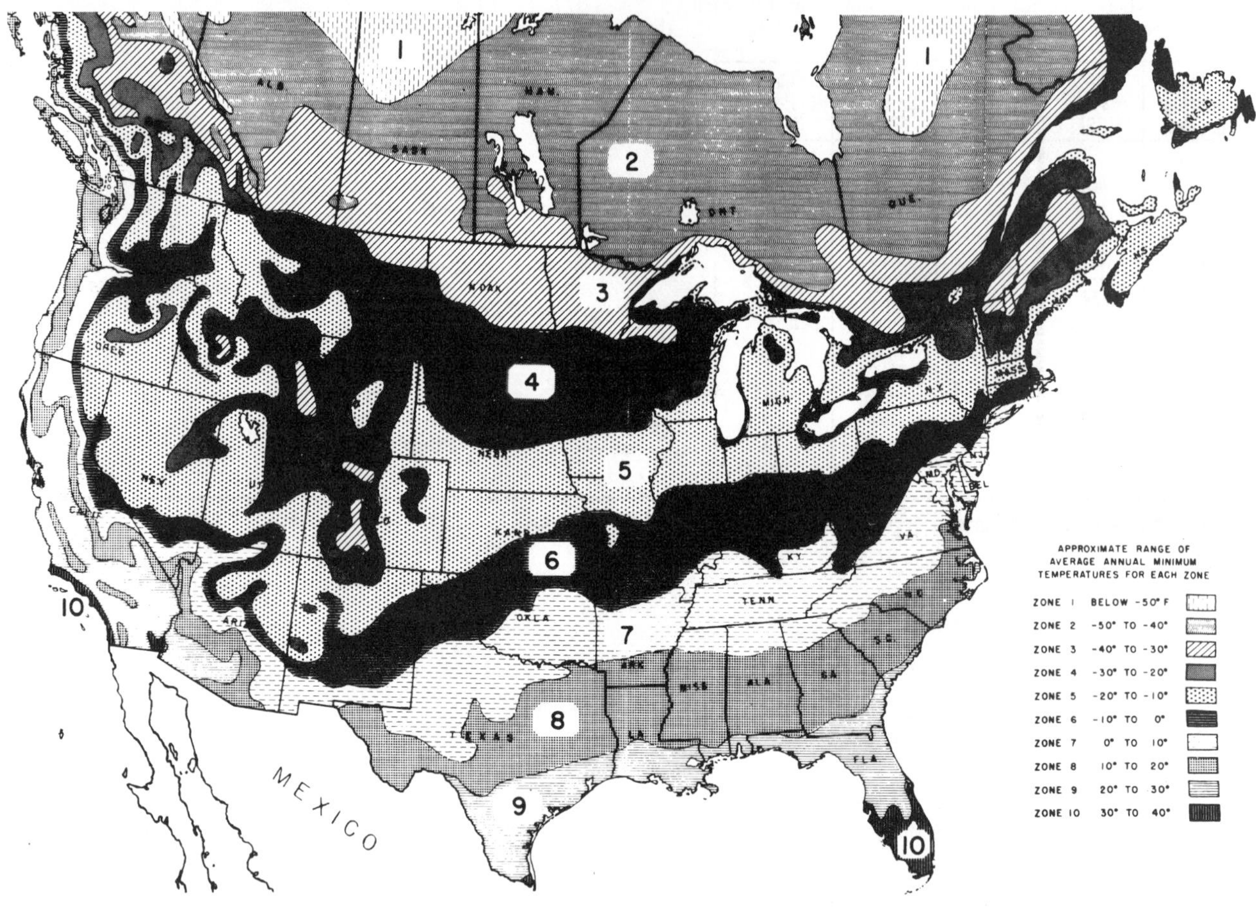

R331113-80 Piping Designations

There are several systems currently in use to describe pipe and fittings. The following paragraphs will help to identify and clarify classifications of piping systems used for water distribution.

Piping may be classified by schedule. Piping schedules include 5S, 10S, 10, 20, 30, Standard, 40, 60, Extra Strong, 80, 100, 120, 140, 160 and Double Extra Strong. These schedules are dependent upon the pipe wall thickness. The wall thickness of a particular schedule may vary with pipe size.

Ductile iron pipe for water distribution is classified by Pressure Classes such as Class 150, 200, 250, 300 and 350. These classes are actually the rated water working pressure of the pipe in pounds per square inch (psi). The pipe in these pressure classes is designed to withstand the rated water working pressure plus a surge allowance of 100 psi.

The American Water Works Association (AWWA) provides standards for various types of **plastic pipe.** C-900 is the specification for polyvinyl chloride (PVC) piping used for water distribution in sizes ranging from 4″ through 12″. C-901 is the specification for polyethylene (PE) pressure pipe, tubing and fittings used for water distribution in sizes ranging from 1/2″ through 3″. C-905 is the specification for PVC piping sizes 14″ and greater.

PVC pressure-rated pipe is identified using the standard dimensional ratio (SDR) method. This method is defined by the American Society for Testing and Materials (ASTM) Standard D 2241. This pipe is available in SDR numbers 64, 41, 32.5, 26, 21, 17, and 13.5. Pipe with an SDR of 64 will have the thinnest wall while pipe with an SDR of 13.5 will have the thickest wall. When the pressure rating (PR) of a pipe is given in psi, it is based on a line supplying water at 73 degrees F.

The National Sanitation Foundation (NSF) seal of approval is applied to products that can be used with potable water. These products have been tested to ANSI/NSF Standard 14.

Valves and strainers are classified by American National Standards Institute (ANSI) Classes. These Classes are 125, 150, 200, 250, 300, 400, 600, 900, 1500 and 2500. Within each class there is an operating pressure range dependent upon temperature. Design parameters should be compared to the appropriate material dependent, pressure-temperature rating chart for accurate valve selection.

R337116-60 Average Transmission Line Material Requirements (Per Mile)

Terrain:		Flat				Rolling				Mountain			
Item		69kV	161kV	161kV	500kV	69kV	161kV	161kV	500kV	69kV	161kV	161kV	500kV
Pole Type	Unit	Wood	Wood	Steel	Steel	Wood	Wood	Steel	Steel	Wood	Wood	Steel	Steel
Conductor: 397,500 – Cir. Mil., 26/7–ACSR													
Structures	Ea.	12[1]				9[2]				7[2]			
Poles	Ea.	12				18				14			
Crossarms	Ea.	24[3]				9[4]				7[4]			
Conductor	Ft.	15,990				15,990				15,990			
Insulators[5]	Ea.	180				135				105			
Ground Wire	Ft.	5,330				10,660				10,660			
Conductor: 636,000 – Cir. Mil., 26/7–ACSR													
Structures	Ea.	13[1]	11[6]			10[2]	9[2]	6[9]		8[2]	8[2]	6[9]	
Excavation	C.Y.	–	–			–	–	120		–	–	120	
Concrete	C.Y.	–	–			–	–	10		–	–	10	
Steel Towers	Tons	–	–			–	–	32		–	–	32	
Poles	Ea.	13	11			20	18	–		16	16	–	
Crossarms	Ea.	26[3]	33[7]			10[4]	9[8]	–		8[4]	8[8]	–	
Conductor	Ft.	15,990	15,990			15,990	15,990	15,990		15,990	15,990	15,990	
Insulators[5]	Ea.	195	165			150	297	297		120	264	297	
Ground Wire	Ft.	5330	5330			10,660	10,660	10,660		10,660	10,660	10,660	
Conductor: 954,000 – Cir. Mil., 45/7–ACSR													
Structures	Ea.	14[1]	12[6]		4[11]	10[2]	9[2]	6[9]	4[11]	8[2]	8[2]	6[9]	4[13]
Excavation	C.Y.	–	–		200	–	–	125	214	–	–	125	233
Concrete	C.Y.	–	–		20	–	–	10	21	–	–	10	21
Steel Towers	Tons	–	–		57	–	–	33	57	–	–	33	63
Poles	Ea.	14	12		–	20	18	–	–	16	16	–	–
Crossarms	Ea.	28[10]	36[7]		–	10[4]	9[8]	–	–	8[4]	8[8]	–	–
Conductor	Ft.	15,990	15,990		47,970[12]	15,990	15,990	15,990	47,970[12]	15,990	15,990	15,990	47,970[12]
Insulators[5]	Ea.	210	180		288	150	297	297	288	120	264	297	576
Ground Wire	Ft.	5330	5330		10,660	10,660	10,660	10,660	10,660	10,660	10,660	10,660	10,660
Conductor: 1,351,500 – Cir. Mil., 45/7–ACSR													
Structures	Ea.			8[14]				8[14]					
Excavation	C.Y.			220				220					
Concrete	C.Y.			28				28					
Steel Towers	Ton			46				46					
Conductor	Ft.			31,680[15]				31,680[15]					
Insulators	Ea.			528				528					
Ground Wire	Ft.			10,660				10,660					

1. Single pole two-arm suspension type construction
2. Two-pole wood H-frame construction
3. 4¾″ x 5¾″ x 8′ and 4¾″ x 5¾″ x 10′ wood crossarm
4. 6″ x 8″ x 26′-0″ wood crossarm
5. 5¾″ x 10″ disc insulator
6. Single pole construction with 3 fiberglass crossarms (5 fog- type insulators per phase)
7. 7′-0″ fiberglass crossarms
8. 6″ x 10″ x 35′-0″ wood crossarm
9. Laced steel tower, single circuit construction
10. 5″ x 7″ x 8′ and 5″ x 7″ x 10′ wood crossarm
11. Laced steel tower, single circuit 500-kV construction
12. Bundled conductor (3 sub-conductors per phase)
13. Laced steel tower, single circuit restrained phases (500-kV)
14. Laced steel tower, double circuit construction
15. Both sides of double circuit strung

Note: To allow for sagging, a mile (5280 Ft.) of transmission line uses 5330 Ft. of conductor per wire (called a wire mile).

R337119-30 Concrete for Conduit Encasement

Table below lists C.Y. of concrete for 100 L.F. of trench. Conduits separation center to center should meet 7.5" (N.E.C.).

Number of Conduits	1	2	3	4	6	8	9	Number of Conduits
Trench Dimension	11.5" x 11.5"	11.5" x 19"	11.5" x 27"	19" x 19"	19" x 27"	19" x 38"	27" x 27"	Trench Dimension
Conduit Diameter 2.0"	3.29	5.39	7.64	8.83	12.51	17.66	17.72	Conduit Diameter 2.0"
2.5"	3.23	5.29	7.49	8.62	12.19	17.23	17.25	2.5"
3.0"	3.15	5.13	7.24	8.29	11.71	16.59	16.52	3.0"
3.5"	3.08	4.97	7.02	7.99	11.26	15.98	15.84	3.5"
4.0"	2.99	4.80	6.76	7.65	10.74	15.30	15.07	4.0"
5.0"	2.78	4.37	6.11	6.78	9.44	13.57	13.12	5.0"
6.0"	2.52	3.84	5.33	5.74	7.87	11.48	10.77	6.0"
	•	• •	• • •	• • • •	• • • • • •	• • • • • • • •	• • • • • • • • •	

Transportation | R3472 | Railway Construction

R347216-10 Single Track R.R. Siding

The costs for a single track RR siding in the Unit Price section include the components shown in the table below.

Description of Component	CSI Line Item	Qty. per L.F. of Track	Unit
Ballast, 1-1/2" crushed stone	02060-150-0320	.667	C.Y.
6" x 8" x 8'-6" Treated timber ties, 22" O.C.	05655-700-1600	.545	Ea.
Tie plates, 2 per tie	05655-750-0300	1.091	Ea.
Track rail	05655-750-1000	2.000	L.F.
Spikes, 6", 4 per tie	05655-750-0200	2.182	Ea.
Splice bars w/ bolts, lock washers & nuts, @ 33' O.C.	05655-750-0100	.061	Pair
Crew B-14 @ 57 L.F./Day		.018	Day

R347216-20 Single Track, Steel Ties, Concrete Bed

The costs for a R.R. siding with steel ties and a concrete bed in the Unit Price section include the components shown in the table below.

Description of Component	CSI Line Item	Qty. per L.F. of Track	Unit
Concrete bed, 9' wide, 10" thick	03310-240-3950	.278	C.Y.
Ties, W6x16 x 6'-6" long, @ 30" O.C.	05120-640-0120	.400	Ea.
Tie plates, 4 per tie	05655-750-0300	1.600	Ea.
Track rail	05655-750-1000	2.000	L.F.
Tie plate bolts, 1", 8 per tie	05655-750-0020	3.200	Ea.
Splice bars w/bolts, lock washers & nuts, @ 33' O.C.	05655-750-0100	.061	Pair
Crew B-14 @ 22 L.F./Day		.045	Day

Change Orders

Change Order Considerations

A Change Order is a written document, usually prepared by the design professional, and signed by the owner, the architect/engineer and the contractor. A change order states the agreement of the parties to: an addition, deletion, or revision in the work; an adjustment in the contract sum, if any; or an adjustment in the contract time, if any. Change orders, or "extras" in the construction process occur after execution of the construction contract and impact architects/engineers, contractors and owners.

Change orders that are properly recognized and managed can ensure orderly, professional and profitable progress for all who are involved in the project. There are many causes for change orders and change order requests. In all cases, change orders or change order requests should be addressed promptly and in a precise and prescribed manner. The following paragraphs include information regarding change order pricing and procedures.

The Causes of Change Orders

Reasons for issuing change orders include:

- Unforeseen field conditions that require a change in the work
- Correction of design discrepancies, errors or omissions in the contract documents
- Owner-requested changes, either by design criteria, scope of work, or project objectives
- Completion date changes for reasons unrelated to the construction process
- Changes in building code interpretations, or other public authority requirements that require a change in the work
- Changes in availability of existing or new materials and products

Procedures

Properly written contract documents must include the correct change order procedures for all parties—owners, design professionals and contractors—to follow in order to avoid costly delays and litigation.

Being "in the right" is not always a sufficient or acceptable defense. The contract provisions requiring notification and documentation must be adhered to within a defined or reasonable time frame.

The appropriate method of handling change orders is by a written proposal and acceptance by all parties involved. Prior to starting work on a project, all parties should identify their authorized agents who may sign and accept change orders, as well as any limits placed on their authority.

Time may be a critical factor when the need for a change arises. For such cases, the contractor might be directed to proceed on a "time and materials" basis, rather than wait for all paperwork to be processed—a delay that could impede progress. In this situation, the contractor must still follow the prescribed change order procedures, including but not limited to, notification and documentation.

All forms used for change orders should be dated and signed by the proper authority. Lack of documentation can be very costly, especially if legal judgments are to be made and if certain field personnel are no longer available. For time and material change orders, the contractor should keep accurate daily records of all labor and material allocated to the change. Forms that can be used to document change order work are available in *Means Forms for Building Construction Professionals.*

Owners or awarding authorities who do considerable and continual building construction (such as the federal government) realize the inevitability of change orders for numerous reasons, both predictable and unpredictable. As a result, the federal government, the American Institute of Architects (AIA), the Engineers Joint Contract Documents Committee (EJCDC) and other contractor, legal and technical organizations have developed standards and procedures to be followed by all parties to achieve contract continuance and timely completion, while being financially fair to all concerned.

In addition to the change order standards put forth by industry associations, there are also many books available on the subject.

Pricing Change Orders

When pricing change orders, regardless of their cause, the most significant factor is *when* the change occurs. The need for a change may be perceived in the field or requested by the architect/engineer *before* any of the actual installation has begun, or may evolve or appear *during* construction when the item of work in question is partially installed. In the latter cases, the original sequence of construction is disrupted, along with all contiguous and supporting systems. Change orders cause the greatest impact when they occur *after* the installation has been completed and must be uncovered, or even replaced. Post-completion changes may be caused by necessary design changes, product failure, or changes in the owner's requirements that are not discovered until the building or the systems begin to function.

Specified procedures of notification and record keeping must be adhered to and enforced regardless of the stage of construction: *before, during,* or *after* installation. Some bidding documents anticipate change orders by requiring that unit prices including overhead and profit percentages—for additional as well as deductible changes—be listed. Generally these unit prices do not fully take into account the ripple effect, or impact on other trades, and should be used for general guidance only.

When pricing change orders, it is important to classify the time frame in which the change occurs. There are two basic time frames for change orders: *pre-installation change orders*, which occur before the start of construction, and *post-installation change orders*, which involve reworking after the original installation. Change orders that occur between these stages may be priced according to the extent of work completed using a combination of techniques developed for pricing *pre-* and *post-installation* changes.

The following factors are the basis for a check list to use when preparing a change order estimate.

Factors To Consider When Pricing Change Orders

As an estimator begins to prepare a change order, the following questions should be reviewed to determine their impact on the final price.

General

- Is the change order work *pre-installation* or *post-installation*?

  Change order work costs vary according to how much of the installation has been completed. Once workers have the project scoped in their mind, even though they have not started, it can be difficult to refocus. Consequently they may spend more than the normal amount of time understanding the change. Also, modifications to work in place such as trimming or refitting usually take more time than was initially estimated. The greater the amount of work in place, the more reluctant workers are to change it. Psychologically they may resent the change and as a result the rework takes longer than normal. Post-installation change order estimates must include demolition of existing work as required to accomplish the change. If the work is performed at a later time, additional obstacles such as building finishes may be present which must be protected. Regardless of whether the change occurs pre-installation or post-installation, attempt to isolate the identifiable factors and price them separately. For example, add shipping costs that may be required pre-installation or any demolition required post-installation. Then analyze the potential impact on productivity of psychological and/or learning curve factors and adjust the output rates accordingly. One approach is to break down the typical workday into segments and quantify the impact on each segment. The following chart may be useful as a guide:

Activities (Productivity) Expressed as Percentages of a Workday

Task	Means Mechanical Cost Data (for New Construction)	Pre-Installation Change Orders	Post-Installation Change Orders
1. Study plans	3%	6%	6%
2. Material procurement	3%	3%	3%
3. Receiving and storing	3%	3%	3%
4. Mobilization	5%	5%	5%
5. Site movement	5%	5%	8%
6. Layout and marking	8%	10%	12%
7. Actual installation	64%	59%	54%
8. Clean-up	3%	3%	3%
9. Breaks—non-productive	6%	6%	6%
Total	100%	100%	100%

Change Order Installation Efficiency

The labor-hours expressed (for new construction) are based on average installation time, using an efficiency level of approximately 60-65%. For change order situations, adjustments to this efficiency level should reflect the daily labor-hour allocation for that particular occurrence.

If any of the specific percentages expressed in the above chart do not apply to a particular project situation, then those percentage points should be reallocated to the appropriate task(s). Example: Using data for new construction, assume there is no new material being utilized. The percentages for Tasks 2 and 3 would therefore be reallocated to other tasks. If the time required for Tasks 2 and 3 can now be applied to installation, we can add the time allocated for *Material Procurement* and *Receiving and Storing* to the *Actual Installation* time for new construction, thereby increasing the Actual Installation percentage.

This chart shows that, due to reduced productivity, labor costs will be higher than those for new construction by 5% to 15% for pre-installation change orders and by 15% to 25% for post-installation change orders. Each job and change order is unique and must be examined individually. Many factors, covered elsewhere in this section, can each have a significant impact on productivity and change order costs. All such factors should be considered in every case.

- Will the change substantially delay the original completion date?

  A significant change in the project may cause the original completion date to be extended. The extended schedule may subject the contractor to new wage rates dictated by relevant labor contracts. Project supervision and other project overhead must also be extended beyond the original completion date. The schedule extension may also put installation into a new weather season. For example, underground piping scheduled for October installation was delayed until January. As a result, frost penetrated the trench area, thereby changing the degree of difficulty of the task. Changes and delays may have a ripple effect throughout the project. This effect must be analyzed and negotiated with the owner.

- What is the net effect of a deduct change order?

  In most cases, change orders resulting in a deduction or credit reflect only bare costs. The contractor may retain the overhead and profit based on the original bid.

Materials

- Will you have to pay more or less for the new material, required by the change order, than you paid for the original purchase?

  The same material prices or discounts will usually apply to materials purchased for change orders as new construction. In some instances, however, the contractor may forfeit the advantages of competitive pricing for change orders. Consider the following example:

  A contractor purchased over $20,000 worth of fan coil units for an installation, and obtained the maximum discount. Some time later it was determined the project required an additional matching unit. The contractor has to purchase this unit from the original supplier to ensure a match. The supplier at this time may not discount the unit because of the small quantity, and the fact that he is no longer in a competitive situation. The impact of quantity on purchase can add between 0% and 25% to material prices and/or subcontractor quotes.

- If materials have been ordered or delivered to the job site, will they be subject to a cancellation charge or restocking fee?

  Check with the supplier to determine if ordered materials are subject to a cancellation charge. Delivered materials not used as result of a change order may be subject to a restocking fee if returned to the supplier. Common restocking charges run between 20% and 40%. Also, delivery charges to return the goods to the supplier must be added.

Labor

- How efficient is the existing crew at the actual installation?

  Is the same crew that performed the initial work going to do the change order? Possibly the change consists of the installation of a unit identical to one already installed; therefore the change should take less time. Be sure to consider this potential productivity increase and modify the productivity rates accordingly.

- If the crew size is increased, what impact will that have on supervision requirements?

  Under most bargaining agreements or management practices, there is a point at which a working foreman is replaced by a nonworking foreman. This replacement increases project overhead by adding a nonproductive worker. If additional workers are added to accelerate the project or to perform changes while maintaining the schedule, be sure to add additional supervision time if warranted. Calculate the hours involved and the additional cost directly if possible.

- What are the other impacts of increased crew size?

  The larger the crew, the greater the potential for productivity to decrease. Some of the factors that cause this productivity loss are: overcrowding (producing restrictive conditions in the working space), and possibly a shortage of any special tools and equipment required. Such factors affect not only the crew working on the elements directly involved in the change order, but other crews whose movement may also be hampered.

  As the crew increases, check its basic composition for changes by the addition or deletion of apprentices or nonworking foreman and quantify the potential effects of equipment shortages or other logistical factors.

- As new crews, unfamiliar with the project, are brought onto the site, how long will it take them to become oriented to the project requirements?

  The orientation time for a new crew to become 100% effective varies with the site and type of project. Orientation is easiest at a new construction site, and most difficult at existing, very restrictive renovation sites. The type of work also affects orientation time. When all elements of the work are exposed, such as concrete or masonry work, orientation is decreased. When the work is concealed or less visible, such as existing electrical systems, orientation takes longer. Usually orientation can be accomplished in one day or less. Costs for added orientation should be itemized and added to the total estimated cost.

- How much actual production can be gained by working overtime?

  Short term overtime can be used effectively to accomplish more work in a day. However, as overtime is scheduled to run beyond several weeks, studies have shown marked decreases in output. The following chart shows the effect of long term overtime on worker efficiency. If the anticipated change requires extended overtime to keep the job on schedule, these factors can be used as a guide to predict the impact on time and cost. Add project overhead, particularly supervision, that may also be incurred.

Days per Week	Hours per Day	Production Efficiency					Payroll Cost Factors	
		1 Week	2 Weeks	3 Weeks	4 Weeks	Average 4 Weeks	@ 1-1/2 Times	@ 2 Times
5	8	100%	100%	100%	100%	100%	100%	100%
	9	100	100	95	90	96.25	105.6	111.1
	10	100	95	90	85	91.25	110.0	120.0
	11	95	90	75	65	81.25	113.6	127.3
	12	90	85	70	60	76.25	116.7	133.3
6	8	100	100	95	90	96.25	108.3	116.7
	9	100	95	90	85	92.50	113.0	125.9
	10	95	90	85	80	87.50	116.7	133.3
	11	95	85	70	65	78.75	119.7	139.4
	12	90	80	65	60	73.75	122.2	144.4
7	8	100	95	85	75	88.75	114.3	128.6
	9	95	90	80	70	83.75	118.3	136.5
	10	90	85	75	65	78.75	121.4	142.9
	11	85	80	65	60	72.50	124.0	148.1
	12	85	75	60	55	68.75	126.2	152.4

Effects of Overtime

Caution: Under many labor agreements, Sundays and holidays are paid at a higher premium than the normal overtime rate.

The use of long-term overtime is counterproductive on almost any construction job; that is, the longer the period of overtime, the lower the actual production rate. Numerous studies have been conducted, and while they have resulted in slightly different numbers, all reach the same conclusion. The figure above tabulates the effects of overtime work on efficiency.

As illustrated, there can be a difference between the *actual* payroll cost per hour and the *effective* cost per hour for overtime work. This is due to the reduced production efficiency with the increase in weekly hours beyond 40. This difference between actual and effective cost results from overtime work over a prolonged period. Short-term overtime work does not result in as great a reduction in efficiency, and in such cases, effective cost may not vary significantly from the actual payroll cost. As the total hours per week are increased on a regular basis, more time is lost because of fatigue, lowered morale, and an increased accident rate.

As an example, assume a project where workers are working 6 days a week, 10 hours per day. From the figure above (based on productivity studies), the average effective productive hours over a four-week period are:

$$0.875 \times 60 = 52.5$$

Depending upon the locale and day of week, overtime hours may be paid at time and a half or double time. For time and a half, the overall (average) *actual* payroll cost (including regular and overtime hours) is determined as follows:

$$\frac{40 \text{ reg. hrs.} + (20 \text{ overtime hrs.} \times 1.5)}{60 \text{ hrs.}} = 1.167$$

Based on 60 hours, the payroll cost per hour will be 116.7% of the normal rate at 40 hours per week. However, because the effective production (efficiency) for 60 hours is reduced to the equivalent of 52.5 hours, the effective cost of overtime is calculated as follows:

For time and a half:

$$\frac{40 \text{ reg. hrs.} + (20 \text{ overtime hrs.} \times 1.5)}{52.5 \text{ hrs.}} = 1.33$$

Installed cost will be 133% of the normal rate (for labor).

Thus, when figuring overtime, the actual cost per unit of work will be higher than the apparent overtime payroll dollar increase, due to the reduced productivity of the longer workweek. These efficiency calculations are true only for those cost factors determined by hours worked. Costs that are applied weekly or monthly, such as equipment rentals, will not be similarly affected.

Equipment

- What equipment is required to complete the change order?

  Change orders may require extending the rental period of equipment already on the job site, or the addition of special equipment brought in to accomplish the change work. In either case, the additional rental charges and operator labor charges must be added.

Summary

The preceding considerations and others you deem appropriate should be analyzed and applied to a change order estimate. The impact of each should be quantified and listed on the estimate to form an audit trail.

Change orders that are properly identified, documented, and managed help to ensure the orderly, professional and profitable progress of the work. They also minimize potential claims or disputes at the end of the project.

Square Foot Costs

Estimating Tips

- The cost figures in this Square Foot Cost section were derived from approximately 11,200 projects contained in the RSMeans database of completed construction projects. They include the contractor's overhead and profit but do not generally include architectural fees or land costs. The figures have been adjusted to January of the current year. New projects are added to our files each year, and outdated projects are discarded. For this reason, certain costs may not show a uniform annual progression. In no case are all subdivisions of a project listed.
- These projects were located throughout the U.S. and reflect a tremendous variation in square foot (S.F.) and cubic foot (C.F.) costs. This is due to differences, not only in labor and material costs, but also in individual owners' requirements. For instance, a bank in a large city would have different features than one in a rural area. This is true of all the different types of buildings analyzed. Therefore, caution should be exercised when using these Square Foot costs. For example, for court houses, costs in the database are local court house costs and will not apply to the larger, more elaborate federal court houses. As a general rule, the projects in the 1/4 column do not include any site work or equipment, while the projects in the 3/4 column may include both equipment and site work. The median figures do not generally include site work.
- None of the figures "go with" any others. All individual cost items were computed and tabulated separately. Thus, the sum of the median figures for Plumbing, HVAC, and Electrical will not normally match the total Mechanical and Electrical costs arrived at by separate analysis and tabulation of the projects.
- Each building was analyzed as to total and component costs and percentages. The figures were arranged in ascending order with the results tabulated as shown. The 1/4 column shows that 25% of the projects had lower costs and 75% had higher. The 3/4 column shows that 75% of the projects had lower costs and 25% had higher. The median column shows that 50% of the projects had lower costs and 50% had higher.
- There are two times when square foot costs are useful. The first is in the conceptual stage when no details are available. Then square foot costs make a useful starting point. The second is after the bids are in and the costs can be worked back into their appropriate units for information purposes. As soon as details become available in the project design, the square foot approach should be discontinued and the project priced as to its particular components. When more precision is required or for estimating the replacement cost of specific buildings, the current edition of *RSMeans Square Foot Costs* should be used.
- When using the figures in this section, it is recommended that the median column be used for preliminary figures if no additional information is available. The median figures, when multiplied by the total city construction cost index figures (see City Cost Indexes) and then multiplied by the project size modifier at the end of this section, should present a fairly accurate base figure, which would then have to be adjusted in view of the estimator's experience, local economic conditions, code requirements, and the owner's particular requirements. There is no need to factor the percentage figures as these should remain constant from city to city. All tabulations mentioning air conditioning had at least partial air conditioning.
- The editors of this book would greatly appreciate receiving cost figures on one or more of your recent projects which would then be included in the averages for next year. All cost figures received will be kept confidential except that they will be averaged with similar projects to arrive at Square Foot cost figures for next year's book. See the last page of the book for details and the discount available for submitting one or more of your projects.

50 17 00 | S.F. Costs

				UNIT COSTS			% OF TOTAL			
			UNIT	1/4	MEDIAN	3/4	1/4	MEDIAN	3/4	
01	0010	**APARTMENTS** Low Rise (1 to 3 story)	S.F.	56.50	71	94				01
	0020	Total project cost	C.F.	5.05	6.70	8.25				
	0100	Site work	S.F.	4.81	6.55	11.55	6.05%	10.55%	14.05%	
	0500	Masonry		1.11	2.57	4.45	1.54%	3.67%	6.35%	
	1500	Finishes		5.95	8.20	10.15	9.05%	10.75%	12.85%	
	1800	Equipment		1.83	2.79	4.14	2.73%	4.03%	5.95%	
	2720	Plumbing		4.38	5.65	7.15	6.65%	8.95%	10.05%	
	2770	Heating, ventilating, air conditioning		2.79	3.44	5.05	4.20%	5.60%	7.60%	
	2900	Electrical		3.25	4.32	5.80	5.20%	6.65%	8.40%	
	3100	Total: Mechanical & Electrical		11.25	14.35	17.90	15.90%	18.05%	23%	
	9000	Per apartment unit, total cost	Apt.	52,500	80,000	118,000				
	9500	Total: Mechanical & Electrical	″	9,900	15,600	20,400				
02	0010	**APARTMENTS** Mid Rise (4 to 7 story)	S.F.	74.50	90	110				02
	0020	Total project costs	C.F.	5.80	8.05	11				
	0100	Site work	S.F.	2.98	5.90	11.65	5.25%	6.70%	9.20%	
	0500	Masonry		4.96	6.85	9.75	5.15%	7.40%	10.50%	
	1500	Finishes		9.40	13.10	15.20	10.55%	13.45%	17.70%	
	1800	Equipment		2.45	3.49	4.48	2.55%	3.48%	4.31%	
	2500	Conveying equipment		1.69	2.08	2.48	2.05%	2.29%	2.69%	
	2720	Plumbing		4.37	7	7.75	5.70%	7.20%	8.95%	
	2900	Electrical		4.92	7	8.20	6.65%	7.20%	8.95%	
	3100	Total: Mechanical & Electrical		15.75	19.70	24.50	18.85%	21%	25%	
	9000	Per apartment unit, total cost	Apt.	84,500	99,500	165,000				
	9500	Total: Mechanical & Electrical	″	15,900	18,400	23,300				
03	0010	**APARTMENTS** High Rise (8 to 24 story)	S.F.	84.50	102	117				03
	0020	Total project costs	C.F.	8.20	10.05	12.20				
	0100	Site work	S.F.	3.07	4.96	6.95	2.58%	4.84%	6.15%	
	0500	Masonry		4.89	8.90	11.05	4.74%	9.65%	11.05%	
	1500	Finishes		9.40	11.75	13.85	9.75%	11.80%	13.70%	
	1800	Equipment		2.72	3.34	4.43	2.78%	3.49%	4.35%	
	2500	Conveying equipment		1.92	2.92	4.16	2.23%	2.78%	3.37%	
	2720	Plumbing		6.25	7.35	10.30	6.80%	7.20%	10.45%	
	2900	Electrical		5.80	7.35	9.90	6.45%	7.65%	8.80%	
	3100	Total: Mechanical & Electrical		17.40	22	26.50	17.95%	22.50%	24.50%	
	9000	Per apartment unit, total cost	Apt.	88,000	97,000	134,500				
	9500	Total: Mechanical & Electrical	″	19,000	21,700	23,000				
04	0010	**AUDITORIUMS**	S.F.	87	119	170				04
	0020	Total project costs	C.F.	5.50	7.65	11.50				
	2720	Plumbing	S.F.	5.55	7.70	9.75	5.85%	7.20%	8.70%	
	2900	Electrical		7.10	9.85	13.20	6.80%	8.95%	11.30%	
	3100	Total: Mechanical & Electrical		13.60	19.30	39	24.50%	30.50%	31.50%	
05	0010	**AUTOMOTIVE SALES**	S.F.	63.50	88	108				05
	0020	Total project costs	C.F.	4.28	5.15	6.65				
	2720	Plumbing	S.F.	2.97	5.15	5.60	2.89%	6.05%	6.50%	
	2770	Heating, ventilating, air conditioning		4.58	7	7.55	4.61%	10%	10.35%	
	2900	Electrical		5.25	8.05	11.15	7.40%	9.95%	12.40%	
	3100	Total: Mechanical & Electrical		14.60	20.50	26	19.15%	20.50%	26%	
06	0010	**BANKS**	S.F.	127	158	200				06
	0020	Total project costs	C.F.	9.10	12.40	16.35				
	0100	Site work	S.F.	14.55	22	33	7.85%	12.95%	17%	
	0500	Masonry		6.65	12.45	23	3.36%	6.95%	10.35%	
	1500	Finishes		11.30	15.45	19.85	5.85%	8.45%	11.25%	
	1800	Equipment		5.10	10.50	22.50	1.34%	5.95%	10.65%	
	2720	Plumbing		4	5.70	8.35	2.82%	3.90%	4.93%	
	2770	Heating, ventilating, air conditioning		7.60	10.15	13.50	4.86%	7.15%	8.50%	
	2900	Electrical		12.05	16.10	21	8.20%	10.20%	12.20%	
	3100	Total: Mechanical & Electrical		29	38	46	16.55%	19.45%	23%	
	3500	See also division 11020 & 11030								

50 17 00 | S.F. Costs

			UNIT	UNIT COSTS			% OF TOTAL		
				1/4	MEDIAN	3/4	1/4	MEDIAN	3/4
13	0010	**CHURCHES**	S.F.	86	109	141			
	0020	Total project costs	C.F.	5.30	6.70	8.85			
	1800	Equipment	S.F.	1.09	2.46	5.20	1.03%	2.24%	4.50%
	2720	Plumbing		3.34	4.67	6.90	3.51%	4.96%	6.25%
	2770	Heating, ventilating, air conditioning		7.80	10.20	14.45	7.50%	10%	12%
	2900	Electrical		7.20	9.90	13.25	7.30%	8.75%	10.90%
	3100	Total: Mechanical & Electrical		22	29	40	18.25%	22%	24%
	3500	See also division 11040							
15	0010	**CLUBS, COUNTRY**	S.F.	92	111	140			
	0020	Total project costs	C.F.	7.40	9.05	12.50			
	2720	Plumbing	S.F.	5.90	8.25	18.80	5.60%	7.90%	10%
	2900	Electrical		7.25	10.35	13.65	7%	8.95%	11%
	3100	Total: Mechanical & Electrical		22	38.50	48	19%	26.50%	29.50%
17	0010	**CLUBS, SOCIAL** Fraternal	S.F.	73.50	105	142			
	0020	Total project costs	C.F.	4.59	6.95	8.30			
	2720	Plumbing	S.F.	4.62	5.75	8.70	5.60%	6.90%	8.55%
	2770	Heating, ventilating, air conditioning		6.65	8.05	10.35	8.20%	9.25%	14.40%
	2900	Electrical		5.85	9.10	11	6.50%	9.50%	10.55%
	3100	Total: Mechanical & Electrical		16.30	31	39.50	21%	23%	23.50%
18	0010	**CLUBS, Y.M.C.A.**	S.F.	92.50	125	152			
	0020	Total project costs	C.F.	4.25	7.10	10.60			
	2720	Plumbing	S.F.	5.85	11.60	13	5.65%	7.60%	10.85%
	2900	Electrical		7	9.55	13.60	6.05%	7.80%	10.20%
	3100	Total: Mechanical & Electrical		26	31.50	35	18.40%	21.50%	28.50%
19	0010	**COLLEGES** Classrooms & Administration	S.F.	103	136	184			
	0020	Total project costs	C.F.	7.45	10.60	16.70			
	0500	Masonry	S.F.	6.85	12.90	15.85	5.65%	8.25%	10.50%
	2720	Plumbing		5.05	10.10	18.15	5.10%	6.60%	8.95%
	2900	Electrical		8.45	12.95	16.10	7.70%	9.85%	12%
	3100	Total: Mechanical & Electrical		31	43.50	51.50	24%	28%	31.50%
21	0010	**COLLEGES** Science, Engineering, Laboratories	S.F.	173	203	247			
	0020	Total project costs	C.F.	9.95	14.50	16.45			
	1800	Equipment	S.F.	9.65	22	24	2%	6.45%	12.65%
	2900	Electrical		14.30	19.50	31	7.10%	9.40%	12.10%
	3100	Total: Mechanical & Electrical		53	63	97.50	28.50%	31.50%	41%
	3500	See also division 11600							
23	0010	**COLLEGES** Student Unions	S.F.	110	155	182			
	0020	Total project costs	C.F.	6.15	8.10	10.40			
	3100	Total: Mechanical & Electrical	S.F.	29	45	53	23.50%	26%	29%
25	0010	**COMMUNITY CENTERS**	″	90	113	152			
	0020	Total project costs	C.F.	5.95	8.50	11.05			
	1800	Equipment	S.F.	2.29	3.90	6.15	1.87%	3.12%	6%
	2720	Plumbing		4.43	7.55	11.05	4.94%	7%	9.10%
	2770	Heating, ventilating, air conditioning		7.30	10.50	14.45	6.95%	10.65%	13.05%
	2900	Electrical		7.75	9.95	14.50	7.35%	9.10%	10.85%
	3100	Total: Mechanical & Electrical		27	32.50	46.50	22.50%	26.50%	32.50%
28	0010	**COURT HOUSES**	S.F.	131	151	176			
	0020	Total project costs	C.F.	10.10	12.20	15.20			
	2720	Plumbing	S.F.	6.25	8.75	12.55	5.95%	7.45%	8.20%
	2900	Electrical		12.90	15.25	18.40	8.55%	9.95%	11.50%
	3100	Total: Mechanical & Electrical		35.50	41	51.50	22.50%	29.50%	30.50%
30	0010	**DEPARTMENT STORES**	S.F.	48.50	66	83			
	0020	Total project costs	C.F.	2.61	3.54	4.59			
	2720	Plumbing	S.F.	1.59	1.92	2.91	1.82%	4.21%	5.90%
	2770	Heating, ventilating, air conditioning		4.43	6.85	10.30	8.20%	9.10%	14.80%

	50 17 00 | S.F. Costs		UNIT	UNIT COSTS 1/4	UNIT COSTS MEDIAN	UNIT COSTS 3/4	% OF TOTAL 1/4	% OF TOTAL MEDIAN	% OF TOTAL 3/4	
30	2900	Electrical	S.F.	5.60	7.65	9.05	9.05%	12.15%	14.95%	30
	3100	Total: Mechanical & Electrical		9.85	12.55	22	13.20%	21.50%	50%	
31	0010	**DORMITORIES** Low Rise (1 to 3 story)	S.F.	83	117	146				31
	0020	Total project costs	C.F.	5.50	8.45	12.65				
	2720	Plumbing	S.F.	5.55	7.45	9.40	8.05%	9%	9.65%	
	2770	Heating, ventilating, air conditioning		5.85	7.05	9.35	4.61%	8.05%	10%	
	2900	Electrical		5.95	8.95	12.45	6.55%	8.90%	9.55%	
	3100	Total: Mechanical & Electrical		31	32.50	34	22%	26%	29%	
	9000	Per bed, total cost	Bed	39,100	43,500	93,000				
32	0010	**DORMITORIES** Mid Rise (4 to 8 story)	S.F.	113	148	181				32
	0020	Total project costs	C.F.	12.50	13.70	16.45				
	2900	Electrical	S.F.	12.05	13.70	17.80	8.20%	10.20%	11.10%	
	3100	Total: Mechanical & Electrical	″	31	35	68	19.50%	30.50%	37.50%	
	9000	Per bed, total cost	Bed	16,100	36,800	76,000				
34	0010	**FACTORIES**	S.F.	43	64	98.50				34
	0020	Total project costs	C.F.	2.75	4.11	6.80				
	0100	Site work	S.F.	4.90	8.95	14.15	6.95%	11.45%	17.95%	
	2720	Plumbing		2.31	4.31	7.10	3.73%	6.05%	8.10%	
	2770	Heating, ventilating, air conditioning		4.50	6.45	8.70	5.25%	8.45%	11.35%	
	2900	Electrical		5.35	8.45	12.85	8.10%	10.50%	14.20%	
	3100	Total: Mechanical & Electrical		12.70	20.50	31	21%	28.50%	35.50%	
36	0010	**FIRE STATIONS**	S.F.	85	116	151				36
	0020	Total project costs	C.F.	4.98	6.85	9.10				
	0500	Masonry	S.F.	12.50	22.50	29.50	8.60%	11.65%	16.45%	
	1140	Roofing		2.78	7.50	8.55	1.90%	4.94%	5.05%	
	1580	Painting		2.28	3.22	3.29	1.37%	1.57%	2.07%	
	1800	Equipment		1.80	2.55	6.15	.90%	2.50%	4.07%	
	2720	Plumbing		5.40	7.95	11.70	5.85%	7.35%	9.50%	
	2770	Heating, ventilating, air conditioning		4.64	7.40	11.50	4.86%	7.25%	9.25%	
	2900	Electrical		6.05	10.40	13.70	6.80%	8.75%	10.65%	
	3100	Total: Mechanical & Electrical		30	35	41.50	19.60%	23%	27%	
37	0010	**FRATERNITY HOUSES** and Sorority Houses	S.F.	85	110	150				37
	0020	Total project costs	C.F.	8.50	8.85	11.35				
	2720	Plumbing	S.F.	6.45	7.35	13.50	6.80%	8%	10.85%	
	2900	Electrical		5.60	12.15	14.85	6.60%	9.90%	10.65%	
	3100	Total: Mechanical & Electrical		15	21.50	26		15.10%	15.90%	
38	0010	**FUNERAL HOMES**	S.F.	90	122	222				38
	0020	Total project costs	C.F.	9.15	10.20	19.60				
	2900	Electrical	S.F.	3.97	7.25	8.60	3.58%	4.44%	5.95%	
	3100	Total: Mechanical & Electrical	″	14.10	21	28.50	12.90%	12.90%	12.90%	
39	0010	**GARAGES, COMMERCIAL** (Service)	S.F.	53	80	109				39
	0020	Total project costs	C.F.	3.42	4.95	7.20				
	1800	Equipment	S.F.	2.87	6.45	10	2.69%	4.62%	6.80%	
	2720	Plumbing		3.52	5.45	9.90	5.45%	7.85%	10.65%	
	2730	Heating & ventilating		4.63	6.30	8.70	5.25%	6.85%	8.20%	
	2900	Electrical		4.90	7.50	10.70	7.15%	9.25%	10.85%	
	3100	Total: Mechanical & Electrical		10.95	20.50	30.50	13.60%	17.40%	27%	
40	0010	**GARAGES, MUNICIPAL** (Repair)	S.F.	74.50	99.50	140				40
	0020	Total project costs	C.F.	4.67	5.90	10.20				
	0500	Masonry	S.F.	7	13.70	21.50	5.60%	9.15%	12.50%	
	2720	Plumbing		3.35	6.45	12.15	3.59%	6.70%	7.95%	
	2730	Heating & ventilating		5.75	8.30	16	6.15%	7.45%	13.50%	
	2900	Electrical		5.55	8.70	12.50	6.65%	8.15%	11.15%	
	3100	Total: Mechanical & Electrical		17.75	35.50	51.50	21.50%	25.50%	28.50%	

50 17 00 | S.F. Costs

		50 17 00 | S.F. Costs	UNIT	UNIT COSTS 1/4	UNIT COSTS MEDIAN	UNIT COSTS 3/4	% OF TOTAL 1/4	% OF TOTAL MEDIAN	% OF TOTAL 3/4	
41	0010	**GARAGES, PARKING**	S.F.	29	42.50	73				41
	0020	Total project costs	C.F.	2.73	3.71	5.40				
	2720	Plumbing	S.F.	.82	1.28	1.97	1.72%	2.70%	3.85%	
	2900	Electrical		1.59	1.96	3.07	4.33%	5.20%	6.30%	
	3100	Total: Mechanical & Electrical	↓	3.26	4.54	5.95	7%	8.90%	11.05%	
	3200									
	9000	Per car, total cost	Car	12,300	15,400	19,700				
43	0010	**GYMNASIUMS**	S.F.	80.50	107	138				43
	0020	Total project costs	C.F.	4.03	5.45	6.70				
	1800	Equipment	S.F.	1.92	3.60	6.90	2.07%	3.35%	6.70%	
	2720	Plumbing		5.10	6.30	7.85	4.95%	6.75%	7.75%	
	2770	Heating, ventilating, air conditioning		5.50	8.40	16.85	5.80%	9.80%	11.10%	
	2900	Electrical		6.15	8.20	10.40	6.60%	8.30%	10.30%	
	3100	Total: Mechanical & Electrical	↓	22	31	36.50	20.50%	26%	29.50%	
	3500	See also division 11480								
46	0010	**HOSPITALS**	S.F.	155	190	288				46
	0020	Total project costs	C.F.	11.75	14.65	21				
	1800	Equipment	S.F.	3.94	7.55	13.05	1.10%	2.68%	5%	
	2720	Plumbing		13.35	18.70	24	7.60%	9.10%	10.85%	
	2770	Heating, ventilating, air conditioning		19.60	25	34	7.80%	12.95%	16.65%	
	2900	Electrical		16.95	22	34	9.85%	11.75%	14%	
	3100	Total: Mechanical & Electrical	↓	47.50	63	103	26.50%	33%	36.50%	
	9000	Per bed or person, total cost	Bed	127,000	203,500	272,000				
	9900	See also division 11700								
48	0010	**HOUSING** For the Elderly	S.F.	76.50	96.50	119				48
	0020	Total project costs	C.F.	5.45	7.55	9.65				
	0100	Site work	S.F.	5.65	8.40	12.10	5.05%	7.90%	12.10%	
	0500	Masonry		2.33	8.70	12.70	1.30%	6.05%	11%	
	1800	Equipment		1.84	2.54	4.04	1.88%	3.23%	4.43%	
	2510	Conveying systems		1.86	2.50	3.38	1.78%	2.20%	2.81%	
	2720	Plumbing		5.65	7.25	9.45	8.15%	9.55%	10.50%	
	2730	Heating, ventilating, air conditioning		2.91	4.12	6.15	3.30%	5.60%	7.25%	
	2900	Electrical		5.70	7.70	9.90	7.30%	8.50%	10.25%	
	3100	Total: Mechanical & Electrical	↓	19.60	23.50	31	18.10%	22.50%	29%	
	9000	Per rental unit, total cost	Unit	71,000	83,000	92,500				
	9500	Total: Mechanical & Electrical	"	15,800	18,200	21,200				
50	0010	**HOUSING** Public (Low Rise)	S.F.	64	89	116				50
	0020	Total project costs	C.F.	5.70	7.10	8.85				
	0100	Site work	S.F.	8.15	11.75	19.05	8.35%	11.75%	16.50%	
	1800	Equipment		1.75	2.85	4.34	2.26%	3.03%	4.24%	
	2720	Plumbing		4.63	6.10	7.75	7.15%	9.05%	11.60%	
	2730	Heating, ventilating, air conditioning		2.33	4.51	4.94	4.26%	6.05%	6.45%	
	2900	Electrical		3.88	5.80	8	5.10%	6.55%	8.25%	
	3100	Total: Mechanical & Electrical	↓	18.40	23.50	26.50	14.50%	17.55%	26.50%	
	9000	Per apartment, total cost	Apt.	70,500	80,000	100,500				
	9500	Total: Mechanical & Electrical	"	15,000	18,500	20,500				
51	0010	**ICE SKATING RINKS**	S.F.	57	128	141				51
	0020	Total project costs	C.F.	4.03	4.12	4.75				
	2720	Plumbing	S.F.	2.05	3.84	3.93	3.12%	3.23%	5.65%	
	2900	Electrical		5.85	9	9.55	6.30%	10.15%	15.05%	
	3100	Total: Mechanical & Electrical	↓	9.80	13.80	17.25	18.95%	18.95%	18.95%	
52	0010	**JAILS**	S.F.	167	216	278				52
	0020	Total project costs	C.F.	15.75	21	26				
	1800	Equipment	S.F.	6.50	19.25	32.50	2.80%	5.55%	11.90%	
	2720	Plumbing		17	21.50	28.50	7%	8.90%	13.35%	
	2770	Heating, ventilating, air conditioning		15.05	20	39	7.50%	9.45%	17.75%	
	2900	Electrical	↓	17.80	23	28.50	8.20%	11.55%	14.70%	

	50 17 00 | S.F. Costs		UNIT	UNIT COSTS 1/4	UNIT COSTS MEDIAN	UNIT COSTS 3/4	% OF TOTAL 1/4	% OF TOTAL MEDIAN	% OF TOTAL 3/4	
52	3100	Total: Mechanical & Electrical	S.F.	44.50	83	98.50	27.50%	30%	34%	52
53	0010	**LIBRARIES**	S.F.	106	132	174				53
	0020	Total project costs	C.F.	7.20	9.05	11.55				
	0500	Masonry	S.F.	8.25	14.60	24.50	5.80%	8%	12.35%	
	1800	Equipment		1.44	3.88	6.05	.41%	1.50%	4.16%	
	2720	Plumbing		3.90	5.65	7.70	3.43%	4.60%	5.70%	
	2770	Heating, ventilating, air conditioning		8.65	14.60	19	7.80%	10.95%	12.80%	
	2900	Electrical		10.75	13.85	17.50	8.35%	10.40%	11.95%	
	3100	Total: Mechanical & Electrical		32	40	50	20.50%	23%	26.50%	
54	0010	**LIVING, ASSISTED**	S.F.	97.50	115	135				54
	0020	Total project costs	C.F.	8.20	9.60	10.90				
	0500	Masonry	S.F.	2.87	3.42	4.02	2.37%	3.16%	3.86%	
	1800	Equipment		2.22	2.58	3.30	2.12%	2.45%	2.66%	
	2720	Plumbing		8.20	10.95	11.35	6.05%	8.15%	10.60%	
	2770	Heating, ventilating, air conditioning		9.70	10.15	11.10	7.95%	9.35%	9.70%	
	2900	Electrical		9.55	10.55	12.20	9%	10%	10.70%	
	3100	Total: Mechanical & Electrical		27.50	32	36	26%	29%	31.50%	
55	0010	**MEDICAL CLINICS**	S.F.	99	122	155				55
	0020	Total project costs	C.F.	7.30	9.85	12.50				
	1800	Equipment	S.F.	2.69	5.65	8.80	1.10%	3.13%	6.35%	
	2720	Plumbing		6.60	9.30	12.45	6.15%	8.40%	10.10%	
	2770	Heating, ventilating, air conditioning		7.85	10.30	15.15	6.65%	8.85%	11.35%	
	2900	Electrical		8.35	12.10	15.80	8.10%	10%	12.20%	
	3100	Total: Mechanical & Electrical		27	37	51	22.50%	27%	33.50%	
	3500	See also division 11700								
57	0010	**MEDICAL OFFICES**	S.F.	93.50	115	141				57
	0020	Total project costs	C.F.	6.95	9.45	12.80				
	1800	Equipment	S.F.	3.25	6.10	8.65	.98%	5.10%	7.05%	
	2720	Plumbing		5.15	7.95	10.75	5.60%	6.80%	8.50%	
	2770	Heating, ventilating, air conditioning		6.25	9.15	11.90	6.15%	8.05%	9.70%	
	2900	Electrical		7.50	10.85	15.20	7.65%	9.80%	11.70%	
	3100	Total: Mechanical & Electrical		20	29	43	19.35%	23%	30.50%	
59	0010	**MOTELS**	S.F.	59	87	113				59
	0020	Total project costs	C.F.	5.25	7.05	11.50				
	2720	Plumbing	S.F.	6	7.60	9.10	9.45%	10.60%	12.55%	
	2770	Heating, ventilating, air conditioning		3.64	5.45	9.75	5.60%	5.60%	10%	
	2900	Electrical		5.60	7.15	9.10	7.45%	9.20%	10.80%	
	3100	Total: Mechanical & Electrical		18.95	23.50	40.50	18.50%	21%	25.50%	
	5000									
	9000	Per rental unit, total cost	Unit	30,000	57,000	61,500				
	9500	Total: Mechanical & Electrical	"	5,850	8,850	10,300				
60	0010	**NURSING HOMES**	S.F.	93.50	119	146				60
	0020	Total project costs	C.F.	7.25	9.10	12.40				
	1800	Equipment	S.F.	2.91	3.86	6.45	2.02%	3.62%	4.99%	
	2720	Plumbing		7.95	12	14.45	8.75%	10.10%	12.70%	
	2770	Heating, ventilating, air conditioning		8.35	12.65	16.80	9.70%	11.45%	11.80%	
	2900	Electrical		9.30	11.50	15.60	9.50%	10.60%	12.50%	
	3100	Total: Mechanical & Electrical		22	30.50	51	26%	29.50%	30.50%	
	9000	Per bed or person, total cost	Bed	41,000	51,500	66,000				
61	0010	**OFFICES** Low Rise (1 to 4 story)	S.F.	77.50	99.50	130				61
	0020	Total project costs	C.F.	5.55	7.65	10.10				
	0100	Site work	S.F.	6.20	10.50	15.35	6.20%	9.70%	13.60%	
	0500	Masonry	"	2.59	5.85	10.60	2.55%	5.40%	8%	
	1800	Equipment	S.F.	.82	1.62	4.42	.77%	1.50%	3.69%	
	2720	Plumbing		2.78	4.30	6.25	3.66%	4.50%	6.10%	
	2770	Heating, ventilating, air conditioning		6.15	8.70	12.55	7.20%	10.40%	11.70%	
	2900	Electrical		6.35	9.05	12.75	7.45%	9.65%	11.40%	

50 17 | Square Foot Costs

50 17 00 | S.F. Costs

			UNIT	UNIT COSTS			% OF TOTAL			
				1/4	MEDIAN	3/4	1/4	MEDIAN	3/4	
61	3100	Total: Mechanical & Electrical	S.F.	16.50	23	34.50	18.25%	22.50%	27%	61
62	0010	**OFFICES** Mid Rise (5 to 10 story)	S.F.	82.50	100	132				62
	0020	Total project costs	C.F.	5.85	7.45	10.55				
	2720	Plumbing	S.F.	2.49	3.86	5.55	2.83%	3.74%	4.50%	
	2770	Heating, ventilating, air conditioning		6.25	8.95	14.30	7.65%	9.40%	11%	
	2900	Electrical		6.10	7.85	10.85	6.35%	7.80%	10%	
	3100	Total: Mechanical & Electrical		15.65	19.95	39	18.95%	21%	27.50%	
63	0010	**OFFICES** High Rise (11 to 20 story)	S.F.	101	128	157				63
	0020	Total project costs	C.F.	7.10	8.85	12.70				
	2900	Electrical	S.F.	6.15	7.50	11.15	5.80%	7.85%	10.50%	
	3100	Total: Mechanical & Electrical	"	19.80	26.50	45	16.90%	23.50%	34%	
64	0010	**POLICE STATIONS**	S.F.	122	156	199				64
	0020	Total project costs	C.F.	9.65	11.85	16.20				
	0500	Masonry	S.F.	11.40	20	26.50	6.70%	10.55%	11.35%	
	1800	Equipment		1.93	8.50	13.45	1.43%	4.07%	6.70%	
	2720	Plumbing		7	13.55	16.85	5.65%	6.90%	10.75%	
	2770	Heating, ventilating, air conditioning		10.60	14.10	19.15	5.85%	10.55%	11.70%	
	2900	Electrical		13.25	19.80	25	9.80%	11.85%	14.80%	
	3100	Total: Mechanical & Electrical		43.50	52.50	71	28.50%	32%	32.50%	
65	0010	**POST OFFICES**	S.F.	96	118	151				65
	0020	Total project costs	C.F.	5.75	7.30	8.65				
	2720	Plumbing	S.F.	4.32	5.35	6.75	4.24%	5.30%	5.60%	
	2770	Heating, ventilating, air conditioning		6.75	8.35	9.25	6.65%	7.15%	9.35%	
	2900	Electrical		7.90	11.15	13.20	7.25%	9%	11%	
	3100	Total: Mechanical & Electrical		23	30	34	16.25%	18.80%	22%	
66	0010	**POWER PLANTS**	S.F.	665	880	1,625				66
	0020	Total project costs	C.F.	18.30	40	85.50				
	2900	Electrical	S.F.	47	99.50	149	9.30%	12.75%	21.50%	
	8100	Total: Mechanical & Electrical	"	117	380	855	32.50%	32.50%	52.50%	
67	0010	**RELIGIOUS EDUCATION**	S.F.	77.50	101	119				67
	0020	Total project costs	C.F.	4.29	6.15	7.70				
	2720	Plumbing	S.F.	3.23	4.58	6.45	4.40%	5.30%	7.10%	
	2770	Heating, ventilating, air conditioning		8.15	9.25	13.05	10.05%	11.45%	12.35%	
	2900	Electrical		6.15	8.65	11.35	7.60%	9.10%	10.35%	
	3100	Total: Mechanical & Electrical		25	32.50	38.50	22%	23%	26%	
69	0010	**RESEARCH** Laboratories and facilities	S.F.	115	166	242				69
	0020	Total project costs	C.F.	9.05	17.40	20.50				
	1800	Equipment	S.F.	5.15	10.10	24.50	.90%	4.58%	8.80%	
	2720	Plumbing		11.60	14.85	24	6.15%	8.30%	10.80%	
	2770	Heating, ventilating, air conditioning		10.40	35	41.50	7.25%	16.50%	17.50%	
	2900	Electrical		13.30	22.50	38	9.45%	11.15%	15.40%	
	3100	Total: Mechanical & Electrical		41	78	111	29.50%	37%	45.50%	
70	0010	**RESTAURANTS**	S.F.	111	143	188				70
	0020	Total project costs	C.F.	9.45	12.35	16.25				
	1800	Equipment	S.F.	7.25	17.80	27	6.10%	13%	15.65%	
	2720	Plumbing		8.90	10.75	14.15	6.10%	8.15%	9%	
	2770	Heating, ventilating, air conditioning		11.30	15.60	20.50	9.20%	12%	12.40%	
	2900	Electrical		11.85	14.60	19	8.35%	10.55%	11.55%	
	3100	Total: Mechanical & Electrical		36.50	38.50	50	19.25%	24%	29.50%	
	9000	Per seat unit, total cost	Seat	4,125	5,475	6,500				
	9500	Total: Mechanical & Electrical	"	1,025	1,375	1,625				
72	0010	**RETAIL STORES**	S.F.	52	70.50	92.50				72
	0020	Total project costs	C.F.	3.55	5.05	7.05				
	2720	Plumbing	S.F.	1.90	3.17	5.40	3.26%	4.60%	6.80%	
	2770	Heating, ventilating, air conditioning		4.10	5.60	8.45	6.75%	8.75%	10.15%	
	2900	Electrical		4.70	6.45	9.30	7.25%	9.90%	11.65%	
	3100	Total: Mechanical & Electrical		12.05	16.05	21.50	17.05%	21%	23.50%	

50 17 00 | S.F. Costs

		50 17 00 | S.F. Costs	UNIT	UNIT COSTS 1/4	UNIT COSTS MEDIAN	UNIT COSTS 3/4	% OF TOTAL 1/4	% OF TOTAL MEDIAN	% OF TOTAL 3/4	
74	0010	**SCHOOLS** Elementary	S.F.	84.50	104	128				74
	0020	Total project costs	C.F.	5.60	7.15	9.25				
	0500	Masonry	S.F.	7.55	12.50	18.65	5.45%	10.65%	14.70%	
	1800	Equipment		2.49	4.23	7.80	1.90%	3.33%	5%	
	2720	Plumbing		4.90	6.95	9.30	5.70%	7.15%	9.35%	
	2730	Heating, ventilating, air conditioning		7.35	11.75	16.40	8.15%	10.80%	15.20%	
	2900	Electrical		7.95	10.40	13.05	8.40%	10.05%	11.70%	
	3100	Total: Mechanical & Electrical		28	35	42.50	24%	27.50%	30%	
	9000	Per pupil, total cost	Ea.	9,825	14,600	43,200				
	9500	Total: Mechanical & Electrical	″	2,775	3,500	12,500				
76	0010	**SCHOOLS** Junior High & Middle	S.F.	86	108	126				76
	0020	Total project costs	C.F.	5.60	7.25	8.10				
	0500	Masonry	S.F.	9.45	13.45	16.20	8%	11.60%	14.30%	
	1800	Equipment		2.81	4.53	7	1.81%	3.26%	5.15%	
	2720	Plumbing		5.60	6.35	8.20	5.50%	6.90%	8.15%	
	2770	Heating, ventilating, air conditioning		6.45	12.40	17.30	8.75%	12.75%	17.45%	
	2900	Electrical		8.50	10.35	13.30	7.90%	9.25%	10.60%	
	3100	Total: Mechanical & Electrical		25.50	36	44	22.50%	25.50%	29.50%	
	9000	Per pupil, total cost	Ea.	11,200	15,100	19,700				
78	0010	**SCHOOLS** Senior High	S.F.	90	110	139				78
	0020	Total project costs	C.F.	5.80	8.15	13.50				
	1800	Equipment	S.F.	2.41	5.70	8.40	1.88%	3.29%	4.80%	
	2720	Plumbing		5.25	7.80	14.10	5.70%	7%	8.35%	
	2770	Heating, ventilating, air conditioning		10.50	12.05	23	8.95%	11.60%	15%	
	2900	Electrical		9.10	11.70	18.75	8.35%	10.10%	12.35%	
	3100	Total: Mechanical & Electrical		31	35.50	60	23%	26.50%	28.50%	
	9000	Per pupil, total cost	Ea.	8,650	17,600	22,000				
80	0010	**SCHOOLS** Vocational	S.F.	74	105	132				80
	0020	Total project costs	C.F.	4.61	6.60	9.15				
	0500	Masonry	S.F.	4.36	10.75	16.45	3.53%	4.61%	10.95%	
	1800	Equipment	″	2.22	3.15	8	1.24%	3.13%	4.68%	
	2720	Plumbing	S.F.	4.75	7.25	10.85	5.40%	6.90%	8.55%	
	2770	Heating, ventilating, air conditioning		6.65	12.35	20.50	8.60%	11.90%	14.65%	
	2900	Electrical		7.75	10.55	14.90	8.45%	10.95%	13.20%	
	3100	Total: Mechanical & Electrical		27	29.50	51	23.50%	29.50%	31%	
	9000	Per pupil, total cost	Ea.	10,300	27,700	41,300				
83	0010	**SPORTS ARENAS**	S.F.	65	86.50	134				83
	0020	Total project costs	C.F.	3.52	6.30	8.15				
	2720	Plumbing	S.F.	3.76	5.70	12.05	4.35%	6.35%	9.40%	
	2770	Heating, ventilating, air conditioning		8.10	9.60	13.30	8.80%	10.20%	13.55%	
	2900	Electrical		6.75	9.15	11.85	8.60%	9.90%	12.25%	
	3100	Total: Mechanical & Electrical		16.80	30	39	21.50%	25%	27.50%	
85	0010	**SUPERMARKETS**	S.F.	60	69.50	84				85
	0020	Total project costs	C.F.	3.34	4.03	6.10				
	2720	Plumbing	S.F.	3.35	4.21	4.91	5.40%	6%	7.45%	
	2770	Heating, ventilating, air conditioning		4.92	6.55	7.95	8.60%	8.65%	9.60%	
	2900	Electrical		7.50	8.60	10.20	10.40%	12.45%	13.60%	
	3100	Total: Mechanical & Electrical		19.25	21	29	20.50%	26.50%	31%	
86	0010	**SWIMMING POOLS**	S.F.	97	163	345				86
	0020	Total project costs	C.F.	7.80	9.70	10.55				
	2720	Plumbing	S.F.	8.95	10.25	14.20	4.80%	9.70%	20.50%	
	2900	Electrical		7.30	11.80	17.20	6.50%	7.25%	7.60%	
	3100	Total: Mechanical & Electrical		17.80	45	61.50	11.15%	14.10%	23.50%	
87	0010	**TELEPHONE EXCHANGES**	S.F.	129	189	239				87
	0020	Total project costs	C.F.	8	12.85	17.70				
	2720	Plumbing	S.F.	5.45	8.60	12.30	4.52%	5.80%	6.90%	
	2770	Heating, ventilating, air conditioning		12.60	25.50	31.50	11.80%	16.05%	18.40%	

	50 17 00 | S.F. Costs		UNIT	UNIT COSTS 1/4	UNIT COSTS MEDIAN	UNIT COSTS 3/4	% OF TOTAL 1/4	% OF TOTAL MEDIAN	% OF TOTAL 3/4	
87	2900	Electrical	S.F.	13.10	21	37	10.90%	14%	17.85%	87
	3100	Total: Mechanical & Electrical	↓	38.50	73.50	104	29.50%	33.50%	44.50%	
91	0010	**THEATERS**	S.F.	81	101	153				91
	0020	Total project costs	C.F.	3.74	5.55	8.15				
	2720	Plumbing	S.F.	2.70	2.93	11.95	2.92%	4.70%	6.80%	
	2770	Heating, ventilating, air conditioning		7.85	9.55	11.80	8%	12.25%	13.40%	
	2900	Electrical		7.10	9.55	19.45	8.05%	9.95%	12.25%	
	3100	Total: Mechanical & Electrical	↓	18.25	27	56	23%	26.50%	27.50%	
94	0010	**TOWN HALLS** City Halls & Municipal Buildings	S.F.	95	121	153				94
	0020	Total project costs	C.F.	8.25	10.15	13.90				
	2720	Plumbing	S.F.	3.77	7.05	13	4.31%	5.95%	7.95%	
	2770	Heating, ventilating, air conditioning		6.85	13.55	19.80	7.05%	9.05%	13.45%	
	2900	Electrical		8.60	12.55	16.75	8.05%	9.50%	12.05%	
	3100	Total: Mechanical & Electrical	↓	29	34	45	22%	26.50%	31%	
97	0010	**WAREHOUSES** And Storage Buildings	S.F.	35.50	51	72				97
	0020	Total project costs	C.F.	1.94	2.77	4.57				
	0100	Site work	S.F.	3.48	6.90	10.45	6.05%	12.95%	19.85%	
	0500	Masonry		2.10	4.79	10.35	3.73%	7.40%	12.30%	
	1800	Equipment		.54	1.17	6.55	.91%	1.82%	5.55%	
	2720	Plumbing		1.12	2.02	3.77	2.90%	4.80%	6.55%	
	2730	Heating, ventilating, air conditioning		1.35	3.62	4.86	2.41%	5%	8.90%	
	2900	Electrical		2.05	3.76	6.25	5.15%	7.20%	10.10%	
	3100	Total: Mechanical & Electrical	↓	5.55	8.55	18.70	12.75%	18.90%	26%	
99	0010	**WAREHOUSE & OFFICES** Combination	S.F.	41.50	55.50	74.50				99
	0020	Total project costs	C.F.	2.12	3.08	4.56				
	1800	Equipment	S.F.	.72	1.39	2.15	.52%	1.21%	2.40%	
	2720	Plumbing		1.60	2.84	4.16	3.74%	4.76%	6.30%	
	2770	Heating, ventilating, air conditioning		2.53	3.96	5.55	5%	5.65%	10.05%	
	2900	Electrical		2.78	4.13	6.50	5.85%	8%	10%	
	3100	Total: Mechanical & Electrical	↓	7.75	11.80	18.85	14.40%	19.95%	24.50%	

Square Foot Project Size Modifier

One factor that affects the S.F. cost of a particular building is the size. In general, for buildings built to the same specifications in the same locality, the larger building will have the lower S.F. cost. This is due mainly to the decreasing contribution of the exterior walls plus the economy of scale usually achievable in larger buildings. The Area Conversion Scale shown below will give a factor to convert costs for the typical size building to an adjusted cost for the particular project.

Example: Determine the cost per S.F. for a 100,000 S.F. Mid-rise apartment building.

$$\frac{\text{Proposed building area} = 100{,}000\text{ S.F.}}{\text{Typical size from below} = 50{,}000\text{ S.F.}} = 2.00$$

Enter Area Conversion scale at 2.0, intersect curve, read horizontally the appropriate cost multiplier of .94. Size adjusted cost becomes .94 x $90.00 = $84.60 based on national average costs.

Note: For Size Factors less than .50, the Cost Multiplier is 1.1
For Size Factors greater than 3.5, the Cost Multiplier is .90

The Square Foot Base Size lists the median costs, most typical project size in our accumulated data and the range in size of the projects.

The Size Factor for your project is determined by dividing your project area in S.F. by the typical project size for the particular Building Type. With this factor, enter the Area Conversion Scale at the appropriate Size Factor and determine the appropriate cost multiplier for your building size.

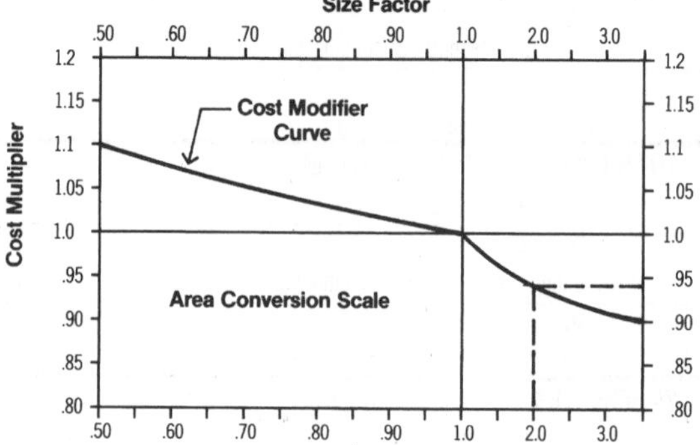

Square Foot Base Size

Building Type	Median Cost per S.F.	Typical Size Gross S.F.	Typical Range Gross S.F.	Building Type	Median Cost per S.F.	Typical Size Gross S.F.	Typical Range Gross S.F.
Apartments, Low Rise	$ 71.00	21,000	9,700 - 37,200	Jails	$216.00	40,000	5,500 - 145,000
Apartments, Mid Rise	90.00	50,000	32,000 - 100,000	Libraries	132.00	12,000	7,000 - 31,000
Apartments, High Rise	102.00	145,000	95,000 - 600,000	Living, Assisted	115.00	32,300	23,500 - 50,300
Auditoriums	119.00	25,000	7,600 - 39,000	Medical Clinics	122.00	7,200	4,200 - 15,700
Auto Sales	88.00	20,000	10,800 - 28,600	Medical Offices	115.00	6,000	4,000 - 15,000
Banks	158.00	4,200	2,500 - 7,500	Motels	87.00	40,000	15,800 - 120,000
Churches	109.00	17,000	2,000 - 42,000	Nursing Homes	119.00	23,000	15,000 - 37,000
Clubs, Country	111.00	6,500	4,500 - 15,000	Offices, Low Rise	99.50	20,000	5,000 - 80,000
Clubs, Social	105.00	10,000	6,000 - 13,500	Offices, Mid Rise	100.00	120,000	20,000 - 300,000
Clubs, YMCA	125.00	28,300	12,800 - 39,400	Offices, High Rise	128.00	260,000	120,000 - 800,000
Colleges (Class)	136.00	50,000	15,000 - 150,000	Police Stations	156.00	10,500	4,000 - 19,000
Colleges (Science Lab)	203.00	45,600	16,600 - 80,000	Post Offices	118.00	12,400	6,800 - 30,000
College (Student Union)	155.00	33,400	16,000 - 85,000	Power Plants	880.00	7,500	1,000 - 20,000
Community Center	113.00	9,400	5,300 - 16,700	Religious Education	101.00	9,000	6,000 - 12,000
Court Houses	151.00	32,400	17,800 - 106,000	Research	166.00	19,000	6,300 - 45,000
Dept. Stores	66.00	90,000	44,000 - 122,000	Restaurants	143.00	4,400	2,800 - 6,000
Dormitories, Low Rise	117.00	25,000	10,000 - 95,000	Retail Stores	70.50	7,200	4,000 - 17,600
Dormitories, Mid Rise	148.00	85,000	20,000 - 200,000	Schools, Elementary	104.00	41,000	24,500 - 55,000
Factories	64.00	26,400	12,900 - 50,000	Schools, Jr. High	108.00	92,000	52,000 - 119,000
Fire Stations	116.00	5,800	4,000 - 8,700	Schools, Sr. High	110.00	101,000	50,500 - 175,000
Fraternity Houses	110.00	12,500	8,200 - 14,800	Schools, Vocational	105.00	37,000	20,500 - 82,000
Funeral Homes	122.00	10,000	4,000 - 20,000	Sports Arenas	86.50	15,000	5,000 - 40,000
Garages, Commercial	80.00	9,300	5,000 - 13,600	Supermarkets	69.50	44,000	12,000 - 60,000
Garages, Municipal	99.50	8,300	4,500 - 12,600	Swimming Pools	163.00	20,000	10,000 - 32,000
Garages, Parking	42.50	163,000	76,400 - 225,300	Telephone Exchange	189.00	4,500	1,200 - 10,600
Gymnasiums	107.00	19,200	11,600 - 41,000	Theaters	101.00	10,500	8,800 - 17,500
Hospitals	190.00	55,000	27,200 - 125,000	Town Halls	121.00	10,800	4,800 - 23,400
House (Elderly)	96.50	37,000	21,000 - 66,000	Warehouses	51.00	25,000	8,000 - 72,000
Housing (Public)	89.00	36,000	14,400 - 74,400	Warehouse & Office	55.50	25,000	8,000 - 72,000
Ice Rinks	128.00	29,000	27,200 - 33,600				

Abbreviations

A	Area Square Feet; Ampere
ABS	Acrylonitrile Butadiene Stryrene; Asbestos Bonded Steel
A.C.	Alternating Current; Air-Conditioning; Asbestos Cement; Plywood Grade A & C
A.C.I.	American Concrete Institute
AD	Plywood, Grade A & D
Addit.	Additional
Adj.	Adjustable
af	Audio-frequency
A.G.A.	American Gas Association
Agg.	Aggregate
A.H.	Ampere Hours
A hr.	Ampere-hour
A.H.U.	Air Handling Unit
A.I.A.	American Institute of Architects
AIC	Ampere Interrupting Capacity
Allow.	Allowance
alt.	Altitude
Alum.	Aluminum
a.m.	Ante Meridiem
Amp.	Ampere
Anod.	Anodized
Approx.	Approximate
Apt.	Apartment
Asb.	Asbestos
A.S.B.C.	American Standard Building Code
Asbe.	Asbestos Worker
A.S.H.R.A.E.	American Society of Heating, Refrig. & AC Engineers
A.S.M.E.	American Society of Mechanical Engineers
A.S.T.M.	American Society for Testing and Materials
Attchmt.	Attachment
Avg.	Average
A.W.G.	American Wire Gauge
AWWA	American Water Works Assoc.
Bbl.	Barrel
B&B	Grade B and Better; Balled & Burlapped
B.&S.	Bell and Spigot
B.&W.	Black and White
b.c.c.	Body-centered Cubic
B.C.Y.	Bank Cubic Yards
BE	Bevel End
B.F.	Board Feet
Bg. cem.	Bag of Cement
BHP	Boiler Horsepower; Brake Horsepower
B.I.	Black Iron
Bit.; Bitum.	Bituminous
Bk.	Backed
Bkrs.	Breakers
Bldg.	Building
Blk.	Block
Bm.	Beam
Boil.	Boilermaker
B.P.M.	Blows per Minute
BR	Bedroom
Brg.	Bearing
Brhe.	Bricklayer Helper
Bric.	Bricklayer
Brk.	Brick
Brng.	Bearing
Brs.	Brass
Brz.	Bronze
Bsn.	Basin
Btr.	Better
BTU	British Thermal Unit
BTUH	BTU per Hour
B.U.R.	Built-up Roofing
BX	Interlocked Armored Cable
c	Conductivity, Copper Sweat
C	Hundred; Centigrade
C/C	Center to Center, Cedar on Cedar
Cab.	Cabinet
Cair.	Air Tool Laborer
Calc	Calculated
Cap.	Capacity
Carp.	Carpenter
C.B.	Circuit Breaker
C.C.A.	Chromate Copper Arsenate
C.C.F.	Hundred Cubic Feet
cd	Candela
cd/sf	Candela per Square Foot
CD	Grade of Plywood Face & Back
CDX	Plywood, Grade C & D, exterior glue
Cefi.	Cement Finisher
Cem.	Cement
CF	Hundred Feet
C.F.	Cubic Feet
CFM	Cubic Feet per Minute
c.g.	Center of Gravity
CHW	Chilled Water; Commercial Hot Water
C.I.	Cast Iron
C.I.P.	Cast in Place
Circ.	Circuit
C.L.	Carload Lot
Clab.	Common Laborer
Clam	Common maintenance laborer
C.L.F.	Hundred Linear Feet
CLF	Current Limiting Fuse
CLP	Cross Linked Polyethylene
cm	Centimeter
CMP	Corr. Metal Pipe
C.M.U.	Concrete Masonry Unit
CN	Change Notice
Col.	Column
CO_2	Carbon Dioxide
Comb.	Combination
Compr.	Compressor
Conc.	Concrete
Cont.	Continuous; Continued
Corr.	Corrugated
Cos	Cosine
Cot	Cotangent
Cov.	Cover
C/P	Cedar on Paneling
CPA	Control Point Adjustment
Cplg.	Coupling
C.P.M.	Critical Path Method
CPVC	Chlorinated Polyvinyl Chloride
C.Pr.	Hundred Pair
CRC	Cold Rolled Channel
Creos.	Creosote
Crpt.	Carpet & Linoleum Layer
CRT	Cathode-ray Tube
CS	Carbon Steel, Constant Shear Bar Joist
Csc	Cosecant
C.S.F.	Hundred Square Feet
CSI	Construction Specifications Institute
C.T.	Current Transformer
CTS	Copper Tube Size
Cu	Copper, Cubic
Cu. Ft.	Cubic Foot
cw	Continuous Wave
C.W.	Cool White; Cold Water
Cwt.	100 Pounds
C.W.X.	Cool White Deluxe
C.Y.	Cubic Yard (27 cubic feet)
C.Y./Hr.	Cubic Yard per Hour
Cyl.	Cylinder
d	Penny (nail size)
D	Deep; Depth; Discharge
Dis.;Disch.	Discharge
Db.	Decibel
Dbl.	Double
DC	Direct Current
DDC	Direct Digital Control
Demob.	Demobilization
d.f.u.	Drainage Fixture Units
D.H.	Double Hung
DHW	Domestic Hot Water
Diag.	Diagonal
Diam.	Diameter
Distrib.	Distribution
Dk.	Deck
D.L.	Dead Load; Diesel
DLH	Deep Long Span Bar Joist
Do.	Ditto
Dp.	Depth
D.P.S.T.	Double Pole, Single Throw
Dr.	Driver
Drink.	Drinking
D.S.	Double Strength
D.S.A.	Double Strength A Grade
D.S.B.	Double Strength B Grade
Dty.	Duty
DWV	Drain Waste Vent
DX	Deluxe White, Direct Expansion
dyn	Dyne
e	Eccentricity
E	Equipment Only; East
Ea.	Each
E.B.	Encased Burial
Econ.	Economy
E.C.Y	Embankment Cubic Yards
EDP	Electronic Data Processing
EIFS	Exterior Insulation Finish System
E.D.R.	Equiv. Direct Radiation
Eq.	Equation
Elec.	Electrician; Electrical
Elev.	Elevator; Elevating
EMT	Electrical Metallic Conduit; Thin Wall Conduit
Eng.	Engine, Engineered
EPDM	Ethylene Propylene Diene Monomer
EPS	Expanded Polystyrene
Eqhv.	Equip. Oper., Heavy
Eqlt.	Equip. Oper., Light
Eqmd.	Equip. Oper., Medium
Eqmm.	Equip. Oper., Master Mechanic
Eqol.	Equip. Oper., Oilers
Equip.	Equipment
ERW	Electric Resistance Welded
E.S.	Energy Saver
Est.	Estimated
esu	Electrostatic Units
E.W.	Each Way
EWT	Entering Water Temperature
Excav.	Excavation
Exp.	Expansion, Exposure
Ext.	Exterior
Extru.	Extrusion
f.	Fiber stress
F	Fahrenheit; Female; Fill
Fab.	Fabricated
FBGS	Fiberglass
F.C.	Footcandles
f.c.c.	Face-centered Cubic
f'c.	Compressive Stress in Concrete; Extreme Compressive Stress
F.E.	Front End
FEP	Fluorinated Ethylene Propylene (Teflon)
F.G.	Flat Grain
F.H.A.	Federal Housing Administration
Fig.	Figure
Fin.	Finished
Fixt.	Fixture
Fl. Oz.	Fluid Ounces
Flr.	Floor
F.M.	Frequency Modulation; Factory Mutual
Fmg.	Framing
Fndtn.	Foundation

Abbreviations

Fori.	Foreman, Inside
Foro.	Foreman, Outside
Fount.	Fountain
FPM	Feet per Minute
FPT	Female Pipe Thread
Fr.	Frame
F.R.	Fire Rating
FRK	Foil Reinforced Kraft
FRP	Fiberglass Reinforced Plastic
FS	Forged Steel
FSC	Cast Body; Cast Switch Box
Ft.	Foot; Feet
Ftng.	Fitting
Ftg.	Footing
Ft. Lb.	Foot Pound
Furn.	Furniture
FVNR	Full Voltage Non-Reversing
FXM	Female by Male
Fy.	Minimum Yield Stress of Steel
g	Gram
G	Gauss
Ga.	Gauge
Gal.	Gallon
Gal./Min.	Gallon per Minute
Galv.	Galvanized
Gen.	General
G.F.I.	Ground Fault Interrupter
Glaz.	Glazier
GPD	Gallons per Day
GPH	Gallons per Hour
GPM	Gallons per Minute
GR	Grade
Gran.	Granular
Grnd.	Ground
H	High; High Strength Bar Joist; Henry
H.C.	High Capacity
H.D.	Heavy Duty; High Density
H.D.O.	High Density Overlaid
Hdr.	Header
Hdwe.	Hardware
Help.	Helper Average
HEPA	High Efficiency Particulate Air Filter
Hg	Mercury
HIC	High Interrupting Capacity
HM	Hollow Metal
H.O.	High Output
Horiz.	Horizontal
H.P.	Horsepower; High Pressure
H.P.F.	High Power Factor
Hr.	Hour
Hrs./Day	Hours per Day
HSC	High Short Circuit
Ht.	Height
Htg.	Heating
Htrs.	Heaters
HVAC	Heating, Ventilation & Air-Conditioning
Hvy.	Heavy
HW	Hot Water
Hyd.;Hydr.	Hydraulic
Hz.	Hertz (cycles)
I.	Moment of Inertia
I.C.	Interrupting Capacity
ID	Inside Diameter
I.D.	Inside Dimension; Identification
I.F.	Inside Frosted
I.M.C.	Intermediate Metal Conduit
In.	Inch
Incan.	Incandescent
Incl.	Included; Including
Int.	Interior
Inst.	Installation
Insul.	Insulation/Insulated
I.P.	Iron Pipe
I.P.S.	Iron Pipe Size
I.P.T.	Iron Pipe Threaded
I.W.	Indirect Waste
J	Joule
J.I.C.	Joint Industrial Council
K	Thousand; Thousand Pounds; Heavy Wall Copper Tubing, Kelvin
K.A.H.	Thousand Amp. Hours
KCMIL	Thousand Circular Mils
KD	Knock Down
K.D.A.T.	Kiln Dried After Treatment
kg	Kilogram
kG	Kilogauss
kgf	Kilogram Force
kHz	Kilohertz
Kip.	1000 Pounds
KJ	Kiljoule
K.L.	Effective Length Factor
K.L.F.	Kips per Linear Foot
Km	Kilometer
K.S.F.	Kips per Square Foot
K.S.I.	Kips per Square Inch
kV	Kilovolt
kVA	Kilovolt Ampere
K.V.A.R.	Kilovar (Reactance)
KW	Kilowatt
KWh	Kilowatt-hour
L	Labor Only; Length; Long; Medium Wall Copper Tubing
Lab.	Labor
lat	Latitude
Lath.	Lather
Lav.	Lavatory
lb.; #	Pound
L.B.	Load Bearing; L Conduit Body
L. & E.	Labor & Equipment
lb./hr.	Pounds per Hour
lb./L.F.	Pounds per Linear Foot
lbf/sq.in.	Pound-force per Square Inch
L.C.L.	Less than Carload Lot
L.C.Y.	Loose Cubic Yard
Ld.	Load
LE	Lead Equivalent
LED	Light Emitting Diode
L.F.	Linear Foot
Lg.	Long; Length; Large
L & H	Light and Heat
LH	Long Span Bar Joist
L.H.	Labor Hours
L.L.	Live Load
L.L.D.	Lamp Lumen Depreciation
lm	Lumen
lm/sf	Lumen per Square Foot
lm/W	Lumen per Watt
L.O.A.	Length Over All
log	Logarithm
L-O-L	Lateralolet
L.P.	Liquefied Petroleum; Low Pressure
L.P.F.	Low Power Factor
LR	Long Radius
L.S.	Lump Sum
Lt.	Light
Lt. Ga.	Light Gauge
L.T.L.	Less than Truckload Lot
Lt. Wt.	Lightweight
L.V.	Low Voltage
M	Thousand; Material; Male; Light Wall Copper Tubing
M^2CA	Meters Squared Contact Area
m/hr; M.H.	Man-hour
mA	Milliampere
Mach.	Machine
Mag. Str.	Magnetic Starter
Maint.	Maintenance
Marb.	Marble Setter
Mat; Mat'l.	Material
Max.	Maximum
MBF	Thousand Board Feet
MBH	Thousand BTU's per hr.
MC	Metal Clad Cable
M.C.F.	Thousand Cubic Feet
M.C.F.M.	Thousand Cubic Feet per Minute
M.C.M.	Thousand Circular Mils
M.C.P.	Motor Circuit Protector
MD	Medium Duty
M.D.O.	Medium Density Overlaid
Med.	Medium
MF	Thousand Feet
M.F.B.M.	Thousand Feet Board Measure
Mfg.	Manufacturing
Mfrs.	Manufacturers
mg	Milligram
MGD	Million Gallons per Day
MGPH	Thousand Gallons per Hour
MH, M.H.	Manhole; Metal Halide; Man-Hour
MHz	Megahertz
Mi.	Mile
MI	Malleable Iron; Mineral Insulated
mm	Millimeter
Mill.	Millwright
Min., min.	Minimum, minute
Misc.	Miscellaneous
ml	Milliliter, Mainline
M.L.F.	Thousand Linear Feet
Mo.	Month
Mobil.	Mobilization
Mog.	Mogul Base
MPH	Miles per Hour
MPT	Male Pipe Thread
MRT	Mile Round Trip
ms	Millisecond
M.S.F.	Thousand Square Feet
Mstz.	Mosaic & Terrazzo Worker
M.S.Y.	Thousand Square Yards
Mtd.	Mounted
Mthe.	Mosaic & Terrazzo Helper
Mtng.	Mounting
Mult.	Multi; Multiply
M.V.A.	Million Volt Amperes
M.V.A.R.	Million Volt Amperes Reactance
MV	Megavolt
MW	Megawatt
MXM	Male by Male
MYD	Thousand Yards
N	Natural; North
nA	Nanoampere
NA	Not Available; Not Applicable
N.B.C.	National Building Code
NC	Normally Closed
N.E.M.A.	National Electrical Manufacturers Assoc.
NEHB	Bolted Circuit Breaker to 600V.
N.L.B.	Non-Load-Bearing
NM	Non-Metallic Cable
nm	Nanometer
No.	Number
NO	Normally Open
N.O.C.	Not Otherwise Classified
Nose.	Nosing
N.P.T.	National Pipe Thread
NQOD	Combination Plug-on/Bolt on Circuit Breaker to 240V.
N.R.C.	Noise Reduction Coefficient
N.R.S.	Non Rising Stem
ns	Nanosecond
nW	Nanowatt
OB	Opposing Blade
OC	On Center
OD	Outside Diameter
O.D.	Outside Dimension
ODS	Overhead Distribution System
O.G.	Ogee
O.H.	Overhead
O&P	Overhead and Profit
Oper.	Operator
Opng.	Opening
Orna.	Ornamental
OSB	Oriented Strand Board

Abbreviations

O.S.&Y.	Outside Screw and Yoke
Ovhd.	Overhead
OWG	Oil, Water or Gas
Oz.	Ounce
P.	Pole; Applied Load; Projection
p.	Page
Pape.	Paperhanger
P.A.P.R.	Powered Air Purifying Respirator
PAR	Parabolic Reflector
Pc., Pcs.	Piece, Pieces
P.C.	Portland Cement; Power Connector
P.C.F.	Pounds per Cubic Foot
P.C.M.	Phase Contrast Microscopy
P.E.	Professional Engineer; Porcelain Enamel; Polyethylene; Plain End
Perf.	Perforated
Ph.	Phase
P.I.	Pressure Injected
Pile.	Pile Driver
Pkg.	Package
Pl.	Plate
Plah.	Plasterer Helper
Plas.	Plasterer
Pluh.	Plumbers Helper
Plum.	Plumber
Ply.	Plywood
p.m.	Post Meridiem
Pntd.	Painted
Pord.	Painter, Ordinary
pp	Pages
PP; PPL	Polypropylene
P.P.M.	Parts per Million
Pr.	Pair
P.E.S.B.	Pre-engineered Steel Building
Prefab.	Prefabricated
Prefin.	Prefinished
Prop.	Propelled
PSF; psf	Pounds per Square Foot
PSI; psi	Pounds per Square Inch
PSIG	Pounds per Square Inch Gauge
PSP	Plastic Sewer Pipe
Pspr.	Painter, Spray
Psst.	Painter, Structural Steel
P.T.	Potential Transformer
P. & T.	Pressure & Temperature
Ptd.	Painted
Ptns.	Partitions
Pu	Ultimate Load
PVC	Polyvinyl Chloride
Pvmt.	Pavement
Pwr.	Power
Q	Quantity Heat Flow
Quan.;Qty.	Quantity
Q.C.	Quick Coupling
r	Radius of Gyration
R	Resistance
R.C.P.	Reinforced Concrete Pipe
Rect.	Rectangle
Reg.	Regular
Reinf.	Reinforced
Req'd.	Required
Res.	Resistant
Resi.	Residential
Rgh.	Rough
RGS	Rigid Galvanized Steel
R.H.W.	Rubber, Heat & Water Resistant; Residential Hot Water
rms	Root Mean Square
Rnd.	Round
Rodm.	Rodman
Rofc.	Roofer, Composition
Rofp.	Roofer, Precast
Rohe.	Roofer Helpers (Composition)
Rots.	Roofer, Tile & Slate
R.O.W.	Right of Way
RPM	Revolutions per Minute
R.S.	Rapid Start

Rsr	Riser
RT	Round Trip
S.	Suction; Single Entrance; South
SC	Screw Cover
SCFM	Standard Cubic Feet per Minute
Scaf.	Scaffold
Sch.; Sched.	Schedule
S.C.R.	Modular Brick
S.D.	Sound Deadening
S.D.R.	Standard Dimension Ratio
S.E.	Surfaced Edge
Sel.	Select
S.E.R.; S.E.U.	Service Entrance Cable
S.F.	Square Foot
S.F.C.A.	Square Foot Contact Area
S.F. Flr.	Square Foot of Floor
S.F.G.	Square Foot of Ground
S.F. Hor.	Square Foot Horizontal
S.F.R.	Square Feet of Radiation
S.F. Shlf.	Square Foot of Shelf
S4S	Surface 4 Sides
Shee.	Sheet Metal Worker
Sin.	Sine
Skwk.	Skilled Worker
SL	Saran Lined
S.L.	Slimline
Sldr.	Solder
SLH	Super Long Span Bar Joist
S.N.	Solid Neutral
S-O-L	Socketolet
sp	Standpipe
S.P.	Static Pressure; Single Pole; Self-Propelled
Spri.	Sprinkler Installer
spwg	Static Pressure Water Gauge
S.P.D.T.	Single Pole, Double Throw
SPF	Spruce Pine Fir
S.P.S.T.	Single Pole, Single Throw
SPT	Standard Pipe Thread
Sq.	Square; 100 Square Feet
Sq. Hd.	Square Head
Sq. In.	Square Inch
S.S.	Single Strength; Stainless Steel
S.S.B.	Single Strength B Grade
sst	Stainless Steel
Sswk.	Structural Steel Worker
Sswl.	Structural Steel Welder
St.;Stl.	Steel
S.T.C.	Sound Transmission Coefficient
Std.	Standard
STK	Select Tight Knot
STP	Standard Temperature & Pressure
Stpi.	Steamfitter, Pipefitter
Str.	Strength; Starter; Straight
Strd.	Stranded
Struct.	Structural
Sty.	Story
Subj.	Subject
Subs.	Subcontractors
Surf.	Surface
Sw.	Switch
Swbd.	Switchboard
S.Y.	Square Yard
Syn.	Synthetic
S.Y.P.	Southern Yellow Pine
Sys.	System
t.	Thickness
T	Temperature; Ton
Tan	Tangent
T.C.	Terra Cotta
T & C	Threaded and Coupled
T.D.	Temperature Difference
Tdd	Telecommunications Device for the Deaf
T.E.M.	Transmission Electron Microscopy
TFE	Tetrafluoroethylene (Teflon)
T. & G.	Tongue & Groove; Tar & Gravel

Th.;Thk.	Thick
Thn.	Thin
Thrded	Threaded
Tilf.	Tile Layer, Floor
Tilh.	Tile Layer, Helper
THHN	Nylon Jacketed Wire
THW.	Insulated Strand Wire
THWN;	Nylon Jacketed Wire
T.L.	Truckload
T.M.	Track Mounted
Tot.	Total
T-O-L	Threadolet
T.S.	Trigger Start
Tr.	Trade
Transf.	Transformer
Trhv.	Truck Driver, Heavy
Trlr	Trailer
Trlt.	Truck Driver, Light
TTY	Teletypewriter
TV	Television
T.W.	Thermoplastic Water Resistant Wire
UCI	Uniform Construction Index
UF	Underground Feeder
UGND	Underground Feeder
U.H.F.	Ultra High Frequency
U.L.	Underwriters Laboratory
Unfin.	Unfinished
URD	Underground Residential Distribution
US	United States
USP	United States Primed
UTP	Unshielded Twisted Pair
V	Volt
V.A.	Volt Amperes
V.C.T.	Vinyl Composition Tile
VAV	Variable Air Volume
VC	Veneer Core
Vent.	Ventilation
Vert.	Vertical
V.F.	Vinyl Faced
V.G.	Vertical Grain
V.H.F.	Very High Frequency
VHO	Very High Output
Vib.	Vibrating
V.L.F.	Vertical Linear Foot
Vol.	Volume
VRP	Vinyl Reinforced Polyester
W	Wire; Watt; Wide; West
w/	With
W.C.	Water Column; Water Closet
W.F.	Wide Flange
W.G.	Water Gauge
Wldg.	Welding
W. Mile	Wire Mile
W-O-L	Weldolet
W.R.	Water Resistant
Wrck.	Wrecker
W.S.P.	Water, Steam, Petroleum
WT., Wt.	Weight
WWF	Welded Wire Fabric
XFER	Transfer
XFMR	Transformer
XHD	Extra Heavy Duty
XHHW; XLPE	Cross-Linked Polyethylene Wire Insulation
XLP	Cross-linked Polyethylene
Y	Wye
yd	Yard
yr	Year
Δ	Delta
%	Percent
~	Approximately
Ø	Phase
@	At
#	Pound; Number
<	Less Than
>	Greater Than

A

B

C

D

E

F

I

J

K

L

M

N

O

P

INDEX

Q

R

S

X

Y

Z

Division Notes

		CREW	DAILY OUTPUT	LABOR-HOURS	UNIT	2006 BARE COSTS MAT.	LABOR	EQUIP.	TOTAL	TOTAL INCL O&P

Division Notes

		CREW	DAILY OUTPUT	LABOR-HOURS	UNIT	2006 BARE COSTS MAT.	LABOR	EQUIP.	TOTAL	TOTAL INCL O&P

Division Notes

		CREW	DAILY OUTPUT	LABOR-HOURS	UNIT	2006 BARE COSTS MAT.	LABOR	EQUIP.	TOTAL	TOTAL INCL O&P

Notes

Notes

Notes

Notes

Reed Construction Data, Inc.

Reed Construction Data, Inc., a leading worldwide provider of total construction information solutions, is comprised of three main product groups designed specifically to help construction professionals advance their businesses with timely, accurate and actionable project, product and cost data. Reed Construction Data is a division of Reed Business Information, a member of the Reed Elsevier plc group of companies.

The Project, Product and Cost & Estimating divisions offer a variety of innovative products and services designed for the full spectrum of design, construction and manufacturing professionals. Through its International companies, Reed Construction Data's reputation for quality construction market data is growing worldwide.

Cost Information
RSMeans, the undisputed market leader and authority on construction costs, publishes current cost and estimating information in annual cost books and on the *Means CostWorks* CD-ROM. RSMeans furnishes the construction industry with a rich library of corresponding reference books and a series of professional seminars that are designed to sharpen professional skills and maximize the effective use of cost estimating and management tools. RSMeans also provides construction cost consulting for owners, manufacturers, designers, and contractors.

Project Data
Reed Construction Data provides complete, accurate and relevant project information through all stages of construction. Customers are supplied industry data through leads, project reports, contact lists, plans and specifications surveys, market penetration analyses and sales evaluation reports. Any of these products can pinpoint a county, look at a state, or cover the country. Data is delivered via paper, e-mail, CD-ROM or the Internet.

Building Product Information
The First Source suite of products is the only integrated building product information system offered to the commercial construction industry for comparing and specifying building products. These print and online resources include *First Source,* CSI's SPEC-DATA™, CSI's MANU-SPEC™, First Source CAD, and Manufacturer Catalogs. Written by industry professionals and organized using CSI's MasterFormat™, construction professionals use this information to make better design decisions. FirstSourceONL.com combines Reed Construction Data's project, product and cost data with news and information from Reed Business Information's *Building Design & Construction* and *Consulting-Specifying Engineer magazines.* This industry-focused site offers easy and unlimited access to vital information for all construction professionals.

International
Reed Construction Data Canada serves the Canadian construction market with reliable and comprehensive project and product information services that cover all facets of construction. Core services include: *BuildSource, BuildSpec, BuildSelect,* product selection and specification tools available in print and on the Internet; Building Reports, a national construction project lead service; CanaData, statistical and forecasting information; *Daily Commercial News,* a construction newspaper reporting on news and projects in Ontario; and *Journal of Commerce,* reporting news in British Columbia and Alberta.

BIMSA/Mexico provides construction project news, product information, cost data, seminars and consulting services to construction professionals in Mexico. Its subsidiary, PRISMA, provides job costing software.

Byggfakta Scandinavia AB, founded in 1936, is the parent company for the leaders of customized construction market data for Denmark, Estonia, Finland, Norway and Sweden. Each company fully covers the local construction market and provides information across several platforms including via subscription, on an ad-hoc basis, electronically and on paper.

Cordell Building Information Services, with its complete range of project and cost and estimating services, is Australia's specialist in the construction information industry. Cordell provides in-depth and historical information on all aspects of construction projects and estimation, including several customized reports, construction and sales leads and detailed cost information.

For more information, please visit our Web site at www.reedconstructiondata.com.

Reed Construction Data, Inc., Corporate Office
30 Technology Parkway South
Norcross, GA 30092-2912
(800) 322-6996
(800) 895-8661 (fax)
info@reedbusiness.com
www.reedconstructiondata.com

Means Project Cost Report

By filling out this report, your project data will contribute to the database that supports the Means Project Cost Square Foot Data. When you fill out this form, Means will provide a 30% discount off one of the Means products advertised in the following pages. Please complete the form including all items where you have cost data, and all the items marked (✔).

$30.00 Discount per product for each report you submit.

Project Description (No remodeling projects, please.)

✔ Building Use (Office, School...) ____________

✔ Address (City, State) ____________

____

✔ Frame (Wood, Steel...) ____________

✔ Exterior Wall (Brick, Tilt-up...) ____________

✔ Basement: (check one) ☐ Full ☐ Partial ☐ None

✔ Number Stories ____________

✔ Floor-to-Floor Height ____________

% Air Conditioned ____________ Tons ____________

Comments ____________

Total Project Cost $ ____________

Owner ____________

Architect ____________

General Contractor ____________

✔ Bid Date ____________

Typical Bay Size ____________

✔ Labor Force: _____ % Union _____ % Non-Union

✔ Project Description (Circle one number in each line)

1. Economy 2. Average 3. Custom 4. Luxury
1. Square 2. Rectangular 3. Irregular 4. Very Irregular

A	✔	General Conditions	$
B	✔	Site Work	$
C	✔	Concrete	$
D	✔	Masonry	$
E	✔	Metals	$
F	✔	Wood & Plastics	$
G	✔	Thermal & Moisture Protection	
GR		Roofing & Flashing	$
H	✔	Doors and Windows	$
J	✔	Finishes	$
JP		Painting & Wall Covering	$

K	✔	Specialties	$
L	✔	Equipment	$
M	✔	Furnishings	$
N	✔	Special Construction	$
P	✔	Conveying Systems	$
Q	✔	Mechanical	$
QP		Plumbing	$
QB		HVAC	$
R	✔	Electrical	$
S	✔	Mech./Elec. Combined	$

Please specify the Means product you wish to receive. Complete the address information.

Product Name ____________

Product Number ____________

Your Name ____________

Title ____________

Company ____________

☐ Company

☐ Home Street Address ____________

City, State, Zip ____________

Return by mail or Fax 888-492-6770.

Method of Payment:

Credit Card # ____________

Expiration Date ____________

Check ____________

Purchase Order ____________

Reed Construction Data/RSMeans
Square Foot Costs Department
P.O. Box 800
Kingston, MA 02364-9988

For more information visit Means Web site at www.rsmeans.com

Reed Construction Data/RSMeans . . . a tradition of excellence in Construction Cost Information and Services since 1942.

Table of Contents

Book Selection Guide

The following table provides definitive information on the content of each cost data publication. The number of lines of data provided in each unit price or assemblies division, as well as the number of reference tables and crews is listed for each book. The presence of other elements such as an historical cost index, city cost indexes, square foot models or cross-referenced index is also indicated. You can use the table to help select the Means' book that has the quantity and type of information you most need in your work.

Unit Cost Divisions	Building Construction Costs	Mechanical	Electrical	Repair & Remodel.	Square Foot	Site Work Landsc.	Assemblies	Interior	Concrete Masonry	Open Shop	Heavy Construc.	Light Commercial	Facil. Construc.	Plumbing	Western Construction Costs	Resi-dential
1	554	273	358	465		496		297	465	553	502	210	979	325	553	161
2	2560	1345	460	1457		8234		543	1383	2466	5287	695	4835	1593	2543	768
3	1393	112	99	721		1250		191	1774	1388	1395	229	1315	81	1388	262
4	854	18	0	666		660		576	1075	830	597	432	1076	0	840	352
5	1876	239	193	965		817		967	738	1844	1077	835	1856	316	1816	759
6	1487	82	78	1465		450		1385	317	1477	607	1603	1574	47	1824	1734
7	1252	159	71	1240		472		485	401	1251	355	963	1306	169	1252	760
8	1893	28	0	1965		313		1682	642	1870	51	1329	2095	0	1852	1263
9	1595	47	0	1425		232		1688	360	1547	157	1337	1811	47	1586	1217
10	871	47	25	508		193		710	170	873	0	404	917	233	871	217
11	1018	322	169	502		137		813	44	925	107	218	1173	291	925	104
12	313	0	0	46		213		1422	27	304	0	65	1442	0	304	63
13	1317	1016	385	659		451		884	75	1292	340	484	1931	929	1183	194
14	345	36	0	258		36		292	0	344	30	12	362	35	343	6
15	1976	13019	636	1752		1542		1192	58	1985	1729	1225	10787	9669	2005	829
16	1279	470	10114	1002		739		1108	55	1297	760	1083	9885	415	1228	554
Totals	**20583**	**17213**	**12588**	**15096**		**16235**		**14235**	**7584**	**20246**	**12994**	**11124**	**43344**	**14150**	**20513**	**9243**

Assembly Divisions	Building Construction Costs	Mechanical	Elec-trical	Repair & Remodel.	Square Foot	Site Work Landsc.	Assemblies	Interior	Concrete Masonry	Open Shop	Heavy Construc.	Light Commercial	Facil. Construc.	Plumbing	Western Construction Costs	Asm Div	Resi-dential
A	612	19	0	192	150	540	612	0	550		542	149	24	0		1	374
B	5584	0	0	809	2479	0	5584	333	1914		0	2023	144	0		2	217
C	1208	0	0	635	853	0	1208	1555	132		0	757	238	0		3	588
D	2392	1014	780	693	1776	0	2392	752	0		0	1262	1010	890		4	867
E	294	0	0	85	257	0	294	5	0		0	257	5	0		5	393
F	124	0	0	0	123	0	124	0	0		0	123	3	0		6	357
G	584	465	172	330	111	1844	584	0	471		423	110	113	559		7	299
																8	760
																9	80
																10	0
																11	0
																12	0
Totals	**10798**	**1498**	**952**	**2744**	**5749**	**2384**	**10798**	**2645**	**3067**	**0**	**965**	**4681**	**1537**	**1449**			**3935**

Reference Section	Building Construction Costs	Mechanical	Electrical	Repair & Remodel.	Square Foot	Site Work Landsc.	Assemblies	Interior	Concrete Masonry	Open Shop	Heavy Construc.	Light Commercial	Facil. Construc.	Plumbing	Western Construction Costs	Resi-dential
Tables	237	237	237	237	4	237	219	237	237	237	237	237	237	237	237	35
Models					105							46				32
Crews	452	452	452	434		452		452	452	431	452	431	434	452	452	431
City Cost Indexes	yes	yes	yes	yes	yes	yes	yes	yes	yes	yes	yes	yes	yes	yes	yes	yes
Historical Cost Indexes	yes	yes	yes	yes	yes	yes	yes	yes	yes	yes	yes	yes	yes	yes	yes	no

Means Building Construction Cost Data 2006

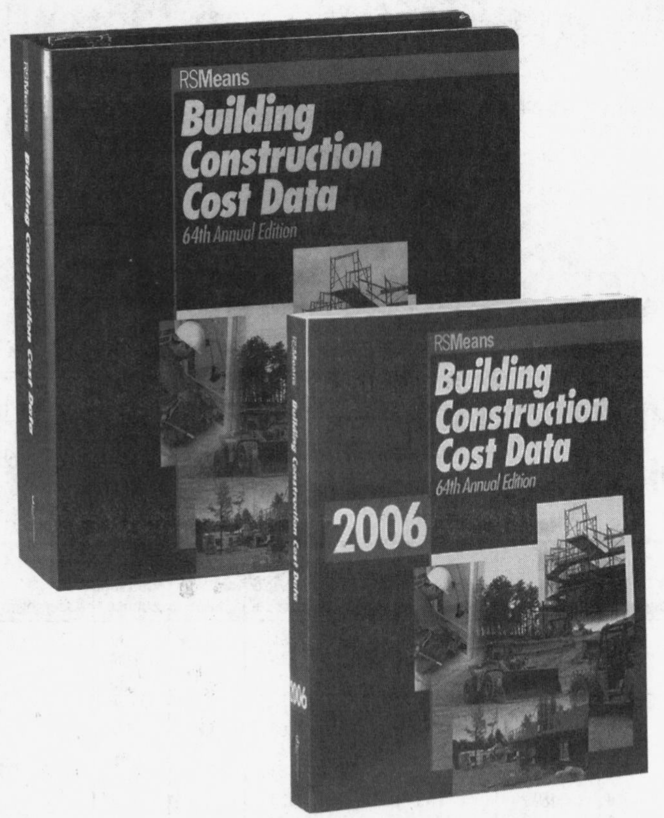

Available in Both Softbound and Looseleaf Editions

The "Bible" of the industry comes in the standard softcover edition or the looseleaf edition.

Many customers enjoy the convenience and flexibility of the looseleaf binder, which increases the usefulness of Means *Building Construction Cost Data 2006* by making it easy to add and remove pages. You can insert your own cost information pages, so everything is in one place. Copying pages for faxing is easier also. Whichever edition you prefer, softbound or the convenient looseleaf edition, you'll be eligible to receive *The Change Notice* FREE. Current subscribers receive *The Change Notice* via e-mail.

$157.95 per copy, Looseleaf
Catalog No. 61016

Means Building Construction Cost Data 2006

Offers you unchallenged unit price reliability in an easy-to-use arrangement. Whether used for complete, finished estimates or for periodic checks, it supplies more cost facts better and faster than any comparable source. Over 23,000 unit prices for 2006. The City Cost Indexes cover over 930 areas, for indexing to any project location in North America. Order and get *The Change Notice* FREE. You'll have year-long access to the Means Estimating **HOTLINE** FREE with your subscription. Expert assistance when using Means data is just a phone call away.

$126.95 per copy
Over 700 pages, illustrated, available Oct. 2005
Catalog No. 60016

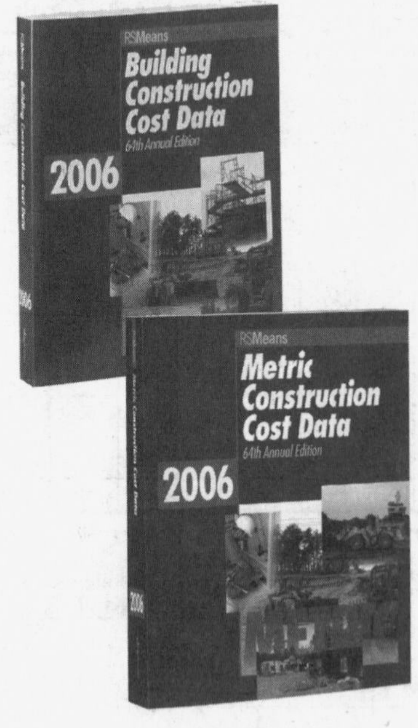

Means Metric Construction Cost Data 2006

An 880+ page compendium of all the data from both the 2006 *Building Construction Cost Data* AND the *Heavy Construction Cost Data*, in **metric** format! Access all of this vital information from one complete source. It contains more than 600 pages of unit costs and 40 pages of assemblies costs. The Reference Section contains over 200 pages of tables, charts and other estimating aids. A great way to stay in step with today's construction trends and rapidly changing costs.

$149.95 per copy
Over 800 pages, illus., available Dec. 2005
Catalog No. 63016

For more information visit Means Web site at www.rsmeans.com

Annual Cost Guides

Means Mechanical Cost Data 2006

• HVAC Controls

Total unit and systems price guidance for mechanical construction... materials, parts, fittings, and complete labor cost information. Includes prices for piping, heating, air conditioning, ventilation, and all related construction.

Plus new 2006 unit costs for:

- Over 2500 installed HVAC/controls assemblies
- "On Site" Location Factors for over 930 cities and towns in the U.S. and Canada
- Crews, labor, and equipment

$126.95 per copy
Over 600 pages, illustrated, available Oct. 2005
Catalog No. 60026

Means Plumbing Cost Data 2006

Comprehensive unit prices and assemblies for plumbing, irrigation systems, commercial and residential fire protection, point-of-use water heaters, and the latest approved materials. This publication and its companion, Means *Mechanical Cost Data*, provide full-range cost estimating coverage for all the mechanical trades.

$126.95 per copy
Over 550 pages, illustrated, available Oct. 2005
Catalog No. 60216

Means Electrical Cost Data 2006

Pricing information for every part of electrical cost planning. More than 15,000 unit and systems costs with design tables; clear specifications and drawings; engineering guides and illustrated estimating procedures; complete labor-hour and materials costs for better scheduling and procurement; and the latest electrical products and construction methods.

- A variety of special electrical systems including cathodic protection
- Costs for maintenance, demolition, HVAC/mechanical, specialties, equipment, and more

$126.95 per copy
Over 450 pages, illustrated, available Oct. 2005
Catalog No. 60036

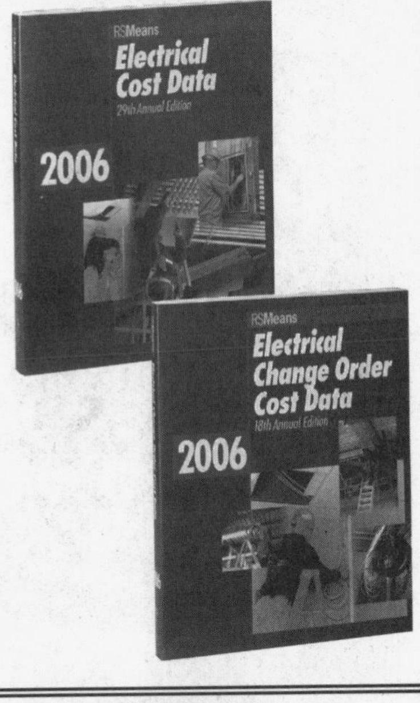

Means Electrical Change Order Cost Data 2006

Means *Electrical Change Order Cost Data* 2006 provides you with electrical unit prices exclusively for pricing change orders based on the recent, direct experience of contractors and suppliers. Analyze and check your own change order estimates against the experience others have had doing the same work. It also covers productivity analysis and change order cost justifications. With useful information for calculating the effects of change orders and dealing with their administration.

$126.95 per copy
Over 450 pages, available Oct. 2005
Catalog No. 60236

Means Square Foot Costs 2006

It's Accurate and Easy To Use!

- **Updated 2006 price information**, based on nationwide figures from suppliers, estimators, labor experts, and contractors
- "How-to-Use" sections, with **clear examples** of commercial, residential, industrial, and institutional structures
- Realistic graphics, offering true-to-life illustrations of building projects
- Extensive information on using square foot cost data, including sample estimates and alternate pricing methods

$137.95 per copy
Over 450 pages, illustrated, available Nov. 2005
Catalog No. 60056

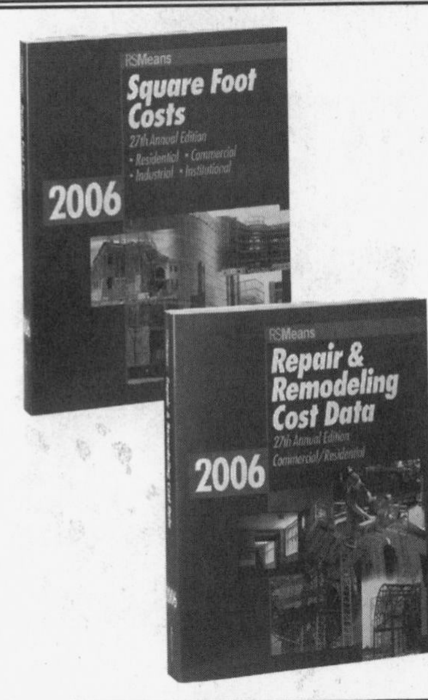

Means Repair & Remodeling Cost Data 2006

Commercial/Residential

Use this valuable tool to estimate commercial and residential renovation and remodeling.

Includes: New costs for hundreds of unique methods, materials, and conditions that only come up in repair and remodeling. PLUS:

- Unit costs for over 16,000 construction components
- Installed costs for over 90 assemblies
- Costs for 300+ construction crews
- Over 930 "On Site" localization factors for the U.S. and Canada.

$107.95 per copy
Over 650 pages, illustrated, available Nov. 2005
Catalog No. 60046

Annual Cost Guides

For more information visit Means Web site at www.rsmeans.com

Means Facilities Construction Cost Data 2006

For the maintenance and construction of commercial, industrial, municipal, and institutional properties. Costs are shown for new and remodeling construction and are broken down into materials, labor, equipment, overhead, and profit. Special emphasis is given to sections on mechanical, electrical, furnishings, site work, building maintenance, finish work, and demolition.

More than 45,000 unit costs, plus assemblies and reference sections are included.

$299.95 per copy
Over 1200 pages, illustrated, available Nov. 2005
Catalog No. 60206

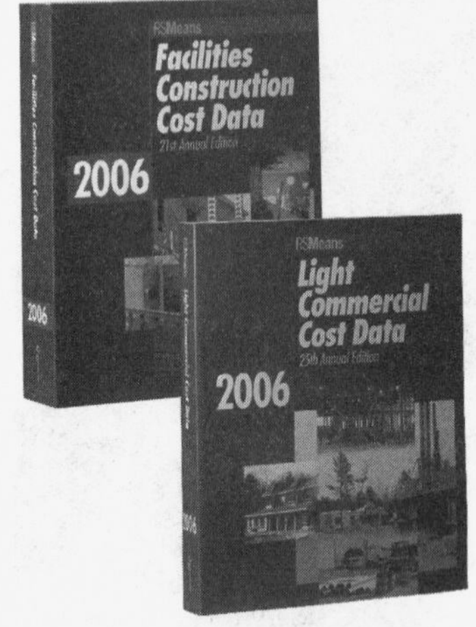

Means Light Commercial Cost Data 2006

Specifically addresses the light commercial market, which is an increasingly specialized niche in the industry. Aids you, the owner/designer/contractor, in preparing all types of estimates, from budgets to detailed bids. Includes new advances in methods and materials.

Assemblies section allows you to evaluate alternatives in the early stages of design/planning.

Over 13,000 unit costs for 2006 ensure you have the prices you need... when you need them.

$107.95 per copy
Over 650 pages, illustrated, available Nov. 2005
Catalog No. 60186

Means Residential Cost Data 2006

Contains square foot costs for 30 basic home models with the look of today, plus hundreds of custom additions and modifications you can quote right off the page. With costs for the 100 residential systems you're most likely to use in the year ahead. Complete with blank estimating forms, sample estimates and step-by-step instructions.

$107.95 per copy
Over 600 pages, illustrated, available Oct. 2005
Catalog No. 60176

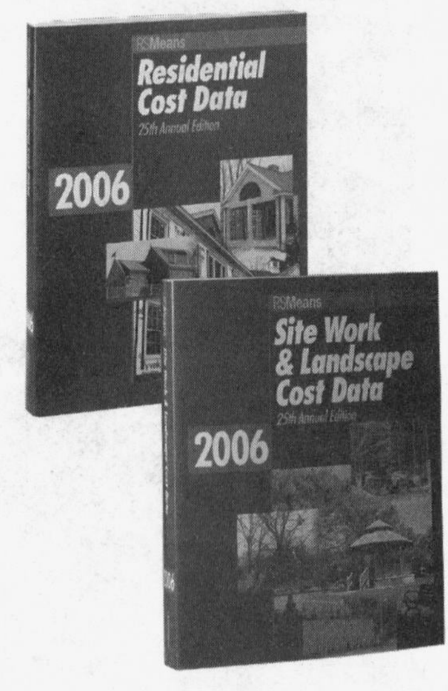

Means Site Work & Landscape Cost Data 2006

Means *Site Work & Landscape Cost Data* 2006 is organized to assist you in all your estimating needs. Hundreds of fact-filled pages help you make accurate cost estimates efficiently.

Updated for 2006!

- Demolition features—including ceilings, doors, electrical, flooring, HVAC, millwork, plumbing, roofing, walls, and windows
- State-of-the-art segmental retaining walls
- Flywheel trenching costs and details
- Updated wells section
- Landscape materials, flowers, shrubs, and trees

$126.95 per copy
Over 600 pages, illustrated, available Nov. 2005
Catalog No. 60286

Means Assemblies Cost Data 2006

Means *Assemblies Cost Data* 2006 takes the guesswork out of preliminary or conceptual estimates. Now you don't have to try to calculate the assembled cost by working up individual components costs. We've done all the work for you.

Presents detailed illustrations, descriptions, specifications and costs for every conceivable building assembly—240 types in all—arranged in the easy-to-use UNIFORMAT II system. Each illustrated "assembled" cost includes a complete grouping of materials and associated installation costs, including the installing contractor's overhead and profit.

$206.95 per copy
Over 600 pages, illustrated, available Oct. 2005
Catalog No. 60066

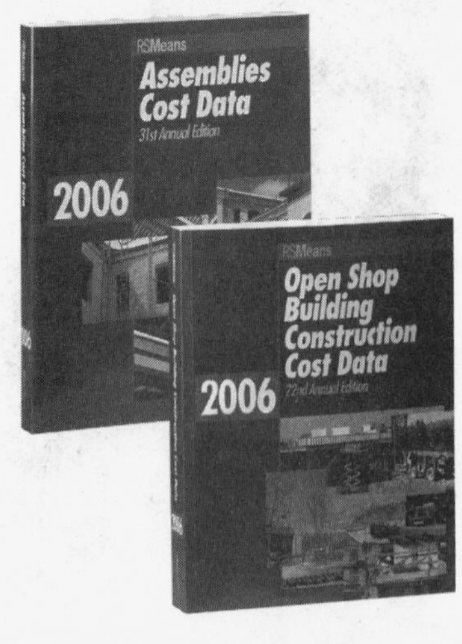

Means Open Shop Building Construction Cost Data 2006

The latest costs for accurate budgeting and estimating of new commercial and residential construction... renovation work... change orders... cost engineering.

Means *Open Shop BCCD* will assist you to:

- Develop benchmark prices for change orders
- Plug gaps in preliminary estimates, and budgets
- Estimate complex projects
- Substantiate invoices on contracts
- Price ADA-related renovations

$126.95 per copy
Over 700 pages, illustrated, available Dec. 2005
Catalog No. 60156

For more information visit Means Web site at www.rsmeans.com

Annual Cost Guides

Means Building Construction Cost Data 2006

Western Edition

This regional edition provides more precise cost information for western North America. Labor rates are based on union rates from 13 western states and western Canada. Included are western practices and materials not found in our national edition: tilt-up concrete walls, glu-lam structural systems, specialized timber construction, seismic restraints, and landscape and irrigation systems.

$126.95 per copy
Over 650 pages, illustrated, available Dec. 2005
Catalog No. 60226

Means Heavy Construction Cost Data 2006

A comprehensive guide to heavy construction costs. Includes costs for highly specialized projects such as tunnels, dams, highways, airports, and waterways. Information on different labor rates, equipment, and material costs is included. Features unit price costs, systems costs, and numerous reference tables for costs and design. Valuable not only to contractors and civil engineers, but also to government agencies and city/town engineers.

$126.95 per copy, Over 450 pages, illustrated, available Nov. 2005
Catalog No. 60166

Means Construction Cost Indexes 2006

Who knows what 2006 holds? What materials and labor costs will change unexpectedly? By how much?

- Breakdowns for 316 major cities
- National averages for 30 key cities
- Expanded five major city indexes
- Historical construction cost indexes

$272.00 per year
$68.00 individual quarters
Catalog No. 60146 A,B,C,D

Means Interior Cost Data 2006

Provides you with prices and guidance needed to make accurate interior work estimates. Contains costs on materials, equipment, hardware, custom installations, furnishings, labor costs... every cost factor for new and remodel commercial and industrial interior construction, including updated information on office furnishings, plus more than 50 reference tables. For contractors, facility managers, and owners.

$126.95 per copy, Over 600 pages, illustrated, available Oct. 2005
Catalog No. 60096

Means Concrete & Masonry Cost Data 2006

Provides you with cost facts for virtually all concrete/masonry estimating needs, from complicated formwork to various sizes and face finishes of brick and block, all in great detail. The comprehensive unit cost section contains more than 8,500 selected entries. Also contains an assemblies cost section, and a detailed reference section that supplements the cost data.

$114.95 per copy
Over 450 pages, illustrated, available Dec. 2005
Catalog No. 60116

Means Labor Rates for the Construction Industry 2006

Complete information for estimating labor costs, making comparisons, and negotiating wage rates by trade for over 300 U.S. and Canadian cities. With 46 construction trades listed by local union number in each city, and historical wage rates included for comparison. No similar book is available through the trade.

Each city chart lists the county and is alphabetically arranged with handy visual flip tabs for quick reference.

$274.95 per copy
Over 300 pages, available Dec. 2005
Catalog No. 60126

Means Facilities Maintenance & Repair Cost Data 2006

Means *Facilities Maintenance & Repair Cost Data* gives you a complete system to manage and plan your facility repair and maintenance costs and budget efficiently. Guidelines for auditing a facility and developing an annual maintenance plan. Budgeting is included, along with reference tables on cost and management and information on frequency and productivity of maintenance operations.

The only nationally recognized source of maintenance and repair costs. Developed in cooperation with the Army Corps of Engineers.

$274.95 per copy
Over 600 pages, illustrated, available Dec. 2005
Catalog No. 60306

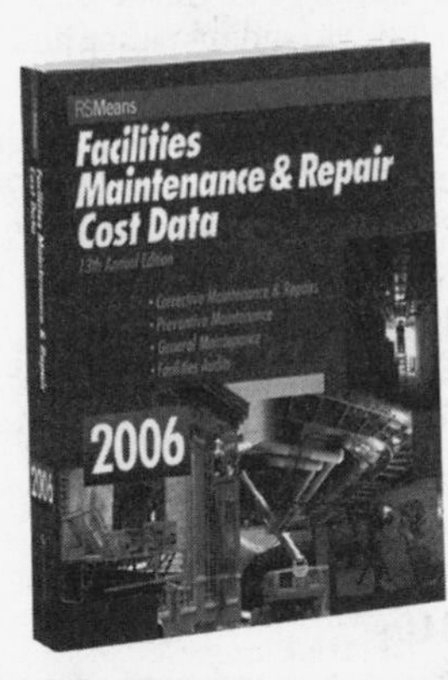

For more information
visit Means Web site
at www.rsmeans.com

Builder's Essentials: Estimating Building Costs for the Residential & Light Commercial Contractor

By Wayne J. DelPico

Step-by-step estimating methods for residential and light commercial contractors. Includes a detailed look at every construction specialty—explaining all the components, takeoff units, and labor needed for well-organized, complete estimates. Covers:

- Correctly interpreting plans and specifications
- Developing accurate and complete labor and material costs
- Understanding direct and indirect overhead costs... and accounting for time-sensitive costs
- Using historical cost data to generate new project budgets
- Allocating the right amount for profit and contingencies

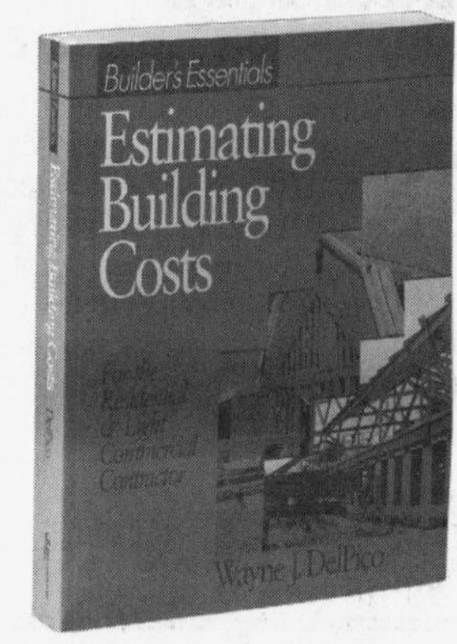

$29.95 per copy
Over 400 pages, illustrated, Softcover
Catalog No. 67343

Green Building: Project Planning & Cost Estimating

By RSMeans and Contributing Authors

Written by a team of leading experts in sustainable design, this book is a complete guide to planning and estimating green building projects, a growing trend in building design and construction. It explains:

- All the different criteria for "green-ness"
- What criteria your building needs to meet to get a LEED, ENERGYSTAR®, or other recognized rating for green buildings
- How the project team works differently on a green versus a traditional building project
- How to select and specify green products
- How to evaluate the cost and value of green products versus conventional ones—not only for their initial installation cost, but their cost over time in maintenance and operation

Features an extensive Green Building Cost Data section

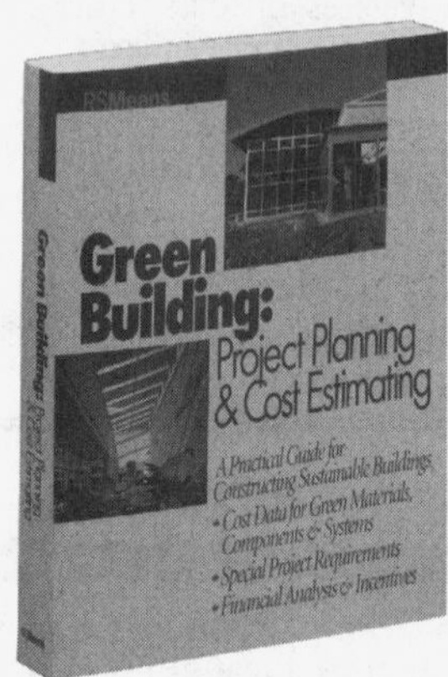

$89.95 per copy
350 pages, illustrated, Hardcover
Catalog No. 67338

Means ADA Compliance Pricing Guide

New Second Edition

By Adaptive Environments and RSMeans

Completely updated and revised to the new 2004 *Americans with Disabilities Act Accessibility Guidelines*, this book features more than 70 of the most commonly needed modifications for ADA compliance—their design requirements, suggestions, and final cost. Projects range from installing ramps and walkways, widening doorways and entryways, and installing and refitting elevators, to relocating light switches and signage, and remodeling bathrooms and kitchens. Also provided are:

- Detailed cost estimates for budgeting modification projects, including estimates for each of 260 alternates.
- An assembly estimate for every project, with detailed cost breakdown including materials, labor-hours, and contractor's overhead.
- 3,000 additional ADA compliance-related unit cost line items.
- Costs that are easily adjusted to over 900 cities and towns.

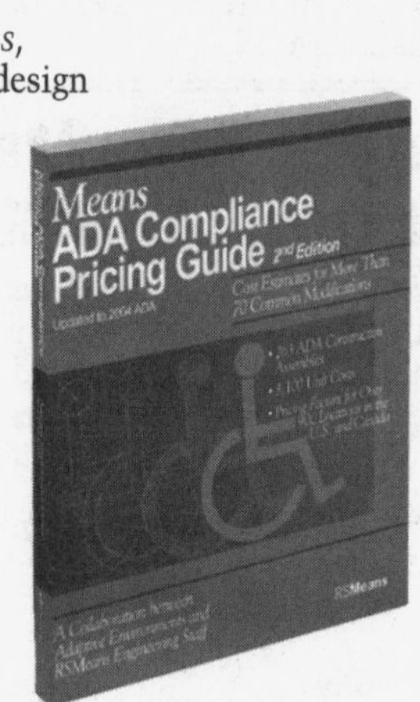

$79.95 per copy
Over 350 pages, illustrated, Softcover
Catalog No. 67310A

For more information
visit Means Web site
at www.rsmeans.com

Reference Books

Value Engineering: Practical Applications . . . For Design, Construction, Maintenance & Operations

By Alphonse Dell'Isola, PE

A tool for immediate application—for engineers, architects, facility managers, owners, and contractors. Includes: making the case for VE—the management briefing, integrating VE into planning, budgeting, and design, conducting life cycle costing, using VE methodology in design review and consultant selection, case studies, VE workbook, and a life cycle costing program on disk.

$79.95 per copy
Over 450 pages, illustrated, Softcover
Catalog No. 67319

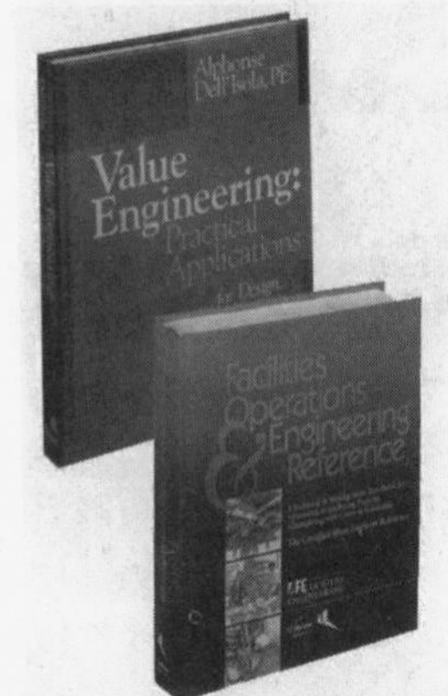

Facilities Operations & Engineering Reference

By the Association for Facilities Engineering and RSMeans

An all-in-one technical referance for planning and managing facility projects and solving day-to-day operations problems. Selected as the official Certified Plant Engineer reference, this handbook covers financial analysis, maintenance, HVAC and energy efficiency, and more.

$109.95 per copy
Over 700 pages, illustrated, Hardcover
Catalog No. 67318

The Building Professional's Guide to Contract Documents

3rd Edition

By Waller S. Poage, AIA, CSI, CVS

A comprehensive reference for owners, design professionals, contractors, and students.

- Structure your documents for maximum efficiency.
- Effectively communicate construction requirements.
- Understand the roles and responsibilities of construction professionals.
- Improve methods of project delivery.

$32.48 per copy, 400 pages
Diagrams and construction forms, Hardcover
Catalog No. 67261A

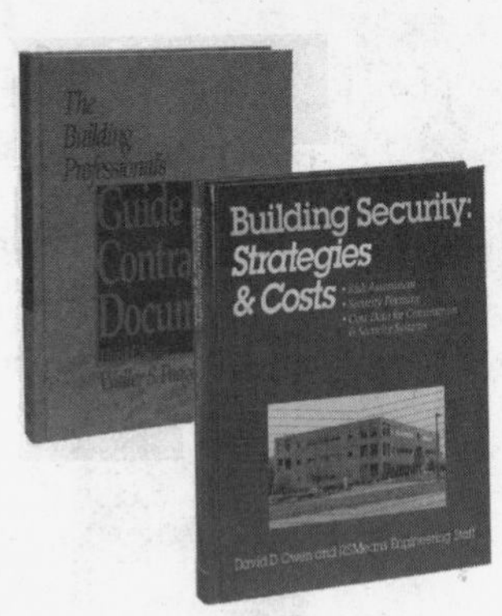

Building Security: Strategies & Costs

By David Owen

This comprehensive resource will help you evaluate your facility's security needs, and design and budget for the materials and devices needed to fulfill them.

Includes over 130 pages of Means cost data for installation of security systems and materials, plus a review of more than 50 security devices and construction solutions.

$44.98 per copy
350 pages, illustrated, Hardcover
Catalog No. 67339

Cost Planning & Estimating for Facilities Maintenance

In this unique book, a team of facilities management authorities shares their expertise at:

- Evaluating and budgeting maintenance operations
- Maintaining and repairing key building components
- Applying Means *Facilities Maintenance & Repair Cost Data* to your estimating

Covers special maintenance requirements of the ten major building types.

$89.95 per copy
Over 475 pages, Hardcover
Catalog No. 67314

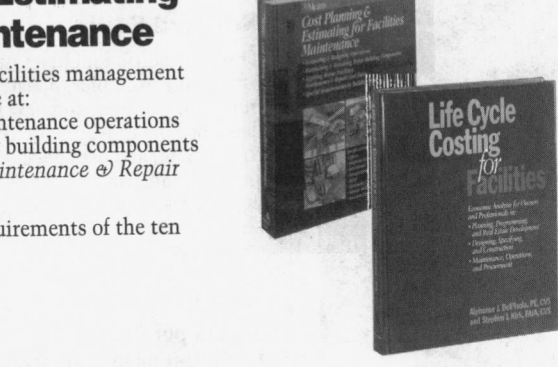

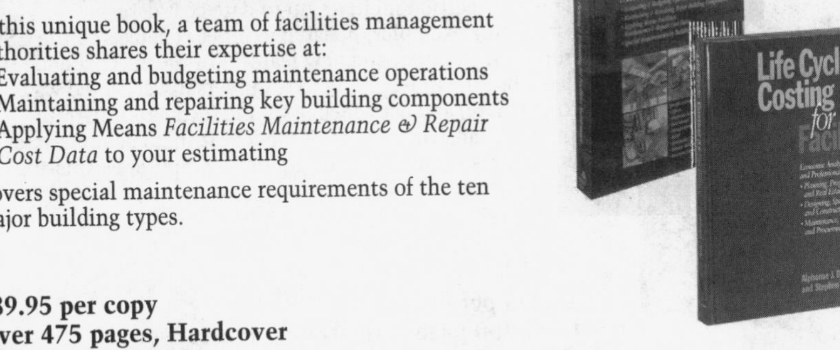

Life Cycle Costing for Facilities

By Alphonse Dell'Isola and Dr. Steven Kirk

Guidance for achieving higher quality design and construction projects at lower costs! Cost-cutting efforts often sacrifice quality to yield the cheapest product. Life cycle costing enables building designers and owners to achieve both. The authors of this book show how LCC can work for a variety of projects – from roads to HVAC upgrades to different types of buildings.

$99.95 per copy
450 pages, Hardcover
Catalog No. 67341

Building & Renovating Schools

This all-inclusive guide covers every step of the school construction process—from initial planning, needs assessment, and design, right through moving into the new facility. A must-have resource for anyone concerned with new school construction or renovation, including architects and engineers, contractors and project managers, facility managers, school administrators and school board members, building committees, community leaders, and anyone else who wants to ensure that the project meets schools' needs in a cost-effective, timely manner.
With square foot cost models for elementary, middle, and high school facilities and real-life case studies of recently completed school projects.

The contributors to this book—architects, construction project managers, contractors, and estimators who specialize in school construction—provide start-to-finish, expert guidance on the process.

$99.95 per copy
Over 425 pages, Hardcover
Catalog No. 67342

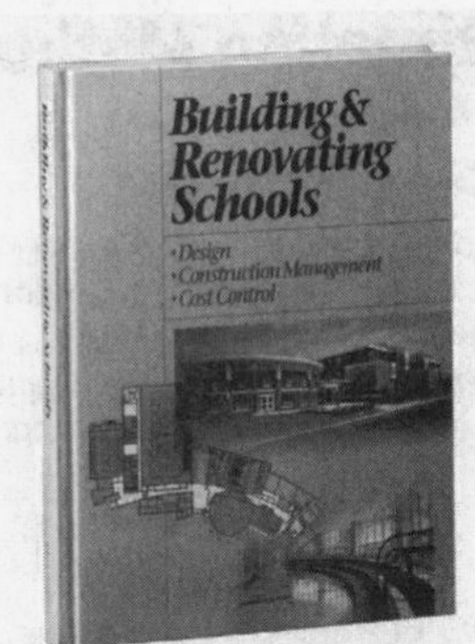

Interior Home Improvement Costs, New 9th Edition

Estimates for the most popular remodeling and repair projects—from small, do-it-yourself jobs to major renovations and new construction. Includes: Kitchens & Baths; New Living Space from your Attic, Basement, or Garage; New Floors, Paint, and Wallpaper; Tearing Out or Building New Walls; Closets, Stairs, and Fireplaces; New Energy-Saving Improvements, Home Theaters, and More!

$24.95 per copy
250 pages, illustrated, Softcover
Catalog No. 67308E

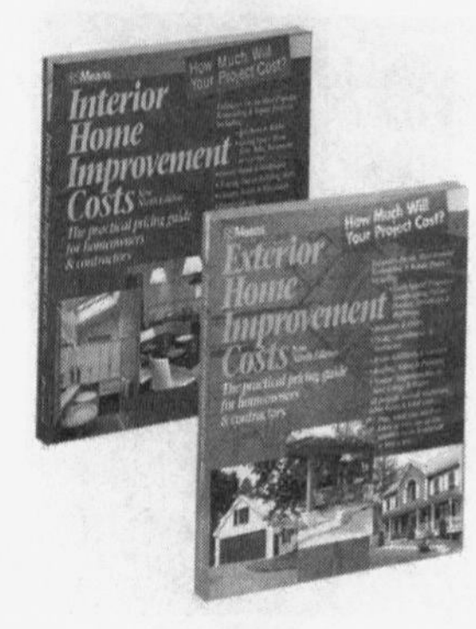

Exterior Home Improvement Costs New 9th Edition

Estimates for the most popular remodeling and repair projects—from small, do-it-yourself jobs, to major renovations and new construction. Includes: Curb Appeal Projects—Landscaping, Patios, Porches, Driveways, and Walkways; New Windows and Doors; Decks, Greenhouses, and Sunrooms; Room Additions and Garages; Roofing, Siding, and Painting; "Green" Improvements to Save Energy & Water.

$24.95 per copy
Over 275 pages, illustrated, Softcover
Catalog No. 67309E

Builder's Essentials: Plan Reading & Material Takeoff

By Wayne J. DelPico

For Residential and Light Commercial Construction

A valuable tool for understanding plans and specs, and accurately calculating material quantities. Step-by-step instructions and takeoff procedures based on a full set of working drawings.

$35.95 per copy
Over 420 pages, Softcover
Catalog No. 67307

Builder's Essentials: Framing & Rough Carpentry, 2nd Edition

By Scot Simpson

Develop and improve your skills with easy-to-follow instructions and illustrations. Learn proven techniques for framing walls, floors, roofs, stairs, doors, and windows. Updated guidance on standards, building codes, safety requirements, and more. Also available in Spanish!

$24.95 per copy
Over 150 pages, Softcover
Catalog No. 67298A
Spanish Catalog No. 67298AS

Builder's Essentials: Best Business Practices for Builders & Remodelers

An Easy-to-Use Checklist System

By Thomas N. Frisby

A comprehensive guide covering all aspects of running a construction business, with more than 40 user-friendly checklists. Provides expert guidance on: increasing your revenue and keeping more of your profit, planning for long-term growth, keeping good employees, and managing subcontractors.

$29.95 per copy
Over 220 pages, Softcover
Catalog No. 67329

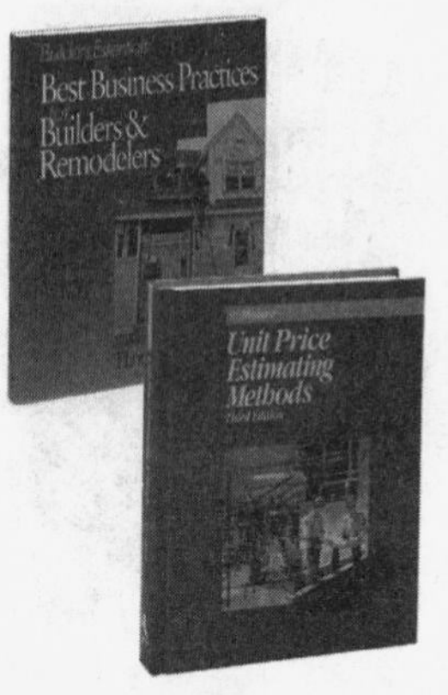

Unit Price Estimating Methods

New 3rd Edition

This new edition includes up-to-date cost data and estimating examples, updated to reflect changes to the CSI numbering system and new features of Means cost data. It describes the most productive, universally accepted ways to estimate, and uses checklists and forms to illustrate shortcuts and timesavers. A model estimate demonstrates procedures. A new chapter explores computer estimating alternatives.

$59.95 per copy
Over 350 pages, illustrated, Hardcover
Catalog No. 67303A

Means Landscape Estimating Methods

4th Edition

By Sylvia H. Fee

This revised edition offers expert guidance for preparing accurate estimates for new landscape construction and grounds maintenance. Includes a complete project estimate featuring the latest equipment and methods, and chapters on *Life Cycle Costing* and *Landscape Maintenance Estimating*.

$62.95 per copy
Over 300 pages, illustrated, Hardcover
Catalog No. 67295B

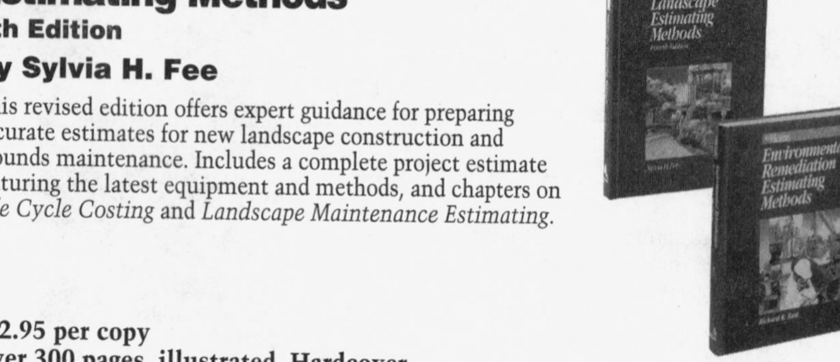

Means Environmental Remediation Estimating Methods, 2nd Edition

By Richard R. Rast

Guidelines for estimating 50 standard remediation technologies. Use it to prepare preliminary budgets, develop estimates, compare costs and solutions, estimate liability, review quotes, negotiate settlements.

A valuable support tool for Means Environmental Remediation Unit Price and Assemblies books.

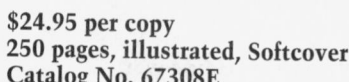

$99.95 per copy
Over 750 pages, illustrated, Hardcover
Catalog No. 64777A

For more information
visit Means Web site
at www.rsmeans.com

Reference Books

Means Illustrated Construction Dictionary, Condensed, 2nd Edition

By RSMeans

Recognized in the industry as the best resource of its kind. This essential tool has been further enhanced with updates to existing terms and the addition of hundreds of new terms and illustrations—in keeping with recent developments. For contractors, architects, insurance and real estate personnel, homeowners, and anyone who needs quick, clear definitions for construction terms.

$59.95 per copy
Over 500 pages, Softcover
Catalog No. 67282A

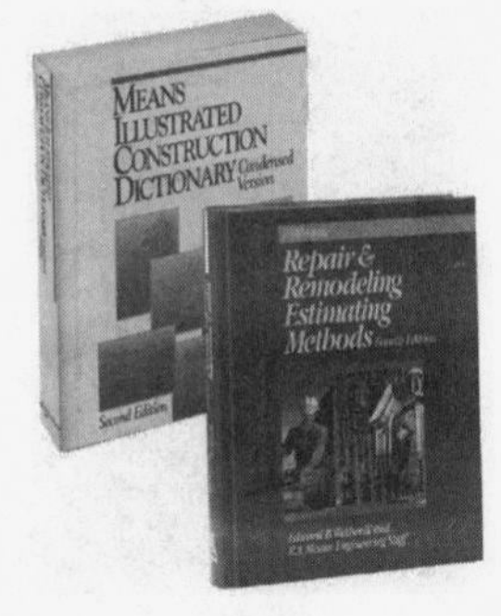

Means Repair & Remodeling Estimating New 4th Edition

By Edward B. Wetherill & RSMeans

This important reference focuses on the unique problems of estimating renovations of existing structures and helps you determine the true costs of remodeling through careful evaluation of architectural details and a site visit.

New section on disaster restoration costs.

$69.95 per copy
Over 450 pages, illustrated, Hardcover
Catalog No. 67265B

Facilities Planning & Relocation

New, lower price and user-friendly format.

By David D. Owen

A complete system for planning space needs and managing relocations. Includes step-by-step manual, over 50 forms, and extensive reference section on materials and furnishings.

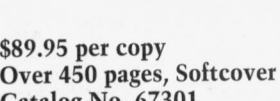

$89.95 per copy
Over 450 pages, Softcover
Catalog No. 67301

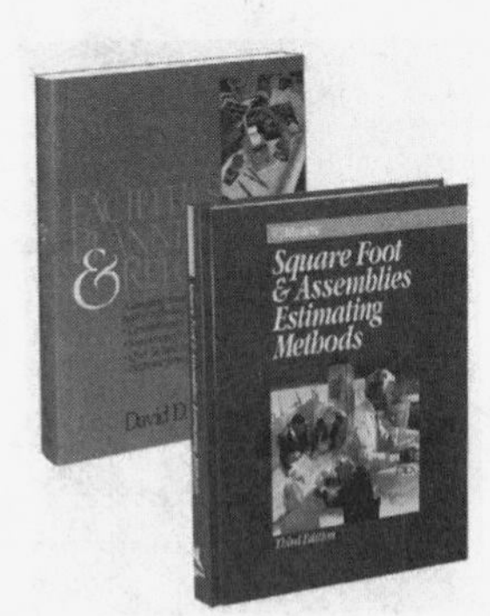

Means Square Foot & Assemblies Estimating Methods, 3rd Edition

Develop realistic square foot and assemblies costs for budgeting and construction funding. The new edition features updated guidance on square foot and assemblies estimating using UNIFORMAT II. An essential reference for anyone who performs conceptual estimates.

$69.95 per copy
Over 300 pages, illustrated, Hardcover
Catalog No. 67145B

Means Electrical Estimating Methods 3rd Edition

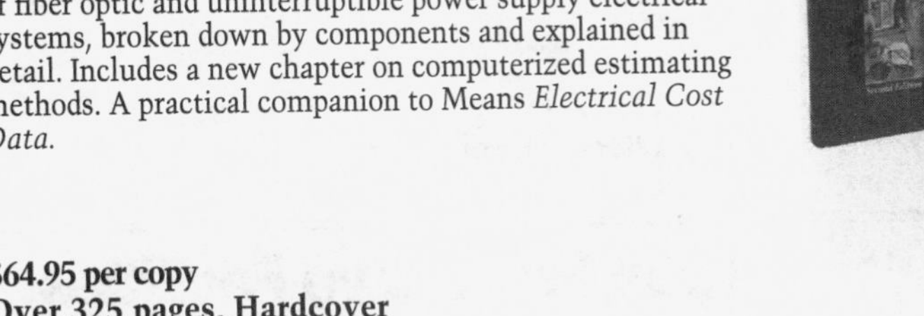

Expanded new edition includes sample estimates and cost information in keeping with the latest version of the CSI MasterFormat and UNIFORMAT II. Complete coverage of fiber optic and uninterruptible power supply electrical systems, broken down by components and explained in detail. Includes a new chapter on computerized estimating methods. A practical companion to Means *Electrical Cost Data*.

$64.95 per copy
Over 325 pages, Hardcover
Catalog No. 67230B

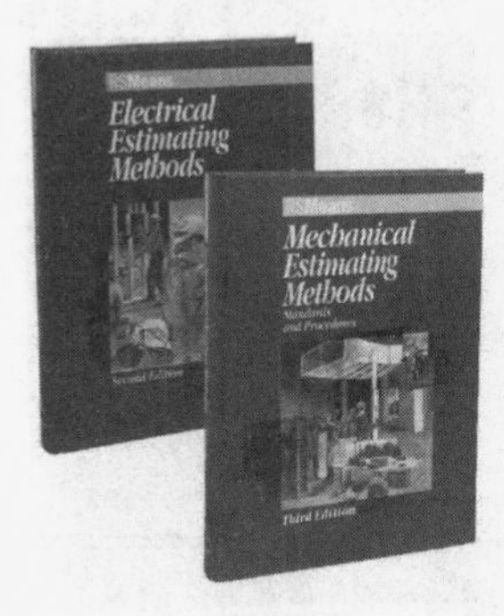

Means Mechanical Estimating Methods 3rd Edition

This guide assists you in making a review of plans, specs, and bid packages with suggestions for takeoff procedures, listings, substitutions and pre-bid scheduling. Includes suggestions for budgeting labor and equipment usage. Compares materials and construction methods to allow you to select the best option.

$64.95 per copy
Over 350 pages, illustrated, Hardcover
Catalog No. 67294A

Means Spanish/English Construction Dictionary

By RSMeans, The International Conference of Building Officials (ICBO), and Rolf Jensen & Associates (RJA)

Designed to facilitate communication among Spanish- and English-speaking construction personnel—improving performance and job-site safety. Features the most common words and phrases used in the construction industry, with easy-to-follow pronunciations. Includes extensive building systems and tools illustrations.

$22.95 per copy
250 pages, illustrated, Softcover
Catalog No. 67327

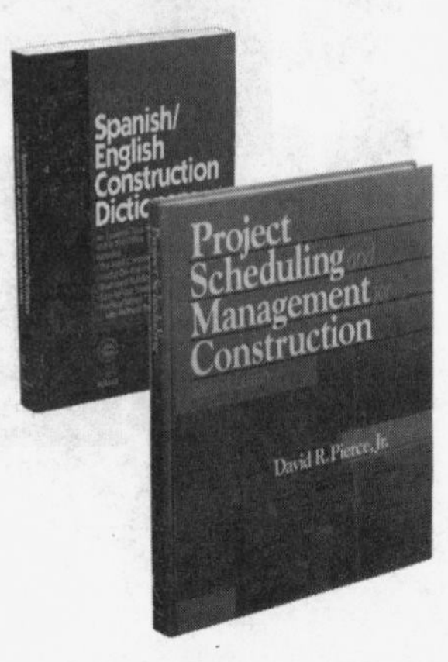

Project Scheduling & Management for Construction

New 3rd Edition
By David R. Pierce, Jr.

A comprehensive yet easy-to-follow guide to construction project scheduling and control—from vital project management principles through the latest scheduling, tracking, and controlling techniques. The author is a leading authority on scheduling with years of field and teaching experience at leading academic institutions. Spend a few hours with this book and come away with a solid understanding of this essential management topic.

$64.95 per copy
Over 300 pages, illustrated, Hardcover
Catalog No. 67247B

Reference Books

For more information visit Means Web site at www.rsmeans.com

Concrete Repair and Maintenance Illustrated

By Peter Emmons

Hundreds of illustrations show users how to analyze, repair, clean, and maintain concrete structures for optimal performance and cost effectiveness. From parking garages to roads and bridges to structural concrete, this comprehensive book describes the causes, effects, and remedies for concrete wear and failure. Invaluable for planning jobs, selecting materials, and training employees, this book is a must-have for concrete specialists, general contractors, facility managers, civil and structural engineers, and architects.

$34.98 per copy
300 pages, illustrated, Softcover
Catalog No. 67146

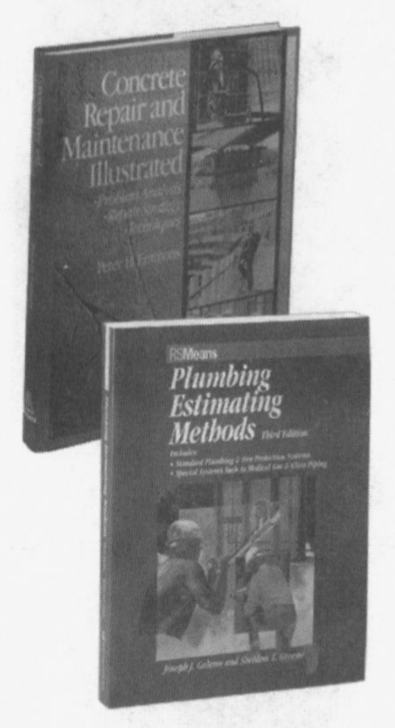

Plumbing Estimating Methods 3rd Edition

By Joseph Galeno and Sheldon Greene

Updated and revised! This practical guide walks you through a plumbing estimate, from basic materials and installation methods through change order analysis. *Plumbing Estimating Methods* covers residential, commercial, industrial, and medical systems, and features sample takeoff and estimate forms and detailed illustrations of systems and components.

$29.98 per copy
330+ pages, Softcover
Catalog No. 67283B

Building Spec Homes Profitably

By Kenneth V. Johnson

Professional guidance from a veteran spec home builder… This complete system covers all aspects of spec home building—from market research and site selection to financing and design to scheduling and supervision, and more!

You'll learn to reduce risk and ensure profits in spec home building, no matter what the economic climate.

$29.95 per copy
200 pages, Softcover
Catalog No. 67312

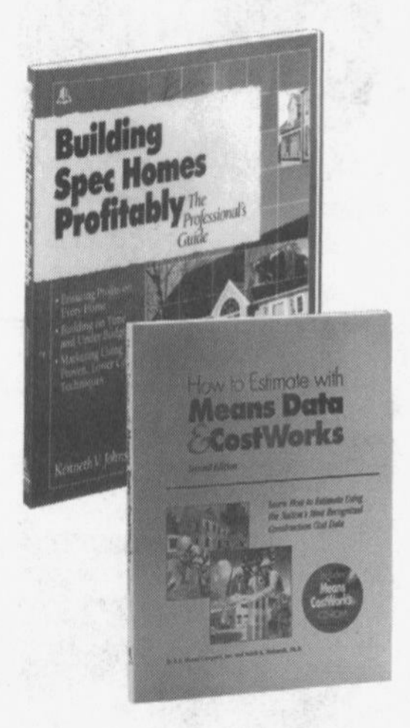

How to Estimate with Means Data & CostWorks

New 2nd Edition

By RSMeans and Saleh Mubarak

Learn estimating techniques using Means cost data. Includes an instructional version of Means *CostWorks* CD–ROM with sample building plans. The step-by-step guide takes you through all the major construction items. Over 300 sample estimating problems are included.

$59.95 per copy
Over 190 pages, Softcover
Catalog No. 67324A

Means Productivity Standards for Construction

3rd Edition

By RSMeans

Completely updated, with well over 3,000 new work items added to cover topics in current demand! This comprehensive reference for the productivity of construction crews, equipment, and labor includes precise descriptions of each construction task organized according to the Construction Specifications Institute's MasterFormat.

$49.98 per copy
800+ pages, Hardcover
Catalog No. 67236A

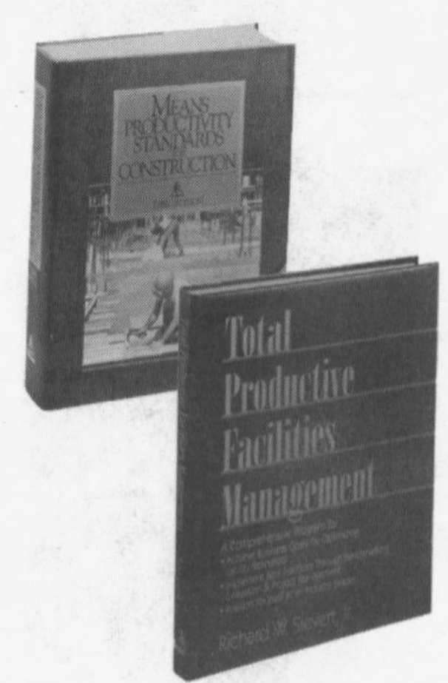

Total Productive Facilities Management

By Richard W. Sievert, Jr.

Today, facilities are viewed as strategic resources... elevating the facility manager to the role of asset manager supporting the organization's overall business goals. Now, Richard Sievert Jr., in this well-articulated guidebook, sets forth a new operational standard for the facility manager's emerging role... a comprehensive program for managing facilities as a true profit center.

$29.98 per copy
275 pages, Softcover
Catalog No. 67321

For more information visit Means Web site at www.rsmeans.com

Reference Books

Preventive Maintenance Guidelines for School Facilities

By John C. Maciha

A complete PM program for K-12 schools that ensures sustained security, safety, property integrity, user satisfaction, and reasonable ongoing expenditures.

Includes schedules for weekly, monthly, semiannual, and annual maintenance with hard copy and electronic forms.

$149.95 per copy
Over 225 pages, Hardcover
Catalog No. 67326

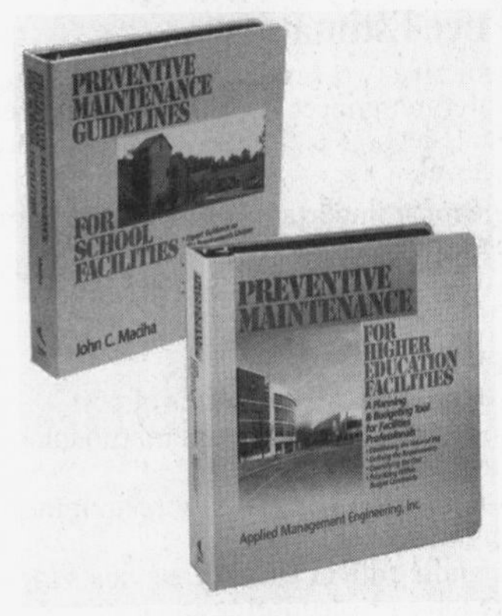

Preventive Maintenance for Higher Education Facilities

By Applied Management Engineering, Inc.

An easy-to-use system to help facilities professionals establish the value of PM, and to develop and budget for an appropriate PM program for their college or university. Features interactive campus building models typical of those found in different-sized higher education facilities, and PM checklists linked to each piece of equipment or system in hard copy and electronic format.

$149.95 per copy
150 pages, Hardcover
Catalog No. 67337

Historic Preservation: Project Planning & Estimating

By Swanke Hayden Connell Architects

Expert guidance on managing historic restoration, rehabilitation, and preservation building projects and determining and controlling their costs. Includes:

- How to determine whether a structure qualifies as historic
- Where to obtain funding and other assistance
- How to evaluate and repair more than 75 historic building materials

$99.95 per copy
Over 675 pages, Hardcover
Catalog No. 67323

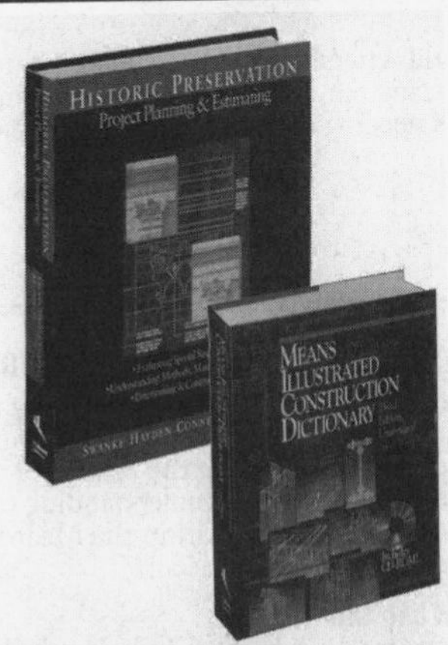

Means Illustrated Construction Dictionary

Unabridged 3rd Edition

Long regarded as the Industry's finest, the Means Illustrated Construction Dictionary is now even better. With the addition of over 1,000 new terms and hundreds of new illustrations, it is the clear choice for the most comprehensive and current information. The companion CD-ROM that comes with this new edition adds many extra features: larger graphics, expanded definitions, and links to both CSI MasterFormat numbers and product information.

$99.95 per copy
Over 790 pages, illustrated, Hardcover
Catalog No. 67292A

Designing & Building with the IBC, 2nd Edition

By Rolf Jensen & Associates, Inc.

This updated comprehensive guide helps building professionals make the transition to the 2003 International Building Code®. Includes a side-by-side code comparison of the IBC 2003 to the IBC 2000 and the three primary model codes, a quick-find index, and professional code commentary. With illustrations, abbreviations key, and an extensive Resource section.

$99.95 per copy
Over 875 pages, Softcover
Catalog No. 67328A

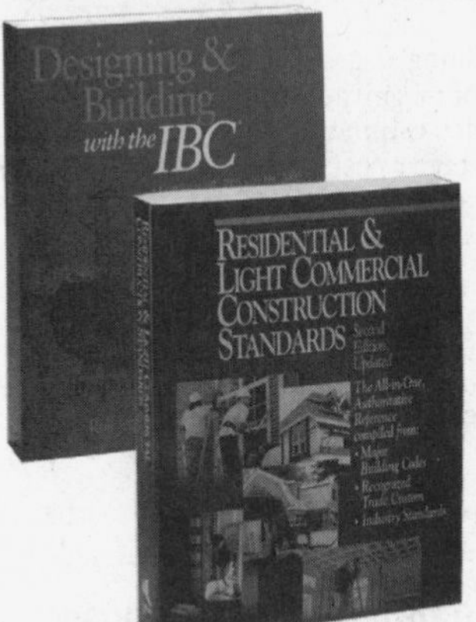

Residential & Light Commercial Construction Standards, 2nd Edition

By RSMeans and Contributing Authors

For contractors, subcontractors, owners, developers, architects, engineers, attorneys, and insurance personnel, this book provides authoritative requirements and recommendations compiled from the nation's leading professional associations, industry publications, and building code organizations.

$59.95 per copy
600 pages, illustrated, Softcover
Catalog No. 67322A

Builder's Essentials: Advanced Framing Methods

By Scot Simpson

A highly illustrated, "framer-friendly" approach to advanced framing elements. Provides expert, but easy to interpret, instruction for laying out and framing complex walls, roofs, and stairs, and special requirements for earthquake and hurricane protection. Also helps bring framers up to date on the latest building code changes, and provides tips on the lead framer's role and responsibilities, how to prepare for a job, and how to get the crew started.

$24.95 per copy
250 pages, illustrated, Softcover
Catalog No. 67330

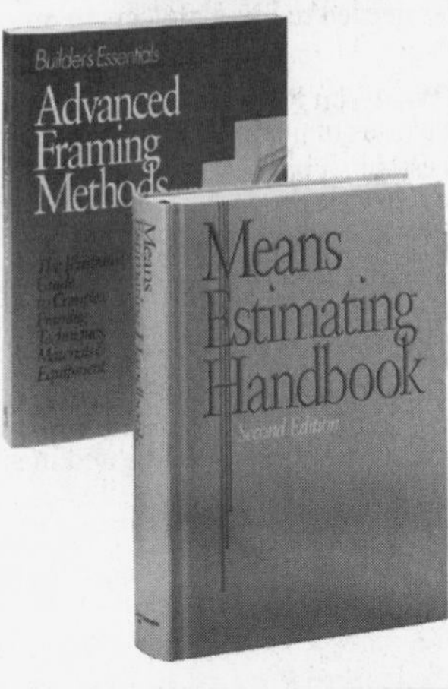

Means Estimating Handbook

2nd Edition

By RSMeans

Updated new Second Edition answers virtually any estimating technical question—all organized by CSI MasterFormat. This comprehensive reference covers the full spectrum of technical data required to estimate construction costs. The book includes information on sizing, productivity, equipment requirements, code-mandated specifications, design standards, and engineering factors.

$99.95 per copy
Over 900 pages, Hardcover
Catalog No. 67276A

Seminars

Means CostWorks Training

This one-day seminar course has been designed with the intention of assisting both new and existing users to become more familiar with the *CostWorks* program. The class is broken into two unique sections: (1) A one-half day presentation on the function of each icon; and each student will be shown how to use the software to develop a cost estimate. (2) Hands-on estimating exercises that will ensure that each student thoroughly understands how to use *CostWorks*. You must bring your own laptop computer to this course.

CostWorks Benefits/Features:
- Estimate in your own spreadsheet format
- Power of Means National Database
- Database automatically regionalized
- Save time with keyword searches
- Save time by establishing common estimate items in "Bookmark" files
- Customize your spreadsheet template
- Hot key to Product Manufacturers' listings and specs
- Merge capability for networking environments
- View crews and assembly components
- AutoSave capability
- Enhanced sorting capability

Unit Price Estimating

This interactive two-day seminar teaches attendees how to interpret project information and process it into final, detailed estimates with the greatest accuracy level.

The single most important credential an estimator can take to the job is the ability to visualize construction in the mind's eye, and thereby estimate accurately.

Some Of What You'll Learn:
- Interpreting the design in terms of cost
- The most detailed, time-tested methodology for accurate "pricing"
- Key cost drivers—material, labor, equipment, staging, and subcontracts
- Understanding direct and indirect costs for accurate job cost accounting and change order management

Who Should Attend: Corporate and government estimators and purchasers, architects, engineers... and others needing to produce accurate project estimates.

Square Foot and Assemblies Cost Estimating

This two-day course teaches attendees how to quickly deliver accurate square foot estimates using limited budget and design information.

Some Of What You'll Learn:
- How square foot costing gets the estimate done faster
- Taking advantage of a "systems" or "assemblies" format
- The Means "building assemblies/square foot cost approach"
- How to create a very reliable preliminary and systems estimate using bare-bones design information

Who Should Attend: Facilities managers, facilities engineers, estimators, planners, developers, construction finance professionals... and others needing to make quick, accurate construction cost estimates at commercial, government, educational, and medical facilities.

Repair and Remodeling Estimating

This two-day seminar emphasizes all the underlying considerations unique to repair/remodeling estimating and presents the correct methods for generating accurate, reliable R&R project costs using the unit price and assemblies methods.

Some Of What You'll Learn:
- Estimating considerations—like labor-hours, building code compliance, working within existing structures, purchasing materials in smaller quantities, unforeseen deficiencies
- Identify problems and provide solutions to estimating building alterations
- Rules for factoring in minimum labor costs, accurate productivity estimates and allowances for project contingencies
- R&R estimating examples are calculated using unit prices and assemblies data

Who Should Attend: Facilities managers, plant engineers, architects, contractors, estimators, builders... and others who are concerned with the proper preparation and/or evaluation of repair and remodeling estimates.

Mechanical and Electrical Estimating

This two-day course teaches attendees how to prepare more accurate and complete mechanical/electrical estimates, avoiding the pitfalls of omission and double-counting, while understanding the composition and rationale within the Means Mechanical/Electrical database.

Some Of What You'll Learn:
- The unique way mechanical and electrical systems are interrelated
- M&E estimates, conceptual, planning, budgeting, and bidding stages
- Order of magnitude, square foot, assemblies, and unit price estimating
- Comparative cost analysis of equipment and design alternatives

Who Should Attend: Architects, engineers, facilities managers, mechanical and electrical contractors... and others needing a highly reliable method for developing, understanding, and evaluating mechanical and electrical contracts.

Plan Reading and Material Takeoff

This two-day program teaches attendees to read and understand construction documents and to use them in the preparation of material takeoffs.

Some of What You'll Learn:
- Skills necessary to read and understand typical contract documents—blueprints and specifications
- Details and symbols used by architects and engineers
- Construction specifications' importance in conjunction with blueprints
- Accurate takeoff of construction materials and industry-accepted takeoff methods

Who Should Attend: Facilities managers, construction supervisors, office managers... and others responsible for the execution and administration of a construction project including government, medical, commercial, educational, or retail facilities.

Facilities Maintenance and Repair Estimating

This two-day course teaches attendees how to plan, budget, and estimate the cost of ongoing and preventive maintenance and repair for existing buildings and grounds.

Some Of What You'll Learn:
- The most financially favorable maintenance, repair, and replacement scheduling and estimating
- Auditing and value engineering facilities
- Preventive planning and facilities upgrading
- Determining both in-house and contract-out service costs; annual, asset-protecting M&R plan

Who Should Attend: Facility managers, maintenance supervisors, buildings and grounds superintendents, plant managers, planners, estimators... and others involved in facilities planning and budgeting.

Scheduling and Project Management

This two-day course teaches attendees the most current and proven scheduling and management techniques needed to bring projects in on time and on budget.

Some Of What You'll Learn:
- Crucial phases of planning and scheduling
- How to establish project priorities and develop realistic schedules and management techniques
- Critical Path and Precedence Methods
- Special emphasis on cost control

Who Should Attend: Construction project managers, supervisors, engineers, estimators, contractors... and others who want to improve their project planning, scheduling, and management skills.

Advanced Project Management

This two-day seminar will teach you how to effectively manage and control the entire design-build process and allow you to take home tangible skills that will be immediately applicable on existing projects.

Some Of What You'll Learn:
- Value engineering, bonding, fast-tracking, and bid package creation
- How estimates and schedules can be integrated to provide advanced project management tools
- Cost engineering, quality control, productivity measurement, and improvement
- Front loading a project and predicting its cash flow

Who Should Attend: Owners, project managers, architectural and engineering managers, construction managers, contractors... and anyone else who is responsible for the timely design and completion of construction projects.

For more information visit Means Web site at www.rsmeans.com

Seminars

2006 Means Seminar Schedule

Location	Dates
Las Vegas, NV	March
Washington, DC	April
Phoenix, AZ	April
Denver, CO	May
San Francisco, CA	June
Philadelphia, PA	June
Washington, DC	September
Dallas, TX	September
Las Vegas, NV	October
Orlando, FL	November
Atlantic City, NJ	November
San Diego, CA	December

Note: Call for exact dates and details.

Registration Information

Register Early... Save up to $100! Register 30 days before the start date of a seminar and save $100 off your total fee. *Note: This discount can be applied only once per order. It cannot be applied to team discount registrations or any other special offer.*

How to Register Register by phone today! Means' toll-free number for making reservations is: **1-800-334-3509.**

Individual Seminar Registration Fee $895. Individual CostWorks Training Registration Fee $349. To register by mail, complete the registration form and return with your full fee to: Seminar Division, Reed Construction Data, RSMeans Seminars, 63 Smiths Lane, Kingston, MA 02364.

Federal Government Pricing All federal government employees save 25% off regular seminar price. Other promotional discounts cannot be combined with Federal Government discount.

Team Discount Program Two to four seminar registrations: Call for pricing.

Multiple Course Discounts When signing up for two or more courses, call for pricing.

Refund Policy Cancellations will be accepted up to ten days prior to the seminar start. There are no refunds for cancellations received later than ten working days prior to the first day of the seminar. A $150 processing fee will be applied for all cancellations. Written notice of cancellation is required. Substitutions can be made at any time before the session starts. **No-shows are subject to the full seminar fee.**

AACE Approved Courses The RSMeans Construction Estimating and Management Seminars described and offered to you here have each been approved for 14 hours (1.4 recertification credits) of credit by the AACE International Certification Board toward meeting the continuing education requirements for recertification as a Certified Cost Engineer/Certified Cost Consultant.

AIA Continuing Education We are registered with the AIA Continuing Education System (AIA/CES) and are committed to developing quality learning activities in accordance with the CES criteria. RSMeans seminars meet the AIA/CES criteria for Quality Level 2. AIA members will receive (14) learning units (LUs) for each two-day RSMeans Course.

NASBA CPE Sponsor Credits We are part of the National Registry of CPE Sponsors. Attendees may be eligible for (16) CPE credits.

Daily Course Schedule The first day of each seminar session begins at 8:30 A.M. and ends at 4:30 P.M. The second day is 8:00 A.M.–4:00 P.M. Participants are urged to bring a hand-held calculator since many actual problems will be worked out in each session.

Continental Breakfast Your registration includes the cost of a continental breakfast, a morning coffee break, and an afternoon break. These informal segments will allow you to discuss topics of mutual interest with other members of the seminar. (You are free to make your own lunch and dinner arrangements.)

Hotel/Transportation Arrangements RSMeans has arranged to hold a block of rooms at each hotel hosting a seminar. To take advantage of special group rates when making your reservation, be sure to mention that you are attending the Means Seminar. You are, of course, free to stay at the lodging place of your choice. **(Hotel reservations and transportation arrangements should be made directly by seminar attendees.)**

Important Class sizes are limited, so please register as soon as possible.

Note: *Pricing subject to change.*

Registration Form **Call 1-800-334-3509 to register or FAX 1-800-632-6732. Visit our Web site www.rsmeans.com**

Please register the following people for the Means Construction Seminars as shown here. Full payment or deposit is enclosed, and we understand that we must make our own hotel reservations if overnight stays are necessary.

☐ Full payment of $____________________enclosed.

☐ Bill me

Name of Registrant(s)
(To appear on certificate of completion)

____

____

P.O. #: ______________________________
GOVERNMENT AGENCIES MUST SUPPLY PURCHASE ORDER NUMBER

Firm Name ______________________________

Address ______________________________

City/State/Zip ______________________________

Telephone No. Fax No. ______________________________

E-mail Address ______________________________

Charge our registration(s) to: ☐ MasterCard ☐ VISA ☐ American Express ☐ Discover

Account No. ____________________ Exp. Date __________

Cardholder's Signature ______________________________

Seminar Name City Dates

____

____

Please mail check to: Seminar Division, Reed Construction Data, RSMeans Seminars, 63 Smiths Lane, P.O. Box 800, Kingston, MA 02364 USA

MeansData™

CONSTRUCTION COSTS FOR SOFTWARE APPLICATIONS

Your construction estimating software is only as good as your cost data.

A proven construction cost database is a mandatory part of any estimating package. The following list of software providers can offer you MeansData™ as an added feature for their estimating systems. See the table below for what types of products and services they offer (match their numbers). Visit online at **www.rsmeans.com/demosource/** for more information and free demos. Or call their numbers listed below.

1. **3D International**
   713-871-7000
   venegas@3di.com
2. **4Clicks-Solutions, LLC**
   719-574-7721
   mbrown@4clicks-solutions.com
3. **Aepco, Inc.**
   301-670-4642
   blueworks@aepco.com
4. **Applied Flow Technology**
   800-589-4943
   info@aft.com
5. **ArenaSoft Estimating**
   888-370-8806
   info@arenasoft.com
6. **BSD - Building Systems Design, Inc.**
   888-273-7638
   bsd@bsdsoftlink.com
7. **CMS - Computerized Micro Solutions**
   800-255-7407
   cms@proest.com
8. **Corecon Technologies, Inc.**
   714-895-7222
   sales@corecon.com
9. **CorVet Systems**
   301-622-9069
   sales@corvetsys.com
10. **Estimating Systems, Inc.**
    800-967-8572
    esipulsar@adelphia.net
11. **MC² - Management Computer**
    800-225-5622
    vkeys@mc2-ice.com
12. **Maximus Asset Solutions**
    800-659-9001
    assetsolutions@maximus.com
13. **Shaw Beneco Enterprises, Inc.**
    877-719-4748
    inquire@beneco.com
14. **Timberline Software Corp.**
    800-628-6583
    product.info@timberline.com
15. **US Cost, Inc.**
    800-372-4003
    sales@uscost.com
16. **Vanderweil Facility Advisors**
    617-451-5100
    info@VFA.com
17. **WinEstimator, Inc.**
    800-950-2374
    sales@winest.com

TYPE	1	2	3	4	5	6	7	8	9	10	11	12	13	14	15	16	17
BID					•	•	•		•		•			•			•
Estimating		•			•	•	•	•	•	•	•		•	•	•		•
DOC/JOC/SABER		•			•	•			•	•		•	•	•			•
ID/IQ		•							•	•		•	•				•
Asset Mgmt.												•				•	•
Facility Mgmt.	•		•									•				•	
Project Mgmt.	•	•						•	•				•	•			
TAKE-OFF					•		•	•	•		•			•	•		
EARTHWORK								•			•						
Pipe Flow				•													
HVAC/Plumbing					•		•		•								
Roofing					•		•		•								
Design	•				•								•	•			•
Other Offers Links:																	
Accounting/HR		•			•								•	•			
Scheduling					•									•	•		•
CAD														•			•
PDA														•	•		•
Lt. Versions		•					•							•	•		•
Consulting	•	•			•		•		•			•		•	•	•	•
Training		•			•	•	•	•	•	•		•	•	•	•	•	•

Qualified re-seller applications now being accepted. Call Carol Polio, Ext. 5107.

FOR MORE INFORMATION
CALL 1-800-448-8182, EXT. 5107 OR FAX 1-800-632-6732

New Titles

Preventive Maintenance for Multi-Family Housing

by John C. Maciha

With 24/7 occupancy, multi-family buildings can be some of the toughest to maintain. Prepared by one of the nation's leading experts on multi-family housing, Preventive Maintenance for Multi-Family Housing puts easy-to-use guidelines right at your fingertips for the what, when, why, and how much of multi-family preventive maintenance.

This complete PM system for apartment and condominium communities features expert guidance, checklists for buildings and grounds maintenance tasks and their frequencies, a reusable wall chart to track maintenance, and a dedicated Web site featuring customizable electronic forms. A must-have for anyone involved with multi-family housing maintenance and upkeep.

$89.95 per copy
225 pages
Catalog No. 67346

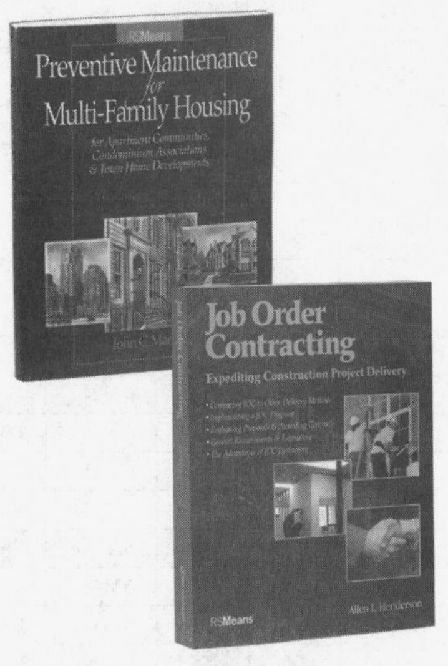

Job Order Contracting

Expediting Construction Project Delivery
by Allen Henderson

Expert guidance to help you implement JOC—fast becoming the preferred project delivery method for repair and renovation, minor new construction, and maintenance projects in the public sector and in many states and municipalities. The author, a leading JOC expert and practitioner, shows how to:

- Establish a JOC program
- Evaluate proposals and award contracts
- Handle general requirements and estimating
- Partner for maximum benefits

This book will quickly bring you up to speed on JOC (also known as Delivery Order Contracting and SABER) so you can achieve faster project starts with fewer delays, cost overruns, and quality disputes.

$89.95 oer copy
192 pages, illustrated, Hardcover
Catalog No. 67348

Kitchen & Bath Project Costs: Planning & Estimating Successful Projects

Project estimates for 35 of the most popular kitchen and bath renovations... from replacing a single fixture to whole-room remodels. Each estimate includes:

- All materials needed for the project
- Labor-hours to install (and demolish/remove) each item
- Subcontractor costs for certain trades and services
- An allocation for overhead and profit

PLUS! Takeoff and pricing worksheets—forms you can photocopy or access electronically from the book's Web site; alternate materials—unit costs for different finishes and fixtures, location factors—easy multipliers to adjust the costs to your location; and expert guidance on estimating methods, project design, contracts, marketing, working with homeowners, and tips for each of the estimated projects.

$29.95 per copy
Over 175 pages
Catalog No. 67347

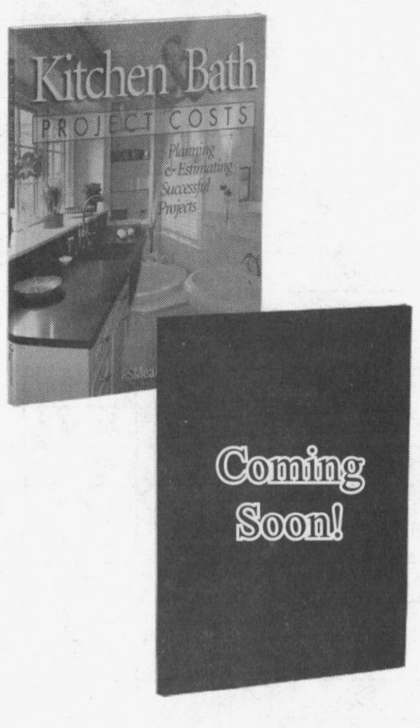

Home Addition & Renovation Costs

New! This essential home remodeling reference gives you 35 project estimates and guidance for some of the most popular home renovation and addition projects... from opening up a simple interior wall to adding an entire second story. Each estimate includes a floor plan, color photos, and detailed costs. Use the project estimates as backup for pricing, to check your own estimates, or as a cost reference for preliminary discussion with homeowners.

Includes:

- Case studies—with creative solutions and design ideas.
- Alternate materials costs—so you can match the estimates to the particulars of your projects.
- Location factors—easy multipliers to adjust the book's costs to your own location.

$29.95 per copy
Over 200 pages, illustrated, Softcover
Catalog No. 67349

The Practice of Cost Segregation Analysis

by Bruce A. Desrosiers and Wayne J. DelPico

This expert guide walks you through the practice of cost segregations analysis, which enables property owners to defer taxes and benefit from "accelerated cost recovery" through depreciation deductions on assets that are properly identified and classified. Cost segregation practice requires knowledge of both tax law and the construction process. In this book, the authors share their expertise in these areas with tax and accounting professionals, cost segregation consultants, facility owners, architects, and general contractors—providing guidance on major aspects of a professional, defensible cost segregation study.

With a glossary of terms, sample cost segregation estimates for various building types, key information resources, and updates via a dedicated Web site, this book is a critical resource for anyone involved in cost segregation analysis.

$99.95 per copy
Over 225 pages
Catalog No. 67345

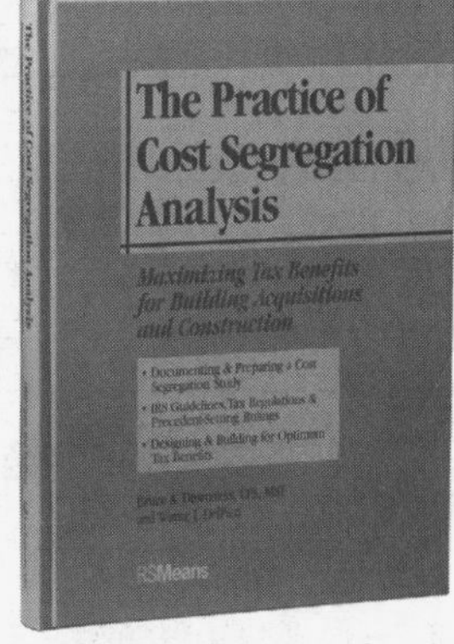

ORDER TOLL FREE 1-800-334-3509
OR FAX 1-800-632-6732.

2006 Order Form

Qty.	Book No.	COST ESTIMATING BOOKS	Unit Price	Total
	60066	Assemblies Cost Data 2006	$206.95	
	60016	Building Construction Cost Data 2006	126.95	
	61016	Building Const. Cost Data–Looseleaf Ed. 2006	157.95	
	60226	Building Const. Cost Data–Western Ed. 2006	126.95	
	60116	Concrete & Masonry Cost Data 2006	114.95	
	60146	Construction Cost Indexes 2006	272.00	
	60146A	Construction Cost Index–January 2006	68.00	
	60146B	Construction Cost Index–April 2006	68.00	
	60146C	Construction Cost Index–July 2006	68.00	
	60146D	Construction Cost Index–October 2006	68.00	
	60346	Contr. Pricing Guide: Resid. R & R Costs 2006	39.95	
	60336	Contr. Pricing Guide: Resid. Detailed 2006	39.95	
	60326	Contr. Pricing Guide: Resid. Sq. Ft. 2006	39.95	
	60236	Electrical Change Order Cost Data 2006	126.95	
	60036	Electrical Cost Data 2006	126.95	
	60206	Facilities Construction Cost Data 2006	299.95	
	60306	Facilities Maintenance & Repair Cost Data 2006	274.95	
	60166	Heavy Construction Cost Data 2006	126.95	
	60096	Interior Cost Data 2006	126.95	
	60126	Labor Rates for the Const. Industry 2006	274.95	
	60186	Light Commercial Cost Data 2006	107.95	
	60026	Mechanical Cost Data 2006	126.95	
	63016	Metric Construction Cost Data 2006	149.95	
	60156	Open Shop Building Const. Cost Data 2006	126.95	
	60216	Plumbing Cost Data 2006	126.95	
	60046	Repair and Remodeling Cost Data 2006	107.95	
	60176	Residential Cost Data 2006	107.95	
	60286	Site Work & Landscape Cost Data 2006	126.95	
	60056	Square Foot Costs 2006	137.95	
		REFERENCE BOOKS		
	67310A	ADA Compliance Pricing Guide, 2nd Ed.	79.95	
	67273	Basics for Builders: How to Survive and Prosper	34.95	
	67330	Bldrs Essentials: Adv. Framing Methods	24.95	
	67329	Bldrs Essentials: Best Bus. Practices for Bldrs	29.95	
	67298A	Bldrs Essentials: Framing/Carpentry 2nd Ed.	24.95	
	67298AS	Bldrs Essentials: Framing/Carpentry Spanish	24.95	
	67307	Bldrs Essentials: Plan Reading & Takeoff	35.95	
	67261A	Bldg. Prof. Guide to Contract Documents 3rd Ed.	32.48	
	67342	Building & Renovating Schools	99.95	
	67339	Building Security: Strategies & Costs	44.98	
	67312	Building Spec Homes Profitably	29.95	
	67146	Concrete Repair & Maintenance Illustrated	34.98	
	67314	Cost Planning & Est. for Facil. Maint.	89.95	
	67317A	Cyberplaces: The Internet Guide 2nd Ed.	59.95	
	67328A	Designing & Building with the IBC, 2nd Ed.	99.95	
	67230B	Electrical Estimating Methods 3rd Ed.	64.95	
	64777A	Environmental Remediation Est. Methods 2nd Ed.	$ 99.95	
	67343	Estimating Bldg. Costs for Resi. & Lt. Comm.	29.95	
	67160	Estimating for Contractors	17.98	
	67276A	Estimating Handbook 2nd Ed.	99.95	
	67249	Facilities Maintenance Management	86.95	
	67246	Facilities Maintenance Standards	79.95	

Qty.	Book No.	REFERENCE BOOKS (Cont.)	Unit Price	Total
	67318	Facilities Operations & Engineering Reference	109.95	
	67301	Facilities Planning & Relocation	89.95	
	67231	Forms for Building Const. Professional	47.48	
	67260	Fundamentals of the Construction Process	25.98	
	67338	Green Building: Proj. Planning & Cost Est.	89.95	
	67323	Historic Preservation: Proj. Planning & Est.	99.95	
	67308E	Home Improvement Costs–Int. Projects 9th Ed.	24.95	
	67309E	Home Improvement Costs–Ext. Projects 9th Ed.	24.95	
	67324A	How to Est. w/Means Data & CostWorks 2nd Ed.	59.95	
	67304	How to Estimate with Metric Units	9.98	
	67306	HVAC: Design Criteria, Options, Select. 2nd Ed.	84.95	
	67281	HVAC Systems Evaluation	84.95	
	67282A	Illustrated Construction Dictionary, Condensed	59.95	
	67292A	Illustrated Construction Dictionary, w/CD-ROM	99.95	
	67348	Job Order Contracting	89.95	
	67347	Kitchen & Bath Project Costs	29.95	
	67295B	Landscape Estimating 4th Ed.	62.95	
	67341	Life Cycle Costing	99.95	
	67302	Managing Construction Purchasing	19.98	
	67294A	Mechanical Estimating 3rd Ed.	64.95	
	67283B	Plumbing Estimating Methods 3rd Ed.	29.98	
	67345	Pract. of Cost Segregation Analysis	99.95	
	67326	Preventive Maint. Guidelines for School Facil.	149.95	
	67337	Preventive Maint. for Higher Education Facilities	149.95	
	67346	Preventive Maint. for Multi-Family Housing	89.95	
	67236A	Productivity Standards for Constr.–3rd Ed.	49.98	
	67247B	Project Scheduling & Management for Constr. 3rd Ed.	64.95	
	67265B	Repair & Remodeling Estimating 4th Ed.	69.95	
	67322A	Resi. & Light Commercial Const. Stds. 2nd Ed.	59.95	
	67327	Spanish/English Construction Dictionary	22.95	
	67145B	Sq. Ft. & Assem. Estimating Methods 3rd Ed.	69.95	
	67287	Successful Estimating Methods	32.48	
	67313	Successful Interior Projects	24.98	
	67233	Superintending for Contractors	35.95	
	67321	Total Productive Facilities Management	29.98	
	67284	Understanding Building Automation Systems	29.98	
	67303A	Unit Price Estimating Methods 3rd Ed.	59.95	
	67319	Value Engineering: Practical Applications	79.95	
		MA residents add 5% state sales tax		
		Shipping & Handling**		
		Total (U.S. Funds)*		

Prices are subject to change and are for U.S. delivery only. *Canadian customers may call for current prices. **Shipping & handling charges: Add 7% of total order for check and credit card payments. Add 9% of total order for invoiced orders.

Send Order To: **ADDV-1001**

Name (Please Print) ______________________

Company ______________________

☐ **Company**

☐ **Home** Address ______________________

____

City/State/Zip ______________________

Phone # ____________ P.O. # ____________

(Must accompany all orders being billed)

Mail To: **RSMeans** P.O. Box 800, Kingston, MA 02364-0800